Third edition

Biological science

D.J. Taylor B.Sc., Ph.D., C.Biol., F.I.Biol.
Director of Continuing Education, Strode's Sixth Form College, Egham

N.P.O. Green B.Sc., C.Biol., M.I.Biol.
Headmaster, St George's College, Buenos Aires, Argentina

G.W. Stout B.Sc., M.A., M.Ed., C.Biol., F.I.Biol.
Headmaster, International School, Mafikeng, South Africa

Editor
R. Soper B.Sc., C.Biol., F.I.Biol.
Formerly Vice-Principal and Head of Science, Collyers Sixth Form College, Horsham

CAMBRIDGE
UNIVERSITY PRESS

PUBLISHED BY THE PRESS SYNDICATE OF THE UNIVERSITY OF CAMBRIDGE
The Pitt Building, Trumpington Street, Cambridge CB2 1RP, United Kingdom

CAMBRIDGE UNIVERSITY PRESS
The Edinburgh Building, Cambridge CB2 2RU, United Kingdom
40 West 20th Street, New York, NY 10011-4211, USA
10 Stamford Road, Oakleigh, Melbourne 3166, Australia

First published 1984
Second edition 1990
Third edition 1997
Low price edition 1998

Printed in the United Kingdom at the University Press, Cambridge

9 8 7 6 5 4 3

A catalogue record for this book is available from the British Library

ISBN 0 521 63923 9 low price paperback

Cambridge University Press is part of the University of Cambridge. It is a charitable enterprise dedicated to printing and publishing for the advancement of knowledge throughout the world. In recognition of the fact that there are many parts of the world where people have difficulty in purchasing their own copies of the best international textbooks, Cambridge University Press is now publishing a series of Low Priced Editions. These are reprints of internationally respected books from Cambridge University Press, specially printed and priced for the benefit of students and teachers in selected countries.

Contents

Preface to the third edition

Since its publication in 1984, *Biological Science* has become established as one of the most comprehensive and authoritative A level Biology texts. It has remained one of our aims in writing the third edition to maintain its reputation as an up-to-date and comprehensive resource for current A level syllabuses.

In recent years there have been significant changes in content and format of syllabuses, with modular courses becoming important alternatives to 'linear' courses, and a new agreed subject core for Biology being established by SCAA in 1993, which has subsequently been revised in 1997. A typical modern syllabus is now composed of a core containing the agreed basis of the subject, with options that develop depth and range of experience in more specialist areas. Typically these options also emphasise the social, ethical and applied aspects of the subject, and emphasise the growing importance of biological sciences in the modern world.

The revision for the third edition has been far more comprehensive than that carried out for the second edition, with many substantial as well as more subtle changes to the text, diagrams, photographs and tables. Much new material has been written, and material which is no longer relevant has been removed. In addition, some of the material in the appendices has been removed and placed in the relevant chapters.

In recognition of the importance and popularity of certain topics, particularly in option areas, three completely new chapters have been added. These provide comprehensive coverage of Microbiology and biotechnology (chapter 12), Human health and disease (chapter 15) and Applied genetics (chapter 25). In addition, there is far more extensive coverage of human nutrition in chapter 10 and human reproduction in chapter 21 in line with present syllabuses. Where relevant, the ethical and social implications of these topics are also discussed. A wider range of topical issues is also included in the Ecology chapter (chapter 10).

In line with the changing emphasis of syllabuses, Variety of life has been condensed from three chapters into one (chapter 2), with examples relevant to current syllabuses being chosen. The chapter includes a new introductory discussion on classification and use of keys. Other chapters have been updated where necessary. Physiological topics throughout the book have, in particular, been modified in the light of new knowledge as well as to match syllabus requirements. The text also takes into account the trend towards a greater focus on higher plants and humans.

In addition to the changes described, a major effort has been made to make the text suitable for a wider range of students. Consideration has been given to reducing unnecessary complexity, especially in the use of language. Particular care has been taken with the introduction to each topic. Some sections have been reorganised, subheadings added, and greater use made of numbered lists and bulleted points. It is hoped that these changes will improve the readability whilst retaining the rigour and depth of the text.

Revision of this edition of the book has largely been carried out by Dennis Taylor during a sabbatical from Strode's College. As in the second edition, the ecology chapters (10 and 11) have been revised by Rosalind Taylor of Kingston University. The new chapter on Health and disease was mostly written by Roland Soper. Academic referees have checked new text with the aim of making it as factually correct as possible. Nevertheless, in an undertaking this large, errors and inaccuracies are difficult to avoid completely, and the authors are always grateful for notification of any that are spotted.

Acknowledgements

The authors and publisher would like to acknowledge the many friends, colleagues, students and advisers who have helped in the production of *Biological Science*.

In particular, we wish to thank:
Dr R. Batt, Dr I. Benton, Dr Claudia Berek, Professor R.J. Berry, Dr A.C. Blake, Dr John C. Bowman, Dr John Brookfield, Mr R. Brown, Dr Stuart Brown, Dr Fred Burke, Mr Richard Carter, Dr Norman R. Cohen, Dr I. Côte, Dr K.J.R. Edwards, Mr Malcolm Emery, Mr Nick Fagents, Dr James T. Fitzsimons, Dr John Gay, Dr Brij L. Gupta, Vivienne Hambleton, Dr David E. Hanke, Dr R.N. Hardy, Reverend J.R. Hargreaves, Dr S.A. Henderson, Mr Michael J. Hook, Mr Colin S. Hutchinson, Illustra Design Ltd, Dr Alick Jones, Mrs Sue Kearsey, Dr Simon P. Maddrell FRS, Professor Aubrey Manning, Dr Chris L. Mason, Mrs Ruth Miller, Dr David C. Moore, A.G. Morgan, Dr Rodney Mulvey, Dr David Secher, Dr John M. Squire, Professor James F. Sutcliffe, Stephen Tomkins, Dr Eric R. Turner, Dr Paul Wheater, Dr Brian E.J. Wheeler, Dr Michael Wheeler.

The authors are particularly indebted to Mrs Adrienne Oxley, who patiently and skilfully organised the pretesting of all the practical exercises. Her perseverance has produced exercises that teachers, pupils and laboratory technicians can depend upon.

However, the authors accept full responsibility for the final content of these books.

Finally, the authors wish to express their thanks to their wives and families for the constant support and encouragement shown throughout the preparation and publication of these books.

We also wish to thank the following for permission to use their illustrations, tables and questions.
Figures: 2.2*a*, 2.37*c*, 2.38*b*, 2.40*a*, 2.40*b*, 2.46, 2.66*b*, 2.66*e*, 8.3 Heather Angel/Biofotos; 2.2*b* Stephen Krasemann/NHPA; 2.2*c* Gerard Lacz/NHPA; 2.6*b*, 5.3, 5.8 Andrew Syred 1995/Microscopix; 2.6*c*, 2.6*d*, 2.7, 2.17*b*, 2.18*b*, 2.25*a*, 2.25*c*, 2.26*a*, 2.27*b*, 2.32*b*, 2.37*d*, 2.48*e*, 2.48*f*, 2.48*g*, 2.66*c*, 2.66*d*, 5.1*b*, 5.13, 5.25, 5.28, 5.30, 5.31, 5.35, 6.3*e*, 6.3*f*, 6.4*a*, 6.4*b*, 6.5*d*, 6.6*e*, 6.7*b*, 6.9*c*, 6.9*d*, 6.10*b*, 6.12*b*, 6.12*c*, 6.12*e*, 6.13*b*, 6.13*d*, 6.15*b*, 6.16*c*, 6.16*d*, 6.22, 6.25, 6.29, 7.3, 7.4*a*, 7.4*b*, 7.6, 8.10*b*, 8.17, 8.19, 8.21*b*, 8.21*e*, 8.21*f*, 9.11*a*, 9.20*a*, 9.20*b*, 9.22*a*, 9.23, 9.33*a*, 9.33*b*, Biophoto Associates; 2.9 Professor Stanley Cohen/Science Photo Library (SPL); 2.12 Dr L. Caro/SPL; 2.18*c* Jurgen Dielenscheider/Holt Sudios International; 2.19*b* B. Heggeler/Biozentrum, University of Basel/SPL; 2.24 NIBSC/SPL; 2.27*a* Andrew Syred 1993/Microscopix; 2.37*b* Roy Edwards; 2.53 R. Umesh Chandron, TDR, WHO/SPL; 2.62*b*, 2.62*c* Shell International Petroleum Co.; 2.62*d* Stephen Dalton/NHPA; 3.1*b*, 3.1*c*, 3.11, 3.17*b* Andrew Lambert; 3.34*b*, 3.34*e* Sir John Kendrew; 3.34*d* Dr Arthur Lesk/SPL; 3.41 Dr J.M. Squire; 3.45 Professor M.H.F. Wilkins, Biophysics Department, King's College, London; 4.4*d* Clive Freeman, The Royal Institution/SPL; 5.5*a*, 5.5*b* A.M. Page, Royal Holloway College, London; 5.6 R. Maisonneuve, Publiphoto Diffusion/SPL; 5.12 Dr Glenn Decker, School of Medicine, John Hopkins University; 5.24 Don Fawcett/SPL; 5.29, 6.14*b*, 6.17*b*, 6.17*c*, 6.18*c*, 6.19*b* 6.20, 6.21, 6.23, 6.24, 6.26*a*, 6.31*a*, 8.16*b*, 8.21*a*, 9.12*e* Dr Paul Wheater; 5.33 Dr Klaus Weber; 6.3*d* Rothamsted Experimental Station; 6.5*c*, 6.6*d*, 6.12*f*, 7.12, 11.2, 11.3, 11.10 Centre for Cell and Tissue Research, York; 6.14*c*, 6.15*c* Life Science Images; 6.18*d* Mr P. Crosby, Department of Biology, University of York; 7.2 Andrew Mounter/Planet Earth Pictures; 7.8, 7.21*b*, 7.23 Dr A.D. Greenwood; 7.21*a* C.C. Black (1971) *Plant Physiology*, **47**, 15–23, with permission of the publisher; 8.1*a* R.L. Mathews/Planet Earth Pictures; 8.1*b* Nick Greaves/Planet Earth Pictures; 8.6*a* Kim Taylor/Bruce Coleman Ltd; 8.6*c*, 8.6*d* Dr Brad Amos/SPL; 8.7*b* Claude Nuridsany & Marie Perennou/SPL; 8.8 Alan Weaving/Ardea; 8.13*a* Charles Day; 8.13*b* King's College School of Medicine and Dentistry, London; 8.15*a*, 8.15*b*, 8.15*c*, 8.15*d* Dr C.A. Saxton, Unilever Research; 8.16*a* Dr L.M. Beidler/SPL; 8.18*b* Mehav Kulyk/SPL; 8.28, 9.35 National Medical Slide Bank; 8.30*a*, 8.30*b* Peter Menzel/SPL; 9.11*b*, 9.12*g* Dr Brij L. Gupta, Department of Zoology, Cambridge; 9.12*f* Bill Longcore/SPL; 9.13 E.F. van Bruggen, State University of Groningen; 9.20*c* Prof. P. Motta, Department of Anatomy, University La Sapienza, Rome/SPL; 9.22*b*, B. Siegwart, P. Gehr, J. Gil & E.R. Wiebel (1971) *Respir. Physiol.*, **13**, 141–59; 9.25 G.M.Hughes (1973) *The Vertebrate Lung*, Oxford Biology Readers, no.59; 9.35 National Medical Slide Bank; 10.16 Dr Martyn Waller; 10.20 Mark Mattock/Planet Earth Pictures; 10.27 Herbert Giradet/Panos Pictures; 10.30 Nick Garbutt/Planet Earth Pictures; 10.37 W. J. Allen/Chilworth Media Associates; 11.1, 11.13 Graham Page, Kingston University; 11.6 John Edward Leigh; 11.7 Nigel Luckhurst; 12.2 Simon Fraser/SPL; 12.4, 12.14*b* Hank Morgan/SPL; 12.5 National Dairy Council; 12.11*a*, 12.27 Andrew Syred/SPL; 12.11*b* National Institute for Research in Dairying, Reading; 12.12 Robert Longuehaye, NIBSC/SPL; 12.14*a*, 12.15 James Holmes/Celltech Ltd/SPL; 12.18 CEPHAS/Stuart Boreham; 12.19 Ricardo Arias, Latin Stock/SPL; 12.21 John Birdsall; 12.22 E. A. Rathbun & N. J. Brewin, John Innes Centre, Norwich; 12.23 Prof. David Hall/SPL; 12.24 David Hall/Panos Pictures; 12.25 Steve McCutcheon/FLPA; 12.26 Gist-Brocades; 12.31 Hattie Young/SPL.
Tables: 3.1 with permission of Plenum Publishing Corporation, copyright Plenum Publishing Corporation; 8.8, 8.9, 8.10 reproduced by permission of the Controller of Her Majesty's Stationery Office; 10.1 copyright © 1971 by W.B. Saunders Company, reprinted by permission of Holt, Rinehart & Winston, CBS Publishing; 11.5, 11.6 by permission of Griffin & George.
Questions: 10.14, 10.16 Open University Foundation Course (S100) Unit 20, copyright © 1971, Open University Press.
Figures: 13.11, 13.14, 13.16*b*, 13.16*c*, 13.17*b*, 13.25*a*, 13.25*b*, 14.3*b*, 14.6, 14.7, 14.11, 14.14*a*, 14.16, 15.7, 17.14*a*, 17.56*a*, 17.56*b*, 18.16*a*, 18.16*b*, 19.11*a*, 19.20, 20.3, 20.15*a*, 20.15*b*, 20.15*c*, 20.24*a*, 20.24*b*, 21.1*c*, 21.23*a*, 21.23*b*, 21.29, 21.42, 21.50*a*, 21.50*b*, 21.50*f*, 22.25*a*, 22.25*b*, 22.29, 23.1, 23.3, 23.7*a–f*, 23.12*a–j*, 24.15, 25.27 Biophoto Associates; 13.12*a* Claus Meyer/Science Photo Library (SPL); 13.12*b* John Lee/Planet Earth Pictures; 13.16*d*, 22.16 Centre for Cell and Tissue Research, York; 13.22 Anderson & Cronshaw (1970) *Planta* **91**, 173–80; 13.25*c* Dr Martin Zimmermann, Harvard University; 13.27 Professor B.E.S. Gunning (1977) *Science Progress* **64**, 539–68, Blackwell Scientific Publications Ltd; 14.1a, 21.36 Dr Paul Wheater; 14.1*b* K.R. Porter/SPL; 14.3*a* Life Science Images; 14.4*b* Professors P.M. Motta & G. Macchiarelli/SPL; 14.4*c*, 14.4*d*, 15.19*a*, 15.19*b* SPL; 14.35 CNRI/SPL; 14.38*b* Ken Eward/SPL;

14.40*a*, 17.8 University of Zurich-Irchel/Nature & Science AG, FL-Vaduz; 14.40*b* BSIP PIR/SPL; 15.4 Unicef/Betty Press; 15.9 Andy Crump, TDR, WHO/SPL; 15.13 Vivien Fifield; 15.15*c*, 21.41 Biophoto Associates/SPL; 15.16, 21.52, 25.21 National Medical Slide Bank; 15.17 D. Phillips/SPL; 15.20*a* Philippe Plailly/SPL; 15.20*b* Scott Camazine/SPL; 15.23 National Institute of Health/SPL; 15.24 Dr Tony Brain/Spl; 15.25 Princess Margaret Rose Orthopaedic Hospital/SPL; 16.15 Dr B.E. Juniper; 16.17 T. Swarbrick, *Harnessing the hormone*, Grower Publications Ltd; 16.19 Long Ashton Research Station; 16.23 Centre Nationale de la Recherche Scientifique, *Regulateurs naturels de la croissance vegetale* (1964); 16.26 Dr Peter Evans, Southampton University; 16.32 Professor Anton Lang (1957) *Proc. Natl. Acad. Sci. USA* **43**, 709–17; 17.10, 17.14*b* Don Fawcett/SPL; 17.22, 17.33*b*, 20.17 Manfred Kage/SPL; 17.25, 21.50*d* Garry Watson/SPL; 17.27*a*, 17.27*b* Natural History Museum, London; 17.43 Profs. P.M. Motta & A. Caggiati/SPL; 17.56*d* Dr L. Orci, University of Geneva/SPL; 17.58 Daniel Heuchlin/NHPA; 17.61 Niall Rankin/FLPA; 17.68 Caroline E.G. Tutin; 18.18 P.G. Munro, Biopolymer Group, Imperial College; 18.19 A. Freundlich, Biopolymer Group, Imperial College; 18.24 Dr J. Squire, Biopolymer Group, Imperial College; 19.7 Dr R. Clark & M. Goff/SPL; 19.9 19.10 Michael & Patricia Fogden; 19.17*a* W. Higgs/ GSF Picture Library; 19.17*b* William S. Paton/Planet Earth Pictures; 19.17*c* Pete Oxford/Planet Earth Pictures; 20.2*a* E.H. Mercer (1959) *Proc. Roy. Soc. Lond. B* **150** 216–36; 20.31, 21.11*a–f*, GSF Picture Library; 21.10 Dr J. Gurdon (1977) *Proc. Roy. Soc. Lond. B* **198** 211–47; 21.13 Sinclair Stammers/SPL; 21.14 Horticultural Research Institute; 21.26 Hermann Eisenbeiss; 21.28 Howard Jones; 21.46*a* David Scharf/SPL; 21.46*b* Dr Everett Anderson/SPL; 21.50*c*, 21.50*g*, 21.50*h* Petit Format/Nestle/SPL; 21.50*e* Keith/Custom Medical Stock Photo/SPL; 22.29 Bettina Cirone/SPL; 23.8 M. Hirons/GSF Picture Library; 23.9, 24.26 ARC Poultry Research Centre; 23.13, 23.14 Dr S.A. Henderson, Department of Genetics, University of Cambridge; 23.28*a* O.L. Miller Jr & B.A. Hamkalo, Visualization of bacterial genes in action, *Science* **169** 392–5, 24 July 1970, copyright © 1970 by the American Association for the Advancement of Science; 24.30 John Birdsall Photography; 25.4 J.C. Revy/SPL; 25.10 British Diabetic Association; 25.12 John Frost Historical Newspaper Service; 25.13*a*, 25.14, 25.15, 26.9*b* Nigel Cattlin/Holt Studios International; 25.16 M. Baret, RAPHO/SPL; 25.17 Philippe Plailly/Eurelios/SPL; 25.18 PPL Pharmaceuticals; 25.20 British Union for the Abolition of Vivisection; 25.25 Cystic Fibrosis Trust; 25.28 Hattie Young/SPL; 25.32 Saturn Stills/SPL; 25.34 Klaus Gulbrandsen/SPL; 25.35 David Parker/SPL; 25.37 Cellmark Diagnostics; 26.3 D.R.B. Booth/GSF Picture Library; 26.7 Charles & Sandra Hood/Bruce Coleman Ltd; 26.8*b* Heather Angel; 26.9*a* Werner Layer/Bruce Coleman Ltd; 26.17 M.P.L. Fogden/Bruce Coleman Ltd; 27.5*a*, 27.5*b* AGPM; 27.6 Semences Nickerson, France; 27.7 D.F. Jones, Connecticut Agricultural Experiment Station; 27.9, 27.13 John Haywood; 27.10*a* S.E. Davis; 27.10*b* Kim Taylor/Bruce Coleman Ltd; 27.12 M.A. Tribe, I. Tallan & M.R. Erant (1978) *Case Studies in Genetics*, CUP.

Chapter One

Introduction to the subject

Biology (*bios*, life; *logos*, knowledge) is a science devoted to the study of living organisms. Science has progressed by breaking down complex subjects of study into their component parts so that today there are numerous branches of biology, some of which are shown in fig 1.1. This principle is often called the 'reductionist' principle and, carried to its logical conclusions, it has focused attention on the most elementary forms of matter in living and non-living systems. This approach to study seeks fundamental understanding by looking at parts rather than the whole. An opposing approach, based on the 'vitalist' principle, considers that 'life' is something special and unique, and maintains that life cannot be explained solely in terms of the laws of physics and chemistry, having properties which are special to the system as a whole. The aim of biology must ultimately be to explain the living world in terms of scientific principles, although appreciating that organisms behave in ways which often seem beyond the capabilities of their component parts. Certainly the consciousness of living organisms cannot be described in terms of physics and chemistry even though the neurophysiologist can describe the working of the single neurone in physicochemical terms. Consciousness may be the collective working of millions of neurones and their electrochemical states, but as yet we have no real concept of the chemical nature of thought and ideas. Nor do we understand completely how living organisms originated and evolved. There have been many attempts to answer this questions from theological to biological and chapters 23–27 attempt to put the different viewpoints, but with the emphasis on the possible biological explanations.

Thus we are reduced to the position that we cannot define precisely what life is nor whence it came. All that we can do is to describe the observable phenomena that distinguish living matter from non-living matter. These are as follows.

Nutrition

All living organisms need food, which is used as a source of energy, and materials for the processes of life, such as growth. Only two sources of energy are used by living organisms, namely light and chemical energy. Those organisms specialised for using light energy carry out photosynthesis and contain pigments, including chlorophyll, which absorb light. They include plants, algae

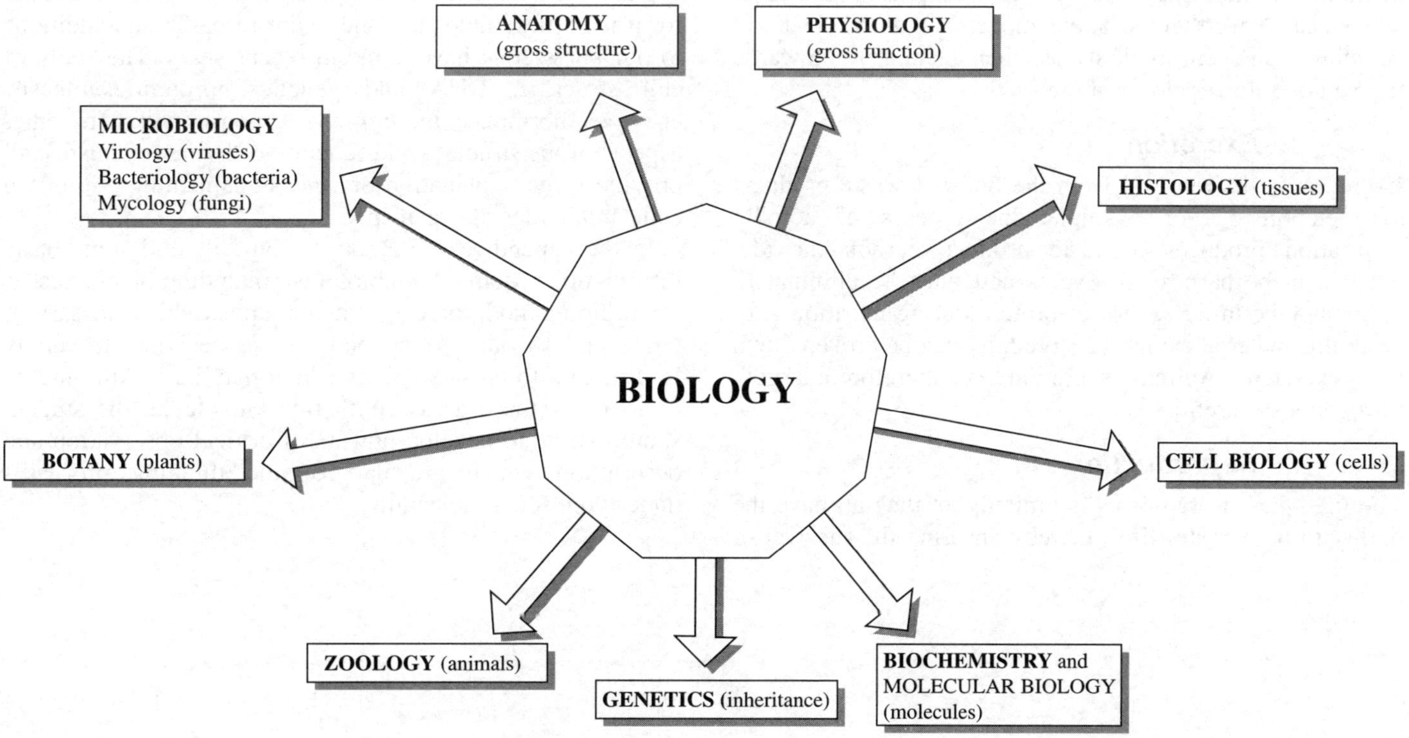

and some simple organisms including bacteria. These organisms which use chemical energy must obtain it from other living organisms. They include animals and fungi. Different methods of nutrition are responsible for some of the most fundamental differences between organisms.

Respiration

All life processes require energy and much of the food obtained by nutrition is used as a source of this energy. The energy is released during the breakdown of certain energy-rich compounds in the process of respiration. The energy released is stored in molecules of adenosine triphosphate (ATP). This compound had been found to occur in all living cells and is sometimes referred to as the 'universal energy carrier'.

Irritability

Living organisms have the ability to respond to changes in both the internal and external environments and thus ensure that they maximise their chances of survival. For example, the blood vessels in the skin of a mammal dilate (increase in diameter) in response to a rise in body temperature, and the consequent heat loss brings about a return to the optimum temperature of the body. A green plant on a window sill in a room grows towards light coming through the window, thus ensuring maximum exposure to light for photosynthesis.

Movement

Some living organisms, such as animals and some bacteria, have the ability to move from place to place, that is they locomote. This is necessary in order for them to obtain their food, unlike other organisms, such as plants, which can manufacture their own food from raw materials obtained in one place. Nevertheless, some movement of whole body structures can occur in plants, as when a leaf grows towards the Sun or a flower closes at night.

Excretion

Excretion is the removal from the body of waste products of metabolism. For example, the process of aerobic respiration produces a waste product, carbon dioxide, which can be harmful in excess and must be eliminated. Animals take in an excess of protein during nutrition and, since this material cannot be stored, it must be broken down and excreted. Animal excretion is, therefore, largely nitrogenous excretion.

Reproduction

The life span of organisms is limited, but they all have the ability to perpetuate 'life', thereby ensuring the survival of the species. The resulting offspring have the same general characteristics as the parents, whether such individuals are produced by asexual or sexual reproduction. The 'reductionist' search for the explanation of this inheritance has revealed the existence of molecules, known as nucleic acids (deoxyribonucleic acid, DNA, and ribonucleic acid, RNA), which contain the coded information passed between organisms from one generation to the next.

Growth

Non-living objects, such as a crystal or a stalagmite, grow by the addition of new material to their outside surface. Living organisms, however, grow from within, using food that they obtain from nutrition. The molecules are formed into new living material.

These seven characteristics can be observed to a greater or lesser extent in all living organisms. They are the *observable* outcome of the all-important property of living material, namely the extraction, conversion and use of energy from the environment. In addition, living material is able to maintain and even increase its own energy content. In contrast to this, dead organic matter tends to disintegrate as a result of the chemical and physical forces of the environment. In order to maintain themselves and prevent this disintegration, organisms have an inbuilt *self-regulating* system to ensure that there is no net energy loss. This control is referred to as homeostasis and operates at all levels of biological organisation, from the molecular level to the community level.

The characteristics of life outlined above are dealt with in detail in this book. Many of the chapters extend the explanations in terms of physical and chemical concepts, for it is in these fields that the major research and additions to our knowledge have come in recent years. The study of cell structure, DNA and genetics, protein synthesis, enzymes, hormones, the immune response, and many other aspects of the structure and function of living organisms, all provide some explanation of what is happening within the cells and bodies of organisms.

In the appendices, in Book 2, you will find some basic information required by a biologist, including biochemistry, scientific method, the experimental approach, a glossary of terms and so on. The appendices are designed to supply information to those students which may be lacking in one or more of these areas. With this knowledge the student should strive to develop powers of critical observation and description which are part of the thinking processes underlying scientific enquiry.

Chapter Two

Variety of life

2.1 Classification

2.1.1 Why classify?

If you have ever watched a child playing with coloured sweets or sorting out stamps, football cards or other collectable items, you may have witnessed one of our most basic instincts in operation, the desire to sort out things into groups. This is an act of classification. **Classification** is grouping things together on the basis of features they have in common. The science of classification is called **taxonomy**. Why do we classify? Some biologists suggest that one reason *why* we classify things is because it has survival advantage. If our senses are besieged by an overwhelming number of different stimuli, we can begin to cope and make sense of things by classifying them. Our first classifications may go wrong; for example some small children may call anything with four legs a dog. But gradually we develop a system that enables us to cope with the complexity of the world.

Something like one-and-a-half million different kinds of living organisms have been discovered on this planet, and it has been estimated that there may be 10–100 million kinds. Not surprisingly therefore, there are records of our attempts to classify these organisms as far back as we can trace. The classifications differ according to the uses to which they are put. The ancient Chinese, for example, organised the animal kingdom into a number of groups, some of which may seem odd to us today, such as fabulous ones, stray dogs, those that have broken a flower vase and those that resemble flies at a distance. More obvious classifications might be into poisonous and edible plants or flying and non-flying animals. As we shall see, modern systems of classification often emphasise our ideas of evolutionary relationships between organism.

As we learn more about living organisms, our classifications are modified, but it is important to realise that there is no single perfect classification. They are all designed for our own convenience.

2.1.2 Taxonomy

The science of taxonomy has two branches, the naming of organisms, or **nomenclature**, and the placing of organisms into groups, or **systematics**. The latter is done on the basis of their similarities and differences.

Biological nomenclature is based on the **binomial system** pioneered by the work of the Swedish naturalist Carl Linnaeus (1707–78). In this system each organism has two Latin names: a **generic** name beginning with a capital letter and a **specific** name beginning with a lower case letter. For example, humans are named *Homo sapiens*; the genus is *Homo* and the species is *sapiens*. Italics are used to indicate Latin names. Alternatively, the words can be underlined, e.g. Homo sapiens. You should try to remember to do this when you use the Latin name of an organism. The genus may be abbreviated to one letter, e.g. *H. sapiens*. The Latin name is internationally agreed and avoids the confusion of local variations in common names. For example, in Britain, the plant *Caltha palustris* has at least 90 common names, including marsh marigold, kingcup, golden cup, brave celandine, grandfather's button and butter-flower. The puma, *Felis concolor*, has more than 20 common names. Additional confusion with English names comes when a single common name refers to more than one species. For example, there are more than 100 different plant species known as raspberries.

2.1.3 The taxonomic hierarchy

Linnaeus eventually extended the binomial system to include more groups than just genus and species. These he arranged in a hierarchy with the largest group, the **kingdom**, at the top of the hierarchy. The groups he proposed are still used today and, in descending order of size, are:

kingdom*
phylum – introduced by Haeckel late nineteenth century
class*
order*
family – introduced in Linnaeus' lifetime
genus*
species*

* introduced by Linnaeus

An actual example from the classification of the animal kingdom is shown in fig 2.1. You can see that each group, or taxon, may contain a number of groups (taxa) lower in the hierarchy. For example, the subphylum Vertebrata contains six classes and the genus *Homo* three species, two of which are extinct. Each group possesses features unique to that group. These are described as diagnostic features.

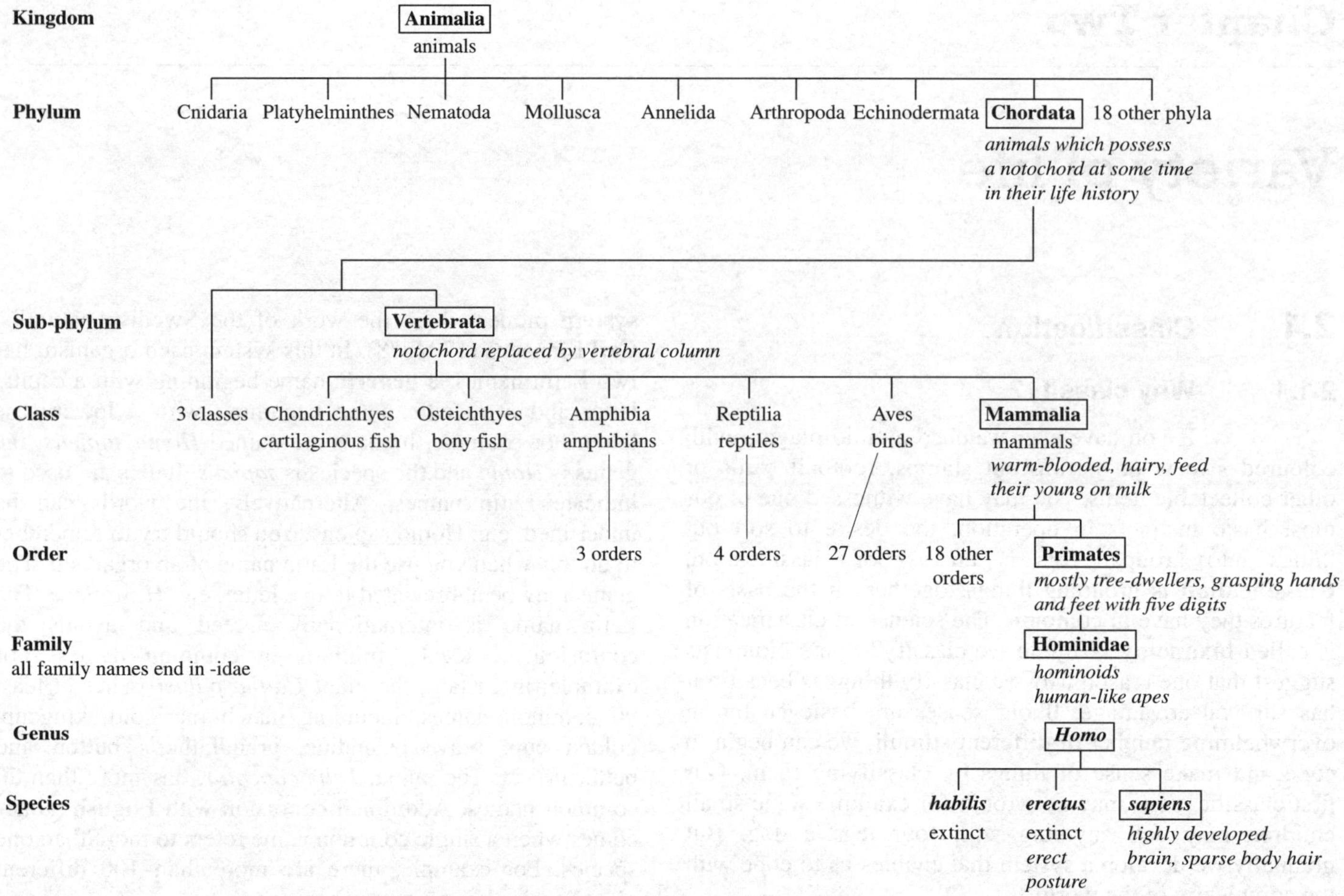

Fig 2.1 *Example of a hierarchy of taxonomic groups. Not all the groups of animals are shown. Note the use of Latin names, although most groups also have common English names.*

For example, only mammals (order Mammalia) possess hair, so hair is a diagnostic feature of mammals. Mammals, however, share with birds, reptiles, amphibians and fish all the diagnostic features of the preceding group in the hierarchy, namely the vertebrates.

Groups may be further subdivided into subgroups, such as subphylum Vertebrata (fig 2.1) or put together into supergroups, such as superclass, if it is convenient to do so. The hierarchies are constructed for our convenience, and they are frequently changed or modified.

2.1.4 Species

The term species has a more precise definition than the other levels in the hierarchy. It can be defined as a **group of closely related organisms which are capable of interbreeding to produce fertile offspring**. Occasionally two organisms which are genetically closely related can interbreed to produce *infertile* offspring. A cross (hybrid) between a donkey and a horse, for example, produces a mule, which is infertile. By definition, therefore, a donkey and a horse should be described as different species.

Mules show some of the advantages of both types of parent (**hybrid vigour**).

There are exceptions to the rule about fertile offspring. For example lions and tigers are considered as different species. If a male tiger mates with a female lion they can have fertile offspring (*tiglons*) although the offspring of female tigers and male lions (*ligers*) are *not* fertile. Normally tigers are forest dwellers and lions are plains dwellers, so they are ecologically isolated. Breeding has only been observed in captivity.

Each species possesses its own distinct structural, behavioural and ecological characteristics (fig 2.2) (see also chapter 27). As we progress up the taxonomic hierarchy, the number of similarities between the members of each group decreases. For example, members of the same genus have more characteristics in common than members of the same family or order.

As we have seen, a precise definition of a species is not really possible. This is not surprising because species can change (evolve) over time. According to the theory of natural selection this process takes place by survival of the fittest, in other words those best adapted to the environment. If the environment changes, then individuals which are better adapted will be selected, and over many generations the species will gradually change. If different populations of the same species become isolated from each

Fig 2.2 (a) Canis familiaris, *the domesticated dog. All breeds of dog are capable of interbreeding and are therefore placed in the same species.*
(b) Canis latrans, *the coyote, a common carnivore and scavenger of North America.*
(c) Canis lupus, *the grey wolf, distributed widely in the northern hemisphere where its range overlaps with the other two species. Coyotes and wolves have been known to mate successfully with dogs, producing fertile offspring. This illustrates the difficulty of deciding what exactly constitutes a species. It is often even more difficult to be precise with larger groups such as genus and order. All canines are placed in the order Carnivora.*

other, for example by ecological or physical barriers such as oceans or mountain ranges, then the different populations may evolve in different ways until they cease to be capable of interbreeding. They become different species.

In some cases there are not necessarily sharp genetic boundaries between one species and another. For example, the herring gull and the lesser black-backed gull are described as different species because they show physical and behavioural differences and do not normally interbreed. However, they occasionally nest in the same place and a few mixed breeding pairs do occur (chapter 27).

2.1.5 Artificial and natural classification

There are two types of classification, artificial and natural. An **artificial classification** is based on one or a few easily observed characteristics, and is usually designed for a practical purpose with an emphasis on convenience and simplicity. The ancient Chinese system already mentioned is an artificial classification. Linnaeus included all worm-like organisms in a single group, the Vermes. This included a wide range of animals, from simple nematode worms and earthworms to snakes. This was an artificial classification because it did not take account of important natural relationships, such as the fact that snakes have backbones and earthworms do not. Snakes have more in common with other vertebrates than with worms. An example of an artificial classification of fish could be to group them as freshwater fish, brackish-water fish and marine fish on the basis of their environment. This would be convenient for the purpose of investigating their mechanisms of osmoregulation. Similarly, all microscopic organisms are known as microorganisms (section 2.2), a convenient group for the purposes of study but not a natural group.

A **natural classification** tries to use natural relationships between organisms. It considers more evidence than artificial classifications, including internal as well as external features. Similarities of embryology, morphology, anatomy, physiology, biochemistry, cell structure and behaviour are all relevant. Most classifications in use today are natural and phylogenetic. A **phylogenetic classification** is one based on evolutionary relationships. In such a system organisms belonging to the same groups are believed to have a common ancestor. The phylogeny (evolutionary history) of a group can be shown by means of a 'family tree', as in fig 2.3.

Another way to classify organisms is to use a **phenetic classification**. This is an attempt to avoid the problem of establishing evolutionary relationships, which can be very difficult and very controversial, especially if there is little or no fossil evidence. The word 'phenetic' comes from the Greek *phainomenon*, 'that which is seen'. This classification is based solely on observable characteristics (phenetic similarity) and all characters used are considered of equal importance. All features of an organism can be considered, the more the better, and they do not necessarily have to be of evolutionary significance. Masses of data are collected and the degrees of similarities between different organisms are calculated, usually by computer because the calculations are extremely complex. The use of computers in taxonomy is known as **numerical taxonomy**. Phenetic classifications often resemble phylogenetic classifications, but they are not constructed with this in view.

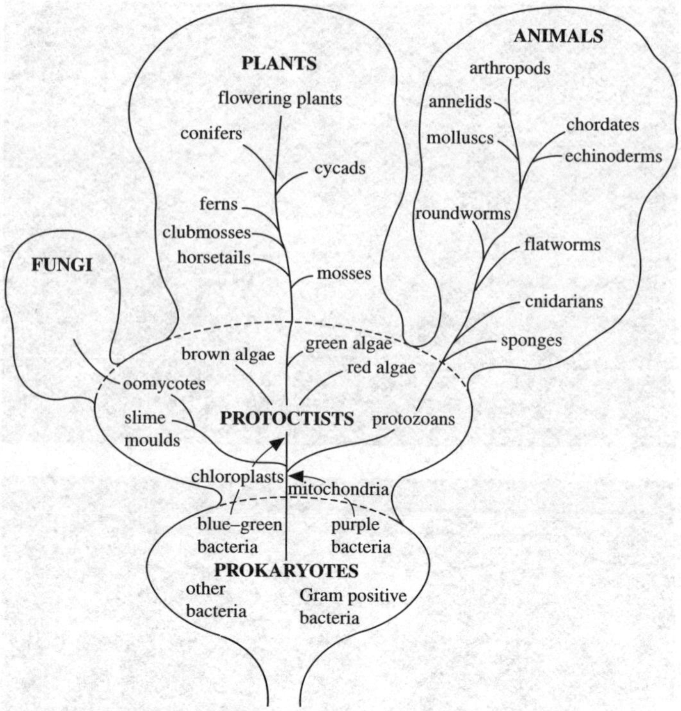

Fig 2.3 *An evolutionary tree of life, including the five kingdoms of Margulis and Schwartz (section 2.2). The lengths of the lines are not related to time.*

2.1.6 Specimen identification and keys

A **key** is a convenient method of enabling a biologist to identify an organism. It involves listing the observable characteristics of the organism and matching them with those features which are diagnostic of a particular group. Most of the characteristics used in identification are based on easily observable features such as shape, colour and numbers of appendages, segments and so on. Hence identification is *artificial* and *phenetic* since it relies purely on the appearance (phenotype) of the organism. Despite this, most diagnostic keys enable organisms to be identified into a group which is part of a natural phylogenetic hierarchical classification system.

There are various types of diagnostic keys, but the simplest is called a **dichotomous key**. This is made up of pairs of statements called **leads**, numbered 1, 2, 3 and so on, where each lead deals with a particular observable characteristic. The paired statements of each lead should be *contrasting* and *mutually exclusive* and, by considering these in order, a large group of organisms may be broken down into progressively smaller groups until the unknown organism is identified.

The characteristics used in keys should be readily observable morphological features, and may be **qualitative**, such as shape of abdomen and colour, or **quantitative**, such as number of hairs and length of stem. Either may be used, but the characteristic must be constant for that species and not subject to variation as a result of environmental influences. In this respect size and colour are often bad examples to use since both can be influenced by the environment, the season, the age or state of the organism at the time of identification. Characteristics chosen should, if possible, exist in two or more states. For example, the characteristic 'stem shape' may exist in one of the two states, 'round' or 'square'.

After each statement there is a number referring to the next lead to be considered, if the statement matches the specimen. For example in the simple key to the cultivated members of the plant family Leguminosae (which includes peas and beans) shown in table 2.1, if the specimen has been keyed as far as lead 5 and it possesses branches without thorns or spines, the next lead to consider is lead 7, and so on.

2.2 Five kingdoms

Until relatively recently it was generally agreed that all organisms should be placed in just two kingdoms, the animal and plant kingdoms. The basic difference between animals and plants was that animals fed on organic material (are **heterotrophic**) whereas plants synthesised their own organic requirements from inorganic compounds (are **autotrophic**). More precisely, a heterotrophic organism is one which has an organic source of carbon, and an autotrophic organism is one which has an inorganic source of carbon, namely carbon dioxide. Animals typically search for their food and so show locomotion. For this they require a nervous system for coordination in the more complex animals, whereas plants are stationary and do not show locomotion or possess a nervous system.

Table 2.1 Extract of key to cultivated Leguminosae.

1	Woody trees and shrubs	2
	Herbaceous and annual plants	15
2	Climbing	3
	Non-climbing	4
3	Flowers bright red	Lobster claw
	Flowers mauve, sometimes white, forming sprays	Wisteria
4	Flowers all or partly yellow	5
	Flowers not yellow	8
5	Branches with thorns and spines	6
	Branches without thorns and spines	7
6	Leaves absent, plant spiny all over	Gorse
	Leaves present on young shoots, spines on older branches	Needlewhin
7	Young stem square, leaves small with three leaflets	Broom
	Stems not square, leaves longer than 2.5 cm	9
8	etc.	

However, this classification ignored the fact that all cellular organisms seem to fall into two natural groups, now known as prokaryotes and eukaryotes. These two groups are fundamentally different. The terms *prokaryote* and *eukaryote* refer to differences in the location of the DNA (the genetic material). In **prokaryotes** the DNA is not enclosed by nuclear membranes and lies free in the cytoplasm. The cells therefore lack true nuclei (*pro*, before; *karyon*, nucleus). The cells of **eukaryotes**, however, do contain true nuclei (*eu*, true). Eukaryotes evolved from prokaryotes.

Classifying all organisms as animals or plants presented other difficulties. For example, fungi are heterotrophic but non-motile, so should they be classified as animals or plants? Such problems have been solved by accepting that there should be more than two kingdoms. In 1982, Margulis and Schwartz proposed a system which used five kingdoms, the Prokaryotae and four eukaryote kingdoms (fig 2.4). This system has been widely accepted and is currently recommended by the Institute of Biology. The eukaryotes

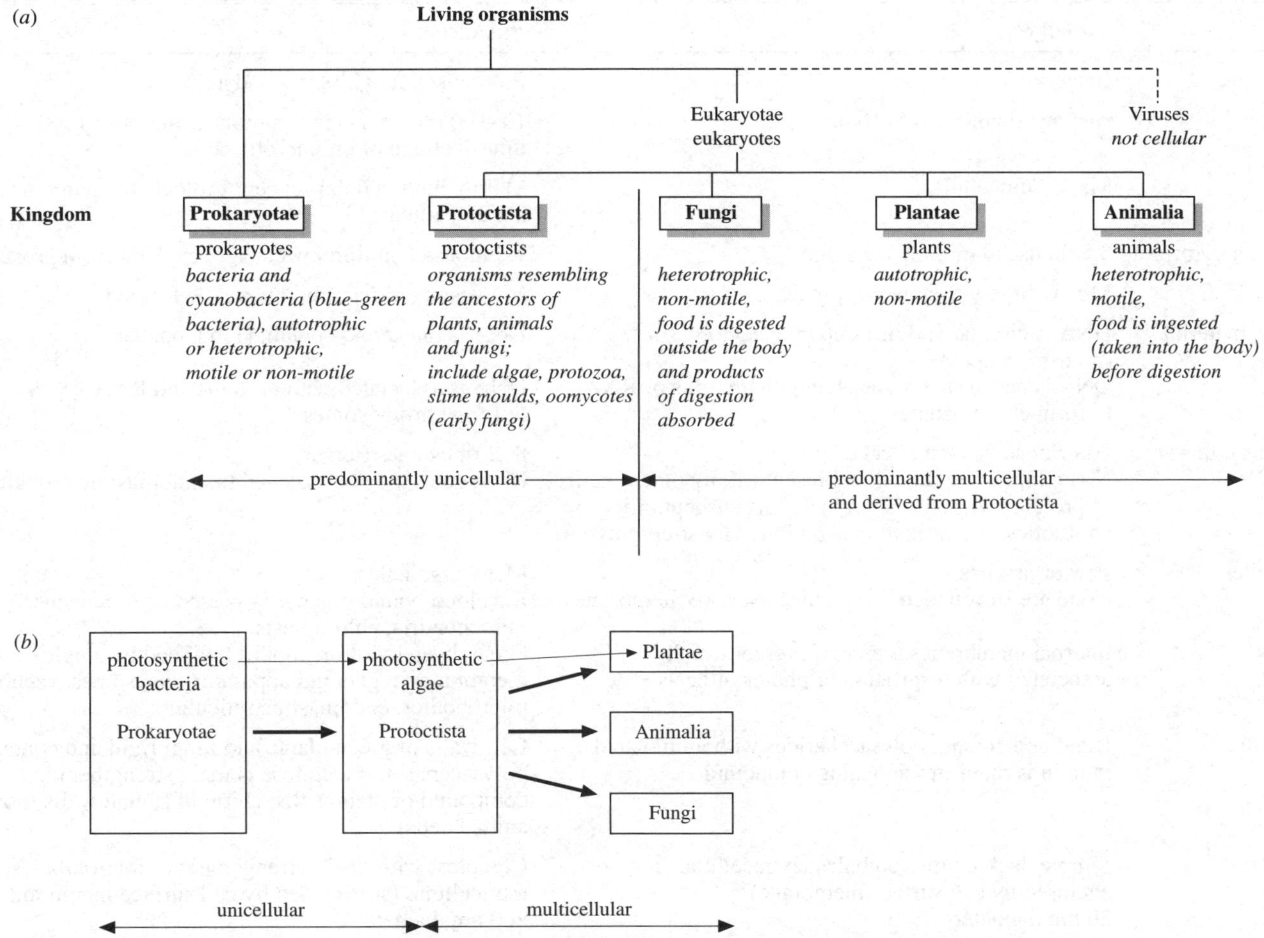

Fig 2.4 (a) *The five kingdom classification of living organisms, according to Margulis and Schwartz. Some of the chief characteristics of the kingdoms are shown. Viruses do not fit neatly into any classification of living organisms because they have a very simple, non-cellular structure and cannot exist independently of other organisms.*
(b) *Evolutionary relationships between the five kingdoms. The diagram shows the trend towards multicellular organisms, the first appearing among the protoctists.*

can be regarded as belonging to a superkingdom, the Eukaryotae. The most controversial group is the protoctists because it is probably an unnatural group. This is discussed later in section 2.6.

One group of 'organisms' that does not fit neatly into any classification scheme is the viruses. Viruses are extremely small particles consisting only of a piece of genetic material (DNA or RNA) in a protective coat of protein. They do not have a cellular structure, unlike all other organisms, and can only reproduce by invading living cells. Their origins are discussed in section 2.4 and they are shown as an extra group in fig 2.4*a*.

Although it is not a natural grouping, the smallest organisms are often collectively known as **microorganisms** or **microbes**. These include the bacteria (prokaryotes), viruses, fungi and protoctists. The grouping is a useful one for practical reasons because the techniques used in their study are often similar. For example, a microscope is needed to see them and sterile (aseptic) techniques are needed for culturing them. The study of microorganisms is a branch of biology known as **microbiology**. Microorganisms have become increasingly important in the areas of biochemistry, genetics, agriculture and medicine, and are the basis of an important section of industry known as **biotechnology**. They will be discussed in chapter 12. Microorganisms, such as bacteria and fungi, are also important ecologically as decomposers (section 10.3.2).

2.3 Prokaryotes

The kingdom Prokaryotae is made up of organisms commonly known as bacteria. They are the most ancient group of organisms, having appeared about 3500 million years ago, and are the smallest organisms with a cellular structure. Their characteristics are summarised in table 2.2. They are mainly single cells, although the blue-green bacteria (Cyanobacteria) may form single rows of cells

Table 2.2 Major differences between prokaryotes and eukaryotes.

Feature	*Prokaryote*	*Eukaryote*
Organisms	Bacteria	Protoctists, fungi, plants and animals
Cell size	Average diameter 0.5–10 μm	10–100 μm diameter common; commonly 1000–10 000 times volume of prokaryotic cells
Form	Mainly unicellular	Mainly multicellular (except Protoctista, many of which are unicellular)
Evolutionary origin	3.5 thousand million years ago	1.2 thousand million years ago, evolved from prokaryotes
Cell division	Mostly binary fission, no spindle	Mitosis, meiosis, or both; spindle formed
Genetic material	DNA is circular and lies free in the cytoplasm (no true nucleus) DNA is naked (not associated with proteins or RNA to form chromosomes)	DNA is linear and contained in a nucleus. DNA is associated with proteins and RNA to form chromosomes
Protein synthesis	70S ribosomes (smaller) No endoplasmic reticulum present (Many other details of protein synthesis differ, including susceptibility to antibiotics, e.g. prokaryotes inhibited by streptomycin)	80S ribosomes (larger) Ribosomes may be attached to endoplasmic reticulum
Organelles	Few organelles None are surrounded by an envelope (two membranes) Internal membranes scarce; if present usually associated with respiration or photosynthesis	Many organelles Envelope-bound organelles present, e.g. nucleus, mitochondria, chloroplasts Great diversity of organelles bounded by single membranes, e.g. Golgi apparatus, lysosomes, vacuoles, microbodies, endoplasmic reticulum
Cell walls	Rigid and contain polysaccharides with amino acids; murein is main strengthening compound	Cell walls of green plants and fungi rigid and contain polysaccharides; cellulose is main strengthening compound of plant walls, chitin of fungal walls (none in animal cells)
Flagella	Simple, lacking microtubules; extracellular (not enclosed by cell surface membrane) 20 nm diameter	Complex, with '9+2' arrangement of microtubules; intracellular (surrounded by cell surface membrane) 200 nm diameter
Respiration	Mesosomes in bacteria, except cytoplasmic membranes in blue-green bacteria	Mitochondria for aerobic respiration
Photosynthesis	No chloroplasts; takes place on membranes which show no stacking	Chloroplasts containing membranes which are usually stacked into lamellae or grana
Nitrogen fixation	Some have the ability	None have the ability

called filaments. Some bacteria stick together in characteristic patterns, forming chains or clusters like bunches of grapes (fig 2.10), but the cells are totally independent of each other. Individual bacterial cells can only be seen with the aid of a microscope, which is why they are known as microorganisms. The study of bacteria is called **bacteriology** and is an important branch of microbiology.

Bacteria range in length from about 0.1 to 10 µm. Their average diameter is about 1 µm, enough room for 200 average-sized globular protein molecules (of 5 nm diameter) to fit across the cell. Such a molecule in solution can diffuse about 60 µm per second; thus no special transport mechanisms are needed for these organisms.

Bacteria occupy many environments, such as soil, dust, water, air, in and on animals and plants. Some are found in hot springs where temperatures may reach 78 °C or higher. Others can survive very low temperatures and periods of freezing in ice. Some have been found in deep cracks in the ocean floor, at very high pressures and temperatures of 360 °C. They form the starting point of unique food chains in these areas of the ocean.

Numbers of bacteria are enormous; one gram of fertile soil is estimated to contain 2.5 thousand million; 1 cm^3 of fresh milk may contain more than 3000 million. Together with fungi their activities are vital to all other organisms because they cause the decay of organic material and the subsequent recycling of nutrients. In addition, they are of increasing importance to humans, not only because some cause disease, but because their very diverse biochemistry can be used in many biotechnological processes. Their importance is discussed further in chapter 12.

2.3.1 Structure

Fig 2.5 shows the structure of a generalised bacterium, a typical prokaryotic cell. Figs 2.6*a–d* show a common rod-shaped bacterium, *Escherichia coli*, which lives in the gut of humans and other vertebrates. It is normally completely harmless. Its presence in water can be used as a very useful indicator of contamination by faeces. *E. coli* has been studied more than any other bacterium and is one of the few organisms whose entire genetic code has been determined. Note how little structure is visible in the cell of *E. coli* compared with a eukaryotic cell (figs 5.5 and 5.6). Fig 2.7 shows another rod-shaped bacterium which, unlike *E. coli*, possesses flagella.

Cell wall

The bacterial cell wall is strong and rigid due to the presence of **murein**, a molecule that consists of parallel

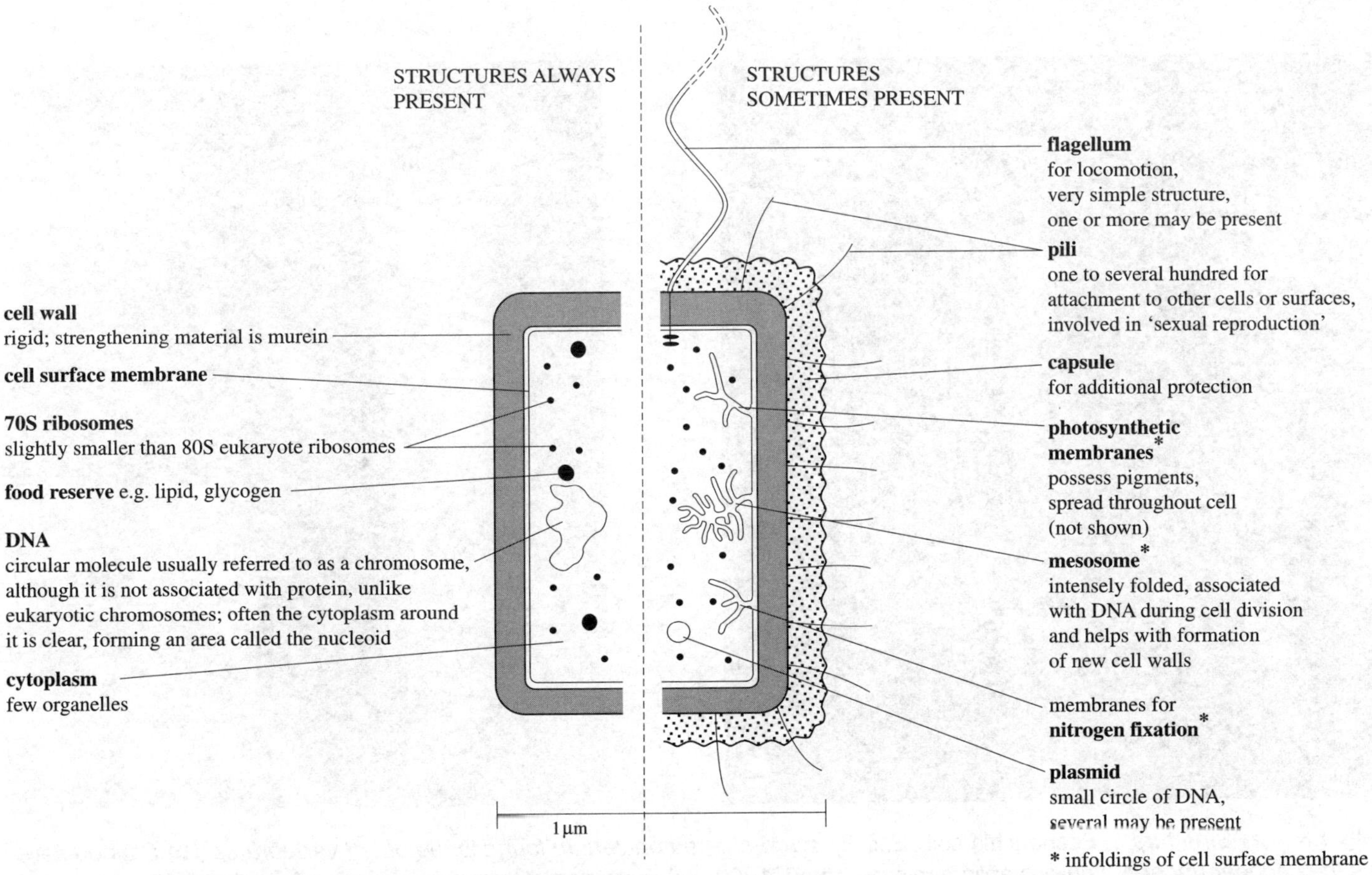

Fig 2.5 *Structure of a generalised rod-shaped bacterium (a typical prokaryote cell). The cell contains little structure compared with a eukaryotic cell.*

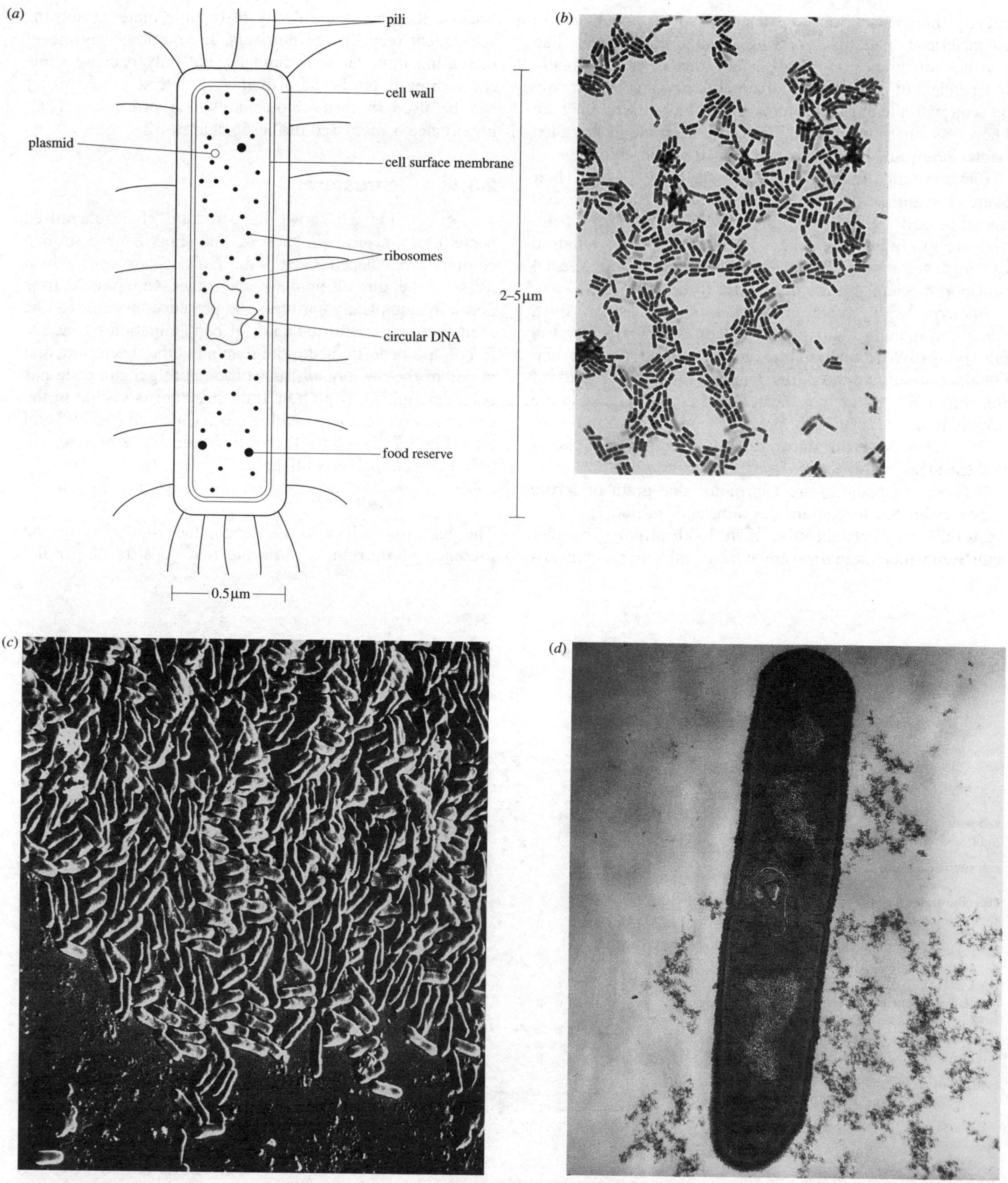

Fig 2.6 (a) *Structure of* Escherichia coli *(*E. coli*), a rod-shaped bacterium found in the gut of vertebrates.* (b) *Stained cells as they appear under a high-power light microscope (× 1000).* (c) *Scanning electron micrograph of a colony of* E. coli. (d) *Transmission electron micrograph of a section of a dividing cell of* E. coli *(× 50 000). This type of division is known as binary fission. The light areas contain DNA. The region containing the DNA is often referred to as the nucleoid.*

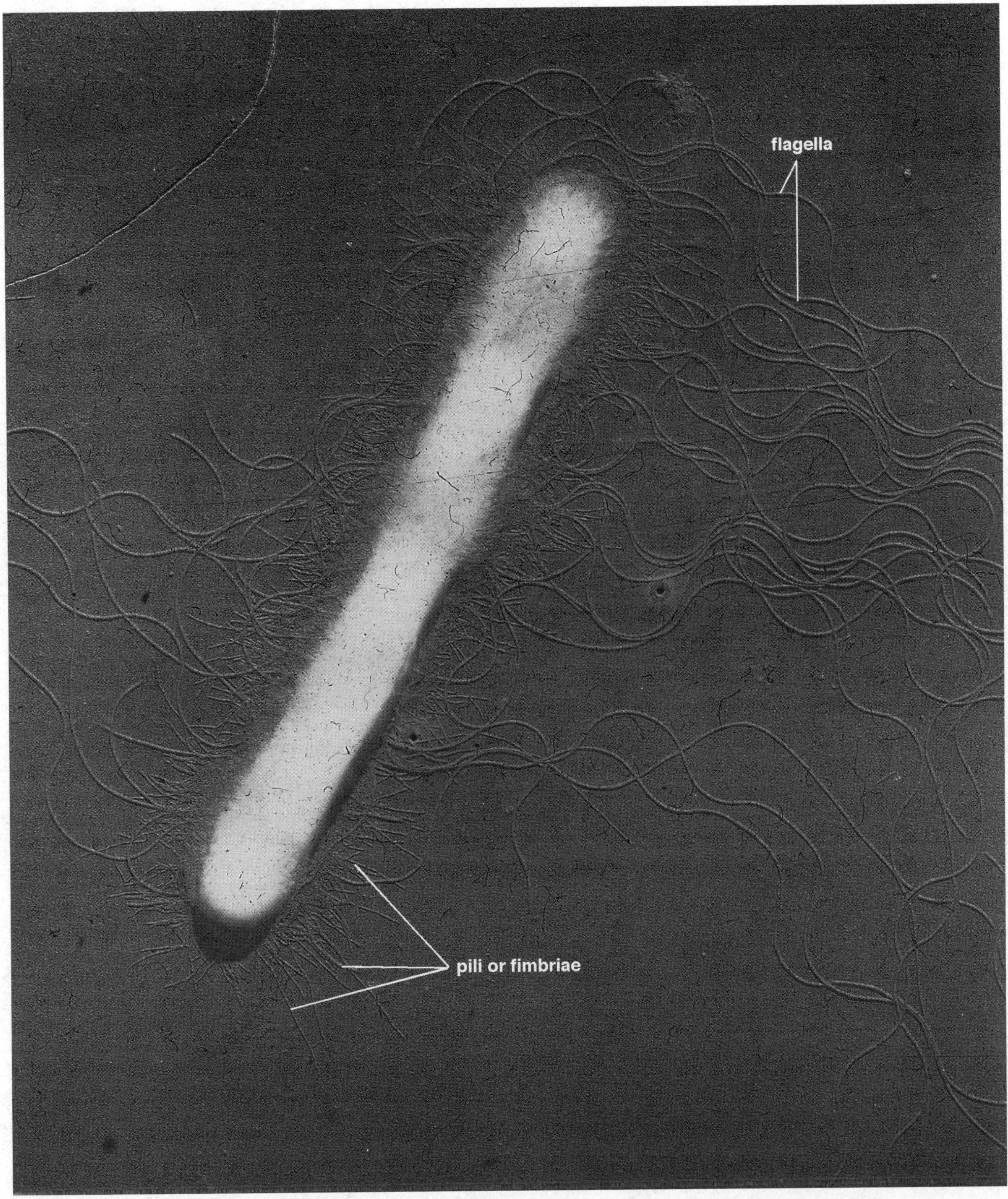

Fig 2.7 *Transmission electron micrograph of a rod-shaped bacterium to show shape, wall, pili and long wavy flagella (× 28 000). The specimen was sprayed with a heavy metal which is opaque to electrons. Sheltered areas remain uncoated, forming an electron-transparent shadow. The photograph is published as a negative to make the shadows black. The technique is known as shadowing and is useful for showing the surface structure of small objects.*

polysaccharide chains cross-linked at regular intervals by short chains of amino acids. Each cell is thus surrounded by a net-like sac which is really one huge molecule. The wall prevents the cell from bursting when it absorbs water (as a result of osmosis). Tiny pores allow the passage of water, ions and small molecules.

In 1884 a Danish biologist, Christian Gram, developed a stain which revealed that bacteria can be divided into two natural groups. We now know that this is due to differences in their wall structure. Some bacteria stain with Gram's stain and are called **Gram positive**, others do not and are called **Gram negative**. A practical exercise involving Gram staining is described in section 12.9.2.

In Gram positive bacteria, such as *Staphylococcus, Bacillus* and *Lactobacillus*, the murein net is filled with other components, mainly polysaccharides and proteins, to form a relatively thick wall. The walls of Gram negative bacteria, such as *Salmonella, E. coli* and *Azotobacter*, are thinner but more complex (fig 2.8). Their murein layer is coated on the outside with a smooth, thin, membrane-like layer of lipids and polysaccharides. This protects them from **lysozyme**, an antibacterial enzyme found in tears, saliva and other body fluids and egg white. Lysozyme digests the polysaccharide backbone of murein. The wall is thus punctured and lysis (osmotic swelling and bursting) of the cell can occur. The same outer layer also gives resistance to penicillin, which attacks Gram positive bacteria by interfering with the cross-linking in the murein of growing cells so making the walls weaker and more likely to burst when water enters by osmosis.

Cell surface membrane, mesosomes and photosynthetic membranes

Like all cells, the living material of bacterial cells is surrounded by a partially permeable membrane. The structure and functions of the cell surface membrane are similar to those in eukaryotic cells (section 5.9). It is also the site of some respiratory enzymes. In addition, in some bacteria it forms mesosomes and/or photosynthetic membranes.

Mesosomes are infoldings of the cell surface membrane (fig 2.5). They appear to be associated with DNA during cell division, organising the separation of the two daughter molecules of DNA after replication and helping in the formation of new cross-walls between the daughter cells.

Among photosynthetic bacteria, sac-like, tubular or sheet-like infoldings of the cell surface membrane contain the photosynthetic pigments, always including bacteriochlorophyll. Similar membranes are associated with nitrogen fixation.

Genetic material (bacterial 'chromosome')

Bacterial DNA is a single circular molecule of about 5 million base pairs and of length 1 mm (much longer than the cell). The total DNA (the genome), and hence the amount of information it contains, is much less than that of a eukaryotic cell: typically it contains several thousand genes, about 500 times fewer than a human cell. (See also table 2.2 and fig 2.5.)

Ribosomes

Ribosomes are the sites of protein synthesis (see table 2.2 and fig 2.5).

Capsules

Capsules are slimy or gummy secretions of certain bacteria which show up clearly after negative staining (when the background, rather than the specimen, is stained). In some cases these secretions unite bacteria into colonies. They also enable bacteria to stick to surfaces such as teeth, mud and rocks, and offer useful additional protection to the bacteria. For example capsulate strains of pneumococci grow in their human hosts causing pneumonia, whereas non-capsulate strains are easily attacked and destroyed by phagocytes, and are therefore harmless.

Spores

Some bacteria, mainly of the genera *Clostridium* and *Bacillus*, form endospores (spores produced inside cells). They are thick-walled, long-lived and extremely resistant, particularly to heat, drought and short-wave radiations. Their position in the cell is variable and is of importance in recognition and classification (see fig 2.10).

Flagella (singular flagellum)

Many bacteria are motile due to the presence of one or more flagella. The flagellum is a simple hollow cylinder of identical protein molecules. It is rigid and wave-shaped (fig 2.7). It propels the cell along by rotating at the base, providing a corkscrew-like motion rather than a beat. Examples of bacteria with flagella are *Rhizobium* (one flagellum) and *Azotobacter* (many flagella), both of which are involved in the nitrogen cycle.

Motile bacteria can move in response to certain stimuli, that is show tactic movements. For example, aerobic bacteria will swim towards oxygen (positive aerotaxis) and motile photosynthetic bacteria will swim towards light (positive phototaxis).

Flagella are most easily seen with the electron microscope using the technique of metal shadowing (fig 2.7).

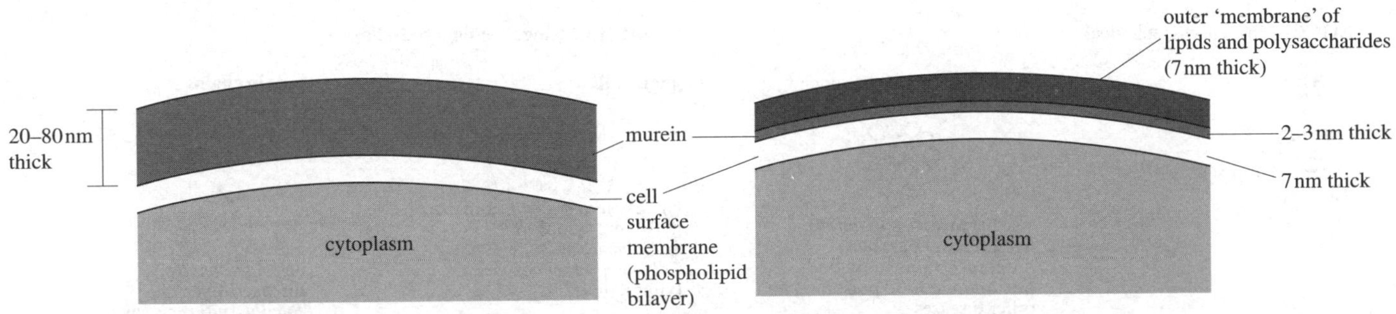

Fig 2.8 *Cell wall structure in Gram positive (left) and Gram negative (right) bacteria. The Gram stain is easily washed out of the thin murein layer of Gram negative bacteria during the decolourising procedure of Gram staining.*

Pili (singular pilus)

Projecting from the walls of some Gram negative bacteria are numerous fine protein rods called **pili** or **fimbriae** (fig 2.7). They are shorter and thinner than flagella and are concerned with attachment to specific cells or surfaces. Various types occur, but of particular interest is the **F pilus**. This is involved in sexual reproduction (section 2.3.3).

Plasmids

In addition to the single circular DNA molecule found in all bacteria, some species also contain one or more plasmids (fig 2.9). A plasmid is a small, self-replicating circle of extra DNA. It possesses only a few genes, which generally give extra survival advantage. Some confer resistance to antibiotics. For example, some staphylococci contain a plasmid which includes a gene for the enzyme penicillinase. This breaks down penicillin, thus making the bacteria resistant to penicillin. The spread of such genes by conjugation (see reproduction) has important implications in medicine. Other plasmid genes are known which:

- confer resistance to disinfectants
- cause disease
- are responsible for the fermentation of milk to cheese by lactic acid bacteria
- confer ability to use complex chemicals as food, such as hydrocarbons, with potential applications in clearing oil spills and producing protein from petroleum.

2.3.2 Cell shape

Bacterial shape is an important aid to classification. The four main shapes found are illustrated in fig 2.10. Examples of both useful and harmful bacteria are given.

2.3.3 Reproduction

Growth of individuals and asexual reproduction

Bacteria have a large surface area to volume ratio and can therefore gain food sufficiently rapidly from their environment by diffusion and active transport mechanisms. Therefore, providing conditions are suitable, they can grow very rapidly. Important environmental factors affecting

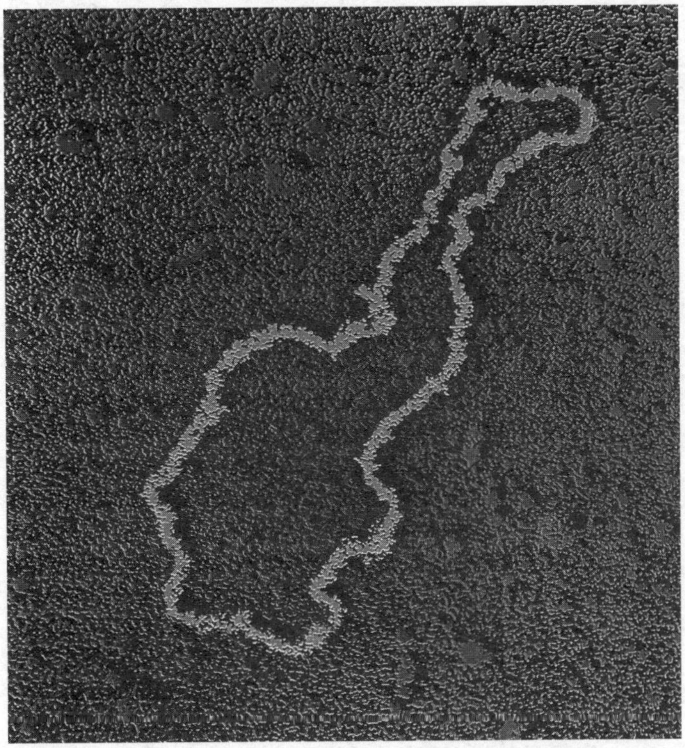

Fig 2.9 *Transmission electron micrograph of a plasmid of bacterial DNA.*

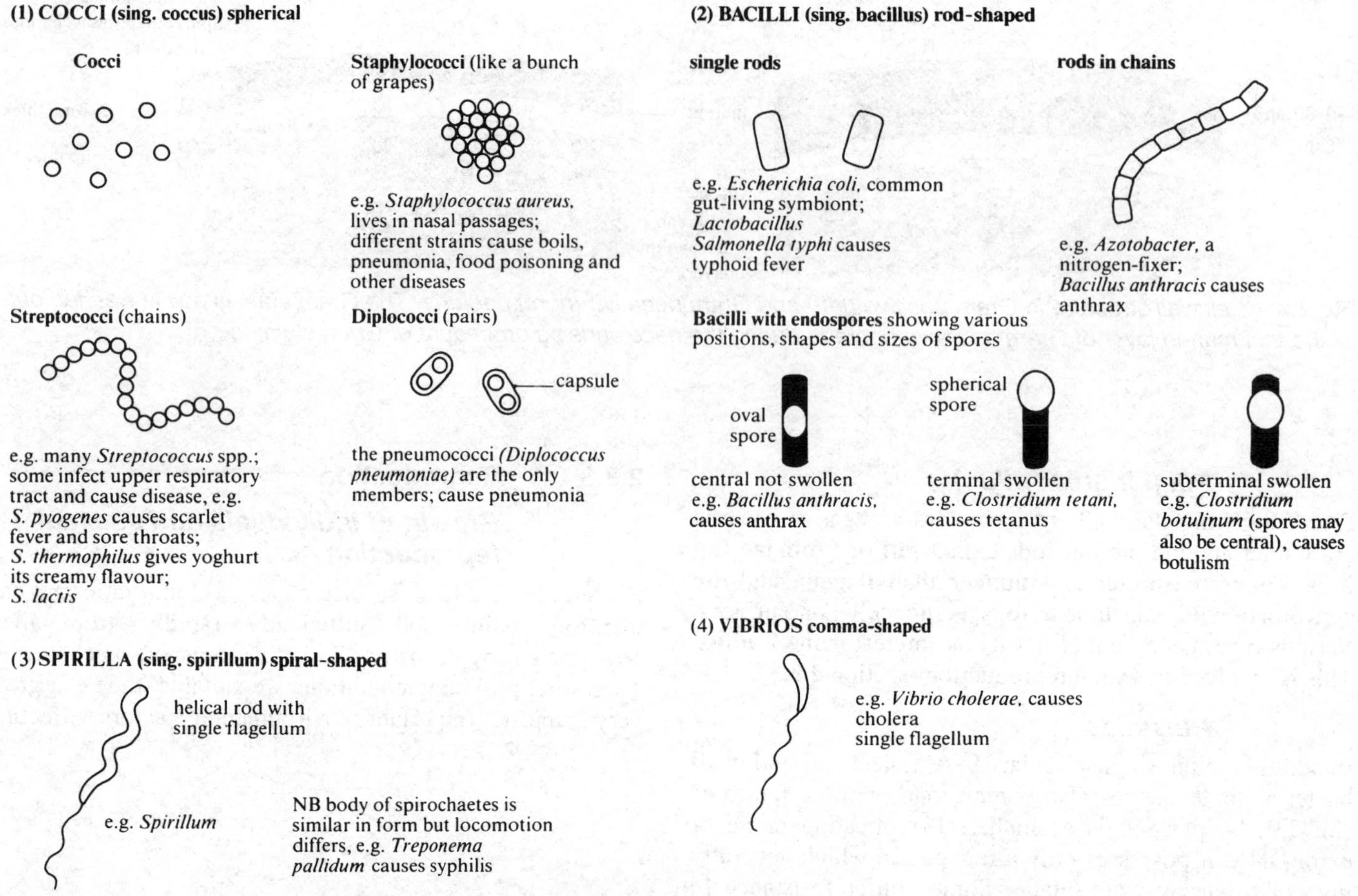

Fig 2.10 *Forms of bacteria, illustrated by some common useful and harmful types.*

growth are temperature, nutrient availability, pH and ionic concentrations. Oxygen must also be present for obligate aerobes and absent for obligate anaerobes.

On reaching a certain size, dictated by the nucleus to cytoplasm ratio, bacteria reproduce asexually by binary fission, that is by division into two identical daughter cells (fig 2.11). Cell division is preceded by replication of the DNA and while this is being copied it may be held in position by a mesosome (figs 2.5 and 2.6*c*). The mesosome may also be attached to the new cross-walls that are laid down between the daughter cells, and plays some role in the synthesis of cell wall material. In the fastest growing bacteria such divisions may occur as often as every 20 minutes.

Sexual reproduction

In 1946 it was discovered that bacteria can take part in a primitive form of sexual reproduction. Gametes are not involved, but the essential feature of sexual reproduction, namely the exchange of genetic material, does take place and is called **genetic recombination**. The process was discovered using *E. coli* as follows. Normally *E. coli* can make all of its own amino acids, given a supply of glucose and mineral salts. Random mutations were induced by exposure to radiation and two particular mutants selected. One could not make biotin (a vitamin) or the amino acid methionine. Another could not make the amino acids threonine and leucine. About 10^8 cells of each mutant were mixed and cultured on media lacking all four growth factors. Theoretically, none of the cells should have grown, but a few hundred colonies developed, each from one original bacterium, and these were shown to possess genes for making all four growth factors. Exchange of genetic information had therefore occurred, but no chemical responsible could be isolated. Eventually it was shown with the electron microscope that direct cell-to-cell contact, that is **conjugation**, can occur in *E. coli* (fig 2.12).

Conjugation therefore involves transfer of DNA between cells in direct contact. One cell acts as the donor ('male') and the other as the recipient ('female'). The ability to serve as a donor is determined by genes in a special type of plasmid called the sex factor, or **F factor** (F for fertility). This codes for the protein of a special type of pilus, the **F pilus** or sex pilus. This enables cells to come into contact. The pilus is hollow and it is believed that the DNA passes through the pilus from the donor (F^+) to the recipient (F^-). The process is described in fig 2.13.

Note that the donor retains the F factor and the recipient also becomes F^+. The process is slow, so the F^- cell can

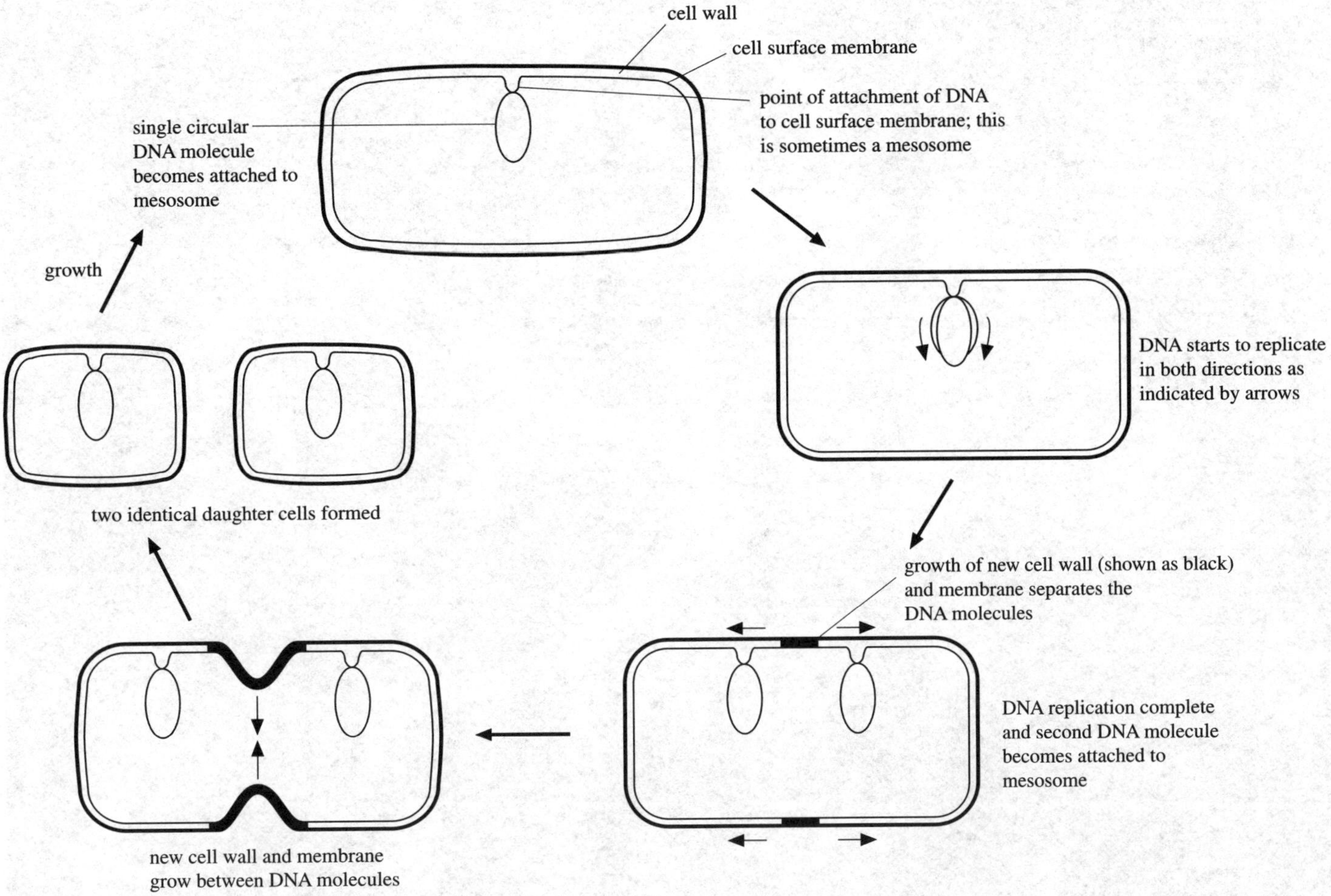

Fig 2.11 *Binary fission in a bacterium, e.g.* E. coli.

replicate by binary fission one or several times before the process is complete, thus maintaining F^- cells in the population.

The F factor is particularly important because in a few cases, about 1 in 100 000, it becomes integrated with the rest of the DNA in the host cell. In such cases, the process of conjugation involves transfer of not only the F factor, but also the rest of the DNA. This takes about 90 min and separation may occur before exchange is complete. Such strains consistently donate all or large portions of their DNA and are called **Hfr strain**s (H = high, f = frequency, r = recombination), because the donor DNA can recombine with the recipient DNA.

2.3.4 Nutrition

Nutrition is the process of acquiring energy and materials. Living organisms can be grouped on the basis of their source of energy or source of carbon, the latter being the most fundamental material required for growth. Only two forms of energy can be used by living organisms to synthesise their organic requirements, namely light and chemical energy. Those that use light are known as **phototrophs** and those that use chemical energy are called **chemotrophs**. Phototrophs carry out photosynthesis.

As already noted, organisms can also be described as autotrophic or heterotrophic, depending on whether their source of carbon is inorganic (carbon dioxide) or organic respectively. Thus four nutritional categories can occur, as shown in table 2.3. There are examples of bacteria in all four categories. The largest group is the chemoheterotrophic bacteria.

Chemoheterotrophic bacteria

These bacteria obtain energy from chemicals in their food. They use an enormous range of chemicals. There are three main groups, namely saprotrophs, mutualists and parasites.

A **saprotroph** is an organism that obtains its food from dead and decaying matter. The saprotroph secretes enzymes onto the organic matter to digest it. Thus digestion is outside of the organism. Soluble products of digestion are absorbed and assimilated within the body of the saprotroph.

Saprotrophic bacteria and fungi constitute the **decomposers** and are essential in bringing about decay and recycling of nutrients. They produce humus from animal and plant remains, but also cause decay of materials useful to humans, especially food. Their importance in the biosphere is stressed in chapter 10.

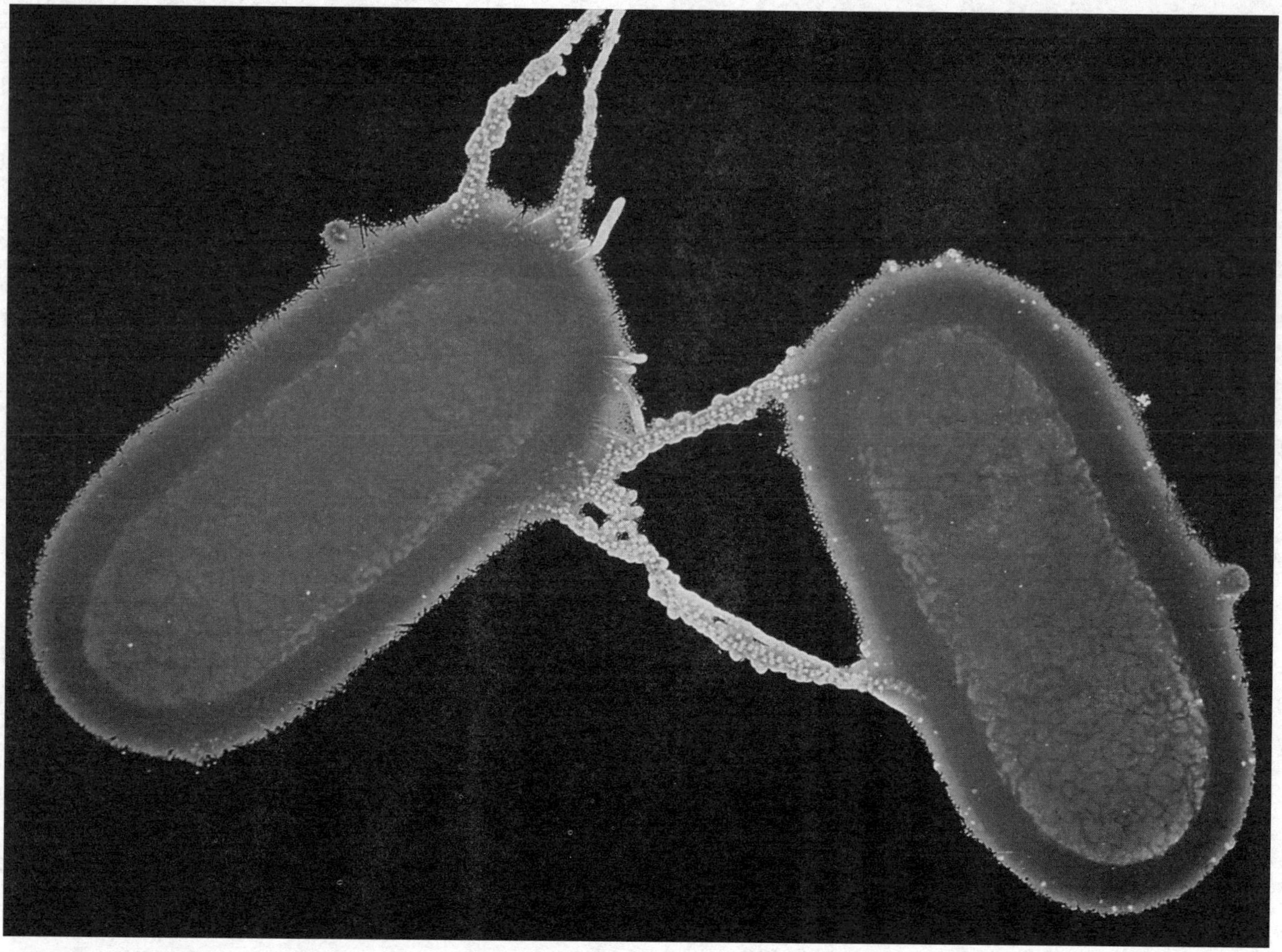

Fig 2.12 *Conjugating bacteria, one 'male' (left) with two 'females' (× 19 475). The second 'female' cell is beyond the top of the photograph.*

Mutualism (or symbiosis) is the name given to any form of close relationship between two living organisms in which both partners benefit. Examples of bacterial mutualists are *Rhizobium*, a nitrogen-fixer living in the root nodules of legumes such as pea and clover, and *Escherichia coli*, which inhabits the gut of humans and probably contributes vitamins of the B and K groups.

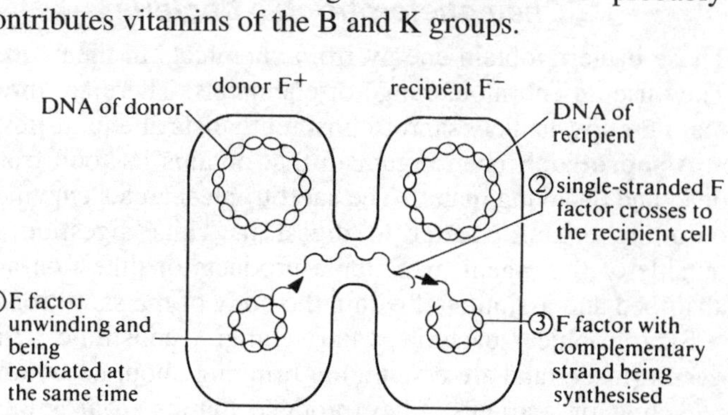

Fig 2.13 *Conjugation between two bacterial cells. 1, 2 and 3 represent successive stages in transfer of the F factor.*

A **parasite** is an organism that lives in or on another organism, the **host**, from which it obtains its food and, usually, shelter. The host is usually of a different species and suffers harm from the parasite. Parasites which cause disease are called **pathogens**. Some examples are given in fig 2.10 and in chapter 15. Some parasites can only survive and grow in living cells and are called **obligate parasites**. Others can infect a host, bring about its death and then live saprotrophically on the remains; these are called **facultative parasites**. It is a characteristic of parasites that they are very exacting in their nutritional requirements, needing 'accessory growth factors' that they cannot manufacture for themselves but can only find in other living cells.

Photoautotrophic bacteria

Cyanobacteria, or blue-green bacteria, are examples of photoautotrophic bacteria. Algae and plants are also photoautotrophic. They all carry out photosynthesis and use carbon dioxide as a source of carbon (table 2.3). The process of photosynthesis first evolved in bacteria, possibly in blue-green bacteria. As we shall see later, the

Table 2.3 The four nutritional categories of living organisms, according to sources of energy and carbon. Examples are given of bacteria in each category. Plants are photoautotrophic. Fungi and animals are chemoheterotrophic.

		CARBON SOURCE	
		autotrophic source of carbon is inorganic (carbon dioxide)	**heterotrophic** source of carbon is organic
ENERGY SOURCE	**phototrophic** (photosynthetic) light energy used	**photoautotrophic** e.g. blue-green bacteria	**photoheterotrophic** e.g. purple non-sulphur bacteria
	chemotrophic (chemosynthetic) chemical energy used	**chemoautotrophic** e.g. *Nitrosomonas* and *Nitrobacter*, nitrifying bacteria involved in the nitrogen cycle	**chemoheterotrophic** most bacteria – all the saprotrophs, parasites and mutualists (symbionts)

chloroplasts of algae and plants are thought to be descendants of what were once free-living photosynthetic bacteria that invaded heterotrophic cells (section 2.6.1).

Blue-green bacteria are common in surface layers of both fresh water and sea water and are also found as gelatinous mat-like growths on shaded soil, rocks, mud, wood and some living organisms. Most blue-green bacteria are single cells but some are linked to form filaments sheathed in mucus, e.g. *Anabaena* and *Spirulina*. They differ from most bacteria, and resemble algae and plants, in producing oxygen from water during photosynthesis. Fig 2.14 shows the structure of *Anabaena*, a typical blue-green bacterium. Photosynthetic membranes characteristically run throughout the cytoplasm and it is here that the photosynthetic pigments are located. The pigments include chlorophyll *a*, again resembling algae and plants, as well as a characteristic blue-green pigment called phycocyanin. The cells of blue-green bacteria tend to be larger than those of other bacteria. The fact that blue-green bacteria produce oxygen in photosynthesis, have photosynthetic membranes running through the cell and contain chlorophyll *a*, indicate that they may be evolutionary links between the rest of the bacteria and eukaryotes.

Some blue-green bacteria, such as *Anabaena*, have the ability to **fix** nitrogen, that is to convert nitrogen gas from the air to ammonia which can then be used in synthesis of amino acids, proteins and other nitrogen-containing organic compounds. This is done in special cells called **heterocysts** which develop when there is a nitrogen shortage. These cells export the nitrogen compounds to neighbouring cells in exchange for other nutrients such as carbohydrate.

Chemoautotrophic bacteria

These are more commonly known as **chemosynthetic** bacteria. They use carbon dioxide as a source of carbon but obtain their energy from chemical reactions. The energy is obtained by oxidising inorganic materials such as ammonia and nitrite. Some are important members of the nitrogen cycle, carrying out a process called **nitrification.** This takes place in two stages. Firstly ammonia is oxidised to nitrite with a release of energy. This is carried out, for example, by *Nitrosomonas*. Secondly nitrite is oxidised to nitrate with the release of more energy. This is done, for example, by *Nitrobacter.*

(1) $$NH_4^+ \xrightarrow{\text{oxygen}} NO_2^- + \text{energy}$$

(2) $$NO_2^- \xrightarrow{\text{oxygen}} NO_3^- + \text{energy}$$

The importance of nitrification is discussed with the nitrogen cycle in section 10.4.1.

2.3.5 Population growth in bacteria

2.1 Consider the situation where a single bacterium is placed in a nutrient medium under optimal growth conditions. Assuming it, and its descendants, divide every 20 min, copy table 2.4 and complete it.

Using the data from your table, draw graphs of number of bacteria (graph A) and $\log_{10}$ number of bacteria (graph B) on the vertical axes against time (horizontal axis). What do you notice about the shapes of the graphs?

The kind of growth shown in table 2.4 is known as **logarithmic**, **exponential** or **geometric**. The numbers form an **exponential series**. This can be explained by reference to line C in table 2.4 where the number of bacteria is expressed as a power of 2. The power can be called the logarithm or exponent of 2. The logarithms or exponents form a linearly increasing series 0, 1, 2, 3, etc., corresponding with the number of generations.

Returning to table 2.4, the numbers in line A could be converted to logarithms to the base 2 as follows:

A Number of bacteria	*1*	*2*	*4*	*8*	*16*	*32*	*64*	*128*	*256*	*512*	*1024*
D Log_2 number of bacteria	0	1	2	3	4	5	6	7	8	9	10

Compare line C with line D. However, it is conventional to use logarithms to the base 10, as in line B. Thus 1 is 10^0, 2 is $10^{0.3}$, 4 is $10^{0.6}$, etc.

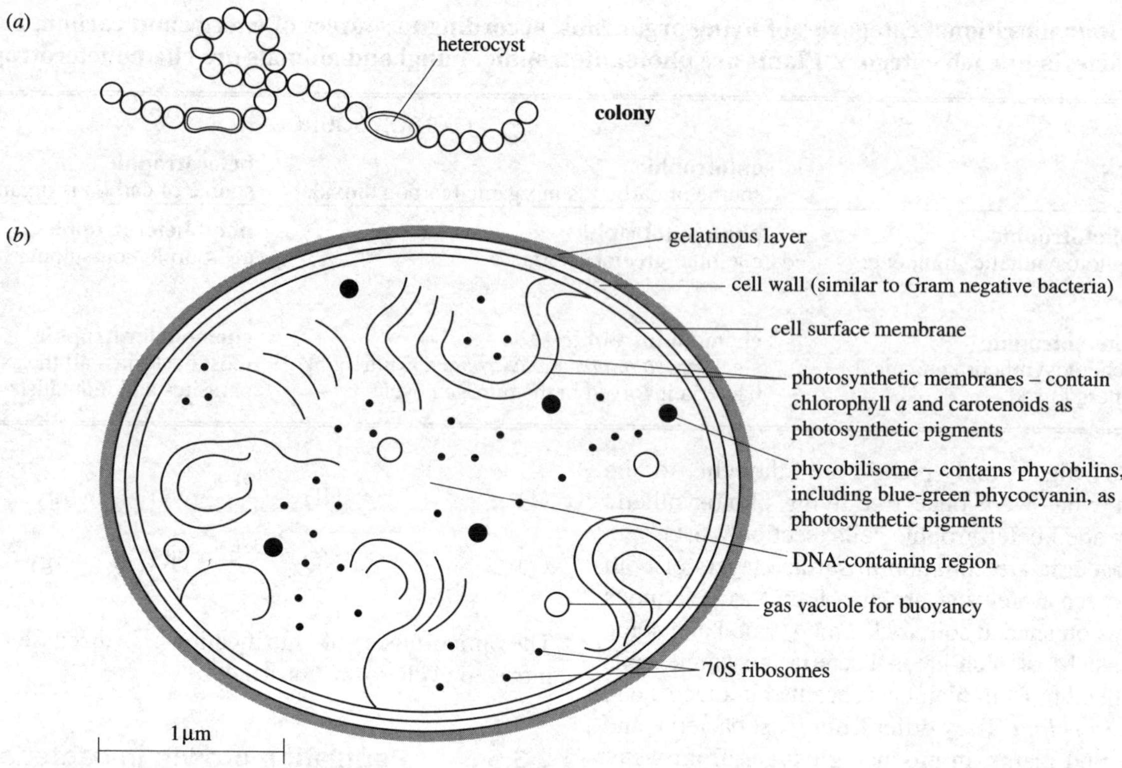

Fig 2.14 *Structure of a blue-green bacterium,* Anabaena. *Cells may occur singly* (b), *or in colonies* (a).

Table 2.4 Growth of a model population of bacteria.

Time (in units of 20 min)	*0*	*1*	*2*	*3*	*4*	*5*	*6*	*7*	*8*	*9*	*10*
A Number of bacteria											
B Log_{10} number of bacteria (to one decimal place)											
C Number of bacteria expressed as power of 2											

The curve in graph A is known as a **logarithmic** or **exponential curve**. Such growth curves can be converted to straight lines by plotting the logarithms of growth against time. Under ideal conditions, then, bacterial growth is theoretically exponential. This mathematical model of bacterial growth can be compared with the growth of a real population. Fig 2.15 shows such growth. The growth curve shows four distinct phases.

- During the **lag phase** the bacteria are adapting to their new environment and growth has not yet achieved its maximum rate. The bacteria may, for example, be synthesising new enzymes to digest the particular spectrum of nutrients available in the new medium.
- The **log phase** is the phase when growth is proceeding at its maximum rate, closely approaching a logarithmic increase in numbers when the growth curve would be a straight line.
- Eventually growth of the colony begins to slow down and it starts to enter the **stationary phase** where growth rate is zero, and there is much greater competition for resources. Rate of production of new cells is slower and may cease altogether. Any increase in the number of cells is offset by the death of other cells, so that the number of living cells remains constant. This phase is a result of several factors, including exhaustion of essential nutrients, accumulation of toxic waste products of metabolism and possibly, if the bacteria are aerobic, depletion of oxygen.

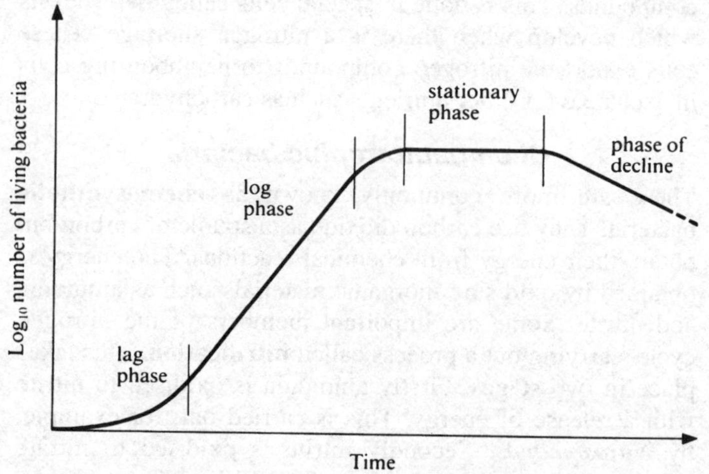

Fig 2.15 *Typical growth curve of a bacterial population.*

Table 2.5 Culture of bacteria at 30 °C.

	Number of cells in millions	
Time/h	*living*	*living and dead*
0	9	10
1	10	11
2	11	12
5	18	20
10	400	450
12	550	620
15	550	700
20	550	850
30	550	950
35	225	950
45	30	950

- During the final phase, the **phase of decline**, the death rate increases and cells stop multiplying. Methods of counting bacteria are described in the practical work in chapter 12.

2.4 Viruses

2.4.1 Discovery

In 1852, the Russian botanist D. I. Ivanovsky prepared an infectious extract from tobacco plants that were suffering from mosaic disease. When the extract was passed through a filter able to prevent the passage of bacteria, the filtered fluid was still infectious. In 1898 the Dutchman Beijerink coined the name 'virus' (Latin for poison) to describe the infectious nature of certain filtered plant fluids. Although progress was made in isolating highly purified samples of viruses and in identifying them chemically as nucleoproteins (nucleic acids combined with proteins), the particles still proved elusive and mysterious because they were too small to be seen with the light microscope. As a result, they were among the first biological structures to be studied when the electron microscope was developed in the 1930s.

2.4.2 Characteristics

Viruses have the following characteristics.

- They are the smallest living organisms.
- They do not have a cellular structure.
- They can only reproduce by invading living cells. Therefore they are all parasitic. They are **obligate endoparasites**, meaning that they can only live parasitically inside other cells. Most cause disease.
- They have a simple structure, consisting of a small piece of nucleic acid, either DNA or RNA, surrounded by a protein or lipoprotein coat.

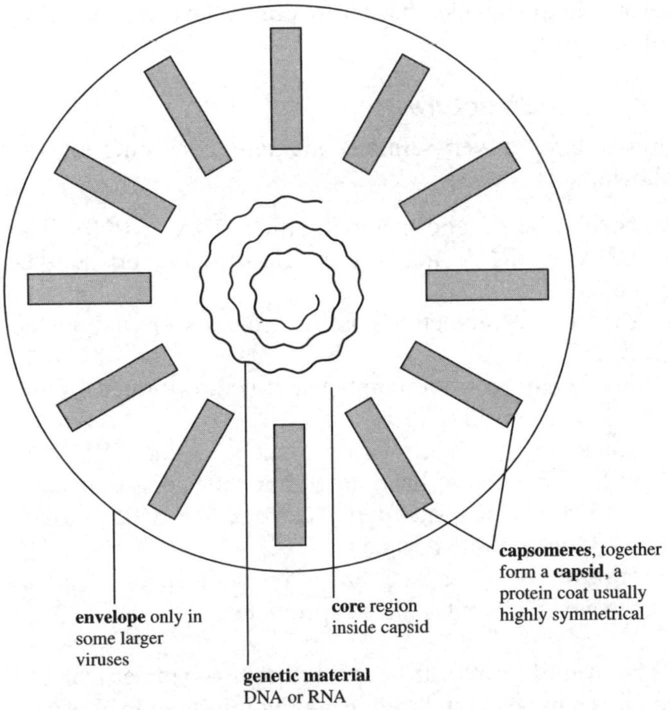

Fig 2.16 *A generalised virus.*

- They are on the boundary between what we regard as living and non-living.
- Each type of virus will recognise and infect only certain types of cell. In other words, viruses are highly specific to their hosts.

These characteristics will now be examined in more detail.

Size

Viruses are the smallest living organisms, ranging in size from about 20–300 nm; on average they are about 50 times smaller than bacteria. They cannot be seen with the light microscope and they pass through filters which retain bacteria.

Origin

The question is often posed, 'Are viruses living?'. If, to be defined as living, a structure must possess genetic material (DNA or RNA), and be capable of reproducing itself, then the answer must be that viruses are living. If to be living demands a cellular structure then the answer is that they are not. It should also be noted that viruses are not capable of reproducing outside the host cell.

We can understand viruses much better if we understand their evolutionary origins. It is suspected, though not proven, that viruses are pieces of genetic material that have 'escaped' from prokaryote and eukaryote cells and have the potential to replicate themselves when they get back into a cell environment. A virus survives in a purely inert state outside cells, but has the set of instructions (genetic code) necessary to re-enter a particular type of cell and instruct it to make many identical copies of itself. It is therefore reasonable to suppose that viruses must have evolved after cells evolved.

Structure

Viruses have a very simple structure consisting of the following:

- **core** – the genetic material, either DNA or RNA. The DNA or RNA may be single-stranded or double-stranded.
- **capsid** – a protective coat of protein surrounding the core.
- **nucleocapsid** – the combined structure formed by the core and capsid.
- **envelope** – a few viruses, such as the HIV and influenza viruses, have an additional lipoprotein layer around the capsid derived from the cell surface membrane of the host cell.
- **capsomeres** – capsids are often built up of identical repeating subunits called capsomeres.

The overall form of the capsid is highly symmetrical and the virus can be crystallised, enabling information about its structure to be obtained by X-ray crystallography as well as electron microscopy. Once the subunits of a virus have been made by the host, they can self-assemble into a virus. Fig 2.16 shows a simplified, generalised structure of a virus.

Certain types of symmetry are common among capsids, notably polyhedral and helical symmetry. A polyhedron is a many-sided figure. The most common polyhedral form in viruses is the icosahedron, which has 20 triangular faces with 12 corners and 30 edges. Fig 2.17*a* shows a regular icosahedron, and fig 2.17*b* the herpes virus, which has 162 capsomeres arranged into an icosahedron.

Helical symmetry is well illustrated by the tobacco mosaic virus (TMV), an RNA virus (fig 2.18*a* and *b*). Here the capsid is made up of 2130 identical protein capsomeres. TMV was the first virus to be isolated in a pure state. It causes a mottled yellowing of tobacco leaves called leaf mosaic (fig 2.18*c*). The virus can spread extremely rapidly, either mechanically if infected plants, or plant parts, come into contact with healthy plants, or even as airborne particles such as the smoke of cigarettes made from contaminated leaves.

Viruses that attack bacteria form a group called **bacteriophages**, or simply phages. Some of these have a distinct icosahedral head, with a tail showing helical symmetry (fig 2.19). Fig 2.20 shows simplified diagrams of some viruses and their relative sizes, and summarises their structures.

2.4.3 Life cycle of a bacteriophage

The life cycle of typical bacteriophage is shown in fig 2.21. *E. coli* is a typical host and can be attacked by at least seven strains of phage, known as T1 to T7. A T-even phage (for instance T2) is illustrated in figs 2.19*a* and *b* and 2.22.

(*a*)

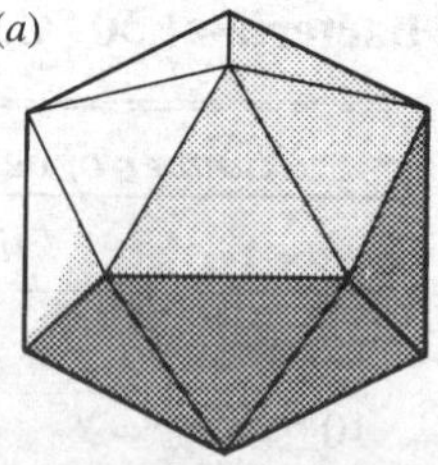

(*b*)

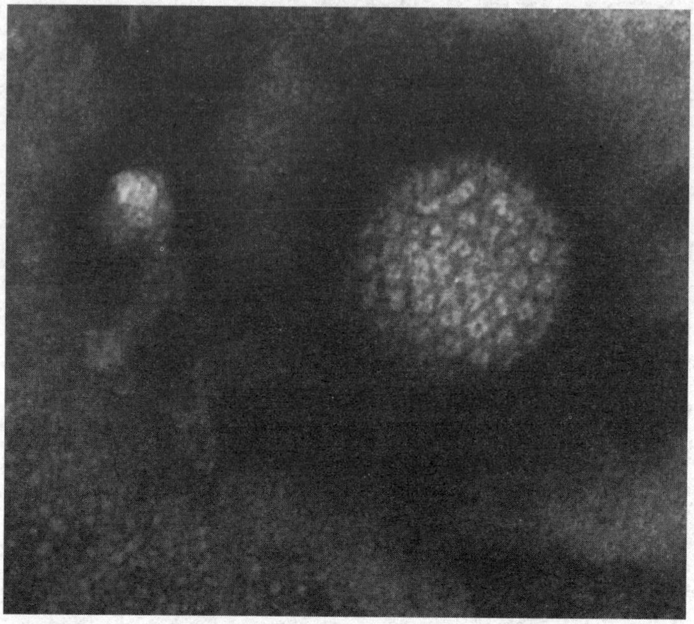

Fig 2.17 *(a) Solid model of an icosahedron. (b) Electron micrograph of a negatively stained herpes simplex virus. Negative staining stains the background, not the specimen. Note how the detailed structure of the virus is revealed. Individual capsomeres are just visible where the stain has penetrated between them.*

The life cycle is the same in principle for all phages. Some complete the life cycle without a break. Such life cycles are called **lytic** cycles. However, some phages, such as lamda phage, insert their DNA into the host DNA and remain dormant for many generations. Each time the host cell divides the phage DNA is copied with the host cell DNA. This dormant stage of the phage is called the **prophage**. Eventually it is activated again and completes its life cycle, causing death of the host cell in the usual way. Such phages are described as **lysogenic**.

2.4.4 Viruses as agents of disease

Viruses can also infect eukaryotic cells and, as in prokaryotic cells, each has its own specific host. For example, TMV will attack only tobacco plants. Between them, viruses cause a wide range of diseases among plants, animals and fungi. Diseases of humans caused by viruses include measles, German measles (rubella), chickenpox, influenza, herpes and AIDS.

Viruses cause many different diseases in almost every other kind of organism.

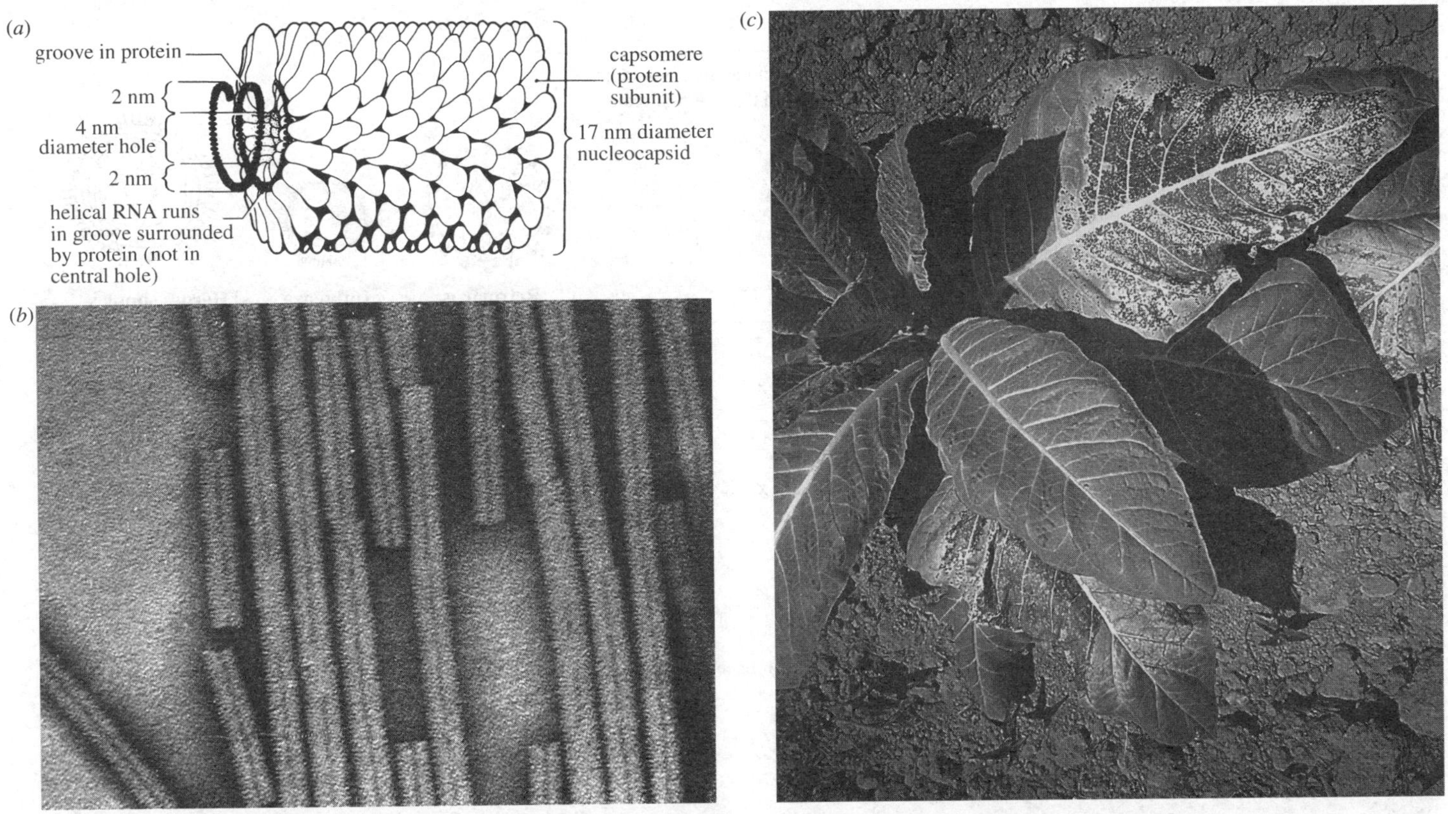

Fig 2.18 (a) *Structure of tobacco mosaic virus (TMV) showing helical symmetry of the capsid. Only part of the rod-shaped virus is shown. The drawing is based on X-ray diffraction, biochemical and electron microscope data.*
(b) *Electron micrograph of a negatively stained tobacco mosaic virus (× 800 000). The capsid (coat) is made of 2130 identical protein capsomeres.* (c) *Tobacco plant infected with TMV. Note the characteristic mosaic pattern on the leaves where tissue is dying.*

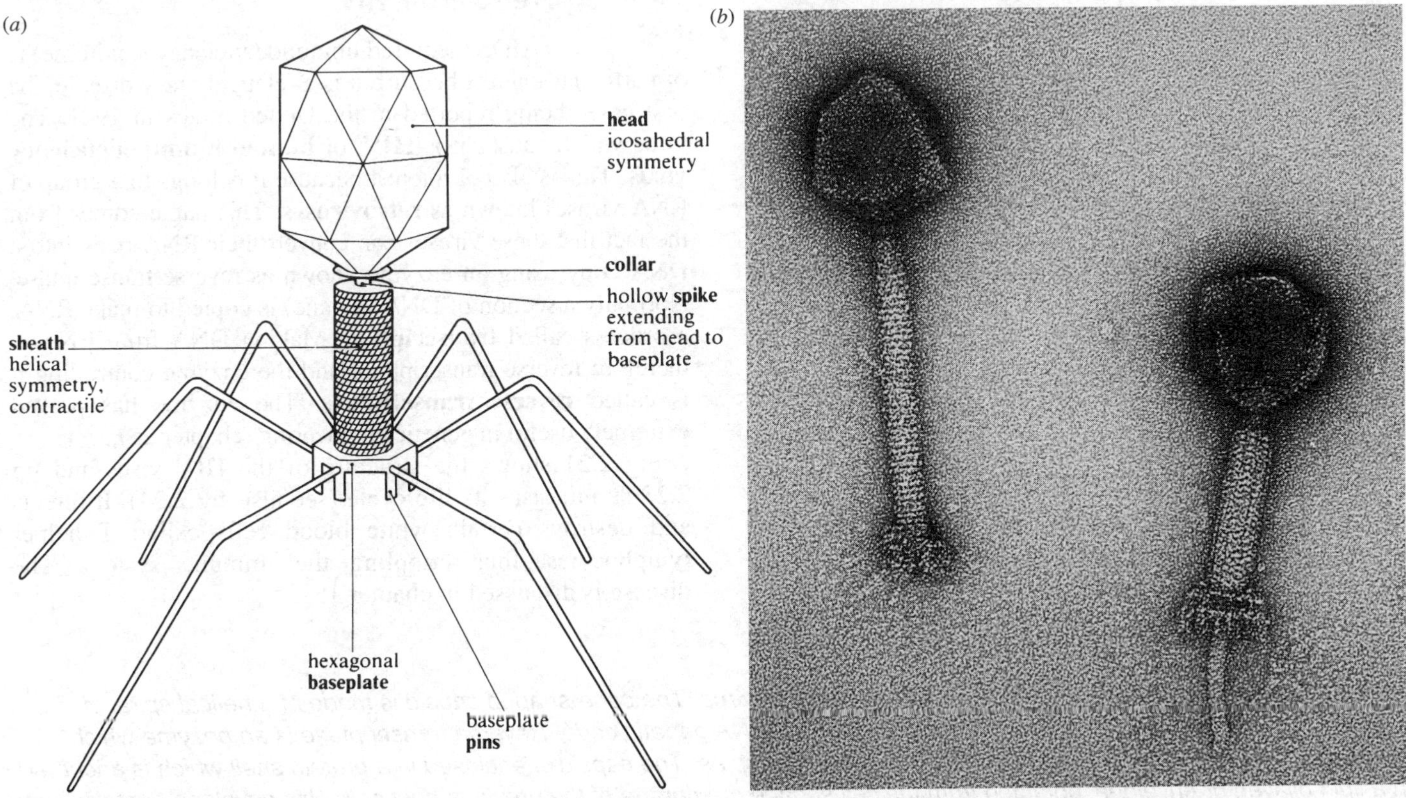

Fig 2.19 (a) *Structure of a T2 bacteriophage.* (b) *Electron micrograph of negatively stained bacteriophages.*

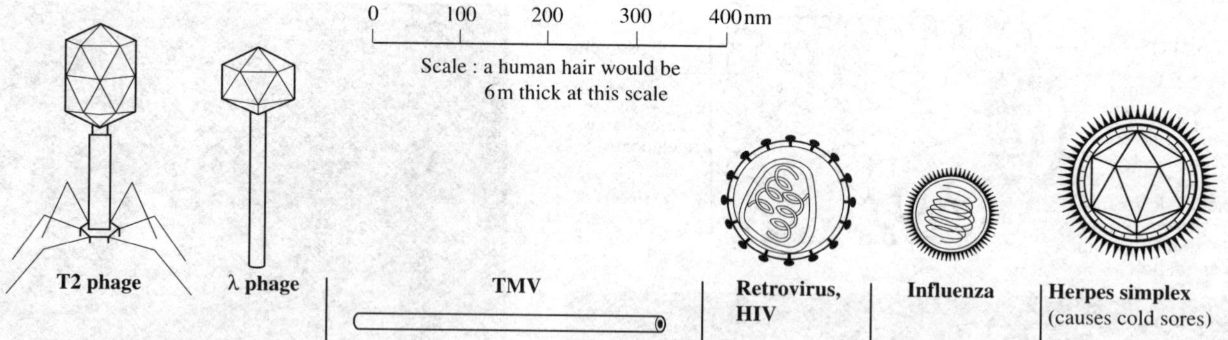

	T2 phage λ phage	TMV	Retrovirus, HIV	Influenza	Herpes simplex (causes cold sores)
envelope	✓	✗	✓	✓	✓
double-stranded DNA	✗	✗	✗	✗	✓
single-stranded RNA	✓	✓	✓	✓	✗
type of organism infected	bacterium	plant (tobacco)	animal (human)	animal (human)	animal (human)
symmetry of capsid	icosahedral head helical tail	helical	helical	helical	icosahedral

Fig 2.20 *Some simplified diagrams of viruses showing different sizes and symmetry. The T2 phage is shown with its tail fibres released prior to infection; the λ (lambda) phage does not have tail fibres.*

2.4.5 Structure and life cycle of a retrovirus, HIV

AIDS (acquired immunodeficiency syndrome) is of particular interest because it is a relatively new disease, the first cases being reported in the United States in 1981. The virus which causes it is **HIV**, or **human immunodeficiency virus**. This is also of interest because it belongs to a group of RNA viruses known as **retroviruses**. This name comes from the fact that these viruses can convert their RNA back into a DNA copy using an enzyme known as reverse transcriptase. Normally a section of DNA (a gene) is copied to make RNA, a process called **transcription**. Making DNA from RNA is therefore reverse transcription, and the enzyme controlling it is called **reverse transcriptase**. The enzyme has proved extremely useful in genetic engineering (chapter 25).

Fig 2.21 shows the structure of the HIV virus and fig 2.23 summarises its life cycle (see also fig 2.24). It infects and destroys certain white blood cells called T helper lymphocytes, thus crippling the immune system. The disease is discussed in chapter 15.

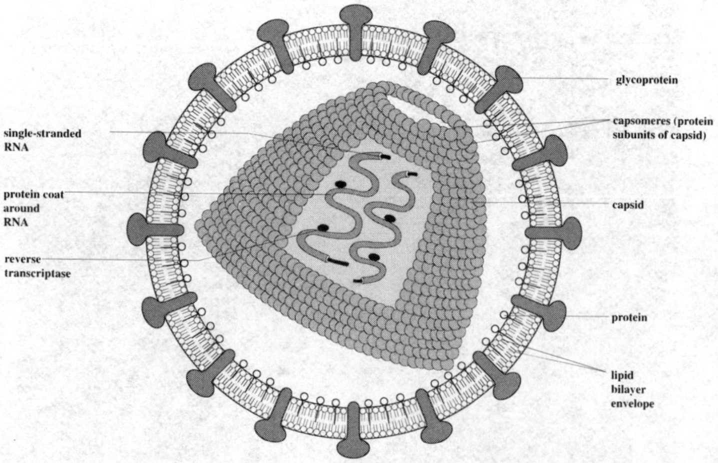

Fig 2.21 *Structure of the HIV virus, an example of a retrovirus. The cone-shaped capsid is made of a helical spiral of capsomeres. It is cut open to reveal the two copies of the RNA genetic code. Reverse transcriptase is an enzyme which converts single-stranded RNA into double-stranded DNA copies. The capsid is enclosed in a protein shell which is anchored in a lipid bilayer, or envelope, obtained from the cell surface membrane of the previous host cell. This envelope contains viral glycoproteins which bind specifically to helper T-cell receptors, enabling the virus to enter its host.*

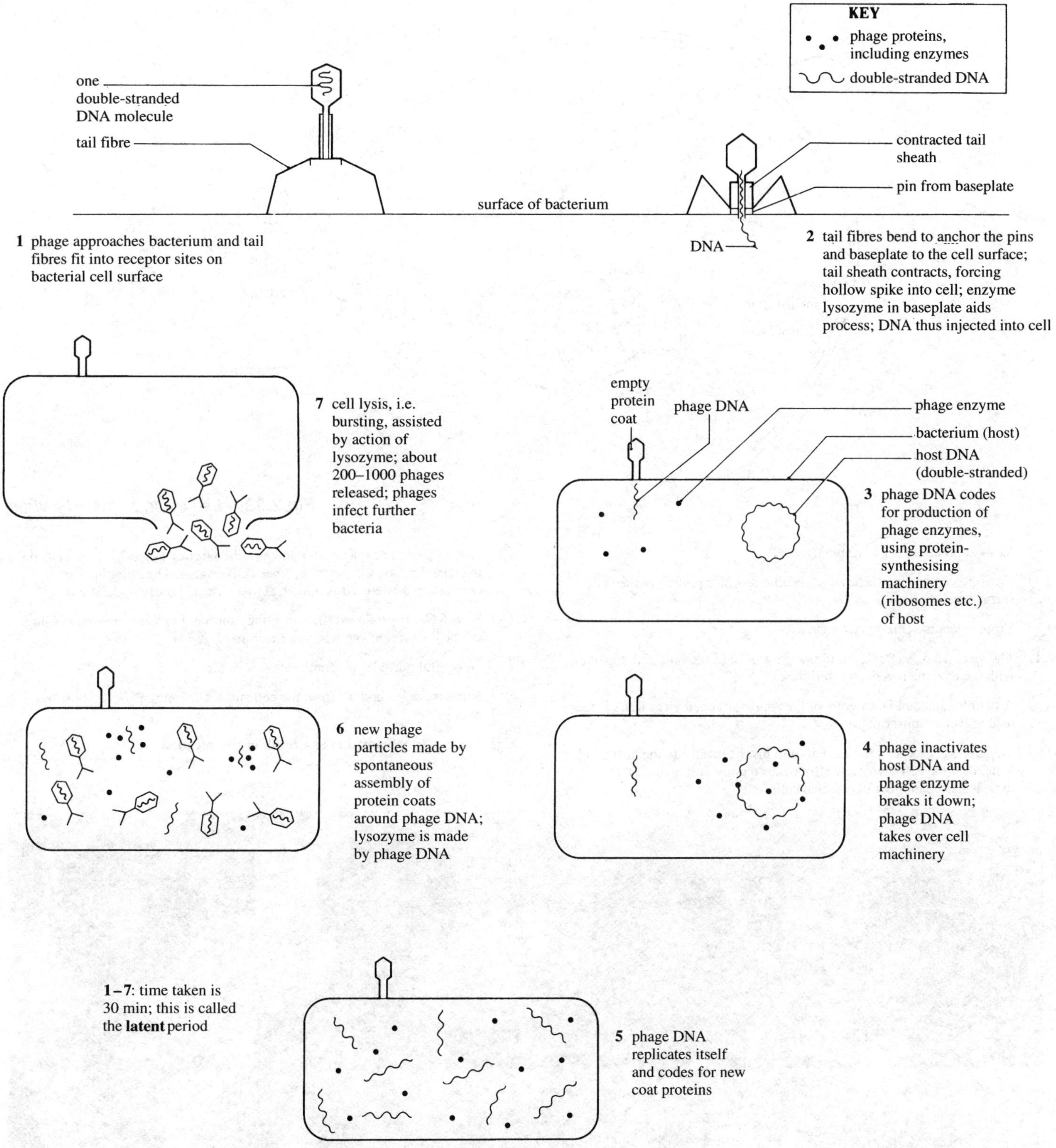

Fig 2.22 *Life cycle of a bacteriophage.*

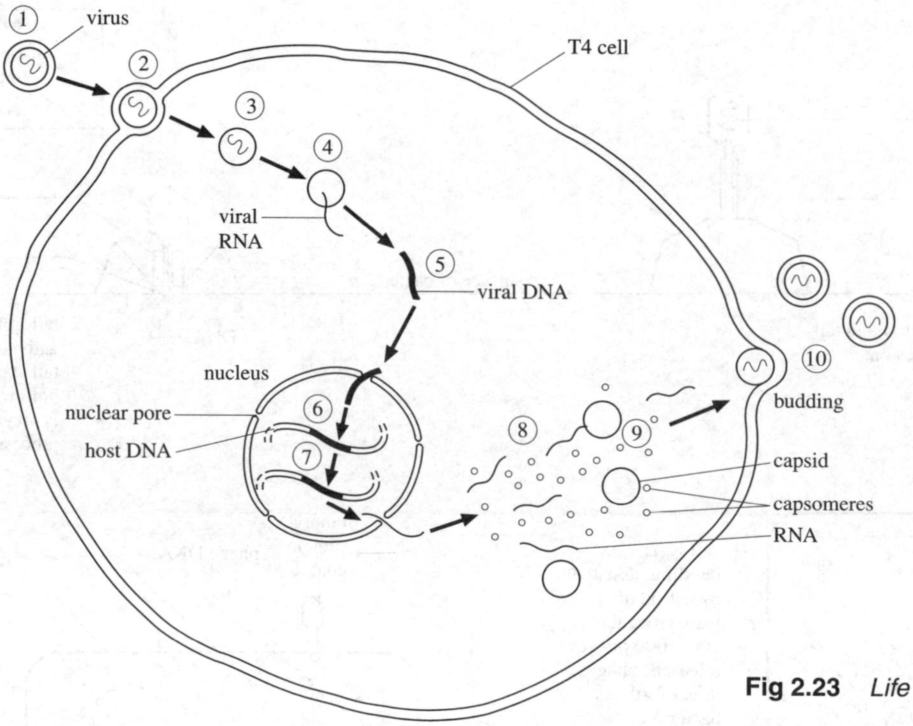

Fig 2.23 *Life cycle of the HIV virus.*

① Virus approaches a T4 lymphocyte cell.

② Virus glycoprotein attaches to a specific receptor protein in the cell surface membrane.

③ Virus enters the cell by endocytosis.

④ The viral RNA is released into the cytoplasm of the host cell, together with the enzyme reverse transcriptase.

⑤ A double-stranded DNA copy of the single-stranded virus RNA is made using reverse transcriptase.

⑥ The DNA copy enters the nucleus and inserts itself into the host DNA. Whenever the cell divides, it also makes a copy of the viral DNA, increasing the number of infected cells.

⑦ After a period of inactivity known as the **latency period**, which lasts on average 5 years, the virus becomes active again. The stimulus for converting a latent virus into an active virus is poorly understood.

⑧ New RNA is produced (transcription) and viral proteins are made using the host's protein synthesising machinery.

⑨ New viral particles assemble.

⑩ Virus particles bud off from the cell surface membrane of the host by exocytosis.

⑪ The cell eventually dies as a result of the infection.

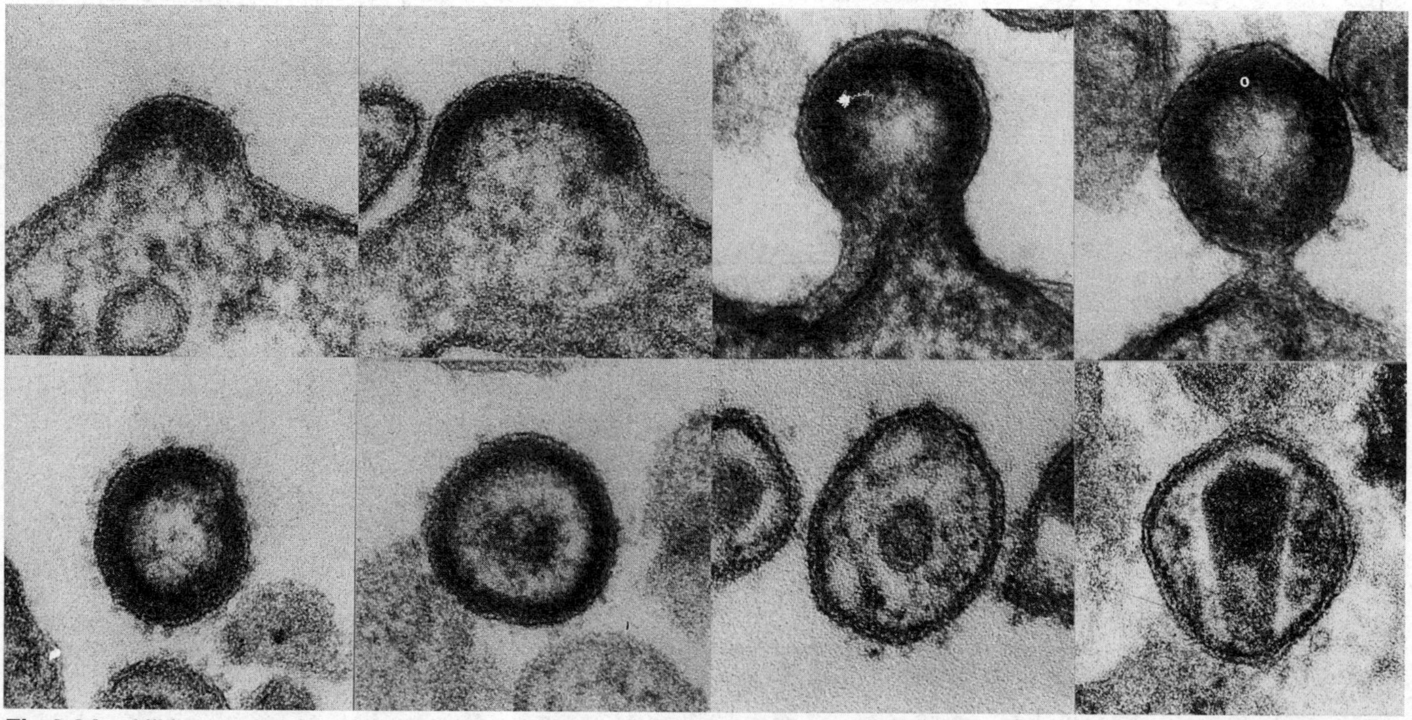

Fig 2.24 *HIV emerging from infected cell.*

2.5 Kingdom Fungi

The fungi are a large and successful group of organisms of about 80 000 named species. They range in size from the unicellular yeasts to the large toadstools, puffballs and stinkhorns, and occupy a very wide range of habitats, both aquatic and terrestrial. They are also of major importance for the essential role that they play in the biosphere, and for the way in which they have been exploited by humans for economic and medical purposes.

Fungi include the numerous moulds which grow on damp organic matter (such as bread, leather, decaying vegetation and dead fish), the unicellular yeasts which are abundant on the sugary surfaces of ripe fruits and many parasites of plants. The latter cause some economically important diseases of crops, such as mildews, smuts and rusts. A few fungi are parasites of animals, but these are less significant in this respect than bacteria.

The study of fungi is called mycology (*mykes*, mushroom). It forms a branch of microbiology because many of the handling techniques used, such as sterilising and culturing procedures, are the same as those used with bacteria (see chapter 12).

2.5.1 Classification and characteristics of Fungi

As discussed in section 2.2, fungi are eukaryotes that lack chlorophyll, and are therefore heterotrophic, like animals. However, they have rigid cell walls and are non-motile, like plants. In the past, they were regarded as plants, but modern classifications, such as that shown in fig 2.4, place them in a separate kingdom. Their classification and characteristics are summarised in table 2.6. The two largest and most advanced groups are the Ascomycota and the Basidiomycota. Structure and nutrition of fungi are discussed in more detail below.

> **2.2** Using those features of the kingdom Fungi given in table 2.6, prepare a table of differences between fungi and typical plant cells.

Table 2.6 Classification and characteristics of fungi.

Kingdom Fungi

General characteristics
Heterotrophic nutrition because they lack chlorophyll and are therefore non-photosynthetic. They can be parasites, saprotrophs or mutualists. Nutrition is absorptive; digestion takes place outside the body and nutrients are absorbed directly. Digestion does not take place inside the body, unlike animals.
Rigid cell walls containing chitin as the fibrillar material. Chitin is a nitrogen-containing polysaccharide, very similar in structure to cellulose. Like cellulose it has high tensile strength. It therefore gives shape to the hyphae and prevents osmotic bursting of cells.
Body is usually a mycelium, a network of fine tubular filaments called hyphae. These may be septate (have cross-walls), e.g. *Penicillium*, or aseptate (no cross-walls), e.g. *Mucor*.
If carbohydrate is stored, it is usually as glycogen, not starch
Reproduce by means of spores
Non-motile

Phylum Zygomycota	*Phylum Ascomycota*	*Phylum Basidiomycota*
Asexual reproduction by conidia or sporangia containing spores	Asexual reproduction by conidia. No sporangia	Asexual reproduction by formation of spores. Not common
Non-septate hyphae and large well-developed branching mycelium	Septate hyphae	Septate hyphae
e.g. *Rhizopus stolonifer*, common bread mould, a saprotroph *Mucor*, common moulds, saprotroph	e.g. *Penicillium* and *Aspergillus*, saprotrophic moulds *Saccharomyces* (yeast), unicellular saprotrophs *Erysiphe,* obligate parasites causing powdery mildews, e.g. of barley	e.g. *Agaricus campestris*, field mushroom, saprotroph

2.5.2 Structure

The body structure of the fungi is unique. It consists of a mass of fine, tubular branching threads called **hyphae** (singular hypha), the whole mass being called a **mycelium**. Each hypha has a thin rigid wall whose chief component is chitin, a nitrogen-containing polysaccharide which is also found as a structural component in the exoskeletons of arthropods (section 2.8.6). The hyphae are not divided into true cells. Instead, the protoplasm is either continuous or interrupted at intervals by cross-walls called **septa** which divide the hyphae into compartments similar to cells. Unlike normal cell walls their formation is not a consequence of nuclear division, and a pore normally remains at their centre allowing protoplasm to flow between compartments. Each compartment may contain one, two or more nuclei, which are distributed at more or less regular intervals along the hyphae. Hyphae having cross-walls are called **septate**, as in *Penicillium* (fig 2.25). Hyphae lacking cross-walls are called **non-septate** (**aseptate**) as in *Mucor* (fig 2.26).

Within the cytoplasm the usual eukaryote organelles are found, such as mitochondria, Golgi apparatus, endoplasmic reticulum, ribosomes and vacuoles. In the older parts, vacuoles are large and cytoplasm is confined to a thin peripheral layer. Sometimes hyphae aggregate to form more solid structures such as the spore-producing bodies of the mushrooms. The yeasts are unusual in being unicellular fungi and therefore lack the typical hypha structure, e.g. *Saccharomyces* (fig 2.27).

Penicillium, *Mucor* and *Rhizopus* are known as moulds. They are widespread saprotrophs, that is they feed on dead organic matter. They are convenient fungi to study because they are easy to grow in culture and show the typical hyphal growth of fungi.

Penicillium species form blue, green and sometimes yellow moulds, common, for example, on bread and decaying fruit. The mycelium forms a circular colony of small diameter with septate hyphae and the spores give colour to the colony (fig 2.25*a*). *Penicillium* reproduces asexually by means of spores called **conidia**. These are found at the tips of special hyphae called **conidiophores** (fig 2.25*b* and *c*). They are not enclosed in a sporangium, but are naked and free to be dispersed as soon as they mature. The structure of the hyphae is shown in fig 2.25*d*. The economic importance of *Penicillium* is discussed in section 12.11.1.

Mucor is a genus which includes a number of well-known moulds. It is common in soil and may also be found growing on bread. It forms more or less circular colonies when grown on agar. Its hyphae are aseptate and profusely branching (fig 2.26*b*). It produces spores in spherical **sporangia** borne on very long stalks known as **sporangiophores** (fig 2.26*a* and *b*). These are numerous in the more mature parts of the mycelium and resemble a collection of pins; hence *Mucor* is often referred to as pin mould. Sporangia are clearly visible using the low power of a microscope. Mucor grows rapidly and can cover a petri dish in 3 days at 20 °C. Internally, its hyphae have the same typical eukaryotic structure as *Penicillium* (fig 2.25*d*) except that the hyphae of *Mucor* lack cross-walls. *Rhizopus* is very similar to *Mucor*. Some hyphae, called *stolons*, are arch-shaped and produce tufts of short, root-like hyphae at their tips. Two or more sporangiophores grow from the same point, unlike *Mucor* where sporangiophores occur singly.

2.3 What is the purpose of the sporangiophores?

Yeasts are unicellular, saprotrophic fungi. They occur widely in nature and are particularly common on the sugary surfaces of ripe fruits. The bloom on grapes, for example, is due to yeast. The fermentation (anaerobic respiration) of the sugars by yeast produces alcohol, a fact made use of by humans for thousands of years and which forms the basis of the wine and brewing industries. Under appropriate conditions yeast cells multiply rapidly by budding, a form of asexual reproduction (fig 2.27*a*). Yeast cells show the usual eukaryotic features (fig 2.27*b* and *c*).

2.5.3 Nutrition

Fungi are heterotrophic, that is they require an organic source of carbon. In addition, they require a source of nitrogen, usually organic such as amino acids; inorganic ions such as K^+ and Mg^{2+}; trace elements such as Fe, Zn and Cu; and organic growth factors such as vitamins. The exact range of nutrients required, and hence substrates on which they are found, is variable. The nutrition of fungi can be described as absorptive because they absorb nutrients directly from outside their bodies. This is in contrast to animals, which normally ingest food and then digest it within their bodies before absorption takes place. With fungi, digestion, if necessary, is performed by the fungus secreting enzymes out of its body on to its food.

Fungi obtain their nutrients as saprotrophs, parasites or mutualists. In this respect they are like most bacteria.

(a)

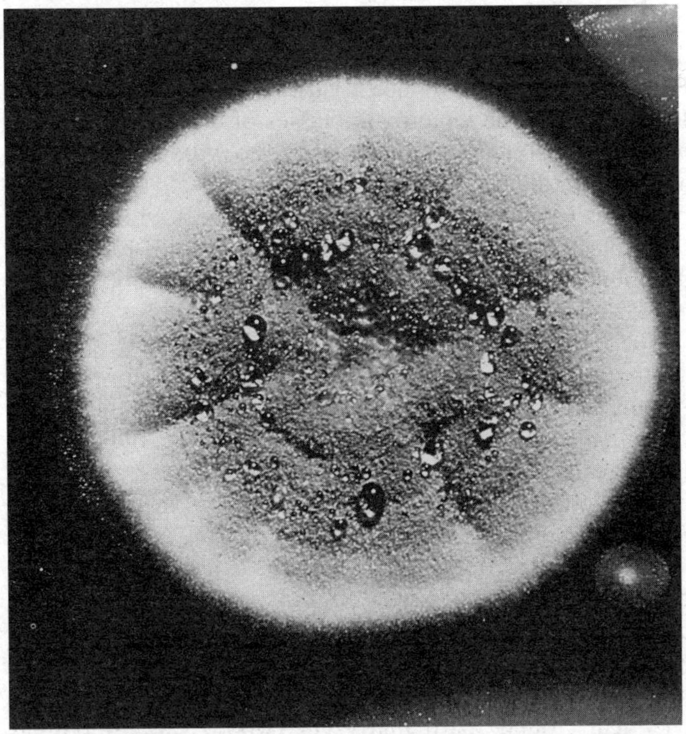

(b)

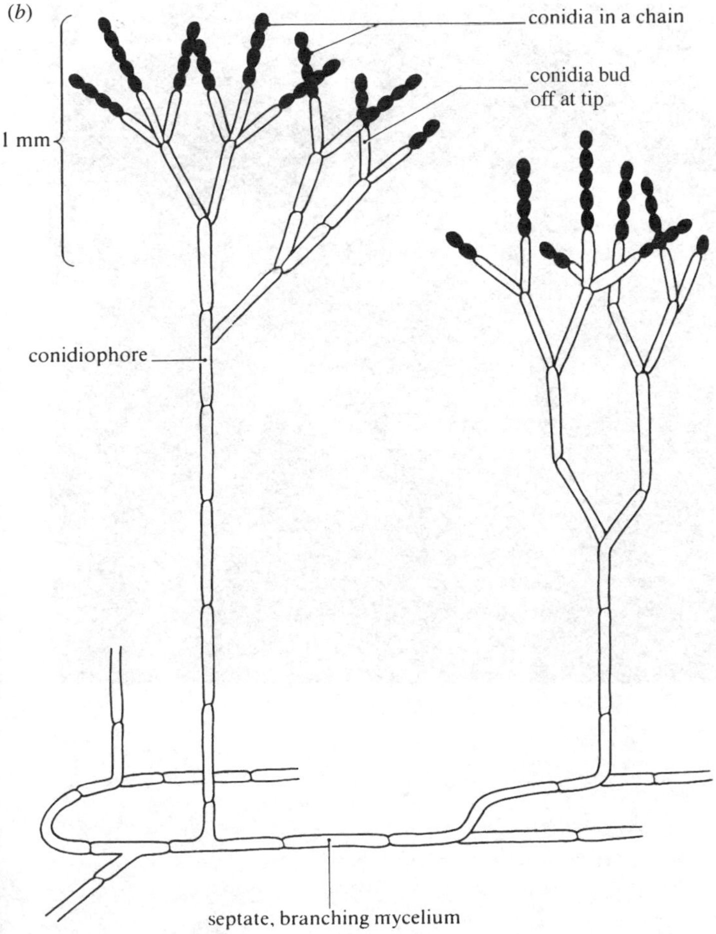

(c)

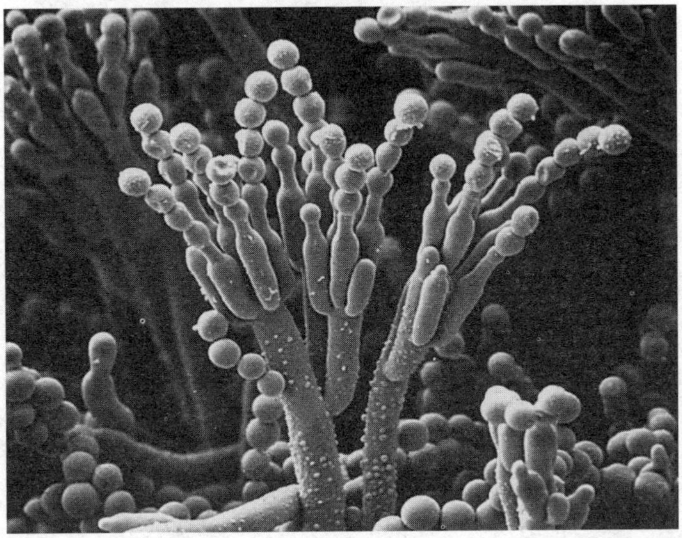

Fig 2.25 (a) Penicillium *growing on nutrient agar in a petri dish. It typically produces a relatively small circular mycelium. The young outer edge of the mycelium appears white, whereas the mature central portion appears darker where coloured spores have been produced.*
(b) Penicillium *showing asexual reproduction. It has a characteristic brush-like arrangement of conidia.*
(c) *Scanning electron micrograph of conidiophore and conidia (spores) of* Penicillium. (d) *LS hypha showing fine structure visible with electron microscope.*

(d)

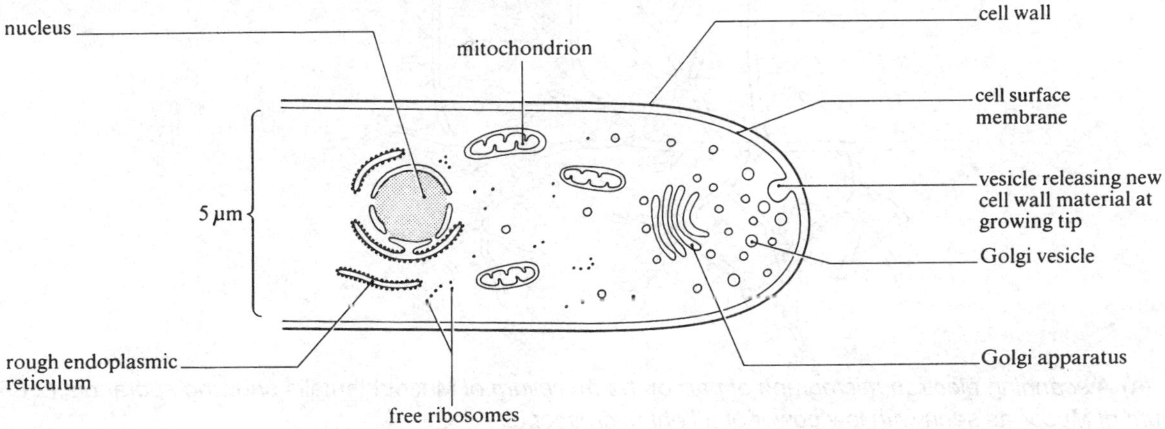

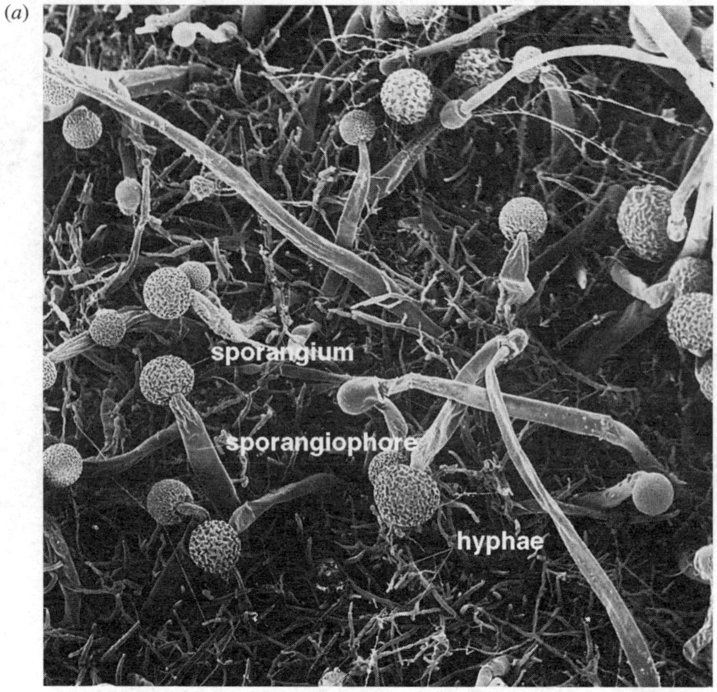

Saprotrophs

Saprotrophic organisms feed on dead organic material. Fungal saprotrophs produce a variety of digestive enzymes. If they secrete the three main classes of digestive enzymes, namely carbohydrases (digest carbohydrates) such as amylases (digest starch), lipases (digest lipids) and proteases (digest proteins), they can utilise a wide range of substrates. For example the *Penicillium* species form green and blue moulds on substrates such as soil, damp leather, bread and decaying fruit.

The hyphae of saprotrophic fungi are usually chemotropic, that is they grow towards certain substrates in response to chemicals diffusing from these substrates.

Fungal saprotrophs usually produce large numbers of light, resistant spores. This allows efficient dispersal to other food sources. Examples are *Mucor, Rhizopus* and *Penicillium*.

Saprotrophic fungi and bacteria together form the decomposers which are essential in the recycling of nutrients. Especially important are the few that secrete the enzymes cellulase and lignase, which break down cellulose and lignin respectively. Cellulose and lignin (a complex chemical found particularly in wood) are important structural components of plant cell walls, and the rotting of wood and other plant remains is achieved partly by decomposers secreting cellulase and lignase.

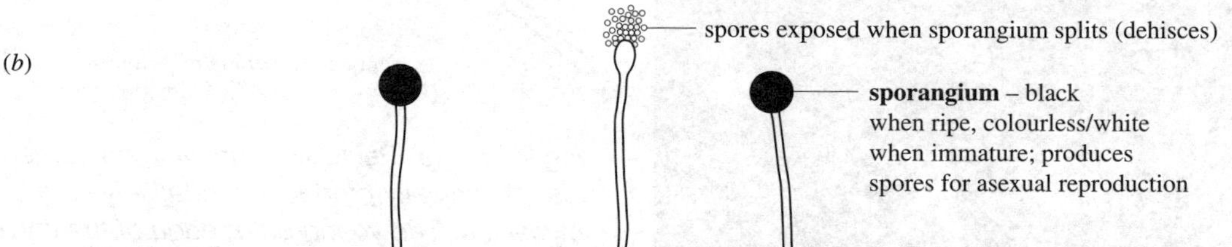

Fig 2.26 (a) *A scanning electron micrograph of part of the mycelium of* Mucor hiemalis *showing sporangia (× 85).* (b) *Mycelium of* Mucor *as seen with low power of a light microscope.*

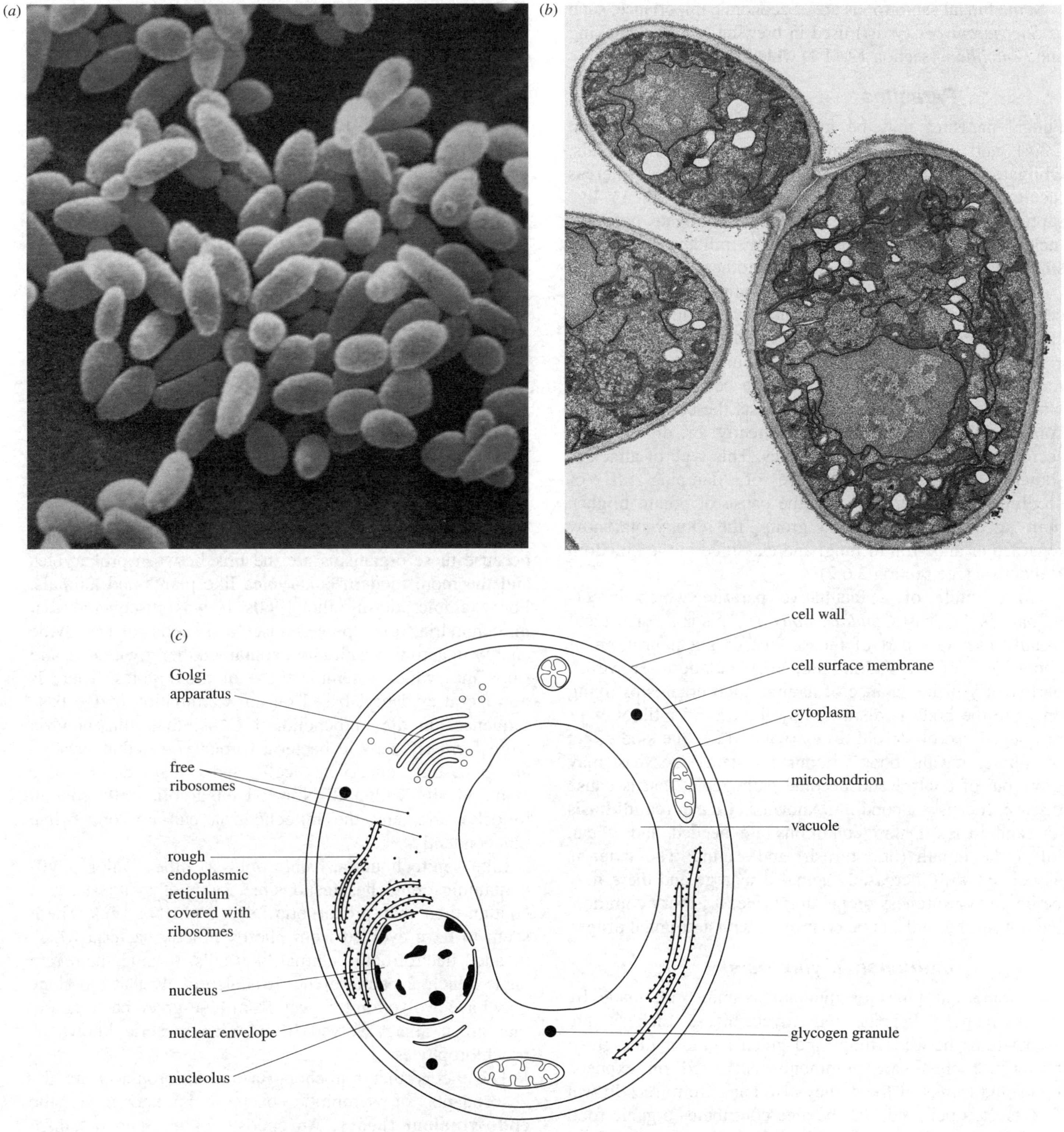

Fig 2.27 *Structure of yeast (*Saccharomyces*). (a) Budding cells as seen with a light microscope (× 4000). (b) Yeast cell seen with a transmission electron microscope (× 10 000). (c) Diagram of fine structure visible with an electron microscope.*

Some fungal saprotrophs are of economic importance, such as *Saccharomyces* (yeast) used in brewing and breadmaking and *Penicillium* (section 12.11.1) which is used in medicine.

Parasites

Fungal parasites may be facultative or obligate (section 2.3.4), and more commonly attack plants than animals. Obligate parasites do not normally kill their hosts, whereas facultative parasites frequently do and then live saprotrophically off the dead remains. Obligate parasites include the powdery mildews, downy mildews, rusts and smuts, which attack cereals and many other crops.

Once inside the plant, hyphae normally grow between cells. Facultative parasites commonly produce enzymes called pectinases which digest the middle lamellae between cells and cause 'soft rot' of the tissue, reducing it to a mush. Subsequently they may invade cells and kill them with the aid of cellulase which digests the cell walls. Cell constituents may be absorbed directly or digested by secretion of further fungal enzymes. This type of attack is shown by *Pythium* (the cause of 'damping off' of seedlings) and *Phytophthora* (the cause of potato blight), both of which belong to a group, the Oomycota, now regarded as ancestral to fungi and classified in the kingdom Protoctista (see section 2.6.2).

An example of a facultative parasite which infects humans is the yeast *Candida albicans*. This is a normal and usually harmless part of the surface or gut microflora of about 5% of the adult human population. However, particularly if the balance of natural microorganisms living on or in the body is disturbed by the use of antibiotics or prolonged use of steroid drugs (which have the side-effect of suppressing the body's immune system), the yeast may grow out of control and become pathogenic, that is cause disease. It causes a condition known as **thrush** (**candidiasis** or candidosis). Damp conditions are needed and it can infect the mouth (oral thrush) and vagina. The latter is associated with increased vaginal discharge and there may be itching or soreness on passing urine. It is very common, but not serious and can be controlled with antifungal drugs.

Mutualism (symbiosis)

Two important types of mutualistic union are made by fungi, namely lichens and mycorrhizas. Lichens are associations between fungi and green algae or blue-green bacteria. Lichens are commonly encrusted on exposed rocks and trunks of trees; they also hang from trees in wet forests. It is believed that the alga contributes organic food from photosynthesis, while the fungus is protected from high light intensity and is able to absorb water and mineral salts. The fungus can also conserve water, enabling some lichens to grow in dry conditions where no plants exist.

A mycorrhiza is an association between a fungus and a plant root. The fungus absorbs mineral salts and water which pass to the plant, and in return receives organic products of photosynthesis. Mycorrhizas are considered in more detail in section 7.10.2.

2.6 Kingdom Protoctista

(Greek *protos*, very first; *ktistos*, to establish)

2.6.1 Classification and characteristics of protoctists

In section 2.2 it was noted that in the five kingdom classification of Margulis and Schwartz (fig 2.4), the Protoctista is probably the most controversial group because it is the least natural. It is really a collection of all the eukaryotic organisms that do not fit neatly into the other three eukaryote kingdoms. Many are unicellular.

The Protoctista contains eukaryotes that are generally regarded as identical or similar to the ancestors of modern plants, animals and fungi (fig 2.28). It includes organisms which resemble early plants (algae), early animals (protozoa) and early fungi (Oomycota). It also includes a group known as the slime moulds which produce spores like fungi but can creep slowly over surfaces and are therefore motile like animals. The earliest eukaryotes were probably unicellular organisms which moved by beating flagella.

The group is fascinating to those interested in evolution because these organisms are the link between prokaryotes and the more modern eukaryotes like plants and animals. For example, during the 1960s it was discovered that mitochondria, the 'powerhouses' of cells that provide energy in aerobic respiration, contained their own DNA and ribosomes which resemble those of prokaryotes. There is now good evidence, based on an examination of the base sequences in the mitochondrial DNA, that mitochondria were formerly aerobic bacteria (prokaryotes) that invaded an ancestral eukaryote cell and 'learned' to live symbiotically within it. Now all eukaryotic cells contain mitochondria, and the mitochondria can no longer live independently.

Like mitochondria, chloroplasts, the chlorophyll-containing organelles responsible for photosynthesis, also contain their own prokaryotic DNA and ribosomes. These seem to have evolved from photosynthetic bacteria which invaded heterotrophic animal-like cells, turning them into algae which are autotrophic. It is also likely that red algae may have evolved in this way from blue-green bacteria and that green algae evolved from green bacteria known as prochlorophytes.

The theory that mitochondria and chloroplasts are the descendants of symbiotic bacteria is known as the **endosymbiont theory**. An endosymbiont is an organism that lives symbiotically *inside* (endo-) another organism.

2.6.2 Phylum Oomycota

Oomycotes are close relations of the fungi and have a similar structure, but are now regarded as a more ancient group. Their cell walls contain cellulose, not chitin, as the strengthening material. Their hyphae are aseptate. In

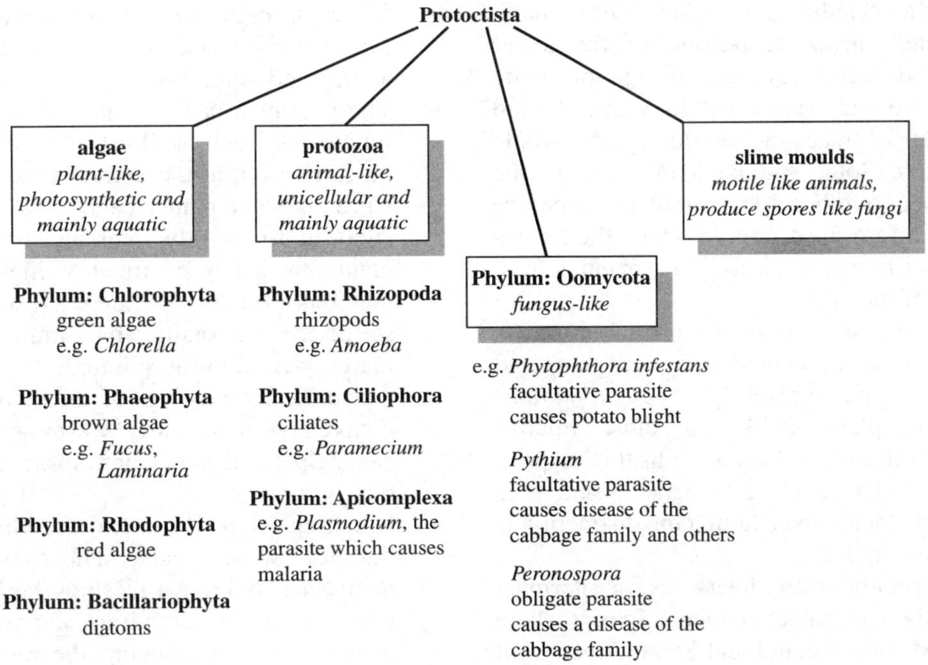

Fig 2.28 *The main groups of Protoctista and some examples of the phyla and genera they contain. Not all groups or phyla are shown.*

this phylum are a number of pathogenic organisms, including the downy mildews. One of these, *Phytophthora infestans*, will be studied as an example of a parasite which is generally described as obligate. *Peronospora*, an obligate parasite, will also be mentioned for comparison. Finally, *Pythium* will be examined as a typical example of a facultative parasite. An obligate parasite is one which can only survive and grow in living cells whereas facultative parasites typically bring about the deaths of their hosts before living saprotrophically on the remains.

Phytophthora infestans

Phytophthora infestans is a pathogen of economic importance because it parasitises potato crops, causing a potentially devastating disease known as potato blight. It does not grow independently of its host and in this respect resembles obligate parasites. It is similar in its structure and mode of attack to another member of the Oomycota, *Peronospora*, which is a common, but less serious, disease of wallflower, cabbage and other members of the plant family Cruciferae.

The *Phytophthora* mycelium overwinters in potato tubers and grows up to the leaves in spring. Blight is usually first noticed in the leaves in August.

A mycelium of branched, aseptate hyphae spreads through the intercellular spaces of the leaves, giving off branched **haustoria** which push into the mesophyll cells and absorb nutrients from them (fig 2.29). Haustoria are typical of obligate parasites. They are specialised penetration and absorption devices. Each is a modified hyphal outgrowth with a large surface area which pushes into cells without breaking their cell surface membranes and without killing them. In warm, humid conditions the mycelium produces long, slender structures called **sporangiophores** which emerge from the lower surface of the leaf through stomata or wounds. These branch and give rise to **sporangia** (fig 2.29). In warm conditions sporangia may behave as spores, being blown or splashed by raindrops on to other plants, where further infection takes place. A hypha emerges from the sporangium and penetrates the plant through a stoma, lenticel or wound. In cool conditions, the sporangium contents may divide to form swimming spores (a primitive feature) which, when released, swim in surface films of moisture. They may encyst until conditions are suitable once more for hyphal growth, then produce new infections.

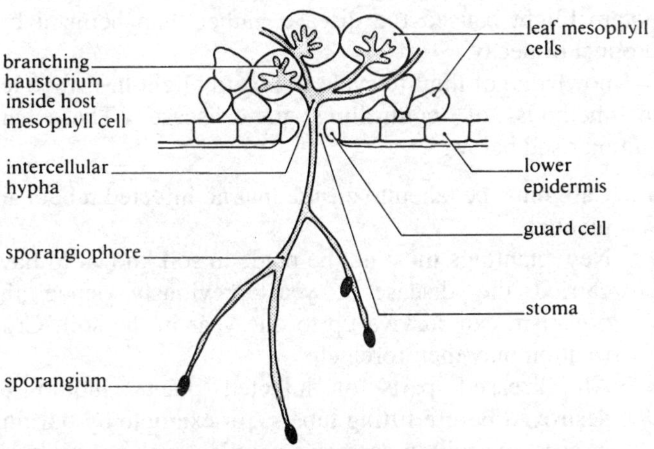

Fig 2.29 Phytophthora infestans *growing in a diseased potato leaf, with sporangiophores emerging from the underside of the leaf.*

Diseased plants show individual leaflets with small, brown, dead, 'blighted' areas. Inspection of the lower surface of an infected leaflet reveals a fringe of white sporangiophores around the dead area. In warm, humid conditions, the dead area spreads rapidly through the whole leaf and into the stem. Some sporangia may fall to the ground and infect potato tubers. Here infection spreads very rapidly, causing a form of dry rot in which the tissues are discoloured a rusty brown in an irregular manner from the skin to the centre of the tuber.

First the base and then the rest of the plant becomes a putrid mass as the dead areas become secondarily infected with decomposing bacteria (saprotrophs). *Phytophthora* thus kills the whole plant, unlike its close relative *Peronospora* which is an obligate parasite. In this respect, *Phytophthora* is not a typical obligate parasite and it is sometimes described as facultative, though the distinction is perhaps not worth stressing here.

The organism normally overwinters as a dormant mycelium within lightly infected potato tubers. Except where the potato is native (Mexico, Central and South America) it is thought that the organism rarely reproduces sexually, unlike *Peronospora*, but under laboratory conditions it can be induced to do so. Like *Peronospora*, it produces a resistant resting spore. It is the result of fusion between an antheridium (male) and an oogonium (female), and a thick-walled spore is produced. This can remain dormant in the soil over winter and cause infection in the following year.

In the past, *Phytophthora* epidemics have had serious consequences. The disease is thought to have been accidentally introduced into Europe from America in the late 1830s and caused a series of epidemics that totally destroyed the potato crop in Ireland in 1845 and in subsequent years. Widespread famine resulted and many starved to death, victims as much of complex economic and political influences as of the disease. Many Irish families emigrated to North America as a result.

The disease is also of interest because in 1845 Berkeley provided the first clear demonstration that microorganisms cause disease by showing that the organism associated with potato blight caused the disease, rather than being a by-product of decay.

Knowledge of the life cycle of potato blight has since led to methods of controlling the disease. These are summarised below.

- Care must be taken to ensure that no infected tubers are planted.
- New plantings must not be made in soil known to have carried the disease a year previously, since the organism can survive up to one year in the soil. Crop rotation may therefore help.
- All diseased parts of infected plants should be destroyed before lifting tubers, for example by burning or spraying with a corrosive solution such as sulphuric acid. This is because tubers can be infected from decaying haulms (stems) and aerial parts.
- Since the organism can overwinter in unlifted tubers, care must be taken to ensure that all tubers are lifted in an infected field.
- The organism can be attacked with copper-containing fungicides, such as Bordeaux mixture. Spraying must be carried out at the correct time to prevent an attack, since infected plants cannot be saved. It is usual to spray at fortnightly intervals, from the time that the plants are a few centimetres high until they are well matured. Tubers intended as seed potatoes can be sterilised externally by immersion in a dilute mercury(II) chloride solution.
- Accurate monitoring of meteorological conditions, coupled with an early warning system for farmers, can help to decide when spraying should be carried out.
- Breeding for resistance to the blight has been carried out for some years. The wild potato, *Solanum demissum*, is known to show high resistance and has been used in breeding experiments. One great obstruction to producing the required immunity lies in the fact that the organism exists in many strains and no potato has been found to be resistant to all of them. New strains of the organism may appear as new strains of potato are introduced. This is a familiar problem in plant pathology and emphasises the need for conservation of the wild ancestors of our modern crop plants as sources of genes for disease resistance.

Pythium

Unlike *Phytophthora, Pythium* is a relatively unspecialised parasite, attacking a great variety of plants and causing a soft rot. It causes 'damping off' in seedlings. It needs damp conditions since it produces swimming spores during asexual reproduction. It can grow on the living plant or on its dead remains, so is a facultative parasite. It can also live saprotrophically in wet soil. It produces extracellular enzymes which help it attack and kill its host rapidly. The first enzymes produced are pectinases which diffuse ahead of the growing fungus and digest the pectin in the middle lamellae which hold the cells together. As a result the plant tissue dissolves into a mush (soft rot). The plant collapses. Later other enzymes are produced which digest the contents of the plant cells, but it does not produce haustoria, unlike *Phytophthora*. Products of digestion are absorbed by the hyphae which grow between the cells.

Damping-off of seedlings is due to destruction of the first shoot as it appears above the soil. Watery spots first appear on the stem at soil level. As these darken, the stem collapses. It can be a serious problem in horticulture, forestry and agriculture. Members of the cabbage family (crucifers) are particularly susceptible, especially when the seedlings are grown in crowded conditions.

2.6.3 Algae

The algae form a large group of protoctistans of great biological importance and significance to humans. No single characteristic is diagnostic. They are best thought of as photosynthetic eukaryotes that evolved in, and have remained in, water. A few algae have escaped to live successfully on land, but unlike plants, which evolved on land, these are insignificant in number compared with those in the oceans and fresh water. The bodies of algae lack true stems, roots and leaves. Such a relatively undifferentiated body is called a **thallus**.

The algae fall naturally into distinct groups, chiefly on the basis of their photosynthetic pigments. These groups are given the status of phyla in modern classifications. Only four of the phyla are shown in fig 2.28. Characteristics of the algae and of two of the main phyla are shown in table 2.7. Two examples of algae, namely *Chlorella* (phylum Chlorophyta) and *Fucus* (phylum Phaeophyta) are examined in more detail below.

2.6.4 Phylum Chlorophyta (green algae)

Chlorella

Chlorella is a unicellular, non-motile green alga. Its structure is shown in fig 2.30. Its habitat is freshwater ponds and ditches. It is easily cultured and has been used as an experimental organism in research on photosynthesis (section 7.6) as well as being investigated as an alternative source of food (single cell protein, section 12.12.3).

2.6.5 Phylum Phaeophyta (brown algae)

Fucus

Fucus is a relatively large and complex brown alga. Its body is a thallus which is differentiated into a stipe, holdfast and fronds (note these are not true stem, roots and leaves). It is a marine alga, common on rocky shores off the British coast. It is well adapted to the relatively harsh conditions of the shore, where it is alternately exposed and covered by the tides.

There are three common species and these are often found at three different levels, or zones, on the shore, a phenomenon called **zonation** (section 10.6.4). They are principally zoned according to their ability to withstand exposure to air. Their chief recognition features and positions on the shore are noted below.

F. spiralis (flat wrack) – towards high tide mark. If suspended, the thallus adopts a slight spiral twist.

F. serratus (common, serrated or saw wrack) – middle zone. Edge of the thallus is serrated.

F. vesiculosus (bladder wrack) – towards low tide mark. Possesses air bladders for buoyancy. The external features of *F. vesiculosus* are shown in fig 2.31.

Adaptations to environment. Before discussing the adaptations of *Fucus* to its environment, some mention must be made of the nature of this environment, which is relatively hostile. Being intertidal, the different species are subjected to varying degrees of exposure to air when the tide recedes. Therefore they must be protected against drying out. Temperatures may change rapidly, as when a cold sea advances into a hot rock pool. Salinity is

Table 2.7 Classification and characteristics of two of the main groups of algae.

Algae	
General characteristics Almost all are specialised for an aquatic existence Great range of size and form, including unicellular, filamentous, colonial and thalloid forms. A thallus is a body which is not differentiated into true roots, stems and leaves and lacks a true vascular system (xylem and phloem). It is often flat Photosynthetic, eukaryotic	
Phylum Chlorophyta ('green algae')	*Phylum Phaeophyta* ('brown algae')
Dominant photosynthetic pigment is chlorophyll; therefore green in appearance. Chlorophylls *a* and *b* present (as in plants)	* Dominant photosynthetic pigment is brown and called fucoxanthin. Chlorophylls *a* and *c* present
Store carbohydrate as starch (insoluble)	* Store carbohydrate as soluble laminarin and mannitol. Also store fat
Mostly freshwater	Nearly all marine (three freshwater genera only)
Large range of types, e.g. unicellular, filamentous, colonial, thalloid	Filamentous or thalloid, often large
e.g. *Chlorella,* a unicellular, non-motile alga *Chlamydomonas,* a unicellular, motile alga *Spirogyra,* a filamentous alga *Ulva,* a thalloid, marine alga	e.g. *Fucus,* a thalloid, marine alga *Laminaria,* large thalloid, marine alga; one of the kelps

* a diagnostic feature.

another factor to which the organism has adapted, and this may increase in an evaporating rock pool, or decrease during rain. The surge and tug of the tide, and the pounding of waves, are additional factors which demand mechanical strength if they are to be withstood. Large waves can pick up stones and cause great damage as they crash down.

Morphological adaptations (overall structure). The thallus is firmly anchored by a holdfast (fig 2.31). This forms an intimate association with its substrate, usually rock, and is extremely difficult to dislodge. In fact, the rock often breaks before the holdfast.

The thallus shows dichotomous branching (branching into two at each branch point). This minimises resistance to the flow of water which can pass between the branches. The thallus is also tough but non-rigid and its midrib is strong and flexible.

F. vesiculosus possesses air bladders for buoyancy, thus holding its fronds up near the surface for maximum interception of light for photosynthesis.

Chloroplasts are mainly located in the surface layers for maximum exposure to light for photosynthesis.

Physiological adaptations. The dominant photosynthetic pigment is the brown pigment **fucoxanthin**. This is an adaptation for photosynthesising under water because fucoxanthin strongly absorbs blue light, which penetrates water much further than longer wavelengths such as red light.

The thallus secretes large quantities of mucilage which fills spaces within the body and exudes on to its surface. This helps to prevent desiccation by retaining water.

The solute potential of the cells is higher (less negative) than that of sea water, so water is not lost by osmosis.

Reproductive adaptations. Release of gametes is synchronised with the tides. At low tide the thallus dries and squeezes the sex organs, which are protected by mucilage, out of the conceptacles. As the tide advances, the walls of the sex organs dissolve and release the gametes.

The male gametes are motile and chemotactic, attracted by a chemical secretion of the female gametes.

The zygote develops immediately after fertilisation, minimising the risk of being swept out to sea.

2.6.6 Protozoa

Like the algae, the protozoans form a large group of protoctistans. They are unicellular, animal-like cells with heterotrophic nutrition. There are over 50 000 known species and they are found in all environments where water is present. Most are free-living and there are various methods of locomotion. Some, however, are parasites, including one (*Plasmodium*) which causes the disease which is estimated to have killed more humans than any other, namely malaria. It is still one of the world's worst killers.

A free-living protozoan, *Paramecium*, is chosen for study in this section as being typical of this level of organisation. *Plasmodium* is studied in Book 2 in chapter 15 on disease.

2.6.7 Phylum Ciliophora (ciliates)

Ciliates are a type of protozoan (fig 2.28). They have the following characteristics:

- unicellular, heterotrophic;
- possession of **cilia**, fine hairs which beat and cause movement of water, either for locomotion or feeding;
- a definite shape due to the presence of a thin, flexible outer region of cytoplasm, called the **pellicle**, which is covered by the cell surface membrane;
- a complex cell structure with a macronucleus and a micronucleus.

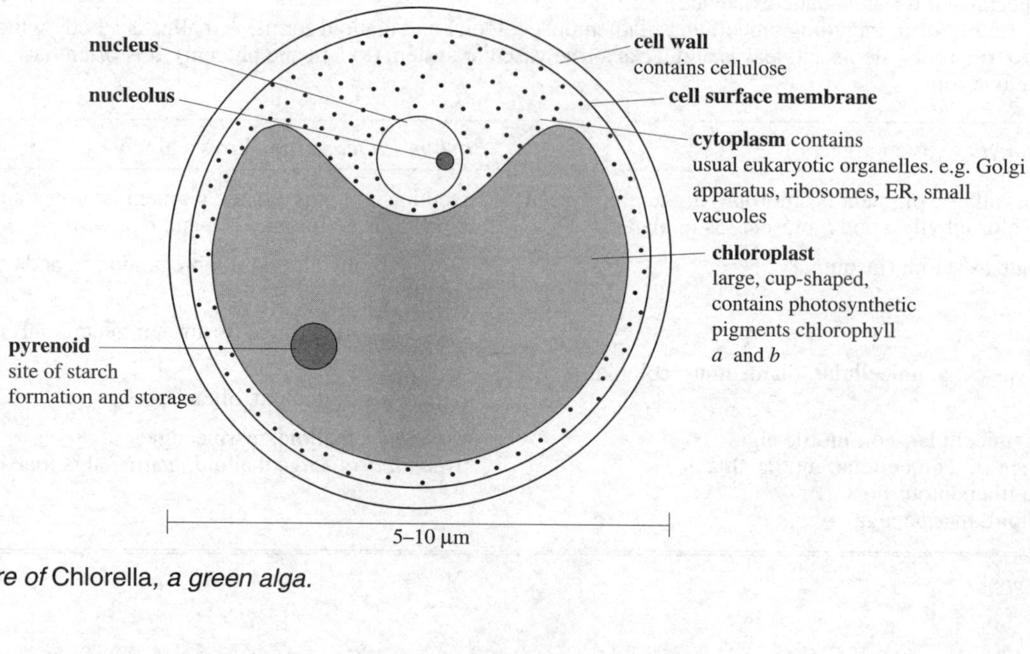

Fig 2.30 *Structure of* Chlorella, *a green alga.*

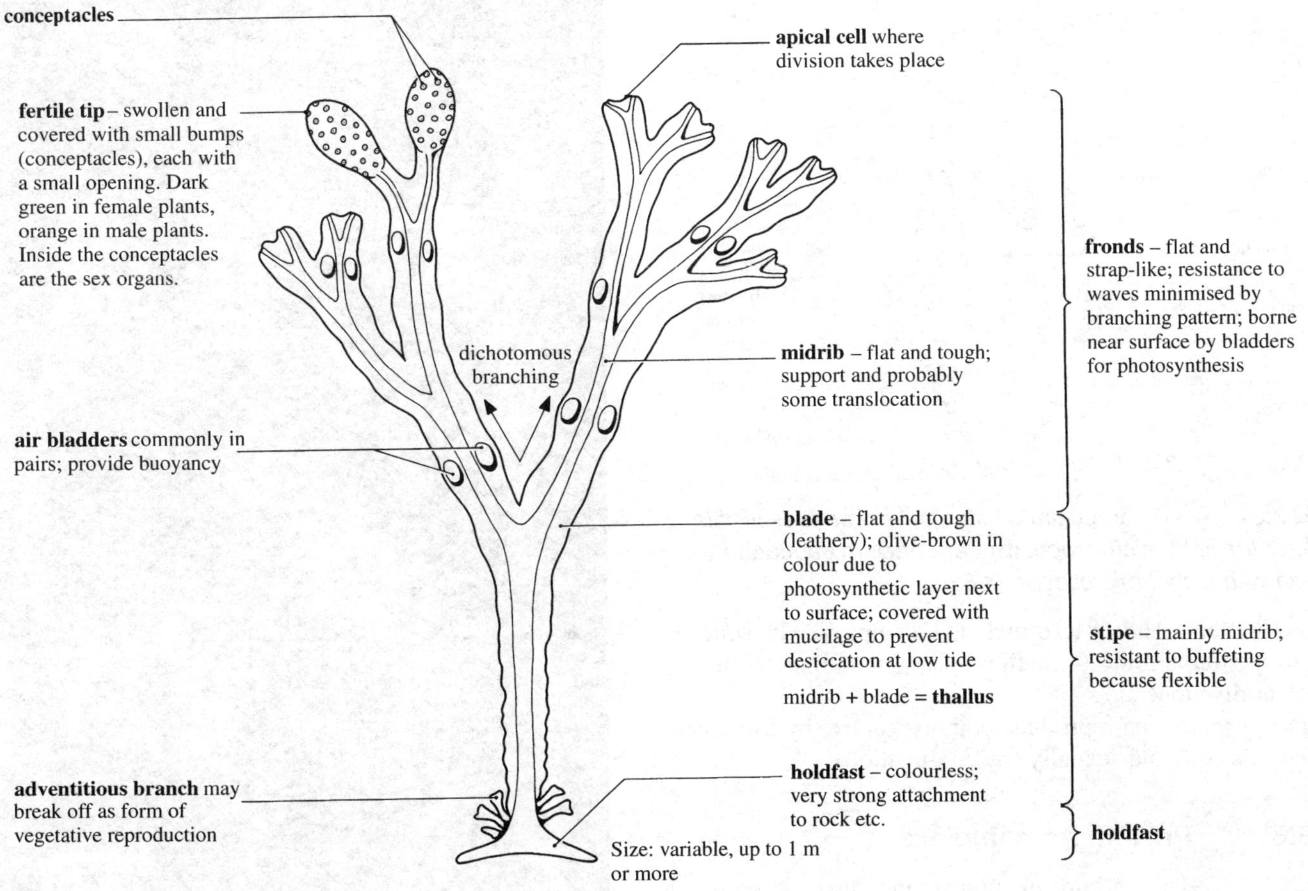

Fig 2.31 *External features of* Fucus vesiculosus*, with notes on structure, particularly adaptations to environment.*

A common example of a ciliate is *Paramecium*. It lives in stagnant water, or slow-flowing fresh water that contains decaying organic matter. Fig 2.32 shows the complex cell structure typical of the ciliates. The complexity of the cell is explained by the fact that it has to perform all the functions of a whole organism, such as feeding, osmoregulation and locomotion. The body shape is characteristic, being blunt at the front (anterior) end and tapered at the back (posterior). Cilia occur in pairs. They run in rows diagonally across the body, causing the body to rotate as they beat and move the cell forward. Between the cilia are holes leading into chambers called **trichocysts**. From these chambers, sharply tipped fine threads can be discharged which are probably used for anchorage during feeding.

Beneath the pellicle is a layer of **ectoplasm**, a clear, firm cytoplasm in the form of a gel. **Basal bodies** (identical to centrioles) are found here. They are the structures from which cilia are formed. There is also a network of fine fibres running between the basal bodies which may be involved in coordinating the beat of the cilia.

The bulk of the cytoplasm is in the form of **endoplasm**, which exists in a more liquid state than the ectoplasm. Here most of the organelles are found. The **oral groove** is a shallow groove found on the ventral (lower) surface near the front of the organism. It tapers back into a narrow tube-like **gullet** at the end of which the endoplasm is exposed to form a 'mouth' or **cytostome**. Both the oral groove and gullet are lined with cilia which beat and cause a current of water to flow towards the cytostome, carrying food particles such as bacteria in suspension. The food particles are ingested into a food vacuole formed by the endoplasm (endocytosis). The vacuoles follow a distinct pathway through the endoplasm, finishing at the **cytoproct** or anal pore, where undigested material is egested (exocytosis). During their movement through the cytoplasm, lysosomes add digestive enzymes to the vacuoles and products of digestion are absorbed into the surrounding cytoplasm.

Two fixed **contractile vacuoles** are present in the endoplasm (fig 2.32). They are responsible for osmoregulation, that is the maintenance of a constant water potential inside the cell (chapter 20). As a result of living in fresh water, water constantly enters the cell by osmosis. This water has to be pumped out by an energy-consuming active transport mechanism to prevent the cell from bursting. Around each contractile vacuole a number of canals radiate outwards and collect water before emptying it into the main vacuole.

The cell contains two nuclei. The larger, bean-shaped **macronucleus** is polyploid (has more than two sets of chromosomes). It controls metabolic activities apart from

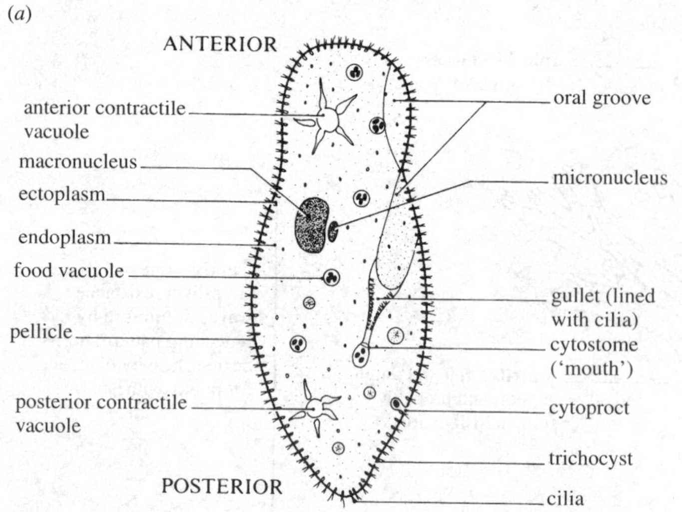

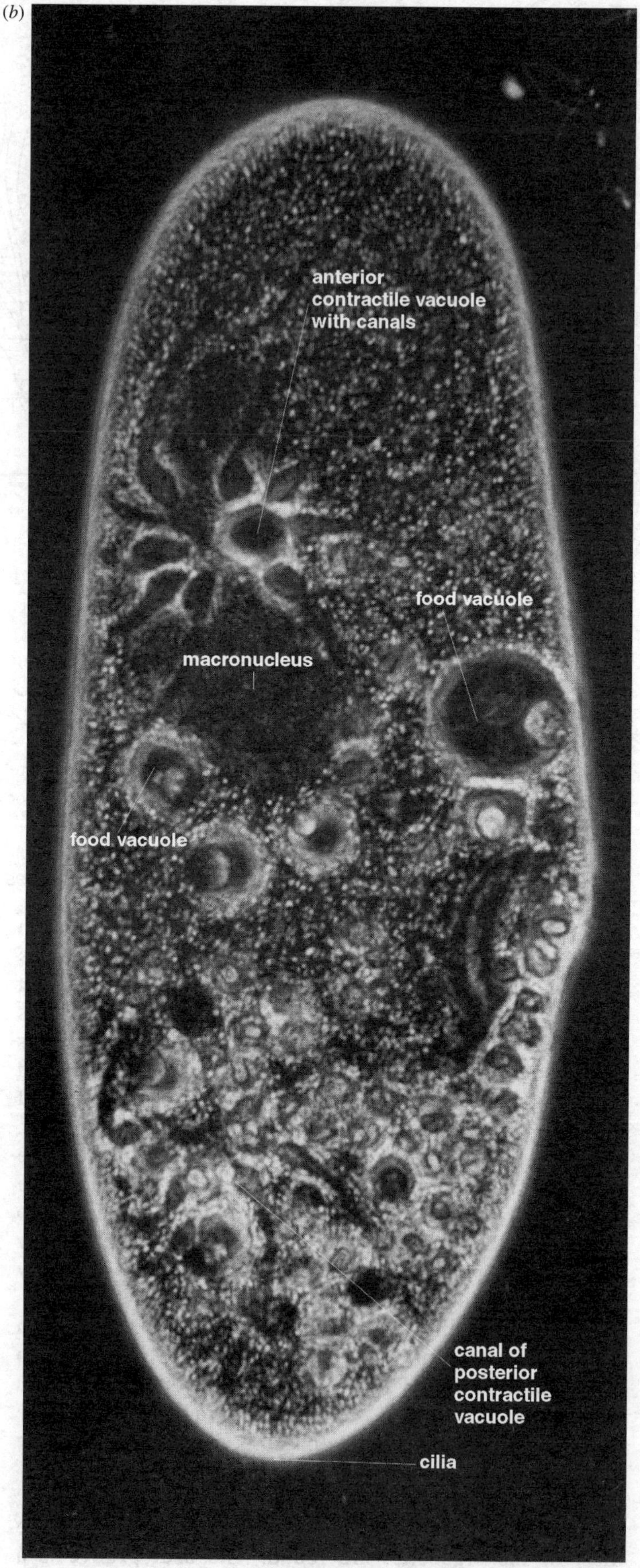

Fig 2.32 (a) Paramecium caudatum – *structures visible under the light microscope.* (b) Paramecium caudatum *as seen with a light microscope (x 832).*

reproduction. The **micronucleus** is diploid. It controls reproduction and the formation of new macronuclei during nuclear division.

Paramecium can reproduce both asexually (by transverse binary fission) and sexually (by conjugation).

2.6.8 Phylum Apicomplexa

This group of protozoans also possesses a pellicle, giving the cell a definite shape. Most, however, possess no special structures for locomotion and have limited movement. Their most distinguishing characteristic is the production of spores during asexual and sexual reproduction. An example is the parasite *Plasmodium* which causes malaria in humans and is discussed in chapter 15.

2.7 Kingdom Plantae

Although life probably began on this planet about 3.5 thousand million years ago, it was not until about 420 million years ago that the first organisms colonised the land. These were the earliest plants. Plants are autotrophic eukaryotes which have become adapted for life on land. The only other autotrophic eukaryotes are algae, which are specialised for life in water. Remember, **autotrophic** means that the organism has an inorganic source of carbon, that is carbon dioxide. Nutrition involves acquiring energy as well as carbon (see section 2.3.4) and plants are photoautotrophic, meaning that their source of energy is light. Their method of nutrition is more commonly referred to as **photosynthesis**.

The story of plant evolution is mainly of gradually improving adaptation to life on land. It is this story which will be one of the main themes in our study of plants. The classification of those plants which we shall consider in this book is shown in fig 2.33, together with a summary of some of the main trends in plant evolution which relate to adaptation to life on land and which will also be studied in this section.

2.7.1 Phylum Bryophyta (liverworts and mosses)

Bryophytes are the simplest land plants. They are thought to have evolved from green algae. The phylum contains two main classes, the Hepaticae, or liverworts, and the Musci, or mosses. Neither group is particularly well adapted for life on land and they are mainly confined to damp, shady places. The classification and characteristics of the bryophytes are summarised in table 2.8.

Bryophytes are small simple plants, with strengthening and conducting tissues absent or poorly developed. There is no true vascular tissue (xylem or phloem). They lack true roots, being anchored by thin filaments called **rhizoids** which grow from the stem. Water and mineral salts can be absorbed by the whole surface of the plant, including the rhizoids, so that the latter are mainly for anchorage, unlike true roots. (*True* roots also possess vascular tissue, as do *true* stems and leaves.) The plant surface lacks a cuticle, or has only a delicate one, and so there is no barrier against loss (or entry) of water. Nevertheless, most bryophytes have adapted to survive periods of dryness using mechanisms that are not fully understood. For example, it has been shown that the well-known xerophytic moss *Grimmia pulvinata* can survive total dryness for longer than a year at 20 °C. Recovery is rapid as soon as water becomes available.

Alternation of generations

In common with all plants and some advanced algae, such as *Laminaria*, bryophytes show alternation of generations. Two types of organism, a haploid gametophyte generation and a diploid sporophyte generation, alternate in the life cycle, summarised in fig 2.34. The haploid generation is called the **gametophyte** (*gameto*, gamete; *phyton*, plant) because it undergoes sexual reproduction to produce gametes. Production of gametes involves mitosis, so the gametes are also **haploid**. The gametes fuse to form a **diploid** zygote which grows into the next generation, the diploid sporophyte generation. It is called the **sporophyte** because it undergoes asexual reproduction to produce spores. Production of spores involves meiosis, so that there is a return to the haploid condition. The haploid spores give rise to the gametophyte generation. One of the two generations is always more

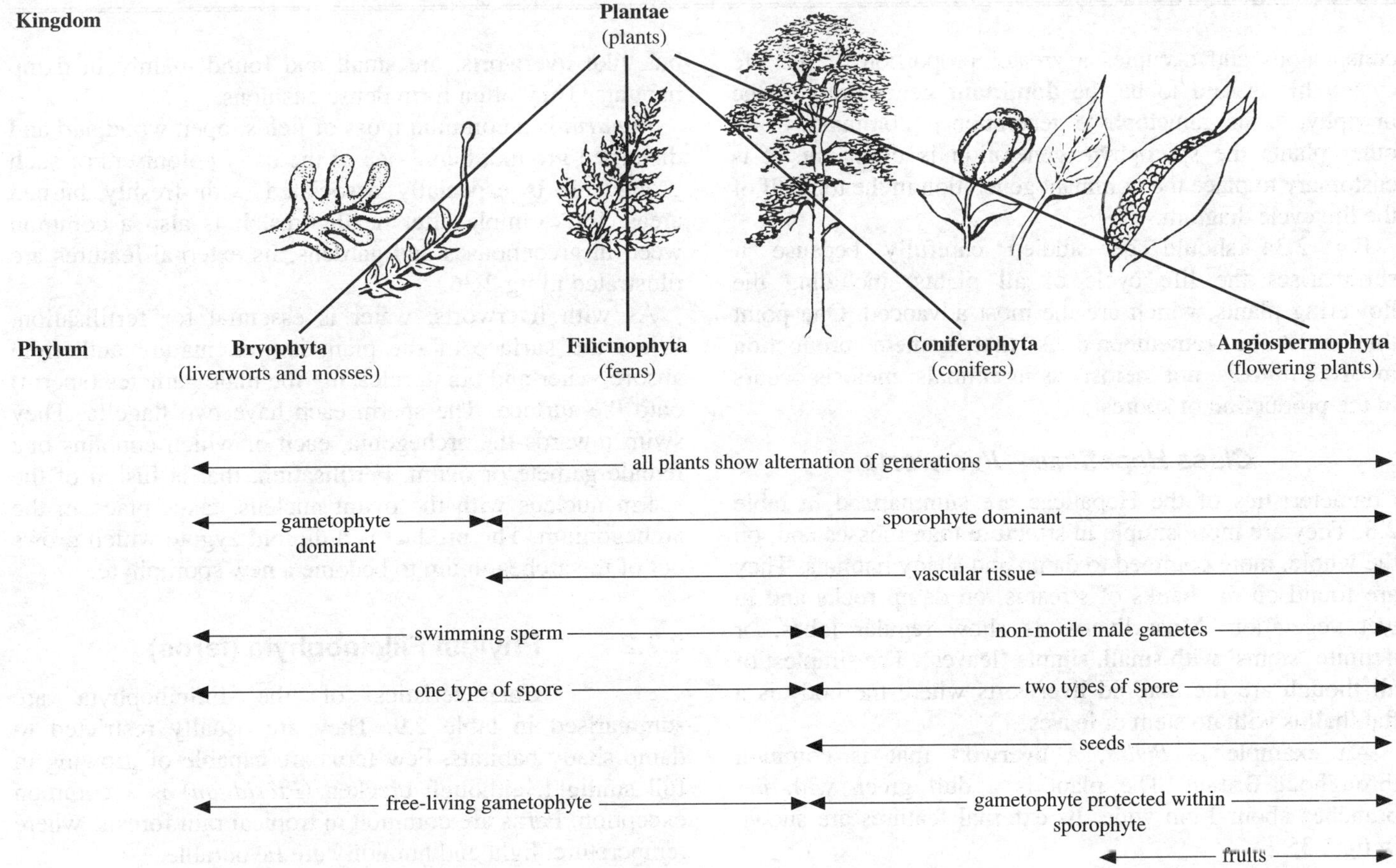

Fig 2.33 *Classification of plants and some of the main trends in plant evolution.*

Table 2.8 Classification and characteristics of the phylum Bryophyta (bryophytes).

Phylum Bryophyta	
General characteristics Alternation of generations in which the gametophyte generation is dominant No vascular tissue, that is no xylem or phloem Body is a thallus, or differentiated into simple 'leaves' and 'stems' No true roots, stems or leaves: the gametophyte is anchored by filamentous rhizoids Sporophyte is attached to, and is dependent upon, the gametophyte for its nutrition Spores are produced by the sporophyte in a spore capsule on the end of a slender stalk above the gametophyte Live mainly in damp, shady places	
Class Hepaticae (liverworts)	*Class Musci* (mosses)
Gametophyte is a flattened structure that varies from being a thallus (rare) to 'leafy' with a stem (majority), with intermediate lobed types	Gametophyte 'leafy' with a stem
'Leaves' (of leafy types) in three ranks along the stem	'Leaves' spirally arranged
Rhizoids unicellular	Rhizoids multicellular
Capsule of sporophyte splits into four valves for spore dispersal: elaters aid dispersal	Capsule of sporophyte has an elaborate mechanism of spore dispersal, dependent on dry conditions and involving teeth or pores
e.g. *Pellia,* a thallose liverwort *Marchantia,* a thallose liverwort, with antheridia and archegonia on stalked structures above the thallus *Lophocolea,* a leafy liverwort, common on rotting wood	e.g. *Funaria* *Mnium*, a common woodland moss similar in appearance to *Funaria* *Sphagnum,* bog moss: forms peat in wet acid habitats (bogs)

conspicuous and occupies a greater proportion of the life cycle; this is said to be the **dominant** generation. In the bryophytes, the gametophyte generation is dominant. In all other plants the sporophyte generation is dominant. It is customary to place the dominant generation in the top half of the life cycle diagram.

Fig 2.34 should be studied carefully because it summarises the life cycle of all plants, including the flowering plants, which are the most advanced. One point that must be remembered is that gamete production involves mitosis, not meiosis as in animals; meiosis occurs in the production of spores.

Class Hepaticae – liverworts

Characteristics of the Hepaticae are summarised in table 2.8. They are more simple in structure than mosses and, on the whole, more confined to damp and shady habitats. They are found on the banks of streams, on damp rocks and in wet vegetation. Most liverworts show regular lobes, or definite 'stems' with small, simple 'leaves'. The simplest of all though are the thalloid liverworts where the body is a flat thallus with no stem or leaves.

An example is *Pellia*, a liverwort that is common throughout Britain. The plant is a dull green with flat branches about 1 cm wide. Its external features are shown in fig 2.35.

Class Musci – mosses

Characteristics of the Musci are summarised in table 2.8. They have a more differentiated structure than liverworts but, like liverworts, are small and found mainly in damp habitats. They often form dense cushions.

Funaria is a common moss of fields, open woodland and disturbed ground, being one of the early colonisers of such ground. It is especially associated with freshly burned areas, for example after heath fires. It is also a common weed in greenhouses and gardens. Its external features are illustrated in fig 2.36.

As with liverworts, water is essential for fertilisation. When the surface of the plant is wet, mature antheridia absorb water and burst, releasing the male gametes (sperm) onto the surface. The sperm each have two flagella. They swim towards the archegonia, each of which contains one female gamete or ovum. Fertilisation, that is fusion of the sperm nucleus with the ovum nucleus, takes place in the archegonium. The product is a diploid zygote which grows out of the archegonium to become a new sporophyte.

2.7.2 Phylum Filicinophyta (ferns)

Characteristics of the Filicinophyta are summarised in table 2.9. They are usually restricted to damp shady habitats. Few ferns are capable of growing in full sunlight, although bracken (*Pteridium*) is a common exception. Ferns are common in tropical rain forests, where temperature, light and humidity are favourable.

Ferns are vascular plants, meaning that they contain vascular tissue. **Vascular tissue** is made up of xylem and phloem. These tissues are concerned with translocation (transport) of water and nutrients round the plant body.

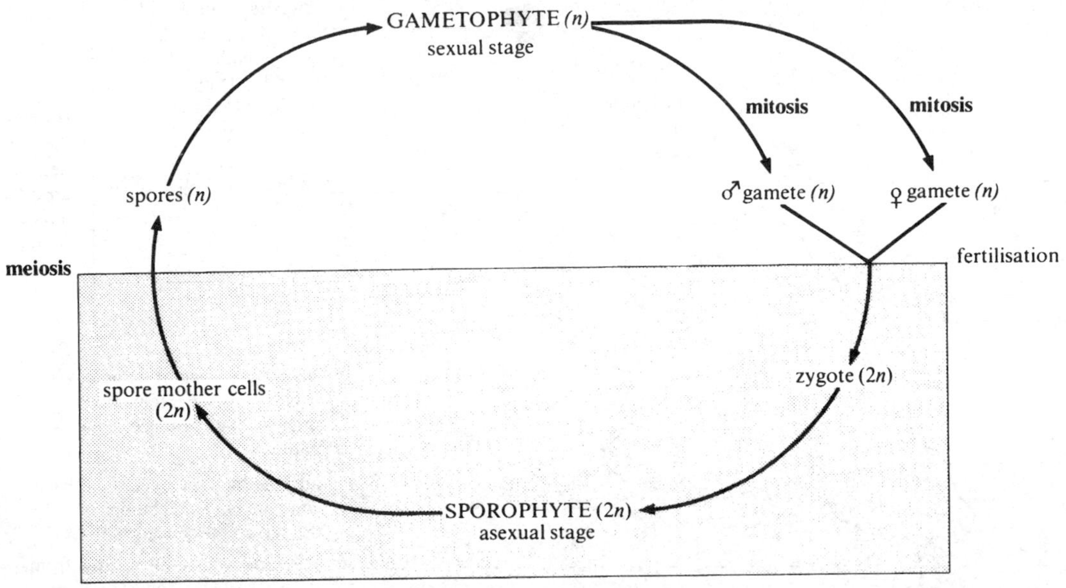

Fig 2.34 *Generalised life cycle of a plant showing alternation of generations. Note the haploid stages (n) and diploid stages (2n). The gametophyte is always haploid and always produces gametes by mitosis. The sporophyte is always diploid and always produces spores by meiosis.*

Xylem carries mainly water and mineral salts, whereas **phloem** carries mainly organic solutes in solution such as sugars. Vascular tissue is a major evolutionary advance compared with the simple conducting cells of some bryophytes and algae. It is found only in the sporophyte generation, and is one reason why the sporophyte generation becomes conspicuous in all vascular plants.

Vascular tissue has two important properties. First, it forms a **transport system**, conducting food and water around the multicellular body, thus allowing the development of large, complex bodies. Secondly, these bodies can be **supported** because xylem, apart from being a conducting tissue, contains lignified cells of great strength and rigidity. Another lignified tissue, sclerenchyma, also develops in vascular plants and supplements the mechanical role of xylem (section 6.2.1).

The sporophyte generation possesses true roots, stems and leaves. Roots penetrate the soil with the result that water and dissolved nutrients can be obtained more easily. Xylem conducts it to other parts of the plant.

Once plant bodies could achieve support above the ground, there must have been competition for light, so there would be a tendency for taller forms to evolve. Ferns and tree-ferns were the dominant vegetation for about 70 million years, from the Devonian to the Permian eras. After this conifers and, later, flowering plants largely replaced them (see the geological time scale in the appendix in Book 2).

Despite these advances in adapting to a land environment, which are associated with the sporophyte generation, in ferns there remains a major problem with the gametophyte. This is even smaller and more susceptible to desiccation (drying) than the bryophyte gametophyte. It is called a **prothallus**, and produces sperm which must swim to reach the female gametes, as is the case in bryophytes.

Table 2.9 Characteristics of the phylum Filicinophyta (ferns).

Phylum Filicinophyta (ferns)
General characteristics
Alternation of generations in which the sporophyte is dominant
Gametophyte is reduced to a small, simple prothallus
Vascular tissue present (xylem and phloem) in sporophyte: sporophyte therefore has true roots, stems and leaves
Leaves relatively large and called fronds
Spores produced in sporangia which are usually in clusters called sori
e.g. *Dryopteris filix-mas* (male fern)
Pteridium (bracken)

The male fern (*Dryopteris filix-mas*)

Dryopteris filix-mas is probably the most common British fern and is found in damp woods, hedgerows and other shady places throughout the country. The fronds (leaves) of the sporophyte may reach a metre or more in height and grow from a thick horizontal stem, or **rhizome**. This bears **adventitious roots**. Branches from the main stem may eventually break away and give rise to separate plants, a form of vegetative reproduction. The bases of the fronds are covered with dry brown scales called **ramenta** that protect the young leaves from frost or drought. The young leaves show a characteristic tightly rolled structure. The ramenta gradually become smaller and less dense up the main axis of the frond. This axis is called the **rachis**, and the leaflets either side are called the pinnae. The small rounded subdivisions of the pinnae are called **pinnules**. The external features of the sporophyte of *Dryopteris filix-mas* are

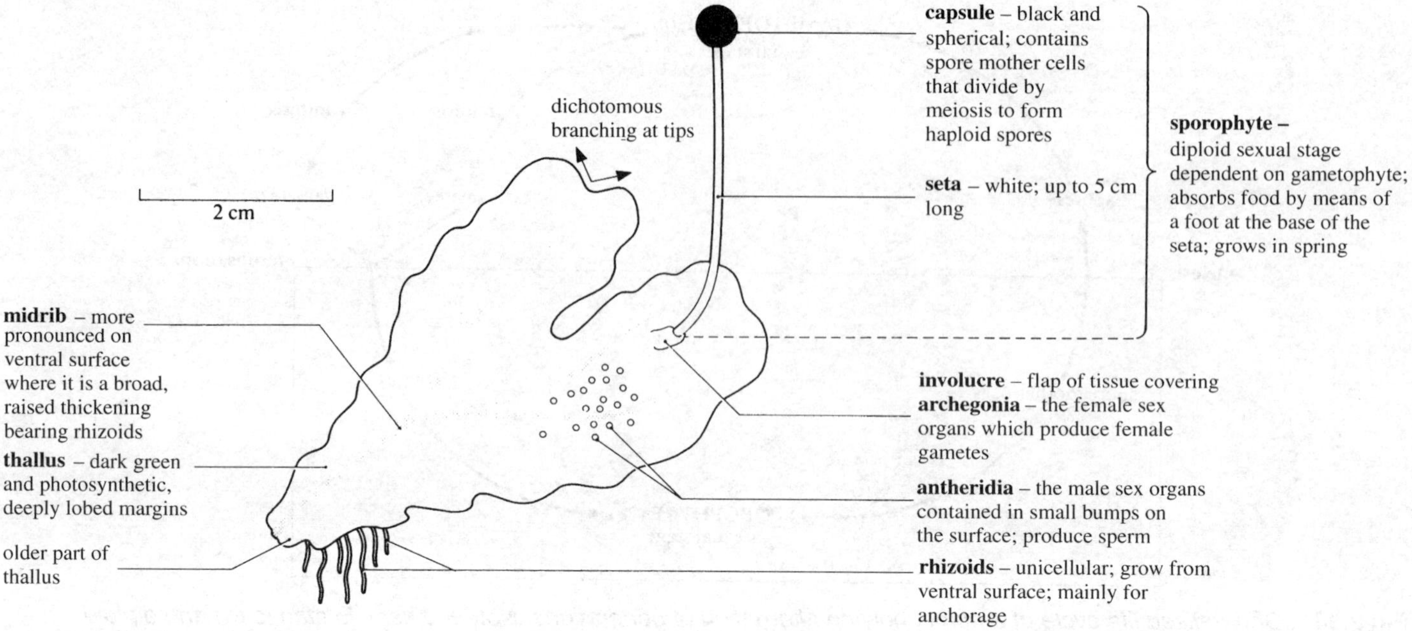

Fig 2.35 *External features of* Pellia, *a liverwort. The gametophyte is shown with the dependent sporophyte generation attached.*

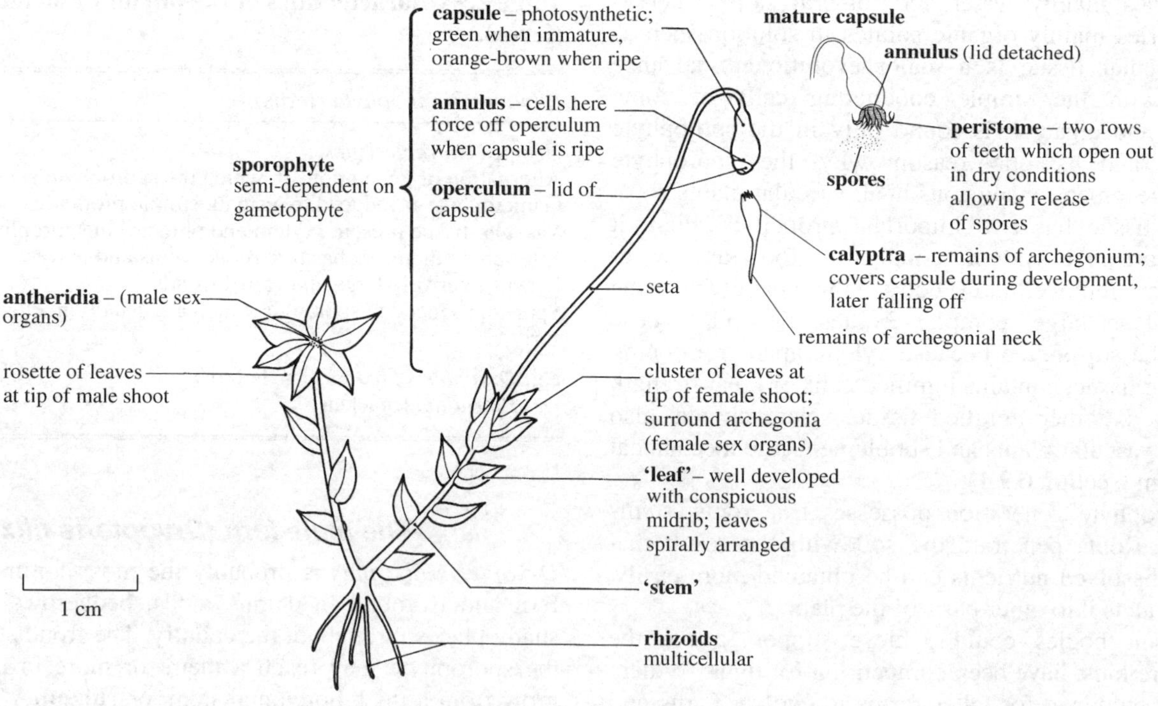

Fig 2.36 *External features of* Funaria, *a moss. The gametophyte is shown with the semi-dependent sporophyte generation attached.*

shown in fig 2.37. The gametophyte is shown in fig 2.38.

Spores are produced during late summer in structures called **sporangia**. Sporangia develop in clusters called **sori** on the undersides of pinnules (fig 2.37*c*, *d* and *e*). Each sorus has a protective covering called an **indusium**. Inside each sporangium diploid spore mother cells divide by meiosis to produce haploid spores. When mature, the indusium shrivels and drops off, and the exposed sporangium walls begin to dry out. Eventually the wall ruptures and spores are catapulted from the sporangium (fig 2.37*e*).

The spores germinate to form the gametophyte generation. The gametophyte is a thin heart-shaped plate of cells about 1 cm in diameter (fig 2.38). It is green and photosynthetic and is anchored by unicellular rhizoids to the soil. This delicate prothallus lacks a cuticle and is prone to drying out, so can only survive in damp conditions.

The gametophyte (prothallus) produces simple antheridia and archegonia on its lower surface. These sex organs protect the gametes within them. Gametes are produced by mitosis of gamete mother cells, the antheridia producing sperm and each archegonium an ovum, as in the bryophytes. Each sperm has flagella. When ripe, and conditions are wet, each antheridium releases its sperm, which swim through a film of water towards the archegonia. The product of fertilisation is a diploid zygote. Note that fertilisation is still dependent on water as in the bryophytes.

The zygote grows into the sporophyte generation. The young embryo absorbs nutrients from the gametophyte until its own roots and leaves can take over the role of nutrition (fig 2.38*b*). The gametophyte soon withers and dies.

2.7.3 Seed-bearing plants

The most successful group of plants have seeds. They probably have their origin among extinct seed-producing members of the ferns and their close relatives. Classification and characteristics of the seed-bearing plants are summarised in table 2.10.

Table 2.10 shows the two main groups of seed-bearing plants, the **conifers** and **angiosperms**. The latter are commonly known as the **flowering plants**. In conifers, ovules (later seeds) are located on the surfaces of specialised scale leaves called ovuliferous scales. These are arranged in cones. In angiosperms, ovules, and therefore seeds, are enclosed, giving more protection.

2.7.4 Phylum Coniferophyta (conifers)

Characteristics of the Coniferophyta are summarised in table 2.10.

Conifers are a successful group of plants of worldwide distribution, accounting for about one-third of the world's forests. They are trees or shrubs, mostly evergreen, with needle-like leaves. Most of the species are found at higher altitudes and further north than any other trees. Conifers are commercially important as 'softwoods', being used not only for timber but for resins, turpentine and wood pulp. They include pines, larches (which are deciduous), firs, spruces and cedars. A typical conifer is *Pinus sylvestris*, the Scots pine.

Pinus sylvestris is found throughout central and northern Europe, Russia and North America. It is native to Scotland, though it has been introduced elsewhere in Britain. It is planted for timber and ornament, being a stately, attractive tree up to 36 m in height with a characteristic pink to orange-brown flaking bark. It grows most commonly on sandy or poor mountain soils and consequently the root system is often shallow and spreading. Its external features are illustrated in fig 2.39.

Each year a whorl of lateral buds around the stem grows out into a whorl of branches. The roughly conical appearance of *Pinus* and other conifers is due to the transition from whorls of shorter (younger) branches at the tops to longer (older) branches lower down. The latter usually die and drop off as the tree grows, leaving the mature trees bare for some distance up their trunks (fig 2.39).

The main branches and trunk continue growth from year to year by the activity of an apical bud. They are said to show **unlimited growth**. They have spirally arranged scale

Table 2.10 Classification and characteristics of the seed-bearing plants.

Seed-bearing plants	
General characteristics Sporophyte is the dominant generation; gametophyte generation is severely reduced Sporophyte produces two types of spores (in other words, it is **heterosporous**). The two types are microspores and megaspores; microspore = pollen grain, megaspore = embryo sac The embryo sac (megaspore) remains completely enclosed in the ovule (megasporangium); a fertilised ovule is a seed* Water is not needed for sexual reproduction because male gametes do not swim (except in a few primitive members); they are conveyed to the ovum by a pollen tube to effect fertilisation* Complex vascular tissues in roots, stems and leaves	
Phylum Coniferophyta (conifers)	*Phylum Angiospermophyta* (flowering plants)
Usually produce cones on which sporangia, spores and seeds develop	Produce flowers in which sporangia, spores and seeds develop*
Seeds are not enclosed in an ovary. They lie on the surface of specialised leaves called ovuliferous scales in structures called cones*	Seeds are enclosed in an ovary*
No fruit because no ovary	After fertilisation, the ovary develops into a fruit*
	Classes Dicotyledoneae and Monocotyledoneae (see table 2.11)

*diagnostic feature.

(a)

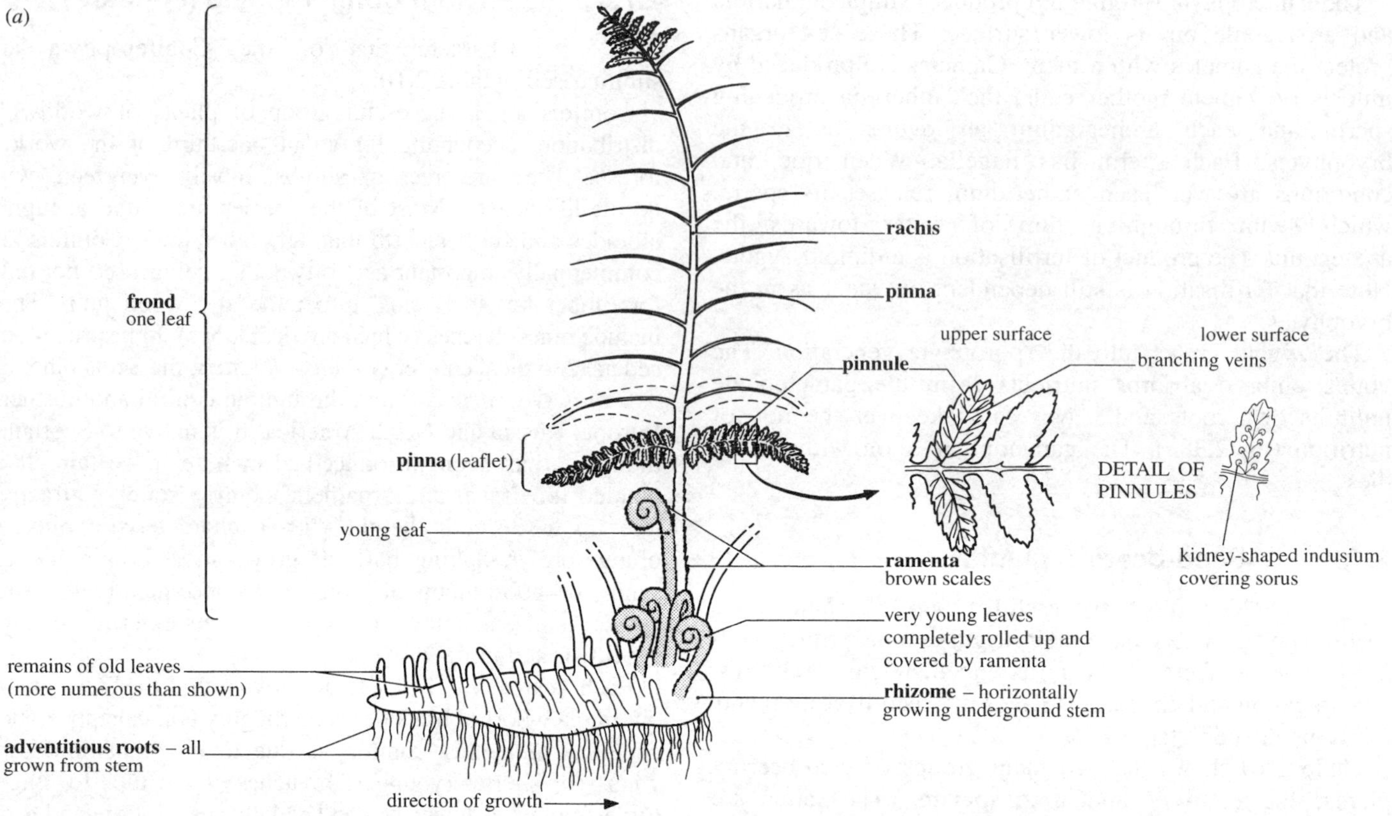

(b)

(c)

Fig 2.37 *External features of the sporophyte generation of Dryopteris filix-mas, the male fern.* (a) *Diagram with details of one pair of pinnae; others have the same structure.* (b) *The fronds.* (c) *Underside of frond showing sori (some covered with indusium).* (d) *and* (e) *opposite.*

(d)

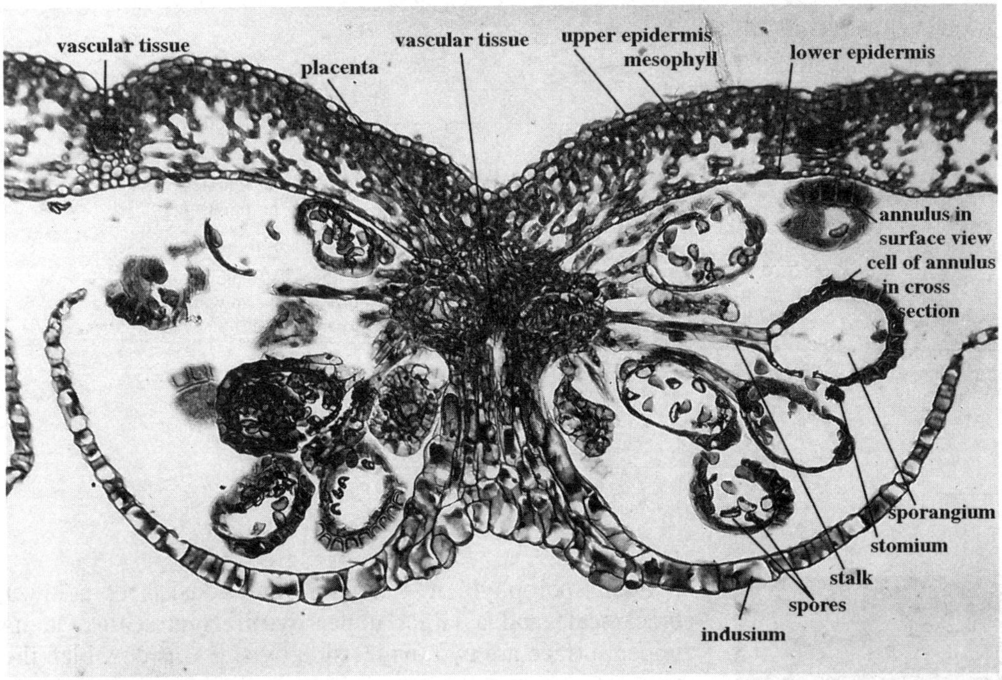

Fig 2.37 (cont.) (d) *LS of a sorus as seen with a light microscope.* (e) *LS of sorus, details of sporangium and spore dispersal.*

(e)

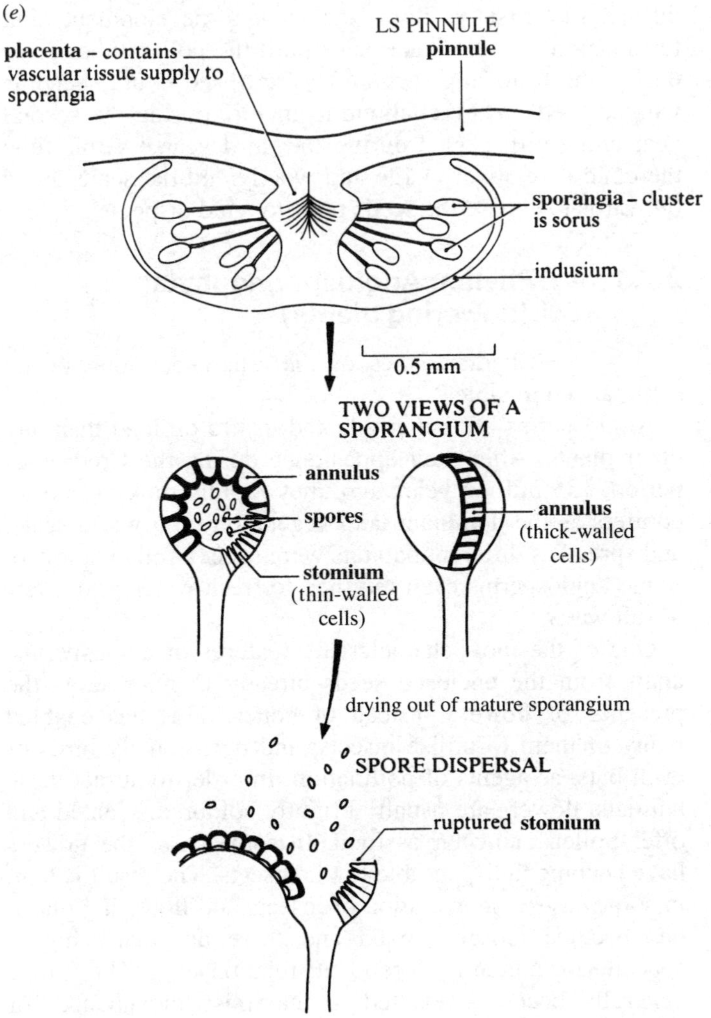

leaves, in the axils of which are buds that develop into very short branches (2–3 mm) called **dwarf shoots**. These are shoots of **limited growth** and at their tips grow two leaves. Once the shoot has grown, the scale leaf at its base drops off leaving a scar. The leaves are needle-like, reducing the surface area available for the loss of water. They are also covered with a thick, waxy cuticle and have sunken stomata, further adaptations for conserving water. These xeromorphic features ensure that the tree does not lose too much water from its evergreen leaves during cold seasons, when water may be frozen or difficult to absorb from the soil. After two to three years the dwarf shoots and leaves drop off together, leaving a further scar.

The tree is the sporophyte generation. In spring, male and female cones are produced on the same tree. The male cones are about 0.5 cm in diameter, rounded and found in clusters behind the apical buds at the bases of new shoots. They develop in the axils of scale leaves in the place of dwarf shoots. Female cones arise in the axils of scale leaves at the tips of new strong shoots, at some distance from the male cones and in a more scattered arrangement. Since they take three years to complete growth and development, they are of various sizes, ranging from about 0.5–6 cm on a given tree. They are green when young, becoming brown or reddish-brown in their second year. Both male and female cones consist of spirally arranged, closely packed sporophylls (modified leaves) around a central axis (fig 2.39).

Each sporophyll of a male cone has two **microsporangia** or **pollen sacs** on its lower surface. Inside each pollen sac, meiosis takes place to form haploid **pollen grains** or **microspores**. These contain the male gametes. Each grain has two large air sacs to aid in wind dispersal. During May the cones become yellow in appearance as they release clouds of pollen. At the end of the summer they wither and drop off.

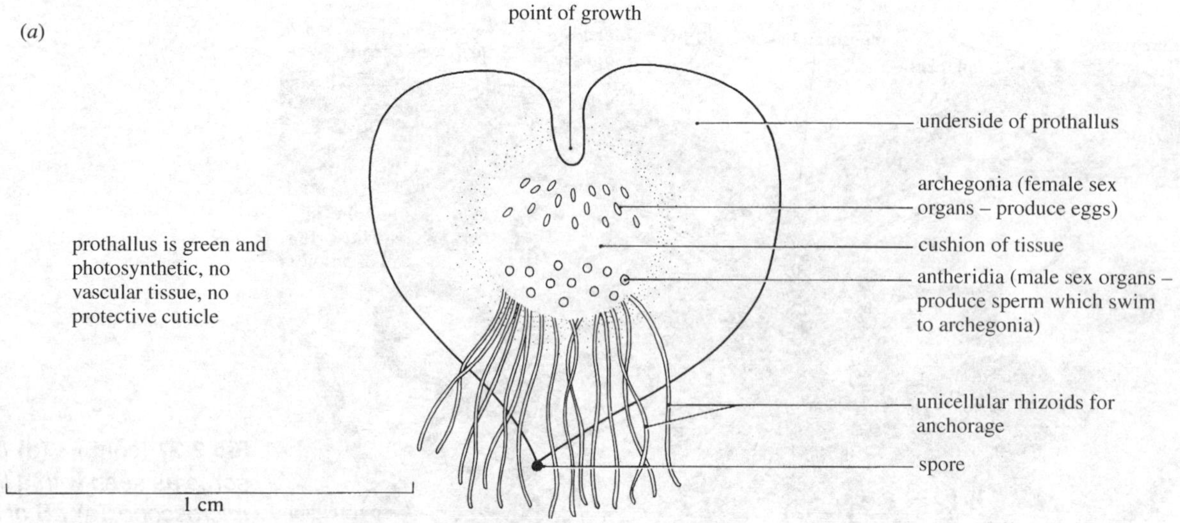

Fig 2.38 *External features of the gametophyte generation of* Dryopteris, *known as the prothallus.* (a) *The prothallus is green and photosynthetic. It has no vascular tissue and no protective cuticle.* (b) *Prothallus with the first frond of the next sporophyte generation growing from it. At first the sporophyte is dependent on the gametophyte for water and minerals but it soon becomes an independent plant and the gametophyte dies.*

Each sporophyll of a female cone consists of a lower bract scale and a larger upper **ovuliferous scale**. On its upper surface are two ovules side by side, inside which the female gametes are produced. Pollination takes place during the first year of the cone's development, but fertilisation does not take place until the pollen tubes grow during the following spring. The fertilised ovules become winged seeds. They continue to mature during the second year and are dispersed during the third year. By this time the cone is relatively large and woody and the scales bend outwards to expose the seeds prior to wind dispersal.

2.7.5 Phylum Angiospermophyta (flowering plants)

Characteristics of the Angiospermophyta are summarised in table 2.10.

Angiosperms are better adapted to life on land than any other plants. After their appearance during the Cretaceous period, 135 million years ago, they rapidly took over from conifers as the dominant land vegetation on a world scale, and spread as different habitats were successfully exploited. Some angiosperms even returned to fresh water, and a few to salt water.

One of the most characteristic features of angiosperms, apart from the enclosed seeds already mentioned, is the presence of **flowers** instead of cones. This has enabled many of them to utilise insects, and occasionally birds or even bats, as agents of pollination. In order to attract these animals, flowers are usually brightly coloured, scented and offer pollen or nectar as food. In some cases the flowers have become indispensable to the insects. The result is that, in some cases, the evolution of insects and flowering plants has become closely linked and there are many highly specialised, mutually dependent, relationships. The flower generally becomes adapted to maximise the chances of pollen transfer by the insect and the process is therefore more reliable than wind pollination. Insect-pollinated plants need not, therefore, produce as much pollen as wind-

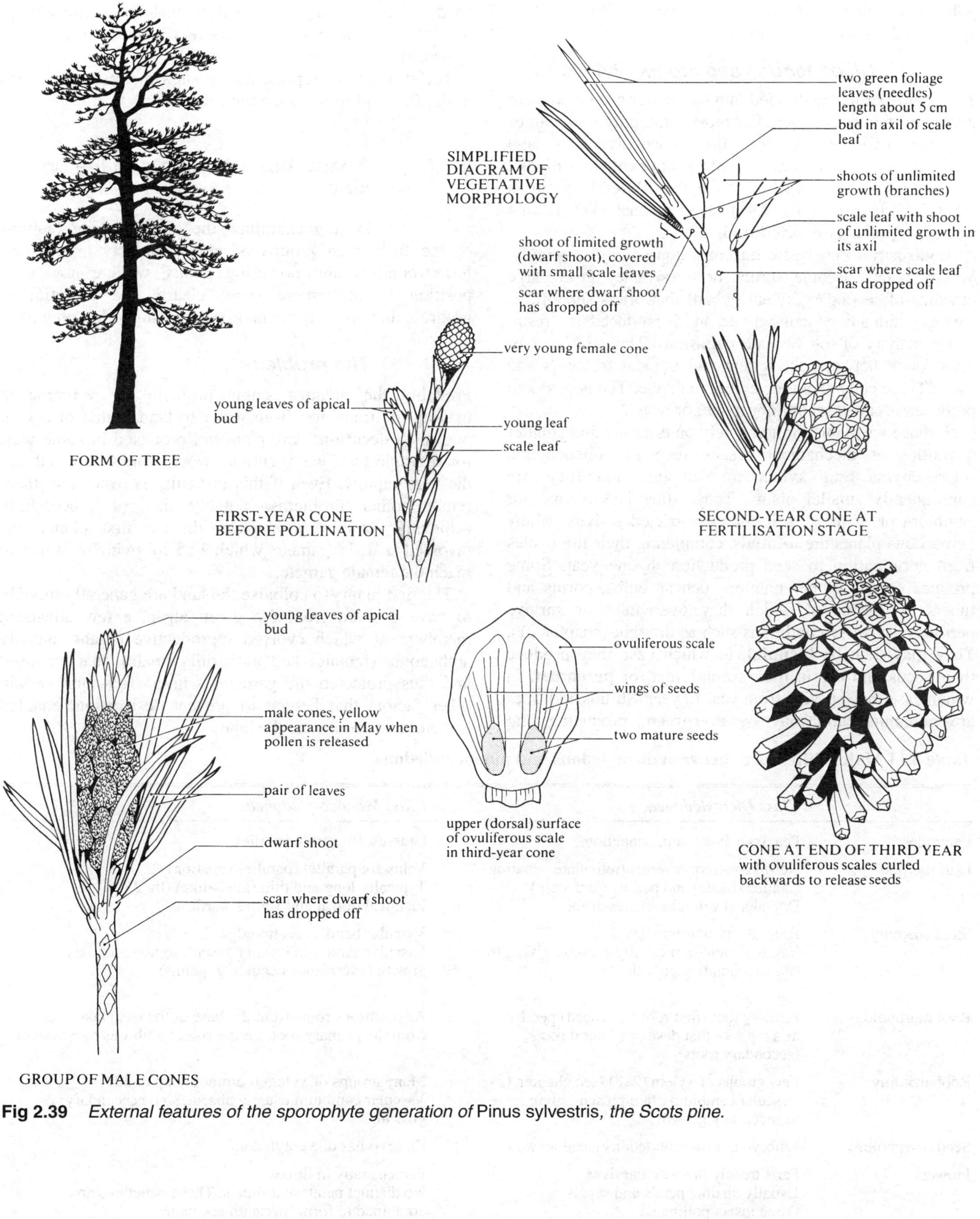

Fig 2.39 *External features of the sporophyte generation of* Pinus sylvestris, *the Scots pine.*

pollinated plants. Nevertheless, many flowers are specialised for wind pollination.

Dicotyledons and monocotyledons

The angiosperms are divided into two major groups that are given the status of classes. The most commonly used names for the two groups are the monocotyledons and dicotyledons, usually abbreviated to **monocots** and **dicots**. A summary of the ways in which they differ is given in table 2.11 (see also fig 2.40). The modern view is that monocots probably evolved from dicots.

Angiosperms may be **herbaceous** (non-woody) or **woody**. Woody plants become shrubs or trees. They grow large amounts of secondary xylem (wood) that offers support, as well as being a conducting tissue, and is produced as a result of the activity of the vascular **cambium**. This is a layer of cells found between the xylem and phloem in stems and roots. These cells retain the ability to divide. The new xylem produced is called **secondary xylem** or wood.

Herbaceous plants, or herbs, rely on turgidity and smaller quantities of mechanical tissues such as collenchyma, sclerenchyma and xylem for support, and they are consequently smaller plants. They either lack a vascular cambium or, if present, it shows restricted activity. Many herbaceous plants are **annuals**, completing their life cycles from germination to seed production in one year. Some produce organs of perennation such as bulbs, corms and tubers by means of which they overwinter or survive periods of adverse conditions such as drought (chapter 21). They may then be **biennial**, in which case they produce their seeds and die in their second year, or **perennial**, in which case they survive from year to year. Shrubs and trees are perennial, and may be **evergreen**, producing and shedding leaves all year round so that leaves are always present, or **deciduous**, shedding leaves in seasons of cold or drought.

The structure of representative angiosperms is described in figs 2.41–44 to illustrate their diversity.

2.7.6 Adaptations of plants to life on land

Having examined the distinguishing features of the four main groups of plants, namely bryophytes, ferns, conifers and flowering plants, we are now in a position to understand more clearly the evolutionary progress that plants have made on adapting to life on land.

The problem

Probably the greatest single problem to overcome in making the transition from water to land is that of drying out, or **desiccation**. Any plant not protected in some way, for example by a waxy cuticle, would tend to dry out and die very rapidly. Even if this difficulty is overcome, there remain other problems, notably that of successfully achieving sexual reproduction. In the first plants this involved a male gamete which had to swim in water to reach the female gamete.

The first plants to colonise the land are generally thought to have evolved from the green algae, a few advanced members of which evolved reproductive organs, namely archegonia (female) and antheridia (male), that enclosed and thus protected the gametes within. This, and certain other factors that helped to prevent desiccation, enabled some of them to invade the land.

Table 2.11 Major differences between dicotyledons and monocotyledons.

	Class Dicotyledoneae	*Class Monocotyledoneae*
Examples	Pea, rose, buttercup, dandelion	Grasses, iris, orchids, lilies
Leaf morphology	Net-like pattern of veins (reticulate venation) Lamina (blade) and petiole (leaf stalk) Dorsal and ventral surfaces differ	Veins are parallel (parallel venation) Typically long and thin (grass-like) (fig 2.40) Identical dorsal and ventral surfaces
Stem anatomy	Ring of vascular bundles Vascular cambium usually present, giving to rise to secondary growth	Vascular bundles scattered Vascular cambium usually absent, so no secondary growth (exceptions occur, e.g. palms)
Root morphology	Primary root (first root from seed) persists as a tap root that develops lateral roots (secondary roots)	Adventitious roots from the base of the stem take over from the primary root, giving rise to a fibrous root system
Root anatomy	Few groups of xylem (2–8) (see chapter 13) Vascular cambium often present, giving rise to secondary growth	Many groups of xylem (commonly up to 30) Vascular cambium usually absent, so no secondary growth
Seed morphology	Embryo has two cotyledons (seed leaves)	Embryo has one cotyledon
Flowers	Parts mainly in fours and fives Usually distinct petals and sepals Often insect pollinated	Parts usually in threes No distinct petals and sepals. These structures are combined to form 'perianth segments' Often wind pollinated

(a)

(b)

Fig 2.40 *Structure of* (a) *a monocotyledonous leaf and* (b) *a dicotyledonous leaf.*

One of the main evolutionary trends in plants is their gradually increasing independence from water.

The main problems associated with the transition from an aquatic to a terrestrial environment are summarised below.

- **Desiccation.** Air is a drying medium and water is essential for life for many reasons (section 3.1.2). Means of obtaining water and conserving it are required.
- **Reproduction.** Delicate sex cells must be protected and motile male gametes (sperm) require water if they are to reach the female gametes.
- **Support.** Air, unlike water, offers no support to the plant body.
- **Nutrition.** Plants require light and carbon dioxide for photosynthesis, so at least part of the body must be above ground. Minerals and water, however, are at ground level or below ground, and to make efficient use of these, part of the plant must grow below ground in darkness.
- **Gaseous exchange.** For photosynthesis and respiration, carbon dioxide and oxygen must be exchanged with the atmosphere rather than a surrounding solution.
- **Environmental variables.** Water, particularly large bodies of water like lakes and oceans, provides a very constant environment. A terrestrial environment, however, is much more subject to changes in important factors such as temperature, light intensity, ionic concentration and pH.

Liverworts and mosses

Mosses are well adapted to a terrestrial environment in their mode of spore dispersal, which depends on the drying out of the capsule and the dispersal of small, light spores by wind. However, they still show a great reliance on water for the following reasons.

- They are still dependent on water for reproduction because sperms must swim to the archegonia. They are adapted to release their sperms when water is available since only then do the antheridia burst. They are partly adapted to land because the gametes develop in protective structures, the antheridia and archegonia.
- There are no special supportive structures, so the plants are restricted in upward growth.
- They are dependent on availability of water and mineral salts close to or at the surface of the soil, because they have no roots to penetrate the substrate. However, rhizoids are present for anchorage, an adaptation to a solid substratum.

2.4 Liverworts and mosses have sometimes been described as the amphibians of the plant world. Briefly explain why this should be.

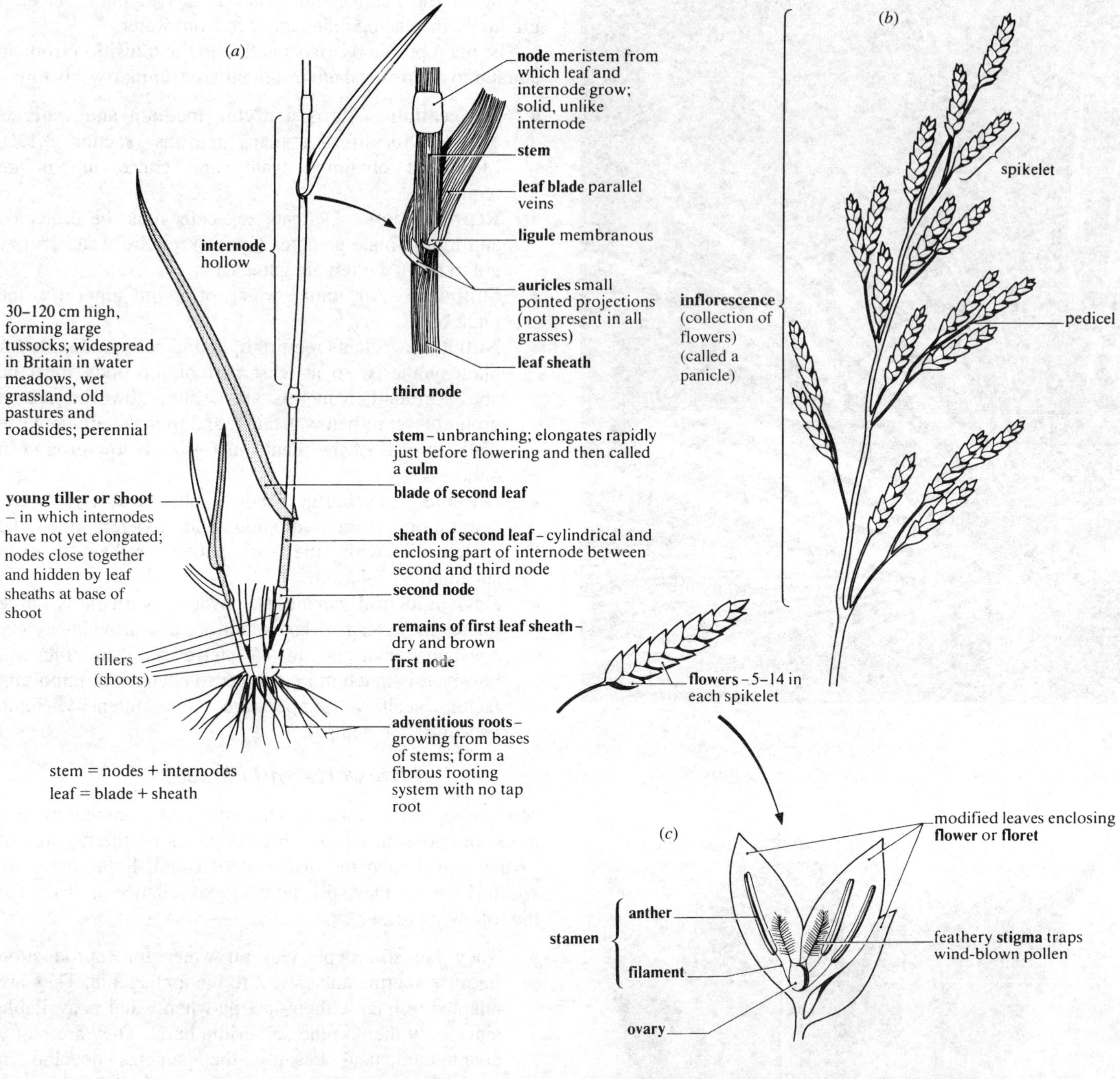

Fig 2.41 *Vegetative and floral morphology of the grass meadow fescue (Festuca pratensis), an herbaceous monocotyledon. The second leaves are shaded. Leaves are typically in two rows, alternating on opposite sides of the stem.* (a) *Vegetative morphology.* (b) *Floral morphology – the inflorescence.* (c) *Detail of one open flower or floret: two small petal-like structures (lodicules) which enclose the ovary have been omitted.*

Ferns

2.5 How are ferns better adapted to life on land than liverworts or mosses?

2.6 In what main respects are mosses, liverworts and ferns poorly adapted to life on land?

Seed-bearing plants – conifers and flowering plants

One of the main problems for plants living on land is the vulnerability of the gametophyte generation. For example, in ferns the gametophyte is a delicate prothallus and it produces male gametes, or sperm, dependent on water for swimming. In seed plants, however, the gametophyte generation is protected and very much reduced.

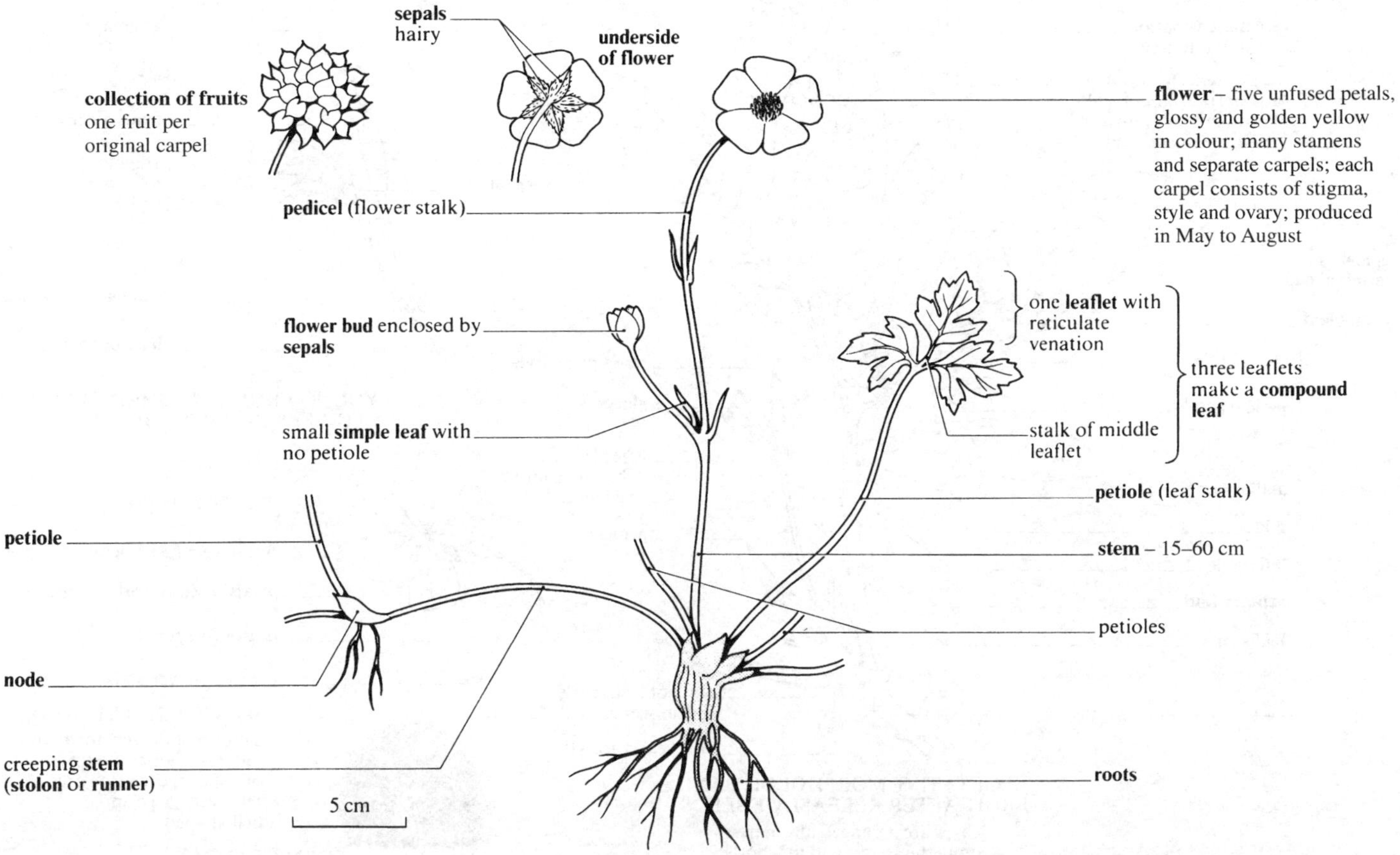

Fig 2.42 *Vegetative and floral morphology of the creeping buttercup (*Ranunculus repens*), an herbaceous dicotyledon. It is a common perennial plant throughout Britain, found in wet fields, woods, gardens and on waste ground.*

Three important advances have been made by seed plants, first the development of two types of spore (heterospory), secondly the development of non-swimming gametes and thirdly the development of seeds.

Heterospory and non-swimming male gametes. An important evolutionary advance was made when certain ferns and their close relatives developed two types of spore. This is known as **heterospory**, and the plants are described as heterosporous. *All* seed-bearing plants are heterosporous. They produce large spores called **megaspores** in one type of sporangium (megasporangium) and small spores called **microspores** in another type of sporangium (microsporangium). When spores grow they form gametophytes (fig 2.34). Megaspores produce female gametophytes and microspores produce male gametophytes. In seed-bearing plants the gametophytes produced by megaspores and microspores are very small and never released from the spores. Thus the gametophytes are protected from desiccation, an evolutionary advance. However, sperms from the male gametophyte still have to travel to the female gametophyte. This is made easier by dispersal of the microspores. Being very small they can be produced in large numbers and blown away from the parent sporophyte by wind. By chance they can be brought into closer proximity to the megaspores, which in seed plants stay attached to the parent sporophyte (fig 2.45). This is the basis of **pollination** in seed plants, where microspores are in fact the pollen grains. Inside the pollen grains male gametes form.

In seed plants another evolutionary advance has occurred. The male gametes no longer have to swim to the female gametes because seed plants have evolved pollen tubes. These grow from the pollen grains to the female gametes and deliver the male gametes. There are no longer any swimming sperm, just male nuclei.

Thus for the first time, plants evolved a mechanism for fertilisation which was not dependent on water. This is one of the main reasons why seed plants are so much more successful than other plants at exploiting dry land. Pollination was originally achieved by wind, a fairly haphazard process involving large wastage of pollen. However, early in the evolution of seed plants flying insects appeared (in the Carboniferous era about 300 million years ago) bringing the possibility of more efficient pollination by insects. The flowering plants have exploited this method to a high degree, although conifers are still wind-pollinated.

Seeds. In the early heterosporous plants, megaspores were released from the parent sporophyte like the microspores. However, in the seed plants megaspores are retained on the parent plant within the megasporangium.

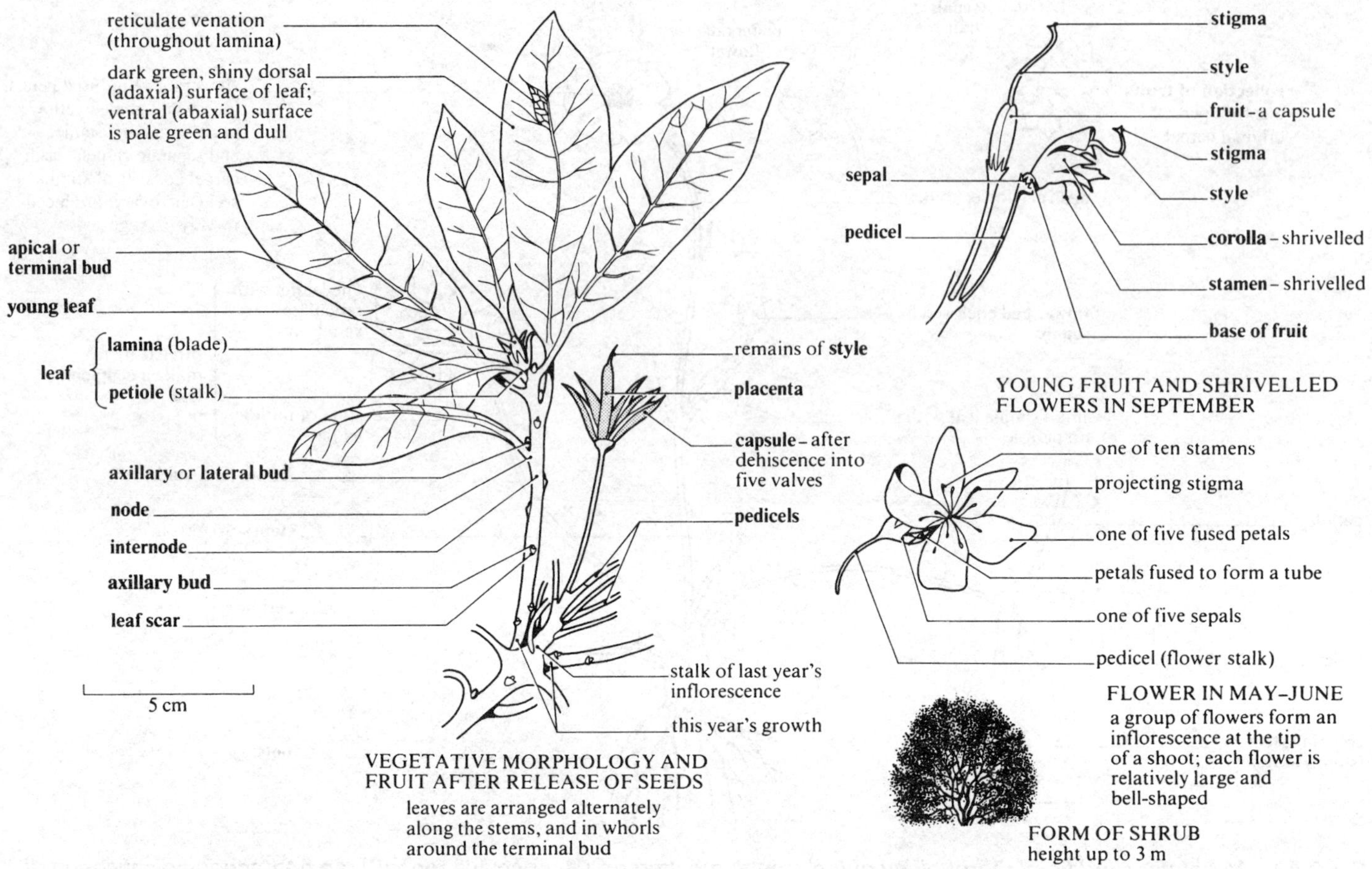

Fig 2.43 *Vegetative and floral morphology of the wild rhododendron,* R. ponticum, *an evergreen dicotyledonous shrub. It is commonly planted in woods and gardens. Originally introduced, it has become naturalised, favouring acid soils (sandy or peaty) on heaths and in woods.*

This is known as an **ovule** in seed plants (fig 2.45). The ovule contains the female gamete. Once this is fertilised the ovule is known as the **seed**. Thus a seed is a fertilised ovule. The ovule/seed brings the following advantages.

- The female gametophyte is protected by the ovule. It is totally dependent upon the parent sporophyte and is not susceptible to desiccation as would be a free-living gametophyte.
- After fertilisation it develops a food store, supplied by the parent sporophyte plant to which it is still attached. The food will be used by the developing zygote (the next sporophyte generation) at germination.
- The seed is specialised to resist adverse conditions and can remain dormant until conditions are suitable for germination.
- The seed may be modified to facilitate dispersal from the parent gametophyte.

The seed is a complex structure because it contains cells from three generations, a parent sporophyte, a female gametophyte and the embryo of the next sporophyte generation. All the essentials for life are supplied by the parent sporophyte and it is not until the seed is mature, containing a food store and an embryo sporophyte, that it is dispersed from the parent sporophyte.

2.7 The chances of survival and development of wind-blown pollen grains (microspores) are much less than those of spores of *Dryopteris*. Why?

2.8 Suggest reasons for the fact that megaspores are large and microspores are small.

2.7.7 Summary of adaptations of seed-bearing plants to life on land

The major advantages that seed plants have over other plants are as follows.

- The gametophyte generation is very reduced. It is always protected inside a sporophyte, which is well adapted for life on land, and is totally dependent on the sporophyte. In other plants the gametophyte is susceptible to drying out.
- Fertilisation is not dependent on water. The male gametes are non-motile and carried within pollen

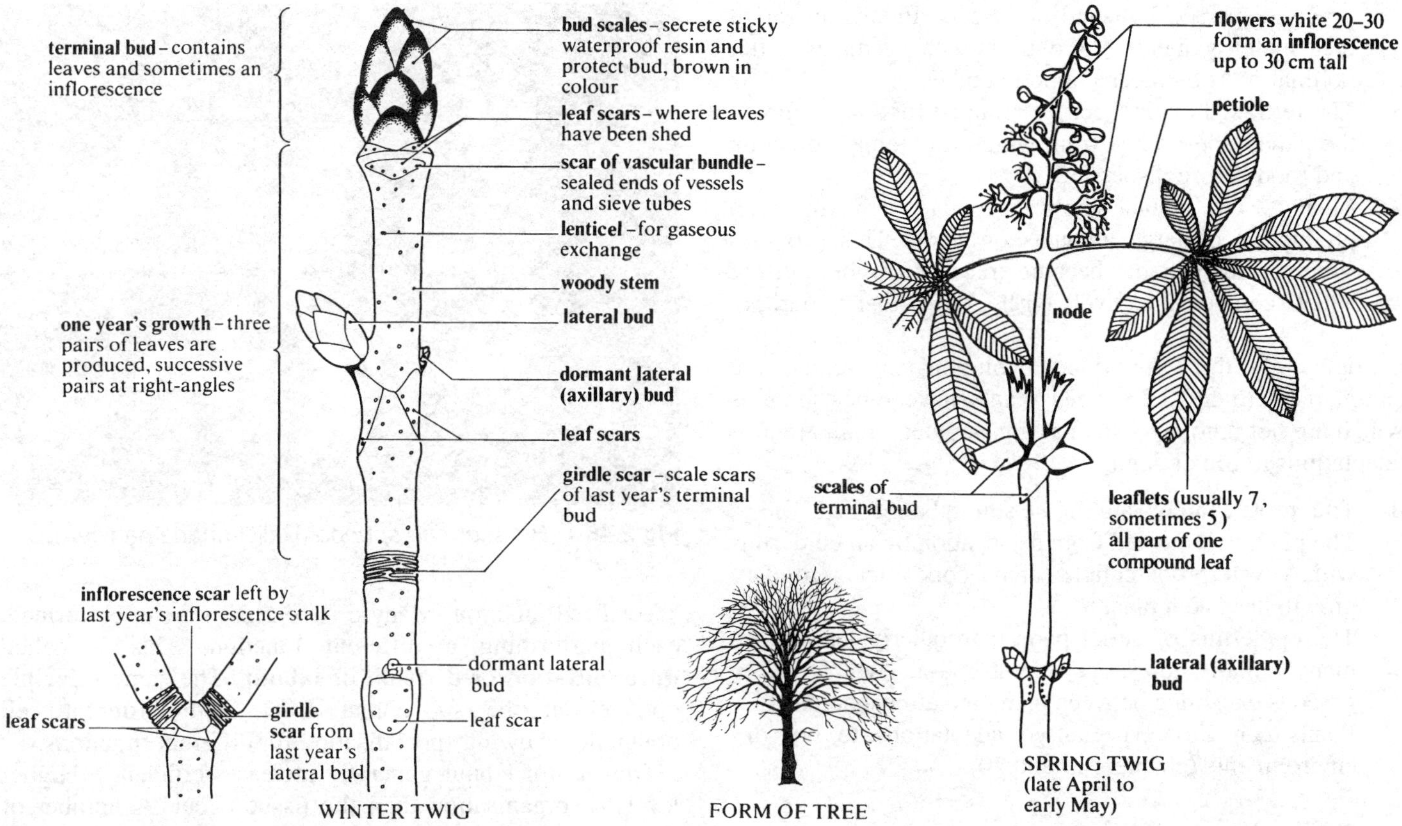

Fig 2.44 *Vegetative and floral morphology of the horse chestnut (*Aesculus hippocastanum*), a deciduous dicotyledonous tree. The tree may reach 30 m or more in height.*

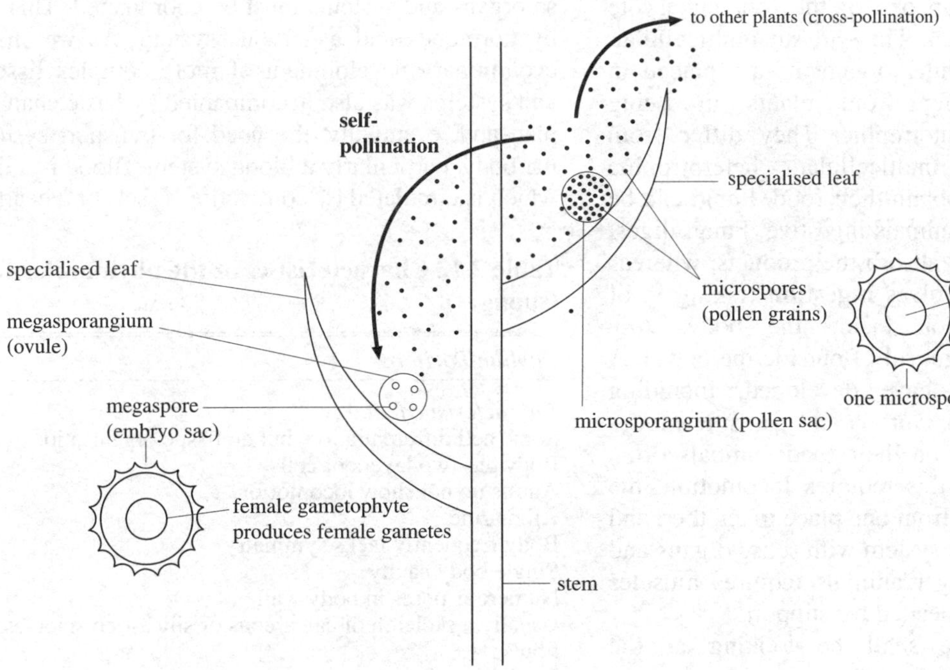

Fig 2.45 *Diagram illustrating the principles of heterospory and pollination.*

grains dispersed by wind or insects. Final transfer of the male gametes to the female gametes after pollination is by means of pollen tubes.
- The fertilised ovule (seed) is retained for some time on the parent sporophyte from which it obtains protection and food before dispersal.
- Many seed plants show secondary growth with production of large amounts of wood. This provides support. Such plants become trees or shrubs and are able to compete effectively for light and other resources.

Some of the important evolutionary trends are summarised in fig 2.33. Seed plants have other features which are not unique to them as a group but which are also adaptations to life on land.

- True roots enable water in the soil to be reached.
- The plant is protected from desiccation by an epidermis with a waterproof cuticle (or by cork after secondary growth has taken place).
- The epidermis of aerial parts, particularly leaves, has many small holes, called **stomata**, which allow gaseous exchange between plant and atmosphere.
- Plants can show specialised adaptations to hot dry environments (chapters 19 and 20).

2.8 Kingdom Animalia (animals)

2.8.1 Evolutionary trends

Animals make up one of the four eukaryote kingdoms, as shown in fig 2.4. They are all multicellular, since the animal-like unicellular organisms are placed in the Protoctista. They differ from plants in being heterotrophic rather than autotrophic. They differ from fungi, which are also multicellular, heterotrophic eukaryotes, in the way they obtain their food. Fungi can be described as absorptive and animals ingestive. Fungi digest food outside their bodies and absorb the products, whereas animal nutrition typically involves **ingestion** (taking in of food) followed by digestion inside the body. Any undigested food is **egested** (got rid of outside the body). A number of feeding habits have developed, including carnivorous, herbivorous, omnivorous and parasitic modes of life. Whereas fungi grow on their food, animals often have to seek it. If they do, this requires locomotion, the ability of the animal to move from one place to another, and this in turn requires a nervous system with sense organs and effectors. Locomotion of larger animals requires muscles and a skeleton, which is also needed for support.

In studying animals, we shall be looking at the evolutionary trends which have led to more and more complex levels of organisation within their bodies. One group of animals, the sponges (fig 2.46), do not form true tissues (table 2.12), but in all other animals tissues are formed. A **tissue** is a group of cells, often similar in structure and origin, operating together to perform a specialised function. Many different tissues can be formed, each performing a different function. This is called **differentiation** or **division of labour**. The same principle operates at the subcellular level, with different cell organelles showing specialisation for different functions.

Fig 2.46 *Breadcrumb sponge* (Halichondria panicea).

Division of labour generally increases efficiency. Higher levels of organisation than the tissue occur. A number of tissues working together form an **organ**, such as the stomach, and a group of organs working together forms a **system**, such as the digestive system. The various systems together make the organism.

Just as the activity of cells is coordinated within a tissue, so organs and systems must be coordinated. This is achieved by hormones and a nervous system. As we shall see, the evolutionary development of more complex tissues, organs and systems was also accompanied by basic changes in body plan and, eventually, the need for transport systems within the body, particularly a blood system. Blood is a liquid tissue which is circulated by contractile vessels or a heart.

Table 2.12 Characteristics of the phylum Porifera (sponges).

Phylum Porifera
Characteristic features
Some cell differentiation, but no tissue organisation
Body has two layers of cells
Adults do not show locomotion
All marine
Body frequently lacks symmetry
Single body cavity
Numerous pores in body wall
Usually a skeleton of calcareous or silica-rich spicules, or horny fibres
No differentiated nervous system
Asexual reproduction by budding
All are hermaphrodite
Great regenerative power
'Dead-end' phylum – it has not given rise to any other group of organisms

Representatives of the main animal phyla will now be studied to examine some of the more important of these developments.

2.8.2 Phylum Cnidaria (cnidarians)

Classification and characteristics of the Cnidaria (the C is not pronounced), are summarised in table 2.13. The cnidarians include jellyfish, sea anemones and corals.

Diploblastic

The body plan is relatively simple, consisting of two layers of cells, an outer **ectoderm** and an inner **endoderm** as shown in fig 2.47. This is known as the diploblastic level of organisation. The ectodermal cells face outwards into the environment and the endodermal cells face inwards into the **enteron**, a cavity with a single opening to the environment, the 'mouth'. Feeding is by means of tentacles arranged around the mouth. Both ingestion (taking in food) and egestion (getting rid of undigested food) take place through this opening. There is some specialisation of cells, so it can be argued that the tissue level of organisation has been achieved. For example, batteries of stinging cells known as **nematoblasts** occur in the ectoderm of the tentacles. These can discharge threads of three types which can penetrate, cling to or kill the prey. The ectoderm also contains sensory cells, which connect with nerve cells, forming a communication network through the mesogloea. Cells containing contractile, muscle-like fibrils allow movement of the body and tentacles, and locomotion in jellyfish. In the endoderm are cells specialised for various aspects of digestion and absorption.

Radial symmetry

Cnidarians are radially symmetrical, meaning that they can be cut in half across any diameter and the two halves will be identical (like a cake). Radial symmetry tends to be associated with organisms which do not show locomotion (see also echinoderms, section 2.8.8). Most animals show bilateral symmetry (associated with locomotion) which not only gives a more compact and streamlined shape but allows greater specialisation of body parts.

Polyps, medusae and polymorphism

Two basic body types occur in the cnidarians, the polyp and the medusa (see fig 2.47). The **polyp** is cylindrical and **sessile**. A sessile organism is one that remains attached to a surface such as a rock throughout its life and shows little or no locomotion. The **medusa** is umbrella-shaped and free-swimming or floating. The two types sometimes alternate in the life cycle, in which case the medusa can act as a dispersal stage (see *Obelia* below). In this situation polyps reproduce asexually by budding off medusae, and the medusae reproduce sexually to produce larvae which develop into polyps. Individual polyps within colonies can also vary in form. For example, they may be specialised for feeding or for asexual reproduction (see *Obelia* below). The situation in which individuals of a species exist in two or more different forms is known as **polymorphism**.

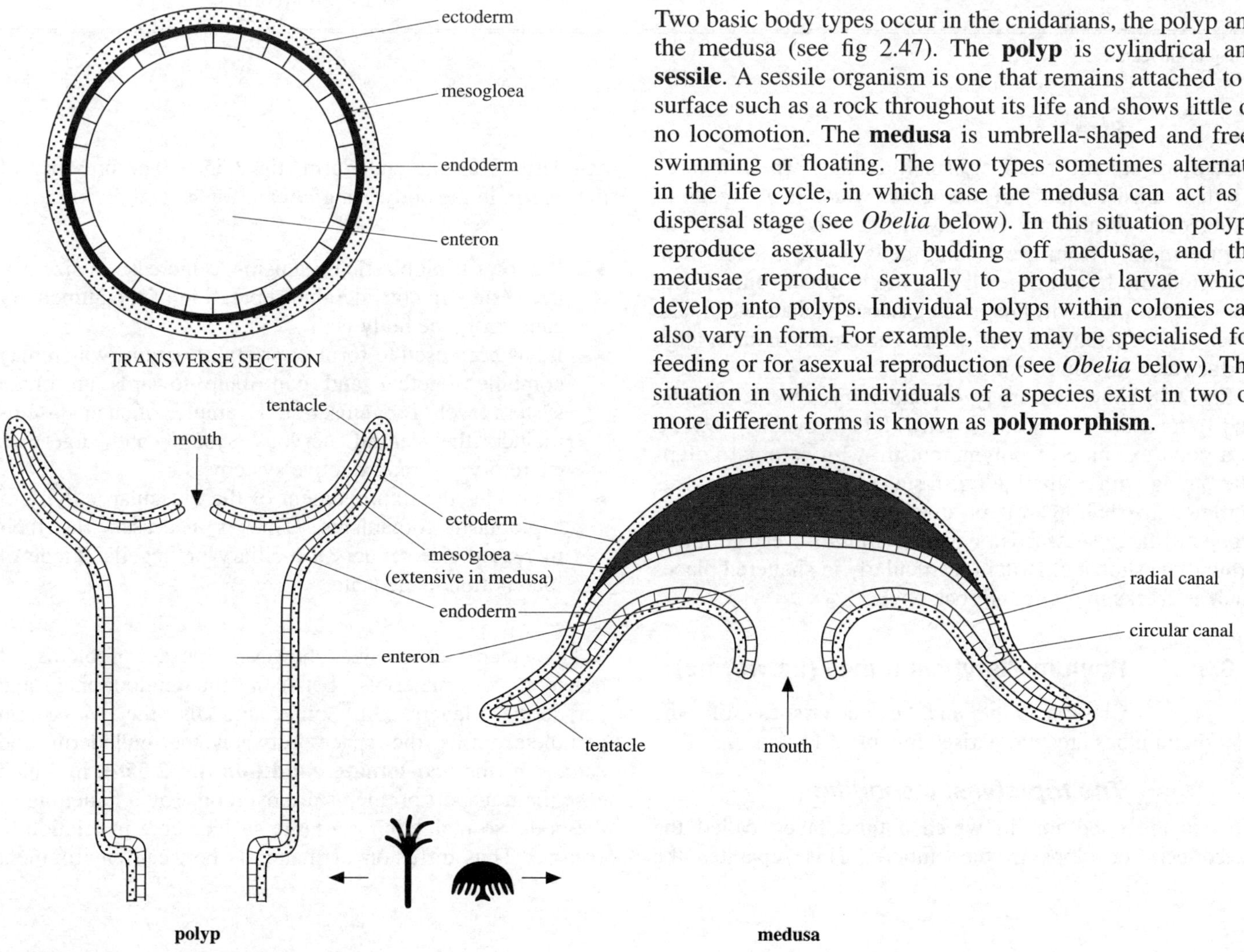

Fig 2.47 *Body plan of cnidarians.*

Table 2.13 Classification and characteristics of the phylum Cnidaria (cnidarians).

Phylum Cnidaria

Characteristic features
Diploblastic animals: body wall composed of two layers of cells, an outer **ectoderm** and an inner **endoderm**; these layers are separated by a structureless, gelatinous layer of **mesogloea** which may contain cells that have migrated from other layers*
Tissue level of organisation achieved
Radial symmetry
Body is basically sac-shaped with a single opening, the 'mouth', for ingestion and egestion. The single cavity within the sac is called the **enteron** and is where digestion takes place
Two structural types, **polyps** and **medusae**. Polyps are sessile (stay in one place) and may be solitary, e.g. *Hydra,* or colonial, e.g. *Obelia.* Medusae are free-swimming and solitary*
Polymorphism exhibited, that is individuals have specialised shapes with different functions – a form of division of labour

*Class Hydrozoa (*hydroids*)*	*Class Scyphozoa* (jellyfish*)*	*Class Anthozoa* (corals and sea anemones*)*
Polyp dominant in life cycle	Small polyp sometimes present as a larval stage	Polyp only – more complex than those of the Hydrozoa
Medusa simple	Large highly organised medusa dominant in life cycle	No medusa
Polyps solitary or colonial		Polyps solitary (anemones, some corals) or colonial (most corals)
Nematoblasts (stinging cells)	Nematoblasts	Nematoblasts
e.g. *Hydra* (no medusa phase) *Obelia*	e.g. *Aurelia* (jellyfish)	e.g. *Actinia* (beadlet anemone) *Madrepora* (coral)

* diagnostic features.

Size

Cnidarians are still relatively small animals. The few large jellyfish consist mainly of mesogloea which is not made of living cells. With only two layers of cells, nutrients can diffuse rapidly from the feeding cells in the endoderm to the ectoderm. In addition, all cells are in direct contact with the water of the environment, so gaseous exchange can take place very efficiently by diffusion. The organism has a large surface area to volume ratio.

Fig 2.48 shows a variety of cnidarians. *Obelia, Aurelia* and *Actinia* are all marine species (live in the sea). *Obelia* is a good example of polymorphism, with colonial polyps alternating with a small jellyfish stage in the life cycle. It is common in shallow waters attached to rocks, shells, the fronds of large seaweeds or piers. *Actinia* is very common around the shores of Britain, particularly in sheltered places such as cracks in rocks and rock pools.

2.8.3 Phylum Platyhelminthes (flatworms)

Classification and characteristics of the Platyhelminthes are summarised in table 2.14.

The triploblastic condition

This is the condition in which a third layer, called the **mesoderm** develops in the embryo. This separates the ectoderm from the endoderm (fig 2.55*a*). The presence of mesoderm in the body is significant for several reasons.

- It allows triploblastic organisms to increase in size and this results in considerable separation of the alimentary canal from the body wall.
- It has been used to form a variety of organs, which may combine together and contribute towards an organ system level of organisation. Examples of such systems include the central nervous system and digestive, excretory and reproductive systems.
- It enables the improvement of the muscular activity of triploblastic organisms. This is necessary as their increased size renders the ciliary or flagellar mode of locomotion inadequate.

This increase in size, however, poses problems of transport of materials between the endodermal and ectodermal layers. In some animals the mesoderm completely fills the space between the endoderm and ectoderm (the **acoelomate condition**, fig 2.55*a*), in which case the transport problems are overcome by a flattening of the body, so maintaining a large surface area in relation to volume. Thus diffusion of materials between environment

Fig 2.48 *A variety of cnidarians.* (a) Hydra *– a freshwater, solitary cnidarian.* (b) Obelia *– a marine, colonial cnidarian with two types of polyp and a medusoid form.* (c) Aurelia *– a jellyfish.* (d) Actinia *– a sea anemone.* (e) Aurelia *medusa.* (f) Obelia *colony.* (g) Hydra. (e)–(g) overleaf.

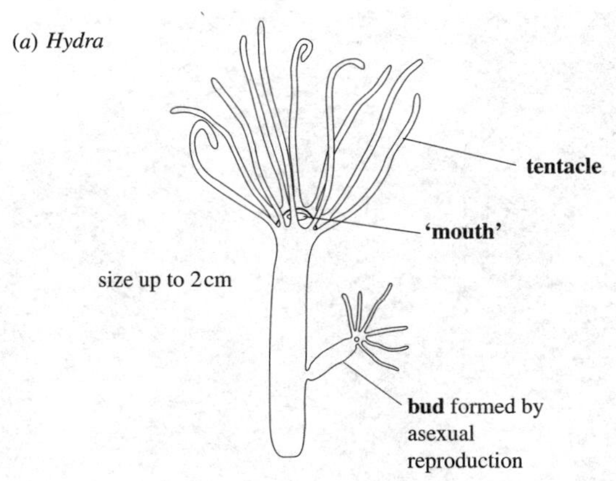

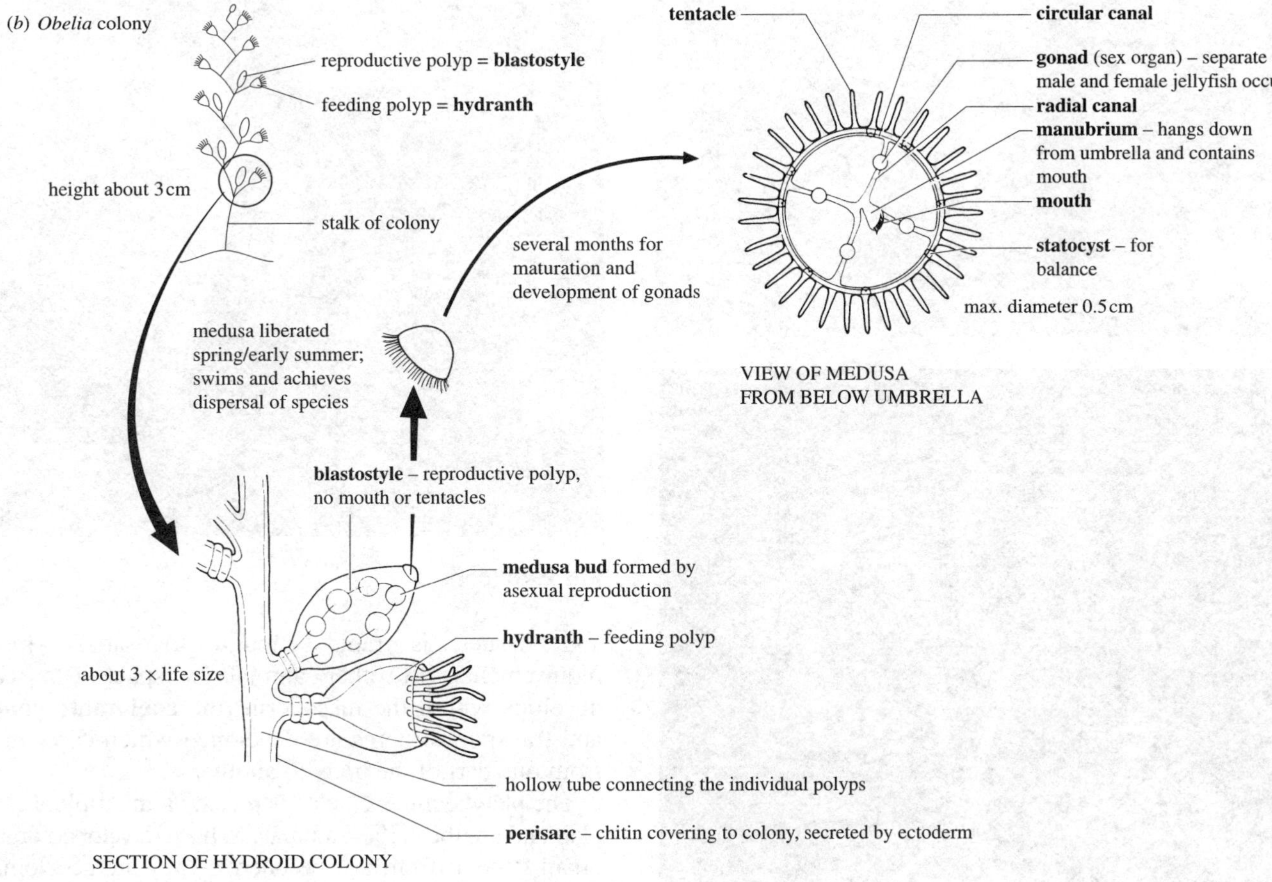

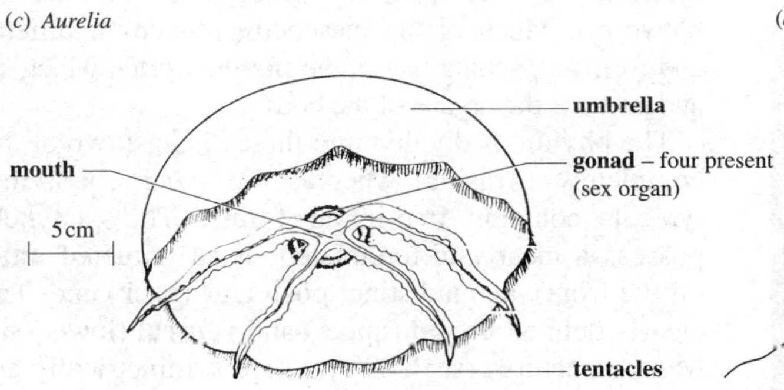

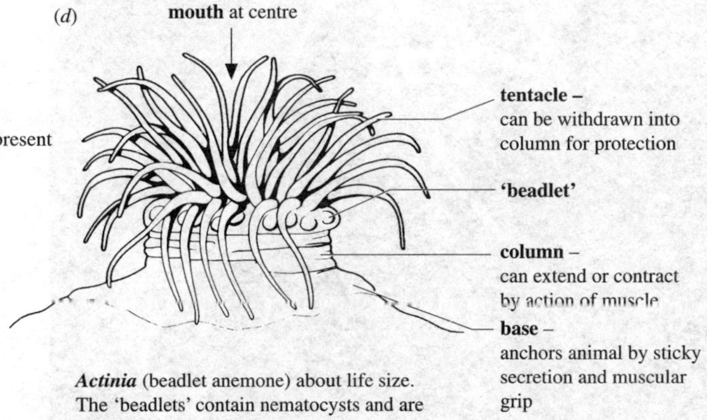

Actinia (beadlet anemone) about life size. The 'beadlets' contain nematocysts and are bright blue in colour. The rest of the animal is a dark red or olive green colour.

(e)

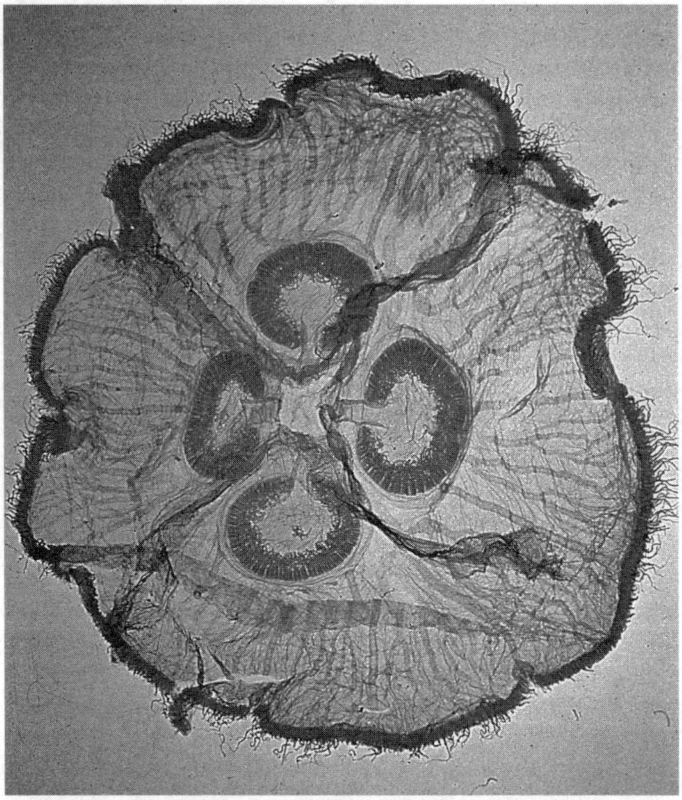

(f)

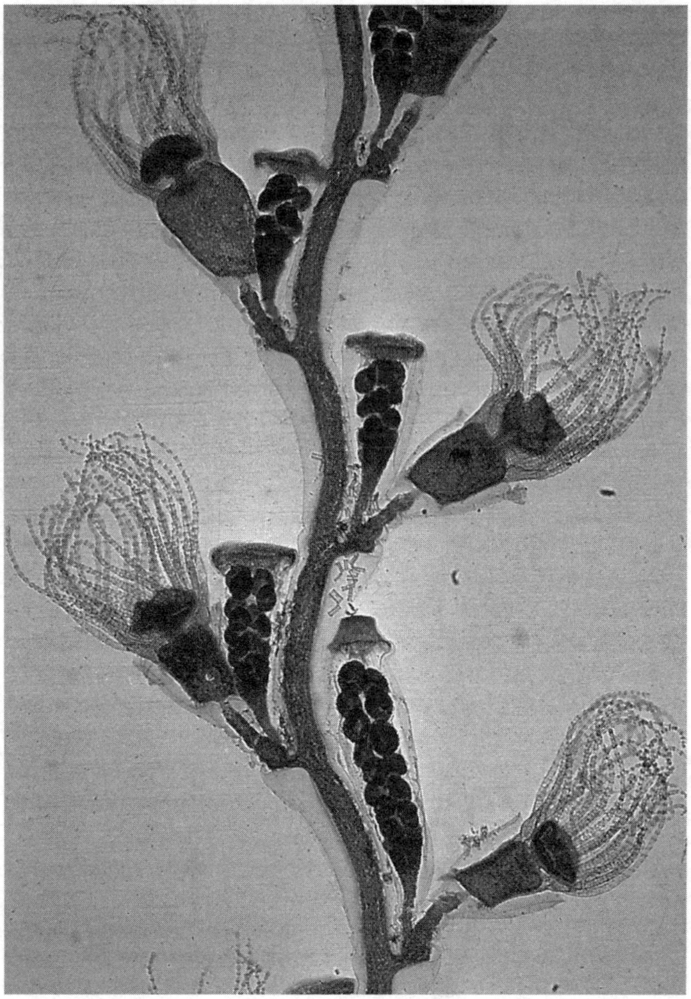

(g)

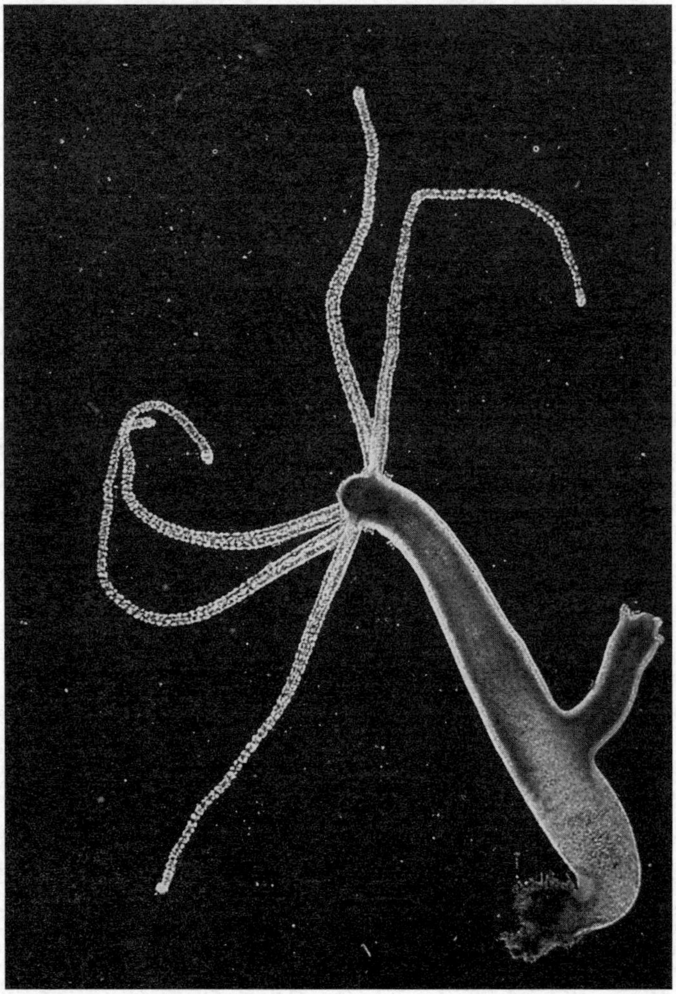

Fig 2.48 (cont.)

and tissues is rapid enough to satisfy metabolic requirements. In other animals a space (the **coelom**) develops within the mesoderm (the **coelomate condition**) and transport systems are developed which carry materials from one part of the body to another.

The platyhelminths are constructed on the triploblastic body plan and are the earliest animals to have developed organs and organ systems from the mesoderm. They are acoelomate and therefore have flat bodies – hence their common name of flatworms. Much of the mesoderm remains undifferentiated and forms a packing tissue, the **mesenchyme**, which supports and protects the organs of the body.

The phylum is divided into three classes; two of these are completely parasitic, whereas the other class, the most typical, contains free-living forms. The platyhelminths possess a clearly differentiated 'head' situated **anteriorly** (at the front), and a distinct **posterior** (back) end. There are clearly defined **dorsal** (upper) and **ventral** (lower) surfaces. Many structures (such as eyes) are symmetrically arranged on the right- and left-hand sides of the body. Such organisation, where the right side is approximately the mirror image of the left and where there is a distinct anterior end, is called **bilateral symmetry**.

Table 2.14 Classification and characteristics of the phylum Platyhelminthes (flatworms).

Phylum Platyhelminthes

Characteristic features
Triploblastic
Bilateral symmetry
Unsegmented (like nematode worms, unlike annelid worms)
Acoelomate
Flattened shape
Mouth but no anus

Class Turbellaria (turbellarians)	*Class Trematoda* (flukes)	*Class Cestoda* (tapeworms)
Free living; aquatic	Endoparasitic (live inside host) or ectoparasitic (live on outer surface of host)	Endoparasitic (live inside host)
Delicate, soft body	Leaf-like shape	Elongated body divided into proglottides which are able to break off
Suckers rarely present	Usually ventral sucker in addition to sucker on 'head' for attachment to host	Suckers and hooks on 'head' (scolex) for attachment to host
Outer suface covered with cilia for locomotion; cuticle absent	Thick cuticle with spines (protection); no cilia in adult (locomotion not needed because not parasitic)	Thick cuticle (protection); no cilia in adult
Enteron present	Enteron present	No enteron (no digestion required – absorbs predigested food from host)
Sense organs in adult	Sense organs only in free-living larval stages	Sense organs only in free-living larval stages
e.g. *Planaria*	e.g. *Fasciola* (liver fluke) *Schistosoma* (blood fluke) – cause of schistosomiasis (bilharzia) in many tropical countries	e.g. *Taenia* (tapeworm)

No transport system has developed, because in the basic body structure all parts are in close proximity to food and oxygen supplies. All platyhelminths are thin and flat, providing a large surface area to volume ratio for gaseous exchange. Many forms possess a much-branched gut, which ramifies throughout the body to facilitate absorption of food materials. In addition, excretory material is collected from all parts by a branched system of excretory tubes.

Class Turbellaria

Planaria is a free-living, carnivorous flatworm found in freshwater streams and ponds. It remains under stones during the day, emerging only at night to feed. It is black in colour and can measure up to 15 mm in length. It has an elongated, extremely flattened body, with a relatively broad anterior 'head' possessing a pair of eyes in the dorsal surface, and a posterior end that is clearly tapered. *Planaria* is bilaterally symmetrical, a body design associated with an active mode of life (fig 2.49).

There is a single gut opening, the mouth, which is located on its ventral surface towards the posterior end of the body. *Planaria* feeds on small worms, crustacea and on the dead bodies of larger organisms.

Class Trematoda

Fasciola hepatica, the liver fluke (fig 2.50), belongs to the class Trematoda, which is one of the major groups of parasites in the animal kingdom. It is **endoparasitic**, meaning it lives inside its host. It lives in the bile ducts of sheep, its most important, or **primary**, **host**. Other primary hosts are cattle and, occasionally, humans.

Many differences exist between *Fasciola* and the free-living *Planaria*. These differences can be attributed to the adaptations that *Fasciola* has evolved in order to survive as an endoparasite. Associated with its parasitic mode of life is a complex life history, involving three larval stages (the miracidium, redia and cercaria), which provide opportunities for increasing its numbers during the life cycle. The large numbers of offspring produced in this way help to offset the high mortality rate that inevitably occurs during infection of new hosts. For a part of its life history *Fasciola* infests a **secondary host**, the freshwater snail (*Limnea*), in which some of its larval stages are able to live and multiply.

Each stage in the life history of *Fasciola* shows structural, physiological and reproductive adaptations suited to its mode of life. Some of these are listed below and shown in fig 2.50. The life cycle is summarised in fig 2.51.

Adult fluke. The body is thin and flat and attaches to the lining of the bile duct. The body wall protects the fluke against the host's enzymes. The gland cells situated here also secrete material which protects the parasite against the host's antitoxins.

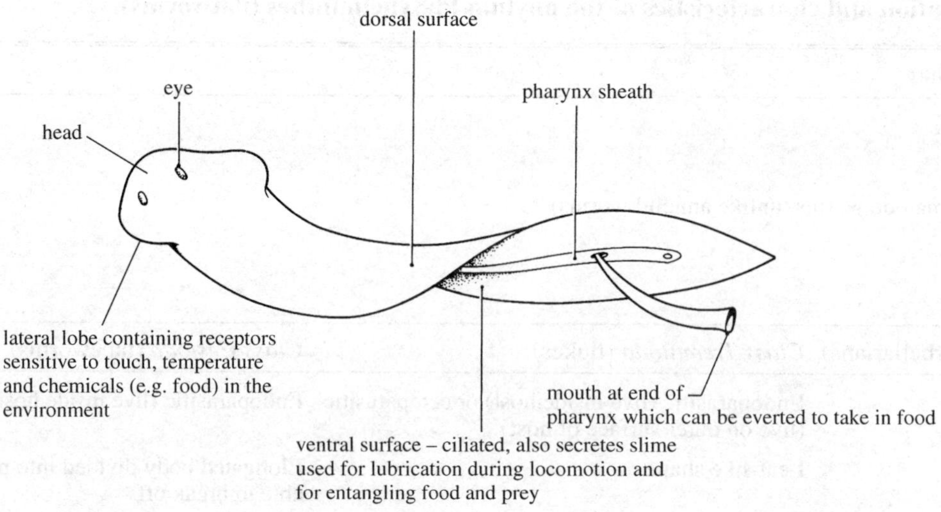

Fig 2.49 Planaria *showing external features.*

A **hermaphrodite** (male and female sex organs in the same organism) reproductive system ensures that self-fertilisation or cross-fertilisation can occur. The fluke can survive anaerobically if there is a shortage of oxygen.

Miracidium. This is the first of the larval stages of *Fasciola* (fig 2.51). Its main function is to find the secondary host, for which it needs sense organs and the ability to move. It also produces more larvae (sporocysts – see below). It has a ciliated epidermis which allows it to swim in water or in moisture on vegetation.

The miracidium is attracted to its secondary host, the freshwater snail, by chemotaxis (locomotion in a particular direction in response to a chemical stimulus). It attaches at its anterior end to the snail's foot, and a gland secretes protein-digesting enzymes onto the surface of the snail to help in the penetration of the host's tissues. Penetration is further helped by muscle cells which help the larva to wriggle through the tissues of the host. In this way it migrates to the digestive glands. There are special germ cells present inside the miracidium which give rise to the next larval forms.

Sporocyst. The function of this stage is to increase numbers to compensate for wastage of larvae that do not find hosts. It is an immobile, closed sac containing germ cells, which multiply to form many rediae, the next larval stage.

Redia. This is a multiplication and feeding stage. It has a muscular pharynx to suck in fluids and tissues from its host. Muscle cells aid locomotion of the larva. Germ cells multiply into more rediae, or into cercariae. There is a pore for the escape of the new rediae or the cercariae.

Cercaria. This bears many features in common with the adult fluke, which include oral and ventral suckers for anchorage to suitable substrates such as grass. There is also a tail to assist in locomotion through water or moisture on vegetation. Glands are present which secrete a cyst wall (fig 2.51). The encysted cercaria undergoes no further development until it is swallowed by a sheep. It has considerable powers of resistance to low temperatures, but is susceptible to desiccation.

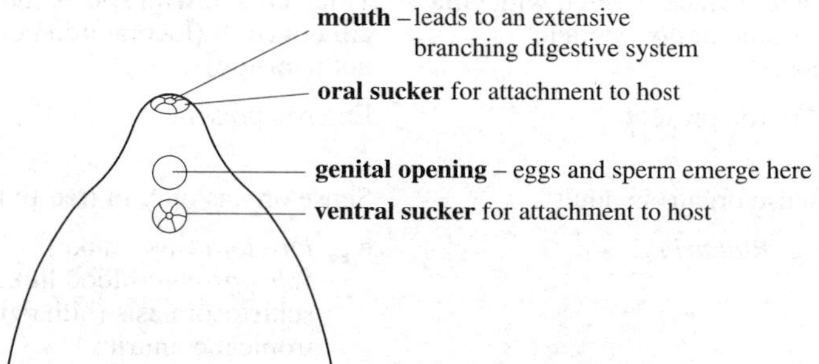

Fig 2.50 *External features of* Fasciola hepatica, *the liver fluke. The fluke lives in the bile duct of the host.*

Limnea is an amphibious snail inhabiting ponds, muddy tracks and damp vegetation. It is able to withstand adverse conditions. Therefore the sporocyst and redia stages of

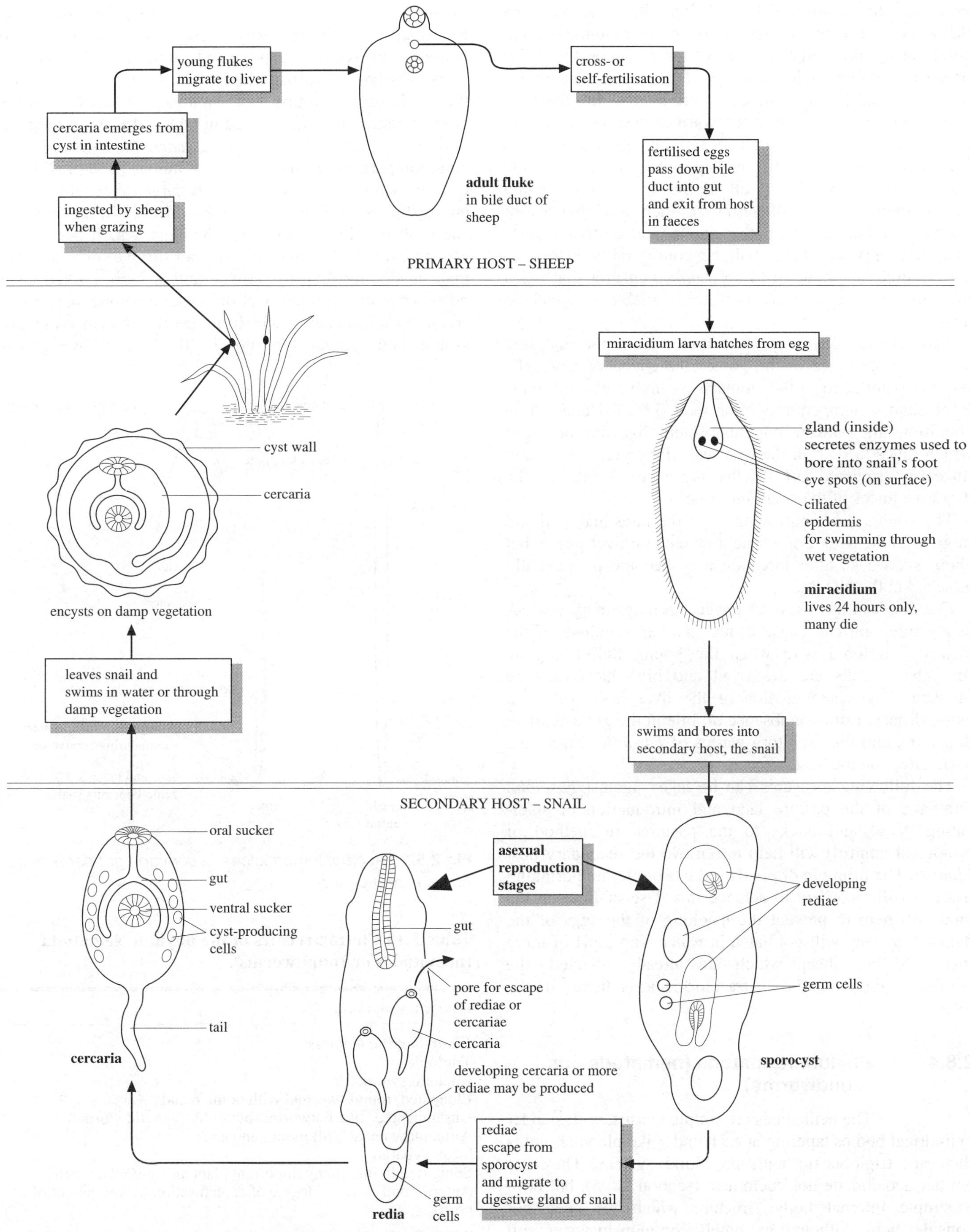

Fig 2.51 *Life cycle of* Fasciola hepatica*, the liver fluke.*

Fasciola's life history, which develop within the snail, are themselves directly protected from such unfavourable conditions. Indeed, in conditions of low temperature, rediae produce daughter rediae instead of cercariae. The rediae remain within the snail and can overwinter within the host, only producing cercariae when warmer weather returns in the spring. *Limnea* is also a very rapid breeder. It has been estimated that one snail may produce up to 160 000 offspring in 12 weeks. If all of these offspring contain developmental stages of *Fasciola,* then the chances of cercariae escaping from the snails and entering new, uninfected primary hosts will be considerably increased. The amphibious mode of life of *Limnea* ensures that when the cercariae escape there is water available in which to disperse.

The release of young adult flukes from the encysted cercaria takes place in the gut of the sheep or cow. The process is initiated in the stomach by high carbon dioxide levels and a temperature of around 39 °C. Under these conditions the parasite releases protein-digesting enzymes which digest a hole in the cyst wall. Emergence of young flukes is triggered off by the presence of bile in the digestive juices of the small intestine.

The young flukes burrow through the intestinal wall and migrate to the liver. For a time they feed on liver tissue, but about six weeks after infection they become permanently attached in the bile ducts.

Fasciola can have several effects on its primary host. A heavy infection can cause death. Liver metabolism of the host is interfered with when the young flukes migrate through it. Cells are destroyed and bile ducts may be blocked; large-scale erosion of the liver (liver rot) will cause dropsy. Little, or absence of, bile in the gut can affect digestion, and the excretory wastes of *Fasciola* can have a toxic effect on the host.

The following measures can be taken against *Fasciola.* Drainage of the pasture land and introduction of snail-eating geese and ducks to the pastures (a method of biological control) will help to remove the secondary host *Limnea*. The filling in of ponds and use of elevated drinking troughs will also help to achieve this. Use of lime on the land will help to prevent the hatching of the eggs of the parasite, as they will not hatch in water with a pH of more than 7.5. For sheep which are already infected, the administration of carbon tetrachloride kills flukes in the liver.

2.8.4 Phylum Nematoda (nematodes or roundworms)

The nematodes are simple worms, with slender cylindrical bodies tapering at each end. Like platyhelminths they are triploblastic with no blood system. They are neither acoelomate nor coelomate (section 2.8.5), but have a simple internal body structure which we need not consider here. Although extremely common in water, soil (as many as 100 000 million per hectare), and a wide range of other habitats, these worms are mostly microscopic or too small to be seen easily. New species are constantly being discovered; one appears to be unique to German beer mats. The roundworms, flukes and tapeworms form the three largest groups of animal parasites. Their characteristics are summarised in table 2.15. An example is *Ascaris* (fig 2.52). One species, *Ascaris lumbricoides*, is a common parasite of the intestine of humans and pigs. It is a creamy white colour and relatively large, about the size of an earthworm (up to 20 cm long). Heavy infections can cause obstructions in the gut. The eggs, which pass out in the faeces, are very resistant and can survive for years. The male is smaller than the female and is more curved at its posterior end. Another well-known nematode parasite is *Wuchereria bancroftii*, which infects the human lymphatic system and causes elephantiasis (fig 2.53). The legs can

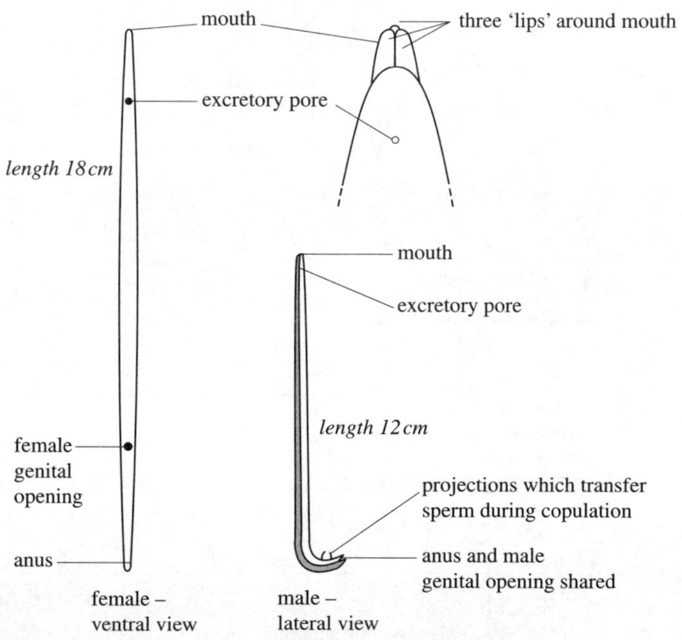

Fig 2.52 Ascaris lumbricoides, *a common gut parasite of humans and pigs.*

Table 2.15 Characteristics of the phylum Nematoda (nematodes or roundworms).

Phylum Nematoda
Characteristic features Triploblastic Bilateral symmetry Elongated, round 'worms' with pointed ends* Unsegmented (like flatworms, but unlike annelid worms) Alimentary canal with mouth and anus Sexes separate Some free living, many important plant and animal parasites Anterior end shows a degree of cephalisation (development of a head)

* diagnostic feature.

swell enormously, resembling those of an elephant. Nematodes can also attack plants, including a wide variety of crops.

2.8.5 Phylum Annelida (annelids or segmented worms)

Classification and characteristics of the annelids are summarised in table 2.16. A variety of annelids is illustrated in fig 2.54. *Nereis* and *Lumbricus* are described in more detail later. Annelids are coelomates.

The coelomate body plan

It has been seen in the platyhelminths that the mesoderm completely fills the space between the ectodermal and endodermal layers and forms a solid middle layer. Such a condition, without a coelom, is said to be **acoelomate** (fig 2.55*a*).

In the annelids and later animal groups, an extensive internal space or body cavity is developed, called a **coelom**. This arises as a split in the mesoderm during development of the embryo. The cavity formed is filled with coelomic fluid, and splits the mesoderm into two layers, the **somatic mesoderm** to the outside and the **splanchnic mesoderm** to the inside (fig 2.55*b*). The somatic mesoderm is attached to the ectoderm and these together form the body wall. The splanchnic mesoderm is attached to the endoderm to form the muscular wall of the gut. The coelom thus separates the body wall from the gut wall.

The majority of the mesoderm which lines the coelom develops into muscle; that of the body wall aids locomotion of the whole animal, whilst that of the gut causes peristalsis of food. Transport of materials between the gut wall and the body wall (and vice versa) is achieved by a well-developed blood vascular system. Note that the gut cavity is in the endoderm. The lining of the coelom is called the **peritoneum**. Portions of the peritoneum which connect the gut wall to the body wall across the coelom are called **mesenteries**. Any organs, such as reproductive or excretory organs, which project into the coelom are bounded by peritoneum (fig 2.55*b*).

Biological significance of the coelom

- Because the coelom separates the gut from the body wall, muscular movements of the body wall, associated with locomotion, can be separated from muscular movements of the gut wall, which move food through the gut (peristalsis) and help to churn the food. Greater powers of locomotion result, and different parts of the gut can become differentiated for different functions, for example a stomach for churning food. This allows a greater variety of diets
- The coelom also provides a cavity in which organs can grow, develop and function independently of other organs.
- Increasing size and complexity are possible. These bring about additional problems of transport and coordination. For example, food is digested in the gut, but the cells of the body wall are separated from the gut by the coelom. Similarly, the gut is some distance from the body surface where gaseous exchange takes place. As body size increases the problems get greater, as already noted at the beginning of section 2.8.3, and some kind of transport system becomes necessary. All coelomates possess a blood vascular system. A **vascular system** is a system of tubes. **Blood** is a liquid tissue which is circulated round the body by the pumping action of muscle in the walls of contractile vessels or a heart. These contain valves to maintain a one-way flow.

  Greater complexity also requires more complex coordination and therefore a more elaborate nervous system. **Cephalisation** (development of a head) is part of this trend. (See also size and surface area:volume ratio below.)
- The coelom performs an additional specialised function in annelids. Here it acts as a **hydrostatic skeleton**, in other words a fluid skeleton. Skeletons serve three main functions, namely support, protection and locomotion. Being a liquid, coelomic fluid is incompressible.

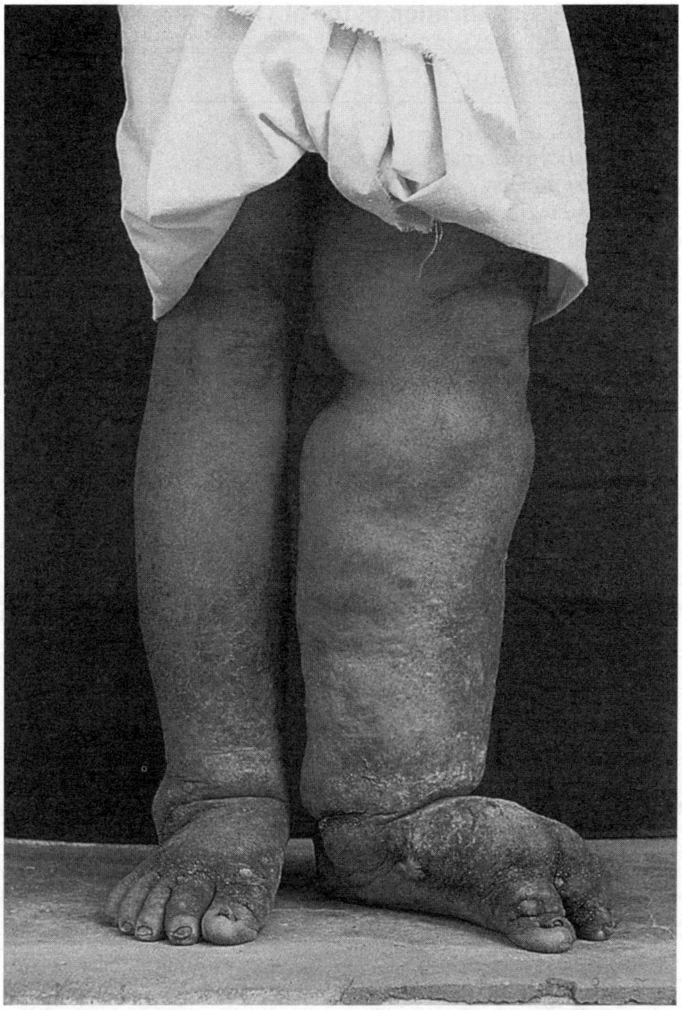

Fig 2.53 *Person suffering from elephantiasis.*

Table 2.16 Classification of the phylum Annelida (annelids or segmented worms).

Phylum Annelida

Characteristic features
Triploblastic, coelomate
Bilateral symmetry
Metameric segmentation
Prostomium, a lip-like extension of the first segment situated above the mouth
Definite cuticle (outer covering)
Chaetae, hair-like structures made of chitin and arranged segmentally (except leeches)*

Class Polychaeta (polychaetes or bristleworms)	*Class Oligochaeta* (oligochaetes or earthworms)	*Class Hirudinea* (leeches)
Marine	Inhabit fresh water or damp earth	Ectoparasites with anterior and posterior suckers
Distinct head	No distinct head	No distinct head
Chaetae numerous on lateral extensions of the body called parapodia*	Few chaetae – in pairs or single, no parapodia*	Small fixed number of segments, no chaetae or parapodia*
No clitellum	Clitellum or 'saddle' which secretes a cocoon in which the eggs are deposited	No clitellum
e.g. *Arenicola* (lugworm) *Nereis* (ragworm)	e.g. *Lumbricus* (earthworm)	e.g. *Hirudo* (leech)

*diagnostic features.

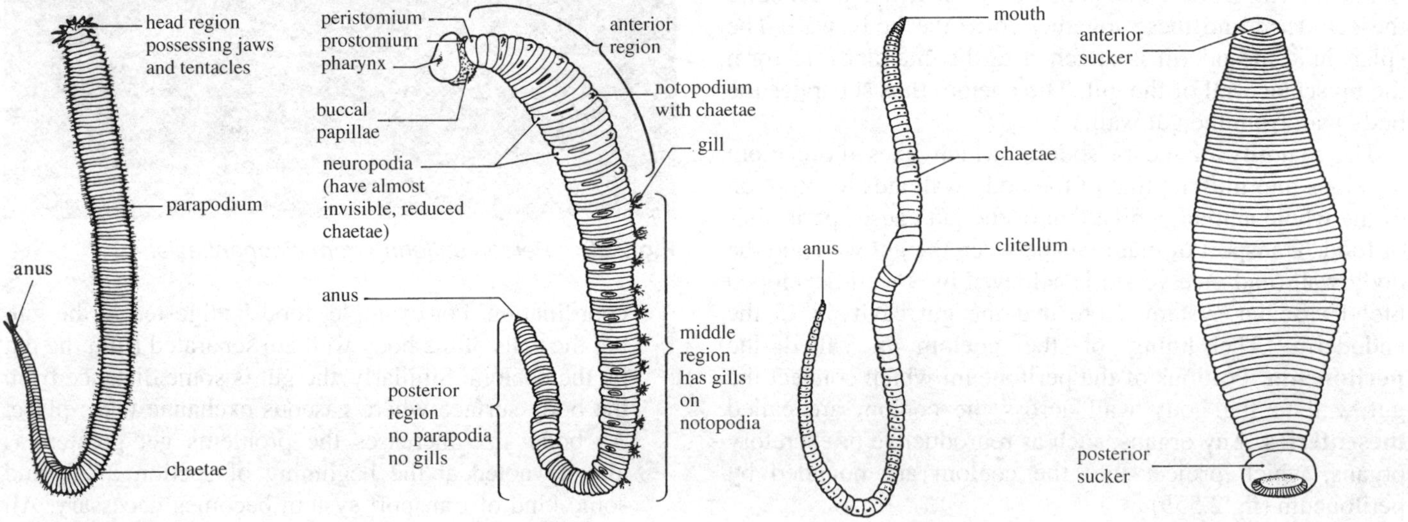

Class Polychaeta *Nereis* (ragworm)

Class Polychaeta *Arenicola* (lugworm)
The true segments have notopodia. Several rings correspond to one true segment.

Class Oligochaeta *Lumbricus terrestris* (earthworm)

Class Hirudinea (leeches) *Hirudo*

Fig 2.54 *A variety of annelids.*

Contraction of muscles can therefore change the shape of the worm but not its volume. During locomotion parts of the body alternately become longer and thinner, then shorter and fatter as different sets of muscles exert pressure on the coelomic fluid. Protection is provided by the ability of fluid to dissipate external forces rapidly and equally in all directions.

- Coelomic fluid may help to circulate food, waste materials and respiratory gases, although this function is mainly carried out by the blood vascular system.

Metameric segmentation

Another evolutionary advance which took place amongst the coelomates was **metameric segmentation**. This is the division of the body transversely into a number of similar parts or segments. Thus a series of similar segments occurs

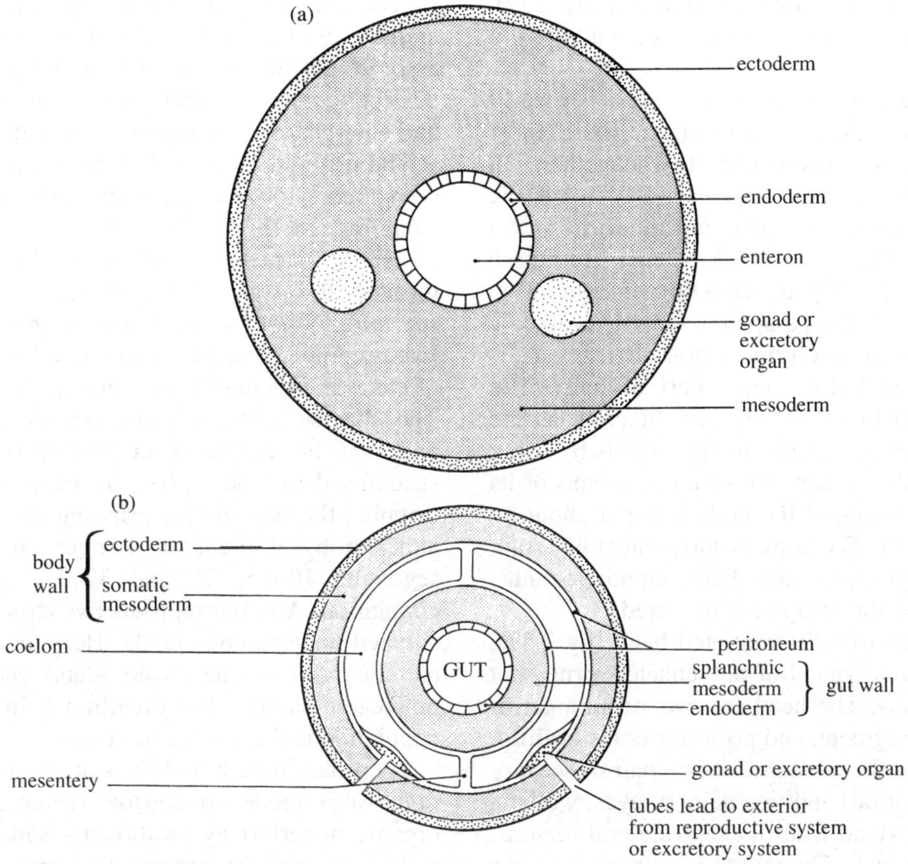

Fig 2.55 (a) *Transverse section of a generalised acoelomate.* (b) *TS generalised coelomate.*

along the length of the body. It originates in the mesoderm but usually affects both mesodermal and ectodermal regions of the body.

Metameric segmentation is most clearly seen in annelids, where the subdivisions may appear externally as constrictions of the body surface. Internally the segments are separated from each other by layers (septa) extending across the coelom. Each segment contains its own blocks of muscle, blood vessels, nerve cells and, in some groups, reproductive organs. However, the segments are not totally independent even in the annelids. The nervous system and blood system, in particular, must also run the length of the body.

Once segmentation and a basic plan for each segment was established, it was possible for evolutionary changes to take place within individual or small groups of segments, and for greater specialisation and division of labour of parts of the body to take place. This occurs in a number of ways. Different functions may be carried out in different segments; fusion of segments can occur, as happens in cephalisation where several segments fuse to form a head; or even loss of some segments, as shown by arthropods. Arthropods have fewer segments and, as we shall see in section 2.8.6, external segmentation is often less obvious, as in the cephalothorax of crustaceans. Internally, however, segmentation in arthropods is almost as clear as in annelids. In the chordates (section 2.8.9) external segmentation is lost, but certain systems still show clear segmentation, for example the muscle blocks in embryos, and the spinal nerves.

Size and surface area:volume ratio

Organisms with a relatively large surface area:volume ratio can rely on diffusion to satisfy their transport needs. Oxygen, nutrients and waste products such as carbon dioxide can diffuse sufficiently rapidly for the organism to survive without a special transport system. Diffusion, however, is only effective over very short distances. As size of a regular figure increases, volume increases faster than surface area so the surface area:volume ratio decreases. This can be shown most easily with a cube, as in fig 2.56, but the same principle applies to spheres, cells and whole living organisms. If the living organism is flattened as volume increases, its surface area can remain high, as occurs in platyhelminths where diffusion is still adequate to satisfy transport needs. However, coelomate animals cannot escape the need for specialised gaseous exchange and transport systems.

Class Polychaeta

Arenicola, the lugworm, lives in burrows in sand or soft mud in the intertidal zone or below low tide (fig 2.54).

Nereis, the ragworm, is a cylindrical bristleworm (fig 2.57). It lives in estuaries under stones, or in mud or muddy sand burrows.

The segmented nature of the body of *Nereis* is clearly visible externally. All segments, apart from those most anterior and posterior, are very similar to each other. On either side of each segment is a projection, called the **parapodium**. It consists of an upper **notopodium** and a lower **neuropodium** (fig 2.57). From each of these structures two hair-like tufts of **chaetae** emerge. Two additional outgrowths of the parapodia are noticeable: a dorsal and a ventral cirrus. A **cirrus** (plural cirri) is a hair-like structure (from the Latin *cirrus*, curl of hair). The parapodia have a good blood supply and function as the animal's gaseous exchange surface. *Nereis* crawls by using its parapodia like paddles. It can also swim by means of its parapodia and lateral flexing of the body brought about by muscles in the body wall. The body is surrounded by a thin cuticle. The alimentary canal runs from mouth to anus. Prey is swallowed when the pharynx is retracted.

Nereis possesses a clearly differentiated head (fig 2.57); this clear cephalisation is typical of polychaete worms, but unlike the other annelids. The head consists of an anterior **prostomium**, the first segment, and posterior **peristomium**, the second segment. On the prostomium is a pair of sensory tentacles (dorsal in position) and two pairs of eyes, whilst a pair of fleshy 'palps' extend from its ventrolateral regions. These are sensitive to touch. The mouth is situated between the two headparts, and on the peristomium are four pairs of long, flexible hairs or cirri. These are also sensitive to touch and have chemoreceptors which are sensitive to various chemicals, giving the animal the equivalent of a sense of taste and smell.

Class Oligochaeta

Lumbricus, an earthworm, is an elongated, cylindrical organism, approximately 12–18 cm in length (fig 2.58). The anterior end of the body is tapered, whilst the posterior end is dorsoventrally flattened. Despite being a terrestrial animal, it has not fully overcome all the problems associated with life on land. In order to protect itself from desiccation it lives underground in burrows in damp soil, and emerges only at night to feed and reproduce.

The differences in body form exhibited by *Lumbricus* as compared with *Nereis* are the result of its adaptation to a subterranean life. The body is streamlined with no projecting structures which might impede its passage through the soil. The prostomium is a small, rounded structure without sensory appendages, overlying the mouth. Each segment, except the first and last, possesses four pairs of chaetae, two positioned ventrally and two ventrolaterally. The chaetae protrude from sacs located in the body wall and can be extended or withdrawn by the action of specialised muscles. They are used during locomotion for gripping the soil and for gripping the sides of the burrow if attacked by a predator. Longer chaetae are present on segments 10–15, 26 and 32–37, and are used during copulation. Another reproductive structure, the **clitellum**, is situated on segments 32–37. Here the epidermis is dorsally and laterally swollen with gland cells that form a very noticeable saddle. The clitellum helps in the processes of copulation and cocoon formation.

There is a mouth and anus at opposite ends of the body. *Lumbricus* feeds on detritus (fragments of decomposing organic material) by swallowing soil. The majority of the soil passes straight through the worm, much of it eventually being deposited as 'worm casts' on the surface of the ground.

Secretions of coelomic fluid via dorsal pores, and mucus from mucous glands in the epidermis (skin), keep the worm's thin cuticle moist. This helps prevent desiccation, improves gaseous exchange and also acts as a lubricant for movement through the soil. It is here that gaseous exchange occurs by diffusion, a process that is helped by the presence of networks of looped blood capillaries in the epidermal layer.

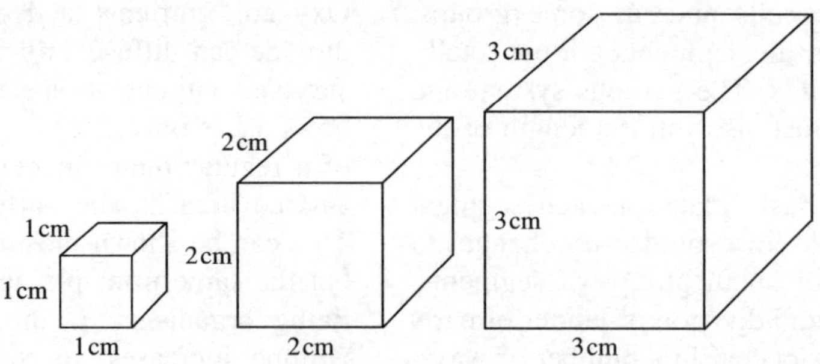

SA	$6 cm^2$	$24 cm^2$	$54 cm^2$
volume	$1 cm^3$	$8 cm^3$	$27 cm^3$
SA : volume ratio	6 : 1	3 : 1	2 : 1

SA = surface area

Fig 2.56 *Effect of increasing size on surface area:volume ratio.*

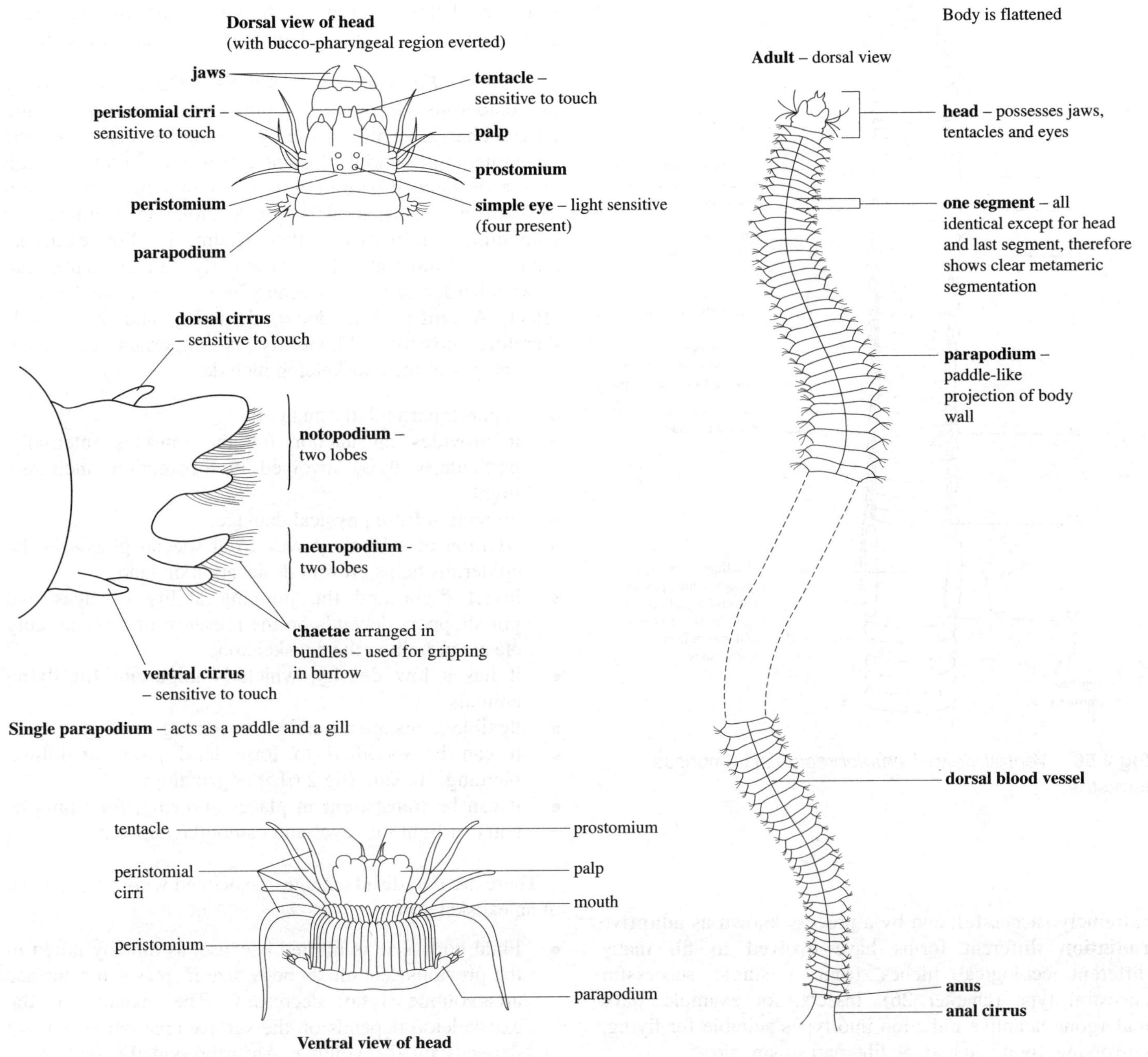

Fig 2.57 *Nereis, the ragworm.*

There is a pair of excretory and osmoregulatory tubes, called nephridia, in every segment except the first three and the last one. They open on to the surface of the worm via pores.

The reproductive system and behaviour of earthworms is very complex. This is due to their terrestrial mode of life and the necessity to avoid desiccation of gametes and fertilised eggs. *Lumbricus* is hermaphrodite (has male and female sex organs). Contact between worms is infrequent, but when it does occur, because they are hermaphrodite, any two worms of the same species are able to copulate. The worms exchange sperm and both are fertilised.

The sex organs are grouped at the anterior end of each worm. The external features associated with reproduction are shown in fig 2.58. Mating and subsequent laying of fertilised eggs in cocoons is a complicated process during which the worms line up facing in opposite directions.

2.8.6 Phylum Arthropoda (arthropods)

Classification and characteristics of the arthropods are summarised in table 2.17. The phylum Arthropoda contains more species than any other phylum. More than three-quarters of all known species are arthropods, insects alone accounting for more than half the known species. They have exploited every type of habitat on land and in water.

The basic body plan of the arthropods has been

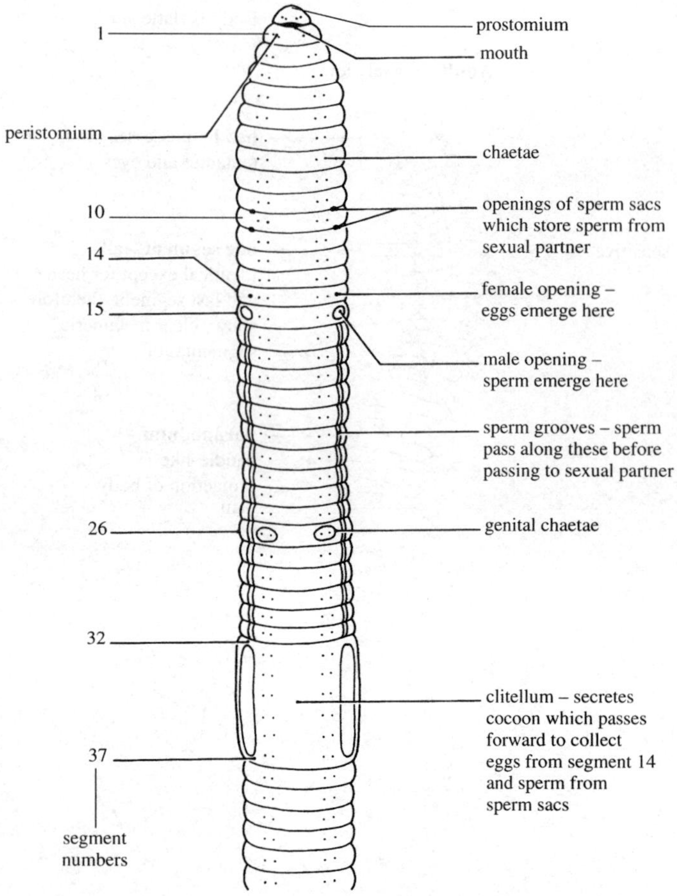

Fig 2.58 *Ventral view of anterior region of* Lumbricus terrestris.

extremely successful, and by a process known as **adaptive radiation** different forms have evolved to fill many different ecological niches from a single successful ancestral type (chapter 26). Insects, for example, have undergone adaptive radiation into types suitable for flying, burrowing, living an aquatic life, parasitism, etc.

The arthropod body design can be regarded as based on the segmented body plan of annelids. It shows clearly how metameric segmentation can be exploited. Ancestral arthropods possessed a series of similar simple appendages along the length of their bodies. These probably served a variety of purposes such as gaseous exchange, food gathering, locomotion and detection of stimuli. In the modern arthropods different segments have tended to become more specialised than in the annelids, so that more elaborate and specialised appendages appear as a result, with greater division of labour in the body. Segmentation is still visible externally but there tend to be fewer segments than in the annelids.

The other major characteristics of the arthropods are discussed below, and these, together with the advanced segmentation noted above, help to explain their success.

Exoskeleton or cuticle. This is secreted by the epidermis. It contains chitin, a nitrogen-containing polysaccharide which strongly resembles cellulose, the strengthening material of plant cell walls. Chitin has high tensile strength (it is difficult to break by pulling from both ends). The properties of the exoskeleton can be altered by combining chitin with other chemicals. For example, addition of mineral salts, particularly calcium salts, can make it harder, as in crustaceans. Protein can have the same effect. A range of hardness, flexibility and stiffness is therefore possible. Flexibility is important at joints. Advantages of the exoskeleton include:

- support, particularly on land;
- it provides an anchor for the muscles internally, particularly those involved in locomotion, including flight;
- protection from physical damage;
- addition of a layer of wax from special glands in the epidermis helps prevent desiccation on land;
- insect flight and the jumping ability of fleas and grasshoppers depends on the presence of a remarkably elastic protein in the exoskeleton;
- it has a low density, which is important for flying animals;
- flexible joints are possible between segments;
- it can be modified to form hard jaws for biting, piercing, sucking (fig 2.61*b*) or grinding;
- it can be transparent in places allowing, for example, entry of light into eyes and camouflage in water.

There are two disadvantages associated with the presence of an exoskeleton.

- Final body size is limited because, as already noted in the previous section, as body size increases the surface area:volume ratio decreases. The extent of the exoskeleton depends on the surface area whereas mass depends on the volume. An arthropod the size of an elephant would either not be able to support its own weight, or the exoskeleton would have to be so massive it would not be able to move. (The other important restriction on size of insects is their breathing mechanism which works mainly by diffusion through tubes called tracheae. The largest living insects are stick insects, of about 30 cm in length, and the larger beetles, such as the hercules beetle which weighs up to 100 g, the size of a mouse.)
- It restricts growth, so periodic moulting (ecdysis) is required if the animal is to grow. However, the arthropod is very vulnerable to attack by predators at this period, and generally seeks the protection of shelter before undergoing the process.

Table 2.17 Classification of the phylum Arthropoda (arthropods).

Phylum Arthropoda

Characteristic features
Triploblastic, coelomate
Metameric segmentation, bilateral symmetry
Exoskeleton* of chitin and sometimes calcareous matter; may be rigid, stiff or flexible
Each segment typically bears a pair of jointed appendages used for locomotion or feeding or sensory purposes*
Coelom much reduced, main body cavity a haemocoel

Superclass Crustacea (crustaceans)**	*Class Insecta* (insects)	*Class Chilopoda* (centipedes)	*Class Diplopoda* (millipedes)	*Class Arachnida* (arachnids)
Mainly aquatic	Mainly terrestrial	Mainly terrestrial	Terrestrial	Terrestrial
Cephalothorax (head and thorax not distinctly separate)	Well-defined head, thorax abdomen	Clearly defined head Other body segments all similar	Clearly defined head Other body segments all similar	Cephalothorax (head and thorax not distinctly separate); thorax separated from abdomen by a narrow waist-like constriction
Two pars of antennae	One pair of antennae	One pair of antennae	One pair of antennae	No antennae
At least three pairs of mouthparts	Usually three pairs of mouthparts	One pair of mouthparts (jaws)	One pair of mouthparts (jaws)	No true mouthparts but one pair of appendages used in capturing prey and one pair of sensory palps
Pair of compound eyes raised on stalks	Pair of compound eyes and simple eyes	Eyes simple, compound or absent	Eyes simple, compound or absent	Simple eyes only (no compound eyes)
Appendages often modified for swimming, as mainly aquatic; number of legs variable, sometimes 10	Three pairs of legs on thorax, one pair per segment. Usually one or two pairs of wings on thorax (on second and/or third segments)	Numerous legs, all identical, one pair per segment	Numerous legs, all identical, two pairs per segment	Four pairs of walking legs (segments 4–7)
Larval form occurs	Life cycle commonly involves metamorphosis either 'complete' or 'incomplete', with a larval stage	No larval form	No larval form	No larval form
Typical gaseous exchange by gills – outgrowths of the body wall or limbs	No gills in adult Gaseous exchange by tracheae (tubes inside body	Gaseous exchange by tracheae	Gaseous exchange by tracheae	Gaseous exchange by 'lung' books or 'gill' books or tracheae
e.g. *Daphnia* (water-flea) *Astacus* (crayfish) Also barnacles, prawns, crabs, lobsters, woodlice	e.g. *Periplaneta* (cockroach) *Apis* (bee) *Pieris* (white butterfly) Also bugs, beetles, fleas, wasps, flies, dragonflies, termites, grasshoppers, earwigs	Mainly carnivorous e.g. *Lithobius* (centipede)	Mainly herbivorous e.g. *Iulus* (millipede)	e.g. *Scorpio* (scorpion) *Epeira* (web-spinning spider) Also mites, ticks

* diagnostic features.
** This superclass contains many classes.

Jointed appendages. The word 'arthropod' means, literally, 'jointed foot'. Jointed appendages are one of the most obvious characteristics of arthropods. They are used for a wide variety of functions such as feeding, locomotion and sensory purposes (fig 2.59).

Haemocoel. In arthropods and molluscs, the coelom is almost completely obliterated during development by another cavity called the **haemocoel** (fig 2.60). It develops from the cavities of the blood vascular system and is therefore filled with blood. The blood is

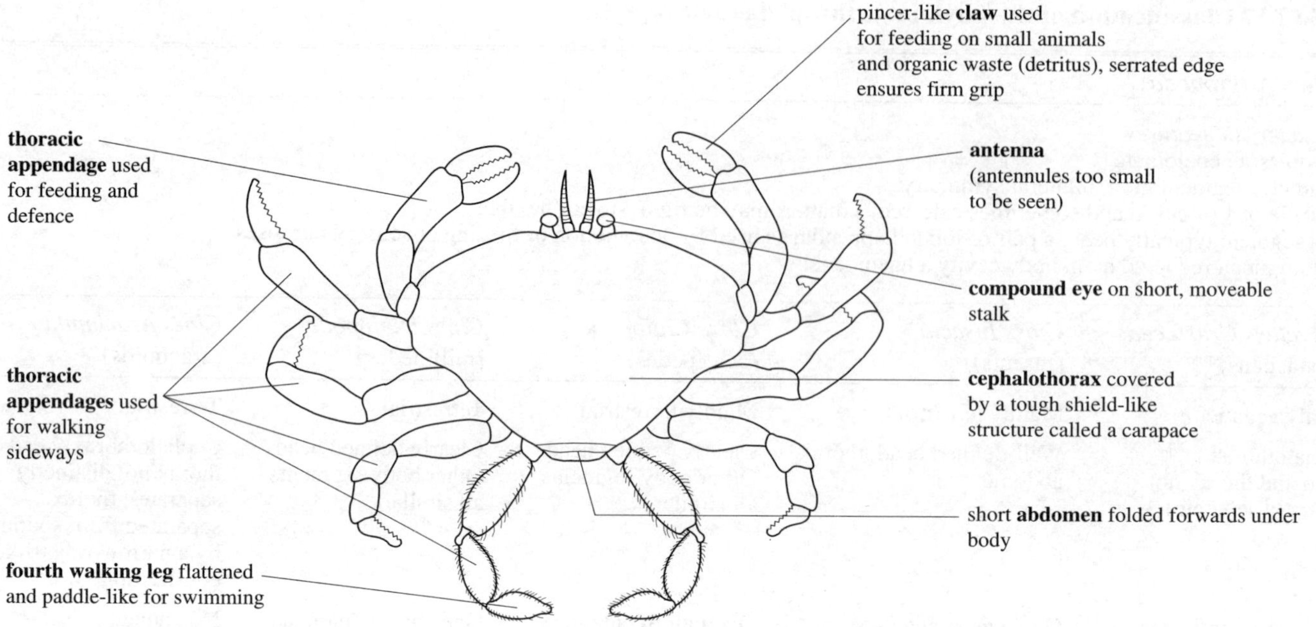

Fig 2.59 *A crustacean* Carcinus maenas, *the shore crab. Dorsal view. Common on rocky shores and beaches. Separate sexes occur. The head is fused to the thorax to form a cephalothorax. The first three pairs of thoracic appendages, called maxillipeds, are involved in feeding but are not visible from the dorsal surface. Note that the appendages are jointed.*

generally circulated in the haemocoel, and through several attached blood vessels. The major organs are bathed in blood. The coelom still exists but is small and confined to the cavities of excretory organs and the reproductive ducts. The high blood volume to body volume in arthropods enables them to maintain a high metabolic rate, allowing them to be very active animals. The danger of blood loss from injury, though, is high.

Specialisation of body parts. The division of labour, which is much more pronounced in arthropods than annelids, has contributed to the development of distinct regions of the body, namely the **head** and, in many cases, a **thorax** and **abdomen**. The head possesses sensory receptors, such as eyes and antennae, as well as feeding appendages. In bilaterally symmetrical animals the front end (the head) is the first part to come into contact with new environments. Thus the front end becomes specialised. The brain is much larger than in annelids and cephalisation much more pronounced.

Flight. Insects have developed flight which greatly increases opportunities for finding food and escaping predators (fig 2.61*a*).

Insect life histories

Life histories of insects are very variable and often highly complex. In many, a process called **metamorphosis** (*meta*, change; *morphe*, form) occurs. This is a change of form or structure of the animal during the course of its life cycle.

In the more primitive insect groups the larval stages often resemble the adult (**imago**) during development. Each successive larval form (called a **nymph** or **instar**) usually looks more and more like the adult. This form of development is termed **hemimetabolous** or **incomplete metamorphosis**. The nymph possesses adaptive features which enable it to live in a different habitat and eat different food from that of the adult. This avoids competition for food between juvenile and adult. The locust shows this type of life cycle.

In groups which evolved later, the larval stages are quite distinct from the adult. The final larval moult produces a pupa, inside which metamorphosis produces the adult

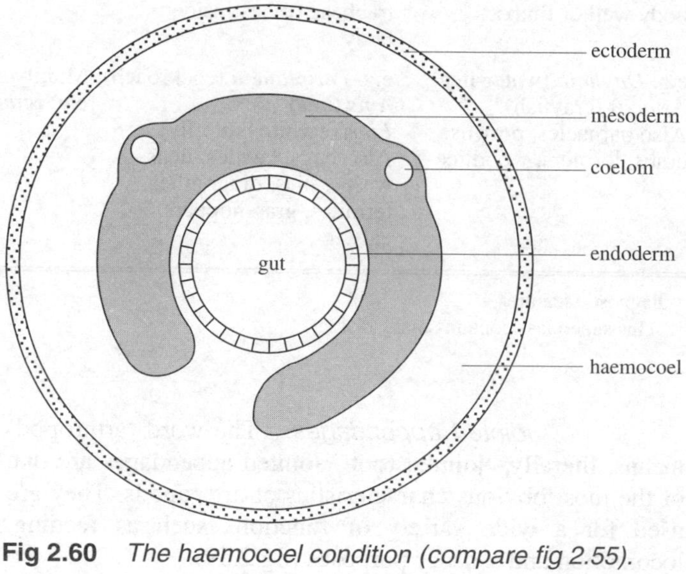

Fig 2.60 *The haemocoel condition (compare fig 2.55).*

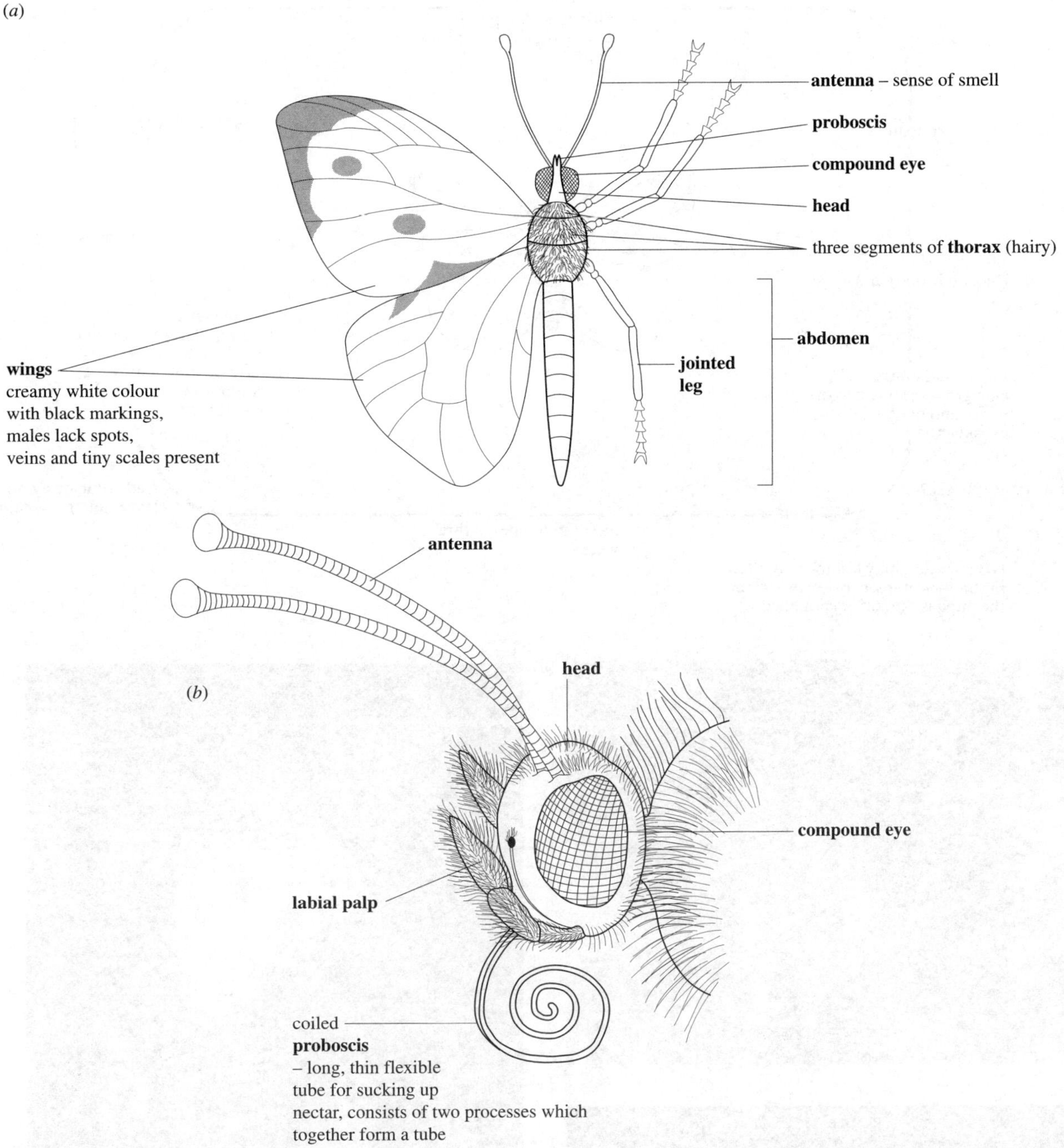

Fig 2.61 (a) *An insect,* Pieris brassicae*, the large (cabbage) white butterfly. Dorsal view, with wings shown on one side and legs the other. Wings are attached to the second and third thoracic segments, legs to all three thoracic segments. Pairs of spiracles, holes leading to the tracheae (respiratory tubes) are present on the first thoracic segment and the first eight abdominal segments.*
(b) *Detail of head of* Pieris.

tissues, using components from the degenerating larval tissues. This is called **holometabolous** or **complete metamorphosis**. An example of this type of life cycle is shown in fig 2.62.

Metamorphosis enables the juvenile and adult forms to live in different habitats and exploit different sources of food, that is to occupy different ecological niches. This reduces competition between juveniles and adults. For instance, dragonfly nymphs prey upon aquatic insects and exchange gases via gills whereas the adults attack terrestrial insects, live in air and exchange gases via tracheae. Also, lepidopteran (butterfly and moth) larvae generally feed on foliage and possess chewing mouthparts, whereas the adults drink nectar and have sucking mouthparts.

Once wings have fully developed, moulting is no longer possible. This restricts growth, so metamorphosis therefore allows the immature stages to provide the feeding and growing stages of the insect's life cycle.

(a)

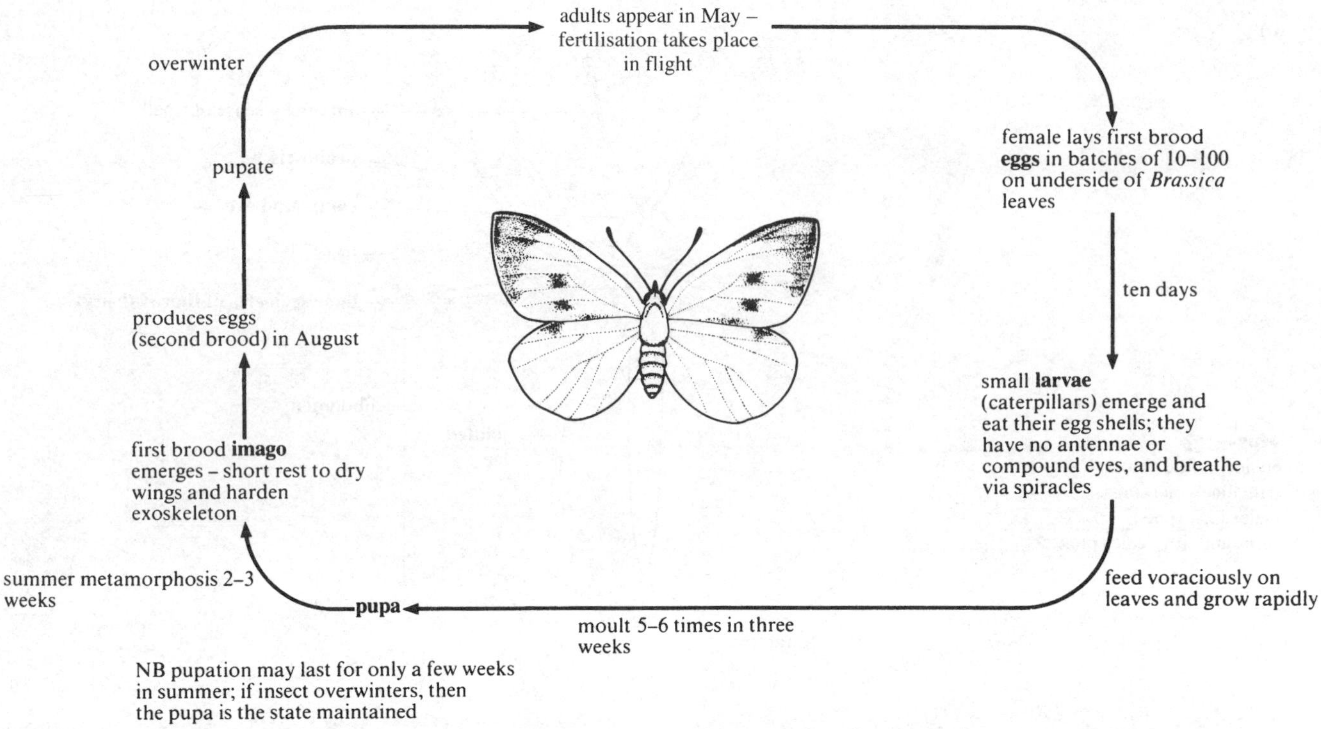

(b)

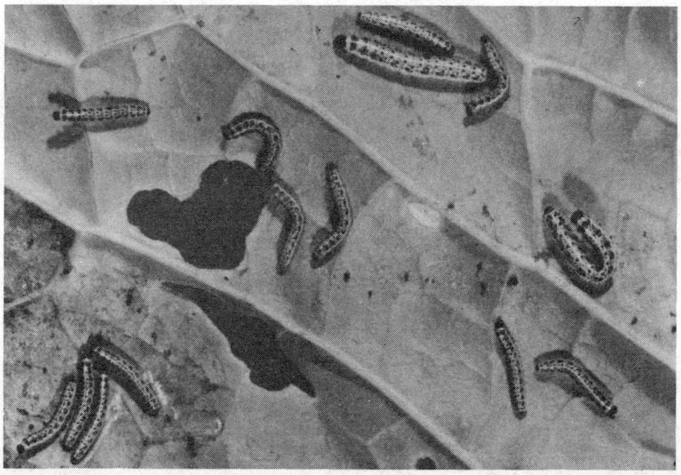

(c)

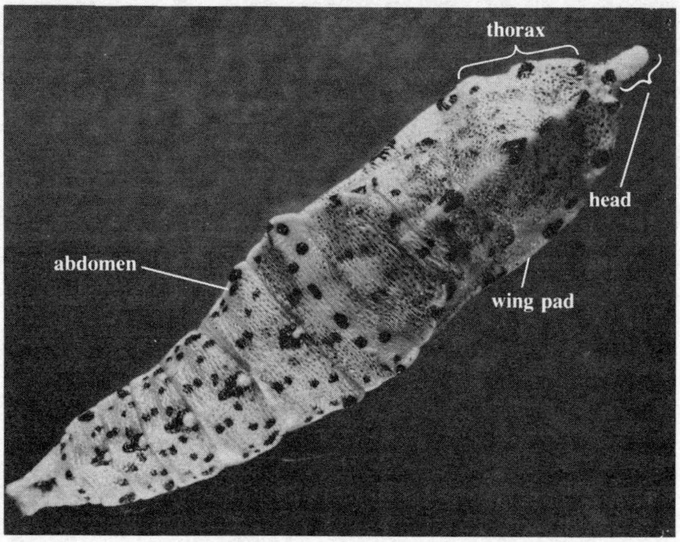

(d)

Fig 2.62 *Life cycle of cabbage white butterfly (*Pieris brassicae*), a member of the insect order Lepidoptera (butterflies and moths);* (a) *diagram of life cycle, an example of complete metamorphosis,* (b) *larvae (caterpillars),* (c) *pupa (chrysalis),* (d) *female imago (adult feeding from a buddleia bush).*

2.8.7 Phylum Mollusca (molluscs)

Classification characteristics of the Mollusca are summarised in table 2.18 (only three of the six classes are shown). The phylum consists of a diverse group of organisms which include slow-moving snails and slugs, relatively sedentary bivalves, such as clams, and highly active cephalopods (fig 2.63). With over 80 000 living species and 35 000 fossil species, the phylum is second only in size to the Arthropoda. One of the molluscs, the giant squid, is the largest non-vertebrate animal, weighing several tonnes and measuring 16 m in length.

The formation of a protective shell, and use of gills or lungs for gaseous exchange, has enabled molluscs to colonise aquatic and terrestrial environments and thus occupy a wide range of ecological niches. However, a shell can be handicap to locomotion, and some of the more active molluscs show a reduction or loss of the shell.

Fig 2.63 *A variety of molluscs.*

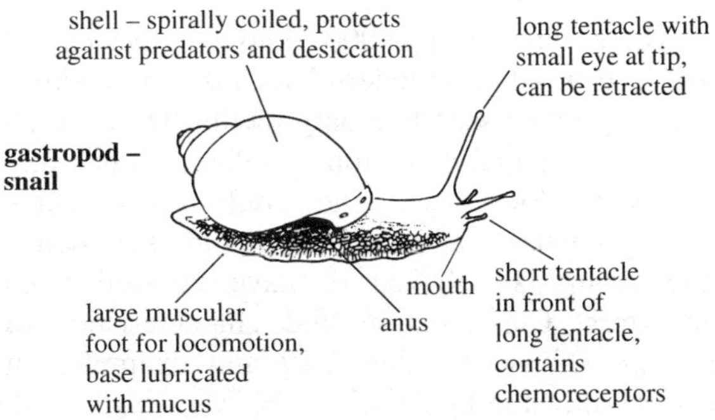

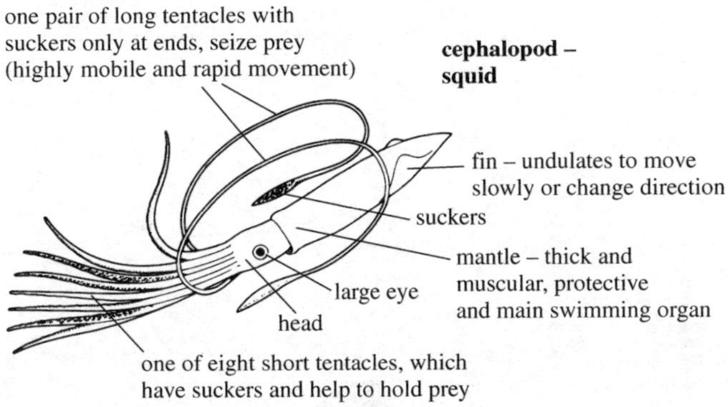

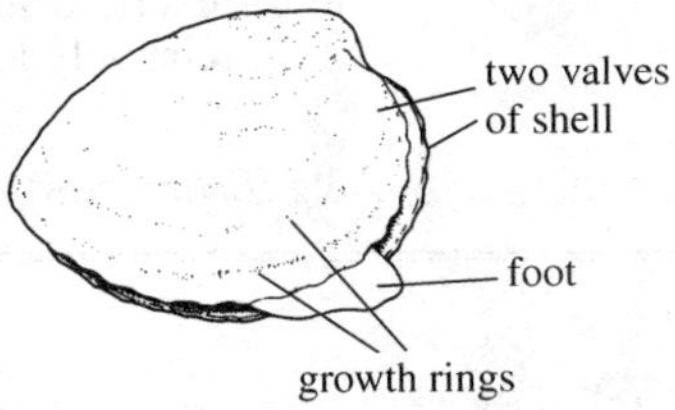

Table 2.18 Classification and characteristics of the phylum Mollusca (molluscs).

Phylum Mollusca		
Characteristic features Unsegmented, triploblastic coelomates Usually bilaterally symmetrical Body soft and fleshy and divided into a head, ventral muscular foot and dorsal visceral hump* Over the hump the skin (mantle) secretes a calcareous shell Main body cavity is a haemocoel No limbs		
Class Gastropoda (gastropods)	*Class Pelycopoda* (bivalves)	*Class Cephalopoda* (cephalopods)
Terrestrial, marine and freshwater	Aquatic	Aquatic. Largest and most complex molluscs
Asymmetrical	Bilateral symmetry	Bilateral symmetry
Shell of one piece, usually coiled due to rotation of hump during growth	Shell consists of two hinged halves called valves (hence the term 'bivalve'. Body enclosed by the valves and laterally compressed)	Shell often reduced and internal or wholly absent
Large flat foot used in locomotion	Foot reduced in size and often used for burrowing in sand or mud	Adapted for fast swimming. Foot modified to form part of head and tentacles
Head, eyes and sensory tentacles	Head greatly reduced in size, tentacles absent	Head highly developed with tentacles with suckers, and well-developed eyes
Radula, a rasping tongue-like structure used in feeding	Filter feeder	Radula and horny beak
Anus is anterior	Anus is posterior	Anus is posterior
e.g. *Helix aspersa* (land snail) *Patella* (limpet) *Buccinum* (whelk) *Limax* (slug)	e.g. *Mytilus edulis* (marine mussel) *Ostrea* (oyster)	e.g. *Sepia officinalis* (cuttlefish) *Loligo* (squid) *Octopus vulgaris* (octopus)

* diagnostic features.

2.8.8 Phylum Echinodermata (echinoderms)

There are over 6000 known species of echinoderms. The word 'echinoderm' means 'spiny skin'; most have hard spiny or wart-like outgrowths. They are all marine, and are largely bottom-dwellers inhabiting shorelines and shallow seas. The adult forms show pentamerous symmetry (a form of radial symmetry) although the group is believed to have evolved from bilateral ancestors. Classification and characteristics of echinoderms are shown in table 2.19 and examples of echinoderms are shown in fig 2.64.

Table 2.19 Classification and characteristics of the phylum Echinodermata (echinoderms).

Phylum Echinodermata

Characteristic features
Triploblastic, coelomate
All marine
Adult shows five-way (pentamerous) radial symmetry*
Tube feet for locomotion*
Calcareous exoskeleton
No head. Mouth generally on lower (oral) surface of body, anus on upper (aboral) surface

Class Stelleroidea (starfish)	*Class Echinoidea* (sea urchins)
Star-shaped, flattened	Globular
Arms not sharply separate from disc	Does not possess arms
Few calcareous plates in body wall; movable spines	Numerous calcareous plates in body wall, attached to each other to form a rigid structure; relatively long movable spines
e.g. *Asterias*	e.g. *Echinocardium*

* diagnostic feature.

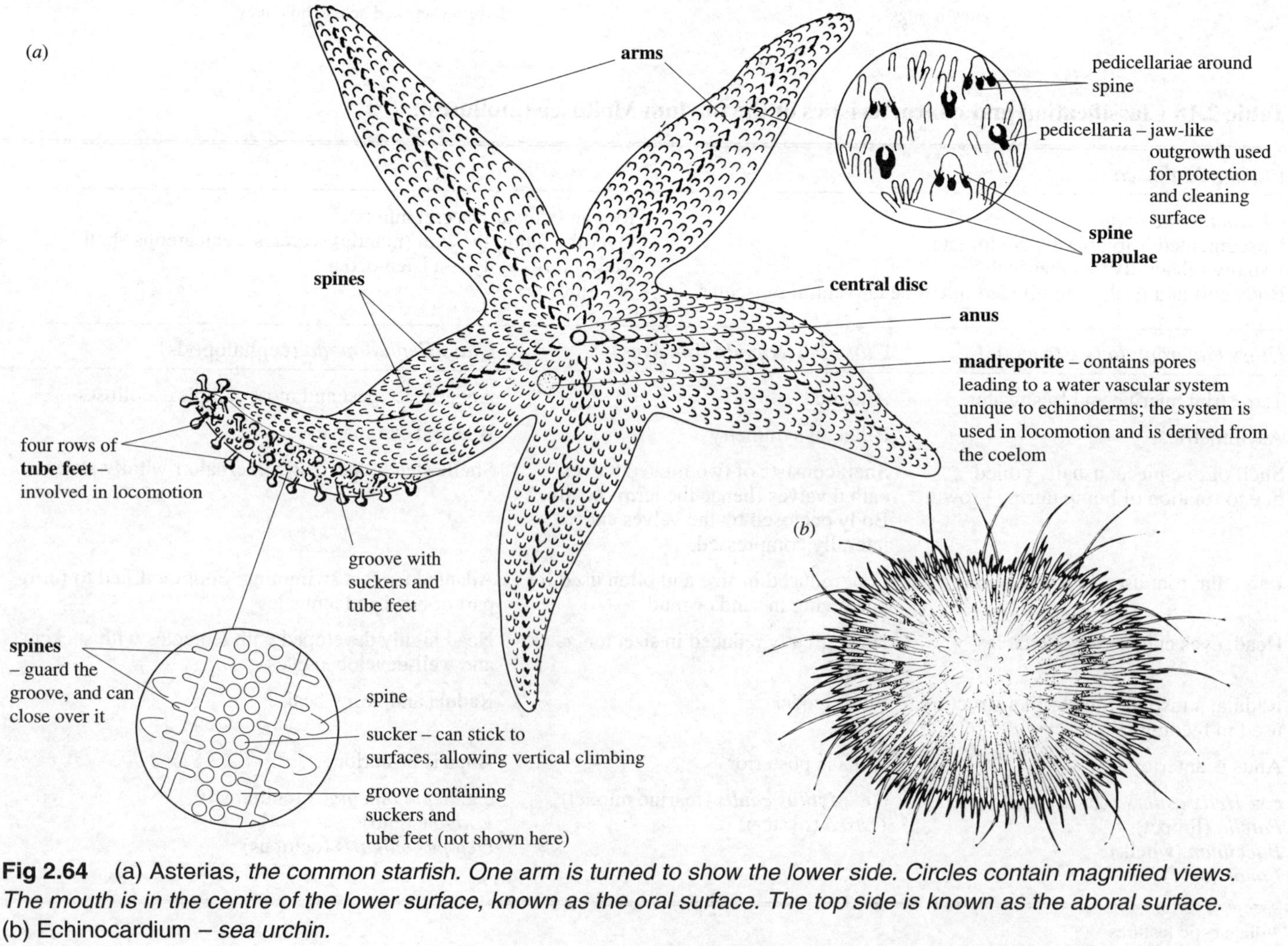

Fig 2.64 (a) Asterias, *the common starfish. One arm is turned to show the lower side. Circles contain magnified views. The mouth is in the centre of the lower surface, known as the oral surface. The top side is known as the aboral surface.* (b) Echinocardium – *sea urchin.*

2.8.9 Phylum Chordata (chordates)

Classification and characteristics of the Chordata are summarised in table 2.20. The chief distinguishing feature of the chordates is the presence of a dorsal, longitudinally running rod, the **notochord**. It lies between the dorsal nerve tube and the gut (fig 2.65). It increases internal support and locomotory power and probably evolved originally in the swimming larval forms of chordate ancestors. It still exists in something like its original form in the few remaining groups of non-vertebrate chordates, but in most chordates it is mainly replaced during embryonic development by the bony vertebrae of the **vertebral column**, or **backbone**. Animals possessing a vertebral column are known as **vertebrates**; all other animals are **non-vertebrates**. We shall only consider the vertebrate chordates here. They are such an important natural group that they are given the status of a subphylum (Vertebrata).

The basic body plan of chordates is shown in figs 2.65*a* and *b*. Representative examples of vertebrates are shown in fig 2.66.

Table 2.21 provides a summary of the key features of evolutionary significance in the different animal phyla.

Table 2.20 Classification and characteristics of the phylum Chordata (chordates).

Phylum Chordata
Characteristic features Notochord present at some stage in the life history. This is a flexible rod of tightly packed, vacuolated cells held together with a firm sheath* Triploblastic coelomate Bilateral symmetry Pharyngeal (visceral) clefts present (slits in the pharynx)* Dorsal, hollow nerve cord* Segmental muscle blocks (myotomes) on either side of the body Post-anal tail (tail starts posterior to anus)* Limbs formed from more than one body segment*
Subphylum Vertebrata (vertebrates)
Characteristic features Notochord replaced in adult by a vertebral column (backbone), a series of vertebrae made either of bone or cartilage* Well-developed central nervous system including brain.* Skull protects the brain Internal skeleton Pharyngeal clefts (gill slits), few in number Two pairs of fins or limbs.* These are attached to the rest of the skeleton by girdles, pectoral and pelvic*

Table 2.20 (cont.)

Class Chondrichthyes (cartilaginous fish)	*Class Osteichthyes* (bony fish)	*Class Amphibia* (amphibians)	*Class Reptilia* (reptiles)	*Class Aves* (birds)	*Class Mammalia* (mammals)
Skin with placoid (tooth-like) scales	Skin with cycloid scales (thin, round and made of bone)	Soft moist skin can be used for gaseous exchange to supplement lungs. No scales	Dry scaly skin with horny scales	Skin bears feathers, legs have scales	Skin bears hair with two types of glands, sebaceous and sweat
Cartilaginous skeleton	Bony skeleton	Bony	Bony	Bony	Bony
Paired, fleshy pectoral and pelvic fins Asymmetric tail fin helps prevent sinking (no air bladder or swim bladder for buoyancy)	Paired pectoral and pelvic fins supported by bony rays, giving greater manoeuvrability. Symmetrical tail fin	Two pairs pentadactyl limbs	Two pairs pentadactyl limbs usually present	Two pairs pentadactyl limbs, front pair form wings	Two pairs pentadactyl limbs
Visceral clefts present as separate gill openings; five pairs	Visceral clefts present as separate gill openings, but covered by a bony flap (operculum), four pairs	Visceral clefts present in aquatic larva (tadpole) only, lungs in adult, which is usually terrestrial	Visceral clefts never develop gills	Visceral clefts never develop gills	Visceral clefts never develop gills
		Metamorphosis from larva to adult in life cycle			
No external ear	No external ear	No external ear	No external ear	No external ear	External ear (in addition to middle and inner ear)
Eggs produced, internal fertilisation	Eggs produced, external fertilisaton	Eggs produced, external fertilisation. Adults must return to water for reproduction	Fertilised yolky eggs laid on land or eggs retained until hatching. Eggs have a leathery skin. Internal fertilisation	As reptiles but eggs in calcareous shells, internal fertilisation.	Only two genera lay eggs, the spiny anteater and the duck-billed platypus. Embryo develops in mother. Mother has mammary glands which produce milk for the newborn. Internal fertilisation
			Teeth lost and beak developed		Different types of teeth for different functions
					Muscular diaphragm between thorax and abdomen
Poikilothermic ('cold-blooded')	Poikilothermic	Poikilothermic	Poikilothermic	Homeothermic ('warm-blooded')	Homeothermic
e.g. *Scyliorhinus* (dogfish) Also sharks, skates and rays	e.g. *Clupea* (herring)	e.g. *Rana* (frog) *Bufo* (toad) Also newts and salamanders	e.g. *Natrix* (grass snake) *Crocodylus* (crocodile) Also lizards, alligators, turtles, tortoises. Dinosaurs were reptiles	e.g. *Columba* (pigeon) *Aquila* (eagle)	e.g. *Homo* (human) *Canis* (dog)

*diagnostic features.

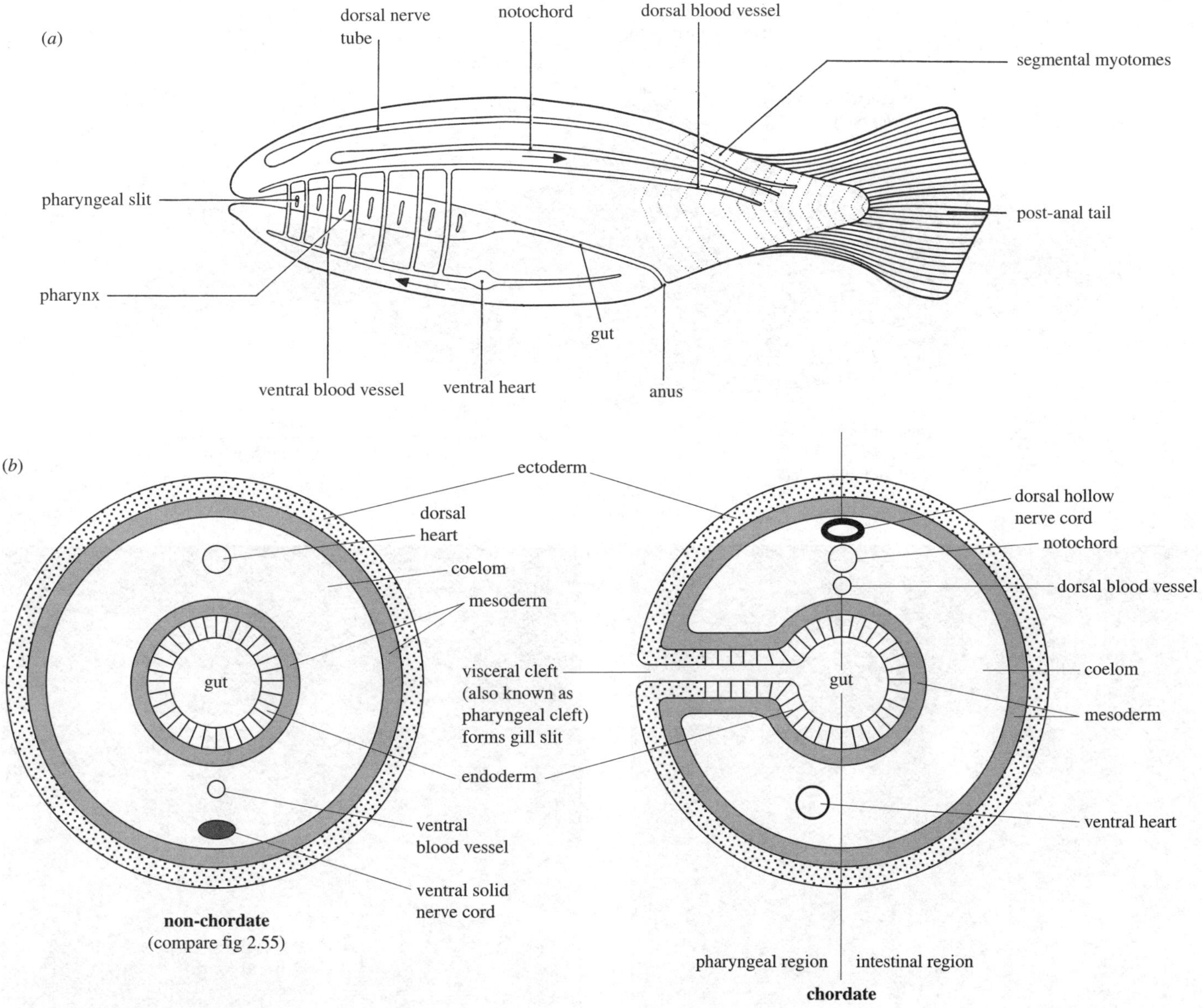

Fig 2.65 (a) *Diagram showing basic chordate body plan.*
(b) *Transverse sections of a non-chordate coelomate and a chordate for comparison.*

Table 2.21 Summary of some key features of evolutionary significance in animal phyla.

Phylum	*Radial symmetry*	*Diploblastic*	*Bilateral symmetry*	*Triploblastic*	*Acoelomate*	*Coelomate*	*Haemocoel*	*Metameric segmentation*
Cnidaria	●	●	×	×	–	–	–	–
Platyhelminthes	×	×	●	●	●	×	×	×
Nematoda	×	×	●	●	body cavity develops differently		×	×
Annelida	×	×	●	●	×	●	×	●
Arthropoda	×	×	●	●	×	●	●	●
Mollusca	×	×	●	●	×	●	●	×
Echinodermata	●	×	×	●	×	●	×	×
Chordata	×	×	●	●	×	●	×	●

Fig 2.66 (a) Clupea harengus, *the herring, a bony fish. The paired pectoral and pelvic fins control pitch (swimming up or down). The single dorsal and ventral fins control roll (leaning to sides) and yaw (turning).* (b) *European tree frog (*Hyla arborea*).* (c) *Lizard.* (d) *Blue tit (*Parus caeruleus*).* (e) *Dormouse (*Muscardinus avellinarius*).*

Chapter Three

Chemicals of life

3.1 Introduction to biochemistry

Biochemistry is the study of the chemicals of living organisms. It has been closely associated with the great expansion in biological knowledge that has taken place during the twentieth century. Its importance lies in the fundamental understanding it gives us of the way in which biological systems work. This finds application in fields like agriculture, with development of pesticides, herbicides and so on; medicine, including the whole pharmaceutical industry; fermentation industries with their vast range of useful products, including baked products; and food and nutrition, including dietetics, food production and preservation. Many of the exciting new developments in biology, like genetic engineering, biotechnology, 'designer' proteins and molecular approaches to genetic disease, are underpinned by an understanding of biochemistry.

Biochemistry is also one of the great unifying themes in biology. At this level, what is often striking about living organisms is not so much their differences as their similarities.

3.1.1 Elements found in living organisms

The Earth's crust contains approximately 100 chemical elements and yet only 16 of these are essential for life. These 16 are listed in table 3.1. The four most common elements in living organisms are, in order, hydrogen, carbon, oxygen and nitrogen. These account for more than 99% of the mass and numbers of atoms found in all living organisms. The four most common elements in the Earth's crust, however, are oxygen, silicon, aluminium, and sodium. The biological importance of hydrogen, oxygen, nitrogen and carbon is largely due to their having valencies of 1, 2, 3 and 4 respectively and their ability to form more stable covalent bonds than any other elements with these valencies. The appendix contains a summary of some basic chemistry, including covalent bonding (appendix, Book 2) which you might find useful to revise before reading this chapter.

The importance of carbon

It is sometimes said that life on our planet is based on carbon. Carbon is an element that is found in all organic molecules. The term organic, meaning living, was used originally because it was thought that only living things could make organic compounds. This myth was dispelled in 1828 when the German chemist Wöhler synthesised the organic molecule urea from inorganic starting materials. This made scientists realise that there was no 'magic' or special life force needed to make the biochemicals of life, and today we have even reached the point where we could in theory synthesise DNA, the genetic material, from inorganic starting materials, and hence 'create' life.

But why is carbon so important? Carbon forms strong covalent bonds, in other words it shares electrons, with other elements. It forms four such bonds, that is it has a valency of four. A simple example is methane, whose **molecular formula** is CH_4 and whose structural formula is shown in fig 3.1. (See also appendix 1, Book 2.) Box 3.1 should be consulted for help with writing formulae.

Table 3.1 The elements found in living organisms.

Chief elements of organic molecules	*Ions*	*Trace elements*	
H hydrogen	Na^+ sodium	Mn manganese	B boron
C carbon	Mg^{2+} magnesium	Fe iron	Al aluminium
N nitrogen	Cl^- chlorine	Co cobalt	Si silicon
O oxygen	K^+ potassium	Cu copper	V vanadium
P phosphorus	Ca^{2+} calcium	Zn zinc	Mo molybdenum
S sulphur			I iodine

Elements in each column are arranged in order of atomic mass, not abundance. Those in the first three columns are found in all organisms.
(Based on A. L. Lehninger, *Biochemistry*, Worth. N.Y. 1970).

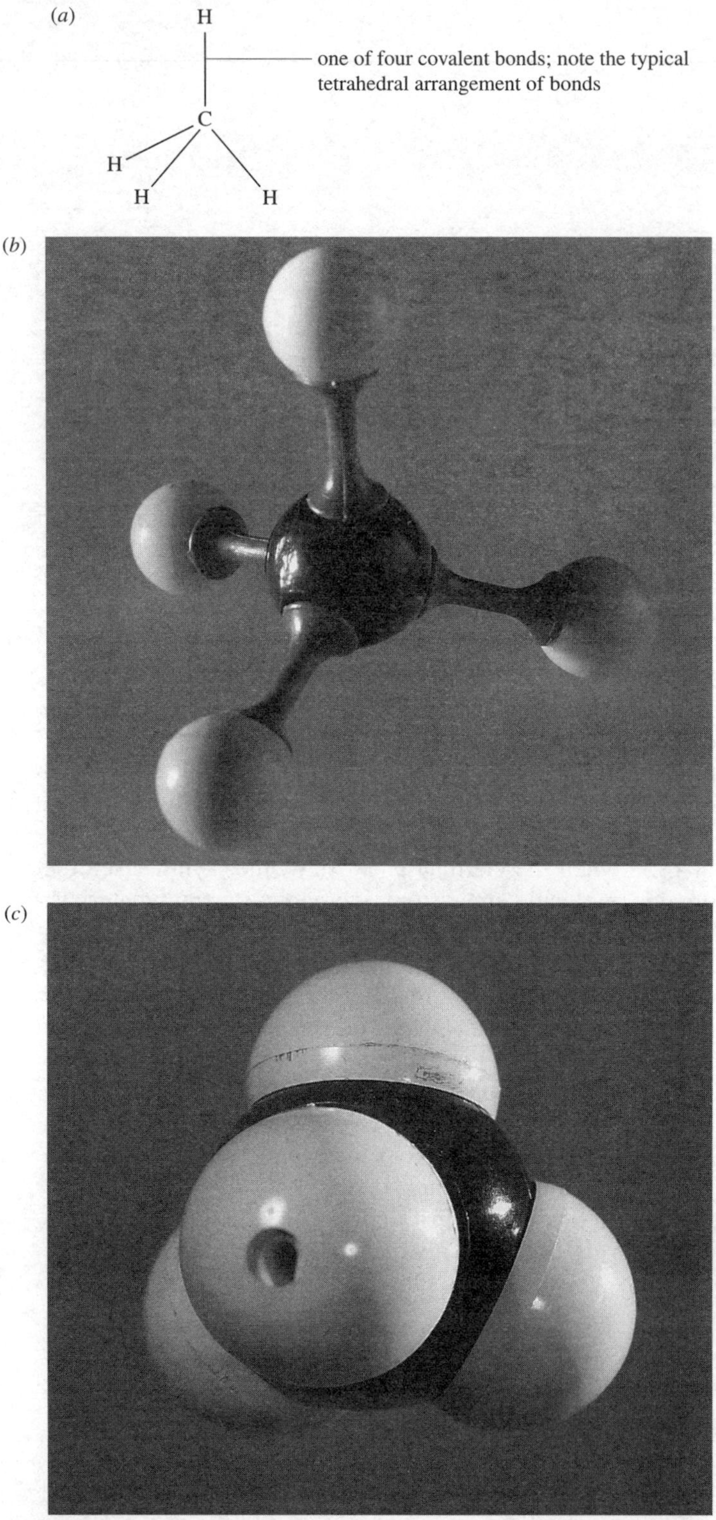

Fig 3.1 *Structure of methane, a simple organic molecule.* (a) *Structural formula. Carbon typically forms four covalent bonds when combining with other elements or with other carbon atoms.* (b) *and* (c) *show two ways of representing molecules as models.* (b) *A ball-and-stick model shows the arrangement of bonds clearly. In this case the tetrahedral arrangement of C bonds is shown.* (c) *A space-filling model gives a more realistic appearance, showing how close the atoms really are.*

> ### Box 3.1 Writing formulae
> Structural formulae are commonly simplified so that emphasis can be placed on the more important chemical groups. A simple example is ethanoic acid, shown in fig 3.2. The simplified version of the structural formula omits the carbon atoms and any hydrogen atoms joined directly to carbon atoms. We know the valency of carbon is 4, so it is possible to deduce the location of missing carbon atoms. The molecular formula can be represented as $CH_3.COOH$. This could also be written as $C_2H_4O_2$ but the former version is much more useful because it gives information about the relative positions of groups present. These, in turn, determine the properties of the molecules.
>
> 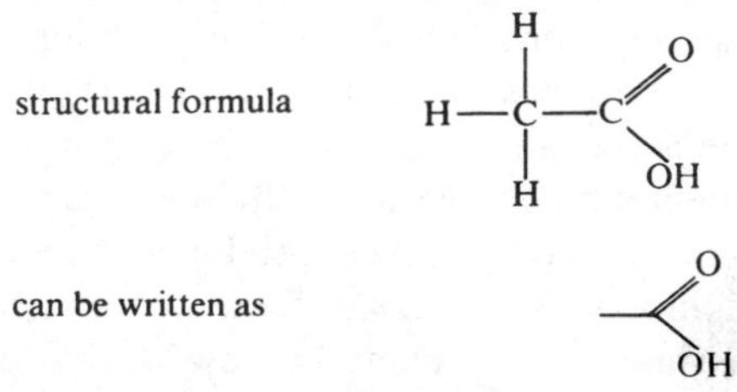
>
>
> **Fig 3.2** *Two ways of representing the structural formula of ethanoic acid, CH_3COOH.*

> **3.1** From what you have read, what is the difference between molecular and structural formulae?

The explanation for the importance of carbon lies in the way carbon atoms can join to each other, forming either chains or rings as shown in fig 3.3. These chains and rings are the skeletons of organic molecules and hence of life itself. They are very stable because the covalent bonds linking the carbon atoms together are strong. Atoms or particular groups of atoms of other elements (referred to simply as **groups**) can be attached at various positions to

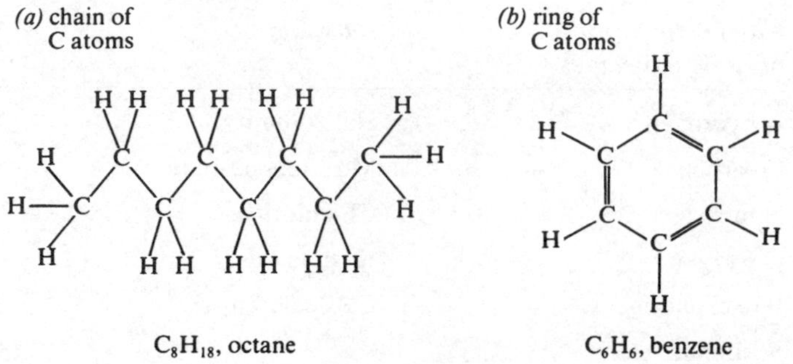

Fig 3.3 *Examples of chain and ring structures formed by C–C bonds.*

Table 3.2 Some common chemical groups found in organic molecules.

Aldehyde group	–C(–H)=O or –CHO
Keto group	–C– over C=O over –C–
Hydroxyl group	–OH
Carboxyl group	–COOH or –C(=O)–O–H
Carbonyl group	>C=O (this group is part of the aldehyde, keto- and carboxyl groups)
Amino group	–N(–H)–H
Sulphydryl group	–S–H
Phosphate group	–O–P(–O–H)(=O)–O–H

the carbon skeleton. Some common groups are shown in table 3.2. Each group has its own particular properties. For example, the carboxyl group, –COOH, is responsible for the acidic nature of fatty acids and amino acids.

Notice the use of shorthand methods to show chemical formulae in table 3.2.

Thus the group C(–H)=O is written as –CHO.

> **3.2** Fig 3.3 shows structural formulae of octane and benzene. Draw simplified versions of the structural formulae of (*a*) octane and (*b*) benzene, using the convention described in fig 3.2 in box 3.1.

Multiple bonds. Note also in table 3.2 and in fig 3.3 that carbon can form double bonds with itself, C=C. In fact, carbon, oxygen and nitrogen can all form strong multiple bonds:

double bonds: >C=C< >C=O >C=N–

triple bonds: –C≡C– –C≡N
(rare in nature)

Compounds containing double (=) or triple (≡) carbon–carbon bonds are called **unsaturated**. In a **saturated** carbon compound, all carbon–carbon bonds are single.

> **3.3** Draw the structural formula for the unsaturated organic compound ethene, C_2H_4.

Summary. The important chemical properties of carbon are:

- it is a relatively small atom with a low mass,
- it has the ability to form four strong, stable covalent bonds,
- it has the ability to form carbon–carbon bonds, thus building up large carbon skeletons with ring and/or chain structures,
- it has the ability to form multiple covalent bonds with other carbon atoms, oxygen and nitrogen.

This unique combination of features is responsible for the enormous variety of organic molecules. Variation occurs in three major ways:

- **size**, determined by the number of carbon atoms in the skeleton,
- **chemistry**, determined by the elements and chemical groups attached to the carbon atoms, and how saturated the carbon skeleton is,
- **shape**, determined by geometry, that is angles of the bonds.

3.1.2 Biological molecules

Living organisms are made of a limited number of types of atom, shown in table 3.1, which combine to form molecules, the building blocks of life. These molecules vary enormously in size from simple molecules such as carbon dioxide and water to macromolecules (giant molecules) such as proteins. The smaller molecules are soluble, easily transported and frequently enter into the general chemical activity of cells known as metabolism. Larger molecules tend to be used for storage or for structural purposes, and some can be described as 'informational' molecules, concerned with carrying genetic information (DNA and RNA) and the expression of that information (proteins).

Of the smaller molecules, water is the most abundant, typically making up between 60–95% of the fresh mass of an organism. Certain simple organic molecules are also found in all living organisms; these are shown in fig 3.4 and act as building blocks for the larger molecules. They are the kinds of molecules which biologists speculate could have been made in the 'primeval soup' of chemicals which is thought to have existed in the early history of the planet, before life itself appeared (chapter 26). These simple organic molecules are made in turn from even simpler, inorganic molecules, notably carbon dioxide, nitrogen and water.

Importance of water

Without water, life could not exist on this planet. It is important for two reasons. Firstly it is a vital chemical

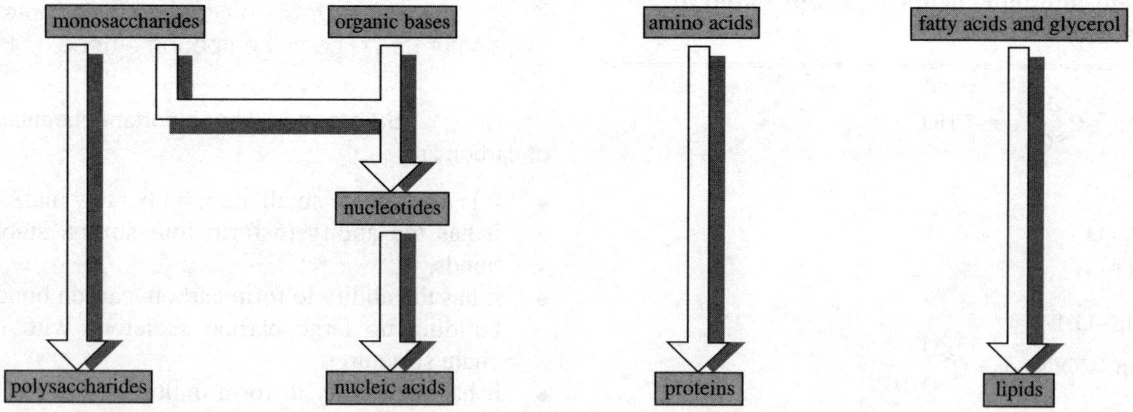

Fig 3.4 *The building blocks of life.*

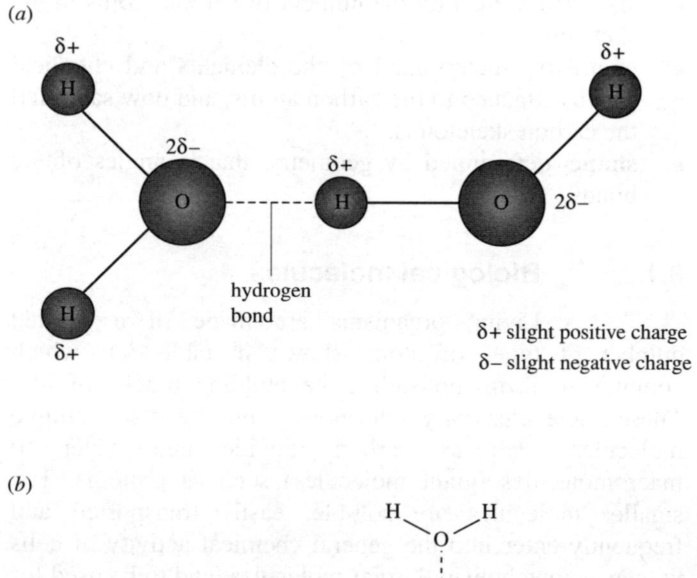

Fig 3.5 *Hydrogen bonding between water molecules.* (a) *Two water molecules attracted to each other by hydrogen bonding.* (b) *Network of hydrogen-bonded water molecules held together by hydrogen bonding. Such networks are constantly forming and breaking in liquid water.*

constituent of living cells, and secondly it provides an environment for those organisms that live in water. It is worthwhile, then, looking at some of its chemical and physical properties.

These properties are rather unusual and due mostly to its small size, its polarity and to hydrogen-bonding between its molecules. Polarity is an uneven charge distribution within a molecule. In water, one part, or pole, of the molecule is slightly positive and the other slightly negative. This is known as a **dipole**. It occurs because the oxygen atom has greater electron-attracting power than the hydrogen atoms. As a result the oxygen atom tends to attract the single electrons of the hydrogen atoms. Electrons are negatively charged, so giving the oxygen atom a slightly negative charge relative to the hydrogen atom.

Water molecules therefore have a weak attraction for each other, with opposite charges coming together and causing them to behave as if they were 'sticky', like magnets (fig 3.5*a*). These attractions are not as strong as normal ionic or covalent bonds and are called **hydrogen bonds**. They are constantly being formed, broken and re-formed in water (fig 3.5*b*). Although individually weak, their collective effect is responsible for many of the unusual physical properties of water. With these features in mind, some of the biologically significant properties of water can be examined.

Biological significance of water

Solvent properties. Water is an excellent solvent for polar substances. These include ionic substances like salts, which contain charged particles (ions), and some non-ionic substances like sugars that contain polar groups (slightly charged) such as the slightly negative hydroxyl group (–OH). On contact with water, the ions and the polar groups are surrounded by water molecules which separate (dissociate) the ions or molecules from each other. This is what happens when a substance dissolves in water (fig 3.6).

Once a substance is in solution its molecules or ions can move about freely, thus making it more chemically reactive than if it were solid. Thus the majority of the cell's

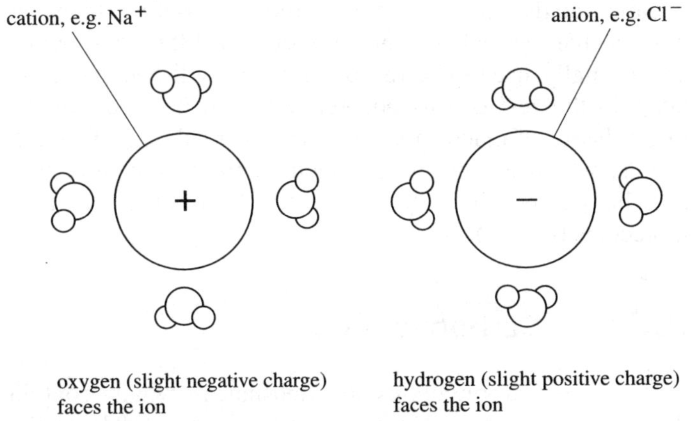

Fig 3.6 *Distribution of water molecules around ions in a solution. Note that the more negatively charged oxygen atom of water faces inwards to the cation but outwards from the anion. Water molecules have the effect of separating the ions because they collectively have a stronger attraction for the ions than the ions have for each other. If the ions were not separated they would form a solid crystal, for example sodium chloride crystals (common salt). Water has the effect of dissolving the salt.*

chemical reactions take place in aqueous solutions. By contrast, non-polar molecules, such as lipids, are repelled by water and usually group together in its presence, that is non-polar molecules are **hydrophobic** (water-hating). Such hydrophobic interactions are important in the formation of membranes and help to determine the three-dimensional structure of many protein molecules, nucleic acids and other cell structures.

Water's solvent properties also mean that it acts as a transport medium, as in the blood, lymphatic and excretory systems, the alimentary canal and in xylem and phloem.

High heat capacity. The heat capacity of water is the amount of heat required to raise the temperature of 1 kg of water by 1 °C. Water has a high heat capacity. This means that a large increase in heat energy results in a relatively small rise in temperature. This is because much of the energy is used in breaking the hydrogen bonds (overcoming the 'stickiness') which restrict the movement of the molecules.

Temperature changes within water are minimised as a result of its high heat capacity. Biochemical processes therefore operate over a smaller temperature range, proceeding at more constant rates and are less likely to be inhibited by extremes of temperature. Water also provides a very constant external environment for many cells and organisms.

High heat of vaporisation. Latent heat of vaporisation is a measure of the heat energy required to vaporise a liquid, that is to overcome the attractive forces between its molecules so that they can escape as a gas. A relatively large amount of energy is needed to vaporise water (make it evaporate or boil away into gas). This is due to the hydrogen bonding. As a result, water has an unusually high boiling point for such a small molecule.

The energy transferred to water molecules to allow them to vaporise results in a loss of energy from their surroundings, that is cooling takes place. This is made use of in the sweating and panting of mammals, the opening of the mouth of some reptiles, such as crocodiles, in sunshine, and may be important in cooling transpiring leaves. The high heat of vaporisation means that a large amount of heat can be lost with minimal loss of water from the body.

High heat of fusion. Latent heat of fusion is a measure of the heat energy required to melt a solid, in this case ice. With its high heat capacity, water requires relatively large amounts of heat energy to thaw it. Conversely, liquid water must lose a relatively large amount of heat energy to freeze. Contents of cells and their environments are therefore less likely to freeze. Ice crystals are particularly damaging if they develop inside cells.

Density and freezing properties. The density of water decreases below 4 °C and ice therefore tends to float. It is the only substance whose solid form is less dense than its liquid form.

Since ice floats, it forms at the surface first and the bottom last. If ponds froze from the bottom upwards, freshwater life could not exist in temperate or arctic climates. Ice insulates the water below it, thus increasing the chances of survival of organisms in the water. This is important in cold climates and cold seasons, and must have been particularly so in the past, such as during Ice Ages. Also, the ice thaws more rapidly by being at the surface. The fact that water below 4 °C tends to rise also helps to maintain circulation in large bodies of water. This may result in nutrient cycling and colonisation of water to greater depths.

High surface tension and cohesion. Cohesion is the force whereby individual molecules stick together. At the surface of a liquid, a force called surface tension exists between the molecules as a result of cohesive forces between the molecules. These cause the surface of the liquid to occupy the least possible surface area (ideally a sphere). Water has a higher surface tension than any other liquid. The high cohesion of water molecules is important in cells and in translocation of water through xylem in plants (chapter 13). At a less fundamental level, many small organisms rely on surface tension to settle on water or to skate over its surface.

Water as a reagent. Water is biologically significant as an essential metabolite, that is it participates in the chemical reactions of metabolism. In particular, it is used as a source of hydrogen in photosynthesis (section 7.6) and is used in hydrolysis reactions.

Some of the biologically important functions of water are summarised in table 3.3.

Table 3.3 Some biologically important functions of water.

All organisms
Structure – high water content of cells (70–95% typical)
Solvent and medium for diffusion
Reagent in hydrolysis
Support for aquatic organisms
Fertilisation by swimming gametes
Dispersal of seeds, gametes and larval stages of aquatic organisms, and seeds of some terrestrial species e.g. coconut
Plants
Osmosis and *turgidity* (important in many ways, such as growth (cell enlargement), support, guard cell mechanism)
Reagent in photosynthesis
Transpiration
Translocation of inorganic ions and organic compounds
Germination of seeds – swelling and breaking open of the testa and further development
Animals
Transport in blood vascular system, lymphatic system, excretory system
Osmoregulation
Cooling by evaporation, such as sweating, panting
Lubrication, as in joints
Support – hydrostatic skeleton of e.g. annelid worms
Protection, for example lachrymal fluid (tears), mucus
Migration in ocean currents

3.1.3 Macromolecules

Simpler organic molecules often associate to form larger molecules. A **macromolecule** is a giant molecule made from many repeating units. Molecules built like this are known as **polymers**. The individual units are known as **monomers**. The units are joined by a chemical process known as **condensation**, which means removal of water. They can be broken down again by the opposite process, **hydrolysis**, or addition of water. There are three important types of macromolecule in biology, namely **polysaccharides**, **proteins** and **nucleic acids** and their constituent monomers are **monosaccharides, amino acids** and **nucleotides** respectively.

Macromolecules account for over 90% of the dry mass of cells. Polysaccharides tend to be used for food storage or structural purposes, whereas nucleic acids and proteins can be regarded as 'informational' molecules. This means that the *sequence* of subunits is important in proteins and nucleic acids and is much more variable than in polysaccharides, where only one or two different subunits are normally used. The reasons for this will become clear later. In the rest of this chapter, we shall be studying the three classes of macromolecules and their subunits in detail. Lipids, which are generally much smaller molecules, will also be studied since they are made from simple organic molecules (fig 3.4).

3.2 Carbohydrates

Carbohydrates are substances which contain the elements carbon, hydrogen and oxygen and have the general formula $C_x(H_2O)_y$, where x and y are variable numbers; their name (hydrate of carbon) is derived from the fact that hydrogen and oxygen are present in the same proportions as in water, namely two hydrogen atoms to one oxygen atom. In addition, they have the following properties:

- all are aldehydes or ketones,
- all contain several hydroxyl groups.

Their chemistry is determined by these groups. For example, aldehydes are very easily oxidised and hence are powerful reducing agents. The structures of these groups are shown in table 3.2.

Carbohydrates are divided into three main classes, monosaccharides, disaccharides and polysaccharides, as shown in fig 3.7.

3.2.1 Monosaccharides

Monosaccharides are single sugar units. Their general formula is $(CH_2O)_n$ and some of their properties are shown in fig 3.7. They are classified according to the number of carbon atoms as trioses (3C), tetroses (4C), pentoses (5C), hexoses (6C) and heptoses (7C). Of these, pentoses and hexoses are the most common.

> **3.4** What would be the molecular formula of pentoses and hexoses?

The chief functions of the common monosaccharides are summarised in table 3.4. It will be seen from table 3.4 that monosaccharides are important as energy sources and as building blocks for the synthesis of larger molecules.

Aldoses and ketoses

In monosaccharides, all the carbon atoms except one have a hydroxyl group attached. The remaining carbon atom is either part of an aldehyde group, in which case the

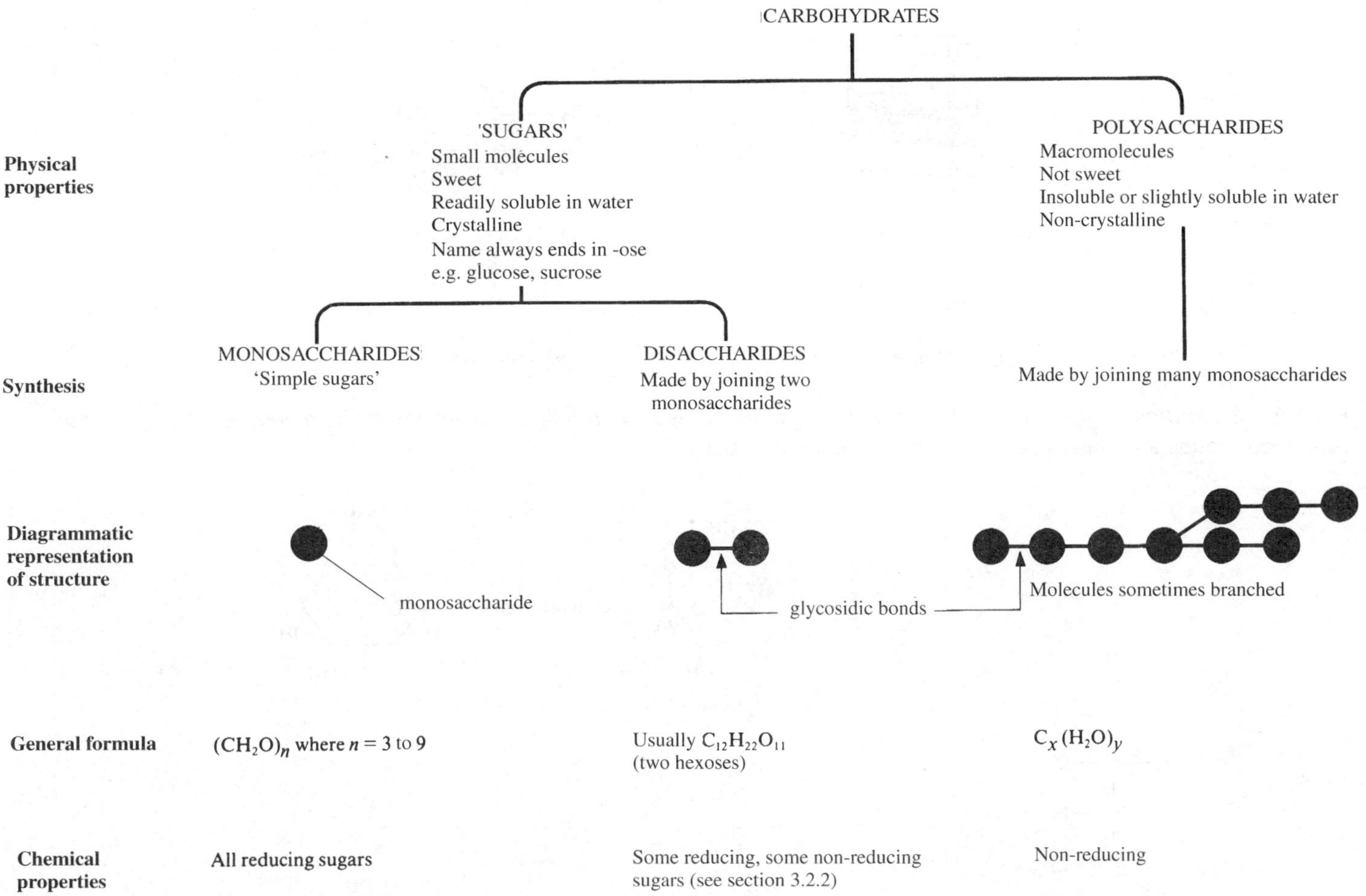

Fig 3.7 *Classification of carbohydrates. Note that monosaccharides and disaccharides are both referred to as sugars. They share certain properties, such as sweetness of taste.*

Table 3.4 Chief functions of monosaccharides.

Trioses $C_3H_6O_3$ e.g. glyceraldehyde, dihydroxyacetone
Intermediates in respiration (see glycolysis), photosynthesis (see dark reactions) and other branches of carbohydrate metabolism

Pentoses $C_5H_{10}O_5$ e.g. ribose, deoxyribose, ribulose

- Synthesis of nucleic acids; ribose is a constituent of RNA, deoxyribose of DNA
- Synthesis of some coenzymes, e.g. ribose is used in the synthesis of NAD and NADP
- Synthesis of ATP requires ribose
- Ribulose bisphosphate is the CO_2 acceptor in photosynthesis and is made from the 5C sugar ribulose

Hexoses $C_6H_{12}O_6$ e.g. glucose, fructose, galactose

- Source of energy when oxidised in respiration; glucose is the most common respiratory substrate and the most common monosaccharide
- Synthesis of disaccharides; two monosaccharide units can link together to form a disaccharide
- Synthesis of polysaccharides; glucose is particularly important in this role

monosaccharide is called an **aldose** or **aldo sugar**, or is part of a keto group, when it is called a **ketose** or **keto sugar**. Thus all monosaccharides are aldoses or ketoses. The two simplest monosaccharides are the trioses glyceraldehyde and dihydroxyacetone. Glyceraldehyde has an aldehyde group and dihydroxyacetone has a keto group (fig 3.8). In general, aldoses, such as ribose and glucose, are more common than ketoses, such as ribulose and fructose.

> **3.5** If you have little experience of chemistry, it might be useful to answer the following with reference to fig 3.8.
> (*a*) What is the valency of each element?
> (*b*) What is the total number of each type of atom? Does it conform with the molecular formula?
> (*c*) How many hydroxyl groups does each molecule contain? Could this have been predicted knowing they were trioses?

A suitable monosaccharide to study in more detail is glucose, the most common monosaccharide. It is a hexose, and therefore has the formula $C_6H_{12}O_6$. Its structure is shown in fig 3.9.

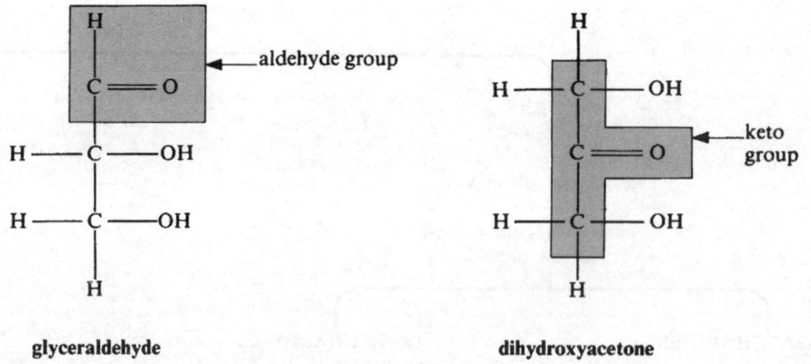

Fig 3.8 *Structures of glyceraldehyde and dihydroxyacetone. Note carefully the positions of the aldehyde and keto groups. Aldehyde groups are always at the end of the chain of C atoms.*

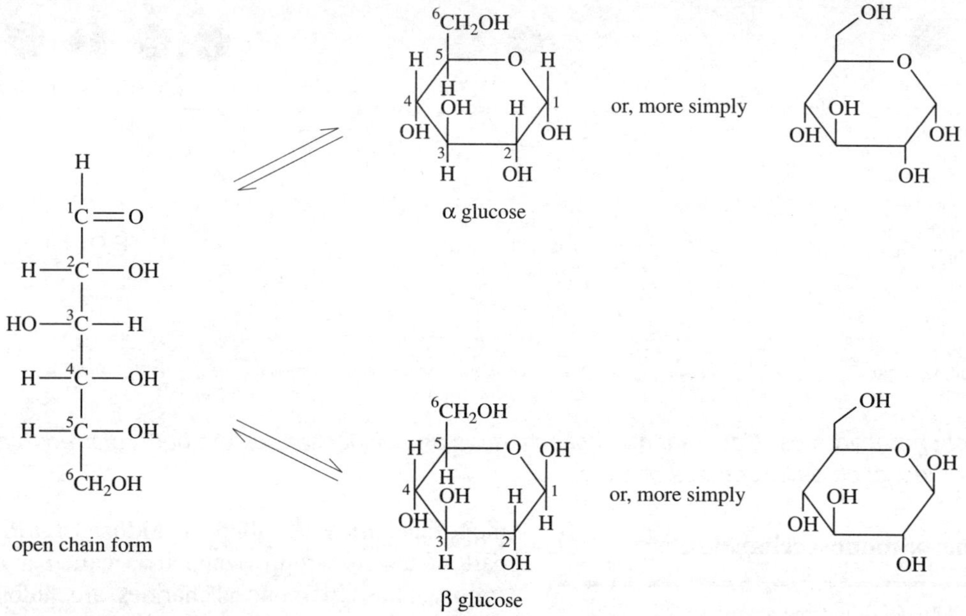

Fig 3.9 *Structure of the open chain and α and β ring forms of glucose. The three forms exist in equilibrium in aqueous solution, with 0.02% open chain, 36% α glucose and 64% β glucose.*

Open chain and ring forms

Fig 3.9 shows glucose as having either an 'open chain' or ring structure. The open chain form can be straight, but because of the bond angles between carbon atoms it is possible for sugars with five and six carbon atoms to bend and form stable ring structures. In hexoses like glucose, the first carbon atom combines with the oxygen atom on carbon atom number five to give a six-membered ring, as shown in fig 3.9. Note that oxygen is part of the ring and that one carbon atom, carbon atom number 6, sticks up out of the ring. In pentoses, the first carbon atom joins with the oxygen atom on the fourth carbon atom to give a five-membered ring, as shown in fig 3.10.

The ring structures of pentoses and hexoses are the usual forms, with only a small proportion of the molecules existing in the open chain form at any one time. The ring structure is the form used to make disaccharides and polysaccharides.

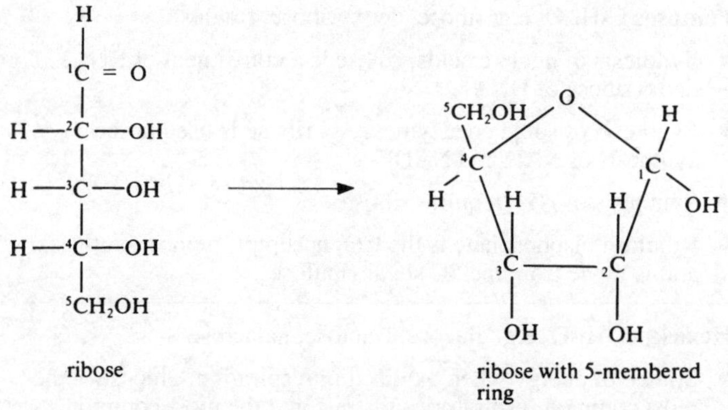

Fig 3.10 *Open chain and ring forms of ribose.*

Alpha (α) and beta (β) isomers

Fig 3.9 shows that glucose can exist in two possible ring forms, known as the **alpha** (α) and **beta** (β) **forms**. The

Fig 3.11 *Space-filling models of α and β glucose.*

hydroxyl group on carbon atom 1 can project below the ring (α glucose) or above the ring (β glucose). Molecules like this which have the same chemical formula but with different structures are said to be isomers of each other. Fig 3.11 shows space-filling models of the two isomers. At any given moment in a glucose solution, some of the molecules will be in the open chain form and some in the ring form. This is more stable and therefore more common. A glucose molecule can switch spontaneously from the open chain form to either of the two ring forms and back again. Overall an equilibrium is reached where the proportions of the different forms remain constant (fig 3.9).

As stated above, only the ring form can be used to make disaccharides and polysaccharides. Despite the relatively small difference in structure between α and β glucose, there are important consequences. Later we shall see that α glucose is used to make the polysaccharide starch and β glucose the polysaccharide cellulose, molecules which have very different properties.

3.2.2 Disaccharides

Fig 3.7 summarises some of the properties of disaccharides. They are formed when two monosaccharides, usually hexoses, combine by means of a chemical reaction known as a **condensation**. This means removal of water, and is shown in fig 3.12:

$$C_6H_{12}O_6 + C_6H_{12}O_6 \underset{\text{hydrolysis}}{\overset{\text{condensation}}{\rightleftharpoons}} C_{12}H_{22}O_{11} + H_2O$$

The bond formed between two monosaccharides as a result of condensation is called a **glycosidic bond** and it normally forms between carbon atoms 1 and 4 of neighbouring units (a 1,4 bond or 1,4 linkage). The process can be repeated many times to build up the giant molecules of polysaccharides (fig 3.12). The monosaccharide units are called residues once they have been linked. Thus a maltose molecule contains two glucose residues.

The most common disaccharides are maltose, lactose and sucrose:

maltose = glucose + glucose
lactose = glucose + galactose
sucrose = glucose + fructose

Maltose occurs mainly as a breakdown product during digestion of starch by enzymes called amylases. This commonly occurs in animals and in germinating seeds. The latter is made use of in brewing beer when barley grain is used as the source of starch. Germination of the barley is stimulated and this results in the conversion of the starch to maltose, a process known as malting. The maltose is then fermented by yeast to alcohol. This involves conversion of maltose to glucose by the action of the enzyme maltase, a process which also occurs in animals during digestion.

Lactose, or milk sugar, is found exclusively in milk and is an important energy source for young mammals. It can only be digested slowly, so gives a slow steady release of energy.

Sucrose, or cane sugar, is the most abundant disaccharide in nature. It is most commonly found in plants, where it is transported in large quantities through phloem tissue. It makes a good transport sugar because it is very soluble, and can therefore be moved efficiently in high concentrations. It is also relatively unreactive chemically. This means it tends not to enter into general metabolism on its way from one place to another. It is sometimes stored for the same reasons. It is obtained commercially from sugar cane and sugar beet and is the 'sugar' we normally buy in shops.

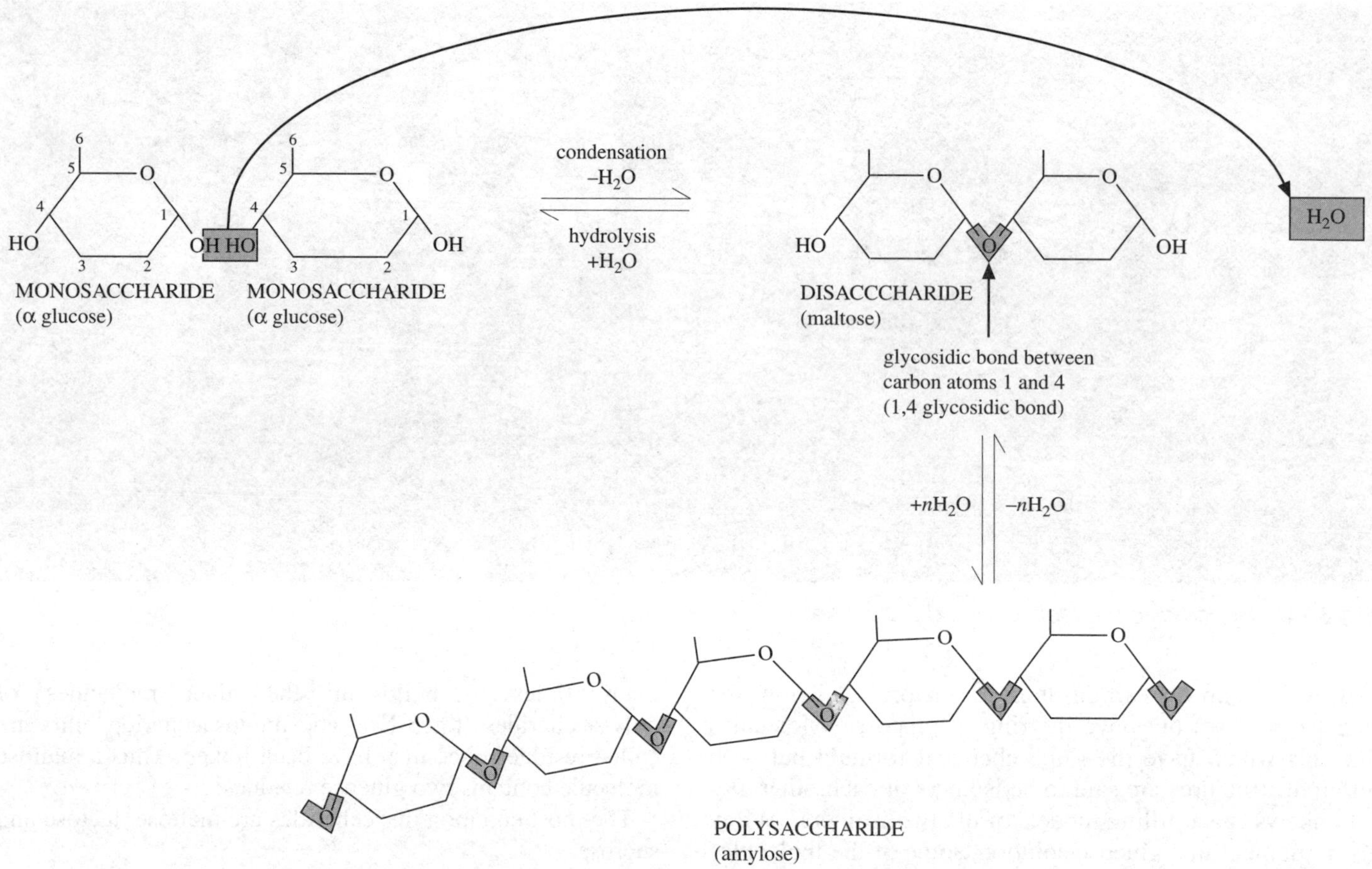

Fig 3.12 *Formation of a disaccharide and a polysaccharide from monosaccharides. In the example illustrated, α glucose is used to make maltose, then starch.*

Reducing sugars

All monosaccharides and some disaccharides, including maltose and lactose, are reducing sugars, meaning that they can carry out a type of chemical reaction known as **reduction**. Sucrose is the only common non-reducing sugar. Two common tests for reducing sugars, Benedict's test and Fehling's test (section 3.7) make use of the ability of these sugars to reduce copper from a valency of 2 to a valency of 1. Both tests involve use of an alkaline solution of copper(II) sulphate ($CuSO_4$) which is reduced to insoluble copper(I) oxide (Cu_2O).

Ionic equation: $Cu^{2+} + e^- \longrightarrow Cu^+$
blue solution — brick-red precipitate

3.2.3 Polysaccharides

Fig 3.7 summarises some of the properties of polysaccharides. They function chiefly as food and energy stores (for example starch and glycogen) and as structural materials (for example cellulose). They are convenient storage molecules for several reasons: their large size makes them more or less insoluble in water, so they exert no osmotic or chemical influence in the cell; they fold into compact shapes (see below) and they are easily converted to sugars by hydrolysis when required.

As we have already seen, polysaccharides are polymers of monosaccharides.

Starch

Starch is a polymer of α glucose (fig 3.12). It is a major fuel store in plants, but is absent from animals where the equivalent is glycogen (see below). It can easily be converted back to glucose for use in respiration. In germinating seeds the glucose may also be used to make cellulose and other materials needed for growth.

Starch has two components, **amylose** and **amylopectin**. Amylose has a straight chain structure consisting of several thousand glucose residues joined by 1,4 bonds as shown in fig 3.12. These bonds cause the chain to coil helically into a more compact shape. Amylopectin is also compact as it has many branches, formed by 1,6 glycosidic bonds (fig 3.13). It has up to twice as many glucose residues as amylose. A suspension of amylose in water gives a blue-black colour with iodine–potassium iodide solution, whereas a suspension of amylopectin gives a red-violet colour. This forms the basis of the test for starch (section 3.7).

Starch molecules accumulate to form starch grains. These are visible in many plant cells, notably in the

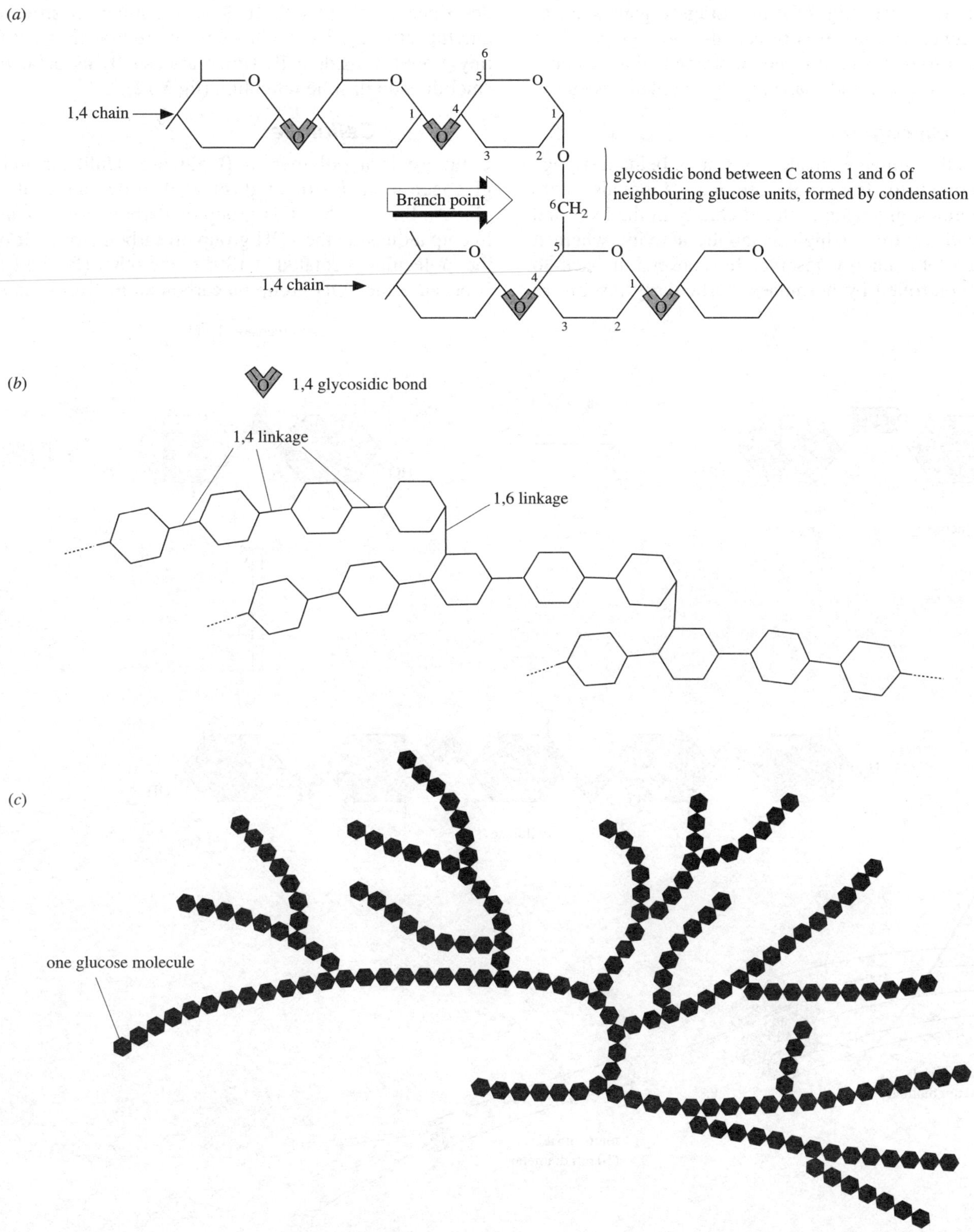

Fig 3.13 *Structure of amylopectin and glycogen at three levels of 'magnification'.* (a) *formation of one branch,* (b) *several branches,* (c) *highly branched structure of whole molecule. The 1,6 linkages cause branching and the 1,4 linkages cause the chains to turn and coil.*

chloroplasts of leaves (fig 7.6), in storage organs such as the potato tuber, and in seeds of cereals and legumes. The grains appear to be made of layers of starch and are usually of a characteristic size and shape for a given plant species.

Glycogen

Glycogen is the animal equivalent of starch, being a storage polysaccharide made from α glucose; many fungi also store it. In vertebrates, glycogen is stored chiefly in the liver and muscles, both centres of high metabolic activity, where it provides a useful energy reserve. Its conversion back to glucose is controlled by hormones, particularly insulin, as described in chapter 9. It is very similar in structure to amylopectin (fig 3.13), but shows more branching. It forms tiny granules inside cells which are usually associated with smooth endoplasmic reticulum (fig 5.12).

Cellulose

Cellulose is a polymer of β glucose. Unlike starch and glycogen it has a structural role. When two molecules of β glucose line up, the –OH group on carbon atom 1 can only line up alongside the –OH group on carbon atom 4 if one of the molecules is rotated at 180° to the other (fig 3.14). This is because the –OH group on carbon atom 1 projects below

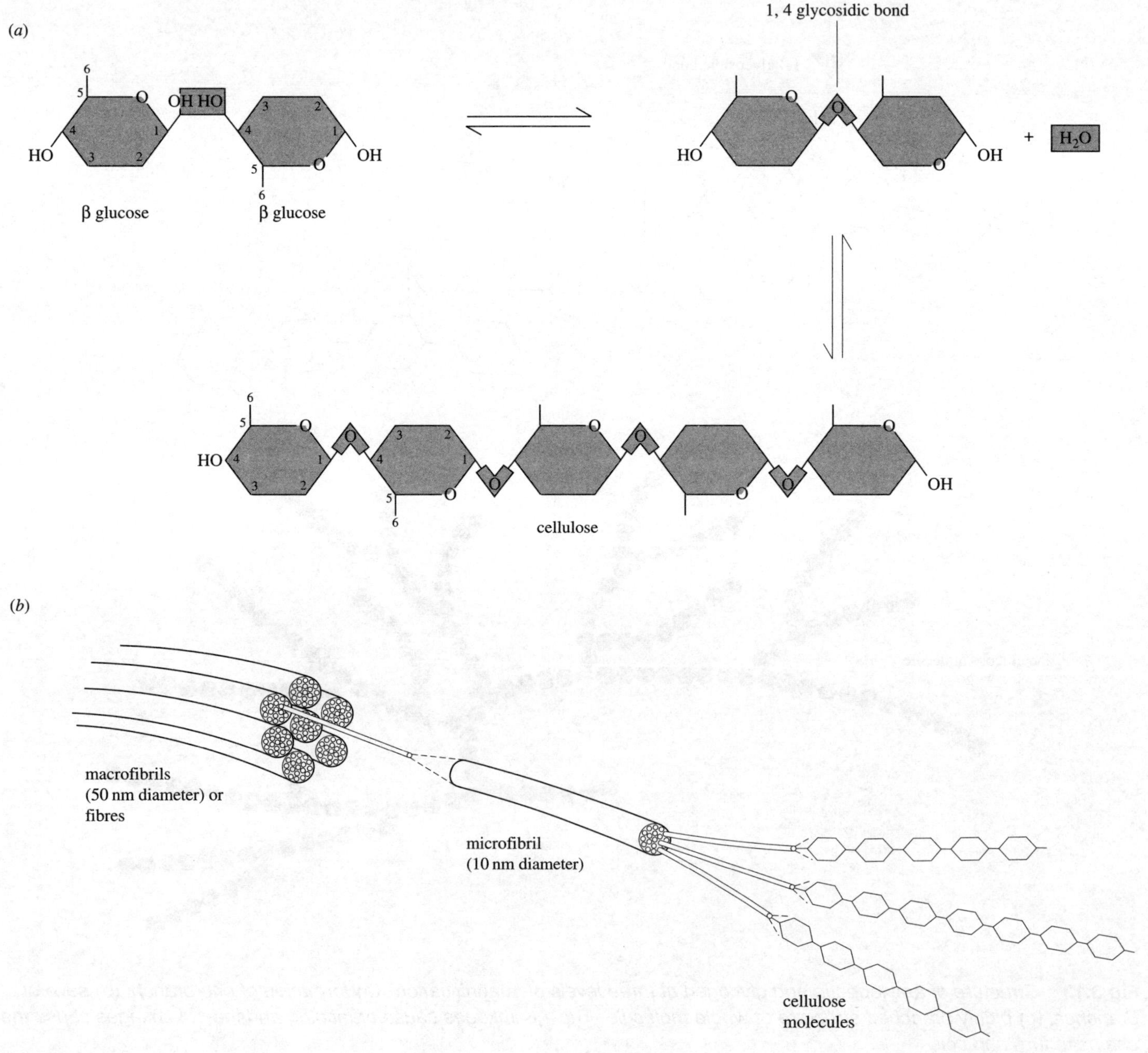

Fig 3.14 *Structure of cellulose.* (a) *Formation of cellulose from β glucose. Note that the β glucose molecules are rotated at 180° to each other so that OH groups on carbon atoms 1 and 4 can come alongside each other during condensation and formation of glycosidic bonds.* (b) *Packing of cellulose molecules into microfibrils and macrofibrils (fibres).*

the ring and the –OH group on carbon atom 4 projects above the ring. This rotation of successive residues is the underlying reason why cellulose has a different structure to starch.

About 50% of the carbon found in plants is in cellulose and it is the most abundant organic molecule on Earth. It is virtually confined to plants, although it is found in some nonvertebrate animals and ancestral fungi. Its abundance is due to its being a structural component of all plant cell walls, making up about 20–40% of the wall on average. The structure of the molecule reveals its suitability for this role. It consists of long chains of glucose residues with about 10 000 residues per chain (fig 3.14*a*). The β 1,4 linkages make the chains straight in contrast to starch where α 1,4 linkages cause the chains to be curved. Hydroxyl groups (–OH) project outwards from each chain in all directions and form hydrogen bonds with neighbouring chains. This cross-linking binds the chains rigidly together. The chains associate in groups of about 60 to 70 to form microfibrils, which are arranged in larger bundles to form macrofibrils (fig 3.14*b*). These have tremendous tensile strength (some idea of this strength can be obtained by trying to break a piece of cotton by pulling on both ends – cotton is almost pure cellulose). In cell walls the macrofibrils are arranged in several layers, in a glue-like matrix made of other polysaccharides as described in section 5.10.10 and shown in fig 5.35. This gives added strength.

Plant cells are therefore wrapped in several layers of cellulose. This prevents the cells from bursting when water enters by osmosis and also helps to determine the shapes of cells, since the direction in which they expand depends on the way the layers are arranged. As a cell inflates with water, pressure develops inside it and the cell becomes turgid. Turgid cells help support plants which lack wood. Despite their combined strength, the layers are fully permeable to water and solutes, an important property in the functioning of plant cells.

Apart from being a structural compound, cellulose is an important food source for some animals, bacteria and fungi. The enzyme cellulase, which catalyses the digestion of cellulose to glucose, is relatively rare in nature and most animals, including humans, cannot utilise cellulose despite its being an abundant and potentially valuable source of glucose. Ruminant mammals like the cow, however, have bacteria living symbiotically in their guts which digest cellulose. The abundance of cellulose and its relatively slow rate of breakdown in nature have ecological implications because it means that substantial quantities of carbon are 'locked up' in this substance, and carbon is one of the chief materials required by living organisms. Commercially, cellulose is extremely important. It is used, for example to make cotton goods and is a constituent of paper and Sellotape.

3.2.4 Compounds closely related to polysaccharides

Chitin

Chitin is closely related to cellulose in structure and function, being a structural polysaccharide (fig 3.15). It occurs in some fungi, where its fibrous nature contributes to cell wall structure, and in some animal groups, particularly the arthropods where it forms an essential part of the exoskeleton. Structurally it is identical to cellulose except that the hydroxyl (–OH) group at carbon atom 2 is replaced by –NH.CO.CH_3. It forms bundles of long parallel chains like cellulose.

Murein

Murein is a polysaccharide which acts as the strengthening material of bacterial cell walls (section 2.31). It is similar in structure to chitin, containing nitrogen like chitin.

3.6 What structural features of carbohydrates account for the fact that a wide variety of polysaccharides exists?

3.3 Lipids

Lipids are sometimes classified loosely as those water-insoluble organic substances which can be extracted from cells by organic solvents such as ether, chloroform and benzene. They cannot be defined precisely because their chemistry is so variable, but we can say that true lipids are formed by condensation reactions between fatty acids and an alcohol.

3.7 What is a condensation reaction?

3.3.1 Constituents of lipids

Fatty acids

Fatty acids contain the acidic group –COOH (the carboxyl group) and are so named because some of the larger molecules in the series occur in fats. They have the general formula R.COOH where R is hydrogen or a group such as

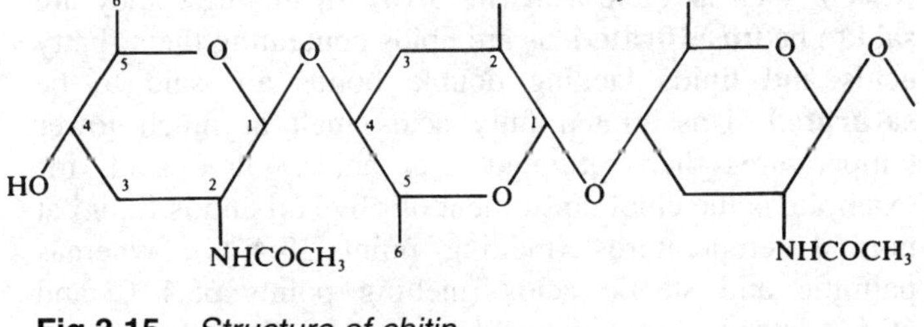

Fig 3.15 *Structure of chitin.*

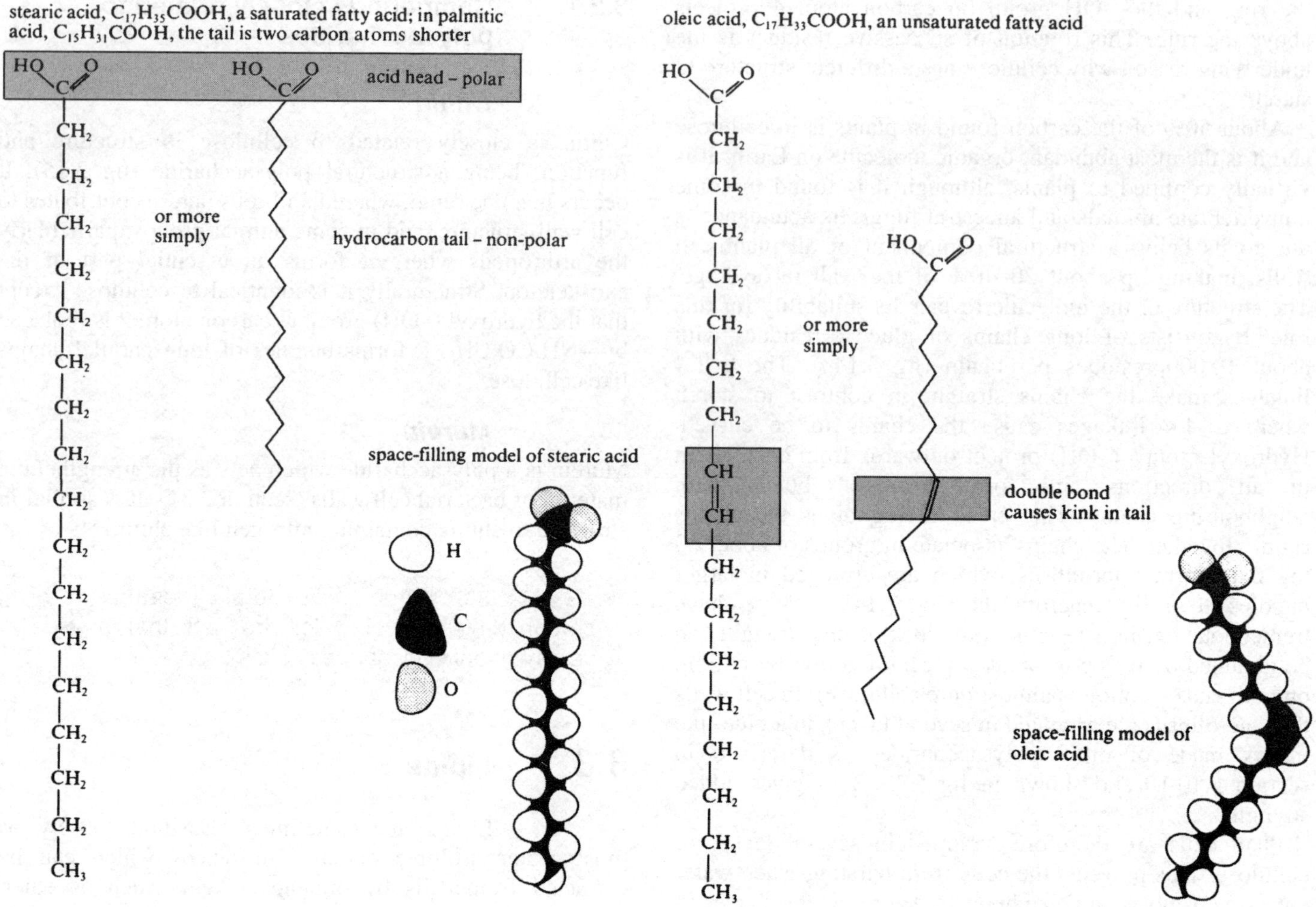

Fig 3.16 *Some examples of common fatty acids.*

$-CH_3$, $-C_2H_5$, and so on, increasing by $-CH_2$ for each subsequent member of the series. There are usually many carbon atoms in the fatty acids used to make lipids. Most naturally occurring fatty acids have an even number of carbon atoms between 14 and 22 (most commonly 16 or 18). The most common fatty acids are shown in fig 3.16. Note the characteristically long chain of carbon and hydrogen atoms forming a **hydrocarbon tail**. Many of the properties of lipids are determined by these tails, including their insolubility in water. The tails are said to be **hydrophobic**, meaning water-hating (*hydro*, water; *phobos*, fear).

Fatty acids sometimes contain one or more double bonds (C=C), such as oleic acid (fig 3.16). In this case they are said to be **unsaturated**, as are lipids containing them. Fatty acids and lipids lacking double bonds are said to be **saturated**. Unsaturated fatty acids melt at much lower temperatures than saturated fatty acids. Oleic acid, for example, is the chief constituent of olive oil and is liquid at normal temperatures (melting point 13.4 °C), whereas palmitic and stearic acids (melting points 63.1 °C and 69.6 °C respectively) are solid at normal body temperatures.

3.8 Cells of poikilothermic ('cold-blooded') animals usually have a higher proportion of unsaturated fatty acids than homoiothermic ('warm-blooded') animals. Can you account for this?

Alcohols

Most lipids are **triglycerides**. These are made from the alcohol glycerol (fig 3.17).

3.3.2 Formation of a lipid

Glycerol has three hydroxyl (–OH) groups, all of which can condense with a fatty acid. Usually all three undergo condensation reactions as shown in fig 3.17, and the lipid formed is therefore called a **triglyceride**.

3.3.3 Properties and functions of triglycerides

Triglycerides are the commonest lipids in nature and are further classified as fats or oils, according to whether they are solid (fats) or liquid (oils) at 20 °C. The higher the proportion of unsaturated fatty acids, the more likely they are to be liquid at a given temperature.

> **3.9** Tristearin and triolein are both lipids. Which is more likely to be an oil?

Triglycerides are non-polar. In other words, there is no uneven distribution of charge within the molecule. This means that they do not form hydrogen bonds with water molecules and therefore do not dissolve in water – they are hydrophobic. They are less dense than water and therefore float. Their tails vary in length according to the particular fatty acids used. Tristearin is an example of a lipid. It is made from three stearic acid molecules (fig 3.16), so its *tails* are 17 carbon atoms long. A space-filling model of tristearin is shown in fig 3.17.

A major function of lipids is to act as energy stores. They have a higher calorific value than carbohydrates, that is a given mass of lipid will yield more energy on oxidation than an equal mass of carbohydrate. This is because lipids have a higher proportion of hydrogen and an almost insignificant proportion of oxygen compared with carbohydrates.

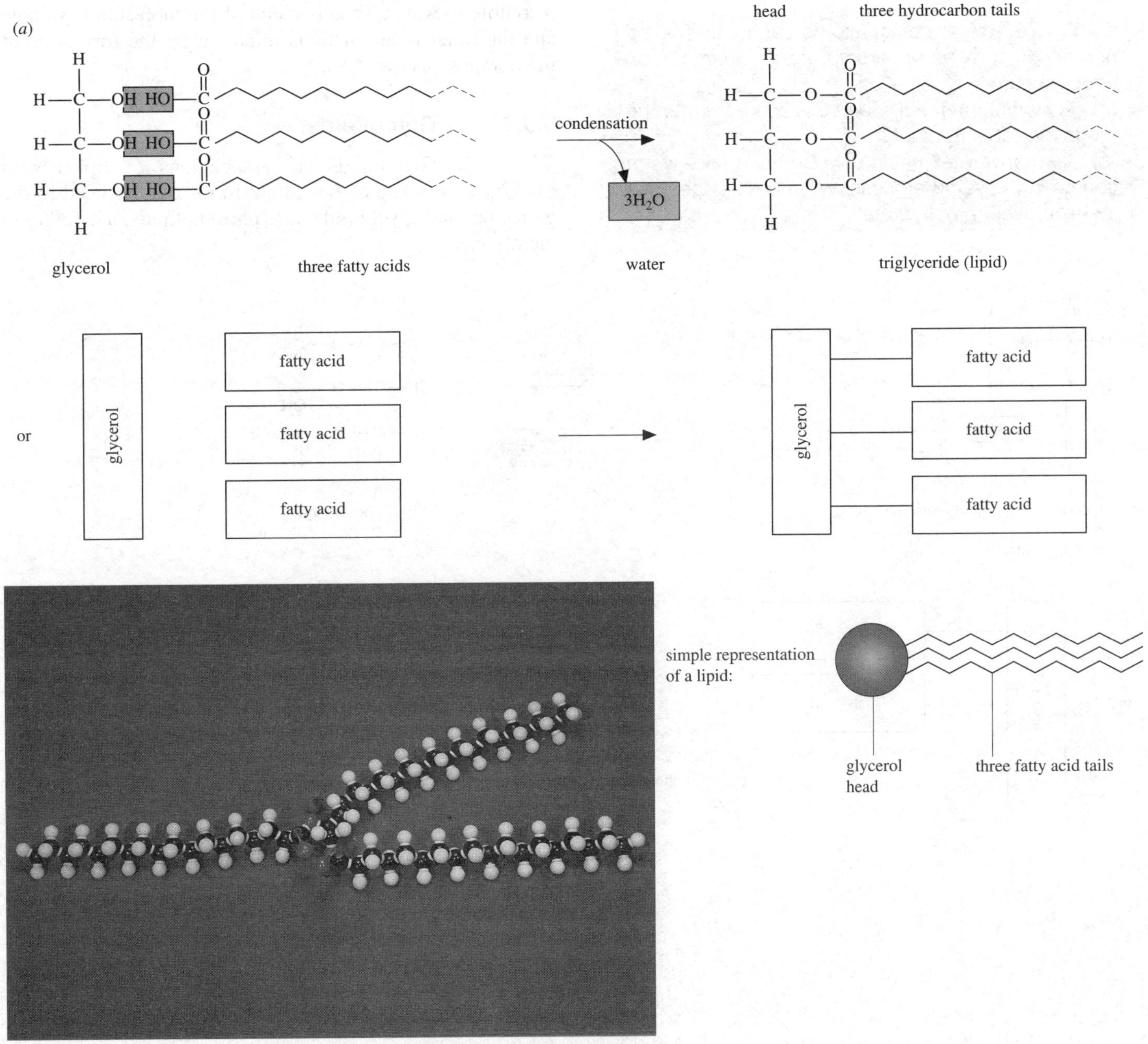

Fig 3.17 *(a) Formation of a triglyceride from glycerol and three fatty acids by three condensation reactions. (b) A space-filling model of tristearin, an example of a triglyceride.*

Animals store extra fat when hibernating, and fat is also found below the dermis of the skin of vertebrates where it serves as an insulator. Here it is extensive in mammals living in cold climates, particularly in the form of blubber in aquatic mammals such as whales, where it also contributes to buoyancy. Plants usually store oils rather than fats. Seeds, fruits and chloroplasts are often rich in oils and some seeds are commercial sources of oils, for example the coconut, castor bean, soyabean and sunflower seed. When fats are oxidised, water is a product. This metabolic water can be very useful to some desert animals, such as the kangaroo rat, which stores fat for this purpose (chapter 20).

3.10 A camel stores fat in the hump primarily as a water source rather than as an energy source.

(*a*) By what metabolic process would water be made available from fat?

(*b*) Carbohydrates could also be used as a water source in the same process. What advantage does fat have over carbohydrate?

3.3.4 Phospholipids

Phospholipids are lipids containing a **phosphate group**. The commonest type is formed when one of the three –OH groups of glycerol combines with phosphoric acid instead of a fatty acid (fig 3.18). The other two –OH groups combine with fatty acids as in the formation of a triglyceride.

The molecule consists of a phosphate head, which is shaded in fig 3.18, with two hydrocarbon tails from the two fatty acids. The phosphate head carries an electrical charge and is therefore soluble in water; in other words it is **hydrophilic**, or water-loving. The tails, however, are still insoluble in water. Thus one end of the molecule is soluble and the other is not. This is important in the formation of membranes (section 5.9).

3.3.5 Glycolipids

Glycolipids are associations of lipids with carbohydrates. The carbohydrate forms a polar head to the molecule, and glycolipids, like phospholipids, are found in membranes.

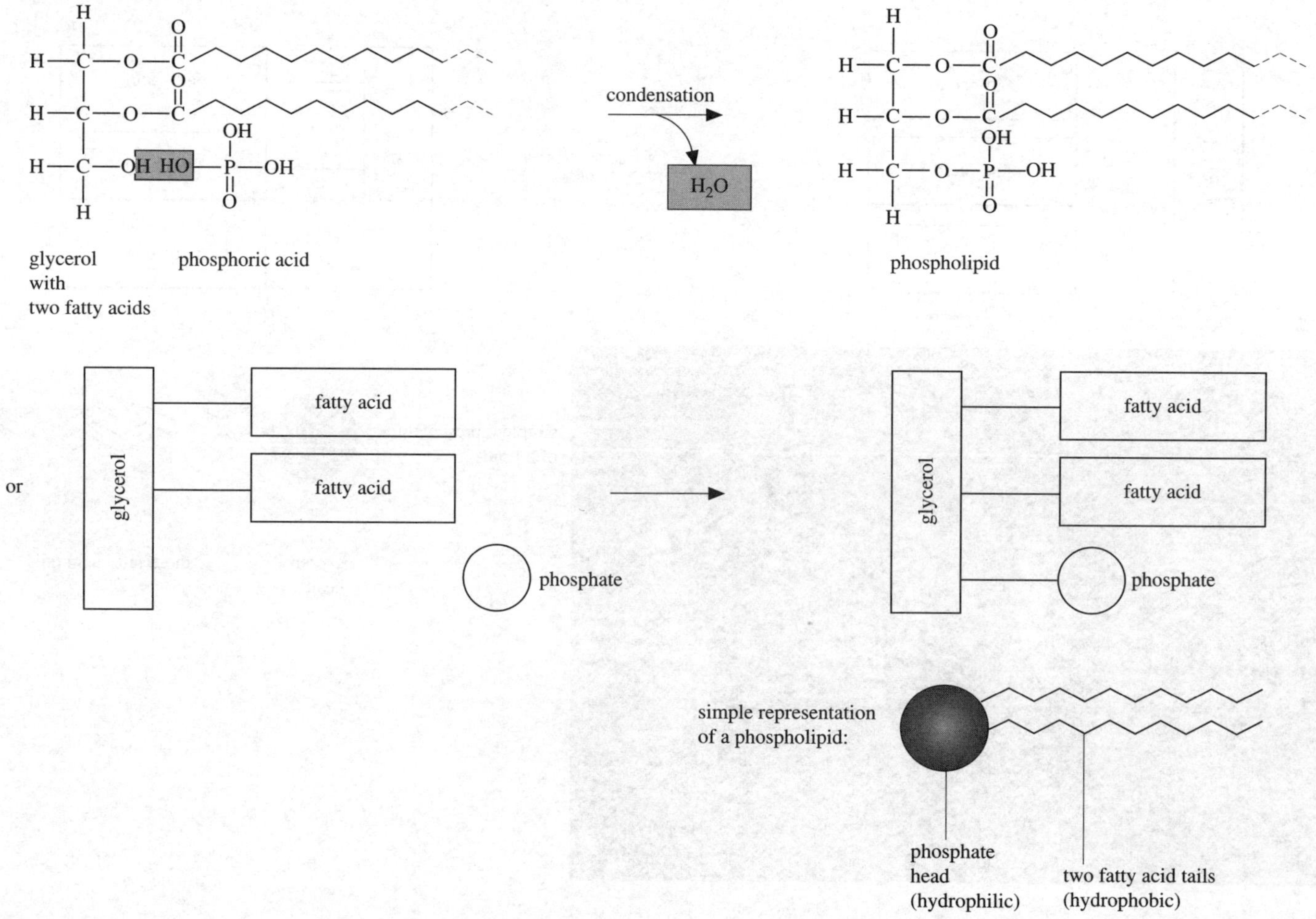

Fig 3.18 *Formation of a phospholipid. The structure can be shown simply as a head with two tails. The phosphate head is strongly polarised and is therefore water soluble (hydrophilic), unlike the non-polar tails (hydrophobic). This is a biologically important property in membranes.*

3.4 Amino acids

Amino acids are the basic units from which proteins are made. Over 170 amino acids are currently known to occur in cells and tissues. Of these, only 20 are commonly found in proteins. These are listed in table 3.5.

Plants are able to make all the amino acids they require from simpler substances. However, animals are unable to synthesise all that they need, and therefore must obtain some 'ready-made' amino acids directly from their diet. These are termed **essential** amino acids. Animals can make the other amino acids they require from these. Note that the essential amino acids are only described as 'essential' because they cannot be synthesised; the non-essential amino acids are just as necessary to make proteins.

3.4.1 Structure and range of amino acids

The general formula of an amino acid is shown in fig 3.19. There is a central carbon atom, known as the α carbon atom, to which is always attached an acidic carboxyl group, **–COOH**, a basic amino group, **$-NH_2$** and a **hydrogen** atom. The fourth position is the only variable part of the molecule. The group here is known as the **R group**. This group gives each amino acid its uniqueness. Table 3.5 shows the names, three-letter abbreviations and R groups of the 20 commonly occurring amino acids.

The simplest amino acid, and therefore the easiest one to learn as an example, is **glycine**, where R is simply hydrogen (fig 3.20*a*). When R is $-CH_3$, the amino acid **alanine** is formed (fig 3.20*b*).

Rare amino acids

A small number of rare amino acids occur in the proteins of organisms. They are made from some of the common amino acids. For example, hydroxyproline is made from proline, and is found in the protein collagen; hydroxylysine is made from lysine, and is also found in collagen.

There is no DNA code for the rare amino acids, and they are made from their parent amino acids *after* they have been incorporated into a protein.

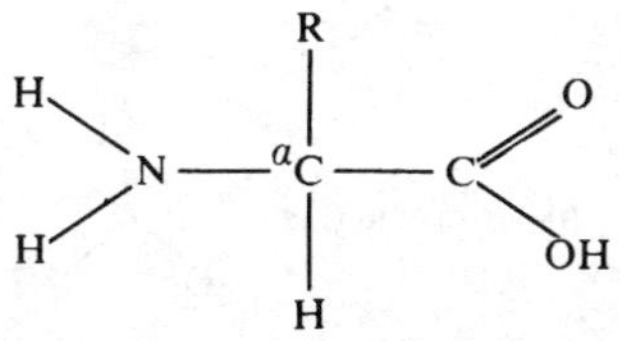

Fig 3.19 *General formula of an amino acid.*

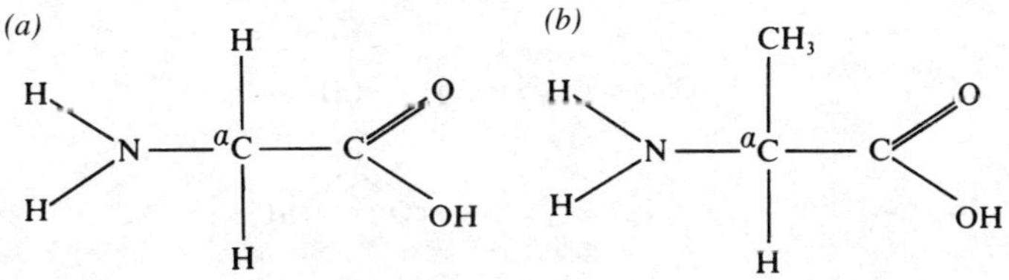

Fig 3.20 *(a) Glycine. (b) Alanine.*

Non-protein amino acids

Over 150 of these are known to occur, either free or in a combined form in cells, but never in proteins. For example, GABA (γ aminobutyric acid) is virtually unique to the nervous system. It is an inhibitory neurotransmitter, important in the brain.

3.4.2 Amino acids are amphoteric

Molecules like amino acids which contain both an acid and a basic part are described as **amphoteric**. They exist mainly as ions and can carry both a positive charge on the basic part and a negative charge on the acid part. Such ions are described as dipolar and are called zwitterions (fig 3.21). This explains the fact that amino acids and proteins can be made to move in an electrical field, as when they are separated by electrophoresis (appendix 1, Book 2). The charge on the amino acid can be affected by changes in its environment. For example, making a solution more acid would increase the concentration of hydrogen ions, and hence positive charges, and these would tend to cancel out the negative charges on the amino acid.

3.4.3 Bonds used in protein structure

Amino acids combine to form proteins. They are joined together by a type of bond known as a **peptide bond**. Once these bonds have been formed, however, the protein typically folds into a particular shape as a result of four other types of bond, namely ionic bonds, disulphide bonds, hydrogen bonds, and hydrophobic interactions. Studying these bonds helps us to understand the structure and behaviour of proteins.

Peptide bond

This is formed when a water molecule is eliminated during a reaction between the amino group of one amino acid and the carboxyl group of another. Elimination of water is known as **condensation** and the bond formed is a covalent bond called a **peptide bond** (fig 3.22). The compound formed is a **dipeptide**. It possesses a free amino group at one end, and a free carboxyl group at the other. This enables further combination between the dipeptide and other amino acids. If many amino acids are joined together in this way, a **polypeptide** is formed (fig 3.23).

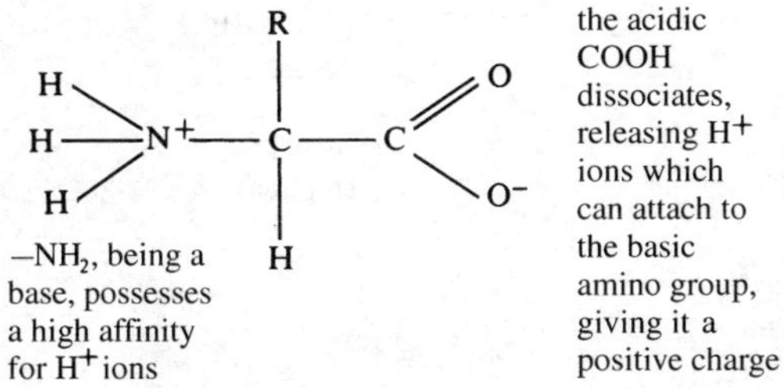

Fig 3.21 *Neutral zwitterion form of an amino acid.*

Table 3.5 The 20 common amino acids found in proteins. Only the R groups are shown. The box represents the constant part of the molecule, shown fully only in alanine. The amino acids are arranged into four categories. All are neutral except for the acidic and basic ones. The two acidic one have an extra COOH group. The three basic ones contain extra nitrogen. Two amino acids contain sulphur, an extra element. Those labelled essential are essential in the diet of humans. Three amino acids are described as aromatic, having rings of carbon atoms.

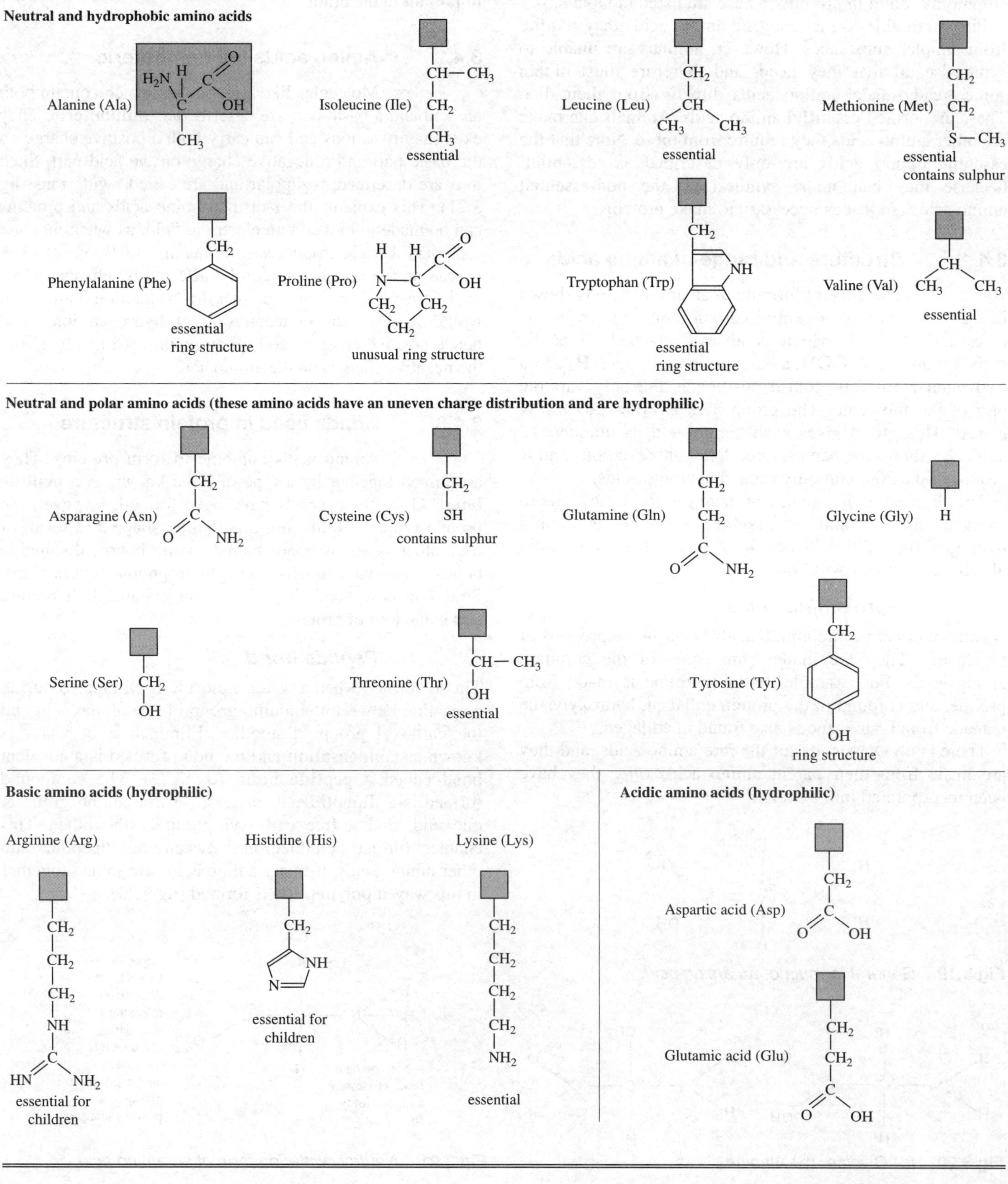

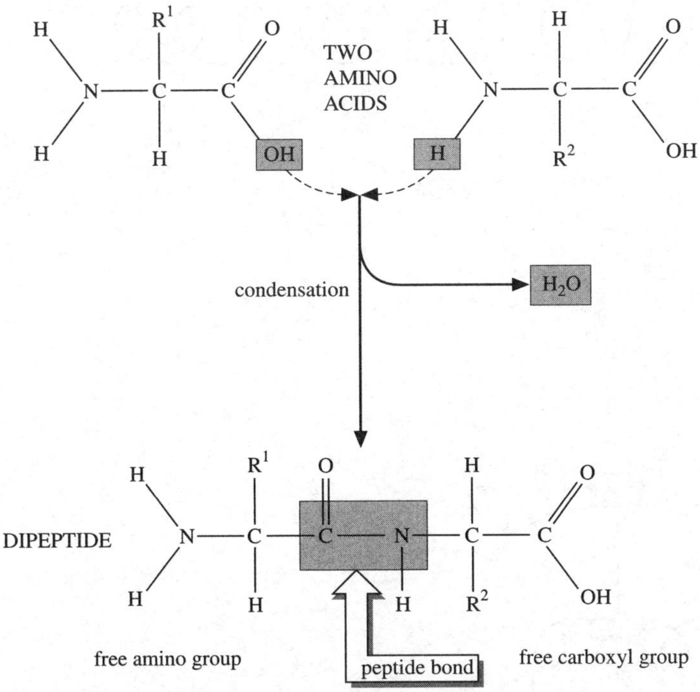

Fig 3.22 *Formation of a dipeptide by a condensation reaction between two amino acids. Condensation is removal of water.*

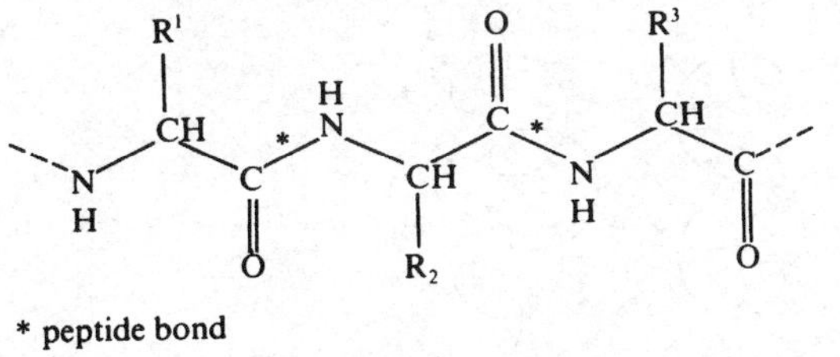

Fig 3.23 *Part of a polypeptide showing the joining of three amino acids.*

> **3.11** Write down the structural formula of the tripeptide formed by alanine, glycine and serine joined together in that order.

Ionic bond

Acidic and basic R groups exist in an ionised (charged) state at certain pHs. Acidic R groups are negatively charged and basic R groups are positively charged. They can therefore be attracted to each other, forming ionic bonds (fig 3.24). In an aqueous environment this bond is much weaker than a covalent bond and can be broken by changing the pH of the medium. This helps to explain the disruptive effect that changes in pH can have on protein structure (section 3.5.4). For example, adding acid to milk makes it curdle because the ionic bonds in casein (milk protein) are broken and the protein ceases to be soluble.

Disulphide bond

The amino acid cysteine contains a sulphydryl group, –SH, in its R group. If two molecules of cysteine line up

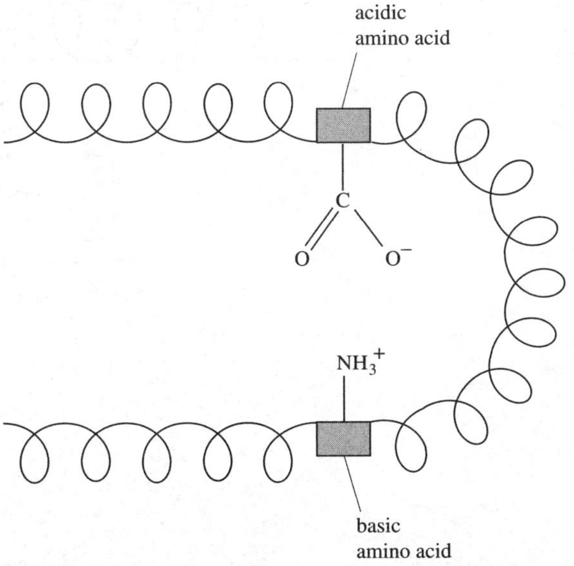

Fig 3.24 *Ionic bond formation. The polypeptide chain of the protein is represented by a coiled spring. The relevant amino acids are shown as boxes with side groups (R groups) attached.*

alongside each other, neighbouring sulphydryl groups can be oxidised and form a disulphide bond (fig 3.25). Disulphide bonds may be formed between different chains of amino acids (see insulin, fig 3.28) or between different parts of the same chain (fig 3.29). In the latter case the disulphide bonds make the molecule fold into a particular shape. They are strong and not easily broken.

Hydrogen bond

Hydrogen bonds have already been discussed in the section on water (section 3.1.2). When hydrogen is part of an OH or NH group it becomes slightly positively charged (electropositive). This is because the electrons that are shared, and which are negatively charged, are attracted more towards the O or N atoms. The hydrogen may then be attracted towards a neighbouring electronegative oxygen or nitrogen atom, such as the O of a C=O group or the N of an NH group (fig 3.26). C=O and NH groups occur along the length of polypeptide chains, as shown in fig 3.23, and they can interact to produce regular shapes such as the α helix discussed later. The hydrogen bond is weak, but as its occurrence is frequent, the total effect makes a considerable contribution towards molecular stability, as in the structure of the α helix (fig 3.30) and silk.

Hydrophobic interactions

Some R groups are non-polar and therefore hydrophobic, such as those on the amino acids tyrosine and valine (table 3.5). If a polypeptide chain contains a number of these groups and is in an aqueous environment, the chain will tend to fold so that the maximum number of hydrophobic groups come into close contact and exclude water (fig 3.27). This is how many globular proteins fold up. The hydrophobic groups tend to point inwards towards

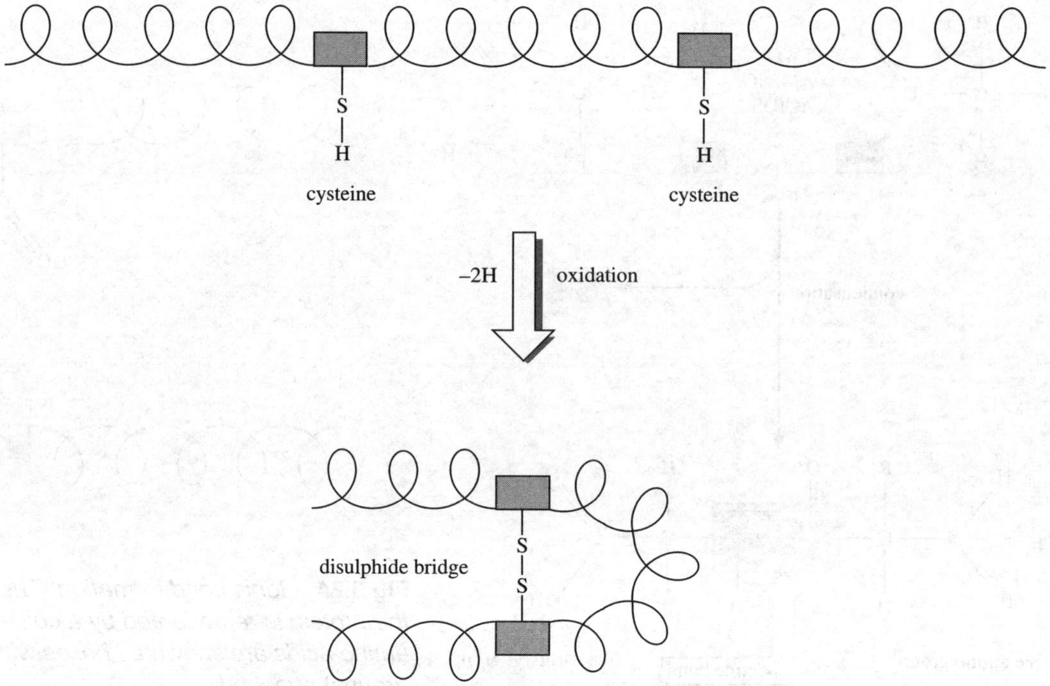

Fig 3.25 *Formation of a disulphide bond between the sulphydryl groups of two cysteines.*

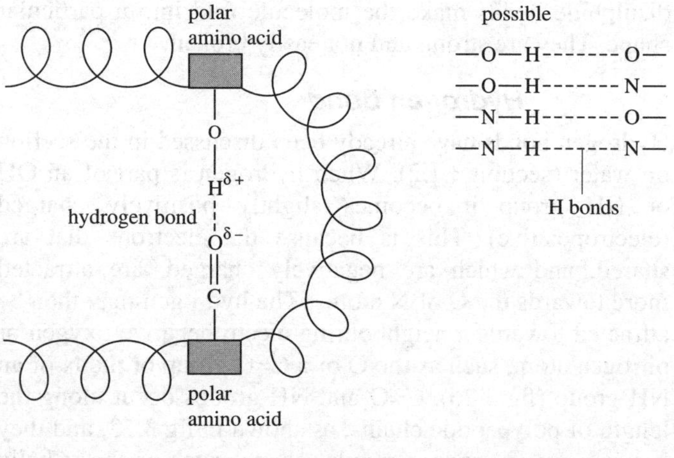

Fig 3.26 *Formation of a hydrogen bond. Hydrogen bonds may form between R groups of polar amino acids or between the C=O and N–H groups either side of peptide bonds (for example, see α helix and β sheet in next section).*

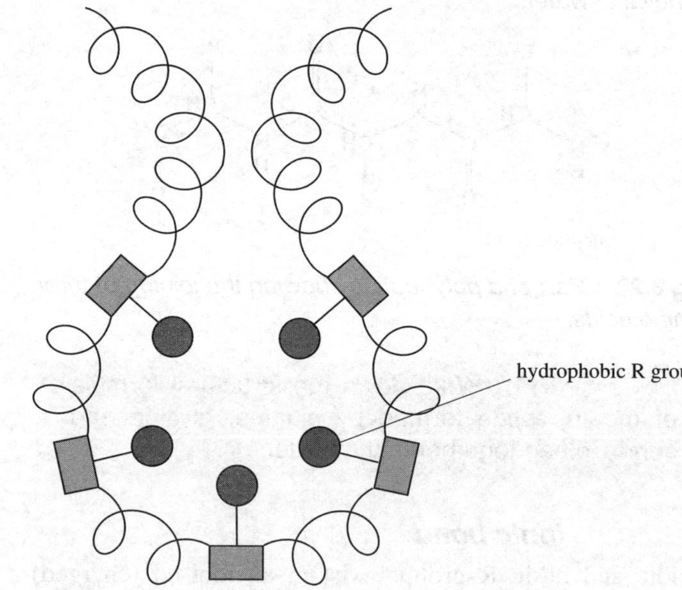

Fig 3.27 *Hydrophobic interactions between hydrophobic R groups. These create a hydrophobic region within the molecule which excludes water.*

the centre of the roughly spherical molecule while the hydrophilic groups face outwards into the aqueous environment, making the protein soluble. In the same way membrane proteins can have hydrophobic regions inside the membrane alongside the hydrophobic tails of the phospholipids, while the hydrophilic regions face outwards alongside the hydrophilic phosphate heads of the phospholipids (fig 5.15).

3.5 Proteins

Proteins are made from amino acids and therefore always contain the elements carbon, hydrogen, oxygen and nitrogen, and in some cases sulphur. Some proteins form complexes with other molecules containing phosphorus, iron, zinc and copper. Proteins are macromolecules of high M_r (relative formula mass or molecular mass), typically between several thousands and several millions, consisting of chains of amino acids. They are polymers and amino acids are the monomers. There are 20 different amino acids which are commonly found in naturally occurring proteins. The potential variety of proteins is unlimited because the sequence of amino acids in each protein is specific for that protein (chapter 23) and is genetically controlled by the DNA of the cell in which it is made. Proteins are the most abundant organic molecules to be found in cells and form over 50% of their total dry mass. They are an essential component of the diet of animals and may be converted to both fat and carbohydrate by the cells. Their diversity enables them to display a great range of structural and metabolic activities within the organism.

3.5.1 Size of protein molecules

Simple peptides containing two, three or four amino acid residues are called di-, tri- and tetrapeptides respectively. Polypeptides are chains of many amino acid residues (up to several thousand – table 3.6). A protein may possess one or more polypeptide chains.

3.5.2 Classification of proteins

Because of the complexity of protein molecules and their diversity of function, it is very difficult to classify them in a single, well-defined fashion. Three alternative methods are given in tables 3.7, 3.8 and 3.9.

3.5.3 Structure of proteins

Each protein possesses a characteristic three-dimensional shape, its **conformation**. There are four separate levels of structure and organisation as follows.

Primary structure

The **primary structure** is the sequence of amino acids in a polypeptide chain. The first person to work out the complete amino acid sequence of a protein was Fred Sanger, working at the Cavendish laboratory in Cambridge, where Watson and Crick also determined the structure of DNA. He worked with the hormone insulin, the smallest protein he could find. It took ten years and the results were published in 1953 (fig 3.28). Max Perutz, another great molecular biologist of the Cavendish, recalls 'it caused a sensation, because it proved

Table 3.6 Sizes of some proteins.

Protein	M_r *(molecular mass)*	*Number of amino acids*	*Number of polypeptide chains*
Ribonuclease	12 640	124	1
Lysozyme	13 930	129	1
Myoglobin	16 890	153	1
Haemoglobin	64 500	574	4
TMV (tobacco mosaic virus)	about 400 000 000	about 336 500	2130

The largest protein complexes are found in viruses where M_rs of over 400 000 000 are commonly found.

Table 3.7 Classification of proteins according to structure.

Type	*Nature*	*Function*
Fibrous 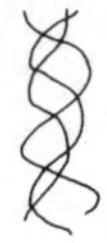	Secondary structure most important (little or no tertiary structure) Insoluble in water Physically tough Long parallel polypeptide chains cross-linked at intervals forming long fibres or sheets	Perform structural functions in cells and organisms e.g. **collagen** (tendons, bone, connective tissue), **myosin** (in muscle), **silk** (spiders' webs), **keratin** (hair, horn, nails, feathers)
Globular 	Tertiary structure most important Polypeptide chains tightly folded to form spherical shape Easily soluble	Form enzymes, antibodies and some hormones, e.g. **insulin** Other important roles
Intermediate	Fibrous but soluble	e.g. **fibrinogen** – forms insoluble fibrin when blood clots

Table 3.8 Classification of proteins according to composition.

Proteins	
Simple	*Conjugated*
Only amino acids form their structure (see tables 3.7 and 3.9)	Complex compounds consisting of globular proteins and tightly-bound non-protein material; the non-protein material is called a **prosthetic** group

Conjugated proteins

Name	*Prosthetic group*	*Location*
Phosphoprotein	Phosphoric acid	Casein of milk Vitellin of egg yolk
Glycoprotein	Carbohydrate	Membrane structure Mucin (component of saliva)
Nucleoprotein	Nucleic acid	Component of viruses Chromosomes Ribosome structure
Chromoprotein	Pigment	Haemoglobin – haem (iron-containing pigment) Phytochrome (plant pigment) Cytochrome (respiratory pigment)
Lipoprotein	Lipid	Membrane structure Lipid transported in blood as lipoprotein
Flavoprotein	FAD (flavine adenine dinucleotide, see section 9.3.5)	Important in electron transport chain in respiration
Metal proteins	Metal	E.g. nitrate reductase, the enzyme in plants which converts nitrate to nitrite

Table 3.9 Protein classification according to function. Proteins are also important in membranes where they function as enzymes, receptor sites and transport sites.

Type	*Examples*	*Occurrence/function*
Structural	Collagen	Component of connective tissue, bone, tendons, cartilage
	Keratin	Skin, feathers, nails, hair, horn
	Elastin	Elastic connective tissue (ligaments)
	Viral coat proteins	'Wraps up' nucleic acid of virus
Enzymes	Trypsin	Catalyses hydrolysis of protein
	Ribulose bisphosphate carboxylase	Catalyses carboxylation (addition of CO_2) of ribulose bisphosphate in photosynthesis
	Glutamine synthetase	Catalyses synthesis of the amino acid glutamine from glutamic acid + ammonia
Hormones	Insulin Gucagon	Help to regulate glucose metabolism
	ACTH	Stimulates growth and activity of the adrenal cortex
Respiratory pigment	Haemoglobin	Transports O_2 in vertebrate blood
	Myoglobin	Stores O_2 in muscles
Transport	Serum albumin	Transport of fatty acids and lipids in blood
Protective	Antibodies	Form complexes with foreign proteins
	Fibrinogen	Forms fibrin in blood clotting
	Thrombin	Involved in blood clotting mechanism
Contractile	Myosin	Moving filaments in myofibrils of muscle
	Actin	Stationary filaments in myofibrils of muscle
Storage	Ovalbumin	Egg white protein
	Casein	Milk protein
Toxins	Snake venom	Enzymes
	Diphtheria toxin	Toxin made by diphtheria bacteria

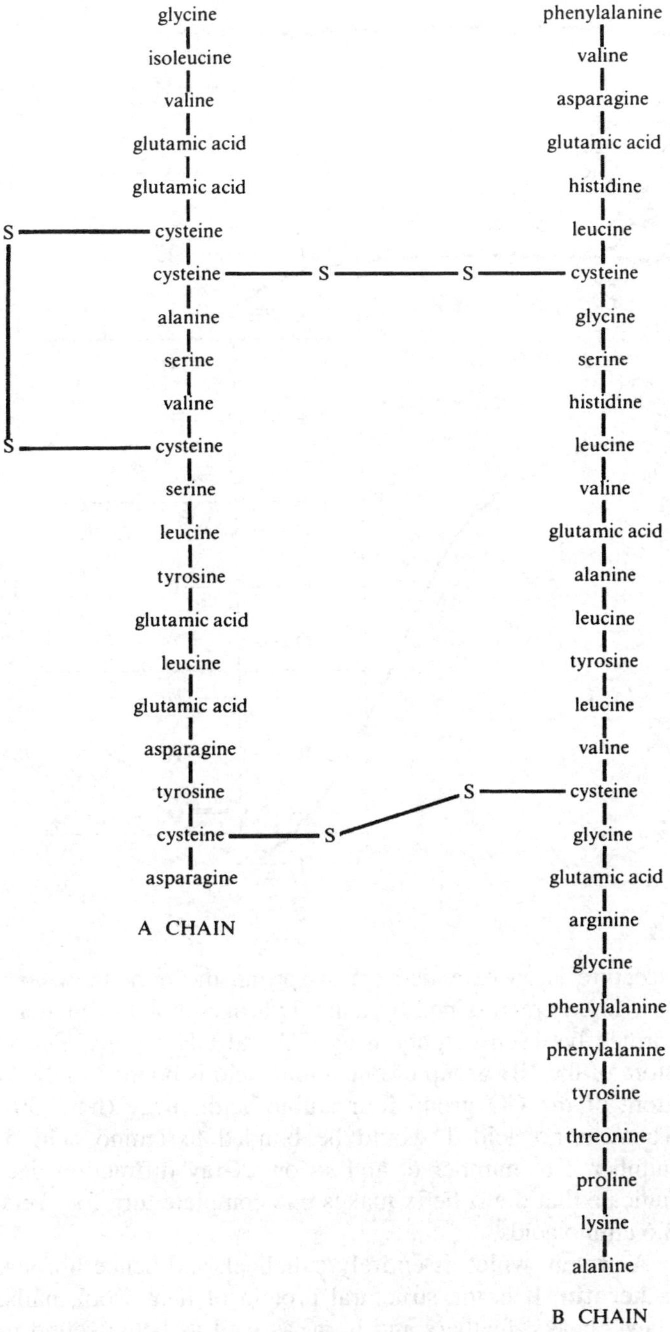

Fig 3.28 *Primary structure (sequence of amino acids) of insulin. The molecule consists of two polypeptide chains held together by two disulphide bridges.*

for the first time that protein has a specific arrangement of amino acids along its chain.' Sanger was awarded the Nobel prize for his work in 1958 (and has since won a second for work on nucleic acids). Insulin is a protein of 51 amino acids. It is made of two polypeptide chains held together by disulphide bridges.

Today, much of the amino acid sequencing is accomplished by machine and well over one hundred thousand protein primary structures are known. Another example, lysozyme is shown in fig 3.29. Table 3.6 shows the number of amino acids in some proteins.

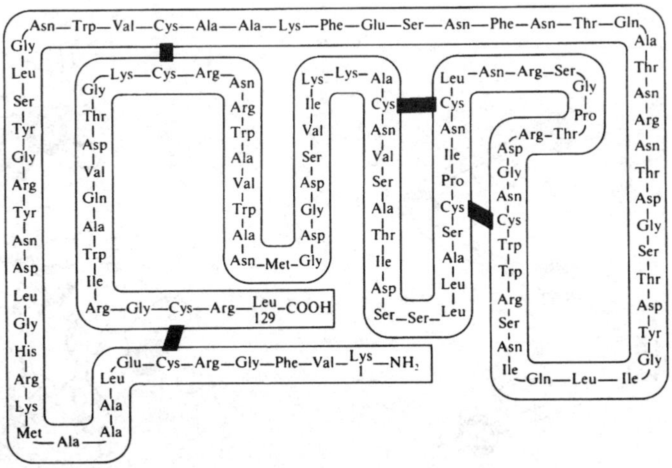

Fig 3.29 *The primary structure of lysozyme. Lysozyme is an enzyme that is found in many tissues and secretions of the human body, in plants, and in the whites of eggs. Its function is to catalyse the breakdown of the cell walls of bacteria. The molecule consists of a single polypeptide chain of 129 amino acid residues. There are four intrachain disulphide bridges.*

There are thousands of different proteins in the human body, all composed of different arrangements of the 20 fundamental amino acids. The sequence of amino acids of a protein dictates its biological function. In turn, this sequence is strictly controlled by the sequence of bases in DNA (chapter 23). Substitution of just a single amino acid can cause a major alteration in a protein's function, as in the condition of sickle cell anaemia (chapter 24). Analysis of amino acid sequences of similar proteins from different species is of interest, because it offers evidence about the possible evolutionary relationships between different species. This is dealt with in chapter 26.

3.12 (*a*) Let the letters A and B represent two different amino acids. Write down the sequences of all the possible tripeptides that could be made with just these two amino acids.
(*b*) From your answer to (*a*), what is the formula for calculating the number of different tripeptides that can be formed from two different amino acids?
(*c*) How many polypeptides, 100 amino acids in length, could be formed from two different amino acids?
(*d*) How many polypeptides, 100 amino acids in length (a modest length for a protein), could be made using all 20 common amino acids?
(*e*) How many peptides/polypeptides could be made (any length) from all 20 common amino acids?

Secondary structure

In addition to the primary structure there is a specific **secondary structure**. The most common secondary

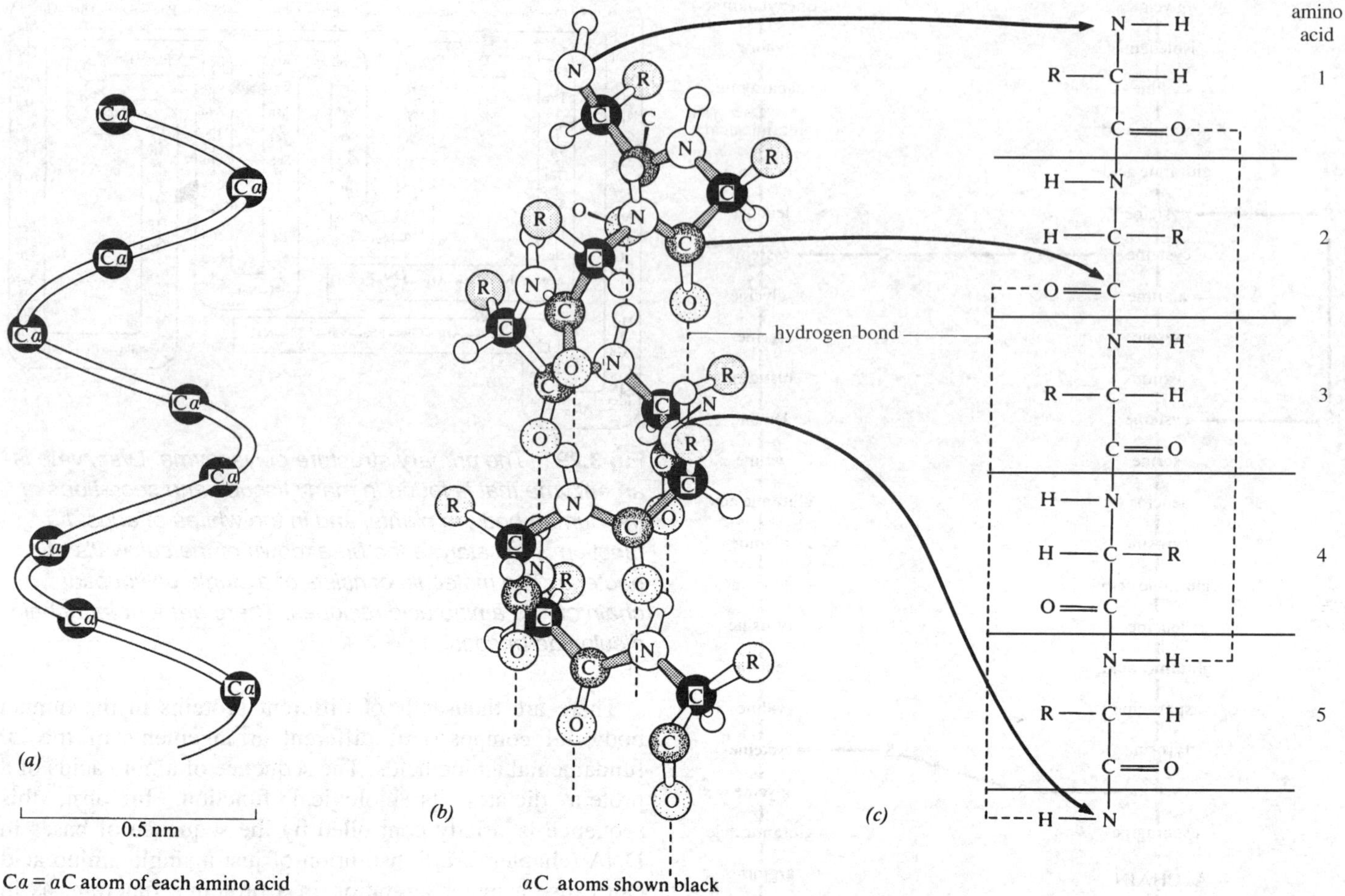

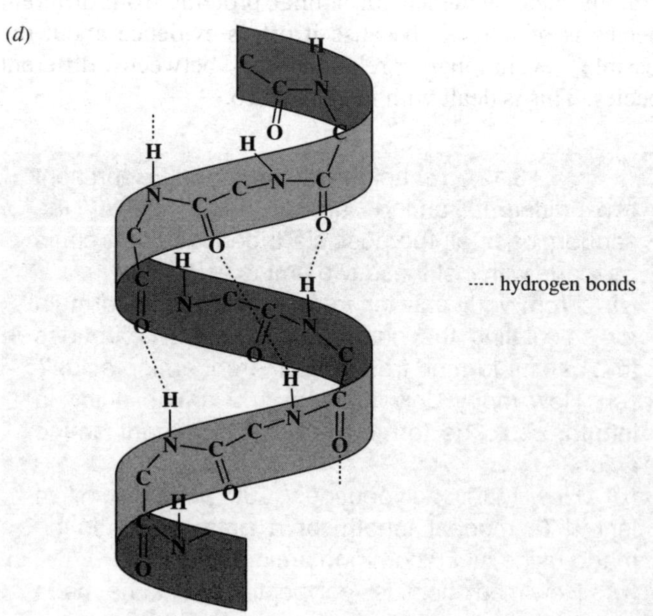

Fig 3.30 *Structure of the α helix.* (a) *Only the α C atoms are shown. A line joining them describes an α helix.* (b) *The entire α helix.* (c) *Part of the α helix straightened out. Hydrogen bonds hold the helix in place.* (d) *α helix shown as a ribbon.*

structure is an extended spiral spring, the α helix, whose structure is maintained by many hydrogen bonds which are formed between neighbouring CO and NH groups. The H atom of the NH group of one amino acid is bonded to the O atom of the CO group four amino acids away (fig 3.30). Thus amino acid 1 would be bonded to amino acid 5, number 2 to number 6, and so on. X-ray diffraction data indicate that the α helix makes one complete turn for every 3.6 amino acids.

A protein which is entirely α-helical, and hence fibrous, is **keratin**. It is the structural protein of hair, wool, nails, claws, beaks, feathers and horn, as well as being found in vertebrate skin. Its hardness and stretchability vary with the degree of cross-linking by disulphide bridges between neighbouring chains.

Theoretically, all CO and NH groups can participate in hydrogen bonding as described, so the α helix is a very stable, and hence a common, structure. α-helical regions are rigid and rod-like. Most proteins are globular molecules in which there are also regions of β sheet (see below) and irregular structure. The fact that they are not entirely α-helical is due mainly to interference with hydrogen bonding by certain R groups, the occurrence of disulphide bridges between different parts of the same chain and the inability of the amino acid proline to make hydrogen bonds.

Another type of secondary structure is the **β-pleated sheet**. The protein that makes silk, namely **fibroin**, is entirely in this form. It is the protein used by silkworms when spinning their cocoon threads. It is made up of a number of adjacent chains which are more extended than the α helices. They are arranged in a parallel fashion, either running in the same direction or in opposite directions as in fig 3.31. They are joined together by hydrogen bonds formed between the C=O and NH groups of one chain and the NH and C=O groups of adjacent chains. Again, all NH and C=O groups are involved in hydrogen bonding, so the structure is very stable and rigid. The whole structure is known as a **β-pleated sheet**. The β sheet of silk has a high tensile strength (cannot be stretched), but the arrangement of the polypeptides makes the silk very supple. In globular proteins a single polypeptide chain commonly folds back on itself several times to form regions of β-pleated sheet.

Yet another arrangement is seen in the fibrous protein collagen, another structural protein like keratin and silk, and one that possesses great tensile strength. Here three polypeptide chains are wound around each other like the strands of a rope to form a **triple helix**. There are about 1000 amino acid residues in each chain, and the complete triple-helix compound is called **tropocollagen**. Each chain is itself in the form of a loose helix (not an α helix) (fig 3.32). The three strands or chains are held together by hydrogen bonds. Many triple helices can lie parallel to form fibrils. They are joined by covalent bonds between neighbouring chains. Fibrils in turn unite to form fibres. The large-scale structure of collagen is therefore built up in stages, as with cellulose. The fact that the protein is extremely resistant to stretching is an essential part of its functioning, for example in tendons, bone, skin, teeth and connective tissue. Proteins which exist entirely in the form of helical coils, such as keratin and collagen, are exceptional.

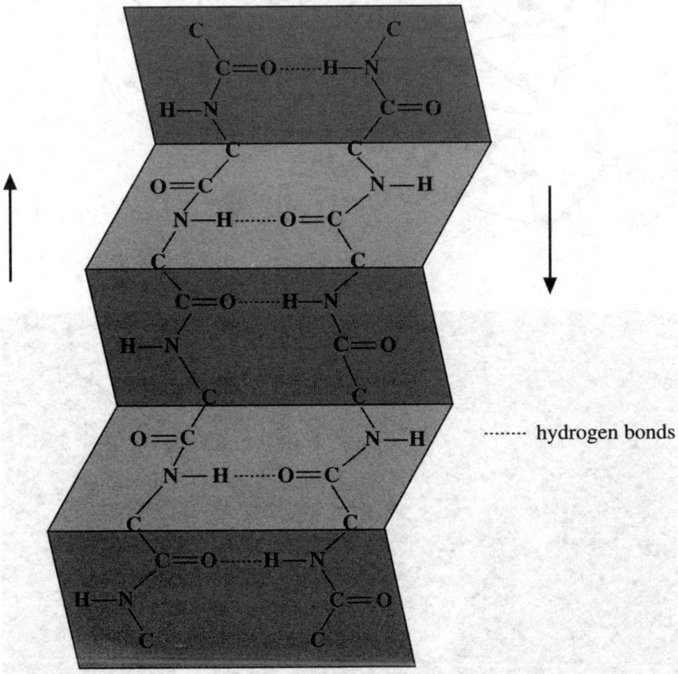

Fig 3.31 *β-pleated sheet. The chains are held parallel to each other by the hydrogen bonds that form between the NH and CO groups. The side groups (R) are not shown but would project above and below the plane of the sheet.*

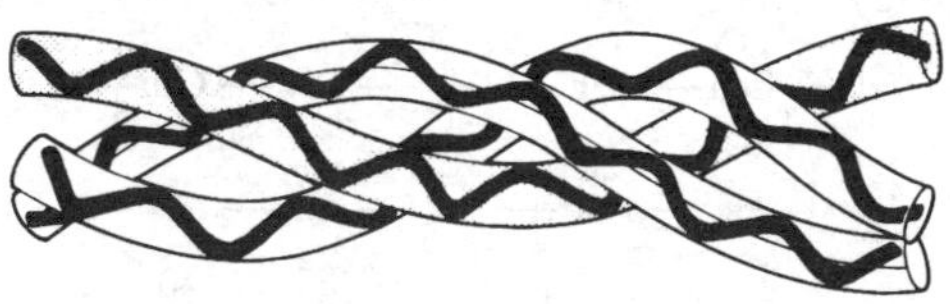

Fig 3.32 *Collagen triple-helix structure.*

Tertiary structure

Usually the polypeptide chain bends and folds extensively, forming a precise, compact 'globular' shape. This is the protein's tertiary structure and it is maintained by the interaction of the four types of bond already discussed, namely ionic, hydrogen and disulphide bonds as well as hydrophobic interactions (fig 3.33). The latter are quantitatively the most important and occur when the protein folds so as to shield hydrophobic side groups from the aqueous surroundings, at the same time exposing hydrophilic side chains, as described above.

The tertiary structure of a protein can be determined by X-ray crystallography. By early 1959, and after many years' work, John Kendrew and Max Perutz had built the first atomic model of myoglobin showing secondary and tertiary structures using this technique (fig 3.34). They received the Nobel prize for their work in 1962:

primary structure – single polypeptide chain of 153 amino acids, the sequence was elucidated in the early 1960s;

secondary structure – about 75% of the chain is α-helical (8 helical sections);

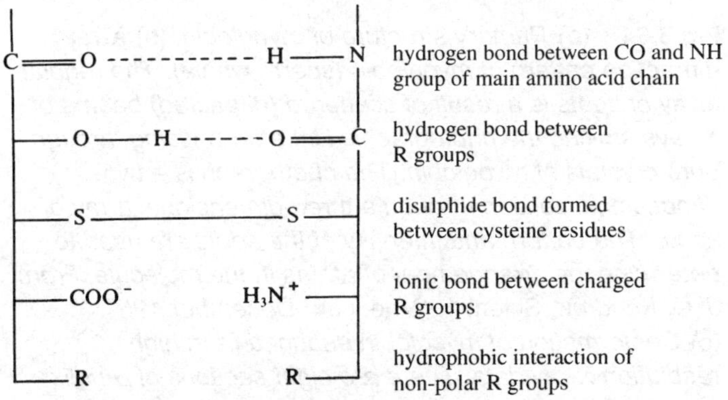

Fig 3.33 *Summary of types of bond stabilising secondary and tertiary structures of proteins. Hydrophobic interactions (associations of non-polar molecules or parts of molecules) to exclude water molecules in the aqueous environment of the cell are particularly important in maintaining structure, as in membranes.*

(a)

10
H_2N— Val — Leu — Ser — Glu — Gly — Glu — Trp — Gln — Leu — Val — Leu — His — Val — Tyr — Ala — Lys — Val —

20, 30
Glu — Ala — Asp — Val — Ala — Gly — His — Gly — Gln — Asp — Ile — Leu — Ile — Arg — Leu — Phe — Lys —

40, 50
Ser — His — Pro — Glu — Thr — Leu — Glu — Lys — Phe — Asp — Arg — Phe — Lys — His — Leu — Lys — Thr —

60
Glu — Ala — Glu — Met — Lys — Ala — Ser — Glu — Asp — Leu — Lys — Gly — His — His — Glu — Ala — Glu —

70, 80
Leu — Thr — Ala — Leu — Gly — Ala — Ile — Leu — Lys — Lys — Gly — His — His — Glu — Ala — Glu —

90, 100
Leu — Lys — Pro — Leu — Ala — Gln — Ser — His — Ala — Thr — Lys — His — Lys — Ile — Pro — Ile — Lys —

110
Tyr — Leu — Glu — Phe — Ile — Ser — Glu — Ala — Ile — Ile — His — Val — Leu — His — Ser — Arg — His —

120, 130
Pro — Gly — Asn — Phe — Gly — Ala — Asp — Ala — Gln — Gly — Ala — Met — Asn — Lys — Ala — Leu — Glu —

140, 150
Leu — Phe — Arg — Lys — Asp — Ile — Ala — Ala — Lys — Tyr — Lys — Glu — Leu — Gly — Tyr — Gln — Gly — COOH

(b)

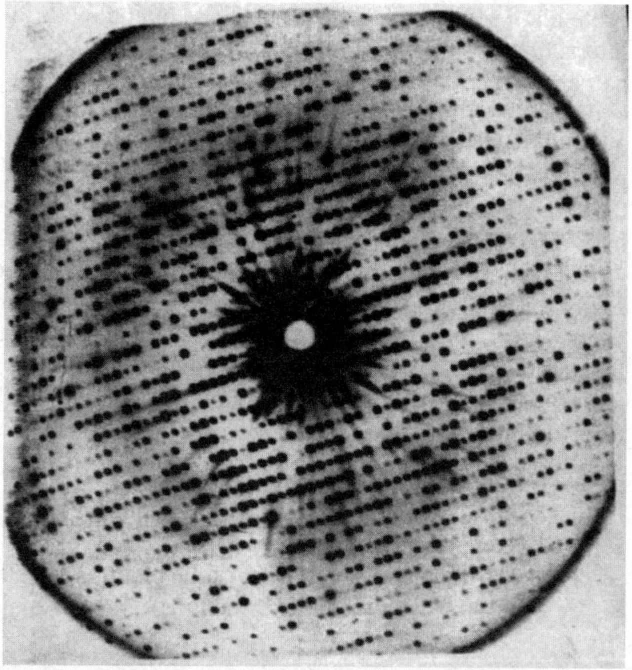

(c)

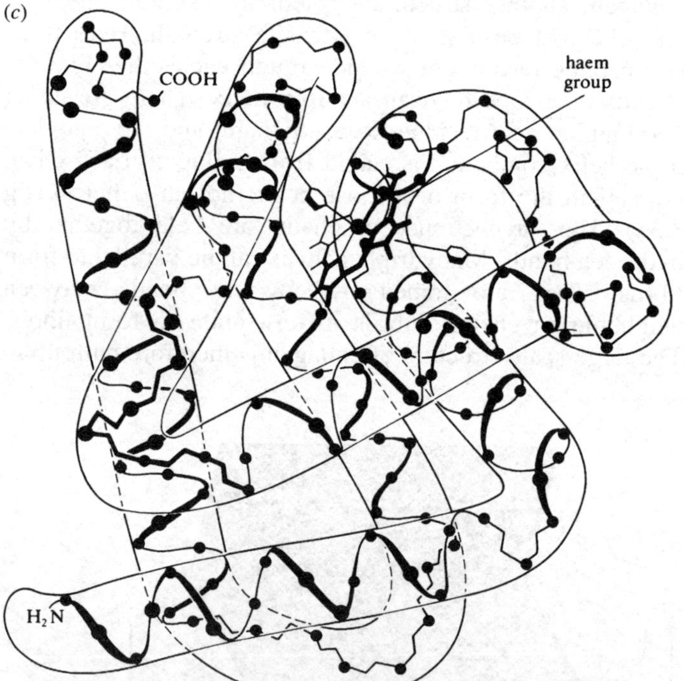

(d)

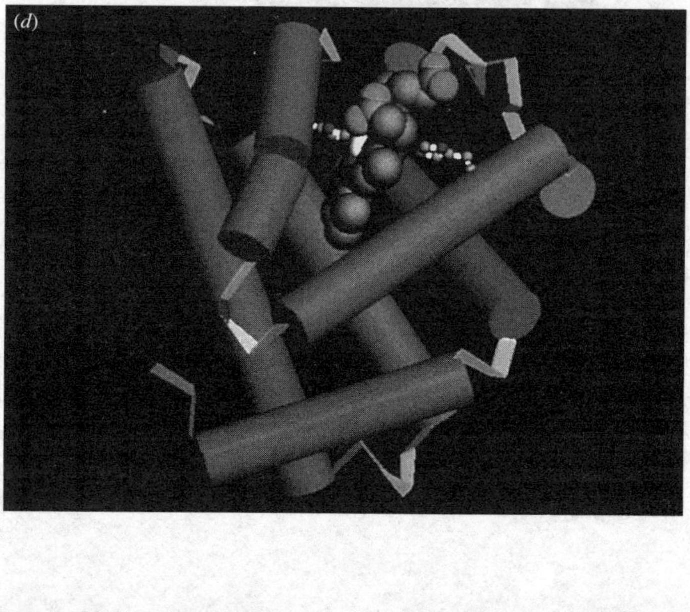

Fig 3.34 (a) *Primary structure of myoglobin.* (b) *X-ray diffraction pattern of myoglobin (sperm whale). The regular array of spots is a result of scattered (diffracted) beams of X-rays striking the photographic film after passing through pure crystals of myoglobin. The photograph is a two-dimensional section through a three-dimensional array of spots. The pattern and intensity of the spots are used to determine the arrangement of atoms in the molecule. From J. C. Kendrew,* Scientific American, *December 1961.*
(c) *Conformation of myoglobin deduced from high resolution X-ray data. There are eight sections of α helix surrounding a haem group which forms a flat disc.*
(d) *Another way of representing the three-dimensional structure of proteins is to show α helices as cylinders, as in this computer-generated diagram of myoglobin. Note the eight α-helical regions. The haem group is shown in the form of a space-filling model held in place by two amino acids (small white balls).*

(e)

Fig 3.34 (cont.) (e) *Ball-and-stick model of myoglobin.*

tertiary structure	–	non-uniform folding of the α-helical chain into a compact shape;
prosthetic group	–	haem group (contains iron).

Myoglobin is formed in muscle where its function is to store oxygen. Oxygen combines with the haem group as in haemoglobin. The haem group gives muscle its red appearance. Further information about the functions of myoglobin can be found in chapter 14. The elucidation of tertiary structure is still very time-consuming. Use of computers and other techniques to predict tertiary structure, based on knowledge of primary and secondary structures, is a fast-growing area of molecular biology. From this follows the possibility of designing proteins with particular shapes for particular functions, with important applications in industry and medicine.

Fig 3.34 shows several ways of representing the three-dimensional structure of a protein. Figure 3.35 shows a further method. The precise shapes formed are essential for the functioning of those proteins with tertiary structure. This is well illustrated by enzymes, as shown in section 4.1.2.

Quaternary structure

Many highly complex proteins consist of more than one polypeptide chain. The separate chains are held together by hydrophobic interactions and hydrogen and ionic bonds. Their precise arrangement is known as the **quaternary structure**. Haemoglobin shows such a structure. It is the red oxygen-carrying pigment found in the red blood cells of vertebrates. It consists of four separate polypeptide chains of two types, namely two α chains and two β chains. These resemble myoglobin in structure. The two α chains each contain 141 amino acids, while the two β chains each contain 146 amino acids. The complete structure of haemoglobin was worked out by Kendrew and Perutz and is illustrated in fig 3.36.

As is typical of globular proteins, its hydrophobic side chains point inwards to the centre of the molecule, and its hydrophilic side chains face outwards, making it soluble in water. A mutation which causes one of the hydrophilic amino acids to be replaced by a hydrophobic amino acid, thereby reducing its solubility, is responsible for the disease sickle cell anaemia (chapter 24).

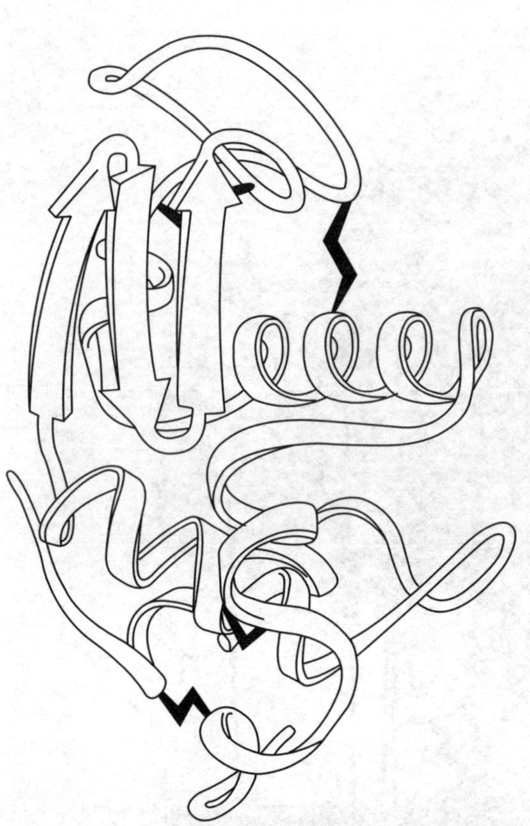

Fig 3.35 *Tertiary structure of lysozyme. The arrows represent a region of β sheet. α helices are shown as regular coils. The rest of the molecule is represented as a ribbon and the four disulphide bridges as black zig-zags (compare fig 3.29).*

Fig 3.36 *Structure of haemoglobin. The molecule consists of four chains: two α chains and two β chains. Each chain carries a haem to which one molecule of oxygen binds. The assembly of a protein from separate polypeptide chains is an example of quaternary structure.*

The protein coats of some viruses, such as the tobacco mosaic virus, are composed of many polypeptide chains arranged in a highly ordered fashion (fig 2.14).

3.5.4 Denaturation and renaturation of proteins

Denaturation is the loss of the specific three-dimensional shape of a protein molecule. The change may be temporary or permanent, but the amino acid sequence of the protein remains unaffected. If denaturation occurs, the molecule unfolds and can no longer perform its normal biological function. A number of agents may cause denaturation as follows.

Heat or radiation. e.g. infra-red or ultra-violet light. Kinetic energy is supplied to the protein causing its atoms to vibrate violently, so disrupting the weak hydrogen and ionic bonds. Coagulation of the protein then occurs.

Strong acids and alkalis and high concentrations of salts. Ionic bonds are disrupted and the protein is coagulated. Breakage of peptide bonds may occur if the protein is allowed to remain mixed with the reagent for a long period of time.

Heavy metals. The positively charged ions of heavy metals (cations) form strong bonds with the negatively charged carboxyl groups on the R groups of proteins and often disrupt ionic bonds. They also reduce the protein's electrical polarity (its overall charge) and thus increase its insolubility. This causes the protein to precipitate out of solution.

Organic solvents and detergents. These reagents disrupt hydrophobic interactions and form bonds with hydrophobic (non-polar) groups. This in turn causes the disruption of hydrogen bonding. When alcohol is used as a disinfectant it functions to denature the protein of any bacteria present.

Renaturation

Sometimes a protein will spontaneously refold into its original structure after denaturation, providing conditions are suitable. This is called **renaturation**, and is good evidence that tertiary structure can be determined purely by primary structure and that biological structures can spontaneously assemble according to a few general principles.

3.6 DNA and RNA, the nucleic acids

Nucleic acids, like proteins, are essential for life. They form the genetic material of all living organisms, including the simplest virus. The term 'nucleic acid' comes from the fact that they are found mainly in the nucleus. Stains for nucleic acids show up nuclei very clearly under the light microscope.

Discovery of the structure of DNA (deoxyribonucleic acid), one of the two types of nucleic acid, represents one of the outstanding milestones in biology because it finally solved the problem of how living organisms store the information needed to control their activities and pass this information on to subsequent generations. Fig 3.4 shows that nucleic acids are made up of units called **nucleotides**. These are arranged to form extremely long molecules known as **polynucleotides**. Thus, to understand their structure, it is necessary first to study the structure of the nucleotide.

3.6.1 Structure of nucleotides

A nucleotide has three components, a 5-carbon sugar, a nitrogenous base and phosphoric acid.

Sugar. The sugar has five carbon atoms; therefore it is a pentose. There are two types of nucleic acids, depending on the pentose they contain. Those containing ribose are called ribonucleic acids or RNA and those containing deoxyribose (ribose with an oyxgen atom removed from carbon atom 2) are called deoxyribonucleic acids or DNA (fig 3.37).

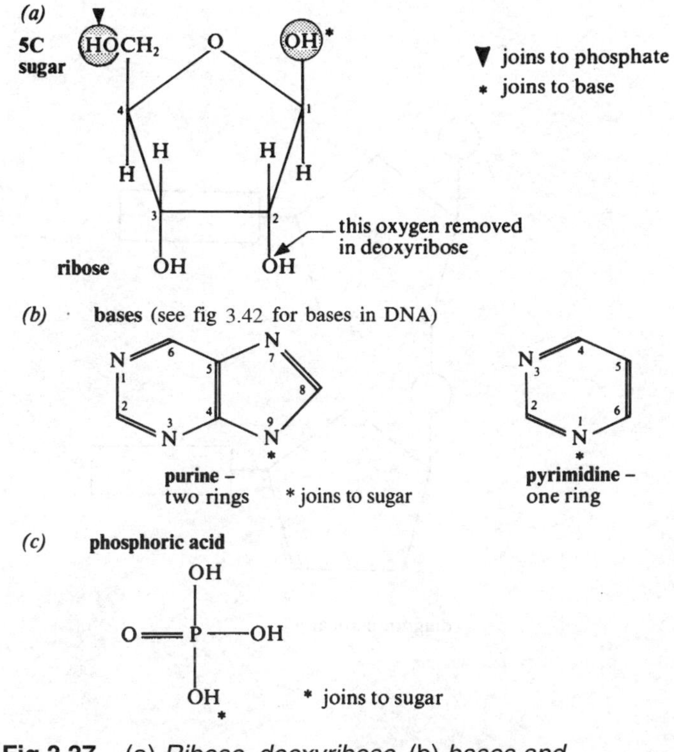

Fig 3.37 (a) *Ribose, deoxyribose,* (b) *bases and* (c) *phosphoric acid, the components of nucleotides.*

Bases. Each nucleic acid contains four different bases, two derived from purine and two from pyrimidine. The nitrogen in the rings gives the molecules their basic nature. The two purines are adenine (A) and guanine (G). The two pyrimidines are thymine (T) and cytosine (C) in DNA, with uracil (U) in place of thymine in RNA. Thymine is chemically very similar to uracil (it is 5-methyl uracil, that is uracil with a methyl group, $-CH_3$, on carbon atom 5). Purines have two rings and pyrimidines have one ring in their structure.

Note that the bases are commonly represented by their initial letters A, G, T, U and C.

Phosphoric acid (fig 3.37). This gives nucleic acids their acid character.

Fig 3.38 shows how the sugar, base and phosphoric acid combine to form a nucleotide. The combination of a sugar with a base occurs with the elimination of water and therefore is a condensation reaction. A nucleotide is formed by further condensation with phosphoric acid.

Different nucleotides are formed according to the sugars and bases used.

Nucleotides are not only used as building blocks for nucleic acids, but they form several important coenzymes, including adenosine triphosphate (ATP), cyclic AMP, coenzyme A, nicotinamide adenine dinucleotide (NAD) and its phosphate NADP, and flavine adenine dinucleotide (FAD) (section 4.5.3).

3.6.2 Formation of dinucleotides and polynucleotides

Two nucleotides join to form a **dinucleotide** by condensation between the phosphate group of one with the sugar of the other, as shown in fig 3.39. The process is repeated up to several million times to make a polynucleotide. An unbranched sugar–phosphate backbone is thus formed, as shown in fig 3.40.

3.6.3 Structure of DNA

Like proteins, polynucleotides can be regarded as having a primary structure, which is the sequence of nucleotides, and a three-dimensional structure. Interest in the structure of DNA intensified when it was realised in the early part of this century that it might be the genetic material. Evidence for this is presented in chapter 23.

By the early 1950s the Nobel prize-winning chemist Linus Pauling of the USA had worked out the α-helical structure which is common to many fibrous proteins, and was applying himself to the problem of the structure of DNA, which evidence suggested was also a fibrous molecule. At the same time Maurice Wilkins and Rosalind Franklin of King's College, London were tackling the same problem using the technique of X-ray crystallography. This involved the difficult and time-consuming process of preparing pure fibres of the salt of DNA through which

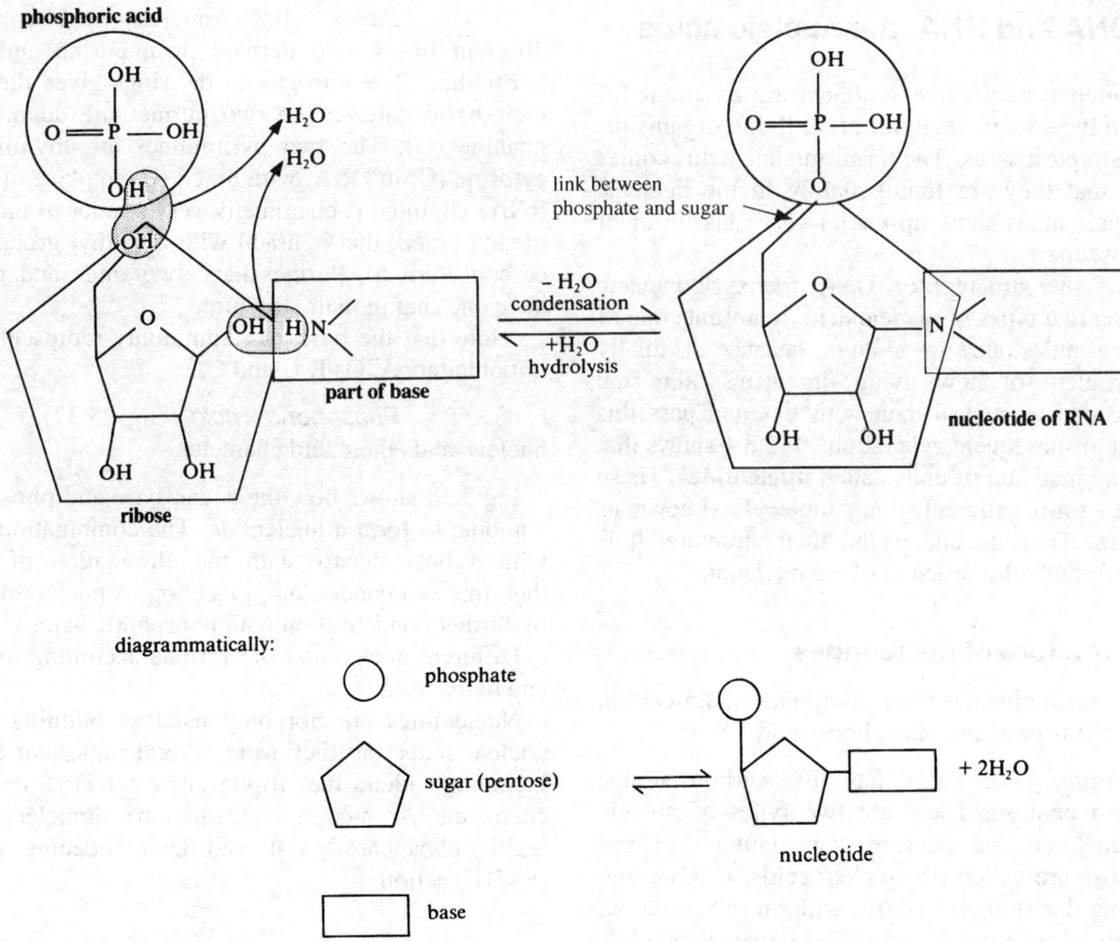

Fig 3.38 *Formation of a nucleotide.*

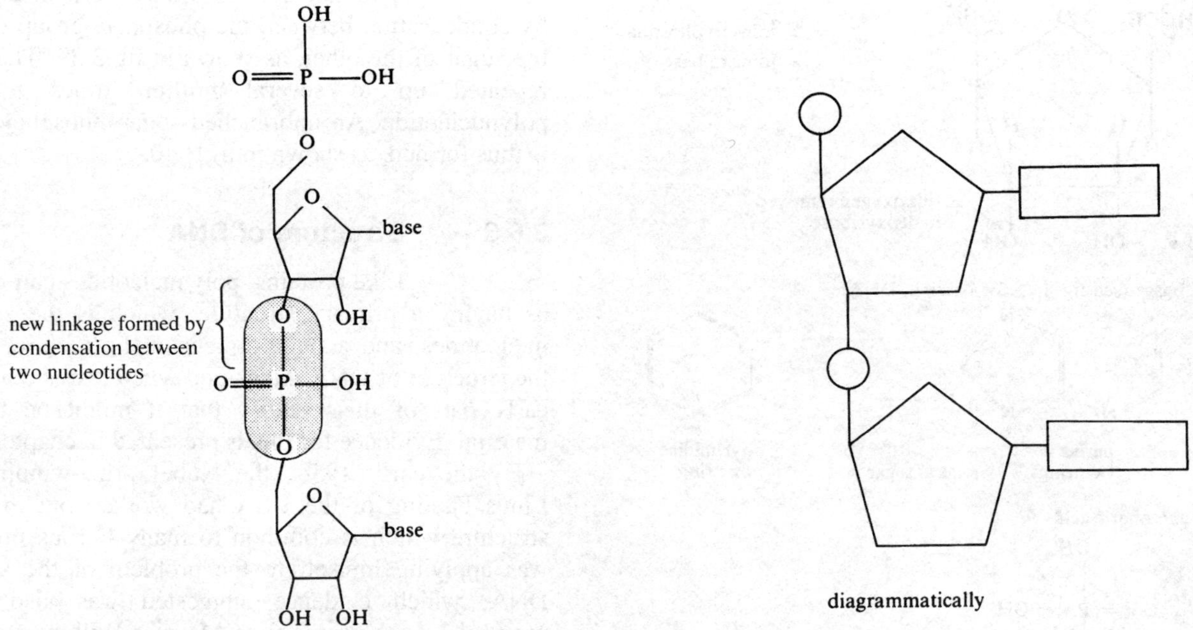

Fig 3.39 *Structure of a dinucleotide.*

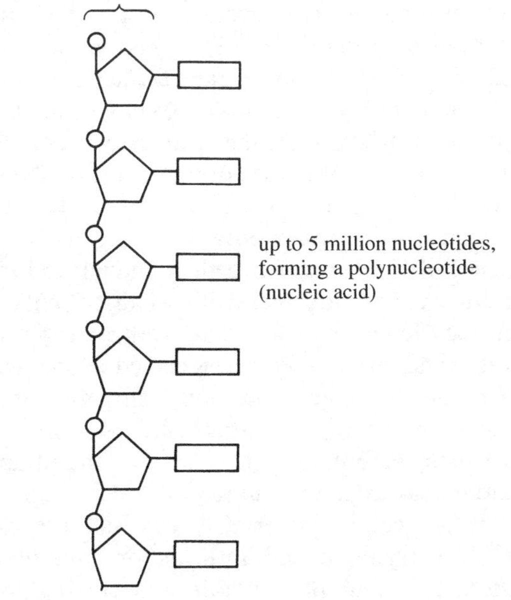

Fig 3.40 *Formation of a polynucleotide.*

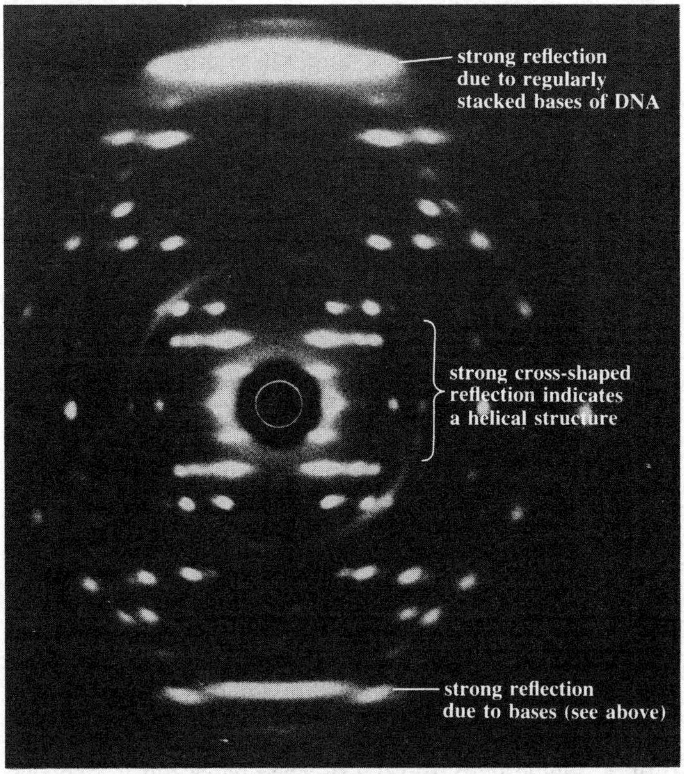

Fig 3.41 *X-ray diffraction photograph of a fibre of DNA. This is the kind of pattern from which the double helical structure was originally deduced (photograph by courtesy of Dr J. M. Squire).*

X-rays could be passed in order to obtain complex X-ray diffraction patterns (fig 3.41). These reveal the gross structure of the molecule but are not as detailed as those from pure crystals of proteins.

Meanwhile, James Watson and Francis Crick of the Cavendish Laboratory in Cambridge had chosen what was to prove the successful approach. Using all the chemical and physical information they could gather, they began building scale models of polynucleotides in the hope that a convincing structure would emerge. Watson's book *The Double Helix* provides a fascinating insight into their work.

Two lines of evidence proved crucial. Firstly, they were in regular communication with Wilkins and had access to the X-ray diffraction data, against which they were able to test their models. These data strongly suggested a helical structure (fig 3.41) with regularity at a spacing of 0.34 nm along its axis. Secondly, they realised the significance of some evidence published by Erwin Chargaff in 1951 concerning the ratio of the different bases found in DNA. Although important, the significance of this had been overlooked. Table 3.10 shows some of Chargaff's data, and supporting data obtained since.

3.13 Examine the table. What does it reveal about the ratios of the different bases?

Watson and Crick had been exploring the idea that there may be two helical chains of polynucleotides in DNA, held together by pairing of bases between neighbouring chains. The bases would be held together by hydrogen bonds.

Table 3.10 Relative amounts of bases in DNA from various organisms.

Source of DNA	*Adenine*	*Guanine*	*Thymine*	*Cytosine*
Human	30.9	19.9	29.4	19.8
Sheep	29.3	21.4	28.3	21.0
Hen	28.8	20.5	29.2	21.5
Turtle	29.7	22.0	27.9	21.3
Salmon	29.7	20.8	29.1	20.4
Sea urchin	32.8	17.7	32.1	17.3
Locust	29.3	20.5	29.3	20.7
Wheat	27.3	22.7	27.1	22.8
Yeast	31.3	18.7	32.9	17.1
Escherichia coli (a bacterium)	24.7	26.0	23.6	25.7
ϕX174 bacteriophage (a virus)	24.6	24.1	32.7	18.5

Amounts are in molar proportions on a percentage basis.

3.14 If this model were correct, can you predict from Chargaff's data which base combines with which in each pair?

Fig 3.42 shows how the base pairs are joined by hydrogen bonds. Adenine pairs with thymine, and guanine with cytosine; the adenine–thymine pair has two hydrogen bonds. Watson tried pairing the bases in this way, and recalls 'my morale skyrocketed, for I suspected that we now had the answer to the riddle of why the number of purine residues exactly equalled the number of pyrimidine residues'.* He noticed the neat way in which the bases fit and that the overall size and shape of the base pairs was identical, both being three rings wide (fig 3.42). Hydrogen bonding between other combinations of bases, while possible, is much weaker. The way was finally open to building the now commonly accepted model of DNA, whose structure is summarised in figs 3.43–45.

Features of the DNA molecule

Watson and Crick showed that DNA consists of two polynucleotide chains. Each chain forms a right-handed helical spiral and the two chains coil around each other to form a double helix (fig 3.43). The chains run in opposite directions, that is are **antiparallel**. Each chain has a sugar–phosphate backbone with bases which project at right-angles and hydrogen bond with the bases of the opposite chain across the double helix (fig 3.44). The sugar–phosphate backbones are clearly seen in a space-filling model of DNA (fig 3.45). The width between the two backbones is constant and equal to the width of a base pair, that is the width of a purine plus a pyrimidine. Two purines would be too large, and two pyrimidines too small, to span the gap between the two chains. Along the axis of the molecule the base pairs are 0.34 nm apart, accounting for the regularity indicated by X-ray diffraction. A complete turn of the double helix comprises 3.4 nm, or ten base pairs. There is no restriction on the sequence of bases in one chain, but because of the rules of base pairing, the sequence in one chain determines that in the other. The two chains are thus said to be **complementary**.

* From *The Double Helix*, James D. Watson, Weidenfeld & Nicolson, 1968.

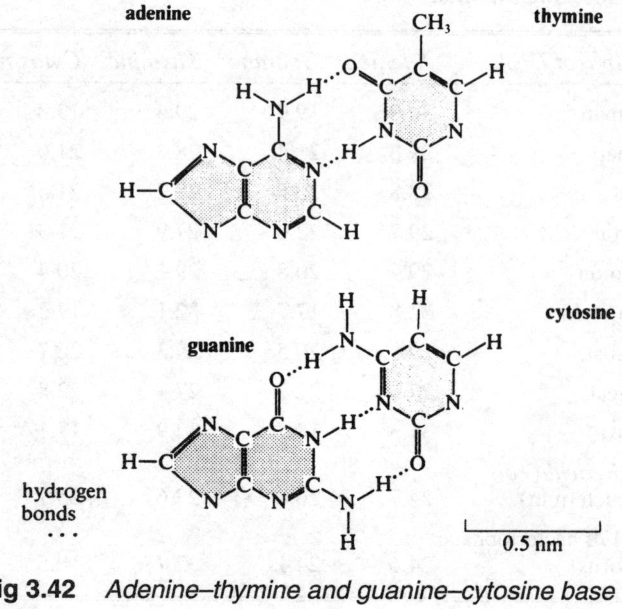

Fig 3.42 *Adenine–thymine and guanine–cytosine base pairs.*

Watson and Crick published their model in 1953 in the journal *Nature* and, together with Maurice Wilkins, were awarded the Nobel Prize for their work in 1962, the same year that Kendrew and Perutz received Nobel prizes for their work on the three-dimensional structure of proteins, also based on X-ray crystallography. Rosalind Franklin died of cancer before the prizes were awarded and Nobel prizes are not awarded posthumously.

To act as genetic material, the structure had to be capable of carrying coded information and of accurate replication. Its suitability for this was not overlooked by Watson and Crick who, with masterly understatement near the end of their paper said 'It has not escaped our notice that the specific pairing we have postulated immediately suggests a possible copying mechanism for the genetic material.' *In a second paper that year they discussed the genetic implications of the structure, and these are dealt with in chapter 23. This discovery, in which structure was shown to be so clearly related to function even at the molecular level, gave great impetus to the science of molecular biology.

3.6.4 Structure of RNA

RNA is normally single stranded, unlike DNA. Certain forms of RNA do assume complex structures, notably transfer RNA (tRNA) and ribosomal RNA (rRNA). Another form is messenger RNA (mRNA). These are involved in protein synthesis and are discussed in chapter 23.

* Watson, J. D. & Crick, F.H.C. (1953) *Nature* 171, 737.

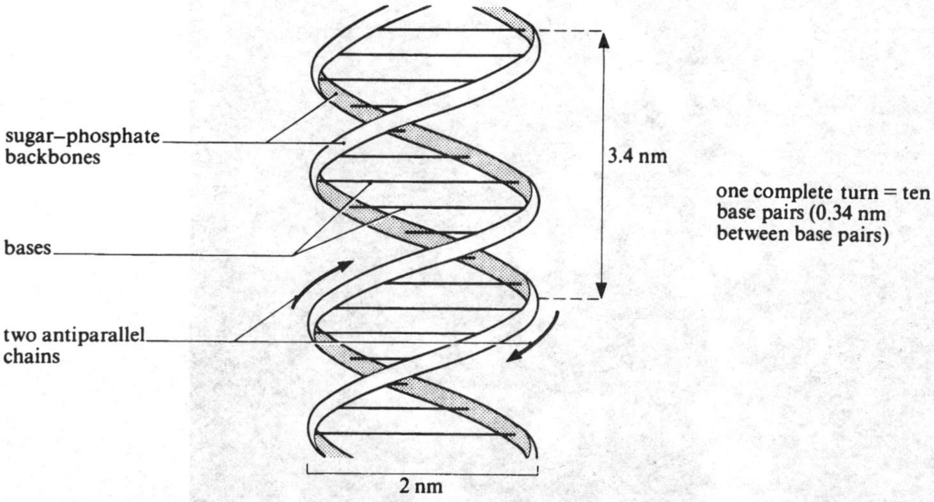

Fig 3.43 *Diagrammatic structure of DNA.*

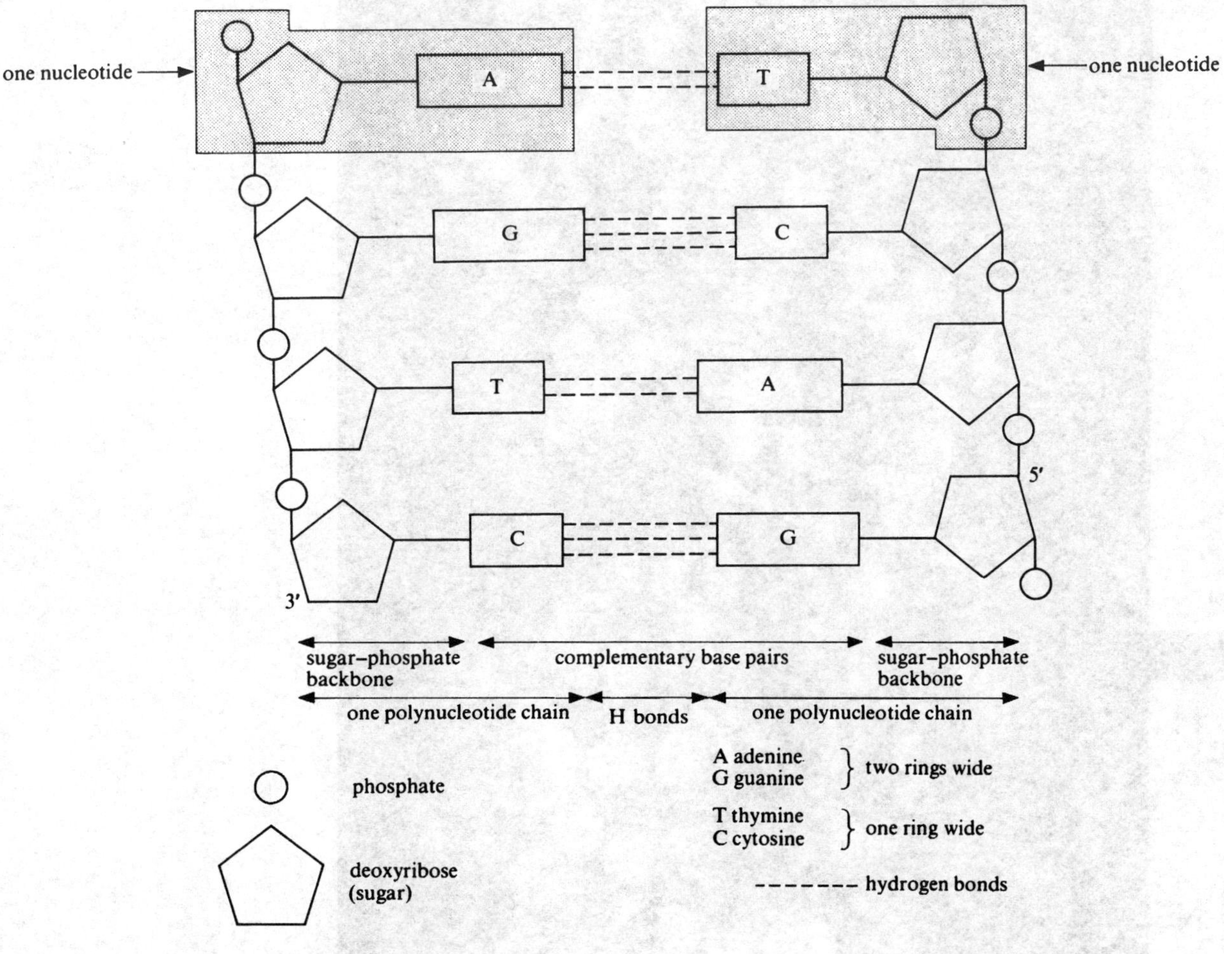

Fig 3.44 *DNA – diagrammatic structure of straightened chains.*

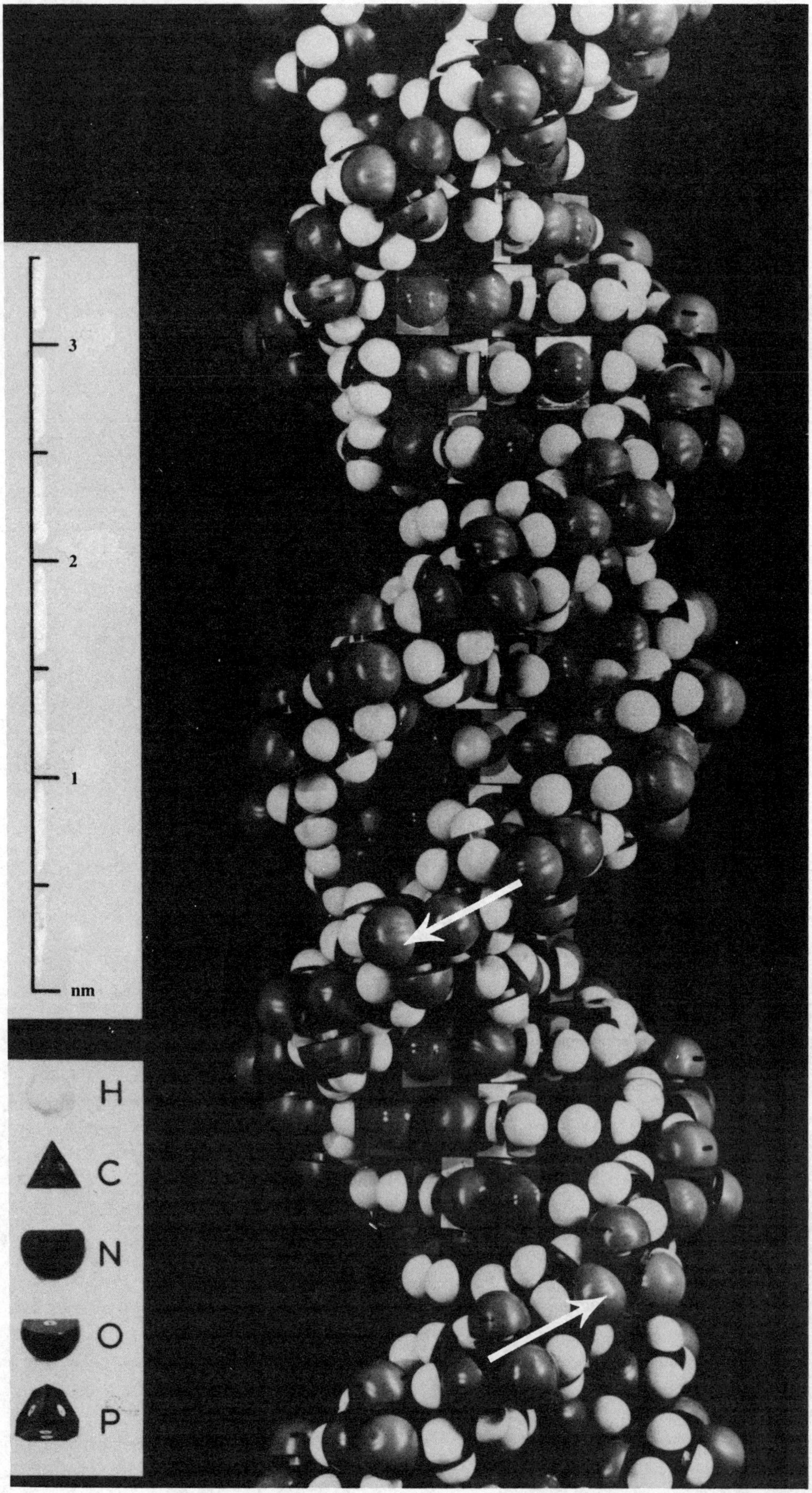

Fig 3.45 *Space-filling model of DNA. Arrows indicate the directions of the two antiparallel sugar–phosphate backbones.*

3.7 Identification of biochemicals

In this section are described some simple tests which can be carried out to identify some of the common biochemicals. More elaborate techniques exist for identifying and separating biochemicals. Of particular importance are chromatography and electrophoresis and these are described in the appendix in Book 2.

It is recommended that you first familiarise yourself with the following tests by using pure samples of the chemicals being tested. Once the techniques have been mastered and familiarity with the colour changes obtained, extracts from various tissues can be studied.

Experiment 3.1: Identification of biochemicals in pure form

NB Any heating that has to be done in the following tests should be carried out in a water bath at the boiling point of water. Direct heating of test-tubes should not take place.

Materials

pH paper
test-tubes
test-tube rack
Bunsen burner
teat pipettes
spatula
1 cm^3 syringe
iodine/potassium iodide solution
Benedict's reagent
dilute sulphuric acid
sodium hydrogencarbonate
Sudan III
Millon's reagent
5% potassium hydroxide solution
1% copper sulphate solution
DCPIP (dichlorophenolindophenol) solution
1% starch solution (cornflour is a recommended source)
1% glucose solution
1% sucrose solution (Analar sucrose must be used to avoid contamination with a reducing sugar)
olive oil or corn oil
absolute ethanol
egg albumen
1% lactose solution
1% fructose solution

Carbohydrates

Reducing sugars. The reducing sugars include all monosaccharides, such as glucose and fructose, and some disaccharides, such as maltose. The only common *non*-reducing sugar is sucrose. Use 0.1–1% sugar solutions.

Test	*Observation*	*Basis of test*
Benedict's test		
Add 2 cm^3 of a solution of the reducing sugar to a test-tube. Add an equal volume of Benedict's solution. Shake and bring gently to the boil.	The initial blue colouration of the mixture turns green, then yellowish and may finally form a brick-red precipitate.	Benedict's solution contains copper sulphate. Reducing sugars reduce soluble blue copper sulphate, containing copper(II) ions (Cu^{2+}) to insoluble red-brown copper oxide containing copper(I). The latter is seen as a precipitate.

Additional information

The test is **semi-quantitative**, that is a rough estimation of the amount of reducing sugar present will be possible. The final precipitate will appear green to yellow to orange to red-brown with increasing amounts of reducing sugar. (The initial yellow colour blends with the blue of the copper sulphate solution to give the green colouration.)

Test	Observation	Basis of test
Fehling's test		
Add 2 cm^3 of a solution of the reducing sugar to a test-tube. Add 1 cm^3 of Fehling's solution A and 1 cm^3 of Fehling's solution B. Shake and bring to the boil.	The initial blue colouration of the mixture turns green to yellow and finally a brick-red precipitate is formed.	As Benedict's test.
Additional information		
Not as convenient as Benedict's test because Fehling's solutions A and B have to be kept separate until the test. Also it is not as sensitive.		

Non-reducing sugars. The most common non-reducing sugar is sucrose, a disaccharide. If reducing sugars have been shown to be absent (negative result in above test) a brick-red precipitate in the test below indicates the presence of a non-reducing sugar. If reducing sugars have been shown to be present, a heavier precipitate will be observed in the following test than with the reducing test if non-reducing sugar is also present.

Test	Observation	Basis of test
Add 2 cm^3 of sucrose solution to a test-tube. Add 1 cm^3 dilute hydrochloric acid. Boil for one minute. Carefully neutralise with sodium hydrogencarbonate (check with pH paper) – care is required because effervescence occurs. Carry out Benedict's test.	As Benedict's test.	A disaccharide can be hydrolysed to its monosaccharide constituents by boiling with dilute hydrochloric acid. Sucrose is hydrolysed to glucose and fructose, both of which are reducing sugars and give the reducing sugar result with the Benedict's test.

Starch. This is only slightly soluble in water, in which it forms a colloidal suspension. It can be tested in suspension or as a solid.

Test	Observation	Basis of test
Iodine/potassium iodide test		
Add 2 cm^3 1% starch solution to a test-tube. Add a few drops of I_2/KI solution. Alternatively add the latter to the solid form of starch.	A blue-black colouration.	A polyiodide complex is formed with starch.

Cellulose and lignin. See appendix A2 (staining) in Book 2.

Lipids

Lipids include oils (such as corn oil and olive oil), fats and waxes.

Test	*Observation*	*Basis of test*
Sudan III		
Sudan III is a red dye. Add 2 cm^3 oil to 2 cm^3 of water in a test-tube. Add a few drops of Sudan III and shake	A red-stained oil layer separates on the surface of the water, which remains uncoloured.	Fat globules are stained red and are less dense than water.
Emulsion test		
Add 2 cm^3 fat or oil to a test-tube containing 2 cm^3 of absolute ethanol. Dissolve the lipid by shaking vigorously. Add an equal volume of cold water.	A cloudy white suspension	Lipids are immiscible with water. Adding water to a solution of the lipid in alcohol results in an emulsion of tiny lipid droplets in the water which reflect light and give a white, opalescent appearance.
Grease spot test		
Rub a drop of the sample into a piece of paper. Allow time for any water to evaporate. Gentle warming will speed up the process.	A permanent transparent spot on the paper.	

Proteins

A suitable protein for these tests is egg albumen.

Test	*Observation*	*Basis of test*
Millon's test		
Add 2 cm^3 protein solution or suspension to a test-tube. Add 1 cm^3 Millon's reagent and boil. NB Millon's reagent is poisonous: take care!	A white precipitate forms which coagulates on heating and turns red or salmon pink.	Millon's reagent contains mercury acidified with nitric acid, giving mercury(II) nitrate and nitrite. The amino acid tyrosine contains a phenol group which reacts to give a red mercury(II) complex. This is a reaction given by all phenolics and is not specific for proteins. Protein usually coagulates on boiling, thus appearing solid. The only common protein lacking tyrosine likely to be used is gelatin.
Biuret test		
Add 2 cm^3 protein solution to a test-tube. Add an equal volume of 5% potassium hydroxide solution and mix. Add two drops of 1% copper sulphate solution and mix. No heating is required.	A mauve or purple colour develops slowly.	A test for peptide bonds. In the presence of dilute copper sulphate in alkaline solution, nitrogen atoms in the peptide chain form a purple complex with copper(II) ions (Cu^{2+}). Biuret is a compound derived from urea which also contains the –CONH– group and gives a positive result.

Vitamin C (ascorbic acid)

This test can be conducted on a quantitative basis if required, in which case the volumes given below must be measured accurately. A suitable source of vitamin C is a 50/50 mix of fresh orange or lemon juice with distilled water. Vitamin C tablets may also be purchased.

Test	*Observation*	*Basis of test*
Using 0.1% ascorbic acid solution as a standard. Add 1 cm^3 of DCPIP solution to a test-tube. Fill a 1 cm^3 syringe with 0.1% ascorbic acid. Add the acid to the DCPIP drop by drop, stirring gently with the syringe needle. Do not shake.* Add until the blue colour of the dye just disappears. Note the volume of ascorbic acid solution used.	Blue colour of dye disappears to leave a colourless solution.	DCPIP is a blue dye which is reduced to a colourless compound by ascorbic acid, a strong reducing agent.

* Shaking the solution would result in oxidation of the ascorbic acid by oxygen in the air. The effects of shaking and of boiling could be investigated.

DNA

See Feulgen's stain (appendix table A2.2 in Book 2).

3.15 How could you determine the concentration of ascorbic acid in an unknown sample?

3.16 You are provided with three sugar solutions. One contains glucose, one a mixture of glucose and sucrose, and one sucrose.
(*a*) How could you identify each solution?
(*b*) Supposing that the apparatus were available, and time permitted, briefly discuss any further experiments you could perform to confirm your results.

3.17 How would you make 100 cm^3 of a 10% glucose solution?

3.18 Starting with stock solutions of 10% glucose and 2% sucrose how would you make 100 cm^3 of a mixture of final concentration 1% sucrose and 1% glucose?

Experiment 3.2: Identification of biochemicals in tissues

A biochemist is often faced with the problem of wanting to identify chemicals (qualitative analysis) or to measure their amounts (quantitative analysis) in living tissue. Sometimes the chemical can be tested for directly, but often some kind of extraction and purification process must first be embarked upon.

A convenient exercise is to take a range of common foods and plant material and to test for the range of biochemicals listed in experiment 3.1. An extraction procedure is designed where possible to give a clear, colourless solution for testing, and you should note the rationale behind the procedures so that you could design your own if necessary.

Materials

- As for experiment 3.1 up to DCPIP solution
- pestle and mortar
- microscope
- slides and cover-slips
- razor blade
- watch glass
- Schultz's solution
- phloroglucinol + conc. hydrochloric acid
- potato tuber
- apple
- cotton wool
- woody stem
- seeds/nuts
- soaked peas
- beans

Microscopic examination of thin sections of tissue

Suitable for: Visible storage products, particularly starch grains, such as potato tuber.

As above with appropriate staining or other chemical testing

Suitable for: Reducing sugars – Mount in a few drops of Benedict's reagent, heat gently to boiling; add water if necessary to prevent drying.
Starch – Mount section in dilute iodine/potassium iodide solution.
Protein – Mount in a few drops of Millon's reagent, heat gently to boiling; add water if necessary to prevent drying.
Oil and fat – Stain material, such as seed, with Sudan III and wash with water and/or 70% ethanol. Then section and mount.
Cellulose, lignin, etc. – see appendix A2.4.2 for staining.

Testing a clear, aqueous solution

Decolourise tissue if necessary: Pigments may interfere with colour tests but can usually be removed with an organic solvent such as 80% ethanol or 80% propanone. *Care must be taken to avoid naked flames.* However, remember these solvents may also remove lipids and soluble sugars. *Suitable for:* Removing chlorophyll from leaves.

Homogenise (grind) material: Sugars and proteins – Small pieces of solid material can be ground with a small quantity of water using a pestle and mortar or a food mixer. The ground material should be squeezed through several layers of pre-moistened fine muslin or nylon and/or filtered or centrifuged to remove solid material. This may be unnecessary if a fairly colourless, fine suspension is obtained. The clear solution can be tested as usual, with further dilution if necessary. The solid residue may also be tested if appropriate.

Lipids – Grind material, transfer to a test-tube and boil. Lipids will escape as oil droplets. Perform the Sudan III test. Alternatively take thin shavings of nuts or other foods, including coloured foods, and do the emulsion test.

Suitable for:

fruit, such as apple, orange	(vitamin C, sugars)
nuts	(oils)
castor oil seed	(oil)
pea seed	(protein)
pine kernels	(protein and oil)
potato	(starch, vitamin C)
egg	(protein)

Subdivision of the above materials, such as into seeds, flesh, skin and juice, may be possible.

Chapter Four

Enzymes

Enzymes can be defined as biological catalysts. A **catalyst** is a substance which speeds up a chemical reaction but remains unchanged itself at the end. Enzymes are *biological* catalysts because they are protein molecules made by living cells. A typical human cell contains several thousand enzymes. They are used to catalyse a vast number of chemical reactions at temperatures suitable for living organisms, that is between approximately 5 and 40 °C. High temperatures would be needed, as well as marked changes in other conditions, if the same speeds of reaction were to be achieved outside the organism. These would be lethal to a living cell. Enzymes are vitally important because in their absence reactions in the cell would be too slow to sustain life.

The chemical (or chemicals) which an enzyme works on is called its **substrate**. An enzyme combines with its substrate to form a short-lived enzyme/substrate complex. This proximity of the enzyme with the substrate in the complex greatly increases the chances of a reaction occurring. Once a reaction has occurred, the complex breaks up into **products** and enzyme. The enzyme remains unchanged at the end of the reaction and is free to interact again with more substrate.

$$\text{substrate} + \text{enzyme} \rightleftharpoons \text{enzyme/substrate complex} \rightleftharpoons \text{enzyme/ product complex} \rightleftharpoons \text{enzyme} + \text{product(s)}$$

or

$$E + S \rightleftharpoons ES \rightleftharpoons EP \rightleftharpoons E + P$$

Anabolism and catabolism

The sum total of all the chemical reactions going on in cells is known as **metabolism**. Metabolism can be divided into two types, namely anabolism and catabolism. These two types of activity often take place in different parts of the cell. **Catabolic reactions** involve the breakdown of molecules and usually release energy. They often involve oxidation or hydrolysis. **Anabolic reactions** involve the synthesis of molecules and usually require energy. They often involve condensation. All these reactions are catalysed by enzymes. An example of an enzyme involved in anabolism is glutamine synthetase, which catalyses the synthesis of the amino acid glutamine from glutamic acid and ammonia:

$$\text{glutamic acid} + \text{ammonia} + \text{ATP} \xrightarrow[\text{synthetase}]{\text{glutamine}} \text{glutamine} + \text{water} + \text{ADP} + P_i$$

(ATP is adenosine triphosphate, ADP is adenosine diphosphate and P_i is inorganic phosphate.) An example of an enzyme involved in catabolism is amylase:

$$\text{starch} + \text{water} \xrightarrow{\text{amylase}} \text{maltose}$$

Metabolic pathways

Commonly, a number of enzymes are used in sequence to convert one substance into one or several products via a series of intermediate compounds. The chain of reactions is referred to as a **metabolic pathway**. Many such pathways are going on at the same time in the cell. The reactions proceed in a controlled manner due to the specific nature of enzymes. A single enzyme generally will catalyse only a single reaction. Thus enzymes serve to control the chemical reactions that occur within cells and ensure that they proceed at an efficient rate.

4.1 Properties of enzymes

Enzymes possess the following major properties.

- All are globular proteins.
- Being proteins, they are coded for by DNA.
- They are catalysts (see above).
- Their presence does not alter the nature or properties of the end product(s) of the reaction.
- They are very efficient. In other words, a very small amount of catalyst brings about the change of a large amount of substrate. For example, one molecule of the enzyme catalase can catalyse the decomposition of about 600 thousand molecules per second of hydrogen peroxide to water and oxygen at body temperature. The efficiency of catalase compared with an inorganic catalyst such as manganese dioxide can easily be demonstrated by adding them separately to hydrogen peroxide and observing the rate of oxygen evolution. A good source of catalase is liver. An average enzyme undergoes about 1000 reactions per second. Without a catalyst at all, reaction rates would be millions of times slower.
- They are highly specific, that is an enzyme will generally catalyse only a single reaction. Catalase, for example, will only catalyse the decomposition of hydrogen peroxide.

- The catalysed reaction is reversible.
- Their activity is affected by pH, temperature, substrate concentration and enzyme concentration. These factors are considered in section 4.3.
- Enzymes lower the activation energy of the reactions they catalyse (see section 4.1.1).
- Enzymes possess active sites where the reaction takes place. These sites have specific shapes (see section 4.1.2).

4.1.1 Activation energy

Consider a mixture of petrol and oxygen maintained at room temperature. Although a reaction between the two substances is thermodynamically possible, it does not occur unless energy is applied to it, such as a simple spark. The same is true of a match. The chemicals in the match head are capable of reacting with an overall release of energy. However, a little energy must be put in to get the reaction started (heat energy generated by friction on the matchbox). This energy is called the **activation energy**. It is the energy required to make the substances react. Enzymes, by functioning as catalysts, serve to reduce the activation energy required for a chemical reaction to take place (fig 4.1). They speed up the overall rate without altering, to any great extent, the temperature at which it occurs.

4.1.2 Mechanism of enzyme action

Enzymes are very specific and it was suggested by Fischer in 1890 that this was because the enzyme had a particular shape into which the substrate or substrates fit exactly. This is often referred to as the '**lock and key**' hypothesis, where the substrate is imagined being like a **key** whose shape is complementary to the enzyme or **lock**. This is shown diagrammatically in fig 4.2. The site where the substrate binds in the enzyme is known as the **active site** and it is this which has the specific shape.

Most enzymes are far larger molecules than the substrates they act on and the active site is usually only a very small portion of the enzyme, between 3 and 12 amino acids. The remaining amino acids, which make up the bulk of the

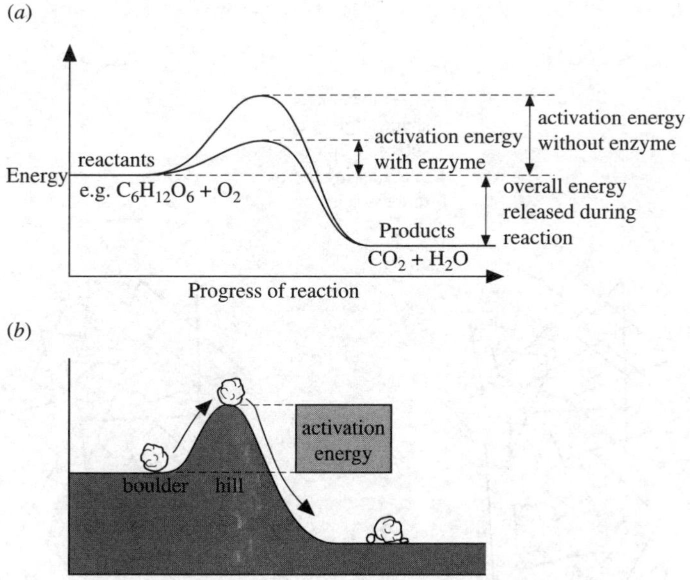

Fig 4.1 *Activation energy. (a) Activation energy for an enzyme-catalysed and an uncatalysed reaction. (b) An analogous situation where rolling a boulder down a hill will result in release of more energy than is used to start it rolling.*

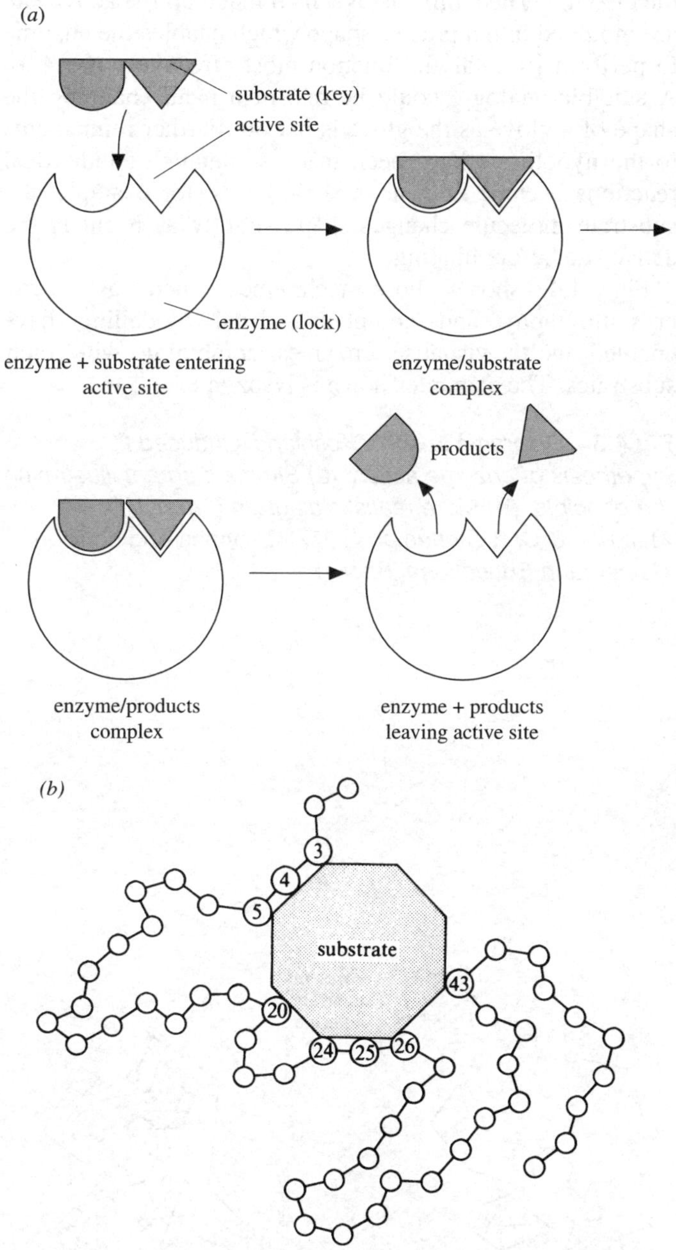

Fig 4.2 *(a) Fischer's 'lock and key' hypothesis of enzyme action. (b) A more realistic diagrammatic representation of an enzyme–substrate complex. The positions of the amino acids of the active site are numbered according to their position in the primary structure of the enzyme.*

enzyme, function to maintain the correct globular shape of the molecule which, as will be explained below, is important if the active site is to function at the maximum rate.

Once formed, the products no longer fit into the active site and escape into the surrounding medium, leaving the **active site** free to receive further substrate molecules.

In 1959 Koshland suggested a modification to the 'lock and key' model known as the **'induced fit' hypothesis**. Working from evidence that suggested that some enzymes and their active sites were physically rather more flexible structures than previously described, he proposed that the active site could be modified as the substrate interacts with the enzyme. The amino acids which make up the active site are moulded into a precise shape which enables the enzyme to perform its catalytic function most effectively (fig 4.3). A suitable analogy would be that of a hand changing the shape of a glove as the glove is put on. Further refinements to the hypothesis have been made as details of individual reactions became known. In some cases, for example, the substrate molecule changes shape slightly as it enters the active site before binding.

Fig 4.4 shows how techniques such as X-ray crystallography and computer-assisted modelling have enabled us to visualise enzymes combining with their substrates. The example shown is lysozyme.

4.2 The rate of enzyme reactions

The rate of an enzyme reaction is measured by the amount of substrate changed, or amount of product formed, during a period of time.

The rate is determined by measuring the slope of the tangent to the curve in the initial stage of the reaction (shown as (*a*) in fig 4.5). The steeper the slope, the greater is the rate. If activity is measured over a period of time, the rate of reaction usually falls, most commonly as a result of a fall in substrate concentration (see next section).

4.3 Factors affecting the rate of enzyme reactions

When investigating the effect of a given factor on the rate of an enzyme-controlled reaction, all other factors should be kept **constant** and at **optimum levels** wherever possible. Initial rates only should be measured, as explained above.

(*a*)

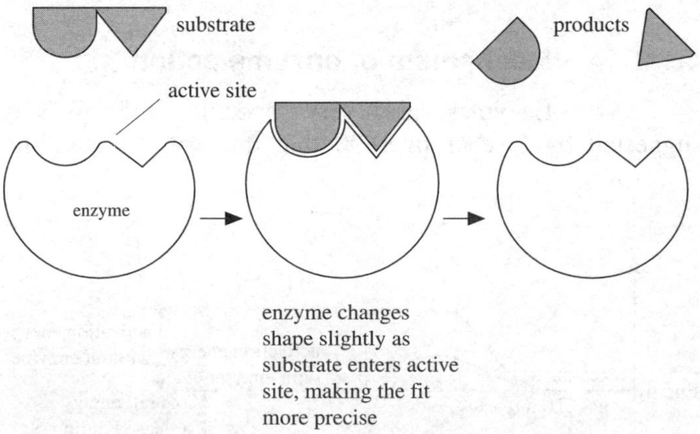

(*b*)

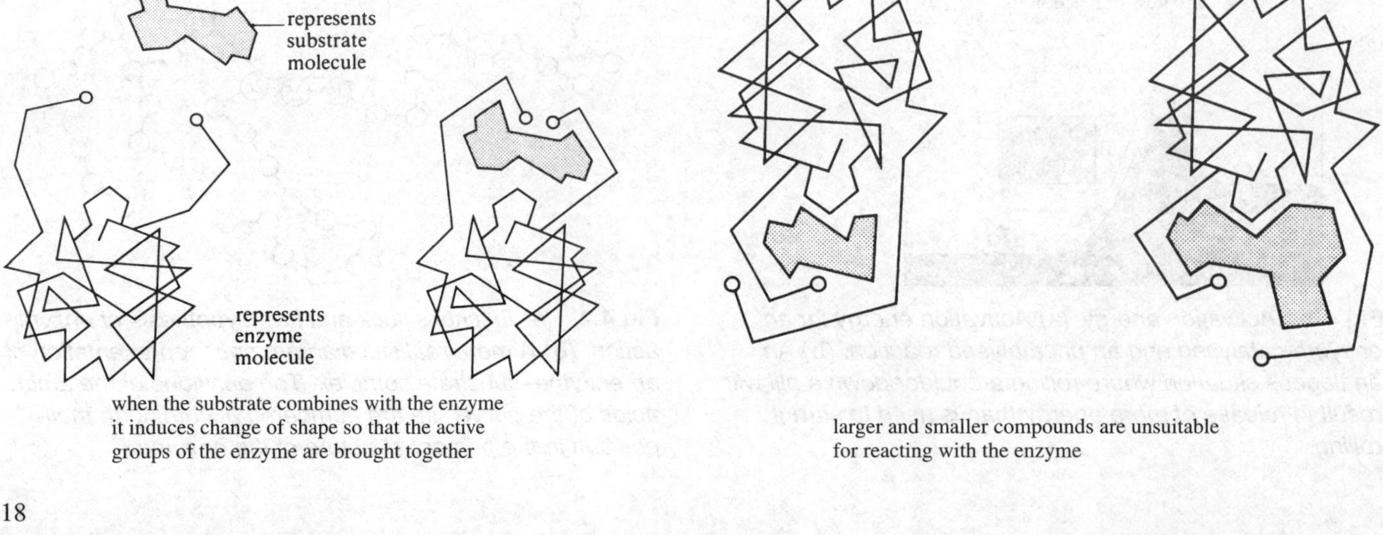

Fig 4.3 *Diagrams to show Koshland's induced fit hypothesis of enzyme action.* (a) *Simple diagram illustrating the principle.* (b) *More realistic diagram (From J. C. Marsden & C. F. Stoneman (1977)* Enzymes and equilibria, *Heinemann Educational Books).*

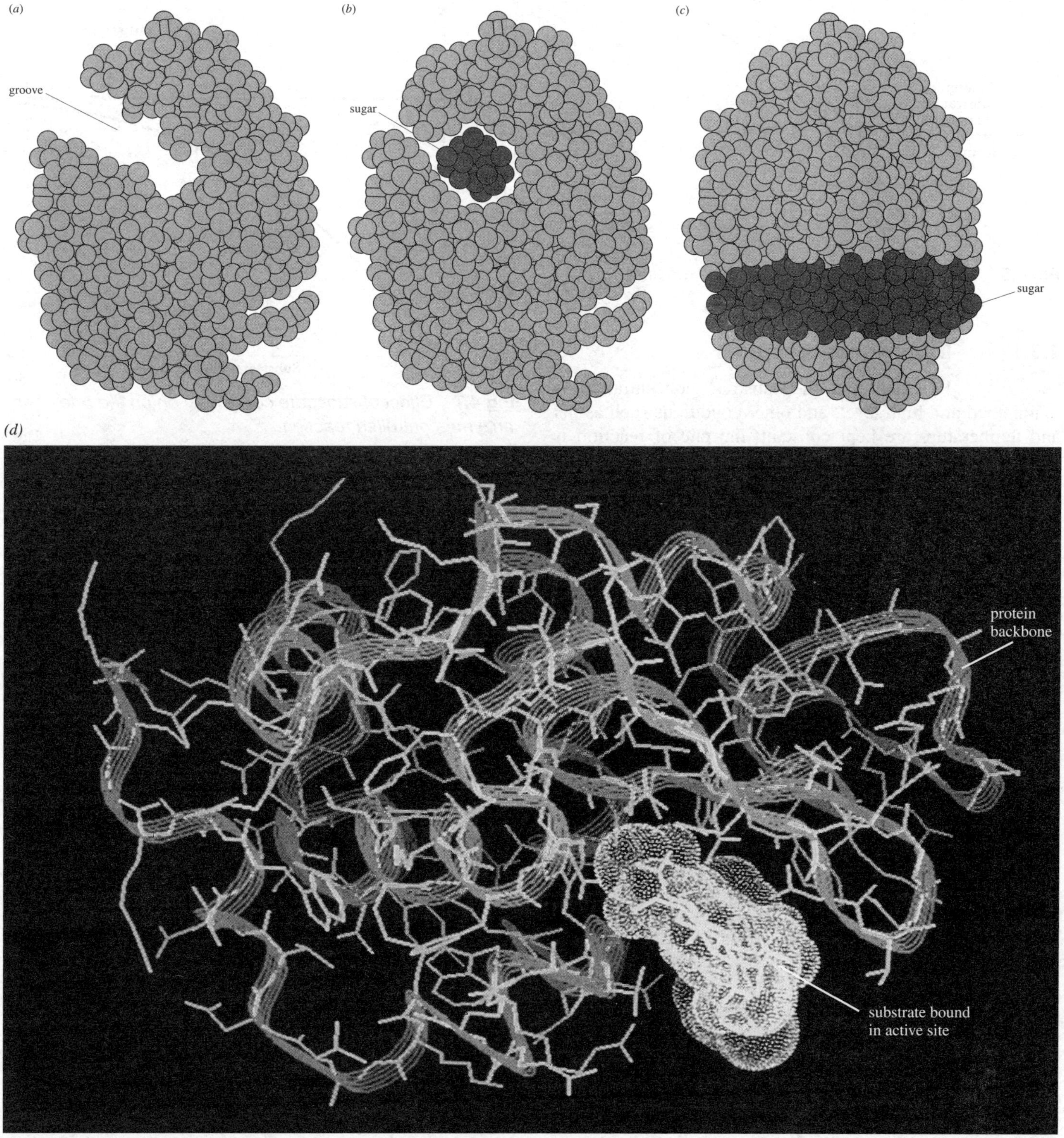

Fig 4.4 *How the enzyme lysozyme works. These computer-generated models show the tertiary structure of lysozyme before and after binding its substrate.* (a) *Side view of the groove-shaped active site which runs across the molecule.* (b) *Side view of active site with substrate molecule in place. Note the slight change in shape of the enzyme which has resulted from substrate binding. This is an example of 'induced fit', first proposed by Koshland in 1959. The substrate is a short chain of sugars which slides neatly into the groove and is split by the enzyme. These sugar chains are found in bacterial cell walls and their breakdown by the enzyme results in bacterial death – the cell explodes as a result of osmosis because the cell wall loses its rigidity. Lysozyme is a common enzyme found in human tears, saliva and mucus as a protective enzyme.* (c) *Front view of active site with substrate molecule in place.* (d) *Computer-generated model of lysozyme showing the substrate bound in the active site.*

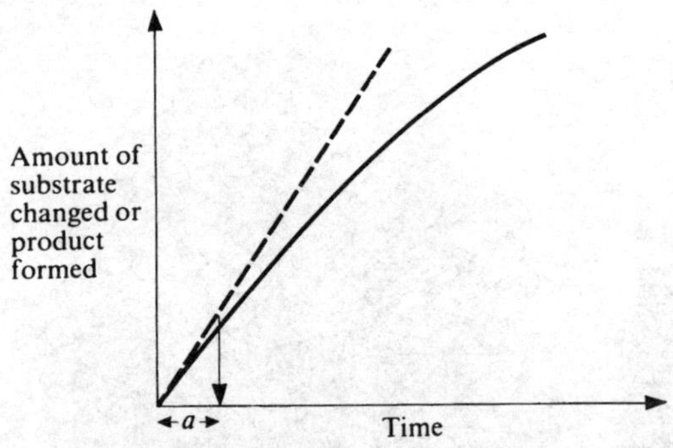

Fig 4.5 *The rate of an enzyme-controlled reaction.*

4.3.1 Enzyme concentration

Provided that the substrate concentration is maintained at a high level, and other conditions such as pH and temperature are kept constant, the rate of reaction is proportional to the enzyme concentration (fig 4.6). Normally reactions are catalysed by enzyme concentrations which are much lower than substrate concentrations. Thus as the enzyme concentration is increased, so will be the rate of the enzyme reaction.

4.3.2 Substrate concentration

For a given enzyme concentration, the rate of an enzyme reaction increases with increasing substrate concentration (fig 4.7). The theoretical maximum rate (V_{max}) is never quite obtained, but there comes a point when any further increase in substrate concentration produces no significant change in reaction rate. This is because at high substrate concentrations the active sites of the enzyme molecules at any given moment are virtually saturated with substrate. Thus any extra substrate has to wait until the enzyme/substrate complex has released the products before it may itself enter the active site of the enzyme.

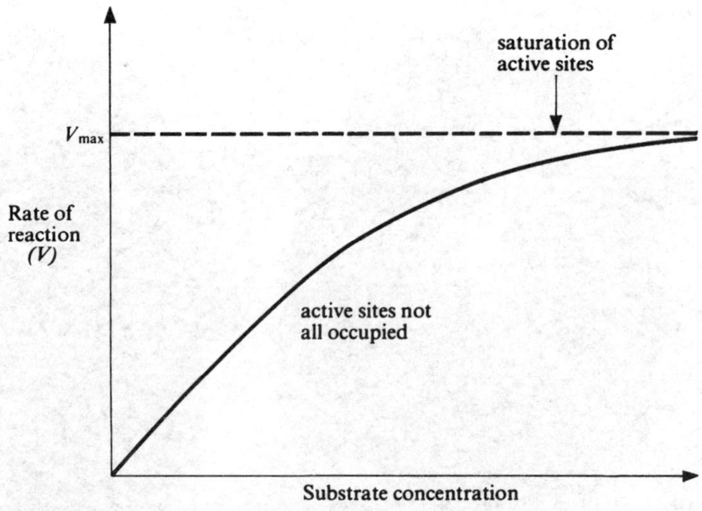

Fig 4.7 *Effect of substrate concentration on the rate of an enzyme-controlled reaction.*

4.3.3 Temperature

Heating increases molecular motion. Thus the molecules of the substrate and enzyme move more quickly and chances of their bumping into each other are increased. As a result there is a greater probability of a reaction occurring. The temperature that promotes maximum activity is referred to as the optimum temperature. If the temperature is increased above this level, then a decrease in the rate of the reaction occurs despite the increasing frequency of collisions. This is because the secondary and tertiary structures of the enzyme have been disrupted, and the enzyme is said to be denatured (fig 4.8). In effect, the enzyme unfolds and the precise structure of the active site is gradually lost. The bonds which are most sensitive to temperature change are hydrogen bonds and hydrophobic interactions.

Most mammalian enzymes have a temperature optimum of about 37–40 °C, but enzymes with higher optima exist. For

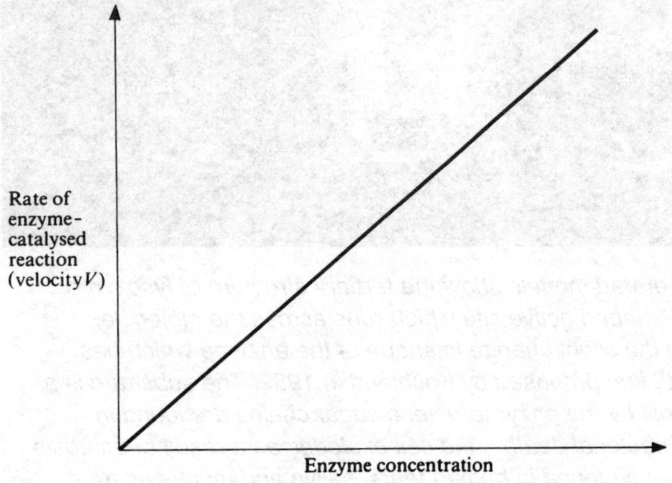

Fig 4.6 *Relationship between enzyme concentration and the rate of an enzyme-controlled reaction.*

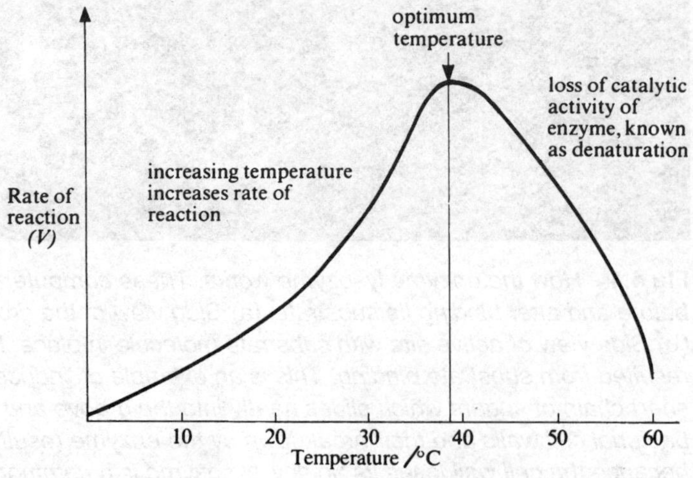

Fig 4.8 *Effect of temperature on the rate of an enzyme-controlled reaction.*

example, the enzymes of bacteria living in hot springs may have an optimum temperature of 70 °C or higher. Such enzymes have been used in biological washing powders for high temperature washes. If temperature is reduced to near or below freezing point, enzymes are **inactivated**, not denatured. They will regain their catalytic influence when higher temperatures are restored.

Today techniques of quick-freezing food are in widespread use as a means of preserving food for long periods. This not only prevents growth and multiplication of microorganisms, but also deactivates their digestive enzymes thus making it impossible for them to decompose food. The natural enzymes in the food itself are also inactivated. However, once frozen, it is necessary to keep the food at subzero temperatures until it is to be prepared for consumption.

4.1 Study fig 4.9 carefully and comment on the shapes of the curves given for the enzyme reaction at different temperatures.

Temperature coefficient, Q_{10}

The effect of temperature on the rate of a reaction can be expressed as the temperature coefficient, Q_{10}.

$$Q_{10} = \text{rate of reaction at } (x + 10)\ °\text{C}/\text{rate of reaction at } x\ °\text{C}$$

Over a range of 0–40 °C, Q_{10} for an enzyme-controlled reaction is 2. In other words, the rate of an enzyme-controlled reaction is doubled for every rise of 10 °C.

4.3.4 pH

Under conditions of constant temperature, every enzyme functions most efficiently over a particular pH range. Often this is a narrow range. The optimum pH is that at which the maximum rate of reaction occurs (fig 4.10 and table 4.1). When the pH is altered above or below this value, the rate of enzyme activity diminishes. As pH decreases, acidity increases and the concentration of H^+ ions increases. This increases the number of positive charges in the medium. Changes in pH alter the ionic charge of the acidic and basic groups and therefore disrupt the ionic bonding that helps to maintain the specific shape of the enzyme (section 3.5.3). Thus the pH change leads to an alteration of enzyme shape, including its active site. If extremes of pH are encountered by an enzyme, then it will be denatured.

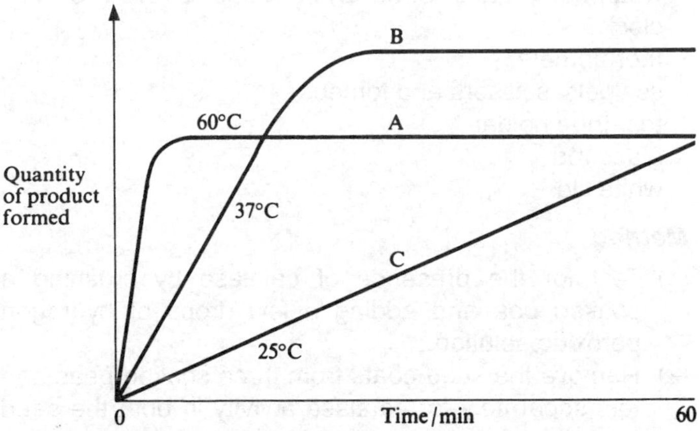

Fig 4.9 *Time course of an enzyme reaction at various temperatures.*

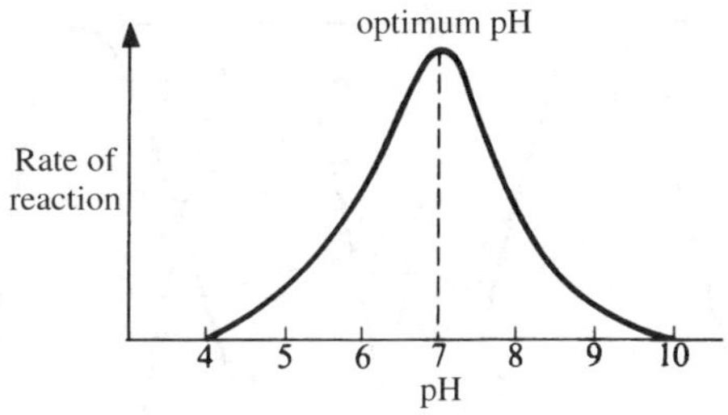

Fig 4.10 *Effect of pH on the rate of an enzyme-controlled reaction.*

Table 4.1 Optimum pH values for some enzymes.

Enzyme	*Optimum pH*
Pepsin*	2.00
Sucrase	4.50
Enterokinase	5.50
Salivary amylase	4.80
Catalase	7.60
Chymotrypsin	7.00–8.00
Pancreatic lipase	9.00
Arginase	9.70

* found in stomach with hydrochloric acid.

4.2 (*a*) In fig 4.11, what is the optimum pH for the activity of enzyme B?
(*b*) Give an example of an enzyme which could be represented by (i) activity curve A, (ii) activity curve B.
(*c*) Why does the enzyme activity of C decrease at pH values between 8 and 9?
(*d*) Why is pH control important in cells?
(*e*) 1 cm^3 of a catalase solution was added to hydrogen peroxide solution at different pH values and the time taken to collect 10 cm^3 of oxygen was measured. The results are given below.

pH of solution	*Time to collect gas/min*
4.00	20.00
5.00	12.50
6.00	10.00
7.00	13.60
8.00	17.40

Draw a graph of these results and comment on them.

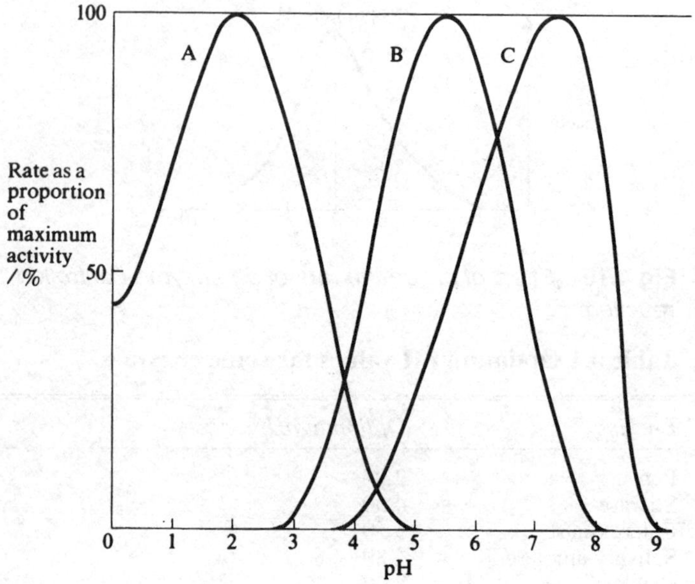

Fig 4.11 *The effect of pH on the activity of three enzymes, A, B and C.*

4.3.5 Practical work

Experiment 4.1: To determine the effect of enzyme concentration on the hydrolysis of sucrose by sucrase (invertase)

Materials

2% sucrose solution
1%, 0.75%, 0.5% sucrase (invertase) solutions
Benedict's reagent
12 test-tubes and rack
water baths at 38 °C and 100 °C
glass rods
stopclock
distilled water
labels
Bunsen burner

Method

(1) Add 2 cm^3 of clear blue Benedict's reagent to 2 cm^3 of clear colourless 1% sucrase solution. Heat the mixture in the water bath maintained at 100 °C for 5 min (Benedict's test).
(2) Repeat (1) using 2 cm^3 of clear colourless 2% sucrose solution and then 2 cm^3 of distilled water.
(3) Boil 5 cm^3 1% sucrase solution.
(4) Take 8 clean, dry test-tubes, label 1–8, and add 1 cm^3 Benedict's reagent to each.
(5) Add 5 cm^3 of 2% sucrose solution to a test-tube labelled S and place in the water bath maintained at 38 °C throughout the experiment.
(6) Add 5 cm^3 of 1% sucrase solution to a test-tube labelled E and place in the water bath at 38 °C.
(7) Leave both test-tubes and contents in the water bath for 5 min to allow them to equilibrate with their surroundings.
(8) Add the enzyme solution to the sucrose solution, invert the test-tube to thoroughly mix the two solutions.
(9) Immediately start the stopclock and replace the tube containing the reaction mixture in the water bath.
(10) Throughout the experiment agitate the mixture continuously to ensure thorough mixing.
(11) After 30 s of incubation remove 1 cm^3 of mixture and place in test-tube 1.
(12) Repeat this procedure every 30 s placing the samples in tubes 2–8 in turn.
(13) Heat tubes 1–8 in the water bath at 100 °C for 5 min. Note the time when the first positive reducing sugar test is obtained indicated by a brick-red precipitate.
(14) Repeat the experiment using the boiled enzyme from (3).
(15) Repeat the entire sequence/experiment twice using the 0.75% and 0.5% sucrase solutions.
(16) Record your observations and comment on your results.

Experiment 4.2: To investigate the distribution of catalase in a soaked pea, and to determine the effect of different temperatures on its activity

Catalase is an enzyme which catalyses the decomposition of hydrogen peroxide, liberating oxygen gas as shown by effervescence:

$$2H_2O_2 \xrightarrow{\text{catalase}} 2H_2O + O_2$$

Hydrogen peroxide is a toxic by-product of metabolism in certain plant and animal cells, and is efficiently removed by catalase, which is one of the fastest acting enzymes known.

Materials

a supply of soaked peas
hydrogen peroxide solution
test-tubes and rack
water baths at 40 °C, 60 °C, 70 °C, 80 °C and 100 °C
clock
thermometer
scalpels, scissors and forceps
test-tube holder
glass rod
white tile

Method

(1) Test for the presence of catalase by crushing a soaked pea and adding a few drops of hydrogen peroxide solution.
(2) Remove the seed coats from three soaked peas and test separately for catalase activity in both the seed coats and the cotyledons.
(3) Place two test-tubes containing distilled water in a water bath at 40 °C.

(4) Boil three whole peas in a test-tube and then place the boiled peas in one of the tubes in the water bath.
(5) Place three whole unboiled peas in the other test-tube in the water bath.
(6) Allow enough time for the peas to reach the temperature of the water bath (at least 10 min).
(7) Test each pea for catalase activity.
(8) Repeat the experiment at 50, 60, 70, 80 and 100 °C.
(9) Record your observations and comment on your results.

Experiment 4.3: To investigate the effect of different pH values on enzyme activity

Materials

Benedict's reagent
buffer solutions at pH 3, 5, 7, 9, 11
1% starch solution
water bath at 38 °C
Bunsen burner
asbestos mat
test-tube holder, test-tubes and rack
5 cm^3 graduated pipettes
thermometer
stopclock
distilled water
stock solution of salivary amylase (such as contained in saliva)

Method

(1) Rinse out the mouth with 5 cm^3 of distilled water and spit this out.
(2) Swill 10 cm^3 of distilled water round the mouth for 1 min and then collect this liquid.
(3) Make up the volume of salivary amylase to 40 cm^3 with distilled water.
(4) Test the salivary amylase, starch and buffer solutions for the presence of reducing sugar using Benedict's reagent.
(5) Label a test-tube pH 3 and add 2 cm^3 of starch solution.
(6) Add 2 cm^3 of buffer solution pH 3 to the same test-tube and mix the two solutions thoroughly.
(7) Boil at least 4 cm^3 of enzyme solution and place 4 cm^3 in a labelled test-tube.
(8) Add 4 cm^3 of unboiled enzyme solution to another labelled test-tube and place all three test-tubes in the water bath and allow the solutions to reach 38 °C (approximately 1 min).
(9) Place a small quantity of Benedict's reagent in each of 11 test-tubes and label them 1–11.
The following three stages must be carried out very quickly.
(10) When the solutions in the water bath have equilibrated, add the buffered starch solution to the unboiled enzyme solution.
(11) Mix the two solutions thoroughly by inverting the test-tube and replace the tube in the water bath.
(12) Start the stopclock and immediately remove a small quantity of reaction mixture (approximately the same volume as the Benedict's reagent) and place it in the test-tube labelled 1.
(13) Throughout the experiment the mixture must be shaken vigorously.
(14) After one minute of incubation remove a second, approximately equal, volume of the mixture and place it in test-tube 2.
(15) Repeat the removal of similar-sized samples of mixture at minute intervals for a further 9 min and place in test-tubes 3–11.
(16) Perform Benedict's tests on test-tubes 1–11 and note the time of incubation at which a positive result (a brick-red precipitate) is first achieved.
(17) Repeat the experiment using the boiled enzyme solution from (7).
(18) Repeat the entire experiment using each of the other buffer solutions.
(19) Plot a graph of time taken for hydrolysis to occur against pH and comment on your results.

4.4 Enzyme inhibition

A variety of small molecules exists which can reduce the rate of an enzyme-controlled reaction. They are called enzyme inhibitors. It is important to realise that inhibition is a normal part of the regulation of enzyme activity within cells. Many drugs and poisons also act as enzyme inhibitors. Inhibition may be competitive or non-competitive. Non-competitive inhibition may be reversible or non-reversible.

4.4.1 Competitive inhibition

This occurs when a compound has a structure which is sufficiently similar to that of the normal substrate to be able to fit into the active site. Normally it does not take part in a reaction but while it remains there it prevents the true substrate from entering the active site. The genuine substrate and the inhibitor therefore **compete** for a position in the active site, and this form of inhibition is called **competitive inhibition**. A characteristic feature of competitive inhibition is that if the substrate concentration is increased, the rate of reaction increases.

4.3 Why should the rate of reaction increase under these conditions?

A diagram illustrating the principle of competitive inhibition, and an example, is shown in fig 4.12.

The knowledge of competitive inhibition helps us to understand the effect of a group of antibiotics known as

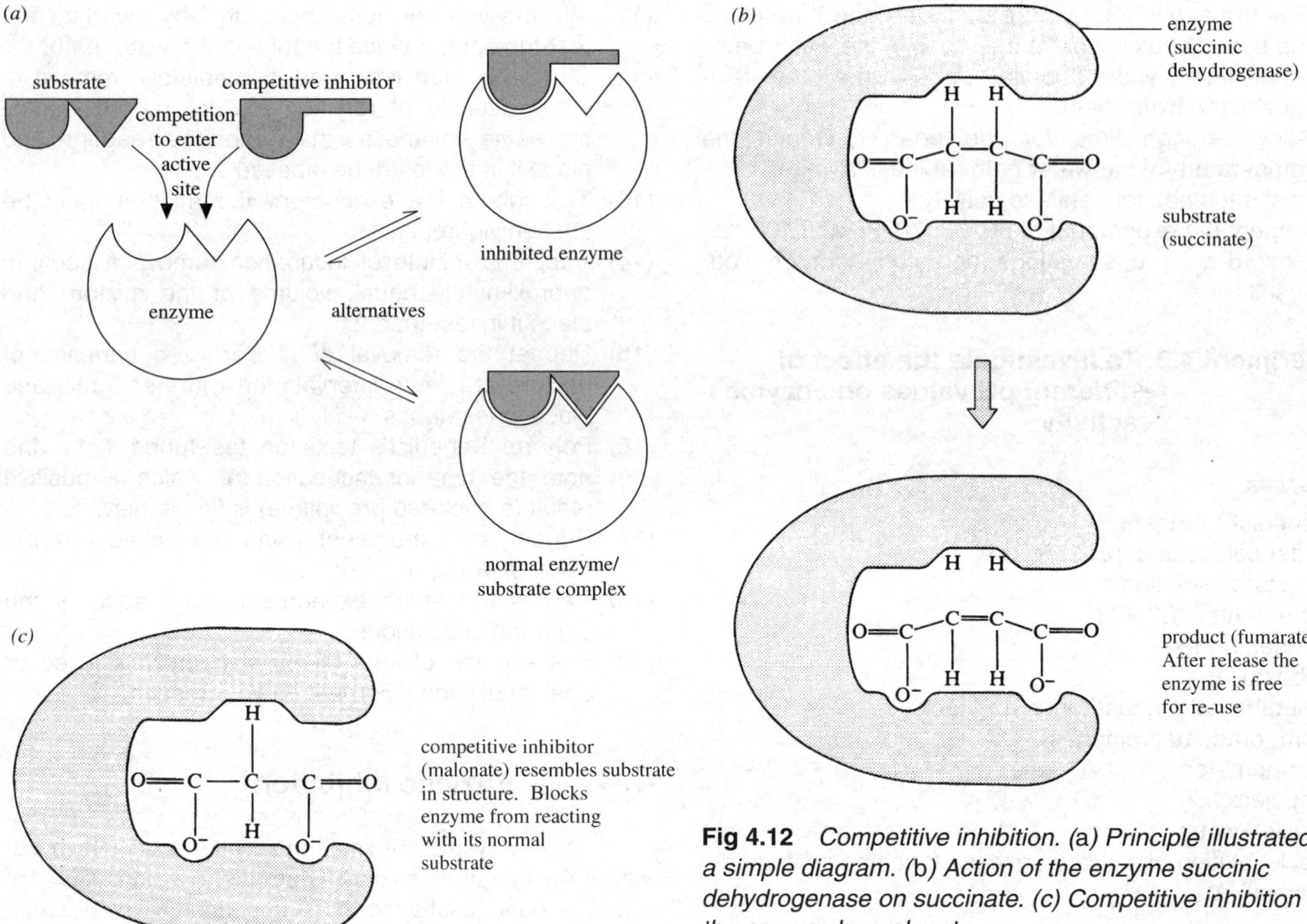

Fig 4.12 *Competitive inhibition. (a) Principle illustrated by a simple diagram. (b) Action of the enzyme succinic dehydrogenase on succinate. (c) Competitive inhibition of the enzyme by malonate.*

sulphonamides. Antibiotics destroy infectious microorganisms without damaging host tissues. Sulphonamides were the first antibiotics to be used and were developed during the 1930s. During the Second World War they were used extensively to prevent the spread of microbial infection in wounds. They are similar in structure to PAB (para-aminobenzoate), a substance essential for the growth of many pathogenic (disease-causing) bacteria. The bacteria require PAB for the production of folic acid, an important enzyme cofactor. Sulphonamides act by inhibiting an enzyme needed for the synthesis of folic acid from PAB.

Animal cells are insensitive to sulphonamides even though they require folic acid for some reactions. This is because they use pre-formed folic acid and do not possess the necessary metabolic pathway for making it.

4.4.2 Non-competitive reversible inhibition

This type of inhibitor has no structural similarity to the substrate and combines with the enzyme at a point other than its active site. It does not affect the ability of the substrate to bind with the enzyme, but it makes it impossible for catalysis to take place. The rate of reaction decreases with increasing inhibitor concentration. When inhibitor saturation is reached, the rate of the reaction will be almost nil. It is a characteristic of this type of inhibition that an increase in substrate concentration does not affect the rate of reaction, unlike with competitive inhibition.

4.4.3 Non-competitive irreversible inhibition

Some chemicals cause irreversible inhibition of enzymes. Two examples will be given.

Very small concentrations of chemical reagents such as the heavy metal ions mercury (Hg^{2+}), silver (Ag^{+}) and arsenic (As^{+}), or certain iodine-containing compounds completely inhibit some enzymes. They combine permanently with sulphydryl (–SH) groups (fig 4.13). These may be in the active site or elsewhere. Either way, the change in structure of the enzyme makes it ineffective as a catalyst. The change may cause the protein of the enzyme molecule to precipitate.

Another example of irreversible inhibition is provided by the nerve gas DFP (diisopropylfluorophosphate) designed for use in warfare. It combines with the amino acid serine at the active site of the enzyme acetylcholinesterase. This enzyme deactivates the neurotransmitter substance

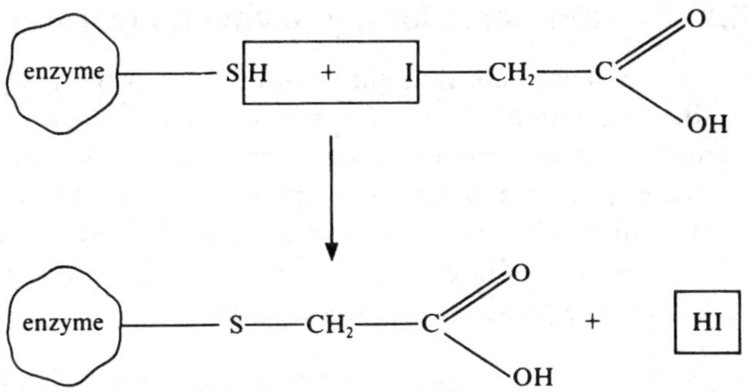

Fig 4.13 *Irreversible inhibition of an enzyme by iodoacetic acid. The iodine reacts with sulphydryl groups.*

acetylcholine. Neurotransmitters are needed to continue the passage of nerve impulses from one nerve cell to another across a synaptic gap (chapter 17). When the impulse has been transmitted, acetylcholinesterase functions to deactivate acetylcholine almost immediately by breaking it down. If acetylcholinesterase is inhibited, acetylcholine accumulates and nerve impulses cannot be stopped, causing prolonged muscle contraction. Paralysis occurs and death may result since the respiratory muscles are among those affected. Some insecticides currently in use, including those known as organophosphates (such as parathion), have a similar effect on insects, and can also cause harm to the nervous and muscular systems of humans who are overexposed to them.

> **4.4** Suggest why it is that substrate concentration has no effect on non-competitive inhibition.

4.4.4 Allosteric enzymes

One of the commonest ways of regulating metabolic pathways in cells is by means of allosteric enzymes. These are enzymes which are 'designed' to change shape (*allo*, different; *steric*, shape). They are regulated by compounds which act as non-competitive inhibitors. These compounds bind to the enzyme at specific sites well away from the active site. They modify enzyme activity by causing a reversible change in the structure of the enzyme's active site. This in turn affects the ability of the substrate to bind to the enzyme (unlike non-competitive reversible inhibition (section 4.4.2)). Compounds of this nature are called **allosteric inhibitors**. Allosteric inhibition is shown in fig 4.14.

An example of this is provided by one of the reactions of glycolysis, the series of reactions that forms the first part of cell respiration. The purpose of cell respiration is to produce ATP. When ATP is at a high concentration, it inhibits one of the enzymes of glycolysis allosterically. However, when cell metabolism increases and more ATP is used up, the overall concentration of ATP decreases and the pathway once again comes into operation because the inhibitor, ATP, has been removed. This is also an example of end-product inhibition.

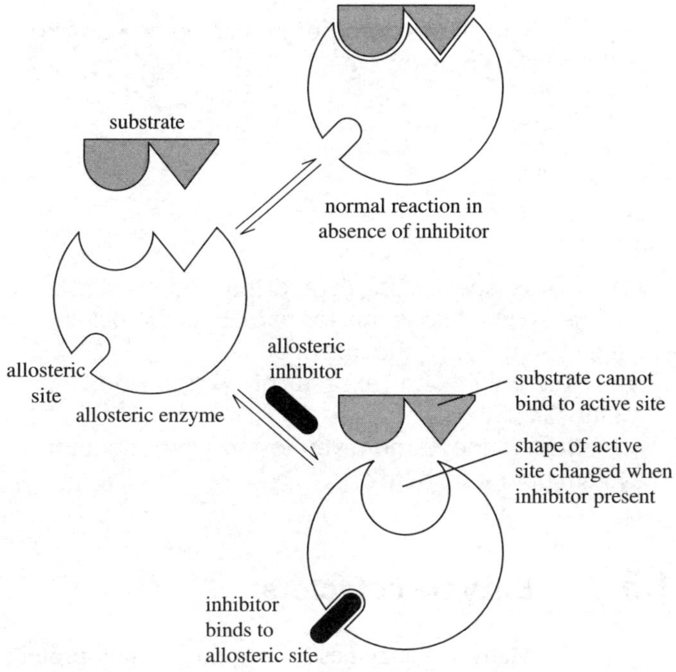

Fig 4.14 *Allosteric inhibition.*

End-product inhibition (negative feedback inhibition)

When the end product of a metabolic pathway begins to accumulate, it may act as an allosteric inhibitor of the enzyme controlling the first step of the pathway. Thus the product starts to switch off its own production as it builds up. The process is self-regulatory. As the product is used up, its production is switched back on again. This is called end-product inhibition and is an example of a **negative feedback** mechanism (fig 4.15) (see also chapter 19).

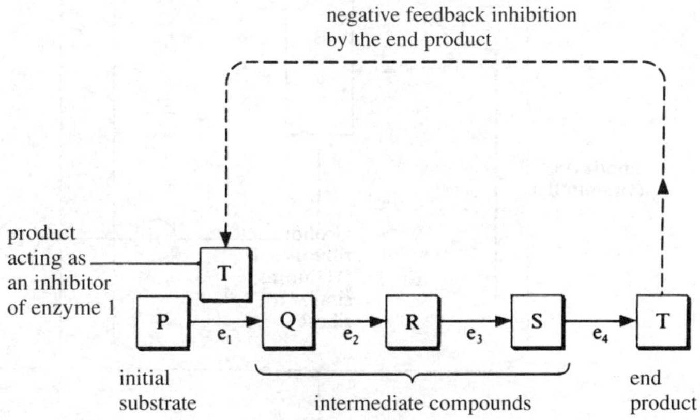

Fig. 4.15 *End-product inhibition. e_1–e_4 are different enzymes of a metabolic pathway.*

4.5 Consider the branching metabolic pathway shown below:

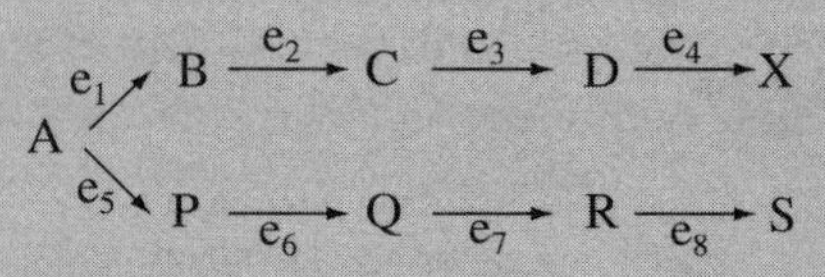

(*a*) If e_1 is specific for A, and the end product X inhibits e_1, what does this tell you about the binding sites of A and X on the enzyme?
(*b*) How might an excess of X regulate the metabolic pathway?
(*c*) What is the name given to the type of control system operating here?

4.5 Enzyme cofactors

Many enzymes require non-protein components called cofactors for their efficient activity. They were discovered as substances that had to be present for enzyme activity, even though, unlike enzymes, they were stable at relatively high temperatures. Cofactors may vary from simple inorganic ions to complex organic molecules, and may either remain unchanged at the end of a reaction or be regenerated by a later process. There are three recognised types of cofactor: inorganic ions, prosthetic groups and coenzymes, which will be examined in the following sections.

4.5.1 Inorganic ions (enzyme activators)

These are thought to mould either the enzyme or the substrate into a shape that allows an enzyme/substrate complex to be formed, hence increasing the chances of a reaction occurring between them and therefore increasing the rate of reaction catalysed by that particular enzyme. For example, salivary amylase activity is increased in the presence of chloride ions.

4.5.2 Prosthetic groups (for example FAD, haem)

If the cofactor is tightly bound to the enzyme on a permanent basis it is known as a **prosthetic group** (from the Greek *prosthesis*, meaning 'addition'). Prosthetic groups are organic molecules. They assist the catalytic function of their enzymes, as in flavine adenine dinucleotide (FAD). This contains riboflavin (vitamin B_2), the function of which is to accept hydrogen (fig 4.16). FAD is concerned with cell oxidation pathways and is part of the respiratory chain in respiration (chapter 9).

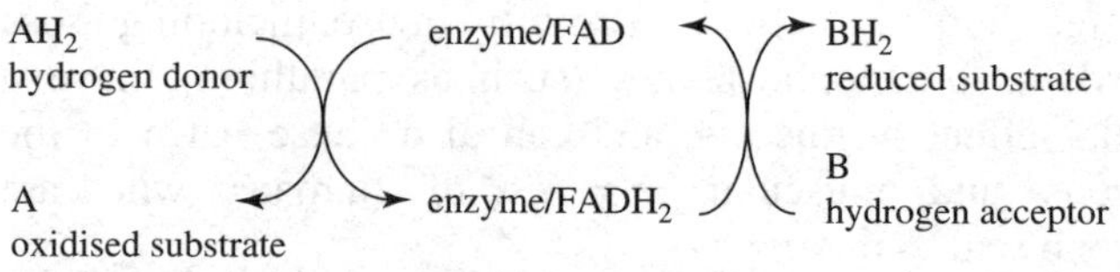

Net effect: 2H transferred from A to B. One enzyme acts as a link between A and B. Both AH_2 and B fit into the active site and FAD passes H_2 from one to the other.

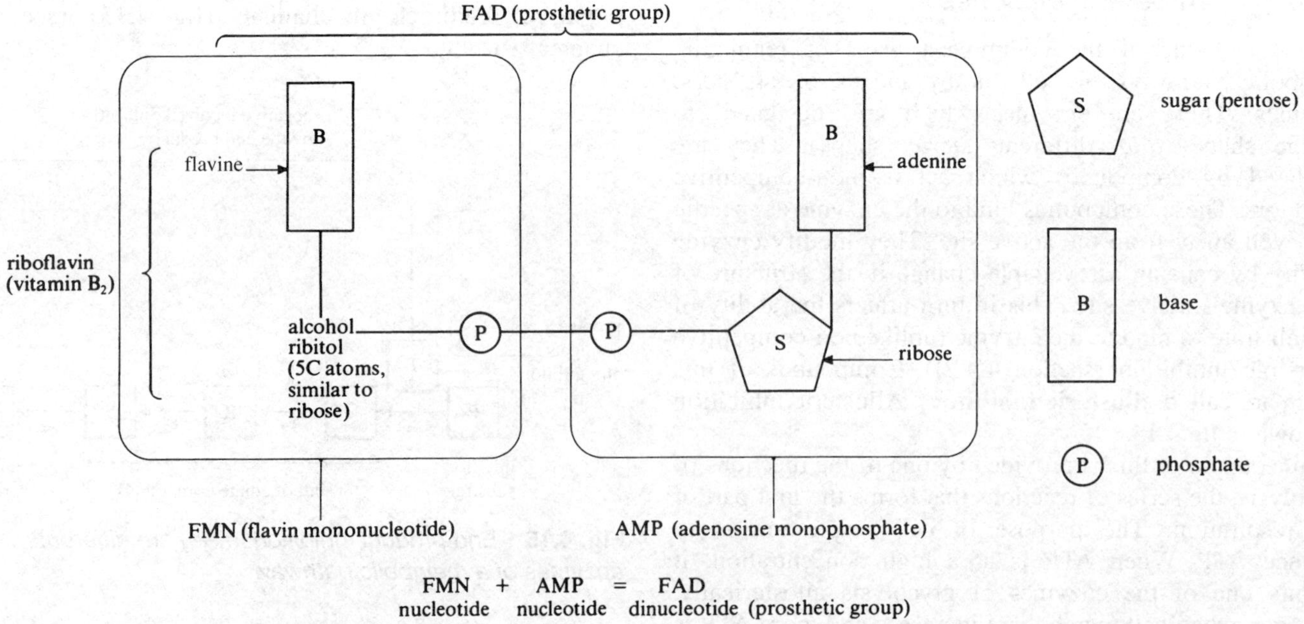

Fig 4.16 *Structure of FAD (flavine adenine dinucleotide), a prosthetic group. FAD is a dinucleotide, formed by the joining of two nucleotides. Nucleotides consist of a sugar, base and phosphate. The two nucleotides in this case are FMN and AMP. Note that riboflavin (vitamin B_2) is part of the structure of one of the nucleotides (FMN), explaining the need for this vitamin in the diet.*

Haem

Haem is an iron-containing prosthetic group. It has the shape of a flat ring (a '**porphyrin ring**' as is found in chlorophyll) with an iron atom at its centre. It has a number of biologically important functions.

Electron carrier. Haem is the prosthetic group of cytochromes (see respiratory chain, chapter 9), where it acts as an electron carrier. In accepting electrons the iron is reduced to Fe(II); in handing on electrons it is oxidised to Fe(III). In other words it takes part in oxidation/reduction reactions by reversible changes in the valency of the iron.

Oxygen carrier. Haemoglobin and myoglobin are oxygen-carrying proteins that contain haem groups. Here the iron remains in the reduced, Fe(II) form (see chapter 14).

Other enzymes. Haem is found in catalases and peroxidases, which catalyse the decomposition of hydrogen peroxide into water and oxygen. It is also found in a number of other enzymes.

4.5.3 Coenzymes (for example NAD, NADP, coenzyme A, ATP)

Like prosthetic groups, coenzymes are organic molecules which act as cofactors, but unlike prosthetic groups they do not remain attached to the enzyme between reactions. All coenzymes are derived from vitamins.

NAD (nicotinamide adenine dinucleotide) (fig 4.17)

This is derived from the vitamin nicotinic acid (niacin) and can exist in both a reduced and an oxidised form. In the oxidised state it functions as a hydrogen acceptor.

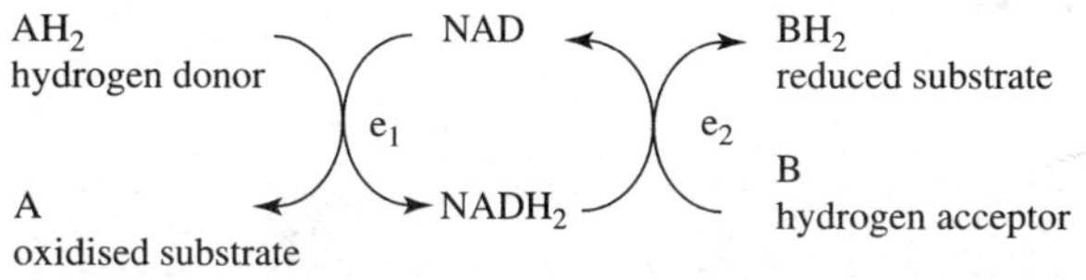

where e_1 and e_2 are two different dehydrogenase enzymes.

Net effect: 2H transferred from A to B. Here the coenzyme acts as a link between two different enzyme systems e_1 and e_2.

4.6 Summarise the characteristic properties of enzymes.

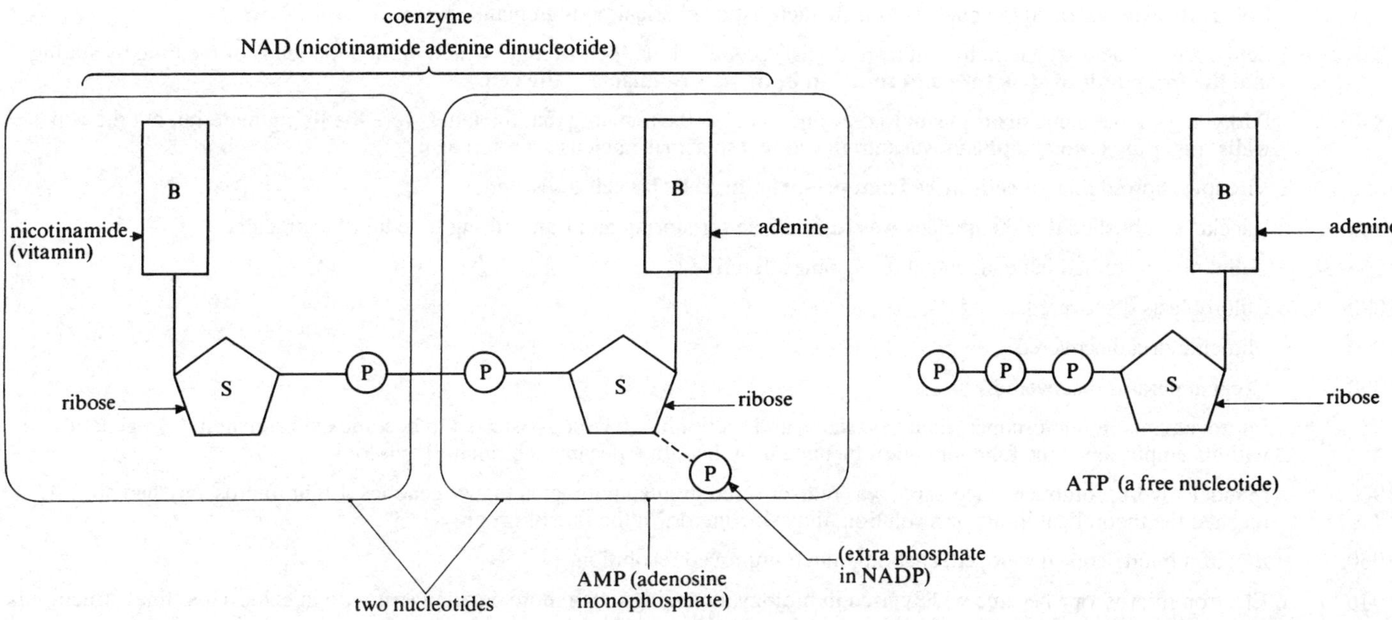

Fig 4.17 *Structure of the coenzyme NAD (nicotinamide adenine dinucleotide) and NADP (NAD with an extra phosphate). NAD and NADP are dinucleotides (see fig 4.16). Note that nicotinamide, also known as niacin, is a vitamin and is part of the structure of one of the nucleotides. AMP is closely related to ATP, the latter having two more phosphate groups. ATP is the energy carrier in cells and is made during cell respiration.*

Chapter Five

Cells

5.1 The cell concept

One of the most important concepts in biology is that **the basic unit of structure and function in living organisms is the cell**. This is known as the **cell theory** and was proposed jointly by two scientists, namely Schleiden, a Belgian botanist, in 1838 and Schwann, a German zoologist, in 1839. The discovery was due to the fact that during the nineteenth century there were dramatic improvements in the quality of lenses for use in microscopy and this in turn led to great interest in the structure of living organisms. Other important discoveries were made during the century, as shown in table 5.1. An essential part of the cell theory is the idea, first proposed in 1855, that **new cells only come from pre-existing cells.**

5.1.1 Why cells?

A cell can be thought of as a bag of chemicals which is capable of surviving and replicating itself. Inside the bag, the chemicals differ in various ways from those outside the bag. Without a barrier between the bag and the

Table 5.1 Some historically important events in cell biology.

1590	Jansen invented the **compound microscope**, which combines two lenses for greater magnification.
1665	Robert Hooke, using an improved compound microscope, examined cork and used the term **'cell'** to describe its basic units. He thought the cells were empty and the walls were the living material.
1650–1700	Antony van Leeuwenhoeck, using a good quality simple lens (mag. × 200), observed nuclei and unicellular organisms, including bacteria. In 1676, bacteria were described for the first time as **'animalcules'**.
1700–1800	Further descriptions and drawings published, mainly of plant tissues, although the microscope was generally used as a toy.
1827	Dolland dramatically improved the quality of lenses. This was followed by a rapid spread of interest in microscopy.
1831–3*	Robert Brown described the nucleus as a characteristic spherical body in plant cells.
1838–9*	Schleiden (a botanist) and Schwann (a zoologist) produced the **'cell theory'** which unified the ideas of the time by stating that **the basic unit of structure and function in living organisms is the cell**.
1840*	Purkyne gave the name **protoplasm** to the contents of cells, realising that the latter were the living material, not the cell walls. Later the term **cytoplasm** was introduced (cytoplasm + nucleus = protoplasm).
1855*	Virchow showed that **all cells arise from pre-existing cells by cell division**.
1866	Haeckel established that the nucleus was responsible for storing and transmitting hereditary characters.
1866–88	Cell division studied in detail and chromosomes described.
1880–3	Chloroplasts discovered.
1890	Mitochondria discovered.
1898	Golgi apparatus discovered.
1887–1990	Improvements in microscopes, fixatives, stains and sectioning. Cytology† started to become experimental. Cytogenetics‡, with its emphasis on the functioning of the nucleus in heredity, became a branch of cytology.
1900	Mendel's work, forgotten since 1865, was rediscovered giving an impetus to cytogenetics. Light microscopy had almost reached the theoretical limits of resolution, thus slowing down the rate of progress.
1930s	Electron microscope developed, enabling much improved resolution.
1946 to present	Electron microscope became widely used in biology, revealing much more detailed structure in cells. This 'fine' structure is called **ultrastructure**.

* significant events in the development of the cell concept. † cytology – the study of cells, especially by microscopy.
‡ cytogenetics – the linking of cytology with genetics, mainly relating structure and behaviour of chromosomes during cell division to results from breeding experiments.

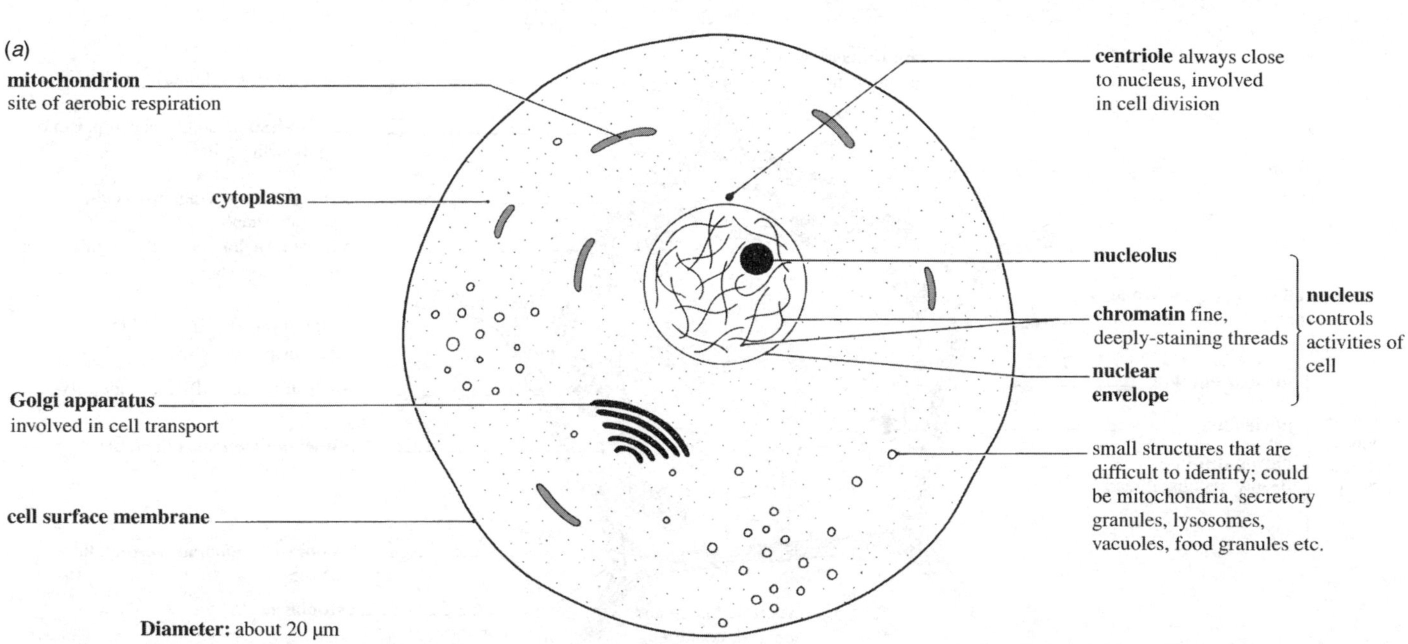

environment the chemicals would mix freely by diffusion and the differences could not be maintained. Life could not exist. The barrier is a very thin membrane. This acts rather like a border control between two countries, controlling the traffic of molecules into and out of the cell. *All* living cells are surrounded by a membrane. It is known as a **cell surface membrane** to distinguish it from any membranes that occur inside the cell. The way in which it controls exchange between the cell and its environment is discussed in section 5.9.8.

5.2 Cells as seen with the light microscope

By the end of the nineteenth century most of the structures visible with a light microscope had been discovered. (A light microscope is one which uses light as a source of radiation.) The cell could be described as a small unit of living protoplasm, always surrounded by a cell surface membrane and sometimes, as in the case of plants, by a non-living wall. The most conspicuous structure in the cell is the **nucleus**, which contains a deeply staining material known as **chromatin** (meaning coloured material). This is the loosely coiled form of chromosomes. Chromosomes appear as thread-like structures just before nuclear division. They contain **DNA**, the genetic material. DNA controls the cell's activities and can replicate itself so that new cells can form.

Figs 5.1*a* and 5.2*a* show the structure of generalised animal and plant cells as seen with a light microscope. (A generalised cell is one which shows *all* the typical features found in a cell.) The only structures shown which had not been discovered by the end of the nineteenth century are lysosomes. Examples of particular types of cell are shown in figs 5.1*b* and 5.2*b* – see also fig 6.16, epithelial cells from the small intestine, and fig 6.2, plant parenchyma cell.

(*b*)

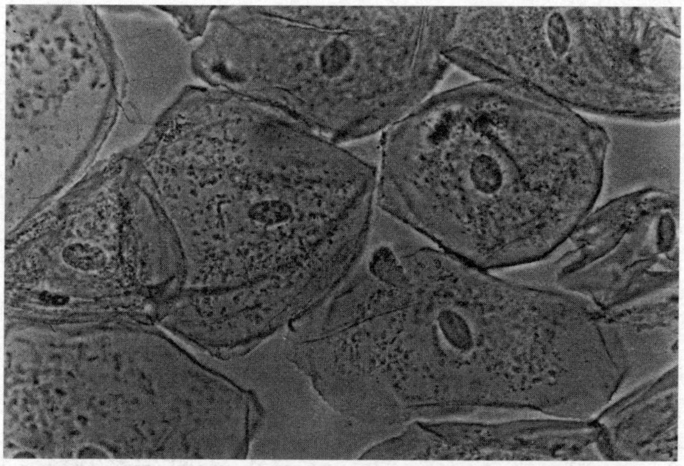

Fig 5.1 (a) *Generalised animal cell as seen with a light microscope.* (b) *Cells from the lining of the human cheek showing typical characteristics of an animal cell. Each cell contains a central nucleus surrounded by cytoplasm containing many organelles such as mitochondria (x400).*

The living material between the nucleus and the cell surface membrane is known as **cytoplasm**. This contains a variety of organelles. An **organelle** is a **distinct part of a cell which has a particular structure and function**. The organelles are described later (figs 5.12 and 5.13, and section 5.10). The only organelle found in animal cells which is absent from plant cells is the centriole. Basically, plant cells are very similar to animal cells but have more structures. The chief differences are the presence in plant cells of:

- a relatively rigid **cell wall** outside the cell surface membrane; pores containing fine threads known as **plasmodesmata** link the cytoplasm of neighbouring cells through the cell walls;

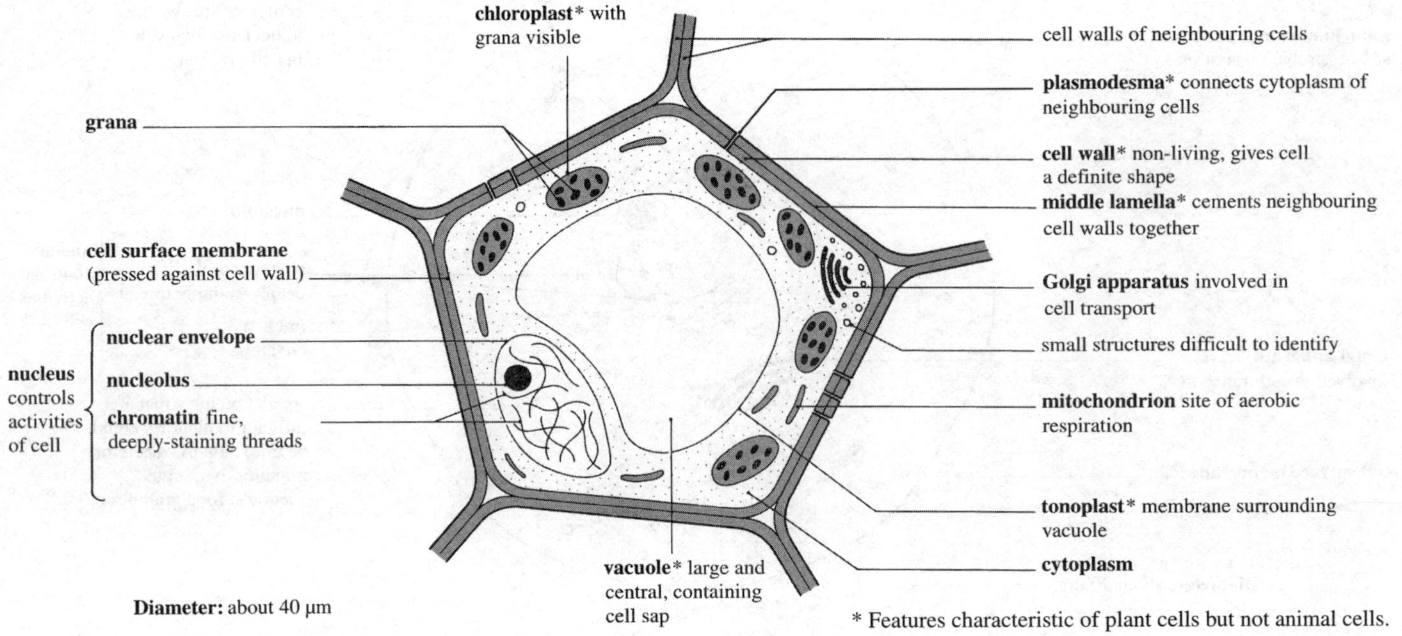

Fig 5.2 *Generalised plant cell as seen with a light microscope.*

- **chloroplasts** in photosynthetic plant cells;
- a **large central vacuole**: animal cells may have small vacuoles such as phagocytic vacuoles (figs 5.22 and 5.31).

Use of the light microscope is described at the end of this chapter (section 5.11).

5.3 Prokaryotes and eukaryotes

As described in chapter 2, there are two fundamentally different types of cell, the prokaryote cell and the eukaryote cell. In prokaryotes the DNA lies free in the cytoplasm in a region known as the **nucleoid**. There is no true nucleus. In eukaryotes the DNA is found inside a nucleus, a structure which is surrounded by two membranes, the **nuclear envelope**. The DNA is also associated with protein to form **chromosomes**. The differences between prokaryote and eukaryote cells are summarised in chapter 2 (table 2.2 and section 2.3).

5.4 Compartments and division of labour

Eukaryotic cells are far larger and more complex than prokaryotic cells. They contain many organelles. The eukaryotic cell has often been compared to a factory where, although different machines and people have different jobs, all are working together with one purpose. Efficiency is improved by '**division of labour**', the sharing out of jobs. In the cell each organelle has its own role, involving its own specialised structure and chemistry. The mitochondrion, for example, is the powerhouse of cells, providing energy in the form of ATP from the specialised reactions of respiration. It has a particular structure which enables it to do this efficiently. The cell as a whole is, in effect, divided up into compartments. This **compartmentation** is often achieved by membranes. Most organelles are surrounded by membranes so that, just as the cell surface membrane controls exchange between the cell and its environment, each membrane-bound organelle can have its own particular unique set of chemicals and chemical reactions. The electron microscope reveals even more structural organisation than the light microscope, as we shall see in section 5.8.

5.5 Units of measurement

Before proceeding further, it is useful to remind yourself of just how small cells are, and of the units of measurement which we have to use in describing them. Table 5.2 summarises the most useful units of measurement. Fig 5.3 shows bacteria on the surface of the sharp end of a pin, which measures about 100 μm in diameter. (The symbol μm is a micrometre.) Between 50 and 100 μm is about the lowest limit of what is visible unaided with the human eye. A very fine human hair is about 30 μm in diameter. Eukaryotic cells vary greatly in size (the largest algal cell is 50 mm in diameter!) but the average animal cell is about 20 μm in diameter compared with about 40 μm for an average plant cell. The average mitochondrion and bacterium are about 1 μm in diameter (a convenient reference size to try to remember). The smallest cell organelles, ribosomes, are about 20 nm in diameter; a DNA molecule is 2 nm in diameter and the smallest atom, a hydrogen atom, is 0.04 nm in diameter.

Table 5.2 Units of measurement used in cell biology.

Unit	*Symbol*	*Fraction of a metre*	*Also*
millimetre	mm	one thousandth, 10^{-3} m	
micrometre	μm*	one millionth, 10^{-6} m	one thousandth of a millimetre
nanometre	nm	one thousand millionth, 10^{-9} m	one thousandth of a micrometre

* μ is the Greek letter mu.
The metre, symbol m, is the internationally agreed basic unit of length.

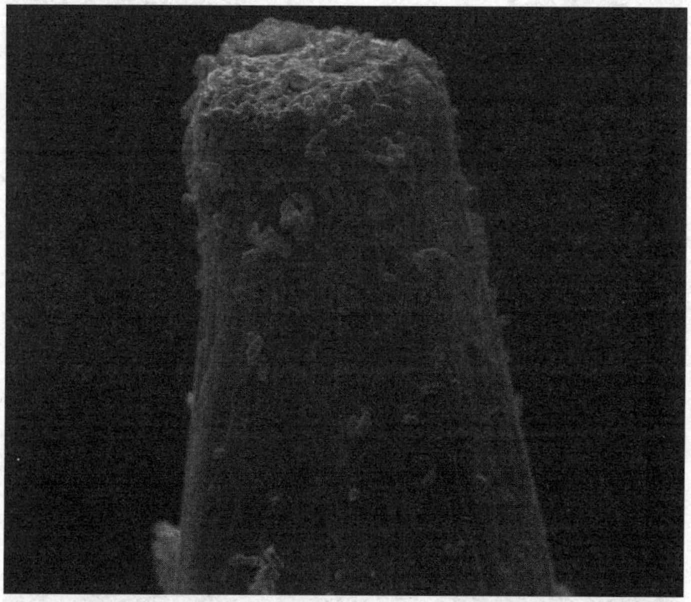

Fig 5.3 *Scanning electron micrograph of bacterial cells on the point of a pin.*

5.6 Electron microscopy

5.6.1 Electron microscopes

By the early 1900s progress on understanding cell structure was limited by the fact that no matter how good the quality of a light microscope, its maximum magnification is limited to about ×1500. This is due to the nature of light itself. Light is a form of electromagnetic radiation, as shown in fig 5.4. One of the properties of light is that it can behave like a series of waves. Wavelengths between about 400 nm (violet) and about 700 nm (red) cause a response in the human eye. However, there is a continuous spectrum of radiation of different wavelengths known as the **electromagnetic spectrum**, of which visible light is only a small part (fig 5.4). All the waves travel at the speed of light, but the shorter the wavelength, the greater the energy that the waves carry. It is not possible to view objects smaller than half the wavelength of the radiation used by the viewer. This is because the object has to be large enough to interfere with the passage of the waves. The smallest object we can view using visible light is therefore about 200 nm in diameter, half the size of the wavelength of violet light. Remembering the dimensions we noted in the previous section, it becomes clear why structures like mitochondria (1 μm, or 1000 nm) are only just visible as granules inside cells. Some organelles, such as ribosomes, are completely invisible using a light microscope. If we want to know more about the detailed structure of mitochondria, ribosomes and the other cell components, then light microscopes are inadequate.

With these limitations in mind, scientists deliberately set out to invent a different kind of microscope, one which used radiation of a much shorter wavelength. At one stage, an X-ray microscope was built, but the best choice turned out to be the electron microscope. Here the radiation used

Fig 5.4 *The electromagnetic spectrum. Visible light makes up only a small part of the spectrum. (Waves are not drawn to scale.)*

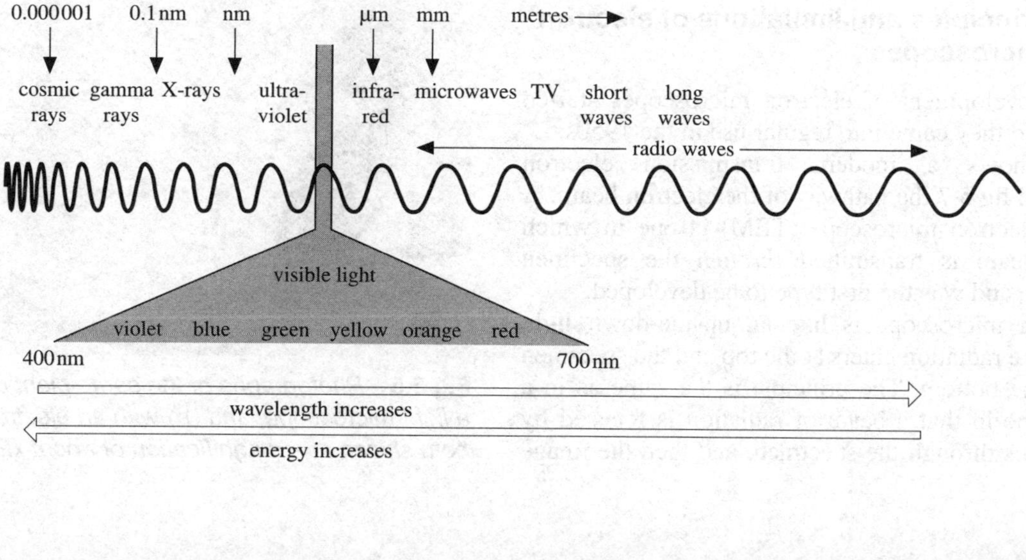

is electrons. Electrons are negatively charged particles which can be found in orbit around the nuclei of atoms. Under certain circumstances they can behave as waves. They have two great advantages over light. Firstly, they have extremely short wavelengths, about the same as X-rays (fig 5.4). Secondly, because they are negatively charged, a beam of electrons can easily be focused through a specimen using electromagnets. This involves bending the beam of electrons just as glass lenses are used to bend light.

With the electron microscope, magnifications up to ×250 000 are commonly obtained with biological material. With some materials even higher magnifications are possible, and even individual atoms have now been seen.

5.6.2 Resolution and magnification

The ability to distinguish between two separate objects is known as **resolution**. If two separate objects cannot be resolved, they will be seen as one object. Resolution is not the same as **magnification**. If you take a photograph and keep on magnifying it (enlarging it), you will not eventually be able to see atoms. Magnification can be increased, *but* the resolution of the photograph stays the same. We enlarge photographs in order to see them more clearly, but if we go too far the picture breaks up into separate blurred dots. Another way of explaining the difference between resolution and magnification is to study a picture of a cell taken with a light microscope and an electron microscope at the same magnification (fig 5.5). Note the difference in clarity between the two pictures in fig 5.5. The resolution is much greater in the electron microscope. The shorter wavelengths of electrons are said to have greater resolving power than those of light.

The resolution of an electron microscope is about 0.5 nm in practice, compared with 200 nm for the light microscope. This does not mean that electron microscopes are better. They are used for different jobs. Light microscopes are still important for getting an overall view of cells or tissues, and preparation of material for them is much quicker and easier. They can also be used to view living material, which is not possible with an electron microscope.

5.6.3 Principles and limitations of electron microscopes

Development of electron microscopes started in the 1930s and they came into regular use in the 1950s.

Fig 5.6 shows a modern transmission electron microscope and fig 5.7 the pathway of the electron beam. A transmission electron microscope (TEM) is one in which the electron beam is transmitted *through* the specimen before viewing, and was the first type to be developed.

The electron microscope is like an upside-down light microscope. The radiation enters at the top and the specimen is viewed at the bottom. The principle is the same as in a light microscope in that a beam of radiation is focused by condenser lenses through the specimen, and then the image

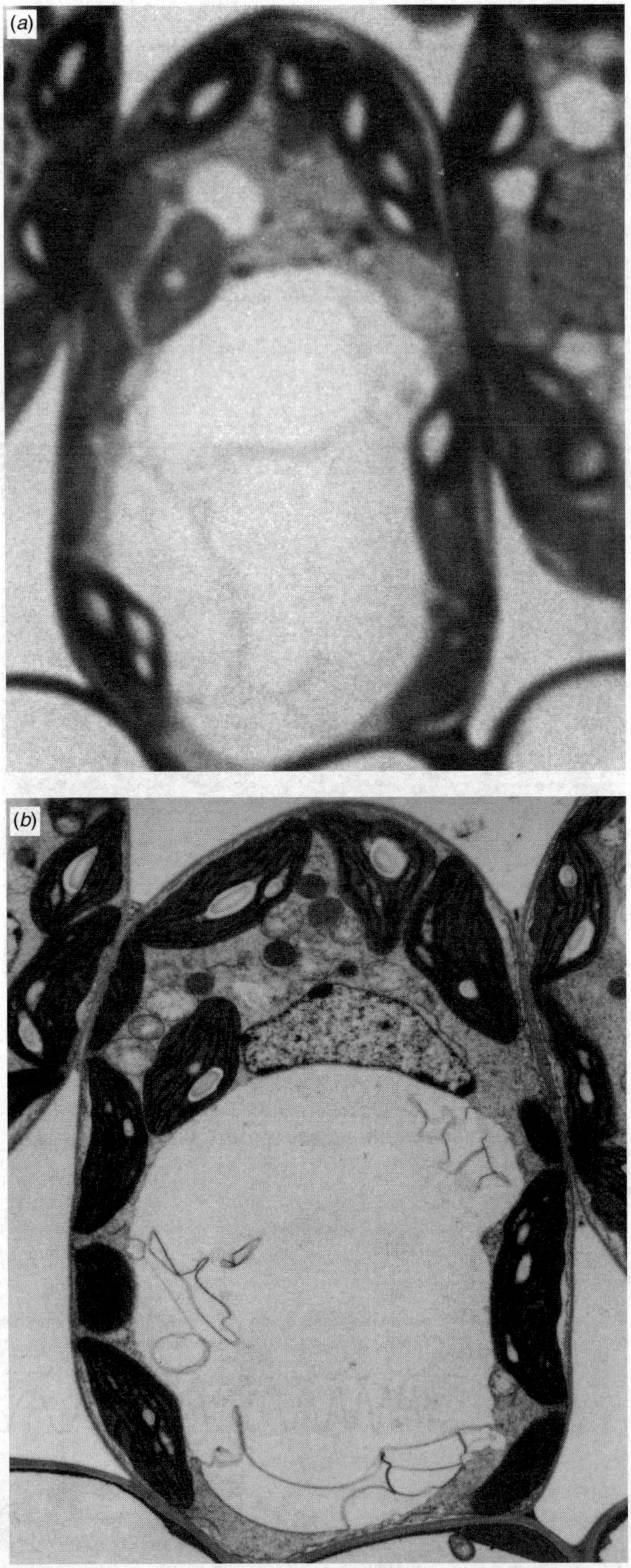

Fig 5.5 *Photographs of the same plant cells seen* (a) *with a light microscope, and* (b) *with an electron microscope, both shown at a magnification of about x2500.*

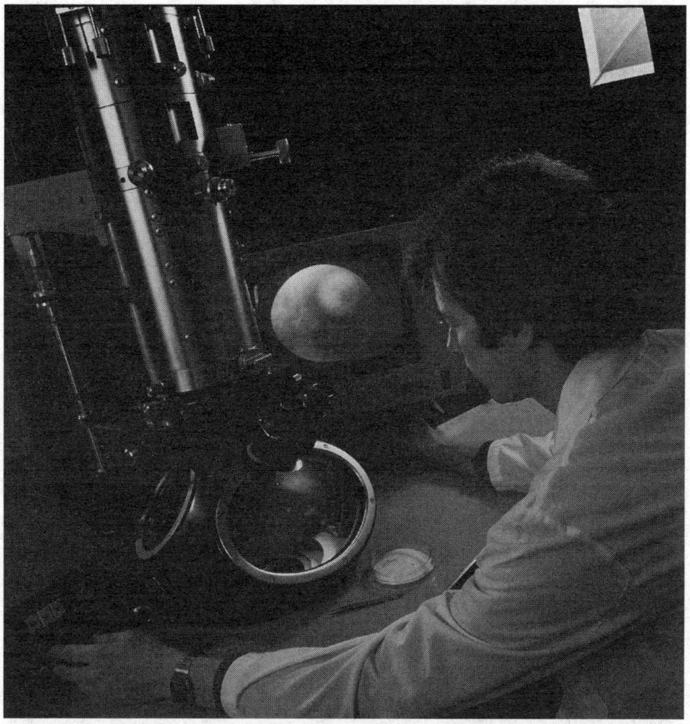

Fig 5.6 *A modern transmission electron microscope.*

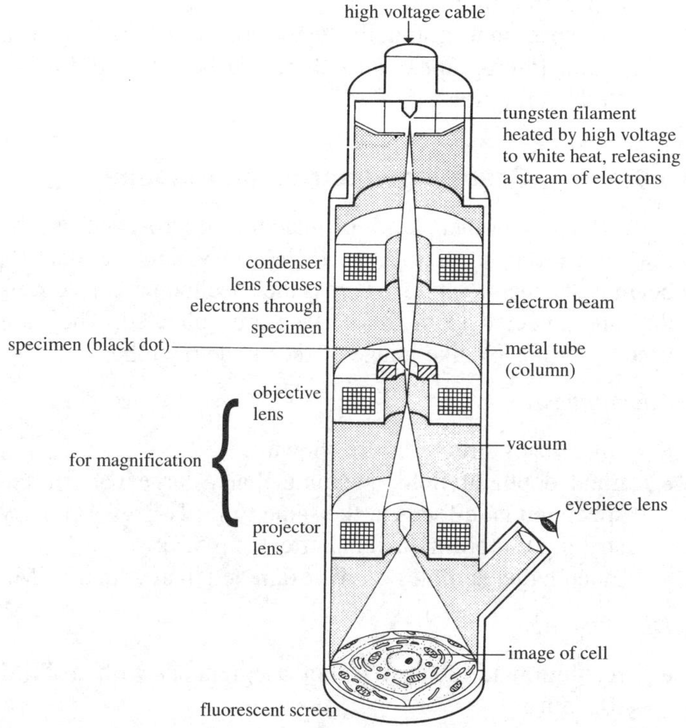

Fig 5.7 *Pathway of the electron beam in the transmission electron microscope.*

is magnified by further lenses. Table 5.3 summarises some of the similarities and differences. A high voltage, such as 50 000 V, is passed through a tungsten filament, like the filament of a light bulb, at the top of the column. The white hot filament releases a stream of electrons, kicked out of their orbits by the high voltage. Electromagnets focus the beam. The inside of the column has to be kept under a high vacuum, otherwise the electrons would collide with air molecules and be scattered. Only very thin sections (slices) of material or very small particles can be observed, because electrons are easily absorbed by larger objects. Those parts of the specimen which are more dense absorb electrons and appear blacker in the final picture. Density differences can be made greater by using stains which contain heavy metals such as lead and uranium.

Electrons cannot be seen with the human eye, so the image is made visible by shining the electrons on to a fluorescent screen. This gives a black-and-white picture. The screen can be lifted out of the way to enable the electrons to pass on to a photographic film so that a permanent record can be obtained of any interesting features. A photograph taken with an electron microscope is called an **electron micrograph**.

Advantage:

- high resolution (0.5 nm in practice).

Disadvantages:

- the specimen must be dead because it is viewed in a vacuum;
- it is difficult to be sure that the specimen resembles a living cell in all its details because preservation and

Table 5.3 Comparison of light and electron microscopes.

	Transmission electron microscope	*Light microscope*
Radiation source	electrons	light
Wavelength	about 0.005 nm	400–700 nm
Max. resolution in practice	0.5 nm	200 nm
Max. useful magnification	×250 000 (on screen)	×1500
Lenses	electromagnets	glass
Specimen	nonliving, dehydrated, relatively small or thin	living or nonliving
	supported on a small copper grid in a vacuum	usually supported on a glass slide
Common stains	contain heavy metals to reflect electrons	coloured dyes
Image	black and white	usually coloured

staining may change or damage the structure;
- expensive to buy and run;
- preparation of material is time-consuming and requires expert training;
- the specimen gradually deteriorates in the electron beam. Photographs must therefore be taken if further study is required.

5.6.4 Scanning electron microscope

Another kind of electron microscope is the scanning electron microscope (SEM) in which the electron beam is scanned to and fro across the specimen and electrons that are reflected from the surface are collected. They are used to form a TV-like image on a cathode ray tube.

Advantages:

- surfaces of structures are shown;
- great depth of field, meaning that a large part of the specimen is in focus at the same time. This gives a very striking three-dimensional effect (fig 5.8);
- much larger samples can be examined than with a TEM.

Disadvantage:

- resolution (5–20 nm) is not as great as with a TEM (0.5 nm).

5.7 Cell fractionation

Microscopy makes a very valuable contribution to our understanding of cells. However, other techniques are also needed if the functions of organelles are to be studied. A common approach to studying function is to isolate a particular cell organelle from other cell components and try to make it perform its normal functions in a test tube. A common procedure is to grind up (homogenise) cells in a suitable medium (with correct pH, ionic composition and temperature). This can be done with a homogeniser (food mixer). The mixture is then centrifuged. The faster the rotation of the centrifuge, the smaller the particles which will be sedimented. Fig 5.9 illustrates the principle. A series of increasing speeds can be used. After each speed, the supernatant (the liquid above the pellet) can be drawn off and recentrifuged. A series of pellets containing cell organelles of smaller and smaller size can therefore be obtained. This is known as **differential centrifugation**. The high speeds require a special centrifuge known as an **ultracentrifuge**.

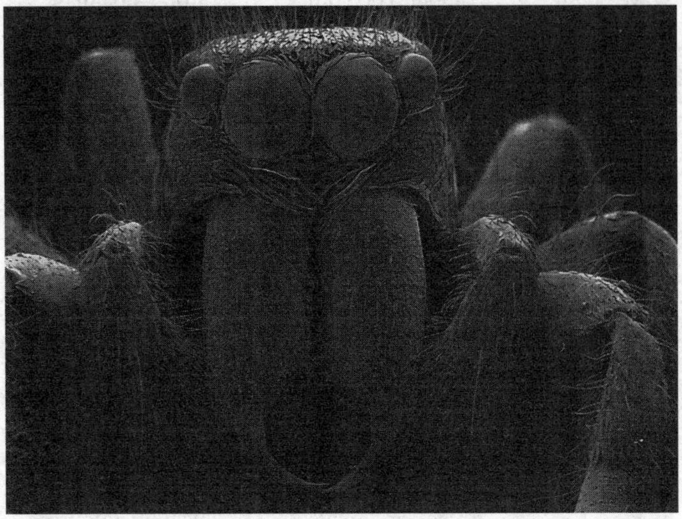

Fig 5.8 *Scanning EM of front view of a spider.*

5.8 Ultrastructure of animal and plant cells

The fine structure of the cell as seen with the electron microscope is known as **ultrastructure**. Figs 5.10 and 5.11 show diagrams of generalised animal and plant cells as seen with the electron microscope and figs 5.12 and 5.13 show actual electron micrographs (pictures taken with an electron microscope), together with summaries of the structures and functions of the different parts seen.

5.1 With reference to figs 5.1, 5.2, 5.10 and 5.11, what additional structures are revealed by the electron microscope compared with the light microscope?

5.2 With reference to figs 5.1, 5.2, 5.10 and 5.11, what structures are found (*a*) in plant cells but not in animal cells, and (*b*) in animal cells but not in plant cells?

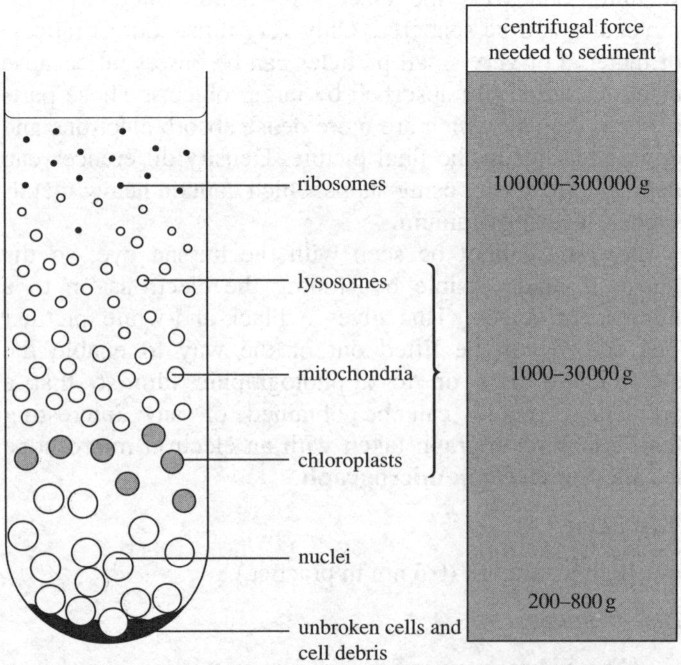

Fig 5.9 *Fractionation of cell components by centrifugation. Centrifugal forces are measured in g (number of times gravity).*

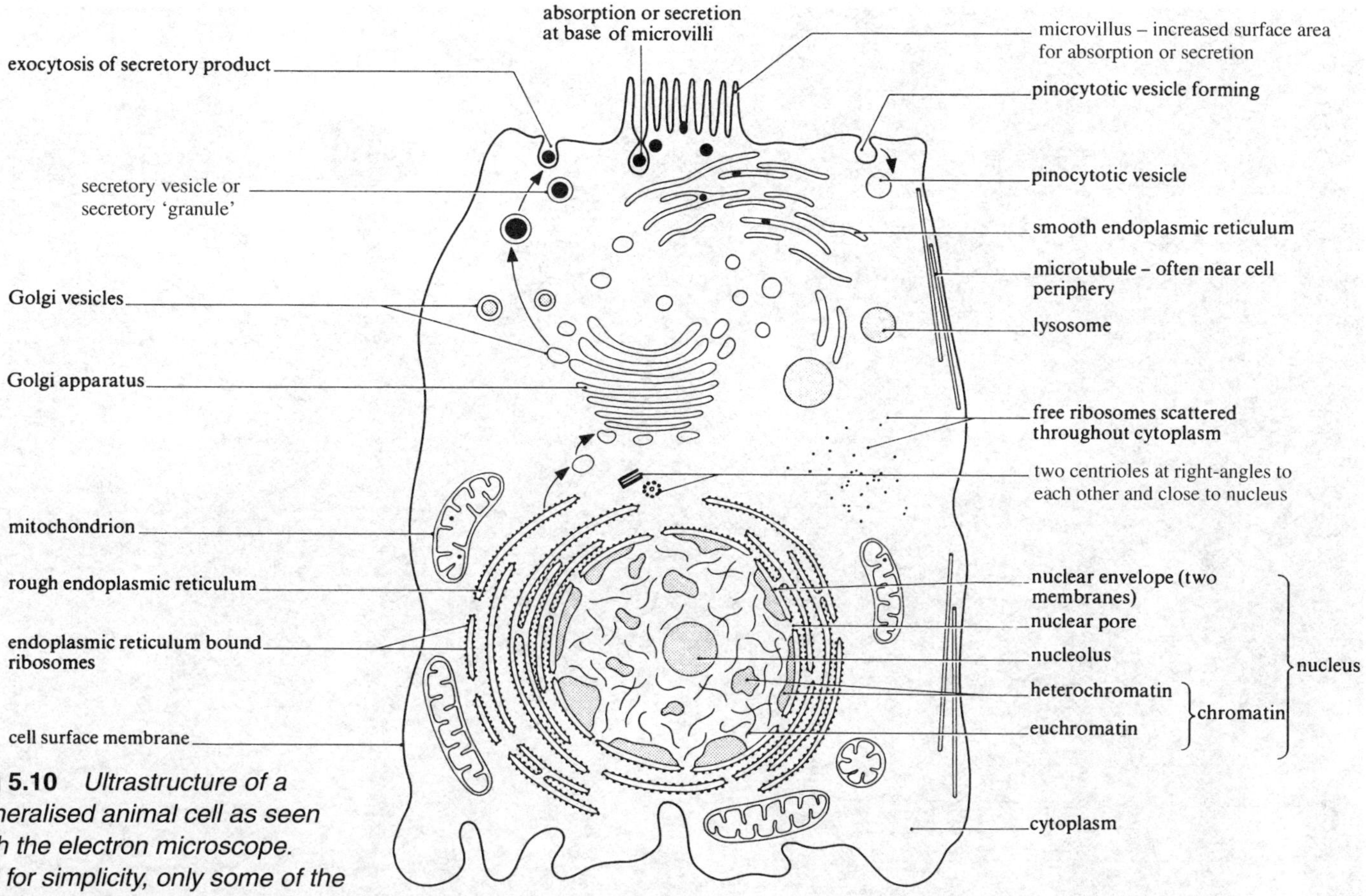

Fig 5.10 *Ultrastructure of a generalised animal cell as seen with the electron microscope. NB for simplicity, only some of the rough endoplasmic reticulum is shown covered with ribosomes. Similarly only some of the free ribosomes are shown.*

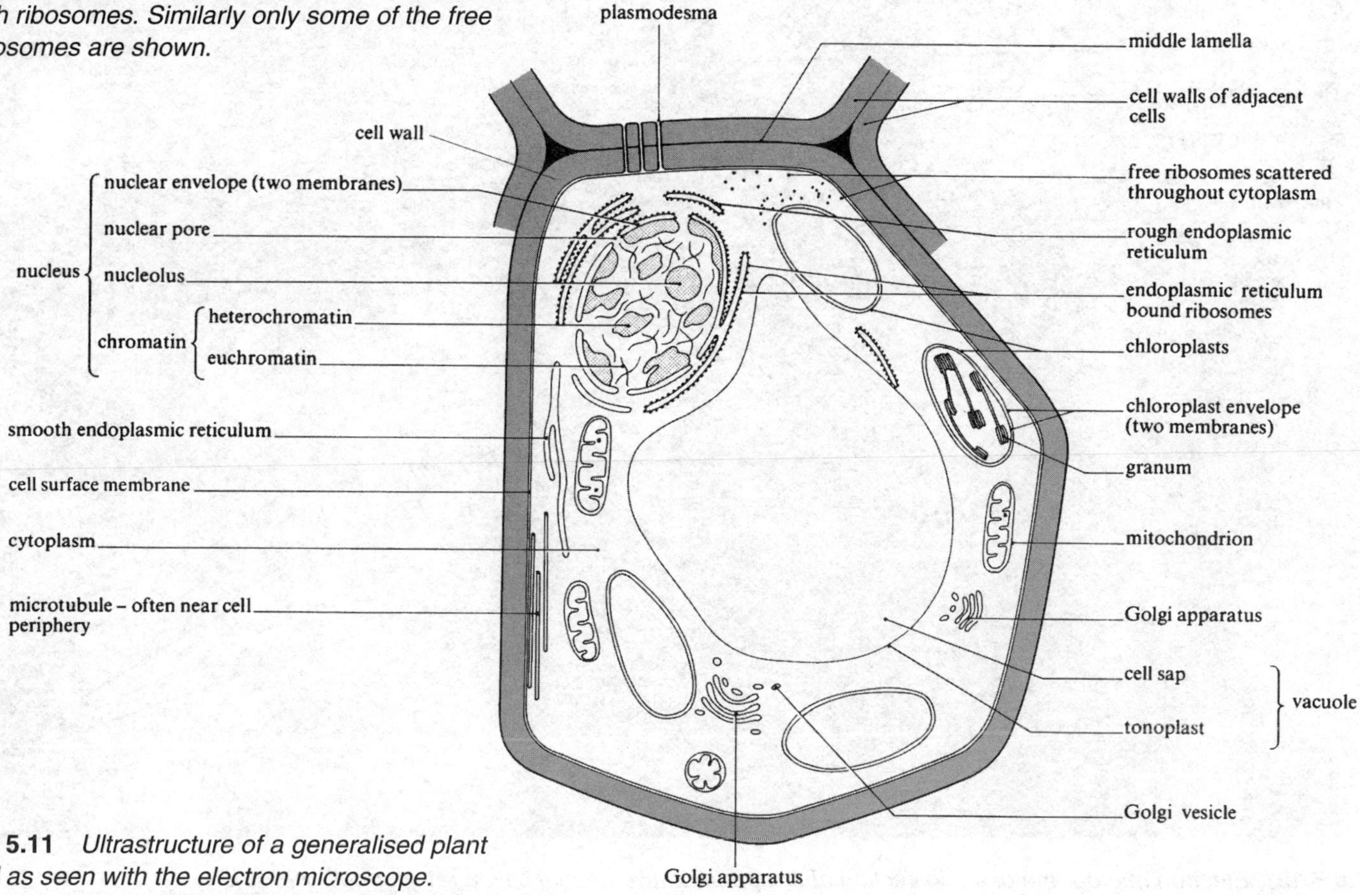

Fig 5.11 *Ultrastructure of a generalised plant cell as seen with the electron microscope.*

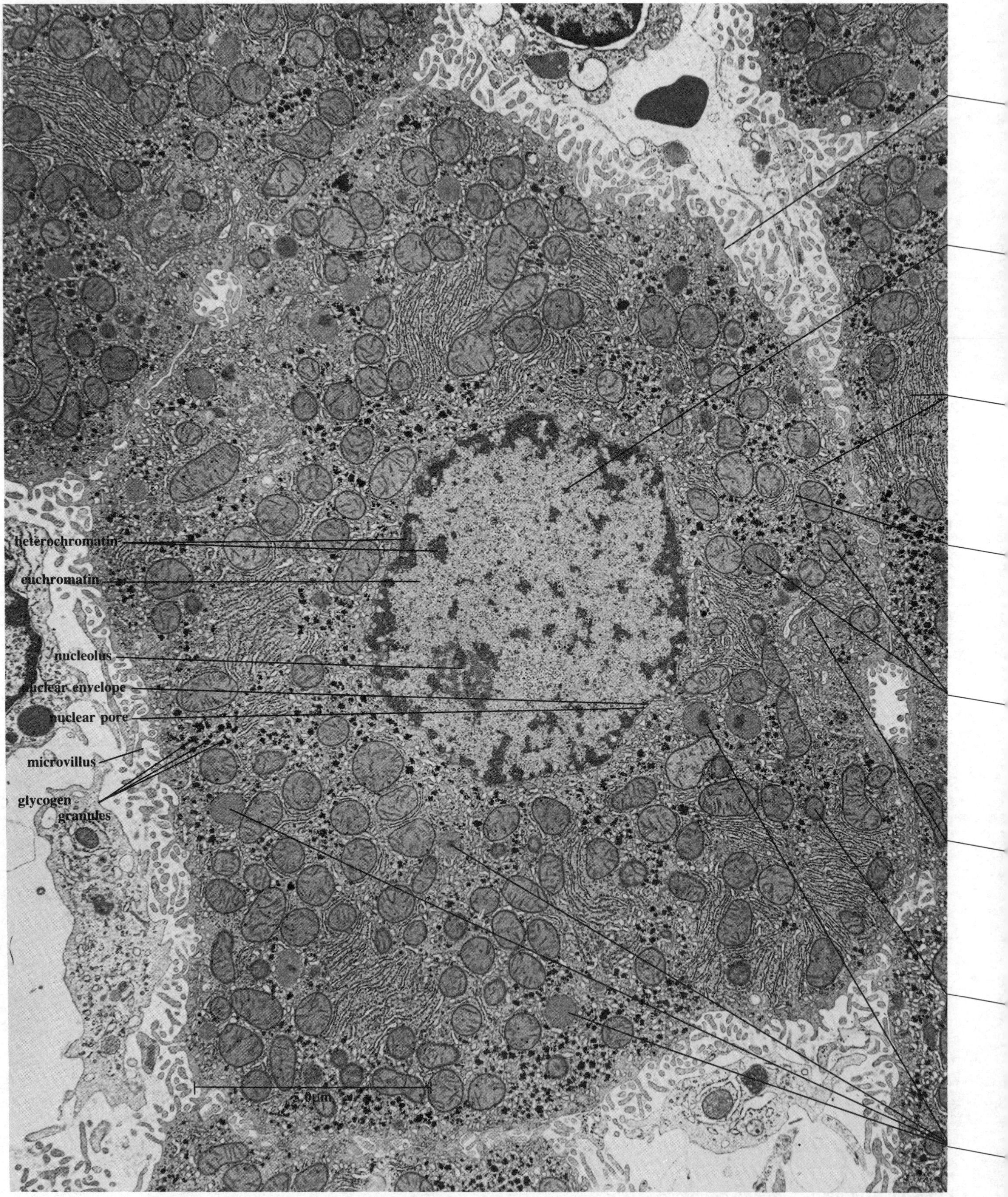

Fig 5.12 *Electron micrograph of a thin section of a representative animal cell, a rat liver cell (x9600).*

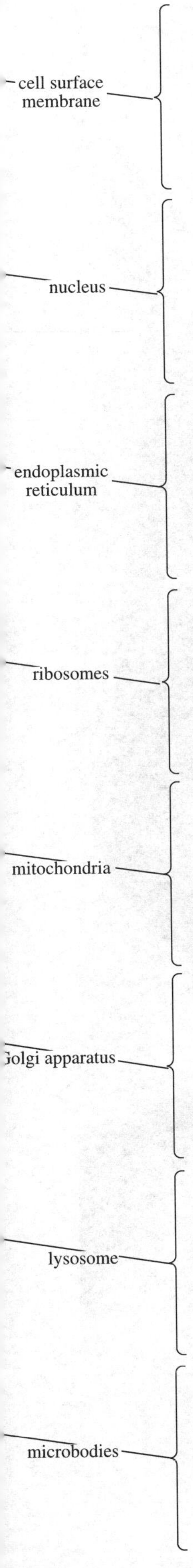

Diagram	*Structure*	*Functions*
Cell surface membrane 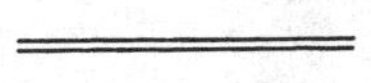	'Trilaminar' appearance (3 layers), a pale layer sandwiched between 2 dark layers	A partially permeable barrier controlling exchange between the cell and its environment
Nucleus 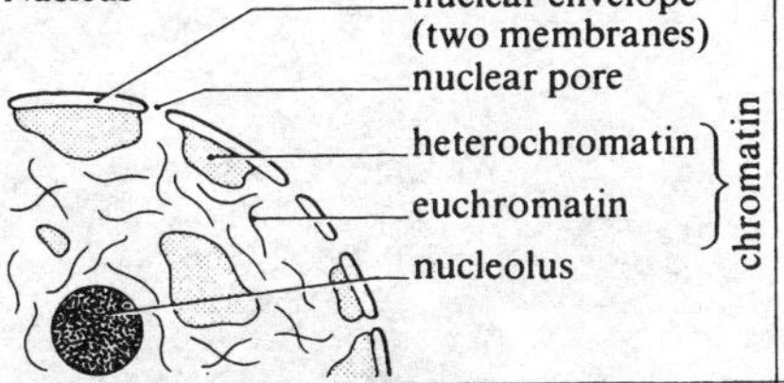	Largest cell organelle, enclosed by an **envelope** of two membranes that is perforated by **nuclear pores**. It contains **chromatin** which is the extended form taken by chromosomes during interphase. It also contains a **nucleolus**.	**Chromosomes** contain DNA, the molecule of inheritance. DNA is organised into genes which control all the activities of the cell. Nuclear division is the basis of cell replication, and hence reproduction. The **nucleolus** manufactures ribosomes.
Endoplasmic reticulum (ER) 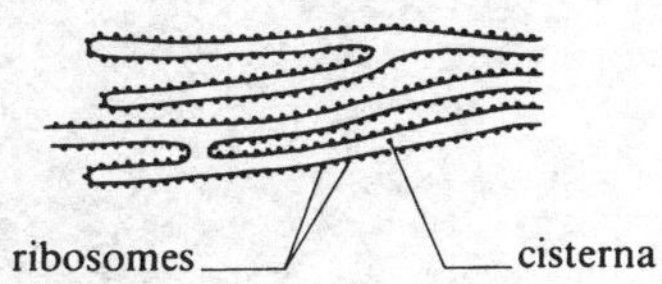	A system of flattened, membrane-bounded sacs called **cisternae**, forming tubes and sheets. It is continuous with the outer membrane of the nuclear envelope.	If ribosomes are found on its surface it is called **rough ER**, and transports proteins made by the ribosomes through the cisternae. **Smooth ER** (no ribosomes) is a site of lipid and steroid synthesis.
Ribosomes 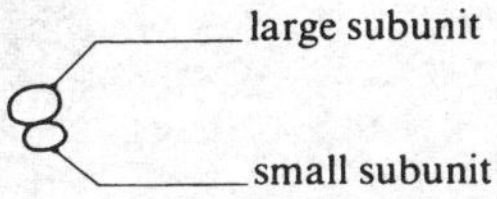	Very small organelles consisting of a large and a small subunit. They are made of roughly equal parts of protein and RNA. Slightly smaller ribosomes are found in mitochondria (and chloroplasts in plants).	Protein synthesis. They are either bound to the ER or lie free in the cytoplasm. They may form **polysomes** (polyribosomes), collections of ribosomes strung along messenger RNA.
Mitochondria (sing. **mitochondrion**) 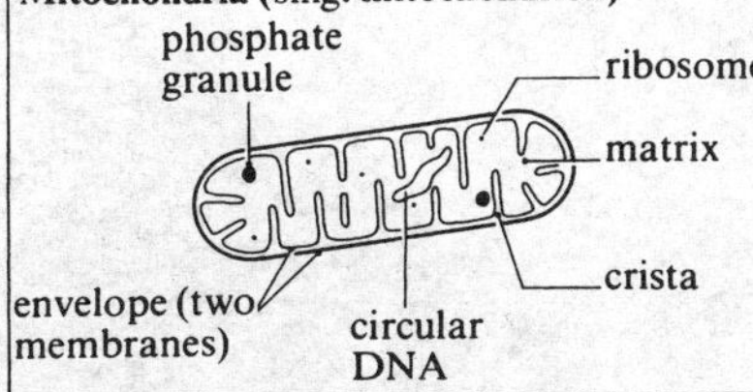	Surrounded by an envelope of two membranes, the inner being folded to form **cristae**. Contains a **matrix** with a few ribosomes, a circular DNA molecule and phosphate granules.	In aerobic respiration cristae are the sites of oxidative phosphorylation and electron transport, and the matrix is the site of Krebs cycle enzymes.
Golgi apparatus 	A stack of flattened, membrane-bounded sacs, called **cisternae**, continuously being formed at one end of the stack and budded off as vesicles at the other.	Internal **processing** and **transport** system. Processing of many cell materials takes place in the cisternae, e.g. proteins from the ER. Golgi vesicles transport the materials to other parts of the cell or to the cell surface membrane for secretion. **Makes lysosomes.**
Lysosomes 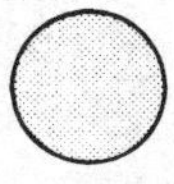	A simple spherical sac bounded by a single membrane and containing digestive (hydrolytic) enzymes. No internal structure visible.	Many functions, all concerned with breakdown of structures or molecules. For example, get rid of old organelles, digest bacteria taken in by phagocytosis.
Microbodies 	A roughly spherical organelle bounded by a single membrane. Its contents appear finely granular except for occasional striking crystalloid or filamentous deposits.	All contain catalase, an enzyme that breaks down hydrogen peroxide. All are associated with oxidation reactions. In plants, are the site of the glyoxylate cycle.

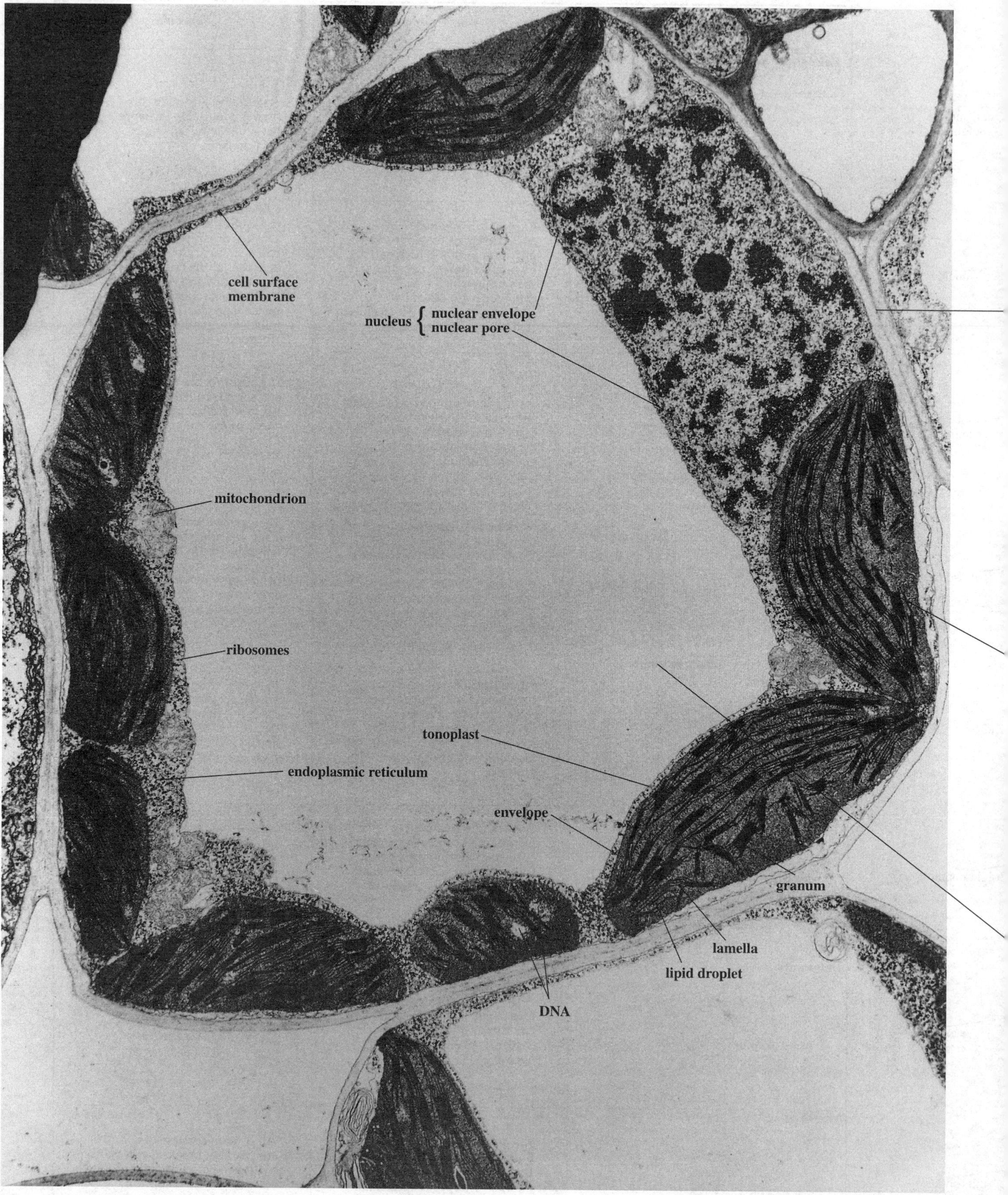

Fig 5.13 *Electron micrograph of a thin section of a representative plant cell, a leaf mesophyll cell (x15 000).*

cell wall

chloroplast

large central vacuole

Diagram	*Structure*	*Functions*
Cell wall, middle lamella, plasmodesmata (sing. **plasmodesma**) 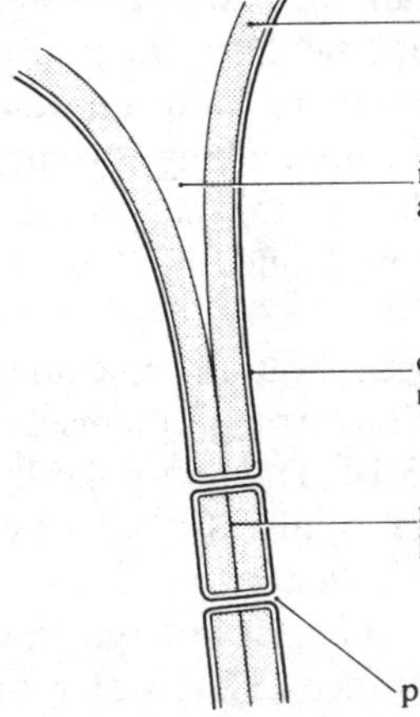	A rigid cell wall surrounding the cell, consisting of cellulose microfibrils running through a matrix of other complex polysaccharides. May be secondarily thickened in some cells.	Provides mechanical support and protection. It allows a pressure potential to be developed which aids in support. It prevents osmotic bursting of the cell. It is a pathway for movement of water and mineral salts. Various modifications, such as lignification, for specialised functions.
	Thin layer of pectic substances (calcium and magnesium pectates).	Cements neighbouring cells together.
Detail of plasmodesma 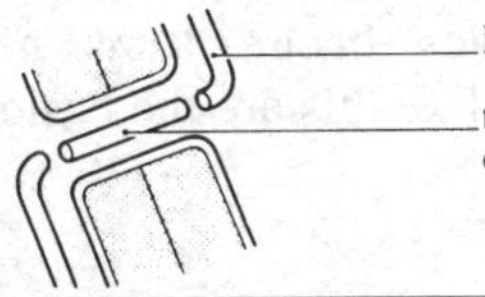	A fine cytoplasmic thread linking the cytoplasm of two neighbouring cells through a fine pore in the cell walls. The pore is lined with the cell surface membrane and has a central tubular core, often associated at each end with ER.	Enables a continuous system of cytoplasm, the **symplast**, to be formed between neighbouring cells for transport of substances between cells.
Chloroplast 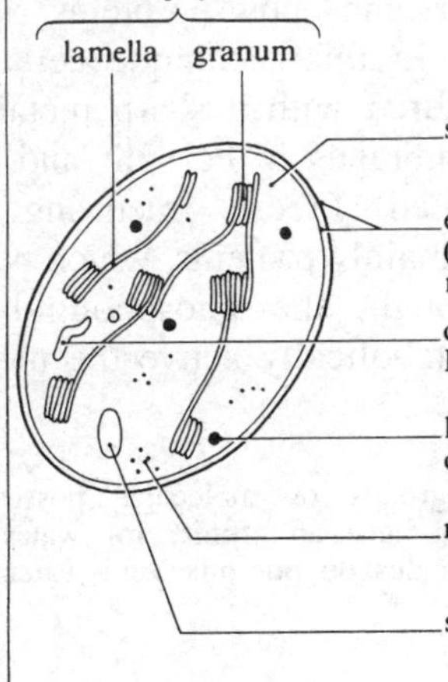	Large plastid containing chlorophyll and carrying out **photosynthesis**. It is surrounded by an envelope of two membranes and contains a gel-like **stroma** through which runs a system of membranes that are stacked in places to form **grana**. It may store starch. The stroma also contains ribosomes, a circular DNA molecule and lipid droplets.	It is the organelle in which photosynthesis takes place, producing sugars from carbon dioxide and water using light energy trapped by chlorophyll. Light energy is converted to chemical energy
Large central vacuole (Smaller vacuoles may occur in plant and animal cells such as food vacuoles, contractile vacuoles.)	A sac bounded by a single membrane called the **tonoplast**. It contains **cell sap**, a concentrated solution of various substances, such as mineral salts, sugars, pigments, organic acids and enzymes. Typically large in mature cells.	Storage of various substances including waste products. It makes an important contribution to the osmotic properties of the cell. Sometimes it functions as a lysosome.

5.9 Cell membranes

Cell membranes are important for a number of reasons. They separate the contents of cells from their external environments, controlling exchange of materials such as nutrients and waste products between the two. They also enable separate compartments to be formed *inside* cells in which specialised metabolic processes such as photosynthesis and aerobic respiration can take place. Chemical reactions, such as the light reactions of photosynthesis in chloroplasts, sometimes take place on the surface of the membranes themselves. Membranes also act as receptor sites for recognising hormones, neurotransmitters and other chemicals, either from the external environment or from other parts of the organism. An understanding of their properties is essential to an understanding of cell function.

5.9.1 Membranes are partially permeable

It has been known since the turn of the century that cell membranes do not behave simply like semi-permeable membranes that allow only the passage of water and other small molecules such as gases. Instead they are better described as **partially permeable**, since other substances such as glucose, amino acids, fatty acids, glycerol and ions can diffuse slowly through them. They also exert a measure of active control over which substances they allow through.

5.9.2 Membranes contain proteins and lipids

Early work showed that organic solvents, such as alcohol, ether and chloroform, penetrate membranes even more rapidly than water. This suggested that membranes have non-polar portions and contain lipids. This was later confirmed by chemical analysis which showed that membranes are made almost entirely of proteins and lipids. The proteins are discussed later. The most common lipids are phospholipids.

5.9.3 Phospholipids

The structure of phospholipids is described in chapter 3 (fig 3.18). Each phospholipid molecule consists of a polar* head containing phosphate, and two non-polar hydrocarbon tails from the fatty acids used to make the molecule. **Polar** means there is an uneven distribution of charge within the molecule, making it soluble in water. The phospholipid molecule is unusual because the head is hydrophilic (water-loving) and the tails are hydrophobic (water-hating).

If a thin layer of phospholipid molecules is spread over the surface of water, they arrange themselves into a single layer, as shown in fig 5.14. The non-polar hydrophobic tails project out of the water, whilst the polar hydrophilic heads lie in the surface of the water.

If the phospholipid is present in large enough amounts to more than cover the surface of the water, or if it is shaken up with the water, particles known as **micelles** are formed, in which hydrophobic tails project inwards away from the water as shown in fig 5.15. Figs 5.15*b* and 5.15*c* show an arrangement known as a **bilayer** in which two layers of phospholipid molecules occur. It is now known that phospholipid bilayers like this are the basic structure of cell membranes.

5.9.4 Proteins

A technique known as freeze fracturing has helped us to understand how proteins fit into the phospholipid bilayer. In this technique cells are rapidly frozen and then fractured with a sharp metal blade. The technique allows membranes to be split and the surfaces inside to be examined. Freeze fracturing reveals the presence of particles, mainly proteins, which penetrate into, and often right through, the phospholipid bilayer. In general, the more metabolically active the membrane, the

*Remember that polar groups or molecules possess an uneven distribution of charge and have an affinity for water (hydrophilic); non-polar groups or molecules do not mix with water (hydrophobic) (section 3.1.2).

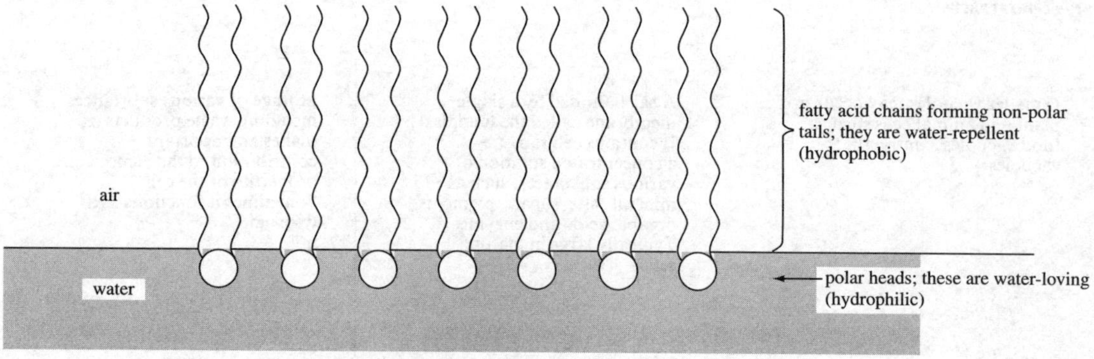

Fig 5.14 *Single layer of phospholipid molecules on the surface of water.*

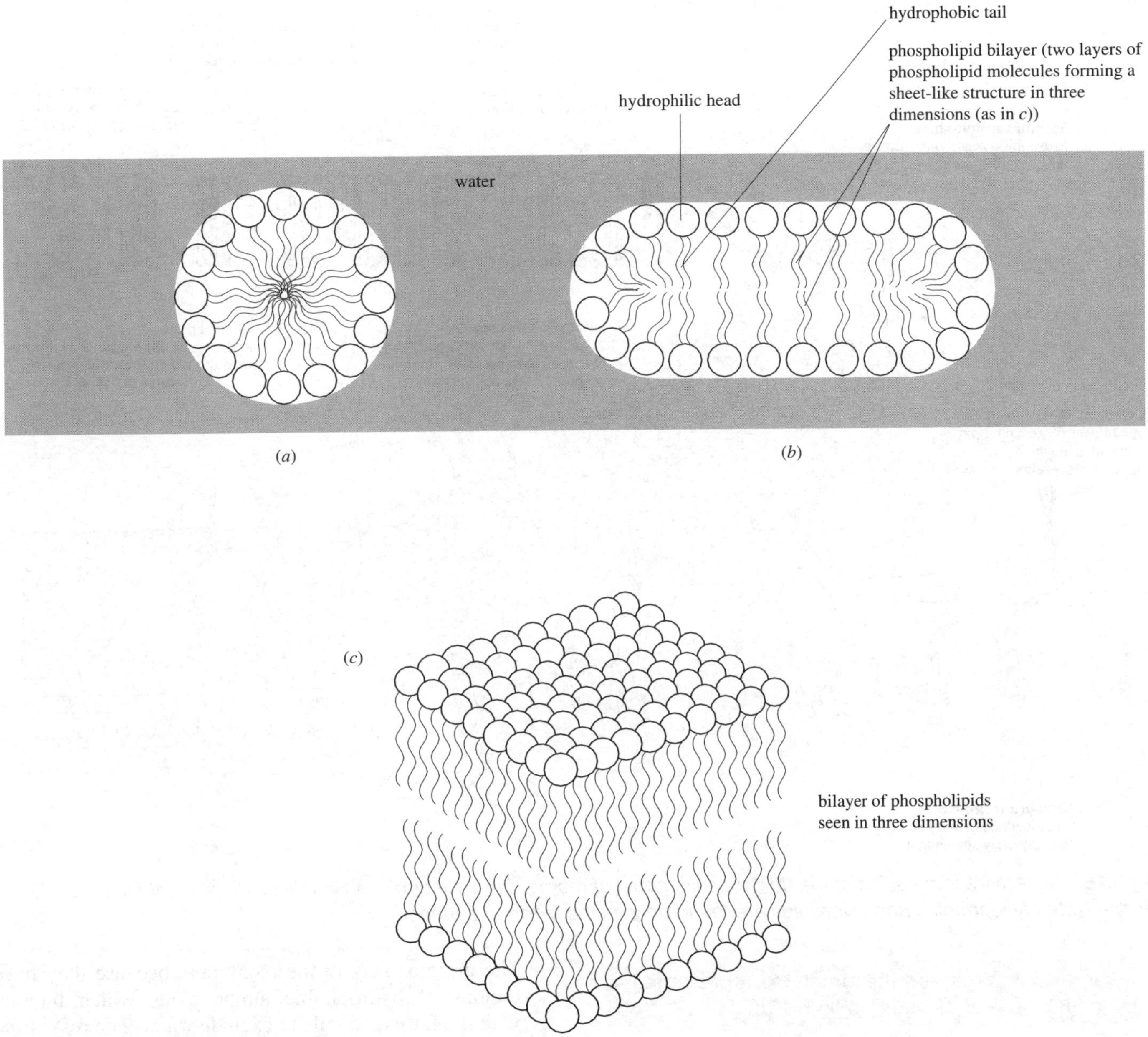

Fig 5.15 *Sections through* (a) *a spherical micelle and* (b) *two layers (a bilayer) of phospholipids.* (c) *Sheet-like bilayer seen in three dimensions.*

more protein particles that are found; chloroplast membranes (75% protein) have many particles (fig 7.12), whereas the metabolically inert myelin sheath (18% protein) has none. The inner and outer sides of membranes also differ in their particle distribution.

5.9.5 Glycolipids and cholesterol

Membranes also contain glycolipids and cholesterol. Glycolipids are lipids combined with carbohydrate. Like phospholipids, they have polar heads and non-polar tails. Cholesterol is closely related to lipids and is slightly polar at one end.

5.9.6 The fluid mosaic model of membrane structure

In 1972, Singer and Nicolson put forward the **'fluid mosaic' model** of membrane structure in which protein molecules float about in a fluid phospholipid bilayer. The scattered protein molecules resemble a mosaic but, since the phospholipid bilayer is fluid, the proteins form a fluid mosaic pattern. The thin sheet-like structure surrounds the cell like the skin of a soap bubble, swirling in much the same way. It has the consistency of olive oil. An artist's impression of the model and a simplified diagram are shown in fig 5.16.

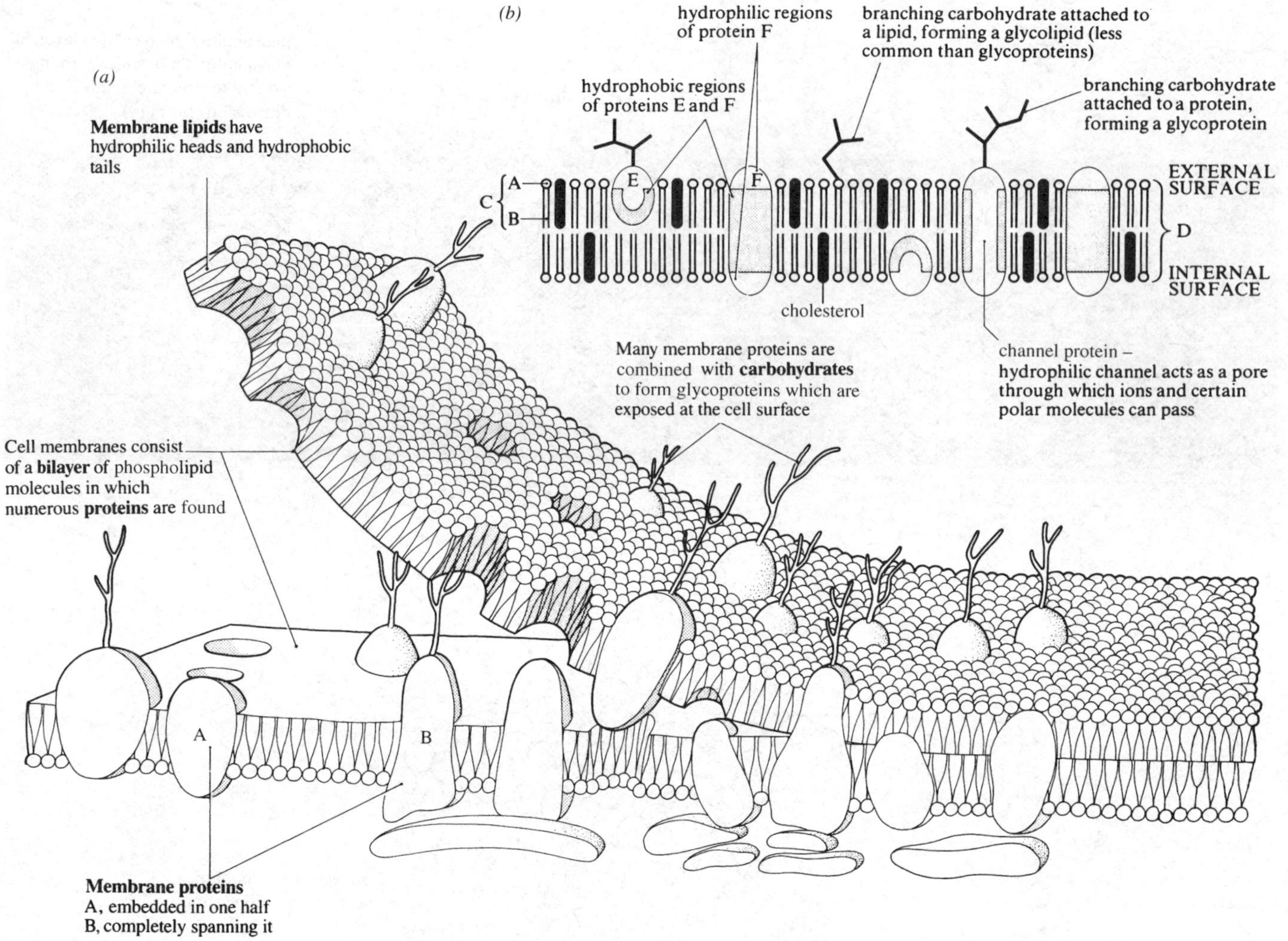

Fig 5.16 (a) *Artist's impression of the fluid mosaic model of membrane structure.* (b) *Simplified diagram of fluid mosaic model. Note glycoproteins and glycolipids are found only on the external surface.*

> **5.3** What are the structures represented by the labels A, B, C and D in fig 5.16*b*?

The membrane has the following features.

- It is about 7 nm thick.
- The basic structure is a phospholipid bilayer.
- The hydrophilic phosphate heads of the phospholipids face outwards into the aqueous environment inside and outside the cell.
- The hydrocarbon tails face inwards and create a hydrophobic interior.
- The phospholipids are fluid and move about rapidly by diffusion in their own layers.
- Some of the fatty acid tails are saturated and some are unsaturated (see chapter 3, fig 3.17). Unsaturated tails are bent and fit together more loosely. Therefore the more unsaturated the tails are, the more fluid the membrane is.
- Most protein molecules float about in the phospholipid bilayer forming a fluid mosaic pattern.
- The proteins stay in the membrane because they have regions of hydrophobic amino acids which interact with the fatty acid tails to exclude water. The rest of the protein is hydrophilic and faces into the cell or out into the external environment, both of which are aqueous.
- Some proteins penetrate only part of the way into the membrane while others penetrate all the way through.
- Some proteins and lipids have short branching carbohydrate chains like antennae, forming glycoproteins and glycolipids respectively.
- Membranes also contain cholesterol. Like unsaturated fatty acids, cholesterol disturbs the close packing of phospholipids and keeps them more fluid. This can be important for organisms living at low temperatures when membranes can solidify. Cholesterol also increases flexibility and stability of membranes. Without it, membranes break up.
- The two sides of a membrane can differ in composition and function.

5.9.7 Functions of membranes

The phospholipid bilayer provides the basic structure of membranes. It also restricts entry and exit of polar molecules and ions. The other molecules have a variety of functions.

- **Channel proteins and carrier proteins** – these are involved in the selective transport of polar molecules and ions across the membrane (see facilitated diffusion and active transport in section 5.9.8).
- **Enzymes** – proteins sometimes act as enzymes, for example the microvilli on epithelial cells lining some parts of the gut contain digestive enzymes in their cell surface membranes.
- **Receptor molecules** – proteins have very specific shapes as discussed in chapters 3 and 4. This makes them ideal as receptor molecules for chemical signalling between cells. For example, hormones are chemical messengers which circulate in the blood but only bind to specific target cells which have the correct receptor sites. Neurotransmitters, the chemicals which enable nerve impulses to pass from one nerve cell to the next, also fit into specific receptor proteins in nerve cells.
- **Antigens** – these act as cell identity markers or 'name tags'. They are glycoproteins, that is proteins with branching carbohydrate side chains like 'antennae'. There is an enormous number of possible shapes to these side chains, so each type of cell can have its own specific markers. This enables cells to recognise other cells, and to behave in an organised way, for example during development of tissues and organs in multicellular organisms. It also means that foreign antigens can be recognised and attacked by the immune system.
- **Glycolipids** also have branching carbohydrate side chains and are involved in cell–cell recognition. They may act as receptor sites for chemical signals. With glycoproteins they are also involved in sticking the correct cells together in tissues.
- **Energy transfer** – in photosynthesis and respiration proteins take part in the energy transfer systems that exist in the membranes of chloroplasts and mitochondria respectively.
- **Cholesterol** acts like a plug, reducing even further the escape or entry of polar molecules through the membrane.

5.9.8 Transport across the cell surface membrane

Cell membranes are only about 7 nm wide but they present barriers to the movement of ions and molecules, particularly polar (water-soluble) molecules such as glucose and amino acids that are repelled by the non-polar, hydrophobic lipids of membranes. This prevents the aqueous contents of the cell from escaping. However, transport across membranes must still occur for a number of reasons, for example:

- to obtain nutrients;
- to excrete waste substances;
- to secrete useful substances;
- to generate the ionic gradients essential for nervous and muscular activity;
- to maintain a suitable pH and ionic concentration within the cell for enzyme activity.

In the following account, movement across the cell surface membrane will be discussed, although similar movements occur across the membranes of cell organelles within cells. There are four basic mechanisms, namely **diffusion, osmosis, active transport** and **bulk transport (endocytosis** or **exocytosis**). The first two processes are passive, that is they do not require the expenditure of energy by the cell; the latter two are active, energy-consuming processes.

Diffusion and facilitated diffusion

Diffusion is the movement of molecules or ions from a region of their high concentration to a region of their low concentration down a diffusion gradient. The process is passive, that is it does not require energy and happens spontaneously. For example, if a bottle of perfume were opened in a closed room, the perfume would eventually spread by diffusion until an equilibrium was reached where the perfume was evenly spread throughout the room. This occurs by the random motion of molecules which is due to their kinetic energy (energy of movement). Each type of molecule moves down its own diffusion gradient independently of other molecules. For example, oxygen diffuses from the lungs into the blood while at the same time carbon dioxide diffuses in the opposite direction.

Three factors in particular affect the rate of diffusion.

(1) The steepness of the diffusion gradient, or difference in concentration between point A and point B: the steeper the gradient, the faster the rate of diffusion. It is an advantage for cells to maintain steep diffusion gradients if rapid transport is required. This can be achieved in the lungs, for example, by speeding up the flow of blood through the lungs or by breathing faster.

(2) The greater the surface area of a membrane through which diffusion is taking place, the greater the rate of diffusion. The larger the cell, assuming it is roughly spherical, the smaller its surface area in relation to its volume. This places a limit on cell size. For example, a very large aerobic cell could not obtain oxygen fast enough to satisfy its needs if it relied on diffusion alone. Microvilli increase the surface area of animal cells for absorption purposes.

(3) Rate of diffusion decreases rapidly with distance (it falls in proportion to the square of the distance). Diffusion is therefore only effective over very short distances. This is another factor which limits cell size. Cells rely on diffusion for internal transport of

molecules so most are no larger than 50 μm in diameter, with no part of the cell more than 25 μm from the cell surface. An amino acid molecule, for example, can travel a few micrometres in several seconds but would take several days to diffuse a few centimetres. It is also essential that membranes are thin so that molecules or ions can cross them rapidly.

The factors affecting the rate at which molecules cross cell membranes by diffusion are summarised in Fick's law. This states that the rate is proportional to:

$$\frac{\text{surface area of membrane} \times \text{difference in concentration across the membrane}}{\text{thickness of membrane}}$$

We can now consider which molecules cross membranes by diffusion. The respiratory gases oxygen and carbon dioxide diffuse rapidly through membranes. Water molecules, although very polar, are small enough to pass between the hydrophobic phospholipid molecules without interference. However, ions and larger polar molecules such as amino acids, sugars, fatty acids and glycerol are repelled by the hydrophobic region of the membrane and diffuse across extremely slowly. Other mechanisms are required for these substances.

Some ions and polar molecules can diffuse through special transport proteins called **channel proteins** and **carrier proteins**. These contain water-filled hydrophilic channels or pores whose shape is specific for a particular ion or molecule. Alternatively several proteins combine, forming a channel between them. Diffusion can occur through the channel in either direction. Since diffusion would not be possible without the protein or proteins, the process is known as **facilitated diffusion**. Transport proteins which allow the passage of ions are called ion channels. Ion channels are usually 'gated', which means they can exist in open or closed states. Gated ion channels are important in the conduction of nerve impulses.

Channel proteins have a *fixed* shape (fig 5.16*b*). It has been shown that the disease cystic fibrosis is caused by a fault in a protein which acts as a chloride ion channel. **Carrier proteins** undergo rapid *changes* in shape, up to 100 cycles per second. They exist in two forms, known as the 'ping' and 'pong' states. Fig 5.17 shows how they work. The binding site faces outwards in one state and into the cell in the other state. The higher the concentration of the solute molecule or ion, the greater its chance of binding. If, as in the case of glucose in the example shown in fig 5.17, there is a higher concentration outside the cell, there will be a net movement of the solute into the cell. Glucose can enter red blood cells in this way. The movement has all the characteristics of diffusion although it is facilitated by the protein. Another example is the movement of chloride and hydrogencarbonate ions into and out of blood cells during the 'chloride shift'. This is one way in which cells achieve partial and selective permeability.

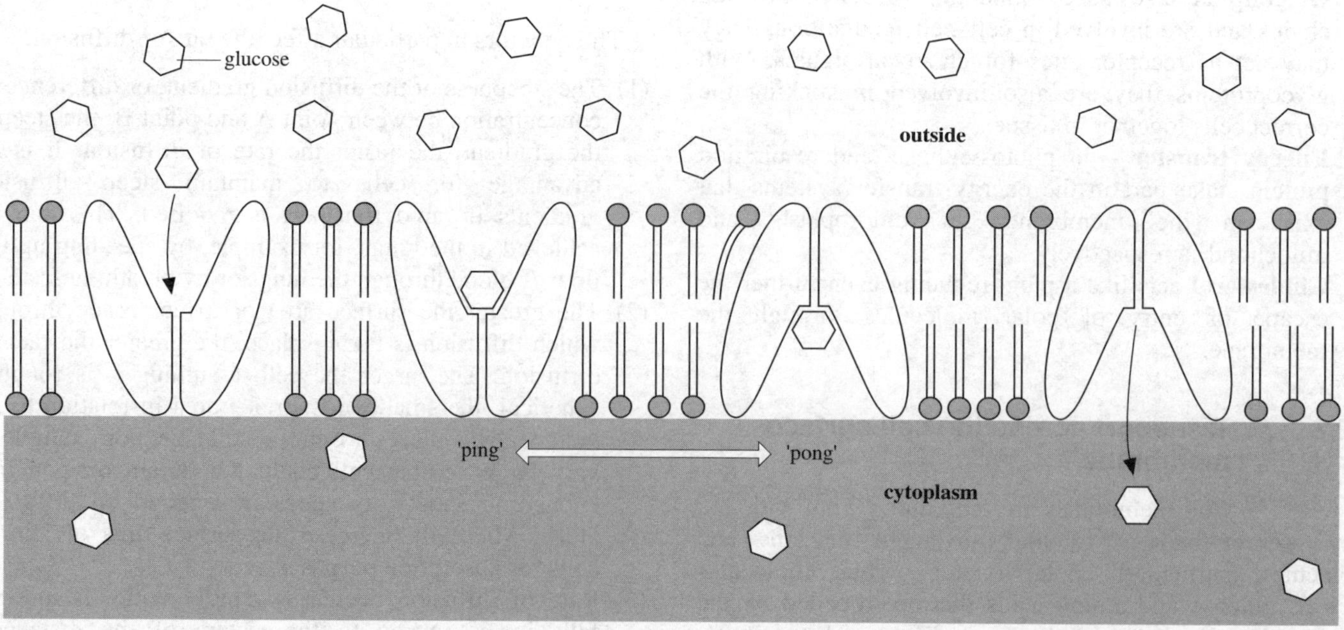

Fig 5.17 *Facilitated diffusion through a carrier protein. The protein alternates rapidly between two different shapes, 'ping' and 'pong'. A higher concentration of glucose (hexagons) is shown outside the cell resulting in a net movement of glucose into the cell down a diffusion gradient.*

Osmosis

Osmosis is the passage of water molecules from a region of their high concentration to a region of their low concentration through a partially permeable membrane. It is best regarded as a form of diffusion in which only water molecules move. Consider the situation in fig 5.18. The solute molecules are too large to pass through the pores in the membrane, so equilibrium can only be achieved by the movement of water molecules. Solution A has the higher concentration of *water*, so there will be a net movement of water from A to B by osmosis. At equilibrium there will be no further net movement of water. The tendency of water molecules to move from one place to another is measured as the **water potential**, represented by the symbol Ψ (Greek letter psi). Water *always* moves from a region of higher water potential to one of lower water potential. Solute molecules reduce Ψ (in effect, they dilute the water!). The extent by which they lower Ψ is known as the **solute potential**, given the symbol Ψ_s. The effects of different solutions on red blood cells are shown in fig 5.19. Osmosis in plant cells is considered in chapter 13.

> **5.4** In fig 5.18, which solution has (*a*) the higher concentration of water molecules, (*b*) the higher concentration of solute molecules, (*c*) the higher water potential, (*d*) the more negative solute potential?
>
> (*e*) Which of the following two values of water potential is the higher, −2000 kPa or −1000 kPa?

Active transport

Active transport is the **energy-consuming transport** of molecules or ions across a membrane **against** a concentration gradient. Energy is required because the substance must be moved **against** its natural tendency to diffuse in the opposite direction. Movement is usually in one direction only, unlike diffusion which is reversible. The energy is supplied in the form of a molecule known as ATP, which is an energy carrier made in respiration. Without respiration, active transport is therefore impossible.

The major ions inside cells and in their environments are sodium ions (Na^+), potassium ions (K^+) and chloride (Cl^-)

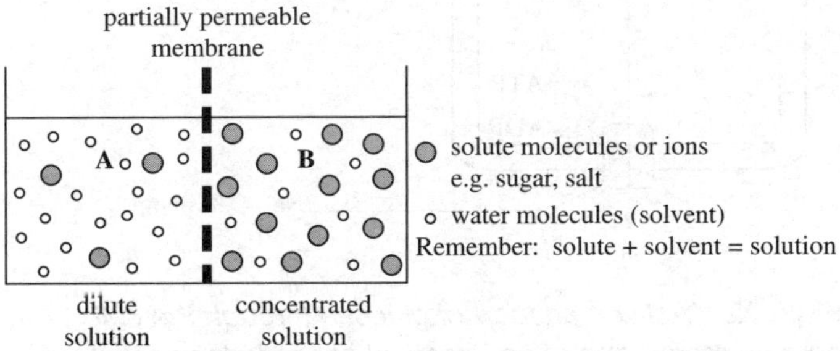

Fig 5.18 *Two solutions separated by a partially permeable membrane.*

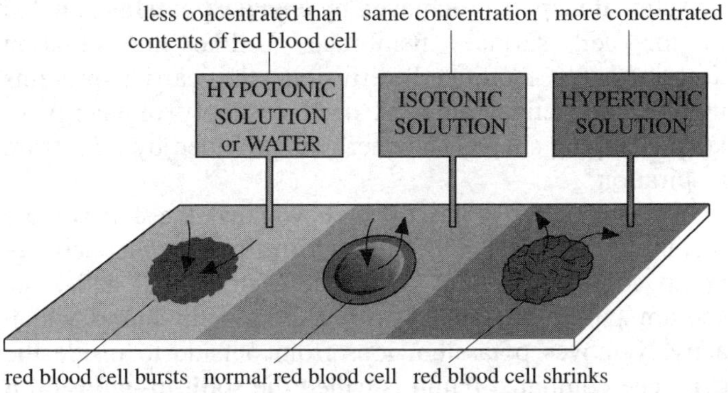

Fig 5.19 *Effects of different solutions on red blood cells. In a hypotonic solution, the solution has a higher water potential than the contents of the red cell. Water therefore enters by osmosis and the cell bursts, dispersing the cell contents. A hypertonic solution has a lower water potential than the cell contents, so water leaves the cell by osmosis and the cell shrinks. In an isotonic solution, water potential of the cell equals that of the external solution and no net movement of water occurs. The cell remains normal. Blood plasma must be kept isotonic to red blood cells and other body cells.*

ions. Look at fig 5.20 which shows that the concentrations of these ions are very different inside and outside a human red blood cell. For example, like most cells, there is a much higher potassium content inside than outside, and a higher concentration of potassium inside than sodium.

If respiration of the red blood cells is inhibited, for example with cyanide, the ionic composition of the cells gradually changes until it comes into equilibrium with the plasma. This suggests that the ions can diffuse passively through the cell surface membrane of the cells, but that normally respiration supplies the energy for active transport to maintain the concentrations shown in fig 5.20. In other words, sodium is actively pumped out of the cell and potassium is actively pumped in.

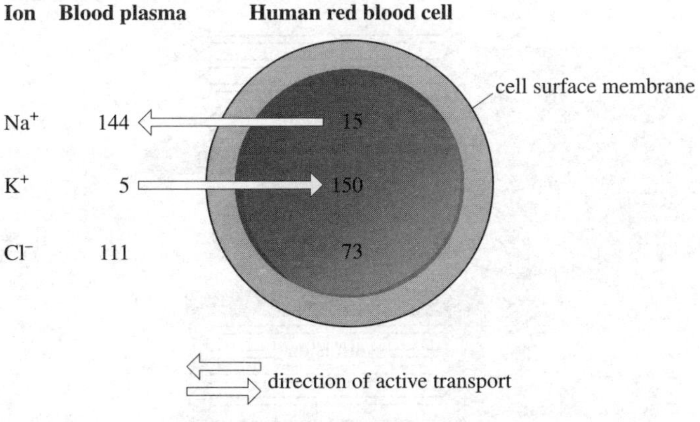

Fig 5.20 *Concentrations of Na^+, K^+ and Cl^- ions in red blood cells and their environment (measured as mM).*

Active transport is achieved by **carrier proteins** situated in the cell surface membrane. Unlike the situation described for facilitated diffusion, the carrier proteins involved in active transport need a supply of energy to keep changing shape. The energy is provided by ATP from respiration.

In recent years it has been shown that the cell surface membranes of most cells have **sodium pumps** that actively pump sodium ions out of the cell. In animal cells, the sodium pump is coupled with a potassium pump which actively moves potassium ions from outside to inside the cell. The combined pump is called the **sodium–potassium pump** (Na^+–K^+ pump). Since this pump is a common feature of almost all animal cells and has a number of important functions, it provides a good example of active transport. Its importance is revealed by the fact that more than a third of the ATP consumed by a resting animal is used to pump sodium and potassium.

The pump is a carrier protein which spans the membrane from one side to the other (fig 5.21). On the inside it accepts sodium and ATP, while on the outside it accepts potassium. The transfer of sodium and potassium across the membrane is brought about by changes in the shape of the protein. Note that for every $2K^+$ taken into the cell, $3Na^+$ are removed. Thus a potential difference is built up across the membrane, with the inside of the cell being negative. This tends to restrict the entry of negatively charged ions (anions) such as chloride. This explains why the chloride concentration inside red blood cells is less than outside (fig 5.20) despite the fact that chloride can diffuse in and out by facilitated diffusion. Similarly, positively charged ions (cations) tend to be attracted into cells. Thus both concentration and charge are important in deciding the direction in which ions cross membranes.

The pump is essential in controlling the osmotic balance of animal cells (osmoregulation). If the pump is inhibited, the cell swells and bursts because a build-up of sodium ions results in excess water entering the cells by osmosis. This explains why bacteria, fungi and plants, which have cell walls, do not need the pump. The pump is also important in maintaining electrical activity in nerve and muscle cells and in driving active transport of some other substances such as sugars and amino acids. Also, high concentrations of potassium are needed inside cells for protein synthesis, glycolysis, photosynthesis and other vital processes.

5.5 Explain the following observations.

(*a*) When K^+ ions are removed from the medium surrounding red blood cells, entry of sodium into the cells and exit of potassium increase dramatically.

(*b*) If ATP is introduced into cells, the exit of Na^+ is stimulated.

Active transport is important in many other situations. It is particularly associated with epithelial cells, as in the gut lining and kidney tubules (nephrons), because these are active in secretion and absorption.

Active transport in the intestine. When the products of digestion are absorbed in the small intestine they must pass through the epithelial cells lining the gut wall. After that, glucose, amino acids and salts pass through the cells of the blood capillary walls, into the blood and then travel to the liver. Soon after feeding, relatively high concentrations of digested foods are found in the gut and absorption is partly a result of diffusion. However, this is very slow and must be supplemented by active transport. Such active transport is coupled to a sodium–potassium pump as shown in fig 5.22.

As sodium is pumped out by the sodium–potassium pump, so it tends to diffuse back in. Situated in the

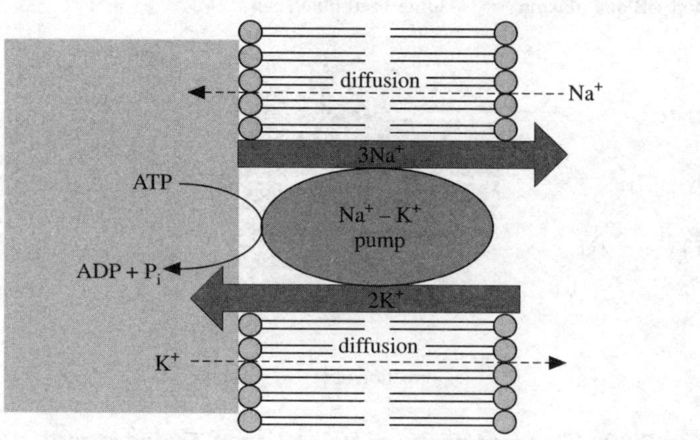

Fig 5.21 *Sodium–potassium pump.*

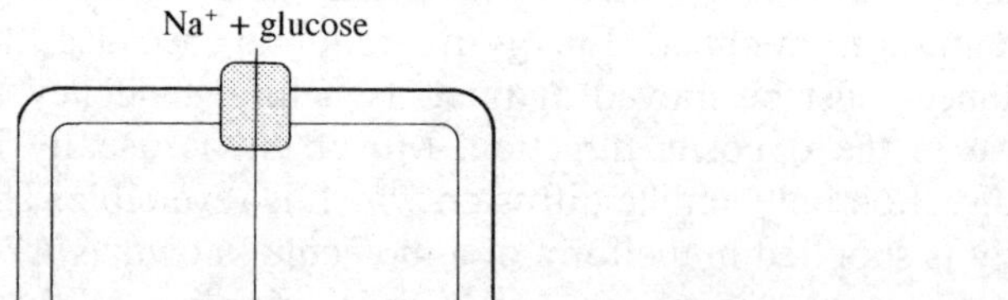

Fig 5.22 *Active transport of glucose through the cell surface membrane of an intestinal cell or kidney cell. (Based on Fig 36-12, L. Stryer (1981) Biochemistry, 2nd ed., Freeman.)*

membrane is a carrier protein which requires both sodium and glucose to function. These are transported together by **facilitated diffusion** into the cell. A similar sodium–amino-acid carrier protein operates in the active transport of amino acids into cells, the active part of the process being the pumping back of sodium ions.

Active transport in nerve cells and muscle cells. In nerve cells and muscle cells a sodium–potassium pump is responsible for the development of a potential difference, called the **resting potential**, across the cell surface membrane (see conduction of nerve impulses, chapter 17, and muscle contraction, chapter 18).

In muscle cells another important carrier protein is the calcium pump. Muscle cells contain a specialised form of endoplasmic reticulum called sarcoplasmic reticulum. Calcium is pumped into it from the surrounding cytoplasm by the calcium pump. Muscle contraction is triggered by the sudden release of the calcium ions in response to a nerve impulse.

Active transport in the kidney. Active transport of glucose and sodium occurs from the proximal convoluted tubules of the kidney and the kidney cortex actively transports sodium. These processes are described more fully in chapter 20.

Active transport in plants. An example of active transport in plants is the active loading of sugar into phloem ready for transport round the plant (chapter 13). This is particularly important in leaves.

Endocytosis and exocytosis

Endocytosis and exocytosis are active processes involving the bulk transport of materials through membranes, either into cells (endocytosis) or out of cells (exocytosis) (fig 5.23).

Endocytosis occurs by an infolding or extension of the cell surface membrane to form a vesicle* or vacuole*. It is of two types.

*Vacuole – fluid-filled, membrane-bound sac; vesicle – small vacuole.

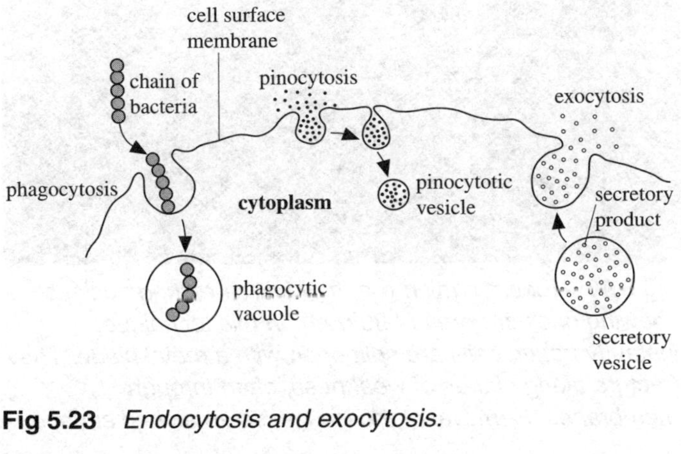

Fig 5.23 *Endocytosis and exocytosis.*

- **Phagocytosis** ('cell eating') – material taken up is in solid form. Cells specialising in the process are called **phagocytes** and are said to be **phagocytic**. For example, some white blood cells take up bacteria by phagocytosis. The sac formed during uptake is called a **phagocytic vacuole**.
- **Pinocytosis** ('cell drinking') – material taken up is in liquid form. Vesicles formed are often extremely small, in which case the process is known as **micropinocytosis** and the vesicles as **micropinocytotic vesicles**. Pinocytosis is used by the human egg cell to take up nutrients from the surrounding follicle cells. In the thyroid gland, the hormone thyroxine is stored as thyroglobulin in hollow structures called follicles. When required, thyroglobulin is taken up by pinocytosis by the follicle cells and then converted to thyroxine for release into the blood. Pinocytosis is very common in both animal and plant cells (fig 5.10).

Exocytosis is the reverse process of endocytosis. Waste materials may be removed from cells, such as solid, undigested remains from phagocytic vacuoles, or useful materials may be secreted. Secretion of enzymes from the pancreas is achieved in this way (fig 5.29). Plant cells use exocytosis to export the materials needed to form cell walls (fig 5.30).

5.10 Structures found in cells

5.10.1 The nucleus

Nuclei are found in all eukaryotic cells, the only common exceptions being mature phloem sieve tube elements and mature red blood cells of mammals. In some protozoa, such as *Paramecium*, two nuclei exist, a micronucleus and a meganucleus. Normally, however, cells contain only one nucleus. Nuclei are conspicuous because they are the largest cell organelles, and they were the first to be described by light microscopists. They are typically about 10 μm in diameter.

The nucleus is vitally important because it controls the cell's activities. This is because it contains the genetic (hereditary) information in the form of DNA. The DNA can also replicate itself which can be followed by nuclear division, thus ensuring that the daughter nuclei also contain DNA. Nuclear division precedes cell division, producing two daughter nuclei with exactly the same DNA content as the parent nucleus. The nucleus is surrounded by a nuclear envelope and contains chromatin and one or more nucleoli.

Nuclear envelope and nuclear pores

What looks like a single membrane around the nucleus in the light microscope is actually a **nuclear envelope** composed of two membranes. The outer membrane is continuous with the endoplasmic reticulum (ER) as shown in figs 5.10, 5.11 and 5.24, and like the ER it may be

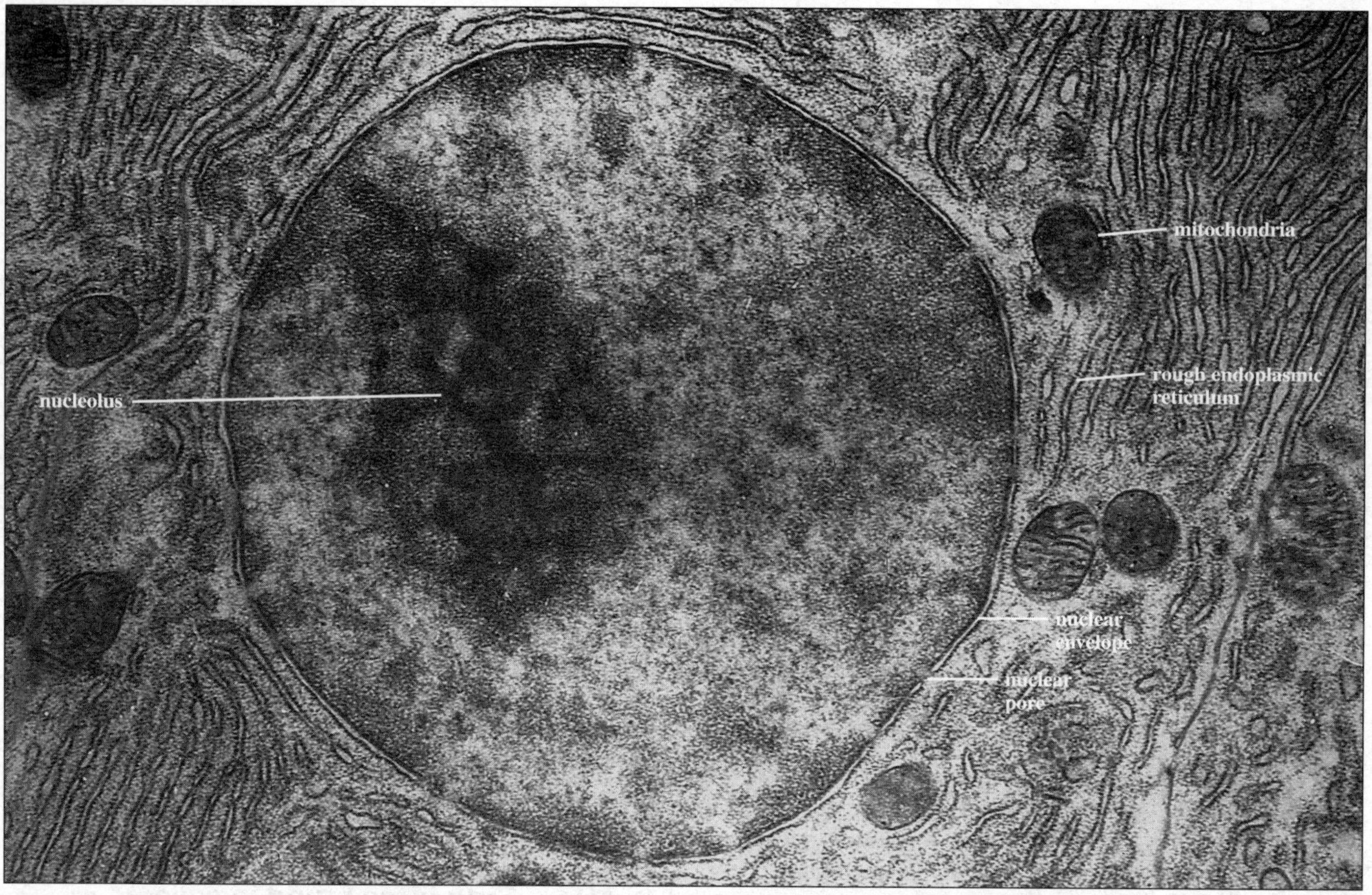

Fig 5.24 *Transmission EM of a nucleus. Note, in reality, the endoplasmic reticulum is continuous with the nuclear envelope.*

covered with ribosomes carrying out protein synthesis. The nuclear envelope is perforated by nuclear pores (fig 5.24) and these are particularly well revealed by freeze etching (fig 5.25). Nuclear pores allow exchange of substances between the nucleus and the cytoplasm. For example, the messenger RNA (mRNA) and newly made ribosomes can leave, and molecules needed for the manufacture of ribosomes and DNA (proteins and nucleotides) and the regulation of DNA (e.g. some hormones) can enter. The pore has a definite structure formed by fusion of the outer and inner membranes of the envelope. This controls the passage of molecules through the pore.

Chromatin

Chromatin is composed mainly of coils of DNA bound to basic proteins called **histones**. The DNA is so long (in humans there is an average of about one metre's length in each nucleus!) that it has to be packaged in an organised manner or it would get tangled like an unravelled ball of string. DNA is wound around the histones which form bead-like structures called nucleosomes, and these in turn are regularly packed in the chromatin.

The term chromatin means 'coloured material' and refers to the fact that this material is easily stained for viewing with the microscope. During nuclear division chromatin

Fig 5.25 *Electron micrograph of freeze-etched nucleus showing nuclear pores (x 30 000). In this technique, instantly frozen cells are split open with a metal blade. They fracture along planes of weakness, often through membranes. Removal of the ice leaves an etched surface.*

stains more intensely and becomes more conspicuous because it condenses into more tightly coiled threads called **chromosomes**. During interphase (the period between nuclear divisions) some of it becomes looser and more scattered. However, some remains tightly coiled and continues to stain intensely. This is called **heterochromatin** and is seen as characteristic dark patches, usually occurring near the nuclear envelope (figs 5.10–13 and fig 5.24). The remaining, loosely coiled chromatin is called **euchromatin**. This is thought to contain the DNA which is genetically active during interphase.

Nucleolus

The **nucleolus** appears as a rounded, darkly stained structure inside the nucleus (figs 5.12 and 5.24). Its function is to make ribosomes. One or more nucleoli may be present. It stains intensely because of the large amounts of DNA and RNA it contains. RNA is a molecule similar to DNA which is copied from DNA. The densely staining core of the nucleolus is made up of the DNA from one or several chromosomes. This contains many copies of the genes that code for the RNA needed to make ribosomes (ribosomal RNA or rRNA). During nuclear division nucleoli seem to disappear, but this is because the DNA disperses. They reassemble after nuclear division.

Around the central core of the nucleolus is a less dense region where ribosomal RNA is beginning to be folded and combined with proteins to make ribosomes. The partly assembled ribosomes move out through the nuclear pores into the cytoplasm, where assembly is completed.

5.10.2 Cytoplasm

The living contents of eukaryotic cells are divided into nucleus and cytoplasm, the two together forming the protoplasm. Cytoplasm is an aqueous (water-containing) substance containing a variety of cell organelles and other structures such as insoluble waste or storage products.

The soluble part of the cytoplasm forms the 'background material' or **'ground substance'** between the cell organelles. It contains a skeleton of very fine fibres (section 5.10.7) but otherwise appears transparent and structureless in the electron microscope. It is about 90% water and forms a solution which contains all the fundamental biochemicals of life. Some of these are ions and small molecules in true solution, such as salts, sugars, amino acids, fatty acids, nucleotides, vitamins and dissolved gases. Others are large molecules such as proteins which form colloidal solutions. A **colloidal solution** is one in which the solute molecules are relatively large (see appendix 1 in Book 2). It may be a sol (runny) or a gel (jelly-like); often the outer regions of cytoplasm are more gel-like.

Apart from acting as a store of vital chemicals, the ground substance is the site of certain metabolic pathways, an important example being glycolysis.

When *living* cytoplasm is examined, great activity is usually seen as cell organelles such as mitochondria move about.

5.10.3 Endoplasmic reticulum (ER)

One of the most important discoveries to be made when the electron microscope was introduced was the occurrence of a system of membranes running through the cytoplasm of all eukaryotic cells. This network, or reticulum, of membranes was named the **endoplasmic reticulum** (ER) and, although it is often extensive, it cannot be seen with a light microscope. The membranes are often covered with small particles called **ribosomes**.

In sections, the ER typically appears as pairs of parallel lines (membranes) running through the cytoplasm, as shown in fig 5.24. Occasionally though, a section will glance through the surface of these membranes and show that, in three dimensions, the ER is usually sheet-like rather than tubular. A possible three-dimensional structure is shown in fig 5.26. The ER consists of flattened, membrane-bound sacs called **cisternae**. These may be covered with ribosomes, forming **rough ER**, or ribosomes may be absent, forming **smooth ER**, which is usually more tubular.

Rough ER is concerned with the **transport of proteins** which are made by ribosomes on its surface. Details of protein synthesis are given in chapter 23. The growing protein, which consists of a chain of amino acids, is bound to the ribosome until its synthesis is complete. A receptor in the membrane of the ER provides a channel through which the protein can pass into the ER once it has been made.

The protein is now transported through the cisternae, usually being extensively modified en route. For example, it may be converted into a glycoprotein. The protein commonly travels to the Golgi apparatus from where it can be secreted from the cell or passed on to other organelles in the same cell. The enzymes in lysosomes have followed this route (see also fig 5.29).

One of the chief functions of smooth ER is **lipid synthesis**. For example, in the epithelium of the intestine the smooth ER makes lipids from fatty acids and glycerol absorbed from the gut, and passes them on to the Golgi apparatus for export. Smooth ER also makes **steroids**, which are a type of lipid. Some steroids are hormones, such as the corticosteroids made in the adrenal cortex and the sex hormones such as testosterone and oestrogen. In muscle cells a specialised form of smooth ER, called **sarcoplasmic reticulum**, is present.

5.10.4 Ribosomes

Ribosomes are tiny organelles, about 20 nm in diameter, that are found in large numbers throughout the cytoplasm of living cells, both prokaryotic and eukaryotic. A typical bacterial cell contains about 10 000 ribosomes, while eukaryotic cells possess many more times this number. They are the sites of protein synthesis.

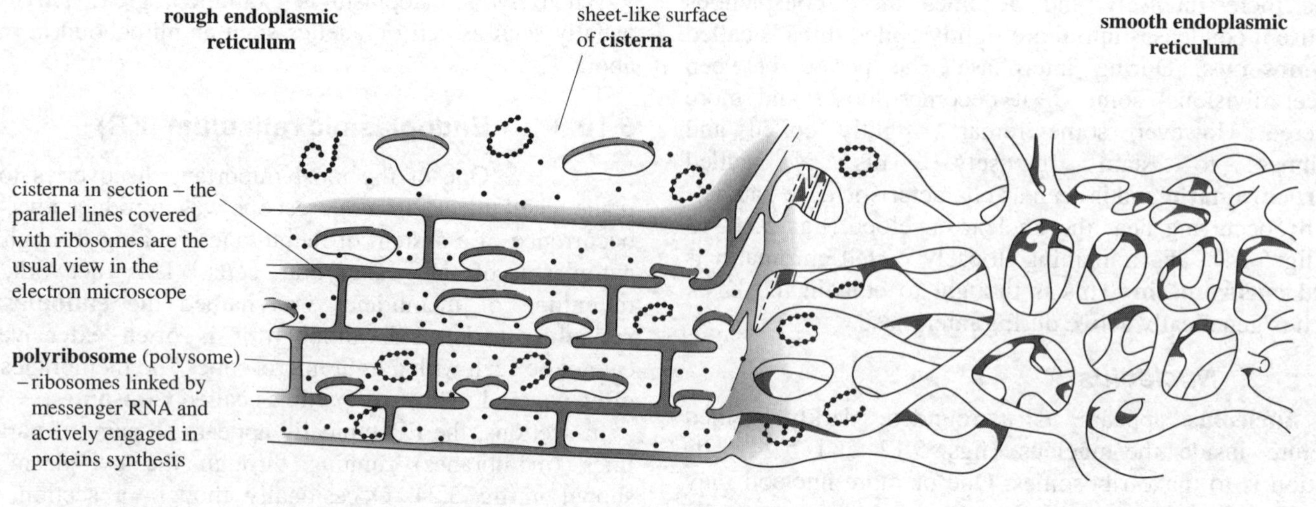

Fig 5.26 *Three-dimensional model of endoplasmic reticulum.*

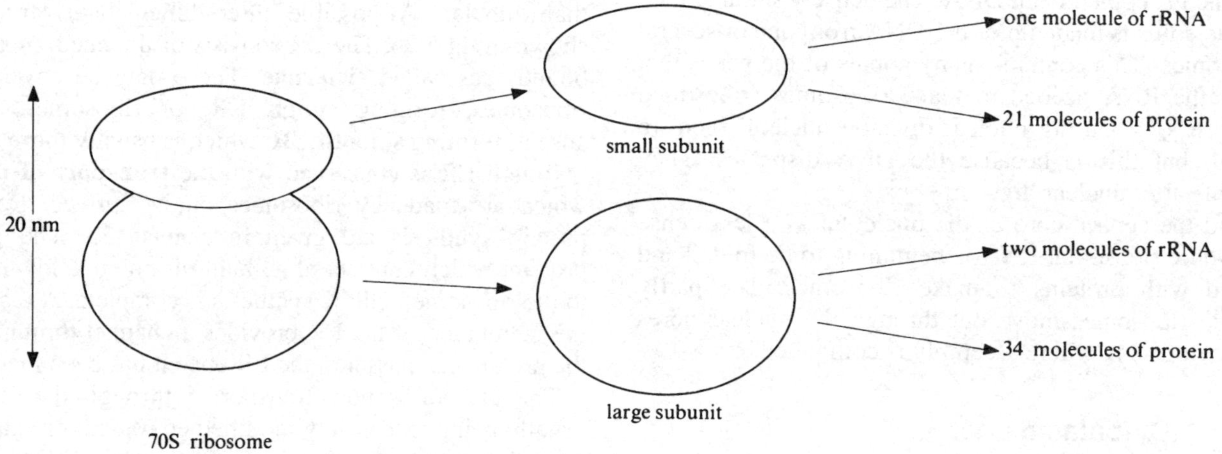

Fig 5.27 *Structure of a 70S ribosome. (The subunits of 80S ribosomes possess more proteins and the large subunit possesses three rRNA molecules.)*

Each ribosome consists of two subunits, one large and one small as shown in fig 5.27. Being so small they are the last organelles to be sedimented in a centrifuge, requiring a force of 100 000 times gravity for 1–2 h (fig 5.9). Sedimentation has revealed two basic types of ribosome, called 70S* and 80S ribosomes. The 70S ribosomes are found in prokaryotes and the slightly larger 80S ribosomes occur in the cytoplasm of eukaryotes. Chloroplasts and mitochondria contain 70S ribosomes, revealing their prokaryotic origins (section 7.4.1).

Ribosomes are made of roughly equal amounts of RNA and protein. The RNA is termed **ribosomal RNA (rRNA)** and is made in the nucleoli. The distribution of rRNA molecules and protein molecules is given in fig 5.27. Together these molecules form a complex three-dimensional structure.

*S = Svedberg unit. This is related to the rate of sedimentation in a centrifuge; the greater the S number, the greater the rate of sedimentation.

During protein synthesis at ribosomes, amino acids are joined together one by one to form polypeptide chains. The process is described in detail in chapter 23. The ribosome acts as a binding site where the molecules involved can be precisely positioned relative to each other. These molecules include messenger RNA (mRNA), which carries the genetic instructions from the nucleus, transfer RNA (tRNA), which brings the required amino acids to the ribosome, and the growing polypeptide chain. In addition there are chain initiation, elongation and termination factors. The process is so complex that it could not occur efficiently, if at all, without the ribosome.

Two populations of ribosomes can be seen in eukaryotic cells, namely free and ER-bound ribosomes (figs 5.24 and 5.26). All of the ribosomes have an identical structure but some are bound to the ER by the proteins that they are making. Such proteins are usually secreted. An example of a protein made by free ribosomes is haemoglobin in young red blood cells.

During protein synthesis, the ribosome moves along the thread-like mRNA molecule. Rather than one ribosome at a time passing along the RNA, the process is carried out more efficiently by a number of ribosomes moving along the mRNA at the same time, like beads on a string. The resulting chains of ribosomes are called **polyribosomes** or **polysomes**. They form characteristic whorled patterns on the ER as shown in fig 5.26.

5.10.5 Golgi apparatus

The Golgi apparatus was discovered by Camillo Golgi in 1898, using special staining techniques. However, its structure was only revealed by electron microscopy. It is found in virtually all eukaryotic cells and consists of a stack of flattened, membrane-bound sacs called **cisternae**, together with a system of associated vesicles (small sacs) called **Golgi vesicles**. It is difficult to build up a three-dimensional picture of the Golgi apparatus from thin sections, but it is believed that a complex system of interconnected tubules is formed around the central stack, as shown in fig 5.28.

At one end of the stack new cisternae are constantly being formed by vesicles from the smooth ER. This 'outer' or 'forming' face is convex, whilst the other end is the concave 'inner' or 'maturing' face where the cisternae break up into vesicles again. The whole stack consists of a number of cisternae moving from the outer face to the inner face.

The function of the Golgi apparatus is to transport and chemically modify the materials contained within it. It is particularly important in secretory cells, a good example being provided by the pancreas. Here specialised cells secrete the digestive enzymes of the pancreatic juice into the pancreatic duct, along which they pass to the duodenum. Fig 5.29*a* is an electron micrograph of such cells, and fig 5.29*b* is a diagram showing the secretion pathway.

(*a*)

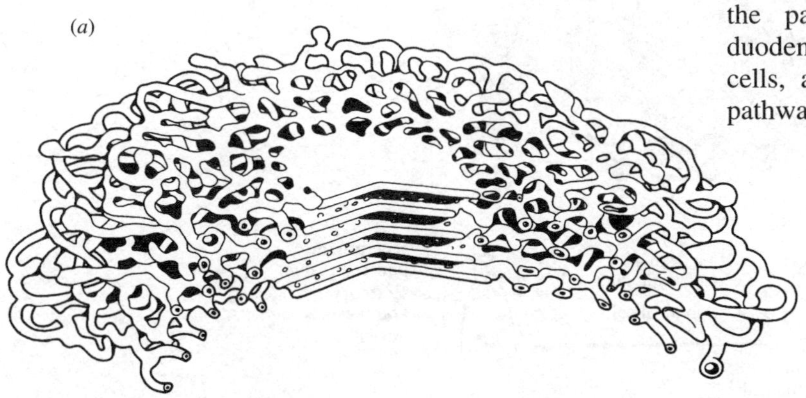

Fig 5.28 (a) *The three-dimensional structure of the Golgi apparatus.* (b) *Transmission electron micrograph showing two Golgi apparatuses. The left-hand one shows a vertical section. The right-hand one shows the topmost cisterna viewed from above (x50 000).*

(*b*)

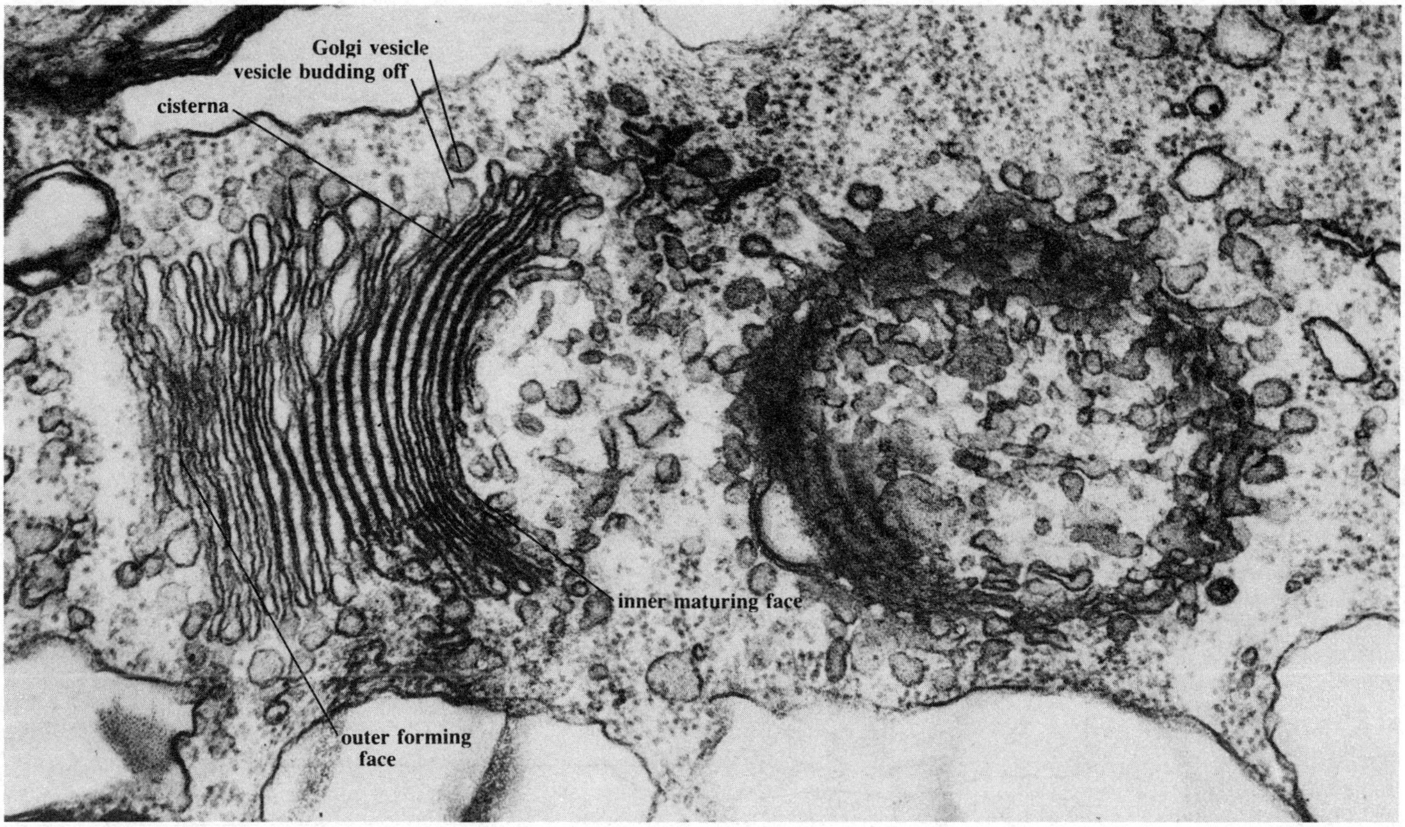

(a)

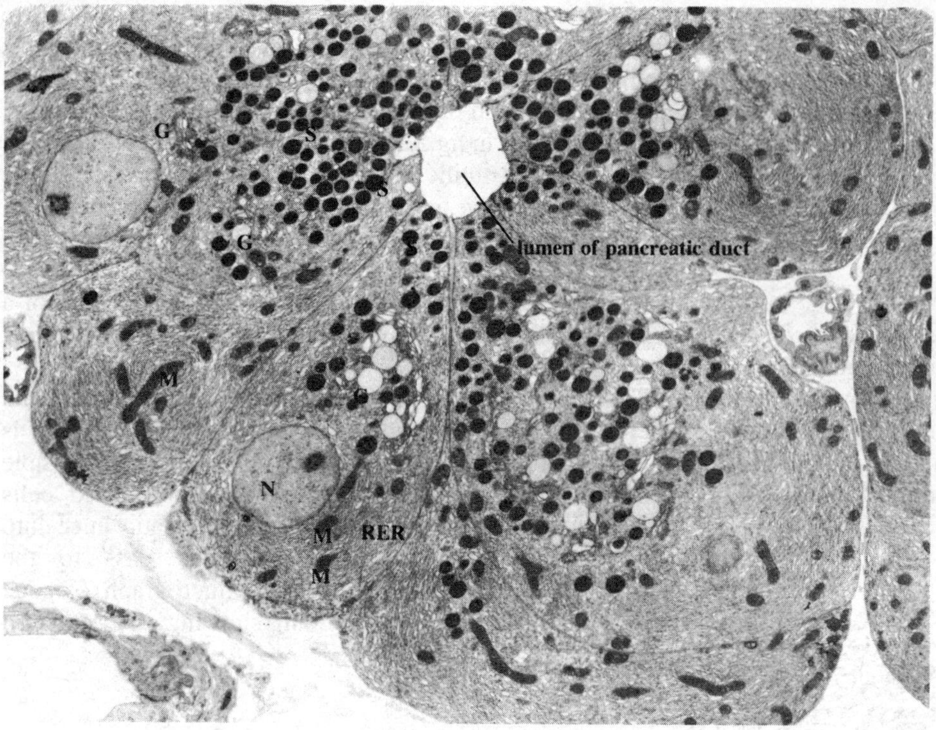

Fig 5.29 (a) *Electron micrograph of a group of enzyme-secreting pancreatic cells (x10 400). N, nucleus; M, mitochondrion; G, Golgi apparatus; S, secretory granules; RER, rough endoplasmic reticulum.* (below) (b) *Diagram of the synthesis and secretion of a protein (an enzyme).*

(b)

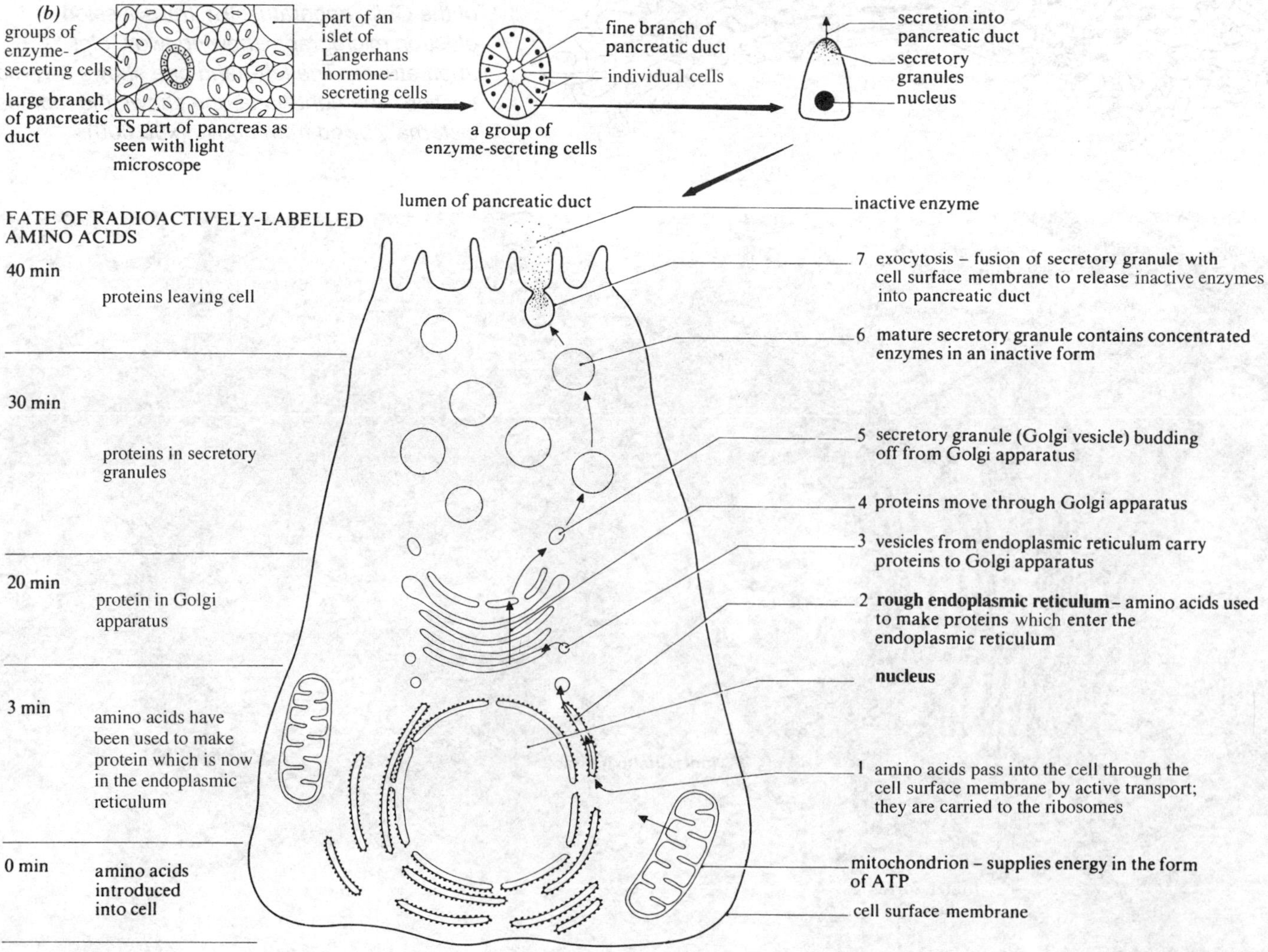

Details of the pathway have been confirmed by using radioactively labelled amino acids. These get used by the cell to make protein. Because they are radioactive they can be traced as they pass through different cell organelles. This is done by homogenising samples of tissue at different times after supplying the amino acids, separating the cell organelles by centrifugation (fig 5.9) and finding which organelles contain the highest proportion of the radioactivity. After concentration in the Golgi apparatus, the protein is carried in Golgi vesicles to the cell surface membrane. The final stage in the pathway is secretion of the inactive enzyme by reverse pinocytosis. The digestive enzymes secreted by the pancreas are synthesised in an inactive form so that they do not attack and destroy the cells that make them. An example is trypsinogen which is converted to active trypsin in the duodenum.

In general, proteins received by the Golgi apparatus from the ER have had short carbohydrate chains added to make them glycoproteins (like the membrane proteins shown in fig 5.16). These carbohydrate 'antennae' can be remodelled in the Golgi apparatus, possibly to become markers that direct the proteins to their correct destinations. However, the exact details of how the Golgi apparatus sorts and directs molecules are unknown.

The Golgi apparatus is also sometimes involved in the secretion of carbohydrates, an example being provided by the synthesis of new cell walls by plants. Fig 5.30 shows the intense activity which goes on at the 'cell plate', the region between two newly formed daughter nuclei where the new cell wall is laid down after nuclear division.

Golgi vesicles are steered into position at the cell plate by microtubules (section 5.10.7). The membranes of the vesicles become the new cell surface membranes of the daughter cells, while their contents contribute to the middle lamella and new cell walls. Cellulose is added separately and involves microtubules, not the Golgi apparatus.

Secretion of enzymes by pancreatic cells and the formation of new plant cell walls are good examples of division of labour, where many cell organelles combine to perform one function.

The goblet cells of the respiratory pathway and gut secrete a glycoprotein called mucin which forms mucus in solution. Mucin is released by the Golgi apparatus. The Golgi apparatus in the leaf glands of some insectivorous plants such as sundews secretes a sticky slime and enzymes which trap and digest insects. The slime, wax, gum and mucilage secretions of many cells are released by the Golgi apparatus.

A second important function of the Golgi apparatus, in addition to secretion, is the **formation of lysosomes** which is described below.

5.10.6 Lysosomes

Lysosomes are found in most eukaryotic cells. They are surrounded by a single membrane and are simple sacs that contain digestive enzymes, such as proteases, nucleases and lipases which break down proteins, nucleic acids and lipids respectively. Such enzymes carry out hydrolysis reactions (adding water) and work best in an acid environment. The contents of the lysosome are therefore acidic. The enzymes have to be kept apart from the rest of the cell or they would destroy it. In animal cells, lysosomes are usually spherical and 0.2–0.5 μm in diameter (fig 5.31).

In plant cells the large central vacuoles may act as lysosomes, although bodies similar to the lysosomes of animal cells are sometimes seen in the cytoplasm.

The enzymes contained within lysosomes are synthesised on rough ER and transported to the Golgi apparatus. Golgi vesicles containing the processed enzymes later bud off to form the lysosomes. Lysosomes have a number of functions, summarised below and in fig 5.32.

Digestion of material taken in by endocytosis (route 1, fig 5.32)

The process of endocytosis is explained in section 5.9.8. Lysosomes may fuse with the vesicles or vacuoles formed by endocytosis, releasing their enzymes into the vacuole and digesting the material inside. This material might be taken in for food, as in the food vacuoles of some protozoans, or for defensive purposes, as in the case of phagocytic vacuoles formed by white blood cells when engulfing bacteria (fig 5.23). The products of digestion are absorbed and assimilated by the cytoplasm of the cell leaving undigested remains. The vacuole usually migrates to the cell surface membrane and releases its contents (exocytosis).

An interesting example of the role of lysosomes occurs in the thyroid gland. In section 5.9.8 the uptake of thyroglobulin by pinocytosis was described. The pinocytic vesicles so formed fuse with lysosomes, and the thyroglobulin is hydrolysed by lysosomal enzymes to produce the active hormone thyroxine. The lysosome then fuses with the cell surface membrane, thus secreting the hormone into the blood.

Autophagy (route 2, fig 5.32)

Autophagy is the process by which unwanted structures within the cell are engulfed and digested within lysosomes. They are first enclosed by a single membrane, usually derived from smooth ER, and this structure then fuses with a lysosome to form an **'autophagic vacuole'**, in which the unwanted material is digested. This is part of the normal turnover of cytoplasmic organelles, old ones being replaced by new ones.

Release of enzymes outside the cell (exocytosis) (route 3, fig 5.32)

Sometimes the enzymes of lysosomes are released from the cell. This occurs during the replacement of cartilage by bone during development. Similarly, bone may be broken down during the remodelling of bone that can occur in response to injury, new stresses and so on. Sperm contain a

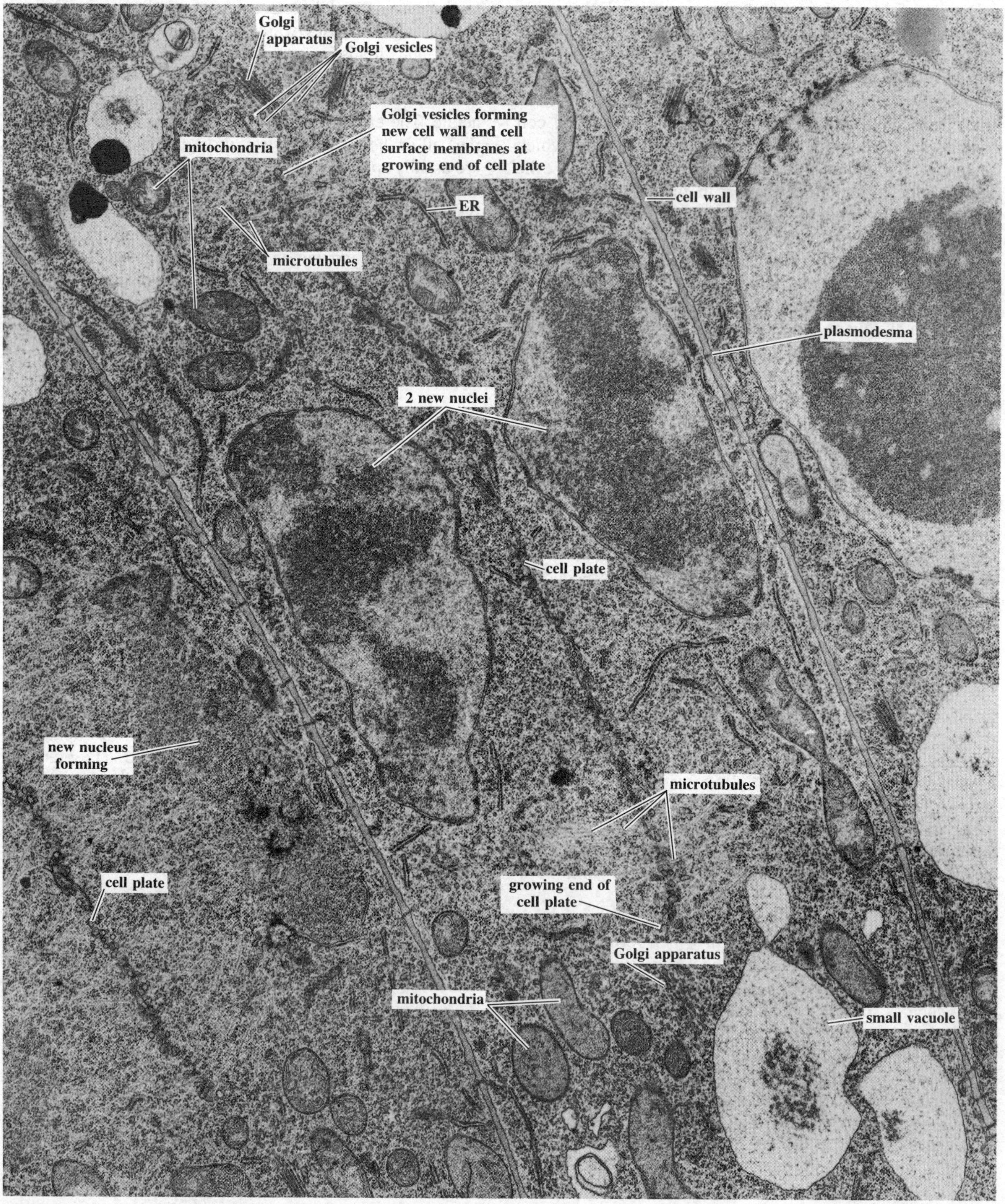

Fig 5.30 *Electron micrograph of a dividing plant cell. The dividing cell runs diagonally across the picture. Each of the two new cells is forming a cell wall from the centre of the cell outwards to the edge. The growing wall is known as the cell plate because it looks like a plate in three dimensions. Note the close association of Golgi apparatus and microtubules with the growing end of the cell plate (x15 000).*

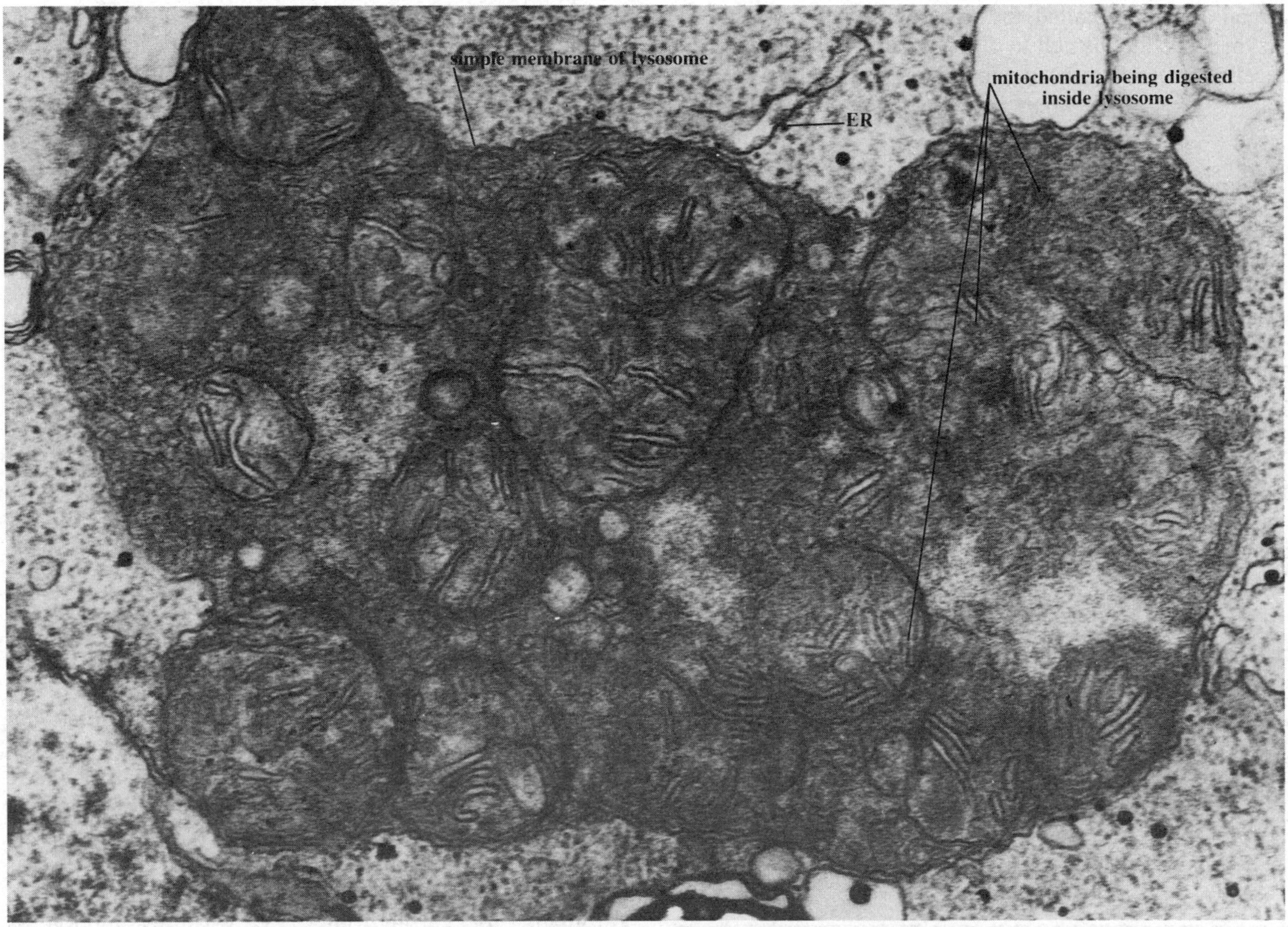

Fig 5.31 *Electron micrograph of a lysosome which has engulfed some old mitochondria and is digesting them (x90 750).*

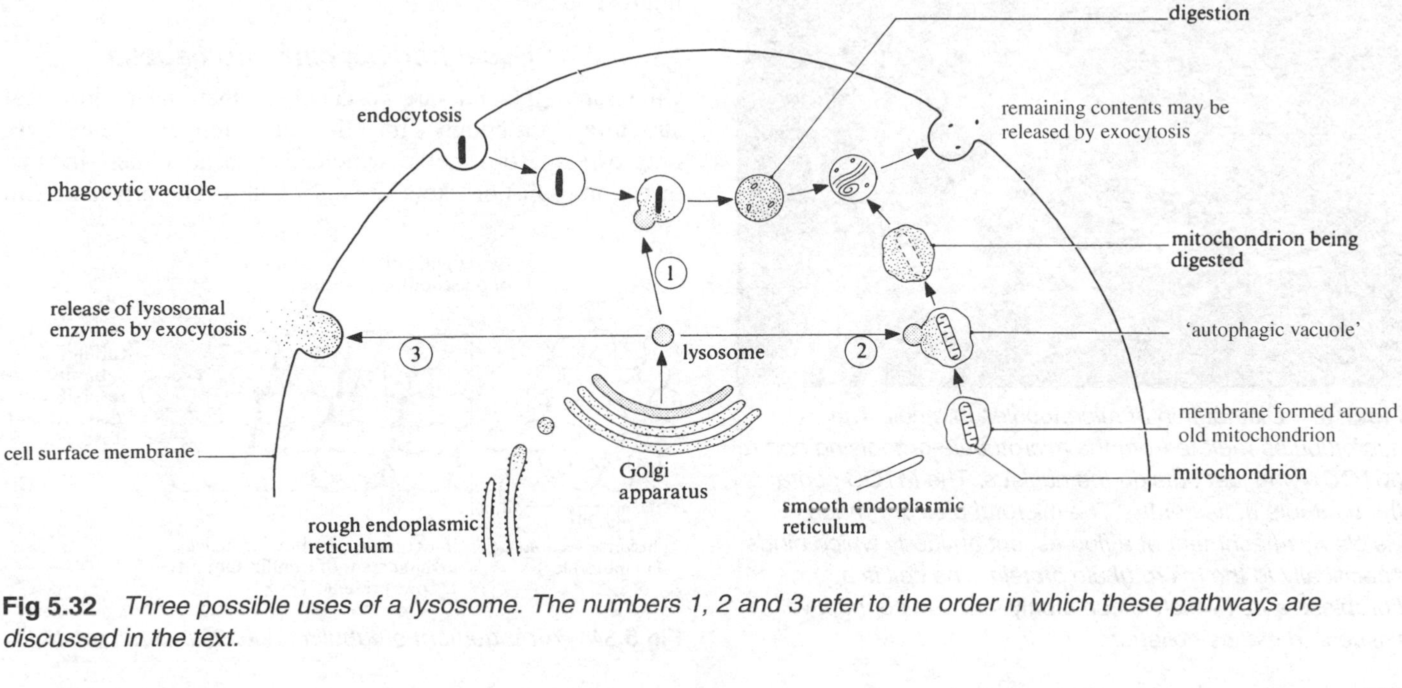

Fig 5.32 *Three possible uses of a lysosome. The numbers 1, 2 and 3 refer to the order in which these pathways are discussed in the text.*

special lysosome called the **acrosome**. This releases its enzymes outside the cell to digest a path through the layers of cells surrounding the egg just before fertilisation.

Autolysis

Autolysis is the **self-digestion** of a cell by releasing the contents of lysosomes within the cell. In such circumstances lysosomes have sometimes been aptly named 'suicide bags'. Autolysis is a normal event in some differentiation processes and may occur throughout a tissue, as when a tadpole tail is reabsorbed during metamorphosis. Another example occurs in the uterus. During pregnancy the uterus grows much larger to accommodate the growing baby. After birth, it gradually returns to its normal size by self-digestion (autophagy) of many of the cells. Autolysis also occurs in muscles which are not exercised! It also occurs after cells die and is one reason why food deteriorates unless refrigerated. Sometimes it occurs as a result of certain lysosomal diseases.

5.10.7 Microtubules

Using electron microscopy, structure has been revealed in the apparently structureless 'background' or ground substance of the cytoplasm. Networks of fibrous protein structures have been shown to exist in all eukaryotic cells. These are known collectively as the **cytoskeleton**. The fibres are of at least three types: **microtubules**, **microfilaments** and **intermediate filaments**. Together, they are concerned with movement, the ability of cells to maintain their shapes, and with various other activities such as endocytosis and exocytosis. Only microtubules will be considered here.

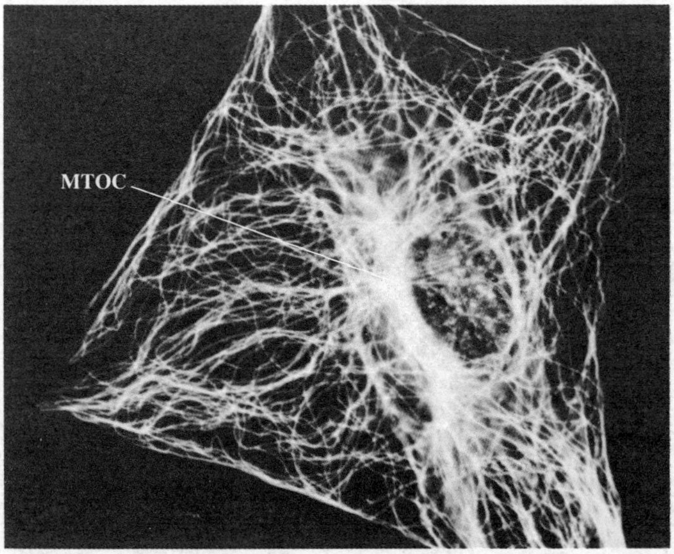

Fig 5.33 *Distribution of microtubules in a cell. The microtubules radiate from the microtubule-organising centre (MTOC) lying just outside the nucleus. The MTOC contains the centriole at its centre. The microtubules are made visible by attachment of a fluorescent antibody which binds specifically to the microtubule protein. The cell is a fibroblast, a cell which is commonly found in connective tissue and makes collagen.*

Nearly all eukaryotic cells contain microtubules (fig 5.33). They are very fine, unbranched, hollow tubes. They have an external diameter of about 24 nm and walls about 5 nm thick made up of helically arranged subunits of a protein called tubulin, as shown in fig 5.34. Their typical appearance in electron micrographs is shown in fig 5.30. Growth of microtubules occurs at one end by addition of tubulin subunits. It apparently requires a template to start and certain very small ring-like structures that have been isolated from cells, and which consist of tubulin subunits, appear to serve this function. In intact cells, centrioles also serve this function and are therefore sometimes known as microtubule-organising centres, or MTOCs. Centrioles contain short microtubules.

Microtubules are involved in a number of cell processes, some of which are described below.

Centrioles and nuclear division

Centrioles are small hollow cylinders (about 0.3–0.5 μm long and about 0.2 μm in diameter) that occur in pairs in most animal cells. Each contains nine triplets of microtubules. At the beginning of nuclear division the centrioles replicate themselves and the two pairs migrate to opposite poles of the spindle, the structure on which the chromosomes line up. The spindle itself is made of microtubules ('spindle fibres') using centrioles as MTOCs. The microtubules control separation of chromatids or chromosomes by a sliding motion as described in chapter 23. Cells of higher plants lack centrioles, although they do produce spindles during nuclear division. The cells are thought to contain smaller MTOCs that are not easily visible even with the electron microscope.

Basal bodies, cilia and flagella

Cilia and flagella are organelles that have identical structures, although flagella are longer. They are outgrowths from cells which can beat either in one direction (cilia) or like a wave (flagella). The organelle can

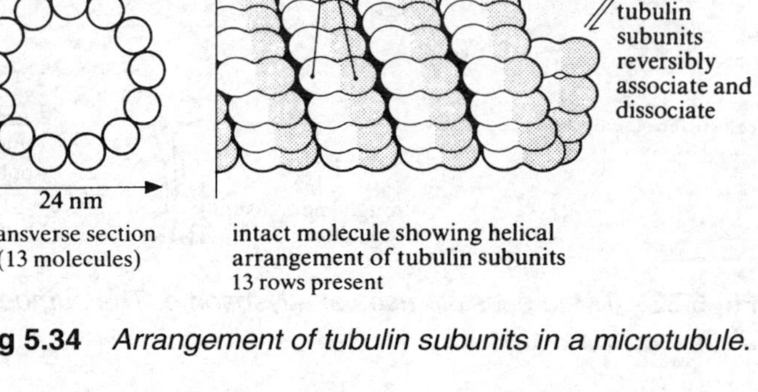

Fig 5.34 *Arrangement of tubulin subunits in a microtubule.*

be used for locomotion of single cells or to move fluids over the surfaces of cells, as cilia do when they move mucus through the respiratory tract. At the base of every cilium and flagellum is a **basal body**. Basal bodies are identical in structure to centrioles and are probably made by replication of centrioles. Like centrioles, they also seem to act as MTOCs because cilia and flagella contain a characteristic '9 + 2' arrangement of microtubules (chapter 18).

In cilia and flagella, microtubules undergo sliding motions which are responsible for the beating movements. Further details are given in chapter 18. Note that bacterial flagella are simpler than eukaryotic flagella and do not have basal bodies.

Intracellular transport

Microtubules are involved in the movements of other cell organelles such as Golgi vesicles, an example being the guiding of Golgi vesicles to the cell plate shown in fig 5.30. There is a constant traffic in cells of Golgi vesicles, of vesicles from the ER to the Golgi apparatus, lysosomes, mitochondria and other organelles. Such movements stop if the microtubule system is disrupted.

Cytoskeleton

Microtubules also have a structural role in cells, their long, fairly rigid, tube-like structure acting like a skeleton in forming part of the '**cytoskeleton**'. They help to determine and maintain the shapes of cells during development, often being found in a zone just beneath the cell surface membrane (figs 5.10 and 5.11). Animal cells in which microtubules are disrupted change to a spherical shape. In plant cells the alignment of microtubules exactly corresponds with the alignment of cellulose fibres during formation of the cell wall, thus helping to determine cell shape.

5.10.8 Microvilli

Microvilli are finger-like extensions of the cell surface membrane of some animal cells (figs 5.10 and 5.12). They increase the surface area by as much as 25 times and are particularly numerous on cells specialised for absorption, such as epithelial cells in the intestine and kidney nephrons. The increase in surface area also improves the efficiency of digestion in the gut because certain digestive enzymes are attached to their surface (see section 8.3.8).

The microvilli can just be seen with a light microscope as a fringe across the top of the cell called a **brush border**.

Each microvillus contains bundles of actin and myosin filaments. Actin and myosin are the proteins found in muscle which cause muscle contraction. At the base of the microvillus actin and other filaments join up with filaments from neighbouring microvilli to form a network of filaments. The system as a whole allows the microvilli to remain upright and retain their shape, while still allowing their movement through the interactions of actin and myosin (similar to muscle contraction).

5.10.9 Mitochondria

Mitochondria are found in all aerobic eukaryotic cells and their structure and function are briefly summarised in fig 5.12. Their chief function is aerobic respiration and they are described in detail in section 9.3.

5.10.10 Cell walls

Plant cells, like those of prokaryotes and fungi, are surrounded by a relatively rigid wall which is secreted by the living cell (the protoplast) within. Plant cell walls differ in chemical composition from those of the prokaryotes and the fungi (table 2.2). The wall laid down during cell division of plants is called the **primary wall**. This may later be thickened to become a **secondary wall**. Formation of the primary wall is described in this section and an early stage of wall formation is shown in fig 5.30.

Structure of the cell wall

The primary wall consists of cellulose fibrils running through a **matrix** of other polysaccharides. Cellulose is a polysaccharide whose chemical structure is described in section 3.2.3. It has a high tensile strength which approaches that of steel. The matrix consists of polysaccharides which are usually divided for convenience into **pectins** and **hemicelluloses**. **Pectins** are acidic and have a relatively high solubility. The **middle lamella** that holds neighbouring cell walls together is composed of sticky, gel-like magnesium and calcium salts (pectates) of pectins.

Hemicelluloses are a mixed group of alkali-soluble polysaccharides. Like cellulose they form chain-like molecules, but the chains are less organised, shorter and more branched.

Cell walls are hydrated and 60–70% of their mass is usually water. Water can move freely through free space in the cell wall.

In some cells, such as leaf mesophyll cells, the primary wall remains the only wall. In most, however, extra layers of cellulose are laid down on the inside surface of the primary wall (the outside surface of the cell surface membrane), thus building up a secondary wall. Within each layer of the secondary wall the cellulose fibres are usually orientated at the same angle, but the fibres of different layers are orientated at different angles, forming a strong cross-ply structure. This is shown in fig 5.35.

Some cells, such as xylem vessel elements and sclerenchyma, undergo extensive **lignification** whereby lignin, a complex polymer (not a polysaccharide), is deposited in all the cellulose layers. In some cells, such as protoxylem, the lignification is only partial. In others it is complete, apart from 'pits' where groups of plasmodesmata were originally present in the primary wall (section 6.1.3 and fig 6.8). Lignin cements and anchors cellulose fibres together. It acts as a very hard and rigid matrix, giving the cell wall extra tensile and particularly compressional strength which prevents buckling. It is the main supporting material of trees. It also protects cells from physical and chemical damage. Together with cellulose, which remains in the wall, it is responsible for the unique characteristics of wood as a construction material.

Box 5A Composite materials

Mechanically strong materials, like cell walls, in which more than one component is present are known as **composite materials** and they are generally stronger than any of their components in isolation. Fibre–matrix systems are used widely in engineering and a study of their properties is an important branch of both modern engineering and biology. The matrix transfers stress to the fibres, which have a high tensile strength. The matrix also improves resistance to compression and shear. An example of a composite material traditionally used in engineering is reinforced concrete in which a concrete matrix is reinforced in various ways, such as by steel rods. More modern and lighter structural materials are fibreglass and carbon fibre in which a plastic matrix is reinforced with glass or carbon fibres. Other biological composites include wood, bone, cartilage and arthropod exoskeletons, which are rigid, and connective tissue and skin, which are flexible.

Fig 5.35 *Electron micrograph of layers from the wall of the green seaweed* Chaetomorpha melagonium *showing cellulose microfibrils about 20 nm wide; the contrast is due to shadowing with a platinum/gold alloy.*

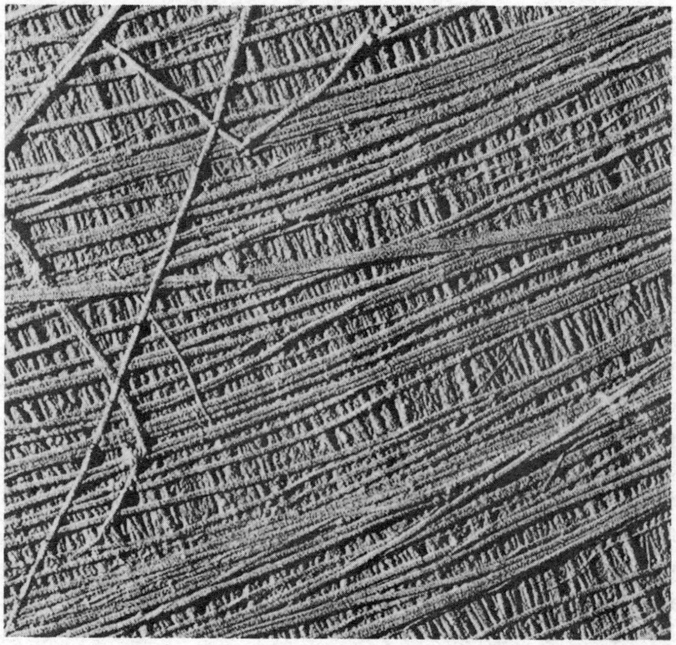

Functions of the cell wall

The main functions of plant cell walls are summarised below.

(1) Mechanical strength and skeletal support is provided for individual cells and for the plant as a whole. Extensive lignification increases strength in some walls (small amounts are present in most walls) and is particularly important in trees and shrubs.
(2) Cell walls are fairly rigid and resistant to expansion and therefore allow development of turgidity when water enters the cell by osmosis. This contributes to the support of all plants and is the main source of support in herbaceous plants and organs, such as leaves, which do not undergo secondary growth. The cell wall also prevents the cell from bursting when exposed to a dilute solution.
(3) Orientation of cellulose microfibrils limits and helps to control cell growth and shape because the cell's ability to stretch is determined by their arrangement. If, for example, cellulose microfibrils form hoops in a transverse direction around the cell, the cell will stretch, as it fills with water by osmosis, in a longitudinal direction.
(4) The system of interconnected cell walls (the **apoplast**) is a major pathway of movement for water and dissolved mineral salts (chapter 13). The walls are held together by middle lamellae. The cell walls also possess minute pores through which structures called **plasmodesmata** can pass, forming living connections between cells and allowing all the protoplasts to be linked in a system called the **symplast** (chapter 13).
(5) Cell walls develop a coating of waxy cutin, the cuticle, on exposed epidermal surfaces reducing water loss and risk of infection. Cork cell walls undergo impregnation with suberin which serves a similar function after secondary growth.
(6) The walls of xylem vessels and sieve tubes are adapted for long-distance translocation of materials through the cells, as explained in chapters 6 and 13.
(7) The cell walls of root endodermal cells are impregnated with suberin that forms a barrier to water movement (chapter 13).
(8) Some cell walls are modified as food reserves, as in storage of hemicelluloses in some seeds.
(9) The cell walls of transfer cells develop an increased surface area and the consequent increase in surface area of the cell surface membrane increases the efficiency of transfer by active transport (chapter 13).

5.10.11 Plasmodesmata

Plasmodesmata (singular *plasmodesma*) are living connections between neighbouring plant cells which run through very fine pores in the walls (fig 5.13). The cell surface membranes of neighbouring cells are continuous and line the pores. Running through the centre of each pore is smooth endoplasmic reticulum. Communication and coordination between plant cells is therefore made easier since molecules and ions do not have to cross a cell surface membrane. Movement is, however, regulated. Viruses can take advantage of these pores and can spread through plasmodesmata.

Sieve plate pores of phloem sieve tubes are formed from plasmodesmata (chapter 6).

5.10.12 Vacuoles

A vacuole is a fluid-filled sac bounded by a single membrane. Animal cells contain relatively small vacuoles, such as phagocytic vacuoles, food vacuoles, autophagic vacuoles and contractile vacuoles. However, plant cells, notably mature parenchyma cells, have a large central vacuole surrounded by a membrane called the **tonoplast** (fig 5.11). The fluid they contain is called **cell sap**. It is a concentrated solution of mineral salts, sugars, organic acids, oxygen, carbon dioxide, pigments and some waste and 'secondary' products of metabolism. The functions of vacuoles are summarised below.

(1) Water generally enters the concentrated cell sap by osmosis through the partially permeable tonoplast. As a result a pressure builds up within the cell and the cytoplasm is pushed against the cell wall. Osmotic uptake of water is important in cell expansion during cell growth, as well as in the normal water relations of plants.

(2) The vacuole sometimes contains pigments in solution. These include **anthocyanins**, which are red, blue and purple, and other related compounds which are shades of yellow and ivory. They are largely responsible for the colours in flowers (for example in roses, violets and *Dahlia*), fruits, buds and leaves. In the latter case they contribute to autumn shades, together with the photosynthetic pigments of chloroplasts. They are important in attracting insects, birds and other animals for pollination and seed dispersal.

(3) Plant vacuoles sometimes contain hydrolytic enzymes and act as lysosomes. After cell death the tonoplast, like all membranes, loses its partial permeability and the enzymes escape causing autolysis.

(4) Waste products and certain secondary products of plant metabolism may accumulate in vacuoles. For example, crystals of waste calcium oxalate are sometimes observed. Secondary products like alkaloids and tannins may offer protection from consumption by herbivores. Latex, a milky liquid, may accumulate in vacuoles, as in dandelion stems. The latex of the rubber tree contains the chemicals needed for rubber synthesis, and the latex of the opium poppy contains alkaloids such as morphine from which heroin is obtained.

(5) Some of the dissolved substances act as food reserves, which can be utilised by the cytoplasm when necessary, for example sucrose and mineral salts.

5.10.13 Chloroplasts

Chloroplasts contain chlorophyll and carotenoid pigments and carry out photosynthesis. They are found mainly in leaves and are described in section 7.4.1.

5.11 Use of the hand lens and microscope

5.11.1 The hand lens

This is a convex lens mounted in a frame. The frame may be small (pocket lens), or much larger for aiding dissection (tripod lens). The hand lens should be held close to the eye and the object brought towards the lens until an enlarged image can be seen. If a drawing is to be made, then the magnification of the drawing in relation to the size of the object must be calculated.

$$\text{drawing magnification} = \frac{\text{linear dimension of the drawing}}{\text{linear dimension of the object}}$$

For example: $\frac{\text{6 cm wide drawing}}{\text{2 cm wide object}} = \text{magnification} \times 3$

5.11.2 The light microscope

The **compound microscope** uses the magnifying powers of two convex lenses to produce a magnified image of a very small object.

Magnification

The magnification of the object is the multiple of eyepiece and objective lens magnifications (table 5.4).

Table 5.4 Magnification of the microscope.

Objective lens	*Eyepiece lens*	*Magnification of the object*
×10	×6	×60
×40	×6	×240
×10	×10	×100
×40	×10	×400

Parts of the microscope

Examine a microscope and identify the parts shown in fig 5.36. The microscope is an expensive instrument and should be handled carefully, paying attention to the following.

- Keep the instrument in a box (or under a cover) when not in use in order to maintain it in a dust-free state.
- Lift the microscope by its supporting arm only. Also support it from below with one hand. Place it on a bench gently to avoid unnecessary jarring.
- The lenses must be kept clean by wiping with a lens tissue (see below).

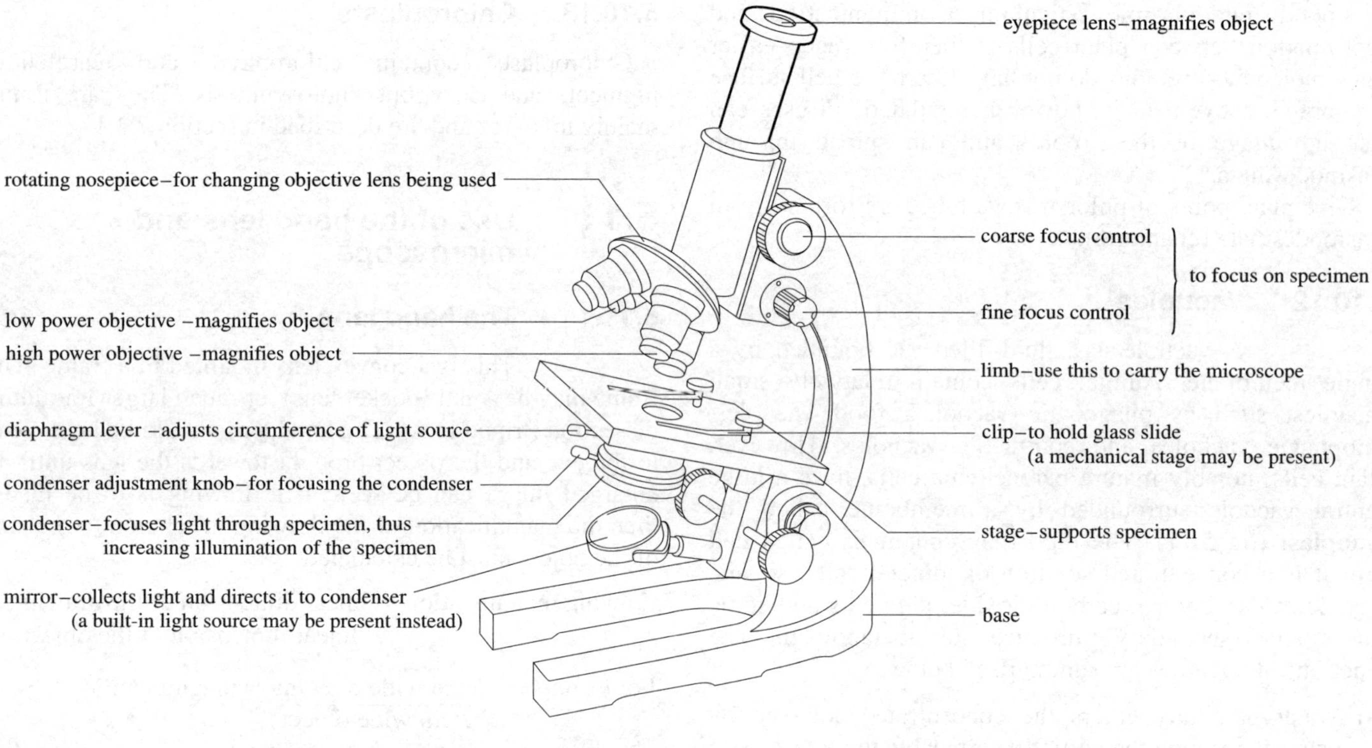

Path of light

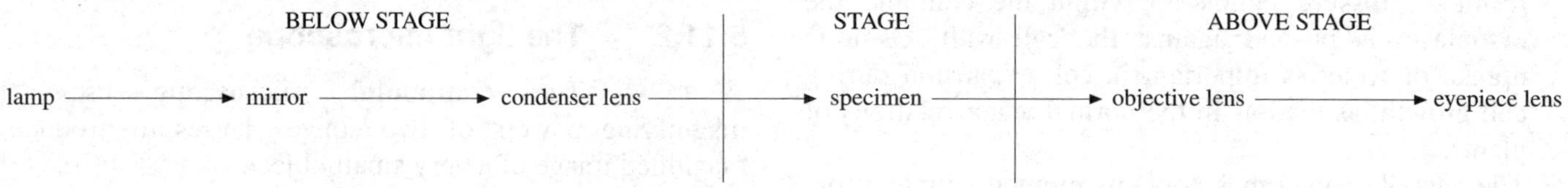

Fig 5.36 *A compound light microscope.*

Cleaning lenses and slides

Dirt (dust, grit, grease etc.) is a common problem, and lenses and slides need regular cleaning. They can be cleaned before use, but if, during use, dirt is a problem, try to isolate the cause. For example, look down the eyepiece lens and try moving the slide or rotating the eyepiece lens to see if the dirt moves in either case. If it does not, the problem could be a dirty objective lens, mirror or lamp surface of condenser lens.

Remove conspicuous dust or grit by blowing, and then sweeping gently with a brush or piece of lens tissue. Huff on the slide or lens to dampen the surface, then polish clean and dry with a lens tissue. (This is specially prepared tissue with all traces of wood fibre removed, which could cause scratching of the lenses. Do not use ordinary tissue. Nappy liners are suitable.)

If necessary, the eyepiece lens can be removed for cleaning and the objective lenses can be unscrewed. Replace them immediately after cleaning to prevent dust from entering the microscope.

Remember:

- handle prepared slides at the ends or edges – do not place fingers on cover-slips.
- never touch lenses with your fingers.
- keep the stage clean.
- keep microscopes and slides covered when not in use.

Adjustment of the microscope for low power work

- Place the microscope on the bench and sit behind it in a comfortable position. Arrange other equipment around it. The object on the stage must be illuminated and this can be done with light from a built-in light source, from a window or a bench lamp. Light from the latter two sources shines on the understage mirror so that the curved surface reflects it up through the hole in the stage. The flat mirror is used to shine light through a sub-stage condenser, if there is one fitted. Light should be evenly spread across the field of view.
- Using the coarse adjustment screw, rack up the tube and turn the nosepiece until the lowest power objective (usually ×10 or 16 mm) *clicks* into line with the microscope tube. The magnifying power of the lens is usually inscribed on its barrel.
- Place the slide to be examined on the microscope stage so that the specimen is over the middle of the hole in the stage, and light can be seen passing through it.

- Viewing the stage and the slide *from the side*, rack down the low power objective using the coarse focus adjustment until the low power objective is about 5 mm from the slide.
- Looking through the microscope, slowly rack up by means of the coarse adjustment until the object is in focus.
- The microscope must **always** be focused **upwards, never downwards.** It is very easy to pass through the plane of focus when looking through the microscope and focusing downwards, and as a result damage the slide.
- Keep both eyes open and use each eye in turn.

Common problems are focusing on dust or dirt on surface of the cover-slip, and having the objective lens too far from the slide.

Moving the specimen

Note the direction in which the specimen moves when the slide is moved to the *left*, to the *right*, *away* and *towards* you. This knowledge will help you locate the specimen or follow a moving one. If a calibrated mechanical stage is present, this can be used as a reference to locate part of the specimen on future occasions.

Adjustment of the microscope for high power work

- High power work needs artificial light for sufficient illumination. Use a bench lamp or microscope lamp with an opal bulb. If a filament bulb is used, it will be necessary to place a sheet of paper between the bulb and the microscope. Swing the mirror so that the flat surface is uppermost and the light is thrown up into the microscope.
- To focus the condenser, leave the slide on the stage. Rack up the sub-stage condenser (fig 5.36) until within 5 mm of the stage. Look down the microscope and rack up the coarse adjustment until the object comes into focus. Now adjust the focus of the condenser until the image of the lamp is just superimposed on the slide. Put the condenser just out of focus so that the lamp image disappears. The lighting should now be at its optimum. Incorporated with the condenser is the **diaphragm**. This adjusts the opening through which light passes and the aperture should be as wide as possible. The definition will then be at its best.
- Turn the nosepiece until the high power objective lens (×40 or 4 mm) clicks into place. If focus has already been achieved under lower power, the nosepiece should automatically bring the high power lens into approximate focus. Adjust carefully using the fine adjustment and always focusing upwards.
- If the focus is still not correct after moving the higher power objective lens then use the following procedure. Look at the stage from the side, lower the tube until the objective lens is almost touching the slide. Watch the reflection of the objective lens in the slide and then aim to make the lens and its image almost meet.
- Look into the microscope and rack up slowly using the fine adjustment until the object is in focus.

Oil immersion

For higher magnification than normal high power work (×400) an oil immersion lens can be used. The light-gathering properties of the lens are greatly increased by placing a fluid between the objective lens and the cover-slip. The fluid must have the same refractive index as the lens itself, so the fluid used is generally cedarwood oil.

- Place the slide on the stage and focus as normal for high power work. Replace the objective lens with an oil immersion lens.
- Place a drop of cedarwood oil on top of the glass cover-slip over the top of the object to be examined.
- Focus the object again under low power, then swing in the oil immersion objective lens so that the tip is in contact with the oil.
- Look down the microscope and very carefully adjust the lens using the fine adjustment. Remember that at the plane of focus the lens is only 1 mm from the cover-slip of the slide.
- After use, clean the oil from the lens with soft tissue.

Measurement of size with the microscope (micrometry)

It is obviously important to be able to make accurate measurements of the real sizes of structures seen with the microscope. Measurement of microscopic objects is called **micrometry**. This can be done using specially designed scales, or micrometers. One scale is placed in the eyepiece lens (the **eyepiece micrometer** or **eyepiece graticule**) and one on the stage (the **stage micrometer**) (fig 5.37). The micrometers have equally spaced divisions. The size of these is not relevant for the eyepiece graticule, but is a known value for the stage micrometer.

Be careful not to get fingerprints on the micrometers. Handle the eyepiece micrometer by the edges only, so that it does not get scratched.

Before using the eyepiece micrometer to measure a particular structure, you will have to find out the real width of each unit on the scale at each magnification. In other words you will have to calibrate the micrometer. This can be done by replacing the specimen with the stage micrometer, and using this to measure the eyepiece units at each magnification.

- Unscrew the top lens of the eyepiece. Check that the eyepiece micrometer is the correct way up, and insert it in the eyepiece. It will lie flat on a ledge inside the eyepiece. Replace the top lens.
- Place the stage micrometer on the stage. Arrange it so that light is passing through the scale (the location of the scale should be clear if viewed carefully with the eye).

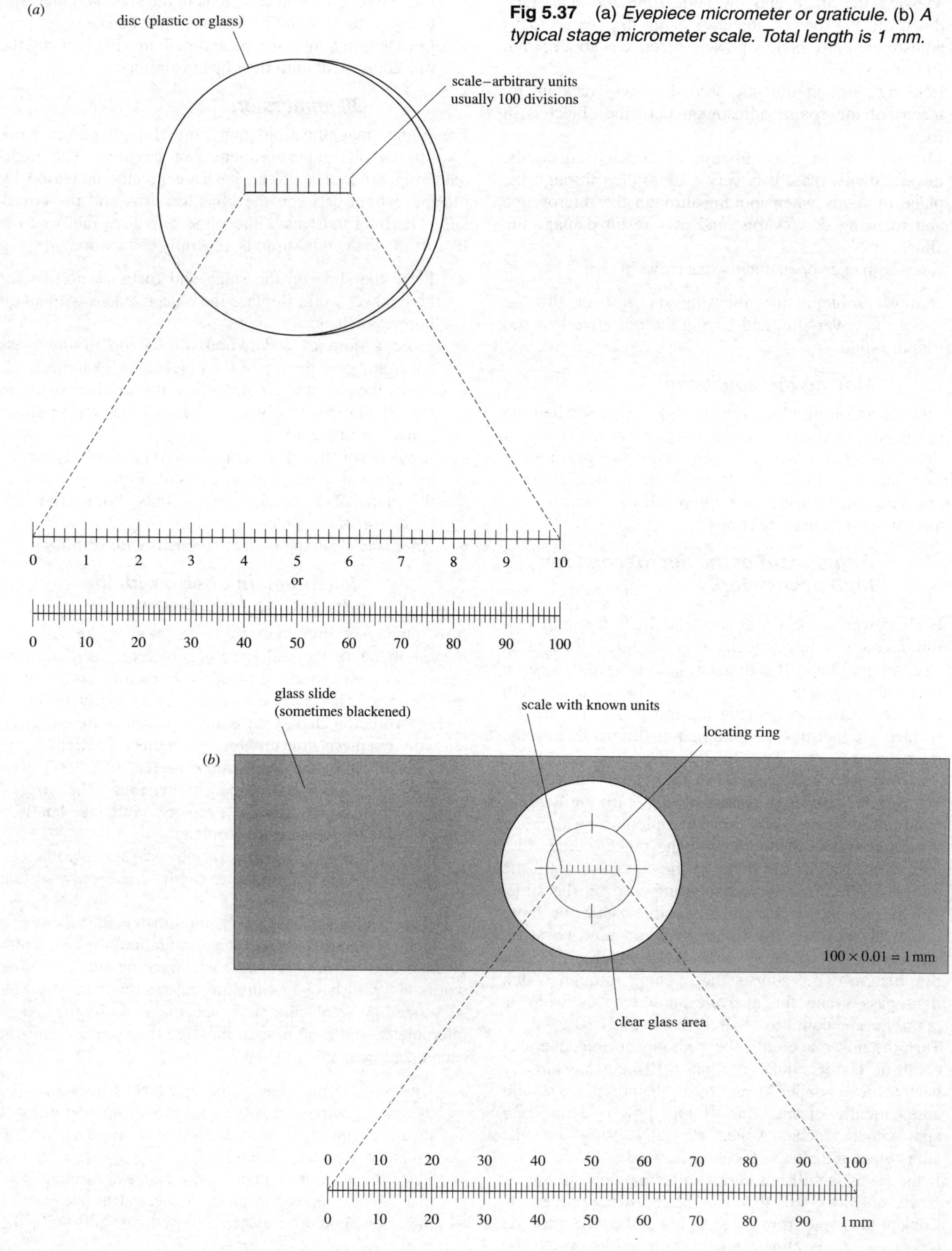

Fig 5.37 (a) *Eyepiece micrometer or graticule.* (b) *A typical stage micrometer scale. Total length is 1 mm.*

- Click the low power objective into position and focus on the stage micrometer scale.
- Rotate the eyepiece lens to line up the eyepiece scale parallel with the stage micrometer scale, and adjust the position of the stage micrometer so that the two scales are lined up next to each other and one can be read off against the other.
- Find out, as accurately as you can, how many stage micrometer divisions correspond to a known number of eyepiece units. (Choose as many units as possible for greatest accuracy.)
- The real width of the stage micrometer divisions will be 0.1 mm or 0.01 mm, which will be written on the slide. Using this information you can calculate the width of one eyepiece unit for this objective lens.
- Repeat the procedure for the other objective lenses that you intend to use for measuring (and any further eyepiece lenses, if necessary). Each eyepiece micrometer needs to be calibrated only once for a particular set of lenses and microscope.
- When the stage micrometer is replaced with a specimen, any part of the specimen can be measured in eyepiece units. Always rotate the eyepiece lens to line up the scale alongside the part to be measured.

It is easier to convert eyepiece units to real measurements if a graph is used. The vertical axis should represent eyepiece units, e.g. 100, and the horizontal axis should represent the actual distance in millimetres (see fig 5.38). Plot the actual length of 1 and 100 eyepiece units for each magnification, joining the two points with a straight line. Plot one graph for each magnification between the same axes, using appropriate scales for the horizontal axis. You can now read off the value in millimetres of any given number of eyepiece units for any magnification. It is useful to keep the graph with the microscope.

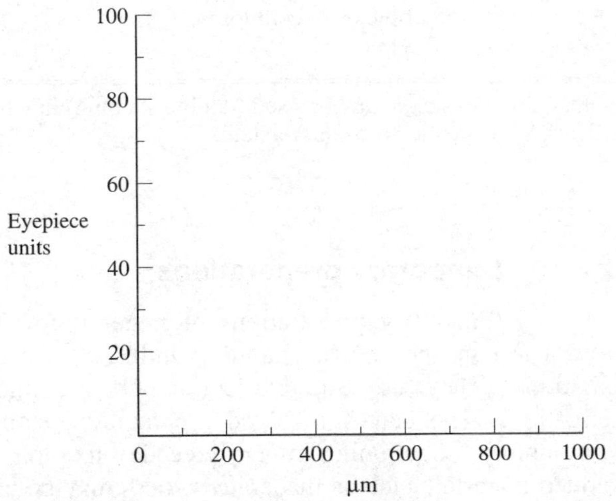

Fig 5.38 *Graph for converting eyepiece units to μm.*

5.12 Microscope techniques

5.12.1 Preparation of material for the microscope

Biological specimens may be examined in a living or preserved form. In the latter case material can be sectioned for closer examination and treated with a wide variety of stains to reveal and identify different structures. Preparations of freshly killed material may be temporary or permanent.

5.12.2 Permanent preparations

Fixation

This is the preservation of material in a life-like condition. Tissues must be killed rapidly and this is best achieved with small pieces of living material. The chemical used is called a **fixative**. By this method the original shape and structure are maintained and the tissue hardens so that thin sections can be cut.

Dehydration

Dehydration, or removal of water, is done to prepare the material for infiltration with an embedding medium or mounting medium (see below) with which water will not mix. Also bacterial decay would eventually occur if water were present. For preservation of fine detail, dehydration should be gradual and carried out by a series of increasingly concentrated ethanol/water or propanone (acetone)/water mixtures, finishing in 'absolute' (pure) ethanol or propanone.

Clearing

Alcohol does not mix with some of the common embedding and mounting media. Where this is the case, it is replaced with a medium (**clearing agent**) which does mix, such as xylol. This also makes the material transparent.

Embedding

Very thin sections can be cut if the material is embedded in a supporting medium. For light microscopy, embedding involves impregnating the material with molten wax which is then allowed to set. A harder material (plastic or resin) must be used for electron microscopy because thinner sections are required, and so more rigid support is needed during cutting.

Sectioning

Most pieces of material are too thick to allow sufficient light to pass through for microscopic investigation. It is usually necessary to cut very thin slices of the material (**sections**) and this may be done with a razor or a microtome. Hand-sectioning can be carried out with a razor which must be of shaving sharpness. For ordinary work, sections should be 8–12 μm thick. The tissue must be held firmly between two pieces of elder pith. The razor is wetted with some of the liquid in which the tissue is stored and the cut is made through pith and tissue, keeping the razor horizontal and drawing it towards the body with a long oblique sliding movement. Cut several sections fairly rapidly. Be content with the thinnest sections showing representative portions of the tissue.

Embedded tissues can be sectioned with a **microtome.** For light microscopy, sections a few micrometres thick can be cut from wax-embedded tissues using a steel knife. The **ultramicrotome** is used for cutting extremely thin sections (20–100 nm) for electron microscopy, and has a diamond or glass knife.

The need for embedding can be avoided in light microscopy by using a **freezing microtome** which keeps the specimen frozen, and therefore rigid, during cutting.

Staining

Most biological structures are transparent, so that some means of obtaining contrast between different structures must be employed. The most common method is staining. Some of the stains used in light microscopy are shown in table 5.5.

Certain stains when used in low concentrations are non-toxic to living tissue and can therefore be used on living material. These are called **vital stains**, for example methylene blue and neutral red.

To stain wax-embedded sections, the wax is dissolved away and the material partially rehydrated before staining.

Mounting

For light microscopy, the final stained sections are 'mounted' on a glass slide in a resinous medium which will exclude air and protect them indefinitely, such as Canada balsam or euparol. The mounted specimen is covered with a glass cover-slip.

The sequence of events described above is typical for preparations of thin sections for permanent preparations. However, two common variations in the order of events are:

(*a*) if hand-cut sections of fresh material are used, sectioning precedes fixation;
(*b*) staining may follow fixation and be carried out at the appropriate stage of the dehydration sequence, for example a stain dissolved in 50% ethanol would be used after dehydration in 50% ethanol.

The above procedures are similar in principle for both light and electron microscopy, although details differ as outlined in table 5.6.

Table 5.5 Common stains for plant and animal tissues.

Stain	*Final colour*	*Suitable for:*
Permanent stains		
aniline blue (cotton blue) in lactophenol	blue	fungal hyphae and spores
borax carmine	pink	nuclei; particularly for whole mounts (large pieces) of animal material, e.g. *Obelia* colony
eosin	pink	cytoplasm (see haematoxylin)
	red	cellulose
Feulgen's stain	red/ purple	DNA; particularly useful for showing chromosomes during cell division
haematoxylin	blue	nuclei; mainly used for sections of animal tissue with eosin as counterstain* for cytoplasm; also for smears
Leishman's stain	red-pink	blood cells
	blue	white blood cell nuclei
light green or fast green	green	cytoplasm and cellulose (see safranin)
methylene blue	blue	nuclei (0.125% methylene blue in 0.75% NaCl solution suitable as a vital stain)
safranin	red	nuclei; lignin and suberin of plants; mainly used for sections of plant tissue with light green as counterstain* for cytoplasm
Temporary stains		
aniline hydrochloride or aniline sulphate	yellow	lignin
iodine solution	blue-black	starch
phloroglucinol + conc. HCl	red	lignin
Schultz's solution (chlor-zinc-iodine)	yellow	lignin, cutin, suberin, protein
	blue	starch
	blue or violet	cellulose

*counterstain: two stains may be used (double staining) in which case the second is called the counterstain.

5.12.3 Temporary preparations

Temporary preparations of material for light microscopy can be made rapidly, unlike permanent preparations. They are suitable for quick preliminary investigations. The stages involved are fixation, staining and mounting. Sectioning may precede fixation, or macerated material, such as macerated wood, may be used. Fresh material may be hand-sectioned with a razor directly into 70% alcohol as a fixative. For staining and mounting, a

Table 5.6 Differences in preparation of material for light and electron microscopes.

Treatment	*For light microscopes*	*For electron microscopes*
Fixation	As for electron microscopy, or for example 99 parts ethanol: 1 part glacial ethanoic acid ('alcohol/acetic'), or 70% ethanol (but this causes shrinkage and damage to delicate structures)	Glutaraldehyde or mixture of glutaraldehyde and osmic acid (OsO_4) is often used. OsO_4 also stains lipids, and hence membranes, black. Smaller pieces of material fixed for more rapid and better preservation of fine structure
Dehydration	Ethanol or propanone series	Ethanol or propanone series
Embedding	Wax	Resin (e.g. araldite, epon) or plastic
Sectioning	Metal knife	Only diamond or glass knives are sharp enough to cut the ultrathin sections required
	Microtome used	Ultramicrotome used
	Sections are few micrometres thick	Sections 20–100 nm thick
Staining	Coloured dyes (reflect visible light)	Heavy metals, e.g. compounds of osmium, uranium, lead (reflect electrons)

number of temporary stains may be used; some suitable for plant materials are shown in table 5.5. In each case the material should be placed on a clean glass slide (wiped clean with alcohol) and a few drops of stain added. In the case of phloroglucinol, one drop of concentrated hydrochloric acid is also added. The specimen is then covered with a thin glass cover-slip to exclude air and dust, and to protect high power microscope objectives (fig 5.39). If the specimens begin to dry out, or if it is known that prolonged examination (longer than 10 min) is required, specimens should be mounted in glycerine after staining.

5.13 Recording by biological drawing

Purpose

- To provide a record of work for future reference.
- To encourage you to study more fully and accurately the specimen that you are investigating.
- To aid memory of what you see by actively recording.

Principles

- Drawing paper of suitable quality must be used. It must be capable of standing some rubbing out of incorrect pencil lines.

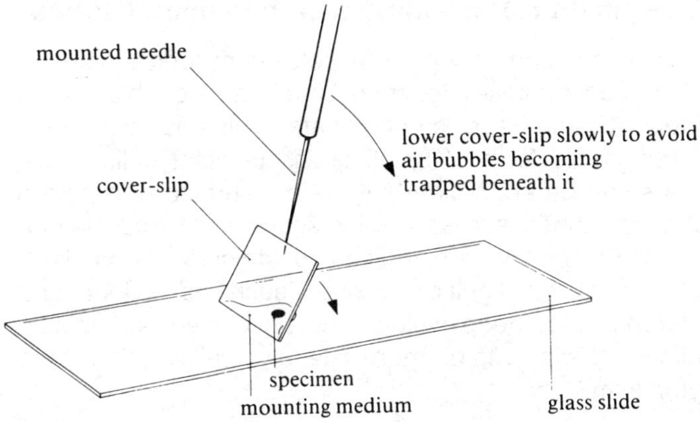

Fig 5.39 *Mounting a specimen and lowering a cover-slip on a glass slide.*

- Pencils should be sharp and of HB quality. No coloured pencils should be used.
- Drawing must be:
  (*a*) large enough – the greater the number of parts the larger the drawing. The drawing should normally occupy more than half the space available.
  (*b*) accurate – relative proportions of the various parts of the specimen should be observed and drawn carefully. If the subject has several similar parts, draw a small portion accurately.
  (*c*) drawn with lines sharp and clear – each line should be considered and then drawn without removing the pencil from the paper. Shading and colour should be avoided.
  (*d*) labelled – these should be as complete as possible with label lines that do not cross; space labels around the figure. Notes on observations can be added to the labels, such as intensity or colour of staining, shrinkage of tissue, unusual or typical features.
- Make two drawings if necessary:
  (*a*) a simple drawing of the main features, and
  (*b*) details of small parts only. For example, a lower power plan of a plant section should outline the different tissues but not include the individual cells. High power detail of the cells in a representative portion of the plan can then be drawn separately.
- Draw what you see and not what you think you should see, and certainly not a textbook copy.
- Every drawing should have a title, magnification, viewpoint of the specimen (such as TS, LS and so on) and explanatory notes. These should be placed in a standard position, such as the top right-hand corner of the page. A scale line should also be drawn where possible.
- Label lines should finish exactly at the structure named. They should be drawn with a ruler and should not cross each other. Labels should be arranged neatly around the drawing (lined up vertically where possible). All relevant structures should be labelled.

Experiment 5.1: Staining starch in plant tissues

A dilute solution of iodine in potassium iodide solution (I_2/KI) can be used to stain starch a deep blue-black colour. Starch is a carbohydrate which is commonly stored in the form of small grains in plant cells. Such grains can be seen clearly if the cut surface of a potato tuber is rubbed across a microscope slide and stained with I_2/KI solution. In addition to staining starch, I_2/KI solution stains lignified tissue (such as xylem and sclerenchyma) deep yellow, and unlignified tissue pale yellow. Nuclei stain more deeply yellow than the cytoplasm.

Materials

microscope and light source
eyepiece micrometer and stage micrometer
clean microscope slide and cover-slip
small brush for transferring sections
lens tissue
plain paper
mounted needle
teat pipette (if stain is not in a dropper bottle)
filter paper
iodine in potassium iodide solution
5% glycerol solution in dropper bottle
section of plant organ, e.g. stem of white deadnettle (*Lamium album*)

I_2/KI solution

Approx. 1 g iodine, 2 g potassium iodide, dissolved in 300 cm^3 water. Concentration may be increased by using less water. Observe standard safety precautions when handling iodine.

Thin, fresh sections may be cut in the conventional way using a razor and supporting the stem between two pieces of pith. They should be cut into 70% alcohol.

Procedure

(1) Using a small brush, transfer a section of the plant organ to be studied to the centre of a microscope slide.

(2) Add two drops of I_2/KI solution, and one drop of 5% glycerol to slow down the rate of evaporation and stop the specimen drying out too quickly.

(3) Add a cover-slip as shown in fig 5.39.

(4) Examine the specimen carefully with a microscope. Note particularly the distribution of the darkly stained starch grains. Note also that any air bubbles will appear to have a thick black outline and are apparently empty. Ignore air bubbles.

(5) Draw a low power plan of your section of white deadnettle. Remember, the aim is to outline accurately the area occupied by each tissue. The following notes will also help.
- White deadnettle has a special supporting tissue called **collenchyma** in each corner of its stem.
- The vascular bundles are variable in size. Extra xylem and phloem *may* grow in between the vascular bundles, eventually making a continuous circle of vascular tissue.

(6) Label your low power plan and indicate the linear magnification of your drawing.

(7) State on the map which tissues contain starch. There is one layer of cells which is particularly abundant in starch grains. This is called the **starch sheath**. Draw this on your map and label it. Below your drawing, describe the *distribution* of starch grains within these cells.

(8) A biologist put forward the hypothesis that the average size of parenchyma cells in the cortex of the white deadnettle stem is greater near the centre of the stem than near the outside.

Make appropriate measurements with the eyepiece graticule in the innermost and outermost regions of the cortex to test this hypothesis. Describe in detail the method you used.

Record your results (measured in eyepiece units) in an appropriate manner and show your calculations.

Convert your measurements to micrometres by calibrating your eyepiece micrometer.

Do your measurements support the hypothesis?

Chapter Six

Histology

Histology is the study of tissues. All multicellular organisms possess groups of cells of similar structure and function assembled together to form tissues. A **tissue** can be defined as a group of physically linked cells and associated intercellular substances that is specialised for a particular function or functions. The cells of a tissue generally share a similar origin in the embryo. Tissues improve the efficiency with which the body functions by allowing division of labour, that is sharing of tasks, with each tissue being specialised for a particular job.

Higher levels of organisation than the tissue occur, particularly in animals. A number of tissues working together as a functional unit is called an **organ**, for example the stomach or the heart. In animals, organs form parts of even larger functional units known as **systems**, for example the digestive system (pancreas, liver, stomach, duodenum and so on) and the vascular system (heart and blood vessels).

The cells of a tissue may be all of one type, for example parenchyma, collenchyma and cork in plants and squamous epithelium in animals. Alternatively, the tissues may contain a mixture of different cell types, as in xylem and phloem in plants and some connective tissues in animals.

The study of tissue structure and function relies heavily on light microscopy and the associated techniques of preserving, staining and sectioning material. These techniques are described in sections 5.11 and 5.12.

In this chapter histology is studied at the level of detail which can be seen with the light microscope. In some cases, though, reference is made to structure as revealed by the electron microscope in order to provide greater clarification. In relating structure to function in tissues it is important to bear in mind the three-dimensional structures of the cells and their relationship to one another. This kind of information is usually 'pieced together' by examining material in thin section, most commonly in transverse section (TS) and longitudinal section (LS). Neither type of section alone can give all the information required, but a combination of the two can often reveal the necessary information. Some cells, such as xylem vessels and tracheids in plants, can easily be examined whole by macerating the tissues. This involves the breakdown of soft tissues leaving behind the harder, lignified xylem vessels, tracheids and fibres.*

Plant tissues can be divided into two groups:

- one type of cell
  - parenchyma — section 6.1.1
  - collenchyma — section 6.1.2
  - sclerenchyma — section 6.1.3
- more than one type of cell
  - xylem — section 6.2.1
  - phloem — section 6.2.2

Animal tissues are divided into four groups:

- epithelial — section 6.3
- connective, including areolar tissue, fibrous tissue, adipose tissue, cartilage, bone — section 6.4
- muscle — section 6.5
- nervous tissue — section 6.6

Table 6.1 shows the characteristic features, functions and distribution of plant tissues. Fig 6.1 will help you to visualise the parts of the plant referred to when different tissues are being discussed.

6.1 Simple plant tissues – tissues consisting of one type of cell

6.1.1 Parenchyma

Structure

The structure of parenchyma is shown in fig 6.2. The cells may be roughly spherical or elongated.

Functions and distribution

- The cells are unspecialised and act as **packing tissue** between more specialised tissues, as in the central pith of stems and outer cortex of stems and roots (fig 6.1). They form a large part of the bulk of the young plant.
- The osmotic properties of parenchyma cells are important because, when turgid, they become tightly packed and provide **support** for the organs in which they are found. This is particular important in the stems of herbaceous plants where they form the main means of support. During periods of water shortage the cells of such plants lose water and this results in the plants wilting.

* The structure of some plant tissue is dealt with elsewhere in this book. More detailed structure of phloem is given in chapter 13 where its structure is related to its function in translocation. Development of plant tissues from meristematic cells is discussed in chapter 22, together with secondary growth and the structure of wood (secondary xylem) and cork.

Table 6.1 Characteristic features, functions and distribution of plant tissues.*

Tissue	*Main functions*	*Living or dead*	*Wall material*	*Cell shape*	*Distribution*
Parenchyma	Packing tissue. Support in herbaceous plants. Metabolically active. Intercellular air spaces allow gaseous exchange. Food storage. Transport of materials through cells or cell walls.	Living	Cellulose, pectins and hemicelluloses	Roughly spherical to elongated	Cortex, pith, medullary rays in wood and packing tissue in xylem and phloem
Modified parenchyma					
(*a*) epidermis	Protection from desiccation and infection. Hairs and glands may have additional functions.	Living	Cellulose, pectins and hemicelluloses, and covering of cutin	Elongated and flattened	Single layer of cells covering entire primary plant body
(*b*) mesophyll	Photosynthesis (contains chloroplasts). Storage of starch.	Living	Cellulose, pectins and hemicelluloses	Roughly spherical, irregular (spongy) or column-shaped (palisade) depending on location	Between the upper and lower epidermis of leaves
(*c*) endodermis	Selective barrier to movement of water and mineral salts (between cortex and xylem) in roots. Starch sheath with possible role in gravity response in stems.	Living	Cellulose, pectins and hemicelluloses, and deposits of suberin	As epidermis	Around vascular tissue (innermost layer of cortex)
(*d*) pericycle	In roots it retains meristematic activity producing lateral roots and contributing to secondary growth if this occurs.	Living	Cellulose, pectins and hemicelluloses	As parenchyma	In roots between central vascular tissue and endodermis
NB The pericycle in the stem is made of sclerenchyma and has a different origin.					
Collenchyma	Support (a mechanical function)	Living	Cellulose, pectins and hemicelluloses	Elongated and polygonal with tapering ends	Outer regions of cortex, e.g. angles of stems, midrib of leaves
Sclerenchyma					
(*a*) fibres	Support (purely mechanical)	Dead	Mainly lignin. Cellulose, pectins and hemicelluloses also present	Elongated and polygonal with tapering interlocking ends	Outer regions of cortex, pericycle of stems, xylem and phloem
(*b*) sclereids	Support or mechanical protection	Dead	As fibres	Roughly spherical or irregular	Cortex, pith, phloem, shells and stones of fruits, seed coats
Xylem	Mixture of living and dead cells. Xylem also contains fibres and parenchyma which are as previously described.				
tracheids and vessels	Translocation of water and mineral salts. Support.	Dead	Mainly lignin. Cellulose, pectins and hemicelluloses also present.	Elongated and tubular	Vascular system
Phloem	Mixture of living and dead cells. Phloem also contains fibres and sclereids which are as previously described.				
(*a*) sieve tubes	Translocation of organic solutes (food)	Living	Cellulose, pectins and hemicelluloses	Elongated and tubular	Vascular system
(*b*) companion cells	Work in association with sieve tubes	Living	Cellulose, pectins and hemicelluloses	Elongated and narrow	Vascular system

* Tissues associated with secondary growth, such as wood and cork, are described in chapter 22.

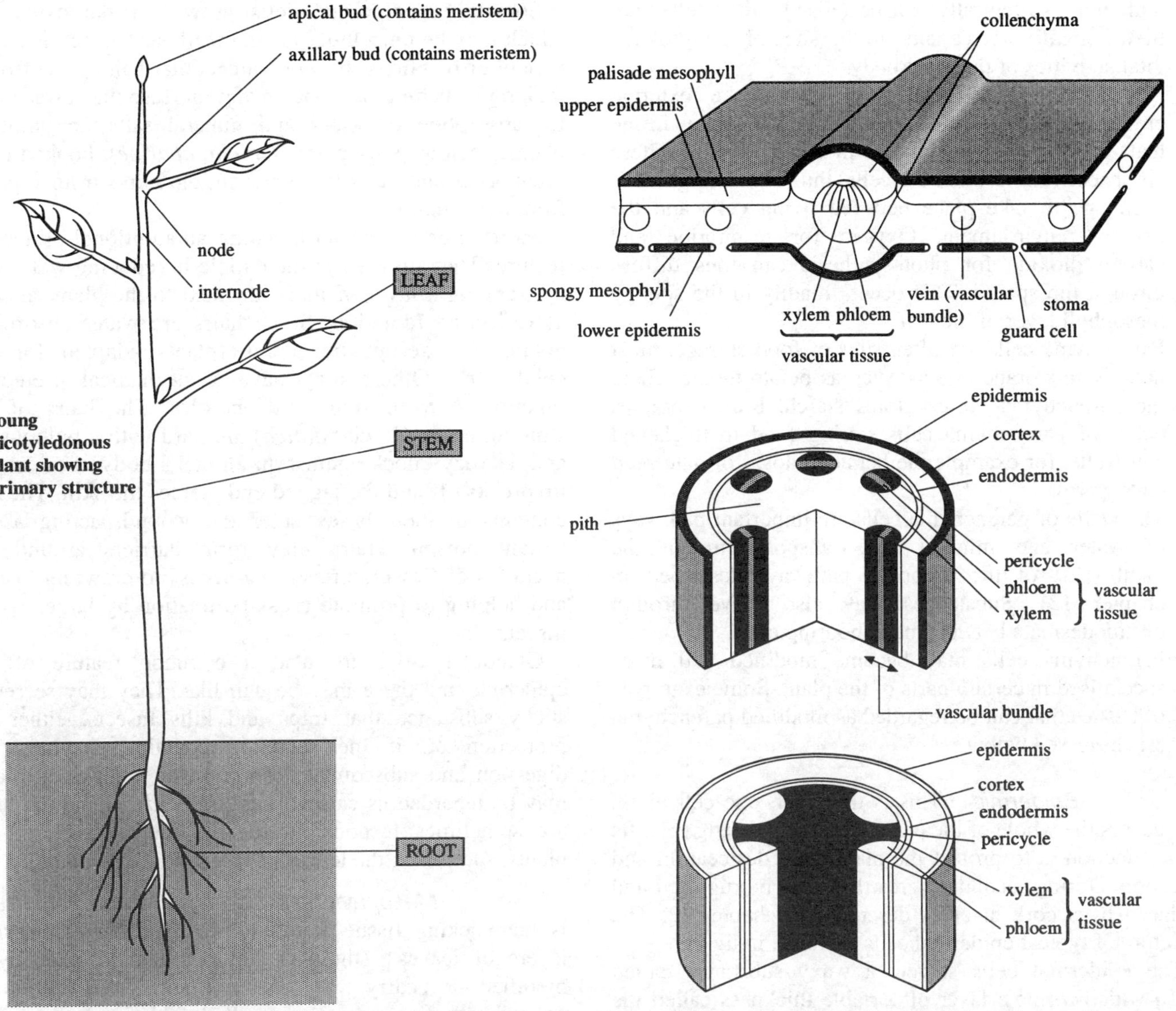

Fig 6.1 *Young dicotyledonous plant showing main features of primary structure of leaf, stem and root. The regions named are referred to during discussions of plant tissues.*

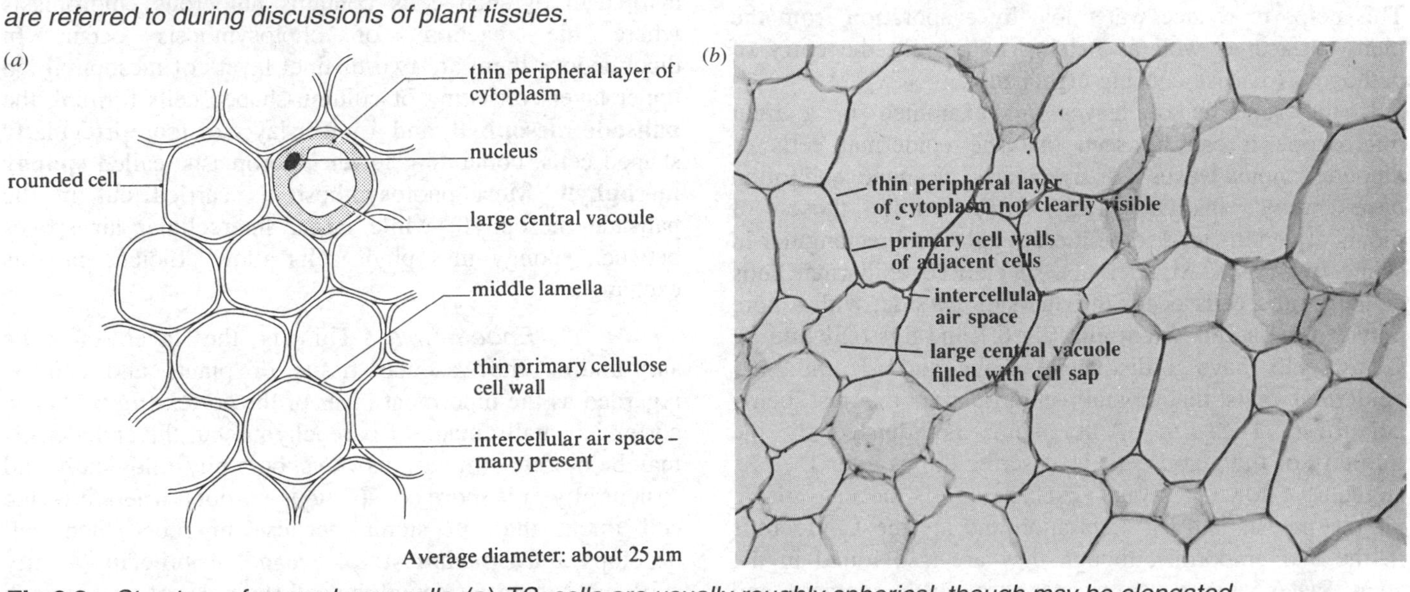

Fig 6.2 *Structure of parenchyma cells.* (a) *TS, cells are usually roughly spherical, though may be elongated.* (b) *TS* Helianthus *stem pith. Pith is the packing and supporting tissue found in the centre of the stems of dicotyledons.*

- Although structurally unspecialised, the cells are **metabolically active** and are the sites of many of the vital activities of the plant body.
- A system of air spaces runs from the external environment where they open as stomata (pores in the leaf) or lenticels (special slits in woody stems). These air spaces run between the cells, thus allowing gaseous exchange to take place between living cells and the external environment. Oxygen for respiration and carbon dioxide for photosynthesis can thus diffuse through the spaces. This occurs readily in the spongy mesophyll layer of the leaf.
- Parenchyma cells are often sites of food storage, most notably in storage organs, such as potato tubers where the parenchyma cortex stores starch. Rare examples occur of parenchyma cells storing food in thickened cell walls, for example the hemicelluloses of date seed endosperm.
- The walls of parenchyma cells are important pathways of water and mineral salt transport through the plant (part of the 'apoplast pathway' described in chapter 13). Substances may also move through plasmodesmata between neighbouring cells.
- Parenchyma cells may become modified and more specialised in certain parts of the plant. Some examples of tissues that can be regarded as modified parenchyma are discussed below.

Epidermis. This is the layer, one cell thick, that covers the whole of the primary plant body (fig 6.1). Its basic function is to protect the plant from desiccation and infection. During secondary growth it may be ruptured and replaced by a cork layer as described in chapter 22. The structure of typical epidermal cells is shown in fig 6.3.

The epidermal cells secrete a waxy substance called **cutin** which forms a layer of variable thickness called the **cuticle** within and on the outer surface of the cell walls. This helps to reduce water loss by evaporation from the plant surface as well as helping to prevent the entry of pathogens (disease-causing organisms).

If the surfaces of leaves are examined in a light microscope it can be seen that the epidermal cells of dicotyledonous leaves are irregularly arranged and often have wavy margins (fig 6.3*b*) while those of monocotyledons tend to be more regular and rectangular in shape (fig 6.3*c*). At intervals, specialised epidermal cells called **guard cells** occur in pairs side by side, with a pore between them called a **stoma** (fig 6.1 and figs 6.3*b* and *c*). Guard cells have a distinctive shape and are the only epidermal cells that contain chloroplasts, the rest being colourless. The size of the stoma is adjusted by the turgidity of the guard cells as described in chapter 13. The stomata allow gaseous exchange to occur during photosynthesis and for respiration and are most numerous in the leaf epidermis, though they are also found in the stem. Water vapour also escapes through the stomata, and this is part of the process called transpiration.

Sometimes epidermal cells grow hair-like extensions which may be unicellular or multicellular and serve a wide variety of functions. In roots, unicellular hairs grow from a region just behind the root tip and increase the surface area for absorption of water and mineral salts. In climbing plants, such as goosegrass (*Galium aparine*), hooked hairs often occur and function to prevent the stems from slipping from their supports.

More often epidermal hairs are an additional protective feature. They may assist the cuticle in reducing water loss by trapping a layer of moist air next to the plant, as well as reflecting radiation. Some hairs are water absorbing, notably on xerophytic plants (plants adapted for dry conditions). Others may have a mechanical protective function as with short, stiff bristles. The hairs of the stinging nettle (*Urtica dioica*) are hard with a bulbous tip and, as they knock against an animal's body, their fragile tip breaks off and the jagged end pierces the skin. The cell contents at their bases enter the wound, acting as an irritant poison. Hairs may form barriers around the nectaries of flowers preventing access to crawling insects and helping to promote cross-pollination by larger flying insects.

Glandular cells are also a common feature of the epidermis and these may be hair-like. They may secrete a sticky substance that traps and kills insects, either for protection or, if the secretion contains enzymes, for digestion and subsequent absorption of food. Such plants may be regarded as carnivorous (fig 6.3*d*). Glandular hairs are sometimes responsible for the scents given off by plants, such as on the leaves of lavender (*Lavendula*).

Mesophyll (see also figs 7.3 and 7.4). This is the packing tissue found between the two epidermal layers of leaves (fig 6.1) and consists of parenchyma modified to carry out photosynthesis. Photosynthetic parenchyma is sometimes called **chlorenchyma**. The cytoplasm of such cells contains numerous chloroplasts where the reactions of photosynthesis occur. In dicotyledons there are two distinct layers of mesophyll: an upper layer consisting of column-shaped cells forming the **palisade mesophyll**, and a lower layer of more irregularly shaped cells, containing fewer chloroplasts, called **spongy mesophyll**. Most photosynthesis is carried out in the palisade mesophyll, while larger intercellular air spaces between spongy mesophyll cells allow efficient gaseous exchange.

Endodermis. This is the layer of cells surrounding the vascular tissue of plants and can be regarded as the innermost layer of the cortex (fig 6.1). The cortex is usually made of parenchyma, but the endodermis may be modified in various ways, both physiologically and structurally. It is more conspicuous in roots, where it is one cell thick, than in stems because in roots each cell develops a **Casparian strip**, a band of **suberin** (a fatty substance) that runs round the cell (fig 6.4). At a later stage further thickenings of the wall may take place. The

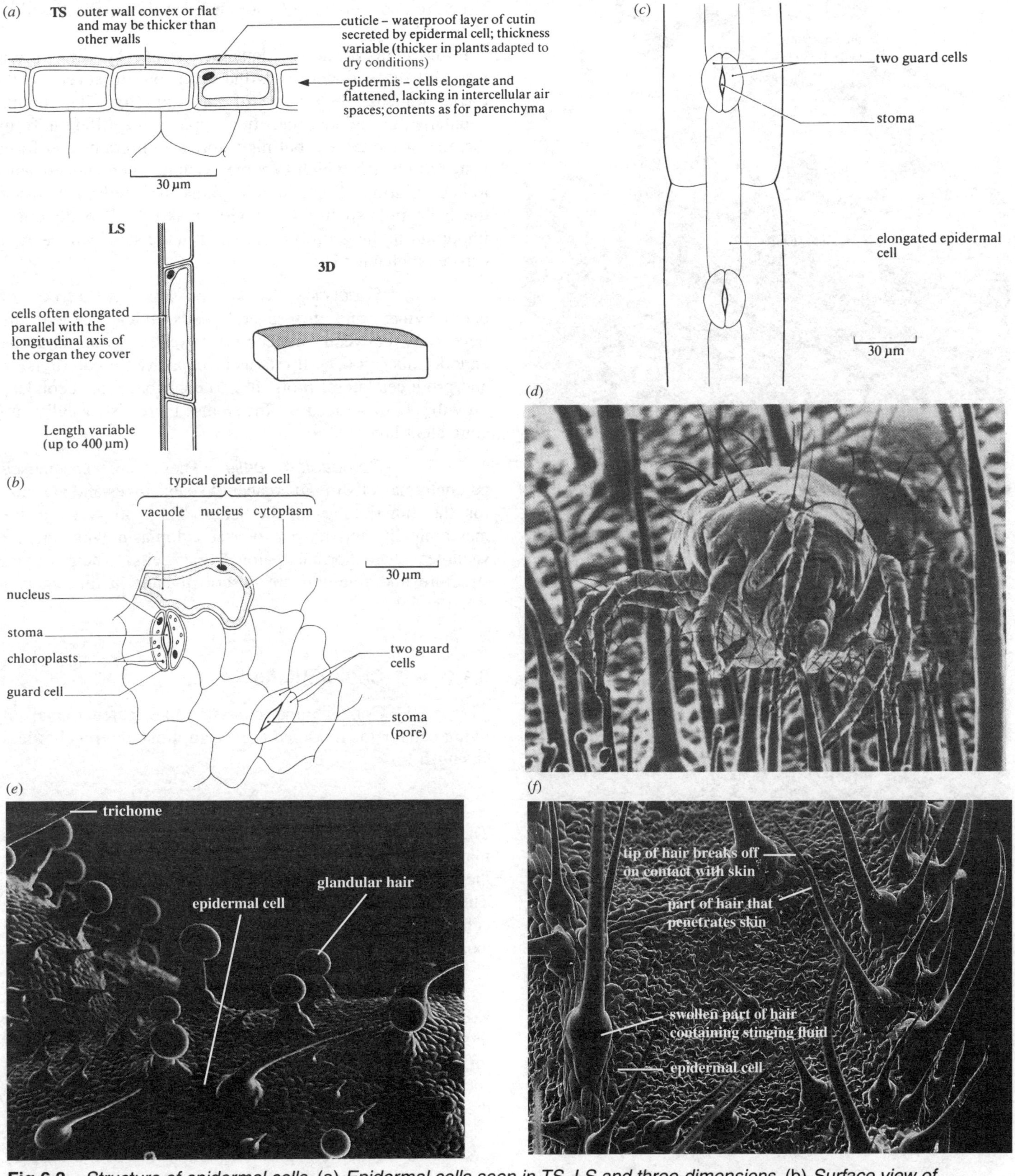

Fig 6.3 *Structure of epidermal cells.* (a) *Epidermal cells seen in TS, LS and three-dimensions.* (b) *Surface view of dicotyledon leaf epidermis.* (c) *Surface view of monocotyledon leaf epidermis.* (d) *Spider mite trapped and killed by the hair glands of a potato leaf. An enzyme capable of digesting animal matter has been found in one type of glandular hair in the potato, so the potato could be regarded as a carnivorous plant. Many other plants not normally thought of as carnivorous may have similar abilities.* (e) *Young leaf of* Cannabis sativa *with glandular hairs and trichomes.*
(f) *Leaf surface of* Urtica dioica *(stinging nettle).*

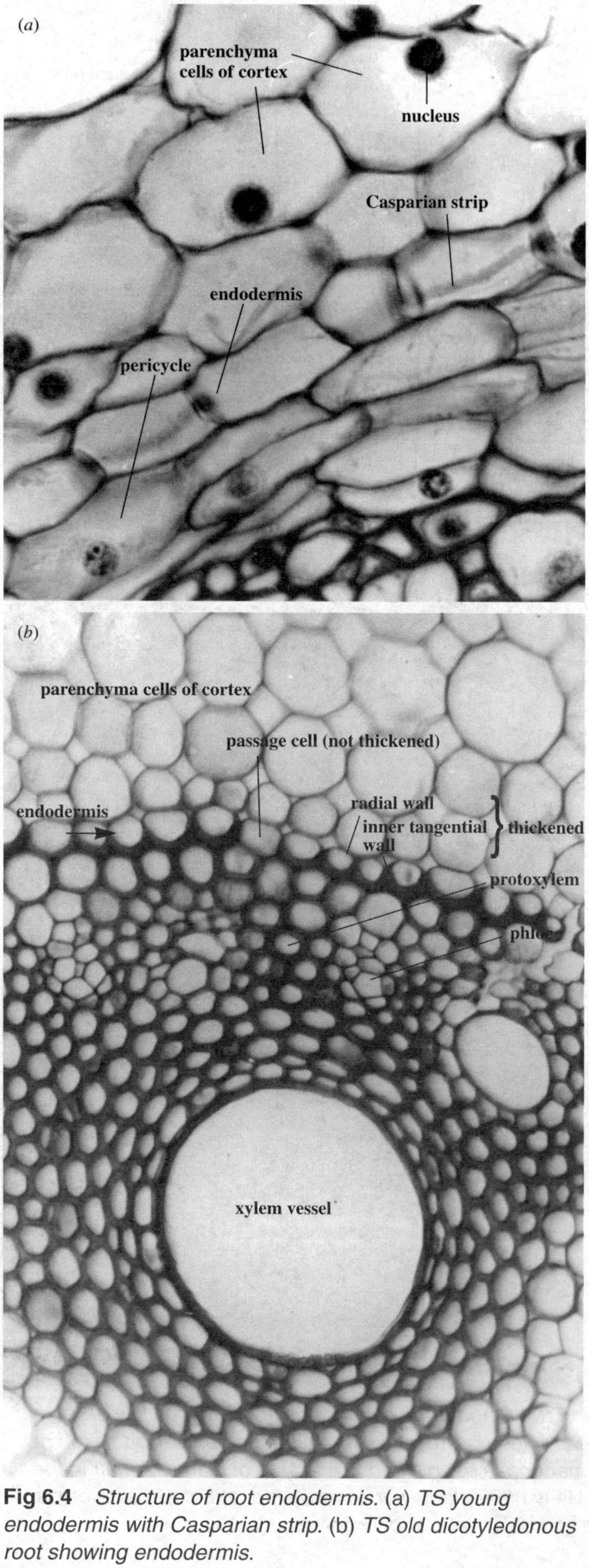

Fig 6.4 *Structure of root endodermis.* (a) *TS young endodermis with Casparian strip.* (b) *TS old dicotyledonous root showing endodermis.*

structure and function of root endodermis are discussed in chapter 13.

In the stems of dicotyledons the vascular bundles form a ring and the endodermis is the layer, one to several cells thick, immediately outside of this ring (fig 6.1). In this situation the endodermis often appears no different from the rest of the cortex, but may store starch grains and form a **starch sheath** which becomes visible when stained with iodine solution. These starch grains may sediment inside the cells in response to gravity, making the endodermis important in the geotropic response in the same way as root cap cells (chapter 16).

Pericycle. Roots possess a layer of parenchyma, one to several cells thick, called the **pericycle**, between the central vascular tissue and the endodermis (fig 6.1). It retains its capacity for cell division and produces lateral roots. It also contributes to secondary growth if this occurs. In stems there is usually no equivalent layer.

Companion cells. These are specialised parenchyma cells found adjacent to sieve tubes and are vital for the functioning of the latter. They are very active metabolically and have a denser cytoplasm with smaller vacuoles than normal parenchyma cells. Their origin, structure and function are described later in this chapter (section 6.2.2).

6.1.2 Collenchyma

Collenchyma consists, like parenchyma, of living cells but is modified to give **support** and **mechanical strength**.

Structure

The structure of collenchyma is shown in fig 6.5. It shows many of the features of parenchyma but is characterised by the deposition of extra cellulose at the corners of the cells. The deposition occurs after the formation of the primary cell wall. The cells also elongate parallel to the longitudinal axis of the organ in which they are found.

Function and distribution

Collenchyma is a **mechanical** tissue, providing support for those organs in which it is found. It is particularly important in young plants, herbaceous plants and in organs such as leaves where secondary growth does not occur. In these situations it is an important strengthening tissue supplementing the effects of turgid parenchyma. It is the first of the strengthening tissues to develop in the primary plant body and, because it is living, can grow and stretch without imposing limitations on the growth of other cells around it.

In stems and petioles its value in support is increased by its location towards the periphery of the organ. It is often found just below the epidermis in the outer region of the

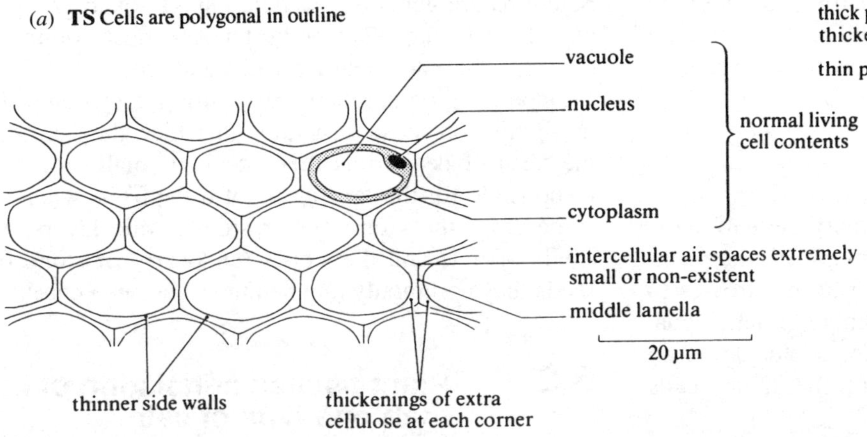

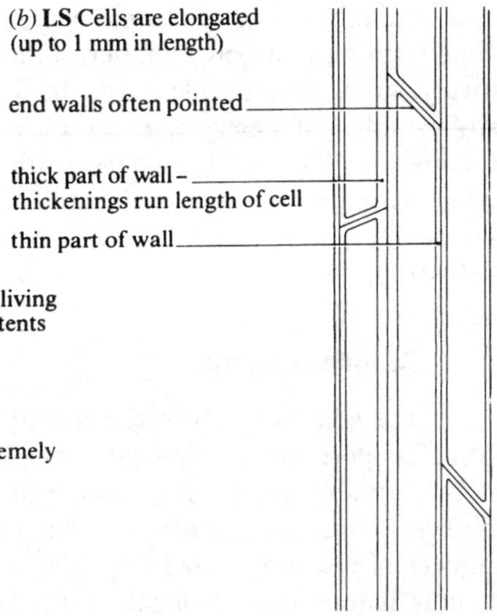

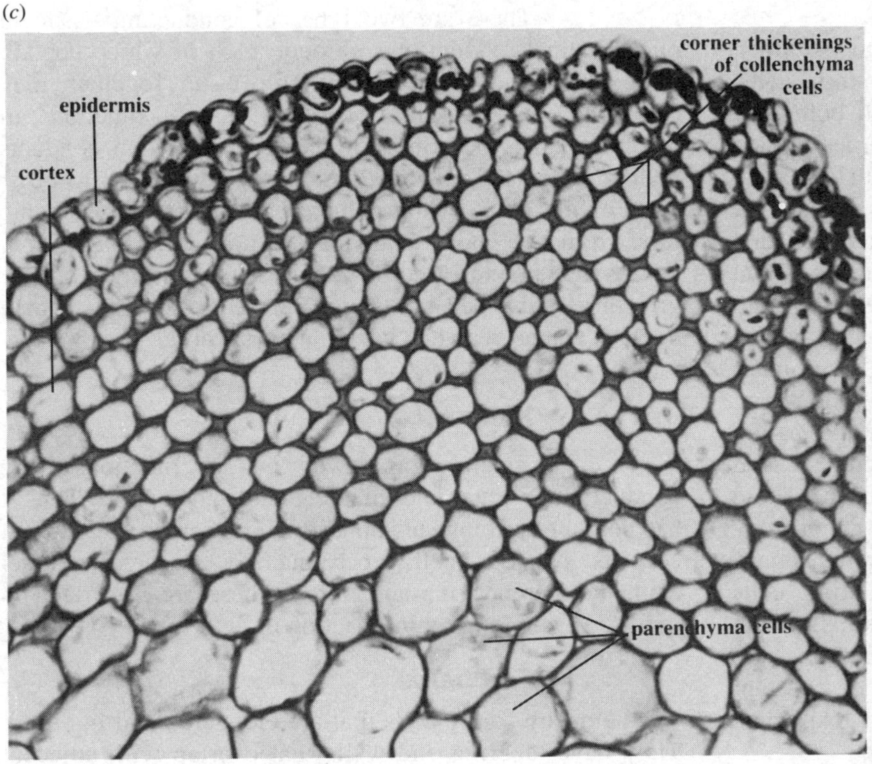

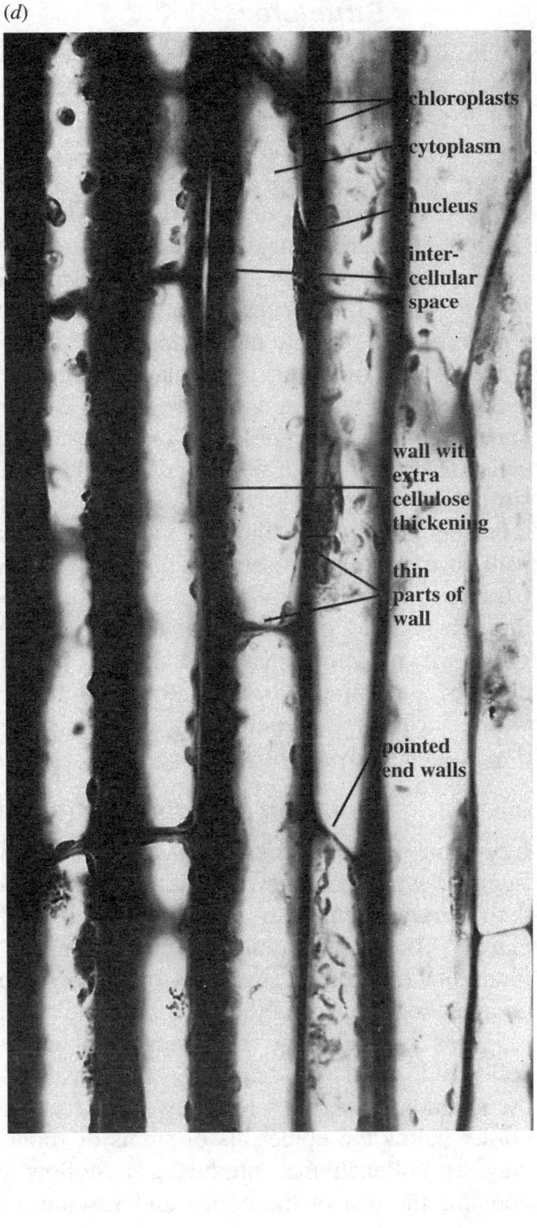

Fig 6.5 *Structure of collenchyma cells.* (a) *TS, cells are polygonal in outlino.* (b) *LS, oclls arc clongated (up to 1 mm in length).* (c) *TS collenchyma from* Helianthus *stem.* (d) *LS collenchyma from* Helianthus *stem.*

cortex and gradually merges into parenchyma towards the inside, thus forming a hollow cylinder in three dimensions. Alternatively, strengthening ridges may be formed, as along the fleshy petioles of celery (*Apium graveolus*) and the angular stems of plants such as dead-nettle (*Lamium*). In dicotyledonous leaves it appears as solid masses running the length of the midrib, providing support for the vascular bundles (fig 6.1).

6.1.3 Sclerenchyma

The sole function of **sclerenchyma** is to assist in providing support and mechanical strength for the plant. Its distribution within the plant is related to the stresses to which different organs are subjected. Unlike collenchyma, the mature cells are dead and incapable of elongation so they do not mature until elongation of the living cells around them is complete.

Structure

There are two types of sclerenchyma cell, namely **fibres**, which are elongated cells, and **sclereids** or **stone cells**, which are usually roughly spherical, although both may vary considerably in size and shape. Their structures are shown in figs 6.6 and 6.7 respectively. In both cases the primary cell wall is heavily thickened with deposits of **lignin**, a hard substance with great tensile and compressional strength. A high tensile strength means that it does not break easily on stretching, and a high compressional strength means that it does not buckle easily.

Deposition of lignin takes place in and on the primary cellulose cell wall and, as the walls thicken, the living contents of the cells are lost with the result that the mature cells are dead. In both fibres and sclereids structures called **simple pits** appear in the walls as they thicken. These represent areas where lignin is not deposited on the primary wall owing to the presence of groups of **plasmodesmata** (strands of cytoplasm that connect neighbouring cells through minute pores in the adjacent cell walls). Each group of plasmodesmata forms one pit. The pits are described as simple because they are tubes of constant width. Their development is best explained diagrammatically as shown in fig 6.8.

Function and distribution of fibres

Individual sclerenchyma fibres are strong owing to their lignified walls. Collectively their strength is enhanced by their arrangement into strands or sheets of tissue that extend for considerable distances in a longitudinal direction. In addition, the ends of the cells interlock with one another, increasing their combined strength.

Fibres are found in the pericycle of stems, forming a solid rod of tissue 'capping' the vascular bundles of dicotyledons (see fig 6.1). They often form a layer in the cortex below the epidermis of stems or roots, in the same way as collenchyma, producing a hollow cylinder that contains the rest of the cortex and vascular tissues. Fibres also occur in both xylem and phloem, either individually or in groups, as described in section 6.2.

Function and distribution of sclereids

Sclereids are generally scattered singly or in groups almost anywhere in the plant body, but are most common in the cortex, pith, phloem and in fruits and seeds.

Depending on numbers and position, they confer firmness or rigidity on those structures in which they are found. In the flesh of pear fruits they occur in small groups and are responsible for the 'grittiness' of these fruits when eaten. In some cases they form very resilient, solid layers, as in the shells of nuts and the stones (endocarp) of stone fruits. In seeds they commonly toughen the testa (seed coat).

6.2 Plant tissues consisting of more than one type of cell

There are two types of conducting tissue in plants, namely **xylem** and **phloem**, both of which contain more than one type of cell (fig 6.1). Together they constitute the **vascular tissue** whose function in translocation is described in chapter 13. Xylem conducts mainly water and mineral salts from the roots up to other parts of the plant, while phloem conducts mainly organic food from the leaves both up and down the plant. Both tissues may be increased in amount as a result of secondary growth as described in chapter 22. Secondary xylem may become extensive, when it is known as **wood**.

6.2.1 Xylem

Xylem has two major functions, the conduction of water and mineral salts, and support. Thus it has both a physiological and a structural role in the plant. It consists of four cell types, namely tracheids, vessel elements, parenchyma and fibres. These are illustrated in transverse and longitudinal section in fig 6.9.

Tracheids

Tracheids are single cells that are elongated and lignified. They have tapering end walls that overlap with adjacent tracheids in the same way as sclerenchyma fibres. Thus they have mechanical strength and give support to the plant. They are dead with empty lumens when mature. Tracheids represent the original, primitive water-conducting cells of vascular plants and are the only cells found in the xylem of the more ancestral vascular plants. They have given rise, in other plants, to xylem fibres and vessels which are described later. Despite their ancestral nature, they obviously function efficiently because conifers, most of which are trees, rely exclusively on tracheids to conduct water from the roots to the aerial parts. Water can pass through the empty lumens without being obstructed by living contents. It passes from tracheid to tracheid through the pits via the 'pit

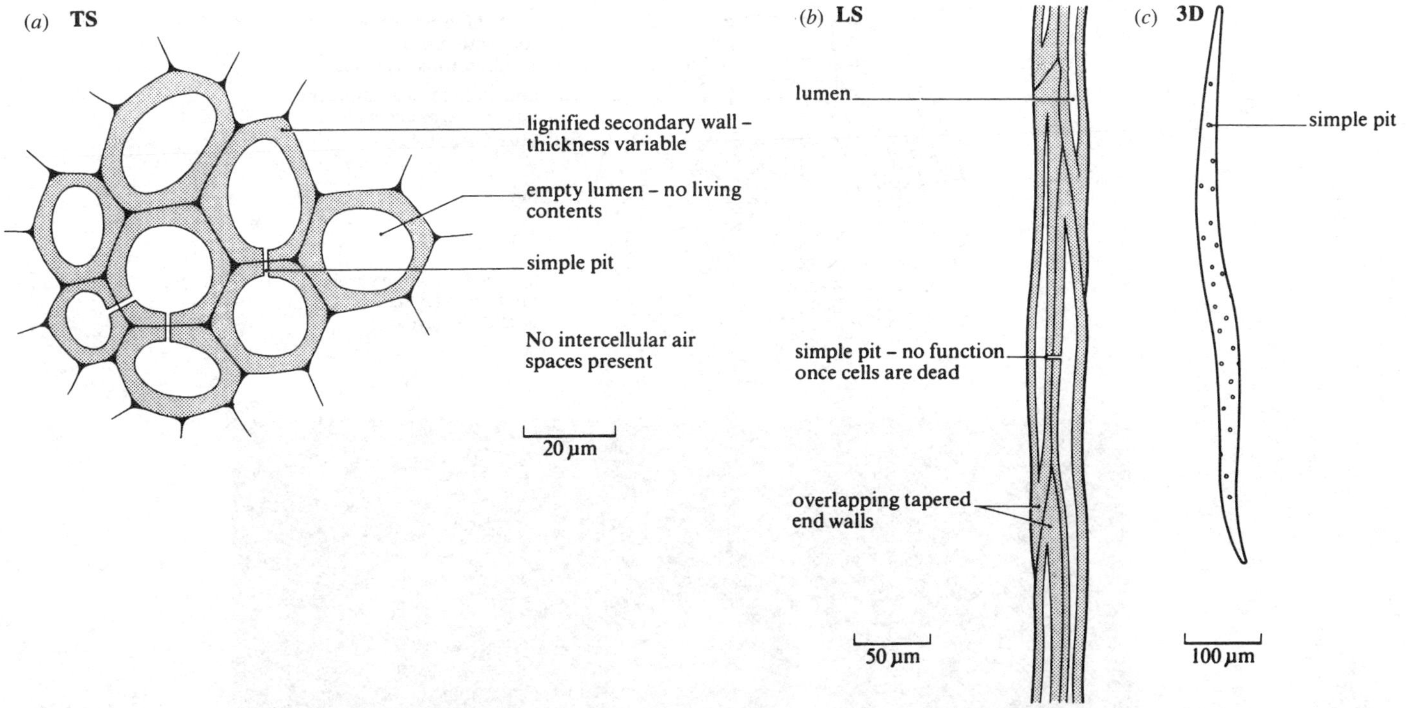

(d)

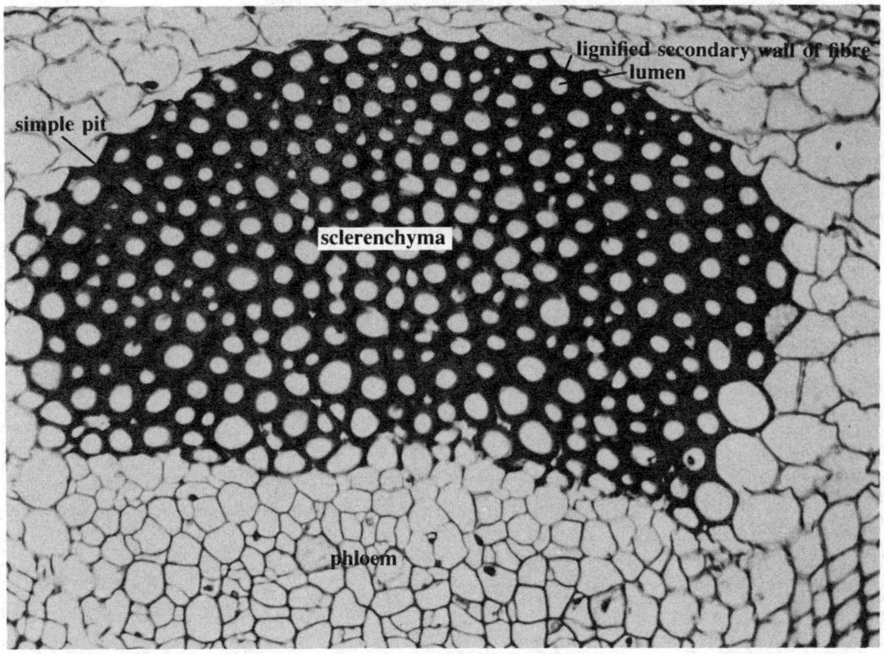

(e)

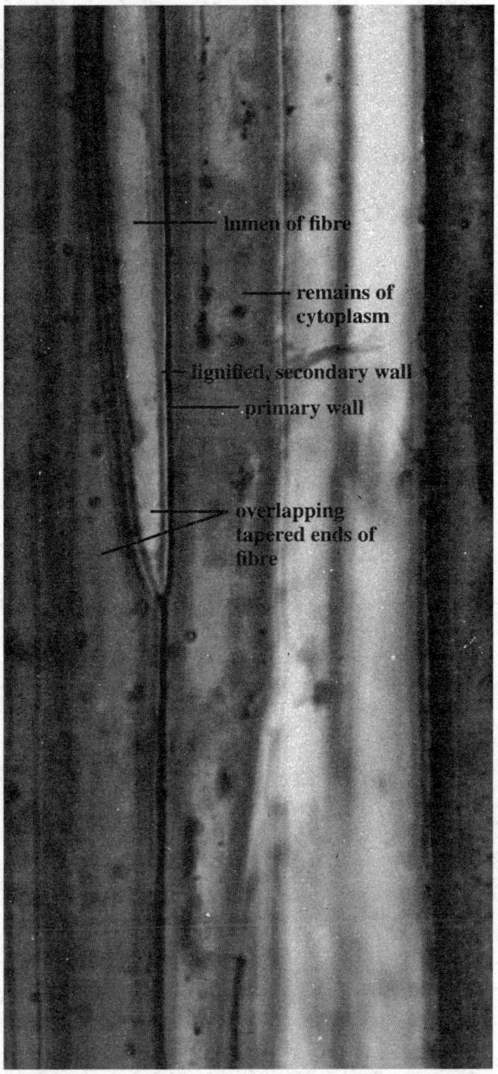

Fig 6.6 *Structure of sclerenchyma cells.* (a) *TS, cells are polygonal in outline.* (b) *LS, cells are elongated (length very variable, commonly >1 mm, up to 250 mm reported).* (c) *Three-dimensional appearance.* (d) *TS sclerenchyma from* Helianthus *stem.* (e) *LS sclerenchyma from* Helianthus *stem.*

(a)

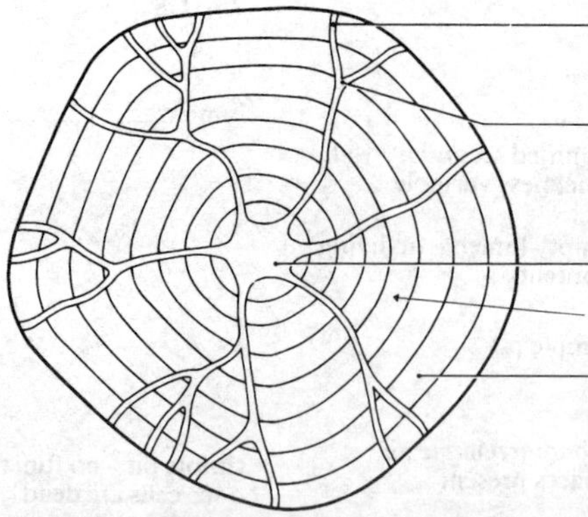

(b)

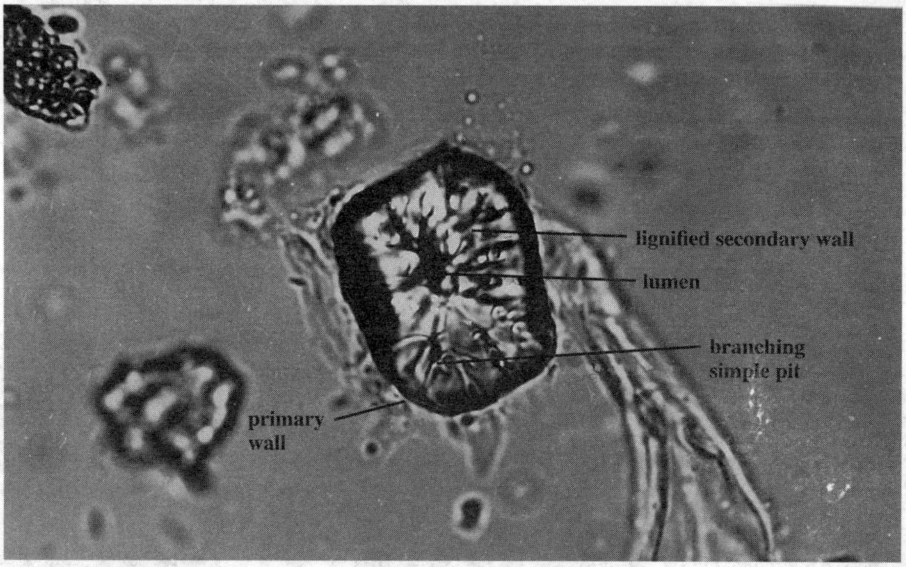

Fig 6.7 *Structure of sclerenchyma sclereids.* (a) *TS or LS, cells are isodiametric.* (b) *Entire sclereid from macerated flesh of pear fruit (×400).*

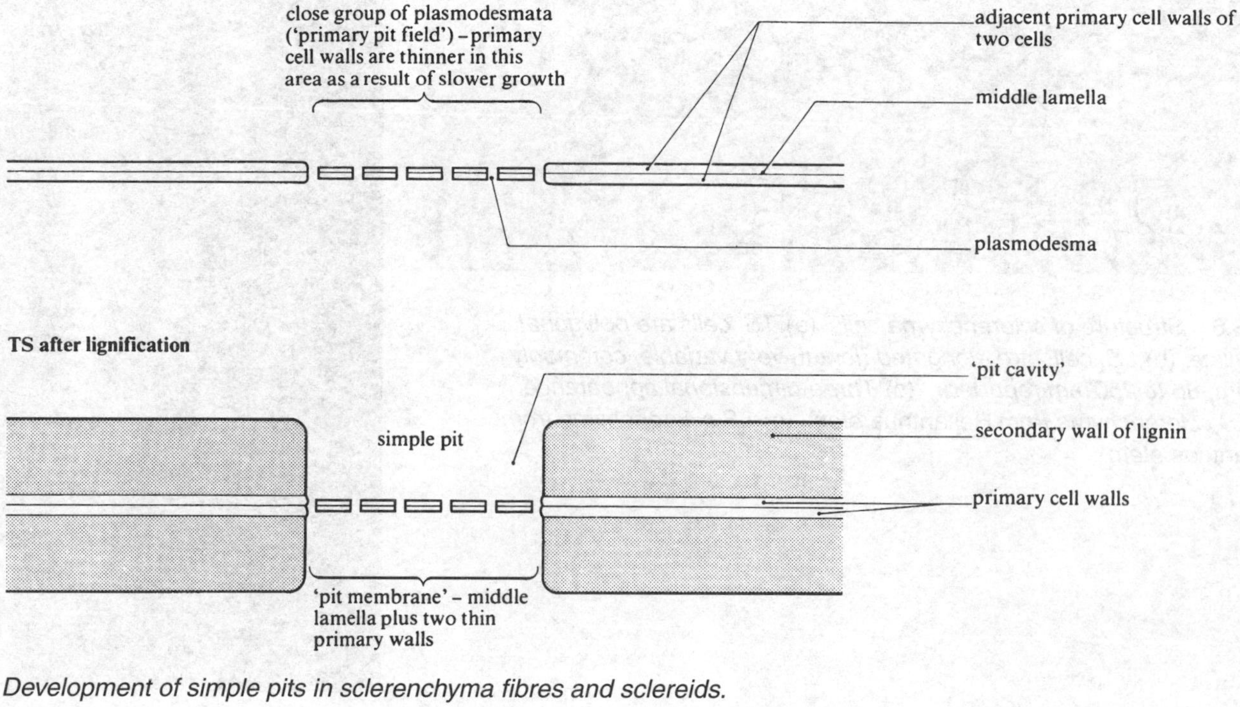

Fig 6.8 *Development of simple pits in sclerenchyma fibres and sclereids.*

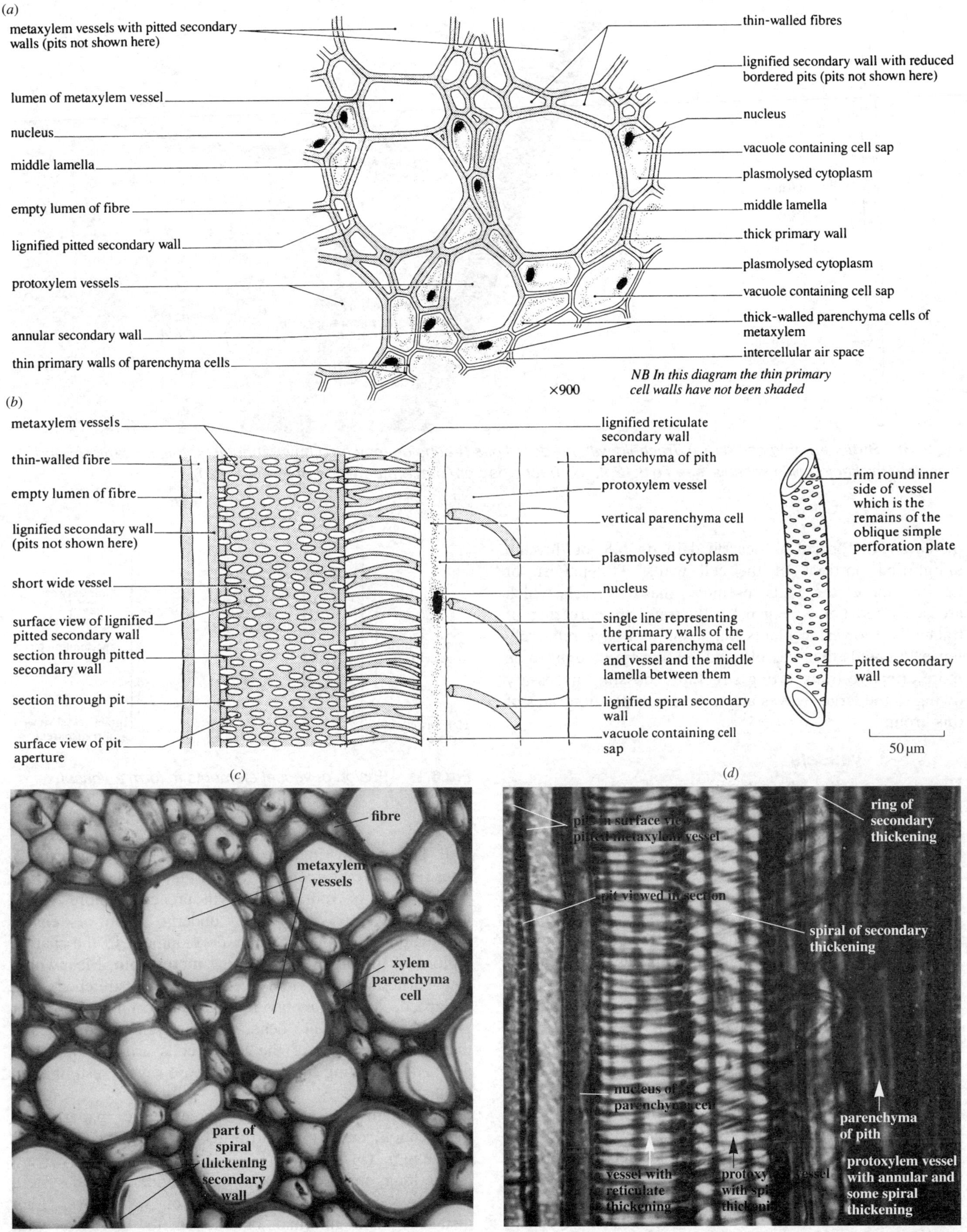

Fig 6.9 *Structure of primary xylem.* (a) *TS.* (b) *LS.* (c) *TS primary xylem from* Helianthus *stem.* (d) *LS primary xylem from* Helianthus *stem.*

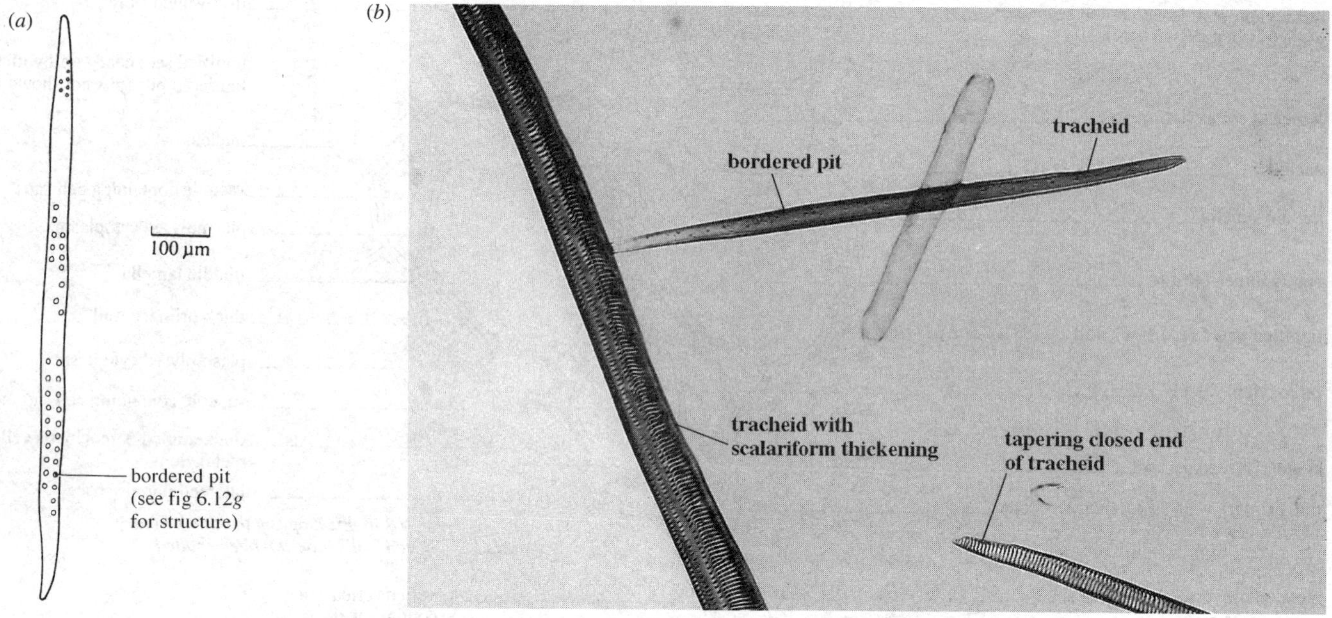

Fig 6.10 *Structure of tracheids.* (a) *Tracheid with bordered pits (tracheids may also have annular, spiral, scalariform and reticulate thickening, like vessels, see fig 6.12g).* (b) *Tracheids from macerated wood of* Pinus *(×120).*

membranes', formed as described in fig 6.8, or through unlignified portions of the cell walls. The pattern of lignification of the walls resembles that of vessels which are described below. Fig 6.10 illustrates the structure of tracheids. Flowering plants (angiosperms) have relatively fewer tracheids than vessels, and vessels are thought to be more effective transporting structures, possibly necessary owing to the larger leaves and higher transpiration rates of this group.

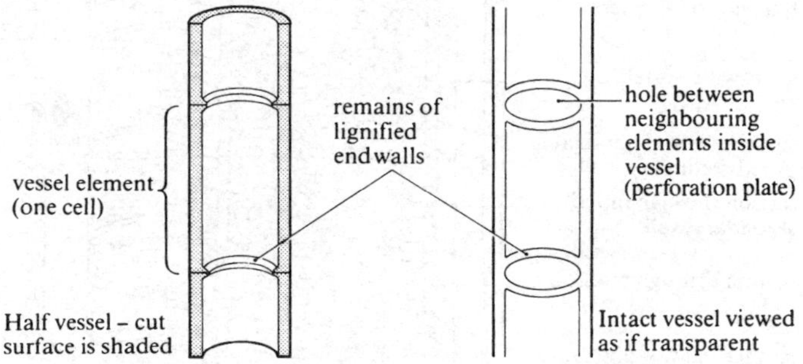

Fig 6.11 *Fusion of vessel elements to form a vessel.*

Vessels

Vessels are the characteristic conducting units of angiosperm xylem. They are very long, tubular structures formed by the fusion of several cells end to end in a row. Each of the cells forming a xylem vessel is equivalent to a tracheid and is called a **vessel element**. However, vessel elements are shorter and wider than tracheids. The first xylem to appear in the growing plant is called **primary xylem** and develops in the root and shoot apices. Differentiated xylem vessel elements appear in rows at the edges of the procambial strands. A vessel is formed when the neighbouring vessel elements of a given row fuse as a result of their end walls breaking down. A series of rims is left around the inner side of the vessel marking the remains of the end walls. The fusion of elements is shown in fig 6.11.

Protoxylem and metaxylem

The first vessels form the **protoxylem**, located in the part of the apex, just behind the apical meristem, where elongation of surrounding cells is still occurring. Mature protoxylem vessels can be stretched as surrounding cells elongate because lignin is not deposited over the entire cellulose wall, but only in rings or in spirals as shown in fig 6.12. These act as reinforcement for the tubes during elongation of the stem or root. As growth proceeds, more xylem vessels develop and these undergo more extensive lignification, completing their development in the mature regions of the organ and forming **metaxylem**. Meanwhile, the earliest protoxylem vessels have stretched and collapsed. Mature metaxylem vessels cannot stretch or grow because they are dead, rigid, fully lignified tubes. If they developed before the living cells around them had finished elongating they would impose severe restraints on elongation.

Metaxylem vessels show three basic patterns of lignification, namely scalariform, reticulate and pitted, as shown in fig 6.12.

The long, empty tubes of xylem provide an ideal system for translocating large quantities of water over long distances with minimal obstruction to flow. As with tracheids, water can pass from vessel to vessel through pits

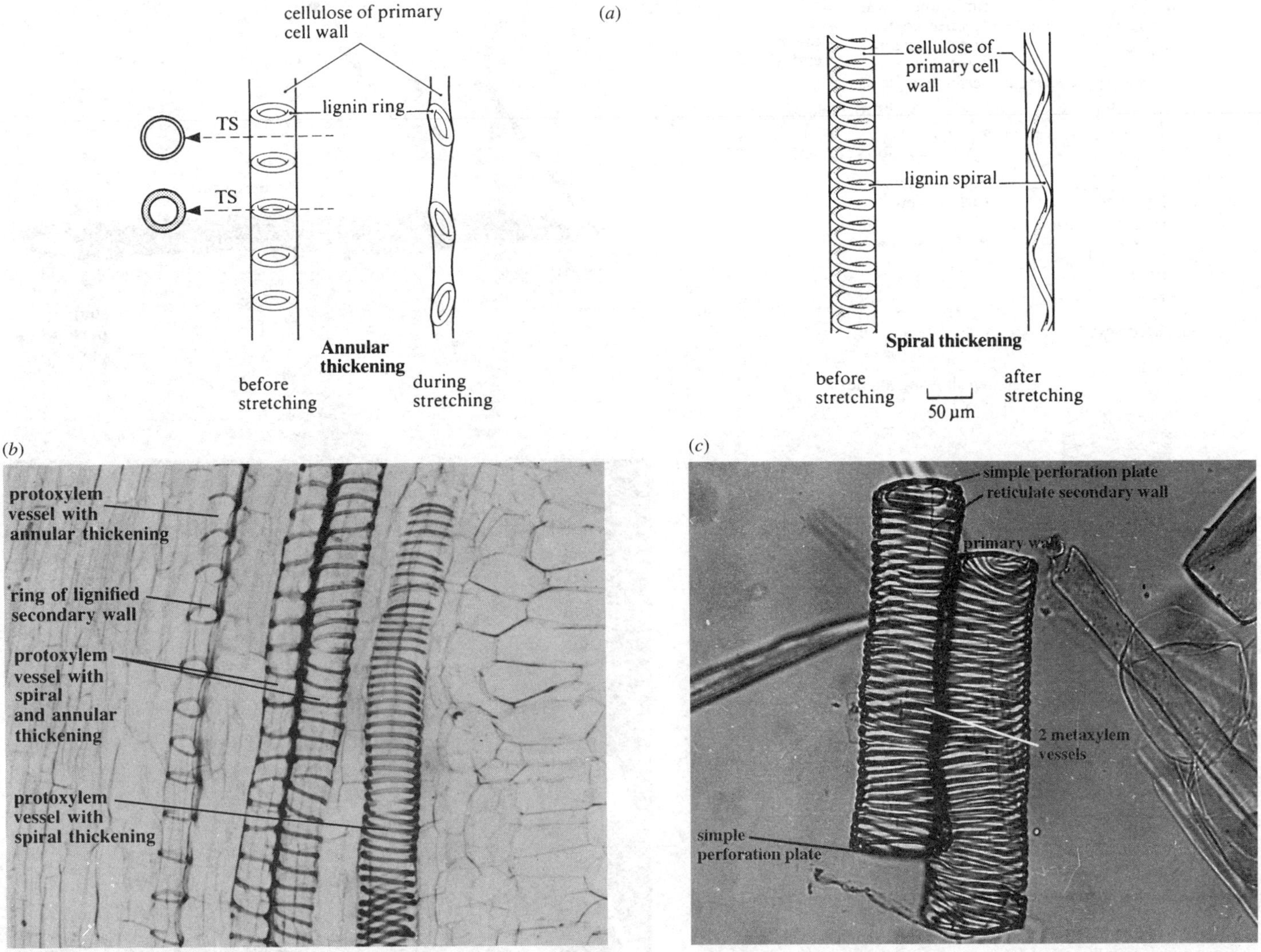

Fig 6.12 *Structure of protoxylem and metaxylem vessels.* (a) *Protoxylem vessels.* (b) *Micrograph of annular and spiral protoxylem vessels.* (c) *Micrograph of metaxylem reticulate vessels from macerated wood.*

or through unlignified portions of the cell wall. The walls also have high tensile strength, being lignified, which is another important feature because it prevents tubes collapsing when conducting water under tension (section 14.4).

The second main function of xylem, namely support, is also fulfilled by the collection of lignified tubes. In the primary plant body the distribution of xylem in the roots is central, helping to withstand the tugging strains of the aerial parts as they bend or lean over. In the stems the vascular bundles are arranged either peripherally in a ring, as in dicotyledons, or scattered, as in monocotyledons, so that in both cases separate rods of xylem run through the stem and provide some support. The supporting function becomes much more important if secondary growth takes place. During this process extensive growth of secondary xylem occurs which supports the large structure of trees and shrubs, taking over from collenchyma and sclerenchyma as the chief mechanical tissue. The nature and extent of the thickness is modified to some extent by the stresses received by the growing plant, so that reinforcement growth can occur and give maximum support.

Xylem parenchyma

Xylem parenchyma occurs in both primary and secondary xylem but it is more extensive and assumes greater importance in the latter. It has thin cellulose cell walls and living contents, as is typical of parenchyma.

Two systems of parenchyma exist in secondary xylem, derived from meristematic cells called ray initials and fusiform initials, as described in chapter 21. The ray parenchyma is the more extensive. It forms radial sheets of tissue called **medullary rays** which maintain a living link through the wood between the pith and cortex. Its functions include food storage, deposition of tannins, crystals and so on, radial transport of food and water, and gaseous exchange through the intercellular spaces.

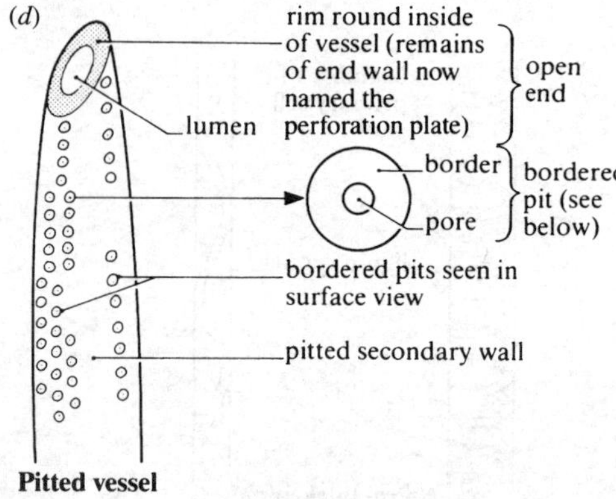

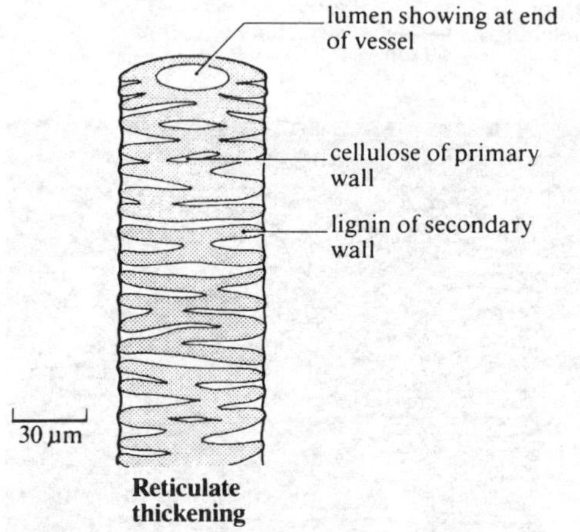

Size is very variable; the longest are several metres in length, though commonly several centimetres long.

Scalariform thickening is similar to reticulate but with fewer interconnections between the bars of thickening. It is less commonly seen. It usually grades into reticulate thickening by progressive lignification

Fig 6.12 *(cont).* (d) *Pitted and reticulate metaxylem vessels.* (e) *Micrograph of metaxylem pitted vessel from macerated wood.* (f) *Scanning electron micrograph of metaxylem vessels (×18 000). Appearance of these vessels in TS will vary according to which part of the vessel is sectioned as indicated in the diagram of extreme left vessel in part* (a). (g) *TS bordered pit to show structure.*

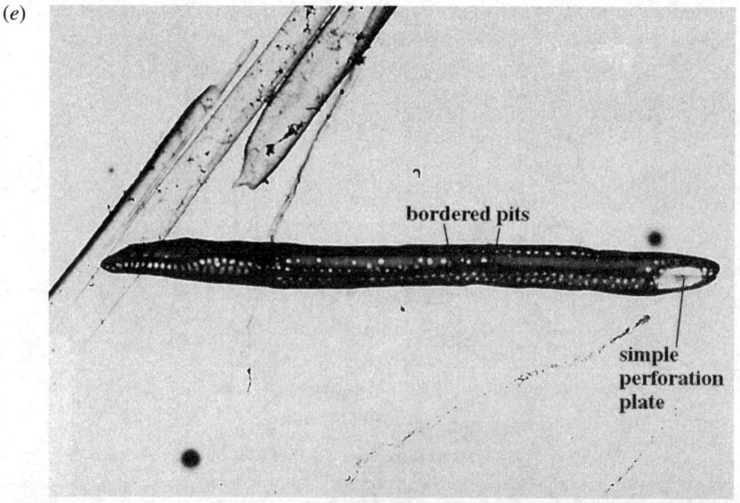

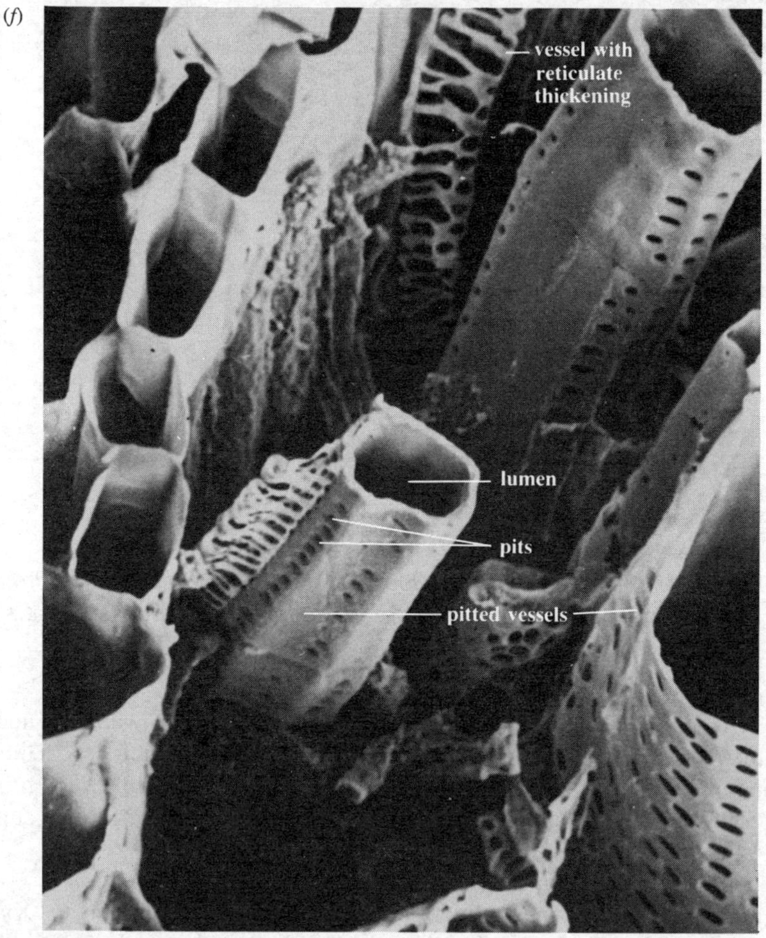

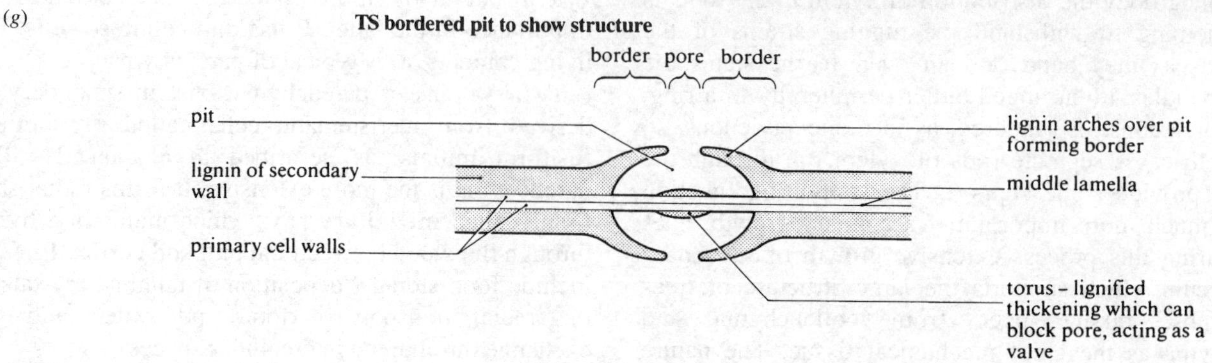

Fusiform initials normally give rise to xylem vessels or phloem sieve tubes and companion cells, but occasionally they give rise to parenchyma cells. These form vertical rows of parenchyma in the secondary xylem.

Xylem fibres

Xylem fibres, like xylem vessels, are thought to have originated from tracheids. They are shorter and narrower than tracheids and have much thicker walls, but they have pits similar to those in tracheids and are often difficult to distinguish from them in section because intermediate cell types occur. Xylem fibres closely resemble the sclerenchyma fibres already described, having overlapping end walls. Since they do not conduct water they can have much thicker walls and narrower lumens than xylem vessels and are therefore stronger and confer additional mechanical strength to the xylem.

6.2.2 Phloem

Phloem resembles xylem in possessing tubular structures modified for translocation. However, the tubes are composed of living cells with cytoplasm and have no mechanical function. There are five cell types in the phloem, namely sieve tube elements, companion cells, parenchyma, fibres and sclereids.

Sieve tubes and companion cells

Sieve tubes are the long tube-like structures that translocate solutions of organic solutes like sucrose throughout the plant. They are formed by the end-to-end fusion of cells called **sieve tube elements** or **sieve elements**. Rows of these cells can be seen developing from the procambial strands of apical meristems where primary phloem develops, together with primary xylem, in vascular bundles.

The first phloem formed is called **protophloem** and, like protoxylem, it is produced in the zone of elongation of the growing root or stem. As the tissues around it grow and elongate, it becomes stretched and much of it eventually collapses and becomes non-functional. Meanwhile, however, more phloem continues to be produced and the phloem that matures after elongation has ceased is called **metaphloem**.

Sieve tube elements have a very distinctive structure. Their walls are made of cellulose and pectic substances, like parenchyma cells, but their nuclei degenerate and are lost as they mature and the cytoplasm becomes confined to a thin layer around the periphery of the cell. Although they lack nuclei, the sieve elements remain living but are dependent on the adjacent companion cells which develop from the same original meristematic cell. The two cells together form a functional unit, the companion cell having dense, very active cytoplasm. The detailed structure of the cells is revealed by the electron microscope and is described in chapter 13.

A conspicuous and characteristic feature of sieve tubes that is visible in the light microscope is the **sieve plate**. This is derived from the two adjoining end walls of neighbouring sieve elements. Originally plasmodesmata run through the walls but the canals enlarge to form pores, making the walls look like a sieve and allowing a flow of solution from one element to the next. Thus sieve tubes are spanned at intervals by sieve plates that mark successive sieve elements. The structure of sieve tubes, companion cells and phloem parenchyma as seen with the electron microscope is shown in fig 6.13.

Secondary phloem, which develops from the vascular cambium like secondary xylem, appears similar in structure to primary phloem except that it is crossed by bands of lignified fibres and medullary rays of parenchyma (chapter 22). It is much less extensive than secondary xylem and is constantly being replaced as described in chapter 22.

Phloem parenchyma, fibres and sclereids

Phloem parenchyma and fibres are found in dicotyledons but not in monocotyledons. Phloem parenchyma has the same structure as parenchyma elsewhere, though the cells are generally elongated. In secondary phloem, parenchyma occurs in medullary rays and vertical strands as already described for xylem parenchyma. Phloem parenchyma and xylem parenchyma have the same functions.

Phloem fibres are exactly similar to the sclerenchyma fibres already described. They occur occasionally in the primary phloem, but more frequently in the secondary phloem of dicotyledons. In secondary phloem they form vertically running bands of cells. Since the secondary phloem is subject to stretching as growth continues, the sclerenchyma probably helps to resist this pressure.

Sclereids occur frequently in phloem, especially in older phloem.

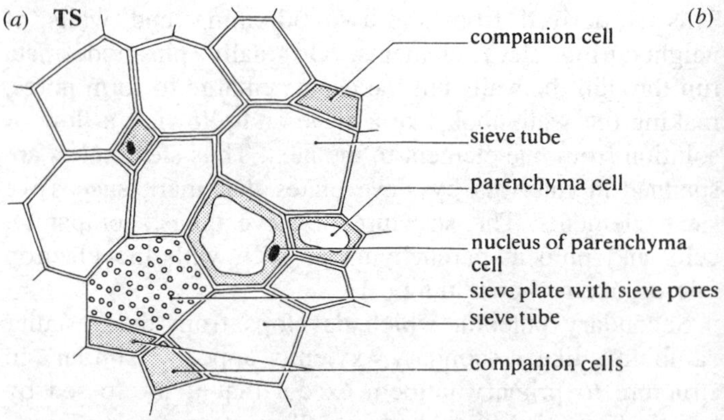

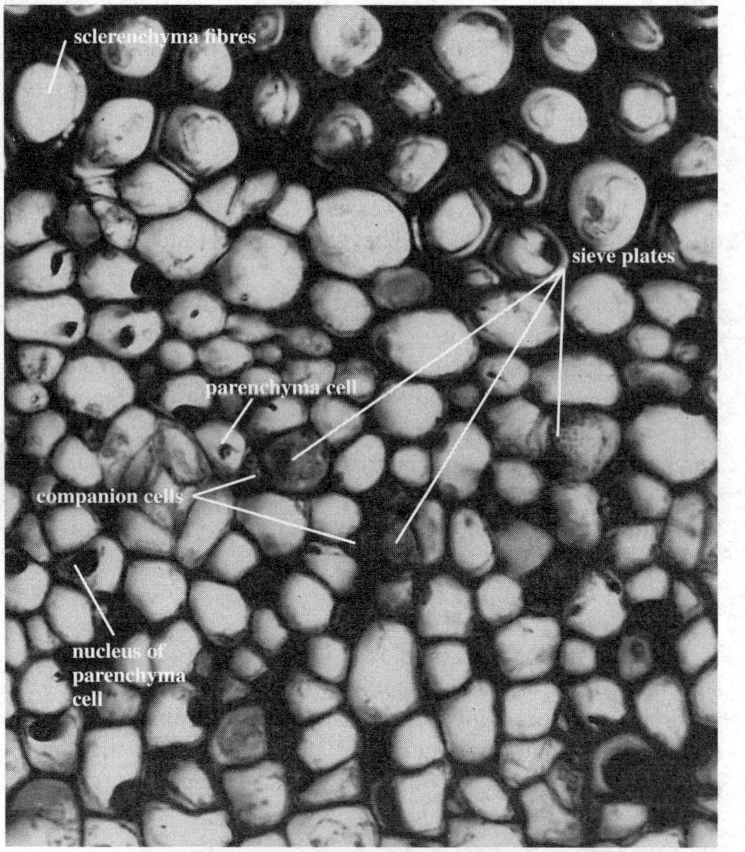

Fig 6.13 *Structure of phloem.* (a) *Diagram of TS.* (b) *Micrograph of TS of primary phloem of* Helianthus *stem (×450).* (c) *Diagram of LS.* (d) *Micrograph of LS of primary phloem of* Cucurbita *stem (×432).*

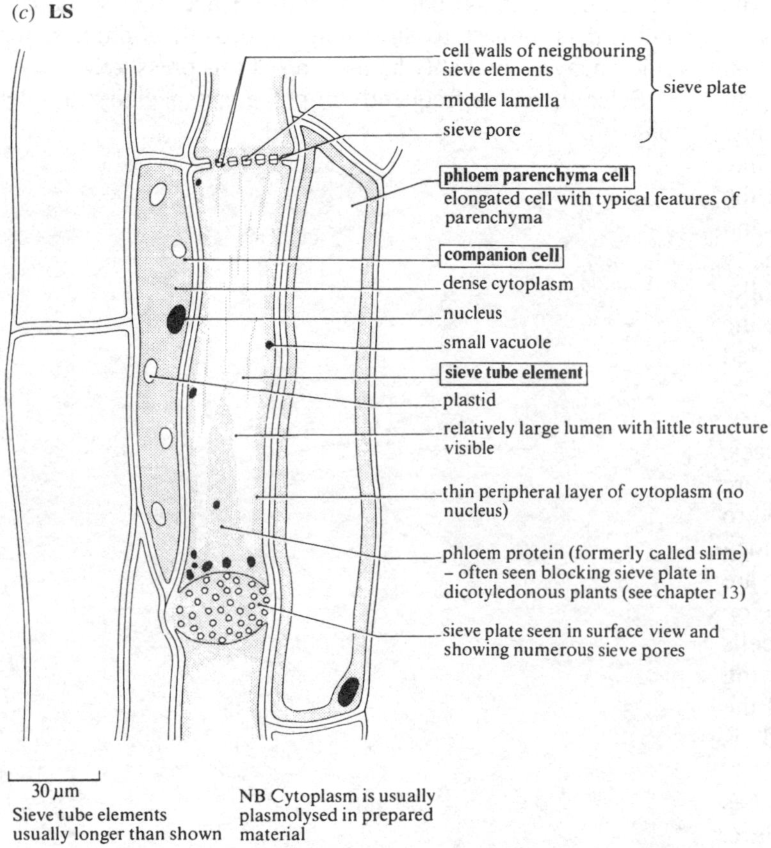

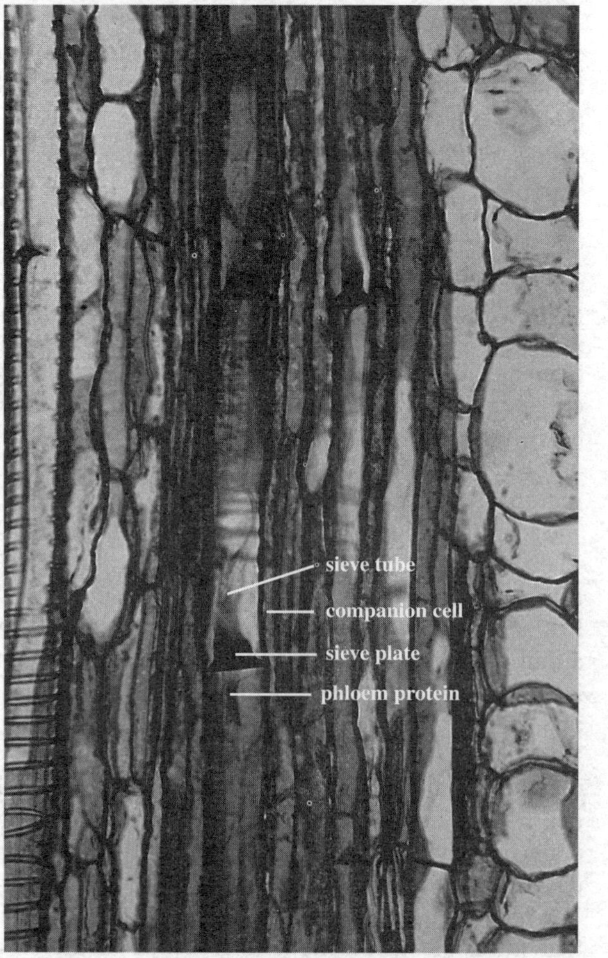

6.3 Animal epithelial tissue

Epithelial tissue is arranged in single or multilayered sheets and covers the internal and external **surfaces** of the body of an organism.

Epithelial cells are held together by small amounts of a carbohydrate-based cementing substance and by special junctions between cells. The bottom layer of cells rests on a **basement membrane** composed of a network of fibres, which include collagen, in a matrix. (The term membrane here should not be confused with the cell membranes discussed in chapter 5 – the term simply means a thin layer.) It is not a barrier to diffusion. As epithelial cells are not supplied with blood vessels, they rely on diffusion of oxygen and nutrients from lymph vessels which run through nearby intercellular spaces. Nerve endings may occur in the epithelium.

Epithelial tissue functions to protect underlying structures from injury through abrasion or pressure, and from infection. Stress is combated by the tissue becoming thickened and keratinised, and where cells are worn off due to constant friction the epithelium shows a very rapid rate of cell division so that lost cells are quickly replaced. The free surface of the epithelium often has a specialised structure and may be absorptive, secretory or excretory in function, or bear sensory cells and nerve endings specialised for reception of stimuli.

Epithelial tissues are classified according to the number of cell layers and the shape of the individual cells, as shown in table 6.2. In many areas of the body the different cell types intermix and the epithelia cannot be classified into distinct types.

6.3.1 Simple epithelia

Squamous epithelium

The cells are thin and flattened. They are so thin that the nucleus causes a bulge (fig 6.14). The edges of squamous

Table 6.2 Classification of epithelial tissues.

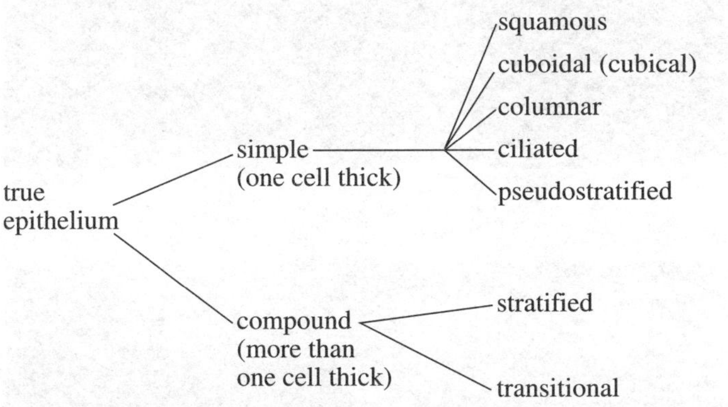

(*c*)

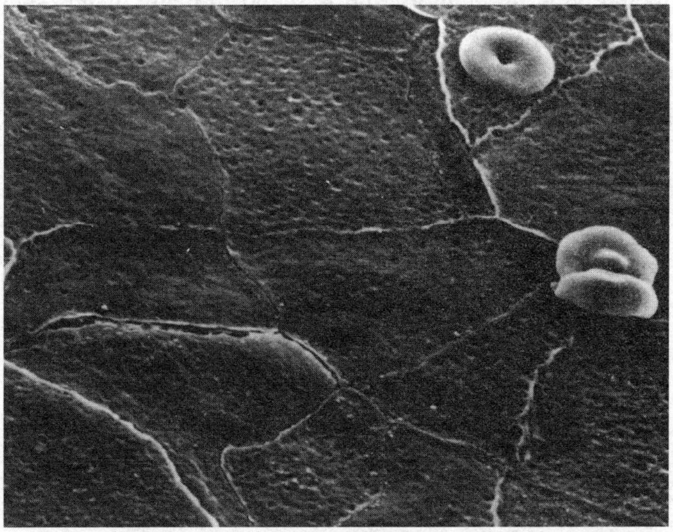

(*a*) (*b*)

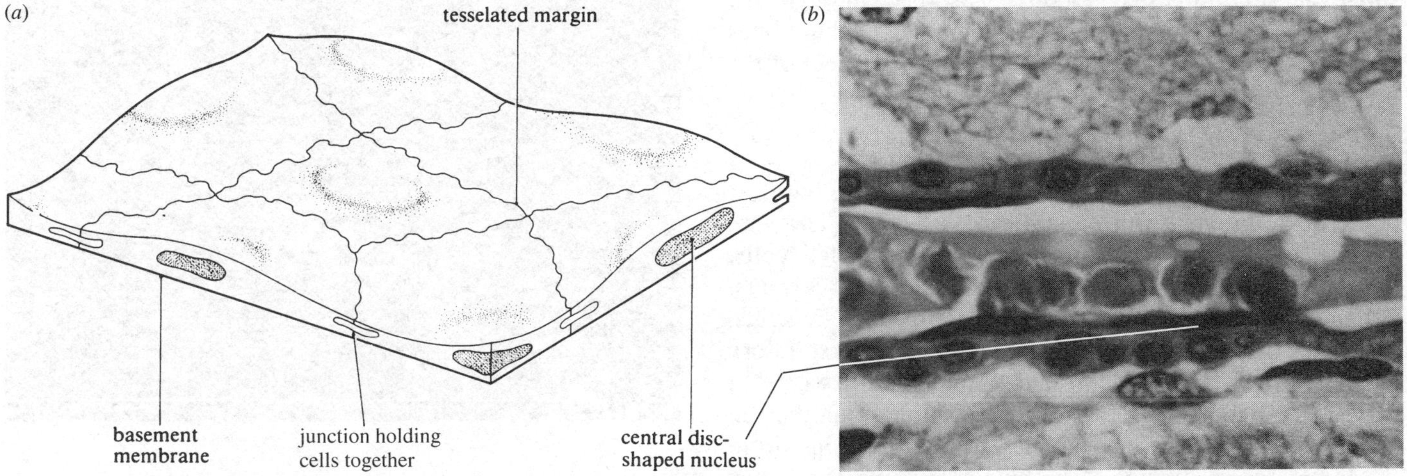

Fig 6.14 *Simple squamous epithelium: (a) diagram; (b) photomicrograph (small blood vessel); (c) electron micrograph.*

cells are irregular as can be seen clearly in surface view. There are special junctions between neighbouring cells which help to bind them firmly together. Squamous epithelium occurs in areas such as the renal capsules of the kidney, the alveoli of the lungs and the blood capillary walls, where its thinness permits diffusion of materials through it. In blood vessels it is referred to as the **endothelium** (*endo*, inside). It also provides smooth linings to hollow structures such as blood vessels and the chambers of the heart, where it allows the relatively friction-free passage of fluids through them.

Cuboidal epithelium

This is the least specialised of all epithelia. The cells are roughly cube-shaped and possess a central spherical nucleus (fig 6.15). When viewed from the surface the cells are either pentagonal or hexagonal in outline. They form the lining of many ducts, such as the salivary and pancreatic ducts, and line the proximal and distal convoluted tubules and collecting ducts of the kidney where they are non-secretory. Cuboidal epithelium in other parts of the body is secretory and is found in many glands such as the salivary, sweat and thyroid glands.

Columnar epithelium

These cells are tall and quite narrow, thus providing more cytoplasm per unit area of epithelium (fig 6.16). Each cell possesses a nucleus situated at its basal end. Goblet cells, which secrete mucus, are often interspersed among the epithelial cells and the epithelium may be secretory and/or absorptive in function. There is frequently a conspicuous striated border or brush border of **microvilli** at the free surface end of each cell. This increases the surface area of the cell for absorption and secretion. Columnar epithelium lines the stomach, where mucus secreted by goblet cells protects the stomach lining from the acidic contents of the stomach and from digestion by enzymes. It also lines the intestine where mucus again protects it from self-digestion and at the same time lubricates the passage of food. In the small intestine digested food is absorbed through the epithelium into the bloodstream. Columnar epithelium lines and protects many kidney ducts, and is a component of the thyroid gland.

Ciliated epithelium

Cells of this tissue are usually columnar in shape but bear numerous cilia at their free surfaces (fig 6.17). They are always associated with mucus-secreting goblet cells, producing fluids in which the cilia set up currents. Ciliated epithelium lines the insides of the oviducts, ventricles of the brain, the spinal canal and the respiratory passages (trachea, bronchi and bronchioles), where it serves to move materials from one location to another. In the respiratory tract, for example, cilia waft mucus up to the throat for swallowing. The mucus traps bacteria, dust and other small particles, preventing it from reaching the lungs.

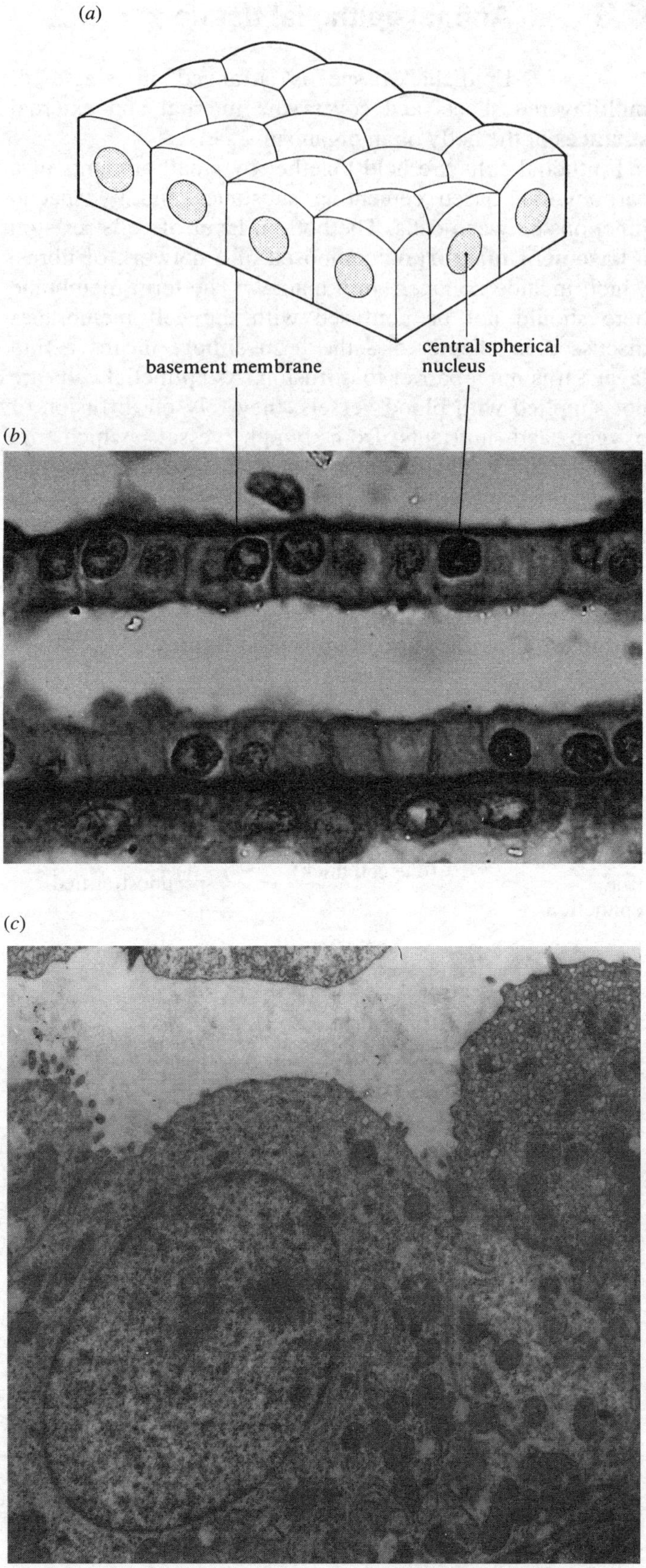

Fig 6.15 *Cuboidal epithelium:* (a) *diagram;* (b) *photomicrograph (kidney);* (c) *electron micrograph.*

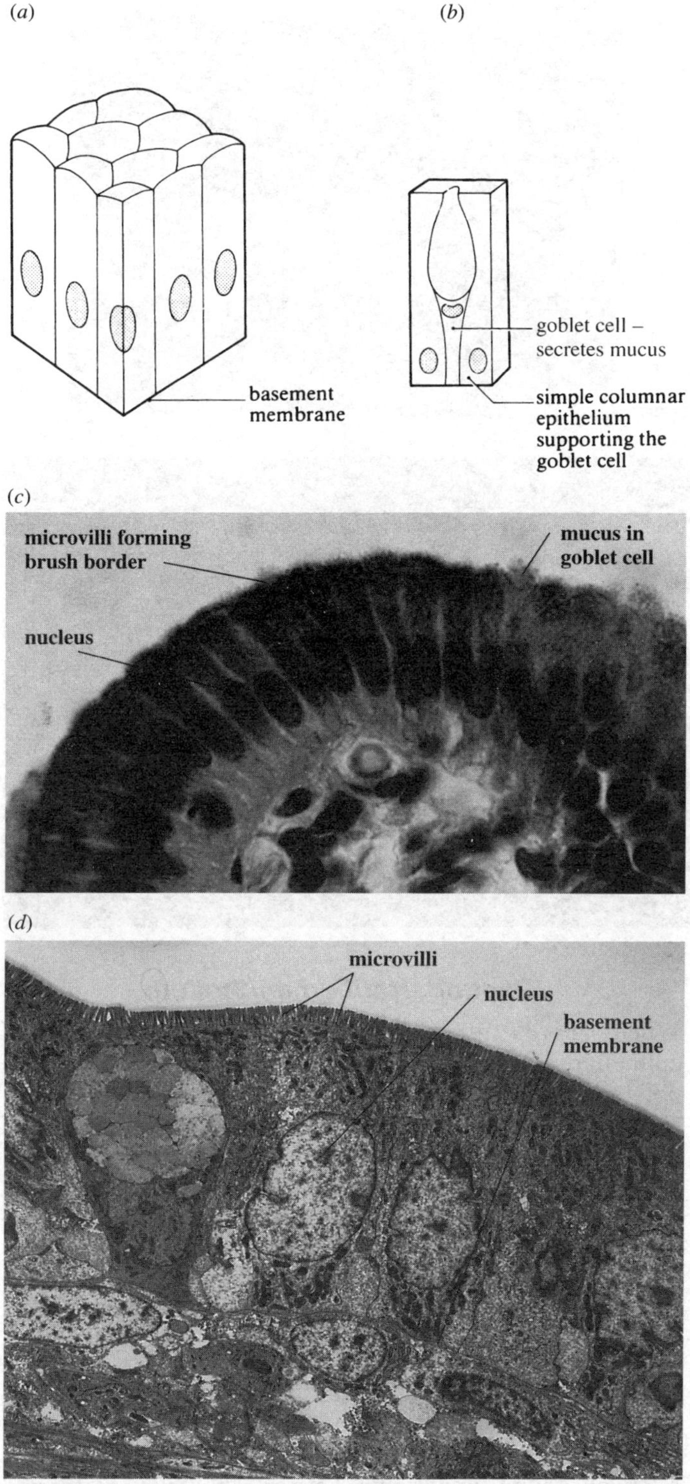

Fig 6.16 (a) *Columnar epithelium;* (b) *showing goblet cell;* (c) *photomicrograph (ileum);* (d) *electron micrograph (ileum).*

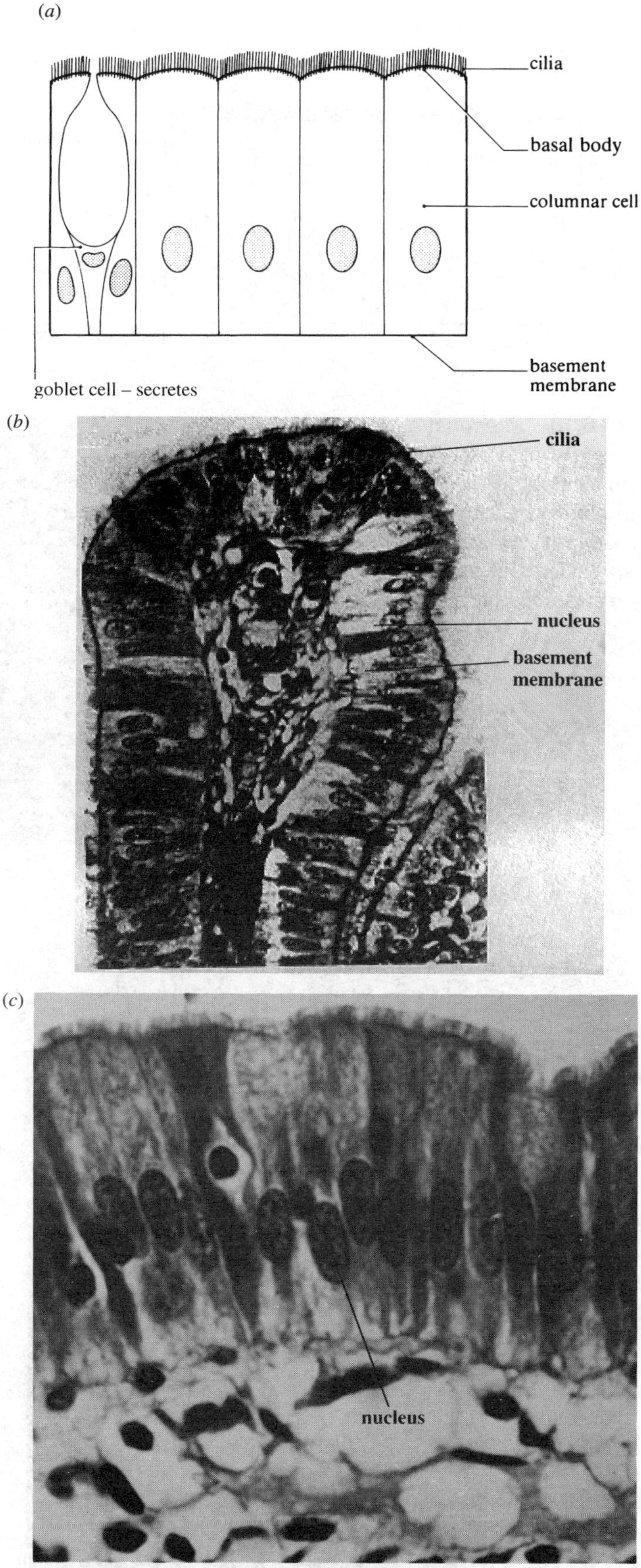

Fig 6.17 *Ciliated columnar epithelium:* (a) *diagram;* (b) *photomicrograph (oviduct);* (c) *photomicrograph (trachea).*

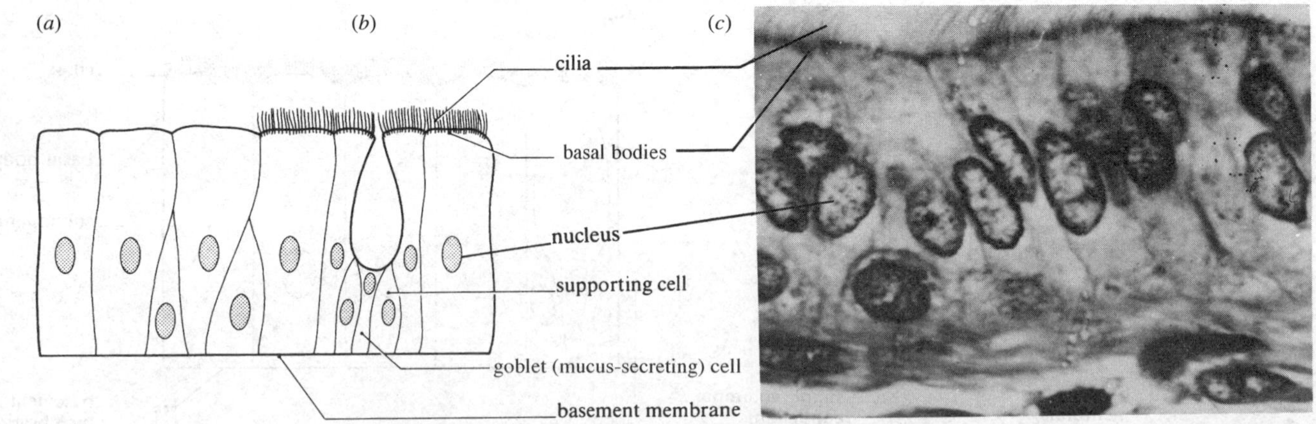

Fig 6.18 *Pseudostratified epithelium:* (a) *columnar;* (b) *ciliated;* (c) *photomicrograph of respiratory, ciliated epithelium;* (d) *scanning electron micrograph of cilia.*

(d)

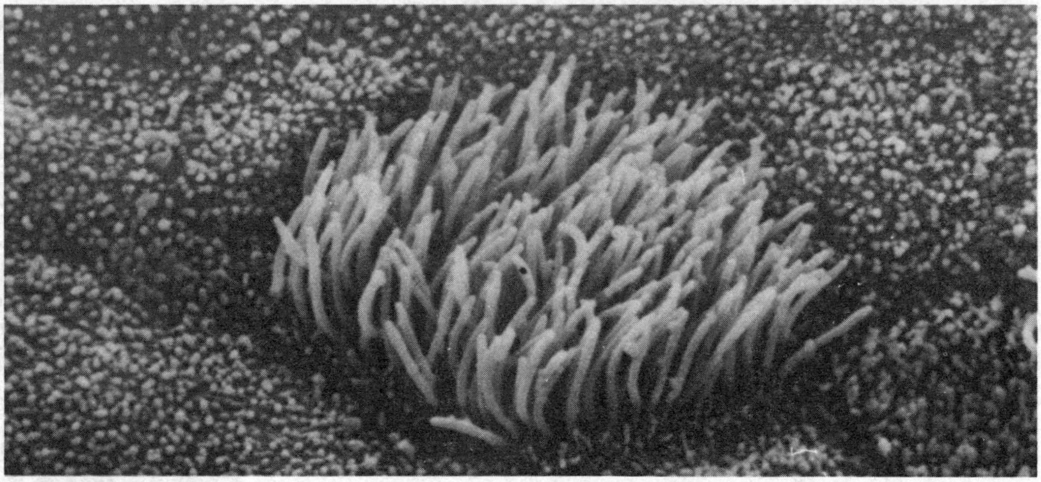

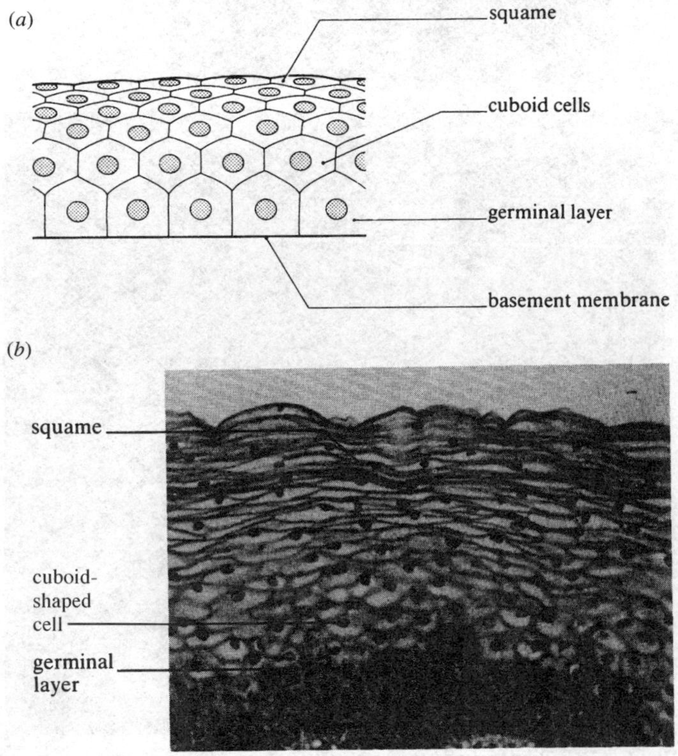

Fig 6.19 *Stratified squamous epithelium:* (a) *diagram;* (b) *photomicrograph (vagina).*

Pseudostratified epithelium

When viewed in section the nuclei of this type of epithelium appear to be at several different levels because not all the cells reach the free surface (fig 6.18). Nevertheless the epithelium is still only one layer of cells thick with each cell attached to the basement membrane. This epithelium is found lining the urinary tract and the respiratory passages (trachea, bronchi and bronchioles, where it is ciliated and columnar).

6.3.2 Compound epithelia

Stratified epithelium

This tissue is made up of a number of layers of cells. It is therefore thicker than simple epithelium and forms a relatively tough, impervious barrier. The cells are formed by mitotic divisions of a germinal layer which rests on the basement membrane (fig 6.19). The first-formed cells are cuboid in shape, but as they are pushed outwards towards the free surface of the tissue they become flattened. In this condition the cells are called **squames**. They may remain uncornified, as in the oesophagus, where the epithelium protects the underlying tissues against mechanical damage by friction with food just swallowed. In other areas of the body the squames may be transformed into a dead horny layer of **keratin** which eventually flakes away. In this condition the epithelium is said to be **cornified**, and is found in particular abundance on external skin surfaces, lining the buccal cavity (mouth) and the vagina, where it affords protection against abrasion.

According to the shape of the cells which make up the stratified epithelium, it may be termed stratified squamous (located in parts of the oesophagus), stratified cuboidal (in the sweat gland ducts), stratified columnar (in the mammary gland ducts), and stratified transitional (in the bladder).

Transitional epithelium

This is often regarded as a modified type of stratified epithelium. It consists of 3–4 layers of cells all of similar size and shape except at the free surface where they are more flattened (fig 6.20). The surface cells do not slough off, and all cells are able to modify their shape when placed under differing conditions. This property is important in locations where structures are subjected to considerable stretching, such as the bladder, ureter and the pelvic region of the kidney. The thickness of the tissue also prevents urine escaping into the surrounding tissues.

6.3.3 Glandular epithelia

Amongst the epithelial cells there may be individual glandular cells, such as the **goblet cells**, or aggregates of glandular cells forming a **multicellular gland**. An epithelium containing many goblet cells is called a mucous membrane.

6.4 Animal connective tissue

Connective tissue is the major **supporting tissue** of the body. It includes the skeletal tissues, bone and cartilage, and in addition it binds other tissues together. Connective tissue also forms sheaths like bags around the organs of the body, separating them so that they do not interfere with each other's activities, as well as surrounding and protecting blood vessels and nerves where they enter or leave organs. Connective tissue is a composite material made up of a variety of cells. It contains several types of **fibre** which are non-living products of the cells, and a fluid or semi-fluid background material or **matrix** between the cells.

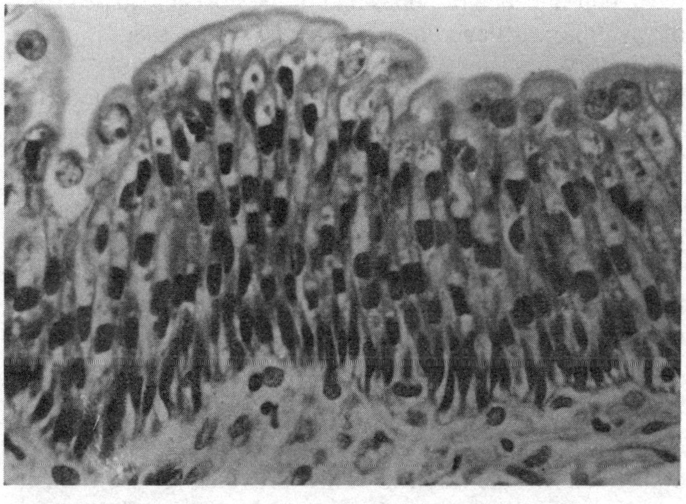

Fig 6.20 *Transitional epithelium (bladder).*

The cells are usually widely separated from each other. An extensive blood supply runs through the tissue in some parts of the body, as in the dermis of the skin, but this is primarily concerned with supplying other structures, such as epithelium, with oxygen and nutrients rather than the connective tissue itself. Connective tissue may be subdivided into a number of types as indicated in table 6.3.

This tissue fulfils many functions other than packing and binding other structures together, such as providing protection against wounding or bacterial invasion (areolar tissue), insulation of the body against heat loss (adipose tissue), providing a supportive framework for the body (cartilage and bone) and producing blood cells.

6.4.1 Areolar, fibrous connective and adipose tissues

Areolar tissue is shown in fig 6.21, fibrous connective tissue in figs 6.22 and 6.23, and adipose tissue in fig 6.24.

6.4.2 Skeletal tissues

Cartilage

Cartilage is a connective tissue consisting of cells embedded in a matrix of **chondrin**. The matrix is deposited by cells called **chondroblasts** and possesses many fine fibrils mostly made up of collagen. Eventually the chondroblasts become enclosed in spaces called **lacunae**. In this condition they are termed **chondrocytes**. The margin of a piece of cartilage is enclosed by a dense layer of cells and fibrils called the **perichondrium**. From here new chondroblasts are produced, which are constantly added to the internal matrix of the cartilage.

Cartilage is a hard but flexible tissue. It is highly adapted to resist any strains that are placed upon it. The matrix is compressible and elastic and is able to absorb mechanical shocks, such as frequently occur between the surfaces of bones where they meet. The collagen fibrils resist any tension which may be imposed on the tissue.

Three types of cartilage are recognisable. For each type the organic components of the matrix are quite distinct.

Hyaline cartilage (fig 6.25) 'Hyaline' means glassy or shiny. The matrix is a semi-transparent material through which fine collagen fibrils run. The chondrocytes near the periphery are flattened in shape, whereas those situated internally are angular. Each chondrocyte is contained in a space called a **lacuna**, and

Table 6.3 Types of connective tissue

- connective tissue
  - loose connective tissue — areolar (fig 6.21)
    - *Cells widely dispersed in semi-fluid matrix with scattered fibres*
  - fibrous connective tissue
    - *More fibres than cells*
    - white fibrous – collagen fibres dominate (fig 6.22)
    - yellow elastic – elastin fibres dominate (fig 6.23)
  - adipose – fat cells dominate (fig 6.24)
  - dentine (ivory) – similar to bone but harder (higher mineral content), found between enamel and pulp cavity in teeth
  - **skeletal**
    - **cartilage**
      - **hyaline** (fig 6.25)
      - yellow elastic
      - white fibrous
    - **bone**
      - spongy
      - membrane
      - **compact** (fig 6.26)
  - blood cell-making tissue — red and white blood cells

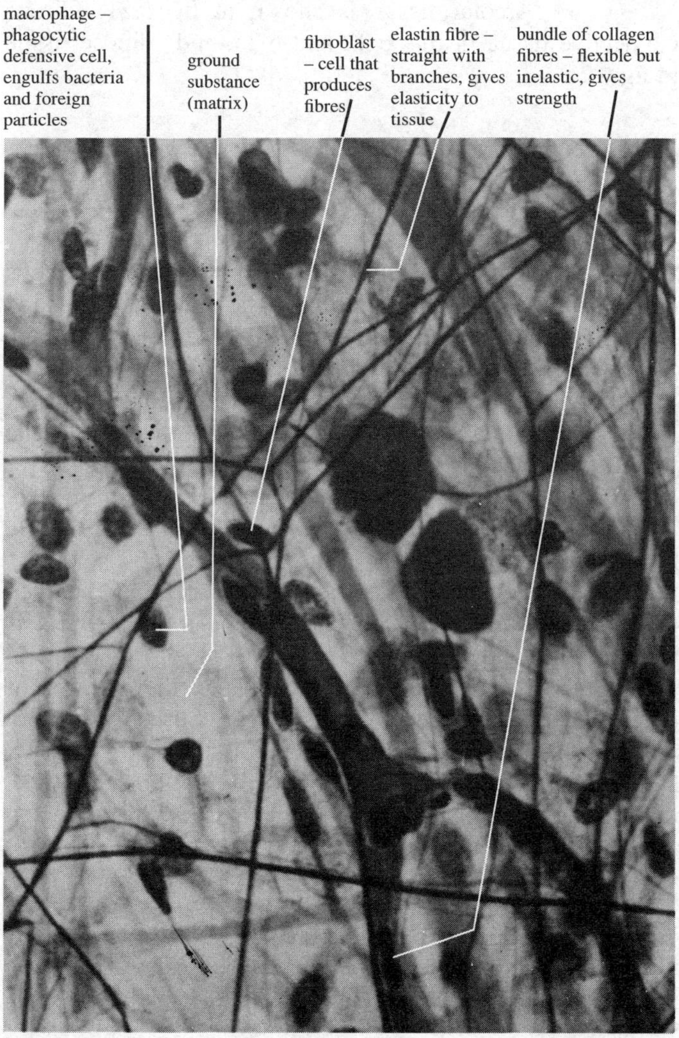

Fig 6.21 *Loose areolar tissue. Cells are widely dispersed. Other cells not shown include fat cells (see adipose tissue), cells involved in response to injury (mast cells), and antibody-producing cells (plasma cells). Areolar tissue is found around all the organs of the body. The fibres are scattered randomly through the matrix.*

each lacuna may enclose one, two, four or eight chondrocytes.

Unlike bone, no processes extend from the lacunae into the matrix, neither are there blood vessels in this area. All exchange of materials between the chondrocytes and the matrix occurs by diffusion.

Hyaline cartilage is an elastic, compressible tissue located at the ends of bones and in the nose. C-shaped rings of hyaline cartilage keep open the air passages of the respiratory system (trachea, bronchi and larger bronchioles). It also forms the skeleton of cartilaginous fish such as sharks, and forms the embryonic skeleton in bony vertebrates.

Yellow elastic cartilage. The matrix is semi-opaque and contains a network of yellow elastic fibres. They confer greater elasticity and flexibility than is found in hyaline cartilage, and permit the tissue to recover its shape quickly after distortion. Examples of its occurrence are in the external ear and the epiglottis.

White fibrous cartilage. This contains large numbers of bundles of densely packed white collagen fibres embedded in the matrix. This provides greater tensile strength than hyaline cartilage, as well as a small degree of

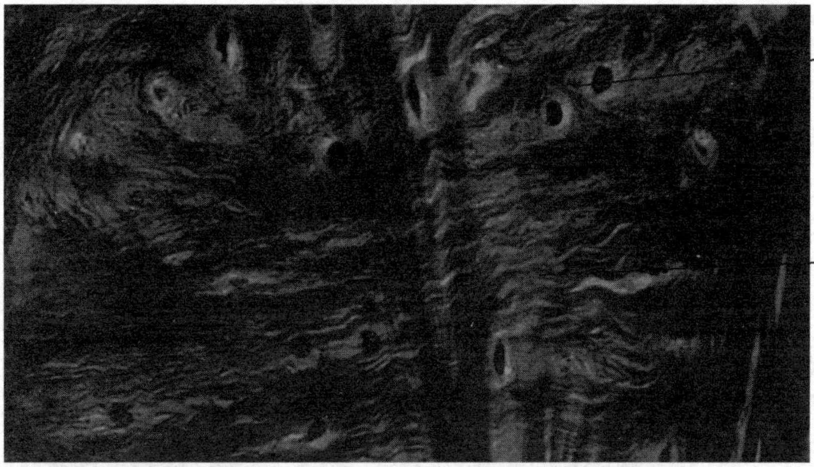

Fig 6.22 *White fibrous tissue. This is composed mainly of parallel bundles of collagen fibres. It is strong and flexible but inelastic. Tendons are almost pure white fibres. They transmit the pull of muscle to bone. The tissue is also found in ligaments, the outer surface of the eye (sclerotic coat and cornea), and other places where strength is required.*

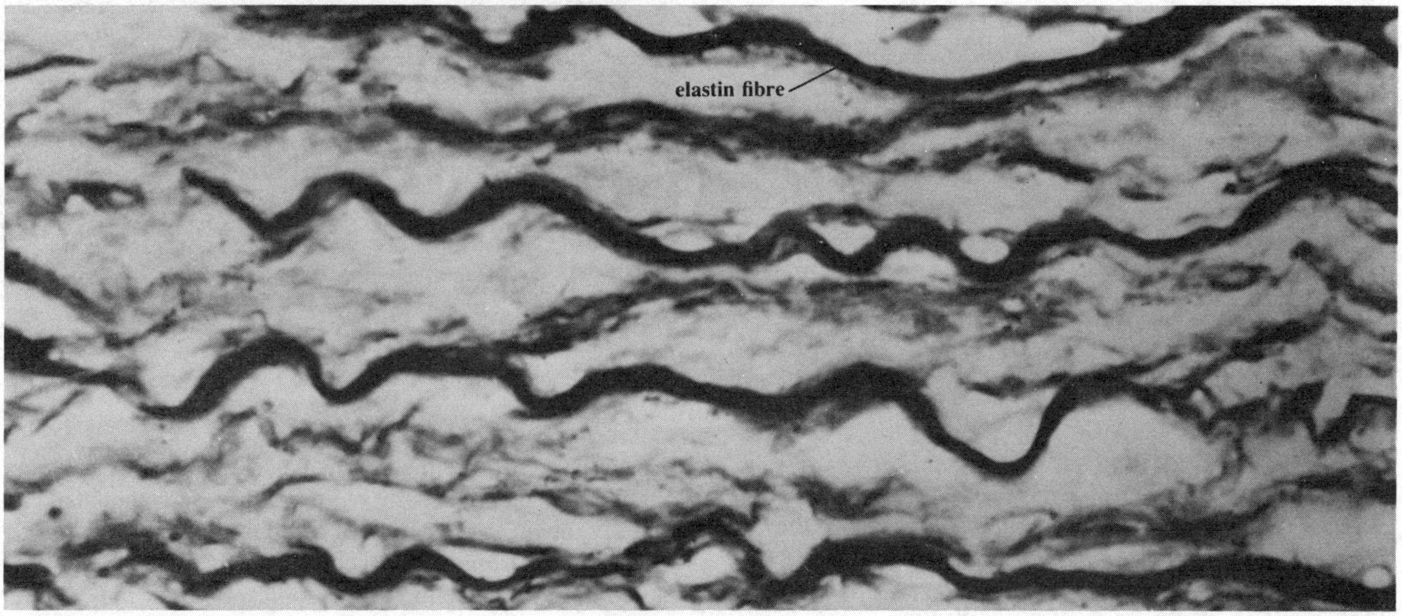

Fig 6.23 *Yellow elastic tissue contains a network of elastin fibres. Elastin is an elastic protein. This tissue is common in elastic structures such as walls of arteries and alveoli of lungs. Also in ligaments.*

flexibility. White fibrous cartilage is located as discs between adjacent vertebrae (intervertebral discs) where it provides a cushioning effect. It is also found in the symphysis pubis (the region between the two pubic bones of the pelvis) and the ligamentous capsules surrounding joints.

Bone

Bone is the most abundant of all animal skeletal materials, and provides **support**, **protection** and some **metabolic functions**. The cells are embedded in a firm, calcified matrix. About 30% of the matrix is composed of organic material, consisting chiefly of collagen fibres and glycoproteins, whilst 70% is inorganic bone salts. The chief inorganic constituent of bone is needle-like crystals of hydroxyapatite, $Ca_{10}(PO_4)_6(OH)_2$, a form of calcium phosphate. Sodium, magnesium, potassium, chloride, fluoride, hydrogencarbonate and citrate ions are also present in variable amounts.

Bone cells, called **osteoblasts**, are contained in lacunae (spaces) which are present throughout the matrix. They lay down the inorganic components of bone. Fine canals containing cytoplasm connect the lacunae to each other and blood vessels passing through them provide the means by which osteoblasts exchange materials.

The structure of bone is specially designed to withstand the compression strains falling upon it and to resist tension.

Bone resorption and reconstruction processes enable a particular bone to adapt its structure to meet any change in the mechanical requirements of the animal during its

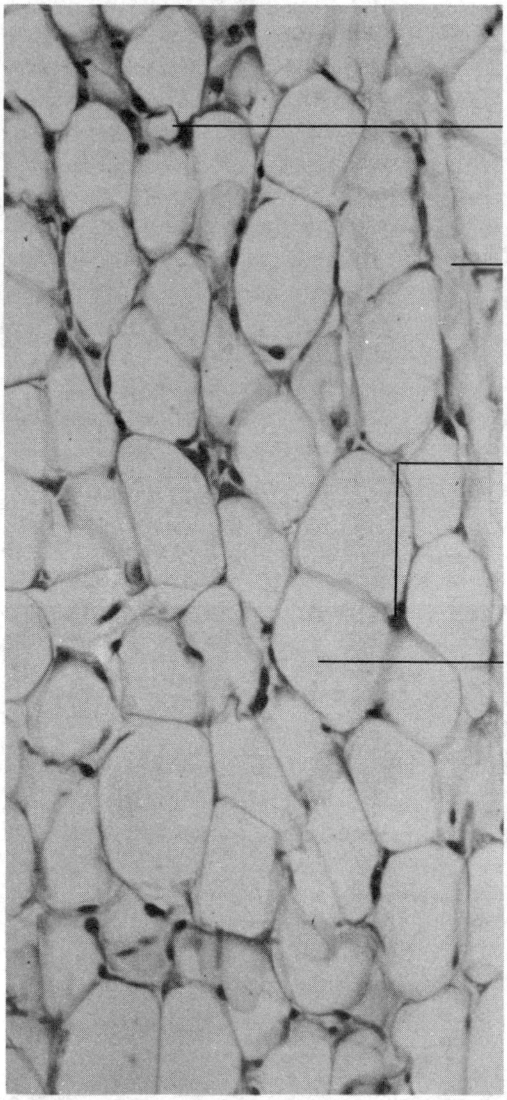

Fig 6.24 *Adipose tissue. This tissue is common in the dermis of the skin and around the kidneys and heart. It acts as an energy store, a shock absorber and can insulate against heat loss.*

development. Calcium and phosphate may be released into the blood as needed, under the control of two hormones, **parathormone** and **calcitonin** (chapter 17).

Compact or dense bone (fig 6.26). Compact bone is used in the growth of long bones (limb bones) and forms the long shaft of the bone between the two swollen ends. A transverse section of compact bone shows it to consist of numerous cylinders, each surrounding a central **Haversian canal**. One such cylinder plus its canal is termed a **Haversian system** or **osteon**. Each cylinder is itself made up of a set of concentric layers called **lamellae** which are cylindrical, an arrangement which increases strength.

Between the lamellae are numerous lacunae (spaces) containing living bone cells called **osteoblasts**. Each cell is capable of bone deposition. As osteoblasts mature they become less active and contain reduced quantities of cell

Fig 6.25 *Hyaline cartilage.*

organelles. They are then known as **osteocytes**. If structural changes in the bone are required they are activated and quickly regain the structure of osteoblasts.

Radiating from each lacuna are many fine channels called **canaliculi** containing cytoplasm which may link up with the central Haversian canal, with other lacunae or pass from one lamella to another.

An artery and a vein run through each Haversian canal, and capillaries branch from here and pass via the canaliculi to the lacunae of that particular Haversian system. They allow the passage of nutrients, respiratory gases and metabolic waste towards and away from the cells. A Haversian canal also contains a lymph vessel and nerve fibres. Transverse Haversian canals communicate with the marrow cavity in the centre of the whole bone and also interconnect with the longitudinal Haversian canals. These contain larger blood vessels and are not encircled by concentric lamellae.

The matrix of compact bone is composed of collagen, manufactured by the osteoblasts, and hydroxyapatite together with quantities of magnesium, sodium, carbonates and nitrates. The combination of organic with inorganic material produces a structure of great strength. The lamellae are laid down in a manner that is suited to the

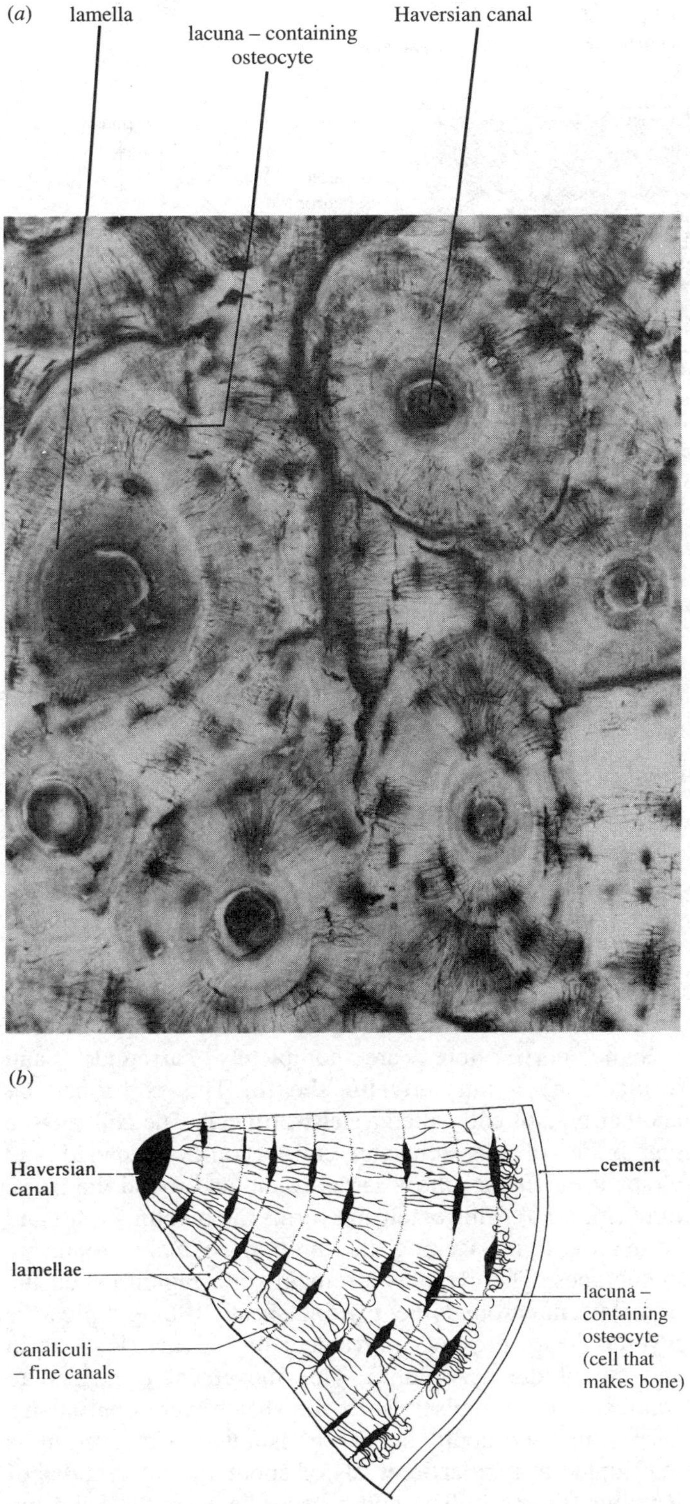

Fig 6.26 (a) *Part of a transverse section of a long bone.* (b) *TS Haversian system. The system forms a cylinder in three dimensions. The presence of large numbers of lamellae (layers) within each cylinder provides the bone with great strength despite its light weight.*

forces acting upon the bone, and the load that has to be carried.

Covering the bone is a layer of dense connective tissue called the **periosteum**. Bundles of collagen fibres from the periosteum pierce the bone, providing an intimate connection between the underlying bone and periosteum and acting as a firm base for tendon insertions. The inner region of the periosteum has blood vessels and forms a layer which contains cells that can develop into osteoblasts.

Spongy bone. Spongy bone consists of a meshwork of thin, interconnecting bony struts called **trabeculae**. Its matrix contains less inorganic material (60–65%) than compact bone. The organic material is primarily composed of collagen fibres. The spaces between the trabeculae are filled with soft marrow tissue.

The trabeculae are orientated in the direction in which the bone is stressed. This enables the bone to withstand tension and compression forces effectively whilst at the same time keeping the weight of the bone to a minimum.

Spongy bone occurs in the embryo, growing organisms, and the swollen ends of long bones.

6.5 Muscle tissue

Muscle tissue makes up 40% of a mammal's body weight. It consists of highly specialised contractile cells or fibres held together by connective tissue. Three types of muscle are present in the body, namely **striated** (voluntary or skeletal), **smooth** (unstriated or involuntary) and **cardiac** (heart) muscle. Further details of striated muscle can be found in chapter 18 and of cardiac muscle in chapter 14.

6.6 Nervous tissue

Nervous tissue contains densely packed nerve cells called **neurones** (or **neurons**), which are specialised for conduction of nerve impulses. Among other cells present are receptor cells and Schwann cells (see below). Nervous tissue is frequently enclosed by connective tissue which contains blood vessels.

6.6.1 Neurones

These are the functional units of the nervous system. Neurones are capable of transmitting electrical impulses, and this provides the means of communication between **receptors**, the cells or organs which receive stimuli, and **effectors**, the tissues or organs which react to stimuli, such as muscles or glands (fig 6.27). Neurones which conduct impulses towards the central nervous system (the brain and spinal cord) are called **sensory neurones**, whilst **motor neurones** conduct impulses away from the central nervous system. **Interneurones** frequently connect

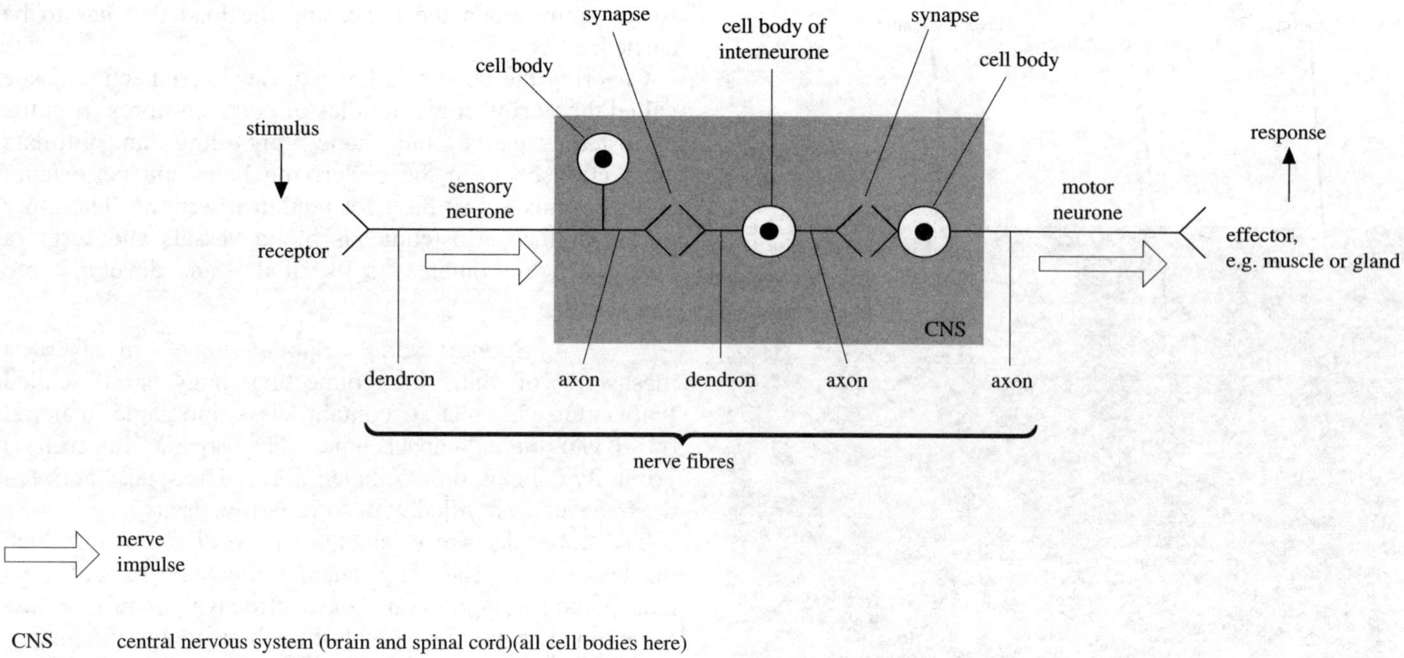

Fig 6.27 *Simplified diagram of a nervous pathway. Alternative names of the neurones are:*
sensory neurone*, afferent neurone;*
motor neurone*, efferent neurone, effector neurone;*
interneurone*, intermediate neurone, association neurone, internuncial neurone, relay neurone.*
Use only one name for each!

sensory neurones with motor neurones. The structure of these neurones, and their alternative names, is shown in fig 6.28.

Each neurone possesses a **cell body** (fig 6.28), which contains a nucleus, most of the cell's other organelles and a variable number of **nerve fibres** extending from it. **Nissl's granules**, which are groups of ribosomes and rough ER associated with protein synthesis, and Golgi apparatus are present in the cell body (fig 6.29).

Nerve fibres which conduct impulses towards the cell body are called **dendrons** (fig 6.27). They are small, relatively wide, and break up into fine terminal branches called **dendrites** (*dendro*, tree). Nerve fibres which conduct impulses away from the cell body are termed **axons**, they are thinner than dendrites and may be several metres long.

The end of an axon breaks up into many fine branches with swollen endings called **synaptic knobs**. These do not join directly to the next nerve cell in the pathway. Instead, there is a tiny gap across which a chemical called a **neurotransmitter** must pass in order to stimulate the next nerve cell (or effector). The neurotransmitter is released from the synaptic knob in response to a nerve impulse travelling along the axon. The sites of these gaps are called **synapses** (fig 6.27).

Some nerve fibres are completely surrounded and insulated by a fatty **myelin sheath**. This is formed by another type of cell called a **Schwann cell**. The cell surface membrane of the Schwann cell becomes extended and wraps itself like a roll of carpet round and round the nerve fibre (fig 6.30). This extension forms the myelin sheath and is mainly lipid, lacking the protein normally found in membranes. The cytoplasm remains in a region called the **neurilemma** around the myelin sheath. Being lipid, the myelin sheath prevents movement of Na^+ and K^+ ions in and out of the nerve fibre. This movement is needed to conduct nerve impulses, so if the sheath were continuous, nerve impulses could not be transmitted. However, it is interrupted at regular intervals of about 1 mm by **nodes of Ranvier** (fig 6.28). The nodes occur between the Schwann cells, and one Schwann cell nucleus is visible in the sheath between each successive pair of nodes.

Nerve fibres with a myelin sheath are described as **myelinated**, for example spinal nerves. Some nerve fibres are non-myelinated. These lack nodes of Ranvier and are only partially surrounded by Schwann cells. Certain diseases involve destruction of myelin sheaths, notably multiple sclerosis and Tay–Sachs disease.

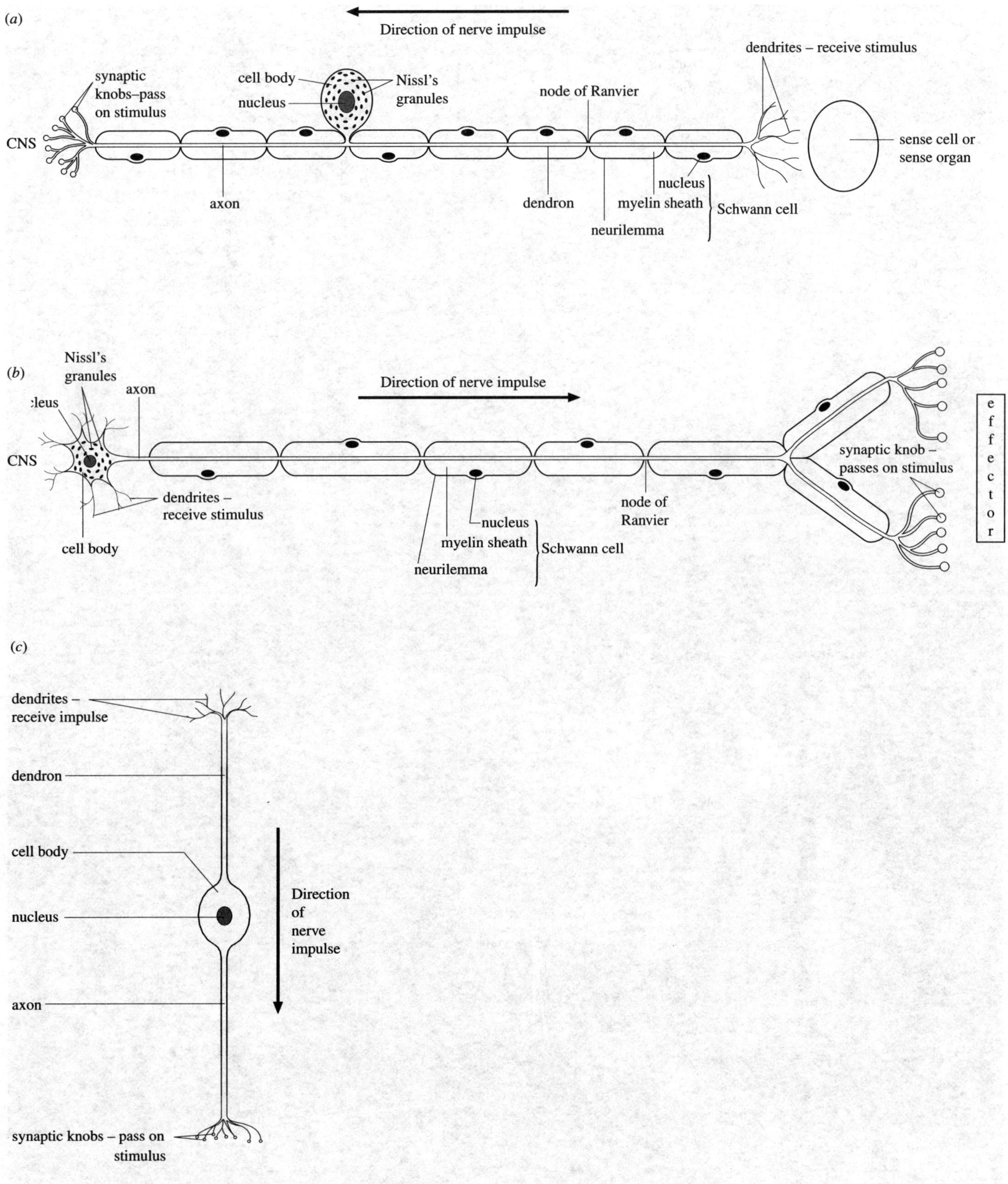

Fig 6.28 *Diagrams of* (a) *sensory neurone,* (b) *motor neurone,* (c) *interneurone.*

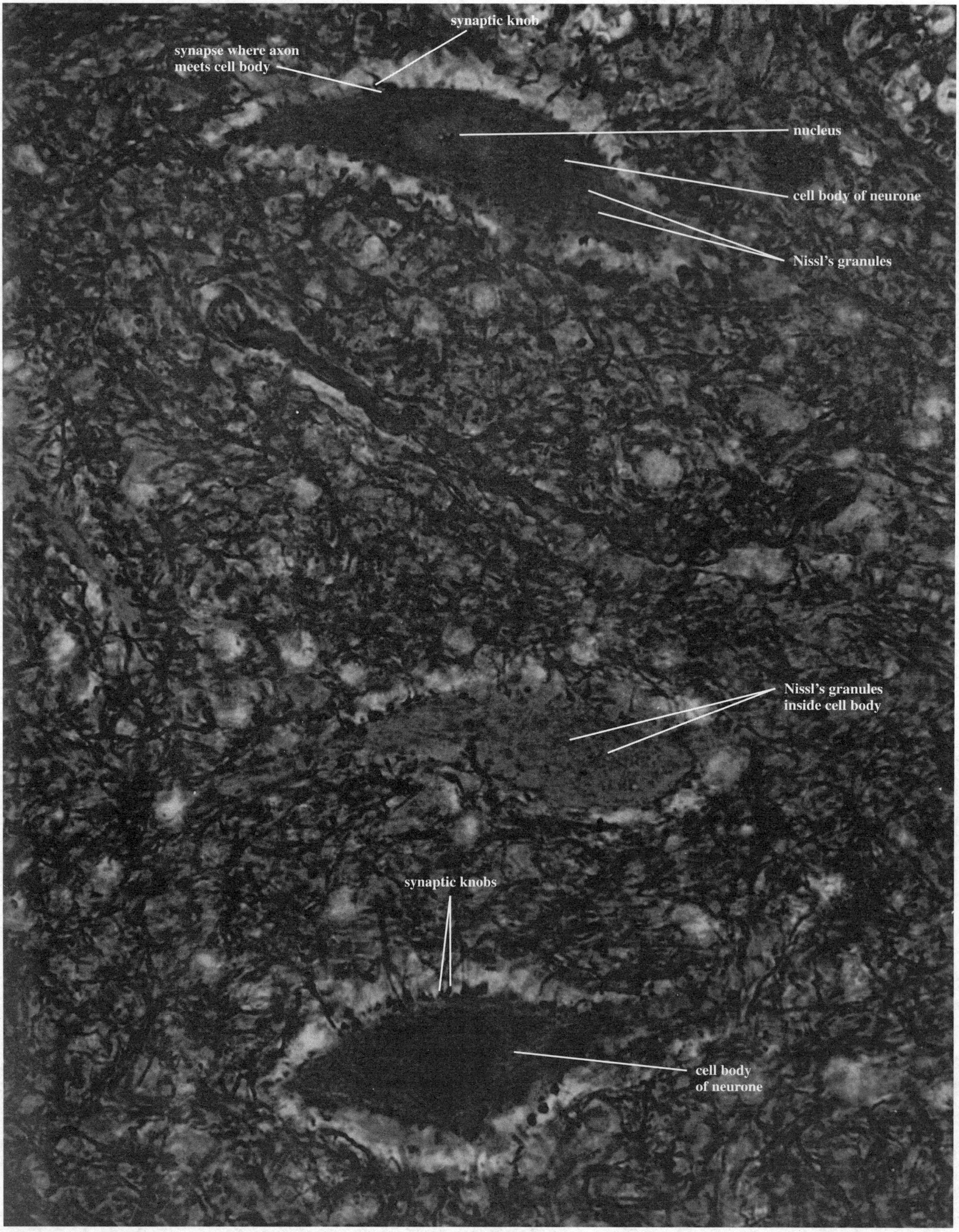

Fig 6.29 *Cell bodies of neurones with synapses and Nissl's granules.*

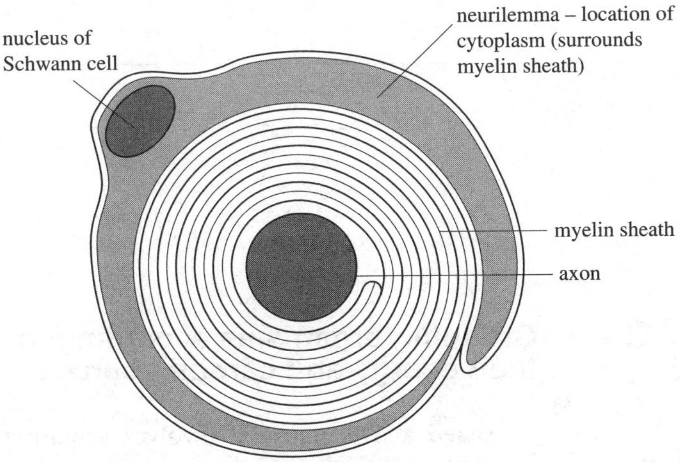

Fig 6.30 *TS myelinated nerve fibre.*

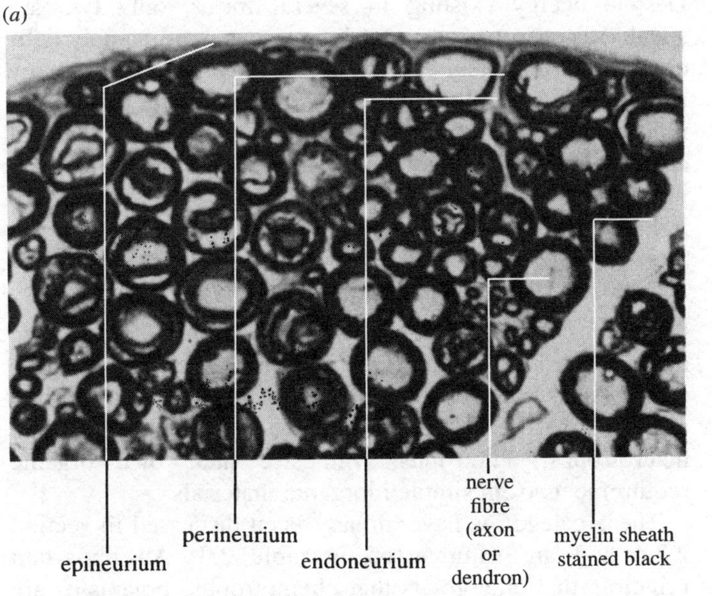

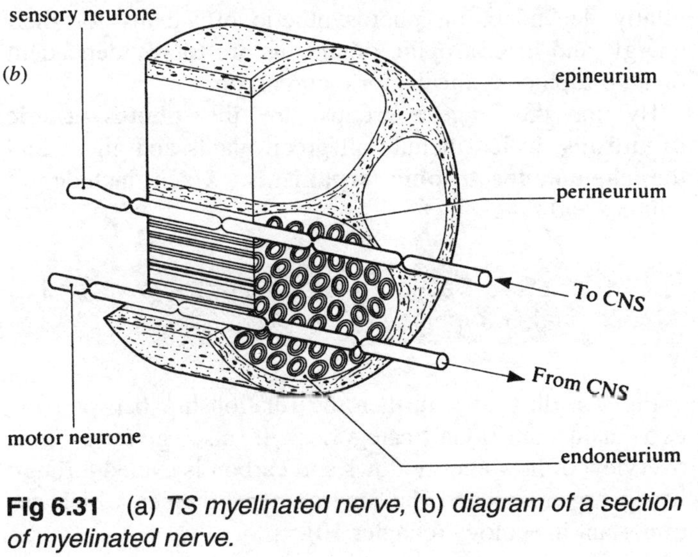

Fig 6.31 (a) *TS myelinated nerve,* (b) *diagram of a section of myelinated nerve.*

6.6.2 Nerves

These consist of bundles of nerve fibres in a connective tissue sheath called the **epineurium**. Inward extensions of the epineurium, called the **perineurium**, divide the fibres into smaller bundles, whilst each fibre is itself surrounded by connective tissue called the **endoneurium** (fig 6.31). Nerves are classified according to the direction in which they convey nerve impulses. Sensory or afferent nerves, such as the optic and auditory nerves, convey impulses *to* the central nervous system, whilst efferent or motor nerves conduct impulses *away from* the central nervous system. Mixed nerves convey impulses in both directions (for example all spinal nerves).

The conduction of nervous impulses is discussed in chapter 17.

Chapter Seven

Autotrophic nutrition

Nutrition is the process of *acquiring* energy and materials. This is the theme of chapters 7 and 8. In chapter 9 **respiration** is considered, the process in which organisms *release* energy from the energy-rich compounds acquired by nutrition.

7.1 Why do living organisms need energy?

Energy may be defined as the **capacity to do work**. All living organisms may be regarded as working machines which require a continuous supply of energy in order to keep working, and so to stay alive. Energy can neither be created nor destroyed (**the law of conservation of energy**). It may occur in various forms, such as light, chemical, heat, electrical, mechanical and sound, and energy can be transferred from one form to another. A simple example would be striking a match, where, in the matchhead, chemical energy is transferred to heat, light and sound energy.

Some common examples of the use of energy in living organisms are:

- synthesis of substances for growth and repair, for example protein synthesis;
- active transport of substances into and out of cells against diffusion gradients, for example the sodium–potassium pump (section 5.9.8);
- phagocytosis, pinocytosis, and exocytosis (section 5.9.8);
- electrical transmission of nerve impulses;
- mechanical contraction of muscles and beating of cilia and flagella;
- heat energy released from respiration used to maintain a constant body temperature in birds and mammals;
- bioluminescence, that is the production of light by living organisms such as fireflies, glow-worms and some deep sea animals;
- electrical discharge, as in the electric eel.

The role of ATP as the energy carrier in cells is described in section 9.2.

7.2 Grouping organisms according to their energy and carbon sources

As stated above, nutrition involves acquiring both **energy** and **materials**. Carbon is the most fundamental material required by living organisms (section 3.1). Living organisms can be grouped on the basis of their source of **energy** or source of **carbon**.

Energy source

Despite energy existing in several forms, only two are suitable as energy sources for living organisms, namely **light energy** and **chemical energy**. Organisms using light energy are described as **photosynthetic** or **phototrophic** (*photos*, light; *trophos*, nourishment), while those using chemical energy are described as **chemotrophic**. Phototrophs contain pigments, including some form of chlorophyll, which absorb light energy and convert it to chemical energy.

Carbon source

Organisms which have an inorganic source of carbon, namely carbon dioxide, are described as **autotrophic** (*autos*, self) and those having an organic source of carbon are described as **heterotrophic** (*heteros*, other). Unlike heterotrophs, autotrophs synthesise their own organic requirements from simple inorganic materials.

These categories have already been discussed in section 2.3.4 and are summarised in table 2.3. An important principle that emerges is that chemotrophic organisms are totally dependent on photosynthetic organisms for their energy, and heterotrophic organisms are totally dependent on autotrophic organisms for their carbon.

By far the largest groups are the **photosynthetic organisms**, which include all green plants and algae, and the **chemoheterotrophic organisms**, which include all animals and fungi.

7.1 Define photoautotrophic and chemoheterotrophic.

Fig 7.1 illustrates further the relationship between the two main nutritional categories. It also gives a brief overview of how energy flows and carbon is cycled through living organisms and the environment, themes which are important in ecology (chapter 10).

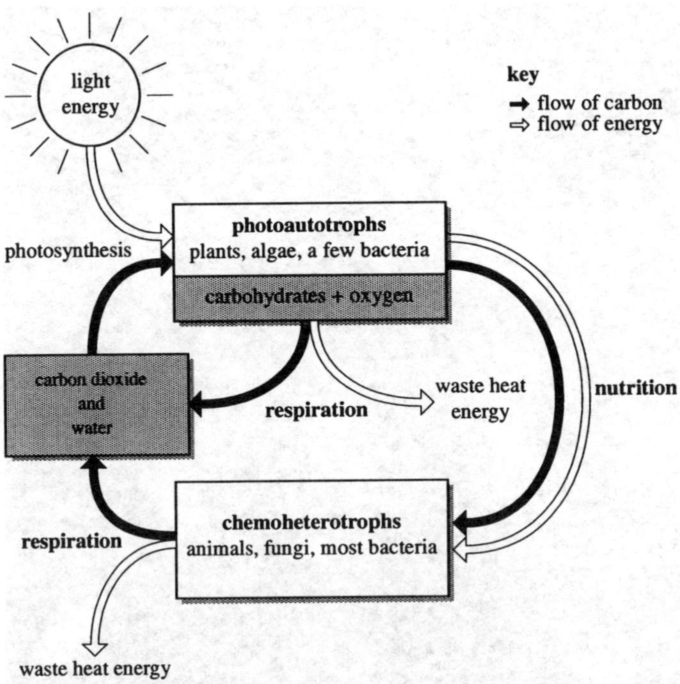

Fig 7.1 *Flow of energy (open arrows) and cycling of carbon (black arrows) through photoautotrophs and chemoheterotrophs, and balance between photosynthesis and respiration. Light energy is converted to chemical energy in photosynthesis and used, together with carbon dioxide and water, in the synthesis of organic materials from inorganic materials. Organic materials are the energy and carbon source for chemoheterotrophs. Energy and carbon dioxide are released again in the process of respiration (carried out by all living organisms). Every energy conversion is accompanied by some loss of energy as heat, which is waste energy.*

7.3 Importance of photosynthesis

Almost all life on Earth depends on photosynthesis, either directly or, as in the case of animals, indirectly. Photosynthesis makes both carbon and energy available to living organisms and produces the oxygen in the atmosphere which is vital for all aerobic forms of life. Humans also depend on photosynthesis for the energy-containing fossil fuels which have developed over millions of years.

Of the total amount of solar radiation intercepted by our planet, about half reaches its surface after absorption, reflection and scattering in the atmosphere. Of this, only about 50% is of the right wavelength to stimulate photosynthesis and, although estimates vary, it is likely that only about 0.2% of this is used in actual plant production (about 0.5% of the energy actually reaching plants). From this small fraction of the available energy virtually all life is sustained. About 40% of all photosynthesis is carried out by tiny algae, known as phytoplankton, which live in the oceans.

7.4 The structure of the leaf

In flowering plants the major photosynthetic organ is the leaf. As with all living organs, structure and function are closely linked. From the equation for photosynthesis

$$\underset{\text{carbon dioxide}}{CO_2} + \underset{\text{water}}{H_2O} \xrightarrow[\text{chlorophyll}]{\text{sunlight}} \underset{\text{carbohydrate}}{(CH_2O)_n} + \underset{\text{oxygen}}{O_2}$$

it can be deduced that first the leaf requires a source of carbon dioxide and water, secondly it must contain chlorophyll and be adapted to receive sunlight, thirdly oxygen will escape as a waste product and finally the useful product, carbohydrate, will have to be exported to other parts of the plant or stored. In its structure the leaf is highly adapted to satisfy these requirements. Fig 7.2 shows the external structure of a leaf. Fig 7.3 shows a labelled photomicrograph of a leaf section which will help you in interpreting sections of dicotyledonous leaves. Fig 7.4 shows high power detail of a single palisade mesophyll cell. Fig 7.5 is a simplified drawing of a vertical section through a dicotyledonous leaf. (Advice on drawing from a light microscope is given in section 5.13.) The epidermises of different leaf types are shown in fig 6.3 and details of stomatal structure and function are dealt with in chapter 14.

The structure and function of different tissues in a dicotyledonous leaf are summarised in table 7.1.

> **7.2** Make a list of the ways in which the structure of the leaf contributes to its successful functioning.

A final point to note is the arrangement of the leaves for minimal overlapping. Such leaf mosaics are particularly noticeable in some plants, such as ivy.

7.4.1 Chloroplasts

In eukaryotes, photosynthesis takes place in organelles called chloroplasts. They are found in the cytoplasm in numbers varying from one (as in the unicellular alga *Chlorella*) to about 100 (palisade mesophyll cells). They are about 3–10 μm (average 5 μm) in diameter, and so are visible with a light microscope (figs 7.3 and 7.4). Chloroplasts are surrounded by two membranes, which form the **chloroplast envelope**. They always contain **chlorophyll** and **other photosynthetic pigments** located on a system of membranes. The membranes run through a ground substance, or **stroma**. Their detailed structure is revealed by electron microscopy. Figs. 5.11 and 7.4 show the typical appearance of chloroplasts in a leaf mesophyll cell as seen at low power in

Fig 7.2 *External structure of a dicotyledonous leaf.*

Table 7.1 Relationship between structure and function in a dicotyledonous leaf.

Tissue	*Structure*	*Function*
Upper and lower epidermis	One cell thick. Flattened cells lacking chloroplasts. External walls covered with a cuticle of cutin (waxy substance). Contains stomata (pores) which are normally confined to, or more numerous in, the lower epidermis. Each stoma is surrounded by a pair of guard cells.	Protective. Cutin is waterproof and protects from desiccation and infection. Stomata are sites of gaseous exchange with the environment. Their size is regulated by guard cells, special epidermal cells containing chloroplasts.
Palisade mesophyll	Column-shaped ('palisade') cells with numerous chloroplasts in a thin layer of cytoplasm.	Main photosynthetic tissue. Chloroplasts may move towards light.
Spongy mesophyll	Irregularly shaped cells fitting together loosely to leave large air spaces.	Photosynthetic, but fewer chloroplasts than palisade cells. Gaseous exchange can occur through the large air spaces via stomata. Stores starch.
Vascular tissue	Extensive finely branching network through the leaf.	Conducts water and mineral salts to the leaf in xylem. Removes products of photosynthesis (mainly sucrose) in phloem. Provides a supporting skeleton to the lamina, aided, by collenchyma of the midrib, turgidity of the mesophyll cells, and sometimes sclerenchyma.

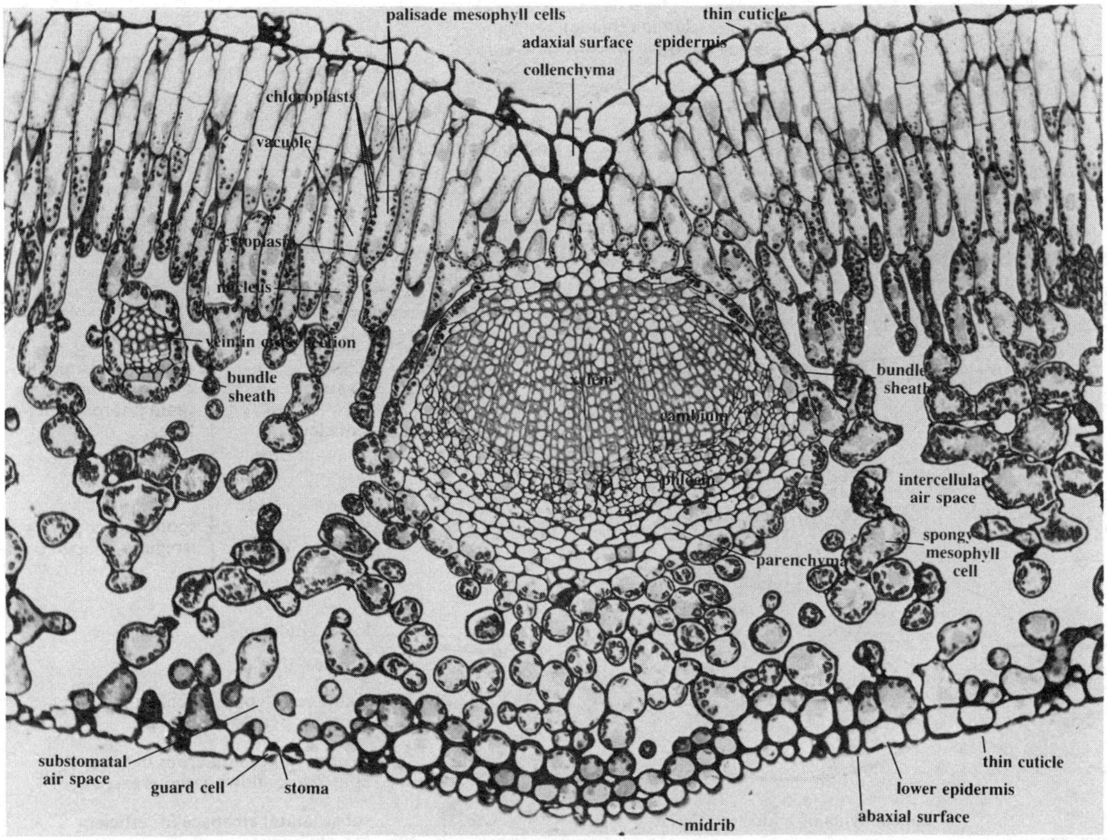

Fig 7.3 *TS lamina and midrib of a privet leaf (*Ligustrum*), a typical dicotyledon.*

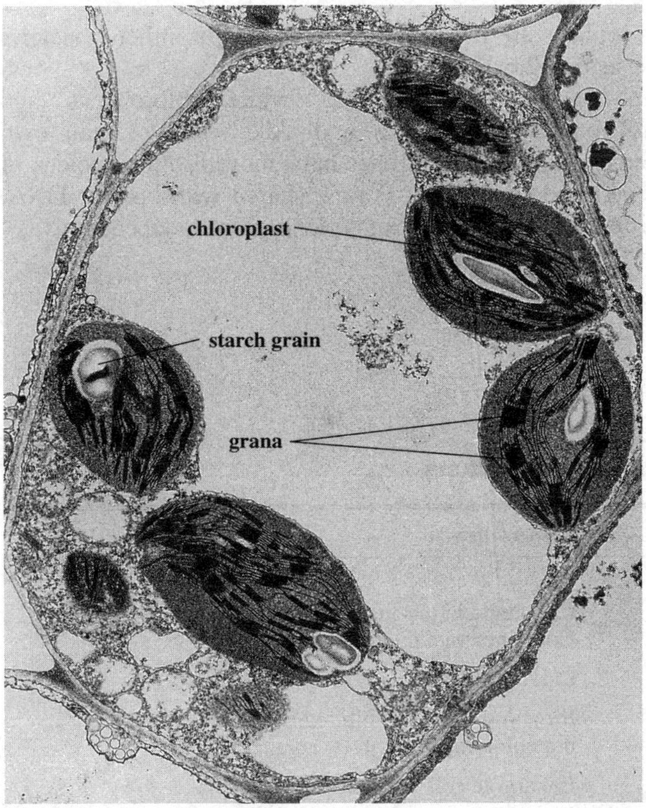

Fig 7.4 *Electron micrograph of a palisade mesophyll cell (x3000).*

the electron microscope. Figs 7.6 and 7.8 show electron micrographs and fig 7.7 a diagram of a chloroplast, illustrating the membrane system. The membrane system is the site of the **light-dependent reactions** in photosynthesis (section 7.6.2). The membranes are covered with chlorophyll and other pigments, enzymes and electron carriers. The system consists of many flattened, fluid-filled sacs called **thylakoids** which form stacks called **grana** at intervals, with lamellae (layers) between the grana. Each granum resembles a pile of coins and the lamellae are often sheet-like (fig 7.8). Grana are just visible under the light microscope as grains.

The stroma is the site of the **light-independent reactions** of photosynthesis (section 7.6.3). The structure is gel-like, containing soluble enzymes, particularly those of the Calvin cycle, and other chemicals such as sugars and organic acids. Excess carbohydrate from photosynthesis is sometimes seen stored as grains of starch. Spherical lipid droplets are often associated with the membranes. They become larger as membranes break down during ageing, presumably accumulating lipids from the membranes.

Protein-synthesising machinery and the endosymbiont theory

An interesting feature of chloroplasts, apart from photosynthesis, is their protein-synthesising machinery. During the 1960s it was shown that both chloroplasts and mitochondria contain DNA and ribosomes. This led to the

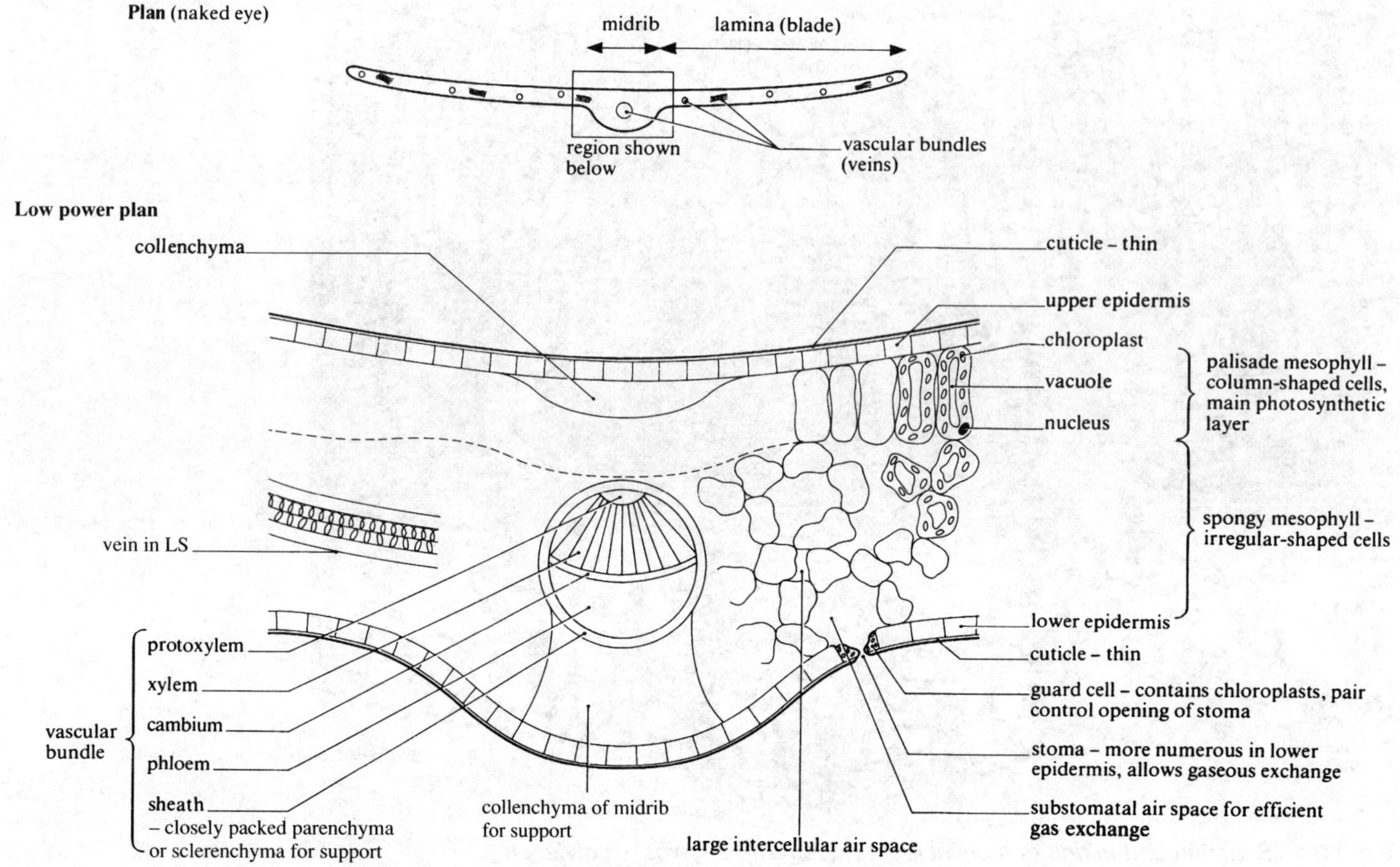

Fig 7.5 *Diagrammatic transverse section of a typical dicotyledon leaf.*

suggestion that they might represent prokaryotic organisms which invaded eukaryotic cells at an early stage in the history of life. Thus the organelles represent an extreme form of symbiosis, a theory known as the **endosymbiont theory**. Some of the evidence for this is presented in table 7.2.

Photosynthetic bacteria (prokaryotes) do not contain chloroplasts. Instead their photosynthetic pigments are located in membranes distributed throughout the cytoplasm. Thus the whole cell is similar to one chloroplast, and is approximately the same size. It is now believed that chloroplasts are the descendants of photosynthetic bacteria (see section 2.6.1).

It has been shown that, while chloroplasts and mitochondria do code for and make some of their own proteins, some of their genes have moved to the nucleus of the cell and so the task is now shared with nuclear DNA. This explains why they can no longer live independently.

Table 7.2 Comparison of prokaryotes, chloroplasts and mitochondria with eukaryotes.

	Prokaryotes, chloroplasts and mitochondria	*Eukaryotes*
DNA	Circular Not contained in chromosomes Not contained in nucleus	Linear Contained in chromosomes Contained in a nucleus
Ribosomes	Smaller (70S)	Larger (80S)
Sensitivity to antibiotics	Protein synthesis inhibited by chloramphenicol, not cycloheximide	Protein synthesis inhibited by cycloheximide, not chloramphenicol
Average diameter	Prokaryote cell: 0.5–10 μm Chloroplast: 1–10 μm Mitochondrion: 1 μm	Eukaryote cell: 10–100 μm

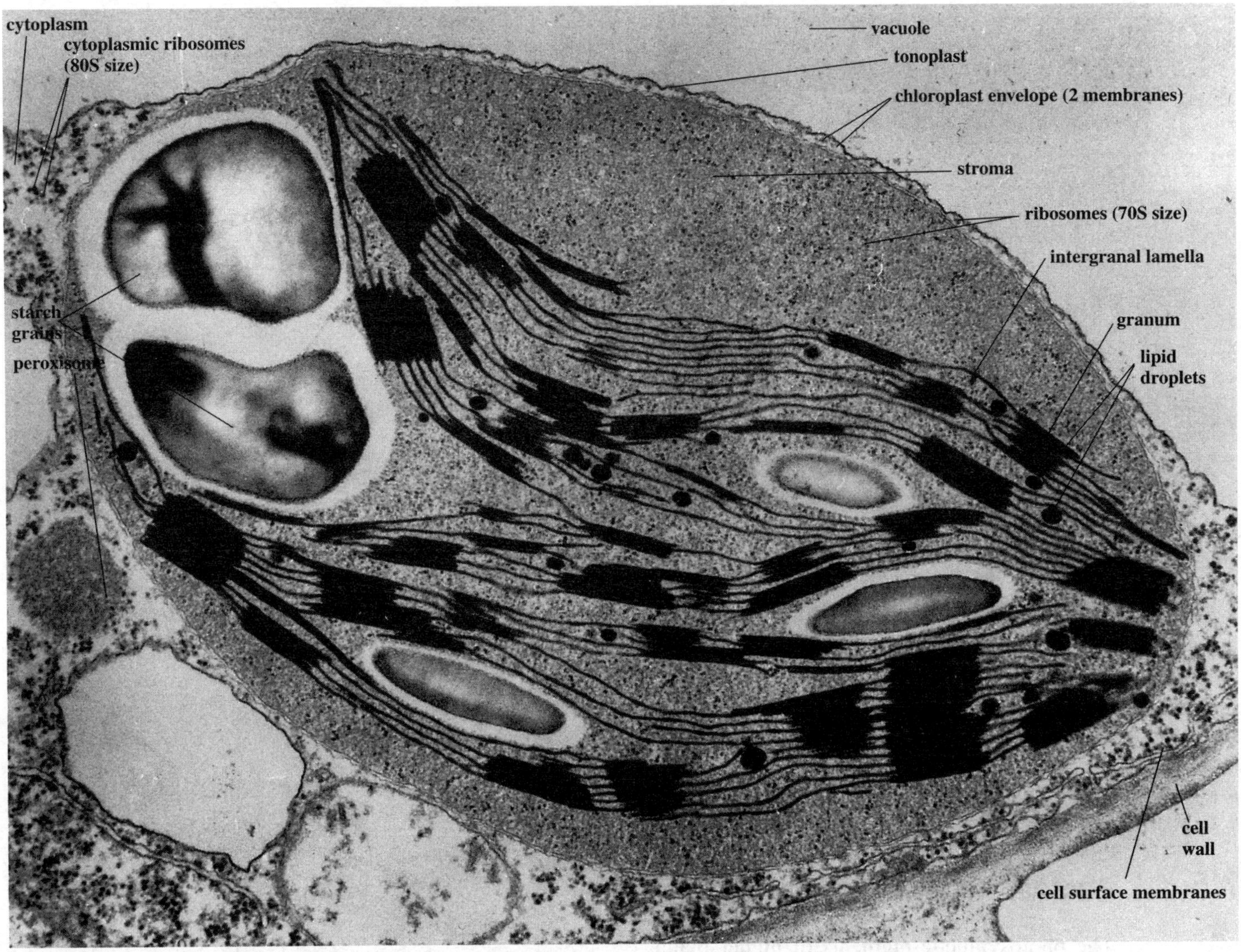

Fig 7.6 *Electron micrograph of a chloroplast (×15 800).*

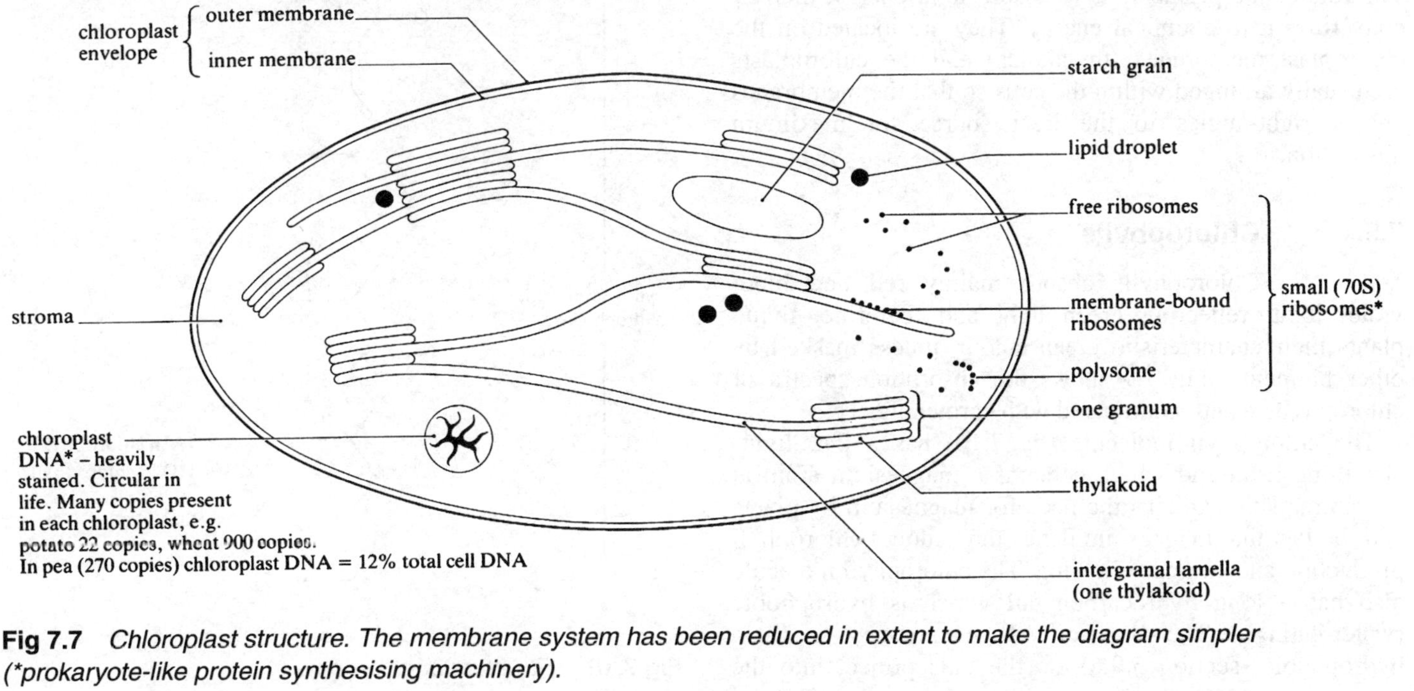

Fig 7.7 *Chloroplast structure. The membrane system has been reduced in extent to make the diagram simpler (*prokaryote-like protein synthesising machinery).*

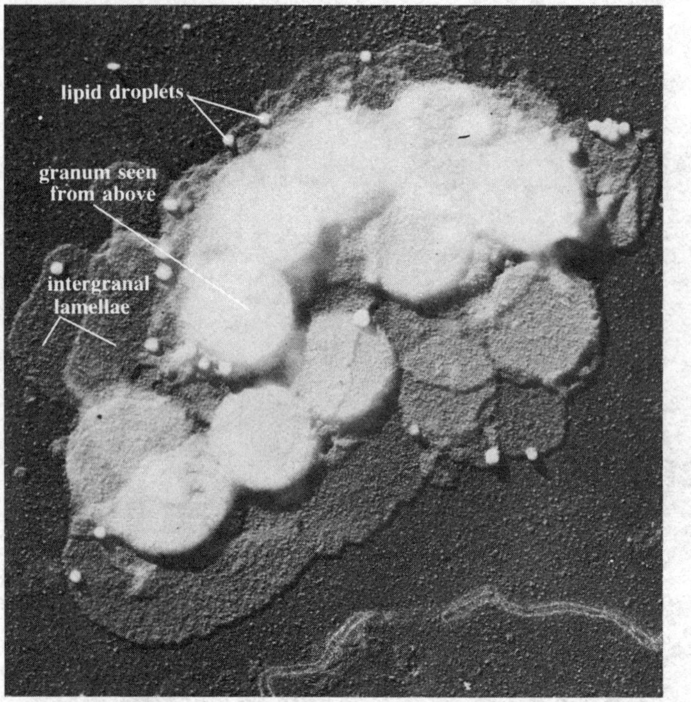

Fig 7.8 Scanning electron micrograph of a 'stripped' chloroplast (a chloroplast whose outer envelope has been removed) looking down from above on the lamellae and grana which can be seen in three dimensions. Note that the lamellae are sheet-like and interconnect the grana. The preparation is a shadowed replica.

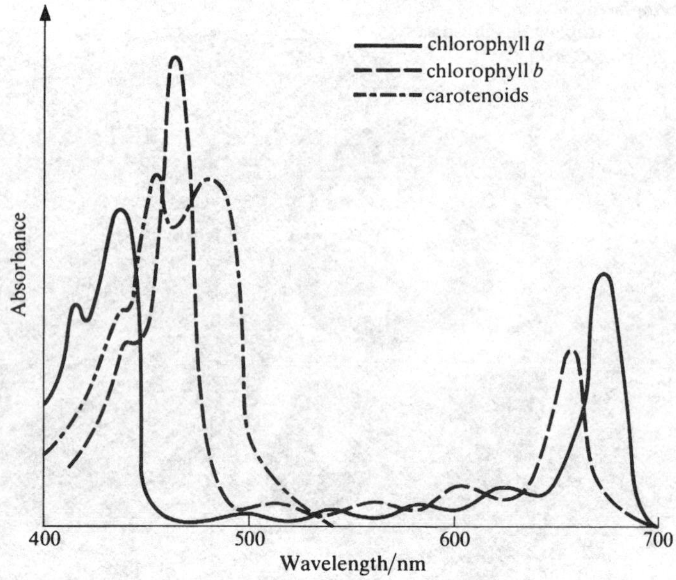

Fig 7.9 *Absorption spectra of chlorophylls* a *and* b*, and carotenoids. An absorption spectrum shows the amounts of light absorbed by a pigment at different wavelengths.*

7.5 Photosynthetic pigments

The photosynthetic pigments of higher plants fall into two classes, the **chlorophylls** and **carotenoids**. The role of the pigments is to absorb light energy, thereby converting it to chemical energy. They are located on the chloroplast membranes (thylakoids) and the chloroplasts are usually arranged within the cells so that the membranes are at right-angles to the light source for maximum absorption.

7.5.1 Chlorophylls

Chlorophylls absorb mainly red and blue-violet light, reflecting green light and therefore giving plants their characteristic green colour, unless masked by other pigments. Fig 7.9 shows the absorption spectra of chlorophylls *a* and *b* compared with carotenoids.

The chlorophyll molecule (fig 7.10) has a flat, light-absorbing head end which contains a magnesium atom at its centre. This explains the need for magnesium by plants and the fact that magnesium deficiency reduces chlorophyll production and causes yellowing. The chlorophyll molecule also has a long hydrocarbon tail which is hydrophobic (water-hating). The interior of membranes is also hydrophobic (section 5.9.6), so the tails project into the

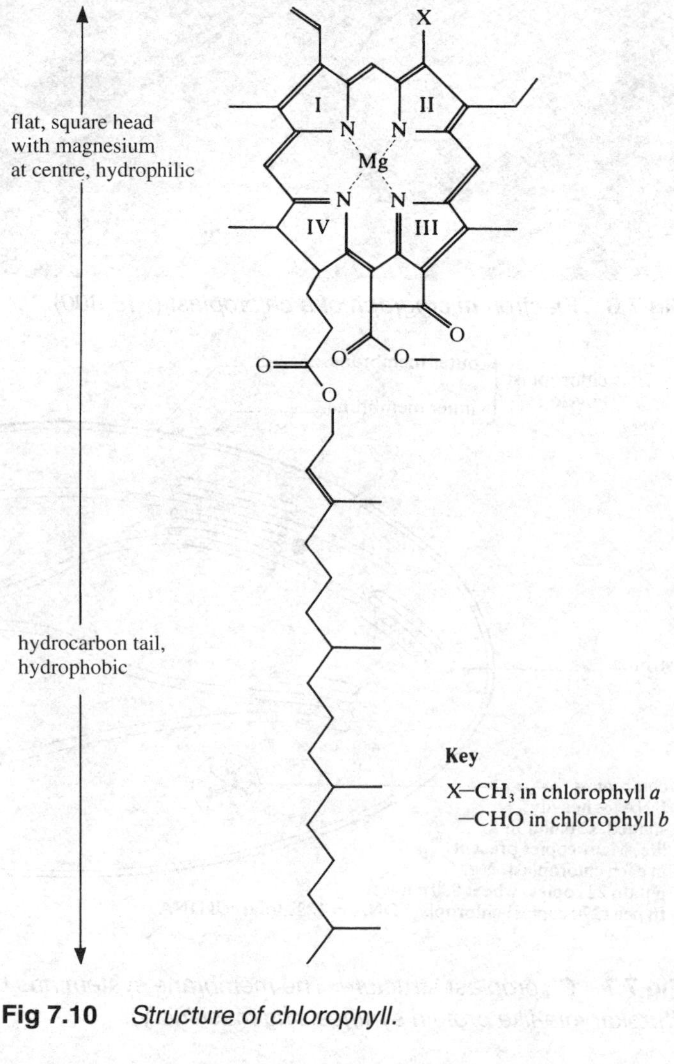

Fig 7.10 *Structure of chlorophyll.*

thylakoid membranes and act like anchors. The heads are hydrophilic and lie flat on the membrane surfaces like solar panels. Different chlorophylls have different side-chains on the head and this modifies their absorption spectra, increasing the range of wavelengths of light absorbed.

Chlorophyll *a* is the most abundant photosynthetic pigment. It exists in several forms, depending on its arrangement in the membrane. Each form differs slightly in its red absorption peak; for example, the peak may be at 670 nm, 680 nm, 690 nm or 700 nm.

7.3 How does the absorption spectrum of chlorophyll *a* differ from that of chlorophyll *b*?

7.5.2 Carotenoids

Carotenoids are yellow, orange, red or brown pigments that absorb strongly in the blue-violet range. They are called **accessory pigments** because they pass the light energy they absorb on to chlorophyll. Carotenoids have three absorption peaks in the blue-violet range of the spectrum (fig 7.9) and, apart from acting as accessory pigments, they may also protect chlorophylls from excess light and from oxidation by oxygen produced in photosynthesis. They are usually masked by the green chlorophylls but can be seen in leaves before leaf-fall because chlorophylls break down first. They are also found in some flowers and fruits where the bright colours attract insects, birds and other animals for pollination or dispersal; for example the red skin of the tomato is due to a carotene.

Carotenoids are of two types, carotenes and xanthophylls. The most widespread and important carotene is β-carotene, which is familiar as the orange pigment of carrots. Vertebrates are able to break the molecule into two during digestion to form two molecules of vitamin A.

7.5.3 Absorption and action spectra

When investigating a process such as photosynthesis that is activated by light, it is important to establish the action spectrum for the process and to use this to try to identify the pigments involved. An **action spectrum** is a graph showing the effectiveness of different wavelengths of light in stimulating the process being investigated. An **absorption spectrum** is a graph of the relative amounts of light absorbed at different wavelengths by a pigment. An action spectrum for photosynthesis is shown in fig 7.11, together with an absorption spectrum for the combined photosynthetic pigments. Note the close similarity, which indicates that the pigments, chlorophylls in particular, are those responsible for absorption of light in photosynthesis.

7.5.4 Excitation of chlorophyll by light

When a molecule of chlorophyll or other photosynthetic pigment absorbs light it is said to become **excited**. The energy from the light is used to boost

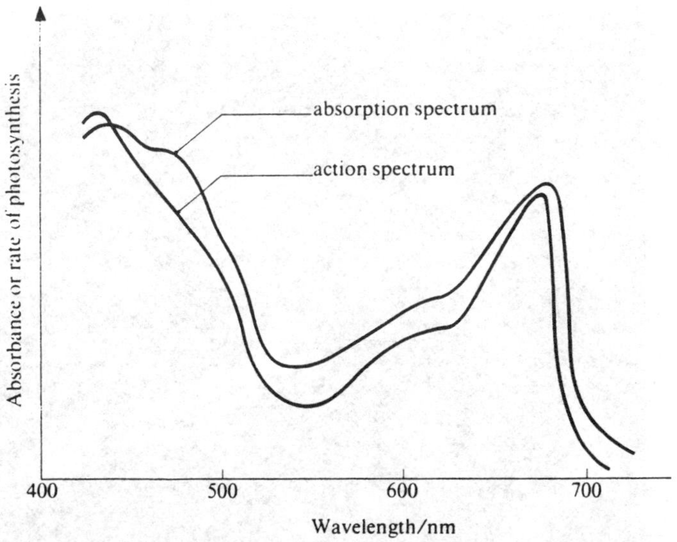

Fig 7.11 *Action spectrum for photosynthesis compared with absorption spectrum of photosynthetic pigments.*

electrons to a higher energy level. The energy of the light is now 'trapped' in the chlorophyll and has been transferred to chemical energy. This excited state is unstable and the molecule will tend to return to its unexcited state. For example, if a solution of chlorophyll has light shone through it and is then observed in darkness, it can be seen to fluoresce. This is because the extra energy of excitation is converted to light of a longer wavelength (less energy) and some waste heat energy. The excited electrons return to their original lower energy state. In the living plant the energy that is released can be passed to another chlorophyll molecule, as explained later. Alternatively, the excited electron itself may pass from the chlorophyll molecule to another molecule called an **electron acceptor**. As electrons have a negative charge, this will leave a positively charged 'hole' in the chlorophyll molecule.

$$\underset{\text{(reduced form)}}{\text{chlorophyll}} \xrightarrow{\text{light energy}} \underset{\text{(oxidised form)}}{\text{chlorophyll}^+} + \underset{\text{electron}}{e^-}$$

Loss of electrons is known as **oxidation** and gaining electrons is **reduction**. Chlorophyll is therefore oxidised and the electron acceptor reduced. Chlorophyll replaces its electrons by removing low energy electrons from another molecule described as an **electron donor**.

The first stages in the process of photosynthesis involve both the movement of energy and of excited electrons between molecules within photosystems, as described below.

7.5.5 Photosystems

The chlorophyll and accessory pigment molecules are located in two types of photosystem, known as **photosystems I** and **II** (**PSI** and **PSII**). These photosystems are visible as particles in the thylakoid membranes, as shown in fig 7.12. Each contains an **antenna complex**, or **light-harvesting complex**, of pigment molecules.

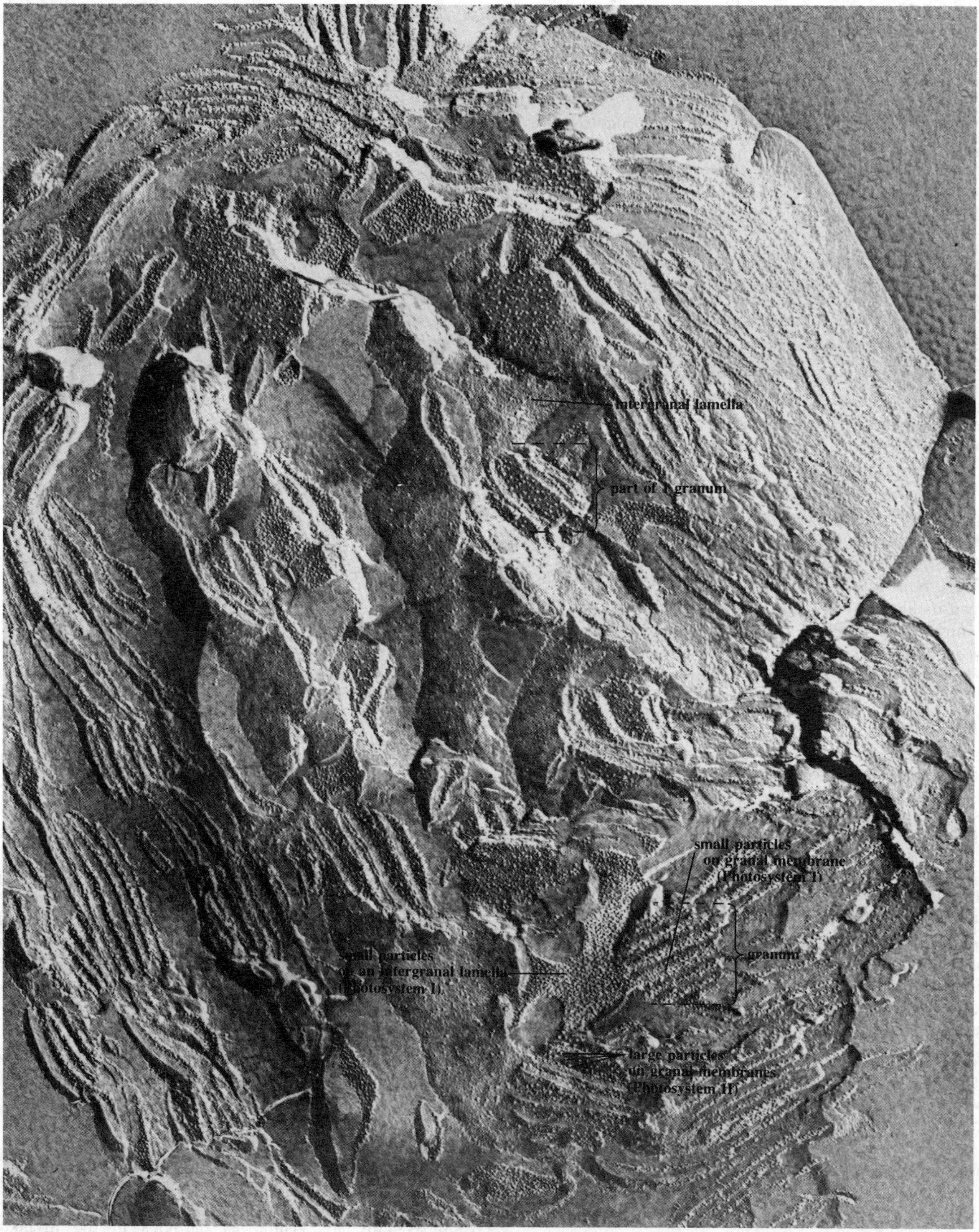

Fig 7.12 *Freeze-fractured isolated thylakoids of a chloroplast. The surfaces of fractured membranes are visible. Note the particles on the membranes. These are the photosystems. PSI particles are smaller than PSII particles.*

The light-harvesting complex contains 200–300 pigment molecules and collects light energy as shown in fig 7.13. Different pigments collect light of different wavelengths, making the process more efficient. All the energy is transferred from molecule to molecule, and finally to a specialised form of chlorophyll *a* known as P700 in PSI and P680 in PSII. P stands for pigment; their absorption peaks are at wavelengths of 700 nm and 680 nm respectively (both red light).

The chlorophylls P700 and P680 become 'excited' by the energy they absorb and release high energy electrons as described above. The fate of these electrons will be examined in section 7.6.2. We can now look at the overall process of photosynthesis.

7.6 Biochemistry of photosynthesis

A commonly used equation for photosynthesis is

$$\underset{\text{carbon dioxide}}{6CO_2} + \underset{\text{water}}{6H_2O} \xrightarrow[\text{chlorophyll}]{\text{light energy}} \underset{\text{sugar e.g. glucose}}{C_6H_{12}O_6} + \underset{\text{oxygen}}{6O_2}$$

This is useful for showing the formation of one molecule of sugar, but it should be realised that it is an overall summary of events. A better summary is

$$CO_2 + H_2O \xrightarrow[\text{chlorophyll}]{\text{light energy}} [CH_2O] + O_2$$

CH_2O does not exist as such, but represents a carbohydrate.

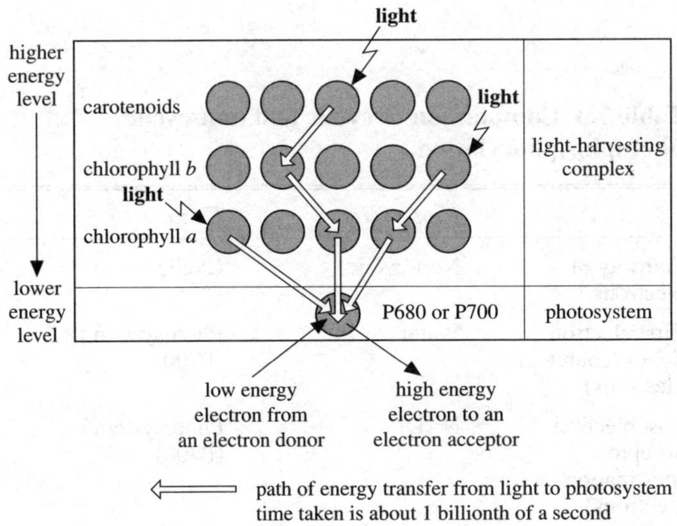

Fig 7.13 *Diagrammatic representation of a photosystem. Light may be absorbed by any of the pigment molecules in the light harvesting complex. The photosystem is located in the thylakoids. The two types of photosystem are visible as two different sized particles in fig 7.12.*

7.6.1 Source of oxygen

Looking at the equation above, a key question is whether the oxygen produced comes from carbon dioxide or water. The most obvious answer would seem to be carbon dioxide, so that the remaining carbon would be added to water to make carbohydrate. With the use of isotopes (appendix 1 in Book 2) in biology during the 1940s it became possible to answer the question directly.

The common isotope of oxygen has a mass number of 16 and is therefore represented as ^{16}O (8 protons, 8 neutrons). A rare isotope has a mass number of 18 (^{18}O). This is stable, but its greater mass can be detected with a mass spectrometer, an important analytical instrument which distinguishes between different atoms and molecules according to their masses. In 1941 an experiment was carried out which produced results summarised in the following equation

$$CO_2 + H_2{}^{18}O \longrightarrow [CH_2O] + {}^{18}O_2$$

The source of oxygen was thus shown to be water. The equation shows two atoms of oxygen coming from one molecule of water. So the balanced equation should be

$$CO_2 + 2H_2O \xrightarrow[\text{chlorophyll}]{\text{light energy}} [CH_2O] + O_2 + H_2O$$

This is the most accurate summary of photosynthesis and provides the extra information that water is produced, as well as used, in photosynthesis. This experiment provided a profound insight into the nature of photosynthesis because it showed that it takes place in two stages, the first of which involves acquiring hydrogen by splitting water into hydrogen and oxygen. This requires energy which must be provided by light (hence the process used to be called **photolysis**: *photos*, light; *lysis*, splitting). Oxygen is released as a waste product. In the second stage, hydrogen combines with carbon dioxide to produce carbohydrate. Addition of hydrogen is an example of a type of chemical reaction called **reduction** (appendix 1).

The fact that photosynthesis is a two-stage process was first established in the 1920s and 1930s. In the first stage the reactions require light and this stage was therefore called the **light reaction**. The second stage did not require light, and was called the **dark reaction**, although it takes place in light! These reactions are now more accurately referred to as the **light-dependent** and **light-independent** reactions. It is now known that the **light-dependent reactions take place on the chloroplast membranes** and the **light-independent reactions in the chloroplast stroma**.

Having established that photosynthesis proceeds by light-dependent reactions followed by light-independent reactions, it remained in the 1950s to discover the nature of these reactions.

7.6.2 Light-dependent reactions

We have seen that what the plant is doing in photosynthesis is making sugar using carbon dioxide and hydrogen (from water). This requires energy. The energy and the hydrogen are supplied by the light-dependent reactions which make ATP (adenosine triphosphate), an energy carrier, and reduced NADP.

ATP is the energy carrier of cells. Its structure is described in section 9.2.1 and its significance is summarised at the end of section 9.2.2. NADP (nicotinamide adenine dinucleotide phosphate) is a type of molecule known as a **hydrogen carrier**. It works in the same way as NAD. The structure of NAD and NADP is described in fig 4.17, and their roles as hydrogen carriers are described in section 4.2.3. It will help you if you read these references first.

ATP is made when energy is used to bond another phosphate to ADP, a process called **phosphorylation** (section 9.5.4). In photosynthesis the energy is supplied by light and the process is therefore called **photophosphorylation**. Reduced NADP is made from NADP in a process called **reduction**. The hydrogen comes from water. This also requires energy which is provided by light. The role of ATP and reduced NADP is simply to carry the energy and hydrogen into the light-independent reactions which follow.

We have seen already in section 7.5.5 that when light shines on photosystems I and II, high energy electrons are released by the chlorophyll molecules in the photosystems. It is the energy from these electrons that is used in the making of ATP and reduced NADP. The mechanism is shown in fig 7.14. The diagram contains a lot of information so should be studied carefully. Note that the vertical axis represents the energy level of the electrons.

The process depends on a flow of electrons from P680 and P700. Light provides the energy that causes this flow. Remember the equation

$$\underset{\text{(reduced form)}}{\text{chlorophyll}} \xrightarrow{\text{light energy}} \underset{\text{(oxidised form)}}{\text{chlorophyll}^+} + \underset{\text{electron}}{e^-}$$

First, an electron from P680 or P700 is boosted to a higher energy level, that is it acquires excitation energy. Instead of falling back into the photosystem and losing its energy it is captured by an electron acceptor (X or Y in fig 7.14). This represents the important conversion of light energy to chemical energy. The electron acceptor is thus reduced and a positively charged (oxidised) chlorophyll molecule is left in the photosystem. The electron then travels downhill, in energy terms, from one electron acceptor to another in a series of oxidation–reduction (redox) reactions. The energy lost during this electron flow is **'coupled'** to the formation of ATP. The pathway followed by the electron can be **cyclic**, returning to where it began, or **non-cyclic**, ending at NADP. When electrons are added to NADP it is changed to reduced NADP.

Non-cyclic photophosphorylation

Excited electrons from P680 (PSII) and P700 (PSI) reduce electron acceptors X and Y respectively, so that P680 and P700 become positively charged (oxidised). The electron donor which provides the replacement electrons for P680 is water. Water is split, releasing electrons which enter P680. Hydrogen ions and oxygen are also released. Oxygen escapes as a waste product (fig 7.14*a*).

Electrons flow from X along a chain of electron carriers, losing a little energy each time they move from one carrier to the next. Eventually they fill the positive holes left in P700. The energy from this flow is coupled to ATP production. Electrons also pass downhill in energy terms from Y to NADP along a chain of electron carriers and combine with hydrogen ions from water to form reduced NADP.

Cyclic photophosphorylation

In cyclic photophosphorylation, electrons from Y are recycled back to P700 via another chain of electron carriers. As the electrons pass down the chain their excitation energy is coupled to ATP production just as in non-cyclic photophosphorylation.

Table 7.3 shows the differences between cyclic and non-cyclic photophosphorylation.

The overall equation for non-cyclic photophosphorylation is

$$H_2O + NADP^+ + 2ADP + 2P_i \xrightarrow[\text{chlorophyll}]{\text{light energy}} {}^1/_2O_2 + NADPH + H^+ + 2ATP$$

Extra ATP can be made via cyclic photophosphorylation. The efficiency of energy conversion in the light-dependent reactions is high and estimated at about 39%.

Table 7.3 Comparison of cyclic and non-cyclic photophosphorylation.

	Non-cyclic	*Cyclic*
Pathway of electrons	Non-cyclic	Cyclic
First electron donor (source of electrons)	Water	Photosystem I (P700)
Last electron acceptor (destination of electrons)	NADP	Photosystem I (P700)
Products	Useful: ATP, reduced NADP Waste: O_2	Useful: ATP only
Photosystems involved	I and II	I only

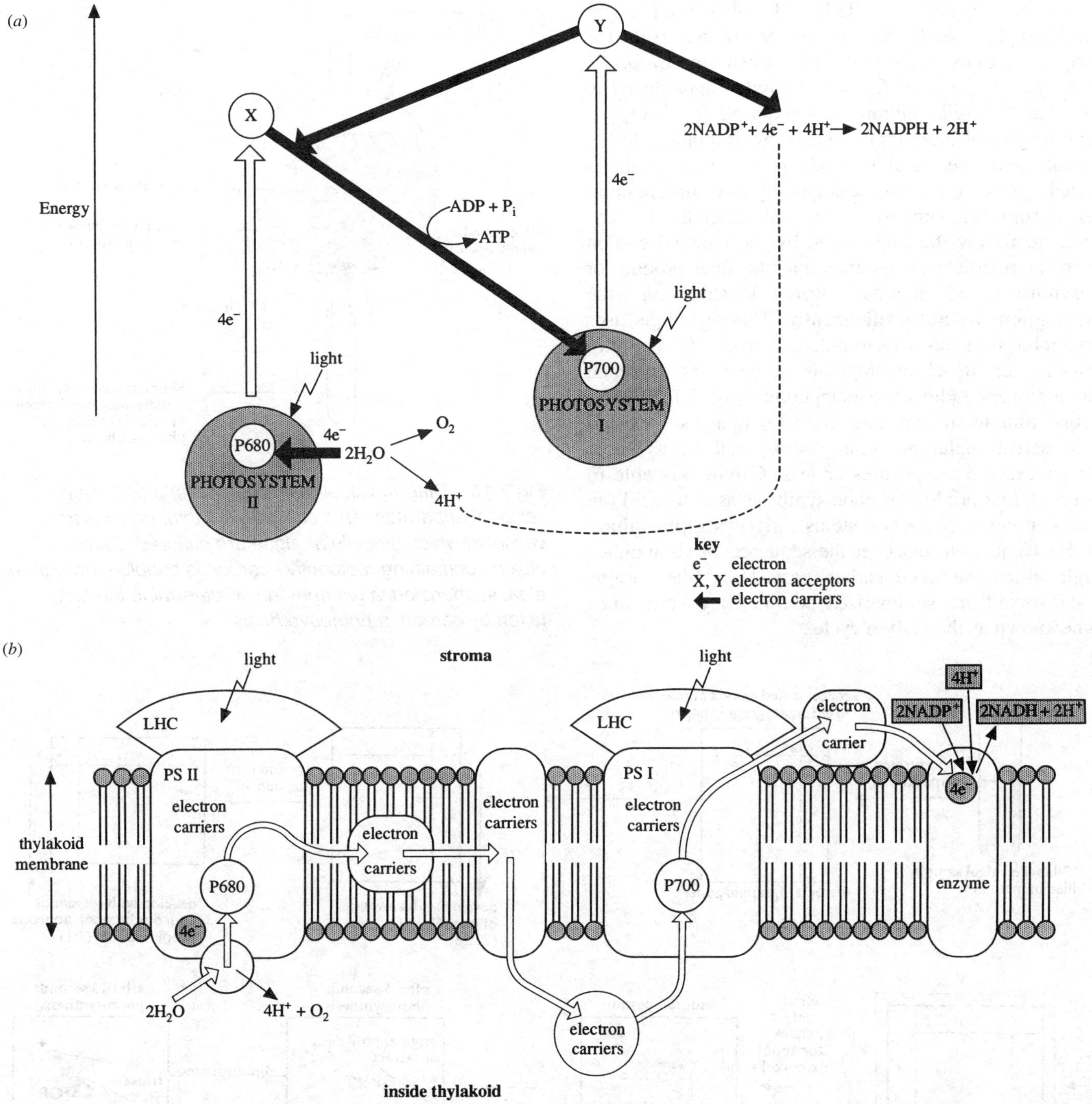

Fig 7.14 (a) *Electron flow in cyclic and non-cyclic photophosphorylation. Broad arrows represent the flow of electrons. As the electrons pass along the carriers they lose energy (see the vertical scale). The fall of electrons from a higher to a lower energy level can be used to drive the manufacture of ATP. Movement of four electrons is shown because two molecules of water are needed to release one molecule of oxygen, at the same time releasing four electrons.* (b) *The relationship between the electron flow and electron carriers in the cell surface membrane.*

7.6.3 Light-independent reactions

The light-independent (or dark) reactions, which take place in the stroma of the chloroplast, do not require light and use the energy (ATP) and reducing power (reduced NADP) produced by the light-dependent reactions to reduce carbon dioxide. The reactions are controlled by enzymes and their sequence was determined by Calvin, Benson and Bassham of the USA during the period 1946–53, work for which Calvin was awarded the Nobel prize in 1961.

Calvin's experiments

Calvin's work was based on use of the radioactive isotope of carbon, ^{14}C (half-life 5570 years, see appendix 1) which only became available in 1945. He also used paper chromatography, which was a relatively new but neglected technique. Cultures of the unicellular green alga *Chlorella* were grown in the now famous 'lollipop' apparatus (fig 7.15). The *Chlorella* culture was exposed to $^{14}CO_2$ for varying lengths of time, rapidly killed by dropping into hot methanol, and the soluble products of photosynthesis extracted, concentrated and separated by **two-dimensional paper chromatography** (fig 7.16 and appendix 1). The aim was to follow the route taken by the labelled carbon through intermediate compounds into the final product of photosynthesis. Compounds were located on the chromatograms by **autoradiography**. This technique uses photographic film sensitive to radiation from ^{14}C. The film is placed over the chromatograms. It becomes darkened where it covers radioactive compounds (fig 7.16). After only one minute of exposure to $^{14}CO_2$ many sugars and organic acids, including amino acids, had been made. However, using 5 s exposures or less, Calvin was able to identify the first product of photosynthesis as a 3C acid (an acid containing three carbon atoms), **glycerate phosphate** (GP). He went on to discover the sequence of compounds through which the fixed carbon passed and the various stages involved are summarised below. They have since become known as the **Calvin cycle** .

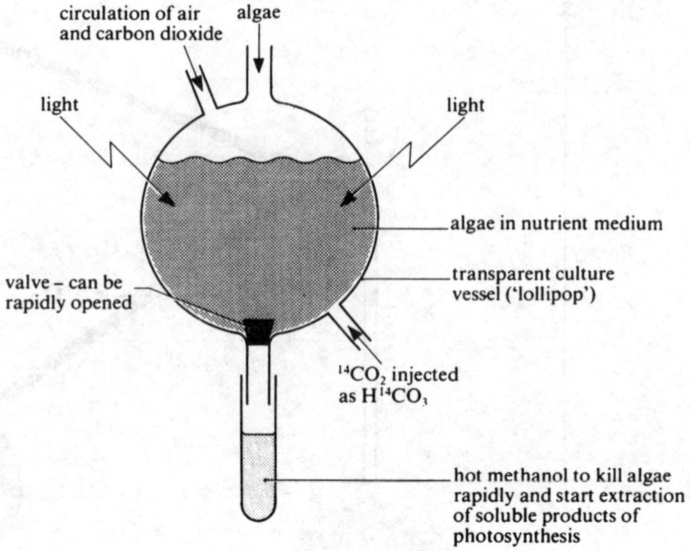

Fig 7.15 *Diagram illustrating the principle of Calvin's 'lollipop' apparatus. This comprises a thin, transparent vessel in which unicellular algae are cultured. Carbon dioxide containing radioactive carbon is bubbled through the algal suspension in experiments to determine the path taken by carbon in photosynthesis.*

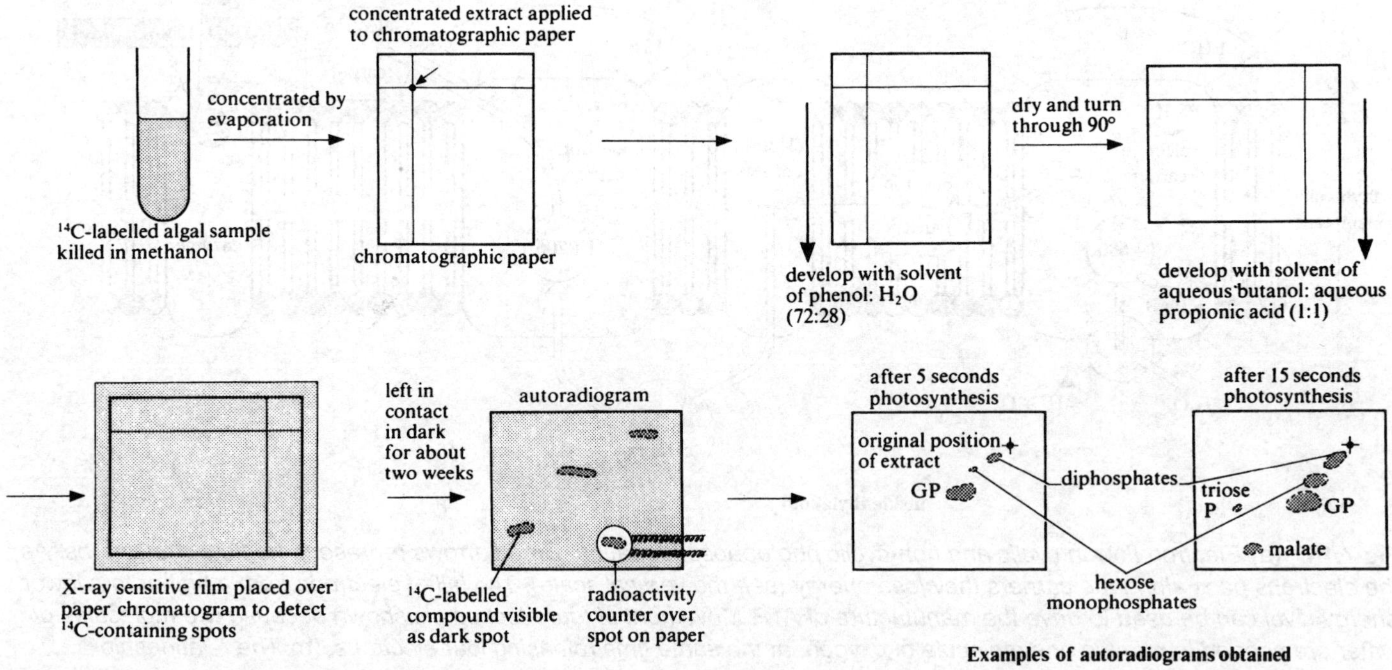

Fig 7.16 *Detection of the products of photosynthesis in algae after brief periods of illumination in the presence of radioactive carbon dioxide, $^{14}CO_2$. Paper chromatography is used to separate the products. Their positions on the paper enable them to be identified. The positions are found by exposing the paper to photographic film. Radiation from the radioactive products blackens the film.*

7.4 What is the advantage of using a radioactive isotope with a long half-life in biological experiments?

7.5 What advantage might be gained by using *Chlorella* rather than a flowering plant?

7.6 Why was the 'lollipop' vessel thin in section rather than spherical?

Stages in carbon pathway

Acceptance of carbon dioxide (carbon dioxide fixation).

$$\text{RuBP} + CO_2 + H_2O \xrightarrow{\text{RuBP carboxylase}} 2\text{GP}$$

RuBP (ribulose bisphosphate) 5C sugar → 2GP (glycerate phosphate) 3C acid

first product of photosynthesis

The carbon dioxide acceptor is a 5C sugar (a pentose), **ribulose bisphosphate** RuBP (ribulose with two phosphate groups). Addition of carbon dioxide to a compound is called **carboxylation**; the enzyme involved is a **carboxylase**. The 6C product is unstable and breaks down immediately to two molecules of **glycerate phosphate** (**GP**). This is the first product of photosynthesis. The enzyme ribulose bisphosphate carboxylase is present in large amounts in the chloroplast stroma, and is in fact the world's most common protein.

Reduction phase.

$$\underset{\text{3C acid}}{\text{GP}} \xrightarrow[\text{NADPH} + H^+ \;\rightarrow\; \text{NADP}^+ + H_2O]{\text{ATP} \;\rightarrow\; \text{ADP} + P_i} \underset{\text{3C sugar}}{\text{TP}}$$

GP is glycerate phosphate, a 3C **acid**. It contains the acidic carboxyl group (—COOH). TP is triose phosphate (glyceraldehyde phosphate), a 3C **sugar**. It contains an aldehyde group (—CHO).

The reducing power of reduced NADP and energy of ATP are used to remove oxygen from GP (reduction). The reaction takes place in two stages, the first using some of the ATP produced in the light-dependent reactions and the second using all the reduced NADP produced in these reactions. The overall effect is to reduce a carboxylic acid group (—COOH) to an aldehyde group (—CHO). The product is a 3C sugar phosphate (a triose phosphate), that is a sugar with a phosphate group attached. This contains more chemical energy than GP, and is the first carbohydrate made in photosynthesis.

Regeneration of the carbon dioxide acceptor, RuBP. Some of the triose phosphate (TP) has to be used to regenerate the ribulose bisphosphate consumed in the first reaction. This process involves a complex cycle, containing 3, 4, 5, 6 and 7C sugar phosphates. It is here that the remaining ATP is used. Fig 7.17 provides a summary of the light-independent reactions. In it the Calvin cycle is represented as a 'black box' into which carbon dioxide and water are fed and TP emerges. The diagram shows that the remaining ATP is used to convert ribulose phosphate (RuP) to ribulose bisphosphate (RuBP), but details of the complex series of reactions are not shown.

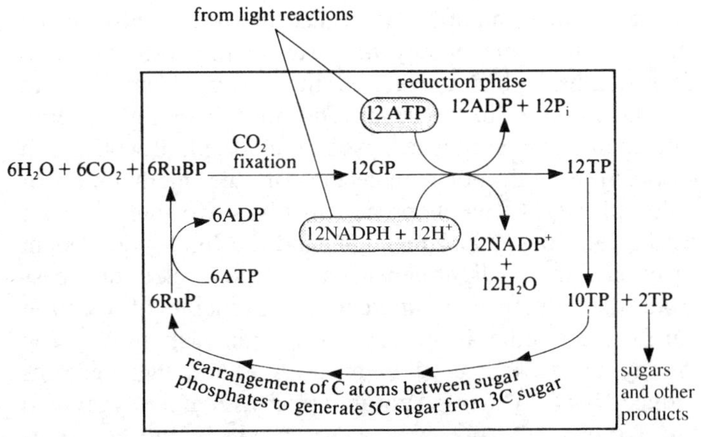

Fig 7.17 *Summary of the light-independent reactions of photosynthesis (Calvin cycle). RuBP, ribulose bisphosphate; RuP, ribulose phosphate; GP, glycerate-3-phosphate; TP, triose phosphate. Note that 10 × TP are used to make 6 × RuP, the 5C sugar ribulose phosphate. TP contains 3 carbon atoms. Therefore 10 × TP contains 30 carbon atoms. 6 × RuP also contains 30 carbon atoms (6 × 5C).*

The overall equation for the light-independent reactions is

$$6H_2O + 6CO_2 \xrightarrow[12\text{NADPH} + 12H^+ \;\rightarrow\; 12\text{NADP}^+ + 12H_2O]{18\text{ATP} \;\rightarrow\; 18\text{ADP} + 18P_i} 2\text{TP}$$

The important point to note is that six molecules of carbon dioxide have been used to make two molecules of a 3C sugar, triose phosphate. The equation can be simplified by dividing by six

$$H_2O + CO_2 \xrightarrow[2\text{NADPH} + 2H^+ \;\rightarrow\; 2\text{NADP}^+ + 2H_2O]{3\text{ATP} \;\rightarrow\; 3\text{ADP} + 3P_i} [CH_2O] + 2H_2O$$

7.7 Redraw fig 7.17 showing only the numbers of C atoms involved; for example 6RuBP = 6 × 5C.

7.6.4 Summary of photosynthesis

Sugar is made during photosynthesis from carbon dioxide and water. The energy for the reaction comes from light and is stored within sugar which is an energy-rich molecule. It is important to realise that the energy in the sugar is the same energy that was in the form of light. Photosynthesis has transferred the energy from light to chemical energy in the sugar. The transfer of light energy into chemical energy is achieved by chlorophyll working in association with other molecules in the membranes of chloroplasts. The whole process involves two stages known as the light-dependent reactions and the light-independent reactions. In the light-dependent reactions electrons pass from one electron to another along chains of electron carriers, dropping in energy level with each move. The energy released is used to make ATP and the electrons reduce NADP. Water is split into hydrogen and oxygen and the oxygen is lost as a waste product. In the light-independent reactions the hydrogen (attached to NADP) and the ATP are used to reduce carbon dioxide to a 3C sugar. The overall equation is

$$CO_2 + H_2O \xrightarrow[\text{chlorophyll}]{\text{light energy}} [CH_2O] + O_2$$

Further summary notes are given in table 7.4.

7.7 Metabolism of glycerate phosphate and triose phosphate

Although triose phosphate is the end-product of the Calvin cycle, it does not accumulate in large quantities since it is immediately converted to other products. The most familiar of these are glucose, sucrose and starch, but fats, fatty acids and amino acids are also made rapidly. Strictly speaking, photosynthesis can be regarded as complete once triose phosphate is made, because subsequent reactions can also occur in non-photosynthetic organisms like animals and fungi. However, it is important to show here how glycerate phosphate and triose phosphate can be used in the synthesis of all the basic food requirements of plants. Fig 7.18 summarises some of the main pathways involved and shows what a central position the reactions of glycolysis and the Krebs cycle have in metabolism. These two pathways are discussed in chapter 9. Both glycerate phosphate and triose phosphate are intermediates in glycolysis.

Synthesis of carbohydrates

Carbohydrates are synthesised in a process which is, in effect, a reversal of glycolysis. The two most common carbohydrate products are sucrose and starch. Sucrose is the form in which carbohydrate is exported from the leaf in the phloem (chapter 13). Starch is a storage product and is the most easily detected product of photosynthesis.

Table 7.4 Summary of photosynthesis.

	Light-dependent reactions	*Light-independent reactions*
Location in chloroplasts	Thylakoids	Stroma
Reactions	Require light. Light energy causes the flow of electrons from electron 'donors' to electron 'acceptors', along a non-cyclic or a cyclic pathway. Two photosystems, I and II, are involved. These contain chlorophylls which emit electrons when they absorb light energy. Water acts as an electron donor to the non-cyclic pathway. Electron flow results in production of ATP (photophos-phorylation) and reduced NADP.	Do not require light. Carbon dioxide is fixed when it is accepted by a 5C compound ribulose bisphosphate (RuBP), to form two molecules of a 3C compound glycerate phosphate (GP), the first product of photosynthesis. A series of reactions occurs called the Calvin cycle in which the carbon dioxide acceptor RuBP is regenerated and GP is reduced to a sugar. (See also fig 7.17.)
Overall equation	$2H_2O + 2NADP^+ \xrightarrow[\text{chlorophyll}]{\text{light}} O_2 + 2NADPH + 2H^+$ also $ADP + P_i \longrightarrow ATP$ (variable amount)	$CO_2 + H_2O \longrightarrow [CH_2O]$ (3ATP → $3ADP + 3P_i$; $2NADPH + 2H^+$ → $2NADP^+ + 2H_2O$)
Results	Light energy is converted to chemical energy in ATP and reduced NADP. Water is split into hydrogen and oxygen. Hydrogen is carried to reduced NADP and oxygen is a waste product.	Carbon dioxide is reduced to carbon compounds such as carbohydrates, using the chemical energy in ATP and hydrogen in reduced NADP.
Summary (omitting ATP and NADP)	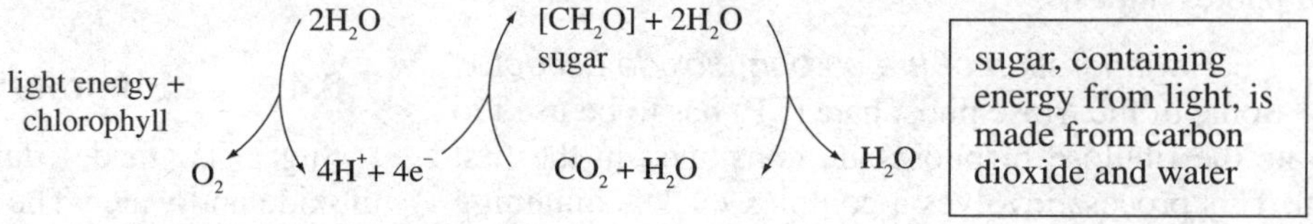	

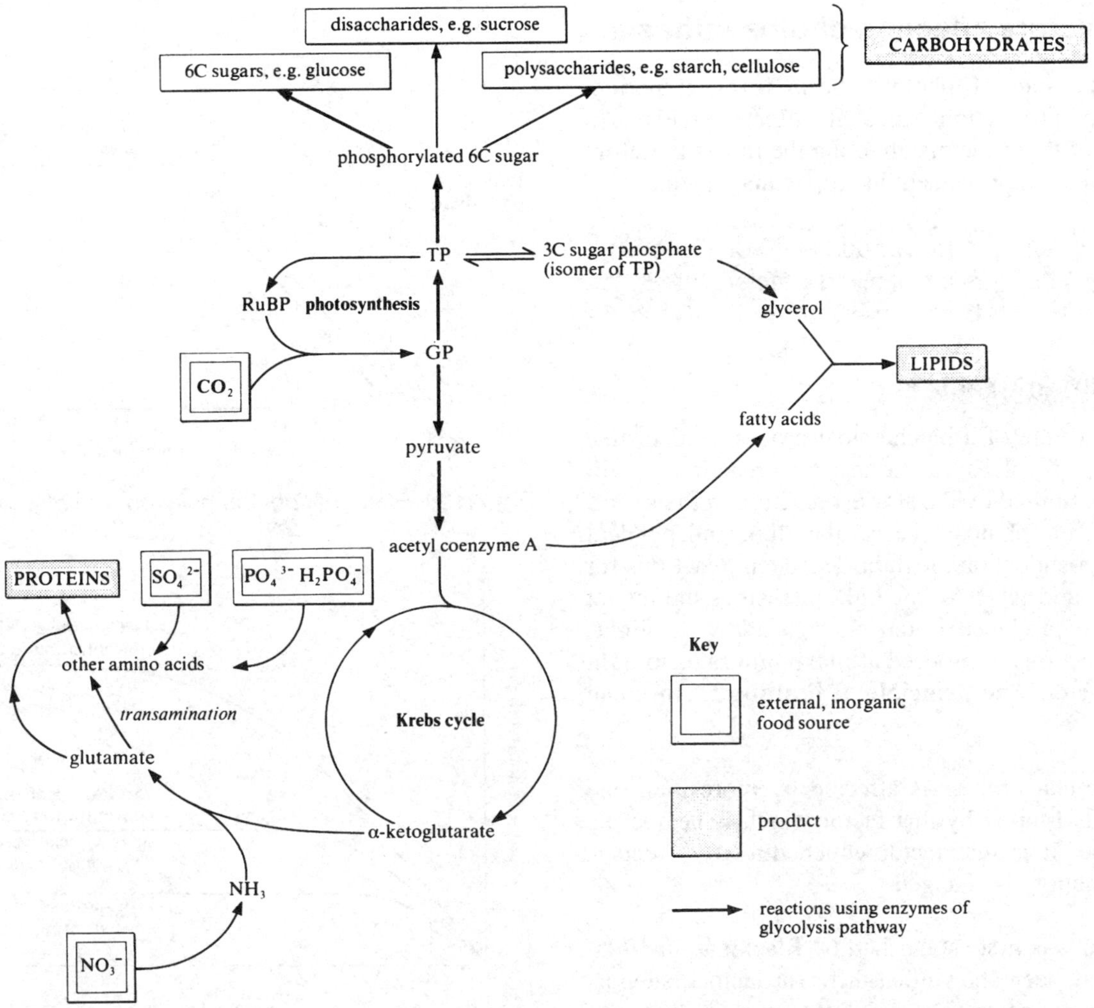

Fig 7.18 *Metabolism of GP and TP showing the relationship between photosynthesis and synthesis of food in plants. Main pathways only are shown. Some intermediate steps are omitted.*

Synthesis of lipids

Lipids are made from glycerol and fatty acids (section 3.3). **Glycerol** is made from triose phosphate. To make fatty acids, glycerate phosphate enters the glycolysis pathway and is converted to an acetyl group which is added to coenzyme A to form acetyl coenzyme A. The acetyl groups are converted to **fatty acids** in both cytoplasm and chloroplasts (not in mitochondria, where *breakdown* of fatty acids occurs).

Synthesis of proteins

Glycerate phosphate and triose phosphate contain the elements carbon, hydrogen and oxygen. Nitrogen and sulphur are also needed if amino acids and hence proteins are to be made. Plants obtain these elements from the soil, or surrounding water if aquatic. Nitrogen is taken up as nitrates or ammonia, and sulphur is taken up as sulphates.

Glycerate phosphate is first converted to one of the acids of the Krebs cycle via acetylcoenzyme A (fig 7.18). Synthesis of the amino acid is summarised below.

(1) NO_3^- (nitrate from roots) $\xrightarrow[\text{nitrate reductase}]{\text{reduction}}$ NO_2^- (nitrite) $\xrightarrow[\text{nitrite reductase}]{\text{reduction}}$ NH_3

(2) NH_3 (ammonia) + Krebs cycle acid ⟶ amino acid

For example,

NH_3 + α-ketoglutarate + reduced NADP $\xrightarrow{\text{transaminase}}$ glutamate + NADP

Reaction (2) is the major route of entry of ammonia into amino acids. By a process called **transamination** other amino acids can be made by transferring the amino group (—NH_2) from one acid to another. For example,

glutamate (amino acid) + oxaloacetate (a Krebs cycle acid) $\xrightarrow{\text{transaminase}}$ α-ketoglutarate (a Krebs cycle acid) + aspartate (amino acid)

Other synthetic pathways for amino acids also occur. Some amino acids are made in the chloroplasts. About one-third of the carbon fixed and about two-thirds of the nitrogen taken up by plants are commonly used directly to make amino acids.

7.8 Factors affecting photosynthesis

The rate of photosynthesis is an important factor in crop production since it affects yields. An understanding of those factors affecting the rate is therefore likely to lead to an improvement in crop management.

> **7.8** From the equation of photosynthesis what factors are likely to affect its rate?

7.8.1 Limiting factors

The rate of a biochemical process which, like photosynthesis, involves a series of reactions, will theoretically be limited by the slowest reaction in the series. For example, in photosynthesis the light-independent reactions are dependent on the light-dependent reactions for reduced NADP and ATP. At low light intensities the rate at which these are produced is too slow to allow the light-independent reactions to proceed at maximum rate, so light is a limiting factor. **The principle of limiting factors** can be stated thus:

when a chemical process is affected by more than one factor, its rate is limited by that factor which is nearest its minimum value: it is that factor which directly affects a process if its quantity is changed.

The principle was first established by Blackman in 1905. Since then it has been shown that different factors, such as carbon dioxide concentration and light intensity, interact and can be limiting at the same time, although one is often the major factor. Consider one of these factors, light intensity, by studying fig 7.19 and trying to answer the following questions.

> **7.9** In fig. 7.19
> (*a*) what is the limiting factor in region A?
> (*b*) what is represented by the curve at B and C?
> (*c*) what does point D represent on the curve?
> (*d*) what does point E represent on the curve?

Fig 7.20 shows the results from four experiments in which the same experiment is repeated at different temperatures and carbon dioxide concentrations.

> **7.10** In fig 7.20 what do the points X, Y and Z represent on the three curves?

In fig 7.20, experiments 1–4 show that once light intensity is no longer limiting, both temperature and carbon dioxide concentration can become limiting. Enzyme-controlled reactions like the light-independent reactions of photosynthesis are sensitive to temperature; thus an increase in temperature from 15 °C to 25 °C results in an increased rate of photosynthesis (compare experiments 2 and 1, or 4 and 3) providing light is not a limiting factor. Carbon dioxide concentration can also be a limiting factor in the light-independent reactions (compare experiments 2 and 4, or 1 and 3). In experiment 2, for example, temperature *and* carbon dioxide concentration are limiting, and an increase in either results in increased photosynthetic rate.

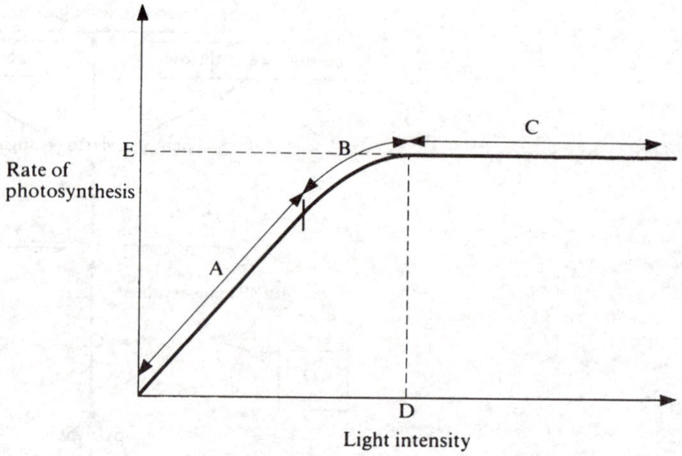

Fig 7.19 *Effect of light intensity on rate of photosynthesis.*

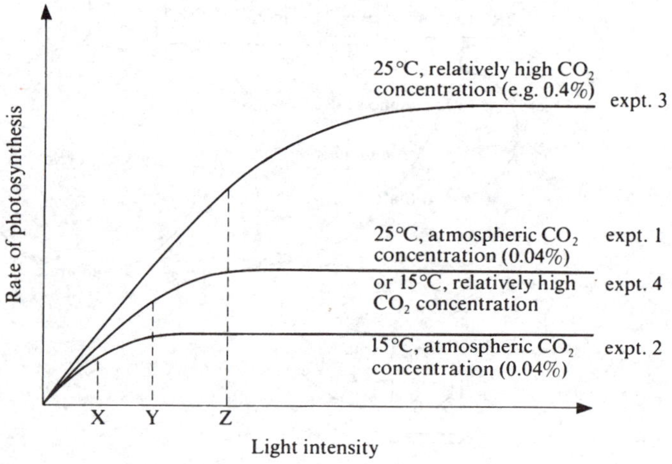

Fig 7.20 *Effect of various factors on rate of photosynthesis.*

7.8.2 Reaction rate graphs

The chief external factors affecting rate of photosynthesis are light intensity, carbon dioxide concentration and temperature. Graphs representing their effects all have the form of fig 7.19, with external factors plotted on the horizontal axis. All show an initial linear increase in photosynthetic rate where the factor being investigated is limiting, followed by a decrease in the rate of increase and stabilising of rate as another factor, or factors, becomes limiting.

In the following it is assumed that factors other than the one under discussion are optimal.

Light intensity

In low light intensities the rate of photosynthesis increases linearly with increasing light intensity (fig 7.19). Gradually the rate of increase falls off as the other factors become limiting. Illumination on a clear summer's day is about 100 000 lux (10 000 ft candles), whereas light saturation for photosynthesis is reached at about 10 000 lux. Therefore, except for shaded plants, light is not normally a major limiting factor. Very high light intensities may bleach chlorophyll and slow down photosynthesis, but plants normally exposed to such conditions are usually protected by devices such as thick cuticles and hairy leaves.

Carbon dioxide concentration

Carbon dioxide is needed in the light-independent reactions where it is used to make sugar. Under normal conditions, carbon dioxide is the major limiting factor in photosynthesis. Its concentration in the atmosphere varies between 0.03% and 0.04%, but increases in photosynthetic rate can be achieved by increasing this percentage (see experiment 3, fig 7.20). The short-term optimum is about 0.5%, but this can be damaging over long periods; then the optimum is about 0.1%. This has led to some greenhouse crops, such as tomatoes, being grown in carbon dioxide-enriched atmospheres. At the moment there is much interest in a group of plants which are capable of removing the available carbon dioxide from the atmosphere more efficiently, hence achieving greater yields. These 'C_4' plants are discussed in section 7.9.

Temperature

The light-independent reactions and, to a certain extent, the light-dependent reactions are enzyme-controlled and therefore temperature-sensitive. For temperate plants the optimum temperature is usually about 25 °C. The rate of reaction doubles for every 10 °C rise up to about 35 °C, although other factors mean that the plant grows better at 25 °C.

> **7.11** Why should the rate decrease at higher temperatures?

Chlorophyll concentration

Chlorophyll concentration is not normally a limiting factor, but reduction in chlorophyll levels can be induced by several factors, including disease (such as mildews, rusts and virus diseases), mineral deficiency (section 7.10.1) and normal ageing processes (**senescence**). If the leaf becomes yellow it is said to be **chlorotic**, the yellowing process being called **chlorosis**. Chlorotic spots are thus often a symptom of disease or mineral deficiency. Iron, magnesium and nitrogen are required during chlorophyll synthesis (the latter two elements being part of its structure) and are therefore particularly important minerals. Potassium is also important. Lack of light can also cause chlorosis since light is needed for the final stage of chlorophyll synthesis.

Specific inhibitors

An obvious way of killing a plant is to inhibit photosynthesis, and various herbicides have been introduced to do this. A notable example is DCMU (dichlorophenyl dimethyl urea) which short-circuits non-cyclic electron flow in chloroplasts and thus inhibits the light-dependent reactions. DCMU has been useful in research on the light-dependent reactions.

Two other factors which are important when growing crops and which have more general effects on plant growth and photosynthesis are water supply and pollution.

Water

Water is a raw material in photosynthesis, but so many cell processes are affected by lack of water that it is impossible to measure the direct effect of water on photosynthesis. Nevertheless, by studying the yields (amounts of organic matter synthesised) of water-deficient plants, it can be shown that periods of temporary wilting can lead to severe yield losses. Even slight water deficiency, with no visible effects, might significantly reduce crop yields. The reasons are complex and not fully understood. One obvious factor is that plants usually close their stomata in response to wilting and this would prevent access of carbon dioxide for photosynthesis. Abscisic acid, a growth inhibitor, has also been shown to accumulate in water-deficient leaves of some species.

Pollution

Low levels of certain gases of industrial origin, notably ozone and sulphur dioxide, are very damaging to the leaves of some plants, although the exact reasons are still being investigated. It is estimated, for example, that cereal crop losses as high as 15% may occur in badly polluted areas, particularly during dry summers. Lichens are very sensitive to sulphur dioxide. Soot can block stomata and reduce the transparency of the leaf epidermis.

> **7.12** Suggest some habitats or natural circumstances in which (*a*) light intensity, and (*b*) temperature might be limiting factors in photosynthesis.

7.9 C_4 photosynthesis

In 1965 it was shown that the first products of photosynthesis in sugarcane, a tropical plant, appeared to be acids containing four carbon atoms (malic, oxaloacetic and aspartic) rather than the 3C acid GP of most temperate plants. Many plants, mostly tropical and subtropical and some of great economic importance, have since been identified in which the same is true and these are called **C_4 plants**. Examples are maize, sorghum, sugarcane and millet. Plants in which the 3C acid GP is the first product of photosynthesis are called **C_3 plants**. It is the biochemistry of these plants which has been described so far.

In 1966, two Australian workers, Hatch and Slack showed that C_4 plants were far more efficient at taking up carbon dioxide than C_3 plants: they could remove carbon dioxide from an experimental atmosphere down to 0.1 parts per million (ppm) compared with the 50–100 ppm of temperate plants. Hatch and Slack discovered a new carbon pathway in C_4 plants which is now called the **Hatch–Slack pathway**. The process in a typical C_4 plant, maize, will be described.

C_4 plants possess a characteristic leaf anatomy in which two rings of cells are found around each of the vascular bundles. The inner ring, or **bundle sheath cells**, contains chloroplasts which differ in form from those in the **mesophyll cells** in the outer ring. The chloroplasts in the plants are therefore described as **dimorphic** (two forms). Figs 7.21 *a* and *b* illustrate this so-called **'Kranz' anatomy** (Kranz means crown or halo, referring to the two distinct rings of cells). The biochemical pathway that takes place in these cells is summarised below and in fig 7.22.

7.9.1 Hatch–Slack pathway

The Hatch–Slack pathway is a pathway for transporting carbon dioxide and hydrogen from mesophyll cells to bundle sheath cells. Once in the bundle sheath cells, the carbon dioxide is released again and normal C_3 photosynthesis occurs (fig 7.22).

Acceptance of carbon dioxide (carbon dioxide fixation) in mesophyll cells

Carbon dioxide is fixed in the **cytoplasm** of the mesophyll cells as shown below:

$$\underset{\substack{\text{(phosphoenolpyruvate)}\\ 3C}}{\text{PEP}} + CO_2 \xrightarrow{\text{PEP carboxylase}} \underset{4C}{\text{oxaloacetate}}$$

The carbon-dioxide-acceptor is phosphoenolpyruvate (PEP) instead of the RuBP of C_3 plants, and the enzyme is PEP carboxylase instead of RuBP carboxylase. PEP carboxylase is much more efficient than the enzyme of C_3 plants for two reasons. Firstly, it has a much higher affinity for carbon dioxide, and secondly it is not competitively inhibited by oxygen. Oxaloacetate is converted to malate, a 4C acid.

Malate shunt

Malate is shunted through plasmodesmata in the cell walls to the chloroplasts of the bundle sheath cells, where it is converted to pyruvate, a 3C acid, by having carbon dioxide and hydrogen removed. The hydrogen is used to reduce NADP (fig 7.22). Note that in the mesophyll cells carbon dioxide and hydrogen are added and that in the bundle

(*a*)

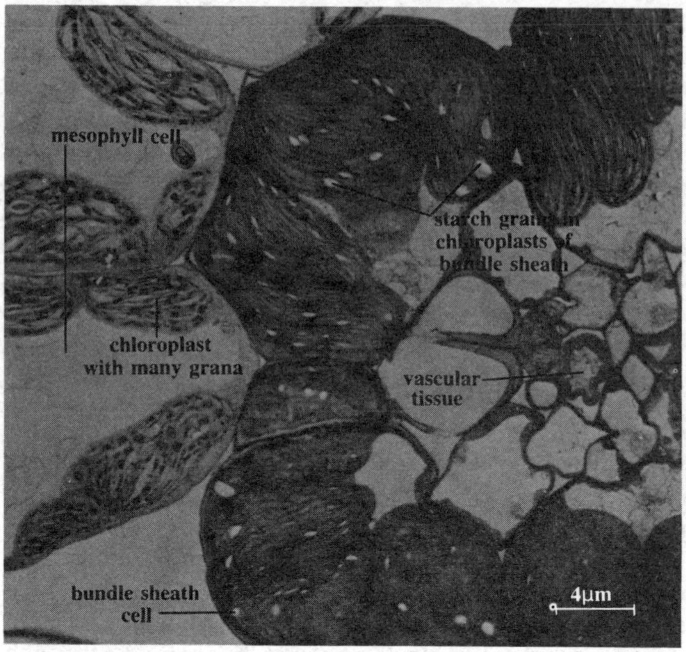

(*b*)

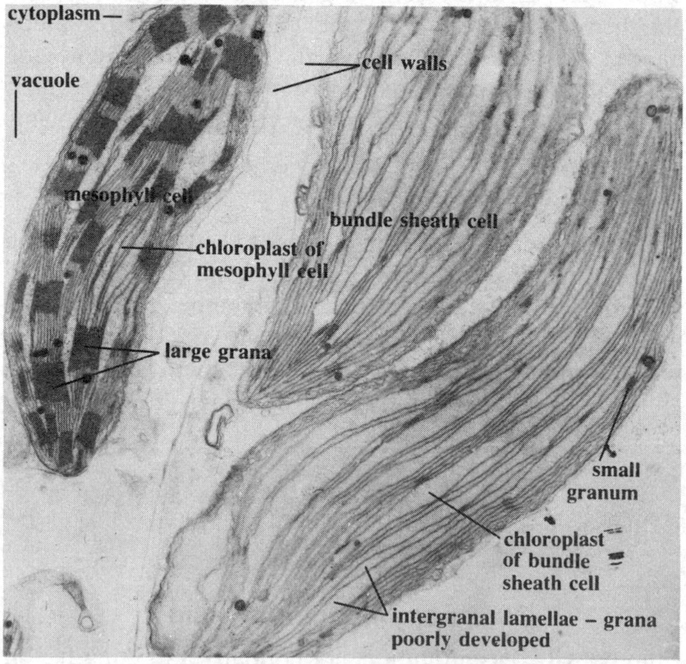

Fig 7.21 *'Kranz' anatomy, characteristic of C_4 plants.* (a) *Section of a leaf of crabgrass to show the difference between bundle sheath chloroplasts and mesophyll chloroplasts. Grana in the bundle sheath chloroplasts are only rudimentary, whereas they are prominent in chloroplasts of the mesophyll. Starch grains are present in both. (Magnification ×4000).* (b) *Electron micrograph of maize leaf showing two types of chloroplasts found in bundle sheath cells and mesophyll cells (×9900).*

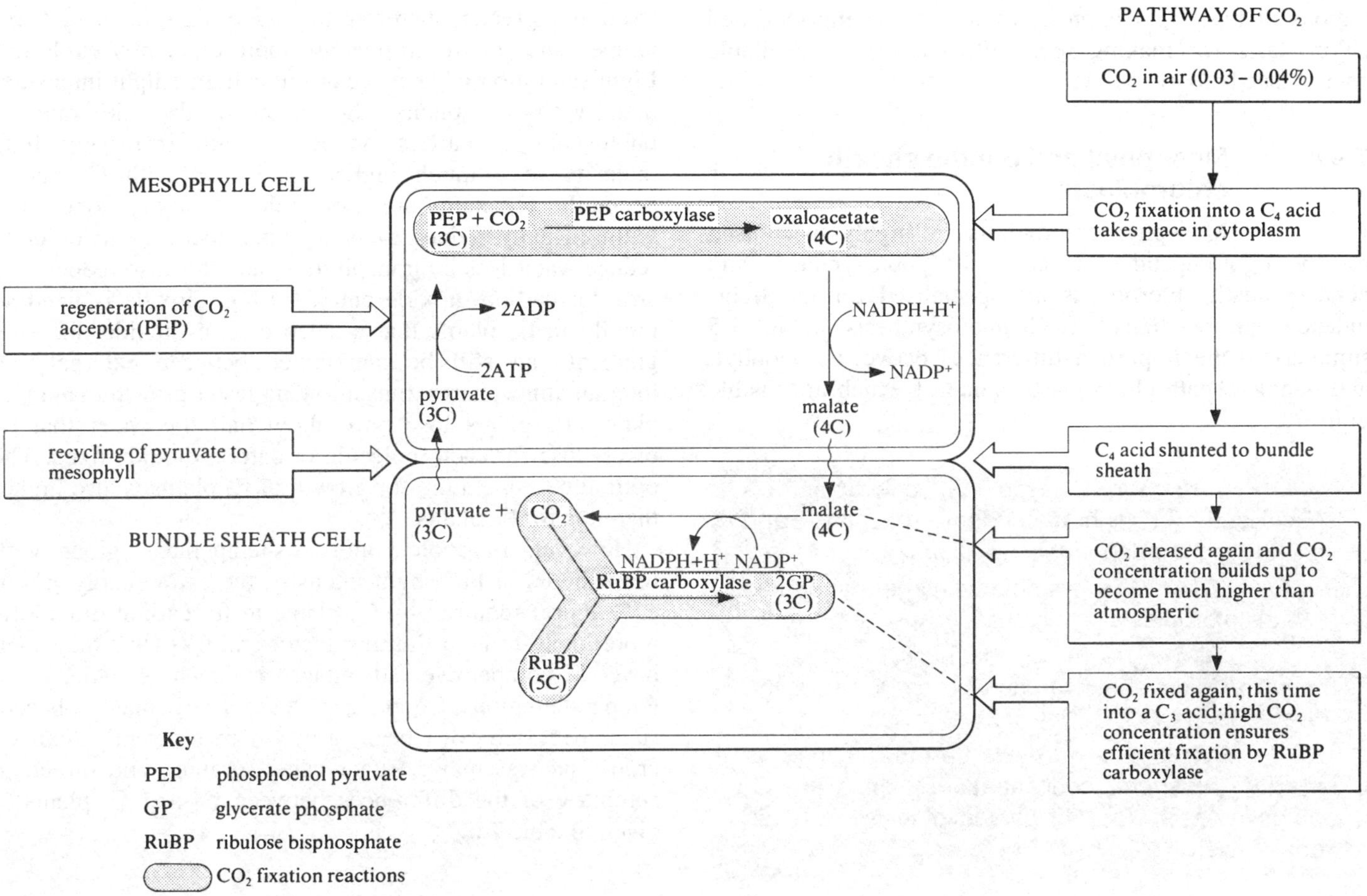

Fig 7.22 *Simplified outline of C_4 pathway coupled with C_3 fixation of carbon dioxide. Transport of carbon dioxide from air to bundle sheath is shown, together with final fixation of carbon dioxide into the C_3 acid GP.*

sheath cells they are removed again. The overall effect of this is to move carbon dioxide and hydrogen from the mesophyll cells to the bundle sheath cells.

Regeneration of the carbon dioxide acceptor

Pyruvate is returned to the mesophyll cells and is used to regenerate PEP by the addition of phosphate from ATP. This requires the energy from two high energy phosphate bonds.

7.9.2 Net result of C_4 pathway

The net result of the C_4 pathway is the use of two high energy phosphate bonds from ATP to transport carbon dioxide and hydrogen from the mesophyll cells to the chloroplasts of the bundle sheath cells. Since the transport requires energy from ATP, it can be regarded as a pumping mechanism.

7.9.3 Refixation of carbon dioxide in the bundle sheath cells

Carbon dioxide and reduced NADP are produced, as well as pyruvate, in the bundle sheath chloroplasts (see malate shunt above). The carbon dioxide is then refixed by RuBP carboxylase in the C_3 pathway, and the reduced NADP used to reduce GP to sugar (section 7.6.3).

Since every carbon dioxide molecule has had to be fixed twice, the energy requirement for C_4 photosynthesis is roughly double that for C_3 photosynthesis. At first sight then, the transport of carbon dioxide and hydrogen by the C_4 pathway seems pointless. However, fixation using PEP carboxylase in the mesophyll is so efficient that a high concentration of carbon dioxide accumulates in the bundle sheath. This means that the RuBP carboxylase works at an advantage compared with the same enzyme in C_3 plants, where carbon dioxide is at atmospheric concentration. There are two reasons for this: firstly, like any enzyme it works much more efficiently at high substrate concentrations; secondly, oxygen is competitively excluded from the enzyme by carbon dioxide.

The main advantage of C_4 photosynthesis, therefore, is that it improves the efficiency of carbon dioxide fixation. It is an addition rather than an alternative to the C_3 pathway. As a result, C_4 plants are photosynthetically more efficient because the rate of carbon dioxide fixation is normally the limiting factor in photosynthesis. C_4 plants consume more energy by using the C_4 pathway, but energy is not normally the limiting factor in photosynthesis, and C_4 plants grow in regions of high light intensity as well as having modified chloroplasts for making more efficient use of available energy (see below).

7.9.4 Mesophyll and bundle sheath chloroplasts

Mesophyll chloroplasts are highly specialised for the light-dependent reactions of photosynthesis and bundle sheath chloroplasts are specialised for the light-independent reactions of C_3 photosynthesis. Table 7.5 summarises the important differences between mesophyll and bundle sheath chloroplasts, some of which are visible in fig 7.21.

7.13 Which type of chloroplast is specialised for light-dependent reactions and which for light-independent reactions?

7.14 Why is it an advantage that bundle sheath chloroplasts lack grana?

7.15 The malate shunt is, in effect, a carbon dioxide and hydrogen pump. What is the advantage of this?

7.16 What would be the effect of lowering oxygen concentrations on (*a*) C_3 photosynthesis, (*b*) C_4 photosynthesis? Explain your answers.

7.9.5 Significance of the C_4 pathway

The C_4 pathway is thought to be more recently evolved than the C_3 pathway, and involves a superior carbon-dioxide-fixing mechanism. Thus C_4 plants increase in dry mass more rapidly than C_3 plants and are more efficient crop plants in certain parts of the world (see below).

C_4 plants have evolved chiefly in the drier regions of the subtropics and tropics, for which they are adapted in two major ways. First, their maximum rate of carbon dioxide fixation is greater; therefore the higher light intensities and temperatures of the tropics are more efficiently exploited. Light saturation takes place at much higher light intensities than with C_3 plants. In other words, the rate of photosynthesis carries on rising with increasing light intensity to a much higher level than with C_3 plants. Secondly, C_4 plants are more tolerant of dry conditions. Plants usually reduce their stomatal apertures in order to reduce water loss by transpiration, and this also reduces the area for carbon dioxide entry. Carbon dioxide is fixed so rapidly in C_4 plants that a steep carbon dioxide diffusion gradient can still be maintained between external and internal atmospheres, thus allowing faster growth than in C_3 plants. C_4 plants lose only about half the water that C_3 plants lose for each molecule of carbon dioxide fixed. The optimum temperature for growth of C_4 plants is also higher than it is for C_3 plants.

However, in cooler, moister, temperate regions with fewer hours of high light intensity, the extra energy (about 15% more) required by C_4 plants to fix carbon dioxide is more likely to be a limiting factor and C_3 plants may even have a competitive advantage in such situations. In temperate regions, C_3 crops, such as wheat, potato, tobacco, sugar beet and soya bean, grow more efficiently than C_4 crops, such as maize, sugar cane, sorghum and millet. A summary of the differences between C_3 and C_4 plants is given in table 7.6.

Table 7.5 Differences between mesophyll and bundle sheath chloroplasts in C_4 plants.

Mesophyll chloroplasts	*Bundle sheath chloroplasts*
Large grana Therefore light-dependent reactions favoured, so plenty of ATP, reduced NADP and O_2 generated.	No grana (or very few and small) Therefore light-dependent reactions occur at very low rate, so little reduced NADP, ATP or O_2 generated
Virtually no RuBP carboxylase so no CO_2 fixation (CO_2 fixation occurs in cytoplasm by PEP carboxylase)	High concentration of RuBP carboxylase so CO_2 fixation occurs as in C_3 plants but more efficiently
Little starch	Abundant starch grains

7.10 Mineral nutrition of plants and animals

Autotrophic nutrition involves not only the synthesis of carbohydrates from carbon dioxide and water, but also the subsequent use of minerals like nitrates, sulphates and phosphates to make other organic substances that are needed, such as proteins and nucleic acids. Heterotrophic organisms like animals also require certain minerals to supplement their organic food. In many cases the same nutrients are required and for the same reasons, so it is convenient to consider the whole area of mineral nutrition as a bridge between autotrophic nutrition (chapter 7) and heterotrophic nutrition (chapter 8).

A nutritional element essential for the successful growth and reproduction of an organism is called an **essential element**. The major essential elements for life are carbon, hydrogen, oxygen, nitrogen, sulphur, phosphorus, potassium, sodium, magnesium, calcium and chlorine. In addition, certain elements, the **trace elements**, are essential in trace amounts (a few parts per million). Of these, all organisms require manganese, iron, cobalt, copper and zinc; some also require combinations of molybdenum, vanadium, chromium and other heavy metals, as well as boron, silicon, fluorine and iodine (see table 3.1). All except carbon, hydrogen and oxygen are taken up as minerals from soil or water by green plants. The mechanism of uptake is discussed in chapter 13.

Table 7.6 Comparison of C_3 and C_4 plants.

	C_3 plants	*C_4 plants*	
Representative species	Most crop plants, e.g. cereals, tobacco, beans	Maize, sugarcane	
Light intensity for maximum rate of photosynthesis	10 000–30 000 foot candles	Not saturated at 10^5 lux	
Effect of temperature rise from 25 °C to 35 °C	No change in rate or lower rate	50% greater at 35°C	
Point at which no more CO_2 can be taken up	40–60 ppm CO_2	Around zero ppm CO_2	
Water loss per g dry mass produced	450–950	250–350	
Carbon dioxide fixation	Occurs once	Occurs twice, first in mesophyll cells, then in bundle sheath cells	
		Mesophyll cells	**Bundle sheath cells**
Carbon dioxide acceptor	RuBP, a 5C compound	PEP, a 3C compound	RuBP
Carbon dioxide-fixing enzyme	RuBP carboxylase, which is inefficient	PEP carboxylase which is very efficient	RuBP carboxylase, working efficiently because carbon dioxide concentration is high
First product of photosynthesis	A C_3 acid, GP	A C_4 acid, e.g. oxaloacetate	
Leaf anatomy	Only one type of chloroplast	'Kranz' anatomy, i.e. two types of cell, each with its own type of chloroplast	
Efficiency	Less efficient photosynthesis than C_4 plants. Yields usually much lower.	More efficient photosynthesis than C_3 plants but use more energy. Yields usually much higher.	

For heterotrophic organisms (animals and fungi) the trace elements (inorganic) are sometimes grouped with vitamins (organic) as **micronutrients**, since both are required in trace amounts and have similar fundamental roles in cell metabolism, often as enzyme cofactors. Vitamins are considered in chapter 8. Autotrophic organisms synthesise their own vitamins. The other essential elements are called **macronutrients**. Deficiency of any of the nutrients mentioned can lead to **deficiency diseases**.

Some examples of the functions of the major minerals are given in table 7.7. A study of the table will reveal that mineral elements are taken up by plants as separate ions, either anions (negatively charged) or cations (positively charged). This is also true of trace elements, though their ions are not shown in the table.

Animals do not obtain all their essential elements in the form of minerals. Much of their nitrogen, for example, is ingested in the form of proteins.

A balance of trace elements is essential for soil fertility. Extreme cases are known of plants thriving in areas of high metal contamination, such as on spoilage tips from mines or over natural mineral deposits, and such plants can prove toxic to grazing animals. On the other hand, they can be useful to humans if they help to cover former unsightly areas.

7.10.1 Mineral element deficiencies

It is not always easy, or possible, to isolate the effects of individual minerals. In plants, for example, chlorosis (lack of chlorophyll) can be caused by lack of magnesium or iron, both having different roles in chlorophyll synthesis (tables 7.7 and 7.8). A common deficiency disease of sheep and cattle called scour, which causes diarrhoea, is due to copper deficiency induced by high levels of molybdenum in the pastures. Different effects may occur in different organisms; lack of manganese, for example, causes grey speck in oats, marsh spot in beans and poor flavour in oats.

The close interaction and varied effects of mineral elements are due to their fundamental effects on cell metabolism. However, it is possible by various means, such as experimentally manipulating mineral uptake, to show that specific sets of symptoms are associated with deficiencies of certain elements. Such knowledge is of importance in both medicine and agriculture because deficiency diseases are common worldwide, both in humans and in their crops and animals.

Table 7.7 Some mineral elements that are essential macronutrients and examples of their uses in living organisms.

MACRONUTRIENTS *Element and symbol*	*Taken up by plants as*	*General importance*	*Common deficiency diseases or symptoms* *Plants*	*Humans*	*Common food source for humans*
Nitrogen, N	Nitrate, NO_3^- Ammonium, NH_4^+	Synthesis of proteins, nucleic acids and many other organic compounds, e.g. coenzymes and chlorophyll.	Stunted growth and strong chlorosis, particularly of older leaves	Kwashiorkor due to lack of protein	Protein, e.g. lean meat, fish and milk.
Phosphorus, P	Phosphate, PO_4^{3-} Orthophosphate, $H_2PO_4^-$	Synthesis of nucleic acids, ATP and some proteins. Also, phosphate is a constituent of bone and enamel. Phospholipids in membranes.	Stunted growth, particularly of roots		Milk is rich in phosphorus
Potassium, K	K^+	Mainly associated with membrane function, e.g. conduction of nervous impulses, maintaining electrical potentials across membranes, Na^+/K^+ pump in active transport across membranes, anion/cation and osmotic balance. Cofactor in photosynthesis and respiration (glycolysis). Common in cell sap of plant vacuoles.	Yellow and brown leaf margins and premature death	Rarely deficient	Vegetables, e.g. brussels sprouts (= buds), and meat
Sulphur, S	Sulphate, SO_4^{2-}	Synthesis of proteins (e.g. keratin) and many other organic compounds, e.g. coenzyme A.	Chlorosis, e.g. 'tea-yellow' of tea		Protein, e.g. lean meat, fish and milk
Sodium, Na	Na^+	Similar to potassium, but usually present in lower concentrations. Often exchanged for potassium.		Muscular cramps	Table salt (sodium chloride) and bacon
Chlorine, Cl	Chloride, Cl^-	Similar to Na^+ and K^+, e.g. anion/cation and osmotic balance. Involved in 'chloride shift' during carbon dioxide transport in blood. Constituent of hydrochloric acid in gastric juice.		Muscular cramps	Table salt and bacon
Magnesium, Mg	Mg^{2+}	Part of structure of chlorophyll. Bone and tooth structure. Cofactor for many enzymes, e.g. ATPase.	Chlorosis		Vegetables and most other foods
Calcium, Ca	Ca^{2+}	Formation of middle lamella (calcium pectate) between plant cell walls and normal cell wall development. Constituent of bone, enamel and shells. Activates ATPase during muscular contraction. Blood clotting.	Stunted growth	Poor skeletal growth, possibly leading to rickets	Milk, hard water

Micronutrients

Table 7.8 Some essential trace mineral elements and examples of their uses in living organisms.

TRACE ELEMENTS – all cations except boron, fluorine and iodine			*Common deficiency diseases or symptoms*		
Element and symbol	*Substance containing*	*Examples of functions*	*Plants*	*Humans*	*Common food source for humans*
Manganese, Mn	Phosphatases (transfer PO_4 groups)	Bone development (a 'growth factor')	Leaf-flecking, e.g. 'grey-speck' in oats	Poor bone development	Vegetables and most other foods
	Decarboxylases, Dehydrogenases	Oxidation of fatty acids, respiration, photosynthesis			
Iron, Fe	Haem group in: haemoglobin and myoglobin	Oxygen carriers		Anaemia	Liver and red meat, some vegetables, e.g. spinach
	Cytochromes	Electron carriers, e.g. respiration, photosynthesis			
	Catalase and peroxidases	Break down H_2O_2			
	Intermediate in chlorophyll synthesis	Chlorophyll synthesis	Strong chlorosis, particularly in young leaves		
Cobalt, Co	Vitamin B_{12}	Red blood cell development		Pernicious anaemia	Liver and red meat (as vitamin B_{12}).
Copper, Cu	Cytochrome oxidase	Last electron carrier in respiratory chain – oxygen converted to water	Dieback of shoots		Most foods
	Plastocyanin	Electron carrier in photosynthesis			
	Tyrosinase	Melanin production		Albinism	
Zinc, Zn	Alcohol dehydrogenase	Anaerobic respiration in plants (alcohol fermentation)	'Mottle leaf' of *Citrus*		
	Carbonic anhydrase	Carbon dioxide transport in vertebrate blood	Malformed leaves, e.g. 'sickle leaf' of cocoa		Most foods
	Carboxypeptidase	Hydrolysis of peptide bonds in protein digestion			
Molybdenum, Mo	Nitrate reductase	Reduction of nitrate to nitrite during amino acid synthesis in plants	Slight retardation of growth; 'scald' disease of beans		Most foods
	Nitrogenase	Nitrogen fixation (prokaryotes)			
Boron, B	—	Plants only. Normal cell division in meristems. Mobilisation of nutrients?	Abnormal growth and death of shoot tips, 'heart-rot' of beet; 'stem-crack' of celery	Not needed	
Fluorine, F	Associated with calcium as calcium fluoride in animals	Component of tooth enamel and bone		Dental decay more rapid	Milk, drinking water in some areas
Iodine, I	Thyroxine (Probably not required by plants)	Hormone controlling basal metabolic rate		Goitre; cretinism in children	Seafoods, salt

Experiments on plants were done in the late nineteenth and early twentieth century, particularly by German botanists, using the now classic water culture or sand culture techniques. In these experiments, plants are grown in prepared culture solutions of known composition. Many economically important plant deficiency diseases are now catalogued with the aid of colour photography, enabling rapid diagnosis.

7.10.2 Special methods for obtaining essential elements

Insectivorous plants

Insectivorous or carnivorous plants are green plants which are specially adapted for trapping and digesting small animals, particularly insects. In this way they supplement their normal autotrophic nutrition (photosynthesis) with a form of heterotrophic nutrition. Such plants typically live in nitrogen-poor habitats, and use the animals principally as a source of nitrogen. Having lured the insect with colour, scent or sweet secretions, the plant traps it in some way and then secretes enzymes and carries out extracellular digestion. The products, notably amino acids are absorbed and assimilated.

Some of the plants are interesting for the elaborate nature of their trap mechanisms, notably the Venus fly trap (*Dionaea muscipula*), pitcher plants (*Nepenthes*) and sundews (*Drosera*). *Drosera* is one of the few British examples, most being tropical or subtropical. It is found on the wetter heaths and moors which are typically acid, mineral-deficient habitats. The details of the various trap mechanisms are outside the scope of this book.

Mycorrhizas

A mycorrhiza is a mutualistic (symbiotic) association between a fungus and a plant root. It is likely that the great majority of land plants enter into this kind of relationship with soil fungi. They are of great significance because they are probably the major route of entry of mineral nutrients into roots. The fungus receives organic nutrients, mainly carbohydrates and vitamins, from the plant and in return absorbs mineral salts (particularly phosphate, ammonium, potassium and nitrate) and water, which can pass to the plant root. Generally only young roots are infected. Root hair production either ceases or is greatly reduced on infection. A network of hyphae spreads through the surrounding soil, covering a much larger surface area than the root could, even with root hairs. It has been suggested that plants of the same species, or even different species may often be interconnected with mycorrhizas, a concept which could radically alter our view of natural ecosystems.

Two groups of mycorrhizas occur, the ectotrophic and endotrophic mycorrhizas. **Ectotrophic mycorrhizas** form a sheath around the root and penetrate the air spaces between the cells in the cortex, but do not enter cells. An extensive intercellular net is formed. They are found mainly in forest trees such as conifers, beech, oak and many others, and involve fungi of the mushroom group. Their 'fruiting bodies' (mushrooms) are commonly seen near the trees.

Endotrophic mycorrhizas occur in virtually all other plants. Like ectotrophic mycorrhizas, they also form an intercellular network and extend into the soil, but they appear to penetrate cells (although in fact they do not break through the cell surface membranes of the root cells).

As we learn more about mycorrhizas, it is likely that the knowledge will be applied with advantage to agriculture, forestry and land reclamation.

Root nodules

Nitrogen fixation in root nodules of leguminous plants is discussed in chapter 13. The bacteria which inhabit the nodules stimulate growth and division of the root parenchyma cells resulting in the swelling or nodule.

7.11 Experimental work

Experiment 7.1: Investigating the Hill reaction

The Hill reaction

In 1939 Robert Hill, working in Cambridge, discovered that isolated chloroplasts were capable of releasing oxygen in the presence of an oxidising agent (electron acceptor). This has since been called the Hill reaction. A number of chemicals substitute for the naturally occurring electron acceptor NADP, one of which is the blue dye DCPIP (2,6 dichlorophenolindophenol) that turns colourless when reduced:

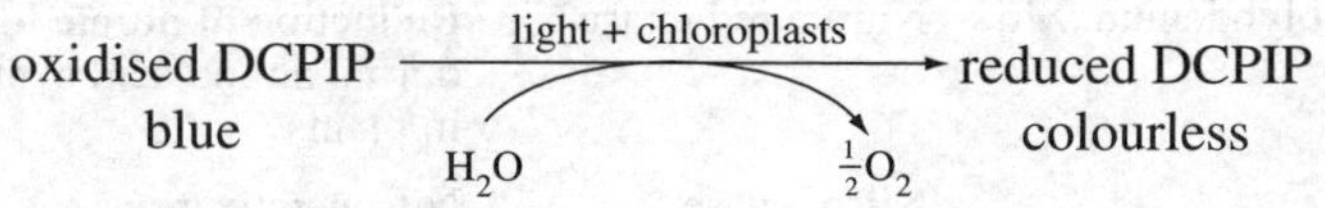

Isolation of chloroplasts

Materials

spinach, lettuce or cabbage leaves
scissors
cold pestle and mortar (or blender or food mixer)
muslin or nylon
filter funnel
centrifuge and centrifuge tubes
ice–water–salt bath
glass rod

Solutions (see notes)

0.05 M; phosphate buffer solution, pH 7.0

isolation medium
DCPIP solution (reaction medium)

Method

Chloroplasts can be isolated by grinding spinach, lettuce or cabbage leaves in a cold medium of suitable osmotic and ionic strength and pH, such as 0.4 M sucrose, 0.01 M KCl and 0.05 M phosphate buffer, pH 7.0. Solutions and apparatus must be kept cold during the isolation procedure if biochemical activity is to be preserved. The operation should also be performed as rapidly as possible, so study the method carefully and assemble the apparatus first.

Sufficient chloroplasts can be isolated using this method to supply several groups of students, if it is not practicable for all groups to prepare their own.

(1) Cut three small spinach, lettuce or cabbage leaves into small pieces with scissors, avoiding midribs and petioles. Place in a cold mortar or blender containing 20 cm^3 of cold isolation medium (scale up quantities for blender if necessary).
(2) Grind vigorously and rapidly (or blend for about 10 s).
(3) Place four layers of muslin or nylon in a funnel and wet with cold isolation medium.
(4) Filter the homogenate through the funnel and collect in pre-cooled centrifuge tubes supported in an ice–water–salt bath. Gather the edges of the muslin and wring thoroughly into the tubes.
(5) Ensure that each centrifuge tube contains about the same volume of filtrate.
(6) If your bench centrifuge has a fixed speed, spin the filtrate for 2–5 min (a small pellet is required, but the time taken should be minimal).

    If a bench centrifuge with variable speed is available, spin the filtrate at 100–200 times gravity for 1–2 min. Respin the supernatant (the liquid above the sediment) at 1000–2000 times gravity for up to 5 min (sufficient time to get a small chloroplast pellet).
(7) Pour away the supernatant. Resuspend the pellet of one centrifuge tube in about 2 cm^3 of isolation medium using a glass rod. Transfer the suspension from this tube to the second centrifuge tube and resuspend the pellet in that tube. (Alternatively, if more than one student group is to be supplied, use 2 cm^3 in each tube and use one tube per group.)
(8) Store this chloroplast suspension in an ice–water–salt bath and use as soon as possible.

The Hill reaction

The chloroplast suspension can now be used to study the Hill reaction. The DCPIP solution should be used at room temperature.

Prepare the following tubes:

(1) 0.5 cm^3 chloroplast suspension + 5 cm^3 DCPIP solution. Leave in a bright light.
(2) 0.5 cm^3 isolation medium + 5 cm^3 DCPIP solution. Leave in a bright light.
(3) 0.5 cm^3 chloroplast suspension + 5 cm^3 DCPIP solution. Place immediately in darkness.
(4) It is useful to add 0.5 cm^3 chloroplast suspension to 5 cm^3 distilled water as a colour standard, showing what the final colour will be if the DCPIP is reduced.

Record your observations after 15–20 min.

If a colorimeter is available, the progress of the reaction can be followed by measuring the decrease in absorbance of the dye as it changes from the blue oxidised to the colourless reduced state. Prepare the mixtures given above for tubes (2) to (4) in colorimeter sample tubes. Insert a red (or yellow) filter and set the colorimeter at zero absorbance using tube (4) as a blank. Then set up tube (1) and immediately take a reading from this tube and return it to the light. Take further readings at 30 s intervals. Plot the rate of the reaction graphically. Once reduction is complete, take a reading from tube (3). Tube (2) can be checked for reduction of dye by first setting the colorimeter at zero with a blank of isolation medium. Ideally the time for complete reduction is about 10 min.

Notes

Prepare the solutions as follows.

0.05 M phosphate buffer solution, pH 7.0

$Na_2HPO_4.12H_2O$	4.48 g (0.025 M)
KH_2PO_4	1.70 g (0.025 M)

Make up to 500 cm^3 with distilled water and store in a refrigerator at 0–4 °C.

Isolation medium

sucrose	34.23 g (0.4 M)
KCl	0.19 g (0.01 M)

Dissolve in phosphate buffer solution at room temperature and make up to 250 cm^3 with the buffer solution. Store in a refrigerator at 0–4 °C.

DCPIP solution (reaction medium)

DCPIP	0.007–0.01 g (10^{-4} M approx.)
KCl	0.93 g (0.05 M)

Dissolve in phosphate buffer solution at room temperature and make up to 250 cm^3. Store in a refrigerator at 0–4 °C. Use at room temperature.

(NB Potassium chloride is a cofactor for the Hill reaction.)

7.17 What change, if any, did you observe in tube (1)?

7.18 What was the purpose of tubes (2) and (3)?

7.19 What other organelles apart from chloroplasts might you expect in the chloroplast suspension?

7.20 What evidence have you that these were not involved in the reduction of the dye?

7.21 Why was the isolation medium kept cold?

7.22 Why was the isolation medium buffered?

7.23 What was (*a*) the electron donor, and (*b*) the electron acceptor, in the Hill reaction?

7.24 During the Hill reaction, DCPIP acts between X and PSI as shown in (fig 7.14*a*) and oxygen is evolved. Does the Hill reaction involve cyclic or non-cyclic photophosphorylation, or both? Give your reasons.

7.25 Fig 7.23 shows the appearance of the chloroplasts after being used in the experiment. The photograph demonstrates the consequences of transferring chloroplasts from the hypertonic isolation medium containing sucrose to the hypotonic reaction medium.

(*a*) How do the chloroplasts in fig 7.23 differ in appearance from normal chloroplasts?

(*b*) Can you explain why transferring the chloroplasts to a medium lacking sucrose should bring about this change?

(*c*) Why was this change desirable before carrying out the Hill reaction?

7.26 What significance do you think the discovery of the Hill reaction might have had on the understanding of the photosynthetic process?

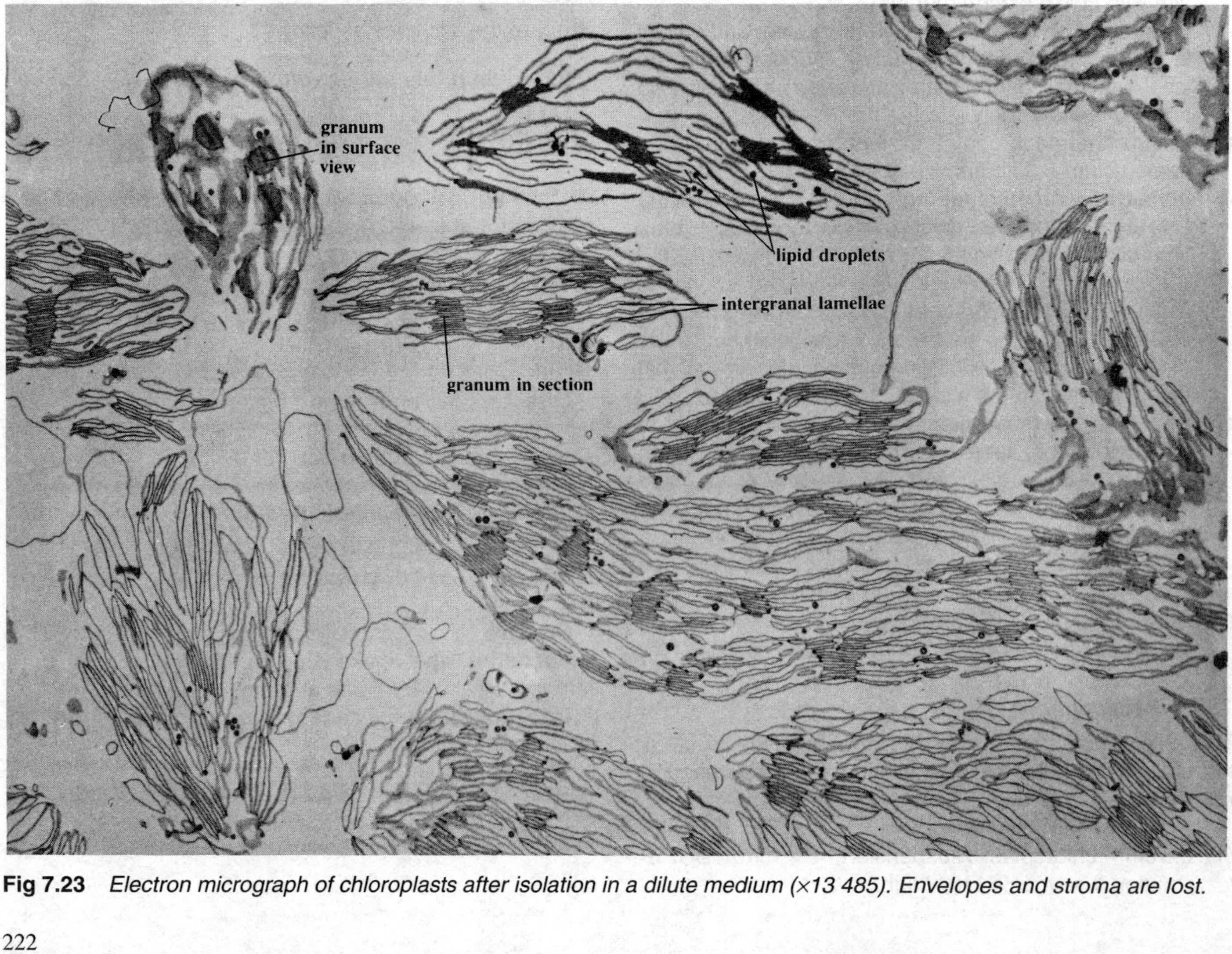

Fig 7.23 *Electron micrograph of chloroplasts after isolation in a dilute medium (×13 485). Envelopes and stroma are lost.*

Experiments: To investigate conditions required for, and products of, photosynthesis

As an indication that photosynthesis has occurred some production of the process can be identified. The first product is glycerate phosphate which is rapidly converted to a number of compounds, including sugars and then starch. The latter can be tested for very easily and can be taken as an indication that photosynthesis has occurred, providing that the precaution is taken of starting the experiment with a destarched leaf or plant.

Destarching a plant

A plant can be destarched by leaving it in the dark for 24–48 h. It is advisable to check that destarching is complete before attempting the following experiment.

> **7.27** Why does this result in destarching?

Experiment 7.2: To test a leaf for starch

Materials

leaf to be tested	hot water bath
test tube	90% ethanol
forceps	iodine/potassium iodide solution
white tile	

Method

Starch can be detected using iodine/potassium iodide solution (I_2/KI) but the leaf must first be decolourised because the green colour of the chlorophyll masks the colour change. This is achieved by placing the leaf in a test-tube of boiling 90% ethanol in a water bath for as long as necessary (naked flames must be avoided because ethanol is highly inflammable).

The decolourised leaf is rinsed in hot water to remove any ethanol and soften the tissues, spread on a white tile, and iodine solution poured on its surface. The red-brown solution stains any starch-containing parts of the leaf blue-black.

Experiment 7.3: To investigate the need for carbon dioxide

Materials

destarched leafy plant such as potted geranium (*Pelargonium*)	cotton wool
	starch test materials
	250 cm^3 conical flask
	clamp and clamp stand
light source such as a bench lamp	lime water
	20% potassium hydroxide solution

Method

Fig 7.24 illustrates a suitable procedure for investigating the need for carbon dioxide. The plant should be left for several hours in the light before testing the relevant leaves for starch.

> **7.28** Describe the conditions to which you would subject the control leaf.

A more satisfactory experiment showing the use of carbon dioxide is one involving the uptake of $^{14}CO_2$ (radioactively labelled carbon dioxide) into sugars and other compounds.

7.11.1 Measuring rates of photosynthesis

> **7.29** From the equation for photosynthesis, what changes in the substances taken up and produced might be used to measure the rate of photosynthesis?

In section 7.8 certain external factors (such as light intensity, carbon dioxide concentration and temperature) were shown to affect the rate of photosynthesis. When a particular factor is being investigated, it is essential that other factors are kept constant and, if possible, at optimum levels so that no other factor is limiting.

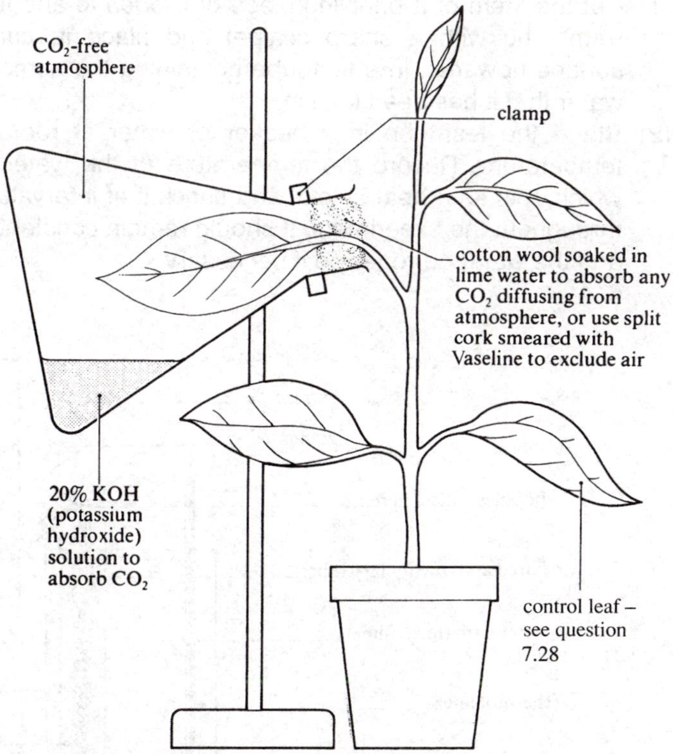

Fig 7.24 *Investigating the need for carbon dioxide in photosynthesis.*

The rate of oxygen evolution

Measuring the rate of oxygen evolution from a water plant is the simplest way to measure the rate of photosynthesis.

Experiment 7.4: To investigate the effect of light intensity on the rate of photosynthesis

Materials

apparatus for collecting gas as shown in fig 7.25	metre rule
test-tube	stopclock
400 cm^3 beaker	light source such as bench lamp
thermometer	Canadian pondweed (*Elodea*), previously well illuminated for several hours
mercury vapour lamp or projector lamp	detergent (washing-up liquid)
sodium hydrogen-carbonate	

Method

It is advisable to use *Elodea* that has been well illuminated and is known to be photosynthesising actively. The addition of 2–10 g of sodium hydrogencarbonate to each dm^3 of pond water may stimulate photosynthesis if there are no obvious signs of bubbles being produced (this increases carbon dioxide availability). The water could also be aerated for an hour before the experiment.

(1) Cut the stem of a bubbling piece of *Elodea* to about 5 cm long with a sharp scalpel and place it, cut surface upwards, in a test-tube containing the same water that it has been kept in.

(2) Stand the test-tube in a beaker of water at room temperature. Record the temperature of the water, which acts as a heat shield, and check it at intervals throughout the experiment. It should remain constant and the water be renewed if necessary.

(3) Fill the apparatus with tap water, ensuring that no air bubbles are trapped in it and push the plunger well in to the end of the syringe (fig 7.25).

(4) Darken the laboratory. Place a bright light source 5 cm from the plant.

(5) Allow the plant to adjust to the light intensity (equilibrate) for 2–3 min. Ensure that the rate of bubbling is adequate (such as more than 10 bubbles per minute). A trace of detergent is sometimes sufficient to lower the surface tension to allow freer escape of bubbles.

(6) Position the *Elodea* so that its bubbles are collected in the capillary tube of the apparatus. Start timing.

(7) Collect a suitable volume of gas in a known period (for example 5–10 min). Measure the length of the bubble by drawing it slowly along the capillary tube by means of a syringe. The bubble can thus be positioned along the scale.

(8) Draw the bubble into the plastic tube connector where it will not interfere with subsequent measurements and repeat the procedure at increasing distances between the light source and *Elodea*, such as 10, 15, 20, 30, 40 and 80 cm. In each case allow time for the plant to equilibrate. The following three measurements are required under each condition: (*a*) the distance between plant and light source, (*b*) the time taken to collect the gas, and (*c*) the length of the gas bubbles collected (this measurement is directly proportional to volume and is used as a measurement of volume).

Results

The intensity of light falling on a given object is inversely proportional to the square of the distance from the source. In other words, doubling the distance between the weed and the lamp does not halve the light intensity received by the weed, but quarters it.

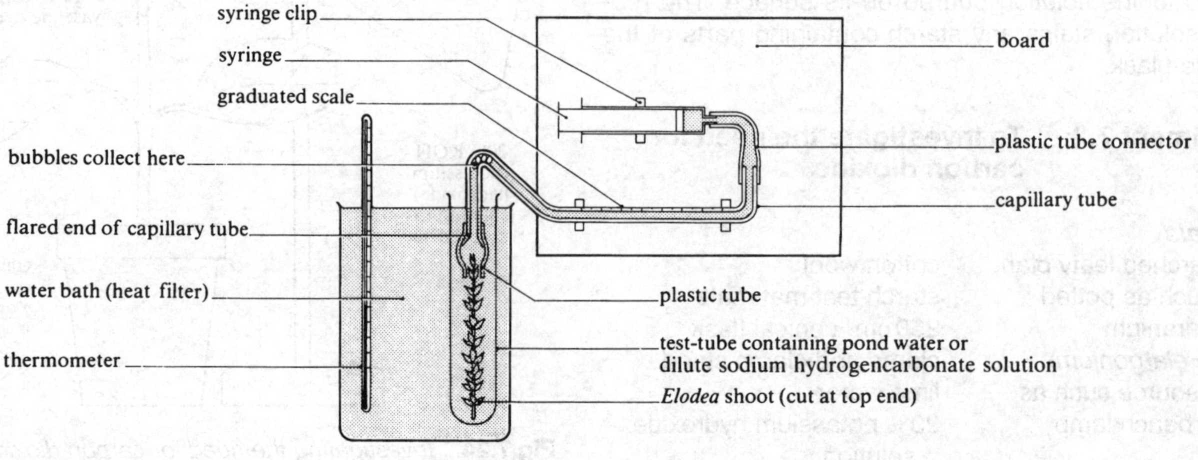

Fig 7.25 *Apparatus for measuring the rate of oxygen evolution by a water plant during photosynthesis.*

$LI \propto 1/d^2$

where LI is the light intensity and d is the distance between object and light source. Plot a graph with rate of photosynthesis on the vertical axis (as length of gas bubble per unit time) and LI on the horizontal axis (as $1/d^2$ or, more conveniently, $1000/d^2$).

7.30 (*a*) State the relationship between gas production and light intensity demonstrated by your results.

(*b*) Why was the laboratory darkened and the temperature kept constant?

7.31 What are the main sources of inaccuracy in this experiment?

7.32 If the gas is collected and analysed it is found *not* to be pure oxygen. Can you account for this?

7.33 Why is it advisable to aerate the water before beginning the experiment?

If a simpler, quicker, though slightly less accurate method is required, the rate of oxygen evolution can be determined by counting the number of bubbles evolved from the cut end of a stem of *Elodea* in a given time period. This can be just as satisfactory, but errors may occur through variations in bubble size. This problem is less likely to arise if a trace of detergent is added to lower the surface tension (see (5) above). The *Elodea* can be anchored to the bottom of the tube with plasticine if necessary.

7.12 Compensation points

Photosynthesis results in uptake of carbon dioxide and evolution of oxygen. At the same time respiration uses oxygen and produces carbon dioxide. If light intensity is gradually increased from zero, the rate of photosynthesis gradually increases accordingly (fig 7.20). There will come a point, therefore, when photosynthesis and respiration exactly balance each other, with no net exchange of oxygen and carbon dioxide. This is called the **compensation point**, or more precisely the **light compensation point**, that is the light intensity at which net gaseous exchange is zero.

Since carbon dioxide concentration affects the rate of photosynthesis there also exists a **carbon dioxide compensation point**. This is the carbon dioxide concentration at which net gaseous exchange is zero for a given light intensity. The higher the carbon dioxide concentration, up to about 0.1% (1000 ppm, parts per million), the faster the rate of photosynthesis. For most temperate plants the carbon dioxide compensation point, beyond which photosynthesis exceeds respiration, is 50–100 ppm, assuming light is not a limiting factor. Atmospheric carbon dioxide concentrations are normally in the range 300–400 ppm, and therefore under normal circumstances of light and atmospheric conditions this point is always exceeded.

Experiment 7.5: To investigate gaseous exchange in leaves

Materials

- four test-tubes thoroughly cleaned and fitted with rubber bungs
- forceps
- test-tube rack
- 2 cm³ syringe
- aluminium foil
- unbleached cotton wool
- no. 12 cork borer
- water bath with test-tube clamps
- bench lamp
- freshly picked leaves
- hydrogencarbonate indicator

The hydrogencarbonate indicator (bicarbonate indicator) solution should be freshly equilibrated with the atmosphere by bubbling fresh air through it until cherry red. Hydrogencarbonate indicator is supplied as a concentrated solution and must be diluted by a factor of ten for experimental use. To equilibrate with atmospheric carbon dioxide, air from *outside* the laboratory should be pumped through the solution. A suitable method is to place the solution in a clear glass wash-bottle to which a tube is attached whose free end is hung from a window. A filter pump is then used to bubble air through the solution until there is no further colour change. The colour of the indicator at this stage is a deep red but will appear orange-red in the test-tubes. Time must be allowed for this procedure before the start of the experiment (100 cm³ of indicator will need to be aerated for at least 20 min).

Method

(1) Label four test-tubes A, B, C and D.
(2) Rinse the four tubes and 2 cm³ syringe with a little of the indicator solution.
(3) Add 2 cm³ of the indicator solution to each tube by means of the syringe. Avoid putting fingers over the ends of the tubes since the acid in sweat will affect the indicator. Also avoid breathing over the open ends of the tubes.
(4) Cover the outside of the tubes A and C with aluminium foil.
(5) Set up the tubes as shown in fig 7.26, using two leaf discs per tube cut from a fresh leaf with a number 12 cork borer.
(6) Arrange the tubes in such a way that they are equally illuminated by a bench lamp.
(7) Place a heat filter in the form of a glass tank of water between the tubes and the light source to prevent a rise in temperature during the experiment.

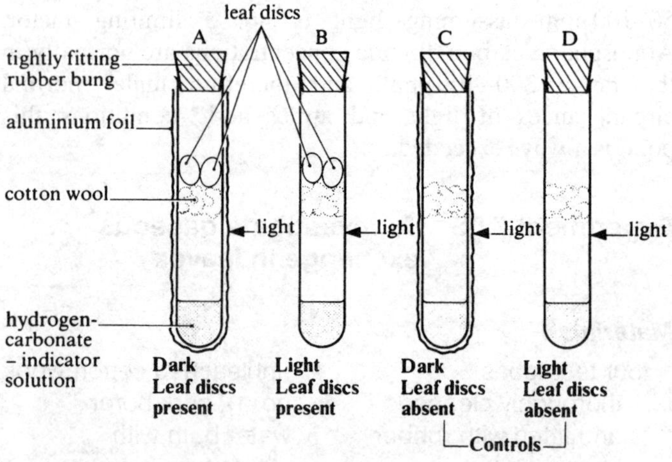

Fig 7.26 *Experiment to investigate gas exchange in leaf discs.*

Alternatively, the tubes can be clamped in a water bath.

(8) Note the colour of the indicator in each tube.
(9) At intervals shake the tubes gently and leave for at least 2 h, preferably overnight. Record the final colour of the indicator in each tube as seen against a white background.

Results

Results can be interpreted using the following guide to colour changes.

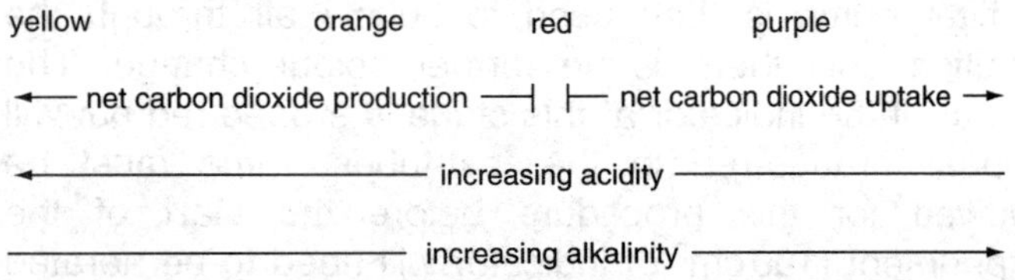

If conditions become more acidic, this can be assumed to be the result of carbon dioxide being produced and dissolving in the indicator solution. If conditions become less acidic, this indicates a lowering of carbon dioxide concentration.

7.34 What can you conclude from your results and why were the controls necessary?

7.35 What is the name given to the equilibrium point where there is no further net uptake or production of carbon dioxide by the leaf discs in tube B?

Modifications of this experiment

(1) **Comparing rates of photosynthesis**. By using leaf discs as described, rather than whole leaves, comparative studies may be carried out using different light intensities, or, for example, old and young leaves on the same plant, yellow and green areas of variegated leaves, leaves of different species (such as a C_3 and a C_4 plant – see C_4 photosynthesis). To compare the rates of photosynthesis, colours of the indicator solutions can be compared during or at the end of the experiment as appropriate. If light intensity is investigated, a mercury vapour lamp should be used. An interesting comparison can be made between shade-loving plants, such as enchanter's nightshade (*Circaea lutetiana*), and other species to determine whether the former are capable of photosynthesising at lower light intensities (that is they have lower light compensation points).

(2) **Using water plants instead of leaf discs**. Water plants such as *Elodea* may be used, providing they are washed well in distilled water to remove traces of dirt and pond water in order to minimise any contribution from microorganisms. The plants should be placed directly in sufficient indicator solution to cover them. The solution has little effect on the plants during the course of the experiment.

Chapter Eight

Heterotrophic nutrition

As noted at the beginning of chapter 7, nutrition is the process of acquiring energy and materials for cell metabolism, including the maintenance and repair of cells, and growth. **Heterotrophic organisms**, or heterotrophs, are organisms that feed on an **organic source of carbon** (fig 8.1). It would be useful, if you have not already done so, to read sections 7.1 and 7.2 at this stage.

The survival of heterotrophs is dependent either directly or indirectly on the activities of autotrophs. All animals and fungi and the majority of bacteria are heterotrophic (table 2.3). The great majority obtain their energy from their food and these will be studied in this chapter. A few bacteria, however, are able to use light energy to synthesise their organic requirements from other organic raw materials. These are called **photoheterotrophs** (table 2.3).

Fig 8.1 (a) *Zebra eating grass. The grass contains energy obtainod from sunlight, and carbon obtained from carbon dioxide, during the process of photosynthesis (autotrophic nutrition). The zebra obtains its energy and carbon from the grass (heterotrophic nutrition).*
(b) *Lion eating zebra. The lion is a carnivore, the zebra a herbivore.*

The way in which heterotrophs obtain their food varies considerably. However, the way in which it is processed into a usable form within the body is very similar in most of them. It involves the following processes:

- **digestion** – reducing large complex food molecules into simpler soluble ones;
- **absorption** – taking the soluble molecules from the region of digestion into the tissues of the organism;
- **assimilation** – using the absorbed nutrients for a particular purpose.

For convenience, the main forms of heterotrophic nutrition may be classified as **holozoic**, **saprotrophic** (or saprophytic), **mutualistic** and **parasitic**, although some overlap between groups may occur. These will be studied in section 8.1.

8.1 Forms of heterotrophic nutrition

8.1.1 Holozoic nutrition

The term **holozoic** is applied mainly to free-living animals which have a specialised digestive tract, the **alimentary canal**. Most animals are holozoic.

The characteristic processes involved in holozoic nutrition are as follows.

- **Ingestion** is the taking in of food.
- **Digestion** is the breakdown of large organic molecules into smaller, simpler soluble molecules. Often two types of digestion occur. **Mechanical digestion** involves mechanical breakdown of the food, for example by teeth. **Chemical digestion** involves the activity of enzymes. The type of chemical process these enzymes catalyse during digestion is **hydrolysis**. Digestion may be either **extracellular** (outside the cell) or **intracellular** (inside the cell).
- **Absorption** is the uptake of the soluble molecules from the digestive region, across a membrane and into the body tissue proper. The food may pass directly into cells or first pass into the bloodstream to be transported to other regions of the body.

- **Assimilation** is using the absorbed molecules to provide either energy or materials to be incorporated into the body.
- **Egestion** is the elimination from the body of undigested waste food materials.

Animals which feed on plants are called **herbivores**, those that feed on other animals are **carnivores**, and those that eat a mixed diet of animal and vegetable matter are termed **omnivores**. Some animals take in food in the form of relatively small particles (microphagous feeders), for example earthworms and filter feeders like mussels. Some ingest food in liquid form (fluid feeders) such as aphids, butterflies and mosquitoes. Some take in food in the form of relatively large pieces (macrophagous feeders), for example *Hydra* and sea anemones, which use tentacles to catch their prey, and large carnivores such as sharks.

8.1.2 Saprotrophic nutrition (*sapros*, rotten; *trophos*, feeder)

Organisms which feed on dead or decaying organic matter are called **saprotrophs**. Other terms sometimes used which mean the same thing are saprophytes (saprophytic nutrition) and saprobionts (saprobiotic nutrition). Many fungi and bacteria are saprotrophs, for example the fungi *Mucor*, *Rhizopus* and yeast. Saprotrophs secrete enzymes onto their food where it is digested. The soluble end-products of this extracellular digestion are then absorbed and assimilated by the saprotroph. Saprotrophs feed on the dead organic remains of plants and animals and contribute to the removal of such remains by decomposing it. Many of the simple substances formed are not used by the saprotrophs themselves but are absorbed by plants. In this way the activity of saprotrophs provides important links in nutrient cycles by making possible the return of vital chemical elements from the dead bodies of organisms to living ones.

The saprotrophic nutrition of Mucor *and* Rhizopus

Mucor and *Rhizopus* are common fungi known as pin moulds. They are often found growing on bread, although they can also live in soil. The structure of *Mucor* is described in section 2.7.2. *Rhizopus* has a very similar structure and is even more common. Both are easy to culture in a laboratory. Their hyphae penetrate the food on which they grow and secrete hydrolysing enzymes from their tips. This results in extracellular digestion as shown in fig 8.2. Carbohydrase and protease enzymes carry out the extracellular digestion of starch to glucose and protein to amino acids respectively. The thin, much-branched nature of the mycelium of *Mucor* and *Rhizopus* ensures that there is a large surface area for absorption. Glucose is used during respiration to provide energy for the organism's metabolic activities whilst glucose and amino acids are used for growth and repair. Excess glucose is converted to glycogen and fat, and excess amino acids to protein granules for storage in the cytoplasm.

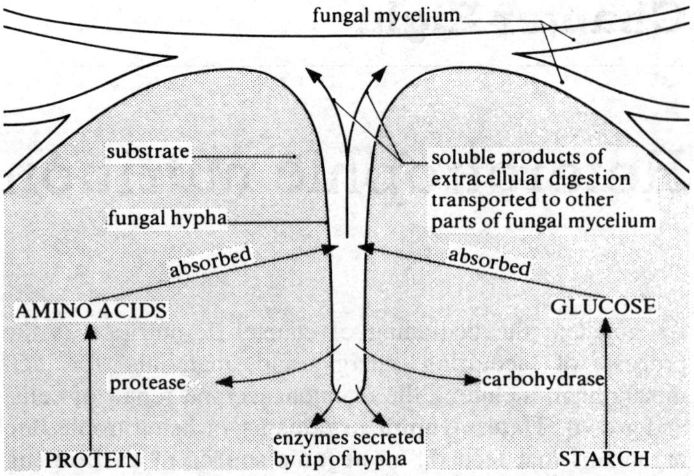

Fig 8.2 *Extracellular digestion and absorption in* Mucor *and* Rhizopus.

> **8.1** Briefly describe the ways in which *Mucor* and *Rhizopus* are important to humans.

8.1.3 Symbiosis: mutualism, parasitism and commensalism

The term **symbiosis** means literally 'living together' (*syn*, with; *bios*, life). It was introduced by the German scientist de Bary in 1879 who described it as 'the living together of dissimilarly named organisms'; in other words, it is an association between two or more organisms of different species. Since de Bary's time the term symbiosis has been restricted by many biologists to meaning a close relationship between two or more organisms of different species **in which all partners benefit**.

Since the 1970s symbiosis has become more important as a topic in biology. For example, we now know that the great majority of plants obtain their minerals with the assistance of fungi, and that much nitrogen fixation is carried out by symbiotic bacteria; knowledge of rumen fermentation, involving symbiotic organisms, is of importance in increasing cattle productivity. At the same time biologists have become increasingly aware that the exact degree of closeness, benefit or harm in a relationship is very variable. Most modern biologists therefore prefer to use something like de Bary's original definition of symbiosis, a move approved by the Society for Experimental Biology in 1975*.

The following definitions will therefore be used in this book. Emphasis is placed on whether the relationship is beneficial or not to both partners.

Symbiosis – the living together in close association of two or more organisms of different species. Many associations involve three or more partners. Nutrition is commonly involved. Three common types of symbiotic relationship are:

- **mutualism** – in which both partners benefit;
- **parasitism** – in which one partner benefits and causes harm to the other;
- **commensalism** – in which one partner benefits but the other receives no harm or benefit.

Mutualism

Mutualism is a close association between two living organisms of different species which is beneficial to both partners. For example, the sea anemone *Calliactis* attaches itself to a shell used by a hermit crab (fig 8.3). The anemone obtains nourishment from the scraps of food left by the crab, and is transported from place to place when the crab moves. The crab is camouflaged by the anemone and may also be protected by the stinging cells in its tentacles. It seems that the anemone is unable to survive unless attached to the crab's shell, and, if the anemone is removed, the crab will seek another anemone and actually place it on its shell.

Herbivorous ruminants contain a vast number of cellulose-digesting bacteria and ciliates (see section 8.6.2). These can only survive in the anaerobic conditions found in a ruminant's alimentary canal. Here the bacteria and ciliates feed on the cellulose contained in the host's diet, converting it into simple compounds which the ruminant is then able to further digest, absorb and assimilate itself.

An important example of mutualism is the formation of root nodules by *Rhizobium*, a bacterium. This is described in section 7.10.2. Other examples are mycorrhizas (section 7.10.2) and endosymbiosis (section 8.1.3).

Fig 8.3 *Sea anemones attached to a whelk shell inhabited by a hermit crab.*

* References: SEB symposia XXIX, *Symbiosis*, CUP (1975) D.H. Jennings & D.L. Lee (eds,); G.H. Harper, (1985) 'Teaching symbiosis' *J. Biol. Ed.* 19 (3), 219–23; D.C. Smith & A.E. Douglas (1987) *The Biology of Symbiosis*, Arnold.

Parasitism (para, *beside;* sitos, *food)*

Parasitism is a close association between two living organisms of different species which is beneficial to one (the **parasite**) and harmful to the other (the **host**). The parasite obtains food from the host and, generally, shelter. A successful parasite is able to live with the host without causing it any great harm. The degree of benefit or harm may be difficult to establish.

Parasites which live on the outer surface of a host are termed **ectoparasites** (for example ticks, fleas and leeches). Such organisms do not always live a fully parasitic existence. Those that live within a host are **endoparasites**, such as *Plasmodium* (a protozoan that causes malaria (chapter 15), the tapeworm *Taenia*, and the liver fluke *Fasciola*. If the organism has to live parasitically at all times, it is said to be an **obligate** parasite, for example the fungus-like organism *Phytophthora* which causes potato blight (section 2.8.2). Facultative parasites are fungi that can feed either parasitically or saprotrophically, for example *Candida*, which causes thrush in humans (section 2.7.3) and *Pythium*, a fungus-like organism which causes 'damping-off' of seedlings (section 2.8.2). Sometimes facultative parasites kill their hosts and then live saprotrophically on the dead remains, as does *Pythium*.

Parasites are highly specialised, possessing numerous adaptations, many of which are associated with their host and its mode of life. This is particularly well illustrated by the tapeworm (*Taenia*) which is specialised for life in the gut, and the liver fluke (*Fasciola*) which lives in the bile duct. The life cycle of the liver fluke is described in section 2.10.3 and fig 2.5.1.

Like the liver fluke, *Taenia* belongs to the group of animals known as flatworms. It shows many adaptations to its mode of life compared with a free-living flatworm such as *Planaria* (section 2.10.3). Some of these adaptations are shown in fig 8.4 and some relating to feeding are discussed below.

Unlike free-living flatworms, the tapeworm has no gut or feeding structures of its own since it can absorb pre-digested food through its cuticle. (The large surface area to volume ratio of flatworms means that no special internal transport system, such as a blood system, is necessary as materials can diffuse rapidly to all parts of the body.) Tapeworms need no special sense organs like eyes since they live in a dark, constant environment and do not need to move around to obtain food. No special locomotory organs are present (free-living flatworms have simple eyes and show a gliding motion caused by cilia). The nervous system of tapeworms is therefore relatively poorly developed compared with free-living flatworms. The tapeworm can also withstand the low oxygen levels of the gut and respire anaerobically.

Table 8.1 shows some of the structural, physiological and reproductive modifications used by various parasites. Microorganisms which cause disease may be regarded as parasites (section 8.1.3).

Fig 8.4 *Structure of an adult tapeworm (*Taenia*).*

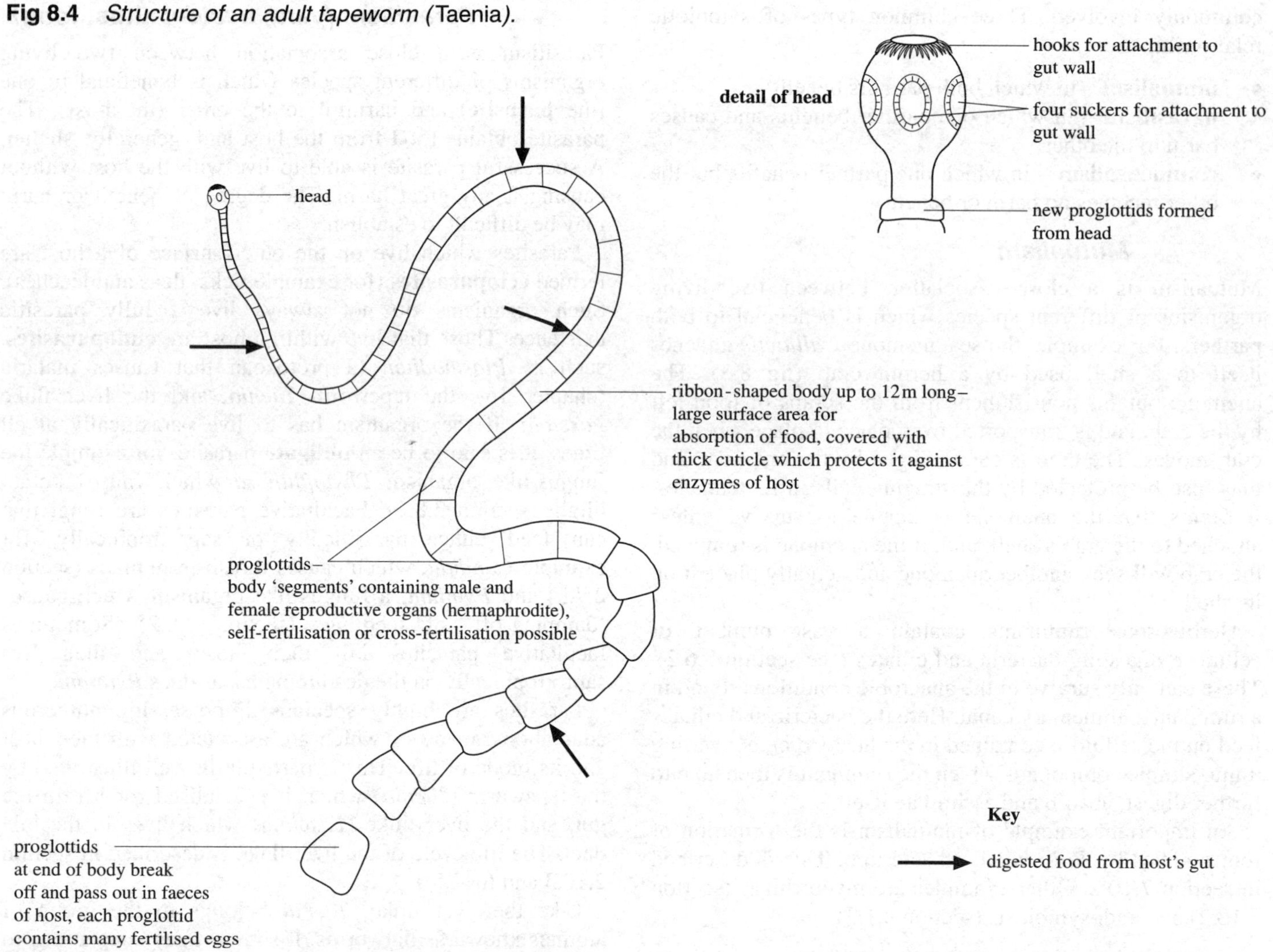

8.2 List the structural, physiological and reproductive features that make *Fasciola* (liver fluke) a successful parasite.

Commensalism (*com-, together; mensa, table*)

Commensalism is a close association between two living organisms of different species which is beneficial to one (the commensal) and does not affect the other (the host). Commensalism means literally 'eating at the same table' and is used to describe symbiotic relationships which do not fit conveniently into the mutualism and parasitism categories. For example, the colonial hydrozoan *Hydractinia* attaches itself to whelk shells inhabited by hermit crabs. It obtains nourishment from the scraps of food left by the crab after it has eaten. In this particular case the crab is totally unaffected by the association. An orchid or lichen (the commensal) growing on a tree (the host) would be another example.

8.2 Feeding mechanisms in a range of animals

8.2.1 Filter feeding

Filter feeders strain small particles of organic matter from water. Many molluscs feed in this way. An example is *Mytilus edulis*, the common mussel, which is found attached to rocks and stones in shallow coastal waters (fig 8.5). It belongs to a group of molluscs known as bivalves. These have a shell with two halves, or valves, which are hinged together. Inside the shell are two large gills, one on each side. The gills are covered with fine beating hairs called **cilia**. The movement of the cilia causes a current of water to enter the animal via one tube (the **inhalant siphon**) and leave via another tube (the **exhalant siphon**) (fig 8.5). The water which enters contains the food of the mussels, such as microscopic protozoa and algae. Numerous secretory cells scattered among the cilia produce streams of sticky mucus which trap the food particles. The trapped food is then swept by special bands of cilia towards the mouth which is located near the front end of the 'gill'. Ciliated structures surround the mouth and sort out the food

Table 8.1 Some structural, physiological and reproductive specialisations of parasites.

	Type of modification	*Examples*
Structural	Absence or degeneration of feeding and locomotory organs – characteristic of gut parasites.	*Fasciola* (liver fluke), *Taenia* (tapeworm)
	Highly specialised mouthparts as in fluid feeders.	*Pulex* (flea), *Aphis* (aphid)
	Development of haustoria in some parasitic green plants.	*Cuscuta* (dodder) (a flowering plant belonging to the family Convolvulaceae which does not possess chlorophyll and parasitises a variety of green plants)
	Boring devices to enter host.	nematode worms
	Attachment organs such as hooks or suckers.	*Taenia, Hirudo* (leech), *Fasciola*
	Outer covering resistant to attack by enzymes.	*Taenia, Fasciola*
	Reduction of sense organs associated with the constancy of the parasite's environment.	*Taenia*
Physiological	Enzyme production to digest host tissue external to parasite.	fungi, *Plasmodium* (a protozoan which infects mammals and birds, and in the case of humans causes malaria)
	Anticoagulant production in blood feeders.	*Pulex, Hirudo*
	Chemosensitivity in order to reach the optimum location in the host's body.	*Plasmodium*
	Production of digestive enzymes to aid penetration into host.	*Cuscuta*
	Ability to respire adequately in anaerobic conditions.	gut parasites
Reproductive	Hermaphrodite condition allowing self-fertilisation, if necessary.	*Taenia, Fasciola*
	Enormous numbers of reproductive bodies, i.e. eggs, cysts and spores.	*Taenia, Fasciola*
	Resistance of reproductive bodies when external to the host.	*Phytophthora* (potato blight)
	Employment of specialised reproductive phases in the life cycle.	*Fasciola*
	Use of secondary hosts as vectors.	*Taenia, Fasciola, Plasmodium*

Fig 8.5 *Filter feeding in the mussel (*Mytilus edulis*).*

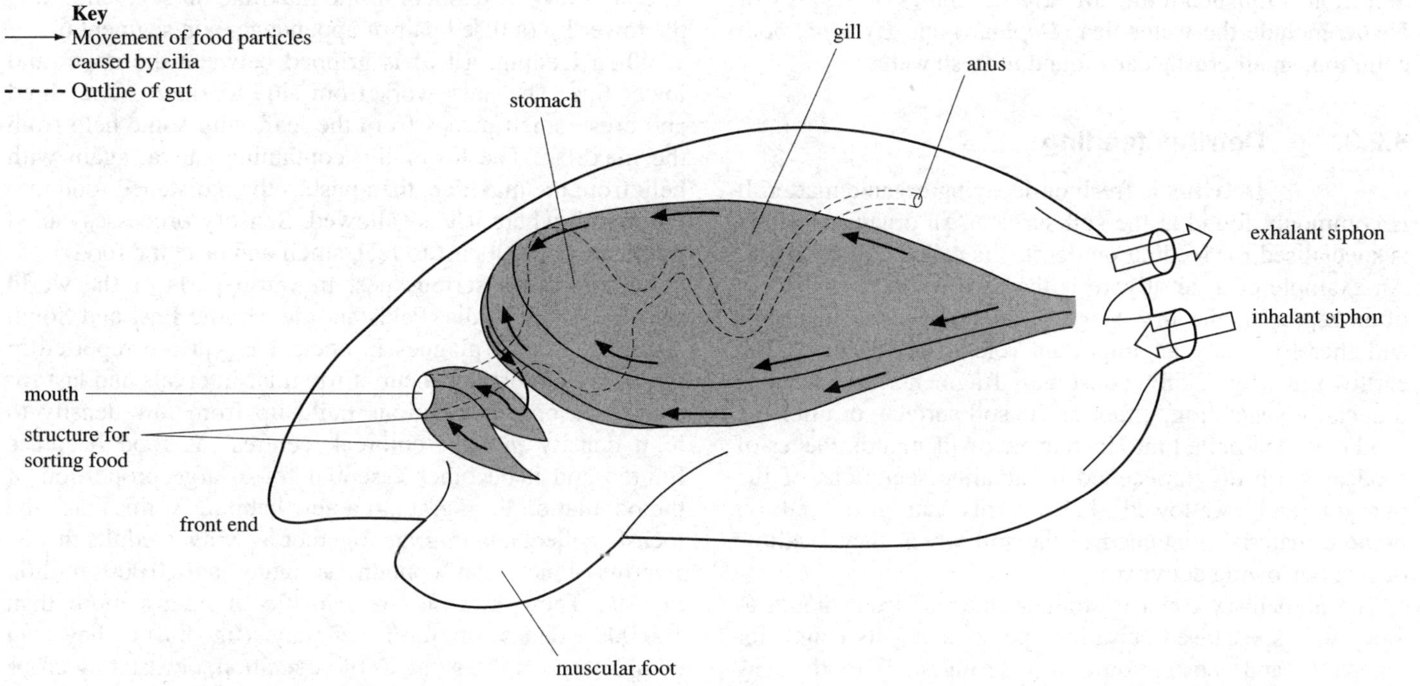

particles to some extent before they enter the mouth. The alimentary canal of the mussel consists of a stomach and short intestine which ends at the anus near the exhalant siphon.

8.2.2 Feeding with tentacles

The phylum Cnidaria contains animals which have a simple structure (section 2.10.2). Cnidarians include jellyfish, sea anemones and the freshwater *Hydra*. They are all carnivores and possess tentacles for capturing food. The tentacles surround the mouth as shown in figs 2.47 and 2.48. There is no true gut, only a simple sac called an **enteron**, with one opening, the mouth. Food is placed in the mouth (**ingestion**) and any undigested remains eventually leave through the mouth (**egestion**).

Along the outside surface of the tentacles are batteries of stinging cells called **nematoblasts**. These cells have projecting hair-like triggers which, when stimulated, release their contents explosively (fig. 8.6). Usually, they have to be stimulated in two ways at the same time, for example by touch and exposure to appropriate chemicals ('scent'). Many types of nematoblasts exist. Some have barbs which penetrate the prey in the initial explosive discharge. Others have minor hooks that cling to the prey, and long hollow threads that release poison (the sting of a jellyfish) which can paralyse and even kill the prey. Sometimes the threads are sticky and loop round the prey, entangling it. The animal then grasps the prey with its tentacles, which pass it to the mouth for ingestion. The mouth opens widely and the prey enters the internal cavity (enteron) for the first, extracellular phase of digestion. Once the food has been reduced to small fragments, these are engulfed by phagocytosis into cells lining the enteron and digestion is finished intracellularly. Examples of the prey of *Hydra* include the water flea (*Daphnia*) and *Cyclops*, both common, small crustaceans found in fresh water.

8.2.3 Detritus feeding

Detritus is fresh or decaying organic matter. It is commonly found at the soil surface. An organism which is specialised for feeding on detritus is called a **detritivore**. An example of a detritivore is the earthworm. Detritivores often represent the first stage in recycling dead materials and therefore play an important role in ecosystems. The earthworm (fig 2.58) consumes fragments of detritus, especially vegetation, either at the soil surface, or after the food has been pulled into its burrow by its mouth. Pieces of food are torn off, moistened by alkaline secretions of the pharynx and swallowed. Earthworms can also feed on organic material contained in the soil which they swallow during burrowing activity.

The alimentary canal is straight and runs from mouth to anus. It is specialised at various points along its length for digestion and absorption of the ingested food. Any undigested material is egested from the anus as 'worm casts'.

Charles Darwin was the first person to point out how valuable earthworms are in maintaining soil structure and fertility. Fertile soil may contain over two million earthworms per hectare. By grinding soil as it passes through their guts, and depositing it at the surface as worm casts, earthworms break it down into fine particles, improving its texture and keeping it turning over (an estimated 50 tonnes per hectare per year). Nitrogenous waste included in the worm casts also adds nutrients which can be used by plants. Excess calcium is also got rid of as calcium salts and these reduce soil acidity, generally improving conditions for plant growth. The burrows of the worms help to aerate the soil, improving drainage and allowing more oxygen to reach plant roots. By pulling down surface detritus into their burrows they increase the overall rate of decomposition and increase productivity of an ecosystem.

8.2.4 Biting and chewing mouthparts

Many insects are herbivorous and have mouthparts for biting and chewing vegetation. Many are pests because they attack crops. The locust is an important example. The mouthparts of the locust are complex (fig 8.7). To understand the arrangement of the mouthparts it helps to know a little about the construction of the insect body as a whole. The body is made up of a series of segments (section 2.10.6). Each segment can bear a pair of jointed appendages. These can be modified in the head region for feeding. The locust uses appendages on segments 4, 5 and 6 of the head. These surround the mouth which is located on the lower surface of the head. Fig 8.7 shows that, in succession, there is an upper lip (**labrum**), a pair of jaws (**mandibles**), a pair of **maxillae** and a lower lip (**labium**). The jaws are on segment 4, the maxillae on segment 5 and the lower lip (a fused pair of appendages) on segment 6.

When feeding, a leaf is gripped between the upper and lower lips. The jaws work from side to side to cut, shred and crush small pieces from the leaf, with some help from the maxillae. The lower lip, containing saliva, again with help from the maxillae, then pushes the moistened food into the mouth where it is swallowed. Sensory processes called **palps** allow the locust to feel, smell and taste the food.

Locusts are a serious pest in warm parts of the world such as Africa, India, Pakistan, the Middle East and South America. Locust plagues in ancient Egypt are reported in the Bible. Outbreaks occur at irregular intervals and last for a few seasons. Populations build up from low density to high density in key 'outbreak centres' as food becomes limited and it becomes essential for a large proportion of the population to migrate to a new habitat. At this stage the locusts collect and migrate together as winged adults in vast swarms which can contain as many as 10 000 million locusts. These swarms are capable of eating more than 100 000 tonnes of food per day (fig 8.8). They can completely strip large areas of vegetation, devastating crops and causing famine. International cooperation means that

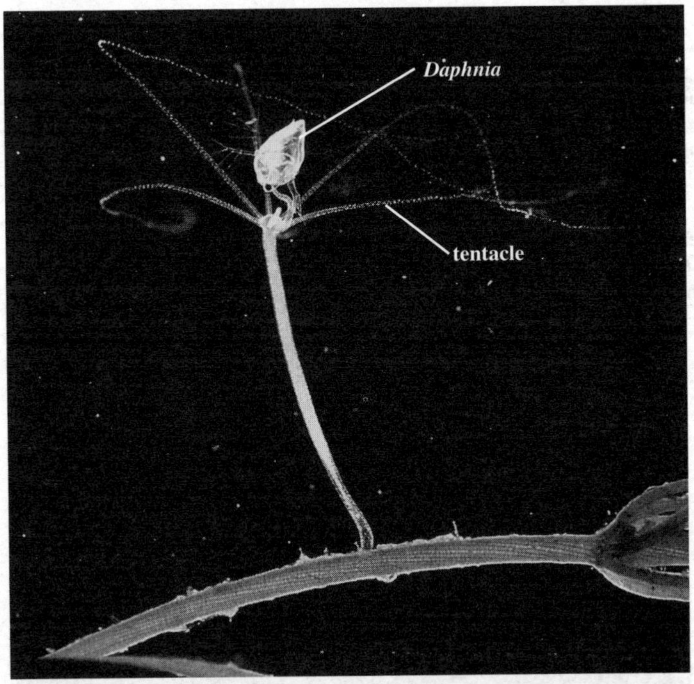

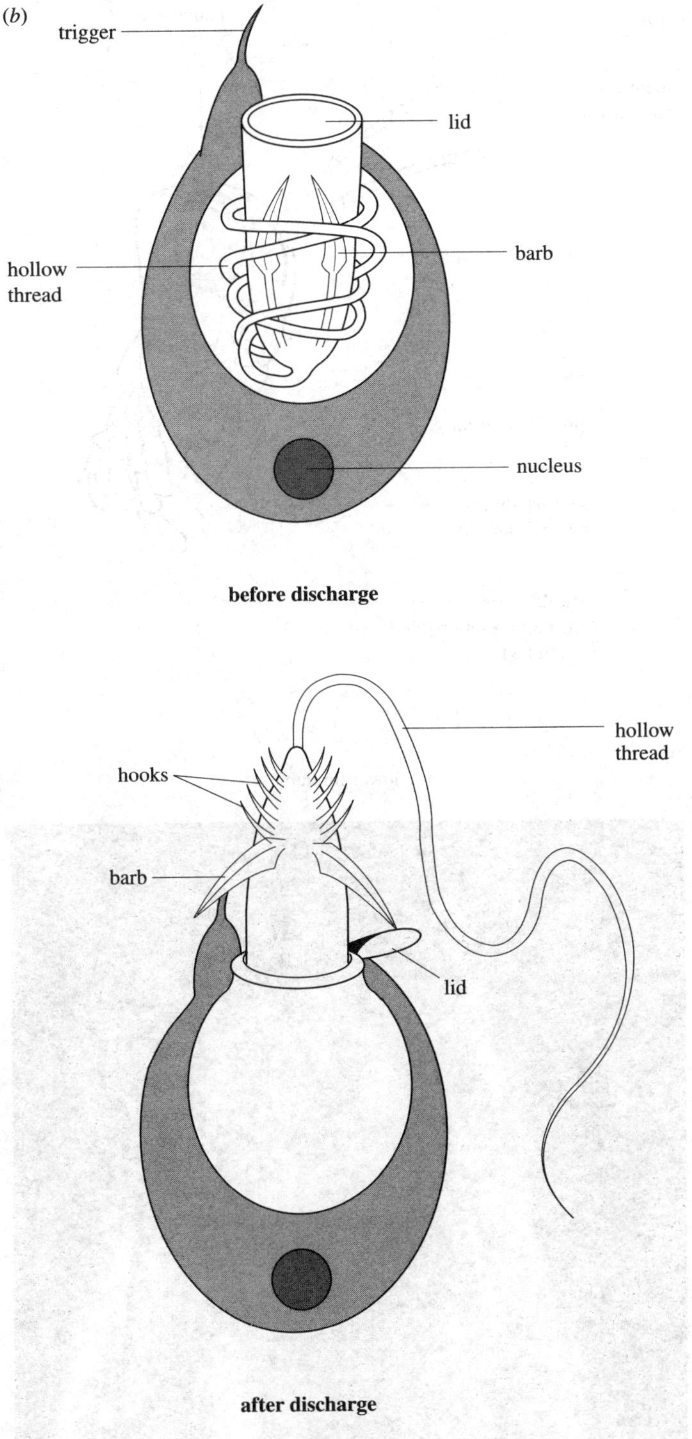

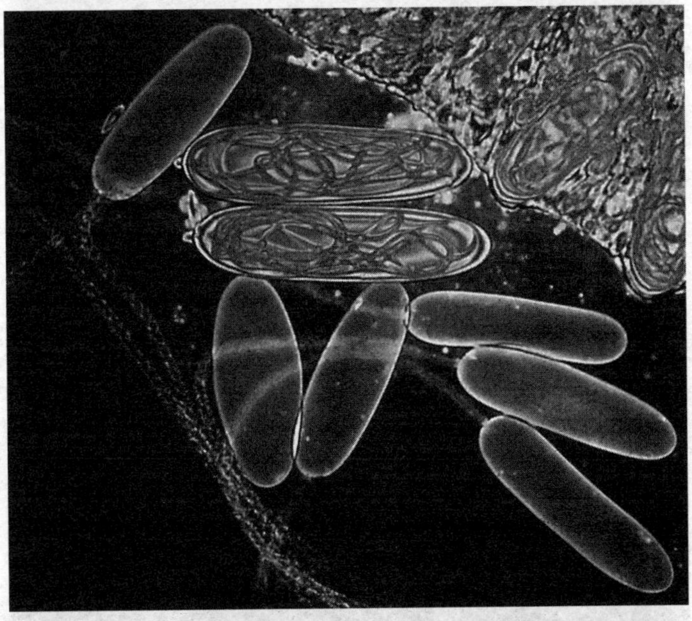

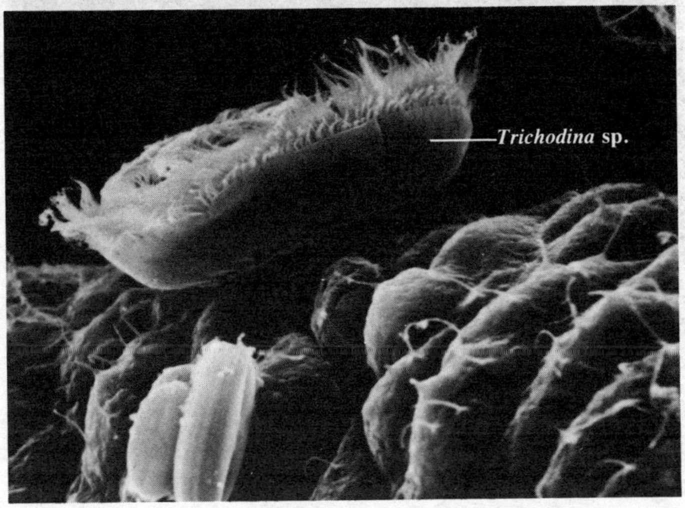

Fig 8.6 (a) Hydra *capturing a water flea (*Daphnia*).* (b) *Nematoblast before and after discharge.* (c) *Light micrograph of a group of nematocysts of the sea anemone* Rhodactis rhodostoma. *All but two of the nematocysts have discharged their stinging threads.* (d) *Scanning electron micrograph of* Trichodina *trapped on the tentacles of a hydra.*

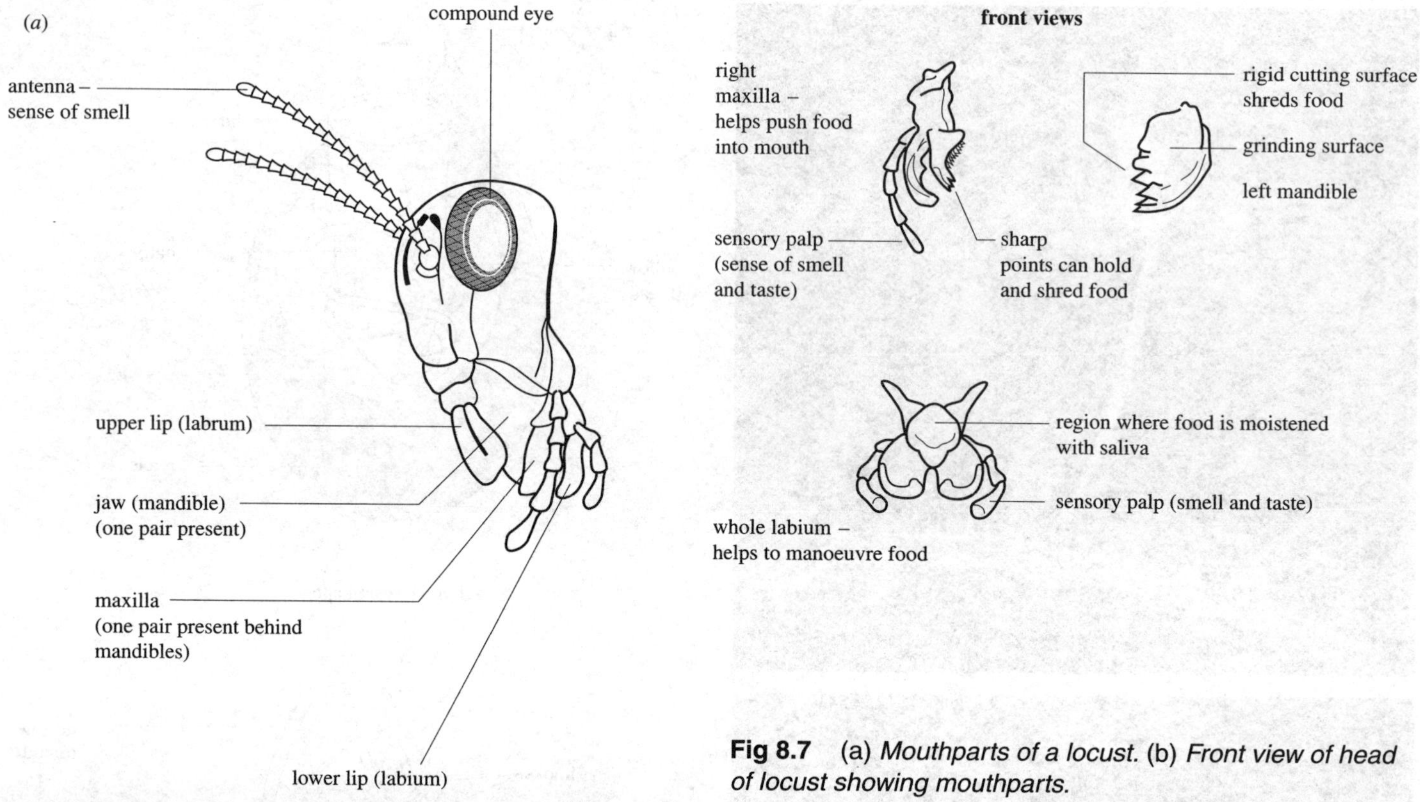

Fig 8.7 (a) *Mouthparts of a locust.* (b) *Front view of head of locust showing mouthparts.*

populations are now constantly monitored. This, and studies of their migratory behaviour, means that swift responses can now be made to minimise damage. Places most at risk can be identified and sprayed with insecticide before the arrival of the locusts. Swarms can be tracked and destroyed with the aid of aeroplanes and land vehicles.

Fig 8.8 *Swarming locusts.*

8.2.5 Fluid feeding

Some insects feed on fluids using specialised mouthparts for **sucking**, for example butterflies, or **piercing and sucking**, for example aphids and mosquitoes.

Sucking

Butterflies, such as the cabbage white butterfly (fig 2.60), possess a feeding device called a **proboscis**. This is formed from the two **maxillae** (which, as noted above, have a completely different structure and function in the locust). Part of each maxilla is greatly elongated and C-shaped in cross section (fig 8.9). The two C-shapes fit together precisely to form the walls of a long tube, the proboscis. Mandibles (jaws) are absent, unlike the locust, and sensory palps are less well developed than in the locust. At rest, the proboscis is coiled up under the head. When feeding, a reflex causes contraction of muscles inside each half of the proboscis which uncoil and straighten the proboscis.

Butterflies feed on nectar from flowers. The proboscis is extended into the corolla (petals) of a flower and its tip placed directly on the nectar, which is a dilute solution of sugar. Muscles in the pharynx (fig 8.9) then begin to contract, causing the nectar to be sucked into the mouth of the insect. It is frequently the case that the depth of the corolla tube of the flower corresponds to the length of the butterfly's proboscis, so that each species of butterfly tends to visit one or a few species of flower only. The butterfly often acts as a vital agent of pollination, so these species often depend on each other for their survival, an example of mutualism.

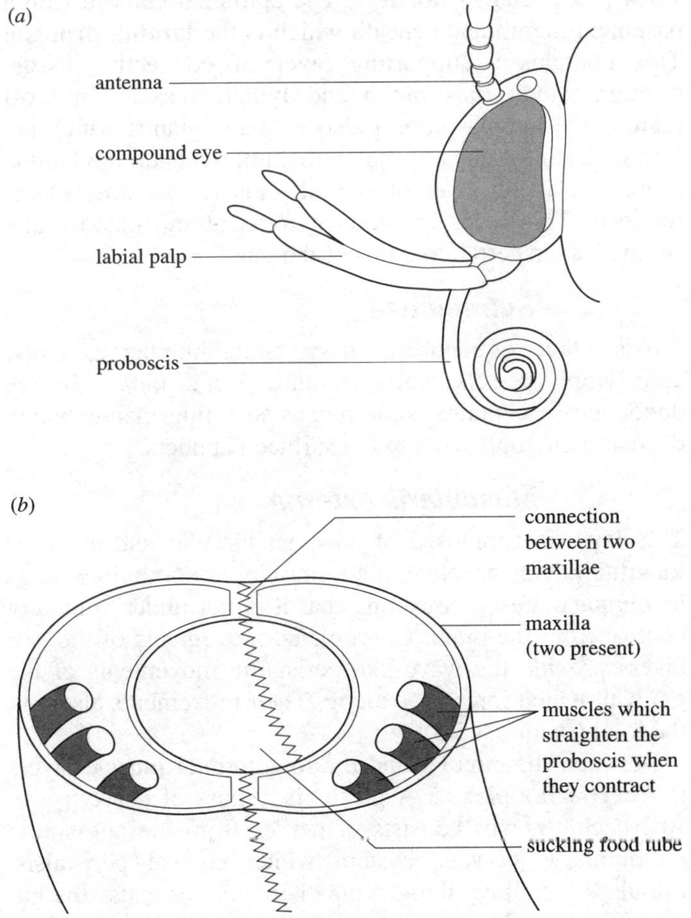

Fig 8.9 (a) *Mouthparts of a butterfly.* (b) *Detail of the proboscis, shown cut transversely.*

Piercing and sucking

Aphids (greenfly) feed on plant juices from leaves and stems. Like the butterfly, the aphid has specialised elongated mouthparts that form a proboscis, but these have to be modified for piercing the plant tissue as well as sucking. They have to penetrate the sieve tubes of the phloem, which are the long tubes which conduct high concentrations of the sugar sucrose and other nutrients through the plant.

In the aphid, the maxillae fit together to form a sharp piercing tube called the **stylet** (fig 8.10). This is enclosed in a sheath formed by an elongated labium (lower lip) and mandibles. The stylet can be pushed through the plant tissues and into the sieve tube. The aphid's work is now done because the contents of the sieve tube are under pressure and are forced up the stylet tube into its gut.

Aphids have been used by scientists to analyse the contents of sieve tubes. After anaesthetising the feeding aphid, its body can be cut off and the stylet continues to exude juice from the sieve tubes. Aphids are a serious pest in agriculture and horticulture. They attack most crops, important examples being cereals, beans, potatoes, fruit trees and bushes, and cotton. They are a common pest of greenhouses. Both biological and chemical controls are used. Ladybirds, for example, are an important predator. However, spraying with insecticides can kill ladybirds and other insect predators, as well as aphids, and populations of aphids have sometimes benefited from the use of insecticides. As well as being important crop pests in their own right, aphids can act as vectors in spreading virus disease from one plant to another.

8.3 The alimentary canal in humans

Digestion and absorption occurs in the alimentary canal, or more plainly the gut, which runs from the mouth to the anus. As the gut wall is continuous with the outside surface of the body, the food in the gut is considered to be 'outside' the body. Food can only be absorbed into the body after it has been **ingested** and broken down physically by the teeth and muscles of the gut wall (**mechanical digestion**), and chemically by its enzymes into molecules of a suitably small size to be absorbed through the gut wall (**chemical digestion**). From here the nutrients enter the blood or lymph and are delivered to the cells of the body tissues where they undergo **assimilation**. Undigested food is **egested** through the anus. A reminder of the stages of nutrition might be useful at this stage (section 8.1.1).

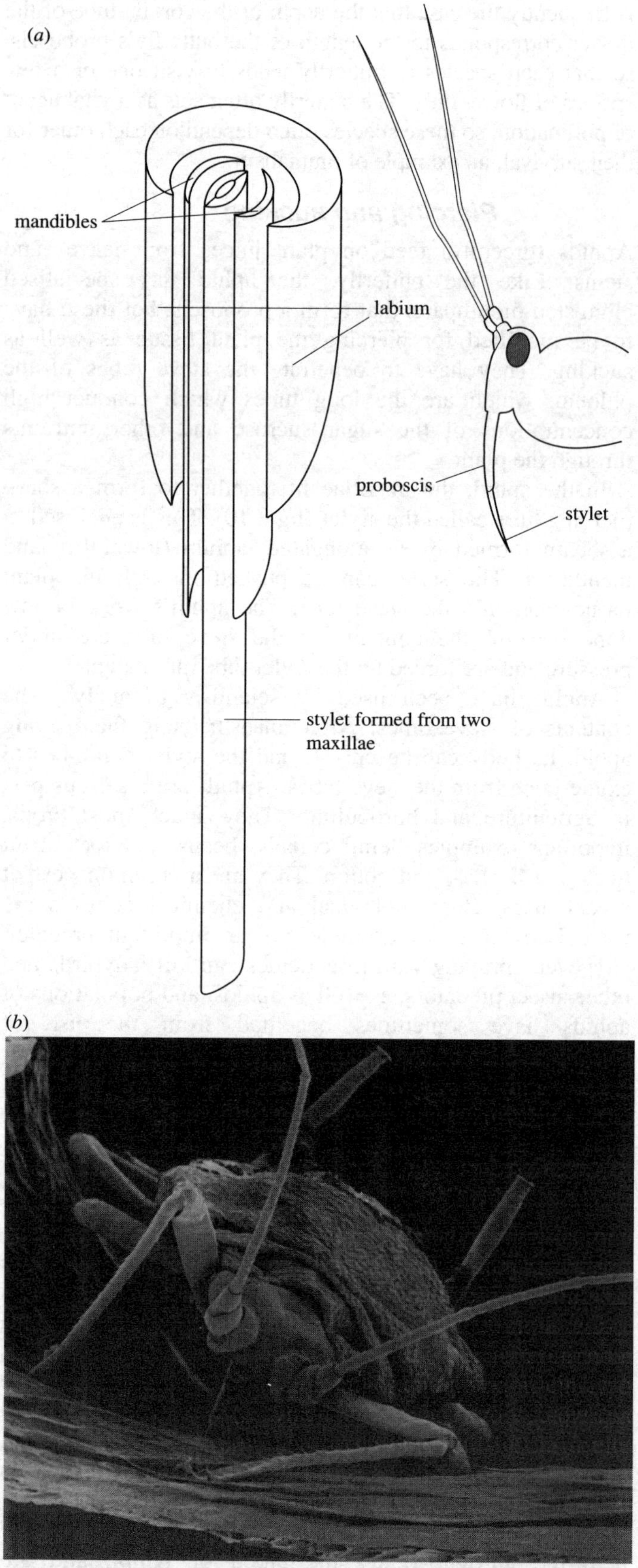

Fig 8.10 (a) *Aphid mouthparts (based on figs 28.15a and 28.16a, p152,* Introduction to Biology, *5th ed., D.G. Mackean, John Murray (1973)).* (b) *Scanning electron micrograph of aphid showing stylet piercing a leaf.*

The gut is specialised into different regions, each designed to carry out a different role in the overall processes of digestion and absorption. These regions and a summary of their functions are shown in fig 8.11.

8.3.1 Generalised structure of the human gut

Although each different region of the gut possesses its own special characteristics, all have a basic common structure as shown in fig 8.12. This consists of four distinct layers: the mucosa, submucosa, muscularis externa and serosa.

Mucosa

This is the innermost layer of the gut and has three layers, the **epithelium**, **lamina propria** and **muscularis mucosa**. It is the major absorbing and secreting layer.

The **epithelium** secretes large quantities of mucus which lubricates the food, helping its passage through the gut. It also prevents digestion of the gut wall by its own enzymes. Some of the epithelial cells have microvilli on their free surfaces. These contain enzymes embedded in their cell surface membranes. The microvilli form a border which is just visible as a fuzzy line in the light microscope and is known as a **'brush border'**. The epithelial cells rest on a basement membrane beneath which is the **lamina propria**. This contains a supporting layer of connective tissue, through which runs blood and lymph vessels. In most regions the lamina propria also contains glands which are formed by infoldings of the epithelium. Outside the lamina propria is a thin layer of smooth muscle, the **muscularis mucosa**. This helps to produce folds of the mucosa and submucosa in certain regions of the gut.

Submucosa

This is a layer of connective tissue containing nerves, blood and lymph vessels, collagen and elastic fibres. In the duodenum it contains some mucus-secreting glands which deposit their contents onto the surface via ducts.

Muscularis externa

This layer is composed of an inner circular and an outer longitudinal layer of smooth muscle. Smooth muscle is involuntary muscle, meaning that it is not under voluntary control from the brain. Coordinated movements of the two layers provide the wave-like peristaltic movements of the gut wall which force food along. Their movements also mix the food (section 8.3.5).

Between the circular and the longitudinal muscle layers is **Auerbach's plexus**. A plexus is a mass of nerve tissue. Auerbach's plexus consists of nerves from the autonomic (involuntary) nervous system which control peristalsis. Impulses travelling along sympathetic nerves cause the gut muscles to relax and the sphincters to close, whilst impulses travelling via the parasympathetic nerves stimulate the gut wall to contract and sphincters to open.

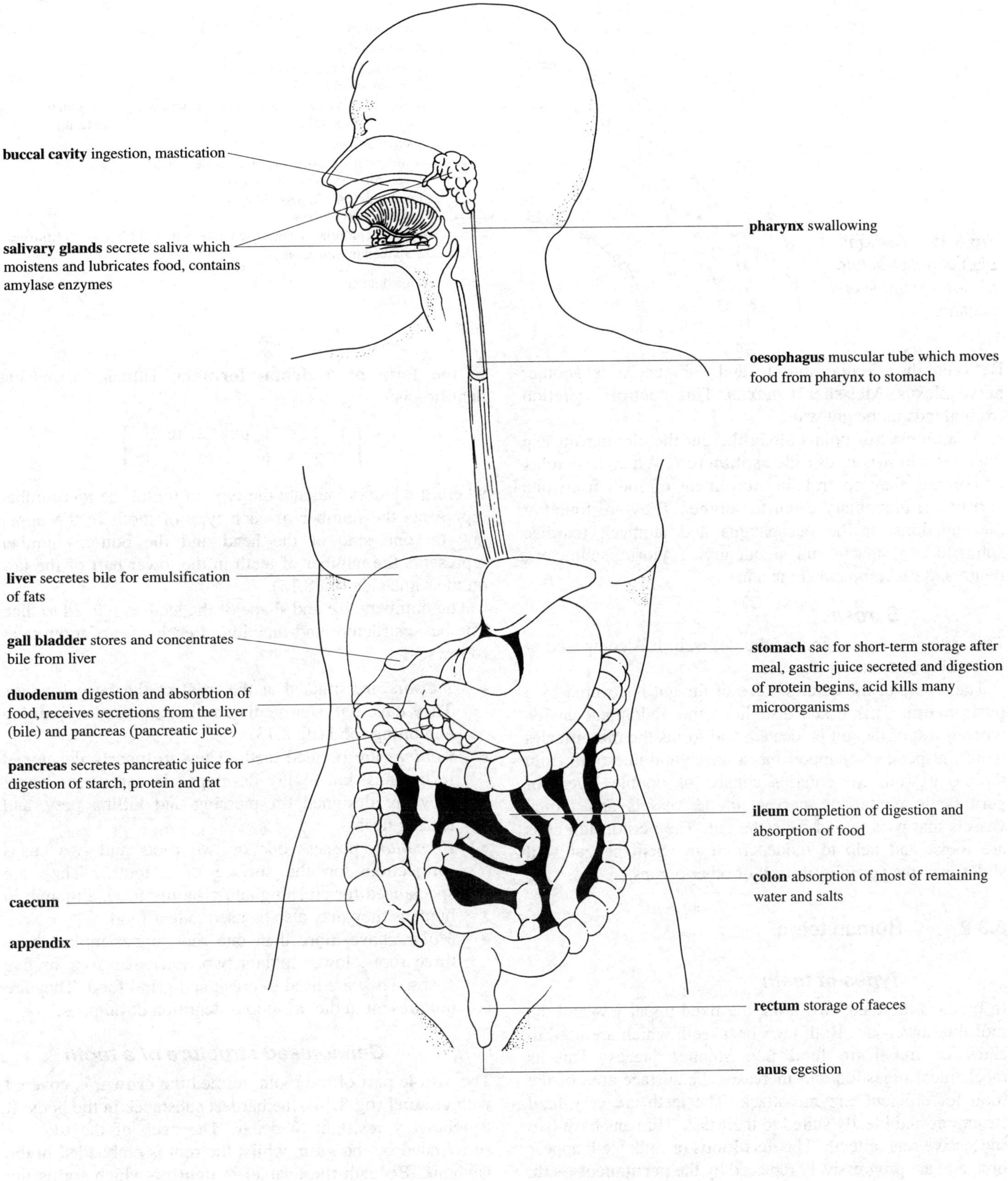

Fig 8.11 *General layout of the organs of the human digestive system and their functions.*

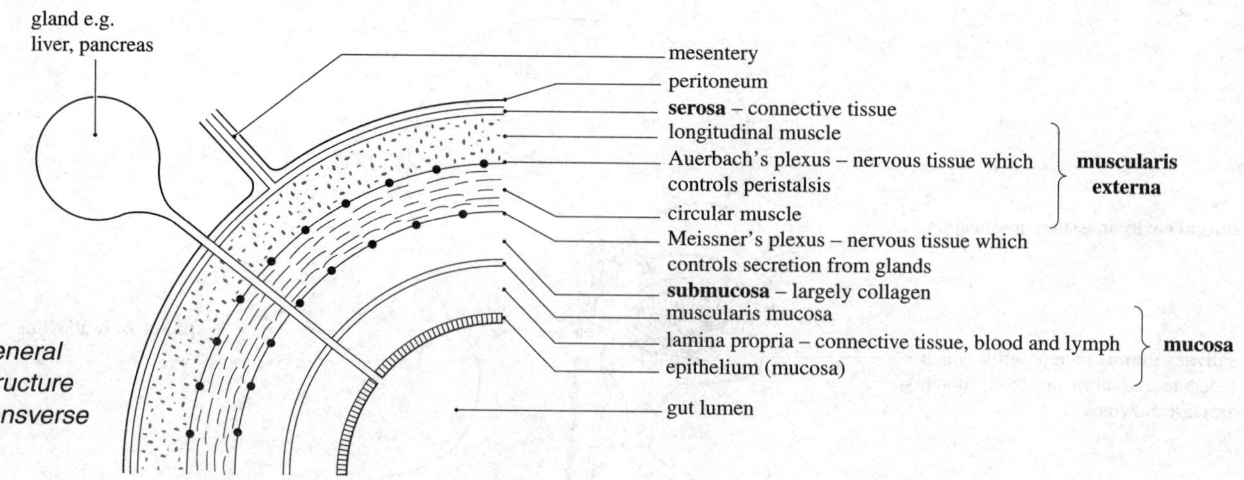

Fig 8.12 *General plan of gut structure as seen in transverse section.*

Between the circular muscle and submucosa is another nerve plexus, **Meissner's plexus**. This controls secretion from glands in the gut wall.

At a number of points along the gut the circular muscle thickens into structures called **sphincters**. When these relax or contract they control the movement of food from one part of the alimentary canal to another. They are found at the junctions of the oesophagus and stomach (cardiac sphincter), stomach and duodenum (pyloric sphincter), ileum and caecum, and at the anus.

Serosa

This is the outermost coat of the gut wall. It is composed of loose fibrous connective tissue.

The whole of the outer surface of the gut is covered by a **peritoneum**. This tissue also lines the abdominal cavity, where most of the gut is located, and forms the **mesenteries** which suspend and support the stomach and intestines from the body wall. Mesenteries consist of double layers of peritoneum containing nerves, blood vessels and lymph vessels that pass to and from the gut. The peritoneum cells are moist and help to reduce friction when the gut wall slides over the portions of itself or other organs.

8.3.2 Human teeth

Types of teeth

In humans there are two jaws, the fixed upper jaw and the movable lower jaw. Both jaws bear teeth which are used to chew or **masticate** food into smaller pieces. This is mechanical digestion and increases the surface area of the food for efficient enzyme attack. The teeth are very hard structures and ideally suited to their task. Humans have two successive sets of teeth. The **deciduous** or milk teeth appear first, and are progressively replaced by the **permanent teeth**. Human teeth have different shapes and sizes and possess uneven biting surfaces. Humans possess up to 32 permanent teeth, consisting of eight incisors (i), four canines (c), eight premolars (pm) and up to 12 molars (m). The arrangement of the teeth can be conveniently expressed in the form of a **dental formula**. Human permanent dentition is:

$$2\left[\mathrm{i}\,\frac{2}{2}\;\mathrm{c}\,\frac{1}{1}\;\mathrm{pm}\,\frac{2}{2}\;\mathrm{m}\,\frac{3}{3}\right]$$

where the letters indicate the type of tooth, the top number represents the number of each type of tooth in the upper jaw on one side of the head and the bottom number represents the number of teeth in the lower part of the jaw on the same side (fig 8.13).

The number, size and shape of the teeth is related to diet. The basic structure and function of each type of tooth is as follows.

- *Incisors* are situated at the front of the buccal cavity. They have flat, sharp edges which are used for cutting and biting food (fig 8.13).
- *Canines* are pointed teeth. They are poorly developed in humans, but highly developed in carnivores where they are designed for piercing and killing prey, and tearing flesh.
- *Premolars* possess one or two roots and two cusps (projections on the surface of a tooth). They are specialised for crushing and grinding food, although in humans they may also be used to tear food.
- *Molars* have more than one root; upper molars have three roots, lower molars two. Each has four or five cusps. They are used to crush and grind food. They are not present in the deciduous dentition of humans.

Generalised structure of a tooth

The visible part of the tooth, termed the **crown**, is covered with **enamel** (fig 8.14), the hardest substance in the body. It is relatively resistant to decay. The neck of the tooth is surrounded by the **gum**, whilst the root is embedded in the jawbone. Beneath the enamel is **dentine** which forms the bulk of the tooth. Though tough, it is not as hard as enamel or as resistant to decay. It contains numerous small canals (canaliculi) containing cytoplasmic extensions of **odontoblasts**, the dentine-producing cells. The **pulp cavity** contains odontoblasts, sensory endings of nerves, and blood

(*a*)

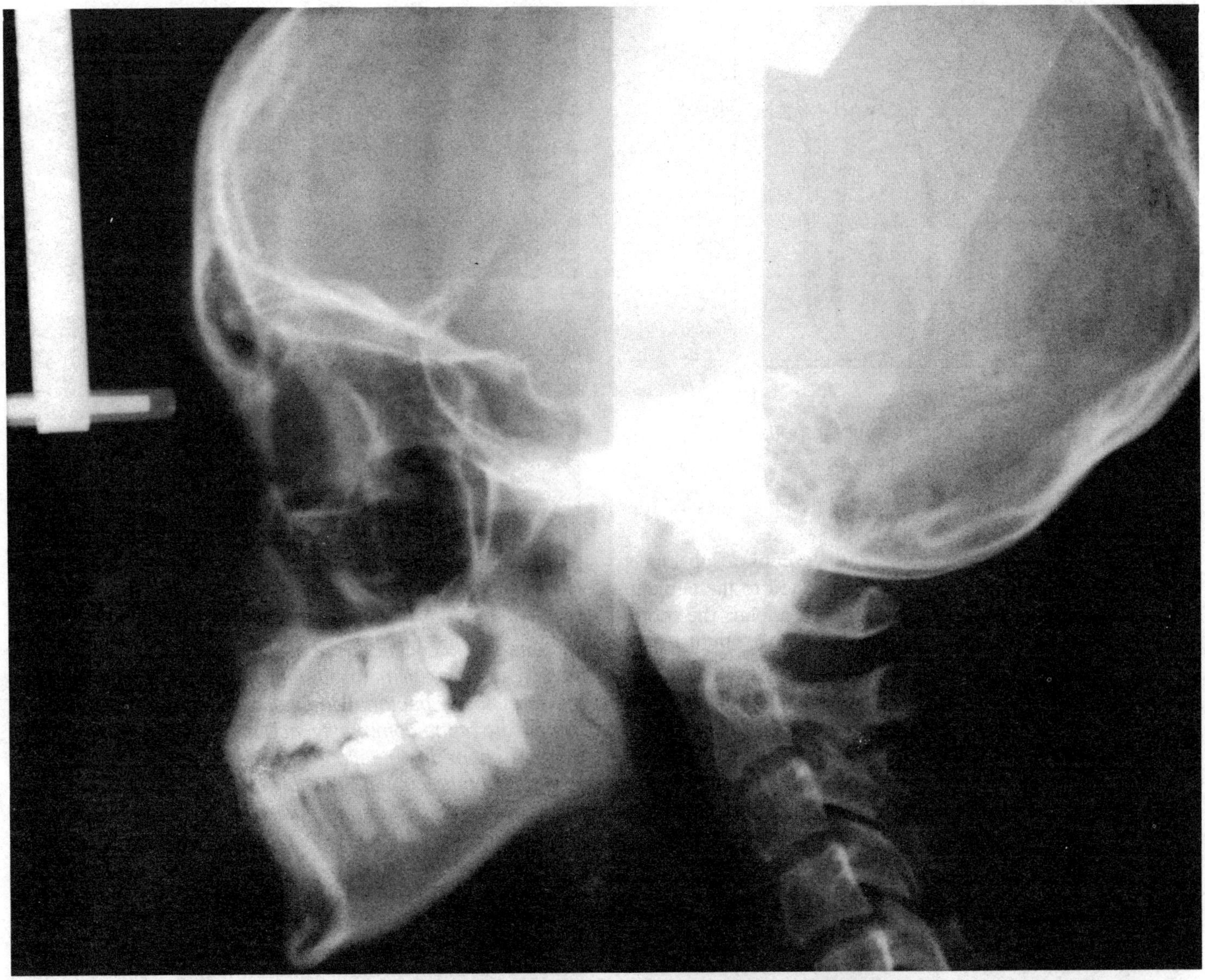

(*b*)

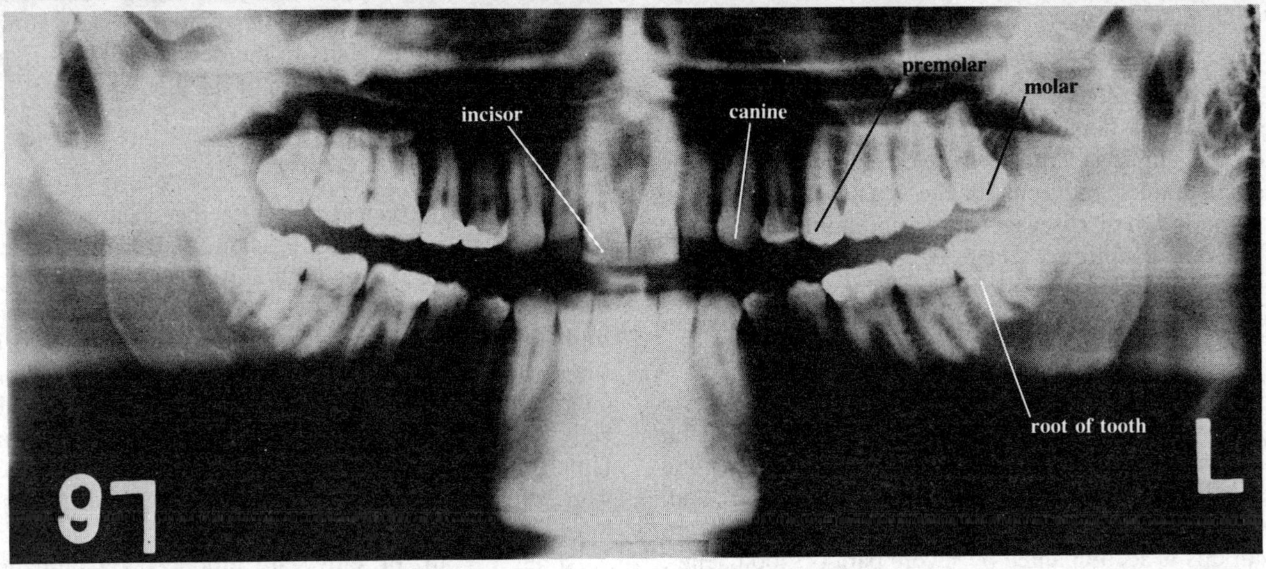

Fig 8.13 *(a) X-ray of side of human head to show permanent dentition on one side.*
(b) X-ray of head from the front to show a complete permanent dentition. Dental formula $2\left[i\,\frac{2}{2}\ c\,\frac{1}{1}\ pm\,\frac{2}{2}\ m\,\frac{3}{3}\right]$

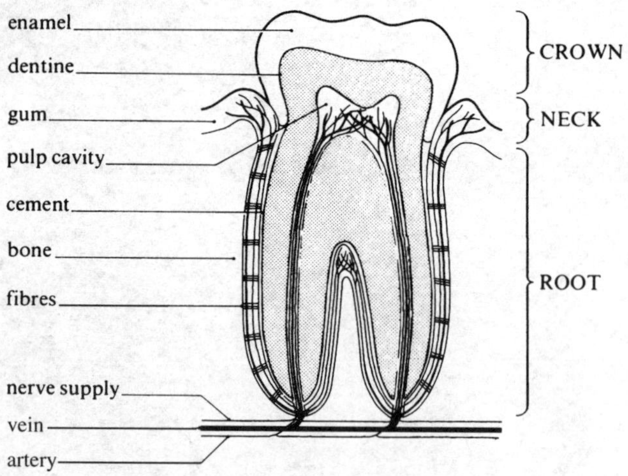

Fig 8.14 *Vertical section of a premolar tooth.*

vessels which deliver nutrients to the living tissues of the tooth and remove their waste products.

The root of the tooth is covered with **cement**, a substance similar to bone. Numerous **fibres**, connected to the cement at one end and the jawbone at the other, anchor the tooth firmly in place. However, it is still able to move slightly and this reduces the chances of it being sheared off during chewing.

Dental disease

Two major dental diseases exist, periodontal disease and dental caries. Both are caused by **plaque**, which is a mixture of bacteria and salivary materials. If allowed to accumulate, the bacteria cause inflammation of the gums (**periodontal disease**). Plaque also combines with certain chemicals in the saliva which make it harden and calcify to form deposits of calculus which cannot be removed by brushing. Some of the bacteria in plaque convert sugar into acid which starts the process of **dental caries** (fig 8.15).

Periodontal disease. This is a disease of the gums caused by microorganisms that are normally present in the mouth in dental plaque, especially in the areas between the gums and the teeth. Neglect of oral hygiene creates favourable conditions for the spread of this disease. At first periodontal disease causes inflammation of the gums. If this condition, which is generally painless, is allowed to continue, the inflammation may spread to the root of the tooth and destroy the fibres which anchor it in place. Eventually the tooth becomes loose and may have to be extracted.

Dental caries. The microorganisms in dental plaque convert sugar in the mouth to acid. Initially the enamel is slowly and painlessly dissolved by the acid. However, when the dentine and pulp of the tooth are attacked, this is accompanied by severe pain or 'toothache', and the possible loss of teeth. Several factors contribute to the spread of dental caries. They include prolonged exposure to sugary foodstuffs, disturbance of saliva composition, lack of oral hygiene and low levels of fluoride in drinking water. Prevention of dental caries may be helped by adding fluoride to drinking water, fluoridation of some foods such as milk, children taking fluoride tablets, brushing teeth with fluoridated toothpaste, good oral hygiene and regular visits to the dentist and oral hygienist, and care with the composition of the diet.

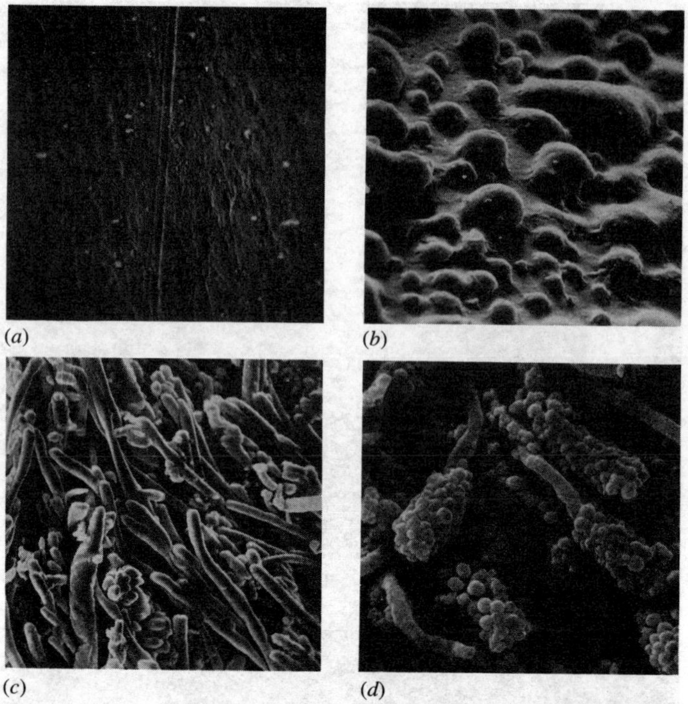

Fig 8.15 *Development of dental plaque.* (a) *Spherical bacteria called cocci deposit as 'pioneer' species and then multiply to form a film.* (b) *Organisms embedded in a matrix of secretions of bacterial and salivary origin.* (c) *The complexity of the community increases and rod- and filament-shaped populations appear.* (d) *In the 'climax' community many unusual associations between different populations can be seen, including 'corn cob' arrangements.*

8.3.3 Buccal cavity

The buccal cavity is the chamber just inside the mouth in which food is chewed. During chewing the muscular tongue moves food around the mouth and mixes and moistens it with saliva. The tongue possesses **taste buds** (fig 8.16)) that contain receptors sensitive to sweet, salty, sour and bitter substances. A simple (inborn) or conditioned (learned) reflex results in stimulation of the salivary glands to secrete saliva. The eye and the olfactory (smell) receptors in the nose are also important receptors in triggering reflexes that bring about salivation (see section 8.4.1).

About 1.5 dm^3 of saliva are produced by humans each day by the **salivary glands**. Saliva is a watery secretion containing the enzymes salivary **amylase** and **lysozyme**. It also contains **mucus** and various **mineral salts**, including chloride ions which speed up the activity of the enzymes. The mucus moistens and lubricates food and makes it easier to swallow. Salivary amylase begins the digestion of starch, first to shorter polysaccharides, and then to the disaccharide maltose. Lysozyme helps to kill bacteria, which are potentially harmful, by catalysing the breakdown of their cell walls (section 5.10.6). Eventually the semi-solid, partially digested food particles are stuck together and moulded into a **bolus** (or pellet) by the tongue, which then pushes it towards the pharynx. From here, as a result of a reflex action, it is swallowed into the oesophagus via the pharynx.

(*a*)

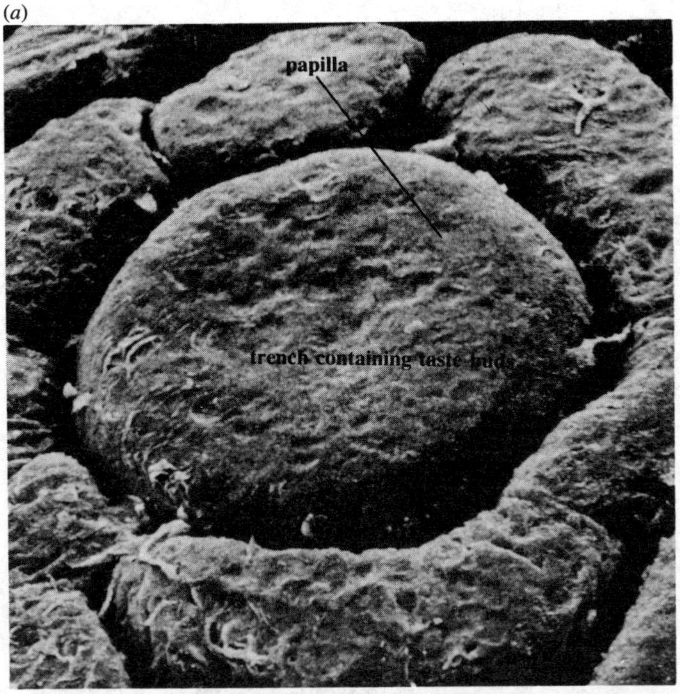

Fig 8.16 (a) *Scanning electron micrograph of the tongue surface of a three-week-old puppy. The taste buds are in the trenches surrounding the surface papillae.* (b) *VS taste buds in the tongue.*

8.3.4 Oesophagus

This is a narrow muscular tube lined by stratified squamous epithelium (see section 6.3.2) containing mucus glands (fig. 8.17). In humans it is about 25 cm long and quickly conveys food and fluids by peristalsis from the pharynx to the stomach.

8.3.5 Peristalsis

Food is pushed through the gut by the muscles of the muscularis externa, the outer longitudinal and inner circular layers of muscle in the gut wall (fig 8.12). Behind the bolus (pellet of food) the circular muscles contract, squeezing and constricting the gut. In front of the food, the longitudinal muscles contract, shortening this section of the gut and pulling it past the advancing bolus (fig 8.18).

Other types of movement are possible in the stomach and small intestine which ensure stirring and mixing of food.

(*b*)

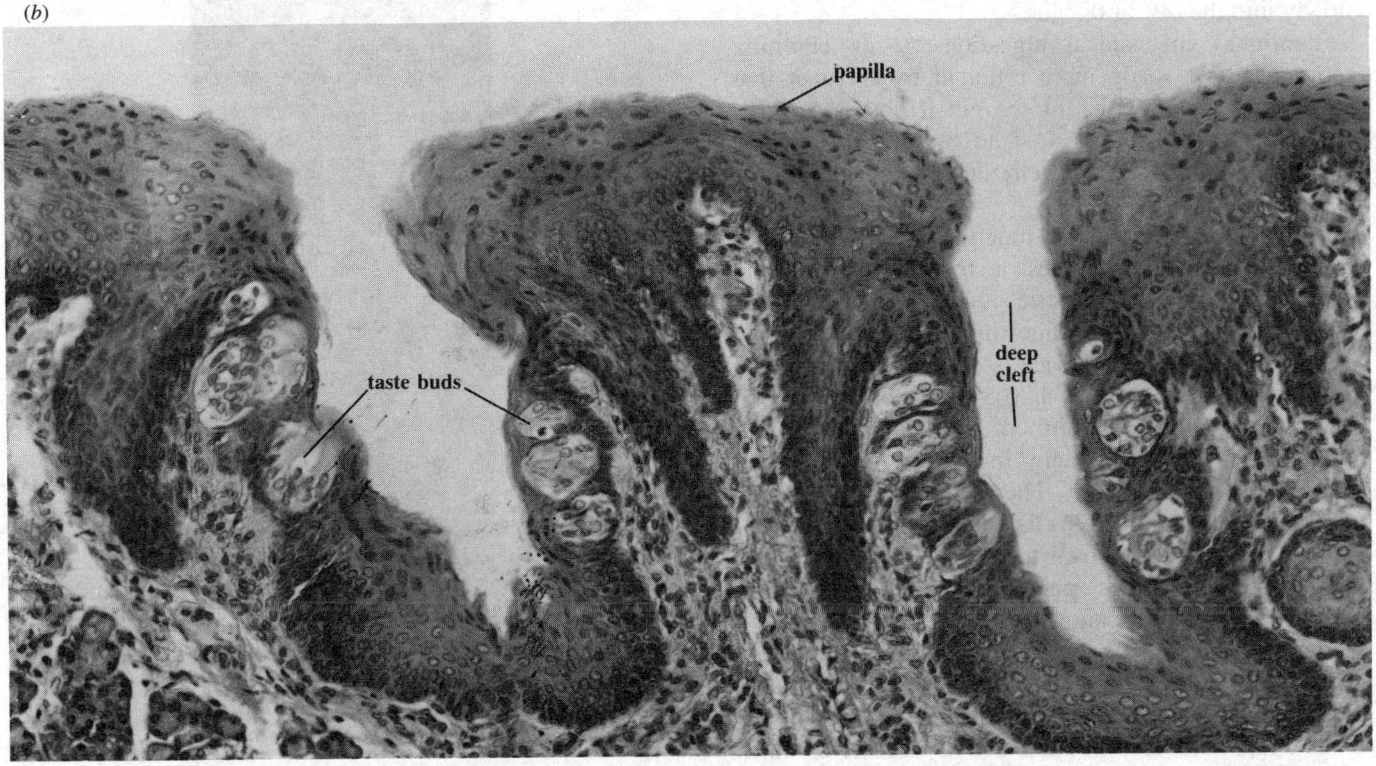

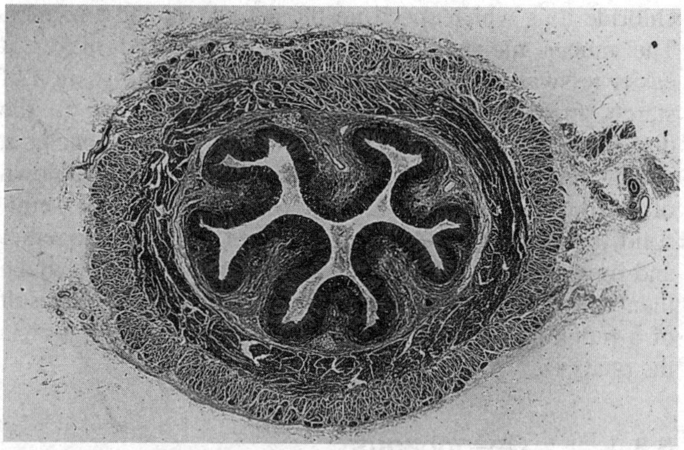

Fig 8.17 *TS human oesophagus.*

Strictly speaking these are not described as peristaltic movements. For example, the gut may be constricted in several parts at once (segmental movements) and sometimes sections of the intestine suddenly and rapidly shorten, throwing food from one end to the other and thus thoroughly mixing it.

8.3.6 Stomach

The stomach in humans is situated below the diaphragm and on the left side of the abdominal cavity (fig 8.11). It is a muscular bag which can stretch to take in food. When unstretched the stomach wall lies in folds, but when fully distended it can hold nearly 5 dm^3 of food and the folds disappear. It has a number of functions.

- It stores food temporarily after meals, releasing food slowly into the rest of the gut.
- It continues mechanical digestion by its churning action. This is made more efficient by the fact that unlike the other regions of the gut it possesses three layers of smooth muscle instead of two, namely the outer longitudinal, middle circular and inner oblique layers.
- The thick mucosa contains mucus-secreting epithelial cells. The mucus provides a barrier between the stomach mucosa and gastric juice (see below) and prevents the stomach self-digesting.
- The main part of the stomach is dotted with numerous gastric pits (figs 8.19 and 8.20). These lead into long, tubular gastric glands formed by infoldings of the epithelium. The glands are lined with cells which secrete the gastric juice. There are two specialised types of cell, the **parietal cells** and the **chief cells**.

Chief cells (also known as zymogen cells) secrete the inactive enzymes **pepsinogen** and **prorennin**. (Inactive enzymes are known as zymogens).

Parietal cells (also known as oxyntic cells) secrete a dilute solution of hydrochloric acid which itself has a number of functions. It makes the stomach contents pH 1–2·5, ideal for the optimum activity of the stomach

(a)

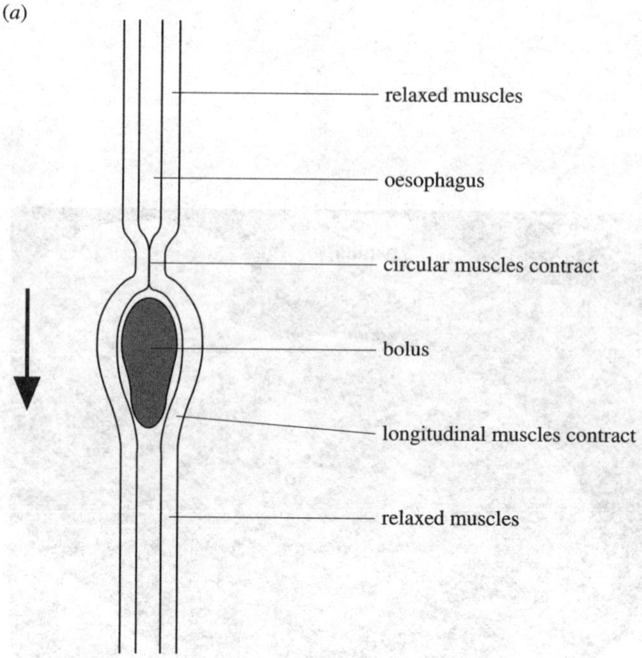

(b)

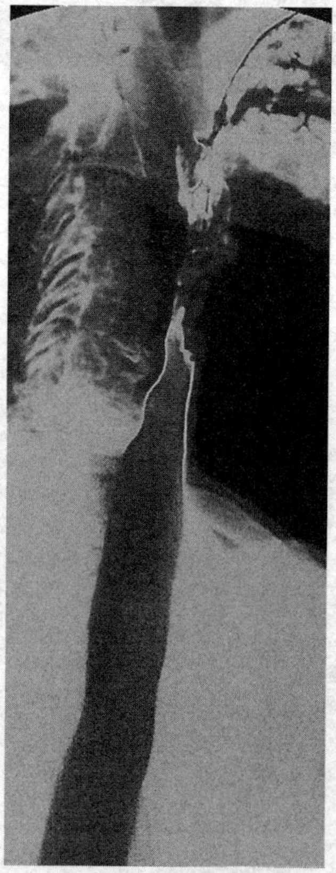

Fig 8.18 (a) *Diagram of peristalsis in the oesophagus.* (b) *X-ray of peristalsis in the human oesophagus. The patient has just swallowed a barium meal, which is opaque to X-rays. The muscles in the oesophagus above the meal are constricted, forcing the meal down towards the stomach.*

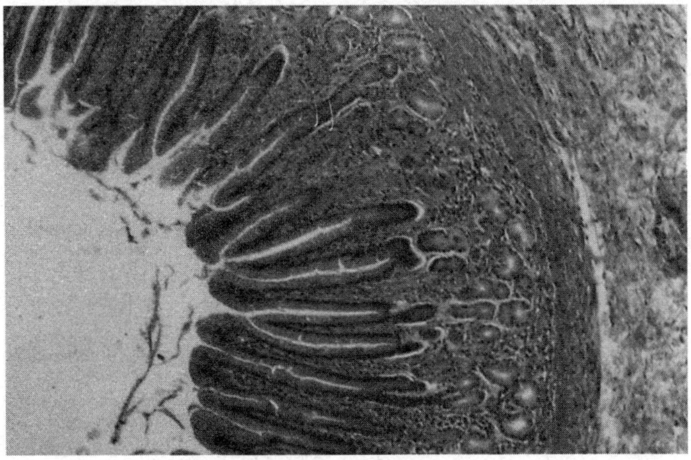

Fig 8.19 *TS gastric pits of a mammal.*

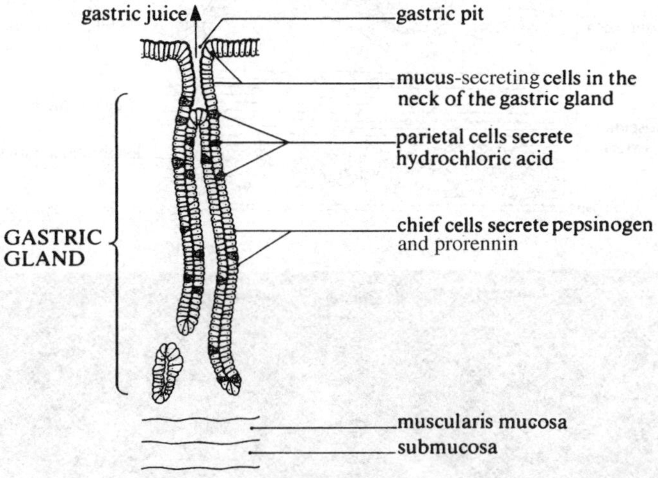

Fig 8.20 *VS stomach wall showing gastric gland. Cells which secrete the hormone gastrin are also present in the gland.*

enzymes. The acid kills many bacteria, thus acting as a defence mechanism. It denatures many proteins: their tertiary structure is altered, making them unfold and so easier to digest. This is particularly important for fibrous proteins, some of which are very common, such as collagen which is found extensively in animal connective tissues. The acid also loosens fibrous and cellular components of tissue. It converts pepsinogen and prorennin to their active forms pepsin and rennin and begins the hydrolysis of sucrose to glucose and fructose.

- Pepsin hydrolyses protein into smaller polypeptides. Rennin coagulates casein, the soluble protein of milk, into the insoluble calcium salt of casein in the presence of calcium ions. This calcium salt is then digested by pepsin.

> **8.3** Why is it necessary for pepsin to be secreted in an inactive state?

- The stomach also contains endocrine cells which secrete the hormone gastrin. This is discussed in section 8.4.

The cardiac sphincter, at the junction between the oesophagus and the stomach, and the pyloric sphincter, at the junction between the stomach and the duodenum, prevent the uncontrolled exit of food from the stomach. Both act as valves and retain food in the stomach for periods of up to four hours. Relaxation of the pyloric sphincter releases small quantities of the food into the duodenum at regular intervals.

The muscles of the stomach wall thoroughly mix up the food with gastric juice and eventually convert it into a semi-liquid mass called **chyme**. Gradually the stomach squirts the chyme into the duodenum through the relaxed, ring-shaped pyloric sphincter.

8.3.7 Structure of the small intestine

The first part of the small intestine is the **duodenum**. It is short, about 25 cm long, and the pancreatic and bile ducts open into it. The duodenum leads on to the **ileum** which is about 3 m long in a living body (it relaxes and gets much longer after death) (figs 8.11 and 8.21). The submucosa and mucosa together are folded (fig 8.21*a*). In addition, the mucosa possesses numerous finger-like projections called **villi**, whose walls are richly supplied with blood capillaries and lymph vessels and contain smooth muscle (fig 8.21*d* and *e*). They are able to contract and relax continuously, thus bringing themselves into close contact with the food in the small intestine. The individual cells on the surface of the villi possess tiny microvilli on their free surfaces (figs 8.21*f*, 6.16 and section 5.10.8). Between them these features greatly increase the surface area of the small intestine (table 8.2).

> **8.4** (*a*) List the features of the small intestine which increase its surface area.
> (*b*) Why is this an advantage?

At the base of the villi the epithelium folds inwards in places to form narrow tubes called **crypts of Lieberkühn** (fig 8.21*d*). It is here that new epithelial cells are made to replace those which are constantly being shed from the villi (the average life of these cells is about five days). The cells in the crypts also secrete **intestinal juice**, a slightly alkaline fluid which contains water and mucus and helps to increase the volume of fluid in the gut. **Paneth cells** at the base of the crypts secrete lysozyme, the antibacterial enzyme already mentioned in saliva.

Throughout the small intestine, special epithelial cells called **goblet cells** secrete mucus, whose function has been described in section 8.3.1 (mucosa). The duodenum also secretes an alkaline fluid which helps to neutralise the acid

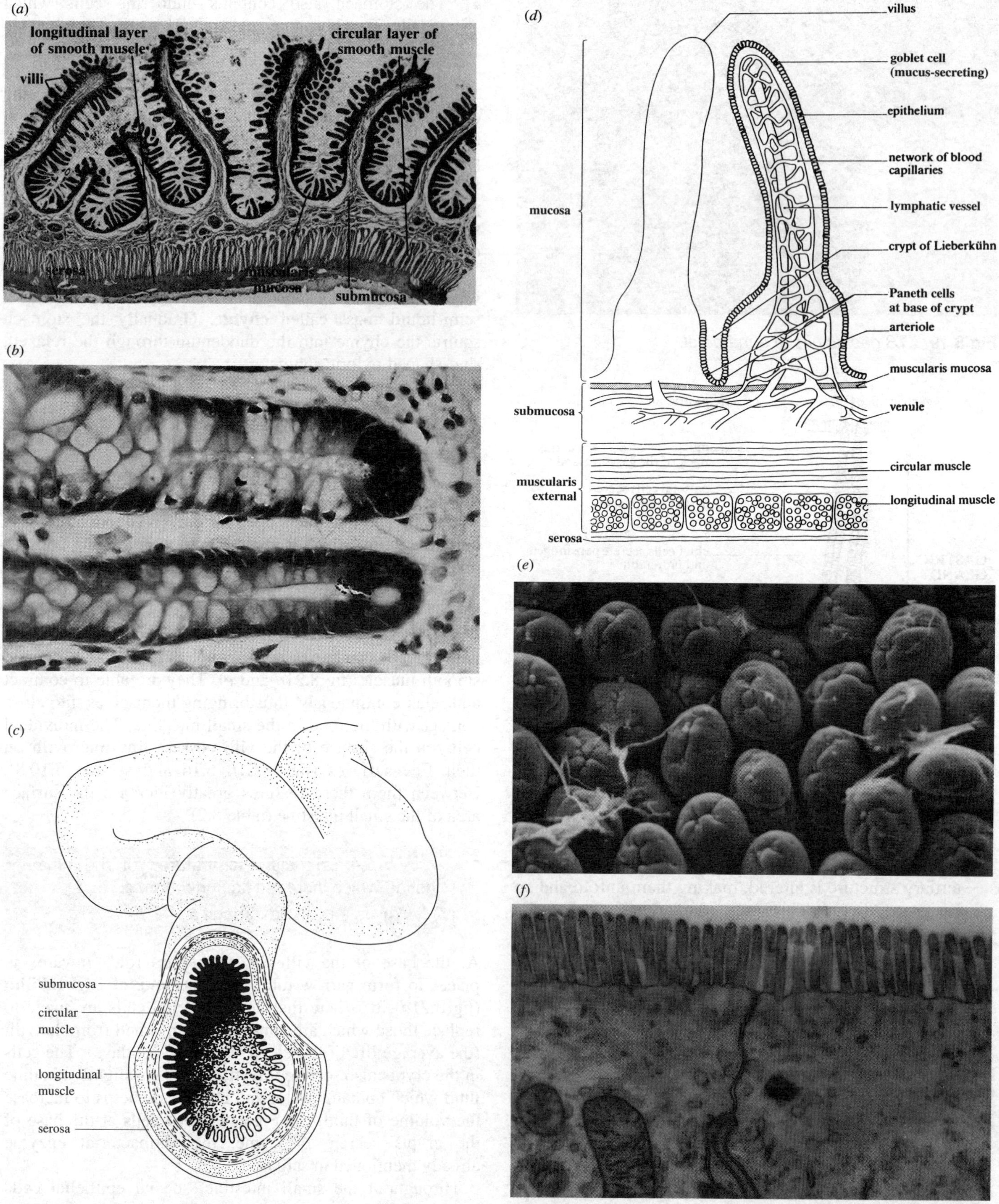

Fig 8.21 (a) *Vertical section through ileum.* (b) *Crypts of Lieberkühn in the ileum.* (c) *Artist's impression of a cross-section of the small intestine.* (d) *Diagram of TS ileum showing a villus.* (e) *Scanning electron micrograph showing villi on surface of small intestine (×200).* (f) *Electron micrograph of epithelial cells showing microvilli (×18 000)*

Table 8.2 Structural features which increase the surface area of the small intestine.

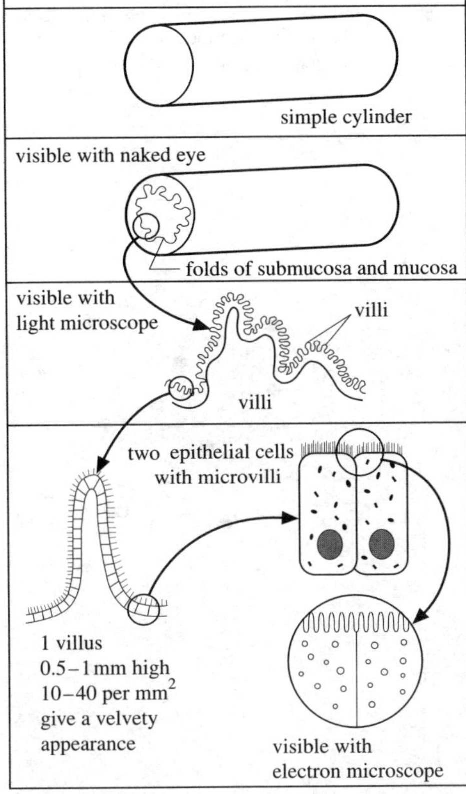

Structure	*Increase in surface area relative to simple cylinder*
simple cylinder	×1
visible with naked eye; folds of submucosa and mucosa	×3
visible with light microscope; villi	×30
two epithelial cells with microvilli; 1 villus 0.5–1 mm high 10–40 per mm² give a velvety appearance; visible with electron microscope	×600

of the stomach and provide an optimum pH of 7–8 for the enzymes of the small intestine.

> **8.5** What would happen to the activity of the intestinal enzymes if the pH in the small intestine remained at 2?

8.3.8 Digestion by enzymes in the small intestine

The general pattern of carbohydrate, protein and lipid digestion is shown in fig 8.22. All the enzymes involved in digestion in the small intestine, apart from those made by the pancreas, are bound to the cell surface membranes of the microvilli of the epithelium (fig 8.21*f*) or located within the epithelial cells. It is at these sites that the final hydrolysis of disaccharides, dipeptides and some tripeptides occurs (fig 8.23). The end-products are monosaccharides and amino acids respectively. A full list of the enzymes involved can be found in table 8.3.

In addition to its own set of enzymes, the small intestine receives alkaline pancreatic juice and bile from the pancreas and liver respectively. **Bile** is produced by liver cells and stored in the gall bladder. It contains a mixture of salts (bile salts) which, in the small intestine, act as natural detergents, reducing the surface tension of fat globules and emulsifying them into droplets, so increasing their total surface area. (This process is called emulsification.) These small droplets are then acted upon more efficiently by the enzyme lipase. Further information about the composition of bile is given in chapter 18.

The pancreas is a large gland located next to the stomach (fig 8.11). Within the pancreas are groups of cells which produce a variety of digestive enzymes that are poured into the duodenum via the pancreatic duct (fig 5.29). They include:

- **amylase** to convert amylose to maltose;
- **lipase** to convert lipids (fats and oils) to fatty acids and glycerol;
- **trypsinogen** which, when converted to **trypsin** by **enterokinase** from the microvilli, digests proteins into smaller polypeptides and more trypsinogen into trypsin;
- **chymotrypsinogen** which is converted to chymotrypsin that digests proteins to amino acids;
- **carboxypeptidases** that converts peptides to amino acids.

A summary of the enzymes secreted by the human gut and their action is given in table 8.3.

8.3.9 Absorption of food in the ileum

Absorption of the end-products of digestion occurs through the villi of the ileum. The structure of the villus is ideally suited for this function as can be seen in figs 8.21*d*, *e*, and *f*. **Monosaccharides**, **dipeptides** and **amino acids** are absorbed either by diffusion or active transport into the blood capillaries (fig 8.23 and fig 5.22).

> **8.6** Suggest one advantage of using active transport in the absorption of monosaccharides, dipeptides and amino acids.

From the villi the blood capillaries join to form the hepatic portal vein which delivers the absorbed food to the liver.

Fatty acids and **glycerol** diffuse into the columnar epithelial cells of the villi. Here they are reconverted into lipids. Proteins present in the epithelial cells coat the lipid molecules to form lipoprotein droplets called **chylomicrons**. These pass out of the epithelial cells by exocytosis and into lymphatic vessels in the villi (fig 8.21*d*). They make the lymph in the lymphatic vessels appear white, so the vessels are sometimes called lacteal vessels (lacteal meaning milky). The chylomicrons are carried by lymph in the lymphatic system to veins near the heart where they enter the liquid part of the blood, the plasma. An enzyme in the blood plasma then hydrolyses the lipids back to fatty acids and glycerol in which form they are taken up by cells. They may be used in respiration or stored as fat in the liver, muscles, mesenteries or below the skin.

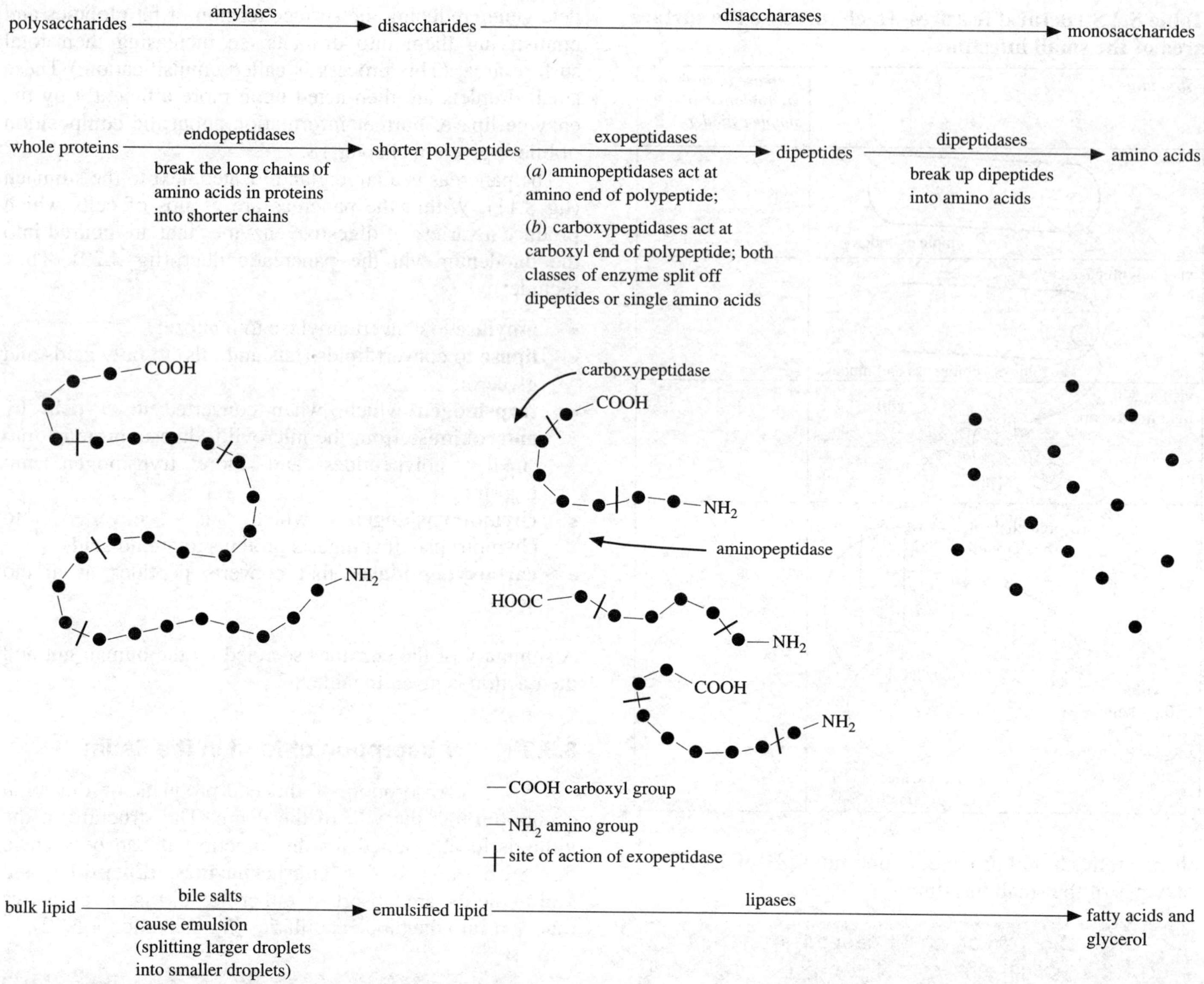

Fig 8.22 *General pattern of enzyme digestion in the human gut.*

Inorganic salts, **vitamins** and **water** are also absorbed in the small intestine.

The sphincter muscle between the ileum and the caecum opens and closes from time to time to allow small amounts of material from the ileum to enter the large intestine.

8.3.10 Large intestine

No digestion takes place in the large intestine or **colon**. Most of the fluids (about 90%) and salts in the gut are absorbed in the small intestine. The colon and caecum remove about 90% of any remaining liquid. Some metabolic waste and inorganic substances, notably calcium and iron, in excess in the body are excreted in the large intestine as salts. Epithelial cells secrete mucus which lubricates the solidifying undigested food remains known as **faeces**. Many symbiotic bacteria present in the large intestine synthesise amino acids and some vitamins, especially vitamin K, which are absorbed into the bloodstream.

In humans the appendix is a blind-ended pouch leading from the caecum and possesses no known function. It is, however, of great significance in herbivores (section 8.6.2). The bulk of the faeces consists of dead bacteria, cellulose and other plant fibres, dead epithelial cells, mucus, cholesterol, bile pigment derivatives and water. Faeces can remain in the colon for 36 h before being passed on to the rectum where they are stored briefly before egestion through the anus. Two sphincters surround the anus, an internal one of smooth (involuntary) muscle and under the control of the autonomic (involuntary) nervous system, and an outer one of striated (voluntary) muscle controlled by the voluntary nervous system.

Table 8.4 summarises the differences in structure between the major regions of the alimentary canal in humans.

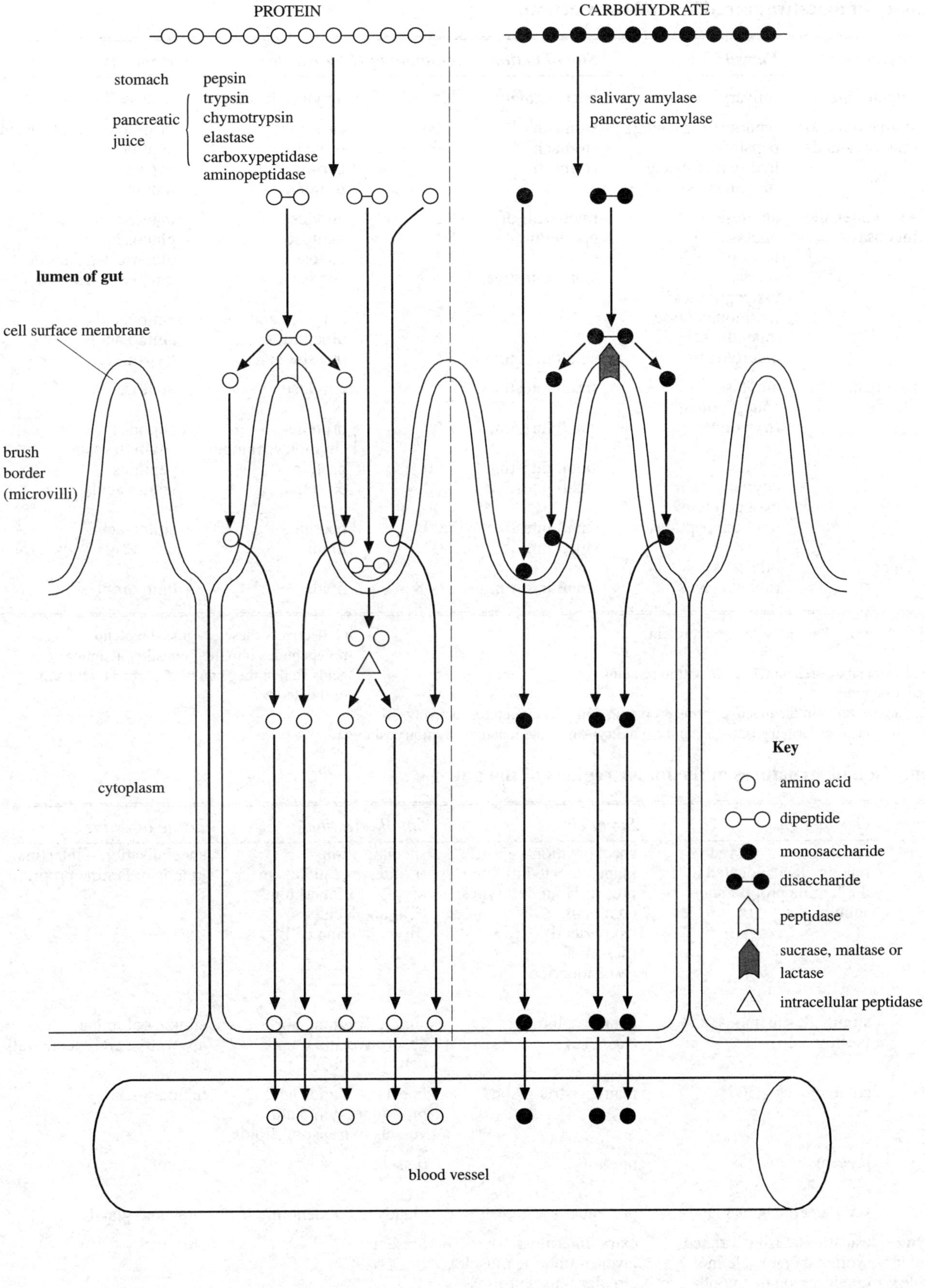

Fig 8.23 *Diagram of an epithelial cell in the ileum with microvilli. The left half shows the final phase of protein digestion, with the absorption of amino acids. The right half shows the corresponding processes for carbohydrates.*

Table 8.3 Summary of digestive secretions and their action.

Secretion	*Source*	*Enzymes*	*Site of action*	*Optimum pH*	*Substrate*	*Products*
saliva	salivary glands	salivary amylase	buccal cavity	6.5–7.5	amylose in starch	maltose
gastric juice	stomach mucosa (gastric glands)	rennin** (in young)	stomach	2.00	casein	insoluble salt of casein
		pepsin**	stomach	2.00	proteins	peptides
		hydrochloric acid (not an enzyme)	stomach	–	pepsinogen prorennin	pepsin rennin
membrane-bound enzymes in microvilli of small intestine	small intestine mucosa	amylase	microvilli of epithelium of small intestine	8.5	amylose	maltose
		maltase		8.5	maltose	glucose
		lactase		8.5	lactose	glucose + galactose
		sucrase		8.5	sucrose	glucose + fructose
		exopeptidases* (aminopeptidase		8.5	peptides and	aminoacids
		dipeptidase)		8.5	dipeptides	amino acids
		enterokinase	small intestine	8.5	trypsinogen	trypsin
pancreatic juice	pancreatic glands	amylase	small intestine	7.00	amylose	maltose
		endopeptidases* (trypsin**	small intestine	7.00	{ proteins { chymotrypsinogen	peptides chymotrypsin
		elastase	small intestine	7.00	proteins	peptides
		chymotrypsin**)	small intestine	7.00	proteins	amino acids
		exopeptidase* (carboxypeptidase)	small intestine	7.00	peptides	amino acids
		lipase	small intestine	7.00	lipids	fatty acids + glycerol
bile	liver	bile salts (not enzymes)	small intestine	7.6–8.6	lipids	lipid droplets

*Exopeptidases split off terminal amino acids from proteins (polypeptides)
Endopeptidases break bonds between amino acids within proteins thus producing smaller peptides

Collectively these enzymes break up polypeptides into their constituent amino acids so that they can be absorbed by the villi of the ileum

**Rennin and pepsin are secreted in the inactive forms prorennin and pepsinogen respectively
Trypsin is secreted in the inactive form trypsinogen, and chymotrypsin as the inactive chymotrypsinogen

Table 8.4 Comparison of structures of the major regions of the gut.

Layer	*Oesophagus*	*Stomach*	*Small intestine*	*Large intestine*
	specialisation – a few mucus glands located in the lamina propria and submucosa	specialisation – gastric glands located in lamina propria, four cell types: (i) mucus (ii) parietal (iii) chief (iv) endocrine	specialisation – (i) intestinal glands in crypts of Lieberkühn (ii) Paneth cells (iii) endocrine cells	specialisation – intestinal glands in lamina propria
mucosa				
(a) epithelium	stratified, squamous (see fig 6.19)	simple columnar (fig 6.16)	simple columnar, absorptive and mucus cells (fig 6.16)	simple columnar, absorptive and mucus cells (fig 6.16)
(b) lamina propria	some mucus glands	many gastric glands	intestinal glands and prominent lymphatic vessels to transport lipids	tubular glands
(c) muscularis mucosa	present	present	present	present
submucosa	some deep mucus glands	present	glands in duodenum	intestinal glands
muscularis externa (inner circular, outer longitudinal muscle)	transitional from striated (voluntary) muscle in upper region to smooth (involuntary) muscle in lower region	extra innermost layer of oblique muscle; circular muscle forms cardiac and pyloric sphincters	present	present
serosa	present	present	present	incomplete

8.4 Nervous and hormonal control of digestive secretions

Secretion of digestive enzymes and other substances, such as hydrochloric acid, is an energy-consuming process. It would be extremely wasteful of both energy and materials if the body carried on producing them in the absence of food. Instead, the bulk of digestive juice is produced only when there is digestive work to be done. The overall control of digestive activity is coordinated and regulated, and this control involves both the nervous system and the endocrine system (the system of hormone-producing glands). This control is discussed in the following sections.

8.4.1 Saliva

Secretion of saliva into the buccal cavity from the salivary glands is controlled by two types of reflex action. First, a simple unconditional (inborn) reflex occurs when food is present in the buccal cavity. Contact with the taste buds of the tongue (fig 8.16) stimulates receptors sensitive to sweet, salty, sour and bitter tastes. Sensory neurones carry nerve impulses from these receptors to the brain. From there, nerve impulses travel along motor neurones to the salivary glands, the effectors, which are stimulated to secrete saliva. Reflexes which pass through the brain are known as cranial reflexes. Secondly, there are the conditioned reflexes of seeing, smelling or thinking of food. If you relax and think of lemon juice dripping onto your tongue it will probably make you start to produce saliva. A **conditioned reflex** is one which has been learned through experience. A well-known example is the experiment carried out by I.P. Pavlov, in which a bell was rung every time he fed some dogs. Eventually the dogs would salivate at the sound of the bell, even when there was no food present. They were said to have been **conditioned**. The reflex is cranial and operates in the same way as the simple reflex described above. The eye, the ear and the olfactory (smell) receptors in the nose are the important receptors.

8.4.2 Gastric juice

Secretion of gastric juice occurs in three phases. The first is the **nervous phase**. The presence of food in the buccal cavity and its swallowing trigger reflex nerve impulses which pass along the vagus nerve from the brain to the stomach. The sight, smell, taste and even the thought of food can trigger the same reflex. The gastric glands of the stomach are stimulated to secrete gastric juice. This takes place before the food has reached the stomach and therefore prepares it to receive food. The nervous phase of gastric secretion lasts for approximately one hour.

The second phase is the **gastric phase** which takes place in the stomach. It involves both nervous and hormonal control. Stretching of the stomach by the food it contains stimulates stretch receptors in the wall of the stomach. These send nerve impulses to Meissner's plexus (see fig 8.11) in the submucosa, which in turn sends nerve impulses to the gastric glands, stimulating the flow of gastric juice. Stretching of the stomach and the presence of food also stimulates special endocrine cells in the mucosa to secrete the hormone **gastrin**. This reaches the gastric glands by way of the bloodstream and stimulates them to produce gastric juice rich in hydrochloric acid for about four hours.

The third phase is the **intestinal phase** which takes place in the small intestine. When acidified chyme enters and makes contact with the walls of the duodenum, it triggers both nervous and hormonal responses. Receptors in the small intestine are stimulated by the presence of food, but the reflexes, which pass through the brain, *inhibit* secretion of gastric juice and slow the release of chyme from the stomach. This prevents too much food being released into the small intestine at once. In addition, the mucosa of the duodenum produces two hormones, **cholecystokinin (CCK)** and **secretin**. (CCK may also be known as pancreozymin, but use only one of these names! CCK is easier and more widely used.) The two hormones are taken in the bloodstream to the stomach, pancreas and the liver. In the stomach secretin inhibits secretion of gastric juice and CCK inhibits stomach emptying.

8.4.3 Pancreatic juice and bile

Secretin and CCK are produced in the duodenum when acidified chyme enters it from the stomach (see above). Secretin is produced in response to the acid, whereas partially digested fats and protein stimulate CCK production. Both hormones are important regulators of the production of pancreatic juice and bile. Secretin is, in effect, an anti-acid hormone. It stimulates the production of hydrogencarbonate ions in the pancreas and the liver, making the pancreatic juice and the bile more alkaline as a result. This helps to neutralise the acid from the stomach. CCK stimulates synthesis of digestive enzymes by the pancreas and the contraction of the gall bladder to release bile into the duodenum (see fig 8.11 for the location of the pancreas, liver and gall bladder). Bile is made in the liver, but stored and concentrated in the gall bladder. It has a pH of 7.6–8.6.

The secretion of bile and pancreatic juice is also stimulated by nervous reflexes. During the nervous and gastric phases of gastric digestion (see section 8.4.2) the vagus nerve also stimulates the liver to secrete bile and the pancreas to secrete enzymes.

Table 8.5 summarises the endocrine control of the various secretions of the alimentary canal and its associated organs.

8.5 The fate of the absorbed food materials – a summary

Monosaccharides and amino acids are both absorbed into blood vessels in the villi and passed to the liver in the hepatic portal vein. Most of the glucose is stored here or in muscle as glycogen and fats, though some leaves the liver in the hepatic vein to be distributed round the body where it is needed for oxidation during respiration or for use in other functions. Between meals, if the body requires more energy, glycogen in the liver can be reconverted to glucose and transported by the blood to those tissues in need.

Amino acids are used for the synthesis of proteins. The functions of proteins are summarised in table 3.9. They are used particularly for growth and repair, being some of the main constituents of protoplasm. Enzymes and some hormones are proteins. Surplus amino acids cannot be stored and are deaminated in the liver. Their amino (NH_2) groups are removed and converted to urea which is taken in the blood to the kidneys and excreted in the urine. The remainder of the amino acid molecule is converted to glycogen and stored.

Absorbed fats bypass the liver by entering the lymphatic system and being released into veins near the heart. Fats represent the major energy store of the body. Normally, however, glucose is in adequate supply and the fats are not required for energy production. In this case they are stored in adipose tissue below the skin, around the heart and kidneys and in the mesenteries. Some fat is incorporated into cell membranes as phospholipids.

Further details of most of these processes will be found elsewhere in this book.

8.6 Herbivores

8.6.1 Teeth

The dentition of herbivores is closely related to their feeding habits and diet. The sheep is a suitable example. A sheep eats grass. Its dental formula is:

$$2\left[\mathrm{i}\,\frac{0}{3}\ \mathrm{c}\,\frac{0}{1}\ \mathrm{pm}\,\frac{3}{2}\ \mathrm{m}\,\frac{3}{3}\right]$$

Upper incisors and canines are absent. In their place is a horny pad against which the chisel-shaped lower incisors and canines bite when the sheep is cropping grass. Between the front and cheek teeth is a large gap, the **diastema**, which provides space for the tongue to manipulate the cropped grass in such a way that grass being chewed is kept apart from that which is freshly gathered.

The cheek teeth possess broad grinding surfaces. The surface area of the upper teeth is further increased by being folded into a W-shape and that of the lower teeth by being folded into an M-shape. The ridges of the teeth are composed of hard enamel whilst the troughs are of dentine. The jaw joint is very loose and allows forward, backward and sideways movement. During chewing the lower jaw moves from side to side, with the W-shaped ridges of the upper cheek teeth fitting closely into the grooves of the M-shaped lower teeth as they grind the grass. The masseter muscle, which provides the power for grinding, is large and the temporal muscle, which is used for biting, is small (fig 8.24). (In contrast, the temporal muscle is large in carnivores, where a powerful bite is needed, and the masseter muscle is small.)

8.6.2 Cellulose digestion in ruminants

A **ruminant** is an animal which has a complicated digestive system in which the 'stomach' typically has several chambers. Among the ruminant animals are deer, giraffe, antelope, cattle, sheep and goats. The first chamber of the stomach is called the **rumen** (fig 8.25). This acts as a fermentation chamber where food, mixed with saliva, undergoes fermentation by mutualistic (symbiotic) microorganisms such as bacteria, protozoans and fungi. Many of these produce cellulases which digest cellulose. Their presence is absolutely essential to the ruminant which is unable to manufacture cellulase itself. The end-products of fermentation are carboxylic acids

Table 8.5 Summary of hormonal control of the secretions of the gut and its associated organs.

Hormone	*Site of production*	*Main stimulus for secretion*	*Target organ*	*Response*
gastrin	stomach mucosa	stretching of stomach by food	stomach	increased secretion of HCl
cholecystokinin (CCK)	mucosa of duodenum	fatty food and protein in the duodenum	pancreas	increased secretion of pancreatic juice rich in enzymes
			gall bladder	contraction of gall bladder to release bile
secretin	mucosa of duodenum	acid chyme in the duodenum	pancreas	increased flow of hydrogen-carbonate in pancreatic juice
			liver	synthesis of bile rich in hydrogen-carbonate
			stomach	inhibits secretion of gastric juice

(particularly ethanoic, propanoic and butanoic acids), carbon dioxide and methane. The acids are absorbed by the host, which uses them as a major source of energy in respiration. In return the microorganisms obtain their energy requirements through the chemical reactions of fermentation, and have an ideal temperature in which to live.

The partially digested food, the 'cud', is passed to the second chamber, the **reticulum**, where it is formed into pellets. It is then regurgitated and thoroughly rechewed. This is called rumination or 'chewing the cud'. The food is then reswallowed and undergoes further fermentation. Eventually the partially digested food is passed through the first three chambers of the gut to reach the **abomasum** which corresponds to the stomach in humans (fig 8.25). From here onwards food undergoes digestion by the usual mammalian digestive enzymes.

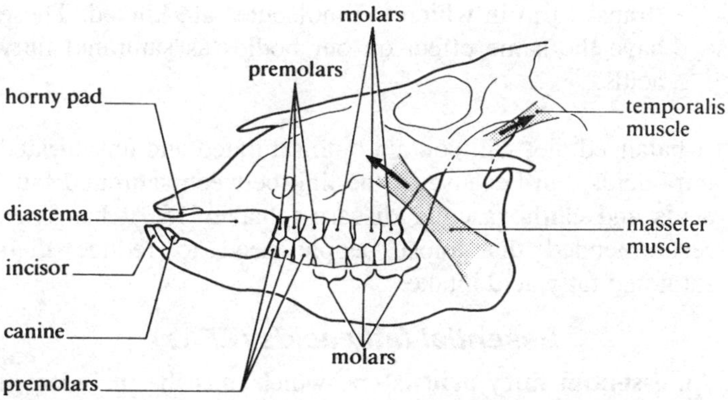

Fig 8.24 *Jaws, dentition and related muscles of the sheep.*

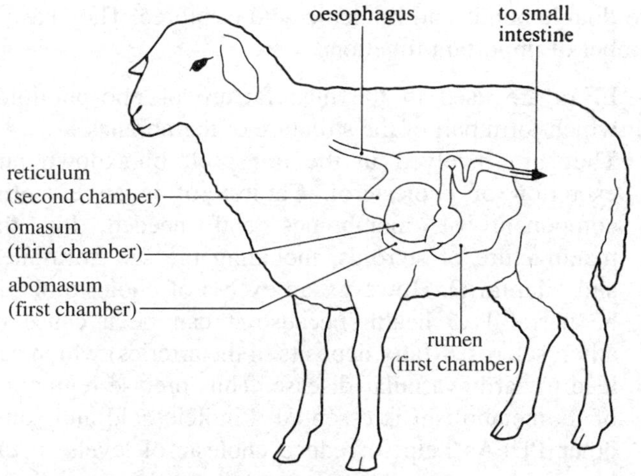

Fig 8.25 *Complex arrangement of chambers preceding the small intestine in a ruminant.*

8.7 Nutrition in humans

Good nutrition has sometimes been described as the key to health. It follows that the opposite, namely malnutrition, can cause ill-health, or disease. **Malnutrition** is, literally, bad nutrition. We tend to associate it with undernutrition, or eating too little, but in its broad sense malnutrition can equally refer to overeating (overnutrition).

Undernutrition not only kills directly by starvation, but indirectly by lowering the resistance of the body to disease. This type of malnutrition is mainly associated with developing countries. In contrast, overnutrition is a leading cause of premature death in developed countries. It can be responsible for, or contribute to, diseases such as coronary heart disease and strokes. Both good nutrition and malnutrition will be examined in this section.

8.7.1 Nutrition, nutrients, food and diet

Nutrition is the acquiring of energy and raw materials needed to maintain life. The two basic nutritional requirements are energy and building materials. These are supplied in the form of chemicals known as **nutrients**. They may be organic, such as carbohydrates, lipids, proteins and vitamins, or inorganic, such as mineral salts. We eat **food** that largely contains these nutrients. **Diet** is the quantity and nature of the food we eat, that is which nutrients and how much of each.

8.7.2 Balanced diet

A **balanced diet** is one which contains the correct proportions and quantity of the various nutrients, water and dietary fibre required to maintain health. Broadly speaking, carbohydrates and fats are needed for energy, proteins for growth and repair processes, and vitamins and minerals for 'protection' of good health and prevention of deficiency diseases.

The ideal diet will vary greatly in different individuals, depending on factors such as their sex, age, activity, body size and the temperature of their external environment (less food is eaten per individual in warm climates).

8.7.3 Water

Water has a wide variety of functions in living cells. It is not therefore regarded simply as a nutrient. Water makes up 65–70% of the total body weight. The importance of water is clear from the fact that a human deprived of water will live for only a few days, whereas it is possible to survive for more than 60 days without food.

8.7.4 Dietary fibre

Dietary fibre is a complex mixture of indigestible compounds derived mainly from plant cell walls. It consists mainly of polysaccharides, particularly cellulose fibres. Its bulk stimulates the movement of food through the gut. There is evidence that fibre helps to reduce blood cholesterol levels, and the risk of bowel cancer and gall stones.

8.7.5 Energy

About 80–85% of the energy content of the average adult diet is provided by carbohydrates and lipids, 15–20% by proteins (up to 5% may be provided by alcohol). Carbohydrates and lipids normally provide the energy actually used. It is needed for three main reasons. These are:

- to maintain the **basal metabolic rate** (**BMR**), that is the activities of the body at rest. This includes the energy for growth when it occurs;
- physical exercise (muscle contraction);
- generation of heat to maintain the body at 37 °C.

BMR accounts for the largest proportion of the energy used.

> **8.7** Why does a mouse require a larger number of joules per unit weight than a human?
>
> **8.8** It has been calculated that 1 g glucose combines with 774 cm^3 oxygen releasing 15.8 kJ heat, and 1 g long-chain fatty acid combines with 2012 cm^3 oxygen releasing 39.4 kJ heat. Why does 1 g of fatty acid release more than twice as much heat as 1 g glucose?

Energy units

Energy units used are joules. In the past calories were used.

4.18 joules = 1 calorie
1000 calories = 1 kilocalorie (kcal) = 1 Cal
1000 joules = 1 kJ (kilojoule) = 1 Joule
1000 kJ = 1 MJ (megajoule)

8.7.6 Carbohydrates

Carbohydrates include sugars and starch (a polysaccharide). Carbohydrates are a major energy source, but sugars are also building blocks for more complex molecules such as nucleic acids, nucleotides (e.g. ATP, NAD) and glycogen.

8.7.7 Lipids (fats and oils)

Lipids include fats, which are solid at room temperature, and oils, which are liquid at room temperature. Like carbohydrates, lipids are a major energy source. Fatty tissues form a convenient long-term energy store in the body. Fats in the diet may also be a source of fat-soluble vitamins (A, D, E and K).

The following are some key facts about lipids.

- The lipids found in foods are mainly triglycerides made from glycerol and three fatty acids.
- There is a large variety of fatty acids in food, but all are either saturated or unsaturated.
- **Saturated** fatty acids have no double bonds (they are saturated with hydrogen). Fats are rich in saturated fatty acids. They have been linked with cardiovascular disease (chapter 15).
- **Unsaturated** fatty acids have one or more double bonds. Oils are rich in unsaturated fatty acids.
- Unsaturated fatty acids may be **monounsaturated** (one double bond) or **polyunsaturated** (more than one double bond).
- Unsaturated fatty acids can be converted to saturated fatty acids by adding hydrogen, a process known as **hydrogenation**. In this way oils can be hardened to make margarines.
- Most natural fatty acids exist in the 'cis' form where the molecules are fairly straight. When unsaturated fatty acids are partially hydrogenated some of the remaining unsaturated fatty acids are converted to the 'trans' form in which the molecules are kinked. These have the same effect on our bodies as saturated fatty acids.

A balanced diet will contain both saturated and unsaturated fatty acids, but because of the link between saturated fatty acids and cardiovascular disease (chapter 15) it has been recommended that most people need to reduce their saturated fatty acid intake.

Essential fatty acids (EFAs)

An **essential fatty acid** is one which must be included in the diet because it cannot be made in the body. Without it, ill-health would result. Only two are strictly essential, namely linoleic acid and alpha linolenic acid. (Note spelling carefully.) They are both polyunsaturated fatty acids (PUFAs) and exist in the cis form. Linoleic acid has two double bonds and linolenic acid has three. They have a number of important functions.

- EFAs are used in the manufacture of phospholipids which form part of the structure of membranes.
- They are involved in the transport, breakdown and excretion of cholesterol. Cholesterol is an essential component of membranes and needed for the manufacture of steroids, including the sex hormones, and vitamin D. However, an excess of cholesterol can be harmful to health because it can be a cause of atherosclerosis (fatty deposits in the arteries) which can lead to cardiovascular disease. Thus precise regulation of its metabolism is essential. Linoleic acid and some other PUFAs help to reduce cholesterol levels in the blood, whereas saturated fatty acids tend to raise them. The best dietary advice is to reduce saturated fatty acid intake.

- Linolenic acid may inhibit the blood clotting associated with atherosclerosis and reduces risk of heart attack in people who have already had one heart attack.
- EFAs are needed to make certain other fatty acids which are physiologically important, such as prostaglandins which are a group of fatty acids that have a wide range of effects on the body. For example, they influence the action of certain hormones, can stimulate inflammatory responses and regulate blood flow to some organs. They are involved in the birth process and have been used in the morning-after pill as an anti-progesterone drug to prevent implantation of a fertilised egg in the uterus.
- Linolenic acid is one of the fatty acids needed for normal development and functioning of the retina and brain.

Deficiency of EFAs is rare since large reserves usually exist in the body fat and daily intake is usually more than adequate.

8.7.8 Proteins

Proteins are used mainly for growth and repair. They have many functions (section 3.5). They can also be used as a source of energy if the diet is deficient in carbohydrate and fat.

Proteins are made of amino acids. There are 20 different amino acids commonly used to make protein, and like fatty acids they can be divided into two types, essential and non-essential.

An **essential amino acid** is one which must be included in the diet because either it cannot be made in the body at all, or it is made too slowly to meet needs. Ill-health would result from insufficient intake. Eight of the 20 amino acids are essential for adult humans, and ten for infants. Non-essential amino acids can be made from essential amino acids in the body. Proteins which are rich in essential amino acids are called first class or high quality proteins. They are most commonly animal proteins, as found in milk and its products, meat, fish and eggs, but soya protein is a useful source of first class protein for vegetarians. Other proteins are referred to as second class or low quality proteins.

8.7.9 Vitamins

Vitamins are organic compounds required in small quantities for good health. They cannot be made within the body, so must be present in the diet.

If a given vitamin is lacking, a characteristic set of symptoms will develop known as a **deficiency disease**. Table 8.6 shows some of the sources and functions of the main vitamins and the deficiency diseases caused by a lack of them. Vitamins A, D, E and K are fat-soluble and the rest are water-soluble.

Vitamin A (retinol)

The proper chemical name of vitamin A is **retinol**. It is found in food of animal origin. The orange pigment carotene, familiar in carrots, and similar pigments called carotenes, are found widely in plants and can be converted to vitamin A during digestion. The structure of carotenes and vitamin A is particularly well adapted for light absorption, both in plants in the form of carotenes and in animals where vitamin A is converted to the light-absorbing molecule **retinal**. The three groups of animals which possess eyes (molluscs, arthropods and vertebrates) all use retinal as the light-absorbing part of their photoreceptor molecules. Light brings about a relatively large change in the structure of retinal, sufficient to trigger the generation of a nerve impulse.

Vitamin A is also needed for healthy skin and other epithelial (surface) tissues and is required by young children for growth.

Deficiency disease. A deficiency of vitamin A affects the rods (which react to light intensity) in the eye much more than the cones (which react to colour) and leads at first to a condition known as **'night blindness'**. This is poor adaptation to conditions of low light intensity when vision is mainly dependent on the rods. Night blindness is caused by a deficiency of retinal in the rods. Eventually the rods themselves become damaged. At the same time the conjunctiva and the cornea become drier and uncomfortable. This can lead to a condition known as **xerophthalmia** (*xero*, dry; *ophthalmia*, eye) with ulcers occurring on the cornea leading to blindness (**keratomalacia**).

Young children are particularly susceptible to vitamin A deficiency because it also reduces growth. Prolonged deficiency can lead to death. The condition is still common in some developing countries and is the most common cause of blindness in children. About 3 million children under the age of 10 years are blind as a result. A person with a healthy diet could be expected to have up to two years' supply in the liver, where it is stored. Average daily consumption in Britain is about twice that needed.

Excess vitamin A. Rare cases have been reported of vitamin A liver poisoning resulting from intake of excess amounts, often from prolonged consumption of large amounts of vitamin pills. Bone damage, hair loss, double vision, vomiting and other problems may also occur. High intakes (in excess of 3300 μg per day) during pregnancy may cause birth defects. In the UK, pregnant women are generally advised not to take vitamin supplements containing vitamin A unless advised to do so at antenatal classes or by a doctor.

Regular intakes should not exceed 6000 μg per day for adolescents, 7500 μg per day for adult women and 9000 μg per day for adult men.

Table 8.6 Sources and functions of the main vitamins required in the human diet and deficiency diseases caused by a lack of them.

Name of vitamin and its designated letter	*Principal sources*	*Function*	*Deficiency diseases and symptoms*
Fat-soluble vitamins			
A (Retinol)	Fish-liver oil, liver, milk and derivatives, carrots, spinach, watercress	Controls normal epithelial structure and growth. Used to make retinal, which is essential for the formation of the visual pigment rhodopsin. Aids 'night vision'	Skin becomes dry, cornea bcomes dry and mucous membranes degenerate. Poor 'night vision'. Serious deficiency results in complete night blindness (xerophthalmia). Permanent blindness (keratomalacia) may occur if the vitamin is not present in the diet.
D (Calciferol)	Fish-liver oil, egg yolk, dairy products, margarine, made by the action of sunlight on a cholesterol-like compound in the skin	Controls calcium absorption from the gut, and concerned with calcium metabolism. Important in bone and tooth formation. Aids absorption of phosphorus.	**Rickets** – this is the failure of growing bones to calcify. Bow legs are a common feature in young children and knock knees in older ones. Deformation of the pelvic bones in adolescent girls can occur which may lead to complications when they give birth. **Osteomalacia** – an adult condition where the bones are painful and spontaneous fractures may occur.
E (Tocopherol)	Wheat germ, brown flour, liver, green vegetables	In rats, it affects muscles and the reproductive system and prevents breakdown of red blood cells. Function in humans is still unknown.	Can cause sterility in rats. Muscular dystrophy. **Anaemia** – increased breakdown of red blood cells.
K (Phylloquinone)	Spinach, cabbage, brussels sprouts, synthesised by bacteria in the intestine	Essential for final stage of prothrombin synthesis in the liver. Therefore it is a necessary factor for the blood-clotting mechanism.	Mild deficiency leads to a prolonged blood-clotting time. Serious deficiency means blood fails to clot at all.
Water-soluble vitamins			
B_1 (Thiamin)	Wheat or rice germ, yeast extract, wholemeal flour, liver, kidney, heart	Acts as a coenzyme for decarboxylation in respiration, especially in Krebs cycle.	**Beriberi** – nervous system affected. Muscles become weak and painful. Paralysis can occur. Heart failure. Oedema (tissues swollen with fluid). Children's growth is reduced. Keto acids, e.g. pyruvic acid, accumulate in the blood.
B_2 (Riboflavin)	Yeast extract, liver, eggs, milk, cheese	Forms part of the prosthetic group of flavoproteins which are used in electron transport.	Tongue sores. Sores at the corners of the mouth.
B_6 (Pyridoxine)	Eggs, liver, kidney, whole grains, vegetables, fish	Converted to a coenzyme for amino acid and fatty acid metabolism	Depression and irritability. Anaemia. Diarrhoea. Dermatitis.
B_5 (Pantothenic acid)	In most foods	Forms part of coenzyme A molecule which is involved in activation of carboxylic acids in cell metabolism.	Poor nerve/muscle coordination. Fatigue. Muscle cramp.
B_3 (Nicotinic acid (niacin) or pp)	Meat, wholemeal bread, yeast extract, liver	Essential component of the coenzymes NAD, NADP which are hydrogen acceptors for a range of dehydrogenase enzymes. Also a part of coenzyme A.	**Pellagra** – skin lesions, rashes. Diarrhoea.
B_{12} (Cyanocobalamin)	Meat, milk, eggs, fish, cheese	RNA nucleoprotein synthesis. Prevents pernicious anaemia.	Pernicious anaemia.
Folic acid (M or Bc)	Liver, white fish, green vegetables	Formation of red blood corpuscles. Synthesis of nucleoproteins.	**Anaemia** – particularly in women during pregnancy.

Table 8.6 *(cont.)*

Name of vitamin and its designated letter	*Principal sources*	*Function*	*Deficiency diseases and symptoms*
H (Biotin)	Yeast, liver, kidney, egg white, synthesis by intestinal bacteria	Used as a coenzyme for a number of carboxylation reactions. Involved in protein synthesis and transamination.	Dermatitis. Muscle pains.
C (Ascorbic acid)	Citrus fruits, green vegetables, potatoes, tomatoes, other fruits e.g. blackcurrants	Concerned with the metabolism of connective tissue and the production of strong skin. Essential for collagen fibre synthesis.	**Scurvy** – skin of gums becomes weak and bleeds. Wounds fail to heal. Connective tissue fibres fail to form. Anaemia. Heart failure.

Vitamin D (calciferol)

For most people enough vitamin D can be made by the action of sunlight on the skin. The light-absorbing molecule found in the skin is made from cholesterol. The active part of the light is ultraviolet (UV) light. There is almost no UV radiation of the suitable wavelength of 280–310 nm in Britain from the end of October to the end of March. Normally, stores of vitamin D build up in the liver in the summer months and provide the body's requirements during the rest of the year.

Most naturally occurring foods are low in vitamin D. Oily fish, such as mackerel, sardines and herring, fish-liver oils and egg yolk are exceptions. These days vitamin D is added as a supplement to some foods, such as margarine (to which it must be added by law) and breakfast cereals.

Vitamin D is converted through reactions first in the liver and then in the kidney (and placenta in pregnant women) to an active form which promotes calcium and phosphate absorption from the intestine. The manufacture of active vitamin D is closely linked with calcium levels in the blood and increases rapidly if calcium levels drop. The active form of vitamin D also affects deposition of calcium and phosphate in bone (bone deposition) and the removal of these from bone (bone reabsorption). In an adult, bone deposition and reabsorption are normally kept in balance so that total bone mass remains constant. A constant remodelling of bones goes on which allows them to adjust in strength and shape to stresses.

In pregnancy and lactation there is extra demand for calcium for the growth of the baby. Vitamin D content of breast milk is relatively low so, where exposure to sunlight is low, supplements may be recommended.

Deficiency disease. A deficiency of vitamin D is particularly damaging in childhood because the skeleton is growing rapidly. For this reason a dietary supply is recommended for children up to the age of 4 years. The deficiency disease is called **rickets**, and is caused by too little calcium and phosphate being added to the bones. This makes the bones too weak and soft to support the weight of the body, causing bowing of the legs and bending of the spine.

In adults vitamin D deficiency leads to a condition known as **osteomalacia** (*osteon*, bone; *malakia*, softness). Here the bones are weakened and soften, becoming less mineralised with calcium and phosphate. Those most at risk are people who receive little exposure to sunlight. In Britain the two groups most at risk are the elderly and the Asian community. Many elderly people lack mobility and spend much of their time indoors. For cultural reasons the Asian diet may be relatively low in calcium as well as vitamin D. The traditional clothing of Asian women leaves little skin exposed to sunlight and darker skins also filter out UV light more effectively than pale skins. Asian children are also particularly at risk. Incidence of rickets in Britain has risen among some poor inner city communities, particularly among Asian communities.

Excess vitamin D. Excess intake of vitamin D can lead to excess calcium uptake. If the excess cannot all be excreted in the urine it may become deposited in the kidneys where it can cause damage. It is more dangerous for infants than for adults.

8.7.10 Minerals

Minerals are inorganic and needed for a wide range of functions. Those needed in the human diet are shown in table 8.7. Seven minerals are needed in only trace amounts for good health, and these are known as **trace elements**. They include manganese, copper, zinc and iodine. Only tiny quantities of trace elements are required in the daily diet. Examples of the roles of minerals are given in table 8.7.

8.7.11 Milk

The only food that most mammals receive during the first weeks of their lives is milk. It provides an almost complete diet during this stage of their development, containing carbohydrate, protein, fat, minerals (especially calcium, magnesium, phosphorus and potassium) and a variety of vitamins. The one major element that milk lacks is iron, a constituent of haemoglobin in blood. However,

Table 8.7 The essential mineral ions needed in a balanced diet and their functions.

Major minerals in order of total amount in body	*Examples of functions*
Calcium	synthesis of bones and teeth
Phosphorus /phosphate	75% is combined with calcium in bones and teeth
	synthesis of nucleic acids (DNA and RNA)
	synthesis of ATP
	synthesis of phospholipids in membranes
Sulphur	mainly present as part of the amino acids cysteine and methionine
Potassium	needed with sodium to maintain electrical potential across cell membranes
	conduction of nerve impulses
Sodium	important constituent of fluid outside cells (tissue fluid)
	helps maintain water balance
	conduction of nerve impulses
	needed with potassium to maintain electrical potential across cell membranes
Chlorine	as sodium (not as important for nerve impulses)
	also a constituent of hydrochloric acid in the stomach
Magnesium	synthesis of bones and teeth
Iron	constituent of the haem group in haemoglobin and myoglobin
Trace elements in order of total amount in body	
Fluorine	associated with structure of bones and teeth, increasing resistance to decay
Zinc	constituent of bone and some enzymes
Copper	constituent of cytochrome oxidase, an electron acceptor in respiration
Iodine	synthesis of the hormone thyroxine
Manganese	constituent of some enzymes involved in respiration and the development of bone
Chromium	involved in the use of glucose
Cobalt	part of vitamin B_{12}

this problem is overcome by the embryo gaining iron from its mother and storing it in its body before birth. It stores enough to allow development until it begins to ingest solid food.

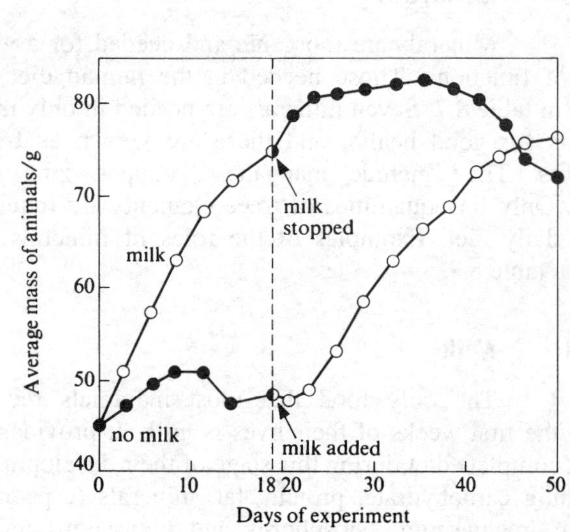

Fig 8.26 *Gowland Hopkins' experiment on feeding milk to rats.*

8.9 Early this century, Frederick Gowland Hopkins in Cambridge performed a famous experiment where he took two sets of eight young rats and fed both on a diet of pure casein (a milk protein), starch, sucrose, lard, inorganic salts and water. The first set also received $3\,cm^3$ of milk per day for the first 18 days. On day 18, the extra milk was denied the first set, but given to the second set of rats instead. The result of the experiment is shown in fig 8.26.

(a) What hypothesis can you deduce from the graph?

(b) Support your answer with comments.

(c) Why is a diet of milk only inadequate for an adult?

8.8 Recommended intakes of nutrients and reference values

The first set of recommendations for human nutritional requirements was made by the League of Nations in 1937. So during the Second World War the government in Britain was able to plan a food policy on a scientific basis. The government later set up a Committee on Medical Aspects of Food Policy (COMA).

In 1979 COMA published tables which summarised the **Recommended Daily Amounts** (**RDAs**) of food energy and nutrients for particular groups of the population. They took account of age, sex, level of activity, and whether a woman is pregnant or lactating (producing milk) and represented the needs of most people within each group, *including those with relatively high needs.*

8.8.1 Dietary Reference Values (DRVs)

In 1987 COMA began reviewing the RDAs, and published another report in 1991. This time they considered 40 nutrients in detail, compared with the previous 10. They also argued that the term RDA was often being misinterpreted as a recommended ideal or a minimum requirement which everyone should be trying to achieve for a healthy life.

> **8.10** Explain why it is a misinterpretation of the term RDA to assume that it is an ideal which everyone should try to achieve.

The term Recommended Daily Amount (RDA) was therefore changed to **Reference Nutrient Intake** (**RNI**) and two further terms were introduced. The two new terms are **Estimated Average Requirement** (**EAR**) and **Lower Reference Nutrient Intake** (**LRNI**). The three terms now used are known collectively as **Dietary Reference Values** (**DRVs**). In tables, DRVs are expressed as amounts per day (tables 8.8 and 8.9). The terms are defined below. Fig 8.27 shows how DRVs relate to the population as a whole. It is assumed that the requirements for energy or any given nutrient within a population is represented by a normal distribution curve (a bell-shaped curve whose mean value occurs with the highest frequency).

- **EAR Estimated Average Requirement**
  This value is given for energy (table 8.8), protein, vitamins and minerals. It represents the estimated

Table 8.8 Estimated Average Requirements (EARs) for energy in the UK (per day).

Age range	*Males*		*Females*	
	MJ	*kcal*	*MJ*	*kcal*
0–3 months (formula fed)	2.28	545	2.16	515
4–6 months	2.89	690	2.69	645
7–9 months	3.44	825	3.20	765
10–12 months	3.85	920	3.61	865
1–3 years	5.15	1230	4.86	1165
4–6 years	7.16	1715	6.46	1545
7–10 years	8.24	1970	7.28	1740
11–14 years	9.27	2220	7.92	1845
15–18 years	11.51	2755	8.83	2110
19–50 years	10.60	2550	8.10	1940
51–59 years	10.60	2550	8.00	1900
60–64 years	9.93	2380	7.99	1900
65–74 years	9.71	2330	7.96	1900
75+ years	8.77	2100	7.61	1810
Pregnant			+0.80[a]	+200[a]
Lactating:				
1 month			+1.90	+450
2 months			+2.20	+530
3 months			+2.40	+570
4–6 months			+2.00	+480
>6 months			+1.00	+240

[a] last three months only.

From *Manual of Nutrition*, Reference Book 342 (HMSO) 10th ed. (1995) table 24, p.68.

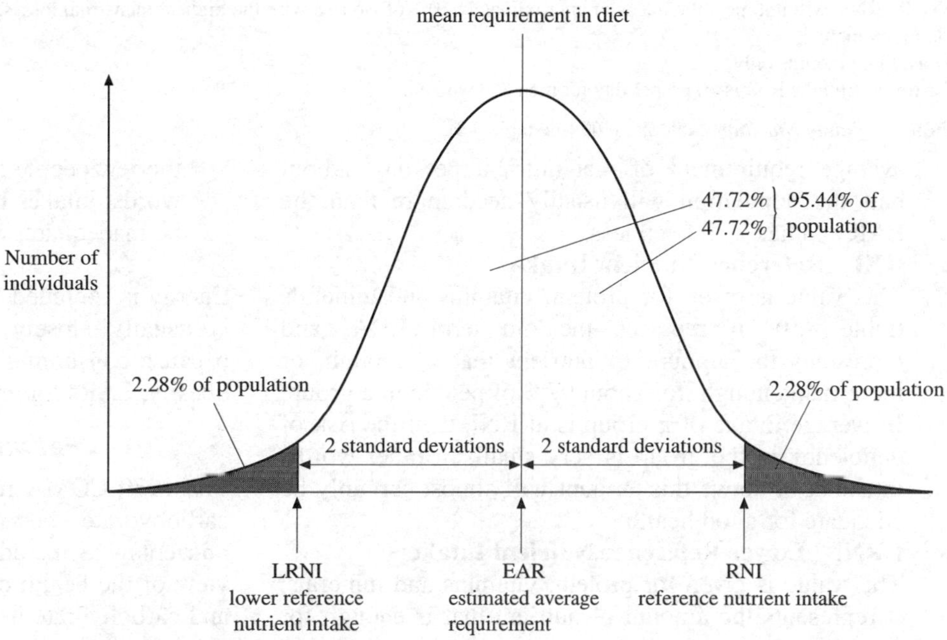

Fig 8.27 *The relationship between LRNI, EAR and RNI in a population.*

Table 8.9 Reference Nutrient Intakes (RNIs) for selected nutrients for the UK (per day).

Age range	*Protein (g)*	*Calcium (mg)*	*Iron (mg)*	*Zinc (mg)*	*Vitamin A (µg)*	*Thiamin (mg)*	*Vitamin B_6*[a] *(mg*[a]*)*	*Folic acid (µg)*	*Vitamin C (mg)*	*Vitamin D (µg)*
0–3 months (formula fed)	12.5	525	1.7	4.0	350	0.2	0.2	50	25	8.5
4–6 months	12.7	525	4.3	4.0	350	0.2	0.2	50	25	8.5
7–9 months	13.7	525	7.8	5.0	350	0.2	0.3	50	25	7
10–12 months	14.9	525	7.8	5.0	350	0.3	0.4	50	25	7
1–3 years	14.5	350	6.9	5.0	400	0.5	0.7	70	30	7
4–6 years	19.7	450	6.1	6.5	500	0.7	0.9	100	30	–
7–10 years	28.3	550	8.7	7.0	500	0.7	1.0	150	30	–
Males										
11–14 years	42.1	1000	11.3	9.0	600	0.9	1.2	200	35	–
15–18 years	55.2	1000	11.3	9.5	700	1.1	1.5	200	40	–
19–50 years	55.5	700	8.7	9.5	700	1.0	1.4	200	40	–
50+ years	53.3	700	8.7	9.5	700	0.9	1.4	200	40	*
Females										
11–14 years	41.2	800	14.8[b]	9.0	600	0.7	1.0	200	35	–
15–18 years	45.0	800	14.8[b]	7.0	600	0.8	1.2	200	40	–
19–50 years	45.0	700	14.8[b]	7.0	600	0.8	1.2	200	40	–
50+ years	46.5	700	8.7	7.0	600	0.8	1.2	200	40	*
Pregnant	+6.0	[c]	[c]	[c]	+100	+0.1[d]	[c]	+100	+10	10
Lactating:										
0–4 months	+11.0	+550	[c]	+6.0	+350	+0.2	[c]	+60	+30	10
4 months	+8.0	+550	[c]	+2.5	+350	+0.2	[c]	+60	+30	10

[a] Based on protein providing 14.7% of the EAR for energy.
[b] These RNIs will not meet the needs of approximately 10% of women with the highest menstrual losses, who may need iron supplements.
[c] No increment.
[d] Last three months only.
* After age 65 the RNI is 10 µg per day for men and women.

From *Manual of Nutrition*, table 25, p.69 (see table 8.8).

average requirement of that nutrient per day. About half the population will usually need more than the EAR and half will need less.

- **RNI Reference Nutrient Intake**
  This value is given for protein, vitamins and minerals (table 8.9). It replaces the old term RDA, and represents the amount of nutrient that is enough, or more than enough, for about 97% of people in a group. If average intake of a group is at RNI, then the risk of deficiency in the group is very small. In other words, intake at or above this amount will almost certainly be adequate for good health.
- **LRNI Lower Reference Nutrient Intake**
  This value is given for protein, vitamins and minerals. It represents the amount of nutrient that is enough for the few people in a group who have low needs. In other words, intakes below this amount will almost certainly be inadequate, with harmful consequences for health.

Energy is excluded from RNIs and LRNIs because appetite is usually closely related to the energy required, and because consuming more energy than required can lead to obesity. EARs for energy are shown in table 8.8.

Fat and carbohydrate

The 1979 COMA report did not include RDAs for fat and carbohydrate because the contribution of these two nutrients was included in the RDA for energy. However, in view of the health concern over the relative amounts of fat and carbohydrate in the diet, and the relative proportions of saturated and unsaturated fats, the 1991 report included recommendations which would help to guide those who plan diets. Table 8.10 shows the energy intake of the average British diet at the time the report was published, and the DRVs. DRVs are not expressed as RNI, LRNI or EAR because, apart from the essential fatty acids, carbohydrates and fats have not been shown to be essential nutrients. Instead, another DRV has been used, namely **recommended percentage of daily energy intake** (average for the whole population).

8.11 Compare the average intakes in 1990 with the DRVs (table 8.10). What are the main ways in which it is recommended that diet should change?
8.12 Suggest two problems that might be encountered in trying to keep an accurate record of an individual's diet.

Note that DRVs are *not recommendations for intakes by individuals or particular groups*. They are reference values which can be used in a variety of ways. However, there is still insufficient data to be able to calculate DRVs with great confidence.

8.8.2 Uses of DRVs

DRVs can be used for a variety of purposes and different DRVs are appropriate in different circumstances.

- **Planning diets**
  When health professionals such as dieticians plan diets for groups, individuals or institutions, or when the government makes national recommendations, it is safest to use the RNIs as the recommended values.

8.13 Explain why this is so.

- **Assessing the diet of individuals**
  One of the least suitable uses of DRVs is in *assessing* the diet of an individual. Assessing diet means judging whether the existing diet is suitable; it is *not* the same as planning a diet. DRVs can be used only as a rough guide to the adequacy of a person's diet.

8.14 Explain why this is the case based on the discussion of DRVs so far.

Table 8.10 Fat, fatty acids and carbohydrates in the diet of British adults – intake in 1990 and DRVs.

Nutrient	*Average intake in 1990 (g/day)*	*Approximate % of daily energy intake*	*DRV (recommended % of daily energy intake) (1991)*
Total fat (glycerol + fatty acids)	87.8	40%	33%
Fatty acids:			
cis-polyunsaturated	13.3	6%	6%
cis-monounsaturated	26.7	12%	12%
saturated	36.5	16%	10% (achieved by only about 5% of adults)
trans-unsaturated	4.8	2%	2%
Total fatty acids	81.3	36%	30%
Total carbohydrate	232	40%	47%
Sugars (excluding those in cell walls and milk)	60	10%	10%
Starch (+ cell wall and milk sugars)	170	30%	37%
Dietary fibre	11.6	11.6 g/day	18 g/ day
Energy	8.6 MJ per day (2061 kcal per day)		
Total energy (actual averages)			
Fat and carbohydrate (excluding dietary fibre)		80%	
Protein		15%	
Alcohol*		5%	
	Total	100%	

* Alcohol contains energy. If the diet contains no alcohol, all the recommendations in the table would be slightly higher.
Data on g/day from Gregory, J., Foster, K., Tyler, H. & Wiseman, M., *The Dietary and Nutritional Survey of British Adults*, HMSO, (1990). Data on sugars and starch have been changed by the author to be comparable with DRVs.

- **Assessing the diets of particular groups**
  The larger the group, the more likely that errors and variations among individuals will be averaged out, and therefore the more accurate will be estimates of mean nutrient intakes of that group. Estimates can then be made of the risks of deficiency within the group.
- **Food labelling**
  RDAs have been used on food labels. Thus the consumer might be informed that 100 g of the food contains one-quarter (or 25%) of the recommended daily amount of protein. This is more useful to the average consumer than being informed that the food contains 15 g of protein per 100 g.

> **8.15** State one disadvantage of providing RDAs on food labels.

  COMA suggested that a system of labelling based on EARs would be more useful than using RDAs. Since EAR is an indication of the average amount of food required rather than the upper limit, this would reduce attempts to reach inappropriately high consumption levels.
- **General use**
  DRVs can also be of more general use. They have relevance to agricultural policy. They may also be used by economists and sociologists in tackling some of the wider issues in society, such as poverty.

> **8.16** How might DRVs be of use to a sociologist investigating poverty?

8.8.3 Effect of age, sex and activity on DRVs

Nutritional requirements vary with age, sex, and activity. DRVs have been set for the two sexes and for different age groups as shown in tables 8.8 and 8.9. They are based on average body weights for the particular group, and apply to healthy people.

Energy (table 8.8)

Energy is needed to maintain the basal metabolic rate (BMR), that is the activities of the body at rest. BMR accounts for a large proportion of the energy used and can be estimated accurately. BMR for a 65 kg man is about 7.56 MJ per day. For a 55 kg woman it is about 5.98 MJ per day.

Any movement requires extra energy because it involves muscular activity, and muscle contraction requires energy. The more physically active a person, the more energy they consume. It is useful to express the amount of energy needed for a given activity as a multiple of BMR. This is known as the **physical activity ratio (PAR)**. For example the PAR for walking is 4, meaning that walking requires four times as much energy as BMR.

Physical activity varies with both occupation and leisure. Occupations are traditionally described as follows. PARs are averaged over the whole working day.

- sedentary, meaning non-active or sitting: PAR is 1.7.
- moderately active: PAR is 2.2 in women and 2.7 in men.
- very active: PAR is 2.3+ in women and 3.0+ in men.

These days leisure activities are at least as important as occupation in determining energy requirements and should, therefore, be taken into account. It has been suggested that the following categories are used: sedentary (PAR 2), moderately active (PAR 3) and very active (PAR 4).

Men have proportionately more muscle and less fat than women. They also weigh more on average (growing, maintaining and moving a heavier body requires more energy). The average energy expenditure of men is therefore higher (table 8.8).

For women during the first six months of pregnancy, reductions in physical activity and metabolic rate compensate for the extra energy needed for growth of the fetus and deposition of fat ready for lactation (producing milk for breastfeeding). It is only in the final three months that extra energy is needed (table 8.8). This is needed for growth of the baby and to build up a store of about 2 kg of fat in the mother in preparation for lactation. Breastfeeding requires extra energy because all the baby's energy comes from the milk and the baby continues to grow rapidly after birth.

Protein

RNIs which take into account age and sex are shown in table 8.9. RNIs for infants and children are higher than for adults on a body weight for weight basis, because protein is needed for growth and maintenance. RNIs are the same, on a weight for weight basis, for all adults (age 19+).

In practice, actual average intakes of protein by adults in the UK are higher than the RNIs. Men consume on average 84 g of protein per day, and women 64 g.

The higher average muscle:fat ratio and average body weight of males compared with females means that males require extra protein from the age of 11 (table 8.9).

There is no need to consume extra protein if there is increased activity because it is not usually used as a source of energy. However, if extra body mass occurs as a result of muscle development, a corresponding increase in protein intake may be justified.

Extra protein is required during pregnancy (table 8.9) to allow for growth of the fetus and extra tissue in the mother, such as uterus, placenta, blood and breasts. Breast milk contains the protein needed for growth of the baby so, during lactation, the mother continues to require extra protein.

Minerals

During pregnancy and lactation, particular attention should be paid to making sure that the mother's diet contains sufficient iron, calcium, vitamin C, vitamin D and folic acid (table 8.9). These are needed for making the baby's haemoglobin, bones, teeth and muscle. A good mixed diet will normally satisfy these requirements. Extra folic acid is sometimes given to pregnant women to reduce the risk of spina bifida in the baby.

8.9 Malnutrition

Malnutrition in its broad sense means bad nutrition, and can be applied to both undernutrition and overnutrition (overeating). In the following sections four examples of malnutrition will be examined. Two of these are typical of developed countries, namely anorexia nervosa and obesity. Two are typical of developing countries, namely starvation or general undernutrition, and protein deficiency. Two examples of vitamin deficiency diseases, caused by lack of vitamin A and vitamin D, are discussed in section 8.7.9.

8.9.1 Anorexia nervosa

Anorexia nervosa is sometimes referred to as 'slimmer's disease'. It has become more common in the last 30 years and is associated mainly with affluent western societies, perhaps because such societies tend to stress that 'thin is beautiful'. The term anorexia nervosa means literally 'loss of appetite through nervous causes', but, strictly speaking, the condition is not caused by a loss of appetite because the typical victim will be extremely hungry. Despite this there is a constant fear of putting on weight and this overcomes the desire to eat.

Anorexia is associated mainly with young women (only 10% of cases are men), often at the beginning of adolescence. It most commonly starts as a result of dieting. Gradually the dieting becomes more and more exaggerated and the woman eats less and less until the fear of putting on weight becomes an obsession. At this stage, psychologically, the woman may still be thinking of herself as overweight even if she is becoming severely underweight and beginning to show signs of starvation.

Physically the body returns to a pre-adolescent state, for example menstruation may cease. Also soft, downy hair may grow around the edges of the face and over the shoulders, as in marasmus (section 8.9.4). Starvation may lead to emaciation, which is an extremely thin body, and other symptons typical of marasmus may appear. The body is usually deprived particularly of carbohydrate and fat, which are the foods that diets focus on. The body may therefore start to use protein as a source of energy. Muscles contain a high proportion of the body's protein, so muscles and other body tissues start to waste away. Other side effects include increased incidence of constipation, low blood pressure, tooth decay (dental caries) and susceptibility to infection. Vitamin and mineral deficiencies may occur. Severe cases may be fatal.

A full discussion of the psychology of the condition and its treatment is outside the scope of this book. However, it is very important for anorexics to recognise that they have a problem and to want to seek help. Self-help groups exist and counselling by experts is available. Voluntary organisations exist which can advise on the best approach. Sometimes a period in hospital is necessary.

8.9.2 Obesity

Obesity is the most common nutritional disorder in Britain today and is just as great a problem in most European countries and in North America. A survey of British adults in 1991 showed that 13% of men and 15% of women were obese. A person is described as obese if they weigh at least 20% more than the average for someone of their height and overall frame size (fig. 8.28). It can be measured more scientifically by measuring the amount of fat in the body compared with total body weight. For young men fat is about 12% of normal body weight and for young women about 26%. A figure of more than 20% for men and 30% for women could be considered as indicating obesity.

Obesity is caused when energy input as a result of eating is greater than energy used. Extra fat, carbohydrate, protein or alcohol can be converted into body fat. This does not necessarily mean that the person eats a lot. Surveys have shown that many obese people eat no more than thin people. Other factors such as little physical exercise may be important. There are exceptions. One man who weighed 365 kg (57 stone) used to eat 15 chickens at a time. The amount of excess food needed to produce obesity may be relatively small. Over time regular small excesses accumulate and cause obesity. Apart from quantity of food, the nature of the food may also be a factor. High energy foods such as carbohydrates and fats are more likely to cause problems.

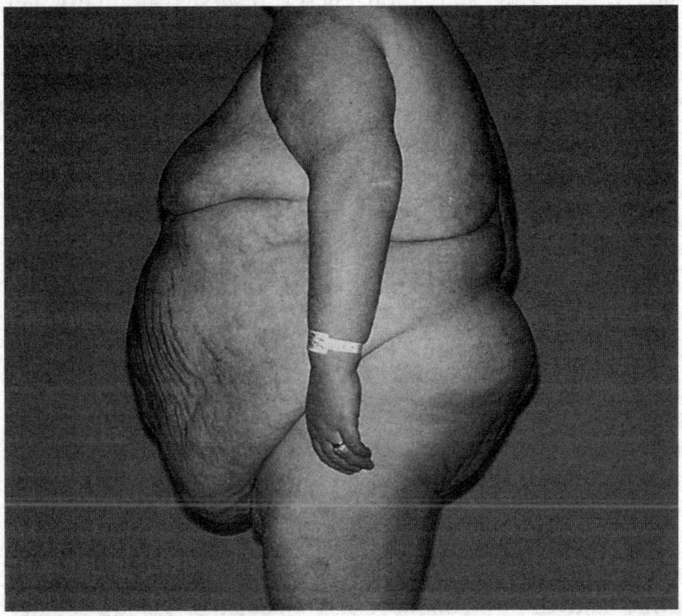

Fig 8.28 *Obese person.*

Obesity often runs in families and there is probably some genetic predisposition in some sufferers. Occasionally it is due to a physiological disorder, particularly involving the hypothalamus or underactivity of the thyroid gland.

Cardiovascular disease is more common in obese people because blood cholesterol levels are typically high and high blood pressure is more common. There is a greater tendency for atherosclerosis in the coronary arteries, with more heart attacks in middle life as a result. There is also an increased incidence of varicose veins.

The extra load on the skeleton commonly causes mechanical difficulties. Flat feet, osteoarthritis of joints, slipped disc and back problems are more common. Hernias are also more common. Movement becomes restricted and difficult and is often slow and awkward. Accidents may happen as a result. Diabetes and some cancers (e.g. gall bladder) are more common in obese people.

Obesity can create emotional problems. Children may become figures of fun and subject to teasing and bullying. Adults are constantly reminded of the association between attractiveness and a slim body by advertising.

Not surprisingly, life expectancy is reduced as a result of obesity. Insurance companies have calculated that a man of 45 who is 10 kg overweight reduces his life expectancy by 25%. The risks are slightly less for women.

To combat obesity, the energy content of the diet must be reduced to the point where energy output is greater than energy input. This should continue until normal weight is restored. This involves counselling the person so that they understand the nature of the problem and can start a low energy diet. A programme of physical exercise may also be recommended.

8.9.3 Starvation and general undereating

The body must maintain a supply of energy for survival. During starvation energy reserves gradually get used up until death results. The first reserve to be used is the carbohydrate glycogen which is stored in the liver and muscles. This supplies energy only for about half a day in the absence of food. Fat stores are then used. In an average person fat can supply enough energy for about 50 days. Fat is broken down in the liver to release fatty acids which are then used instead of glucose in cell respiration. However, ketones made from the fatty acids also tend to build up in the blood, causing a condition called **ketosis** and making the blood acidic. One ketone produced is acetone. It is made in small quantities but can be smelt on the breath and is a sign of ketosis.

For about the first week of fasting, muscle protein is also used as a source of energy. It is converted to glucose, a process called gluconeogenesis. Use of protein then more or less ceases until the fat begins to run out. The renewed use of protein represents the final phase of starvation before death (fig 8.29). Body tissues such as muscle begin to waste away and the person becomes emaciated. Death usually occurs when about half the body's proteins have been used. Complete starvation leads to death in 40 to 60 days.

Children who are undernourished over an extended period of time have stunted growth, and are usually thin. They may develop marasmus (section 8.9.4). It is only relatively recently that the problem of undernourishment has been largely overcome in developed countries, but there are still many problems remaining in providing sufficient food in developing countries.

8.9.4 Protein deficiency, kwashiorkor and marasmus

Protein deficiency can arise in two basic ways. Firstly it may occur when the diet contains sufficient energy, but not enough protein. This occurs in some parts of Africa where the staple food (the food that makes up the bulk of the diet) is corn meal (maize), yam or cassava, all of which are starchy and therefore energy-rich, but deficient in protein in some way. Corn meal lacks one of the essential amino acids, tryptophan, without which proteins cannot be made. Protein deficiency is not common in wheat-growing areas.

A second cause of protein deficiency is lack of sufficient energy in the diet. In this situation the body's own protein is used as a source of energy, as explained for starvation.

Kwashiorkor

In both types of protein-deficiency, a disease called **kwashiorkor** can develop. This term was first adopted in 1935 from the Ghanaian word meaning 'the disease of the child removed from the breast by the birth of the next one'. Switching the child from a milk diet to a starchy diet results in protein deficiency.

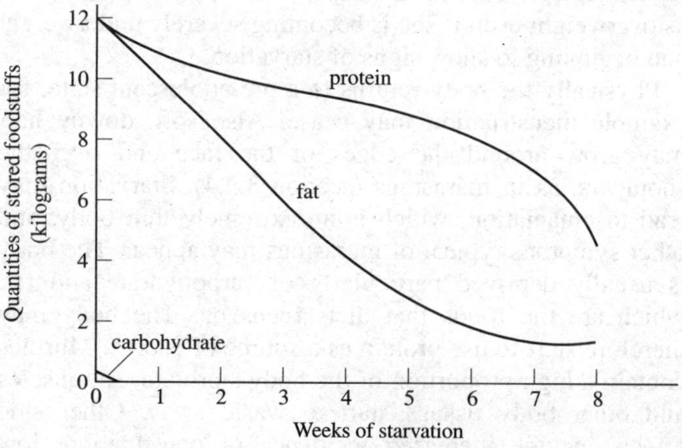

Fig 8.29 *Effect of starvation on the food stores of the body. (From* Textbook of Medical Physiology, *9th ed., A.C. Guyton & J.E. Hall (1996), W.B. Saunders & Co.)*

The characteristic appearance of a child suffering from kwashiorkor can be seen in fig 8.30*a*. Signs and symptoms are:

- the hair changes, becoming thin, straight, sparse and easily removed. It loses pigment and may become white or red.
- the lower cheeks develop large swellings, giving a characteristic 'moon-faced' appearance.
- swollen abdomen due to accumulation of gases and distention of the small intestine caused by abnormal growth of bacteria.
- oedema. This is swelling of the body tissues with fluid, particularly noticeable in the feet and lower legs (later the hands). It is caused by a reduction in blood plasma protein. Water potential of the blood therefore increases and water moves from the blood into the tissue fluid, causing swelling.
- thin muscles, underweight and reduced growth, particularly height. Mental development is also slower.
- skin lesions which cause a 'flaky paint' or 'crazy paving' appearance of the skin. The skin becomes rough. Wound healing is delayed. Jaundice may occur.
- little interest in surroundings and irritability. Babies avoid eye contact, even with their mothers. They may cry continuously and often do not respond to pain or comfort.
- fatty liver. Biochemical changes cause accumulation of fat in the liver which can cause permanent damage.
- vitamin deficiency diseases may be associated with the condition, particularly those due to lack of vitamins A and D.
- reduced resistance to infection.

Kwashiorkor is often fatal.

Marasmus

Another common condition caused by undernutrition is **marasmus**. Originally this was thought to be due to a lack of energy in the diet, but it may not be as clear cut as this, as discussed below. The appearance of a child suffering from some symptoms of marasmus is shown in fig 8.30*b* and can be compared with the child suffering from kwashiorkor. Signs and symptoms are:

- wizened and shrunken features, giving the face the appearance of an old man's face, with sunken eyes.
- thin muscles, thin arms and legs, and low body fat.
- hair is not affected.
- no oedema.
- very underweight. One definition of marasmus states that the child should be more than 60% underweight for its age.
- reduced resistance to infection.
- vitamin deficiency as with kwashiorkor.

As we have learned more about malnutrition caused by undereating, the distinctions between the causes of marasmus and kwashiorkor have become less and less clear. Different children in the same family have developed the two different conditions while feeding on the same diet, and sometimes a child develops marasmus after kwashiorkor. The child in fig 8.30*b* shows the thin limbs associated with marasmus and also the swollen abdomen associated with kwashiorkor. There is a tendency to refer to both conditions now simply as malnutrition, or **protein–energy malnutrition** (**PEM**). This always involves stunted growth and reduced resistance to infectious diseases.

(*a*)

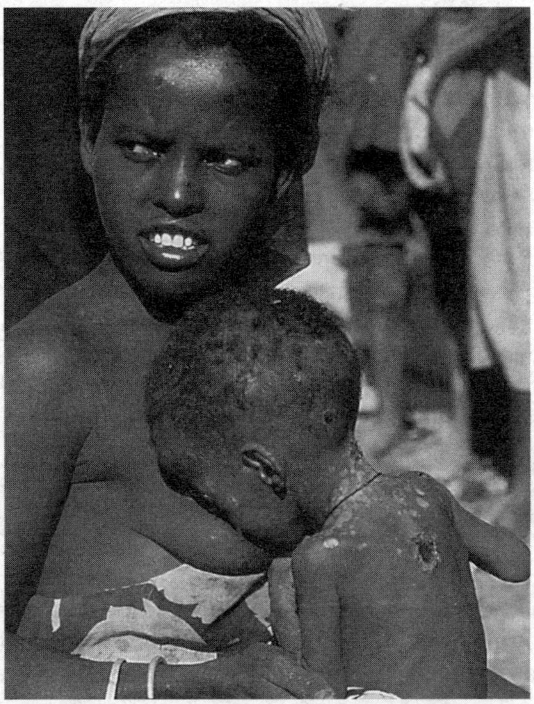

(*b*)

Fig 8.30 (a) *Child showing thin hair and skin lesions associated with kwashiorkor.* (b) *Child showing symptoms of marasmus and kwashiorkor.*

Chapter Nine

Energy utilisation

At the beginning of chapter 7 we considered what energy is and why living organisms need energy. From the examples given, we can see that energy exists in various forms, such as chemical, electrical, mechanical, light and heat energy. You should be able to give examples of the uses of these different forms of energy in living organisms. (If you can't, look back at section 7.1.)

Energy is an important theme in biology. All systems, from cells to ecosystems, require energy so that they can work. Just as a cell would soon die if its energy supply were cut off (cyanide can do this effectively to aerobic cells, as you will see later), so an ecosystem would soon collapse without a constant input of energy from the Sun.

9.1 Explain why animals are dependent on light energy.

Acquiring the energy needed by an individual is one of the two functions of nutrition, the other being to acquire the materials needed to build and repair cells. However the energy in the food has to be made available to cells in a usable form. This is the role of respiration which is the theme of this chapter. The relationship between autotrophic nutrition, heterotrophic nutrition and respiration is shown in fig 7.1. Another way of showing the energy transfer between the environment and cells is given in fig 9.1.

9.1 What is respiration?

Respiration is the process by which chemical energy in organic molecules is released by oxidation. This energy is then made available to living cells in the form of ATP. The biochemical process which occurs within cells is called **cell respiration**. If it requires oxygen, it is described as **aerobic respiration**; if the process takes place in the absence of oxygen, it is described as **anaerobic respiration**.

The organic molecules most commonly used as substrates in cell respiration are carbohydrates, for example glucose, or fats. They are broken down gradually by a series of enzyme-controlled reactions. Each releases a small

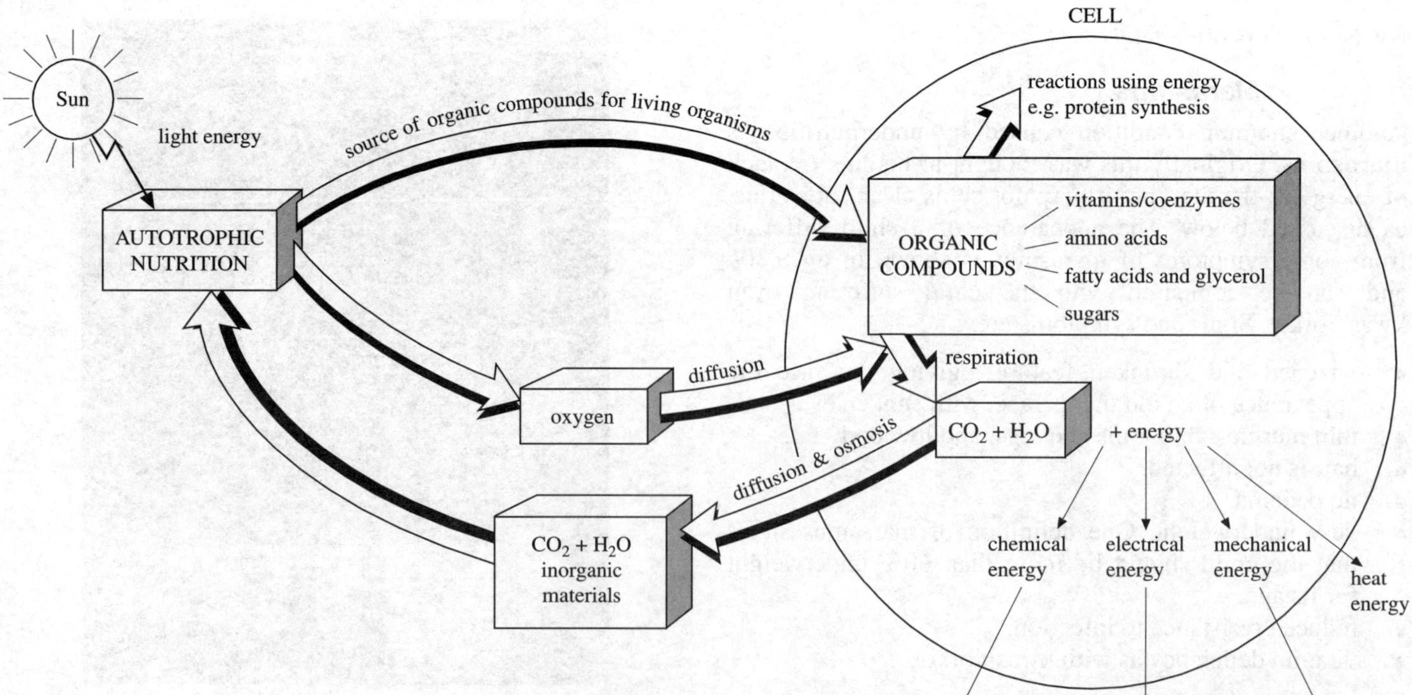

Fig 9.1 *Flow of energy from the Sun to a cell, and finally to waste heat energy. Cycling of carbon is also shown. Nutrition and respiration are responsible.*

amount of energy, some of which is transferred to molecules of a chemical called **adenosine triphosphate** (**ATP**). The rest of the energy is lost as heat. ATP is the energy carrier of cells. The energy in the ATP can then be used when required in reactions in the cell which require energy.

Cell respiration should not be confused with gaseous exchange, which is the process of acquiring oxygen from, and getting rid of carbon dioxide into, the environment. Gaseous exchange may involve organs or structures with specialised surfaces for the efficient exchange of gases, such as lungs and gills (section 9.4).

9.2 ATP

9.2.1 Structure of ATP

Two ways of representing the structure of ATP are shown in fig 9.2.

ATP is adenosine triphosphate, ADP is adenosine diphosphate and AMP is adenosine monophosphate.

AMP is a nucleotide. This is best shown in fig 9.2*a*. A **nucleotide** is made up of a sugar, base and phosphate (section 5.6.1). In AMP the sugar is ribose and the base is adenine. ATP has two extra phosphate groups, making a total of three.

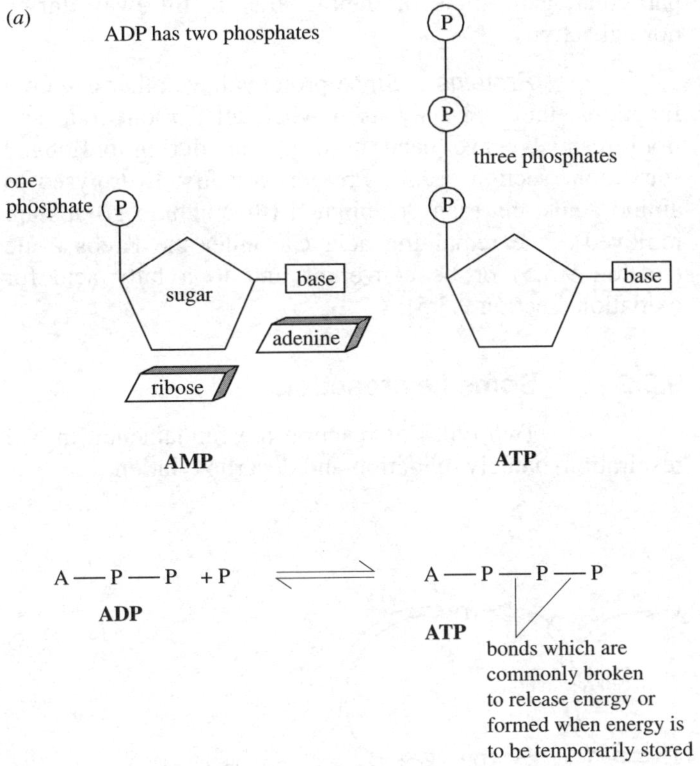

Fig 9.2 *Two ways of representing the structure of ATP.*

9.2.2 Significance of ATP

ATP can be converted to ADP and inorganic phosphate (P_i). This reaction releases energy.

$$ATP + H_2O \xrightarrow{\text{hydrolysis}} ADP + P_i + \text{energy}$$
$$(30.6\,\text{kJ per mole of ATP})$$

The reaction requires water and is described as a **hydrolysis** reaction. This is a common type of biochemical reaction and was encountered many times in chapter 3. The third phosphate group is split from the ATP, and this phosphate remains in the cell in inorganic form. The reaction releases 30.6 kJ for every mole of ATP that is hydrolysed. (A mole is a convenient known quantity of a substance.)

ADP and phosphate can be converted back to ATP, but this requires 30.6 kJ of energy per mole of ATP.

$$30.6\,\text{kJ} + ADP + P_i \xrightarrow{\text{condensation}} ATP + H_2O$$

This reaction releases water and is known as **condensation**. Adding phosphate to ADP is known as **phosphorylation**. Putting the two reactions together:

$$ATP + H_2O \underset{\text{condensation}}{\overset{\text{hydrolysis}}{\rightleftharpoons}} ADP + P_i + 30.6\,\text{kJ per mole of ATP}$$

The enzyme which catalyses this reaction is known as **ATPase.**

All cells require energy, as already explained, and all cell in every kind of organism use ATP as their source of energy when performing their work. ATP is therefore known as the **'universal energy carrier'** or the **'energy currency'** of cells. A useful analogy is the battery. Think of all the uses to which you can put a battery. How could you get light energy, mechanical energy, sound energy, electrical energy, from it? The convenience of a battery is that the same source of energy, the battery, can be used in a wide range of ways to perform work – it is merely 'plugged in' to the right piece of apparatus. ATP does the same in cells. It can be used to make muscles contract, make nerves function, drive active transport and synthesise proteins, and is used in all the other work the cell has to do by being 'plugged in' to the correct piece of cell machinery.

The analogy can be extended further because batteries have to be made in the first place and some, like ATP, are rechargeable. When batteries are made in a factory, energy has to be used in their manufacture. Likewise ATP is made using energy, in this case from the oxidation of organic molecules during respiration. (Since the energy to add the phosphate to the ADP (phosphorylation) comes from oxidation, the process is known as oxidative phosphorylation. In photosynthesis ATP is made using energy from light – photophosphorylation, section 7.6.2). In the manufacture of ATP, the equivalent of the factory is the mitochondrion, a specialised site where the chemical assembly lines for efficiently producing most of the ATP in aerobic respiration are located. Finally there is the recharging of the battery once its energy has been used. Once ATP has been converted to ADP and P_i and released this energy, the ADP and P_i can be rapidly converted back to ATP by re-entering the process of respiration and receiving more energy from the oxidation of more organic molecules.

The actual amount of ATP in the cell at any one time is surprisingly small. ATP should therefore be thought of, not as a store of energy, but as a carrier of energy. Long-term storage of energy takes place in molecules like fats and glycogen. The cell is very sensitive to the amount of ATP present. As soon as the rate at which it is being used goes up, the rate of respiration goes up as well to keep up the supply.

The role played by ATP as the link between cell respiration and the cell's energy requirements is summarised in fig 9.3. This diagram looks simple but it summarises a very important principle.

We can say that the overall function of respiration is to make ATP.

9.2 Copy fig 9.3 and add an extra part to the left to show how the Sun's energy enters glucose.

Summary

- To make ATP from ADP and inorganic phosphate requires 30.6 kJ of energy per mole.
- ATP is found in all living cells and is therefore known as the universal energy carrier. No other carriers are used. This makes things simpler – it reduces the amount of cell machinery needed and is more efficient and economical.
- ATP is mobile and can carry energy to energy-consuming processes anywhere in the cell.
- ATP can release energy quickly. Only one chemical reaction, hydrolysis, is required.
- The rate at which ATP can be re-formed from ADP and inorganic phosphate (the rate of respiration) can be varied quickly according to demand.
- ATP is made during respiration using chemical energy from the oxidation of organic molecules such as glucose, and during photosynthesis using light energy from the Sun. Making ATP from ADP and inorganic phosphate is a phosphorylation reaction. When the energy to carry out the phosphorylation comes from oxidation it is called oxidative phosphorylation (this happens in respiration) and when the energy comes from light it is called photophosphorylation (this happens in photosynthesis).

9.3 Cell respiration

9.3.1 Respiratory substrates

Respiration involves oxidation of an organic compound, the respiratory substrate. Carbohydrates, fats and proteins may all be used.

Carbohydrates. When these are available they are usually used first by most cells. Polysaccharides such as starch (in plants) and glycogen (in animals and fungi) are hydrolysed to monosaccharides before they enter the respiratory pathway.

Lipids (fats or oils). Lipids are used mainly when carbohydrate reserves have been exhausted. They are first converted to glycerol and fatty acids. Fatty acids are energy-rich and some cells such as skeletal muscle cells, in particular, gain some of their energy in this way during normal activity.

Proteins. Since proteins have other essential functions, they are only used when all carbohydrate and lipid reserves have been used up, as during prolonged starvation (section 8.9.3). Proteins are first hydrolysed to amino acids and then deaminated (their amino groups are removed). The remaining acid can enter the Krebs cycle (section 9.3.5) or be converted first to a fatty acid for oxidation (section 9.3.5).

9.3.2 Some key reactions

Two types of reaction are fundamental in cell respiration, namely oxidation and decarboxylation.

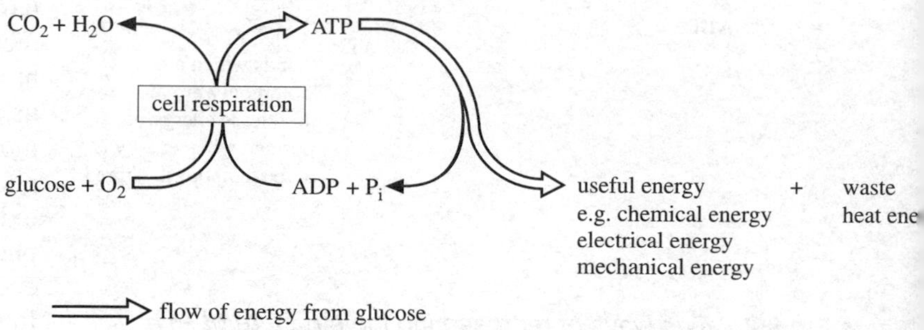

Fig 9.3 *Flow of energy from glucose through ATP to useful work. Once in ATP the energy can be used to do work in the cell, before being lost as heat. ATP is constantly being made in respiration and used in cell reactions. It continuously recycles in cells.*

Oxidation

This may occur in three ways during cell respiration.

(1) Adding oxygen.

$$A + O_2 \longrightarrow AO_2$$

(2) Removal of hydrogen (dehydrogenation). During aerobic respiration glucose is oxidised by a series of **dehydrogenations**. At each dehydrogenation, hydrogen is removed and used to reduce a coenzyme, known as a hydrogen carrier:

AH_2	$+ B$	$\xrightarrow{\text{dehydrogenase}}$	A	$+ BH_2$
reduced respiratory substrate	coenzyme (hydrogen carrier)		oxidised respiratory substrate	reduced coenzyme

Most of these oxidations occur in the mitochondrion, where the usual hydrogen carrier is NAD (nicotinamide adenine dinucleotide):

$$NAD + 2H \longrightarrow NADH_2$$

or, more accurately,

$$NAD^+ + 2H \longrightarrow NADH + H^+$$

NADH (reduced NAD) is later reoxidised, releasing energy, as explained later (section 9.3.5). Enzymes that catalyse dehydrogenation reactions are called **dehydrogenases**. Gradually all the hydrogen is removed from glucose and added to hydrogen carriers. This hydrogen will then be oxidised to water, using oxygen and releasing the energy needed to make ATP. You may have witnessed the energy which can be released from the oxidation (burning) of hydrogen, when a lighted taper is introduced into a test tube of hydrogen. A small explosion, heard as a pop, occurs. In the cell the same amount of energy is released but in a series of small steps known as the respiratory chain.

(3) Removal of electrons.

For example $\quad Fe^{2+} \longrightarrow Fe^{3+} + e^-$

Electrons can be transferred from one compound to another, like hydrogen in the reactions described in (2) above. The compounds involved are called electron carriers. This occurs in mitochondria (section 9.3.5).

Decarboxylation

This is the removal of carbon from a compound by using the carbon to make carbon dioxide. Glucose contains six carbon atoms, as well as hydrogen and oxygen. Since only the hydrogen is needed (see (2) above) carbon is removed by decarboxylation and the carbon dioxide released as a waste product in aerobic respiration.

9.3.3 An overview of cell respiration

Before studying the details of cell respiration it is useful to have an overview of the process. Fig 9.4 outlines the stages of aerobic and anaerobic respiration. Note that there is one aerobic pathway and two anaerobic pathways. Note also that the first stage in all these pathways is **glycolysis**.

9.3.4 Glycolysis

Glycolysis is the oxidation of glucose to pyruvate. One glucose molecule (six carbon atoms, or 6C) is broken down into two molecules of pyruvate (3C) as shown in fig 9.5. It occurs in the cytoplasm of cells, not in the mitochondria, and does not require the presence of oxygen. The process may be divided into three stages:

- **phosphorylation of the sugar** – this 'activates' the sugar, making it more reactive. The process *uses* some ATP. Bearing in mind that the whole point of respiration is to *make* ATP, this may seem unfortunate, but it can be regarded as an investment which allows ATP-producing reactions to occur later.
- **lysis** – the phosphorylated 6C sugar is split into two 3C sugar phosphates. This is the origin of the term glycolysis, which means 'sugar splitting'. The sugar phosphates are isomers of each other. One is converted to the other before continuing, giving two identical 3C sugar phosphates.
- **oxidation by dehydrogenation** – each 3C sugar phosphate is converted to pyruvate. This involves a dehydrogenation, making a reduced NAD molecule, and production of two ATP molecules. The process happens twice, once for each 3C sugar phosphate molecule, so two reduced NAD and four ATP molecules are made (fig 9.5).

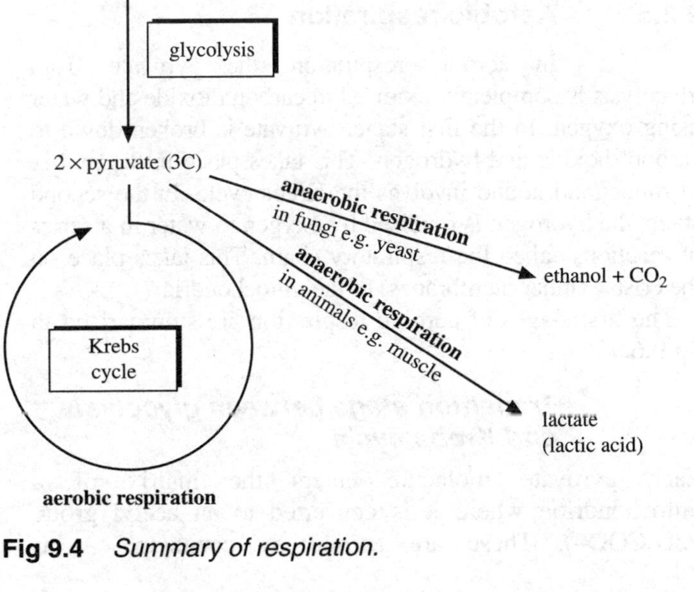

Fig 9.4 *Summary of respiration.*

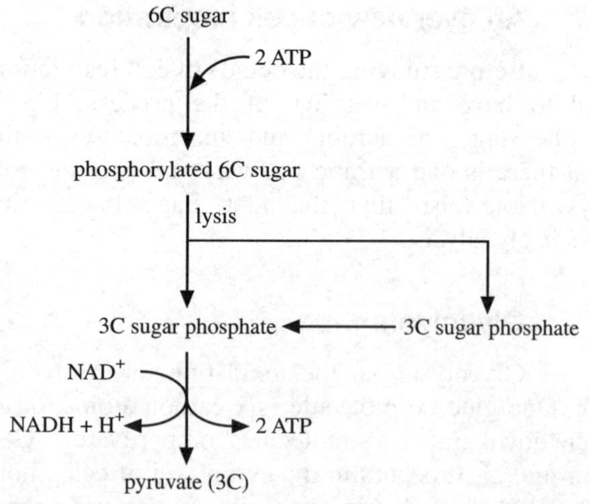

Fig 9.5 *Glycolysis in outline. Only the three key stages are shown.*

Overall two ATP molecules are used for phosphorylation reactions in the first stage, whilst four ATP molecules are produced in the third stage. Therefore there is a net gain of two ATP molecules. Four hydrogen atoms are also released and added to NAD. Their fate will be discussed later. The overall reaction is:

$$\text{glucose} + 2\text{NAD}^+ \longrightarrow 2\text{pyruvate} + 2\text{ATP} + 2\text{NADH} + 2\text{H}^+$$

The input and output of materials during glycolysis is shown in table 9.1.

In the respiration of lipids, the glycerol is easily converted to a 3C sugar phosphate which can join glycolysis. This conversion uses one ATP, but yields three ATPs.

The fate of pyruvate depends on the availability of oxygen in the cell. If it is present, pyruvate will enter a mitochondrion and be completely oxidised into carbon dioxide and water (**aerobic respiration**). If oxygen is unavailable, pyruvate will be converted into ethanol or lactate (**anaerobic respiration**) (fig 9.4).

9.3.5 Aerobic respiration

In aerobic respiration, the pyruvate from glycolysis is completely oxidised to carbon dioxide and water using oxygen. In the first stage, pyruvate is broken down to carbon dioxide and hydrogen. This takes place in the matrix of mitochondria and involves the Krebs cycle. In the second stage, the hydrogen is oxidised by oxygen to water in a series of reactions called the respiratory chain. This takes place on the cristae (inner membranes) of the mitochondria.

The first stages of aerobic respiration are summarised in fig 9.6.

Transition stage between glycolysis and Krebs cycle

Each pyruvate molecule enters the matrix of a mitochondrion where it is converted to an acetyl group (CH_3COO-). These are carried by coenzyme A as

Table 9.1 Budget for glycolysis. ADP, P_i and H_2O are ignored.

Total input	*Total output*
1 molecule of glucose (6C)	2 molecules of pyruvate (2 × 3C)
2 ATP	4 ATP
$2 \times NAD^+$	$2 \times (NADH + H^+)$
Overall 'profit': $2ATP + 2(NADH + H^+)$	

acetylcoenzyme A (acetylcoA). Acetyl groups have two carbon atoms (2C), so the conversion of pyruvate (3C) to acetyl (2C) involves loss of carbon. This is lost as carbon dioxide in a decarboxylation reaction. A dehydrogenation also occurs, so that reduced NAD is made from NAD.

Krebs cycle

The details of this cycle were worked out by Sir Hans Krebs in the 1930s. (It is also known as the TCA cycle (tricarboxylic acid cycle) and the citric acid cycle because of the acids it contains. The acids exist as salts in the cell, so their names end with -ate. Thus citric acid is referred to as citrate).

The Krebs cycle takes place in the matrix of the mitochondrion. A summary is shown in fig 9.6. Acetyl groups (2C) enter the cycle by combining with a 4C compound, oxaloacetate, to form a 6C compound, citrate. As the acetyl groups pass round the cycle, the two carbon atoms are lost in carbon dioxide in two decarboxylation reactions, and the hydrogen is added to hydrogen carriers in four dehydrogenation reactions, resulting in a total of three reduced NAD and one reduced FAD molecules. One molecule of ATP is also made directly for every turn of the

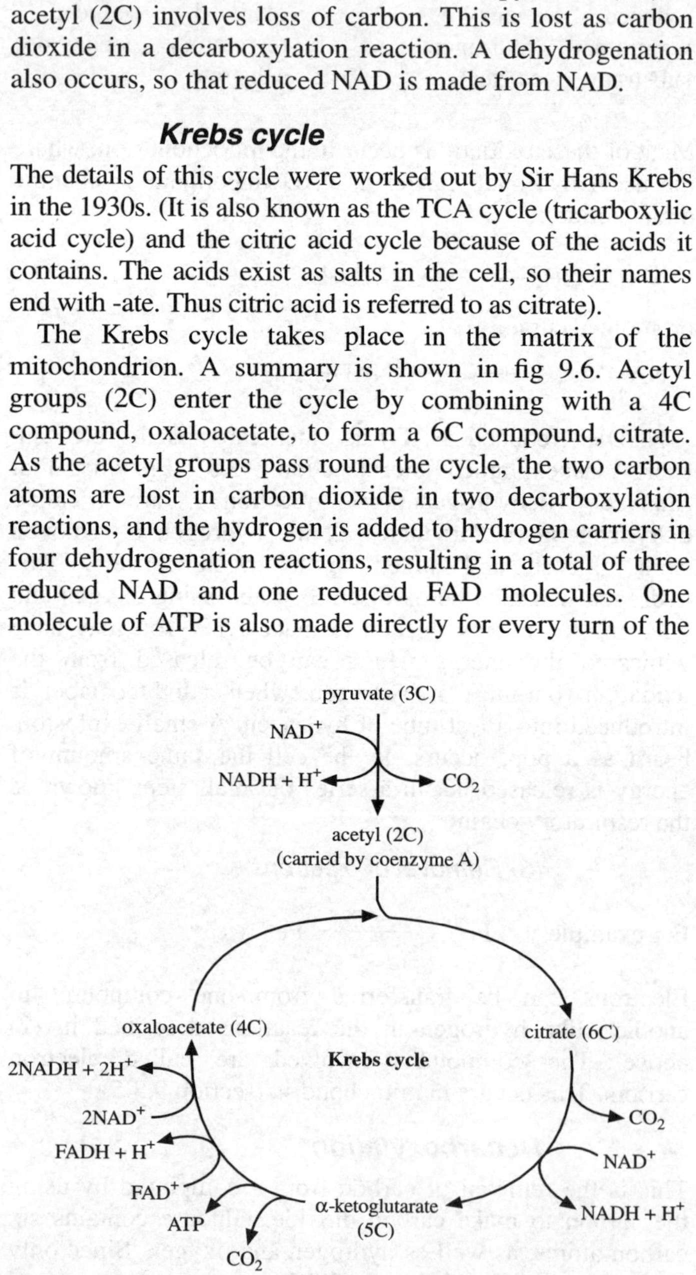

Fig 9.6 *Simplified diagram of the Krebs cycle. The link reaction between glycolysis and the Krebs cycle (pyruvate to acetylcoenzyme A) is also shown. This diagram joins to fig 9.5.*

cycle. (Remember that two acetyl groups were made from one glucose molecule, so two turns of the cycle occur per glucose molecule used.) Oxaloacetate is regenerated at the end of the cycle, ready to accept another acetyl group.

The overall budget for aerobic respiration so far is summarised in table 9.2.

The overall equation is:

$$C_6H_{12}O_6 + 6H_2O \longrightarrow 6CO_2 + 4ATP + 12H_2$$

The hydrogen is on the hydrogen carriers NAD and FAD. All the hydrogen from the original glucose is now on hydrogen carriers. All the carbon is lost in carbon dioxide. (You may be puzzled to note from the equation that six molecules of water have also been used. Water is needed as a source of oxygen in decarboxylation reactions – some of the oxygen in the carbon dioxide comes from this water. This is a detail which you can ignore.)

The respiratory chain and oxidative phosphorylation

The hydrogens (carried as 10 reduced NAD and two reduced FAD molecules) now move to the inner membrane of the mitochondrion. This has folds called **cristae** which increase its surface area (fig 9.12). Hydrogen is a fuel. As already explained, it can be oxidised to water using oxygen, thus releasing energy:

$$2H_2 + O_2 \longrightarrow 2H_2O + \text{energy}$$

Some of the energy is used to make ATP from ADP and inorganic phosphate, in the process of oxidative phosphorylation which was mentioned in section 9.2.2. The energy is not released all in one reaction, but in a series of smaller steps, some of which release enough energy to make ATP. This series of reactions is known as the **respiratory chain**. The respiratory chain is a series of hydrogen and electron carriers ending with oxygen. Hydrogen or electrons are passed from one carrier to the next, moving downhill in energy terms, until they reach oxygen, which is reduced to water as a result. At each transfer some energy is released and in some of the transfers this is coupled to the formation of ATP (fig 9.7 – follow the arrowheads carefully in this figure). The caption to the figure explains the process in more detail. The final stage involves cytochrome oxidase which contains copper. This stage can be specifically inhibited by cyanide (or carbon monoxide). Cyanide combines with the copper and prevents oxygen combining with it.

Fig 9.7 shows that for each reduced NAD that enters the respiratory chain, 3ATP can be made as the hydrogen or electrons flow to oxygen. However, for each reduced FAD, only 2ATP are made because reduced FAD enters the chain at a lower energy level.

The overall budget for the respiratory chain is shown in table 9.3.

The overall equation for the respiratory chain is:

$$12H_2 + 6O_2 \longrightarrow 12H_2O + 34ATP$$

Combining the two equations, we have:

$$(1)\ C_6H_{12}O_6 + 6H_2O \xrightarrow[\text{Krebs cycle}]{\text{glycolysis}} 6CO_2 + 12H_2 + 4ATP$$

$$(2)\ 12H_2 + 6O_2 \xrightarrow{\text{respiratory chain}} 12H_2O + 34ATP$$

Add (1) and (2):

$$C_6H_{12}O_6 + 6O_2 \longrightarrow 6CO_2 + 6H_2O + 38ATP$$

Thus 38 molecules of ATP are produced for every glucose molecule oxidised in aerobic respiration.

9.3 What is the *precise* role of oxygen in respiration?

A summary of aerobic respiration is provided in fig 9.8.

Oxidation of fatty acids

When lipids are used as the respiratory substrate, each fatty acid molecule which is released from lipid is oxidised by a process which involves 2C acetyl fragments being split off from the acid so that the long fatty acid molecule is shortened 2C atoms at a time. Each acetyl group is carried by coenzyme A, forming acetylcoenzyme A. This can enter

Table 9.2 Overall budget for aerobic respiration of one glucose molecule. Remember that two turns of the Krebs cycle take place per glucose molecule.

	CO_2	*ATP*	*NADH + H^+*	*FADH + H^+*
Glycolysis	–	2	2	–
Pyruvate → Acetyl coA	2	–	2	–
Krebs cycle	4	2	6	2
TOTAL	$6CO_2$	4ATP	$10(NADH + H^+)^*$	$2(FADH + H^+)^*$

*enter respiratory chain in cristae of mitochondria.

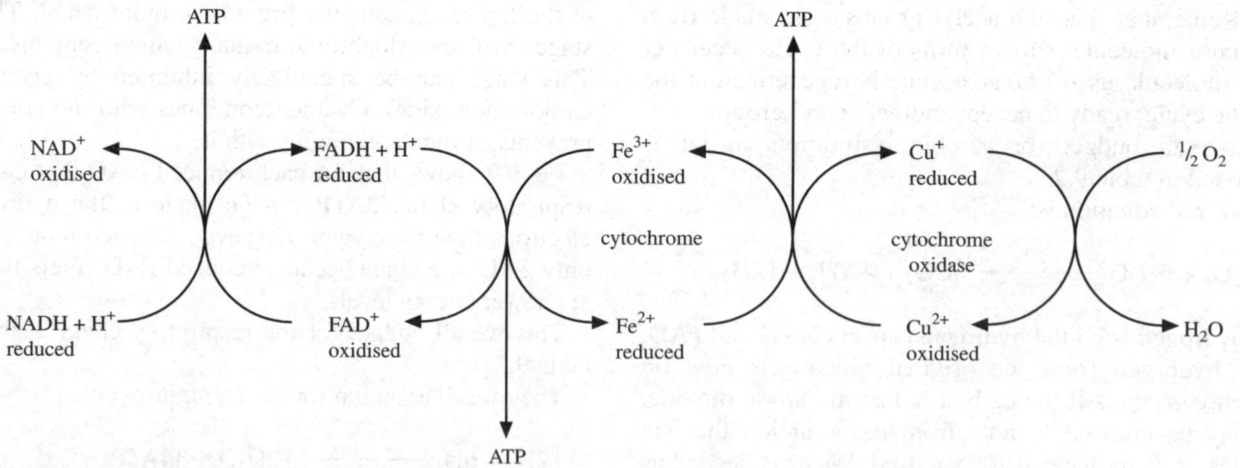

Fig 9.7 Simplified diagram of the respiratory chain. Hydrogen passes from reduced NAD to FAD. Hydrogen then splits into hydrogen ion (H^+) and electrons. Electrons flow from reduced FAD to iron (Fe), copper (Cu), and oxygen, where they are re-united with the hydrogen ions to form water. (Addition of an electron is reduction, loss of an electron is oxidation (section 9.3.2).) The iron is part of a haem group in a protein molecule called a cytochrome. Like haemoglobin, which is also an iron-containing protein, cytochrome is coloured (pink). Copper is part of a group of proteins known collectively as cytochrome oxidase. Cytochromes carry electrons rather than hydrogen.

Table 9.3 Budget for the respiratory chain. Each reduced NAD molecule results in production of three ATP and the release of hydrogen which combines with oxygen to form water, H_2O. Therefore, 10 reduced NAD molecules result in the production of 30ATP, $10H_2O$ and use of 10 oxygen atoms, or five molecules of oxygen. Each reduced FAD molecule results in the production of two ATP.

Entering	*Produced*	*Used*
$12H_2$ in the form of $10NADH + H^+$ and	$30ATP + 10H_2O$	$5O_2$
$2FADH + H^+$	$4ATP + 2H_2O$	O_2
TOTAL	$34ATP + 12H_2O$	$6O_2$

$12H_2O$ are made, but $6H_2O$ are used earlier in respiration (see equation at end of section 9.3.5). Therefore there is an overall production of $6H_2O$.

Krebs cycle as usual. A great deal of energy is released from each fatty acid molecule thus oxidised, for instance 147ATP per molecule of stearic acid. Not surprisingly, therefore, fatty acids are important energy sources, contributing, for example, at least half the normal energy requirements of heart muscle, resting skeletal muscle, liver and kidneys.

9.3.6 Anaerobic respiration

Anaerobic respiration is often referred to as fermentation. A variety of microorganisms use anaerobic respiration as their major source of ATP. Some bacteria are actually killed by normal atmospheric levels of oxygen and have to live where there is no oxygen. They are termed **obligate anaerobes** (for example *Clostridium tetani* which causes tetanus).

Other organisms, such as yeasts and gut parasites (such as tapeworms), can exist whether oxygen is available or not. These can survive on anaerobic respiration when they have to, although respire aerobically when possible, and are called **facultative anaerobes**. Also some cells that are temporarily short of oxygen (such as muscle cells) are able to respire anaerobically.

Like aerobic respiration, the first part of anaerobic respiration is glycolysis. This makes pyruvate from glucose and produces a 'profit' of two ATP and two reduced NAD molecules (table 9.1). In aerobic respiration the hydrogen on the reduced NAD is oxidised to water with release of energy, but this is not possible in anaerobic respiration because oxygen is absent. Instead, the hydrogen is added back to the pyruvate, and its potential for releasing energy is therefore wasted. We shall look at how this happens in animals and fungi. The details are shown below.

Anaerobic respiration in fungi, e.g. yeast

pyruvate ⟶ ethanal + CO_2
enzyme: pyruvate decarboxylase

ethanal + NADH + H^+ ⟶ ethanol + NAD^+
enzyme: alcohol dehydrogenase

Overall: pyruvate ⟶ ethanol + CO_2

No more ATP is made. Ethanol is the alcohol found in alcoholic drinks. This process is known as **alcoholic fermentation** and its occurrence in yeast is made use of in the manufacture of beer, wine and other alcoholic drinks. Production of carbon dioxide by yeast is used in bread making, to make dough rise. Ethanol is a waste product that

Fig 9.8 *Summary of aerobic respiration.*

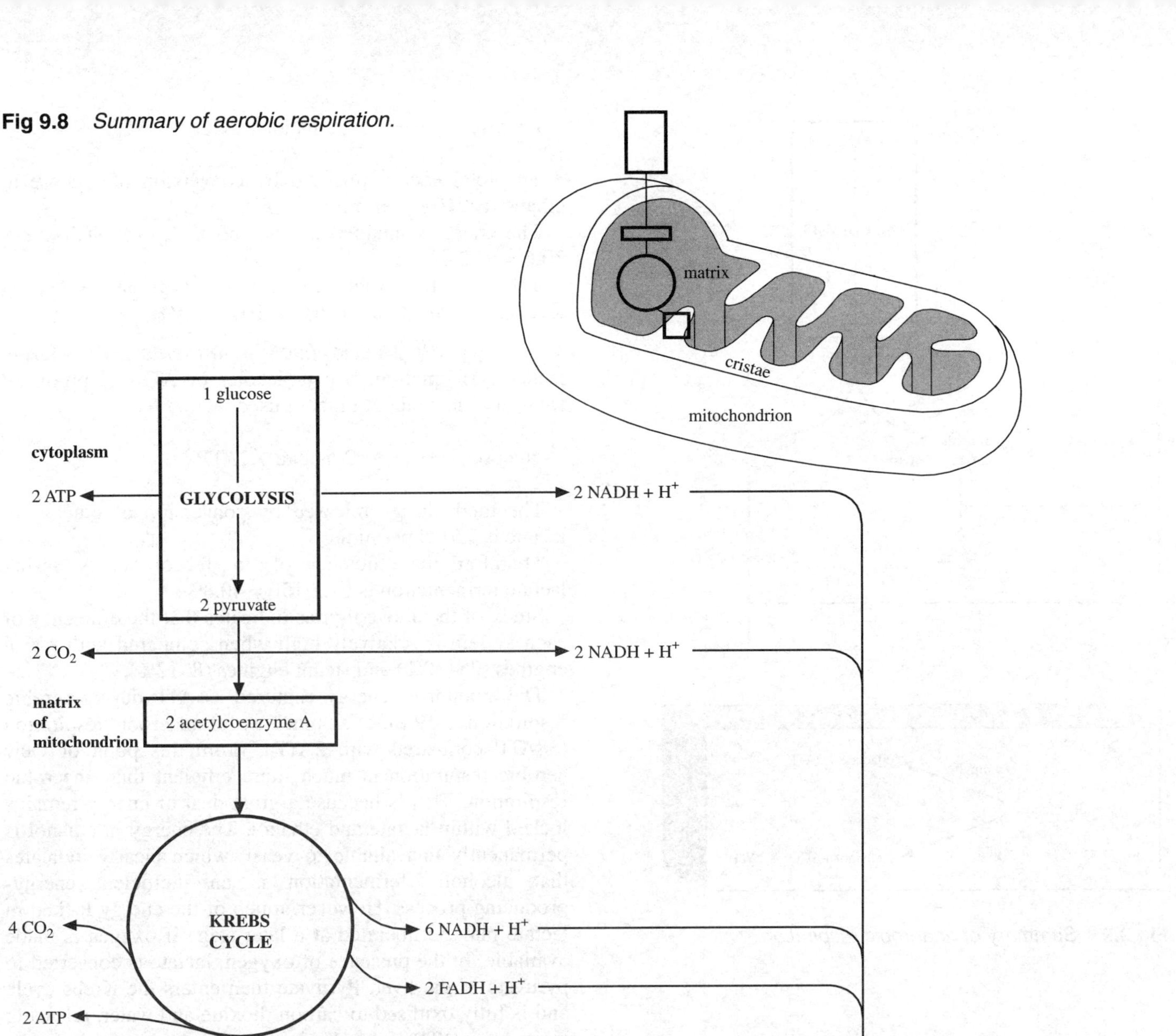

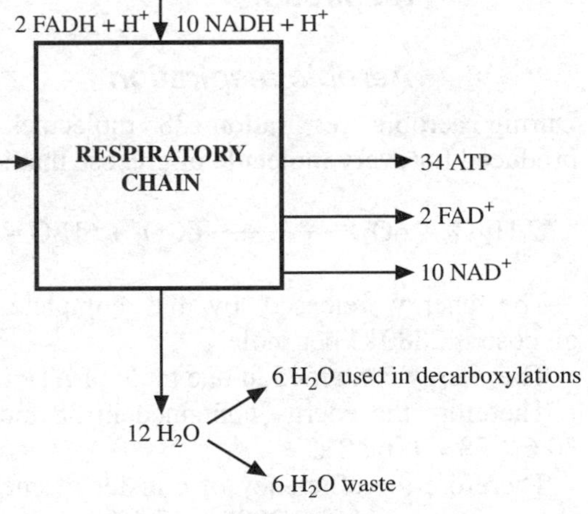

still contains a lot of energy (for example, it is used to make gasohol, a fuel that is used for cars in Brazil). The energy in ethanol is unavailable in the absence of oxygen.

Overall two molecules of ATP are produced per glucose molecule.

Anaerobic respiration in animals, e.g. muscle tissue

pyruvate + NADH + H$^+$ $\rightarrow$ lactate + NAD$^+$

enzyme: lactate dehydrogenase

Note that no carbon dioxide is produced, unlike in fungi. Also, alcohol is not made. Instead, the product is lactate (lactic acid) whose build-up in muscles contributes to the sensation of fatigue and can contribute to cramp. The build-up of an oxygen debt during vigorous exercise is discussed in section 9.3.8.

As with anaerobic respiration in fungi, overall only two molecules of ATP are produced per glucose molecule and the waste product, lactate, still contains a lot of energy.

A summary of anaerobic respiration is given in fig 9.9.

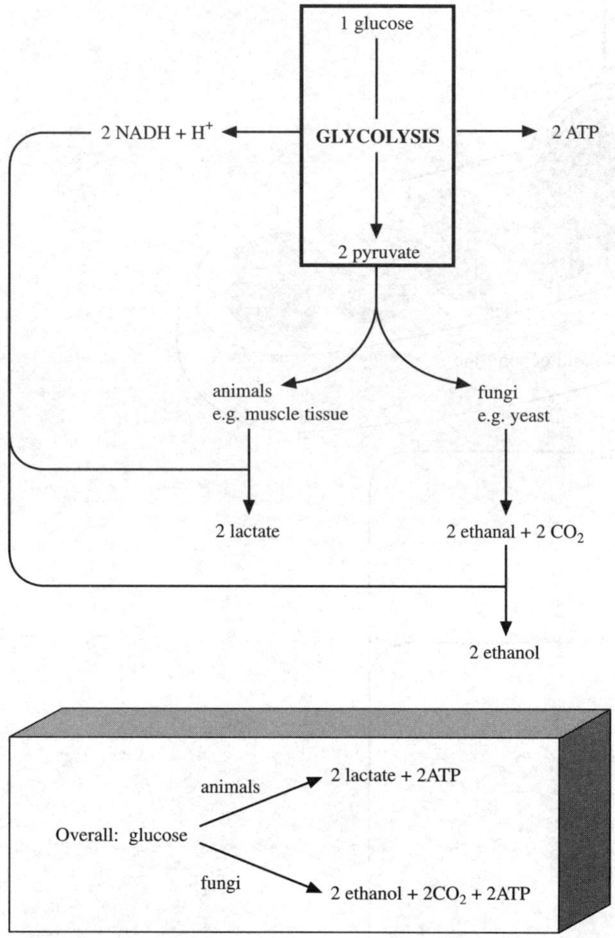

Fig 9.9 *Summary of anaerobic respiration.*

9.3.7 Efficiency of aerobic and anaerobic respiration

Aerobic respiration

During aerobic respiration 38 molecules of ATP are produced for every molecule of glucose that is oxidised.

$$C_6H_{12}O_6 + 6O_2 \longrightarrow 6CO_2 + 6H_2O + 38ATP$$

The energy released by the complete oxidation of glucose is 2880 kJ per mole.

The energy contained in one mole of ATP is 30.6 kJ.

Therefore the energy contained in 38 moles of ATP is $30.6 \times 38 = 1162.8$ kJ.

Therefore the efficiency of transfer of energy in aerobic respiration is = 1162.8/2880 = 40.4%.

Anaerobic respiration

(1) Yeast (alcoholic fermentation). During alcoholic fermentation, two molecules of ATP are produced for every molecule of glucose used.

$$\text{glucose} \longrightarrow 2\,\text{ethanol} + 2CO_2 + 2ATP$$

The total energy released by conversion of glucose to ethanol is 210 kJ per mole.

The energy contained in two molecules of ATP is $2 \times 30.6 = 61.2$ kJ.

Therefore the efficiency of transfer of energy during alcoholic fermentation is 61.2/210 = 29.1%.

(2) Muscle (lactate fermentation). During lactate fermentation, two molecules of ATP are produced for every molecule of glucose used.

$$\text{glucose} \longrightarrow 2\,\text{lactate} + 2ATP$$

The total energy released by conversion of glucose to lactate is 150 kJ per mole.

Therefore the efficiency of transfer of energy during lactate fermentation is 61.2/150 = 40.8%.

Study of the above figures indicates that the efficiency of each system is relatively high when compared with petrol engines (25–30%) and steam engines (8–12%).

The amount of energy captured as ATP during aerobic respiration is 19 times as much as for anaerobic respiration (38ATP compared with 2ATP). From this point of view aerobic respiration is much more efficient than anaerobic respiration. This is because a great deal of energy remains locked within lactate and ethanol. The energy in ethanol is permanently unavailable to yeast, which clearly indicates that alcoholic fermentation is an inefficient energy-producing process. However, much of the energy locked in lactate may be liberated at a later stage if oxygen is made available. In the presence of oxygen, lactate is converted to pyruvate in the liver. Pyruvate then enters the Krebs cycle and is fully oxidised to carbon dioxide and water, releasing many more ATP molecules in the process. Alternatively the pyruvate can be converted back to glucose by the reverse of glycolysis, using energy from ATP.

9.3.8 Oxygen debt and the immediate effects of exercise

There is only a small store of ATP in cells. Normally the body can replace the ATP as fast as it is used, but if a sudden change is made from rest to exercise, it takes a little while to adjust. The body has evolved mechanisms by which it can supply energy to the muscles at the required rate until the rate of aerobic respiration has increased sufficiently. One of these mechanisms is anaerobic respiration. It is important to realise that it is a supplement to aerobic respiration (an extra) rather than an alternative.

Fig 9.10 shows a graph of oxygen uptake during and immediately after a period of exercise. To satisfy the energy demands of the exercise aerobically, 3 dm^3 of oxygen per minute must be supplied (see graph). In the example shown, this is not achieved until six minutes after exercise began. The **oxygen deficit** (the amount of oxygen that was

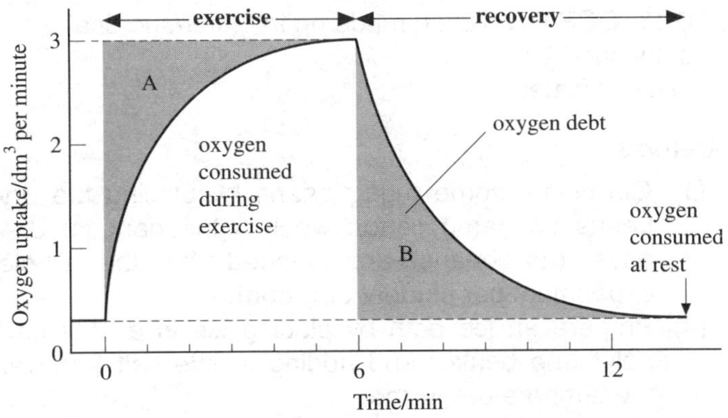

Fig 9.10 *Oxygen uptake during exercise and recovery. This shows the principle of oxygen debt.*

needed, but not supplied from outside the body by breathing), is shown in region A in the graph. During that first six minutes various mechanisms would be used to maintain the supply of energy as described below.

> **9.4** The rate of uptake of oxygen increases immediately exercise starts. How is the supply of oxygen from outside the body to the cells increased during exercise?

Oxygen reserves

The body has reserve oxygen capacity. This includes extracting more oxygen than usual from the lungs, the body fluids and haemoglobin. Also in the muscles, oxygen is stored in a molecule known as **myoglobin**. This is very similar to haemoglobin (fig. 3.36) and, like haemoglobin, can combine reversibly with oxygen. However, it does not release its oxygen until oxygen levels become very low and haemoglobin has already released most of its oxygen. It is therefore only used at times of high demand.

Creatine phosphate system

The amount of ATP in muscles is sufficient only for about three seconds of maximal muscle contraction. Creatine phosphate is another chemical like ATP which contains a phosphate group that can be removed to release energy. It releases enough energy to make ATP from ADP and P_i. Muscle cells have 2–4 times as much creatine phosphate as ATP, so energy can very rapidly be switched to ATP from creatine phosphate when needed. ATP and creatine phosphate together can provide maximum muscle power for about 8–10 seconds.

Anaerobic respiration – the glycogen–lactic acid system

Although anaerobic respiration only produces two molecules of ATP per glucose molecule compared with 38ATP for aerobic respiration, anaerobic respiration can operate much faster than aerobic respiration. In fact, its rate of ATP production is about 2.5 times faster than that of aerobic respiration (five molecules of ATP produced for every two produced by aerobic respiration in a given time period). Anaerobic respiration can therefore supply energy rapidly. It uses glycogen stored in the muscle as a source of glucose. It can provide enough energy for about 90 seconds of maximum muscle activity.

All these systems are improved by regular exercise.

Overall we can see that the creatine phosphate and anaerobic systems supply energy rapidly, but only for short periods of time. The aerobic system supplies energy for an unlimited time, providing the supply of raw materials is adequate. Sports or activities which rely on short explosive bursts of activity, such as a 100 m race or weightlifting, use mainly the creatine phosphate system. Additional energy from anaerobic respiration could be used in a 200 m race. Most of the energy in a 400 m race would come from anaerobic respiration, and sports like tennis, squash and soccer would rely almost entirely on anaerobic respiration during active phases of the game. Endurance sports such as marathon running, jogging and cross-country skiing are almost entirely aerobic.

After exercise, the graph in fig 9.10 does not return immediately to the resting requirement of $0.25\,dm^3$ of oxygen per minute. Instead, the body continues to breathe in and use extra oxygen. The amount of this oxygen is shown in region B of the graph and is known as the **oxygen debt**. It is used for the following.

- It replaces the oxygen reserves in the body, including restoring normal levels of oxygen in the lungs, tissue fluids, haemoglobin and myoglobin.
- It restores creatine phosphate – after exercise the creatine combines with the phosphate again, using energy from aerobic respiration.
- Replacement of oxygen reserves and creatine phosphate takes place rapidly, which explains the steep downward part of the curve in the first few minutes of recovery (fig 9.10). The slow part of the recovery is the removal of lactic acid resulting from anaerobic respiration. This first diffuses into the blood where it is carried away from the muscles, mainly to the liver. Here it is reconverted to pyruvate and reduced NAD. Some of the pyruvate enters the normal aerobic pathway through the Krebs cycle to be oxidised, yielding ATP. This ATP can be used to convert the rest of the pyruvate back to glucose by the reverse of glycolysis (up to 75% of the pyruvate). The heart can also convert lactic acid back to pyruvate and reduced NAD as an extra energy source during heavy exercise.

> **9.5** Why does the lactic acid level in the blood continue to rise *after* exercise when anaerobic respiration has ceased?

9.3.9 Fermentation in industry

Fermentation processes are commercial or experimental processes in which microorganisms are cultured in containers, called **fermenters** or **bioreactors**. They are an important part of modern industry and are discussed in chapter 12.

Experiment 9.1: To investigate the oxidation of a Krebs cycle intermediate

The most efficient way of releasing energy from a substrate and storing this for future use is by a series of smaller reactions, each one reversible and enzyme-controlled. One of the intermediate reactions involved is the oxidation, by removal of hydrogen, of succinate to fumarate.

There are substances which accept such hydrogen atoms and, doing so, change colour. One example is 2,6 dichlorophenolindophenol (DCPIP). It is blue in its oxidised form but loses its colour when reduced.

If the coloured form of DCPIP is decolourised by a tissue extract, one explanation could be that it has accepted hydrogen atoms from succinate. If the rate of decolourisation increased when succinate was added this would tend to confirm the hypothesis that DCPIP was a hydrogen acceptor of atoms from succinate.

As most living processes are governed by enzymes, these must be present before the oxidation will occur. The enzyme succinate dehydrogenase reduces succinate and further experiments could reveal the presence of the enzyme. In this experiment mitochondria are isolated from germinating mung bean seedlings and a suspension of these used as a source of enzyme. It is essential to carry out the extraction as quickly as possible. Once cells are disrupted, further metabolism is short-lived.

The experiment is divided into two parts. The first part consists of the extraction of the enzyme required and the second uses the extracted enzyme to oxidise succinic acid. DCPIP is used to indicate that a reaction has or has not occurred.

Ideally all the apparatus concerned with the first part of this experiment (the preparation of the enzyme extract) should be placed in a refrigerator for at least one hour before it is required for the experiment.

Materials

4 centrifuge tubes (capacity 15 cm^3)
2 glass rods
2 × 10 cm^3 graduated pipettes
2 × 1000 cm^3 beakers (polythene preferably)
ice
salt
mung beans
test-tubes and rack
1 cm^3 graduated pipette
stopclock
Solutions (see notes)
buffer/sucrose solution
buffer/sucrose + succinate (succinic acid) solution
0.1% DCPIP (solution made up in buffer/sucrose solution)
distilled water

Method

(1) Germinate some mung beans by placing the dry beans on damp cotton wool in the dark for 3–4 days (24 beans are needed for the whole experiment per student or group).
(2) Prepare an ice bath by placing ice in a 1000 cm^3 polythene beaker and adding a little salt to lower the temperature further.
(3) Place the flask containing buffer/sucrose solution and two centrifuge tubes in the ice bucket.
(4) Take 12 mung beans and remove their testas and radicles.
(5) Place six beans in each centrifuge tube.
(6) Add 1 cm^3 of buffer/sucrose solution which does not contain succinate to each tube.
(7) Crush the beans thoroughly using a cold glass rod, keeping the tubes in the ice bucket.
(8) Add a further 10 cm^3 of buffer/sucrose solution to each centrifuge tube.
(9) Place the centrifuge tubes on opposite sides of the centrifuge head and spin the tubes at maximum speed for 3 min.
(10) Place the centrifuge tubes back in the ice bucket.
(11) Pipette 15 cm^3 of distilled water into a test-tube and mark the position of the meniscus.
(12) Pour off the distilled water and carefully fill the tube to the mark with supernatant from the centrifuge tubes.
(13) The next step must be carried out very quickly: add 0.5 cm^3 of DCPIP solution to the reaction tube and mix the contents by placing a thumb over the end of the tube and inverting the tube.
(14) Start the stopclock as the solutions are mixing.
(15) Note the colour of the solution after 20 min.
(16) Repeat the entire experiment using the buffer/sucrose solution containing succinate.

The experiment can be monitored colorimetrically. This is done as follows.

(1) Using a red filter, switch on the colorimeter and allow it to warm up for 5 min.
(2) Add 0.5 cm^3 of DCPIP solution to 15 cm^3 of supernatant as before.
(3) Mix the solutions and start the stopclock.
(4) Place the tube in the colorimeter and adjust the needle to 0% transmission.
(5) Take readings after 1, 2, 5, 10 and 20 min.
(6) Repeat the experiment using buffer/sucrose containing succinate.
(7) Plot a graph of percentage transmission (vertical axis) against time.
(8) Draw your own conclusions from the results that you obtain.

Notes on how to make solutions

Buffer/sucrose solution (100 cm^3)

disodium hydrogen phosphate (Na_2HPO_4)	0.76 g
potassium dihydrogen phosphate (KH_2PO_4)	0.18 g
sucrose	13.60 g
magnesium sulphate	0.10 g

Buffer/sucrose + succinic acid (100 cm^3)

As for buffer/sucrose, plus

succinate	1.36 g
sodium hydrogencarbonate	1.68 g

The best method for making up these solutions is to make up enough buffer/sucrose solution for both halves of the experiment (solutions are made up in distilled water). Divide the solution into two and add succinate and sodium hydrogencarbonate (in the correct concentration) to one half.

There will be effervescence when succinate and sodium hydrogencarbonate are added to the buffer/sucrose solution. The solution should be shaken well to get rid of as much carbon dioxide as possible as this could affect the experiment.

DCPIP solution

Use 0.1 g of dichlorophenolindophenol in 10 cm^3 of buffer/sucrose solution (without succinate for both experiments). The solid does not dissolve very well and so after thorough mixing the suspension should be filtered.

9.3.10 Mitochondria

Mitochondria are present in all eukaryotic cells and are the major sites of aerobic respiration within cells. They were first seen as granules in muscle cells in 1850.

The number of mitochondria per cell varies considerably and depends on the type of organism and nature of the cell. Cells with high energy requirements possess large numbers of mitochondria (for example, liver cells contain upwards of 1000 mitochondria) whilst less active cells possess far fewer. Mitochondrial shape and size are also tremendously variable. They may be spiral, spherical, elongated, cup-shaped and even branched, and are usually larger in active cells than in less active ones. Their length ranges from 1.5–10 μm, and width from 0.25–1.00 μm, but their diameter does not exceed 1 μm.

9.6 Why should the diameter of mitochondria remain fairly constant when the length is so variable?

Mitochondria are able to change shape, and some are able to move to areas in the cell where a lot of activity is taking place. This provides the cell with a large concentration of mitochondria in areas where ATP need is greater. Other mitochondria have a more fixed position (as in insect flight muscle, fig 9.11).

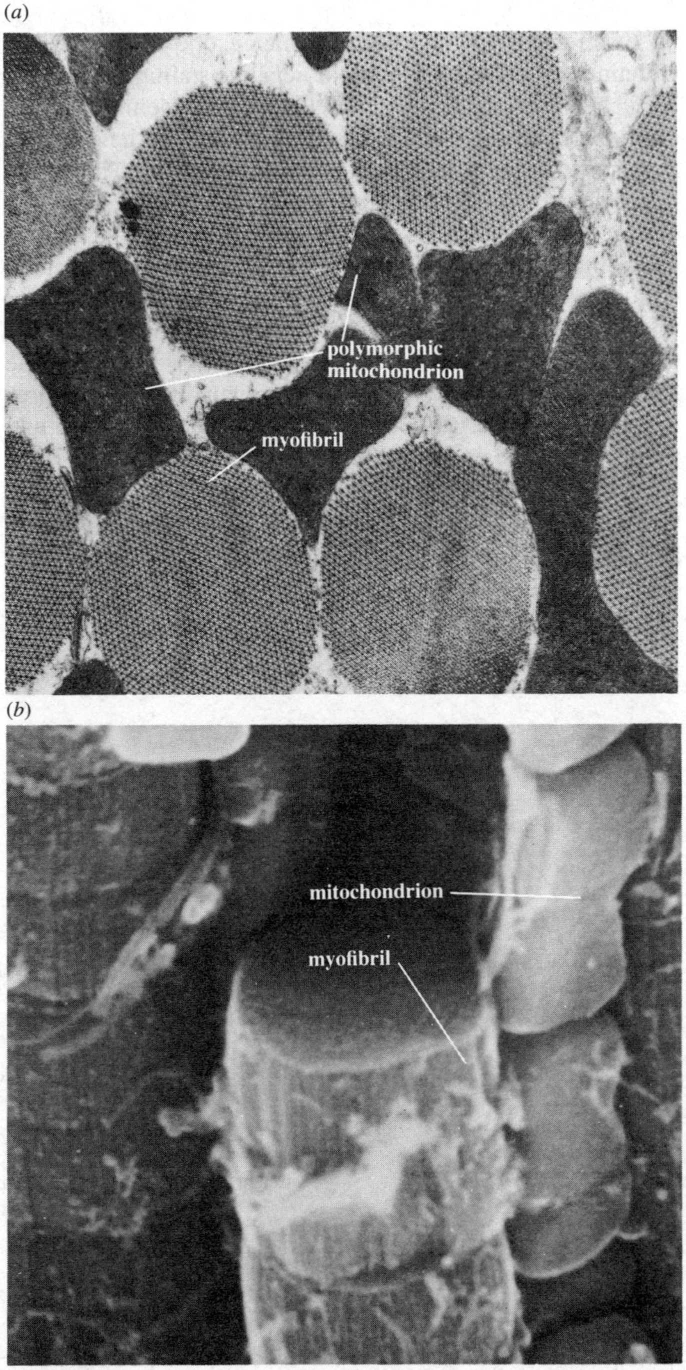

Fig 9.11 (a) *Transmission electron micrograph, and* (b) *scanning electron micrograph of the flight muscle of a house fly. Each muscle fibril (myofibril) is surrounded by polymorphic mitochondria (mitochondria with variable shape). Myofibrils are the part of the muscle that contracts, a process that requires a lot of energy.*

Structure of mitochondria

Mitochondria can be extracted from cells by cell homogenisation and ultracentrifugation techniques as described in chapter 5 (section 5.7). Once isolated they may be examined with an electron microscope using various techniques such as sectioning or negative staining.

Each mitochondrion is bounded by two membranes (an envelope), the outer one being separated from the inner by a narrow space, the intermembranal space. The inner membrane is folded inwards into a number of shelf-like **cristae** (fig 9.12). The cristae increase its surface area, providing space for the components of the respiratory chain which are located in the membrane. Active transport mechanisms are responsible for the movement of ADP and ATP across it. Negative staining techniques which stain the space around structures rather than the structures themselves (fig 9.12*e*) show the presence of **elementary particles** on the matrix side of the inner membrane. Each particle consists of a head piece, stalk and base (fig 9.12*c* and *d*). Whilst the photograph (fig 9.12*f*) suggests that the particles stick out from the membrane into the matrix, it is known that this is an artefact produced by the method of preparation, and that the particles are tucked into the membrane. The head piece is associated with ATP synthesis and is the coupling enzyme, **ATPase**, which links ('couples') the making of ATP to the respiratory chain. At the base of the particle, and extending through the inner membrane, are the components of the respiratory chain itself. The mitochondrial matrix contains most of the enzymes of the Krebs cycle. Fatty acid oxidation also takes place here. In addition, mitochondrial DNA, RNA and 70S ribosomes are present.

9.7 Name four chemical substances which are involved in respiration which would enter the mitochondrion from the cytoplasm and four which would leave.

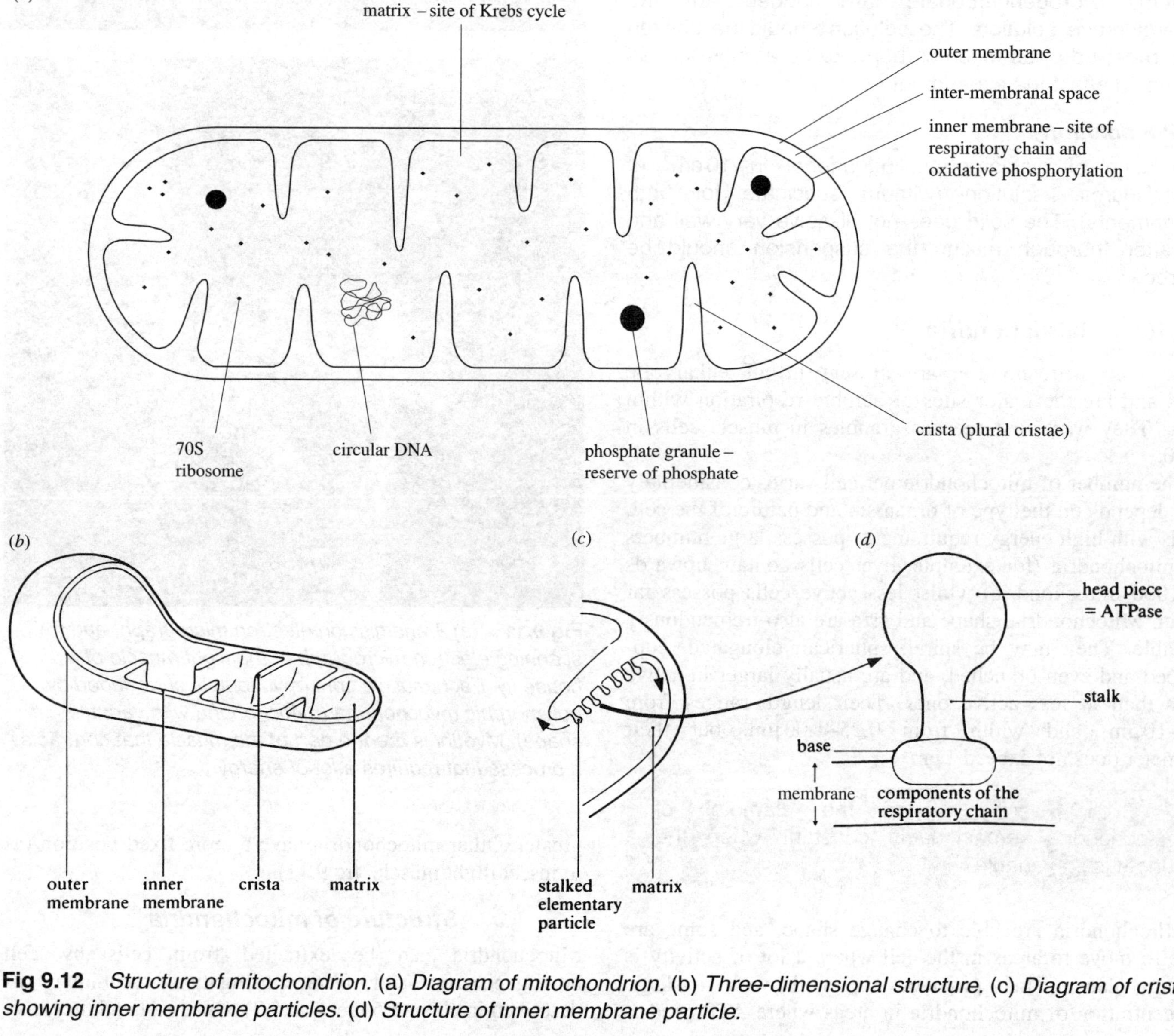

Fig 9.12 *Structure of mitochondrion.* (a) *Diagram of mitochondrion.* (b) *Three-dimensional structure.* (c) *Diagram of crista showing inner membrane particles.* (d) *Structure of inner membrane particle.*

(e)

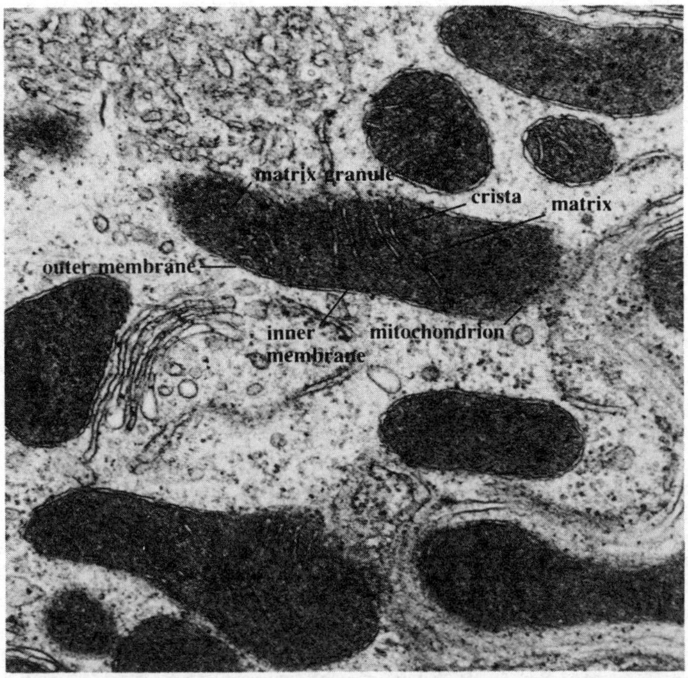

(f)

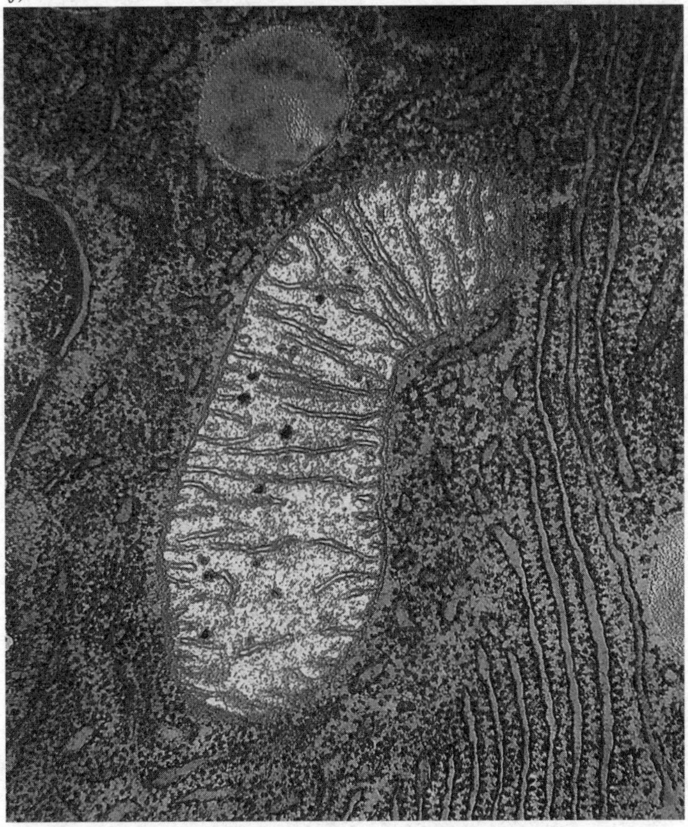

(g)

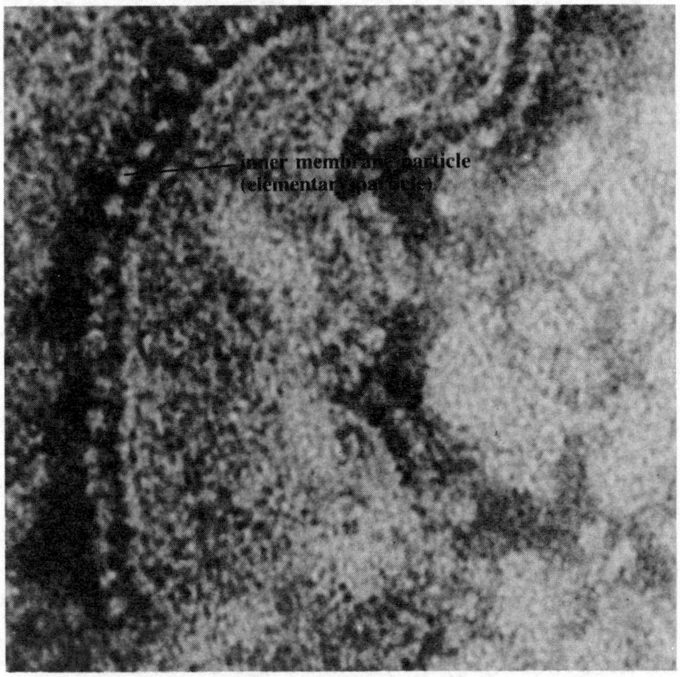

Fig. 9.12 (cont.) (e) *Low power electron micrograph of mitochondria.* (f) *High power electron micrograph of mitochondrion.* (g) *Transmission electron micrograph of negatively stained inner membrane particles from osmotically disrupted mitochondria of the house-fly.*

Evolution of mitochondria – the endosymbiont theory

Mitochondria contain circular DNA (fig 9.13) and 70S ribosomes like bacteria (prokaryotes). This and other evidence suggests that mitochondria, like chloroplasts, were once free-living bacteria. They are believed to have invaded an ancestral eukaryote cell and entered into a successful mutualistic (symbiotic) union with it. (See section 2.6.1 for further information.)

9.4 Gaseous exchange

Gaseous exchange refers to the exchange of respiratory gases between the cells of the organism and the environment. Aerobic organisms require oxygen for respiration whilst both aerobic and most anaerobic organisms must get rid of carbon dioxide, a waste product of respiration. The area where gaseous exchange with the environment actually takes place is called the **respiratory surface**. Gaseous exchange takes place in all organisms by the physical process of **diffusion**. In order for this to occur effectively the respiratory surface must have the following properties.

- It must be **permeable**, so that gases can pass through.
- It must be **thin**, because diffusion is only efficient over distances of 1 mm or less.
- It should possess a **large surface area** so that sufficient amounts of gases are able to be exchanged according to the organism's need.
- It should possess a **good blood supply** (and sometimes a good ventilation mechanism) in those organisms (larger animals) which use blood as a transport

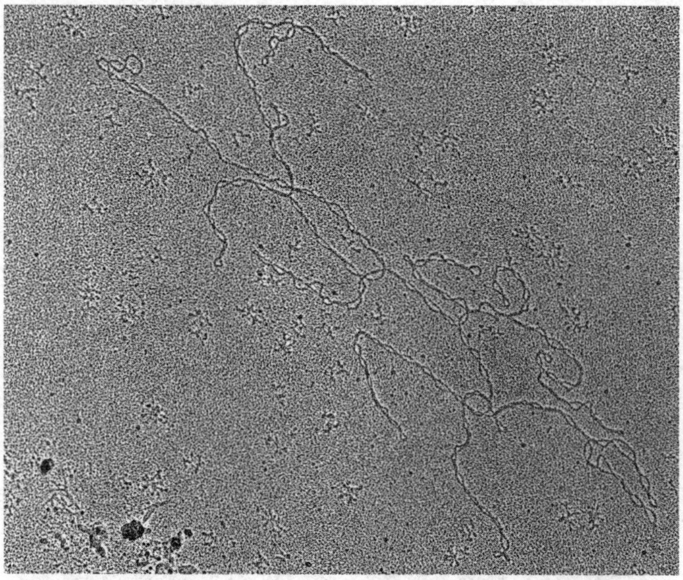

Fig 9.13 *Electron micrograph of mitochondrial DNA from the brewer's yeast* Saccharomyces carlsbergensis. *The molecule is a 'supercoiled' circle of DNA with a circumference of 26 μm. It is made up of some 75 000 nucleotides. It codes for some, but not all, of the mitochondrial proteins. Nuclear DNA contains the remaining genes needed to control mitochondria.*

medium. This helps to maintain a steep diffusion gradient, that is a large difference in concentration, across the respiratory surface.

Fick's law provides a way of considering how the maximum rate of diffusion of respiratory gases is achieved in the examples we shall study. From the law we can predict that the diffusion rate of a respiratory gas through a respiratory surface is proportional to:

surface area* × difference in concentration**/thickness*

* of the respiratory surface
** of the gas either side of the membrane

A similar relationship with membranes has already been noted in section 5.9.8.

Organisms acquire their oxygen either direct from the atmosphere or from oxygen dissolved in water. There are marked differences in the oxygen content of air and water. A given volume of air contains far more oxygen (21% by volume) than an equal volume of water (0.8% by volume). Therefore it follows that an aquatic organism such as a fish must pass a correspondingly much greater volume of water over its gaseous exchange surface than an air-breathing vertebrate passes air in order to obtain sufficient oxygen for its metabolic needs. This requires a different ventilation mechanism. Water is also more than 700 times denser than air and 100 times more viscous. It therefore requires more energy to pass water over the respiratory surface. Oxygen diffuses about 1 000 times slower through water than through air, which means that it is harder to maintain a steep concentration gradient across the respiratory surface. It is not surprising then that fish are unable to achieve the higher metabolic rates of animals with lungs.

9.4.1 A unicellular organism, e.g. *Amoeba*

Amoeba is a single-celled organism which belongs to a group commonly known as protozoans in the kingdom Protoctista. Its shape varies as it moves. It measures less than 1 mm in diameter and possesses a large surface area to volume ratio, typical of unicellular organisms. The respiratory surface is the cell surface membrane. Diffusion of gases therefore occurs over the whole surface of the animal, and is fast enough to satisfy its metabolic needs. Oxygen enters the cell and carbon dioxide leaves, each down its own diffusion gradient. Diffusion across membranes is described in section 5.9.8.

9.4.2 The need for special respiratory structures and pigments

As animals increase in size, so their surface area to volume ratio decreases. This makes simple diffusion over the body surface inadequate to supply oxygen to any cells of the organism that are not in direct contact with the surrounding medium. Also the increased metabolic activity of many of these larger animals increases their rate of oxygen consumption.

In order to cope with the increased demand, certain regions of the body have developed into specialised respiratory surfaces. Different organisms possess different types of gaseous exchange surface. Each has evolved to work efficiently in a particular environment. They can be classified as shown in fig 9.14. Generally they have a large surface area and are often associated with a transport system, the blood vascular system. The possession of a transport system provides a link between the respiratory surface and all the other tissues of the organism, and enables oxygen and carbon dioxide to be continuously exchanged between the respiratory surface and cells. The presence of a respiratory pigment in the blood further increases the efficiency of the blood's oxygen-carrying capacity (see below). In addition there may be special ventilation movements which assist in ensuring a rapid exchange of gases between the animal and the surrounding environment by maintaining steep diffusion gradients.

Respiratory pigments

Respiratory pigments are coloured molecules which act as oxygen carriers by binding reversibly to oxygen. All known respiratory pigments contain a coloured non-protein portion, such as haem in haemoglobin, linked to a protein molecule. Haemoglobin is red. At high oxygen concentrations, the pigment combines with oxygen, whereas at low oxygen concentrations the oxygen is quickly released. Blood that contains any form of

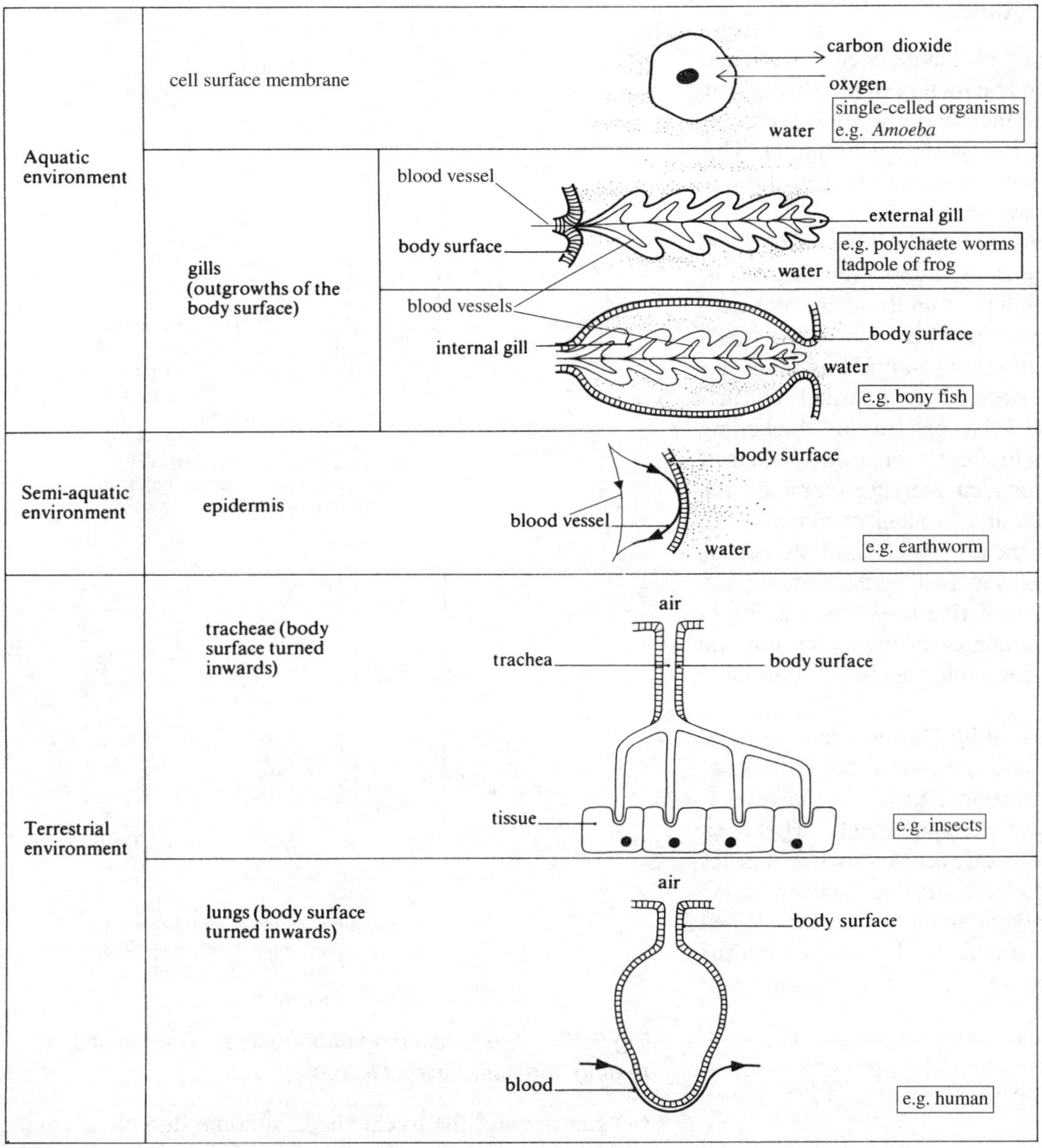

Fig 9.14 *Some types of respiratory surface in animals.*

respiratory pigment is a more efficient oxygen carrier than blood without one. This is because the pigment permits far greater amounts of oxygen to be taken up and transported. In mammals and other vertebrates the pigment is haemoglobin which is enclosed in red blood cells.

A more detailed account of the transport of oxygen by haemoglobin can be found in chapter 14.

9.4.3 Segmented worms, e.g. earthworm

The earthworm belongs to a group of animals known as segmented worms. There are no special organs designed for gaseous exchange in the earthworm. Gas exchange takes place by diffusion over the whole body surface. Special organs are unnecessary as the cylindrical shape of the worms gives a high surface area to volume ratio, and their relative inactivity means that there is only a low rate of oxygen consumption.

Segmented worms do, however, possess a blood vascular system, unlike some simpler animals and unicellular organisms. This contains the respiratory pigment haemoglobin in solution. Pumping activity by the major blood vessels circulates blood and dissolved gases round the body and maintains steep diffusion gradients.

The earthworm lives on land and keeps its thin skin (cuticle) moist by glandular secretions from the epidermis. Looped blood capillaries are present in the epidermis immediately below the cuticle (fig 9.14). The distance between body surface and blood vessels is small enough to enable rapid diffusion of oxygen into the blood. Earthworms have little protection against desiccation (drying out) and therefore they tend to stay in moist conditions.

9.4.4 Insects, e.g. locust

In insects gaseous exchange occurs by means of a system of tubes called the **tracheal system**. This system allows oxygen to diffuse from the outside air directly to the tissues, without the need for transportation by blood. This is much faster than diffusion of dissolved oxygen through the tissues and allows higher metabolic rates.

Pairs of holes called **spiracles**, found on the second and third thoracic segments, and first eight abdominal segments, lead into air-filled sacs. Extending from these are branched tubes called **tracheae** (singular **trachea**) (fig 9.15). Each trachea secretes a thin layer of strong, supporting chitinous material around its outer surface. This is usually further strengthened by spiral or circular patterns of thickening which maintain an open pipeline even when the lumen of the trachea is subjected to reduced pressure (compare the cartilage hoops in the trachea and bronchi of humans). In each body segment the tracheae branch into numerous smaller tubes called tracheoles which spread among the insect tissues, and in the more active ones, such as flight muscle, end within cells. The degree of branching may be adjusted according to the metabolic needs of individual tissues.

Tracheoles lack a chitinous lining. At rest they are filled with watery fluid (fig 9.16) and the rate at which oxygen diffuses through them, and carbon dioxide in the reverse direction, satisfies the insect's requirements. However, during exercise, increased metabolic activity by the muscles leads to accumulation of products such as lactic acid, so making the tissue's solute potential more negative. When this occurs the water in the tracheoles leaves by osmosis into the tissues, causing more air and therefore more oxygen to enter the tracheoles and come into close contact with the tissues just at the time when it is required.

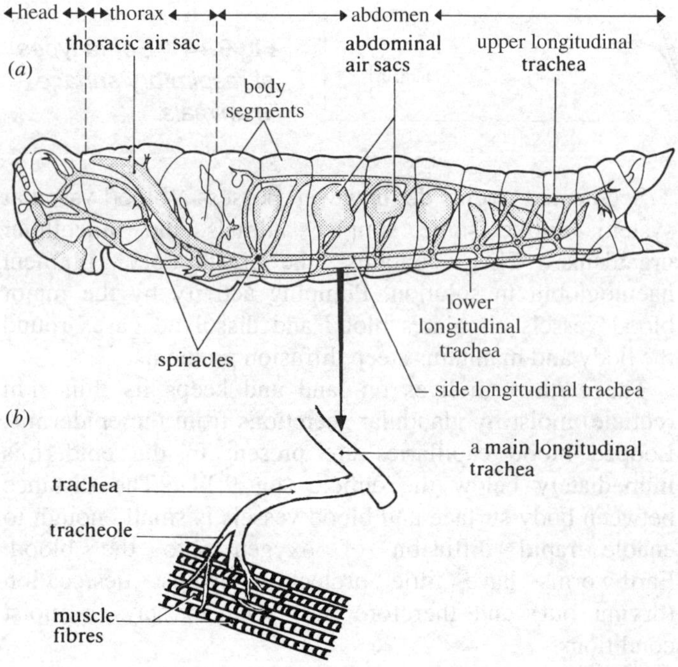

Fig 9.15 (a) *Tracheal system of locust.* (b) *Structure of insect trachea.*

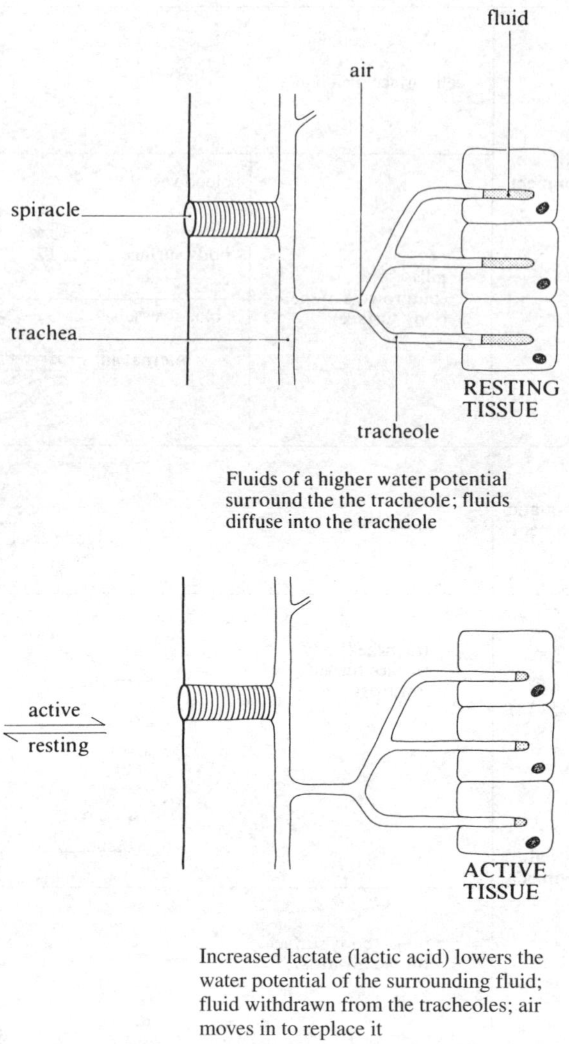

Fig 9.16 *The functioning of tracheoles – conditions in resting and active insect tissues.*

The overall flow of air in and out of the insect is regulated by a valve mechanism at the spiracles. Each spiracle is controlled by a system of valves operated by tiny muscles. It also has hairs around its edges which prevent foreign bodies entering and reduce loss of water vapour. The size of the aperture is adjusted according to the level of carbon dioxide in the body.

Increased activity leads to increased carbon dioxide production. This is detected by chemoreceptors and the spiracles are opened accordingly. Ventilation (breathing) movements by the body may also be triggered by the same stimulus, notably in larger insects such as the locust. Muscles contract and flatten the insect body, decreasing the volume of the tracheal system, thus forcing air out (expiration). Inspiration (intake of air) is achieved passively when the elastic nature of the body segments returns them to their original shape.

There is evidence to suggest that the thoracic and abdominal spiracles open and close alternately, and that this, in conjunction with ventilation movements, provides a one-way flow of air through the animal, with air entering in through the thorax and out via the abdomen.

Even though the tracheal system is a highly effective means of gaseous exchange, in most insects it relies entirely on diffusion of oxygen through the body. Since this can only occur efficiently across small distances, it imposes severe limitations on the size that insects can attain. Diffusion is only effective over distances of up to 1 cm; therefore, even though some stick insects may be up to 30 cm in length, no insects can be more than 2 cm broad!

9.4.5 Bony fish, e.g. herring

Fish possess **gill slits** in the wall of the pharyngeal region of the gut (the region between the buccal cavity and oesophagus). These connect with the outside environment, water. The tissue between the slits forms supports known as branchial arches or **gill arches**. In bony fish there are four pairs of gill arches separating five pairs of gill slits (fig 9.17*a*). Each gill is made up of two rows of **gill filaments** arranged in the shape of a V (fig 9.17*b*). The filaments possess **lamellae** (fig 9.17*c*), thin plates which have a rich supply of blood capillaries. These plates greatly increase the surface area of the respiratory surface. The barrier between blood and water is only several cells thick so diffusion between the two is rapid.

(*a*)

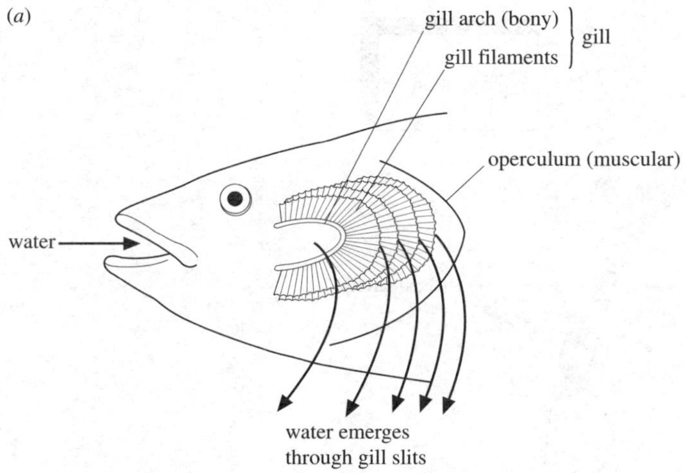

(*b*)

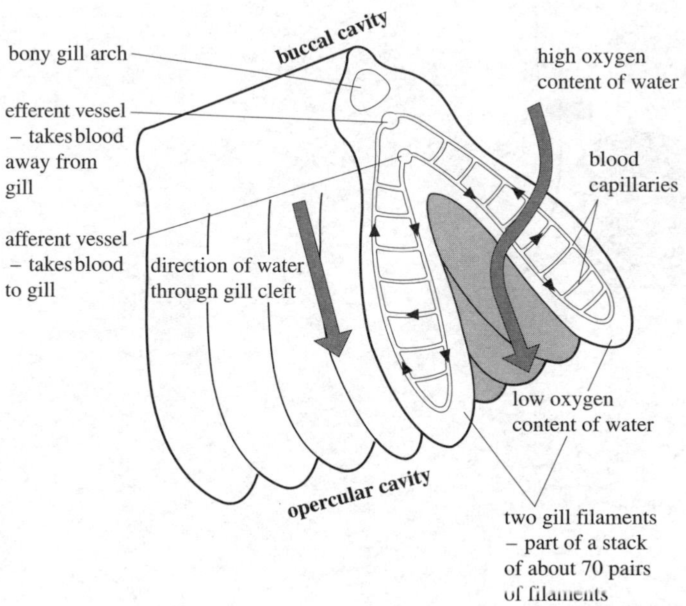

(*c*)

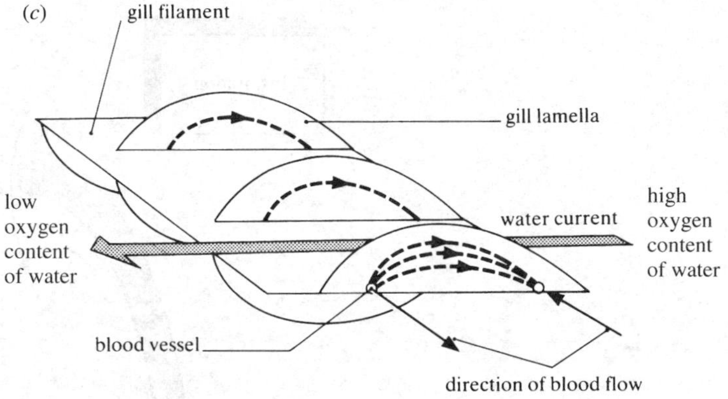

(*d*)

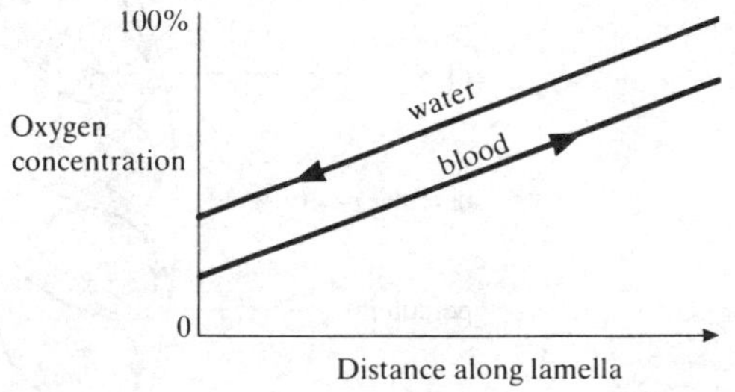

Fig 9.17 (a) *External view of the gills of a bony fish. The operculum is shown as transparent for convenience.* (b) *Part of a stack of gill filaments.* (c) *Detail of part of one gill filament showing the lamellae which increase the surface area.* (d) *Variation in the oxygen concentration of blood and water as they pass across a gill lamella.*

A moveable gill cover, the **operculum**, which is reinforced with thin layers of bone, encloses and protects the gills. There is a space between the inside surface of the operculum and the gills called the **opercular cavity** (fig 9.18). This plays a part in the fish's ventilation mechanism. The operculum can shut tight against the side of the fish or open, controlling movement of water in and out of the opercular cavity like a valve.

During inspiration the buccal cavity expands, and this decreases the pressure within, causing water to be drawn in through the mouth. At the same time, the outside water pressure presses the posterior (rear) end of the operculum shut, preventing entry of water from this region. However, also active at this time are muscles in the operculum which contract, causing the opercular cavity to be enlarged. The pressure in the opercular cavity is less than that in the buccal cavity and hence water is drawn from the buccal cavity over the gills into the opercular cavity. Therefore gaseous exchange is able to continue even when the fish is taking in a fresh supply of water.

When expiration takes place the mouth closes, as does the entrance to the oesophagus, and the floor of the buccal cavity is raised. This forces water over the gills, through the gill slits, and into the opercular cavity where the increased pressure forces open the posterior end of the operculum. Water therefore leaves to the outside environment. The coordinated activity of the buccal cavity and the opercular muscles ensures that a continuous flow of water passes over

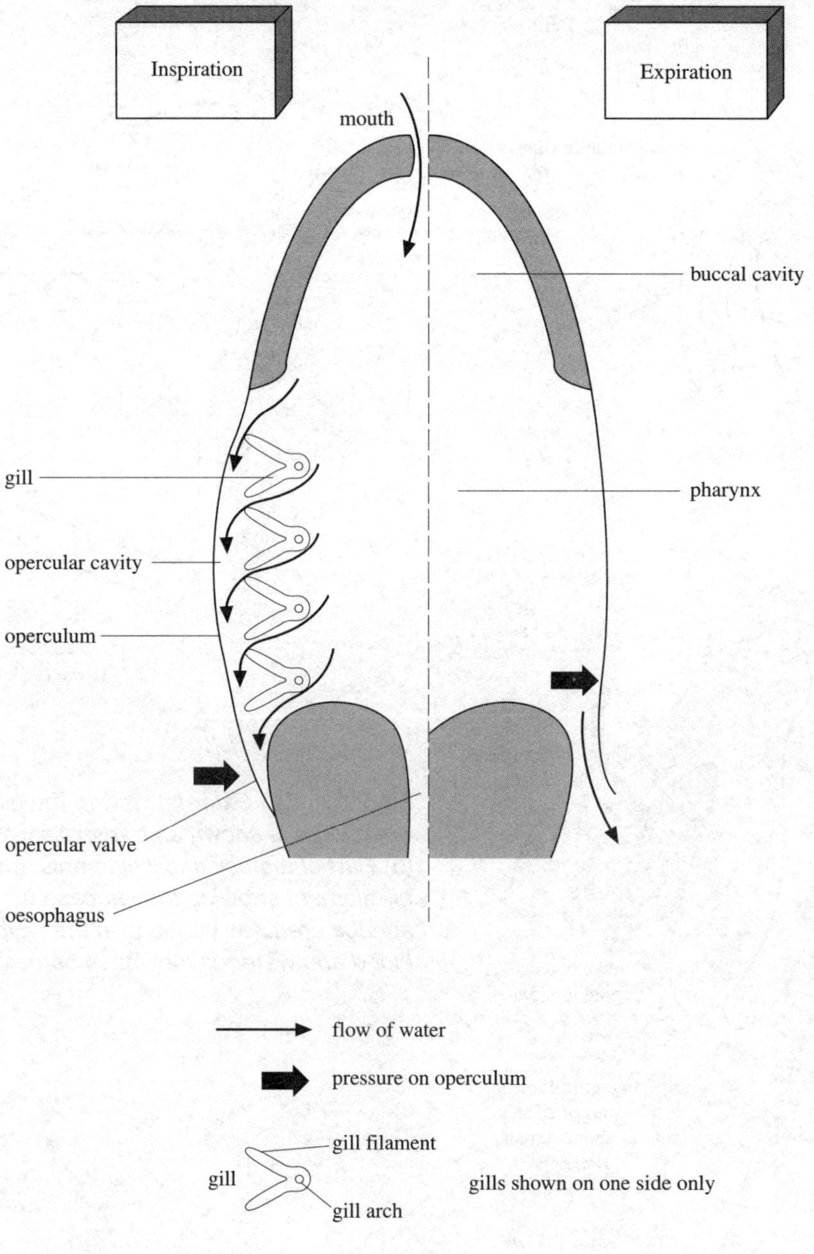

Fig 9.18 *Ventilation mechanism of a bony fish.*

the gills for most of the time, thus maintaining a high concentration of oxygen in water near the gills and low concentration of carbon dioxide.

Neighbouring gill filaments overlap at their tips, providing resistance to water flow. This slows down the passage of water over the gill lamellae, thus increasing the time available for gaseous exchange to take place. The blood in the lamellae flows in the opposite direction to that of the water (fig 9.17*c*). This is called a **countercurrent system** (fig 9.17*d*), and is a more efficient arrangement than a parallel current system in which the two fluids would travel in the same direction. A countercurrent system ensures that blood will constantly meet water with a relatively higher concentration of dissolved oxygen in it, and that a concentration gradient will be maintained between blood and water throughout the entire length of the filament and across each lamella. In this way bony fish are able to extract 80% of the oxygen in water.

9.8 Try to explain why blood flowing in the same direction as the water current (a parallel current system) would be a relatively inefficient mechanism for exchange of gases.

9.5 Gaseous exchange in a mammal

9.5.1 Structure of the respiratory system

The respiratory surface of a mammal consists of many air sacs called **alveoli** inside a pair of **lungs**. The lungs are situated next to the heart in the thoracic cavity and are connected to the atmosphere by tubes (fig 9.19). Air passes into the lungs through these tubes. Twelve pairs of bony **ribs** surround and protect the lungs and heart. **Intercostal muscles** are attached to the ribs, and a large **diaphragm** separates the thorax from the abdomen. These are involved in the ventilation mechanism, as described in section 9.5.4.

Air enters the body through two nostrils, each of which possesses a border of large hairs which trap particles in the air and filter them out of the system. While passing through the nasal passages the air is warmed and moistened and its odour detected. Air passes from the nasal passages, through the pharynx and into the trachea. This is a tube which lies in front of the oesophagus and extends into the thoracic cavity. The wall of the tube is strengthened and held open by horizontally arranged C-shaped bands of cartilage. The open section of the C is next to the oesophagus (fig 9.19). The cartilage prevents collapse of the tube during inspiration (breathing in). The cartilage can be seen in a section of the trachea (fig 9.20).

At its lower end the trachea splits into two **bronchi**. Within the lungs each bronchus subdivides many times into

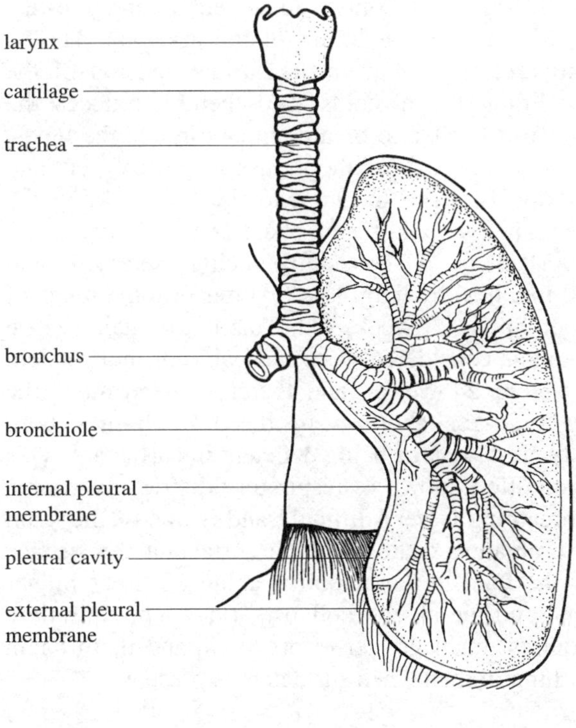

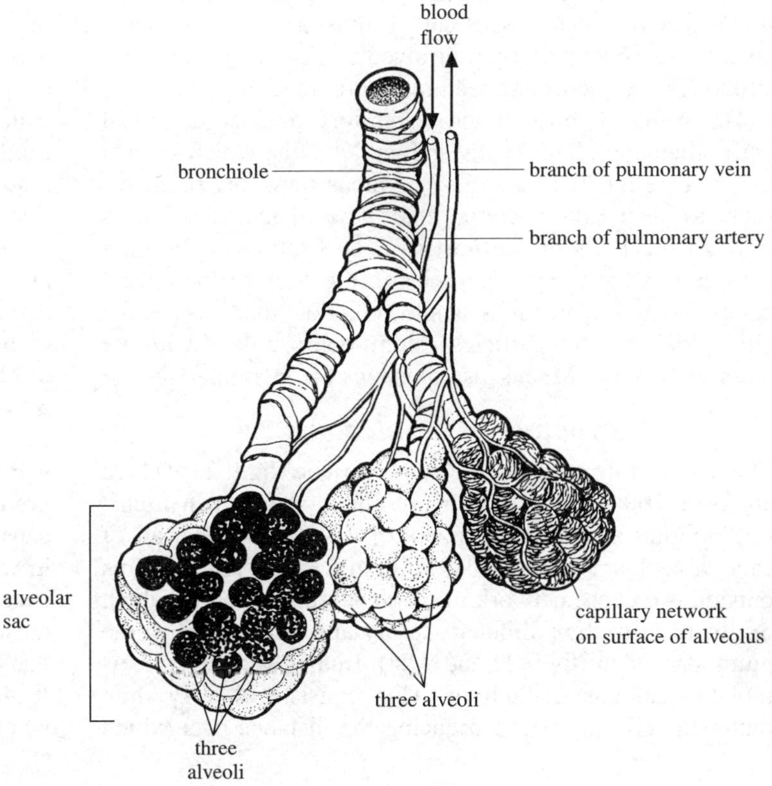

Fig 9.19 *Human trachea and lungs.*

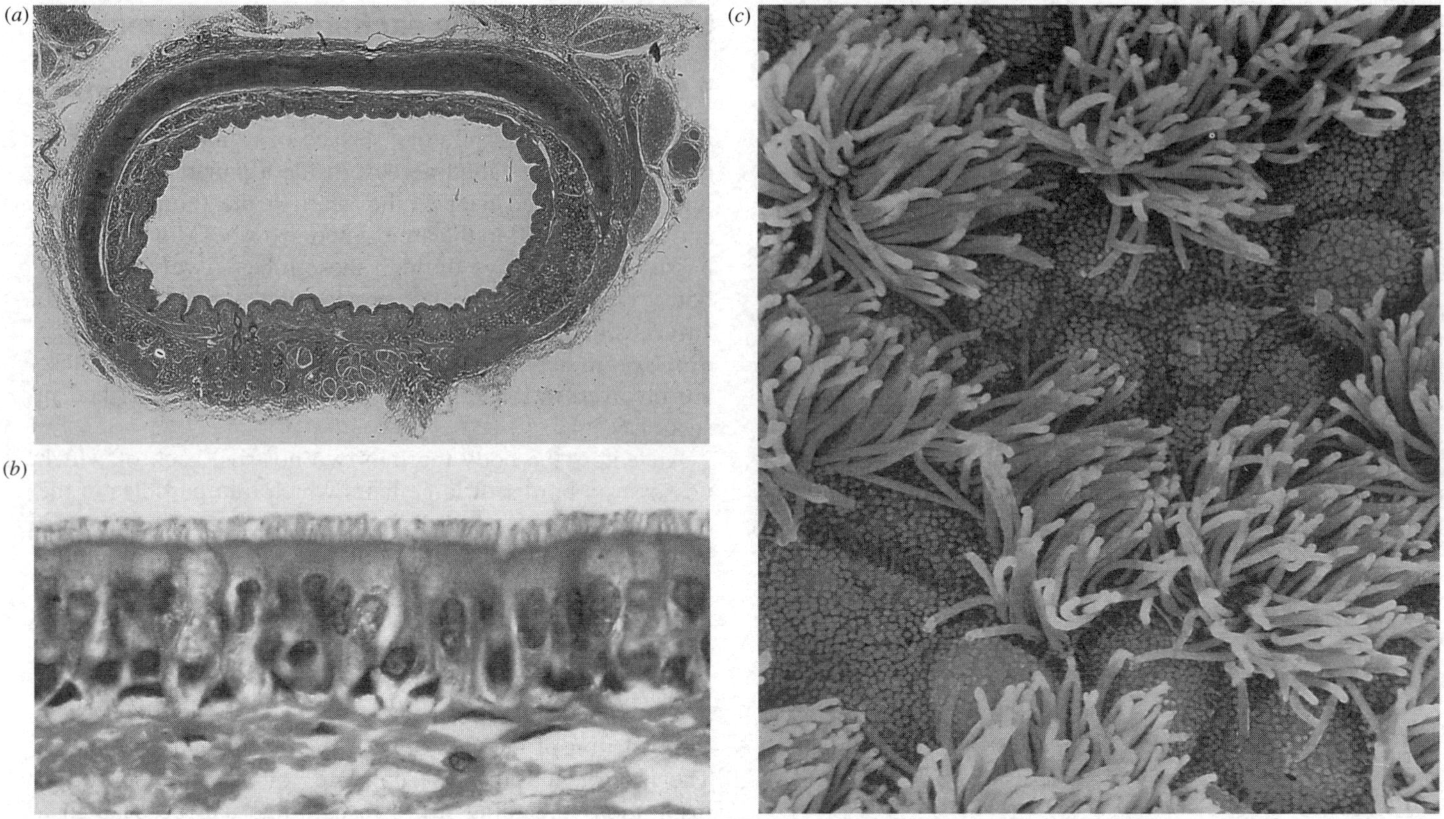

Fig 9.20 (a) *TS trachea seen at low power with a light microscope.* (b) *High power, showing ciliated epithelium and goblet cells.* (c) *Scanning electron micrograph of the surface of ciliated epithelium.*

much smaller tubes called **bronchioles**. These in turn branch into finer and finer tubes, ending with the **alveolar ducts** which lead into sacs called **alveolar sacs**. Into each alveolar sac opens a group of alveoli. A summary of these structures and their main features is provided in fig 9.21.

The walls of most of the respiratory passage are lined with ciliated epithelial cells and goblet cells, which secrete mucus (see figs 9.20 and 6.17). Mucus traps any particles, such as dust and bacteria, that have managed to pass through the hairs of the nostrils. The beating of the cilia then carries the trapped particles to the back of the buccal cavity where the mucus is swallowed. Note that it is not the cilia which trap the particles – don't confuse them with the hairs in the nose. Mucus also moistens the incoming air.

Structure of the alveolus

The alveoli form the gas exchange surface (fig 9.22). There are over 700 million alveoli present in the human lungs, representing a total surface area of 70–90 m^2. The wall of each alveolus is only 0.0001 mm thick (0.1 μm). On its outside is a dense network of blood capillaries, all of which originate from the pulmonary artery and rejoin to form the pulmonary vein (fig 9.23 and 9.24). Lining each alveolus is moist squamous epithelium. This consists of very thin, flattened cells (fig 6.14), reducing the distance over which diffusion must occur (fig 9.24). Collagen and elastic fibres are also present. The latter allow the alveoli to expand and recoil easily during breathing.

Special cells in the alveolus wall secrete a detergent-like chemical on to the inside lining of the alveolus. This is called a **surfactant**. It lowers the surface tension of the fluid layer lining the alveolus, and thereby reduces the amount of effort needed to breathe in and inflate the lungs. Surfactant also speeds up the transport of oxygen and carbon dioxide between the air and the liquid lining the alveolus and helps to kill any bacteria which reach the alveoli. Surfactant is constantly being secreted and reabsorbed in a healthy lung. It is first made in the lungs of a fetus when about 23 weeks old. This is the main reason why the fetus is considered to be incapable of independent existence before 24 weeks, and therefore determines the legal age limit for abortion in the UK. Babies born prematurely are at risk of being deficient in surfactant. This causes a condition known as respiratory distress syndrome in which breathing is very difficult, and is one of the main causes of premature death. Without surfactant the surface tension of the fluid in the alveoli is about 10 times higher than normal and the alveoli collapse after each expiration. It also requires much greater effort to expand them again when breathing in than when surfactant is present.

Fig 9.21 *Summary of the histology of the respiratory pathway.*

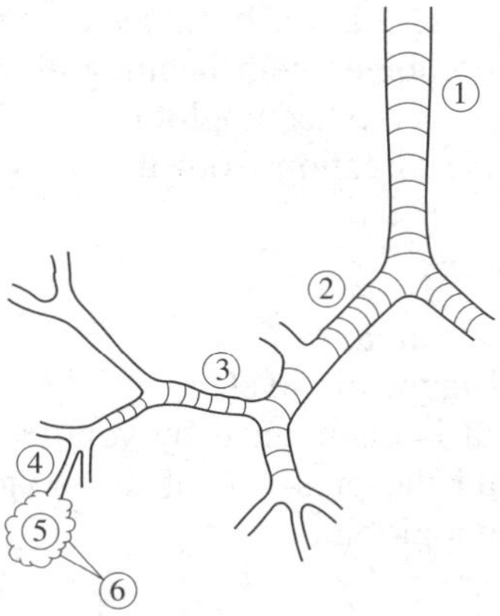

	Cartilage	Ciliated epithelium with goblet cells	Smooth muscle	Connective tissue with elastic fibres and collagen
① trachea	✓	✓	✓	✓
② two bronchi	✓	✓	✓	✓
③ bronchioles	cartilage gradually lost	✓	✓	✓
④ alveolar duct	no cartilage	no cilia no goblet cells	✓	✓
⑤ alveolar sac	no cartilage	no cilia no goblet cells	✓	✓
⑥ alveoli	no cartilage	no cilia no goblet cells squamous epithelium (thin flattened cells) with liquid surfactant on inner surface and blood capillaries on outer surface	(✓) very little	✓

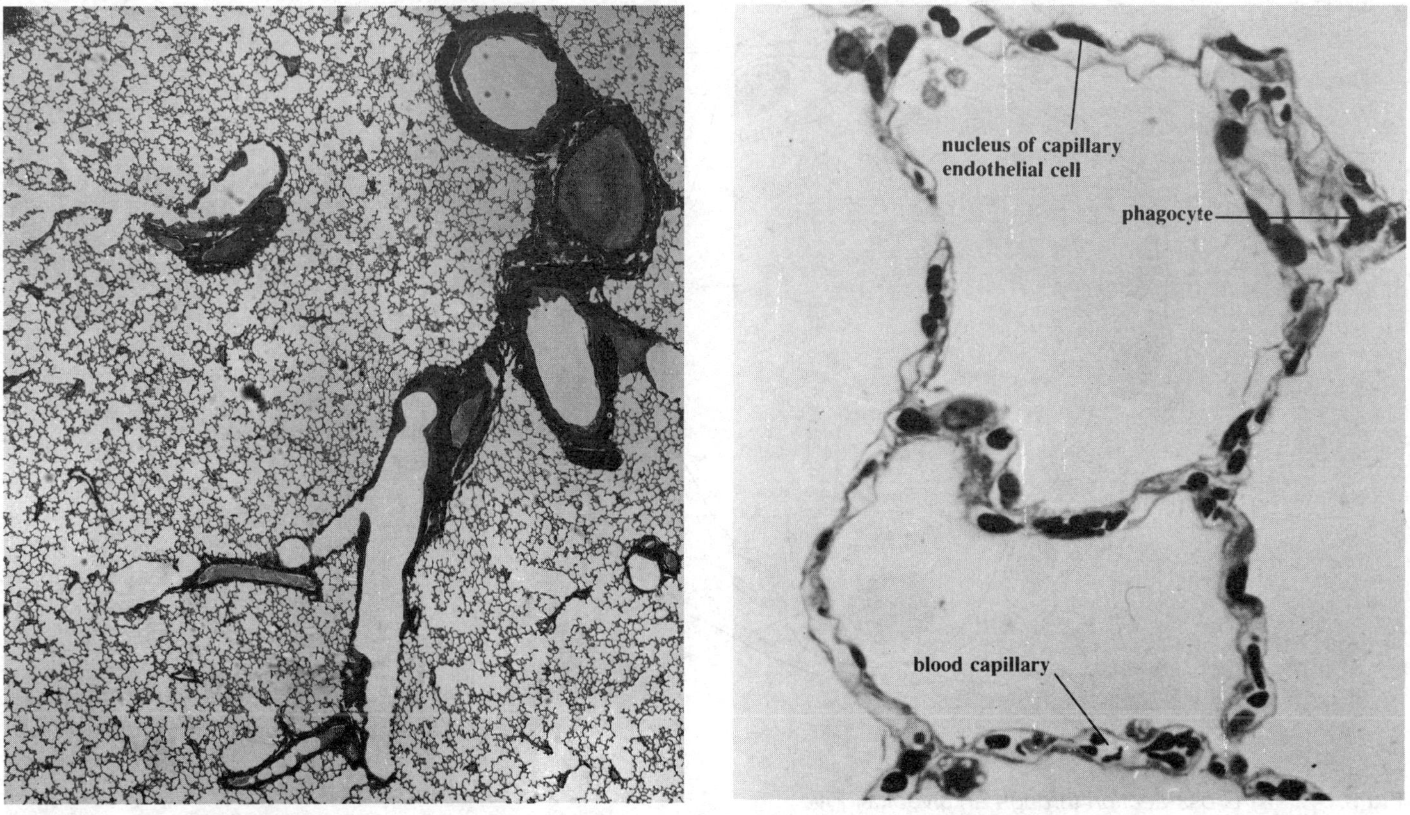

Fig 9.22 (a) *Human lung tissue as seen with a light microscope at low power.*

(b) *High power view of alveoli.*

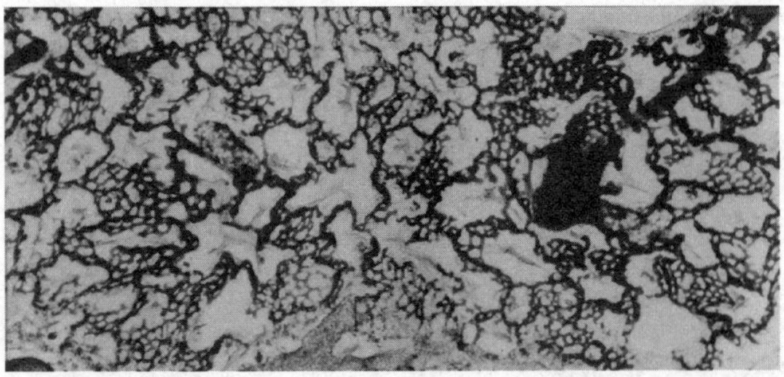

Fig 9.23 *Human lung injected to show blood vessels.*

9.5.2 Gaseous exchange at the alveolus

Oxygen diffuses across the thin barrier represented by the epithelium of the alveolus and the endothelium of the capillary (fig 9.24*b*). It passes first into the blood plasma and then combines with haemoglobin in the red blood cells to form oxyhaemoglobin. Carbon dioxide diffuses in the reverse direction from the blood to the alveoli.

Diffusion is efficient because:

- alveoli have a large surface area,
- the gases have a short distance to travel,
- a steep diffusion gradient is maintained by ventilation, a good blood supply and the presence of an oxygen-carrying compound, haemoglobin,
- surfactant is present.

9.9 How many times must a molecule of oxygen diffuse across a cell surface membrane in passing from the inside of an alveolus to haemoglobin?

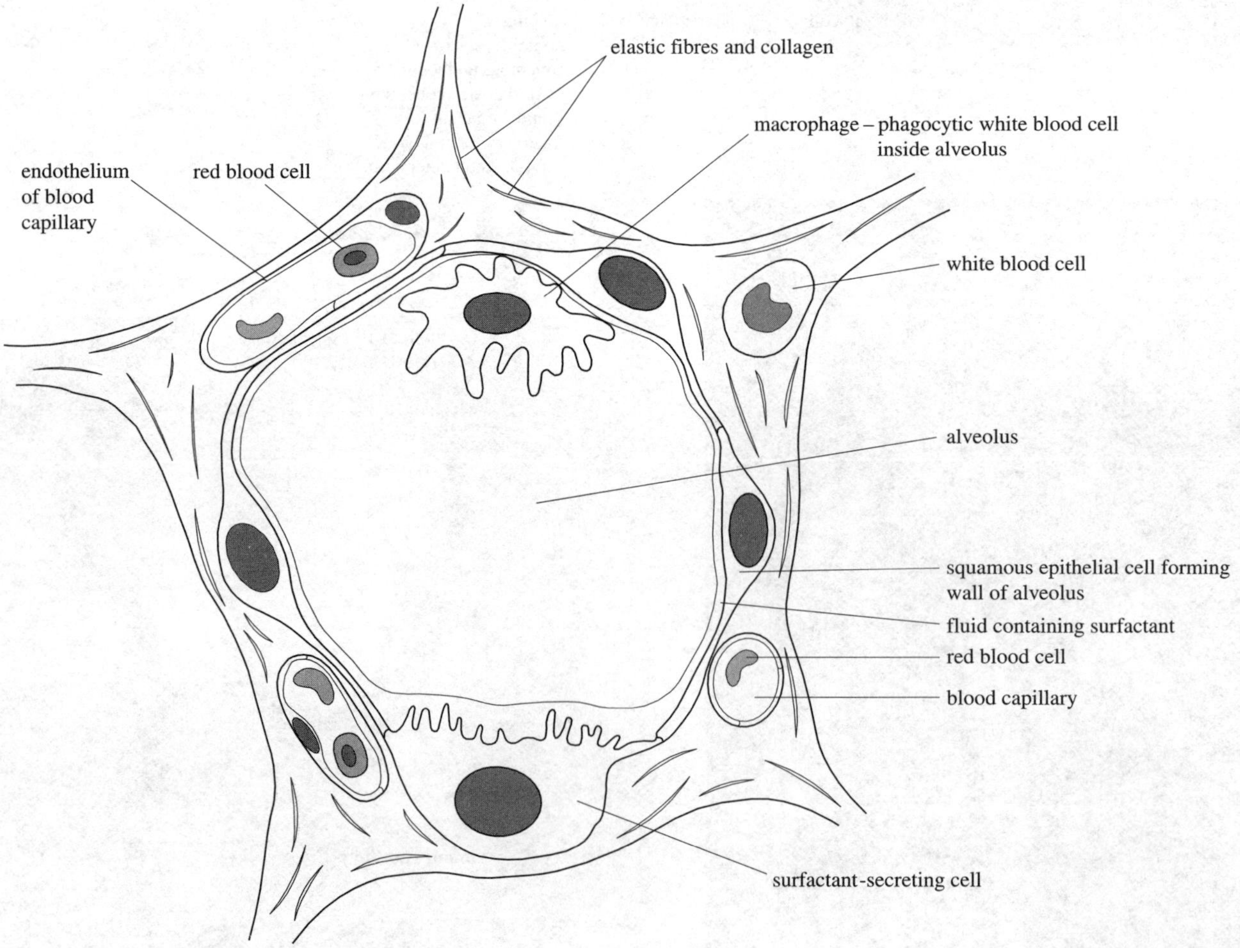

Fig 9.24 (a) *Cross-section through an alveolus. Five neighbouring alveoli are also partly shown, together with the structures between the alveoli.*

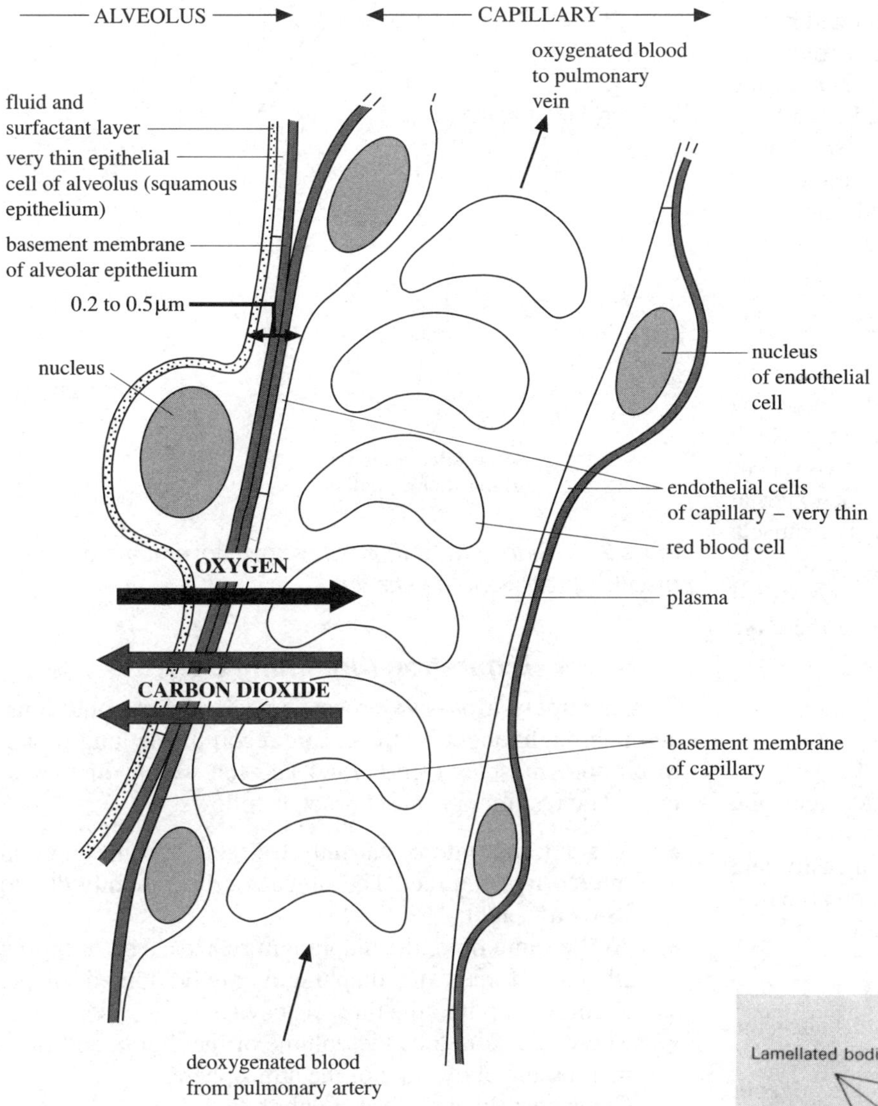

Fig 9.24 (b) *Relationship between an alveolus and a capillary.*

The diameter of the blood capillaries is smaller than the diameter of the red blood cells passing through them. This means that the red blood cells have to be squeezed through the capillaries by blood pressure. During this process they bend into an umbrella shape (fig 9.25), exposing more of their surface area to the surface of the alveolus and allowing greater uptake of oxygen. Progress of the cells is also relatively slow, thus increasing the time available for gaseous exchange to take place. When blood leaves the alveolus it possesses the same concentration of oxygen and carbon dioxide as the air in the alveoli.

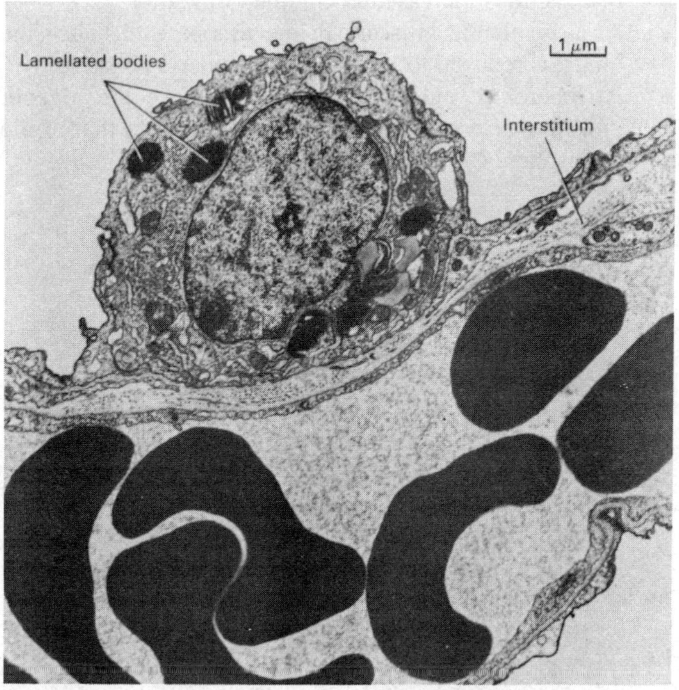

Fig 9.25 *Electron micrograph of a section through a capillary on the surface of an alveolus (dog lung). Red blood cells can be seen as dark objects in the lower part of the photograph*

9.5.3 Pleural cavity

Each lung is surrounded by a pleural cavity. This is a space lined by two tough, flexible, transparent pleural membranes (pleura). These protect the lungs, stop them leaking air into the thoracic cavity and reduce friction between the lungs and the wall of the thorax. The inner membrane is in contact with the lungs, whilst the outer membrane lines the walls of the thorax and diaphragm. The pleural cavity contains pleural fluid which is secreted by the membranes. This lubricates the pleura, thus reducing friction as the membranes rub against each other during breathing movements. The cavity is air-tight and its pressure stays at 3–4 mmHg lower than that in the lungs. This negative pressure is maintained during inspiration and helps the alveoli to inflate and the lungs to fill any extra available space provided by the expanding thorax.

9.5.4 The mechanism of ventilation (breathing)

Air is passed in and out of the lungs by movements of the intercostal and diaphragm muscles which alter the volume of the thoracic cavity. There are two sets of intercostal muscles between each pair of ribs. The external intercostals are outside the internal intercostals. The muscle fibres run diagonally but in opposite directions in the two sets of muscles (fig 9.26). The diaphragm consists of circular and radial muscle fibres arranged around the edge of a circular inelastic sheet of white fibres (collagen).

Inspiration (breathing in)

This is an active process.

- The external intercostal muscles contract and the internal intercostals relax.
- This pulls the rib cage up and out. (You can easily feel this by placing a hand on your chest as you breathe in.)
- At the same time, the diaphragm muscles contract.
- This flattens the diaphragm.
- Both actions increase the volume of the thorax.
- As a result the pressure in the thorax, and hence the lungs, is reduced to less than atmospheric pressure.
- Air therefore enters the lungs, inflating the alveoli, until the air pressure in the lungs is equal to that of the atmosphere (fig 9.27).

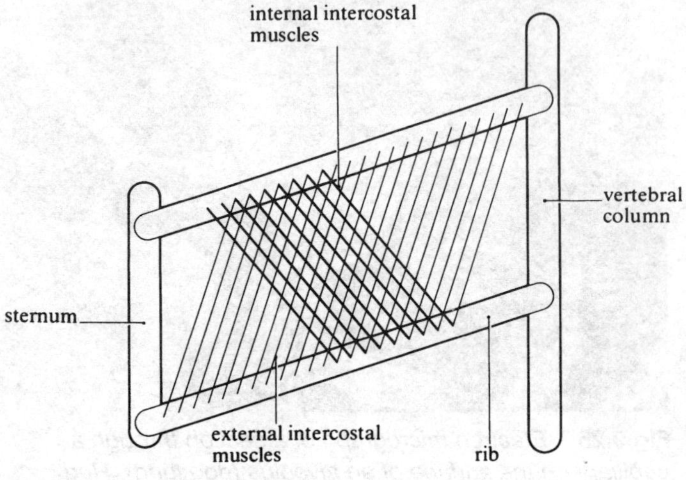

Fig 9.26 *Diagrammatic representation of the position of the intercostal muscles.*

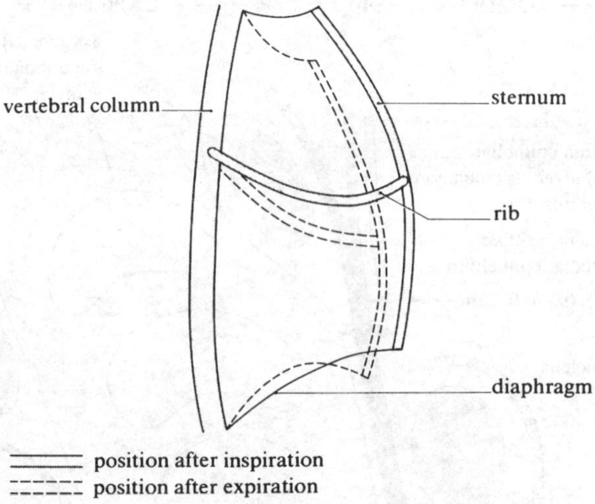

Fig 9.27 *Side view of thorax to show movements during breathing (only one rib shown).*

Expiration (breathing out)

This is largely a passive process under resting conditions, and is brought about by the elastic recoil of the lung tissue, respiratory muscles (see below) and the weight of the rib cage. The respiratory muscles act as follows.

- The external intercostal muscles relax and the internal intercostals contract. The rib cage drops, mainly due to its own weight.
- At the same time, the diaphragm relaxes. The dropping rib cage forces the diaphragm into a domed shape, pushing it up into the thoracic cavity.
- These events reduce the volume of the thorax and raise its pressure above that of the atmosphere.
- Consequently air is forced out of the lungs.

Under conditions of exercise, forced breathing occurs. When this happens additional muscles are brought into action and expiration becomes a much more active, energy-consuming process. The internal intercostals contract more strongly and move the ribs vigorously downwards. The abdominal muscles also contract strongly, causing more active upward movement of the diaphragm. This also happens when sneezing or coughing.

9.5.5 Control of ventilation

Normally we are not conscious of our breathing because it is controlled involuntarily. We can take over some voluntary control, and this will be discussed after considering involuntary control.

Involuntary control of breathing is carried out by a **breathing centre** located in the medulla of the brain (the medulla is part of the hindbrain) (fig 9.28). The ventral (lower) portion of the breathing centre acts to increase the rate and depth of inspiration and is called the **inspiratory centre**. The dorsal (top) and lateral (side) portions inhibit

Fig 9.28 *Nervous control of breathing.*

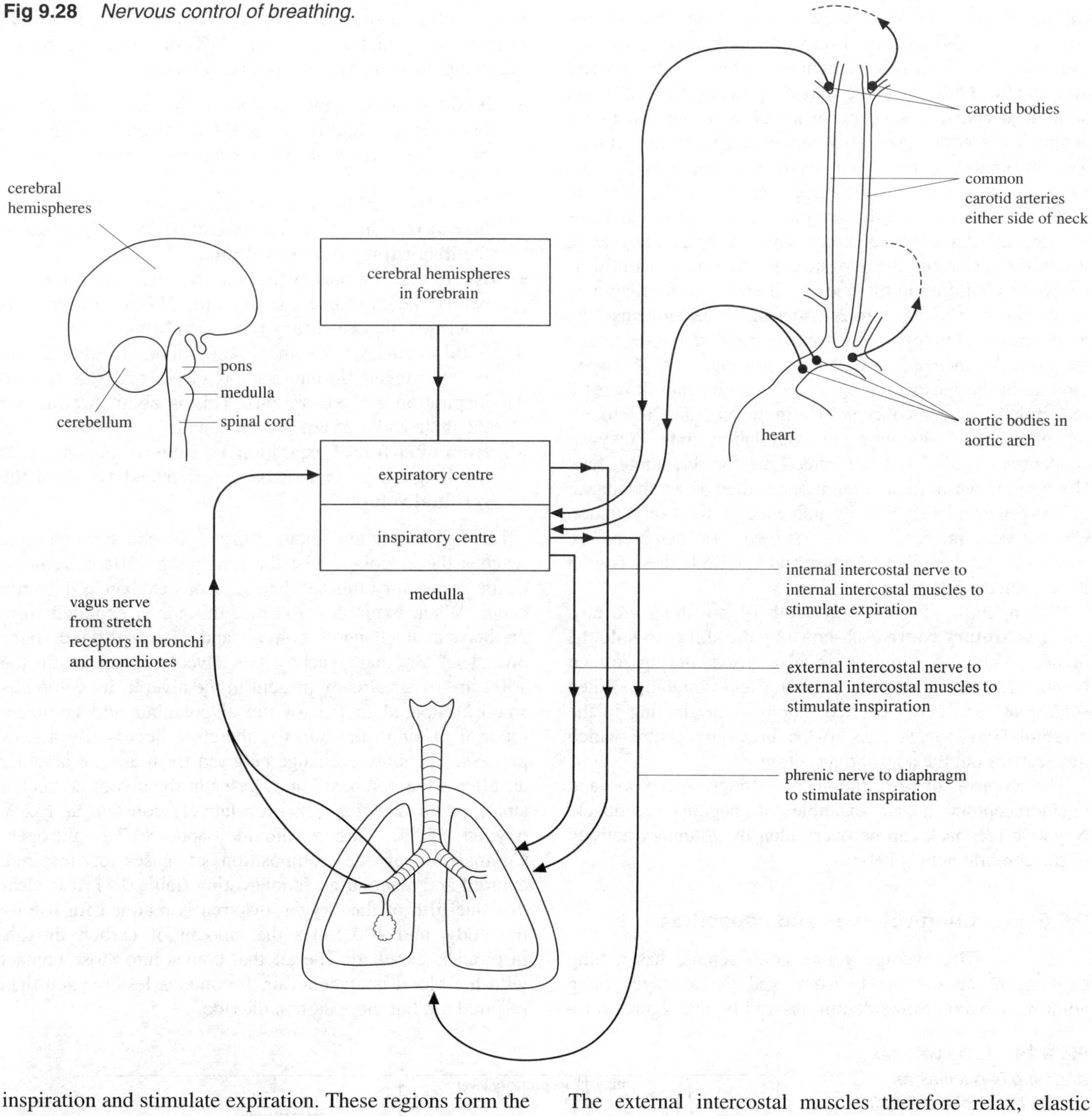

inspiration and stimulate expiration. These regions form the **expiratory centre**. The breathing centre communicates with the intercostal muscles by way of the **intercostal nerves** and with the diaphragm by way of the **phrenic nerves**. The bronchial tree (the mass of bronchioles and the bronchi) is connected to the brain by the **vagus nerve** (fig 9.28). Rhythmic nerve impulses to the diaphragm and intercostal muscles bring about ventilation movements.

Inspiration inflates the lungs and, as they inflate, stretch receptors (also known as proprioceptors) located in the bronchial tree are stimulated to send more and more nerve impulses via the vagus nerve to the expiratory centre. This temporarily inhibits the inspiratory centre and inspiration. The external intercostal muscles therefore relax, elastic recoil of the lung tissues occurs and expiration takes place. After this has occurred, the bronchial tree is no longer stretched and the stretch receptors no longer stimulated. Therefore the expiratory centre becomes inactive and inspiration can begin again. The whole cycle is repeated rhythmically throughout the life of the organism. Forcible expiration can be achieved by contraction of the internal intercostal muscles.

A basic rhythm of breathing is maintained by the medulla even if all nervous input is cut. However, under normal circumstances various stimuli modify this basic rhythm. The main stimulus that controls breathing is the concentration of carbon dioxide in the blood, rather than the oxygen concentration. When carbon dioxide levels increase (as, for example, during exercise) **chemoreceptors** in the **carotid** and **aortic bodies** of the blood system (fig 9.28) are stimulated and they send nerve impulses to the inspiratory centre. The medulla itself also contains such chemoreceptors. The inspiratory centre then sends out impulses via the external intercostal and phrenic nerves to the external intercostal muscles and diaphragm causing them to increase the rate at which they contract. This increases the rate of inspiration. Carbon dioxide quickly becomes harmful if allowed to build up in the body. It dissolves to form an acid which can start to denature enzymes and other proteins. The body has therefore evolved extremely rapid responses to any increase. An increase of 0.25% in concentration of carbon dioxide in the air can double the ventilation rate. It takes a reduction in oxygen concentration in the air from 20% to 5% to produce a doubling in ventilation rate. Oxygen concentration also has an effect on the breathing rate. However, under normal circumstances there is an abundance of oxygen available, and its influence is relatively minor. Chemoreceptors sensitive to oxygen concentration are located in the medulla and aortic and carotid bodies, as with the carbon dioxide receptors.

Within limits, the rate and depth of breathing are also under **voluntary control** as shown by the ability to hold the breath. Voluntary control is also used during forced breathing, speech, singing, sneezing and coughing. When such control is being exerted, impulses originating in the cerebral hemispheres pass to the breathing centre which then carries out the appropriate action.

The control of inspiration by stretch receptors and chemoreceptors is an example of negative feedback. Negative feedback can be overridden by voluntary activity of the cerebral hemispheres.

9.5.6 Lung volumes and capacities

The average young adult female has a lung capacity of approximately 4 dm^3 and the average young adult male 5 dm^3. Some definitions and typical figures for a young adult male are given below (see also fig 9.29). Figures for females are about 80% of those for males, reflecting differences in overall body mass.

- **Tidal volume** is the volume of gas exchanged during one breath in and out. It is about 450 cm^3 during quiet breathing. After maximal exercise it rises to about 3 dm^3.
- If after normal inspiration the male continues to inhale, he can take in an extra 1500 cm^3 of air. This is called the **inspiratory reserve volume**.
- If after a normal expiration the male continues to exhale, he can force out an extra 1500 cm^3 of air. This is termed the **expiratory reserve volume**.
- **Vital capacity** is the maximum volume of air that can be exchanged during one breath in and out (forced inspiration and expiration). This is about 5.7 dm^3 for the male and 4.25 dm^3 for the female.
- Even after forced expiration 1500 cm^3 of air remain in the lungs. This cannot be expelled and is called the **residual volume**.

During inspiration about 300 cm^3 of the tidal volume reaches the alveoli, whilst the remaining 150 cm^3 remains in the respiratory tubes, where gaseous exchange does not occur. When expiration follows, this air is expelled from the body as unchanged room air and is termed **dead space air**. The air that reaches the alveoli mixes with the 1500 cm^3 of air already present in the alveoli. Its volume is small compared to that of the alveolar air and complete renewal of air in the lungs is therefore necessarily a slow process. The slow exchange between fresh air and alveolar air affects the composition of gases in the alveoli to such a small extent that they remain relatively constant at 13.8% oxygen, 5.5% carbon dioxide and 80.7% nitrogen. Comparison of the composition of gases of inspired, expired and alveolar air is interesting (table 9.4). It is clear that one-fifth of the oxygen inspired is retained for use by the body, and 100 times the amount of carbon dioxide inspired is expelled. The air that comes into close contact with the blood is alveolar air. It contains less oxygen than inspired air, but more carbon dioxide.

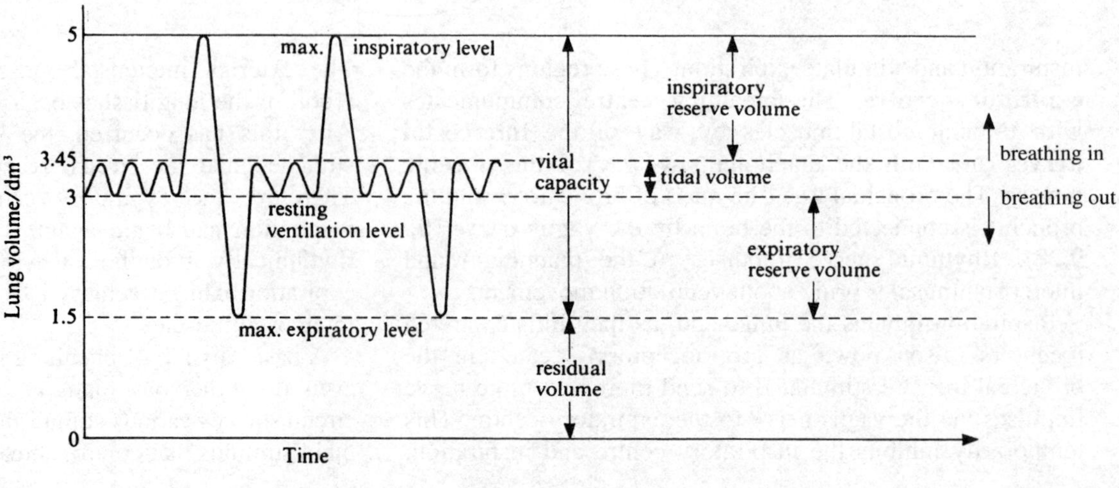

Fig 9.29 *Lung volumes and lung capacities as shown by a spirometer. It is normal, as in this case, to publish graphs upside down compared with an actual trace from a spirometer. When using a spirometer the graph goes down while breathing in (air being removed from the lid and lid dropping) and up while breathing out (lid filling and rising).*

Table 9.4 Percentage composition by volume of gases in inspired, alveolar and expired air.

Gas	*Inspired air*	*Alveolar air*	*Expired air*
Oxygen	20.95	13.8	16.4
Carbon dioxide	0.04	5.5	4.0
Nitrogen	79.01	80.7	79.6

9.5.7 Using a spirometer to measure respiratory activity

An instrument commonly used in schools, laboratories and hospitals for measuring the volume of air which enters and leaves the lungs is the spirometer. Essentially it consists of an air-filled lid (like a box with no bottom) with a capacity of six dm^3 or more, which fits inside a box of water. The lid has an entry tube and an exit tube which come together at a mouthpiece which is placed in the subject's mouth during use. A valve ensures that the air breathed out enters the lid through one tube and leaves through the other. On its return the carbon dioxide can be removed by soda-lime in a canister. (A gradually increasing concentration of carbon dioxide in the spirometer being re-breathed would be dangerous.) Medical grade oxygen may also be used to fill the lid. The air in the subject's lungs and the air in the spirometer is therefore a sealed system.

The lid is counterbalanced so that when gas is passed in or out, it rises or falls accordingly. When the subject breathes out, the lid fills and is raised; when he or she breathes in, the lid drops. A pen attached to the lid writes on a piece of graph paper attached to a slowly rotating drum (**kymograph**), recording the up and down movements of the lid (fig 9.29).

Detailed instructions for the operation of the spirometer are supplied by the manufacturer and will not be dealt with here. However, it is important to be able to analyse the tracings recorded by the spirometer, and to understand what information can be derived from them.

From spirometer tracings the metabolic rate, respiratory quotient, tidal volume, rate of breathing and consumption of oxygen can be measured.

The breathing rate (or ventilation rate) is calculated as the number of breaths taken per minute. Pulmonary ventilation (PV) is expressed in terms of the breathing rate multiplied by the tidal volume:

$$\text{PV} = \text{breathing rate} \times \text{tidal volume}$$

For example, if breathing rate is 15 breaths per minute and tidal volume is $400\,cm^3$, then PV = $15 \times 400\,cm^3$ = $6000\,cm^3$ per minute (that is $6\,dm^3$ of air will be exchanged between subject and outside environment each minute).

9.10 The volume of air exchanged in the alveoli is in fact less than that of the pulmonary ventilation. Suggest why this is the case.

Some further typical figures are as follows:

breathing rate at rest: 15 breaths per minute
breathing rate after maximal exercise: 40–50 breaths per minute
PV at rest: $6.75\,dm^3$ per minute
PV after maximal exercise (tidal volume $3\,dm^3$ per minute $\times$ breathing rate 45 per minute) = $135\,dm^3$ per minute
oxygen consumption at rest: 0.25–0.4 dm^3 per minute
oxygen consumption after maximal exercise: $3.6\,dm^3$ per minute (up to $5.1\,dm^3$ per minute for a marathon runner)

Measuring the metabolic rate of an organism

As respiration is directly involved with most metabolic activities within the body, its measurement gives a relatively accurate indication of metabolic activity. The metabolic rate can be calculated by measuring the rate of oxygen consumption. As the oxygen in the spirometer is used, the lid gradually gets lower, assuming carbon dioxide is being absorbed and not returned to the spirometer. The rate of oxygen consumption is measured as the overall drop in the lid over a given time period (see question 9.16, section 9.5.9).

9.11 Consider fig 9.30. It can be seen that the smaller the mammal the higher its metabolic rate. Why is this so?

9.12 How can you compare the metabolic rates of mammals of different size?

9.5.8 The basal metabolic rate (BMR)

The BMR of an organism is the minimum rate of energy conversion required just to stay alive during complete rest or sleep. Before the BMR of human subjects is measured they undergo a standardised rest period of 12–18 h physical and mental relaxation. No meal is eaten

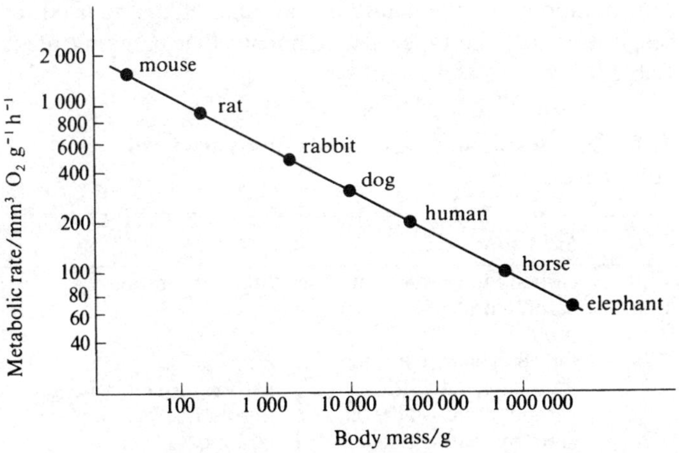

Fig 9.30 *Metabolic rate of animals, calculated per gram body mass, plotted on logarithmic coordinates.*

during this time. This ensures that the alimentary canal is empty before measurements are taken. The BMR varies with age, sex, size and state of health of the individual and is clearly correlated with body surface area to volume ratio.

9.5.9 Respiratory quotient (RQ)

Consider the equation:

$$C_6H_{12}O_6 + 6O_2 \longrightarrow 6CO_2 + 6H_2O + \text{energy}$$

From this it is quite clear that in a given time the volume of carbon dioxide produced during respiration of carbohydrate is equal to the volume of oxygen consumed (remember, one mole of any gas occupies the same volume under the same conditions of temperature and pressure). The ratio of CO_2:O_2 is called the **respiratory quotient**, and for respiration using carbohydrate its value is 1, that is

RQ = volume of CO_2 evolved/volume of O_2 absorbed (from direct observations)

Or

RQ = moles or molecules of CO_2 evolved/moles or molecules of O_2 evolved (from equations)

Therefore, from the above equation,

RQ = $CO_2/O_2 = 6/6 = 1$.

9.13 The equation for respiration of the fat tripalmitin is:

$$2C_{51}H_{98}O_6 + 145O_2 \rightarrow 102CO_2 + 98H_2O$$

What is the RQ for tripalmitin?

9.14 What is the RQ when glucose is respired anaerobically to ethanol and carbon dioxide?

Determination of respiratory quotients can yield valuable information about the nature of the substrate being used for respiration and the type of metabolism that is taking place (table 9.5).

Table 9.5 Respiratory quotients of a variety of substrates.

RQ	*Substrate*	
>1.0	carbohydrates plus some anaerobic respiration	
1.0	carbohydrates	
0.9	protein	
0.7	fat, such as tripalmitin	
0.5	fat associated with carbohydrate synthesis	carbon dioxide released during respiration is being put to other uses and therefore not released from the body
0.3	carbohydrate with associated organic acid synthesis	

9.15 Why is the usual RQ for humans between 0.7 and 1.0?

9.16 From the spirometer trace given in fig 9.31 calculate

(*a*) breathing rate;
(*b*) tidal volume;
(*c*) pulmonary ventilation;
(*d*) oxygen consumption.

Experiment 9.2: Use of a respirometer to measure oxygen uptake in small terrestrial nonvertebrates such as woodlice

The oxygen uptake by the nonvertebrates in this experiment is measured using a manometer. Fig 9.32 shows the apparatus which is used. It is called a **respirometer** (don't confuse this with a spirometer).

In respiration, oxygen is taken up and carbon dioxide is given off. So to ensure that the manometer is recording oxygen consumption alone, soda-lime is incorporated in the apparatus in order to absorb the carbon dioxide evolved. The apparatus can be used to investigate the effect of temperature on oxygen uptake. A water bath is used in order to keep the temperature of the atmosphere surrounding the organisms constant whilst readings are being taken.

Materials

- manometer
- manometer fluid
- 1 cm^3 syringe
- 2 boiling tubes
- 2 pieces of zinc gauze (to fit the diameter of a boiling tube)
- glass beads (or any equivalent non-absorbent material of equal volume to the nonvertebrates)
- clamps and stands
- water bath
- thermometer
- stopclock
- graph paper
- small nonvertebrates such as woodlice or blowfly larvae
- soda-lime

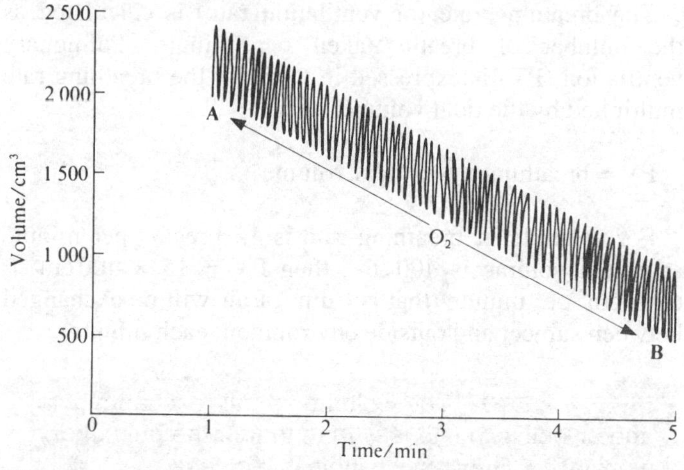

Fig 9.31 *Spirometer tracing of subject at rest.*

Fig 9.32 *Apparatus used in the investigation of oxygen uptake in small terrestrial nonvertebrates.*

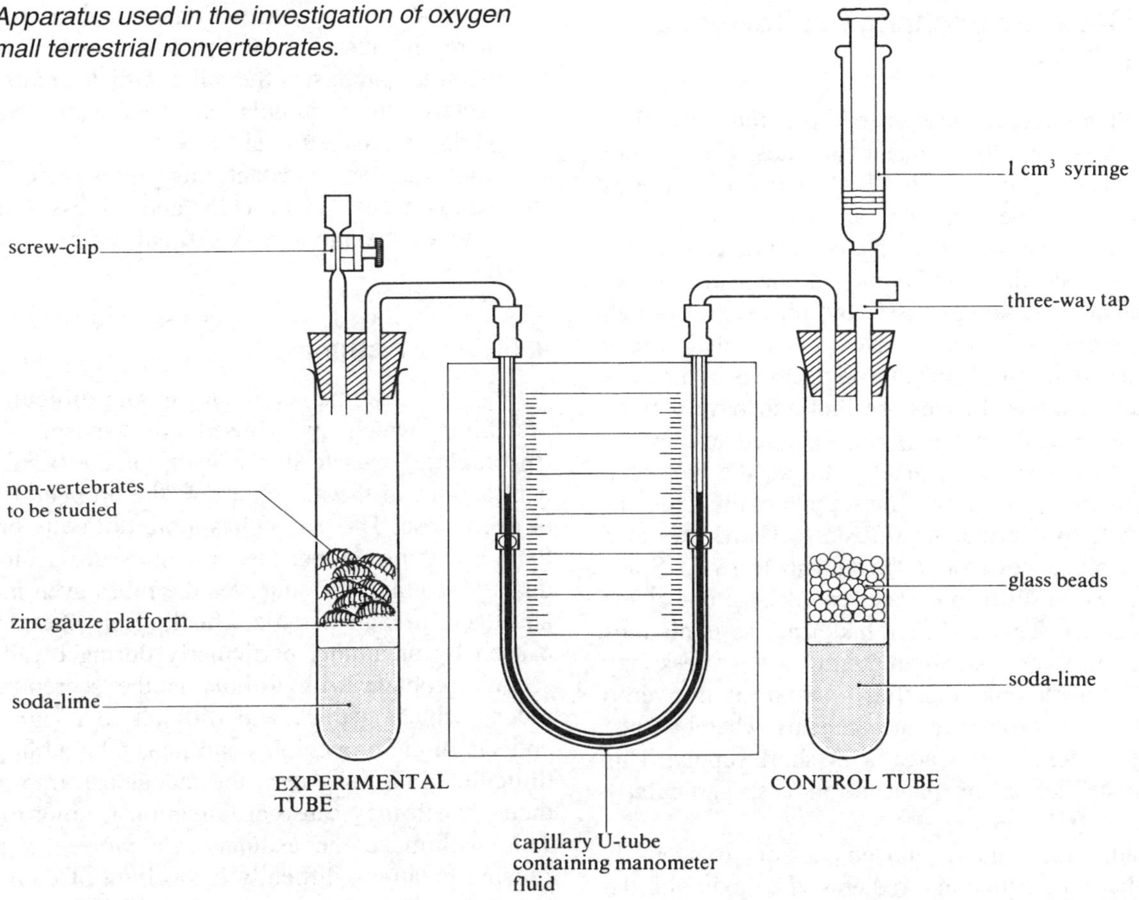

Method

(1) Half fill a manometer with fluid and connect a 1 cm^3 syringe to the three-way tap attached to one arm of the manometer.

(2) Place equal volumes of soda-lime in the bottom of each of two boiling tubes and then place a zinc-gauge platform 1 cm above the soda-lime.

(3) Place some nonvertebrates in one boiling tube (experimental) and an equal volume of glass beads in the other tube (control). The animals must not come into contact with the soda-lime and so the platform must be an absolute barrier between the animals and the soda-lime.

(4) Connect the manometer to the two boiling tubes as shown in fig 9.32, and adjust the three-way tap and screw-clip so that the apparatus is open to the atmosphere.

(5) Clamp the apparatus so that the boiling tubes are in a water bath at 20 °C and leave the apparatus at this temperature with the taps open for at least 15 min.

(6) Close the tap and screw-clip, note the position of the manometer fluid and start the stopclock.

(7) At regular intervals, read off the position of the manometer fluid against the scale.

(8) At the end of the experiment open the tap and screw-clip again.

(9) Plot a graph of the change in fluid level against time.

(10) Calculate the rate of oxygen uptake.

(11) Repeat the experiment several times over a range of temperatures, such as 20, 25, 30, 35 and 40 °C.

(12) Plot a graph of rate of oxygen consumption against temperature.

Notes

(1) The fluid that is used in the manometer can be dyed water, oil or mercury. The less dense the fluid, the greater the displacement in the manometer.

(2) In order to measure the change in the manometer fluid levels, a scale must be attached to the manometer. This can be done by attaching the manometer U-tube to a piece of hardboard on which a scale or graph paper has been glued. Alternatively an adhesive metric scale can be attached to the arm of the manometer itself. The scale is available from Philip Harris Ltd.

(3) Before any readings are taken in the experiment, the apparatus must be checked to ensure that it is air-tight. This can be done by pushing air into the apparatus using the syringe, causing the manometer fluid to be displaced. The tap should then be used to close off the apparatus to the atmosphere and, if the apparatus is air-tight, the difference in levels of fluid should not decrease.

9.6 Gaseous exchange in flowering plants

Plants require less energy per unit mass than animals as they possess lower metabolic rates. They do not therefore need to maintain the high rates of gaseous exchange of the more complex animals, and rely on diffusion through spaces between the cells (intercellular air spaces). No special ventilation mechanisms exist. Flowering plants exchange gases by diffusion through pores called **stomata** in their leaves and on their green stems, or if the stems are woody, through cracks in the bark or slits called **lenticels**. Leaves are thin and have a large surface area, as seen in chapter 7 (fig 7.5), and so they are the main sites of gaseous exchange. Inside the leaf of a dicotyledon, there is a spongy mesophyll with large air spaces which allow efficient diffusion. There are also specially large air spaces around the stomata (fig 7.5). Since the system relies on diffusion, water can diffuse out of the plant just as easily. This can be a handicap, as is obvious when plants wilt. Even a small amount of water stress may reduce plant growth (and therefore yield if it is a crop plant). Plants have protective mechanisms whereby they can close their stomata if water is in short supply. This depends on the action of plant hormones, particularly abscisic acid.

Once inside the plant, movement of oxygen is determined by the diffusion gradients that exist in the intercellular air spaces. In this way oxygen travels towards the cells and dissolves in the surface moisture of their walls. From here it passes by diffusion into the cells themselves. Carbon dioxide leaves the plant by the same pathway but in the reverse direction.

The whole situation becomes more complex in photosynthesising plants. Here oxygen is also produced by the chloroplasts as a waste product of photosynthesis. The oxygen may be used up immediately in respiration by mitochondria contained in the same cell, and waste carbon dioxide from respiration may be used by the chloroplasts for photosynthesis.

> **9.17** (*a*) Construct a table showing the major differences between photosynthesis and aerobic respiration.
> (*b*) Make a list of similarities (including biochemical similarities), between photosynthesis and aerobic respiration.

9.7 Respiratory disease and ill health

9.7.1 Short-term effects of smoking on ventilation and gaseous exchange

Smoking has some relatively short-term effects on breathing and gaseous exchange, as well as long-term effects discussed in the following sections (9.7.2–9.7.5):

- nicotine causes constriction of the finer bronchioles, increasing resistance to the flow of air.
- nicotine paralyses the cilia which remove dirt and bacteria; the accumulation of extra material in the air passages can restrict air flow.
- smoke acts as an irritant; this causes secretion of excess mucus from goblet cells and excess fluid into the airways, making it more difficult for air to pass through them.

9.7.2 Asthma

Asthma is a form of difficult or heavy breathing which is caused by spasms of smooth (involuntary) muscle in the walls of the bronchioles. The contractions of the muscle cause the bronchioles to narrow or even close. The person has more difficulty breathing out than breathing in because the pressure from the lungs during breathing out squeezes the tubes even more. During an attack, a characteristic whistling or wheezing sound is caused by breathing, particularly during breathing out. A second problem with asthma is the secretion of excess mucus which is thick and difficult to cough away. This collects in the bronchioles and makes breathing even more difficult. In the long term the mucus can trap bacteria and their growth may cause an infection leading to bronchitis. This will make the asthma even worse. A third factor causing breathing difficulty is swelling of the lining of the respiratory pathway.

The cause of asthma is an over-reaction to one of a variety of possible stimuli. These stimuli normally have no or little effect on the air passages of people who do not suffer from the condition. Commonly, there is an allergic response to substances like pollen, household dust (which contains many possible causes such as mites and the spores of moulds), a particular food, or feathers from a pillow. Emotional disturbance may also provoke an attack. Other triggers include cold air, exercise and smoking. The worrying increase in asthma among children in urban environments is thought to be related to increases in atmospheric pollution from traffic. Certain particles, as well as gases, emitted from exhaust pipes may be responsible, and legal limits on such emissions are gradually becoming stricter in industrialised countries. Asthma, though, is a complex problem and more research is needed to establish who are most at risk and why. Anti-inflammatory drugs are prescribed to help control the problem.

9.7.3 Emphysema

Emphysema is caused by the gradual breakdown of the thin walls of the alveoli. Gradually the air spaces become larger (fig 9.33) and the total surface area for gaseous exchange decreases. The most obvious sign is therefore increasing breathlessness. In advanced cases, patients find it difficult even to walk across a room, and in extreme cases may be confined to a chair unable even to

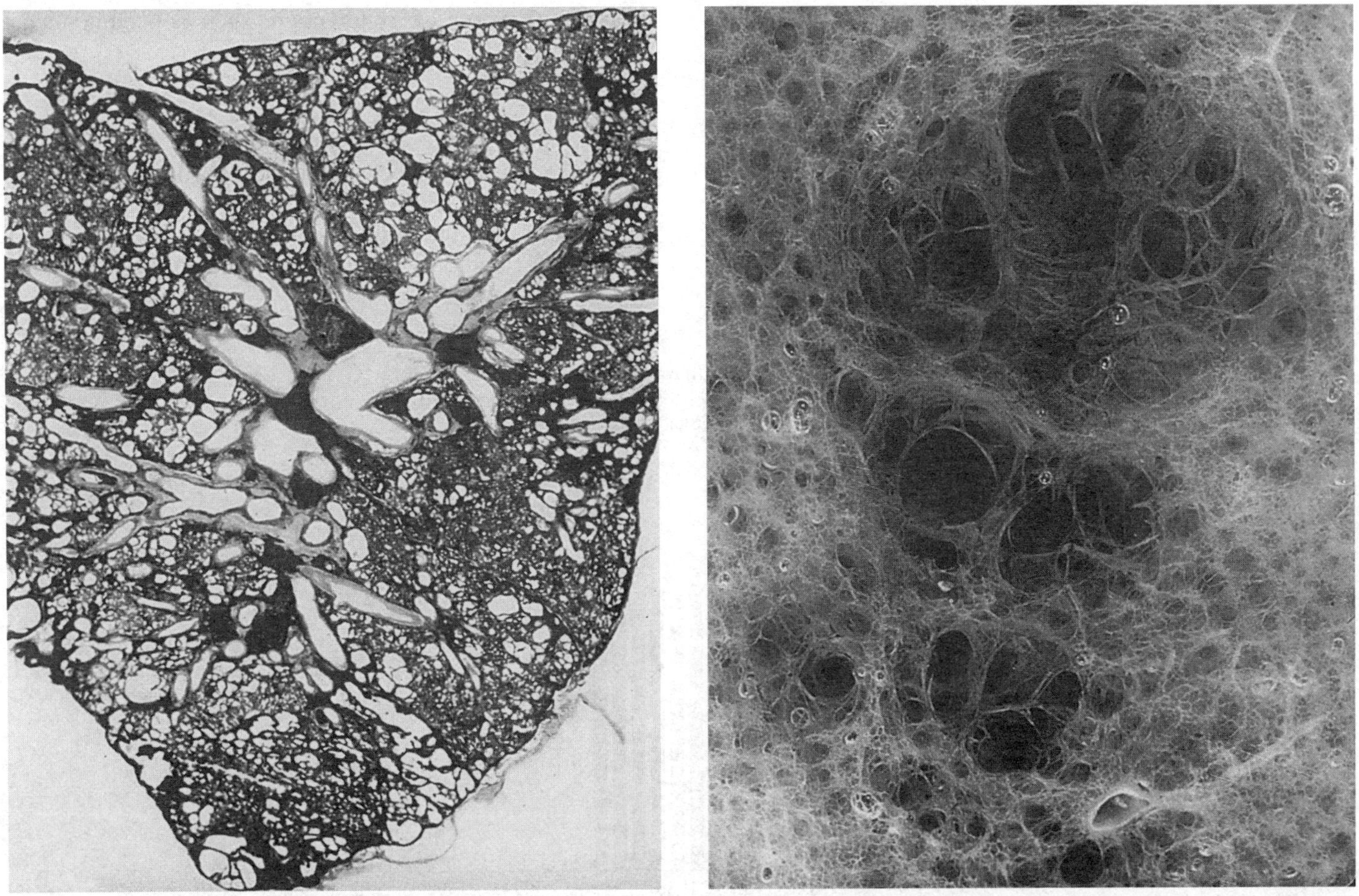

Fig 9.33 (a) *Section of lung from a patient who suffered from emphysema. Note that there are fewer alveoli than usual and the air spaces are larger (compare fig 9.22).* (b) *Close-up view of lung tissue from emphysema lung.*

raise an arm without a noticeable increase in breathing stress. The lungs also lose elasticity, so that it becomes more difficult to exhale air. A lot of air remains in the lung during expiration. This is easily demonstrated by measuring 'forced expiratory volume' (FEV), the amount of air that can be forcibly exhaled in a given time (usually one second) after a maximum inspiration. FEV also declines in asthma sufferers. Breathing out becomes a more conscious activity – the patient is forced to think about it and work to achieve it. Inflammation and narrowing of the finer bronchioles also occurs.

The root cause of emphysema is long-term irritation of the lungs, most commonly by cigarette smoke, air pollution or industrial dust. The damage is caused by both chemical and physical factors. Chemicals in cigarette smoke, for example, disturb the normal balance between breakdown and replacement of elastic tissue. They inhibit enzymes which prevent the breakdown of elastic tissue in the walls of the alveoli and also inhibit repair processes. White blood cells in the lungs secrete protein-digesting enzymes in response to the increasing stress and these also break down the walls of the alveoli. Heavy coughing associated with bronchitis (see below) may also physically damage the walls of the alveoli.

Emphysema is very common in Britain. About 1 person in 100 suffers from it and it is 10 times more common in men than women. Death is typically preceded by years of increasing disability.

9.7.4 Bronchitis

Bronchitis is inflammation of the lining of the air passages and may be chronic or acute. A **chronic** disease is one with a gradual onset and of long duration. An **acute** condition flares up quickly and dies down in a relatively short space of time. **Acute bronchitis** usually lasts a few days only and is a side-effect of an infection like a cold.

Chronic bronchitis is a much more serious and common problem. Britain has the highest death rate in the world from this disease (one person in 2000 every year, about 30 000 people per year). About 1 million others are affected, three times as many men as women. Death rate is six times higher among smokers. It is often associated with emphysema, especially in its late stages. Like emphysema, it is most commonly caused by smoking and to a lesser extent by air pollution. The same sensation of breathlessness occurs, due to reduced gaseous exchange.

The tars in cigarette smoke are the chemicals which are mainly responsible for the inflammation. One symptom of the disease is secretion of excess mucus from the goblet cells in response to the irritation. Smoking destroys or paralyses the cilia which normally sweep away the mucus. The main sign is therefore a cough in which the excess mucus is coughed up as a thick and greenish-yellow sputum. Coughing and breathlessness increase as the disease progresses and the more damaged the system becomes, the more likely infections such as pneumonia are to occur.

9.7.5 Lung cancer

Lung cancer is the most common form of cancer in men in the UK (breast cancer is the most common for women). It is the third most common cause of death in the UK after coronary heart disease and strokes (fig 9.34).

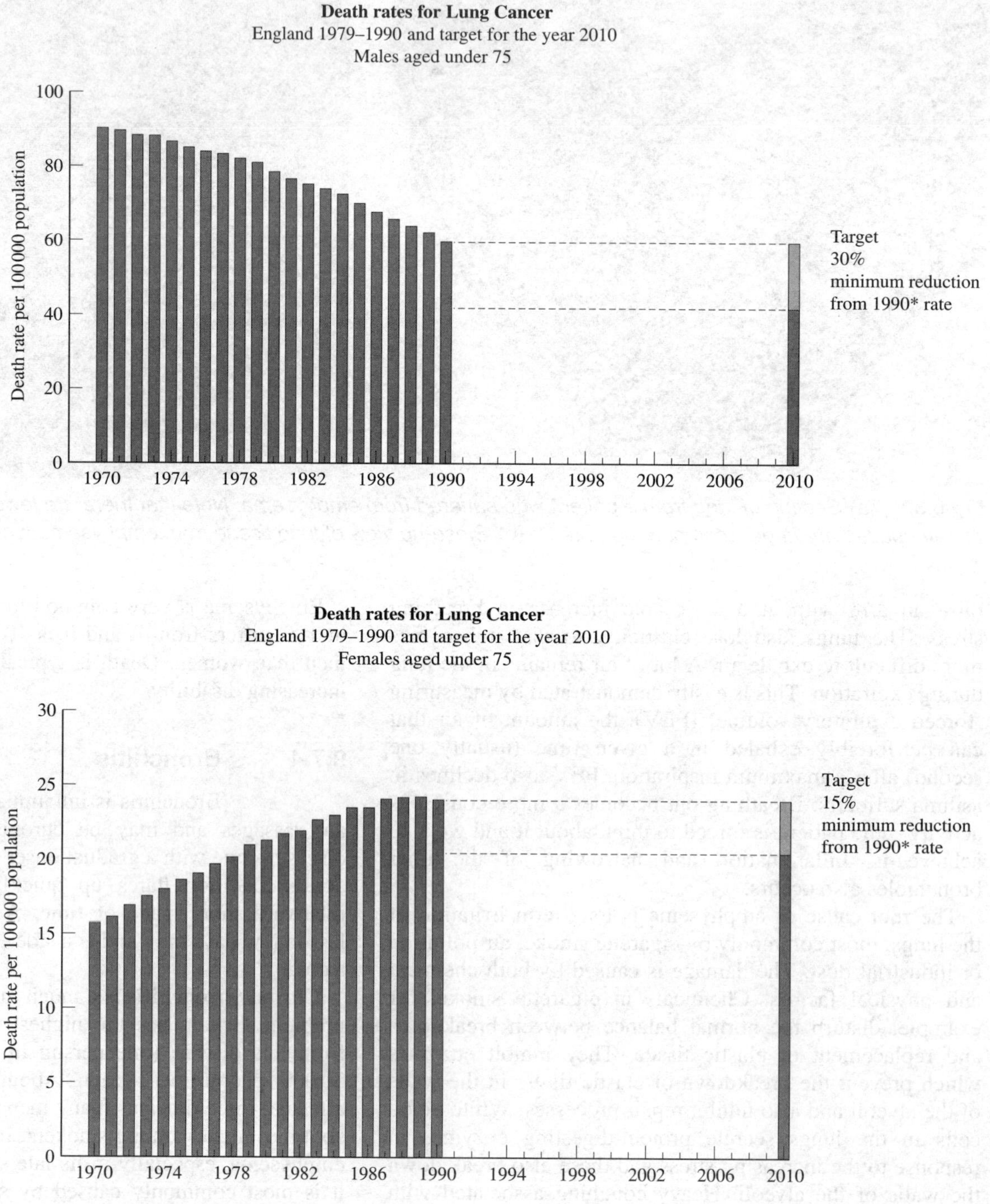

Fig 9.34 *Death rates from lung cancer.*

It accounted for about 1 in 18 deaths (5–6%) in Britain during the 1980s. Cancer is caused by cells dividing repeatedly out of control. They cease to respond to the normal signals around them and form unspecialised masses of cells called **tumours** (fig 9.35). Sometimes these cells break away from the original site and invade other tissues of the body to start secondary tumours. Lung cancer usually starts in the epithelium of the bronchioles, so-called bronchial carcinoma. It then usually spreads throughout the lungs. It is caused almost exclusively by smoking (99.7% of those who die from lung cancer are smokers). The tars in the smoke are responsible. They contain chemicals which cause cancer ('carcinogens'). The irritation causes thickening of the epithelium by extra cell division and it may be this that triggers the cancer. As a result of campaigns against smoking, there has been a decline in deaths from lung cancer (fig 9.34).

9.7.6 Effects of ageing on the respiratory system

There is a decline in the efficiency of the respiratory system with ageing. There is a gradual loss of elastic tissue and the chest wall becomes less capable of expansion. These changes show up as a reduction in vital capacity (the maximum volume of air that can be expired after a maximum inspiration). This may decrease by as much as 35% by the age of 70. All other aspects of function decline in performance, notably the action of cilia and the protective activity of white blood cells. This leaves the system more prone to diseases like pneumonia, bronchitis and emphysema.

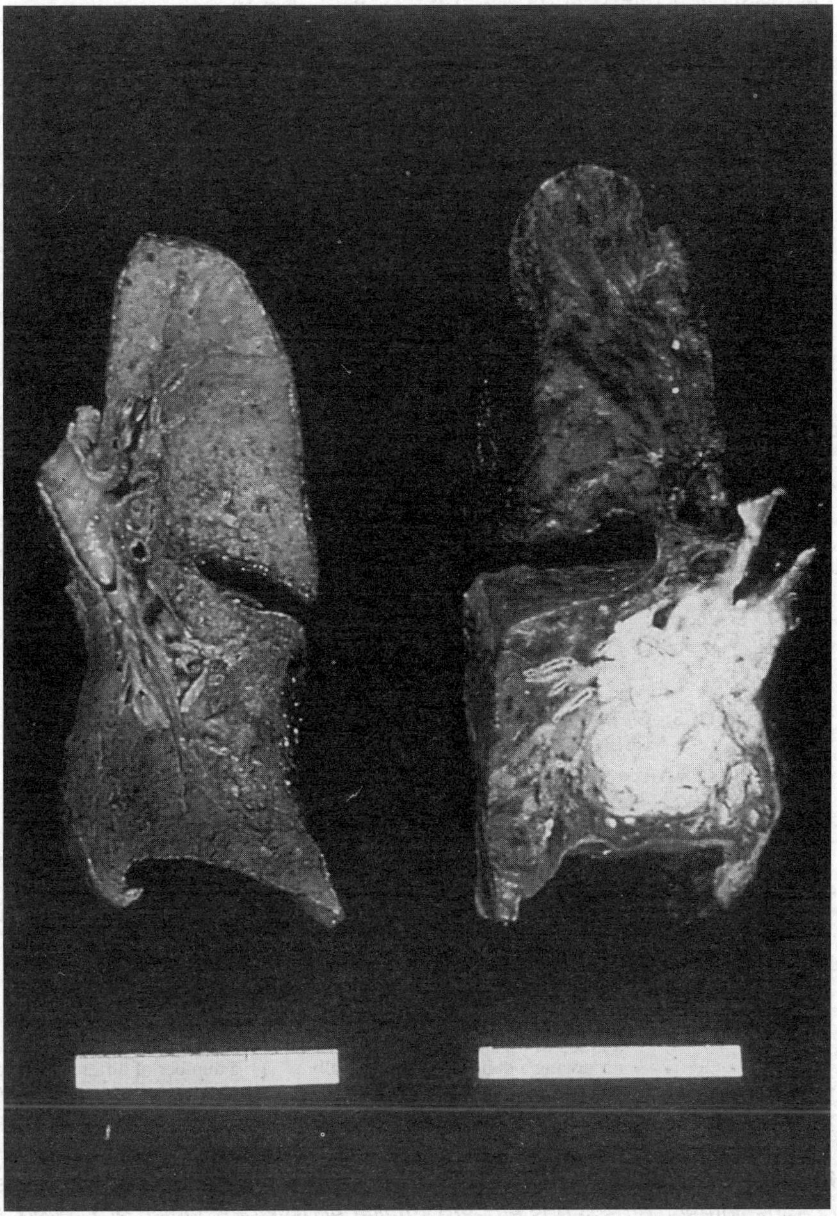

Fig 9.35 *The effect of lung cancer. The lung on the left shows the normal healthy condition. The one on the right shows a large white cancerous growth in the lower part.*

Chapter Ten

Organisms and the environment

Ecology is the study of the relationships of living organisms with each other and their non-living or physical surroundings. Ecological studies give us the scientific foundations for our understanding of agriculture, forestry and fisheries. Ecology also gives us the basis for predicting, preventing and remedying pollution. It helps us to understand the likely consequences of massive environmental intervention, as in the construction of dams or diversion of rivers, and provides the rationale underpinning biological conservation.

The relationship of ecology to other branches of biology is summarised in fig 10.1, which shows that living organisms can be studied at different levels of organisation. Ecology spans the right-hand portion of the diagram, which includes individual organisms, populations and communities. Ecologists regard these as the living part (**biotic component**) of a system called the **ecosystem**. This also includes a non-living part (the **abiotic component**) which contains matter and energy. Populations, communities and ecosystems are terms which have precise meanings in ecology. They are defined in fig 10.1. The different ecosystems together form the **biosphere**, or **ecosphere**, which includes all living organisms and the physical environments with which they interact. Thus, the oceans, land surface and lower parts of the atmosphere all form part of the biosphere.

10.1 Approaches to ecology

The holistic approach (one in which a whole picture is built up that is more important than its parts) is the distinctive characteristic of ecological science. Ecologists must simultaneously consider all of the factors interacting in a particular place. The sheer scope of this task presents problems, and in practice most ecologists adopt one of several main approaches when undertaking a new investigation. These approaches are described below.

- *Ecosystem approach* This approach uses studies which focus on the **exchange of energy and matter** between living and non-living components of the system. The functional relationships between organisms (such as feeding) and between organisms and their environment are emphasised, rather than species composition and identification of rarities.
- *Community approach* Community ecology (synecology) is mainly concerned with the biotic components of ecosystems. One important aspect of community studies is the concept of **succession** and **climax communities** (section 10.6.1).
- *Population approach* Modern population studies (autecology) are concerned with the characteristic mathematical forms of the growth, maintenance and

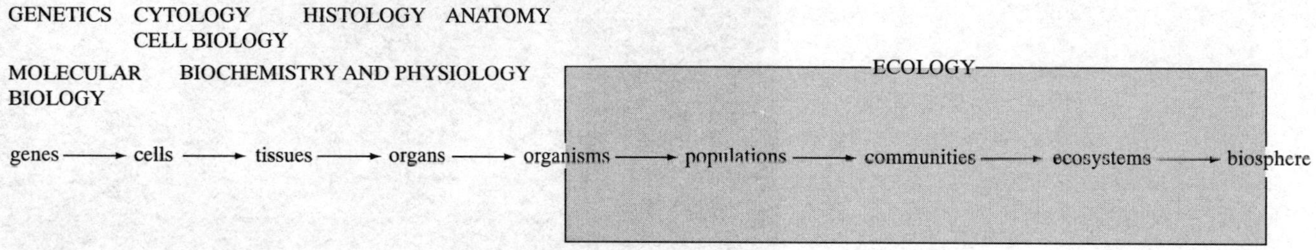

Definitions

population: a group of organisms of one species occupying a defined area and usually isolated to some degree from other similar groups

community: any group of organisms belonging to a number of different species that coexist in the same habitat or area and interact through trophic and spatial relationships

ecosystem: a community of organisms and their physical environment interacting as an ecological unit

Fig 10.1 *Levels of organisation from genes to ecosystems. The whole planet, the Earth, operates as an ecosystem. The oceans, forests, grasslands etc. are smaller ecosystems which are linked, by energy flow and exchange of materials, to form the overall planetary ecosystem.*

decline of species populations. They provide the scientific basis for understanding 'outbreaks' of pests and disease in agriculture and medicine, and help us to predict the critical numbers of individuals needed for continued survival of a species. Traditional autecology focuses on the environmental relationships of a particular species. It examines how characteristics such as an organism's morphology, behaviour, food preferences and so on are linked with its habitat, distribution and evolutionary history.

- *Habitat approach* **Habitat** is a spatial concept. It describes the typical environment of a particular organism, population, community or ecosystem (section 10.5). Particular locations within the same overall habitat may have their own special conditions and are sometimes referred to as **microhabitats**, such as the bark of a rotting log in an oakwood. The habitat approach is also convenient for studying those characteristics of the physical environment which are intimately linked with plants and animals, such as soils, moisture and light.
- *Evolutionary or historical approach* By studying how ecosystems, communities, populations and habitats have changed over time we can gain insights into why these changes occurred. This gives us a good basis for predicting the likely nature of future change. **Evolutionary ecology** views the changes since life evolved. It examines how events, such as the formation of a mountain barrier, have influenced the form and distribution of species and taxa. It answers questions such as 'why are kangaroos only found in Australia?' and 'why are tropical rain forests so species diverse?' It helps us to understand the triggers for extinction and speciation and at a detailed level to understand why species have a particular size or form, and reproductive strategy. **Palaeoecology** applies our modern knowledge of ecosystems to the study of fossil organisms. It attempts to reconstruct past ecosystems and, in particular, to see how ecosystems and communities functioned before humans became a major influence. **Historical ecology** is concerned with change since the developing technology and culture of humans made human activity a major influence on ecological systems. Understanding the human impact is a vital aspect of conservation management. Distinguishing human-induced and natural biosphere change is an important part of this process which may have economic significance. For example, is acidification a natural ecosystem process or is it wholly due to industrial pollutants and thus avoidable if we modify human behaviour?

These approaches to ecology interact and overlap. However, they provide a useful framework for study. In this chapter it is not possible to consider all in equal depth. We will focus on ecosystems, communities and populations.

10.2 Ecosystems

10.2.1 Definitions and key concepts

Ecosystems are made up of **biotic** and **abiotic** components. The organisms which comprise the biotic component are collectively known as the **community**.

Key points about ecosystems

- **There is a close association between living (biotic) and non-living (abiotic) components.** Both affect each other and are equally important for the ecosystem.
- **Ecosystems can be studied at any level of organisation.** For example you can apply ecosystem principles to a:

  puddle → pond → lake → sea → ocean → planet

  You can also study ecosystems over different time periods. A puddle is best studied over hours or days, whereas the ecological relationships in a lake ecosystem may only become fully apparent over many years.
- **All organisms and all features of the physical environment are necessary for the system to be maintained and flourish.** The tendency for a system to maintain a stable state, a balance between the contributions of all its parts, is known as **homeostasis** (self-regulation). Changes move the system away from equilibrium. Small changes will normally be countered by **feedback** processes within the system and the original equilibrium will be restored. Large changes can alter the state of the system in a major way and move it well away from its original equilibrium point. This happens when the feedback mechanisms cannot respond rapidly enough to maintain the original equilibrium. A new equilibrium will be reached but the original ecosystem may be radically changed or altered.

  Changes are not necessarily detrimental; for example the development of scrubland or forest from initial bare ground during ecological succession (section 10.6.1). However, there are many examples where human manipulation of ecosystems has produced unexpected and unwished-for side effects, such as problems with insect pests in simplified agricultural ecosystems. Also, changes may not be immediately obvious. For example, we have only recently understood the link between waste products of fuel combustion and acid damage to lakes and forests (section 10.8.1).

10.2.2 Overall structure of ecosystems

The general structure of an ecosystem can be seen in fig 10.2 which shows, in a simplified way, the overall structure of a terrestrial and an aquatic ecosystem.

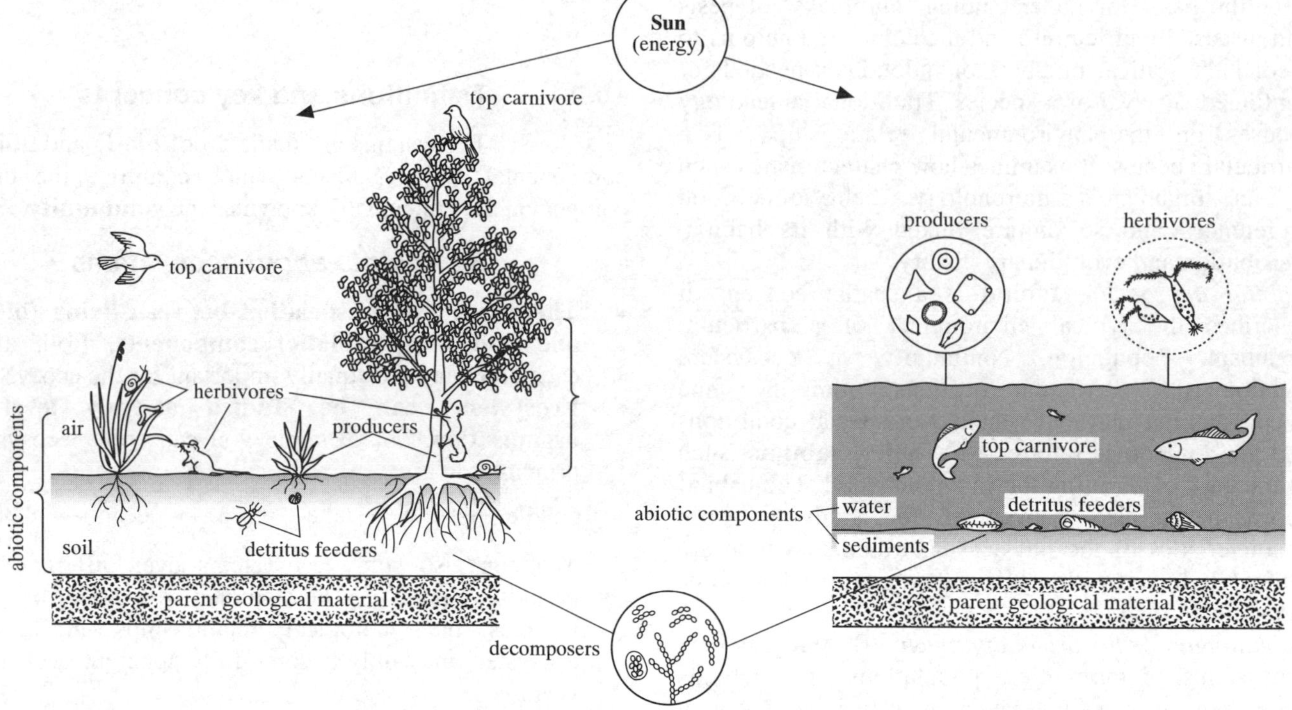

Notes	*Land*	*Water*
producers	trees, herbs, grasses	phytoplankton
primary consumers	small animals e.g. field mouse, squirrel, caterpillar	zooplankton
secondary consumers (top carnivores)	bird of prey	large fish
detritivores	soil non-vertebrates	bottom-living non-vertebrates
decomposers	bacteria and fungi	bacteria and fungi

Fig 10.2 *A simple schematic comparison of terrestrial and aquatic ecosystems (see also section 10.3.2). (Modified from E. P. Odum (1975)* Ecology, *2nd ed., Holt, Rinehart & Wilson.)*

The most striking point is the similarity in ecosystem structure found in these two very different environments. The biotic component of all ecosystems can usefully be subdivided into autotrophic and heterotrophic organisms. Heterotrophs are dependent on autotrophs for their existence. This point is fundamental to our understanding of **food chains** and **food webs** and the movement of energy and nutrients in ecosystems (sections 10.3 and 10.4).

The non-living or abiotic component of an ecosystem includes soil, water and climate. Soil and water contain a mixture of inorganic and organic nutrients. Soil derives most of its physical and chemical properties from its geological parent material. Similarly water quality and salinity are influenced by bedrock and basal sediments and by the soils and rocks of the surrounding area. Climate includes such environmental variables as light, temperature, humidity and rain or snow, which are important influences on the types of living organisms that can flourish in an ecosystem.

The essence of ecosystem studies, however, lies in understanding how connections between the different organisms and their abiotic environment work. **Energy flow** and **biogeochemical cycling** are the important functional links between the different ecosystem components.

10.2.3 Energy flow and biogeochemical cycling

Energy may be defined as the capacity to do work and living organisms can be likened to machines in that they require energy to keep working and stay alive. The energy that powers most ecosystems is ultimately derived from the Sun.* Solar energy is captured by photoautotrophs

* Recently deep sea exploration has found areas near hot vents linked to underwater volcanoes which are unexpectedly rich in life. Organisms include previously unknown species. Sunlight cannot reach the ocean depths and life in the deep waters is often sparse, being dependent on organic debris from surface illuminated waters as a food source. In these hot vents, however, it appears that autotrophic bacteria are the primary food suppliers for some unique tube worms and clams. These bacteria feed on hydrogen sulphide released from the hot vents and are therefore chemoautotrophs.

in photosynthesis. Autotrophs in turn form the food source or potential chemical energy supply for all other organisms in the ecosystem.

The chemicals found in living organisms are derived originally from the abiotic components of ecosystems, such as soil, water and air, to which they eventually return by way of the decomposition of the waste products or dead bodies of organisms. Bacteria and fungi bring about decomposition, obtaining energy from the waste products and dead organisms in the process. Thus a constant cycling of the chemical materials needed by living organisms occurs within an ecosystem. Since both living and non-living parts of the ecosystem are involved in these chemical cycles they are called **biogeochemical cycles**.

The energy to drive these cycles is also supplied by the Sun because it drives the Earth's weather systems and regulates Earth surface climates. Temperature, wind speed and direction, evaporation and rainfall, all ultimately depend on the input of solar energy. Rates of erosion and weathering, the breakdown of rocks to fine particles, in turn reflect climatic conditions. Thus the supply of abiotic nutrient elements is also ultimately dependent on solar energy.

While the chemicals in ecosystems are constantly recycled and used again, some of the energy transferred within the ecosystem is changed into forms which cannot be used again by the system, mainly heat energy. To maintain the ecosystem, frequent and regular inputs of solar energy are needed. Thus we make an important distinction between chemical elements which are *recycled* and energy which is said to *flow* through ecosystems. This is illustrated in fig 10.3.

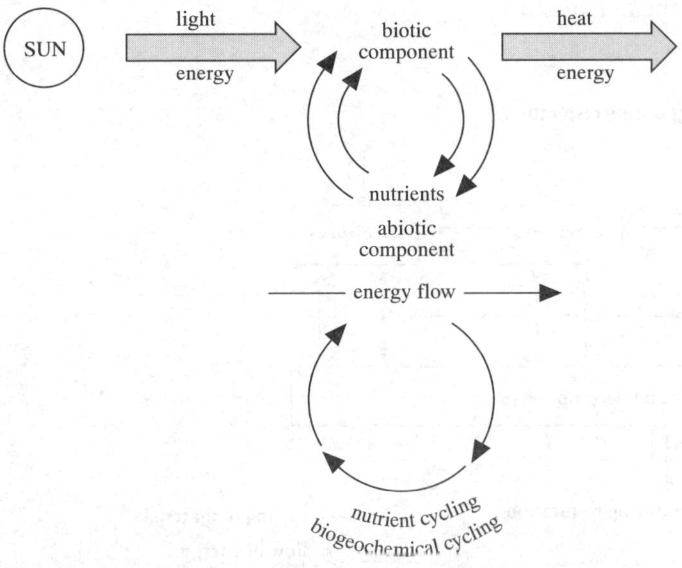

Fig 10.3 *Schematic summary of an ecosystem showing energy flow and nutrient or biogeochemical cycling.*

10.3 Ecosystems and energy flow

The study of energy flow through ecosystems is called energetics. The SI unit of energy is the joule, though the traditional unit, the calorie, is still often used. Both units are defined in table 10.1, which also includes references to the energy content of representative foods and organisms, and to daily food requirements of representative organisms.

10.1 Why are the figures for energy content in table 10.1 quoted for dry mass rather than fresh (wet) mass?

10.2 Account for the large difference in daily energy requirements of humans and small birds or mammals on a weight for weight basis.

Table 10.1 Units of energy and energy content of some living organisms and biological molecules.

Energy units		
calorie (cal) or gram calorie	–	the amount of heat (or energy) needed to raise the temperature of one gram of water through 1 °C (14.5 °C to 15.5 °C)
kilocalorie (kcal or Cal)	–	1000 cal
joule (J)	–	10^7 ergs: 1 erg is the amount of work done when 1 newton moves through 1 metre (1 newton (N) is a unit of force) Alternatively 981 ergs is the work done in raising a gram weight against the force of gravity to a height of 1 cm
kilojoule (kJ)		1000 J
		[1 J = 0.239 cal 1 cal = 4.186 J]

Energy content *(averages or approximations)*	*joules per gram dry mass (energy value)*
carbohydrate	16.7
protein	20.9
lipid	38.5
terrestrial plants	18.8
algae	20.5
non-vertebrates	12.6
(excl. insects)	22.6
vertebrates	23.4
(differences between these groups of organisms are due partly to different mineral contents)	

Daily food requirements	*kJ per kg live body mass*
humans	167 (about 12 500 kJ day^{-1} for a 70 kg adult)
small bird or mammal	4186
insect	2093

Based on data from table 3.1, Odum, E.P. (1971) *Fundamentals of Ecology*, 3rd. ed. Saunders.

10.3.1 The Sun as a source of energy

As we have seen the Sun is usually the ultimate source of energy in ecosystems. Of the Sun's energy which reaches the Earth, about 40% is reflected immediately from the clouds, dust in the atmosphere and the Earth's surface without having any heating effect. A further 15% is absorbed and converted to heat energy in the atmosphere, particularly by ozone in the stratosphere, and by water vapour. The ozone layer absorbs almost all short-wave ultraviolet radiation which is important because such radiation is hazardous to exposed living material. The remaining 45% of incoming energy penetrates to the Earth's surface. This represents an average of about 5×10^6 kJ m^{-2} yr^{-1}, though the actual amount for a given locality varies with latitude and local features such as aspect and climate. Just under half the radiation striking the Earth's surface is in the **photosynthetically active range (PAR)**, the visible wavelengths. However, even under optimum conditions only a very small proportion, about 5% of incoming radiation (or 10% PAR) is converted in photosynthesis into **gross primary productivity (GPP)**. A more typical figure for good conditions is 1% of total radiation (2% PAR) while the biosphere average is about 0.2% of total incident radiation. **Net primary productivity (NPP)** (the net gain of organic material in photosynthesis after allowing for losses due to respiration) varies between 50 and 80% of gross primary productivity (see section 10.3.5).

As a global average the energy fixed by Earth's green plants is only 0.1% of that received by the Earth. Terrestrial systems, which cover the 30% of the Earth which is not covered by oceans, fix half the total sunlight captured. Cultivated crops achieve higher rates of GPP and NPP during their short growing periods, but so far it has proved impossible to achieve higher rates of photosynthetic fixation on a sustained basis under normal field conditions.

10.3.2 Energy transfers: food chains and trophic levels

The energy-containing organic molecules produced by autotrophic organisms are the source of food (materials and energy) for heterotrophic organisms. These animals may in turn be eaten by other animals, and in this way energy is transferred through a series of organisms, each feeding on the preceding organism and providing raw materials and energy for the next organism. Such a sequence is called a **food chain**, or the **grazing link** in the ecosystem. Each stage of the food chain is known as a **trophic level** (*trophos*, food). The first trophic level is occupied by the autotrophic organisms, called **producers**. The organisms of the second trophic level are usually called **primary consumers** while those of the third level are called **secondary consumers**, and so on.

There are usually four or five trophic levels, and seldom more than six. This is partly because at every feeding stage some energy is wasted from the chain of animals feeding on each other. This point is also evident from figs 10.4 and 10.10 and relates to the discussion in section 10.3.5. Recent work has, however, suggested that factors other than energy loss may also be important in limiting the length of food chains. The availability of sufficient food of the preferred types and territorial space may also restrict the numbers of end-of-chain organisms and thus the length of food chains.

It is estimated that in some ecosystems as much as 80% of primary production is not eaten by primary consumers. Instead, on death, plant material is consumed in various ways by detrital feeders and decomposer organisms. Food chains in which most primary production is decomposed or consumed as detritus are termed **detrital food chains**. Tropical moist forests are a noteworthy example of ecosystems where the detrital link is more important than the grazing link.

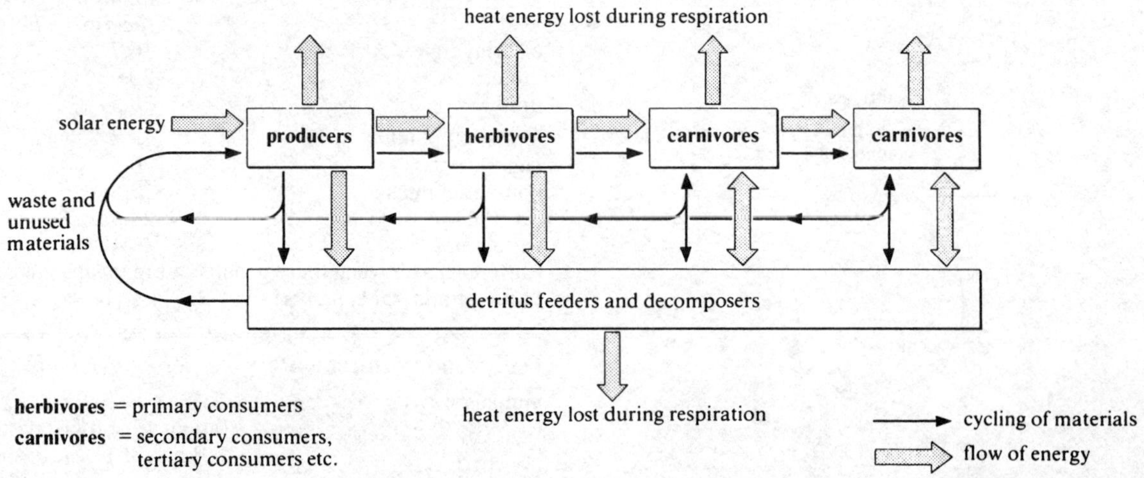

Fig 10.4 *Flow of energy and cycling of materials through a typical food chain. Note that a two-way exchange is possible between carnivores and detritus feeders/decomposers. The latter feed on dead carnivores; carnivores may eat living detritus feeders/decomposers.*

Producers

The producers are autotrophic organisms, and are typically green plants and algae. Some bacteria, such as blue-green bacteria, also photosynthesise and are thus also producers. Microscopic algae and blue-green bacteria are the main producers in aquatic ecosystems and are known as **phytoplankton**. These contrast with terrestrial ecosystems in which larger plants dominate, such as the grasses characteristic of savannah, steppe and many agricultural ecosystems and the conifer and broadleaf trees of forest ecosystems.

Primary consumers

Primary consumers feed on producers. They are therefore also called **herbivores**. Some primary consumers do not eat the producer but live as plant parasites, such as aphids, some fungi and even other plants (e.g. broomrape, *Orobanche*, an unusual plant with no chlorophyll). More common are plants such as mistletoe, which parasitises its host tree for nutrients but has its own chlorophyll system and thus is also a producer.

On land, herbivores typically include insects, reptiles, birds and mammals. In aquatic ecosystems (freshwater and marine) herbivores are typically small crustaceans, such as water fleas, crab larvae and barnacles, and molluscs which include bivalves such as mussels and clams. Most are filter-feeders and extract the producers from the water as described in section 8.2.1. Together with protozoans, many of them form the **zooplankton**, the small or microscopic drifting animals which feed on the phytoplankton. Life in the oceans and lakes is almost totally dependent on planktonic organisms which are found at the beginning of virtually all food chains.

Secondary and tertiary and other top consumers

Secondary consumers feed on herbivores and are thereby flesh eaters or carnivores. Tertiary and other higher order consumers that feed on the secondary (or tertiary as appropriate) consumers are also carnivores.

Secondary and tertiary consumers may be:

- **predators**, which hunt, capture and kill their prey;
- **carrion feeders**, which feed on corpses;
- **parasites** which do not eat their prey but feed off the host organism while it continues to live.

Two examples of predator food chains are given below.

plant (such as leaves) → slug → frog → grass snake → stoat

rosebush sap → aphid → ladybird → spider → insectivorous bird → hawk

Typically carnivores become larger and fewer in number at each successive trophic level. Parasite food chains are very different; the parasites get smaller at successive trophic levels and typically increase in number.

> **10.3** Give example food chains for any major habitat you have investigated, e.g. marine, freshwater, woodland, grassland, etc.

Decomposers and detrivores (detrital food chains)

When organisms die their bodies form a source of energy and raw materials (nutrients) for other organisms. Similarly waste materials passed from the bodies of living organisms are also a source of energy and nutrients. These materials are not wasted by ecosystems. They form the food for many other organisms referred to as **decomposers** and **detritivores**.

Decomposers are microorganisms, mainly fungi and bacteria, which live as saprotrophs on **dead organic matter (DOM)**. They secrete digestive enzymes onto dead or waste material and absorb the products of digestion. **Detritivores** feed on small fragments of decomposing or dead material termed detritus. Many small animals in both terrestrial and aquatic ecosystems are detrital feeders. Examples include ragworms in estuarine environments, sludgeworms in fresh waters and on land earthworms, woodlice and very small animals such as mites and springtails. Methods for isolating and examining some of these organisms are given in section 11.2.

Detritivores may be fed upon by carnivores which may then be consumed by other carnivores, thus building up a food chain based on detritus. These are termed **detrital food chains** to distinguish them from the **grazing food chains** we have already described where living primary producers form the base of the food chain. Two typical detritus food chains of woodlands are:

leaf litter →	earthworms → *Lumbricus* spp.	blackbird → *Turdus nerula*	sparrowhawk *Accipiter nisus*
dead animal →	blowflies → and blowfly maggots (larvae) *Calliphora vomitoria*	common frog → *Rana temporaria*	grass snake *Natrix natrix*

Rates of decomposition vary with substrate and climate. The organic matter of animal urine, faeces and corpses may be consumed within a matter of weeks, whereas fallen trees and branches may take many years to decompose. Essential to the breakdown of wood (and other plant material) is the action of fungi which produce cellulase, softening the wood and allowing small animals to penetrate and ingest material. Decomposition is most rapid in warm and moist environments, such as tropical rainforest, but takes place slowly in cool and/or dry conditions. The virtual absence of litter from the rainforest floor and the low content of humus in rainforest soils by comparison with the conspicuous litter layer and significant humus content of soils in temperate oakwoods or beechwoods reflect this point. This has important implications for human use of these systems (see section 10.9.5).

10.3.3 Food webs

In food chains each organism is shown as feeding on only one other type of organism. However, the feeding relationships within an ecosystem are usually far more complex than this. Most organisms feed on more than one other organism. Some feed in both grazing and detrital food chains. This is particularly true of carnivores at the higher trophic levels. Many carnivores have highly varied diets and operate as secondary, tertiary, quaternary and higher consumers. Some animals, including humans, feed on organisms at all trophic levels; plants, animals and fungi. These organisms are called **omnivores**.

Grazing and detrital food chains interlink in a complex manner. An earthworm, for example, may feed as both a herbivore (fine rootlets) and a detritivore. Waste products and dead bodies from every trophic level form the raw materials for detrital food chains. This mesh of interlinking food chains that characterises the real world is called a **food web**. Fig 10.5 illustrates woodland and freshwater food webs. Only some of the many possible interrelationships can be shown on such diagrams and it is usual to include only one or two carnivores at the highest level. Such diagrams illustrate the feeding relationships among organisms in an ecosystem and provide a basis for more quantitative studies of energy flow and exchange of material through the biotic component of ecosystems.

10.3.4 Ecological pyramids

The first pyramid diagrams were prepared by Charles Elton in the 1920s. His pyramids were based on field observations of the numbers of animals in different size classes. He did not include primary producers, nor distinguish detritivores and decomposers in his model. However, Elton observed that predators were typically larger than their prey and realised that this relationship was quite specialised for particular sizes of predator and prey. In the 1940s, the American ecologist Raymond Lindeman suggested that Elton's idea could be adapted to a trophic model, that is one based directly on the feeding levels of organisms irrespective of their sizes. However, while it is a straight-forward matter to assign animals to different size classes, it is much more difficult to identify their trophic level. We can do this only in the most general terms.

> **10.4** Using the information from fig 10.5*a* and section 10.3.2 on consumer organisms, identify and write out food chains for a hawk at trophic levels 3, 4, 5 and 6.

Feeding relationships and the efficiency of energy transfer through the biotic component of ecosystems have traditionally been summarised in pyramid diagrams. These give an apparently simple and fundamental basis for comparing:

- different ecosystems;
- seasonal variation within a particular ecosystem;
- change in an ecosystem.

Three types of pyramid have been used. These are:

- pyramids of numbers, based on counting the numbers of organisms at each trophic level;
- pyramids of biomass, which note the weight (usually dry weight) of organisms at each trophic level;
- pyramids of energy, which monitor the energy content of the organisms at each trophic level.

Energy pyramids are considered the most important since they deal directly with the fundamentals of food chains, the flow of energy.

Pyramids of numbers

In a numbers pyramid based on trophic levels, the organisms of a given area are first counted and then grouped into their trophic levels (as best one can). When this is done, a progressive decrease in the number of organisms at each successive level is often found. For diagrammatic purposes the number of organisms in a given trophic level can be represented as a rectangle whose length (or area) is proportional to the number of organisms in a given area (or volume if aquatic). An idealised numbers pyramid is shown in fig 10.6*a*.

Although the data needed to construct pyramids of numbers may be relatively easy to collect by straightforward sampling techniques, there are a number of complications associated with their use. Three important problems are as follows.

- Producers vary greatly in size, but a single grass plant or alga, for example, is given the same status as a single tree. This explains why a true pyramid shape is often not obtained (see fig 10.6*b*). Parasitic food chains may also give inverted pyramids.
- The range of numbers is so great that it is often difficult to draw the pyramids to scale. Although logarithmic scales may be used, their interpretation needs care.
- The trophic level of an organism may be difficult to ascertain.

> **10.5** What changes would you expect to occur to the pyramid of numbers shown in fig 10.6*b* during the midwinter season?

Pyramids of biomass

Some of the disadvantages of using pyramids of numbers can be overcome by using a **pyramid of biomass** in which the total mass of the organisms (**biomass**) is estimated for each trophic level. Such estimates involve weighing representative individuals, as well as recording numbers, and so are more laborious and expensive in terms of time

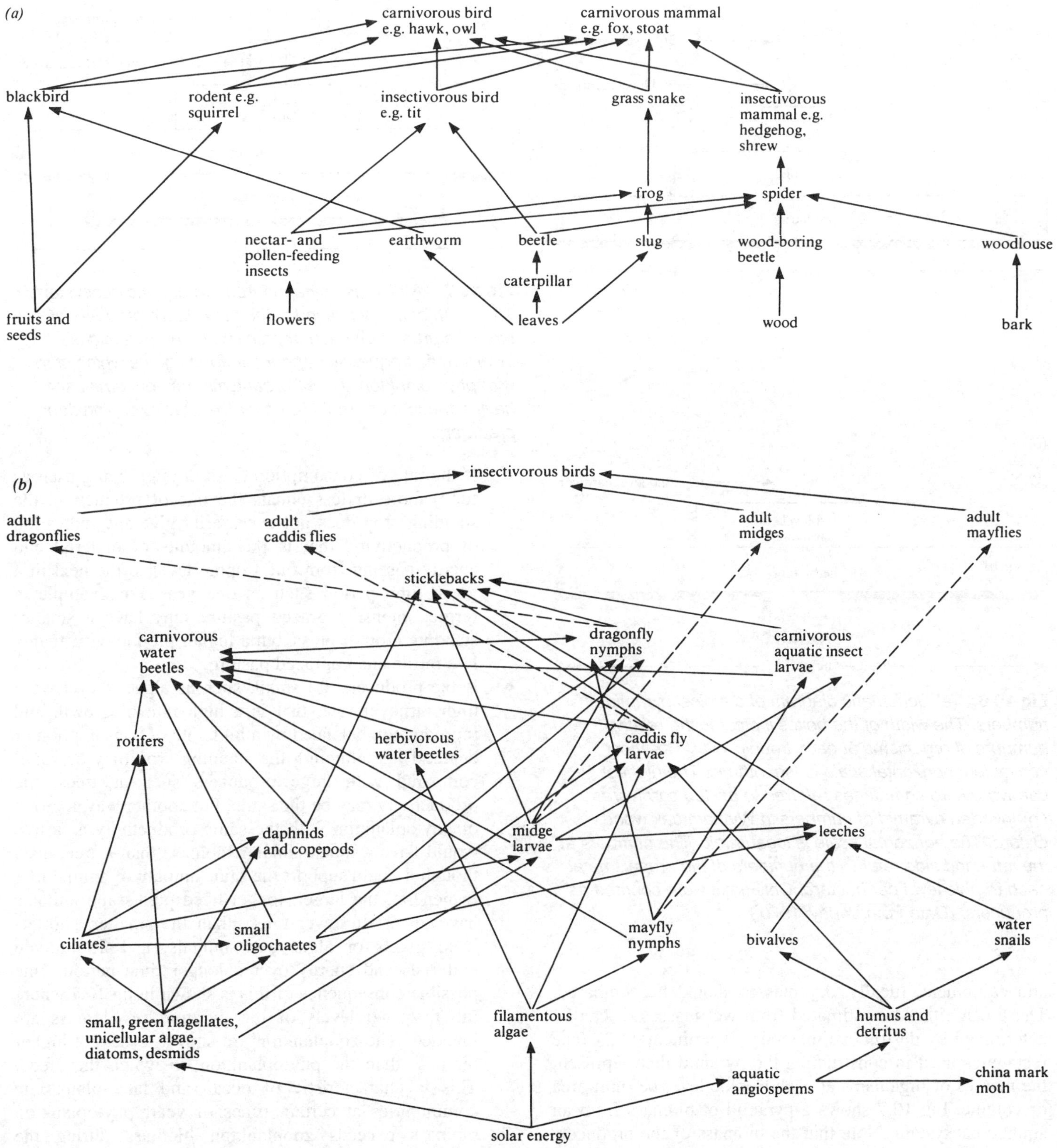

Fig 10.5 *(a) Feeding relationships in a woodland, forming a food web.* (b) *Food web of a freshwater habitat (based on Popham (1955)* Some aspects of life in fresh water, *Heinemann).*

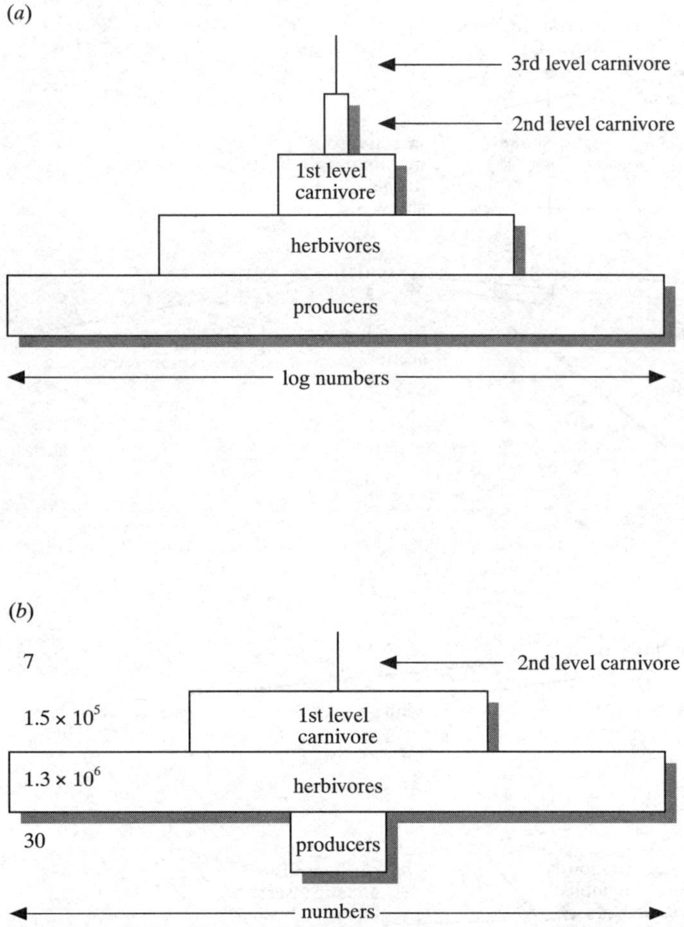

Fig 10.6 *(a) Schematic diagram of a typical pyramid of numbers. The width of the boxes indicates the relative numbers of organisms at each trophic level. The most convenient horizontal scale is logarithmic. The highest carnivores are sometimes referred to as 'top carnivores'. (b) Inverted pyramid of numbers in Wytham oak wood, Oxford. The horizontal scale is logarithmic. The numbers at the left-hand side refer to the numbers of the organisms at each trophic level per hectare. Only oaks were counted as producers. (Data from Varley, 1970.)*

and equipment. Ideally, dry masses should be compared. These can either be estimated from wet masses or can be determined by destructive methods (experiment 11.2). The rectangles used in constructing the pyramid then represent the masses of organisms at each trophic level per unit area or volume. Fig 10.7 shows a pyramid of biomass from an aquatic ecosystem. Note that the biomass of the producers (phytoplankton) is smaller than the primary consumers (zooplankton), but the pyramid becomes a typical shape above this level. The biomass at the time of sampling, in other words at a given moment in time, is known as the **standing biomass** or **standing crop biomass**. It is important to realise that this figure gives no indication of the *rate* of production (**productivity**) or consumption of biomass. This can be misleading in two ways.

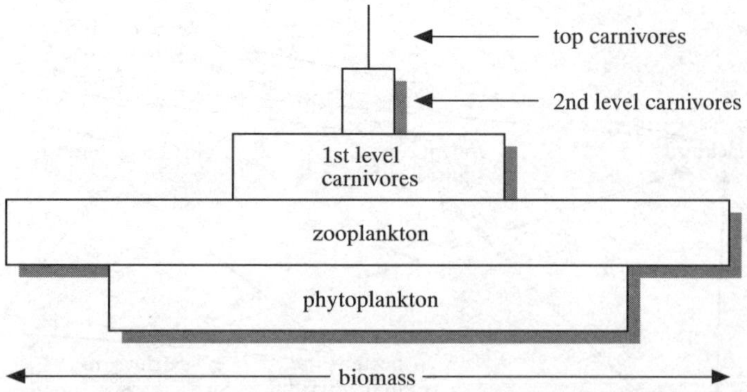

Fig 10.7 *A biomass pyramid from an aquatic ecosystem. The width of the boxes indicates the relative amounts of biomass present at each trophic level. The pyramid is inverted, as frequently happens in food chains which start with phytoplankton. These organisms are very small and have a much more rapid turnover than their zooplankton predators.*

- If the rate of consumption (loss through being used as food) more or less equals the rate of production, the standing crop does not necessarily give any indication of productivity, that is the amounts of material and energy passing from one trophic level to the next in a given time period such as one year. For example, a fertile, intensely grazed pasture may have a smaller standing crop of grass, but a higher productivity, than a less fertile and ungrazed pasture.
- If the producers are small, such as algae, they have a high turnover rate, that is a high rate of growth and reproduction balanced by a high rate of consumption or death. Thus, although the standing crop may be small compared with large producers such as trees, the productivity may be the same. Put another way, a group of phytoplankton with the same productivity as a tree would have a much smaller biomass than a tree, even though it could support the same amount of animal life. In general, the larger, longer-lived plants and animals have lower 'turnover rates' than the smaller, shorter-lived plants or algae and animals, and accumulate materials and energy over a longer time period. One possible consequence of this is shown in fig 10.7 where the first two levels of the pyramid of biomass are inverted. The zooplankton are shown to have a higher biomass than the phytoplankton on which they feed. This is characteristic of ocean and lake planktonic communities at certain times of year; phytoplankton biomass exceeds zooplankton biomass during the spring 'bloom', but at other times the reverse is often true. Such apparent anomalies are avoided by using pyramids of energy as described below.

These differences may also highlight useful information. For example, persistence of an algal bloom and a broad-based biomass pyramid in an aquatic ecosystem may indicate the onset of eutrophication (section 10.8.2).

Similarly, the frequent inversion of marine pyramids suggests that for these systems harvesting plant or algal material rather than animal biomass is not the sensible strategy it appears to be for many terrestrial ecosystems.

10.6 Fig 10.8 shows the standing crop biomass of producers and primary consumers in a lake throughout the year, as well as certain environmental variables.
(a) In what months could an inverted pyramid of biomass be obtained?
(b) What factors account for (i) the spring rise in phytoplankton production, (ii) the rapid decline in phytoplankton in the summer, (iii) the increase in phytoplankton in the autumn, (iv) the decrease in phytoplankton during the winter?

Pyramids of energy

The most fundamental and ideal way of representing relationships between organisms in different trophic levels is by means of a pyramid of energy. This has a number of advantages.

- It takes into account the *rate* of production, in contrast to pyramids of numbers and biomass which depict the standing states of organisms at a particular moment in time. Each bar of a pyramid of energy represents the amount of energy per unit area or volume that flows through that trophic level in a given time period. In fig 10.9 a pyramid of energy for an aquatic ecosystem is shown.

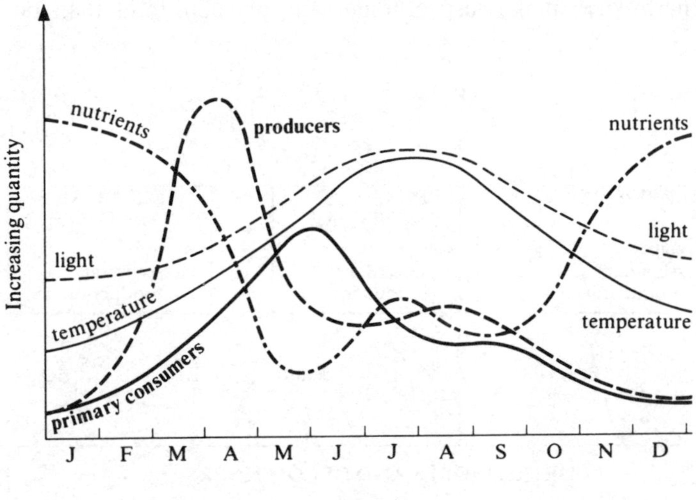

Fig 10.8 *Changes in standing crop biomass of producers and primary consumers and in certain environmental variables in a lake during one year. (From M.A. Tribe, M.R. Erant and R.K. Snook (1974)* Ecological principles, Basic Biology Course 4, *CUP.)*

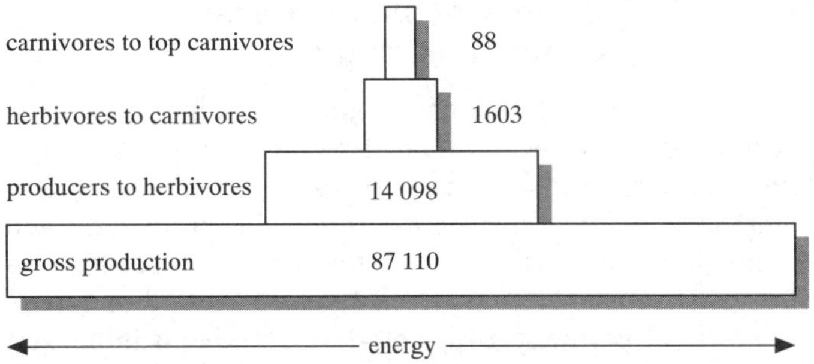

Fig 10.9 *An energy pyramid for Silver Springs, Florida. The figures represent energy flow in kJ m^{-2} yr^{-1}. (From E.P. Odum (1971)* Fundamentals of ecology, *3rd edition, W.B. Saunders.)*

- Weight for weight, two species do not necessarily have the same energy content, as table 10.1 indicates. Comparisons based on biomass may therefore be misleading.
- Apart from allowing different ecosystems to be compared, the relative importance of populations within one ecosystem can be compared and inverted pyramids are not obtained.
- Input of solar energy can be added as an extra rectangle at the base of the pyramid of energy.

Although pyramids of energy are sometimes considered the most useful of the three types of ecological pyramid, they are the most difficult to obtain data for because they require even more measurements than pyramids of biomass. One extra piece of information needed is the energy values for given masses of organisms. This requires combustion of representative samples. In practice, pyramids of biomass can sometimes be converted to pyramids of energy with reasonable accuracy, based on previous experiments.

Problems with ecological pyramids

- The most fundamental problem is that of identifying an organism's trophic level. As discussed above, many organisms feed at several trophic levels.
- Some ecologists also consider that assigning all plant material to the producer level may be inappropriate. Many plants have organs such as tubers or produce fruits and seeds which do not contain chlorophyll. These structures are plant products rather than primary producing photosynthetic organs. Many herbivores cannot digest chlorophyll; others are highly selective in their diet, eating only seeds or fruits or nectar. Many ecologists think that the trophic level model described by ecological pyramids should be adapted to take account of these important differences.
- Another difficulty is that DOM is often omitted from pyramid diagrams. Yet, as we have already noted, as much as 80% of all energy fixed by producers may not be eaten by consumers but by detritivores or is used by decomposers.

10.3.5 Efficiency of energy transfer: production ecology

The study of productivity is known as **production ecology**, and involves the study of energy flow through ecosystems. Energy enters the biotic component of the ecosystem through the producers, and the rate at which this energy is stored by them in the form of organic substances which can be used as food materials is known as **primary productivity**. This is an important parameter to measure as it determines the total energy flow through the biotic component of the ecosystem, and hence the amount (biomass) of life which the ecosystem can support.

> **10.7** In considering the primary productivity of an ecosystem, which groups of organisms other than plants make a contribution?

As mentioned in section 10.3.1, the amount of the Sun's radiation intercepted by the Earth's surface varies with latitude and with details of location such as aspect and altitude. The amount intercepted by plants also varies with light quality and the organisation and amount of vegetation cover. In Britain, incident radiation on plants averages about 1×10^6 kJ m^{-2} yr^{-1}. Of this, as much as 95–99% is immediately lost from the plant by reflection, radiation or heat of evaporation. The remaining 1–5% of incoming radiation is absorbed by the chlorophyll and used in the production of organic molecules. The rate at which this chemical energy is stored by plants is known as **gross primary productivity (GPP)**. Between 20–25% of the GPP is used by the plant in simultaneous respiration and photorespiration, leaving a net gain known as the **net primary productivity (NPP)** which is stored in the plant. It is this energy which is potentially available to the next trophic level.

When herbivores and carnivores consume other organisms, food (materials and energy) is thereby transferred from one trophic level to the next. Not all of the energy and materials available in the food is used by the consumer organisms for production. Some energy is lost as heat in respiration. Other losses occur in the organic waste products of metabolism and are **excreted**. Some food materials remain undigested and are lost immediately in the process of **egestion**.

food consumed = growth + respiration + egesta + excreta

These waste products are used as food sources by decomposers and detritivores.

Some of these terms can be measured easily in domestic animals or in laboratory studies of wild animals. Growth is measured as increase in biomass, or better as increase in energy value of the body, with time. Faeces and excreta can be collected, weighed and subtracted from the mass of food consumed to determine food retained and used for growth and respiration.

The energy remaining in heterotrophs after losses through egestion, excretion and respiration is available for production, that is growth, repair and reproduction. Production by heterotrophs is called **secondary production** (whatever the trophic level).

Fig 10.10 shows clearly that energy is lost at every stage in the food chain and the length of the food chain is obviously limited by the extent of these losses. The proportion of energy lost in the first transfer of energy from solar energy received to net primary production is high. Subsequent transfers are much more efficient. The average efficiency of transfer from plants to herbivores is about 10% and from animal to animal is about 20%. In general, herbivores make less efficient use of their food than do

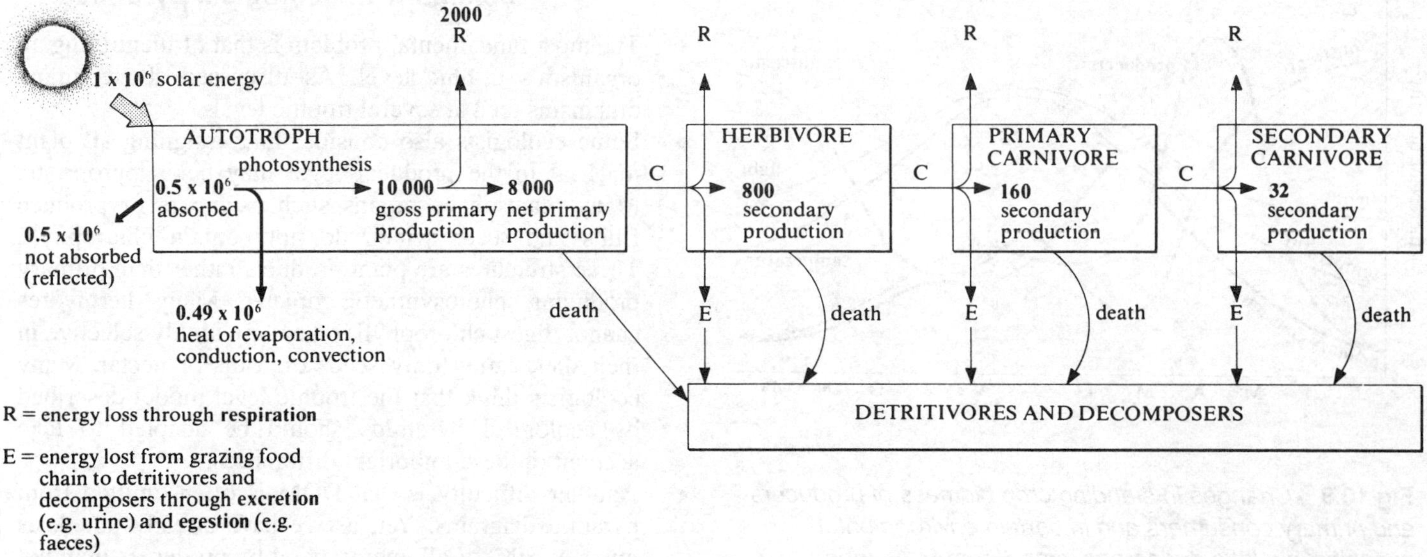

Fig 10.10 *Energy flow through a grazing chain, such as a grazed pasture. Figures represent kJ m^{-2} yr^{-1}.*

carnivores because plants contain a high proportion of cellulose and sometimes wood which are relatively indigestible and therefore unavailable as energy sources for most herbivores.

Energy lost in respiration cannot be transferred to other living organisms. However, the energy lost from a food chain in the form of excreta and egesta is not lost to the ecosystem because it is transferred to detritivores and decomposers. Similarly, any dead organisms, fallen leaves, twigs and branches and so on will start detritus and decomposer food chains. Detrital pathways are often complex and are less well understood than the conventionally described grazing pathways. Nevertheless they are just as important and, in terms of energy flow, frequently more important than grazing pathways.

The proportion of net primary production flowing directly into detritus and decomposer food chains varies from one system to another. In a forest ecosystem most of the primary production enters the detrital rather than the grazing pathway, with the result that the litter and humus on the forest floor is the centre of much of the consumer activity, even though the organisms involved are mostly inconspicuous. However, in an ocean ecosystem or an intensively grazed pasture more than half the net primary production may enter the grazing food chain. Most intensive agricultural systems ignore the potential value of detritus-based food production.

The figures quoted in this section have been on an annual basis. If the ecosystem is stable, that is not changing such as during succession (section 10.6.1), the total biomass at the end of the year will be the same as at the beginning. All the energy that went into primary production will then have passed through the various trophic levels and none retained as net production. Quite often, however, an ecosystem will be in a process of change. A young forest, for example, would retain some of the energy input in the form of increased biomass at the end of the year. A year is a useful period over which to express productivity because it takes into account seasonal variations where these exist. For example, primary productivity is usually greater in the part of the year when new plant or algal growth commences and secondary production increases later.

One of the reasons for studying energy flow through ecosystems is that it has important implications for the way in which humans obtain their own food and energy requirements. It opens the way to analysing agricultural systems for their efficiency, and suggests where improvements can be made. Since energy is lost at each trophic level, it is clear that, for omnivores like humans, eating plants is a more efficient way of extracting energy from a system (table 10.2). However, in suggesting improved methods for providing food, other factors must be considered. For example, animal protein is generally a better source of the essential amino acids, though some pulse crops, such as soyabean, are richer sources than most plants. Also animal protein is more easily digested, since the tough plant cell walls must first be broken down before the plant protein is released. Finally, there are many ecosystems where animals can concentrate food from large areas where it would be difficult to grow or harvest plant crops. Examples are grazing on poor quality pasture land, such as by sheep in Britain, reindeer in Scotland and Scandinavia and eland in East Africa, or taking fish from aquatic ecosystems.

Table 10.2 Outputs of agricultural food chains in UK.

Food chain	*Example*	*Energy yield of food to humans* (kJ × 10^3 ha^{-1})	*Protein yield of food to humans* (kg ha^{-1} yr^{-1})
(*a*) Cultivated plant crop ⟶ humans	Monocultures of wheat and barley	7800–11 000	42
(*b*) Cultivated plant crop ⟶ livestock ⟶ humans	Barley-fed beef and bacon pigs	745–1423	10–15
(*c*) Intensive grassland ⟶ livestock ⟶ humans	Intensive beef herd on carefully managed pasture		
	Meat	339	4
	Milk	3813	46
(*d*) Grassland and crops ⟶ livestock ⟶ humans	Mixed dairy farm		
	Milk	1356	17

Data from Duckham, A.N. & Mansfield, G.B. (1970) *Farming systems of the world*, Chatto and Windus.

10.4 Biogeochemical cycles – the cycling of matter

Biogeochemical cycling is the other major feature of ecosystems (along with energy flow). For many elements in an ecosystem, a cycle can be drawn which summarises the movement of the element through the living components of the ecosystem. During the cycle the element may be combined within complex organic molecules. These are later broken down in decomposition to simpler organic and inorganic forms which can be used again to make the living material of living organisms. As well as this actively **cycling pool** of an element, all cycles have a larger **reservoir pool** which is usually abiotic. Exchanges between the reservoir and active cycling pools are typically limited and often slow processes, for example the chemical weathering of phosphate rock, and fixation by lightning of nitrogen into nitrates during thunderstorms.

An understanding of biogeochemical cycling and maintenance of effective cycling is important. Human activity generally speeds movement of material through the cycles and may fundamentally upset the balance of cycles. This may lead to build-up of material at one point in the cycle, in other words, **pollution** (section 10.8).

The biogeochemical cycles for nitrogen and carbon are summarised in figs 10.11 and 10.12. Hydrogen, which has vital importance in photosynthesis, cycles in the water or hydrological cycle as shown in fig 10.13.

10.4.1 The nitrogen cycle

Nitrogen in the atmosphere (N_2) is very inert, and it takes a lot of energy to split the bonds in the nitrogen molecule so that it can form other compounds, such as nitrites and nitrates. However, nitrogen is an essential component of biological molecules such as proteins and DNA. The only organisms capable of splitting the nitrogen molecule are a few bacteria. They use it to form nitrites or nitrates, a process known as **nitrogen fixation** (fig 10.11). This is the major way in which nitrogen enters the biotic component of an ecosystem.

Nitrogen fixation

Nitrogen fixation is an energy-consuming process because the two nitrogen atoms of the nitrogen molecule must first be separated. Nitrogen-fixers achieve this using an enzyme, nitrogenase, and energy from ATP. Non-enzymic separation requires the much greater energy of industrial processes or of ionising events in the atmosphere, such as lightning and cosmic radiation.

Nitrogen is so important for soil fertility, and the demand for food production so great, that colossal amounts of ammonia are produced industrially each year to be used mainly for nitrogenous fertilisers such as ammonium nitrate (NH_4NO_3) and urea ($CO(NH_2)_2$).

The amounts of nitrogen fixed commercially are now approximately equivalent to the amounts fixed naturally. We are still relatively ignorant of the likely effects of this gradual accumulation of fixed nitrogen on the biosphere. There is no new counterbalancing removal mechanism taking industrially fixed nitrogen back to the atmospheric reservoir pool.

A relatively small amount of fixed nitrogen (5–10%) is formed by ionising events in the atmosphere. The resulting nitrogen oxides dissolve in rain, forming nitrates.

The legumes, such as clover, soyabean, lucerne and pea, are probably the greatest natural source of fixed nitrogen. Their roots possess characteristic swellings called **nodules** which are caused by colonies of nitrogen-fixing bacilli (genus *Rhizobium*) living within the cells. The relationship is mutualistic because the plant gains fixed nitrogen in the form of ammonia from the bacteria and, in return, the bacteria gain energy and certain nutrients, such as carbohydrates, from the plants. In a given area legumes can contribute as much as 100 times more fixed nitrogen than free-living bacteria. It is not surprising, therefore, that they are frequently used to add nitrogen to the soil, especially since they have the added benefit of making good fodder crops.

> **10.8** Farmers often say that legumes are 'hard on the soil', meaning that they place a large demand on soil minerals. Why should this be so?

All nitrogen-fixers incorporate nitrogen into ammonia, but this is immediately used to make organic compounds, mainly proteins.

Decay and nitrification

Most plants depend on a supply of nitrate from the soil for their nitrate source. Animals in turn depend directly or indirectly on plants for their nitrogen supply. Fig 10.11 shows how nitrates are recycled from proteins in dead organisms by saprotrophic bacteria and fungi. The sequence from proteins to nitrate is a series of oxidations, requiring oxygen and involving aerobic bacteria. Proteins are decomposed via amino acids to ammonia when an organism dies. Animal wastes and excreta are similarly decomposed. Chemosynthetic bacteria (*Nitrosomonas* and *Nitrobacter*) then oxidise ammonia to nitrate, a process called **nitrification**.

> **10.9** In which of the nutritional categories would you place bacteria and fungi which are decomposers?

Denitrification

Nitrification can be reversed by denitrifying bacteria (**denitrification**) whose activities can therefore reduce soil fertility. They only do this under anaerobic conditions, when nitrate is used instead of oxygen as an oxidising agent (electron acceptor) for the oxidation of organic compounds.

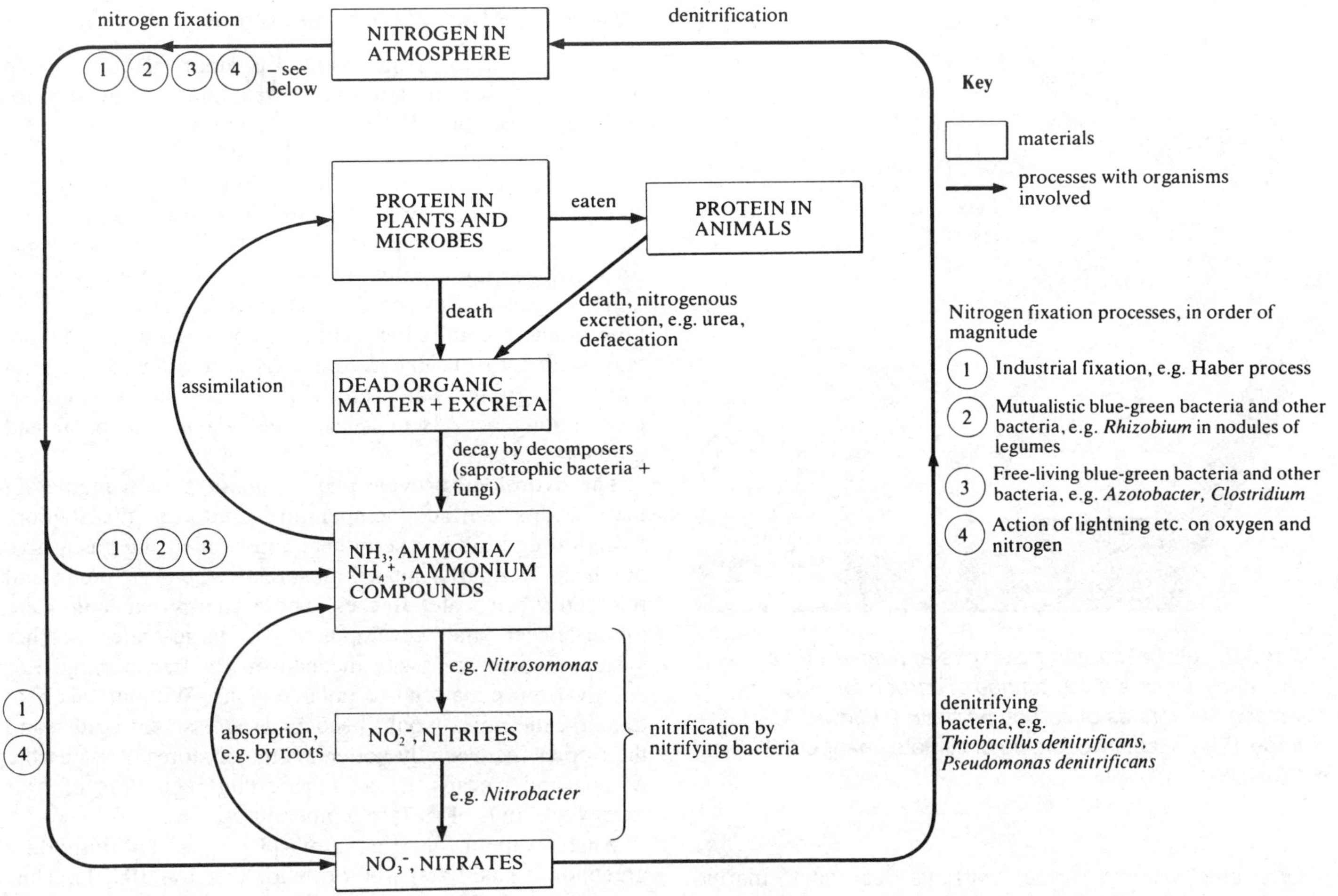

Fig 10.11 *Nitrogen cycle. 79% by volume of the atmosphere is nitrogen. This is the main nitrogen reservoir.*

Nitrate itself is reduced. The bacteria are therefore **facultative aerobes**.

> **10.10** What natural areas or situations might favour denitrification?
>
> **10.11** Why should good drainage and ploughing increase soil fertility?

10.4.2 The carbon cycle

The main carbon store is the estimated 75 million billion tonnes in the Earth's rocks. A further 5000 billion tonnes are found in fossil fuel reserves: coal, oil, gas and peat, and about 150 billion tonnes are held in the uppermost ocean bed sediments. These carbon sources are not normally available to living organisms; they represent the main carbon reservoirs. Of more importance to living organisms are the exchange or cycling pools of carbon shown in fig 10.12.

The main carbon source for living organisms is carbon dioxide present in the atmosphere or dissolved in surface waters. In photosynthesis green plants, algae and blue-green bacteria convert carbon dioxide to simple carbohydrates, the building blocks for all other organic molecules. This conversion of carbon dioxide in photosynthesis, and the counterbalancing release of carbon dioxide in respiration, is an important mechanism helping to maintain the balance of the natural carbon cycle. However, not all carbon dioxide fixed is returned to the atmosphere by respiration. In anaerobic environments, such as waterlogged soils, or at the bottom of still waters with poor illumination, decomposition is very slow and organic matter accumulates. These accumulating peat deposits and organic sediments may in the very long term generate new 'fossil' fuel deposits.

In oceans, the main removal mechanisms taking carbon dioxide from the atmosphere are photosynthesis, mostly by phytoplankton, and dissolving in surface waters. Much of this carbon dioxide is quickly released back into the atmosphere, either directly from the water or by respiration. However, as in terrestrial ecosystems, some carbon dioxide is locked away for a longer time, such as when cool surface

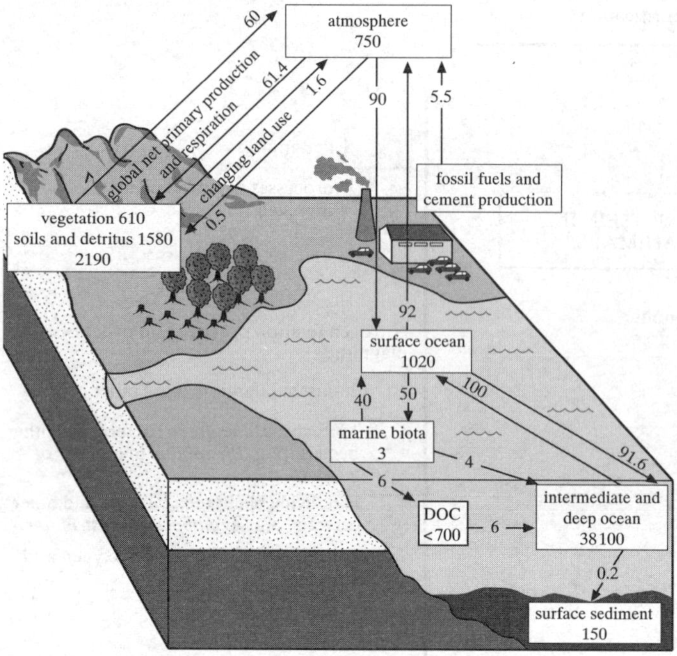

Fig 10.12 *Global carbon reservoirs and flows at the present day. Units are 10^9 tonnes of carbon for reservoir sizes and 10^9 tonnes of carbon per year for flows.* (Climate change 1994, Radiative forcing of climate change *(1995) IPCC/CUP.)*

waters sink below warmer currents, or when marine organisms form carbonate shells that later form carbonate rocks such as limestone.

In practice rates of carbon exchange between the active cycling pools can vary from year to year depending on climate. The balance is also affected by human activities, notably by changing land-use as in deforestation or reforestation and in human use of fossil fuels and in cement manufacture. Records show that human activity has led to a marked build-up of carbon dioxide in the atmosphere since the industrial revolution (section 10.8.1).

The effect of human activities in increasing the rate of release of carbon dioxide from long-term stores, such as fossil fuels and carbonates used in cement manufacture, and how this might affect the Earth's climate and ecosystems is the subject of much current research and debate (section 10.8.1). The general feeling is that we may be risking major environmental change by adding carbon dioxide to the atmosphere at the present rates. Efforts are now being made by governments to reduce carbon dioxide emissions by combustion in industry, and to reduce the rate of use of fossil fuels by finding alternative sources of energy, such as solar or wind power.

10.4.3 The hydrological cycle

Water is an essential component of all living organisms. It is important as a solvent and as a medium for chemical reactions. The oceans are the main reservoir for the hydrological cycle, containing 97% of all water on Earth (fig 10.13). Evaporation from the oceans and subsequent condensation and precipitation is the source of all fresh water for terrestrial and freshwater organisms. Fresh water may rapidly evaporate into the atmosphere or quickly return to the oceans via surface run-off and river flow. Some precipitation, especially in areas of vegetation, may soak into the underlying soils and enter longer-term storage as ground water. Fresh water may also be stored as ice for long periods in icecaps and glaciers of polar and mountain areas.

The hydrological cycle plays a major part in regulating the Earth's surface temperature and its distribution. Evaporating liquids take in heat and condensing gases give out heat. Similarly, heat is absorbed when ice melts and released when water freezes. These energy exchanges are important in the development of large-scale weather systems which are a key mechanism for transporting heat energy from equatorial to polar regions. Without this heat transfer, the poles would become progressively colder and the tropics increasingly hotter. As major stores of water, the oceans and icecaps are also important regulators of heat energy and thus of surface temperatures.

Water vapour in the atmosphere is an important greenhouse gas, like carbon dioxide (section 10.8.1). Thus water vapour also has an important influence on surface temperatures of the Earth. Understanding how interactions between the carbon and water cycles operate and, in turn, influence surface temperatures of the Earth, is an important challenge for current research. This knowledge would help us to predict and minimise the likely impact of human activity on the greenhouse effect.

10.5 Factors influencing environments and habitats

The environment in which an organism lives will be determined by a range of physical or abiotic factors such as light, heat and moisture (table 10.3). However, the real details of an individual's environment may be much more precise than this. For example, for a beetle living on a fallen tree trunk it is the details of temperature and moisture *on* the log which matter. However, an organism's habitat is not solely a matter of *physical conditions*. Its environment may be modified, or even mainly determined, by *other living organisms* (biotic factors). No matter how good the abiotic conditions, an organism may not thrive if this is also

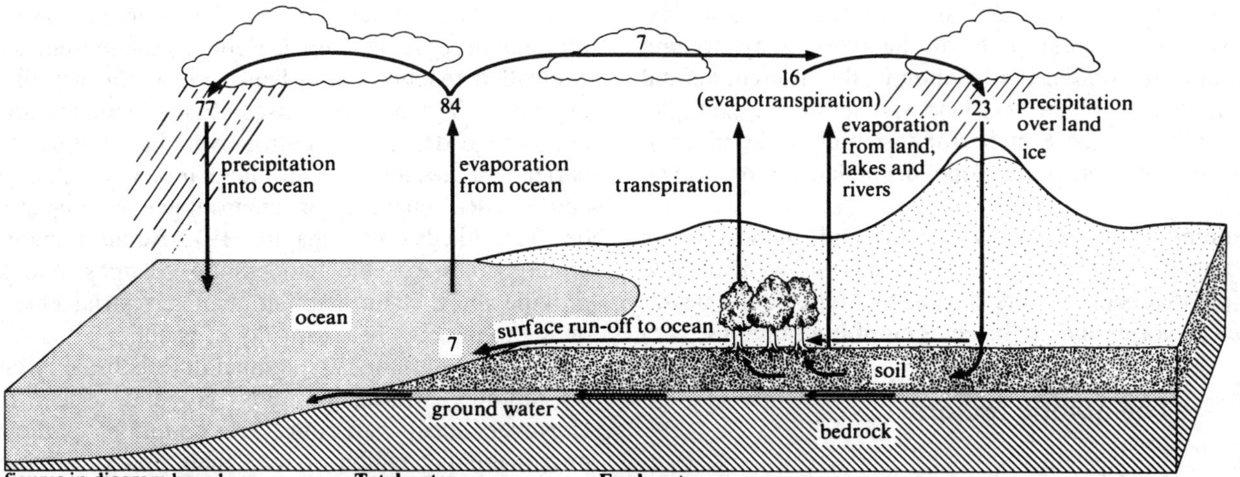

figures in diagram based on mean annual global precipitation of 100 units

Total water
oceans 97%
fresh water 3%

Fresh water
ice sheets and glaciers 75%
ground water 25% } to nearest whole %
lakes 0.3%
soil moisture 0.06%
atmosphere 0.035%
rivers 0.03%

Fig 10.13 *The hydrological cycle and water storage. (Based on R.J. Chorley & P. Haggett eds. (1967)* Physical and information models in geography, *Methuen.)*

Table 10.3 Factors influencing the distribution of organisms.

Abiotic factors		*Biotic factors*		*Combined factors*	
Factor	*Main examples*	*Factor*	*Main examples*	*Factor*	*Main examples*
light	quantity quality duration	competition predation mutualism		soil type	texture organic content soil air soil water
heat	temperature extremes seasonality	human activity	pesticides fire domestication of plants and animals	fire	
water	salinity, nutrients rain, snow, hail, dew humidity water currents and pressure		land use change e.g. in agriculture and dam construction		
atmosphere	gaseous content: e.g. carbon dioxide, oxygen air currents and air pressure weather systems		deforestation pollution e.g. linked with use of fossil fuels		
topography	altitude aspect gradient		contaminated land		
		(You can probably think of many other examples.)			

the perfect location for its main competitors and predators or lacks its preferred food or nesting materials and so on. An organism's habitat is thus determined by both abiotic and biotic factors.

A related, useful and important concept is the **ecological niche** which combines the ideas of spatial habitat with the functional relationships of the organism. An organism's niche thus describes both its location and 'job' (e.g. pollinator, decomposer, predator and so forth) within a particular community or ecosystem. To understand more completely why an organism not only exists but flourishes in a particular place, you must study its absolute physical and biotic constraints, which will determine its **potential niche** (where it could occur), and its preferences and behaviour, which will determine the more restricted range of its **realised niche** (where it does occur).

If two species occupy the same niche they will generally compete with one another until one is displaced (see also section 10.7.5). Similar habitats have similar ecological niches and in different parts of the world may contain morphologically similar but taxonomically different animal and plant species. Open grassland and scrub, for example, will typically provide a niche for fast running herbivores, but these may be horses, antelope, bison, kangaroos, and so on.

A summary of the main abiotic and biotic factors influencing the distribution of organisms is given in table 10.3. Note that some factors such as soil type do not neatly fit these categories and are listed as **combined factors**.

10.5.1 The abiotic factors

Light

As the source of energy for photosynthesis light is essential for life, but it also influences living organisms in many other ways. The intensity, quality (wavelength or colour), and duration (photoperiod) of light can all have different effects.

Light intensity is affected by the angle of incidence of the Sun's rays to the surface of the Earth. This varies with latitude, season, time of day and aspect of slope. Photoperiod or daylength is a more or less constant 12 hours at the equator but at higher latitudes it varies seasonally. Plants and animals of higher latitudes typically show photoperiodic responses that synchronise their activities with the seasons, such as flowering and germination of plants (chapter 16), migration, hibernation and reproduction of animals (chapter 17). Light quality is important to plants. Only light at certain wavelengths can be used by chlorophyll. In aquatic systems some seaweeds, such as red algae, have different light intercepting pigments and can thus survive in locations where green algae would find light quality limiting.

The need for light by plants has an important effect on the structure of communities. Aquatic plants are confined to surface layers of water, and in terrestrial ecosystems competition for light favours certain strategies such as gaining height through growing tall or climbing, and increasing leaf surface area. In woodland this results in stratification (fig 10.15).

> **10.12** Identify the various ways in which light affects the activity of organisms.

Temperature

The main source of heat is the Sun's radiation. Geothermal sources are important only in a minority of habitats, such as the growth of bacteria in hot springs. A given organism will survive only within a certain temperature range for which it is metabolically and structurally adapted. If the temperature of a living cell falls far below freezing, the cell is usually physically damaged and killed by the formation of ice crystals. At the other extreme, if temperatures are too high, proteins become denatured. Between the extremes enzyme-controlled reactions, and hence metabolic activity, double in rate with every 10°C rise. Most organisms are able to exert some degree of control over their temperature by a variety of responses and adaptations so that extremes and sudden changes of environmental temperature can be 'smoothed out' (chapter 19). Aquatic environments undergo less extreme temperature changes, and therefore provide more stable habitats than terrestrial environments owing to the high heat capacity of water.

As with light intensity, temperature is broadly dependent on latitude, season, time of day and aspect of slope. However, local variations are common, particularly in microhabitats which have their own microclimates. Vegetation often has some microclimatic effect on temperature, as in forests, or on a smaller scale within individual clumps of plants or the shelter of leaves and buds of individual plants.

Moisture and salinity

Water is essential for life and is one of the major limiting factors in terrestrial ecosystems. It is precipitated from the atmosphere as rain, snow, sleet, hail or dew. There is a continuous cycling of water, the **hydrological cycle**, which governs water availability over land surfaces. Terrestrial plants mainly absorb water from the soil. Rapid drainage, low rainfall and high evaporation, or a combination of these factors, can result in dry soils, whereas the opposite extremes can lead to permanent waterlogging.

Plants can be classified according to their ability to tolerate water shortage as xerophytes (high tolerance), mesophytes (medium tolerance) and hydrophytes (low tolerance/water-adapted). Some of the xeromorphic adaptations are summarised in table 10.4 (see chapters 14 and 20 for a fuller discussion). Similarly, terrestrial animals show adaptations for gaining and conserving water, particularly in dry habitats (see chapter 20 and table 10.4).

Aquatic organisms also have problems with water regulation (chapter 20). Salinity of water is important, as can be seen in the differences between freshwater and marine species. Relatively few plants and animals can withstand large fluctuations in salinity, such as are found in estuaries or salt marshes. One example is the snail *Hydrobia ulvae* which can survive a range of salinities from 50–1600 mmol dm^{-3} of sodium chloride. Salinity may also be important in terrestrial habitats; if evaporation exceeds precipitation, soils may become saline. This is a serious problem in some irrigated areas.

Atmosphere

The atmosphere is a major part of the ecosphere. Like the oceans, it is constantly circulating. This is a mass flow phenomenon. The energy which drives these circulations comes from the Sun.

On a large scale, atmospheric circulation is important in the distribution of water vapour because this can be picked

Table 10.4 Adaptations of plants and animals to dry conditions.

	Examples
Reducing water loss	
Leaves reduced to needles or spines	Cactaceae, Euphorbiaceae (spurges), conifers
Sunken stomata	*Pinus, Ammophila*
Leaf rolls into cylinder	*Ammophila*
Thick waxy cuticle	Leaves of most xerophytes; insects
Swollen stem with large volume to surface area ratio	Cactaceae and Euphorbiaceae
Hairy leaves	Many alpine plants
Leaf-shedding in drought	*Fouquieria splendens* (ocotillo or candle plant)
Stomata open at night, close during day	Crassulaceae (stonecrops)
Efficient carbon dioxide fixation at night with partial opening of stomata	C_4 plants, e.g. *Zea mais*
Uric acid as nitrogenous waste	Insects, birds and some reptiles
Long loop of Henlé in kidneys	Desert mammals, e.g. camel, desert rat
Tissues tolerant of high temperatures, reducing sweating or transpiration	Many desert plants, camel
Burrowing behaviour	Many small desert mammals, e.g. desert rat
Spiracles covered with flaps	Many insects
Increasing water uptake	
Extensive shallow root system and deep roots	Some Cactaceae, e.g. *Opuntia,* and Euphorbiaceae
Long roots	Many alpine plants, e.g. *Leontopodium alpinum* (edelweiss)
Burrow for water	Termites
Water storage	
In mucilaginous cells and cell walls	Cactaceae and Euphorbiaceae
In specialised bladder	Desert frog
As fat (water is a product of oxidation)	Desert rat
Physiological tolerance of water loss	
Apparent dehydration can occur without death	Some epiphytic ferns and clubmosses, many bryophytes and lichens, *Carex physiodes* (sedge)
Loss of high proportion of body mass with rapid recovery when water is available	*Lumbricus terrestris* (70% loss of body mass), camel (30% loss)
Evasion	
Pass unfavourable season as seed	Californian poppy
Pass unfavourable season as bulbs or tubers	Some lilies
Seed dispersal, with some reaching more favourable conditions	
Escape behaviour	Soil organisms, e.g. mites, earthworm
Aestivation in mucus-containing sheath	Earthworm, lungfish

up locally (by evaporation), carried in moving air masses and deposited in a distant location (by precipitation). Similarly pollutant gases, such as sulphur dioxide, released into the atmosphere at industrial locations may be deposited in solution in rainfall many miles from the source (section 10.8.1). The pattern of atmospheric circulation will thus affect the distribution of pollutants and their eventual precipitation in rainfall.

Wind can interact with other environmental variables to affect growth of vegetation, particularly trees in exposed places, where they may become stunted and distorted on their windward sides. Wind is also important in increasing evaporation and transpiration under conditions of low humidity.

Dispersal of spores, seeds and so on, through the atmosphere, aided by wind, increases the spread of non-motile organisms like plants, fungi and some bacteria. Winds may also influence the dispersal or migration of flying animals.

Topography

The influence of topography is intimately connected with the other abiotic factors since it can strongly influence local climate and soil development. The main topographic factor is altitude. Higher altitudes are associated with lower average temperatures and a greater diurnal temperature range, higher precipitation (including snow), increased wind speeds, more intense radiation, lower atmospheric pressures, all of which have an influence on plant and animal life. As a result, vertical **zonations** are common, as shown in fig 10.14 (see also section 10.6.4).

Mountain chains can act as climatic barriers. As air rises over mountains it cools and precipitation tends to occur. Thus a rain 'shadow' occurs on the leeward side of the mountains where air is drier and precipitation is less. This affects the ecosystems. Mountains also act as barriers to dispersal and migration and may play important roles as isolating mechanisms in the process of speciation, as described in chapter 26.

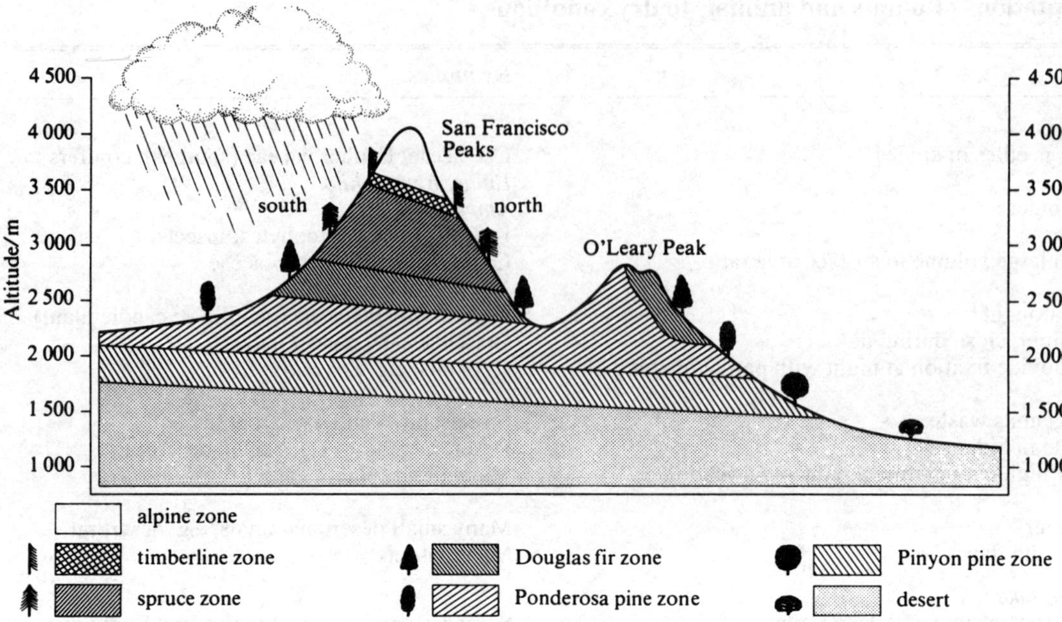

Fig 10.14 *Zonation of vegetation on San Francisco Peaks, Arizona, as viewed from the south east. (From Merriam (1890). Redrawn and modified from W.D. Billings (1972)* Plants, man and the ecosystem, *2nd ed. Macmillan.)*

Another important topographic factor is **aspect**. In the northern hemisphere south-facing slopes receive more sunlight, and therefore higher light intensities and temperatures, than valley bottoms and north-facing slopes (the reverse being true in the southern hemisphere). This has striking effects on the natural vegetation and on land use by humans.

Steepness of slope (**gradient**) is a third topographic factor. Steep slopes generally suffer from faster drainage and run-off, and the soils are therefore thinner and drier, with more xeromorphic vegetation. At slope angles in excess of 35° soil and vegetation are typically unable to develop, and screes of loose material form.

10.5.2 Soils

Soil is a complex mixture of organic and inorganic materials. It is the medium for plant growth. Understanding the characteristics of soil, in particular knowing how to maintain good quality soil and how to improve difficult soils, is of importance for successful agriculture. It thus has considerable economic significance.

Soils have four main constituents. These are:

- the mineral matter;
- the organic matter;
- the soil air;
- the soil water. (This may be better described as the '**soil solution**' since it is not pure water but contains dissolved ions.)

The soil mineral matter

Soils are composed of mineral fragments of widely varying size: gravel and stones, and fine earth. In practice the fine earth is commonly further subdivided into clays, silts and sands. The relative proportions of the main particle size groups determine soil texture.

Soil texture importantly influences drainage, nutrient retention and the soil temperature regime. It is thus important agriculturally. Medium- and fine-textured soils, such as clays, clay loams and silt loams, are generally more suitable for plant growth because they have the most satisfactory nutrient and water retention. Sandy soils are faster draining and lose nutrients through leaching, but may be advantageous in obtaining early crops as the surface dries more rapidly than that of a clay soil in early spring, resulting in a warmer soil. Stone content of the soil (particles >2 mm) may also have importance agriculturally since it will affect wear and tear on agricultural implements, and will modify the drainage characteristics of fine earth. Generally, as the stone content of a soil increases, its water-holding capacity decreases.

The soil organic matter

The organic matter content of a soil, although only a small percentage by volume, is very important. It is a major source of the key soil nutrients phosphorus, nitrogen and sulphur; it increases the water-holding capacity of the soil; it encourages the formation of aggregates in the soil and thus is particularly useful in clay rich soils; and it is a source of energy and materials for microorganisms. Soil organic matter comprises two main constituents, namely DOM and living organisms.

Humus is organic material formed from the partial decomposition of DOM. Much of the soil humus is present not as individual molecules but as complexes with inorganic molecules, especially the clay minerals. These clay–humus complexes are fundamental to almost every physical, chemical and biological process that occurs in soils. For example they influence water and nutrient retention and affect crumb structure and aeration.

Earthworms are particularly important soil organisms. They feed on DOM and at the same time ingest large amounts of mineral particles. Because they migrate between the upper and lower layers of the soil, earthworms are important mixing agents in the soil. They also open up channels for air and water and so help to improve soil texture and related properties. Earthworms thrive best in neutral to acid soils and are rarely found where the pH is less than 4.5.

The soil air

The soil air occupies the pore spaces within the soils. It is readily displaced by the soil water. In a waterlogged soil there is no soil air. The soil air differs from air in the outside atmosphere in several important ways. It contains more carbon dioxide and water vapour but has less oxygen.

The soil water

The soil water or the **soil solution** contains dissolved chemicals that have been produced by weathering and by mineralisation in the process of decomposition. There is a continuous movement of solutes (dissolved chemicals) from the mineral and organic components of the soil into the soil solution and from the soil solution into plants.

10.5.3 Biotic factors

The biotic factors which affect the survival and distribution of an organism include:

- intraspecific factors – those which occur between members of the same species, such as competition for food and territory (section 10.7.4).
- interspecific factors – those which occur between members of different species, such as predator–prey interactions, host–parasite interactions, competitive exclusion and resource partitioning (section 10.7.5).
- other detailed mutualistic relationships such as species-specific pollinators and fruit dispersal agents.

However, as shown in table 10.3, humans have become the predominant biotic influence on the distribution and success of other species as discussed in sections 10.8 and 10.9.

10.6 Community ecology

A **community** is a group of interacting organisms or species populations living together in a particular place. It represents the living (biotic) part of an ecosystem and could be seen as a dynamic unit with a web of energy flow and an exchange or cycling of matter as described in sections 10.3 and 10.4. Community ecology focuses on the interactions between the community members. This means understanding competition, predator–prey relationships (including herbivory) and mutualism (section 10.7.5).

Community structure is often very diverse. A woodland community, for example, will generate many different niches and microhabitats, and supply many different food sources, as a result of its vertical structure (fig 10.15) and spatial variation beneath and between the mature tree canopies.

10.6.1 Primary and secondary succession

All biotic communities are continually changing. They change in response to external factors, such as changing climates, or as a result of internal factors caused by the organisms themselves, such as accumulation of DOM. The time scales of change are highly variable. So long as the abiotic factors remain relatively constant, the biotic community will develop through time from an initial bare rock or open water start point to a **climax community**. The climax community is considered to be the most complex, diverse and productive community a given area can *sustain*. It may vary seasonally or fluctuate in a minor way, but it is essentially stable unless some catastrophic intervention occurs. A volcanic eruption or human-induced fire or deforestation would be a typical catastrophic event. The climax community would be destroyed but DOM and chance survivors would remain. These would start a new sequence of changes until a fairly stable community was once again established.

The change from bare rock or open water is rapid, especially in the initial stages, and follows a series of recognisable and hence predictable stages. This process is called **succession**. Individual successions are known as **seres** and the developmental phases are called **seral stages**. A succession developing on newly emerged land or water is termed a **primary succession**. A succession developing following a fire or similar major disruption to an established community is called a **secondary succession**.

Opportunities for primary succession are relatively uncommon. Examples would be land or lakes emerging during glacier retreat, or a new island created by volcanic activity as occurred when Surtsey appeared off Iceland in 1963. Secondary successions are much more common. Scrub invasion on the lowland heaths and chalk downlands of southern England are widespread examples of secondary succession (fig 10.16). Here the disruption to the natural succession was caused by forest clearance by humans, and has subsequently been reinforced by centuries of grazing activity by sheep and cattle. A community where human intervention has led to the long-term establishment of a community very different from the original climax is termed a **plagioclimax**. The succession is said to have been **deflected**. If pressures such as grazing and burning management which caused the succession to be deflected are removed then, inevitably, a renewed succession will occur. This is an important consideration for the conservation management of these communities.

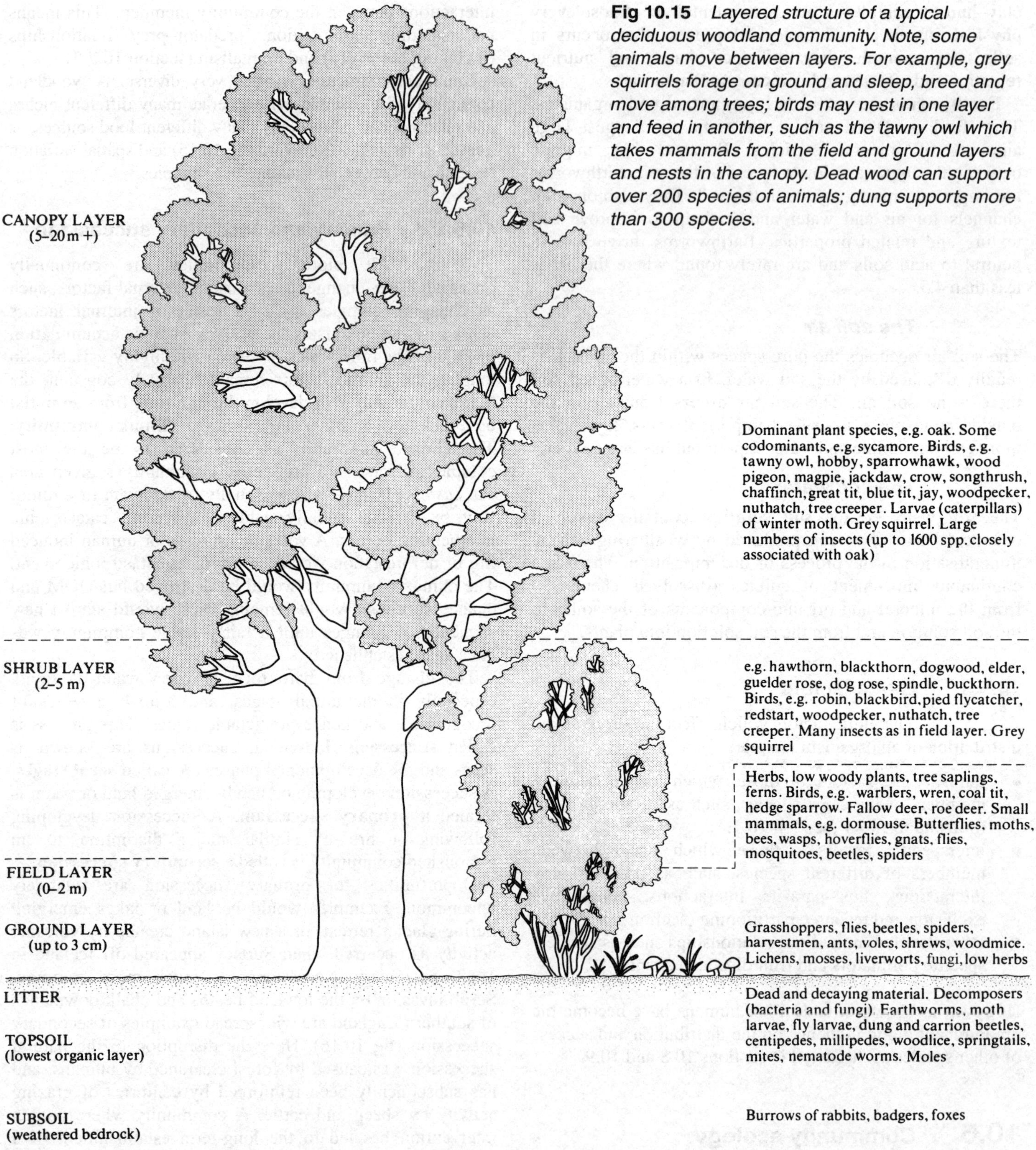

Fig 10.15 *Layered structure of a typical deciduous woodland community. Note, some animals move between layers. For example, grey squirrels forage on ground and sleep, breed and move among trees; birds may nest in one layer and feed in another, such as the tawny owl which takes mammals from the field and ground layers and nests in the canopy. Dead wood can support over 200 species of animals; dung supports more than 300 species.*

10.6.2 The process of succession

The concept of succession is most easily illustrated for a bare rock surface. The initial rock surface presents a hostile environment for most living organisms. Early colonisers will need to be autotrophs which can tolerate these hostile conditions. They will also need good dispersal mechanisms to ensure arrival at the site. Typical organisms are lichens, algae and blue-green bacteria.

Fig 10.16 *Scrub invasion of lowland heath in south-east England.*

Lichens, for example, colonise rock surfaces. They exude chemicals which help to break down the rock surface, at the same time supplying themselves with essential inorganic materials. When they die, these primary colonisers provide DOM which may accumulate in the small depressions and cracks in the rock. This debris is a good food source for decomposers and marks the beginning of soil formation and accumulation of readily available nutrients for plant growth. A series of increasingly demanding plant species can then colonise the area. First arrivals and early colonisers will be those with good seed dispersal mechanisms and a wide range of tolerance. We call these species **opportunists** or **generalists**. Each organism in turn contributes to the developing biomass (living and dead) of the community. This paves the way for later arrivals which will typically be more competitive longer-living species. These characteristically show a narrower range of tolerance to environmental conditions.

During a succession more and more of the available nutrients become locked up in the biomass of the community, with a consequent decrease in nutrients in the abiotic component of the ecosystem. The amount of detritus produced also increases and detritus feeders take over from grazers as the main primary consumers. Appropriate changes in food webs occur and detritus becomes the main source of nutrients. These and other trends that can be expected to occur in successions are summarised in table 10.5.

Succession is a good example of organisms modifying their environment and thereby creating opportunities for other organisms. Of course, as the plant community develops so new niches for herbivores, secondary consumers, detrital feeders and decomposers emerge until a complex biotic community and ecosystem develops. A schematic summary of this process is shown in fig 10.17.

> **10.13** What factors are likely to affect the number and diversity of species reaching an area?

In the latter stages of succession, biotic interactions become more important in forming the detailed community structure. The variety of living organisms tends to increase, so inevitably their interactions become more complex (section 10.7.5). The tropical rainforest communities, which are among the longest established climax communities, are renowned for their species richness and the extreme complexity of their biotic interactions.

The outline given above is an idealised picture of succession for a terrestrial community. In real life the situation may be less clear cut. Much will depend on available spores and seed supplies. This is especially true in secondary succession where colonisers will also reflect surviving seed stores and vegetative organs in cleared or abandoned land.

The climax community is often described as having one **dominant** or several **co-dominant** species. The term dominance is rather subjective but normally refers to those species with the greatest collective biomass or productivity, although physical size of individuals is also usually considered important.

Table 10.5 Summary of changes in an ecosystem during a typical secondary succession.

Characteristic	*Stage of ecosystem development* *Immature (early)*	 *Mature (late)*
Gross production/community respiration (P/R ratio)	high (>1)	approaches 1
Food chains	linear, mainly grazing	web-like, mainly detritus-feeding
Total organic matter (or biomass)	small	large
Species diversity	low	high*
Structure of community	simple	complex (stratification, many microhabitats)
Niche specialisation	broad	narrow
Size of organism	small	large
Strategies adopted by species	opportunist	specialist

* Most plant and some animal examples of succession show a peak of species diversity before climax.

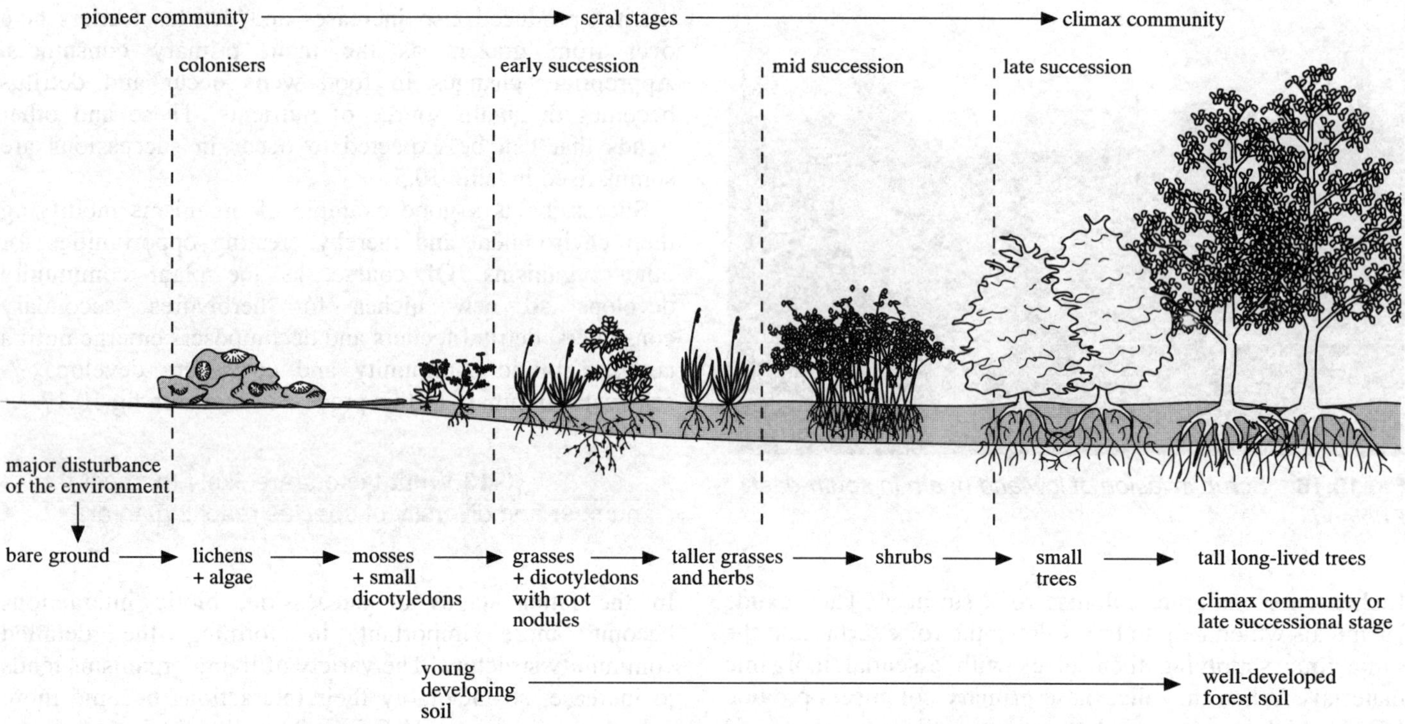

Fig 10.17 *Schematic summary of a sere showing characteristic vegetational change during succession.*

The time scales of succession are very difficult to specify. Initial changes may be rapid but subsequent rates of change are often slow. For example on the skeletal soils of moraines left by retreating glaciers a well-developed soil and plant community has been observed to occur in 30–70 years. Whether this represents the final climax stage is much more difficult to judge. Indeed some ecologists argue that since climate is always changing, biotic communities too must always be changing in sympathy with the new climatic conditions. The optimum stable climax community must therefore also be changeable. In other words as human observers of this process we are much better informed about the early stages of succession than about changes that may occur after 200 or 300 years, the lifespan of some trees which in many terrestrial environments are the major determinants of climax community structure.

10.6.3 Applying succession principles to land restoration

In practical terms, the speed with which successional change takes place can be very important. Mining wastes, motorway verges and railway embankments are just three examples where colonisation by vegetation is an important means of stabilisation. In the case of mine waste, land reclamation and landscape improvement are also important objectives. Some wastes are quickly colonised. China clay waste has been observed to establish woody vegetation in very short time periods, such as 10–20 years. It also quickly establishes a self-sustaining nitrogen cycle essential to the maintenance of tree cover. This latter point is important. It means that very little, if any, human help is needed to revegetate these areas. Commonly an initial seeding will include a mix of grasses and nitrogen-fixing legumes, to enable a vegetation cover to establish quickly, but little subsequent treatment is needed.

Other wastes such as colliery spoil are more difficult to colonise. These wastes are often very acidic and may also be contaminated with heavy metals. Relying on natural succession would take a very long time since few organisms can tolerate these conditions. We can help succession in these circumstances in several ways. Lime can be added to reduce acidity and top soil (very expensive) or some organic material may be added. This material will improve water retention by the waste. It will also bind toxic heavy metals and help to buffer extremes of pH. The organic material also provides a food source for earthworms. Earthworm activity will increase the aeration and mixing of the soil. Organic amendments are often readily and cheaply available as the waste products of other industries, for example sewage sludge, finely shredded domestic waste and farmyard manure.

10.6.4 Zonation

Within a community at any one time species may be spatially distributed according to variations of the physical environment. This is called **zonation**. A good example is the zonation of seaweeds and marine animals that occurs on rocky shores from low-tide to high-tide level and into the splash zone. Physical conditions vary through these zones, notably length of exposure to air between successive tides, and each zone is occupied by species adapted for its particular conditions. Zonation on a seashore is described and illustrated in section 11.4. Another good example of zonation is the vertical zonation that occurs on mountains with increasing altitude (shown in fig 10.14). Superficially zonations may resemble succession, but it is important to recognise the basic difference, namely that with zonation the species vary in space (spatially), whereas with succession the species vary in time (temporally).

10.7 Population ecology

A **population** may be defined as a group of organisms of one species occupying a particular place and usually isolated to some extent from other similar groups by geography or topography. The roe deer of a particular woodland, the frogs of a particular valley or trout of a particular lake are all examples of ecological populations. Population studies are not just about numbers of a given species living in a given area at a given moment in time. More importantly we want to know

- how populations grow;
- how populations are maintained;
- and how and why populations decline.

The study of how and why population size changes over time is called **population dynamics**. Typically, it examines the characteristics of a group of organisms, such as their density, natality (birth rate), mortality (death rate), survivorship, age structure, migration and form of growth of the population. Population interactions such as competition, predation and parasitism not only regulate growth of a given population but influence the structure of communities.

10.7.1 Birth rate (natality) and death rate (mortality)

Population size may increase as a result of **immigration** from neighbouring populations, or by reproduction of individuals within the population. For mammals, rate of reproduction is expressed as **birth rate** or **natality**, the number of young produced per female per unit time (usually per year).

Population size may decrease as a result of **emigration** or death (**mortality**). In population biology, mortality strictly means *rate* of death and may be expressed in terms of per cent, or numbers per thousand, dying per year.

10.7.2 Survivorship curves

The percentage of individuals that die before reaching reproductive age (**pre-reproductive mortality**) is one of the chief factors affecting population size, and for a given species is much more variable than birth rate. Many populations remain more or less the same size year after year. In these cases, an average of only two offspring from each male–female pair must survive to reproductive age.

10.14 The data below give the average number of fertilised eggs produced in their lifetime by females of different organisms (fecundity).

oyster	100×10^6	mouse	50
cod	9×10^6	dogfish	20
plaice	35×10^4	penguin	8
salmon	10×10^4	elephant	5
stickleback	5×10^2	Victorian Englishwoman	10
winter moth	200		

(a) If each population remains stable in numbers, how many fertilised eggs from each female must, on average, survive?

(b) For each species write down the number of fertilised eggs which must die before becoming adult if the population remains stable. Then express this number as a percentage of the total number produced (this gives the pre-reproductive mortality).

(c) Try to suggest why the fecundities of the stickleback and dogfish are so much lower than for the other fish in the table.

Source of question: Open University, S100, Unit 20, *Species and Populations*, p. 26.

If we start with a population of newborn individuals and the numbers of survivors is plotted against time, a **survivorship curve** is obtained. On the vertical axis, actual numbers of survivors may be plotted, or percentage survival:

$$\frac{\text{number of survivors}}{\text{number in original population}} \times 100\%$$

Different species have characteristic survivorship curves. Some representative examples are shown in fig 10.18.

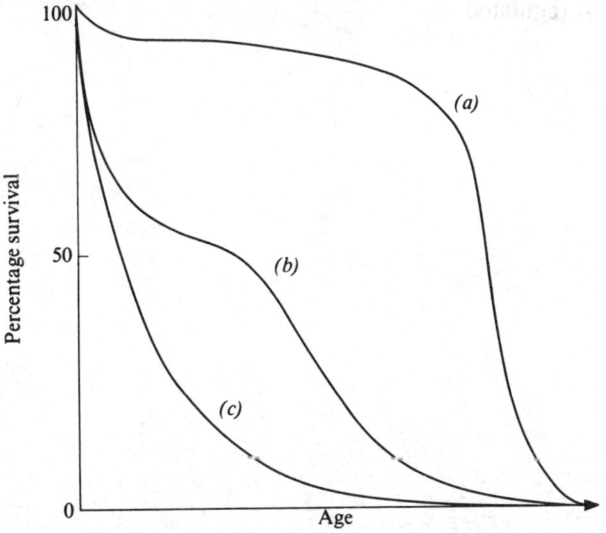

Fig 10.18 *Three types of survivorship curve. Curves (a), (b) and (c) are explained in the text.*

Most animals and plants exhibit a phenomenon called senescence or ageing, manifested as a declining vigour with increasing age beyond maturity. Once senescence begins, there is increasing likelihood of death occurring within a given time period. The immediate cause of death can vary, but the underlying cause is a reduced resistance to external factors such as disease.

Curve *(a)* in fig 10.18 shows an almost ideal curve for a population in which senescence is the major factor affecting mortality. A curve like (*a*) would also be obtained for an annual crop plant such as wheat, where all the plants in a given field survive well early in life and then senesce simultaneously.

Curve (*b*) is for a population with a high mortality rate early in life, such as might occur for mountain sheep or for humans in a country in which starvation and disease are prevalent. **Curve** (*c*) shows the kind of smooth curve that would be obtained if there was a constant mortality rate throughout life (50% per unit time). Such a curve is obtained if chance is the major factor influencing mortality and the organisms die out before senescence becomes evident. Some animal populations show survivorship curves which approximate closely to this model curve, for example *Hydra*, where there is no special risk attached to being young. Most nonvertebrates and plants show a curve similar to (*c*), but with high juvenile mortality superimposed so that the initial part of the curve descends even more steeply.

10.15 Which population, (*a*) or (*b*), would need the higher reproductive rate to maintain a stable population? Explain your choice.

By plotting survivorship curves of species it is possible to determine the mortality rates of individuals of different ages and hence to determine at which ages they are most vulnerable. By identifying the factors causing death at these ages, an understanding can be gained of how population size is regulated.

10.16 The following figures apply to sockeye salmon in a Canadian river system. Each female salmon lays 3200 eggs in a gravelly shallow in the river in autumn. 640 fry (young fish derived from these eggs) enter a lake near the shallow in the following spring. 64 smolts (older fish survivors from the fry) leave the lake one year later and migrate to the sea. Two adult fish (survivors of these smolts) return to the spawning grounds $2\frac{1}{2}$ years later; they spawn and then die. Calculate the percentage mortalities for sockeye salmon for each of the following periods:

(*a*) from laying eggs to movement of fry into the lake six months later;
(*b*) from entering the lake as fry to leaving the lake as smolts 12 months later;
(*c*) from leaving the lake as smolts to returning to the spawning grounds as adult salmon 30 months later.

Draw a survivorship curve for the sockeye salmon in this river system (plot percentage survival against age). What is the pre-reproductive mortality for these salmon?

Source of question: Open University, S100, Unit 20, *Species and Populations*, p. 71.

10.7.3 Population growth and growth curves

Populations grow and decline in characteristic ways. The size of population increase will be determined by the **reproductive potential** of the organisms concerned and by **environmental resistance**. The maximum reproductive potential is the rate of reproduction given unlimited environmental resources. It will vary according to the age structure of a population and will be influenced by male:female ratios.

Environmental resistance means the sum total of limiting factors, both biotic and abiotic, which act together to prevent the maximum reproductive potential from being realised. It includes external factors such as predation, food supply, heat, light and space, and internal regulatory mechanisms such as intraspecific competition and behavioural adaptations. Strong feedback links exist between all these factors, for example intraspecific competition arises in response to some resource (such as space) which is in limiting supply.

The balance between biotic potential and environmental resistance defines the **carrying capacity** for a particular organism with a given set of environmental resources.

Growth curves

Two basic forms of growth curve can be identified, the J-shaped growth curve and the S-shaped or sigmoidal growth curve.

The **S-shaped** or **sigmoidal growth curve** describes a situation in which, in a new environment, the population density of an organism increases slowly initially, as it adapts to new conditions and establishes itself, then increases rapidly, approaching an exponential growth rate. It then shows a declining rate of increase until a zero population growth rate is achieved where rate of reproduction (natality) equals rate of death (mortality) (fig 10.19*a*). The slowing rate of population growth results from increasing competition for essential resources, such as food or nesting materials. The decline in growth rate continues until eventually feedback in terms of increased mortality and reproduction failures (fewer matings, stress-induced abortion) reduces population growth rate to zero.

This type of population growth is said to be **density-dependent** since, for a given set of resources, growth rate depends on the numbers present in the population. The point of stabilisation or zero growth rate is the maximum **carrying capacity** of the given environment for the organism concerned.

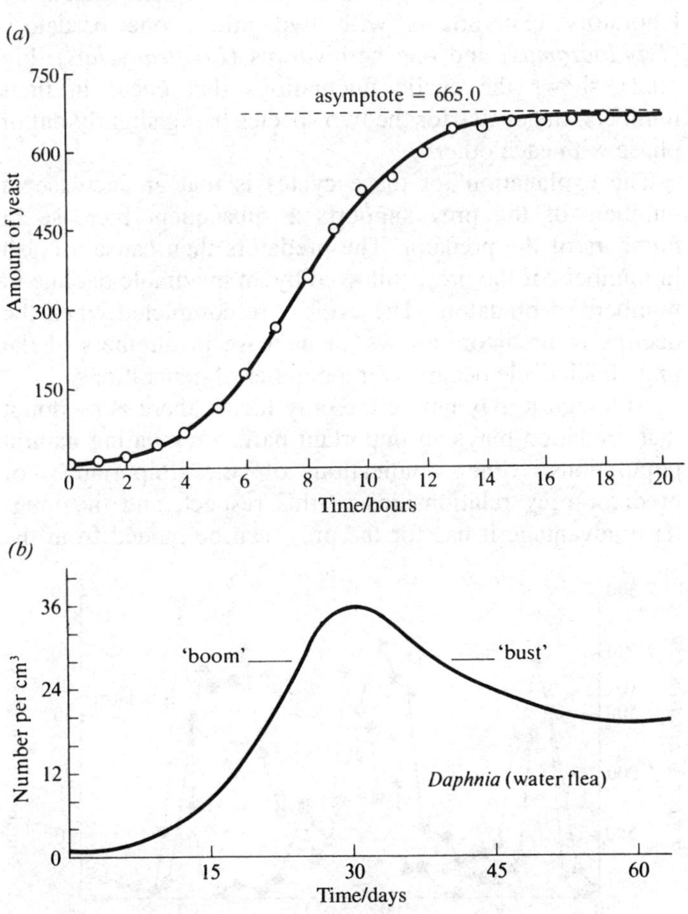

Fig 10.19 *Types of population growth curve. (a) The sigmoidal growth curve (S-shaped curve). The growth of yeast in a culture. A simple case of the sigmoid growth form in which environmental resistance (in this case detrimental factors produced by the organisms themselves) is linearly proportional to density. (b) The diagram shows the curve of water fleas grown in culture. These curves are sometimes called 'boom and bust' curves.*

Many populations of microorganisms, plants and animals, under both laboratory and field conditions broadly follow this basic sigmoidal pattern. A useful example is when a fresh culture medium is inoculated with bacteria (fig 12.8). Phytoplankton in lakes and oceans may show sigmoidal growth in spring, as do insects such as flour beetles or mites introduced into a new habitat with abundant food and no predators.

The **J-shaped growth curve** describes a situation in which, after the initial establishment phase (lag phase), population growth continues in an exponential form until stopped *abruptly*, as environmental resistance becomes *suddenly* effective (fig 10.19*b*). Growth is said to be **density-independent** since regulation of growth rate is not tied to the population density until the final crash. The crash may be triggered by factors such as seasonality or the end of a breeding phase, either of the organism itself or of an important prey species. It may be associated with a particular stage in the life cycle, such as seed production, or it may be induced by human intervention as when an insecticide is used to control an insect pest population. Following the crash, such populations typically show a fluctuating recovery pattern giving the 'boom and bust' cycles characteristic of some insect species and associated with algal blooms.

10.7.4 Factors within species which affect population size

Once a population has finished its initial growth phase there usually continues to be fluctuations in population size over time. Important influences are likely to be variations in climatic conditions (such as temperature), food supply and predation. Sometimes fluctuations are regular and may be called **cycles**.

Population size may change as a result of changes in birth rate or death rate, or possibly both. Factors which bring about these changes often become more effective as population density increases. Hence they may be described as density-dependent factors. Food shortage and increased predation are two factors which sometimes operate in this way. They have direct effects on mortality for obvious reasons. Two regulatory mechanisms which affect birth rate are **territorial behaviour** and the physical effects of **overcrowding**.

Territorial behaviour

Territorial behaviour or **territoriality** occurs in a wide range of animals, including certain fish, reptiles, birds, mammals and social insects. It has been particularly well studied in bird populations. Either the male bird, or both the male and the female, of a pair may establish a breeding territory which they will defend against intruders of the same species. The song of the bird and often a visual display, such as that of the robin's red breast, are means of asserting territorial claims, and intruders usually retreat, sometimes after a brief 'ritual fight' in which neither

Fig 10.20 *Display of territoriality.*

competitor is seriously damaged. There is little or no overlap between neighbouring territories of the same species and, in areas where the territory includes the food of the species, it will contain sufficient food to support the birds and their young (fig 10.20). As population sizes grow, territories usually become smaller and able to support fewer new birds. In extreme cases, some birds may be unable to establish territories and therefore fail to breed. Regulation is therefore due to spatial interactions.

Overcrowding

Another factor which may affect birth rate is overcrowding. Laboratory experiments with rats show that when a certain high population density is reached, birth rate is greatly reduced even if there is no food shortage. Various hormonal changes occur which affect reproductive behaviour in a number of ways; for example failure to copulate, infertility, number of abortions and eating of young by the parents all increase, and parental care decreases. The young abandon the nest at an earlier age, with consequent reduction in chances of survival. There is also an increase in aggressive behaviour. Changes like these have been demonstrated for a number of mammals and could operate under natural circumstances outside the laboratory. Natural populations of voles, for example, show a similar kind of regulation.

Another density-dependent factor that may affect the population size of a species is dispersal. For example, at high aphid densities not only does the rate of reproduction of the aphids decrease but a higher proportion develop wings and leave the plant on which they are feeding.

Some factors regulating population sizes such as climate have been regarded as density-independent, but it is inevitable that they interact with other factors that are density-dependent. Thus, although the terms are useful, it is better to avoid stressing the difference between density-dependent and density-independent factors because complex interactions may occur.

10.7.5 Factors between species which affect population size

It is seldom possible to confine studies of population dynamics to single species. A number of well-recognised types of interaction may occur between populations of different species. These are termed **interspecific interactions**. Populations from different trophic levels may also interact, as, for example, in the cases of predator–prey relationships and host–parasite relationships. Other relationships exist, some of which are subtle and complex, including some mutualistic relationships where both partners benefit. At a given trophic level there may be **interspecific competition**, that is competition between members of different species for available resources such as food and space. This can result in resource partitioning.

Predator–prey relationships

A commonly used and simple model of predator–prey relationships is one that has been well illustrated by laboratory experiments with two mites, one predatory (*Typhlodromus*) and one herbivorous (*Eotetranychus*). Fig 10.21 shows the cyclic fluctuations that occur in their numbers, the cycles for the two species being slightly out of phase with each other.

The explanation for these cycles is that an increase in numbers of the prey supports a subsequent increase in numbers of the predator. The predators then cause a crash in numbers of the prey, followed by an inevitable decline in numbers of predators. The cycles are completed when the decline in predators allows an increase in numbers of the prey. Each cycle occurs over a number of generations.

Although it may not be the only factor, there is no doubt that predation plays an important part in regulating natural populations. Some indication of the importance of predator–prey relationships in this respect, and the long-term advantage it has for the prey, can be gained from the

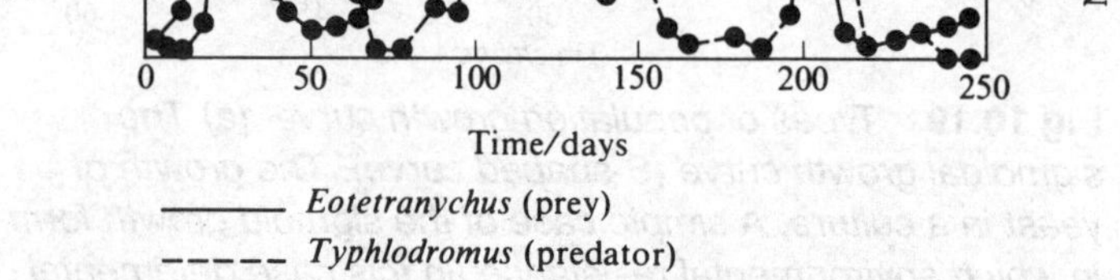

Fig 10.21 *Oscillations in the populations of the predatory mite* Typhlodromus *and its prey, the plant mite* Eotetranychus, *in a laboratory habitat. (From M.K. Sands (1978)* Problems in ecology, *Mills & Boon Limited.)*

fate of a population of deer on the Kaibab Plateau in Arizona. In 1906 the area was declared a wildlife refuge, and in order to protect the deer from their predators, a culling programme of the deers' main predators, such as pumas, wolves and coyotes, was planned for the next 30 years. Until 1906 the deer population had remained stable at about 4000, but subsequently a 'population explosion' occurred, as shown in fig 10.22, with the result that the carrying capacity of the rangeland, thought to be about 30 000 deer, was exceeded. The population grew rapidly to an estimated 100 000 by 1924, following a J-shaped growth curve instead of its normal sigmoidal curve. The deer population then crashed due to starvation and disease. Meanwhile the rangeland had been seriously damaged by overgrazing and did not recover to its 1906 level. As a result the carrying capacity of the area dropped to 10 000 deer.

Host–parasite relationships

In some of the cases studied host and parasite populations show similar out-of-phase cycles to those described above, notably when insects parasitise other insects.

Interspecific competition

Competition may occur between populations within an ecosystem for any of the available resources, such as food, space, light or shelter. If two species occur at the same trophic level then they are likely to compete with each other for food if they eat the same prey. Specialisation by one or both species over a period of time may lead to resource partitioning, so that they come to occupy separate niches within the trophic level and minimise the extent of competition (see below). Alternatively, if the competitors occupy the same niche, or strongly overlapping niches, an equilibrium situation may be reached in which neither succeeds as well as it would in the absence of the competitor, or one of the competitors declines in numbers to the point of extinction. The latter phenomenon is known as **competitive exclusion**.

Interspecific competition is difficult to study in wild populations but some classic work on laboratory populations was done by the Russian biologist Gause in 1934 who worked on competition between several species of *Paramecium*. Some of his results are shown in fig 10.23.

10.17 With reference to fig 10.23,
(*a*) what type of population growth curve is shown by the two species when grown in isolation?
(*b*) what resources might the two species be competing for in the mixed culture?
(*c*) what factors give *P. aurelia* a competitive advantage over *P. caudatum*?

When the two species are cultured together *P. aurelia* has a competitive advantage over *P. caudatum* for gaining food and after five days the numbers of *P. caudatum* start to decrease until, after about 20 days, the species has become 'extinct', that is it has been competitively excluded. *P. aurelia* takes longer to reach the stationary phase of growth than when grown in isolation, so is also affected adversely by the competition, even though it is more successful than its competitor. This helps to explain the selection pressure for competitors to adapt to separate niches. Under natural circumstances, the less successful competitor rarely becomes extinct, but merely becomes rare.

The **competitive exclusion principle** (or **Gausian exclusion principle**) has since been confirmed in further animal experiments. Competitive exclusions have also been shown to occur in plant populations, such as in mixed cultures of duckweed (*Lemna*) species.

The study of natural populations is made more complex by the larger number of interacting populations and by the fact that the environmental variables such as temperature, moisture and food supply cannot be controlled.

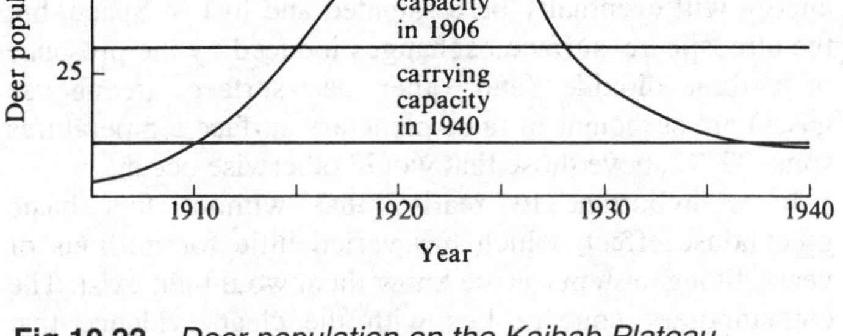

Fig 10.22 *Deer population on the Kaibab Plateau following eradication of predators.*

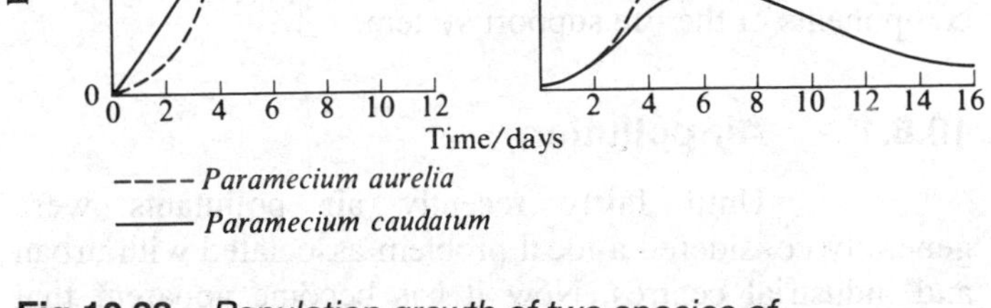

Fig 10.23 *Population growth of two species of* Paramecium, *(a) cultured separately, (b) cultured together.*

One form of competition which has attracted a lot of interest involves complex organic molecules that are produced by plants and animals which affect the growth of other living organisms. They include antibiotics and growth inhibitors, such as penicillin which is produced by the fungus *Penicillium* and has antibiotic properties against Gram positive bacteria (section 2.2.2). Chemical competition among microorganisms is very intense and the relationships between them very complex.

Resource partitioning

The more specialised an organism becomes for a particular niche, the less chance there is of direct competition. So, in long-established communities which are species-rich, **resource partitioning** (specialisation by different species to make use of different resources) leads to less competition and a more stable community structure.

Resource partitioning may take several forms, for example:

- specialisation of morphology and behaviour for different foods, such as the beaks of birds which may be modified for picking up insects, drilling holes, cracking nuts, tearing flesh and so on;
- vertical separation (stratification, see fig 10.15), such as canopy dwellers and forest floor dwellers;
- horizontal separation, such as the occupation of different microhabitats.

However, despite a tendency for each species to evolve its own particular niche, some direct competition between species for available resources will still occur.

10.8 Human impacts on ecosystems

Until the last ten thousand years or so, living systems evolved in response to changes in the abiotic environment, unaffected by human activity. Since the development of agriculture and technology, an increasing human impact on the environment has occurred. In the last two centuries especially, widespread industrialisation has led to potentially damaging environmental pollution.

Pollution may be defined as the release into the environment of substances or energy in such quantities and for such duration that they cause harm to people or other organisms or their environment. Pollution can affect all aspects of environment, human-made and natural, abiotic and biotic, and may be readily transferred between components of the life support system.

10.8.1 Air pollution

Until fairly recently air pollutants were generally considered a local problem associated with urban and industrial centres. Now it has become apparent that pollutants may be transported long distances in the air, causing adverse effects in environments far removed from the source of emission. Air pollution and its control is thus a global issue demanding international cooperation. Important atmospheric pollutants include gases such as chlorofluorocarbons (CFCs), sulphur dioxide (SO_2), hydrocarbons (HCs) and the oxides of nitrogen (NO_x). Increasing levels of natural gases, such as carbon dioxide, in the atmosphere as a result of human activity can also be considered as a form of pollution.

Gases occurring naturally in the atmosphere may be affected seriously by pollution, such as the depletion of ozone (O_3) high in the atmosphere. Paradoxically in some locations ozone is occurring with increasing frequency as a ground-level air pollutant. These raised levels damage many crop plants and, in association with hydrocarbons and NO_x pollutants, are an important constituent of photochemical smog and may generate a direct human health hazard. Dust, noise, waste heat, radioactivity and electromagnetic pulses may also pollute the atmosphere.

Carbon dioxide and the greenhouse effect

The main exchange pathways in the carbon cycle were summarised in fig 10.12. Much of the carbon is locked away in complex inorganic forms, such as carbonates in rocks, and in complex organic forms in fossil fuels and biomass. The release of some of this stored carbon into the atmosphere, as carbon dioxide, has increased in recent times with the burning of large amounts of fossil fuels and with large-scale deforestation (section 10.8.3).

10.18 How does deforestation increase atmospheric carbon dioxide?

Carbon dioxide is normally present in the lower atmosphere, the troposphere, in very small amounts, about 300 ppm or 0.03% by volume. Its importance lies in its contribution to the planetary greenhouse effect. Carbon dioxide is transparent to incoming short-wave radiation from the Sun, but absorbs strongly the long-wave radiation which the Earth re-radiates into Space. It therefore 'traps' outgoing radiation, warming the lower atmosphere which in turn radiates energy back to the surface of the Earth. Ultimately, of course, any given 'package' of incoming energy will eventually be dissipated and lost to Space, but the atmosphere–surface exchanges induced by the presence of carbon dioxide (and other near-surface greenhouse gases) are sufficient to raise planetary surface temperatures some 32 °C above those that would otherwise occur.

It is important to realise that without this basic greenhouse effect, which has varied little for millions of years, living systems as we know them would not exist. The contemporary concern lies with the clear evidence that carbon dioxide levels (and those of other greenhouse gases, notably carbon monoxide, methane and chlorofluorocarbons (CFCs)), are rising at a rate unprecedented in recent Earth history and their increased presence may lead to an increasingly warmer surface environment (fig 10.24), namely an **enhanced greenhouse effect**. This may in turn lead to increased evaporation and a greater atmospheric water vapour content. Since water vapour also acts as a powerful long-wave absorber, this may further increase surface temperatures. A resulting rise in surface temperatures would cause changes in the distribution pattern and intensity of the major planetary weather systems which would profoundly affect human activities and the distribution of organisms.

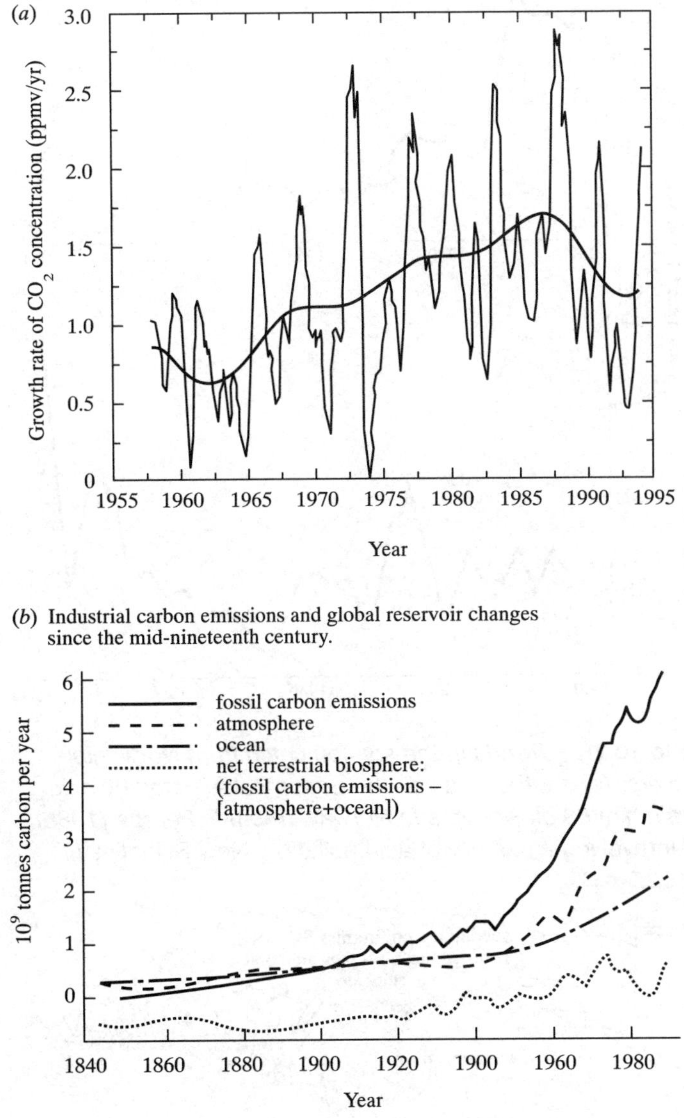

Fig 10.24 *(a) Growth rate of carbon dioxide concentration since 1958 (from the Mauna Loa record). The smooth curve shows the same data filtered to suppress any variations on a time scale less than 10 years. Currently atmospheric carbon dioxide concentration is about 355 ppmv.*
(b) Industrial carbon emissions and global reservoir changes since the mid-nineteenth century. The calculation implies that the terrestrial biosphere was a net source of carbon before 1940 (negative values) and has been a net sink since about 1960. (From Climate Change *(1994) IPCC Scientific Assessment, WMO/UNEP, CUP.)*

In 1988 an Intergovernmental Panel on Climate Change (IPCC) was established to coordinate scientific information and research on the likely causes and consequences of increased levels of carbon dioxide and other greenhouse gases and to identify effective remedies. The Earth Summit in 1992 attempted to secure international agreement on targets for CO_2 emissions for all countries. We cannot be certain whether enhanced greenhouse warming will happen though some scientists consider that is has already begun. Nevertheless international governmental action to control pollutant greenhouse gases is a welcome trend.

Ozone depletion

The atmosphere provides a thermal blanket and radiation shield to the Earth. In the upper atmosphere, 15–50 km above the Earth, oxygen and ozone absorb much of the incoming short-wave radiation which is mostly very harmful to living organisms, in particular in damaging their genetic material. Although ultraviolet radiation of certain wavelengths has a beneficial effect in the production of vitamin D (section 8.7.9), overexposure to strong sunlight is known to increase the risk of skin cancer. Furthermore radiation absorption by ozone high in the atmosphere warms these higher levels and creates a deep temperature inversion layer (where the highest temperatures are at greatest altitude). This effectively limits the movement of air in the atmosphere by convection. Any change or weakening of this inversion layer would profoundly alter global weather patterns and hence Earth surface climates.

Ozone is produced high in the atmosphere by the action of sunlight on oxygen molecules. Chlorofluorocarbons are a group of chemicals, including carbon tetrachloride and chloroform, which are commonly used as solvents, aerosol propellants and refrigerator coolants. They are not readily broken down in the atmosphere and may contribute to increased 'greenhouse' warming. Eventually they diffuse into the stratosphere. High in the atmosphere they are broken down by sunlight, releasing chlorine and fluorine. These react with ozone and break it down into oxygen faster than it can be reformed from oxygen into ozone. CFC pollution thus shifts the oxygen–ozone equilibrium.

In 1987 a seasonal, but complete, depletion of the ozone layer occurred above Antarctica for the first time and, in the 1990s, ozone thinning over the Arctic has been regularly observed. Scientists are not certain why this is happening. A possible explanation may be that increased cooling of the stratosphere is taking place due to the enhanced greenhouse effect which is trapping more radiated heat near Earth's surface. This may in turn mean the Arctic stratosphere is more frequently cold enough to enable formation of stratospheric ice clouds. Rapid ozone depletion then takes place. A worrying outcome may be that in spite of worldwide efforts to reduce emissions of CFCs the stratospheric ozone layer may not, as previously predicted, recover quickly in the next century. Recovery may also depend on controlling the enhanced greenhouse effect.

Acid rain

Acid rain is neither a simple nor a single phenomenon. The acid gases sulphur dioxide (SO_2) and oxides of nitrogen (NO_x) are produced by burning fossil fuels. Incomplete combustion of these fuels also releases hydrocarbons. These may have effects as dry gases or they may be washed out of the atmosphere to produce acid precipitation in rain and snow (fig 10.25). The most industrialised areas of the world, such as the eastern USA, western Europe, north-east China and Japan, have all experienced rainfall with a pH well below 4.0.

Acid rainfall (pH <5) often causes major changes in ecosystems and damage to buildings. This often happens in countries bordering those which are major sources of pollutants. Norway and Sweden, for example, receive acid rain as a result of air pollutants emitted in the UK and industrial centres of Europe which are transported by prevailing high-level winds. Acid rainfall in central Sweden and southern Norway has affected salmon and trout fisheries (fig 10.26) and damaged forests. Tree injury associated with acid pollution is now widespread in Europe (fig 10.27) and evidence for damage to beech and yew has been recorded in Britain.

Acid rain leaches magnesium and calcium from soils and from damaged leaves. Eventually aluminium, manganese and heavy metals such as iron and cadmium come into solution and may reach toxic concentrations, causing damage to tree roots and the breakdown of mycorrhizas. This decreases the capacity of the tree to take up water and nutrients. Disease induced by mineral deficiencies becomes common, a situation made worse by dry conditions.

Cures, such as adding lime to lakes (Sweden) and forests (W. Germany) can only be viewed as temporary stop-gaps. The remedy lies in reducing the release of pollutant gases. Attention has been focused on reducing sulphur dioxide emissions since these have significant and clearly identifiable industrial sources, most notably coal-fired electricity generators. Furthermore the desulphurisation technology is available and effective, though costly. In the long term, however, it may be equally important to reduce hydrocarbon and nitrogen oxide emissions.

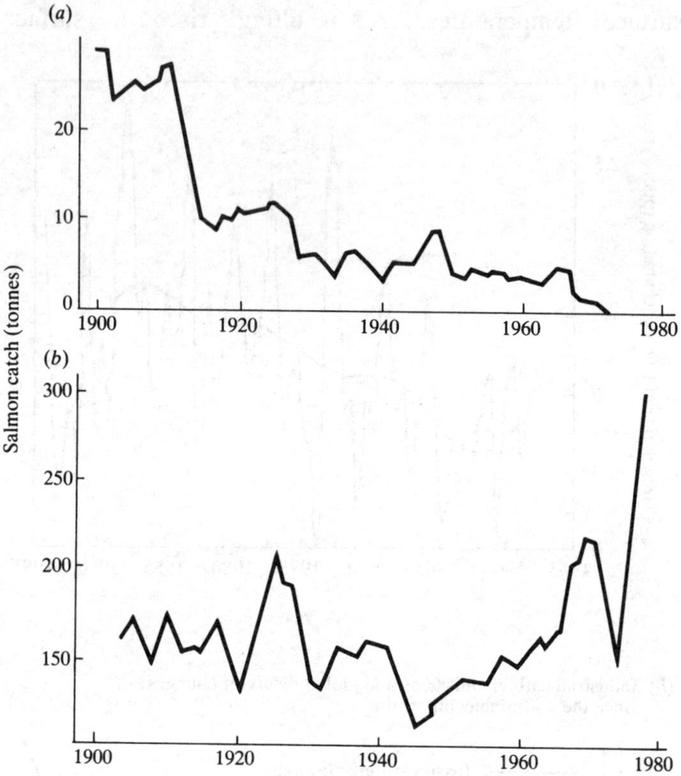

Fig 10.26 *Trends in the salmon catch from Norwegian rivers; (a) southern rivers, the area most affected by acid rain, (b) 68 other rivers in Norway. (From F. Pearce (1986)* Unravelling a century of acid pollution, New Scientist *11, 1527, p.33.)*

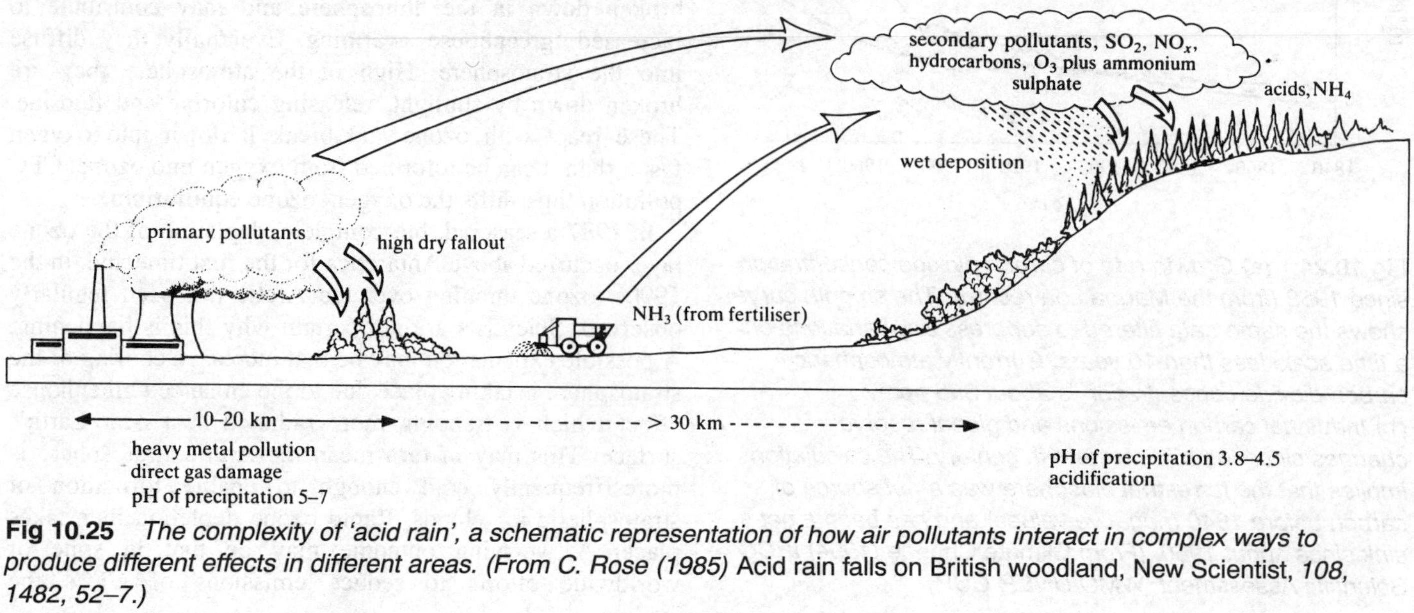

Fig 10.25 *The complexity of 'acid rain', a schematic representation of how air pollutants interact in complex ways to produce different effects in different areas. (From C. Rose (1985)* Acid rain falls on British woodland, New Scientist, *108, 1482, 52–7.)*

Fig 10.27 *Acid dieback in a German forest.*

10.8.2 Water pollution

Until recently water pollution has been a relatively local problem of the developed world. **Eutrophication** is the most common problem, where inland waters and rivers are polluted with nitrogen and phosphorus run-off from fertilisers used in intensive agriculture and discharge of phosphate-rich sewage effluents. Such problems are increasingly occurring on a worldwide basis and now affect marine as well as freshwater ecosystems.

Sewage from coastal settlements discharges, sometimes untreated, into coastal waters where it generates a direct health hazard for recreational bathers as well as marine organisms. Land drainage from urban areas, industrial and waste disposal sites is often contaminated with heavy metals or hydrocarbons. Biological concentration of heavy metals in marine food chains may give lethal doses, as occurred following the industrial discharge of mercury into coastal waters at Minimata in Japan. Here, concentrations in fish led to the deaths of many humans and other animal predators. At sub-lethal levels heavy metals and contaminants such as pesticide and oil derivatives may lower resistance to disease.

During the last decade measures to control and eventually stop toxic waste dumping and incineration at sea have been introduced by countries bordering the North Sea in an attempt to reduce pollution and damage to this ecosystem.

Another major problem is caused by excessive soil erosion on the land surface. This increases the silt load of rivers and coastal waters which may beneficially enrich fisheries. However, it can also be destructive. For example, it is leading to coral reef destruction in the Australian Great Barrier Reef as a result of deforestation on the mainland.

Other important forms of water pollution include thermal pollution and oil pollution.

Eutrophication

Eutrophication means nutrient enrichment. Over a long time period, typically several thousand years, lake ecosystems classically show a natural progression from an **oligotrophic** (few nutrients) to a **eutrophic** or even **dystrophic** (rich in nutrients) state (table 10.6). In the twentieth century, however, rapid eutrophication has occurred in many lakes, in semi-enclosed seas such as the Baltic, Mediterranean and Black Sea, and in river systems worldwide. This is due to human activity.

The main factors that cause eutrophication are heavy use of nitrogen fertilisers on agricultural land and the increased discharge of phosphates from sewage works. The phosphate problem reflects not only a larger human population but also the modern tendency for more people to live in urban areas and the development of mains sewage systems.

Eutrophication generates acute economic as well as ecological problems. Good quality water resources are important for many industrial processes, vital for human and livestock drinking water supplies, essential for commercial and recreational fisheries and necessary for the maintenance of recreational amenities and navigation routes on major waterways (table 10.7).

Nitrates and particularly phosphates are the nutrients most commonly limiting primary productivity in aquatic ecosystems. Additional nitrate and phosphate, therefore, favours an increase in rapidly growing competitive planktonic species. As it takes longer for consumer organisms to increase in number in response to environmental change, this means that not all the increased primary production is eaten by the consumer organisms. Instead the excess material enters the decomposition pathway. Breakdown to simple inorganic nutrients is an oxygen-demanding process. Dissolved oxygen levels may be reduced below those necessary for the successful growth

Table 10.6 The general characteristics of oligotrophic and eutrophic lakes.

	Oligotrophic	*Eutrophic*
Depth	deeper	shallower
Summer oxygen in hypolimnion	present	absent
Algae and blue-green bacteria	high species diversity, with low density and productivity, often dominated by green algae	low species diversity with high density and productivity, often dominated by blue-green bacteria
Blooms	rare	frequent
Plant nutrient flux	low	high
Animal production	low	high
Fish	salmonids (e.g. trout, char) and coregonids (whitefish) often dominant	coarse fish (e.g. perch, roach, carp) often dominant

(Source: C.F. Mason (1981) *Biology of freshwater pollution*, Longman.)

Note

In the classic mode of natural lake eutrophication a newly formed deep lake (a classic situation would be following retreat of an ice sheet) contains few nutrients since there has been no opportunity for weathering and sediment removal from the surrounding catchment. Primary and secondary productivity are hence low, the waters are clear, and oxygen status is good throughout.

With time, as weathering proceeds, nutrient status increases, primary and secondary productivity rise, organic and inorganic sediments accumulate and the lake becomes shallower. The more productive waters are less clear and the hypolimnion (see fig 10.29) may become seasonally oxygen-depleted.

A **dystrophic** lake is one which receives large quantities of organic matter from terrestrial plants giving the water a brown colouration. Such lakes typically have peat-filled margins and may develop into peat bogs.

Table 10.7 The main effects of eutrophication on the receiving ecosystem and the problems for human societies associated with these effects.

Effects

(1) Species diversity decreases and the dominant biota change
(2) Plant, algal and animal biomass increase
(3) Turbidity increases
(4) Rate of sedimentation increases, shortening the life span of the lake
(5) Anoxic conditions may develop

Problems

(1) Treatment of drinking water may be difficult and the supply may have an unacceptable taste or odour
(2) The water may be injurious to health
(3) The amenity value of the water may decrease
(4) Increased vegetation may impede water flow and navigation
(5) Commercially important species (such as salmonids and coregonids) may disappear

(Source: C.F. Mason (1981) *Biology of freshwater pollution*, Longman.)

and reproduction of other species. In extreme cases, the death of fish and other species and their subsequent decomposition can impose a further oxygen demand, making the situation increasingly worse. This may not just be a problem for the immediately affected area. Zones of oxygen depletion on an otherwise unaffected river system may be sufficient to disrupt breeding in migratory species such as salmon and eels.

Deoxygenation of a river caused by organic wastes is a slow process so that the point of maximum deoxygenation may occur some distance downstream of a discharge (fig 10.28). In the River Thames in 1967, at low flow in autumn, minimum oxygen conditions prevailed for 40 km downstream of London Bridge, whereas in spring, with high flow, only 12 km had minimum oxygen. During the last 30 years major efforts have been made to clean up the River Thames. Oxygen depletion on this scale no longer occurs and fish are found throughout the course of the river.

10.19 List the factors that will determine the degree of deoxygenation.

In lake ecosystems the problem of eutrophic oxygen depletion may be made worse by seasonal water **stratification**, when the water forms layers with different temperatures. In temperate regions thermal stratification typically establishes in early summer (fig 10.29). This occurs for two main reasons.

- The Sun heats the surface water. Warmer water is less dense so it remains in the top layer of the lake (**epilimnion**). Heat transfer to deeper layers can only take place by conduction which is a slow process in water.
- Rivers and streams draining into the lake are shallow by comparison. Their water will have become warmed throughout its depth. This warmer water, being also lighter and less dense, mixes only with the surface waters of the lake further raising its temperature by comparison with the deeper water (the **hypolimnion**).

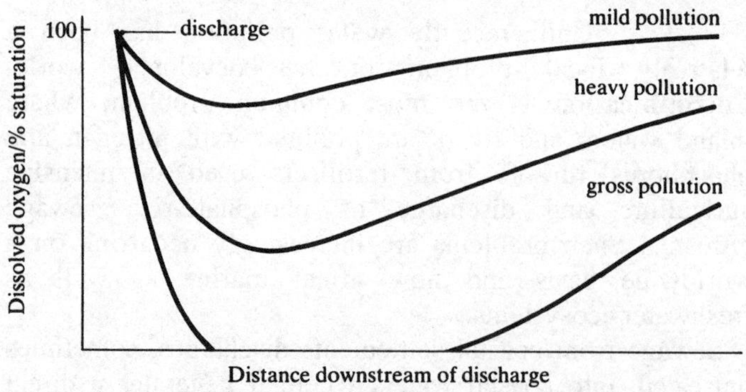

Fig 10.28 *A typical 'oxygen sag' curve; the effect of organic discharge on the oxygen content of river water. (From C.F. Mason (1981)* Biology of fresh water pollution, *Longman.)*

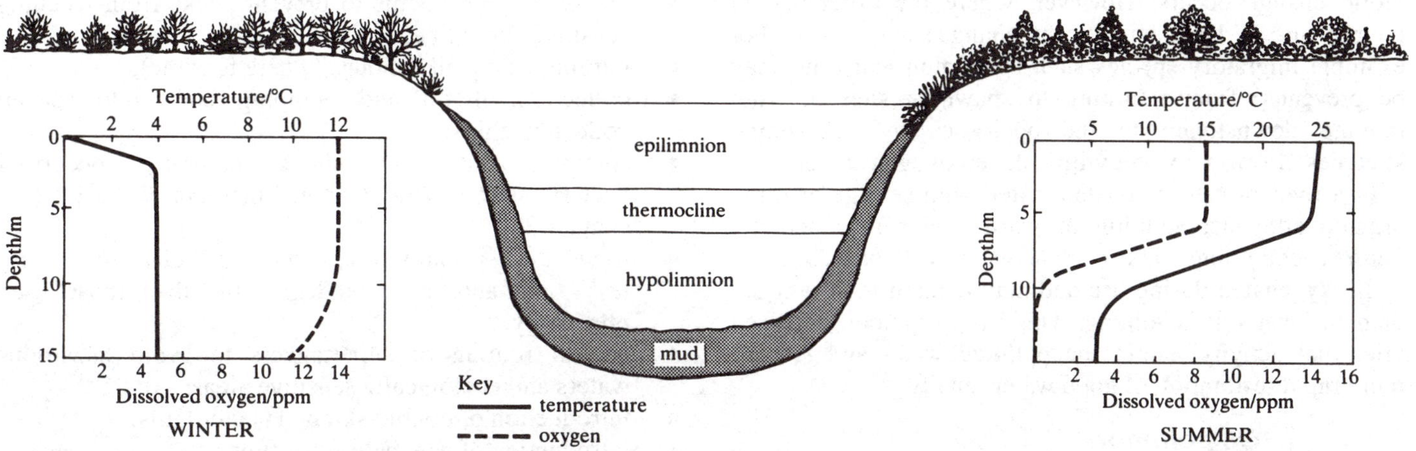

Fig 10.29 *Thermal stratification in a mid-latitude lake (Linsley ponds, Conn. USA). In summer, a warm oxygen-rich circulating layer of water, the epilimnion, is separated from cold oxygen-poor hypolimnion waters by a broad zone of rapid temperature change called the thermocline. A similar gradient in oxygen also occurs across the thermocline. (Modified from E.P. Odum (1971)* Fundamentals of ecology, *Saunders.)*

This has important consequences for lake ecosystems. In particular it affects the supply of oxygen to the deep waters.

Oxygen supplies in lake waters come from three main sources. These are:

- photosynthesis which requires light and is therefore most rapid in the surface waters;
- diffusion from the atmosphere;
- oxygen in stream and river waters draining into the lake.

These sources all primarily enrich the surface waters. Oxygen in the deeper water will depend on effective diffusion from above or extreme turbulence linked with storm events. Storminess and good mixing of the water is more characteristic of the winter season. So once the thermal stratification pattern has become established life in the deep waters will be dependent on the oxygen supplies locked away in spring.

In a healthy lake ecosystem most primary producers are eaten. Very few die and form food materials for detritivores and decomposers. However, if increased phytoplankton production occurs in the surface waters, encouraged by the warmth and increased nutrient status of the epilimnion or by eutrophication, then the excess primary producers not consumed by herbivores will fall into the hypolimnion when they die. Here their decomposition makes an additional oxygen demand on a restricted oxygen resource. Providing sufficient oxygen is present to meet this extra demand, as well as the needs of existing fauna, no major problem will arise. However, if there is insufficient oxygen, sudden and catastrophic fish kills may result in late summer when oxygen supplies approach exhaustion.

Monitoring eutrophication

Changes associated with eutrophication can be monitored biologically and chemically. This gives the opportunity for remedial action before catastrophic ecosystem damage occurs. Changes in phytoplankton species present may help to indicate eutrophication, for example blue-green bacterial blooms are common. Eutrophic waters characteristically show high abundance and low species diversity of phytoplankton.

A useful chemical indicator of eutrophication is the **biochemical oxygen demand (BOD)** (Experiment 11.9), which measures the rate of oxygen depletion by organisms. It is assumed this primarily reflects microorganism activity in decomposing organic matter present in waters. (As discussed, organic matter typically increases as waters become nutrient-enriched.) However, oxygen consumption by algae will inevitably also be included in the test. In practice this is normally not important, but in some cases it may account for up to 50% of the total BOD. BOD is thus an approximate rather than precise guide to water quality. It is most useful when used in conjunction with other water quality indicators.

> **10.20** What are the advantages and disadvantages of biological rather than chemical monitoring of eutrophication?

Thermal pollution

Thermal pollution can be a problem in rivers and coastal areas. It most typically arises from the use of water as a coolant in industrial processes, such as in electricity generating plants. Water discharging from electricity stations may be warmer than that taken in. Warm water has a lower dissolved oxygen content than cold water. At the same time the increased warmth raises the metabolic rate of organisms in the water and thus increases their oxygen demand. With small temperature differences no major biotic change occurs. However, where the difference in temperature is large, significant changes may result. For example migratory species such as salmon and trout may be prevented from returning to spawning sites or from moving downstream to the open sea by intervening stretches of warmer water with reduced oxygen content.

In a river or estuary contaminated with sewage or other organic effluents, addition of warm water may actually *improve* conditions. The coolant water will have become well oxygenated during use and may contain more oxygen than the water it is joining. This may promote increased microbial activity within the polluted water and lead to more rapid restoration of good water quality.

Oil pollution

Oil pollution is a major hazard for marine and coastal environments. The main sources of marine oil pollution are:

- damage to oil tanker ships through collision with other ships, explosion, or wrecking;
- seepage from offshore installations;
- flushing of tanker holds.

Each year about 10 million tonnes of crude oil is spilt into the Earth's oceans from incidents of this type.

Crude oil kills seaweeds, molluscs and crustaceans when washed onto rocky shores. Marine mammals may also be affected by oil spills, for example when their fur is oiled. However, usually the worst casualties of major oil spills are fish-eating birds because oil penetrates and mats the plumage making flight impossible; heat insulation is lost and birds may die of hypothermia; buoyancy is also reduced causing birds to drown; attempts to preen leads to ingestion of oil and gut irritation. Phytoplankton, however, do not appear to be greatly affected by the oil, though shading may temporarily restrict light penetration and reduce photosynthesis.

The long-term damage to ecosystems caused by major oil spills is minimal. Recovery is more rapid and complete when oil is left to disperse naturally. Bacterial decomposition aided by break-up of the slick by wind and wave action can mean complete recovery in 3 or 4 years in warm or temperate climates. In colder environments, such as the Alaskan coast affected by the *Exxon Valdez* spill in 1989, effects are more persistent since bacterial activity is slower. Use of **dispersants** may accelerate break up of oil slicks, but the agents used often cause lasting damage because detergents are toxic and are not readily biodegradable.

Studies have shown, however, that the less spectacular but persistent spills and leakages, linked for example with coastal refineries and oil terminals, have a more damaging effect on marine and coastal ecosystems than the well-publicised major incidents.

Techniques for treating and preventing oil pollution have improved. These include:

- use of floating booms to prevent slicks from reaching sensitive shore lines;
- burning heavy oil residues (where feasible);
- collection of oil and pumping back into special collection ships;
- spraying onto oil slicks naturally occurring bacteria such as *Pseudomonas* that can digest oil (see chapter 25);
- use of new specially designed oil spill cleaners that are less toxic and more biodegradable than those used previously;
- careful routing of supertankers to avoid hazardous waters and ecologically sensitive areas;
- introduction of double-skinned tanker hulls;
- introduction of new ballast systems.

10.8.3 Degradation of terrestrial ecosystems

Since prehistoric times, humans have manipulated ecosystems in their search for food, shelter, fuel and other resources and by the impact of discarded waste materials. The first hunter–gatherer communities had minimal impact but, as human societies became increasingly settled, numerous and technically advanced, their impact on Earth's ecosystems increased dramatically. For example, deforestation in Britain, which began in Neolithic times, expanded rapidly to the point where at the start of the twentieth century only 3% of the land surface remained wooded. Deforestation on a global scale is now an issue of great concern.

The destruction of woodlands and forests is usually in response to the need for more land for growing crops and rearing livestock, and results in the creation of completely new, managed ecosystems. Mismanagement of these ecosystems has created other problems, such as soil erosion, desertification and adverse effects associated with the widespread use of new synthetic organic compounds as pesticides.

Global deforestation

Forests are the natural climax vegetation of many parts of the world covering, until recent years, a third of the land surface. Today temperate forests are not significantly decreasing in area though in the past large scale deforestation has occurred. In fact major replanting and conservation initiatives are underway in many countries. Tropical forests on the other hand are declining at a rate that will decrease their 1950 extent (15% of global land surface) to 300 million hectares (7% of global land surface) by the year 2000 AD. It has been estimated that twelve million hectares of forest, an area the size of England, are disappearing annually and a further ten million hectares are being degraded by removal of good timber species, inappropriate management and inattention to conservation needs.

Where human population densities are low, tropical forest clearance for agriculture using traditional shifting cultivation methods does little lasting damage. A small area is felled and the timber burnt leaving a clearing for cultivation. Ash from the burn provides nutrients for the crops and the burn destroys pests. Crops planted in the ash enriched soil are initially very productive, though, as the nutrient supply from the ash becomes exhausted, crop productivity declines. At the same time rainforest understorey species will begin to invade the clearing. After 3 or 4 years the plot is abandoned as it is easier and more rewarding to clear and fire a new area than to try to maintain and cultivate the existing plot. The original plot will then quite quickly regenerate to mature forest, typically taking 30 to 40 years to establish a mature rainforest canopy.

Re-establishment of mature vegetation is important. In tropical moist forests the main biomass store and the main nutrient store is in the above-ground vegetation. Tropical forest soils often lack any significant nutrient store. The main nutrient supply for the vegetation comes from recycled nutrients. Litter is rapidly decomposed and its nutrients are recycled within five or six weeks or less. The trees have a thick surface root mat and the fine rootlets absorb 99.9% of all inorganic nutrients percolating down through them. Many trees are legumes and have root nodules with nitrogen-fixing bacteria; others commonly have mycorrhizas which, in their feeding process, transfer nutrients from decomposing litter direct to the tree roots. Maintaining an unbroken cycle is important to maintaining fertility. This is quite easy in shifting cultivation, where the small-scale clearances resemble the natural gaps which occur from time to time in the forest canopy. It is very difficult to maintain fertility when large-scale clearance takes place.

Small gap size clearances and long fallow periods between felling produce a sustainable forest agriculture. If the vegetation is felled again before reaching maturity, then the ash produced will be less abundant and the nutrient supply for new crops will be reduced. The clearing will be less productive and will be abandoned more quickly, increasing demands on the surrounding forested areas.

Fig 10.30 *Villager planting maniok in a clearing in Madagascar rainforest produced by slash and burn.*

As human population size increases, larger areas of land may be cultivated to supply food. In addition, fuel wood gathering will lead to further deforestation. Changing land use may make the situation much worse, such as clearance of large areas for the growth of cash crops or for ranching. In some forests commercial logging for tropical hardwoods such as teak and mahogany is another major cause of deforestation.

Loss of forests is serious for many reasons.

- There is a loss of traditionally harvested products such as timber, poles, twine, fuel-wood, honey, fruit, game animals and herbs, that at one time supplied local people with their needs.
- The demand for softwood timber (for building), pulp wood (for making paper) and tropical hardwood (for furniture) is rising globally. Long-term supplies are threatened.
- Forests are often on uplands and on watersheds. Here they catch large amounts of rain. An intact forest canopy softens the impact of intense tropical rainfall in many ways. It releases large amounts of water back to the atmosphere in evaporation and transpiration and channels water gently through the vegetation to the soil. Infiltration to the soil is high and there is a long delay before water percolates through to the streams and rivers. If the forest canopy is removed, the tropical soil surface bakes hard in the intense heat. Rainfall cannot easily penetrate the surface and is rapidly lost from the area in surface run-off. Regeneration is difficult and, at the same time a flood hazard is created in the plains below. For example, in Bangladesh in the summer of 1988, flooding occurred on an unprecedented scale affecting most of the country, largely due to the deforestation in the mountains to the north in India and Nepal.
- More rapid run-off of rainwater results in soil erosion. This can remove the topsoil and leave ground unsuitable for growing crops. It can also lead to silting of reservoirs which reduces their useful life, whilst harbours and estuaries must be continually dredged to keep them open.
- Deforestation increases global carbon dioxide (see section 10.8.1) which may have long-term effects on the global climate.
- Forests have species-rich and diverse wildlife communities. Their destruction will lead to innumerable extinctions of little-known forms of life with the consequent loss of genetic variety and potential resources. Tropical forests have already given us anti-cancer and anti-malaria drugs and scientists are actively investigating tropical moist forest plants for drugs to control HIV and many other diseases.

Soil erosion and desertification

Deforestation is not the only cause of soil erosion. Mismanagement of farmland and grassland may lead to a rapid loss of soil. Soil formation is a slow process. Hilly areas with steep slopes which are regularly cultivated and also have high rainfall are especially susceptible to soil erosion. Parts of south-east Asia have long-established systems of terracing that have proved very effective in holding soil. These are in areas where forests have been cleared. Principal soil conservation measures are terrace cultivation, contour ploughing and making ridges to stop run-off of rain. Annually, five million hectares of farmed land are coming out of crop production, worldwide, because of erosion losses. Grasslands that are overgrazed by livestock frequently lose the plant cover that holds the topsoil. Plants are eaten to their roots and die, with the result that water running freely across the land surface causes sheet erosion, carrying off the topsoil. Channelled rainwater forms gullies which cut deep into the land surface. About seven million hectares of grazing land are lost this way each year and much of this will become desert.

Deserts may form naturally, for example when rains persistently fail in semi-arid areas, but their creation can be accelerated by human activity in a process called **desertification**. This is a general term for the degradation of dry land areas so that formerly productive land becomes useless. It usually results from:

- overgrazing by livestock,
- overcultivation,
- deforestation (as mentioned above),
- poor irrigation practices.

Overgrazing reduces the sparse vegetation cover, and trampling by livestock may break up the soil. Erosion of top soil by wind or during flash floods commonly results. Overcultivation can lead to the removal of nutrients and humus from the soil, leading again to sparse vegetation cover and the increased risk of erosion.

In irrigated areas, waterlogging and salinisation are the main problems. **Waterlogging** occurs when the water table lies close to the soil surface. Excess water use during irrigation may increase and extend the problem. This is a particular problem when the cultivated crops are intolerant of waterlogging, such as wheat and cotton. **Salinisation** is the increase in the concentration of soluble salts in the soil. This occurs when high rates of evaporation draw water and salts to the surface of the soil by capillary action. It also happens when saline irrigation water is used on soils which are not very permeable, and when deep ground water sources, which are naturally more salty than most other 'fresh' water, are used for irrigation. Few crops can tolerate a salt concentration of above 0.5–1.0%, so salinisation can be economically very important.

Restoration of desertified land is a complex problem. It may require the exclusion of people and their livestock from the area to allow surface vegetation to recover. Other practices may be used to reduce soil removal by flooding, such as the planting of trees, terracing on slopes or fencing to prevent soil loss. Installation of good drainage and efficient leaching systems can reduce the problems of waterlogging or salinisation. However, this can be administratively, socially, economically and politically difficult. Unless the socio-economic factors that trigger the use of dry lands are resolved, it is unlikely that local projects to alleviate the problems of land degradation will make a significant impact on desertification.

10.8.4 Pesticides and the environment

Pesticides are chemical substances used by humans to control pests. The term pesticide is an all-embracing word for **herbicides** (which kill plants), **insecticides** (which kill insects), **fungicides** (which act on fungi), and so on. Most pesticides are poisons and aim to kill the target species, but the term also includes chemosterilants (chemicals causing sterility) and growth inhibitors.

In Britain pesticide use is mainly associated with agriculture and horticulture, though pesticides are also widely used in food storage and to protect wood, wool and other natural products. In many countries pesticides are used in forestry and they are also used extensively to control human and animal disease vectors (as in malaria, chapter 15).

Ecological characteristics of pesticides

The important ecological characteristics of pesticides are toxicity, persistence and specificity.

Toxicity. For a particular species toxicity is commonly defined by the **lethal dose 50** (LD_{50}). This is the single dose which kills half an experimental laboratory population. In the field, when the organism is subjected to additional environmental stresses, a higher proportion may die. Nevertheless, by definition, some survive. (In short the aim in agriculture is to reduce crop injury to an acceptable level, this point being largely a question of economics). Unfortunately the survivors form the basis of a resistant pest population and, in organisms such as insects with a rapid life cycle, resistance and pest resurgence are common problems.

Persistence. This is the length of time that the pesticide remains in the environment, including within organisms, without being broken down. An example of a persistent insecticide is the organochlorine DDT which was commonly used between the 1940s and 1960s.

Though persistence is generally an undesirable quality (particularly on food crops) in some instances, for example in the control of animal parasites and soil-borne diseases, some degree of persistence is an important practical and economic requirement. However, long-term persistence can be very damaging. For example, in the mid-1960s DDT was detected in the livers of penguins in the Antarctic, a habitat far from areas where DDT might have been used.

Toxicity and persistence are linked in that a lethal but non-persistent chemical may in the long run do less damage than a sublethal persistent chemical. This is because the latter has more opportunity for incorporation into food chains where it may be metabolised to a more toxic form or more typically accumulate to toxic concentrations in predators at the top of the chain (see question 10.21).

Pesticide poisoning has had devastating effects on some top carnivores, most notably birds. The peregrine falcon, for example, disappeared completely from the eastern USA as a result of DDT poisoning. Birds are especially vulnerable because DDT induces hormonal changes that affect calcium metabolism and result in the production of thinner egg shells with a consequently high loss of eggs through breakage.

DDT is now banned in most developed countries such as Britain and the USA. However, it is relatively cheap to produce and is still considered a good option for certain tasks, such as malaria control. When considering whether to use pesticides it is often a case of choosing the lesser of two evils. For example, DDT has completely eradicated malaria in many parts of the world.

Specificity. Pesticides vary in their specificity, that is the range of organisms they affect. DDT is an example of a broad-spectrum pesticide, because it seriously affects many different kinds of animals. Narrow-spectrum pesticides only affect a restricted range of organisms, for example pirimicarb only kills aphids and flies but does not affect beetles and most other insects. Likewise dalapon kills monocotyledonous plants but does not affect dicotyledons, whereas phenoxyacetic acid weedkillers eliminate dicotyledons but do not affect monocotyledons.

Use of broad-spectrum pesticides can lead to **pest resurgence**, that is when numbers of the pest after treatment increase to *more than before* the treatment. This is because the pesticide not only kills the pest, but also the predators of the pest.

A good example of this was seen in a study of the use of DDT to control cabbage white butterfly, *Pieris rapae*, on brussels sprouts. Initial pesticide application gave good control but subsequently numbers of butterfly larvae *exceeded* those in an unsprayed control area (fig 10.31). This effect was even more pronounced following repeat applications of DDT to 'control' the new infestation. Examination of the crop ecosystem showed that pesticide concentrations on the leaves were rapidly reduced due to subsequent growth of existing and new leaves. However, levels in the soil remained high, especially if crop residues were ploughed in. Thus eggs deposited by adults entering the crop from surrounding areas after spraying were little affected by the pesticide, but the main larval predators, the soil-dwelling ground beetle *Harpalus rufipes* and the harvestman *Phalangium opilio*, showed reduced numbers and survivors fed less frequently. Larval predation was thus significantly reduced and larval numbers exceeded pre-spraying levels. Further applications of DDT merely made this situation worse (fig 10.31). Predatory species are often more badly affected by pesticide use than the pest species. This is because they occur in lower numbers than their prey and therefore the population is more vulnerable and recovers more slowly.

(*a*) 1964 (after one application of DDT, 6 July 1964)

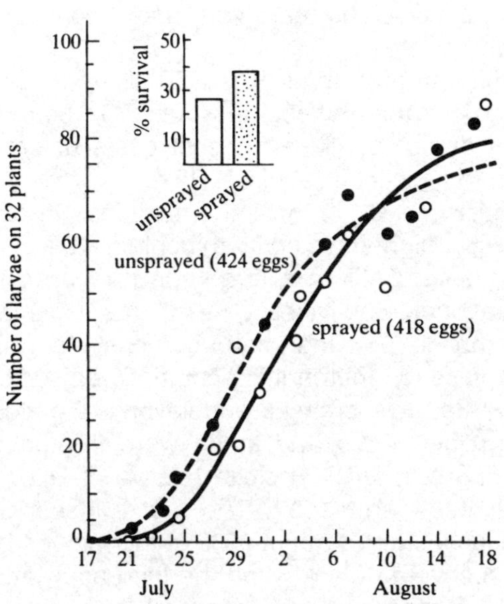

(*b*) 1965 (after three applications of DDT, 6 July, 20 August 1964 and 28 June 1965)

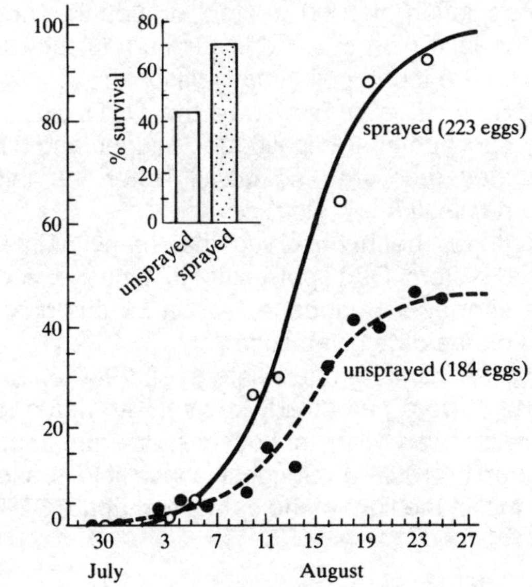

Fig 10.31 *An illustration of the differential effects of DDT on crop and soil fauna.* Pieris *lives on crops; spraying with DDT to control it is effective for only a very short period in the first year (*a*). As the soil-living predators of* Pieris *are affected by residuals in the soil,* Pieris *increases markedly after repeat spraying (*b*). (From J.D. Dempster (1968) The control of* Pieris rapae *with DDT,* J. Appl. Ecol. *5, 451–62.)*

10.21 Fig 10.32 shows the amount of DDT at different levels in a food chain, the data for which were collected in the USA.

(*a*) If the concentration of DDT in the water surrounding the algae was 0.02 ppm, what was the final concentration factor for DDT in passing **from water** into (i) producers, (ii) small fish, (iii) large fish, (iv) the top carnivore.

(*b*) What conclusions can you draw from your answers to (*a*)?

(*c*) At which trophic level (i) is DDT likely to have the most marked effect, (ii) would DDT be most easily detected, (iii) are insect pests of crops found (a typical target of DDT)?

(*d*) Suggest ways in which penguins in the Antarctic might have come to contain DDT.

(*e*) Clear Lake, California, is a large lake used for recreational activities such as fishing. Disturbance of the natural ecosystem by eutrophication (nutrient enrichment, see section 10.8.2) led to increased populations of midges during the 1940s and these were treated by spraying with DDD, a close relative of DDT, in 1949, 1954 and 1957. The first and second applications killed about 99% of the midges but they recovered quickly and the third application had little effect on the population.

Analysis of small fish from the lake showed levels of 1–200 ppm of DDD in the flesh eaten by humans, and 40–2500 ppm in fatty tissues. A population of 1000 western grebes that bred at the lake died out and levels of 1600 ppm of DDT were found in their fatty tissues.

(i) Suggest a reason why the DDD did not succeed in eradicating the midges and why they recovered so quickly after the third application.

(ii) It has been observed that many animals die from DDT poisoning in times of food shortage. Suggest a reason for this based on the data given so far.

(*f*) In Great Britain, the winters of 1946–7 and 1962–3 were particularly severe. The death toll of birds was high in both winters, but much higher in 1962–3. Suggest a possible reason for this in the light of the data given about DDT.

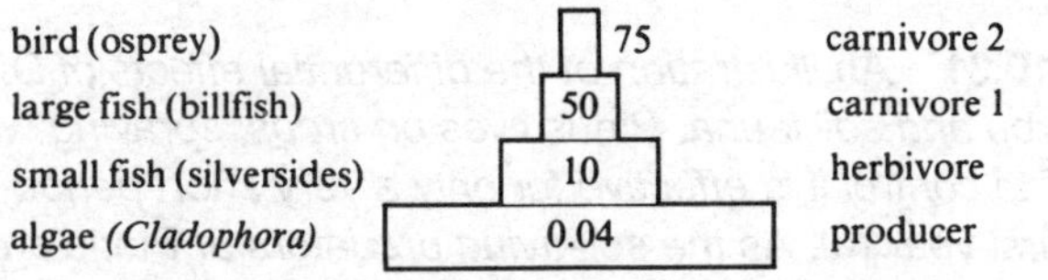

Fig 10.32 *Biomass and amounts of DDT at different trophic levels in a food chain. Figures represent amounts of DDT in parts per million (ppm).*

Long-term consequences of pesticide exposure, even at low doses, and possible synergistic links with other contaminants or disease vectors are little known due to the relative newness of most pesticides. There is mounting concern that the 'harmless' traces of pesticide metabolites left as residues on food, though not directly toxic and certainly not lethal, may, nevertheless, lower disease resistance or be biologically accumulated to significant levels. Pesticide residues in the North Sea are thought by many scientists to be linked with the rapid spread of viral disease in the common seal population during summer 1988.

The general effect of pesticide use is to reduce species diversity. It also tends to increase productivity at lower levels of ecosystems and lessen productivity at higher levels. The effects on decomposer organisms are poorly understood and the implications of all these for nutrient cycling and soil fertility need further study. Fig 10.33 summarises the main ways in which pesticides affect ecosystems. You should consider the implications of these changes.

Problems such as those outlined, especially resistance, resurgence and health risk to the human population, have led to more extensive consideration of alternative control techniques. The main alternatives, **biological control** and **integrated control** (more carefully targeted use of pesticides linked with biological control), are now briefly considered.

Biological control of pests

Biological control of pests has traditionally meant regulation by natural enemies: predators, parasites and pathogens. As such it represents a form of population management preventing unchecked exponential growth of pest species (section 10.7.3, growth curves). Some scientists take a wider view and include other techniques, such as genetic manipulation. **Cultural control** methods such as crop rotation, tillage, mixed cropping, removal of crop residues and adjustment of harvest or sowing times to favour crop or natural enemies rather than pest may also be considered biological control.

Classic biological control has been most successfully applied to introduced species which may lack natural controls, either biological or physical in a new environment. The control of cottony cushion scale, *Icerya purchasi*, on newly established citrus plantations in California in the late-nineteenth century is the first truly successful example of scientifically planned biological control. This pest was introduced with nursery stock from its native Australia. Field searches in Australia identified two natural enemies, a parasitic fly *Cryptochetum iceryae* and a predatory ladybeetle, *Rodolia cardinalis*, commonly known as Vedalia. These were taken to California and, after careful study and successful controlled release in canvas tents, were released into the wild. Both parasite and predator spread rapidly and effective and persistent control was achieved within months.

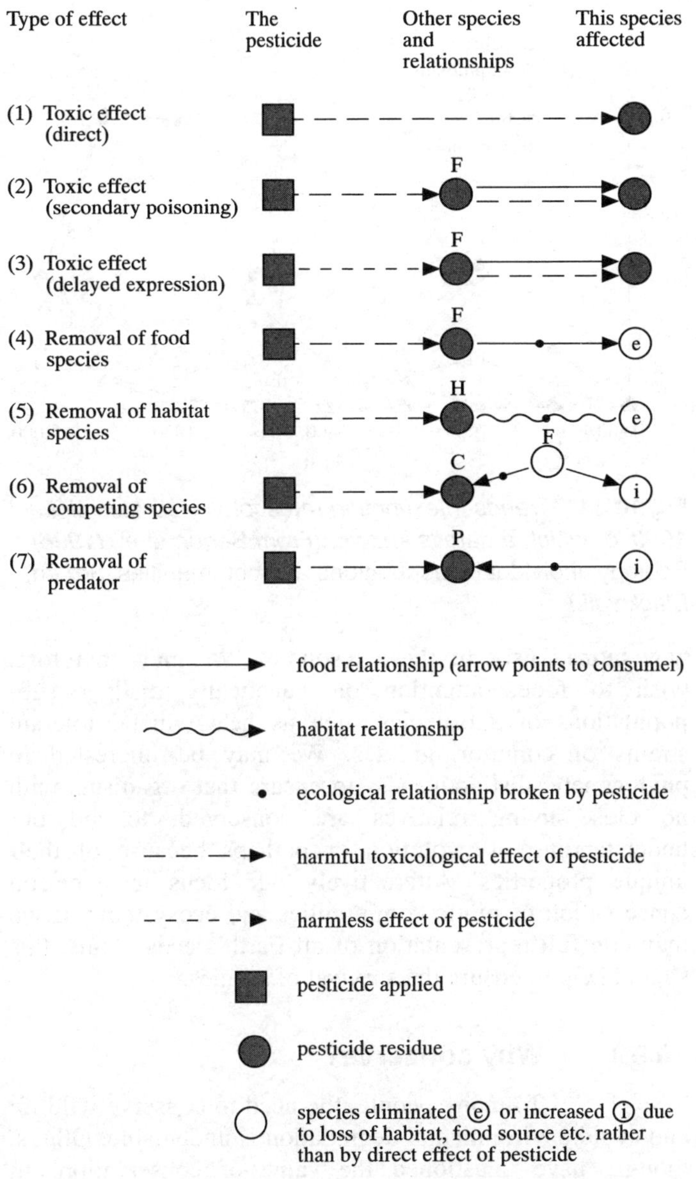

Fig 10.33 *The main ways in which pesticides affect ecosystems. C, competing species; F, food species; H, habitat species; P, predator species. (Modified from N.W. Moore (1967)* A synopsis of the pesticide problem, *Advances in Ecological Research, J.B. Cragg (ed.) pp. 75–126, Blackwell.)*

Success is not always so immediate. Careful matching of climatic conditions and monitoring of interactions with native species is essential. In attempts to control the walnut aphid, *Chromaphis juglandicola*, in California a parasitic wasp, *Trioxys pallidus*, was introduced from Cannes in France. It was moderately successful in coastal areas but in the hot interior, the main walnut-growing area, the wasp died out after one season. Ten years later after extensive searching in similar hot and dry environments a new strain of *Trioxys* was introduced from Iran. It successfully overwintered and within a year over 50 000 square miles were cleared of the pest and parasitisation exceeded 90%.

Biological control is widely used in commercial glasshouses which are enclosed environments. Soil sterilisation in winter kills all beneficial predatory controls. New plants introduced the following spring typically bring in some pests. With no natural enemies present, populations grow rapidly unless pesticides are used or natural enemy control is quickly re-established. Release of predators, once pest populations have reached a large enough number to provide sufficient food for the predators, can be very effective in keeping pest numbers to a low level where damage to the crop is minimum.

All biological control requires careful analysis of the ecosystem into which it is being introduced. In particular, where predatory control is planned it is essential to identify all the likely prey organisms. Unwished-for consequences may otherwise result. For example the mongoose which was introduced to Jamaica to control the black rat showed a preference for the existing natural enemies of the rat. It also severely reduced some bird populations and caused the extinction of several reptiles.

Integrated control of pests

Integrated control is pest population management which combines and integrates biological and chemical controls in a sensitive way. Pesticides are used as necessary in a manner least disruptive to complementary biological control. It aims to keep pest populations below the level of economic injury, or even prevent their development, while causing minimum harm and disruption to a crop (agroecosystem) or 'natural' ecosystem and especially the beneficial natural enemies of the crop or host species.

One example of this is to use more-selective chemicals, such as the aphicide pirimicarb. Its use has enabled development of an integrated control programme for peach–potato aphid, *Myzus persicae*, and for red spider mite, *Tetranychus* sp., on glasshouse chrysanthemums in Britain. Both were formerly controlled by organophosphate insecticides which are relatively unselective and to which the aphid was developing resistance.

For economic reasons (research and development costs versus the smaller market for selective chemicals) another approach that is used is the better targeted use of broad-spectrum chemicals. Targeting may be improved spatially and by better timing of pesticide applications. The development of pheromones (chapter 17) has greatly aided spatial targeting. Sex attractants can be used to lure insects to some other control measure, such as a pesticide or chemosterilant. Alternatively, a pheromone might be used to aggregate a population in an area to be treated with a pesticide. In this way much less pesticide is used, with minimal effects on non-target species and the wider environment. Another promising use of pheromones alone is to inhibit behavioural responses such as mating by saturating the atmosphere with the appropriate pheromone (very low concentrations will do this). Insects become habituated to the constant stimulus and the appropriate reaction is suppressed.

Frequency of pesticide application can be reduced by careful timing to cause maximum damage to the pest (such as at mating) while minimising effects on associated species. This requires close study of the life histories of all the species concerned and a clear understanding of the ecosystems involved.

10.9 Conservation

Conservation is about maintaining the biosphere. It means taking action to avoid species decline and extinction and permanent detrimental change to the environment. To achieve conservation we need to understand how the biosphere functions or, more immediately, how organisms and environment interact. In other words we must apply our knowledge of ecology and the related environmental sciences.

Successful conservation, however, is not just a matter of science. It also requires public and governmental support. For this reason conservationists often stress the utilitarian benefits of successful conservation.

Extinction

Extinction, as the fossil record shows, has always been part of the story of life on Earth. Since the emergence of humans, however, many new causes of extinction have arisen which are directly attributable to human use of biosphere resources. Extinction due to overhunting probably began this trend, whereas pollution of land, air and water (section 10.8), the widespread use of pesticides, habitat fragmentation and loss, and agricultural intensification are among the prime causes of species rarity and extinction during the last 400 years. As fig 10.34 shows there is considerable evidence that rates of extinction have accelerated since the agricultural and industrial revolutions.

Research shows that extinction rates for birds and mammals are presently between 10 and 100 times greater than the 'natural' background rate. Many more species are rare and their long-term survival is also in doubt. Rare species are typically **genetically eroded**, which means that their gene pool is reduced and with it their capacity to adapt to change in their environment. Chance adversity or environmental change can easily bring about the extinction of a rare species.

Biodiversity

The most obvious definition of **biodiversity** is the variety of species on Earth. Biodiversity can, however, be viewed at much smaller and larger scales. We may wish to conserve as much genetic diversity as possible within each species as insurance against future environmental change or new human uses of these resources. We may, therefore, wish to focus attention on genetically distinct sub-populations or subspecies such as heavy metal tolerant strains of common grasses. We may be interested in phylogenetic distinctiveness to ensure that organisms with no close living relatives are conserved, to aid our understanding of evolution or perhaps because of their unique properties. Alternatively our focus may be on conservation of biotic communities and ecosystems. If we maintain full representation of all Earth's ecosystems, this should help to ensure the survival of species.

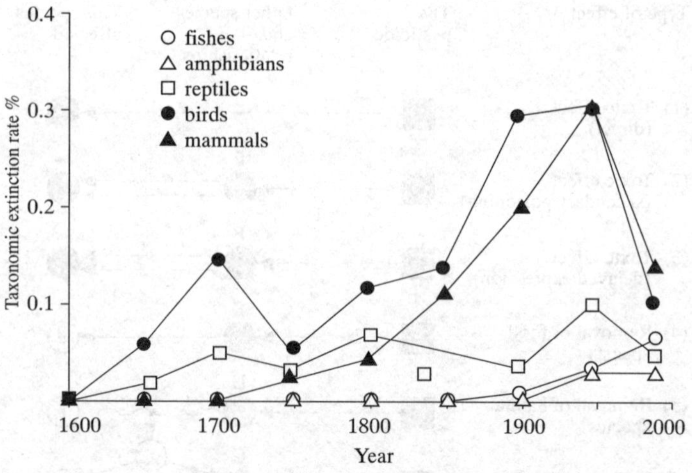

Fig 10.34 *Trends in extinction rates for vertebrates since 1600 for which a date is known. (From Begon* et al. *(1996)* Ecology: individuals, populations and communities, *3rd ed, Blackwell.)*

10.9.1 Why conserve?

To many people, the need to conserve wildlife and avoid environmental degradation is undeniable. Others, though, have questioned the value of conservation. In particular the justification for financial investment in species and countryside conservation has been challenged. Policy decisions are overwhelmingly matters of economics. Yet standard economics has failed to assign more than nominal value to ecological resources.

Developing a new ecological or 'green' economics is an important contemporary challenge. This may be relatively straightforward for species with obvious resource value such as crop plants or plants with known medicinal use. It is much more difficult for species with only indirect or potential value, for example as a tourist attraction or as part of a landscape in a National Park or as an essential but unattractive decomposer organism.

Most justifications for nature conservation centre on the human benefits that will accrue from maintaining a full range of biodiversity and avoiding environmental degradation. These can broadly be summarised as aesthetic, utilitarian, and ecological or scientific reasons for conservation. To these we must add one further category, ethical reasons for conservation.

Ethical reasons. Cultural tradition, religious beliefs, political persuasion and other similar concepts all shape our attitudes to nature. Some people argue very strongly that nature does not exist simply for humans to transform and modify as they please for their own purposes. They assert that all living species have a right to coexist with us on Earth, and we have no right to cause the extinction or to diminish the quality of life of any organism. Sometimes linked with this idea is the concept of **custodianship** which challenges us to pass on to future generations all the diversity of life and quality of environment that we, ourselves, inherited. Implicit in custodianship, however, is the notion of beneficial human inheritance, whereas the ethical argument accepts no justification of human benefit or need for guardianship.

Aesthetic reasons. Humans derive pleasure from natural environments and the presence of other living organisms. This can be hard to measure objectively but is undeniably true, as shown by the numerous local, national and international organisations that exist worldwide to promote wildlife and countryside conservation. Our appreciation of nature permeates art, design, literature and music and influences our recreational pursuits. Some people consider that contact with nature is essential for human well-being.

Utilitarian reasons. Wildlife contributes to our immediate needs in many fundamental ways. Obvious examples are agriculture and forestry and fisheries. As well as direct crop benefits, we make use of pollinating insects and beneficial predators in pest control. Many plant species have important medical uses; animal studies have enabled us to understand and treat human diseases more effectively. Numerous industrial processes depend on plant and animal materials and, increasingly, microorganisms are being used industrially, for example to concentrate valuable metals from low-grade metal ores.

Ecological or scientific reasons. It is not just the direct benefits of species and their genetic resources that are important to humans. Our well-being also depends on the maintenance of a fully functional biosphere. In particular maintenance of balanced biogeochemical cycles is vital to the avoidance of pollution, as discussed for acidification and eutrophication in sections 10.8.1 and 10.8.2, and to the regulation of Earth's climatic systems as discussed in section 10.8.1. As you will have read in the discussion of deforestation and desertification (section 10.8.3), loss of vegetation cover can have profound effects on soil erosion, lead to siltation of rivers and coasts and may even result in changes in rainfall and climate patterns.

10.9.2 Conservation of genetic diversity

Rare and endangered species

Human pressures on the biosphere have caused the extinction of many species. The woolly mammoth, dodo, great auk and Bengal tiger are all examples of species made extinct by humans. It is estimated that, at present, one species is lost every day.

If we are to prevent the continued rapid extinction of species due to human activity we must:

- identify the species at greatest risk;
- investigate why they have become vulnerable;
- attempt to remedy the problem.

The IUCN publishes detailed lists of species at risk of extinction in a series known as the **Red Data Books**. Four categories of risk are identified.

Rare	Species with small populations either restricted geographically with localised habitats or with widely scattered individuals. These species are at risk of becoming more rare, but they are not in immediate danger of extinction.
Vulnerable	These are species under threat of or actually declining in number, or species which have been seriously depleted in the past and have not yet recovered.
Endangered	Species with low population numbers that are in considerable danger of becoming extinct.
Extinct	Species which cannot be found in areas they recently inhabited nor in other likely habitats.

The Red Data Books for vertebrates list all known species in these four categories. This is impossible for plants since it is estimated that over 10% of all known plants, up to 60 000 species, are either rare or in danger of genetic erosion or extinction in the next 30 to 40 years.

We can help to prevent extinctions by:

- protecting and restoring habitats;
- establishing game parks, national parks, nature reserves and similar protected areas;
- controlling and reducing the impacts of modern intensive agriculture;
- reducing the use of bio-poisons such as pesticides;
- restricting trade in endangered species;
- providing refuges and assisted breeding programmes for endangered species, for example in zoos and botanic gardens;
- establishing sperm banks and seed stores to maintain the full range of genetic diversity of species.

The giant panda, *Ailuropoda melanoleuca*, symbol of the Worldwide Fund for Nature (WWF), is an example of an endangered species whose extinction has so far been prevented by a combination of habitat restoration measures and a captive breeding programme.

Giant pandas are found in eastern Tibet and southwest China. At one time the species was very much endangered since its habitat, of bamboo forest, was being encroached upon increasingly by the human population. Giant pandas feed almost exclusively on arrow bamboo, a plant that goes through cycles of abundance and scarcity. In Sichuan province in 1983 large areas of forest suddenly died back and at least 59 pandas died of starvation. Without human help this endangered species might have become extinct altogether.

By the early 1990s sufficient studies had been made of this species to begin to understand its physiology and ecology in the wild. It is now easy to see how ill-equipped giant pandas are to cope with change. Pandas live solitary lives and only come together for breeding. Adults spend 95% of their waking hours feeding on bamboo, which is a poor diet, but they will rarely eat anything else in preference. Breeding captive pandas has been successful at Wolong in Sichuan province. Use has been made of the latest techniques in reproduction technology, such as artificial insemination and sperm banking. Since interest in its survival has developed and forest reserves for it have been made, the giant panda population has increased to nearly one thousand.

Genetic resources for human use

Only 30 of the 250 000 known higher plant species currently account for 95% of human nutrition (fig 10.35). In developed countries it is common for just a few varieties of these species to be used. For example, half of all the wheat grown in the Canadian wheatlands is of one variety, 'Neepawa'. The danger of this is that if major physical or biotic environmental change occurs, the species or varieties in use will not continue to thrive. Problems of pest attack are an obvious hazard of monocultural practice. This was clearly shown by the catastrophic spread of potato blight and resulting crop failure that led to widespread famine in Ireland in the 1840s.

The solution to problems on this large scale is to find a wild relative that is naturally resistant to the pest, or better adapted to the new climatic situation. The desired genes from the new stock can then be transferred into the crop cultivar by a careful programme of crossbreeding. Similarly, disease resistance in stock animals can be improved by incorporating genes from wild relatives.

Such options clearly will not continue to be possible if human activity eliminates wild relatives of important crop and animal species. Conservation in seed stores, sperm banks, field gene banks and cryopreservation have obvious importance for domesticated plant and animal species. Maintaining genetic diversity in non-domesticated species is also important for humans, since some wild species have become important sources of drugs used in medicine. Aspirin was originally derived from the leaves of a species of willow, *Salix alba*, whilst the rosy periwinkle,

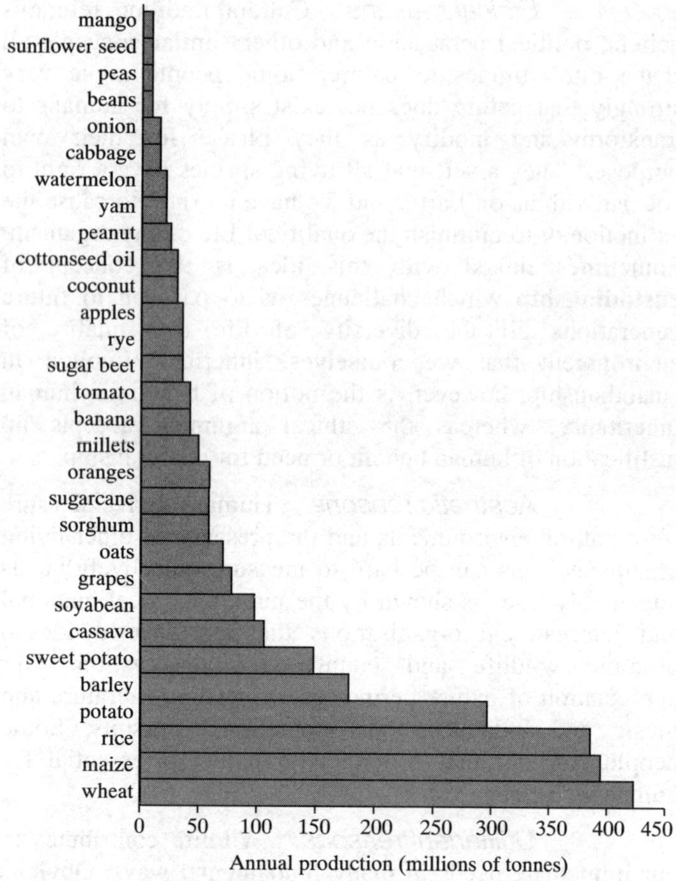

Fig 10.35 *Annual production of the major crop plants. (From O. Sattaur (1989)* The shrinking gene pool, New Scientist, *29.7.89, 37–41.)*

Catharanthus roseus, has yielded potent anti-cancer drugs for the treatment of Hodgkin's disease and some types of lymphatic cancer. Recently a variant of this plant was found in the West Indies that had the capacity to produce ten times the yield of this valuable anti-cancer drug.

In 1982 the world value of plant-based prescription drugs was estimated at $40 billion. These drugs came from just 41 species out of 5000 plants tested. Given a world total of 250 000 known plant species, it seems quite probable that many useful plant-based drugs have yet to be discovered. Realising that potential obviously depends on conserving the maximum plant genetic diversity possible.

Botanic gardens

Plants are best conserved **in situ**, that is where they occur naturally. In situ conservation can achieve maintenance of a large number of individuals for minimum management effort and cost. Species protected in their natural habitat can continue to evolve alongside their pollinators, symbionts, competitors and predators. This will ensure that they retain their 'fitness' and adaptiveness.

However, where habitats are particularly fragmented and vulnerable to exploitation in situ conservation may not be feasible. This is true for many tropical species where rates of habitat loss are very high. Ex situ conservation in botanic gardens and special tree collections, called arboreta, is a possible solution to this problem.

Worldwide there are about 1500 botanic gardens. Most are located in North America, Europe and the former Soviet Republics, whereas plant diversity is greatest in the tropics and subtropics. This poses obvious problems, such as differing day length and temperature regimes, if the aim is not merely to preserve examples of living specimens but to ensure reproduction and long-term survival of the species. These difficulties may be overcome by combining the cultivation expertise of botanic gardens with the more innovative methods of germ plasm conservation based on establishment of seed banks, field gene banks and cryopreservation (see below).

Seed banks

Seed banks are a convenient and space-saving method of conserving germ plasm. Originally focused on food crops and their wild relatives, they now include stores of many endangered wild species.

The seeds of many species can remain dormant for thousands of years if kept at low humidity (5–10%) and low temperature (–20 °C). Seeds that can be stored in this way are termed **orthodox** and include major crop plants such as cereals, soyabean, cotton and many vegetable species. One difficulty is ensuring that the seed remains viable. At Wakefield House, the major seed store for the Royal Botanic Gardens at Kew in Britain, all seed is X-rayed before storage to check an embryo is present. Subsamples are regularly tested for germination and if viability falls below acceptable levels, about 85%, new seed for storage is generated by growing the stored seed and harvesting the seed. This can prove a challenging and costly exercise. Even when successful, the enforced inbreeding may lead to a loss of vigour in the stored seed.

Some seeds, called **recalcitrant**, are damaged by drying and cannot be conserved for long periods in a seed bank. About 20% of all species, and 70% of tropical species, have recalcitrant seeds, including several commercially important species such as cocoa, rubber and tea. Other plants like potatoes, that mainly reproduce vegetatively, also require an alternative long-term conservation strategy to seed stores. For these species deep frozen germ plasm, rather than whole seed storage, and field gene banks are possible alternatives.

Cryopreservation

Cryopreservation involves storage of cells from embryos and shoot tips in liquid nitrogen at –196 °C. This stops all metabolic processes, which means that the material can be preserved indefinitely assuming no mechanical failure of the storage system occurs. The main biological drawback of this approach is that germ plasm is evolutionarily frozen. It is no longer subject to adaptation processes and may therefore prove biologically 'unfit' after prolonged storage so that it loses its value for reintroduction of species to the wild or for breeding into modern crop cultivars.

Field gene banks

Field gene banks are permanent living plant collections, not just of trees, as in arboreta, but also of savannah grasses, varieties of wheat, rice, cotton and others. A small plot or strip of each variety of a particular species is grown. For example the International Cocoa Genebank in Trinidad, which specialises in cocoa types from Latin America, grows 16 trees of each of 2500 types of *Theobroma cacao* that it holds. These plots give useful information on plant requirements, especially for germination and seed production, which is also vital companion information for seed banks. The main drawbacks are the space required and the inevitable susceptibility to pests and natural disasters such as fire and flood.

In 1989 a Botanic Gardens Conservation Strategy was drawn up by a group of major international agencies concerned with conservation. This aims to ensure integration of local, national and global conservation schemes and to standardise procedures for ex situ conservation, including the exchange and duplication of material and data between major collections. The Strategy also advises botanic gardens how they can stop illegal trade in endangered species.

Zoos

Zoos were originally established as collections of animals for curiosity's sake. Conservation, involving captive breeding, is a relatively recent aim. The aim of captive breeding programmes is to preserve the genetic stocks of threatened species so that they can be reintroduced into the wild when conditions permit. Therefore a complementary programme remedying conditions in the natural habitat that caused the species to become endangered is also needed.

Most zoos have only a few individuals of each species. To avoid inbreeding and weakening of stock, cooperation with other zoos is essential. The International Species Inventory System (ISIS) based at Minnesota Zoo in the USA coordinates advice on which individuals to exchange for breeding.

Transporting would-be suitors around the world is expensive and has no guarantee of successful breeding as was most famously demonstrated by giant pandas at London Zoo. Artificial insemination is one solution to this problem. In captive wild animals, sperm can be collected from anaesthetised males using a probe which electrically stimulates the genitals. This approach can also be used for sperm collection from animals living in the wild, so that the gene pool of captive breeding lines can be increased without further depleting wild stocks.

In practice many captive breeding programmes have proved so successful that, in zoos, overpopulation problems have resulted, especially where return to the wild has not been possible. Overpopulation occurs because life in captivity poses fewer threats, whereas mortality in the wild is high. Lions, for example, rarely live more than 7 years in the wild but, in zoos, a 20-year lifespan is common.

Reducing birth rates is rarely an appropriate solution. An ageing, less viable population results, and continuity of important learning patterns achieved by copying adult role models, such as mothering, is broken.

Reintroductions pose many problems and have achieved limited success so far, often because the problems that originally threatened the species still remain. The story of the Hawaiian goose illustrates this point well. In 1949 the wild population was reduced to 12 birds. Captive breeding programmes have since ensured its survival and to date over 3000 birds have been released back into the wild. Yet a viable wild population has not been re-established since the introduced predators, that caused the original problem, still remain.

Nevertheless reintroduction is an essential aim. Captive breeding without rapid reintroduction into the wild will inevitably lead to changed characteristics in the captive population, such as docility. Equally, learned behaviour such as finding food, establishing territories, wariness of enemies, especially humans, would be lost, reducing the prospects for long-term survival and successful reproduction of captive bred animals in the wild. As in the case of plants, successful animal conservation requires close collaboration between ex situ and in situ conservation efforts.

10.22 How can sperm banks contribute to the conservation of a species such as the African rhinoceros, populations of which are becoming increasingly isolated and unable to meet each other?

10.9.3 A case study in species conservation: the African elephant

The main herds of non-domesticated elephants are found in Africa (*Loxodonta africana africana* (bush elephant) and *L. africana cyclotis* (forest elephant)). They have no serious natural predators, yet their numbers are in decline, dropping from 1.2 million in 1981 to 623 000 in 1989. Humans are the main threat to the elephants' survival. They compete with the elephants for land for forestry, agriculture and settlement, destroying their habitat. They kill elephants which threaten crops or property and, above all, kill elephants for their ivory.

It is not just the absolute decline in elephant numbers that is causing concern. Changing population structure caused by illegal poaching is also a problem. Poachers select the animals with the largest tusks. Males, which have a typical tusk weight of 9.3 kg compared with 4.7 kg for cows and mature calves, have been preferentially eliminated. As a result, by 1987 cows and calves were the main source of poached ivory in parts of East Africa. One tonne of ivory represented 113 dead elephants compared with 54 in 1979 and the mainly female harvest was estimated to cause the deaths of an additional 55 calves orphaned too young to survive. Surveys in Ambroseli National Park in Kenya in 1988 showed that just 22% of the population was male; data from Mikimu Reserve in Tanzania showed only 0.4% of the population was male. These data suggest that the long-term prospects of elephant populations seem very poor. Some researchers have predicted their extinction by the year 2010 unless effective conservation action is taken.

Elephants are an important **keystone species** (one on which many others depend) in the ecology of the African savannah. A medium-density herd will generate a wide diversity of habitats and promote a varied community of associated grazing and browsing game animals (table 10.8). In Kenya the revenue from tourism related to game viewing is ten times greater than the estimated income from sales of poached ivory and benefits many more people. Ensuring the elephants' survival has clear utilitarian benefits, as well as other conservation benefits.

Since 1989 the African elephant has been protected by a total ban on ivory sales under international legislation known as CITES (Convention on International Trade in Endangered Species). Not all African states support this ban. Southern African countries of Zimbabwe, Botswana, Malawi, Zambia and South Africa objected, since they already have successful sustainable management

Table 10.8 Schematic summary of the relationship between changing elephant population density and associated plant and animal communities.

Density of elephant population	*low*	*medium*	*high*
Plants	mainly woody plants low species diversity	mixed trees and grasses high species diversity and good abundance of each species	mainly open grassland a few species dominant
Animals	browsers, e.g. giraffe, impala, Grant's gazelle	mixed browsers and grazers	grazers, e.g. zebra, wildebeest, Thompson's gazelle, buffalo
Level of protection	outside Game/ National Park	Park boundary area	central Park area

programmes. Elephant populations in these countries have a good age and sex structure and are stable, or in some cases rising and requiring culling to maintain rangeland. Sustainably managed herds generate income from sale of ivory, meat and hide, and at the same time encourage tourism. Elephants are regarded as a valuable resource, generating jobs and income which is used for development projects. Local support for maintenance of elephant herds is therefore strong and poaching is considered antisocial.

However, some people consider that a total ban on ivory sales is the only way to prevent extinction. They argue that we must educate people worldwide so that they no longer want carved ivory products and compare the ivory trade with the fur trade: it is no longer acceptable to wear fur coats made from the skins of exotic and rare species. Public disapproval has killed demand for fur products and helped to save many endangered species.

The debate continues: while ivory remains available from sustainably managed herds it is difficult to enforce a ban elsewhere. The pro-ivory lobby point out that the black rhino has continued to decline in spite of maximum protection under CITES. The anti-ivory group argue that illegally obtained ivory is marketable only because the legal trade continues. They claim that as herds elsewhere diminish, illegal poaching of sustainably managed herds will inevitably take place.

10.9.4 Planning for the future

Knowledge from ecological studies of how we are changing our environment is forcing us to look critically at how we use the Earth's resources. We need to reduce pollution and conserve resources while achieving a good standard of living for all people. The Earth Summit held at Rio de Janeiro in 1992 was an important milestone in raising political and public awareness of environmental concerns and the need to promote **sustainable development**. This has been defined as 'meeting the needs of the present without compromising the ability of future generations to meet their own needs'.

The Earth Summit identified 27 principles for environmental and social development into the twenty-first century. Together these initiatives are known as **Agenda 21**. What is important is how the general public, business community, industry and politicians, respond to Agenda 21. One target set in Rio acknowledged the importance of local community participation. Local communities in each participating country were charged with consulting the people they represent to formulate a **Local Agenda 21** by 1996.

10.23 Has your local authority identified a Local Agenda 21?

What local conservation measures are proposed?

You may expect to find new schemes for collection, separation and recycling of waste from households and local businesses.

How are local environment action groups contributing to Local Agenda 21.

What is happening in your school or college?

Is your home a 'green' household?

Some resources, such as food crops, are easily renewed in short time scales. Regular, typically annual, harvests are produced. We call these **renewable resources**. Providing we do not catastrophically damage our environment, as in desertification, these resources are potentially always available. Other resources, notably mineral resources and fossil fuels, were formed thousands or millions of years ago. They cannot be replenished on human time-scales. These resources are called **non-renewable**. For example, it is estimated that in 70 years almost all the known oil deposits that are recoverable with current technology will be used up unless we reduce our rates of consumption. Similarly, natural gas reserves will be exhausted in 150 years unless we change our pattern of use.

10.9.5 Sustainable use of plant and animal resources

The discussions of deforestation (section 10.8.3) and African elephants (section 10.9.3) show how valuable resources can be diminished or destroyed by over-exploitation due to ignorance or greed. In contrast, management based on ecological knowledge brings continued benefits, though short-term profits may be less. The attempt to regulate sea fishing using an agreed international quota system is one example of this. Long-lived species and ecosystems such as elephants and forests also require sensitive management if future stocks are to be maintained. Productive and sustainable management of woodland ecosystems is a long-established concern of human populations.

Overfishing

A state of chronic overfishing now exists in the world's oceans. Fig 10.36 shows landings of sea fish since 1950. Today nine of the world's 17 major fishing grounds show a steep decline in size of catch and four are fished out. In 1992 the cod fisheries off Newfoundland in Canada were closed indefinitely. In the North Sea mackerel have declined by more than 50% since the 1960s, while the herring industry which was closed between 1977 and 1982 has never recovered its former abundance. The world's largest fishery, anchovies off the coasts of Peru and Chile, has collapsed completely.

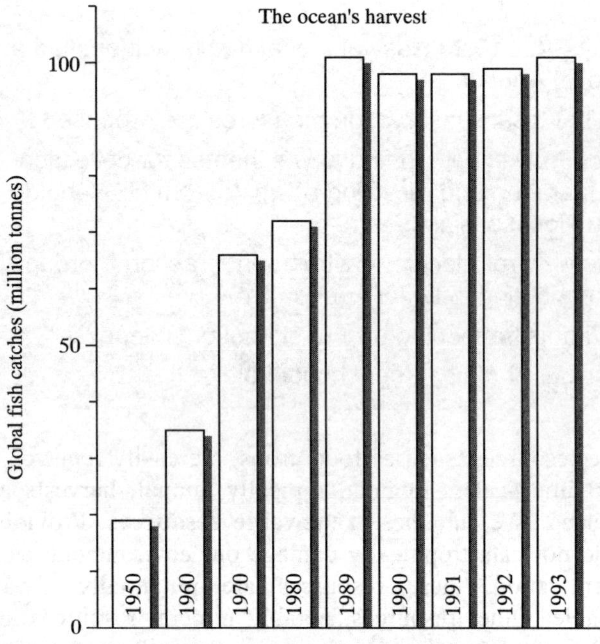

Fig 10.36 *Global fish catches 1950–93. (From F. Pearce* New Scientist, 2016, *10th February 1996 p. 4.)*

Overfishing is the root cause of these problems. Signs of overfishing were clearly evident in the 1960s, yet during the subsequent decades the recorded global marine fish catch rose by nearly 50%. These increased landings reflected the development of large 'factory' fishing fleets (that is ships equipped not only to freeze but to completely process the catch at sea) and the targeting of one fish species after another. In other words, as one fishery or fish species became exhausted another species would be substituted.

Heavy fishing produces a population of mainly young small individuals because fish are caught as soon as they reach a catchable size. These young fish would make rapid growth if left longer in the sea and would soon reach a more valuable size, giving heavier landings of better quality fish per unit fishing effort, and yielding a better profit margin. However, individual fishermen are tempted to achieve weight quotas and maintain income by catching more rather than fewer fish. If this reaches the point where fish are caught before spawning, so that the reproductive capacity of the stock is seriously impaired, then a catastrophic reduction in numbers occurs and local extinction may result.

Fisheries quotas are intended to safeguard fish stocks. However, although scientists use quite sophisticated population ecology models on which to base their recommendations, there are many problems with the quota system. These include:

- acquiring reliable data with which to model the fish populations;
- allowing for climate change;
- persuading governments to accept scientific recommendations;
- policing international agreements;
- accommodating the needs of traditional local fishing communities as well as major commercial fleets.

Tropical rainforests

Some ecosystems are very difficult to manage in a way that provides sustainable resources. In parts of Amazonia agricultural success following rainforest clearance has been very transient. Even schemes with substantial external financial support and agricultural expertise have been abandoned after 10 or 15 years. Weed invasion, pest and disease problems and, above all, declining soil fertility make projects no longer economically viable. Mature rainforest rarely re-establishes on these sites. In the intervening period soil nutrients are lost and soil structure is impaired; no forest seed store remains and, even when dispersal brings seeds of mature rainforest species, germination is poor in the exposed conditions. (Rainforest species require the high humidity found beneath a mature canopy for successful germination.) At best, a secondary forest of lower productivity than the original rainforest is established, in which only very low intensity shifting cultivation can be practised. Intensive agriculture is clearly not a sustainable use of these areas. It wastes resources since the secondary forest is less productive than the original. At the same time many mature forest species are permanently lost. Sustainable use of tropical moist forests which also allows economic and social development of the region concerned is a major contemporary challenge for the world community.

Traditional woodland management systems in Britain

For centuries Britain's woodlands were an important source of fuel and timber resources. Management practices evolved to ensure their continued regeneration and productivity. The broadleaved woodland remnants we seek to conserve today reflect this management history, and are in most cases very different from the original forest cover of Britain. These woodlands include many species in semi-natural communities that will die out if we do not continue to manage them in a traditional manner.

There are very few woodlands that are primary forest remnants and even these have been modified by human activities. Most broadleaved woodlands are secondary, having developed from formerly cleared sites. 'Ancient woodlands' are those that date from before 1600 AD and have the highest conservation value. They often contain great species diversity, particularly of slow-growing, long-lived organisms such as lichens, and they bear the marks of human management back to very early times.

Most woodlands were managed in one of two traditional systems as wood pastures or coppice. **Wood pastures** were common on poorer soils and livestock grazing, as well as timber produce, was important in these woodlands. Grazing pressure restricted regeneration, a problem overcome by **pollarding**, the cutting of the main stem above the browse line (c. 2 m) to encourage new growth of lateral branches which could be periodically harvested. Over centuries the more palatable species, such as hazel, have been lost and browse-tolerant species, such as holly and hawthorn, have become dominant in the understorey. The future of many of these woodlands depends on regeneration by the control of grazing pressure.

On richer soils the **coppice system** predominated. Understorey species, such as ash and hazel, were cut close to ground level on a 5–20 year cycle depending on the rate of regrowth and the products required. The larger coppice poles were used to provide materials for house building, furniture and fencing; lighter materials were used for basket weaving, sheep bundles or wattle (to hold wall plaster) or were used in thatching. Woods provided much fuel and very little was wasted. Occasional **standard** trees were allowed to grow to near full maturity. These were felled on a longer time scale and sawn into planks or fashioned into beams. This management pattern of coppice with standards resulted in periodic expanses of flowers on the woodland floor. Such woods have a species-rich ground flora, with many rare and historically interesting plants.

Conserving ancient woodland helps to retain the rich plant and animal communities of the primary forests, many members of which are now rare.

Fig 10.37 *Wood pasture in the New Forest.*

Hedgerows

Many people consider that hedgerows have high value to wildlife conservation. Hedges were used in Saxon times to divide and mark out land plots and to contain livestock. Over the centuries hedgerows have increased their species richness. On average, one new species of woody shrub is added to every 27 m length of hedgerow every hundred years. Ancient hedgerows have therefore the greatest diversity of wildlife and some are the only places in an arable area where plants typical of ancient woodland are found.

Between 1950 and 1980 many thousands of kilometres of hedgerow were lost in lowland Britain to make fields larger for arable farming and to make ploughing and harvesting by large machines easier. This also removed what farmers perceived as a reservoir of weeds, pests and diseases. Hedgerows, however, undoubtedly conserve many species of animals which act as predators of many plant pests. They reduce wind speeds, providing important areas of livestock shelter in their lee and helping to reduce soil erosion. Hedgerows also provide a refuge and home for song birds, game birds and other woodland species and give shelter to beneficial pollinating insects. They can also act as 'corridors' between isolated broadleaved woodlands in agricultural areas. Small mammal species, such as the dormouse, which do not readily colonise new areas may benefit from the habitat continuity that hedgerows can provide. Plants with only localised dispersal mechanisms may also benefit from these hedgerow 'corridors'.

In some parts of Britain farmers can now get grants to re-establish hedgerows. However, a newly planted hedge, although quickly providing the benefits of shelter, will be much less diverse in species than the long-established hedgerows removed during the 1960s.

10.9.6 Waste management and recycling

Human societies generate large amounts of waste. Many wastes contain valuable materials that can be reclaimed and re-used, reducing the need to exploit new supplies of a resource. Reclamation and recycling of scrap metals rather than mining new materials is an obvious example. The energy requirements and pollution associated with recycling are often much less than those caused by mining and ore processing methods. For example, in Europe the steel industry re-uses scrap metal, so achieving energy savings up to 50%; aluminium recycling achieves 95% in energy savings. Also, many businesses are reducing packaging on products in an attempt to save energy costs in production and reduce waste and pollution.

Recycling also beneficially reduces the total volume of waste requiring disposal. In Britain most household and municipal waste is sent to landfill sites. These are rapidly becoming filled, especially in the densely populated south-east. In 1996, as part of its response to Agenda 21, the UK government introduced a new landfill tax designed to encourage business and consumers to dispose of less waste and to recover more of the waste which is produced using recycling and waste-to-energy incineration schemes (where the energy generated is used for other purposes).

At present in Britain between 1% and 8% of municipal waste is recycled, whereas the government target for recycled household materials is 25% by the year 2000.

The difficulties of recycling include:

- the need for effective separation of wastes. For example, one piece of Pyrex in a consignment of glass can render the recycled product worthless.
- the fluctuating price received for materials saved for recycling. For example, the price for waste paper in 1996 was only $\frac{1}{3}$ the price paid in 1995.
- the need for easily accessible recycling points for householders. Most waste collection authorities rely on the 'bring system', where the householder takes recyclables to a collection point which restricts those who can participate. More recently authorities have experimented with kerbside collection schemes for different wastes in an attempt to increase householder participation rates.

Much municipal waste is biodegradable. During this process methane, a highly flammable and important greenhouse gas, is produced. Using carefully designed recovery systems the methane can be collected and used for generating electricity. A landfill site at Mucking in Essex which serves several London boroughs generates 3.8 MW of electricity, sufficient to run the plant and supply a small community of 30 000 people. At the same time the potential fire hazard associated with uncontrolled methane release is reduced, and the carbon dioxide released as waste is a far less active greenhouse gas than the original methane.

In the future waste-to-energy incineration plants are likely to become the main method of waste disposal. Such energy production from waste beneficially reduces demands on fossil fuel reserves as well as reducing demand for landfill.

Sewage processing uses decay microorganisms, normally present in the soil and fresh water, to break down human wastes. If these organic rich wastes were discharged to the receiving waters without pre-treatment their inevitable subsequent decomposition would lead to rapid and extreme deoxygenation in the receiving water. Sewage works employ the natural decomposition processes of the microorganisms and produce, as byproducts, nutrient-rich dried sludge and methane. Providing it has not become contaminated with heavy metals (such as by lead from petroleum wastes) the sludge may be used on the land as fertiliser. Sewage sludge is also often used in reclamation schemes since it improves 'soil' structure as well as providing vital missing nutrients (section 10.5.2). The methane (biogas) that is also produced during decomposition can be burnt to provide the heating and electricity for the sewage processing works and, in some cases, small local communities or businesses.

New sources of energy

In Britain most of our energy is provided from the burning of fossil fuels which are non-renewable. About 10% of our electrical energy is generated from nuclear power, which apart from its inherent dangers and the problems of nuclear waste disposal and decommissioning of obsolete nuclear plants, is also dependent on a non-renewable resource, uranium. An awareness of the rate of depletion of fossil fuels and pollution caused by their use, coupled with unease about nuclear power, has prompted many people to think creatively about alternative sources of energy.

Globally, biomass energy (wood, charcoal, crop residues, dung and other organic materials) is the most important fuel energy source aside from fossil fuel. This source supplies 14% of the world's energy needs and is the main source of energy in the developing world, supplying 35% of needs. Biotechnology can also be used to produce alternative sources of energy based on naturally produced organic materials. For example, Brazil has large industrial plants for converting sugar, from cane, into ethanol (gasohol) which can be used to fuel cars. In North America cereals such as maize, wheat and barley are fermented to produce a variety of products including fuel alcohol. Sugar beet widely grown in northern Europe could similarly be processed for fuel alcohol. Fuel alcohol is potentially a cleaner source of energy than fossil fuels since it is sulphur-free and does not require a lead additive.

New and renewable sources of energy come in many forms. Solar, wind, hydro, tides and waves, biomass technologies, geothermal sources and others present an almost limitless potential if we can find suitable ways to harness them.

Hydro power is the most widely used form of renewable energy. It currently produces energy equivalent to 500 000 MW worldwide, supplying about 23% of the world's electricity. Most remaining potential hydro resources are concentrated in developing countries. Massive schemes, such as the Itaipu dam in Brazil, produce cheap electricity but distribution systems are often poor, limiting the benefits of the schemes. High environmental and social costs typically include:

- displacement of people as environmental refugees;
- flooding of valuable farmland and forests;
- increased incidence of water-borne diseases such as bilharzia;
- disrupted river flow;
- increased soil erosion.

Silting and acidification problems often limit the productive life of these schemes, sometimes to as little as 30 years.

Most renewable energy resources, however, have a less damaging environmental impact than fossil fuels and nuclear power, though many are not suitable for large-scale centralised power generation. The resources themselves, such as waves, wind and sunshine, are too dispersed to make it easy to generate large amounts of power. They are also often unreliable, for example because of cloudiness or lack of wind, so for essential energy supplies a back-up capacity is needed.

Renewable energy resources can, however, play a useful part in an integrated power generation system for greater energy needs. For example, wind-generated electricity already makes a small contribution to Britain's energy needs and is widely used in other countries. In California, the use of solar energy for space and water heating has become a regular feature of modern house design. It has been also proposed that solar energy should provide the estimated 10 MW of power needed to supply the Olympic Games in Sydney, Australia in the year 2000. Tidal power has been successfully harnessed at La Rance, France. However, though several designs have been proposed for using wave power to generate electricity none, as yet, has proved commercially successful.

Renewable energy sources are ideally suited to supplying small-scale dispersed power needs, such as in remote island communities, mountain regions and semi-arid areas. They are particularly useful in developing countries where sunshine is plentiful, providing energy for local industries, schools, health clinics, water pumping stations and many other small-scale projects.

10.9.7 Conservation agencies in Britain

There are many individuals, groups and organisations concerned with conservation. Some operate locally, others have national or even international agendas. All contribute in different ways to changing the way we think about and act towards the environment.

Non-governmental organisations (NGOs)

In Britain membership of environmental organisations such as the National Trust (NT), the Royal Society for the Protection of Birds (RSPB) and the various Wildlife Trusts has doubled since 1980. Every county and major urban area in Britain has a Wildlife Trust which owns and manages small nature reserves. These Trusts, which are largely staffed by volunteers, provide protection for locally important conservation sites and involve many of their members in wildlife recording and fund-raising. The 48 County Wildlife Trusts collectively own or manage over 2000 nature reserves. They are linked with the more recently formed 50 Urban Wildlife Groups in a Wildlife Trusts Partnership under the umbrella of the Royal Society for Nature Conservation (RSNC). Their junior club 'WATCH' has promoted many nationwide educational schemes typically linked with industrial sponsors. Recent examples include an ozone monitoring project sponsored by Volvo and a three-year Riverwatch project sponsored by National Power. The RSPB owns 126 generally quite large reserves. Though primarily aimed at conservation of birds, these sites also importantly conserve a wide variety of other wildlife as well. The RSPB junior membership is the Young Ornithologist's Club.

Other conservation charities, such as the NT, are not solely concerned with wildlife and landscape conservation. Nevertheless the NT owns 340 **Sites of Special Scientific Interest** (**SSSI**) and is an important coastal and countryside land owner. Groups, such as the Wildfowl Trust (ducks, geese and swans), the Rare Breeds Trust (conserving domesticated animal varieties) and the Woodland Trust, are more specific in their concerns. Organisations such as Friends of the Earth (FoE) and the Councils for the Protection of Rural England, Scotland and Wales (CPRE, CPRS, CPRW) are concerned with political lobbying and direct action at a wider level. Greenpeace is an important international agency renowned for its controversial, sometimes confrontational but often very effective, direct action. The Flora and Fauna Preservation Society is an international voluntary agency based in Britain which specialises in work promoting species conservation and protection. Much more widely known and also based in Britain is WWF, the Worldwide Fund for Nature. WWF raises large sums of money in support of species and habitat conservation and is particularly active on the international scene. It also sponsors a wide range of environmental education activities.

Statutory Conservation Organisations

Although the voluntary movement is important in Britain, there are also many statutory (government and state-funded) organisations concerned with conservation and the environment. The main nature and countryside conservation agencies were established immediately following the Second World War. These were the Nature Conservancy and the National Parks Commission. Both organisations have subsequently experienced several reorganisations in an attempt to clarify their role and widen their achievements. The Nature Conservancy (since 1973 the Nature Conservancy Council (NCC)) was responsible for establishing National Nature Reserves (NNRs) intended to conserve the best examples of Britain's wildlife and habitats. Over 200 had been designated in mainland Britain by 1990. The Conservancy also identified SSSIs. Today these form an important countrywide series of about 6000 protected locations.

The National Parks Commission was charged with setting up Britain's National Parks. It had limited success since it had no land-owning powers. The Commission's role was to persuade, cajole and advise local authorities and county councils to act. Ten parks were designated in the 1950s in England and Wales, mainly in upland and coastal locations. The Norfolk and Suffolk Broads finally achieved National Park status in 1988.

In 1968 the Countryside Commission (CC) replaced the National Parks Commission. The Countryside Commission deals with the wider countryside as well as the National Parks. It is responsible for designating Areas of Outstanding Natural Beauty (AONBs), the definition of Heritage Coasts and the establishment of long-distance footpaths. It advises local and regional planning authorities on countryside matters, taking a particular interest in urban fringe areas. It has encouraged the setting up of Country Parks, whose recreational and educational emphasis has relieved pressure on the more-sensitive and scientifically important sites managed by the former NCC (see below).

Since 1987 government funds have also been available for **Environmentally Sensitive Areas** (ESAs) giving support to farmers to maintain traditional agricultural practices in countryside that might otherwise be spoilt by modern farming methods. The South Downs, which were rejected as a National Park in the 1950s, is included in this scheme. Since 1990, to reduce the levels of overproduction of food in Europe, farmers have been required to 'set-aside' land from agricultural production. Properly managed, this scheme offers another new opportunity for diversifying the countryside and promoting wildlife conservation.

Other significant statutory organisations are the Forestry Commission, which was established in 1919, and the National Rivers Authority (NRA), which was set up in 1989 and, since April 1996 has been linked with the former Pollution Inspectorate in the new Environment Agency. The Forestry Commission is the statutory body responsible for timber production. It is required to take account of conservation interests on its estates and co-operates with both statutory and voluntary nature conservation organisations. Although much criticised in the past for extensive use of regimented plantings of non-native conifers, the Forestry Commission today takes account of scenic, wildlife and recreational issues. Native hardwoods are now also used on its estates which have been important in re-establishing woodland habitats in Britain. The National Rivers Authority (now the Environment Agency) has responsibility, in England and Wales, for water quality and conservation where this is related to matters of water abstraction, flood control, water quality and similar issues.

In 1990 the British government passed major new environmental legislation. This brought pollution control, environmental assessment and wildlife and countryside conservation into one legislative framework. It included major reorganisation of the Britain's nature conservation agencies to form separate, national organisations for England, Scotland and Wales. In England, English Nature replaced the NCC but remained separate from the Countryside Commission. In Wales and Scotland, however, the national branches of the NCC were amalgamated with the Welsh and Scottish Countryside Commissions to give single unified wildlife and countryside conservation agencies. These are the Countryside Council for Wales and Scottish National Heritage. A new umbrella agency, the Joint Nature Conservation Committee (JNCC) which includes representatives from all these national groups, was formed to deal with issues such as management of major estuaries like the Severn and Solway Firth which require co-operation between the national bodies.

European Environmental initiatives are increasingly influencing nature conservation in Britain. An important current development is the European Habitats Directive. The aim of this directive is to promote the maintenance of biodiversity. Each member state in the European Union has to identify Sites of Community Interest. A final list of key European nature conservation sites will then be agreed in consultation with the European Union. Termed Special Areas for Conservation (SACs), these sites will be strongly protected under European law. The year 2004 is the target date for full establishment of this scheme which will be known as **Natura 2000**.

Chapter Eleven

Quantitative ecology

The principles of ecology, as outlined in the previous chapter, are based on qualitative and quantitative data obtained from studies carried out on animals, plants, microorganisms and the abiotic environment. This chapter deals with both qualitative and quantitative aspects of ecological investigation, and presents a general introduction to some of the methods and techniques of obtaining, presenting and analysing data relating to the abiotic and biotic environments.

Before attempting any ecological investigation it is essential to identify the exact aims and objectives of the study and the degree of accuracy required. These, in turn will clarify the methods and techniques to be employed and will ensure that the data collected are relevant to the study and are adequate to form a basis for valid conclusions. In many cases, this simplifies the methods and techniques and reduces the time, money, resources and effort needed for the study. However, it must be stressed that investigations frequently have to be modified in the light of problems encountered during the investigation.

11.1 Methods of measuring environmental factors

The main environmental factors which must be studied in order to complement biotic analyses are soil factors, water, topography and climatic factors, such as humidity, temperature, light and wind. Many of the methods used to measure environmental factors are included below in experiments. Other methods of quantitative study are described in outline only.

11.1.1 Soil factors

Soils vary considerably in structure and chemical composition. In order to obtain a basic idea of the structure or profile of the soil, a pit is dug so that a clean-cut vertical section of the soil can be seen. The various thicknesses of clearly differentiated bands (horizons), shown in terms of colour and texture, can be measured directly, and samples removed from these horizons and used for the various analyses described below.

Alternatively a soil auger, which is an elongated cork-screw implement, is screwed into the ground to the desired depth and then removed (fig 11.1). Soil trapped in the threads of the screw at various levels is removed into separate bags for subsequent analysis. When using this method of obtaining a soil sample, it is important to keep a record of the level each part of the sample occupied in the ground. This information should be recorded on the relevant bag.

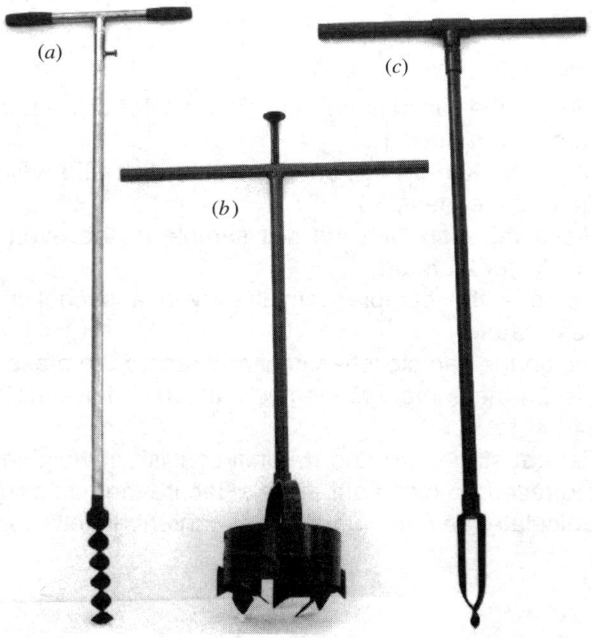

Fig 11.1 *(a) Simple screw auger, (b) post-hole auger and (c) 'Dutch' auger.*

Notes

(*a*) Soil samples should not be stored in polythene bags for long periods before analysis because they tend to 'sweat'. The humidity and temperature changes will affect microorganisms present and will therefore affect the pH of the sample and the form in which nutrient elements are held in the soil.

(*b*) Screw auger sampling is most useful as an indicator of soil profile and hence soil type. Though sufficient material can be taken this way for pH testing and field estimation of soil texture, samples from a soil pit, or taken with a Dutch or post-hole borer (fig 11.1), are preferable to screw auger samples for accurate tests of water content, organic matter content, soil organisms and so on. It is particularly difficult to take reliable uncontaminated samples in sufficient bulk from coarse-textured soils using a screw auger.

Experiment 11.1: To investigate the water content of a soil sample

Materials

about 80 g soil
aluminium foil pie dish
balance accurate to 0.1 g
thermostatically controlled oven
thermometer reading up to 150 °C
desiccator
tongs

Method

(1) Weigh the aluminium foil dish while still empty. Record the mass (*a*).
(2) Add a broken-up soil sample to the dish and weigh. Record the mass (*b*).
(3) Place the dish with the soil sample in the oven at 110 °C for 24 hours.
(4) Remove the sample from the oven and cool in a desiccator.
(5) Weigh the sample when cool and record the mass.
(6) Return the sample to the oven at 110 °C for a further 24 hours.
(7) Repeat stages (4) and (5) until consistent weighings are recorded (constant mass). Record the mass (*c*).
(8) Calculate the percentage water content as follows:

$$\frac{b-c}{b-a} \times 100\%$$

(9) Retain the soil sample in the desiccator for experiment 11.2.

Notes

The value obtained in the experiment is the percentage total water present. This amount will depend upon recent rainfall. Alternative estimates of water content include field capacity and available water. The **field capacity** is the amount of water retained in the soil after excess water has drained off under the influence of gravity. To obtain this value the soil in the field should be flooded until surface water persists for several minutes, 48 hours before the sample is removed for investigation. The **available water** is the water which is available to be taken up by plants, and may be estimated by drying the weighed sample to constant mass at room temperature. The difference between wet mass and dry mass is the amount of available water present.

Experiment 11.2: To investigate the organic (humus) content of a soil sample

Materials

dried soil sample from experiment 11.1 in desiccator
crucible and lid
tripod, bunsen burner, heat proof mat, fireclay triangle
desiccator
tongs

Method

(1) Heat the crucible and lid strongly in the bunsen flame to remove all traces of moisture. Place in the desiccator to cool. Weigh and record the mass (*a*).
(2) Add the dried soil sample (kept from the previous experiment) from the desiccator and weigh. Record the mass (*b*).
(3) Heat the soil sample in the crucible, covered with the lid, to red-heat for 1 h to burn off all the organic matter. Allow to cool for 10 min and remove to the desiccator.
(4) Weigh the crucible and sample when cool.
(5) Repeat (3) and (4) until constant mass is recorded. Record the mass (*c*).
(6) Calculate the percentage organic content as follows:

$$\frac{b-c}{b-a} \times 100\%$$

(7) Repeat the experiment on soil samples taken from different areas to demonstrate variations in organic content.

Note

The percentage organic content obtained in this experiment is relative to dried soil and not to fresh (wet) soil. The organic content of a soil may be quoted as a percentage of fresh (wet) soil using the data obtained in experiment 11.1.

> **11.1** 60 g of a fresh sample of soil produced the following data on analysis. After repeatedly heating at 110 °C and cooling in a desiccator, consistent readings of dry mass of 45 g were obtained. The dry soil was heated repeatedly to red-heat in a crucible, cooled in a desiccator and weighed. The mass was now found to be 30 g. Calculate the water content and organic content of the fresh soil sample.

Experiment 11.3: To investigate the air content of a soil sample

Materials

tin can of volume about 200 cm^3
500 cm^3 beaker
water
100 cm^3 measuring cylinder
chinagraph pencil
drill
metal seeker

Method

(1) Place the empty can, open end uppermost, into the $500\,cm^3$ beaker and fill the beaker with water above the level of the can. Mark the water level in the beaker.
(2) Carefully remove the can containing the water and measure this volume of water in a measuring cylinder. Record the volume (*a*). The water level in the beaker will fall by an amount corresponding to the volume of water in the can.
(3) Perforate the base of the can using a drill, making about eight small holes.
(4) Push the open end of the can into soil from which the surface vegetation has been removed until soil begins to come through the perforations. Gently dig out the can, turn it over and remove soil from the surface until it is level with the top of can.
(5) Place the can of soil, with open end uppermost, gently back into the beaker of water and loosen soil in the can with seeker to allow air to escape.
(6) The water level in the beaker will be lower than the original level because water will be used to replace the air which was present in the soil.
(7) Add water to the beaker from a full $100\,cm^3$ measuring cylinder until the original level is restored. Record volume of water added (*b*).
(8) The percentage air content of the soil sample can be determined as follows:

$$\frac{b}{a} \times 100\%.$$

(9) Repeat the experiment on soil samples from different areas.

Experiment 11.4: To investigate the approximate relative proportions of solid particles (soil texture) in a soil sample

Materials

$500\,cm^3$ measuring cylinder
$100\,cm^3$ soil sample
$300\,cm^3$ water

Method

(1) Add the soil sample to the measuring cylinder and cover with water.
(2) Shake the contents vigorously.
(3) Allow the mixture to settle out, according to density and surface area of particles, for 48 hours.
(4) Measure the volume of the various fractions of soil sample.

Results

A gradation of soil components is seen. Organic matter floats at the surface of the water, some clay particles remain in suspension, larger clay particles settle out as a layer on top of sand and stones which are layered according to their sizes.

Experiment 11.5: To investigate the pH of a soil sample

Materials

long test-tube (145 mm) and bung
test-tube rack
barium sulphate
BDH universal indicator solution and colour chart
soil sample
spatula
distilled water
$10\,cm^3$ pipette

Method

(1) Add about 1 cm of soil to the test-tube and 1 cm of barium sulphate, which ensures flocculation of clay particles that remain in suspension.
(2) Add $10\,cm^3$ of distilled water and $5\,cm^3$ of BDH universal indicator solution. Seal the test-tube with the bung. Shake vigorously and allow contents to settle for 5 min.
(3) Compare the colour of liquid in the test-tube with the colours on the BDH reference colour chart and read off the corresponding pH.
(4) Repeat the experiment on soil samples from different areas.

Note

pH is one of the most useful measurements which can be made on a soil. Although a simple measurement, it is a product of many interacting factors and is likely to be a good guide to nutrient status and to types of plants (and therefore animals) that flourish. Acid soils tend to be less nutrient-rich (poorer cation-holding capacity).

11.1.2 Water factors

Water, like soil, is an important medium for life. This section outlines some of the basic practical methods used in monitoring the physical and chemical water properties that are of vital importance to living organisms.

Experiment 11.6: To investigate the pH of a water sample

Materials

universal indicator test paper or pH meter
water sample

Method

Either

(1) Dip a piece of universal indicator test paper into the water sample and compare the colour produced with the colour chart. Read off the pH value.

or

(2) Rinse the probe of the pH meter with distilled water, dip it into the water sample and read off the pH value. (This method is more precise, but the meter must be accurately calibrated using prepared solutions of known pH before the experiment begins.) Rinse the probe with distilled water before returning it to buffer solution for storage.

Experiment 11.7: To investigate the chloride content of a water sample (giving a rough estimate of salinity)

Materials

water sample
10 cm^3 pipette
burette
distilled water in a wash bottle
3 conical flasks
white tile
potassium chromate indicator
50 cm^3 silver nitrate solution (2.73 g 100 cm^{-3})

Method

(1) Place 10 cm^3 of the water sample into a conical flask and add two drops of potassium chromate indicator solution.
(2) Titrate silver nitrate solution from the burette, shaking the conical flask constantly.
(3) The end-point of the titration is given by a reddening of the silver chloride precipitate.
(4) Repeat the titration on a further two 10 cm^3 water samples. Calculate the mean volume of silver nitrate used.
(5) The volume of silver nitrate solutions used is approximately equal to the chloride content of the water sample (in g dm^{-3}).

Experiment 11.8: To investigate the dissolved oxygen content of a water sample

The technique described here is the Winkler method which gives an accurate measure of oxygen content but requires many reagents. A simpler but less accurate method is described in Nuffield Advanced Science, *Biological Science.*

Miniaturised 'field kit' versions of the Winkler method are now available from several suppliers (e.g. Hanna dissolved oxygen test kit).

Materials

10 cm^3 of alkaline iodide solution (3.3 g NaOH, 2.0 g KI in 10 cm^3 distilled water) (CARE)
10 cm^3 of manganese chloride solution (4.0 g $MnCl_2$ in 10 cm^3 distilled water)
5 cm^3 of concentrated hydrochloric acid (CARE)
starch solution (as indicator)
distilled water in a wash bottle
0.01 M sodium thiosulphate solution (see point (8) in method)
3 × 5 cm^3 graduated pipettes
burette
white tile
3 conical flasks
250 cm^3 water sample in glass bottle with ground glass stopper

Method

(1) Collect the water sample carefully without splashing and stopper the sample bottle under water to prevent entry of air bubbles.
(2) Add 2 cm^3 of manganese chloride solution and 2 cm^3 of alkaline iodide solution to the sample using pipettes whose tips are placed at the bottom of the sample bottle. The heavier salt solutions will displace an equal volume of water from the top of the sample bottle. Replace the stopper carefully (the bottle should be completely filled by the sample). Shake well to mix reagents throughout the water sample. A complex precipitate of manganic-oxide-hydroxide will form in direct proportion to the amount of oxygen present in the sample. The sample may now be set aside (e.g. transported back to a laboratory before continuing the analysis).
(3) Add 2 cm^3 of concentrated hydrochloric acid and stopper the bottle so that no air bubbles are trapped. Shake the bottle thoroughly to dissolve the precipitate. This leaves a solution of iodine in an excess of potassium iodide. The iodine formed is directly proportional to the oxygen originally present in the water sample. The dissolved oxygen is now fixed and exposure to air will not affect the result.
(4) Remove a 50 cm^3 sample of this solution and place it in a conical flask. Titrate with 0.01 M sodium thiosulphate solution from the burette as follows:
(*a*) add thiosulphate solution whilst shaking the flask until the yellow colour becomes pale;
(*b*) add three drops of starch solution and continue to titrate and shake until the blue-black colouration of the starch disappears.
Record the volume of thiosulphate used.
(5) Repeat stage (4) with two further 50 cm^3 samples of water and obtain the mean volume used (x).
(6) Using these solutions, 1 cm^3 of 0.01 M thiosulphate solution corresponds to 0.056 cm^3 of oxygen at standard temperature and pressure (STP).

(7) Calculate the concentration of oxygen per dm^3 of water using the following formula:

oxygen in $cm^3 dm^{-3} = 0.056 \times x \times 1000/50$ at STP

where x = volume of thiosulphate solution required for the titration of 50 cm^3 of samples.

(8) In comparative studies for water pollution work and estimating BOD, dissolved oxygen levels are commonly expressed in mg dm^{-3}. Calculation of the final result is simpler if a working solution of 0.0125 M sodium thiosulphate is used. Then 1 cm^3 sodium thiosulphate solution is equivalent to 0.1 mg oxygen.

(*a*) Prepare a stock solution of 0.1 M sodium thiosulphate. To do this, dissolve 24.82 g $Na_2S_2O_3.5H_2O$ (sodium thiosulphate) in distilled water. Add a pellet of NaOH (sodium hydroxide) and dilute to 1 dm^3. Store in a brown bottle. This solution may be kept for two or three weeks.

(*b*) Prepare, as needed, a working solution of 0.0125 M sodium thiosulphate. To do this take 125 cm^3 of the stock solution and dilute to 1 dm^3 ($\times 8$ dilution).

(*c*) Carry out the method following the procedure outlined above but using 0.0125 M sodium thiosulphate in step (4):

$$\text{mg } O_2 \text{ in 1 dm}^3 \text{ sample} = x \times 0.1 \times 1000/50 \quad \text{or } x \times 2$$

where x = mean volume of 0.0125 M thiosulphate solution required for the titration of 50 cm^3 of sample.

(9) It may be useful to compare the actual oxygen content with the potential maximum value, the **saturation level**. This will be especially relevant if you are sampling a stream or river at different seasons of the year, in other words under different temperature conditions. The amount of oxygen that can be held in solution is temperature-dependent. Thus to estimate the percentage saturation one further measurement, the temperature of the water sample at collection, is needed. This can be easily achieved using a simple mercury thermometer. By reference to table 11.1 the percentage oxygen saturation value for the water can easily be calculated:

$$\frac{\text{mg } O_2 \text{ dm}^{-3} \text{ present in sample tested}}{\text{mg } O_2 \text{ dm}^{-3} \text{ held at saturation}} \times 100\%.$$

Notes

(1) It is quite common to use 25 cm^3 water samples, thus saving on reagent, with appropriate adjustment of the final calculation (mg $O_2 dm^{-3} = x \times 4$).

(2) It is essential to dissolve all the precipitate present, since this contains all the oxygen from the water. It may sometimes be necessary to add extra acid to achieve this.

(3) It is important to add enough manganese chloride and alkaline iodide solution to ensure all oxygen is trapped in the precipitate. In practice any convenient size of glass-stoppered bottle can be used, providing these two reagents are added in equal amounts and in the approximate ratio of 1 cm^3 manganese

Table 11.1 The solubility of oxygen in water.

Temperature/ °C	*Saturation value for O_2/ mg dm^{-3}*	*Adjustment for saline waters/ mg dm^{-3}*
0	14.63	0.0925
1	14.23	0.0890
2	13.84	0.0857
3	13.46	0.0827
4	13.11	0.0798
5	12.77	0.0771
6	12.45	0.0745
7	12.13	0.0720
8	11.84	0.0697
9	11.55	0.0675
10	11.28	0.0653
11	11.02	0.0633
12	10.77	0.0614
13	10.53	0.0595
14	10.29	0.0577
15	10.07	0.0559
16	9.86	0.0543
17	9.65	0.0527
18	9.46	0.0511
19	9.27	0.0496
20	9.08	0.0481
21	8.91	0.0467
22	8.74	0.0453
23	8.57	0.0440
24	8.42	0.0427
25	8.26	0.0415
26	8.12	0.0404
27	7.97	0.0393
28	7.84	0.0382
29	7.70	0.0372
30	7.57	0.0362

Notes: The solubility of oxygen in water varies with temperature, atmospheric pressure and the concentration of dissolved salts. In saline waters saturation values will be lower and suitable adjustment must be made. The data show the necessary adjustment in mg dm^{-3} for each 1 part per thousand (ppt) change in salinity.

Data are based on work by Montgomery, Thorn and Cockburn at the Water Pollution Research Laboratory. It has been reproduced from Klein, L., (1966) *River Pollution*, Vol. 3, Butterworth.

chloride and 1 cm^3 alkaline iodide to 100 cm^3 of water sample.

(4) When collecting samples in the field the bottles should be rinsed at least three times with the water to be tested before collecting the test sample. The bottle should be pointing upstream so that water flows in easily without splashing. Bottles should be thoroughly cleaned before use, if possible including acid rinsing.

(5) Waterproof gloves (e.g. washing-up gloves) should always be worn when taking samples from rivers, streams and ponds which may be polluted. In shallow rivers and streams, samples should be taken mid-stream; deeper water samples may be taken from bridges or using a boat, as suitable. Appropriate safety procedures should always be followed.

Experiment 11.9 To investigate the biochemical oxygen demand (BOD) of a water sample

A dissolved oxygen test measures the current oxygen status of a stream or river. This is a useful starting point. However, dissolved oxygen content can vary considerably from day to day due to a range of other environmental factors such as sunlight and windiness. Of more fundamental significance is the rate at which oxygen is being used by organisms present in the water. If water contains large amounts of organic wastes, then the rate of microorganism activity (effectively decomposition) may be high and the water may rapidly become depleted of oxygen. This will have important consequences for continued microorganism activity and the lives of other aerobic organisms.

Materials

500 cm^3 – 1 dm^3 water sample and

either

reagents and glassware as described for the Winkler method in experiment 11.8 above

or

an appropriately calibrated oxygen electrode

Method

Pre-checks

(1) If necessary adjust the sample pH to a range of 6.5–8.5 (to optimise microorganism activity).

(2) If the oxygen content of the sample is known to be very low (e.g. you have already measured the dissolved oxygen) the sample should be oxygenated for 5–10 min. This is important since the test measures the rate of oxygen consumption and microorganism activity. The results will be misleading if there is an insufficient initial oxygen supply.

(3) If high organic contamination is suspected, prepare sample dilutions (see note at end of method) before incubating. Remember to check that the BOD of the dilution water itself is negligible. To do this incubate dilution water in the same way as samples. If necessary (that is if there is a significant decrease in dissolved oxygen) adjust results for oxygen loss in dilution water controls as well as for dilution factor itself.

Test procedure

(1) Place portions of the sample (dilute if necessary) into three glass-stoppered bottles of 125 cm^3 or 250 cm^3 capacity. Pour carefully to avoid trapping air bubbles. Ensure bottles are completely full.

(2) Immediately determine the oxygen content of one bottle (express as mg dm^{-3}).

(3) Incubate the remaining two bottles *in the dark* (no photosynthesis) at a standard temperature (20 °C) or the temperature of the original sample for 1–5 days. The standard procedure is to incubate in darkness at 20 °C for five days.

(4) Determine the oxygen content of the incubated bottles (mg dm^{-3}).

(5) Subtract the mean value for the incubated samples from the original sample. This gives the sample BOD in mg dm^{-3}, unless the sample was diluted before incubation. In this case use the following formula:

$$\text{BOD} = (x - y)(a + 1)\ \text{mg dm}^{-3}$$

where x is the initial dissolved oxygen in mg dm^{-3}, y is the mean final dissolved oxygen in mg dm^{-3}, and a is the volume(s) of dilution water to 1 volume of sample.

Note

River water does not usually require dilution. A badly polluted stream or pond might require up to four parts dilution water to one part sample. Such contaminated water is a health risk and requires great care in handling and is best avoided for student class work. Tap water was formerly commonly used for dilution but high chlorination now often makes this unsuitable. Synthetic dilution water is preferable (distilled or deionised water with appropriate chemicals added). Advice on the preparation of synthetic dilution waters is given in H.L. Golterman, R.S. Clymo & M.A.M. Ohnstad (1978) *Methods for physical and chemical analysis of fresh waters,* IBP Handbook No. 8, Blackwell Scientific Publications, 2nd edition.

Any samples absorbing more than 6 mg dm^{-3} oxygen or having a final dissolved oxygen content less than 40% saturation should be diluted.

In some cases a considerable part of the BOD may be due to oxidation of ammonia. If wished this nitrification can be inhibited by adding 1 cm^3 of 0.5 g dm^{-3} solution of allylthiourea to each sample. For a fuller discussion see Golterman *et al.* as cited above.

Water current

The simplest method of measuring water current is to record the time taken for a floating object to cover a known distance. In order to eliminate the effects of wind, it is preferable to use an object which is mainly submerged. Alternatively an L-shaped tube 50 cm high, 10 cm long and 2 cm in diameter can be placed in a stream with the short end facing upstream. By measuring the height to which water rises in the long limb, the velocity of the current can be measured using the formula:

$$v = \sqrt{(2hg)}$$

where v is the speed of the current (cm s^{-1}), g is the acceleration due to gravity (981 cm s^{-2}) and h is the height of the column (cm).

11.1.3 Climatic factors

Some simple approaches to the measurement of key atmospheric and climatic characteristics are outlined below.

Humidity

The relative humidity of air is a measure of the moisture content of air relative to air fully saturated with water vapour. It varies with temperature, since air expands on heating and can hold more water vapour. Relative humidity is measured by a **whirling hygrometer**, which is a wet and a dry bulb thermometer mounted on a wooden frame resembling a football rattle (fig 11.2). It is whirled around until both thermometers give constant temperature readings. These temperatures are then examined in hygrometer tables, or on a specially calibrated slide rule supplied with the whirling hygrometer kit, and the corresponding relative humidity read off. Dew-point values (the temperature at which the same air sample would become saturated) can also be estimated from these data.

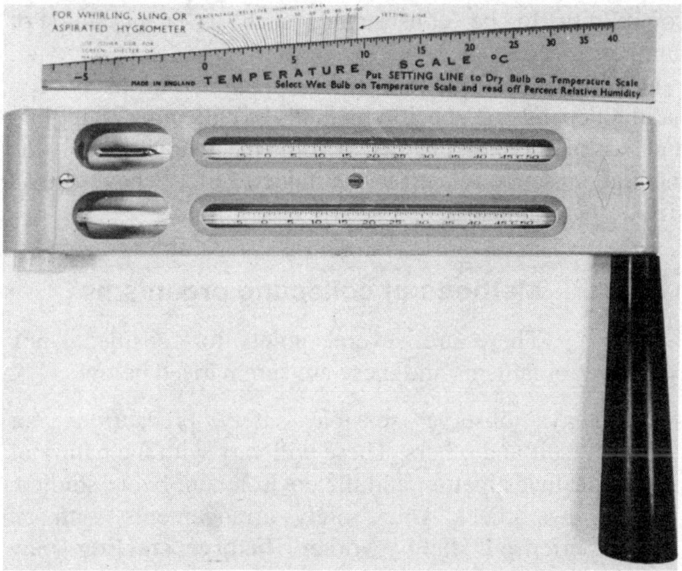

Fig 11.2 *Whirling hygrometer.*

Temperature

Air, water and soil temperatures can be measured using a mercury thermometer, but measurements of temperature at a point in time provide little real information of ecological significance. It is the range of temperatures over a period of time which has more significance in ecological studies. Hence sophisticated time-based recordings of temperature are normally used, or the maximum and minimum temperatures recorded using a maximum–minimum thermometer.

Temperatures in microhabitats and inaccessible habitats, such as the centre of a tree, are measured using a thermistor (fig 11.3). This is an electrical device which can be miniaturised to fit into the tip of a ballpoint pen and whose resistance varies with temperature. By measuring the resistance of the thermistor, and comparing this with previous temperature-calibrated resistances, the environmental temperature can be obtained.

The temperature extremes of microhabitats (microclimates) are also useful in ecological studies since they can often explain the disappearance of a particular species from an area, such as frost-sensitive plants.

Light

Light varies in intensity, duration and quality (wavelength). Measurements of all three aspects are required to provide the information relevant to ecological study, and specialised techniques are required to record them. For practical purposes some indication of intensity related to particular areas is generally required, so that the incident light in

Fig 11.3 *A thermistor in use.*

different areas can be compared. For this purpose an ordinary photographic exposure meter is adequate. Light intensities over a given period of time can be recorded using Ozalid papers which have a cumulative sensitivity to light.

Wind speed and direction

The wind speed in a habitat at a given point in time is not as ecologically significant as the degree of exposure to wind experienced by the habitat. In this respect wind frequency, intensity and direction are all important. However, for most practical purposes a simple wind-gauge, indicating the direction of the wind, and a simple anemometer (fig 11.4), indicating wind speed, are adequate for comparing features of wind in different habitats.

11.2 Biotic analysis

In analysing the organisms living in a given habitat (the biotic component of the ecosystem) the community structure must be determined in terms of species present in the habitat and numbers within each population. It is obviously impractical to attempt to find and count all the members of a given species, and so sampling techniques have to be devised which will give indications of species present and their numbers. Generally speaking, the more accurate the results required the more time-consuming the method, so it is necessary to be clear about objectives. Also, if possible, non-destructive techniques should be used.

In all cases reliable methods of sampling (recording

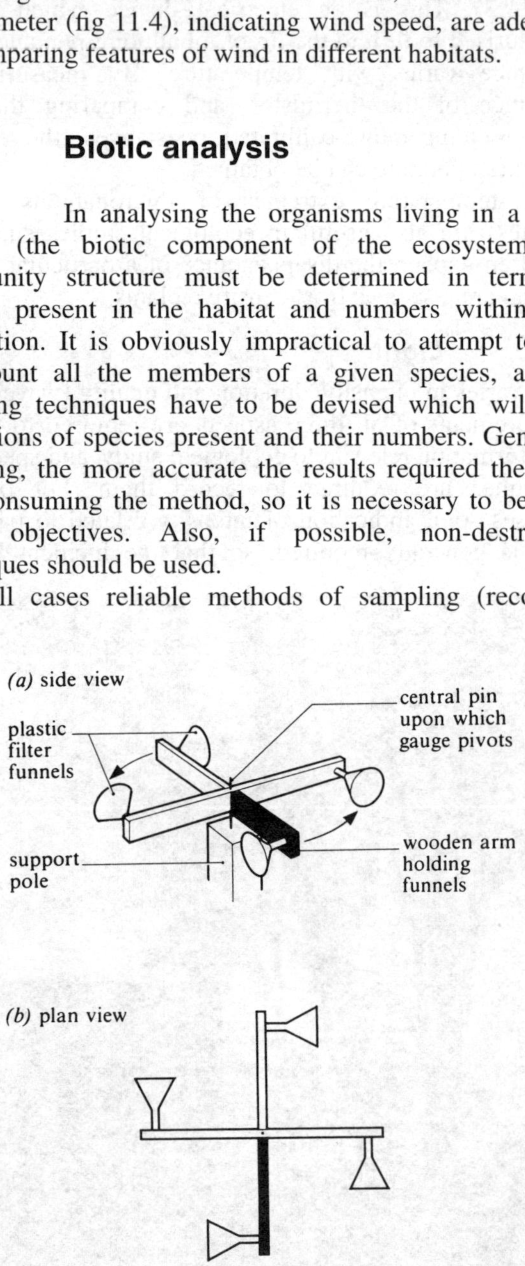

Fig 11.4 *Simple anemometer which may be used to determine wind speed in terms of the rate of turning of the wooden arm painted black.*

and/or collecting) organisms are required, and it is safe to say that 'no stone should be left unturned' (providing it is replaced!) since organisms will occupy almost all available microhabitats. For example, at first sight a square metre of grass, soil, sand, rocky shore or stream bed may not appear to support many species, but closer examination, involving hand-sorting the soil, grass and weeds, turning over stones and examining roots, stems, flowers and fruits of plants and holdfasts of seaweeds, will reveal many more species.

In recording data, as many species as possible (plant and animal) should be identified in the field, using keys if necessary. Only if the species are obviously common locally and not known to be rarities should they be collected. Over-collection can have serious effects on local communities. In the case of collected animals, attempts should be made to keep them alive and to release them in a similar microhabitat to that in which they were collected. It is necessary to identify the organisms as accurately as possible, that is to the level of species. This cannot always be done, but it should be possible to identify them at least as far as class, order or family. Identification of specimens depends upon familiarity with keys. The principles of classification, key construction and details of how to use a key are described in appendix 3 in Book 2.

A list of all the species in the habitat gives some indication of the diversity of structure of the community, the **species richness** or **diversity**. (There are various numerical ways of expressing species richness using mathematical formulae. The numerical value is called the **diversity index** but details of this will not be considered here.)

These data provide information enabling possible food chains and food webs to be constructed, but are inadequate in providing information related to quantitative aspects of the community. The extent of the diversity is only fully revealed when the numbers of organisms within each species, that is the population sizes, are determined. This information enables a more detailed picture of the community to be constructed, such as a pyramid of numbers (section 10.3.4).

Obtaining the qualitative and quantitative data of a habitat depends on specific methods of collecting, sampling and estimating organisms within the habitat, and the method chosen is related to the mode of life, behaviour and size of the organism.

11.2.1 Methods of collecting organisms

There are several points to consider when collecting organisms and these are summarised below.

(1) Always observe sensible safety precautions for yourself and others. These will vary with the nature of your investigation and the environment being studied. Always check your safety arrangements with an experienced field worker **before starting** any ecological project work.

(2) Observe the Countryside Code at all times.
(3) Respect the environments, communities and ecosystems in which you are working. Be sure you cause no damage nor permanent change.
(4) Always obtain permission from the landowner before beginning an ecological study in an area.
(5) Consult the local Wildlife Trust, university, college or the statutory countryside conservation organisations such as English Nature about where and what you are to study and what you may collect.
(6) Never remove organisms from their habitat or destroy them unnecessarily.
(7) Leave the habitat as undisturbed as possible, for example replace stones, turf, logs and so on to their original positions.
(8) Where it is necessary to remove organisms from the habitat for identification, take as few as possible and, if practicable, return them to the habitat.
(9) Keep specimens separate when removing them to the laboratory for identification, to prevent contamination or being eaten by predators, for example do not put ragworm and crabs in the same collecting vessel. Useful collecting equipment includes jam jars, Kilner jars, polythene bottles, specimen tubes and polythene bags.
(10) Always record as much information as possible concerning the topography of the habitat and climate at the time of collection as the information may have a bearing on what is collected:

    (*a*) nature of rock or substratum (grass, mud, soil etc.);
    (*b*) nature of aspect (for example flat, south-facing, angle of slope etc.);
    (*c*) drainage;
    (*d*) soil, mud or sand profile;
    (*e*) temperature of substrate, water and air;
    (*f*) substrate or water pH;
    (*g*) cloud cover and rainfall;
    (*h*) relative humidity of air;
    (*i*) light intensity (such as shaded or open, possibly a meter reading);
    (*j*) wind speed and direction (such as still, gentle breeze, gale, south-west);
    (*k*) time of day and date.

An example of how some of these features may be recorded is shown in table 11.2.

There are a variety of methods of collecting specimens. A summary of methods and their applications is shown in table 11.3 and in figs 11.5–11.10. Specimens should be collected from traps at regular intervals, identified, counted and, where possible, released. In the case of pitfall traps it should be realised that if natural predators and prey are collected it is probable that the prey will not be present when the trap is emptied. Where this is believed to be happening, 70% alcohol should be placed in the trap to kill the organisms as they fall in. Imagination and ingenuity are required in collecting specimens.

Table 11.2 Field booking sheet for recording edaphic, physiographic and climatic features.

Area .... Grid reference .... Date ....

(1) **Underlying rock**
(2) **Substratum/soil**
    (*a*) surface feature..............................
    (*b*) depth of horizon A..............................
    (*c*) " " " B..............................
    (*d*) " " " C..............................
    (*e*) pH..............................
    (*f*) temperature..............................

(3) **Topography**
    (*a*) aspect, direction......... angle..............................
    (*b*) height above sea level..............................
    (*c*) relief..............................
    (*d*) drainage..............................
    (*e*) land use..............................
    (*f*) high or low water, time......... height..............................

(4) **Climate**
    (*a*) air temperature, range..............................
    (*b*) rainfall..............................
    (*c*) cloud cover/sunlight..............................
    (*d*) relative humidity..............................
    (*e*) wind direction..............................
    (*f*) wind speed..............................
    (*g*) light intensity (horizontal), N....., S....., E....., W............
    (*h*) time of day..............................

Generally, sites where specimens are collected are not randomly chosen and consequently the results obtained from the collections must be interpreted in the light of biased selection of collecting site. Whilst this may not affect the species of organism collected, so that community structure will be accurately represented, it is likely to give biased indications of numbers present. For example, the use of baits and lures to attract organisms to sticky traps, pitfall traps and mammal traps will influence the results, and conclusions based on quantitative data will reflect this bias. Therefore in discussion of results it is necessary to state clearly that bias exists.

11.2.2 Methods of sampling an area

In order to standardise the sites where abiotic and biotic aspects of ecosystems are investigated, transects and/or quadrats are commonly used and collecting and sampling is often confined to the area of the transect or quadrat.

Line transect. This may be used to sample a uniform area but is particularly useful where it is suspected that there is a transition in habitats and populations through an area (fig 11.10). For example, a tape or string running along the ground in a straight line between two poles indicates the position of the transect and sampling is rigorously confined to species actually touching the line.

Table 11.3 A summary of various methods used to collect organisms.

Collecting method	*Structure and function*	*Organisms collected*
beating tray	A fabric sheet of known area is attached to a collapsible frame and held under a branch which is beaten with a stick or shaken. Organisms fall onto the sheet and are removed using a pooter (see later notes).	non-flying insects, larval stages, spiders
kite net	A muslin net is attached to a handle and swept through the air. Organisms become trapped in the net. All netting techniques must be standardised to ensure uniformity of sampling, e.g. eight, figure-of-eight sweeps per examination of the net.	flying insects
sweep net	A nylon net is attached to a steel handle and swept through grass, bushes, ponds or streams.	insects, crustaceans
plankton net	A bolting silk net is attached to a metal hoop and rope harness and towed through the water. A small jar is attached to the rear of the net to collect specimens.	plankton
sticky trap	Black treacle and sugar are boiled together and smeared onto a sheet of thick polythene which is then attached to a piece of chipboard with drawing pins. This can be hung in various situations and at various heights. Jam and beer can be added to the sticky substances to act as attractants.	flying insects
pitfall trap	A jam jar or tin is buried in the soil with the rim level with ground level. This is best placed where the ground falls away from rim level to prevent water entering the jar. A piece of slate supported on three stones acts as a lid preventing rainwater from entering. The trap can be baited with either sweet foods such as jam or decaying meat. Traps should be regularly cleared (fig 11.5).	walking/crawling insects, myriapods, spiders, crustaceans
light trap	A mercury vapour light trap attracts flying organisms which hit baffles and fall down into the base and become trapped in cardboard egg boxes or crumpled-up paper. Cotton wool soaked in chloroform is added before examining the contents to anaesthetise or kill the organisms (fig 11.6).	night-flying insects, particularly moths and caddis flies
mammal trap	A Longworth mammal trap (fig 11.7) is left in a runway and filled with bedding material. Bait, e.g. grain or dried fruit, can be left outside and inside the trap. The trap can be left unset for some time until organisms become accustomed to it and then set. Animals are captured alive so the trap must be visited regularly. Some animals may remain 'trap shy' and never enter it, whereas others become 'trap happy' and visit it regularly. These two patterns can present problems when using the technique to estimate population sizes.	shrews, voles and mice
kick sampling	This is used for collecting in running fresh water. An open sweep or plankton net is held vertically downstream of the area being sampled by turning over stones and scraping off organisms which are then swept into the net. Alternatively the area being sampled is agitated by kicking and stamping so that organisms are displaced vertically and swept into the net by the current.	aquatic insects and crustaceans
pooter	This is used to collect small insects from beating trees or directly off vegetation for closer examination and/or counting (fig 11.11).	aphids, small insects and spiders
hand-sorting	Samples of soil or vegetation, e.g. grass, leaf litter, pond and seaweed, are placed at one end of a tray and small amounts of material are systematically examined between the fingers, specimens are removed to a collecting jar and sorted material passed to the other end of the tray. The sample is then examined as it is moved back to the original end of the tray.	mites, enchytraeid worms, insect larvae and small insects
extractions	5 cm^3 of 4% formaldehyde are added to 50 cm^3 of water and used to water a square metre of lawn or grassland. Earthworms are driven out from their burrows and collected and immediately washed in water to remove the formaldehyde.	earthworms
flotation	Add a known mass of soil to a beaker of saturated salt solution, stir vigorously for several minutes and allow soil to settle. Organisms float to surface in dense salt solution. Pour off surface layer of fluid into a petri dish and examine under binocular microscope. Remove all specimens into another petri dish containing 70% alcohol to kill and fix the specimens. Mount each specimen separately in glycerine on a microscope slide, cover with a cover-slip and identify under binocular microscope or low power of compound microscope.	mites, insects, eggs, cocoon, larval and pupal stages
Tullgren funnel (dry extraction)	Many soil and leaf litter-dwelling organisms move away from a source of heat and towards moister conditions. A soil or leaf litter sample is placed in the sieve about 25 cm below a 25 W bulb in a metal reflector (fig 11.8). Every two hours the bulb is moved 5 cm nearer to the sample until the bulb is 5 cm from the soil sample. The apparatus is left for a total of 24 h. All small arthropods move downwards and drop through the metal gauze into the alcohol beneath.	small arthropods e.g. millipedes, centipedes, mites, springtails and collembola
Baermann funnel (wet extraction)	A soil sample is placed in a muslin bag, submerged in a funnel containing water and suspended 25 cm from a 60 W bulb in a metal reflector (fig 11.9). The apparatus is left for 24 h. The water and the gentle heating encourage organisms to leave the sample, move out into the water and sink to the base of the funnel. They are removed at intervals by opening the clip in the apparatus and allowing them to fall into the alcohol.	small arthropods, enchytraeid worms and nematodes

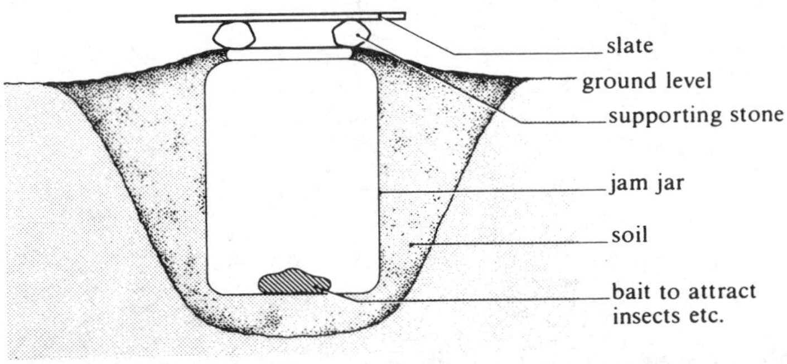

Fig 11.5 *Simple pitfall trap made by sinking a jam jar into the soil.*

Fig 11.6 *Mercury vapour lamp in use attracting insects.*

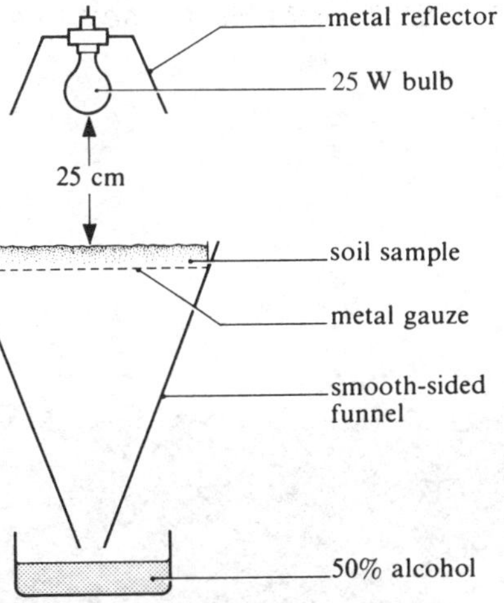

Fig 11.8 *Tullgren funnel.*

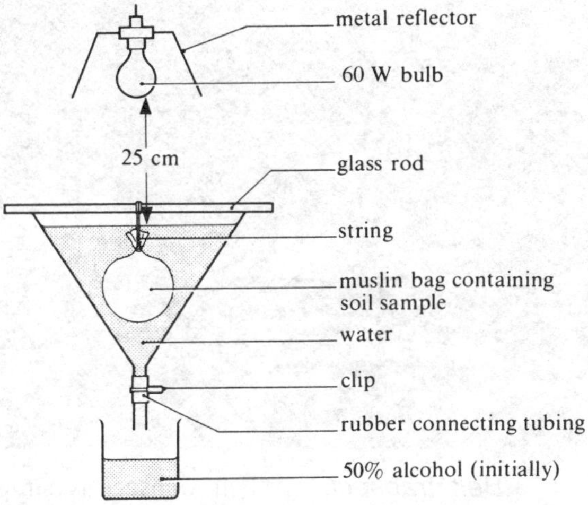

Fig. 11.9 *Baermann funnel.*

Fig 11.7 *(below) Longworth mammal trap.*

Fig 11.10 *The position of a line transect a cross a rocky shore.*

Belt transect. A belt transect is simply a strip of chosen width through the habitat, made by setting up two line transects, say 0.5 m or 1 m apart, between which species are recorded. An easier method of obtaining both qualitative and quantitative data from a belt transect is to use a quadrat frame in conjunction with a line transect.

Height variations recorded along line or belt transects produce a profile of the transect, sometimes known as a **profile transect**, which can be used when presenting data (fig 11.21).

A decision over which type of transect to use depends on the qualitative and quantitative nature of the investigation, the degree of accuracy required, the nature of the organisms present, the size of the area to be investigated and the time available. Over a short distance a line transect might be used and a continuous record kept of each plant species lying immediately beneath it. Alternatively, over a longer distance the species present every metre, or other suitable distance along the transect, may be recorded.

Quadrat. A quadrat frame is typically a metal or wooden frame, preferably collapsible to facilitate carrying, which forms a square of known area, such as $0.25\,m^2$ or $1\,m^2$ (fig 11.12). The size of quadrat used will depend on the organisms being studied. A $0.25\,m^2$ (flexible*) quadrat would be suitable for a study of lichens on trees, but a $10\,m^2$ or $20\,m^2$ quadrat** would be needed for an investigation of a woodland.

The quadrat is placed to one side of a line transect and sampling carried out. It is then moved along the line transect to different positions. Both the species present within the frame and the numbers or abundance (section 11.2.3) of these may be recorded, depending upon the nature of the investigation. In all cases the method of recording the species must be consistent, for example all species partially or completely visible within the quadrat are listed. The structure of the quadrat frame can be modified according to the demands of the investigation. For example, it can be divided by string or wire into convenient sections to assist in counting or estimating numbers or abundance of the species (fig 11.12). This is particularly useful when studying a habitat supporting several species of plants.

* A small flexible quadrat can be made by marking out the quadrat dimensions, in waterproof ink, onto a clear plastic sheet.

** Instant portable quadrats, of any size, can be made using tent pegs or canes to fix the corner posts and strong nylon string, looped at appropriate intervals (10 m or 20 m) to mark the quadrat sides. The string should be mounted on a dispenser board for tangle-free assembly and dismantling. If needed, matching lengths of dividing strings can be mounted on separate dispenser boards or, for smaller quadrat sizes, they can be wound separately onto the quadrat dispenser (fig 11.13).

Fig 11.11 *Pooter in use collecting small non-vertebrates.*

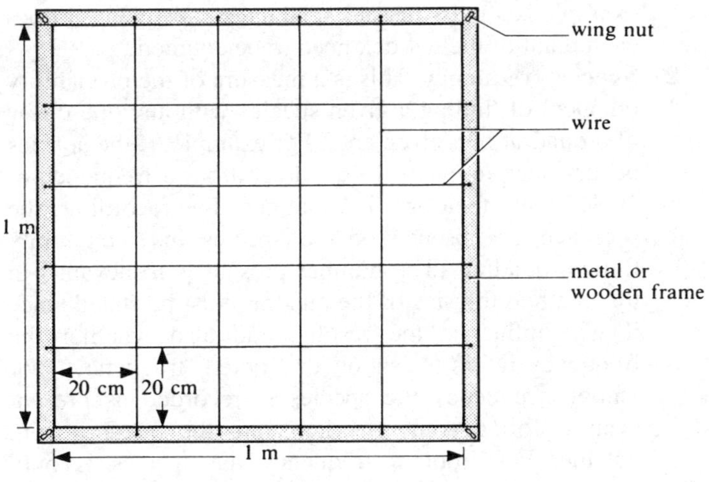

Fig 11.12 *Quadrat frame (1 m^2) with wire sub-quadrats (each 400 cm^2) forming a graduated quadrat.*

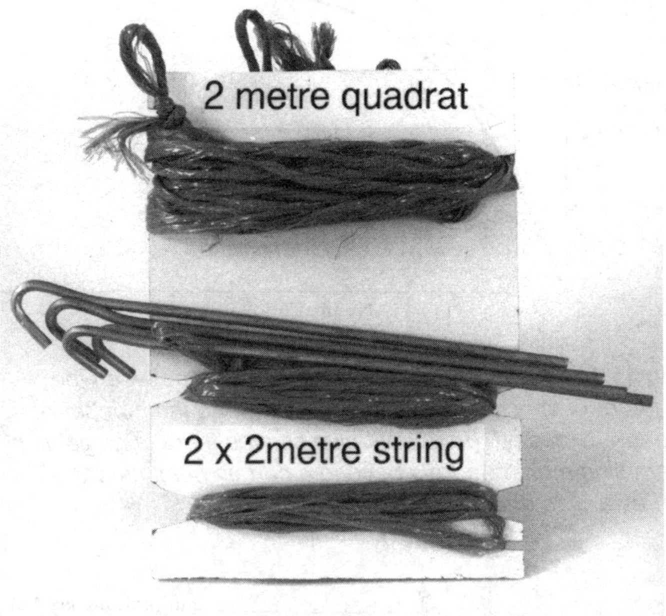

Fig 11.13 *Quadrat dispenser.*

A quadrat may be used without a transect when studying an apparently uniform habitat. In this case the quadrat is used randomly. One fairly random and traditionally used sampling technique is to fling a robust quadrat over the shoulder and record the species within it wherever it falls. This is repeated several times so that a representative sample of the area is covered. This obviously needs care and the resultant sample may still be biased, for example according to the investigator's throwing skills. An alternative, and sounder, approach is to choose a sampling point using a table of random numbers or random numbers generated on an suitable pocket calculator. Each pair of random numbers can be used to identify a random coordinate, the **sampling point**, on an imaginary grid laid over the area. The sides of the grid may be marked by measuring tapes. Alternatively random number pairs may be used to plot a random walk using one number to determine the distance walked and the second to indicate the direction of the walk.

Investigations have shown that in a uniform habitat there comes a point beyond which analysing the species within a quadrat becomes unnecessary, as it does not increase the number of different species recorded. This relationship is shown in fig 11.14. As a rule of thumb, once five quadrats have failed to show any new species it may be assumed that no further species will be found. However, when an assumption such as this is made it must be stated in the ecological report as it may affect the reliability of the results.

Pin frame (point quadrat). This is a frame bearing a number of holes through which a 'pin', such as a knitting needle can be passed (fig 11.15). It is particularly useful with transect studies of overgrown habitats where

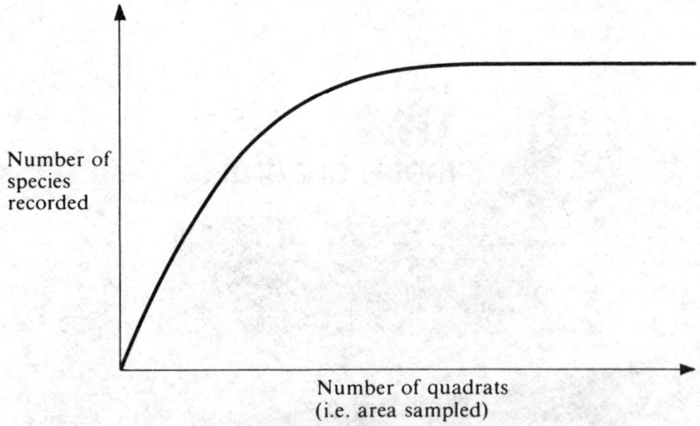

Fig 11.14 *Graph showing the relationship between the number of species recorded in an area and the number of quadrats studied. (In quantitative studies there is no point in sampling more quadrats beyond a certain point as it is unrewarding and uneconomical of time.)*

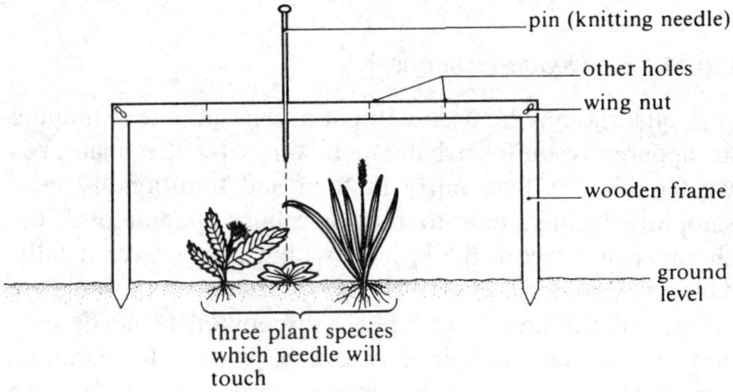

Fig 11.15 *Pin frame or point quadrat.*

several plant species may overlap. All species touched by the pin as it descends to the ground are recorded for each of the holes.

Permanent quadrat. In long-term ecological investigations involving the study of community change (succession) or seasonal changes, a permanent quadrat or transect is used. Metal pegs and nylon rope are used to mark out an area of ground. Periodic samples of abiotic and biotic factors can be taken and the results presented in such a way as to reveal trends and changes and possible factors accounting for, or associated with, these changes.

11.2.3 Methods of estimating population size

In all studies in quantitative ecology it is essential to be able to estimate, with a reasonable degree of accuracy, the number of organisms within a given area of ground or volume of water or air. In most cases this is equivalent to estimating the population size, and the methods employed are determined by the size and mode of life of organisms involved and the size of the area under investigation. The numbers of plants and sessile or slow-moving animals in a small area may be counted directly, or their percentage cover or abundance estimated, whereas indirect methods may be required for fast-moving organisms in large open areas. In habitats where organisms are difficult to observe, because of their behaviour and mode of life, it is necessary to estimate numbers of organisms using either the removal method or the capture–recapture method. Methods of estimating populations may be either objective or subjective.

Objective methods

The use of quadrats, direct observation and photography are known as direct counting methods, whereas the removal and capture–recapture techniques are indirect counting methods.

Quadrat method. If the number of organisms within a number of quadrats, representing a known fraction of the total area, are determined, an estimate of the total numbers in the whole area can be obtained by simple multiplication. This method provides a means of calculating three aspects of species distribution.

(1) **Species density.** This is the number of individuals of a given species in a given area. It is obtained by counting the number of organisms in randomly thrown quadrats. The method has the advantages of being accurate, enabling different areas and different species to be compared, and providing an absolute measure of abundance. The disadvantages are that it is time-consuming and requires individuals to be defined, for example is a grass tussock counted as one plant or does each plant of the tussock need to be counted?

(2) **Species frequency.** This is a measure of the probability (chance) of finding a given species with any one throw of a quadrat in a given area. For example, if the species occurs once in every ten quadrats it has a frequency of 10%. This measure is obtained by recording the presence or absence of the species in a randomly thrown quadrat. (The number present is irrelevant.) In this method the size of the quadrat must be stated since it will influence the results, and also whether the frequency refers to 'shoot' or 'rooted' frequency. (For 'shoot' frequency the species is recorded as present even if foliage only overlaps into the quadrat from outside. For 'rooted' frequency the species is only recorded as present if it is actually rooted in the quadrat.) This method has the advantage of being quick and easy and useful in certain large-scale ecosystems such as woodland. The disadvantages are that quadrat size, plant size and spatial distribution (that is random, uniform or clumped) (see appendix section A2.8, Book 2) all affect the species frequency.

11.2 What is the species frequency of the species recorded in 86 quadrats out of 200 studied?

(3) **Species cover.** This is a measure of the proportion of ground occupied by the species and gives an estimate of the area covered by the species as a percentage of the total area. It is obtained either by observing the species covering the ground at a number of random points, by the subjective estimate of percentage of quadrat coverage or by the use of a pin frame (fig 11.15). This is a useful method for estimating plant species, especially grasses, where individuals are hard to count and are not as important as cover. However it has the disadvantages of being slow and tedious.

11.3 If a pin frame containing ten pins was used ten times and 36 units were recorded for plant X, what is the percentage cover of X?

Direct observation. Direct counting is not only applicable to sessile or slow-moving animals but also to many larger mobile organisms such as deer, wild ponies and lions, and wood pigeons and bats as they leave their roost.

Photography. It is possible to obtain population sizes of larger mammals and sea birds which congregate in open spaces by direct counting from aerial photographs.

Removal method. The removal method is very suitable for estimating numbers of small organisms, particularly insects, within a known area of grassland or volume of water. Using a net in some form of standard sweep, the number of animals captured is recorded and the animals kept. This procedure is repeated a further three times and the gradually reducing numbers recorded. A graph is plotted of number of animals captured per sample against the previous cumulative number of animals captured. By extrapolating the line of the graph to the point at which no further animals would be captured (that is number in sample = 0) the total population may be estimated, as shown in fig 11.16.

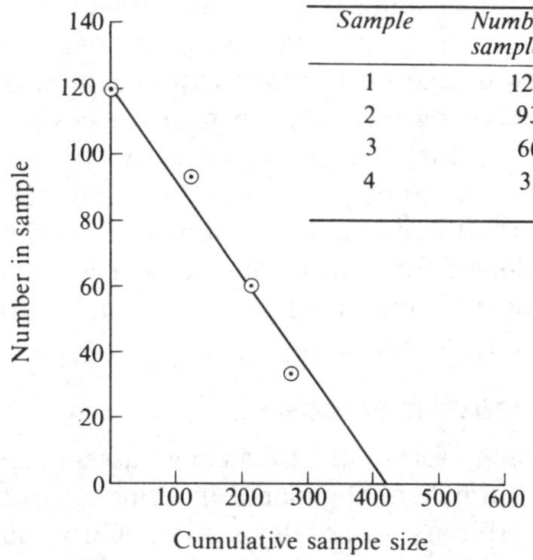

Sample	*Number in sample*	*Cumulative sample size*
1	120	0
2	93	120
3	60	213
4	35	273

Fig 11.16 *Graph of number in sample against cumulative sample size. Extrapolation of the line to the point when the number in the sample equals zero gives an estimate of the number in the population.*

Capture–recapture method. This method involves capturing the organism, marking it in some way without causing it any damage, and replacing it so that it can resume a normal role in the population. For example, fish are netted and their operculum tagged with aluminium discs; birds are netted and rings attached to their legs; small mammals may be tagged by dyes, or by clipping the fur in a distinctive pattern; and arthropods are marked with paint. In all cases some form of coding may be adopted so that individual organisms are identified. Having trapped, counted and marked a representative sample of the population, the individuals are released in the same area. At a later stage the population is trapped again and counted, and the population size estimated using the expression below:

$$\text{Estimated total population} = \frac{\text{number of organisms in initial sample} \times \text{number of organisms in second sample}}{\text{number of marked organisms recaptured}}$$

This estimate of population size is called the **Lincoln index**. It relies on a number of assumptions which are summarised below.

(1) Organisms mix randomly within the population. (This does not always apply since some organisms live in colonies, troops or shoals.)
(2) Sufficient time must elapse between capture and recapture to allow random mixing. The less mobile the species the longer the time lapse must be.
(3) It is only applicable to populations whose movement is restricted geographically.
(4) Organisms disperse evenly within the geographical area of the population.
(5) Changes in population size as a result of immigration, emigration, births and deaths are negligible.
(6) Marking does not hinder the movement of the organisms or make them conspicuous to predators.

11.4 In an attempt to estimate the number of trout in a small lake, 625 trout were netted, marked and released. One week later 873 trout were netted and of these 129 had been marked. What was the estimated size of the population?

Where plants and small animals, such as barnacles, are concerned, direct counting becomes very tedious and, depending upon the degree of accuracy required from the study, may be replaced by estimating percentage cover or abundance within a quadrat frame. In the early stages of estimation it is advisable to use a graduated quadrat frame (fig 11.12) to increase the accuracy of estimation. Various schemes may be adopted for representing percentage cover or abundance, some being totally subjective, others partially, or completely, objective.

Subjective methods

These involve some form of frequency assessment, frequency scale or estimate of abundance in terms of cover. For example, an arbitrary scale devised by Crisp and Southward for limpets on a rocky shore uses the following letters, frequencies and percentages.

A	abundant	>50%
C	common	10–50%
F	frequent	1–10%
O	occasional	<1%
R	rare	present – only a few found in 30 min searching

These assessments and scales are arbitrary and the frequencies can be adjusted to varying percentage values, for example in a particular study, abundant may represent >90%. The value of using the five categories above is that they can be applied to methods of presenting data, such as in constructing kite diagrams, as described in appendix section A2.7.3 in Book 2. The major disadvantage of this method is that it is subjective and tends to rate small species with poor cover lower than conspicuous species, flowering species and species occurring in clumps.

Table 11.4 Five-point scale for water pollution studies using presence and absence indicator species.

Level of pollution	*Oxygen concentration*	*Indicator organisms*
(A) clean water or very low pollution levels	high	stonefly nymph mayfly nymph
(B) low pollution levels	↓	caddis fly larva freshwater shrimp
(C) high pollution levels	↓	water louse bloodworm
(D) very high pollution levels	low	sludgeworm rat tailed maggot
(E) extreme pollution levels	no oxygen	no apparent life

This point scheme is used in the Philip Harris and Griffin water pollution study packs. The Philip Harris scheme includes colour photographs of key indicator species. The Griffin package has an excellent series of black and white drawings of a wider range of indicator organisms as well as procedures for calculating the Trent Biotic Index and other simple pollution indicator tests.

11.2.4 Biotic indices

In some situations organisms make good indicators of environmental quality. Examples are the use of lichens as air pollution indicators and the use of non-vertebrates to monitor river pollution. Diatoms and plants can also be used as biological indicators of water quality.

Assessment of water quality using non-vertebrates is now in widespread use. Techniques rely on the assessment of non-vertebrates found in the stream sediment and the presence and absence of key taxa or families. Several different schemes have been proposed from the simple five-point scale shown in table 11.4 to the Biological Monitoring Working Party (BMWP) scores which are now routinely used in the water industry in Britain. The example which follows shows how to estimate the Trent Biotic Index (TBI), the first scheme used in the water industry and essentially the basis for subsequent methods.

Method for estimating the Trent Biotic Index (TBI)

Principles. The approach is based on a two-way classification which looks at the number of non-vertebrate taxa present from certain defined groups in relation to the presence of six key indicator organisms. Each group denotes the limit of identification which can be reached for a given set of organisms without needing to use lengthy identification techniques. Thus molluscs and crustacea are identified to species level whereas mayflies are distinguished to genus level only (see table 11.5). The mayfly *Baetis rhodani* is an important exception which is separately identified in the scheme since it is more pollution tolerant than other mayfly species. It is grouped with the caddis fly larvae in terms of pollution tolerance.

An example of the two-way approach used in the Trent Biotic Index is shown in table 11.5. The combination of the total number of groups present and the highest (most pollution sensitive) indicator organism present gives a score ranging from 0, very polluted, to 10, very clean with many groups present and several species of stoneflies. In the example highlighted several species of caddis fly larvae (Trichoptera) are the highest indicator species present. Overall, representatives of 7 groups were found and thus a TBI of 6 is recorded. If only one species of caddis fly larvae was present then the TBI would be 5.

Materials.

waterproof footwear (waders or wellingtons)
waterproof gloves
white sample tray
stiff brush
handlens (×10)
sweep net (see table 11.3)
identification key

Method. Kick sample the river bed to disturb bottom-living non-vertebrates. Use the sweep net downstream of the disturbed area to capture the organisms.

Table 11.5 The Trent Biotic Index.

Summary table

The maximum value is 10. Biotic indices are effectively marks out of ten with zero representing virtually lifeless heavily polluted waters.

	Indicator species		*Total number of groups present* 0–1	2–5	6–10	11–15	16+
			Trent Biotic Index				
Clean	Plecoptera nymph present (stoneflies)	More than one species	–	7	8	9	10
		One species only	–	6	7	8	9
Organisms in order of tendency to disappear as degree of pollution increases	Ephemeroptera nymph present (mayflies)	More than one species*	–	6	7	8	9
		One species only*	–	5	6	7	8
	Trichoptera larvae present (caddisflies)	More than one species**	–	5	6	7	8
		One species only**	4	4	5	6	7
	Gammarus present (freshwater shrimps)	All above species absent	3	4	5	6	7
	Asellus present (water louse, water skaters)	All above species absent	2	3	4	5	6
	Tubificid worms and/or red chironomid larvae present	All above species absent	1	2	3	4	–
	All above types absent	Some organisms such as *Eristalis tenax* not requiring dissolved oxygen may be present	0	1	2	–	–
Heavily polluted							

* *Baetis rhodani* excluded

** *Baetis rhodani* (Ephemeroptera) is counted in this section for the purpose of classification.

Groups used in calculating the TBI

The term 'group', for the purposes of calculating the TBI, means any one of the organisms included in the following list.

Each known species of Platyhelminthes (flatworms).
Annelida (worms) excluding genus *Nais*.
Genus *Nais* (worms).
Each known species of Hirudinae (leeches).

Each known species of Mollusca (snails).
Each known species of Crustacea (*Asellus*, shrimps).
Each known species of Plecoptera (stonefly).
Each known genus of Ephemeroptera (mayfly) excluding *Baetis rhodani*.
Baetis rhodani (mayfly).
Each family of Trichoptera (caddisfly).
Each species of Neuroptera (alderfly).
Family Chironomidae (midge larvae) except *Chironomus thummi*.
Chironomus thummi (blood worms).
Family Simulidae (blackfly larvae).
Each known species of other fly larvae.
Each known species of Coleoptera (beetles and beetle larvae).
Each known species of Hydracarina (watermites).

Procedure (see also text)

(1) Sort each sample, separating the animals according to group (see above groups list). Count the total number of groups present.
(2) Note which indicator species are present, starting from the top of the list.
(3) To find the Trent Biotic Index, take the highest indicator species, e.g. caddis fly, *Trichoptera*, and work along the line. Note from the top the group number and read off the Trent Biotic Index.

e.g.	Highest indicator animal	Trichoptera
	Number of indicator species	more than one
	Total numbers of groups	7
	Trent Biotic Index	6

(Source: The Griffin Pollution test kit: handbook for users.)

Empty the contents into the white tray for identification. Record all the groups and key indicator organisms present. Return organisms to the river when identification is complete. Use table 11.5 to establish the Trent Biotic Index.

It is important to standardise the kick time if you are comparing rivers or different locations on the same river. The recommended kick sample time is 3 minutes. You should also investigate beneath stones and boulders and use the brush to disturb and remove organisms from these locations. A one-minute search is standard.

The TBI has been criticised since it takes no account of the relative abundance of organisms. This can lead to misclassification of sites. For example, it can often happen that a single stonefly is swept downstream to a polluted location following heavy rain. It may have poor survival prospects but if you are sampling soon after the rain event your results will be distorted by its chance presence.

The TBI has also been criticised for its taxonomic inconsistency. In other words it requires identification of some organisms to species level, such as *B. rhodani*, whereas others are not distinguished beyond genus or family level.

Alternative approaches have been devised to overcome these criticisms. The Biological Monitoring Working Party was set up to identify a rigorous universally applicable scheme for Britain. Its recommended method is now widely used. Similar schemes are used in other countries. Estimating BMWP scores requires much more taxonomic knowledge than the simpler Trent Biotic Index. Full discussion of the BMWP method is beyond the scope of this text.

11.3 Ecological research projects and investigations

Ecological projects are broadly concerned with studying either the organisms in an area (**community ecology** or **synecology**) or a single species (**autecology**). In both cases it is necessary to spend time reading about and discussing the project, so as to clarify the aims, nature and extent of the project. All investigations should include problems which have to be solved or hypotheses to be tested.

The aims of the project should be stated clearly and should include both general and specific aims. For example:

(1) to develop and encourage an attitude of curiosity and enquiry;
(2) to develop the ability to plan an investigation, construct hypotheses and design experiments;
(3) to develop the ability to formulate questions and collect relevant qualitative and quantitative data to answer these;
(4) to develop practical and observational skills including the use of apparatus and biological keys;
(5) to develop the ability to record data accurately;
(6) to develop the ability to apply existing knowledge to the interpretation of data;
(7) to develop a critical attitude to data, assessment of their validity and conclusions based on them;
(8) to develop the ability to communicate biological information by means of tables, graphs and the spoken and written word;
(9) to develop an appreciation of organisms and the importance of conservation;
(10) to develop an understanding of the interrelationships between organisms, between organisms and their environment, and the dynamic aspects of ecology. This can be extended to include very specific aims, such as those described for autecological investigations (section 11.5).

11.3.1 Writing up the project or investigations

Irrespective of the quality of both the investigation and the data obtained, the project or investigation is of little use to other scientists until it is presented as a report and this should take the following form.

(1) **Introduction:** including the idea, the problems, hypotheses and aims (what you set out to do and why).
(2) **Method:** the strategy of the project (what you did (was done), where and how it was done, including all practical details of apparatus and techniques employed both in the field and in the laboratory).
(3) **Results and observations:** tabulated data, graphs, histograms, profiles, presence–absence graphs, kite diagrams and any other relevant and realistic way of representing data and relationships clearly and concisely.
(4) **Discussion of results:** this involves an analysis of the results, preferably quantitative if possible, tentative conclusions based on data presented and references to already published material.
(5) **Discussion of significance of conclusions:** criticisms of the techniques employed, sources of error and suggestions for further study.
(6) **List of references consulted.**

11.4 A synecological investigation

A synecological investigation involves studying the abiotic and biotic elements associated with a natural community (biotic element of ecosystem) found in a particular defined geographical area (or ecosystem) such as an oak woodland or a rocky shore, which may contain several plant and animal species and possibly several habitats. In such an investigation it is necessary to carry out the following exercises:

(1) map the area and habitat(s) in plan view and, if necessary, in profile;
(2) identify the species and estimate the number of each species present;
(3) measure (possibly collect and analyse) the abiotic factors within the habitat(s).

The overall aim of such an investigation is to determine the qualitative and quantitative relationships between the plant and animal populations within the area being studied and the possible interactions between these and soil, topographic and climatic factors. Given this information, it is possible to explain the nature and extent of the factors governing the number and distribution of organisms in terms of a food web and, depending upon the sophistication of the investigation, pyramids of numbers, standing crop biomass and energy.

11.4.1 Mapping an area

Plan view

The following simple method is designed primarily for mapping a small area, such as a grassland 10 m × 10 m or a small pond, but can be used on a large scale, for example to map the whole rocky shore of a bay.

(1) Select the approximate area to study and stretch a measuring tape along one side of the area. This marks the base line AY (fig 11.17).
(2) From the base line measure the perpendicular distance to certain natural landmarks within the area or marker poles showing the limit of the study area. Record these measurements.
(3) Transfer the measurement of AY and the various perpendicular distances to a sheet of squared paper using a suitable scale.
(4) Using the base line and measured distances to perpendicular landmarks drawn in (3) above as a guide, complete the map freehand.
(5) If the area is relatively small divide the actual base line AY into an equal number of sections and from these lay out perpendicular string line transects. Repeat the procedure using the extreme left transect AF as a new base line to produce a string grid, as shown in fig 11.18. Draw these grid lines on the map and label them using A, B, C etc. along one edge and 1, 2, 3 etc. along the other edge.
(6) Mark the positions of obvious structural and vegetational zones.
(7) Using a quadrat frame, pin frame or sweep net, depending on the area, systematically sample the area from, say, left to right and record the species present and their numbers or abundance.
(8) If the area is extremely large and a qualitative and quantitative study is required, belt transects spaced out at set intervals across the area, and set at right-angles to any suspected zonation, can be used in conjunction with random quadrats to sample the area at particular points called **stations**. If the area has no obvious zonation then a number of random quadrat samples should be used (and their positions noted) rather than a transect approach. Direct measurement of the abiotic features of the environment should be made as frequently as possible or samples removed for subsequent analysis.

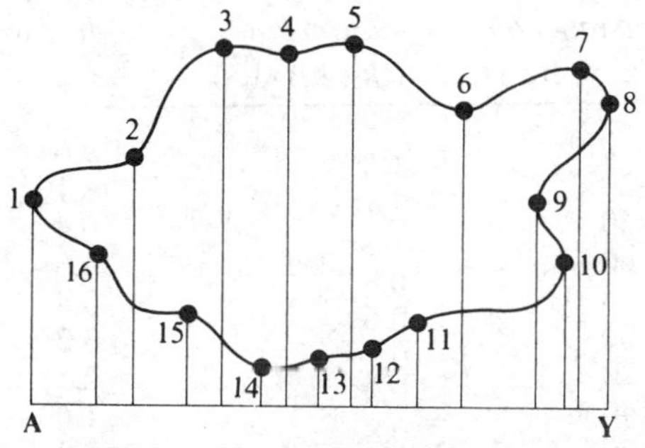

Fig 11.17 *A suggested method of mapping the significant aspects of an area, such as a small, irregularly shaped pond.*

Plotting a height profile

In some areas, distribution of organisms may be influenced by a factor related to height, such as on a rocky shore. Here the length of time each part of the shore is exposed due to the vertical motion of tides is height dependent. In such cases it is necessary to produce a height profile showing how the height along the transect varies, as from high to low water marks in the rocky shore example used below. At each point (station) along the transect where the community is sampled, the height should be obtained accurately by the use of a surveying theodolite and measuring points. Over short distances a simple home-made levelling device attached to a reference pole, and a graduated pole can achieve relatively accurate results as described below (fig 11.19).

(1) Attach the levelling device at a convenient height (h_1), such as 1.5 m, on the measuring pole.

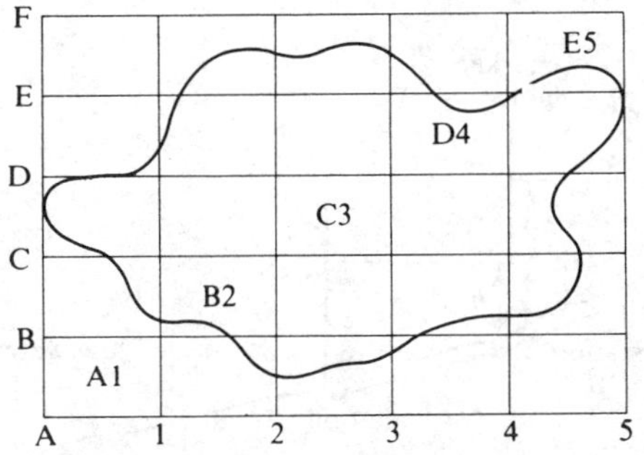

Fig 11.18 *Map of the area under investigation showing the various sub-sections, for example A1–E5, obtained by the use of a string grid. These provide reference areas for subsequent study.*

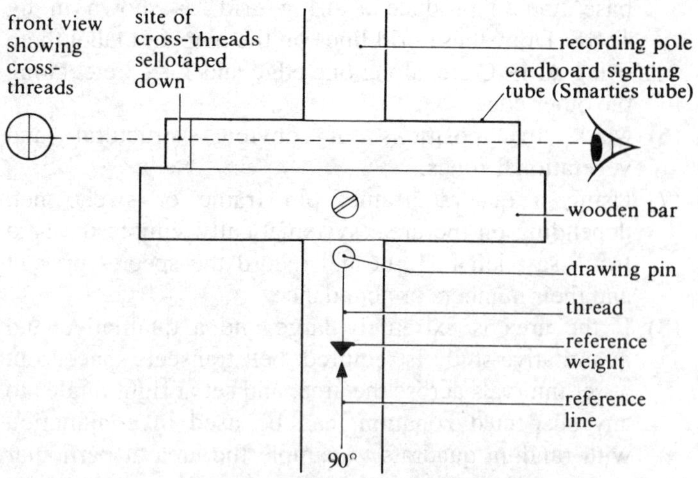

Fig 11.19 *A simple home-made levelling device attached to a reference pole. The position of the pole is adjusted until the sighting bar is shown to be horizontal by the thread indicating 90° on the reference point on the pole. Holding the pole steady, the observer looks along the sighting tube and indicates to the person holding the graduated pole the corresponding level position on their pole, as shown by the cross-wire sights. This height is recorded.*

(2) Set out a line transect from high water mark to the water's edge.

(3) Set up the reference pole at a specific point, such as high water mark, on the transect and the marker pole at a known distance (x) further down the shore. Mark these positions on the transect and label them A and B. Keep to one side of the transect line whilst taking readings to avoid trampling on the specimens to be studied.

(4) When the wooden sighting bar is horizontal, look along the sighting tube identifying the point where the centre of the cross-threads 'hits' the marker pole. The exact position of this point is then located by the person holding the marker pole and the height (h_2) recorded. The height difference between the stations is equal to $h_2 - h_1$.

(5) Move the reference pole to station B and the marker pole a known distance (x_1) to station C. Repeat stages (3) and (4) and record the new height (h_3) (fig 11.20).

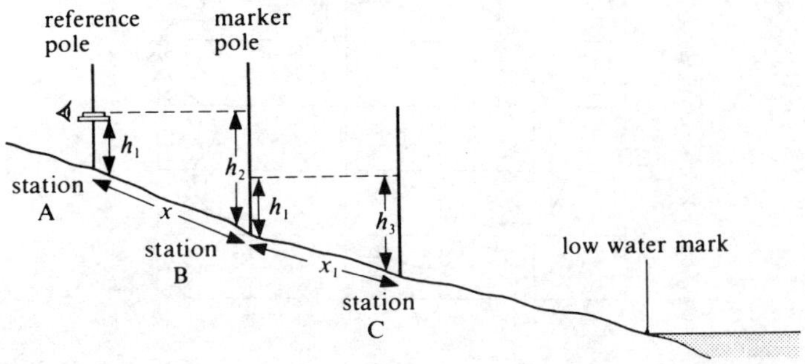

Fig 11.20 *Methods of obtaining heights and horizontal distances of stations above low water mark.*

(6) Continue to obtain readings h_4, h_5 and so on, distances x_2, x_3 and so on and stations D, E and so on, to the water's edge at low water. Record all distances as shown in table 11.6, and calculate the heights and horizontal distances of the stations above the low water mark.

(7) Transfer these data to a scale representation of the shore profile and mark on the positions of the stations (fig 11.21).

11.4.2 Identifying and estimating the number of each species present

Line and belt transects, frame quadrats and pin frames are used to sample systematically the area as described in section 11.2.2. Specimens are identified using a key and the number of organisms are either counted directly or estimated as described in section 11.2.3.

11.4.3 Recording and representing data

Data should be recorded directly they are obtained using some form of field booking sheet. In the case of synecological investigations of marine habitats the information shown on the booking sheets illustrated in tables 11.7 and 11.8 has proved successful. These sheets are best attached to a clipboard, completed in pencil and kept in a large polythene bag to protect them from rain. Once all the data have been collected they must be represented in some suitably efficient diagrammatic form that will highlight relationships between organisms and/or the nature of the environment. Methods of representing data are given in appendix section A2.7 in Book 2 and include presence–absence graphs, kite diagrams, and histograms. Trophic pyramids are described in section 10.3.4. Some examples of the use of all four methods of representation are included in figs 11.22–11.24.

Table 11.6 Horizontal and vertical distances recorded at stations A–K on a rocky shore. Northumberland 1968.

Station	*Horizontal distance/m* (x, x_1 etc.)	*Height between stations/m* ((h_2–h_1) etc.)	*Height above low water/m* ((h_2–h_1) etc.)
A	0		9.6
B	20	1.5	8.1
C	40	1.7	6.4
D	60	1.8	4.6
E	80	0.8	3.8
F	100	0.6	3.2
G	120	0.7	2.5
H	140	0.9	1.6
I	160	0.8	0.8
J	180	0.4	0.4
K	200	0.4	0

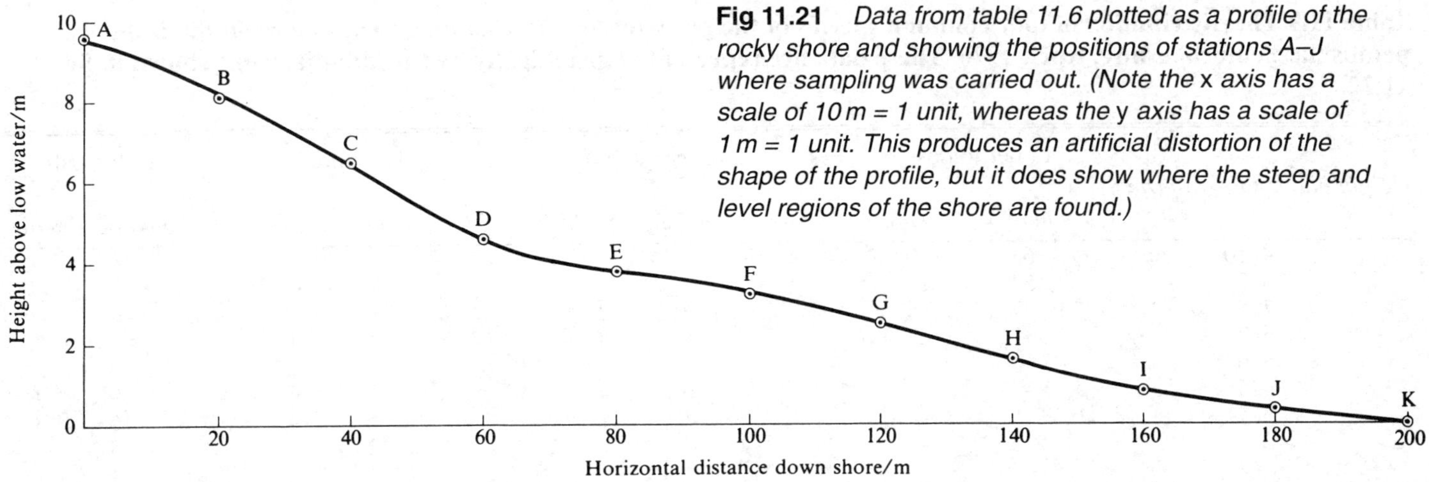

Fig 11.21 *Data from table 11.6 plotted as a profile of the rocky shore and showing the positions of stations A–J where sampling was carried out. (Note the x axis has a scale of 10 m = 1 unit, whereas the y axis has a scale of 1 m = 1 unit. This produces an artificial distortion of the shape of the profile, but it does show where the steep and level regions of the shore are found.)*

Table 11.7 Suggested format of field booking sheet.

Field Booking Sheet – Marine Ecology

(1) Name of site and grid reference . . . .
(2) Nature of profile (rocky, sandy, muddy, dune) . . . .
(3) Sketch map of area showing area(s) of study/position(s) of transects.

(4) Special features (exposure, aspect, etc.) . . . .
(5) Date . . . .
(6) Weather, conditions . . . .
(7) Tide data:
  predicted high water . . . .
  " low water . . . .
  observed high water . . . .
  " low water . . . .
  predicted tidal range . . . .
(8) Notes and key to recorded data (abundance scales – % cover, reference height of level, e.g. h_1 etc.)

Table 11.8 Data required on a booking sheet for investigating the synecology of marine habitats.

Station name	Horizontal distance from origin/m	Level reading (h_2)/m etc.	Change in height (h_2–h_1 etc.)/m	Height above low water/m	Time exposed	Time covered	ANIMALS (species and abundance)	PLANTS (species and abundance)	NOTES

Table 11.9 The distribution of four common species of the periwinkle, *Littorina* on a rocky shore on the Dale peninsula, Pembrokeshire. April 1976. These data are represented graphically by the kite-diagrams shown in fig 11.22.

Height above low water/m	L. neritoides		L. saxatalis		L. littorea		L. littoralis	
	number	*scale*	*number*	*scale*	*number*	*scale*	*number*	*scale*
9–10	63	10	–	–	–	–	–	–
8–9	54	10	–	–	–	–	–	–
7–8	7	4	3	2	–	–	–	–
6–7	–	–	8	4	1	1	–	–
5–6	–	–	17	8	3	2	2	2
4–5	–	–	6	4	13	6	9	4
3–4	–	–	1	1	6	4	16	8
2–3	–	–	–	–	2	2	5	4
1–2	–	–	–	–	1	1	1	1
0–1	–	–	–	–	–	–	–	–

Abundance scale: ⩾20 = 10; 19–15 = 8; 14–10 = 6, 9–5 = 4; 4–2 = 2; 1 = 1.

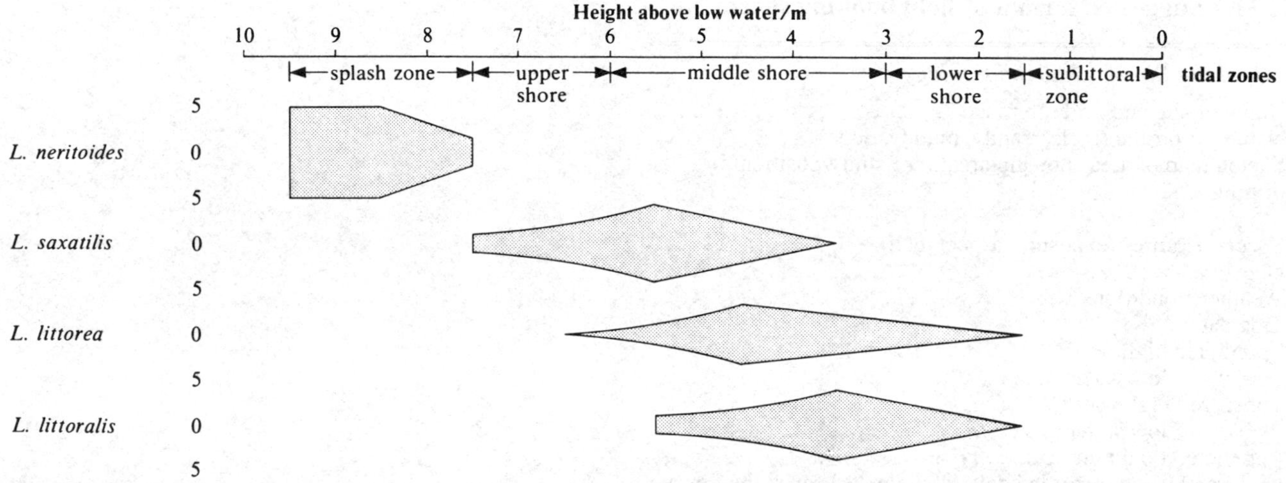

Fig 11.22 *Kite-diagrams showing the frequency and distribution of four common species of periwinkle,* Littorina, *on a rocky shore on the Dale peninsula, Pembrokeshire, April 1976. See table 11.8 for data.*

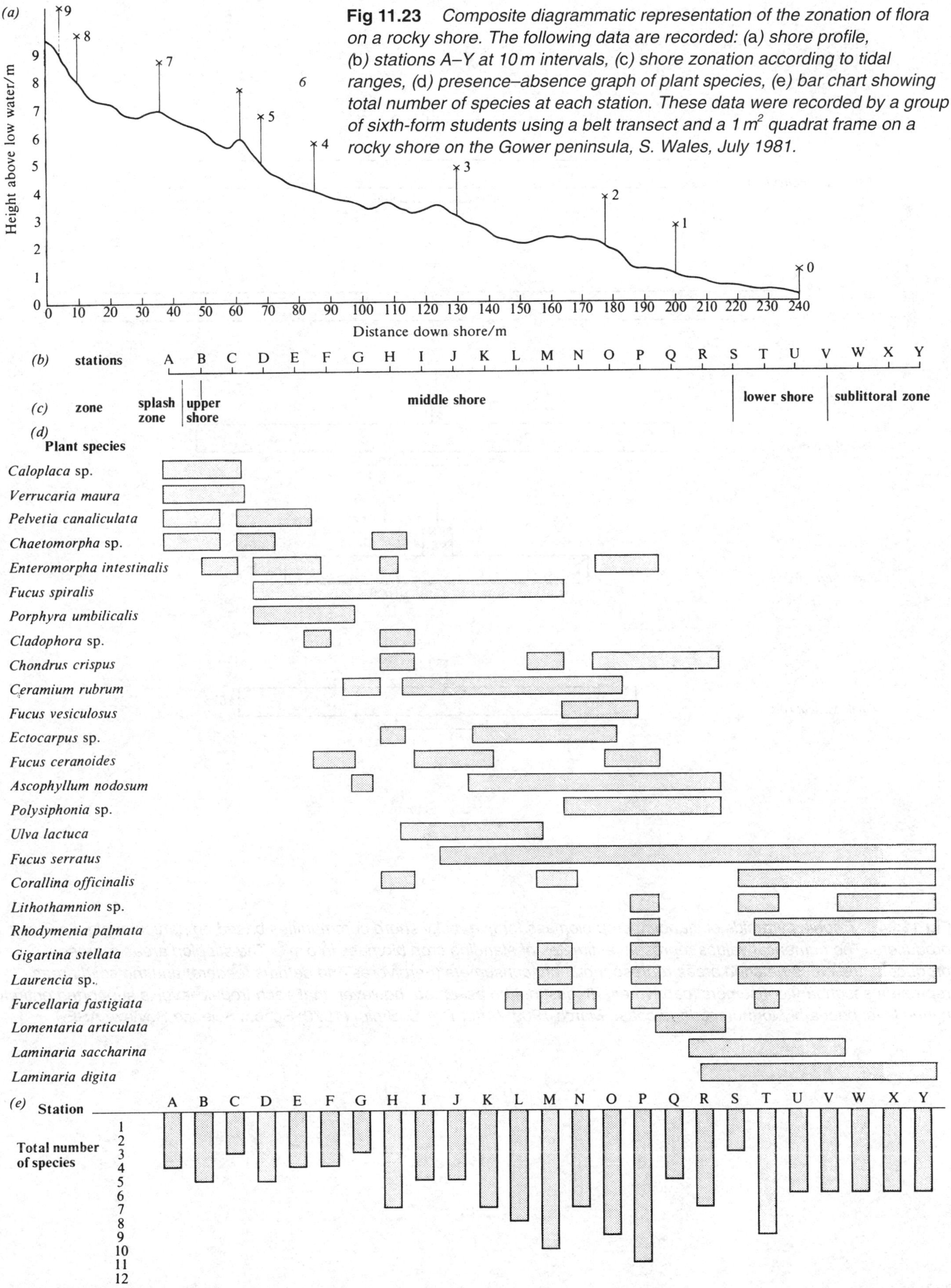

Fig 11.23 *Composite diagrammatic representation of the zonation of flora on a rocky shore. The following data are recorded: (a) shore profile, (b) stations A–Y at 10 m intervals, (c) shore zonation according to tidal ranges, (d) presence–absence graph of plant species, (e) bar chart showing total number of species at each station. These data were recorded by a group of sixth-form students using a belt transect and a 1 m² quadrat frame on a rocky shore on the Gower peninsula, S. Wales, July 1981.*

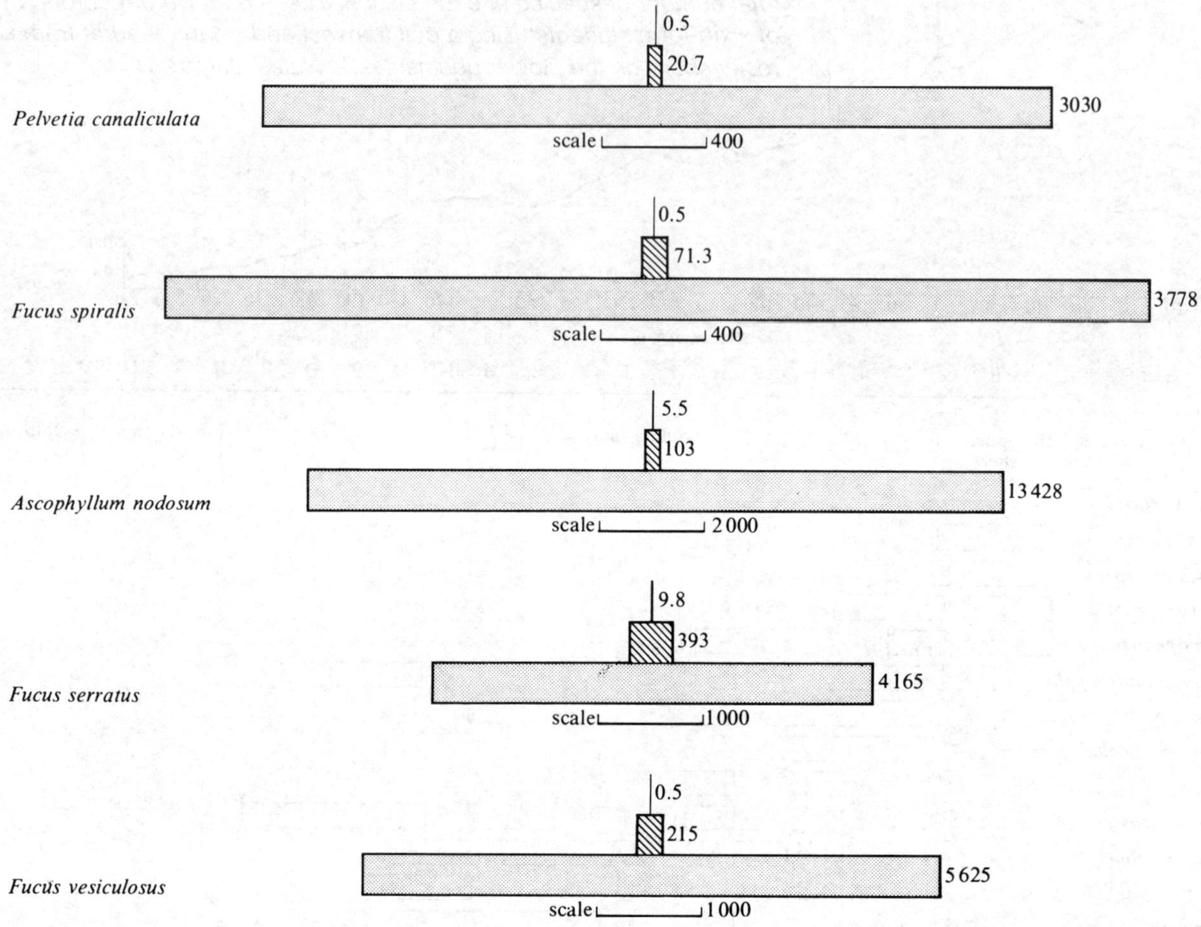

Fig 11.24 *Trophic pyramids of standing crop biomass for five rocky shore communities based on particular algal primary producers. The numerical values represent estimates of standing crop biomass in g m^{-2}. The stippled areas represent producers, the cross-hatched areas represent primary consumers (herbivores and detritus feeders) and the solid areas represent secondary consumers (carnivores). It must not be assumed, however, that each trophic level is supported entirely by the level beneath. Northumberland coast, March 1969. (After D.A.S. Smith (1970)* School Science Review, *ASE.)*

11.4.4 Collecting and analysing abiotic factors

The amount of time spent on this stage of a synecological investigation will depend upon the nature of the area being studied. It is more applicable to areas where soil factors predominate, such as woodland, grassland and salt marsh, than to a rocky shore. The abiotic factors to be studied and the methods of study are described in section 11.1.

11.5 An autecological investigation

An autecological investigation involves studying all the ecological factors related to a single plant or animal species throughout its life cycle. The aim of the investigation is to describe as precisely as possible the ecological niche of the species. The species selected for study should be one which is both common and locally available. Initially the investigation should concern itself with undertaking extensive background reading on the species selected. During the reading, notes should be made on all aspects of the biology of the species and also on opportunities for practical work. This may involve either repeating investigations carried out by others or developing new investigations to be undertaken as part of the current study.

A straightforward approach to an autecological study is to prepare a comprehensive list of the questions which must be answered in order to reveal all there is to know about the species under investigation. The study should be undertaken as rigorously as possible and treated as a research project. Therefore it must involve some measure of original investigation including observation, measurement and experimentation. It must not simply be a report based on knowledge gleaned from reading books, journals and magazines. The species under investigation should be studied over a period of at least one full year.

A guide to the sorts of questions to be asked in the investigations of an **animal** is given below.

(1) **Classification.** What is the name of the species? What other groups of organisms does it resemble most closely? What are the similarities and differences between related species? What is its full taxonomic description?

(2) **Habitat.** Where is it found? What are the characteristic abiotic features of the area? How do these factors change over the course of a year?

(3) **Structure.** What is its adult structure? What are its characteristic external features? What are its dimensions and mass?

(4) **Movement.** How does it move from place to place? Which parts of the organism are involved in the movement and what are the functions of these parts? How are these parts adapted to the environment?

(5) **Nutrition.** What are the food sources of the organism? When does the organism feed? How much food is eaten? How is the food captured and ingested? What special features assist ingestion? Are there any unusual features of digestion and absorption.

(6) **Respiration.** Where is the gaseous exchange surface? How does gaseous exchange occur? How much oxygen is required by the organism?

(7) **Excretion.** What are the waste products of metabolism? How are these removed from the organism? What special organs of excretion are present?

(8) **Reproduction.** Are the sexes separate? What visible external differences are there between the sexes? Does any form of courtship occur? Does the organism defend a territory? How does mating occur? When does mating occur? How often does mating occur? How many gametes are produced? Where does fertilisation occur?

(9) **Life cycle.** How long does development take? What degree of parental care is shown? Are there larval stages? When do adults become sexually mature? What is the typical lifespan of an individual of the species?

(10) **Behaviour.** How does the organism receive stimuli? To which stimuli does the organism mainly respond? How are the major sense organs adapted to the mode of life of the organism? To what extent does learning occur? How does the organism react to other members of the same species? How does the species react to unfavourable weather conditions? How does the organism communicate?

(11) **Ecology.** How many organisms occur in the population? What other organisms live in the same habitat? How are the various species distributed within the habitat? How is the species related to other species in the same habitat in terms of position in food chains and food webs? Is the organism a host, parasite or symbiont? What is the ecological niche of the species?

Similarly the sorts of questions to be asked in the investigation of a **flowering plant** are given below.

(1) **Classification.** What is the name of the species? What subspecies, varieties and ecotypes of the species exist? What are the similarities and differences between closely related species? What is its full taxonomic description?

(2) **Habitat.**
(*a*) *Edaphic factors* – What is the parent rock type? What type of soil profile is shown? How thick are the various horizons? What is the percentage water content (field capacity) of the soil? What is the percentage organic content of the soil? What is the mineral composition of the soil? What is the pH of the soil?

What is the height and seasonal variation of the water table in relation to the life history and distribution of the species?

(*b*) *Climatic factors* – What are the extremes and mean temperatures in the habitats? What is the annual rainfall in the habitats? What is the mean relative humidity of the air in the habitats? What is the direction of the prevailing wind? How much light is received by the plant?

(*c*) *Topographical factors* – To which direction is the species normally exposed? Does the species appear to prefer exposed or sheltered sites? Does the species appear to prefer sloping or flat habitats? Does altitude appear to affect the distribution of the species?

(3) **Structure.** How extensive is the root system? What form does the root system take? How does the stem branch? How many leaves are carried on each branch? What shapes are the leaves? What variations in length and breadth exist between the leaves? How tall does the plant grow?

(4) **Physiology.** What pigments are present in the leaves and petals? Which surface of the leaf has the highest transpiration rate? What effect has darkness on transpiration rate? Do diurnal changes in water content of leaves occur?

(5) **Reproduction.**

(*a*) *Flower* – How many flowers are produced per plant? How many and of what shape and size are the sepals, petals, anthers, carpels or pistil? What variation in petal colour exists? What pigments are present in the leaves? When does flowering begin? How long is the flowering period? How does pollination occur? What adaptations to insect or wind pollination are shown?

(*b*) *Fruits and seeds* – How are the fruits formed? What is the structure of the fruit? How many seeds are produced per flower? How are the fruits and seeds dispersed? How far are fruits and seeds dispersed?

(*c*) *Perennation* – How does vegetative propagation occur? What are the organs of perennation? At what rate does the species colonise an area?

(6) **Life cycle.** What type of seed is produced? What conditions are required for germination? When do the seeds germinate? What percentage of seeds germinate? Which form of germination occurs? At what rate does the shoot system develop? What is the extent of growth in terms of space and time? (Why do some of the seedlings not become mature?)

(7) **Ecology.** Does the species grow as solitary plants or in patches? What size are the patches? Which species grow in the same habitat? What degree of competition exists between the species being studied and other species? Is the species a parasite, host or symbiont? How is the species related to animals in terms of position in the food web? Does the species offer protection or shelter to animals? If so, which animals and how is this provided? What is the ecological niche of the species?

Fungi, algae, mosses, liverworts or conifers may be used in autecological studies and questions above may be modified as appropriate to the species under investigation.

Chapter Twelve

Microbiology and biotechnology

What is microbiology?

Microbiology is the study of microorganisms, those organisms which are so small that they can only be seen clearly, if at all, with the use of a microscope. They include bacteria, viruses, fungi and protoctists such as protozoa and microscopic algae. The classification and some of the chief characteristics of these organisms have been described in chapter 2.

Microorganisms show great diversity and have enormous potential for exploitation by humans. They grow and multiply rapidly given suitable conditions, and use and make a huge range of chemicals. It is this versatility which makes them so useful. They can even be genetically engineered to make further useful products, such as human insulin. Our exploitation of microorganisms is probably only in its infancy, but already it makes an important contribution to our industrial production. One word has come to symbolise the use of microorganisms and other biological systems, namely biotechnology.

What is biotechnology?

Biotechnology has been defined as the application of organisms, biological systems or biological processes to manufacturing and servicing industries. It does not refer simply to microorganisms, although microorganisms play an important role. In fact, use of any biological process in a manufacturing process could be regarded as biotechnology. This would include, for example, genetic engineering and cloning of plants in agriculture, horticulture and forestry. Examples of biotechnology are shown in fig 12.1.

Biotechnology provides both **products** and **services**. Examples of products are alcohol in the brewing industry and newer products such as human insulin from genetically engineered bacteria. Examples of services are treatment of sewage or detection of pollution using a biosensor. Here the process rather than an end-product is important.

Biotechnology could be defined more broadly as the use of other living organisms for the benefit of humans. It would then include the breeding and improvement of domesticated animals such as cattle or pigs, and crops such as wheat and potatoes. New techniques of genetic engineering are particularly relevant here because they give us a way of introducing new desirable factors into living organisms much more precisely and quickly than the breeding programmes of the past.

In this chapter we shall examine some of the principles of large-scale manufacture in biotechnology and look at some examples of biotechnology. The new technologies will continue to change our society just as the old ones have done. Inevitably they will raise new social and ethical issues, and some of these are discussed at the end of chapter 25 on applied genetics.

12.1 Requirements for growth

Scientists first began trying to grow bacteria and fungi under controlled conditions in the mid-nineteenth century. The two great pioneers working on bacteria were Louis Pasteur, working in Paris, and Robert Koch, working in Berlin. These and other scientists realised the importance of being able to grow pure cultures of the organisms they were trying to study. This meant somehow developing techniques by which different microorganisms could be isolated from each other. In addition, suitable nutrients and environmental conditions for growing the microorganisms had to be developed. Some microorganisms, particularly those that cause disease and therefore live parasitically, have complex requirements; others have relatively simple requirements.

12.1.1 Essential nutrients

Raw materials are needed for growth, maintenance, and multiplication. In addition, a source of energy is needed. When microorganisms are grown, nutrients are supplied in a suitable medium known as the **nutrient medium**. The following are typical components of nutrient media.

- **A source of carbon** – Most bacteria and all fungi and protozoans are heterotrophic, meaning that they need an *organic* source of carbon (see section 2.5.4 and table 2.3). This is often provided in the form of glucose or the salt of an organic acid such as sodium ethanoate (acetate). However, as a group, bacteria in particular can use a vast range of organic compounds as a source of carbon, including fatty acids, alcohols, proteins, hydrocarbons and methane. Some soil bacteria and fungi, and some bacteria that live in the guts of herbivores such as ruminants, can digest cellulose and use it as a carbon source. All disease-causing bacteria are heterotrophic.

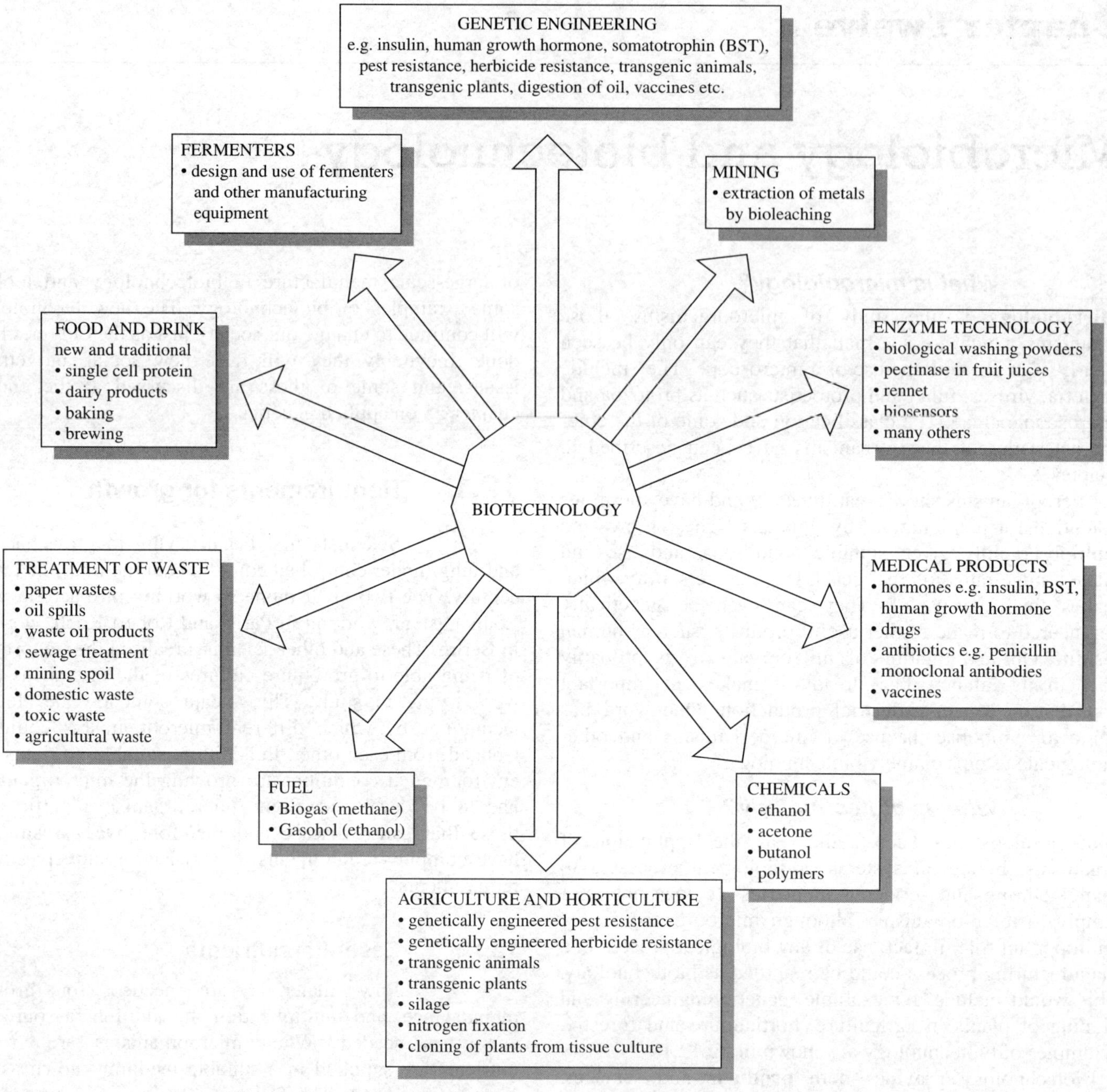

Fig 12.1 *Some of the applications of biotechnology. Categories are not rigid and may overlap. Genetic engineering is an important technique in biotechnology. It can be involved in any of the categories shown in order to 'improve' microorganisms, plants and animals. It will become increasingly important in the twenty-first century.*

All the algae and some bacteria, such as the blue-green bacteria, are autotrophic, meaning that their source of carbon is carbon dioxide. The algae are photosynthetic, but some of the bacteria are photosynthetic and some are chemosynthetic (table 2.3).

- **A source of nitrogen** – This may be organic, for example amino acids, peptides and proteins, or inorganic, for example ammonium salts and nitrates. Amino acids are often supplied in the form of partially digested proteins called peptones.
- **Growth factors** or **vitamins** are sometimes needed. Growth factors are the equivalent of the vitamins needed by animals, and many *are* vitamins. They are organic compounds which are essential for growth and

are needed in very small amounts. They include some of the B vitamins (thiamine or B_1, riboflavine or B_2, niacin or B_3, and B_6) as well as folic acid and para-amino benzoic acid. Trace amounts only are needed for healthy growth. Other organic compounds may also be important, such as purines and pyrimidines.

Microorganisms vary in their ability to make their own growth factors from simpler substances. If their needs are complex, the medium used to grow them in a laboratory is often derived from natural substances in which they normally grow, such as blood, soil, meat extract or yeast extract.

- **Mineral salts** – Those most commonly needed are the positive ions calcium, potassium, sodium, iron and magnesium, and the negative ions chloride, phosphate (a possible source of phosphorus) and sulphate (a possible source of sulphur). As mentioned above, nitrogen may be supplied as ammonium or nitrate. The needs of algae are similar to those of plants (see tables 7.7 and 7.8).

> **12.1** Give one example of the role of each of the following: (i) iron and phosphate in heterotrophic bacteria, (ii) nitrate and magnesium in autotrophic bacteria. (Use table 7.7 for help if you need it.)

- **A source of energy** – The need for energy by living cells is discussed at the beginning of chapters 7 and 9. The energy may be supplied as chemical energy or light energy. If it uses chemical energy, the organism is described as chemotrophic; if it uses light energy, it is described as phototrophic or photosynthetic (table 2.3). Photosynthetic microorganisms include some bacteria, such as the blue-green bacteria, and algae. When chemical energy is needed it is usually supplied as a sugar such as glucose.
- **Water** – Although this is not strictly a nutrient, it is required by all living cells. In general bacteria need more moisture than yeasts, and yeasts need more than moulds.

An example of a relatively simple culture medium is given in table 12.1.

> **12.2** What is each of the ingredients in the medium in table 12.1 needed for?

12.1.2 Environmental variables

As well as nutrients, environmental variables have to be carefully controlled. All microorganisms will grow best at a particular temperature and pH, and they may be sensitive to factors such as oxygen concentration and light intensity. **Optimum conditions** are the conditions which give the shortest generation time. Each species of microorganism will be adapted for a particular environment.

Temperature

Microorganisms differ widely in their temperature requirements, so temperature has an important effect on which microorganisms grow in a particular environment. The effect of temperature is best understood by considering its effects on enzymes. All cells contain enzymes, and the rate at which enzymes work, and therefore the rate of cell reactions and growth, is affected by temperature. A given microorganism will have an optimum temperature at which it grows best and a temperature range outside of which it will not grow. The upper limit is usually the point at which certain vital enzymes begin to denature. The optimum temperature ranges from 0 °C to 5 °C for some ocean-dwelling bacteria at or near the poles, to 70 °C or higher for bacteria living in hot springs or rotting vegetation such as compost heaps or silage where temperatures can be relatively high. There are bacteria which live in or close to volcanic vents in the ocean floor which thrive in temperatures of 118 °C.

Microorganisms can be placed in four categories according to their temperature requirements. The temperatures quoted below are only a general guide because there is a continuous range of types.

- **thermophiles** – optimum temperature for growth is above 45 °C, e.g. some species of *Lactobacillus* (found in milk).
- **mesophiles** – optimum temperature for growth is 25–45 °C. For bacteria living in mammals the optimum temperature is about 37 °C with a maximum of 42–43 °C, e.g. *Escherichia coli* (*E. coli*), a bacterium which is common in the human gut. Yeast (a fungus) is also mesophilic.
- **psychrophiles** – optimum temperature for growth is below 15 °C and they cannot grow above 20 °C. These can be a problem in refrigerated food.
- **psychrotrophs** – optimum temperature above 15 °C but can grow at much lower temperatures (5 °C or lower), e.g. some *Candida* yeasts, some moulds. These organisms are important in the spoilage of refrigerated dairy products.

pH

As with temperature, each species has its own minimum, maximum and optimum pH at which it will grow. For most bacteria the optimum is around pH 7 (neutral). Microorganisms that thrive in acid conditions, such as the

Table 12.1 A relatively simple medium used for growth of *Escherichia coli*, a bacterium which is common in the human gut.

		Concentration/ g dm^{-3}
K_2HPO_4	dipotassium hydrogen phosphate	7.0
KH_2PO_4	potassium dihydrogen phosphate	2.0
$(NH_4)_2SO_4$	ammonium sulphate	1.0
$MgSO_4$	magnesium sulphate	0.1
$CaCl_2$	calcium chloride	0.02
glucose		10.0
distilled water to 1 dm^3		

bacteria that produce vinegar, are described as **acidophiles**, while those that thrive in alkaline conditions are described as **alkalophiles**. Most yeasts are capable of growing in acid conditions, with an optimum pH of 4.5–5.0 being typical. Moulds prefer slightly acid conditions.

Oxygen concentration

Oxygen is needed for aerobic respiration. Microorganisms differ in their oxygen requirements as follows.

- **Obligate aerobes** – Microorganisms that can only live if oxygen is present, e.g. *Mycobacterium tuberculosis*, the bacterium that causes tuberculosis, moulds, e.g. *Penicillium*, algae and most protozoans. Most bacteria and fungi are obligate aerobes.
- **Facultative aerobes** – These use oxygen if it is present but can survive anaerobically in its absence, e.g. *E. coli* and many other bacteria, yeasts and some protozoans.
- **Obligate anaerobes** – Microorganisms which survive only in anaerobic conditions such as mud or the rumen, e.g. *Clostridium tetani* which lives in soil and is the cause of tetanus, and *Clostridium botulinum* which can be found in damaged tins of food rich in protein such as meat, and causes botulism, an often fatal form of food poisoning.
- **Microaerophiles** – These microorganisms grow best at concentrations of oxygen much lower than air, e.g. *Lactobacillus* in milk.

12.3 If bacteria of types (i) to (iv) below were mixed with nutrient agar in four separate test tubes, show by means of dots the distribution you would expect to find: (i) aerobic, (ii) anaerobic, (iii) facultative aerobic, (iv) microaerophilic.

Ionic and osmotic balance

When a cell is in a solution which is more concentrated than its contents, it usually loses water by osmosis. However, halophiles are organisms which are specialised for living in high salt concentrations, for example in the Salt Lakes of America, the Dead Sea in Israel and in salt marshes. An example of a halophile is *Halobacterium*. In contrast, if protozoans did not have a contractile vacuole which collects excess water and removes it, they would swell and burst in dilute media because they do not have a cell wall (section 2.3.1). The correct balance of ions is also important.

12.2 Culture media

A **medium** (plural *media*) is a solid or liquid preparation containing nutrients for the culture (growth) of microorganisms, animal cells or plant tissue cultures. A **culture** is a collection of microbial cells growing on or in a medium.

12.2.1 Solid and liquid media

Microorganisms may be cultured in a **solid** medium or a **liquid** medium (a **broth**).

Solid media

Solid media are particularly suitable for bacteria and fungi and are prepared by mixing the liquid nutrient solution with a gelling agent, usually **agar**, at a concentration of about 1–2%, thus producing **nutrient agar**. Agar is an extract from red algae. At a concentration of 1–2% agar melts at 95–100 °C and sets at about 44 °C. It can be sterilised by heating first (section 12.2.4) and then cooling. Microorganisms can either be added to the surface of the agar after setting or, if they are not damaged by temperatures of around 44 °C, can be added just before setting so that they can be distributed throughout the medium. Agar has the advantage that it is transparent and, being a complex polysaccharide, is not easily broken down by microorganisms. Examples of the uses of solid and liquid media are given in the following sections. Plant tissue cultures are sometimes grown on solid media (chapter 22).

Liquid media

Liquid media are often useful for measuring population growth (section 12.5). They may be placed in a test tube, stoppered by a plug of cotton wool or a metal cap, or in a glass, screw-capped bottle such as a universal bottle (McCartney bottle, fig 12.3) which holds about 25 cm^3, enough for one agar plate. The medium must be sterilised before it can be used for growth of a cell culture. Adding a small quantity of cells to the medium is called **inoculation**. After inoculation, the medium is left in a thermostatically controlled incubator at the optimum growth temperature. Cells grown are evenly spread throughout the medium.

If large volumes of medium are used, they should be

agitated to ensure mixing and to prevent the cells from settling out. Mechanical shakers or a magnetic stirrer can be used. Large volumes should also have sterile air passed through them in order to maintain the oxygen concentration throughout the medium. Commercial filters or use of glasswool or non-absorbent cottonwool can be used to filter the air and sterilise it. Air is introduced through a sparger, a device with many small holes to give fine bubbles. It is fitted to the end of a tube leading to the bottom of the culture vessel. Liquid cultures may be grown as batch cultures or continuous cultures (section 12.10).

12.2.2 Enrichment and selective media

An **enrichment medium** is a medium in which substances are added to meet the requirements of certain microorganisms in preference to others. As a result, certain microorganisms grow better than others. Starting with a mixed culture, often taken from the 'wild', one or a few types eventually come to dominate. Using a suitable medium, it is therefore possible to selectively grow the microorganism wanted. Conditions such as temperature, pH, carbon source and energy source are modified until those are found that favour the chosen organism as much as possible. For example, the Salmonella bacteria that cause typhoid (*Salmonella typhi*) can be selected for and later positively identified by growing a sample of food or faeces in a medium containing a selenium compound (selenium is an element with similar properties to sulphur). Normal gut bacteria such as *E. coli* which might be present in a faecal sample are inhibited by the selenium. Enrichment culture is one of the most powerful tools available to the microbiologist for isolating microorganisms.

A **selective medium** is one in which one or more substances are added which inhibit the growth of all but one or a few organisms. Examples include the addition of penicillin to a culture to select for those organisms resistant to it, or the selection of hybridised cells during the production of monoclonal antibodies (section 12.11.2).

> **12.4** How would you set about trying to isolate from the soil an organism which could use atmospheric nitrogen as its only source of nitrogen (a nitrogen-fixing organism)?

12.2.3 Indicator media

These are media which contain an indicator which enables colonies of one organism to be visibly distinguished from those of another. For example, certain bacteria can break down ('ferment') the sugar lactose to an acid. In a medium known as MacConkey's agar, which contains the pH indicator neutral red and bile salts, colonies of lactose fermenters will appear red and non-lactose fermenters colourless or pale pink. This is useful, for example, in detecting whether water is contaminated by sewage. Such water contains gut bacteria such as *E. coli* which ferment lactose. Similarly, if *Salmonella* were present in a sample of gut bacteria it would form colourless colonies in MacConkey's agar.

12.2.4 Preparing media

Dried media can be bought commercially, already containing agar. They are normally soaked for 15 minutes in water and then autoclaved in bottles or flasks for 15 minutes at 121 °C to sterilise. The medium mixes and dissolves during autoclaving. Inside an autoclave, water is boiled under pressure. The autoclave has a lid with a safety valve to let out the steam when the required pressure has been reached. The higher the pressure, the higher the boiling point of the water. It is used to sterilise solutions and equipment such as glassware. In order to kill all bacteria and their resistant spores, a pressure of 103 kPa (15 pounds per square inch) for 15 or 20 minutes is normally recommended. At this pressure, the temperature inside the autoclave will reach 121 °C.

12.3 Aseptic technique

Aseptic technique is using sterilised equipment and solutions and preventing their contamination while in use. Bacteria and fungal spores are abundant in most environments, including laboratories. The microbiologist uses a range of special techniques and apparatus which are designed to prevent contamination of nutrient media. Although a 'nuisance', these quickly become part of the daily routine of being a microbiologist. You will only be able to take basic precautions in school or college laboratories. Special laboratories are needed for routine microbiological work. These will have easily cleaned surfaces and special enclosed benches which receive filtered sterile air (fig 12.2).

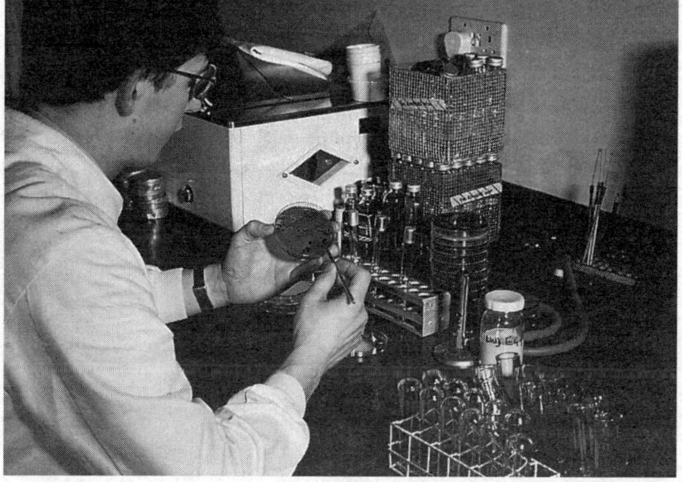

Fig 12.2 *A modern microbiology laboratory with a laminar flow bench in use.*

12.3.1 Pouring plates

This is one of the most basic and useful of microbiological techniques. A **plate** is a petri dish containing nutrient agar. Petri dishes are specially designed shallow circular dishes made of glass or plastic. They are used for growth of bacteria, fungi or tissues on a solid nutrient medium. They are typically about 9 cm in diameter. Glass dishes can be re-used after autoclaving. Plastic dishes are disposable and are usually autoclaved to destroy the culture. They melt in the process. They are bought in sealed packs which have been sterilised, usually by radiation with gamma rays. The lid prevents contamination, but gas molecules can diffuse between the inside of the dish and the environment through microscopic irregularities where the base meets the lid. Oxygen can therefore reach the culture and carbon dioxide can escape.

The procedure for pouring molten nutrient agar into a petri dish ('pouring a plate') is shown in fig 12.3. It is assumed that the agar has been prepared in small bottles (McCartney bottles).

12.4 Methods of inoculation

The introduction of a small number of microorganisms into a nutrient medium is called **inoculation**. Aseptic technique must be used to avoid contamination. The procedure differs for solid and liquid media.

12.4.1 Inoculating a solid medium

Streak plate or dilution plate

The technique is illustrated in fig 12.4. It is suitable for isolating pure colonies of bacteria from a mixture of bacteria. An inoculating loop is used, which must first be 'flamed', as shown in fig 12.4*a*, to sterilise it. The wire loop is then used to lift a thin film of liquid suspension, or a small amount of solid material, containing the microorganisms being investigated from the previous culture or source of the microorganisms. The loop is gently stroked across the surface of the medium in a series of sets of streaks. The plate is partly rotated between each set of streaks so that each set spreads out the bacteria from the previous set of streaks, diluting them until they are separated into single bacteria (fig 12.4). (Don't expect to see anything along the final streaks until after incubation!) The streaks can be made very rapidly once the technique has been practised.

Bacteria from natural habitats like soil, milk and water can be isolated in this way. In the case of a solid medium like soil, it may be easier to suspend a sample in water, or to grow a sample in liquid medium first. Pasteurised milk is a safe source for routine work. Safety regulations and guidelines should always be consulted before carrying out experiments with bacteria or fungi to minimise the risk of harmful organisms being cultivated.

Spread plate

This technique is illustrated in fig 12.5. It is suitable for inoculating solid media with cells from a liquid suspension. It is used when determining the number of living cells (the 'viable count') in a sample after serial dilutions (see section 12.6.1). It can also be used to produce a continuous 'lawn' of microorganisms over the surface of the agar by using a heavy inoculation. This is convenient for testing the activity of inhibitors, such as antibiotics or disinfectants, which can be added to wells cut in the agar or to filter paper discs added to the surface of the agar. The inhibitor can diffuse through the agar, causing a zone of inhibition around the wells or filter paper which is visible after incubation. The diameter of the zone can be used as a measure of the degree of inhibition.

Pour plate

This is an alternative to the spread plate method for inoculating with cells from a liquid suspension and for counting living cells. Larger numbers can be counted, up to 1000 colonies per plate, because the cells grow throughout the medium, not just on its surface. Smaller colonies result (fig 12.6).

A known, small volume (up to 0.5 cm^3) of cell suspension is added to a suitable volume (about 15–20 cm^3) of sterilised molten nutrient agar in a small bottle which has been cooled to about 45 to 50 °C in a water bath. The lid of the bottle should be removed and the mouth of the bottle flamed as shown in figs 12.3*a* and *b* before adding the cell suspension. Mix the cell suspension thoroughly with the nutrient agar in the bottle by rotating (not shaking) the bottle back and forth between the palms of the hands as shown in fig 12.6*a*. Then pour the mixture into a sterile petri dish as shown in fig 12.3*c*. Label the base of the dish and incubate. The final appearance of the plate is shown in fig 12.6*b*.

(*a*)

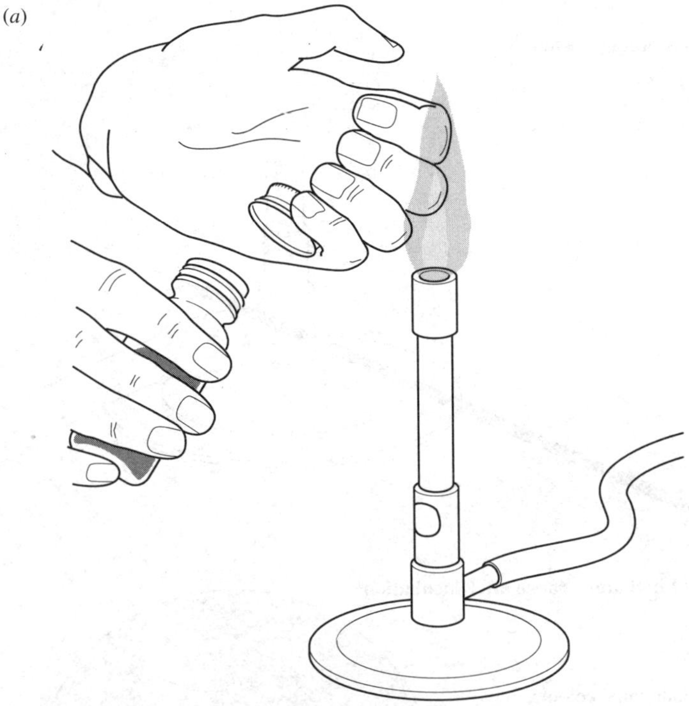

Allow liquid nutrient agar to cool to about 50 °C (just cool enough for comfortable handling) after removal from autoclave. Unscrew cap of bottle with little finger as shown. It can be held by the little finger during the procedure to avoid placing it on the bench where it could become contaminated.

(*b*)

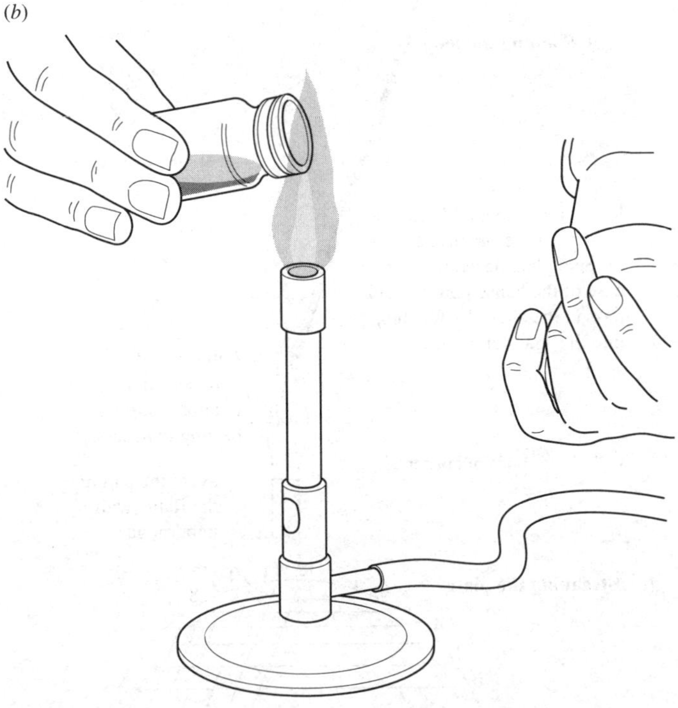

'Flame' mouth of bottle for a few seconds. This is done to produce an upward flow of air from the bottle so that any organisms in the area will not fall into the bottle. (It is not done to kill microorganisms – the bottle would have to be heated for too long to achieve this effect.)

(*c*)

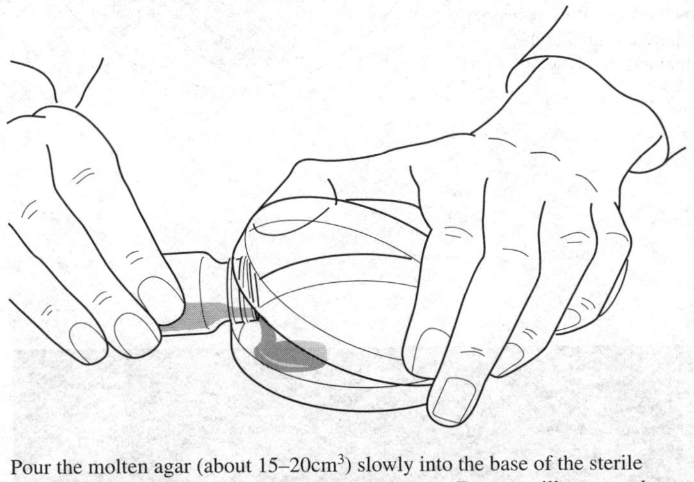

Pour the molten agar (about 15–20cm^3) slowly into the base of the sterile petri dish, lifting the lid only as much as necessary. Do not spill agar on the edges of the dish (if so, start again). Replace the lid and allow the agar to set.

(*d*)

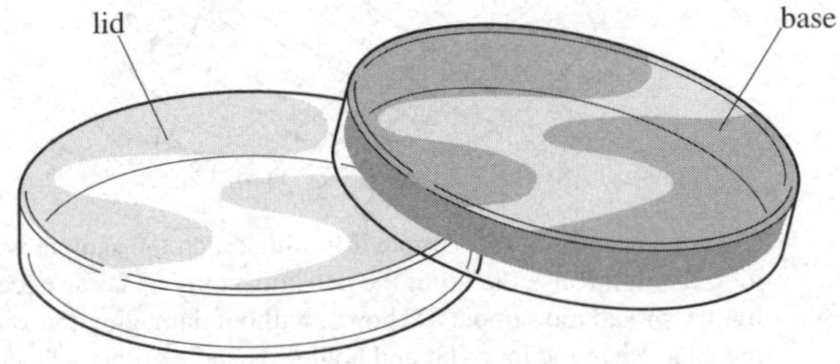

plates are dried with base resting on lid as shown

drying several plates

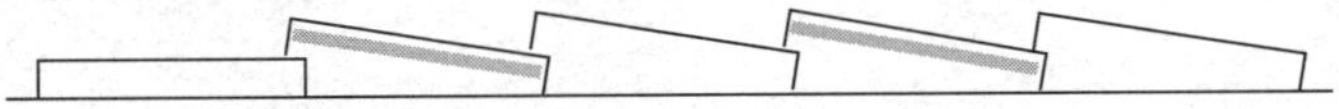

If the plate is to be used for 'streaking' or 'spreading' the surface of the agar and lid can be dried once the agar is set. They are best dried upside down in an incubator at about 37 °C, and arranged as shown to minimise the risk of contamination (slight in these circumstances).

Fig 12.3 *Procedure for pouring a plate.*

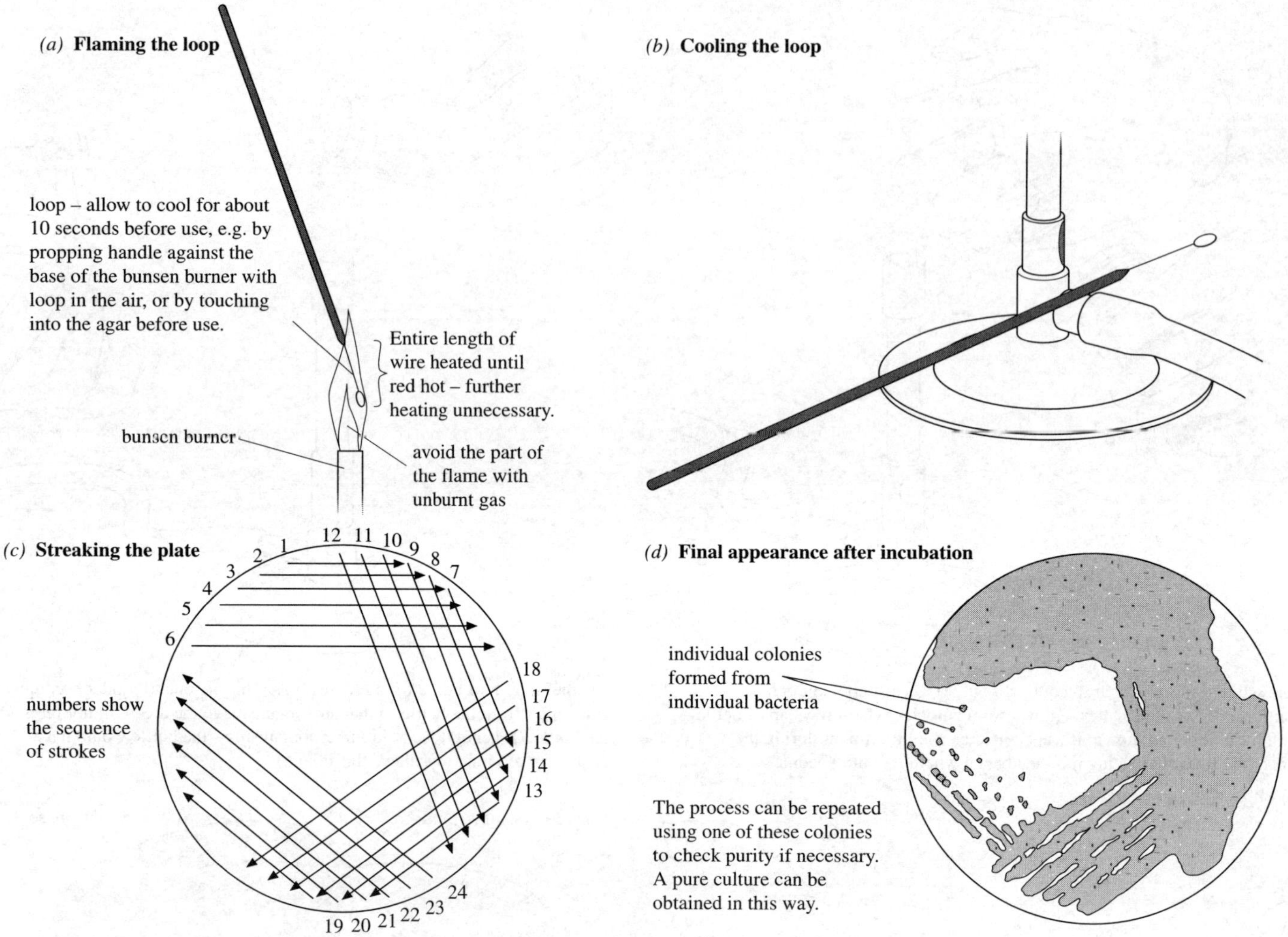

Collect a loopful of the sample if liquid. Touch the sample with the loop if solid. Raise the lid of the petri dish only as far as necessary and lightly spread the sample as shown, without damaging the surface of the agar. Keep the loop flat and hold the handle at the balance point near the centre. Each line represents one stroke with the wire loop. Flame the loop and cool between each set of six lines. Flame again at the end. Label the base of the plate*. Incubate upside down to prevent any condensation falling onto the cultures.

* **Labelling the plate**
- date
- initials
- contents

If you use mirror-writing you will be able to read it through the lid

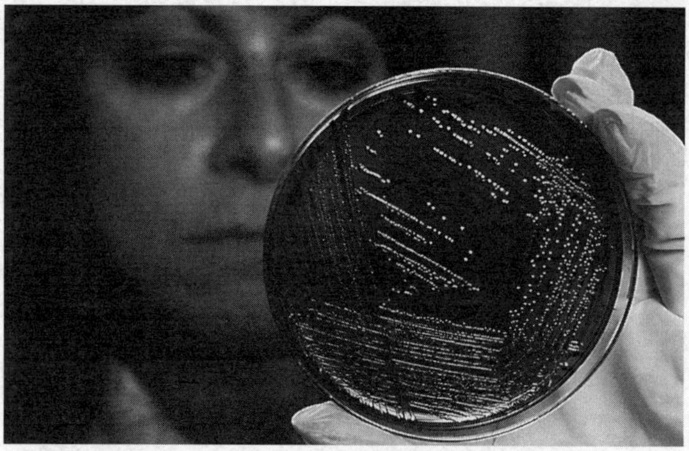

Fig 12.4 *Streak or dilution culture.*

(*a*) Stand the spreader in a small beaker of alcohol.

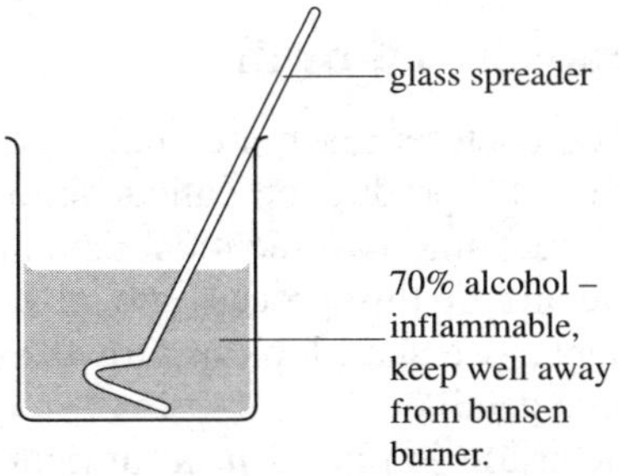

(*b*) Transfer a known, small volume (up to $0.5\,\mathrm{cm}^3$) of cell suspension to the surface of the nutrient agar in a petri dish. A teat on the end of the pipette helps to control delivery.

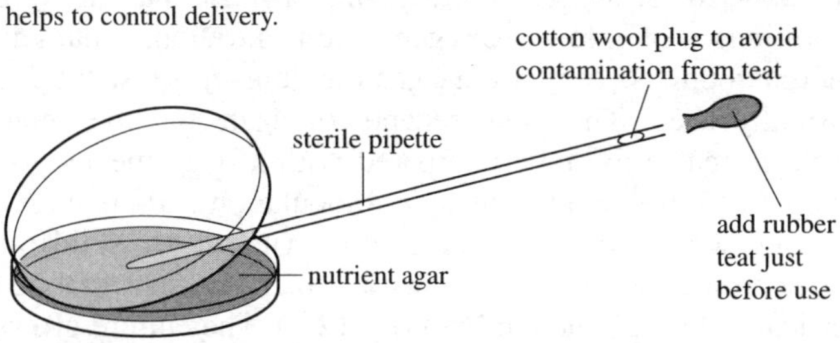

(*c*) Remove the spreader from the alcohol and allow excess alcohol to drain from it.

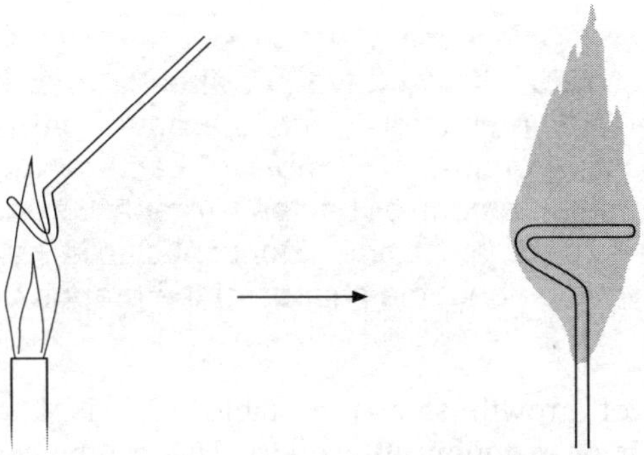

Ignite the alcohol by passing the spreader through a bunsen flame. Do not leave in the bunsen flame or it may crack. (Alcohol burns at a much lower temperature than that of the bunsen flame.) After the flame has disappeared, allow to cool for about 10 seconds. Store in the beaker after use.

(*d*) Push the liquid over the surface of the agar using the spreader. Rotate the plate at the same time to ensure an even coverage. Label the base of the plate and incubate.

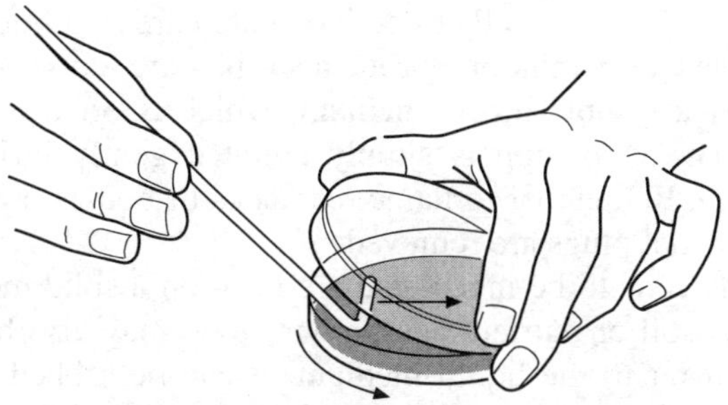

(*e*) **Final appearance after incubation**

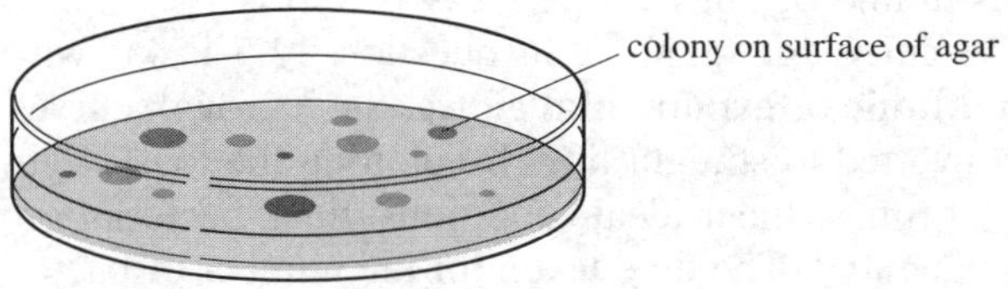

Count the number of colonies. This corresponds to the number of bacteria originally inoculated.

Fig 12.5 *Spread culture (above).*

(*a*)

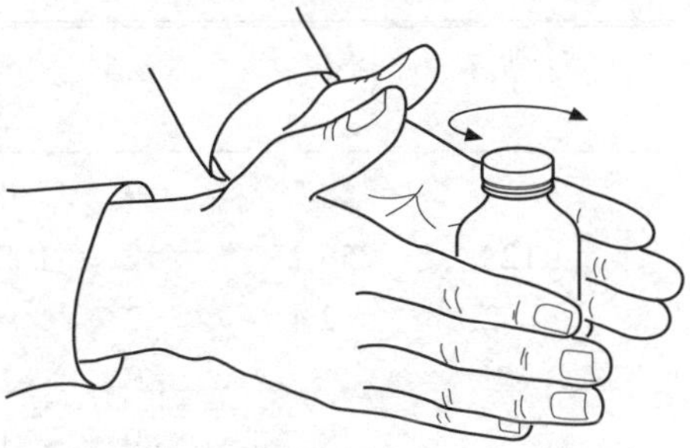

(*b*)

Fig 12.6 *Pour plate.*

Stab culture

This is used for anaerobic organisms or those that thrive in conditions of low oxygen concentration, namely microaerophiles. A nutrient agar medium in a test tube is normally used. The greater depth of agar, and the small surface area in the tube compared with a dish, means less oxygen can diffuse into the agar. Inoculation is done with a straight wire (no loop). A sample (liquid or solid) is collected at the tip of the wire which is then stabbed vertically through the medium (fig 12.7). The culture grows out from the stab line.

12.4.2 Inoculating a liquid medium

If the cells to be inoculated are in a liquid, for example water, milk or a broth, a sterile wire loop is used to transfer a sample to the medium, which is often in a test tube. The wire loop is simply agitated gently inside the medium. Remember to flame the necks of bottles if caps or cotton wool plugs are removed.

If the cells to be inoculated are in or on a solid medium, such as soil or nutrient agar, a wire loop may also be used for transfer to the liquid medium. It can be rubbed on the inside surface of the vessel containing the liquid medium to ensure successful transfer.

In both cases the tube containing the liquid medium can be tapped afterwards to help mix the culture.

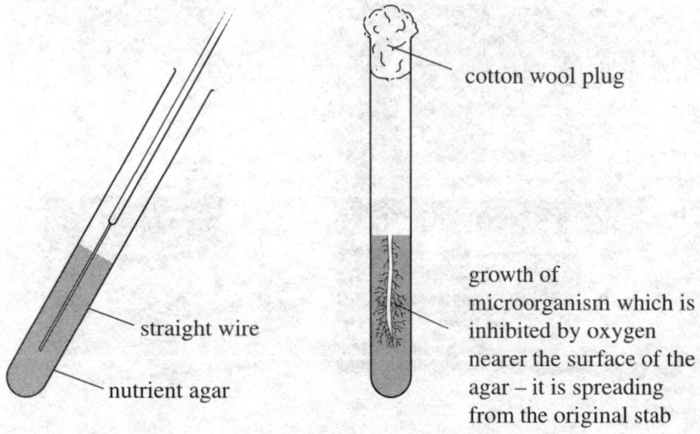

Fig 12.7 *Stab culture.*

12.5 Bacterial growth

12.5.1 Population growth

When bacteria reach a certain size they divide by a form of asexual reproduction called **binary fission** in which the cell divides into two identical daughter cells (fig 2.11). Further details of this process are given in section 2.5.3. In this chapter we shall look at growth of the whole population in more detail.

If a single bacterium is placed in a nutrient medium in optimum growth conditions, it, and its descendants, should divide every 30 minutes, as shown in table 12.2.

> **12.5** If you have not already answered question 2.1 in chapter 2, use the data from table 12.2 to draw graphs of number of bacteria (graph A) and $\log_{10}$ number of bacteria (graph B) on the vertical axes against time (horizontal axis). What do you notice about the shapes of the graphs?

The kind of growth shown in table 12.2 is known as **logarithmic** or **exponential growth**. This can be explained by looking at line C in table 12.2 where the number of bacteria is expressed as a power of 2. The power can be called the logarithm (log) or exponent. The logs or exponents form a linearly increasing series 0, 1, 2, 3, etc., corresponding with the number of divisions.

The numbers in line A of table 12.2 can be converted to logs to the base 2 as shown in line D. (Compare line C with line D.) However, it is conventional to use logs to the base 10, as in line B. Thus 1 is 10^0, 2 is $10^{0.3}$, 4 is $10^{0.6}$, etc.

The curve for graph A in question 12.5 is known as a **logarithmic** or **exponential curve**. Such growth curves can be converted to straight lines by plotting the logs of growth against time. Under ideal conditions, then, bacterial growth is exponential. The time taken for the number of bacteria to double is constant during exponential growth. This is

Table 12.2 Growth of a model population of bacteria.

	Time /h	*0*	*0.5*	*1*	*1.5*	*2*	*2.5*	*3*	*3.5*	*4*	*4.5*	*5*
	*	0	1	2	3	4	5	6	7	8	9	10
A	**	1	2	4	8	16	32	64	128	256	512	1024
B	***	0.0	0.3	0.6	0.9	1.2	1.5	1.8	2.1	2.4	2.7	3.0
C	****	2^0	2^1	2^2	2^3	2^4	2^5	2^6	2^7	2^8	2^9	2^{10}
D	*****	0	1	2	3	4	5	6	7	8	9	10

* Number of divisions
** Number of bacteria
*** $\log_{10}$ number of bacteria
**** Number of bacteria expressed as power of 2
***** $\log_2$ number of bacteria

known as the **doubling time** or the **generation time** and can be calculated from the graph.

The ideal model of bacterial growth can be compared with the growth of a real population in a closed vessel (with no external changes such as renewal of nutrients). Fig 12.8 shows such growth. Note that two curves are shown, one for the total number of bacteria, including all those that have died. In practice, this is an easier number to determine (section 12.6.2). A more useful graph to study, though, is the number of living bacteria, known as the **viable count** (section 12.6.1). This curve has four distinct phases. The first is the **lag phase** in which the bacteria are adapting to their new environment and growth has not yet achieved its maximum rate. The bacteria may, for example, be synthesising new enzymes to digest the particular range of nutrients available in the new medium.

The next phase is the **log phase** when growth is proceeding at its maximum rate, closely approaching an ideal logarithmic increase in numbers when the growth curve would be a straight line. During this phase the doubling time is constant and at its shortest. Eventually growth of the colony begins to slow down, doubling time starts to increase, and it starts to enter the **stationary phase** where growth rate of the population is zero and there is much greater competition for resources. Rate of production of new cells is slower and may cease altogether. Any increase in the number of cells is offset by the death of other cells, so that the number of living cells remains constant. This phase is a result of several factors, including exhaustion of essential nutrients, accumulation of toxic waste products, such as alcohol, and possibly, if the bacteria are aerobic, shortage of oxygen. Changes in pH sometimes occur which also slow down growth.

During the final phase, the **phase of decline**, the death rate increases and becomes greater than the rate of multiplication. Eventually the cells stop multiplying altogether. Methods of counting bacteria are described in the next section.

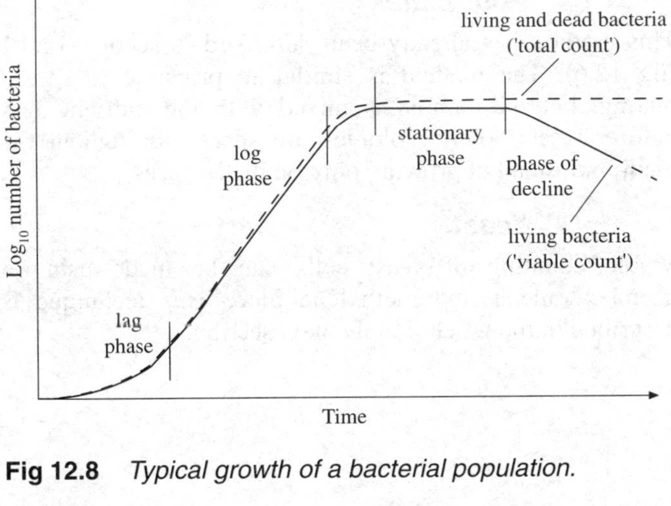

Fig 12.8 *Typical growth of a bacterial population.*

12.6 What would be the shape of the graph obtained if a sample of bacteria was taken from the culture at the beginning of the stationary phase, inoculated into fresh medium and the bacterial population growth measured?

12.7 A culture of bacteria was set up in a nutrient solution and kept at 30 °C. From time 0, and at the times indicated in table 12.3, a count was made of the number of bacterial cells in the culture.

Using the data in table 12.3, draw graphs to show what happened. Explain, using your graphs, what you think was affecting the changes in cell numbers.

12.8 What is the shortest doubling time (generation time) of the bacteria in question 12.7?

Similar principles to these can be applied to the growth of any population and have even been applied to human populations. Any population can, in theory, achieve exponential growth if the doubling time remains constant. However, limiting factors always come into operation sooner or later and a study of these factors is the basis of population ecology (section 10.7).

12.5.2 Diauxic growth

Diauxic means literally that the growth has two phases. It occurs when a bacterium is in a medium containing two different and alternative nutrients. It will often use one in preference to the other until the preferred one is exhausted. It will then switch to the alternative. However, growth usually slows noticeably while the change is being achieved. An example is *E. coli*, the common gut

Table 12.3 Culture of bacteria at 30 °C.

	Number of cells in millions	
Time/h	*living*	*living and dead*
0	9	10
1	10	11
2	11	12
5	18	20
10	400	450
12	550	620
15	550	700
20	550	850
30	550	950
35	225	950
45	30	950

bacterium. This can use glucose or lactose as an energy and carbon source. If both are present, the glucose is used first and then growth slows while enzymes are made which can digest the lactose.

12.5.3 Production of primary and secondary metabolic products

Primary metabolites are products of metabolism needed for growth and are vital for survival.

Secondary metabolites are products of metabolism which are not needed for growth and are not vital for survival. However, they do serve useful functions and are often protective or inhibitory to other competing microorganisms. Some are toxic to animals; they may therefore be a form of chemical warfare. They are often not produced during the most active periods of growth. Once growth slows, spare materials become available for their manufacture. Some important antibiotics are secondary metabolites.

12.6 Measuring population growth of bacteria and fungi

The typical growth curve of a population of bacteria was examined in the previous section. A similar growth curve could be expected for yeast, a unicellular fungus, or the growth of any population.

When measuring the growth of a population of bacteria or yeast, we can do it directly by counting the number of cells, or indirectly by measuring some indication of the number of cells, such as the cloudiness of a solution, or production of a gas. It is usual to inoculate a small sample of the microorganism into a sterilised nutrient medium and to place the culture in an incubator at the optimum temperature for growth. Other conditions should be as close as possible to optimum (section 12.1). Growth can be measured from the time of inoculation.

It is good practice in all scientific investigations to set up replicates and controls where possible and appropriate. Some of the techniques for measuring population growth require practice and are not very precise even in the hands of experts. It is therefore wise, if possible, to set up at least two samples (one replicate) for each treatment. Controls with no microorganisms added to the nutrient medium can test whether your technique is genuinely aseptic. All the techniques described improve with experience and it is advisable to practice them first if they are to be used for project work.

Two types of cell count are possible, namely viable counts and total counts. The **viable count** is the total of *living* cells only. The **total count** is the total number of cells, *living* and *dead*, and is often easier to measure.

12.6.1 Viable counts

It may be important to know the number of living cells. For example, the effectiveness of pasteurising milk in killing certain bacteria could be measured by making viable counts before and after pasteurisation. In an industrial process it is only the living cells which will be contributing to the process, so here again it is important to know the number of living cells in the culture. Viable counts are made using spread plates or pour plates, though viable yeast cells can also be counted using an adaptation of haemocytometry.

Spread plates

This method has already been described in section 12.4.1 (fig 12.5). A known, small volume of sample is added to the nutrient agar in a petri dish. One problem with the technique is that traces of the sample are bound to remain on the spreader and in the pipette, so the final count will be a slight underestimate. However, this is usually not important.

The method relies on the principle that each bacterium will grow into a single colony after a suitable time period, such as two days. The number of bacteria in the original added sample is therefore equal to the number of colonies after incubation. Only those colonies with about 100 000 or more individuals are visible to the naked eye. Typically a colony contains several million bacteria.

Providing the sample does not have too few or too many bacteria, the number of colonies can be counted easily although, for safety reasons, lids of petri dishes should not be removed during counting. It is usual to prepare a dilution series, so that the ideal number of colonies is obtained from one of the series. A dilution series is a series of samples of different known dilutions (fig 12.9). Once a suitable dilution has been found, further replicates can be carried out at this dilution to improve accuracy and reliability. In experiment 12.3 in section 12.9.2 the number of bacteria in milk samples is estimated using this method. Some limitations of the method are discussed at the end of this section.

Pour plates

This method has already been described in section 12.4.1 (fig 12.6). The method is similar in principle to spread plating, but the sample is mixed with the nutrient agar before it sets so the colonies are spread throughout the medium instead of growing only on the surface.

Yeast

Viable counting of yeast cells can be made using a haemocytometer and methylene blue. This technique is described in more detail in the next section.

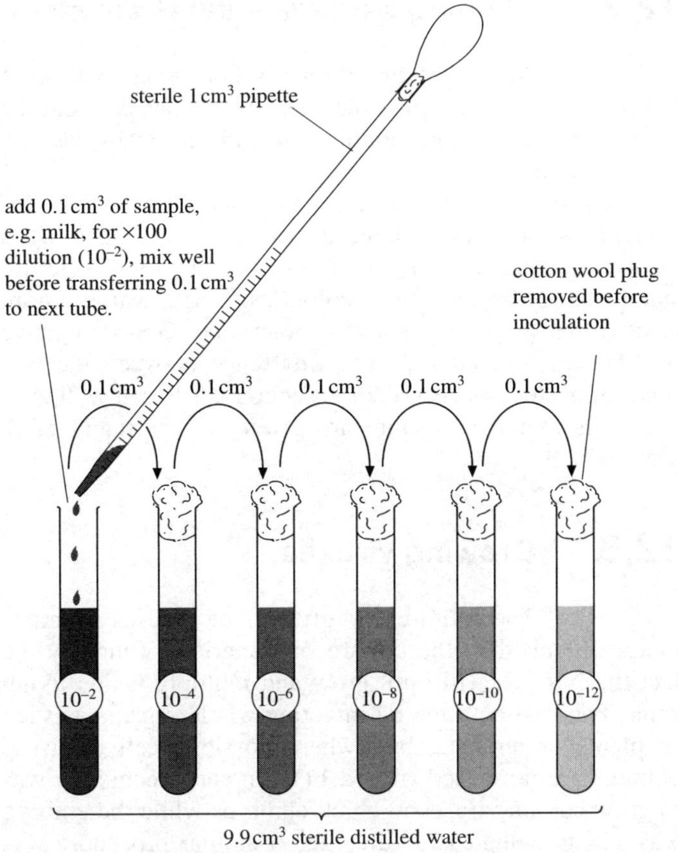

Fig 12.9 *Making a serial dilution.*

Problems associated with viable counting

There are a number of problems associated with viable counting.

- Some bacteria form chains or groups of cells, for example streptococci and staphylococci (fig 2.10). Each group of cells will give rise to only one colony. The results of viable counts are therefore sometimes expressed as numbers of colony-forming units (cfu) rather than number of bacteria.
- If more than one type of bacterium is present, as for example in a soil, milk or water sample, conditions will not favour all types equally. Some will therefore grow more rapidly than others, giving variable numbers of visible colonies.

12.6.2 Total counts

Haemocytometer

The number of bacteria or yeast cells in a liquid culture, such as a broth culture, can be counted directly using a microscope. The technique is easier for yeast because the cells are much larger. With bacteria, an oil immersion lens is required (see section 5.11.2).

A special counting slide called a **haemocytometer slide** is most commonly used (fig 12.10). The technique is known

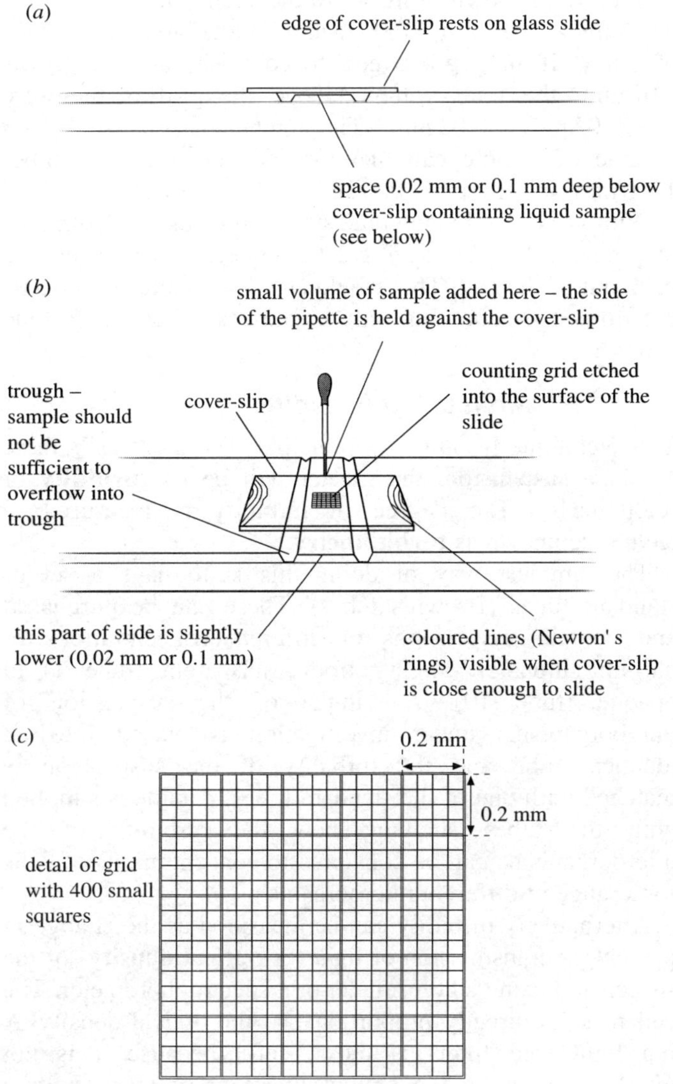

Fig 12.10 *Haemocytometry.*

as **haemocytometry**. To ensure close enough contact between slide and cover-slip, the cover-slip should be pressed down firmly either side of the counting chamber (but don't break the cover-slip!). It should be moved slightly until coloured lines appear as indicated in fig 12.10. The liquid sample in the pipette is drawn under the cover-slip by capillary action. The slide should be left for 2 to 30 minutes to allow cells to settle, making counting easier. The microscope is focused on the grid and the number of cells in the entire grid, or a representative sample, can be counted. At least 600 cells should be counted. Where cells lie on a boundary, they can be judged as in the square on two sides (e.g. upper and right boundaries) and out of the square on the other two sides, rather than trying to count half cells. Each small square is a known area and the depth of liquid below the cover-slip is constant (usually either 0.02 mm or 0.1 mm). The volume of each square can therefore be calculated, and the number of cells in a given volume can be estimated. The sample should be diluted if there are more than about 30 cells in some of the squares. Remember to allow for any dilution in the calculations.

In the grid shown in fig 12.10 the small squares are each 0.05 mm × 0.05 mm = 0.0025 mm^2, a total area for the grid of 1 mm^2. If the gap between the cover-slip and the grid is 0.02 mm, the total volume above the grid is therefore 1 × 0.02 mm = 0.02 mm^3. The number of cells in a given volume of sample can therefore be counted. Remember 1000 mm^3 = 1 cm^3 and 1000 cm^3 = 1 dm^3 (1 litre).

With yeast, methylene blue stain can be used. This stains only dead cells blue. Living cells actively transport the dye back out of the cell. This would enable a viable count to be determined by counting only colourless or very pale blue cells.

Measuring turbidity

This technique is simple in principle. The more cells there are in a suspension, the greater will be its **turbidity** or 'cloudiness'. The degree of turbidity is measured, a technique known as **turbidimetry**.

The simplest way of doing this is to have a set of standard tubes (Brown's tubes). These can be purchased and contain suspensions of different concentrations of barium sulphate, ranging from transparent (tube 1) to opaque (tube 10). A sample of the suspension of microorganisms under investigation is placed into an identical tube and the turbidity of the suspension is matched with one of the standard tubes. A table is supplied with the tubes showing how the turbidity of the microorganisms can be converted to concentrations of cells for a range of different microorganisms.

Alternatively turbidity can be measured as the change in percentage transmission of light (or **optical density**) of the suspension using a colorimeter or a spectrophotometer. The cell mass is directly proportional to the optical density. A red light (red filter) is most useful because it is not interfered with by the yellowish colour of many culture media.

Measuring turbidity is likely to be subject to errors if cells grow in clumps like some bacteria (see viable counting). Also, the most accurate estimates are obtained from population densities that are not too high or too low. Dilution of the sample may be necessary for an ideal density of cells (10^6–10^{10} cells per cm^3).

12.6.3 Non-counting methods

Various other methods for measuring the growth of microorganisms can be devised. For example, with fungi, the growth in diameter of the mycelium can be measured over time. This would be suitable, for example, if the effect of temperature on fungal growth were being investigated, or the effect of an inhibitory substance in the medium. If a fungus is growing in a liquid medium, samples of the culture could be filtered or centrifuged at suitable intervals and the fresh or dry mass of the mycelium measured.

12.7 Staining bacteria – the Gram stain

Although direct microscopic examination of living bacteria is possible using a phase contrast microscope, it is more common to kill and stain bacteria before examination.

One stain which is important in the identification of bacteria is the Gram stain. It was first developed by a Danish physician, Christian Gram, in 1884. Before staining, all bacteria are colourless. Afterwards Gram positive bacteria are stained violet and Gram negative bacteria are stained red. The difference between the two types of bacteria is described in section 2.5.1 (cell wall).

Details of the procedure are given in experiment 12.2, section 12.9.

12.8 Growing viruses

The culture, or growth, of viruses is made more difficult than the culture of bacteria or fungi by the fact that viruses will only grow and multiply inside living cells. This can be done by infecting whole organisms such as plants or animals, but, where possible, cell or tissue cultures are now used (fig 12.11). An early technique was to grow certain viruses in chick embryos while the embryo was still growing inside the egg. A similar procedure was used to grow viruses for mumps and influenza vaccines in the amniotic fluid surrounding the chick embryo.

Fig 12.11 *Cell cultures for growing viruses in petri dishes, incubated in a sterile hot room.*

Cell cultures are a suspension of cells in a liquid medium and tissue cultures are small pieces of plant or animal organ grown in liquid or on a solid medium. In plant tissue culture, meristems are usually used, that is regions such as the root and shoot tips where cells are actively dividing. The material can be taken from a suitably infected plant or inoculated with the appropriate virus when in culture. Animal tissue culture is usually set up by growing individual cells to form a single layer of cells over the surface of a glass container (a **monolayer culture**). The polio virus was originally grown in such cultures derived from monkey kidney. Human cells from the lung, amnion or human cancers have also been used in tissue culture work.

12.9 Experimental work

The following practical work is designed to cover some of the basic microbiological techniques associated with bacteriology, using milk as a relatively safe source of bacteria. Milk is a useful food source for bacteria as well as mammals, and certain bacteria are characteristically associated with it.

12.9.1 Bacterial content of milk

Bacteria inevitably enter milk during milking and handling, even under the most hygienic conditions. Milking is normally followed immediately by cooling to retard bacterial growth. The untreated (raw) milk is pasteurised, a heat process intended to kill pathogenic bacteria, though many non-pathogenic bacteria survive. Bacteria present are:

15–30 °C *Streptococcus lactis* (Gram +) dominates, together with many other streptococci (Gram +) and coryneform (club-shaped) bacteria (for example *Microbacterium*, *Brevibacterium*) which resemble *Lactobacillus* but may have swollen ends to the rods.
Streptococcus lactis grows well at 10 °C but growth ceases at >40 °C.

30–40 °C *Lactobacillus* (Gram +) and gut-living bacilli (Gram –) dominate, such as *E. coli*.

Streptococcus lactis and *Lactobacillus* are lactic acid bacteria. They produce lactic acid during fermentation (anaerobic respiration) of lactose (milk sugar) and the accumulating acid causes souring of milk. Colonies of *S. lactis* and *Lactobacillus* are relatively small (maximum diameter of a few millimetres on a culture) and never coloured, appearing chalky white. *S. lactis* forms smooth-textured colonies with entire edges. If finely divided calcium carbonate is included in the nutrient agar, streptococci show clear zones around each colony where lactic acid dissolves the calcium carbonate. Streptococci are responsible for the normal souring of milk. They are spherical and form chains (fig 12.12). Lactobacilli are rods which tend to stick together in long chains (fig 12.12). Colonies may have a rough surface texture with irregular edges.

A number of other bacteria may be found in milk, including the gut-living rod *Alcaligenes* (Gram –) found singly or in chains. It may be recognised on MacConkey's agar by a yellowish (alkaline) zone around each colony.

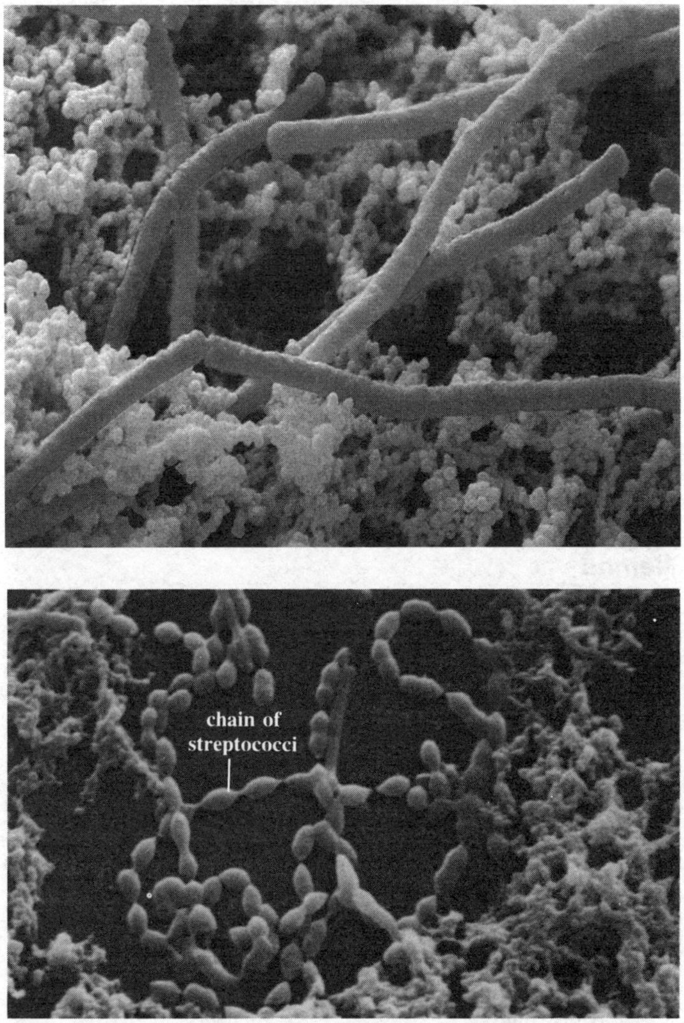

Fig 12.12 *(a) Scanning electron micrograph of* Lactobacillus bulgaricus, *a rod-shaped Gram positive bacterium which causes souring of milk. (b) Light micrograph of chains of streptococci, recovered from milk by filtration.*

12.9.2 Bacteriology experiments

The following three experiments are exercises in the use of microbiological techniques, all including aseptic technique. The first experiment is to culture milk bacteria and involves pouring agar plates and producing a streak culture. The second experiment is to use Gram-staining bacteria from experiment 12.1 for examination with a light microscope. The third involves the counting of bacterial colonies using the technique of serial dilution.

Experiment 12.1: To investigate the bacterial content of fresh and stale milk

The aims of the experiment are to determine the effect of leaving milk unrefrigerated for 24 hours and why milk becomes stale. Milk is almost a complete food for humans and the experiments show that it is also a good culture medium for certain bacteria.

Materials

4 sterile nutrient agar plates
inoculating loop
bunsen burner
indelible marker or wax pencil
fresh pasteurised milk
stale milk (milk left at room temperature for 24 h)
incubator set at 35 °C

Method

(1) Place the inoculating loop in the bunsen burner flame until the loop is red hot (fig 12.4).
(2) Allow the loop to cool and then dip it into a sample of fresh, well shaken milk.
(3) Lift the lid of a sterile agar plate slightly with the other hand and lightly spread the contents of the inoculating loop over the surface of the agar as described in fig 12.4.
(4) Close the lid of the plate and return the loop to the bunsen burner flame until red hot.
(5) Label the base of the plate with an indelible marker (or wax pencil).
(6) Repeat with a second plate and another sample of fresh milk.
(7) Flame the loop again and, having allowed it to cool, dip it into a sample of stale milk.
(8) Spread the contents of the loop over the surface of a third plate and then close the lid.
(9) Label the base of the plate with an indelible marker.
(10) Repeat with a fourth plate and a second sample of stale milk.
(11) Place the four plates in an incubator at 35 °C for about three days. They should be placed upside down to prevent condensation falling onto the cultures. After incubation, the two halves of each plate should be taped together for safety reasons.
(12) Record the appearance of the colonies and compare with the description in section 12.9.1.

Notes

(1) Students may pour their own plates, if McCartney tubes of sterile molten nutrient agar are supplied.
(2) The particular streaking technique used progressively reduces the number of bacteria in each streak. It is suitable in situations where large numbers of bacteria are present, as in milk, and is normally used to isolate pure colonies of bacteria from mixed cultures.
(3) The plates can be placed in a refrigerator after incubation until required. This prevents further bacterial growth.
(4) When the plates are no longer required they should be placed in a disposable autoclave bag and autoclaved for 15 min before final disposal.
(5) Other experiments using milk could be performed. The experiment above is the simplest. The effect of refrigeration could be studied. Also, if samples of raw (unpasteurised) milk could be obtained (for instance direct from a dairy farm) the effect of the process of pasteurisation on the bacterial content of milk could be studied. Pasteurisation of milk can be accomplished by placing raw milk in a sterile test-tube plugged with cotton wool and heating at 63 °C for 35 min in a water bath. A third variation would be to incubate some plates at 10 °C instead of 35 °C. This lower temperature favours growth of *Streptococcus lactis* compared with *Lactobacillus*.

Experiment 12.2: To stain bacteria for examination with a light microscope

Although direct microscopic examination of living bacteria is possible using a phase contrast microscope, it is more common to kill and stain bacteria before examination.

One stain which is important in the identification of bacteria is the Gram stain. Before staining, all bacteria are colourless. Afterwards **Gram positive** bacteria are stained violet and **Gram negative** bacteria are red. The difference between the two types of bacteria is described in section 2.5.1 (cell wall).

Materials

basic stain = crystal violet (0.5% aqueous)
mordant = Lugol's iodine
decolouriser = acetone–alcohol (50:50 acetone: absolute alcohol)
counterstain = safranin (1% aqueous)
wire loop
bunsen burner
glass slides scrupulously clean (wipe with alcohol)
forceps
staining rack set up over sink or dish
distilled water in wash bottle
blotting paper
immersion oil and microscope with oil immersion lens

Method

(Stages 1 to 6 should take about 5 min.) (Based on *Bacteriology*, J. Humphries, John Murray, 1974.)

(1) **Prepare a smear** of bacteria on the slide as follows. Flame a wire loop and cool. Place a loopful or two of tap water on the centre of a clean slide. Touch the wire loop lightly on a selected bacterial colony from the experiment above, opening the lid of the plate a

minimal amount for safety reasons. Transfer the bacteria to the slide and gently mix with the water. Spread the bacteria over the slide, using the loop, to cover an area about 3 × 1 cm. Flame the loop again. It is important to achieve the correct thickness of the smear. It should appear only faintly opalescent and is more usually too thick than too thin. It should also be of even thickness. Allow the smear to become perfectly dry in air (a few minutes).

(2) **Fix the bacteria.** Holding the slide with forceps, pass it horizontally just over a yellow bunsen flame three times. It is important that it is not overheated and should feel comfortable against the skin after each passage over the flame. Fixing kills the bacteria by coagulating the cytoplasm and also makes them stick to the slide.

(3) **Stain the bacteria.** Staining is likely to soil the bench so should be done on a rack over a sink or dish. A rack can be made by arranging two glass or metal rods across the sink or dish 5 cm apart and absolutely horizontal. If supported on plasticine they are easily adjusted. Flood the slide with crystal violet stain. Leave for 30 s. This makes all bacteria violet.

(4) Wash off with Lugol's iodine; flood with Lugol's iodine and leave for 30 s. Wash off the iodine with distilled water from a wash bottle. The iodine fixes the stain more permanently into the cells.

(5) Flood the slide with acetone–alcohol until no more colour is seen to come off (about 3 s); *immediately* wash with water to prevent excessive decolourisation. Repeat if necessary (only experience will show how much washing is needed). This decolourises Gram negative bacteria. Gram positive bacteria stay violet.

(6) Flood the slide with safranin and leave for 1 min. Wash off the stain with water. Gently dry the slide between sheets of clean blotting paper and allow to dry finally in air. Safranin is described as a counterstain. It is used after the crystal violet to stain any Gram negative bacteria red.

(7) Apply a drop of immersion oil and examine under the oil immersion lens (section 5.11.2).

Results

Are your observations in agreement with the description given in section 12.9.1 of the bacterial content of milk?

Experiment 12.3: To compare the numbers of bacteria present in fresh and stale milk

If a single bacterium is placed on nutrient agar it will grow to form a colony which is easily seen with the naked eye, unlike the original bacterium. This can be made use of when counting bacteria.

After sterilising the apparatus, the first part of the experiment involves the technique of serial dilution. The numbers of bacteria in milk are vast, so counting can be made more manageable by diluting by a known factor and taking a small sample of known volume. A series of dilutions is prepared. In the second part of the experiment samples of each dilution are cultured and the one giving the most suitable number of colonies (a reasonably large number but with no overlap of colonies) when grown on agar is used to calculate the number of bacteria in a given volume of milk.

Materials

6 sterile nutrient agar plates
8 1 cm^3 graduated pipettes
10 cm^3 graduated pipette
6 test-tubes and test-tube rack
cotton wool
oven at 160 °C
indelible marker
bunsen burner
100 cm^3 distilled water
fresh milk
stale milk
70% alcohol
aluminium foil
glass spreader

Sterilisation of apparatus

(1) Place cotton wool plugs in each of six test-tubes and cover plugs loosely with aluminium foil.

(2) Place a small piece of cotton wool in the top of each of eight 1 cm^3 graduated pipettes and the 10 cm^3 graduated pipette and wrap each pipette separately in aluminium foil.

(3) Place the test-tubes and pipettes in a hot air oven at 160 °C for 60 min (bottles of media and water should not be sterilised in an oven).

(4) Allow all apparatus to cool before use.

Serial dilution of milk and inoculation of agar plates

(1) Label the six sterile plugged test-tubes F1, F2, F3, S1, S2 and S3, and remove the aluminium foil covers from the plugs.

(2) Label the base of each of six sterile nutrient agar plates F1, F2, F3, S1, S2 and S3.

(3) Transfer 9.9 cm^3 of sterile distilled water to each of the six test-tubes using the following technique.
(*a*) Remove the cotton wool plug from the flask containing sterile distilled water using the little finger and fourth finger of one hand.
(*b*) Whilst holding the plug, draw up 9.9 cm^3 of water using the sterile 10 cm^3 graduated pipette held in the other hand.
(*c*) Replace the plug.
(*d*) Remove the plug from the first test-tube using the same method as in (*a*).
(*e*) Transfer 9.9 cm^3 of water to the test-tube.
(*f*) Replace the plug.
(*g*) Repeat for the five remaining test-tubes.

(4) Shake the sample of fresh milk and transfer 0.1 cm^3 of this milk using a sterile 1 cm^3 pipette to tube F1, removing and replacing the plug as before. This gives a ×100 dilution.
(5) Shake the tube gently to ensure thorough mixing.
(6) Using a fresh pipette, transfer 0.1 cm^3 from tube F1 to the sterile plate labelled F1, lifting the lid by a minimal amount.
(7) Dip the glass spreader in 70% alcohol, allow excess alcohol to drip off and then hold the spreader vertically in a bunsen burner flame.
(8) Cool the spreader and spread the sample milk over the surface of the plate.
(9) Re-sterilise the spreader.
(10) Using the same pipette as in point (6), transfer 0.1 cm^3 from tube F1 to tube F2, removing and replacing the bungs as before.
(11) Shake the tube F2 to ensure thorough mixing. This gives a ×10 000 dilution.
(12) Repeat from (10)–(11), using F3 for F2. This gives a ×1 000 000 dilution. Repeat (6)–(9) using F3 for F1.
(13) Repeat the serial dilution technique using the sample of stale milk and prepare plates S1, S2 and S3.
(15) Incubate the six plates upside down at 35 °C for about three days.
(16) The lids of the plates should then be taped down to avoid the risk of pathogens being spread.
(17) Examine the plates for bacterial growth. Count the numbers of individual colonies where practical. Record results in the form of a table and use them to calculate the number of bacteria in 1 cm^3 of undiluted milk.

Notes

See notes (3) and (4) at the end of experiment 12.1.

12.9.3 Practical work with fungi

The methods for handling fungi are in many cases the same as those for bacteria, being standard microbiological techniques. Many saprotrophic fungi, like bacteria, can be cultured on nutrient agar and, if pure cultures are required, the sterile techniques described in sections 12.3 and 12.4 should be used. Common fungi suitable for culture in this way are *Mucor*, *Rhizopus* and *Penicillium*. A suitable culture medium is a 2% malt agar prepared in petri dishes. Selected fungi can be isolated from mixed cultures grown by chance contamination of substrates such as bread, fruit and other moist foods. Spores can be transferred and added to the culture medium by a sterile mounted needle. Cultures can be conveniently examined with low power stereoscopic microscopes.

12.10 Large scale production

You have read about, and may have attempted, some of the practical aspects of working with microorganisms in the previous sections. Moving from a laboratory scale to an industrial scale presents new problems for biotechnologists. The problems may involve any area of science, including engineering, chemistry and biology. Economic, social and sometimes ethical issues also start to be important factors in decision-making. In this section we shall consider some of the practical aspects of large-scale production before looking at specific examples and related issues in later sections.

Microorganisms are particularly suitable for industrial processes for the following reasons:

- they have simple nutritional requirements;
- growth conditions can be controlled very precisely in fermenters (large vessels in which the microorganisms are grown);
- they have fast growth rates;
- reactions can be carried out at lower temperatures than conventional industrial procedures – energy costs are therefore lower;
- they produce higher yields and have higher specificity than conventional processes;
- a wide range of chemicals can be used and produced;
- some complex chemicals, such as hormones and antibiotics, can be manufactured which are difficult to produce by other methods, and specific isomers (such as L-amino acids) can be produced;
- the genetics of microorganisms are relatively simple and techniques for genetic manipulation are continually advancing.

However, there are specialist techniques, such as aseptic techniques and complex methods of separation, which can make the process more technically demanding.

12.10.1 An overview

Fig 12.13 is a flow diagram showing some of the key stages which are typically involved in setting up a biotechnological process.

12.10.2 Screening

We know that microorganisms show great diversity in the chemical reactions they carry out and the products they can form. However, only a tiny fraction of their potential has been exploited. There is a constant search for potentially useful microorganisms, particularly by commercial companies such as drugs companies. Microorganisms are cultured from all over the world, and from many different environments, in the hope of discovering new commercially useful products or more efficient means of producing existing products. Very often

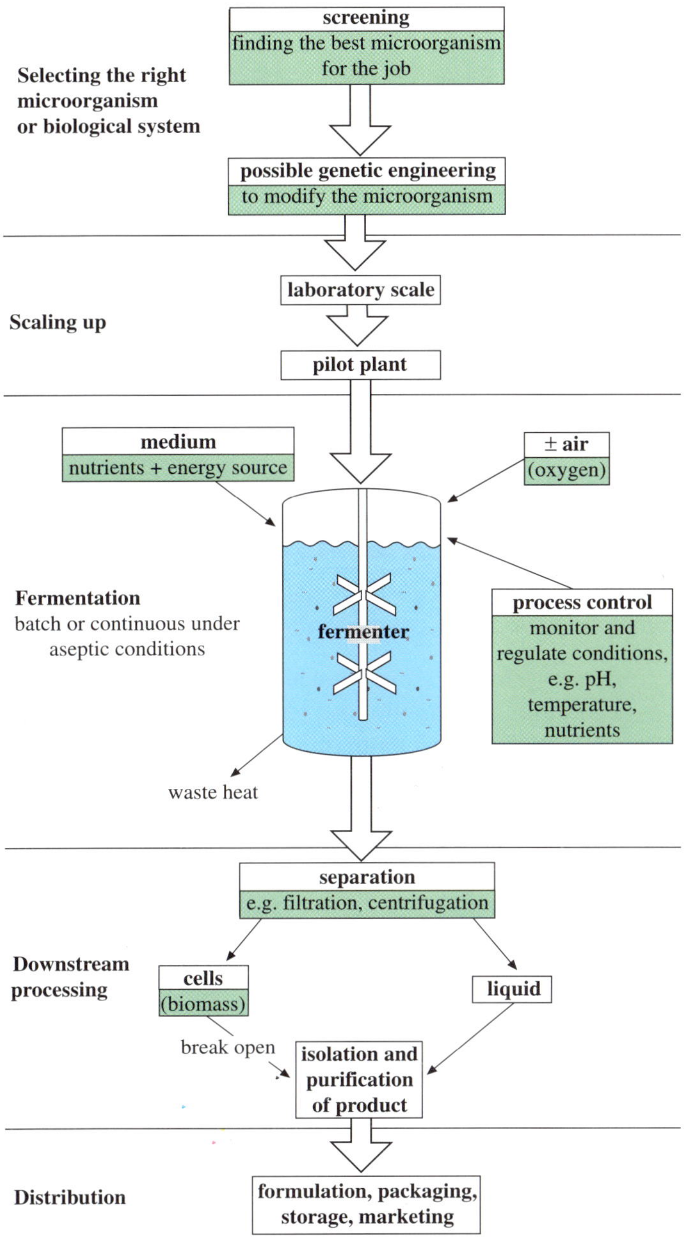

Fig 12.13 *Overview of a biotechnological process.*

this work is purely empirical, meaning that chance plays a large part in any discovery. Testing microorganisms in this way is known as **screening**. A good example is the constant screening that takes place for new antibiotics. The first antibiotic was discovered in 1928 by Alexander Fleming and named penicillin after the fungus that produces it, *Penicillium*. Natural antibiotics are chemicals which are made by microorganisms and which kill or inhibit the growth of other microorganisms. Since 1928 more than 5000 different antibiotics have been isolated from microorganisms, including a range of different penicillins with slightly different structures and activities. Most of the antibiotics discovered are unsuitable for medical use, mainly because they are too toxic. However, *Streptomyces* species have been a particularly rich source of different antibiotics, including streptomycin.

Antibiotics are used to treat bacterial or fungal diseases of humans and domesticated animals. Some also inhibit growth of cancer. They appear to be products of secondary metabolism (section 12.5.3). With systematic screening there is always the hope of finding a new 'wonder-drug' or a microorganism which will improve on the performance of one currently in use.

12.10.3 Scaling up

Any new biotechnological manufacturing process must first be tried out on a laboratory scale. After initial investigations using ordinary laboratory apparatus it is usual to make a 'pilot plant'. This involves use of a relatively small fermenter, anything from 2 to 200 dm^3 (fig 12.14). The **fermenter** is the tank or vessel in which the process will be carried out. Optimum nutrient and physical conditions for maximum yields must be determined.

New factors come into play when the process has to be scaled up from pilot production to full-scale, involving thousands of dm^3 (fig 12.15). Sometimes a successful transition is not possible. Some of the important factors are as follows.

- A major problem is maintaining aseptic conditions. It is easy to contaminate both inputs and outputs to the main fermenter. Engineering techniques have improved with experience and a good example of success with aseptic technique was the manufacture of Pruteen (section 12.12.3).
- Physical factors, such as mixing and aerating the media and getting rid of waste heat, create the biggest problems in moving from one scale to another. Chemical engineers are needed to solve these problems. Some of these problems are discussed below.
- To supply enough oxygen in large-scale cultures, air must be forced through the medium because the simple agitation used at the laboratory scale is inadequate. Small bubbles are more effective than large bubbles, so a **sparger** is used (a tube with small holes). The mixture may also be stirred. **Baffles** in the vessel walls increase turbulence and improve the efficiency with which bubbles dissolve in the medium since they take longer to reach the top of the fermenter.
- Anti-foaming agents are required to reduce the foaming caused by stirring and aeration.
- Heat is produced by the activity of microorganisms and large-scale production. Cooling water must be circulated around the fermenter.
- It is more difficult to keep conditions constant, such as supply of nutrients, pH and oxygen concentration, throughout the medium on a large scale. Sophisticated monitoring devices and control processes are needed. Overcoming these difficulties, for example by stirring the medium, may have other undesirable consequences such as affecting the distribution of organisms in the culture. In the manufacture of Pruteen, for example, the

Fig 12.14 *(a) Start-up flask of mammalian cell culture used for the commercial production of monoclonal antibodies. The hybridoma cells formed in the flask are later multiplied in fermenters to produce the antibodies. (b) Pilot-scale commercial fermentation unit in a biotechnology laboratory.*

yields originally obtained were significantly lower than predicted from the laboratory and pilot plant scales. It was discovered that because the methanol, which was the only carbon and energy source, was being added at one point only, the circulating bacteria were going through successive 'feast and famine' periods as they rotated once every several minutes. Yields were raised to those predicted by adding the methanol at various points around the fermenter.

Fig 12.15 *A large-scale fermenter (2000 dm^3) used in commercial production of monoclonal antibodies.*

- Microorganisms may change their metabolism according to conditions, and precise control of conditions is much more difficult on a large scale. Any small changes in the product may have harmful effects on people or animals using it.
- The more complex the equipment and procedures, the more opportunity there is for faults or accidents to occur. These may have economically disastrous consequences and have to be considered when deciding whether the operation is economically worthwhile.
- Large quantities of water must be supplied, sterilised and disposed of.
- Because of the large quantities involved, raw materials should be readily available, stable, and easy to handle and store. Chemical changes and microbial contamination of stored material must be minimal. They must be economic because cost is a major concern.
- Sometimes the strains of microorganisms selected for use in fermentation processes can easily revert by mutation to lower yielding strains which grow more rapidly. This is more likely as the scale of manufacture increases.
- Handling powders in bulk can be a health hazard to workers.
- Materials used in construction of the fermenter must be corrosion-resistant, to prevent traces of metal contamination. They should also be non-toxic to microorganisms and must be able to resist repeated sterilisation with high pressure steam.
- Extra hazards are posed by large-scale fermentation, especially of genetically modified organisms. These are due to the scale of the operation and the opportunity for a greater degree of exposure to an organism and its biologically active products compared to laboratory scale work.

12.10.4 Fermenter design and use

Fermenters, also known as **bioreactors**, are chambers in which microorganisms are cultured in a liquid or solid medium. The processes which take place in fermenters are referred to as **fermentations**. The term fermentation originally applied only to anaerobic processes but is now used more broadly to include all the processes, whether aerobic or anaerobic. Fig 12.16 shows a typical fermenter. It is a complex diagram so you will need to spend some time studying the different parts. Bear in mind the list of scale-up problems identified in the previous section. The contents of most fermenters are stirred during operation, but this is not always the case, as with the now discontinued production of the single cell protein Pruteen by ICI, where air introduced at high velocity at the bottom of the vessel was used to achieve mixing. The **product** is either the cells themselves (biomass) or some useful cell product. *All* operations must be carried out under sterile conditions to avoid contamination of the culture. In addition, it must be possible to keep all inlets and outlets of the fermenter sterile. The fermenter and the medium used are sterilised before use, either together or separately. Stock cultures of the organism to be used in the fermentation are kept in an inactive form (for example stored frozen). A sample is re-activated, grown up to sufficient bulk using aseptic techniques (**scale-up**), and then added to the fermenter, a process known as **inoculation**. Once inside the fermenter, the organism grows and multiplies, using the nutrient medium.

The fermenter is usually made of high grade stainless steel so that it does not corrode and leak toxic metal salts into the medium. All equipment, materials and air used must be sterile. Equipment is sterilised by steam under pressure. All surfaces must be accessible to the steam and as smooth and polished as possible to avoid cracks or rough surfaces where microorganisms can collect. The medium may be sterilised before inoculation by passing steam through the cooling coils. Air is sterilised by filtration.

For safety reasons air leaving the fermenter may also need to be sterilised, for example to prevent escape of genetically engineered microorganisms.

12.10.5 Batch, fed-batch and continuous culture

Two basic types of fermentation are possible, **batch fermentation** (or **closed system**) and **continuous culture** (or **open system**). In the more common batch fermentation the conditions are set up and not changed from outside once the fermentation starts; for example, no more nutrients are added. This is why the process is described as a closed system. The process is stopped once sufficient product has been formed. The contents of the fermenter are then removed, the product isolated, the microorganism discarded and the fermenter is cleaned and set up for a fresh batch.

Continuous culture involves continuous long-term operation over many weeks, during which nutrient medium is added as fast as it is used, and the overflow is harvested. Although this is closer to the situation in natural environments, such as the gut, continuous culture has found only limited application. However, it is used for the production of single cell protein where a large biomass of cells is required. ICI, for example, used to produce the single cell bacterial protein (SCP) Pruteen from the bacterium *Methylophilus methylotrophus* (section 12.12.3). The fermenter was run continuously for as long as 100 days and could produce 150 tonnes per day.

Although continuous culture is still not commonly used, increasing use is being made of **'fed-batch' culture**. This is a compromise between the two systems. With fed-batch culture more control of the process is involved than with batch culture. The period of growth is extended by adding nutrients at low concentrations during the fermentation, and not adding all the nutrients at the beginning. One advantage of this is that the rate of growth can be regulated to keep pace with the rate at which oxygen can be supplied. The process is commonly used to produce yeast cells for the baking industry. If the yeast has too much sugar it starts to respire anaerobically and produce alcohol at the expense of biomass. The process is also used in the production of penicillin (section 12.11.1).

The advantages and disadvantages of batch and continuous culture are summarised in table 12.4.

12.10.6 Downstream processing

Downstream processing is the name given to the stage after fermentation when the desired product is recovered and purified. Usually the contents of the fermenter are first separated into a liquid component and a solid component which contains the cells. This is usually done by filtration or centrifugation. The liquid may contain the desired product in solution or it may be the cells or some product inside the cells that is needed.

A whole range of biochemical separation and purification techniques is available, such as drying, chromatography, solvent extraction and distillation. As an indication of the importance of downstream processing, it involves over 90% of the 200 staff employed by Eli Lilly in their human insulin plant. The techniques used in downstream processing during the production of penicillin and mycoprotein are considered later in sections 12.11.1 and 12.12.3 respectively.

12.11 Medical products

One of the most economically successful ways of using the technology of large-scale fermentation is to manufacture medical products. Such products are effective in relatively small amounts and command relatively high prices. There is the additional advantage of an obvious

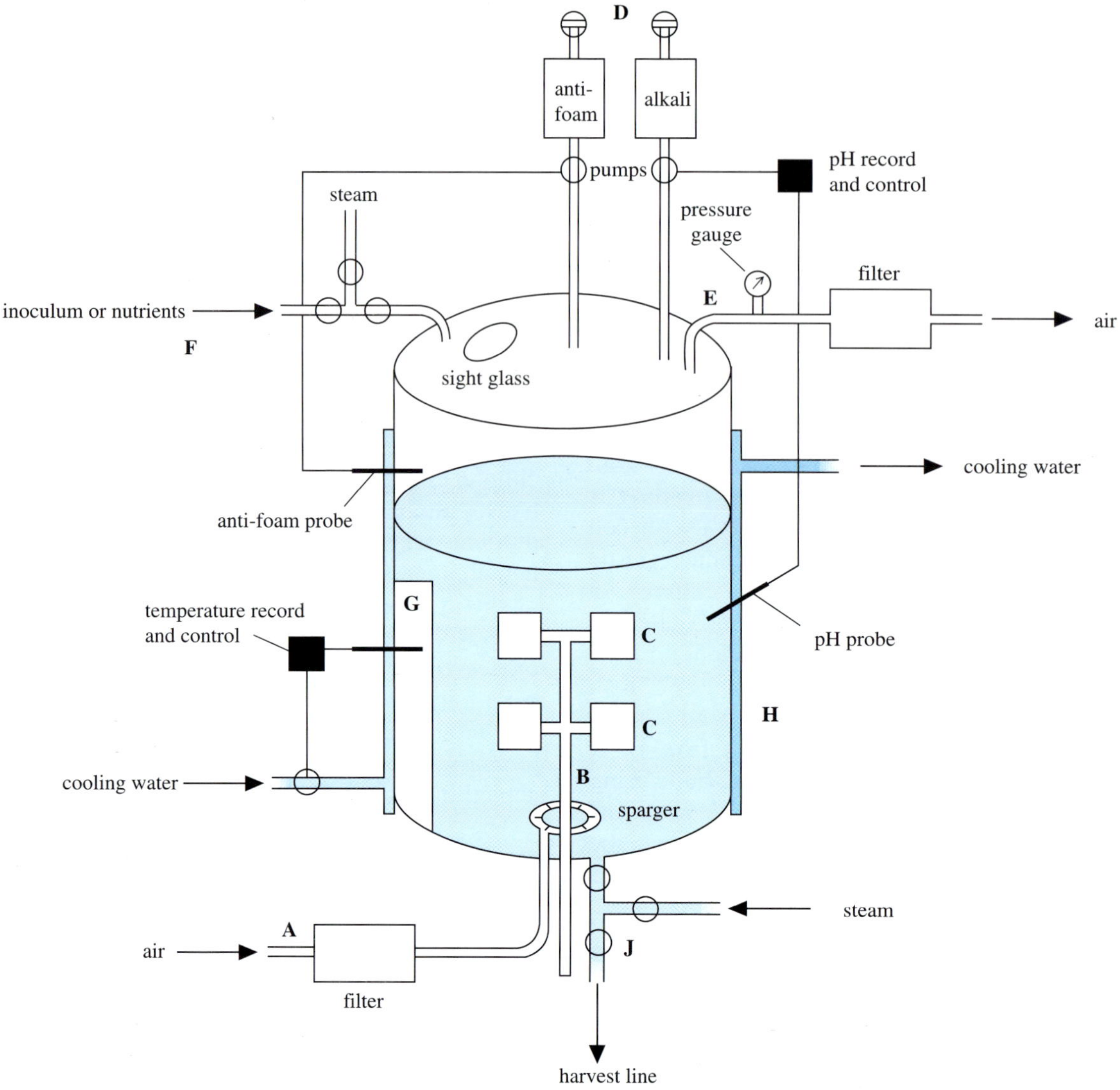

Fig 12.16 *A typical fermenter. Size of vessel is very variable, ranging from 1 dm^3 (experimental) to 500 000 dm^3 (commercial production). Shape and material used in construction are also variable, although cylindrical stainless steel vessels are common.*

Key:

A *Air inlet – most fermentations are aerobic, requiring large volumes of sterile air. The sparger is a specially designed part that releases air bubbles to help the mixing process, provide oxygen for aerobic respiration and aid release of volatile waste products.*

B *Stirrer shaft – present in most fermenters. Agitation increases the rate at which oxygen dissolves; maintains diffusion gradients of oxygen and nutrients into cells and products out of cells; prevents clumping of cells or mycelia of fungi; promotes heat exchange between medium and cooling surfaces. Shaft bearings must be strong and sterile.*

C *Stirrer paddles – usually flat and vertical.*

D *Alkali and anti-foam inlets - these are connected to pH and anti-foam probes which monitor the contents of the fermenter. Aeration and agitation generate foam, particularly from proteins, and prevent escape of contents through the exhaust. When foam touches the anti-foam probe, it completes an electrical circuit which activates a pump connected to a supply of anti-foaming agent. Other probes are connected by electrical circuits to other 'effectors'. Alkali is added if (as is usual) acidity increases during fermentation, so that pH is kept constant.*

E *Exhaust – contents of fermenter are under pressure, so a pressure gauge and safety valve are attached.*

F *Top of fermenter has a part where medium and inoculum (microorganism) can be added, and which gives access for cleaning.*

G *Baffle – vertical fin on inside wall, helps to increase agitation and reduce swirling of contents as culture is rotated. This improves efficiency of oxygen transfer.*

H *Cooling jacket – reduces temperature because the culture generates heat during growth.*

J *Harvest – samples may be taken here during fermentation so that the process can be monitored.*

Table 12.4 Advantages and disadvantages of batch and continuous culture.

	Advantages	*Disadvantages*
Batch culture	Suitable for production of secondary metabolites whose production is not associated with growth, e.g. antibiotics. Can use strains which are too unstable for continuous culture. Easier to set up and run than continuous culture. Fermenters are less specialised and may be used for a greater variety of processes, depending on demand.	Turnaround time between batches can be prolonged, wasting possible production time. Environment changes in the fermenter as the fermentation progresses. Nutrients get used up and products build up. Heat output, acid or alkali production, oxygen consumption increase in rate as growth progresses. Therefore, conditions gradually become unfavourable and growth rate gradually declines. See graph fig 12.8.
Continuous culture	Gives more control. Aim is to keep environmental conditions constant. Nutrients are replaced as fast as they are used and products are removed as fast as they are made. Productivity is greater because there is no turnaround time (continuous process). Therefore more cost effective in some situations. Optimum (maximum or exponential) growth rate is maintained once achieved. Smaller fermenters are needed because higher yields are obtained. More suitable for production of biomass. Also used for production of ethanol whose synthesis is proportional to rate of growth. Demand for labour is more regular.	Greater risk of contamination, although good engineering design can solve this problem. Control is more complex. When used for brewing gives greater yields but has given flavour problems.

social benefit. Some of the most successful products are considered in the following sections.

12.11.1 Penicillin production

Production of the antibiotic penicillin is a good example of the use of fed-batch culture for the production of a secondary metabolite.

> **12.9** What is meant by (*a*) antibiotic, (*b*) fed-batch culture, (*c*) secondary metabolite?

Since its discovery in 1928 (see section 12.10.2) penicillin has probably saved millions of lives. It was first used on a large scale during the Second World War where it was mainly used to cure troops suffering from the sexually transmitted disease gonorrhoea. Its use against more life-threatening diseases, such as pneumonia, followed when sufficient supplies were available. The death rate from pneumonia was about 30% before its use. Penicillin acts by inhibiting cell wall synthesis in some bacteria, particularly Gram positive bacteria (section 2.5.1, cell wall). Only growing cells are killed. Some bacteria contain plasmids which make them resistant to penicillin (section 2.5.1, plasmids). There are now several hundred penicillins, all with the same basic structure but with different side chains on the molecule. Some are synthetic, some semi-synthetic and some are natural. They have different specifications, in other words they vary in their effectiveness against different bacteria. This provides the incentive for looking for alternative forms of the antibiotic.

The original screening process for a suitable strain of the fungus *Penicillium*, from which penicillin is obtained, was carried out at the beginning of the Second World War in the USA, although Fleming discovered penicillin in England (in *Penicillium notatum*). The search was started by a scientist from Oxford University, Sir Howard Florey, who was helped by the greater resources of the Americans. Cultures and soil samples from all round the world were tested, but the highest yields were obtained from a strain of *Penicillium chrysogenum* found growing on a mouldy melon purchased locally. Since then yields have increased about 2000-fold due to selection of high-yielding mutant strains and better culturing, extraction and purification techniques.

The composition of the medium used in the fermentation is critical. The preferred carbon source now is glucose. Glucose promotes growth but tends to inhibit production of the enzymes needed for penicillin synthesis. It is usual, therefore, to stimulate rapid growth of the fungus for the first 30 to 40 hours, and then to add glucose at low concentrations, either in regular doses, or as a continuous feed (hence it is a fed-batch culture). Nitrogen is supplied in the form of a cheap protein such as soyabean flour or fishmeal. Other nutrients added are phosphate, calcium carbonate and phenylacetic acid which increases yields because it is used to help make the penicillin molecule. Since penicillin is a secondary metabolite, its production starts after the initial rapid growth phase. The primary metabolites of the first phase of growth include carbon dioxide and ethanol from respiration.

The culture is started from spores of the fungus. About 3–5 tonnes of mycelium is eventually used to inoculate a 50 000 dm^3 fermenter. Fermenters up to 200 000 dm^3 have been built for penicillin production because the demand is so huge. About half a tonne of glucose is used per day. Production is very sensitive to temperature, which can rise by as much as 2 °C per hour, so temperature is very carefully controlled. pH is also carefully controlled between 6.5 and 7.0. The process is allowed to run for up to 15 days.

Downstream processing involves removal of the mycelium by filtration. The penicillin remains behind in solution in the liquid part of the medium. It is extracted by a series of solvents. With each extraction some of the impurities are removed, so the penicillin gets more and more pure until it is left dissolved in water in a pure state. The water is then removed by vacuum evaporation and the penicillin crystallises as a sodium or potassium salt (penicillin itself is a weak acid).

12.11.2 Monoclonal antibodies

Humans have an immune system which is able to respond to the invasion of certain foreign molecules by making antibodies (chapter 14). Molecules which stimulate the formation of antibodies are called **antigens**. They are usually proteins or glycoproteins. Each antigen stimulates the production of a specific antibody which matches it exactly and is able to combine with it and bring about its destruction. Thus if, for example, the body is invaded by a particular type of bacterium, the antigens in the cell surface membranes of the bacterial cells will be recognised as foreign by the immune system, and antibodies specific to those antigens will be made. Antibodies are made by special lymphocytes (a type of white blood cell) called B cells.

In the 1970s César Milstein and Georges Köhler working in Cambridge, were looking for a way to make pure antibodies of one type only. Up to that time antibodies were prepared from the blood of animals that had been deliberately exposed to the relevant antigen. However, the final product was still impure and contained hundreds of different antibodies. Milstein and Köhler solved the problem by developing a technique for producing monoclonal antibodies, work for which they were awarded the Nobel prize in 1984. **Monoclonal** means belonging to one clone. Each type of antibody is made by one type of B cell which clones itself, in other words multiplies to make many identical copies of itself, in response to a particular antigen. In theory, if a particular type of B cell could be isolated and cultured, large quantities of a single pure antibody could be collected. Since antibodies produced in this way are all from one clone, they are described as monoclonal. However, B cells survive only a few days in a culture medium. Milstein and Köhler solved this problem by fusing B cells with cancer cells, which are immortal, to form **hybridoma cells**. The hybridoma cells continue to multiply and can be cloned so that large quantities of antibody can be produced.

Production of monoclonal antibodies

The relevant antigen is injected into an experimental animal, usually a mouse. After allowing time for an immune response to take place, lymphocytes are removed from the spleen of the animal and mixed with a special type of cancer cell in a suitable culture medium. Fusion of some of the lymphocytes with the cancer cells is stimulated by adding polyethylene glycol to the medium. Conditions are arranged so that only hybridised cells will survive. It is then possible to select the ones making the relevant antibodies from all the others, and to culture these separately. The cells continue to multiply indefinitely and are a constant source of pure antibody.

Uses of monoclonal antibodies

Monoclonal antibodies are specific for a particular antigen. They can be used to 'find' or identify that antigen and possibly to destroy it. A large number of applications have been found and are being developed. So far, they are used mainly for diagnostic purposes.

Pregnancy testing. Nearly one-third of the 150 diagnostic monoclonal antibodies currently in use are for pregnancy testing. As soon as the embryo reaches the uterus (within four days of conception) it begins to grow into the wall of the uterus, a process called **implantation**. It is vital that the embryo sends a signal to the woman's ovaries at this stage so that they can respond appropriately. If the signal is received, the ovaries continue to produce the hormones that maintain the lining of the uterus, and the woman will not have a period and lose the embryo. The signal to the ovaries is a hormone produced by finger-like extensions of the chorion, a layer of cells which grows out from the embryo into the uterus wall. The name of the hormone is human chorionic gonadotrophin (HCG for short – chapter 21). It circulates in the mother's blood to reach the ovaries. By the time her period should normally have started (about 14 days after conception) HCG has built up to a high enough level to be detectable in the urine. It is a glycoprotein, so monoclonal antibodies can be produced against it by the method described above. Modern pregnancy tests work by detecting HCG in a sample of the woman's urine using these monoclonal antibodies. The procedure is described in fig 12.17. The kit can be bought in chemists for testing at home and gives a result in five minutes. Even earlier detection of pregnancy is possible if a doctor tests the blood.

Fig 12.17 *Use of home pregnancy kit.*

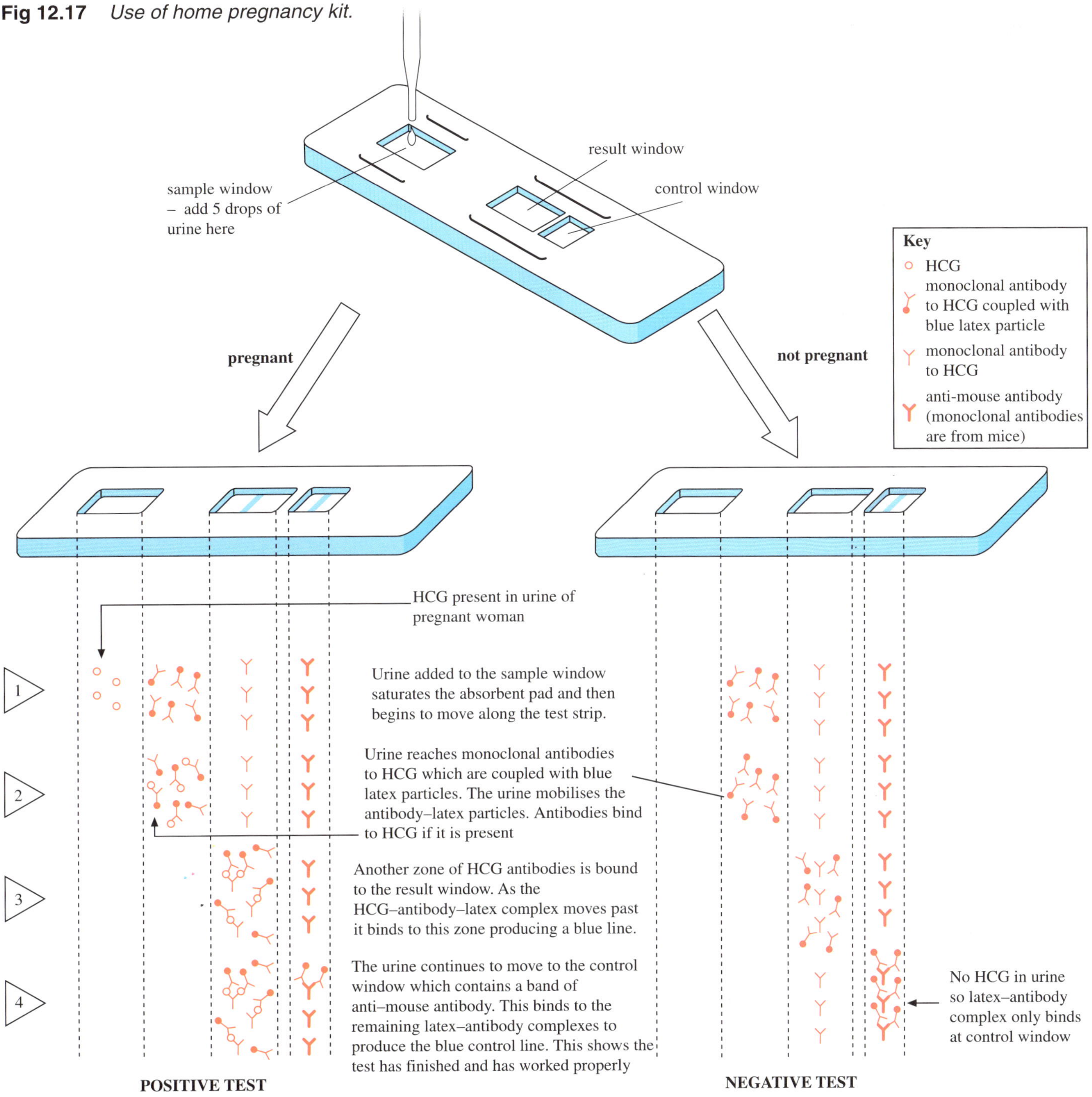

> **12.10** Suggest three advantages of being able to conduct a pregnancy test at home.

Diagnosis of disease. One of the most common sexually transmitted diseases (STD) is *Chlamydia*. It is a small Gram negative bacterium which is unusual in that it grows inside cells. Symptoms of infection are very mild and are sometimes hard to distinguish from gonorrhoea, another common STD. Both diseases can cause pelvic inflammatory disease in women if they spread through the uterus and into the Fallopian tubes. This causes pain and discomfort and may cause infertility. *Chlamydia* was particularly difficult to diagnose until monoclonal antibodies became available. Use of monoclonals has made tests for both diseases much more rapid and reliable. Results can be obtained within 15–20 minutes rather than involving laborious procedures in hospitals which take several days.

Another use of monoclonal antibodies is in a diagnostic kit for streptococcal throat infections, which is available for use by local doctors, enabling immediate diagnosis and treatment. This might previously have had to wait several days for tests to be completed in a hospital laboratory.

Monoclonal antibodies have also been developed that can distinguish between the closely related herpes virus 1, which causes cold sores on the lips, and herpes virus 2, which causes genital infections. About 10–20% of these genital infections, however, are caused by herpes virus 1. Since the recommended treatment differs for the two viruses, it is important to distinguish between them. Again, the test takes only about 15–20 minutes.

One exciting aspect of diagnosis with monoclonal antibodies is the research being carried out to try to detect cancers much earlier than is now possible. Leukaemias and lymphomas are both cancers of white blood cells (lymphocytes) which are sometimes difficult to distinguish between. Early precise diagnosis with monoclonals which recognise the different antigens on these cells is now becoming possible. The importance of this is that early treatment of cancer greatly increases the chances of survival.

Work is also proceeding on early diagnosis of the most common cancers such as lung, breast, colon and rectal cancers. At present, samples of body fluids such as blood and fluids from around the lungs are tested most frequently. However, techniques for identifying directly tumours inside the body are also being developed. One method is to add a radioactive isotope, such as iodine-131, to a monoclonal antibody which is specific for an antigen associated with a cancer cell. The antibody will seek out these antigens in the body and accumulate at cells possessing them. The location of these cells can then be detected from outside the body using special equipment sensitive to the radiation from the isotope. Eventually such targetting may be used in treating the cancer because the radiation could be used to kill the cancer cells (see below).

Treatment of disease – magic bullets. It is hoped that eventually monoclonal antibodies will be able to be used in the *treatment* of disease, not just in its diagnosis. The way they seek out and attach themselves to specific targets such as cancer cells has led to their being called 'magic bullets'. If a radioactive isotope or a toxic drug is attached to the antibody, the hope is that this would destroy the target cell. One problem is that the antigens most characteristic of cancer cells are also found on some normal cells, so these would also be killed. Another is that the antibody attaches to the *surface* of the cancer cell but does not enter it to deliver the toxin. However, some progress is being made on this work.

Preventing rejection of transplants. The main problem with transplants is that the body's immune system recognises the new organ as foreign and attacks it. One way around this problem is to try to suppress the immune system of the patient. A monoclonal antibody has been developed which is very effective at preventing rejection of transplanted kidneys. It reacts with an antigen found on all T cells. T cells are a type of lymphocyte which normally attack virus-infected cells and cancer-causing cells. However, they are also involved in the rejection of transplanted organs (chapter 14). A monoclonal antibody is more effective than the usual drugs which are used to suppress the immune system. It also only suppresses the T cells rather than the whole immune system which the drugs affect, leaving the patient with more protection against disease.

Tissue typing for transplants. To reduce the risk of rejection, before carrying out a transplant, a donor who is as compatible as possible must be found. This means that the donor's antigens must be as close a match as possible to those of the recipient. The more unlike the antigens are, the greater the chance of rejection. Monoclonal antibodies can be used to find out the types of antigens present in the donor and increase the accuracy of matching.

12.11.3 Insulin and human growth hormone

Both these products can now be produced from genetically engineered bacteria. Although genetic engineering can be regarded as a branch of biotechnology, it is dealt with in chapter 25 with other applications of genetics. Insulin and human growth hormone are discussed in chapter 25.

12.12 Food and drink

Some of the oldest biotechnologies are concerned with the making of foods and drinks. The making of bread, cheese, yogurt, vinegar and alcoholic drinks, such as beer and wine, date back thousands of years to prehistoric times. There are many references to wine in the Old Testament of the Bible, the Romans drank beer and wine, and as long ago as 4000 BC the ancient Egyptians are known to have made beer and used wild yeasts to raise dough in the making of bread. By some definitions, the breeding of animals and plants for food can also be regarded as biotechnology, and agriculture dates back more than 10 000 years.

In this section we shall not only be looking at modern approaches to some of these old technologies but also at some of the newer methods for producing food.

12.12.1 Yeast fermentation – bread, beer and wine

Yeast is a unicellular fungus. Wild yeasts are common and it is not difficult to imagine how the raising of dough and the fermentation of sugars to produce alcohol could have been discovered, even though the actual cause of the process could not have been understood. It was Louis Pasteur, in the nineteenth century, who first showed that fermentation was due to the activity of microorganisms.

Bread

Bread is made from flour obtained by grinding cereal grain, usually wheat. The flour is mainly starch (the white part of the grain), which is part of the food store normally used at germination of the grain. Enzymes present in the grain partially digest the starch to sugars like maltose and glucose. Amylases from fungi, which digest starch, can be added to increase the sugar content. Yeast uses sugars as a source of energy in respiration. Both aerobic and anaerobic respiration result in production of carbon dioxide gas. When making bread, bubbles of the gas are trapped inside the warm dough causing it to rise. This stage is called **proving**. Strains of the yeast *Saccharomyces cerevisiae* are selected for their high production of carbon dioxide. Alcohol is also a product of anaerobic respiration but this evaporates during the baking process that follows proving.

Beer

The oldest fermentation industry is that of brewing. Beer is brewed from barley grain which has been partially germinated to convert its starch store to sugars such as maltose, a process known as **malting**. Gibberellins are used to speed up this process and to control it more precisely (chapter 16). Also, more enzymes, such as amylase, are sometimes added to increase the amount of sugar produced. This in turn leads to more alcohol being produced. Nitrogen is also present from the digestion of proteins.

The grain is then killed by slow roasting to about 80 °C, although 'malts' for lager production are roasted at lower temperatures. The exact conditions affect flavour and colour. The grain is then crushed between rollers and added to hot water to extract the sugars. The liquid obtained is called **wort**. Hops are added for their bitter flavour and antimicrobial properties. The mash is boiled to concentrate it and the hops are removed before cooling the mash to a suitable temperature for fermentation. It is added to a large batch fermenter where it is inoculated with brewer's yeast, usually from a previous batch (fig 12.18). Two commonly used species are *Saccharomyces cerevisiae* and *Saccharomyces carlsbergensis*. The latter is used to make lager. Fermentation is anaerobic. During this process sugar is converted to alcohol and carbon dioxide as in baking. After 2–5 days, the alcohol reaches a final concentration of 3.5–8% (mostly 3.5–5%). Traditionally, lager yeasts grow at the bottom of the fermenter and are called **bottom fermenters**, whereas beer yeasts float at the top and are called **top fermenters**. However, newer types of beer yeast are bottom fermenters, enabling the same type of fermenter to be used for both processes. After fermentation the beer is separated from the yeast.

Wine

The source of sugar for the fermentation, which lasts several days, is grapes. Different grape varieties and different yeasts are responsible for the different flavours of wines. Red wines get their colour from the skin of the grapes used. White wines are made from grapes which have usually had their skins removed. Commercially selected strains of yeast are now used in preference to wild yeasts because they are more reliable. Fermentation starts when the grapes are crushed to form the 'must'. A second fermentation by lactic acid bacteria may take place during which malic acid is converted to lactic acid and carbon dioxide. This reduces the acidity of the wine.

Fig 12.18 *Reading the temperature of a brew of beer in a brewery.*

12.12.2 Lactic acid fermentation – dairy products

A range of products, known as dairy products, can be made by the fermentation of milk. Milk contains lactic acid bacteria which break down lactose, the sugar found in milk, to lactic acid during anaerobic respiration. These bacteria may be killed during pasteurisation, so it is usual to add them back again if fermentation is to be carried out. Different products are obtained depending on the conditions under which fermentation takes place, the additives used and the exact composition and source of the milk. Final textures and flavours are very variable, as can be seen from the range of products which include all the cheeses, butter, yogurt, sour cream and fromage frais (fig 12.19).

Yogurt

Unlike cheese, which is made from part of the milk (see below), yogurt is made by fermenting the whole milk. Originally it was thinner and more acidic, but in western countries it has been modified to suit national tastes. Since it is already acidic (pH 3.7–4.3), fruits can be added for flavour.

A variety of milks may be used, either alone or mixed, including whole, semi-skimmed, skimmed, evaporated and dried milks. Each combination will give a particular type of yogurt. Low-fat varieties have become increasingly popular. A flow diagram of the manufacturing process is shown in fig 12.20.

Fig 12.19 *Range of dairy products.*

All milk in the UK is heat treated (pasteurised), usually at 72 °C for 15 seconds. This is needed to destroy any pathogenic (disease-causing) microorganisms. These are not commonly present, but milk is an ideal medium for the growth of microorganisms and contamination by harmful bacteria such as *Salmonella*, or fungi, is always a possibility. In addition, some bacteria which are not harmful nevertheless affect the quality of milk if allowed to grow in it. Extra heat treatment is required for yogurt manufacture for the reasons shown in fig 12.20. In order to replace the lactic acid bacteria killed by the heat, a 'starter culture' is added. This is a culture of selected bacteria which gives a controlled and predictable fermentation. The starter normally contains *Lactobacillus bulgaricus*, which produces lactic acid and ethanal, a characteristic yogurt flavour, and *Streptococcus thermophilus* which adds a characteristic creamy flavour. These bacteria can be seen in roughly equal proportions if Gram staining of a sample of yogurt is carried out (section 12.9.2). Both types are Gram positive and will therefore appear purple. If you would like to try counting the bacteria, you should find something like 10^8 per gram of each of the two types mentioned.

Cheese

To make cheese the milk must first be separated into curds and whey. This is done by inoculating the milk in carefully controlled, hygienic conditions with a starter culture containing the required microorganisms. As fermentation proceeds, lactic acid builds up and starts to sour the milk, causing the soluble milk protein casein to coagulate (solidify). This is what happens when milk 'curdles'. Coagulation is increased and controlled more precisely by adding rennet which contains the enzyme rennin. Traditionally this is obtained from calves' stomachs, but increasingly similar proteinase enzymes are being obtained from certain fungi or from genetically engineered bacteria. The advantage of these newer sources is that they can be used to make vegetarian and kosher cheeses. The solid part of the milk, mainly protein and fat, is known as **curds** and the liquid is **whey**. The curds are separated and pressed into

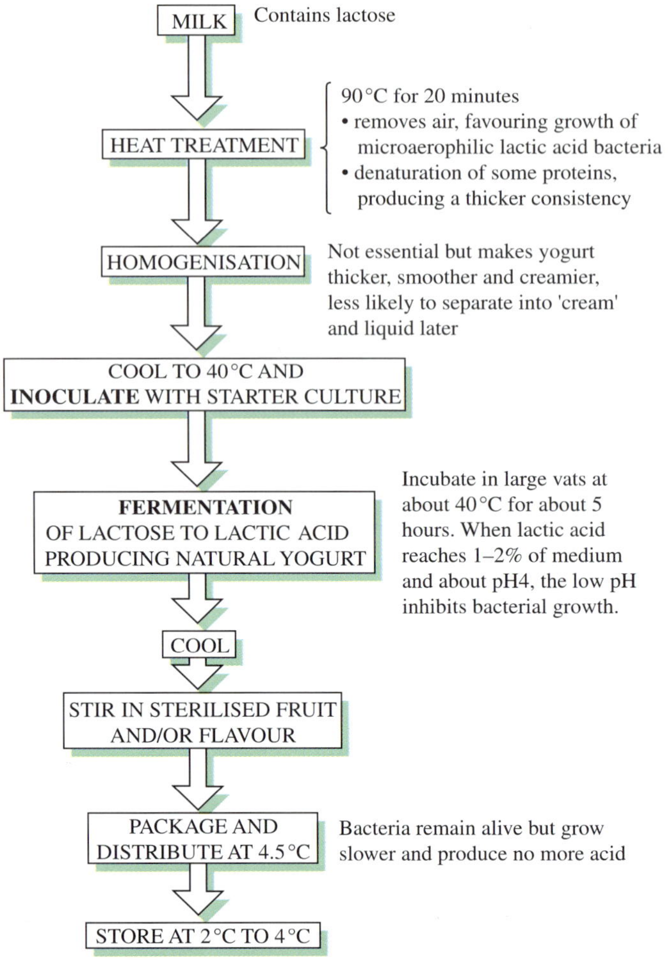

Fig 12.20 *Flow diagram of yogurt manufacture.*

the desired shape and drained. The whey can be used as a food source for yeasts which in turn can be processed and added to cattle feed. The product is rich in protein and certain vitamins.

Maturing or ripening of the cheese then adds its characteristic flavour and texture. This is done by adding further microorganisms, either bacteria (e.g. Cheddar), or mould fungi (e.g. Roquefort and other blue cheeses) or a combination of the two (e.g. Camembert). All these processes are done under controlled conditions which vary for different cheeses. Some cheeses, such as cottage cheese and cream cheese, are unripened.

Butter

Butter is made from the cream of fresh milk and must by law contain at least 80% fat. The cream is first pasteurised and later it is churned.

12.12.3 Single cell protein (SCP)

A relatively new food source is 'single cell protein' (SCP). Its production began in the late 1960s. The term refers to protein obtained from the large-scale growth of microorganisms such as bacteria, yeasts and other fungi, and algae. The protein may be used for human consumption or animal feed. It may also be a useful source of minerals, vitamins, fat and carbohydrate. In theory this should release proteins, such as soyabean meal and grain, which are currently used in animal feeds, for human consumption. However, there are agricultural surpluses in the western world which mean that this has not happened, as we shall see. Although there is a global shortage of protein, the shortage is in developing countries which cannot afford this technology.

There are several advantages in using microorganisms as a food source.

- They occupy less room than conventional crops and animals.
- They grow much more rapidly.
- They can grow on a wide range of cheap or waste products of agriculture and industry, e.g. petroleum products, methane, methanol, ethanol, sugar, molasses, cheese, whey, and waste from pulp and paper mills. They may therefore have the secondary advantage that they can help to recycle materials and clean up waste.
- There are fewer ethical issues associated with their exploitation and no animal rights issues.
- They are more easily modified by genetic engineering.
- They have a relatively high protein content.
- They are independent of climate and do not occupy large areas of land.

One of the first major SCP products was Pruteen, whose production was an excellent example of good engineering design in a continuous fermentation process. Pruteen was marketed as an animal feed additive. Early hopes for SCP were dampened by changes in the economic and political aspects of its production. Problems were as follows.

- A rise in oil prices in the late 1970s greatly increased the cost of production since the process, including production of methanol, is very energy-demanding.
- In the developed countries such as the USA and Europe where SCP was to be made, agricultural surpluses became common, particularly of rival high-protein products like grain, soya bean and dairy products.
- Developing countries, where protein is scarce, could not afford the investment necessary to set up SCP production themselves, and did not have the necessary expertise.
- There was an increase in production and reduction in price of competitive animal feed additives like soya bean, fishmeal, and gluten from maize. The latter is a byproduct of biotechnological fuel production.

Pruteen was made by ICI. After its failure as a commercial product, the production plant was dismantled, but ICI made use of the expertise gained by collaborating with Rank, Hovis, McDougall to produce a protein from a fungus called *Fusarium*. This mycoprotein (*myco-* refers to fungus) is unusual in being used for human consumption. It is marketed as Quorn, and its manufacture is described later in this section.

Pruteen

In 1980 ICI built a huge fermenter at Billingham on the north bank of the Tees in north-east England. It cost £40 million, had a capacity of 1500 cubic metres (1.5 million dm^3) and was 60 m tall. It had a potential to produce 70 000 tonnes of SCP per year. The organism used was a bacterium, *Methylophilus methylotrophus*, and its organic carbon and energy source was the cheap and readily available methanol. The process was aerobic. A summary of the process shows the raw materials required and the products made:

$$\begin{array}{c}\text{organic carbon (methanol)} + \\ \text{nitrogen} + \text{mineral salts} + \text{oxygen}\end{array} \longrightarrow \begin{array}{c}\text{SCP (biomass)} + CO_2 + \\ H_2O + \text{heat energy}\end{array}$$

The waste carbon dioxide was bottled and sold.

Screening for a suitable bacterium involved looking for one with a rapid growth rate capable of growing on a relatively cheap source of carbon and energy, heat-tolerant due to the heat released during the process, non-pathogenic to any other organisms and with a high percentage of protein in its dry mass. The bacterium chosen, *M. methylotrophus*, was found locally and has a generation time (doubling time) of 2–5 hours. Nitrogen was added as ammonium salts, and additional minerals that were needed included phosphorus, calcium and potassium. Methanol was obtained from natural gas on the same industrial site at Billingham. It produces no harmful by-products. Temperature was carefully monitored and kept within the range 30–40 °C, and pH was kept at 6.7. The fermenter could be run continuously for several months.

Pruteen manufacture is a classic example of continuous batch culture and the basic principles of manufacture are as described in section 12.10.5. The fermenter was tall and narrow, making aeration and cooling easier. It had a unique air-lift system to aerate the medium. Compressed air was introduced at the bottom of the fermenter and the rising bubbles mixed the contents (rather than using paddles). 6000 tonnes were produced per month. 2 tonnes of methanol produced 1 tonne of dried Pruteen, which is 72% protein and 8% moisture. It was rich in essential amino acids, had a high vitamin content and was twice as nutritious as soyabean meal. It was used as an animal feed.

Being a continuous culture, bacteria were harvested continuously once they had reached their maximum growth rate (exponential growth). Some of the contents of the fermenter were removed, treated with a chemical which caused the bacteria to clump together and then centrifuged. The separated cells were then spray-dried and the useful components of the liquid recycled to the fermenter. After drying, the bacterial cells were ground up to improve digestibility and the pH and mineral balance adjusted before packaging.

Mycoprotein

Fungi are another possible source of SCP. Yeast is used for both animal and human consumption (see yeast extract below) in processes similar to those for bacteria. Moulds are also used. These have the typical fungal body described in section 2.5.2 which consists of a mass of fine thread-like hyphae called a mycelium. A good example of the use of moulds is the manufacture of mycoprotein (Quorn). In 1985 Rank, Hovis, McDougall jointly formed a company with ICI, called Marlow Foods, to produce and market mycoprotein under the trade name of Quorn. The fungus used is *Fusarium graminearum*. It was originally isolated in the early 1960s from soil in a field near Marlow in Buckinghamshire, hence the name of the company. Product development was started in 1964. It was passed as safe for human consumption and launched in 1986, the first commercial product being a savoury pie sold by J. Sainsbury. Although originally used as an ingredient of manufactured foods such as pies and curries, Quorn pieces became available for home cooking in 1990 and minced Quorn in 1992. It is given a mild savoury flavour and has been a commercial success.

The doubling time of the fungus in culture is 5.5 hours, which is slower than bacteria, and it uses glucose as a carbon and energy source. This comes from any convenient cheap source of starch such as corn, wheat, rice, potato or molasses. It produces 0.5 kg dry biomass per kilogram of sugar used. Fungi have the advantage that they can be grown at a pH which is acidic enough to inhibit the growth of bacteria, thus reducing the risk of contamination. *Fusarium* is grown at 30 °C in continuous culture. Ammonium salts are the nitrogen source and mineral salts are also added to maintain growth, as with bacteria. Sterilisation of all materials used, and aeration and cooling of the medium are required as described for previous processes. Agitation of the medium is achieved by a special aeration mechanism because the hyphae would tend to become tangled up with stirrers and not be evenly distributed inside the fermenter. This mechanism is called an air-lift fermenter because the culture is circulated continuously, once every 2 minutes, by the air through a vertically elongated loop about 40 m high. The fermenter contains about 40 000 dm^3. As it is a continuous process, the product is harvested and fresh medium added continuously to keep the volume constant.

Eukaryotic cells contain a higher proportion of nucleic acid than prokaryotic cells, so the mycoprotein product contains a significant amount of nucleic acid (5–15% dry mass). This is mostly RNA and must be reduced because consumption of more than 2 g a day in the human diet can lead to gout and kidney stones. It is removed by heating the culture to 64 °C for 20–30 minutes in a separate steam-heated vessel. This inactivates fungal proteases (so the protein is not digested) but allows natural RNAase enzymes to break down the RNA. RNA is reduced to about 1%, well below the recommended limit of 2% set by the World Health Organisation.

An advantage of using fungi is that the mycelium is easier to separate from the medium than are bacterial cells. Filtration is sufficient and centrifugation is not needed. Filtering and drying leaves a thin flexible sheet of Quorn. At this stage it looks and tastes a little like raw pastry. Vegetable flavours and a little egg white are added. It is then sliced, diced or shredded for use. The fungus is already fibrous (one reason why it was chosen), so it is easy to give it the texture of meat.

For human consumption, factors other than just economics are important. Very strict safety guidelines must be adhered to, and the nutritional value of the food has to be acceptable. Long-term studies (10 or more years) with 11 species of animals, including rats, pigs and cows, for up to four generations have shown no long-term harmful effects. Human trials were also carried out before its release on the mass market. The final product has several health advantages over meat. Its composition compared with some typical animal protein sources is shown in table 12.5. It is cholesterol-free and high in fibre, unlike meat. It is low in fat and 'calories' (energy) with a good polyunsaturated:saturated fatty acid ratio (see section 8.7). It is also a good source of vitamin B_{12} and zinc, which are often lacking in the diets of vegetarians.

There are also psychological barriers to overcome in adopting new foods. Presentation, including packaging and advertising, is important (fig 12.21). Odour, colour, taste and texture must all be carefully planned. It was decided to market Quorn originally as a meat substitute. It can easily be woven into fibres which successfully mimic the texture of meat, and it can be flavoured to taste like chicken or, less commonly, beef.

Fig 12.21 *Food containing Quorn.*

Table 12.5 Typical composition of Quorn mycoprotein compared with traditional animal protein sources.

Component per 100 g	*Myco-protein (cooked)*	*Raw lean beef*	*Cooked stewing steak*	*Roast chicken, meat only*	*Cheddar cheese*	*Fresh cod*	*Grilled beef sausage*
Protein	12.3	20.3	30.9	24.8	26.0	17.4	13.0
Fat	3.2	4.6	11.0	5.4	34.4	0.7	17.3
Dietary fibre	4.8	0	0	0	0	0	0
Cholesterol	0	59	82	76	70	50	40
Energy (kJ)	355	514	932	621	1708	318	1104
Ratio PUFAs to SFAs*	2.5	0.1	0.1	0.5	0.2	2.2	0.1

* PUFA = polyunsaturated fatty acids – these are beneficial to health (section 8.7.7)
SFAs = saturated fatty acids

12.11 In advertising, the source of Quorn is usually referred to as a 'natural, tiny plant' or as a 'tiny relative of the mushroom'. Suggest why this is so.

Yeast extract

The left-over yeast from brewing can be used in a number of ways. One example is the manufacture of whisky which, like beer, depends on fermentation of sugars from germinating barley. Whisky, however, is distilled from the fermented malt. This requires boiling and kills the yeast. Yeast produced during brewing can be used as a replacement.

Yeast cells are rich in B vitamins, particularly niacin (B_6), riboflavin (B_2), thiamin (B_1), folic acid and B_{12}. They can be dried and made into vitamin-rich tablets or converted into products like Marmite. This involves heating the yeast in large vats to 50 °C and adding salt to encourage the process of autolysis (section 5.10.6). **Autolysis** is self-digestion and is carried out by enzymes in dying cells. The products are filtered and centrifuged to remove cell walls and then concentrated into a thick paste. Vegetable extract is added to Marmite.

An alternative to autolysis is **hydrolysis** with hydrochloric acid. This is later neutralised with sodium hydroxide. The product is also used in a whole variety of foods as a meaty, salty flavouring, for example in crisps, hamburgers, soups, sauces and gravy powders.

SCP from photosynthetic organisms

Both photosynthetic bacteria, such as blue-green bacteria, and algae are used to make SCP. *Spirulina* is an example of a blue-green bacterium that is used. It was made into cakes by the Aztecs and is the main food of flamingoes in the lakes of the Rift Valley in East Africa. *Spirulina* has been grown on a small scale in Mexico and Hawaii for sale in health food shops. It has a very high protein content and a high growth rate. It can be skimmed off from pond or lake surfaces. *Chlorella* is a unicellular alga which is commonly used. In Japan and Taiwan dried *Chlorella* is sold as health food.

The nutritional and environmental requirements of photosynthetic organisms used in SCP production have been discussed in sections 12.1.1 and 12.1.2. The fact that they all need light and most need carbon dioxide as a carbon source are the obvious differences compared with fungi and non-photosynthetic bacteria. They are generally cultured in non-sterile conditions in warm, mineral-rich open ponds and are usually in a mixed culture (one that contains different species).

Growing algae on sewage serves the dual purpose of purifying the sewage and producing SCP for animal feed. This has been done in Israel.

12.13 Agriculture

This section deals with some of the ways in which production of food using biotechnology can be extended to agriculture.

12.13.1 Genetic engineering

Some of the most important applications of genetic engineering are in agriculture. As already discussed, genetic engineering can be regarded as an aspect of biotechnology. Examples of its use in agriculture which are considered in this book are production of somatotrophin (BST), pesticide and herbicide resistance, and transgenic plants and animals. These are all discussed in chapter 25 which deals with the practical applications of genetics.

12.13.2 Silage

The making of silage is a traditional anaerobic fermentation process carried out on farms. It preserves the nutritive value of grass for winter feed for farm animals, particularly dairy cattle. Grass has a natural population of lactic acid bacteria such as *Lactobacillus* on its surface. After harvest it is finely chopped and can be loaded into a large 'bin' called a silo or silage clamp or, more recently, it is sometimes stacked in a large bale wrapped in black plastic. The bacteria use the natural sugars in grass, such as glucose, fructose and sucrose, as an energy source in anaerobic respiration (traditional fermentation) and convert them to lactic acid. Other fermentation products are also formed which add flavour, but lactic acid is the main product. It reduces the pH to about 4, which is low enough to inhibit the activity of decomposing bacteria which would otherwise completely rot the grass and lower its nutritive value. Temperatures in fermenting grass can become quite high, so the bacteria must be heat resistant. Contamination with *Listeria* bacteria must be avoided because this can lead to blood poisoning or stillbirth of farm animals. Providing fermentation is active and the pH is less than 5.5 this should not be a problem.

Improved reliability and quality of silage production is now possible by adding commercially prepared inoculants of fast-growing lactic acid bacteria. These are applied as powders or sprays to the newly harvested grass. ICI produce three products in fermenters at Billingham, named Ecosyl, Ecohay and Ecobale. Increased productivity of the animals is reported to occur, partly because the conversion to lactic acid is more efficient and partly because the silage tastes better and therefore the cattle eat more.

Fig 12.22 *The effect of* Rhizobium *inoculant on pea seedlings. The pea seedling on the left has its roots growing under sterile conditions in a flask. The plant has been fed with a mineral salts medium lacking any source of nitrogen. Consequently, it shows poor growth and yellow leaves. The roots of the similar pea seedling on the right were inoculated with* Rhizobium, *which converts nitrogen gas in the atmosphere into forms which the plant can use. Consequently this plant is healthy and shows no sign of nitrogen deficiency.*

12.13.3 Nitrogen fixation

Crops of the legume family (Papilionaceae), which includes peas, beans, clover and alfalfa, have been used for centuries in crop rotation schemes because their roots add nitrogen to the soil. In the nineteenth century it was discovered that the characteristic swellings on the roots of these plants, root nodules, contain bacteria which can 'fix' nitrogen from the air and convert it to nitrate. This process is described in more detail in section 10.4.1. The bacteria are various strains of *Rhizobium* (fig 12.22). Individual strains are adapted to infect different legume species. The bacteria normally infect the roots as they grow in the natural habitat where *Rhizobium* occurs naturally. However, for situations where seed is to be grown in places where the bacteria may not exist, for example if the seed is for export, a technique has been developed for 'inoculating' the seed with the bacterium. Cultures of the required *Rhizobium* strain are grown in fermenters and afterwards added to a suitable medium, commonly sterile peat. This can be added to the soil as the seeds are planted, for example by being added to the seed drill. Inoculated seed is not needed where the crop has been grown regularly because *Rhizobium* can survive in the soil for many years using simple organic compounds and ammonium or nitrate as a nitrogen source. The bacterium only fixes nitrogen when in plant roots.

Examples of seeds that have been exported are clover to Australia, soya beans to N. America and alfalfa to Europe. Using such crops in rotation reduces the need for industrially produced nitrogen fertiliser. The advantages of this are discussed in section 10.8.2.

12.14 Fuel from biomass – new energy sources

Biomass is a traditional fuel in the form of products like coal, gas, oil, wood, peat, and dried animal dung. Some of these products are becoming scarcer and more expensive. Many new methods are being explored for using living organisms and biological processes as sources of fuel. **Artificial photosynthesis**, producing hydrogen gas as a fuel from water, is a long-term possibility. Another basic approach is to change the energy trapped in biomass into another form which can be used as fuel. Among the raw materials being investigated are **waste materials** such as animal manures, sewage sludge, domestic wastes, food wastes, paper wastes, spoilt crops, sugarcane tops and molasses. Various **crops** (such as maize, sugarcane, sugar beet) and water plants (such as kelps and water hyacinth) might also be used. Two processes currently dominate, namely production of biogas (methane) by bacteria, and production of ethanol by yeast. Both processes are anaerobic.

12.14.1 Biogas

Overall equation:

	$C_6H_{12}O_6$	$\longrightarrow$	$3CH_4$	+	$3CO_2$
	glucose		methane		carbon dioxide
energy value:	$16\,kJ\,g^{-1}$		$56\,kJ\,g^{-1}$		

Biogas is about 54–70% methane. Most of the rest is carbon dioxide, with traces of nitrogen, hydrogen and other gases. (Natural gas is about 80% methane.) A mixture of microorganisms is used in the fermentation, including a group of bacteria called **methanogens**, e.g. *Methanobacterium*, which can produce methane from carbon dioxide and hydrogen. These are **archaebacteria**, an ancient group of organisms closely related to the true bacteria. A wide range of waste materials or plant products can be used as a substrate for fermentation (see above). In the USA the water hyacinth, a vigorous plant which can block canals and waterways, has been used. The process is ideal for small-scale use and therefore for local fuel use. This is now common in India and China for example (fig 12.23).

The manure from one cow in one year can be converted to an amount of methane which is equivalent to over 227 litres (dm^3) of petrol. For example, 0.5 kg of cow manure could generate enough gas to cook a family's meals for a day. In China, over 18 million family-scale digesters have been built. The gas is typically used for cooking, lighting, tractor or car fuel and for running electricity generators.

On a larger scale, the gas can be produced as a by-product of landfill, sewage or waste from factories such as sugar factories and distilleries. It can be used to drive electricity generators in sewage works and waste treatment plants. In Britain, rubbish could be a major source of methane, with up to 20 dm^3 of gas being obtained per kilogram of rubbish. At the moment the gas is collected from landfill sites by sinking pipes into the compacted rubbish and sucking out the gas.

Bearing in mind the growing shortage of landfill sites and the nuisance they cause, it may be worth developing processes for *fermenting* products like paper and cardboard, even if the fuel produced is no cheaper than conventional fuels. The economics are usually more favourable in developing countries which lack their own fossil fuel reserves and have dwindling supplies of timber. Raw sewage and dried animal dung can also be used in a fermenter. The fuel value of fermented dung is six times greater than that of dried dung.

Fig 12.23 *Biogas digester in India. Animal manure is put into the digester, where it decomposes and gives off methane gas.*

12.14.2 Ethanol

Overall equation:

	$C_6H_{12}O_6$	$\longrightarrow$	$2C_2H_5OH$	+	$2CO_2$
	glucose		ethanol		
energy value:	$16\,kJ\,g^{-1}$		$30\,kJ\,g^{-1}$		

Ethanol has been produced successfully for use as a fuel in Brazil since 1975. Sugarcane is the starting material. It is crushed by rollers after harvest to extract the juice. Sucrose is extracted from the juice as a commercial product, but this leaves a syrup called **molasses** which contains glucose and fructose. The molasses can be used as material for fermentation by the yeast *Saccharomyces cerevisiae*. Ethanol is distilled to separate it from the other fermentation products. The dry fibrous material left behind when the sugarcane is crushed can be used as fuel in the distillery.

Some Brazilian cars are adapted to run on pure alcohol, although the ethanol is mixed with a little petrol to stop people drinking it. Over 11 000 million dm^3 were produced in 1985. The process started to save Brazil money when oil prices rose in 1983, and since then oil consumption has been cut by 20%. Some cars in Brazil and in the USA can run on mixtures of alcohol and petrol called **gasohol** (fig 12.24). In the USA the starting biomass is starch from maize. Over 2280 million dm^3 per year were being produced in the mid 1980s.

Although the Brazilian scheme seems to have been a success, not all is straightforward. Debate continues about the most economic use for molasses and whether other crops might be a better source of carbohydrate. For example, molasses could be used instead of oil as a raw material in the plastics industry.

Fig 12.24 *An ethanol-powered car filling up with Alcool (gasohol) in Brazil.*

12.15 Microbiological mining

Useful metals such as copper, iron, uranium, gold, lead, nickel and cobalt are found naturally as minerals, otherwise known as ores. Where these ores are sufficiently concentrated they can be mined and the metals extracted. It is only recently that the important role and potential of microorganisms in the extraction process has been realised. Copper will be used as an example to illustrate the principles involved.

Copper was one of the first metals used by humans. Bronze, which is an alloy of copper and tin, was first made over 5000 years ago, leading to the 'Bronze Age' when bronze was valued for its strength and cutting power as well as its decorative value. Copper naturally occurs mainly as copper sulphides. For example, more than 50% of the world's supply comes from copper pyrites, $CuFeS_2$, which contains iron as well as sulphur. Extraction of copper from the ore is difficult. It has long been known, though, that copper can be recovered from water which has drained through rocks containing the copper ores. This process of metal **leaching** is now known to be due to the action of bacteria. The bacteria convert insoluble metal compounds to soluble metal compounds such as copper sulphate, from which copper can be more easily extracted.

The bacterium chiefly responsible for the leaching of metal sulphides was identified in 1947 as *Thiobacillus ferrooxidans*. Other important bacteria are *T. thiooxidans* and *Leptospirillum ferrooxidans*. These bacteria thrive in acid conditions and can often work at high temperatures. They obtain their energy by oxidising inorganic substances. *T. ferrooxidans*, for example, obtains energy by oxidising Fe^{2+} in ores to Fe^{3+} and reduced forms of sulphur, like the sulphides in ores, to sulphuric acid. *T. ferrooxidans* is autotrophic, so can be classified as chemoautotrophic or chemosynthetic (table 2.3).

Bacterial leaching is now used throughout the world as an additional technique for extracting metals from ores, mainly for copper and uranium (fig 12.25). Suitable combinations of bacteria are used, each making its own unique contribution. More than 10% of the copper extracted in the USA in 1983, worth over 300 million dollars, was recovered using this method. Advantages of bacterial leaching include the following.

Fig 12.25 *An open pit mine in Rio Tinto, Spain, which produces 2.3 million tonnes per year of gold, silver and copper.*

- Lower grade ores can be exploited commercially. The expense of extracting by conventional methods meant that only ores particularly rich in the metal could be used, so there was much loss of potential product in the mining areas.
- Deep mining can be avoided by washing out the metal using bacterial leaching. The rock is fractured first by explosive charges and the leaching solution is then pumped in. The leaching solution containing the soluble metal compounds is recovered by pumping from wells sunk into the rock. This saves the cost, danger and environmental damage of deep mining which brings large quantities of rock to the surface, creating waste tips.
- Traditional methods of extracting the copper from the ore use high temperatures. They consume fossil fuels and therefore contribute to pollution such as acid rain

as well as being expensive. (Bacterial leaching might also be used in future to clean fossil fuels by leaching out sulphur compounds.)

- Uncontrolled leaching from the waste of mines has led to pollution of surrounding waters with heavy metals. This can be avoided by controlled leaching with recovery of the metal. It is estimated that more than 33 thousand million kilograms of copper are located in mine dumps in the western USA alone. A good location for a dump is a valley so that it can be sprayed with water from a river. The water, containing soluble leached metal, can then be collected at a dam downstream and pumped through a metal recovery plant. Cleaned water can be recycled to the dump if desired.
- Improvements are being attempted by genetically engineering the bacteria, particularly *T. ferrooxidans*.

12.16 Enzyme technology

This section will be concerned with the use in biotechnology of enzymes which have been separated from cells.

Enzymes are the biological catalysts which make possible the organised chemical activity of cells. Although we have made use of them for thousands of years, we did not begin to understand how they work until the late nineteenth century. We now know that they are complex protein molecules with specific three-dimensional shapes and that their structure is coded for by DNA. The number of possible overall shapes is infinite.

To the industrialist, enzymes have two major attractions. Firstly, because of their variety, they have the potential to catalyse a vast range of industrially important chemical reactions. Secondly, they are much more efficient and specific than the inorganic catalysts commonly used. As a result, they may achieve at normal temperatures and pressures what might otherwise require extremely high temperatures and pressures. For example, one of the world's largest industrial processes is the Haber process in which ammonia (NH_3) is produced from nitrogen and hydrogen gases at a temperature of 500 °C and at high pressures. Nitrogen-fixing bacteria, however, can make ammonia from atmospheric nitrogen and hydrogen at room temperature and normal atmospheric pressure using enzymes, with ATP as an energy source. If the technology could be devised to do this with enzymes, great energy savings would be made. Another advantage is that, because of their specificity, enzymes also give purer products, which is important in the pharmaceutical, food and agricultural industries.

There are disadvantages however, mainly relating to the instability of proteins once they are removed from the cell environment. They are easily denatured by temperature and pH change, and by the organic solvents which often have to be used in industrial processes. They may also be inhibited by products of the reaction. Two other problems are that they are expensive to produce and that they must be taken from 'safe' (for example, non-pathogenic) organisms if they are to be used to make products for use in animals, including humans. Less than 200 enzymes are used out of the roughly 2500 which have been isolated and described so far (which in turn is only about 10% of those found in nature). Most come from only 11 species of fungi, four species of yeasts and eight species of bacteria.

In the future there are exciting opportunities opening up for re-designing enzymes. One possibility is to deliberately change individual amino acids by changing the genes that code for the enzymes. As we learn more about the rules governing the way that proteins fold into their specific three-dimensional shapes, we may even start to be able to make completely new *designer enzymes*. This is known as **protein engineering**. Also, the more we learn about the ways in which enzymes work, the more likely it will be that we can design non-protein or part-protein catalysts which are far more stable than normal enzymes. This is probably the direction most industrialists would favour. Of more immediate benefit is the search for natural enzymes which offer improved alternatives to the processes currently used. All this requires the investment of large sums of money in research and development.

Pharmaceutical companies, cheese-makers, brewers and distillers, wine producers, detergent manufacturers, textiles manufacturers, fruit juice producers and many others now use enzymes in manufacturing processes. Worldwide sales of enzymes now exceed 1 billion dollars per year. Table 12.6 gives some idea of the range of uses of enzymes.

12.16.1 Source of enzymes

Microorganisms are preferred to plants and animals as a source of enzymes because:

- they have high growth rates;
- they can be grown economically in bulk in fermenters under controlled conditions;
- they carry out a wide range of chemical reactions;
- they can be genetically engineered relatively easily, and mutant varieties can relatively easily be produced, to improve performance;
- they have simple nutritional requirements;
- they can be grown on cheap, often waste, substrates;
- production rate can be altered to suit demand;
- they produce a lot of extracellular enzymes; because these leave the cell, they are easier to recover and purify.

12.16.2 Why isolate enzymes?

Since the enzymes used come mainly from microorganisms, it might be argued that it would be simpler to use whole cells in fermenters rather than isolate the enzymes first. Where more than one chemical reaction is

Table 12.6 Some common industrial uses of enzymes.

Application	*Enzymes used*	*Uses*
Biological detergents	Primarily proteases, produced in an extracellular form by bacteria	Used for pre-soak or main wash, break down protein stains on clothes; also used in dishwashers to remove food residues
	Amylases	Remove starch stains from clothes; also used in dishwashers to remove resistant starch residues
	Cellulase	Softens and brightens colour of cotton fabrics
Brewing industry	**Enzymes produced from barley** during mashing stage of beer production	Digest starch to sugars and proteins to amino acids and peptides for use by yeast in growth and alcohol production
	Enzymes from microorganisms	
	Amylases	Digest polysaccharides in the malt
	Proteases	Digest proteins in the malt; remove cloudiness during cold storage
	β-glucanase	Digests polysaccharides in beer and prevents cloudiness
Baby foods	Trypsin	To predigest baby foods
Dairy industry	Rennet (rennin) from stomachs of young calves	Used to coagulate protein in manufacture of cheese
Leather industry	Proteases	Removal of hair from hides; also makes leather more pliable
Paper industry	Amylases	Degrade starch to smaller molecules which have lower viscosity and are used for sizing (filling in spaces between cellulose fibres to make a smoother product) and coating paper
Photographic industry	Proteases	Dissolve gelatin off scrap film allowing recovery of silver present

Modified from Table 5.2 *Biotechnology*, 2nd ed., John E. Smith, New Studies In Biology (1988), Edward Arnold, and *Enzymes, their nature and role*, Wiseman & Gould, Hutchinson Educational.

required to make a product, and therefore more than one enzyme, this is what is normally done. However, using purified enzymes rather than whole microorganisms can be an advantage for several reasons.

- Building and running fermenters in which to grow microorganisms is expensive.
- Fermentations must be carried out in aseptic conditions.
- Aeration and mixing of whole cells is energy-consuming.
- Products in fermenters are in an impure form.
- Wastes from fermenters have to be disposed of and may cause pollution.

Manufacturing processes carried out with purified enzymes involve the use of specially designed 'reactors'. The design of these is outside the scope of this book, but they are much simpler than fermenters.

12.16.3 Producing isolated enzymes

When the source of the enzyme is a microorganism, the microorganism is first grown in bulk in a fermenter as discussed earlier in this chapter. Glucose is usually avoided as a nutrient because it inhibits the production of many useful enzymes. Batch fermentations are used, except for the production of glucose isomerase from *Bacillus coagulans*, which is a continuous fermentation. Many enzymes are secreted from the cells as extracellular enzymes. These can be purified more easily than those which remain inside the cells. In the latter case, the cells are separated from the liquid contents of the fermenter and then broken open by various means to release their contents. Purification is then carried out as far as is necessary. Some enzymes do not need to be completely pure.

The enzymes may be used as they are, for example in fruit juice production, meat tenderisation and in washing powders, or may be immobilised before use (section 12.16.7).

12.16.4 Fruit juice production

During the manufacture of fruit juices the fruit must first be crushed. Like any other part of the plant, fruits are made up of cells with cell walls. These contain cellulose and other complex polysaccharides called hemicelluloses. Cell walls are very tough and difficult to break open. In order to improve yields and the quality of the product, cellulases and hemicellulases are added during the crushing stage to digest the cellulose and hemicelluloses in the walls, making them more 'soluble' and ensuring more complete disintegration of the tissues. The enzymes are selected to work at low pH because fruit juice is acidic. Their pH optimum is about 4–5.

Plant cells are held together by other complex polysaccharides called pectins. These are the sticky compounds which cause jam to set. They tend to be converted into water-soluble pectins during the storage and processing of fruits, and are therefore present in fruit juices in solution even if the 'bits', the unbroken cells and cell wall debris, are removed. At low temperatures, the pectins start to come out of solution and form a colloidal suspension (a suspension of tiny particles which do not settle). This gives the drink a cloudy appearance. This is an unattractive feature to some consumers, particularly in the USA and Britain, although it should, perhaps, be regarded more as a sign of high fruit content. If enzymes called pectinases are added to the juice they partially digest the pectins to smaller polysaccharides and sugars which remain in solution even at low temperatures. The pectinases therefore clarify the drink (fig 12.26). The drink is then described as 'chill-proofed'. The source of the pectinases is bacteria.

The same principle is applied to beers and wine, although the haze in beer is due to protein and tannins rather than polysaccharides. Proteases such as pepsin and papain are added to beer to break down the proteins and reduce the haze when it is chilled.

12.16.5 Meat tenderisation

Meat is mainly muscle protein. Muscles are bundles of protein fibres wrapped in blocks by connective tissue. Connective tissue also contains structural proteins, particularly collagen and elastin. Collagen forms 'white fibres', which are tough and non-elastic; elastin forms 'yellow fibres' which are elastic in nature. The meat can be made easier to chew, in other words more tender, by predigesting some of these connective tissue proteins and some of the muscle fibres. The fibres become shorter and more easily separated and the meat therefore breaks down more easily.

Enzymes that break down proteins are called proteases and the one most commonly used for meat tenderisation is **papain**. It comes from the sap of the papaya plant. Meat tenderisers containing this enzyme can be purchased for home use. The meat should be left 'marinading' in the juice

(a)

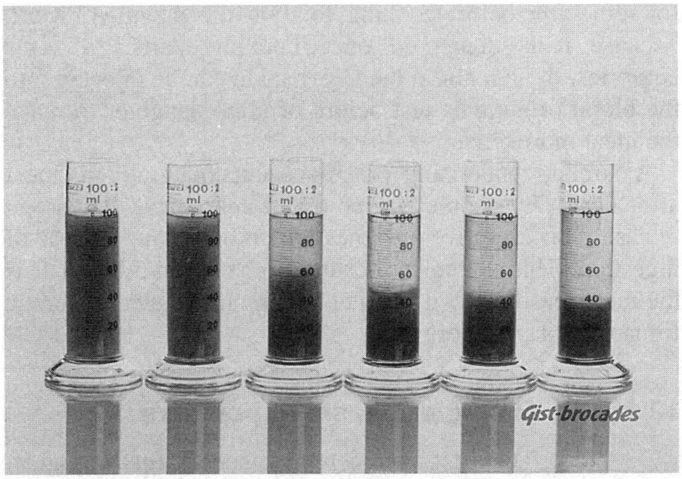

(b)

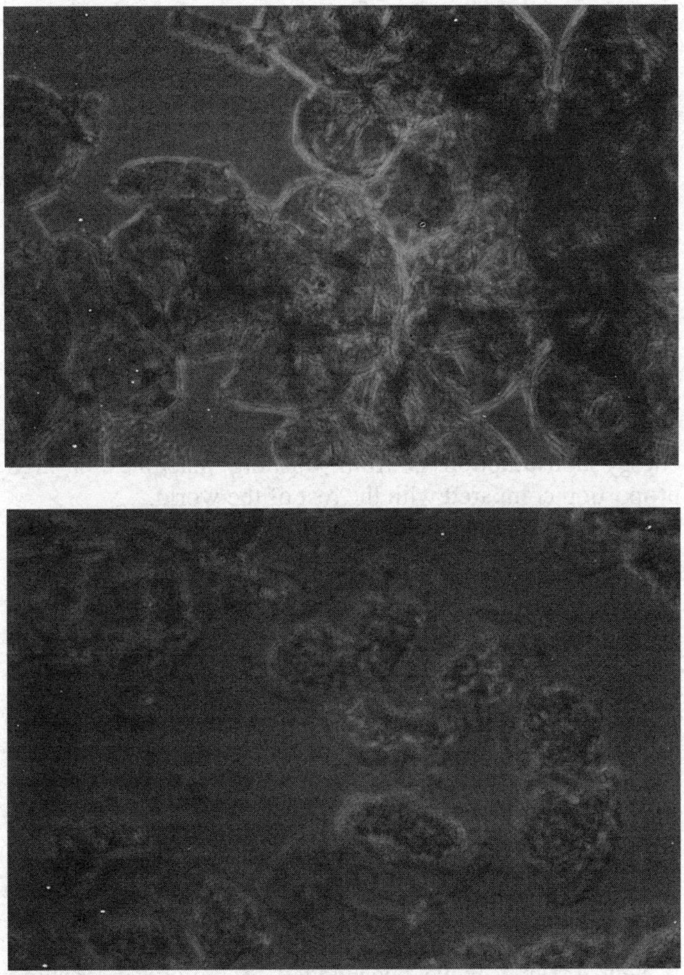

Fig 12.26 *(a) A series of flasks showing the effect of increasing pectinase dosage on apple juice. The two cloudy flasks on the left have no enzyme. The others, from left to right, contain 0.5, 1.0, 1.5 and 2.0 grams per 100 dm*^{3}*.*
(b) Apple pulp before (middle) and after (bottom) the action of pectinase. You can see that the pectinases (Rapidase) are removing the pectin coating of the apple cells, making them more fragile and less viscous.

for some time before cooking, to allow the enzyme to work, because it is denatured once cooking starts. In some countries, though not in the UK, papain can be injected into the blood of animals just before or after slaughter to make the meat more tender.

A similar tenderising process occurs naturally in meat after death. Lysosomes in the dying cells break down and release their digestive enzymes which begin the process of digestion. This is known as autolysis (section 5.10.6). It is the main reason why it is usual to hang meat in cold storage for several days before use.

12.16.6 Biological washing powders

The first commercially successful biological washing powders were introduced in the mid 1960s. They contained proteases (protein-digesting enzymes) which are particularly useful for removing stains of biological origin, such as blood, grass and egg, formed from protein and other materials. These are dislodged by digestion of the protein. Lower temperature washes are possible, saving energy, although the enzymes used have been selected to function over a wide range of temperatures (10–90 °C, optimum about 55 °C), having been obtained originally from thermophilic (heat-loving) bacteria. In washing powders, proteases must also function in alkaline conditions (pH 9–10) and in the presence of the high levels of phosphate found in some detergents. They are particularly useful for pre-wash and soak use and for low temperature washes. In western Europe 25% of domestic energy consumption is from washing machines, a high proportion compared with the rest of the world.

Soon after the introduction of biological washing powders, health problems began to occur, particularly among the workers in the factories making the detergents. Some people developed allergies to the powders, and lung irritation and respiratory disorders were experienced after breathing in fine detergent dust from the air. This led to the withdrawal in the early 1970s of the powders from sale in the USA, although they remained in use in Europe. The product was re-formulated to overcome these problems (fig 12.27), an inert wax being added to the powder to make it safer on contact and less likely to be airborne. Liquids were also developed. They have now been re-introduced in the USA and by the mid 1990s made up 15% of the market there compared with 85% in western Europe.

New enzymes have been added to biological washing powders to improve effectiveness. Cellulase digests the loose cellulose microfibrils from damaged cotton fibres, making cotton fabrics brighter, softer and smoother. Enzymes are also added that digest carbohydrates, for example amylases that digest starches. One problem is that proteases can digest some other enzymes since enzymes are proteins! This has made finding suitable enzymes more difficult. Even more recently a suitable lipase has been found in a fungus which is active at normal temperatures.

Fig 12.27 *SEM of biological washing powder, showing encapsulation of granules. Some granules are partially opened to show the enzymes inside.*

The relevant gene has been transferred by genetic engineering into another fungus which is more suitable for growing in a fermenter. Lipases digest lipids, in other words fat, oil and grease stains.

An organism which is commonly used as a source of enzymes for washing powders is the bacterium *Bacillus*. Several species are used. The enzymes produced are extracellular, making it relatively easy to collect and purify them.

12.16.7 Immobilised enzymes

Advantages of immobilised enzymes

As already discussed in the introduction to this section, commercial use of enzymes is limited by a number of factors. Two important ones are the instability of enzymes and their high cost. The cost can be greatly reduced by **immobilising** the enzyme. This means binding it to, or trapping it in, a solid support which can be recovered easily from the reaction mixture during downstream processing. The enzyme can then be re-used which greatly reduces the cost of the process.

Another advantage of immobilisation is that it sometimes makes the enzyme more stable, possibly by restricting its ability to change shape and denature as a result of changes in pH, temperature, and solvents. For example, glucose isomerase is stable for almost a year at 65 °C when immobilised, whereas it denatures within a few hours at 45 °C when in solution.

Immobilising the enzyme can also mean that continuous (open) production can be achieved more easily by passing the reactants over the enzyme and collecting the product at the end.

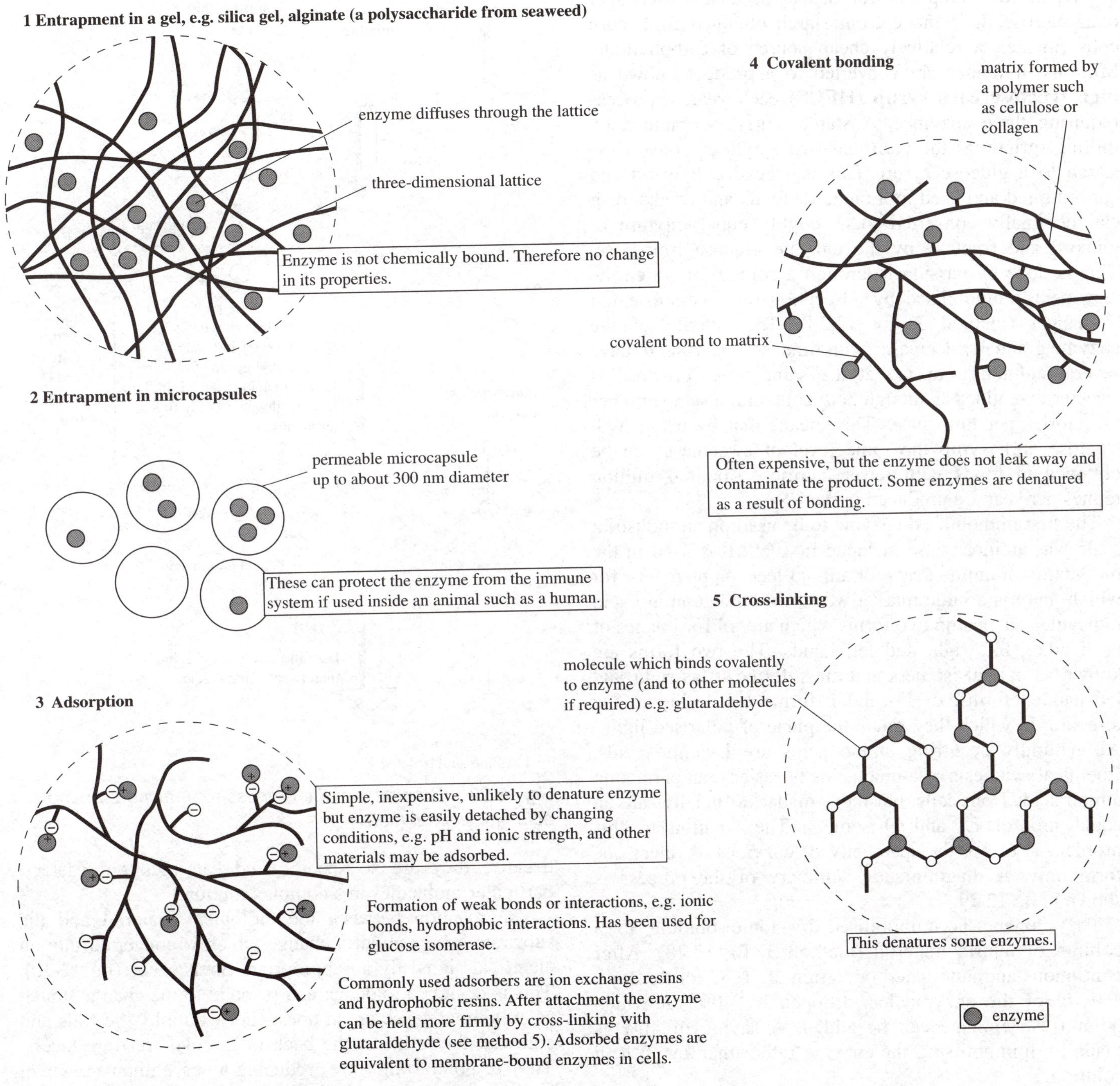

Fig 12.28 *Methods for immobilising enzymes.*

Methods for immobilising enzymes

There are various ways of immobilising enzymes, as shown in fig 12.28. They involve either trapping the enzyme (**entrapment**) or attaching it to a fixed structure or matrix. Entrapment has the advantage that the enzyme remains in its natural state. However, it is difficult for large molecules to reach the enzyme.

Entrapment in beads of alginate is easily demonstrated in a class experiment and is a commonly used method in industry. A solution containing the enzyme and sodium alginate is dripped into a solution of calcium chloride. The droplets start to gel immediately they contact the calcium chloride and perfectly spherical beads of gel, with the enzyme entrapped inside, are formed. The gel can be further stabilised with polyacrylamide for long-term use. It can also be prepared in the form of sheets if supported by a woven cloth.

Applications of immobilised enzymes

The best example of a successful process involving immobilised enzymes is the production of high fructose corn syrup. This is widely used as a sweetener in the USA and Japan, for example in fruit drinks, because it is cheaper than sucrose. It is made from starch obtained from corn cobs (maize), a relatively cheap source of carbohydrate. Millions of tonnes are converted to a product known as **high fructose corn syrup (HFCS)** each year, a process requiring three enzymes. A starch slurry is obtained by milling (grinding) the corn and two amylases convert the starch to a glucose syrup. This can be decolourised and concentrated and used in a range of foods and drinks, or it can be finally converted to a roughly equal mixture of glucose and fructose by the enzyme glucose isomerase. This is done by passing it through a column in which the enzyme is immobilised by adsorption on a cellulose ion exchanger (method 3, fig 12.28). The activity of the enzyme gradually decreases with time, so it is usual to have several columns working at the same time. Fructose is sweeter than glucose, though both contain the same number of calories per unit mass. This means that by using high fructose corn syrup the same level of sweetness can be obtained in foods with fewer calories. About 4 million tonnes per year are produced in the USA.

The first immobilised enzyme to be used on an industrial scale was aminoacylase in Japan in 1969. It is used in the production of amino acids for animal feed supplements, for which there is a large market worldwide. Each amino acid molecule can exist in two forms which are mirror images of each other, like right and left hands. The two forms are known as optical isomers and are referred to as right and left-handed forms, or D- and L-forms (according to the direction in which they rotate the plane of polarised light). All naturally occurring amino acids are L-amino acids. Chemical synthesis of amino acids is easier than extracting amino acids from cells, but those made 'artificially' are an equal mix of L- and D-isomers. The solution to this problem is to use the specificity of enzymes to select one form only. A diagrammatic summary of the process is shown in fig 12.29.

The enzyme is immobilised by ionic binding to a column of matrix material (method 3, fig 12.28). After continuous and automated operation at 50 °C for 30 days, activity of the enzyme has dropped by 40%, at which point fresh enzyme can be added. A saving of 40% is made by immobilising the enzyme rather than using it in solution.

Another use of immobilised enzymes is in the manufacture of semisynthetic penicillins from natural penicillins. The immobilised enzyme chemically modifies one of the side groups on the basic penicillin structure which increases the antibiotic activity of the penicillins.

12.17 Biosensors

A **biosensor** is an electronic monitoring device which uses a biological material, such as a cell, an enzyme or an antibody, to detect or measure a chemical compound. Enzymes and antibodies are particularly useful because they are so specific and can pick out (detect) particular molecules in a complex mixture.

The reaction between the biological material and the substrate brings about a change which is converted into an electrical signal by an appropriate **transducer** (fig 12.30). This is designed to detect and respond to the change, much as an animal sense organ does. (For example, the rods and cones in the retina at the back of the eye are transducers. They respond to light by producing a nerve impulse, which is an electrical event.) The electrical signal in the biosensor is amplified to give some form of read-out, such as a digital display or print-out. Many types of change may occur, such as release of heat, light, a change in pH or mass, a flow of electrons or production of a new chemical.

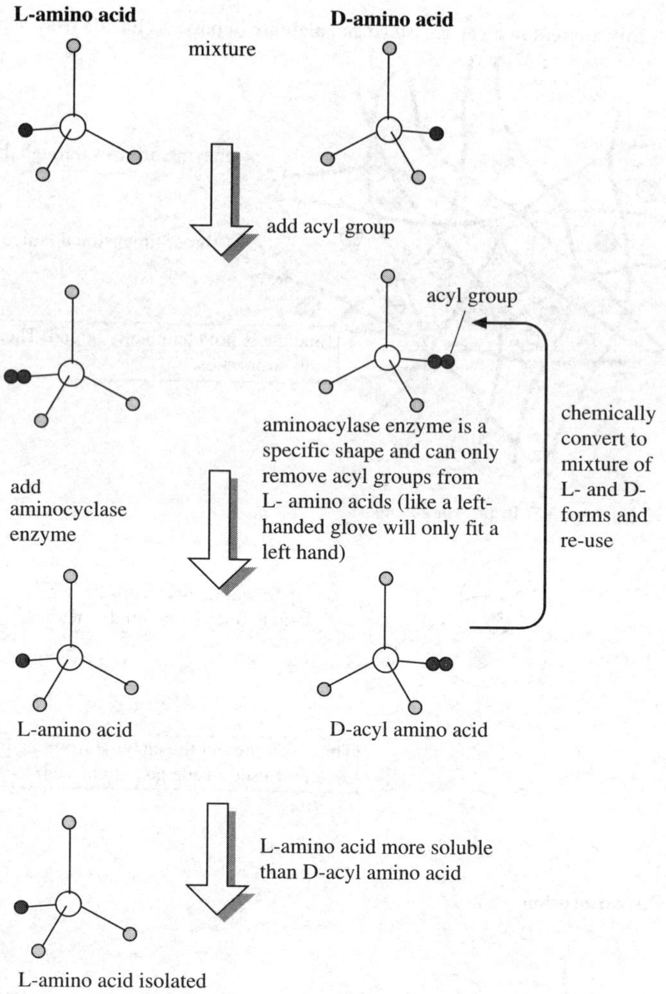

Fig 12.29 *Summary of the process for making L-amino acids for animal feed.*

12.17.1 Advantages and problems with biosensors

The main advantages are:

- they are specific – complex mixtures can be analysed for a particular chemical without the need for purification;

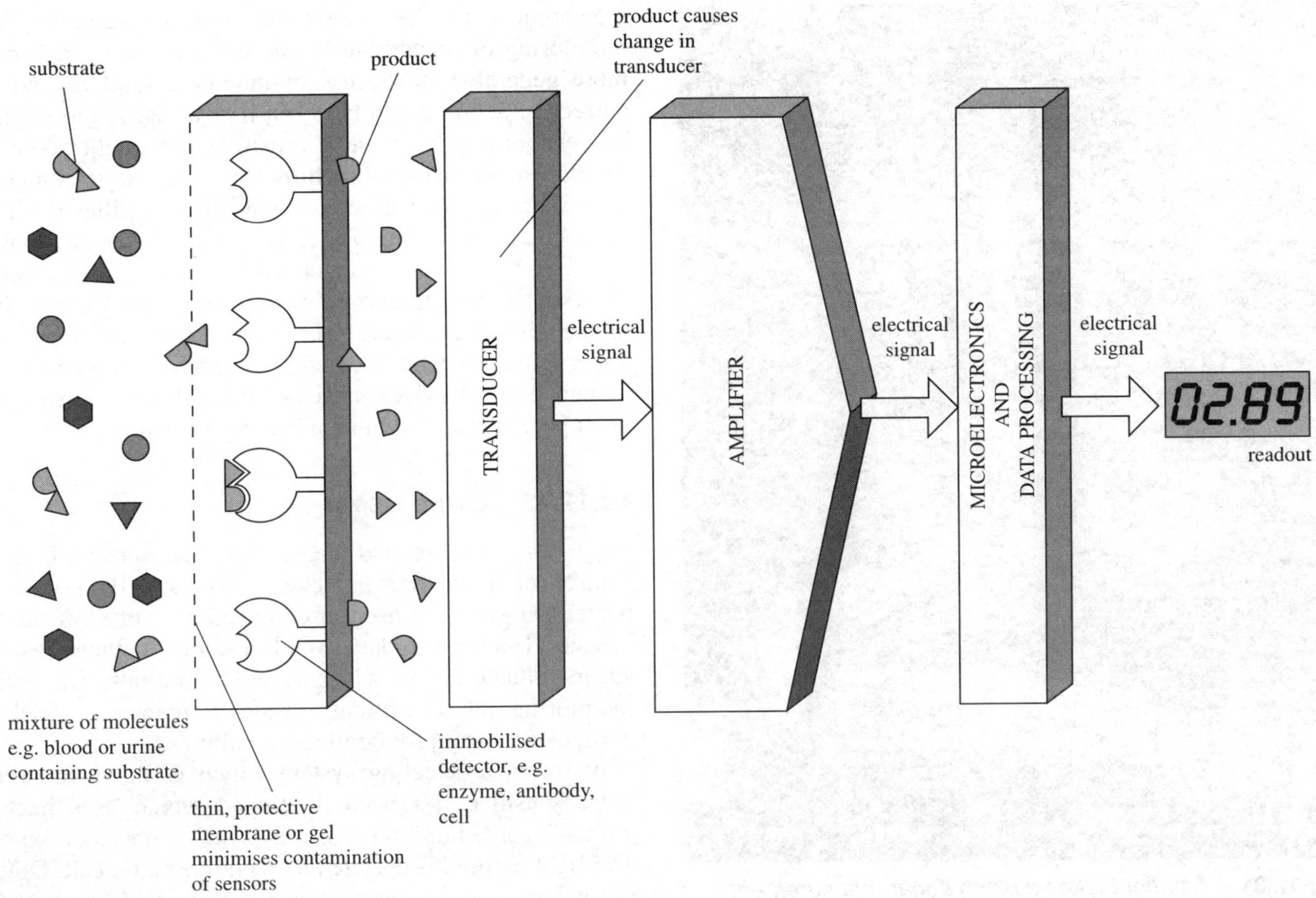

Fig 12.30 *A biosensor.*

- they are very sensitive, so only very small samples are needed and very low concentrations can be detected;
- they are rapid in response;
- they are safe to use;
- they are accurate;
- they can be made very small;
- they can be mass produced.

The main problems are:

- they are not very robust and need careful handling;
- they are not very stable;
- they are not sterilisable.

12.17.2 Monitoring blood glucose

An example of a commonly used biosensor is the one developed for detecting glucose in the blood of diabetics (fig 12.31). This contains the enzyme **glucose oxidase** in an immobilised form. The enzyme oxidises glucose in the blood to release electrons. These are collected and converted into an electrical current. The current generated is proportional to the *amount* of glucose present. It is extremely sensitive and can measure the glucose concentration in a single drop of blood and display the result within 20 seconds.

It is hoped that eventually it will be possible to implant such devices in blood vessels in the skin of diabetics, allowing them to monitor more accurately their insulin requirements. If this is linked to a minipump so that insulin is automatically released when needed, the diabetic will have, in effect, an automatic pancreas. This fine control would reduce the common secondary effects of diabetes, such as eye and kidney damage, suffered by some diabetics as a result of the relatively crude peaks and troughs of insulin concentration obtained with occasional injections.

12.17.3 Medical applications

The largest application of biosensors at present is in medicine. Enzymes are being used increasingly for routine automatic analysis of body fluids for metabolites, drugs and hormones. They are particularly useful for clinical diagnosis. Using biosensors reduces the risk of errors in diagnosis and also reduces costs once the biosensors are mass produced. Less time and less expertise is needed. It allows GPs to do tests in their surgeries without involving hospital laboratories. This saves money and avoids the need for patients to make return visits to receive a diagnosis. Treatment can also be started quicker. Another advantage is that there is less chance of a sample being mishandled, lost or contaminated. This would also be particularly useful in testing for use of drugs in sport. Kits

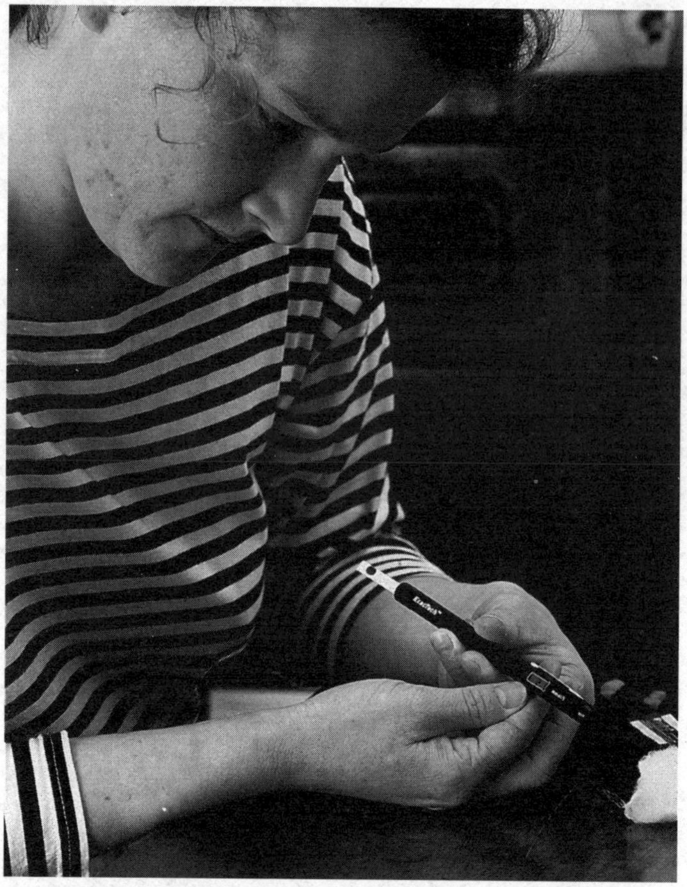

Fig 12.31 *A patient takes a blood sugar measurement using an Exactech biosensor. The patient places a drop of blood from a pin prick onto the lighter of the two panels at the end of the disposable strip (seen projecting from the left). An electronic readout then gives the blood sugar measurement, allowing the user to calculate the correct dose of insulin to be taken.*

are already in use by police and doctors for detecting small amounts of drugs in humans.

Home diagnosis kits may become possible since many people would like to have this facility. However, care would be needed to make counselling available if tests for serious conditions became possible.

Biosensors are already in use which enable critical metabolites to be monitored during surgery. Such monitoring of metabolite levels in the body could become more general with the use of miniature implants, so that corrective action could be taken if any changes took place. Biosensors which are more sensitive and smaller could be developed by using '**biochips**'. Just as large computers have been reduced in size by the introduction of silicon microchips, so further size reduction of biosensors may be possible by using semiconducting organic molecules in place of silicon. Electrical signals could pass along these molecules and electrical circuits could be just one molecule wide. Biochips would be small enough to implant in the human body. Devices such as artificial sense organs and heart pacemakers might then become possible.

12.17.4 Other uses

The second largest use of biosensors is in the control of industrial processes. Living cells (yeasts or bacteria) have been used in conjunction with electrodes to measure L-amino acids, alcohol, phenols, methane, various sugars, ethanoic acid (vinegar) and antibiotics. They allow monitoring of conditions inside fermenters, which is particularly useful for continuous cultures.

An oxygen-detecting system which is about 100 times more sensitive has been developed which uses bacteria which become luminous when exposed to traces of oxygen. The light emitted is detected by a photoelectric cell. Optical biosensors are generating a lot of interest. One reason is that remote sensing may be possible in hazardous environments by using fibre optics.

Many enzyme reactions produce heat. Thermal biosensors can detect temperature changes as small as 0.0001 °C. They can be used to detect the presence of lactic acid.

Future applications are expected in the fields of agriculture, veterinary science, defence (detection of nerve gases, toxins and explosives), and the environment (mainly detection of pollution). Annual growth in the use of biosensors is 30% or over in all these cases. The markets, though, are still relatively small, worth less than £50 million in 1992. This is due partly to the disadvantages noted above.

Answers and discussion

Chapter 2

2.1

Time (in units of 20 min)	*0*	*1*	*2*	*3*	*4*	*5*	*6*	*7*	*8*	*9*	*10*
A Number of bacteria	1	2	4	8	16	32	64	128	256	512	1024
B Log_{10} number of bacteria	0.0	0.3	0.6	0.9	1.2	1.5	1.8	2.1	2.4	2.7	3.0
C Number of bacteria expressed as power of 2	2^0	2^1	2^2	2^3	2^4	2^5	2^6	2^7	2^8	2^9	2^{10}

Graph A increases in steepness as time progresses. Graph B (a logarithmic plot) is a straight line (increases linearly with time). See fig 2.1(ans).

2.2 Consult tables 2.6 and 2.7, chapter 2.

2.3 The sporangiophores bear the sporangia above the main mycelium so that the spores are more likely to catch air currents and be dispersed.

2.4 Amphibians, like liverworts and mosses, are only partially adapted to life on land, having bodies which easily lose water, and they still rely on water for sexual reproduction. Both groups of organisms are thought by some scientists to represent intermediate stages in the evolution towards more advanced forms which are better adapted to life on land.

2.5 The sporophyte has become adapted for life on land although the gametophyte is still dependent on water for swimming gametes.

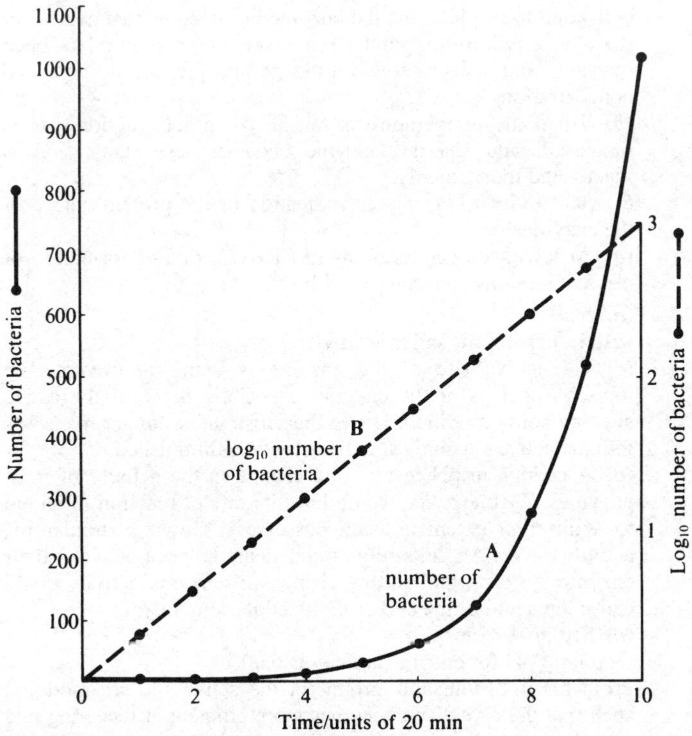

Fig 2.1(ans) *Growth of a model population of bacteria as plotted on arithmetic and logarithmic scales.*

The sporophyte generation has true vascular tissue and true roots, stem and leaves with which to exploit the land environment more successfully.
The sporophyte is the dominant generation, the life of the gametophyte being short.
The mature sporophyte is no longer dependent on the gametophyte.

2.6 Sexual reproduction is dependent on water since it involves free-swimming sperm.
The gametophyte thallus is susceptible to desiccation.
The plants are often relatively intolerant of high light intensities.

2.7 The *Dryopteris* spore can develop wherever it falls, providing conditions are moist and fertile. Pollen grains must reach the female parts of the sporophyte.

2.8 The megaspore is large because it must contain sufficient food reserves to support the female gametophyte and subsequent development of the embryo sporophyte until the latter becomes self-supporting. Microspores, by being small, can be produced economically in large numbers and are light enough to be carried by air currents, thus increasing the chances of the male gametes that they contain reaching the female parts of the plants.

Chapter 3

3.1 The molecular formula shows the number of each type of atom. The structural formula shows the arrangement of the atoms relative to each other. Note that angles of bonds can also be shown; see figs 3.3 and 3.5, for example.

3.2 (*a*)  C_8H_{18}, octane

(*b*) 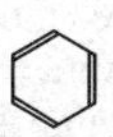 C_6H_6, benzene

3.3 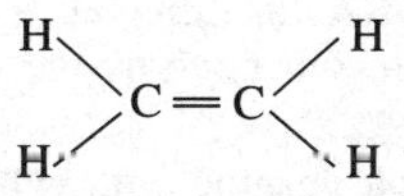

3.4 pentose $C_5H_{10}O_5$ hexose $C_6H_{12}O_6$

3.5 (*a*) Valency of C = 4, O = 2, H = 1.
(*b*) Molecular formula is $C_3H_6O_3$ in both cases; the compounds are therefore trioses.

(*c*) Each contains two hydroxyl groups. This could have been predicted, since it has already been explained that in monosaccharides all the carbon atoms except one have a hydroxyl group attached.

3.6 The principal sources of variation are as follows.
(*a*) Both pentoses and hexoses can be used to make polysaccharides, although normally only one type of monosaccharide is used in each type of polysaccharide.
(*b*) Two types of linkage, 1,4 and 1,6, are common between sugar units. Thus branching can occur.
(*c*) Lengths of chains and branches, and extent of branching can vary enormously.
(*d*) α- and β-forms of monosaccharides are important. (Compare starch and cellulose).
(*e*) Sugars may be ketoses or aldoses.
(*f*) The high chemical reactivity of sugars (aldehyde, ketone and hydroxyl groups) means that they are very reactive molecules.

3.7 One that occurs when two compounds are joined by the elimination of a water molecule.

3.8 Body temperatures of poikilothermic animals become lower in cold environments. Lipids rich in unsaturated fatty acids (which have low melting points) generally remain liquid at temperatures lower (usually 5°C or lower) than those rich in saturated fatty acids. This may be necessary if the lipid is to maintain its function, such as a constituent of membranes.

3.9 Triolein because it contains three *unsaturated* oleic acid molecules. Tristearin is a fat, triolein an oil.

3.10 (*a*) Cell respiration (internal or tissue respiration). Fat undergoes oxidation.
(*b*) Only the hydrogen part of carbohydrate and fat molecules yields water on oxidation ($2H_2 + O_2 \longrightarrow 2H_2O$) and fats contain relatively more hydrogen than carbohydrates on a weight basis (nearly twice as much).

3.11

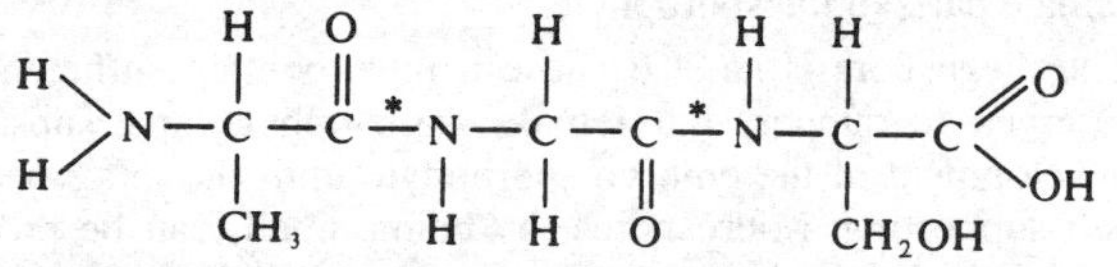

*peptide bond.

3.12 (*a*) AAA AAB ABA ABB
BAA BAB BBA BBB
(*b*) $2^3 = 8$
(*c*) $2^{100} = 1.27 \times 10^{30}$
(*d*) $20^{100} = 1.27 \times 10^{130}$ This is much larger than the number of atoms in the Universe (estimated at about 10^{100})! Thus, there is effectively an infinite potential for variation among protein structures.
(*e*) 20^n where *n* is the number of amino acids in the molecule.

3.13 The outstanding feature is that the ratio of adenine to thymine is always about 1.0, and so is the ratio of guanine to cytosine. In other words, the number of adenine molecules equals the number of thymine molecules and guanine = cytosine. Note also that the number of purine residues (adenine + guanine) therefore equals the number of pyrimidine residues (thymine + cytosine). Also revealed is the fact that the DNAs of different organisms have different base compositions, in other words the ratio of A:G or T:C is variable.

3.14 Adenine must pair with thymine and guanine with cytosine to account for the observed base ratios.

3.15 Compare the volume of the unknown sample needed to reduce the dye with the volume of 0.1% ascorbic acid solution needed in the standard described.

Percentage ascorbic acid in unknown sample =

$$\frac{\text{volume 0.1\% ascorbic acid used in standard}}{\text{volume of unknown sample used}} \times \frac{0.1}{100}$$

3.16 (*a*) Carry out Benedict's test on all three solutions. The sucrose solution would not give a brick-red precipitate on boiling. The glucose and glucose/sucrose solutions could be distinguished by pre-treating both as for hydrolysis (see non-reducing sugar test) and repeating Benedict's test. The glucose/sucrose mix will now show a greater amount of reducing sugar. (In practice, different dilutions of the solutions may have to be tried for convincing results. 0.05% glucose solution, 0.5% sucrose solution and a mixture of equal volumes of 0.1% glucose solution and 1.0% sucrose solution are suitable.)
(*b*) (i) Paper chromatography or thin-layer chromatography.
(ii) Effect on plane-polarised light using a polarimeter (both sucrose and glucose are dextro-rotatory, but sucrose produces a greater degree of rotation than glucose).
(iii) Sucrose is converted to reducing sugars (glucose + fructose) by the enzyme sucrase (invertase). The reaction may either be followed using a polarimeter or by Benedict's test.

3.17 Dissolve 10 g glucose in distilled water and make up to 100 cm^3. (Do not add 10 g glucose to 100 cm^3 distilled water because the final volume would be greater than 100 cm^3.)

3.18 Add 10 cm^3 of 10% glucose to 50 cm^3 of 2% sucrose solution and make up to 100 cm^3 with distilled water.

Chapter 4

4.1 (*a*) Initially the reactions A and B are fast and a lot of product is formed. Later, product formation levels off and there is no further increase. This may be because (i) all substrate has been converted to product, (ii) the enzyme has become inactivated, or (iii) the equilibrium point of a reversible reaction has been reached, and substrate and product are present in balanced concentrations.
(*b*) When the temperature is raised, (i) initial reaction rate is increased, and (ii) the enzyme becomes less stable and is inactivated more rapidly.
(*c*) Sensitivity to heat is an indication of the protein nature of the enzymes.
(*d*) At lower temperatures (as in curve C) rate of formation of product remains constant over 1 h.

4.2 (*a*) 5.50
(*b*) (i) pepsin, (ii) salivary amylase
(*c*) The active site of the enzyme is being destroyed. The ionisable groups of the enzyme, especially those of the active site, are being modified. Hence the substrate no longer fits easily into the active site and catalytic activity is diminished.
(*d*) A change in pH results in a change in the activity of most enzymes. Each enzyme would have its rate of reaction modified to a different extent as each possesses its own particular pH activity curve. All cells rely on a delicate balance between their enzyme systems, and so any changes in enzyme activity could cause the death of the cell or multicellular organism.
(*e*) See fig 4.2 (ans).
Optimum pH for enzyme activity is 6.00.
From pH 4–6, ionisable groups of the active site are modified such that the active site becomes more efficient at receiving and complexing with its substrate. The reverse is true when the pH changes from 6–8.

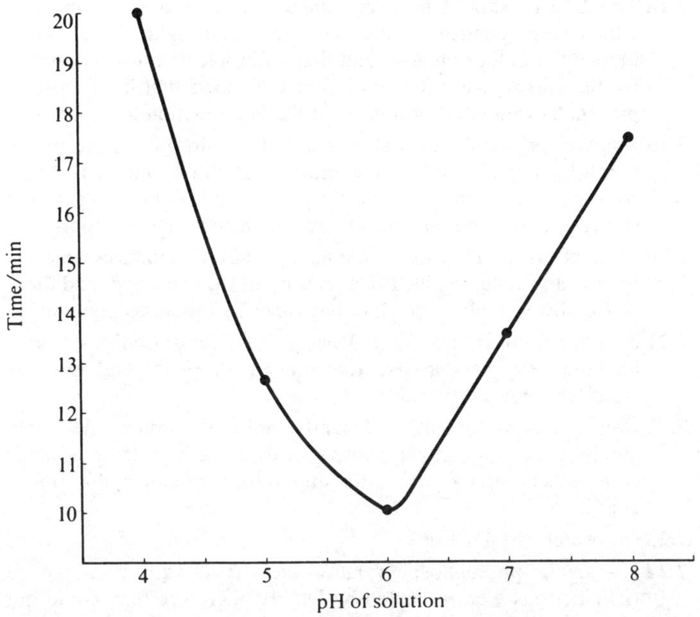

Fig 4.2(ans) *Activity of catalase on hydrogen peroxide at varying pH.*

4.3 Increasing the substrate concentration increases the probability of substrate molecules fitting into the active sites rather than inhibitor molecules.

4.4 Increased substrate concentration has no effect on the overall rate as there is no competition for the active site.

4.5 (*a*) The two sites are located on different parts of the enzyme, an active site for binding with the substrate and an allosteric site for binding with X.
(*b*) (i) X could act as an allosteric inhibitor of e_1 and therefore only permit production of S along the A–S pathway. This situation could remain until the surplus of X had been used up.
(ii) X could act as an allosteric inhibitor of e_5, again enhancing the production of S at the expense of X.
(*c*) Feedback inhibition.

4.6 (1) All are proteins and synthesised within living organisms.
(2) They catalyse chemical reactions by lowering the activation energy required to start the reaction.
(3) Only small amounts of enzyme are needed to catalyse reactions.
(4) At the end of a reaction the enzymes are unchanged.
(5) Each enzyme is specific and possesses an active site where enzyme and substrate combine temporarily to form an enzyme/substrate complex before products are released.
(6) Enzymes work best at an optimum pH and optimum temperature.
(7) Being proteins, enzymes are denatured by extremes of pH and temperature.

Chapter 5

5.1 Endoplasmic reticulum, ribosomes, microtubules, microvilli (visible as a 'brush border' in the light microscope).
In addition, small structures that are difficult to identify with certainty using a light microscope can be easily identified with the electron microscope, such as lysosomes and mitochondria.

5.2 (*a*) Cell wall with middle lamella and plasmodesmata, chloroplasts, large central vacuole (animal cells do possess small vacuoles, such as food vacuoles, contractile vacuoles)
(*b*) Centrioles, microvilli. (Pinocytotic vesicles are more commonly seen in animal cells.)

5.3 A: polar head of phospholipid (hydrophilic)
B: non-polar hydrocarbon tails of phospholipid (hydrophobic)
C: phospholipid
D: phospholipid layer

5.4 (*a*) A, (*b*) B, (*c*) A, (*d*) B, (*e*) –1000 kPa.

5.5 (*a*) A Na^+/K^+ pump operates whereby pumping out of Na^+ is linked to pumping in of K^+. Without K^+, no pumping out of Na^+ can occur, so Na^+ accumulates within the cells by diffusion and K^+ leaves the cells by diffusion.
(*b*) ATP is a source of energy for active transport of Na^+ ions.

Chapter 7

7.1 Photoautotrophic organisms use light energy from the Sun as an energy source for synthesising organic compounds from inorganic materials, with carbon dioxide as a source of carbon. Chemoheterotrophic organisms use organic compounds which are synthesised from pre-existing organic sources of carbon, using energy from chemical reactions.

7.2 **Overall form and position**
Large surface area to volume ratio for maximum interception of light and efficient gaseous exchange.
Blade often held at right-angles to incident light, particularly in dicotyledons.
Stomata
Pores in the leaf allow gaseous exchange. Carbon dioxide needed for photosynthesis, with oxygen a waste product.
In dicotyledons, stomata are located mainly in the shady lower epidermis, thus minimising loss of water vapour in transpiration.
Guard cells
Regulate opening of stomata (ensure stomata open only in light when photosynthesis occurs).
Mesophyll
Contains special organelles for photosynthesis, the chloroplasts, containing chlorophyll.
In dicotyledons, palisade mesophylls cells, with more chloroplasts, are located near the upper surface of the leaf for maximum interception of light. Length of the cells increases the chance for light absorption.
Chloroplasts are located near the periphery of the cell for maximum absorption of light and easier gas exchange with intercellular spaces.
Chloroplasts may be phototaxic (that is move within the cell towards light).
In dicotyledons, spongy mesophyll has large intercellular spaces for efficient gaseous exchange.
Vascular system
Supplies water, a reagent in photosynthesis; also mineral salts.
Removes the products of photosynthesis.
Supporting skeleton provided together with collenchyma and sclerenchyma.

7.3 Chlorophyll *a* absorption in red light is about twice that of chlorophyll *b* and the absorption peak is at a slightly longer wavelength (lower energy). Absorption in the blue is lower and shifted to a slightly shorter wavelength (higher energy). Note that only very slight differences in chemical structure cause these differences.

7.4 If an isotope has a shorter half-life (for example 11C, 20.5 min) it rapidly decays to the point at which it is undetectable, thus severely restricting its usefulness in biological experiments, which often take hours or days to complete.

7.5 Photosynthesis in *Chlorella* and higher plants is biochemically similar so that *Chlorella* was used for the following reasons:
(1) *Chlorella* culture is virtually a chloroplast culture since a large volume of every cell is occupied by a single chloroplast;
(2) greater uniformity of growth can be achieved;
(3) the cells are very rapidly exposed to radioactive carbon dioxide and also quickly killed, so handling techniques are easier.

7.6 For maximum illumination of algae.

7.7

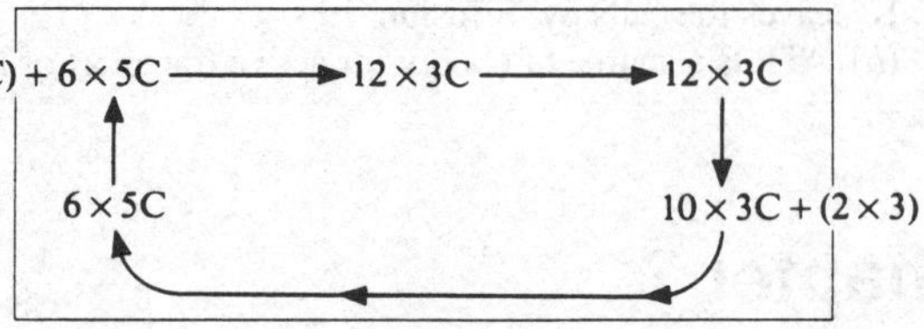

This emphasises the cyclic flow of carbon; the complexity of Calvin's cycle is due mainly to the difficulty of converting 10 × 3C into 6 × 5C.

7.8 Availability of carbon dioxide, water, light and chlorophyll.

7.9 (*a*) In region A light intensity is the limiting factor.
(*b*) B: some factor other than light intensity is becoming the limiting factor. In region B, both light intensity and the other factor(s) are limiting. C: light intensity is no longer a limiting factor.
(*c*) D: the 'saturation point' for light intensity under these conditions, that is the point beyond which an increase in light intensity will cause no further increase in the rate of photosynthesis.
(*d*) E: the maximum rate of photosynthesis attainable under the conditions of the experiment.

7.10 X, Y and Z are the points at which light ceases to be the major limiting factor in the four experiments. Up to these points there is a linear relationship between light intensity and rate of photosynthesis.

7.11 Enzymes would start to become denatured.

7.12 Some likely situations would be (*a*) in a shaded community such as a wood; dawn and twilight in a warm climate; (*b*) a bright winter's day.

7.13 Mesophyll chloroplasts for light reactions, bundle sheath chloroplasts for dark reactions.

7.14 Oxygen would compete with carbon dioxide for the active site of RuBP carboxylase.

7.15 **Carbon dioxide pump.** By acting as a carbon dioxide pump, the malate shunt increases carbon dioxide concentration in the bundle sheath cells, thus increasing the efficiency with which RuBP carboxylase works.
Hydrogen pump. Malate carries hydrogen from $NADPH_2$ in the mesophyll to NADP in the bundle sheath cells, where $NADPH_2$ is regenerated. The advantage is that $NADPH_2$ is generated by the efficient light reaction in the mesophyll chloroplasts and can be used as reducing power in the Calvin cycle of bundle sheath chloroplasts, whose own synthesis of $NADPH_2$ is limited.

7.16 (*a*) Lowering oxygen concentrations stimulates C_3 photosynthesis because it reduces competition between oxygen and carbon dioxide for the active site of RuBP carboxylase.
(*b*) Lowering oxygen concentration does not affect C_4 photosynthesis because PEP carboxylase does not accept oxygen.

7.17 The dark blue colour of the dye should disappear as it is reduced, leaving the green of the chloroplasts.

7.18 The DCPIP should have remained blue in tubes (2) and (3), which were controls. Tube (2) shows that light alone cannot induce the colour change, and that chloroplasts must be present for the Hill reaction to occur. Tube (3) shows that light must be present as well as chloroplasts for the Hill reaction to occur.

7.19 The two organelles closest in size to the chloroplasts are nuclei (slightly larger) and mitochondria (slightly smaller). More rigorous differential centrifugation or density gradient centrifugation would be necessary to isolate pure chloroplasts.

7.20 Indirect evidence suggests that nuclei and mitochondria were not involved in reducing DCPIP because light was needed, and these organelles lack chlorophyll or any other conspicuous pigment.

7.21 To reduce enzyme activity. During homogenisation destructive enzymes may be released from other parts of the cell, such as from lysosomes or vacuoles.

7.22 Cell reactions operate efficiently only at certain pHs; any significant change in pH, caused for example by release of acids from other parts of the cell, might have affected chloroplast activity.

7.23 (*a*) water (*b*) DCPIP

7.24 Non-cyclic photophosphorylation only: (i) oxygen was evolved (ii) electrons were accepted by DCPIP, therefore they could not recycle into PSI.

7.25 (*a*) The chloroplasts lack chloroplast envelopes (bounding membranes) and stroma. Only the internal membrane system remains.
(*b*) The medium lacking sucrose was hypotonic to the chloroplasts. Without the protection of the cell walls, broken during homogenisation, chloroplasts absorb water by osmosis, swell and burst. The stroma dissolves, leaving only membranes.
(*c*) The change was desirable because bursting the chloroplasts allows more efficient access of DCPIP to the membranes where the Hill reaction is located.

7.26 The discovery of the Hill reaction was a landmark for several reasons:
(1) it showed that oxygen evolution could occur without reduction of carbon dioxide, providing evidence for separate light and dark reactions and the splitting of water;
(2) it showed that chloroplasts could carry out a light-driven reduction of an electron acceptor;
(3) it gave biochemical evidence that the light reaction of photosynthesis was entirely located in the chloroplast.

7.27 The plant continues to use sugars in the dark, for example for respiration. Photosynthesis ceases in the dark, so as sugars are depleted starch reserves are converted to sugars, including sucrose which travels from the leaves to other parts of the plant.

7.28 It should be placed in an identical flask but with water replacing potassium hydroxide solution. Unsoaked cotton wool should secure the leaf stalk. (The stalk itself could be surface-treated with lime water to check whether possible injury here could affect photosynthesis.)

7.29 Rates of carbon dioxide uptake, oxygen production and carbohydrate production could be used. Rate of increase in the dry mass of leaves may also be measured. This is particularly suitable for crop plants over a growing season when relatively large samples may be taken. An experiment for measuring carbon dioxide uptake is described in experiment 7.5.

7.30 (*a*) The rate of gas production is directly proportional to *LI* up to a *LI* of *x* units. At this point light saturation began to occur and this was complete at *y* units (*x* and *y* values depend on experimental conditions). Thereafter some factor other than light was limiting the rate of gas production.
(*b*) The laboratory was darkened to avoid extra light which could have stimulated extra photosynthesis. Temperature was kept constant because this also affects the rate of photosynthesis.

7.31 (*a*) Temperature may vary as the lamp heats the water (this should be avoided by the water bath).
(*b*) The carbon dioxide concentration of the water may vary during the experiment, especially if potassium hydrogencarbonate was added earlier.
(*c*) Any stray light which is admitted to the laboratory will affect photosynthesis.

7.32 As the bubble of oxygen rises through the water, some of the dissolved nitrogen will come out of solution and enter the bubble, and some of the oxygen will dissolve. This exchange is due to the different partial pressures (concentrations) of oxygen and nitrogen in the bubble and the water, there being a tendency for them to come to equilibrium with time. Traces of water vapour and carbon dioxide will also be present in the collected gas. Once the gas has been collected, it will tend to come into equilibrium with atmospheric air by diffusion of gases through the water.

7.33 The amount of oxygen produced by photosynthesis in the experiment must all be collected. If the water is not saturated with air, some of the oxygen released in photosynthesis will dissolve in the water and reduce the amount recorded.

7.34 Specimen results are given in the following table.

Time/h	*Colour of indicator*			
	tube A	*tube B*	*tube C*	*tube D*
0	red	red	red	red
18	yellow	purple	red	red

The control tubes, C and D, were necessary to prove that any changes that took place in tubes A and B were due to the presence of leaves. In tube A conditions became more acidic as a result of carbon dioxide being produced during respiration. Photosynthesis did not take place in the absence of light. In tube B conditions became less acidic, indicating a net uptake of carbon dioxide. The carbon dioxide produced by respiration was used in photosynthesis, together with that already in the air inside the leaf and dissolved in the indicator solution. The rate of photosynthesis was greater than the rate of respiration.

7.35 The carbon dioxide compensation point. At this point rate of photosynthesis equals rate of respiration.

Chapter 8

8.1 (1) Decompose organic matter and therefore help recycling of elements from dead to living organisms.
(2) Render food unfit for human consumption (such as make bread mouldy).
(3) In the Far East *Mucor* has been used to produce alcohol. A mixture of *Mucor* and yeast was added to rice. *Mucor* converted the rice to sugars which the yeast then converted to alcohol.

8.2 See section 2.10.3 and table 8.1.

8.3 Active pepsin would digest cells that produce it, there being no mucus barrier within the gastric glands.

8.4 (*a*) The folds of the wall of the small intestine, villi and microvilli.
(*b*) It increases tremendously the surface area for secretion and absorption and makes it very efficient at these processes.

8.5 Rate of enzyme activity would decrease or stop as the enzymes would be denatured by the low pH.

8.6 It ensures that even if the soluble food molecules are in concentrations lower than those already in the blood, they will still pass into the blood.

8.7 Because of the continual heat loss from the relatively larger body surface of the mouse.

8.8 Fats are much richer in hydrogen than carbohydrates. As most of the energy that is released in the body arises by the oxidation of hydrogen to water, so fats liberate more heat than carbohydrates.

8.9 (*a*) Certain 'factors' (now known as vitamins) are needed in small amounts in the diet, which are essential for healthy growth and development.
(*b*) The growth 'factors' must be contained in the 3 cm^3 rations of milk provided for the rats, which confirms that only minute amounts are required. When the milk was stopped, growth was quickly curtailed. Rats without milk did grow initially, therefore they must have had a small store of vitamins in their body initially.
(*c*) It is deficient in iron, vitamin B and roughage.

8.10 The RDA quoted for a particular group is not the *average* requirement for that group, but a level which would cover the needs of almost everyone in that group. Many people regard it as a minimum desirable intake. For most people, though, the RDA is much more than they need. Its misuse could result in an average person over-eating.

8.11 The contribution of fat to energy intake should decrease, from 40% to 33%. This should be achieved by cutting down intake of saturated fatty acids from 16% to 10% of energy intake. There should be an equivalent rise in carbohydrate in the form of starch, and cell wall sugars from dietary fibre. (Milk contains saturated fats.)

8.12 There are many possible answers, e.g. the individual may make an error in estimating amount of intake; day-to-day and long-term variations in diet occur; food composition tables are not perfectly accurate and are based on assumptions such as the amount of fat in any meat consumed.

8.13 The risk of deficiency would be very small for any individual. Most individuals, though, would be eating more than they need.

8.14 It is difficult to measure accurately the existing intake of an individual. An individual's intake may vary over time.
If an individual's consumption is between the LRNI and the RNI one can only state that the nearer the RNI, the less likelihood of a deficiency. Unless there are physical signs or symptoms of deficiency it is impossible to state whether the diet is inadequate.

8.15 The consumer might believe that this is the average requirement and that he or she should try to consume the full RDA, whereas most people do not need this amount.

8.16 Purchasing food takes up a high proportion of the income of lower income groups. Such groups might be more at risk from nutrient deficiencies and assessment and planning of diet would be particularly important in such a situation.

Chapter 9

9.1 Light energy is needed for photosynthesis. Photosynthetic organisms (mainly plants and algae) are at the beginning of almost all food chains. Animals are therefore dependent on plants, either directly or indirectly, for their energy and materials.

9.2 See fig 9.2 (ans).

9.3 Oxygen is the final hydrogen acceptor in the respiratory chain.

9.4 Oxygen supply is increased by increased rate and depth of breathing, and an increase in the power and rate of heart beat.

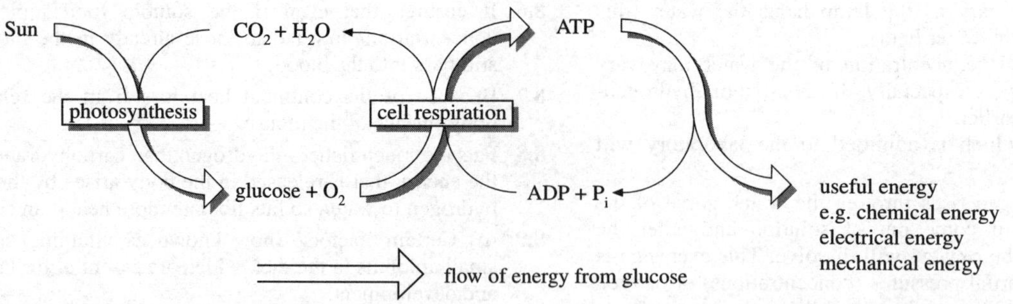

Fig 9.2(ans)

9.5 Lactic acid is being removed from the muscle by the blood, to be taken to the liver.

9.6 For rapid diffusion of molecules between cytoplasm and mitochondrion.

9.7

Entering	*Leaving*
pyruvate	
oxygen	carbon dioxide
reduced hydrogen carrier, e.g. reduced NAD	oxidised hydrogen carrier, e.g. NAD
ADP	ATP
phosphate	water

9.8 When blood and water first meet, the concentration gradient of oxygen between them will be great. However, as blood and water flow along together the gradient will decrease until blood shows a percentage saturation for oxygen equal to that of the water. This would be well below the blood's maximum possible saturation point and therefore inefficient (fig 9.8 (ans)).

9.9 Five – into epithelial cell lining alveolus, out of this cell, into endothelial cell of blood capillary, out of this cell, and into red cell.

9.10 Because not all the air exchanged reaches the alveoli. Some is in the bronchioles, bronchi and trachea ('dead space air').

9.11 Smaller mammals have a large surface area to volume ratio from which heat can be lost and therefore must use up more oxygen in order to maintain a constant body temperature.

9.12 By relating oxygen consumed per gram of body weight in unit time.

9.13 RQ = $\frac{CO_2}{O_2} = \frac{102}{145} = 0.70$. An RQ of 0.7 is typical of fats (lipids).

9.14 RQ = $\frac{CO_2}{O_2} = \frac{2}{0} = \infty$ (infinity).

This is typical of anaerobic respiration. If anaerobic and aerobic respiration are occurring at the same time, very high RQ values are obtained.

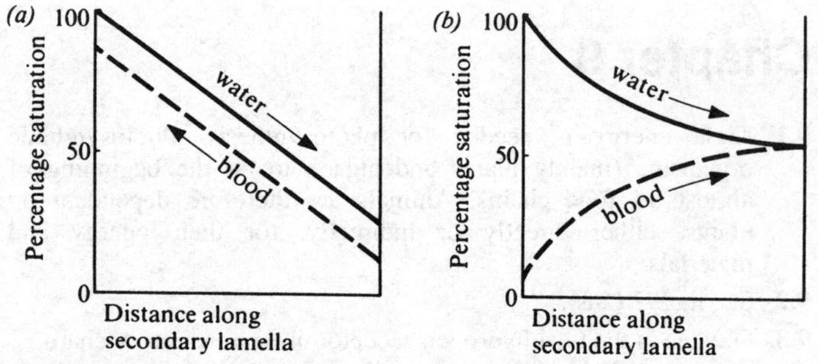

Fig 9.8(ans) (a) *Counterflow of water and blood.* (b) *Parallel flow of water and blood.*

9.15 Because humans generally respire a mixture of carbohydrate and fat.

9.16 (*a*) Breathing rate is about 17 breaths min^{-1}.
(*b*) Tidal volume is 450 cm^3 (average).
(*c*) Pulmonary ventilation is $17 \times 450\ cm^3 = 7.65\ dm^3\ min^{-1}$.
(*d*) Oxygen uptake is given by the slope of the line AB. Therefore oxygen consumption is 1500 cm^3 in 4 min = 375 $cm^3\ min^{-1}$.

9.17 (*a*)

Aerobic respiration	*Photosynthesis*
This is a catabolic process and results in the breakdown of carbohydrate molecules into simple inorganic compounds	An anabolic process which results in the synthesis of carbohydrate molecules from simple inorganic compounds.
Energy is incorporated into ATP for immediate use.	Energy is accumulated and stored in carbohydrate. Some ATP is formed.
Oxygen is used up.	Oxygen is released.
Carbon dioxide and water are released.	Carbon dioxide and water are used up.
The process results in a decrease in dry mass.	Results in an increase in dry mass.
In eukaryotes the process occurs in mitochondria.	In eukaryotes the process occurs in chloroplasts.
Takes place continuously throughout the lifetime of all cells, and is independent of chlorophyll and light.	Occurs only in cells possessing chlorophyll and only in the presence of light.

(*b*) *List of similarities between photosynthesis and aerobic respiration*
Both are energy-converting processes.
Both require mechanisms for exchange of carbon dioxide and oxygen.
Both require special organelles in eukaryotes, that is mitochondria for respiration and chloroplasts for photosynthesis; mitochondria and chloroplasts resemble prokaryotic organisms in possessing circular DNA and a prokaryote-type protein-synthesising system.
The light reactions of photosynthesis resemble cell respiration in the following ways:
(i) phosphorylation occurs (that is synthesis of ATP from ADP and P);
(ii) this is coupled to flow of electrons along a chain of electron carriers;
(iii) the electron carriers must be organised on membranes for coupling to take place; these are cristae in mitonchondria and thylakoids in chloroplasts.

Chapter 10

10.1 Dry mass is used because the water content of food samples or organisms may vary and water contributes no energy.

10.2 Small birds or mammals have a much higher surface area to volume ratio than humans and therefore lose body heat relatively more rapidly. Since small mammals and birds are endothermic ('warm-blooded') like humans they must consume relatively more energy to maintain body heat. (Birds also have a higher metabolic rate and body temperature than mammals.)

10.3 Example for grassland:

grass ⟶ sheep ⟶ human.
Festuca ovina,
sheep's fescue

10.4 seed – blackbird – hawk; trophic level 3, T3
leaf litter – earthworm – blackbird – hawk, T4
leaf – caterpillar – beetle – insectivorous bird – hawk, T5
rose bush (sap) – aphid – ladybird – spider – insectivorous bird – hawk, T6

10.5 Since the main primary producers are trees, their numbers would not be changed during the winter season. However herbivores dependent on the leaves, flowers and fruits for their food sources would be greatly reduced in number since in a mid-latitude deciduous woodland their food source would not be available in winter. It is likely that in winter the pyramid of numbers for this woodland would no longer be inverted; certainly any inversion between trophic levels 1 and 2 would be greatly reduced. In winter detritus pathways are more important than grazing food chains.

10.6 (*a*) May, June and July

(*b*) (i) Increase in light intensity and duration, and increase in temperature coupled with the availability of nutrients. Photosynthesis and growth are therefore favoured.

(ii) Grazing by primary consumers, such as zooplankton, and decrease in production due to depletion of nutrients. (The latter is due to the dead remains of producers sinking through the lake to colder, non-circulating water.)

(iii) Decline in numbers of zooplankton. Increase in nutrients (circulation of nutrients improves in the autumn as the surface layers of water cool and mix more freely with the colder, deeper layers). Temperature and light are still favourable.

(iv) Light and temperature unfavourable for photosynthesis and growth.

10.7 Blue-green bacteria and some other bacteria are also photosynthetic (they are prokaryotes, not plants). Chemosynthetic bacteria are also autotrophic (section 7.2) and therefore make a contribution to primary productivity. The total contribution of all these organisms is small compared with autotrophic eukaryotes (photosynthetic protoctista and plants).

10.8 Mutualistic bacteria in the root nodules of legumes fix nitrogen which leads to increased growth and thus to increased demand for other minerals, notably potassium and phosphorus. (However, ploughing-in of legumes is sometimes done, thus keeping the minerals in the soil.)

10.9 Chemoheterotrophic. They can be classified further as saprotrophic.

10.10 Anywhere there is insufficient oxygen for decomposition of all accumulating organic matter, such as bogs, aquatic sediments like mud deposits, arctic tundra, deeper zones of soil and waterlogged soils.

10.11 Both increase aeration and hence oxygen content of soil. This stimulates decomposition and nitrification. It also inhibits denitrification, oxygen being used instead of nitrate.

10.12 *Photosynthesis* (see chapter 7 and section 10.3)
On average 1–5% of the radiation incident on plants is used in photosynthesis
Source of energy for rest of food chain
Light is also needed for chlorophyll synthesis

Transpiration (see chapter 13)
About 75% of the radiation incident on plants is wasted in causing water to evaporate thereby causing transpiration
Important implications for water conservation

Photoperiodism (see chapters 16 and 17)
Important for synchrony of plant and animal behaviour (particularly reproduction) with seasons

Movement (see chapters 16 and 18)
Phototropism and photonasty in plants: important for reaching light
Phototaxic movements of animals and unicellular plants; important for locating suitable habitat

Vision in animals (see chapter 17)
One of the major senses

Other roles
Synthesis of vitamin D in humans
Prolonged exposure to ultra-violet damaging, particularly to animals, therefore pigmentation, avoidance behaviour, etc.

10.13 Geographical barriers, such as oceans; ecological barriers, such as unfavourable habitats separating areas of favourable habitats; distance over which dispersal must operate; air and water currents: size and nature of invasion areas

10.14 (*a*) Two eggs from each female must, on average, survive.

(*b*)

	Number of fertilised eggs that must die for stable population	*Pre-reproductive mortality*
oyster	$(100 \times 10^6) - 2$	>99.9%
codfish	$(9 \times 10^6) - 2$	>99.9%
plaice	$(35 \times 10^4) - 2$	>99.9%
salmon	$(10 \times 10^4) - 2$	>99.9%
stickleback	498	498/500= 99.6%
winter moth	198	99.0%
mouse	48	96.0%
dogfish	18	90.0%
penguin	6	75.0%
elephant	3	60.0%
Victorian Englishwoman	8	80.0%

(*c*) The stickleback and dogfish give birth to live young, that is they are viviparous. Therefore fewer eggs need to be produced owing to the greater degree of parental involvement in the development of offspring. Also, the female parent could not physically support any greater numbers of offspring.

10.15 Population (*b*), since a high percentage of individuals would die before reproductive age is reached. Population (*a*) would have to combine its high survival rate with low reproductive rate to maintain a stable population size.

10.16 (*a*) Out of 3200 eggs, 640 survive, so 2560 die – a mortality of 80%.

(*b*) Out of 640 fry, 64 survive, so 576 die – a mortality of 90%.

(*c*) Out of 64 smolts, 2 survive, so 62 die – a mortality of about 97%.

The total pre-reproductive mortality for salmon is 3198 out of 3200 = 99.97% (see fig 10.16 (ans)).

10.17 (*a*) A sigmoid (S-shaped) growth curve.

(*b*) Food and space. Food is more likely in this case.

(*c*) Faster reproductive rate. More efficient feeding. Greater resistance to toxic waste products, either of *Paramecium* or of bacteria growing in the same culture (*P. aurelia* has been shown to be more resistant than *P. caudatum*). Production of a poison or growth inhibitor (allelopathy). Predation.

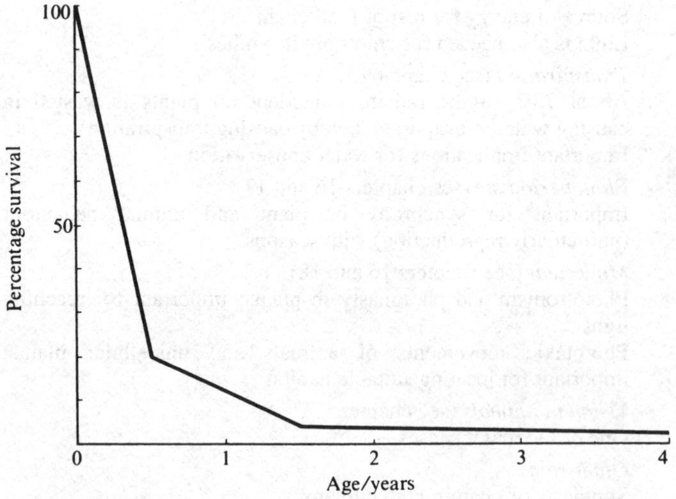

Fig 10.16(ans) *Graph showing pre-reproductive mortality for salmon.*

10.18 (*a*) Deforestation reduces the total world volume of photosynthetic material and thus reduces consumption of atmospheric carbon dioxide in photosynthesis.
(*b*) Removal of the tree canopy exposes the forest floor to sunlight and warmer temperatures. In forests or woodlands with significant litter and soil humus contents this exposure will favour accelerated rates of decomposition and carbon dioxide release.

10.19 BOD of discharge
BOD of receiving water
Nature of organic material
Total organic load of the river
Temperature
Extent of aeration from atmosphere (varies with wind etc.)
Dissolved oxygen in stream
Numbers and type of bacteria in effluent and stream
Ammonia content of effluent

10.20 Organisms live in their environment all the time. Their presence (or absence) therefore reflects the suitability of that environment for their living requirements at all times. A short-lived but severe pollution incident occurring at night would be reflected by the absence of sensitive organisms long after visible and chemical evidence of the pollution incident had disappeared. Biological indicators can therefore be a more sensitive and representative reflection of environmental conditions. 24 hour continuous chemical monitoring can be done but is not usually a viable routine practice for most water courses. This is especially true of small rivers and streams and remote areas. It also requires much time-consuming and, in the long term, expensive laboratory analysis. The main drawbacks of biological methods are that they require reasonable expertise at identification and will also be affected by seasonal factors.

10.21 (*a*) (i) × 2 (ii) × 500 (iii) × 2500 (iv) × 3750
(*b*) DDT is subject to progressive concentration as it passes along the food chain. This suggests that it is a persistent chemical, not easily broken down, and that it is stored rather than metabolised in living organisms. (In fact, it remains active for 10–15 years in soil.)
(*c*) (i) and (ii) 4th trophic level (top carnivore) (iii) 2nd trophic level (herbivore)
(*d*) DDT has spread all over the world as a result of two factors. First, it is carried at very low concentrations in water. If it is washed off agricultural land and into rivers some of it reaches the sea and becomes concentrated in marine food chains. Penguins feed on fish and are part of these food chains. Secondly, DDT can be carried in the atmosphere, both because it is volatile and because it is sprayed as a dust which can be carried by wind systems over large distances.
(*e*) (i) A small proportion of the original midges were resistant to DDD and these were not killed by the spraying procedure. Between sprays their numbers increased and after successive sprays they continued to breed and eventually constituted the greater part of the population. In other words the population had undergone intensive selection pressure (see chapter 26).
(ii) The data given suggest that DDD (and therefore DDT) is stored predominantly in fatty tissues. (This is because DDD and DDT are soluble in fat rather than water.) During times of food shortage, fat is mobilised and used so that the DDD or DDT accumulated over a long period is released into the bloodstream in relatively high concentrations.
(*f*) It has been suggested that the high death toll of birds in the winter of 1962–3 compared with 1946–7 was due to the additional effects of DDT mobilisation from fatty tissues. In 1946–7 the use of DDT was limited: in the late 1950s and early 1960s its use was widespread.

10.23 A population with a common gene pool will slowly evolve over time. If the population is very small its members may become inbred and lose vigour. In the recent past the black rhino has been hunted, for its horn, to near extinction. Each local population is now a tiny fraction of the original population and physically isolated. Such animals will be increasingly inbred. Outbreeding strengthens genetic diversity and, where populations are very small, this may be important for the health of the animals. Semen may be collected from anaesthetised wild or captive males and used to inseminate anaesthetised captive females at oestrus (the time of ovulation). Such artificial insemination technology makes it less necessary to move the animals themselves yet allows the spread of genes. It also makes possible the storage of genetic material (by cryopreservation – deep-frozen semen) in the event of a lack of males, in a local population, where females are still found.

Chapter 11

11.1 Fresh mass of soil = 60 g
dry mass of soil = 45 g
therefore mass of water = 60 g – 45 g = 15 g

therefore percentage water content of fresh soil $= \frac{15}{60} \times 100 = 25\%$

Dry mass of soil = 45 g
dry mass of soil after combustion = 30 g
therefore mass of organic material = 15 g

therefore percentage organic content of fresh soil $= \frac{15}{60} \times 100 = 25\%$

11.2 43%.

11.3 36%.

11.4 4230.

Chapter 12

12.1 (i) iron – found in cytochromes which are electron carriers in respiration

phosphate – synthesis of nucleic acids, ATP, phospholipids in membranes

(ii) nitrate – source of nitrogen for synthesis of proteins, nucleic acids and many other organic molecules

magnesium – part of structure of chlorophyll (bacteriochlorophyll) and cofactor for many enzymes, eg AT Pase

12.2 K_2HPO_4 and KH_2PO_4 – source of K and P. (These also act as buffers, tending to resist changes in pH caused by products of bacterial growth.)

$(NH_4)_2SO_4$ – source of N and S.

$MgSO_4$ – source of Mg and S.

$CaCl_2$ – source of Ca and Cl.

Glucose – source of C and energy.

12.3 See fig 12.3(ans).

12.4 Prepare a medium which is free from any nitrogen-containing compounds but which contains all other nutrients needed for growth. Inoculate with soil, place in contact with nitrogen and incubate under sterile conditions. The only organisms which will be able to grow and multiply will be nitrogen fixers.

12.5 Graph A increases in steepness as time progresses. Graph B (a logarithmic plot) is a straight line (increases linearly with time). See fig 2.1(ans).

12.6 The graph would be a typical growth curve for a bacterial population (see fig 12.8) except that there would be no lag phase because the bacteria are already adapted for that medium.

12.7 See fig 12.7(ans). Factors responsible for the changes are discussed in section 12.1. The difference in the growth curve of living bacteria compared with living and dead bacteria is due to the following:

(*a*) a few cells die during lag and log phases;

(*b*) during the stationary phase the combined total of living plus dead cells continues to increase slowly for some time since some cells are still reproducing;

(*c*) during the phase of decline the combined total of living plus dead cells remains constant, though many are dying.

12.8 Doubling time is between 2.5 and 3 hours.

12.9 (*a*) See section 12.10.2.

(*b*) See section 12.10.5.

(*c*) See section 12.5.3.

12.10 Any of the following: confidentiality – the user is the first to know if she is pregnant; a rapid result is obtained; can be used from the first day a period is due because it is very sensitive; simple, giving confidence that the test has been performed correctly.

12.11 Fungi may have a bad image. Eating fungi may seem unappealing or even dangerous to some consumers. Emphasis on 'natural' and 'plants' is reassuring. Reference to familiar food like mushrooms is also reassuring.

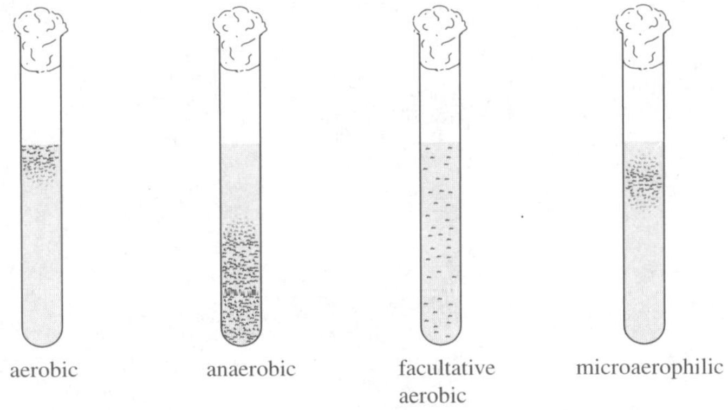

Fig 12.3(ans) *Distribution of different bacteria.*

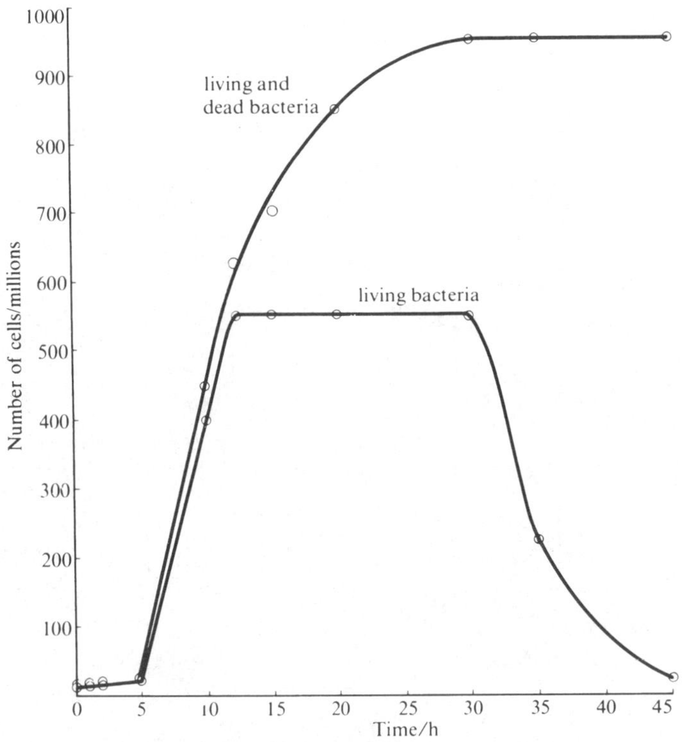

Fig 12.7(ans) *Growth of a bacterial population.*

Chapter Thirteen

Transport in plants

It was explained in section 5.9.8 that the exchange of substances between individual cells and their environments takes place by the passive processes of diffusion and osmosis, and the active processes of active transport and endocytosis or exocytosis. Within cells substances generally move by diffusion, but active processes, such as cytoplasmic streaming, can also occur. Over short distances these means of transport are rapid and efficient and in unicellular organisms, and multicellular organisms which have a large surface area to volume ratio, they are efficient enough for special transport systems to be unnecessary. For example, respiratory gases are exchanged by diffusion between the body surface and the environment in small organisms such as the earthworm.

In larger and more complex organisms, cells may be too widely separated from each other and from their external environments for these processes to be adequate. Specialised long-distance transport systems which can move substances more rapidly become necessary. Materials are generally moved by a mass flow system, **mass flow** being the bulk transport of materials from one point to another as a result of a pressure difference between the two points. With mass flow all the materials are swept along together at similar speeds, as in a river, whereas in diffusion molecules move independently of each other according to their own diffusion gradients. Some of the mass flow systems of plants and animals are summarised in table 13.1.

Note that animals are able to use the power of muscle contraction to force liquids or gases from one place to another, as when the heart pumps blood round the body. Plants on the other hand have to rely only on mechanisms such as evaporation, active transport and osmosis as will be seen later.

Both animals and plants have vascular systems. A **vascular system** is one which has tubes which are full of fluid being transported from one place to another. In animals the blood system is a vascular system. In plants the xylem and phloem form vascular systems. These systems require a source of energy to function. In the case of movement through xylem, the energy required comes directly from the Sun. Where mass flow occurs through specialised transport systems, such as phloem in plants and respiratory systems in animals, these systems are linked with specialised exchange systems whose function is to maintain concentration gradients between the transport system and the cells it serves.

Table 13.2 summarises the main groups of substances that move through plants and gives details of their movement.

The movement of substances through the conducting, or vascular, tissues of plants is called **translocation**. In vascular plants the vascular tissues are highly specialised and are called **xylem** and **phloem**. Xylem translocates mainly water and mineral salts (as well as some organic

Table 13.1 Some mass flow systems of animals and plants.

Mass flow system	*Material(s) moved*	*Driving force*
Plants		
vascular system:		
xylem (chapter 13)	mainly water and mineral salts	transpiration and root pressure
phloem (chapter 13)	mainly organic food, e.g. sucrose	active transport and osmosis
Animals		
alimentary system (chapter 8)	food and water	muscles of alimentary canal
respiratory system (chapter 9)	air or water	respiratory muscles
blood vascular system (chapter 14)	blood	heart or contractile blood vessels
lymphatic system (chapter 14)	lymph	general muscular activity in the body

Table 13.2 Movement of substances through plants.

	Uptake	*Transport*	*Elimination*
Water	osmosis into root	mass flow through xylem	diffusion (transpiration) through stomata (also small loss from cuticle and lenticels)
Solutes	diffusion or active transport into root	mass flow through xylem (mainly inorganic solutes) or phloem (mainly organic solutes)	shedding of leaves, bark, fruits and seeds; otherwise retained until death or passed to next generation in embryo of seed
Gases*	diffusion through stomata, lenticels, epidermis	diffusion through intercellular spaces and through cells	diffusion through stomata, lenticels, epidermis

* Movement of gases is considered in further detail in chapter 9.

nitrogen and hormones) from the roots to the aerial parts of the plant. Phloem translocates a variety of organic and inorganic solutes, mainly from the leaves or storage organs to other parts of the plant.

The study of translocation has important economic applications. For example, it is useful to know how herbicides, fungicides, growth regulators and nutrients enter plants, and the routes that they take through plants, in order to know how best to apply them and to judge possible effects that they might have. Also, plant pathogens such as fungi, bacteria and viruses are sometimes translocated, and such knowledge could influence treatment or preventive measures. In the 1960s, for example, a new group of fungicides was introduced which were described as **systemic** because they were translocated throughout plants. They provide longer-term, and more thorough, protection from important diseases like mildews.

13.1 Plant water relations

13.1.1 Osmosis

An understanding of plant water relations depends upon an understanding of osmosis and diffusion, which are explained in section 7.9.8. You should read this section if you have not already done so. It is pointed out in section 7.9.8 that osmosis can be regarded as a special kind of diffusion in which water molecules are the only molecules diffusing. This is due to the presence of a partially permeable membrane which does not allow the passage of solute particles. **Osmosis** is the movement of water molecules from a region of their high concentration (a dilute solution) to a region of their low concentration (a more concentrated solution) through a partially permeable membrane.

13.1.2 Terms used

In 1988 the Institute of Biology recommended the use of the term **water potential** to describe water movement through membranes. The two main factors affecting the water potential of plant cells are solute concentration and the pressure generated when water enters and inflates plant cells. These are expressed in the terms **solute potential** and **pressure potential** respectively. All of these terms are explained below.

13.1.3 Water potential (symbol ψ, the Greek letter psi)*

Water potential is a fundamental term derived from thermodynamics. Water molecules possess kinetic energy, which means that in liquid or gaseous form they move about rapidly and randomly from one location to another. The greater the concentration of water molecules in a system, the greater the total kinetic energy of water molecules in that system and the higher its so-called water potential. Pure water therefore has the highest water potential. If two systems containing water are in contact (such as soil and atmosphere, or cell and solution) the random movements of water molecules will result in the net movement of water molecules from the system with the higher water potential (higher energy) to the system with the lower water potential (lower energy) until the concentration of water molecules in both systems is equal. This is, in effect, diffusion involving water molecules.

Water potential is usually expressed in pressure units by biologists (such as pascals; 1 pascal = 1 newton per m^2).** Pure water has the highest water potential, and by convention this is set at zero.

* Technically, ψ means potential and water potential should be represented ψ_w. Since in living systems the solvent we are concerned with is always water, it is simpler to use ψ and assume the w.

** Pressure was formerly measured in atmospheres (atm), but now pascals are used (Pa).

1 Pa	=	$1\,N\,m^{-2}$		(N = newton)		
1 bar	=	0.987 atm	=	10^5 Pa	=	100 kPa
1 atm	=	1.0132 bar	=	1.0132×10^5 Pa		
1000 kPa	=	1 MPa				

Note the following main points:

- pure water has the maximum water potential, which by definition is zero;
- water *always* moves from a region of higher ψ to one of lower ψ;
- all solutions have lower water potentials than pure water and therefore have negative values of ψ (at atmospheric pressure and a defined temperature);
- osmosis can be defined as the movement of water molecules from a region of higher water potential to a region of lower water potential through a partially permeable membrane.

Advantage of using water potential

Water potential can be regarded as the tendency of water molecules to move from one place to another. The higher (less negative) the water potential the greater tendency to leave a system. If two systems are in contact, water will move from the system with the higher water potential to the one with the lower water potential. The two systems do not necessarily have to be separated by a membrane.

Using the term water potential, the tendency for water to move between any two systems can therefore be measured, not just from cell to cell in a plant, but also, for example, from soil to root or from leaf to air. Water can be said to move through a plant down a gradient of water potential from soil to air. The steeper the gradient, the faster the flow of water along it.

13.1.4 Solute potential, ψ_s

The effect of dissolving solute molecules in pure water is to reduce the concentration of water molecules and hence to lower the water potential. All solutions therefore have lower water potentials than pure water. The amount of this lowering is known as the **solute potential**. In other words, solute potential is a measure of the change in water potential of a system due to the presence of solute molecules. ψ_s is always negative. The more solute molecules present, the lower (more negative) is ψ_s (see also section 7.9.8). For a solution, $\psi = \psi_s$.

13.1.5 Pressure potential, ψ_p

If pressure is applied to pure water or a solution, its water potential increases. This is because the pressure is tending to force the water from one place to another. Such a situation may occur in living cells. For example, when water enters plant cells by osmosis, pressure may build up inside the cell making the cell turgid and increasing the pressure potential (section 13.1.8). Also, water potential of blood plasma is raised to a positive value by the high blood pressure in the glomerulus of the kidney. Pressure potential is usually positive, but in certain circumstances, as in xylem when water is under tension (negative pressure) it may be negative.

Summary

Water potential is affected by both solute potential and pressure potential, and the following equation summarises the relationship between the two terms.

$$\underset{\text{water potential}}{\psi} = \underset{\text{solute potential}}{\psi_s} + \underset{\text{pressure potential}}{\psi_p}$$

Solute potential is negative and pressure potential is usually positive.

13.1.6 Movement of water between solutions by osmosis

The terms mentioned above can only be used with confidence if they are properly understood. Question 5.4 on page 145 in Book 1 can be a useful test of your understanding of ψ and ψ_s. ψ_p is dealt with below.

13.1.7 Osmosis and plant cells

The partially permeable membranes of importance in the water relations of plant cells are shown in fig 13.1. The cell wall is usually freely permeable to substances in solution, so is not important in osmosis. The cell contains a large central vacuole whose contents, the cell sap, contribute to the solute potential of the cell. The

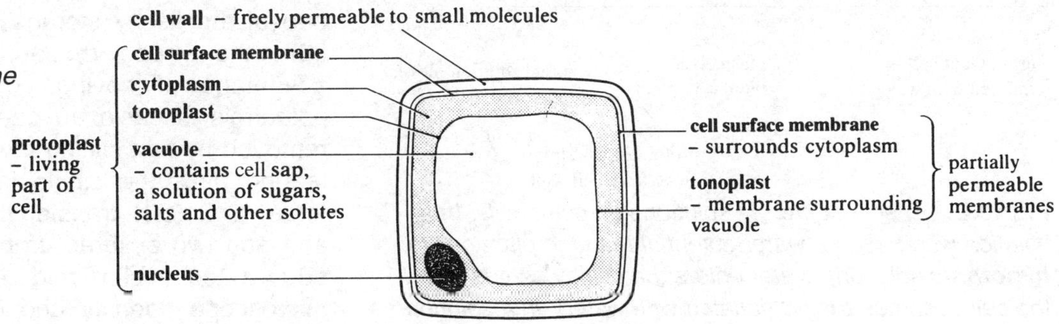

Fig 13.1 *Partially permeable membranes of a typical plant cell. The cell surface membrane would normally be pressed tightly up against the cell wall*

two important membranes are the cell surface membrane and the tonoplast.

If a plant cell is in contact with a solution of lower water potential than its own contents (such as the concentrated sugar solution in Experiment 13.1 below), then water leaves the cell by osmosis through the cell surface membrane (fig 13.2). Water is lost first from the cytoplasm and then from the vacuole through the tonoplast. The protoplast, that is the living contents of the cell surrounded by the cell wall, shrinks and eventually pulls away from the cell wall. This process is called **plasmolysis**, and the cell is said to be **plasmolysed**. The point at which plasmolysis is just about to happen is called **incipient plasmolysis**. At incipient plasmolysis the protoplast has just ceased to exert any pressure against the cell wall, so the cell is **flaccid**. Water will continue to leave the protoplast until its contents have the same water potential as the external solution. No further shrinkage then occurs.

> **13.1** What occupies the space between the cell wall and the shrunken protoplast in plasmolysed cells?

The process of plasmolysis is usually reversible without permanent damage to the cell. If a plasmolysed cell is placed in pure water or a solution of higher water potential than the contents of the cell, water enters the cell by osmosis (fig 13.2). As the volume of the protoplast increases it begins to exert pressure against the cell wall and stretches it. The wall is strong and relatively rigid, so the pressure inside the cell rises rapidly. The pressure is called the **pressure potential** (ψ_p). As the pressure potential of the cell increases, due to water entering by osmosis, the cell becomes **turgid**. Full turgidity, that is maximum ψ_p, is achieved when a cell is placed in pure water.

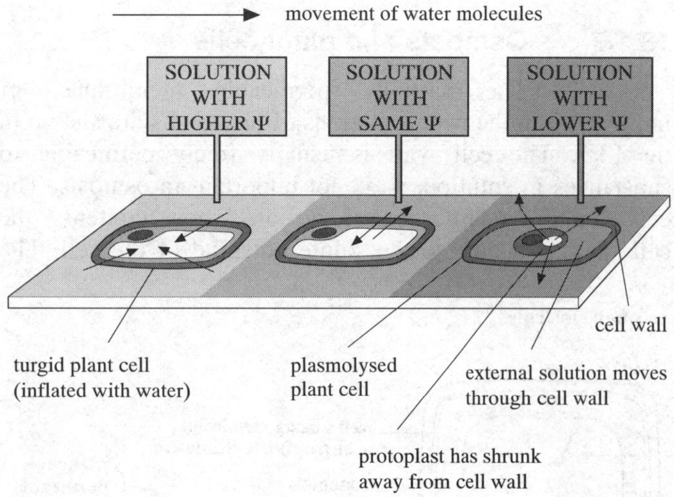

Fig 13.2 *Effect of different solutions on plant cells. In a solution with a higher water potential than the plant cell (a* ***hypotonic*** *solution), water enters the cell by osmosis and the cell becomes* ***turgid*** *(inflated with water). In a solution with a lower water potential (****hypertonic*** *solution), water leaves the cell by osmosis and the living part of the cell, the protoplast, shrinks. In a solution of equal water potential (****isotonic*** *solution), the cell remains normal.*

When the tendency for water to enter a cell is exactly balanced by pressure potential, the amount of water leaving the cell equals that entering the cell. There is no further net uptake of water and the cell is now in equilibrium with the surrounding solution. The contents of the cell are still likely to be of lower solute potential than the external solution because only a small amount of water is needed to raise the pressure potential to the equilibrium point, and this is not sufficient to dilute the cell contents significantly. Pressure potential therefore accounts for the fact that at equilibrium the solute potential of a plant cell can still be lower (its contents more concentrated) than that of the external solution.

Pressure potential is a real pressure rather than a potential one, and can only develop to any extent if a cell wall is present. Animal cells have no cell wall and the cell surface membrane is too delicate to prevent the cell expanding and bursting in a solution of higher water potential. Animal cells must therefore be protected by osmoregulation (chapter 20).

> **13.2** What is the ψ_p of a flaccid cell?
>
> **13.3** Which organisms, apart from plants, possess cell walls?

Experiment 13.1: To investigate osmosis in living plant cells

Materials

- onion bulb or young rhubarb epidermis
- microscope
- 2 slides and cover-slips
- scalpel and forceps
- distilled water
- 1 M sucrose solution
- 2 teat pipettes
- filter paper

Method

Remove a strip of epidermis from the inner surface of one of the fleshy storage leaves of the onion bulb, or from the young rhubarb petiole. Rhubarb has the advantage of having coloured cell sap, but onion epidermis is easier to peel off. The epidermis can be removed by first slitting it with a scalpel, then lifting and tearing back the single layer of cells with fingers or forceps. Quickly transfer the epidermal strip to a slide and add two or three drops of distilled water. Carefully add a cover-slip and examine the cells with a microscope. Identify and draw a few epidermal cells. Repeat using another strip of epidermis and 1 M sucrose solution instead of distilled water. Observe the strip over a period of 15 min and draw any changes observed in one or more representative cells at high power. The possibility of reversing the process observed can be investigated by irrigating with distilled water under the cover-slip to wash away the sucrose solution. Use filter paper to absorb any excess liquid.

Results

Fig 13.3 shows the appearance of onion epidermal cells left in 1 M sucrose solution for varying lengths of time.

13.1.8 Movement of water between cells by osmosis

Consider the situation in fig 13.4, in which two plant cells possessing different water potentials are in contact.

> **13.4** (*a*) Which cell has the higher (less negative) water potential?
> (*b*) In which direction will water move by osmosis?
> (*c*) At equilibrium the two cells will have the same water potential, which will be the average of the two, namely –1000 kPa. Assuming that ψ_s does not change significantly, what would be ψ_p at equilibrium in cell A and cell B?

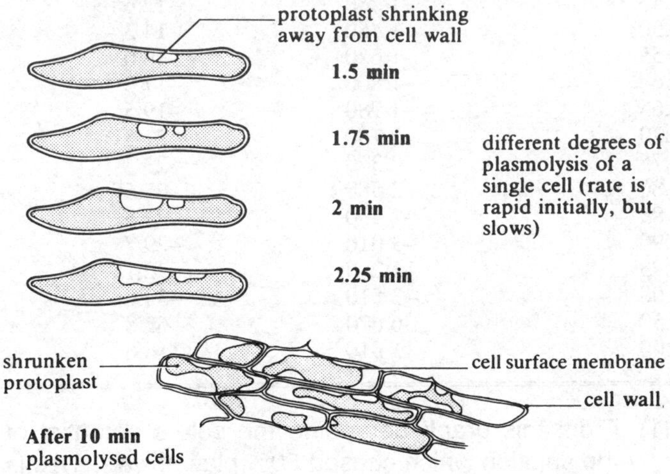

Fig 13.3 *Appearance of onion epidermal cells during plasmolysis. Strips of epidermal cells were left in 1 M sucrose solution for varying lengths of time.*

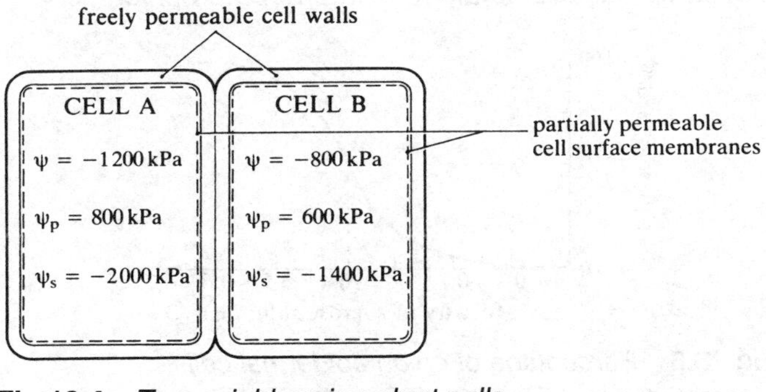

Fig 13.4 *Two neighbouring plant cells.*

13.1.9 Effect of heat and alcohol on membranes

The partial permeability of cell membranes can be destroyed by certain chemicals and treatments, such as ethanol and high temperatures. The membranes are still present but behave as if holes had been punched through them and they no longer provide a barrier to the passage of large molecules such as sucrose. High temperature and alcohols denature membrane proteins. Alcohols at high concentrations can also dissolve lipids.

Experiment 13.2: To determine the mean solute potential of the cell sap in a sample of plant cells using the method of incipient plasmolysis

There are several methods available for determining the solute potential of plant cells, but the most convenient is that of incipient plasmolysis. It makes use of the following relationships:

(1) ψ of a cell = $\psi_s + \psi_p$; ψ of a solution = ψ_s.
(2) $\psi^{cell} = \psi^{solution}$ when the two are in equilibrium.

Samples of the tissue being investigated are allowed to come to equilibrium in a range of solutions of different concentrations (water potentials) and the aim is to find which solution causes incipient plasmolysis, that is shrinkage of the protoplasts to the point where they just begin to pull away from the cell walls. At this point pressure potential is zero since no pressure is exerted by the protoplasts against the cell walls, so $\psi^{cell} = \psi_s^{cell} = \psi^{solution} = \psi_s^{solution}$ (from (1) and (2) above). In other words, the solution causing incipient plasmolysis has the same solute potential as the cell sap.

In practice, solute potential varies between cells in the same tissue and so some plasmolyse in more dilute solutions than others. Incipient plasmolysis is said to have been reached when 50% of the cells have plasmolysed. At this point 50% of the cells are unplasmolysed and the average cell can be said to be at incipient plasmolysis. The solute potential obtained is a mean value for the tissue.

Materials

onion bulb or rhubarb petiole
6 petri dishes
6 test-tubes
test-tube rack
labels or wax pencil
2 × 10 cm^3 or 25 cm^3 graduated pipettes
2 × 100 cm^3 beakers
brush (fine paintbrush)
distilled water
1 M sucrose solution
fine forceps
Pasteur pipettes

slides and cover-slips
microscope
graph paper
razor blade or sharp scalpel

Method

(An alternative method using beetroot is described after this first method.)

(1) Label six petri dishes and six test-tubes appropriately for each of the following sucrose solutions: 0.3 M, 0.35 M, 0.4 M, 0.45 M, 0.5 M and 0.6 M.

(2) Using a 10 cm^3 or 25 cm^3 graduated pipette, a beaker of distilled water and a beaker of 1 M sucrose solution, make up 20 cm^3 of sucrose solution of the required concentration in each test-tube. Table 13.3 shows the amounts used.

(3) Make sure that the solutions are mixed thoroughly by shaking. This is very important. Add the solutions to the appropriate petri dishes.

(4) **Onion**. Remove one of the fleshy storage leaves of an onion. While it is still attached to the leaf, cut the inner epidermis into six squares of approximately 5 mm side using a razor blade or scalpel. Remove each of the six squares using the fine forceps and immediately place one square of tissue into each petri dish. Agitate each dish gently to ensure that the tissue is completely immersed and washed with the sucrose solution. Leave for about 20 min.

Rhubarb. Score the outer epidermis into six squares of approximately 5 mm side and remove the epidermis as described for onion.

(5) Remove the tissue from the 0.60 M solution and, using a brush, mount it on a slide in sucrose solution of the same concentration. Add a cover-slip and examine with a microscope.

(6) Select a suitable area of cells using lower power. Switch to a medium or high power objective and move the slide through the selected area, recording the state (plasmolysed or unplasmolysed) of the first 100 cells viewed. Cells in which there is any sign of the protoplast pulling away from the cell wall should be counted as plasmolysed.

(7) Repeat for all other squares of tissue, mounting them in their respective solution.

(8) From the total number of cells counted and number plasmolysed, determine the percentage of plasmolysed cells for each solution. Plot a graph of percentage of plasmolysed cells (vertical axis) against molarity of sucrose solution (horizontal axis).

(9) Read off from the graph the molarity of the sucrose solution which causes 50% of the cells to plasmolyse.

(10) Plot a graph of solute potential (vertical axis) against molarity of sucrose solution (horizontal axis) using the data provided in table 13.4.

Table 13.3 Sucrose dilution table for experiment 13.2.

Concentration of sucrose solution	*Volume of distilled water/cm^3*	*Volume of 1 M sucrose solution/cm^3*
0.30 M	14	6
0.35 M	13	7
0.40 M	12	8
0.45 M	11	9
0.50 M	10	10
0.60 M	8	12

Table 13.4 Solute potentials of given sucrose solutions at 20 °C.

Concentration of sucrose solution (molarity)	*Solute potential/kPa*	*Solute potential/atm*
0.05	−130	−1.3
0.10	−260	−2.6
0.15	−410	−4.0
0.20	−540	−5.3
0.25	−680	−6.7
0.30	−820	−8.1
0.35	−970	−9.6
0.40	−1 120	−11.1
0.45	−1 280	−12.6
0.50	−1 450	−14.3
0.55	−1 620	−16.0
0.60	−1 800	−17.8
0.65	−1 980	−19.5
0.70	−2 180	−21.5
0.75	−2 370	−23.3
0.80	−2 580	−25.5
0.85	−2 790	−27.5
0.90	−3 010	−29.7
0.95	−3 250	−32.1
1.00	−3 510	−34.6
1.50	−6 670	−65.8
2.00	−11 810	−116.6

(11) From this graph determine the solute potential of the solution which caused 50% plasmolysis. This is equal to the mean solute potential of the cell sap.

Results

A typical graph for onion epidermis is shown in fig 13.5. Similar results are obtained using rhubarb epidermis.

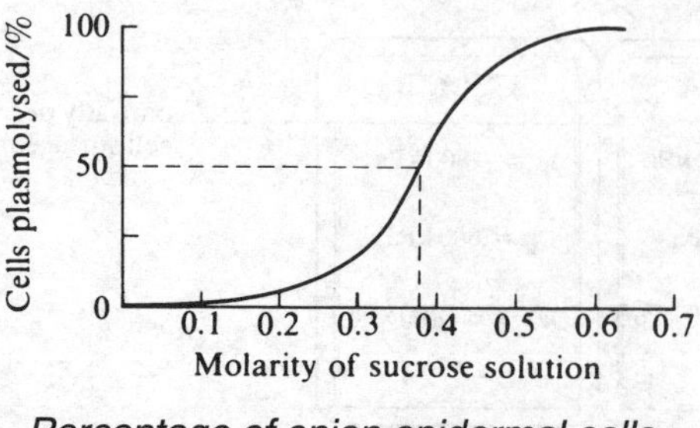

Fig 13.5 *Percentage of onion epidermal cells plasmolysed in different concentrations of sucrose solution.*

13.5 What is the solute potential of onion epidermal cells if 50% of the cells were plasmolysed in 0.38 M sucrose solution?

Use of beetroot tissue

Beetroot is a less convenient material to use, but a combination of this experiment with experiment 13.3 would enable an estimation of the pressure potential of beetroot cells to be made, although it should be pointed out that different beetroots may have different water potentials and solute potentials. Beetroots normally have a lower solute potential than onion or rhubarb because they have more sugar and inorganic salts in their vacuoles.

Modifications to the method are as follows:

(1)–(3) As for onion and rhubarb experiment above, except use solutions of the following concentrations: 0.4 M, 0.45 M, 0.5 M, 0.55 M, 0.6 M and 0.7 M.

(4) Cut a rectangular 'chip' of beetroot with square ends of approximately 5 × 5 mm. Thin sections (maximum 0.5 mm thick) should be cut from the end of this chip using a razor blade. The thinner the sections, the easier it is to count plasmolysed cells. The coloured sap enables easy detection of plasmolysis. The sections could be cut immediately before the practical class and kept in distilled water. Add several sections of beetroot tissue to each sucrose solution, and leave for about 30 min. Meanwhile examine similar sections, mounted in distilled water, with a microscope to become familiar with the appearance of the unplasmolysed cells. The margins of the sections are likely to be thinner and easier to examine. Some damaged cells may be colourless, and some small cells near vascular tissue may be seen. These can be ignored in subsequent counts.

(5)–(11) As before, starting with tissue from 0.7 M solution.

Results

A set of results obtained for beetroot is given in table 13.5.

13.6 What is the mean solute potential of the beetroot cells used in this experiment? (You will need to draw a graph to determine this.)

Table 13.5 Percentage of beetroot cells plasmolysed in different concentrations of sucrose solution.

Molarity of sucrose solution	*Percentage of plasmolysed cells**
0.30	2.5
0.40	3.5
0.45	13.5
0.50	74.0
0.55	100.0
0.60	100.0

* Sample size 200 cells.

Experiment 13.3: To determine the water potential of a plant tissue

Water potential is a measure of the tendency of water molecules to pass from one place to another. The principle in this experiment is to discover a solution, of known water potential, in which the tissue being examined neither gains nor loses water. Samples of the tissue are allowed to come to equilibrium in a range of solutions of different concentrations and the solution which induces neither an increase nor a decrease in mass or volume of the tissue has the same water potential as the tissue. The method described below relies on volume rather than mass changes.

Materials

fresh potato tuber or fresh beetroot
6 petri dishes
5 test-tubes
test-tube rack
labels or wax pencil
2 × 10 cm^3 or 25 cm^3 graduated pipettes
tile
distilled water
1 M sucrose solution
scalpel or knife
2 × 100 cm^3 beakers
graph paper

Method

(1) Label six petri dishes appropriately, one for each of the following: distilled water, 0.1 M, 0.25 M, 0.5 M, 0.75 M and 1.0 M sucrose solutions. Label five test-tubes appropriately, one for each of the sucrose solutions.
(2) Using a graduated pipette, a beaker of distilled water and a beaker of 1 M sucrose solution, make up 20 cm^3 of sucrose solution of the required concentration in each test-tube. A dilution table is useful as described in experiment 13.2 (table 13.3).
(3) Shake the tubes to mix the solutions thoroughly.
(4) Pour the solutions into the appropriate petri dishes. Add 20 cm^3 of distilled water to the sixth petri dish.

(5) Place the petri dishes on graph paper, making sure their lower surfaces are dry.
(6) Using the knife or scalpel, cut 12 rectangular strips of tissue approximately 2 mm thick, 5 mm wide and as long as possible (about 5 cm) from a slice of tissue (2 mm thick) taken from the middle of a large beetroot or potato. It is important to work quickly to avoid loss of water through evaporation as this would lower the water potential of the tissue.
(7) Completely immerse at least two strips in each petri dish and immediately measure their lengths against the graph paper seen through the bottoms of the dishes. Agitate the contents of each dish to wash the strips.
(8) Leave in covered petri dishes for at least 1 h, preferably 24 h.
(9) Measure the lengths again, and calculate the mean percentage change in length. Plot a graph of the mean percentage change in length (vertical axis) against the molarity of the sucrose solution (horizontal axis). Changes in length are proportional to changes in volume.
(10) Read off from the graph the molarity of the sucrose solution which causes no change in length.
(11) Plot a graph of solute potential (vertical axis) against molarity of sucrose solution (horizontal axis) using the data provided in table 13.4.
(12) From this graph, determine the solute potential of the solution which caused no change in length. The water potential of the tissue is determined according to the following:

$$\psi^{\text{cell}} = \psi^{\text{external solution}} = \psi_s.$$

(13) If beetroot has been used and its solute potential determined from experiment 13.2, calculate the pressure potential from:

$$\psi = \psi_s + \psi_p.$$

Table 13.6 Lengths of beetroot strips left in distilled water or different concentrations of sucrose solution for 24 hours.

Molarity of sucrose solution	*Length of beetroot strip at start/cm* 1	2	3	*Length of beetroot strip after 24 h/cm* 1	2	3
0.00 (distilled water)	4.8	5.0	5.3	5.0	5.3	5.6
0.10	5.1	4.8	4.9	5.3	4.9	5.1
0.20	5.1	4.9	4.9	5.2	4.9	5.0
0.25	5.2	4.8	5.0	5.2	4.9	5.0
0.30	4.9	4.9	5.0	4.9	5.0	5.1
0.40	4.9	5.0	4.8	4.9	5.0	4.8
0.50	5.0	4.8	5.1	4.8	4.7	5.0
0.60	4.8	5.0	5.0	4.6	4.9	4.9
0.75	4.9	4.9	5.0	4.6	4.7	4.8
0.90	4.9	5.0	4.9	4.5	4.7	4.7
1.00	4.8	4.9	4.9	4.7	4.6	4.4
1.50	4.9	4.9	4.9	4.5	4.1	4.5

Results

More accurate results are likely to be obtained by pooling class results. See table 13.6 for specimen results.

13.7 What is the mean water potential of beetroot cells from the data in table 13.6? (You will need to determine mean percentage changes in length and draw graphs.)

13.8 Why are at least two strips of tissue added to each dish?

13.9 Why are the petri dishes covered when left?

13.10 If the solute potential of beetroot cells is –1400 kPa and their water potential is –950 kPa, what is their mean pressure potential?

13.11 Consider the experiment illustrated in fig 13.6 in which the hollow inflorescence stalk (scape) of a dandelion (*Taraxacum officinale*) is first cut longitudinally into four strips 3 cm long and the pieces then immersed in distilled water or sucrose solutions of different concentrations.
(*a*) Why did cutting the scape longitudinally result in immediate curling back of the cut strips?
(*b*) Why did scape B bend further outwards in distilled water?
(*c*) Why did scape C bend inwards in concentrated sucrose solution?
(*d*) Why did scape A retain the same curvature in dilute sucrose solution?
(*e*) Which of the following could be determined for scape cells using this method: solute potential, water potential or pressure potential? Design an experiment to determine the relevant value, giving full experimental details.

13.12 The red colour of beetroot is contained in the cell vacuoles. Using this information, design experiments to investigate the effects of heat and ethanol on the partial permeability of beetroot cell membranes.

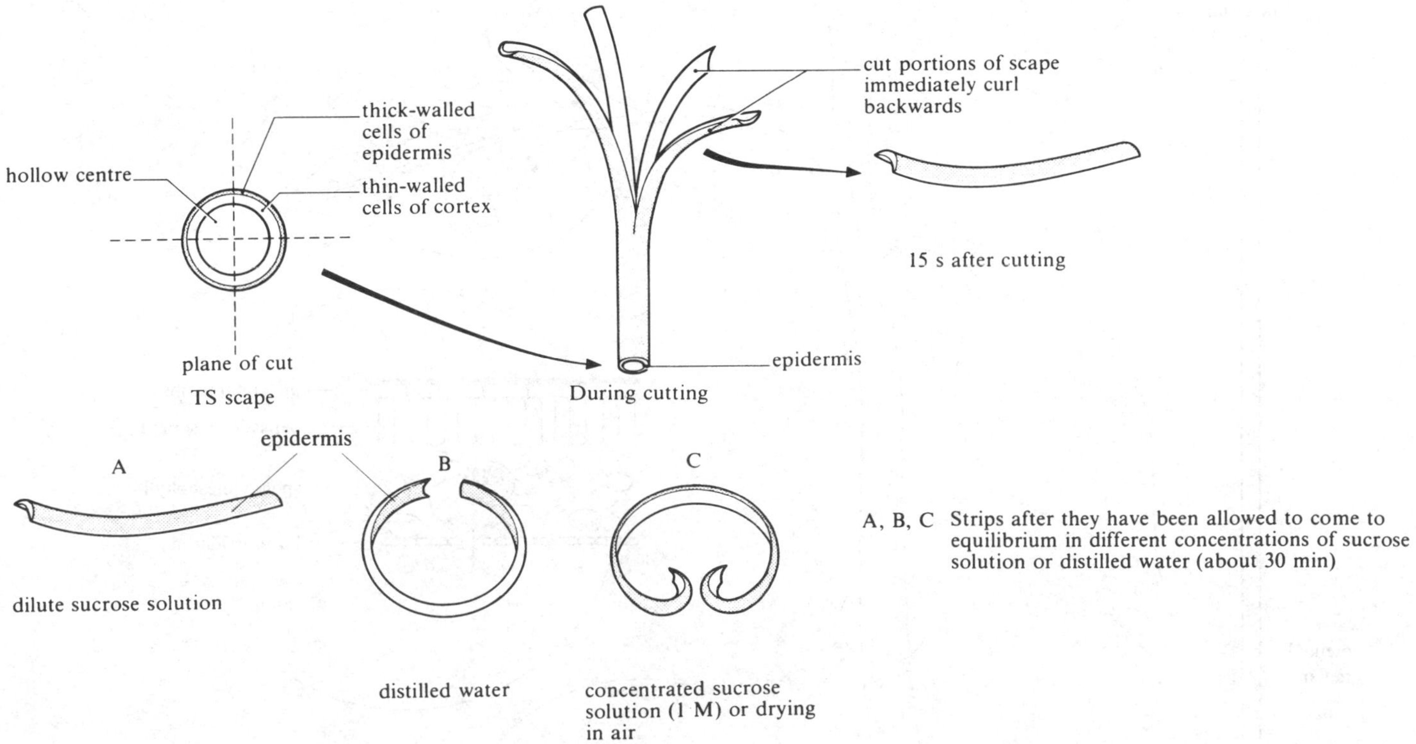

Fig 13.6 *Experiment on dandelion scapes. Investigation of the effects of distilled water and sucrose solutions on the curvature of strips of dandelion scape.*

13.2 Movement of water through the flowering plant

Water in the plant is in direct contact with water in the soil and with water vapour in the air around the plant. It has already been stated that water moves from higher to lower water potentials. Plant physiologists therefore think of water as moving through plants from a region of higher water potential in the soil to a region of lower water potential in the atmosphere, down a gradient of water potentials (fig 13.7). The water potential of moderately dry air is much lower than that of the plant so there is a great tendency for water to leave the plant.

Most of the water entering the plant does so through the root hairs. It travels across the root cortex to the xylem, ascends in the xylem to the leaves and is lost by evaporation from the surface of the mesophyll cells before diffusing out through the stomata. Loss of water from the surface of a plant is called **transpiration**, and the flow of water from the roots to the stomata forms the **transpiration stream**. It is estimated that more than 99% of the water absorbed by the average plant is lost.

13.3 Transpiration and movement of water through the leaf

Water normally leaves the plant as water vapour. The change from a liquid state to a vapour state requires the addition of energy which is provided by the Sun, and it is this energy that maintains the flow of water through the entire plant. Transpiration may occur from the following three sites.

- **Stomata:** by evaporation of water from cells and diffusion of the water vapour through stomata, the pores found in the epidermis of leaves and green stems. (About 90% of the water is lost this way.)
- **Cuticle:** by evaporation of water from the outer walls of epidermal cells through the waxy cuticle covering the epidermis of leaves and stems. (About 10% of the water lost, varying with thickness of cuticle.)
- **Lenticels:** by evaporation of water through lenticels. These are small slits in the stems and bark of trees for gas exchange. (Minute proportions, although this is the main method of water loss from deciduous trees after leaf fall.)

The quantities of water lost by transpiration can be very large. A herbaceous plant, such as cotton or sunflower, can lose between 1–2 dm^3 of water per day, and a large oak tree may lose more than 600 dm^3 per day.

Water is brought to the leaf in the xylem vessels. The structure of vessels is described in section 6.2.1. The xylem is part of the vascular bundles which spread to form a fine branching network throughout the leaf. The branches end in one or a few xylem vessels that possess little lignification. Water can therefore escape easily through their cellulose walls to the mesophyll cells of the leaf. Fig 13.8 shows the

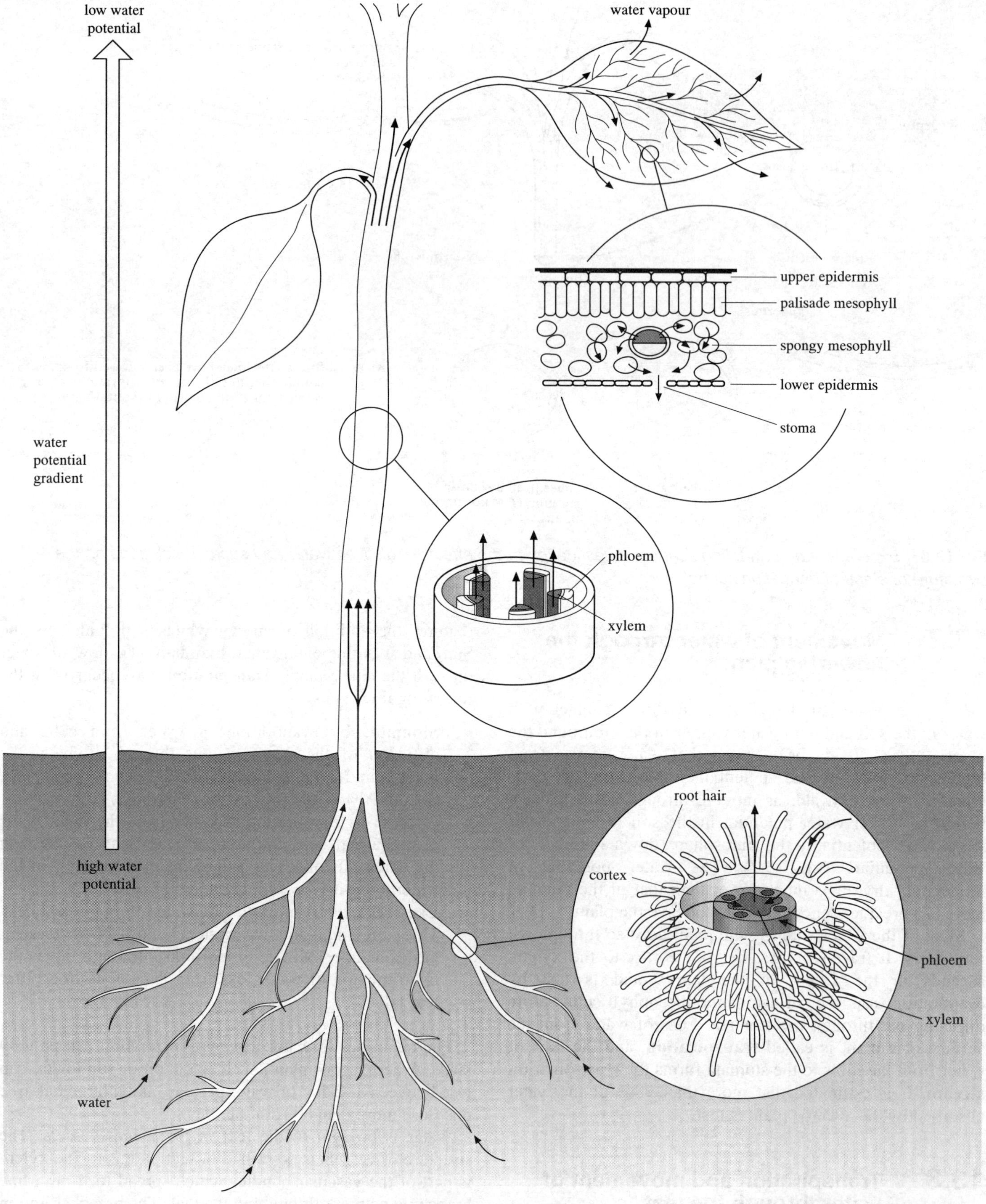

Fig 13.7 *Movement of water through a plant. Water in the soil is at a high water potential and water in the atmosphere is at a low water potential. Water moves from high to low potential down a gradient through the plant. The gradient is maintained by solar energy and evaporation of water from the surface of the plant (transpiration).*

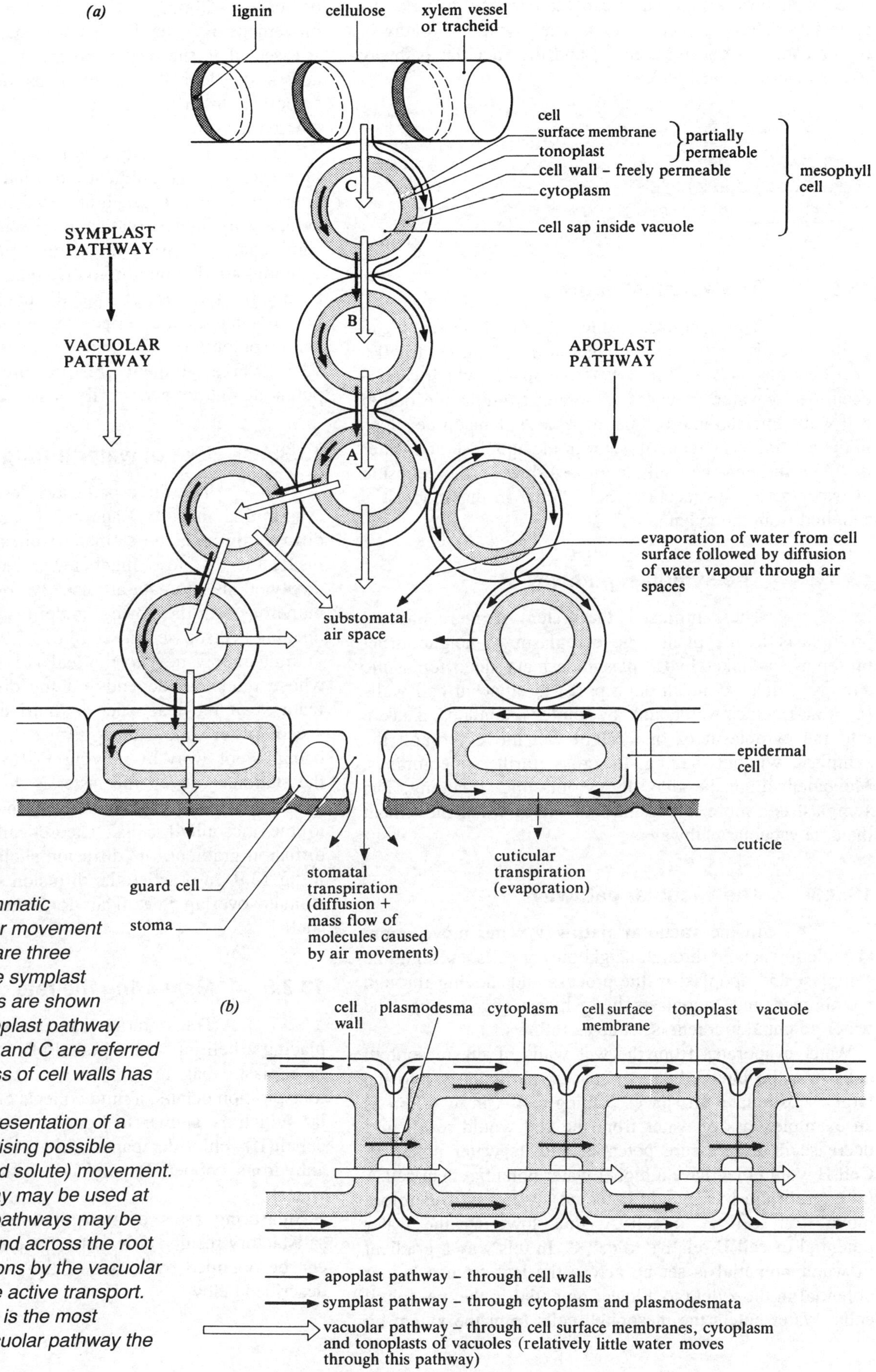

Fig 13.8 *(a) Diagrammatic representation of water movement through a leaf. There are three possible pathways: the symplast and vacuolar pathways are shown to the left, and the apoplast pathway to the right. Cells A, B and C are referred to in the text. Thickness of cell walls has been exaggerated.*
(b) Diagrammatic representation of a group of cells summarising possible pathways of water (and solute) movement. More than one pathway may be used at the same time. Such pathways may be used across the leaf and across the root cortex. Movement of ions by the vacuolar pathway would involve active transport. The apoplast pathway is the most important, and the vacuolar pathway the least important.

three pathways which water can then follow, namely the apoplast pathway (cell walls), the symplast pathway (cytoplasm and plasmodesmata) and the vacuolar pathway (from vacuole to vacuole).

13.13 Why does transpiration occur mainly through leaves and not so much through the cuticle and lenticels?

13.3.1 The apoplast pathway

The **apoplast** is the system of adjacent cell walls which is continuous throughout the plant. Up to 50% of a cellulose cell wall may be 'free space' which can be occupied by water. As water evaporates from the mesophyll cell walls into the intercellular air spaces, tension develops in the continuous stream of water in the apoplast, and water is drawn through the walls in a mass flow by the cohesion of water molecules (section 13.4). Water in the apoplast is supplied from the xylem.

13.3.2 The symplast pathway

The **symplast** is the system of interconnected protoplasts in the plant. The cytoplasm of neighbouring protoplasts is linked by the plasmodesmata, the cytoplasmic strands which extend through pores in adjacent cell walls (fig 13.8*b*). Once water, and any solutes it contains, is taken into the cytoplasm of one cell it can move through the symplast without having to cross further membranes. Movement might be aided by cytoplasmic streaming. The symplast is a more important pathway of water movement than the vacuolar pathway.

13.3.3 The vacuolar pathway

In the **vacuolar pathway** water moves from vacuole to vacuole through neighbouring cells, crossing the symplast and apoplast in the process and moving through membranes and tonoplasts by osmosis (fig 13.8*b*). The water potential gradient is set up as follows.

Water evaporates from the wet walls of the mesophyll cells into the intercellular air spaces, particularly into the larger substomatal air spaces. Taking cell A in fig 13.8*a* as an example, loss of water from the cell would result in a decrease in its pressure potential and its water potential. Cell B would then have a higher water potential than cell A (at equilibrium they would be equal). Water will therefore move from cell B to cell A, thus lowering the water potential of cell B relative to cell C. In this way a gradient of water potential is set up across the leaf from a higher potential in the xylem to a lower potential in the mesophyll cells. Water enters the mesophyll cells from the xylem by osmosis. Although it is convenient to describe the movement of water in a step-by-step fashion, it should be stressed that the water potential gradient that develops across the leaf is a continuous one, and water moves smoothly down the gradient as a liquid would in moving along a wick.

It is sometimes imagined that water moves across the leaf in response to a gradient of solute potentials. However, although a water potential gradient exists, there is no evidence to suggest that solute potentials of the cells differ significantly from one another. Differences in water potential are due mainly to differences in pressure potential (remember that loss of a small amount of water from a cell has a much greater effect on pressure potential than on solute potential). The same applies to the root (section 13.5) where gradients of pressure potential and water potential, but not necessarily of solute potential, exist.

13.3.4 Exit of water through stomata

The three pathways described end with water evaporating into air spaces. From here water vapour diffuses through the stomata, from a high water potential inside the leaf to a much lower one outside the leaf. In dicotyledons, stomata are usually confined to, or are more numerous in, the lower epidermis. Control of stomatal opening is discussed in section 13.3.9.

Immediately next to the leaf is a layer of stationary air whose thickness depends on the dimensions and surface features of the leaf, such as hairiness, and also on wind speed. Water vapour must diffuse through this layer before being swept away by moving air (mass flow). The thinner the stationary layer, the faster is the rate of transpiration. There is a diffusion gradient from the stationary layer back to the mesophyll cells. Theoretically each stoma has a diffusion gradient, or 'diffusion shell' around it, as shown in fig 13.9. In practice the diffusion shells of neighbouring stomata overlap in still air to form one overall diffusion shell.

13.3.5 Measuring the rate of transpiration

Transpiration can easily be demonstrated by placing a bell jar over a potted plant with the pot enclosed in a plastic bag to prevent water loss from the soil. As transpiration occurs, a fluid collects on the inside of the bell jar which is shown to contain water when tested with cobalt(II) chloride paper (blue to pink in water) or anhydrous copper(II) sulphate crystals (white to blue in water).

Measuring rates of transpiration can be difficult, but satisfactory results, at least for the purposes of comparison, can be obtained by means of the two simple experiments described below.

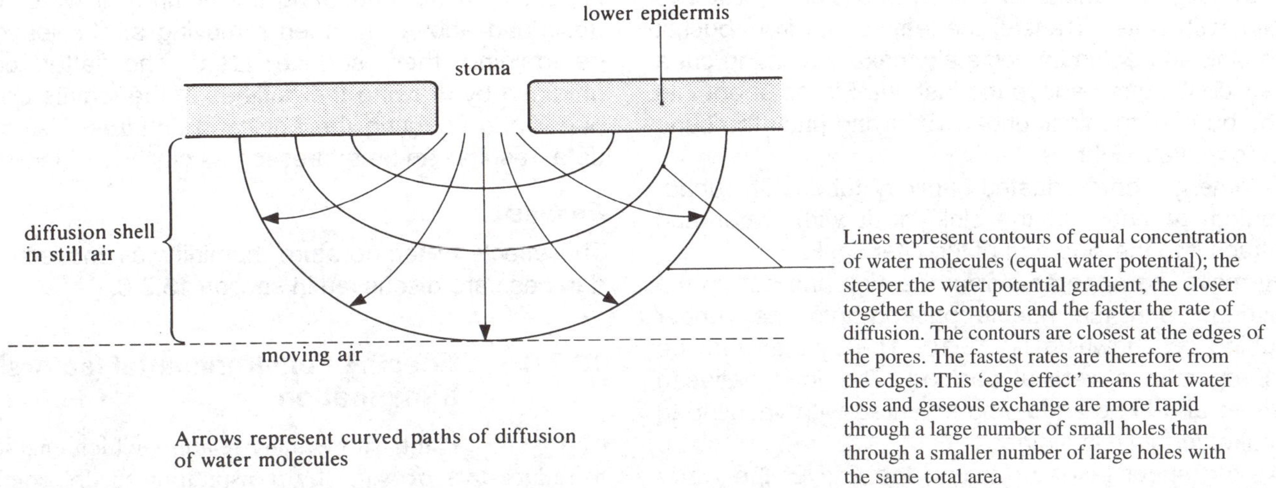

Fig 13.9 *Diffusion of water molecules from a stoma.*

Experiment 13.4: To investigate and measure factors affecting rate of transpiration using a potometer

A **potometer** is a piece of apparatus designed to measure the rate of water uptake by a cut shoot or young seedling. It does not measure transpiration directly, but since most of the water taken up is lost by transpiration, the two processes are closely related. Potometers are available commercially, but a simple version may be set up as shown in fig 13.10.

Materials

potometer (fig 13.10: conical filter flask, short rubber tubing, rubber bung with a single hole, hypodermic syringe and needle, graduated capillary tube)
large black polythene bag
large transparent polythene bag
small electric fan
retort stand and clamp
stop clock
thermometer
vaseline (petroleum jelly)
leafy shoot such as lilac
bucket

Method

(1) Select a suitable leafy plant, cut off the shoot and immerse the cut end immediately in a bucket of water to minimise the risk of air being drawn into the xylem. Immediately cut the shoot again under water, with a slanting cut, a few centimetres above the original cut. The stem must be thick enough to fit tightly into the bung of the potometer.

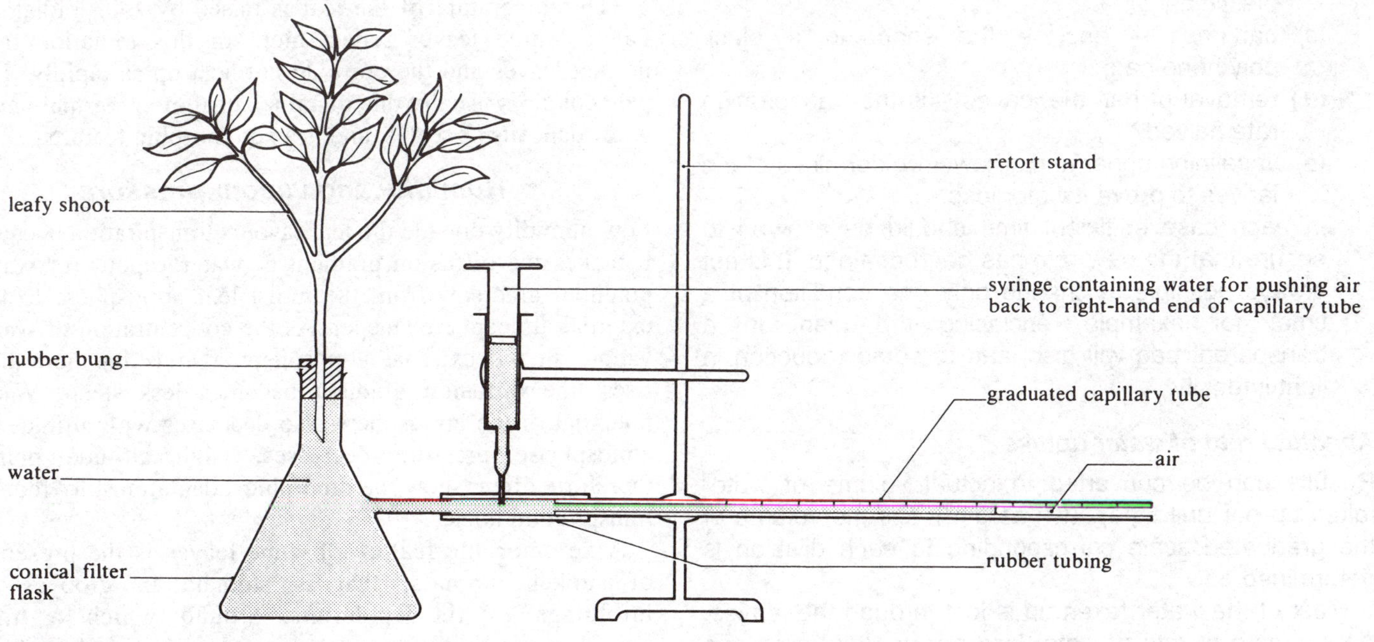

Fig 13.10 *A simple potometer.*

(2) Submerge a conical filter flask in a sink of water to fill it with water. Transfer the leafy shoot from bucket to sink and again immediately make a slanting cut a few centimetres above the last cut. Fit the shoot into the bung of the flask under water and push the bung in to make a tight fit.
(3) Submerge the graduated capillary tube, with rubber tubing attached, in the sink, fill it with water and attach it to the side arm of the filter flask.
(4) Remove the apparatus from the sink and set up the syringe with the needle pushed into the rubber tubing as shown in fig 13.10. The syringe can be clamped in a vertical position. The joint between shoot and bung should be smeared with vaseline to make certain it is airtight.
(5) As the shoot takes up water, the end of the water column in the capillary tube can be seen to move. It may be returned to the open end of the tube by pushing in water from the syringe. Allow the shoot to equilibrate for 5 min whilst regularly replacing the water taken up.
(6) Measure the time taken for the water column to move a given distance along the capillary tube and express the rate of water uptake in convenient units, such as cm min^{-1}. A number of readings should be taken, to ensure that the rate is fairly constant, and the mean result calculated. The temperature of the air around the plant should be noted.
(7) Each time the air bubble reaches the end of the graduated section of the tube return it to its original position with the syringe.
(8) The effects of some of the following factors on rate of uptake of water could be investigated:
   (*a*) wind – use a small electric fan (do not strongly buffet the leaves or the stomata will close);
   (*b*) humidity – enclose the shoot in a transparent plastic bag;
   (*c*) darkness – enclose the shoot in a black polythene bag;
   (*d*) removal of half the leaves – is the transpiration rate halved?
   (*e*) vaselining upper and/or lower epidermises of the leaves to prevent water loss.

   In each case sufficient time should be allowed to ensure that the new rate has been attained. It is not always possible to change only one condition at a time; for example, enclosing the plant in a transparent bag will also lead to some reduction in light intensity.

Absolute rate of water uptake

Results can be converted to actual volume of water taken up per unit time, such as $cm^3 h^{-1}$, if the volume of the graduated scale corresponding to each division is determined.

Most of the water taken up is lost through the leaves. An estimate of rate of water loss per unit leaf area can be obtained by measuring the volume of water lost as described above and then removing all the leaves and determining their surface area. The latter can be obtained by drawing the outlines of the leaves on graph paper and counting the enclosed squares. Using these data, results can be expressed as $cm^3 h^{-1} m^{-2}$ leaf area.

Results

The effects of temperature, humidity, air movement and darkness are discussed in section 13.3.6.

13.3.6 Effects of environmental factors on transpiration

Plants show many features which enable them to reduce loss of water by transpiration in dry conditions. Such features are described as **xeromorphic**. Plants growing in dry habitats and subjected to drought are called **xerophytes** and possess many xeromorphic features which are described in more detail in chapter 20. Plants growing under conditions in which there is normally an adequate water supply are called **mesophytes**, but nevertheless can show some xeromorphic features.

Temperature

The external factor which has the greatest effect on transpiration is temperature. The higher the temperature, the greater the rate of evaporation of water from mesophyll cells and the greater the saturation of the leaf atmosphere with water vapour. At the same time, a rise in temperature lowers the relative humidity of the air outside the leaf. Both events result in a steeper concentration gradient of water molecules from leaf atmosphere to external atmosphere. The steeper this gradient, the faster is the rate of diffusion. Alternatively, it can be said that water potential increases inside the leaf while decreasing outside the leaf.

The temperature of the leaf is raised by solar radiation. Pale-coloured leaves reflect more of this radiation than normal leaves and therefore do not heat up as rapidly. The pale colour is usually due to a thick coat of epidermal hairs, waxy deposits or scales, and is a xeromorphic feature.

Humidity and vapour pressure

Low humidity outside the leaf favours transpiration because it makes the diffusion gradient of water vapour (or water potential gradient) from the moist leaf atmosphere to the external atmosphere steeper. As the concentration of water vapour in the external atmosphere, that is the humidity, rises, the diffusion gradient becomes less steep. Water potential of the atmosphere also decreases with altitude as atmospheric pressure decreases. High altitude plants therefore often show xeromorphic adaptations to reduce transpiration rates.

A xeromorphic feature of some leaves is the presence of sunken stomata, that is stomata in grooves or infoldings of the epidermis, around which a high humidity can build up and reduce transpiration losses. In some cases the whole leaf may roll up enclosing a humid atmosphere, such as in *Ammophila* (marram grass) (fig 13.11). A coat of epidermal hairs or scales will tend to trap a layer of still moist air next to the leaf, thus reducing transpiration.

Air movement

In still air a layer of highly saturated air builds up around the leaf, reducing the steepness of the diffusion gradient between the atmosphere inside the leaf and the external atmosphere. Any air movement will tend to sweep away this layer. Thus windy conditions result in increased transpiration rates, the increase being most pronounced at low wind speeds. High winds may result in closing of the stomata, stopping transpiration.

Hairs and scales trap still air as described above, tending to reduce transpiration rates.

Light

Light affects transpiration because stomata usually open in the light and close in darkness. At night, therefore, only small amounts of water are lost (through the cuticle or lenticels). As stomata open in the morning, transpiration rates increase.

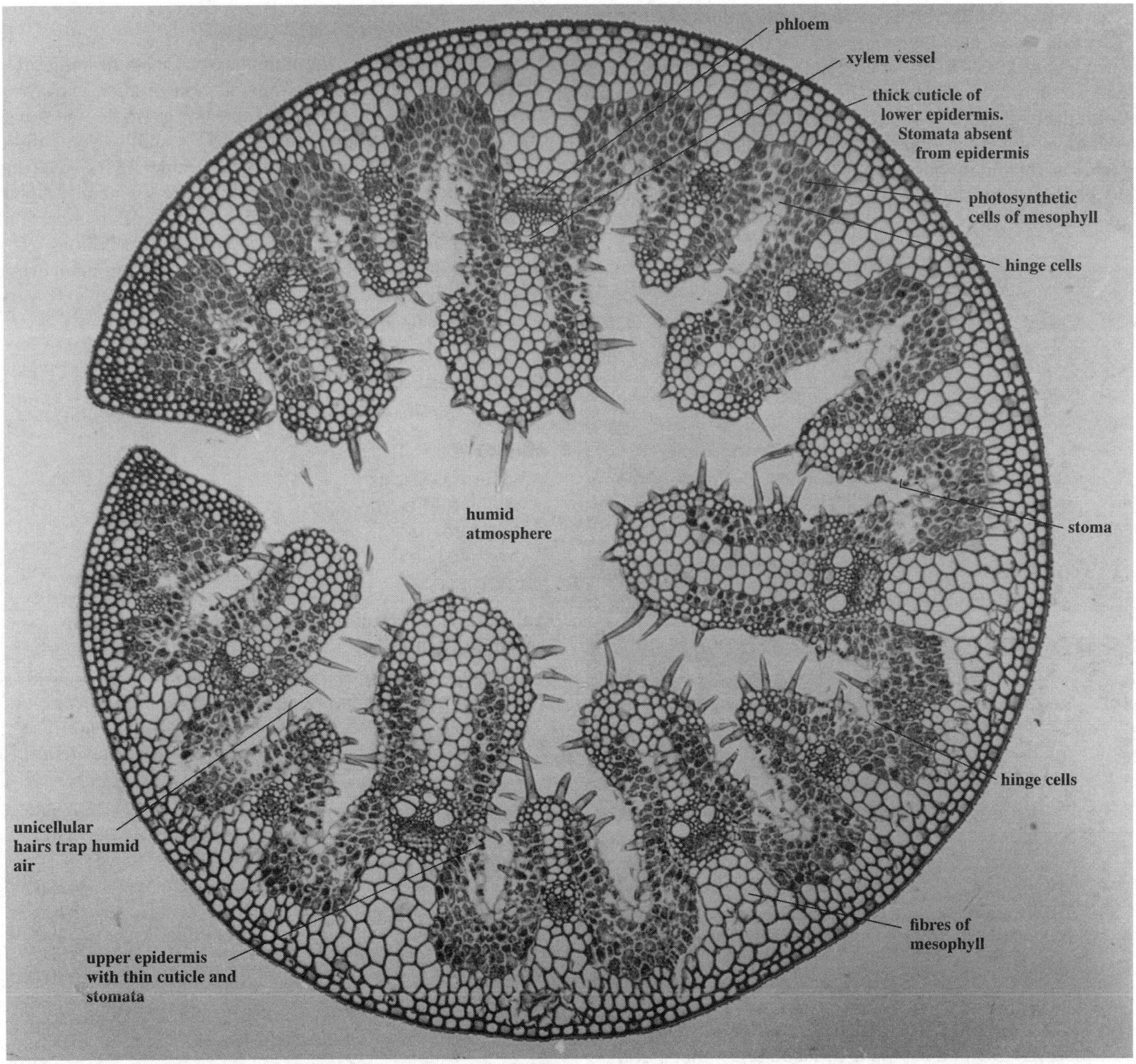

Fig 13.11 *A transverse section of the xeromorphic leaf of* Ammophila *(marram grass) to show distribution of tissues. The leaf is shown in the rolled condition.*

13.3.7 Effect of plant or internal factors on the rate of transpiration

The effects of some xeromorphic adaptations on transpiration rates have been considered above. Further examples of the ways in which such 'internal' as opposed to 'external' (environmental) factors can operate are given below.

Leaf surface area and surface area to volume ratio

Transpiration of a plant increases with its total leaf surface area, and with leaf surface area to volume ratio. Reduction of leaf surface is achieved when leaves are reduced to needles, such as in *Pinus* and other conifers, or to spines, as in cacti. There may also be a reduction in size in dry conditions. The shedding of leaves in dry or cold seasons by deciduous plants is a xeromorphic adaptation. In cold seasons water may be unavailable through being frozen in the soil.

Surface area to volume ratio can be reduced by using the stem as the main photosynthetic organ, as in cacti. Fig 13.12 shows the characteristic reduction in leaf surface area of succulents like cacti.

(*a*)

(*b*)

Fig 13.12 *(a) The cactus* Opuntia *has thick fleshy stems to conserve water, no leaves to reduce transpiration, and spines to protect it from grazing. (b)* Sempervivum *is a typical succulent with fleshy leaves for storing water.*

Cuticle

The cuticle is a layer which is secreted by the epidermis. It consists of a fatty substance called **cutin** which is relatively waterproof. In general, the thicker the cuticle the lower the rate of transpiration through it. Where it is thin, as in ferns, 30–45% of the transpiration losses can be through it.

The upper surfaces of dicotyledonous leaves, which are exposed to direct sunlight and are less protected from air currents than the lower surfaces, often possess thicker cuticles than the lower surfaces. Increased wax deposits on leaves can virtually eliminate cuticular transpiration. Also, waxy leaves are usually shiny and so reflect more solar radiation.

Stomata

In general, the greater the number of stomata per unit area, the greater is the rate of stomatal transpiration; however, their distribution is also important. For example, the lower surfaces of dicotyledonous leaves usually have more stomata than their upper surfaces (table 13.7), whereas monocotyledonous leaves, which are generally held vertically rather than horizontally, have similar upper and lower surfaces with similar stomatal distributions (see maize and oat, table 13.7). On average, fewer stomata occur in plants adapted to dry conditions. The number may vary within the same species as a result.

Experiment 13.5: To investigate stomatal distribution

Materials

clear nail varnish
slides and cover-slips
fine forceps
fresh fully expanded leaves
microscope

Table 13.7 Stomatal densities in the leaves of some common plants.

Plant	*Number of stomata/cm^{-2}* *upper epidermis*	*lower epidermis*
Monocotyledons		
maize (*Zea mais*)	5 200	6 800
oat (*Avena sativa*)	2 500	2 300
Dicotyledons		
apple (*Malus* spp.)	0	29 400
bean (*Phaseolus vulgaris*)	4 000	28 100
cabbage (*Brassica* spp.)	14 100	22 600
lucerne (*Medicago sativa*)	16 900	13 800
Nasturtium	0	13 000
oak (*Quercus* spp.)	0	45 000
potato (*Solanum tuberosum*)	5 100	16 100
tomato (*Lycopersicon esculentum*)	1 200	13 000

Based on Weier, T. E., Stocking, C. R. and Barbour, M. G. (1970) *Botany, an Introduction to Plant Biology*, 4th ed., John Wiley & Sons, p. 192.

Method

A convenient means of examining stomatal distribution is to make a replica of the leaf surface using clear nail varnish. Spread a thin layer of the nail varnish over the leaf using the brush in the bottle. Allow it to dry, then peel off the thin replica with fine forceps, lay it on a slide and add a cover-slip. It may be mounted in water for convenience. Examine with a microscope. Count the number of stomata in a given field of view and repeat several times in different areas. Obtain a mean value. Determine the area of the field of view of the microscope by measuring the diameter with a calibrated slide or transparent rule and using the formula πr^2 for the area, where r is the radius and $\pi = 3.142$. The number of stomata per square centimetre can then be calculated.

Compare the densities of stomata in upper and lower epidermises of the same leaf, and in different species. Is there any correlation between stomatal densities and the habitats of plants?

13.14 Examine fig 13.13. Describe and explain the relationships between the three variables shown.

13.3.8 Functions of transpiration

Transpiration has been described as a 'necessary evil' because it is inevitable, but potentially harmful. It happens because of the existence of wet cell walls from which evaporation occurs. Water vapour escapes mainly through the stomata which are needed for gaseous exchange between the plant and its environment. Gaseous exchange is essential to obtain the carbon dioxide needed for photosynthesis. (Because there is much more oxygen than carbon dioxide in the atmosphere, plants can get all the oxygen they need even with stomata shut. They therefore respire in the dark just as efficiently as they do in the light.) If there was no cuticle, stomata would be unnecessary and gaseous exchange would be even more efficient. However, loss of water could not then be controlled. The cuticle reduces water loss and further control is exercised by the stomata, which in most plants are highly sensitive to water stress and close under conditions of drought. They also usually close during the night when photosynthesis ceases. Loss of water can lead to wilting, serious desiccation, and often death of a plant. There is good evidence that even mild water stress results in reduced growth rate and causes economic losses in crops through reductions in yield.

Although it is inevitable, it is worth asking whether there might be some advantages associated with transpiration. Two possibilities are as follows.

- The evaporation of water from mesophyll cells that accompanies transpiration requires energy and therefore results in cooling of the leaves in the same way that sweating cools the skin of mammals. This is sometimes important in direct sunlight when leaves absorb large amounts of energy and experience rises in temperature which, under extreme conditions, can inhibit photosynthesis. However, it is unlikely that the cooling effect is of significance under normal conditions. Plants that live in hot climates usually have other means of avoiding heat stress.
- It has been suggested that the transpiration stream is necessary to distribute mineral salts throughout the plant, since these move with the water. Whilst this may be true, it seems probable that very low transpiration rates would be sufficient. For example, mineral salt supply to leaves is just as great at night, when transpiration is low, as during the day because the xylem sap is more concentrated at night. Uptake of mineral salts from the soil is largely independent of the transpiration stream.

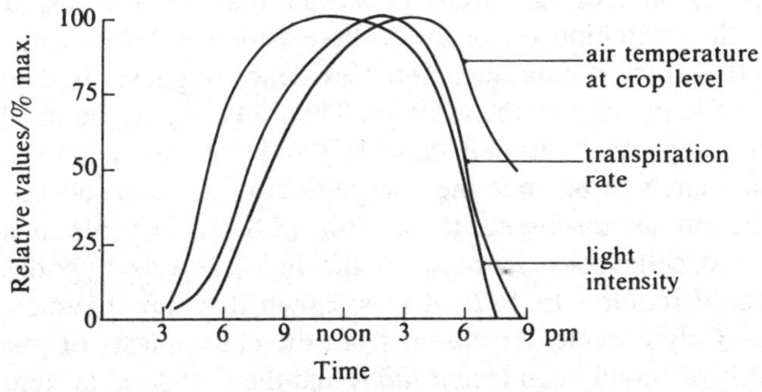

Fig 13.13 *Relationship between light intensity, air temperature and transpiration rate from lucerne leaves. (From data by L. J. Briggs & H. L. Shantz (1916)* J. Agr. Res., *5, 583–649; cited by A. C. Leopold (1964)* Plant growth and development, *p. 396. McGraw-Hill.)*

13.3.9 Stomata – structure and mechanism of opening and closing

Stomata are pores in the epidermis through which gaseous exchange takes place. They are found mainly in leaves, but also in stems. Each stoma is surrounded by two guard cells which, unlike the other epidermal cells, possess chloroplasts. The guard cells control the size of the stoma by changes in their turgidity. The appearance of guard cells and stomata is well revealed by the scanning electron microscope, as shown in fig 13.14.

The appearance of epidermal cells, guard cells and stomata in surface view, as seen with the light microscope, is dealt with in section 6.1. Fig 13.15 is a diagram of section through a typical stoma, and shows that the guard cell walls are unevenly thickened. The wall furthest from the pore (called the dorsal wall) is thinner than the wall next to the pore (the ventral wall). Also, the cellulose

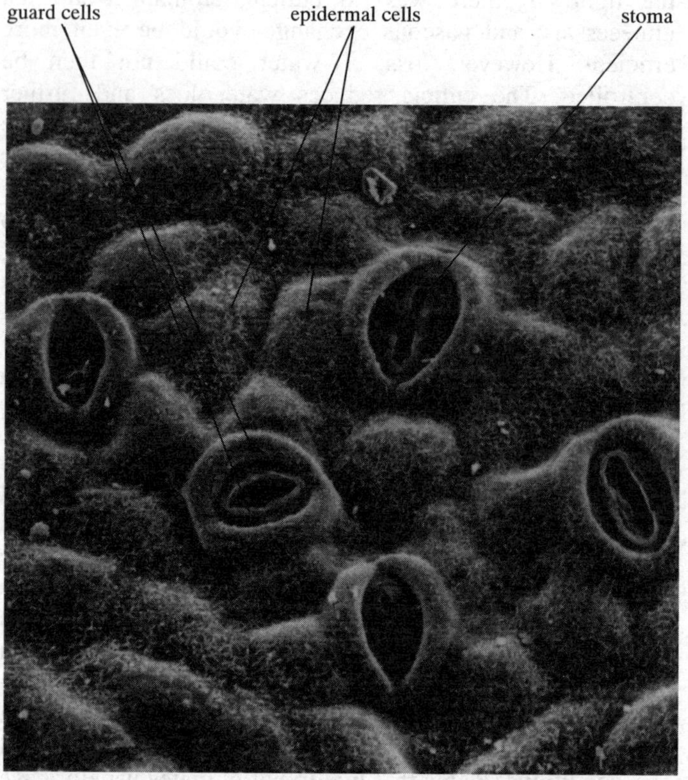

Fig 13.14 *Scanning electron micrograph of stomata on the lower surface of a leaf.*

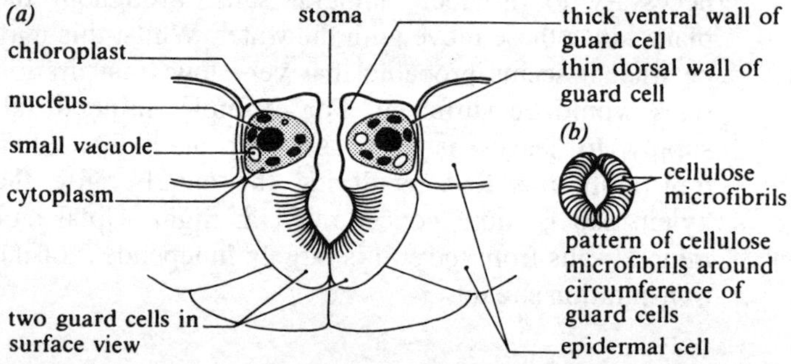

Fig 13.15 *(a) Vertical section through a stoma, showing also part of the lower surface of the leaf. (b) Pattern of cellulose microfibrils in guard cell walls.*

microfibrils that make up the walls are arranged so that the ventral wall is less elastic than the dorsal wall. Some of the cellulose microfibrils form hoops around the sausage-shaped guard cells as shown in fig 13.15*b*. These hoops are inelastic. As the cells inflate with water, that is become turgid, the hoops tend to stop the cells increasing in diameter (getting fatter) so they can only expand by increasing in length. Because the ends of the guard cells are joined, and also because the thin dorsal walls stretch more easily than the thick ventral walls, each cell becomes semicircular in shape (fig 13.15). Thus a hole, the stoma, appears between the guard cells. The same effect can be obtained by inflating a sausage-shaped balloon which has had a piece of adhesive tape stuck along one side to mimic the non-elastic ventral wall of the guard cell. The tape can also be wound loosely round the balloon to mimic the hoops.

When the guard cells lose water and turgidity, the pore closes. The question remains as to how the turgidity changes are brought about.

A traditional hypothesis, the 'starch–sugar hypothesis', suggested that an increase in sugar concentration in guard cells during the day led to their solute potential becoming more negative, resulting in entry of water by osmosis. However, sugar has never been shown to build up in guard cells to the extent necessary to cause the observed changes in solute potential. It has now been shown that potassium ions and associated negative ions accumulate in guard cells during the day in response to light and are sufficient to account for the observed changes. There is still doubt about which negative ions balance the potassium. In some species studied large quantities of organic acid ions accumulate, such as malate. At the same time the starch grains that appear in guard cell chloroplasts in darkness decrease in size. The reason is that starch is converted to malate and that this requires blue light. A possible route is:

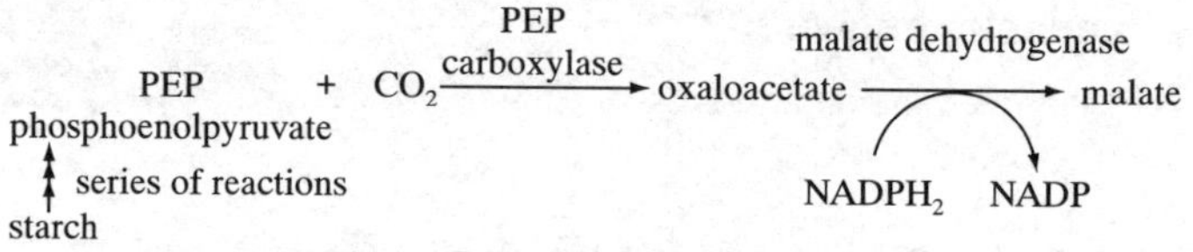

(Compare C_4 photosynthesis, section 7.9.)

Some plants, such as onion, have no starch in their guard cells. Here malate does not accumulate during stomatal opening and chloride ions (Cl^-), are taken up with the positive ions.

In darkness, potassium ions (K^+) move out of the guard cells into surrounding epidermal cells. The water potential of the guard cells increases as a result and water moves out of the cells. The loss of pressure makes the guard cells change shape again and the stoma closes.

Certain questions remain to be answered. For example, what causes the potassium ions to enter the guard cells in the light, and what function is served by the chloroplasts apart from storing starch? Potassium may enter in response to the switching on of an ATPase which is located in the cell surface membrane. Some evidence suggests that blue light may activate the ATPase. The ATPase may be needed to pump out hydrogen ions (H^+) and potassium ions may then enter to balance the charge (a similar pump occurs in phloem as discussed in section 13.8.4). The pH inside guard cells does decrease in the light, as this hypothesis would require. In 1979 it was shown that the enzymes of the Calvin cycle are absent from the chloroplasts of guard cells of broad bean (*Vicia faba*) and the thylakoid system is poorly developed, although chlorophyll is present. Normal C_3 photosynthesis therefore cannot occur and starch cannot be made by this route. This might explain why starch is made at night rather than during the day as in normal photosynthetic cells.

13.4 Ascent of water in the xylem

Xylem in flowering plants contains two types of water-transporting cell, the tracheid and the vessel, whose structures as seen in the light microscope are discussed in section 6.2.1 together with the appearance of vessels as seen with the scanning electron microscope (fig 6.12). The structure of the secondary xylem (wood) is dealt with in chapter 22. Xylem, together with phloem, forms the vascular or conducting tissue of higher plants. Vascular tissue consists of bundles of tubes called **vascular bundles** whose structure and arrangement in the primary stems of dicotyledonous plants (dicots) is shown in fig 13.16.

> **13.15** In the dicot stem, what is the overall shape of the following tissues in three dimensions: (*a*) epidermis, (*b*) xylem, (*c*) pericycle and (*d*) pith?

The fact that water can move up the xylem may be demonstrated by placing the cut end of a shoot in a dilute solution of a dye such as eosin. The dye rises in the xylem and spreads through the network of veins in the leaves. Sectioning and examination with a light microscope reveals the stain to be in the xylem.

Better evidence that xylem conducts water is given by 'ringing' experiments. These were among earlier experiments done before radioactive isotopes made the tracing of substances through living organisms much easier. In one type of ringing experiment an outer ring of bark, including phloem, is removed and, in the short term, this does not affect the upward movement of water. However, lifting a flap of bark, removing a section of xylem, and replacing the flap of bark leads to rapid wilting. Water therefore moves in the xylem.

Any theory for water movement up the xylem has to account for the following observations.

- Xylem vessels are narrow dead tubes ranging in diameter from 0.01 mm in 'summer wood' to about 0.2 mm in 'spring wood'.
- Large quantities of water are carried at relatively high speeds, up to $8\,m\,h^{-1}$ being recorded in tall trees and commonly in other plants at $1\,m\,h^{-1}$.
- To move water through such tubes to the height of a tall tree requires pressures of around 4000 kPa. The tallest trees, the giant sequoias or redwoods of California and *Eucalyptus* trees of Australia, can reach heights greater than 100 m. Water will rise in fine capillary tubes due to its high surface tension, a phenomenon called **capillarity**, but could rise only about 3 m in even the finest xylem vessels by this method.

The **cohesion–tension theory** of water movement adequately accounts for these observations. According to this theory, evaporation of water from the cells of a leaf is responsible for raising water from the roots. Evaporation results in a reduced water potential in the cells next to the xylem as described in section 13.3. Water therefore enters these cells from the xylem sap which has a higher water potential, passing through the moist cellulose cell walls of the xylem vessels at the ends of the veins, as shown in fig 13.8.

The xylem vessels are full of water and, as water leaves them, a tension is set up in the columns of water. This is transmitted back down the stem all the way to the root by **cohesion** of water molecules. Water molecules have high cohesion, that is they tend to 'stick' to each other, because, being polar, they are electrically attracted to each other and are held together by hydrogen bonding (section 3.1.2). They also tend to stick to the vessel walls, a force called **adhesion**. The high cohesion of water molecules means that a relatively large tension is required to break a column of water, that is a water column has a high tensile strength. The tension in the xylem vessels builds up to a force capable of pulling the whole column of water upwards by mass flow, and water enters the base of the columns in the roots from neighbouring root cells. It is essential that the xylem walls should be rigid if they are not to buckle inwards, as happens when sucking up a soggy straw. Lignin provides this rigidity. Evidence that the contents of xylem vessels are under high tension comes from measuring daily changes in the diameters of tree trunks using an instrument called a dendrogram. The minimum diameters are recorded during daylight hours when transpiration rates are highest. The minute shrinkage of each xylem vessel under tension combines to give a measurable shrinkage in diameter of the whole trunk.

Estimates of the tensile strength of a column of xylem sap vary from about 3000–30 000 kPa, the lower estimates being the more recent. Water potentials of the order required to generate enough tension to raise water, about –4000 kPa, have been recorded in leaves, and it seems likely that xylem sap has the required tensile strength to withstand this tension, though there may be a tendency for the columns to break, particularly in vessels of relatively large diameter.

Critics of the theory point to the fact that any break in a column of sap should stop its flow, the vessel tending to fill with air and water vapour, a process known as **cavitation**. Shaking, bending and shortage of water can all cause cavitation. It is well known that the water content of tree trunks gradually decreases during summer as the wood becomes filled with air. This is made use of in the lumber industry because such wood floats more easily. However, breaks in water columns do not greatly affect water flow rates. The explanation may be that water flows from one vessel to another, or by-passes air-locks by moving through neighbouring parenchyma cells and their walls. Also, it is calculated that only a small proportion of the vessels need be functional at any one time to account for the observed flow rates. In some trees and shrubs water moves only

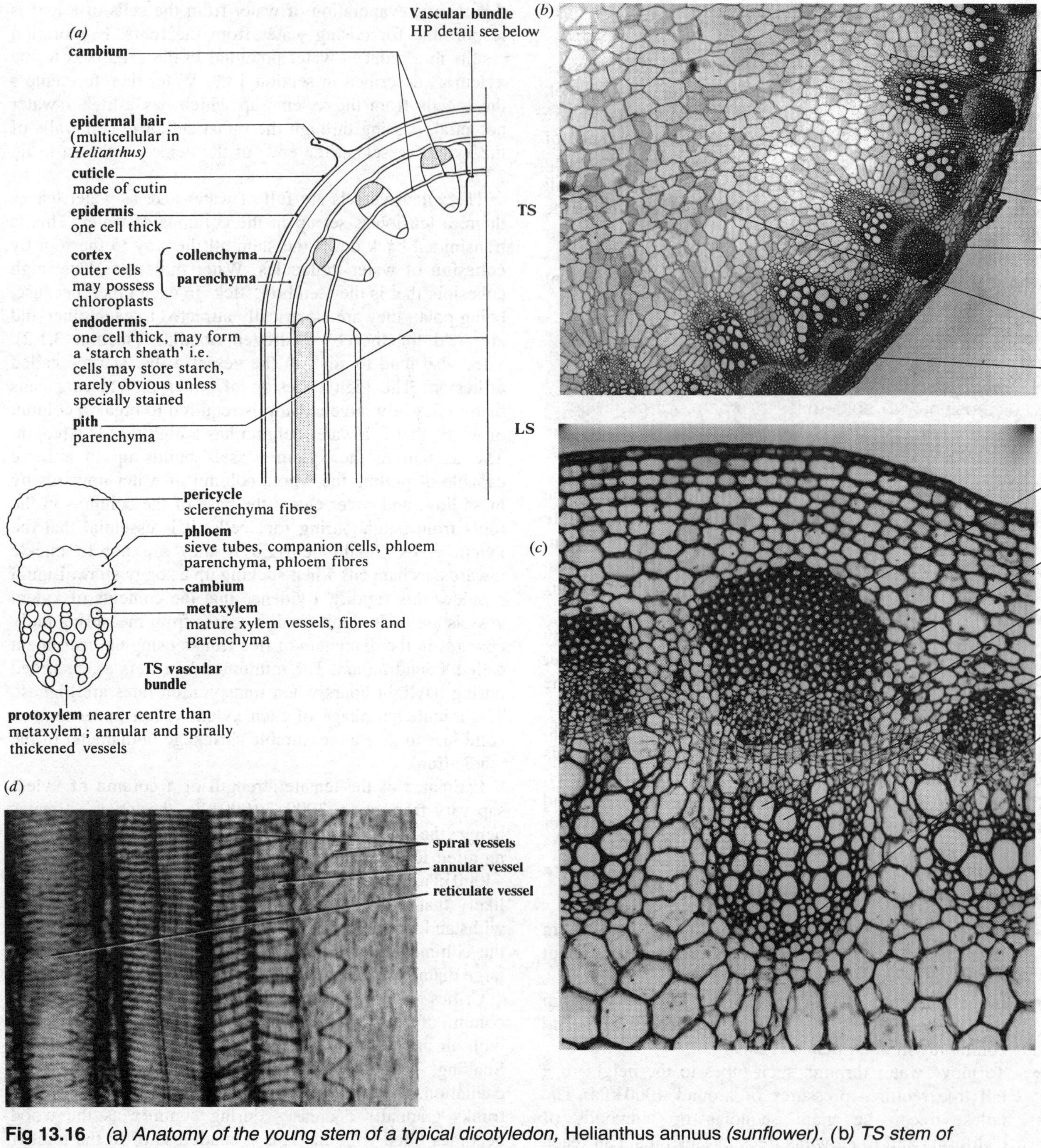

Fig 13.16 *(a) Anatomy of the young stem of a typical dicotyledon,* Helianthus annuus *(sunflower). (b) TS stem of* Helianthus *as seen with low power of a light microscope. (c) High power view of a TS of a vascular bundle from a stem of* Helianthus. *(d) LS of a stem of* Helianthus.

through the younger outer wood, which is therefore called **sapwood**. In oak and ash, for example, water moves mainly through the vessels of the current year, the rest of the sapwood acting as a water reserve. New vessels are added throughout the growing season, mostly early in the season when flow rates are higher.

A second force involved in water movement up the xylem is **root pressure**. This can be observed and measured when a freshly cut root stump continues to exude sap from its xylem vessels. The process is inhibited by respiratory inhibitors such as cyanide, lack of oxygen and low temperatures. The mechanism probably depends on active secretion of salts or other solutes into the xylem sap, thus lowering its water potential. Water then moves into the xylem by osmosis from neighbouring root cells.

The positive hydrostatic pressure of around 100–200 kPa (exceptionally 800 kPa) that is generated by root pressure is usually not sufficient alone to account for water movement up the xylem, but it is no doubt a contributing factor in many plants. It can be sufficient, however, in slowly transpiring herbaceous plants, when it can cause guttation. **Guttation** is the loss of water as drops of liquid from the surface of a plant (as opposed to vapour in transpiration). It is favoured by the same conditions that favour low transpiration rates, including dim light and high humidity. It is common in many rainforest species and is frequently seen at the tips of the leaves of young grass seedlings.

> **13.16** Summarise the properties of xylem which make it suitable for the long-distance transport of water and solutes.

13.5 Uptake of water by roots

The primary structure of a typical dicot root is shown in fig 13.17.

Water is absorbed mainly, but not exclusively, by the younger parts of roots in the regions of the root hairs. As a root grows through the soil, new root hairs develop a short distance behind the zone of elongation and older hairs die. These hairs are tubular extensions of epidermal cells (fig 13.17) and greatly increase the available surface area for uptake of water and mineral salts. They form a very intimate relationship with soil particles.

Fig 13.18*a* is a diagrammatic representation of the pathway taken by water across a root. A water potential gradient exists across the root from higher potential in the epidermis to lower potential in the cells adjacent to the xylem. This gradient is maintained in two ways:

- by water moving up the xylem, as described, setting up tension in the xylem and thus lowering the water potential of its sap;

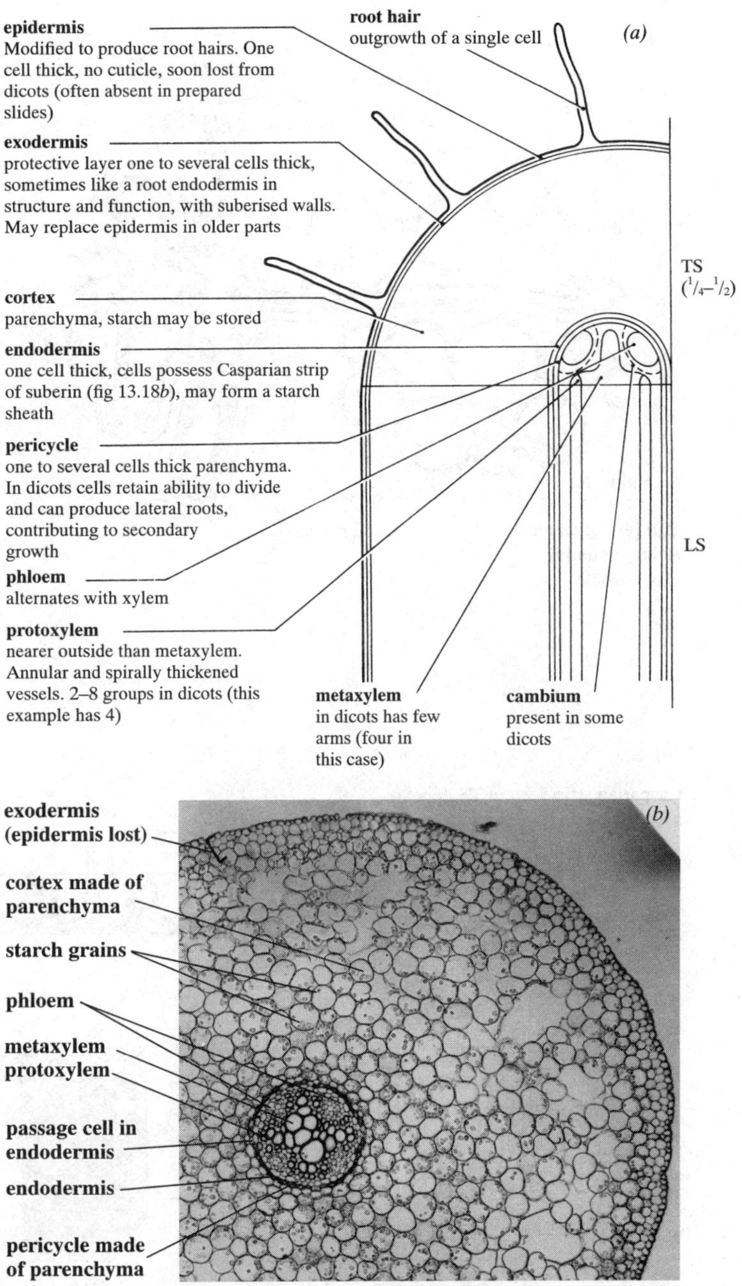

Fig 13.17 *(a) Anatomy of a young root of a typical dicotyledon,* Ranunculus *(buttercup). (b) Low power view of a TS of the root of* Ranunculus *as seen with a light*

- the xylem sap has a more negative (lower) solute potential than the dilute soil solution.

Water moves across the root by pathways similar to those in the leaf, namely apoplast, symplast and vacuolar pathways.

13.5.1 Symplast and vacuolar pathways

As water moves up the xylem in the root, it is replaced by water from neighbouring parenchyma cells, such as cell A in fig 13.18*a*. As water leaves cell A, the

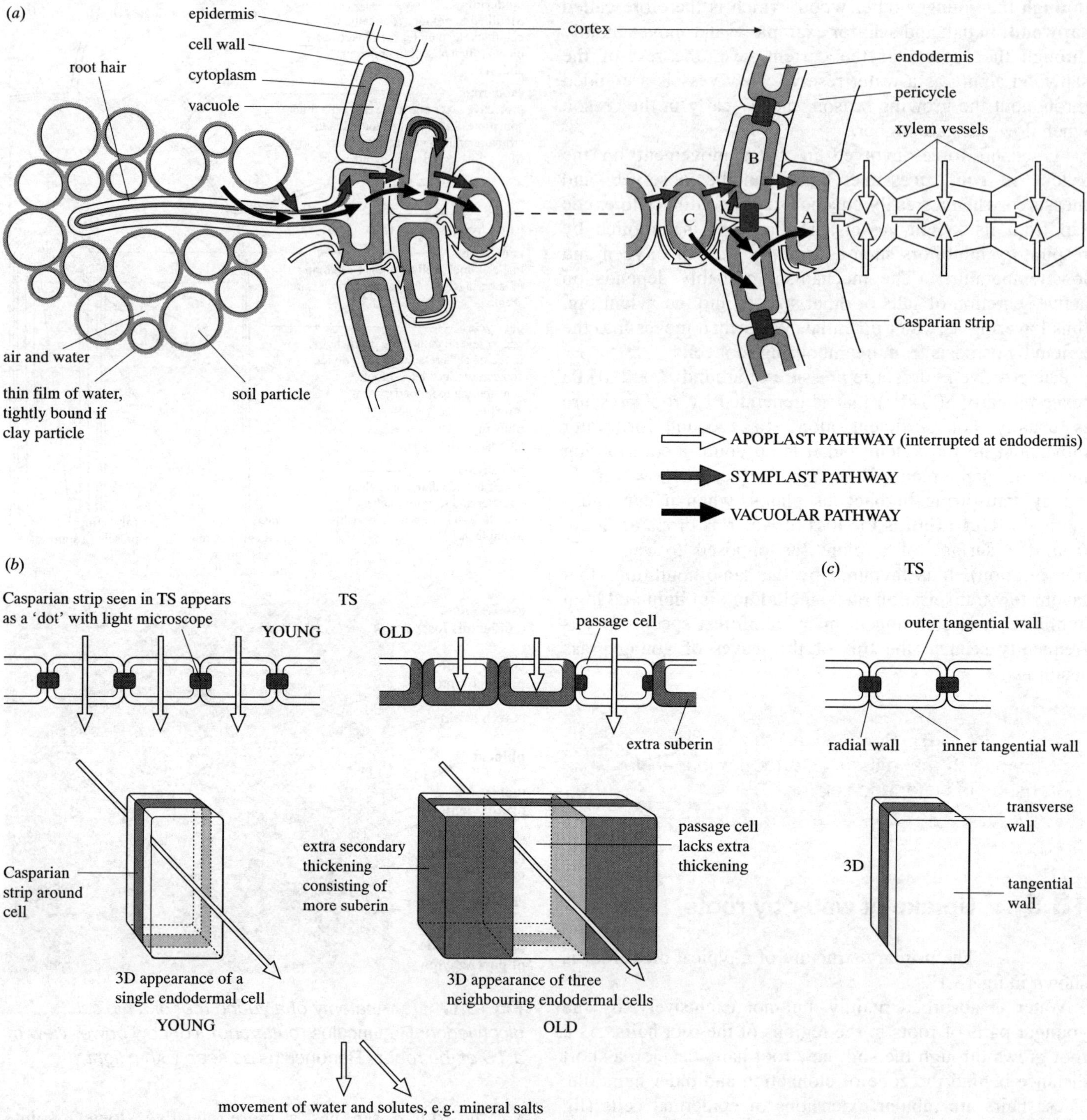

Fig 13.18 *(a) Diagrammatic representation of water and ion movement across a root shown in TS. Thickness of cell walls is exaggerated for clarity. Cells A, B and C are referred to in the text. The apoplast pathway is of greatest importance for both water and solutes. The symplast pathway is less important, except at the endodermis. Movement along the vacuolar pathway is negligible. (b) Structure and function of root endodermis showing Casparian strip in young endodermal cells and deposition of extra suberin in older endodermal cells, with exception of 'passage cells'. (c) Naming of walls. The transverse and radial walls are anticlinal (at right-angles to the surface of the root) and the tangential wall is periclinal (parallel).*

water potential of cell A decreases and water enters it from cell B by osmosis or through the symplast in exactly the same way as described for cells A and B in the leaf (section 13.3.2). Similarly the water potential of cell B then decreases and water enters it from cell C and so on across the root to the epidermis.

The soil solution has a higher water potential than cells of the epidermis including the root hairs. Water therefore enters the root from the soil by osmosis.

> **13.17** Arrange the following in order of ψ: soil solution, xylem sap, cell A, cell B, cell C, root hair cell. (Use the symbol > to mean greater than.)

13.5.2 Apoplast pathway

The apoplast pathway operates in much the same way as in the leaf (section 13.3.1). However, there is one important difference. When water moving through spaces in the cell walls reaches the endodermis its progress is stopped by a waterproof substance called **suberin** which is deposited in the cell walls in the form of bands called **Casparian strips** (fig 13.18*b*). Therefore water and solutes, particularly salts in the form of ions, must pass through the cell surface membrane and into the living part (cytoplasm) of the cells of the endodermis. In this way the cells of the endodermis can control and regulate the movement of solutes through to the xylem. Such control is necessary as a protective measure against the entry of toxic substances, harmful disease-causing bacteria and fungi, and so on. As roots get older the extent of suberin in the endodermis often increases, as shown in fig 13.18*b*. This blocks the normal exit of water and mineral salts from the cell (see fig 13.18*b*). However, plasmodesmata may stay as pores in the cell walls, and 'passage cells' in which no extra thickening occurs also remain to allow water and solutes to pass through to the xylem. The relative importance of apoplast, symplast and vacuolar pathways is not known.

13.6 Uptake of mineral salts and their transport across roots

As part of their nutrition, plants require certain mineral elements in addition to the carbohydrates made in photosynthesis. The uses of these elements are described in table 7.7. In plants, minerals are taken up from the soil or surrounding water by roots. Uptake is greatest in the region of the root hairs. The involvement of mycorrhizas is discussed in section 7.10.2.

In attempting to explain the uptake and movement of mineral ions, the following facts should be remembered.

- Mineral elements exist in the form of salts. Salts are made up of ions and in solution the ions can separate ('dissociate') and move about freely.
- Ions can cross membranes in a number of ways. One method is active transport. This requires energy in the form of ATP (made during respiration) and can lead to an accumulation of ions against a concentration gradient (section 5.9.8).
- There is a continuous system of cell walls, the apoplast, extending inwards from the epidermis of the root. Water, and any solutes it contains, enters the apoplast from the soil.

Fig 13.19 shows the uptake of potassium ions by young cereal roots which had previously been thoroughly washed in pure water. After 90 minutes the respiratory inhibitor potassium cyanide was added to the solutions.

> **13.18** (*a*) Describe the uptake of potassium ions at 0 °C and 25 °C.
> (*b*) Suggest why potassium cyanide (KCN) has the effects it does.
> (*c*) Suggest why the roots were thoroughly washed before placing them in a solution containing potassium ions.

Fig 13.19 shows that there are two distinct phases of uptake. The first phase lasts for about 10–20 minutes. Uptake during this phase is relatively rapid. Potassium ions come into contact with the epidermis of the root and start to move through the cell walls of the apoplast pathway, either by mass flow in the transpiration stream or by diffusion. The results show that this phase is more or less independent of temperature since it occurs just as rapidly at 0 °C. It is a passive process.

The second phase is temperature dependent and does not occur at 0 °C when the rate of metabolism and respiration is very low. Its inhibition by KCN shows that it is dependent on respiration. In fact, during this phase, the potassium ions are being taken into the cells of the roots across cell surface membranes by active transport.

Similar results to those in fig 13.19 can be obtained with isolated tissues. Those of storage organs, such as carrot, are commonly used. The data shown in fig 13.20 confirm the inhibition of respiration by potassium cyanide.

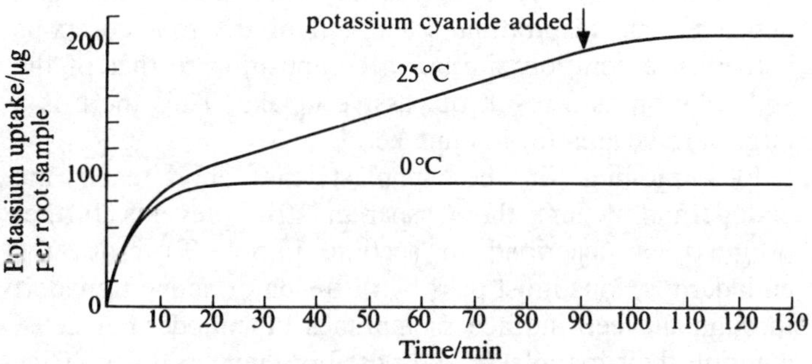

Fig 13.19 *Absorption of potassium ions by young cereal plants in an aerated solution.*

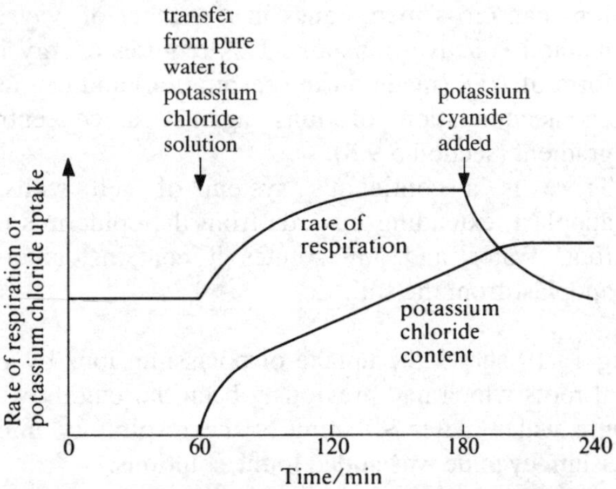

Fig 13.20 *Rate of respiration and uptake of potassium chloride by carrot discs. (Based on data by Robertson & Turner (1945).)*

To summarise so far, the uptake of ions by roots is a combination of:

- **passive uptake**, where ions move by mass flow and diffusion through the apoplast;
- **active uptake**, or **active transport**, in which ions can be taken up into cells against a concentration gradient using energy from respiration.

> **13.19** Fig 13.20 shows that the rate of respiration of carrot discs increases when they are transferred from pure water to potassium chloride solution. From the results shown, account for this increase.
>
> **13.20** Why does the rise in potassium chloride content stop when KCN is added?
>
> **13.21** In an experiment similar to that described in fig 13.19, but involving phosphate uptake, 16% of the phosphate taken up by barley roots over a short period could be washed out after transferring to pure water again. Explain.
>
> **13.22** Could ions reach the xylem entirely by means of the apoplast pathway?

Active transport is selective and dependent on respiration, whereas diffusion is non-selective and not dependent on respiration. Each cell of the root cortex is bathed in a solution similar in composition to that of the soil solution as a result of passive uptake. Thus there is a large surface area for ion uptake.

Ions moving in the apoplast can only reach the endodermis, where the Casparian strip prevents further progress as described in section 13.5.2. To cross the endodermis, ions must pass by diffusion or active transport through the cell surface membranes of endodermal cells, entering their cytoplasm and possibly their vacuoles. Thus the plant monitors and controls which types of ions eventually reach the xylem.

> **13.23** How could you demonstrate, using a radioactive ion and autoradiography, that the endodermis is a barrier to the movement of ions through cell walls?

Ions can also move through the symplast pathway. Once they are taken into the cytoplasm of one cell, they can move through the symplast without having to cross further membranes. The symplast extends from the epidermis right through to the xylem. Fig 13.18*a* summarises the possible ways in which ions can cross the root.

The final stage in the movement of mineral salts across the root is the release of ions into the xylem. To achieve this, ions must leave living cells at some stage, crossing back through a cell surface membrane. This could be by diffusion or active transport.

13.7 Translocation of mineral salts through plants

The pathway of mineral salts across the root to the xylem, described above, is the first stage in their translocation. Once in the xylem, they move by mass flow throughout the plant in the transpiration stream. Movement of mineral elements in the xylem can be demonstrated by ringing experiments like those already described, in which removal of tissues external to the xylem, such as phloem, has no effect on upward movement of ions. Analysis of the xylem sap also reveals that although some of the nitrogen travels as inorganic nitrate or ammonium ions, much of it is carried in the organic form of amino acids and related compounds. Some conversion of these ions to amino acids must therefore take place in the roots. Similarly, small amounts of phosphorus and sulphur are carried as organic compounds.

Thus, although xylem and phloem are traditionally regarded as conducting inorganic and organic materials respectively, the distinction is not clear-cut.

The chief **sinks**, that is regions of use, for mineral elements are the growing regions of the plant, such as the apical and lateral meristems, young leaves, developing fruits and flowers, and storage organs.

13.8 Translocation of organic solutes in phloem

Not all parts of a plant are photosynthetic. The main photosynthetic organs are the leaves. For those parts such as roots which are some distance from the sites of photosynthesis, there is a need for a transport system to circulate the products of photosynthesis. In vascular plants phloem is the tissue which carries products of photosynthesis away from the leaves to other parts. Fig

13.21 summarises the relationship between photosynthetic cells which produce organic food, and non-photosynthetic cells which receive the food. If you study fig 13.21 you will see that organic solutes must be able to move up and down in the same plant. This contrasts with movement in the xylem, which is only upwards. Note also that storage organs act either as sources (losing food) or as sinks (gaining food) at different times.

Typically, about 90% of the total solute carried in the phloem is the carbohydrate sucrose, a disaccharide. This is a relatively inactive and highly soluble sugar, playing little direct role in metabolism and so making an ideal transport sugar since it is unlikely to be used in transit. Once at its destination it can be converted back to the more active monosaccharides, glucose and fructose. Its high solubility means that it can be present in very high concentrations, up to 25% mass to volume in the phloem of plants such as sugarcane.

Phloem also carries certain mineral elements in various forms, particularly nitrogen and sulphur in the form of amino acids, phosphorus in the form of inorganic phosphate ions and sugar phosphates, and potassium ions. Small amounts of vitamins, growth substances such as auxins and gibberellins, viruses and other components may also be present.

Evidence for the circulation of carbon within the plant can be obtained by supplying leaves with carbon dioxide containing the radioactive isotope ^{14}C. Radioactive carbon dioxide is used in photosynthesis and ^{14}C passes into an organic compound such as sucrose. Its movement around the plant can then be traced by various techniques used to locate radioactive isotopes, such as autoradiography, holding a Geiger counter next to the plant surface, or extracting the isotope from different parts. Eventually, both phloem and xylem will be involved in the circulation of carbon. For example, carbon in the form of sucrose may reach the roots and be used in the manufacture of amino acids from nitrates. The amino acids, containing the carbon, can then travel up the shoots in the xylem.

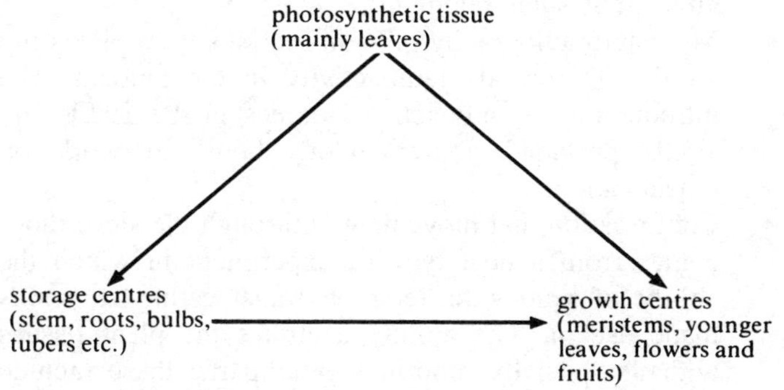

Fig 13.21 *Movement of organic solutes in a green plant.*

13.8.1 Features of phloem translocation

Before considering possible mechanisms for phloem translocation it is useful to list some outstanding facts which any hypothesis has to account for.

- **The quantity of material moved can be very large.** It is estimated, for example, that as much as 250 kg of sugar can be conducted down the trunk of a large tree during a growing season.
- **The rate of flow is high**, commonly 20–100 cm h^{-1}. Maximum rates in excess of 600 cm h^{-1} have been recorded.
- **The distances travelled can be very large.** The tallest trees, such as *Eucalyptus*, may be over 100 m tall. The leaves of *Eucalyptus* trees are located mainly near the top of the trunk, so assimilates must travel the length of the stem and often a considerable distance through the roots.
- **The amount of phloem is not great.** In a tree trunk, the functional phloem tissue is a layer only about the thickness of a postcard around the circumference. It forms the innermost layer of the bark of woody stems and roots. The older phloem, which is nearer to the outside, becomes stretched and dies as the plant grows and its circumference increases.
- **Movement is through tubes called sieve tubes which are very fine**, not more than 30 μm in diameter. This is comparable with a very fine human hair. At regular intervals the tubes are spanned by sieve plates with pores of even smaller diameter. The smaller the diameter of the tubes and pores, the greater is their resistance to the passage of fluid, and the greater the force required to move it. Pressure inside sieve tubes is high.
- Apart from sieve plates, sieve tubes have other structural features which must be taken into account (see next section).

> **13.24** How many sieve plates per metre would be encountered by a sucrose molecule moving through a sieve tube whose sieve tube elements were 400 μm long? (The sieve tube elements are the individual cells that make up the tube. They have sieve plates at each end.)

13.8.2 Structure of sieve tubes

The structure of phloem as seen with the light microscope is described in section 6.2.2. Phloem contains tubes called sieve tubes which are made from cells called sieve tube elements (or sieve elements). The elements are arranged end to end and fuse together to form the sieve tubes. Each element is separated from the next by a sieve plate, which contains pores to allow the flow of liquid from one element to the next.

In contrast to xylem vessels, which are dead empty tubes with few, if any, internal obstructions, phloem sieve tubes are living and do apparently contain obstructions to the flow of solution, namely the sieve plates and, to a lesser extent, the cytoplasm. Fig 13.22 is an electron micrograph of a mature sieve tube element, and fig 13.23 is a diagram showing the main features of sieve tube elements and their neighbouring companion cells.

During development of a sieve tube element its nucleus degenerates, making it an unusual example of a living cell with no nucleus; in this respect it is like mammalian red blood cells. At the same time many other important changes take place, the results of which are shown in fig 13.23. The cell walls at each end of the element develop into sieve plates. These are formed when the plasmodesmata of the end walls enlarge greatly to form sieve pores. A surface view of a sieve plate is shown in fig 6.13. The effect of all the changes is to leave a tube-like structure lined by a very narrow layer of living cytoplasm. This is surrounded by a cell surface membrane.

Closely associated with each sieve tube element are one or more companion cells. These come from the same parent cell as the neighbouring sieve tube element. Companion cells have dense cytoplasm with small vacuoles, and the usual cell organelles. They are metabolically very active, as indicated by their numerous mitochondria and ribosomes (fig 13.23). They have a very close relationship with sieve tube elements and are essential for their survival.

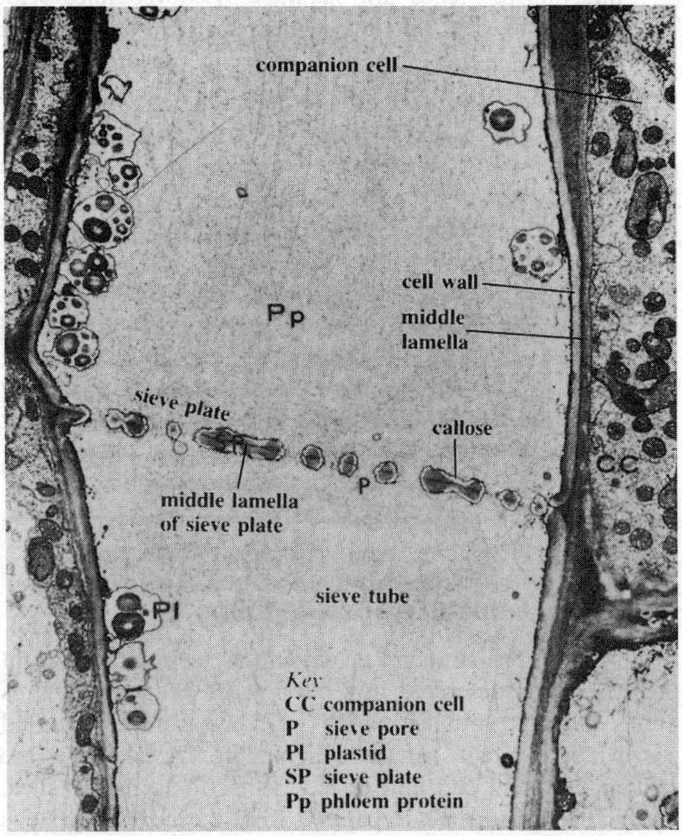

Fig 13.22 *Electron micrograph of a mature sieve tube element.*

In some plants sieve tube elements develop large quantities of a fibrous protein called **phloem protein** (P-protein). This sometimes forms deposits large enough to be seen with the light microscope. There has been much debate in the past about its function, but it is no longer believed to play a role in translocation.

13.8.3 Evidence for movement in phloem

It is important, especially in view of the discussion so far, to be certain that organic solutes really are carried in the phloem sieve tubes. Several types of experiment have been done, all of which support this belief.

- The earliest evidence for movement of sugars and other compounds in phloem came from ringing experiments, in which a ring of tissue containing phloem was removed from the outer region of the stem, leaving the xylem intact. Malpighi obtained evidence in 1675 for ascent of water in wood and descent of food in 'bark'. He removed rings of bark from trees (bark contains the phloem) and found that the leaves did not wilt, but that growth below the ring was greatly reduced. This is because he had stopped sugars moving down the plant without affecting passage of water upwards.
- Mason and Maskell, working with cotton plants in Trinidad during the 1920s and 1930s, did many ringing experiments, one of which is described in fig 13.24. From the results of the experiments shown in fig 13.24, Mason and Maskell concluded that some sideways exchange of sugars can take place between xylem and phloem when they are in contact and the phloem is interrupted (fig 13.24*a*) but that downward movement occurs in phloem (*b* and *c*).
- In 1945, a non-radioactive isotope of carbon, ^{13}C, was introduced into a plant as $^{13}CO_2$ and detected by mass spectrometry. A ring of phloem was killed with a fine jet of steam and translocation of ^{13}C-labelled sucrose through this section was shown to be prevented. Movement of mineral elements in the xylem is not affected by such treatment.
- Microautoradiography of stem sections from plants fed with $^{14}CO_2$ reveals radioactivity in the phloem. The introduction of radioactive isotopes in the 1930s and 1940s provided a tremendous boost to work on translocation.
- Confirmation that movement is through the sieve tubes comes from a neat type of experiment in which the ability of aphids to feed on translocating sugars is made use of. The aphid penetrates the plant tissues with its specially modified mouthparts; these include extremely fine, tube-like 'stylets', which are pushed slowly through the plant's tissues to the phloem. They can be shown to penetrate individual sieve tubes, as revealed by fig 13.25.

If the aphid is anaesthetised with carbon dioxide and the body removed, leaving the stylets in the plant, the

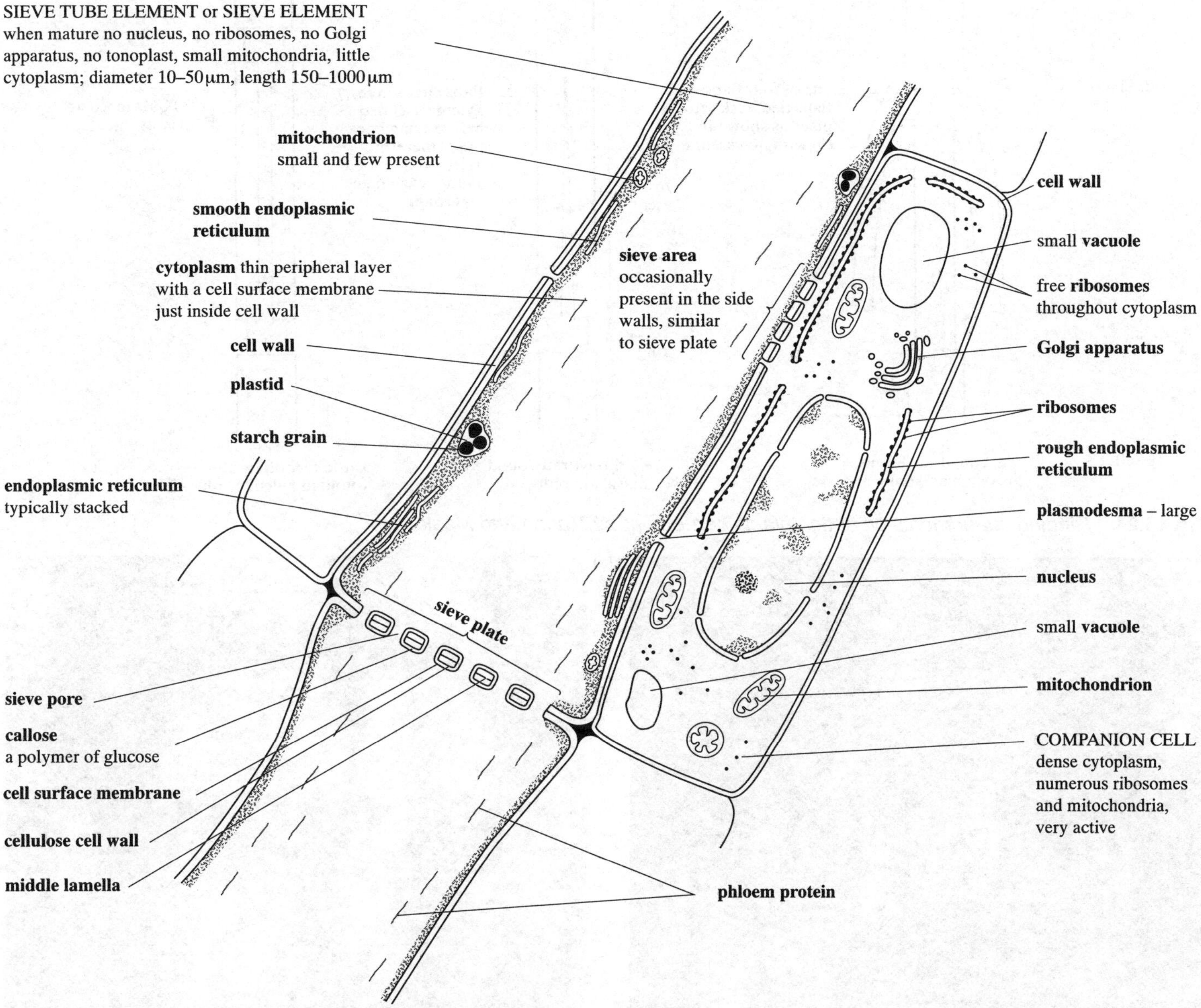

Fig 13.23 *Diagrammatic LS of sieve tube elements and a companion cell as seen with the electron microscope. If the sieve tube is damaged, for example by a grazing animal, more callose is rapidly deposited, blocking the sieve plate and preventing loss of valuable solutes from the sieve tube.*

contents of the sieve tube will continue to be forced up the tube of the mouthparts by the high pressure in the sieve tube, and the oozing fluid can be collected by microcapillary tubes. This technique has found a number of useful applications, for example in estimating rate of flow through sieve tubes (rate of exudation from the tube) and in analysing their contents.

- Finally, improvements in the sensitivity of film used in microautoradiography have enabled precise location of the weakly emitting isotope of hydrogen, tritium (^3H), in sieve tubes rather than other phloem cells. The isotope is supplied as part of an amino acid or sucrose.

Experiments have also established that different materials are carried up and down the phloem at the same time, although it is probable that this bidirectional movement is in neighbouring sieve tubes rather than in the same sieve tube.

13.8.4 Mechanism of translocation in phloem

The facts which any hypothesis must account for are summarised in sections 13.8.1 and 13.8.2, that is large quantities of material move at relatively rapid speeds through very fine sieve tubes. Within the tubes are apparent

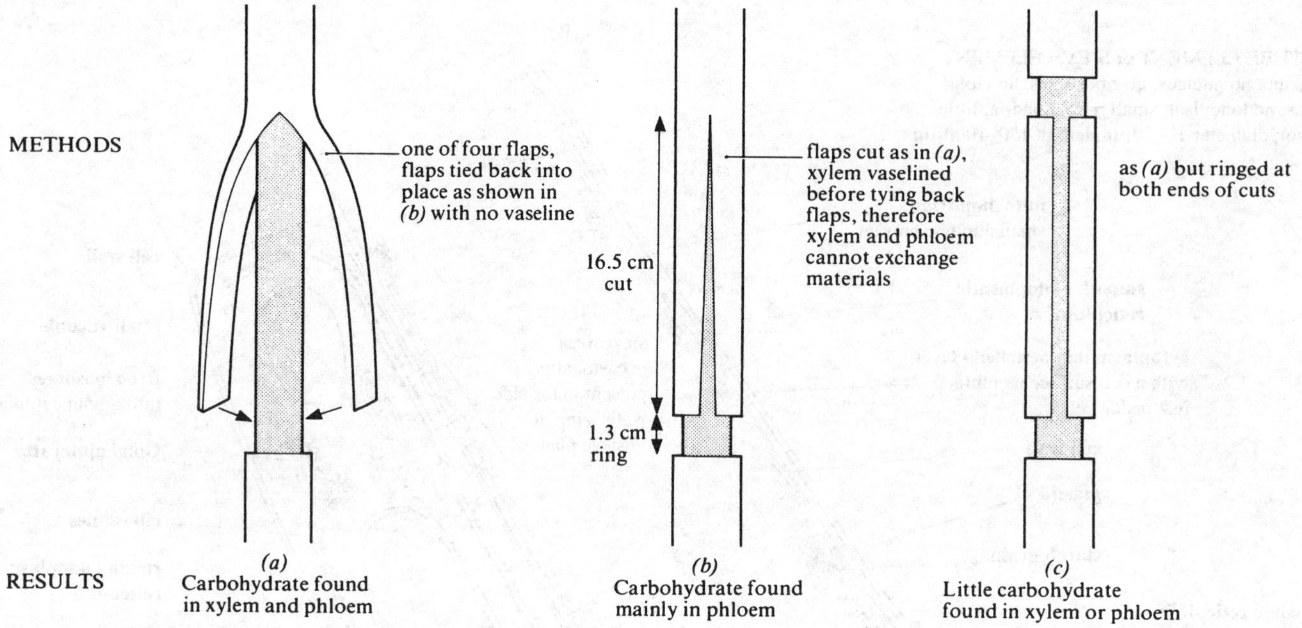

Fig 13.24 *Ringing experiments on cotton plants carried out by Mason and Maskell.*

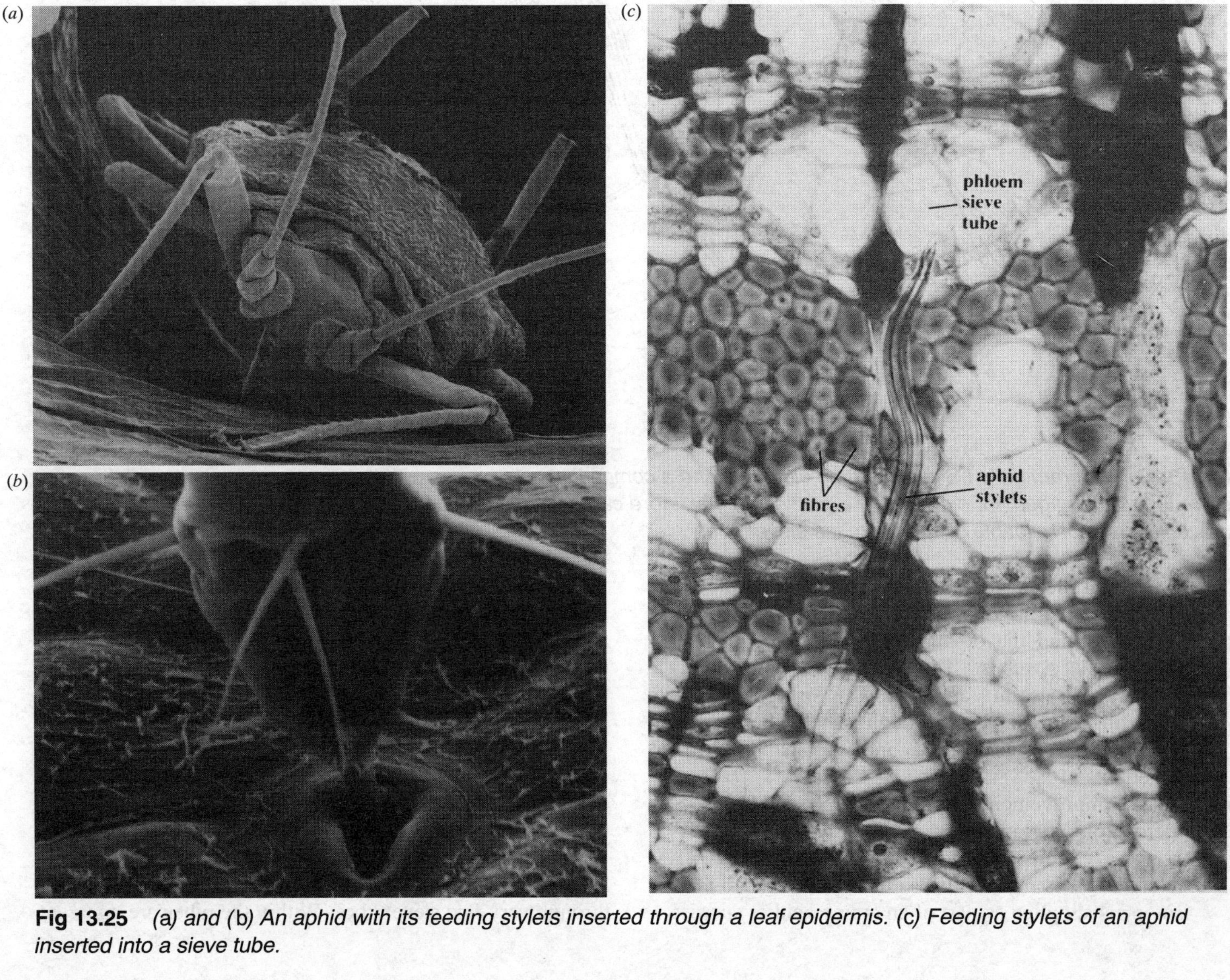

Fig 13.25 *(a) and (b) An aphid with its feeding stylets inserted through a leaf epidermis. (c) Feeding stylets of an aphid inserted into a sieve tube.*

obstructions, the sieve plates, and other structural features such as phloem protein for which no certain roles have been found. Combine these facts with the fact that the system is delicate and easily damaged by interference, and it is not surprising that research workers have found it difficult to establish the mechanism of translocation through sieve tubes.

It is now believed that a mass flow of solution occurs through sieve tubes. Diffusion is far too slow to account for the rates observed. The evidence for mass flow through sieve tubes is summarised below.

- When phloem is cut, sap oozes out, apparently by mass flow. This is sometimes used commercially as a source of sugar by tapping trees. For example the sugar palm exudes 10 dm^3 of sugar-rich sap per day.
- The prolonged exudation of sucrose solution from aphid stylets, as described in section 13.8.3, is evidence of hydrostatic pressure (pumping pressure) in sieve tubes.
- Certain viruses move in the phloem translocation stream, indicating mass flow rather than diffusion since the virus is incapable of locomotion. It cannot diffuse because it is not in solution.

Münch's mass flow hypothesis and the pressure flow hypothesis

In 1930, Münch put forward a purely physical hypothesis to explain how mass flow might be brought about in sieve tubes. It can be illustrated by the model shown in fig 13.26.

In the model there is an initial tendency for water to pass by osmosis into A and C, but the tendency is greater for A because the solution in A is more concentrated than that in C. As water enters A, a pressure (hydrostatic pressure) builds up in the closed system A–B–C, forcing water out of C. Mass flow of solution occurs through B along the pressure gradient that is generated. There is also an osmotic gradient from A to C. Eventually the system comes into equilibrium as water dilutes the contents of A and solutes accumulate at C.

The model can be applied to living plants. The leaves are represented by A. They make sugar during photosynthesis, thus making the ψ_s of the leaf cells more negative. Water, brought to the leaf in xylem (D) enters the leaf cells by osmosis, raising their pressure potential (ψ_p). At the same time, sugars are used in the sinks, such as roots (C), for various purposes including respiration and synthesis of cellulose. This makes the ψ_s of these cells less negative (higher). A pressure gradient exists from leaves to roots, or, in more general terms, from sources to sinks, resulting in mass flow. In the living plant, equilibrium is not reached because solutes are constantly being used at the sinks (C) and made at the sources (A).

The Münch hypothesis is a purely physical explanation and so does not explain why sieve tubes must be living and metabolically active. It also does not explain the observation that leaf cells are capable of loading sieve tubes against a concentration gradient, that is the fact that the ψ_s of sieve tubes is more negative than that of the leaf cells. The hypothesis has therefore been modified to include an active loading mechanism of solutes into the sieve tubes. The osmotic and hydrostatic pressure gradient therefore starts in the tubes rather than in the photosynthetic cells. It is also believed that unloading at the sinks is an active process. The modern version of Münch's hypothesis is known as the **pressure flow hypothesis**.

Loading sieve tubes

It has been shown that the sucrose concentration in sieve tubes in leaves is commonly between 10 and 30%, whereas it forms only about a 0.5% solution in the photosynthetic

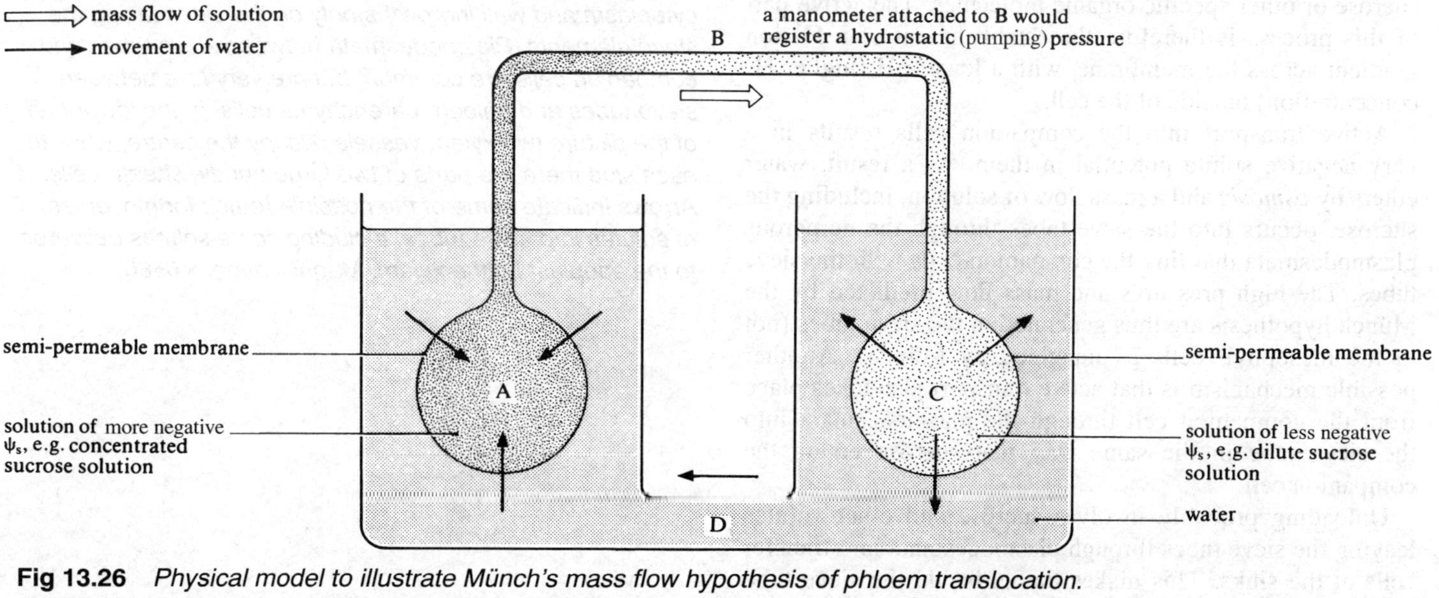

Fig 13.26 *Physical model to illustrate Münch's mass flow hypothesis of phloem translocation. Equivalents in living plant: A: source, such as leaf; B: phloem; C: sink, such as roots, meristems, fruits; D: xylem, cell walls and intercellular spaces.*

cells where it is produced. As stated above, loading of sieve tubes therefore takes place against a concentration gradient. The mechanism of loading has been the subject of much research in recent years. First the organic solutes have to move from the chloroplasts to the phloem tissue, a journey of 3 mm at most. Both symplast and apoplast routes are involved. The symplast route involves travelling through plasmodesmata and the apoplast route involves travelling by diffusion or mass flow in the transpiration stream through the cell walls.

In 1968 a modified type of companion cell, now known as a **transfer cell**, was reported by Gunning and fellow workers. The cells are found next to the sieve tubes as shown in fig 13.27. Transfer cells have numerous internal projections of the cell wall, a result of extra thickening of the wall. This results in an approximately tenfold increase in surface area of the cell surface membrane lining the wall. It is thought that such cells are thus modified for active uptake of solutes from neighbouring cells. Numerous mitochondria in their cytoplasm provide the energy for this. They are not found in all plants, but are common in the pea family and some other families. Even when transfer cells are absent, similar active transport processes are believed to occur.

Active loading of sucrose (and other substances such as amino acids, phosphates, potassium and ammonium ions) into companion cells is thought to be carried out by specific carrier protein molecules in the cell surface membranes of the companion cells. These carrier systems are thought to be similar to those in animal and bacterial cells in which transport of organic molecules is linked with transport of H^+ ions. H^+ ions are pumped out of the cell by a carrier which uses ATP as an energy source (fig 13.28). The H^+ ion gradient thus established represents potential energy. The H^+ ions diffuse rapidly back into the cell by way of specific carrier proteins that only function if they co-transport sucrose or other specific organic molecules. The active part of this process is therefore the establishment of a H^+ ion gradient across the membrane, with a lower pH (higher H^+ concentration) outside of the cell.

Active transport into the companion cells results in a very negative solute potential in them. As a result, water enters by *osmosis* and a mass flow of solution, including the sucrose, occurs into the sieve tubes through the numerous plasmodesmata that link the companion cells with the sieve tubes. The high pressures and mass flow predicted by the Münch hypothesis are thus generated in the sieve tubes (not in the mesophyll cells as imagined by Münch). Another possible mechanism is that active transport also takes place from the companion cell through the plasmodesmata into the sieve tube in the same way that sucrose enters the companion cell.

Unloading probably involves sucrose and other solutes leaving the sieve tubes through plasmodesmata into transfer cells at the sinks. This makes the ψ_s inside the sieve tube less negative, maintaining the pressure gradient between source, where sugar is loaded, and sink, where it is

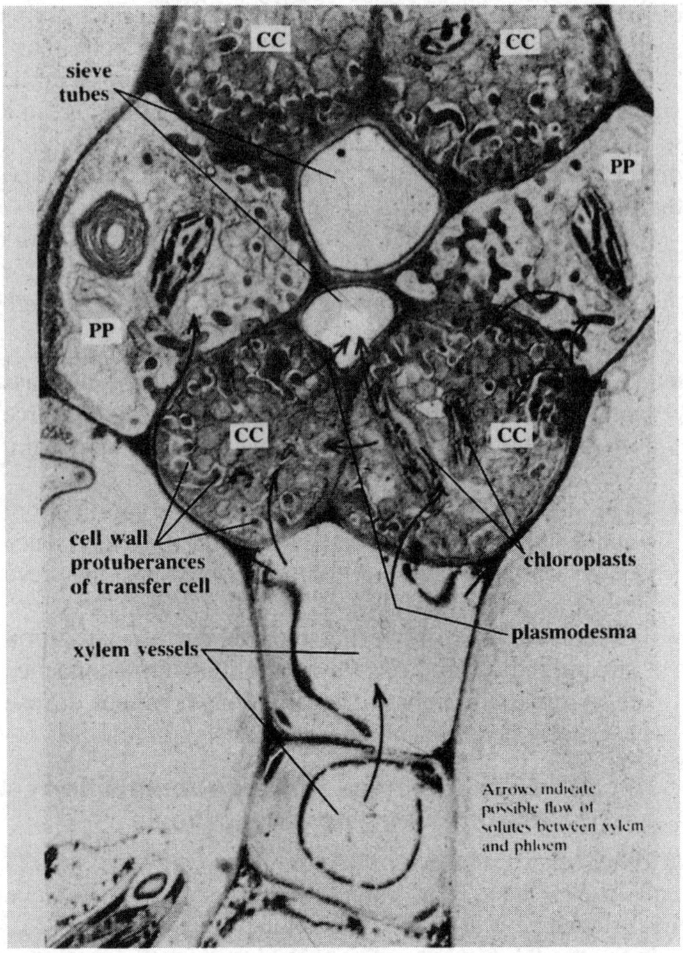

Fig 13.27 *Transverse section of a minor vein in a leaf of* Senecio vulgaris. *The phloem region, in the top half of the picture, shows six cells arranged around two sieve tubes. There are 4 companion cells (cc) modified as transfer cells. These have dense cytoplasm. There are phloem parenchyma cells (pp) to each side. These have less dense cytoplasm and wall ingrowths only on the sides facing the sieve elements. Plasmodesmata between sieve tubes and companion cells are common, but are very rare between sieve tubes and phloem parenchyma cells. In the lower half of the picture two xylem vessels occupy the centre, while to each side there are parts of two large bundle sheath cells. Arrows indicate some of the possible routes for movement of solutes into sieve tubes, including some solutes delivered to the apoplast by the xylem. Magnification ×6560.*

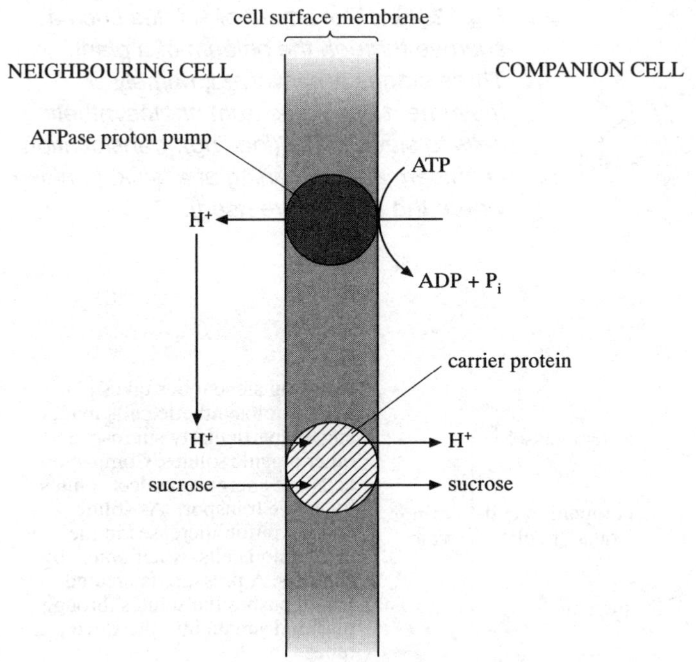

Fig 13.28 *Loading a companion cell with sucrose. H^+ ions are pumped out of the companion cell by a proton pump located in the cell surface membrane. This pump requires energy from ATP and has ATPase activity. Protons pass by facilitated diffusion back into the cell with sucrose through a combined hydrogen ion/sucrose carrier protein.*

unloaded. Unloading may also occur through the cell surface membrane of the sieve tube into the cell walls and into the apoplast.

Critical assessment of the pressure flow hypothesis

- The hypothesis predicts that mass flow will occur through sieve tubes and this seems to occur (the evidence has been given).
- The hypothesis requires the existence of an osmotic gradient and high pressure in the phloem. These have been demonstrated in a number of plants.

  The pressure gradients required to move solutes at the observed rates are relatively high. Assuming that the sieve pores are completely open, a *gradient* of 13 kPa per metre has been theoretically calculated as adequate and until recently it was doubted that such large gradients existed. Several attempts have been made to measure pressure in phloem directly, a very difficult task. Actual pressures varying from 1000–2000 kPa have been reliably recorded in recent years, and the gradient has been measured at 20 kPa per metre. It is therefore likely that the required pressure gradients *are* generated.
- Another criticism of the pressure flow hypothesis has been that it does not explain why sieve tubes should be living as opposed to the dead tubes of xylem. However, only living cells can maintain cell surface membranes, and these are required to prevent leakage of sucrose from the sieve tubes. Recent studies indicate that little metabolic energy is expended by living sieve tubes, suggesting that movement of solutes through them is passive, as predicted by the pressure flow hypothesis.
- Sieve plates are thought to be necessary, despite the resistance to flow which they create, to support the sieve tubes, preventing them from bulging and splitting or exploding outwards with the high internal pressures.
- Contents of the sieve tube are at a pH of about 7.5 to 8.0 (slightly alkaline). This would be expected if hydrogen ions were being pumped out as explained above. The pump would also explain the high concentrations of ATP found in sieve tubes. A summary of the movement of sucrose through the plant is given in fig 13.29.

13.8.5 First-aid mechanisms – a possible role for sieve plates, phloem protein and plastids

One danger faced by plants is damage from being eaten by animals. If sieve tubes are ruptured, leakage of high energy substances such as sucrose would be costly to the plant. Usually, damaged sieve tubes are sealed within minutes by deposition of callose across the sieve plates, blocking the sieve pores. This represents another possible role for sieve plates. It has been suggested that when phloem protein is present, it serves a first-aid function by blocking the sieve pores as soon as the tube is broken. When a sieve tube is cut, release of the pressure inside the tube causes a surge of its contents until they come up against a sieve plate and block it. This prevents escape of the contents of the sieve tube.

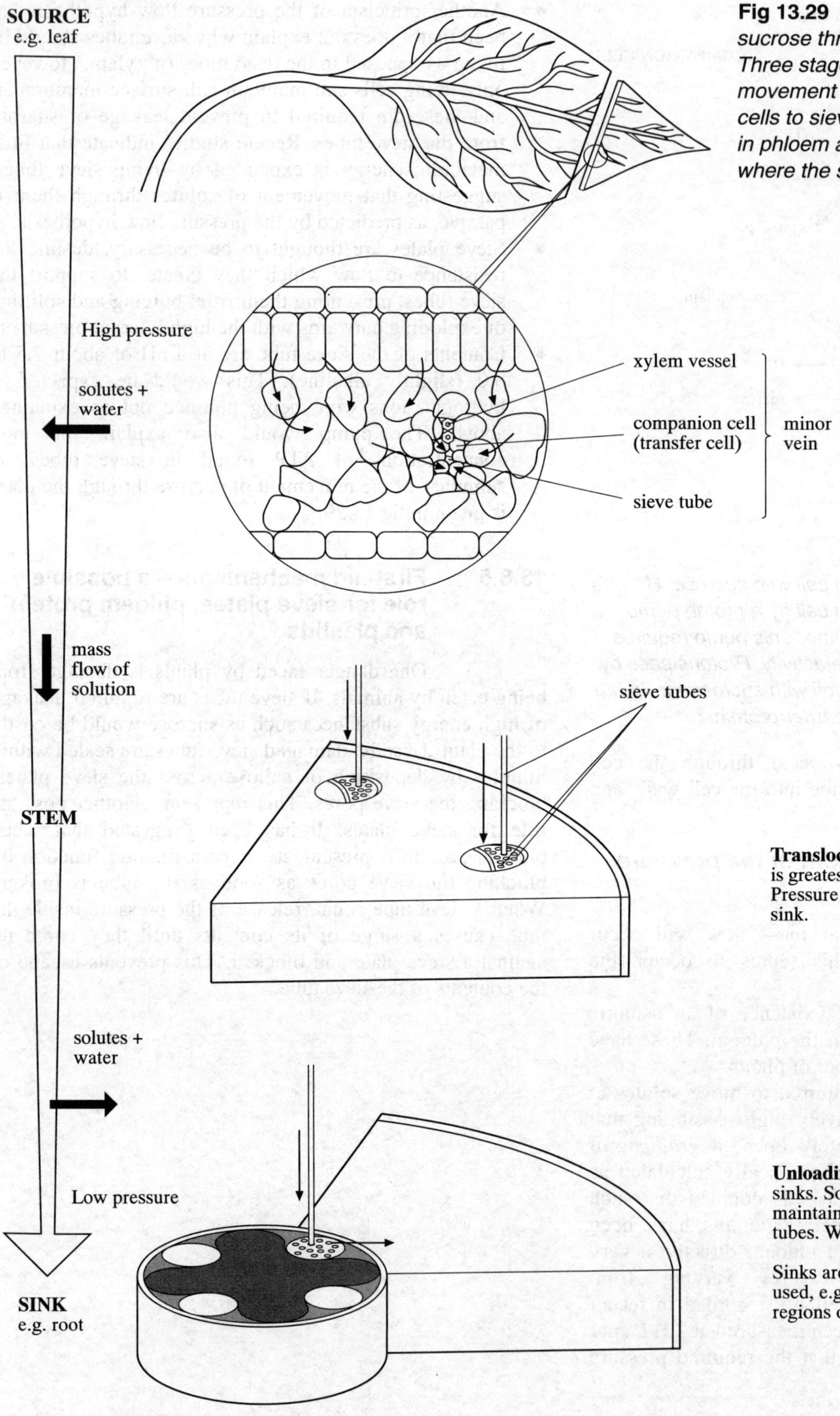

Fig 13.29 *Movement of solutes such as sucrose through the phloem of a plant. Three stages are involved, namely movement of solutes from photosynthetic cells to sieve tubes (loading), translocation in phloem and unloading at a 'sink' (a place where the solutes are used).*

Loading sieve tubes takes place here. Photosynthetic cells make sugars, particularly sucrose, and other organic solutes. Companion cells use energy to collect solutes by active transport. As solute concentration increases in the companion cells, water enters by osmosis. A pressure is created which pushes the solutes through plasmodesmata into the sieve tubes.

Translocation – Pressure inside sieve tubes is greatest at the source and lowest at the sink. Pressure pushes sucrose, etc. from source to sink.

Unloading of the sieve tubes takes place at the sinks. Solute is removed for use, thus maintaining the pressure gradient in the sieve tubes. Water follows.

Sinks are any region where solutes are being used, e.g. roots, fruits, storage organs and regions of growth.

Chapter Fourteen

Transport in animals

The simpler animals such as cnidarians and platyhelminths lack specialised systems for the transport and distribution of materials. The organisms in these groups possess a large surface area to volume ratio, and diffusion of gases over the whole body surface is sufficient for their needs. Internally the distance that materials have to travel is again small enough for them to move by diffusion or cytoplasmic streaming (sections 5.9.8 and 5.10.2).

As organisms increase in size and complexity so the quantity of materials moving in and out of the body increases. The distance that materials have to travel within the body also increases, so that diffusion becomes inadequate as a means for their distribution. Some other method of conveying materials from one part of the organism to another is therefore necessary. This generally takes the form of a mass flow system, as described at the beginning of chapter 13 (see also table 13.1). There are two circulatory systems which rely on mass flow in animals, namely the blood vascular system and the lymphatic system. A vascular system is one which contains fluid-filled vessels involved in transport.

14.1 General characteristics of a blood vascular system

The purpose of a blood vascular system (or blood system for short) is to provide rapid mass flow of materials from one part of the body to another over distances where diffusion would be too slow. On reaching their destination the materials must be able to pass through the walls of the circulatory system into the organs or tissues. Likewise, materials produced by these structures must also be able to enter the circulatory system. In other words, exchange systems are linked to mass flow systems.

Every blood system possesses three distinct characteristics:

- a circulatory fluid, the blood;
- a contractile, pumping device to propel the fluid around the body – this may either be a modified blood vessel or a heart;
- tubes through which the fluid can circulate, the blood vessels.

Two distinct types of blood system are found in animals. They are the open and closed blood systems.

The open blood system (most arthropods, some cephalopod molluscs). Blood is pumped by the heart into an aorta which branches into a number of arteries. These open into a series of blood spaces collectively called the **haemocoel**. In other words, blood does not stay in the blood vessels, hence the term 'open'. Blood under low pressure moves slowly between the tissues, gradually percolating back into the heart through open-ended veins. Distribution of blood to the tissues is poorly controlled.

The closed blood system (echinoderms, most cephalopod molluscs, annelids, vertebrates including humans).

- Blood stays in the blood vessels. It does not come into direct contact with the body tissues.
- Blood is pumped by the heart rapidly around the body under high pressure and back to the heart.
- Distribution of blood to different tissues can be adjusted, depending on demand.
- The only entry and exit to the system is through the walls of the blood vessels.

Blood vessels are named according to their structure and function. Vessels conveying blood away from the heart are called **arteries**. These branch into smaller arteries called **arterioles**. The arterioles divide many times into microscopic **capillaries** which are located between the cells of nearly all the body tissues. It is here that exchange of materials between blood and tissues takes place.

Within the organ or tissue the capillaries reunite to form **venules** which begin the process of returning blood to the heart. The venules join to form **veins**. Veins carry blood back to the heart. The structure of each type of blood vessel is discussed in detail later in section 14.5.

14.2 The development of blood systems in animals

14.2.1 Annelids

Annelids are coelomate animals. The presence of a coelom separates the body wall from the internal organs and gives the advantage of independence of movement of internal structures such as the gut. However, this means there is a need for some form of transport system in the body. A blood system has evolved which connects gut and body wall. The earthworm, for example, has a well-developed blood system in which blood circulates around the body through a system of blood vessels.

Two main blood vessels run the length of the body, one dorsal and one ventral. They are connected by blood vessels in each segment. Near the front of the animal, five pairs of these connecting vessels are contractile and act as pumps. The main blood vessels can also pump blood.

The blood contains haemoglobin dissolved in the plasma rather than being carried in red blood cells. Haemoglobin transports oxygen around the body.

14.2.2 Arthropods

Arthropods have an open blood system (see above). The coelom is drastically reduced and its place taken by the haemocoel. This is a network of blood-filled spaces called **sinuses** in which the internal organs are suspended. Gaseous exchange in most arthropods is achieved by the tracheal system (section 2.8.6), and the blood vascular system is not used for transporting respiratory gases. Arthropod blood is colourless and contains no haemoglobin.

14.2.3 Vertebrates

The blood systems of all vertebrates possess a muscular heart, lying in a ventral position near the front of the animal. The heart is responsible for pumping blood rapidly to all parts of the body. Arteries carry blood away from the heart, and veins carry blood from the body back to the heart. Oxygen is carried by haemoglobin in red blood cells. The human blood system will be studied in detail as an example of both a vertebrate and a mammal. The lymphatic system will also be examined briefly where its functions overlap with those of the blood system.

14.3 Composition of blood

The average adult has about 5 dm^3 of blood. Technically speaking blood is a liquid tissue. It is made up of several types of cell which are found bathed in a fluid matrix called **plasma**. The different types of blood cell can be seen in a blood smear (fig 14.1*a*). The cells make up about 45% by volume of the blood. The other 55% is plasma. If blood is centrifuged, the cells (and platelets, which are really cell fragments) form a red pellet at the bottom of the tube, with the straw-coloured plasma above.

14.1 Why is the pellet red?

Plasma from which the clotting protein fibrinogen has been removed is called **serum**. The pH of the blood is kept between 7.35 and 7.45.

(*a*)

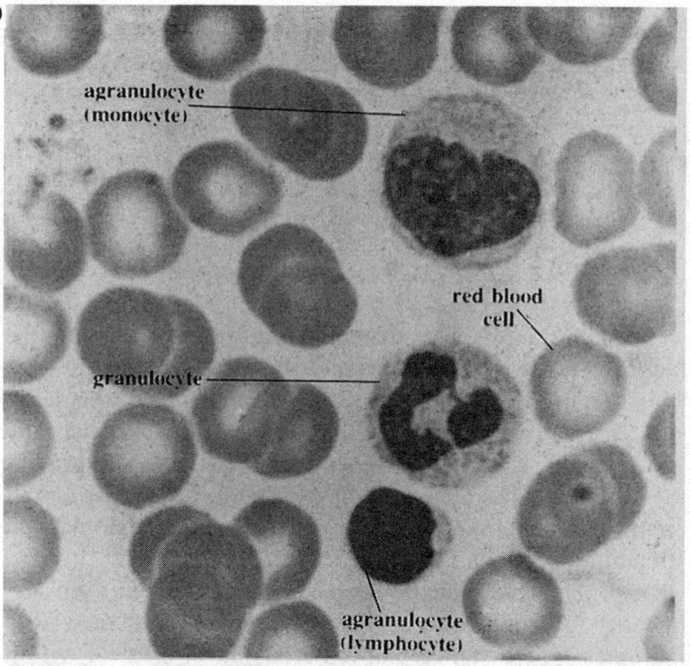

(*b*)

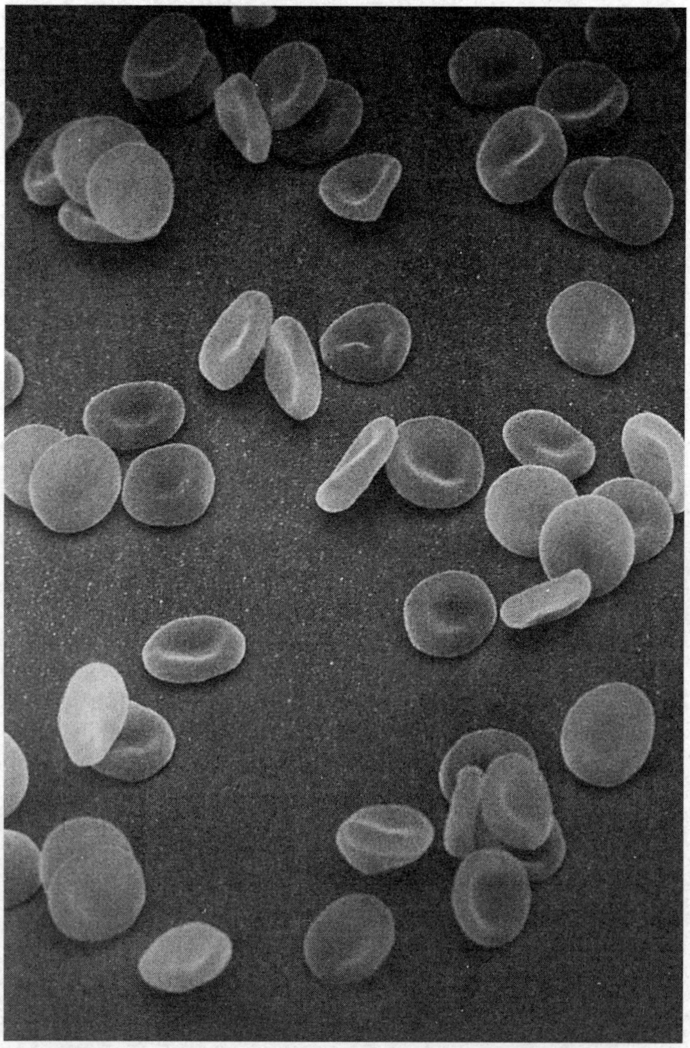

Fig 14.1 *(a) A blood smear showing red blood cells and three types of white cell. (b) A scanning electron micrograph of red blood cells of a mammal.*

In this section we shall look at the structure of blood and briefly consider its functions. In section 14.8 its functions will be examined in more detail.

14.3.1 Plasma

Plasma is a pale straw-coloured liquid. It consists of 90% water and 10% of a variety of substances in solution and suspension. The major components of blood plasma together with a summary of their functions are shown in table 14.1. The most abundant solute is sodium. The functions of plasma are dealt with in more detail in section 14.8.

14.3.2 Blood cells

Red blood cells (erythrocytes)

In humans red blood cells are small cells that lack nuclei when mature. They appear as circular, biconcave discs (table 14.2 and fig 14.1*b*). Their average diameter is 7–8 μm (an average animal cell is about 20 μm in diameter) and they are about 2.2 μm thick. Their particular shape results in a larger surface area to volume ratio than that of a sphere and therefore increases the area which can be used for gaseous exchange. Each cell is very thin, thus allowing efficient diffusion of gases across its surface. The shape makes it very flexible and this property allows it to squeeze through capillaries whose internal diameters are smaller than its own by bending into shapes like umbrella-tops. There are approximately five million red blood cells per mm^3 of blood (one drop of blood has a volume of about 50 mm^3). However this figure varies according to the age, sex and state of health of each individual. They make up about half the volume of blood, giving it an enormous oxygen-carrying capacity (about 20 cm^3 of oxygen per 100 cm^3 of blood).

Red blood cells are packed with **haemoglobin**, the oxygen-carrying protein pigment which gives blood its red colour. The lack of a nucleus makes more room for haemoglobin. (Red blood cells also lack mitochondria which makes more room and also means they have to respire anaerobically. Therefore they do not use up any of the oxygen they carry.) Haemoglobin combines reversibly with oxygen to form oxyhaemoglobin in areas of high oxygen concentration, and releases the oxygen in regions of low oxygen concentration. Red blood cells also contain the enzyme carbonic anhydrase which plays a role in carbon dioxide transport (section 14.8.4).

In the adult, each red blood cell has a relatively short life span of about three months (due to the lack of a nucleus to control repair processes) after which time it is destroyed in the spleen or liver. The protein portion of the red blood cell is broken down into amino acids. The iron of the haem

Table 14.1 Components of blood plasma and their functions.

Component	*Function*
Components maintained at a constant concentration	
Water	Major constituent of blood and lymph. Provides cells of the body with water. Transports many dissolved materials round the body. Regulation of water content helps to regulate blood pressure and blood volume.
Plasma proteins (7–9% of plasma)	
Serum albumins	Very abundant. Bind to and transport calcium. Produced by liver. Contribute to solute potential of the blood.
Serum globulins:	
α-globulins	Produced by liver. Bind and transport the hormone thyroxine, lipids and fat-soluble vitamins (A, D, E and K).
β-globulins	Produced by liver. Bind and transport iron, cholesterol and the fat-soluble vitamins (A, D, E and K).
γ-globulins	Antibodies. Produced by lymphocytes. Important in immune response.
Prothrombin	A protein involved in blood clotting.
Fibrinogen	Produced by liver. Takes part in blood clotting.
Enzymes	Take part in metabolic activities.
Mineral ions	
These include: Na^+, K^+, Ca^{2+}, Mg^{2+}, Cl^-, HCO_3^-, $H_2PO_4^-$, HPO_4^{2-}, PO_4^{3-}, SO_4^{2-}	All help collectively to regulate solute potential and pH levels in the blood. They also have a variety of other functions, e.g. Ca^{2+} may act as a clotting factor.
Components that occur in varying concentrations	
Products of digestion e.g. sugars, fatty acids, glycerol, amino acids. Vitamins. Excretory products e.g. urea. Hormones e.g. insulin, sex hormones, growth hormone.	All are being constantly transported to and from cells within the body.

Table 14.2 Types of cell found in blood (diagrams not drawn to scale).

Component	*Origin*	*Number of cells per mm^3*	*Function*	*Structure*
Red blood cells	bone marrow	5 000 000	transport of oxygen and some carbon dioxide	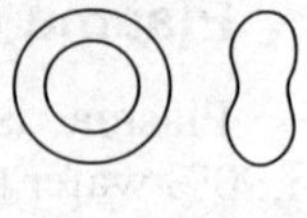
White blood cells	bone marrow			
(*a*) *Granulocytes* (72% of total white blood cell count)				
neutrophils (70%)		4 900	engulf bacteria	
eosinophils (1.5%)	bone marrow	105	allergic responses and anti-histamine properties	
basophils (0.5%)		35	produce histamine and heparin	
(*b*) *Agranulocytes* (28%)				
monocytes (4%)	bone marrow	280	engulf bacteria	
lymphocytes (24%)	bone marrow lymphoid tissue spleen	1 680	production of antibodies	
Platelets	bone marrow	250 000	start blood-clotting mechanism	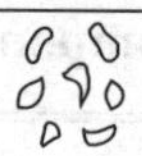

group is extracted and stored in the liver as ferritin (an iron-containing protein). It may be re-used later in the production of further red blood cells or as a component of cytochrome. The remainder of the haem molecule is broken down into two bile pigments, bilirubin (red) and biliverdin (green). These are later excreted by way of the bile into the gut.

Between 2–10 million red blood cells are destroyed and replaced each second in the human body. Each one contains about 250 million molecules of haemoglobin, a large protein, so the rate of protein synthesis is impressive. The rate of destruction and replacement is partly determined by the amount of oxygen in the atmosphere. If the quantity of oxygen being carried in the blood is low, the marrow is stimulated to produce more red blood cells than the liver destroys. This is one of the ways in which we acclimatise to lower oxygen levels at high altitudes, and is made use of by athletes in high altitude training. When the oxygen content of the blood is high, the situation is reversed.

14.2 List the main categories of substances transported by the blood.

White blood cells (leucocytes)

These cells are larger than red blood cells, and are present in much smaller numbers, there being about 7000 per mm^3 of blood. All have nuclei. They play an important role in the body's defence mechanisms against disease. Although they have nuclei, their life span in the bloodstream is normally only a few days. All are capable of a crawling movement known as **amoeboid movement**. This allows them to squeeze through pores in capillary walls to reach the tissues and sites of infection.

White blood cells can only be seen easily with a light microscope if they are stained. Staining shows that there are two main groups of white blood cell, the granulocytes and the agranulocytes, according to whether they show granules in their cytoplasm.

Granulocytes (polymorphonuclear leucocytes) (72%). These are made in the bone marrow but by cells different from those that make red blood cells. Granulocytes can be further subdivided into neutrophils, eosinophils and basophils.

- **Neutrophils** (phagocytes): These make up about 70% of the total number of white cells. They commonly squeeze between the cells of the capillary walls and wander through the intercellular spaces. From here they move to infected areas of the body. They are actively phagocytic and engulf and digest disease-causing bacteria (section 14.8.5).

- **Eosinophils**: These possess cytoplasmic granules which stain red when the dye eosin is applied to them. Generally they represent only 1.5% of the total number of white cells, but their population increases in people with allergic conditions such as asthma or hayfever. They possess anti-histamine properties. The number of eosinophils present in the bloodstream is under the control of hormones produced by the adrenal cortex in response to stress of various kinds.
- **Basophils**: These represent 0.5% of the white blood cell population. The granules in these cells stain blue with basic dyes such as methylene blue. The cells produce heparin, an anti-clotting protein and histamine, a chemical found in damaged tissues which is involved in inflammation. Inflammation stimulates repair of damaged tissues. Overproduction of histamine occurs in some allergies, such as hayfever.

Agranulocytes (mononuclear leucocytes) (28%). These cells possess non-granular cytoplasm and have either an oval or bean-shaped nucleus (fig 14.1*a*). Two main types exist.

- **Monocytes** (4%). These are formed in the bone marrow and have a bean-shaped nucleus. They spend only about 30–40 hours in the blood and then enter the tissues where they become macrophages. **Macrophages** are phagocytic and engulf bacteria and other large particles. They also play a role in the immune system by processing certain antigens, as explained in section 14.9. Together with neutrophils they form a system of phagocytes throughout the body which acts as a first line of defence against infection.
- **Lymphocytes** (24%). These are produced in the thymus gland and lymphoid tissues from cells which originate in the bone marrow. The cells are rounded and possess only a small quantity of cytoplasm. Amoeboid movement is limited. They are also found in lymph and the body tissues. Two types occur, T cells and B cells (section 14.9). They are involved in immune reactions (such as antibody production, graft rejection and in killing tumour cells). The life span of these particular cells can vary from a matter of days to up to ten years or more.

14.3.3 Platelets

Platelets are irregularly shaped, membrane-bound cell fragments, usually lacking nuclei, and are formed from special bone marrow cells. They are about one-quarter the size of red blood cells. They are responsible for starting the process of blood clotting. There are about 250 000 platelets per mm^3 of blood. They survive about 5–9 days before being destroyed by the spleen and liver.

14.4 The circulation

A general plan of the human circulation is shown in fig 14.2*a*. It shows the following features.

- It is a **double circulation**. This means that the blood passes through the heart twice for each circuit of the body. The advantage of this is that the blood can be sent to the lungs to pick up oxygen and then be returned to the heart to be pumped again before travelling round the body. Blood loses pressure going through the capillaries of the lungs, so the pressure is restored and boosted before the blood is circulated to the rest of the body. In vertebrates which have a single circulation this is not possible. In fish, for example, blood from the heart first goes to the gills to collect oxygen, but then continues round the whole body before returning to the heart. This is a **single circulation** (fig 14.2*b*). Only birds and mammals have true double circulations. It is probably no coincidence that only birds and mammals are warm blooded. Warm-bloodedness requires a high metabolic rate and this is only possible if a good supply of oxygen is available for high levels of aerobic respiration. Animals with a high metabolic rate can maintain higher levels of activity than other animals.

  Double circulation is made possible by the heart being divided into two. One half pumps deoxygenated blood to the lungs and the other half pumps oxygenated blood to the rest of the body. In effect we have two hearts which are stuck together and beat at the same time. The beginnings of a double circulation are seen in amphibians, and reptiles have an almost completely divided heart.
- The organs are arranged in parallel rather than in series (fig 14.2*a*). If they were arranged in series, blood would pass from organ A to B to C and so on, losing pressure, oxygen and nutrients at each stage. This would be extremely inefficient. Also, any damage done to a blood vessel linking two organs would interrupt the whole circulation.
- A portal vessel links the gut to the liver (fig 14.2*a*). A **portal vessel** is a blood vessel which links two organs, neither of which is the heart. (A similar system links the hypothalamus with the pituitary gland.) This means that the gut and liver are linked in series, not in parallel, a disadvantage as explained above. However, there is an overriding advantage: blood from the gut is very variable in composition, depending on whether or not it contains digested food products (or other substances such as alcohol). One of the functions of the liver is to monitor blood passing through it and help to maintain a relatively constant composition. For example, the liver can remove excess glucose from blood and store it as glycogen.

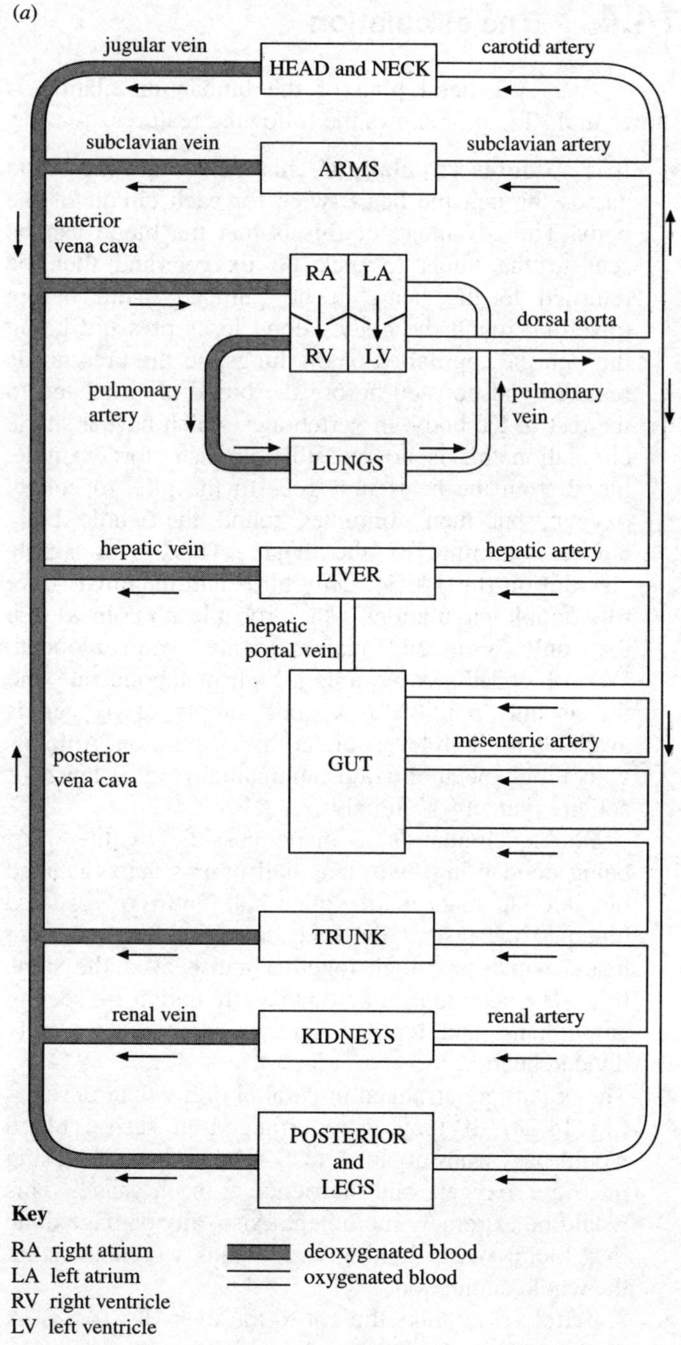

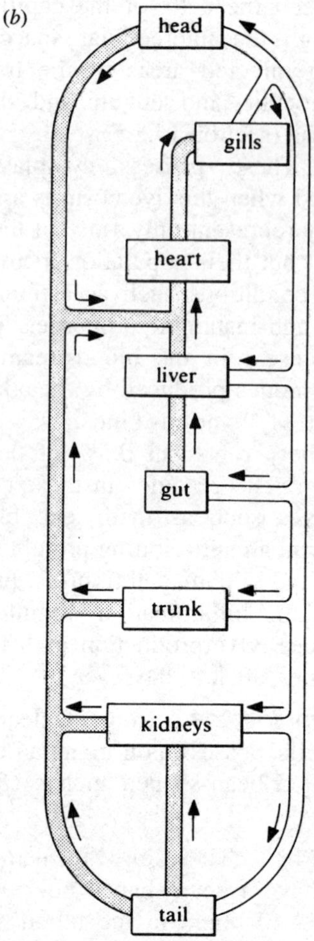

Fig 14.2 *(a) Double circulation of blood in the human body. The main blood vessels are shown. (b) Single circulation of blood in a fish.*

14.5 Blood vessels

14.5.1 General structure

As blood circulates round the body it passes through a series of arteries, capillaries and veins. Basically each artery and vein consists of three layers:

- the **endothelium**, an inner lining of squamous epithelium;
- the **tunica media** ('middle coat'), a middle layer of smooth involuntary muscle and elastic fibres;
- the **tunica externa** ('external coat'), an external layer consisting mainly of inelastic white fibres (collagen fibres).

The structure of these vessels is shown and compared in table 14.3 and fig 14.3. Some of the tissues are also discussed in chapter 6 (see figs 6.14, 6.22 and 6.23).

14.5.2 Arteries

The large arteries near the heart (that is the aorta, subclavians and carotids) must be able to withstand the high pressure of blood leaving the left ventricle of the heart. The walls of these arteries are thick and the middle layer is mainly composed of elastic fibres. This enables them to dilate (increase in diameter, or stretch) but not rupture when the heart contracts and forces blood into them at high pressure. Between heartbeats the arteries undergo elastic recoil and contract, tending to smooth out the flow of blood along their length (fig 14.4).

Table 14.3 Comparison of the structure and function of an artery, capillary and vein (diagrams are not drawn to scale).

Artery	*Vein*	*Capillary*
Transports blood away from the heart.	Transports blood towards the heart.	Link arteries to veins. Site of exchange of materials between blood and tissues.
Tunica media thick and composed of elastic and smooth muscle tissue.	Tunica media relatively thin and only slightly muscular. Few elastic fibres.	No tunica media. Only tissue present is squamous endothelium (section 6.3.1). No elastic fibres.
No semilunar valves (except where leave heart).	Semilunar valves at intervals along the length to prevent backflow of blood.	No semilunar valves.
Pressure of blood is high and has a pulse.	Pressure of blood low and no pulse detectable.	Pressure of blood falling and no pulse detectable.
Blood flow rapid.	Blood flow slow.	Blood flow slowing.
Low blood volume.	Much higher blood volume than capillaries or arteries.	High blood volume.
Blood oxygenated except in pulmonary artery.	Blood deoxygenated except in pulmonary vein.	Mixed oxygenated and deoxygenated blood.

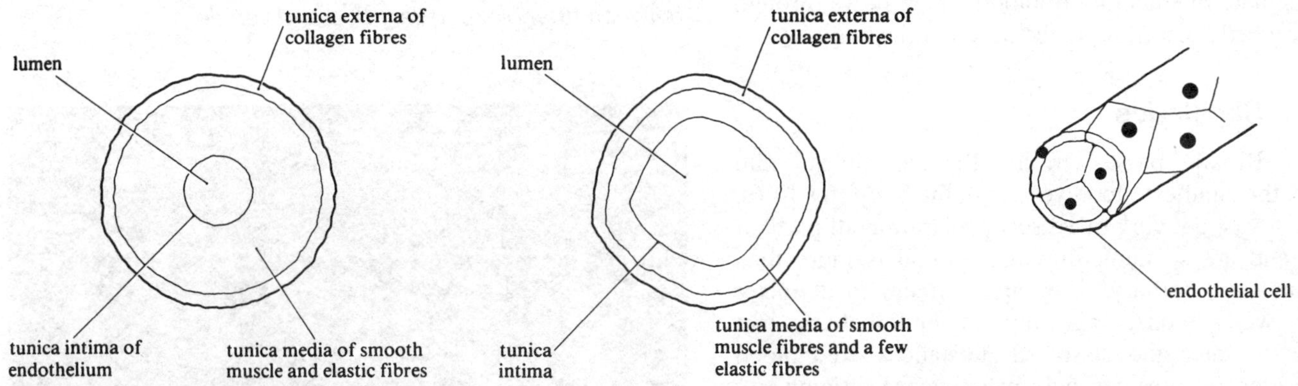

(*a*)

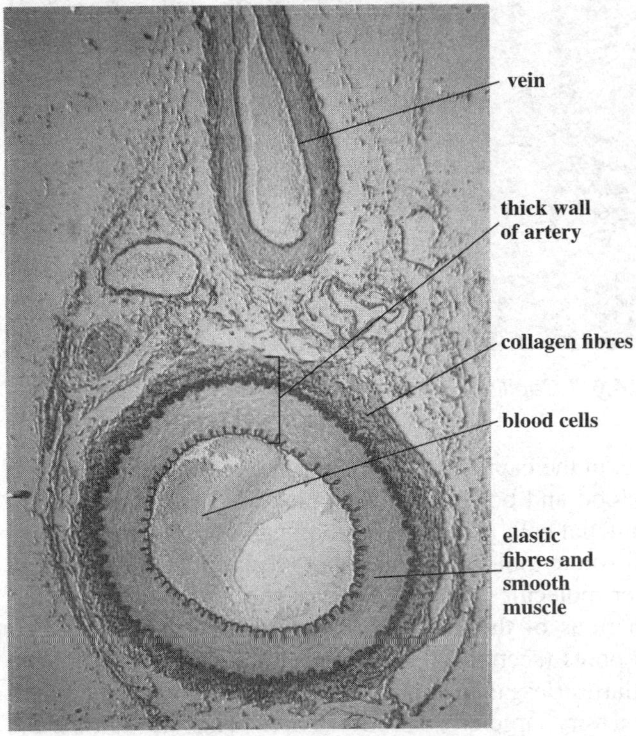

(*b*)

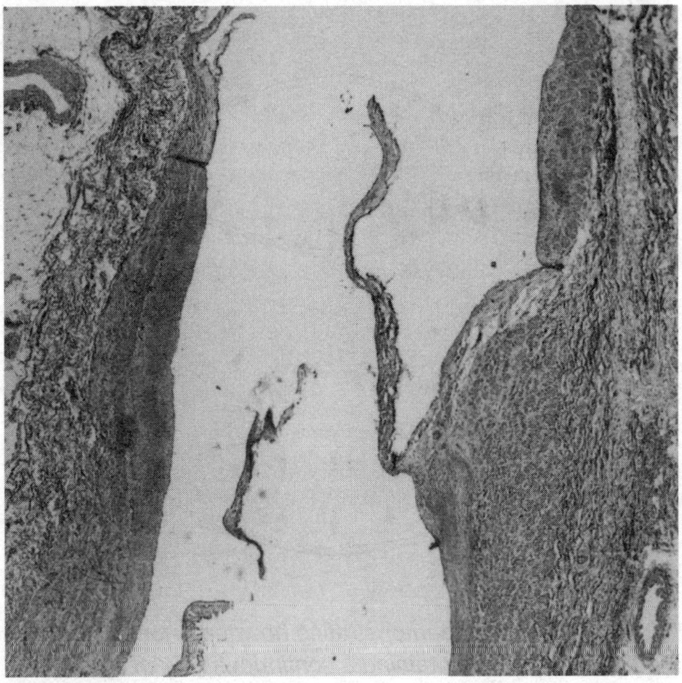

Fig 14.3 *(a) TS of an artery and a vein. (b) LS of a vein showing a valve.*

The arteries further away from the heart have a similar structure but possess relatively more smooth muscle fibres in the middle layer. They are supplied with neurones (nerve cells) from the sympathetic nervous system. Stimulation from this system regulates the diameter of these arteries and this is important in controlling the flow of blood to different parts of the body.

14.5.3 Arterioles

Blood passes from the arteries into small vessels called arterioles. These consist only of endothelium wrapped round by a few muscle fibres at intervals. Many arterioles possess 'sphincters' (fig 14.5) at the point where they enter the capillaries. These are circular muscle fibres which, when they contract, prevent blood from flowing into the capillary network. Also present in certain regions of the body are cross-connections, or 'shunt vessels', which act as short cuts between arterioles and venules (small veins) and serve to regulate the quantity of blood which flows through the capillary beds according to the needs of the body.

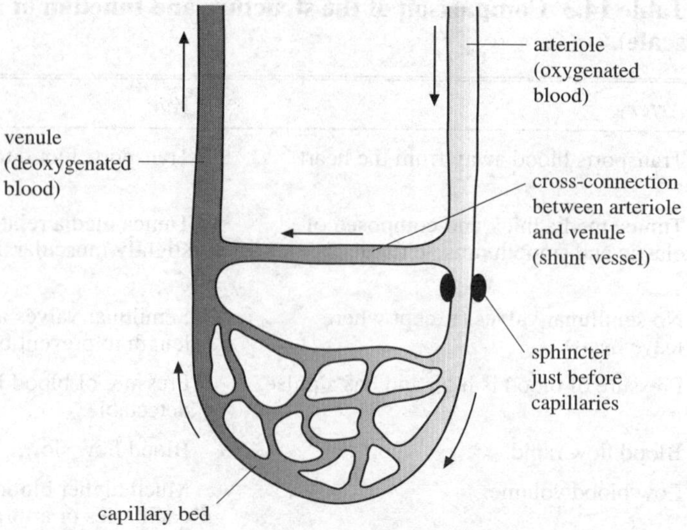

Fig 14.5 *The possible routes that blood may take between arteriole, capillary bed and venule.*

14.5.4 Capillaries

Blood passes from the arterioles into capillaries, the smallest blood vessels in the body (fig 14.6). They form a vast network of vessels penetrating all parts of the body, and are so numerous that no cell is more than 25 μm from any capillary. They are 4–10 μm in diameter and their walls, consisting only of endothelium, are permeable to water and dissolved substances. Red blood cells (diameter 7–8 μm) can only just squeeze through (fig 14.7).

Fig 14.6 *Capillary bed showing arterioles and capillaries.*

It is in the capillaries that exchange of materials between the blood and body cells takes place. Where neighbouring endothelial cells join, a very small gap occurs which allows small solute molecules and ions to leak out of the capillary. Larger molecules such as proteins stay in the blood. In the glomerulus of the kidney the walls of the capillaries also have pores (section 20.5). Blood flows slowly through the capillaries (less than 1 mm per second) because in total they have a very large cross-sectional area. This allows time for exchange.

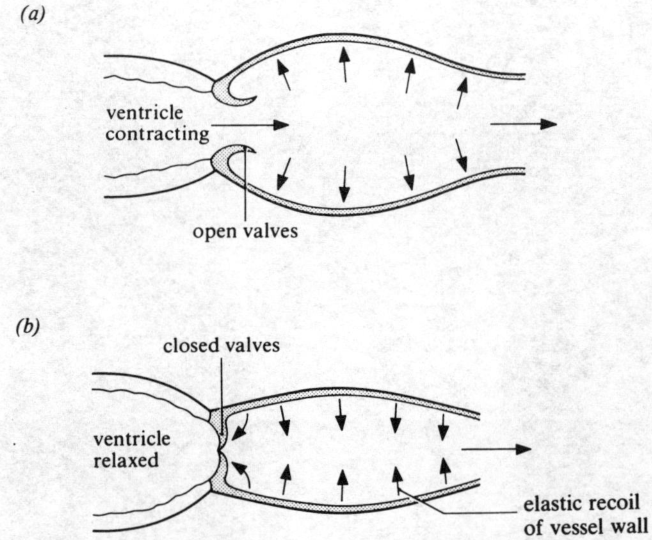

Fig 14.4 *Diagram demonstrating how the arteries near the heart help in maintaining a continuous flow of blood in spite of a discontinuous flow received from the ventricles. (From Clegg & Clegg (2nd ed. 1963)* Biology of the mammal, *Heinemann Medical Books.)*

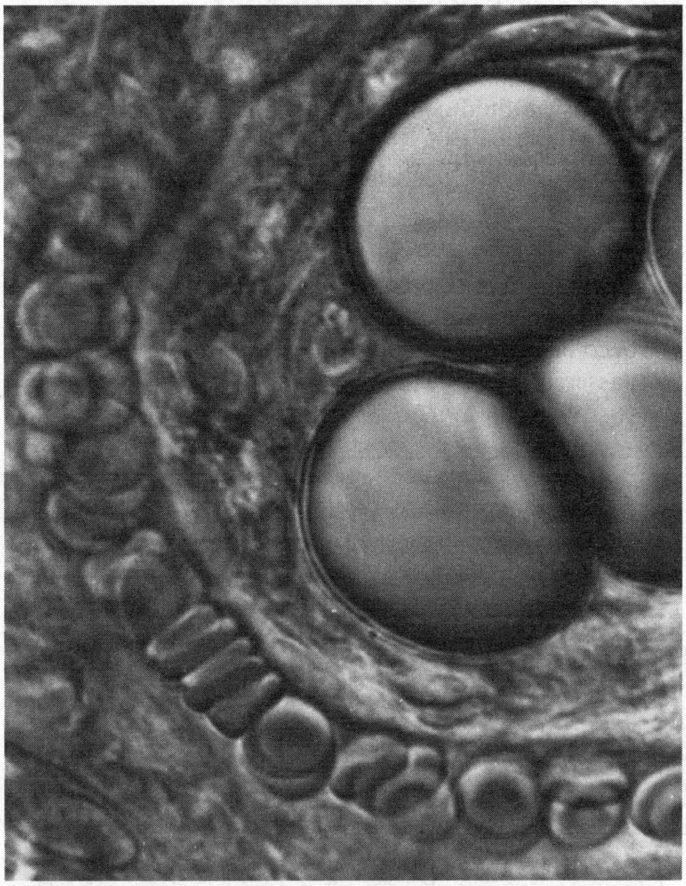

Fig 14.7 *Red blood cells squeezing through a capillary.*

14.5.5 Venules

Blood from the capillary beds drains into venules, whose walls consist of a thin layer of collagen fibres. These are tough and inelastic. They pass the blood into veins which eventually carry it back to the heart.

14.5.6 Veins

A vein possesses less muscle and elastic fibres in its middle layer than an artery and the diameter of its lumen is greater. Semi-lunar valves (fig 14.8) are present,

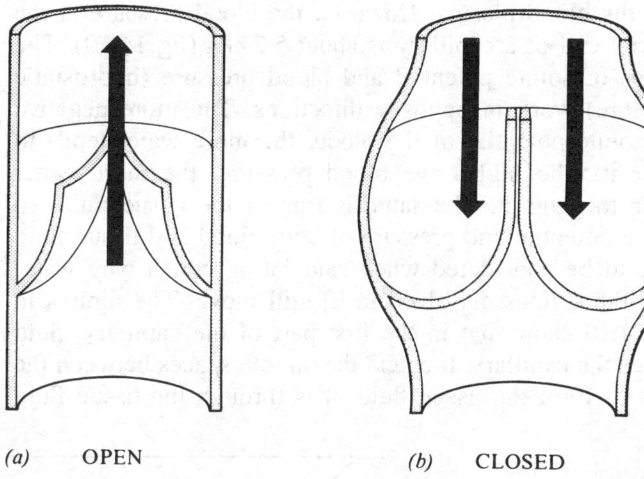

Fig 14.8 *Action of semilunar valve in a vein. (a) Upward pressure of the blood forces the valve open and blood flows towards the heart. (b) Backflow of blood closes the valve, which resembles a pocket or cup when full of blood.*

which are formed from folds of the inner walls of the vein. They function to prevent backflow of blood, thereby maintaining a one-way flow of blood. A number of veins are located between the large muscles of the body (as in the arms and legs). When these muscles contract they exert pressure on the veins and squeeze them flat (fig 14.9). This helps the flow of blood to the heart.

14.6 Formation of tissue fluid

Tissue fluid is formed when blood passes through the capillaries. As mentioned, the capillary walls are permeable to small solute molecules and ions, but not to the red blood cells, platelets and plasma proteins. **Tissue fluid** is therefore a watery liquid which resembles plasma minus its proteins.

The solute potential exerted by solutes in the plasma is about –3.5 kPa and this is far more negative than the solute potential in the tissue fluid. Under these conditions one would normally expect water to flow from the tissue fluid

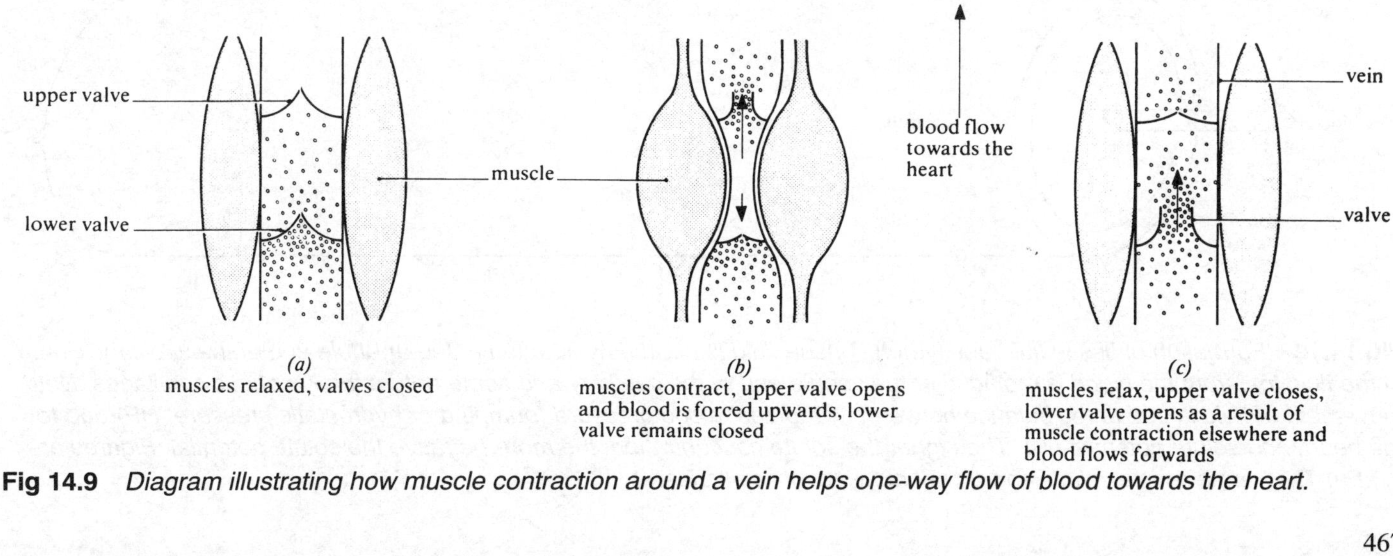

Fig 14.9 *Diagram illustrating how muscle contraction around a vein helps one-way flow of blood towards the heart.*

into the blood plasma. However, the blood pressure at the arterial end of a capillary is about 5.2 kPa (fig 14.10). The forces of solute potential and blood pressure (hydrostatic pressure) work in opposite directions. The more negative the solute potential of the blood, the more water tends to *enter* it. The higher the blood pressure, the more water tends to *leave* it. The same is true of the tissue fluid, so solute potential and pressure of both blood and tissue fluid have to be considered when calculating which way water (and the solutes dissolved in it) will move. The figures in fig 14.10 show that in the first part of the capillary, fluid leaves the capillary. It enters the minute spaces between the cells to form the tissue fluid. It is through the tissue fluid that exchange of materials between blood and tissues occurs.

The blood cannot afford to lose so much fluid constantly. Also, if tissue fluid continued to build up the tissue would swell, a condition called **oedema**. Therefore it is normally returned to the blood at the same rate that it is produced. This occurs in two ways.

- As tissue fluid is forming, protein molecules in solution stay behind in the blood. The blood therefore becomes more concentrated, in other words its solute potential becomes more negative. It also loses pressure as it passes through the narrow capillary, so by the time it

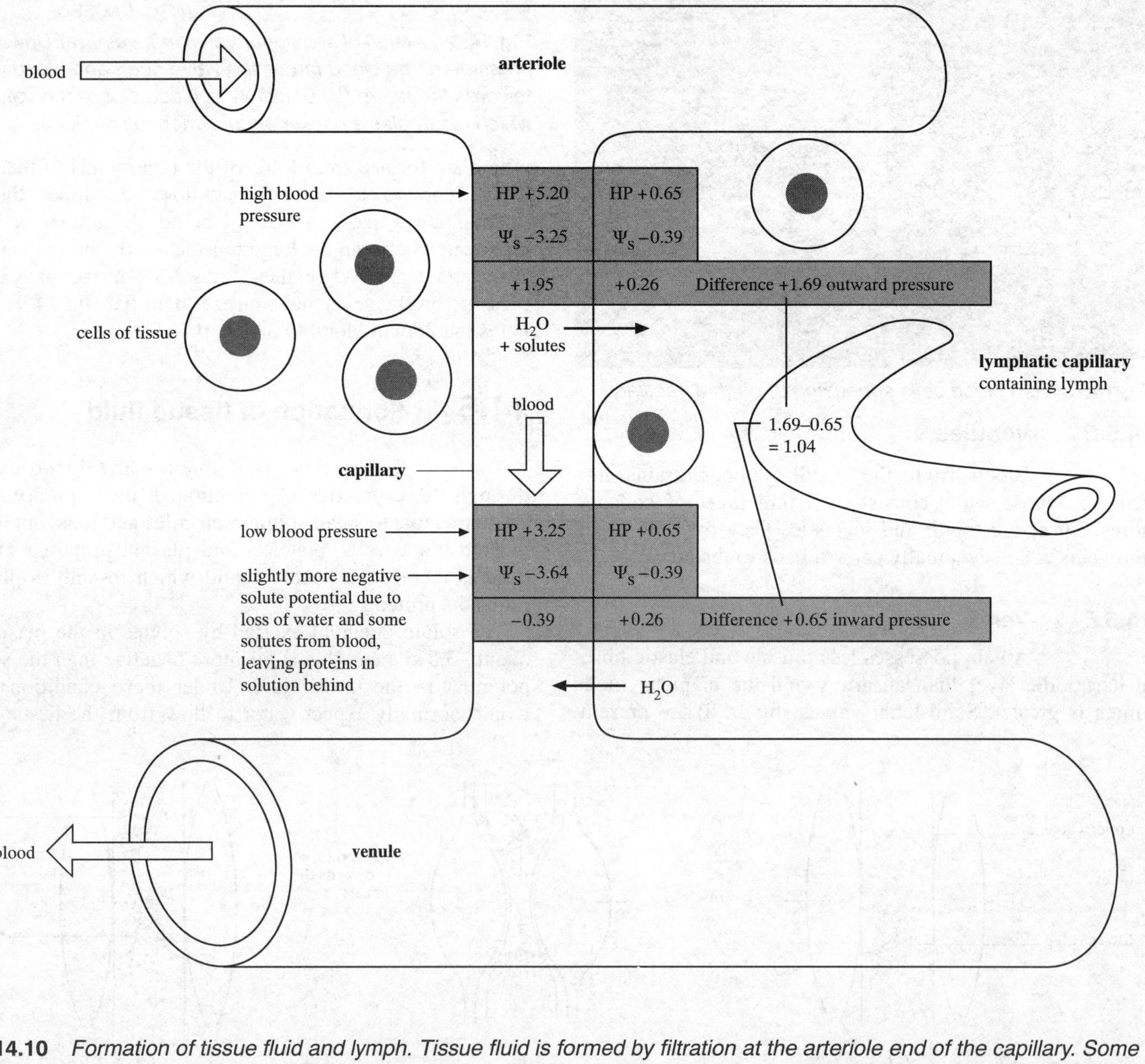

Fig 14.10 *Formation of tissue fluid and lymph. Tissue fluid is formed by filtration at the arteriole end of the capillary. Some of the fluid lost from the blood is replaced at the venule end of the capillary and some enters the lymphatic capillaries. Net movement of fluid is due to the balance between two forces, blood pressure (pumping or hydrostatic pressure, HP) due to the heart, and solute potential (ψ_s). The higher the solute concentration, the more negative the solute potential. Figures are in kPa. They are average values and do not apply to all capillaries.*

reaches the venule end of the capillary, fluid is tending to re-enter the blood capillary, as shown in fig 14.10.

- The rest of the tissue fluid drains into blindly ending lymphatic capillaries, and once inside these the fluid is termed **lymph**. The lymphatic capillaries join to form larger lymphatic vessels. The lymph is moved through the vessels by contraction of the muscles surrounding them, and backflow is prevented by valves present in the major vessels which act in a similar fashion to those found in veins (fig 14.11).

The lymphatic vessels of the body empty the lymph into the blood system at the subclavian veins just after they leave the arms and just before they reach the heart (fig 14.12).

Situated at intervals through the lymphatic system are **lymph glands** or nodes. Lymphocytes, in the course of circulation through the blood and lymph, accumulate in the lymph nodes. They produce antibodies and are an important part of the body's immune system. Phagocytes in the nodes also remove bacteria and foreign particles from the lymph.

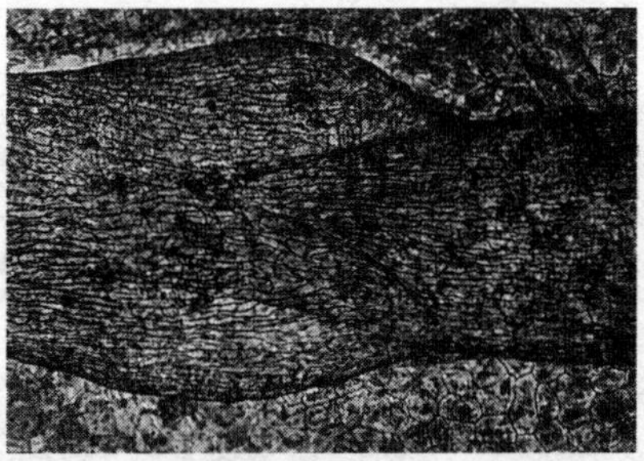

Fig 14.11 *LS through lymph vessel showing an internal valve.*

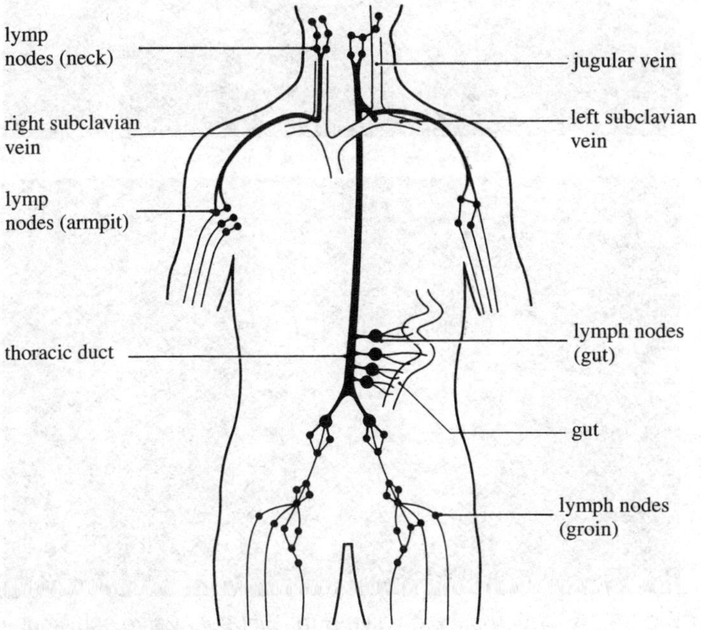

Fig 14.12 *Human lymphatic system. (From E. G. Springthorpe (1973)* An introduction to functional systems in animals, *Longman.)*

14.7 The heart

14.7.1 Structure

The heart is situated between the two lungs and behind the sternum in the thorax. It is surrounded by a tough sac, the **pericardium**, the outer part of which consists of inelastic white fibrous tissue. The inner part is made up of two membranes. The inner membrane is attached to the heart and the outer one is attached to the fibrous tissue. Pericardial fluid is secreted between them and reduces the friction between the heart wall and surrounding tissues when the heart is beating. The inelastic nature of the pericardium as a whole prevents the heart from being overstretched or overfilled with blood.

There are four chambers in the heart, two upper thin-walled **atria** (singular **atrium**) and two lower thick-walled **ventricles** (fig 14.13). The atria receive blood from the veins and pump it to the ventricles which in turn pump it into arteries. The walls of the atria are thin because they only have to pump blood into the ventricles.

The right side of the heart is completely separated from the left. The right side deals with deoxygenated blood and the left side with oxygenated blood. The right atrium receives deoxygenated blood from the general circulation of the body whilst the left atrium receives oxygenated blood from the lungs. The muscular wall of the left ventricle is at least three times as thick as that of the right ventricle. This difference is due to the fact that the right ventricle only has to pump blood to the lungs, which are very near the heart in the thorax, whereas the left ventricle pumps blood all round the body. Therefore the blood entering the aorta from the left ventricle is at a much higher blood pressure (approximately 14.0 kPa) than the blood entering the pulmonary artery (2.1 kPa). The circulation to and from the lungs is called the **pulmonary circulation** and the circulation round the body is called the **systemic circulation.**

> **14.3** What other advantages are there in supplying the pulmonary circulation with blood at a lower pressure than that of the systemic circulation?

As the atria contract they force blood into the ventricles, and rings of muscle which surround the venae cavae and pulmonary veins at their point of entry into the atria contract and close off the veins. This prevents blood returning into the veins. The left atrium is separated from the left ventricle by a **bicuspid** (two-flapped) valve, whilst a **tricuspid** valve separates the right atrium from the right ventricle (fig 14.14*a*). Jointly, these are known as the **atrioventricular valves**. Attached to the ventricle side of the flaps are fibrous cords which in turn attach to conical-shaped papillary muscles which are extensions of the inner

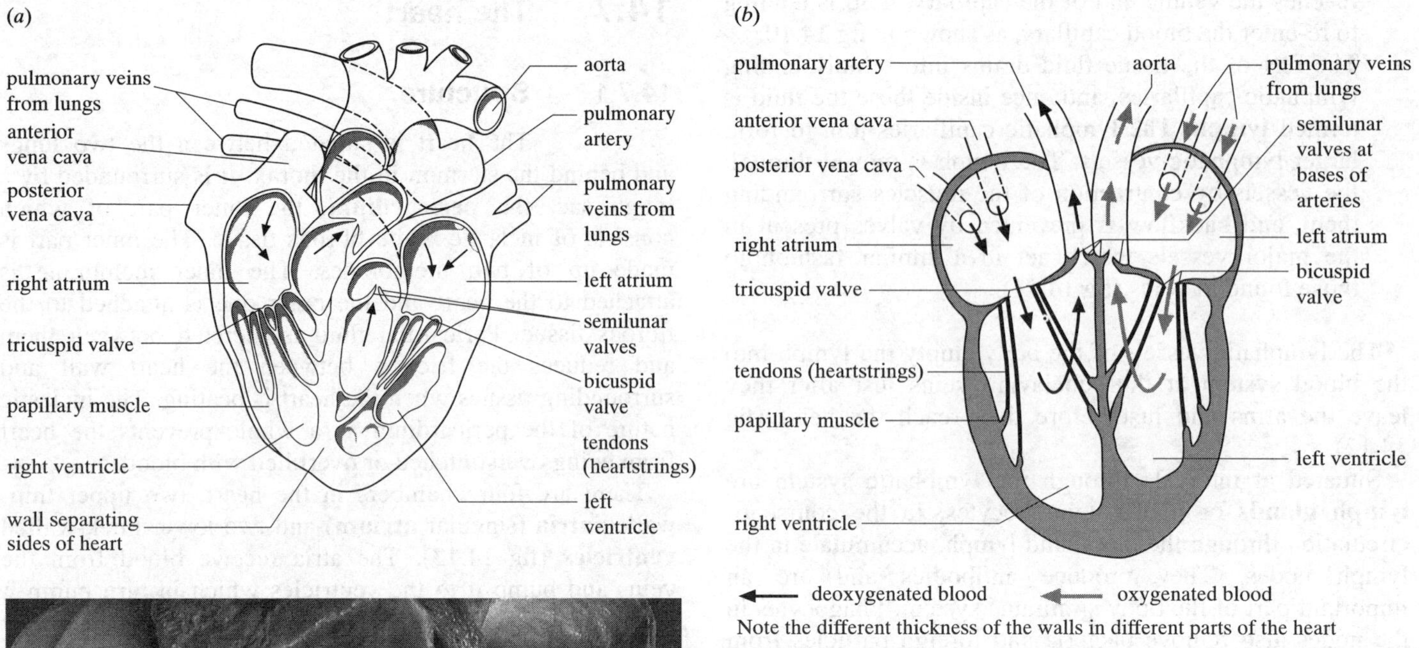

Fig 14.13 *(a) Section through heart. (b) Simplified diagram of the heart.*

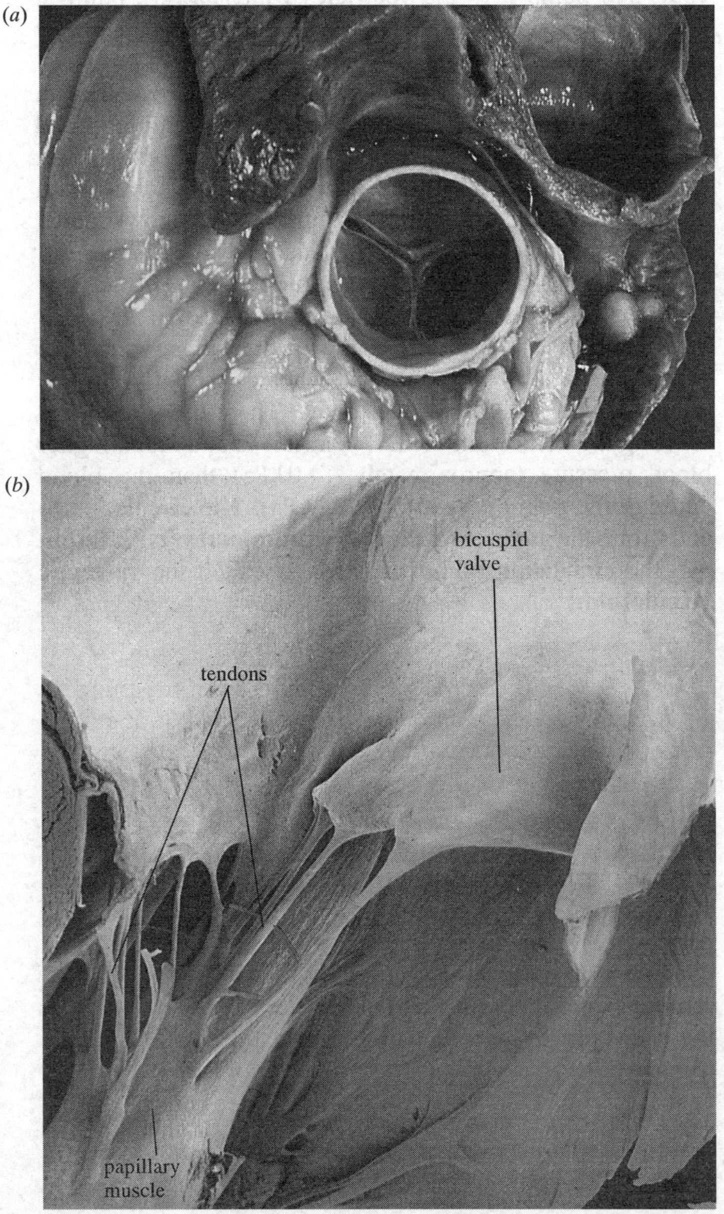

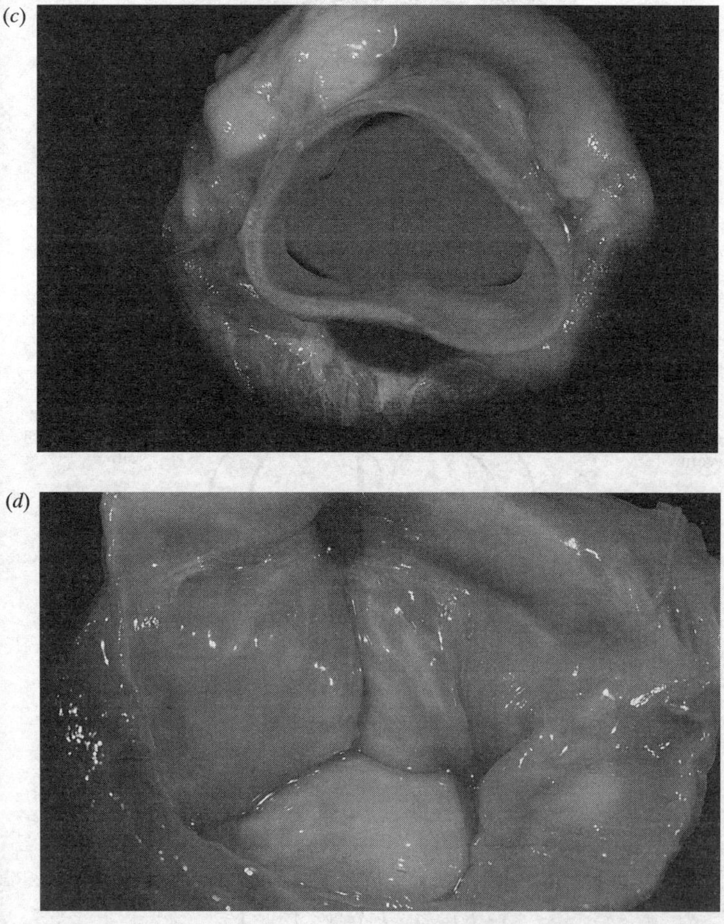

Fig 14.14 *(a) Tricuspid valves. (b) Bicuspid valve showing tendons which join the valve flaps to papillary muscle. (c) Open semilunar valve at the base of pulmonary artery. (d) Closed semilunar valve at the base of the pulmonary artery.*

wall of the ventricles. The atrioventricular valves are pushed open when the atria contract but, when the ventricles contract, the flaps of each valve press tightly closed so preventing return of blood to the atria. At the same time the papillary muscles contract so tightening the fibrous cords. This prevents the valves from being turned inside out. Semilunar valves (pocket valves) (fig 14.14*b*) are found at the points where the pulmonary artery and aorta leave the heart. These prevent blood from getting back into the ventricles.

Just beyond the aortic valve are the openings of the two coronary arteries. These are the only blood vessels which supply oxygenated blood to the walls of the heart.

Structure of cardiac muscle

The walls of the heart are composed of cardiac muscle fibres, connective tissue and tiny blood vessels. Each muscle fibre possesses one or two nuclei and many large mitochondria. Each fibre is made up of many myofibrils. These contain actin and myosin filaments which bring about contraction in the same way as skeletal muscle (section 18.4). They account for the striped appearance of the muscle fibres (striations in figs 14.15 and 14.16). The dark bands known as **intercalated discs** are cell surface membranes separating individual muscle cells. The structure of the membranes is modified to allow ions to diffuse rapidly across them. This allows rapid spread of excitation (action potentials) through the muscle. When one cell becomes excited, the action potential spreads quickly to all the others, so that the whole mass of fibres behaves as one unit. The fibres branch and cross-connect with each other to form a complex net-like arrangement. Cardiac muscle contracts more slowly than skeletal muscle and does not fatigue as easily. No neurones are present in the wall of the heart.

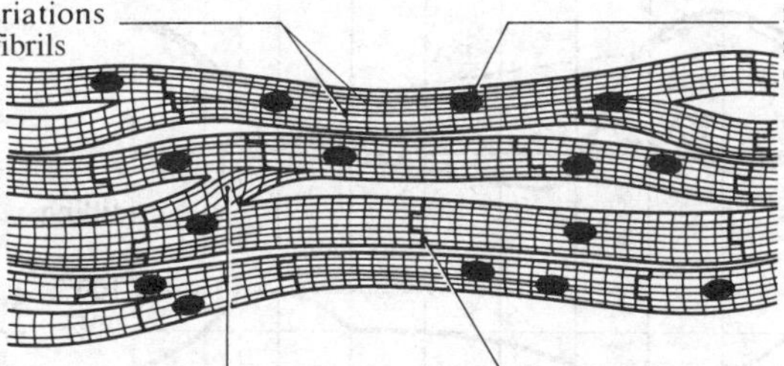

Fig 14.15 *Structure of cardiac muscle.*

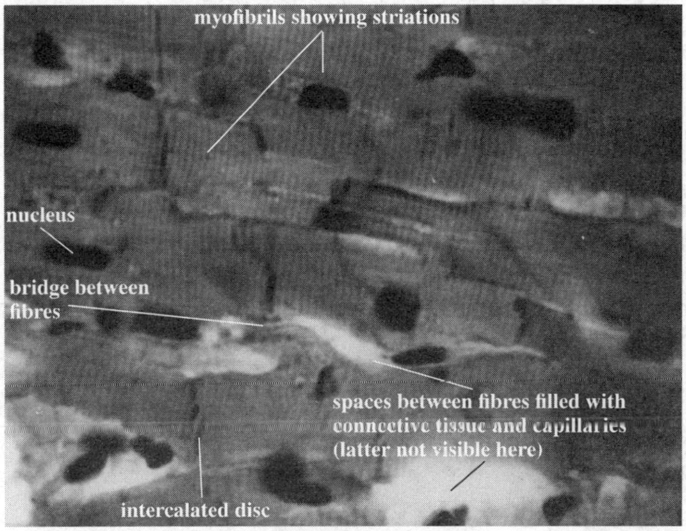

Fig 14.16 *Photograph of cardiac muscle.*

14.7.2 The cardiac cycle

The cardiac cycle refers to the sequence of events which takes place during the completion of one heartbeat. It involves repeated contraction and relaxation of the heart muscle. Contraction is called **systole** and relaxation is called **diastole**. It occurs as follows.

- **Atrial diastole**. During the time when the atria and the ventricles are both relaxed, blood returning to the heart under low pressure in the veins enters the two atria. Oxygenated blood enters the left atrium and deoxygenated blood enters the right atrium. At first the bicuspid and tricuspid valves are closed (fig 14.17*a*) but, as the atria fill with blood, pressure in them rises (fig 14.18). Eventually it becomes greater than that in the relaxed ventricles and the valves are pushed open.
- **Atrial systole**. When atrial diastole ends, the two atria contract simultaneously. This is termed atrial systole and results in blood being pumped into the ventricles (fig 14.17*b*).
- **Ventricular systole**. Almost immediately (about 0.1 to 0.2 seconds later) the ventricles contract. This is called ventricular systole (fig 14.17*c*). When this occurs the pressure in the ventricles rises and closes the atrioventricular valves, preventing blood from returning to the atria. The pressure forces open the semi-lunar valves of the aorta and pulmonary artery and blood enters these vessels. The closing of the atrioventricular valves during ventricular systole produces the first heart sound, described as **'lub'**.
- **Ventricular diastole**. Ventricular systole ends and is followed by ventricular diastole (fig 14.17*d*). The high pressure developed in the aorta and pulmonary artery tends to force some blood back towards the ventricles and this closes the semi-lunar valves of the aorta and pulmonary artery. Hence backflow into the heart is prevented. The closing of the valves causes the second heart sound, **'dub'**. The two heart sounds are therefore:

  ventricular systole = 'lub'
  ventricular diastole = 'dub'

Ventricular systole, and the elastic recoil of the arteries as blood at high pressure is forced through them, causes a pulse. As blood gets further and further away from the heart, the pulse becomes less and less pronounced until, in the capillaries and veins, blood flows evenly (figs 14.19 and

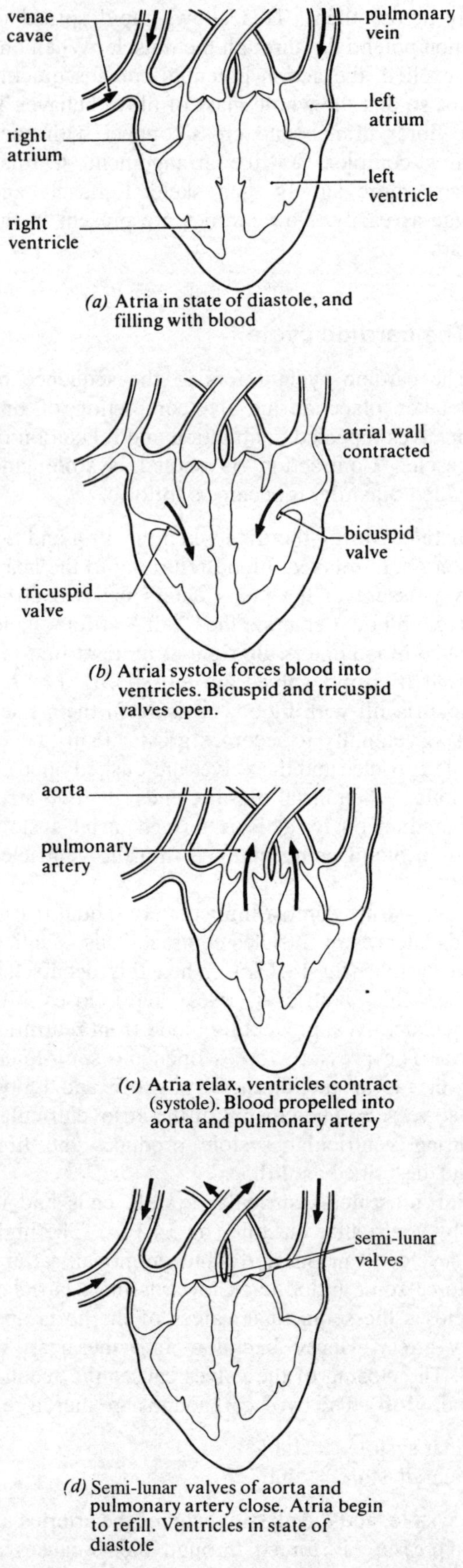

Fig 14.17 *Sequence of heart actions involved in one complete heartbeat – the cardiac cycle.*

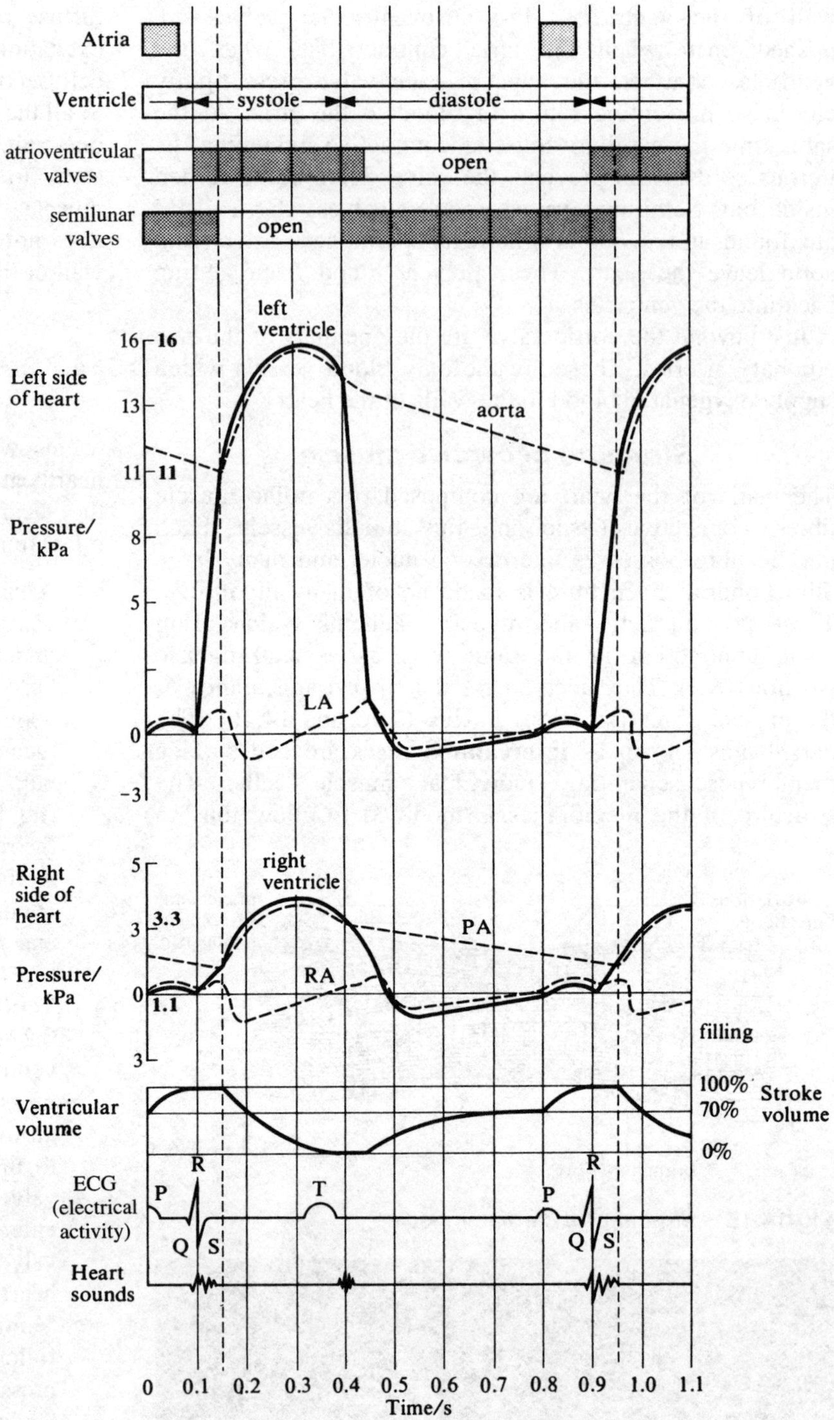

Fig 14.18 *The sequence of events in the cardiac cycle starting with the onset of atrial systole. Part of the next cycle is shown. PA, blood pressure in the pulmonary artery; RA, right atrial pressure; LA, left atrial pressure. (From J. H. Green (1968)* An introduction to human physiology, *Oxford University Press.)*

14.20). One complete heartbeat consists of one systole and one diastole and lasts for about 0.8 s (fig 14.18).

14.7.3 Myogenic stimulation of heart rate

When a heart is removed from a mammal and placed in a well-oxygenated salt solution at 37 °C it will continue to beat rhythmically for a considerable time, without stimuli from the nervous system or hormones. This demonstrates the **myogenic** nature of the stimulation of the heart, that is heart muscle has its own 'built-in' mechanism for bringing about its contraction (*myo*, muscle; *genic*, giving rise to).

The stimulus for contraction of the heart originates in a specific region of the right atrium called the **sino-atrial node** (or **SAN** for short). This is located near the opening of the venae cavae (fig 14.21). It consists of a small number of cardiac muscle fibres and a few nerve endings from the autonomic nervous system (the involuntary part of the nervous system – see next section). The SAN can stimulate the heartbeat on its own, but the rate at which it beats can be varied by stimulation from the autonomic nervous system.

The cells of the SAN slowly become depolarised during atrial diastole. This means that the charge across the membrane is gradually reduced. At a certain point an action potential (section 17.1.1) is set up in the cells (nerve impulses are started in the same way). A wave of excitation similar to a nerve impulse passes across the muscle fibres of the heart as the action potential spreads from the SAN. It causes the muscle fibres to contract. The SAN is known as the **pacemaker** because each wave of excitation begins here and acts as the stimulus for the next wave of excitation.

Once contraction has begun, it spreads through the walls of the atria through the network of cardiac muscle fibres at the rate of 1 m s^{-1}. Both atria contract more or less simultaneously. The atrial muscle fibres are completely separated from those of the ventricles by a layer of connective tissue called the atrio-ventricular septum, except for a region in the right atrium called the **atrio-ventricular node** (**AVN**) (fig 14.21).

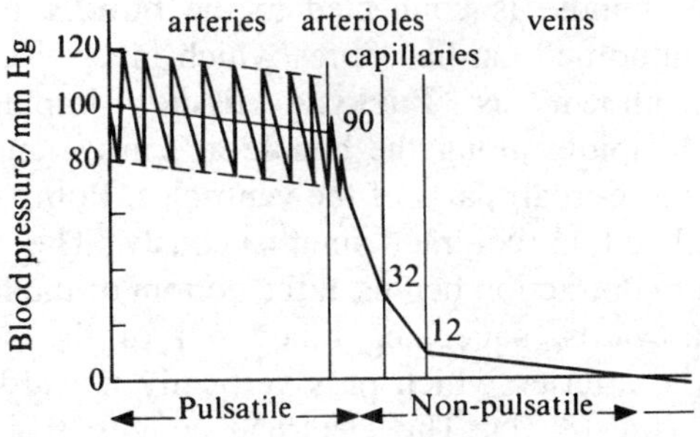

Fig 14.20 *Blood pressure throughout the human circulatory system. (From J. H. Green (1968)* An introduction to human physiology, *Oxford University Press.)*

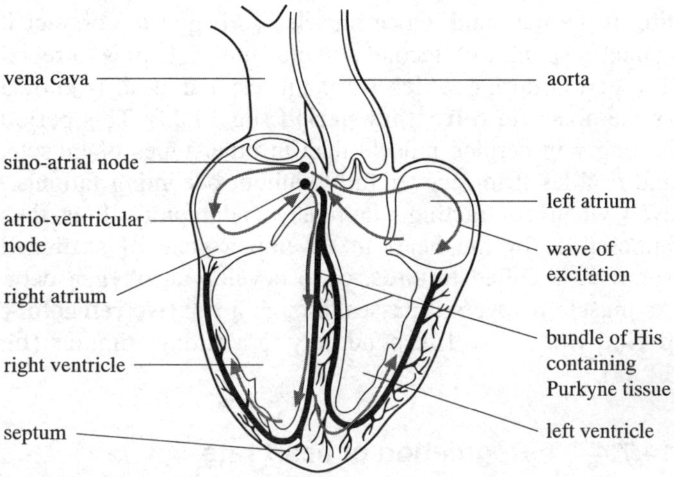

Fig 14.21 *Position of the sino-atrial and atrio-ventricular nodes, and the bundle of His.*

The structure of the AVN is similar to that of the SAN and is connected to a bundle of specialised muscle fibres, the **AV bundle**, which provides the only route for the transmission of the wave of excitation from the atria to the ventricles. There is a delay of approximately 0.15 s in conduction from the SAN to the AVN, which means that atrial systole is completed before ventricular systole begins.

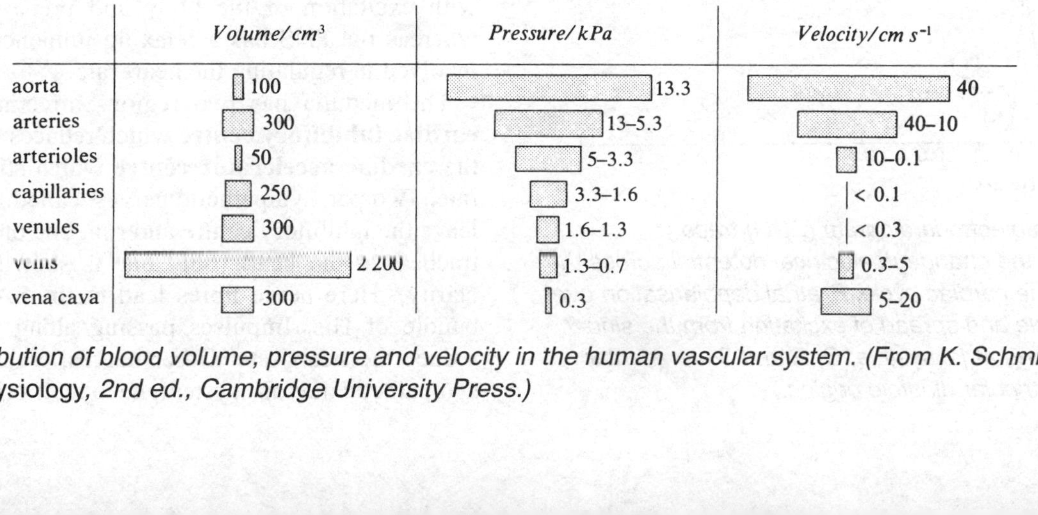

Fig 14.19 *Distribution of blood volume, pressure and velocity in the human vascular system. (From K. Schmidt-Nielsen (1980)* Animal physiology, *2nd ed., Cambridge University Press.)*

The AV bundle is connected to the **bundle of His**, a strand of modified cardiac fibres which gives rise to finer branches known as **Purkyne tissue**. Impulses are conducted rapidly along the bundle at $5\,m\,s^{-1}$, and spread out from there to all parts of the ventricles. Both ventricles are stimulated to contract simultaneously. The wave of ventricular contraction begins at the bottom of the heart and spreads upwards, squeezing blood out of the ventricles towards the arteries which pass vertically upwards out of the heart (fig 14.21). The electrical activity that spreads through the heart during the cardiac cycle can be detected using electrodes placed on the skin and an instrument called an electrocardiogram (ECG). This has medical use since certain heart defects can be detected (fig 14.22).

Certain characteristics of cardiac muscle make it suited to its role of pumping blood round the body throughout the life of the mammal. Once muscle has begun to contract it cannot respond to a second stimulus until it begins to relax. The period during which it cannot respond at all is known as the **absolute refractory period** (fig 14.23). This period is longer in cardiac muscle than in other types of muscle, and enables it to recover fully without becoming fatigued, even when contracting vigorously and rapidly. It is thus impossible for the heart to develop a state of sustained contraction called **tetanus**, or to develop an oxygen debt. As muscle recovers it passes through a **relative refractory period** when it will respond only to a strong stimulus (fig 14.23).

14.7.4 Regulation of heart rate

The basic rate of the heartbeat is controlled by the activity of the SAN as described earlier. Even when removed from the body and placed into an artificial medium the heart will continue to beat rhythmically, although more slowly. In the body, however, the demands on the blood system are constantly changing and the heart rate has to be adjusted accordingly. This is achieved by control systems, one nervous and the other chemical. This is a **homeostatic response** whose overall function is to maintain constant conditions within the bloodstream even though conditions around it are constantly changing.

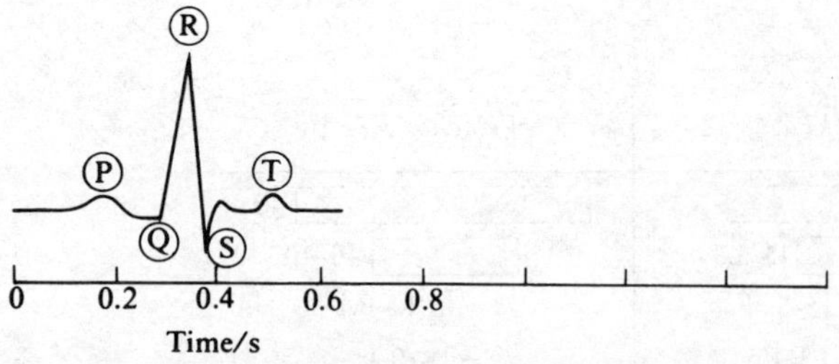

Fig 14.22 *An electrocardiogram (ECG) trace demonstrating the change in electrical potential across the heart during one cardiac cycle. P, atrial depolarisation over the atrial muscle and spread of excitation from the sino-atrial node during atrial systole; Q, R and S, ventricular systole; T, ventricular diastole begins.*

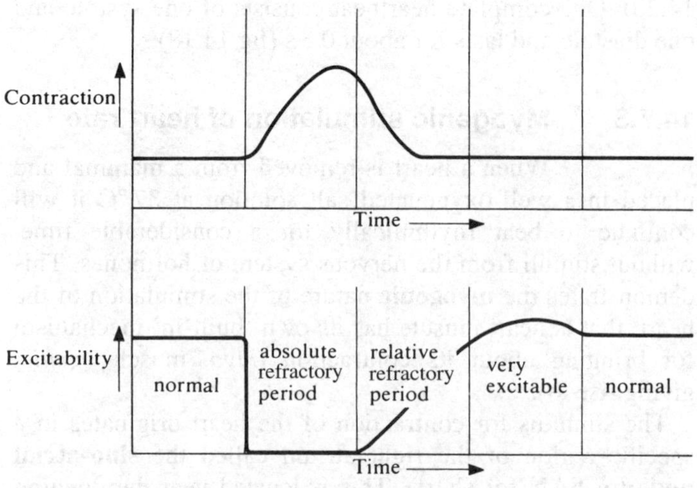

Fig 14.23 *The refractory period of cardiac muscle. The upper figure shows a record of the contraction of the muscle, the lower figure shows the varying excitability of the muscle to stimuli. (From Clegg & Clegg (2nd edition, 1963)* Biology of the mammal, *Heinemann Medical Books.)*

The amount of blood flowing from the heart over a given period of time is known as the **cardiac output** and depends upon the volume of blood pumped out of the heart at each beat, the **stroke volume**, and the **heart rate** (number of beats per minute):

cardiac output = stroke volume × heart rate

It is the cardiac output which is the important variable in supplying blood to the body. One way of controlling cardiac output is to control the heart rate.

Nervous control of heart rate

Overall nervous control of the cardiovascular system (heart and blood vessels) is located in a part of the hindbrain known as the **medulla**. Part of its function is to control heart rate. Certain nerves link the medulla with the heart, as shown in fig 14.24. The nervous system is divided into a voluntary nervous system and an autonomic nervous system which acts automatically and is not under voluntary control. The autonomic nervous system itself is divided into a sympathetic nervous system (SNS) and a parasympathetic nervous system (PNS). The SNS is generally associated with excitation of the body and preparing it for action, whereas the PNS has a relaxing influence. These are both involved in regulating the heart rate.

The medulla has two regions affecting heart rate, the **cardiac inhibitory centre** which reduces the heart rate, and the **cardiac accelerator centre** which stimulates the heart rate. Two parasympathetic nerves, called the **vagus nerves**, leave the inhibitory centre and run, one on either side of the trachea, to the heart (only one is shown in fig 14.24 for clarity). Here nerve fibres lead to the SAN, AVN and the bundle of His. Impulses passing along the vagus nerves reduce the heart rate. Other nerves, which are part of the sympathetic nervous system, have their origin in the cardiac

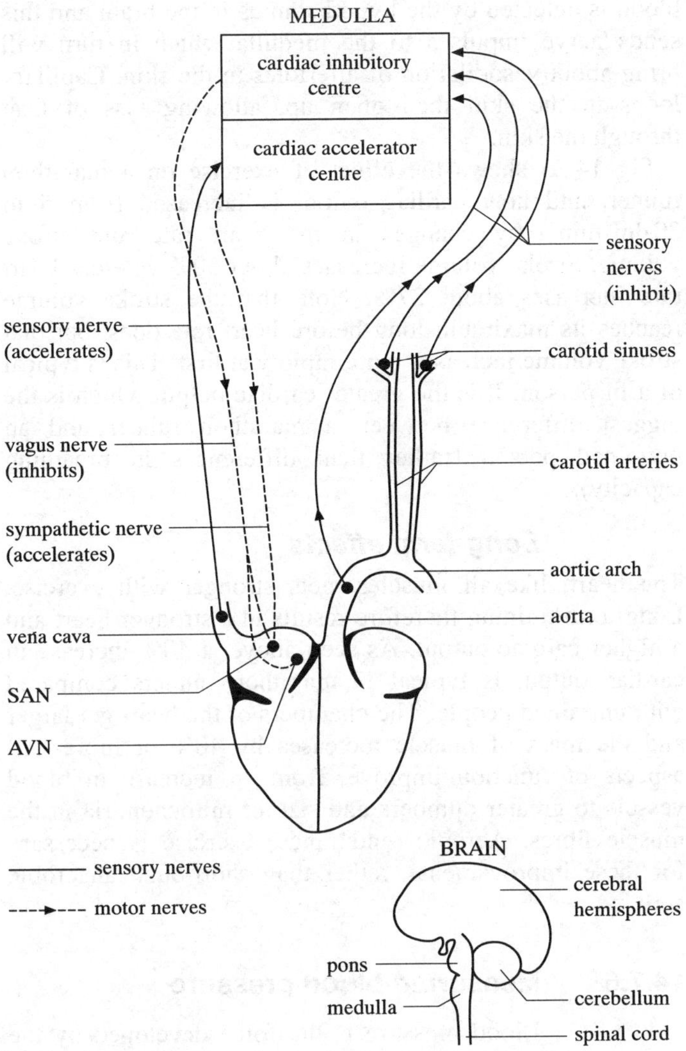

Fig 14.24 *Nervous control of the heart rate. Neurones connect the heart to the cardiac inhibitory and cardiac accelerator centres in the medulla of the hindbrain.*

accelerator region of the medulla. These run parallel to the spinal cord and enter the SAN. Stimulation by these nerves results in an increase in the heart rate. It is the coordinated activity of the inhibitory and accelerator centres in the medulla that controls the heart rate.

Sensory nerve fibres from stretch receptors within the walls of the aortic arch, the carotid sinuses and the vena cava run to the cardiac inhibitory centre in the medulla. Impulses received from the aorta and carotids decrease the heart rate, while those from the vena cava stimulate the accelerator centre which increases the heart rate. As the volume of blood passing to any of these vessels increases so does the stretching of the walls of these vessels. This stimulates the stretch receptors and increases the number of nerve impulses transmitted to the centres in the medulla.

For example, under conditions of intense activity body muscles contract strongly and this increases the rate at which venous blood returns to the heart. Consequently the walls of the vena cava are stretched by large quantities of blood and the heart rate is increased. At the same time the increased blood flow to the heart places the cardiac muscle of the heart under increased pressure. Cardiac muscle responds automatically (no nerves are involved) to this pressure by contracting more strongly during systole and pumping out an increased volume of blood. In other words stroke volume is increased. This relationship between the volume of blood returned to the heart and cardiac output was named after the English physiologist Starling and is known as **Starling's law**.

The increased stroke volume stretches the aorta and carotids which in turn, via stretch reflexes, signal the cardiac inhibitory centre to slow the heart rate. Therefore there is an automatic fail-safe mechanism which serves to prevent the heart from working too fast, and to enable it to adjust its activity in order to cope effectively with the volume of blood passing through it at any given time.

Hormonal control of heart rate

A number of hormones affect heart rate, either directly or indirectly. We shall consider those that have a direct effect first.

The most important of these is **adrenaline**. Adrenaline is secreted by the medulla (middle) of the adrenal glands. The adrenal medulla also secretes smaller amounts of the hormone noradrenaline which has similar effects to adrenaline. Both stimulate the heart, although adrenaline is more effective. Cardiac output and blood pressure are increased by increasing heart rate. The two hormones also have other effects on the body which prepare the body for action (the 'flight or fight' response) as described in section 17.1.2.

Thyroxine, produced by the thyroid gland, raises basal metabolic rate (section 17.6.4). This in turn leads to greater metabolic activity, with greater demand for oxygen and production of more heat. As a result, vasodilation (dilation of the blood vessels) followed by increased blood flow occurs, and this leads in turn to increased cardiac output. Heart rate is also directly stimulated by thyroxine.

Other factors controlling heart rate

A number of other stimuli act directly on cardiac muscle or on the SAN. They are briefly summarised in table 14.4.

Many activities affect the cardiovascular centres in the medulla in some way or other, for example emotions, such as embarrassment, which causes blushing, or anger, which can make the skin turn white, sights and sounds. In such instances sensory impulses are transmitted to the brain where they pass to the cardiovascular centre in the medulla.

Table 14.4 Factors other than nervous and hormonal factors which affect heart rate.

Non-nervous stimulus	*Effect on heart rate*
high pH	decelerates
low pH (e.g. high CO_2 levels, as is the case during active exercise)	accelerates
low temperature	decelerates
high temperature	accelerates
mineral ions	the rate is affected directly or indirectly

14.7.5 Effects of exercise on the cardiovascular system

Short-term effects

During exercise the muscles need an increased blood supply so that oxygen and glucose can be provided for aerobic respiration and waste carbon dioxide and heat can be removed. Blood flow increases dramatically and this is achieved in a number of ways.

During periods of continuous heavy exercise the output of blood from the left ventricle of the heart may increase from the resting condition of 4–6 $dm^3 min^{-1}$ to 15 (untrained female) and 22 (untrained male) $dm^3 min^{-1}$, or even 30 $dm^3 min^{-1}$ for an athlete (this would fill an average bath in about two minutes). This is brought about both by an increased rate of contraction (heart rate) and a more complete emptying of the ventricles (stroke volume). It typically results in a rise in pressure of about 30% in the arteries. Roughly speaking, heart rate can triple and stroke volume can double in response to maximal exercise. In a trained athlete this, combined with vasodilation of blood vessels due to adrenaline and the sympathetic nervous system (see below), can result in a 25-fold increase in blood flow through the muscles about half of which is due to vasodilation and half due to increased blood pressure. Vasodilation also takes place in the heart's own blood vessels and in the lungs.

How is the increased cardiac output brought about? In anticipation of exercise, and during its early phase, the sympathetic nervous system and adrenaline stimulate an increased heart rate. However, during a period of prolonged exercise, the rate is maintained by further nervous and hormonal factors. For example, dilation of the veins in the muscles increases venous return to the heart, which results in increased cardiac output as discussed in section 14.7.4 (see also fig 14.24). Vasoconstriction of arterioles (section 14.7.7) occurs in tissues which are less in need of oxygen, particularly the gut, liver, kidneys and spleen. The effect of a rising carbon dioxide concentration due to exercise is discussed in section 14.7.7.

Eventually vasodilation of skin blood vessels occurs in response to a build-up of heat. The rising temperature of the blood is detected by the hypothalamus in the brain and this sends nerve impulses to the medulla which in turn will bring about vasodilation of arterioles in the skin. Capillary loops in the skin then open up, allowing loss of heat through the skin.

Fig 14.25 shows the effect of exercise on a marathon runner, and how cardiac output is increased from 5 to 30 $dm^3 min^{-1}$ by changes in the heart rate and stroke volume. Stroke volume increases about 50% whereas heart rate increases about 27%. Note that the stroke volume reaches its maximum long before heart rate does, but that stroke volume increases more rapidly at first. This is typical of a fit person. It is the greater cardiac output which is the biggest difference between a marathon runner and an untrained person (rather than differences in breathing capacity).

Long-term effects

The heart, like all muscles, gets stronger with exercise. Long-term training therefore results in a stronger heart and a higher cardiac output. As seen above, a 40% increase in cardiac output is typical in marathon runners compared with untrained people. The chambers of the heart get larger and the mass of muscle increases by 40% or more. All aspects of function improve, from an increase in blood vessels to greater numbers and size of mitochondria in the muscle fibres. Aerobic (endurance) exercise is necessary for these improvements, rather than short-burst anaerobic activity.

14.7.6 Measuring blood pressure

Blood pressure is the force developed by the blood pushing against the walls of the blood vessels. It is usually measured in the brachial artery in the arm by using

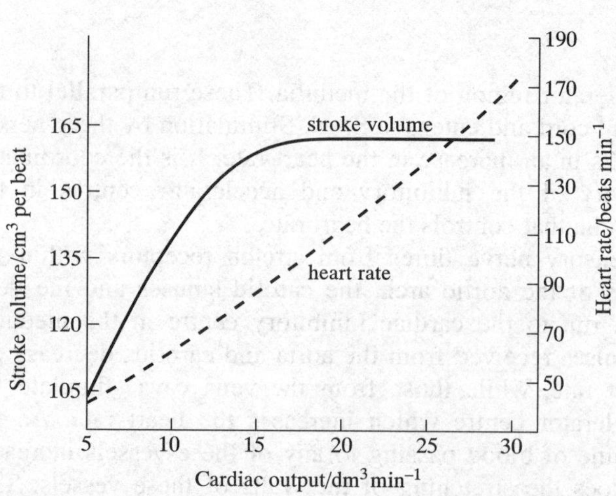

Fig 14.25 *Effects of exercise on a marathon runner. Stroke volume and heart rate at different levels of cardiac output are shown.*

a **sphygmomanometer**. The **systolic** pressure is produced by the contraction of the ventricles and the **diastolic** pressure is the pressure in the arteries when the ventricles relax. Blood pressure is affected by age, sex and state of health. The mean pressures for a healthy young person are about:

systolic: 16 kPa (120 mm Hg)
diastolic: 10 kPA (80 mm Hg).

This is expressed as 120 over 80.

Both pressures are affected by cardiac output and peripheral resistance (resistance to blood flow in the peripheral circulation) and indicate the general state of heart and blood vessels. Conditions which lead to narrowing and hardening of the arteries **(atherosclerosis)** or damage to the kidneys may increase the blood pressure, a condition known as **hypertension**, and impose a strain on the heart and blood vessels. This may lead to a weakening of artery walls and their rupture, or the clogging of narrowed vessels by blood clots **(thrombosis)**. These are very serious if they affect the brain or the heart and lead to cerebral haemorrhage (stroke), cerebral thrombosis or coronary thrombosis (section 15.5).

14.7.7 Regulation of blood pressure

Blood pressure depends on several factors:

- heart rate;
- stroke volume;
- resistance to blood flow by the blood vessels (peripheral resistance);
- strength of heartbeat.

Heart rate and stroke volume have been discussed in the last section.

Resistance to blood flow is altered by contraction **(vasoconstriction)** or relaxation **(vasodilation)** of the smooth muscle in the blood vessel walls, especially those of the arterioles. This peripheral resistance is increased by vasoconstriction but decreased by vasodilation. Increased resistance leads to a rise in blood pressure, whereas a decrease produces a fall in blood pressure. All such activity is controlled by a **vasomotor centre** in the medulla of the hindbrain.

Nerve fibres run from the vasomotor centre to all arterioles in the body. Changes in the diameter of these blood vessels are produced mainly by variation in the activity of constrictor muscles. Dilator muscles play a less important role.

Vasomotor centre activity is regulated by impulses coming from pressure receptors **(baroreceptors)** located in the walls of the aorta and carotid sinuses in the carotid arteries (fig 14.26). Stimulation of parasympathetic nerves in these areas, caused by increased cardiac output, produces vasodilation throughout the body and consequent reduction in blood pressure as well as a slowing of the heart rate. The opposite occurs when blood pressure is low. In this case, a fall in blood pressure increases nerve impulse transmission along sympathetic nerves. This causes body-wide vasoconstriction and a rise in blood pressure.

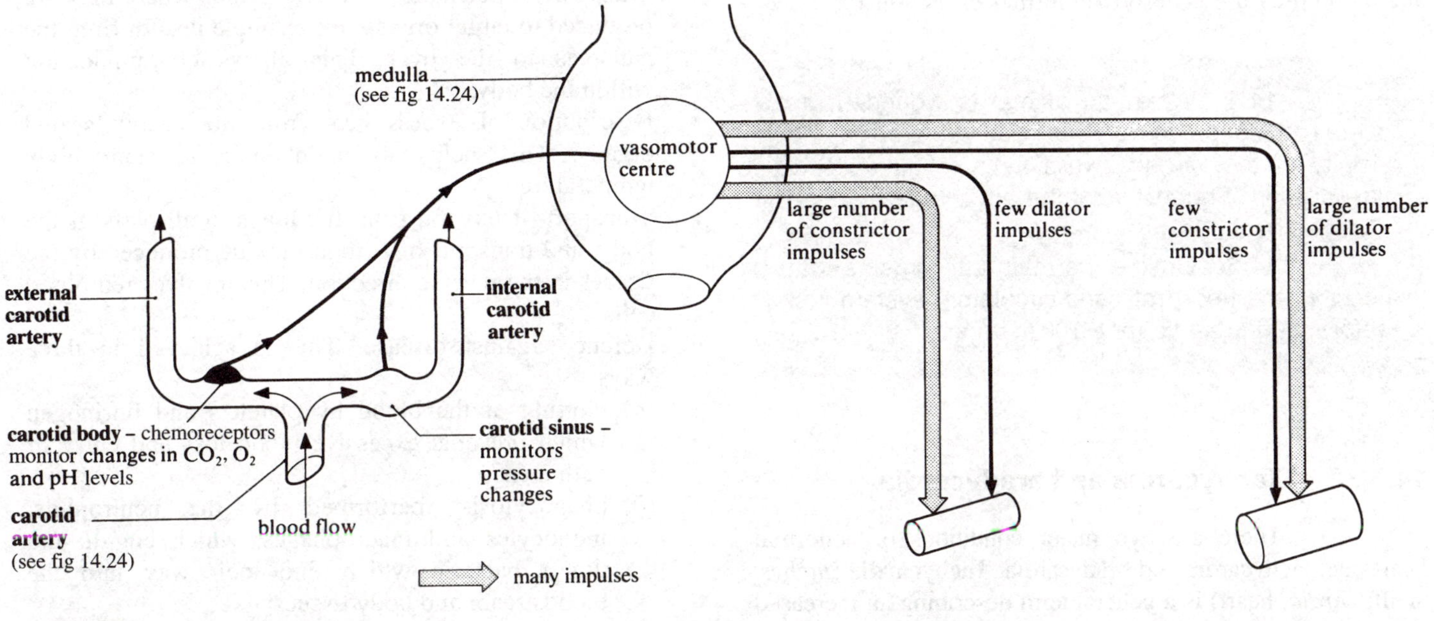

Fig 14.26 *Control of blood pressure. Relationships between the carotid body, carotid sinus, vasomotor centre and general circulatory system are shown.*

Chemical control of the vasomotor centre

Blood arriving at the carotid bodies carrying a high concentration of carbon dioxide stimulates chemoreceptors in these regions to transmit impulses to the vasomotor centre (fig 14.26). Nerve fibres leaving the chemoreceptors link with fibres from the carotid sinus by means of synapses before passing to the vasomotor centre. When the vasomotor centre is stimulated in this way it sends impulses to the blood vessels to vasoconstrict and therefore raises blood pressure. As increased carbon dioxide concentration in the body is usually brought about by increased activity by body tissues, the blood containing the carbon dioxide will be transported more rapidly to the lungs where removal of carbon dioxide in exchange for oxygen can take place more quickly.

Carbon dioxide can also directly affect the behaviour of the smooth muscle of the blood vessel itself. When a tissue suddenly becomes very active, producing a large quantity of carbon dioxide, the carbon dioxide acts directly on the blood vessels in the area and stimulates them to *dilate*. This increases their own blood supply thus allowing more oxygen and glucose to reach the active cells. It must be remembered, however, that when the carbon dioxide leaves this localised area it will have the effect of promoting vasoconstriction elsewhere via vasomotor activity. This is a good illustration of how dynamic and adjustable the control of blood pressure and therefore circulation and distribution of blood can be.

Other stimuli, such as types of emotional stress (for example excitement, pain and annoyance), increase sympathetic activity and therefore blood pressure. Also when the adrenal medulla is stimulated to produce adrenaline by impulses from higher nervous centres this again increases the rate of heartbeat, and therefore raises blood pressure. The significance of the control of heart rate and blood pressure is described further in section 19.1.

14.4 When an animal is wounded, its overall blood pressure rises, but the area in the vicinity of the wound swells as a result of local vasodilation. Suggest what the advantage of these changes might be.

14.5 Outline the main adjustments that occur to the heart rate and circulatory system just before, during and after a 100 m race.

14.7.8 Tachycardia and bradycardia

There are two major conditions of abnormal heart rate, tachycardia and bradycardia. **Tachycardia** (*tachys*, swift; *cardia*, heart) is a general term describing an increased heart rate. It may be caused by a variety of factors including emotional states, such as anxiety, anger and laughter, and overactivity of the thyroid gland. Severe tachycardia is often the result of changes in the electrical activity of the heart. Regions of the heart, other than the SAN, may also contribute to the stimulus for contraction in tachycardia.

A condition known as **bradycardia** (*bradys*, slow) occurs where the heart rate is reduced below the mean level. Long-term training, such as that carried out by athletes, results in an increase by as much as 40% in stroke volume because the heart gets stronger. Therefore, in order to maintain a constant cardiac output at all times, their resting heart rate is reduced.

Underactivity of the thyroid gland, and changes in the electrical activity of the SAN, can also give rise to bradycardia. The regulatory effects of the sympathetic and parasympathetic nervous systems on the SAN tend to counteract temporary conditions of tachycardia and bradycardia and restore normal heart rhythm.

14.8 Functions of mammalian blood

Blood performs many major functions. In the following list, the first four functions are carried out solely by the plasma.

- Transport of soluble organic compounds (digested food) from the small intestine to various parts of the body where they are stored or assimilated (used), and transport from storage areas to places where they are used, such as transport of glucose from the liver to the muscles when glycogen is converted to glucose.
- Transport of soluble excretory materials to organs of excretion. Urea is made in the liver and transported to the kidneys for excretion, and carbon dioxide is made by all cells and taken to the lungs to be excreted.
- Transport of hormones from the glands where they are produced to target organs, for example insulin from the pancreas to the liver. This allows communication within the body.
- Distribution of excess heat from the deeply seated organs. This helps to maintain a constant body temperature.
- Transport of oxygen from the lungs to all parts of the body, and transport of carbon dioxide produced by the tissues in the reverse direction. This involves red blood cells.
- Defence against disease. This is achieved in three ways:
  - (*a*) clotting of the blood by platelets and fibrinogen which prevents excessive blood loss and entry of pathogens;
  - (*b*) phagocytosis, performed by the neutrophils, monocytes and macrophages, which engulf and digest bacteria which find their way into the bloodstream and body tissues;
  - (*c*) immunity, achieved by antibodies and lymphocytes.

- Maintenance of a constant blood solute potential and pH as a result of plasma protein activity. As the plasma proteins and haemoglobin possess both acidic and basic amino acids, they can combine with or release hydrogen ions and so minimise pH changes over a wide range of pH values. In other words, they act as buffers.

14.8.1 Oxygen transport

Haemoglobin, found in the red blood cells, is responsible for the transport of oxygen round the body. Haemoglobin has four polypeptide chains and therefore has a quaternary structure (section 3.5.3). Each polypeptide chain possesses a globin polypeptide chain that is linked to a haem group, which is responsible for the characteristic red colour of the blood. An iron atom (Fe II) is located within each haem group, and each of these can combine with one molecule of oxygen (fig 14.27):

$$\underset{\text{whole haemoglobin molecule}}{Hb + 4O_2} \underset{\text{low } O_2 \text{ concentration}}{\overset{\text{high } O_2 \text{ concentration}}{\rightleftharpoons}} \underset{\text{oxyhaemoglobin}}{HbO_8}$$

(Sometimes the biochemist's standard abbreviation HbO_2 is used to represent oxyhaemoglobin.)

Combination of oxygen with haemoglobin, to form oxyhaemoglobin, occurs under conditions when the concentration of oxygen is high, such as in the lung alveolar capillaries. When the concentration of oxygen is low, as in the capillaries of metabolically active tissues, the bonds holding oxygen to haemoglobin become unstable and oxygen is released. This diffuses in solution into the surrounding cells. Release of oxygen from haemoglobin is called **dissociation**.

The amount of oxygen that can combine with haemoglobin is determined by the oxygen concentration or **partial pressure**. (Partial pressure is a term which is used instead of concentration when referring to gases.) In a mixture of gases, the partial pressure of a given gas, such as oxygen, is that part of the total pressure which is due to that gas. Thus the more oxygen there is in the air, the greater its partial pressure. It is still measured in millimetres of mercury. For example, atmospheric pressure at sea level is 760 mm Hg. Approximately one-fifth of the atmosphere is oxygen, therefore the partial pressure of oxygen in the atmosphere at sea level is $^1/_5 \times 760 = 152$ mm Hg.

Dissociation curves

The greater the concentration, or partial pressure, of oxygen the more saturated haemoglobin becomes with oxygen. The degree to which the haemoglobin is saturated at different oxygen partial pressures can be measured. It might be imagined that a simple linear (straight line) relationship between degree of saturation and oxygen partial pressure will be obtained, but this is not the case, as fig 14.28 shows. The graph is not a straight line, but S-shaped, or **sigmoid**. This curve is called an **oxygen dissociation curve**.

The graph shows that at an oxygen partial pressure of approximately 30 mm Hg only 50% of the haemoglobin is present as oxyhaemoglobin, and at a partial pressure of zero no oxygen is attached to the haemoglobin molecule. 100% saturation is rarely achieved in natural conditions. The S shape of the curve is physiologically important. Over the steep part of the curve, a small *decrease* in the oxygen partial pressure of the environment will bring about a large fall in the percentage saturation of haemoglobin. The oxygen given up by the pigment in such a situation is available to the tissues. It is precisely those oxygen partial pressures which are likely to be found in tissues which bring about this large release of oxygen.

The Bohr effect

In regions with an increased partial pressure of carbon dioxide the oxygen dissociation curve is shifted to the right. This is known as the **Bohr effect** (fig 14.29). This shift is a physiological advantage. If you examine fig 14.29 you will see a vertical line at a partial pressure of about 29 mm Hg (the partial pressure which gives 50% saturation for the middle of the three curves). The three curves in fig 14.29 represent three different carbon dioxide concentrations or

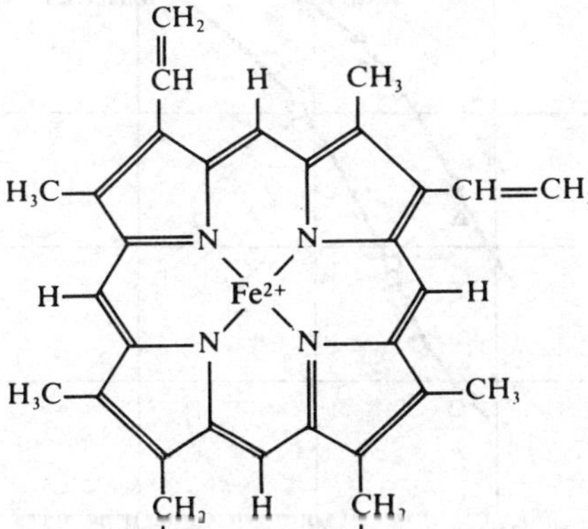

Fig 14.27 *A haem molecule.*

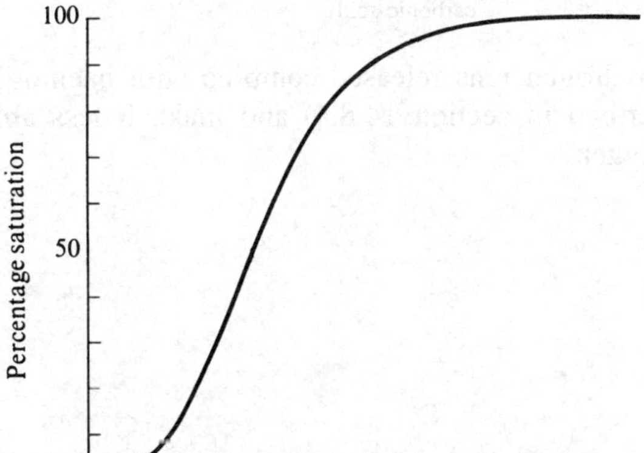

Fig 14.28 *Oxygen dissociation curve of haemoglobin.*

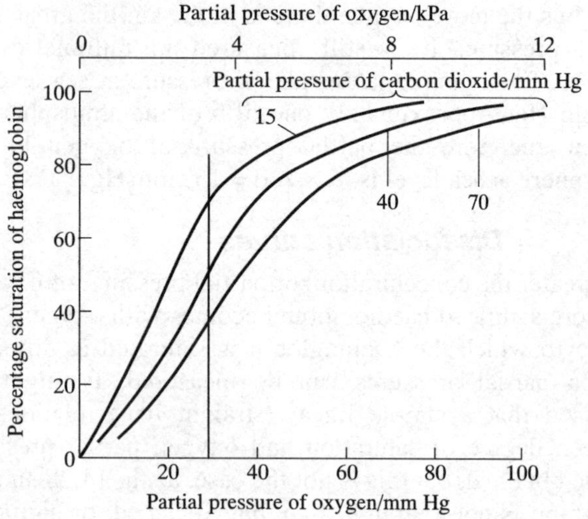

Fig 14.29 *Oxygen dissociation curves of haemoglobin at different partial pressures of carbon dioxide, showing the Bohr effect.*

partial pressures. The curve to the right represents a high partial pressure (70 mm Hg). Note from the vertical line that this results in a lower saturation of the haemoglobin at an oxygen partial pressure of 29 mm Hg. The curve to the left represents a low partial pressure of carbon dioxide (15 mm Hg) and at an oxygen partial pressure of 29 mm Hg this results in a higher saturation of the haemoglobin with oxygen. The effect of increased carbon dioxide is therefore to cause oxygen to be released from the haemoglobin molecule.

Carbon dioxide is a product of respiration. The faster respiration is occurring, the faster it is produced. High levels of respiration are therefore associated with high partial pressures of carbon dioxide. These are the conditions when oxygen is most needed, so it is an advantage that the carbon dioxide makes the haemoglobin release oxygen.

Carbon dioxide has this effect because when it dissolves it forms a weak acid:

$$H_2O + CO_2 \rightleftharpoons \underset{\text{carbonic acid}}{H_2CO_3} \rightleftharpoons H^+ + HCO_3^-$$

The hydrogen ions released combine with haemoglobin (as described in section 14.8.4) and make it less able to carry oxygen.

14.6 (*a*) Temperature also affects the dissociation of haemoglobin. Bearing in mind the effect of carbon dioxide, suggest what the effect of a rise in temperature might be, and explain the physiological advantage of the change.

(*b*) The oxygen dissociation curve is not the same for all animals. For example, compared with humans, the curve for small mammals is displaced to the right. Suggest why this is the case.

14.7 Consider fig 14.30. The oxygen dissociation curve of the fetus is to the left of that of its mother. Suggest why this is so.

14.8 The oxygen dissociation curve of the South American llama, which lives in the High Andes at an altitude of about 5000 m above sea level, is located to the left of most other mammals (fig 14.31). Suggest why this is so.

14.8.2 Myoglobin

Myoglobin is a red pigment very similar in structure to one of the polypeptide chains of haemoglobin. The two proteins have presumably evolved from a common ancestral molecule. It is found in skeletal muscles and is the main reason why meat appears red. It shows a great affinity for oxygen and its oxygen dissociation curve is displaced well to the left of haemoglobin (fig 14.32). In fact it only begins to release oxygen when the partial pressure of oxygen is below 20 mm Hg. In this way it acts as a store of oxygen in resting muscle, only releasing it when supplies of oxyhaemoglobin have been exhausted.

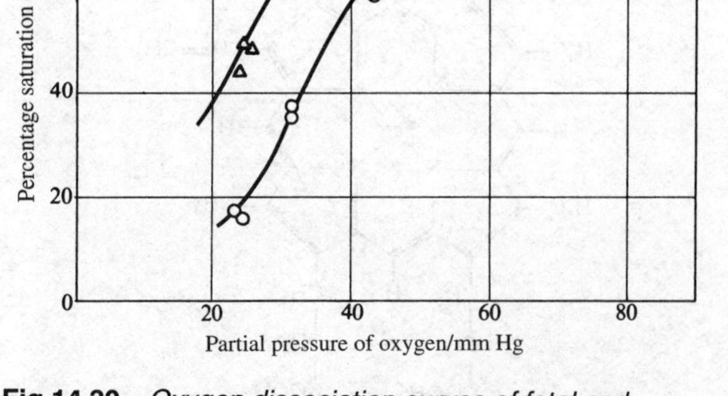

Fig 14.30 *Oxygen dissociation curves of fetal and maternal blood of a goat.*

Fig 14.31 *Oxygen dissociation curves of the llama and other mammals.*

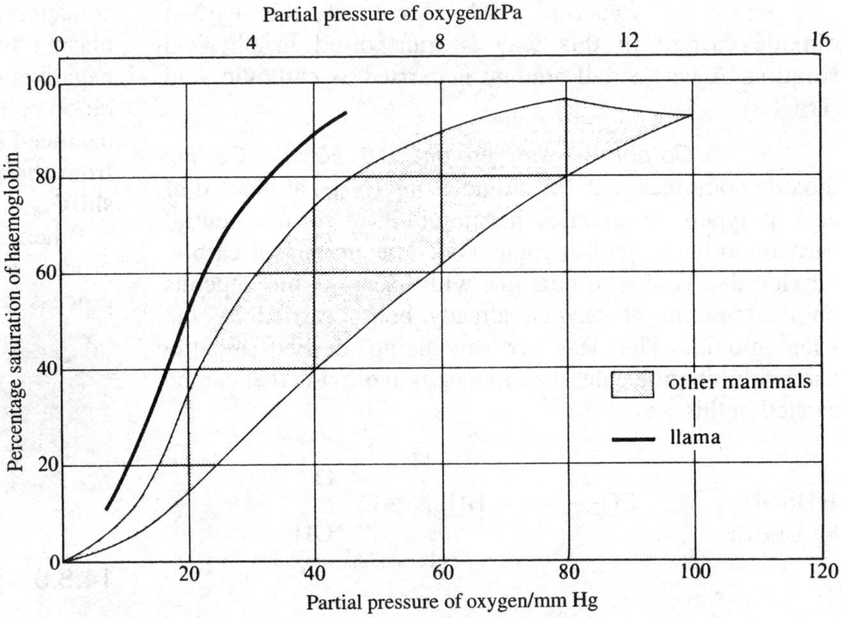

Fig 14.32 *Comparison of haemoglobin and myoglobin oxygen dissociation curves. Myoglobin remains 80% saturated with oxygen until the partial pressure of oxygen falls below 20 mm Hg. This means that myoglobin retains its oxygen in the resting cell but gives it up when vigorous muscle activity uses up the available oxygen supplied by haemoglobin.*

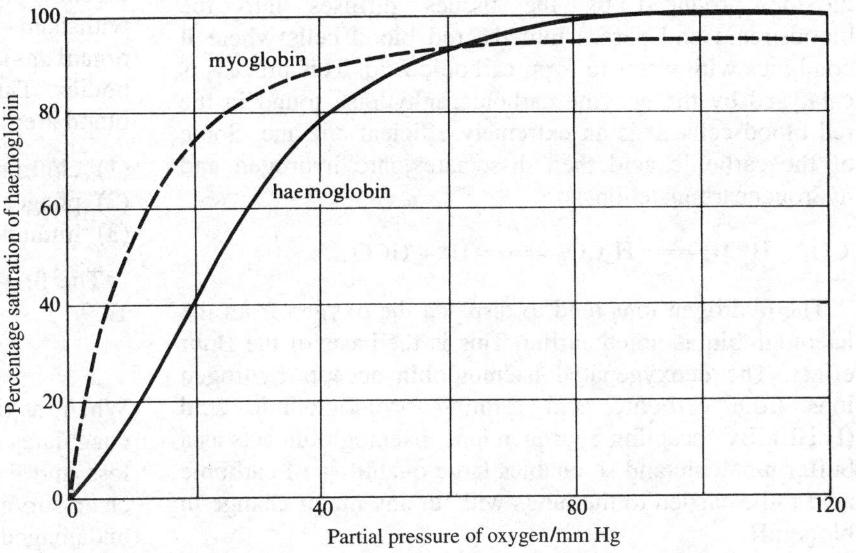

14.8.3 Carbon monoxide and haemoglobin

The affinity of the iron(II) in haemoglobin for carbon monoxide is about 250 times as great as it is for oxygen. Therefore haemoglobin will combine with any carbon monoxide available in preference to oxygen to form a relatively stable compound called **carboxyhaemoglobin**. If this happens oxygen is prevented from combining with haemoglobin, and therefore the transport of oxygen round the body by the blood is no longer possible. In humans collapse follows quickly after exposure to carbon monoxide and unless the victim is removed from the gas, asphyxiation is inevitable. A concentration of about 0.1% in air (about 0.6 mm Hg) can be lethal. Carboxyhaemoglobin is a cherry red colour and this may show in the skin colour of a victim. Removal from the gas must be followed by administering a mixture of almost pure oxygen that contains a small amount of carbon dioxide. The carbon dioxide stimulates the respiratory centre in the medulla. Faster breathing results, which helps to flush the carbon monoxide out of the lungs. Pure oxygen is used because the oxygen partial pressure in air is not sufficient to replace the carbon monoxide attached to the haemoglobin.

The two most common sources of carbon monoxide are car exhaust fumes and tobacco smoke.

14.8.4 Transport of carbon dioxide

Carbon dioxide is carried by the blood in three different ways. It must not be allowed to accumulate in the body because it forms an acid in solution and could lead to fatal changes in blood pH.

In solution (5%). Most of the carbon dioxide carried in this way is transported in physical solution. A very small amount is carried as carbonic acid (H_2CO_3).

Combined with protein (10–20%). Carbon dioxide combines with the amino group (NH_2) at the end of each polypeptide chain of haemoglobin to form a neutral carbamino-haemoglobin compound. The amount of carbon dioxide that is able to combine with haemoglobin depends on the amount of oxygen already being carried by the haemoglobin. The less oxygen being carried by the haemoglobin molecule, the more carbon dioxide that can be carried in this way:

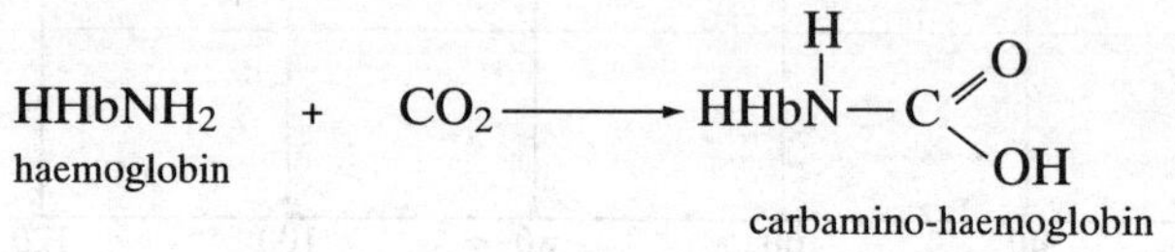

As hydrogencarbonate (85%). Carbon dioxide produced by the tissues diffuses into the bloodstream and passes into the red blood cells where it combines with water to form carbonic acid. This process is catalysed by the enzyme carbonic anhydrase found in the red blood cells. It is an extremely efficient enzyme. Some of the carbonic acid then dissociates into hydrogen and hydrogencarbonate ions:

$$CO_2 + H_2O \rightleftharpoons H_2CO_3 \rightleftharpoons H^+ + HCO_3^-$$

The hydrogen ions tend to displace the oxygen from the haemoglobin as noted earlier. This is the basis of the Bohr effect. The deoxygenated haemoglobin accepts hydrogen ions from carbonic acid forming haemoglobinic acid (H.Hb). By accepting hydrogen ions, haemoglobin acts as a buffer molecule and so enables large quantities of carbonic acid to be carried to the lungs without any major change in blood pH.

The majority of hydrogencarbonate ions formed within the red blood cells diffuse out into the plasma along a concentration gradient and combine with sodium in the plasma to form sodium hydrogencarbonate. The loss of negatively charged hydrogencarbonate ions from the red blood cells leaves them with a more positive charge. This is balanced by chloride ions diffusing into the red blood cells from the plasma. This phenomenon is called the **'chloride shift'**.

When the red blood cells reach the lungs the reverse process occurs and carbon dioxide is released. The whole process is summarised in fig 14.33.

> **14.9** Summarise how carbon dioxide in the blood is expelled as gaseous carbon dioxide by the lungs.

14.8.5 Defensive functions of the blood

We are equipped with a complex system of defence mechanisms whose function is to enable us to withstand attacks by pathogens (disease-causing organisms), and to remove foreign materials from our bodies. Three important defensive mechanisms involving blood are:

(1) clotting of blood } both contributing to
(2) phagocytosis } wound healing
(3) immune response to infection.

The first two are discussed here and the third in section 14.9.

Clotting

When a tissue is wounded, blood flows from it and coagulates to form a blood clot. This prevents further blood loss and entry of pathogenic microorganisms which is of clear survival value. It is just as important that blood in undamaged vessels does not clot. The highly complex series of reactions that takes place in order for coagulation to be achieved serves at the same time to prevent it from

Fig 14.33 *Carbon dioxide transport by the plasma and red blood cell.*

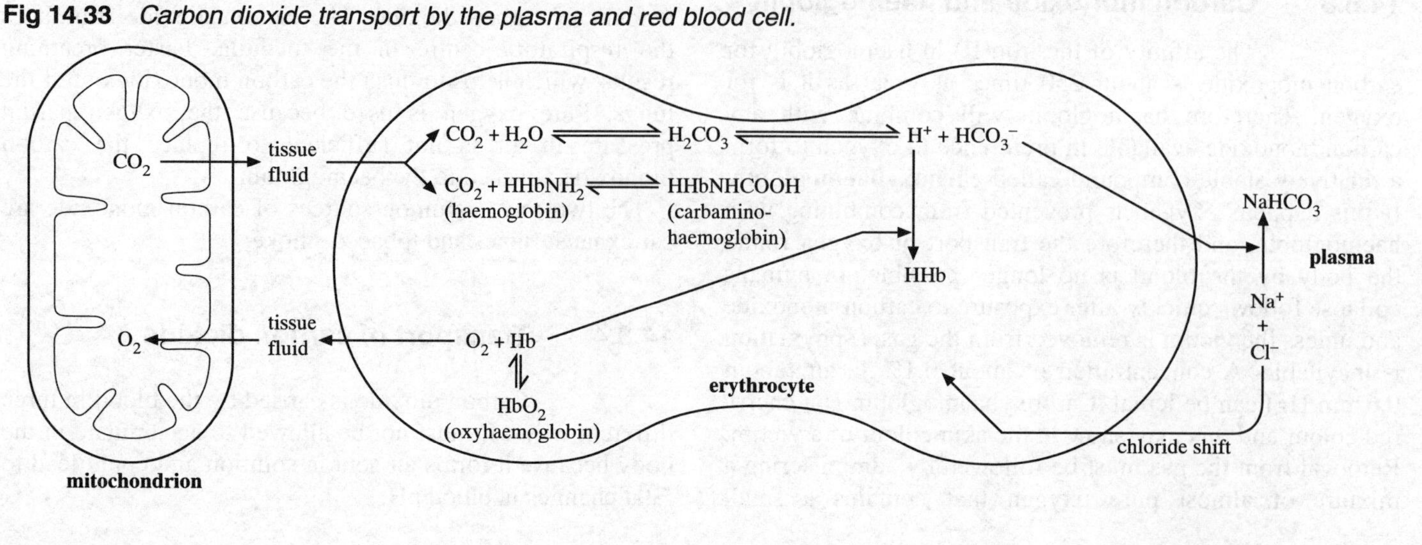

occurring unnecessarily. The whole clotting process depends on at least 12 clotting factors working in harmony with each other. Only the main clotting factors are described in this account. The process is summarised in fig 14.34.

Blood escaping from a skin wound is exposed to the air, and mixes with substances oozing from the damaged cells and ruptured platelets. Among these substances are:

- **thromboplastin**, a lipoprotein which is released from injured tissues;
- **clotting factors VII and X**, which are enzymes found in the plasma;
- **calcium ions**.

Together these substances catalyse the conversion of prothrombin to thrombin. **Prothrombin** is a protein found in the plasma. It is converted to an active enzyme, **thrombin**, which is a protease (a protein-digesting protein). Thrombin hydrolyses fibrinogen, another plasma protein, to **fibrin**. Fibrinogen is a large protein molecule which is soluble. Fibrin, however, is insoluble and fibrous in nature and forms very tangled needle-like fibres. Blood cells become trapped in the meshwork (fig 14.35) and a blood clot is formed. This dries to form a scab which acts to prevent further blood loss and as a mechanical barrier to the entry of the pathogens.

Because the clotting process is so elaborate, it means that the absence or low concentration of any of the essential clotting factors can produce excessive bleeding. For example, if an essential factor necessary for the action of thromboplastin is absent or only present in minute amounts, the individual will bleed profusely from any minor cut. One of the most common conditions associated with excessive bleeding is **haemophilia**. This is caused by a fault in one of two genes. One gene controls production of a protein called factor VIII; a fault in this protein is responsible for haemophilia A which accounts for 85% of haemophilia cases. The other gene controls production of the protein factor IX, a fault in which is responsible for haemophilia B which accounts for the remaining 15% of cases. Both genes are located on the X chromosome. The male carries only one X chromosome, his other sex chromosome being a Y which carries no gene for blood clotting, so the condition is usually seen only in males, where only one faulty chromosome is needed to cause the condition. Two faulty X chromosomes would be needed to cause the condition in a woman. A female with one faulty chromosome will be a **carrier** since the faulty gene is recessive to the normal gene on the other X chromosome (section 24.6.1).

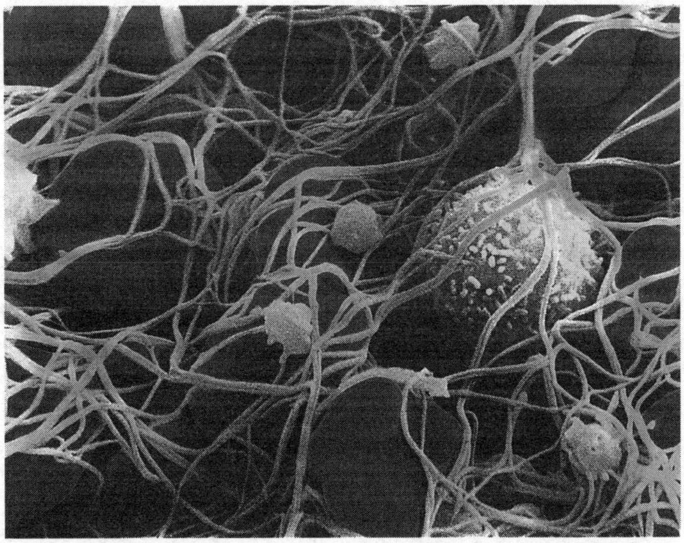

Fig 14.35 *SEM of a blood clot. Red blood cells can be seen trapped in a mesh of fibrous threads, as well as a white blood cell (large spiky cell) and some small platelets.*

Clotting does not occur in undamaged blood vessels because the lining of the vessels is very smooth and does not promote platelet or cell rupture. Also present are substances which actively prevent clotting. One of these is **heparin**, which is present in low concentrations in the plasma and is produced by mast cells found in the connective tissues and the liver. It serves to prevent the conversion of prothrombin into thrombin, and fibrinogen to fibrin and is widely used clinically as an anticoagulant.

If a clot does form within the blood circulation it is called a **thrombus** and leads to a medical condition known as **thrombosis**. This may happen if the endothelium of a

Fig 14.34 *Major stages involved in blood clotting.*

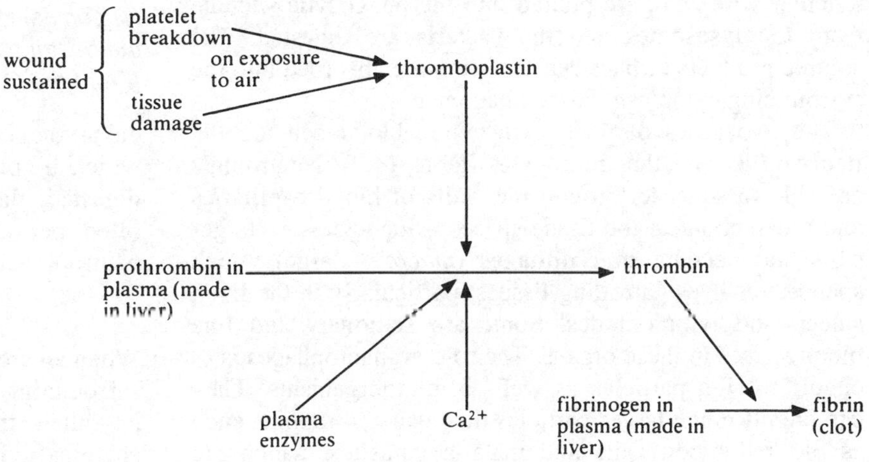

blood vessel is damaged and the roughness of the damaged area promotes platelet breakdown and sets in motion the clotting process. Coronary thrombosis, a thrombus developing in the coronary artery of the heart, is particularly dangerous and can lead to a swift death (section 15.5).

Phagocytosis

The main function of some of the white blood cells is the engulfing of invading microorganisms, and the clearing up of dead cells and debris, such as dust in the lungs. This process is called **phagocytosis** and the cells that carry it out are called **phagocytes**. Phagocytes form the body's first line of defence against attack by microorganisms.

Phagocytes show a kind of movement known as **amoeboid movement**, which means that they can crawl by a flowing movement of their cytoplasm. They are attracted to areas where cell and tissue damage has occurred. The stimulus for this attraction is the release of a variety of chemicals by the damaged blood cells, tissues, blood clotting products and bacteria. Movement towards a source of chemicals is called **chemotaxis**. They are able to recognise invading bacteria. Their ability to do this is enhanced by a group of about 20 proteins known collectively as **complement**. These proteins are also activated by bacterial infection. They have the following effects:

- some attract phagocytes by chemotaxis;
- some are involved in opsonisation (see below);
- some punch holes in the cell surface membranes of bacteria, causing the cells to swell and burst;
- some promote inflammation.

Opsonisation is the coating of bacteria with proteins called opsonins. The **opsonins** are usually a complement protein or an antibody. The process of opsonin recognition is shown in fig 14.36. Phagocytes have receptors in their cell surface membranes which match the opsonins and enable them to recognise, bind to and therefore engulf the bacteria. A phagocytic vacuole is formed (fig 14.37). Small lysosomes fuse with the phagocytic vacuole, forming a phagolysosome. Lysozyme and other hydrolytic enzymes, together with acid, are poured into the phagocytic vacuole from the lysosomes and the bacteria are digested. The soluble products of bacterial digestion are absorbed into the surrounding cytoplasm of the phagocyte.

The two types of phagocytic white blood cell are the **neutrophils** and the **monocytes** (table 14.2). Neutrophils are able to squeeze through the walls of blood capillaries and move about in the tissue spaces. Monocytes are larger cells and become **macrophages** (*macro* – large) which spend their lives patrolling tissues, particularly in the liver, spleen and lymph nodes. Some are stationary and line blood spaces in these organs. The role of macrophages is to engulf foreign particles as well as microorganisms. They can engulf much larger particles than can neutrophils, such as old red blood cells and malaria parasites, which are

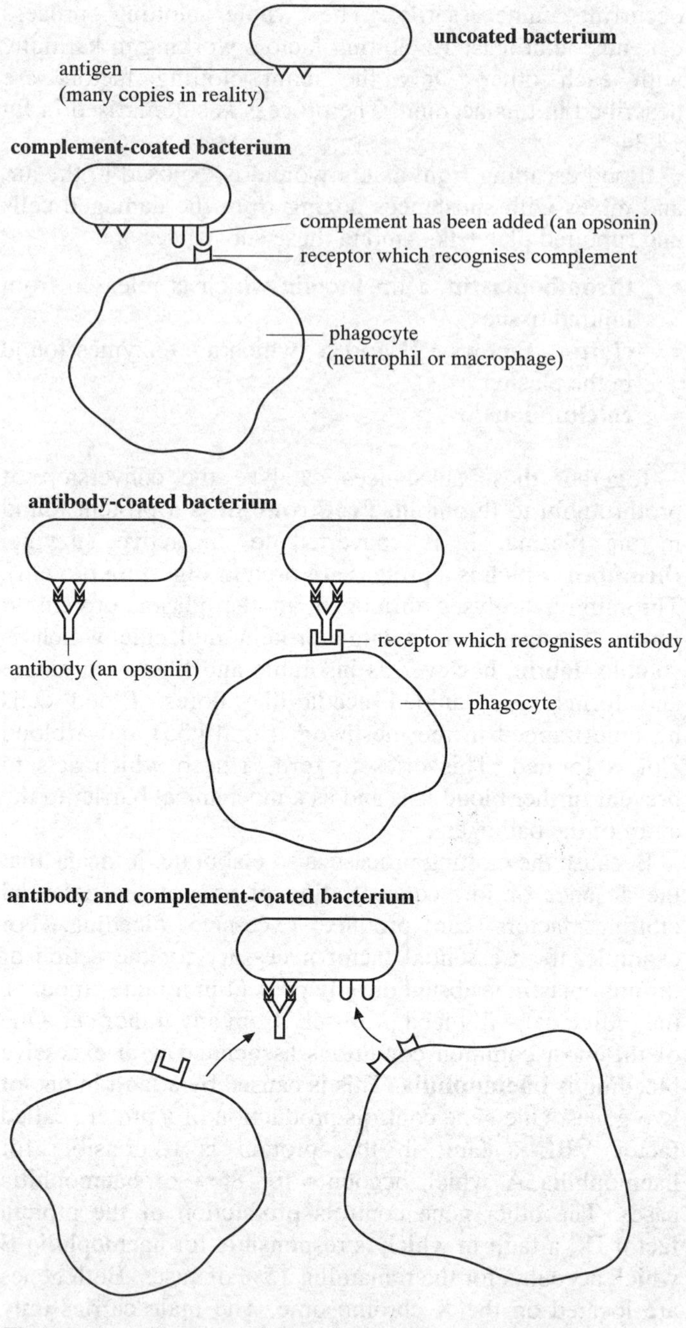

Fig 14.36 *The role of complement and antibodies (opsonins) in coating bacteria (opsonisation). Once coated, the bacteria are recognised by phagocytes and engulfed. (From the Greek word* opson, *meaning seasoning!)*

eukaryote cells and therefore much larger than bacteria which are prokaryotes. If the particles they ingest cannot be digested, they can retain them for long periods of time, often permanently. The macrophages together with the neutrophils form the body's **reticulo-endothelial system**.

Inflammation

When an area of the body is wounded or infected, the tissue surrounding the wound becomes swollen and painful. This is called **inflammation** and is due to the escape of chemicals, including histamine and 5-hydroxytryptamine,

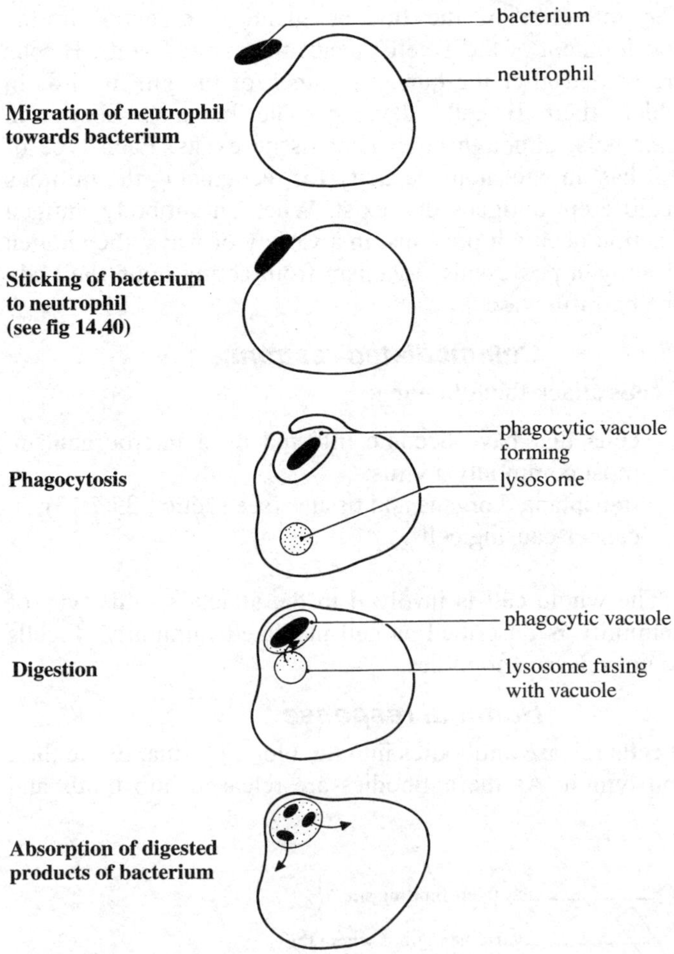

Fig 14.37 *Phagocytosis of a bacterium by a phagocyte.*

from the damaged tissues. Collectively they cause local vasodilation of capillaries. This increases the amount of blood in the area and raises the temperature locally. Leakiness of the capillaries is also increased, permitting escape of plasma and white blood cells into the surrounding tissues and a consequent swelling of the area, a condition known as **oedema**. This plasma contains chemicals, which inhibit the growth of bacteria or kill them, and antibodies and phagocytes, all of which help to combat spread of infection. One of the chemicals is **interferon** which is secreted mainly by macrophages and some other white blood cells if they are exposed to foreign antigens. Interferon makes body cells resistant to infection by viruses. Phagocytes also engulf dead cell debris. Fibrinogen is also present to assist blood clotting if necessary, and the excess tissue fluid tends to dilute and reduce the effect of any potential toxic chemicals.

Wound healing

Towards the end of the inflammatory phase, cells called **fibroblasts** appear and secrete **collagen**. This is a fibrous protein and becomes linked to polysaccharide to form a meshwork of randomly arranged fibrous scar tissue. Vitamin C is important for collagen formation; without it hydroxyl groups cannot be attached to the collagen molecule and so it remains incomplete. After about 14 days the disorganised mass of fibres is reorganised into bundles arranged along the lines of stress of the wound. Numerous small blood vessels begin to spread through the wound. They function to provide oxygen and nutrients for the cells involved in repairing and healing the wound.

Whilst these processes are going on within the wound, the epidermis around it is also engaged in repair and replacement activity. Some epidermal cells migrate into the wound and ingest much of the debris and fibrin of the blood clot which has formed over the wound. When the epidermal cells meet, they unite to form a continuous layer under the scar. When this is complete the scab sloughs off thus exposing the epidermis to the surrounding atmosphere.

Summary of events.

(1) Wound occurs, blood flows.
(2) Clotting process occurs.
(3) Inflammation occurs.
(4) White cells migrate into wound. They absorb foreign matter and bacteria, and remove cell debris.
(5) Fibroblasts enter the wound and synthesise collagen which is built up into scar tissue.
(6) Epidermal cells remove any final debris in the wound, and also begin to dismantle the scar.
(7) Epidermis creates a new skin surface in the area of the wound.
(8) Scab sloughs off.

If the wound is small, phagocytosis is usually sufficient to cope with any invasion by pathogens. However, if there is considerable damage, the immune response of the body is put into action.

14.9 The immune system

Immunity was defined by Sir Macfarlane Burnet as 'the capacity to recognise the intrusion of material foreign to the body and to mobilise cells and cell products to help remove that particular sort of foreign material with greater speed and effectiveness'.

In this section we shall be confining our attention to the **immune response**. This is the second line of defence after the phagocytes which, according to the definition above, are also part of the immune system. The immune response is the production of antibodies in response to antigens. Each antigen is recognised by a specific antibody.

14.9.1 Antibodies, antigens, B cells and T cells

An **antibody** is a molecule that is synthesised by an animal in response to the presence of foreign substances known as antigens. Each antibody is a protein molecule called an **immunoglobulin**. Its structure consists of two heavy chains (H-chains) and two light chains (L-chains) (fig 14.38). The antibody has a constant and variable part, the variable part acting something like a key which specifically fits into a lock. The human body can produce an estimated 100 million different antibodies, recognising all kinds of foreign substances, including many the body has never met. It does this by shuffling different sections of parts of the genes which produce the variable regions (like building different models from a basic set of shapes).

An **antigen** is a molecule which can cause antibody formation. All cells possess antigens in their cell surface membranes which act as markers, enabling cells to 'recognise' each other. Antigens are usually proteins or glycoproteins, that is proteins with a carbohydrate tail, although almost any complex molecule can be antigenic. The body can distinguish its own antigens ('self') from foreign antigens ('non-self') and normally only makes antibodies against non-self antigens. Microorganisms carry antigens on their surfaces.

Two systems of immunity have been developed by mammals, a **cell-mediated immune response** and a **humoral immune response**. The two systems involve the development of two types of lymphocyte, the T and B cells. Both types arise from precursor cells in the bone marrow. The influence of the thymus gland is essential in the development of the T cells, hence the name T cells. B cells are named after the bursa, a branch of the gut in birds in which their B cells develop. The bursa is absent in mammals, although equivalent tissue exists. Each type of cell has an enormous capacity for recognising the millions of different antigens that exist. When an antibody–antigen reaction occurs it prevents, in a variety of ways, the antigen or antigen-possessing organism from acting upon the body in a harmful way.

Cell-mediated response

T cells attack the following:

- cells that have become infected by a microorganism, most commonly a virus;
- transplanted organs and tissues (see section 25.7.13);
- cancer-causing cells.

The whole cell is involved in the attack, so this type of immunity is described as cell-mediated immunity. T cells do not release antibodies.

Humoral response

B cells release antibodies into the blood plasma, tissue fluid and lymph. As the antibodies are released into fluids and

(*a*)

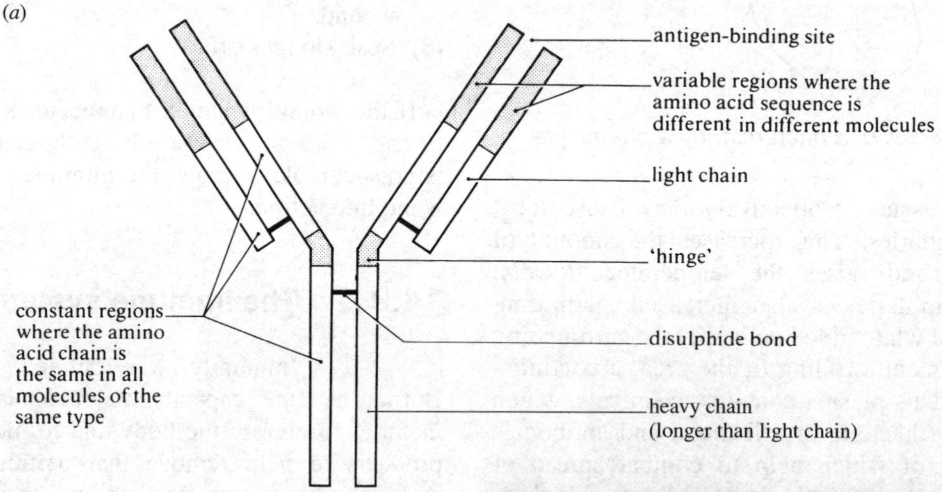

(*b*)

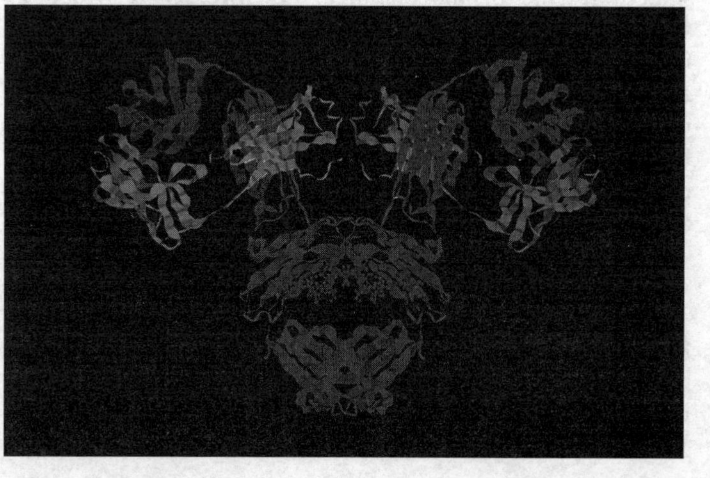

Fig 14.38 *Antibody molecule. (a) Diagram of the basic structure of an antibody molecule. The molecule is made up of four polypeptide chains, two heavy and two light chains. Antigens are bound between the light and heavy chains of the variable regions. (b) Computer-generated picture of antibody structure.*

the attack on the microorganisms takes place in the fluid, this type of immunity is described as humoral ('humor' means fluid). The antibodies of B cells attack bacteria and some viruses.

14.9.2 T cells and the cell-mediated response

The thymus gland is situated in the thorax just above the heart. It begins to function in the embryo and is at its most active at the time of, and just after, birth. After the period of weaning it decreases in size and soon ceases to function.

Evidence that the thymus gland is important in the development of the immune response can be demonstrated by the following experiments.

(1) Removal of the gland from a newborn mouse results in death from a deficiency of lymphocytes in its tissue fluid and blood.
(2) Tissue from an older mouse grafted onto an experimental newborn mouse with the gland removed is unable to recognise and react with antigens.
(3) If the thymus gland is removed from a much older mouse, this mouse suffers no adverse effects.

The stem cells of the bone marrow which give rise to T lymphocytes must pass through the tissue of the thymus gland before they can become fully functional. Here they develop into cells called **thymocytes**. At this stage any cells that recognise 'self' (the body's own antigens) are destroyed, so that the body does not attack itself later. Some of the thymocytes mature into T cells. They leave the thymus gland in the bloodstream. Some stay in the blood and some migrate to the tissue fluid, lymph nodes and other organs such as the spleen.

The cell surface membranes of T cells contain specific receptors with particular shapes, similar to antibodies. However, these receptors do not recognise *whole* antigen molecules, unlike antibodies. They bind only to *fragments* of antigens or other foreign molecules which are presented to them by other cells, often macrophages. Mature T cells possess a T4 molecule (**T4 cells**) or a T8 molecule (**T8 cells**) which give them different functions. T4 cells are known as **helper cells**. The HIV virus, which causes AIDS, infects mainly T-helper cells. There are two types of T8 cells, known as **suppressor cells** and **killer cells** (or cytotoxic cells). Each type of T cell produces a different type of lymphokine. **Lymphokines** are small peptide molecules with various functions which are described briefly below. They are also known as cytokines or interleukins.

T4 cells work in association with macrophages. The macrophage first captures an antigen-carrying organism. It then 'chops off' a piece of the antigen and presents it at its cell surface where it is recognised as a foreign peptide by a T4 cell (one with a matching receptor). The T4 cell then produces large amounts of lymphokines. These have various functions. In particular they:

- stimulate T cells to multiply;
- promote inflammation;
- stimulate B cells to make antibodies.

Killer cells (one type of T8 cell) produce smaller amounts of lymphokines, but kill body cells which have become infected by viruses, and cancer cells. This is done by a chemical attack or by 'punching' holes in the cells. They recognise foreign peptides which are *not* parts of antigens. They can recognise, for example, a stray part of a virus on the outside of an infected cell or a mutant protein produced by a cancer cell. They also attack and gradually destroy transplanted organs (see section 25.7.13).

The activity of all the different types of white blood cell, including phagocytes, is decreased by the lymphokines which are secreted by suppressor cells and stimulated by the lymphokines which are secreted by helper cells. The relative numbers of these two types of cells therefore regulates the whole immune response (fig 14.39).

14.9.3 B cells and the humoral response

The action of B cells is not as complicated. Each B cell has the function of recognising a particular antigen and producing antibodies that will bind to it. The cell surface membrane of B cells contains antigen receptors whose specific shape is identical to the antibodies that that cell can make. All the receptors in the membrane of one cell are identical, so a given cell can only recognise one type of antigen. When it binds to an antigen the cell is activated to clone itself, meaning that it multiplies to form many identical copies of itself. Activation requires the presence of lymphokines secreted by helper T cells as well as antigen, as described above and shown in fig 14.39. It is therefore slightly misleading to think in terms of separate cell-mediated and humoral responses because the two are dependent on each other.

Two types of B cell form, namely **memory cells** and **effector cells** (meaning they carry out the response). Effector cells are also known as plasma cells. These secrete huge numbers of antibody molecules into the blood, tissue fluid and lymph. Effector cells live for only a few days. The memory cells survive for long periods of time and enable a rapid response to be made to any future infection (see next section).

Once an antibody has reacted with an antigen, destruction of the antigen-bearing structure is brought about in a number of different ways. The most common method is to identify the antigen as a target for the action of phagocytes. Phagocytes have receptors that bind to the tail of the Y-shaped antibody (the opposite end to the variable regions).

Fig 14.39 *Summary of the main stages of the cell-mediated and humoral immune responses.*

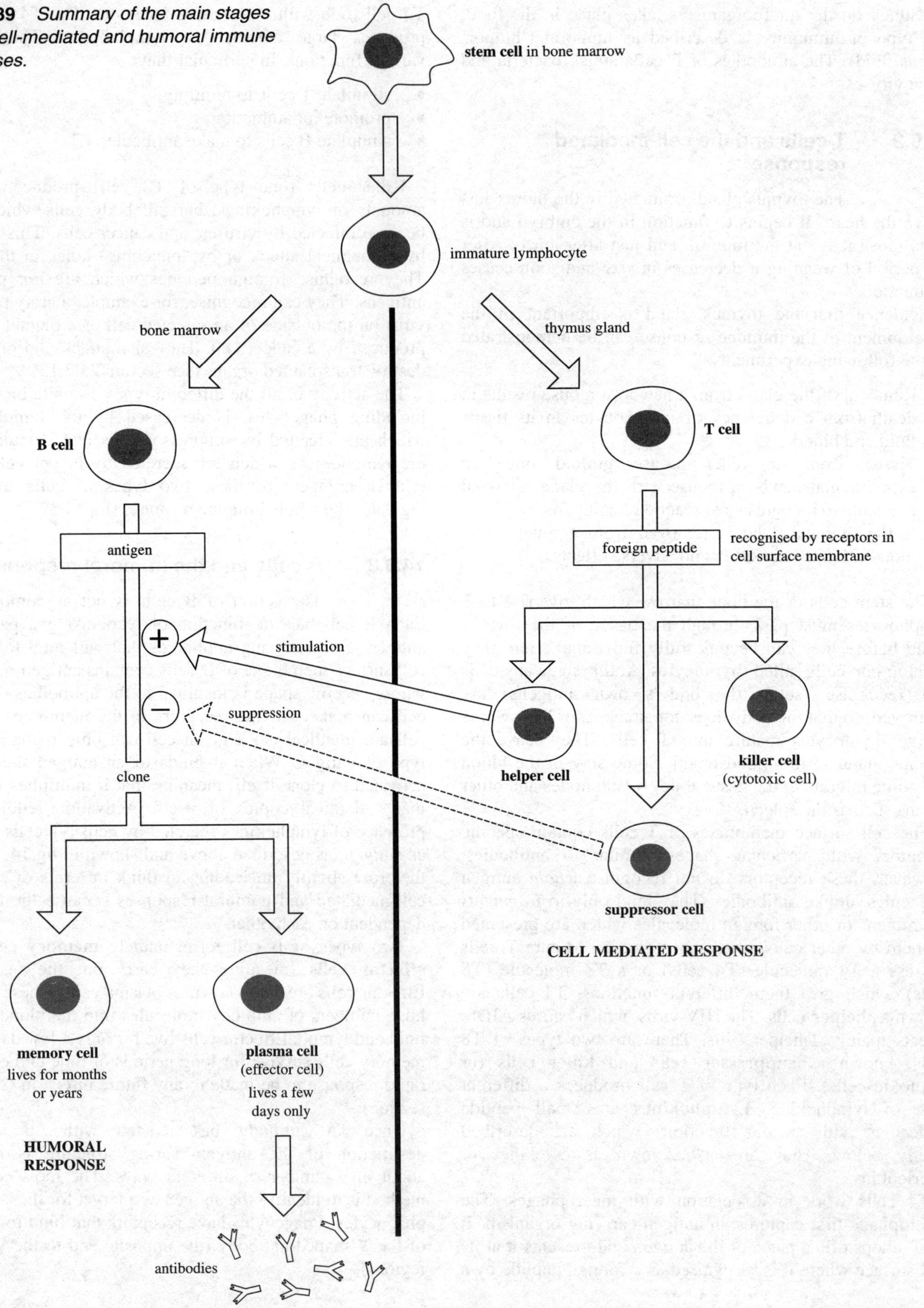

(a)

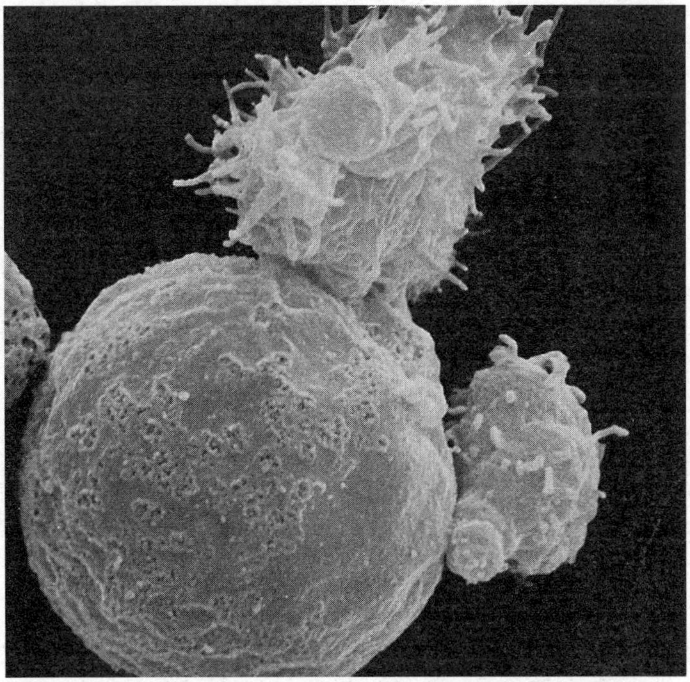

(b)

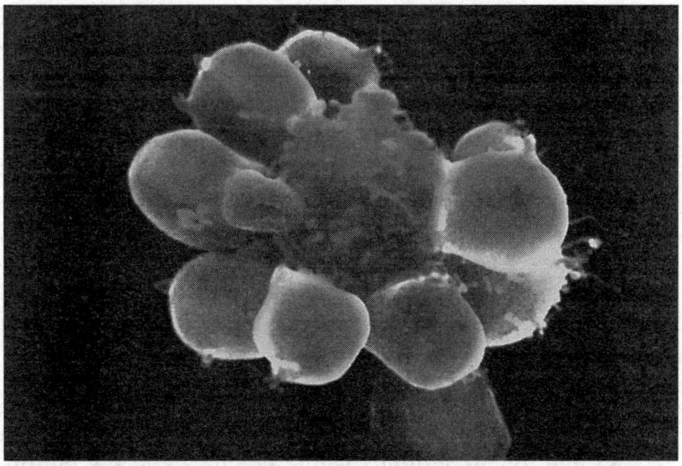

Fig 14.40 *(a) SEM of killer T cells (relatively small cells, top and right of picture) approaching a larger cell, which typically is a tumour cell, a virus-infected cell or a cell from a transplanted organ or tissue.*
(b) SEM of a killer T cell (relatively large cell in centre of picture) stuck to a cluster of foreign red blood cells which it is in the process of destroying.

14.9.4 The immune system has a memory

The memory cells mentioned in the previous section are important if a second infection of an antigen occurs. The population of memory cells is much larger than the original population of B cells from which they came. Therefore the response to the second infection, called the **secondary response**, is much more rapid and is also greater than the **primary response** to the original infection, as shown in fig 14.41. The primary response may not be rapid enough to prevent a person suffering from an infection, but if that person survives, they will rarely suffer from it again because of the greater secondary response. With each exposure, the response gets more efficient. This is the basis of vaccination (and booster doses) (see next section).

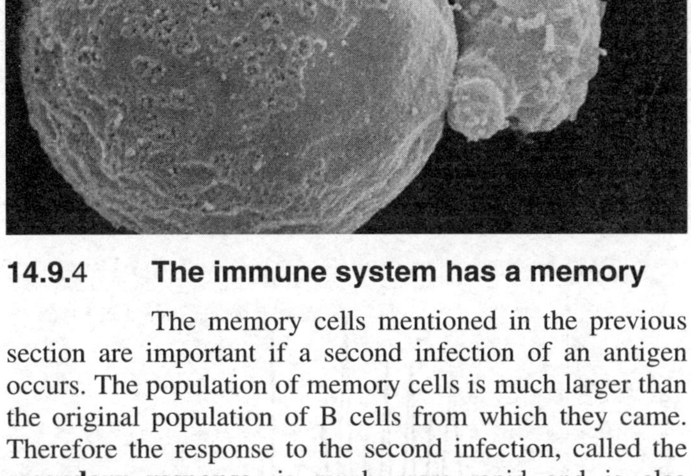

Fig 14.41 *Primary and secondary response to an initial and later dose of antigen. The secondary response is more rapid and intense than the first. IgM and IgG are two different types of antibody (immunoglobulins). IgM is responsible for the primary response. IgG starts the secondary response. (The heavy chain of the antibodies is different.)*

14.9.5 Types of immunity

Immunity may be described as active or passive. Both types may be acquired naturally or artificially. Providing immunity artificially is called **immunisation**.

Natural active immunity

This is the kind of immunity which is obtained as a result of an infection. The body manufactures its own antibodies when exposed to an infectious agent. Because memory cells, produced on exposure to the first infection, are able to stimulate the production of massive quantities of antibody when exposed to the same antigen again, this type of immunity is most effective and generally persists for a long time, sometimes even for life.

Artificial active immunity (vaccination)

This is achieved by injecting (or less commonly administering orally) small amounts of antigen, called the **vaccine**, into the body of an individual. The process is called **vaccination**. If the whole organism is used as the vaccine, it is first made safe by being killed or attenuated (see below). The antigen stimulates the body to manufacture antibodies against the antigen. Often a second, booster injection is given and this stimulates a much quicker production of antibody which is longer lasting and which protects the individual from the disease for a considerable time. Several types of vaccine are currently in use.

- **Toxoids**. Toxins (poisons) produced by tetanus and diphtheria bacteria are detoxified with formaldehyde, yet their antigen properties remain. Therefore vaccination with the toxoid will stimulate antibody production without producing symptoms of the disease.
- **Killed organisms**. Some dead viruses and bacteria are able to provoke a normal antibody response and are used for immunisation purposes. An example is the flu vaccine which contains dead flu viruses.
- **Live vaccines (attenuated organisms)**. An attenuated organism is one which has been 'crippled' in some way so that it cannot cause disease. Often it can only grow and multiply slowly. Attenuation may be achieved by culturing the organisms at higher temperatures than normal, or by adding specific chemicals to the culture medium for long periods of time. Alternatively, the attenuated organism may be a mutant variety with the same antigens but lacking the ability to cause disease. Attenuated vaccines for the bacterial disease tuberculosis (TB), and for measles, mumps, rubella (german measles) and polio are in general use.
- With smallpox, which is now extinct, a live virus vaccine was used. However, the virus in the vaccine was not an attenuated virus but a closely related and harmless form.
- **New vaccines**. Vaccine development made little progress for many years, but new approaches to vaccine design are now possible using modern techniques of molecular biology and genetic engineering. Antigens are very often proteins and proteins are coded for by genes. If the gene for an antigen is transferred into a bacterium, using standard techniques described in chapter 12, the bacterium can be used as a 'factory' for producing large quantities of the antigen for use in vaccines. Cholera, typhoid and hepatitis B vaccines have been prepared in this way. Some vaccines can be made safer in this way, for example the whooping cough vaccine. An alternative approach is to synthesise antigens artificially from amino acids once their amino acid sequences are known.

Vaccination is a common experience in developed countries and is one of the weapons which have helped to reduce the incidence of infectious diseases so dramatically in those countries. Other weapons have been social and environmental improvements such as treatment to clean water for drinking, better nutrition and sewage treatment.

An example of the use of vaccination is the recommendation that all children in the UK should have the MMR vaccine at 2 years of age. This protects against measles, mumps and rubella. Also three doses of another triple vaccine, DTP (diphtheria, tetanus and pertussis (whooping cough)) are recommended at various ages. In some countries vaccination is a legal requirement. Smallpox has been made extinct by vaccination and some childhood diseases such as diphtheria, polio and measles have become extremely rare (fig 14.42). Polio may become

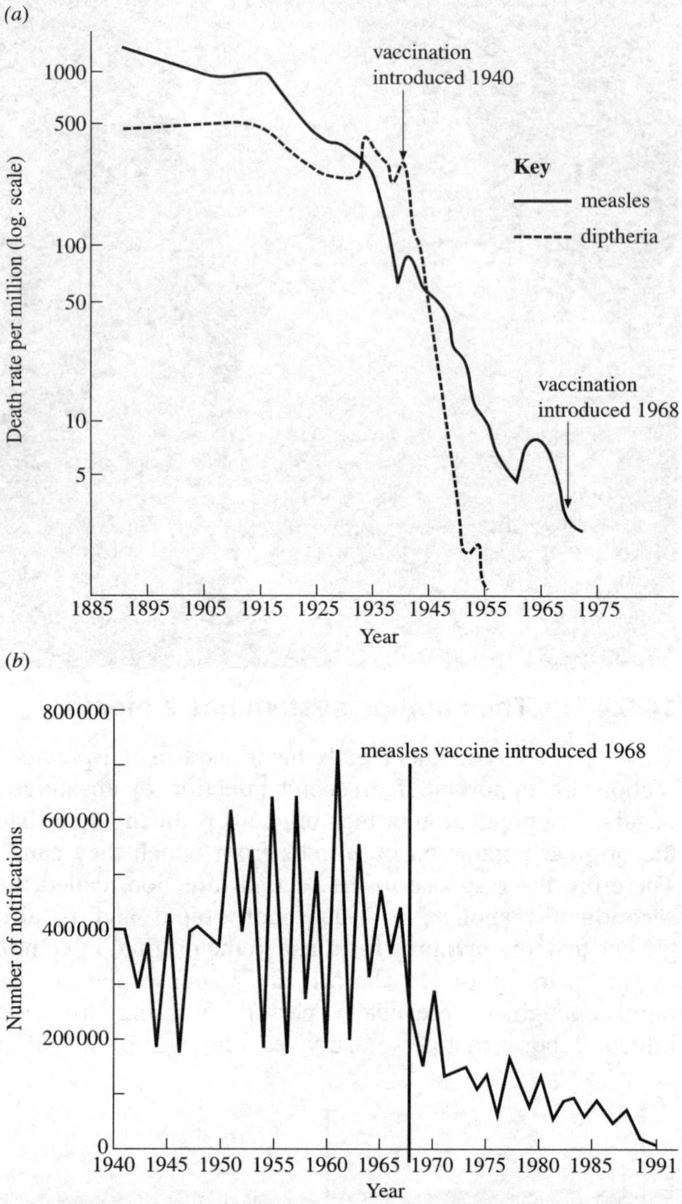

Fig 14.42 *The effect of vaccination on diphtheria and measles. (*a*) Death rate due to diphtheria and measles in children under the age of 15 years in England and Wales from the year 1885 to 1972. (Note that the vertical scale is logarithmic.) Vaccination for diphtheria was introduced in 1940. From being a feared killer of children, it was virtually eradicated within 15 years. (*b*) Measles notifications (reported cases) in England and Wales between the years 1950 and 1991. A measles vaccine was introduced in 1968. The graph shows that measles still tends to occur in minor epidemics in roughly two-year cycles.*
((a) From Registrar General's Statistical Review for England and Wales, *Part 1, Tables Medical, HMSO (1887–1974) HMSO. (b) From* The Health of The Nation and You, *Dept of Health, HMSO (1992) (Source OPCS).)*

extinct in the near future. The World Health Organisation, with support from various organisations like UNICEF and the World Bank, has targetted six major diseases in the developing world in its EPI programme (Expanded Programme on Immunisation). The diseases are diphtheria, whooping cough, tetanus, polio, measles and TB. Although not feared so much any more in developed countries, these are still killer diseases worldwide. Over 80% of children in developing countries now receive vaccination against these diseases. Hepatitis B is also being targetted.

Improvements to some vaccines, such as the flu vaccine, are still needed and there is still no vaccine against some diseases, notably cancers, leprosy, malaria and AIDS, despite intensive research.

Passive immunity

In passive immunity antibodies from one individual are passed into another individual. They give *immediate* protection, unlike active immunity which takes a few days or weeks to build up. However, it only provides protection against infection for a few weeks, for the antibodies are broken down by the body's natural processes, so their numbers slowly fall and protection is lost.

Natural passive immunity

Passive immunity may be gained naturally. For example, antibodies from a mother can cross the placenta and enter her fetus. In this way they provide protection for the baby until its own immune system is fully functional. Passive immunity may also be provided by colostrum, the first secretion of the mammary glands. The baby absorbs the antibodies through its gut.

Artificial passive immunity

Here antibodies which have been formed in one individual are extracted and then injected into the blood of another individual which may or may not be of the same species. They can be used for immediate protection if a person has been, or is likely to be, exposed to a particular disease. For example, specific antibodies used for combating tetanus and diphtheria used to be cultured in horses and injected into humans. Only antibodies of human origin are now used for humans. Antibodies against rabies and some snake venoms are also available. Antibodies against the human rhesus blood group antigen are used for some rhesus negative mothers when carrying rhesus positive babies, as explained in section 14.9.8.

A summary of the different types of immunity is given in table 14.5.

Table 14.5 Summary of different types of immunity.

	Active *antigens received*	*Passive* *antibodies received*
Natural	Natural active e.g. fighting infection, rejecting transplant	Natural passive from mother via milk or placenta
Artificial	Artificial active vaccination (injection of antigens)	Artificial passive injection of antibodies

14.9.6 Monoclonal antibodies

In the 1970s a technique was developed for isolating clones of B cells which produced only one type of antibody. For the first time it became possible to make large quantities of a single pure antibody. Antibodies produced in this way from single clones are called **monoclonal antibodies**. They have a number of important applications. Details of their production and applications are described in section 12.11.2.

14.9.7 Blood groups

When a patient receives a blood transfusion it is vital that they receive blood that is compatible with their own. If it is incompatible, a type of immune response occurs. This is because the donor's red cell membranes possess glycoproteins (known as **agglutinogens**) which act as antigens and react with antibodies (agglutinins) in the recipient's plasma. The result is that the donor's cells are agglutinated (in other words, the cells link or attach to each other when the antigens on their surfaces interact with the antibodies). Two antigens exist, named **A** and **B**. The complementary plasma antibodies are named **a** and **b**, and are present in the plasma all the time; they are not produced in response to the donor's antigen as is the case in the immune reactions already studied. A person with a specific antigen in the red cells does not possess the corresponding antibody in the plasma. For example, anyone with antigen **A** in the red cell membranes has no antibody **a** in the plasma and is classified as having blood group A. If only **B** antigens are present the blood group is B. If both antigens are present the blood group will be AB, and if no antigens are present the blood group is O (table 14.6).

When transfusion occurs it is important to know what will happen to the cells of the donor. If there is a likelihood of them being agglutinated by the recipient's plasma antibodies then transfusion should not take place.

Fig 14.43 indicates the consequences of mixing different blood groups together. Individuals with blood group O are termed **universal donors** because their blood can be given to people with other blood groups. It possesses cells which will not be agglutinated by the recipient's plasma antibodies. Although group O possesses **a** and **b** antibodies,

Table 14.6 Blood groups.

Blood group	*O*	*A*	*B*	*AB*
percentage of population	46%	42%	9%	3%
antigen	–	A	B	A + B
antibody	a + b	b	a	–

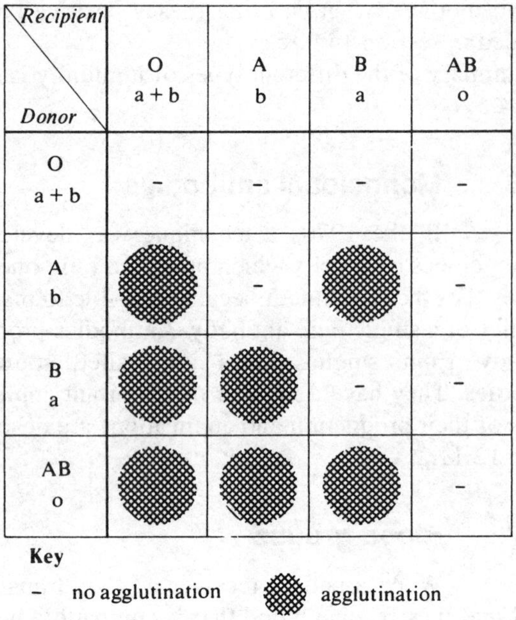

Fig 14.43 *Interactions between human blood groups. Cell antigens are denoted by capital letters, antibodies by small letters.*

when the donated blood forms a relatively small proportion of the total blood volume, there will be very little agglutination of the recipient's cells because the donated plasma is diluted so much by the recipient's blood. However, in major transfusions the blood group must be matched more accurately. Individuals with group AB can receive blood from anyone and are called **universal recipients**. However, they can only donate to blood group AB.

The relationship between blood groups and the major histocompatibility complex (MHC) is discussed in section 25.7.13.

14.9.8 The rhesus factor

Of the total population, 85% possess red cells containing an antigen called the rhesus factor and are termed **rhesus positive**. The remainder of the population lack the rhesus antigen and are therefore described as **rhesus negative**. Rhesus negative blood does not usually contain rhesus antibodies in its plasma. However, if rhesus positive blood enters a rhesus negative individual the recipient responds by manufacturing rhesus antibodies.

The practical importance of this is made obvious when a rhesus negative mother bears a rhesus positive child. The rhesus factor is inherited. (Rhesus positive is dominant and rhesus negative recessive.) During the later stages of the pregnancy, fragments of the rhesus positive red blood cells of the fetus may enter the mother's circulation and cause the mother to produce rhesus antibodies. These can pass across the placenta to the fetus and destroy fetal red cells. Normally the antibodies are not formed in large enough quantities to affect the first-born child. However, subsequent rhesus positive children can suffer destruction of their red cells. A rhesus baby is usually premature, anaemic and jaundiced, and its blood needs to be completely replaced by a transfusion of healthy blood. The condition is known as **haemolytic disease of the newborn**. It can be fatal, especially if the baby is born prematurely as often happens. Although a blood transfusion can now be undertaken whilst the baby is still in the womb, with modern screening methods the problem can be avoided, as explained below.

Protection against the rhesus reaction

(1) If an intravenous injection of anti-rhesus antibodies, called **anti-D**, is given to a rhesus negative mother within 72 hours of her giving birth, sensitisation of the rhesus negative mother by rhesus positive fetal cells is prevented. The anti-rhesus antibodies attach themselves to the rhesus antigens on the fetal cells which are in the mother's circulation and prevent them from being recognised by the mother's antibody forming cells. Hence the antibody process in the mother is not set in motion (fig 14.44). This means of prevention obviously depends on careful screening of all pregnant women. Testing blood groups is part of antenatal care in the UK.

(2) If a rhesus negative mother of blood group O is carrying a rhesus positive child of any blood group other than O, the problem will not arise. This is because if fetal cells enter the mother's circulation, the mother's **a** and **b** antibodies will destroy the blood cells before the mother has time to manufacture anti-rhesus antibodies.

14.9.9 Transplantation

Replacement of diseased tissue or organs by healthy ones is called transplantation and is a technique used increasingly in surgery today. However, when foreign tissue is inserted into or, in the case of skin, onto another individual it is usually rejected by the recipient because it acts as an antigen, stimulating the immune response in the recipient.

Types of transplant

The following terms are used for the different kinds of transplant.

Autograft – tissue grafted from one area to another on the same individual. Rejection is not a problem. This can be used in skin grafting.

Isograft – a graft between two genetically identical individuals such as identical twins. Again, rejection is not a problem.

Allograft – a tissue grafted from one individual to a genetically different individual of the same species.

Xenograft – a graft between individuals of different species such as from pig to human.

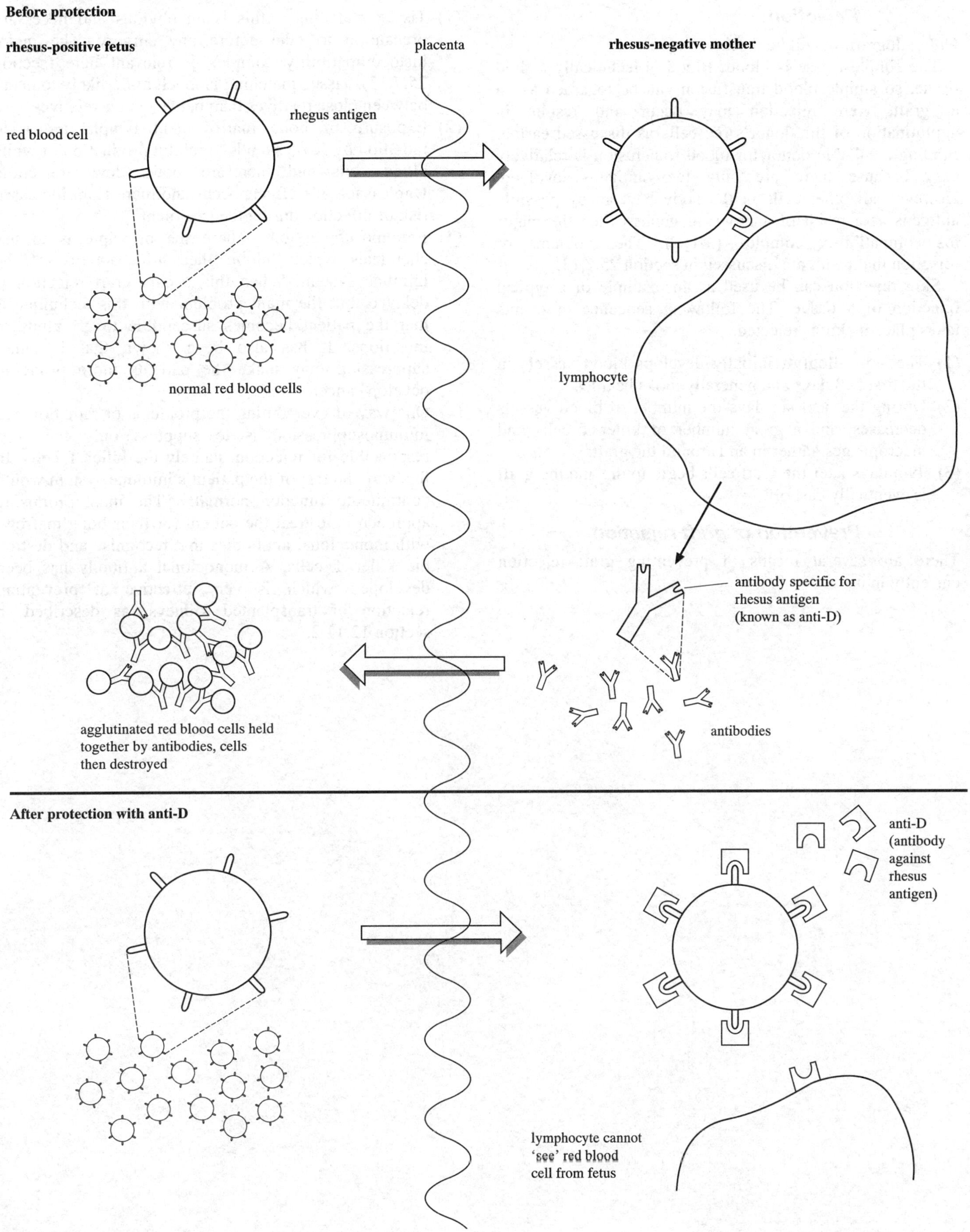

Fig 14.44 *Rhesus disease and its prevention.*

Rejection

Only allografting will be discussed here.

The simplest case is blood. Blood is technically a fluid tissue, so simple blood transfusion can be regarded as an allograft. Here rejection may occur and results in agglutination of the donor's red cells as discussed earlier. Finding a suitable donor for blood transfusion is relatively easy because there are only two antigens involved. However, all other cells in the body have many possible antigens as a result of a group of genes called the major histocompatibility complex (MHC). The problems of rejection that result are discussed in section 25.7.13.

Skin rejection can be used as an example of a typical rejection of a tissue. The following sequence of events takes place if skin is rejected.

(1) The skin allograft initially develops blood vessels in the first 2–3 days and generally looks healthy.
(2) During the next six days the number of blood vessels decreases, and a great number of killer T cells and macrophages gather in and around the graft.
(3) Two days later the graft cells begin to die and the graft is eventually cast off.

Prevention of graft rejection

There are several means of preventing graft rejection currently in use.

(1) Tissue matching – this is an obvious and necessary precaution to take before any surgery. The major histocompatibility complex is relevant here (section 25.7.13). Tissue matching is much more likely to occur between close relatives than between non-relatives.
(2) Exposure of bone marrow and lymph tissues to radiation by X-rays tends to inhibit production of white blood cells and therefore slows down rejection. Unpleasant side-effects occur and there is an increased risk of infection during the treatment.
(3) Immunosuppression – here the principle is to use chemicals which inhibit the entire activity of the immune system. When this occurs graft rejection is delayed, but the main problem with this technique is that the patient becomes susceptible to all kinds of infections. It has also been shown that immunosuppression may make the patients more prone to develop cancer.
(4) One way of overcoming the problems of radiation and immunosuppression is to suppress only the cells responsible for rejection, namely the killer T cells. In this way the rest of the patient's immune system would continue to function normally. The most promising approach is to treat the patient (or their bone marrow) with monoclonal antibodies that recognise and destroy the killer T cells. A monoclonal antibody has been developed which is very effective at preventing rejection of transplanted kidneys, as described in section 12.11.2.

Chapter Fifteen

Health and disease

15.1 What is meant by health and disease?

The World Health Organisation has defined health as 'a state of complete physical, mental and social well-being and not merely the absence of disease or infirmity'. Note that there are three dimensions to this definition, namely physical, mental and social. Ill-health may not necessarily involve disease, and the term *disease* is rather more difficult to define than *health*. One possible definition is that it is a 'bodily disorder', or a 'disordered state of an organ or organism'. This may be suitable when describing a person with tuberculosis, a liver affected by alcohol, or a lung with a tumour, but what of a broken arm? The arm is certainly in a 'disordered state', but it would not normally be described as diseased. In fact, it is arguable that the term 'disorder' is sometimes to be preferred to disease, as perhaps in the case of genetic disorders/diseases. A better understanding of what we mean by disease can be obtained by considering how diseases may be classified.

Classification of disease

It is convenient to classify diseases into the following six main groups.

1 Diseases caused by other living organisms. Disease-causing organisms typically include viruses, bacteria, fungi, protozoans, flatworms and roundworms. These organisms live as parasites in or on the human body and interfere with its normal working. Diseases which are caused by bacteria, viruses and fungi are commonly referred to as **infectious diseases** or **communicable diseases**, such as cholera (caused by a bacterium) and measles (caused by a virus). Diseases caused by other organisms are more commonly referred to as **parasitic diseases**, such as malaria (caused by a protozoan).

2 Diseases that are 'human-induced' or 'self-inflicted'. These diseases are brought by humans on themselves, either as individuals or collectively as a society. They could also be described as **social diseases**. Many are particularly associated with modern industrialised societies and include coronary heart disease, alcoholism, drug abuse, lung cancer, domestic and industrial accidents, industrial diseases such as asbestosis, and pollution-related disorders. The latter include brain damage brought on by lead or mercury poisoning, some cases of asthma and possibly some cases of cancer found in people living near nuclear power stations.

3 Deficiency diseases. These are related to the absence of certain nutrients in the diet. They may be due to the absence of one of the main food groups such as protein which results in kwashiorkor and marasmus (chapter 9). The absence of specific vitamins may result in a number of diseases such as pellagra (vitamin B_1), scurvy (vitamin C) or rickets (vitamin D). Deficiency of minerals in the diet may also result in disease, such as calcium and phosphate deficiency causing rickets or lack of iodine causing thyroid goitre. Deficiency diseases are discussed in chapter 8.

4 Genetic and congenital (present at birth) disorders. These disorders are raising increasing concerns in the medical services and society in general. Examples of genetic disorders are cystic fibrosis, Huntington's disease and Down's syndrome. These and others are discussed in chapter 25. Advances in medical science ensure that many children who would in the past have died in infancy from such disorders are surviving and living to adulthood. This means that as a society we must provide means whereby adults with mental or physical handicaps, often severe, can lead fulfilling lives. On the other side of the coin, genetic screening can increasingly provide the information before birth whereby babies with genetic disorders can be aborted. This raises many social and ethical issues which provoke controversy in society. Some of these issues are discussed in chapters 21 and 25.

5 Ageing and degenerative diseases. Degeneration of the body tissues can also cause disease. For example, weakening of the eye muscles causes long-sightedness in many older people, and diseases of the circulatory system, such as arteriosclerosis (hardening of the arteries) result from ageing. Ageing of the joint and bone tissues often leads to arthritis.

6 Mental illness. Mental illness covers a wide variety of disorders. Examples are schizophrenia, senile dementia and depression. Certain drugs have been developed that control or reduce various forms of mental illness. The treatment of these illnesses changed dramatically during the twentieth century from life long confinement in 'lunatic asylums' to 'care in the community'.

The six groups described above may also be grouped into the following categories.

- **Infectious or communicable disease.** Many diseases are passed on from one organism to another, that is they are transmitted and are thus said to be infectious, contagious, communicable or transmissible. Many of the diseases in group 1 above are infectious and are transmitted by way of droplets of liquid, in the air, food or water, by sexual intercourse or simply by touch. Many are transmitted by way of an intermediate organism called a **vector**, for example malaria by way of the mosquito (section 15.3.3), or bubonic plague by way of the rat flea.
- **Non-infectious disease.** Disease groups 2 to 6 above are described as non-infectious or non-transmissible. It could be argued that group 4 is a special case of transmission because genetic disorders are transmitted from parent to offspring due to the presence or absence of one or more inherited genes.

It is clear from the discussion above that there are no rigid boundaries between the disease categories. For example, many diseases involve a genetic predisposition and therefore an overlap between group 4 above and other categories, such as heart disease (group 2) and some mental illness (group 6). Like health, disease has physical, mental and social dimensions. Some diseases, for example, are associated with poverty, such as tuberculosis, and may therefore be described as having an important social dimension.

15.2 Global distribution of disease

When the distribution of disease worldwide is studied, the most striking feature to emerge is the difference between developed and developing countries. Fig 15.1 shows the number of deaths and the percentage of deaths due to different causes in developed (industrialised) countries and developing (Third World) countries. Infectious and parasitic diseases represent a relatively high proportion of disease in developing countries, whereas circulatory and degenerative diseases are, relatively speaking, the most important in developed countries. The infectious disease measles, for example, is still an important killer disease in many developing countries (table 15.1) although it is declining rapidly as a result of vaccination programmes (section 15.2.1). Measles is a more serious disease than most people think. It can cause pneumonia, blindness, deafness and inflammation of the brain (encephalitis) which in turn may cause brain damage (table 15.3, page 501). It used to be an important killer in developed countries (fig 15.2); 4188 people died of measles in England and Wales in 1930, but only 26 in 1980.

Why have these changes come about and why are there such striking differences between developed and developing countries? The temptation is to suggest that medical intervention is responsible, but fig 14.42 shows that death rates from measles were declining rapidly in England and Wales long before medical intervention. Vaccination against measles was not introduced in Britain until 1968. The same is true of some other infectious diseases which have been killers in the past, such as whooping cough and tuberculosis. The truth is that, in fighting disease, social and economic factors are just as important as medical intervention.

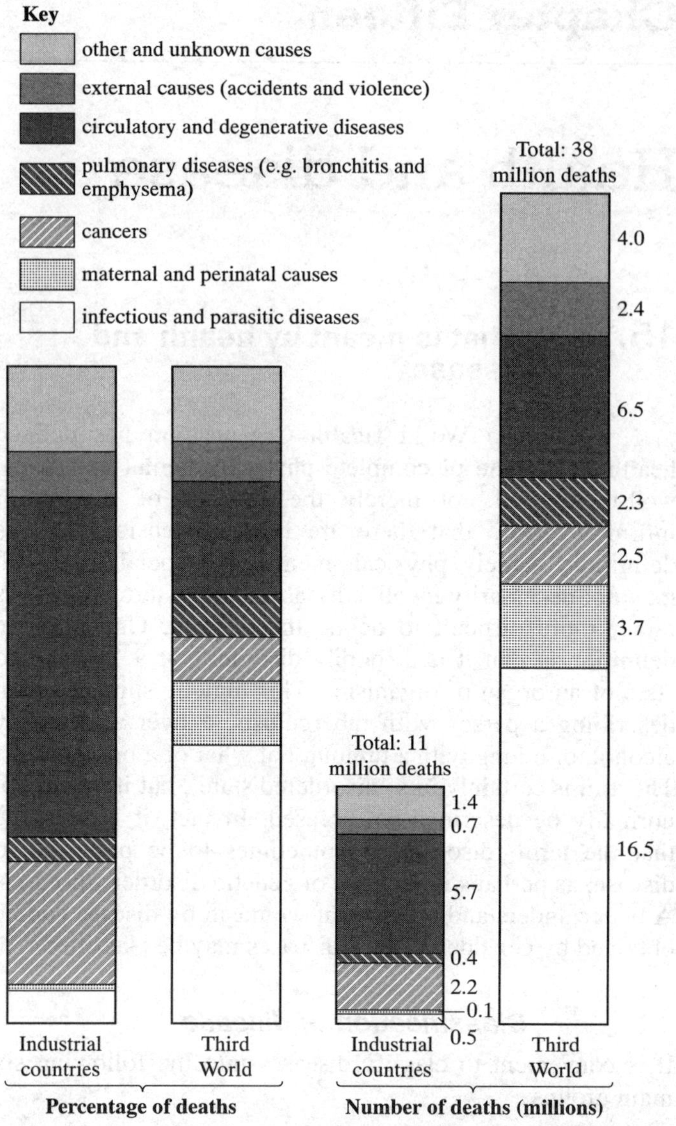

Fig 15.1 *Leading causes of death (mortality) in developed and developing countries in the year 1985. Percentage of deaths are shown to the left and numbers of deaths in millions are shown to the right. (Data from Lopez, A.D., 1993,* Causes of death in the industrialized and the developing countries: estimates for 1985' *in Jamison, D.T. and Mosley, H. (eds)* Disease Control Priorities In Developing Countries, *OUP).*

Some important infectious diseases are transmitted by faecal contamination of water and food. Examples are cholera, diarrhoea, typhoid and dysentery. These declined rapidly in England and Wales after the Public Health Act, 1875 which introduced public hygiene measures for the

Table 15.1 Mortality for selected infectious diseases – world estimates.

Disease	*Deaths per year (mortality)*
Respiratory disease (e.g. pneumonia, bronchitis, influenza, diphtheria)	10 000 000
Diarrhoea	4 300 000
Measles	2 000 000
Malaria	1 500 000
Tetanus	1 200 000
Tuberculosis (TB)	900 000
Hepatitis B	800 000
Pertussis (whooping cough)	600 000
Typhoid	600 000
Schistosomiasis (blood fluke)	250 000
HIV	200 000

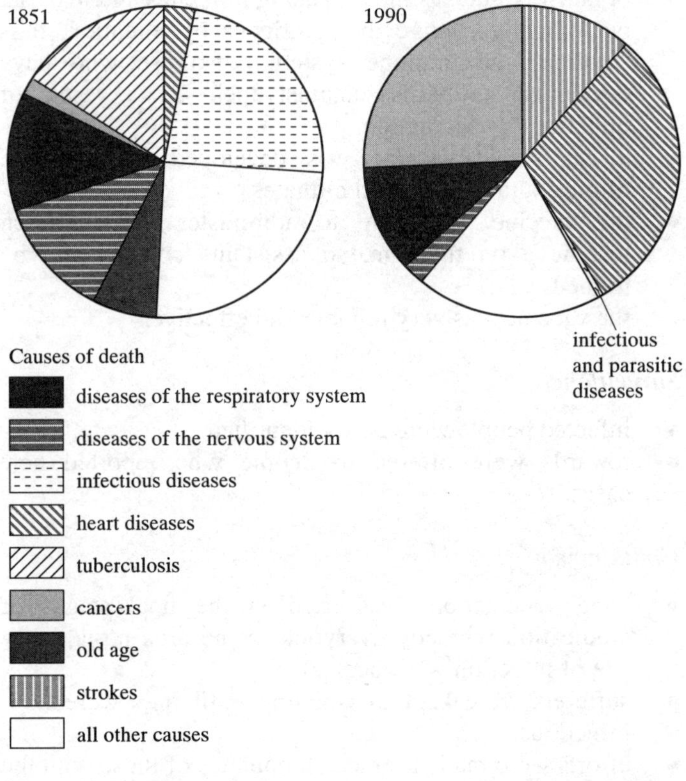

Fig 15.2 *Causes of death in England and Wales, 1851 and 1990. (Data for 1851 from Registrar-General, 1855,* England and Wales Report; *for 1990 from Registrar-General, 1992,* Annual Abstract of Statistics 1992, *HMSO, London, Table 2.20, p37.)*

proper disposal and treatment of sewage, and the purification of water. When considering infectious disease in general, improvements in living standards have been just as important. These increase *resistance* to disease. Particularly important are good nutrition and housing. Tuberculosis, a respiratory infection, began to decline as living standards improved since it is typically spread when people live in close contact, such as several people living in one room, and also when people are malnourished. Measles is a particularly important disease in developing countries because it tends to affect very young children, often under the age of one year before they have been vaccinated, and when the body is less able to fight infection. If the body is weakened by malnutrition, or by other infectious diseases or parasites, it makes measles and many other infectious diseases much more dangerous. Most measles deaths are caused by secondary infections.

To summarise, infectious diseases have been brought under control in the UK and other developed countries by improvements in hygiene, housing and nutrition as well as by direct medical intervention. The latter includes improvements in prevention, such as vaccination, and in treatment, such as the use of antibiotics (considered in more detail later). Overall social, economic and medical factors are all important and all interact in a complex way.

As vaccination programmes are extended to developing countries, so the incidence of infectious disease is declining in them. Measles is one of six major diseases targetted by the World Health Organisation (WHO) for prevention by means of vaccination in its Expanded Programme on Immunisation (section 15.2.1 and fig 15.3). The five others are tetanus, pertussis (whooping cough) and polio (fig 15.3), tuberculosis and diphtheria.

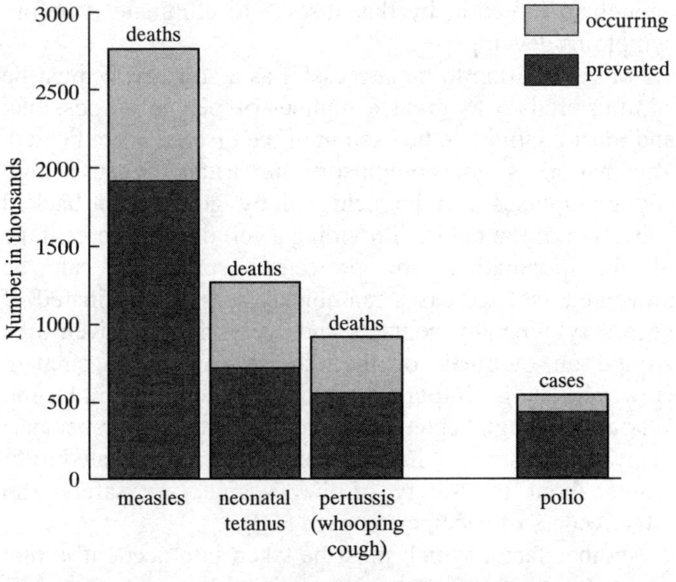

Fig 15.3 *Number of deaths due to measles, neonatal tetanus and pertussis (whooping cough), and number of polio cases occurring and prevented by vaccination in developing countries, 1990. (From* Work of WHO, Biennial Report 1992–3, *(1994), WHO, Geneva.)*

As infectious disease has been virtually eliminated as a major killer in developed countries, so other diseases have taken their place (figs 15.1 and 15.2). Notable among the circulatory diseases are coronary heart disease and strokes. The emergence of circulatory diseases and cancers as leading causes of death in developed countries is partly related to changes in lifestyle. Heart disease and strokes are discussed in section 15.5, and lung cancer has been discussed in section 7.9.5. These diseases are regarded as modern epidemics and will probably be brought under much greater control as societies develop strategies to tackle them. It can be argued that they are self-inflicted. Some major diseases of developed countries however are not related to lifestyle. They are due to an increasing length of life (longevity). As we age, so the body begins to degenerate. A whole range of degenerative diseases exists. Cancers, for example, become more common as we age.

Fig 15.2 compares causes of death in England and Wales for 1851 and 1990. Such data give an indication of the changes that can be expected as countries tackle infectious disease.

15.2.1 Vaccination

Role of vaccination

Over 10 million deaths worldwide per year are due to infectious diseases. To combat this terrible loss of life, vaccination is a powerful weapon and has been one of the great success stories of human medicine. Vaccination is the giving of antigens from a disease-causing organism, either by injection or orally, with the aim of causing the body's immune system to learn to make antibodies against the disease. The body should then be able to respond fast enough to infection by that disease to eliminate it before symptoms develop.

For vaccination to be successful as a strategy, it must be administered to as great a number of people as possible, and must continue to be used until the disease is eradicated. Vaccination is not compulsory in Britain because high enough uptakes can be achieved by government-backed education of the public. Providing a certain high percentage of the population are protected, epidemics can be prevented. Isolated cases can quickly be 'ring vaccinated' if necessary, whereby contacts and everyone in a given area around an outbreak of the disease can be vaccinated. However, it is important not to become complacent. Whooping cough began to increase in Britain again because many people stopped having their children vaccinated after scares about the safety of the vaccine (see safety and effectiveness of vaccines on page 500).

Another factor which must be taken into account is that diseases can spread across international boundaries, so it is important to have, in addition to national policies, a world-wide coordinated approach. This is one of the roles of the World Health Organisation.

The smallpox story

The most successful example of the effectiveness of vaccination is the elimination of smallpox. Up to the late 1960s a combined total of some 15 million cases of smallpox occurred annually in 33 different countries. The World Health Organisation started an eradication programme in 1956 and by 1977 the last case was reported in Somalia, and the disease was effectively extinct. The virus is still kept in secure laboratories in the USA and Russia, although there have been calls for the last samples to be destroyed.

Factors which contributed to the successful eradication of smallpox were:

Vaccination

- the virus did not keep changing its surface antigens, so the vaccine remained highly effective. Some organisms, such as those causing influenza and malaria, occasionally change their antigens by mutation, thus 'fooling' the immune systems of those who have developed antibodies against them as a result of infection or vaccination.
- a heat-stable vaccine was developed for work in tropical and sub-tropical climates.
- the vaccine was easy to administer by a scratch technique on the arm, so assistants could easily be trained.
- the vaccine was very reliable and effective.

Surveillance

- infected people were easily identified.
- rewards were offered to people who reported new cases.

Containment

- 'ring vaccination' was used in the final stages of eradication whereby everyone in the area around any site of infection was vaccinated.
- sufferers were kept in isolation until they were non-infectious.
- efforts were made to trace all contacts of those with the disease.
- international restrictions were made on the travel of those who had not been vaccinated.

All this required great international cooperation and financial support.

Vaccination programmes

Children in the UK can be given a series of vaccines to protect them against a range of diseases (table 15.2). In some countries vaccination is a legal requirement. Vaccination programmes have been particularly successful in virtually eliminating polio (poliomyelitis) and diphtheria (fig 14.42) in developed countries. Polio may soon become extinct

Table 15.2 Immunisation schedule offered to children in the UK. (From Department of Health, *Immunisation against Infectious Diseases* (1992) HMSO.)

	2 months	*3 months*	*4 months*	*1 year*	*4 years*	*10–13 years*	*14 years*	*14–15 years*
Diphtheria	✓	✓	✓	–	✓	–	–	–
Tetanus	✓	✓	✓	–	✓	–	–	✓
Polio	✓	✓	✓	–	✓	–	–	✓
Pertussis (whooping cough)	✓	✓	✓	–	–	–	–	–
Measles	–	–	–	✓1	–	–	–	–
Mumps	–	–	–	✓1	–	–	–	–
Rubella (German measles)	–	–	–	✓1	–	✓2	–	–
Haemophilus influenzae B (bacterium which may cause respiratory infection)	✓4	✓4	✓4	✓4	–	–	–	–
BCG (tuberculosis)	–	–	–	–	–	–	✓3	–

1 MMR vaccine (Measles, Mumps, Rubella) formerly given at 2 years of age, 2 females, 3 susceptibles, 4 one dose at 13 months to 4 years of age.

worldwide. Diphtheria is very rare in the UK, reduced to only 13 cases and no deaths between 1986 and 1991.

By 1984 the World Health Organisation's programme of vaccination, which is targetted against six major diseases (measles, pertussis [whooping cough], tetanus, polio, tuberculosis [TB], diphtheria), had immunised some 50% of the children in the world and this had risen to 80% by the mid-1990s. It has been estimated that this programme prevented more than one million deaths annually between 1974 and 1984. However, despite great progress, by 1990 about 3 million children were still dying each year from these diseases and about 4.6 million were still not fully vaccinated. Measles was killing 1.4 million annually (one every 20 seconds). Another 490 000 died of pertussis annually and 450 000 of TB. The annual cost, one-third of which is given in aid, is about US$ 1.5 thousand million.

The WHO Expanded Programme of Immunisation (EPI) aims to immunise more than 90% of the world's newborn against a number of viral and bacterial diseases by the year 2000. Hepatitis B is now also being targetted and it is hoped to eradicate polio by the year 2000.

There is still a need for new vaccines, for example against malaria, dengue fever, sleeping sickness, worm infections, HIV, leprosy and others. Not only are new vaccines required, but also more effective and safer vaccines than those in use at the present time are needed. For example the vaccine for cholera is only effective in 50% of patients and the duration of immunity is relatively short (see section 15.3.1). The flu vaccine also needs improvement to make it more effective.

Types of vaccine

The different types of vaccine are described in section 14.9.5. There has been much debate about the relative merits of live and killed vaccines. Generally speaking, live vaccines are more effective although in the past they have been more risky. Other factors such as cost, safety, politics and social acceptance can determine whether there is high uptake of a particular vaccine and whether it is successful.

There are many approaches to making and using a vaccine. For example in the UK three different vaccines have been licensed for typhoid vaccination:

(i) a killed whole cell vaccine (no longer available);
(ii) a polysaccharide extract from the capsule of the typhoid bacterium;
(iii) a live attenuated strain of the typhoid bacterium, *Salmonella typhi*.

The second one is the most recent, having been introduced in 1992. It requires the least number of doses and is the most preferred.

The most important issue for developed countries is the safety of the vaccine, whereas in a developing country the question of cost and how to deliver the vaccine are probably of greater importance.

A relatively recent development in the production of vaccines is that of using genetic engineering techniques. Many pathogens cannot be cultured outside their natural host. Thus the conventional approaches to vaccination cannot be used. For example, the microbe causing human syphilis (*Treponema pallidum*) and the bacterium that causes leprosy (*Mycobacterium leprae*) have never been grown in vitro (outside the body). Thus it is not possible to generate live attenuated or inactivated vaccines by culturing techniques. In these examples recombinant DNA technology offers an alternative approach, allowing genes for antigens to be transferred from these organisms to more useful hosts such as *E. coli*, yeast or mammalian cells. These can then be used to produce bulk quantities of antigens for vaccines. For example, the surface antigen gene of hepatitis is simple to identify, clone and express. However, not all protective antigens are as simple to develop.

Safety and effectiveness of vaccines

There is sometimes public controversy regarding the effectiveness of vaccines. Up to 1986, 160 million doses of measles vaccine (a live attenuated virus) had been administered in the UK, with an excellent record of protection. Of the children given vaccination, 5–15% developed fever on the fifth day, lasting several days. One recipient of vaccine in one million developed a disorder of the central nervous system known as **encephalitis**. This can cause extreme concern in parents when these facts appear in newspapers just as vaccination campaigns begin. The chances of this complication from measles vaccination is, in fact, less than the incidence of encephalitis from an unknown origin. Whooping cough (pertussis) vaccine, which contains dead bacteria, sometimes has a rare neurological adverse reaction resulting in convulsions and brain damage. This occurs once in about 100 000 doses and the possibility of permanent brain damage occurs once in 300 000 doses. Both measles and diphtheria vaccines can have local reactions of inflammation and laryngitis. However, deaths from these diseases still occur amongst unvaccinated children and parents must weigh up the information and take the responsibility of deciding whether or not to have their children vaccinated.

15.3 Infectious disease

Infectious diseases are those caused by other living organisms which invade the body and live parasitically. Such organisms are called **pathogens**. Some examples are given in tables 15.3–6. Further details of cholera, tuberculosis, malaria, AIDS, typhoid, paratyphoid and *Salmonella* food poisoning are given in the following sections.

The body's defence mechanisms against invading organisms are discussed in chapter 14 (see section 14.8.5 and 14.9 immune system).

The following technical terms are of use when discussing infectious disease.

aetiology – the study of the cause of disease
epidemiology – the study of all the factors that contribute to the appearance of a particular disease
causative agent – the organism which causes the disease
vector – an organism which carries a disease from one person to another or from an infected animal to a human, e.g. the mosquito is a vector for malaria (it is *not* the causative agent)
incubation period – the period of time between the original infection and the appearance of signs and symptoms
infective period – the time during which a person is capable of passing the disease on to another person
carrier – a person who has been infected but develops no signs or symptoms; the carrier can pass the disease on to another person
notifiable disease – a disease which must be reported by doctors to the health authorities due to its seriousness (e.g. cholera, TB, polio)
epidemic – situation in which a disease spreads rapidly through a large number of people and later 'disappears' again
pandemic – an epidemic which spreads across whole continents
endemic – describes a disease which is always present at a low level in a given population or region
signs – *visible* expression of the disease which can be found by examining a patient, e.g. a rash or a high temperature
symptoms – an indication of a disease which is not detectable by examination and can only be reported by the patient, e.g. a headache, nausea
prevention – measures taken to prevent a person from getting a disease, e.g. vaccination, sewage treatment, hygiene
treatment – measures taken to cure a disease or alleviate symptoms once a person has the disease, e.g. use of antibiotics

15.3.1 Cholera

Cholera is a good example of a waterborne disease. It is endemic in parts of Asia, particularly India. Epidemics occur from time to time in other countries, as in Peru in 1992 which was the first outbreak in South America of the twentieth century. In 1991 more than 16 000 people died worldwide from half a million cases of cholera. Improved treatment has reduced the death rate dramatically, but it is still a serious disease. Until it was discovered how to treat cholera the death toll was enormous. For example half a million New York residents were killed in one epidemic in 1832.

Transmission, signs and symptoms

The organism which causes cholera is a comma-shaped motile bacterium called *Vibrio cholerae*. The main source of infection is water contaminated by faeces from a sufferer of the disease or a 'carrier'. A carrier is an individual infected with vibrios who does not develop the typical symptoms of cholera. It is estimated that only about one

Table 15.3 Some common viral diseases of humans. Diseases are grouped according to method of spread.

Name of disease	*Cause*	*Method of spread*	*Signs and symptoms*	*Type of vaccination*
Influenza	Myxovirus (DNA virus); three types, A, B and C, of varying severity	Droplet infection	Sudden fever with headache, sore throat and muscular aches. Affects epithelia of respiratory passages, trachea and bronchi. Recovery within one week, but after-effects may last a month. Secondary infection of lung tissue by bacteria, leading to pneumonia may occur.	Killed virus: must be of right strain
Common cold	Large variety of viruses, most commonly rhino-virus (RNA virus)	Droplet infection	Nasal and bronchial irritation, resulting in sneezing and coughing. Usually only affects upper respiratory passages. Secondary bacterial infection may occur.	Living or inactivated virus given as intramuscular injection; not very effective because so many different strains of rhinovirus
Smallpox*	Variola virus (DNA virus), a pox virus	Droplet infection (contagion possible via wounds in skin, clothing, bedding, dressings)	High fever and generalised aching. Affects respiratory passages. Rash on the body two days later which spreads over body. Secondary infection by bacteria causes permanent scarring of the skin.	Living attenuated virus applied by scratching skin; no longer carried out because virus is extinct
Mumps	A paramyxovirus (RNA virus)	Droplet infection (or contagion via infected saliva to mouth)	Occurs mainly in children. Fever, followed by swelling of the parotid (salivary) glands on one or both sides, lasting about ten days. Testes, ovaries and pancreas are other organs that may be affected. Inflammation of the testes in male after puberty may cause sterility.	Living attenuated virus
Measles	A paramyxovirus (RNA virus)	Droplet infection	Occurs mainly in children. Sore throat, runny nose, watery eyes, cough and fever. Small, white spots (Koplik's spots) appear inside mouth on wall of cheek. Two days later reddish rash appears on neck at hair line – spreads over body. Recovery one week later, but virus can damage kidneys or brain. Secondary bacterial infections may occur.	Living attenuated virus
German measles (Rubella)	Rubella virus	Droplet infection	Occurs mainly in older children and adults. Affects respiratory passages, lymph nodes in neck, eyes and skin. Slight fever, body rash which disappears after three days. Complications rare, except in women during first four months of pregnancy, when there is 20% chance of blindness, deafness or other serious defects of the baby.	Living attenuated virus; more essential for girls because disease causes complications in pregnancy
Poliomyelitis ('polio')	Poliovirus (a picornavirus) (RNA virus), three strains exist	Droplet infection or via human faeces	Fever, headache and feeling of stiffness in neck and other muscles. Nerve cells to muscles are destroyed causing paralysis and muscle wasting. When breathing muscles are paralysed, an 'iron lung' may be needed. Most cases of paralysis occur in children aged 4–12 years, but adults may also be affected.	Living attenuated virus given orally, usually on sugar lump
Yellow fever	An arbovirus, that is arthropod-borne virus (RNA virus)	Vector – arthropods, e.g. ticks, mosquitoes	Fever, headache, backache, nausea, tenderness in pit of stomach. Affects lining of blood vessels and liver. Fourth day vomit blood and bile (so-called 'black vomit'). Eyes become yellow. Faeces coloured black due to digested blood.	Living attenuated virus (control of vectors also important)
AIDS	HIV virus – a retrovirus (RNA virus)	Blood-borne. Can be spread by sexual intercourse – homo- and heterosexuals	Laboratory evidence of infection, i.e. antibodies to the virus, but few symptoms except swollen glands. AIDS-related complex (ARC) may develop in about 25% of people testing positive to AIDS virus. ARC includes loss of appetite, loss of weight, fevers, persistent dry cough, white spots due to thrush (*Candida albicans*), shingles, lymphoma (cancer of the lymphatic system), pneumonia, cryptosporidiosis (severe diarrhoea), tuberculosis and other diseases resulting from the breakdown of the immune system.	Not available
Hepatitis B	DNA virus	Blood-borne and can be spread by sexual intercourse	Incubation 6 weeks to 6 months. Infects liver. Flu-like symptoms. Jaundice, nausea and severe loss of appetitite	Genetically engineered

*Last recorded natural case in Somalia, October 1977; disease extinct, though virus kept in a few laboratories

Table 15.4 Some common bacterial diseases of humans. Diseases are grouped according to method of spread.

Name of disease	*Cause*	*Method of spread*	*Signs and symptoms*	*Type of vaccination or antibiotic*
Diphtheria	*Corynebacterium diphtheriae* (rod-shaped, Gram +)	Droplet infection	Bacteria grow on moist mucous membranes of upper respiratory tract. Toxin spread by blood to all parts of body. Slight fever, sore throat followed by severe damage to heart, nerve cells and adrenal glands.	Toxoid
Tuberculosis (TB)	*Mycobacterium tuberculosis* (rod-shaped, member of actinomycetes)	Droplet infection. Drinking milk from infected cattle.	Tubercle bacilli may infect many organs, but pulmonary TB of the lungs most common. General weight loss and cough. Sputum may contain blood.	BCG living attenuated bacteria. Must test first to see if already immune. Antibiotics, e.g. streptomycin.
Whooping cough (pertussis)	*Bordetella pertussis* (rod-shaped, Gram –)	Droplet infection	Mainly in young children. Severe coughing bouts, each cough followed by 'whoop' sound at expiration of air through narrowed passages.	Killed bacteria
Gonorrhoea	*Neisseria gonorrhoeae* (coccus, Gram –)	Contagion by sexual contact	Affects mainly mucous membranes of urinogenital tract. In male, burning feeling of discomfort on passing urine, after a yellow discharge. Accompanied by fever, headache and general feeling of illness. Can affect prostate gland or epididymis of testis. Untreated, disorders of joints may follow. In female, no symptoms in genital tract. Infection commonest in urethra and cervix of womb. Infection spreads from here to fallopian tubes which become filled with pus. Sterility results.	Antibiotics, e.g. penicillin, streptomycin
Syphilis	*Treponema pallidum* (a spirochaete)	Contagion by sexual contact	Chronic in nature and widespread in body tissues. Incubation 2–4 weeks. First show is sores or chancres (painless ulcer) on any part of body e.g. vagina, penis, lips, fingers, nipples. Sores heal in 3–8 weeks. Next phase 6–8 weeks later, includes fever, skin rashes. Patient highly infectious. Final phase patient non-infectious. Many tissues damaged e.g. heart disease, insanity, blindness.	Antibiotics, e.g. penicillin
Typhus	*Rickettsia*	Epidemic typhus: vector – louse. Endemic typhus: vector – rat flea. From rat to rat by flea and louse.	After 12–14 days headache, pain in back and limbs. Measles-like rash on armpits, hands and forearm. Delirium sets in, then coma. May affect linings of blood vessels causing clots. Death can result from toxaemia, heart or kidney failure.	Killed bacteria or living non-virulent strain. Antibiotics e.g. tetracyclines, chloramphenicol (control of vectors also important).
Tetanus	*Clostridium tetani* (rod-shaped Gram +)	Wound infection	Toxins cause muscular spasms in the region of mouth and neck. Extend throughout body. Convulsions become so severe patient dies through lack of oxygen.	Toxoid
Botulism	*Clostridium botulinum* In canned or smoked food etc. grows anaerobically.	Eating infected food.	Within 24 hours vomit, constipation, paralysis of muscles and intense thirst. 50% mortality.	Antitoxins can be used to neutralise toxins
Cholera	*Vibrio cholerae* (comma-shaped, Gram –)	Faecal contamination: (a) food- or water-borne material contaminated with faeces from infected person; (b) handling of contaminated objects; (c) vector, e.g. flies moving from human faeces to food.	Bacteria produce toxins causing inflammation of the gut and severe diarrhoea. Fluid loss so intense that diarrhoea is termed 'rice water'. Resulting dehydration and loss of mineral salts can cause death.	Killed bacteria: short-lived protection and not always effective. Genetically engineered vaccine now available. Antibiotics, e.g. tetracyclines, chloramphenicol.
Typhoid fever	*Salmonella typhi* (= *S. typhosa*) (rod-shaped, Gram –)	As cholera	Mild fever, slight abdominal pains. Affects alimentary canal, spreading to lymph and blood, lungs, bone marrow and spleen. Intensity of fever and pain increase and diarrhoea follows. Ulceration and rupture of intestine may follow. Infection spreads to other organs – lungs, bone marrow and spleen. Occurs 2 or 3 weeks after infection.	Polysaccharide extract from the bacterial capsule. Genetically engineered vaccine now available
Bacterial dysentery (bacillary dysentery)	*Shigella dysenteriae* (rod-shaped, Gram –)	As cholera	Bacteria produce toxins in the intestine causing abdominal pain with blood and mucus in diarrhoea. Comes on rapidly 2 or 3 days after swallowing an infecting dose of bacteria.	No vaccine. Antibiotic, e.g. tetracyclines.
Bacterial food poisoning (gastro-enteritis or salmonellosis)	*Salmonella* spp. (rod-shaped, Gram –)	Mainly foodborne – meat from infected animals, mainly poultry and pigs. Also via faecal contamination as with cholera.	Affects alimentary canal. Can be very brief, involving 'diarrhoea and vomiting' and develops in hours. Toxins develop rapidly in the gut and cause symptoms after an infected meal.	No vaccine. Antibiotics, e.g. tetracyclines; usually not necessary and not very effective.

Table 15.5 Diseases of humans caused by protozoa.

Cause	*Disease*	*Transmission*	*Symptoms*	*Control and treatment*
Plasmodium spp.	Malaria	*Anopheles* spp. mosquito bite	After ten days high fever develops. Fever may be continuous, irregular or twice daily.	Destruction of mosquito larvae with oil spray or insecticide. Drainage of breeding places of mosquitoes. Preventive drugs (prophylactics), e.g. chloroquine. Drugs to kill parasites in humans, e.g. primaquine.
Entamoeba histolytica	Amoebiasis (amoebic dysentery)	Uncooked food, unhygienic food preparation, 'carrier' handling food	Diarrhoea with loss of blood in stools, fever, nausea and vomiting. Can lead to death.	Hygienic food handling and preparation. Prevention of spread by flies. No acceptable chemical prophylaxis. Drugs to kill parasites in humans, e.g. metronidazole, diloxanide furoate.
Trypanosoma spp.	Trypanosomiasis (sleeping sickness) also disease of cattle (nagana), transferable between cattle and humans	Tsetse fly bite	Lymph glands enlarge, fever, enlargement of spleen and liver follow. Later parasite invades nervous system; results in sleepiness and muscular spasms.	Tsetse flies live in restricted areas – 'fly belts'. Fly screens on doors and windows, spraying of cattle, moving human settlements to areas cleared of flies. Drugs to kill parasites in humans, e.g. pentamidine.

Table 15.6 Diseases of humans caused by fungi.

Disease	*Cause*	*Transmission*	*Symptoms*	*Control and treatment*
Tinea pedis	'Athlete's foot'	Communal changing and bathing facilities with wet floors	Presence of sodden, peeling and cracked skin between toes. Often persistent in hot summer months.	Disinfection of communal bath and shower floors. Exclusion of infected individuals. Drugs: griseofulvin (antibiotic) taken by mouth.
Tinea capititis	Head ringworm	Highly contagious, direct contact by way of combs, brushes, caps, hats etc.	Small scaly spot with broken hairs. Spot increases in size, covered with greyish scales, thicker at edges forming a distinct margin.	Local application of fungicides as a variety of ointments. Drugs: griseofulvin (antibiotic) taken by mouth.
Candida albicans	Candidiasis (thrush)	Can occur in mouth, vagina, intestine, etc. Infection may arise due to loss of acidity in vagina, for example during pregnancy or as a result of diabetes in women. Infants can be infected in the mouth region at birth.	Local infection of yeast organisms forming fluffy white patches. Red inflamed skin under patch. Severe irritation.	Search for underlying predisposing factor. Drugs used locally as lotions, creams or pessaries (vaginal infection). Drugs: amphotericin.

infected person in 50 develops the disease, the rest being carriers. Thus the faeces containing vibrios are a considerable threat when the 'carriers' are free moving in a society. Drinking contaminated water, or washing food or utensils in it, is the most common means of transmission. Direct contamination of food with faeces as a result of poor hygiene is also possible.

Vibrio cholerae multiplies in the intestine, releasing a powerful toxin which results in violent inflammation of the intestine and production of a watery diarrhoea. The organisms can only multiply and flourish in the human intestine, although they can survive outside of the body.

The main sign of the disease is severe diarrhoea due to irritation of the bowel by toxins from the vibrios. The liquid of the faeces is so profuse and cloudy that it is called 'rice water'. Up to 15 dm^3 of fluid can be lost in a day. Abdominal pain and vomiting are also common. Fever is absent; in fact, the skin feels 'deathly' cold and often damp. Dehydration is rapid and quickly results in death unless rehydration treatment is given.

If the faeces are not collected or disposed of hygienically then one patient can infect a wide area of a community. Each cubic centimetre of diarrhoea will contain up to a thousand million vibrios. Thus a mild case of cholera or a carrier may create a great hazard as they move around.

Treatment and prevention

The prime cause of death from cholera is loss of water with its mineral salts, that is dehydration. Replacement of these is therefore the first task. Where there are epidemics, for example in refugee camps, cheap and immediate therapy is by oral rehydration, that is, by packs of water containing balanced mineral salts and sugar administered by mouth (fig 15.4). This replacement therapy must be 1.5 volumes to each volume of stool (faeces) lost. The fluid may also be administered by a drip feed into a vein.

Various antibiotics, such as tetracyclines and chloramphenicol, are effective at destroying the vibrios and decreasing the diarrhoea. Chloramphenicol is effective against tetracycline-resistant vibrios.

During a devastating epidemic in London in 1849 the physician John Snow demonstrated that cholera was transmitted by way of water by removing the handle of the Broad Street pump (now Broadwick Street). This had been the sole water supply for a poor slum community that was badly affected by cholera. The local epidemic ceased. Authorities were not convinced and in a second epidemic in 1854 Snow showed that cholera-ridden communities were drawing their water from the lower Thames. Populations drawing water from the upper, purer reaches of the Thames were almost cholera-free. In 1875 the Public Health Act was introduced which resulted in proper sewage treatment and water purification, and since 1893 any new cases in Britain have come in from abroad.

Improvements in basic human hygiene in all parts of the industrialising world, including better garbage and waste disposal, and in-house toilets with appropriate water flushing increased the cleanliness of cities. By 1900 life

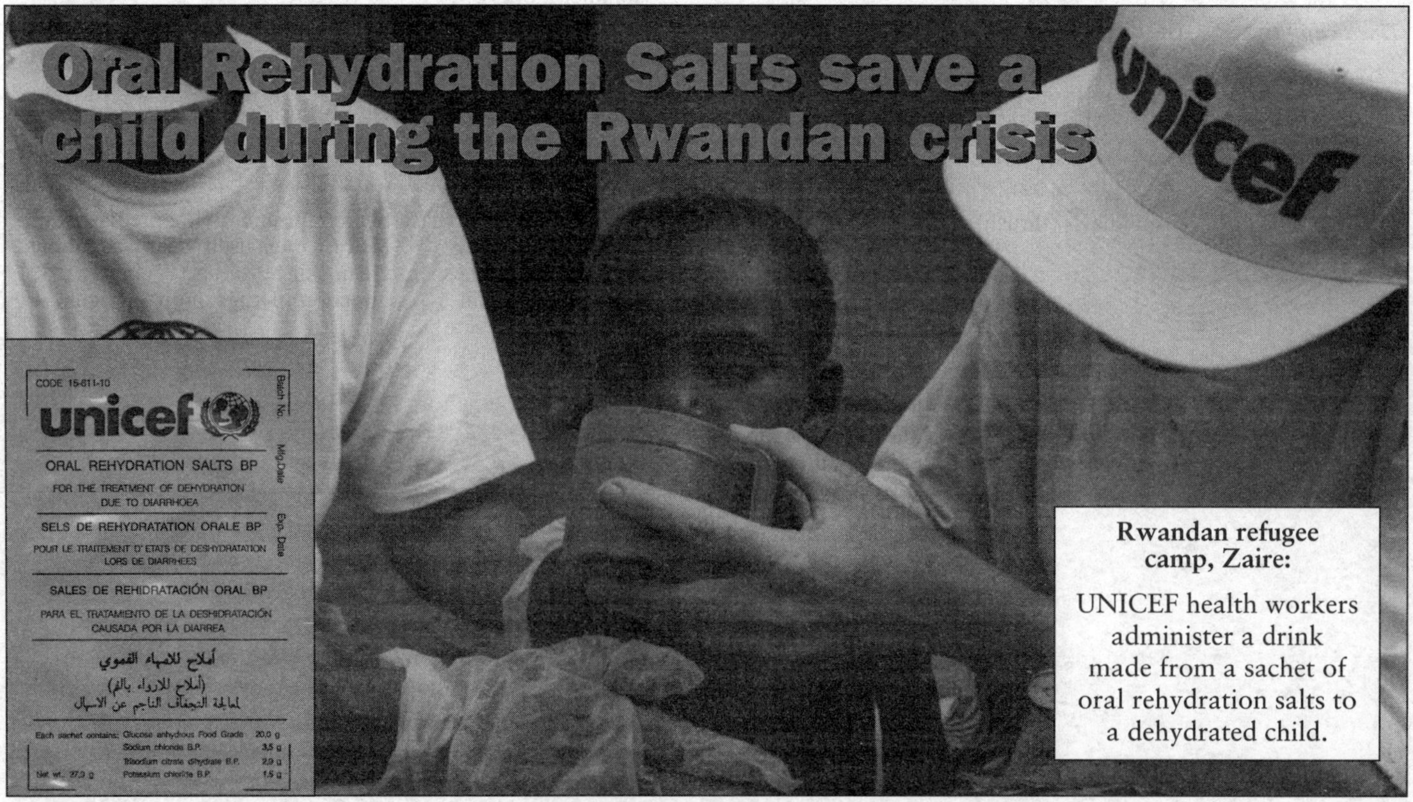

Fig 15.4 *Leaflet from UNICEF which invites donations and points out that a small sachet of ready-mixed rehydration salts costs only 7p. These sachets can be used for a variety of diarrhoeal diseases, not just cholera.*

expectancy in Britain had improved markedly and nearly all infectious diseases were in decline, not just cholera.

Cholera is one of five epidemic diseases which must be notified to the World Health Organisation in Geneva. Reporting these diseases means that prompt international action can be taken to try to prevent their spread. This will include restrictions on travel into or out of the affected area.

The key measures in controlling cholera are:

- provision of clean drinking water.
- proper sewage treatment and sanitation.
- high standards of public and personal hygiene, particularly in relation to food (such as washing hands after defaecation).
- health education.
- **vaccination** – recommended for people visiting areas where cholera is endemic and for those living in such areas. The cholera vaccine contains heat-killed bacteria. It is only about 40–80% effective and protection only lasts for about 3–6 months. A second booster dose will bring a rapid response and will give immediate protection during an epidemic. Work is in progress on a genetically engineered vaccine. An important step has been the identification and cloning of the genes coding for the toxin and its production. One promising approach is to develop a live attenuated strain of the bacterium which has had one of the two toxin genes removed.
- control of flies which act as vectors in transferring faecal material to food.
- isolation of patients and hygienic disposal of faeces and vomit from patients.
- identification of carriers and prevention of their working in the food industry.
- immediate examination of diarrhoeal diseases for bacterial content and effective treatment supplied immediately.
- close contacts should be treated with drugs to kill any possible cholera bacteria, and people in the area should be vaccinated.

In the 1970s it was discovered that the tiny cholera vibrio could live inside algae, resting encysted in a dormant state for months and years. Since that time many new cholera epidemics in India, South America and southern Asian coastal regions have appeared. These have been traced to a new strain of *Vibrio cholerae* which first appeared in Indonesia in 1961. This new, more virulent strain, has been shown to be capable of surviving in sea water and may move around the world as a result, particularly when water is contaminated by sewage. Sewage tends to encourage the growth of algae because it contains nutrients, and resulting 'algal blooms' may also help to transmit the disease.

15.3.2 Tuberculosis (TB)

Historically TB has been one of the world's worst killer diseases. Traces of TB have been found in skeletons of the late Stone Age. Fig 15.5 shows the death rate from tuberculosis in England and Wales from 1838 to 1970. It was one of the world's great killer diseases in the nineteenth century and was the largest single cause of death in England and Wales during that time, accounting for one-fifth of all deaths. By 1990, the numbers of children dying from the disease had been reduced to less than 1 in 1 million in developed countries, but at least 30 million people worldwide still suffered symptoms, 95% of whom were living in developing countries. Up to 3 million a year are currently dying from TB, and up to one-third of the world's population carry the bacteria with no ill effects.

In the early 1990s the WHO declared the disease a global emergency as cases in the developed world, including Britain, began to increase again and resistance to drugs began to grow. In 1992 there were 5802 new cases in England. Worldwide, 8 million new cases are reported every year. In 1993, Dr Kochi, manager of the WHO's TB programme, said 'TB is out of control in many parts of the world. The disease, which is preventable and treatable, has been grossly neglected and no country is immune to it.' Its global distribution is indicated in fig 15.6.

The disease is caused by a fungus-like bacterium called *Mycobacterium tuberculosis*, first discovered by Robert Koch in 1882. It is sometimes referred to as the tubercle bacillus, bacilli being rod-shaped bacteria. The most common form in the UK is pulmonary TB which infects the lungs, although other organs may be affected. Two strains of the bacterium may cause the disease, the human and bovine forms. The latter can be present in cattle and can enter the milk of cows. It is very resistant and can remain

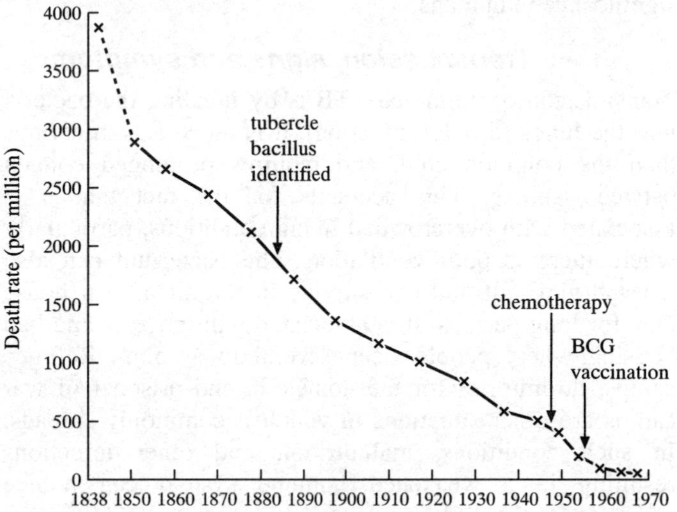

Fig 15.5 *Pulmonary tuberculosis: death rates in England and Wales from 1838 to 1970.*

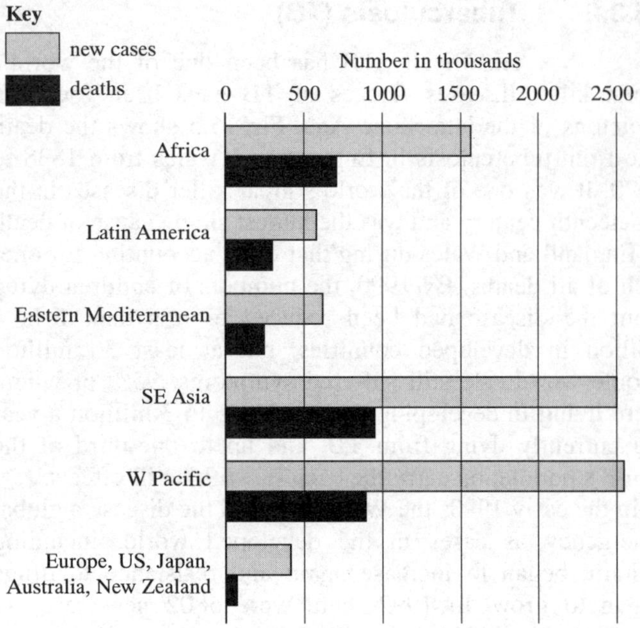

Fig 15.6 *New cases of TB and deaths from TB in different parts of the world in the late 1980s. (Source: World Health Organisation, 1989/90.)*

alive for long periods in milk products. It is a very serious disease of cattle and has also been responsible for a great deal of illness and death in humans in the past, particularly in children. Today, however, all milk in the UK is produced from cows that have been 'tuberculin tested', that is certified free of *Mycobacterium*. The milk also undergoes treatment at bottling plants where it is subjected to pasteurisation, sterilisation or ultra-high temperature. These processes destroy at least 99% of all bacteria, including all pathogens. Thus bovine tuberculosis is no longer of significance in humans.

Transmission, signs and symptoms

Transmission of pulmonary TB is by inhaling the bacteria into the lungs (droplet infection). It is much less infectious than the common cold, and requires prolonged contact between people. This accounts for the fact that it is associated with overcrowded living conditions, particularly where there is poor ventilation. The bacterium can also resist drying out and can survive in the air and in house dust for long periods. It is associated with poverty and bad housing where people sleep several to a room. Refugee camps, dormitories for the homeless, and prisoner of war camps are other situations in which it commonly spreads. In such conditions, malnutrition and other infections resulting in a weakened immune system can reduce resistance to the disease.

Tuberculosis can affect almost any tissue or organ in the body, but disease of the lung is by far the most frequent. It was commonly known as 'consumption' in the past because it consumed the body, causing it to waste away. The outcome of infection by tubercle bacilli depends on a variety of factors. These include the age of the patient, the state of nutrition (which is usually related to social class) and the presence or absence of immunity. Immunity can be acquired by an individual as a result of a previous mild infection or by vaccination (see below).

The disease frequently shows itself by vague symptoms such as loss of appetite, loss of weight and excessive sweating. There are often no symptoms in early tuberculosis and the disease may only be accidentally discovered through a routine X-ray of the lungs (fig 15.7). The disease starts as an inflammation in one lung, which develops into a cavity. Then further cavities develop, spreading into both lungs. As progressive destruction of the lungs occurs the symptoms become more dramatic with coughing, appearance of blood in the sputum, chest pains, shortness of breath, fever and sweating, poor appetite and weight loss.

Treatment and prevention

Effective medical treatment only began in 1947 with the introduction of the antibiotic streptomycin. Mass vaccination did not begin in Britain until 1954. The decline up to this point must have been due mainly to improving social conditions, particularly improved housing. Vaccination accelerated the decline (fig 15.5) and by 1970 the annual death rate in Britain had fallen to 1526.

Vaccination. The development of an effective vaccine against the disease resulted from the work of two French scientists, Albert Calmette and Camille Guérin,

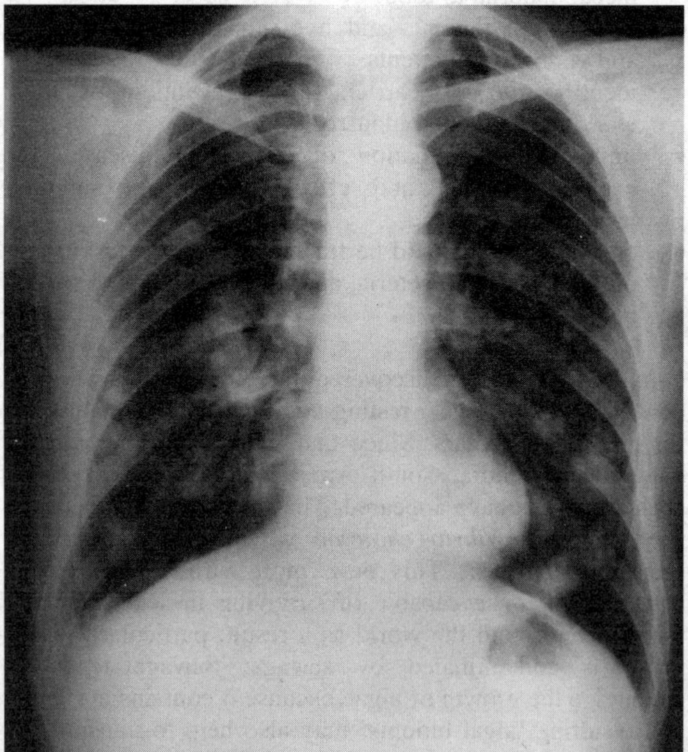

Fig 15.7 *Chest X-ray of TB sufferer. A normal lung would be uniformly dark. The clouded areas show areas of infection.*

hence the name of the vaccine Bacille Calmette-Guérin (BCG). As far back as 1921 they developed attenuated (less virulent) strains which were found to be effective for vaccination. Before treating any individual it is important to check if they are already suffering from TB or have recovered from it. The test is to puncture the skin with a special instrument which has a ring of six short needles (the Heaf test). This introduces a protein called tuberculin, purified from dead tubercle bacilli. In the absence of past or present TB the skin shows no reaction, but if an individual has the disease or has recovered, then the skin swells and reddens at the injection site. This indicates a substantial immunity and no vaccine is offered.

A detailed study of 50 000 healthy children, reported on in 1963, showed that the incidence of TB per 1000 children was 1.91 if unvaccinated, and 0.4 if vaccinated. The benefit of vaccination therefore lasts for a long period of time because the children still had immunity after more than ten years. Today children are vaccinated at age twelve to fourteen years (table 15.2). Tuberculin tests indicate that about 10% of children are positive at this age. These children are given a routine X-ray to ensure that no active pulmonary tuberculosis is present, and very few children have the disease.

Antibiotics. A cure for people already affected by TB did not come until 1943 when the antibiotic streptomycin was discovered. The number of cases started to fall more rapidly after this (fig 15.5) and continued to decline up to the mid-1980s, aided by the introduction of further antibiotics such as rifampicin, isoniazid and others. At that time in Western countries more than 80% of all active TB cases were of people over sixty years of age.

Resurgence of the disease

After 1980, the demographics of the disease shifted in that more and more young people between the ages of twenty five and thirty were developing the disease. Between 1980 and 1986 five different surveys in the USA showed a relationship between the rise of homelessness and surges of TB in young adult populations. It became clear by 1985 that new mutant strains of drug-resistant TB were also present in the population. In 1986 patients with strains of *Mycobacterium* resistant to both isoniazid and rifampicin numbered 0.5% of cases, by 1991 this had risen to 3% and to 6.9% in 1994. The main contributing factors were courses of treatment being too short to kill the bacteria and patients not completing the doses prescribed. The full course of treatment lasts 6–8 months and requires consumption of several tablets a day, every day. This gives a combination of at least three or four antibiotics to reduce the chance of a strain multiplying which is resistant to one of the antibiotics. The problem is made worse by the fact that the patient starts to feel well again after a few weeks. Supervision by health workers is difficult not only in developing countries, but also in large cities such as New York where many sufferers are homeless, and where TB has become a new epidemic.

From the beginning of the AIDS epidemic it was noted that HIV-positive members of the community developed a high rate of tuberculosis (section 15.3.4). Many developing countries took steps to heed a WHO warning regarding this relationship between HIV and TB. Doctors in the USA and most of Western Europe, however, took little notice of these facts for they tended to view the TB risk for HIV patients as a Third World problem. In Africa TB began to spread rapidly and HIV patients did not respond well to the two cheapest antituberculosis drugs, thiacetazone and streptomycin. By 1990 health experts in some African countries were predicting defeat in their efforts to control tuberculosis.

The new strains of drug-resistant bacteria spread rapidly and there were clear interconnections between these new strains and HIV. Patients with AIDS, with its immunodeficiency, were very susceptible to infection, and death rates rose to 90–100% fatality. The percentage increases in TB for different European countries are shown in table 15.7. These are directly related to the increase in drug-resistant strains and HIV infection.

A report in 1996 on TB in Edinburgh during the period 1988–1992 showed the following:

4.1% increase in TB cases recorded among people over 65 years
12.6% increase in TB cases recorded in younger patients

In the elderly, most cases were the result of reactivation of TB caught in childhood or youth. The rise in both age groups was entirely due to the increased resistance to antibiotics of the tubercle bacilli.

Immigration is also associated with an increase in TB. For example areas in Britain with large immigrant populations have shown increases 25% higher than in the indigenous residents.

The World Health Organisation has started to achieve a more successful treatment of TB with its DOTS campaign (directly observed treatment under supervision). The patient is given pills under supervision and watched each time to check the pills are swallowed. This is done over a period of 6 to 8 months and results in a cure of over 85% of cases.

15.3.3 Malaria

Malaria has been one of the world's worst killer diseases throughout recorded human history. Despite attempts to eradicate it, it remains one of the worst diseases in terms of deaths annually, and has actually increased in incidence since the 1970s. About 200 to 300 million new cases occur worldwide each year, and about 1.5 million deaths, over two-thirds of which occur in Africa. It is particularly common in Africa south of the Sahara, and is widespread throughout Asia and Latin America (fig 15.8). It used to be common in Europe and North America. (Oliver Cromwell died of malaria and Sir Walter Raleigh suffered from it.)

Table 15.7 Percentage increase in tuberculosis cases in some European countries during the period 1987–91. (From World Health Organisation, *Press Release June 17*, 1992.) The US reported a 12% increase from 1986–91.

Country	*Time period*	*% increase in TB*
Switzerland	1986–90	33
Denmark	1984–90	31
Italy	1988–90	28
Norway	1988–91	21
Ireland	1988–90	18
Austria	1988–90	17
Finland	1988–90	17
Netherlands	1987–90	9.5
Sweden	1988–90	4.6
United Kingdom	1987–90	2.0
France, Germany, Belgium	1987–91	stable

Malaria provides a good example of how social, economic and biological factors are all important in controlling disease.

Transmission, signs and symptoms

The causative agent of malaria is the protozoan parasite *Plasmodium*. Four species cause malaria, but most cases are caused by one of two species: *Plasmodium vivax* is found in the subtropics; *Plasmodium falciparum* is more common and more lethal, and is found in both the tropics and subtropics. It is the largest single cause of death in Africa.

Malaria is transmitted by female mosquitoes of the genus *Anopheles*. This mosquito is itself a parasite, the females visiting humans for occasional meals of blood. During feeding, infected mosquitoes pass on the malaria parasite from their salivary glands. The mosquito is described as a **vector**.

An immature form of *Plasmodium* (the sporozoite) is injected into the blood of humans by the mosquito. This form disappears from the bloodstream as it enters various cells of the body, particularly the liver. Here it multiplies to produce large numbers of a form (the merozoite) which can infect other liver cells. Finally it leaves the liver and enters red blood cells. Each parasite cell in a red blood cell undergoes further division. The red cell bursts and the released parasite cells can enter other red blood cells. As a result of this extensive division, millions of these parasites can be present in the blood. Some of the parasites transform into male and female forms of the parasite (gametocytes).

If infected red blood cells are now sucked up by a mosquito the parasites survive digestion in the stomach of the mosquito and transform into male and female gametes. Fertilisation occurs and the zygote penetrates the wall of the stomach where it grows to produce a swelling containing immature parasites.

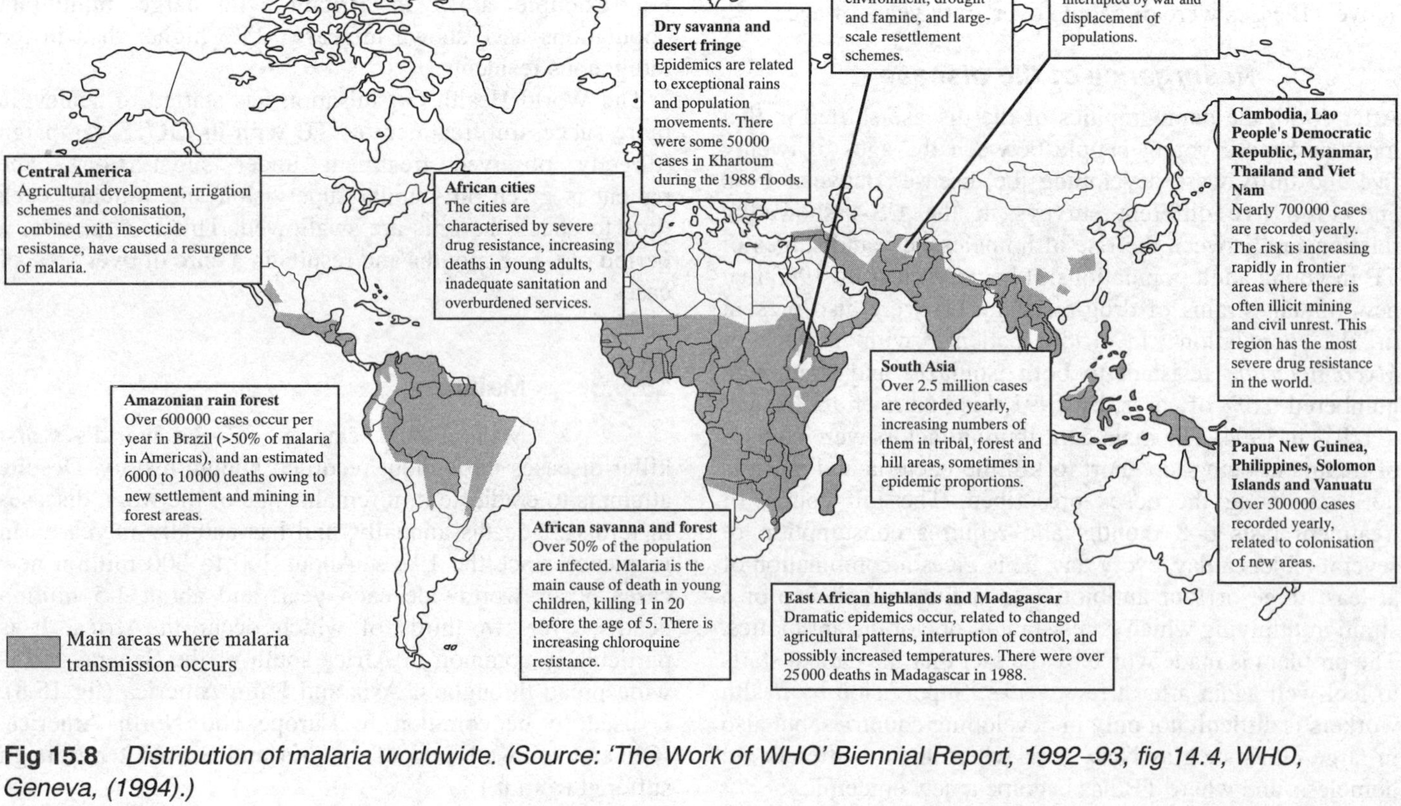

Fig 15.8 *Distribution of malaria worldwide. (Source: 'The Work of WHO' Biennial Report, 1992–93, fig 14.4, WHO, Geneva, (1994).)*

People who have been exposed to infection since birth and who have survived attacks of malaria develop a certain amount of tolerance to the disease. People with no history of previous infection will develop serious disease very rapidly. Ten days after infection a fever develops and the body temperature increases rapidly to 40.6–41.7 °C. The fever may last as long as 12 hours accompanied by headache, generalised aches and nausea. After the fever, sweating starts and then the temperature falls. The area of the abdomen over the spleen is tender.

The symptoms and fever coincide with the multiplication of the parasite when the red blood cells burst open and the parasites are released. The fevers occur every third day in *P. vivax* and *P. falciparum* malarias. The attacks can be complicated, however, as a result of successive infections by mosquitoes.

The infections by *P. falciparum* cause malignant malaria in which the fever is accompanied by other complications. The parasite tends to accumulate in blood vessels in the brain, causing convulsions or coma. Other common complications include kidney failure and pneumonia. Malaria caused by *P. falciparum* can be fatal within two or three days.

Prevention and treatment

Prophylaxis, that is the use of medicines (drugs) to prevent disease, can be used by people entering areas endemic for malaria. The usually accepted prevention is chloroquine or mefloquine which is taken weekly before and during a visit to endemic areas. The dose must also be taken for six weeks after leaving the area. Other synthetic drugs are proguanil hydrochloride and pyrimethamine. The effectiveness of these drugs has declined as the parasite developed resistance. *P. falciparum* is now resistant in most areas where malaria is endemic, such as Latin America, East Africa and S.E. Asia. New drugs are constantly being sought, but pharmaceuticals companies are tending to scale down work on antimalarial drugs because there is little profit to be made from developing countries.

A drug called Fansidar and another called Lariam (mefloquine) are effective against chloroquine-resistant *Plasmodium*. Mefloquine can cause unpleasant side effects such as nausea, dizziness, diarrhoea, vomiting, headache, abdominal pain and occasionally psychiatric disturbance. It was introduced in 1985 but already half the malaria cases in Thailand are resistant to it. A traditional drug, quinine, has come back into favour, but resistance is also growing to this.

Together with the use of drugs, other measures such as clothing (long sleeves and trousers) and mosquito nets at night should be used in mosquito areas (Fig 15.9). These are now often impregnated with new insecticides.

It may be necessary to administer drugs by injection. This then produces a high concentration of the chosen drug to kill the malaria organisms in the blood.

Eradication of malaria

In 1955 the World Health Organisation launched a programme designed to eradicate malaria from the world. Large sums of money were devoted to this project and the methods employed were as follows.

- **Drainage of stagnant water.** The larval stages of the mosquito live in stagnant water, so drainage removes breeding sites. This has had some success. However, the process is expensive and incomplete because rural populations must ensure that ponds, small ditches and even containers holding water are not allowed to provide breeding places for mosquitoes.
- **Destruction of the breeding stages of the mosquito.** The larvae and pupae of mosquitoes obtain their oxygen by means of small tubes which are pushed through the water surface film. Thus any method of blocking these tubes will result in the death of the intermediate life stages of the mosquito. The simplest method is a thin layer of oil spread over the water surface to block the breathing tubes. Petroleum oil sprayed from back packs is used. This method was used effectively in Brazil in 1938 and eliminated *Anopheles gambiae* by 1940.
- **Destruction of the adult mosquitoes.** This is aimed at killing the mosquitoes that enter houses. Thus the indoor surfaces are sprayed with a persistent insecticide. If dwellings are sprayed for three years, the cycle of man–mosquito–man can be disrupted because *P. vivax* and *P. falciparum* eventually die out in infected patients.

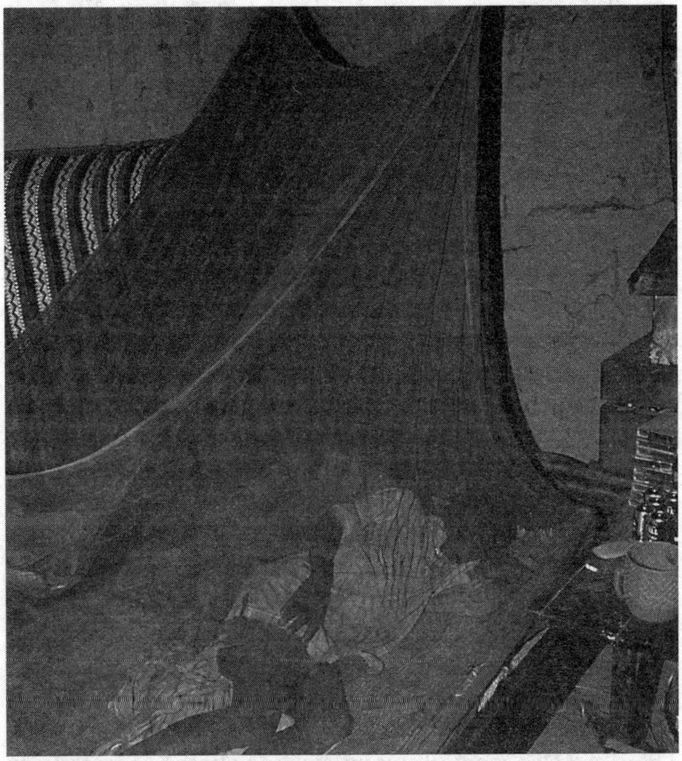

Fig 15.9 *This boy is sleeping under a mosquito net which has been sprayed with insecticide as a protective measure against mosquito bites.*

Such a programme requires tremendous resources in effort and money. In spite of this, very great efforts have eliminated malaria in many areas such as Chile, Europe (e.g. Cyprus, France, Italy, Netherlands), parts of Asia (e.g. Singapore) and N. America. In Africa it is still an important part of the public health programme, but the complete eradication of malaria in the world is still a very distant prospect.

Many factors work against anti-malaria programmes. For example, national boundaries are not barriers to mosquitoes and malaria can travel widely and rapidly. Political and financial difficulties in neighbouring countries, particularly in times of war, make it difficult to sustain a programme of preventive measures. In addition, migration of people into an area, for example looking for work or clearing and developing new land, can result in rapid infection of people not previously exposed to the disease and local epidemics (fig 15.8). This has happened as a result of the exploitation of the Amazon rainforest and in other areas including Madagascar, Ethiopia and Sri Lanka.

Another problem is that mosquitoes have become resistant to pesticides. In 1951 the first DDT-resistant strains of mosquitoes were reported from Greece, Panama and the USA. In many parts of the world spraying no longer prevents transmission of malaria. In addition, the passage of persistent insecticides such as DDT and dieldrin through the food chain became an increasing problem because birds and mammals at the top of the chain were found to have increasing amounts of the insecticides in their bodies (chapter 10). These effects have been found both on land and in the seas. DDT breaks down very slowly in soils and marshes and may be detected up to 30 years after spraying. Thus spraying with DDT and dieldrin has been suspended in many places.

Control of malaria has also been made more difficult because there is an enormous reservoir of the disease in monkeys, birds, rodents and reptiles which are also affected by it.

As a result of all these problems in trying to eradicate malaria, in 1969 WHO gave up the aim of trying to eradicate it and settled instead for a 'control' policy.

Vaccination. In the 1980s and 1990s there was a shift in emphasis to development of a vaccine against malaria. So far, despite intensive efforts, it has proved impossible, although it is a top priority of malaria research. It must be cheap as well as effective if it is to be afforded by developing countries.

The parasite is only vulnerable to attack by antibodies when not inside liver cells or red blood cells. Attempts have been made to develop vaccines against these vulnerable stages using dead or attenuated parasites, but without success. One possible reason is that the antigens of the parasite may change over time, so that antibodies against the original antigens fail to recognise the new antigens. There are also many different existing strains of malaria as well as different species. Genetic engineering to produce large quantities of relevant antigens has been attempted, as well as other approaches. Mapping of the parasite's genes is underway with the 'Malaria Genome Project'. Work is also in progress on human genes responsible for the variable resistance shown to malaria.

A synthetic protein vaccine developed in Colombia started trials in Africa in 1994. It has been shown to be safe and to trigger a strong immune response. Trials are due to be completed in 1998.

15.3.4 AIDS – Acquired Immune Deficiency Syndrome

AIDS is thought to have originated in Central Africa. The HIV virus which causes it then appears to have migrated via Haiti to the USA, and has subsequently been identified in 71 other countries throughout the world. By late 1993 it had infected an estimated 14 million people, with over 3 million estimated cases of AIDS. By late 1995 an estimated 23 million people were infected (fig 15.10), and there were an estimated 3.1 million new infections and 1.5 million deaths. Half the new infections were of women and the majority were aged under 25. By the year 2000, around 40 million people may be infected by HIV. Fig 15.10 shows that the greatest problem is in sub-Saharan Africa, and this is where the greatest increases are being observed. The disease is also spreading rapidly in Asia (notably India, China, Vietnam and Cambodia), Central and E. Europe and the former Soviet Union.

AIDS is a disorder which damages the human body's immune system. It is caused by the HIV virus (Human Immunodeficiency Virus). This is an RNA virus and its structure is shown in fig 2.21 (see also figs 2.23 and 2.24). The virus replicates inside the T4 lymphocytes or 'helper' cells (see section 14.9.2). Thus these cells can no longer 'help' or induce other T cells, called killer cells, to fight invaders. The body's immune system breaks down, leaving the patient exposed to a variety of diseases (see later). It is important to realise, however, that infection with the HIV virus does not necessarily result in AIDS. As with other diseases, some people remain symptomless and are therefore termed 'carriers'.

Transmission, signs and symptoms

The HIV virus can only survive in body fluids and is transmitted by blood or semen. In 90% of cases the transmission is achieved by sexual contact. People can contract the disease as follows.

- **Intimate sexual contact.** The disease was first associated with homosexual communities, notably in American cities such as San Francisco, Los Angeles and Miami where there were high levels of promiscuity amongst homosexuals. Since then it has become clear that transmission can also take place between heterosexuals. It passes from the infected partner to his/her unaffected partner through vaginal or anal intercourse, or oral sex. The risk becomes propor-

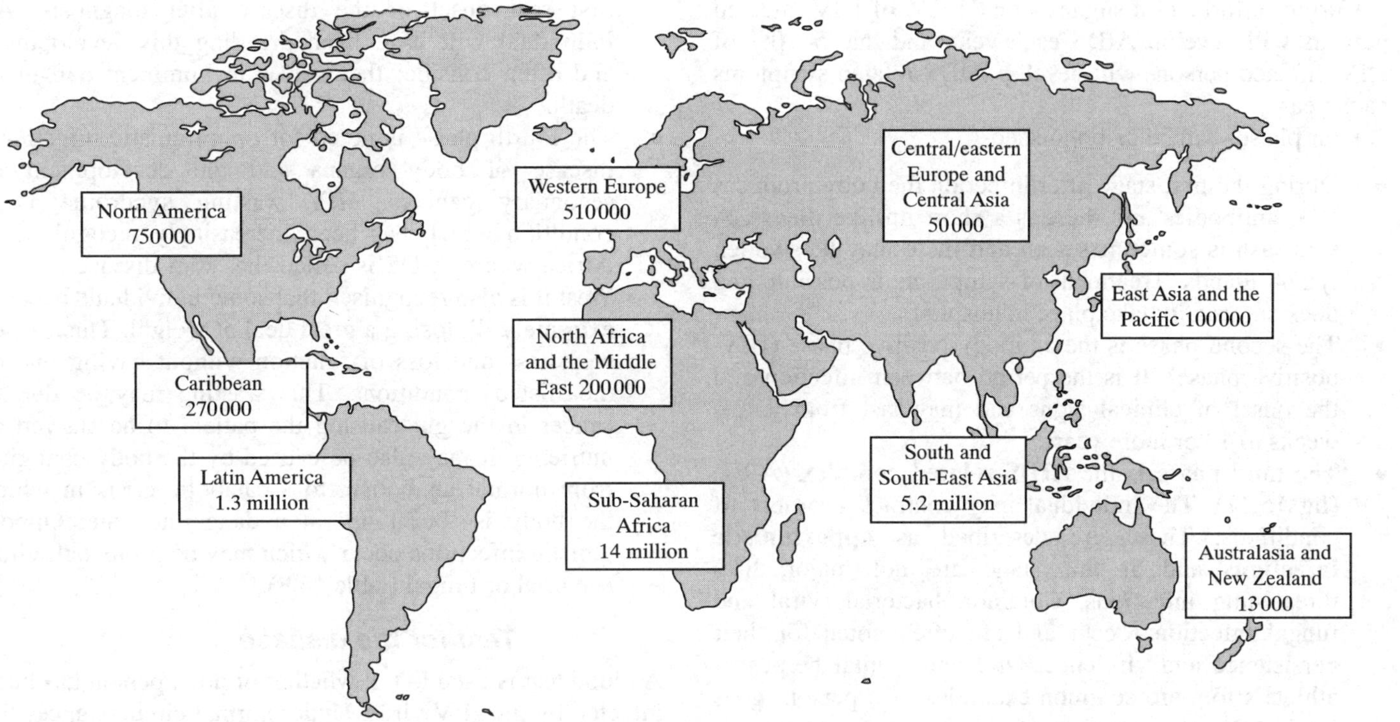

Fig 15.10 *Estimated total numbers of HIV infections in adults in different parts of the world, 1996. (Source: UNAIDS Geneva.)*

tionately greater amongst those who are promiscuous. In America and Europe the disease is still largely confined to homosexuals, but in Africa it is also present in the heterosexual community. There the most rapid spread is amongst prostitutes and this pattern may be occurring in the Western World. In October 1988 it was estimated that 50 000 people in the United Kingdom had the HIV virus in their blood and most of them were homosexuals (85%) who were expected to develop the fatal disease in the next 10–15 years.

- **Infected blood entering the bloodstream.** AIDS can be contracted by intravenous drug users practising self-injection by means of unsterilised needles and syringes. HIV has spread rapidly amongst intravenous drug users. Of the 250 000 heroin addicts in New York, it is officially estimated that 60% have already been infected by the virus. Once in the bloodstream of the drug addict, HIV can be passed on through sexual activity, to other drug users and to the general public.

  Unfortunately the disease can be contracted after being given blood or blood products already infected with HIV. This has happened to some haemophiliacs who have been given factor VIII from infected blood. As a result of extensive screening programmes and other measures, which include heat treatment of blood products to inactivate the virus, the treatment of haemophilia has been prevented from being a means of HIV transmission in countries with adequate medical facilities.

  Close contact between infected and non-infected people through cuts and open wounds has also been known to pass on the virus.
- **From mother to baby.** An infected pregnant woman can pass on the virus to her baby through the placenta, at birth or through breast milk during suckling. The chances of the infection being transmitted from the mother to her baby are currently estimated to be 25–50%.

The virus binds to receptors present in the surface of the T4 lymphocytes. From here it enters the lymphocytes by endocytosis or by fusing with the cell surface membrane and injecting its viral RNA directly into the cell (fig 2.23). The viral RNA is then copied into DNA by the activity of an enzyme called **reverse transcriptase** (section 2.4.5 and fig 2.23). The viral DNA enters the lymphocyte nucleus and becomes incorporated into the cell's own DNA. Thus it becomes a permanent part of the cells of an infected individual. Every time the human cell divides, so does the viral DNA, and thus spread of the viral genes is rapid.

The viral DNA may remain dormant for at least six years, the so-called **latency period**. However, suddenly, for some unknown reason, the lymphocyte begins to make copies of the viral genes in the form of messenger RNA. These then migrate from the nucleus into the lymphocyte cytoplasm and direct the synthesis of viral proteins and RNA. These assemble to form new HIV viruses which leave the lymphocyte by budding out from underneath the cell surface membrane (fig 2.24). The viruses spread and infect many other lymphocytes and brain cells. Eventually the cells in which the virus has multiplied are killed.

Current information suggests that 1–2% of HIV infected persons will develop AIDS each year, and that 5–10% of HIV infected persons will develop AIDS-related symptoms each year.

Four phases can be distinguished.

- During the first stage after infection the body produces HIV antibodies and there is a short flu-like illness. A skin rash is sometimes seen and there may be swollen lymph glands. Treatment of symptoms is possible and does not usually take place in hospital.
- The second phase is the antibody-positive phase (HIV-positive phase). It is the period between infection and the onset of clinical signs and may last from a few weeks to 13 or more years.
- The third phase is the **AIDS-related complex** (ARC) (fig 15.11). The individual may contract a variety of conditions. These are described as **opportunistic infections** and at this stage are not major, life-threatening infections. Common bacterial, viral and fungal infections occur and are often noted for their persistence and virulence. Oral and genital herpes or athlete's foot are common examples. If a person 'goes into' ARC the duration of this type of infection is lengthened, compared with that in a normal healthy person. Loss of weight may be seen at this stage (up to 10% body mass). A significant drop occurs in the number of T helper cells.

  Appropriate nursing is required since this stage is the first real onset of the disease after diagnosis. An individual will have been dreading this development and often consider themselves in imminent danger of death.

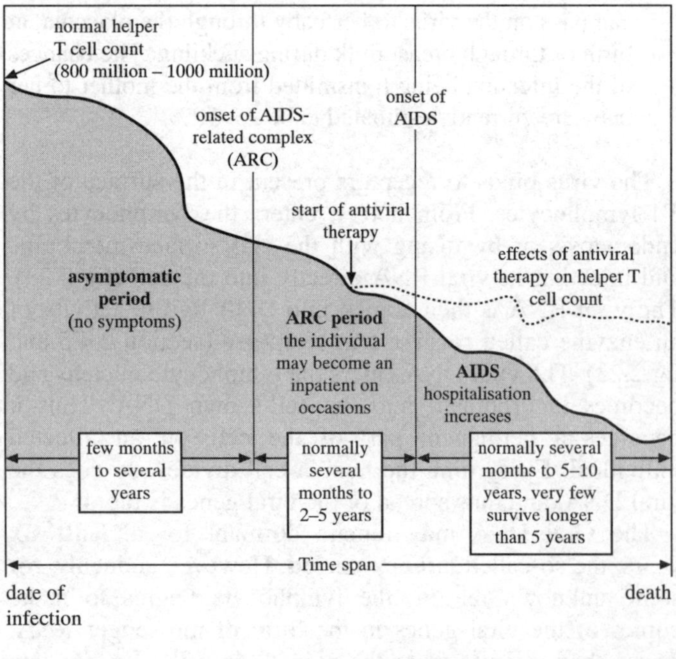

Fig 15.11 *Diagrammatic representation of the progress of HIV-related illness. Many people die when they get the first major opportunistic infection (onset of AIDS). It is very difficult to state when a person is going into the terminal stages of the disease. Many people make remarkable recoveries over a period of months or years.*

- The fourth phase is noted for opportunistic infections, disease of body organs and the development of secondary cancers. HIV wasting syndrome is a condition which has been increasingly recognised in Africa where AIDS is called the 'slim disease'. In the West it is also recognised that some individuals become extremely ill, losing a great deal of weight. This causes weakness and loss of function without having one of the 'listed' conditions. This wasting may be due to cancer in the gut causing the patient to be starved of nutrients. It may also be caused by the body changing from normal anabolism to a catabolic crisis in which the body is 'burnt up' at a dangerous rate. Opportunistic infections occur which may be protozoal, viral, bacterial or fungal (table 15.8).

Test for the disease

A blood test is used to tell whether or not a person has been infected by the HIV virus. Under normal circumstances the immune system reacts to infection by producing antibodies, and when the HIV virus enters the body, anti-HIV antibodies are produced. The blood of the person being tested is added to HIV proteins which have been commercially prepared. If there are anti-HIV antibodies in the blood sample they will bind to the viral proteins and the person is then described as HIV positive. However, if the test proves negative that person may still be infected. This is because it takes up to three months or longer for HIV antibodies to be produced after infection.

Treatment and prevention

AIDS is caused by a virus and while bacteria can be controlled by antibiotics, these are not effective against viruses. Most treatments are therefore limited to relieving symptoms.

Present research on treatment and prevention is concentrating on three areas:

(1) restoring or improving the damaged immune system of victims.
(2) developing drugs that will stop the growth of the virus and also treat the other infections and symptoms that result from HIV infection.
(3) developing a vaccine against the virus.

These three methods are considered below.

Restoring the immune system. The strengthening of the immune system seems a logical step to take to help the body cells to combat the virus. Suppressor T-cells in cultures appear to be able to control the HIV virus. There are also individuals infected with the HIV virus who have no sign of the disease, nor any trace of the virus in the blood. They have in fact shown some

Table 15.8 Secondary infections associated with HIV infection.

Type of infection and cause	*Signs and symptoms*
Protozoal infections	
Pneumonia caused by *Pneumocystis carinii* (PCP or pneumocystosis)	About 60% of individuals contract pneumocystosis as the first AIDS-related infection and it is the most common cause of death in AIDS. Normally the immune system would keep this infection at bay, but in HIV patients life-threatening pneumonia develops. Treatment is by the drug co-trimoxazole.
Cryptosporidiosis caused by a small protozoan in the water supply	In patients whose immune system is damaged the population of protozoa in the gut rises to high levels and causes diarrhoea. There is great loss of fluid and intravenous fluid replacement is needed.
Toxoplasmosis caused by a protozoan (often associated with cats and raw meat)	Infection can cause lesions in the cerebrum of the brain. The patient lapses into paralysis and unconsciousness.
Viral infections (only most common shown)	
Herpes simplex virus	Also associated with ARC
Cytomegalovirus (CMV)	In HIV-related illness causes retinitis when the patient may rapidly become blind.
Bacterial infections	
Tuberculosis (TB)	See section 15.3.2
Salmonellosis	This infection is commonly associated with food poisoning, and is particularly dangerous to AIDS patients who should avoid undercooked foods, particularly eggs and poultry. (See section 15.3.6).
Fungal infections Candidiasis	This is a virulent infection in AIDS patients, and may extend from the mouth down the alimentary tract. (See table 15.6).
Secondary cancers (neoplasms)	Kaposi's sarcoma (KS) A purplish skin cancer was one of the first manifestations of AIDS in the western world. Not seen in all patients, but when present can be extremely disfiguring, especially if on the face. May develop internally and cause obstructions in the gut.
Non-Hodgkin's lymphomas	Seen in much higher numbers among patients with AIDS. Found in the central nervous system, the bone marrow and the gut. Often rapidly fatal.

improvement in their immune systems. In 1987 an experiment showed that blood samples from three individuals with antibodies to the virus in their blood, but no detectable virus, when taken and treated by removing T-suppressor cells began to grow the virus. When the cells were put back into the blood samples the virus was again suppressed. This showed that infection by the virus does not necessarily mean the disease will develop, and that AIDS is an opportunistic infection, causing disease only in someone whose immune system has been severely weakened.

Bone marrow transplants are a proven technique for the treatment of a form of cancer called leukaemia, the success of which depends on the genetic closeness of the donor and the recipient, with ideally identical twins or siblings being used. In theory a similar procedure should be applicable to AIDS patients and should provide the body with an immediate source of immune cells. Unfortunately to date there has been only one notable success using a combination of bone marrow transplant (from an identical twin), antiviral drugs and transfusions of lymphocyte blood cells.

Another line of attack is to use protein substances known as **lymphokines** (from lymphocytes and Greek *kinen* 'to move'). These are produced in small amounts by lymphocytes and move from cell to cell carrying a message through the immune system. The system can then deal with a threat from cancer cells or viruses. The best known is **interferon**, which can be produced by genetic engineering, and is now approved for use against cancer. A specific type of interferon called alpha-interferon has shown some success in causing the regression of Kaposi's sarcoma (a skin cancer associated with AIDS). Interleukin is another immune system activator which has shown some success against Kaposi's sarcoma.

Development of drugs.

- **Azidothymidine** (AZT) has received most attention in the media. Since 1986 there have been numerous trials with modest success and certainly life has been prolonged in a proportion of patients. It does have side-effects such as anaemia (requiring frequent blood transfusions) and there is the problem of supply in relation to an overwhelming demand from patients.
- **Zalcitabine** is a sister compound of AZT that has been shown to block replication of the HIV virus in laboratory cultures. Unfortunately trials have shown the drug to be highly toxic in humans.
- **Glycyrrhizin** is a component of licorice and has been used to treat hepatitis and various allergies. Japanese researchers have shown it to be capable of halting the growth of the HIV virus. No clinical trials have yet been conducted.
- **Ribavirin** has been used against influenza and other viruses in many countries, and against more viruses and in more animals than any other agent. In laboratory studies it has been found to inhibit replication of the AIDS virus and increase the number of T4 cells without damaging the infected cells.

There are a number of other drugs that are being developed and tested. However, in all the work on drug testing a very difficult issue is whether the use of a **placebo** (a comparable drug of useless effect) in drug trials for so lethal a disease as AIDS is proper. Should every AIDS patient who wants an experimental drug get it, even if it has not been proven in long-term trials? AIDS patients who are clamouring for a cure and have nothing to lose, argue that it is immoral and unfair to deny them such medication on demand. In terms of proper research and testing, however, scientists say that there is no other way than to give placebos to some patients and the real drug to the rest. Another factor is that of expense. Millions of pounds would be used up supplying a drug that may not work and in any case would never be useful for the majority of people.

Vaccine. In order to protect people not yet exposed to HIV the ideal solution would be to develop a vaccine. The most difficult part will be to overcome the ability of the virus to change its genetic structure and so alter the proteins on the surface of the virus which the antibodies recognise. Another problem in vaccine production is the matter of safety. Antiviral vaccines are generally made up of live viruses that have been attenuated (weakened), enabling them to protect against, but not cause a disease e.g. vaccines for measles, polio and rubella.

Researchers all over the world are striving to produce an effective vaccine, but as yet no product has been made ready for human trials. The best that vaccine researchers have been able to test so far, in laboratory animals, are a variety of protein antigens from the AIDS virus.

Simply developing a vaccine, even if it were possible, does not end the story, for how to test it safely and prove its effectiveness are formidable obstacles. Furthermore other difficult questions would be who gets it, and at what age? Should vaccination be a condition for employment, marriage, travel, insurance policies? Should it be voluntary or compulsory for high risk groups only, or for the general population? Should young people be allowed vaccination without parental approval? Should vaccination have to be revealed if so requested?

There are other obvious precautions which can be followed in trying to prevent the disease.

- The use of a barrier during intercourse can prevent the virus from infecting through blood or semen. Thus the use of a sheath or condom is recommended. This practice has been encouraged through many advertising campaigns throughout the world. There is evidence that this advice is beginning to be understood and used.
- Restriction to one sexual partner and the absence of promiscuity will also clearly reduce the risk of infection.
- A reduction in the spread of HIV can be brought about by the use of clean needles and syringes by drug addicts. Some health authorities in European countries, such as the Netherlands, provide free sterile needles and syringes for drug users.
- Since October 1985 all blood donated in Britain has been tested for the presence of antibodies to HIV which indicates whether or not the donor is infected. Blood containing these antibodies is not used.
- Education about the disease has an important part to play, particularly in reassuring the public about the real risks. There is no evidence that infection can occur by droplet infection through the nose or mouth, or by casual contact such as shaking hands. Healthcare staff who tend AIDS patients have never contracted the disease in this way.

15.3.5 Typhoid and paratyphoid fevers (*Salmonella typhi* and *S. paratyphi*)

Typhoid has been a great scourge of humans. It has typically been a disease of armies and over-crowded communities and is therefore associated with poverty, starvation and war. Before 1875 typhoid was a widespread disease in the UK, but the Public Health Act of that year brought about change. It led to improvements in sanitation and water supply both of which contributed to prevention of the spread of the disease.

In recent years some two hundred cases of typhoid have been notified to public health authorities in England and Wales, but most of these have been acquired abroad on holiday visits. Unfortunately, typhoid fever is always present in communities in the form of actual cases of the disease or in the form of carriers of the bacteria.

Typhoid and paratyphoid are caused by the bacteria *Salmonella typhi* and *Salmonella paratyphi* respectively. Paratyphoid is very similar to typhoid, but is usually milder in its symptoms.

Transmission, signs and symptoms

The bacteria are derived from the faeces of a sufferer from the disease or from a carrier. The disease is spread by water or from contaminated food. Typhoid spread needs only a small number of organisms, and is therefore described as having **high infectivity**. Paratyphoid, however, like food poisoning (salmonellosis) needs a larger dose of infecting organisms (**low infectivity**). Common sources of infection are listed below.

- Water supplies may be contaminated by human faeces through seepage of sewage into a reservoir, leakage from defective underground sewers or discharge of sewage into a river. Salmonella can live for about one week in sewage, or longer in water which has been diluted with sewage.
- Serious epidemics have also been traced to food. Milk can be contaminated by a carrier or from dirty equipment which has been washed with contaminated water. Outbreaks of typhoid in England, such as Epping, 1931 (260 cases, 8 deaths) and Bournemouth, 1938 (718 cases, 70 deaths) were caused by infected milk supplies. Aberdeen had an outbreak in 1964 caused by contaminated corned beef.
- Shellfish, oysters and mussels are often responsible for transmitting typhoid, due to their filter method of feeding whereby pathogenic organisms are extracted from the surrounding water.

When the infecting organisms have been swallowed they migrate to the lymph glands, where they multiply during a ten-day incubation period. After this time the organisms enter the bloodstream and the patient develops headache and muscular pains. A fever develops reaching its peak after about one week. A faint rash may appear. In the second week diarrhoea develops and the patient has mental confusion. The third week shows the peak of the illness and the patient deteriorates and may die. Other illnesses accompany these symptoms including bronchitis, pneumonia, meningitis and abscesses.

Since typhoid is a comparatively rare disease in the UK it may not be diagnosed readily. If there is a history of recent travel abroad, then any fever accompanied by other symptoms should be considered as a possible indication of typhoid. Blood samples will confirm, after culturing, the presence of antibodies against *S. typhi*.

Treatment and prevention

The disease had a 20% fatality rate before the use of antibiotics. Chloramphenicol and ampicillin are effective and reduce fatality rate to between 1 and 5%.

Any patient with typhoid requires the highest standards of nursing together with isolation and hygienic disposal of faeces. The two most important preventive measures are proper sewage treatment and purification of water supplies. Contamination of food can be reduced by personal hygiene, hygienic measures in the food trade and in the home, and control of flies, which can transfer faecal material to food (see also cholera, section 15.3.1).

Vaccination is effective by the injection of a suspension of the polysaccharide from the capsule of the bacteria. There is often a reaction to the injection with pain and swelling of the arm, and sometimes a fever. However, the vaccination is essential for travellers to parts of the world where sanitation is poor.

15.3.6 Salmonellosis – *Salmonella* and food poisoning

There are many cases of 'food poisoning' which develop within hours after infection. These illnesses are caused by salmonella organisms other than *S. typhi* or *S. paratyphi*, for example *Salmonella enteritidis*. They are referred to as **salmonellosis**. The economic and social costs of food poisoning can be very high. Salmonellosis in England and Wales was estimated to cost between £231 and £331 million in 1988 and 1989 in terms of lost production through sickness-related absence from work, investigation and treatment.

In contrast to typhoid, salmonellosis organisms have low infectivity, in other words large numbers of bacteria are required to cause infection. This is also in contrast to the high infectivity of *E. coli* 157, an increasingly important agent in food poisoning. Other bacteria may also cause food poisoning, for example *Listeria monocytogenes* which is particularly associated with pâté and dairy products such as soft cheeses and causes **listeriosis**. *Staphylococcus aureus*, *Bacillus* and *Clostridium* species are also well-known causative agents which are beyond the scope of this book.

Salmonella enteritidis is particularly associated with eggs and meat from poultry. It was important in the rising incidence of salmonellosis from 1987 and accounted for 53% of salmonellosis in 1989 and 63% in 1991. At that time hen's eggs represented a newly recognised source of salmonellosis.

Fig 15.12 shows the number of cases in Britain of salmonellosis resulting from *Salmonella enteritidis* found in eggs. Despite environmental health regulations designed to eradicate the bacterium, the strain prevalent in poultry has increased since 1988.

Transmission, signs and symptoms

Salmonellosis is a classic foodborne disease. Table 15.9 shows the different types of food that were responsible for salmonella outbreaks during 1989–1991 and shows the

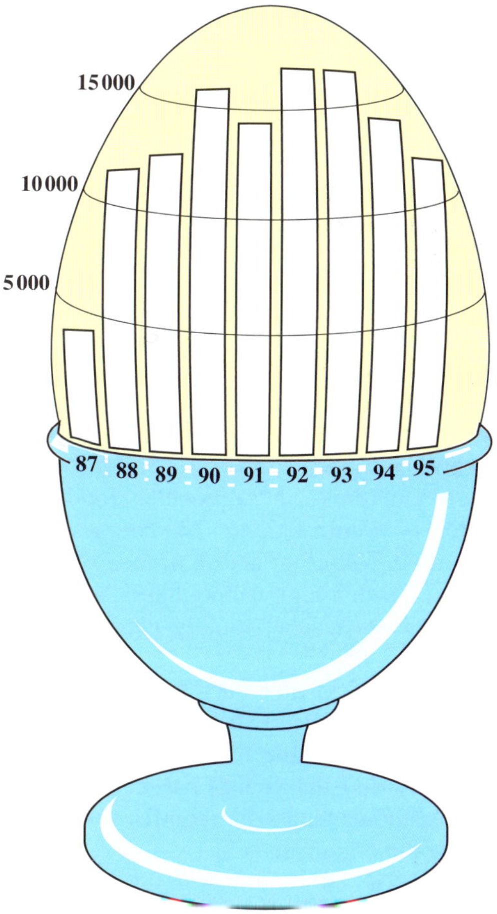

Fig 15.12 *Cases of food poisoning in Britain resulting from* Salmonella enteritidis *found in eggs, 1987–1995. (From* The Times*, 23rd March 1996.)*

Table 15.9 Reported food sources in general outbreaks of salmonella infection 1989–91. (From *Communicable Disease Report* vol.3, Review no. 12, 5 Nov. 1993, Public Health Laboratories Service.)

Food	*Number of outbreaks due to* Salmonella enteritidis	*Total number of salmonella outbreaks*
chicken	16	23
turkey	5	21
beef	3	7
pork or ham	–	5
cold meats	2	7
other meats or pies	8	24
gravy or sauces	1	1
milk	–	3
other dairy products	–	1
eggs	45	50
vegetables or spices	–	1
bakery products	3	7
sweets or puddings	20	21
mixed foods	42	47
other/not stated*	114	174
total	**259**	**392**

*includes 12 outbreaks where infection was presumed to have been contracted abroad

relative importance of *S. enteritidis*. Salmonellosis not only infects humans but also many food animals such as poultry, pigs and cattle from which humans can be directly infected. The faeces of rats, mice and domestic pets may also contaminate human food.

Signs and symptoms of salmonella food poisoning appear suddenly, within 12 to 24 hours after eating contaminated food. Vomiting and diarrhoea occur, accompanied by a fever with a high temperature. Abdominal pain or discomfort, and headache, usually occur. Most people recover within a few days.

Dehydration is a risk, and may lead to complications such as low blood pressure and kidney failure. These complications are responsible for most of the deaths that occur. The elderly and the young, particularly babies, are most at risk. Accurate diagnosis requires a culture of the faeces to isolate the organisms.

Treatment and prevention

The treatment is to replace fluid and salt loss and to use drugs. Antibiotics are effective against salmonella organisms, and include tetracyclines and ampicillin, but they do not shorten the course of acute diarrhoea. Recently drug-resistant strains of *Salmonella* have begun to appear.

There are several methods of prevention.

- **Carriers** Symptomless carriers of the disease can retain the organisms in the faeces for some time and present a problem in society. About 2–5 people per thousand of the general population are thought to be carriers. Known carriers and people who have suffered the disease are not allowed to work in the food industry until samples of their faeces have been shown to be clear of the pathogenic organisms.
- **Hygiene in the home and food trade** Storage, preparation and cooking of food should all be carried out hygienically. Food should be stored cool or refrigerated to minimise bacterial growth. Great care should be taken in kitchens to separate the different preparatory processes, so that raw meat, particularly poultry and eggs do not cross-contaminate other foodstuffs. Cutting boards, dish cloths, dirty kitchen utensils and unwashed hands can all harbour salmonella and bring about transfer from food to food and place to place. Environmental Health Officers regularly inspect restaurants, shops and factories.
- **Thorough cooking** Cooking should be thorough enough to kill all bacteria. One of the greatest dangers is not thawing frozen food sufficiently. In addition, the Chief Medical Officer recommended in 1988 that raw eggs should be avoided and vulnerable groups such as the elderly, sick, babies and pregnant women should consume only eggs which have been cooked until the white and yolk are solid.
- **Meat inspection** by Environmental Health Officers is essential and any animals suffering from salmonellosis should not be used in the food industry. Low levels of salmonella contamination are regarded as inevitable amongst poultry, and only large outbreaks result in slaughter of the birds.
- Food poisoning is a **notifiable disease**. However, many mild cases go unreported. In 1995 food poisoning from all causes was at its highest level since records began in 1949: 80 000 cases were recorded in 1995 compared with 63 000 for 1992. Accurate monitoring of the incidence of food poisoning is essential in helping to decide what preventative measures to take.
- **Control of rodents** whose faeces may contain salmonella.
- **Proper sewage disposal** (see typhoid and cholera).
- **Control of flies** (see typhoid and cholera).
- **Government intervention** In 1989 the government took steps to control the rise in salmonella cases. Testing of poultry flocks became compulsory. More than three million infected laying hens were slaughtered and the Food Safety Act of 1990 raised standards in food production.

15.4 Disinfectants, sterilisation and antiseptics

As we have seen, microorganisms are extremely numerous and widespread and many have the ability to invade humans and other organisms and live parasitically and cause disease. It was not until their existence was established by Louis Pasteur in the nineteenth century that any effort could be directed to destroy them. Up to that time most surgical operations were followed by a period of pain and fever and diseases such as tetanus, septicaemia (blood poisoning) and gangrene. The deeper the surgeon operated into the body, the greater the risk of wound infection. Pasteur showed that microorganisms could be killed or removed by various techniques such as heat, chemicals or filtration.

Joseph Lister, who was appointed Professor of Surgery at the University of Glasgow in 1860, was greatly influenced by the findings of Pasteur and he began to develop chemical methods for the destruction of microorganisms. He began with carbolic acid (phenol) applied to surgical instruments, catgut used for sutures (stitches), ligatures and surgeons' hands to 'clean' whatever came in contact with the wounds of surgery (fig 15.13). He applied carbolic acid to the patient's skin before cutting and to wounds, and also sprayed the air of the operating theatre. His methods soon produced a reduction in infections following operations. The techniques that followed from his antiseptic work were designed to eliminate the microorganisms from the instruments and the body surface of the patient, rather than killing the organisms after they had been introduced.

Carbolic acid is an example of an antiseptic (see below). The concept of antisepsis was followed by asepsis or sterilisation. This involved sterilisation of the patient's skin before cutting and similar sterilisation of instruments, gowns, caps, masks, rubber gloves and catgut, in fact anything that came into contact with the patient during surgery.

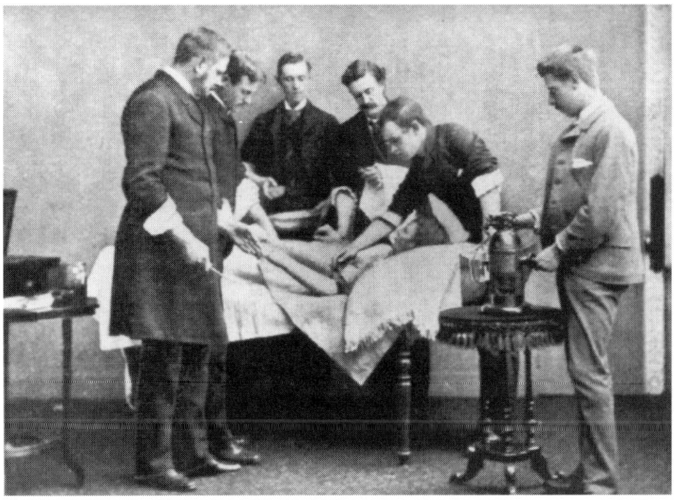

Fig 15.13 *Operation in Lister's day. The figure on the right holds a carbolic spray as pioneered by Lister.*

15.4.1 Antiseptics and disinfectants

Antiseptics and **disinfectants** are chemical substances that are used to destroy microorganisms in order to prevent infection. There are few that will destroy all microorganisms, but certain substances will destroy most of them.

The difference between an antiseptic and a disinfectant is that an antiseptic can be applied more or less safely directly to the human body, whereas disinfectants cannot. Antiseptics are used on living surfaces such as the skin. Disinfectants are used on substances such as working surfaces (kitchen tops, baths, sinks etc.), crockery, cutlery, operating theatres, drains, etc.

Antiseptics and disinfectants are generally prepared as liquids. Their effectiveness varies in that temperature, exposure time, concentration and the presence of organic matter can all limit their action.

The resistance of different microorganisms to given disinfectants or antiseptics varies considerably. For example the polio virus is very resistant, as is the tuberculosis bacterium (tubercle bacillus). Spore forms of bacteria and fungi are extremely resistant to changes in environmental conditions, so it is not surprising that they are also resistant to cold, chemical disinfectants.

There is a wide range of useful disinfectants.

- **Hypochlorites**, in the form of calcium hypochlorite and sodium hypochlorite, form hypochlorous acid and oxygen. The acid rapidly kills microorganisms. Domestos and Milton are commercial examples of this type of disinfectant.
- **Phenol** was one of the first disinfectants to be used by Lister although it is rarely used today. Phenol derivatives are used in combination with detergents for hospital cleaning purposes.
- The alcohols **ethanol** and **isopropanol** are often used as antiseptics, but they are also used as disinfectants for storing boiled syringes. 70% alcohol destroys tubercle bacilli and some viruses but not fungal or bacterial spores. The vapour is inflammable and so can be dangerous. The substances are also expensive.
- **Aldehydes**, in particular formaldehyde, can be used for disinfection and sterilisation of surfaces, but formaldehyde is too much of an irritant for skin use. Its power to destroy decay organisms led to its use as a preservative for dead organic material, although less harmful compounds are often used today.
- **Detergents** added to water increase the 'wettability' of the water and this helps to bring disinfectant molecules onto the surface. Additionally grease is removed and the cleansing action of the detergent removes bacteria and dust. Following with disinfectant allows greater penetration of the disinfectant.
- **Chlorxylenol** (marketed as Dettol) and iodine can be used as disinfectants as well as in more dilute form as antiseptics.

Useful antiseptics are shown in table 15.10.

15.4.2 Sterilisation

This process involves the removal of all life forms from any non-living object or material. Sterilisation is used in many ways to achieve an end result that cannot be obtained by disinfectants or antisepsis. For example liquids often require sterilisation, e.g. water for sterile processes, culture media for growing microorganisms. Solid materials such as instruments, clothing, even 'sterile' rooms and operating theatres all need to receive sterile treatment. There are a number of different methods which can be used.

- **Heat treatment** Usually heating in a dry oven or exposing to steam. Heat is one of the oldest ways of sterilising food in order to make it safe to eat. The amount of heat treatment that is used in commercial food processing depends on the kind of food and the microorganisms that may be found in the food. Only a very short time is required for 'pasteurising' milk above 70 °C and thus eliminating the bacillus causing tuberculosis (section 15.3.2).

  Tinned food is heat treated and may then be stored for long periods. To kill the spores of *Clostridium botulinum* (which causes botulism) requires some 200 minutes at 250 °C, compared with most bacterial spores which only require temperatures of around 120 °C and a much shorter treatment time.
- **Steam treatment** Using an autoclave, simply a steel chamber with a door for loading and unloading apparatus, cultures etc. Steam under pressure is fed into the chamber. The time of exposure to steam depends on the temperature and pressure in the autoclave. The temperature of 120 °C only requires about 12 minutes at a pressure of 150 kN m^{-2} for adequate sterilisation.
- **Radiation** Visible wavelengths have no effect on bacteria other than a warming effect. Shorter wavelengths, such as ultraviolet light, can result in death if the dosage is high enough. Most effective for this purpose are X-rays and γ (gamma) rays which can destroy microorganisms and so are useful for sterilisation. γ radiation from radio-isotopes allows simple and easily controlled sterilisation of disposable hypodermic syringes, scalpels, bandages etc. as well as some types of food.

Table 15.10 Examples of antiseptics and their uses.

Antiseptic	*Use*
phenol (carbolic)	Used by Joseph Lister but is poisonous to humans and is no longer used as an antiseptic. Not as efficient as modern disinfectants.
chlorinated phenol derivatives, e.g. hexa-chlorophene, chlorxylenol ('Dettol')	Skin cleanser (about 100 times more effective than phenol). Not safe for babies. Hexachlorophene prepared in form of lotion or soap. Needs several days of washing with soap to be effective since the chemical accumulates in the skin. Widely used in soap by medical profession. Doctors, nurses and midwives use it for 'scrubbing up' before operations or the dressing of wounds and burns. Dettol can be used in preventing infection before and after childbirth. It has a powerful odour and therefore should not be used on cutlery and crockery.
soap	Skin. Not very powerful.
cetrimide – also a detergent (contains quaternary ammonium salts)	Cleaning minor wounds, and skin around major wounds. Virtually non-toxic to humans. Kills many bacteria as well as some fungi, e.g. *Candida* which causes thrush.
70% ethanol	Not a very good antiseptic but combines antiseptic properties with removal of grease and therefore good cleansing agent for skin.
iodine	Used as potassium iodide in 90% ethanol. Formerly used on wounds but can injure tissues. Still used on intact skin in preparation for surgery. Kills bacteria rapidly, including spores. Stains brown which limits its use.
hydrogen peroxide	An oxidising agent, decomposed by enzymes in blood to release oxygen. Can be used on skin or to clean wounds.
acridine dyes	Wounds
brilliant green, crystal violet	Ringworm

15.4.3 Antibiotics

Antibiotics are chemicals produced by micro-organisms which are capable of destroying or inhibiting the growth of another microorganism (section 12.10.2). The microorganisms that produce antibiotics are mostly bacteria, but a few fungi also produce them. The definition of antibiotic has become looser with time. For example, as we have learned more about the chemical nature of antibiotics, scientists have begun to make more and more *synthetic* antibiotics. Also, chemicals which are active against microorganisms have been isolated from a variety of organisms such as plants, insects and amphibians. The treatment of disease with chemicals is called **chemotherapy**.

There are two types of antibiotic, biostatic and biocidal. **Biostatic** agents inhibit the growth and multiplication of susceptible microorganisms, e.g. chloramphenicol, erythromycin, sulphonamides and tetracyclines. The microorganisms can continue growth and multiplication if the agent is removed. **Biocidal** agents kill microorganisms, e.g. streptomycin, cephalosporins, penicillins and polymyxins. Biocidal antibiotics may become biostatic at lower concentrations. A biocide that kills bacteria is described as **bactericidal**. A biostat that acts on bacteria is described as **bacteriostatic**.

In 1928 Sir Alexander Fleming observed that a contaminating growth of fungus on an agar plate containing a growth of *Staphylococcus aureus* had apparently inhibited the growth of the bacteria. The commercial exploitation of this discovery was not achieved until 1941, when it played a considerable part in the Second World War in terms of the wounded and their infections (section 12.11.1).

The first bactericidal chemical (not strictly an antibiotic since it was synthetic) preceded the work of Fleming, because in 1935 it was announced in Germany that a red dye called prontosil could cure infections by certain bacteria (haemolytic streptococci). The active part of this dye was a substance called **sulphanilamide** and many derivatives of this compound, called **sulphonamides**, have been synthesised in succeeding years. They have been used in the treatment of bacterial infections such as streptococcal and some staphylococcal infections, meningitis, urinary infections and pneumonia. They also provided a means of combating sexually transmitted diseases such as gonorrhoea.

Sulphonamides are bacteriostatic drugs and therefore do not kill bacteria, but simply stop their growth and multiplication. They are competitive inhibitors of enzymes as described in section 4.4.1. The defence mechanisms of the body follow up the action of the drug and remove the bacteria. The bactericidal antibiotics that developed from Fleming's discovery are more powerful and rapid because they actively destroy bacteria. This action becomes extremely important in regions of the body where natural defence mechanisms are poor e.g. central nervous system and joint cavities.

Mechanism of action of antibiotics

Antibiotics are effective for a variety of reasons. Generally speaking they interfere with some aspect of metabolism of the microorganism which is not found in its host. Thus the microorganism suffers, but the host does not. Bacteria have a slightly different protein synthesising machinery and different cell walls to eukaryotes, and these are two obvious points of attack. Table 15.11 summarises some of the mechanisms by which antibiotics work.

Broad-spectrum antibiotics are effective against a broad range of different bacteria. For example, tetracyclines and chloramphenicols are highly active against almost all

Table 15.11 Most commonly used antibiotics.

Antibiotic	*Targetted process*	*Original source*	*Mechanism of action and general points*
penicillins cephalosporins	cell wall synthesis	*Penicillium notatum* (fungus) *Cephalosporin acremonium* (fungus)	In most Gram positive bacteria, inhibit formation of peptide links between molecules in cell wall. Wall bursts open (lysis). Effective only against growing bacteria. Widely used and well tolerated in the body, though some people are allergic.
vancomycin		*Streptomyces* (filamentous bacteria)	
rifampicin (synthetic form of rifamycin group)	transcription (RNA synthesis)	*Streptomyces*	Binds to RNA polymerase in bacteria but not mammals, prevents transcription.
streptomycin chloramphenicol erythromycin tetracyclines	translation (protein synthesis)	*Streptomyces*	Bind to bacterial (70S) ribosomes, not 80S eukaryote ribosomes. (Tetracycline can bind to 80S ribosomes but cannot enter mammalian cells.) Inhibit translation and protein synthesis. Streptomycin useful against *Staphylococcus aureus* (which causes spots and boils, blood poisoning or food poisoning, problems in hospitals especially drug-resistant form). Chloramphenicol only used for serious diseases, e.g. typhoid, or if no suitable alternative, as has serious and sometimes fatal effects on bone marrow. Erythromycin is useful alternative for patients allergic to penicillin.
anthracyclines	DNA replication		Anti-cancer drugs. Inhibit DNA synthesis in all cells but effect most marked in rapidly growing cancer cells.
amphotericin B nystatin	cell membrane function		Used mainly against fungi. Amphotericin binds to ergosterol (only found in fungal membranes) and distorts its shape, opening channels in membrane and affecting normal exchange of ions and molecules.
polymixin		*Bacillus polymixa*	Polymixin used against Gram negative bacteria, potentially dangerous to kidney and nervous system.
interferons	block protein synthesis and cell growth, only attach to cells infected by viruses, prevent modification of viruses inside cells	white blood cells	Active against viruses.

the common Gram positive and Gram negative pathogens. **Narrow-spectrum antibiotics** are effective against a narrow range of bacteria. Penicillins, for example, are ineffective against most Gram negative bacteria, which include some important pathogens such as the bacterium that causes tuberculosis.

Resistance

A major problem that is increasing is that many bacteria have been found to show resistance to antibiotics. The emergence of antibiotic-resistant bacteria is closely linked to the extent that antibiotics are used in humans and items of human diet. Resistant strains may appear rapidly or slowly, according to the amount or type of antibiotic used.

Resistance may occur through exclusion of the antibiotic. For example, Gram negative bacteria are naturally resistant to penicillins because their cell wall has a complex structure which excludes penicillins. In other cases entry of the antibiotic may be slowed down enough to enable the antibiotic to be broken down by enzymes as it arrives in the cell. A bacterium can achieve tetracycline resistance by making a protein which is inserted into the cell surface membrane and pumps out the antibiotic as soon as it enters.

Antibiotics may also be destroyed by enzymes inside the cells being targetted. A well-known and important example is the group of enzymes known as **penicillinases** which hydrolyses and destroys penicillins and cephalosporins. The bacterium therefore has resistance to these antibiotics. Some of the semi-synthetic penicillins are sufficiently different in structure to avoid hydrolysis.

Another problem is that, if the target of the antibiotic changes slightly, it may become inefficient. An example of this is resistance to streptomycin caused by a change in the structure of the ribosome, to which streptomycin normally binds. A change of only one amino acid in one of the ribosomal proteins as a result of random mutation can be sufficient.

An organism may acquire resistance in one of two ways:

- **Mutation** Spontaneous, random mutation can occur at any time. Microorganisms occur in such large numbers that there is a high chance of a resistant individual eventually appearing in the population. As soon as it does, use of the antibiotic to which it is resistant will give it a selective advantage over non-resistant types and it will multiply and eventually become the dominant type.
- **Transfer of resistance** Resistance can spread from one bacterium to another by transfer of the relevant genes. The most common mechanism is **conjugation**, a simple form of sexual reproduction described in section 2.3.3. The resistance genes are often on a plasmid (a small circular piece of DNA) which can replicate and send a copy into another bacterium by conjugation. Other methods of transferring genes also occur. Multiple resistance, that is resistance to two or more antibiotics, is commonly acquired using these methods and this can also be transferred from one bacterial species to another. Multiple resistant *Staphylococcus aureus* (MRSA) is now a common problem in hospitals.

Antibiotics in human food

Concern has been growing about the use of antibiotics in farming. They have been used for the treatment of bovine mastitis or other infections in animals. Antibiotics are also used in feeding young farm stock to encourage growth, employed as food preservatives or to control disease in plants. As a consequence, these antibiotics occur as traces in human diets and then result in hazards to health. The hazards are direct toxic effects, hypersensitive reactions and the production of antibiotic resistance to pathogenic organisms transmissible to humans.

In order to ensure that there are no antibiotic residues in human food derived from animals, at least 48 hours should be allowed between last treatment and time of slaughter.

15.5 Cardiovascular disease

In the developed countries infectious diseases are no longer the major cause of death, as discussed in section 15.2. The average life expectancy is now about 76 years. As people age, however, they become prone to diseases of the heart and blood vessels, and cancers, and these two types of disease account for about two-thirds of all deaths in developed countries.

Cardiovascular disease (disease of the heart and blood vessels) is Britain's biggest killer, accounting for about 40% of all premature deaths. The two major cardiovascular diseases are **coronary heart disease**, which accounted for about 25% of all deaths in Britain in 1990 (30% for males and 23% for females), and **strokes**, responsible for about 10% of all deaths in Britain in 1990. Overall, this is between 300 and 400 people per day and represents about a five-fold increase since the Second World War. It is not surprising then that cardiovascular disease has been called the 'modern epidemic' as it is on a scale comparable to the major infectious diseases of the past.

Unfortunately Britain has one of the highest rates of cardiovascular disease in the world and in 1992 the Government set targets aimed at reducing its incidence (fig 15.14). The figures show targets for those aged under 65. Slightly less ambitious targets were set for the over-65s. The targets acknowledge the fact that, to a large extent, the deaths are avoidable and that it is important to understand their causes and to try to develop more effective strategies to reduce the numbers of deaths.

15.5.1 Atherosclerosis

By far the most common cause of cardiovascular disease is **atherosclerosis**. The process leading to atherosclerosis starts with the deposition of yellow fatty

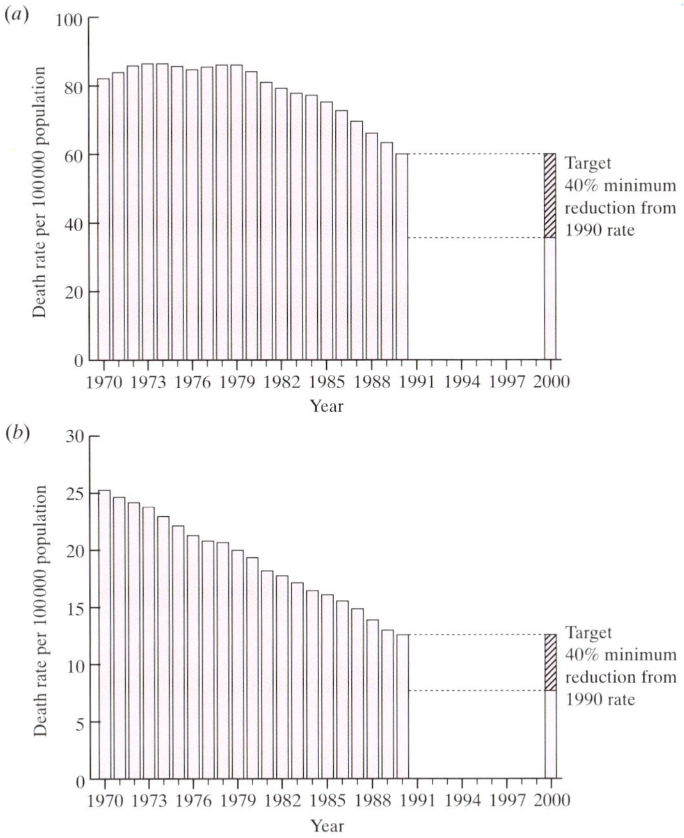

Fig 15.14 *(a) Death rates for coronary heart disease in the UK for people aged under 65 and health target for the year 2000. (b) Death rates for stroke in the UK for people aged under 65 and health target for the year 2000. (From* The Health of the Nation *(1992) HMSO.)*

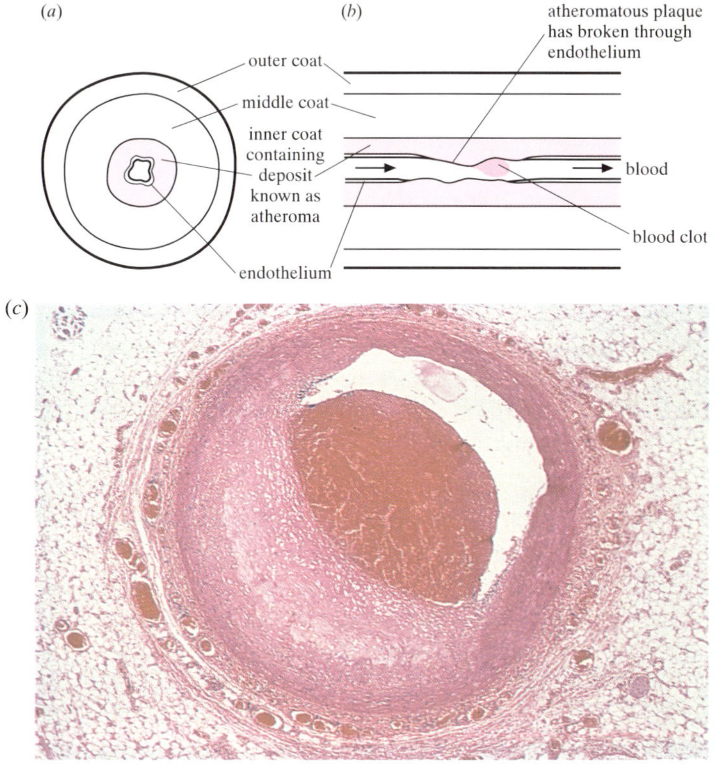

Fig 15.15 *Narrowing of arteries caused by atherosclerosis. (a) TS artery narrowed by atheroma. (b) LS narrowed artery with blood clot forming. (c) Section through human coronary artery showing almost total obstruction by a haemorrhage developing into an atheromatous plaque. This obstruction of the coronary artery led to a heart attack.*

streaks containing a high proportion of cholesterol in the inner coat of arteries. The deposits form beneath the inner lining known as the endothelium (fig 15.15). Later, fibres are deposited in the cholesterol and these often start to calcify and become hard, a process known as **arterio-sclerosis**. The deposits are referred to as **atheromatous plaques**. As a plaque increases in size it protrudes into the lumen of the artery and begins to block it. This commonly occurs in the aorta and coronary arteries which supply the muscle of the heart. If the plaque breaks through the smooth endothelium, its rough surface commonly causes a blood clot to develop. This is called a **thrombus** which may build up until it is large enough to block the artery. If the clot breaks away, it may block an artery at another location. A clot that breaks away like this is called an **embolus**.

The artery wall is made weaker by atheromatous plaques and may stretch as a result. Local stretching is called **aneurysm**. It may rupture, a process known as **haemorrhage**. This is more likely if arteriosclerosis has occurred. Once an artery is blocked, the tissue it supplies will suffer oxygen starvation and will be severely damaged or die. If thrombosis occurs in a coronary artery (**coronary thrombosis**), the heart is damaged and a 'heart attack' may occur. The medical term for a heart attack is **myocardial infarction**. (Myocardial refers to heart muscle; infarction means suffocation due to lack of oxygen.) If thrombosis occurs in the brain (**cerebral thrombosis**) a stroke may occur. Strokes are sometimes referred to as **cerebro-vascular accidents**. They are also caused by cerebral haemorrhage. They usually result in permanent damage to the cerebral hemispheres due to oxygen starvation. The cerebral hemispheres are the conscious part of the brain and control many functions such as speech and motor coordination (section 17.2.4). Both heart attacks and strokes may result in death.

A muscle that is exercised without an adequate blood supply will give rise to pain as a result of cramp. When the heart is involved, such pain is called **angina**. An angina attack may be brought on even by gentle exercise such as climbing stairs. The pain may spread out from the centre of the chest to the neck, jaws, arms and back.

Thus coronary heart disease has two main forms, angina and myocardial infarction (heart attack). A heart attack may be caused by a coronary thrombosis or simply by narrowing of the artery by atherosclerosis until the blood supply is sufficiently restricted. About half-a-million people a year in Britain have heart attacks and about one-third die as a result. Half of these die within one hour. There are now great efforts taken to try to avoid these deaths by carrying special equipment in ambulances and by suitable treatment in hospital casualty units. Drugs can be used to restore normal heart rhythms and a heart which has stopped beating can sometimes be restarted by administration of an electrical shock across the chest wall.

15.5.2 Possible causes of and methods for reducing atherosclerosis and cardiovascular disease

A number of factors are known or believed to be involved in development of atherosclerosis, and hence cardiovascular disease. Some are more important than others and usually several act together to bring it about. The three most important are diet, hypertension (high blood pressure) and smoking.

Diet. Atheroma contains fats and cholesterol, and it has been shown clearly that experimental animals fed on a high fat diet develop diseased arteries. In countries such as Greece and Japan where the average diet is relatively low in fat, cardiovascular disease is much less common. The main problem is caused by saturated fats (section 8.7.7) which cause a rise in blood cholesterol levels. In 1992 the UK government recommended that the average percentage of food energy derived by the population from saturated fatty acids should be reduced from 17% (proportion in 1990) to no more than 11% by 2005 (a 35% reduction). A 12% reduction in total fat was also proposed (from about 40% to 35% of food energy). On the other hand, polyunsaturated fatty acids, found in unsaturated fats, are thought to help reduce cholesterol levels in blood and are therefore beneficial to health.

Hypertension (high blood pressure). Raised blood pressure can considerably increase the chances of developing cardiovascular disease. It has been shown that men under the age of 50 years with a blood pressure of 170/100 are twice as likely to die of coronary heart disease as men with normal blood pressure of about 120/80. High blood pressure itself is associated with a number of different factors, including stress, obesity, smoking, drinking excessive amounts of alcohol and lack of exercise. There is also a genetic predisposition in some people. Some of these factors can obviously be avoided by changes in lifestyle.

Drugs known as β-blockers can be used to reduce hypertension. The hormone adrenaline has an excitatory effect on the body (section 17.6.5). It binds to specific receptors known as α- and β-receptors in its target organs. β-blockers block the β-receptors, thus inhibiting the action of adrenaline and reducing heart rate. The same drugs are used to treat angina because they reduce the need for oxygen in the heart muscle.

Smoking. Heavy smokers are more likely to develop cardiovascular disease. For example, people under the age of 45 years who smoke more than 25 cigarettes a day are 15 times more likely to die of heart disease than non-smokers. Nearly 40% of cardiovascular deaths are due to smoking. Smoking increases atherosclerosis and decreases the ability to remove blood clots that build up at atheromatous plaques.

The effects of smoking are many and complex.

- Carbon monoxide and nicotine are both toxic to the endothelium, the thin lining of the blood vessels, damaging the lining and making penetration by fats and cholesterol easier.
- Carbon monoxide combines with haemoglobin and reduces oxygen transport by about 15% in smokers. Oxygen deficiency is a cause of angina and may induce a heart attack.
- Nicotine increases blood pressure, heart rate and constriction of blood vessels.
- Cigarette smokers produce more fibrinogen, the blood clotting protein, and reduced levels of the enzymes involved in removing blood clots.
- Smoking greatly stimulates the sticking of blood platelets to the surface of the endothelium and these are involved in blood clotting.
- Nicotine has a direct effect on raising blood fat levels.

Smoking is the largest single cause of premature death in Britain. More than one-third of the extra deaths caused by smoking are due to cardiovascular disease. The effects of smoking on the lungs are discussed in chapter 9.

Physical exercise. Many studies have shown that the more a person is physically active at work or during their leisure time, the less chance they have of suffering from cardiovascular disease. The effects of exercise on the cardiovascular system are described in section 14.7.

Gender. Death rates from cardiovascular disease in women are less than half those for men, and women rarely suffer from it before the menopause. There is a protective effect from the female sex hormones and a harmful effect from the male hormone testosterone. After the menopause women show an increase in blood fats and a sharp increase in rates of cardiovascular disease.

Lipids are insoluble in plasma and are therefore carried in a combined form with other molecules. In particular they are commonly combined with proteins to form spherical particles called lipoproteins. The size and density of these particles varies. Two common types are the low density and high density lipoproteins (LDLs and HDLs respectively). HDLs contain 21% cholesterol while LDLs contain 55% cholesterol. Generally speaking, the higher the amount of HDLs and the lower the amount of LDLs in the blood, the better for health. LDLs tend to stick to artery walls and unload their fats, contributing to atheroma. In males, HDL levels tend to drop at puberty and LDL levels gradually increase with age. This is thought to be due to testosterone. Oestrogen on the other hand increases HDLs and therefore tends to protect women between puberty and menopause. They still get heart disease but on average about 10 years later than men.

Heredity. If one parent suffered from premature coronary heart disease then the risk of a man suffering a myocardial infarction is doubled. If both parents suffered, the risk is increased by a factor of five. This indicates that there could be a genetic predisposition to heart disease. It is anticipated that the genes responsible will be identified fairly soon and tests may become available to identify those at risk.

Stress. There is great difficulty in defining stress and quantifying it. However, there is general agreement that psychological and emotional stress is often an important factor in triggering attacks of angina or even a myocardial infarction.

Age. Arteriosclerosis appears to be an inevitable consequence of ageing and this increases the risk of cardiovascular disease (section 15.7).

15.5.3 Treatment of cardiovascular disease

Pacemakers

When there are problems with regularity of heartbeat an artificial pacemaker can be used to gain control over the electrical activity of the heart. A pacemaker has two basic components, a pulse generator containing a power source and one or two pacing leads, each with an electrode on its tip.

Pacemaking can be temporary or permanent. When long-term control of the heart is required, a permanent pacemaker is implanted under the skin. The two most common modes of pacing are

- **demand** this detects the heart's own rhythms and stimulates depolarisation of the heart muscle, and therefore contraction, as necessary;
- **fixed rate** this fires at a predetermined rate, irrespective of the heart's own activity.

The pacemaker is a small, metal unit weighing between 30 and 130 g (fig 15.16). It is powered by a lithium battery with a life-span of up to 15 years. It is implanted in the chest under local anaesthetic.

Heart transplant surgery

In cases of heart disease where all other treatments are inadequate or inappropriate, a heart transplant may be advised. The first heart transplant was carried out in 1967 in South Africa. At first, survival rates were low, but great improvements in drugs used to prevent rejection have resulted in the majority of transplant patients surviving more than five years. The operation itself is relatively simple, with life being sustained by a heart–lung machine during the operation. The difficult task is caring for the patient afterwards, and relatively few institutions are equipped to do this. Although a relatively common procedure, demand greatly exceeds the supply of donor hearts. Attempts to develop artificial hearts have not been successful. The latest proposal to meet the demand is to use hearts from pigs which have been genetically engineered to avoid the potential rejection problems. This not only raises animal rights issues but the possibility of transferring disease from one species to another. This is of serious concern, especially in the light of the transfer of BSE to humans from beef. Other ethical issues are raised by transplant surgery, particularly heart transplants. Choices have to be made about who to treat when there is a scarcity of organs. Should smokers be as entitled to treatment as non-smokers, for example? The high cost of transplant surgery also raises the question of how limited resources are best allocated within the National Health Service.

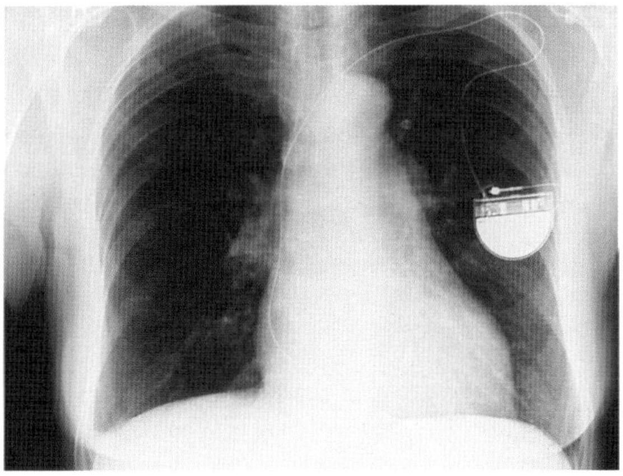

Fig 15.16 *X-ray showing pacemaker in position.*

15.6 Cancer

Cancers caused about 25% of deaths in Britain in 1991 and are the most common cause of death after cardiovascular disease. This is typical of developed countries (section 15.2). Breast cancer is the most common cancer in women and lung cancer in men (table 15.12). Cancer is not a single disease; more than 200 types of cancer are known.

Cancers are a result of uncontrolled cell division. The type of nuclear division involved is mitosis. The problem is caused by mutations or abnormal activation of the genes which control cell division. When the genes are abnormal they are called **oncogenes** (*onkos* means tumour). About 100 of these have been discovered. A single faulty cell may divide to form a clone of identical cells. Eventually an irregular mass of relatively undifferentiated cells called a **tumour** is formed. Tumour cells (fig 15.17) can break away and spread to other parts of the body, particularly in the bloodstream or lymphatic system, causing **secondary tumours** or **metastases.** This process is called **metastasis**. Tumours that spread and eventually cause ill health and death are described as **malignant**. The majority of tumours, such as common warts, do not spread and are described as **benign**.

Table 15.12. The ten most common cancers in men and women in the UK. (From Cancer Research Campaign, Scientific Yearbook 1995–96.)

Men		*Women*	
Site in the body	*% of all cancers*	*Site in the body*	*% of all cancers*
Lung	21	Breast	19
Skin*	14	Skin*	11
Prostate	10	Lung	8
Bladder	6	Colon	6
Colon	6	Stomach	3
Stomach	5	Ovary	3
Rectum	5	Cervix	3
Lymph node**	3	Rectum	3
Oesophagus	2	Uterus	2
Pancreas	2	Bladder	2
Other cancers	26	Other cancers	40

*Non-melanoma, nearly always curable
**Non-Hodgkin's lymphoma

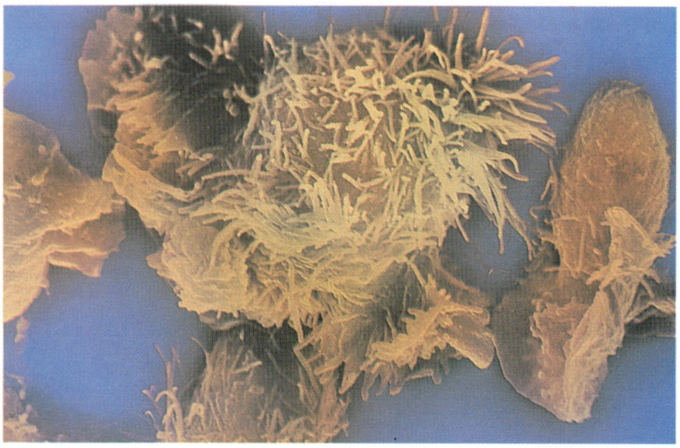

Fig 15.17 *False colour SEM of human cancer cells. They are typically large in size and have a 'hairy' surface which is thought may increase mobility.*

15.6.1 Causes of cancer

Changes in genes are called mutations and any factor bringing about a mutation is called a **mutagen**. An agent which causes cancer is called a **carcinogen**. Most mutated cells are either destroyed by the body's immune system or die with no ill-effect on the body. It is believed that development of a malignant cancer cell involves several steps and is usually caused by more than one factor operating over several years rather than a single factor. More than one mutation to the genes may occur. Up to 20% of the cancer worldwide may be caused by viruses.

Retroviruses. Evidence that cancers are genetic in origin was provided by work with retroviruses. Retroviruses are RNA viruses which, when they invade animal cells, use the enzyme reverse transcriptase (a viral coat protein) to make DNA copies of the viral RNA. The DNA is inserted into the host DNA where it may stay and be replicated for generations of cells. Some retroviruses are harmless. However, HIV is a harmful retrovirus and other retroviruses cause cancer. These contain a gene which alters host cell division genes, switching them on and causing the cell to become malignant. The genes become oncogenes. The advantage to the virus is that the cell makes many copies of itself and therefore of the virus.

DNA viruses. DNA viruses contain DNA as their hereditary material. Some contain their own oncogenes which can cause uncontrolled cell division of host cells. Examples which infect humans are the papilloma viruses which cause warts. Some papilloma viruses have been implicated in some forms of cervical cancer, making this a sexually transmitted disease. The Epstein–Barr virus may cause one form of Burkitt's lymphoma which is common in Africa.

Hereditary predisposition. About 5% of human cancers show a strong genetic predisposition, in other words they tend to run in families. More than 40 types of cancer, including cancer of the breast, ovary and colon, come into this category. The genes responsible may be oncogenes, or genes which lead to failure to kill cancer cells. In most cases other factors are required, but in a few cases, such as retinoblastoma, a single faulty gene is responsible. Retinoblastoma starts in the eye and spreads to the brain, causing death if untreated. It is caused by a dominant gene.

Two breast cancer genes have been identified and named BRCA1 and BRCA2. BRCA1 was cloned in 1994, and codes for a protein involved in transcription. A woman with one of these genes has about an 80% risk of developing breast cancer before the age of 70.

Ionising radiation. This includes X-rays, γ-rays and particles from the decay of radioactive elements. Cancers were caused in workers with X-rays at the beginning of the twentieth century and factory workers painting the dials of watches with a luminous paint containing radioactive radium and thorium. The radiation causes the formation of chemically active and damaging ions inside cells which can break DNA strands or cause mutations. The types of cancer linked with ionising radiation include skin cancer, bone marrow cancer, lung cancer and breast cancer. Medical and dental X-rays also expose patients to ionising radiation.

Ultraviolet light. This is the most common form of carcinogenic radiation and is non-ionising. DNA absorbs ultraviolet light and the energy is used in converting the bases into more reactive forms which react with surrounding molecules. Sunlight contains ultraviolet light and prolonged exposure to it can result in skin cancers, including melanoma which is highly malignant and commonly causes death through secondary brain tumours. Depletion of the ozone layer results in a higher proportion of ultraviolet light reaching the Earth's surface. The brown skin pigment melanin offers some protection.

Radon gas. Radon gas is a natural source of radiation released from certain rocks such as granite. It may accumulate in houses in areas where these rocks are found. It has been linked to the development of leukaemia (cancer of white blood cells), lung, kidney and prostate cancers, although the evidence is inconclusive.

Chemical mutagens. Many chemicals are now recognised as causing cancer. The first example was described in 1775 as soot and coal tar, when chimney sweeps were discovered to develop cancer of the scrotum. Later mineral oils were also found to be carcinogenic, when shale oils were used as a lubricant in the cotton-spinning mills. The workers developed cancers of the abdominal wall where their clothes had been splashed. Workers in the synthetic dye industry in the late nineteenth century developed bladder cancer.

The list of chemical carcinogens has steadily lengthened over the last 90 years and now includes, in addition to the above, inorganic arsenic compounds which produce skin cancer and asbestos products which cause lung cancer. Some food additives (flavours, colourings and stabilisers) have been considered as possible carcinogens because they cause cancers in experimental animals. As a result a number have been withdrawn.

Tobacco smoke contains chemicals responsible for lung cancer (section 9.7). The most important of these are polycyclic hydrocarbons which are converted in the body to carcinogens. Many common foods contain carcinogenic chemicals (table 15.13) although the levels are mostly low.

Table 15.13 Dietary carcinogens. (*From Molecular and Cell Biology*, Stephen L. Wolfe, Wadsworth, 1993).

Food	*Active agent*
alcohol	metabolised to acetaldehyde
basil	estragole
black pepper	safrole, piperine
celery, parsnips, parsley	furocoumarins
coffee, tea, chocolate	caffeine or theobromine*
common mushrooms	hydrazines
foods cooked in gas ovens	nitrosamines
grilled or barbecued beef, chicken, or pork	heterocyclic amines, nitropyrenes, nitrosamines
herbs, herbal teas	pyrrolizidine alkaloids
mould growth in peanut butter, grains, cheese, bread, fruits	aflatoxins and sterigmatocystin
mustard, mustard seed, horseradish	allylisothiocyanate
oil of sassafras	safrole
rhubarb	anthroquinone

*Not directly carcinogenic but promotes activity of carcinogen

15.6.2 Preventing and controlling cancer

There must be constant vigilance in the workplace regarding the possible dangers of exposure to carcinogens. The general public and workers should be made aware of any problems that may exist. Educational campaigns can encourage individuals to pay attention to early signs that may indicate the presence of cancer. Early diagnosis is important for increasing the chances of successful treatment.

Control should be considered as:

- prevention – protection against known carcinogens;
- early diagnosis – including screening programmes;
- treatment – urgent action once diagnosis is made.

Methods of prevention, diagnosis and treatment of cervical, breast and colon cancer are described below. Lung cancer is discussed in section 9.7.

Cervical cancer

England and Wales have one of the highest death rates from cervical cancer in the developed world (fig 15.18). Smoking doubles susceptibility to the disease. It is estimated that an effective screening programme should reduce deaths from the disease by over 80%. Deaths in England are falling gradually, but were still more than 1500 per year in the early 1990s. The Government's target is to reduce that by at least 20% by the year 2000.

The main method chosen to reduce the death rate was to set up an effective screening programme, because early diagnosis can be followed by effective treatment and cure. The screening test is known as the **cervical smear test** and was introduced in the 1960s. A national computerised system known as 'call and recall' was introduced in 1988. Women aged 20–64 are invited for cervical screening every 5 years. A few cells are gently scraped from the cervix and examined with a microscope for signs of abnormality which can lead to malignancy later (fig 15.19). Treatment at this stage is simple, free and can halt progress of the disease.

15.1 What information would you include in an education leaflet about cervical screening for the general public?

Breast cancer

Breast cancer is the commonest form of cancer among women in the developed world. Its incidence has increased slowly since the 1960s (table 15.12). It affects about one

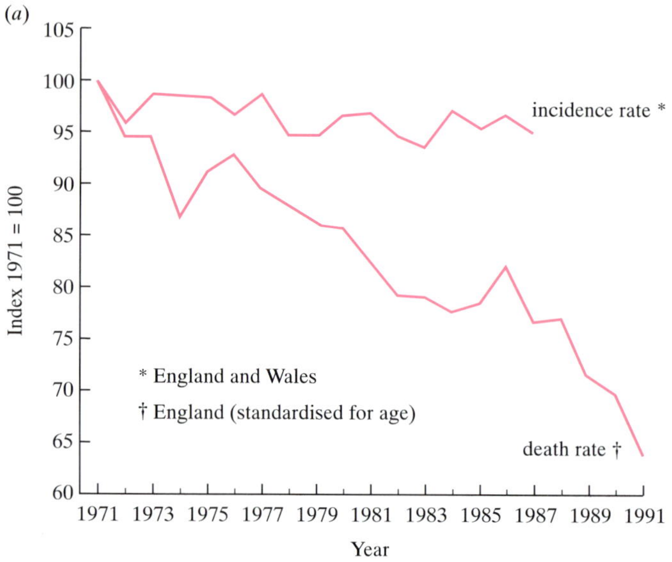

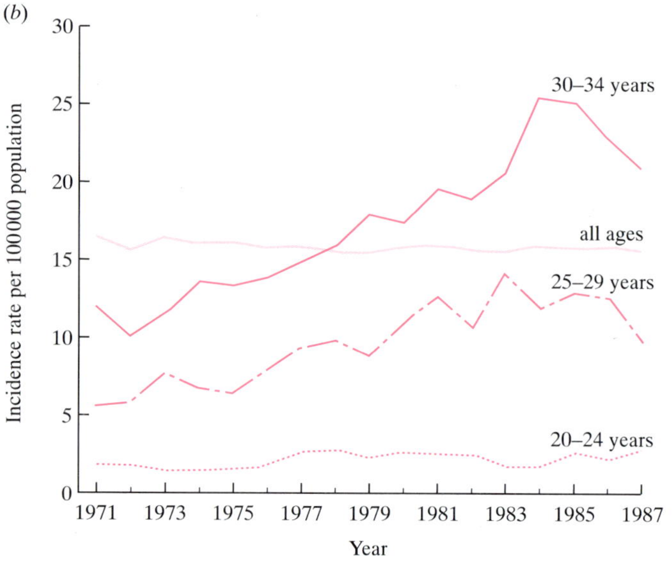

Fig 15.18 *(a) Incidence and death rates for cervical cancer from 1971–91. (b) Incidence rates for cervical cancer by age from 1971–87. (Source: OPCS (ICD 1 80), figs 3 and 4.)*

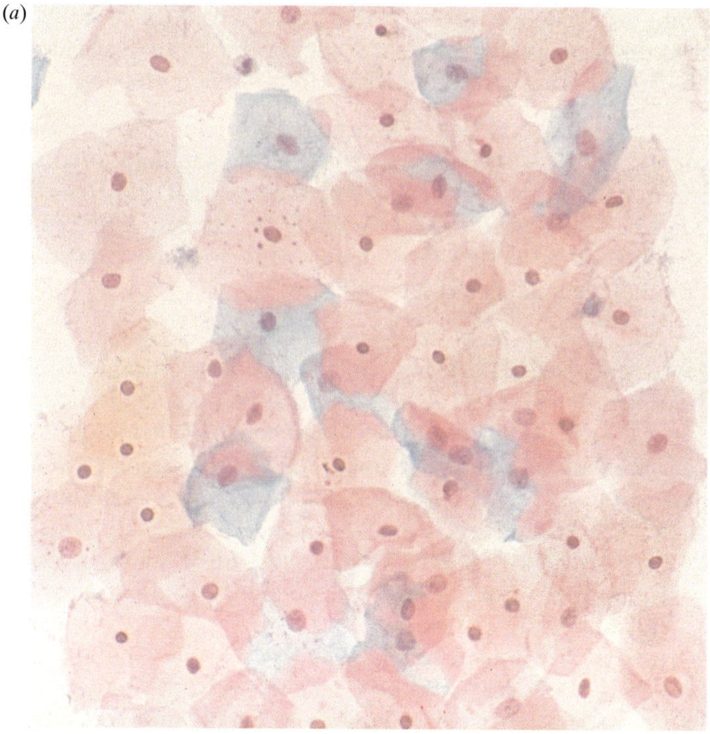

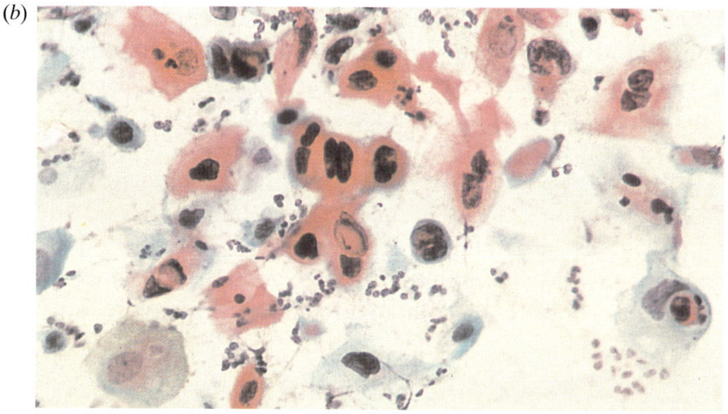

Fig 15.19 *Results of a cervical smear test. (a) Normal epithelial cells stained pink and blue. Note the small nuclei and large amounts of cytoplasm. (b) Malignant epithelial cells showing large, dark stained nuclei, indicating cell division, and very little cytoplasm.*

woman in twelve at some stage in their lives. In 1993 15 000 women died in England of breast cancer. About 90% were aged over 50. Once again, early diagnosis allows scope for effective treatment. A national breast cancer screening programme, using a computerised call and recall system, is used. Women aged 50–64 are invited to be screened every three years. Women of 65 or over may be screened on request. The screening technique is known as **mammography** and involves X-raying the breast (fig 15.20).

With both cervical and breast cancer screening, black and ethnic minority groups show a lower uptake of the service and attempts are being made by the UK Government to encourage more of these women to go for screening. The aim is to reduce deaths by 25% by the year 2000 compared with 1990, saving about 1250 lives per year.

Another method used for early diagnosis, which can be done by the woman herself, is to feel the breast for unusual lumps. Women can be educated for 'breast awareness' and should report any changes to their doctor without delay.

In 5% of cases of breast cancer there is a genetic predisposition. There is debate about the most appropriate response if genetic screening reveals a woman to be carrying the breast cancer genes BRCA1 or BRCA2. Trials with tamoxifen as a prophylactic (preventative measure) are underway and some at-risk women have opted to have their breasts removed surgically rather than risk developing the disease.

The main treatments for breast cancer are:

- surgery to remove the tumour;
- radiotherapy of the breast using X-rays to kill tumour cells;

(a)

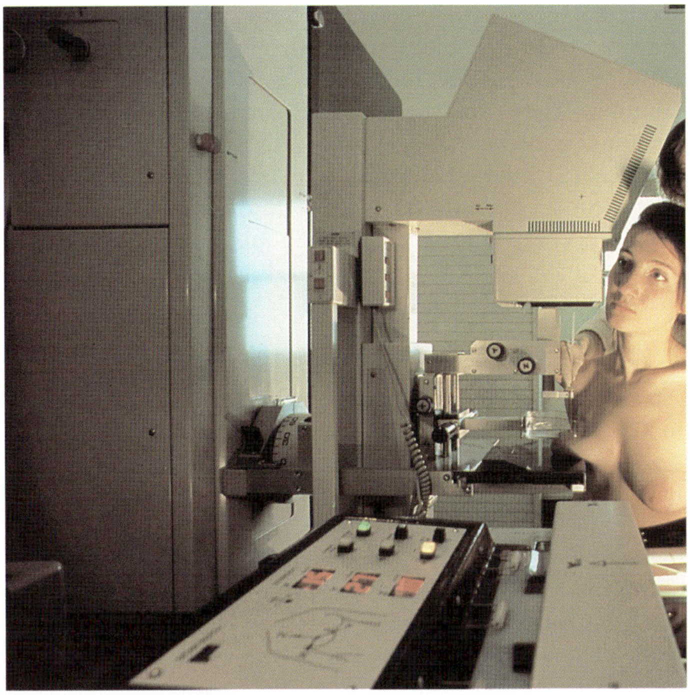

(b)

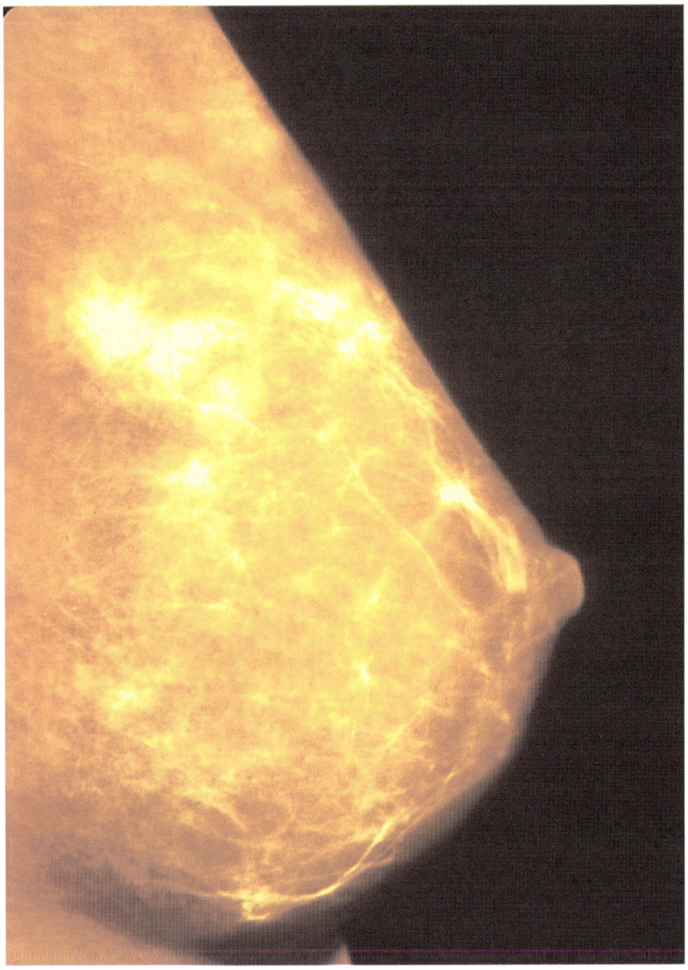

Fig 15.20 *(a) Woman undergoing mammography examination for the early detection of cancerous tumours in the breast. (b) Mammogram of breast in profile showing cancer tumours (bright areas).*

- chemotherapy of the whole body using chemicals which kill dividing cells (both normal and tumour cells);
- chemotherapy using tamoxifen, an anti-oestrogen hormone. Oestrogen can stimulate the growth of breast cancers, therefore growth can be controlled using tamoxifen.

The treatments may be used individually or in combination. The cancer may return some time after treatment. A woman whose breast cancer is diagnosed early has an 80% chance of surviving for 5 or more years after diagnosis.

Colon cancer

The colon is part of the large intestine. Cancer of the large intestine (bowel cancer) is the third most common form of cancer in Britain, accounting for about 14% of cancer deaths. About 1 person in 1200 is diagnosed each year. The condition may be symptomless in its early stages. Commonly though the tumour causes bleeding of the gut wall, leading to anaemia and traces of blood in the faeces. It may partially obstruct the colon affecting bowel movements, causing diarrhoea or constipation. Early diagnosis would therefore include being observant for these signs. The presence of the tumour can be revealed by means of a barium meal, which shows up an obstruction if the gut is X-rayed. Alternatively internal examination by endoscopy can be performed. The tumour can be removed surgically with an 80% chance of a complete cure if diagnosed early. The diseased section of bowel is removed and the two ends of the gut sewn together again.

There is growing evidence that diets relatively high in fat and meat may be associated with a higher occurrence of colon cancer, as well as some other forms of cancer. Reducing consumption of fatty foods and meat, and frequently eating fresh fruit and vegetables, and cereals with a high fibre content, can help to reduce the risk. There is also a genetic predisposition in some people. These people may be identifiable in future by genetic screening techniques and be able to take specific precautions, particularly regular check-ups. Early surgery to remove growths as they appear is relatively straightforward.

15.7 Ageing

After reaching maturity we enter a phase of gradual decline known as **ageing**. The body changes that lead to a decreasing life expectancy with age are known as **senescence**. Senescence is a characteristic of most living organisms. Despite advances in biology we still cannot fully explain the process, although there are many theories.

Humans have long cherished the idea that we may one day be able to prevent or slow down the process of ageing. But is death inevitable? To the biologist, ageing can be viewed as a kind of disease, a malfunctioning of the body systems. Biologists believe that they may be on the verge of being able to intervene in the human ageing process and the prospect of humans living for several hundred years is now being seriously considered. Needless to say, there would be many serious ethical arguments to resolve if this became possible.

Figs 15.21 and 15.22 summarise some of the changes associated with ageing.

15.7.1 Changes in the brain

Unlike most cells of the body, nerve cells either cannot regenerate themselves by dividing to replace dead cells, or do so very slowly. As a result, the number of nerve cells (neurones) in the brain declines with age. The average weight of the brain of a 90 year old is about 10% less than that of a 30 year old. In general, therefore, there is a decline in brain function with age. However, there is conflicting evidence about the degree and nature of decline in the absence of specific conditions such as Alzheimer's disease (see below). Intellectual skills and memory decline overall, but some types of problem solving which depend on experience or creative thinking may improve. Memory loss occurs, but this mainly affects rote learning and recall of specific facts, whereas recall of interconnected facts such as memory of current affairs or the plot of a novel may be unaffected. Accessing information rather than losing information from the brain seems to be the main problem. There is great individual variation.

Senile dementia

Dementia is mental deterioration as a result of physical changes within the brain. Certain intellectual functions, particularly memory, are progressively lost. **Senile dementia** is degeneration of brain cells as a result of the ageing process and typically occurs after the age of 65. Dementia before this age is sometimes referred to as presenile dementia.

Apart from the inability of neurones to replace themselves, another factor affecting the progress of senility is the supply of blood to the brain, which may be reduced as atherosclerosis or arteriosclerosis develops. A stroke or injury to the brain can cause dementia. Other possible causes of dementia-like symptoms are depression, chest infections, low blood sugar levels, hypothermia, alcohol abuse and hypothyroidism. It therefore has no single cause, and is simply the name given to a collection of symptoms rather than a single disease. Alzheimer's disease (see below) is by far the commonest cause of senile dementia, accounting for nearly 80% of cases. Creutzfeldt–Jakob disease ('mad cow disease') is another, rare, cause of dementia.

The earliest symptom of senile dementia is usually memory loss, particularly for recent events. Intellectual functions gradually decline, including understanding and powers of reasoning, resulting in confusion. The individual may lose concentration and interest in life generally and show lack of initiative. Personality may change. The person may often be irritable, emotionally unstable, with sudden extreme changes in mood, for example from tears to laughter. They may become embarrassingly uninhibited, abandon politeness, neglect personal hygiene or become antisocial. In the advanced stages, slowness, stiffness and awkwardness of movement can occur. Towards the end a person may cease to speak, think or move.

Risks increase with age. About 10% of over-65s suffer, and about 20% of over-80s.

Alzheimer's disease

Alzheimer's disease was first described by Alois Alzheimer in 1907. He examined brain tissue from patients dying with dementia. The tissue has a characteristic appearance under the microscope, with protein 'plaques' accumulating outside brain cells and tangled deposits of protein appearing inside the cells. Two key parts of the brain affected are the cortex of the cerebral hemispheres (the conscious part of the brain) and the hippocampus (involved with memory). Both plaques and tangles are caused by accumulation of abnormal proteins, amyloid β-protein in the case of plaques, and an abnormal form of tau protein which is overloaded with phosphate groups in the case of tangles. Normal tau protein is associated with microtubules in the elaborate cytoskeleton of neurones. The amyloid protein gene is found on chromosome 21 and the extra chromosome 21 in sufferers of Down's syndrome results in Alzheimer-like changes in their brains. The brains of Alzheimer's sufferers also shrink from loss of nerve cells (fig 15.23).

Alzheimer's disease is difficult to diagnose because its symptoms are similar to those of other diseases that cause dementia. It usually requires a post-mortem examination of brain tissue to confirm diagnosis. There is a genetic predisposition to the disease in some people, so it tends to run in families. There is also evidence that high levels of aluminium may contribute to the onset of the disease.

Nerve conduction velocity

Fig 15.22 shows a slight but steady decline in the speed of conduction of nerve impulses with age. Defects in synthesis of neurotransmitters may be one reason.

15.7.2 Changes in the locomotory system

The locomotory system includes joints, skeleton, muscles and tendons. Some of the changes in the system can be explained with reference to the fibrous protein collagen, a key constituent of bone and other connective tissues. Tendons are almost pure collagen. Muscle contains little collagen, but the amount increases with age as muscle fibres decline and are replaced by collagen (hence meat from older animals is tougher to eat). The structure of collagen also changes with age, with the fibres becoming thicker and less elastic. This would tend, for example, to make bone more brittle.

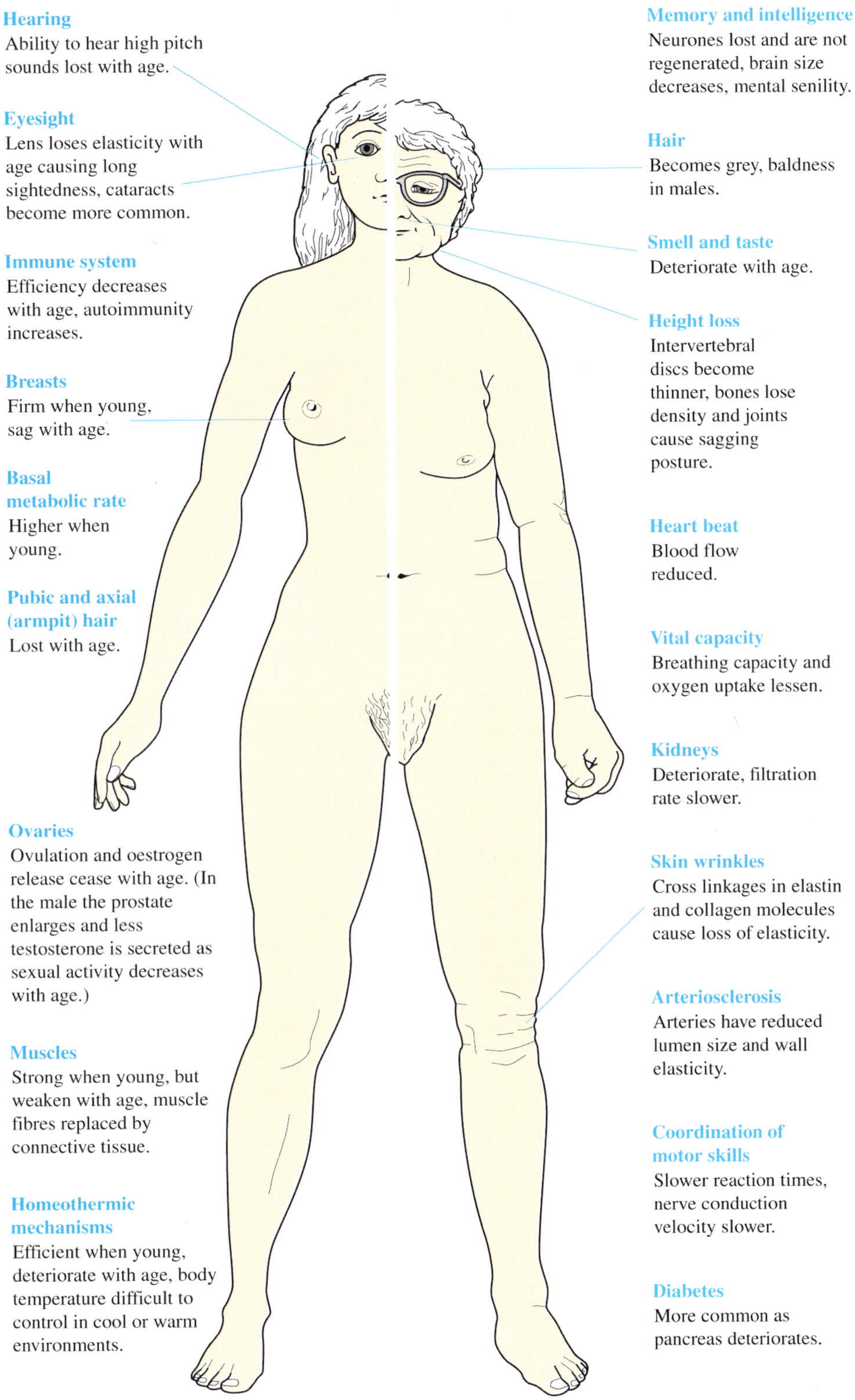

Fig 15.21 *Changes in the body that occur with ageing. (Based on fig 8.1, Philip Gadd (1983)* Individuals and populations, *Cambridge Social Biology Topics, CUP.)*

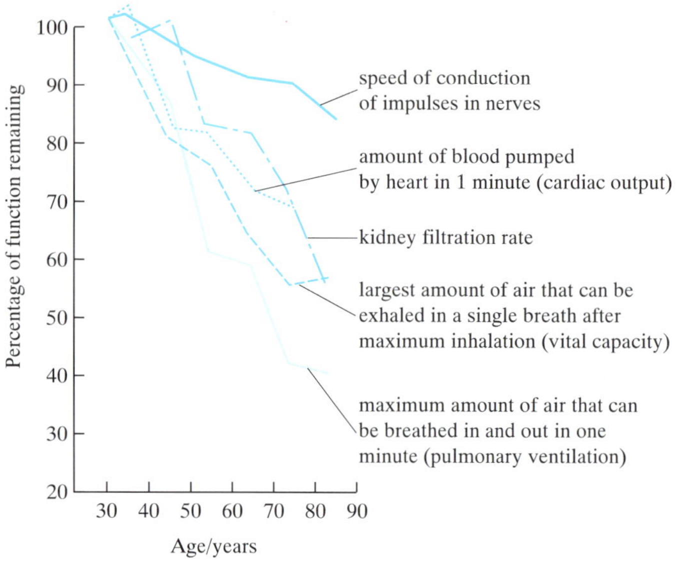

Fig 15.22 *Changes in the body with ageing. (From R. Passmore & J.S. Robson, A Companion to Medical Studies, vol 2, Blackwell Sci, 2nd ed., 1980.)*

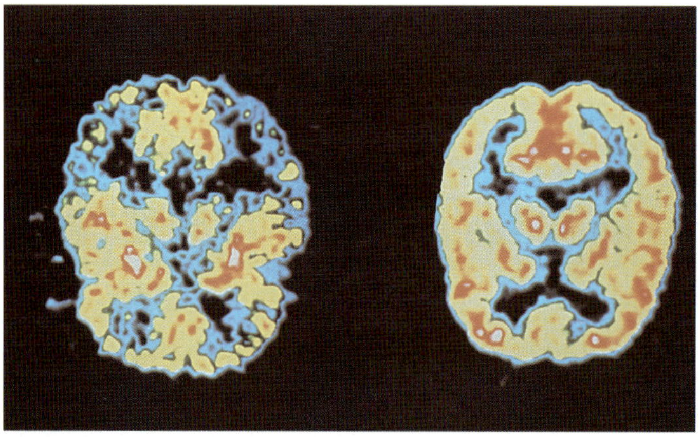

Fig 15.23 *Brain scans of (left) normal human brain and (right) brain of patient suffering from Alzheimer's disease. Areas of red and yellow show high brain activity, and blue and black show areas of low activity. The scan on the right shows reduction in function and blood flow in both sides of the brain which is often seen in Alzheimer's.*

Changes in bones

Bones become thinner, weaker and less flexible with ageing, partly because calcium starts to be lost faster than it is replaced from the age of about 35 years. This happens to everybody to some extent, but in some people it is particularly serious, leading to a condition known as **osteoporosis**.

Osteoporosis affects the whole skeleton but tends to affect the hips, wrists and vertebrae worst of all. Even a minor fall can result in broken bones. Apart from fractures, weakened bones in the spine result in compression of the bones and intervertebral discs and loss of height. The vertebrae may also collapse, causing curvature of the spine. Osteoporosis is most common in elderly women, affecting about 1 in 4 over the age of 60. Women are four times more vulnerable than men. This is because before menopause, which is the cessation of monthly periods, oestrogen helps to maintain bone strength. After menopause, levels of oestrogen decline. Oestrogen is antagonistic (has an opposite effect) to the hormone parathormone which stimulates the raising of blood calcium levels. Therefore in the absence of oestrogen, loss of calcium occurs from the bones, making the bones weaker and more easily fractured (fig 15.24).

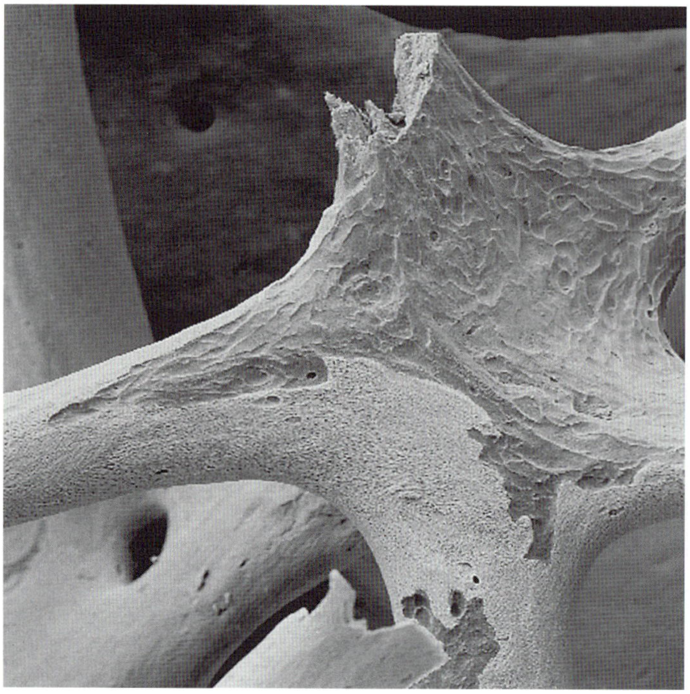

Fig 15.24 *Scanning electron micrograph of the brittle and spongy fractured femur from a patient with osteoporosis. The dark areas show where bone mineral has been reabsorbed by the body, leaving the bone brittle and easy to fracture.*

Hormone replacement therapy (HRT) is recommended for most women after the menopause if they wish to avoid the problem. Oestrogen is taken either in pill form or by implants below the skin. It reduces the incidence of osteoporosis to the same level as in men. The undesirable side effects of long-term oestrogen treatment, such as an increased risk of blood clotting, can be reduced by adding progesterone to the oestrogen. A balanced diet rich in calcium, and regular exercise, also help to keep bones strong. Smoking and drinking alcohol increase the risk of osteoporosis.

Osteoporosis is an important disease in terms of the demands it makes on the health service. An estimated £640 million is spent each year in the UK on treating people with fractures, particularly hip fractures, due to osteoporosis. Only a quarter of those who break a hip stage a full recovery, and a quarter need to be placed permanently in nursing homes.

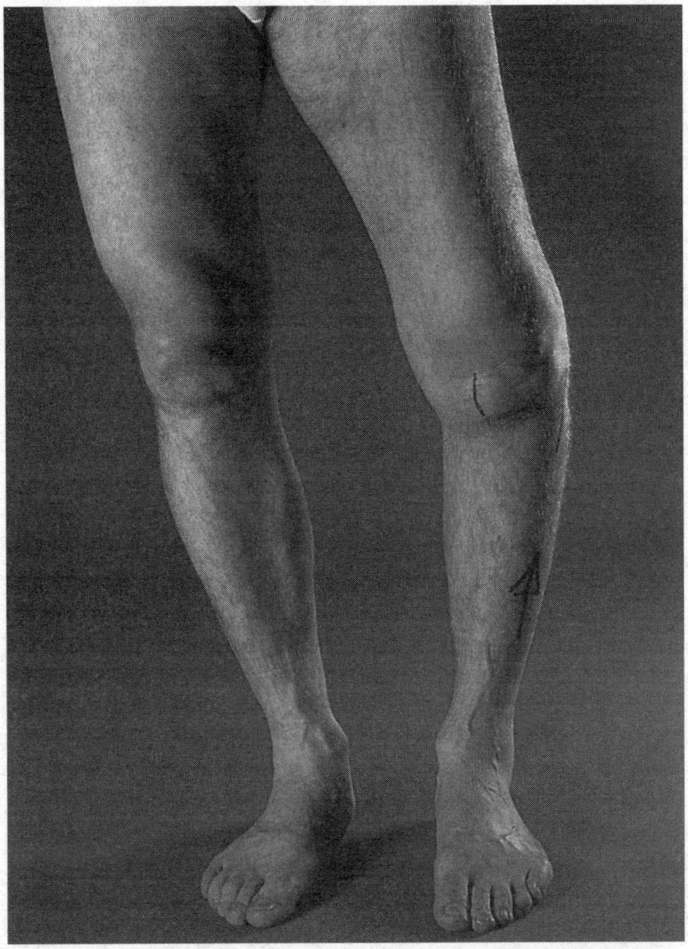

Fig 15.25 *A severe case of osteoarthrosis in the left knee of a man. The picture clearly shows the swelling and deformation of the joint which is associated with severe cases of the condition.*

Changes in joints and osteoarthrosis

The load-bearing joints such as the knees and hips are particularly subject to changes, as are the fingers. Joints in general tend to become stiffer and painful.

As ageing progresses, the smooth, tough cartilage which covers the ends of bones where they articulate with other bones becomes weaker and less extensive. It gradually breaks down. Eventually, in extreme cases, the ends of the bones may become exposed and start to grate against each other. They become thicker and denser and start to wear. This causes pain and stiffness, a condition which used to be known as **osteoarthritis** (fig 15.25). The more modern term is **osteoarthrosis**, because the term arthritis implies inflammation which does not occur.

A less frequent type of joint degeneration known as **rheumatoid arthritis** is commoner in older people than young people. It is an autoimmune disease, in which the body's immune system attacks its own tissues. Affected joints become hot and swollen during attacks. This is a genuine inflammatory response.

Changes in muscles

As a person gets older, muscle fibres tend to be replaced by connective tissue such as collagen, making the muscle weaker. This results in loss of body weight and body strength. Since the heart is a muscle it also becomes weaker.

15.7.3 Changes in the cardiovascular system

Fig 15.22 shows the decline with age in cardiac output (amount of blood pumped by the heart in one minute). By the age of 80 the cardiac output at rest has declined to about 70% of that of a 30-year old. Larger differences are noted after exercise. This could have effects on other body systems.

Two common causes of this decline are atherosclerosis and arteriosclerosis. Atherosclerosis (section 15.5.1) is not strictly speaking an inevitable consequence of ageing, although it takes many years to develop. Arteriosclerosis, however, does appear to be a consequence of ageing. It is commonly known as hardening of the arteries and is due to loss of elastic tissue in the artery walls as we age. Blood pressure rises as a result (hypertension), making haemorrhage (bursting of blood vessels) and thrombosis (blood clotting) more likely. The incidence of strokes is particularly associated with arteriosclerosis.

15.7.4 Changes in the respiratory system

Ageing is associated with a gradual loss of elastic tissue in the lungs and a decline in muscle power which affects the rate and extent of chest expansion. These and other degenerative changes bring about a decline in vital capacity and pulmonary ventilation (gas exchanged per minute) as shown in fig 15.22. This places limits on the amount of oxygen that can be made available to the body and hence the amount of work or exercise that can be carried out in a given time period.

15.8 Respiratory and genetic diseases

Respiratory disease (asthma, emphysema, bronchitis, lung cancer and the effects of smoking tobacco) is discussed in chapter 9. Genetic disease is discussed in chapter 25.

Chapter Sixteen

Coordination and control in plants

Plants, like animals, need some form of internal coordination if their growth and development is to proceed in an orderly fashion, with suitable response to their environment. Unlike animals, plants do not possess nervous systems and rely entirely on chemical coordination. Their responses are therefore slower and they often involve growth. Growth, in turn, can result in movement of an organ. In this chapter plant movements will be examined before studying the various ways in which plants coordinate their activities.

16.1 Plant movements

It is a characteristic of plants that they do not show locomotion (movement of the entire organism). However, movements of individual plant organs are possible and are modified by the sensitivity of the plant to external stimuli. Movements induced by external stimuli fall into two main categories: tropisms (tropic movements) and taxes (tactic movements).

16.1 What is the basic reason for the fact that animals show locomotion whereas plants do not?

16.1.1 Tropisms

A **tropism** is a movement of part of a plant in response to, and directed by, an external stimulus. The movement is almost always a growth movement. Tropic responses are described as positive or negative depending on whether growth is towards or away from the stimulus respectively. Some examples of tropisms are shown in table 16.1.

16.2 Complete a fourth column to table 16.1 to show for each response how it is advantageous to the plant involved.

Phototropism and geotropism will be discussed in more detail later in this chapter (sections 16.2.1 and 16.2.2).

Table 16.1 Examples of tropisms.

Stimulus	*Type of tropism*	*Examples*
light	phototropism	Shoots and coleoptiles positively phototropic Some roots negatively phototropic, e.g. adventitious roots of climbers like ivy
gravity	geotropism	Shoots and coleoptiles negatively geotropic Roots positively geotropic Rhizomes, runners, dicotyledonous leaves **diageotropic*** Lateral roots, stem branches **plagiogeotropic***
chemical	chemotropism	Hyphae of some fungi positively chemotropic, e.g. *Mucor* Pollen tubes positively chemotropic in response to chemical produced at micropyle of ovule
water	hydrotropism (special kind of chemotropism)	Roots and pollen tubes positively hydrotropic
solid surface or touch	haptotropism (thigmotropism)	Tendrils positively haptotropic, e.g. leaves of pea Central tentacles of sundew, an insectivorous plant, positively haptotropic
air (oxygen)	aerotropism (special kind of chemotropism)	Pollen tubes negatively aerotropic

* diageotropism: growth at 90° to gravity, that is horizontal growth. plagiogeotropism: growth at some other angle to gravity, that is not horizontal or directly towards or away from gravity.

16.1.2 Taxes

A **taxis** is a movement of an entire cell or organism (that is locomotion) in response to, and directed by, an external stimulus. As with tropisms they can be described as positive or negative, and can be further classified according to the nature of the stimulus. Note that this kind of movement occurs in a wide range of organisms, not just plants. Examples are given in table 16.2.

Table 16.2 Examples of taxes.

Stimulus	*Taxis*	*Examples*
light	phototaxis	**positive:** *Euglena*, a unicellular alga, swims towards light, chloroplasts move towards light, fruit flies fly towards light **negative**: earthworms, blowfly larvae, woodlice and cockroaches move away from light
chemical	chemotaxis	**positive:** sperms of liverworts, mosses and ferns swim towards substances released by the ovum; motile bacteria move towards various food substances **negative:** mosquitoes avoid insect repellent
air (oxygen)	aerotaxis (special kind of chemotaxis)	**positive:** motile aerobic bacteria move towards oxygen
gravity	geotaxis	**positive:** planula larvae of some cnidarians swim towards sea bed **negative**: ephyra larvae of some cnidarians swim away from sea bed
magnetic field	magnetotaxis	certain motile bacteria
resistance	rheotaxis	**positive:** *Planaria* move against water current, moths and butterflies fly into the wind

16.3 *Euglena* and *Chlamydomonas* are unicellular algae which swim by means of flagella. Both organisms are positively phototactic, meaning that they move towards light. This is an advantage since they are both photosynthetic. Design an experiment to demonstrate the preferred light intensity of *Euglena* or *Chlamydomonas* in phototaxis.

16.4 Fig 16.1 illustrates the distribution of motile bacteria 10 min after being placed under a cover-slip with a filament of a green alga, such as *Spirogyra*. (*a*) Put forward a hypothesis to account for the final distribution of the bacteria. (*b*) How could you check your hypothesis?

16.1.3 Kinesis

Another type of locomotory response is **kinesis**. Since this is virtually confined to the animal kingdom it is discussed in chapter 17 with animal behaviour.

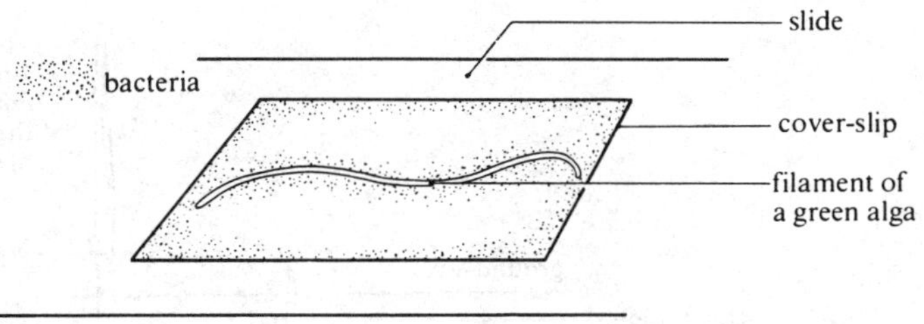

Fig 16.1 *Distribution of motile bacteria on a slide.*

16.2 Plant growth substances

Chemical coordination in animals is controlled by **hormones**. Hormones work at very low concentrations at sites some distance from where they are made. Plants are coordinated by chemicals which do not necessarily move from their sites of synthesis and hence, by definition, should not always be termed hormones. In view of this, and because their effects are usually on some aspect of growth, they are called **growth substances**. It is also important to realise that the precise mechanisms of action of plant growth substances are not yet clear and that they probably do not work in the same way as animal hormones. It should be borne in mind that growth can be divided into the three stages of cell division, cell enlargement and cell differentiation (specialisation), and that these stages have particular locations in plants (section 22.4). The action and distribution of different plant growth substances therefore reflects this. Five major types of growth substance are recognised:

- auxins, usually associated with cell enlargement and differentiation;
- gibberellins, also usually associated with cell enlargement and differentiation;
- cytokinins, associated with cell division;
- abscisic acid, usually associated with dormancy, as with buds;
- ethene (ethylene), often associated with ageing (senescence).

In this chapter, each type of growth substance will first be discussed separately, and then key stages in the life cycle of a plant will be discussed to emphasise the fact that growth substances often interact with each other to achieve their effects.

16.2.1 Auxins and phototropism

Discovery of auxins

The discovery of auxins was the result of investigations into phototropism that began with the experiments of Charles Darwin and his son Frances. Using oat coleoptiles as convenient material (fig 16.2), they showed that the growth of shoots towards light was the result of some 'influence' being transmitted from the shoot tip to the region of growth

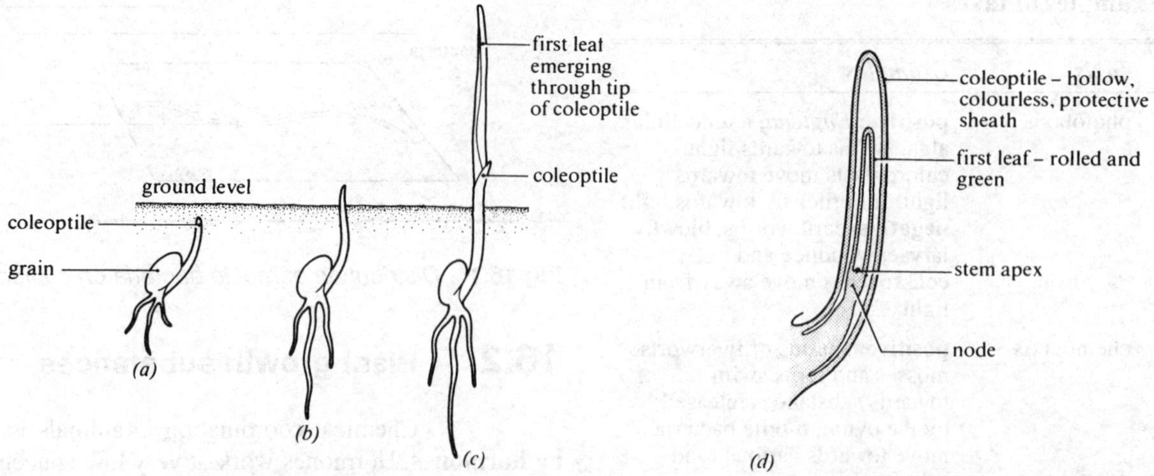

Fig 16.2 *Germination of a typical grass seedling: (a), (b) and (c) stages in germination, (d) section of coleoptile at stage (b).*

behind it. Some of their experiments are summarised in fig 16.3, the diagrams representing results obtained from many seedlings.

> **16.5** *(a)* List carefully the conclusions you could draw from experiments *(a)*–*(d)* in fig 16.3, given that the curvature was due to growth in the region behind the tip.
> *(b)* Why was experiment *(c)* necessary, bearing in mind the result from experiment *(b)*?

If the tropic response is analysed in terms of the following: stimulus → receptor → transmission → effector → response, then the largest gap in our knowledge remains the nature of the transmission. In 1913 the Danish plant physiologist Boysen-Jensen added to our knowledge. Fig 16.4 summarises some of his experiments.

> **16.6** What extra information is provided by Boysen-Jensen's experiments?
>
> **16.7** If these experiments were repeated in uniform light, draw diagrams to show what results you would expect. Give reasons for your answers.

In 1928 the Dutch plant physiologist Went finally proved the existence of a chemical transmitter. His aim had been to intercept and collect the chemical as it passed back from the tip and to demonstrate its effectiveness in a variety of tests. He reasoned that a small diffusing molecule should pass freely into a small block of agar jelly, whose structure is such that relatively large spaces exist between its molecules. Fig 16.5 illustrates some of his experiments.

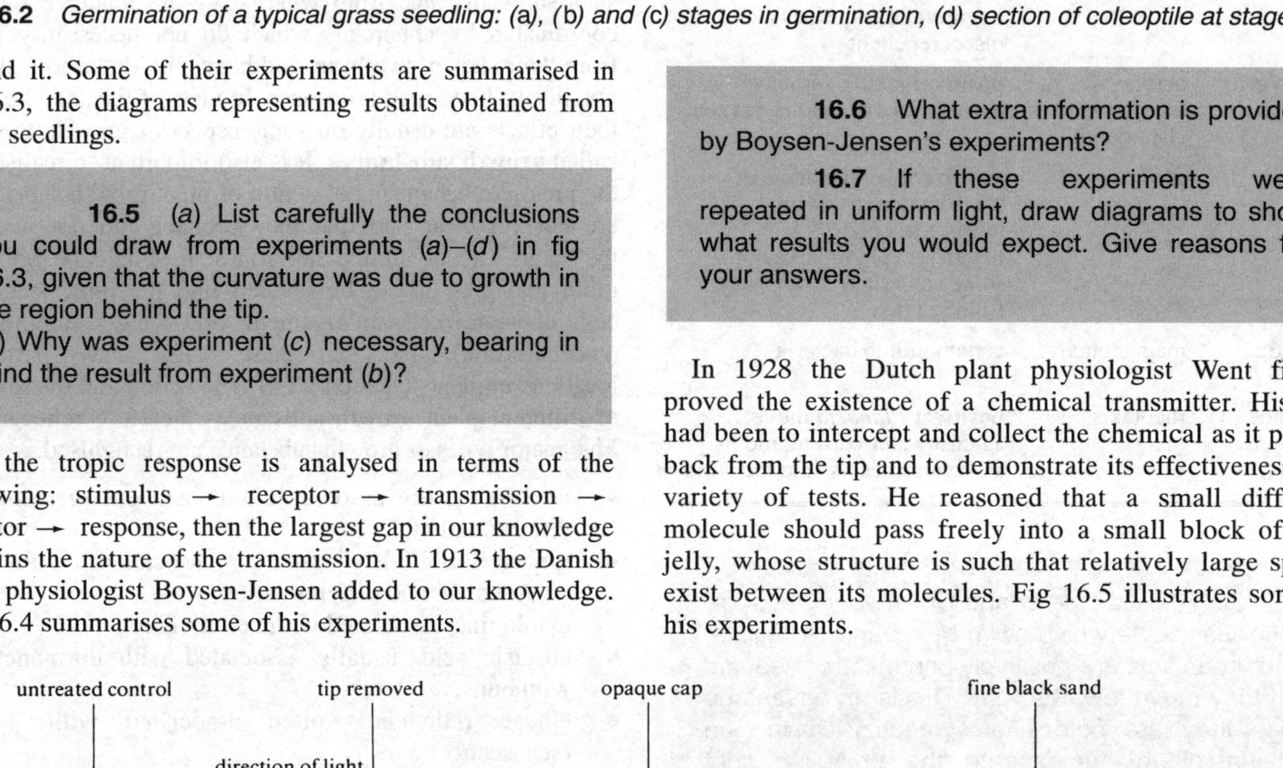

Fig 16.3 *Darwin's experiments on phototropism using oat coleoptiles. (a), (b), (c) and (d) are separate experiments showing treatment (left) and result (right).*

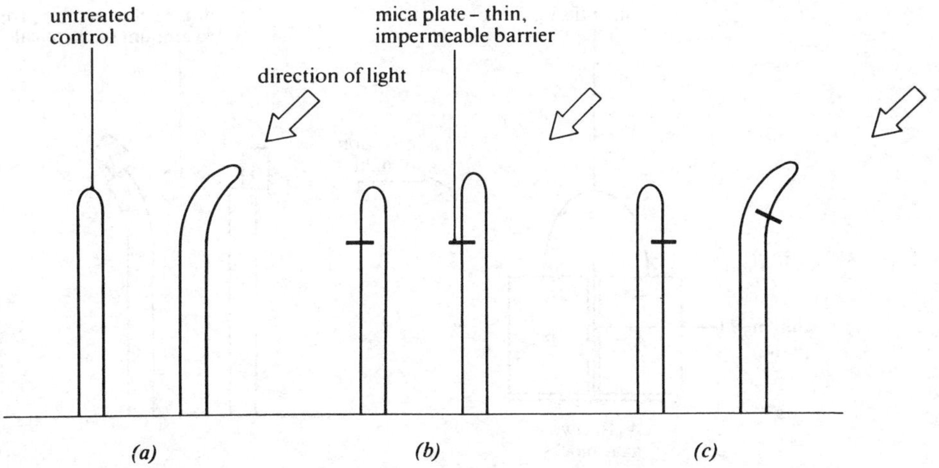

Fig 16.4 *Boysen-Jensen's experiments on phototropism using oat coleoptiles. (a), (b) and (c) are separate experiments showing treatment (left) and result (right).*

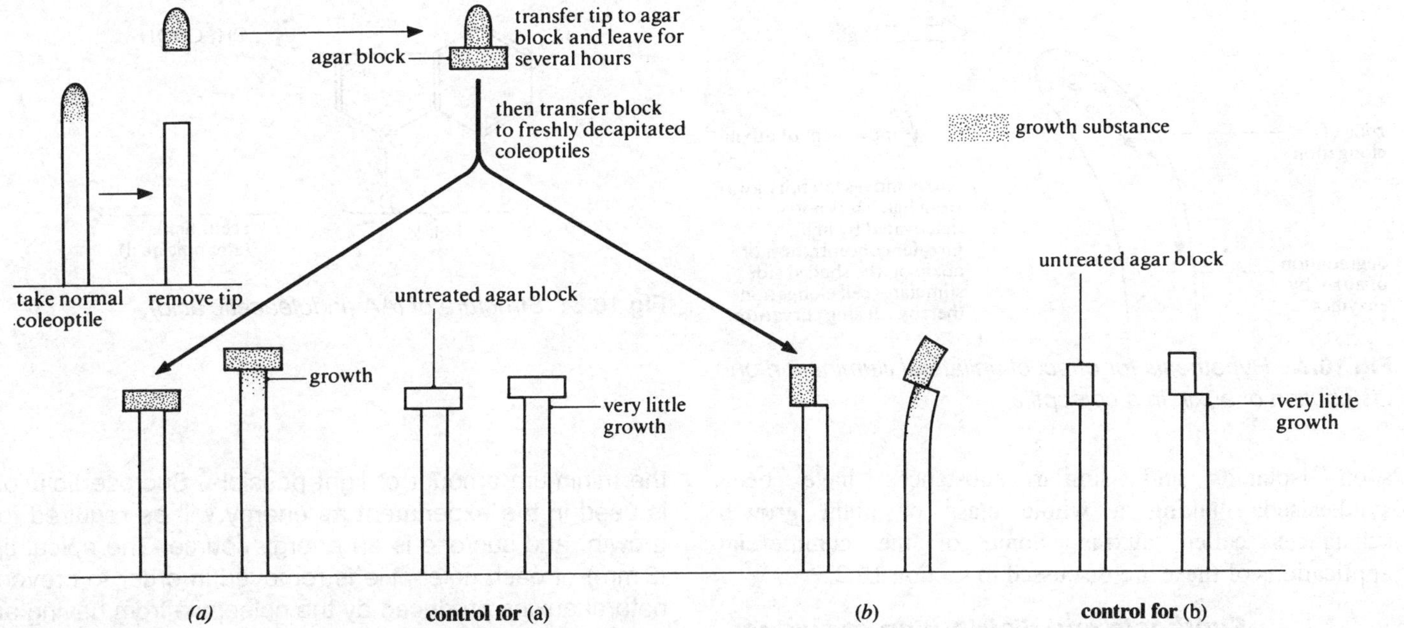

Fig 16.5 *(below) Went's experiments. (a) and (b) are separate experiments showing treatment (left) and result (right). Control experiments are shown alongside. All treatments were carried out in darkness or uniform light.*

> **16.8** What would you conclude from the results shown in fig 16.5?
>
> **16.9** What result would you expect if the treated block had been placed on the right side of the decapitated coleoptile in experiment (*b*)?

A further experiment of note carried out by Went is illustrated in fig 16.6. In control experiments the tip was exposed to uniform light or darkness before transfer of agar blocks, and the degree of curvature induced by blocks **A** and **B** was the same. Unilateral illumination of the tip, however, resulted in unequal distribution of the chemical in blocks **A** and **B** (fig 16.6). Not only does this support the conclusions from Boysen-Jensen's experiments about the effect of light on the distribution of the chemical, but it shows how a test of measuring the amount of the chemical present, that is a bioassay, can be set up. A **bioassay** is an experiment in which the amount of a substance is found by measuring its effects in a biological system. Went showed that the degree of curvature of oat coleoptiles was directly proportional to the concentration of the chemical (at normal physiological levels).

The chemical was subsequently named 'auxin' (from the Greek *auxein*, to increase). In 1934 it was identified as indoleacetic acid (IAA). IAA was soon found to be widely distributed in plants and to be intimately concerned with cell enlargement. Fig 16.7 summarises present beliefs concerning the movement of IAA during unilateral illumination of coleoptiles. It should be pointed out, however, that the coleoptile is the simplest system so far studied and that others appear to be more complex. Also, there is little evidence for the development of auxin gradients in the critical period before the response is measured.

Structure of IAA

The structure of IAA is shown in fig 16.8.

Other chemicals with similar structures and activity were

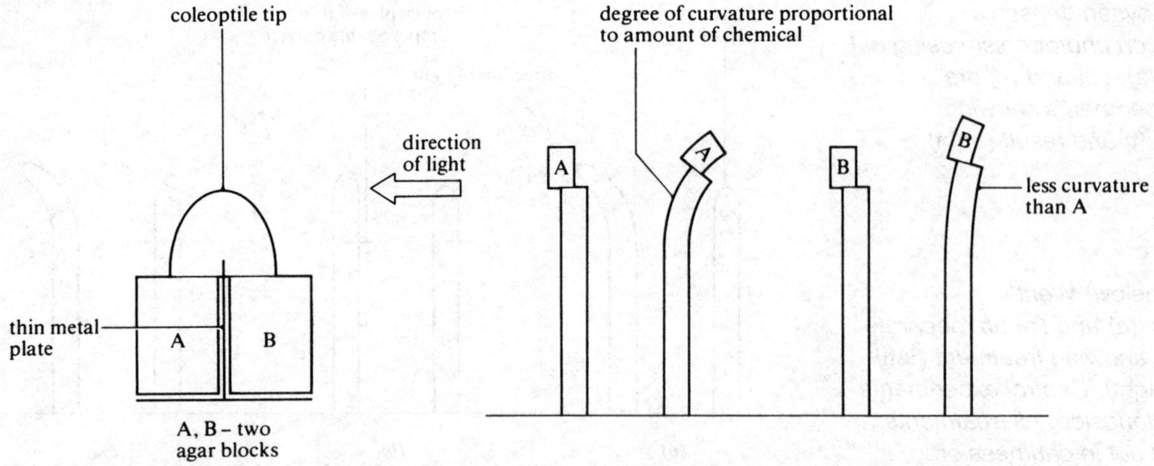

Fig 16.6 *Went's experiment showing effect of unilateral light on distribution of the chemical (auxin).*

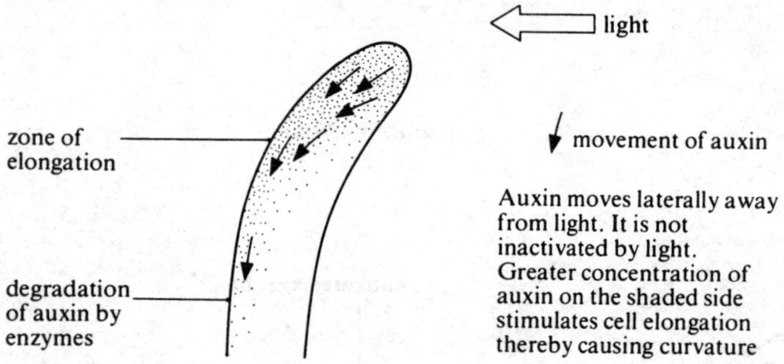

Fig 16.7 *Hypothesis for effect of unilateral illumination on distribution of auxin in a coleoptile.*

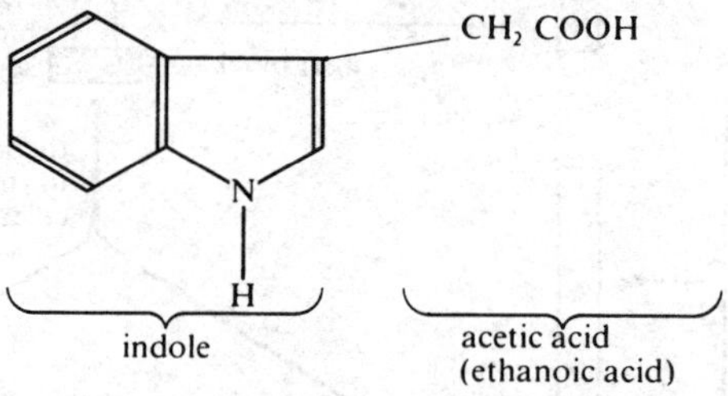

Fig 16.8 *Structure of IAA (indoleacetic acid).*

soon isolated, and similar substances have been synthesised, making a whole class of plant growth substances called auxins. Some of the commercial applications of these are discussed in section 16.2.5.

Synthesis and distribution of auxins

Auxins are made continuously in the shoot apex and young leaves. Movement away from the tip is described as basipetal (from apex to base of the organ) and polar (in one direction only). It moves, apparently by diffusion, from cell to cell and is eventually inactivated and degraded by enzymes. Long-distance transport can also occur via the vascular system (mainly phloem) from shoots to roots. A little auxin is probably made in roots. The effects of different auxin concentrations on shoot growth can be investigated by means of an experiment such as experiment 16.1.

Experiment 16.1: To investigate the effects of indoleacetic acid (IAA) on the growth of oat coleoptiles

The aim of the experiment is to investigate the effect of various concentrations of IAA on growth of oat coleoptiles. Growth is affected by white light and therefore cutting and transferring of coleoptiles during the experiment should be carried out under red light or in the minimum amount of light possible. Sucrose solution is used in the experiment as energy will be required for growth, and sucrose is an energy source. The apical tip (3 mm) of each coleoptile is removed in order to prevent natural auxins produced by the coleoptile from having an effect on growth.

Materials

germinating oat seedlings with coleoptiles at least 1.5 cm long (Soak 100 oat grains in water overnight, place the soaked seeds on damp paper towelling in a dish, cover the dish with aluminium foil and place in the dark to germinate (five days in an incubator at 20 °C). In order to obtain the 60 coleoptiles required for each experiment, at least 100 grains should be soaked to allow for germination failure.)
6 test-tubes in a test-tube rack
6 petri dishes + lids
$5 \times 5\ cm^3$ graduated pipettes
$25\ cm^3$ measuring cylinder or $10\ cm^3$ graduated pipette
coleoptile cutter (fig 16.9)
paint brush
2% sucrose solution
distilled water
stock IAA solution ($1\ g\ dm^{-3}$) IAA is not readily

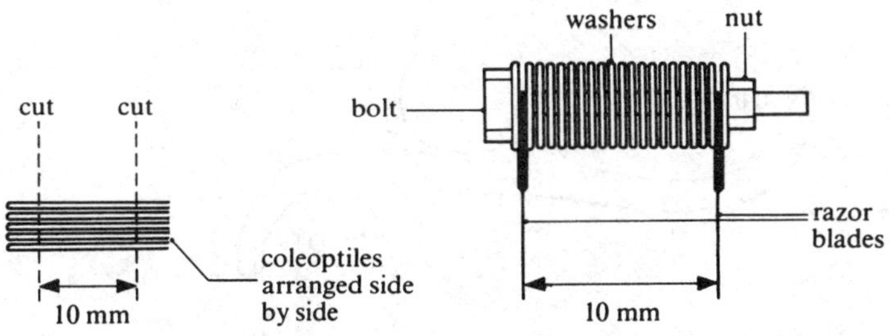

Fig 16.9 *Cutting 10 mm lengths of coleoptiles.*

soluble in water and is therefore first dissolved in ethanol: dissolve 1 g of IAA in 2 cm^3 ethanol and dilute to 900 cm^3 with distilled water. Warm the solution to 80 °C and keep at this temperature for 5 min. Make up to 1 dm^3 with distilled water. Adjust quantities according to final volume required.

Method

(1) Take six test-tubes and six petri dishes and label them **A–F**.
(2) Add 18 cm^3 of 2% sucrose solution to each test-tube.
(3) Using a clean 5 cm^3 pipette, add 2 cm^3 of IAA solution to tube **A** and mix the two solutions thoroughly.
(4) Using a fresh pipette transfer 2 cm^3 of solution from tube **A** to tube **B** and mix the contents of tube **B** thoroughly.
(5) Using a fresh pipette each time, transfer 2 cm^3 from tube **B** to tube **C**, mix, then transfer 2 cm^3 from tube **C** to tube **D**, mix, then transfer 2 cm^3 from tube **D** to tube **E**.
(6) Add 2 cm^3 distilled water to tube **F**.
(7) Transfer the solutions from tubes **A–F** to petri dishes **A–F**.
(8) Take 60 germinated oat seedlings and cut 10 mm lengths of coleoptile, starting about 2 mm back from the tips. Use a double-bladed cutter with the blades held exactly 10 mm apart by a series of washers, two nuts and two bolts (see fig 16.9). If the tips of the coleoptiles are placed in a line, several lengths can be cut simultaneously.
(9) Using a paint brush, transfer 10 lengths of coleoptile to each dish, avoiding cross contamination of the solutions (the greater the number of coleoptiles used, the more statistically valid the results).
(10) Place a lid on each and incubate the dishes at 25 °C for three days in the dark.
(11) Remeasure the lengths of coleoptiles as accurately as possible.
(12) Ignoring the largest and smallest figures for each dish, calculate the mean (average) length.
(13) Plot a graph of mean length (vertical axis) against IAA concentration in parts per million (horizontal axis).

> **16.10** What is the concentration in parts per million (ppm) of IAA in each petri dish (1 g dm^{-3} = 1000 ppm)?

(14) Comment on the results and compare them with fig 16.10. More accurate results can be obtained by combining class results.

16.2.2 Auxins and geotropism

It is a common observation that roots are positively geotropic, that is grow downwards, and shoots are negatively geotropic, that is grow upwards. That gravity is the stimulus responsible can be demonstrated by using a piece of equipment called a **klinostat** (fig 16.11). As the chamber rotates, all parts of the seedling receive, in turn, equal stimulation from gravity. A speed of four revolutions per hour is sufficient to eliminate the one-sided effect of gravity and to cause straight shoot and root growth. A non-rotating control shows the normal response to gravity, with shoot growing up and root growing down. It is important to ensure even illumination during the experiment (or to carry it out in darkness) so that there can be no directional response to light.

The involvement of auxins in geotropism is demonstrated by the experiment shown in fig 16.12, which uses the techniques introduced by Went. Auxin moves out of the horizontally placed coleoptile tip but moves downwards as it does so. The greater auxin concentration on the lower surface of an intact coleoptile would stimulate greater cell elongation here, and hence upward growth.

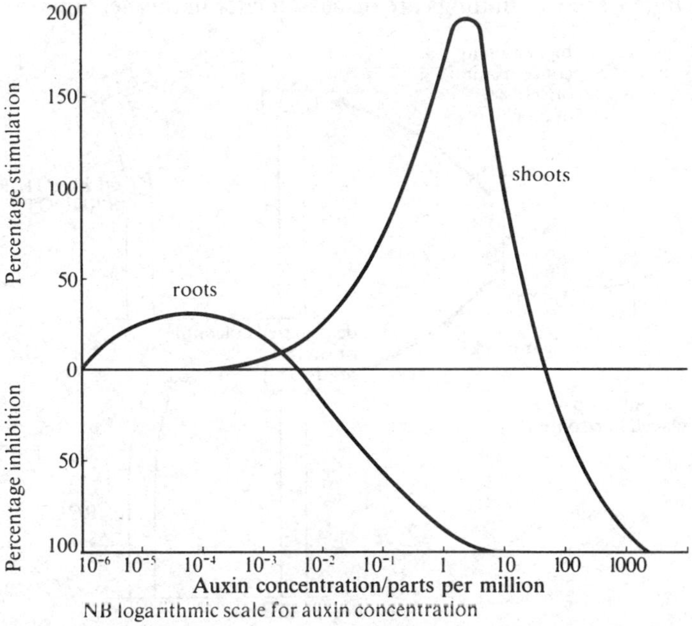

Fig 16.10 *Effect of auxin concentration on growth responses of roots and shoots. Note that concentrations of auxin which stimulate shoot growth inhibit root growth.*

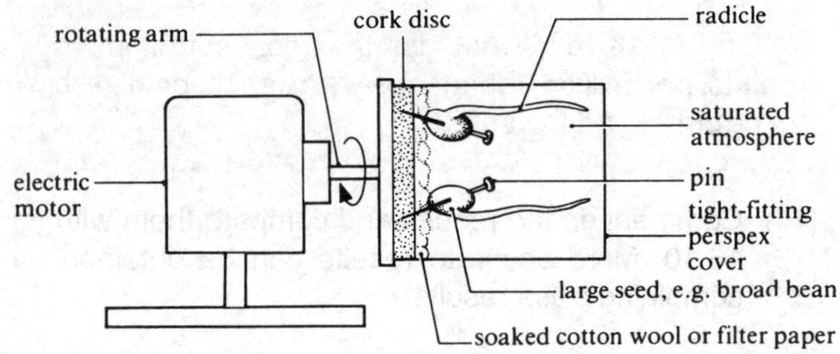

Fig 16.11 *Klinostat showing broad beans after several days growth with rotation.*

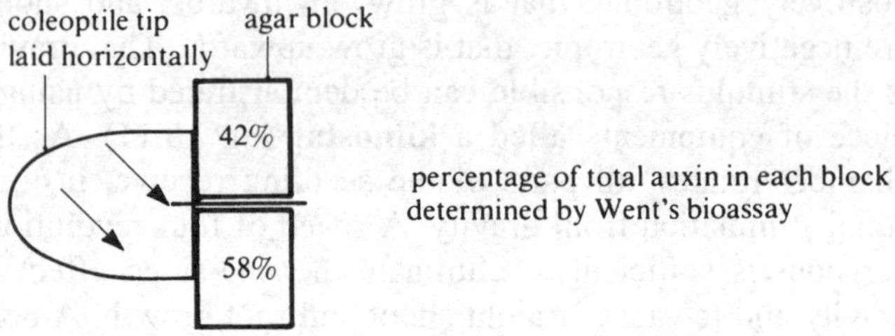

Fig 16.12 *Effect of gravity on distribution of auxin from a horizontal coleoptile tip.*

Decapitation of a root tip removes its sensitivity to gravity, but it is not so easy to demonstrate movement of auxins in roots because very low concentrations are present, and these do not give convincing results in the bioassay described. An interesting result though is obtained from the experiment shown in fig 16.13.

Observations of this type led to the hypothesis summarised in fig 16.14, which suggests that the opposite responses of roots and shoots are due to different sensitivities to auxin. Modifications of the hypothesis in the light of recent findings are discussed later in this section.

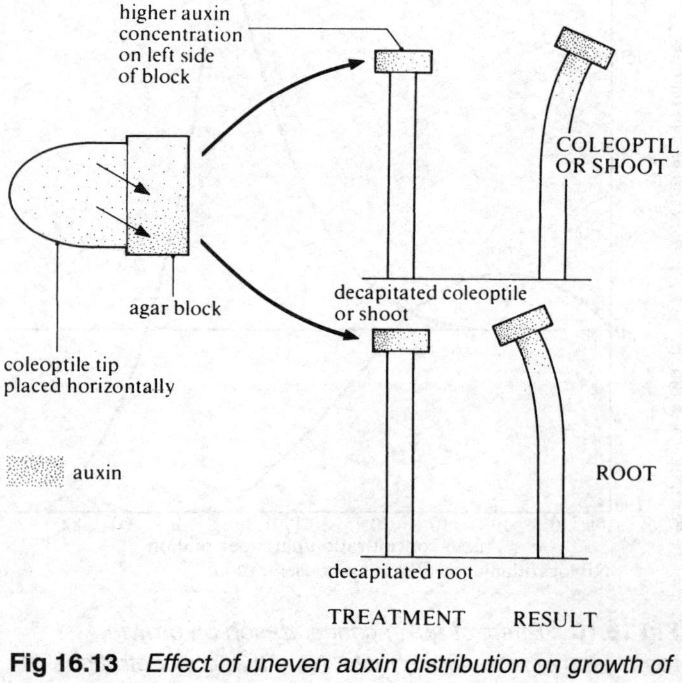

Fig 16.13 *Effect of uneven auxin distribution on growth of decapitated coleoptile and root.*

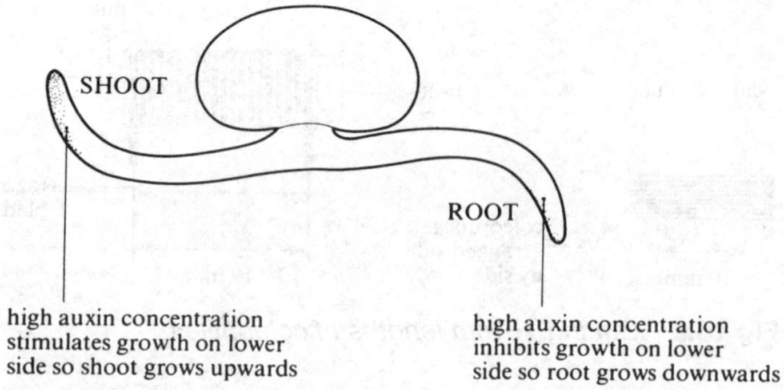

Fig 16.14 *Hypothesis for redistribution of auxin in a horizontally placed seedling.*

The different sensitivity of roots to auxin (fig 16.10) could also explain the negative phototropism shown by some; the higher accumulation of auxin on the shaded side would cause inhibition of growth with the result that cells on the light side would elongate faster and the root would grow away from light. An important aspect of plant growth regulation has thus been revealed. It is not only the nature of the growth substance which is relevant (qualitative control) but the amount of that substance (quantitative control).

The gravity-sensing mechanism

The question now arises as to how the gravity stimulus is detected. Darwin showed that removal of the root cap, the group of large parenchyma cells that protect the root tip as it grows through the soil abolishes the geotropic response. A section of the root cap reveals the presence of large starch grains contained in amyloplasts within the cells (fig 16.15).

It was suggested as long ago as 1900 that these cells act as **statocytes**, that is gravity receptors, and that the starch grains are **statoliths**, structures which move in response to gravity. The so-called **starch–statolith hypothesis** proposes that sedimentation of the starch grains through the cells occurs so that they come to rest on the lower sides of the cells with respect to gravity (fig 16.15). In some unknown way this affects the distribution of growth substances which are known to be produced sometimes in the root apex, sometimes in the root cap and sometimes in both. There is much evidence to support this hypothesis. All plant organs which are sensitive to gravity contain statocytes. They are found, for example, in the vascular bundle sheaths of shoots. Plants from which the starch grains have been removed by certain treatments lose their sensitivity to gravity, but regain it if allowed to make more starch.

> **16.11** How is this mechanism of gravity detection similar to that in animals?

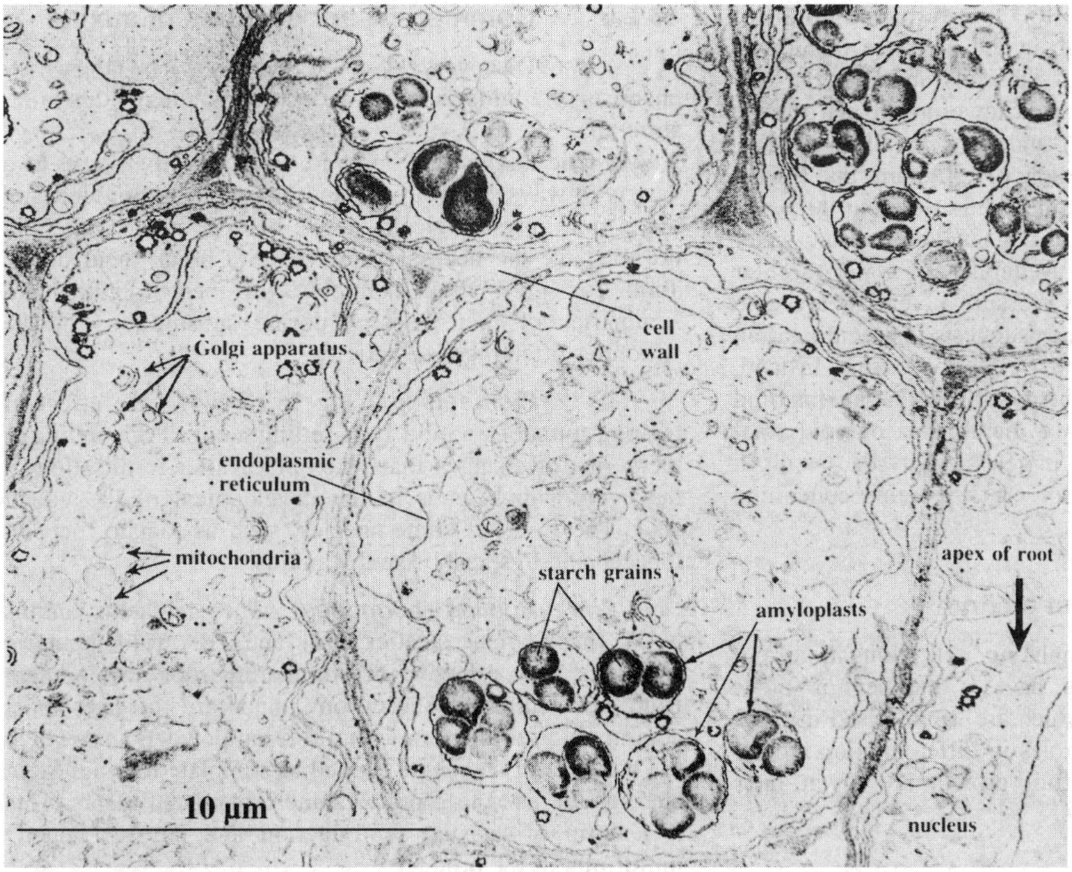

Fig 16.15 *Electron micrograph of section of root cap showing amyloplasts with starch grains located at the bottoms of the cells.*

Modern hypotheses on geotropism

In coleoptiles the gravity response seems to be mediated by auxins as described above, but with most shoots a geotropic response is still obtained if the tip is removed, and there is still doubt as to whether movement of auxin is involved. In roots, auxin redistribution does occur, but probably not dramatically enough to account for the observed changes in growth rates. Transmission of a growth inhibitor from the root cap to the zone of elongation has been shown, but this is not necessarily auxin. Several groups of workers have been unable to find auxin in the root caps of maize seedlings, a common experimental plant. Instead, abscisic acid, a well-known growth inhibitor, has been found. Ethene, another growth inhibitor, could also be involved. Finally, gibberellins (growth promoters) have been found in higher concentrations than normal in the rapidly growing sides of both shoots and roots when they are geotropically stimulated.

16.12 What can you conclude from the experiments shown in fig 16.16? Controls, using untreated agar, showed no curvature. When IAA was used instead of abscisic acid no significant curvature was obtained.

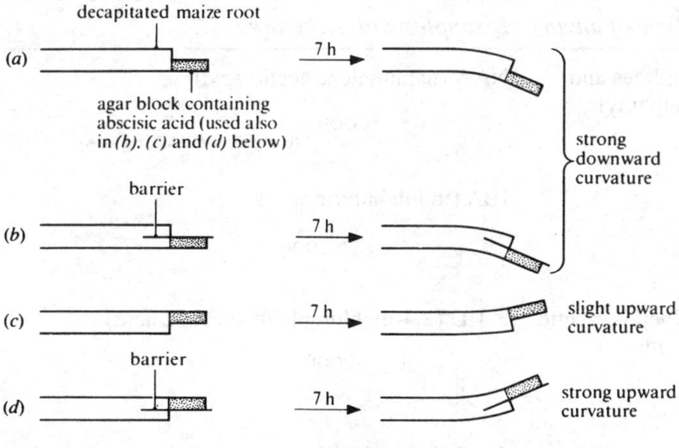

Fig 16.16 *Effect of abscisic acid on geotropic response to decapitated roots. (Based on experiments by Pilet, 1975.)*

16.2.3 Mode of action of auxins

The effect of auxins on cell enlargement is now reasonably well understood. During cell extension the rigid cellulose framework of the cell wall must be loosened. Extension then occurs by a combination of osmotic swelling as water enters the cell and by the laying down of new cell wall material. The orientation of the existing cellulose microfibrils probably helps to determine the direction of extension. 'Wall loosening' is induced by acid conditions, and by auxins. In 1973 four different groups of workers all demonstrated that, in the presence of auxins, hydrogen ion secretion out of the cells and into the cell walls is stimulated. This causes a lowering of pH outside the cell (increase in acidity) and hence wall loosening, possibly by an enzyme with a low pH optimum that breaks bonds in cell wall polysaccharides, thus allowing the walls to stretch more easily. The ability to maintain a low water potential inside the cell, and availability of water to enter the cell and generate a high pressure potential, are also necessary. More recent evidence suggests that acidification of the walls may not be the first effect of auxins. Instead, researchers are investigating the possibility that auxins bind to receptors in the cell surface membranes of epidermal cells and bring about changes in gene activity that result in production of new enzymes or other proteins concerned with growth.

16.2.4 Other effects of auxins

Apart from stimulating cell elongation and hence shoot growth, auxins have a number of other important roles in the plant which are summarised in table 16.4. Further details of their roles in differentiation, apical dominance, abscission and fruit growth are given later under the appropriate headings.

Table 16.3 Commercial applications of auxins.

Type of auxin	*Examples with structures*	*Uses*
Indoles and naphthyls	NAA (naphthalene acetic acid) 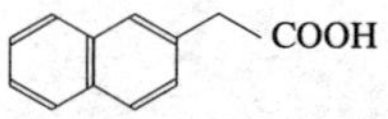 (compare with IAA, fig 16.8) IBA (indolebutyric acid) 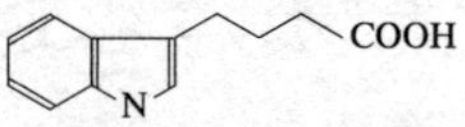	**Fruiting** – help natural fruit set; sometimes cause fruit setting in absence of pollination (parthenocarpy) **Rooting hormone** – promote rooting of cuttings
Phenoxyacetic acids	2,4-D (2,4-dichlorophenoxyacetic acid) 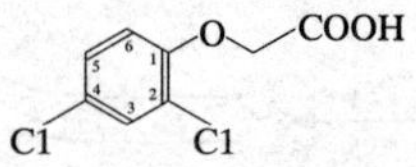 2,4,5-T (2,4,5-trichlorophenoxyacetic acid) as above but extra chlorine atom in 5 position of ring MCPA (2-methyl-4-chlorophenoxyacetic acid) as 2,4-D but methyl group (CH_3) in 2 position of ring instead of chlorine	**Selective weedkillers** – kill broad-leaved species (dicotyledons). Used in cereal crops and on lawns. Also in conifer plantations for scrub clearance (conifers unaffected). 2,4-D/2,4,5-T mix used in Vietnam war by US as the defoliant 'Agent Orange' **Potato storage** – inhibit sprouting of potatoes **Fruiting** – prevent premature fruit drop (retard abscission)
Benzoic acids	2,3,6-trichlorobenzoic acid 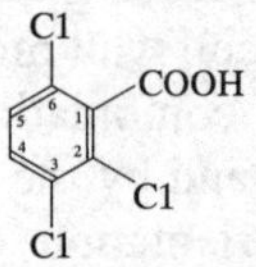 2,4,6-trichlorobenzoic acid as above except chlorine atom in 4 position instead of 3 position of ring	Powerful **weedkillers**. Useful against deep-rooted weeds such as dandelion (*Taraxacum officinale*) and bindweed (*Convolvulus arvensis*)

16.2.5 Commercial applications of auxins

Discovery of IAA led to the synthesis by chemists of a wide range of active compounds with similar structure.

Synthetic auxins have proved commercially useful in a variety of ways. They are cheaper than IAA to produce, and often more physiologically active because plants generally do not have the necessary enzymes to break them down. Table 16.3 gives some examples with their structures and summaries of their uses. Chlorine substitutions in the structures often increase activity.

Fruit setting. Fig 16.17 shows the effect of treating tomatoes with a fruit-setting auxin. Fruit setting is the series of changes that takes place after fertilisation in the ovary which leads to the development of the young fruit. The fruits of some species, such as tomato, pepper, tobacco and figs, can be set by auxins.

Rooting hormones. The synthetic auxins NAA (naphthalene acetic acid) and IBA (indolebutyric acid) are very effective at stimulating root development from stem cuttings. These compounds are the ones most commonly used in commercial rooting powders into which the cut ends of stems are dipped to stimulate rooting. Stem cuttings are used to produce some ornamental plants. This is a form of asexual reproduction and ensures genetic uniformity of the product.

Fig 16.17 *The three large trusses were set by spraying with beta naphthoxyacetic acid. The small one in the bottom left-hand corner was not sprayed and produced only one normal size tomato.*

Weedkillers. A group of auxins known as phenoxyacetic acids (see table 16.3) form effective herbicides. They are relatively cheap to manufacture and very effective. Their main attraction, though, is that they are selective. They affect broad-leaved plants (dicotyledonous plants or dicots) much more than monocotyledonous plants (monocots). Monocots include cereals and other grasses. 2,4-D will therefore kill dicot weeds in cereal crops and lawns. 2,4,5-T was particularly effective against woody perennials such as those found in rough pastureland. It has been banned, however, due to the presence of traces of a dioxin, an extremely toxic chemical which can cause fetal abnormalities, cancer and a particularly severe form of acne known as chloracne.

These weedkillers cause twisted growth and growth overall is reduced, eventually stopping. It is believed that the auxins interfere with the normal expression of genes and that the correct balance of enzymes needed for growth is not achieved. It is not known why they are more effective against dicots than monocots. It is not simply that higher quantities are absorbed by the broader dicot leaves.

Another group of auxins, the benzoic acids, are also useful weedkillers (table 16.3).

Plant growth regulators

It is important to distinguish between naturally occurring **plant growth substances** and synthetic **plant growth regulators** which are often based on the structure of the naturally occurring compounds but are more effective and less easily degraded by the plant and therefore tend to be used commercially, for example:

plant growth substance	auxin (indoleacetic acid)
plant growth regulators	phenoxyacetic acids (2,4-D)
	indolebutyric acid (IBA)
	indolepropionic acid (IPA)

16.2.6 Gibberellins

Discovery of gibberellins

During the 1920s a team of Japanese scientists at the University of Tokyo was investigating a particularly damaging worldwide disease of rice seedlings, caused by the fungus *Gibberella* (now called *Fusarium*). Infected seedlings became tall, spindly and pale and eventually died or gave poor yields. By 1926 a fungal extract had been isolated which induced these symptoms in rice plants. An active compound was crystallised by 1935 and a further two by 1938. These compounds were called **gibberellins**, after the fungus. Language barriers and then the Second World War delayed the initiation of work in the West, but immediately after the war there was competition between British and American groups to isolate these chemicals. In 1954 a British group isolated an active substance which they called **gibberellic acid**. This was the third, and most active, gibberellin (**GA_3**) isolated by the Japanese. Gibberellins were isolated from higher plants during the 1950s, but the chemical structure of GA_3 was not completely worked out until 1959 (fig 16.18). Now more than 50 naturally occurring gibberellins are known, all differing only slightly from GA_3.

Structure of gibberellins

All are **terpenes**, a complex group of plant chemicals related to lipids; all are weak acids and all contain the **gibbane** skeleton (fig 16.18).

gibbane skeleton

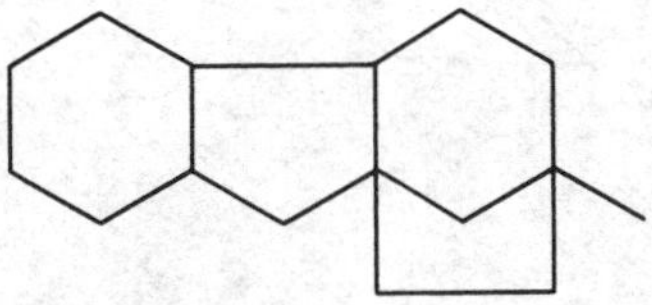

gibberellic acid (GA_3)

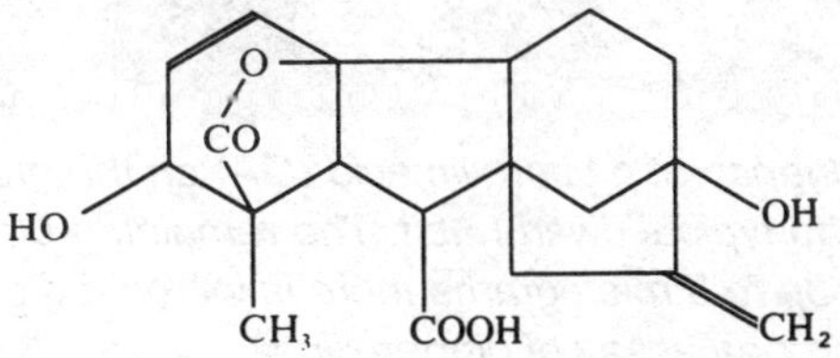

Fig 16.18 *Structures of gibbane skeleton and gibberellic acid (GA_3).*

Synthesis and distribution of gibberellins

Gibberellins are most abundant in young expanding organs, being synthesised particularly in young apical leaves (possibly in chloroplasts), buds, seed and root tips. They migrate after synthesis in a non-polar manner, that is up or down the plant from the leaves. They move in phloem and xylem.

Effects of gibberellins

Like the auxins, the main effect of gibberellins is on stem elongation, mainly by affecting cell elongation. Thus genetically dwarf varieties of peas and maize are restored to normal growth and dwarf beans can be converted into runner beans (fig 16.19). Stem growth of normal plants is promoted. Further information, relating to interaction with auxins, is given in section 16.3.

One of the classic effects of gibberellins, which has been much studied in an attempt to understand their mechanism of action, is the breaking of dormancy of certain seeds, notably of cereals. Germination is triggered by soaking the seed in water. After imbibing water the embryo secretes gibberellin which diffuses to the aleurone layer, stimulating synthesis of several enzymes including α-amylase (fig 16.20). These enzymes catalyse the breakdown of food reserve in the endosperm, and the products of digestion diffuse to the embryo, where they are used in growth.

16.13 (*a*) What is the substrate of α-amylase? (*b*) What is the product of the reaction it catalyses? (*c*) What other enzyme is required to complete digestion of its substrate? (*d*) Why is α-amylase so important in cereal seeds?

16.14 Explain the role of storage proteins in the aleurone layer by reference to fig 16.20.

Experiment 16.2: To test the following two hypotheses, (*a*) that gibberellin can stimulate breakdown of starch in germinating barley grains and (*b*) that gibberellin is produced in the embryo

The presence of amylase in barley seeds can be detected by placing a cut seed on the surface of agar containing starch. If the surface is moist, amylase will diffuse from the seed and catalyse digestion of the starch. Addition of iodine to the agar would stain remaining starch blue-black, revealing a clear 'halo' of digestion around the seed. The size of this circular zone gives a rough indication of how much amylase was present. In practice, sterile handling techniques are an important precaution because contaminating bacteria and fungi may also produce amylase.

Fig 16.19 *The influence of gibberellic acid (GA) on the growth of variety* Meteor *dwarf pea. The plant on the left received no GA and shows the typical dwarf habit. The remaining plants were treated with GA; the dose per plant in micrograms is shown. With doses up to 5 micrograms there is increased growth of the stems with increase in GA dosage. This is the principle of the dwarf pea assay of gibberellins.*

Fig 16.20 *Role of gibberellin in mobilising food reserves of barley grain during breaking of dormancy.*

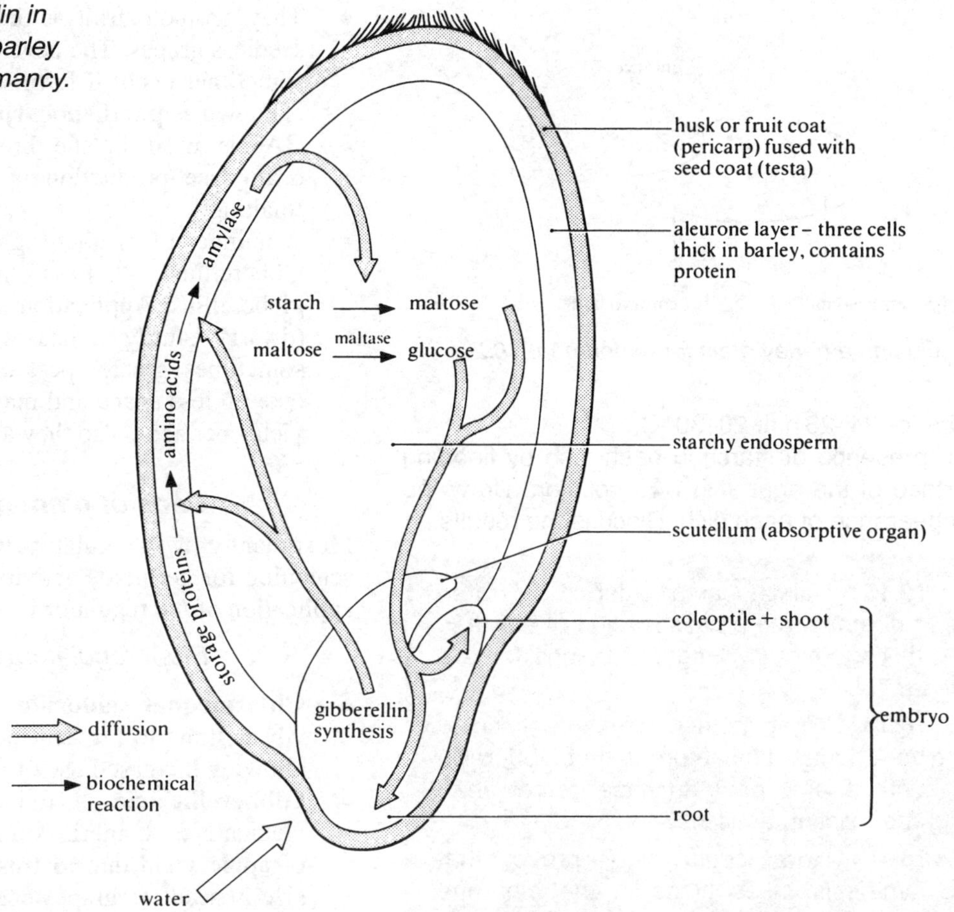

With careful experiment design, further hypotheses may be tested (see, for example, Coppage, J. & Hill, T. A. (1973) *J. Biol. Ed.* **7**, 11–18).

Materials (per student)

white tile
scalpel
forceps
50 cm^3 beaker
labels or chinagraph pencil
iodine/potassium iodide solution
sterile distilled water in sterile flasks (×3)
5% sodium hypochlorite solution or commercial sterilising fluid, e.g. Milton's or 70% alcohol
two starch agar plates (sterile): 1% agar containing 0.5% starch poured to a depth of about 0.25 cm in sterile petri dishes
two starch–gibberellin agar plates: as above but add gibberellin (GA_3) to the agar before autoclaving at the concentration of 1 cm^3 of 0.1% GA_3 solution per 100 cm^3 agar (final concentration 10 ppm GA_3) (Gibberellin does not dissolve readily in water and is best dissolved in ethanol; some of the GA_3 is destroyed during autoclaving but the seeds are sensitive to concentrations as low as 10^{-5} ppm GA_3.)
dehusked barley grains: to dehusk barley grains, soak them in 50% (v/v) aqueous sulphuric acid for 3–4 h and then wash thoroughly (about ten times) in distilled water; violent shaking of the grains in a conical flask removes most of the husks; grains should be used immediately since soaking starts germination. Alternatively, 'embryo' and 'non-embryo' halves can be separated (see below) and stored under dry conditions in a fridge for a maximum of 2–3 days. Wheat grains are naked and do not need dehusking, and may be used as a substitute for barley.

Method

(1) Take two starch agar, and two starch–gibberellin agar plates appropriately labelled +GA or –GA. These have been sterilised. Label one of each type 'embryo' and the other 'non-embryo'.
(2) Cut at least two dry barley grains transversely in half (fig 16.21) on a tile, thus separating into 'embryo' and 'non-embryo' halves.
(3) Sterilise the halves in 5% sodium hypochlorite solution for 5 min. Then wash in three changes of sterile distilled water in sterile flasks.
(4) Using forceps sterilised by rinsing in 70% alcohol, place the halves immediately in the relevant dishes, cut face downwards, with minimal lifting of lids as follows:

–GA embryo half	+GA embryo half	–GA non-embryo half	+GA non-embryo half

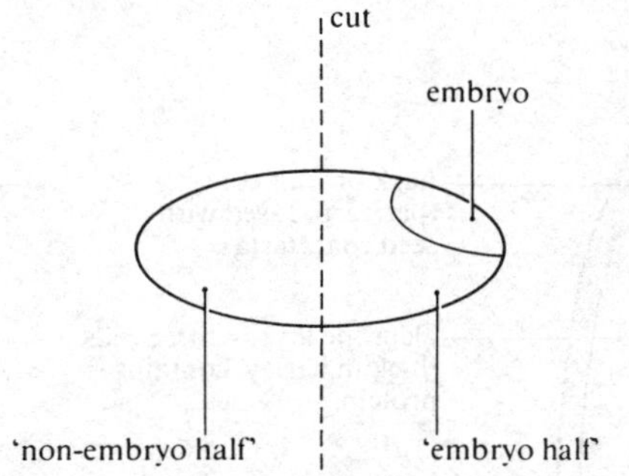

Fig 16.21 *Cutting a barley grain for experiment 16.2.*

(5) Incubate for 24–28 h at 20–30 °C.
(6) Test for presence of starch in each dish by flooding the surface of the agar with I_2/KI solution. Draw the final appearance of each dish. Discuss the results.

16.15 Using starch agar it is often possible to demonstrate an association of amylase activity with fingerprints. Suggest reasons for this association.

16.16 What further experiment could you do, given the facilities, to prove that gibberellin causes synthesis of *new* amylase rather than activating pre-existing amylase?

16.17 How could you prove that amylase synthesis takes place in the aleurone layer?

16.18 How might the effect of gibberellin on barley seeds be used as a bioassay for gibberellin activity?

Other effects of gibberellins

Further effects of gibberellins on flowering, fruit growth and dormancy and their involvement in photoperiodism and vernalisation are discussed later under the appropriate headings. Their effects are summarised in table 16.4.

Mode of action of gibberellins

The mechanism of action of gibberellins remains unclear. In cereal grains GA_3 has been shown to stimulate synthesis of new protein, particularly α-amylase, and is effective in such low concentrations (as little as $10^{-5} \mu g\,cm^{-3}$) that it must be operating at a profound level in cell metabolism, such as the 'switching' on or off of genes which takes place during cell differentiation (section 23.9). No conclusive evidence that this is so has yet been obtained, and higher concentrations are required for its other effects. In cell elongation it is dependent on the presence of auxins.

Commercial applications of gibberellins

Gibberellins are produced commercially from fungal cultures.

- They promote fruit setting and are used for growing seedless grapes. The development of seedless fruits can sometimes occur if fertilisation does not take place and is known as **parthenocarpy**.
- GA_3 is used in the brewing industry to stimulate α-amylase production in barley, and hence promote 'malting'.
- A number of synthetic growth retardants act as 'anti-gibberellins', that is they inhibit the action of gibberellins. Application of these often results in short (dwarf), sturdy plants with deep green leaves and sometimes greater pest and disease resistance. They take up less space and may in the future lead to higher yields per acre; also they are less inclined to blow over.

Uses of plant growth regulators

Most plant growth regulators were discovered as a result of screening for herbicide activity. The rate and timing of the application of the regulator is of critical importance.

Plant regulators increasing yield.

- **Chlormequat chloride** to shorten and stiffen wheat straw to prevent lodging (that is, blowing over) allowing increased use of nitrogenous fertilisers.
- **Gibberellic acid** to increase fruit set of mandarins, clementines, tangerines and pears; to overcome losses of apple yield due to frost damage; to increase berry size in seedless grapes enabling the product to be sold fresh rather than dry as raisins; to overcome low temperature constraints to sugarcane growth in Hawaii.
- **Ethephon** (releases ethene) to increase latex flow in rubber.
- **Glysophosine** to ripen sugarcane.

Plant regulators improving quality.

- **Gibberellic acid** coupled with mechanical thinning to increase berry size of seedless grapes in California.
- **Ethene** to de-green citrus fruits.
- **Gibberellins** to delay ripening and improve storage life of bananas.

Plant regulators increasing the value of crops.

- **Gibberellic acid** to advance or retard maturity of globe artichokes to capture higher prices outside the main production season; to retard ripening of grapefruit to spread harvest and capture higher priced market; to force rhubarb for early production.
- **Maleic hydrazides** to extend storage life of bulbs and root crops.

Growth retardants.

- **Ethephon** (releases ethene) shortening stems of forced daffodils, retarding elongation and stimulating branching in tomatoes, geraniums and roses.

Table 16.4 Roles of plant growth substances in plant growth and development.

Process affected	*Auxins*	*Gibberellins*
Stem growth	**Promote cell enlargement in region behind apex.** Promote cell division in cambium.	**Promote cell enlargement in presence of auxin.** Also promote cell division in apical meristem and cambium. **Promote 'bolting' of some rosette plants.**
Root growth	**Promote at very low concentrations. Inhibitory at higher concentrations,** e.g. geotropism	Usually inactive
Root initiation	**Promote growth of roots from cuttings and calluses.**	Inhibitory
Bud (shoot) initiation	Promote in some calluses but sometimes antagonistic to cytokinins and inhibitory. Sometimes promote in intact plant if apical dominance broken (see below).	Promote in chrysanthemum callus. Sometimes promote in intact plant if apical dominance broken.
Leaf growth*	Inactive	Promote
Fruit growth	**Promote.** Can sometimes induce parthenocarpy.	**Promote.** Can sometimes induce parthenocarpy.
Apical dominance	**Promote, i.e. inhibit lateral bud growth.**	Enhance action of auxins.
Bud dormancy*	Inactive	**Break**
Seed dormancy*	Inactive	**Break,** e.g. cereals, ash
Flowering*	Usually inactive (promote in pineapple)	**Sometimes substitute for red light.** Therefore promote in long-day plants, inhibit in short-day plants.
Leaf senescence	Delay in a few species	Delay in a few species
Fruit ripening	–	–
Abscission	**Inhibit.** Sometimes promote once abscission starts or if applied to plant side of abscission layer.	Inactive
Stomatal mechanism	Inactive	Inactive

Process affected	*Cytokinins*	*Abscisic acid*	*Ethene*
Stem growth	**Promote cell division in apical meristem and cambium** Sometimes inhibit cell expansion	**Inhibitory, notably during physiological stress, e.g. drought, waterlogging**	**Inhibitory, notably during physiological stress**
Root growth	Inactive or inhibit primary root growth	Inhibitory, e.g. geotropism?	Inhibitory, e.g. geotropism?
Root initiation	Inactive or promote lateral root growth	–	–
Bud (shoot) initiation	Promote, e.g. in protonemata of mosses	–	–
Leaf growth*	Promote	–	–
Fruit growth	**Promote.** Can rarely induce parthenocarpy	–	–
Apical dominance	**Antagonistic to auxins, i.e. promote lateral bud growth**	–	–
Bud dormancy*	**Break**	**Promotes,** e.g. sycamore, birch	**Breaks**
Seed dormancy*	**Break**	**Promotes**	–
Flowering*	Usually inactive	Sometimes promote in short-day plants and inhibit in long-day plants (antagonistic to gibberellins)	Promotes in pineapple
Leaf senescence	**Delay**	Sometimes promotes	–
Fruit ripening	–	–	**Promotes**
Abscission	Inactive	**Promotes**	–
Stomatal mechanism	Promotes stomatal opening	**Promotes closing of stomata under conditions of water stress (wilting)**	Inactive?

* Light and temperature are also involved – see photoperiodism and vernalisation.

NB The information presented in this table is generalised. The growth substances do not necessarily always have the effects attributed to them and variation in response between different plants is common. It is best to pay closest attention to the positive effects. Those stressed in bold type are the most important for the A-level student.

- **Piproctanyl chloride** dwarfing ornamentals, mainly chrysanthemums.
- **Dikegulac sodium** retarding growth of hedges and woody ornamentals.

Other targets for plant growth regulators include:

- promotion of root initiation in plant propagation;
- breaking or enforcing dormancy in seed buds and storage organs;
- controlling development of lateral stems;
- increasing resistance to pests and adverse weather conditions, such as drought/pollution;
- controlling size, shape and colour of crops grown for the processed food market;
- suppressing unwanted vegetative growth;
- promoting or delaying flowering/controlling fruit set and ripening.

The benefit to cost ratio of using gibberellic acid to increase fruit setting in mandarin oranges has been reported as 9:1, for the production of seedless grapes 40:1, whilst the use of glysophosine has increased sucrose production in sugarcane by 10–15%. However the use of most plant growth regulators is restricted to commercially small and specialist crops with the exception of antilodging compounds. A number of regulators considered potentially useful during laboratory screening tests proved unreliable for commercial usage. Consequently the main thrust in this field centres on the development of herbicides and pesticides. Further manipulation of crops to human need will probably require a balanced mixture of several plant growth regulators.

16.2.7 Cytokinins

Discovery of cytokinins

During the 1940s and 1950s efforts were made to perfect techniques of plant tissue culture. Such techniques provide an opportunity to study development free from the influence of other parts of the organism, and to study effects of added chemicals. While it proved possible to keep cells alive, it was difficult to stimulate growth. During the period 1954–6 Skoog, working in the USA, found that coconut milk contained an ingredient that promoted cell division in tobacco pith cultures. Coconut milk is a liquid endosperm (food reserve) and evidence that it contained growth substances had already been obtained in the 1940s when it had been studied as a likely source of substances to promote embryo growth. In a search for other active substances a stale sample of DNA happened to show similar activity, although a fresh sample did not. However, autoclaving fresh DNA (heating it under pressure) produced the same effect and the active ingredient was shown to be chemically similar to the base adenine, a constituent of DNA (chapter 3). It was called **kinetin**. The term **kinin** was used by Skoog for substances concerned with the control of cell division. Later the term **cytokinin** was adopted, partly from cytokinesis, meaning cell division, and partly because kinin has an entirely different meaning in zoology (it is a blood polypeptide). The first naturally occurring cytokinin to be chemically identified was from young maize (*Zea mais*) grains in 1963 and hence was called **zeatin**. Note again the chemical similarity to adenine (fig 16.22).

Synthesis and distribution of cytokinins

Cytokinins are most abundant where rapid cell division is occurring, particularly in fruits and seeds where they are associated with embryo growth. Evidence suggests that in mature plants they are frequently made in the roots and move to the shoots in the transpiration stream (in xylem). Cytokinins may be re-exported from leaves via the phloem.

Effects of cytokinins

Cytokinins, by definition, promote cell division. They do so, however, only in the presence of auxins. Gibberellins may also play a role, as in the cambium. Their interaction with other growth substances is discussed later in section 16.3.2.

One of the intriguing properties of cytokinins is their ability to delay the normal process of ageing in leaves. If a leaf is detached from a plant it will normally senesce very rapidly, as indicated by its yellowing and loss of protein, RNA and DNA. However, addition of a spot of kinetin will result in a green island of active tissue in the midst of yellowing tissue. Nutrients are then observed to move to this green island from surrounding cells (fig 16.23).

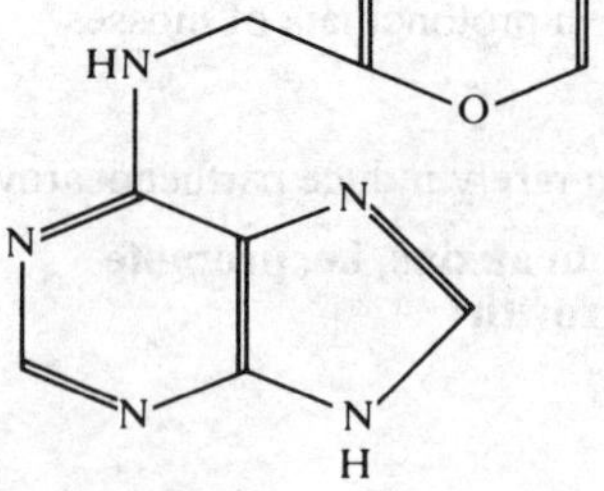

kinetin (6-furfuryl adenine) – a synthetic cytokinin

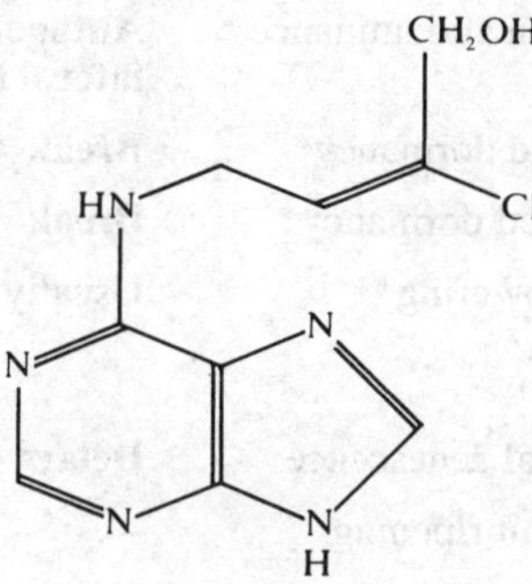

zeatin – a natural cytokinin

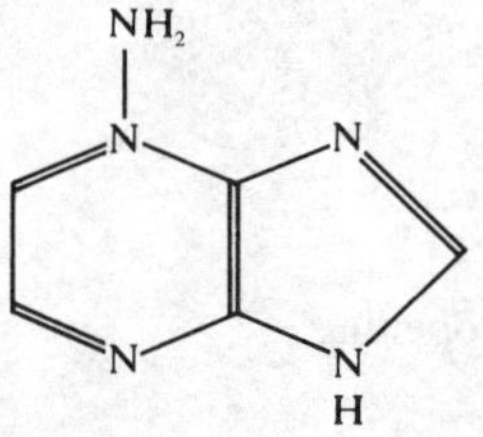

adenine – related to cytokinins

Fig 16.22 *Structures of kinetin, zeatin and adenine.*

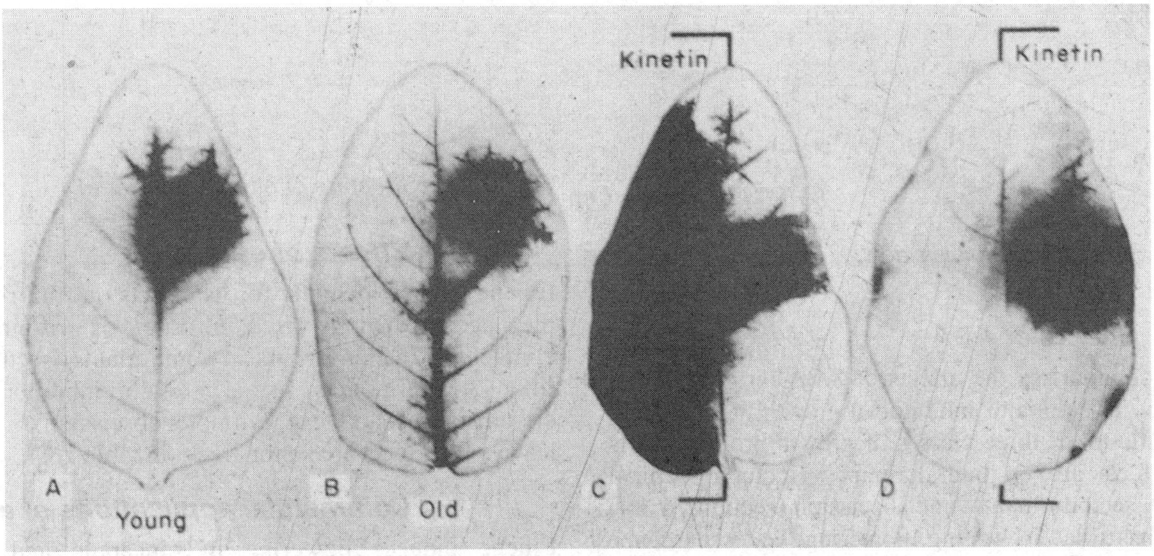

Fig 16.23 *Effect of kinetin upon translocation of an amino acid in tobacco leaves. Radioactive amino acid was supplied as indicated and after a period of translocation the leaves were exposed to photographic film. In the resulting autoradiographs, the areas containing the amino acid appear black.*

> **16.19** Study fig 16.23 and then answer the following.
> (*a*) What difference is there in the fate of applied amino acid between an old leaf and a young leaf?
> (*b*) Why should there be this difference?
> (*c*) What is the effect of kinetin on distribution of radioactive amino acid in old leaves?

Even when kinetin is applied to dying leaves on an intact plant a similar, though less dramatic, effect occurs. It has been shown that levels of natural cytokinins decrease in senescing leaves. A natural programme of senescence may therefore involve movement of cytokinins from older leaves to young leaves via the phloem.

Cytokinins are also implicated in many stages of plant growth and development (table 16.4 and section 16.3).

Mode of action of cytokinins

The similarity of cytokinins to the base adenine, a component of the nucleic acids RNA and DNA, suggests that they may have a fundamental role in nucleic acid metabolism. Some unusual bases, derived from transfer RNA molecules, have been shown to have cytokinin activity, raising the possibility that cytokinins are involved in transfer RNA synthesis. Whether this is true or not, it does not necessarily account for their role as growth substances, and further evidence is still being sought.

Commercial applications of cytokinins

Cytokinins prolong the life of fresh leaf crops such as cabbage and lettuce (delay of senescence) as well as keeping flowers fresh. They can also be used to break the dormancy of some seeds.

16.2.8 Abscisic acid

Discovery of abscisic acid

Plant physiologists have more recently obtained evidence that growth inhibitors, as well as growth promoters like auxins, gibberellins and cytokinins, are important in the normal regulation of growth. It had long been suspected that dormancy was caused by inhibitors when a group at the University of Aberystwyth, led by Wareing, set about trying to find them in the late 1950s. In 1963, an extract from birch leaves was shown to induce dormancy of birch buds. The leaves had been treated with short days to mimic approaching winter. Pure crystals of an active substance were isolated from sycamore leaves in 1964. The substance was called **dormin**. It turned out to be identical to a compound isolated by another group in 1963 from young cotton fruit. This accelerated abscission and was called **abscisin II** (**abscisin I** is a similarly acting, but chemically unrelated and less active compound). In 1967 it was agreed to call the substance **abscisic acid** (**ABA**). It has been found in all groups of plants from mosses upwards and a substance that plays a similar role, lunularic acid, has been found in algae and liverworts.

Structure of abscisic acid

Like the gibberellins, ABA is a terpenoid and has a complex structure (fig 16.24). It is the only growth substance in its class.

Synthesis and distribution of ABA

ABA is made in leaves, stems, fruits and seeds. The fact that isolated chloroplasts can synthesise it again suggests a link with the carotenoid pigments, which are also made in chloroplasts. Like the other growth substances, ABA moves in the vascular system, mainly in the phloem. It also moves from the root cap by diffusion (see geotropism).

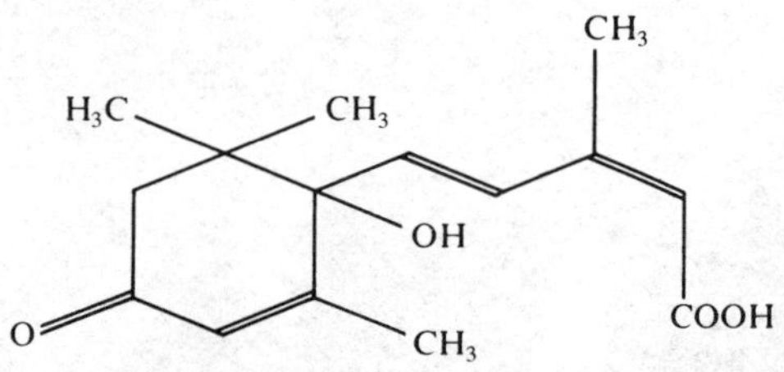

Fig 16.24 *Structure of abscisic acid.*

Effects of ABA

Table 16.4 summarises the effects of ABA on growth and development. It is a major inhibitor of growth in plants and is antagonistic to all three classes of growth promoters. Its classical effects are on bud dormancy (including apical dominance), seed dormancy and abscission (section 16.3.4) but it also has roles in wilting, flowering, leaf senescence and possibly geotropism. It is associated with stress, particularly drought. In wilting tomato leaves, for example, the ABA concentration is 50 times higher than normal and ABA is thought to bring about closure of stomata. High concentrations stop the plant growing altogether.

Mode of action of ABA

This is unknown.

Commercial applications of ABA

ABA can be sprayed on tree crops to regulate fruit drop at the end of the season. This removes the need for picking over a long time-span.

16.2.9 Ethene (ethylene)

Discovery of ethene as a growth substance

It was known in the early 1930s that ethene gas speeded up ripening of citrus fruits and affected plant growth in various ways. Later it was shown that certain ripe fruits, such as bananas, gave off a gas with similar effects. In 1934 yellowing apples were shown to emit ethene and it was subsequently shown to emanate from a wide variety of ripening fruits and other plant organs, particularly from wounded regions. Trace amounts are normal for any organ.

Structure of ethene

The structure of an ethene molecule is shown in fig 16.25.

Synthesis and distribution of ethene

As mentioned above, ethene is made by most or all plant organs. Despite being a gas, it does not generally move freely through the stem of air spaces in the plant because it tends to escape more easily from the plant surface. However, movement of the water-soluble precursor of ethene from waterlogged roots to shoots in the xylem has been demonstrated.

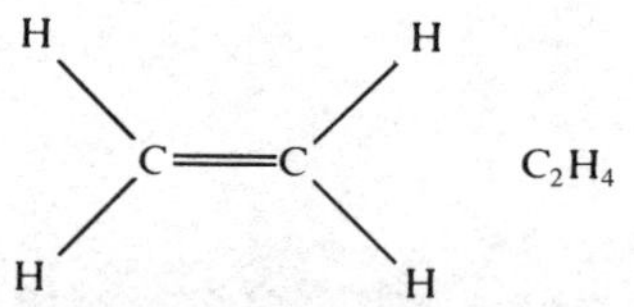

Fig 16.25 *Structure of ethene (ethylene).*

Effects of ethene

Ethene is known chiefly for its effects on fruit ripening and the accompanying rise in rate of respiration (the climacteric) which occurs in some plants (section 16.3.5). Like ABA it acts as a growth inhibitor in some circumstances and can promote abscission of fruits and leaves. Its effects are summarised in table 16.4.

Commercial applications of ethene

Ethene induces flowering in pineapple and stimulates ripening of tomatoes and citrus fruits. Fruits can often be prevented from ripening by storage in an atmosphere lacking oxygen; ripening can subsequently be regulated by application of ethene with oxygen. The commercial compound 'ethephon' breaks down to release ethene in plants and is applied to rubber trees to stimulate the flow of latex.

16.3 Synergism and antagonism

Having studied individual growth substances it has become clear that they generally work by interacting with one another, rather than each controlling its own specific aspect of growth. Two kinds of control emerge. In the first, two or more substances supplement each other's activities. It is often found that their combined effect is much greater than the sum of their separate effects. This is called **synergism** and the substances are said to be **synergistic**. The second kind of control occurs when two substances have opposite effects on the same process, one promoting and the other inhibiting. This is called **antagonism** and the substances are said to be **antagonistic**. Here the balance between the substances determines response.

Some of the better understood phases of plant growth and development can now be studied and the importance of synergism and antagonism demonstrated.

16.3.1 Shoot growth

The effect of gibberellins on elongation of stems, petioles, leaves and hypocotyls is dependent on the presence of auxins.

> **16.20** How could you demonstrate this experimentally?

16.3.2 Cell division and differentiation

Cytokinins promote cell division only in the presence of auxins. Gibberellins sometimes also play a role, as in the cambium when auxins and gibberellins come from nearby buds and leaves. The interaction of cytokinins with other growth substances was demonstrated in the classic experiments of Skoog in the 1950s, already mentioned. His team showed the effect of various concentrations of kinetin and auxin on growth of tobacco pith callus. A high auxin to cytokinin ratio promoted root formation whereas a high kinetin to auxin ratio promoted lateral buds which grew into leafy shoots. Undifferentiated growth would occur if the growth substances were in balance (fig 16.26).

16.3.3 Apical dominance

Apical dominance is the phenomenon whereby the presence of a growing apical bud inhibits growth of lateral buds. It also includes the suppression of later root growth by growth of the main root. Removal of a shoot apex results in lateral bud growth, that is branching. This is made use of in pruning when bushy rather than tall plants are required.

> **16.21** (*a*) What plant growth substance is made in the shoot apex? (*b*) Design an experiment to show whether it is responsible for apical dominance.

It is interesting to note that auxin levels at the lateral buds are often *not* high enough to cause inhibition of growth. Auxins exert their influence in an unknown way, possibly by somehow 'attracting' nutrients to the apex. In cocklebur it appears that the fall in auxin level in the stem after decapitation permits the lateral buds to inactivate the high levels of ABA that they contain. Gibberellins often enhance the response to IAA. Kinetin application to lateral buds, however, often breaks their dormancy, at least temporarily. Kinetin plus IAA causes complete breaking of dormancy. Cytokinins are usually made in the roots and move in the xylem to the shoots. Perhaps, therefore, they are normally transported to wherever auxin is being made, and they combine to promote bud growth.

Apical dominance is a classic example of one part of a plant controlling another via the influence of a growth substance. This is called **correlation** (fig 16.27).

16.3.4 Abscission

Abscission is the organised shedding of part of the plant, usually a leaf, unfertilised flower or fruit. At the base of the organ, in a region called the **abscission zone**, a layer of living cells separates by the breakdown of their middle lamellae, and sometimes breakdown of their cell walls. This forms the **abscission layer** (fig 16.28). Final shedding of the organ occurs when the vascular strands are broken mechanically, such as by the action of wind. A protective layer is formed beneath the abscission layer to prevent infection or desiccation of the scar, and the vascular strand is sealed. In woody species the protective layer is corky, being part of the tissue produced by the cork cambium, namely the periderm (section 22.4).

Abscission of leaves from deciduous trees and shrubs is usually associated with the onset of winter, but in the tropics it often occurs with the onset of a dry season. In both cases it affords protection against possible water

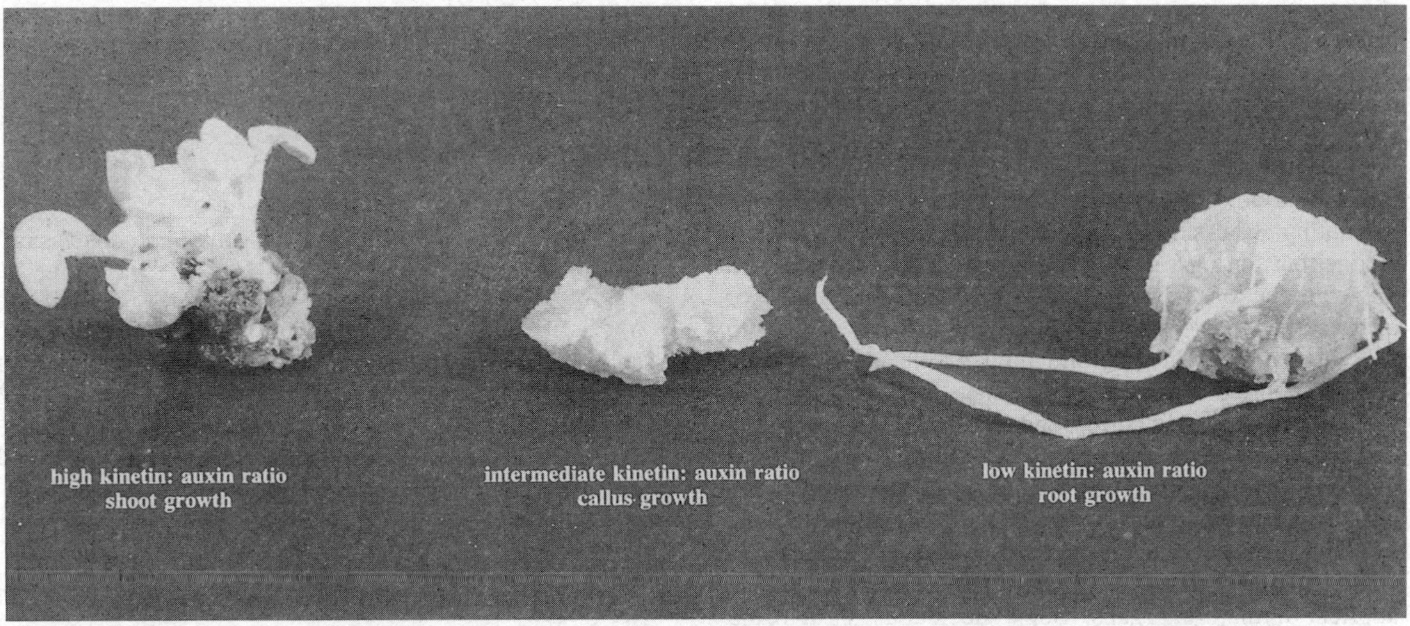

Fig 16.26 *Cultures of tobacco callus. The culture medium in each case contains IAA (2 mg dm^{-3}). The culture 0.2 mg dm^{-3} kinetin (centre) continues growth as a callus; with a lower kinetin addition (0.02 mg dm^{-3}) it initiates roots and with a higher kinetin addition (0.5 mg dm^{-3}) it initiates shoots.*

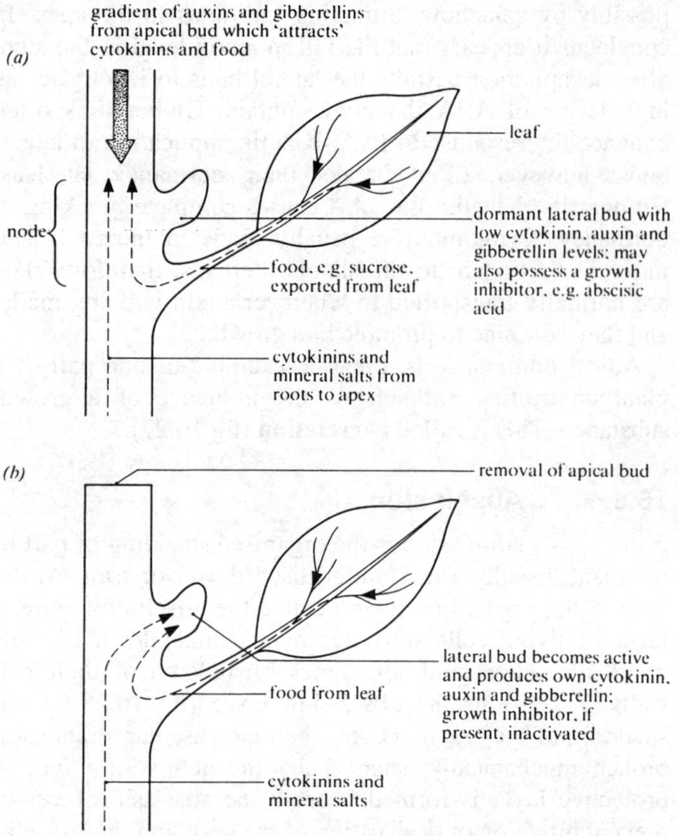

Fig 16.27 *Possible involvement of plant growth substances in apical dominance, (a) in presence of apical bud, (b) after removal of apical bud.*

shortage, leaves being the main organs through which water is lost by transpiration. In winter, for example, soil water may be unavailable through being frozen. In evergreen species, abscission is spread over the whole year and the leaves are usually modified to prevent water loss.

It has been shown that as a leaf approaches abscission, its output of auxin declines. Fig 16.29 summarises the effect of auxin on abscission. It is worth noting that once abscission has been triggered, auxins seem to accelerate the process.

Abscisic acid (ABA) acts antagonistically to auxin by promoting abscission in some fruits. Unripe seeds produce auxins, but during ripening auxin production declines and ABA production may rise. In developing cotton fruits, for example, two peaks of ABA occur. The first corresponds to the 'June drop' when self thinning of the plants occurs: only the aborted immature fruitlets have high ABA levels at this stage. The second corresponds with seed ripening.

There is some doubt as to whether ABA also affects leaf abscission. Applications of high concentrations are effective, but this could be a result of stimulating ethene production. Ethene is produced by senescing leaves and ripening fruits and always stimulates abscission when applied to mature organs. Some deciduous shrubs and trees produce ABA in their leaves just before winter (section 16.5.1) but this may be purely to induce bud dormancy.

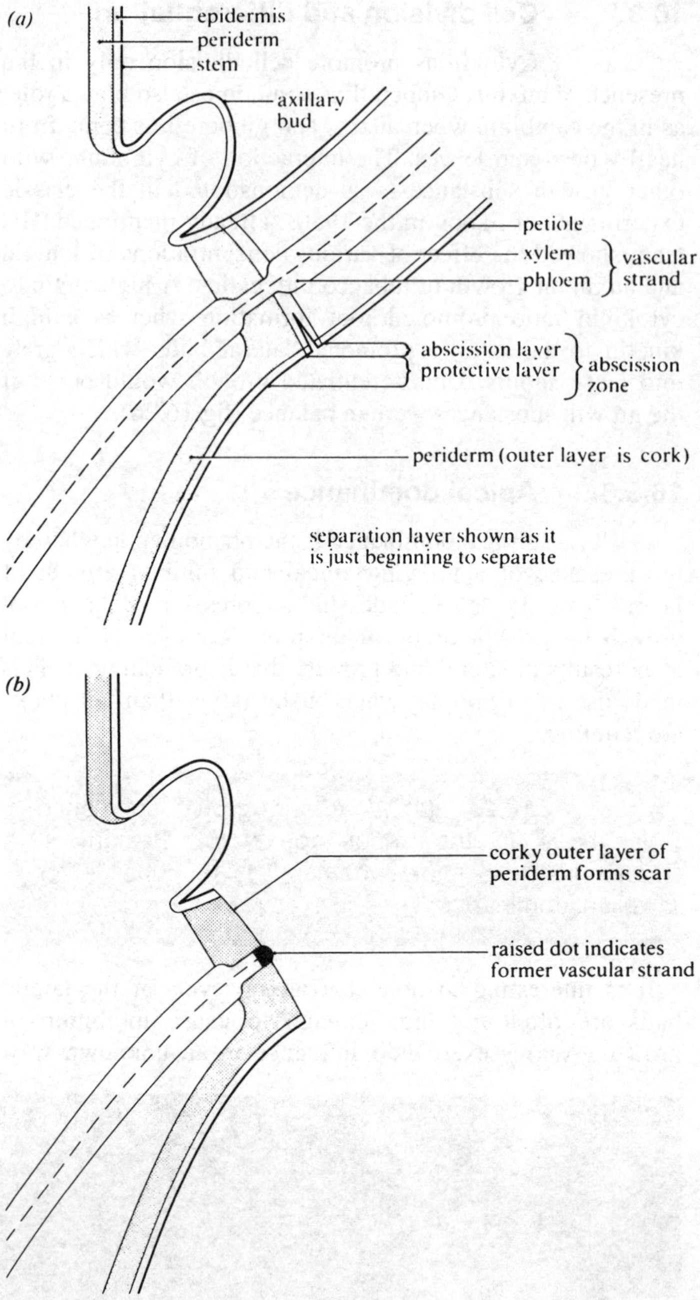

Fig 16.28 *Abscission zone of a leaf, (a) during abscission, (b) after abscission.*

Abscission is of immense horticultural significance because of its involvement in fruit drop. Commercial applications of auxins and ABA reflect this and have been discussed earlier in this chapter.

If flowers are not fertilised they are generally abscised (see 'fruit set' below).

16.3.5 Pollen tube growth, fruit set, fruit development and parthenocarpy

Germinating pollen grains are a rich source of auxins as well as commonly stimulating the tissues of the style and ovary to produce more auxin. This auxin is

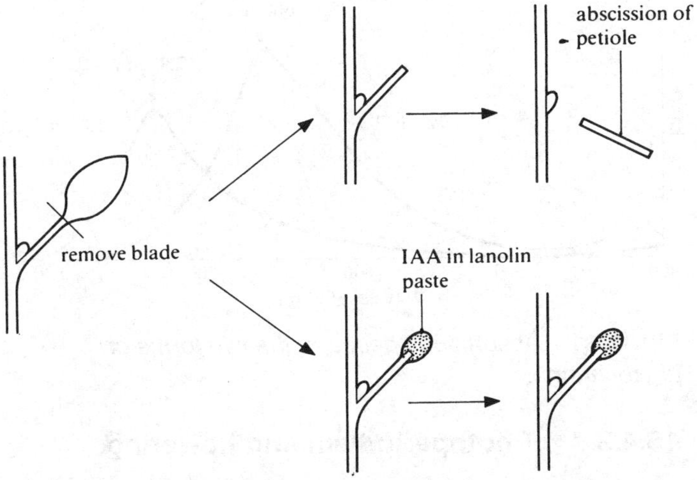

Fig 16.29 *Effect of auxin (IAA) on abscission of a leaf petiole. Removal of the leaf blade leads to abscission of the petiole. IAA substitutes for the presence of the leaf blade.*

necessary for 'fruit set', that is retention of the ovary which becomes the fruit after fertilisation. Without it abscission of the flower normally occurs. After fertilisation, the ovary and the ripe seeds continue to produce auxins which stimulate fruit development.

A few natural examples are known where fruit development proceeds without fertilisation, and therefore without seed development, for example banana, pineapple and some seedless varieties of oranges and grapes. Such development is called **parthenocarpy**. Unusually high auxin levels occur in these ovaries. Parthenocarpy can sometimes be artificially induced by adding auxins, as in tomato, squash and peppers. Seedless pea pods can just as easily be induced. Gibberellins have the same effect in some plants, such as the tomato, including some that are not affected by auxins, for example cherry, apricot and peach. Developing seeds are not only a rich source of auxins and gibberellins, but also of cytokinins (section 16.2.7). These growth substances are mainly associated with development of the embryo and accumulation of food reserves in the seed, and sometimes in the pericarp (fruit wall) from other parts of the plant.

Fruit ripening is really a process of senescence and is often accompanied by a burst of respiratory activity called the **climacteric**. This is associated with ethene production. The subsequent roles of ethene and ABA in fruit abscission were discussed in section 16.3.4.

16.4 Phytochrome and effects of light on plant development

The importance of environmental stimuli to the growth and orientation of plant organs has already been discussed with plant movements. The stimulus which has the widest influence on plant growth is light. Not only does it provide the energy for photosynthesis and influence plant movements, but it directly affects development. The effect of light upon development is called **photomorphogenesis**.

16.4.1 Etiolation

Perhaps the best way to demonstrate the importance of light is to grow a plant in the dark! Such a plant lacks chlorophyll (is chlorotic) and therefore appears white or pale yellow rather than green. The shoot internodes become elongated and thick and the plant is described as **etiolated**. In dicotyledonous plants, the epicotyl or hypocotyl (section 22.4.2) elongates in hypogeal or epigeal germination respectively and the plumule tip is hooked. Dioctyledonous leaves remain small and unexpanded. In monocotyledonous plants, the mesocotyl elongates during germination and the leaves may remain rolled up. In all leaves, chloroplasts fail to develop normal membrane systems and are called **etioplasts**. Plants make less supporting tissue and are fragile and collapse easily. Eventually they use up their food reserves and die unless light is reached for photosynthesis. Yet as soon as the plant is exposed to light, normal growth begins again. The significance of etiolation is that it allows maximum growth in length with minimum use of carbon reserves which, in the absence of light, the plant cannot obtain by photosynthesis.

16.22 How does the morphology (structure) of an etiolated plant suit it for growing through soil?

16.4.2 Discovery of phytochrome

The first stage in any process affected by light must be the absorption of light by a pigment, the so-called **photoreceptor**. The characteristic set of wavelengths of light it absorbs form its absorption spectrum (section 7.3.2); remaining wavelengths are reflected and give the substance a characteristic colour (chlorophyll, for example, absorbs red and blue light and reflects green light).

The seeds of many plants germinate only if exposed to light. In 1937 it was shown that, for lettuce seeds, red light promoted germination but far-red light (longer wavelength) inhibited germination. Borthwick and Hendrick, working at the US Department of Agriculture in the 1950s, plotted an **action spectrum** for the germination response (a spectrum of wavelengths showing their relative effectiveness at stimulating the process). Fig 16.30 shows that the wavelength most effective for germination was about 660 nm (red light) and for inhibition of germination about 730 nm (far-red light).

They also showed that only brief exposures of light were necessary and that the effects of red light were reversed by far-red light and vice versa. Thus the last treatment in an alternating sequence of red/far-red exposures would always

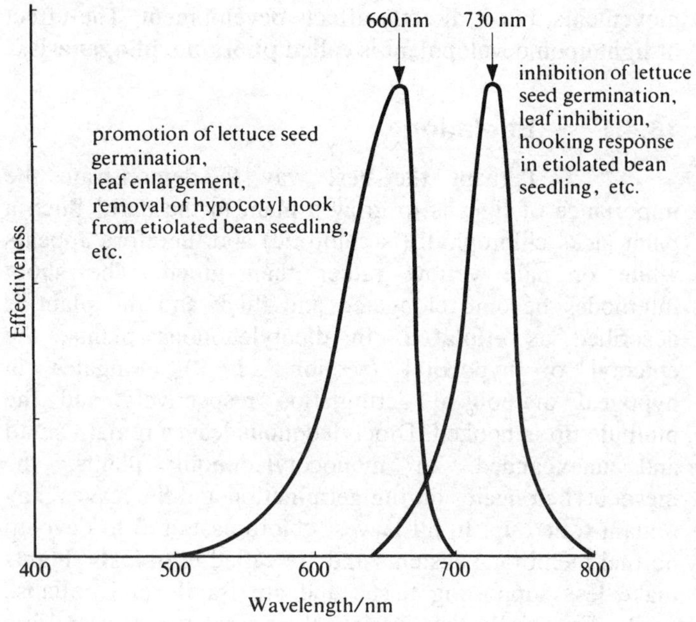

Fig 16.30 *Typical action spectra of a phytochrome-controlled response.*

be the effective one. The US team eventually isolated the pigment responsible in 1960 and called it **phytochrome**. Phytochrome, as they predicted, is a blue-green pigment existing in two interconvertible forms. One form, $\mathbf{P_{FR}}$ or $\mathbf{P_{730}}$ absorbs far-red light and the other, $\mathbf{P_R}$ or $\mathbf{P_{660}}$, absorbs red light. Absorption of light by one form converts it rapidly and reversibly to the other form (within seconds or minutes depending on light intensity):

$$P_R \underset{\text{far-red light (730nm)}}{\overset{\text{red light (660nm)}}{\rightleftharpoons}} P_{FR} \longrightarrow \text{biochemical process}$$

(slowly in darkness: $P_{FR} \rightarrow P_R$)

Normal sunlight contains more red than far-red light, so the $\mathbf{P_{FR}}$ form predominates during the day. This is the physiologically active form, but reverts slowly to the more stable, but inactive, $\mathbf{P_R}$ form at night. Phytochrome was shown to consist of a pigment portion attached to a protein. It is present in minute amounts throughout plants (hence it is not visible, despite its colour) but is particularly concentrated in the growing tips. Its absorption spectrum is shown in fig 16.31.

16.23 What is the difference between an absorption spectrum and an action spectrum?

A number of developmental processes are mediated by low intensities of red light and reversed by far-red light or darkness, showing the involvement of phytochrome (fig 16.30 and table 16.5). The most significant of these is flowering, which is discussed later. The involvement of phytochrome in a process is shown by matching the action spectrum of the response with the absorption spectrum of phytochrome.

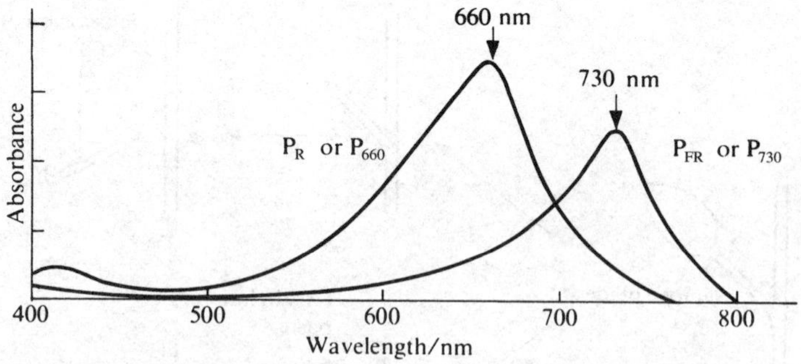

Fig 16.31 *Absorption spectra of the two forms of phytochrome.*

16.4.3 Photoperiodism and flowering

One of the important ways in which light exerts its influence on living organisms is through variations in daylength (**photoperiod**). The further from the equator, where days are almost a constant 12 hours, the greater the seasonal variation in daylength. Thus daylength is an important environmental signal in temperate latitudes where it varies between about 9 and 15 hours during the year. The effects of photoperiod on animals are discussed in section 17.8.5. In plants it is a matter of common observation that phenomena such as flowering, fruit and

Table 16.5 Some phytochrome-controlled responses in plants.

General process affected	*Red light promotes*
Germination*	Germination of some seeds, e.g. some lettuce varieties
	Germination of fern spores
Photomorphogenesis (light-controlled development of form and structure)	Leaf expansion in dicotyledons. Leaf unrolling in grasses (monocotyledons). Chloroplast development (etioplasts converted to chloroplasts: see etiolation). Greening (protochlorophyll converted to chlorophyll). Inhibition of internode growth (including epicotyl, hypocotyl, mesocotyl), i.e. preventing of etiolation. Unhooking of plumule in dicotyledons
Photoperiodism	Stimulates flowering in long-day plants. Inhibits flowering in short-day plants. See flowering

* Experiments designed to investigate the effects of light on seed germination are described by J. W. Hannay in *J. Biol. Ed.* (1967) **1**, 65–73. The variety of lettuce suggested, 'Grand Rapids', is no longer available but some modern varieties could be screened for suitability. Such varieties currently available are 'Dandie', 'Kloek' and 'Kweik'. 'Dandie' is probably the most reliable but is fairly expensive because it is a winter-forcing variety. It is available from Suttons in small packets or from E. W. King & Co., Coggeshall, Essex in 10 g packets (enough for about 250 dishes of 50 seeds each). *Phacelia* seeds make an interesting contrast to lettuce.
NB In this experiment a green leaf can be used as a far-red filter.

seed production, bud and seed dormancy, leaf fall and germination are closely attuned to seasonal influences like daylength and temperature, and that survival of the plant depends on this.

The process involving the most profound change is flowering, when shoot meristems switch from producing leaves and lateral buds to producing flowers. The importance of photoperiod in flowering was discovered as early as 1910 but was first clearly described by Garner and Allard in 1920. They showed that tobacco plants would flower only after exposure to a series of short days. This occurred naturally in autumn, but could be induced by artificially short days of seven hours in a greenhouse in summer. As they examined other plants it became obvious that some required long days for flowering (**long-day plants**, LDPs) and some would flower whatever the photoperiod once mature (**day-neutral plants**).

Additional complications have since been found. For example, some plants are day-neutral at one temperature, but not at another; some require one daylength followed by another; in some, the appropriate daylength only accelerates flowering and is not an absolute requirement.

An important advance in our understanding came when it was shown that it is really the length of the dark period which is critical. Thus short-day plants (SDPs) are really long-night plants. If they are grown in short days, but the long night is interrupted by a short light period, flowering is prevented. Long-day plants will flower in short days if the long night period is interrupted. Short dark interruptions, however, do not cancel the effect of long days. Table 16.6 summarises the three main categories of plant.

Flowering in chrysanthemums

In the classical work of Garner and Allard, chrysanthemums were described as typical short-day plants. However it was later shown by Schwabe that, in addition, an exposure to cold temperatures (1–7 °C in a refrigerator for around three weeks) hastens flower bud formation in both long- and short-day conditions. In the absence of this cold temperature treatment, the plants may remain vegetative even in short days. For the rapid onset of flowering, short-day photoperiodic treatment and warm temperature are required following exposure to a lower temperature (vernalisation, see section 16.5). Evidence suggests that the vernalisation treatment is more effective if given during the dark period. The growing tip has been shown to be the seat of perception of the vernalisation stimulus. Flowering in chrysanthemums can be inhibited by:

- transfer to long-day treatment;
- transfer to low light intensity during short-day treatment;
- application of auxin paste suggesting that hormone balance may be important in inducing flowering.

Table 16.6 Classification of plants according to photoperiodic requirements for flowering.

Short-day plants (SDPs)	*Long-day plants (LDPs)*
e.g. cocklebur (*Xanthium pennsylvanicum*), chrysanthemum, soybean, tobacco, strawberry	e.g. henbane (*Hyoscyamus niger*), snapdragon, cabbage, spring wheat, spring barley
Flowering induced by dark periods longer than a critical length, e.g. cocklebur 8.5 h; tobacco 10–11 h	Flowering induced by dark periods shorter than a critical length, e.g. henbane 13 h
(Under natural conditions equivalent to days shorter than a critical length, e.g. cocklebur 15.5 h; tobacco 13–14 h)	(Under natural conditions equivalent to days longer than a critical length, e.g. henbane 11 h)
Day-neutral plants	
e.g. cucumber, tomato, garden pea, maize, cotton Flowering independent of photoperiod	

NB Tobacco (SDP) and henbane (LDP) both flower in 12–13 h daylength.

It is unusual for short-day plants to require vernalisation. Commercially, chrysanthemum propagation cuttings are taken during January–March when the vernalisation requirement will be met by the prevailing winter temperature.

16.4.4 Quality and quantity of light

The next step is to find the quality (colour) and quantity of light required. Remembering that the cocklebur (SDP) will not flower if its long night is interrupted, experiments revealed that red light was effective in preventing flowering but that far-red light reversed the effect of red light. Therefore phytochrome is the photoreceptor. These experiments were, in fact, part of the programme which led to the discovery of phytochrome. As would be expected, a LDP held in short days is stimulated to flower by a short exposure to red light during the long night. Again this is reversed by far-red light. The last light treatment always determines the response.

In some, though not all, cases low light intensities for a few minutes are effective, again typical of a phytochrome-controlled response. The higher the intensity used, the shorter the exposure time required.

16.24 *Xanthium* (cocklebur) is a short-day plant. Flowering is therefore normally induced by long nights. In experiments the effects of red and red/far-red light interruptions of three consecutive 12 h ('long') nights on subsequent flowering were investigated. Flowering intensity was measured by an index of 'floral stage' from 1 to 8, and the results are shown in table 16.7.

Draw graphs of these results, plotting floral stage on the vertical axis and duration of light on the horizontal axis. (The two graphs can be drawn between the same axes.) Comment on the results.

16.4.5 Perception and transmission of the stimulus

It was shown in the mid-1930s that the light stimulus is perceived by the leaves and not the apex where the flowers are produced.

16.25 How could you demonstrate this with a short-day plant?

In addition to this, a cocklebur plant with just one induced leaf will flower even if the rest of the plant is under non-inductive conditions. This implies that some agent, that is a hormone, must pass from the leaf to the apex to bring about flowering. This concept is supported by the observation that the flowering stimulus can be passed from an induced plant to a non-induced plant by grafting, and that the stimulus is apparently the same for SDP, LDP and day-neutral plants because grafting between these is successful. The hypothetical flowering hormone has been called '**florigen**' but has never been isolated. Indeed, its existence is doubted by some plant physiologists.

Table 16.7 Effect of red/far-red light interruptions of long nights on flowering of cocklebur.

Red light		*Two minutes red light followed by FR light*	
Floral stage	*Duration of red light/s*	*Floral stage*	*Duration of FR light/s*
6.0	0	0.0	0
5.0	5	4.0	12
4.0	10	4.5	15
2.6	20	5.5	25
0.0	30	6.0	50

(After Downs, R. J. (1956) *Plant Physiol.* 31 279–84.)

16.4.6 Mode of action of phytochrome

How, then, does phytochrome exert its control? At the end of a light period it exists in the active P_{FR} form. At the end of a short night its slow transition back to the inactive P_R form, which takes place in darkness, may not be complete. It can be postulated, therefore, that in LDPs P_{FR} promotes flowering and in SDPs it inhibits flowering. Only long nights remove sufficient P_{FR} from the latter to allow flowering to occur. Unfortunately, short exposures to far-red light, which would have the same effect as a long night, cannot completely substitute for long nights, so the full explanation is more complex. Some time factor is also important.

We do know that gibberellins can mimic the effect of red light in some cases. Gibberellic acid (GA_3) promotes flowering in some LDPs, mainly rosette plants like henbane which bolt before flowering. (Bolting is a rapid increase in stem length.) GA_3 also inhibits flowering in some SDPs. Antigibberellins (growth retardants) nullify these effects.

So, does P_{FR} stimulate gibberellin production, and is this the flowering hormone? There are too many exceptions for this to be the case. Abscisic acid inhibits flowering in some LDPs, such as *Lolium*, but induces it in some SDPs, such as strawberry. In short, our understanding of the flowering process is still incomplete.

16.5 Vernalisation and flowering

Some plants, especially biennials and perennials, are stimulated to flower by exposure to low temperatures. This is called **vernalisation**. Here the stimulus is perceived by the mature stem apex, or by the embryo of the seed, but not by the leaves as in photoperiodism. As with photoperiod, vernalisation may be an absolute requirement (such as in henbane) or may simply hasten flowering (as in winter cereals).

Long-day plants (for example cabbage), short-day plants (such as chrysanthemum) and day-neutral plants (such as ragwort) can all require vernalisation. The length of chilling required varies from four days to three months, temperatures around 4 °C generally being most effective. Like the photoperiodic stimulus, the vernalisation stimulus can be transmitted between plants by grafting. In this case the hypothetical hormone involved was called **vernalin**. It has subsequently been discovered that during vernalisation gibberellin levels increase, and application of gibberellins to unvernalised plants can substitute for vernalisation (fig 16.32). It is now believed that 'vernalin' is a gibberellin. It is clear now that photoperiodism and vernalisation serve to synchronise the reproductive behaviour of plants with their environments, ensuring reproduction at favourable times of the year. They also help to ensure that members of the same species flower at the same time and thus encourage cross-pollination and cross-fertilisation, with the attendant advantages of genetic variability.

Fig 16.32 *Carrot plants (var.* Early french forcing*). left: control; centre: maintained at 17 °C but supplied 10 mg of gibberellin daily for 4 weeks; right: plant given vernalising cold treatment (6 weeks). All photographed 8 weeks after completion of cold treatment.*

16.5.1 Photoperiodism and dormancy

The formation of winter buds in temperate trees and shrubs is usually a photoperiodic response to shortening days in the autumn, for example in birch, beech and sycamore. The stimulus is perceived by the leaves and, as mentioned before, abscisic acid (ABA) levels build up. ABA moves to the meristems and inhibits growth. Short days also induce leaf fall from deciduous trees (abscission). Often buds must be chilled before dormancy can be broken (‘**bud-break**’). Similarly some seeds require a cold stimulus (‘**stratification**’) after imbibing water before they will germinate, thus preventing them from germinating prematurely once ripe. Gibberellins can substitute for the cold stimulus, and natural bud-break is accompanied by a rise in gibberellins as well as, in many cases, a fall in ABA. Breaking of bud dormancy in birch and poplar has been shown to coincide with a rise in cytokinins.

Apart from buds and seeds, storage organs are involved in dormancy and again photoperiod is important. For example, short days induce tuber formation in potatoes, whereas long days induce onion bulb formation.

16.26 Some buds remain dormant throughout the summer. What causes dormancy here?

Chapter Seventeen

Coordination and control in animals

Irritability or **sensitivity** is a characteristic feature of all living organisms and refers to their ability to respond to a stimulus. The stimulus is received by a **receptor**. It is transmitted by means of nerves or hormones, and an **effector** brings about a **response**.

Animals, unlike plants, have two different but related systems of coordination: the **nervous system** and the **endocrine system**. These are compared in table 17.1. A good case study of the difference between nervous and hormonal control is the control of digestive secretions in the gut. This is discussed in section 8.4. The two systems have developed in parallel in animals. Plants also have a chemical coordination system equivalent to hormones, so the extra possession of a nervous system in animals is probably related to their need to seek food. This requires sense organs and locomotion, which are controlled by a nervous system.

17.1 The nervous system

The nervous system is made up of highly specialised cells whose function is to

- receive **stimuli** from the environment. In multicellular organisms this is done by modified nerve cells called **receptors**. The structure and function of these is described in section 17.5.
- convert the stimuli into the form of electrical impulses, a process called **transduction**.
- transmit them, often over considerable distances, to other specialised cells called **effectors** which are capable of producing an appropriate **response**. The structure and function of effectors is described briefly in section 17.5 and chapter 18.

Table 17.1 Comparison of nervous and hormonal control in animals.

Nervous control	*Hormonal control*
Electrical and chemical transmission (nerve impulses and chemicals across synapses)	Chemical transmission (hormones) through blood system
Rapid transmission and response	Slower transmission and relatively slow-acting (adrenaline an exception)
Often short-term changes	Often long-term changes
Pathway is specific (through nerve cells)	Pathway not specific (blood around whole body), target is specific
Response often very localised, e.g. one muscle	Response may be very widespread, e.g. growth

Between the receptors and effectors are the conducting cells of the nervous system, the **neurones**. These are the basic structural and functional units of the nervous system and spread throughout the organism forming a complex communication network.

Types of nerve cell – a summary

The structure of the different types of nerve cell are shown and described in section 6.6. The three types described are **sensory neurones**, **motor neurones** and **interneurones** (fig 6.27). Each has a nerve cell body and a nerve fibre. The cell body contains the nucleus. The part of the fibre which conducts nerve impulses away from the cell body is the axon. Axons end in swellings called synaptic knobs. Leading towards the cell body is the part of the fibre known as the dendron which *receives* nerve impulses at fine branches called dendrites. Many nerve fibres are surrounded by a fatty myelin sheath and are said to be myelinated. In these neurones, the dendron and the axon form one fibre, with the cell body on a short branch to one side. Another type of neurone called a **bipolar neurone** will be described in this chapter. A bipolar neurone has two unconnected fibres, a dendron and an axon, which enter and leave at opposite sides of the cell body.

17.1.1 The nerve impulse

Measuring electrical activity in neurones

The fact that information can be transmitted along neurones as electrical impulses and bring about muscle contraction and secretion by glands has been known for over 200 years. Details of the mechanisms, however, were discovered only in the last 50 years, following the discovery of certain axons in squid which have a diameter of approximately 1 mm. These **giant axons** (which are involved in escape responses) have large enough diameters to allow electrodes to be inserted into them and experiments to be carried out.

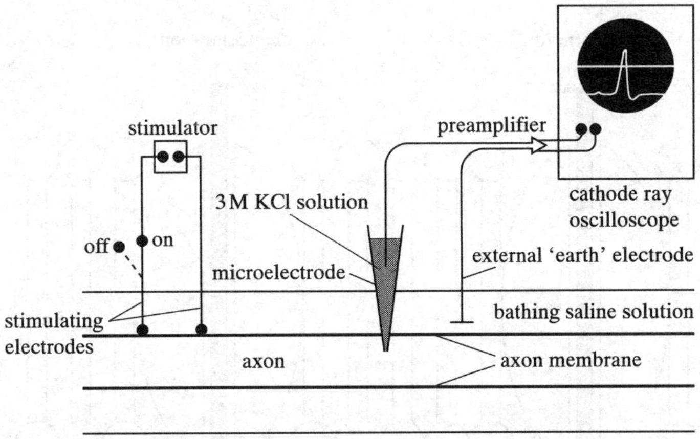

Fig 17.1 *The basic apparatus used to record electrical activity in the axon of an isolated neurone. The stimulator produces a current which generates an action potential in the axon and is detected and recorded using a microelectrode inserted into the axon. The signal is sent to a dual-beam cathode ray oscilloscope.*

The apparatus now used to investigate electrical activity in neurones is shown in fig 17.1. The **microelectrode**, composed of a small glass tube drawn out to a fine point of 0.5 μm diameter, is filled with a solution capable of conducting an electric current such as 3 M potassium chloride. This is inserted into an axon and a second electrode, in the form of a small metal plate, is placed in the saline solution bathing the neurone that is being investigated. Both electrodes are connected by leads to a **preamplifier** to complete the circuit. The preamplifier increases the signal strength in the circuit approximately 1000 times and provides the input to a **dual-beam cathode ray oscilloscope**. All movements of the microelectrode are controlled by a **micromanipulator**, a device with adjusting knobs similar to those of a microscope, which enables delicate control over the position of the tip of the microelectrode.

When the tip of the microelectrode penetrates the axon cell surface membrane, the beams of the oscilloscope separate. The distance between the beams indicates the potential difference between the two electrodes of the circuit and can be measured. This value is called the **resting potential** of the axon (see below). In sensory cells, neurones and muscle cells this value changes with the activity of the cells and hence they are known as **excitable cells**. All other living cells show a similar potential difference across the membrane, known as the **membrane potential**, but in these cells this is constant and so they are known as **non-excitable cells**.

The resting potential

The potential difference (a charge) which exists across the cell surface membrane of all cells is usually, as in the case of nerve cells, negative inside the cell with respect to the outside. The membrane is said to be **polarised**. The potential difference across the membrane at rest is called the **resting potential** and this is about –70 mV (the negative sign indicates that the inside of the cell is negative with respect to the outside). The resting potential is maintained by active transport and passive diffusion of ions. This was discovered by Curtis and Cole in the USA, and Hodgkin and Huxley in England, in the late 1930s when they examined squid axons, as shown in table 17.2.

Table 17.2 Ionic concentrations of extracellular and intracellular fluids in squid axon. (Values given are approximations in mmol kg^{-1} H_2O, data from Hodgkin, 1958.)

Ion	*Extracellular concentration*	*Intracellular concentration*
K^+	20	400
Na^+	460	50
Cl^-	560	100
A^- (organic anions)	0	370

The cytoplasm inside the axon (known as **axoplasm**) has a high concentration of potassium (K^+) ions and a low concentration of sodium (Na^+) ions, in contrast to the fluid outside the axon which has a low concentration of potassium ions and a high concentration of sodium ions. (The distribution of chloride (Cl^-) and other ions is ignored in the following descriptions since it does not play a vital role in the process.) The gradients across the membrane are known as electrochemical gradients. The **electrochemical gradient** of an ion is due to its electrical and chemical properties. The electrical property of an ion is its charge – it will tend to be attracted to an opposite charge and repelled by a similar charge. Its movement is also affected by its concentration in a solution, a chemical property. It will tend to move from a high concentration to a low concentration. Predicting its movement is therefore difficult since it depends on a balance of both charge and concentration.

The resting potential is maintained by the active transport of ions *against* their electrochemical gradients by **sodium/potassium pumps (Na^+/K^+ pumps)**. These are carrier proteins located in the cell surface membrane (see section 5.9.8 active transport and fig 5.21). They are driven by energy supplied by ATP and couple the removal of three sodium ions from the axon with the uptake of two potassium ions as shown in fig 17.2.

The active movement of these ions is opposed by the *passive* diffusion of the ions which constantly pass down their electrochemical gradients through specific ion channel proteins as shown in fig 17.2. The rate of diffusion is determined by the permeability of the axon membrane to the ion. Potassium ions have a membrane permeability which is 20 times greater than that of sodium ions, for reasons explained below. Therefore potassium ion loss from

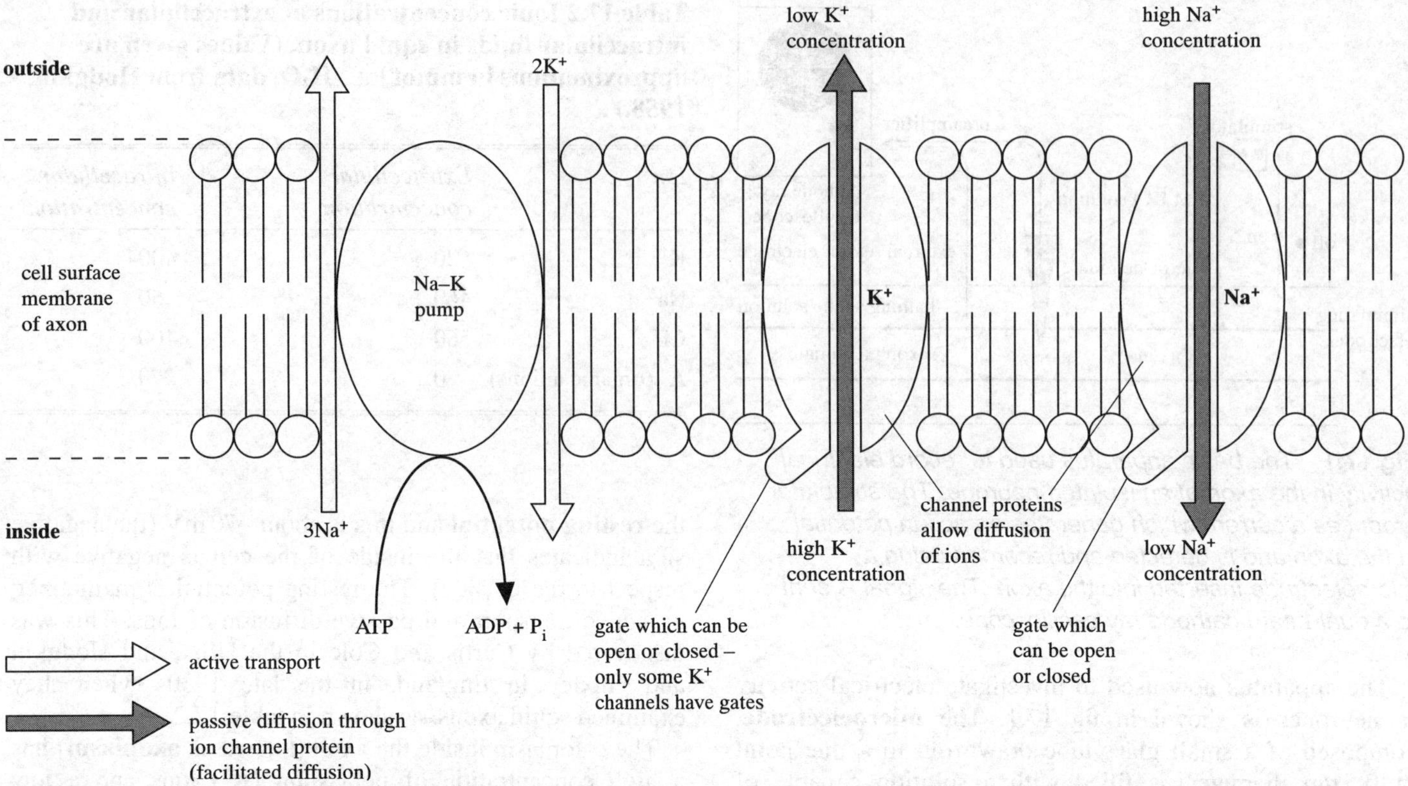

Fig 17.2 *Active and passive ion movements across the cell surface of an axon. The movements are responsible for the generation of a negative potential inside the axon. This is called the resting potential. Active transport takes place through the sodium/potassium pump. Ion channels (proteins) allow the passive movement of ions down their electrochemical gradients. The ion channels are gated, meaning they can be closed. They can be closed in response to a change in voltage across the membrane as happens during the propagation of a nerve impulse. Not all the gates shut at the same time.*

the axon is greater than sodium ion gain. This leads to a net loss of potassium ions from the axon, and the production of a *negative* charge within the axon. The value of the resting potential is largely determined by the potassium ion electrochemical gradient.

We can now turn our attention to how these cells generate nerve impulses. Basically, nerve impulses are due to changes in the permeability of the nerve cell membrane to potassium and sodium ions which lead to changes in the potential difference across the membrane and the formation of 'action potentials'.

The action potential

The experimental stimulation of an axon by an electrical impulse, as shown in fig 17.3, results in a change in the potential across the axon membrane from a negative inside value of about −70 mV to a positive inside value of about +40 mV. This polarity change is called an **action potential** and appears on the cathode ray oscilloscope as a spike, as shown in fig 17.3.

This is generated by a change in the sodium ion channel. This channel, and some of the potassium ion channels, are known as **gated channels**, meaning they can be opened or closed with polypeptide chains called gates. When the gate is closed, the membrane is not very permeable to the ion that normally passes through the gate. Opening the gate increases permeability. Regulation is achieved mainly by the sodium gate which is closed in the resting cell. This helps to explain why the membrane is normally 20 times more permeable to potassium.

An action potential is generated by a sudden and brief opening of the sodium gates. This happens in response to a stimulus which brings about a slight depolarisation, or loss of charge, of the axon membrane. Opening the gates increases the permeability of the axon membrane to sodium ions which enter the axon by diffusion. This increases the number of positive ions inside the axon, which therefore becomes *further* **depolarised**. First the negative resting potential is cancelled out, at which point the membrane

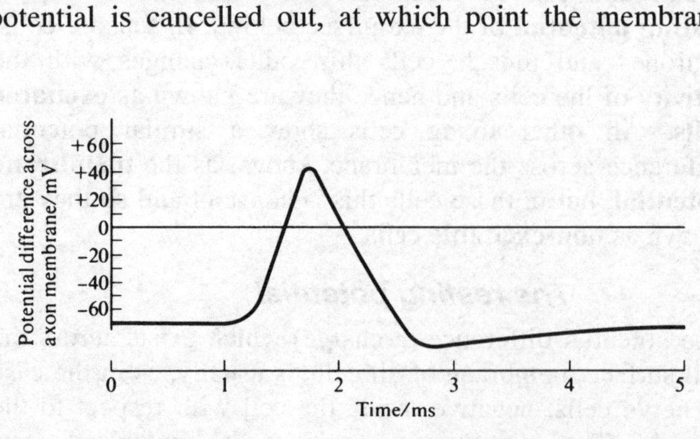

Fig 17.3 *A typical action potential in the axon of a squid.*

is completely depolarised, then it makes the potential difference across the membrane positive. Since the sodium gates are sensitive to depolarisation, the greater the depolarisation, the more gates open. This allows more sodium into the cell, causing greater depolarisation. The two processes therefore reinforce each other. This is called a positive feedback loop. It causes an explosive acceleration in the rate of entry of sodium. (Remember this is taking place within thousandths of a second.) The potential difference peaks at 40 mV, as shown in fig 17.4*a*. This peak corresponds to the maximum concentration of sodium inside the axon. The total depolarisation associated with the action potential has therefore been from –70 mV to + 40 mV, a total of 110 mV. Calculations have revealed that relatively few sodium ions (about one-millionth of the internal sodium ions present, depending upon axon diameter) enter the axon and produce this depolarisation.

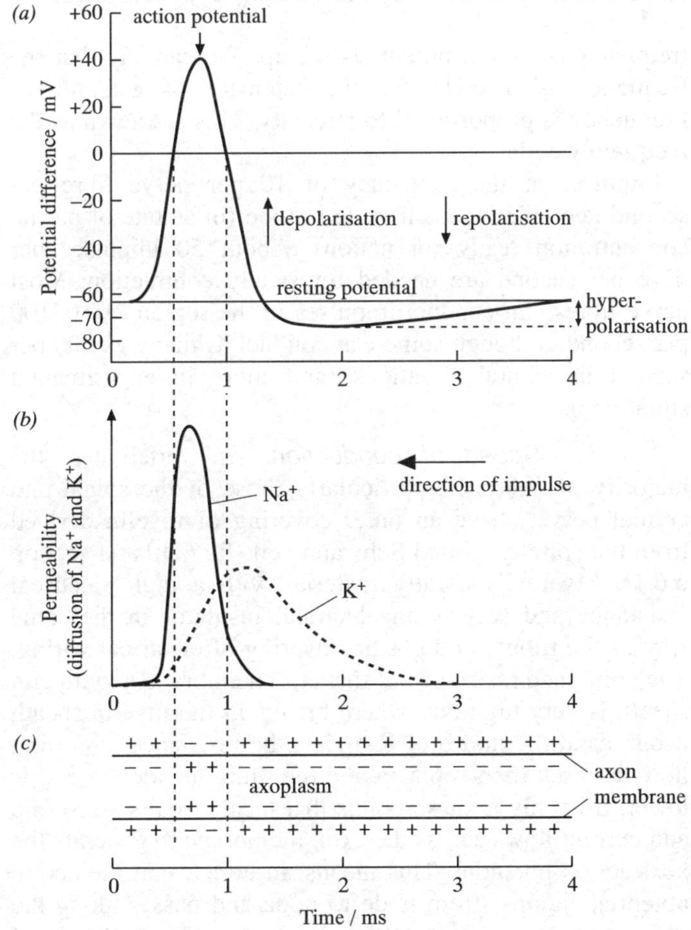

Fig 17.4 *Diagrams to show changes in the axon during the propagation of a nerve impulse: (a) the membrane potential, showing the electrical events associated with the nerve impulse, (b) change in permeability to ions, (c) the net charge across the axon membrane during production of an action potential.*

A fraction of a second after the sodium gates open, depolarisation of the axon membrane causes the potassium gates to open. Potassium therefore diffuses *out* of the cell (fig 17.4*b*). Since potassium is positively charged, this makes the inside of the cell less positive, or more negative, and starts the process of **repolarisation**, or return to the original resting potential.

At the peak of the action potential, the sodium gates start to close again. Sodium permeability therefore declines. The sodium–potassium pump continues to work during this time, so it gradually begins to restore the original resting potential. This repolarisation is shown by the falling phase of the action potential 'spike' (fig 17.4*a*) and results in the membrane potential returning to its original level. In fact, there is a slight overshoot into a more negative potential than the original resting potential. This is called **hyperpolarisation**. It is due to the slight delay in closing all the potassium gates compared with the sodium gates (the potassium graph starts to fall after the sodium graph in fig 17.4*b*). As potassium ions continue to return to the inside of the axon their positive charge restores the normal resting potential.

From the above account it can be seen that whilst the resting potential is determined largely by potassium ions, the action potential is determined largely by sodium ions (fig 17.4).

Nerve impulses travel as action potentials. A nerve impulse is an action potential which passes along an axon as a **wave of depolarisation**. The *outside surface* of the axon is negatively charged at the site of the action potential.

17.1 Give two reasons why there is a sudden influx of sodium ions into the axon following an increase in sodium ion permeability of the axon membrane.

17.2 If the permeability of the axon membrane to sodium ions and potassium ions increased simultaneously, what effect would this have on the action potential?

17.3 Hodgkin and Katz, in 1949, investigated the effect of sodium ions on the production of action potentials in squid axons. Intracellular microelectrodes recorded action potentials from axons bathed in isotonic sea water containing different concentrations of sodium ions. The results are shown in fig 17.5. Which action potentials correspond with axons placed in normal sea water, one-half sea water and one-third sea water? Explain the effect of these solutions on the action potentials.

Features of action potentials

Propagation (conduction) of nerve impulses. A nerve impulse is a **wave of depolarisation** that moves

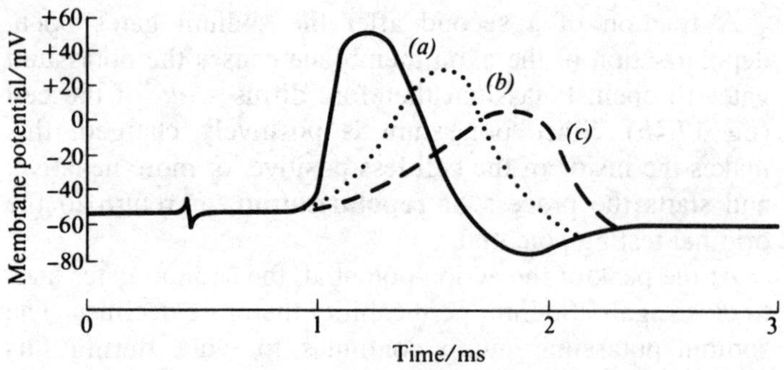

Fig 17.5 *Membrane potentials recorded from squid axons bathed in sea water containing different concentrations of sodium ions.*

along the surface of a nerve cell. Action potentials are **propagated**, that is self generated, along the axon by the effect of sodium ions entering the axon. This creates an area of positive charge, and a flow of current is set up in a **local circuit** between this active area and the negatively charged resting region immediately ahead. The current flow in the local circuit reduces the membrane potential in the resting region and this depolarisation produces an increase in sodium permeability and the development of an action potential in this region. Repeated depolarisations of immediately adjacent regions of the membrane result in the action potential being moved, or propagated, along the axon. Action potentials are capable, in theory, of being transmitted over an infinite distance, that is, they do not lose strength. The reason for this is that the production of an action potential at each point along the axon is a self-generating event resulting from a change in the local concentration of ions. So long as the outside and inside environments of the axon have the necessary differences in ionic concentrations, an action potential at one point will generate another action potential at the next point.

All-or-nothing law and coding for stimulus intensity. The body must have a way of distinguishing between weak and strong stimuli. One possibility would be to have a variable size of action potential depending on the size of the stimulus. In order to investigate whether this occurs, an experiment can be done in which stimuli of increasing intensity are given, and the size of the action potential measured. The results are shown in fig 17.6. They show that the size of the action potential is *not* affected by the size of the stimulus. This is known as the **all-or-nothing law** because the action potential either occurs or does not and its size is constant. In other words, there is a **threshold stimulus intensity** above which the action potential will be triggered whatever the strength of the stimulus.

It has also been shown that the *speed* at which the impulse is transmitted is not affected by the strength of the stimulus. So how *does* the body distinguish between weak and strong stimuli? The answer lies with the frequency of the action potentials (the number in a given time). The stronger the stimulus, within limits, the greater the

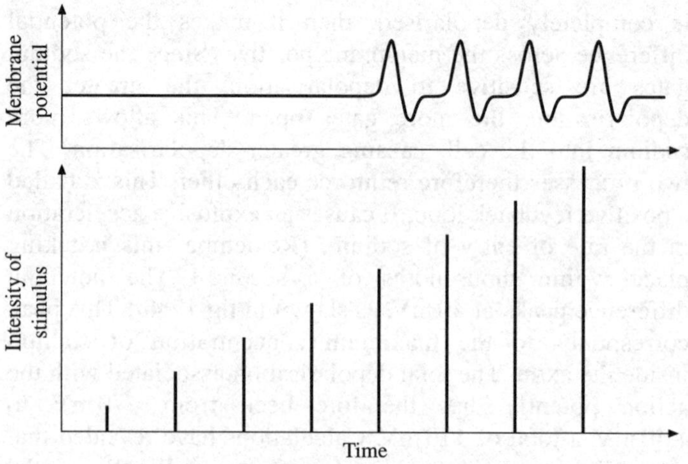

Fig 17.6 *The 'all-or-nothing' response. A certain 'threshold intensity' of stimulus must be reached before the nerve cell responds. It is also possible to see from the figure that the size of the action potential is always the same and is not affected by the intensity of the stimulus.*

frequency of action potentials set up. We can say that the frequency is a code for the intensity of a stimulus. Frequency is proportional to intensity. This is known as the **frequency code**.

Impulses at the frequency of 10 per nerve fibre per second keep the biceps muscle in tone (in a state of partial concentration ready for action). About 50 impulses per fibre per second are needed for steady contraction. Most nerve fibres can conduct impulses in the region of 10–100 per second, although some can conduct as many as 500 per second in natural situations (and more in experimental situations).

Speed of conduction. In vertebrates, the majority of neurones, particularly those of the spinal and cranial nerves, have an outer covering of myelin derived from the spirally wound Schwann cell (fig 6.30 and section 6.6.1). Myelin is a fatty material with a high electrical resistance and acts as an electrical insulator in the same way as the rubber and plastic covering of electrical wiring. The combined resistance of the axon membrane and myelin sheath is very high but, where breaks in the myelin sheath occur, as at the **nodes of Ranvier**, the resistance to current flow between the axoplasm and the fluid outside the cell is lower. It is only at these points that local circuits are set up and current flows across the axon membrane to generate the next action potential. This means, in effect, that the action potential 'jumps' from node to node and passes along the myelinated axon faster than the series of smaller local currents in a non-myelinated axon. This type of conduction is called **saltatory conduction** (*saltare*, to jump) and can lead to conduction speeds of up to $120\,m\,s^{-1}$ (fig 17.7).

In non-myelinated axons (axons lacking a myelin sheath) typical of those found in non-vertebrates, the speed of conduction of the action potential depends on the resistance of the axoplasm. This resistance, in turn, is related to the

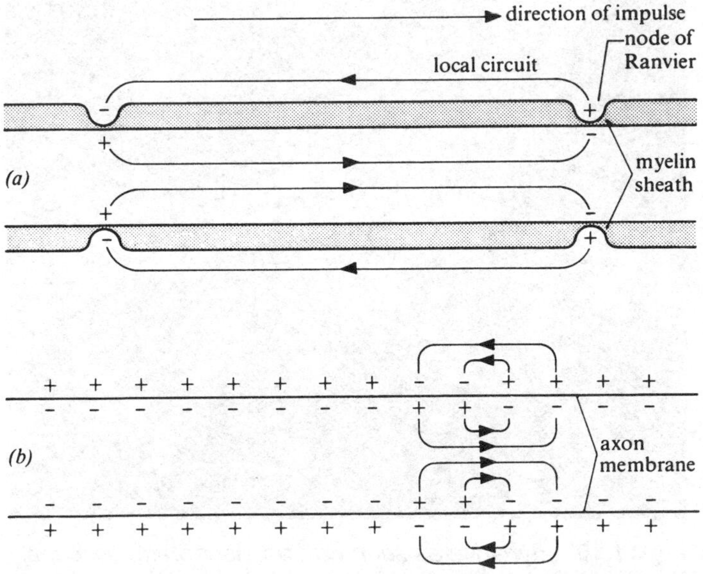

Fig 17.7 Diagrams showing the difference in lengths of the local circuits produced (a) in a myelinated axon and (b) a non-myelinated axon. In (a) conduction is described as saltatory since the action potential effectively 'jumps' from node to node.

diameter of the axon. The smaller the diameter, the greater is the resistance. In the case of fine axons (<0.1 mm) the high resistance of the axoplasm has an effect on the spread of current and reduces the length of the local circuits, so that only the region of the membrane immediately in front of the action potential is involved in the local circuit. These axons conduct impulses at about $0.5\,\mathrm{m\,s^{-1}}$. Giant axons, typical of many annelids, arthropods and molluscs, have a diameter of approximately 1 mm and conduct impulses at velocities up to $100\,\mathrm{m\,s^{-1}}$, which are ideal for conducting information vital for survival.

17.4 Explain, in terms of the resistance of the axoplasm and local circuits, why giant axons conduct impulses at greater velocities than fine axons.

Effect of temperature on speed of conduction. Temperature has an effect on the rate of conduction of nerve impulses and as temperature rises to about 40°C the rate of conduction increases.

17.5 Why do myelinated axons of frog having a diameter of $3.5\,\mu m$ conduct impulses at $30\,\mathrm{m\,s^{-1}}$ whereas axons of the same diameter in cat conduct impulses at $90\,\mathrm{m\,s^{-1}}$?

17.1.2 The synapse

A synapse is the link between one neurone and another. There is no physical contact between one neurone

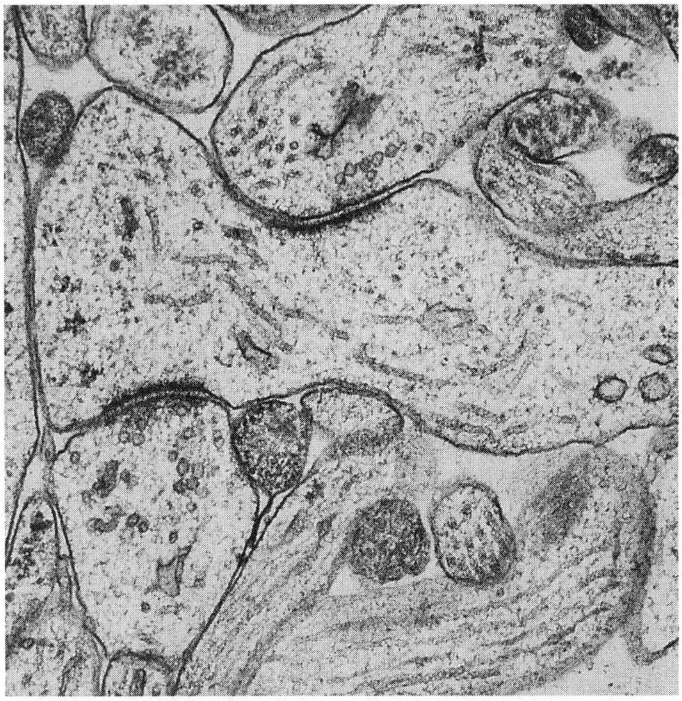

Fig 17.8 Transmission electron micrograph of a motor neurone showing synapses. The synaptic cleft appears dark.

and the next. Instead there is a tiny gap called the **synaptic cleft**. Synapses are usually found between the fine branches at the end of the axon of one neurone and the **dendrites** or **cell body** of another neurone. The number of synapses is usually very large, providing a large surface area for the transfer of information. For example, over 1000 synapses may be found on the dendrites and cell body of a single motor neurone in the spinal cord. Some cells in the brain may receive up to 10 000 synapses (fig 17.8).

Another type of synapse, with a similar function but a different structure, exists between the terminals of a motor neurone and the surface of a muscle fibre. This is called a **neuromuscular junction** and is described later in this section.

Structure of the synapse

The structure of a synapse is shown in figs 17.9 and 17.10. A synapse consists of a swelling at the end of a nerve fibre called a **synaptic knob** lying in close proximity to the membrane of a dendrite. The cytoplasm of the synaptic knob contains numerous mitochondria (for energy) and small **synaptic vesicles** (diameter 50 nm). Each vesicle contains a chemical called a **neurotransmitter** which is responsible for the transmission of the nerve impulse across the synapse. The membrane of the synaptic knob nearest the synapse is thickened and is called the **presynaptic membrane**. The membrane of the dendrite is also thickened and termed the **postsynaptic membrane**. These membranes are separated by a gap of about 20 nm, the **synaptic cleft**. The presynaptic membrane is modified for the attachment of synaptic vesicles and the release of

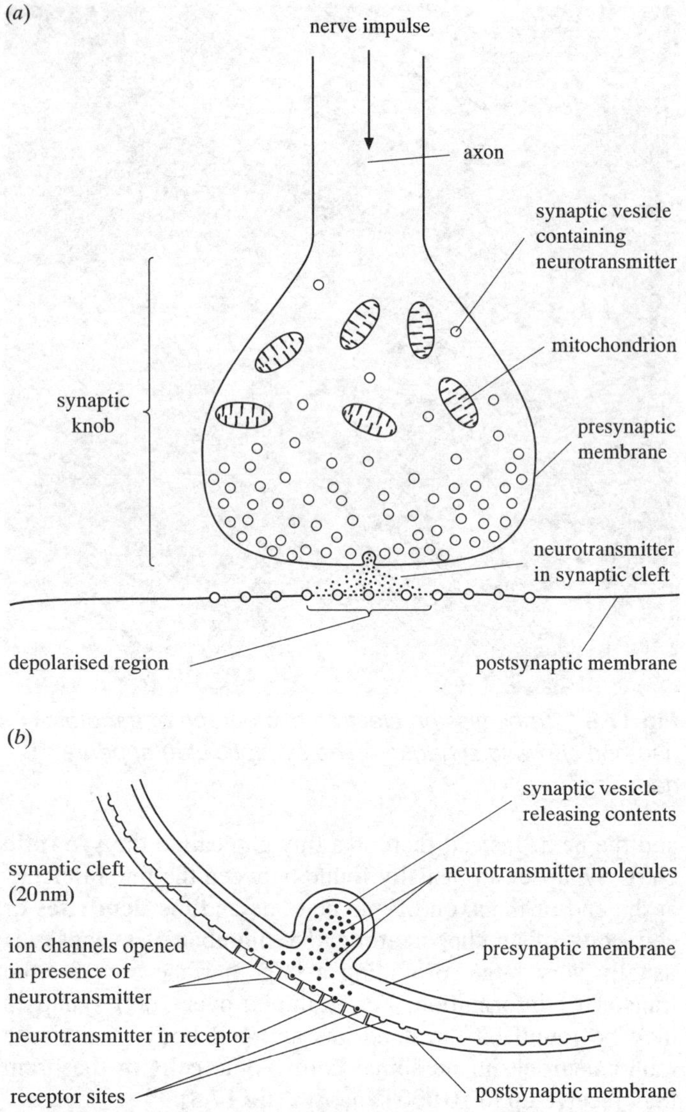

Fig 17.9 *Structure of a synapse.*

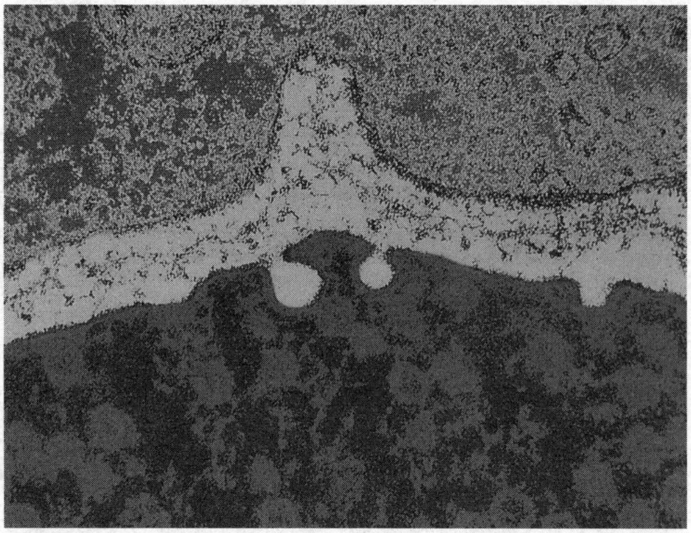

Fig 17.10 *Synapse as seen with an electron microscope.*

transmitter substance into the synaptic cleft. The postsynaptic membrane contains large protein molecules, which act as **receptor sites** for the transmitter substances, and numerous **channels** and **pores**, normally closed, for the movement of ions into the postsynaptic neurone (fig 17.11).

As stated, synaptic vesicles contain a neurotransmitter substance. This is produced either in the cell body of the neurone, from where it passes down the axon to the synaptic knob, or directly in the synaptic knob. In both cases, synthesis of transmitter substances requires enzymes produced by ribosomes in the cell body. In the synaptic knob the transmitter substance is 'packaged' into vesicles and stored ready for release. The two main transmitter substances in vertebrate nervous systems are **acetylcholine (ACh)** and **noradrenaline**, although other substances exist and are described at the end of this section. Acetylcholine is an ammonium compound. It was the first transmitter substance to be isolated (in 1920). Noradrenaline is described in section 17.6.5. Neurones releasing acetylcholine are described as **cholinergic neurones** and those releasing noradrenaline are described as **adrenergic neurones**. Noradrenaline is released by nerves in the sympathetic nervous system whereas acetylcholine is released by almost all other nerves (except some in the brain).

Mechanism of synaptic transmission

In the following account, the example of the neurotransmitter acetylcholine will be used. The arrival of nerve impulses at the synaptic knob depolarises the presynaptic membrane, causing calcium channels to open, increasing the permeability of the membrane to calcium (Ca^{2+}) ions. As the calcium ions rush into the synaptic knob they cause the synaptic vesicles to fuse with the presynaptic membrane, releasing their contents into the synaptic cleft (**exocytosis**). The vesicles then return to the cytoplasm where they are refilled with transmitter substance. Each vesicle contains about 3000 molecules of acetylcholine.

Acetylcholine diffuses across the synaptic cleft, creating a delay of about 0.5 ms, and attaches to a specific receptor site (a protein) on the postsynaptic membrane that recognises the molecular structure of the acetylcholine molecule. The arrival of the acetylcholine causes a change in the shape of the receptor site which results in ion channels opening up in the postsynaptic membrane. Note that in the transmission of the nerve impulse along the axon, ion channels are opened in response to depolarisation, whereas in the postsynaptic membrane of the synapse it is in response to binding of neurotransmitter to receptor proteins.

Entry of sodium ions through the postsynaptic membrane causes **depolarisation** (fig 17.4*a*) of the membrane. This excites the cell, making it more likely to set up a nerve impulse (action potential). Having produced a change in the permeability of the postsynaptic membrane the acetylcholine is immediately removed from the synaptic cleft, by

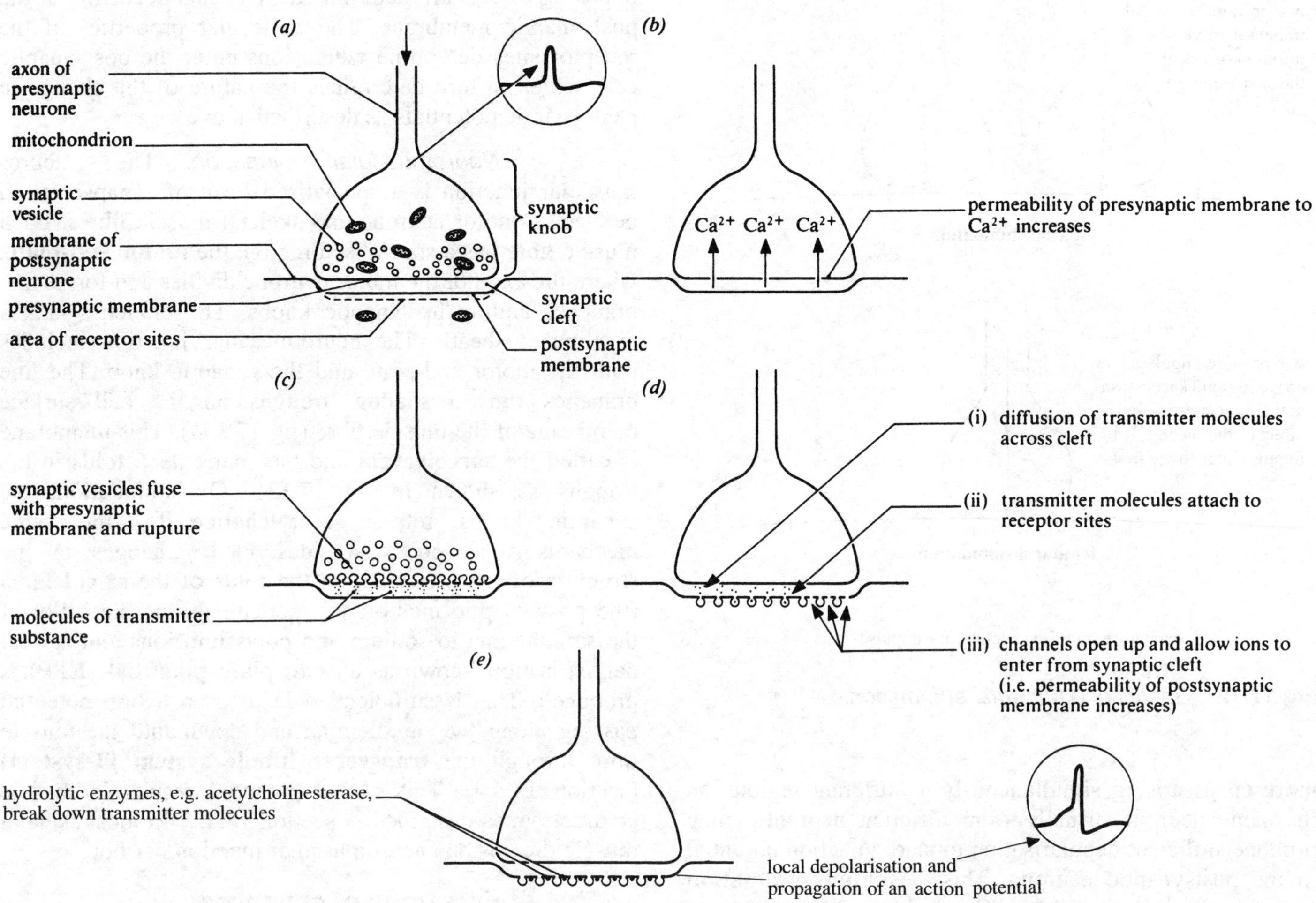

Fig 17.11 *Summary diagrams showing the mechanisms involved in chemical transmission at a synapse, (a) to (e) time sequence.*

the enzyme **acetylcholinesterase (AChE)**, also called **cholinesterase**. The enzyme is situated on the postsynaptic membrane and hydrolyses the acetylcholine to choline. The choline is reabsorbed into the synaptic knob to be recycled into acetylcholine by synthetic pathways in the vesicle (fig 17.11). Some nerve gases and insecticides work by inhibiting AChE as explained in section 4.4.3.

Other neurotransmitters

Some neurotransmitters are inhibitory in their effects. They cause **hyperpolarisation** rather than *depolarisation* of the postsynaptic membrane. Hyperpolarisation makes the inside of the cell even more negative, making it *less* likely to set up a nerve impulse.

There are three possible ways of removing neurotransmitters from the synaptic cleft:

- reabsorption by the presynaptic membrane;
- diffusion out of the cleft;
- hydrolysis by enzymes.

Roles of synapses

Excitatory synapses, spatial summation and temporal summation. At **excitatory** synapses ion-specific channels open up allowing sodium ions to enter and potassium ions to leave down their respective concentration gradients. This leads to a depolarisation in the postsynaptic membrane. The depolarising response is known as an **excitatory postsynaptic potential (epsp)** and the size of this potential is usually small but longer lasting than that of an action potential. The size of epsps fluctuates in steps, suggesting that transmitter substance is released in 'packets' rather than individual molecules. Each 'step' is thought to correspond to the release of transmitter substance from one synaptic vesicle. A single epsp is normally unable to produce sufficient depolarisation to reach the threshold required to propagate an action potential in the postsynaptic neurone. However, it does increase the chance of a further epsp generating an action potential, an effect called **facilitation**. The depolarising effect of several epsps is additive, a phenomenon known as **summation**. Two or

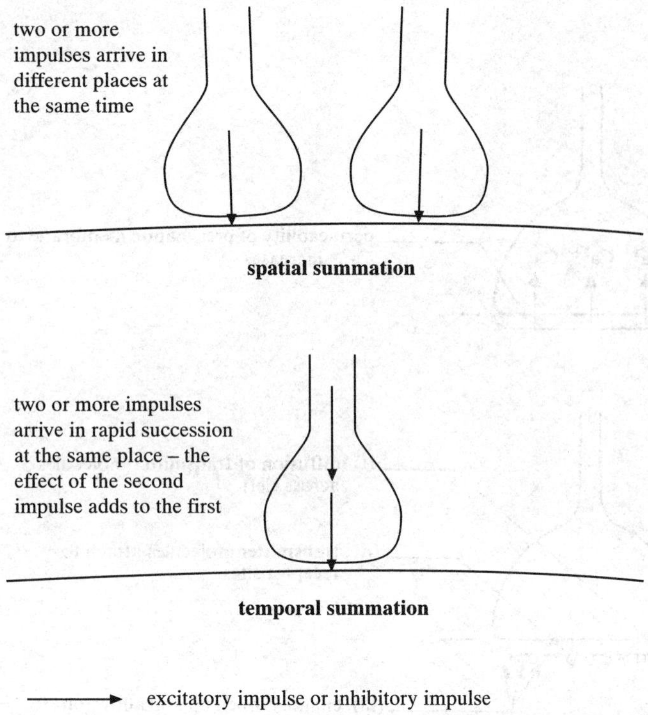

Fig 17.12 *Spatial and temporal summation.*

more epsps arising simultaneously at different regions on the same neurone, usually from different neurones, may produce sufficient depolarisation to start an action potential in the postsynaptic neurone. This is **spatial summation** (related in space) (fig 17.12). The rapid repeated release of transmitter substance from several synaptic vesicles by the *same* synaptic knob, as a result of an intense stimulus, produces individual epsps which are so close together that they summate and give rise to an action potential in the postsynaptic neurone. This is **temporal summation** (summation through time) (fig 17.12). Therefore impulses can be set up in a single postsynaptic neurone as a result of either weak stimulation by several presynaptic neurones or repeated stimulation by one presynaptic neurone.

Inhibitory synapses. At inhibitory synapses the release of transmitter substance increases the permeability of the postsynaptic membrane by opening up ion-specific channels to chloride (Cl^-) and potassium ions. As chloride ions rush in and potassium ions rush out down their concentration gradients, they produce a hyperpolarisation of the membrane known as an **inhibitory postsynaptic potential (ipsp)**. In other words they make the inside of the neurone *more* negative, down to as low as −90 mV. This makes the neurone *less* likely to trigger an action potential.

Transmitter substances themselves are neither inherently excitatory nor inhibitory. For example, acetylcholine has an excitatory effect at most neuromuscular junctions and synapses, but has an inhibitory effect on neuromuscular junctions in cardiac muscle and gut muscle. These opposing effects are determined by events occurring at the postsynaptic membrane. The molecular properties of the receptor sites determine which ions enter the postsynaptic cell, which in turn determines the nature of the change in postsynaptic potentials as described above.

Neuromuscular junction. The neuromuscular junction is a specialised form of synapse found between a motor neurone and skeletal muscle fibres. Each muscle fibre has a specialised region, the **motor end-plate**, where the axon of the motor neurone divides and forms fine branches ending in synaptic knobs. The branches lack a myelinated sheath. The neuromuscular junction includes both the motor end-plate and the synaptic knob. The fine branches run in shallow troughs on the cell surface membrane of the muscle fibre (fig 17.13*a*). This membrane is called the **sarcolemma** and has many deep folds in the troughs as shown in fig 17.13*b*. On stimulation, the synaptic knobs release acetylcholine by the same mechanisms as previously described. Changes in the structure of receptor sites on the folds of the sarcolemma (the postsynaptic membrane) increase the permeability of the sarcolemma to sodium and potassium ions and a local depolarisation known as an **end-plate potential (EPP)** is produced. This is sufficient to lead to an action potential passing along the sarcolemma and down into the muscle fibre through the **transverse tubule system (T-system)** (section 18.4.4). This action potential results in muscle contraction, as described in section 18.4.7. In most skeletal muscle there is one neuromuscular junction per fibre.

Functions of synapses

Since synapses have the effect of slowing down nerve impulses by about 0.5 ms per synapse, it can be assumed that their advantages outweigh this disadvantage. What are the advantages? They are summarised as follows.

- **Unidirectionality**. The release of transmitter substance at the presynaptic membrane, and the location of receptor sites on the postsynaptic membrane, ensure that nerve impulses can pass only in one direction along a given pathway. This gives *precision* to the nervous system, allowing nerve impulses to reach particular destinations.
- **Amplification**. Sufficient acetylcholine is released at a neuromuscular junction by each nerve impulse to excite the postsynaptic membrane to produce a response in the muscle fibre. Thus nerve impulses arriving at the neuromuscular junction, however weak, are adequate to produce a response from the effector, thereby increasing the *sensitivity* of the system.
- **Adaptation and fatigue**. The amount of transmitter substance released by a synapse steadily falls off in response to constant stimulation until the supply of transmitter substance is exhausted and the synapse is described as **fatigued**. Further passage of information along this pathway is not possible until after a period of recovery. The significance of fatigue is that it prevents

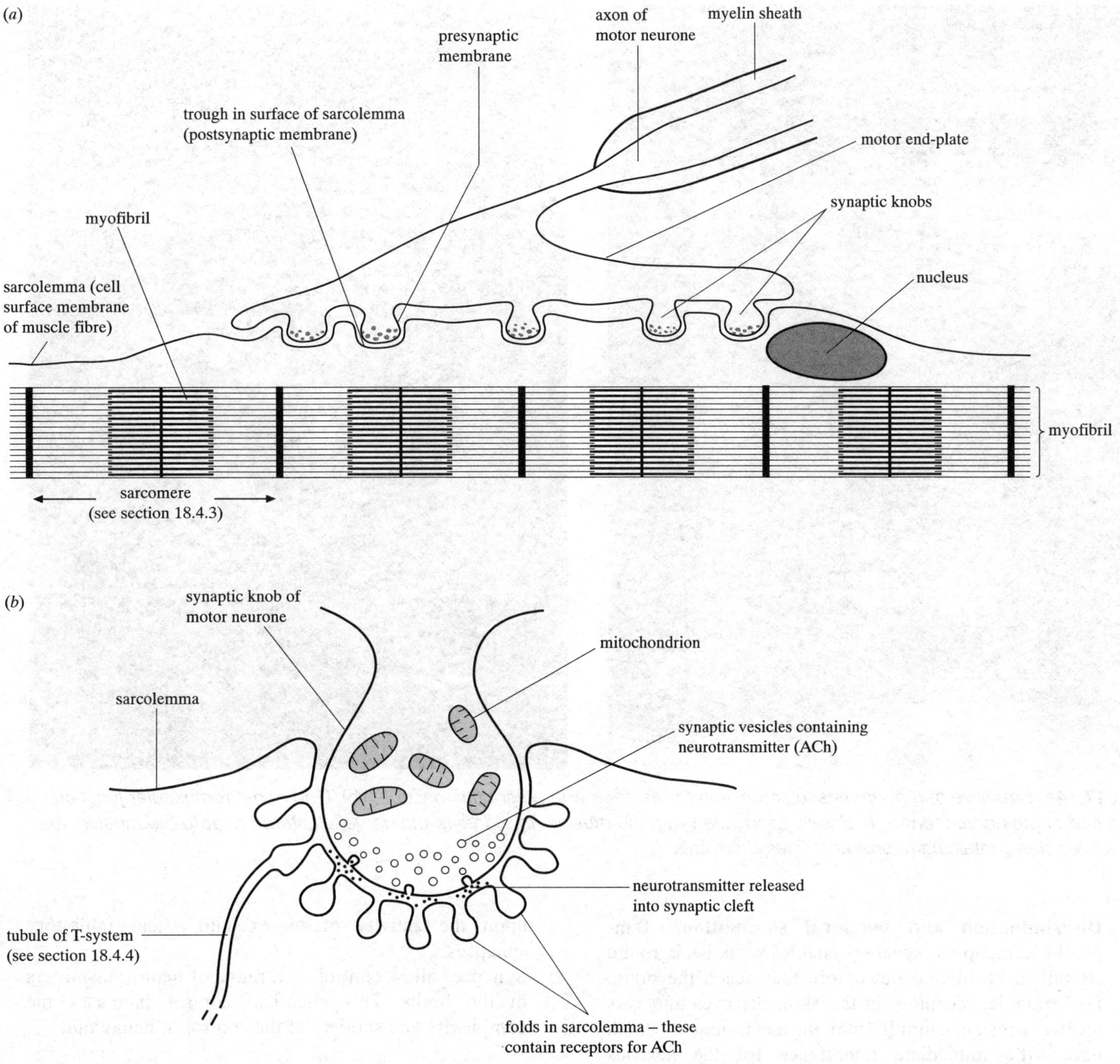

Fig 17.13 *Structure of a neuromuscular junction: (a) detail visible with a light microscope, (b) detail visible with an electron microscope.*

damage to an effector through overstimulation. **Adaptation** also occurs at the level of the receptor and this is described in section 17.4.2.

- **Integration, convergence and spatial summation.** A postsynaptic neurone may receive impulses from a large number of excitatory and inhibitory presynaptic neurones. This is known as **convergence**, and the postsynaptic neurone is able to summate the stimuli from all the presynaptic neurones (a simple example is shown in fig 17.12). This spatial summation enables the synapse to act as a centre for the integration of stimuli from a variety of sources and the production of a coordinated response.
- **Facilitation**. This occurs at some synapses and involves each stimulus leaving the synapse more responsive to the next stimulus. The sensitivity of the synapse has therefore been increased, and subsequent weaker stimuli may cause a reponse. Facilitation is not temporal summation in that it is a chemically controlled response of the postsynaptic membrane and not an electrical summation of postsynaptic membrane potentials.

(a) (b)

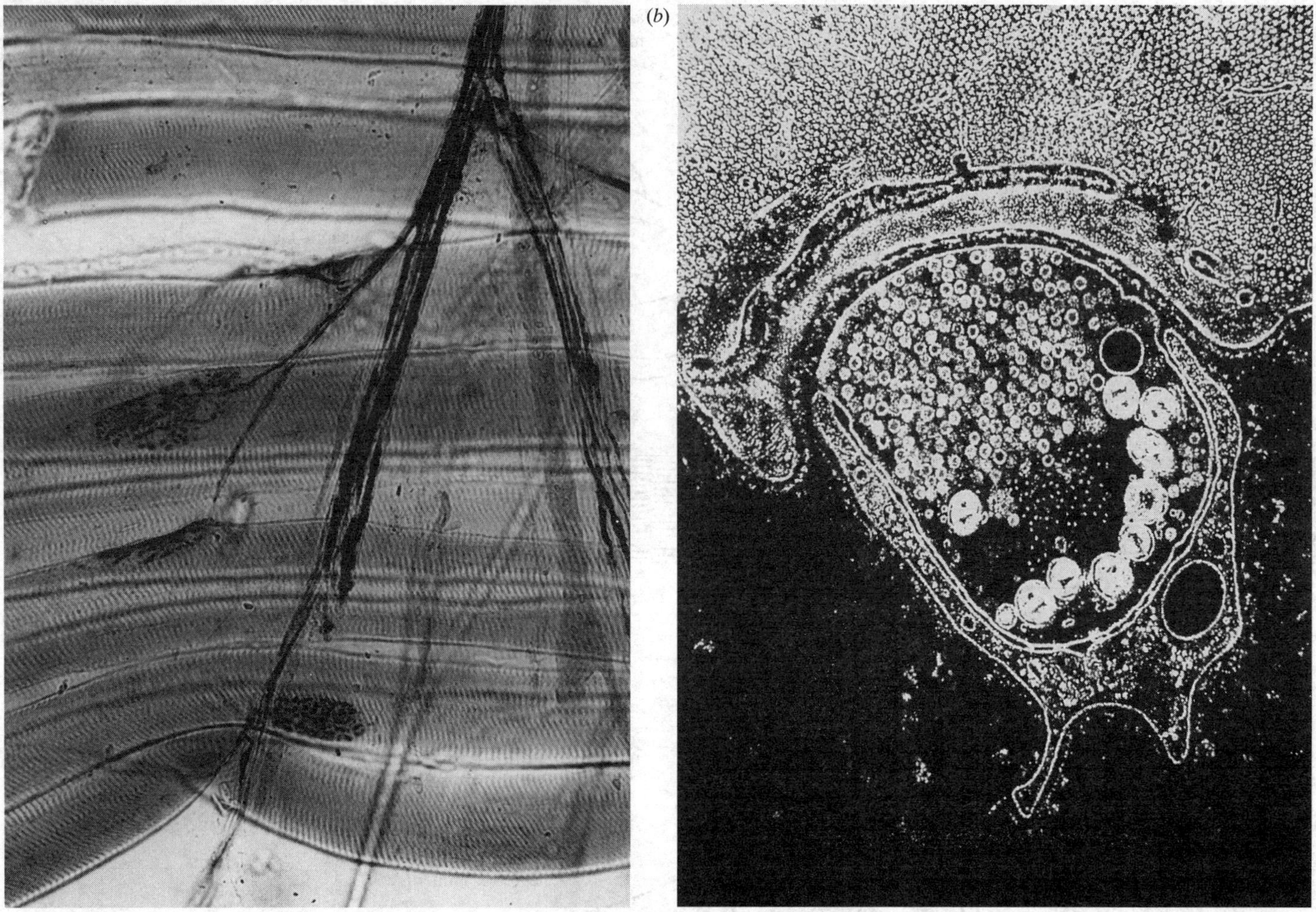

Fig 17.14 *Structure of a neuromuscular junction (a) as seen with a light microscope. (b) TEM of neuromuscular junction. The end of the nerve (centre) is closely associated with the muscle fibre. Inside the nerve, clustered near to the muscle, are small vesicles containing neurotransmitter chemicals.*

- **Discrimination and temporal summation**. Temporal summation at synapses enables weak background stimuli to be filtered out before they reach the brain. For example, receptors in the skin, the eyes and ears receive constant stimuli from the environment which have little immediate importance for the nervous system, such as background noise. Only *changes* in intensity of stimuli are significant to the nervous system. These increase the frequency of stimuli and pass across the synapse to cause a response.
- **Inhibition**. The transmission of information across synapses and neuromuscular junctions may be prevented at the postsynaptic membrane by the activity of certain neurotransmitters or drugs, as described in the next section. Presynaptic inhibition is also possible. It occurs at synaptic knobs that are in close contact with synaptic knobs from inhibitory synapses. Stimulation of these inhibitory synapses reduces the number of synaptic vesicles released by the inhibited synaptic knob. This arrangement enables a given nerve terminal to produce a variable response depending upon the activity of its excitatory and inhibitory synapses.
- Synapses allow control by a range of neurotransmitters in the brain. This chemical control increases the complexity and subtlety of the control of behaviour.

Drugs, neurotransmitters and synapses

Although acetylcholine and noradrenaline are the main neurotransmitters outside the brain, more than 50 natural neurotransmitters have been isolated from the brain, all associated with their own particular nervous pathways. In addition, many other chemicals affect synapses. These may affect the brain, and therefore behaviour, or other parts of the body such as the neuromuscular junction. Some of these substances and their functions are described in table 17.3 and below.

Amino acids. The major excitatory neurotransmitter in the brain is the amino acid **glutamate** (glutamic acid). A number of chemicals mimic glutamate

Table 17.3 Summary of chemical substances which affect the synapse and neuromuscular junction.

Substance	*Site of action*	*Function/effect*
acetylcholine	nervous system	excitation or inhibition
glutamic acid (amino acid)	brain	excitation
glycine (amino acid)	brain and spinal cord postsynaptic membrane	inhibition muscles relax
strychnine	brain and spinal cord postsynaptic membrane blocks glycine receptors	antagonistic to glycine muscles contract but will not relax
GABA (gamma aminobutyric acid) (amino acid)	brain	inhibition
β-endorphins } enkephalins	brain opioid receptors in postsynaptic membrane	block 'pain pathways' and stimulate reward pathways
morphine } **heroin**	brain opioid receptors in postsynaptic membrane	mimic endorphins and enkephalins
noradrenaline (monoamine)	brain and SNS	excitation
dopamine (monoamine)	brain	excitation, control of complex muscular movement, some emotional responses
serotonin	brain	excitation, control of mood, hallucinations
LSD (lysergic acid diethylamide)	brain	mimic serotonin? produce hallucinations
mescaline		
amphetamine	brain and SNS	inhibits monoamine oxidase so increases level of noradrenaline
cocaine	brain and SNS	as amphetamine above (and other effects)
Prozac	brain and SNS	blocks re-uptake of serotonin so increases effect
nicotine	postsynaptic membrane in SNS and PNS	mimics action of ACh on nicotinic receptors
muscarine	postsynaptic membrane in PNS	mimics action of ACh on muscarinic receptors
caffeine	brain	stimulates dopamine pathways
atropine	SNS and PNS	blocks action of ACh on muscarinic receptors
curare	SNS and PNS neuromuscular junction postsynaptic membrane	blocks action of ACh on nicotinic receptors
tetanus toxin	presynaptic membrane	prevents release of inhibiting transmitter substance
botulinum toxin	presynaptic membrane	prevents release of ACh
organophosphorous weedkillers and insecticides, some nerve gases	postsynaptic membrane	inactivates acetylcholine-esterase and prevents breakdown of ACh

Drugs are indicated in bold type.

and bind to glutamate receptors in the brain. One group of glutamate receptors are the major sites of action of the illicit drug known as **angel dust**. (An illicit drug is one which is illegal outside of medical use.)

The amino acid **glycine** is inhibitory, causing chloride channels to open in the postsynaptic membrane and resulting in hyperpolarisation (the contents of the cell become more negative). Glycine is important in the spinal cord where it helps control skeletal movements by making muscles relax (preventing their stimulation). Its importance is revealed by the effect of **strychnine** which blocks the glycine receptors and therefore inhibits the effects of glycine. The slightest stimulation then causes muscle contraction. Victims of strychnine poisoning suffocate through being unable to relax the diaphragm muscles.

The amino acid **GABA** is the most common neurotransmitter in the brain and is also inhibitory. It helps control muscle movement, for example by acting in the cerebellum (a part of the brain discussed in section 17.2.4). A deficiency of GABA produces the uncontrolled movements of Huntington's disease (section 25.7.5). It is the target of antianxiety drugs such as Valium, which enhance its activity.

Opioids, endorphins and enkephalins. Opioids are drugs which are derived from opium, that is obtained from the opium poppy. These drugs have been used for centuries as pain killers. **Morphine** and **heroin** are modern examples. In the brain there are naturally occurring substances which have similar effects, although they are 200 times stronger than morphine. These substances react with the opiate receptor in the brain (discovered in 1973) and are collectively called **endorphins** (meaning endogenously produced morphine-like compounds). Many have been identified and the best known are a group of small peptides known as **enkephalins**, for example **metenkephalin** and **β-endorphin**. The enkephalins are five amino acids long. They reduce pain, influence emotion and are involved with certain types of mental illness. Endorphins and enkephalins act at the postsynaptic membrane and suppress the synaptic activity which would normally lead to the sensation of pain. They are therefore natural pain killers. If used for medical purposes they are addictive and are only briefly effective. A burst of β-endorphin is released naturally during childbirth.

Opioids also produce pleasant sensations and may be associated with internal 'reward' pathways which reinforce certain types of behaviour. Since β-endorphin levels increase during exercise, this may help to explain the 'high' that joggers experience.

Research into these chemicals has opened up new ideas on brain functioning. It also helps to explain the control of pain and healing by such diverse activities as hypnosis, acupuncture and faith healing. Many more chemical substances of this type have yet to be isolated, identified and have their function determined.

Monoamines. **Noradrenaline**, another hormone secreted by the adrenal gland, is a monoamine and a neurotransmitter in the sympathetic nervous system which prepares the body for action (section 17.2.3). It also exists in the brain, increasing alertness and helping to maintain the state of arousal. It enhances our response to new stimuli. Pep pills containing **amphetamine** increase the level of noradrenaline in the brain by inhibiting the action of monoamine oxidase (MAO). This is the enzyme which under normal conditions removes noradrenaline once it has been reabsorbed at the synapses; MAO therefore prevents overstimulation. Another effect of amphetamine is to cause excess dopamine release and therefore stimulate reward pathways (see below). Amphetamine also stimulates the sympathetic nervous system indirectly by raising levels of noradrenaline at sympathetic nerve endings.

Some antidepressants are **monoamine oxidase inhibitors** and, presumably, are effective in treating depression by prolonging the effects of noradrenaline. Clinical depression is an illness which is characterised by fewer nerve impulses being transmitted through the brain. MAO inhibitors promote the activity of all the monoamines and can have undesirable side effects. A newer drug, **Prozac**, specifically blocks the re-uptake of serotonin into presynaptic axons and therefore promotes serotonin effects only (see below). It is now used widely for treatment of depression.

Dopamine, another monoamine, has a structure very similar to that of noradrenaline (fig 17.15). It is the natural neurotransmitter of 'dopamine pathways' in the brain. These are partly concerned with voluntary control of complex muscular movements. A deficiency of dopamine results in Parkinson's disease. Dopamine is also involved in emotional responses in the cerebral cortex and has been linked with schizophrenia. It can also stimulate the 'pleasure' centre of the hypothalamus. Amphetamines trigger the release of dopamine.

Serotonin is associated with control of moods, including depression, elation and mania. It is also involved in the onset of sleep, sensory perception and temperature regulation in the hypothalamus. Fig 17.15 shows that it has some similarity in structure to **LSD (lysergic acid diethylamide)**. LSD is a hallucinogenic drug as is mescaline. They are thought to mimic some of the activities of serotonin. It is common to find similarities in structure between drugs that alter behaviour and neurotransmitters, suggesting that the drugs recognise the same receptors and either block or stimulate the natural effects.

Nicotine. Nicotine mimics the effect of the neurotransmitter acetylcholine (ACh) on certain receptors (not all those that acetylcholine binds to). The receptors concerned are called **nicotinic receptors** and are found in the membranes of postsynaptic cells. They are found in the sympathetic and parasympathetic nervous systems (section 17.2.3). Activation of the receptors leads to depolarisation and therefore excitation of the postsynaptic cell or effector.

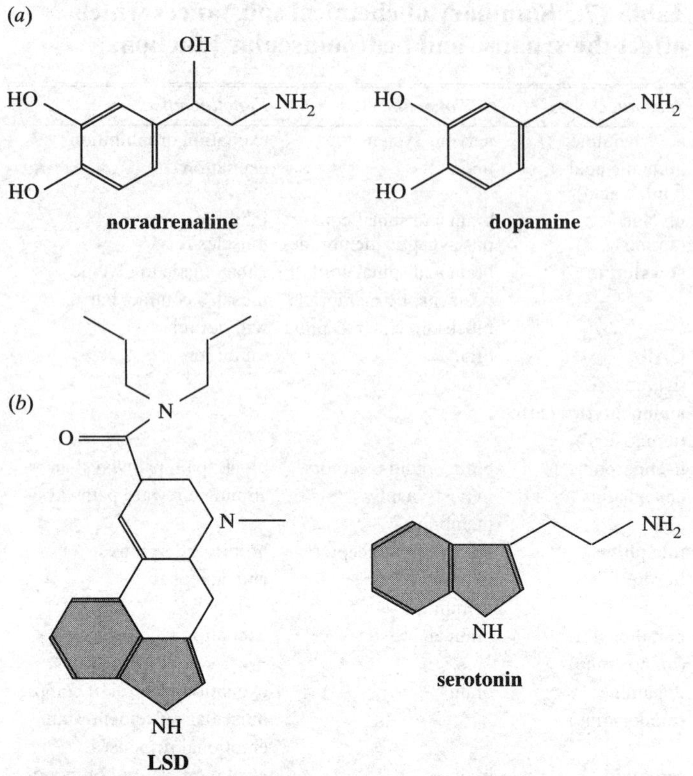

Fig 17.15 *Similarities in structure between (a) the two natural neurotransmitters noradrenaline and dopamine and (b) the drug LSD and the natural neurotransmitter serotonin. Many drugs that affect behaviour, such as LSD, resemble natural neurotransmitters.*

Nicotine causes strong sympathetic vasoconstriction in abdominal organs such as the gut, and in the limbs, but at the same time parasympathetic effects occur, such as increased gastrointestinal activity and, sometimes, slowing of the heart.

Nicotinic receptors are also found at neuromuscular junctions. If nicotine is applied directly to the junction it makes the muscle contract, mimicking acetylcholine. Nicotinic receptors are found in a few regions of the brain, but it is not understood how this relates to the psychological effects of the drug. Many of the receptors stimulate release of dopamine and therefore stimulate reward and pleasure pathways. However, the behavioural effects of nicotine are very subtle compared with some drugs.

Atropine. Cholinergic receptors, that is receptors stimulated by acetylcholine, are of two types, namely nicotinic receptors (mentioned above) and muscarinic receptors. (These receptors are named after the drug muscarine, a poison isolated from certain mushrooms. They are in the parasympathetic nervous system, for example in the heart and gut.) Atropine blocks the action of acetylcholine in the muscarinic receptors. Acetylcholine slows the heart and stimulates peristalsis in the gut, so atropine has the opposite effect because it blocks the receptors, preventing the binding of acetylcholine.

Atropine is obtained from the deadly nightshade (*Atropa belladonna*). The use of the species name belladonna, meaning beautiful lady, refers to the use by women in the Middle Ages of a deadly nightshade extract, containing atropine, as eye drops to make the pupils dilate. This was thought to make women look more attractive to men. The drug inhibits parasympathetic stimulation of the iris, causing the circular muscles of the iris to relax. Today the drug is used for medical purposes. It can be used to dilate the pupils during an eye examination, before a general anaesthetic to inhibit mucous production in the respiratory tract, and to inhibit secretion of stomach acid in patients suffering from gastritis (overproduction of acid).

Curare. Curare specifically blocks the nicotinic receptors of acetylcholine. It therefore has opposite effects to acetylcholine and nicotine at these receptors. The effects of curare are particularly noticeable at neuromuscular junctions where it prevents muscle contraction, causing rapid paralysis throughout the body. The victim dies as a result of being unable to breathe. (Artificial respiration can save a victim if applied until the effect of the drug wears off.) This accounts for its use as a poison to tip arrows by some South American tribes when hunting: it prevents wounded animals escaping into thick undergrowth. Curare and related drugs can be used as a muscle relaxant on patients during surgical operations, which makes operating easier for the surgeon. Breathing is maintained artificially.

Caffeine. Like nicotine, caffeine is a stimulant, although a relatively weak one. It is thought to cause dopamine release in the brain and therefore stimulates reward pathways.

Cocaine. Cocaine is a stimulant, having very similar, though not identical, effects to amphetamine. Cocaine blocks the re-uptake of monoamines into the presynaptic axons. This results in overstimulation of dopamine pathways and other monoamine pathways. It has many effects. It occurs naturally in coca leaves, which are chewed by local people in the highlands of the Andes. Until about 1900 it was added to Coca Cola. As a drug of abuse it is used for the sense of euphoria it promotes, which helps in social situations. It can become addictive, particularly in its pure form **crack**. This leads to social withdrawal, depression and eventually heart and kidney disease. Cocaine has been used medically as a local anaesthetic for the eyes, nose, mouth (during dentistry) and throat, but non-medical use is illegal.

17.2 The parts of the nervous system

The nervous system can be divided into the central nervous system (brain and spinal cord) and the peripheral nervous system. The peripheral nervous system can be divided into the voluntary nervous system, which is under voluntary control from the brain, and the autonomic nervous system which operates automatically (involuntary). This system is divided into the sympathetic nervous system (SNS), which has a mainly excitatory effect on the body, and the parasympathetic nervous system (PNS), which acts antagonistically (oppositely) to the SNS and has a mainly calming influence. These subdivisions are summarised in fig 17.16.

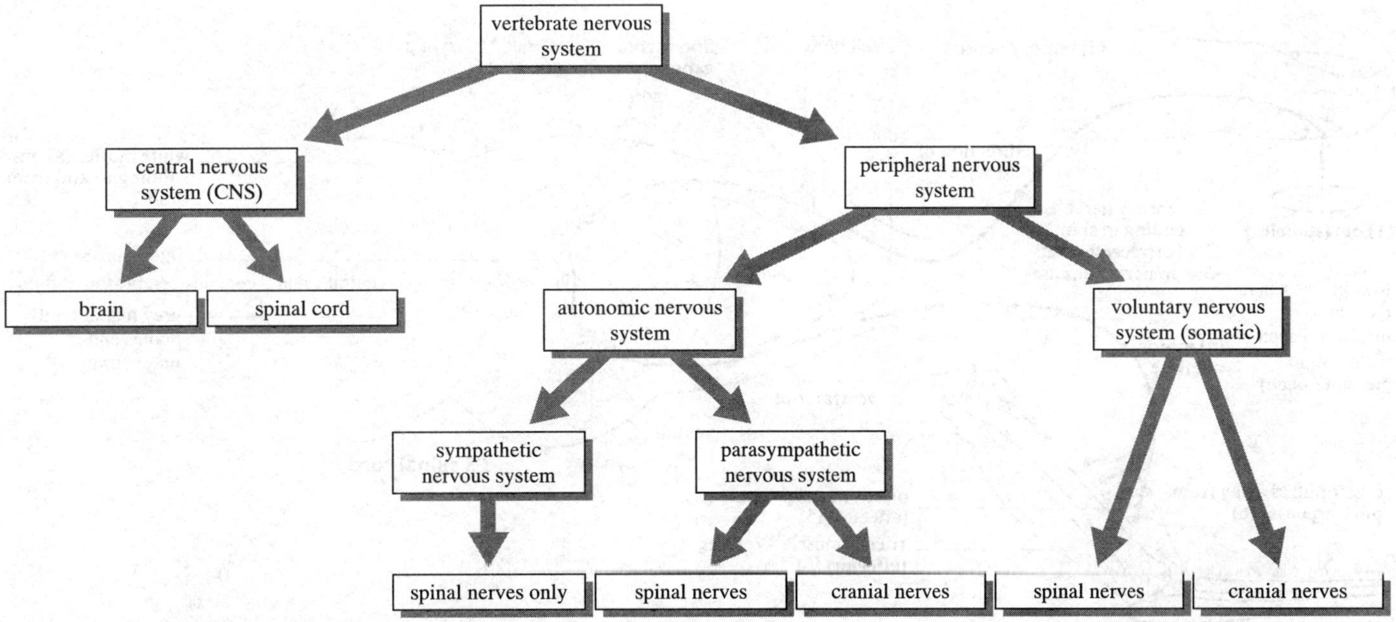

Fig 17.16 *Outline classification of the vertebrate nervous system. (All spinal nerves and some cranial nerves have both sensory and motor neurones.)*

17.2.1 The peripheral nervous system

All the nerves of the body together make up the peripheral nervous system. They all enter or leave the central nervous system, either the spinal cord in the case of **spinal nerves**, or the brain in the case of **cranial nerves**.

Spinal nerves arise from the spinal cord between the vertebrae along most of the length of the spinal cord. They all carry both sensory and motor neurones and are described as **mixed** nerves. Further details of spinal nerves and the spinal cord are given below. **Cranial nerves** arise from the brain and, with one exception (the vagus), supply receptors and effectors of the head. There are 12 pairs of cranial nerves in mammals, numbered I–XII in Roman numerals. Not all cranial nerves are mixed.

Three examples of cranial nerves are:

- *cranial nerve II* – **optic nerve** – this is a *sensory nerve* running from the retina to the brain;
- *cranial nerve III* – **oculomotor nerve** – this is a *motor nerve* running from the brain to the four eye muscles and helps control eye movements;
- *cranial nerve X* – **vagus nerve** – this is a *mixed nerve*. It runs between the brain and the heart, gut and part of the respiratory tract and decreases heart rate, stimulates peristalsis and is concerned with speech and swallowing. It includes an important motor nerve of the autonomic nervous system supplying the heart, bronchi and gut.

17.2.2 Reflex action and reflex arcs

The simplest form of response in the nervous system is **reflex action**. This is a rapid, automatic response to a stimulus which is not under the voluntary control of the brain. It is described as an **involuntary action**. The same stimulus produces the same response every time. The nervous pathway taken by nerve impulses in a reflex action is called the **reflex arc**.

The simplest reflex arcs in humans involve only two neurones, a sensory neurone and a motor neurone. The knee-jerk reflex is an example. Other reflex arcs involve three neurones: a sensory neurone, an interneurone and a motor neurone. One example is the withdrawal reflex associated with pricking a finger on a pin. This is a spinal reflex, meaning that the reflex arc passes through the spinal cord rather than the brain (these are called cranial reflexes). It is illustrated in fig 17.17. Note that there is a one-way system of nerve fibres through the spinal cord. Sensory neurones enter through the dorsal root (not *route*) and motor neurones emerge from the ventral root. All the nerve cell bodies are in the central nervous system (spinal cord). The nerve cell bodies of the sensory neurones are situated in the dorsal root, in a swelling called a ganglion. The nerve cell bodies of the interneurones and motor neurones are in the 'grey matter'. This appears grey because of the extra density of all the nuclei. The white matter around it contains only nerve fibres.

Importance of reflexes

Simple reflex arcs such as the one described allow the body to make automatic involuntary adjustments to changes in the external environment, such as the iris–pupil reflex and balance during locomotion. They also help control the internal environment, such as breathing rate and blood pressure, and prevent damage to the body as in cuts and burns. These help the body to maintain constant conditions, in other words they are involved in homeostasis.

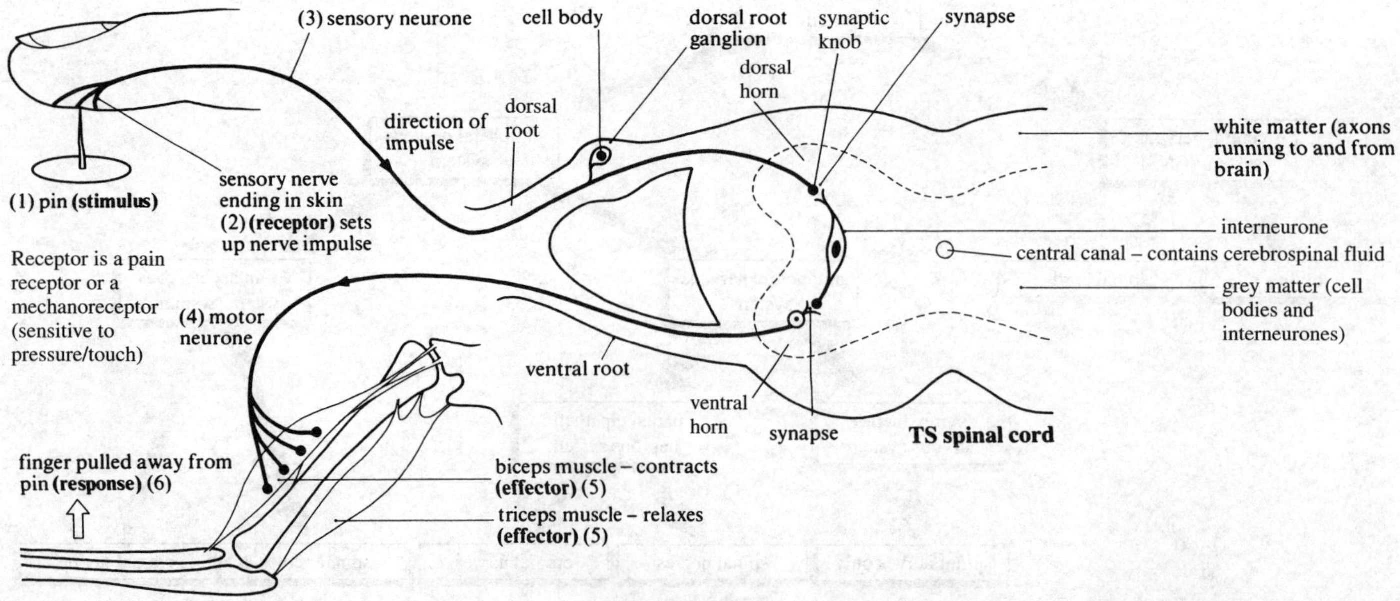

Fig 17.17 *A simplified example of reflex action and a reflex arc. The numbers in brackets refer to the basic structures in any reflex arc. Note the sequence: 1, stimulus; 2, receptor; 3, sensory neurone; 4, motor neurone; 5, effector; 6, response.*

More complex reflexes

In more complex reflexes, the sensory neurone synapses in the spinal cord with a secondary sensory neurone which passes to the brain. The brain identifies this sensory information and can store it for further use or send a motor nerve impulse down a motor neurone to synapse directly with spinal motor neurones and bring about a suitable response at an effector.

Conditioned reflexes

These are forms of reflex actions where the type of response is modified by past experience. These reflexes are coordinated by the brain. **Learning** forms the basis of all conditioned reflexes, such as in toilet training, salivation on the sight and smell of food, and awareness of danger (section 17.9). The first demonstration of conditioned reflexes is a classic example. Pavlov showed that dogs could learn a reflex by ringing a bell every time he gave them food. Eventually they would start to produce saliva when he rang the bell, even if no food was presented.

Many conditioned reflexes are modified versions of more simple reflexes. For example, if an empty metal baking tin is picked up and found to be extremely hot, burning the fingers, it will probably be dropped immediately, whereas a boiling hot, cooked casserole in an expensive dish, equally hot and painful, will probably be put down quickly but gently. The reason for the difference in the response reveals the involvement of conditioning and memory, followed by a conscious decision by the brain. The stimulus in both cases produces impulses passing to the brain in a sensory neurone. When the information reaches the brain it is interpreted and associated with information coming from other sense organs, for example the eyes, concerning the *cause* of the stimulus. The incoming information is compared with stored information concerning the nature and cause of the present stimulus and the likely outcome of allowing the spinal reflex to proceed. In the case of the metal tin, the brain works out that no further damage to either the body or the tin will occur if it is dropped and so initiates impulses in an excitatory motor neurone. This passes back down the spinal cord and synapses with the cell body of the motor neurone of the spinal reflex. Such is the speed of conduction through the pathway described that the impulses from the excitatory motor neurone reach the spinal motor neurone at the same time as impulses from the interneurone. The combined effect of these sends excitatory impulses to the effector muscle along the spinal motor neurone and the tin is dropped.

In the case of the casserole dish, the brain computes that dropping the casserole would probably scald the legs and feet, ruin the meal and break an expensive dish, whereas holding it until it could be put down safely would not cause much more damage to the fingers. If this decision is reached, impulses are initiated which pass down the spinal cord in an **inhibitory** motor neurone. These impulses arrive at the synapse with the spinal motor neurone at the same time as stimulatory impulses from the interneurone, and the latter are cancelled out. No impulses pass along the motor neurone to the muscle effector and the dish is held. Simultaneous brain activity would initiate an alternative muscle response which would result in the dish being put down quickly and safely.

The accounts of reflex arcs and reflex activity given above are, of necessity, simplified generalisations. The whole process of the coordination, integration and control of body functions is much more complex. For example, neurones connect different levels of the spinal cord together, controlling say the arms and legs, so that activity in one can be related to the other whilst at the same time other neurones from the brain achieve overall control.

Another reflex system exists for the control of activities which do not involve voluntary (skeletal) muscle. This is the autonomic nervous system.

17.2.3 The autonomic nervous system

The autonomic nervous system (*autos*, self; *nomos*, governing) is that part of the peripheral nervous system which controls activities inside the body that are normally involuntary, such as heart rate, peristalsis and sweating. It consists of motor neurones passing to the smooth muscles of internal organs. Smooth muscles are involuntary muscles. Most of the activity of the autonomic nervous system is controlled within the spinal cord or brain by reflexes known as **visceral reflexes** and does not involve the conscious control of higher centres of the brain. However, some activities, such as the control of the anal sphincter muscles which control defaecation, and bladder sphincter muscles which control urination (micturition), are also under the conscious control of the brain and control of these has to be learned. It is thought that many other autonomic activities may be able to be controlled by conscious effort and learning: many forms of meditation and relaxation have their roots in the control of autonomic activities, and considerable success has already been achieved in regulating heart rate and reducing blood pressure by conscious control or 'will power'. The overall control of the autonomic nervous system is maintained, however, by centres in the medulla (a part of the hind brain) and hypothalamus (also in the brain) (see section 17.2.4). These receive and integrate sensory information and coordinate this with information from other parts of the nervous system to produce the appropriate response.

The autonomic nervous system is composed of two types of neurone, a **preganglionic** neurone, which leaves the central nervous system in the ventral root before synapsing with several **postganglionic** neurones leading to effectors (fig 17.18).

There are two divisions of the autonomic nervous system: the **sympathetic** (SNS) and the **parasympathetic nervous systems** (PNS). The structure of the two systems differs mainly in the organisation of their neurones and these differences are shown in fig 17.18.

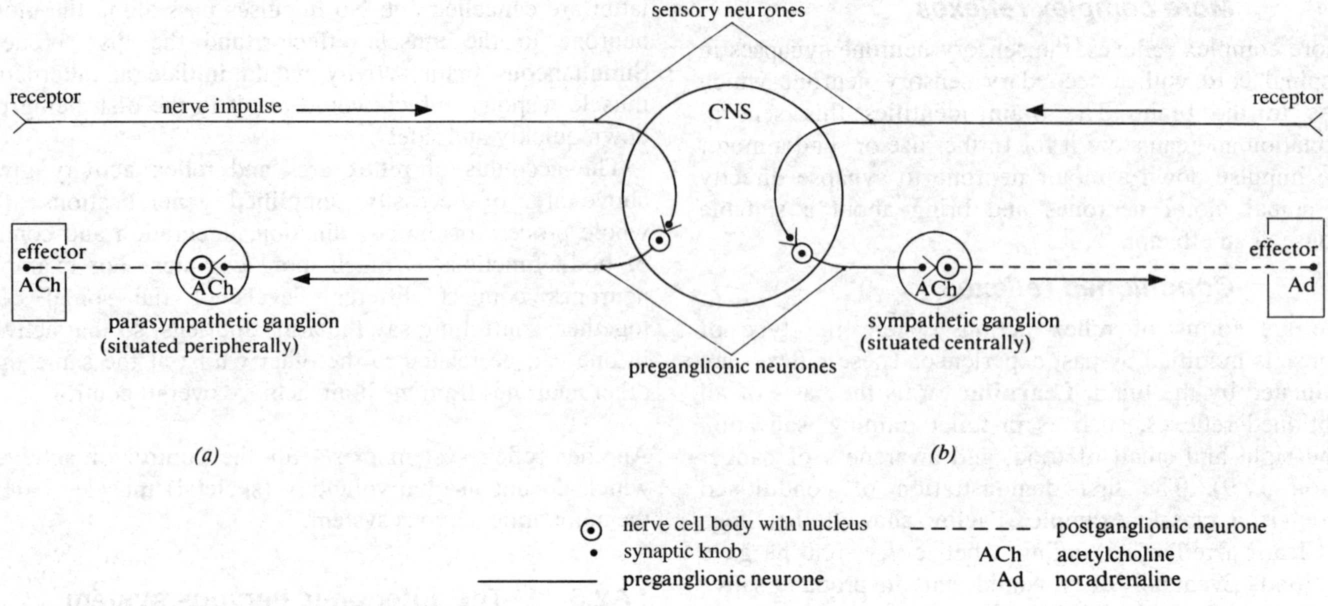

Fig 17.18 *Simplified diagram showing the basic features of (a) the parasympathetic nervous system and (b) the sympathetic nervous system (the sensory neurones are not part of the autonomic nervous system).*

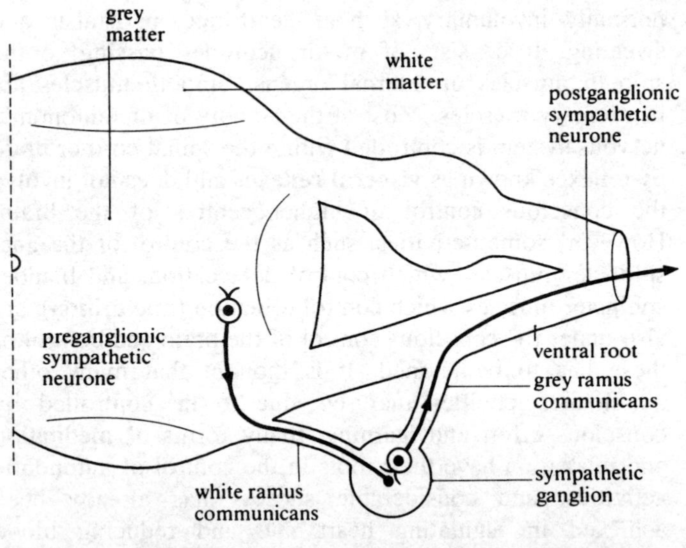

Fig 17.19 *Simplified diagram showing the position of a sympathetic ganglion and its relationship to the spinal cord and spinal nerve.*

In the sympathetic nervous system the synapses and cell bodies of the postganglionic neurones in the trunk region are situated in ganglia (swellings) close to the spinal cord. Each **sympathetic ganglion** is connected to the spinal cord by a **white ramus communicans** and to the spinal nerve by a **grey ramus communicans** as shown in fig 17.19. A chain of connected sympathetic ganglia runs alongside the spinal cord. The ganglia of the parasympathetic nervous system are situated close to, or within, the effector organ (fig 17.18).

Other differences between the two systems include the nature of the chemical transmitter substance released at the postganglionic effector synapse, their general effects on the body and the conditions under which they are active. These differences are summarised in table 17.4.

The sympathetic and parasympathetic nervous systems generally have opposing (antagonistic) effects on organs they supply, and this enables the body to make rapid and precise adjustments of involuntary activities in order to maintain a steady state. For example, an increase in heart rate due to the release of noradrenaline by sympathetic neurones is compensated for by the release of acetylcholine by parasympathetic neurones. This action prevents heart rate becoming excessive and will eventually restore it to its normal level when secretion from both systems balances out. A summary of the antagonistic effects of these systems is shown in table 17.5. A careful study of this table will give you a good understanding of the functions of the two systems. A diagrammatic summary of the structure of the two systems is given in fig 17.20.

17.2.4 The central nervous system

The central nervous system consists of the brain and spinal cord. Like a telephone exchange with ingoing and outgoing wires, it is responsible for the coordination and control of the activity of the nervous system. It develops from an infolding of the outer layer of the embryo known as the ectoderm. This is immediately above the notochord, the long strengthening rod which becomes the backbone in vertebrates. The infolding ectoderm forms a dorsal, hollow neural tube running the length of the animal. The neural tube differentiates during development to form an expanded anterior (front) region, the **brain**, and a long cylindrical **spinal cord**.

Table 17.4 Summary of the differences between the sympathetic and parasympathetic nervous systems.

Feature	*Sympathetic*	*Parasympathetic*
Origin of neurones	Emerge from cranial, thoracic and lumbar regions of CNS	Emerge from cranial and sacral regions of CNS
Position of ganglion	Close to spinal cord	Close to effector
Length of fibres	Short preganglionic fibres Long postganglionic fibres	Long preganglionic fibres Short postganglionic fibres
Number of fibres	Numerous postganglionic fibres	Few postganglionic fibres
Distribution of fibres	Preganglionic fibres cover a wide area	Preganglionic fibres cover a restricted region
Area of influence	Effects diffuse	Effects localised
Transmitter substance	Noradrenaline released at effector	Acetylcholine released at effector
General effects	Increases metabolite levels, e.g. blood sugar Increases metabolic rate Increases rhythmic activities, e.g. heart rate Raises sensory awareness	Decreases metabolite levels, e.g. blood sugar None Decreases rhythmic activities, e.g. heart rate Restores sensory awareness to normal levels
Overall effect	Excitatory homeostatic effect	Inhibitory homeostatic effect
Conditions when active	Dominant during danger, stress and activity; controls reactions to stress	Dominant during rest Controls routine body activities

Table 17.5 Summary of the effects of the sympathetic and parasympathetic nervous systems on the body.

Region	*Sympathetic*	*Parasympathetic*
Head	Dilates pupils None Inhibits secretion of saliva	Constricts pupils Stimulates secretion of tears Stimulates secretion of saliva
Heart	Increases strength and rate of heart beat	Decreases strength and rate of heart beat
Lungs	Dilates bronchi and bronchioles Increases ventilation rate	Constricts bronchi and bronchioles Decreases ventilation rate
Gut	Inhibits peristalsis Inhibits secretion of alimentary juices Contracts anal sphincter muscle	Stimulates peristalsis Stimulates secretion of alimentary juices Inhibits contraction of anal sphincter muscle
Blood	Constricts arterioles to gut and smooth muscle Dilates arterioles to brain and skeletal muscle Increases blood pressure Increases blood volume by contraction of spleen	Maintains steady muscle tone in arterioles to gut, smooth muscle, brain and skeletal muscle, allowing normal blood flow Reduces blood pressure None
Skin	Contracts hair erector muscles (hair 'stands on end') Constricts arterioles in skin of limbs (skin whitens) Increases secretion of sweat	None Dilates arterioles in skin of face (skin reddens) None
Kidney	Decreases output of urine	None
Bladder	Contracts bladder sphincter muscle	Inhibits contraction of bladder sphincter muscles
Penis	Induces ejaculation	Stimulates erection
Glands	Releases adrenaline from adrenal medulla	None

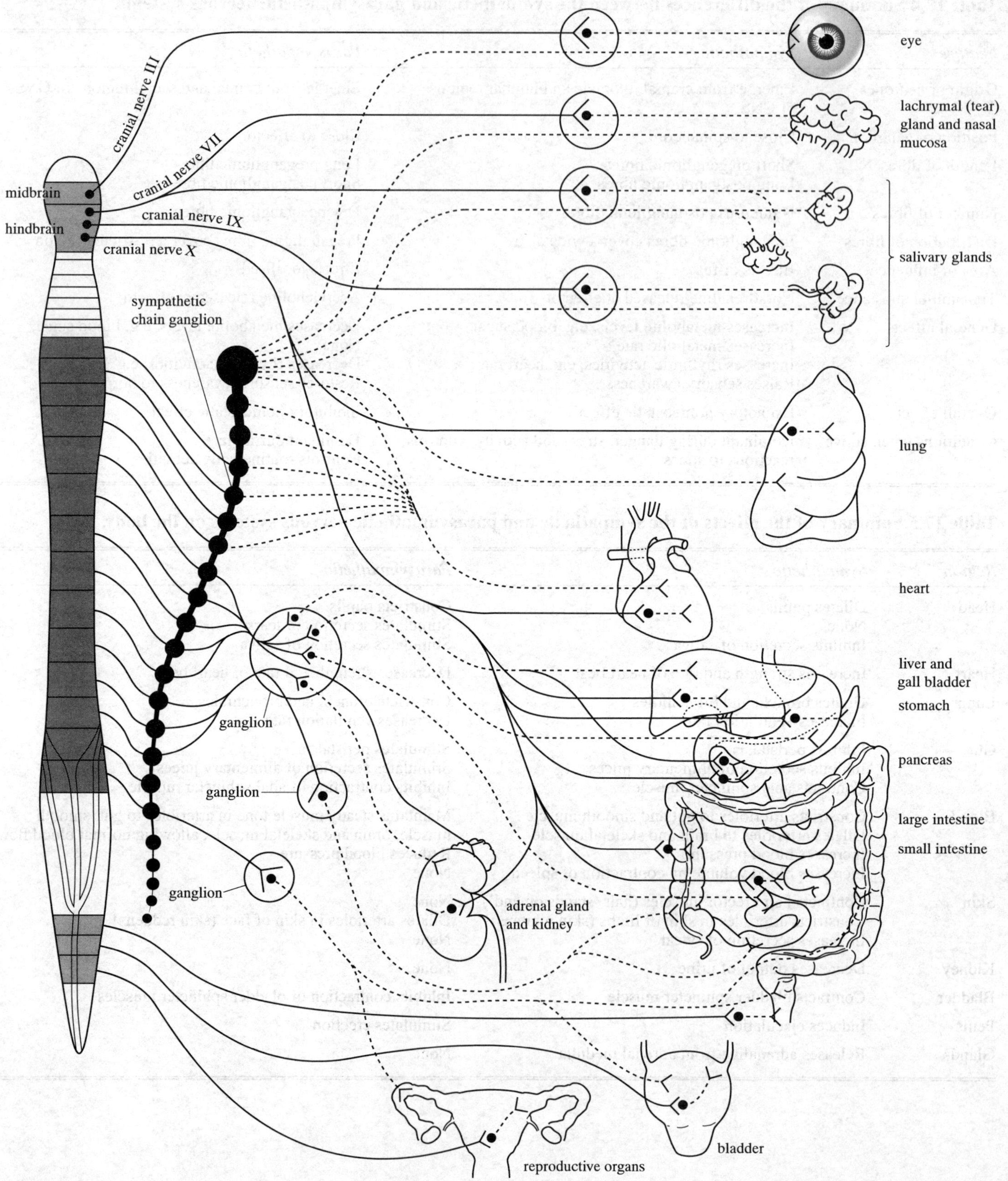

Fig 17.20 *Structure of the sympathetic and parasympathetic nervous systems. Together they form the autonomic nervous system which is the part of the nervous system responsible for involuntary (automatic or reflex) activity. The solid lines represent preganglionic fibres, and the dashed lines represent postganglionic fibres.*

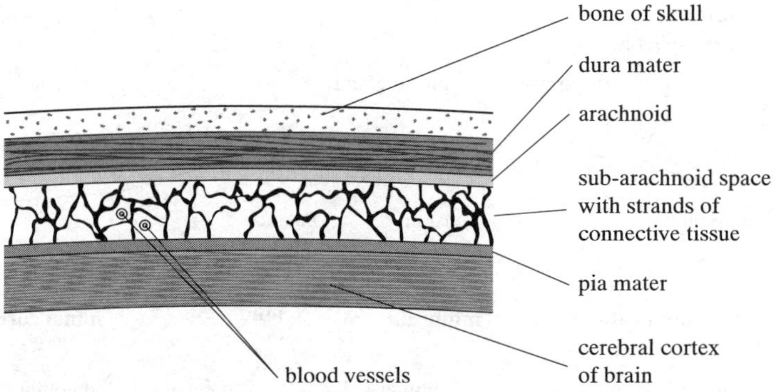

***Fig 17.21** The three meninges of the brain (dura mater, arachnoid and pia mater).*

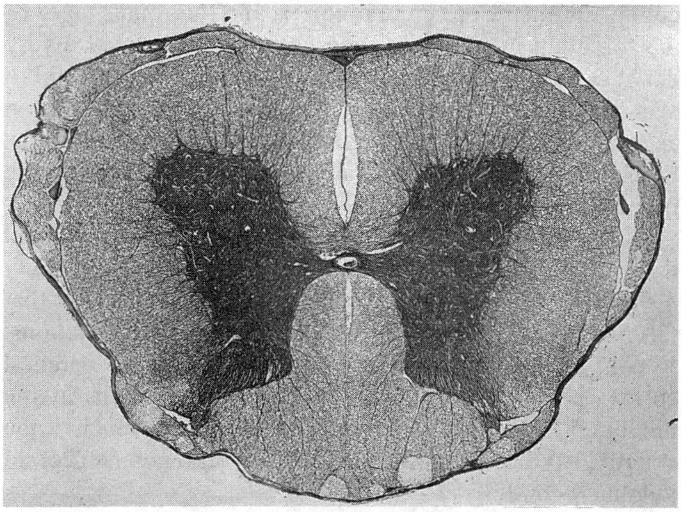

Fig 17.22 *Light micrograph of a cross-section of human spinal cord. The butterfly-shaped grey matter in the middle consists of nerve cells. The surrounding white matter is made up of myelinated nerve fibre bundles. Surrounding the cord are three layers of membranes (meninges).*

The meninges and cerebrospinal fluid

The central nervous system is surrounded by three layers or 'membranes' called **meninges** (fig 17.21) and is completely encased within the protective bones of the skull and vertebral column. The outer membrane forms the tough **dura mater** which is attached to the skull and vertebrae, and the inner membrane forms the thin **pia mater** which lies next to the nervous tissues. Between the two is the **arachnoid** 'membrane'. This includes a space, the **sub-arachnoid space**, strands of connective tissue, blood vessels and **cerebrospinal fluid** (**CSF**). Most of this fluid is contained in the central canal of the spinal cord and continues forward to occupy four expanded cavities within the brain called the **ventricles**. The fluid therefore comes into contact with the outside and inside of the brain, and blood vessels lie within it for the supply of nutrients and oxygen to the nervous tissues and the removal of wastes (see fig 17.21). It also contains lymphocytes to protect against infection. Meningitis is caused by an infection of the meninges. About 100 cm^3 of fluid is present in the CNS and, apart from its nutritive, excretory and defensive functions, it supports the nervous tissues and protects them against mechanical shock. A continual circulation of fluid is maintained by ciliated cells lining the ventricles and central canal.

The spinal cord

The spinal cord (fig 17.22) is a cylinder of nervous tissue running from the base of the brain down the back. It is protected by the vertebrae of the backbone (vertebral column) and the meninges. It has a H-shaped central area of **grey matter**, composed of nerve cell bodies, dendrites and synapses surrounding a central canal which contains cerebrospinal fluid. Around the grey matter is an outer layer of **white matter**, containing nerve fibres whose fatty myelin sheaths give it its characteristic colour. There are 31 pairs of **spinal nerves** and these divide close to the spinal cord to form two branches called the **dorsal root** and **ventral root**. Sensory neurones enter the dorsal root and have their cell bodies in a swelling, the **dorsal root ganglion**, close to the spinal cord. The sensory neurones then enter the **dorsal horn** of the grey matter where they synapse with interneurones. These, in turn, synapse with motor neurones in the **ventral horn** and leave the spinal cord via the ventral root (fig 17.17). Since there are many more interneurones than motor neurones, some integration must occur within the grey matter. Some sensory neurones synapse directly with motor neurones in the ventral horn, as in the familiar knee-jerk reflex. From the thorax region downwards, a **lateral horn** is present between the dorsal and ventral horns which contains the cell bodies of the preganglionic autonomic neurones. The white matter is composed of groups of nerve fibres, running between the grey matter and the brain and providing a means of communication between spinal nerves and the brain. **Ascending tracts** carry sensory information to the brain and **descending tracts** relay motor information to the spinal cord.

To summarise, the functions of the spinal cord include acting as a coordinating centre for simple spinal reflexes, such as the knee-jerk response, and autonomic reflexes, such as contraction of the bladder, and providing a means of communication between spinal nerves and the brain.

The brain

The human brain has been described as the most complex structure in the Universe. Although it has been compared to a computer, it is far superior to any computer yet built. Understanding how it functions remains one of the most challenging problems in biology. Nevertheless, progress has accelerated in recent years.

The brain contains an estimated 100 thousand million nerve cells with more than 1000 miles of nerve fibres per cubic centimetre of cerebral cortex, the thin outer layer of the forebrain in which our consciousness is located. Every nerve cell in the cortex receives, on average, about 1000 to 10 000 connections from other nerve cells, giving an astronomical number of combinations of nerve cells. Somehow, from this 'neural network', we achieve the process of conscious thought.

Although in the early twentieth century it was imagined that the functions of the brain were somehow vaguely distributed throughout its structure, we now know that different parts of the brain have very specific functions. Vision, for example, is located at the back of the cerebral cortex; heart rate is controlled from the medulla in the hindbrain. The cerebral cortex is conscious, but much of the activity of the brain is unconscious and not subject to voluntary control.

In this section we shall look at the main structures of the brain and their functions.

Origins of the brain

It helps to understand the structure of the brain if we examine its origins in simpler vertebrates and in the human embryo. Fig 17.23 shows the structure of a fish brain in section and from above. It shows that the brain is divided into three main regions, the forebrain, midbrain and hindbrain. These three divisions are seen just as clearly in the early embryonic development of the human brain. However, the story of evolution in the vertebrates includes a massive increase in the size of the brain in relation to the rest of the body. This increase in size is due mainly to an increase in size of the forebrain. Birds and mammals have larger forebrains than amphibians and reptiles, with the brains of mammals being largest. In mammals the forebrain has a folded surface and grows back over the midbrain and hindbrain so that, if the brain is viewed from above, the midbrain is no longer visible (fig 17.24*a*).

Of the mammals, primates have the largest forebrains of all relative to body size. Note that size alone is not the only indicator of quality. An average elephant's brain is four times heavier than an average human brain, and the average human male brain is larger than the average human female brain, a piece of evidence used by scientists in the nineteenth century to justify the assertion that men had superior brains (even though they ignored the fact that men generally have a larger body size too). Fig 17.24*b* shows a diagram of the main regions of the human brain as seen in section. Fig 17.25 shows a section through an actual human brain.

Fig 17.23 shows that vision in the fish is located in the midbrain. In humans, the eyes are still connected by the optic nerves to the midbrain, but most of the information continues forward to the forebrain where it is processed and converted to vision. The hindbrain in humans still more or less retains the same functions as in fish, and is divided into cerebellum, pons and medulla. It is with these 'older' parts that we shall start this brief tour of the brain's structure and function.

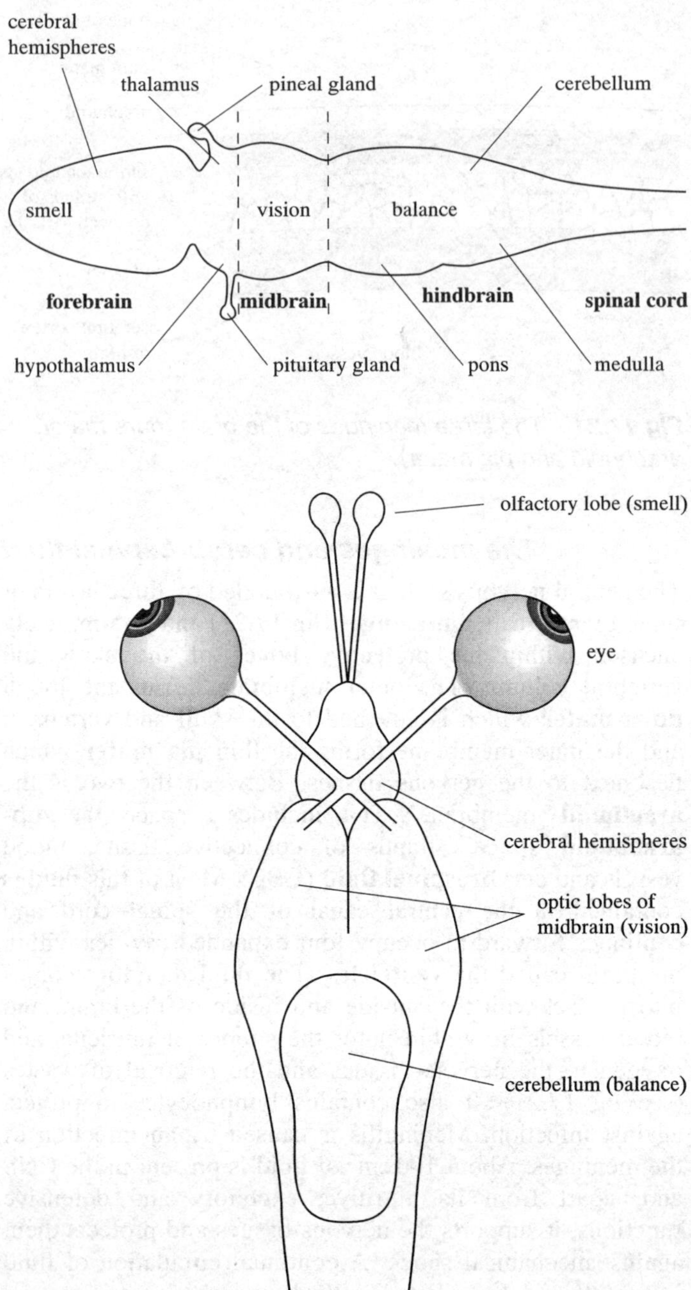

Fig 17.23 *Fish brain as seen from the side and from above.*

Hindbrain

Cerebellum. This is located at the back of the brain under the cerebral hemispheres. It is very folded and has an outer region, the cortex, which contains many nerve fibres and cell bodies. Like the cerebral cortex it appears greyish in colour. The cerebellum has been called the gyroscope of the body because it is concerned with balance. It receives information from the organs of balance in the ears and is concerned with the control and precision of all movements involving voluntary muscle.

Patients whose cerebellum has been damaged at first are not able to walk. They can learn to walk again, but do so

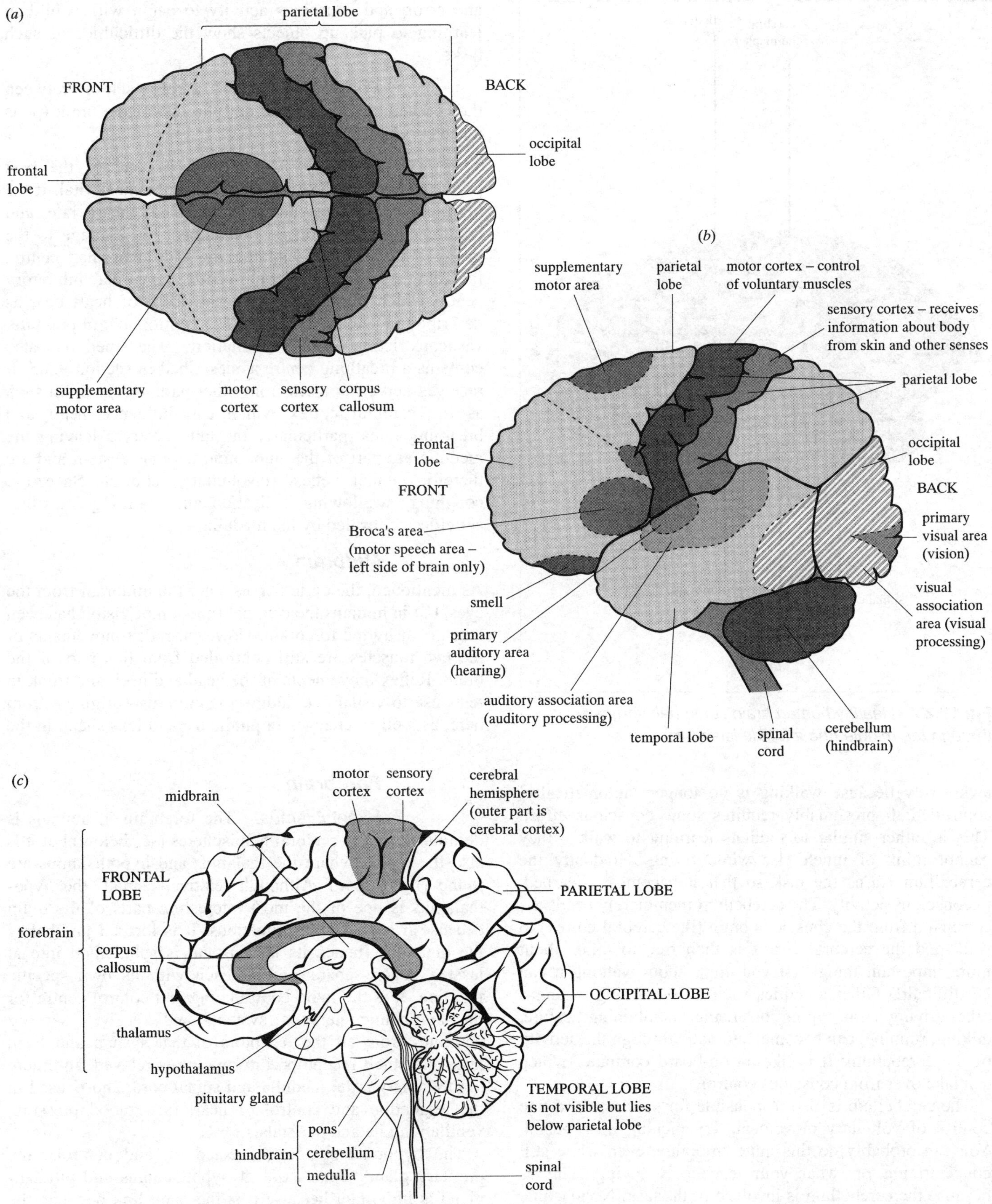

Fig 17.24 *External view of human brain as seen from (a) the top (b) the left side. A section through the brain is shown in (c).*

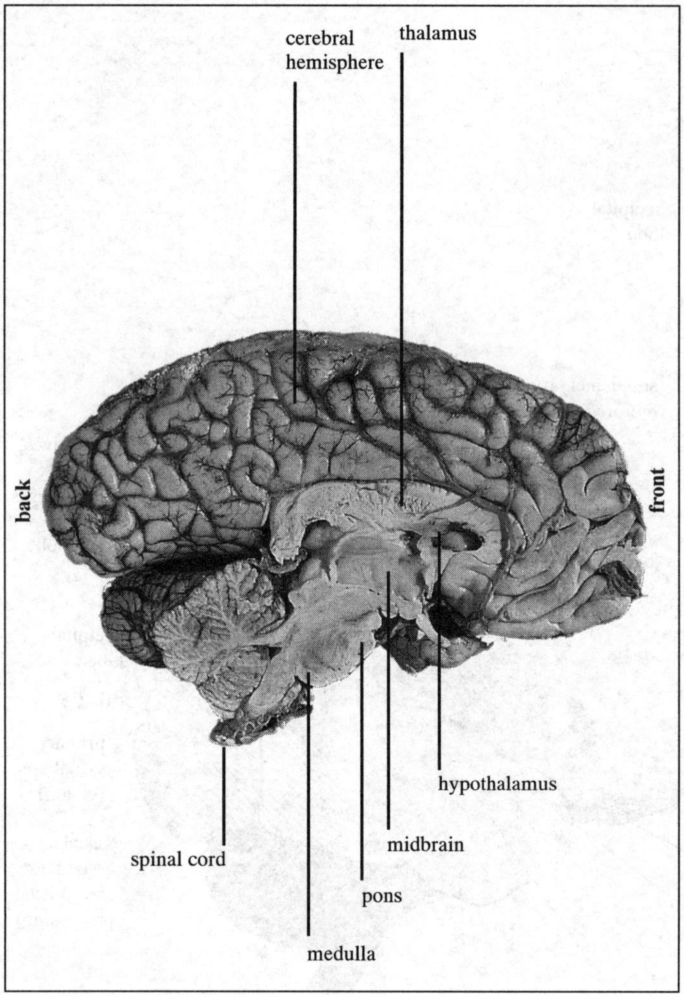

Fig 17.25 *Healthy human brain cut in half vertically through the middle (the left side is shown).*

awkwardly because walking is no longer 'automatically' controlled. It presumably requires some conscious effort. This is rather similar to toddlers learning to walk – they cannot think of much else while trying. Gradually the cerebellum learns the task so that it becomes a learned, unconscious activity. The cerebellum then merely needs the command from the conscious brain (the cerebral cortex) to walk and the cerebral cortex is then free to think about more important things. (If you think about walking it can be difficult!) Other activities such as swimming, riding a bike, driving a car, typing, balancing, maintaining posture, talking, running, can become automatic through the activity of the cerebellum. It is like an on-board computer which can take over from conscious control.

The cerebellum is also responsible for precision and fine control of voluntary movement. Try picking up a pencil. You can probably do this quite smoothly, even while still concentrating on what your teacher is saying. This is because the cerebellum is involved in the task. No machine has yet been made which can do this as smoothly or with as much versatility because it is an extremely difficult task. Once again, this is something that has to be learned by trial and error, and conscious activity to begin with. Children learning to pick up objects show the difficulties of such tasks.

Pons. The pons is a relay station between the cerebellum, spinal cord and the rest of the brain (pons means bridge).

Medulla. The medulla is one of the best protected parts of the brain and one of the most vital. If the brain is cut above the medulla, basic heart rate and breathing rate can still be maintained, but damage to the medulla is fatal. It contains the cardiovascular centre, including a cardiac accelerator centre and cardiac inhibitory centre which regulate the rate and force of heart beat as described in section 14.7.4. It also controls blood pressure, vasoconstriction and vasodilation. The medulla also contains a breathing centre as described in section 9.5.5. It receives nerve impulses from other parts of the brain such as the cerebral cortex which can influence heart and breathing rates, particularly the latter. Nerves leaving the medulla are part of the autonomic nervous system and are therefore under reflex (involuntary) control. Sneezing, coughing, swallowing, salivation and vomiting are other activities controlled by the medulla.

Midbrain

As mentioned, the optic nerves enter the midbrain from the eyes, but in humans the original function of vision has been taken over by the forebrain. However, reflex movements of the eye muscles are still controlled from this part of the brain. Reflex movements of the head and neck and trunk in response to visual and auditory stimuli also originate from here, as well as changes in pupil size and lens shape in the eye.

Forebrain

Hypothalamus. The forebrain in humans is dominated by the cerebral hemispheres (see below) but it is also the region where the thalamus and hypothalamus are found (fig 17.24*c*). Although relatively small, the hypothalamus is one of the most interesting parts of the brain because it has so many functions. It is located just below the thalamus (hence its name), and is subdivided into at least a dozen separate areas, each with its own specific function. It is the main coordinating and control centre for the autonomic nervous system, and receives sensory neurones from all the receptors of that system and from taste and smell receptors. Information is relayed from here to effectors via the medulla and spinal cord, and is used in the regulation and control of heart rate, blood pressure, ventilation rate and peristalsis.

The hypothalamus is connected to, and controls, the pituitary gland. This 'axis' of hypothalamus and pituitary gland is important because it is the main link between the nervous system and the endocrine system and its hormones (see the example of ADH below, and section 17.6.2).

Within the hypothalamus are centres which control

various aspects of mood and emotions, such as aggression, rage, fear and pleasure. Artificial stimulation of these centres with electrodes provides some of the most convincing evidence of how localised brain function can be. For example, stimulation of the aggression centre in a cat makes its back arch, hair stand on end, tail lash, pupils dilate and it starts snarling. It will attack any suitable object, such as a rat or sometimes the experimenter. Interestingly, and probably no coincidence, this centre is close to the fear centre. Fear can easily turn to aggression.

Another centre is the pleasure centre. A rat given the choice of pressing a lever to obtain food or one to stimulate its pleasure centre starves to death for obvious reasons. While the notion of a pleasure centre is probably an oversimplification of what is a very complex area of behaviour. Other centres, however, do have very precise roles. For example, stimulation of the thirst centre will make an animal drink. There are also centres associated with hunger, satiety (a feeling of being 'full' or having eaten enough) and control of body temperature (section 19.5.4). The thirst centre, for example, helps to regulate the solute concentration of the blood and hence the osmotic properties of the blood. A high solute concentration (relatively low water concentration) results in the secretion of the hormone ADH (antidiuretic hormone) through nerve cells to the posterior lobe of the pituitary gland, from where it enters the blood. ADH decreases the volume of urine as explained in section 20.6.

All these centres monitor the blood with the overall purpose of maintaining homeostasis, or constant conditions, within the body. Relatively speaking the hypothalamus receives more blood than any other part of the brain. Using the information it receives, the hypothalamus, in association with the pituitary gland, is one of the major regulators of homeostasis.

Cerebral hemispheres

The outer 2–4 mm of the cerebral hemispheres (or cerebrum) is known as the **cerebral cortex**. It consists of 'grey matter' containing thousands of millions of nerve cells, including their cell bodies. Beneath the cortex is the 'white matter' (in contrast to the spinal cord where the grey matter is on the inside and the white matter on the outside).

The cerebral hemispheres are the site of consciousness, our sense of self, perhaps the most mysterious property of the brain. So far, the nature of consciousness is entirely unexplained. We are not even sure how many other animals show consciousness. Other rather vague properties such as intelligence, reasoning, personality, learning, emotions and the 'will' are located here, as well as some more specific functions, as shown in fig 17.26.

The right hemisphere controls the left side of the body and the left hemisphere controls the right side of the body.

Primary sensory areas receive sensory impulses from receptors in most parts of the body. The sensory areas form localised regions of the cortex associated with certain

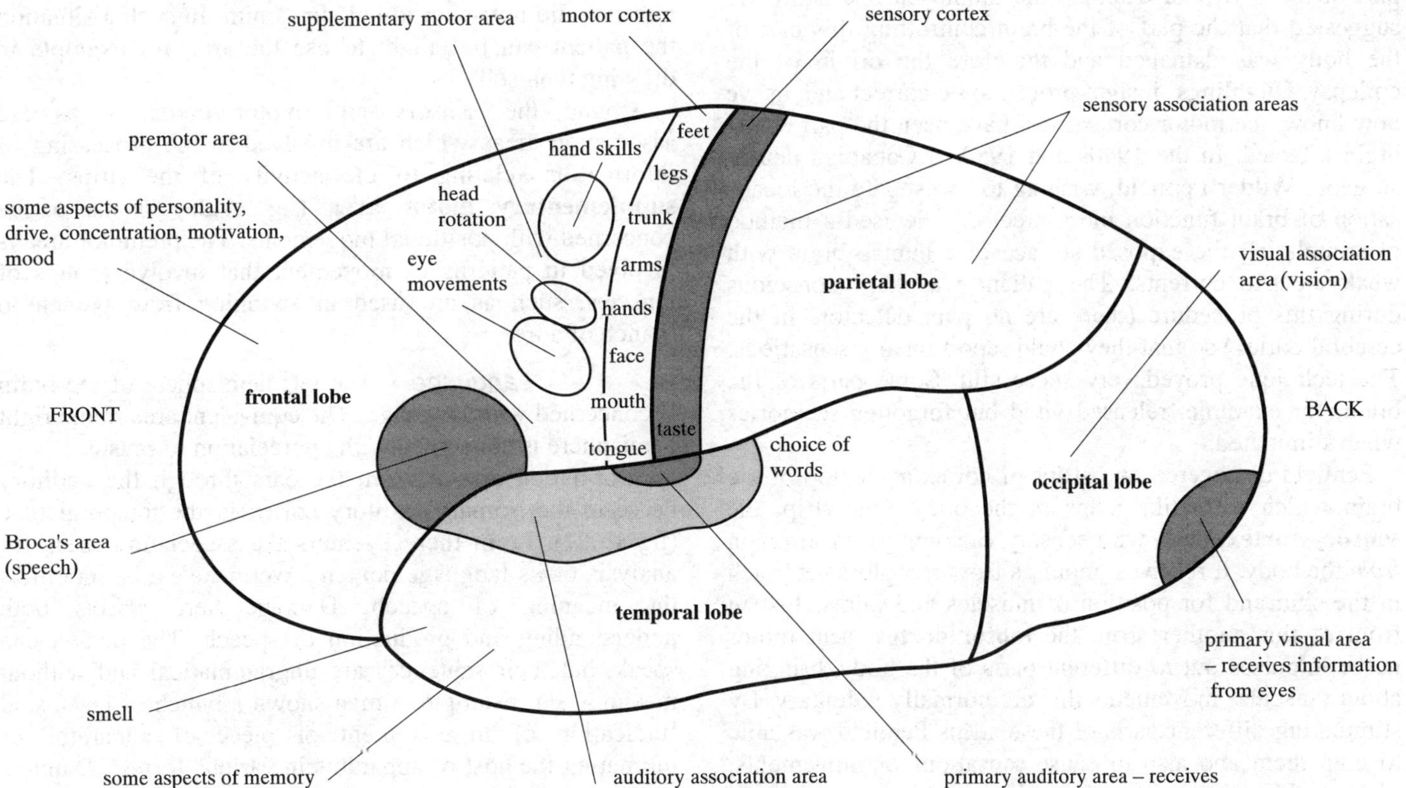

Fig 17.26 *Localisation of function in the cerebral hemispheres. Broca's area is in the left hemisphere in most people.*

senses as shown in fig 17.26. The size of the region is related to the number of receptors in the sensory structure.

Association areas are so named for several reasons. First, they associate incoming sensory information with information stored in the memory, so that the information is 'recognised'. Secondly, the information is associated with information arriving from other receptors. Thirdly, the information is 'interpreted' and given meaning within its present context and, if necessary, a decision is made about the most appropriate response. The association area passes instructions to its associated motor area. Association areas, therefore, are involved in memory, learning and reasoning, and the degree of success of the outcome may loosely be termed **intelligence**. Association areas are usually next to their related sensory area; for example, the **visual association area** is situated immediately next to the visual cortex in the **occipital lobe**. Some association areas may have a restricted function and are linked to other association centres that can further develop the activity. For example, the **auditory association area** only interprets sounds into broad categories which are relayed to more specialised association areas where 'sense' is made of the words.

Motor areas control movement of voluntary muscles, sending out nerve impulses in motor nerves.

The sensory and motor cortex. In the 1860s, a British neurologist, John Hughlings, discovered that when some of his epileptic patients had fits, the uncontrollable movements always started in one particular part of the body, for example the thumb on one hand. He suggested that the part of the brain controlling this part of the body was damaged and therefore the origin of the epilepsy. Hughlings' insight proved to be correct and, as we now know, the motor cortex must have been the part of the brain affected. In the 1940s and 1950s a Canadian neurosurgeon, Wilder Penfield, wishing to investigate the localisation of brain function more precisely, devised a method of stimulating the exposed surface of a human brain with weak electric currents. The patient was kept conscious during this procedure (there are no pain detectors in the cerebral cortex) so that they could report on any sensations. The technique proved very successful. Some parts of the brain, for example, released vivid but forgotten memories when stimulated.

Penfield discovered two strips of cortex at the top of the brain which were like maps of the body. One strip, the **sensory cortex**, dealt with sensory (incoming) information *from* the body. It receives impulses from receptors for touch in the skin and for position of muscles and joints. Just in front of this another strip, the **motor cortex**, sent motor nerve impulses out *to* different parts of the body, bringing about muscular movements that are normally voluntary. By stimulating different parts of these strips Penfield was able to map them and also to cause sensations or movements, like twitching the fingers, almost like working the strings of a puppet.

The maps are unusual in that they show us the body as the brain experiences it. Thus the lips, tongue and hands, which are very sensitive, and subject to very fine control of movement, 'occupy' relatively large areas of the strips, whereas the legs or trunk, for example, occupy relatively small areas. The tongue, for example, is involved in talking, and moves about very quickly and precisely during eating (if you think about it while eating you will bite your tongue). Motor control of the tongue is therefore complex and requires a relatively large part of the motor cortex. The tongue is also very sensitive, as you may have experienced in probing a huge cavern in a damaged tooth, only to discover that it is a very small hole. It therefore sends a relatively large amount of information to the brain, requiring a relatively large part of the sensory cortex.

The sensory and motor maps can each be illustrated graphically with a model called a **homunculus** (meaning 'little man'), as shown in fig 17.27. Fig 17.28 shows maps of the sensory and motor cortex.

Damage to the motor cortex, for example by a stroke, results in paralysis of the part of the body that the damaged part normally controls. Damage to the left hemisphere affects the right side of the body and vice versa. Damage to the sensory cortex results not only in loss of sensation from the relevant body part, but can also lead to loss of the knowledge of its existence and the person behaving as if it did not exist, even if that part can be operated by the motor cortex. For example, a man recovering from an accident complained that an arm had been sewn on to him as he tried to rationalise the experience of having an arm in bed with him that did not seem to belong to him. In such a situation the patient can be taught to use the arm, for example in dressing themselves.

Around the sensory and motor cortex strips are association areas which are involved in the processing of information relating to the activity of the strips. The **supplementary motor area** (fig 17.26) seems to be concerned with positional movements. The premotor area is involved in patterns of movement that involve groups of muscles, such as are used in swinging from branch to branch of a tree.

Language. The left hemisphere of the brain is concerned with language. The equivalent area in the right hemisphere is concerned with appreciation of music.

Information arrives from the ears through the auditory nerve to the primary auditory cortex in the temporal lobe (fig 17.27). From there it enters the association areas for analysis of its language content. Wernicke's area interprets the meaning of speech. Damage here affects both understanding and production of speech. The person can speak, but their sentences are ungrammatical and without meaning. For example, a man shown a bunch of keys, said 'indication of measurement of piece of apparatus or intimating the cost of apparatus in various forms'. Damage to the region between the parietal lobe and temporal lobe (see fig 17.26) can affect the ability to find the right word. Here word associations are common. For example, a man

(a) (b)

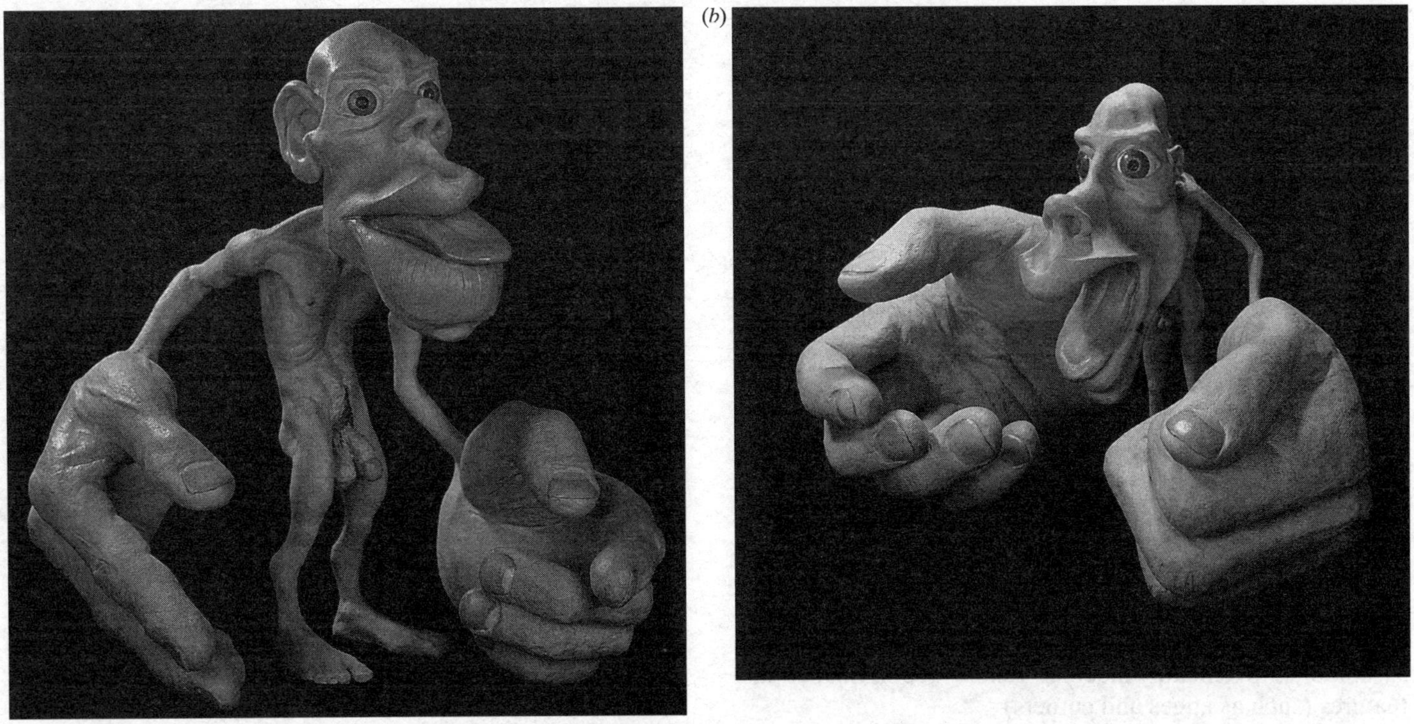

Fig 17.27 *(a) Sensory homunculus: this model shows what a man's body would look like if each part grew in proportion to the area of the cortex of the brain concerned with its sensory perception. (b) Motor homunculus: this model shows what a man's body would look like if each part grew in proportion to the area of the cortex of the brain concerned with its movement.*

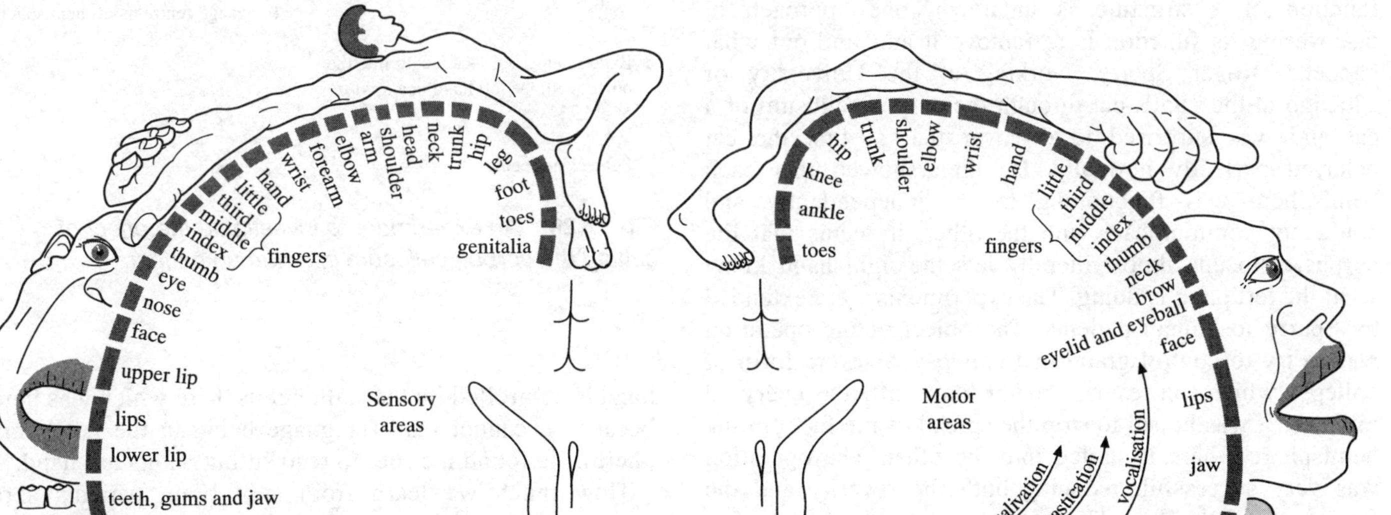

Fig 17.28 *Maps of the sensory and motor cortex.*

shown a nail file said it was a 'pair of scissors'. Dyslexia probably involves the occipital lobe (vision) and the temporal lobe (language).

A motor area associated with language is Broca's area in the frontal lobe, near the area of the motor cortex dealing with the mouth (lips and tongue) (fig 17.26). It is concerned with speech. Nerve impulses from here travel via the premotor area to the muscles of the tongue, pharynx and mouth. Breathing is also controlled during speech. Damage to Broca's area results in impaired speech, although the victim can understand speech perfectly well.

Vision. The primary visual area in the occipital lobe at the back of the head receives information from the eyes. Damage to this would result in blindness, even if the eyes were functioning perfectly. Damage to the association areas, however, affects the processing of vision. For example, a person might be able to see a friend but not recognise them if the program for constructing and recognising faces is damaged. Different association areas in the occipital lobe are concerned with analysis of features such as colour, movement, binocular vision, depth, shapes and features (such as edges and corners).

Corpus callosum

The right and left hemispheres are connected by a thick band of nerve fibres called the corpus callosum. When the function of a structure is unknown, one approach to discovering its function is to remove it and find out what happens. Roger Sperry, working at the University of Chicago in the 1950s cut through the corpus callosum of a cat and was surprised to discover that at first the cat behaved perfectly normally. He then showed that each hemisphere was functioning totally independently and could not communicate with the other. It seems that the corpus callosum almost literally lets the right hand know what the left hand is doing. The experiments were extended by Sperry to human patients. The object of the operation was to try to control grand mal epilepsy, a severe form of epilepsy which can result in major fits as often as every 30 minutes. It was hoped to stop the spread of the fit from the hemisphere where it started into the other. The operation was very successful, reducing both the severity and the frequency of the fits. The patient appeared normal after recovery. However, Sperry was able to show by carefully designed experiments that the two hemispheres were not communicating, as with the cat. One experiment is shown in fig 17.29. Further experiments confirmed that language is located in the left hemisphere.

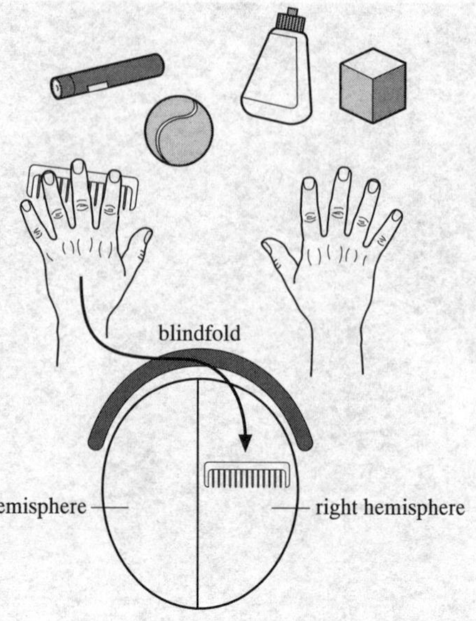

1 Subject is blindfolded.

2 Experimenter places the chosen object in the left hand. (No one must name the object or the right ear will hear and inform the left hemisphere.)

Left hand sends nerve impulses to right hemisphere.

Right hemisphere knows object is a comb.

3 Experimenter asks subject to find the object with the left hand in a mixture of objects.

Subject succeeds.

4 Experimenter asks subject to find th object with right hand (This is controlled by the left hemisphere.)

Subject fails.

5 Experimenter asks subject to say what the object was.
Subject fails.

Only the left hemisphere can speak (language centre in left hemisphere)

Fig 17.29 *An experiment to investigate the effect of cutting the corpus callosum in a human patient.*

Careful observation of such 'split brain' patients has revealed subtle changes in their behaviour. Because the left hemisphere tends to be dominant, the left side of the body (controlled by the right hemisphere) rarely shows spontaneous activity. The patient generally does not respond to stimulation of the left side. One man found that when he (or rather his left hemisphere) tried to read a paper, his left hand tried to throw it on the floor. The left hand is controlled by the right hemisphere which was bored because it cannot read (language being in the left hemisphere). He found it easier to read sitting on his left hand.

How much we learn from split brain patients about normal behaviour is questionable. Such problems as those described should not affect anyone with an intact corpus callosum. Information can pass from one hemisphere to the other and tasks can be shared. However, the importance of the corpus callosum becomes clear, and it raises interesting questions about the nature of identity, especially when we are 'in two minds' about whether to do something.

17.3 Evolution of the nervous system

A study of animal evolution shows a progressive increase in complexity from cnidarians to mammals.

17.3.1 Cnidarians, e.g. *Hydra*

The development of multicellular organisation in the cnidarians led to a separation of stimulus and response, receptor and effector. Specialised cells, the nerve cells, evolved which link receptor to effector. The nervous system of primitive cnidarians, for example *Hydra*, is a **nerve net** or **plexus** composed of a single layer of short neurones in contact throughout the organism. Impulses spread out in all directions from the point of stimulation and at each synapse an impulse is lost. This impulse is used, effectively, to 'charge' the synapse so that subsequent impulses can cross the synapse. This process is called **facilitation** and, since an impulse is lost at each synapse in facilitating (making easier) the passage of the next impulse, the mechanism of conduction is called **decremental conduction**. Nervous conduction in this organism is therefore slow, due to the number of synapses to cross, and restricted because impulses die out as they progress outwards from the stimulus. The system is useful in producing localised responses, say within a tentacle, but of little value to the whole organism unless the stimulus is **intense** or **prolonged**. In most cnidarians, such as jellyfish and sea anemones, in addition to the nerve net there is a system of bipolar neurones arranged in tracts, called **through conduction tracts**, and able to transmit impulses rapidly over considerable distances and without apparent loss. This system enables the organism to make fairly rapid responses of the whole body to harmful stimuli, such as the withdrawal of tentacles, and this foreshadows the aggregation of neurones into nerves seen in higher organisms.

17.3.2 Annelids, e.g. earthworm

In the annelids neurones are associated into nerves. This has resulted in a nervous system with a single **ventral nerve cord** running the entire length of the organism. It consists of one pair of ganglia per segment joined by connecting neurones. Nerves to and from the tissues arise from the ganglia of each segment as shown in fig 18.26.

As a result of the unidirectional method of locomotion, annelids possess a head. This structure is specialised to assist with feeding and, since it is the first part of the body to come into contact with new environmental situations, it contains all the sensory structures necessary to detect stimuli associated with these situations. The increased input of sensory information from these receptors to the nervous system is dealt with by the enlarged anterior end of the nerve cord. This concentration of feeding apparatus, sense organs and nervous tissues into a 'head' region is called **cephalisation**. It should be emphasised though that the term applies to the development of *all* the features associated with the head and not just the nervous tissue.

The annelid nervous system shows all the basic features found in all other non-vertebrate groups. The enlarged anterior region of the nerve cord forms a pair of **cerebral ganglia** situated above the pharynx and linked to the ventral nerve cord around each side of the pharynx.

17.3.3 Arthropods, e.g. insects

In arthropods the basic organisation of the nervous system is almost identical to that of annelids except that the cerebral ganglia overlie the oesophagus. Cerebral ganglia are analogous to the vertebrate brain but do not possess the same degree of control over the entire nervous system seen in vertebrates. For example, removal of the head of a non-vertebrate has very little effect on movement, whereas in vertebrates the brain controls all movement of the body. Non-vertebrate cerebral ganglia, in fact, appear to act as relay centres between receptors and effectors and their role in integration and coordination is limited to a few responses involving hormones, such as the timing of reproductive activities in annelids and the control of moulting in arthropods (section 22.3.3).

17.4 Sensory receptors

The coordinated activity of an organism relies upon a continuous input of information from the internal and external environments. If this information leads to a change in activity or behaviour of the animals, it is a **stimulus**. The specialised region of the body detecting the stimulus is known as a **sensory receptor**.

The simplest and most primitive type of receptor consists of a single sensory neurone which is capable of detecting the stimulus and giving rise to a nerve impulse passing to the central nervous system, for example, skin mechanoreceptors such as the **Pacinian corpuscle** (section 17.5.1).

More complex receptors, known as sense cells, consist of modified epithelial cells able to detect stimuli. These form synaptic connections with their sensory neurones which transmit impulses to the CNS, for example mammalian taste buds.

The most complex receptors are **sense organs** such as the eye and ear. These are composed of a large number of sense cells, sensory neurones and associated accessory structures. In the eye there are two types of sense cells, rods and cones, many connecting neurones and many accessory structures such as the lens and iris.

A classification of receptors based on the type of stimulus detected is shown in table 17.6.

Animals only detect stimuli existing in one of the forms of energy shown in table 17.6. Structures transforming stimulus energy into electrical responses (nerve impulses) in axons are known as **transducers** and, in this respect, receptors act as **biological transducers**.

All receptors transform the energy of the stimulus into an electrical response which initiates nerve impulses in the neurone leaving the receptor. Thus receptors **encode** a variety of stimuli into nerve impulses which pass into the

Table 17.6 Types of receptors and the stimuli detected by them.

Type of receptor	*Type of stimulus energy*	*Nature of stimulus*
photoreceptor	electromagnetic	light
electroreceptor	electromagnetic	electricity
mechanoreceptor	mechanical	sound, touch, pressure, gravity
thermoreceptor	thermal	temperature change
chemoreceptor	chemical	humidity, smell, taste

central nervous system where they are decoded and used to produce the required responses.

17.4.1 The mechanism of transduction

Transduction is the name given to the process by which a receptor converts a stimulus into a nerve impulse. The stimulus may be light, electricity, touch, etc., as shown in table 17.6, but the nerve impulse is always the same in nature.

All sensory cells are excitable cells, and they share with nerve cells and muscle cells the ability to respond to an appropriate stimulus by producing a rapid change in their electrical properties. When not stimulated, sensory cells are able to maintain a resting potential as described in section 17.1.1. They respond to a stimulus by producing a change in membrane potential. Bernard Katz, in 1950, using a specialised stretch receptor known as a muscle spindle, was able to demonstrate the presence of a depolarisation in the immediate region of the sensory nerve ending in the muscle spindle. This localised depolarisation is found only in the sensory cell and is known as the **generator potential**. Subsequent investigations involving intracellular recordings, made by penetrating the membranes of receptor cells in muscle spindles and the mechanoreceptors of the skin (the Pacinian corpuscles), have revealed the following information about transduction:

- the generator potential results from the stimulus producing an increase in the permeability of the sensory cell membrane to sodium and potassium ions which flow down their electrochemical gradients;
- the magnitude of the generator potential varies with the intensity of the stimulus;
- when the generator potential reaches a certain threshold value it gives rise to an action potential (fig 17.30);
- the frequency of the nerve impulses in the sensory axon is directly related to the intensity of the stimulus.

17.4.2 Properties of receptors

There are various ways in which the effectiveness of receptors can be increased. Some of these are described below.

Sensory cells with various thresholds

Some sense organs, such as stretch receptors in muscle, are composed of many sense cells which have a range of thresholds. If a cell has a low threshold it responds to a weak stimulus. As the strength of the stimulus increases the cell can respond by producing an increasing number of impulses in the sensory neurone leaving the cell. However, at a given point saturation occurs, and the frequency of impulses in the neurone cannot be increased. A further increase in intensity of stimulus will then excite sense cells with higher thresholds, and these too will produce a frequency of nerve impulses which is proportional to the intensity of the stimulus. In this way the range of the receptor is extended (fig 17.31).

Adaptation

Most receptors initially respond to a strong constant stimulus by producing a high frequency of impulses in the sensory neurone. The frequency of these impulses gradually declines and this reduction in response, with time, is called **adaptation**. For example, on entering a room you may immediately notice a clock ticking but after a while become unaware of its presence. The rate and extent of adaptation in a receptor cell is related to its function and there are two types, rapidly and slowly adapting receptors.

Rapidly adapting receptors (phasic receptors) respond to changes in stimulus level by producing a high frequency of impulses at the moments when the stimulus is switched 'on' *or* 'off'. For example, the Pacinian corpuscle and other receptors concerned with touch and the detection of sudden changes act in this way. They register *change* in the stimulus.

Slowly adapting receptors (tonic receptors) register a constant stimulus with a slowly decreasing frequency of impulses.

Adaptation is thought to be related to a decrease in the permeability of the receptor membrane to ions due to sustained stimulation. This progressively reduces the size and duration of the generator potential and, when this falls below the threshold level, the sensory neurone ceases to fire.

The advantage of adaptation of sense cells is that it provides animals with precise information about *changes* in the environment. At other times the cells do not send signals, thus preventing overloading of the central nervous system with irrelevant and unmanageable information. This contributes to the overall efficiency and economy of the nervous system and enables it to ignore unchanging background information and to concentrate on monitoring aspects of the environment which have most survival value.

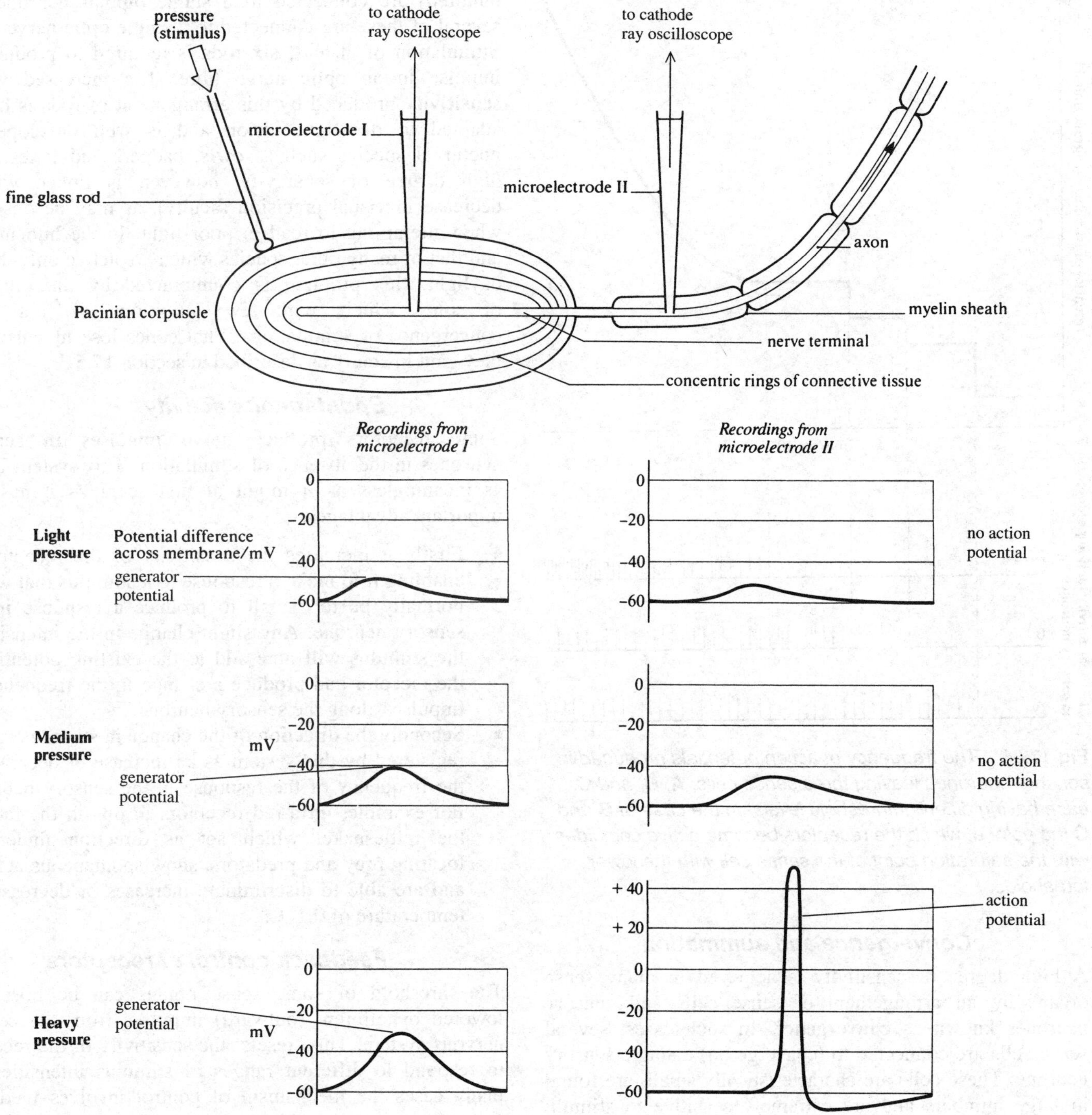

Fig 17.30 *The electrical activity recorded by two microelectrodes and inserted into (I) the axon within a Pacinian corpuscle and (II) the axon of the sensory neurone leaving the corpuscle. As the pressure on the fine glass rod (the stimulus) is increased, the size of the generator potential increases and at a certain threshold triggers an action potential in the sensory neurone.*

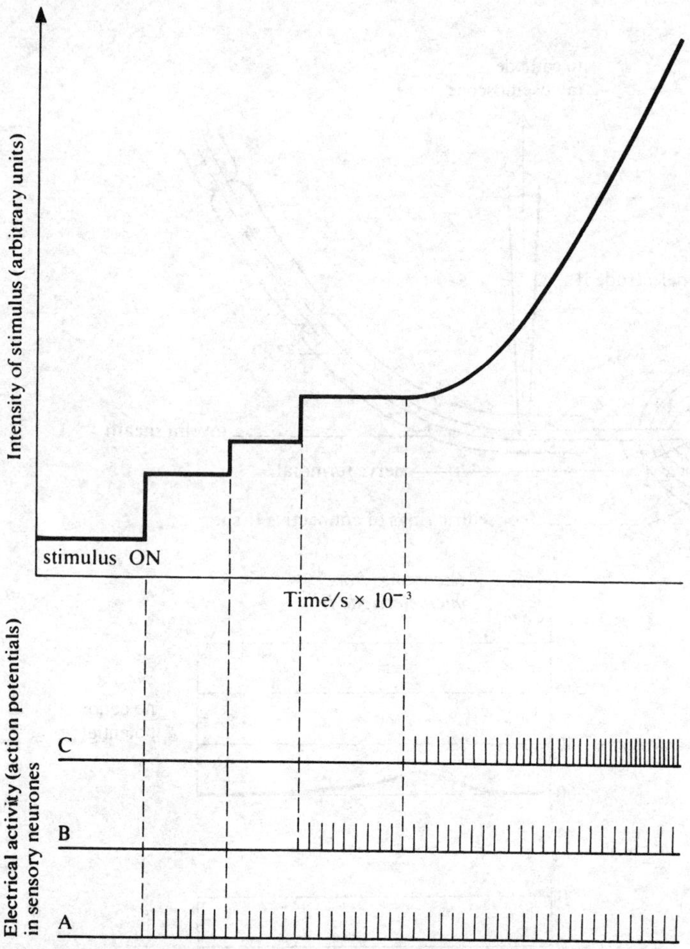

Fig 17.31 *The frequency of action potentials produced in sensory neurones leaving three sense cells, **A**, **B**, and **C**, each having different threshold levels. In the case of **B** and **C** the point at which the receptors become active coincides with the saturation point of the sense cell with the lower threshold.*

Convergence and summation

A high degree of sensitivity is achieved in many sense organs by an arrangement of sense cells and sensory neurones known as **convergence**. In such cases, several sense cells are connected to (converge on) a single sensory neurone. These cells are characteristically small, are found in large numbers and are extremely sensitive to stimuli. Whilst the effect of a stimulus on a single one of these cells would not produce a response in the sensory neurone, the combined effect of the simultaneous stimulation of several cells is cumulative. This cumulative stimulatory effect produced in the sensory neurone is known as **summation** and is similar in function to the summative effect described for synapses in section 17.1.2 and effectors in section 17.5.

A good example of convergence and summation is provided by rod cells of the retina in the eye. Some of these cells are capable of detecting a single quantum of light, but the generator potential produced is inadequate to produce an action potential in a neurone of the optic nerve. However, several rods (ranging from two or three to several hundred) are connected to a single bipolar neurone and several of these are connected to a single optic nerve fibre. Stimulation of at least six rods is required to produce an impulse in an optic nerve fibre. The increased **visual sensitivity** produced by this arrangement of rods is highly adapted to dim-light vision and is well developed in nocturnal species such as owls, badgers and foxes. This high degree of sensitivity, however, is linked with a decrease in visual precision (acuity), as may be observed when attempting to read in poor light. In the human eye, and that of many other species which are active only during daylight, this problem is counteracted by the presence of cones which, with few exceptions, do not show convergence or summation. What cones lose in sensitivity they gain in acuity as described in section 17.5.3.

Spontaneous activity

Some receptors produce nerve impulses in sensory neurones in the absence of stimulation. This system is not as meaningless as it might at first seem as it has two important advantages.

- Firstly it increases the sensitivity of the receptor by enabling it to make a response to a stimulus that would normally be too small to produce a response in the sensory neurone. Any slight change in the intensity of the stimulus will now add to the existing potential in the receptor and produce a change in the frequency of impulses along the sensory neurone.
- Secondly the direction of the change in stimulus can be registered by this system as an increase or decrease in the frequency of the response in the sensory neurone. For example, infra-red receptors in pits in the face of the rattlesnake, which act as direction finders in locating prey and predators, show spontaneous activity and are able to discriminate increases or decreases in temperature of 0.1 °C.

Feedback control of receptors

The threshold of some sense organs can be raised or lowered by efferent (outward) impulses from the central nervous system. This 'resets' the sensitivity of the receptor to respond to different ranges of stimulus intensities. In many cases the mechanism of control involves feedback from the receptors, which produces changes in accessory structures enabling the receptor cells to function over a new range. This occurs, for example, in the iris of the eye.

17.5 Structure and function of receptors

17.5.1 Mechanoreceptors

Mechanoreceptors are considered the most primitive type of receptors and may respond to a range of mechanical stimuli such as touch, pressure, vibration and stretching.

Touch, pressure and vibration

The difference between touch and pressure is one of degree, and the detection of these stimuli depends on the position of the receptors within the skin. Touch receptors are also found in other regions of the body and account for the increased sensitivity in these regions. For example, two stimuli may be resolved by the tip of the tongue when 1 mm apart whereas this is only possible at a distance of 60 mm in the middle of the back.

Specialised sense organs known as **Meissner's corpuscles** are situated immediately beneath the epidermis in the skin and respond to touch (fig 17.32). They consist of a single twisted ending of a neurone enclosed within a fluid-filled capsule. Another type of receptor, the **Pacinian corpuscle** (fig 17.30), is found in the skin, joints, tendons, muscles and gut area, and consists of the ending of a single neurone surrounded by many layers of connective tissue. Pacinian corpuscles respond to pressure. Other mechanoreceptors are also shown in fig 17.32.

All touch and pressure receptors are thought to produce a generator potential as a result of deformation of the receptor membrane leading to an increase in the permeability of the receptor cell to ions.

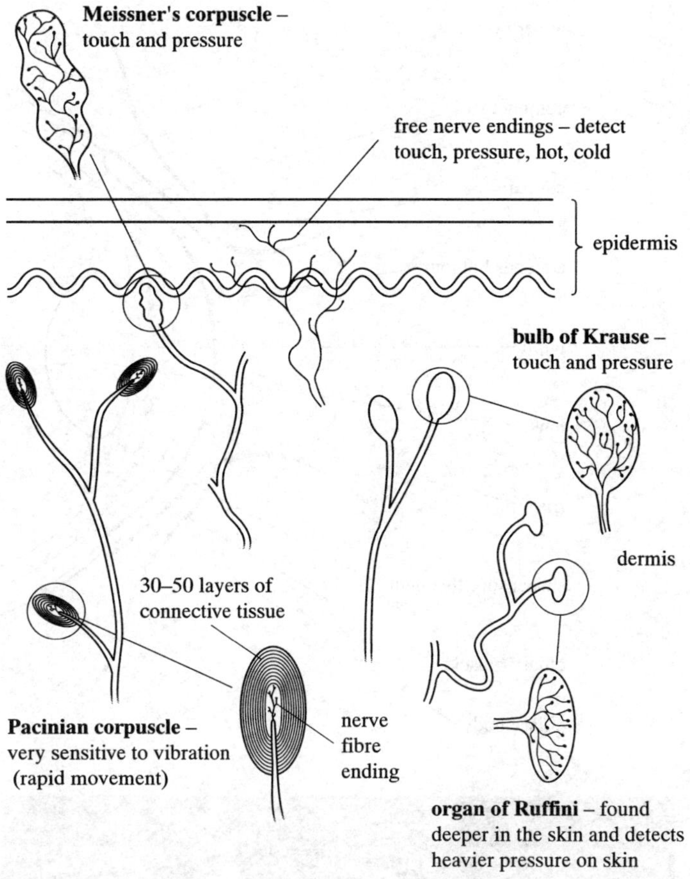

Fig 17.32 *Five types of touch receptor found in the skin.*

Muscle spindle

Proprioreceptors are stretch receptors, sensitive to the position and movement of parts of the body. They respond to changes in the state of contraction of muscles and act as stretch receptors in all activities associated with the control of muscular contraction. Specialised proprioceptors called muscle spindles have been found in the muscles of mammals and other animals. The muscle spindle has three main functions, one static and two dynamic:

- to provide information to the central nervous system on the state and position of muscles and structures attached to them (a static function);
- to initiate reflex contraction of the muscle and return it to its previous length when stimulated by a load (a dynamic function);
- to alter the state of tension in the muscle and reset it to maintain a new length (a dynamic function).

The structure and function of the muscle spindle is described in detail in section 18.4.4.

17.5.2 Thermoreceptors

The many free nerve endings in the skin are the major structures detecting temperature (fig 17.32). There are separate hot and cold receptors.

17.5.3 The eye

The eye is a sense organ which receives light of various wavelengths reflected from objects at varying distances, the **visual field**, and converts it into electrical impulses. Optic nerves transmit these impulses to the brain where an image of remarkable precision is perceived.

Light travels as waves of electromagnetic radiation and the wavelengths perceived by the human eye occupy a narrow range, the **visible spectrum**, from 380–760 nm. Light is a form of energy and is emitted and absorbed in discrete packets called **quanta** or **photons**. The wavelengths of the visible spectrum carry sufficient energy in each quantum of radiation to produce a photochemical response in the sense cells of the eye.

The camera and the eye work on the same basic principles. These are:

- controlling the amount of light entering the structure;
- focusing images of the external world by means of a lens system;
- registering the image on a sensitive surface;
- processing the 'captured' images to produce a pattern which can be 'seen'.

Structure and function of the human eye

The eyes are held in protective bony sockets of the skull called **orbits** by four **rectus** muscles and two **oblique** muscles which control eye movement. Each human eyeball

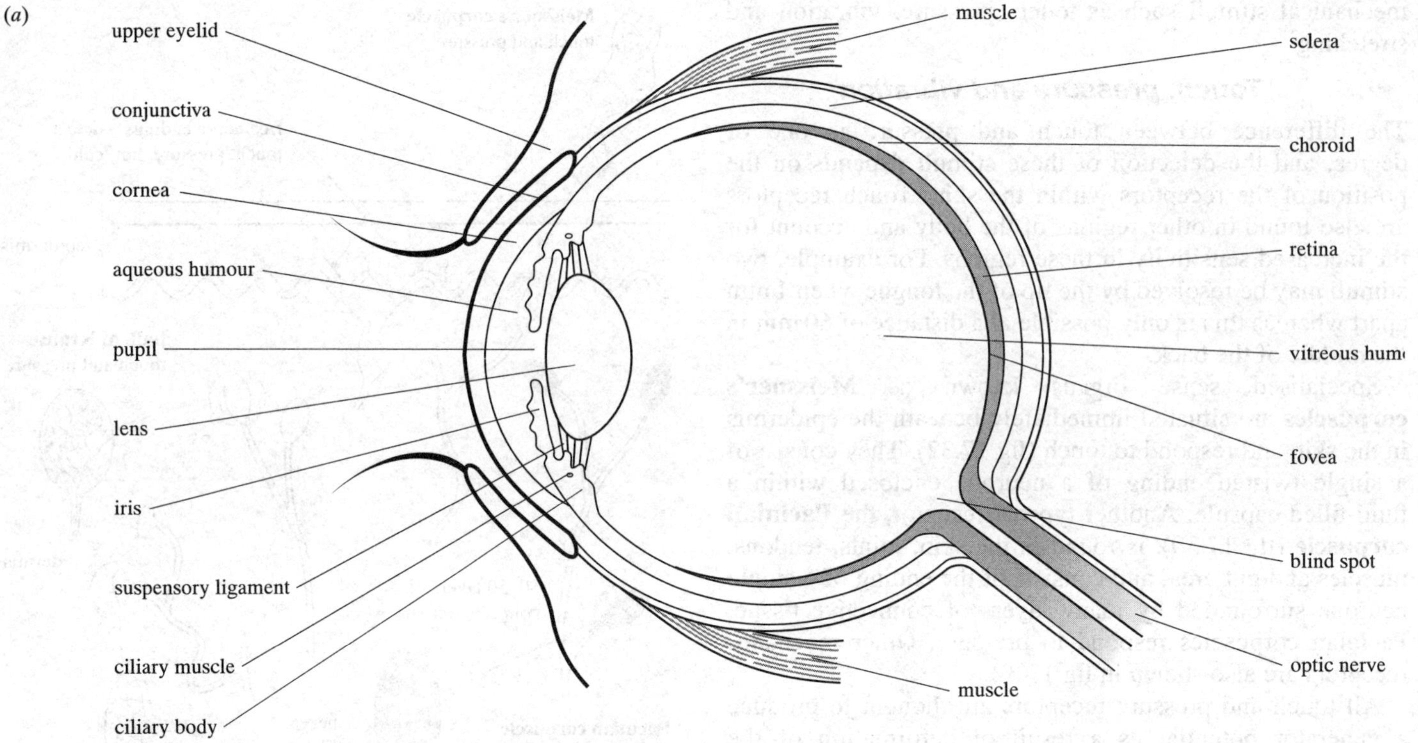

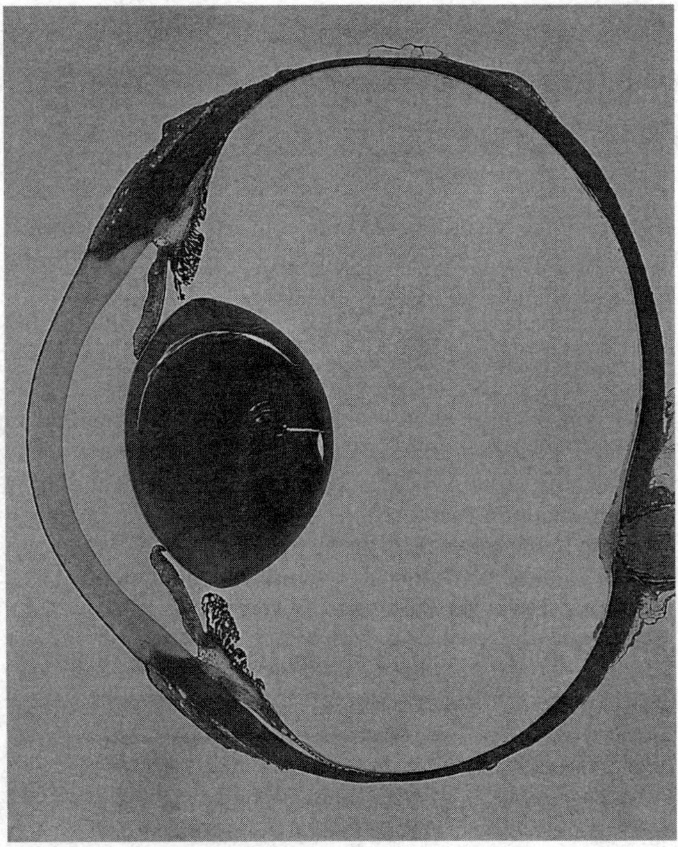

Fig 17.33 *Structure of the human eye: (a) diagram of section through eye, (b) section.*

is about 24 mm in diameter and weighs 6–8 g. Most of the eye is composed of 'accessory structures' concerned with bringing light to the photoreceptor cells which are situated in the innermost layer of the eye, the **retina**. The eye is composed of three concentric layers: the sclera (sclerotic coat) and cornea; the choroid, ciliary body, lens and iris; and the retina (fig 17.33). The eye is supported by the hydrostatic pressure (3.3 kPa) of the aqueous and vitreous humours.

The structure of the human eye is shown in fig 17.33 and brief notes on the function of the various parts are given below.

Sclera – external covering of eye; very tough, containing collagen fibres, protects and maintains shape of eyeball.

Cornea – transparent front part of the sclera; the curved surface acts as the main structure refracting (bending) light towards the retina.

Conjunctiva – thin transparent layer of cells protecting the cornea and continuous with the epithelium of eyelids; the conjunctiva does not cover the part of the cornea over the iris.

Eyelid – protects the cornea from mechanical and chemical damage and the retina from bright light by reflex action.

Choroid – rich in blood vessels supplying the retina, and covered with black pigment cells to prevent reflection of light within the eye.

Ciliary body – at the junction of sclera and cornea; contains tissue, blood vessels and ciliary muscles.

Ciliary muscles – circular sheet of smooth muscle fibres that form bundles of circular and radial muscles which alter the shape of the lens during accommodation.

Suspensory ligament – attaches the ciliary body to the lens.

Lens – transparent, elastic biconvex structure; provides fine adjustment for focusing light on to the retina and separates the aqueous and vitreous humours.

Aqueous humour – clear solution of salts secreted by the ciliary body, finally draining into the blood through a canal.

Iris – circular, muscular diaphragm containing the pigment which gives the eye its colour; it controls the amount of light entering the eye.

Pupil – a hole in the iris; all light enters the eye through this.

Vitreous humour – clear semi-solid substance supporting the eyeball.

Retina – contains the photoreceptor cells, rods and cones, and cell bodies and axons of neurones supplying the optic nerve.

Fovea – most sensitive part of retina, contains cones only; most light rays are focused here. Less than 0.5 mm in diameter.

Optic nerve – bundle of nerve fibres carrying impulses from the retina to the brain.

Blind spot – point where the optic nerve leaves eye; there are no rods or cones here, therefore it is not light-sensitive.

> **17.6** List, in order, the structures through which light passes before striking the retina.

Accommodation

Accommodation is the reflex mechanism by which light rays from an object are brought to focus on the retina. It involves two processes and these will be considered separately.

Reflex adjustment of pupil size. In bright light the circular muscle of the iris diaphragm contracts, the radial muscle relaxes, the pupil becomes smaller and less light enters the eye, preventing damage to the retina (fig 17.34). In poor light the opposite muscular contractions and relaxations occur. The added advantage of reducing the pupil size is that this increases the **depth of focus** of the eye.

Refraction of light rays. Light rays from distant objects (more than about 6 m away) are parallel when they strike the eye. Light rays from near objects (less than 6 m away) are diverging (spreading out) when they strike the eye. In both cases the light rays must be **refracted** or bent to focus on the retina, though refraction must be greater for light from near objects. The normal eye is able to accommodate light from objects from about 25 cm to infinity. Refraction occurs when light passes from one medium into another with a different refractive index, and this occurs at the air–cornea surface and at the lens.

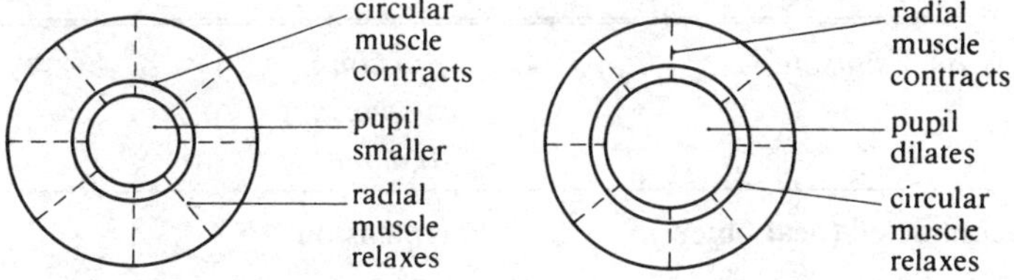

Fig 17.34 *The iris/pupil response to variations in light intensity.*

The degree of refraction at the corneal surface cannot be varied and depends on the angle at which the light strikes the cornea (which, in turn, depends upon the distance of the object from the cornea). Most refraction occurs here, and consequently the function of the lens is to produce the final refraction that brings light to a sharp focus on the retina; this is regulated by the ciliary muscles. The state of contraction of the ciliary muscles changes the tension on the suspensory ligaments. This acts on the natural elasticity of the lens which causes it to change its shape (radius of curvature) and thus the degree of refraction. As the radius of curvature of the lens decreases, it becomes thicker and the amount of refraction increases. The complete relationship between these three structures and refraction is shown in table 17.7. Fig 17.35 shows the changes occurring in the eye during accommodation to light from distant and near objects.

The image produced by the eye on the retina is inverted and reversed but the mental image is perceived in the correct position because the brain learns to accept an inverted reversed image as normal.

Structure of the retina

The photoreceptor cells of the retina face towards the choroid layer. The cells are therefore covered by the cell bodies and axons linking the photoreceptor cells to the brain as shown in fig 17.36.

The retina is composed of three layers of cells each containing a characteristic type of cell. First there is the **photoreceptor layer** (outermost layer) containing the photosensitive cells, the **rods** and **cones**, partially embedded in the pigmented epithelial cells of the choroid. Next is the **intermediate layer** containing bipolar neurones with synapses connecting the photoreceptor layer to the cells of the third layer. Cells called **horizontal** and **amacrine cells** found in this layer enable lateral inhibition to occur (see later). The third layer is the **internal surface layer** containing **ganglion cells** with dendrites in contact with bipolar neurones and axons of the optic nerve.

Structure and function of rods and cones

Rods and cones have an essentially similar structure as shown in fig 17.36, and their photosensitive pigments are attached to the outer surfaces of the membrane in the outer

Table 17.7 Relationship between structures changing the shape of the lens and the degree of refraction.

Ciliary muscle	*Tension in suspensory ligament*	*Shape of lens*	*Refraction*
contracted (near object)	no tension	more curved, thick	increased
relaxed (distant object)	taut	less curved, thin	decreased

segment. They have four similar regions whose structure and function are summarised below.

Outer segment. This is the photosensitive region where light energy is converted into a generator potential. The entire outer segment is composed of flattened membranous vesicles containing the photosensitive pigments. Rods contain 600–1000 of these vesicles stacked up like a pile of coins and they are enclosed by an outer membrane. Cones are so-called because their outer segments are essentially cone-shaped. These have fewer membranous vesicles which are formed by repeated infoldings of the outer membrane.

Constriction. The outer segment is almost separated from the inner segment by an infolding of the outer membrane. The two regions remain in contact by cytoplasm and a pair of cilia which pass between the two. These cilia consist of nine peripheral fibres only, the usual central two being absent. These no longer have a function.

Inner segment. This is an actively metabolic region. It is packed with mitochondria, which make energy available for visual processes, and polyribosomes for the synthesis of proteins involved in the production of the membranous vesicles and visual pigment. The nucleus is situated in this region.

Synaptic region. Here the cells form synapses with bipolar cells. Some bipolar cells may have synapses with several rods. This is called synaptic convergence, and whilst it lowers visual acuity it increases visual sensitivity. This is explained in section 17.4.2 under the heading 'Convergence and summation'. Other bipolar cells link *one* cone to *one* ganglion cell and this gives the cones greater visual acuity than rods. Horizontal cells and amacrine cells link certain numbers of rods together and cones together. This allows a certain amount of processing of visual information to occur before it leaves the retina. For example these cells are involved in lateral inhibition (see later).

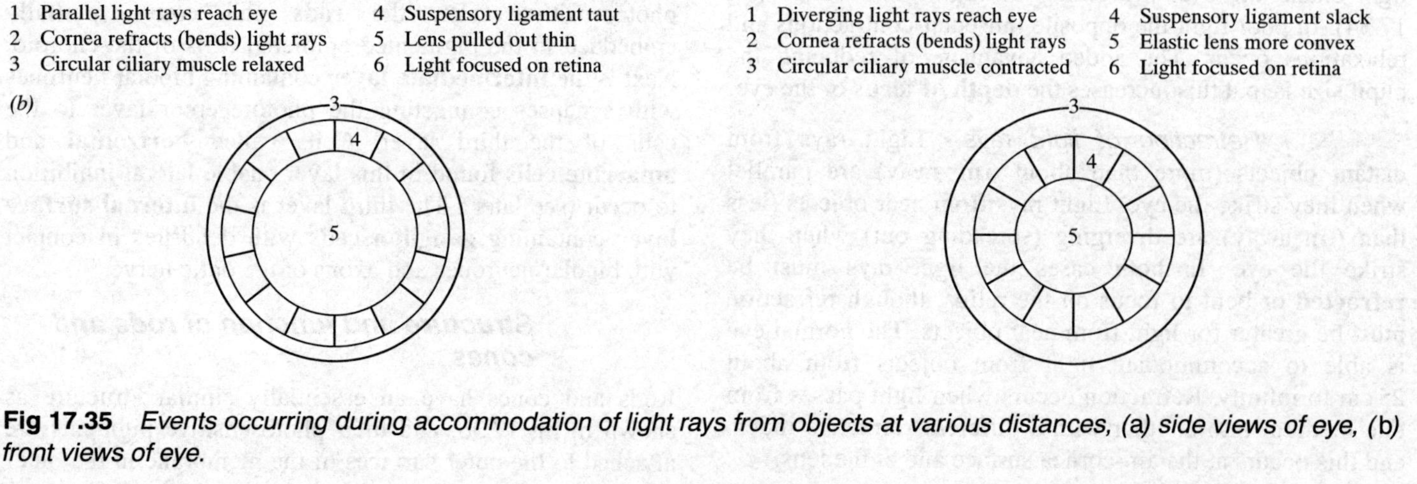

Fig 17.35 *Events occurring during accommodation of light rays from objects at various distances, (a) side views of eye, (b) front views of eye.*

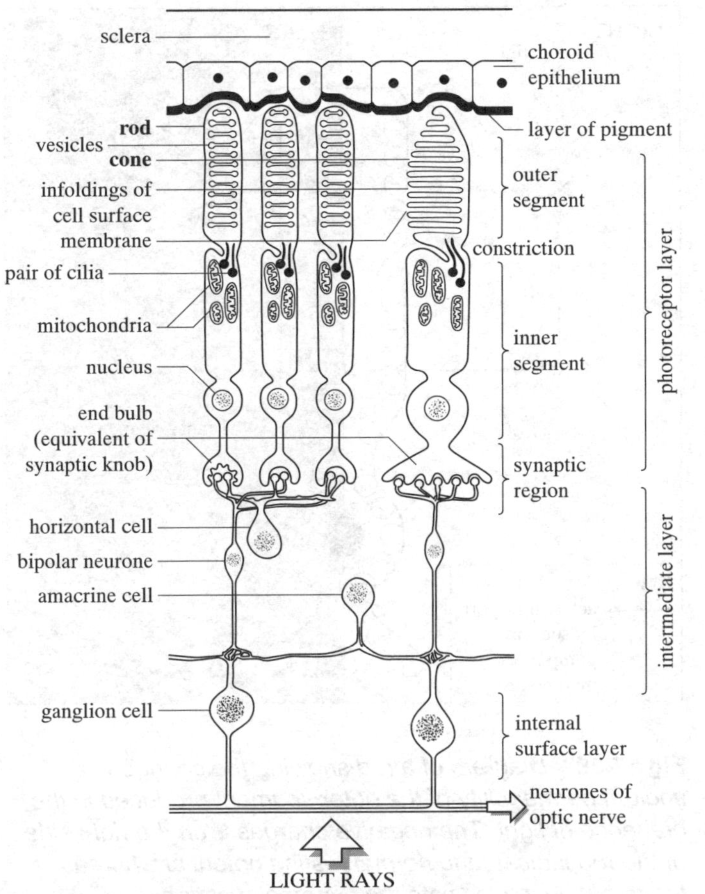

Fig 17.36 *Diagrammatic section through the retina of the eye showing the ultrastructure of a rod and a cone. Connections between the sensory cells and the neurones of the optic nerve are shown in the inner segment. Light rays must pass through the ganglion cells and the intermediate layers before reaching the rods and cones.*

Differences between rods and cones

Rods are more numerous than cones, 120 million as opposed to 6 million, and have a different distribution. The rods are distributed uniformly throughout the retina except at the fovea, where the cones have their greatest concentration (50 000 per square millimetre). Since the cones are tightly packed together at the fovea, this gives them higher visual acuity (see below).

Rods are much more sensitive to light than cones and respond to lower light intensities. Rods only contain one visual pigment and, being unable to discriminate colour, are used principally for night vision. Each cone contains one of three visual pigments which enables them to differentiate colours. Cones are used mainly in daylight.

The rods have a lower visual acuity because they are less tightly packed together and they undergo synaptic convergence, but this latter point gives them increased collective sensitivity required for night vision (see below).

Sensitivity and acuity

Acuity refers to the precision, or sharpness, of the image we see. Thus if one part of the retina could distinguish between two separate, but closely placed points, and another part could not, the former part would have greater acuity. Acuity is greatest at the fovea, which is normally at the centre of our field of vision, and gradually decreases towards the edges of our field of vision. We move our eyes in order to maintain the image of what we are interested in on the fovea. The fovea contains only cones, and about 90% of the cones are located there. Many of the cones in the fovea synapse with only one bipolar neurone, as shown in fig 17.36, and the bipolar neurones synapse with only one ganglion cell. This 1:1 relationship provides maximum possible acuity because each part of the image is being detected by a different cell and there is no 'blurring' or combining of information. Acuity also increases as the number of cones in a given area increases (rather like a photograph where the closer the individual dots that make up the picture, the sharper the picture) so the fact that they are tightly packed together is an advantage. Cones further away from the fovea (and some in it) combine with two or more bipolar neurones, so some acuity is lost.

Overall there are about 120 million rods and 6 million cones in each retina, but only about 1.2 million ganglion cells. Thus convergence occurs (section 17.4.2) with about 105 photoreceptor cells to 1 ganglion cell. The degree of convergence is actually much greater for rods than cones, so they have much less acuity than cones. The greater the convergence, the lower the acuity. However, **sensitivity** to light increases with convergence (see question 17.7 below). Many rods synapse to a single bipolar cell and many bipolar cells synapse to one ganglion cell. In dim light, vision is poorest at the fovea, where most photoreceptor cells are cones, and best at the edges of the field of vision. You may have noticed that you can see more stars on a dark night out of the 'corners' of your eyes. In dim light only the rods are activated, so colour vision is lost as well as acuity.

17.7 Explain why synaptic convergence should increase visual sensitivity.

17.8 Explain in your own words why objects are seen more clearly at night by not looking directly at them.

The mechanism of photoreception

Rods contain the light-sensitive pigment **rhodopsin**, which is attached to the outer surface of the vesicles. Rhodopsin is a molecule formed by the combination of a protein called **scotopsin** with a small light-absorbing molecule called **retinene (retinal)**. Retinene is a carotenoid molecule and is a derivative of vitamin A. It exists in two isomeric forms according to the light conditions as shown in fig 17.37.

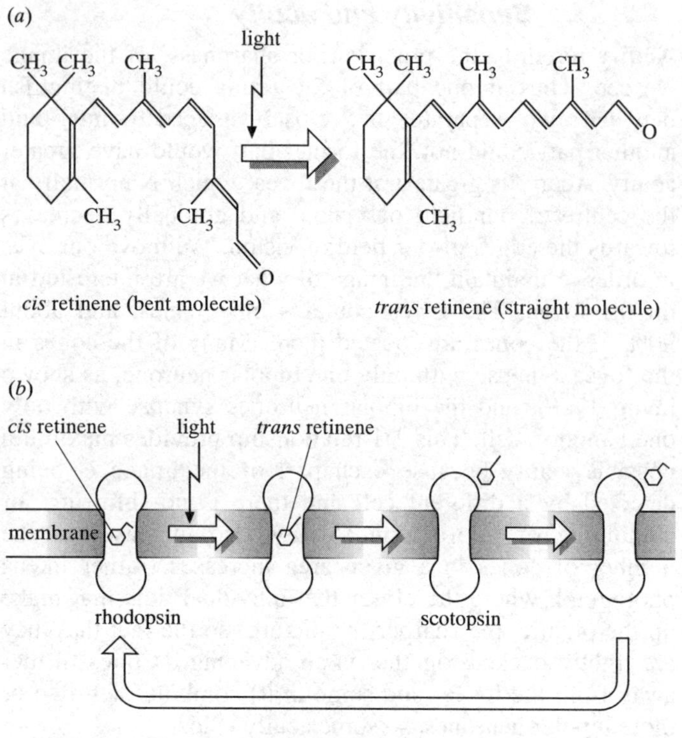

Fig 17.37 *(a) The action of light changes the structure of retinene from the* cis *isomer to the* trans *isomer (*b*) bleaching and regeneration of rhodopsin.*

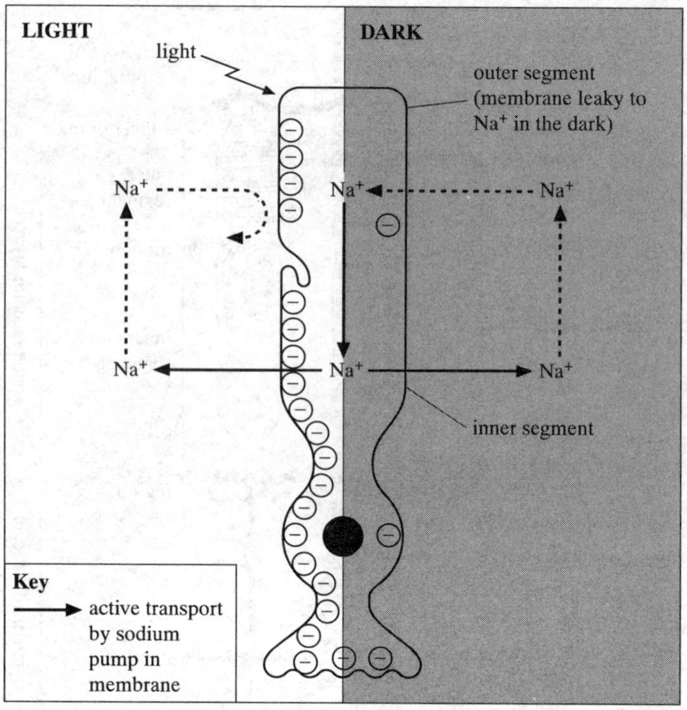

Fig 17.38 *Diagram of a rod showing the changes in sodium permeability of the outer segment produced in the presence of light. The negative charges ⊖ on the right side of the rod indicate the normal resting potential whereas those on the left indicate the hyperpolarisation.*

When a molecule of rhodopsin is exposed to light it is known that one photon of light will produce the above isomeric change. Once the change has happened the rhodopsin breaks down into retinene and scotopsin, a process known as **bleaching**.

$$\text{rhodopsin} \xrightarrow{\text{bleaching}} \text{retinene} + \text{scotopsin}$$

Rhodopsin is reformed immediately in the absence of further light stimulation (and therefore in the dark). *Trans* retinene is first converted into *cis* retinene and then recombined with scotopsin (fig 17.37*b*). This process is called **dark adaptation** and in total darkness it takes about 30 minutes for all rods to adapt and the eyes to achieve maximum sensitivity again.

How does the change in rhodopsin in rods in the light lead to production of an action potential? This involves changes in the outer segment and inner segment membranes of the rod (fig 17.38). In the inner segment of the rod is a sodium pump which continuously pumps out sodium ions. In the dark the outer segment membrane is leaky and allows the sodium back in again by diffusion, reducing the negative charge inside the cell (−40 mV instead of the normal −70 mV of most cells). In the light, however, the permeability of the outer segment membrane to sodium ions decreases whilst the inner segment continues to pump out sodium ions, thus making the inside of the rod more negative (fig 17.38). This causes hyperpolarisation of the rod. This situation is exactly opposite to the effect normally found in sensory receptors where the stimulus produces a depolarisation and not a hyperpolarisation. The hyperpolarisation *reduces* the rate of release of excitatory transmitter substance from the rod which is released at its maximum rate during darkness. The bipolar neurone linked by synapses to the rod cell also responds by producing a hyperpolarisation, but the ganglion cells of the optic nerve supplied by the bipolar neurone respond to this by producing an action potential.

Role of horizontal cells and amacrine cells

Horizontal cells synapse with several bipolar neurones (fig 17.36). They are responsible for a phenomenon called **lateral inhibition**. This increases both sensitivity and acuity of vision. If the cells receive stimuli from two rods which are of equal intensity they cancel out (inhibit) the stimuli. They therefore enhance contrast between areas that are weakly stimulated and areas that are strongly stimulated. This makes features such as edges of objects stand out more clearly. Amacrine cells are stimulated by bipolar neurones and synapse with ganglion cells. They transmit information about changes in the level of illumination.

Table 17.8 Colours of the visible spectrum and approximate ranges of their wavelengths.

Colour	*Wavelength/nm*
red	above 620
orange	590–620
yellow	570–590
green	500–570
blue	440–500
violet	below 440

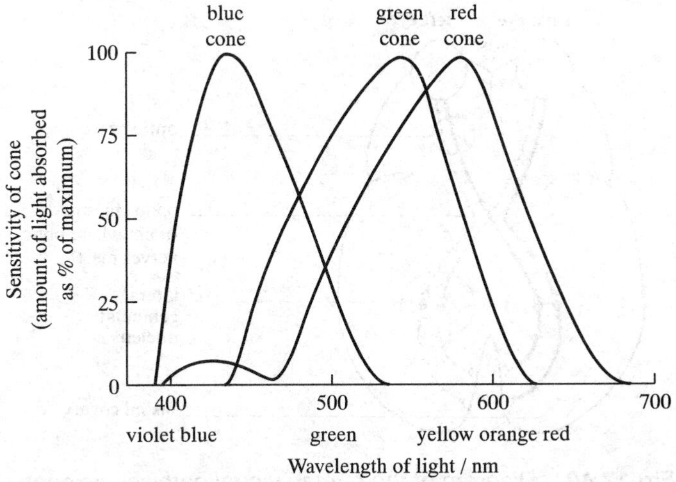

Fig 17.39 *Sensitivities of the three types of cone to light of different wavelengths (colours).*

Colour vision

The human eye absorbs light from all wavelengths of the visible spectrum and perceives these as six colours broadly associated with particular wavelengths as shown in table 17.8. Colour as such does not exist, only different wavelengths of light. Colour is an 'invention' of the brain. There are three types of cone each possessing a different pigment which investigations have shown absorb light of different wavelengths. These are 'red', 'green' and 'blue' cones, which are most sensitive to red, green and blue light respectively. In laboratory tests it has been shown that all the different colours can be seen by using different combinations of pure red, green and blue light.

> **17.9** Using table 17.8 suggest what colour would be seen in the following situation. A person places a green filter (average wavelength 530 nm) over one eye and a red filter (average wavelength 620 nm) over the other eye and looks at an object.

The most generally accepted theory of colour vision is the **trichromacy theory**, which states that different colours are produced by the degree of stimulation of each type of cone. For example, equal stimulation of all cones produces the colour sensation of white. Fig 17.39 shows the sensitivity of the three types of cone to different colours of light. You can see that, although the cones are described as blue, green and red cones, this refers to their maximum sensitivity. They are also less sensitive to a limited range of other colours, and their sensitivities overlap. Different colours are therefore produced by different degrees of stimulation of the three cones. For example, orange light stimulates only green and red cones, blue light stimulates blue cones strongly and green cones weakly, and green light stimulates all three cones.

Although the trichromacy theory explains most of the experimental evidence obtained about colour vision, there are still a few facts that it does not explain. These are outside the scope of this book, but it is clear that further refinement of the theory is necessary.

The initial discrimination of colour occurs in the retina but the final colour perceived involves interpretation by the brain.

Colour-blindness. The complete absence of a particular type of cone or a shortage of one type can lead to various forms of colour-blindness or degrees of 'colour-weakness'. This is the inability to distinguish certain colours. For example, a person lacking red or green cones is 'red–green colour-blind' because they cannot distinguish between red and green, whereas a person with a reduced number of either cones will have difficulty in distinguishing a range of red–green shades. Colour-blindness, or its extent, is determined using test charts, such as the Ishihara test charts, composed of a series of dots of several colours. Some charts bear a number which a person with normal vision can perceive, whilst the colour-blind sufferer sees a different number or no number at all. Colour-blindness is a sex-linked recessive characteristic resulting in the absence of appropriate colour genes in the X chromosome. About 2% of men are red colour-blind, 6% are green colour-blind, but only 0.4% of women show any sign of colour-blindness.

Binocular vision and stereoscopic vision

Binocular vision occurs when the visual fields of both eyes overlap so that the fovea of both eyes are focused on the same object. It provides the basis of **stereoscopic vision**. Stereoscopic vision depends upon the two eyes producing slightly different images on the retina at the same time which the brain interprets as one image. The resolution of the two retinal images produced in stereoscopic vision occurs in the area of the brain called the **visual cortex**.

The more frontally the eyes are situated the greater the overlap between the eyes and the more that can be seen stereoscopically. For example, humans have a total visual field of 180° and a stereoscopic visual field of 140°. The

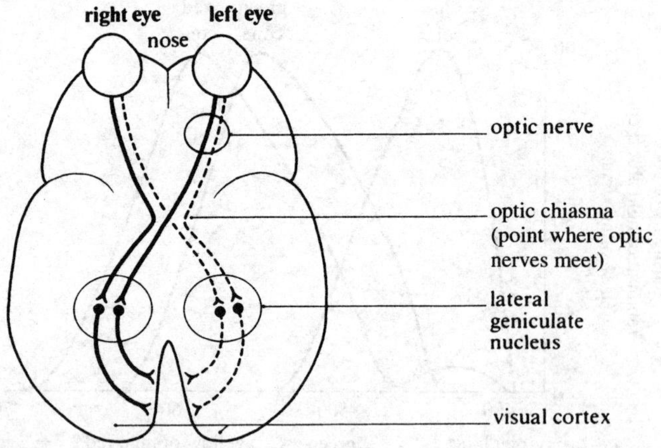

Fig 17.40 *Diagram of the human visual pathway as seen from the* underside of the brain.

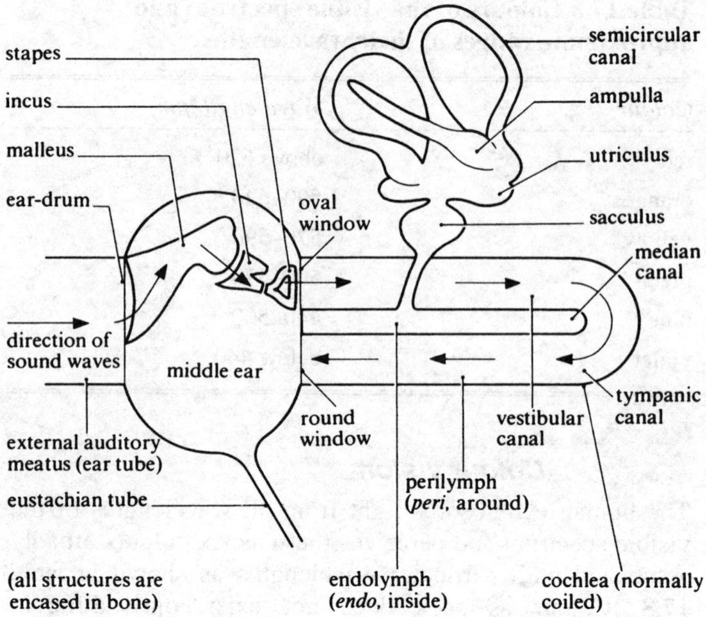

Fig 17.41 *Diagram showing the major structures in the mammalian ear involved in hearing and balance (not to scale).*

horse has laterally placed eyes with a limited forward stereoscopic visual field which it uses for viewing distant objects. For nearer objects the horse turns its head and uses monocular vision to examine details. Frontally placed eyes and centrally situated foveas, producing good visual acuity, are essential for good stereoscopic vision which provides an increased appreciation of size and perception of the depth and distance of objects. Stereoscopic vision is found mainly in predatory animals where it is vital when capturing prey by pouncing or swooping, as shown by members of the cat family, hawks and eagles. Animals which are hunted have laterally placed eyes giving wide visual fields but restricted stereoscopic vision. For instance the rabbit has a total binocular visual field of 360° and frontal stereoscopic vision of 26°.

The role of the brain

Nerve impulses generated in the retina are carried by the million or so neurones of the optic nerve to the primary visual area of the visual cortex, which is situated in the occipital lobe at the back of the brain (fig 17.26). Here each part of the retina, involving perhaps only a few rods and cones, is represented and it is here that the visual input is interpreted and we 'see'. However, what we see has meaning only after reference to other regions of the cortex and the temporal lobes, where previous visual information is stored and used in the analysis and identification of the present visual input (section 17.2.4). In humans, axons from the left side of the retina of both eyes, which see the right visual field, pass to the left visual cortex and axons from the right side of each retina, which see the left visual field, pass to the right visual cortex. The point where the axons from those regions of the retina closest to the nose cross is called the **optic chiasma** and is included in the visual pathway shown in fig 17.40. About 20% of the optic neurones do not pass to the visual cortex but enter the midbrain, where they are involved in reflex control of pupil size and eye movements.

17.5.4 Mammalian ear

The mammalian ear is a sense organ containing receptors sensitive to gravity, movement and sound. Movements and positions of the head relative to gravity are detected by the vestibular apparatus which is composed of the **semicircular canals**, the **utricle** and the **saccule**. All other structures of the ear are involved in receiving, amplifying and transducing sound energy into electrical impulses and producing the sensation of hearing in the auditory regions of the brain.

Structure and function of the ear

The ear consists of three sections specialising in different functions (fig 17.41). The **outer ear** consists of the **pinna**, strengthened by elastic cartilage, which focuses and collects sound waves into the **ear tube** (external auditory meatus); the sound waves cause the **tympanic membrane** (eardrum) to vibrate. In the **middle ear**, vibrations of the tympanic membrane are transmitted across to the membranous **oval window** by movement of the three ear ossicles, the **malleus**, **incus** and **stapes** (hammer, anvil and stirrup). A lever system between these bones and the relative areas of contact of the malleus with the tympanic membrane (60 mm^2) and the stapes with the oval window

($3.2\ mm^2$) amplifies the pressure on the tympanic membrane 22 times. Damage to the tympanic membrane, due to atmospheric pressure changes, is prevented by a connection between the air-filled middle ear and the pharynx, the **Eustachian tube**. Finally there is the **inner ear** which consists of a complex system of canals and cavities within the skull bone which contain a fluid called **perilymph**. Within these canals are membranous sacs filled with **endolymph** and sensory receptors. Auditory receptors are found in the cochlea, and balance receptors are found in the utricle and saccule and the ampullae of the semicircular canals. The perilymph is enclosed by the membranes of the oval window and round window.

The nature of sound

Sound is produced by the vibration of particles within a medium. It travels as waves consisting of alternating regions of high and low pressure and will pass through liquids, solids and gases. The distance between two identical points on adjacent waves is the **wavelength** and this determines the **frequency** of **vibrations** or **pitch** (whether it sounds high or low). The human ear is sensitive to wavelengths between 40 and 20 000 Hz (cycles per second). The audible range of dogs reaches 40 000 Hz, and that of bats 100 000 Hz. Human speech frequencies vary between 500 and 3000 Hz, and sensitivity to high frequencies decreases with age.

Tone depends upon the number of different frequencies making up the sound. For example, a violin and trumpet playing the same note, say middle C, produce the same fundamental frequency of 256 Hz but sound different. This is due to overtones or harmonics produced by the instrument which give it its distinctive quality or **timbre**. The same principle applies to the human voice and gives it its characteristic sound.

The **intensity** (loudness) of a sound depends upon the amplitude of the sound waves produced at the source and is a measure of the energy they contain.

Cochlea and hearing

The cochlea is a spiral canal 35 mm long and subdivided longitudinally by a membranous triangle into three regions as shown in fig 17.42.

Both the vestibular and tympanic canals contain perilymph, and the two canals are connected at the extreme end of the cochlea via a small hole. The median canal contains endolymph. The **basilar membrane** separates the median and tympanic canals and supports sensory hair cells that can be brought into contact with the **tectorial membrane** above. This unit, consisting of basilar membrane, sensory cells and tectorial membrane, is called the **organ of Corti** and is the region where transduction of sound waves into electrical impulses occurs.

Sound waves transmitted from the ear tube to the oval window produce vibrations in the perilymph of the vestibular canal and these are transmitted via **Reissner's membrane** to the endolymph in the median canal. From

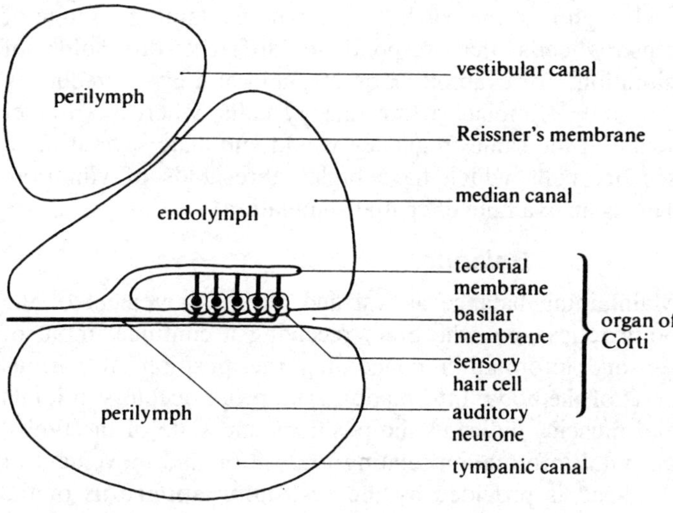

Fig 17.42 *Diagram of a TS cochlea showing the organ of Corti.*

there they are transferred to the basilar membrane and the perilymph in the tympanic canal, and are finally dissipated into the air of the middle ear as vibration of the round window.

The precise mechanism of transduction of pressure waves into nerve impulses is not known but is believed to involve relative movement of the basilar and tectorial membranes. Vibrations of the basilar membrane, induced by pressure waves, push the sensory hairs against the tectorial membrane and force the two membranes to slide past each other. The distortion produced in the sensory hairs due to the shearing forces causes a depolarisation of the sensory cells, the production of generator potentials and the initiation of action potentials in the axons of the auditory nerve.

Pitch and intensity discrimination

The ability to distinguish the pitch of a sound depends upon the frequency of the vibration-producing movement of the basilar membrane and stimulating sensory cells in a specific region of the organ of Corti. These cells supply a particular region of the auditory cortex of the brain where the sensation of pitch is perceived. The basilar membrane becomes broader and more flexible as it passes from the base of the cochlea to its apex and its sensitivity to vibration changes along its length so that only low-frequency sounds can pass to the apex. High-frequency (pitch) sounds stimulate the basilar membrane at the *base* of the cochlea and low frequency sounds stimulate it at the *apical end*. A pure sound, consisting of a single frequency, will only stimulate one small area of the basilar membrane whereas most sounds, containing several frequencies, simultaneously stimulate many regions of the basilar membrane. The auditory cortex integrates the stimuli from these various regions of the basilar membrane and a 'single' blended sound is perceived.

The intensity or loudness of the sound depends upon

each region of the basilar membrane containing a range of sensory cells that respond to different thresholds of vibration. For example, a quiet sound at a given frequency may only stimulate a few sensory cells, whereas a louder sound at the same frequency would stimulate several other sensory cells which have higher thresholds of vibration. This is an example of spatial summation.

Balance

Maintaining balance at rest and during movement of the body relies upon the brain receiving a continual input of sensory information concerning the position of various parts of the body. Information from proprioceptors in joints and muscles indicates the positions and state of the limbs, but vital information relating to position and movement of the head is provided by the **vestibular apparatus** of the ear, the **utricle**, **saccule** and **semicircular canals**.

The basic sensory receptors within these structures consist of cells which have hair-like extensions, **hair cells**, attached to dense structures supported in the **endolymph**. Movement of the head results in deflection of the hairs and production of a generator potential in the hair cells.

Regions of the walls of the utricle and saccule, called **maculae**, contain receptor cells which have their hair-like processes embedded in a gelatinous mass that contains granules of calcium carbonate. This mass is called an **otoconium** (fig 17.43). Otoconia respond to the pull of gravity acting at right-angles to the Earth's surface and are mainly responsible for detecting the direction of movement of the head with respect to gravity.

The utricle responds to vertical movements of the head and the otoconia produce maximum stimulation when pulling the receptor hairs downwards, such as when the body is upside down (fig 17.44).

The saccule responds to lateral (sideways) movement of the head. The hair cells of the saccule are horizontal when the head is upright. Tilting of the head to the left produces a differential response from left and right saccules. The left receives increased stimulation as the otoconia pull downwards on the hairs whereas decreased stimulation

Fig 17.43 *Scanning electron micrograph of the inner ear showing part of the region of the saccule known as the macula. This forms the organ of balance and consists of supporting cells (general background) on which sensory ciliated cells (hair-like structures) are found. When the head tilts sideways, calcium carbonate crystals called otoconia (angular structure between ciliated cells at bottom) are moved by the endolymph and stimulate the ciliated cells.*

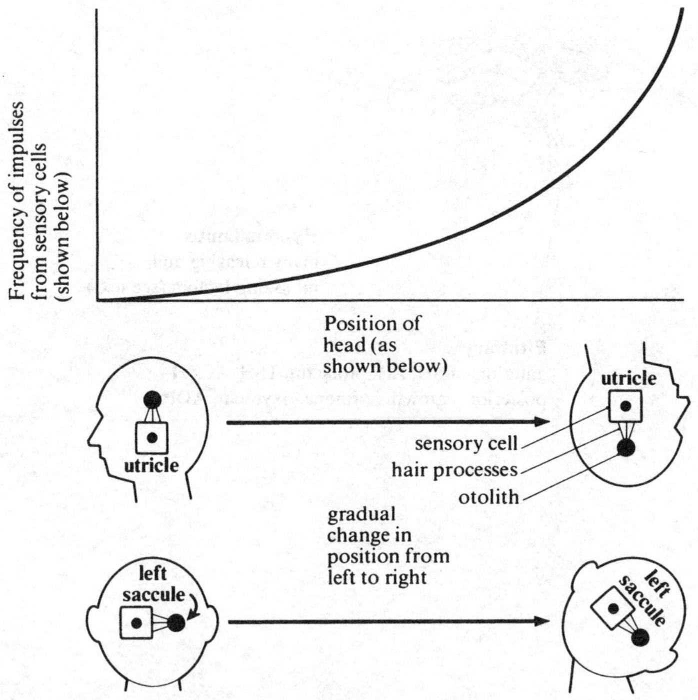

Fig 17.44 *Graph showing the effect of head position on the activity of receptor cells of the utricle and saccule.*

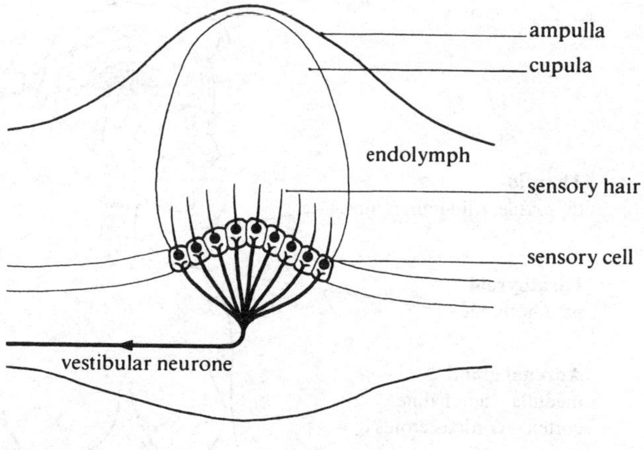

Fig 17.45 *Diagram showing a TS through the ampulla of a semicircular canal.*

occurs on the right. These displacements produce impulses passing to the cerebellum where the orientation of the head is perceived.

The three semicircular canals are arranged in three planes at right-angles to each other and detect the direction and rate of change of position of the head. At the base of each canal is a swelling, the **ampulla** containing a conical gelatinous structure, the **cupula**. This encloses the hair-like projections of the receptor cells. The cupula extends fully across the ampulla (fig 17.45). Rotational movement of the head, semicircular canals and cupula is resisted by the inertia of the endolymph which remains stationary. This produces a relative displacement of the cupula which is bent in the opposite direction to the head movement. The receptor cells respond by producing generator potentials leading to action potentials in the vestibular neurones. The direction and rate of displacement are both detected by the receptor cells. Linear acceleration is detected by both the maculae and cupulae.

17.6 The endocrine system

The endocrine system is made up of a number of glands called endocrine glands. A **gland** is a structure which secretes a specific chemical substance or substances. There are two types of gland in the body, exocrine glands and endocrine glands. An **exocrine gland** is one which secretes its product into a duct, for example the sweat gland which secretes sweat into tubes called sweat ducts that lead to the surface of the skin. An **endocrine gland** has the following characteristics:

- it secretes chemicals called **hormones**;
- it has no duct (a **ductless gland**), instead, the hormone is secreted directly into the bloodstream;
- it has a rich supply of blood with a relatively large number of blood vessels.

Some glands have both endocrine and exocrine functions, such as the pancreas, which secretes the hormones insulin and glucagon, and also secretes pancreatic juice containing digestive enzymes into the pancreatic duct that leads to the gut.

A hormone is a **chemical messenger**. It has the following properties:

- it travels in the blood;
- it has its effect at a site different from the site where it is made, called the **target**, hence the term messenger;
- it fits precisely into receptor molecules in the target like a key in a lock – it is therefore **specific** for a particular target;
- it is a small soluble organic molecule;
- it is effective in low concentrations.

The endocrine and nervous systems function in a coordinated way to maintain a homeostatic state within the body. The two systems are compared in table 17.1. As we shall see, the hypothalamus in the brain is an important link between them (see also section 17.2.4). Despite their obvious differences, both systems share a common feature in the release of chemical substances as a means of communication between cells. It is believed that the two systems originated and developed side by side as the needs of communication became more complex due to the increase in the size and complexity of organisms. In both cases the principal role of the systems is the coordination and control of many of the major physiological activities of organisms.

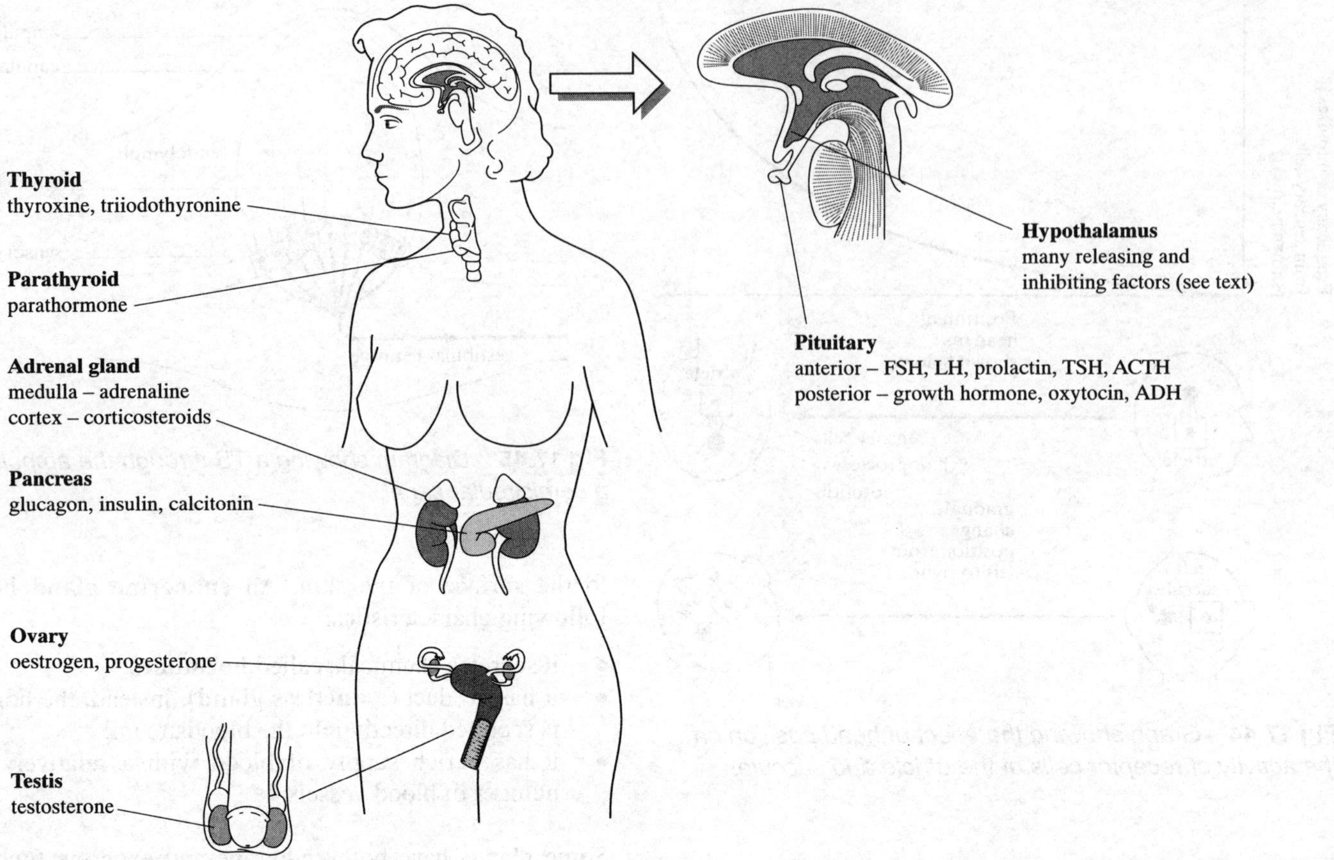

Fig 17.46 *Locations of the main endocrine organs in the human body.*

The major endocrine glands of the body are shown in fig 17.46 and their hormones and effects are summarised in table 17.9.

17.6.1 Mechanisms of hormone action

As summarised in table 17.10, all vertebrate hormones belong to one of four chemical groups:

- peptides and proteins;
- derivatives of amines, such as tyrosine;
- steroids;
- fatty acids.

Release of hormones

The mechanisms controlling the release of hormones by glands are as follows:

- presence of a specific metabolite in the blood, for example excess glucose in the blood causes the release of insulin from the pancreas which lowers the blood glucose level;
- presence of another hormone in the blood, for example many of the hormones released from the anterior pituitary gland are 'stimulating' hormones which cause the release of other hormones from other glands in the body;
- stimulation by neurones from the autonomic nervous system, for example adrenaline and noradrenaline are released from the cells of the adrenal medulla by the arrival of nerve impulses in situations of anxiety, stress and danger.

Negative feedback

In the first two cases above, the timing of hormone release and the amount of hormone released are regulated by feedback control. This is usually negative feedback control (rather than positive feedback control). A good example of negative feedback control is provided by thyroxine (see section 17.6.4).

The cascade effect

Hormones which are released by the presence of another circulating hormone are usually under the control of the hypothalamus and pituitary gland and the final response often involves the secretion of three separate hormones. A good example is the role of ACTH in cortisol production. This is shown in fig 17.47. The mechanism, known as the **cascade effect**, is significant because it enables the effect of the release of a small amount of initial hormone to become amplified (magnified) at each stage in the pathway. Cortisol, the final hormone released in the example in fig 17.47, is an anti-stress hormone secreted by the outer

Table 17.9 Summary of the major human endocrine glands, their functions and the control of their secretion.

Gland	*Hormone*	*Functions*	*Secretion control mechanism*
Hypothalamus	Releasing and inhibiting hormones and factors Posterior pituitary hormones produced here	Control of anterior pituitary hormones	Feedback mechanisms involving metabolite and hormone levels
Posterior pituitary gland	Receives hormones from hypothalamus – no hormones synthesised here Stores and secretes the following:		Feedback mechanisms involving hormones and nervous system
	Oxytocin	Ejection of milk from mammary gland, contraction of uterus during birth	
	Antidiuretic hormone (ADH) (vasopressin)	Reduction of urine secretion by kidney	Blood solute potential
Anterior pituitary gland	Follicle stimulating hormone (FSH)	In male, stimulates spermatogenesis In female, growth of ovarian follicles	Blood oestrogen and testosterone levels via hypothalamus and pituitary gland
	Luteinising hormone (LH)	In male, testosterone secretion	Blood testosterone levels via hypothalamus and pituitary gland
		In female, secretion of oestrogen and progesterone, ovulation and maintenance of corpus luteum	Blood oestrogen levels via hypothalamus and pituitary gland
	Prolactin	Stimulates milk production and secretion	Hypothalamus hormones
	Thyroid stimulating hormone (TSH)	Synthesis and secretion of thyroid hormones, growth of thyroid glands	Blood levels of thyroxine via hypothalamus and pituitary gland
	Adrenocorticotrophic hormone (ACTH or corticotrophin)	Synthesis and secretion of adrenal cortex hormones, growth of gland	Blood ACTH via hypothalamus
	Growth hormone (GH)	Protein synthesis, growth, especially of bones of limbs	Hypothalamus hormones
Parathyroid gland	Parathormone	Increases blood calcium level Decreases blood phosphate level	Blood Ca^{2+} level, and blood PO_4^{3-} level
Thyroid gland	Triiodothyronine (T_3) and thyroxine (T_4)	Regulation of basal metabolic rate, growth and development	TSH
	Calcitonin	Decreases blood calcium level	Blood Ca^{2+} level
Adrenal cortex	Glucocorticoids (cortisol)	Protein breakdown, glucose/glycogen synthesis, adaptation to stress, anti-inflammatory/allergy effects	ACTH
	Mineralocorticoids (aldosterone)	Na^+ retention in kidney, Na^+ and K^+ ratios in extracellular and intracellular fluids, raises blood pressure	Blood Na^+ and K^+ levels and low blood pressure
Adrenal medulla	Adrenaline (epinephrine)	Increases rate and force of heartbeat, constriction of skin and gut capillaries Dilation of arterioles of heart and skeletal muscles, raises blood glucose level	Sympathetic nervous system
	Noradrenaline (norepinephrine)	General constriction of small arteries, raising of blood pressure	Nervous system
Islets of Langerhans	Insulin (beta cells)	Decreases blood glucose level, increases glucose and amino acid uptake and utilisation by cells	Blood glucose and amino acid levels
	Glucagon (alpha cells)	Increases blood glucose level, breakdown of glycogen to glucose in liver	Blood glucose level
Stomach	Gastrin	Secretion of gastric juices	Food in stomach
Duodenum	Secretin	Secretion of pancreatic juice Inhibits gastric secretion	Acidic food in duodenum
	Cholecystokinin (pancreozymin)	Emptying of gall bladder and release of pancreatic juice into duodenum	Fatty acids and amino acids in duodenum
Kidney	Renin	Conversion of angiotensinogen into angiotensin	Blood Na^+ level, decreased blood pressure
Ovary	Oestrogens (17β-oestradiol)	Female secondary sex characteristics, oestrous cycle	FSH and LH
	Progesterone	Gestation, inhibition of ovulation	LH
Corpus luteum	Progesterone and oestrogen	Growth and development of uterus	LH
	Progesterone and oestrogen	Fetal development	Developing fetus
Placenta	Chorionic gonadotrophin	Maintenance of corpus luteum	Developing fetus
	Human placental lactogen	Stimulates mammary growth	Developing fetus
Testis	Testosterone	Male secondary sexual characteristics	LH and FSH

Table 17.10 Summary table showing chemical nature of the major hormones of the body.

Chemical group	*Hormone*	*Major source*
Peptides and proteins	Growth hormone Oxytocin ADH (Vasopressin)	Posterior pituitary gland
	Parathormone	Parathyroid gland
	Calcitonin	Thyroid gland
	Insulin Glucagon	Islets of Langerhans (pancreas)
	Gastrin	Stomach mucosa
	Secretin	Duodenal mucosa
Amines	Adrenaline	Adrenal medulla
	Noradrenaline	Sympathetic nervous system and adrenal medulla
	Thyroxine Triiodothyronine	Thyroid gland
	Releasing and inhibiting hormones and factors of the hypothalamus	Hypothalamus
	Follicle stimulating hormone Luteinising hormone Prolactin Thyroid stimulating hormone Adrenocorticotrophic hormone	Anterior pituitary gland
Steroids	Testosterone	Testis
	Oestrogens Progesterone	Ovary and placenta
	Corticosteroids	Adrenal cortex
Fatty acids	Prostaglandins	Many tissues

region (cortex) of the adrenal glands. It belongs to a group of hormones known as glucocorticoids which help to regulate blood sugar levels at times of stress (see also section 17.6.5).

Effects on target cells

Hormones are very specific and only exert their effects on target cells which possess the particular protein receptors that recognise the hormone. Non-target cells lack these receptors and therefore do not respond to the circulating hormone. Once attached to a receptor, the hormone may exert its effect in a number of ways. Three of the most important are through effects on:

- the cell membrane;
- enzymes located in the cell membrane (second messenger mechanism);
- genes.

Examples of each of these follow.

Cell membrane. Insulin exerts one of its effects by increasing the uptake of glucose into cells. It binds with a receptor site and alters the permeability of the membrane to glucose. Adrenaline works on smooth muscle cells by opening or closing ion channels for sodium or potassium ions or both, changing membrane potentials and either stimulating or inhibiting contraction as a result.

Second messenger mechanism of hormone action. Adrenaline and many peptide hormones bind to receptor sites on the cell membrane but cannot enter the cells themselves. Instead they cause the release of a 'second messenger' which triggers a series of enzyme-controlled reactions. These eventually bring about the hormonal response. In many cases this 'second messenger' is the nucleotide **cyclic AMP** (cyclic adenosine monophosphate), as shown in fig 17.48. The figure also shows that when the hormone binds to the receptor site it activates the receptor protein to become the enzyme **adenyl cyclase**. This converts ATP to cyclic AMP. In general, cyclic AMP can

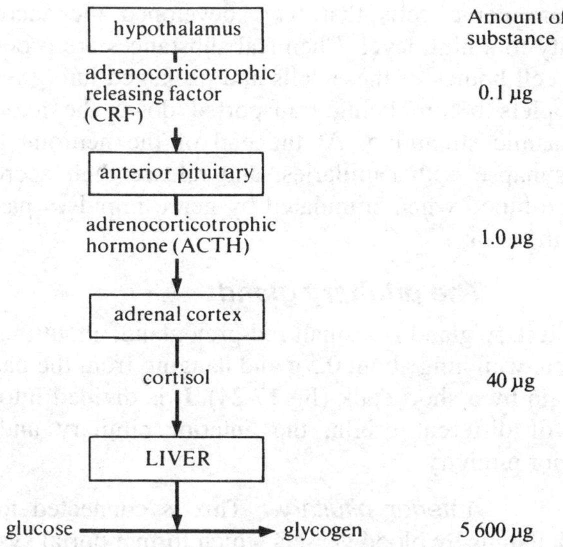

Fig 17.47 *An example of the 'cascade' effect in the control of the conversion of glucose to glycogen as a result of the release of adrenocorticotrophic releasing factor. The total amplification in this example is 56 000 times (data from Bradley, 1976).*

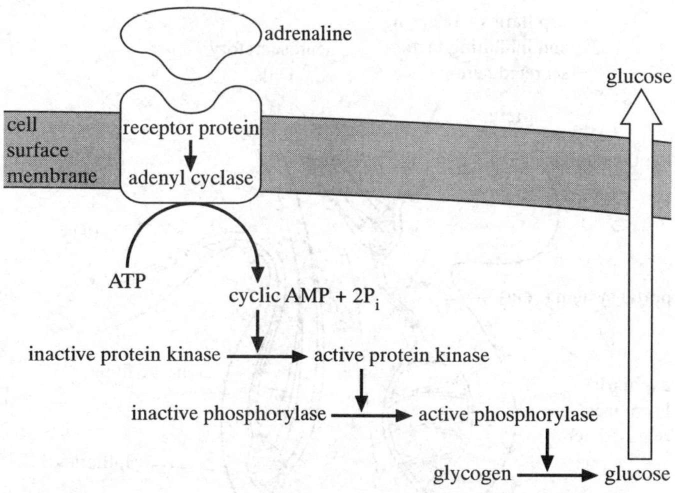

Fig 17.48 *Simplified diagram showing how adrenaline causes the release of glucose from a liver cell. The activation of membrane bound adenyl cyclase produces cyclic AMP which activates enzyme systems leading to the breakdown of glycogen to glucose. Glucose diffuses out of the cell into the bloodstream.*

trigger a wide variety of responses, depending on the particular cell stimulated.

In the case of adrenaline, cyclic AMP activates the enzyme protein kinase which in turn activates a phosphorylase enzyme which is needed to convert glycogen into glucose. The overall effect of adrenaline in this situation is therefore to release glucose. At each stage in the process an amplification occurs because only a few molecules of adenyl cyclase are needed to activate many molecules of protein kinase, and so on. This is the cascade effect.

Other hormones which use cyclic AMP as a second messenger include ADH (section 17.6.2 and 20.6), TSH (section 17.6.4), ACTH (section 17.6.5), glucagon (section 17.6.6), LH and FSH (section 21.6.6), and most releasing hormones from the hypothalamus.

Genes. Steroid hormones (the sex hormones and hormones secreted by the adrenal cortex) pass through the cell surface membrane and bind to a receptor protein in the cytoplasm. The complex formed passes to the cell nucleus where the hormones exert a direct effect upon the chromosomes by switching on genes and stimulating transcription (messenger RNA formation). The messenger RNA enters the cytoplasm and is translated into new proteins, such as enzymes, which carry out a particular function. For example, the hormone thyroxine passes through the cell surface membrane and binds directly to receptor proteins in the chromosomes, switching on certain genes.

In many cases of hormone action hormones appear to exert their effects by influencing enzymes associated with membranes or genetic systems.

17.6.2 The hypothalamus and pituitary gland

As stated earlier, much of the coordination of the body is achieved by the nervous and endocrine systems acting together. The major centres in the body for the coordination of the two systems of control are the hypothalamus and pituitary gland. The hypothalamus plays a dominant role in collecting information from other regions of the brain and from blood vessels passing through it. This information passes to the pituitary gland which, by its secretions, directly or indirectly regulates the activity of all other endocrine glands.

The hypothalamus

The hypothalamus is situated at the base of the forebrain immediately beneath the thalamus and above the pituitary gland (fig 17.24). It contains several distinct regions of nerve cells whose axons terminate on blood capillaries in the hypothalamus and posterior pituitary as shown in fig 17.49. (The pituitary gland can be divided into two parts, the anterior pituitary and the posterior pituitary.) Many physiological activities, such as hunger, thirst, sleep and temperature regulation, are regulated by nervous control through nerve impulses passing from the hypothalamus along neurones of the autonomic nervous system, giving involuntary (reflex) control of these processes. The control

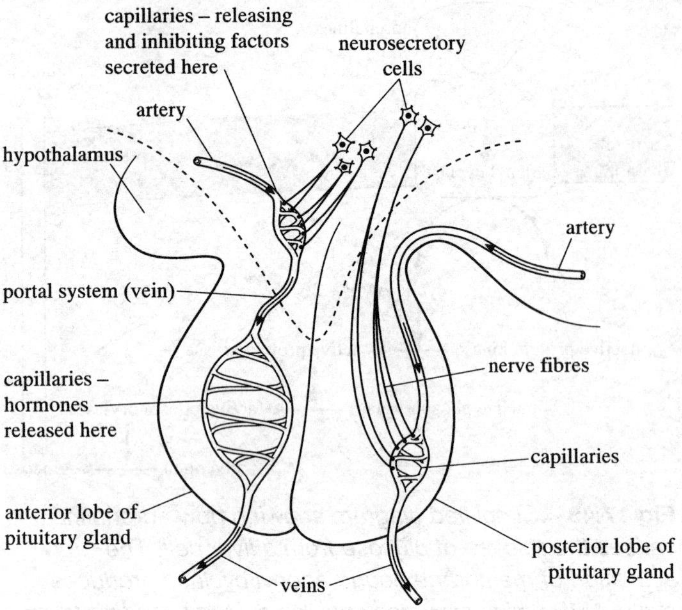

Fig 17.49 *Diagram showing the relationship between neurosecretory cells and blood vessels in the hypothalamus and pituitary gland.*

by the hypothalamus of endocrine secretion, however, lies in the ability of the hypothalamus to monitor metabolite and hormone levels in the blood. A **metabolite** is any molecule that is taking part in metabolism, such as glucose. Information gathered in this way, together with information from almost all parts of the brain, then passes to the pituitary gland, either by the release of 'hormones' into blood vessels which supply the pituitary or by neurones. The information relayed by neurones passes through specialised neurones called **neurosecretory cells.**

All nerve cells release a chemical substance, a neurotransmitter, at their terminal synapse, but neurosecretory cells are nerve cells that have developed the secretory capacity to a high level. Chemical substances are produced in the cell bodies of these cells and packaged into granules or droplets before being transported down the axon by cytoplasmic streaming. At the end of the neurone these cells synapse with capillaries, and release their secretion into the blood when stimulated by nerve impulses passing down the axon.

The pituitary gland

The pituitary gland is a small red-grey gland, about the size of a pea, weighing about 0.5 g and hanging from the base of the brain by a short stalk (fig 17.24). It is divided into two lobes of different origin, the anterior pituitary and the posterior pituitary.

Anterior pituitary. This is connected to the hypothalamus by blood vessels which form a portal system. A **portal system** is one which connects two organs, neither of which is the heart. It has one capillary bed in the hypothalamus and a second in the anterior pituitary.

Nerve terminals from specialised neurosecretory cells release two groups of chemical substances, known as 'releasing factors' and 'inhibiting factors', into the blood capillaries at the hypothalamus end of the portal system. These pass to the pituitary end where they cause the release of six hormones known as **trophic hormones**. A trophic hormone is one which stimulates other endocrine glands to release their hormones. The trophic hormones are produced and stored by the anterior pituitary. These six hormones pass into the blood vessels that leave the pituitary and exert their effects on specific target organs throughout the body, as shown in table 17.11. Growth hormone is considered in more detail in section 22.5.

The release of growth hormone and prolactin (see table 17.11) can both be stimulated and inhibited by the hypothalamus, whereas the release of the other four shown in the table (FSH, LH, TSH and ACTH) is regulated by

Table 17.11 Summary table showing the main hormones of the hypothalamus, the anterior pituitary hormones influenced by them and their target organs.

Hypothalamus hormone	*Anterior pituitary hormone and response*	*Site of action*
Growth hormone releasing factor (GHRF) Growth hormone release-inhibiting hormone (GHRIH)* (somatostatin)	Growth hormone (GH) (see section 22.8.1)	Most tissues
Prolactin releasing factor (PRF) Prolactin inhibiting factor (PIF)	Prolactin Inhibition of prolactin secretion	Ovary and mammary gland
Luteinising hormone releasing hormone (LHRH)*	Follicle stimulating hormone (FSH) Luteinising hormone (LH)	Ovary and testis
Thyrotrophin releasing hormone (TRH)*	Thyroid stimulating hormone (TSH)	Thyroid gland
Adrenocorticotrophin releasing factor (CRF)	Adrenocorticotrophic hormone (ACTH)	Adrenal cortex

*Releasing factors with an established identity are known as 'hormones'. Luteinising hormone releasing hormone is able to stimulate the release of both FSH and LH.

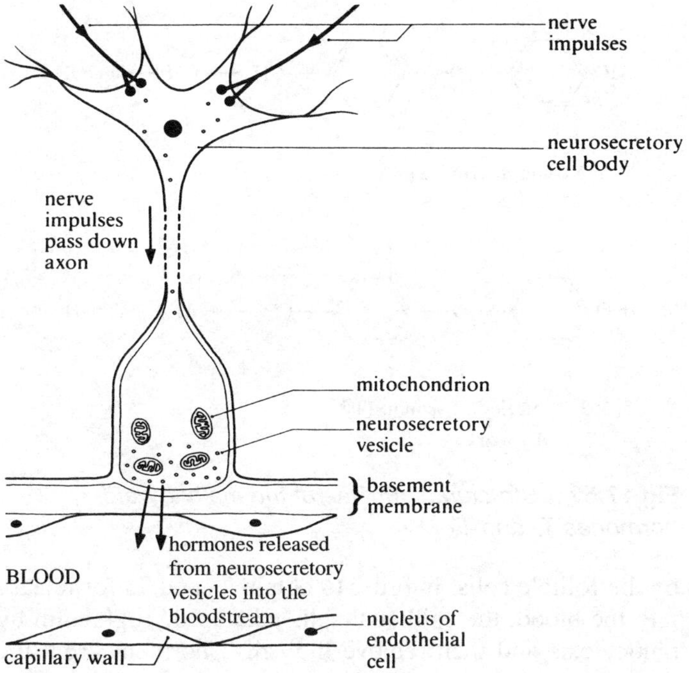

Fig 17.50 *Diagram showing a neurosecretory cell and its attachment to a blood capillary (not drawn to scale).*

negative feedback of hormones from the target glands acting on receptors in the hypothalamus and anterior pituitary. The six pituitary hormones stimulate the release of target gland hormones and, as the levels of these rise, they inhibit the secretion of the hypothalamus and pituitary hormones. When the blood levels of these target hormones fall below a certain level, hypothalamus and pituitary inhibition ceases allowing increased secretion from these glands. This is an example of negative feedback and is a control mechanism described in sections 19.5.4 and 20.6.

Posterior pituitary. This originates as an extension of the brain. It does not synthesise any hormones but stores and releases two hormones, **antidiuretic hormone** (**ADH** or **vasopressin**) and **oxytocin**. Antidiuretic hormone is released in response to a fall in the water content of blood plasma and leads to an increase in the permeability to water of the distal and collecting tubules of the nephron in the kidney so that water is retained in the blood plasma. A reduced volume of concentrated urine is excreted (section 20.6). Oxytocin causes contraction of the uterus during birth and the ejection of milk from the nipple (section 21.3.8).

ADH and oxytocin are produced by neurosecretory cell bodies lying in the hypothalamus and pass down the nerve fibres. These neurosecretory cells are much more specialised than those connected with the secretion of releasing factors and they form structures consisting of a swollen synapse attached to a capillary and surrounded by connective tissue (fig 17.50). Nerve impulses are relayed to the cell bodies of these neurosecretory cells from other

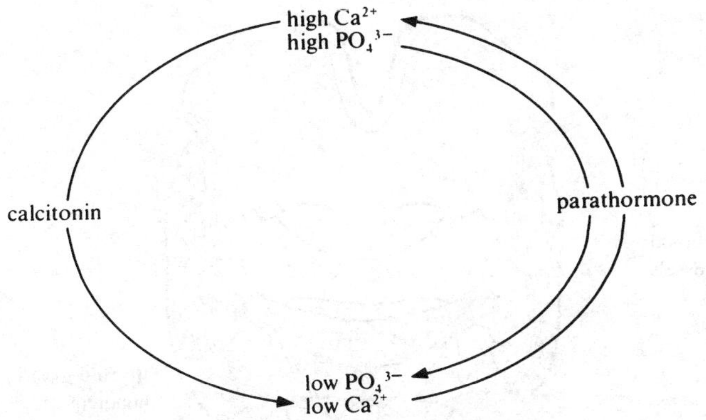

Fig 17.51 *Role of parathormone and calcitonin in the regulating of blood calcium level.*

regions of the brain and transmitted down the axons to the swollen ends of the axons where hormones stored in vesicles are released into the bloodstream and carried to target organs. Since the whole process involves both the nervous system and the endocrine system the response is known as a **neuroendocrine response**. Many neuroendocrine responses result in a type of behaviour pattern known as a **neuroendocrine reflex** and many examples of these are associated with courtship and breeding activity.

17.6.3 Parathyroid glands

In humans there are four small parathyroid glands embedded in the thyroid gland. They produce only one hormone called **parathormone** which is a peptide composed of 84 amino acids. Parathormone and the thyroid hormone **calcitonin** (see below) work antagonistically (in opposite ways) to regulate the plasma calcium and phosphate levels. The release of parathormone increases the plasma calcium level to its normal level and decreases the plasma phosphate level. The activity of the parathyroid glands is controlled by the simple negative feedback mechanism shown in fig 17.51.

Overactivity of the gland reduces the level of calcium in the plasma and tissues due to calcium excretion in urine. This can lead to a state of tetany in which muscles remain contracted. Also the rate of excretion of phosphate is reduced and the level of phosphate ions in the plasma rises.

17.6.4 Thyroid gland

The thyroid gland produces three active hormones, **triiodothyronine** (**T_3**), **thyroxine** (**T_4**) and **calcitonin**. T_3 and T_4 help to regulate metabolic rate, growth and development, whilst calcitonin is involved in the regulation of calcium levels in the blood. Calcitonin is referred to in section 17.6.3 above.

The structure of the thyroid gland

The thyroid gland is a bow-tie shaped structure found in the neck. It has lobes on each side of the trachea and larynx,

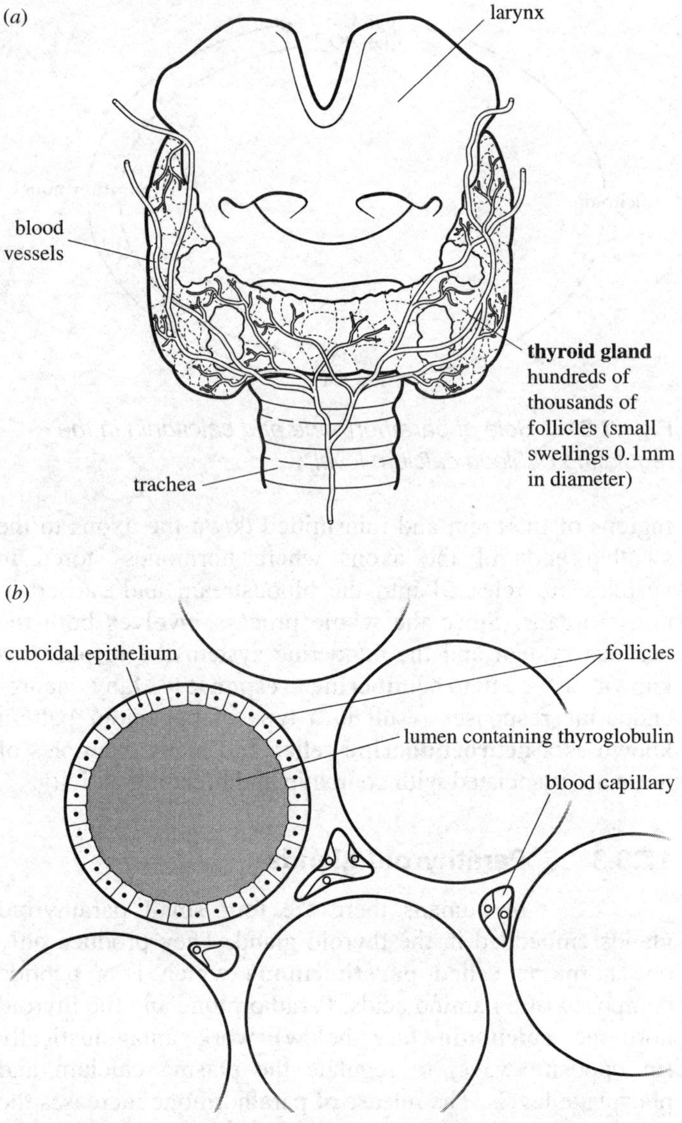

Fig 17.52 *Structure of the thyroid gland: (a) bilobed appearance around the trachea, (b) section through a group of follicles.*

connected by a thin band of tissue (fig 17.52). It is made up of hundreds of thousands of tiny follicles which have a diameter of about 0.1 mm. Each follicle is a hollow sphere whose wall consists of a single layer of cuboidal epithelium. It is these cells that secrete thyroxine and T_3. They become columnar in shape and develop microvilli on their inner surface when the gland is activated by **thyroid stimulating hormone** (**TSH**) from the anterior pituitary gland.

The formation and release of T_3 and T_4

Triiodothyronine contains three iodine atoms in its structure and thyroxine contains four, hence the names T_3 and T_4 (fig 17.53). Iodine is taken up by active transport from the blood plasma in the numerous capillaries surrounding the follicles. It is taken up as iodine ions (I^-) and secreted into the lumen of the follicle. Here it is used to make T_3 and T_4 which are bound to a protein called thyroglobulin formed

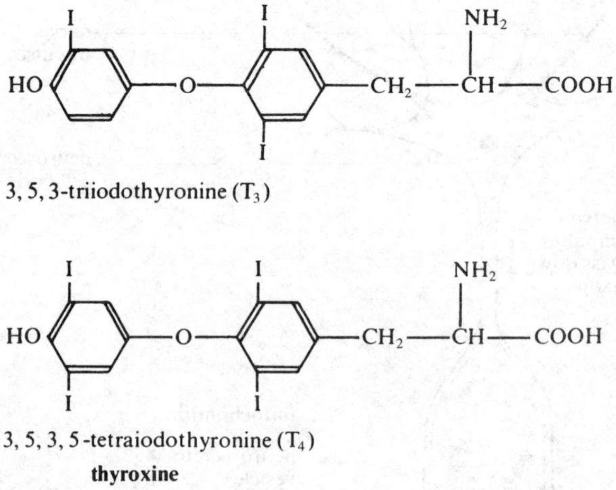

Fig 17.53 *Structural formulae of the main thyroid hormones T_3 and T_4.*

by the follicle cells. In order to obtain T_3 and T_4 for release into the blood, the epithelial cells take up thyroglobulin by pinocytosis and then remove the active hormone from the thyroglobulin. T_3 and T_4 have identical structures apart from the number of iodine atoms and both have the same effect on the body. Since T_4 (thyroxine) is secreted in much larger amounts, the following account will refer to thyroxine only.

The function of thyroxine (T_4)

The main effect of thyroxine is to control the basal metabolic rate (BMR). **Metabolism** is the term used to describe all the chemical reactions going on in the body. The **basal rate** is the rate at rest. Basal metabolic rate is therefore the rate at which oxygen and food are used to release energy and is directly related to the rate of cell respiration. The BMR in humans averages about $160\,kJ\,h^{-1}\,m^{-2}$ body surface and is maintained at a steady state by the action of thyroxine. Thyroxine promotes the breakdown of glucose and fats to provide energy (section 19.5.2). Further effects which lead to an increase in energy release include increases in the uptake of oxygen by the body and the rate of cell respiration in the mitochondria. The number and size of mitochondria are also increased. The overall effect is to increase the rate of cell respiration and therefore the rate of ATP formation and heat production by the tissues. Increased heart rate and cardiac output also occur.

Thyroxine also has important influences on many other body processes. Thyroxine and growth hormone (GH) have a joint stimulatory effect on protein synthesis, leading to an increase in growth rate, particularly of the skeletal system. Unlike GH, thyroxine also stimulates brain development, so a deficiency in childhood can cause mental retardation (see later).

In many of the metabolic processes with which it is involved thyroxine appears to enhance the effects of other hormones such as insulin, adrenaline and glucocorticoids (section 19.5.2).

Control of thyroxine release

The effects of thyroxine are longer lasting than those of most other hormones and hence it is vital that large changes in its secretion are prevented. This is one of the reasons for the storage of T_4 within the gland so that it is readily available for release.

The level of T_4 circulating in the blood controls its release from the thyroid gland by negative feedback mechanisms involving the hypothalamus and anterior pituitary. Fig 17.54 shows that if excess T_4 is present in the blood it switches off its own production by switching off production of **TRH (thyrotrophin releasing hormone)** by the hypothalamus and **TSH (thyroid stimulating hormone)** by the anterior pituitary. This is negative feedback. The general principle is that the product of a series of reactions controls its own production by turning off the pathway when it reaches a certain level. This is comparable with a thermostat where the product, heat, switches off its own production when a certain temperature is reached. If there is too little of the product, its production is switched on again because, in the case of thyroxine, inhibition of TRH and TSH is removed.

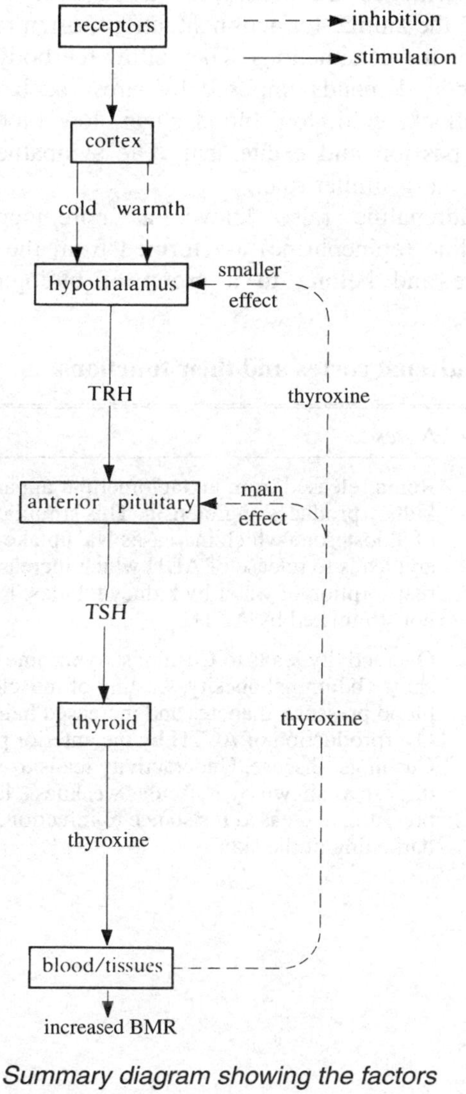

Fig 17.54 *Summary diagram showing the factors regulating thyroxine secretion and leading to homeostatic control of metabolic rate (BMR).*

Superimposed on this feedback system are environmental factors, such as temperature, which influence higher centres of the brain. The temperature centre in the hypothalamus can also be stimulated directly. Prolonged exposure to cold, for example, causes the 'resetting' of the threshold in the pituitary for negative feedback so that release of TRH is stimulated (fig 17.54) and BMR and heat production are raised.

Overactivity of the thyroid gland (hyperthyroidism)

Both over- and underactivity of the thyroid gland can produce a swelling in the neck known as a **goitre**. Overactivity may be due to overproduction of thyroxine from an enlarged thyroid gland. The symptoms are increases in heart rate, ventilation rate and body temperature. The basal metabolic rate may increase by 50% with associated increases in oxygen consumption and heat production. Patients become very nervous and irritable and the hands shake when held out. Extreme hyperthyroidism is termed **thyrotoxicosis** and is associated with increased excitability of cardiac muscle which may lead to heart failure unless treated. This may involve the surgical removal of most of the gland or the destruction of the same amount by administering radioactive iodine.

Underactivity of the thyroid gland (hypothyroidism)

A lack of TSH production by the anterior pituitary, iodine deficiency in the diet or failure of enzyme systems involved in thyroxine production may result in hypothyroidism. If there is a deficiency of thyroxine at birth this will lead to poor growth and mental retardation, a condition known as **cretinism**. If the condition is diagnosed at an early stage thyroxine can be given to restore normal growth and development. Thyroxine deficiency in later life gives rise to a condition known as **myxoedema** and the symptoms are a reduction in metabolic rate accompanied by decreased oxygen consumption, ventilation, heart rate and body temperature. Mental activity and movement become slower and weight increases due to the formation and storage of a semi-fluid material under the skin. This causes the face and eyelids to become puffy, the tongue swells, the skin becomes rough and hair is lost from the scalp and eyebrows. All of these symptoms can be eliminated and the condition treated by taking thyroxine tablets.

17.6.5 Adrenal glands

There are a pair of adrenal (*ad*, to, *renes*, kidneys) glands located one just above each kidney (fig 17.46). Each gland is composed of two types of cells from different origins, and these cells function independently. The outer region is called the **cortex** and forms 80% of the gland. The inner region is called the **medulla** and is closely linked with the nervous system.

Adrenal cortex

The adrenal cortex produces steroid hormones of two types, as shown in table 17.12. All steroids are formed from a molecule called **cholesterol** which the cortex is able to synthesise and take up from the circulation following absorption from the diet. Steroids are lipid-soluble substances which diffuse through cell membranes and attach to cytoplasmic receptor proteins. The complexes formed then migrate into the nucleus where they attach to specific areas of the chromosome and switch on or off certain genes.

The size of the adrenal gland is closely linked to the output of ACTH and the ability to withstand stress. During long periods of stress the size of the gland increases. Investigations into the behaviour of organisms under stress have shown that the output of adrenal hormones increases with the rise in number in the population. In organisms where social hierarchies exist, there is a positive correlation between position in the hierarchy and increased size of the adrenal gland.

Control of cortical hormone release

Mineralocorticoid release is stimulated by the activity of **renin** and **angiotensin** as described in table 17.12. **Glucocorticoids** are secreted in response to **adrenocorticotrophic hormone (ACTH)**. An example of the role of ACTH in regulating the release of the glucocorticoid hormone cortisol is shown in fig 17.47. This is an example of the cascade effect which amplifies the amount of hormone produced.

ACTH is a protein molecule which contains 39 amino acids. It attaches to receptors on the surface of the cortical cells and activates adenyl cyclase to convert ATP to cyclic AMP (section 17.6.1). This activates enzymes known as **protein kinases** which stimulate the conversion of cholesterol to glucocorticoids.

Adrenal medulla

The adrenal medulla forms the centre of the adrenal gland. It is richly supplied with nerves as well as blood vessels. The cells of the medulla are modified neurones of the sympathetic nervous system. When stimulated they secrete adrenaline and noradrenaline in the ratio of 4:1 (section 17.6.1). Noradrenaline is also secreted as a neurotransmitter by synapses of the sympathetic nervous system. The adrenal medulla is not essential to life since its function is to boost the sympathetic nervous system. The actions of these hormones are widespread throughout the body and prepare the animal for action situations, often referred to as 'fight or flight' situations. They allow the body to respond to sudden demands imposed by stress such as exercise, pain, shock, cold, low blood sugar, low blood pressure, anger, passion and excitement. The sympathetic nervous system has a similar role.

Noradrenaline (also known as norepinephrine) and adrenaline (epinephrine) are formed from the amino acid tyrosine and belong to a group of biologically active

Table 17.12 Summary table showing the hormones secreted by the adrenal cortex and their functions.

Hormones	*Function*	*Notes*
Mineralocorticoids e.g. aldosterone	Control water and salt content of body by stimulating cation pumps in membranes to conserve Na^+ and Cl^- and remove K^+. Prevent excessive Na^+ loss in sweat, saliva and urine and maintain osmotic concentration of body fluids at a steady state	Renin released from juxtaglomerular apparatus in kidney produces angiotensin. This stimulates release of aldosterone which increases Na^+ uptake by kidney and leads to release of ADH which increases reabsorption of water by kidney tubules. Release is not stimulated by ACTH
Glucocorticoids e.g. cortisol	(A) Carbohydrate metabolism (1) promote gluconeogenesis (2) promote liver glycogen formation (3) raise blood glucose level (B) Protein metabolism (1) promote breakdown of plasma protein (2) increase availability of amino acids for enzyme synthesis in the liver (C) Other roles (1) prevent inflammatory and allergic reactions (2) decrease antibody production	Overactivity leads to Cushings' syndrome and patients show abdominal obesity, wasting of muscles, high blood pressure, diabetes and increased hair growth. Overproduction of ACTH by the anterior pituitary is Cushings' disease. Underactivity leads to Addison's disease as shown by muscular weakness, low blood pressure, decreased resistance to infection, fatigue and darkening of the skin

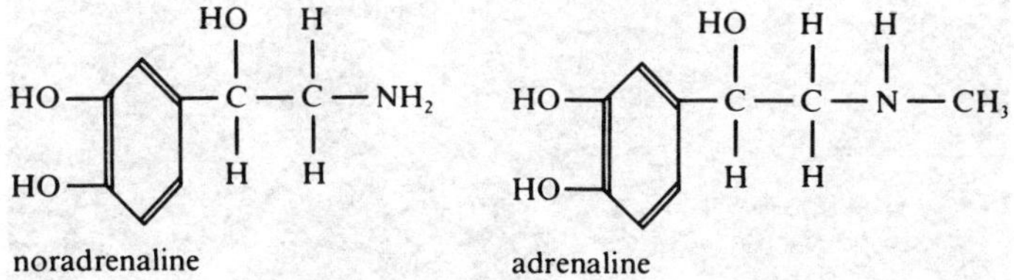

Fig 17.55 *Molecular structure of noradrenaline and adrenaline.*

molecules called **catecholamines** (fig 17.55). The effects of both hormones are basically identical as shown in table 17.13, but they differ in their effects on blood vessels. Noradrenaline causes vasoconstriction of all blood vessels whereas adrenaline causes *vasoconstriction* of blood vessels supplying the skin and gut and *vasodilation* of blood vessels to muscles and the brain. Both of these hormones activate two types of receptor sites on the target tissues known as α and β **adrenergic receptors**. These activate adenyl cyclase to make cyclic AMP and this leads to the specific tissue responses shown in table 17.13. Most organs have both α and β receptors with α receptors appearing to be more receptive to noradrenaline than adrenaline and vice versa in the case of β receptors.

17.6.6 Pancreas

Structure of the pancreas

Sections of the pancreas as seen at low power and high power with a light microscope are shown in fig 17.56*a* and *b*, and a diagrammatic interpretation of the structure is shown in fig 17.56*c*. The pancreas has both exocrine and endocrine functions. The bulk of the gland is made up of cells which surround the numerous branches of the pancreatic duct. Each branch is surrounded by a ring of cells (fig 17.56*c*) called an **acinus** (acinus means grape pip). The acinar cells are exocrine cells. They secrete the enzymes of the pancreatic juice into the pancreatic duct in a way described in section 5.9.8 and shown in fig 5.23. The endocrine cells within the pancreas are described below.

Table 17.13 Physiological effects of noradrenaline and adrenaline.

Dilate pupils of eyes
Cause hair to stand on end
Relax bronchioles thus increasing air flow to lungs
Inhibit peristalsis
Inhibit digestion
Prevent bladder contraction
Increase force and rate of heartbeat
Cause almost general vasoconstriction
Increase blood pressure
Stimulate conversion of liver glycogen to glucose
Decrease sensory threshold
Increase mental awareness

Discovery of insulin

In this section we are concerned with the endocrine function of the pancreas. In 1868 Paul Langerhans, a German scientist who later became a Professor of Pathology, noticed tiny patches of cells about the size of a pinhead in the tissue of the pancreas. Examination with a microscope revealed that the cells were unlike most of the other cells of the pancreas and had a rich supply of blood vessels. They became known as islets (small islands) of Langerhans. Their structure is shown in fig 17.56*c*. They contain a small number of large cells, known as α **cells** (alpha cells), many smaller cells known as β **cells** (beta cells) and blood capillaries.

The function of the islets remained unknown until the twentieth century. Scientists discovered that removing the pancreas from an animal such as a dog resulted in a disease similar to the human disease sugar diabetes, and it was suspected that the pancreas produced a hormone which helped to regulate blood sugar. The hormone was named '**insulin**', meaning island, even before its existence was confirmed. It was isolated by Banting, Best and Macleod in Canada in 1921. Injections of insulin into a dog from which the pancreas had been removed cured it of diabetes. They then succeeded in treating the condition in humans using insulin extracted from the pancreases of animals. Since then millions of lives have been saved worldwide and human insulin can now be prepared from genetically engineered bacteria as described in section 25.1.2.

Two hormones

It is now known that the islets of Langerhans secrete two hormones, insulin and **glucagon**. Both are proteins. Insulin is secreted by the β cells and glucagon by the α cells. These two hormones have antagonistic effects on the glucose level in the blood.

Insulin

Insulin is a small protein composed of 51 amino acids, whose primary structure (fig 5.27) was determined in 1950 by Fred Sanger in Cambridge. It is released in response to a rise in blood glucose level above 90 mg per 100 cm^3 blood. It is carried in the blood plasma (the liquid part of blood) bound to β globulin and has an important effect on every organ of the body, although its main effect is on the liver and muscle. Receptor sites on cell surface membranes bind insulin, and this reaction leads to changes both in cell permeability and the activity of enzyme systems within the cell, the overall effect being to reduce blood sugar:

- increase in the rate of conversion of glucose to glycogen, called **glycogenesis** – this takes place mainly in the liver and muscle (glycogen granules can clearly be seen with an electron microscope, fig 5.12);

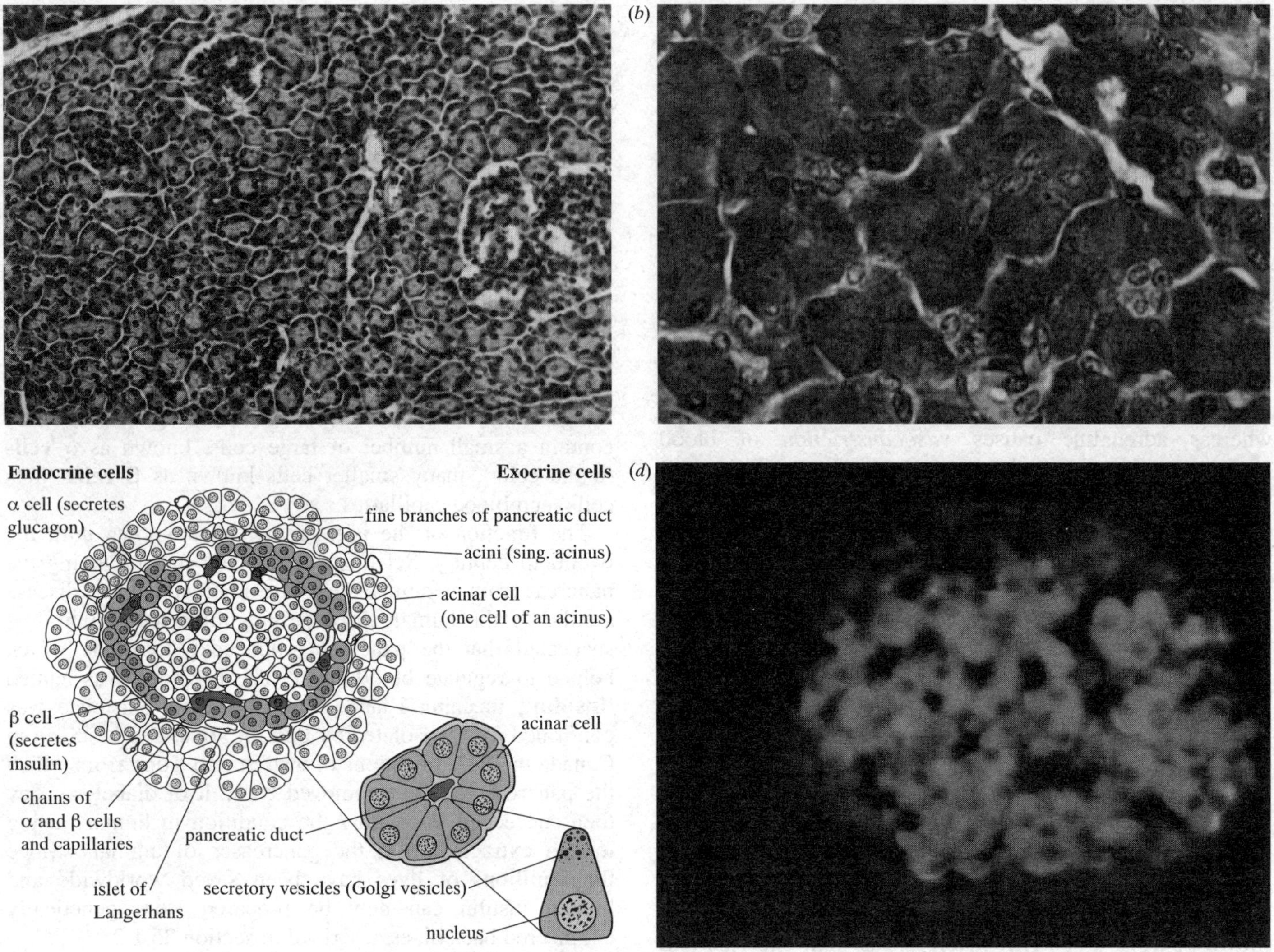

Fig 17.56 *Structure of the pancreas. (a) Low power light micrograph showing the islets of Langerhans surrounded by acinar cells. (b) High power light micrograph showing detail of one islet of Langerhans. (c) Diagram of islet of Langerhans and surrounding acinar cells. (d) Light micrograph of an islet of Langerhans showing central position of insulin-secreting cells (lighter) and the peripheral glucagon-secreting cells (darker).*

- increase in the rate of uptake of glucose by cells, especially skeletal muscle;
- increase in the use of glucose rather than other substances, such as fat, as a source of energy for cell respiration;
- increase in the conversion of glucose into fatty acids and fats, and fat deposition;
- increase in the rate of uptake of amino acids into cells and the rate of protein synthesis;
- decrease in gluconeogenesis (production of glucose).

Regulation of insulin production

Production of insulin is regulated by a negative feedback mechanism. A rise in blood sugar is detected by the β cells in the pancreas, which in response produce more insulin. As insulin levels rise, glucose is removed from the blood by the methods described and, as the blood glucose level gets lower, the β cells reduce output of insulin.

The secretion of insulin is vital to life since it is the only hormone which lowers the blood glucose level. A deficiency in insulin production leads to the metabolic disease known as **diabetes mellitus** in which the blood glucose reaches such a level that it exceeds that which the kidneys can reabsorb, and so is excreted in the urine.

The effects of insulin deficiency and excess are summarised in table 17.14.

Glucagon

Glucagon is a protein composed of 29 amino acids and is released, along with several other hormones, in response to a fall in blood glucose level below normal. This is usually the result of an increase in metabolic demand, for example as a result of exercise. Its role is to increase blood glucose level. Its main target is the liver. Glucagon stimulates the conversion of glycogen to glucose (**glycogenolysis**). It also stimulates the breakdown of proteins and fats to glucose

Table 17.14 Some of the effects of insulin deficiency and excess.

Deficiency	*Excess*
high blood glucose level (hyperglycaemia)	low blood glucose level (hypoglycaemia)
breakdown of muscle tissue	hunger
loss of weight	sweating
tiredness	irritability
	double vision

and conversion of lactic acid to glucose, processes known as **gluconeogenesis** (*gluco*, glucose; *neo*, new; *genesis*, giving rise to) (section 19.6.2).

Receptor sites in the liver cell membrane bind glucagon which activates adenyl cyclase to form cyclic AMP. The action of glucagon in bringing about an increase in glucose is similar to that of adrenaline and, in both cases, the cyclic AMP activates phosphorylase enzymes which stimulate the breakdown of glycogen to glucose as shown in fig 17.47. Glucagon has no effect on *muscle* glycogen. Regulation of glucagon secretion is similar to that of insulin in principle except that the α cells, not the β cells, are involved and they respond to falling blood glucose levels rather than rising levels.

Summary diagrams of the role of insulin and glucagon in controlling blood sugar level are shown in figs 19.4 and 19.22.

17.7 The study of behaviour (ethology)

Behaviour may be defined as the outwardly expressed course of action produced in organisms in response to stimuli from a given situation. The action modifies, in some way, the relationship between the organism and its environment and its adaptive significance is the perpetuation of the species. All living organisms exhibit a variety of forms of behavioural activity determined by the extent to which they are able to respond to stimuli. This varies from the relatively simple action of the growth of a plant stem towards a light source, to the complex sexual behaviour patterns of territory defence, courtship and mating seen in birds and mammals.

Plant behaviour is restricted to movements produced by growth or turgor changes and is stereotyped and predictable. The two main activities associated with plant behaviour are tropisms and taxes and details of these are described in section 16.1.

Animal behaviour is far more complex and diverse than plant behaviour and therefore it is extremely difficult to investigate and account for with any degree of scientific validity. The three main approaches to behavioural studies are the vitalistic, mechanistic and ethological approaches.

Vitalistic approach

This seeks to account for behavioural activities in terms of what animals are seen to do, and attempts to relate this to changes in the environment. It involves the total rejection of any study of the animal outside its natural environment. The technique has its foundations in natural history and has provided a wealth of valuable data, but it is essentially non-scientific since all the observations relate to past events which cannot be tested experimentally.

Mechanistic approach

This is an experimental approach and involves the study of particular aspects of behaviour under controlled conditions in a laboratory. It may be criticised on the grounds of the artificiality of the experimental situation, the nature of the behaviour activities and the way in which the results are interpreted. This technique is, however, used extensively in psychology and was pioneered by Pavlov.

Ethological approach

This is the contemporary approach to behavioural investigations and attempts to explain responses observed in the field in terms of the stimuli eliciting the behaviour. It involves both of the techniques outlined above and was pioneered by Lorenz, von Frisch and Tinbergen.

In all behavioural studies, great care has to be taken in interpreting the results of observations in order to eliminate subjectivity. For example, care must be taken to avoid putting oneself in the place of animal (**anthropocentrism**), or interpreting what is observed in terms of human experience (**anthropomorphism**) or interpreting the cause of the observation in terms of its outcome (**teleology**).

Recent advances in audio-visual technology have assisted the recording of behaviour activities. Infra-red photography has enabled animals to be filmed at night and time-lapse photography and slow-motion cinematography have enabled, respectively, slow-moving activities, such as moulting in insects, and fast-moving activities, such as bird flight, to be recorded and subsequently seen at speeds more suited to analysis by behaviouralists. The use of miniature cassette tape recorders for recording sounds and their subsequent analysis using sound spectrographs and computers has helped in the study of auditory communication between organisms. The movement of organisms is now studied either using implanted miniaturised signal generators, emitting signals that can be followed using direction-finding equipment, or by the use of tracking radar and satellite surveillance. These two techniques are employed successfully in following the migrations of mammals, birds and locusts.

Whatever the approach and techniques used in the investigation of behaviour, the fundamental explanation of behaviour activity must begin with a stimulus, end with a response and include all the stages occurring at various levels of oganisation within the body linking *cause* and *effect*.

Broadly speaking there are two forms of behaviour, **innate behaviour** and **learned behaviour**, but the distinction between the two is not clear-cut and the majority of behavioural responses in higher organisms undoubtedly contain components of both. However, for simplification in this elementary introduction to behaviour, the various aspects of behaviour are considered under these two headings in the next two sections.

17.8 Innate behaviour

Innate behaviour does not involve a single clear-cut category of behaviour, but rather a collection of responses that are predetermined by the inheritance of specific nerve or cytoplasmic pathways in multicellular or single-celled organisms. As a result of these 'built-in' pathways a given stimulus will produce, invariably, the same response. These behaviour patterns have developed and been refined over many generations (**selected**) and their primary adaptive significance lies in their survival value to the species. Another valuable feature of innate behaviour is the economy it places on nerve pathways within multicellular organisms, since it does not make enormous demands on the higher centres of the nervous system.

There is a gradation of complexity associated with patterns of innate behaviour which is related to the complexity of nerve pathways involved in their performance. Innate behaviour patterns include orientations (taxes and kineses), simple reflexes and instincts. The latter are extremely complex and include biological rhythms, territorial behaviour, courtship, mating, aggression, altruism, social hierarchies and social organisation. All plant behaviour is innate.

Taxes

A taxis or **taxic response** is a movement of the whole organism in response to an external directional stimulus. Taxic movements may be towards the stimulus (**positive**, +), away from the stimulus (**negative**, –), or at a particular angle to the stimulus, and are classified according to the nature of the stimulus. Some examples of types of taxes are shown in table 16.2. In some cases organisms are able to move by maintaining a fixed angle relative to the directional stimulus. For example, certain species of ants can follow a path back to their nest by setting a course relative to the Sun's direction. Other organisms orientate themselves so that, for example, their dorsal side is always uppermost. This is called the **dorsal light reaction** and is found in fish such as plaice which maintain their dorsal surface at right-angles to the sky.

Many organisms detect the direction of the stimulus by moving the head, which bears the major sensory receptors, from side to side. This is known as a **klinotaxic response** and enables symmetrically placed receptors on the head, such as photoreceptors, to detect the stimulus. If both receptors are equally stimulated, the organism will move forwards in approximately a straight line. This type of response is shown by *Planaria* moving towards a food source and by blowfly larvae moving away from a light source. In all cases of klinotaxis it is thought that successive stimulation of receptors on each side of the body is necessary in order to provide the 'brain' with a continuous supply of information since there is no long-term 'memory'.

Kineses

A **kinetic response** is a non-directional movement response in which the *rate* of movement is related to the *intensity* of the stimulus and not the *direction* of the stimulus. For example, the direction of movement of the tentacles of *Hydra* in search of food is random and slow, but if saliva, glutathione or water fleas are placed close to the *Hydra* the rate of movement of the tentacles increases.

Both kinetic and taxic responses can be observed through the use of woodlice in a **choice chamber** as described in experiment 17.1.

Experiment 17.1: To investigate orientation behaviour in woodlice by the use of a simple choice chamber.

Materials

old pair of tights	anhydrous calcium chloride
bases of two petri dishes	adhesive tape
Araldite	ten woodlice
hot metal rod	plasticine
cotton wool	

Method

(1) Cut a circle out of an old pair of tights 10 cm in diameter and stretch over the base of an 8.5 cm petri dish. Attach with Araldite, held in place by an elastic band until it sets.
(2) Burn out a 1.0 cm hole in the bottom of this petri dish using a hot metal rod.
(3) Divide the base of another petri dish in half using a plasticine strip 8.5 cm long, 1.4 cm deep and 0.5 cm wide.
(4) Place cotton wool soaked in water in one half of this petri dish and granules of anhydrous calcium chloride in the other half.
(5) Attach the petri dish base prepared in (1) above to the petri dish base prepared in (4) with adhesive tape as shown in fig 17.57.
(6) Introduce ten woodlice into the apparatus through the hole in the upper Petri dish and record the position and number active at 1 min intervals in a table such as table 17.15.
(7) After 20 min plot a graph of numbers present against time for each environment.
(8) Calculate the percentage number of woodlice active in the dry environment for each minute interval and plot on a graph against time.
(9) Explain the nature of the results obtained in terms of kineses and taxes.

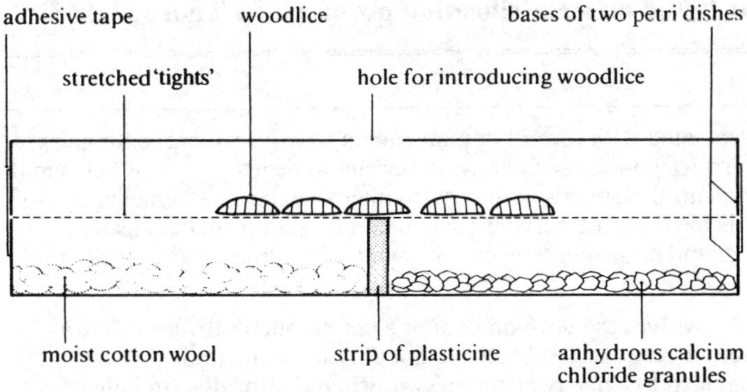

Fig 17.57 *Choice chamber apparatus for investigating orientation behaviour in woodlice.*

In simple experiments of this type the response of organisms to environments with extremes of a given variable can be investigated. Taxic responses are observed by the preference shown by the organisms for a particular environment. For example, woodlice exposed to areas of high and low humidity in a choice chamber congregate in larger numbers in the area of highest humidity, showing them to be **positively hydrotaxic**. More complex experiments can be devised using combinations of variables in order to determine which is strongest in eliciting a final response.

Kinetic responses are observed by recording the activity of woodlice at, say, 20 s intervals, in relation to their position in the choice chamber. Results of such investigations show that when first introduced into the choice chamber at the junction of two environments some woodlice move around whilst others remain stationary. After a short time all the woodlice begin moving and the speed of movement and rate of turning is always greatest in the drier side of the choice chamber than in the humid side. The increased, apparently random, moving and turning of the woodlice on the dry side is believed to indicate an attempt to find optimal conditions and, when these are found, the moving and turning response diminishes. These responses are examples of **orthokinesis**. The woodlice move more slowly on the humid side and consequently usually congregate there. The preference shown for the humid side of the choice chamber indicates a positive taxic response to humidity.

Not all orientation behaviour patterns are rigid, and the response shown by an organism may vary depending upon other factors such as the degree of hunger, thirst, light, dark, heat, cold and humidity.

17.8.1 Simple reflexes in vertebrates

A simple reflex is an involuntary stereotyped response of part of an organism to a given stimulus. It is determined by the presence of an inherited pattern of neurones forming spinal and cranial reflex arcs, and the structure and function of these was described in section 17.2.

In terms of behaviour, simple spinal reflexes are either **flexion** responses, involving withdrawal of a limb from a painful stimulus, or **stretch** responses, involving the balance and posture of the organism. Both of these responses are primarily involuntary and most require no integration or coordination outside that found in the spinal cord. However both types of response may be modified by the brain according to circumstances and in the light of previous experience. When this happens innate and learned behaviour patterns overlap and the reflex action is now described as 'conditioned' as described in table 17.16. Many simple cranial reflexes too, may be conditioned, for example blinking in response to a sudden movement.

17.8.2 Instincts

Instincts are complex, inborn, stereotyped behaviour patterns of immediate adaptive survival value to the organism and are produced in response to sudden changes in the environment. They are unique to each species and differ from simple reflexes in their degree of complexity. Konrad Lorenz, a Nobel prize-winning ethologist, defined instincts as 'unlearned species-specific motor patterns'.

Table 17.15 Specimen arrangement of table of results.

	Humid		*Dry*		
Time/min	*Number present*	*Number active*	*Number present*	*Number active*	*Percentage active dry side*
0	4	1	6	4	40%
1	5	2	5	5	50%
2		etc.		etc.	
3					
–					
–					
–					
20					

Table 17.16 Summary of the major types of learned behaviour based on a classification proposed by Thorpe (1963).

Learned behaviour	*Features of the learned behaviour*
Habituation	Continuous repetition of a stimulus not associated with reward or punishment (reinforcement) extinguishes any response to the stimulus, e.g. birds **learn** to ignore a scarecrow. Important in development of behaviour in young animals in helping to understand neutral elements in the environment, such as movements due to wind, cloud-shadows, wave-action, etc. It is based in the nervous system and is not a form of sensory adaptation since the behaviour is permanent and no response is ever shown to the stimulus after the period of habituation.
Associative learning: Classical conditioning (conditioned reflex)	Based on the research of Pavlov on dogs. It involves the development of a conditioned salivary reflex in which animals **learn** to produce a **conditioned response** (salivation), not only to the natural **unconditioned stimulus** (sight of food) but also to a newly acquired **conditioned stimulus** (ticking of a metronome) which was presented to the dog along with the unconditioned stimulus. Animals learn to **associate** unconditioned stimuli with conditioned stimuli so either produces a response. For example, birds avoid eating black and orange cinnabar moth larvae because of bad taste and avoid all similarly coloured larvae even though they may be nutritious.
Associative learning: Operant conditioning (trial-and-error learning)	Based on the research of Skinner on pigeons. Trial motor activities give rise to responses which are reinforced either by rewarding (positive) or punishment (negative). The **association** of the outcome of a response in terms of reward or punishment increases or decreases respectively future responses. Associative learning efficiency is increased by repetition as shown in investigations carried out on learning in cuttlefish.
Latent learning (exploratory learning)	Not all behavioural activities are apparently directed to satisfying a need or obtaining a reward (i.e. appetitive behaviour). Animals explore new surroundings and **learn** information which may be useful at a later stage (hence latent) and mean the difference between life and death. For example in mice, knowledge of the immediate environment of its burrow may help it escape from a predator. At the time of acquiring this knowledge it had no apparent value. This appears to be the method by which chaffinches learn to sing, as described in section 17.8.2.
Insight learning	Probably the 'highest' form of learning. It does not result from immediate trial-and-error learning but may be based on information previously learned by other behavioural activities. Insight learning is based on advanced perceptual abilities such as thought and reasoning. Kohler's work on chimpanzees suggested 'insight learning': when presented with wooden boxes and bananas too high to reach the chimps stacked up the boxes beneath the bananas and climbed up to get them. Observations revealed that this response appeared to follow a period of 'apparent thought' (previous experience of playing with boxes [latent learning] may have increased the likelihood of the response).
Imprinting	A simple and specialised form of learning occurring during receptive periods in an animal's life. The learned behaviour becomes relatively fixed and resistant to change. Imprinting involves young animals becoming associated with, and identifying themselves with, another organism, usually a parent, or some large object. Lorenz found that goslings and ducklings deprived of their parents would follow him and use him as a substitute parent. 'Pet lambs', bottle-fed, show similar behaviour and this may have a profound and not always desirable effect later in life when the animal finds difficulty in forming normal relationships with others of the species. In the natural situation imprinting has obvious adaptive significance in enabling offspring to acquire rapidly skills possessed by the parents, e.g. learning to fly in birds, and features of the environment, e.g. the 'smell' of the stream in which migratory salmon were hatched and to which they return to spawn.

Instinctive behaviour is predominant in the lives of non-vertebrate animals where, in insects for example, short life cycles prevent modifications in behaviour occurring as a result of trial-and-error learning. Instinctive behaviour in insects and in vertebrates therefore is a 'neuronal economy measure' and provides the organism with a ready-made set of behaviour responses. These responses are handed down from generation to generation and, having successfully undergone the rigorous test of natural selection, clearly have important survival significance.

However, before concluding that instinctive behaviour patterns are completely inflexible as a result of their genetic origin, it must be stressed that this is not so. All aspects of the development of an organism, whether anatomical, biochemical, physiological, ecological or behavioural, are the result of the influence of constantly varying environmental factors acting on a genetic framework. In view of this, no behaviour pattern can be purely instinctive (that is genetic) or purely learned (that is environmental), and any subsequently described behaviour activity, whilst being either superficially instinctive or superficially learned, is influenced by both patterns. Some authorities prefer the terms **species-characteristic behaviour**, in preference to instinctive behaviour, and **individual-characteristic behaviour**, in preference to learned behaviour. But despite this terminology the same principle of genetic and environmental interaction applies. This point was demonstrated very clearly by Professor W. H. Thorpe in his investigations of chaffinch song. He found that chaffinches, whether reared by parent, reared in isolation, or deaf from the time of hatching, all produce sounds clearly identifiable to the human ear as those of a chaffinch. Sound spectrograms show, however, that these are only rudimentary songs, and that chaffinches reared by parents, listening to the songs of their parents and those of other chaffinches in the population, develop identical sound patterns to the older birds, characteristic of the local population. It was apparent that bird songs within a species have 'local dialects'. Songs of deaf birds or those in isolation remain rudimentary, thus demonstrating that the environment can significantly modify an instinctive pattern.

17.8.3 Motivation

The extent and nature of any behavioural response is modified by a variety of factors that are collectively known as **motivation**. For example, the *same* stimulus does not always evoke the *same* response in the *same* organism. The difference is always circumstantial and may be controlled by either internal or external factors. Presenting food to a starved animal will produce a different response from that shown by an animal that has been fed. In between the two extremes responses of varying strengths will be produced depending upon the degree of hunger experienced by the organism. However, if the act of feeding would place a hungry animal in danger of being attacked by a predator, the feeding response would be curbed until the danger passed.

Many behavioural responses associated with reproduction have a motivational element. For example, many female mammals are only receptive to mating attempts by males at certain times of the year. These times coincide with the period of oestrus and have the adaptive significance of ensuring that mating coincides with the optimum time for fertilisation and therefore the production of offspring at the most favourable time of the year. These behavioural patterns are known as **biological rhythms** and are described in section 17.8.5. In many species the degree of motivation, or 'drive', coincides in males and females, but in other species some system of communication between the sexes is essential to express the degree of motivation. In many primate species the timing of oestrus is signalled by a swelling and change of colour of the genital area of the female and this is displayed to the male (fig 17.58). Such behaviour reduces the likelihood of a male attempting to mate at a time when the female is not receptive. The signals used to bring about a change in behaviour are known as **sign stimuli** and, depending upon their origin or function, are classified as motivational, releasing or terminating stimuli.

Motivational stimuli

This type of stimulus may be **external**, for example increasing day length inducing territorial and courtship behaviour in birds, or **internal**, for example depleted food stores in the body during hibernation results in awakening and food seeking. Motivational stimuli provide the 'drive' or 'goal' preparing the organism for activity which may be triggered off by the second type of sign stimulus.

Fig 17.58 *Female chimpanzee signalling to the male that she is sexually receptive.*

Releasing stimuli or 'releasers'

A releaser is either a simple stimulus or a sequence of stimuli produced by a member of a species which evokes a behavioural response in another member of the same species. The term 'releaser' was introduced by Lorenz and its role in behaviour was extensively studied by Tinbergen.

The effect of a releaser was demonstrated during an investigation into feeding in herring gulls. Young herring gulls normally peck at a red spot on the yellow lower mandible of the parent's bill to signal the parent to regurgitate fish which the young then swallow. In a series of controlled experiments, carried out by Tinbergen and Perdeck using cardboard models of adult gulls' heads, they found that the releaser of the begging response was the presence of a contrasting colour on the beak. Such was the strength of the releaser that a pointed stick with alternating coloured bands was able to elicit a greater response than the parent birds, as shown in fig 17.59.

Terminating stimuli

Terminating stimuli, as the name implies, complete the behavioural response and may be external or internal. For example, the external visual stimuli of a successfully completed nest will terminate nest building in birds, whereas the internal satisfaction or 'satiety' accompanying ejaculation in the male will terminate copulation and likewise a full stomach will terminate feeding.

Further examples of sign stimuli for a selection of behavioural mechanisms are discussed in sections 17.8.4–17.8.9.

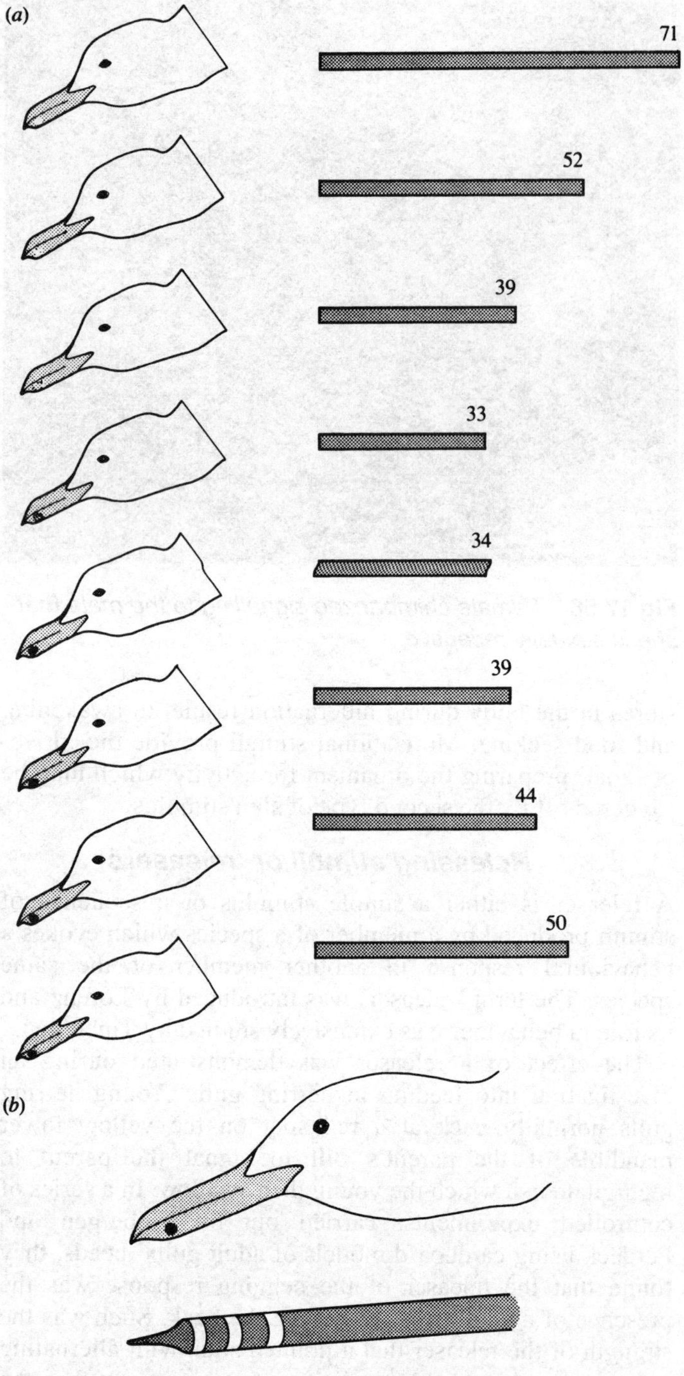

Fig 17.59 *(a) The horizontal bars indicate the number of pecking responses made by herring gull chicks to a series of cardboard models of adult herring gull heads having grey bills and spots of varying shade. (From Hinde, R. A. 1966, after Tinbergen, N., 1951.)*
(b) The artificial bill, coloured red with three white bars evoked 20% more pecks than an accurate three-dimensional model of an adult herring gull head and beak, coloured yellow with a red spot. (From Tinbergen and Perdeck, 1950, Behaviour, *3.1.)*

17.8.4 Innate releasing mechanisms

Lorenz suggested that there must exist a means of filtering out stimuli which are irrelevant from those that are relevant to producing the correct behavioural response. Investigations suggest that this may occur peripherally at the receptors or centrally within the central nervous system. For example, Schneider found that the chemoreceptors on male moth antennae are only sensitive to the sex-attracting chemicals (**pheromones**) produced by the female of that species and not to those of other species. Modifications of Tinbergen and Perdeck's experiments on the herring gull have been carried out, and the results suggest, as Lorenz postulated, that centrally situated neurosecretory mechanisms control the response to sign stimuli.

17.8.5 Biological rhythms

Many behavioural activities occur at regular intervals and are known as biological rhythms or **biorhythms**. Well-known examples of these include the courtship displays and nesting behaviour of birds in the spring and the migration of certain bird species in autumn. The time interval between activities can vary from minutes to years depending on the nature of the activity and the species. For example, the polychaete lugworm *Arenicola marina* lives in a U-shaped burrow in sand or mud and carries out feeding movements every 6–7 min. This cyclical feeding pattern has no apparent external stimulus nor internal physiological motivational stimulus. It appears that the feeding pattern rhythm is regulated by a biological 'clock' mechanism dependent, in this case, on a 'pacemaker' originating in the pharynx and transmitted through the worm by the ventral nerve cord.

Rhythms involving an internal clock or pacemaker are known as **endogenous rhythms**, as opposed to **exogenous rhythms** which are controlled by external factors. Apart from examples such as the feeding behaviour of *Arenicola*, most biological rhythms are a blend of endogenous and exogenous rhythms.

In many cases the major external factor regulating the rhythmic activity is **photoperiod**, the relative lengths of day and night. This is the only factor which can provide a reliable indicator of time of year and is used to 'set the clock'. The exact nature of the clock is unknown but the clockwork mechanism is undoubtedly physiological and may involve both nervous and endocrine systems. The effect of photoperiod has been studied extensively in relation to behaviour in mammals, birds and insects and, whilst it is evidently important in activities such as preparation for hibernation in mammals, migration in birds and diapause in insects, it is not the only external factor regulating biological rhythms. Lunar rhythms, too, can influence activity in certain species, such as the palolo worm of Samoa. This polychaete worm swarms and mates through the whole South Pacific on one day of the year, the first day of the last lunar quarter of the year, on average the

2nd of November. The influence of lunar rhythms on tidal variations is well known, and these are two exogenous factors which have been shown to impose a rhythmic behaviour pattern on the midge *Clunio maritimus*. The larvae of *Clunio* feed on red algae growing at the extreme lower tidal limit, a point only uncovered by the tide twice each lunar month. Under natural conditions these larvae hatch, the adults mate and lay eggs in their two-hour-long life during which they are uncovered by the tide. In laboratory conditions of a constant 12 h light – 12 h dark photoperiod the larvae continued to hatch at about 15-day intervals, demonstrating the apparent existence of an endogenous clock programmed to an approximately semi-lunar rhythm coinciding with the 14.8-day tidal cycle.

The behaviour of many completely terrestrial insects appears to be controlled by endogenous rhythms related to periods of light and dark. For example, *Drosophila* emerge from pupae at dawn whereas cockroaches are most active at the onset of darkness and just before dawn. These regularly occurring biological rhythms, showing a periodicity of about 24 h, are known as **circadian** (*circa*, about; *dies*, day) rhythms or **diurnal** rhythms. In an investigation of the activity of a cockroach (*Periplaneta*) under two different light regimes (12 h light and 12 h darkness for 10 days followed by total darkness of 10 days), the cockroach restricted its activity in the latter regime to a time approximately related to the period of activity associated with the onset of darkness under the former light regime. The results of this investigation are shown in fig 17.60 and indicate that in the absence of an external time-cue the circadian rhythm persisted even though the onset of activity varied by a small amount each day. These results are consistent with the idea that circadian rhythms are controlled by an endogenous mechanism or 'clock', governed or 'set' by exogenous factors.

Circadian rhythms are believed to have many species-specific adaptive significances and one of these involves orientation. Animals such as fish, turtles, birds and some insects which migrate over long distances are believed to use the Sun and stars as a compass. Other animals, such as honeybees, ants and sandhoppers use the Sun as a compass in locating food and their homes. Compass orientation by Sun or Moon is only accurate if organisms using it possess some means of registering time so that allowances can be made for the daily movement of the Sun and Moon. The increasingly familiar concept of 'jetlag' is an example of a situation where the human internal physiological circadian rhythm is out of step with the day-and-night rhythm of the destination.

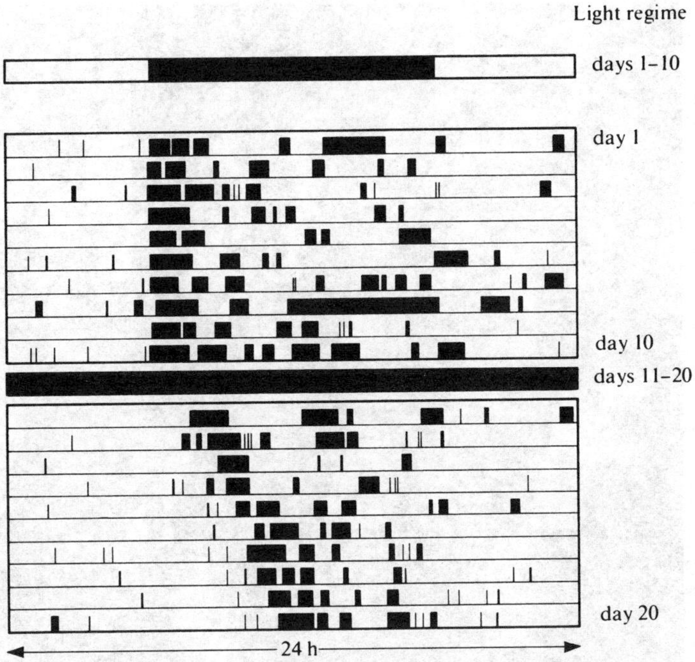

Fig 17.60 *Results of cockroach activity over a 20-day period. During days 1–10 the cockroach received a light regime of 12 hour light : 12 hour dark cycle and between days 11–20 the cockroach was kept in constant darkness as indicated in the figure. The black areas shown for each day represent the time and duration of each 'burst' of activity.*

17.8.6 Territorial behaviour

A territory is an area held and defended by an organism or group of organisms against organisms of the same, or different, species. Territorial behaviour is common in all vertebrates except amphibia, but is rare in non-vertebrates. Research into the nature and function of territoriality has been carried out on birds and groups of primates. In the latter it forms an important part of their social behaviour.

The exact function of territory formation probably varies from species to species, but in all cases it ensures that each mating pair of organisms and their offspring are adequately spaced to receive a share of the available resources, such as food and breeding space. In this way the species achieves optimum utilisation of the habitat. The size of territories occupied by any particular species varies from season to season, according to the availability of environmental resources. Birds of prey and large carnivores have territories several square miles in area in order to provide all their food requirements. Herring gulls and penguins (fig 17.61), however, have territories of only a few square metres, since they move out of their territories to feed and use them for breeding purposes only.

Territories are found, prior to breeding, usually by males. Defence of the area is greatest at the time of breeding and fiercest between males of the same species. There are a variety of behavioural activities associated with territory formation and they involve threat displays between owners of adjacent territories. These threat displays involve certain stimuli which act as releasers. For example, Lack demonstrated that an adult male robin (*Erithacus rubecula*) would attack a stuffed adult male robin displaying a red breast, and a bunch of red breast feathers, but not a stuffed young male robin which did not have a red breast. The level

Fig 17.61 *Aerial photograph showing the territories occupied by penguins.*

of aggression shown by an organism increases towards the centre of the territory. The aggressiveness of males is determined partly by the level of testosterone in the body and this can affect territory size. For example, the territory size of a red grouse can be increased by injecting the bird with testosterone. Fig 17.62 shows the changes in territory size of three red grouse treated in this way. The boundary between adjacent territories represents the point where neighbouring animals show equally strong defence behaviour. Despite the apparent conflict and aggression associated with territory formation, actual fighting, which would be detrimental to the species, is rare and is replaced by threats, gestures and postures. Having obtained a territory, many species, particularly carnivores, proceed to mark out the boundary by leaving a scent trail. This may be done by urinating or rubbing glandular parts of the body against objects called **scent posts** along the boundary of the territory.

Although territorial behaviour involves the sharing of available resources amongst the population there are inevitably some organisms unable to secure and defend a territory. In many bird species, such as grouse, these weaker organisms are relegated to the edges of the habitat where they fail to mate. This appears to be one of the adaptive significances of territoriality as it ensures that only

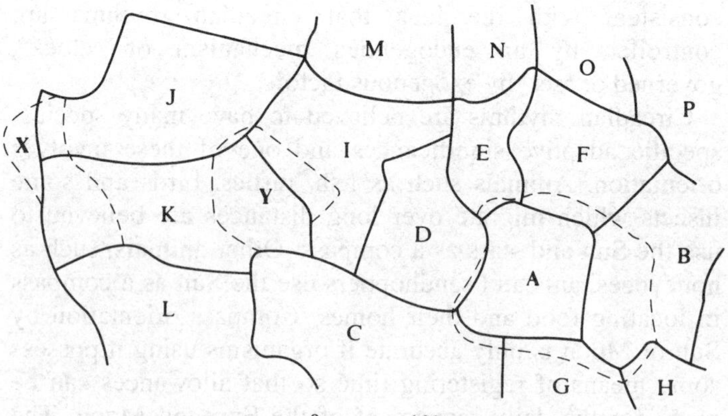

Fig 17.62 *The solid lines indicate the territories of a group of male red grouse. The dotted lines show changes which occurred after birds A, X and Y had received doses of testosterone. Birds X and Y had not previously held territories. (After Watson, A. (1970).* J. Reprod. Fert., *Suppl., 11.3.)*

the 'fittest' find a territory, breed and thus passes on their genes to the next generation. Thus a further function of territorial behaviour is associated with **intraspecific competition** and may act as a means of regulating population size.

17.8.7 Courtship and mating

There are many elaborate and ritualistic species-specific behaviour patterns associated with courtship and mating. In birds, mammals and some fish these two processes often follow the establishment of a territory by the male. Courtship is a complex behaviour pattern designed to stimulate organisms to sexual activity, and is associated with pair formation in those species where both sexes are involved in the rearing of offspring, as in thrushes, or in gregarious mixed-sex groups such as baboons. The majority of these species show rhythmic sexual activity of the type described in section 17.8.5.

Courtship behaviour is controlled primarily by motivational and releasing stimuli and leads to mating which is the culmination of courtship. During mating, the behavioural activities are initiated by releasing stimuli and ended by terminating stimuli associated with the release of gametes by the male.

The motivational stimuli for courtship in most species are external, such as photoperiod, and lead to rising levels of reproductive hormones and the maturation of the gonads. In most species this produces striking changes in the secondary sexual characteristics and other behavioural activities including colouration changes, as in the development of a red belly in male sticklebacks; increase in size of parts of the body, as in the plumage of birds of paradise; mating calls, as in nightingales; postural displays, as in grebes (fig 17.63) and the use of chemical sex attractants, as in butterflies and moths.

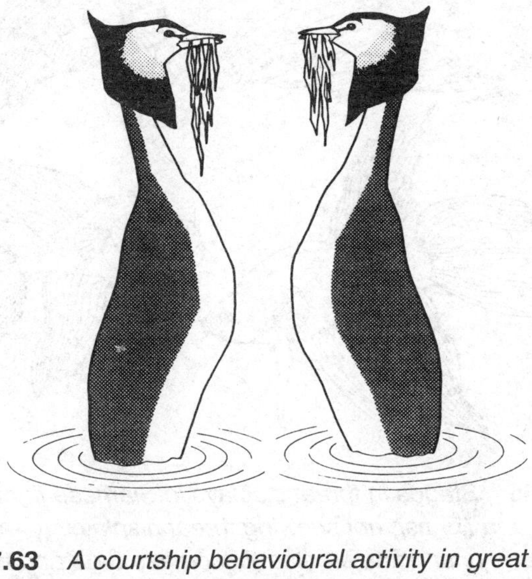

Fig 17.63 *A courtship behavioural activity in great crested grebes. Male and female grebes are shown here presenting nesting material to each other. (After Huxley, J. S. (1914).* Proc. Zool. Soc. Lond., *1914 (2), 491–562.)*

Of the variety of signals used in the courtship to attract members of the opposite sex, sight, sound and smell play important roles. For example, the male fiddler crab, *Uca*, uses a visual display and attracts females by waving an enlarged chela in a bowing movement similar to that of a violinist. The vigour of the movement increases as a female is attracted to the male. Many insects, amphibia, birds and mammals use auditory signals in courtship. Some species of female mosquito attract males by the sounds produced by the frequencies of their wing beats, whilst grasshoppers, crickets and locusts **stridulate**. This involves either rubbing the hindlegs against each other or against the elytron (hardened wing case), or rubbing the elytra together to produce a 'chirping' sound which is species-specific and only produces a response from members of *that* species.

Some species of spiders employ a mechanical means of attracting the opposite sex; male spiders approach the web of a female sitting at the centre of the web and pluck a thread of the web at a species-specific frequency. The plucking 'serenades' the female and reduces her natural aggressive manner so enabling the male to approach and mate her. Unfortunately, if the male 'woos' a female of the wrong species or 'plays the wrong tune' he is attacked and killed!

The secretion and release by organisms of small amounts of chemical substances, leading to specific physiological or behaviour responses in other members of the same species, is used in courtship and mating and, as described later, the regulation of behaviour within social groups. These substances are called **pheromones** and are usually highly volatile compounds of low relative molecular mass. Many of these compounds function as natural sex attractants and the earliest to be identified were **civetone** from the civet cat and **muscone** from the musk deer. Both of these substances are secretions from the anal glands and are used commercially in the preparation of perfume. Mares, cows and bitches secrete pheromones whilst on 'heat'. This is undetectable by human olfactory epithelium but detectable by the males of the species concerned. **Bombykol**, a pheromone released by reversible glands at the tip of the abdomen of unfertilised adult female silk moths, is capable of attracting males of the same species from considerable distances. The olfactory receptors on the antennae of male moths detect the presence of the pheromone molecules in great dilutions and the males make a rheotactic response by flying upwind until they reach the female. Pheromones are used increasingly as a method of **biological control** in insect pest species such as the gypsy moth. In these cases the artificial release of the pheromone **gyplure** attracts males to the source of release where they can be captured and killed. This not only immediately reduces the number of male moths in the population but also, in preventing them from breeding, reduces the size of the next generation.

Pheromones are also used to induce mating as in the case of the queen butterfly *Danaus gilippus*. The pheromone is released by the male and brushed on to the female by a pair

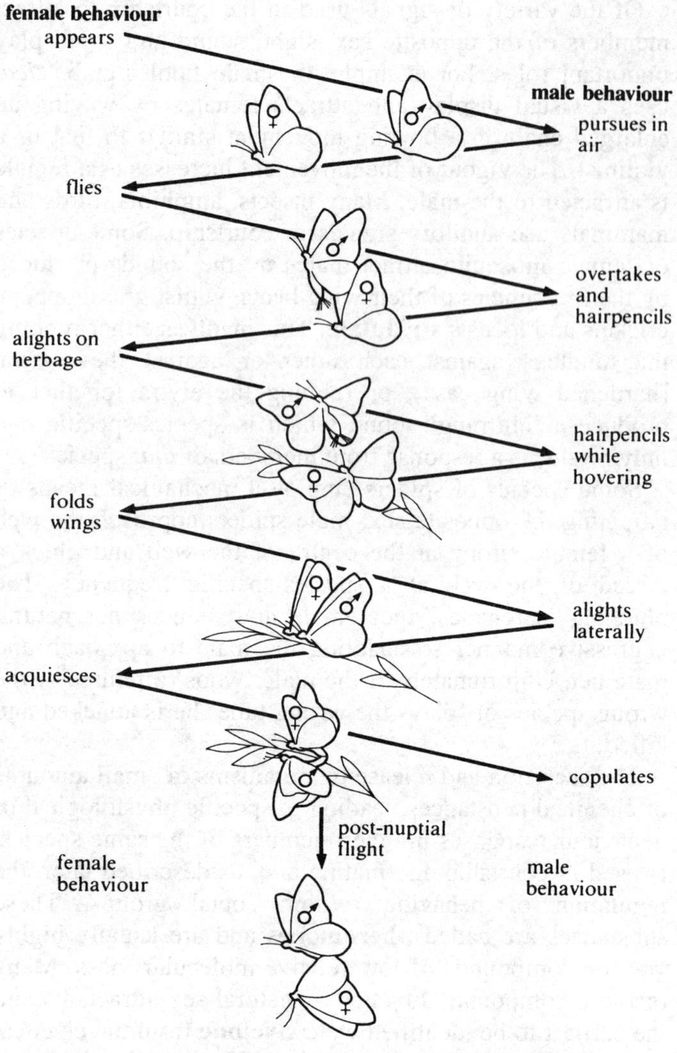

Fig 17.64 *Courtship and copulation in the queen butterfly. (The arrows indicate the stimuli and responses involved in the behavioural activities.) (From Brower, L. P., Brower, J. V. Z. and Cranston, F. P. (1965). Zoologica, 50, 18.)*

of brush-like structures, called hairpencils, everted from the tip of the abdomen. The entire courting and mating sequence is shown in fig 17.64.

Courtship in some species is accompanied by conflict behaviour on the part of one or both sexes. In species where individuals normally live a solitary existence, courtship conflict may be associated with changing attitudes to other members of the species as a result of increasing hormone levels. Other significances of this behaviour may be the tightening of the pair bond between the mating pair and the synchronisation of gonad development so that gametes mature at the same time. In certain species of spider, such as wolf spiders, conflict between male and female only diminishes for the act of copulation which culminates in the female killing the male.

17.8.8 Aggression (agonistic behaviour)

Aggression is a group of behavioural activities including threat postures, rituals and occasionally physical attacks on other organisms, other than those associated with predation. They are usually directed towards members of the same sex and species and have various functions, including the displacement of other animals from an area, usually a territory or a source of food, the defence of a mate or offspring and the establishment of rank in a social hierarchy.

The term 'aggression' is emotive and suggests an existence of unnecessary violence within animal groups; the alternative term '**agonistic**' is preferable. Agonistic behaviour has the adaptive significance of reducing intraspecific conflict and avoiding overt fighting which is not in the best interests of the species. Most species channel their 'aggression' into ritual contests of strength and threat postures which are universally recognised by the species. For example, horned animals such as deer, moose, ibex and chamois may resort to butting contests for which 'ground rules' exist. Only the horns are allowed to clash and they are not used on the exposed and vulnerable flank. Siamese fighting fish, *Betta splendens*, resort to threat postures involving increasing their apparent size as shown in fig 17.65.

The threats issued by two organisms in an agonistic conflict situation are settled invariably by one of the organisms, generally the weaker, backing down and withdrawing from the situation by exhibiting a posture of submission or appeasement. In dogs and wolves an appeasement posture may take the form of the animal lying down on its back or baring its throat to the victor.

Fig 17.65 *Stages in threat displays in Siamese fighting fish. (a) and (b) fish not showing threat displays, (c) and (d) operculum, o, and fins erected to increase their apparent size during threat displays. (From Hinde, R. A., 1970, after Simpson, 1968.)*

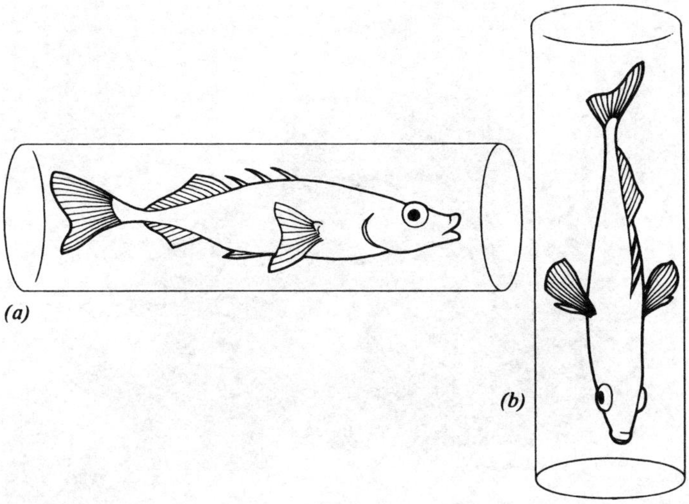

Fig 17.66 *(a) Full threat posture in male stickleback. (b) Reduced threat posture when contained in a tube held vertically. (After Tinbergen, N., 1951.)*

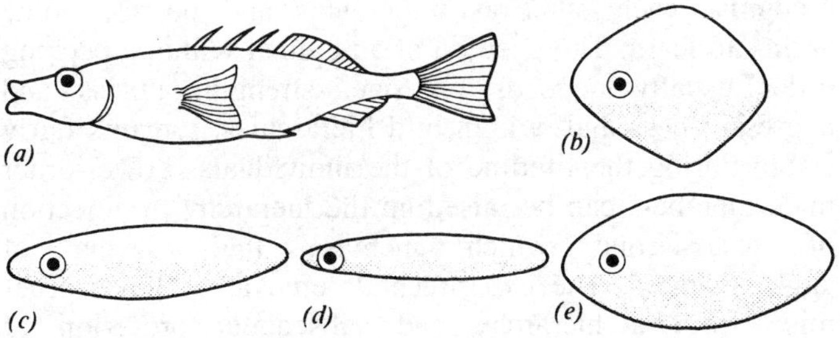

Fig 17.67 *Models used as releasers of aggressive behaviour in male sticklebacks holding a territory. (a) Accurate model not having a red belly does not elicit aggression from male stickleback. (b)–(c) are models of sticklebacks not having an accurate shape but having a red belly. All these models produced aggressive responses from male sticklebacks. (After Tinbergen, 1951.)*

During actual physical contact animals often refrain from using their most effective weapons on another member of the same species. For example, giraffes will fight each other using the short horns on their heads, but in defence against other animals they use their very powerful feet.

For agonistic behavioural activities to be most effective they must be stereotyped for any species and Tinbergen demonstrated several of these during investigations carried out on sticklebacks. In one series of experiments he demonstrated that the effectiveness of threat posture depended on the stickleback assuming a horizontal position with fins and spines outstretched. When trapped in a specimen tube and kept vertical a male stickleback does not have the same threat potential to ward off other sticklebacks as it has when free-swimming or held horizontally (fig 17.66).

In another series of experiments he demonstrated that agonistic behaviour in male sticklebacks defending a territory is triggered off by 'releasers' which can take the form of almost any object whose underside is coloured red. These objects act as mimics of male sticklebacks whose bellies turn red during the breeding season and who appear to the territory holder as a potential threat (fig 17.67).

At times of stress, for example during conflict situations or during courtship and mating, an organism may perform an action which is trivial and irrelevant to the situation. This is known as **displacement activity** and occurs when motivation is high but two conflicting 'releasers' present themselves. For example, one of a pair of birds involved in a territorial dispute may begin nest-building activities, such as pulling up grass, when presented with a choice between fighting or fleeing. Such displacement activities act as an outlet for pent-up activities. Many human activities may be considered displacement activities in certain circumstances, for example fist clenching, fist banging, nail biting, straightening clothes, finger drumming, etc. A similar form of behaviour is called **vacuum activity** which occurs when motivation is high and no releaser presents itself. In this case the normal response is produced but is not directed towards the normal object or situation, and so provides a means of reducing frustration; for example, showing irritation towards someone who is not the cause of the irritation but acts as a substitute.

17.8.9 Social hierarchies

Many species of insects and most vertebrates show a variety of group behavioural activities associated with numbers of individuals living together temporarily or permanently. This is known as **social behaviour** and the coherence and cooperation achieved has the adaptive significance of increasing the efficiency and effectiveness of the species over that of other species. In a social group of this kind a system of communication is essential, and the efficiency of the organisation is further increased by individuals carrying out particular roles within the society. One aspect of social behaviour arising out of these points is the existence of **social hierarchies** or **pecking orders**.

A pecking order is a **dominance hierarchy**. That is to say that animals within the group are arranged according to status. For example, in a group of hens sharing a hen-house a linear order is found in which hen A will peck any other hen in the group, hen B will peck all hens other than A and so on. Position in the hierarchy is usually decided by some agonistic form of behaviour other than fighting. Similar patterns of dominance have been observed in other species of birds and in mice, rats, cows and baboons. The institutional organisation of all human societies is based on a pattern of dominance hierarchy.

Pecking orders exist only where animals are able to

recognise each other as individuals and possess some ability to learn. The position of an animal within a pecking order usually depends on size, strength, fitness and aggressiveness and, within bird hierarchies, remains fairly stable during the lifetime of the individuals. Lower-order male members can be raised up the hierarchy by injection of testosterone which increase their levels of aggressiveness. The experimental removal of lower-order mice from a hierarchy and subsequent provision of unlimited food for them increases their mass, improves their vigour and can raise their position in the hierarchy when reintroduced to the group. Similarly placing lower-order mice into other groups where they are dominant appears to give them a degree of 'self-confidence' (to use an anthropomorphic term) which stays with them when reintroduced to their original groups and results in their rank increasing.

One advantage of pecking order is that it decreases the amount of individual aggression associated with feeding, mate selection and breeding-site selection. Similarly it avoids injury to the stronger animals which might occur if fighting was necessary to establish the hierarchy. Another advantage of pecking order is that it ensures that resources are shared out so that the fittest survive. For example, if a group of 100 hens is provided with sufficient food for only 50 hens, it is preferable, in terms of the species, for 50 hens to be adequately fed and the weaker 50 hens die than for them all to live and receive only half rations, as this might prevent successful breeding. In the short term, social hierarchies increase the genetic vigour of the group by ensuring that the strongest and genetically fittest animals have an advantage when it comes to reproducing.

Social organisation

When animals come together to form a cohesive social group individuals often assume specialised roles, which increases the overall efficiency of the group (fig 17.68). These roles include members specialised or designated for food-finding, reproduction, rearing and defence. Cooperation between members of a society sharing division of labour depends upon stereotyped patterns of behaviour and effective means of communication. These patterns of behaviour and methods of communication vary between species and are vastly different for primate and insect societies. Primate societies are flexible, in that roles are interchangeable between members of the group, whereas in insect societies differences in body structure and reproductive potential affect their role within the society, a feature called **polymorphism**.

Ants, termites and bees are social insects that live in colonies and have an organisation based on a **caste system**. In the honeybee colony there is a single fertile female queen, several thousand sterile female workers and a few hundred fertile male drones. Each type of honeybee has a specific series of roles determined primarily by whether it hatched from a fertilised or an unfertilised egg. Fertilised eggs are diploid and develop into females; unfertilised eggs are haploid and develop into males. Secondly, the type of food provided for female larvae determines whether they will become queens or workers. This food is called **royal jelly** and is one example of the importance of chemical substances in the organisation of the society. Information within the colony is transmitted either by chemical odours and pheromones during the many licking and grooming activities called **trophallaxes**, or by particular forms of visual orientation displays known as **dances**.

Fig 17.68 *Social grooming in adult chimpanzees provides a means of social cohesion within a group.*

Karl von Frisch, a German zoologist and Nobel prize-winner, investigated the nature of these dances using marked worker bees in specially constructed observation hives. Worker bees 'forage' for sources of nectar and communicate the distance and direction of the source to other workers by the nature of a dance generally performed on a vertical comb in the hive. If the distance is less than about 90 m the worker performs a **round dance** as shown in fig 17.69*a* which intimates that the source is less than 90 m from the hive but gives no indication of direction. The **waggle dance** is performed if the source is greater than 90 m, and includes information about distance of source from the hive and its direction relative to the hive and the position of the Sun. The dance involves the worker walking in a figure-of-eight and waggling her abdomen during which, according to von Frisch, the speed of the dance is inversely related to the distance of the food from the hive; the angle made between the two loops of the figure-of-eight and the vertical equals the angle subtended at the hive by the Sun; and the food source and the intensity of the waggles is related to the amount of food at the source (fig 17.69*b*). It is thought that allowances for movement of the Sun are made by the use of an inbuilt 'biological clock' and that bees orientate on cloudy days by substituting polarised light from the Sun for the position of the Sun.

Recent evidence has suggested that bees may use high-frequency sound to communicate sources of food to other

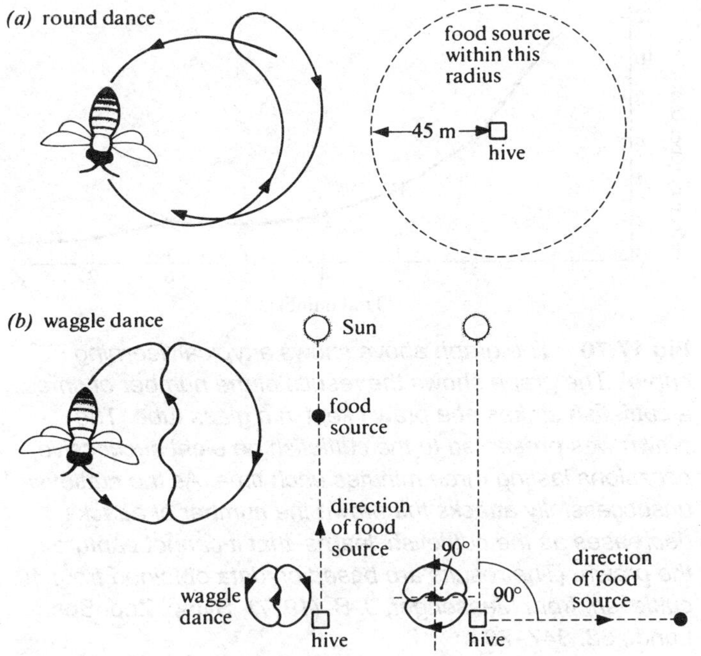

Fig 17.69 *Honeybee dances.*
(a) Round dance is performed when food is less than 90 metres from hive. (b) Waggle dance shows relation between hive, sun and direction of food source.

workers, but whether this is the main means of communication has yet to be demonstrated clearly. This does not, though, invalidate the data and interpretations of von Frisch and may be an associated communication system which augments the visual dance displays. The returning worker may also communicate the type of flower visited by feeding the other workers with some of the nectar collected.

17.8.10 Altruistic behaviour

One area of social behaviour which is not fully understood concerns the way in which certain organisms expend time and energy in caring for other members of the species. This phenomenon is called **altruism** and refers to a form of social behaviour whereby one organism puts itself either at risk or personal disadvantage for the good of other members of the species.

In the case of activities associated with mating and parental care, altruism is not so difficult to comprehend since the action is clearly in the interests of the parents, offspring and species. For example, the female baboon protects and cares for its offspring for almost six years whilst most bird species feed and protect their demanding offspring until they are capable of fending for themselves. What is not so clear is the reason why some organisms give support to organisms which are *not* their offspring, for example, birds and monkeys that call out warnings to others in danger and female monkeys who carry and care for the babies of other monkeys.

One insight into the mechanisms regulating this type of behaviour is seen in the **eusocial** insects such as honeybees, wasps and ants that have a caste system. Here the advantages to the species of division of labour based on a social hierarchy are apparent, but what is slightly obscure is the mechanism by which such behaviour arose.

Here the sterile female workers are prevented, by definition, from producing offspring, yet they spend their lives looking after their brothers and sisters. Investigation of the chromosome composition of the queen, drones and workers show that sisters (queen and workers) are more closely related to each other than mothers are to sons and daughters. This is because the fertile queen is diploid, the sterile worker is diploid and the drone (male) is haploid. Hence by helping their sister (queen) to reproduce they are effectively aiding in the production of queens, workers and drones with a genetic complement closer to their own than if they had offspring of their own. The conferring of a genetic advantage on closely related organisms forms the basis of altruistic behaviour.

Altruistic behaviour is very common amongst primates and varies from the extremes of social protection which exist between members of the same troop (monkeys), through acts of mutual grooming and food sharing (apes) to deliberate acts of self-sacrifice for family, God and country (humans). The extent of the altruistic behaviour appears to be related to close relatives (**kin**) such as offspring and **siblings** (brothers, sisters and cousins) with whom they share certain alleles. Thus the adaptive significance of altruistic behaviour is to increase the frequency of those alleles common both to the donor and recipient(s) of the altruistic behaviour. This behaviour is called **kin selected** and has led to the establishment of altruism because of the way it confers genetic advantages in kin by promoting survival and reproduction within the species.

In situations where the altruistic behaviour is directed generally towards members of the species rather than to close relatives, it is postulated that again the result of the behaviour will enable selection of those alleles responsible for the behaviour to be perpetuated within the benefiting group. This conclusion is backed up by observation which reveals that such general altruistic behaviour is much commoner in species such as the zebra, which live in coherent families, than in species such as the wildebeest where family groupings are uncommon.

Parallel examples of altruistic behaviour occur in humans too, where once again the intensity of the behaviour is related to kinship, and responses are strongest between family members sharing the same alleles.

17.9 Learned behaviour

17.9.1 Memory

Memory is the ability to store and recall the effects of experience and without it learning is not possible. Past experiences, in the form of stimuli and responses, are recorded as a 'memory trace' or **engram** and, since the extent of learning in mammals is proportional to the extent of the cerebral hemispheres, it would appear that these are the site of engram formation and storage.

The nature of the engram is not known and it exists only as a hypothetical concept backed up by conflicting data. Of the two broad areas of thought on the nature of the engram, one is based on changes in neuronal structure and organisation within the central nervous system and the other is based on permanent changes in brain biochemistry.

Histological examination of brain tissue shows the existence of neurones arranged in loops, and this has given rise to the concept of '**reverberating circuits**' as units of the engram. According to this view these circuits are continuously active carrying the memory information. It is doubtful if this activity could last for any length of time, and experiments suggest that memory has greater performance and stability than could be achieved by this mechanism alone. For example, cooling the brains of rats down to 0 °C causes all electrical activity in the nervous system to cease, but on restoring the rats to normal temperatures there is no impairment in memory. However, it is thought that such circuits may play a role in **short-term memory**, that is memory lasting at most for minutes, and in facilitating particular neural pathways. Events associated with short-term memory take longer to be recalled following concussion or amnesia and gradually disappear in old age. Long-term memory is more stable and suggests that some mechanism for permanent change exists in the brain.

Evidence based on the latter observation suggests that memory is a biochemical event involving the synthesis of substances within the brain. Extracts of the 'brains' of trained flatworms or rats injected into untrained flatworms or rats reduce the time taken by the latter organisms to learn the same task as compared with control groups. The active substance in all the experiments appears to be RNA.

Further evidence exists which suggests that the composition of the RNA of neurones changes during learning and that this may result in the synthesis of specific 'memory proteins' associated with the learned behaviour. Investigations have shown that injections of the protein-inhibiting drug, puromycin, also interfere with memory. For example, injecting puromycin into the brains of mice recently trained to choose one direction in a maze destroys their ability to retain this learning, whereas a control group, injected with saline, retained the learned behaviour. In conclusion, it would appear that the nature of memory is far from being clarified, but it seems probable that changes in electrical properties of neurones, the permeability of synaptic membranes, enzyme production associated with synapses and synaptic transmission are all concerned with formation of a 'memory trace'. Certainly it seems that memory is associated closely with events occurring at synapses.

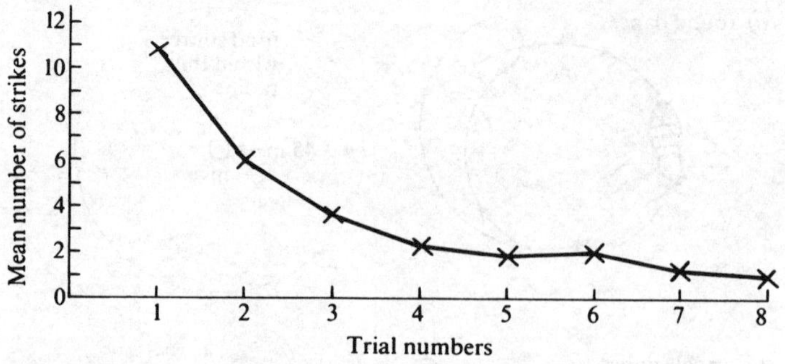

Fig 17.70 *The graph above shows a typical 'learning curve'. The graph shows the results of the number of times a cuttlefish strikes at a prawn kept in a glass tube. The prawn was presented to the cuttlefish on eight successive occasions lasting three minutes each time. As the cuttlefish unsuccessfully attacks the prawn the number of attacks decreases as the cuttlefish 'learns' that it cannot capture the prawn. (The results are based on data obtained from 40 cuttlefish, from Messenger, J. B. (1977),* Symp. Zoo. Soc. Lond., *38, 347–76.)*

17.9.2 Learning

Learning is an adaptive change in individual behaviour as a result of previous experience (fig 17.70). The degree of permanence of newly acquired learned behaviour patterns depends upon memory storing the information gained from the experience. In humans, acquiring or learning 'facts', for example for examinations, may be short-lived whereas the ability to carry out coordinated motor activities such as toilet training, riding a bicycle or swimming, lasts throughout life. Learning is generally thought of in terms of vertebrates, and mammals in particular, but has been demonstrated in all groups of animals except protozoans, cnidarians and echinoderms where neural organisation is absent or poorly developed. Psychologists have attempted to establish general 'laws of learning' but all attempts so far have failed. It would appear that learning is an individual event and occurs in different ways in different species and different contexts.

The classification and features of learned behaviour presented in this chapter are artificial and must be recognised as such. They do, however, cover the spectrum on current thinking on types of learning and are backed up by experimental evidence. A summary of the major types of learned behaviour is given in table 17.16 and is designed to provide only an introduction to the topic of learning.

Chapter Eighteen

Movement and support in animals

Movement and locomotion

Movement can occur at:

- cell level, for instance cytoplasmic streaming and swimming of gametes;
- organ level, such as heartbeat and movement of a limb;
- the level of the organism.

Movement of the whole organism from place to place is termed **locomotion**. Plants show cell and often organ movement, but they do not show locomotion, that is move from place to place in search of food or water. Plant movements are considered in chapter 16.

Whilst a few animals can survive successfully by remaining attached to one place (**sessile**), the vast majority have locomotory systems which presumably evolved to enable them to search for and acquire food. However, even sessile animals exhibit a great degree of mobility of their bodily parts.

Locomotion is used for:

- finding food;
- avoiding capture by predators;
- dispersal;
- finding new and favourable habitats;
- bringing together individuals for reproductive activity.

Locomotion involves coordination between the nervous, muscular and skeletal systems. Muscles used for locomotion are attached to the skeleton and are therefore called **skeletal muscles**. They act as machines, converting chemical energy into mechanical energy. They have the ability to contract and, when they do, they move systems of levers (bones) that make up part of the skeleton. Coordinated movement of the levers enables the animal to move about. The posture of the animal is also maintained by the musculo-skeletal system which is under the overall control of the central nervous system.

Other muscles within the body serve not to move the whole organism but to move materials from place to place within it. Cardiac muscle (section 6.5) of the heart pumps blood round the body, whilst smooth muscle located in the walls of various blood vessels constricts or dilates them to alter blood flow. Smooth muscle in the wall of the gut propels food along by means of peristalsis (section 8.3.5). These are just a few of many such activities which are constantly occurring within the body.

In this chapter we will be primarily concerned with locomotion, and two systems will be discussed in detail, namely the skeletal and muscle systems. This will be followed by a review of the types of locomotion that occur in a few representative organisms.

Support

As animals and plants increased in size through the process of evolution, the need for support became greater. This was particularly true once living organisms left water and colonised land. The skeleton in animals, and various mechanical tissues in plants, contribute to this support. In plants, the relevant mechanical tissues are collenchyma, sclerenchyma, and xylem. In addition, the turgid cells of the parenchyma are important for support. These tissues and their roles are discussed in chapters 6 and 22.

18.1 Skeletal systems

18.1.1 Functions of skeletons

The general functions of a skeleton are as follows.

- **Support.** The vast majority of animals possess some form of supportive structure. Structures of different design are needed for aquatic or terrestrial animals, animals with four legs or two legs, and for those that move over the ground or through the air. However, all skeletons provide a rigid framework for the body and are resistant to compression and tension (stretching) forces. They help to maintain the shape of the body. For terrestrial organisms the skeleton supports the weight of the body against gravitational force and in many cases raises it above the ground. This permits more efficient movement over the ground. Within the body, organs are attached to, and suspended from, the skeleton.
- **Protection.** The skeleton protects the delicate internal organs in those organisms with an exoskeleton (arthropods, section 18.1.3), and parts of the endoskeleton (section 18.1.4) are designed for a similar function. For example, in humans the cranium (skull) protects the brain and the sense organs of sight, smell, balance and hearing; the vertebral column protects the spinal cord, and the ribs and sternum protect the heart, lungs and large blood vessels.

- **Locomotion.** Skeletons composed of rigid material provide a means of attachment for the muscles of the body. Parts of the skeleton operate as levers on which the muscles can pull. When this occurs, movement takes place. Soft-bodied animals rely on muscles acting against body fluids to produce their form of locomotion (section 18.1.2).
- In addition the skeleton may have other functions, such as **making blood cells** and acting as a **store of calcium and phosphate** (see section 18.2.2).

Three major types of skeleton are generally recognised, namely the hydrostatic skeleton, exoskeleton and endoskeleton.

18.1.2 Hydrostatic skeleton

This is characteristic of soft-bodied animals. Here fluid is secreted within the body and surrounded by the muscles of the body wall. The fluid presses against the muscles which in turn are able to contract against the fluid. The muscles are not attached to any structures and thus they can only pull against each other. The combined effect of muscle contraction and fluid pressure serves to maintain the shape and form of the animal. Generally there are two muscle layers, one longitudinal and the other circular. When they act antagonistically against each other locomotion is achieved. If the body is segmented (as in the earthworm) then such pressure is localised and only certain segments will move or change shape. A detailed account of the function of the hydrostatic skeleton in locomotion is given for the earthworm in section 18.5.1.

18.1.3 Exoskeleton

This is a particular characteristic of the arthropods. Epidermal cells secrete a non-cellular exoskeleton, composed mainly of **chitin**. It acts as a hard outer covering to the animal and is made up of a series of plates or tubes covering or surrounding organs. Chitin is very tough, light and flexible. However, it can be strengthened by impregnation with 'tanned' (hardened) proteins, and, particularly in the aquatic crustaceans like crabs, by calcium carbonate. Where flexibility is required, as at the joints between plates or tubes, chitin remains unmodified. This combination of a system of plates and tubes joined together by flexible membranes provides both protection and mobility.

Arthropods are the only non-vertebrate group to possess jointed appendages. The joints are hinges, and the levers on either side are operated by flexor and extensor muscles which are attached to inward projections of the exoskeleton (fig 18.1). Chitin is permeable to water which could lead to desiccation of terrestrial animals like insects. This is prevented by the secretion of a thin waxy layer over the exoskeleton from gland cells in the epidermis. Therefore, the exoskeleton supports and protects the delicate inner parts of the animal and in addition prevents their drying up.

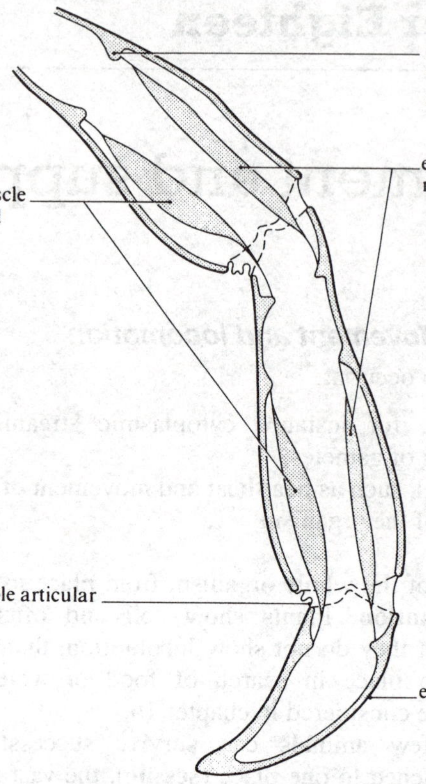

Fig 18.1 *Vertical section of an arthropod limb showing joints and muscles.*

The hollow tubular form of the exoskeleton is very efficient for support and locomotion in small animals, such as most arthropods, and can support a much greater weight without giving way than a solid cylindrical strut (like a bone) *of the same mass*. However, it loses this efficiency when organisms become bigger and their mass increases. As noted in section 2.8.6, as body size increases the surface area to volume ratio decreases. The extent of the exoskeleton depends on surface area. In large animals the exoskeleton would have to increase enormously in weight and thickness to do the same job just as efficiently. The end product would then be very heavy and cumbersome.

Growth takes place by **ecdysis** (moulting) in juvenile stages (larvae and nymphs) in insects and throughout adult life in crustaceans. This involves shedding the exoskeleton, thus exposing a new soft exoskeleton. Whilst still soft, growth takes place and the exoskeleton is extended and moulded into a larger form which often includes a change of shape. The new exoskeleton then hardens. The animal is vulnerable to predators whilst the new exoskeleton is hardening. At this stage the skeleton is unable to support the weight of the animal and movement is virtually impossible. This is less of a problem for aquatic species as their body weight is supported by the water, but aquatic and terrestrial organisms usually hide away during this time in an attempt to decrease their chances of being eaten by predators. Moulting is quite expensive in terms of the energy expenditure involved in building the exoskeleton in the first place and material loss when it is shed.

18.1.4 Endoskeleton

This is an internal skeleton and is the type which is typical of vertebrates (the only other animals with an internal skeleton being certain molluscs such as cuttlefish). In vertebrates the skeleton is:

- made either of cartilage or bone (unlike the chitin of exoskeletons);
- located inside the organism, with muscles 'outside' (in contrast to the exoskeleton where muscles are inside the skeleton);
- made of living tissue and so can grow steadily within the animal – this avoids the necessity for moulting which is typical of animals with exoskeletons;
- jointed, like the exoskeleton, although the joints are more complex. A number of different types of joint exist and bones that form them are maintained in their correct positions by ligaments.

Skeletal design in quadrupeds (four-legged animals) and bipeds (two-legged animals) is essentially the same, but there are slight differences at the shoulder and hip. This is associated with the type of locomotion shown by the animals concerned.

18.2 Skeletal tissues

The vertebrate skeleton is composed either of cartilage or bone. Both tissues provide an internal supporting framework for the body. Only cartilaginous fish (such as dogfish and sharks) possess a completely cartilaginous endoskeleton. All other vertebrates have a bony skeleton in their adult form, but with cartilage also present in certain regions, such as at the joints or between the vertebrae. In the embryo stage the skeleton of bony vertebrates is first laid down as hyaline cartilage (section 6.4.2). This is of great biological significance, as cartilage is capable of internal enlargement, and so different parts of the skeleton are able to grow in proportion to each other during the development of the organism. Bone is different from cartilage in this respect as it can grow only by addition of material to its outer surface.

18.2.1 Cartilage

Three types of cartilage are recognisable: **hyaline cartilage**, **white fibrous cartilage** and **yellow elastic cartilage**. A detailed account of their structure can be found in section 6.4.2. All types consist of a hard matrix penetrated by numerous connective tissue fibres. The matrix is secreted by living cells called **chondroblasts**. These later become enclosed in spaces (lacunae) scattered throughout the matrix. In this condition the cells are termed **chondrocytes**.

Hyaline cartilage is the most common type and is found particularly at the ends of bones articulating (meeting and moving relative to each other) to form joints. Its matrix of chondroitin sulphate is compressible and elastic, and is well able to withstand heavy weight and absorb sharp mechanical shocks such as might take place at joints. Embedded within the matrix are fine collagen fibres which provide resistance to tension (stretching). Dense connective tissue, called the perichondrium, surrounds the outer surface of this cartilage at all places except where it passes into the cavity of a joint.

White fibrous cartilage contains a dense meshwork of collagen fibres and is found as discs between vertebrae and as a component of tendons. It is very strong yet possesses a degree of flexibility. Yellow elastic cartilage possesses many yellow elastic fibres and is located in the external ear, epiglottis and pharynx.

18.2.2 Bone

Bone is a hard, tough connective tissue composed mainly of calcified material. Details of its structure can be found in section 6.4.2. It has a hard matrix (harder than cartilage) which contains living cells. The matrix is designed to resist both compression and tension. The mineral part of the matrix (about 70%) is a form of calcium phosphate and is very resistant to compression (high compressional strength). The organic part of the matrix (about 30%) includes many collagen fibres which are very resistant to tension (high tensile strength). These forces are discussed in more detail in section 18.2.3. The arrangement of the tissue into cylinders, which is described in section 6.4.2, also increases the strength of the whole bone.

When a vertical section of a long bone, such as a femur, is examined microscopically it is seen to be made up of several distinct components. It consists of a hollow shaft or **diaphysis**, with an expanded head or **epiphysis** at each end. Covering the entire bone is a sheath of tough connective tissue, the **periosteum**. The diaphysis is composed of compact bone whilst the epiphyses are composed of spongy bone covered by a thin layer of compact bone (section 6.4.2). The layout of the bony material is designed to withstand compression forces and to give maximum strength to the bone (fig 18.2).

Fatty yellow marrow occupies the marrow cavity of the diaphysis, whilst red marrow is present amongst the bony struts **(trabeculae)** of the epiphyses. Numerous small openings penetrate the surface of the bone, through which nerves and blood vessels cross into the bony tissue and marrow.

Apart from the functions listed at the beginning of section 18.1, a bony skeleton also produces red blood corpuscles and white blood cells. Also it takes part in the maintenance of constant calcium and phosphorus levels in

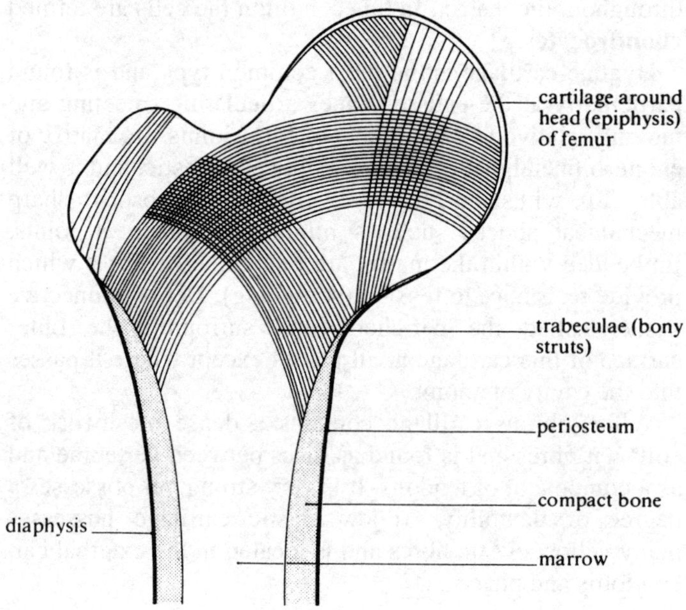

Fig 18.2 *Vertical section of the head of a femur showing arrangement of trabeculae (bony struts) in spongy bone.*

the bloodstream (see chapter 17) by providing a store of calcium and phosphate ions. These can be mobilised by the action of parathyroid and calcitonin hormones of the parathyroid and thyroid glands respectively.

18.2.3 Structure related to function

Using the femur (the thigh bone) as an example, we can now look at how the different tissues associated with a bone are adapted for the functions they perform. The femur is one of the 'long' bones of the body, which are involved in locomotion.

The femur is basically a hollow bone. When a compression force is exerted on one side of the bone, the other side is subjected to tension. Along the central axis of the bone the forces diminish and are neutralised. The material in the centre of the bone consequently does not need to contribute to its strength. Reduction in weight of the bone due to the absence of bone along its central axis is advantageous to the animal as it lightens the weight of the femur without reducing its mechanical efficiency. The perimeter of the femur is composed of compact bone which resists the tension and compressional forces (see section 8.4.4). Spongy bone at the head of the femur is a meshwork of interconnecting bony struts. They maintain the rigidity of the bone but with the minimum of weight.

Cartilage acts as a cushion between two articulating bones. Its matrix can be deformed by compression but will return to its original shape because it possesses good powers of elasticity (extensibility). The cartilage also reduces friction between the smooth, moving articular surfaces.

The tendons consist of inelastic white fibrous tissue and attach the muscles to the femur. The pull of the muscle is concentrated over a small area. Tendons also help to prevent muscle rupture if the muscle is suddenly subjected to a heavy load.

Ligaments are also composed of inelastic white fibrous tissue and connect the femur to other bones articulating with it at joints. They confine the movement of the structures at each joint to a specific direction and therefore promote the efficiency of its operation. Ligaments also strengthen the joint.

18.2.4 Support in vertebrates

Among the first animals to live on land were the amphibians. They evolved from fish and as they migrated from water to land they were faced with the problem of holding their bodies off the ground in the absence of any support by the air. As a consequence, evolution favoured the development of structures which linked together by interlocking projections. Collectively the vertebrae formed a strong but reasonably flexible girder that supported the weight of the body.

The legs of early amphibians splayed out from the sides of their bodies so that the animals were able to drag themselves over the ground. This type of stance and locomotion is also seen in primitive reptiles (fig 18.3*a*). When in motion, most of the muscular energy is used to hold the trunk off the ground. Such is the effort required to maintain this position that the animals spend the majority of their time whilst on land resting their bellies on the ground.

Later, the trend in evolution of the reptiles was towards bringing the limbs into a position beneath the body and raising the body well clear of the ground (fig 18.3*b*). This stance provides greater efficiency in locomotion and means that the weight of the body is transmitted through the four relatively straight limbs. This trend reaches its conclusion in the mammals (fig 18.3*c*).

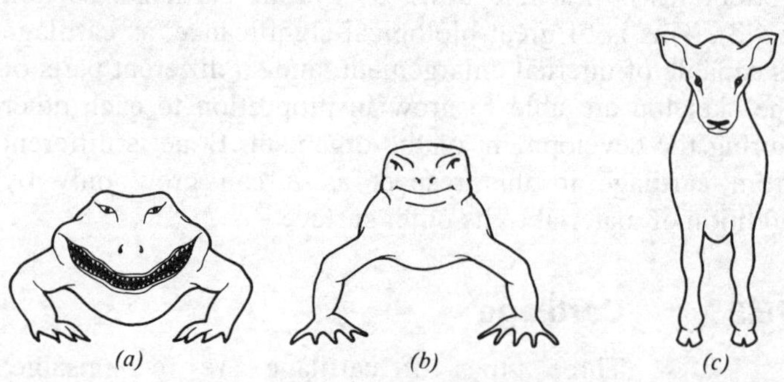

Fig 18.3 *Types of stance in vertebrates:*
(a) a primitive amphibian stance – legs projected laterally from body and then down; (b) modern reptilian stance – intermediate between amphibian and mammals;
(c) mammalian stance – legs project straight down from beneath the body.

Some reptiles and mammals have evolved a **bipedal** gait, walking, running or hopping on their hindlimbs. This releases the forelimbs for developing manipulative skills such as feeding, building and cleaning. A special type of locomotion, called **brachiation** is typical of some monkeys and apes. These animals swing from tree to tree using their long arms and hands to grasp the branches. Other animals that climb and move about in trees are too small to brachiate; instead they jump from branch to branch. The most specialised form of aerial locomotion is true flight. This evolved simultaneously during the Jurassic period in the flying reptiles (pterodactyls) and in the first birds (which were descended from reptiles). The forelimbs were modified and adapted into wings. Flying reptiles eventually became extinct, but birds survived and evolved into many highly varied forms.

18.3 Anatomy of the skeleton of a mammal (the rabbit)

All mammalian skeletons can be divided into two main parts:

- the **axial skeleton**, which consists of the skull, vertebral column, ribs and sternum;
- the **appendicular skeleton**, which consists of two girdles, the pectoral and posterior pelvic girdle, attached to each of which is a pair of limbs.

18.3.1 The axial skeleton

The skull

The **skull** consists of the cranium to which the upper jaw is fused, and a lower jaw which articulates with the cranium. Muscles connect the lower jaw to the skull and cranium. The **cranium** is composed of a number of flattened bones tightly interlocking to form a series of **immovable joints**. Besides enclosing and protecting the brain, it protects the olfactory organs (organs of smell), middle and inner ear and the eyes. At the back end of the cranium are two smooth, rounded protuberances which articulate with the atlas vertebra to form a hinge joint that allows the nodding of the head.

The vertebral column

The vertebral column is the main axis of the body. It consists of a series of bones called **vertebrae**, placed end to end, and separated by **intervertebral discs** made of cartilage (fig 18.4). The vertebrae are held together by ligaments which prevent their dislocation, but permit a

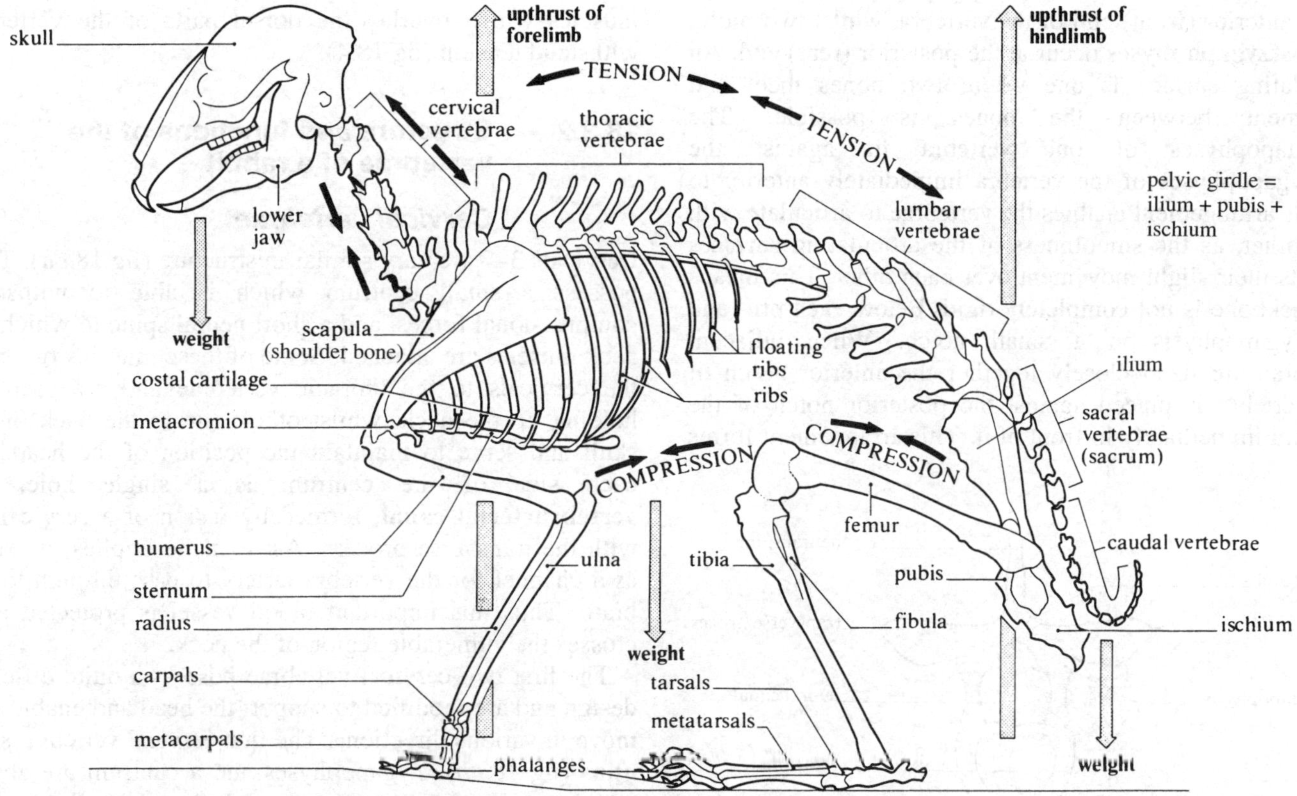

Fig 18.4 *Skeleton of rabbit seen from left side. The vertebrae together resist compression whilst the ligaments and muscles which link one vertebra to another resist tension. The abdominal muscles prevent the weight of the body from forcing the girdles apart.*

Table 18.1 Number and types of vertebrae in a range of mammals.

Types of vertebra	*Region*	*Number of vertebrae* rat	rabbit	cat	cow	human
cervical	neck	7	7	7	7	7
thoracic	chest	13	12–13	13	13	12
lumbar	abdomen	6	6–7	7	6	5
sacral	hip	4	4	3	5	5
caudal	tail	30±	16	18–25	18–20	4

degree of movement, so that the vertebral column as a whole is flexible. The vertebral column also gives protection to the spinal cord. On the vertebrae are numerous projections for the attachment of muscles. When the muscles are active, they may bend the vertebral column ventrally, dorsally or from side to side.

The total number of vertebrae varies in different mammals. Nevertheless, in all mammals five regions of the vertebral column can be distinguished. The number and types of vertebrae in a variety of mammals are given in table 18.1.

Vertebrae from different regions of the vertebral column all have the same basic design. The structure of a typical vertebra is shown in fig 18.5. Note that two facets (articulating surfaces), called **prezygapophyses**, are present at the anterior (front) end of the vertebra, whilst two more, the **postzygapophyses** occur at the posterior (rear) end. An articulating surface is one where two bones meet and movement between the bones is possible. The prezygapophyses of one vertebra fit against the postzygapophyses of the vertebra immediately anterior to it. This arrangement enables the vertebrae to articulate with each other, as the smoothness of the articulating surfaces permits their slight movement over each other. This means the backbone is not completely rigid. Below each pre- and post-zygapophysis is a small notch. When adjacent vertebrae are fixed closely together the anterior notch of one vertebra is placed against the posterior notch of the vertebra immediately in front of it. This arrangement forms

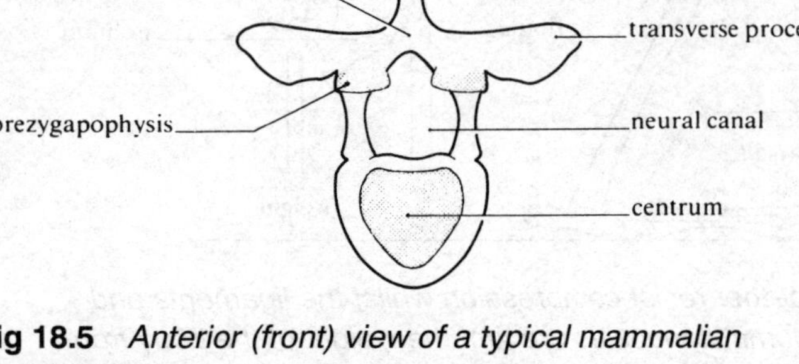

Fig 18.5 *Anterior (front) view of a typical mammalian vertebra.*

a hole through which a spinal nerve can pass. Other structures characteristic of all vertebrae are the **neural spine** and **transverse processes** for muscle attachment. The **centrum** forms a central rigid body to the vertebra over which the **neural arch** encloses the spinal cord.

Whilst there is a great degree of similarity between vertebrae, their design varies in different regions of the vertebral column. This is because of uneven distribution of body weight along the length of the column, and because the vertebrae are modified and adapted to perform those functions which are specific to each region.

When a rabbit stands up, its vertebral column is supported by the fore- and hindlimbs, with the bulk of the body weight suspended between them. The centra of the vertebrae withstand compression, whilst ligaments and muscles which overlay the dorsal parts of the vertebrae withstand tension (fig 18.4).

18.3.2 Structure and functions of the vertebrae of a rabbit

Cervical vertebrae

Cervicals 3–7 are very similar in structure (fig 18.6*a*). They possess a small centrum which is able to withstand compressional forces, and a short neural spine to which the neck muscles are attached. Some of these muscles run from the cervicals to the thoracic vertebrae and are used for holding up the neck, whilst others run to the back of the skull and serve to maintain the position of the head. On each side of the centrum is a single hole, the **vertebrarterial canal**, formed by fusion of a cervical rib with the transverse process. As its name implies, it serves as a channel for the vertebral artery to pass through to the brain. Thus this important blood vessel is protected as it crosses the vulnerable region of the neck.

The first two cervical vertebrae possess a quite different design and are modified to support the head and enable it to move in various directions. The first cervical vertebra is the **atlas** (fig 18.6*b*). Zygapophyses and a centrum are absent and the neural spine is very reduced. On its anterior surface are two concave depressions, the **articular facets** which articulate with the two processes from the skull to form a hinge joint. This supports the skull and permits it to be

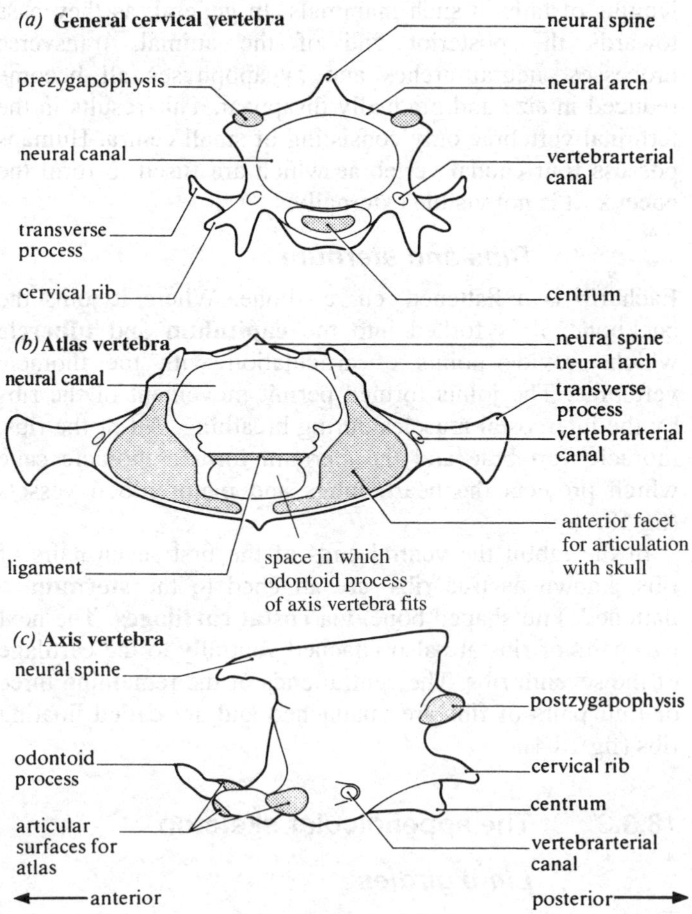

Fig 18.6 *(a) Fifth cervical vertebra of a rabbit, anterior view. Note the characteristic vertebrarterial canal. (b) Anterior view of atlas vertebra of a rabbit. Note absence of centrum and anterior facets. (c) Side view (left) of axis vertebra of a rabbit. Note odontoid process and forwardly projecting neural spine.*

nodded up and down. Wide, flattened transverse processes provide a large surface area for the attachment of those muscles that bring about the nodding action.

The second cervical vertebra is the **axis** (fig 18.6*c*). It possesses a peg-like structure called the **odontoid process** which projects forwards from the centrum. The process is formed by the fusion of the centrum of the atlas to that of the axis, and it fits into the cavity of the atlas below the ligament (fig 18.6*b*) so that it is separated from the neural canal. This arrangement gives a pivot joint which enables the head to be rotated from one side to the other (that is to be shaken). Such activity is brought about by muscles on the left and right sides of the neck. They run forwards from the neural spine of the axis to attach to the transverse processes of the atlas (fig 18.7). No prezygapophyses are present.

Thoracic vertebrae

These possess long, backwardly pointing neural spines and short transverse processes. Also present on the transverse

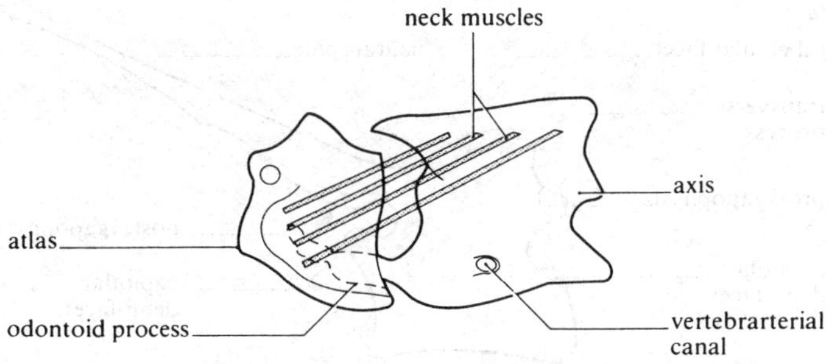

Fig 18.7 *The arrangement of the neck muscles between the atlas and axis vertebrae in rabbit.*

processes are small, rounded projections called **tubercular facets**. Anterior and posterior half or **demi-facets** are present on the sides of the centrum. Both types of facet are for articulation with the ribs (fig 18.8*a*). The end of the rib which joins to a thoracic vertebra branches into two projections, one being called the **capitulum** and the other the **tuberculum**. The tuberculum articulates with the facet of the transverse process whilst the capitulum articulates with two demi-facets of the centrum. Here the arrangement is quite complex. When two thoracic vertebrae are closely applied to each other, the anterior demi-facet of one vertebra fits closely to the posterior demi-facet of the vertebra in front of it to form a common depression. The capitulum of the rib fits into this depression thus effectively articulating with two vertebrae. As a result the thoracic vertebrae serve to support the ribs, but because of the complex arrangement between the vertebrae and ribs, movement between them is strictly limited. Some forward and sideways movement can occur, but in general the thoracic vertebrae are the least flexible of all.

Lumbar vertebrae

The vertebrae of this region are subject to the greatest stress in terms of gravity and locomotion. Not only must they provide rigidity for the body, but they must also permit bending, sideways movement and rotation of the trunk. Therefore, not surprisingly, this is the region where the large muscles of the back are attached and where there are many modifications of the vertebrae. The centrum and neural arch are massive, although the centrum is quite short. This arrangement provides greater flexibility between the lumbar vertebrae. The transverse processes are long and wide. They point forwards and downwards. Extra muscle bearing projections called meta-, ana-, and hypapophyses are present on the vertebrae (fig 18.8*b*). They also interlock with each other and keep the vertebrae in their correct positions relative to each other when this part of the vertebral column is placed under stress.

Sacral vertebrae (sacrum)

The four sacral vertebrae are fused together to form a broad structure, the **sacrum** (fig 18.8*c*). The most anterior sacral vertebrae possess well-developed transverse processes

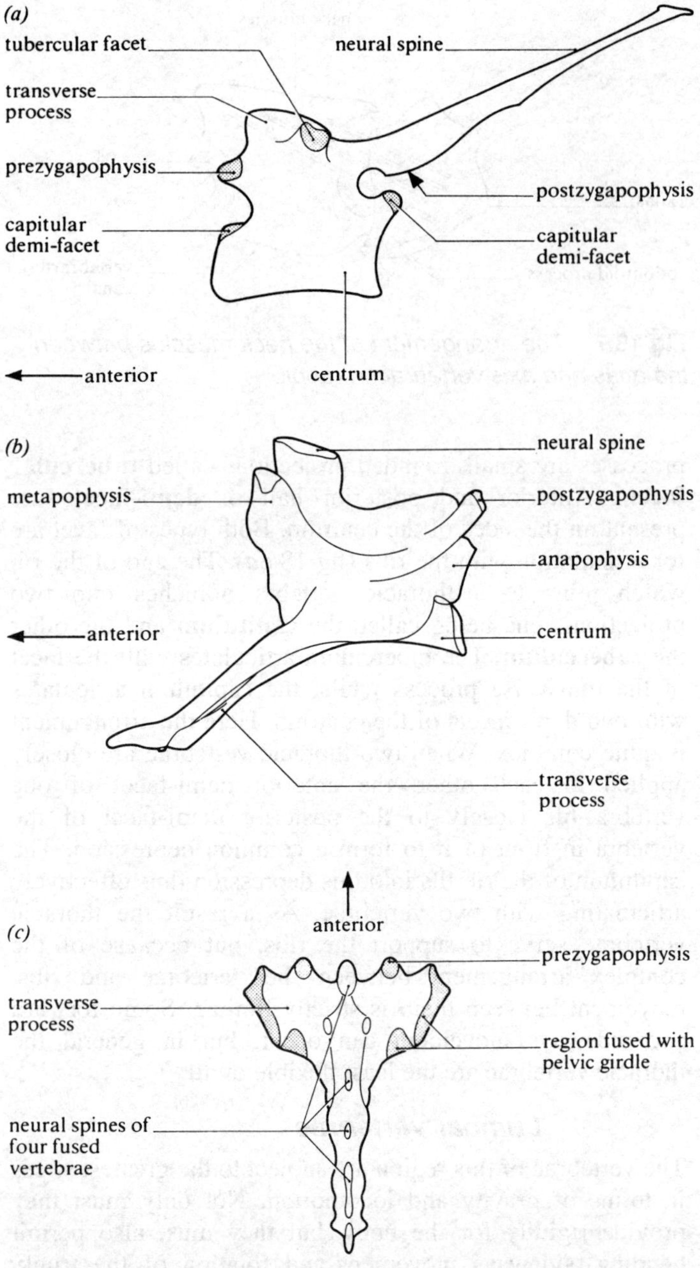

Fig 18.8 *(a) Left side view of thoracic vertebra of a rabbit. Note long neural spine and demi-facets. (b) Lumbar vertebra of a rabbit from left side. No hypapophysis is shown. Where it does occur (first and second lumbar vertebrae) it exists as a small projection from the ventral surface of the centrum. (c) Dorsal view of sacrum of a rabbit.*

which are fused to the pelvic girdle. It is through the sacrum that the weight of the body of a stationary animal is transmitted to the pelvic girdle and the legs. When an animal moves forwards, the thrust developed by the hindlimbs is transmitted via the pelvic girdle through the sacrum to the rest of the axial skeleton.

Caudal vertebrae

The number of caudal vertebrae varies greatly from one mammal to another (table 18.1) and is related to different lengths of tails in such mammals. In general, as they pass towards the posterior end of the animal, transverse processes, neural arches and zygapophyses all become reduced in size and gradually disappear. This results in the terminal vertebrae only consisting of small centra. Humans possess four caudal vertebrae which are fused to form the **coccyx**. It is not visible externally.

Ribs and sternum

Each rib is a flattened, curved bone. Where it joins the backbone, it is forked into the **capitulum** and **tubercle** which provide points of articulation with the thoracic vertebrae. The joints formed permit movement of the ribs by the intercostal muscles during breathing. All of the ribs, thoracic vertebrae and the sternum form a thoracic cage which protects the heart, lungs and major blood vessels (fig 18.9).

In the rabbit the ventral ends of the first seven pairs of ribs, known as true ribs, are attached to the **sternum**, a flattened, kite-shaped bone, via **costal cartilages**. The next two pairs of ribs are also attached ventrally to the cartilage of the seventh ribs. The ventral ends of the remaining three or four pairs of ribs are unattached and are called floating ribs (fig 18.4).

18.3.3 The appendicular skeleton

Limb girdles

These provide a connection between the axial skeleton and the limbs. The width of the pectoral girdle separates the forelimbs, and that of the pelvic girdle the hindlimbs, and both contribute to the stability of the animal. A number of areas are modified for muscle attachment and articulation with the limb bones.

Pectoral girdle. This is composed of two separate halves. Each half consists of the scapula, coracoid process and clavicle. It is not fused to the axial skeleton but flexibly attached to it by ligaments and muscles. This arrangement enables the girdle and its associated limbs to

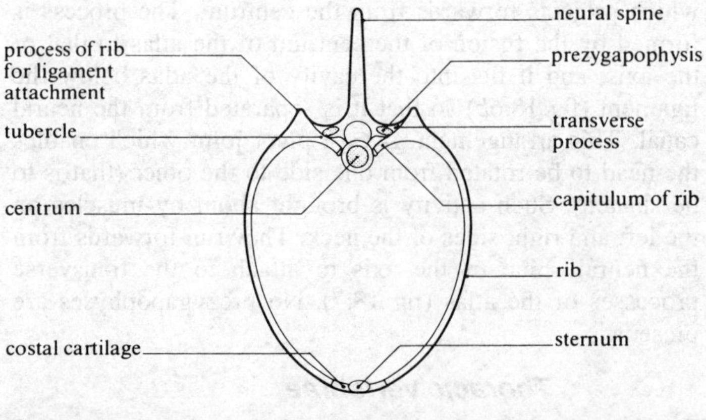

Fig 18.9 *Anterior view of thoracic vertebra of a rabbit attached to a pair of ribs.*

be moved through a great variety of planes of movement and angles. The girdle is strong enough to support the majority of the weight of a quadruped when it is stationary. It also acts as a shock absorber when the animal lands at the end of a jump.

The **scapula** is a flat, triangular-shaped bone which covers a number of the anterior ribs (figs 18.4 and 18.10*a*). At one end is a concave depression, the **glenoid cavity**, which articulates with the head of the humerus to form a ball-and-socket joint. A spine runs along the outer surface of the scapula and, at its free end, close to the glenoid cavity, are two projections, the **acromion** and **metacromion**, which are both used for muscle attachment. The **coracoid process** is all that remains of a small bone, the coracoid, which has fused with the scapula to form a projection above the glenoid cavity.

The **clavicle** is variable in size and shape in different mammals. In humans it is well developed, with one end articulating with the acromion process and the other with the sternum. It is used for muscle attachment and aiding the complex movements of the arms. It is sometimes referred to as the collar bone in humans. Its removal has no serious consequences. In quadrupeds it is much smaller and relatively less important. It forms the 'wishbone' in birds.

> **18.1** What advantages are there to mammals in possessing a flexible connection between the pectoral girdle and vertebral column?

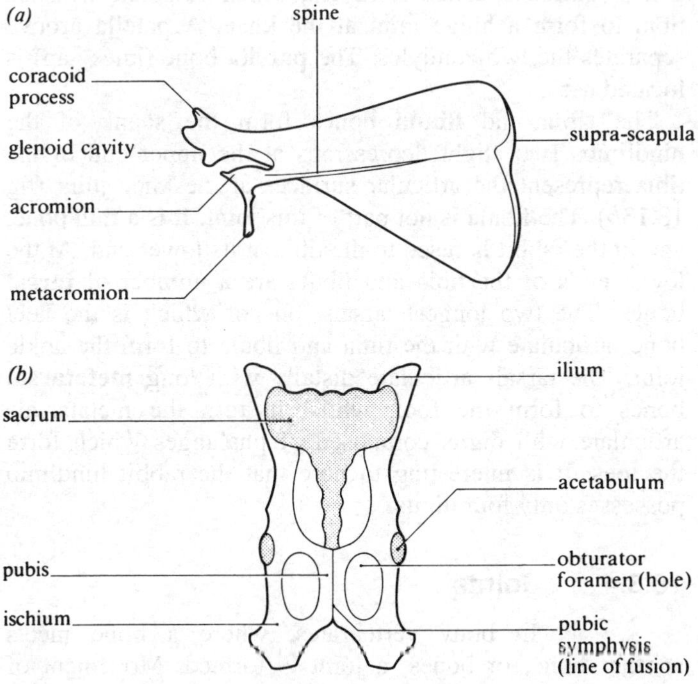

Fig 18.10 *(a) Left scapula of a rabbit. (b) Ventral view of the pelvic girdle of a rabbit. Note how the sacrum is fused to the ilium.*

Pelvic girdle. Like the pectoral girdle, this consists of two halves, each half comprising three bones, the ilium, ischium and pubis (fig 18.4). However, they are fused to each other to form a single structure. The ilium is fused to the sacrum of the vertebral column on each side. On the outer edge of each half is a depression, the **acetabulum**, which articulates with the head of the femur to form the ball-and-socket hip joint (fig 18.10*b*). The ilium is above the acetabulum. Dorsally it possesses a large crest to which the thigh muscles are attached.

Between the ischium and pubis is a large hole, the **obturator foramen**. Except for a small aperture through which blood vessels and nerves pass to the legs, it is covered by a sheet of tough inflexible connective tissue which provides yet another surface for muscle attachment. Such a design could be an adaptation to reduce the weight of the pelvic girdle and so lighten the load that has to be supported by the hind legs.

Ventrally a line of fusion can be seen where the two halves of the pelvic girdle meet. This is the **pubic symphysis**. Flexible cartilage in this region permits a widening of the female's girdle when giving birth.

Limbs

The limbs of all mammals are designed on the same basic plan, that of the **pentadactyl limb**, so named because each limb ends in five digits (fingers or toes) (fig 18.11). There are numerous variations of the general plan, which are adaptations to the different modes of life of different animals. In some cases the number of digits per limb has been reduced during evolution (section 26.7).

Forelimb. The upper part of the forelimb consists of a single bone, the **humerus**. At its upper end is the head which articulates with the glenoid cavity of the scapula. This forms a ball-and-socket joint at the shoulder which allows universal movement. Near the head are two roughened projections, between which is a groove, the

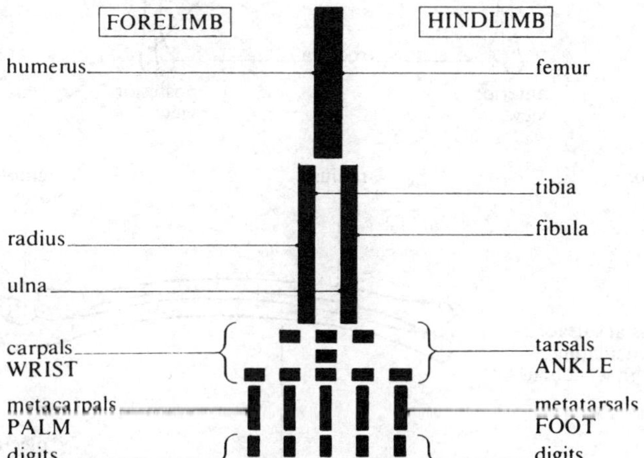

Fig 18.11 *Vertebrate pentadactyl limb.*

bicipital groove. It is along this groove that the tendon of the biceps muscle passes. At its lower end is the **trochlea** which articulates with the forearm to form a hinge joint at the elbow. A hole perforates the humerus just above the trochlea in the rabbit, but is absent in humans. Also visible is the characteristic **deltoid ridge** which runs anteriorly along the upper half of the humerus (fig 18.12*a*).

The lower part of the forelimb, the forearm, is composed of two bones, the **ulna** and **radius**. The ulna is the longer of the two. A notch, the **sigmoid notch**, at its upper end articulates with the trochlea of the humerus. Beyond the elbow joint is a projection, the **olecranon process**. This is an important structure, for when the arm is straightened it prevents any further backward movement of the forearm; hence dislocation does not occur. On the anterior surface of the humerus, above the trochlea, is a hollow into which the radius fits when the arm is bent (fig 18.12*b*).

The radius is a flattened, slightly curved bone which is relatively simple in design. In humans it is not firmly bound to the ulna; muscles are able to rotate the radius about the ulna so that the palm of the hand can be turned downwards or upwards, contributing to human manipulative skills. This freedom of movement is not present in the rabbit where both bones are tightly bound and the palm always faces downwards. However, this is not disadvantageous as the limb is in the best position for burrowing and running.

Distally (further from the main part of the body) the ulna and radius articulate with a number of small **carpal** bones which form the wrist. The carpals articulate with five long **metacarpals** which finally articulate with five **digits**. The first digit on the inside of the limb is composed of two **phalanges** whereas all the others contain three.

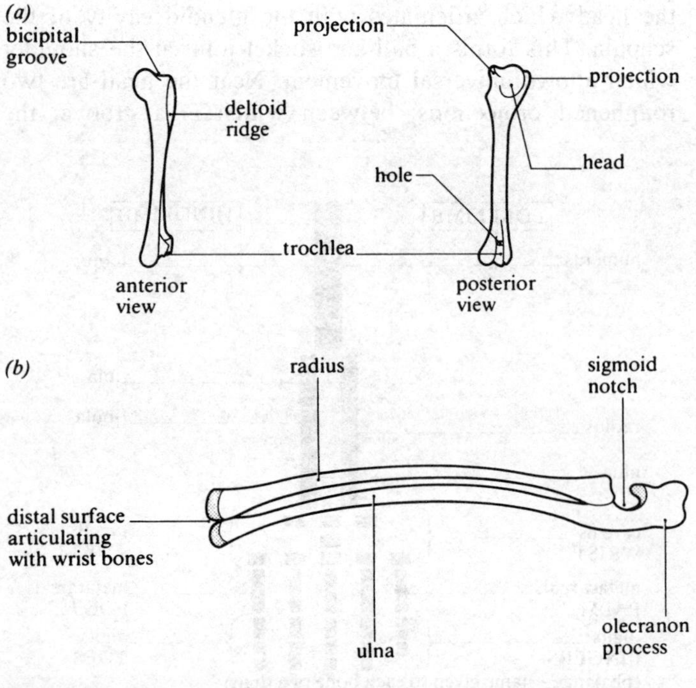

Fig 18.12 *(a) Left humerus of a rabbit, anterior and posterior views. (b) Ulna and radius of a rabbit, side view.*

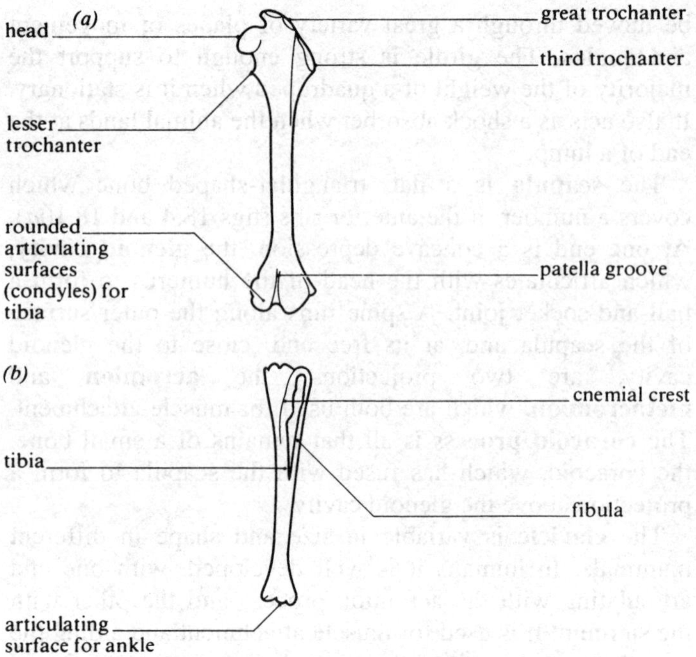

Fig 18.13 *(a) Left femur of a rabbit, anterior view. (b) Anterior view of left tibia and fibula of a rabbit.*

Hindlimb. The upper part of the hindlimb consists of a single bone, the **femur.** At its upper end is a large round head which articulates with the acetabulum of the pelvic girdle to form a ball-and-socket joint at the hip (fig 18.13*a*). Three processes called **trochanters** protrude below the head and provide points of attachment for the thigh muscles. The lower end of the femur has two curved convex surfaces, called **condyles**, which articulate with the tibia to form a hinge joint at the knee. A patella groove separates the two condyles. The **patella** bone (knee cap) is located here.

The **tibia** and **fibula** bones form the shank of the hindlimb. Two slight depressions at the upper end of the tibia represent the articular surfaces at the knee joint (fig 18.13*b*). The fibula is not part of this joint. It is a thin bone, and in the rabbit is fused to the tibia at its lower end. At the lower ends of the tibia and fibula are a number of **tarsal** bones. The two longest tarsals, one of which is the heel bone, articulate with the tibia and fibula to form the ankle joint. The tarsals articulate distally with long **metatarsal** bones to form the foot, whilst in turn the metatarsals articulate with digits composed of phalanges which form the toes. It is interesting to note that the rabbit hindlimb possesses only four digits.

18.3.4 Joints

In bony vertebrates, where a bone meets another bone, or bones, a joint is formed. Movement of parts of the skeleton over each other is only possible if there is a joint between them. A variety of different types of joint exist in the mammalian skeleton. They are summarised in table 18.2.

Table 18.2 A variety of joints in the endoskeleton of a mammal.

Type of joint	*General characteristics*	*Examples*	*Function*
Immovable/suture	A thin layer of fibrous connective tissue exists between the bones, holding them firmly in position	Between bones of skull; between sacrum and ilia of pelvic girdle; between bones of pelvic girdle	Provides strength and support for the body, or protection of delicate structures which cannot withstand any kind of deformation
Partially movable	Bones are separated from each other by cartilaginous pads		
(*a*) Gliding		Joints between vertebrae; wrist and ankle bones	Bones glide over each other to a limited extent. Collectively they provide a wide range of movement and confer strength on the limb.
(*b*) Swivel/rotating/ pivot		Joint between atlas and axis vertebrae	Permits shaking of head from side to side
Freely movable/synovial (fig 18.14)	Articulating bone surfaces are covered with cartilage and separated from each other by a synovial cavity containing synovial fluid		
(*a*) Hinge	Relatively few muscles operate this joint	Elbow, knee and finger joints	Permits movement in one plane. Capable of bearing heavy loads
(*b*) Ball and socket	Variety of muscles attached to the bones of the joint	Shoulder and hip joints	Permits movement in all planes, and some rotation. Unable to bear very heavy loads

Synovial joints (hinge joints and ball-and-socket joints) are similar to each other in design. The end surface of each articulating bone is covered by a smooth layer of hyaline cartilage. Though a living tissue, the cartilage contains no blood vessels or nerves. The nutrients and respiratory gases it requires diffuse from the synovial membrane and fluid. The cartilage serves to reduce friction between the bones during movement. Because of its elastic properties, the cartilage also acts as a shock absorber.

The bones of the joint are held in position by a number of ligaments which collectively form a strong fibrous 'capsule'. They run from one side of the joint to the other and are arranged in such a way as to cope effectively with the particular stresses suffered by the joint. The inner surface of the capsule is lined by a thin, cellular synovial membrane which secretes synovial fluid into the synovial cavity (fig 18.14). Synovial fluid contains mucin, a lubricant for the joint surfaces. The fluid serves to reduce friction between the joint surfaces. The synovial membrane acts as a waterproof seal preventing escape of synovial fluid. Therefore the joint effectively requires no maintenance.

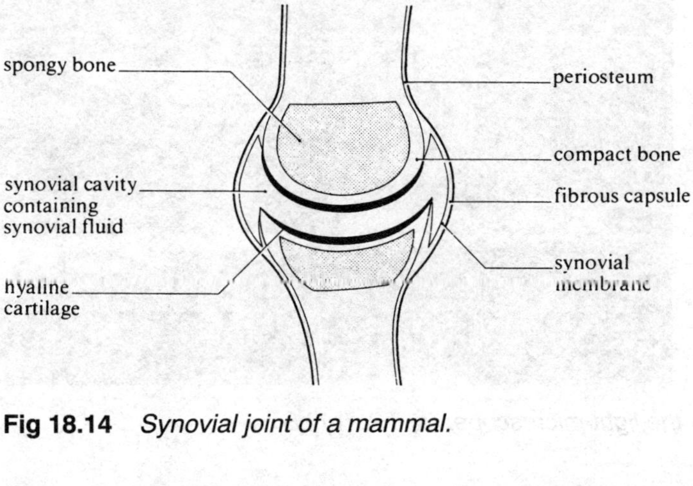

Fig 18.14 *Synovial joint of a mammal.*

18.4 The muscle system

Muscles are composed of many elongated cells called **muscle fibres** which are able to contract and relax. During relaxation muscles can be stretched, but they show elasticity which allows them to regain their original size and shape after being stretched. Muscles are well supplied with blood which brings nutrients and oxygen, and takes away metabolic waste products such as carbon dioxide. The amount of blood arriving at a muscle at any one time can be adjusted according to its need. Each muscle possesses its own nerve supply. Three distinct types of muscle can be identified.

- **Skeletal muscle** (section 18.4.1) (also called striated, striped, voluntary). Muscle which is attached to bone. It is concerned with locomotion, contracts quickly and fatigues quickly. It receives nerves from the voluntary nervous system.
- **Smooth muscle** (also called unstriated, unstriped, involuntary). Muscle which is found in the walls of organs of the body such as the gut and bladder and is concerned with movement of materials through them. It contracts slowly and fatigues slowly. It receives nerves from the autonomic nervous system.
- **Cardiac muscle**. Muscle found only in the heart. It is self-stimulating and does not fatigue. It receives nerves from the autonomic nervous system. Its structure is described in chapter 14.

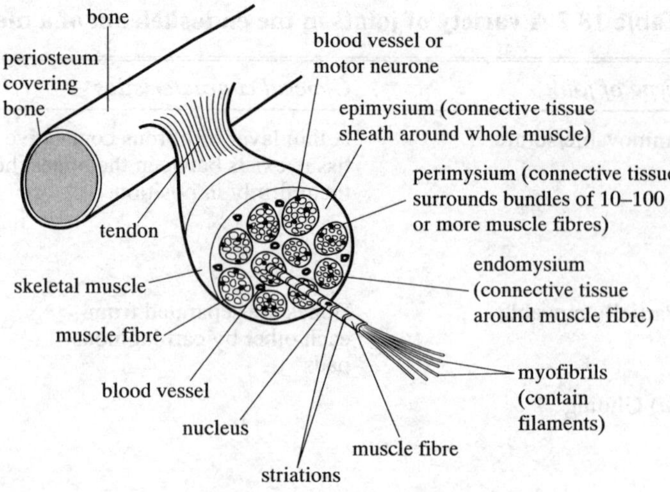

Fig 18.15 *A skeletal muscle showing the layers of connective tissue and other structures.*

18.4.1 Skeletal muscles – gross structure

A skeletal muscle is attached to bone in at least two places, namely the **origin**, a fixed **non-movable** part of the skeleton, and the **insertion**, a movable part of the skeleton. Attachment is by means of tough, relatively inelastic tendons made up almost entirely of collagen (section 3.5). At one end a tendon is continuous with the outer covering of the muscle, while the other end combines with the outer layer of the bone (the periosteum) to form a very firm attachment (fig 18.15).

As muscles can only produce a shortening force (that is contract), it follows that at least two muscles or sets of muscles must be used to move a bone into one position and back again. Pairs of muscles acting in this way are termed **antagonistic** muscles. They may be classified according to the type of movement they bring about. For example, a **flexor** muscle bends a limb by pulling two parts of the skeleton towards each other, such as the biceps which flexes the arm, causing the ulna and radius bones in the lower arm to be pulled upwards towards the humerus in the upper arm. An **extensor** is antagonistic to a flexor. The triceps muscle is antagonistic to the biceps and extends the lower arm. The activity of flexor and extensor muscles in the leg are examined in section 18.6.4.

It is rare that a movement will involve a single pair of antagonistic muscles. Generally, groups of muscles work together to produce a particular individual movement.

18.4.2 Striated muscle – histology

The structure of striated muscle as seen with a light microscope (its histology) is shown in fig 18.16. It is clear from the appearance of the muscle in longitudinal section (or in preparations of whole muscle fibres teased out from the muscle) why it is called striped or striated muscle. The reason for this appearance only becomes clear when an electron microscope is used, as described in the

(*a*)

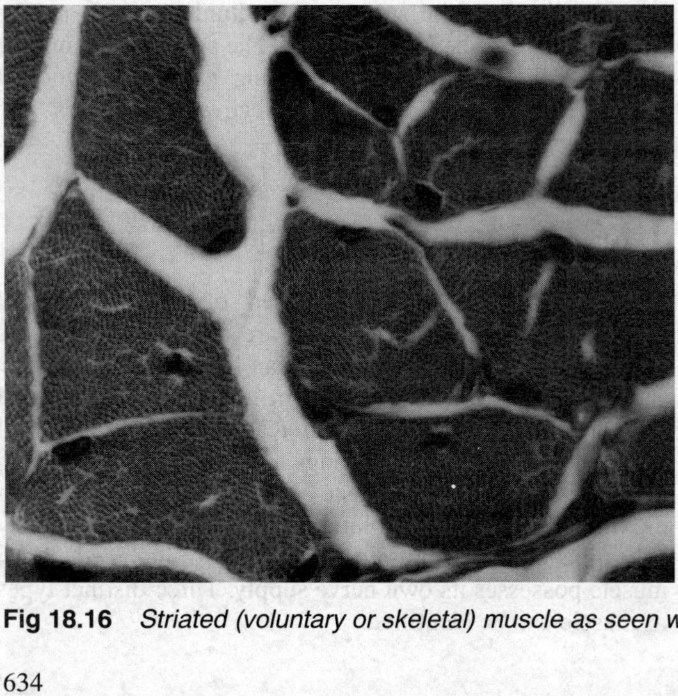

(*b*)

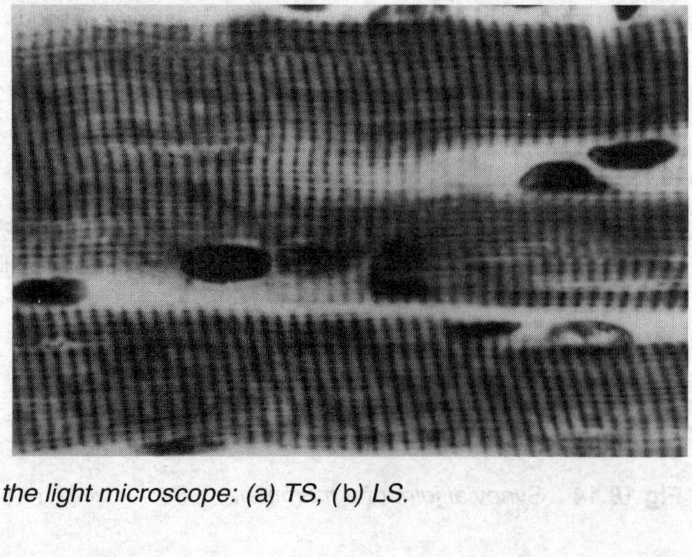

Fig 18.16 *Striated (voluntary or skeletal) muscle as seen with the light microscope: (a) TS, (b) LS.*

next section. The muscle is made up of many cells which are referred to as muscle fibres because they are so long. They can be several centimetres long and between 0.01 and 0.1 mm in diameter (an average cell is about 0.02 mm in diameter). They are cylindrical in shape and arranged parallel to each other. Each muscle fibre has many nuclei, a special arrangement not found in the other types of muscle. The nuclei are located near the surface of each fibre as can be seen in fig 18.16. Bundles of muscle fibres are surrounded by collagen fibres and connective tissue. Collagen also occurs between fibres. Each muscle fibre is surrounded by a cell surface membrane called the **sarcolemma** (*sarco-* means flesh and refers to muscle). This is very similar in structure to a typical cell surface membrane.

Inside the muscle fibres, it is just possible to see with a light microscope that there are numerous thin **myofibrils** (*myo-* also refers to muscle) which possess characteristic cross-striations. The myofibrils line up in parallel with their cross-striations next to each other, forming the stripes that are seen with the microscope.

18.4.3 Striated muscle – ultrastructure

It has only been possible to work out how muscle contracts by studying its ultrastructure (fine structure) with an electron microscope. In the electron microscope the myofibrils are clearly seen. They are about 1 μm across compared with the 100 μm (0.1 mm) of the whole fibre. The structure of a myofibril is shown in fig 18.17. It has a series of dark bands which line up with those of neighbouring myofibrils to form the striations seen in the whole fibre. Close examination of the myofibril shows that it is made up of two types of 'filaments' which run longitudinally. There are thin filaments and thick filaments. The thin filaments are made of a protein called **actin** and the thick filaments are made of a protein called **myosin**. In some places they overlap, like partly interlocking your fingers together. Where they do they produce a darker appearance (as your overlapping fingers would if you shone a light behind them). This explains the dark bands and striations previously mentioned (fig 18.16). The technical term for these bands is **A bands**. The light bands between the A bands are called **I bands**. Only actin filaments are present in the I bands. Closer examination shows another region in the A band where the filaments do not overlap and only myosin filaments are present. This is called the H zone. Finally, running through the middle of each light band is a line called the **Z line**. The overall structure is best described by means of diagrams such as fig 18.17.

It was soon realised that the distance between two Z lines represents the functional unit of muscle. This is the unit which is capable of contraction and is called a **sarcomere**. The myofibril, and therefore the muscle fibre, is made up of thousands of sarcomeres. If it is possible to understand how one sarcomere contracts, then it is possible to understand how the whole muscle contracts.

The banding pattern and the corresponding arrangement of actin and myosin filaments in a sarcomere is shown in fig 18.17. Figs 18.18 and 18.19 show the appearance as seen with an electron microscope.

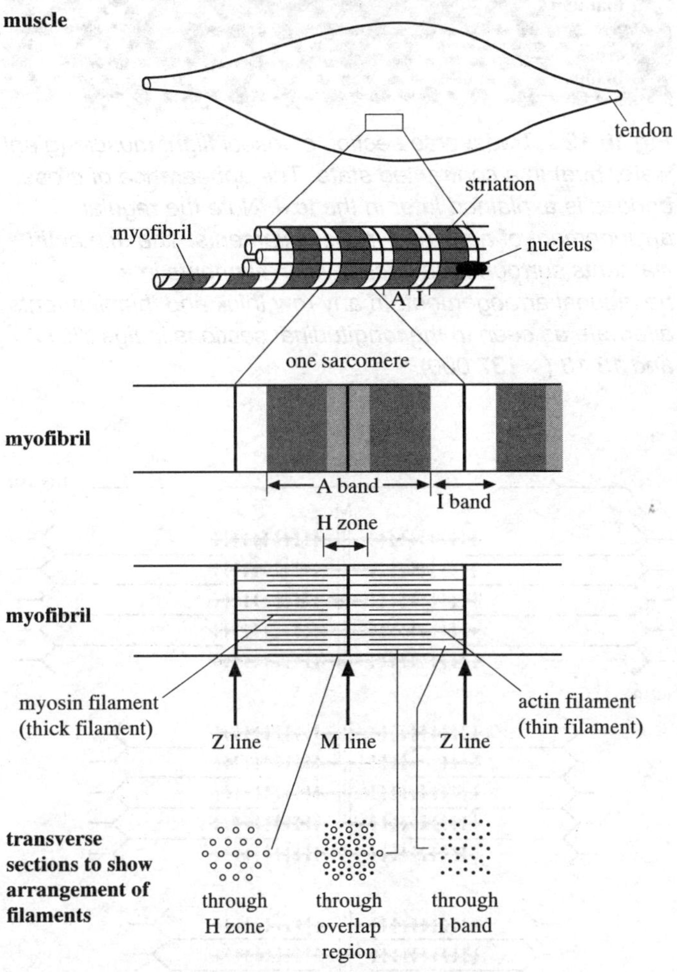

Fig 18.17 *Fine structure of skeletal muscle.*

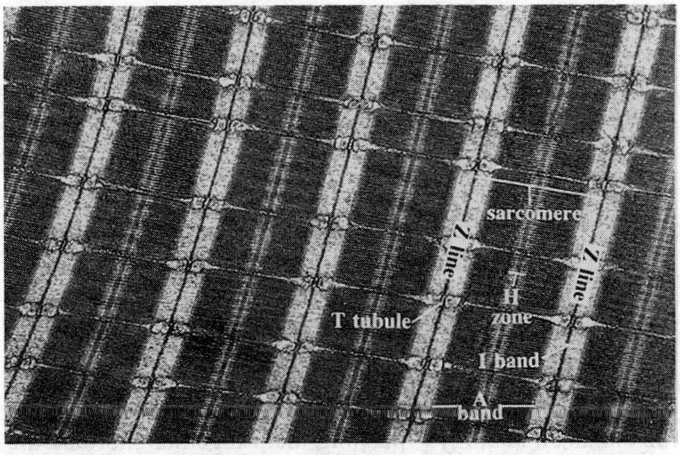

Fig 18.18 *Longitudinal section of fish muscle (roach). Note how the myofibrils, which run across the picture from left to right, are precisely lined up (× 7650).*

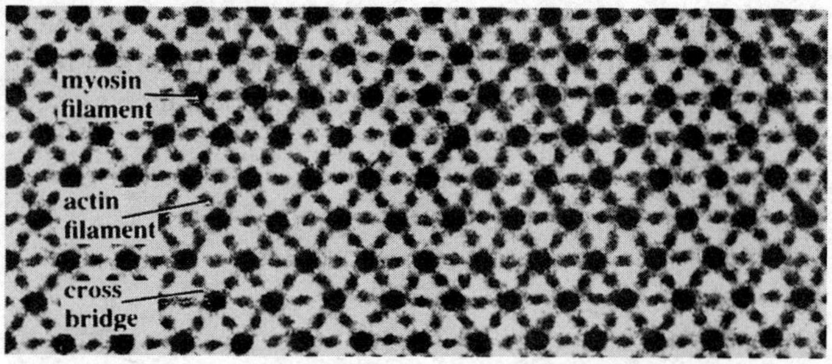

Fig 18.19 *Transverse section of insect flight muscle (giant water bug) in a contracted state. The appearance of cross bridges is explained later in the text. Note the regular arrangement of actin and myosin filaments. The thin actin filaments surround the thick myosin filaments in a hexagonal arrangement. In any row thick and thin filaments alternate as seen in the longitudinal sections in figs 18.17 and 18.18 (×137 000).*

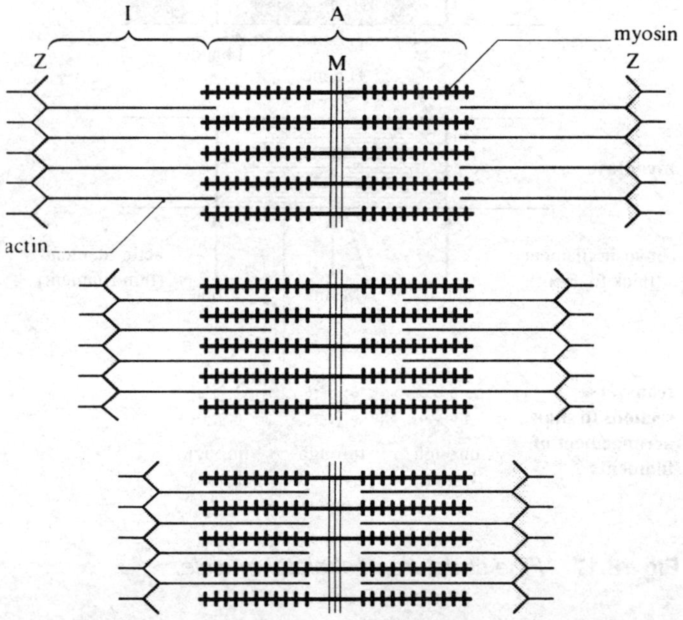

Fig 18.20 *Diagrammatic representation of how a sarcomere contracts by the actin filaments sliding between myosin filaments. Myosin is shown with its myosin heads. These are explained in the text.*

18.4.4 The sliding filament theory of muscle contraction

Once the structure was understood, it was proposed independently by two research teams that muscle contracts by the actin and myosin filaments sliding past each other (fig 18.20) (again, you can use interlocking fingers as an analogy – pushing them together shortens the distance they span, equivalent to muscle contraction by sliding filaments). Nothing actually contracts except the length of the sarcomeres and hence the whole muscle. The two research teams who proposed the hypothesis in 1954 were H. E. Huxley and J. Hanson, and A. F. Huxley and R. Niedergerke. One piece of supporting evidence was the fact that, as muscle contracts, the dark bands (A bands) remain the same length and the light bands (I bands) and the H zones get shorter. This is explained by the hypothesis. Evidence has now confirmed the hypothesis which is known as the **sliding filament theory**.

We can now examine in more detail how the sliding is brought about. To do this, the structures of actin and myosin need to be examined in more detail.

Myosin (thick filaments)

A molecule of myosin consists of two distinct regions, a long rod-shaped region called a myosin rod, and a myosin head which consists of two similar globular parts (fig 18.21*a*). The globular heads appear at intervals along the myosin filaments, projecting from the sides of the filament. Where the actin and myosin filaments overlap, the myosin heads can attach to neighbouring actin filaments. The importance of this will become clear when we deal with the actual contraction mechanism of the sarcomere.

Actin (thin filaments)

Each actin filament is made up of two helical strands of globular actin molecules (G-actin) which twist round each other (fig 18.21*b*). The whole assembly of actin molecules is called F-actin (fibrous actin). It is thought that an ATP molecule is attached to each molecule of G-actin.

Contraction mechanism

An outline of the contraction mechanism is as follows. Where the actin and myosin filaments overlap the myosin heads can attach like 'hooks' to neighbouring actin filaments (F-actin), forming cross bridges. The bridges then move (each like a straight finger bending) to pull the actin filaments past the myosin filaments. Not all the bridges form at the same time (about half are attached at any given time which gives a smoother contraction). The energy for

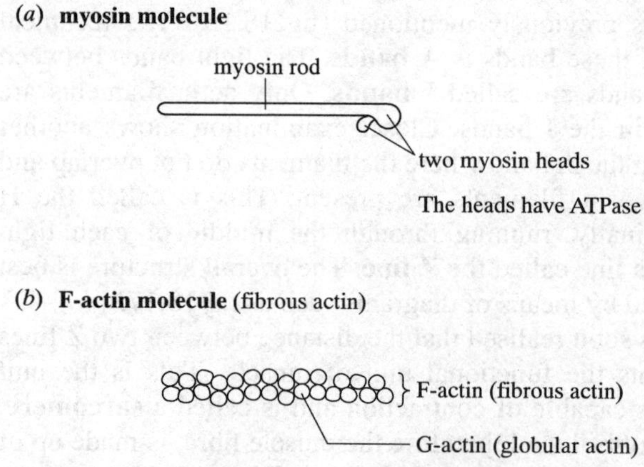

Fig 18.21 *(a) Structure of myosin showing its rod-like shape with two heads. (b) Structure of actin.*

this movement is provided by ATP. For ATP to release its energy it must be hydrolysed by the enzyme ATPase to ADP and phosphate. Each myosin head has ATPase activity. After sliding has occurred, the bridge detaches and changes back to its original shape (like the finger straightening) before re-forming again a little further along the actin filament. The sequence of events is shown in fig 18.22. It is sometimes described as a ratchet mechanism. The bridges form and re-form 50 to 100 times per second, using up ATP rapidly. This explains the need for numerous mitochondria in the muscle fibre, which can supply the ATP as a result of aerobic respiration. The sarcomere can shorten by as much as 30 to 60% of its length.

18.2 What happens to the length of the A band as the sarcomere contracts?

The question still remains as to how the whole process is started and stopped. Calcium ions activate the process (see roles of tropomyosin and troponin below). Calcium ions are located in the sarcoplasmic reticulum (the specialised endoplasmic reticulum of the muscle fibre) which forms swollen areas or vesicles at the Z lines of the sarcomeres, as shown in fig 18.23. Here the vesicles are in contact with tubes formed from the sarcolemma (the cell surface membrane which covers the muscle fibre). The sarcolemma

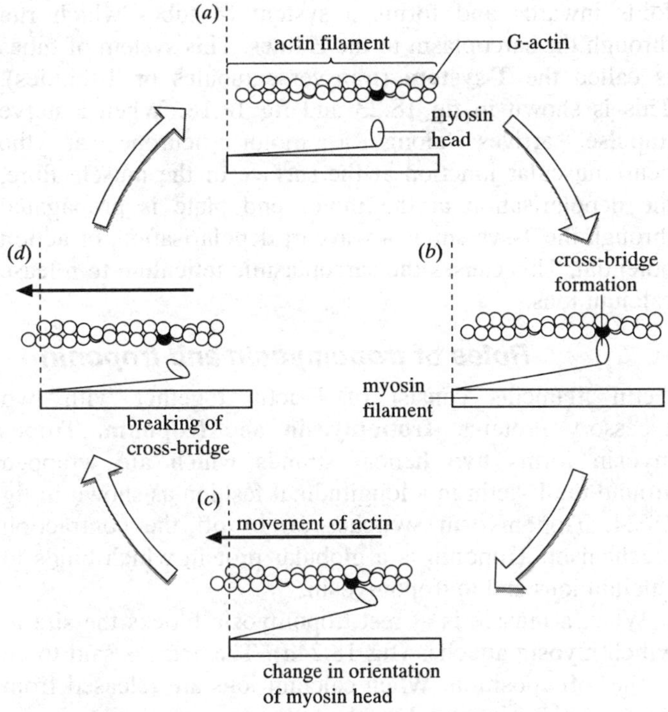

Fig 18.22 *Movement of actin by myosin during muscle contraction. By looking at the dashed line on the left of each diagram, you can see how the actin has moved in (c) and (d).*

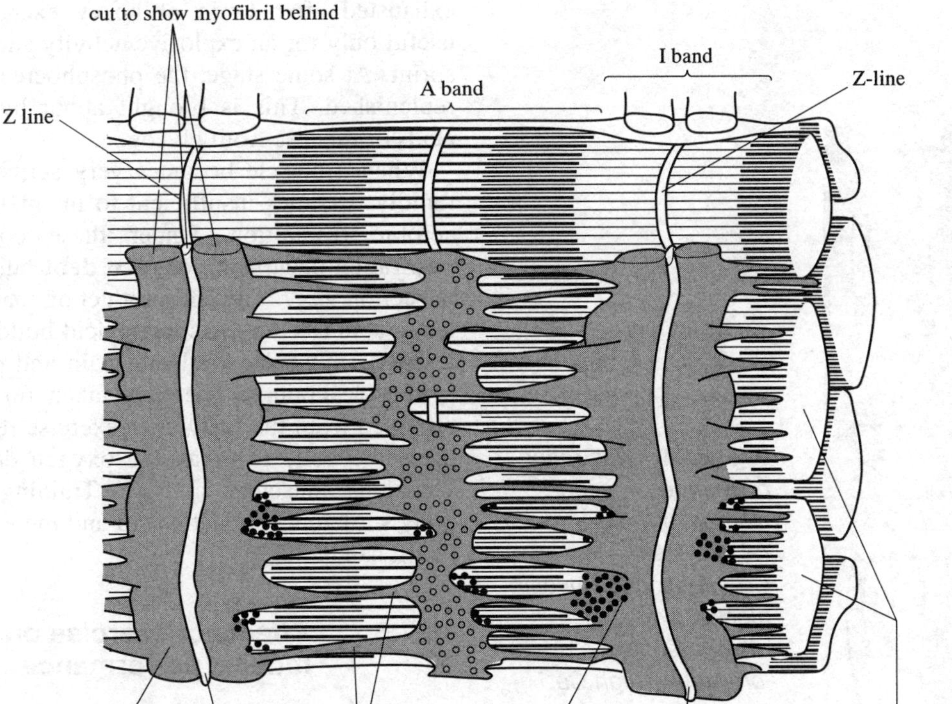

Fig 18.23 *Sarcoplasmic reticulum and T-system.*

folds inwards and forms a system of tubes which run through the sarcoplasm to the Z lines. This system of tubes is called the **T-system** (transverse tubules or T-tubules). This is shown in fig 18.23 and fig 18.18. When a nerve impulse arrives along a motor neurone at the neuromuscular junction at the surface of the muscle fibre, the depolarisation at the motor end plate is propagated through the T-system as a wave of depolarisation, or action potential. This causes the sarcoplasmic reticulum to release calcium ions.

Roles of tropomyosin and troponin

Actin filaments consist of F-actin together with two accessory proteins, **tropomyosin** and **troponin**. Tropomyosin forms two helical strands which are wrapped around the F-actin in a longitudinal fashion as shown in fig 18.24. Tropomyosin switches on or off the contraction mechanism. Troponin is a globular protein which binds to calcium ions and to tropomyosin.

When a muscle is at rest tropomyosin blocks the site to which myosin attaches (fig 18.24*a*). The actin is said to be in the 'off' position. When calcium ions are released from the sarcoplasmic reticulum they bind to troponin causing it, and the tropomyosin which is attached to it, to move away from the myosin binding site (fig 18.24*b*). The actin filament is now in the 'on' position and binds to myosin as shown in fig 18.22. When excitation of the muscle by nerve impulses ceases, calcium ions are pumped by active transport back into the sarcoplasmic reticulum by a calcium pump located in the membrane of the sarcoplasmic reticulum. This also requires ATP. The muscle then relaxes.

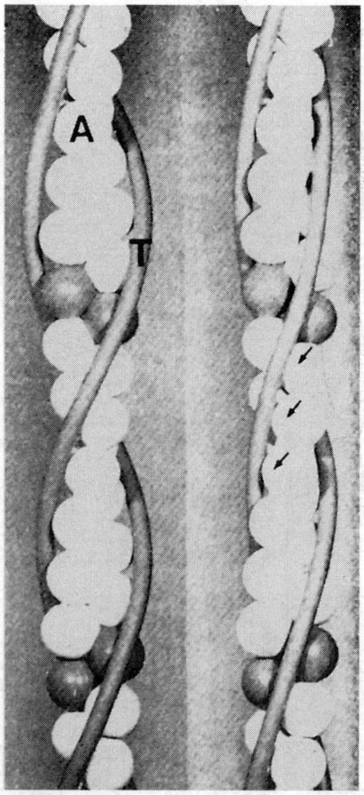

Fig 18.24 *Illustration of changes in actin filament structure (a) 'off' state – low Ca^{2+} level: tropomyosin blocks myosin attachment site. (b) 'on' state – high Ca^{2+} level: tropomyosin moves to expose attachment sites (arrows). A, actin; T, tropomyosin. Troponin is not shown (it lies near the grey balls).*

18.4.5 The energy supply

Within the body the source of energy for muscle contraction is usually glucose obtained from blood or from glycogen stored in the muscle, but may also be fatty acids. When these molecules are oxidised during respiration, ATP is generated.

Normally the oxygen used in aerobic respiration is supplied by haemoglobin. However muscles also have their own store of oxygen because they contain a protein similar to haemoglobin called **myoglobin** (section 14.8.2), which combines with oxygen and only releases it if the rate of its supply from haemoglobin cannot keep up with demand, as in strenuous exercise.

In resting muscle the level of ATP is low. It is soon used up when a muscle contracts, and has to be restored quickly by other processes until the rate of aerobic respiration adjusts to the activity.

One method is to use **phosphocreatine** to generate ATP under anaerobic conditions. This ensures that there is always a constant supply of ATP in the muscle which it can use for immediate contraction. There would only be sufficient phosphocreatine to supply total energy demand for about 5–10 s, and normally about 70% of the store is exhausted after 1 min of heavy exercise. It is therefore useful only for an explosive activity such as a sudden short sprint. At some stage, the phosphocreatine level has to be replenished. This is brought about by oxidation of fatty acids or glucose from glycogen.

When a muscle becomes very active its oxygen supply rapidly becomes insufficient to maintain adequate rates of aerobic respiration. Under these conditions anaerobic respiration occurs. An **oxygen debt** builds up as explained in section 9.3.7. The end-product of anaerobic respiration is lactic acid (lactate). As lactic acid builds up in the muscle, it causes muscular tiredness, pain and possibly contributes to muscle cramps. The time taken for lactate to be fully removed from the body after exercise represents the time it takes the body to repay the oxygen debt incurred during strenuous muscular activity. Training can increase the body's tolerance to lactic acid and increase the oxygen debt that it can build up.

18.4.6 Effects of exercise on muscles and muscle performance

Basic muscle size is inherited, but exercise can increase the size of muscles by up to 60%. This is mainly the result of an increase in the diameter of individual muscle fibres and an increase in their numbers. There is also an increase in the number of myofibrils within each muscle fibre.

Long-term biochemical changes

The number and size of mitochondria increase within the fibres. Processes that take place in mitochondria such as the Krebs cycle, electron transport and oxidation of fatty acids all occur more rapidly.

Stamina training may double the ability of mitochondria to generate ATP. More phosphocreatine, glycogen and fat are stored and, as a result of the presence of more myoglobin, more oxygen is stored. The combined effect of these changes means that the athlete relies less on anaerobic respiration and therefore produces less lactate. There is greater ability to release fatty acids from fat stores for energy liberation. Hence fit people use up more fat during exercise than unfit people.

Long-term increase in muscle strength

Muscle strength is increased only if the muscle is working against a load (resistance) greater than that which it is normally used to. Either intensity or duration of exercise can be increased to achieve this. Muscles working at, or close to, their maximum force of contraction will increase in strength very quickly even if the daily exercise is a matter of minutes. Regular exercise is necessary to maintain strength of muscle. Without it they revert to their former state and become 'out of condition', losing both speed of contraction and strength.

Blood supply to muscles

Regular exercise results in an increase in the number of blood vessels supplying blood to the muscles. This serves to provide a more efficient system for glucose and oxygen transport, and for the removal of waste products of respiration.

During long-term exercise both circulatory and respiratory systems adapt so that any oxygen debt that builds up at the beginning of the bout of exercise can adequately be repaid during the exercise.

The ability to sustain exercise at a persistently high rate is generally dependent on the rate and efficiency at which oxygen can be taken up and used.

Coordination

Exercise improves coordination between pairs of antagonistic muscles, thus enabling more complex and skilful movement. Exercise also improves the speed at which muscles relax as well as contract. If a muscle does not relax rapidly enough, it may be torn by the pulling effect of the opposing muscle.

Muscles can be overstretched due to over-rigorous training, causing straining or tearing of muscle tissue. The likelihood of this happening can be reduced by the use of warming-up exercises. After strenuous exercise muscles are generally shorter and tighter and more prone to injury. Cooling-down exercises, concentrating on flexibility, can help prevent this by gently stretching the muscles.

18.4.7 Slow and fast muscle fibres

There are two major types of skeletal muscle fibre, namely **slow** or **tonic** fibres, and **fast**, **twitch**, or **fast-twitch** fibres. Table 18.3 indicates their structure, location and general properties. Some muscles contain purely slow fibres, some just fast fibres, and some contain mixtures of both.

Together the two types of fibre allow the organism to move about and to maintain posture. The fast fibres allow fast muscle contraction. Predators possess many fast fibres and use them for fast reactions to capture prey. On the other hand, prey species can also react quickly in order to avoid capture by predators. In both cases speed of body movement would influence the survival chances of the organism concerned.

When an animal is still it has to maintain a particular posture. This is achieved by contraction of the slow muscle fibres. They generate a slower, more sustained contraction, whilst at the same time consuming less fuel than the fast muscle fibres.

In humans both types of fibre occur in all muscles, but one or other usually predominates. The functional significance of this is that the predominantly slow muscles are suited to long-term slow contractions, and consequently are found in the extensor muscles, whilst fast muscle fibres dominate in the flexor muscles, which are designed to react at speed. Posture is maintained mainly by extensor muscles.

Fast muscle fibres are sometimes known as white muscle fibres. They contain relatively little of the red pigment myoglobin, which stores oxygen in muscle. Slow muscle fibres are sometimes known as red muscle fibres because they contain much more myoglobin.

18.5 Locomotion in selected non-vertebrates

18.5.1 Locomotion in the earthworm (*Lumbricus terrestris*)

Locomotion is not possible without a skeleton. The skeleton in segmented worms like the earthworm is the coelom, which is a hydrostatic skeleton. Its skeletal role is explained in section 18.1.2. The origin of the coelom is described in section 2.10.5. The coelom is surrounded by the body wall which contains two layers of antagonistic muscles (muscles that have opposite effects), an outer circular layer and an inner longitudinal layer. The circular muscle is divided into separate units along the length of the animal by septa (barriers) between the segments of the body, but the muscle fibres of the longitudinal layer generally extend over several segments. Locomotion is brought about by the coordinated activity of the two muscle layers and of the muscles that control the **chaetae**. These are bristle-like hairs which project from the lower surface of the worm.

Table 18.3 Structure, location and general properties of slow and fast skeletal muscle fibres.

	Slow/tonic muscle fibres	*Fast/twitch muscle fibres*
Structure	Many mitochondria Poorly developed sarcoplasmic reticulum Red – due to presence of myoglobin and cytochrome pigments Low in glycogen content Capillaries in close contact with fibres to allow fast exchange of materials	Few mitochondria Well-developed sarcoplasmic reticulum White – little or no myoglobin or cytochrome pigments Abundance of glycogen granules
Location	Deeply seated inside the limbs	Relatively close to the surface
Innervation	Associated with small nerve fibres (of $5\,\mu m$ diameter). A number of end-plates are distributed along the length of the fibre. This is called multi-terminal innervation. Relatively slow conduction of action potential (2–8 $m\ s^{-1}$) 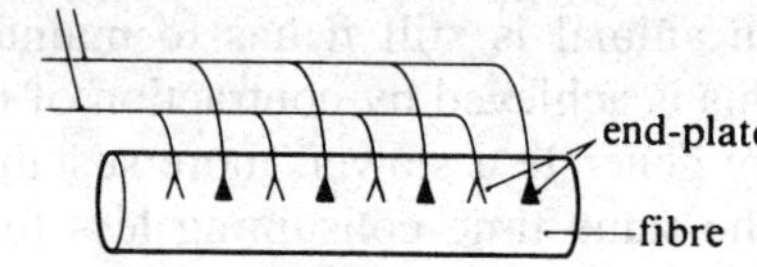 	Associated with large nerve fibres (of 10–20 μm diameter). Usually one or possibly two end-plates per fibre Relatively fast conduction of action potential (8–40 $m\ s^{-1}$) 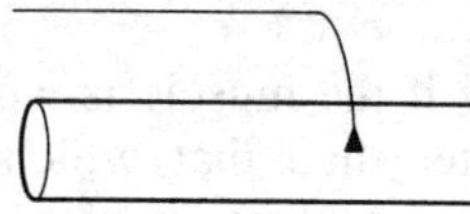
Excitability	Membrane electrically inexcitable. Each impulse causes release of only small amount of acetylcholine. Therefore amount that membrane is depolarised depends on frequency of stimulation	Membrane electrically excitable
Response	Slow graded muscular contraction of long duration. Relaxation process slow (up to 100 times slower than twitch fibre)	Fast contraction (3 times faster than slow fibres) Fatigues quite quickly
Physiological activity	Depend on aerobic respiration for ATP production Many continue to function anaerobically if oxygen is in short supply in which case lactate is formed and an oxygen debt incurred Carbohydrate or fat store mobilised at same rate as respiratory substrate is oxidised Heat transported away from muscle as soon as it is produced Steady state between muscle activity and its needs is set up	Depend on the anaerobic process of glycolysis for ATP supply Oxygen debt quickly built up Glycogen used extensively as respiratory substrate Heat produced is absorbed by the fibres as the circulatory system does not immediately remove it Muscle contraction occurs during a period when the circulatory system has not had time to increase the oxygen supply to the muscle
Function	Enable sustained muscle contractions to occur. This is used for the maintenance of posture by the organism	Immediate, fast muscle contraction is permitted at a time when the circulatory system is still adjusting to the needs of the new level of muscle activity. Therefore of great importance during locomotion.

When an earthworm begins to move forward, contraction of the circular muscles begins at the anterior end of the body and continues, segment by segment, as a wave along the length of the body. This activity exerts pressure on the coelomic fluid in each segment, stretching the relaxed longitudinal muscle and changing the shape of the segments such that they become longer and thinner. This causes the anterior end of the worm to extend forwards. Chaetae, four pairs of which are present in all segments except the first and last, are retracted during the activity of the circular muscles and therefore do not slow down the forward movement (fig 18.25).

While the anterior end of the worm is moving forward, longitudinal muscle in more posterior segments contracts, causing this region of the worm to swell and press against the surrounding soil. Chaetae in this region are extended and help the worm to grip the soil. This is particularly useful during burrowing for the worm can exert a powerful thrust against the surrounding soil particles during forward locomotion.

Contraction of the circular muscle is quickly followed by contraction of the longitudinal muscles throughout the length of the body, and this means that different parts of the worm may either be moving forward (when the circular muscles contract) or static (when the longitudinal muscles contract) at any given movement. The net effect is a smooth

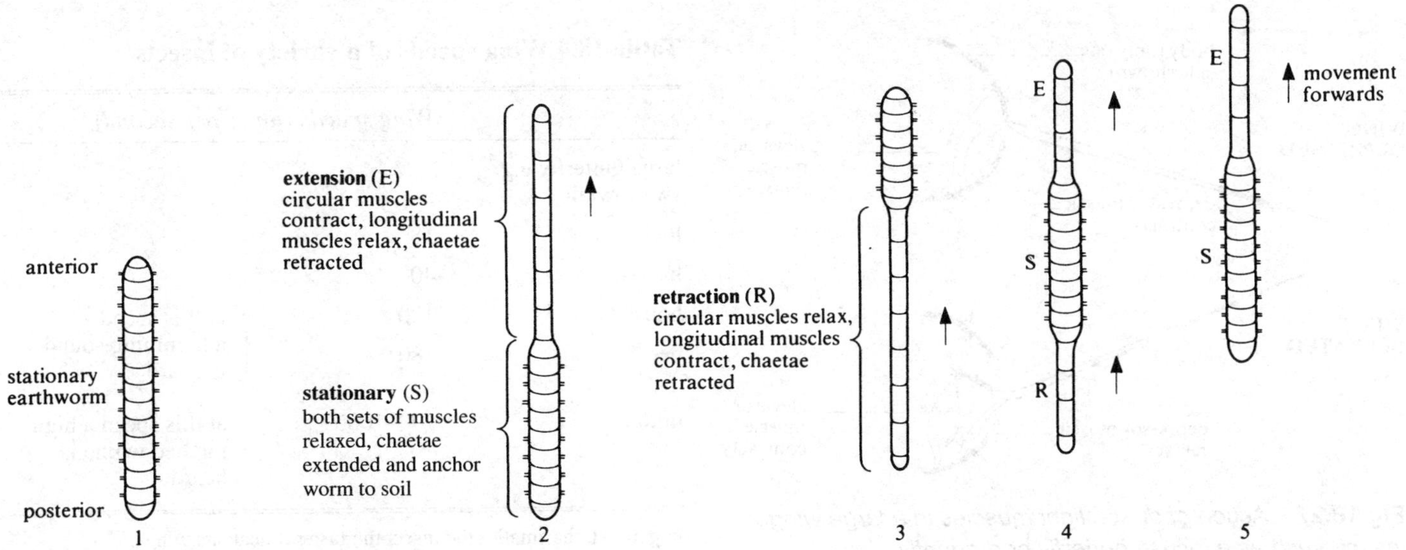

Fig 18.25 *Locomotion in the earthworm.*

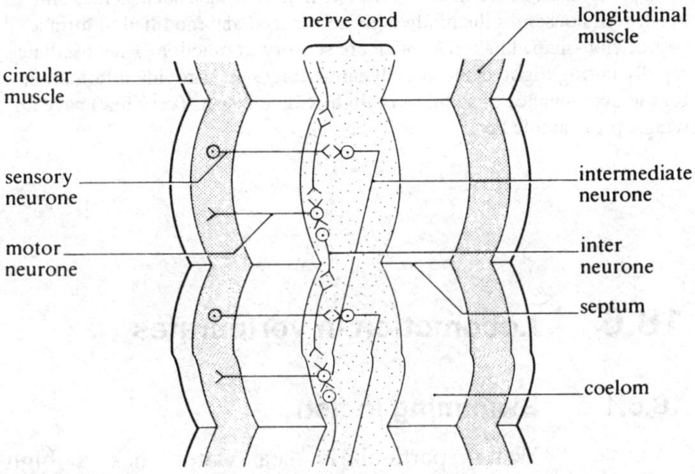

Fig 18.26 *Segmental nerves of longitudinal muscle in the earthworm (dorsal view). A similar arrangement is present in the circular muscle.*

peristaltic wave of activity moving along the length of the worm as it progresses forwards. The worm is also able to crawl backwards by reversing the direction of contraction of the muscles.

Control of muscle contraction is brought about by a complex network of neurones inside the segments. All segments are in contact with longitudinal nerve cord and also possess their own set of segmental nerves. This means that localised control of each segment is possible, as well as control of overall activity of the animal (fig 18.26).

The ventral nerve cord possesses a **giant axon** which runs along its length and conducts impulses along the body. When sensory receptors in the head are stimulated, impulses pass along the giant axon stimulating the longitudinal muscles to contract, thus causing the anterior end to be pulled back from the stimulus. Also present in the nerve cord are two **lateral fibres** which run the length of the worm and carry impulses from the tail to the head. If the tail of the worm is stimulated, impulses pass along the lateral fibres from tail to head and cause the tail to be withdrawn. This is the basis of the worm's escape reaction.

18.5.2 Locomotion in insects

Insects have an exoskeleton. Its role in relation to muscles and locomotion is described in section 18.1.3.

Walking

This is achieved by the coordinated activity of three pairs of legs, one pair being attached to each of the three thoracic segments of the animal. Each leg consists of a series of hollow cylinders whose walls are composed of rigid exoskeletal material (see fig 18.1). The cylinders are linked together by joints and soft pliable membranes. Where the leg joins to the body, a form of ball-and-socket joint occurs, but all other joints in the leg are hinge joints. Bending and straightening of the legs is achieved by antagonistic flexor and extensor muscles attached to the inner surface of the exoskeleton on either side of a joint (fig 18.1).

When an insect begins to walk, three legs remain on the ground to support the animal whilst the other three move forward. The first leg on one side pulls the insect, whilst the third leg of the same side pushes. The second leg on the other side serves as a support for this activity. The process is then repeated but with the role of each trio of limbs reversed.

Many insects possess a pair of claws and a sticky pad at the ends of their legs. The pad consists of minute hollow tubes which secrete a sticky fluid that helps the insect to stick to smooth surfaces. Thus, these insects are able to walk up vertical surfaces as well as upside down.

Flight

The wings of insects are flattened extensions of the exoskeleton and are supported by an intricate system of

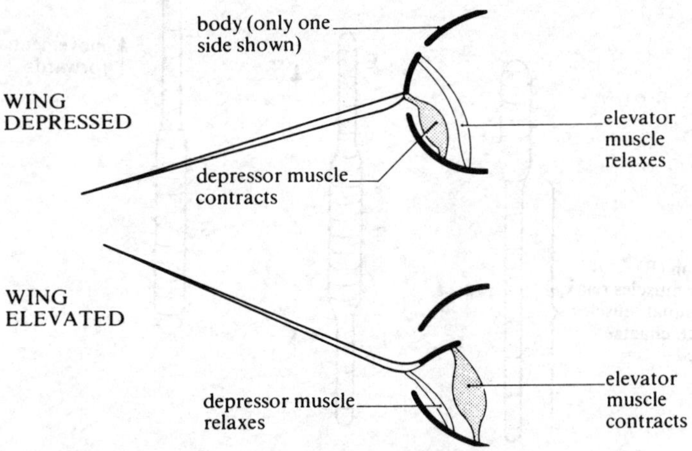

Fig 18.27 *Action of direct flight muscles in a large winged insect, such as a locust, butterfly or dragonfly.*

veins. Their movement is controlled by two main groups of muscles, **direct** and **indirect muscles**. In insects with large wings (such as locusts, butterflies and dragonflies) the muscles are actually attached to the bases of the wings (fig 18.27); these are the direct muscles. They raise and lower the wings as well as controlling the angle of the wing stroke during flight. When the angle of one wing is adjusted with respect to the other, the insect turns in the air. These muscles are also used to fold up the wings when the insect is stationary.

In large, winged insects such as the locust and butterfly the rate of wing beat is between 5–50 beats per second (table 18.4). Here the flight muscles contract each time as a result of a single nerve impulse. Hence impulses are generated at the same rate as the wings beat. Insect flight muscle which responds in this way is called **synchronous muscle**. In the housefly, which has a wing beat frequency of 120–200 beats per second, contraction of the flight muscles is much too fast to be triggered by individual nerve impulses. This muscle is termed **asynchronous** and receives roughly one impulse per 40 wing beats, which is necessary to maintain the muscle in an active state during flight. It can contract further and generate more power than synchronous muscle. Asynchronous muscle can also automatically contract in response to being stretched. This is called the **stretch reflex** and occurs faster than the speed of a nerve impulse.

> **18.3** The sarcoplasmic reticulum of insect flight muscle is modified to increase its surface area by being perforated at intervals. Can you suggest a reason for this?
>
> **18.4** Would synchronous or asynchronous muscle be expected to contain more sarcoplasmic reticulum? Give a reason for your answer.

Table 18.4 Wing speeds of a variety of insects.

	Wing speeds (times per second)	
large butterfly e.g. swallowtail	5	
locust	18	
hawkmoth	40	
housefly	120	at this speed a humming sound is heard
bee	180	
midge	700–1000	at this speed a high pitched whine is heard

In general, the smaller the insect the faster it beats its wings.
NB Some insects have two pairs of wings (such as locusts and dragonflies). In some cases both pairs of wings beat together, as in bees; in others the back pair of wings beats slightly ahead of the front pair, for example locusts. Some insects have one pair of wings, such as flies and beetles. In houseflies the hindwings are reduced and modified to form a pair of club-shaped halteres which are sensory in function. They oscillate rapidly during flight, detect aerodynamic forces and provide information for the maintenance of stability in flight. Some insects (very few) have no wings, for example fleas.

18.6 Locomotion in vertebrates

18.6.1 Swimming in fish

Water, particularly sea water, has a high density, many hundreds of times greater than air. As such it represents a comparatively viscous medium to move through. However, its density is made use of by fish for support and to provide a medium against which the fish can thrust during swimming movements.

Any successful organism shows many adaptive features suited to the environment in which it lives and a fish is no exception. The body of most fish is highly streamlined, being tapered at both ends. This means that water flows readily over the body surface and that drag is reduced to a minimum. Apart from the fins, no other structures project from a fish, and it seems that the faster the fish, the more perfect is the streamlining. The scales of bony fish are moistened by slimy secretions from mucus or oil glands which considerably reduces friction between the fish and the water. The fins are also adaptations for moving efficiently through the water. The median fin (along the midline of the body) and the dorsal and ventral fins, which are unpaired, help to stabilise the fish. The paired pectoral and pelvic fins are used for steering and balancing the animal and the caudal, or tail, fin contributes to the forward movement of the fish through the water. Details of how the fins operate are discussed below.

18.6.2 Propulsion in fish

In animals like the earthworm and insects, whose locomotion was studied in previous sections, the body is clearly divided into segments. Although the vertebrate body is also built on a segmented plan, few systems still show it clearly. However, blocks of muscle, called **myotomes**, located on either side of the vertebral column in fish show a clear segmental pattern. These are responsible for movement of the fish. Each myotome has a zig-zag shape (which you may have noticed if you have eaten fish). The vertebral column is a long, flexible rod and, when myotomes on one side of it contract, it bends easily. It is part of the endoskeleton and is adapted for locomotion in its flexibility. The myotomes contract and relax alternately on each side of the vertebral column, beginning at the front end of the fish and travelling towards its tail. In other words, the blocks on opposite sides of the spinal cord are antagonistic. This activity bends the body of the fish into a series of waves, the number of bends increasing the longer and thinner the fish.

- Very compact fish such as the tunny show little evidence of this wave-like action, with as much as 80% of their forward thrust being achieved purely by the side-to-side lashing of the tail and caudal fin. This locomotion is called **ostraciform**.
- Longer fish, such as the dogfish and the majority of bony fish, exhibit **carangiform locomotion**. Here the posterior half of the fish is thrown into a series of waves.
- **Anguilliform locomotion**, as demonstrated by eels, is where the body is very long and thrown into many waves, so that different parts of the body are moving to the left or to the right at the same time.

These types of movement are shown in fig 18.28.

Forward propulsion is generally achieved by the side-to-side movement of the tail to which is attached the caudal fin. As the fish is thrust forward through the water, the side-to-side movement of the tail tends to make the head swing from side to side in a direction opposite to that of the tail. This is called **lateral drag**. Fortunately, it is not very pronounced because the water tends to resist the movement of the relatively large front end of the body (compared to the tail) and the large surface area of the dorsal median fin also resists sideways movement. Also it requires a much greater force to move the body sideways through the water than to propel it forwards. The magnitude of the force that the tail and caudal fin apply to the water depends on:

- their speed of action;
- their surface area:
- the angle at which they are held with respect to the direction of movement.

18.6.3 Locomotion in a bony fish, the herring

The external features of a herring are shown in fig 2.65. Bony fish possess a structure called the **swim bladder** or air bladder. The swim bladder is a sac that lies between the vertebral column and the gut and provides the fish with 'neutral buoyancy'. This means that the fish has a density equal to that of the surrounding water and therefore does not need to use energy to keep itself from sinking. It can concentrate its efforts on moving through the water.

With the development of a swim bladder, the paired fins could change from their original function of providing lift. In bony fish these fins are much smaller than in cartilaginous fish like sharks and dogfish, and are used instead as stabilisers or brakes, in the latter case being spread vertically at 90° to the body. Each pectoral fin may be used independent of the other pectoral fin, and in this way they act as pivots round which the fish can turn rapidly. When the fish is swimming in a straight line the paired fins are pressed firmly against the sides of the body, thus improving its streamlined shape. The symmetrical shape of the tail fin means that it transmits most of the force it develops against the water in a forward direction rather

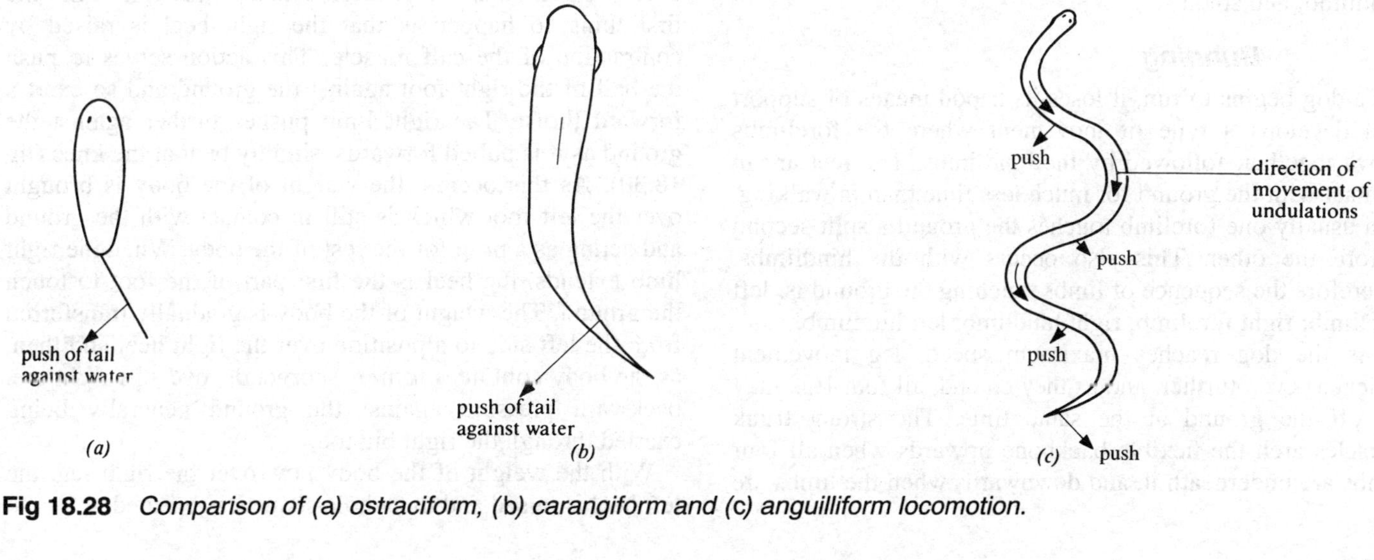

Fig 18.28 *Comparison of (a) ostraciform, (b) carangiform and (c) anguilliform locomotion.*

than up or down. The median fins resist the tendency of the head to swing from side to side in reaction to the tail as already described.

Two types of swim bladder exist.

- **Open swim bladder** (e.g. goldfish, herrings). The bladder is connected to the pharynx by a duct. Air is taken in or expelled from the bladder via the mouth and duct, thus decreasing or increasing the relative density of the fish respectively.
- **Closed swim bladder** (e.g. codfish). The bladder has completely lost its connection with the pharynx. By automatically increasing or decreasing the amount of gas in its bladder, the fish can match the density of the surrounding water and thus preserve 'neutral buoyancy'. Gas is extracted from, or returned to, the blood.

18.5 Summarise the adaptations that fish possess for efficient swimming.

18.6.4 Locomotion in quadrupeds, the dog

Walking

When a dog walks, its vertebral column remains rigid, and forward movement is achieved by the activity of the hindlimbs. They are moved forwards and backwards by alternative contraction of flexor and extensor muscles respectively.

When its extensor muscle contracts, each hindlimb, acting as a lever, extends and exerts a backward force against the ground, thrusting the animal forward and slightly upwards. When the flexor contracts, the limb is lifted clear of the ground and pulled forward. Only one limb is raised at any one time, the other three providing a tripod of support which balances the rest of the body. Beginning with the left forelimb in a stationary dog, the sequence of leg movement is as follows when it walks forward: left forelimb; right hindlimb; right forelimb; left hindlimb; and so on.

Running

As a dog begins to run, it loses its tripod means of support and develops a type of movement where the forelimbs move together, followed by the hindlimbs. The feet are in contact with the ground for much less time than in walking, and usually one forelimb touches the ground a split second before the other. This also occurs with the hindlimbs. Therefore the sequence of limbs touching the ground is: left forelimb; right forelimb; right hindlimb; left hindlimb.

As the dog reaches maximum speed, leg movement quickens even further, and as they extend, all four legs may be off the ground at the same time. The strong trunk muscles arch the flexible backbone upwards when all four limbs are underneath it, and downwards when the limbs are fully extended. In this way the thrust of the limbs is increased and the stride of the dog considerably lengthened, both of which enable the dog to increase its speed (fig 18.29).

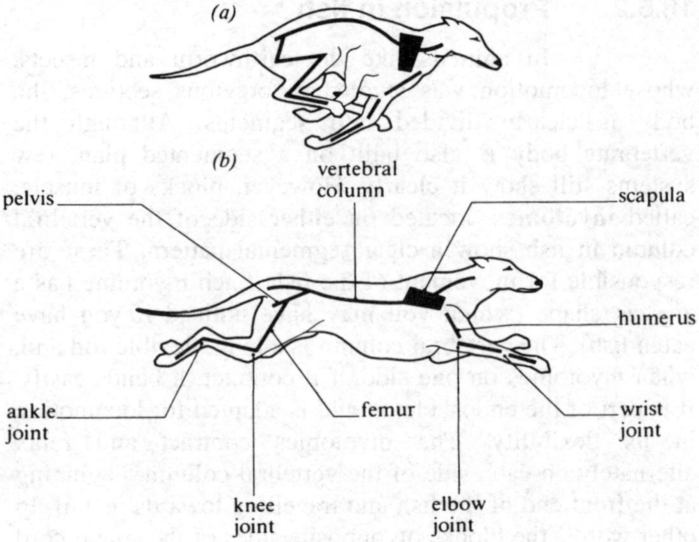

Fig 18.29 *Running sequence in a dog (such as a greyhound). (a) Backbone fully arched and feet immediately under the body. (b) Backbone fully extended and somewhat concave, limbs fully extended.*

18.6.5 Walking in the human

Humans have a bipedal gait, meaning they walk on two legs. You will find it useful to refer to some of the terms introduced in section 18.4.1, namely origin and insertion, flexor and extensor, and antagonistic muscles.

Walking

In the standing position the weight of the body is balanced over two legs. When a stride is taken by the right limb the first thing to happen is that the right heel is raised by contraction of the calf muscle. This action serves to push the ball of the right foot against the ground and so exert a forward thrust. The right limb pushes further against the ground as it is pulled forwards, slightly bent at the knee (fig 18.30). As this occurs, the weight of the body is brought over the left foot which is still in contact with the ground and acting as a prop for the rest of the body. When the right limb extends, the heel is the first part of the foot to touch the ground. The weight of the body is gradually transferred from the left side to a position over the right heel and then, as the body continues to move forwards, over the right toe, backward pressure against the ground generally being exerted through the right big toe.

With the weight of the body now over the right leg, the left heel is raised and the whole sequence repeated.

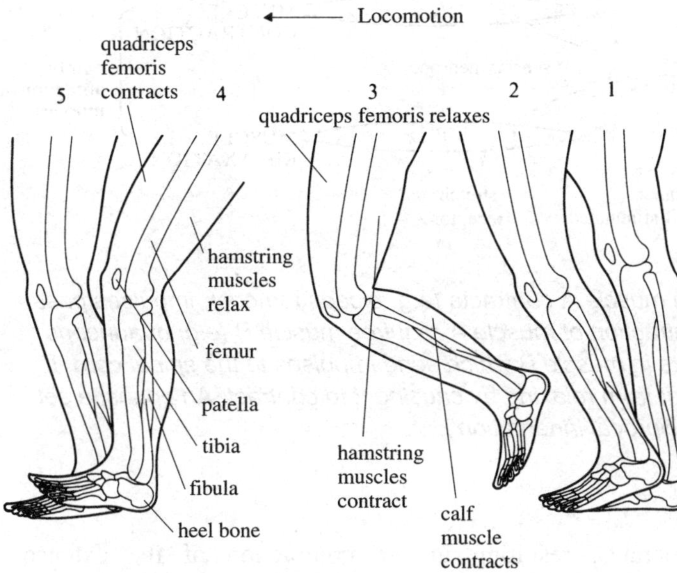

Fig 18.30 *Successive positions of the right leg during a single walking pace.*

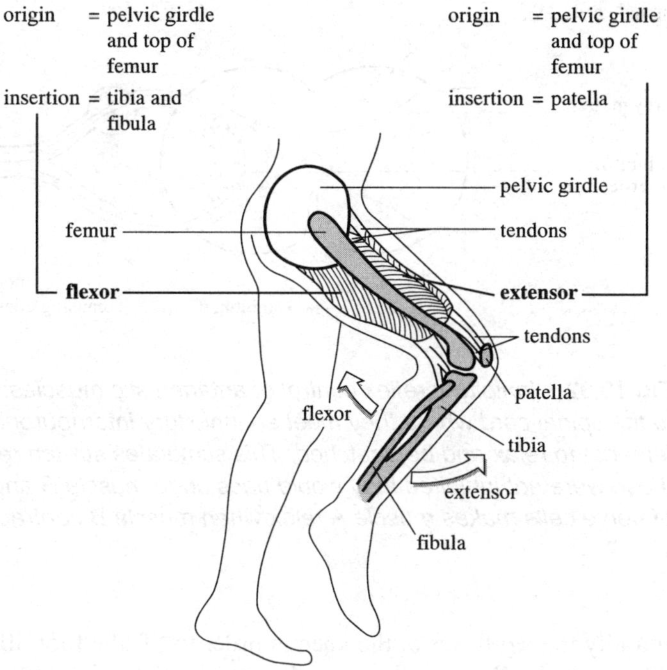

Fig 18.31 *An example of a pair of antagonistic muscles involved in walking.*

Antagonistic muscles

The pattern of muscle activity involved in walking is complex, and involves sets of antagonistic muscles. However, the principles involved can be examined by using the flexing and extending of the leg at the knee as an example. The knee is a hinge joint. The flexors are the hamstring muscles at the back of the upper part of the leg (thigh). An example of one of these is the **biceps femoris**. (This also extends the thigh at the hip joint, which involves pulling it further away from the trunk.) The extensors are in the front of the thigh and the main one is the **quadriceps femoris**. This extends the leg at the knee joint (as well as flexing the thigh at the hip joint, which involves pulling the thigh closer to the trunk of the body). Origins and insertions are as follows and are shown in fig 18.31.

- Biceps femoris: origin = pelvic girdle and top of femur
  insertion = tops of fibula and tibia (bones of lower leg)
- Quadriceps femoris: origin = pelvic girdle and top of femur
  insertion = patella (knee-cap)

The two muscles work antagonistically so that as one contracts the other relaxes. In this way the movements indicated by arrows in fig 18.31 are brought about. The whole process is controlled by nerve impulses from motor neurones supplying the muscle fibres. These are carefully coordinated by reflexes as described below.

Inhibitory reflexes

For a limb, or part of a limb, to be moved backwards and forwards it must be operated by at least two opposing muscles or sets of muscles. When one contracts the other must relax. This is achieved by a simple inhibitory reflex mechanism. Normally, as a muscle begins to stretch, special receptors called **stretch receptors** (or muscle spindles or proprioreceptors) detect the stretching and send nerve impulses to the spinal cord which continue back to the muscle and make it contract sufficiently to resist the stretching. Imagine, for example, putting a weight in one hand: the arm would drop and the biceps muscle in your upper arm would stretch unless the biceps resisted the force by contracting. If its force of contraction exactly balances the effect of the weight, your hand will stay in the same place. This is under reflex control, although voluntary control can take over if necessary. If, however, the lower arm is being extended (moving away from the upper arm) the extensor muscle at the back of the arm is contracting and the biceps is being stretched (extended). It is important to inhibit the normal reflex which would make it contract in this situation.

The sensory neurone which leads from the stretch receptor to the spinal cord also synapses with interneurones in the grey matter of the spinal cord (fig 18.32). When suitably stimulated, these inhibit the motor neurones leading to the antagonistic muscle, which is therefore unable to contract and so remains relaxed.

A good example of this is the mechanism of walking.

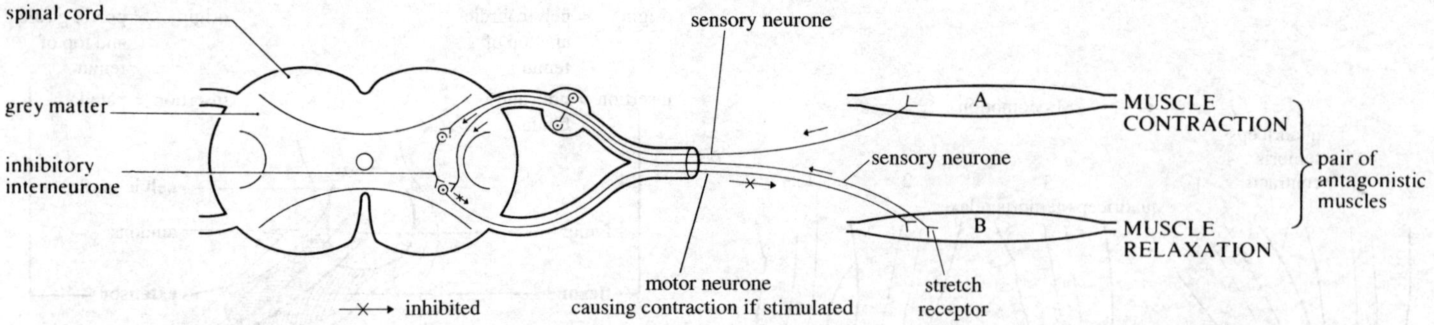

Fig 18.32 *Inhibitory reflex control of antagonistic muscles. When muscle A contracts (e.g. biceps femoris), impulses pass to the spinal cord where they meet an inhibitory interneurone. Contraction of muscle A requires muscle B (e.g. quadriceps femoris) to relax and be stretched. This stimulates stretch receptors in muscle B which send impulses to the spinal cord. If these were not inhibited they would pass on to muscle B and stop it from relaxing by causing it to contract. An opposite set of nerve cells makes muscle A relax when muscle B contracts ('reciprocal innervation').*

Initially the leg flexes at the knee in order to lift the foot off the ground. During flexing, the antagonistic extensor muscles are stretched but are reflexly inhibited from contracting. After flexion, the limb is straightened and the foot is again brought into contact with the ground. With the flexor muscles no longer contracting, inhibition of the extensor muscles ceases and the stretch reflex now operates, resulting in the contraction of the extensor muscles. When the limb is straight, no stretching in the extensor muscle stretch receptor is detected and the stretch reflex ceases. The whole process is then free to be repeated.

18.6 Why do sprinters generally run on their toes?

Chapter Nineteen

Homeostasis

An organism may be thought of as a complex system of chemical processes. These processes are self-regulating and tend to maintain a steady state within an external environment which is liable to change. The ability to maintain a steady state within a constantly changing environment is essential for the survival of living systems. In order to maintain this condition organisms from the simplest to the most complex have developed a variety of structural, physiological and behavioural mechanisms designed to achieve the same end, that is the preservation of a constant internal environment. The maintenance of a constant internal environment is called **homeostasis**. The advantage of a constant internal environment was first pointed out by the French physiologist Claude Bernard in 1857. Throughout his research he had been impressed by the way in which organisms were able to regulate physiological conditions, such as body temperature and water content, maintaining them within fairly narrow limits. This concept of self-regulation leading to physiological stability was summed up by Bernard in the now classic statement, 'La fixité du milieu interieur est la condition de la vie libre.' (The constancy of the internal environment is the condition of the free life.)

Bernard went on to distinguish between the **external environment** in which organisms live, and the **internal environment** in which individual cells live (in mammals, this is tissue fluid). He realised the importance of conditions in the latter being kept stable. For example, mammals are capable of maintaining a constant body temperature despite changes in the external temperature. If it is too cold the mammal may move to warmer or more sheltered conditions (a behavioural response); if this is not possible, self-regulating mechanisms operate to raise the body temperature and prevent further heat loss (a physiological response). All metabolic systems operate most efficiently if maintained within narrow limits close to optimum conditions. So the organism, as a whole, will function more efficiently if its cells are maintained at optimum conditions. Homeostatic mechanisms prevent large fluctuations from the optimum which are caused by changes in external and internal environments.

In 1932 the American physiologist Walter Cannon introduced the term **homeostasis** (*homoios*, same; *stasis*, standing) to describe how Bernard's 'constancy of the internal environment' is maintained. Homeostatic mechanisms maintain the stability of the cell environment and in this way provide the organism with a degree of independence of the environment. The more effective the mechanisms used, the more independent the organism is of the external environment. Independence of the environment can be used as a measure of the 'success' of an organism and, on this basis, more complex organisms such as mammals and flowering plants are seen as 'successful' since they are able to maintain relatively constant levels of activity despite fluctuations in environmental conditions. Such organisms are sometimes referred to as '**regulators**', meaning that they can regulate their own internal environments. They can typically exploit a wider range of environments and habitats than '**non-regulators**', which tend to be confined to environments which are more stable, such as oceans or lakes. Examples of non-regulators are cnidarians and algae, such as seaweeds and phytoplankton.

In order to achieve stability, the activities of organisms need to be regulated at all levels of biological organisation, from the molecular level to the level of the population. This means that organisms must use a range of structural, biochemical, physiological and behavioural mechanisms. In all these respects mammals are better equipped than simple animals like cnidarians to cope with changes in environmental conditions.

The mechanisms of regulation found in living organisms have many features in common with the mechanisms of regulation used in non-living systems, such as machines. The systems of both organisms and machines achieve stability by some form of control. In 1948, Wiener introduced the term **cybernetics** (*cybernos*, steersman) to mean the science of control mechanisms. This science is also commonly referred to as **control theory**. Plant and animal physiologists have used many of the very precise mathematical models of control theory to explain the functioning of biological control systems. Before studying some of the self-regulating mechanisms of living organisms, such as body temperature regulation and blood sugar levels, it is useful to have some understanding of the principles underlying control systems.

19.1 Control systems in biology

The application of control theory to biology has led to a clearer understanding of the different stages involved in the regulation of physiological processes. For example, living systems are now seen to be **open systems**;

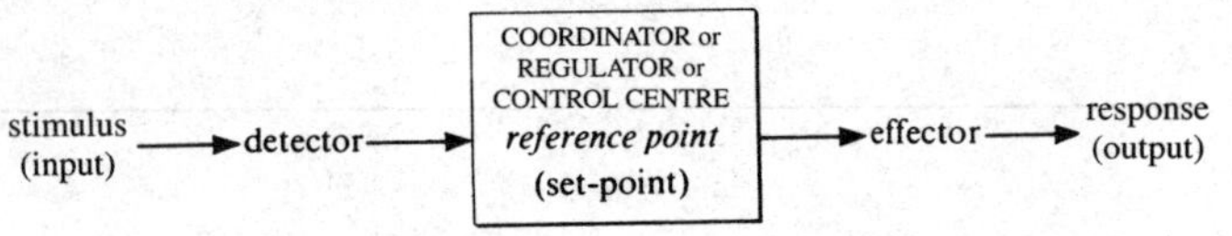

Fig 19.1 *Basic components of a control system.*

that is they require a continuous exchange of matter between the environment and themselves. They are in a steady state with their environment but require a continuous input of energy in order to prevent them coming to equilibrium with the environment. A simple analogy is a fountain of water. It needs a constant input of material (new water) and can only be kept operating by a continuous input of energy. Yet it appears to remain constant, in other words in a steady state, with respect to its environment. The basic components of any control system are summarised in fig 19.1. Various terms may be used for each stage of the process and commonly used alternatives are shown in fig 19.1. The regulator in mammals is either an endocrine gland producing hormones, or part of the nervous system, often the brain or spinal cord.

The efficiency of the control system can be judged by:

- how little change there is from the reference point (or optimum level);
- the speed with which the level is restored after change.

Any change from the set-point activates the control systems and returns conditions towards their optimum level. As conditions return to the optimum, the corrective processes can be switched off, a process known as '**negative feedback**'. Feedback requires the action of the system to be related to a **reference point** or **set-point**, which is the optimum level of the variable being controlled (similar to the temperature set by a thermostat in a heating system). There are two forms of feedback, **negative** and **positive**. Negative feedback is more common.

Negative feedback

Negative feedback is associated with increasing the stability of systems (fig 19.2). If the system is disturbed, the disturbance sets in motion a sequence of events which tends to restore the system to its original state. The principle of negative feedback can be illustrated in terms of the regulation of oven temperature by the use of a thermostat. In the electric oven the control system includes an **effector**

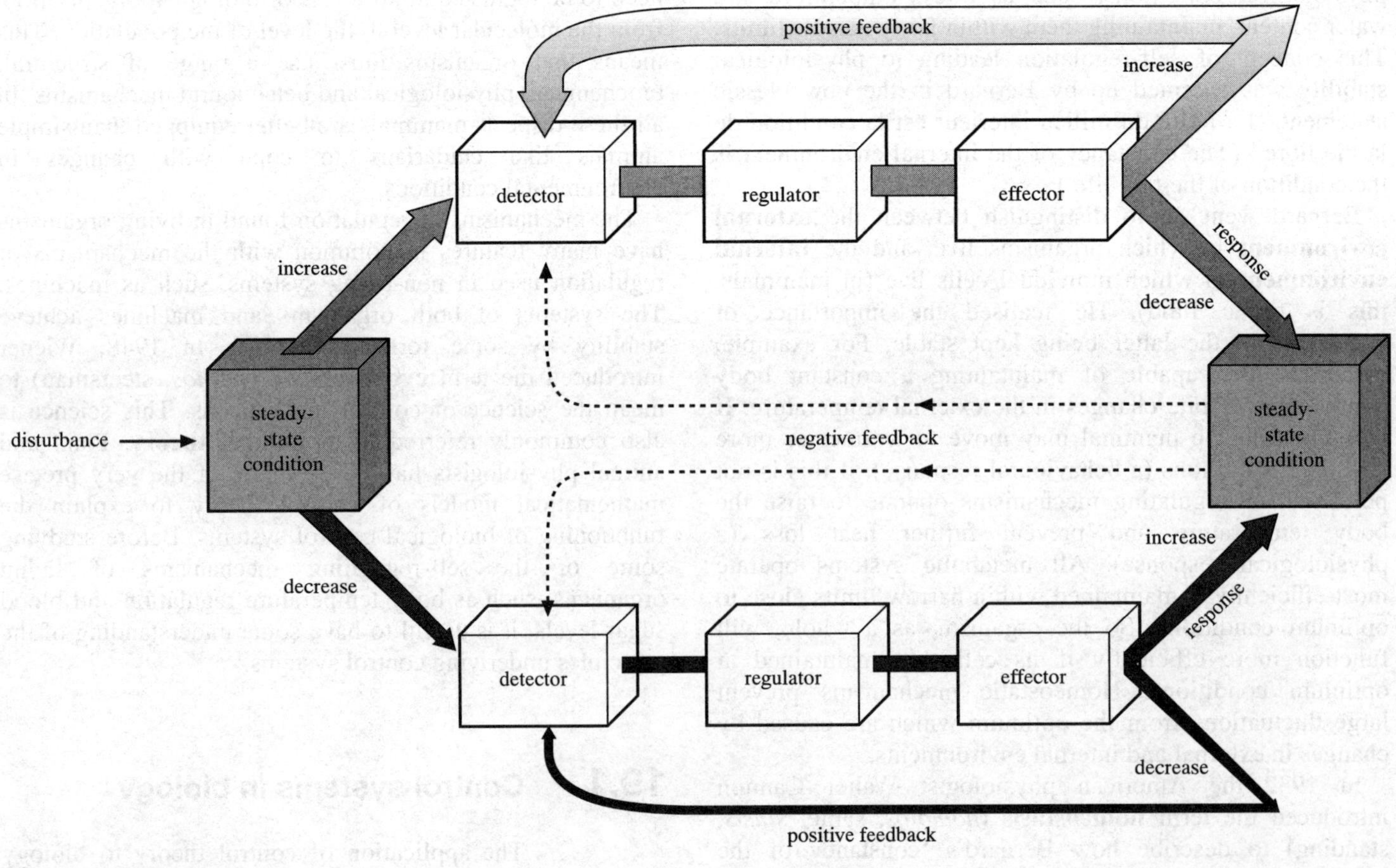

Fig 19.2 *A homeostatic control system. Negative feedback is shown by dotted lines. Negative feedback provides a means of switching off a process.*

(an element through which an electric current (an **input**) flows and which acts as a source of heat), an **output** (the oven temperature) and a thermostat which can be set to a desired level, the **set-point**. The thermostat acts as a **detector** (or **receptor**) and a **regulator**. The **stimulus** is heat. If the thermostat is set to read 150 °C, an electric current will provide a source of heat which will flow until the oven temperature passes the set-point of 150 °C, then the thermostat will cut out and no more heat will be supplied to the oven. When the oven temperature falls below 150 °C the thermostat will cut in again and the electric current will increase the temperature and restore the set-point. In this system the thermostat is functioning as an **error detector** where the error is the difference between the output and the set-point. The error is corrected by the effector (the heating element) being switched on or off. This is an example of a steady-state negative feedback system which is typical of many of the physiological control mechanisms found in organisms.

Examples of biological negative feedback mechanisms include the control of:

- oxygen and carbon dioxide levels in the blood by controlling rate and depth of breathing (section 9.5.5);
- heart rate (section 14.7.4);
- blood pressure (section 14.7.7);
- hormone levels, e.g. thyroxine (section 17.6.4 and fig 19.3), sex hormones (sections 21.6.4 and 21.6.6);
- metabolite levels, e.g. glucose (sections 17.6.6 and 19.2);
- water and ionic balances (section 20.3.5);
- the regulation of pH (section 20.8);
- body temperature (section 19.5).

Fig 19.3 illustrates the role of negative feedback in the control of thyroxine release by the thyroid gland. In this example the **detector** is the hypothalamus, the **regulator** is the pituitary gland and the **effector** is the thyroid gland.

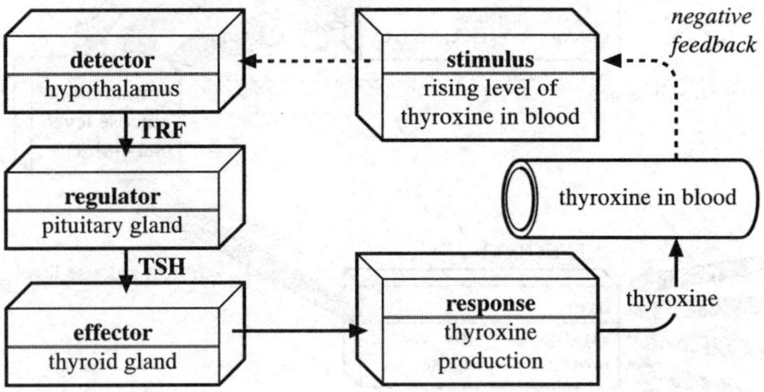

Fig 19.3 *The control of thyroxine production. This is a biological example of a simple control system. Thyroxine is a hormone produced by the thyroid gland. TRF, thyroid releasing factor; TSH, thyroid stimulating hormone; see section 17.6.4*

Positive feedback

Positive feedback is rare in biological systems since it leads to an unstable situation and extreme states. In these situations a disturbance leads to events which increase the disturbance even further (fig 19.2). For example, during the propagation of a nerve impulse, depolarisation of the membrane of the neurone produces an increase in its sodium permeability. As sodium ions pass into the axon through the membrane they cause a further depolarisation which leads to even more sodium ions entering. The rate at which sodium ions enter therefore increases very rapidly and produces an action potential. In this case positive feedback acts as an amplifier of the *response* (depolarisation). Its extent is limited by other mechanisms as described in section 17.1.1. Positive feedback also occurs during labour when the hormone oxytocin stimulates muscular contractions of the uterus, which in turn stimulate the release of more oxytocin (section 21.7.12).

More complex mechanisms

There are several control mechanisms in the body which are more complex than those previously described. Broadly speaking they either involve the use of additional detectors (early-warning systems) or additional effectors ('fail-safe' systems). For example, temperature detectors situated externally and internally enable warm-blooded animals (homeotherms) to maintain an almost constant 'core' body temperature. Temperature receptors in the *skin* act as detectors of changes in the *external* environment. They send impulses to the hypothalamus which acts as a regulator and stimulates corrective measures before any change in *blood* temperature occurs. Other examples of this type of early-warning system include the control of ventilation during exercise, and control of appetite and thirst. Similarly multiple detectors and effectors provide fail-safe mechanisms for many vital processes, such as regulation of arterial blood pressure where stretch receptors in the carotid sinus and aorta, and baroreceptors in the medulla, respond to blood pressure changes and produce responses in various effectors including the heart, blood vessels and kidneys. Failure of any one of these is compensated for by the others.

19.2 Control of blood glucose levels

One of the most important metabolites in the blood is glucose. Its level must be controlled strictly. Glucose is the main respiratory substrate and must be supplied continuously to cells. The brain cells are especially dependent on glucose and are unable to use any other metabolites as an energy source. Lack of glucose results in fainting. The normal level of glucose in the blood is about 90 mg per 100 cm^3 blood, but may vary from 70 mg per 100 cm^3 blood during fasting up to 150 mg per 100 cm^3 blood following a meal. The sources of blood glucose and its relationships with other metabolites are described in section 19.6.2.

The control of blood glucose level is a good example of homeostasis and involves the secretion of at least six hormones and two negative feedback pathways. A rise in blood glucose level (**hyperglycaemia**) stimulates insulin secretion (section 17.6.6) whereas a fall in blood glucose level (**hypoglycaemia**) inhibits insulin secretion and stimulates the secretion of glucagon (section 17.6.6) and other hormones which raise blood glucose levels, such as adrenaline. The control mechanism is summarised in fig 19.4.

19.3 Temperature regulation

All living systems require a continuous supply of heat energy in order to survive.

The major source of heat for all living organisms is the Sun. Solar radiation is converted into heat energy whenever it strikes and is absorbed by a body. The extent of this solar radiation depends upon geographical location, and is a major factor in affecting the climate of a region. This, in turn, determines the presence and abundance of species. For example, organisms inhabit regions where the normal air temperatures vary from –40 °C, as in the Arctic, to 50 °C in desert regions. In some of the latter regions the surface temperature may rise as high as 80 °C. The majority of living organisms exist within confined limits of temperature, say 10–35 °C, but various organisms show adaptations enabling them to exploit habitats at both extremes of temperature (fig 19.5). This is not the only source of heat available to organisms. Solar radiation is used by photosynthetic organisms and this energy becomes locked up in the chemical bonds of organic molecules such as sugars. This provides an internal source of heat energy when released by the reactions of respiration.

Temperature indicates the amount of heat energy in a system and is a major factor determining the rate of chemical reactions in both living and non-living systems. Heat energy increases the rate at which atoms and molecules move and this increases the probability of reactions occurring between them.

19.3.1 The influence of temperature on the growth and distribution of plants

Temperature can act as a limiting factor in the growth and development of plants by influencing the rates of processes like cell division, photosynthesis and other

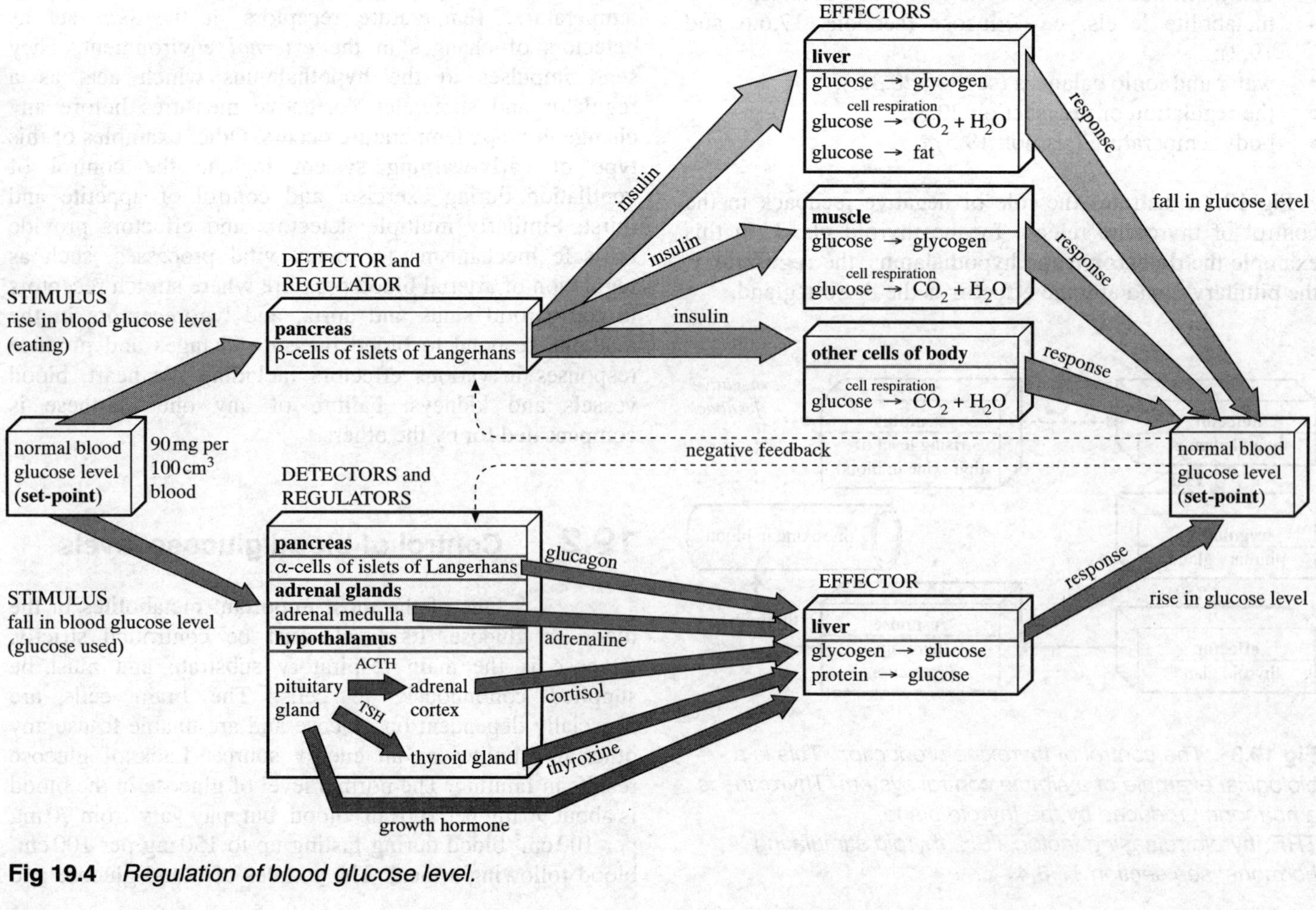

Fig 19.4 *Regulation of blood glucose level.*

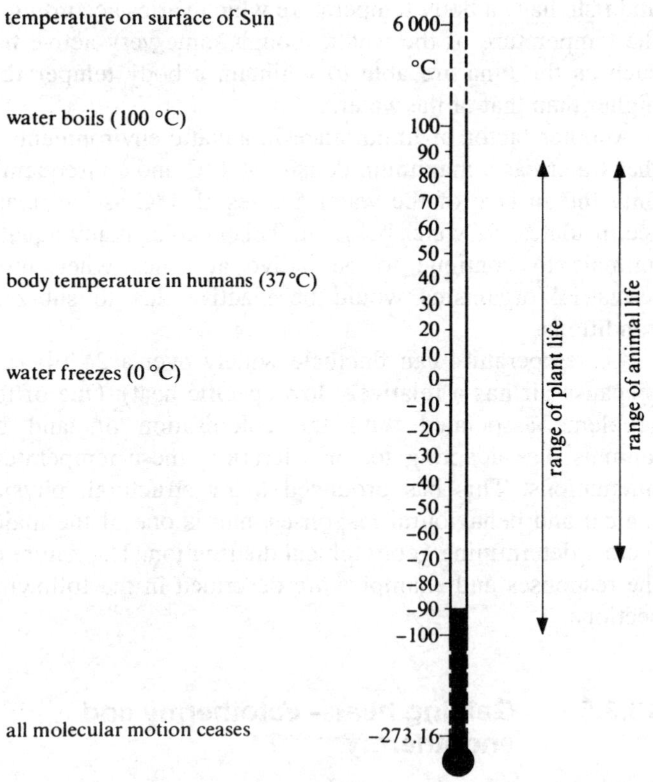

Fig 19.5 *Temperature reference points in living organisms and the ranges of temperature they are able to tolerate.*

metabolic processes. The light-independent reactions of photosynthesis are temperature dependent and lead to the various metabolic pathways described in chapter 7. The rates of photosynthesis, and the accumulation of sufficient food materials to enable the plant to complete its life cycle, are important factors in determining the geographical range of plants.

19.3.2 Adaptations of plants to low temperatures

The flora of northern temperate climates and the tundra show many adaptations that enable plants to take maximum advantage of the short warm summers. For example, the only plant species found are mosses, a few grasses and fast-growing annuals. Plants living in extreme northerly or southerly latitudes are subjected to long periods of adverse conditions, such as low light intensity, low temperatures and frozen soil. In order to survive in these conditions, plants show many structural, physiological and behavioural adaptations. For example, most temperate woody perennials are deciduous and lose their leaves, under the influence of the plant growth regulator substance abscisic acid (ABA), in order to prevent water loss by transpiration during periods when soil water uptake is limited by low temperatures (section 16.2.8). Wind and snow damage is also avoided by the shedding of leaves during these periods, when the rate of photosynthesis would be severely limited by low light intensities, low temperatures and unavailability of water and salts. Throughout these periods the regions of next year's growth, the buds, are protected by scale leaves and are dormant. Their metabolic activity is inhibited by the presence of abscisic acid. Many coniferous species dominate the vegetation of the more temperate regions, particularly in northern latitudes. These species have needle-like leaves which reduce the amount of snow which can accumulate on them in winter and have a thick cuticle to prevent water loss in summer. Many species of annuals have short growing periods and survive the winter by producing resistant seeds or organs of perennation.

Low temperatures are required by many plant species in order to break dormancy. For example, lilac buds develop more quickly after being exposed to low temperatures than to high temperatures. Other examples of the effects of low temperature on plant growth are described in chapter 16 and section 22.4.

19.3.3 Adaptations of plants to high temperatures

In many regions of the world high temperatures are associated with water shortage, and many of the adaptations shown by plants in these regions are related to their ability to resist desiccation.

Plants are unable to escape high temperatures by moving to shaded areas and therefore they have to rely on structural and physiological adaptations to avoid overheating. It is the aerial parts of plants which are exposed to the heating effect of solar radiation, and the largest exposed surface is that of the leaves. Leaves characteristically are thin structures with a large surface area to volume ratio to allow gaseous exchange and absorption of light. This structure is also ideal for preventing damage by excessive heating. A thin leaf has a relatively low heat capacity and therefore will usually assume the temperature of the surroundings. In hot regions a shiny cuticle is secreted by the epidermis which reflects much of the incident light, thus preventing heat being absorbed and overheating the plant. The large surface area contains numerous stomatal openings which permit transpiration. As much as $0.5\,\mathrm{kg\,m^{-2}\,h^{-1}}$ of water may be lost from plants by transpiration in hot, dry weather, which would account for the loss of approximately $350\,\mathrm{W\,m^{-2}}$ in terms of heat energy. This is nearly half the total amount of energy being absorbed. As a result of these mechanisms plants are able to exercise a considerable degree of control over their temperature.

> **19.1** Why do plants suffer permanent physiological damage if exposed to temperatures in excess of 30 °C when the humidity is high?

During hot, sunny days many plants wilt. This occurs when the rate of transpiration is greater than the rate of water uptake. There is an overall loss of turgidity by cells such as parenchyma. **Wilting** may be observed in plants growing in greenhouses in response to the very high temperatures which develop in the leaves, even if adequately supplied with water. In such a situation it may therefore be a mechanism for preventing overheating by reducing the surface area of leaves exposed to direct sunlight. Wilting reduces the yield of plants. Once the temperature begins to fall, the plants recover very quickly even if the degree of wilting looks severe.

Plants living in dry conditions are called **xerophytes** and show many structural adaptations which enable them to survive (section 20.10). In most cases these adaptations are primarily concerned with regulating water loss, although the characteristic needle-shaped leaves permit maximum heat loss. The mechanisms for withstanding high temperatures, on the other hand, are mainly physiological. One physiological mechanism used by plants (not just xerophytes) to avoid wilting in dry conditions is to produce more abscisic acid. This causes stomata to close, thus reducing loss of water by transpiration.

19.3.4 The influence of temperature on the growth and distribution of animals

Temperature influences the metabolic activity of animals in a number of ways:

- the main effect is on the rate of enzyme activity and the rate of movement of atoms and molecules. This directly affects the rate of growth of animals (chapter 22).
- Temperature may also affect the geographical distribution of animals through its influence on plants as primary producers in the food chains. The ecological range of most animal species, with the exception of some insects, birds and mammals which are able to migrate, is determined by the local availability of food.

The variety of responses shown by animals to temperature depends upon:

- the degree of variation in temperature shown by the environment;
- the degree of control the organism has over its own body temperature.

Life is believed to have originated in the marine environment, the environment that poses the fewest temperature problems for living organisms because temperature fluctuations in aquatic environments are slight. The biological significance of this is that aquatic organisms have a relatively stable environment with regard to temperature and therefore do not need the same range of responses to temperature change that is shown by terrestrial animals. Most aquatic organisms, including non-vertebrates and fish, have a body temperature which varies according to the temperature of the water, though some very active fish such as the tuna are able to maintain a body temperature higher than that of the water.

Another factor of importance in aquatic environments is that water has a maximum density at 4 °C and consequently only the surface of the water freezes at 0 °C as ice floats. Ice insulates the water below it. This enables many aquatic animals to continue to be active at times when most terrestrial organisms would be inactive due to sub-zero conditions.

Air temperature can fluctuate widely over a 24 h period (because air has a relatively low specific heat). One of the problems associated with the colonisation of land by animals was adapting to, or tolerating, these temperature fluctuations. This has produced many structural, physiological and behavioural responses, and is one of the major factors determining geographical distribution. The nature of the responses and examples are described in the following sections.

19.3.5 Gaining heat – ectothermy and endothermy

All animals gain heat from two sources:

- the external environment;
- the release of heat energy as a result of chemical reactions within their cells.

The extent to which different groups of animals are able to generate and conserve this heat is variable. All non-vertebrates, fish, amphibia and reptiles are unable to maintain their body temperature within narrow limits using *physiological* mechanisms, though many do so using behavioural mechanisms. Consequently these animals are described as **poikilothermic** (*poikilos*, various; *therme*, heat). Alternatively, because they rely more on heat derived from the environment than metabolic heat in order to raise their body temperature, they are termed **ectothermic** (*ecto*, outside). These animals used to be described as 'cold-blooded' but this is a misleading and inaccurate term.

Birds and mammals are able to maintain a fairly constant body temperature independently of the environmental temperature by using physiological mechanisms. They are described as **homeothermic** (or **homoiothermic** – *homoios*, like) or less correctly as 'warm-blooded'. Homeotherms are relatively independent of external sources of heat and rely on a high metabolic rate to generate heat which must be conserved. Since these animals rely on internal sources of heat they are described as **endothermic** (*endos*, inside). Some poikilotherms may, at times, have temperatures higher than those of homeotherms and, in order to prevent confusion, the terms ectotherm and endotherm are used throughout this text. Fig 19.6 shows the abilities of several vertebrates to regulate their body temperature in various external temperatures.

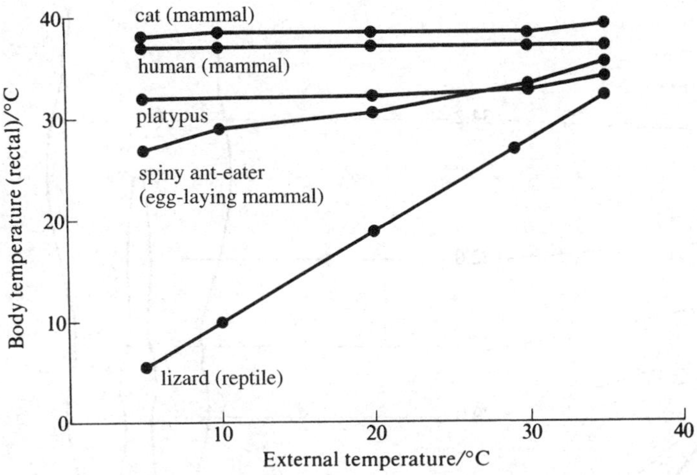

Fig 19.6 *The relationships between external (environmental) and internal (body) temperatures in vertebrates kept for 2 h at the external temperatures indicated.*

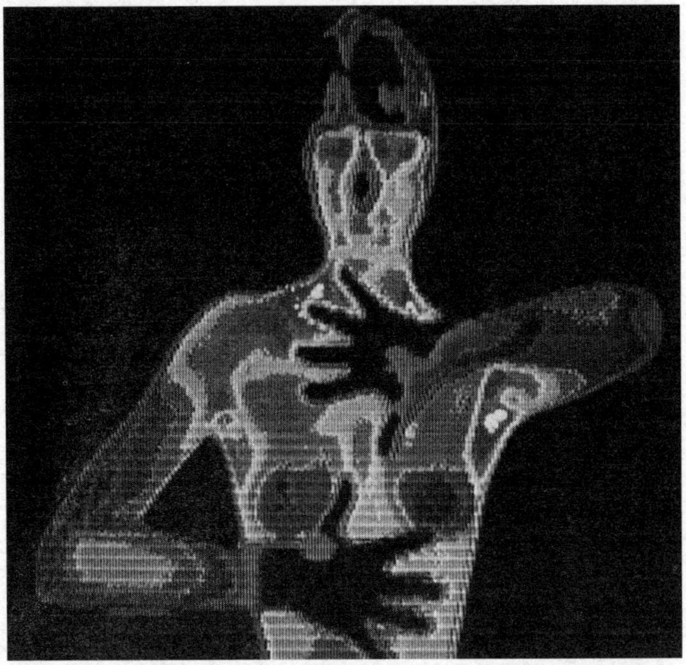

Fig 19.7 *Infra-red photo of a young woman with cold hands. The brighter areas of her body are warmer, whilst her hands show as dark areas.*

19.3.6 Losing heat

Whilst the main method of *gaining* heat differs between ectotherms and endotherms, the methods of heat loss to the environment are the same in both types of organism. These are radiation, convection, conduction and evaporation. In practice, heat can be transferred in either direction by the first three methods depending upon the **thermal gradient** (the direction of the temperature difference from hot to cold) but can only be *lost* from organisms by the latter method.

Radiation. Heat is transferred by infra-red radiation. Infra-red waves are part of the electromagnetic spectrum and are longer than the waves of the visible spectrum. They can be detected by special film sensitive to infra-red radiation, as shown in fig 19.7. Such photos show radiation of heat from the body. Heat loss (or gain) is proportional to the temperature difference between the body and its surroundings. Radiation accounts for about 50% of the total heat loss in humans and provides the main route for controlled heat loss in animals.

Convection. Heat is transferred between organisms and environment by convection currents in the air or water in contact with the surface of an organism. Convection currents are mass movements of air or water molecules. It is due to still air or water next to the skin being warmed, and rising as a result. The rate of heat transfer by this process is linked to the rate of air or water movement. It may be reduced by insulating materials such as feathers or hair in animals or clothing in humans.

Conduction. Conduction is the transfer of energy by physical contact between two bodies, for instance between the organism and the ground or the air. The molecules of the skin and anything it is in contact with are in constant motion due to kinetic energy. The energy of this motion can be transferred from a region of higher temperature to a region of lower temperature. Heat exchange by this means is relatively insignificant for most terrestrial organisms, but may be considerable for aquatic and soil-dwelling organisms. Conduction to physical objects such as a chair is insignificant in humans (about 3%) but conduction to air is more significant, about 15% of heat loss in humans.

Evaporation. Heat is lost from the body surface during the conversion of water to water vapour. The evaporation of 1 cm^3 of water (1 g) requires the loss of 2.45 kJ from the body. It is a limiting factor in the distribution of many plant and animal species. In humans water loss by evaporation takes place continuously through the skin even when a person is not sweating. It also occurs by evaporation from the lungs (lost in expired air). Neither of these can be controlled. However, partial control of heat loss is possible by regulating loss of water by sweating and panting.

19.3.7 Core temperature and surface temperature

When reference is made to body temperature in animal studies it usually refers to the **core temperature**. This is the temperature of the tissues below a level of 2.5 cm beneath the surface of the skin. This temperature is most easily measured by taking the temperature of the rectum (rectal temperature). Temperatures near the surface of the body can vary tremendously depending upon position and external temperature (fig 19.8).

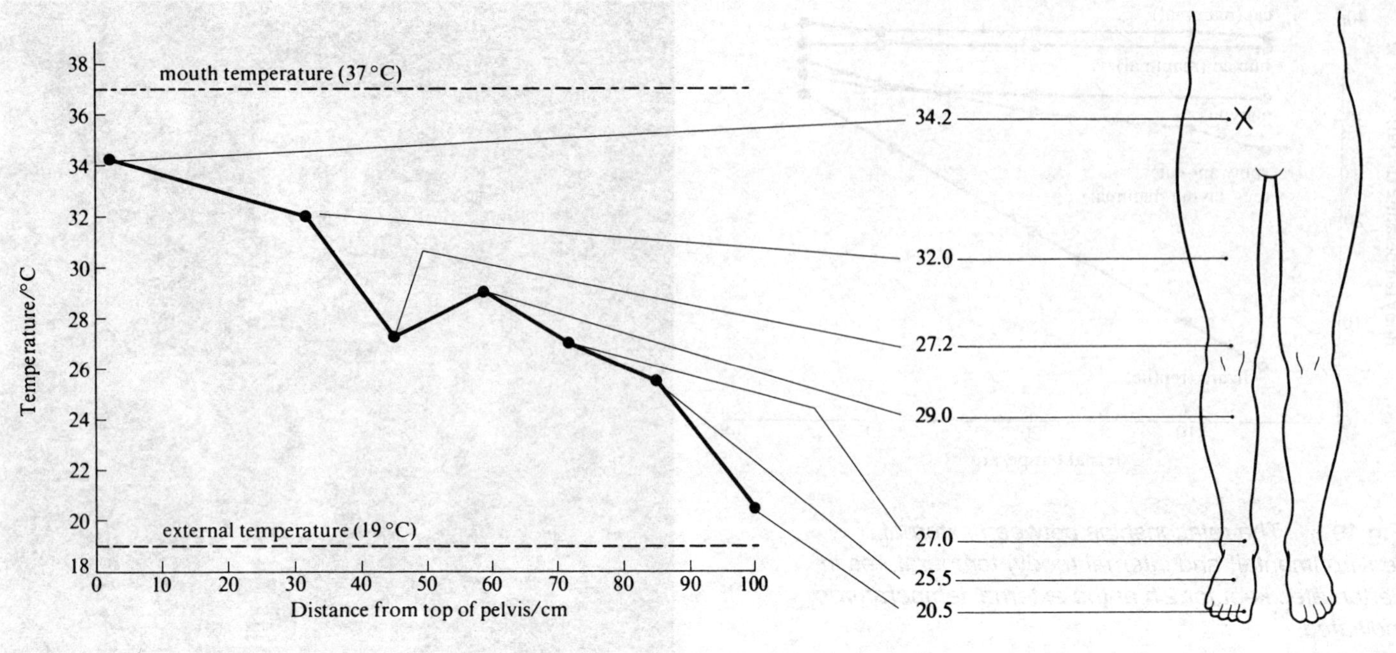

Fig 19.8 *Skin temperatures recorded at the pelvis and on the leg during a period of exposure to an external temperature of 19 °C. The low temperature at the knee cap probably reflects its poor blood supply.*

19.4 Ectothermic animals

The majority of animals are ectothermic, and their activity is determined by the environmental temperature. The metabolic rate of ectotherms is relatively low and they lack physiological mechanisms for conserving heat.

19.4.1 Temperature regulation in aquatic ectotherms

Aquatic ectotherms live within a restricted temperature range, determined by the size of the body of water in which they live. For example, the temperature of a pond can vary considerably throughout the year, whereas that of an ocean may change by only a few degrees. Despite this wide temperature fluctuation in small bodies of water, many insect species have aquatic stages in the life cycle (larvae, pupae or nymphs), since the aquatic temperatures are more stable and less extreme than terrestrial temperatures during the winter months. Mayflies, dragonflies, caddisflies, midges and mosquitoes all have an aquatic stage.

Aquatic non-vertebrates are able to tolerate greater temperature fluctuations than aquatic vertebrates due to their relatively simple physiology. Fish have a higher rate of metabolism than aquatic non-vertebrates but the majority of the extra heat produced as a result is rapidly spread around the body and lost to the environment by conduction through the gills and skin. Consequently, fish usually have a body temperature which is in equilibrium with that of the water. Fish cannot maintain a temperature below that of the water but may in some cases, as in the case of tuna fish, retain heat by means of a countercurrent heat exchange system. This can raise the temperature of the 'red' swimming muscle to about 12 °C above that of the sea water.

19.4.2 Temperature regulation in terrestrial ectotherms

Terrestrial (land-living) ectotherms have to cope with greater temperature fluctuations than those of aquatic ectotherms, but they have the benefit of living at higher environmental temperatures. This allows them to be more active. The relatively poor thermal conductivity of air reduces the rate of heat loss from the organisms whilst water loss by evaporation may be used to cool them. Many species are able to maintain temperatures slightly above or below air temperature and thereby avoid extremes.

Heat is gained and lost by terrestrial ectotherms by behavioural and physiological activities. The main sources of heat gain are the absorption of solar radiation and conduction from the air and the ground. The amount of heat absorbed depends upon:

- the colour of the organism;
- its surface area;
- its position relative to the Sun's rays.

Colour. A species of Australian grasshopper is dark in colour at low temperatures and absorbs solar radiation, thus heating up rapidly. As the temperature rises above a set point, further absorption is reduced by the exoskeleton becoming lighter in colour. This colour change is believed to be a direct response by pigment cells to body temperature. Such organisms are known as **basking heliotherms** (*helios*, sun).

Orientation. Changes in position of the organism relative to the Sun's rays vary the surface area exposed to heating. This practice is common in many terrestrial ectotherms including insects, arachnids, amphibia and reptiles. It is a form of *behavioural* thermoregulation. For example, the desert locust is relatively inactive at 17 °C but by aligning itself at right-angles to the Sun's rays it is able to absorb heat energy. As the air temperature rises to approximately 40 °C it re-orientates itself parallel to the Sun's rays to reduce the exposed surface area. Further increase in temperature, which may prove fatal, is prevented by raising the body off the ground or climbing up vegetation. Air temperature falls off rapidly over short distances above ground level, and these movements enable the locust to find a more favourable microclimate.

Reptiles

Crocodiles, too, regulate their body temperature on land by varying their position relative to the Sun's rays. They also open their mouths to increase heat loss by evaporation. If the temperature becomes too high, they move into the water which is relatively cooler. At night they retreat to water in order to avoid the low temperatures which would be experienced on land. Thermoregulation mechanisms have been studied most extensively in lizards.

Many different species of lizard have been studied and show a variety of responses to different temperatures. Lizards are terrestrial reptiles and exhibit many behavioural activities, as is typical of other ectotherms. Some species, however, use a number of physiological mechanisms to raise and maintain their body temperatures above that of the environment (fig 19.9). Other species are able to keep their body temperature within confined limits by varying their activity and taking advantage of shade or exposure. In both these respects lizards foreshadow many of the mechanisms of homeothermy shown by birds and mammals.

Surface temperatures in desert regions can rise to 70–80 °C during the day and fall to 4 °C at dawn. During the periods of extreme temperatures, most lizards seek refuge by living in burrows or beneath stones. This response and certain physiological responses are shown by the horned lizard (*Phrynosoma*) which inhabits the deserts of the south-west of the USA and Mexico. In addition to burrowing, the horned lizard is able to vary its orientation and colour, and as the temperature becomes high it can also reduce its body surface area by pulling back its ribs. Other responses to high temperatures involve panting, which removes heat by the evaporation of water from the mouth,

Fig 19.9 Varanus *rock monitor basking at its treehole in the sunlight to raise its body temperature.*

Fig 19.10 *Shovel-snouted lizard* (Aporosaura anchietae) *in the Namib desert. In the afternoon when the sand is hot, the lizard lifts opposite pairs of feet alternately so that they can cool in the air. This is known as thermal dancing.*

pharynx and lungs, eye bulging, thermal dancing (fig. 19.10) and the elimination of urine from the cloaca.

The marine iguana (*Amblyrhynchus*), a reptile, normally maintains a temperature of 37 °C as it basks on the rocky shore of the Galapagos Islands, but it needs to spend a considerable time in the sea feeding on seaweed at a temperature of approximately 25 °C. In order to avoid losing heat rapidly when immersed in water the iguana reduces the blood flow between surface and core tissues by slowing its heart rate (bradycardia).

Amphibians

The moist skin of amphibians provides an ideal mechanism to enable heat loss by evaporation. This water loss, however, cannot be regulated physiologically as in mammals. Amphibians lose water immediately on exposure to dry conditions and, whilst this helps heat loss, it leads to dehydration if the amphibian does not find moist shaded conditions where the rate of evaporation is reduced.

19.5 Endothermic animals

Birds and mammals are endothermic and their activity is largely independent of environmental temperature. In order to maintain a constant body temperature, which is normally higher than the air temperature, these organisms need to have a high metabolic rate and an efficient means of controlling heat loss from the body surface. The skin is the organ of the body in contact with the environment and therefore monitors the changing temperatures. The actual regulation of temperature by various metabolic processes is controlled by the hypothalamus of the brain.

19.5.1 Skin structure

The term 'skin' applies to the outer covering of vertebrate animals. The skin is the largest organ of the body and has many different functions. The structure of the skin varies in different vertebrate groups. The basic structure of human skin will be described here (fig 19.11).

The skin is composed of two main layers, the epidermis and dermis. These cover the underlying, or 'subcutaneous' tissue, which contains specialised fat-containing cells known as **adipose tissue**. The thickness of this layer varies according to the region of the body and from person to person.

Epidermis

The cells of this region are separated from the dermis by a basement membrane. The epidermis is composed of many layers of cells that form a stratified epithelium (section 6.3.2). The cells immediately above the basement membrane are cuboidal epithelial cells (section 6.3.1) and form a region known as the **Malpighian layer**. The cells in this layer are constantly dividing by mitosis and replace all the cells of the epidermis as they wear away. The Malpighian layer forms the lower region of the **stratum granulosum**, which is composed of living cells that become flatter as they approach the outer region of the epidermis, the **stratum corneum**. Cells in this region become progressively flattened and synthesise **keratin**, which is a fibrous protein which makes the cells waterproof. As the keratin content of the cells increases, they are said to become 'cornified'. Their nuclei disappear and the cells die. The thickness of the stratum corneum increases in parts of the body where there is considerable friction, such as the ball of the foot and the bases of the fingers. The outer covering of the skin forms a semi-transparent, thin, tough, flexible, waterproof covering pierced by the hair follicles and by pores, which are the openings of the sweat glands. The outermost squamous epithelial cells are continually being shed as a result of friction.

The stratum corneum has become modified in many vertebrates to produce nails, claws, hooves, horns, antlers, scales, feathers and hair. Keratin is the main component of all these structures.

Dermis

The dermis is a dense matrix composed of connective tissue rich in elastic fibres and contains blood capillaries, lymph vessels, muscle fibres, sensory cells, nerve fibres, pigment cells, sweat glands and hair follicles.

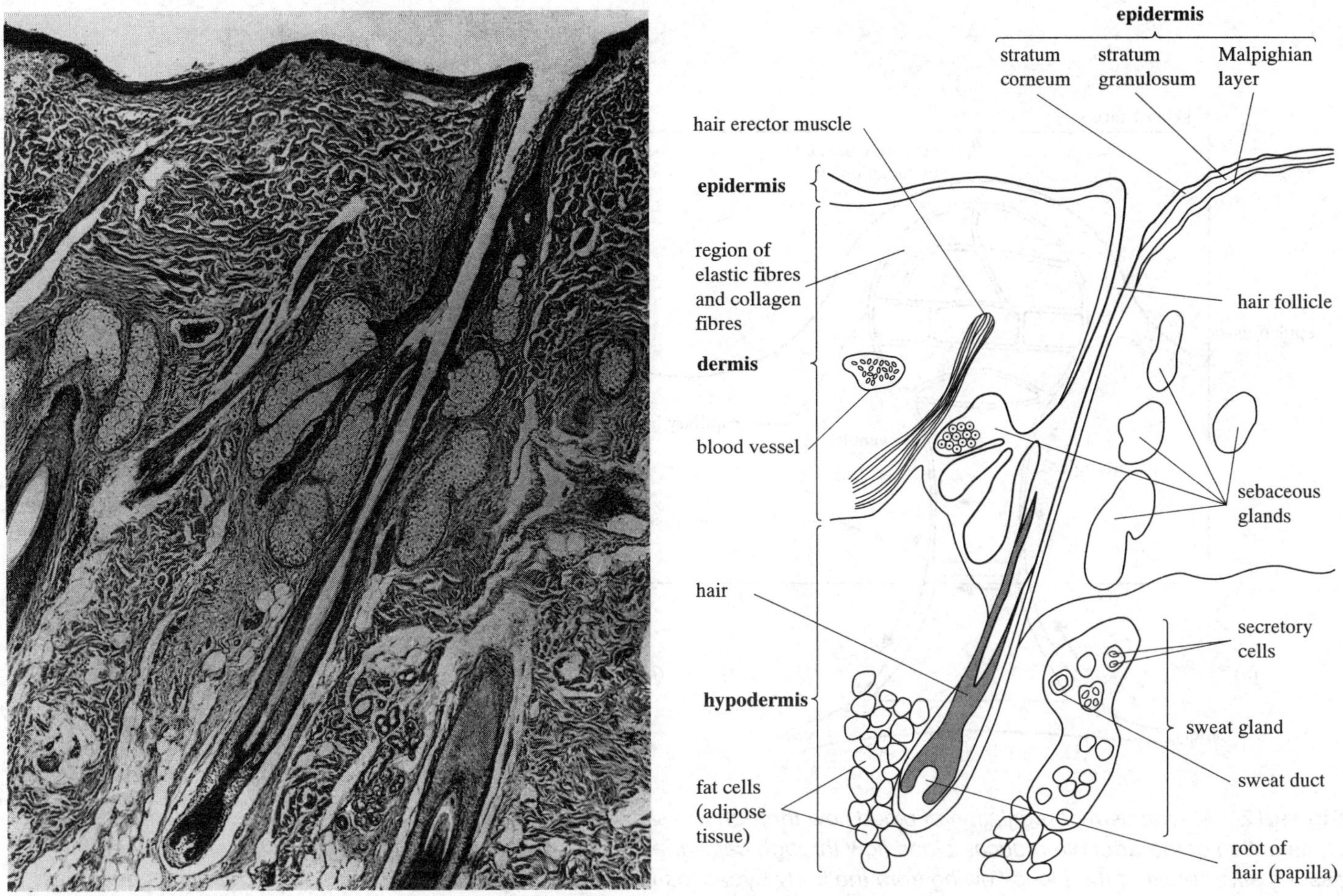

Fig 19.11 *(a) Vertical section through human skin. (b) Diagram of skin structure based on (a).*

Hair follicles are infoldings of the epidermis with a hair root, or **papilla**, at the base from which the hair develops. Hair is composed of cuboid epithelial cells which become cornified by impregnation with keratin. The outer region of the hair contains varying amounts of the pigment melanin which determines hair colour. The centre of the hair may contain air bubbles, and as the number increases with age, and melanin production falls off, the hair becomes white. Blood capillaries supply the growing hair with nourishment and remove waste substances. The upper part of the hair projects beyond the epidermis and is kept supple and waterproof by **sebum**, an oily secretion produced by **sebaceous glands** which open into the hair follicle. Sebum contains fatty acids, waxes and steroids, and spreads along the hair and onto the skin where it keeps the follicle free from dust and bacteria, as well as forming a thin waterproof layer over the skin. This not only prevents water loss from the skin but also prevents water entering the skin.

At the base of the follicle is a smooth muscle, the **hair erector muscle**, which has its origin on the basement membrane and its insertion on the hair follicle. Contraction of this muscle alters the angle between the hair and the skin which results in variation of the amount of air trapped above the skin. This is used as a means of thermoregulation and also as a behavioural response to danger in some vertebrates. When the hair 'stands on end' it increases the apparent size of the organism which may be sufficient to frighten off would-be attackers. The distribution of hair in humans is much more restricted than in other mammals.

Sweat glands are coiled tubular glands situated in the dermis and connected to a sweat duct which opens as a pore on the surface of the skin. They are found over the entire human body, but in some mammals are restricted to the pads of the feet. They are absent in birds. Water, salts and urea are brought to the glands by blood capillaries, and the secretory activity of the glands is controlled by the activity of the sympathetic nerve fibres. There are two types of sweat glands, **eccrine** and **apocrine**. Eccrine glands are the most common (approximately 2.5 million in humans), being found in most regions of the body. Apocrine glands are found in the armpits, around the nipples, the pubic region, hands, feet and anus. These release an odourless fluid which may later produce a strong odour due to bacterial activity in the fluid. Certain zinc and aluminium compounds can inhibit the activity of the glands and destroy bacteria. Most antiperspirants and deodorants contain these compounds.

Blood capillaries are numerous in the dermis and supply

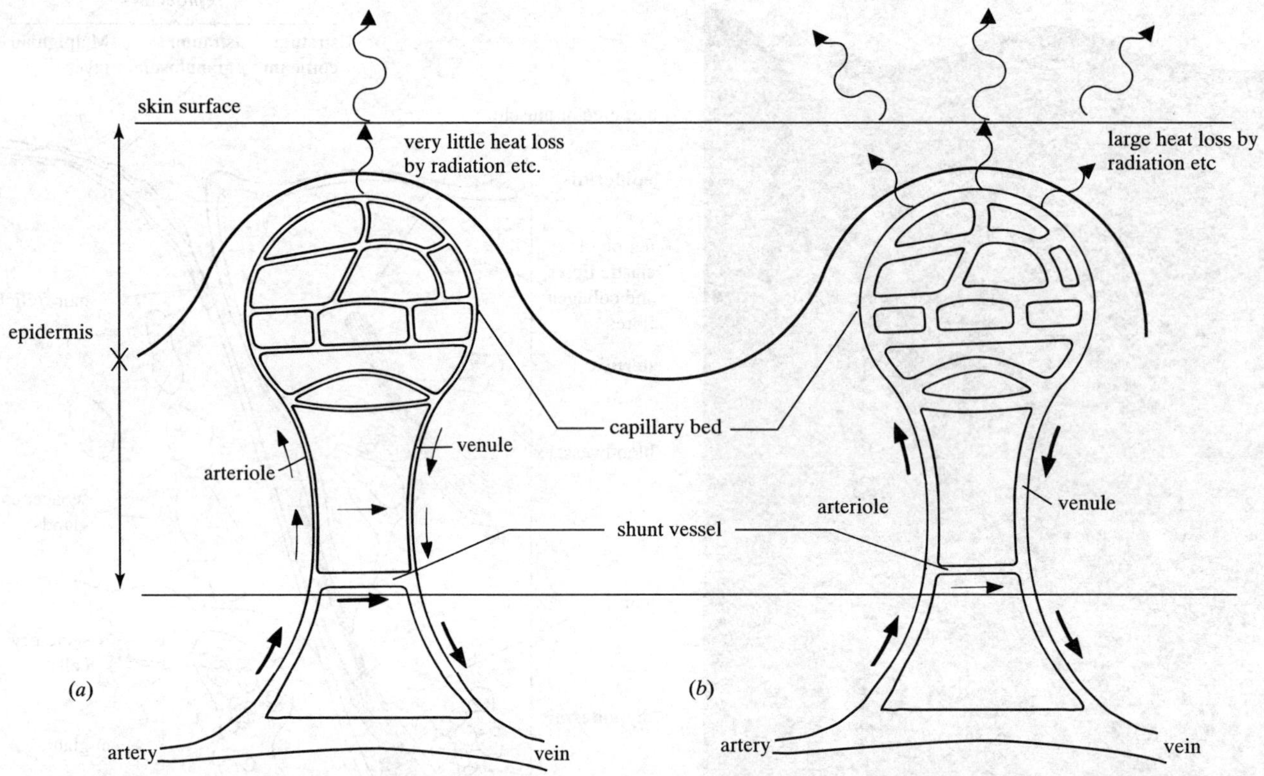

Fig 19.12 *Mechanism of regulation of blood flow through the skin. (a) Blood flow through the skin preventing heat loss. Constriction of the arteriole reduces blood flow through capillaries. Only sufficient blood passes into the skin to keep the tissues alive. Most of the blood flowing from the body bypasses the skin through the shunt vessels and reduces heat loss. (b) Blood flow through the skin increasing heat loss. Dilation of the arteriole increases blood flow through the capillaries. The capillaries dilate due to the rise in blood pressure within them. Heat is lost from the blood by radiation, convection and conduction and blood flow is increased to the sweat glands.*

the various structures already described. Many of the capillaries form loops and have shunts (fig 19.12) which enable the body to vary the amount of blood flowing through the capillaries. This is one of the many ways of regulating body temperature, as described in section 19.5.3.

Motor neurones run to the muscles and glands in the dermis whilst sensory neurones carry nerve impulses from the many sense cells situated in the dermis. These **sense cells** detect heat, cold, touch, pain and pressure. Some of the sense cells are simple and consist of free nerve endings, whereas others, such as the Pacinian corpuscle, are in capsules (fig 17.33).

19.5.2 Sources of heat

The major source of heat in endotherms is energy-releasing biochemical reactions which occur in living cells. The heat energy is released mainly by the breakdown of carbohydrates and fats taken in as part of the diet. Most of it comes from active tissues such as the liver and voluntary (skeletal) muscle. The rate of heat release in a resting organism is known as the **basal metabolic rate** (BMR). This provides a 'base-line' for comparing the energy demands of various activities, and of different organisms (see section 9.5.8). The energy content of food required to meet the demands of the basal metabolic rate for an average-sized male human over a 24-hour period is approximately 8000 kJ. The exact amount per individual depends upon size, age and sex, being slightly higher in males.

The rate of energy release is regulated by factors such as environmental temperature and hormones. The hormone thyroxine, which is released from the thyroid gland, increases the metabolic rate and therefore heat production. Its effects are long term. The hormone adrenaline produces short-term increases in metabolic activity. Other sources of heat energy are initiated by nerve impulses. Repeated stimulation of voluntary muscle by motor neurones produces the shivering response which can increase heat production by up to five times the basal level. During shivering various groups of muscle fibres within a muscle contract and relax out of phase, so that the overall response is an uncoordinated movement. This response may be reinforced by other muscular activity such as rubbing the hands together, stamping the feet and limited forms of exercise. In many mammals there are areas near the thoracic blood vessels which are rich in brown fat cells; stimulation of these by sympathetic neurones causes the rapid oxidation of the numerous fat droplets in the cells in aerobic respiration. The resulting release of energy by the mitochondria in these cells is particularly important for hibernating animals since it helps to rapidly raise the core temperature during waking from hibernation. The hypothalamus is the centre controlling heat production for most of the mechanisms described above. Its role is discussed in more detail in section 19.5.4.

19.5.3 Loss of heat

Heat is lost from endotherms by the four mechanisms described in section 19.3.6, that is conduction, convection, radiation and evaporation. In all cases, the rate of loss depends upon the temperature differences between the body core and the skin, and the skin and the environment. The rate can be increased or decreased depending upon the rate of heat production and the environmental temperature.

There are three factors limiting heat loss, as described below.

The rate of blood flow between the body core and skin

The rate of heat loss from the skin by radiation, convection and conduction depends upon the amount of blood flowing through it. If the blood flow is low, the skin temperature approaches that of the environment whereas, if the flow is increased, the skin temperature then approaches core temperature. The skin of endotherms is rich in blood vessels and blood can flow through it by one of two routes:

- through capillary networks in the dermis;
- through shunt pathways deep in the dermis which link arterioles and venules.

Arterioles have relatively thick muscular walls which can contract or relax to alter the diameter of the vessels and the rate of blood flow through them. The degree of contraction is controlled by sympathetic nerves from the vasomotor centre in the medulla of the hindbrain. This in turn is controlled by nerve impulses received from the thermoregulatory centre in the hypothalamus. The rate of blood flow through the skin in humans can vary from less than 1 cm^3 per minute per 100 g in cold conditions to 100 cm^3 per minute per 100 g in hot conditions, and this can account for an increase in heat loss by a factor of five or six. Constriction of the arterioles forces blood through the low resistance shunt vessels from arteries to veins and the bulk of the blood by-passes the capillaries (fig 19.12). This is a typical response preventing heat loss. Dilation of the arterioles encourages blood flow through the capillary beds and not through the shunt vessels. This increases blood flow through the skin and therefore heat is lost more rapidly.

The rate of sweat production and evaporation from the skin

Sweat is a watery fluid containing between 0.1 and 0.4% of sodium chloride, sodium lactate (from lactic acid) and urea. It is less concentrated than blood plasma and is secreted from tissue fluid by the activity of sweat glands under the control of neurones which are part of the sympathetic nervous system. These neurones come from the hypothalamus. Sweating begins whenever the body temperature rises above its mean value of 36.7 °C. Approximately 900 cm^3 of sweat are lost per day in a temperate climate, but the figure can rise as high as 12 dm^3 per day in very hot dry conditions, providing that there is adequate replacement of water and salts.

19.2 The latent heat of evaporation of sweat is 2.45 kJ cm^{-3}. Calculate the percentage of energy lost by sweating from a heavy manual worker who loses 4 dm^3 per day of sweat and has a daily energy intake of 50 000 kJ.

When sweat evaporates from the skin surface, energy is lost from the body as latent heat of evaporation and this reduces body temperature. The rate of evaporation is reduced by low environmental temperatures, high humidity and lack of wind.

Many mammals have so much hair on their bodies that sweating is restricted to bare areas, for instance pads of the feet of dogs and cats, and the ears of rats. These mammals increase heat loss by licking their bodies and allowing moisture to evaporate, and by panting and losing heat from the moist nasal and buccal cavity. Humans, horses and pigs are able to sweat freely over the entire body surface.

Experiments have now confirmed that sweating only occurs as a result of a rise in core temperature. Experiments on humans and other animals have shown that lowering the core temperature, by swallowing ice water or cooling the carotid blood vessels with an ice pack around the neck, while at the same time exposing the skin to heat, results in a decrease in the rate of sweating. Opposite effects have been recorded by reversing the environmental conditions. Blood from the carotid vessels flows to the hypothalamus and these experiments have indicated its role in thermoregulation. Inserting a thermistor against the eardrum gives an acceptable measure of the temperature in the hypothalamus. The relation between changes in temperature in this region and the skin and rate of evaporation of sweat are shown in fig 19.13. The data come from an experiment carried out on a naked man in a heat-controlled chamber. Examine figure 19.13 and answer the following questions.

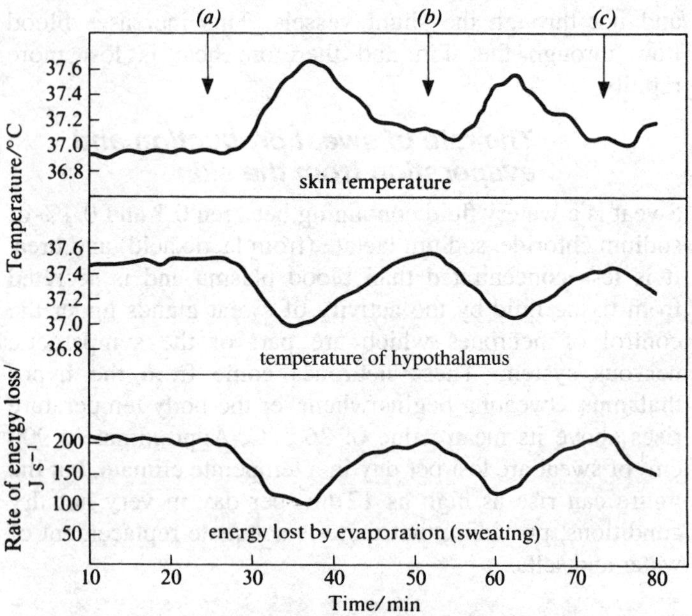

Fig 19.13 *Graphs showing the relation between skin temperature, temperature of the hypothalamus and rate of evaporation for a human in a warm chamber (45 °C). Iced water was swallowed at the points labelled* (a), (b) *and* (c).

19.3 Suggest why iced water was not given until 20 minutes after the start of the experiment.

19.4 Describe the relationship between the temperature of the hypothalamus and the rate of sweating.

19.5 Suggest why the skin temperature rises shortly after the ingestion of iced water.

The amount of insulation between the core and the environment

Insulation for the body is provided by air trapped outside the skin by hair and by fat in the dermis and just below the dermis (subcutaneous fat). In mammals, hair or clothes trap a layer of air, known as stagnant air, between the skin and the environment, and because air is a poor conductor of heat it reduces heat loss. Thick hair is known as fur. The amount of insulation provided by this means depends upon the thickness of the layer of trapped air. Reflex contractions of the hair erector muscles in response to decreasing temperatures increase the angle between the hair and the skin and so more air is trapped. The response is still present in humans but, as there is only a small amount of body hair, it only produces the effect known as 'goose-flesh' or 'goose-pimples'. Humans compensate for the lack of body hair by taking advantage of the insulating effects of clothing. The seasonal accumulation of a thick layer of subcutaneous fat is common in mammals, particularly in those species which do not hibernate and manage to withstand cold temperatures. Aquatic mammals, particularly those living in cold waters, such as the whale, sea-lion, walrus and seal, have a thick layer of fat known as blubber, which effectively insulates them against the cold.

19.5.4 Heat balance and the role of the hypothalamus

If the temperature of a body is to remain constant, then over a period of time the following equation must apply:

heat gained by body = heat lost by body

Endothermic animals are able to balance heat gain and heat loss by generating heat energy internally, and regulating the amount lost (fig 19.14). This is known as **homeothermy**. Any mechanism which has an input and an output and is capable of maintaining a constant value must be regulated by a control system as described earlier in this chapter (see fig 19.1).

Mammals have a well-developed control system involving receptors and effectors and an extremely sensitive control centre, the hypothalamus. The hypothalamus is located in the brain (fig 17.24). It monitors the temperature of the blood flowing through it. This blood is at core temperature (section 19.3.7). If the hypothalamus is to control a constant core temperature, as is the case with endotherms, it is vital that information regarding changes in the external temperature is also transmitted to the hypothalamus. Without such information the body would gain or lose a great deal of heat before changes in core temperature would activate the hypothalamus to take corrective measures. This problem is overcome by having thermoreceptors in the skin. These detect changes in the environmental temperature and send nerve impulses to the hypothalamus before changes in the core temperature take place. There are two types of thermoreceptors, hot and cold, which generate nerve impulses when suitably stimulated. Some pass to the hypothalamus and others to the sensory areas of the cortex, where the sensations associated with temperature (feeling hot or cold) are experienced according to the intensity of stimulation, the duration and the numbers of receptors stimulated. There are estimated to be 150 000 cold receptors and 16 000 heat receptors in humans. This enables the body to make rapid and precise adjustments to maintain a constant core temperature. In the context of control systems, the skin receptors act as disturbance detectors, responding before changes in body temperature take place. Factors bringing about changes in internal temperature, such as metabolic rate or disease, will immediately affect the core temperature and in these situations be detected by thermoreceptors in the hypothalamus. In most cases the activity of both skin and hypothalamus receptors combine to control body temperature.

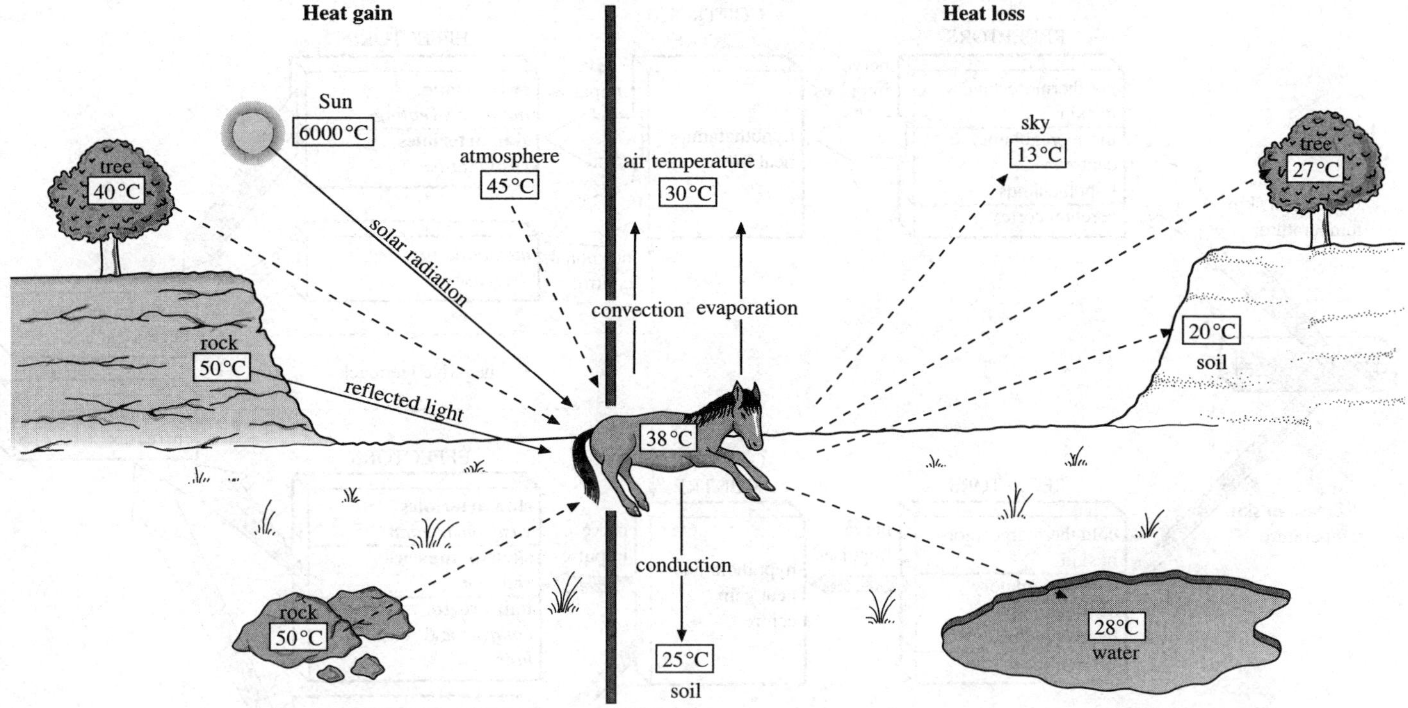

Fig 19.14 *Diagram showing the energy exchanges between a horse, with a body temperature of 38 °C, and the environment on a hot sunny day with an air temperature of 30 °C. The dotted lines represent radiation of heat.*

Heat gain centre and heat loss centre

Investigations into the activity of the hypothalamus have shown that there are two distinct centres concerned with temperature regulation, namely the **heat gain centre** and the **heat loss centre**. The functions of these centres are summarised in table 19.1. A summary of the control of temperature is shown in fig 19.15 and fig 19.16.

Fever

Some diseases produce an increase in core temperature known as **fever** as a result of the 'thermostat' in the hypothalamus being set at a higher temperature. It is believed that certain substances known as **pyrogens**, which may be toxins produced by pathogenic organisms or substances released by white blood cells known as neutrophils, directly affect the hypothalamus and increase the set-point. The raised body temperature stimulates the defence responses of the body and helps the destruction of pathogens. Antipyretic drugs such as aspirin and paracetamol lower the set-point and provide relief from the unpleasant symptoms of fever, but probably slow down the normal defence mechanisms. In cases of extremely high temperature these drugs are valuable in preventing irreversible damage to the brain.

Table 19.1 Functions of the heat loss and heat gain centres of the hypothalamus. These are situated in the anterior and posterior hypothalamus respectively and have antagonistic (opposite) effects.

Anterior hypothalamus (heat loss centre)	*Posterior hypothalamus (heat gain centre)*
Activated by increase in the temperature of the hypothalamus	Activated by nerve impulses from cold receptors in the skin or decrease in temperature of the hypothalamus
Increases vasodilation. Therefore increases heat loss from the skin by radiation, convection and conduction	Increases vasoconstriction. Therefore decreases heat loss from the skin by radiation, convection and conduction
Increases sweating and panting	Inhibits sweating and panting
Decreases metabolic activity	Increases metabolic activity through shivering and release of thyroxine and adrenaline
Decreases thickness of air layer by flattening hair (relaxing hair erector muscles)	Increases thickness of air layer by contraction of hair muscles, making hair stand on end

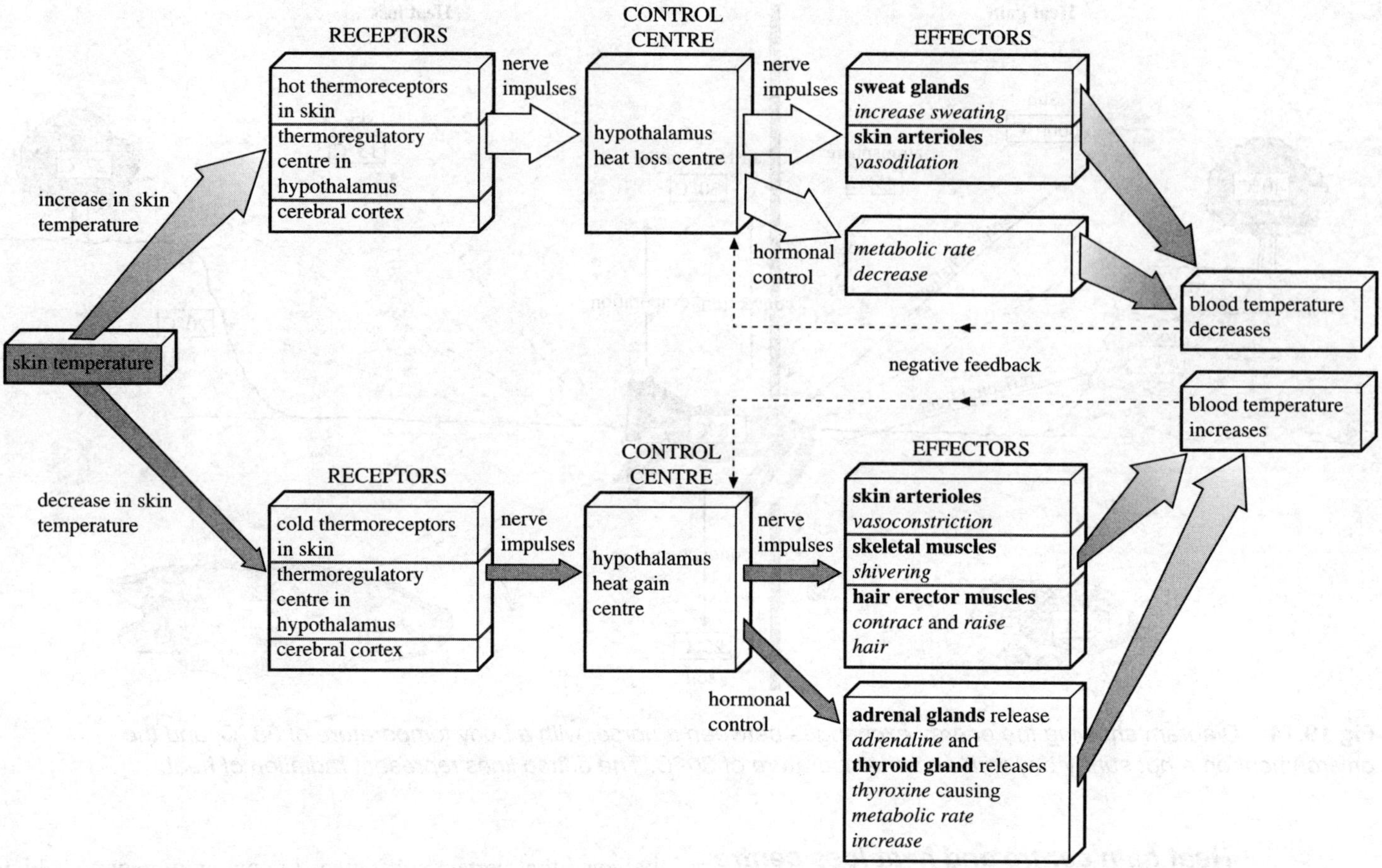

Fig 19.15 *Summary of reflex control of body temperature in a mammal involving the environment, hypothalamus, skin and blood temperature.*

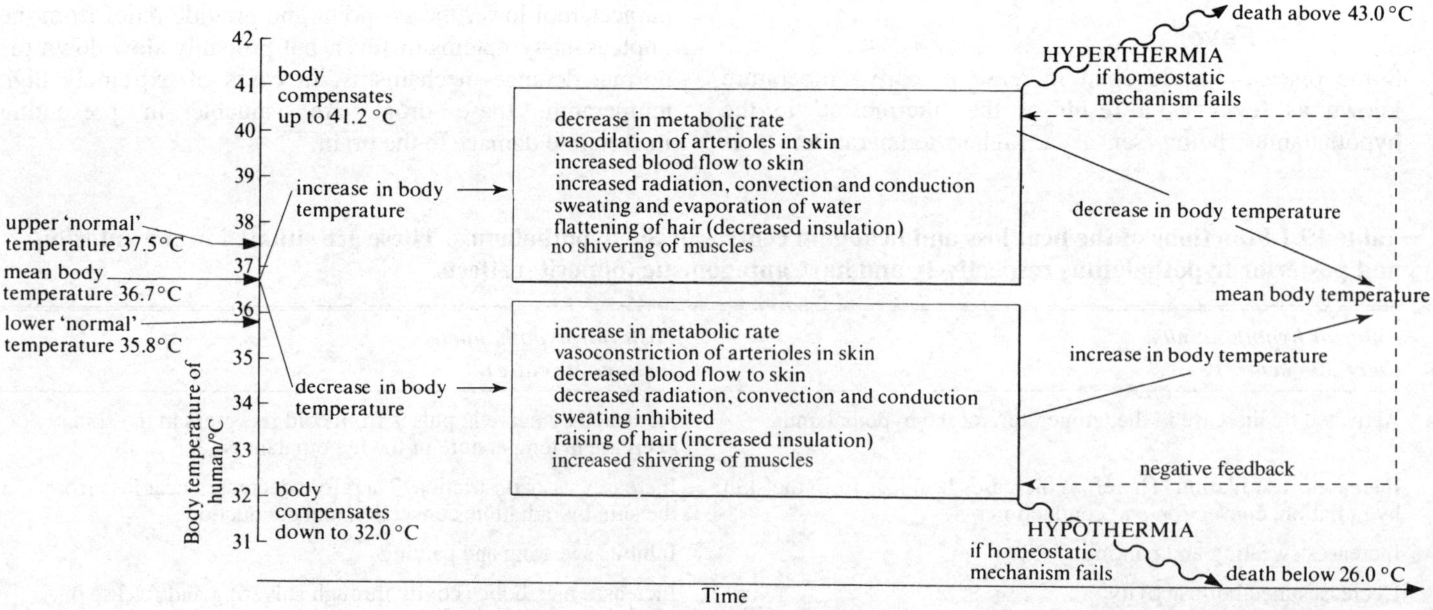

Fig 19.16 *Homeostatic control of human body temperature by physiological means. Behavioural changes also occur, for example adding or removing clothing, huddling in cold conditions to decrease surface area or stretching out in hot conditions to expose more skin.*

19.6 The onset of fever is often accompanied by shivering and a feeling of cold known as chill. Explain these symptoms in terms of the mechanism of control of body temperature.

19.5.5 Adaptations to extreme climates

Bergman's rule

The total heat production of endotherms depends upon the *volume* of the body whilst the rate of heat loss depends upon *surface area*. Volume increases more rapidly than surface area as the size of an animal increases. For this reason animals living in cold regions tend to be large, for example polar bears and whales, whilst animals living in hot climates are generally smaller, for instance insectivorous mammals. This phenomenon is known as **Bergman's rule** and is observed in many species, including the tiger, which decreases in size with distance from the Poles. There are exceptions to this rule, but the organisms concerned have adaptations favouring survival in these regions. For example, small mammals in temperate or arctic regions have a large appetite enabling them to maintain a high metabolic rate. They have small extremities, such as small ears, to reduce heat loss and are forced to hibernate in winter. Large mammals living in hot regions, such as the elephant and hippopotamus, have the opposite problems. The elephant has extremely large ears which are well supplied with blood, and flapping of these ears encourages heat loss by radiation and convection. The hippopotamus lacks sweat glands and adopts a similar behavioural response to temperature as the crocodile, in that it moves between land and water in an attempt to minimise the effects on its body of changes in temperature.

Allen's rule

Species living in colder climates have smaller extremities than related species in warmer climates. This is known as **Allen's rule** and may be seen in closely related species of, for instance, the fox (fig 19.17).

19.5.6 Adaptations to life at low temperatures

Dormancy

All ectotherms and many endotherms are unable to maintain a body temperature that allows normal activity during cold seasons, and they respond by showing some form of dormancy. Some of these responses are quite startling; for example, the larva of an insect parasite *Bracon*, which invades the Canadian wheat sawfly, is able to survive exposure to temperatures lower than –40 °C. This larva accumulates glycerol in its blood. The glycerol acts as an 'antifreeze' and is able to prevent the formation of ice crystals.

(*a*) (*b*) (*c*)

Fig 19.17 *Variation in ear length shown by three species of fox, each of which occupies a different geographical region. This is an example of Allen's rule. (a) The Arctic fox has the shortest ears, (b) the European fox, and (c) the Bat-eared fox from Botswana has the longest ears.*

Heat exchangers

Excessive heat loss to the environment from appendages is prevented in many organisms by the arrangement of blood vessels within the appendages. The arteries carrying blood towards the appendage are surrounded by veins carrying blood back to the body. The warm arterial blood from the body is cooled by the cold venous blood flowing towards the body. Similarly the cold venous blood from the appendage is warmed by the warm arterial blood flowing towards the appendage. Because the blood reaching the appendage is already cooled, the amount of heat lost is considerably reduced. This arrangement is known as a **countercurrent heat exchanger** and is found in the flippers and flukes of seals and whales, in the limbs of birds and mammals and in the blood supply to the testes in mammals (fig 19.18).

The countercurrent exchange principle is used for the transfer of materials other than heat, such as respiratory gases in fish gills and ions in the loop of Henle (sections 9.4.5 and 20.5.6).

Hypothermia

Hypothermia is a reduction in *core* temperature below about 32 °C. The technique of deliberately causing a state of hypothermia can be used in heart surgery since it allows the surgeon to carry out repairs to the heart without the risk of brain damage to the patient. By reducing the body temperature to 15 °C, the metabolic demands of the brain cells are so reduced that blood flow to the brain can be stopped, without any adverse effects, for up to one hour. For operations requiring a longer time than this, a heart–lung machine is used, in addition to hypothermia, to maintain blood circulation in the tissues.

19.5.7 Adaptations to life at high temperatures

Animals living in conditions where the air temperatures are higher than skin temperature gain heat, and the only means of reducing body temperature is by evaporation of water from the body surface. Hot regions may be, in addition, either particularly dry or humid, and this poses additional problems. In hot dry regions, although heat can be lost by the free evaporation of water, animals have the problem of finding adequate supplies of water to replace the water lost. In hot humid regions water is freely available but the humidity gradient between organisms and the environment often prevents evaporation. In this case behavioural mechanisms of temperature control often become more important. These involve taking advantage of the shade and breezes associated with humid forest and jungle habitats.

Bergman's rule

As noted in section 19.5.5, animals living in hot climates tend to be smaller than those living in cold climates. This can be explained by the fact that the amount of heat gained from the environment is approximately proportional to the body surface area. The majority of animals living in deserts, therefore, are small, such as the kangaroo rat (*Dipodomys*), and have fewer problems than larger animals, such as camels. In addition they are able to live in burrows in sand and soil where the microclimate poses fewer problems to life. Adaptations of the kangaroo rat for conserving water are discussed in section 20.3.1.

The camel – a case study

The camel is superbly adapted to a hot, dry climate in the following ways.

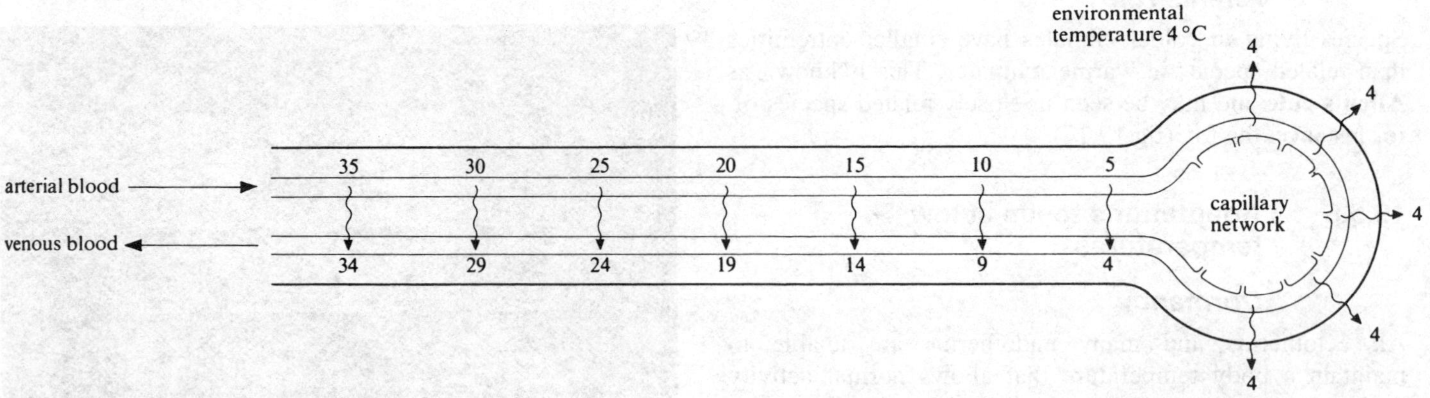

Fig 19.18 *Diagram showing the blood supply between the body of an endotherm with a stable temperature of 35 °C and an appendage such as a leg, in an environment at 4 °C. Heat flows from a warm body to a cool body and the rate of heat loss between the two bodies is proportional to the temperature difference between them. The countercurrent flow shown delivers blood to the capillary network at 5 °C and collects blood from it at 4 °C. The amount of heat lost to the environment is therefore proportional to the temperature difference of 1 °C. Likewise blood returning to the core of the body is only 1 °C cooler than the blood leaving the core. This mechanism prevents the excessive loss of metabolic energy and helps maintain the core temperature at 35 °C.*

- In hot dry conditions, with free access to water, it can regulate its body temperature between about 36 and 38 °C by losing heat through the evaporation of water from the body surface.
- If the camel is deprived of water, as say during a journey across the desert lasting several days, the difference in body temperature between morning and evening steadily increases according to the degree of dehydration. This daily fluctuation can be from 34 °C in the early morning to 41 °C in the late afternoon. By effectively being able to store up heat during the day the camel does not need to lose this heat by the evaporation of water. It functions, in fact, like a storage radiator. In a series of investigations carried out by Schmidt-Nielsen it was found that a 500 kg camel tolerating a 7 °C temperature rise stored approximately 12 000 kJ of heat energy. If this amount of heat were lost by the cooling effect of evaporation, in order to maintain a constant body temperature, it would require the loss of 5 dm^3 of water. Instead the heat is lost by radiation, conduction and convection during the night.
- A second advantage of becoming 'partially ectothermic' during the day is that this reduces the temperature difference between the hot desert air and the camel, and therefore reduces the rate of heat gain.
- The fur of the camel acts as an efficient insulating barrier by reducing heat gain and water loss. In an experiment in which a camel was shorn like a sheep, the water loss increased by 50% over that of a control camel.
- The final significant advantage shown by the camel is its ability to tolerate dehydration. Most mammals cannot tolerate dehydration beyond a loss of body mass of 10–14%, but the camel can survive losses of up to 30%, because it is able to maintain its blood volume even when dehydrated. Heat death as a result of dehydration is due to a fall in volume of the blood. This results in the circulatory system being unable to transfer heat from the body core to the surface quickly enough to prevent overheating.
- Contrary to popular belief, the camel is unable to store water in advance of conditions of water shortage. It may be able to gain water from the oxidation of fat stored in the hump, but there is some doubt about the effectiveness of this.
- The camel is able to drink a vast volume of water in a short space of time to rehydrate the body tissues after a period of severe dehydration. For example, a 325 kg camel is known to have drunk 30 dm^3 of water in less than ten minutes. This is roughly equivalent to a human of average build and weight drinking about 7 dm^3 (12 pints) of water!

19.6 The liver

Apart from the skin, the liver is the largest visceral organ of the body. It is situated just below the diaphragm in the abdomen and makes up 3 to 5% of the body weight. It is basically an organ of homeostasis. It controls many metabolic activities essential for maintaining a constant blood composition. Many of its functions are associated with the metabolism of food brought from the gut. Because of its rich blood supply it also regulates many activities associated with blood and the circulatory system (fig 19.19). Despite the enormous variety of metabolic activities carried out by the liver its structure is relatively simple.

(*a*)

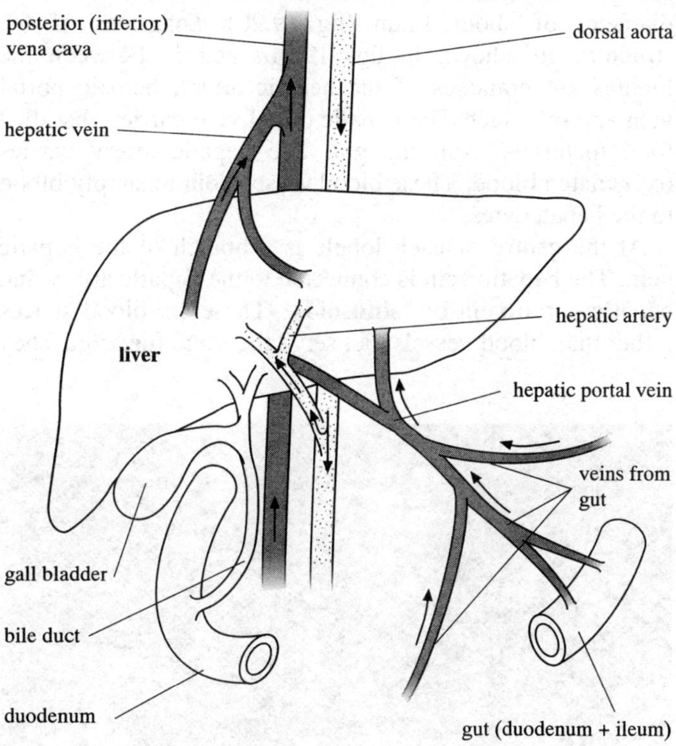

(*b*)

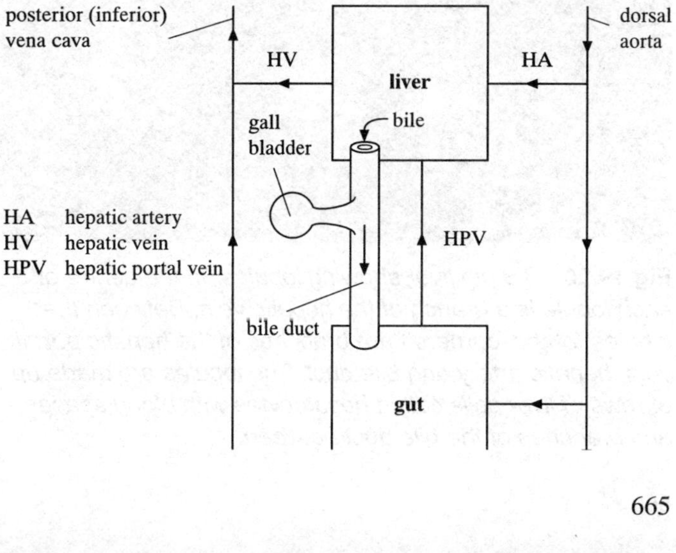

Fig 19.19 *(a) Diagram showing the blood supply to and from the liver and the positions of the bile duct and gut. (b) Map of the blood supply to and from the liver. Movement of bile is also shown.*

19.6.1 The position and structure of the liver

The liver is made up of several lobes and has a variable shape depending upon the amount of blood present within it. It is surrounded by a tough, fibrous capsule.

The cells of the liver are called **hepatocytes**. The only other cells found in the liver are nerve cells and cells associated with blood and lymph vessels. Hepatocytes have prominent nuclei and Golgi apparatus, many mitochondria and lysosomes, and are rich in glycogen granules and fat droplets. They are tightly packed together and, where their surface is in contact with blood vessels, there are microvilli which are used for the exchange of materials between the two.

The pig liver is convenient to study because the functional units of the liver, known as lobules, are more clearly defined than in other mammals. Each lobule has a diameter of about 1 mm (fig 19.20). Diagrams of the structure are shown in figs 19.21*a* and *b*. Between the lobules are branches of the hepatic artery, hepatic portal vein and bile duct. The hepatic portal vein carries absorbed food materials from the gut. The hepatic artery carries oxygenated blood. These blood vessels join to supply blood to the hepatocytes.

At the centre of each lobule is a branch of the hepatic vein. The hepatic vein is connected to the hepatic artery and hepatic portal vein by '**sinusoids**'. These are blood spaces rather than blood vessels, but serve the same function. They radiate like the spokes of a wheel from the centre to the edges of the lobule. Blood flows slowly from the hepatic artery and hepatic portal vein to the hepatic vein past the hepatocytes, which also form rows across the lobule. As blood flows along the sinusoids, exchange of materials takes place between the blood and the hepatocytes, across the microvilli of the hepatocytes which touch the sinusoids. The sinusoids have a lining of thin endothelial cells containing pores with a diameter of up to 10 nm.

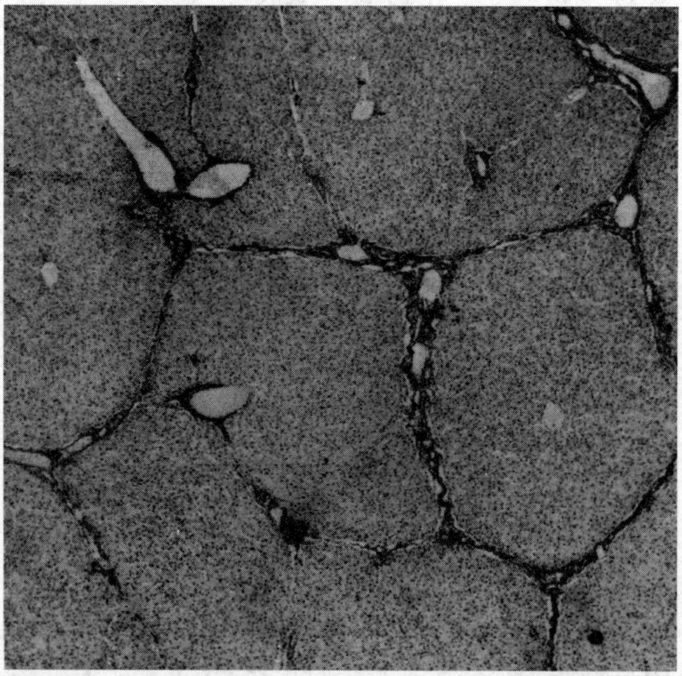

Fig 19.20 *TS pig liver showing lobules. In the centre of each lobule is a branch of the hepatic vein. Between the lobules (at the 'corners') are branches of the hepatic portal vein, hepatic artery and bile duct. The lobules are made up of rows of liver cells called hepatocytes with blood spaces and branches of the bile duct between.*

The sinusoids alternate with **bile canaliculi**. These are small canals which carry bile, made by the hepatocytes which line them, to the branches of the bile duct. Note that the bile flows in the opposite direction to the blood and does not mix with the blood (fig 19.21).

One other type of cell is found in the liver, known as the **Kupffer cell**. Kupffer cells are found attached to the walls of the sinusoids, but are now known to be macrophages (section 14.3.2). They are phagocytic and are involved in the breakdown of old red blood cells and the ingestion of potentially harmful bacteria.

19.6.2 Functions of the liver

It has been estimated that the liver carries out several hundred separate functions involving thousands of different chemical reactions. A vast amount of blood flows through it at any given time (approximately 20% of the total blood volume). The liver and the kidneys between them are the major organs responsible for maintaining the steady state of the blood. All food materials absorbed from the alimentary canal pass directly to the liver where they are stored or converted into some other form as required by the body at that time.

Carbohydrate metabolism

Role of insulin. Sugars such as glucose enter the liver from the gut by the hepatic portal vein, which is the only blood vessel in the body having an extremely variable sugar content. This gives a clue to the role of the liver in carbohydrate metabolism as the organ which maintains the blood glucose level at approximately 90 mg glucose per 100 cm^3 blood. The liver prevents blood glucose levels from varying up and down according to how recently food has been eaten. Low levels of glucose would be particularly damaging because some tissues cannot store glucose, such as the brain. All hexose sugars, including galactose and fructose, are converted to glucose by the liver and stored as the insoluble polysaccharide glycogen. Up to 100 g of glycogen are stored here but more is stored in muscle. The conversion of glucose to glycogen is known as **glycogenesis** and is stimulated by the presence of insulin. This is a hormone produced by the pancreas in response to high blood sugar levels (section 17.6):

$$\text{glucose} \underset{\text{(phosphorylation)}}{\overset{\text{insulin}}{\rightleftharpoons}} \text{glucose phosphate} \underset{\text{(condensation)}}{\rightleftharpoons} \text{glycogen}$$

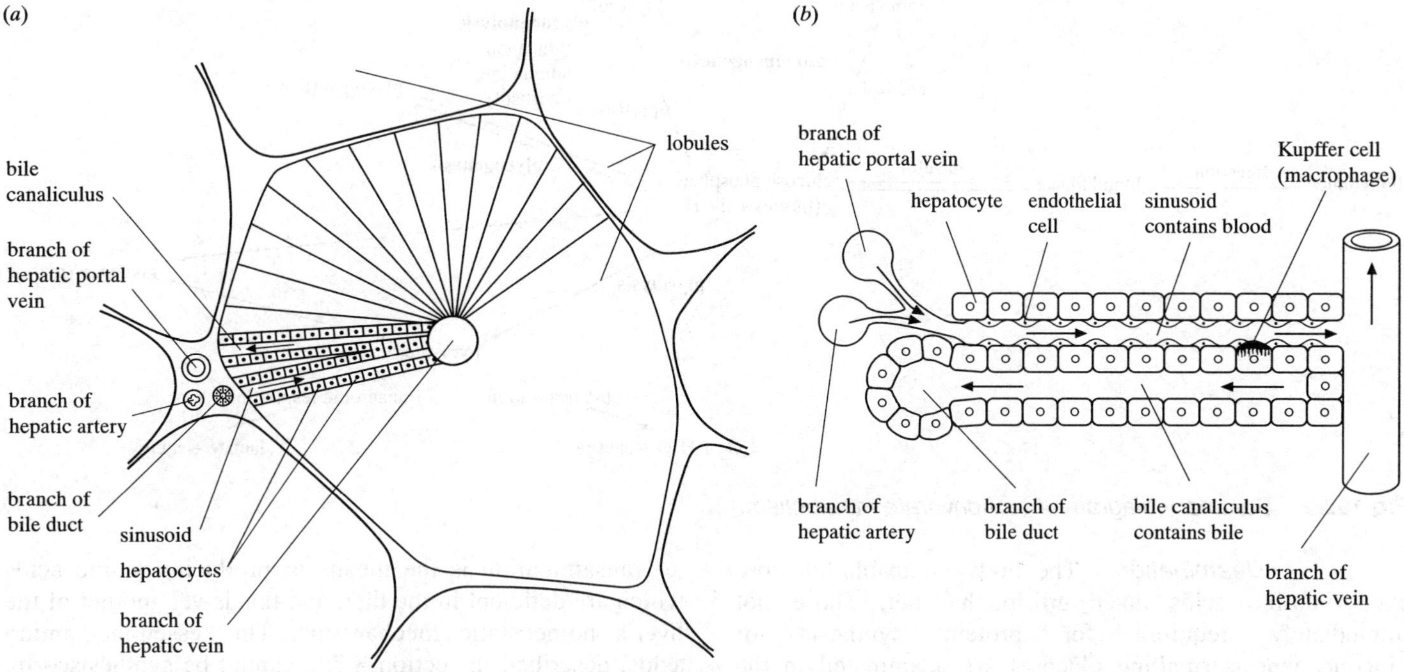

Fig 19.21 *(a) Diagram of a transverse section of a liver lobule. (Arrows indicate flow of blood in sinusoids and flow of bile in canaliculi.) (b) A simplified diagram of part of a liver lobule.*

Role of glucagon. Glycogen is broken down to glucose to prevent the blood glucose level falling below 60 mg per 100 cm^3 blood. This process is called **glycogenolysis** and involves the activation of a phosphorylase enzyme by the hormone glucagon. Glucagon is made in the pancreas and is released when blood sugar levels fall (section 17.6.6). In times of danger, stress or cold this activity is also stimulated by adrenaline, released by the adrenal medulla, and noradrenaline released both by the adrenal medulla and the endings of the sympathetic neurones (section 17.6.5):

$$\underset{\text{(store)}}{\text{glycogen}} \underset{}{\overset{\text{phosphorylase}}{\rightleftharpoons}} \text{glucose phosphate} \rightleftharpoons \underset{\text{(free)}}{\text{glucose}}$$

Lactic acid from anaerobic respiration. Muscle does not convert glycogen directly to glucose as shown above because it does not release glucose into the blood for the rest of the body, unlike the liver. Muscle only breaks down glycogen if it needs glucose for its own respiration, in which case the glycogen is converted to glucose phosphate (as above), but is then converted to pyruvate (glycolysis) which is used to produce ATP during aerobic or anaerobic respiration. Lactic acid (lactate) produced by anaerobic respiration in skeletal muscle can be converted later into glucose and hence glycogen in the liver:

$$\text{lactate} \rightarrow \text{pyruvate} \rightarrow \text{glucose} \rightarrow \text{glycogen}$$

Gluconeogenesis. When the demand for glucose has exhausted the glycogen store in the liver, glucose can be synthesised from non-carbohydrate sources. This is called **gluconeogenesis**. Low blood glucose levels (hypoglycaemia) stimulate the sympathetic nervous system to release adrenaline which helps satisfy immediate demand as described above. Low blood glucose levels also stimulate the hypothalamus to release CRF (section 17.6.5) which in turn releases adrenocorticotrophic hormone (ACTH) from the anterior pituitary gland. This leads to the synthesis and release of increasing amounts of the glucocorticoid hormones, particularly cortisol (also known as hydrocortisone). These stimulate the release of amino acids, glycerol and fatty acids, present in the tissues, into the blood and increase the rate of synthesis of enzymes in the liver which convert amino acids and glycerol into glucose. (Fatty acids are converted into acetyl coenzyme A and used directly in the Krebs cycle.)

Manufacture of fats. Carbohydrate in the body which cannot be used or stored as glycogen is converted into fats and stored.

A summary of carbohydrate metabolism involving the liver, muscles and tissues is shown in fig 19.22.

Protein metabolism

The liver plays an important role in protein metabolism which may be considered under the headings of:

- deamination;
- urea formation;
- transamination;
- plasma protein synthesis.

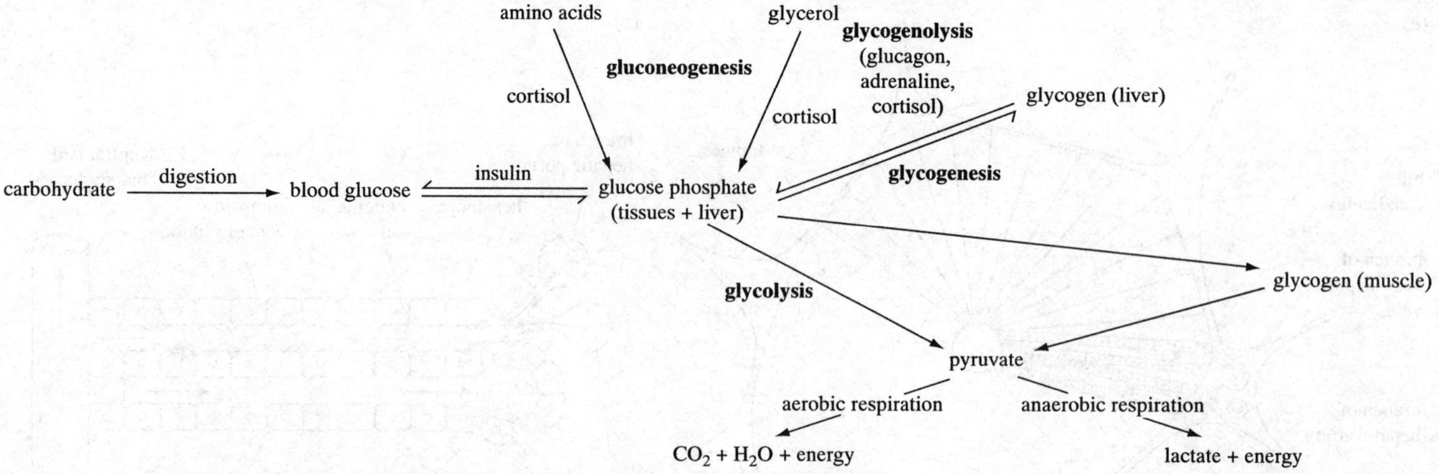

Fig 19.22 *Summary diagram of carbohydrate metabolism.*

Deamination. The body is unable to store excess amino acids taken up in the diet. Those not immediately required for protein synthesis or gluconeogenesis (making glucose) are deaminated in the liver. This process is described in section 20.4.

Transamination. This is the synthesis of amino acids by the transfer of the amino group from an amino acid to another organic acid. The general principle underlying these reactions is the exchange of chemical groups between the amino acid and the other organic acid:

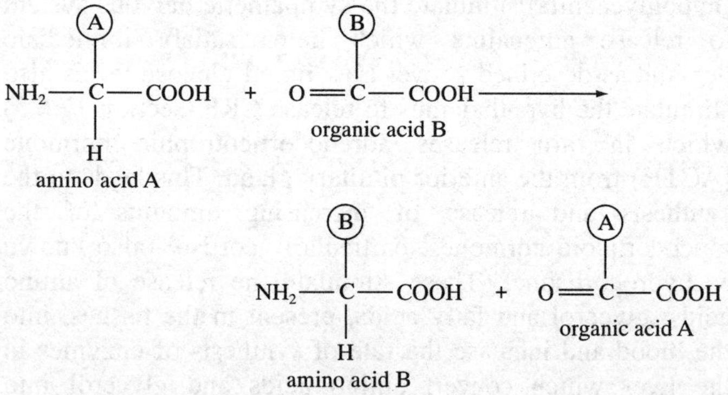

For example, the amino acid, glutamic acid, could be synthesised by the following reactions:

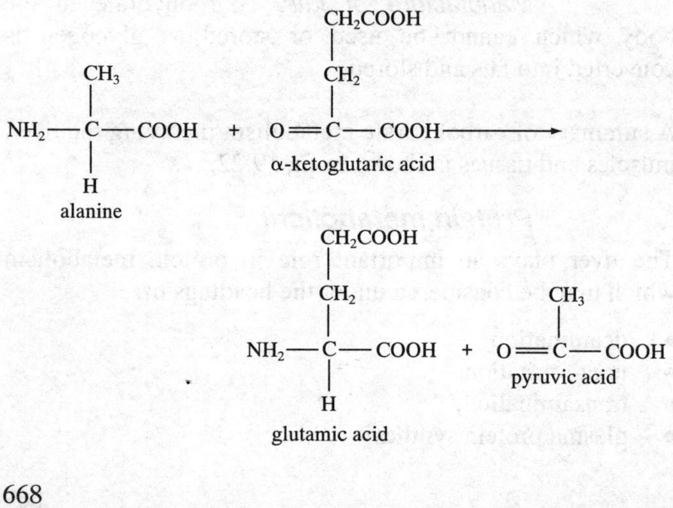

Transamination is the means of producing amino acids which are deficient in the diet, and this is yet another of the liver's homeostatic mechanisms. The 'essential' amino acids, described in section 8.7.8, cannot be synthesised by transamination in the liver and must be obtained from the diet.

Plasma protein production. Plasma proteins are vital components of plasma and the majority of them are synthesised from amino acids in the liver.

- Albumin is the commonest protein and about 4 g per 100 cm^3 is normally present in the blood. It plays an important part in exerting an osmotic potential which opposes the hydrostatic pressure developed in blood vessels. The antagonistic effects of these two factors maintain the balance of fluids inside and outside of blood vessels (section 14.6). Albumins also act as transport molecules within the blood, carrying substances such as calcium, bile, salts, and some steroid hormones.
- Globulins are very large molecules and blood carries about 3.4 g per 100 cm^3. α- and β-globulins transport hormones (including thyroxine and insulin), cholesterol, lipids, iron and the vitamins B_{12}, A, D and K. γ-globulins are antibodies and are produced by lymphocytes and other cells of the immune system (not the liver). They are involved in the immune response (section 14.9). The other main plasma proteins are the blood-clotting factors prothrombin and fibrinogen, and their functions are described in section 14.8.5.

Fat metabolism

The liver is involved in the processing and transport of fats rather than their storage. Liver cells carry out the following functions:

- converting excess carbohydrates to fat;
- removing cholesterol from the blood and breaking it down or, when necessary, synthesising it;

- if glucose is in short supply, the liver can break down fats into fatty acids and glycerol for respiration. Fatty acids are converted to acetyl groups which combine with coenzyme A to form acetyl coenzyme A. This enters Krebs' cycle for oxidation as described in section 9.3.5. Fatty acids can also be exported from the liver after conversion to other chemicals. Glycerol can be converted to glucose as previously described under gluconeogenesis.

Vitamin storage

The main vitamins stored in the liver are the fat-soluble vitamins A, D, E and K. The livers of certain fish, such as cod and halibut, contain high concentrations of vitamins A and D. Vitamin K is a vital factor in blood clotting.

The liver also stores some of the water-soluble vitamins, namely vitamins B and C, especially those of the B group such as nicotinic acid, vitamin B_{12} and folic acid. Vitamin B_{12} and folic acid are required by the bone marrow for the formation of red blood cells and deficiency of these vitamins leads to various degrees of anaemia.

Mineral storage

Those elements required in small amounts such as copper, zinc, cobalt and molybdenum (trace elements) are stored in the liver along with iron and potassium (see breakdown of red blood cells below). Approximately one-thousandth of the dry mass of liver tissue in humans is iron. Most of this iron is temporary and comes from the breakdown of old red blood cells. It is stored here for later use in the manufacture of new red blood cells in the bone marrow.

Storage of blood

The blood vessels leaving the spleen and gut join to form the hepatic portal vein and, together with the blood vessels of the liver, they contain a large volume of blood which acts as a reservoir, although the blood is constantly moving through it. Sympathetic neurones and adrenaline from the adrenal medulla can constrict many of these hepatic vessels and make more blood available to the general circulation. If the blood volume increases, as for example during a blood transfusion, the hepatic veins along with other veins can dilate to take up the excess volume.

Making red blood cells

The liver of the fetus is responsible for red blood cell production but this function is gradually taken over by cells of the bone marrow (section 14.3.2). Once this process is established, the liver takes the opposite role and assists in breaking down red blood cells and haemoglobin.

Breakdown of haemoglobin

Red blood cells have a life span of about 120 days. They are then broken down by the activity of phagocytic macrophage cells in the liver, spleen and bone marrow. The haemoglobin they contain is released and dissolves in the plasma. It is taken up from here by macrophages (special white blood cells) in the liver, spleen and lymph glands. The macrophages in the liver are called Kuppfer cells. Inside the macrophages haemoglobin is broken down into **haem** and **globin**. Globin is the protein part of the molecule and is broken down to its individual amino acids. These can be used according to demand. The iron is removed from haem and the remaining part of the molecule forms a green pigment called **biliverdin**. This is converted to **bilirubin**, which is yellow and a component of bile. The accumulation of bilirubin in the blood is a characteristic symptom of liver disease and produces a yellowing of the skin, a condition known as **jaundice**.

The iron is not wasted. In the blood it can combine with a plasma protein to form a complex called **transferrin**. The iron may then be re-used by cells in the bone marrow to make more haemoglobin. Alternatively it may be stored in the liver. In this case it is taken up by hepatocytes and stored as a compound called **ferritin**.

Bile production

Bile is a viscous, greenish yellow fluid secreted by hepatocytes. Between 500–1000 cm^3 of bile are produced each day and stored and concentrated in the gall bladder. It is composed of about 98% water, 0.8% bile salts, 0.2% bile pigments, 0.7% inorganic salts and 0.6% cholesterol.

Bile is secreted into the duodenum where it is involved in digestion and the absorption of fats and is a means of excretion of bile pigments. The stimulus for its release into the duodenum is the presence of the hormone **cholecystokinin** (CCK), also known as **pancreozymin** (section 8.4.3).

Bile salts are made from the steroid **cholesterol** which is synthesised in hepatocytes. The commonest bile salts are sodium glycocholate and sodium taurocholate. They are secreted with cholesterol and phospholipids in the form of large spherical particles called **micelles**. Molecules of bile salts resemble detergents in having a water-soluble (hydrophilic) end and a lipid-soluble (hydrophobic) end. The cholesterol and phospholipids hold the bile salt molecules together so that all the hydrophobic ends of the molecules point the same way. Inside the gut the hydrophobic ends attach to lipid droplets in the food whilst the other ends are attached to water. This decreases the surface tension of the lipid droplets and enables the lipid molecules to separate, forming an emulsion of smaller droplets. These have an increased surface area for attack by the enzyme pancreatic lipase which converts the lipids into glycerol and fatty acids so that they can then be absorbed from the gut. Bile salts also activate the enzyme lipase, but their action, in all cases, is purely physical. They are not enzymes. Too little bile salt in bile increases the concentration of cholesterol which may precipitate out in the gall bladder or bile duct as cholesterol gall stones. These can block the bile duct and cause severe discomfort.

Bile pigments have no function and are excretory products.

Cholesterol production

Cholesterol is produced by the liver and is used as the starting point for the synthesis of other steroid molecules. The major source of cholesterol is the diet, and many dairy products are rich in cholesterol or fatty acids from which cholesterol can be synthesised. Thyroxine both stimulates cholesterol formation in the liver and increases its rate of excretion in the bile. Excessive amounts of cholesterol in the blood can be harmful. It can be deposited in the walls of arteries, leading to **atherosclerosis** (narrowing of the arteries) and increased risk of the formation of a blood clot which may block blood vessels, a condition known as **thrombosis**. This is often fatal if it occurs in the coronary artery in the wall of the heart (coronary thrombosis or 'heart attack') or brain (cerebral thrombosis or 'stroke'). Although cholesterol is harmful in excess, it is essential to have some in the diet for the reasons stated.

Hormone breakdown

The liver destroys almost all hormones to various extents. Testosterone and aldosterone are rapidly destroyed, whereas insulin, glucagon and gut hormones, female sex hormones, adrenal hormones, ADH and thyroxine are destroyed less rapidly. In this way the liver has a homeostatic effect on the activities of these hormones.

Detoxification

Detoxification means the removal of toxins, or poisons. These are usually naturally occurring compounds which can be toxic if allowed to build up in the body. Detoxification is part of homeostasis and helps to maintain the composition of blood in a steady state. Bacteria and other pathogens are removed from the blood in the sinusoids by Kupffer cells, but the toxins they produce are dealt with by biochemical reactions in the hepatocytes. Toxins are rendered harmless by one or more of the following reactions: oxidation, reduction, methylation (the addition of a $-CH_3$ group) or combination with another organic or inorganic molecule. Following detoxification these substances, now harmless, are excreted by the kidney. The major toxic substance in the blood, though, is ammonia, whose fate is described in section 20.4.

The detoxification process also removes harmful substances taken into the body such as alcohol and nicotine. Alcohol taken in excess over a period of time can result in liver breakdown, such as cirrhosis of the liver in alcoholics. It is removed by the enzyme alcohol dehydrogenase.

Some of the metabolic activities of the liver may be potentially harmful and evidence is growing that certain food additives may be converted into poisonous or carcinogenic (cancer-producing) substances by liver activity. Even the pain killer paracetamol, if taken in excess, is changed into a substance which affects enzyme systems and can cause liver, and other tissue, damage.

Heat production

Evidence is accumulating to show that the widespread belief that the metabolic activity of the liver results in it being a major source of heat production in the body of mammals may be false. Many of the liver's metabolic activities are endothermic and therefore require heat energy rather than release it. Under conditions of extreme cold the hypothalamus will increase the energy-releasing activity of the liver by stimulating the release of the hormones adrenaline and thyroxine. At 'normal' temperatures, however, the liver has been shown to be 'thermally neutral' but is usually 1–2 °C hotter than the rest of the body core.

Fig 19.23 summarises the functions of the liver.

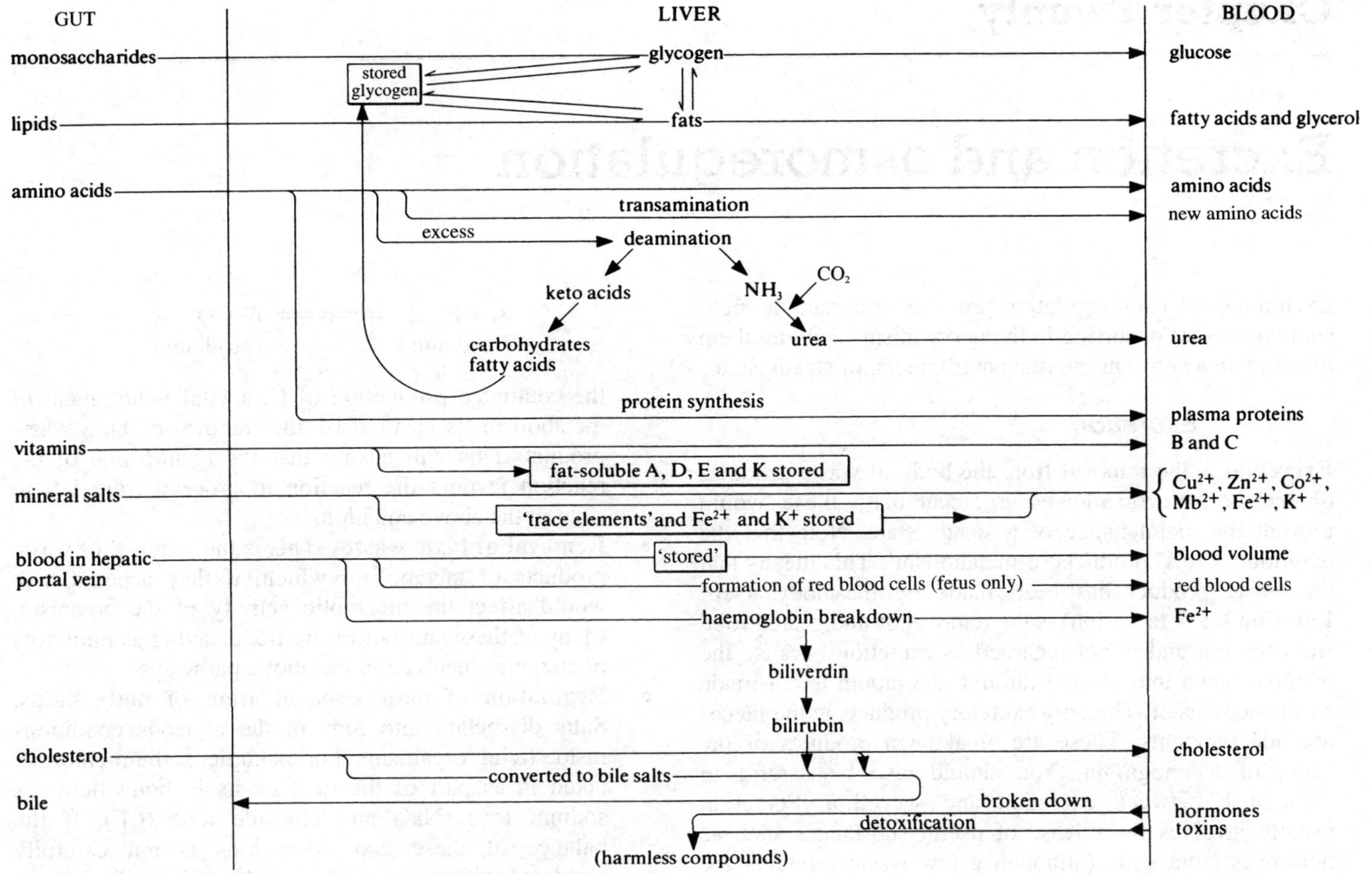

Fig 19.23 *Summary diagram of the functions of the liver.*

Chapter Twenty

Excretion and osmoregulation

Excretion and osmoregulation are two important homeostatic processes occurring in living organisms, helping them to maintain a constant internal environment, or steady state.

Excretion

Excretion is the removal from the body of waste products of metabolism. If allowed to accumulate these would prevent the maintenance of a steady state. Note that the definition states 'products of metabolism'. This means that the waste product has been made by the body itself. **Egestion** (or defaecation) is the removal of undigested food from the gut and is not regarded as excretion because the material taken into the gut through the mouth is not made by the body itself. The only excretory products in the faeces are bile pigments. These are breakdown products of the haem of haemoglobin. You should also be careful to distinguish between excretion and **secretion**. Secretion usually involves the release of useful substances such as hormones from cells (although a few waste products are secreted into the kidney tubules during excretion, as we shall see later).

Osmoregulation

Osmoregulation is the maintenance of constant osmotic conditions in the body. It involves regulation of the water content and solute concentration of body fluids, particularly of sodium, potassium and chloride ions. Body fluids include cell contents as well as fluids outside cells such as blood plasma, tissue fluid and lymph. It is vital that the composition of these fluids should remain constant in order for the cells to work efficiently.

20.1 The significance of excretion and osmoregulation

Excretion and osmoregulation have a number of functions which may be summarised as follows.

- **Removal of unwanted by-products of metabolic pathways**. This is necessary in order to prevent unbalancing the chemical equilibria of reactions. Many metabolic reactions are reversible and the direction of the reaction is determined by the relative concentrations of reactants and products. For example, in the enzyme-catalysed reaction:

$$\underset{\text{(reactants)}}{A + B} \rightleftharpoons \underset{\text{(products)}}{C + D}$$

  the continued production of C, a vital requirement of metabolism, is ensured by the removal of D, a waste product. This will ensure that the *equilibrium* of the reaction favours the reaction to proceed from left to right in the above equation.
- **Removal of toxic wastes**. This is the removal of waste products of metabolism which, if they accumulated, would affect the metabolic activity of the organism. Many of these substances are toxic, acting as inhibitors of enzymes involved in metabolic pathways.
- **Regulation of ionic concentration of body fluids**. Salts dissociate into ions in the aqueous conditions inside living organisms. For example, sodium chloride, taken in as part of the diet, exists in body fluids as sodium ions (Na^+) and chloride ions (Cl^-). If the balance of these and other ions is not carefully regulated within narrow limits, the efficiency of many cell activities is reduced; for instance a reduction in sodium ion concentration leads to a decrease in nervous coordination. Other important ions whose concentrations must be carefully regulated are K^+, Mg^{2+}, Ca^{2+}, Fe^{2+}, H^+, Cl^-, I^-, PO_4^{3-} and HCO_3^-, as they are vital for many metabolic activities including enzyme activity, protein synthesis, production of hormones and respiratory pigments, membrane permeability, electrical activity and muscle contraction. Their effects on water content, solute potential and pH of body fluids are described below.
- **Regulation of water content of body fluids**. The amount of water within the body fluids and its regulation is one of the major problems faced by organisms. The solutions to this problem have produced some of the most important structural and functional adaptations shown by organisms. The mechanisms of obtaining water, preventing water loss and eliminating water are varied, but they are of great importance in maintaining the solute potential and volume of body fluids at a steady state, as described later in this chapter. The solute potential of body fluids depends upon the relative amounts of solute and solvent (water) present. The mechanisms of regulation of solutes and water are known as **osmoregulation**.
- **Regulation of pH**. The nature of pH and methods of its measurement are described in appendix A1.1.5 but

the mechanisms of excreting those ions which have a major influence on pH, such as hydrogen and hydrogencarbonate ions, are considered in this chapter. For example, the pH of urine may vary between 4.5 and 8 in order to maintain the pH of the body fluids at a fairly constant level.

20.1.1 Excretory products

The major excretory products of animals and plants, and their sources, are as follows.

- **Nitrogenous compounds such as urea, ammonia and uric acid**. These come from breakdown of proteins, nucleic acids or excess amino acids. They are discussed in more detail in section 20.2.
- **Oxygen from photosynthesis** in plants, algae and some bacteria. Some of this may be used in respiration.
- **Carbon dioxide from cell respiration**. In autotrophic organisms this may be used as a source of carbon.
- **Bile pigments** from the breakdown of haem in the liver.

20.1.2 Excretory structures

The following structures are used for excretion in different animals:

- the cell surface membrane of unicellular organisms;
- the Malpighian tubules and tracheae of arthropods;
- the kidneys, liver, gills and skin of fish and amphibia;
- the kidneys, liver, lungs and skin of vertebrates.

The cells of organisms having a relatively simple structure are usually in direct contact with the environment and their excretory products are immediately removed by diffusion. As organisms increase in complexity, excretory organs develop to remove excretory products from the body and pass them to the external environment. The most important excretory organs in vertebrates are the skin, lungs, liver and kidney. The roles of the first three only will be described at this stage.

Skin

Water, urea and salts are actively secreted from capillaries in the skin by the tubules of the sweat glands. Sweat is secreted onto the skin where the water evaporates. In this way heat is lost from the body and this helps to regulate the body temperature.

Lungs

Carbon dioxide and water vapour diffuse from the moist surfaces of the lungs, which in mammals are the only excretory organs for carbon dioxide. Some of the water released at the lung surface is metabolic, that is, produced as a waste product of respiration, but its exact origin is not really important in view of the large volume of water contained within the body.

Liver

Considering the many homeostatic roles of the liver described in section 19.6.2 it is not surprising that these include excretion. Bile pigments are excretory products from the breakdown of the haemoglobin of old red blood cells. They pass to the duodenum in the bile for removal from the body along with the faeces, for whose colour they are partly responsible. The most important excretory role of the liver is the formation of urea from excess amino acids (section 20.4).

20.1.3 Excretion in plants

Plants do not have as many problems regarding excretion as do animals. This is because of fundamental differences in physiology and mode of life between animals and plants. Plants are producers and they synthesise all their organic requirements according to demand. For example, plants manufacture only the amount of protein necessary to satisfy immediate demand. There is never an excess of protein and therefore very little excretion of nitrogenous waste substances. If proteins are broken down into amino acids, the latter can be recycled into new proteins. Three of the waste substances produced by certain metabolic activities in plants, that is oxygen, carbon dioxide and water, are raw materials (reactants) for other reactions, and excesses of carbon dioxide and water are used up in this way. Water is also a solvent. The only major gaseous excretory product of plants is oxygen. During light periods the rate of production of oxygen is far greater than the plant's demand for oxygen in respiration and this escapes from plants into the environment by diffusion.

Many organic waste products of plants are stored within dead permanent tissues such as the 'heartwood' or within leaves or bark which are removed periodically. The bulk of most perennial plants is composed of dead tissues into which excretory materials are passed. In this state they have no adverse effects upon the activities of the living tissues. Similarly, many mineral salts, taken up as ions, may accumulate. Organic acids, which might prove harmful to plants, often combine with excess cations and precipitate out as insoluble crystals which can be safely stored in plant cells. For example, calcium ions and sulphate ions are taken up together, but sulphate is used up immediately in amino acid synthesis leaving an excess of calcium ions. These combine with oxalic and pectic acids to form harmless insoluble products such as calcium oxalate and calcium pectate. Substances are not only eliminated through leaf loss but also through petals, fruits and seeds, although this excretory function is not the primary function of their dispersal. Aquatic plants lose most of their metabolic wastes by diffusion directly into the water surrounding them.

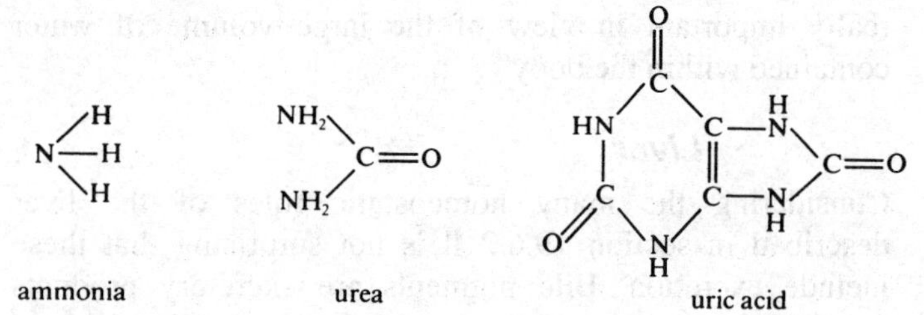

Fig 20.1 *Molecular structure of the three main nitrogenous excretory products.*

20.2 Nitrogenous excretory products and environment

Nitrogenous waste products are produced by the breakdown of proteins, nucleic acids and excess amino acids. The first product of the breakdown of excess amino acids is ammonia. It is produced by removal of the amino group from amino acids, a process called **deamination** (section 20.4). Ammonia may be excreted immediately or converted into the nitrogen-containing compounds urea or uric acid (fig 20.1). The exact nature of the excretory product is determined mainly by the availability of water to the organism (that is, its habitat), and the extent to which the organism controls water loss (table 20.1). The correlation with habitat may be summarised thus:

ammonia	aquatic (water conservation not a problem)
urea	aquatic/terrestrial
uric acid	terrestrial

20.2.1 Ammonia

The major source of ammonia is the deamination of excess amino acids. Ammonia is extremely toxic and must be eliminated. Being very soluble, it can be eliminated from the body rapidly and safely if diluted in a sufficient volume of water. This presents no real problems to organisms which have ready access to water but this applies only to those organisms living in freshwater. It is therefore excreted rapidly as ammonium ions (NH_4^+) in most aquatic organisms, from protozoa to amphibia, before it reaches concentrations which are toxic to the organism.

Marine and terrestrial organisms have an acute problem of gaining or conserving water respectively. Therefore very little is available for the elimination of nitrogenous waste. Table 20.1 reveals that organisms living in these environments have developed alternative means of nitrogen excretion. These involve the development of many anatomical, biochemical, physiological and behavioural mechanisms involving the elimination of nitrogenous waste whilst maintaining the composition of the body fluids at a steady state.

Table 20.1 Summary of the relationship between excretory products and habitat of representative animal groups.

Animal	*Excretory product*	*Habitat*
protozoan	ammonia	aquatic
terrestrial insect	uric acid	terrestrial
freshwater bony fish	ammonia	aquatic
marine bony fish	urea, trimethylamine oxide	aquatic
bird	uric acid	terrestrial
mammal	urea	terrestrial

20.2.2 Urea

Urea is formed in the liver as described in section 20.4. It is much less toxic than ammonia and is the main nitrogenous excretory product in mammals. Its excretion is described in section 20.5.

20.2.3 Uric acid

Uric acid and its salts are ideal excretory products for terrestrial organisms and essential for organisms such as land-living insects and birds which produce shelled eggs. They combine a high nitrogen content with low toxicity and low solubility. They can be stored in cells, tissues and organs without producing any toxic or harmful osmoregulatory effects, and they require very little water for their excretion. As the concentration of uric acid in the tissue rises it settles out as a solid precipitate. The details of uric acid excretion are described in section 20.3.3. Humans excrete small quantities of uric acid but this is produced from the breakdown of nucleic acids and not from the breakdown of proteins. Approximately 1 g of uric acid is excreted in urine per day.

20.3 Nitrogenous excretion and osmoregulation in representative animals

Throughout this review it should be remembered that the environment influences the nature of the excretory product and the process of osmoregulation. The processes of nitrogenous excretion and osmoregulation involve the same body structures. This is because elimination of nitrogenous waste is usually associated with problems of gaining or losing water. The two processes will therefore be considered together.

20.3.1 The effect of environment on osmoregulation

The bodies of many aquatic organisms have a water potential which is higher than that of the environment (the body contents are less concentrated). The body therefore *loses* water by osmosis. The water loss is replaced in various ways, including drinking and eating. Organisms with bodies of lower water potential than the surrounding water (body contents more concentrated) *gain* water by osmosis. In order to minimise these exchanges the organisms often have an impermeable outer covering.

All terrestrial organisms face the problem of water loss from their body fluids to the environment. The body fluids of these organisms are maintained by specialised osmoregulatory/excretory organs, such as Malpighian tubules and kidneys. A balance must be achieved between the amount of water lost and gained. The problems of water balance are described in detail in section 20.6.

Osmoregulation also involves regulation of solute concentrations by the processes of diffusion and active transport.

Adaptations to severe drought

The kangaroo rat (*Dipodomys*) is remarkable among mammals in being able to tolerate drought conditions in the deserts of North America. It flourishes in these conditions by possessing a unique combination of structural, physiological and behavioural adaptations. Water loss by evaporation from the lungs is reduced by exhaling air at a temperature below body temperature. As air is inhaled it gains heat from the nasal passages which are cooled as a result. During exhalation water vapour in the warm air condenses on the nasal passages and is conserved. The kangaroo rat feeds on dry seeds and other dry plant material and does not drink. Water produced by the chemical reactions of respiration, and that present in minute amounts in its food, are its only sources of water. The classic investigations by Knut Schmidt-Nielsen (summarised in table 20.2) revealed the overall water metabolism balance for a kangaroo rat weighing 35 g metabolising 100 g of barley in experimental surroundings at 25 °C and a relative humidity of 20%. Throughout this period the only source of water was the barley grain.

Finally the kangaroo rat avoids excessive evaporative water losses in the wild by spending much of its time in the relatively humid atmosphere of its underground burrow.

Table 20.2 Water metabolism for a kangaroo rat under experimental conditions. The absorbed water was the water present in the food.

Water gains	cm^3	*Water losses*	cm^3
oxidation water from cell respiration	54.0	urine	13.5
absorbed water from food	6.0	faeces	2.6
		evaporation	43.9
Total water gain	60.0	Total water loss	60.0

The other spectacular example of water conservation is the camel, whose physiological adaptations are described in section 19.5.7.

20.3.2 Protozoans

Protozoans are single-celled organisms belonging to the Kingdom Protoctista (section 2.8.4). They are found in freshwater and marine habitats and provide an excellent example of the basic problems of osmoregulation faced by animal cells. An animal cell which is unprotected by any osmoregulatory mechanisms will die if its contents have a different water potential to their environment, as demonstrated by the red blood cell in fig 5.19. A single cell like a protozoan must therefore have an osmoregulatory mechanism. The cell contents are separated from the external environment only by a partially permeable cell surface membrane.

Excretion

Excretion of carbon dioxide and ammonia occurs by diffusion over the entire surface of the cell. This has a relatively large surface area to volume ratio which allows the efficient removal of waste substances.

Osmoregulation in freshwater species

Remember that the more concentrated a solution, the lower its water potential (and solute potential) (section 5.9.8 osmosis). When a cell is in a solution, water moves from the region of higher water potential to the region of lower water potential.

All freshwater species of protozoans have a lower water potential than their surroundings (more concentrated solution in their cells). There is therefore a constant tendency for water to enter the cell by osmosis through the cell surface membrane. The problem is overcome by the presence of organelles known as **contractile vacuoles**. These remove water which enters the cell by osmosis, therefore preventing the cell from increasing in size and bursting. The exact location and structure of the contractile vacuole is extremely variable. In *Amoeba* a contractile vacuole can form anywhere within the cell and release its fluid into the external environment at any point on its outer surface (fig 20.2). In *Paramecium* there are two contractile vacuoles with fixed positions (fig 20.3, see also fig 2.32). The method of functioning, however, appears to be the same in all species. Small vesicles (or canals in some species, such as *Paramecium*) in the cytoplasm fill with fluid from the cytoplasm. At first the fluid has the same composition, and therefore the same water potential, as the cytoplasm (fig 20.4). Most of the ions are then pumped out of the fluid by active transport, using energy in the form of ATP supplied by mitochondria that surround the vesicles. The remaining fluid in the vesicles is mostly water and is loaded into the contractile vacuole by the vesicle. The contractile vacuole gradually fills. Its membrane is almost

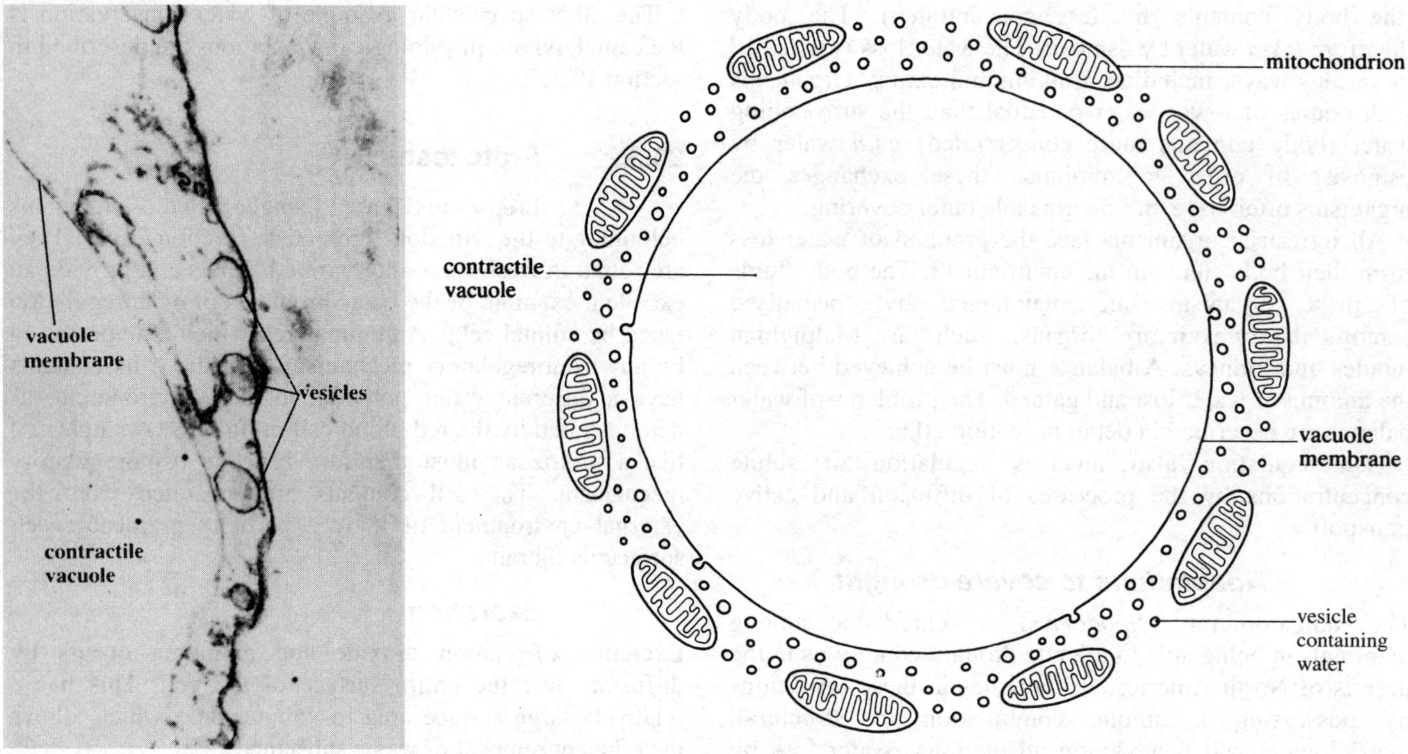

Fig 20.2 *Electron micrograph and diagram of contractile vacuole of* Amoeba*. The tiny vesicles fill with fluid from the cytoplasm. Ions in the fluid are then pumped back into the cytoplasm before the vesicles fuse with the membrane of the contractile vacuole, discharging water into the vacuole.*

completely impermeable to water so, despite its contents being very dilute (high water potential), water cannot escape back into the cytoplasm by osmosis. When it reaches a certain size, the contractile vacuole fuses with the cell surface membrane, contracts suddenly and releases its water.

Osmoregulation in marine species

Many of the marine protozoans do not have functional contractile vacuoles because their cell contents have the same water potential as sea water.

20.3.3 Insects

The great majority of insects are adapted for life on land. One of the major problems of life on land is the prevention of water loss. Adaptations for preventing water loss include:

- an almost impermeable waxy layer covering their exoskeletons which reduces water loss from the body surface;
- the only openings to the body for gaseous exchange are pairs of small holes in certain segments – these are called **spiracles**;
- there are valve-like structures in the spiracles to reduce water loss from the tubes that lead from the spiracles to the cells (the gaseous exchange system described in section 9.4.4);
- the excretory product is semi-solid, not liquid (see below);
- cleidoic eggs, meaning that the embryo develops inside an egg with a relatively impermeable shell that prevents water loss.

The strong exoskeleton is covered by a thin waterproof layer, the epicuticle (0.3 μm thick). Water loss by evaporation is prevented by the way the molecules are organised. A highly organised single layer of lipid molecules is covered by several layers of irregularly arranged lipid molecules. If these wax or grease lipid layers are rubbed by sharp particles, such as sand, the structure is damaged, evaporation rate increases and the insect risks dehydration. Interestingly, as the air temperature around an insect is increased, there is a gradual increase in the rate of evaporation until a particular temperature is exceeded after which the evaporation rate increases rapidly. This point is known as the **transition temperature** (fig 20.5) and it marks the temperature at which the ordered orientation of the single layer of lipid molecules breaks down.

Some insects living on dry food in very dry habitats are able to take up water from the air providing that the relative humidity of the air is above a certain value, such as 90% for the mealworm (*Tenebrio*) and 70% for the house mite (*Dermatophagoides*).

The problem of preventing water loss by excretion is overcome by specialised excretory organs called

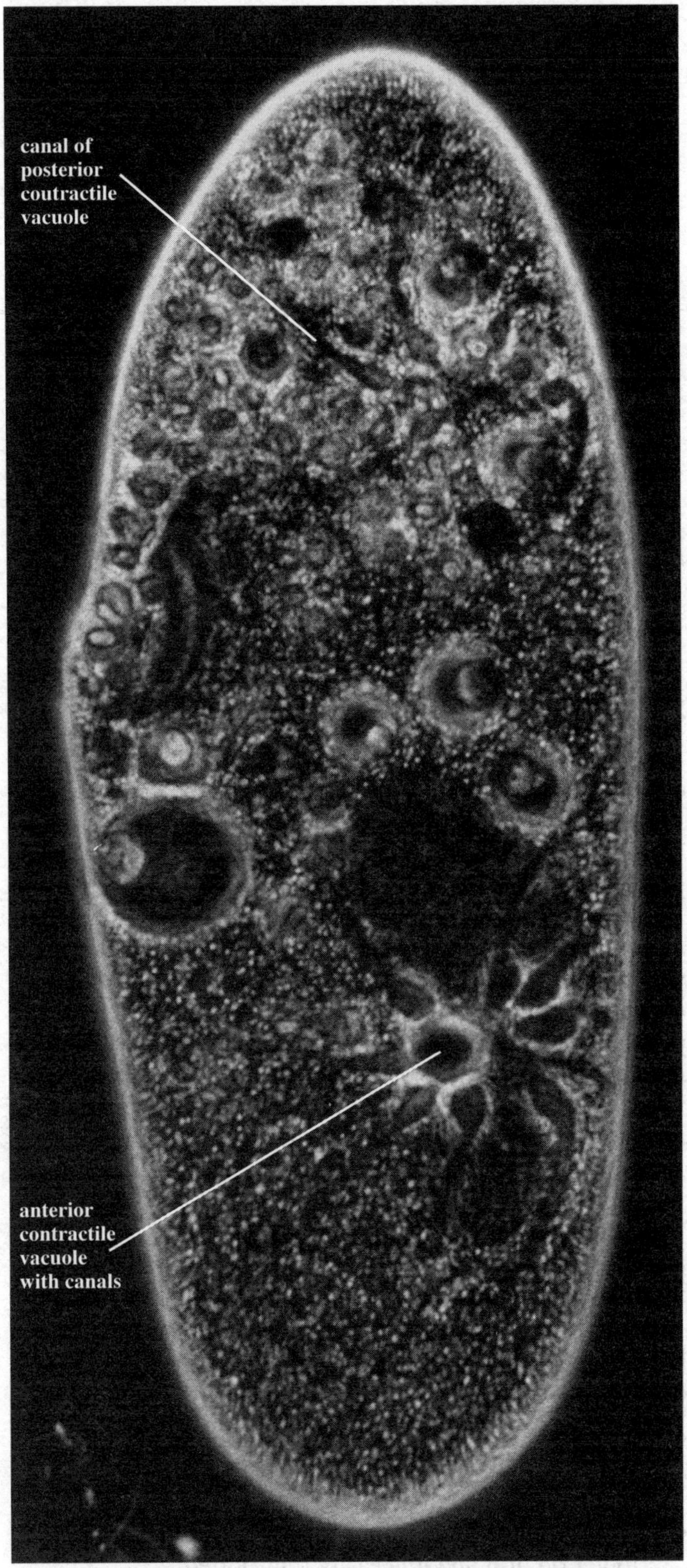

Fig 20.3 *Photomicrograph of the fixed contractile vacuoles of* Paramecium. *A system of canals around each vacuole brings fluid from the cytoplasm to the vacuole. The vacuole itself swells and bursts at regular intervals, releasing water back into the environment each time it bursts. In the photograph, the anterior vacuole is full and the posterior vacuole has recently burst.*

Malpighian tubules which produce and excrete the almost insoluble waste substance **uric acid**.

Malpighian tubules are blind-ending extensions of the hindgut in insects. They lie in the abdomen and are bathed in blood. (The blood system of insects is open, meaning that the blood vessels empty blood into a cavity, the haemocoel, and do not have capillaries. The main organs are bathed in blood.) The number of tubules is variable in insects. Some have a pair and others may have several hundred. *Rhodnius*, a blood-sucking bug, has four tubules as shown in fig 20.6. The tubules of a wasp are shown in fig 20.7. In all cases they open into the hindgut at its junction with the midgut and may be long and slender or short and compact.

The tubule has two distinct regions, an **upper segment** (furthest from the gut), composed of a single layer of cells, and a **lower segment**. The upper segment absorbs fluid from the blood. As the fluid passes along the inside of the tubules, the cells of the lower segment, which have microvilli on their inner surfaces, absorb water and various salts. It is in the lower segment that the nitrogenous waste precipitates out of solution as solid crystals of uric acid. The concentrated contents of the tubule, still fluid, pass into the hindgut or rectum where they mix with waste materials from digestive processes. **Rectal glands** in the wall of the rectum absorb water from the faeces and uric acid suspension until the waste is dry enough for it to be eliminated from the body as pellets.

20.3.4 Freshwater fish

The excretory and osmoregulatory organs of the fish are the gills and kidneys. The gills are in contact with the external environment and are permeable to water, nitrogenous waste and ions. They have a large surface area for efficient exchange of respiratory gases, but this presents a problem when it comes to osmoregulation. This is particularly true when the fish lives in fresh water.

The internal body fluids of freshwater bony fish are more concentrated than their environment. Despite having a relatively impermeable outer covering of scales and mucus, there is a considerable movement of water by osmosis, and loss of ions by diffusion, through the highly permeable gills, which also serve as the organs of excretion for the waste nitrogenous substance ammonia. In order to maintain the body fluids at a steady state, freshwater fish have to lose a large volume of water continually (fig 20.8). They do this by producing a large volume of very dilute urine (of higher water potential than blood) which contains some ammonia and a number of solutes. Up to one-third of the body mass can be lost per day as urine. Ions which are lost from the body fluids are replaced by food and by active uptake from the external environment by special cells in the gills.

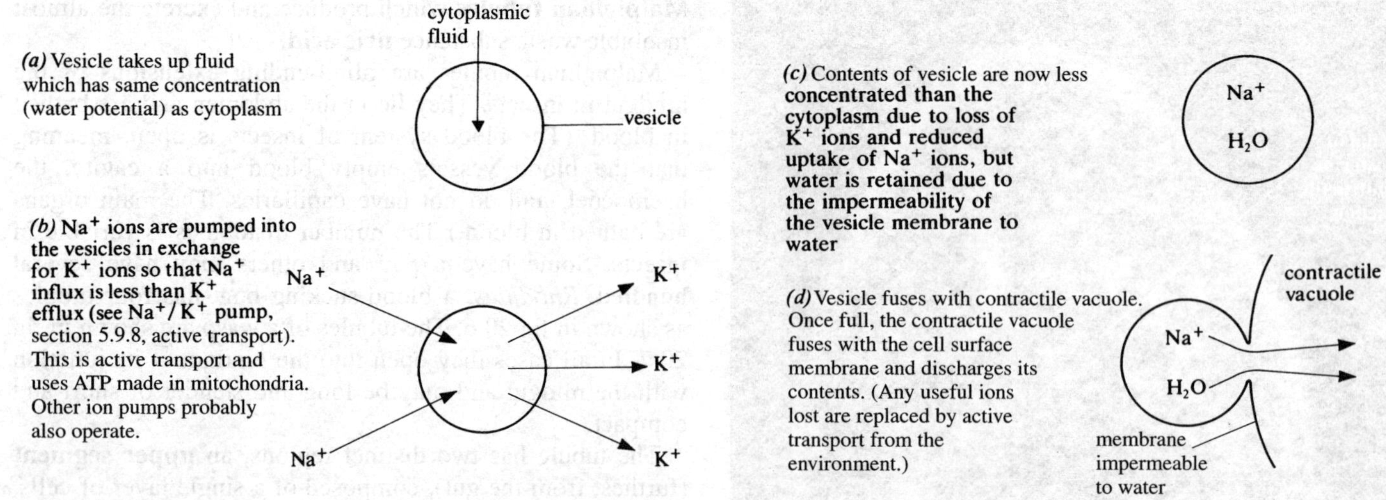

Fig 20.4 *Diagrammatic explanation of a possible mechanism of water uptake by a contractile vacuole.*

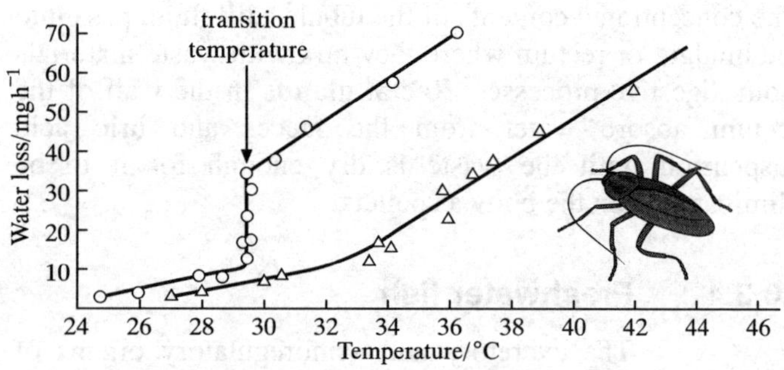

Fig 20.5 *Graph showing the water loss from the exoskeleton of a cockroach at various air temperatures (triangles). The circles indicate water loss plotted against the surface temperature of the exoskeleton. This shows the dramatic increase in water loss at about 29.5 °C, the* ***transition temperature****.*

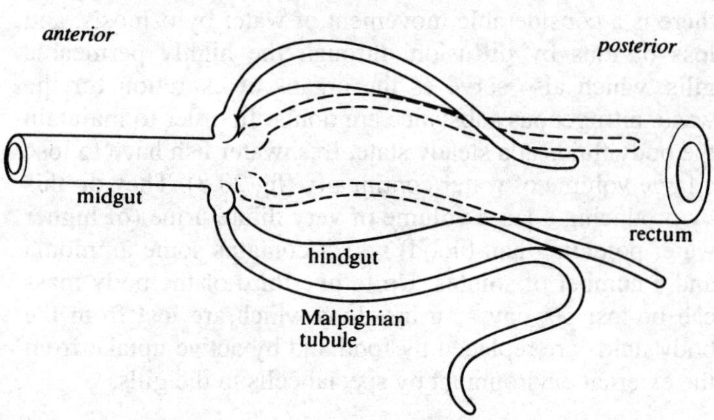

Fig 20.6 *Diagram showing the position of Malpighian tubules in relation to the gut of* Rhodnius*, a bug.*

20.3.5 Summary of water balance

The efficient functioning of animal cells relies on the maintenance of the steady state of the cell contents. Homeostatic exchange of water between cells, tissue fluid, lymph and blood plasma and the environment present problems for both aquatic and terrestrial forms of life. Aquatic organisms gain or lose water by osmosis through all permeable parts of the body surface depending on whether the environment is more dilute or more concentrated than their body. Terrestrial organisms have the problem of losing water and many mechanisms are used to maintain a steady-state water balance, as summarised for insects and mammals in table 20.3. This steady state is achieved by balancing loss and water gain, as shown in fig 20.9.

20.4 Formation of urea in humans

Urea is the nitrogenous waste product of humans and other land-living mammals. The advantages of using urea as a nitrogenous waste product are that it is:

- non-toxic – it can therefore be carried round the body in the blood from the liver where it is made until it is removed by the kidneys;
- very soluble – it does not require a great deal of water to get rid of it and it is easily transported;
- a small molecule – it is therefore easily filtered in the kidneys.

The body is unable to store excess amino acids taken in in the diet. Those not immediately needed for protein synthesis or making sugar must be got rid of. This takes place in the liver in two main stages:

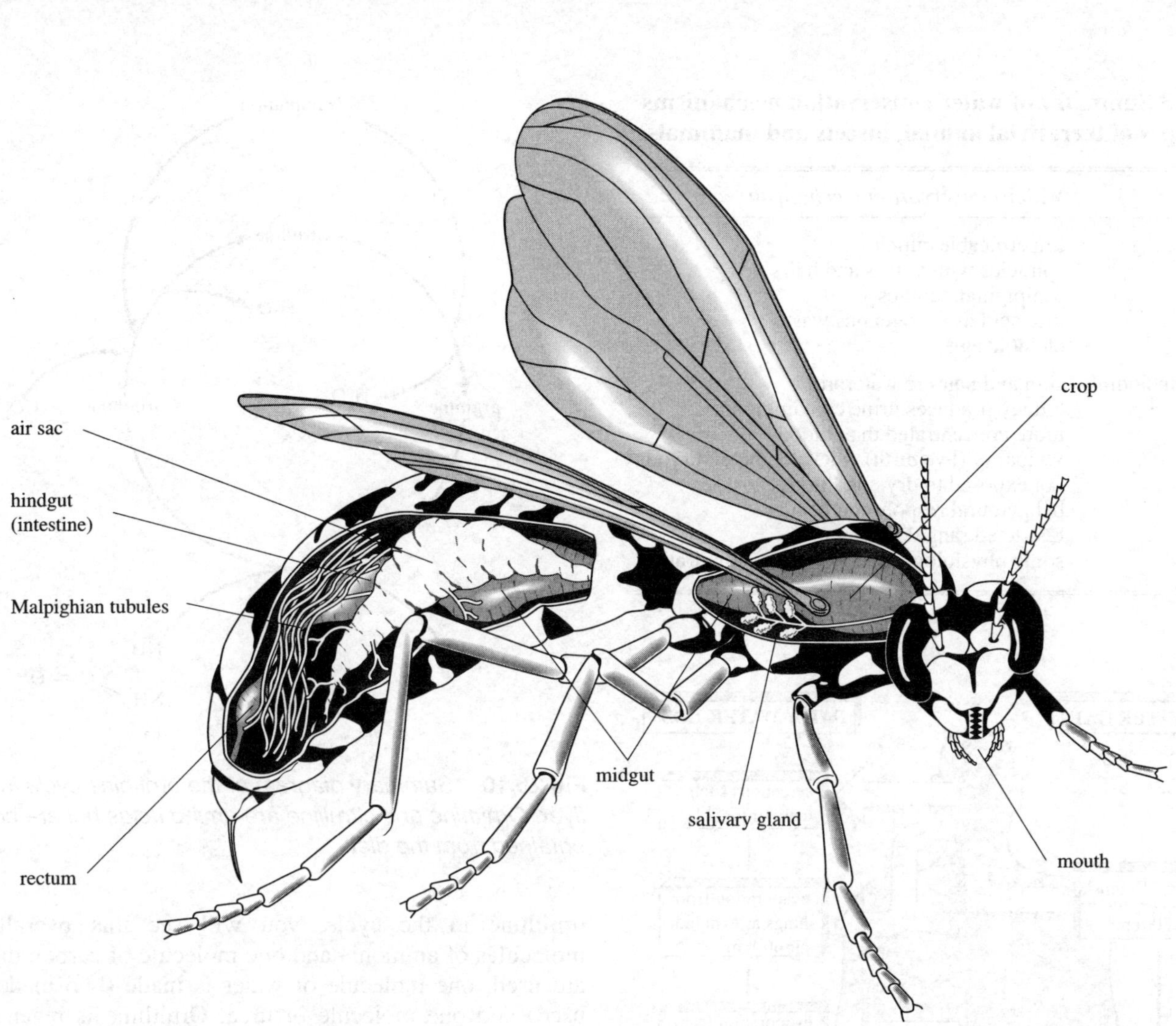

Fig 20.7 *Malpighian tubules of a wasp. The tubules are specialised excretory organs which are extensions of the hind gut.*

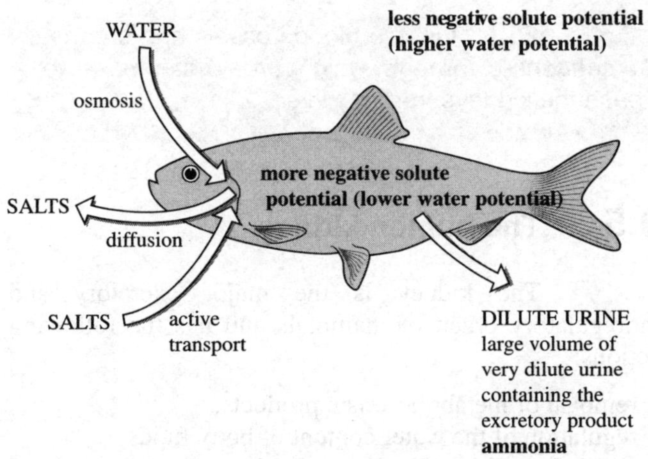

Fig 20.8 *Excretion and osmoregulation in a freshwater bony fish.*

- **deamination** – the amino group is removed from the amino acid and used to make ammonia;
- **detoxification** – ammonia is toxic (poisonous) and so is converted to a harmless product, urea, for transport to the kidneys.

Deamination

The amino acid is oxidised using oxygen. This results in removal of the amino group ($-NH_2$) and leaves an acid. The acid can enter the Krebs cycle and be used as a source of energy in cell respiration. The amino group is converted to ammonia (NH_3) during deamination.

Detoxification

Ammonia is converted into urea in the liver:

$$\underset{\text{ammonia}}{2NH_3} + \underset{\text{carbon dioxide}}{CO_2} \longrightarrow \underset{\text{urea}}{(NH_2)_2C{=}O} + \underset{\text{water}}{H_2O}$$

This occurs by a cyclic reaction known as the **ornithine cycle** which is summarised in fig 20.10. If you start at

Table 20.3 Summary of water conservation mechanisms in two types of terrestrial animal, insects and mammals.

Organism	*Water conservation mechanism*
insect	impermeable cuticle spiracles with valves and hairs Malpighian tubules uric acid as nitrogenous waste cleidoic egg
mammal (including humans)	skin and hair are waterproof kidney produces urine containing urea more concentrated than blood viviparity (live birth) – developing embryo is not exposed to dry conditions behavioural response to heat restricted range of habitats some physiological tolerance to dehydration

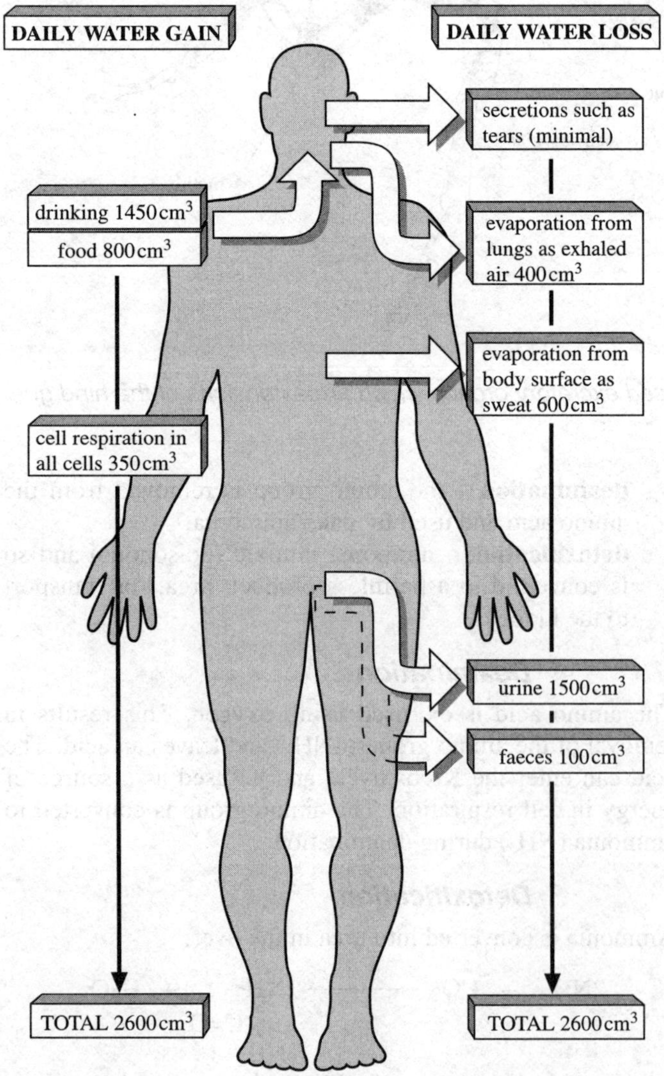

Fig 20.9 *Daily water loss and water gain by the human body.*

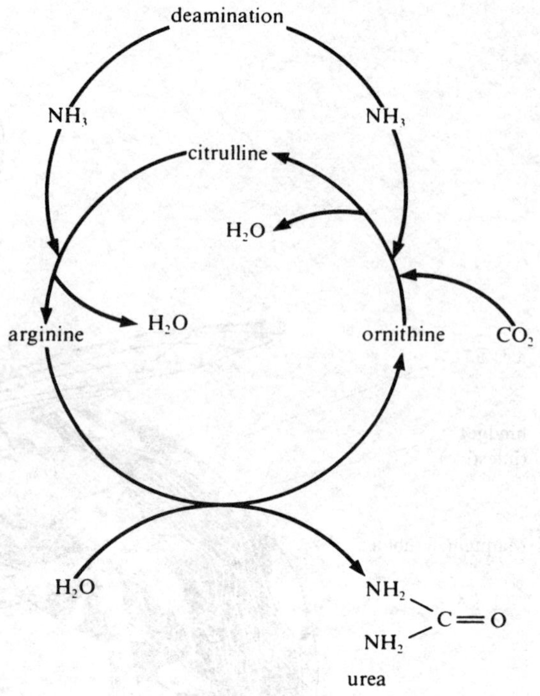

Fig 20.10 *Summary diagram of the ornithine cycle in the liver. Ornithine and citrulline are amino acids but are not obtained from the diet.*

ornithine in the cycle, you will see that overall two molecules of ammonia and one molecule of carbon dioxide are used, one molecule of water is made (two made, one used) and one molecule of urea. Ornithine is regenerated ready for the next cycle.

Urea is transported in the blood plasma from the liver to the kidneys.

20.1 List the blood vessels and organs, in sequence, through which urea must pass to reach the kidneys from the liver.

20.5 The human kidney

The kidney is the major excretory and osmoregulatory organ of mammals and has the following functions:

- removal of metabolic waste products;
- regulation of the water content of body fluids;
- regulation of the pH of body fluids;
- regulation of the chemical composition of body fluids by removal of substances which are in excess of immediate requirements.

The kidney has a rich blood supply and regulates the blood composition at a steady state. It therefore contributes to homeostasis. This ensures that the composition of the tissue fluid is maintained at an optimum level for the cells bathed by it and enables the cells to function efficiently at all times.

20.5.1 Position and structure of kidneys

There are a pair of kidneys in humans situated towards the back of the lower part of the abdominal cavity, on either side of the vertebral column. The left kidney lies slightly above the right.

The kidneys receive blood from the aorta via the **renal arteries**, and the **renal veins** return blood to the posterior (inferior) vena cava. Urine formed in the kidneys passes by a pair of **ureters** to the **bladder** where it is stored until it is released via the **urethra** (fig 20.11). Two muscle sphincters surround the urethra where it leaves the bladder, one of which is under voluntary control. These control release of urine, a process known as urination or micturition.

A transverse section of the kidney shows two distinct regions, an outer **cortex** and an inner **medulla** (fig 20.12). The cortex is covered by fibrous connective tissue, forming a tough capsule. The cortex contains glomeruli, which are just visible to the naked eye, renal corpuscles and parts of the nephrons (see below). The medulla is composed of tubular parts of the nephrons and blood vessels, which together form **renal pyramids**. The apex of each pyramid is called a **papilla**. All the pyramids project into the **pelvis** which leads into the ureter (fig 20.12). A large number of blood vessels run through the kidney and supply a vast network of blood capillaries.

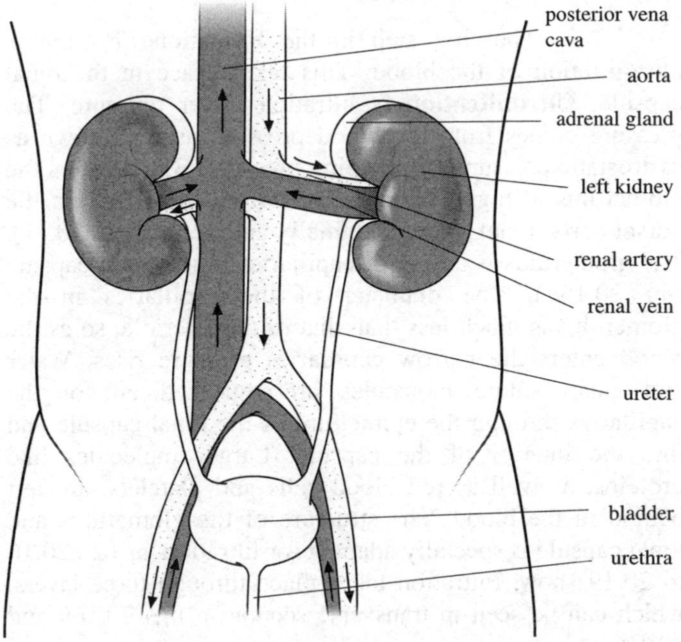

Fig 20.11 *Human excretory system. The size of the organs relative to the outline of the body is exaggerated for clarity.*

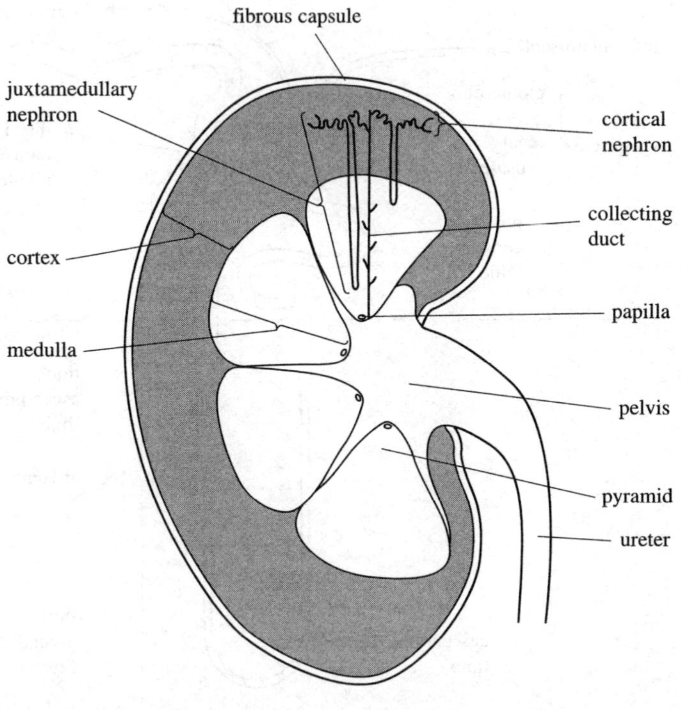

Fig 20.12 *TS through kidney showing the position of two nephrons.*

20.5.2 Nephron – overall structure and blood supply

The basic unit of structure and function of the kidney is the **nephron** (fig 20.13) and its associated blood supply. Each kidney, in a human, contains an estimated one million nephrons each having an approximate length of 3 cm. The total length of tubules in each kidney is about 120 km. This offers an enormous surface area for the exchange of materials. About one fifth of the blood passes through the kidneys for each circuit of the body and about 125 cm^3 of fluid is filtered out of the blood per minute. About 99% of the water is returned to the blood, so only about 1 cm^3 of urine is made per minute, although this varies with factors like drinking.

Each nephron is composed of six regions, each having its own particular structure and function:

(1) renal corpuscle (Malpighian body), composed of renal capsule and glomerulus (the renal capsule is also known as the Bowman's capsule);
(2) proximal convoluted tubule;
(3) descending limb of the loop of Henle;
(4) ascending limb of the loop of Henle;
(5) distal convoluted tubule;
(6) collecting duct.

These regions are shown in fig 20.13.

There are two types of nephrons, cortical nephrons and juxtamedullary nephrons, which differ in their positions in the kidney. **Cortical nephrons** are found in the cortex and have relatively short loops of Henle which just extend into

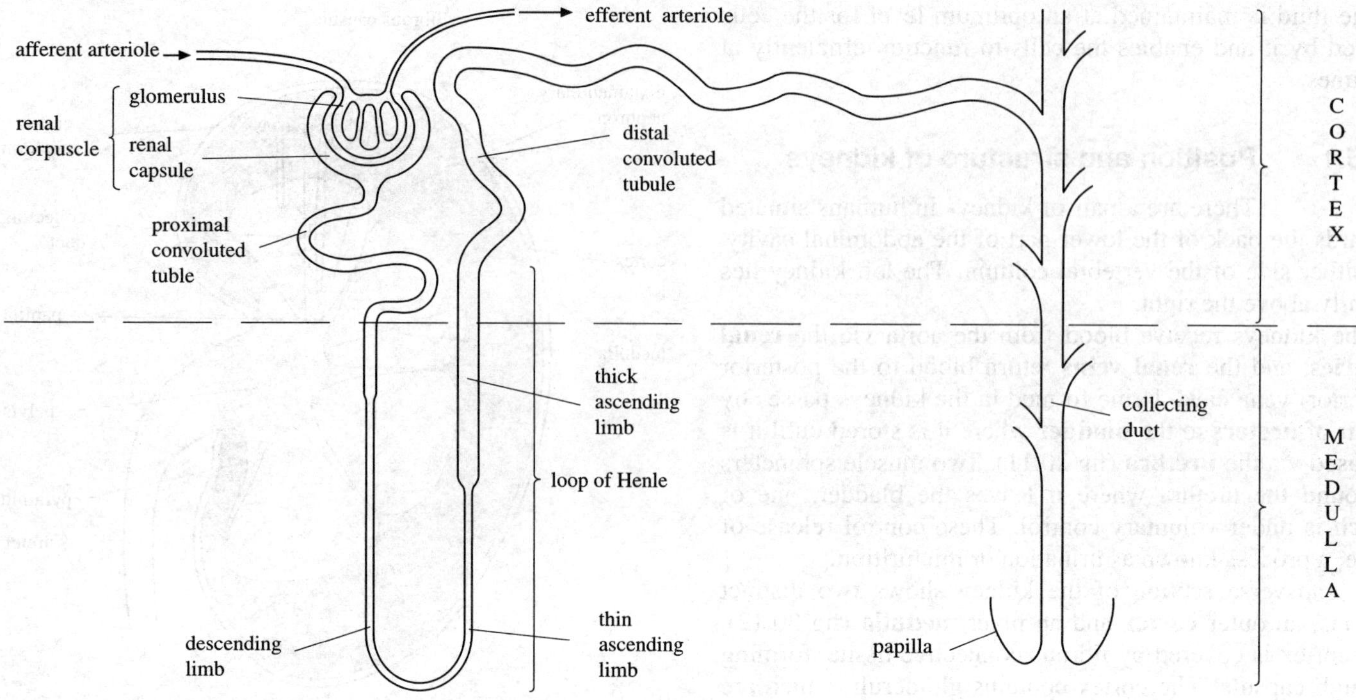

Fig 20.13 *Diagram showing the structure of a nephron. (Not to scale.)*

the medulla. **Juxtamedullary nephrons** have their renal corpuscle close to (= *juxta*) the junction of the cortex and medulla. They have long loops of Henle which extend deep into the medulla (fig 20.14*a*). The two types of nephrons have different uses. Under normal conditions of water availability the cortical nephrons deal with the control of blood volume, whereas, when water is in short supply, increased water retention occurs through the juxtamedullary nephrons.

Blood enters the kidney by the renal artery which branches into finer and finer arteries before entering the glomerulus of a renal corpuscle as an afferent arteriole. (Afferent means *to*, and efferent means *from*.) Filtered blood leaves the glomerulus by an efferent arteriole and flows through a network of capillaries in the cortex which surround the proximal and distal convoluted tubules and the loops of Henle in the medulla (fig 20.14*b*). The capillaries of the vasa recta run parallel to the loops of Henle and the collecting ducts in the medulla. These networks of blood vessels return blood, containing substances which are useful to the body, to the general circulation. Blood flow through the vasa recta is much less than through the capillaries around the proximal and distal convoluted tubules and this enables a water potential (solute potential) gradient to be maintained in the tissue of the medulla, as described later.

20.5.3 Histology of the kidney

Fig 20.15 shows the structure of the cortex and medulla regions of the kidney as seen with a light microscope.

20.5.4 Ultrafiltration

The first step in the formation of urine is ultrafiltration of the blood. This takes place in the renal capsule. **Ultrafiltration** is filtration under pressure. The pressure comes from the blood pressure and is known as hydrostatic pressure, or pumping pressure. Blood enters the glomerulus at high pressure direct from the heart via the dorsal aorta, renal artery and finally an arteriole (fig 20.11). The glomerulus is a knot of capillaries in the renal capsule (fig 20.15*d*). The diameter of the capillaries in the glomerulus is much less than that of the arteriole, so as the blood enters the narrow capillaries pressure rises. Water and small solute molecules are squeezed out of the capillaries through the epithelium of the renal capsule and into the interior of the capsule. Larger molecules like proteins, as well as red blood cells and platelets, are left behind in the blood. The structure of the glomerulus and renal capsule is specially adapted for filtration, as figs 20.16 to 20.19 show. Filtration takes place through three layers, which can be seen in transverse section in figs 20.16*b* and 20.19:

- **Endothelium of the blood capillary** This is very thin and is perforated with thousands of pores of about 10 nm diameter. They occupy up to 30% of the area of

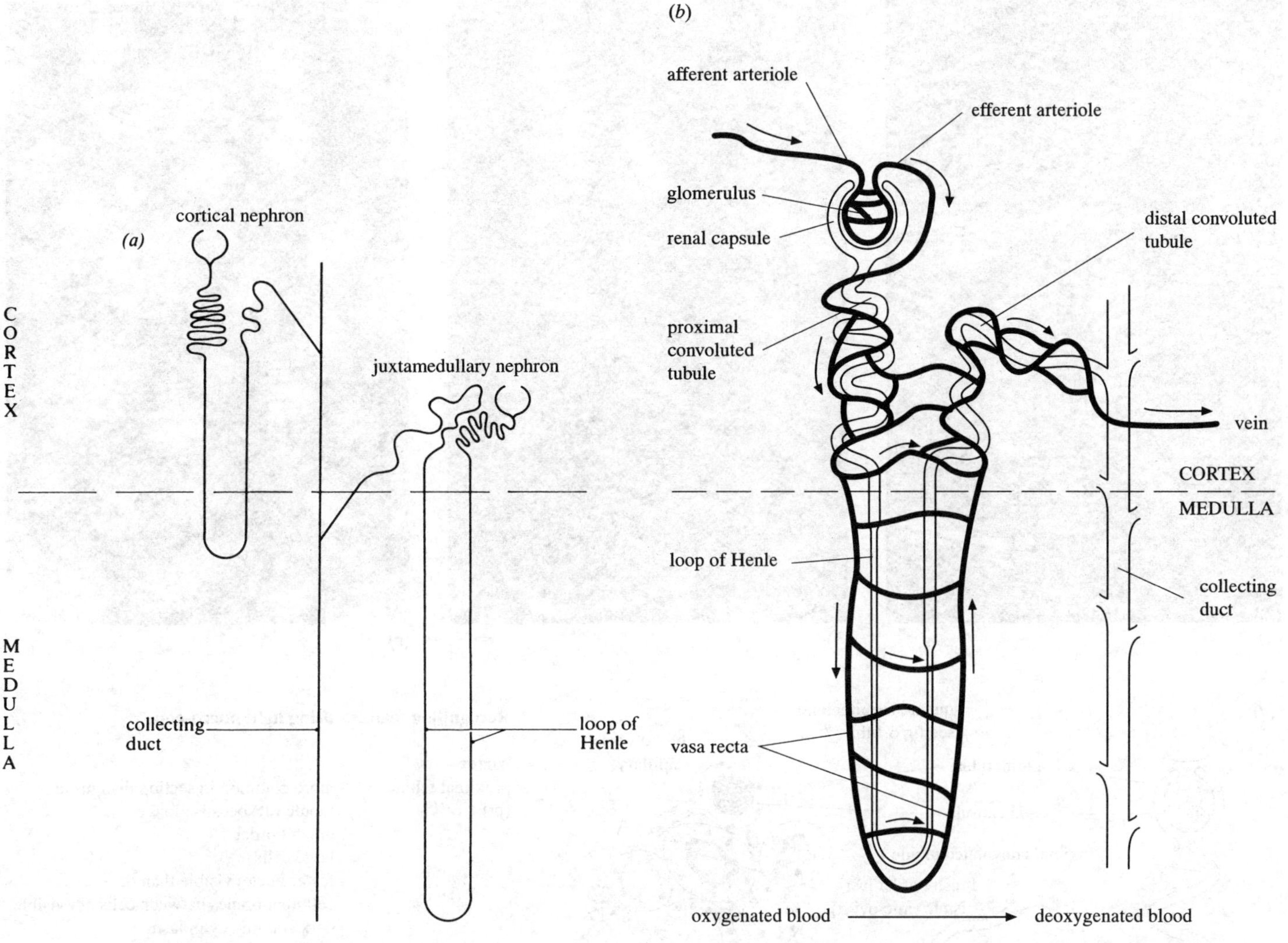

Fig 20.14 *(a) Cortical and juxtamedullary nephrons and (b) juxtamedullary nephron and its associated blood supply.*

the wall. The pores are not a barrier to plasma proteins because they are too large.

- **Basement membrane of the blood capillaries** All epithelial cells rest on a 'basement membrane'. It consists of a meshwork of fibres, including collagen fibres. Water and small solute molecules can pass through spaces between the fibres. Red blood cells and platelets are too large. Protein molecules are too large and are also repelled by negative electrical charges on the fibres.
- **Epithelium of the renal capsule** This is made of cells which are highly modified for filtration, called **podocytes** (pod meaning *foot*). Each cell has many foot-like extensions projecting from its surface. The extensions interlink with extensions from neighbouring cells as shown in figs 20.16*a*, 20.16*b* and 20.17. They fit together loosely, leaving slits called **slit pores** or **filtration slits** about 25 nm wide (fig 20.19). The filtered fluid can pass through these slits.

About 20% of the plasma is filtered into the capsule. Of the three layers, the basement membrane is the main filtration barrier. The filtered fluid in the capsule is called **glomerular filtrate (GF)**. It has a chemical composition similar to that of blood plasma. It contains glucose, amino acids, vitamins, ions, nitrogenous waste (mainly urea, but also some uric acid and creatinine), some hormones and water.

Blood passing from the glomerulus has a lower water potential due to the increased concentration of plasma proteins and a reduced hydrostatic pressure.

Factors affecting the glomerular filtration rate (GFR)

The filtration pressure forcing fluid out of the glomerulus depends not only on the hydrostatic pressure of the blood, but also on the pressure of the glomerular filtrate (fig 20.20). If this equalled the hydrostatic pressure of the blood they would cancel each other out. In fact, the hydrostatic pressure of the glomerular filtrate is much lower than that of the blood, although not zero. Similarly the solute potential either side of the filtration barrier will affect the flow of fluid. Water tends to move from less negative solute potentials to more negative solute potentials (from less

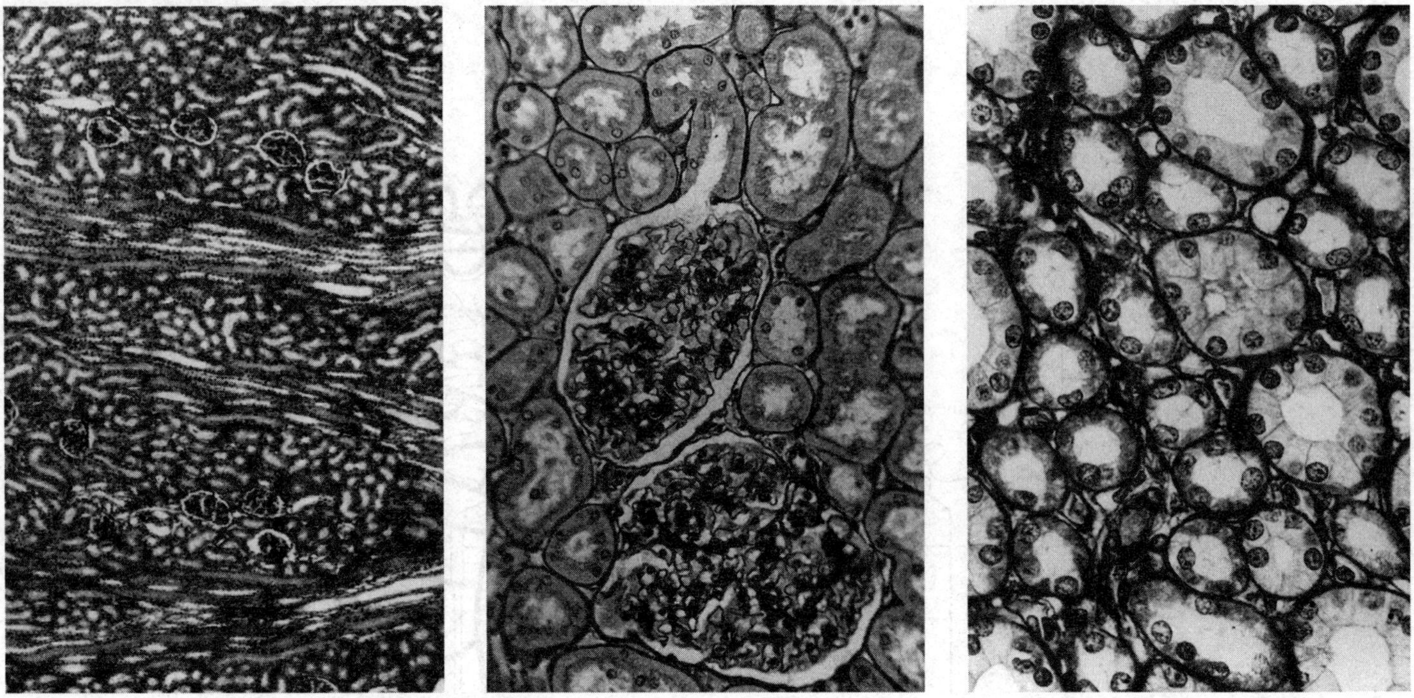

(*a*) (*b*) (*c*)

(*d*)

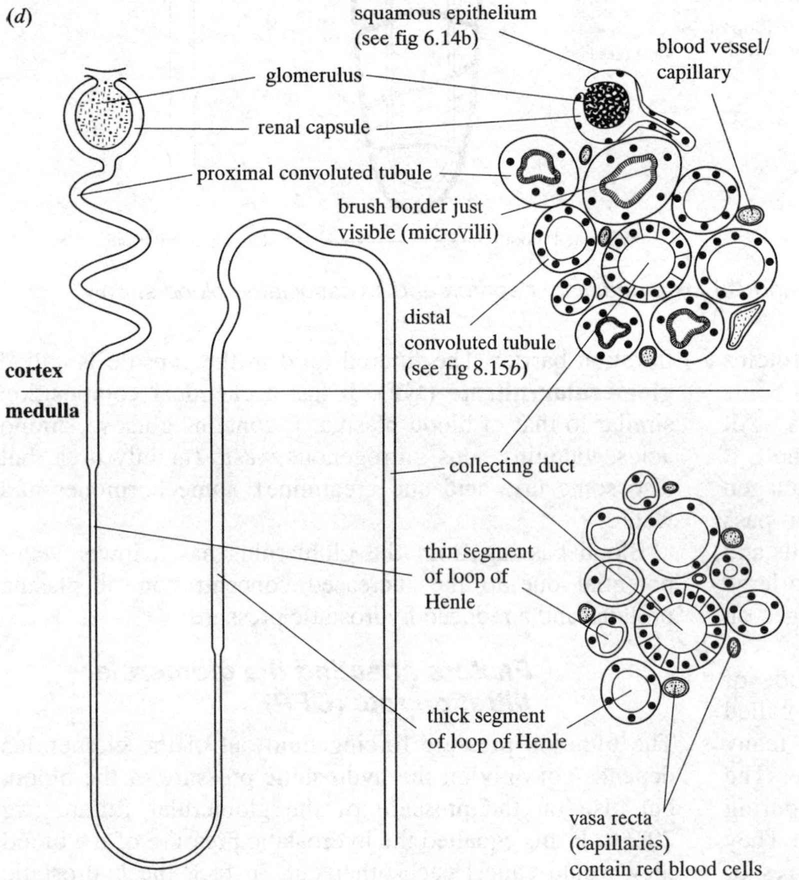

Recognition features using light microscope

cortex	
proximal tubule: (pt)	more common in section than distal tubule (dt) because longer brush border large cells fewer nuclei visible than dt cell membranes between cells not visible dark staining cytoplasm small irregular lumen
distal tubule: (dt)	less common in section than pt no brush border smaller cells than pt more nuclei visible than pt cell membranes between cells not visible pale staining cytoplasm large regular lumen
collecting duct:	greater diameter cell membrane visible pale staining cytoplasm cuboidal/columnar cells large lumen
medulla	
thick segment of loop of Henle:	no cell membrane visible thick walls
thin segment of loop of Henle:	no cell membrane visible thin walls, nuclei causing a bulge

Fig 20.15 *Histology of the kidney. (a) Low power TS cortex showing sections through tubules and glomeruli. (b) High power TS cortex showing section through two glomeruli. (c) TS medulla showing sections through loops of Henle and collecting ducts. (d) Diagram interpreting sections through cortex and medulla.*

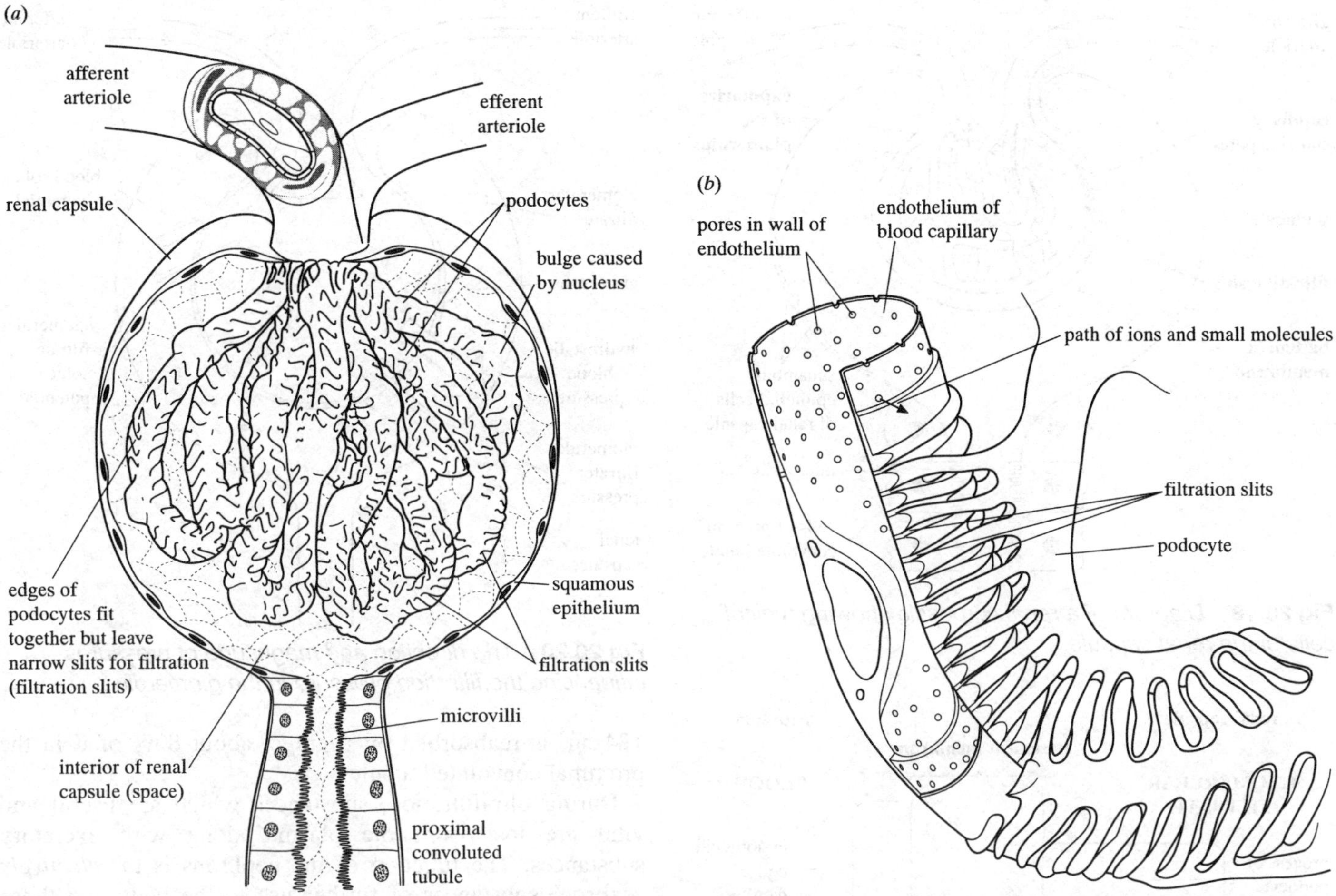

Fig 20.16 *(a) Structure of the renal corpuscle. The upper part shows afferent and efferent arterioles. Special epithelial cells called podocytes cover the outside surfaces of the capillaries of the glomerulus. The capillaries themselves are therefore hidden, although their outline is revealed. (Rather like a glove concealing fingers while revealing their shape. The glove is equivalent to the podocytes and the fingers are equivalent to the capillaries.) The lower part of the diagram shows the start of the proximal convoluted tubule, which has cuboidal epithelial cells with microvilli (a brush border). (Based on L. C. Junqueira & J. Carneiro (1980)* Basic Histology *3rd ed. Lange Medical Publications.) (*b*) Detailed view of podocytes and TS capillary. Note how the podocytes fit together like loosely interlocking fingers, leaving slits through which glomerular filtrate can pass on its way into the renal capsule.*

Fig 20.17 *Scanning electron micrograph of podocytes, ×900. The podocytes have a cell body with numerous braucle of varying size, which restrict the size of molecule able to pass across the membrane.*

concentrated to more concentrated solutions). As the blood flows from the afferent arteriole to the efferent arteriole through the glomerulus it loses water and small solute molecules, but the plasma proteins remain in the blood and increase in concentration by about 20% as a result of the loss of water. This makes the solute potential, and therefore the water potential, of the blood more negative and tends to decrease GFR. When all the forces are taken into account, the GFR is positive and fluid moves from the glomerulus to the renal capsule. The greater the water potential of the blood compared with the glomerular filtrate, the greater the filtration pressure and the GFR.

Filtration rate can be increased by raising blood pressure. It can also be raised by dilating the afferent arterioles (vasodilation) and therefore decreasing the resistance to the flow of blood into the glomerulus. A third regulatory mechanism is to increase the resistance in the efferent arterioles by constricting them (vasoconstriction).

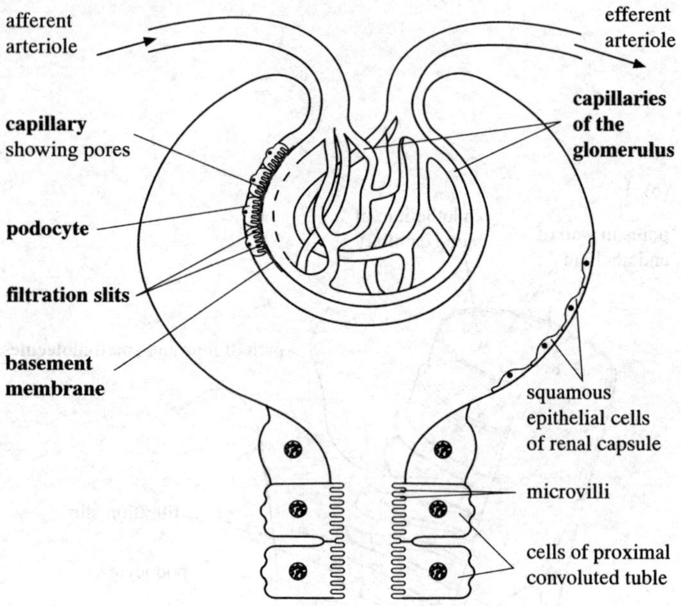

Fig 20.18 *Diagram of a renal corpuscle showing typical cells of the renal capsule.*

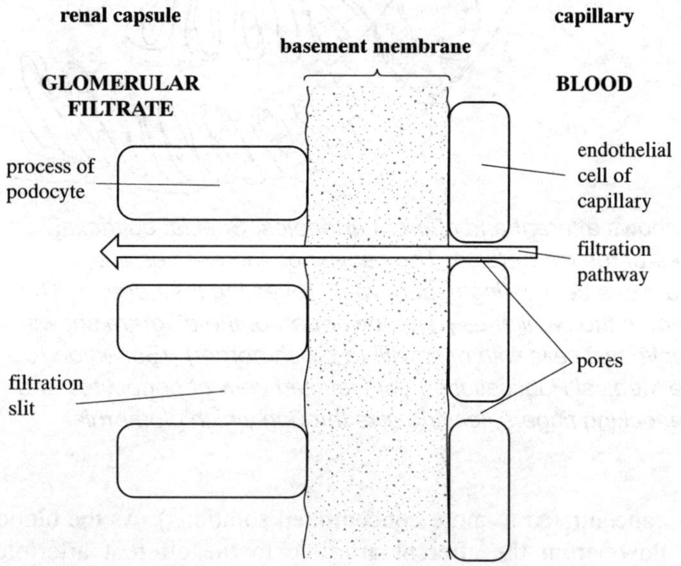

Fig 20.19 *Diagram showing the path taken by fluid (glomerular filtrate) as it passes from the plasma in a glomerular capillary to the lumen of a renal capsule.*

Vasodilation and vasoconstriction are under both nervous and hormonal control.

20.5.5 Selective reabsorption in the proximal convoluted tubule

Ultrafiltration produces about 125 cm^3 of glomerular filtrate per minute in humans. This is equivalent to about 180 dm^3 per day. Since only 1.5 dm^3 of urine is produced each day, a great deal of reabsorption must occur. In fact, of the 125 cm^3 of filtrate produced per minute

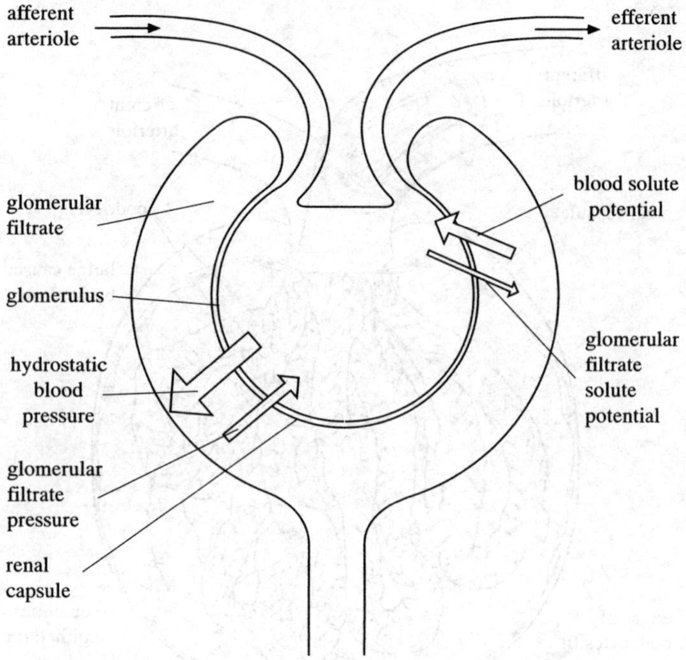

Fig 20.20 *The direction and magnitude of pressures influencing the filtration pressure in the glomerulus.*

124 cm^3 is reabsorbed on average, about 80% of it in the proximal convoluted tubule.

During ultrafiltration, substances which are useful and vital are lost from the plasma along with excretory substances. The function of the nephrons is to *selectively reabsorb* substances of further use to the body and those required to maintain the composition of the body fluids in a steady state. Further waste substances may be added to the tubules by active *secretion* from the blood capillaries surrounding the tubules.

Formation of urine therefore involves three key processes, namely ultrafiltration, selective reabsorption and secretion.

Analysis of fluid in the nephrons

By using extremely fine pipettes it has been possible to remove fluid from different parts of the nephron and to analyse its content in an attempt to find out what effect the different parts of the nephron have on the composition of the fluid. It is also possible to measure the rate of flow using the polysaccharide inulin as a tracer. It is injected into the blood from where it is filtered into the nephron. It is not reabsorbed from, or secreted into, the nephron so, as water is reabsorbed from the nephron, the concentration of inulin increases in proportion to the amount of water reabsorbed. As the amount of water decreases the flow rate of glomerular filtrate decreases.

In the example given in fig 20.21, the flow rate in the renal capsule is set at 100 arbitrary units. This is called the **flow rate index (FRI)**. Flow rates at other points can be compared with this. FRI gives a measure of the amount of water. For example, if it changes from 100 to 40, 60% of the water must have been reabsorbed. Using the figures in

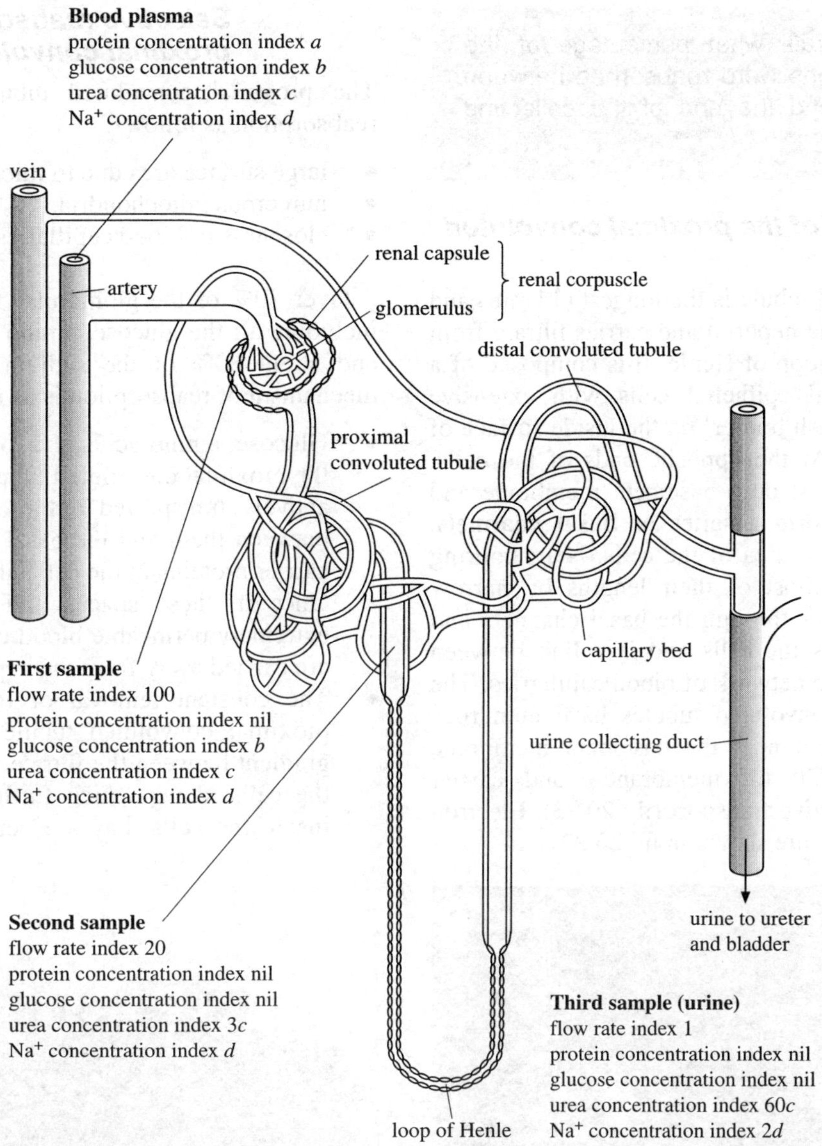

Fig 20.21 *Diagram of a single nephron and part of its blood supply.* (*Based on* Nuffield Advanced Science (Biological Science) Study Guide, *Penguin (1970) p.373.)*

fig 20.21, the amount of reabsorption of different solutes can be calculated as they move through the nephron. Try the following questions. If you cannot answer a question, look up the answer for an explanation and then try the next question before you look at further answers.

20.2 What happens to the concentration of solutes as fluid passes from the blood to the renal capsule? Explain.

20.3 In passing from the renal capsule to the end of the proximal convoluted tubule, the flow rate index changes from 100 to 20. What percentage of water has been reabsorbed back into the blood from the proximal convoluted tubule?

20.4 Only 20%, or one fifth, of the water remains at the end of the proximal convoluted tubule, so the concentration of all the solutes should have increased by 5 times, *unless* the proximal convoluted tubule has had an effect on the solute. Thus urea, for example, should have changed from a concentration index *c* to concentration index 5*c*. However, it has changed to 3*c*, only 3/5 or 60% of that predicted. This means that only 60% of the original urea remains and therefore 40% must have been reabsorbed. What percentage, if any, of the glucose and sodium ions were reabsorbed?

20.5 What changes, if any, took place in the amount of water, sodium ions and urea in the nephron between the end of the proximal convoluted tubule and the end of the collecting duct?

> **20.6** Overall, what percentage of the water and sodium ions were reabsorbed between the renal capsule and the end of the collecting duct?

Structure of the proximal convoluted tubule

The proximal convoluted tubule is the longest (14 mm) and widest (60 μm) part of the nephron and carries filtrate from the renal capsule to the loop of Henle. It is composed of a single layer of cuboidal epithelial cells with extensive microvilli forming a 'brush border' on the inside surface of the tubule (fig 20.22). At the opposite ends of the cells, their outer membranes rest on a basement membrane and are folded inwards to form a series of **basal channels**. These increase the surface area of the cells. Neighbouring cells are separated for most of their lengths by narrow spaces and fluid circulates through the basal channels and spaces. This fluid bathes the cells and is a link between them and the surrounding network of blood capillaries. The cells of the proximal convoluted tubules have numerous mitochondria concentrated near the basement membrane where they provide ATP for membrane-bound carrier molecules involved in active transport (fig 20.23). Electron micrographs of such cells are shown in fig 20.22.

Selective reabsorption in the proximal convoluted tubule

The proximal convoluted tubule cells are adapted for reabsorption as follows:

- large surface area due to microvilli and basal channels;
- numerous mitochondria;
- closeness of blood capillaries.

Over 80% of the glomerular filtrate is reabsorbed here, including all the glucose, amino acids, vitamins, hormones and about 80% of the sodium chloride and water. The mechanism of reabsorption is as follows.

- Glucose, amino acids and ions diffuse into the cells of the proximal convoluted tubule from the filtrate and are actively transported out of the cells into the spaces between them and the basal channels. This is done by carrier proteins in the cell surface membranes.
- Once in these spaces and channels they enter the extremely permeable blood capillaries by diffusion and are carried away from the nephron.
- The constant removal of these substances from the proximal convoluted tubule cells creates a diffusion gradient between the filtrate in the proximal tubule and the cells, down which further substances pass. Once inside the cells they are actively transported into the

(*a*)

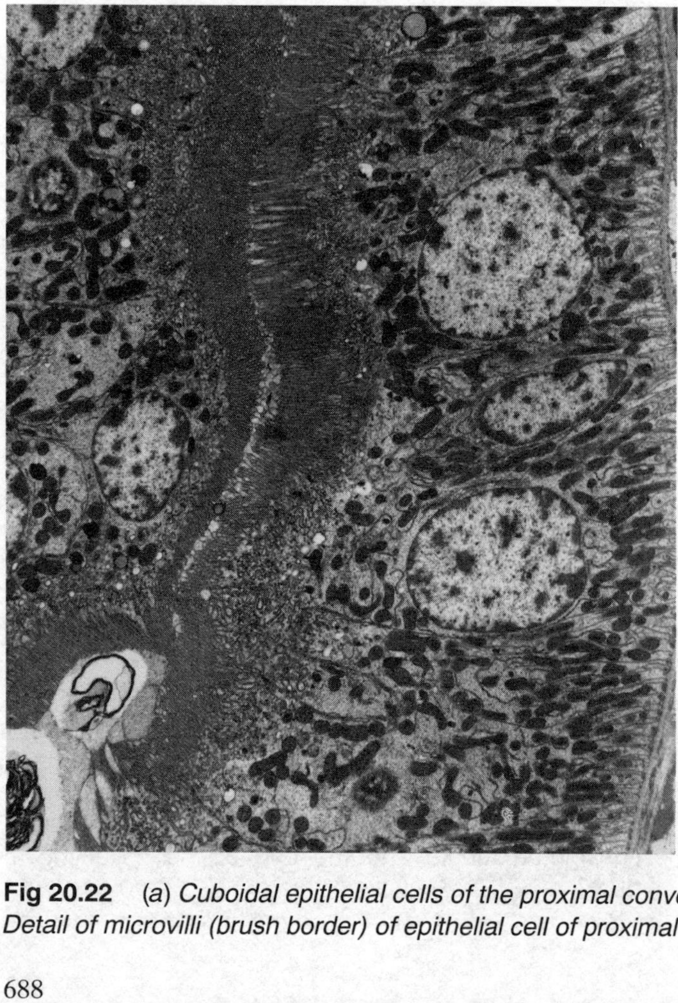

(*b*)

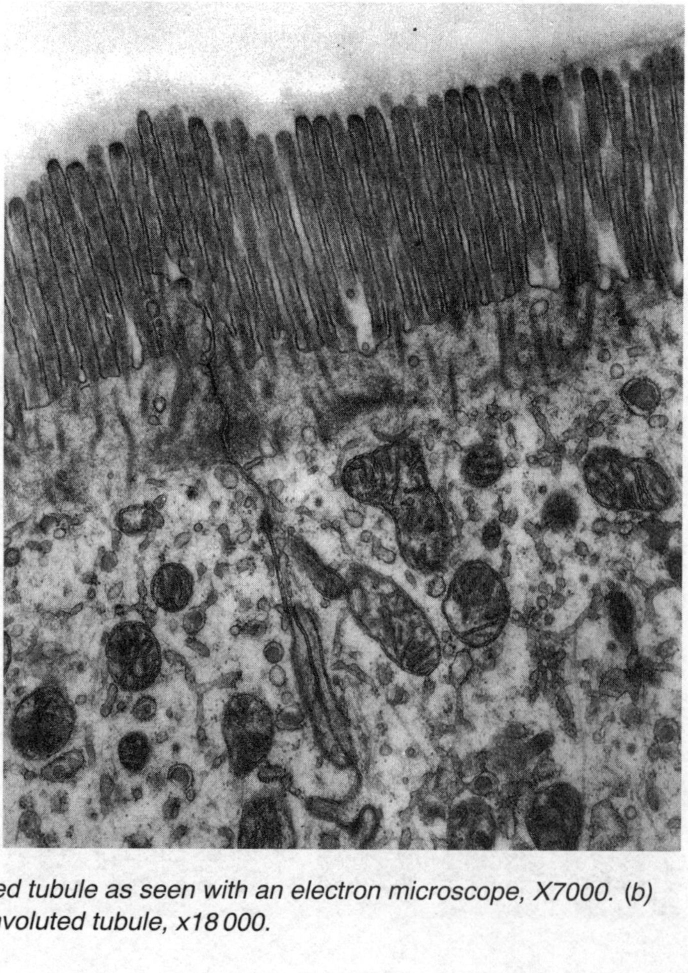

Fig 20.22 *(a) Cuboidal epithelial cells of the proximal convoluted tubule as seen with an electron microscope, X7000. (b) Detail of microvilli (brush border) of epithelial cell of proximal convoluted tubule, x18 000.*

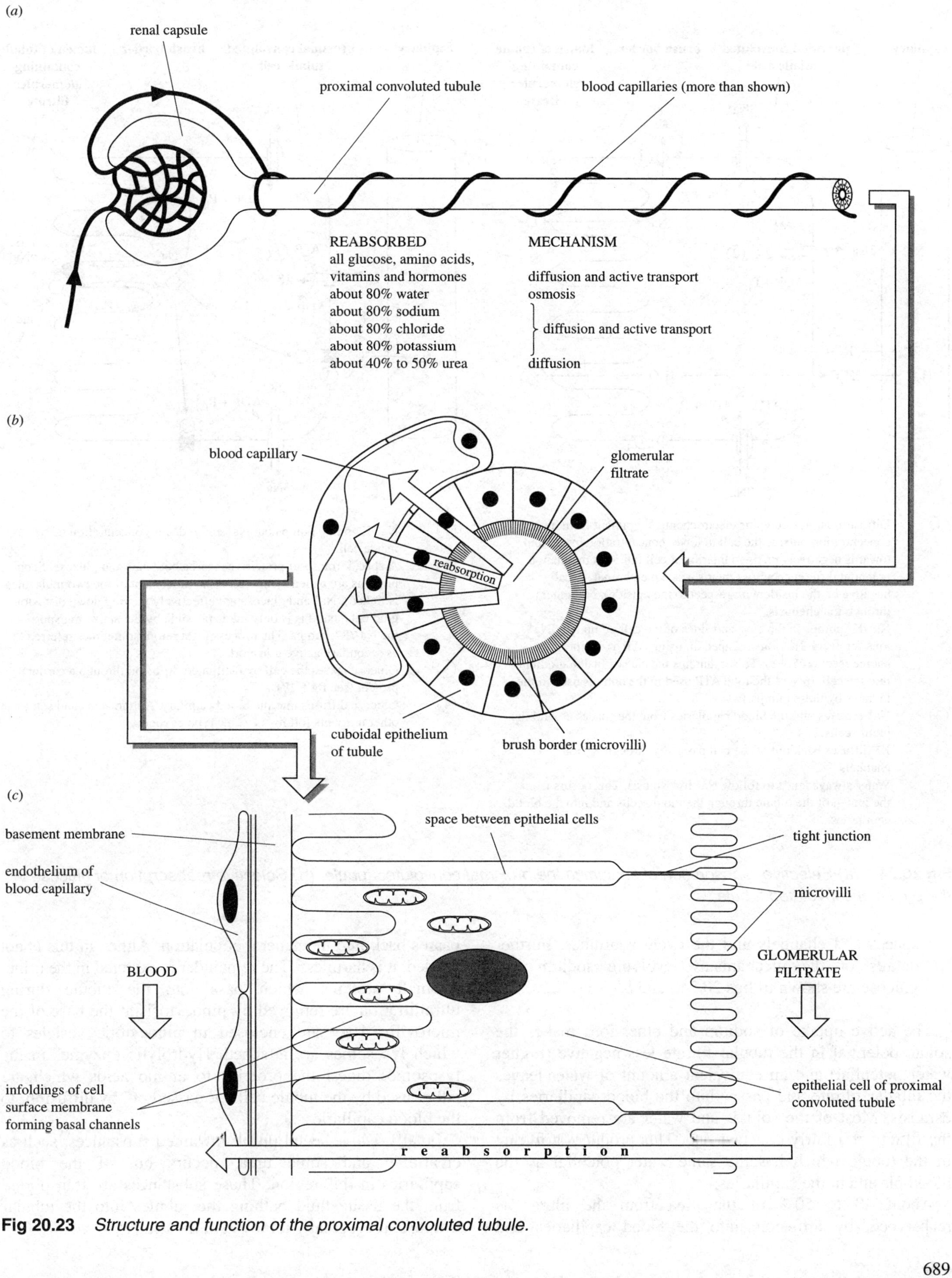

Fig 20.23 *Structure and function of the proximal convoluted tubule.*

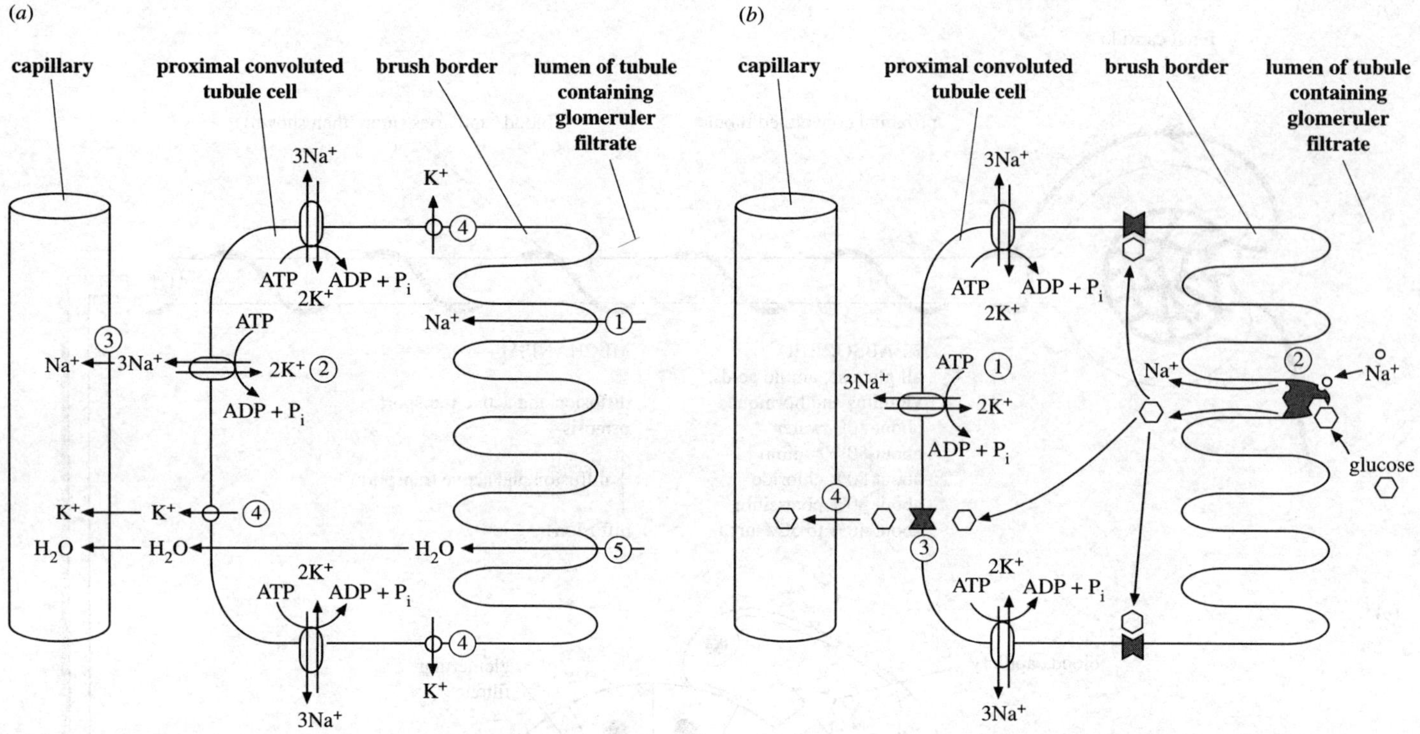

(a)

1. Diffusion of Na^+ down its electrochemical gradient from higher concentration outside the cell to lower concentration inside and towards negatively charged interior of cell (all cells maintain a potential difference across their surface membranes, usually negative on the inside with respect to the outside). Na^+ enters through ion channels.
2. Na^+/K^+ pumps in the base and sides of the cell pump out $3Na^+$ ions for every $2K^+$ ions pumped in, using ATP as an energy source (section 5.9.8). This maintains the Na^+ diffusion gradient into the cell. (6% of the total ATP used in the body is used in the kidneys by these pumps.)
3. Na^+ diffuses into the blood capillaries from the spaces around the tubule cells.
4. K^+ diffuses back out of the cell passively through K^+ ion channels.
5. Water always tends to follow Na^+ by osmosis. This occurs from the lumen of the tubule through the tubule cells and into the blood capillaries.

(b)

1. Na^+/K^+ pump pumps out Na^+ and reduces concentration of Na^+ inside cell.
2. A special transport protein reabsorbs both Na^+ and glucose. Such proteins are called symporters (the movement of the two molecules is linked). Na^+ and glucose are effectively moving down diffusion gradients, but this is only made possible by the active transport of the Na^+/K^+ pumps. The process is therefore sometimes referred to as secondary active transport.
3. Glucose leaves the cell by facilitated diffusion through a carrier protein (see fig 5.17).
4. Glucose diffuses into the blood capillary. Amino acids and some other nutrients follow the same type of route.

Fig 20.24 *(a) Selective reabsorption of sodium in the proximal convoluted tubule. (b) Selective reabsorption of glucose in the proximal convoluted tubule.*

spaces and channels and the cycle continues. Further details of the mechanisms involving sodium and glucose are shown in figs 20.24*a* and *b*.

The active uptake of sodium and other ions makes the solute potential in the tubular filtrate less negative (higher water potential) and an equivalent amount of water leaves the tubular filtrate and passes into the blood capillaries by osmosis. Most of the solutes and water are removed from the filtrate at a fairly constant rate. This produces a filtrate in the tubule which has the same water potential as the blood plasma in the capillaries.

About 40 to 50% of the urea from the filtrate is reabsorbed, by diffusion, into the blood capillaries and passes back into the general circulation. Although this is not needed, it is harmless. The remainder is excreted in the urine.

Small proteins which pass into the tubule during ultrafiltration are removed by pinocytosis at the base of the microvilli. They are enclosed in pinocytotic vesicles to which lysosomes are attached. Hydrolytic enzymes in the lysosomes digest the proteins to amino acids which are either used by the tubule cells or passed on, by diffusion, to the blood capillaries.

Finally, active secretion of unwanted substances, such as creatinine and some urea, occurs out of the blood capillaries in this region. These substances are transported from the tissue fluid bathing the tubules into the tubular filtrate and eventually removed in the urine.

20.5.6 The loop of Henle

The function of the loop of Henle is to conserve water. The longer the loop of Henle, the more concentrated the urine that can be produced. This is a useful adaptation to life on land. Birds and mammals are the only vertebrates which can produce a urine which is more concentrated than the blood and they are the only vertebrates with loops of Henle. The urine of a human can be 4 to 5 times as concentrated as the blood. The drier the natural habitat of an animal, the longer its loop of Henle. For example, the beaver, a semi-aquatic mammal, has a short loop of Henle and produces a large volume of dilute urine, whereas the desert-dwelling kangaroo rat and the jerboa (hopping mouse) have long loops of Henle and produce small volumes of highly concentrated urine. Their urine is 6 to 7 times more concentrated than human urine and they do not need to drink water. They get enough from food and metabolic water produced during cell respiration.

The loop of Henle, together with the capillaries of the vasa recta and collecting duct, creates and maintains an osmotic gradient in the medulla which extends from the cortex to the tips of the pyramids (see fig 20.12).

The gradient extends across the medulla from a less concentrated salt solution at the cortex to a more concentrated salt solution at the tips of the pyramids. Water leaves the nephrons by osmosis in response to this gradient, as will be explained later, making the fluid inside the nephrons, which becomes the urine, more concentrated.

The loop of Henle has three distinct regions, each with its own function. These are:

- the **descending limb** which has thin walls;
- the **thin ascending limb** – this is the lower half of the ascending limb and has thin walls like the descending limb;
- the **thick ascending limb** – this is the upper half of the ascending limb and has thick walls.

The descending limb is highly permeable to water and permeable to most solutes. Its function is to allow substances to diffuse easily through its walls. Both parts of the ascending limb are almost totally impermeable to water. The cells in the thick part can actively reabsorb sodium, chloride, potassium and other ions from the tubule. Normally water would follow by osmosis the movement of these ions into the cells, but this cannot occur because the cells are impermeable to water as stated. The fluid in the ascending limb therefore becomes very dilute by the time it reaches the distal convoluted tubule.

The loop of Henle as a countercurrent multiplier

It has been noted that there is a gradient of salts across the medulla. The gradient is from about 300–1200 mOsm dm^{-3}, but for simplicity this will be referred to as 300 to 1200 units. We can now examine how this is achieved. It is best to imagine a starting situation where the whole loop of Henle is filled with fluid at a concentration of 300 units (the normal concentration of tissue fluid and blood), and that this is in equilibrium with tissue fluid of the same concentration in the surrounding medulla. From this starting point a gradient has to be built up across the medulla. The process can be thought of as starting in the thick ascending limb. This carries out active transport of sodium ions out of the cells of the ascending limb into the tissue fluid of the medulla by means of a sodium–potassium pump (section 5.9.8, active transport). Sodium ions then diffuse from the fluid in the ascending limb into the cells of the ascending limb to replace the sodium ions lost. As they do they pass through a carrier protein which also accepts chloride and potassium ions. These are cotransported into the cells with sodium ions, against their concentration gradients, and diffuse out with sodium ions into the medulla. The whole process is driven by the sodium–potassium pump and results in sodium, potassium and chloride ions accumulating in the tissue fluid of the medulla. The concentration of these ions in the ascending limb *decreases* while the concentration in the medulla *increases* because water cannot leave the ascending limb by osmosis due to its impermeable wall. On the other hand, the descending limb is very permeable to water, and not very permeable to ions. A difference of about 200 units between the ascending limb and the medulla can be maintained by the pump, as shown in fig 20.25.

Box 20.1 The use of osmolarity

The concentrations of particles (ions or molecules) in solutions can be described in terms of solute potential or in terms of osmolarity. It is traditional in animal physiology to use osmolarity rather than solute potential. Because every particle (ion or molecule) in a solution contributes to the solute potential of that solution, osmolarity is used to refer to the total number of moles of all particles in 1 dm^3 of solution. One osmole is 1 mole of any combination of particles. For example, 1 mole of KCl produces a solution which has an osmolarity of 2 osmoles. This is because 1 mole of KCl will dissolve to produce 1 mole of K^+ ions and 1 mole of Cl^- ions, that is 2 moles of particles are present.

Solutions with the same osmolarity have the same solute potential. Osmolarity is measured as osmoles per dm^3 or, more appropriately, as milliosmoles per dm^3 in the case of glomerular filtrate. This is written as mOsm dm^{-3}. Normal blood plasma and tissue fluid osmolarity is about 300 mOsm dm^{-3}. Normal urine is 300–1000 mOsm dm^{-3}. The osmolarity of the tissue fluid in the kidney medulla is about 1200 mOsm dm^{-3}. Seawater is about 1000 mOsm dm^{-3}.

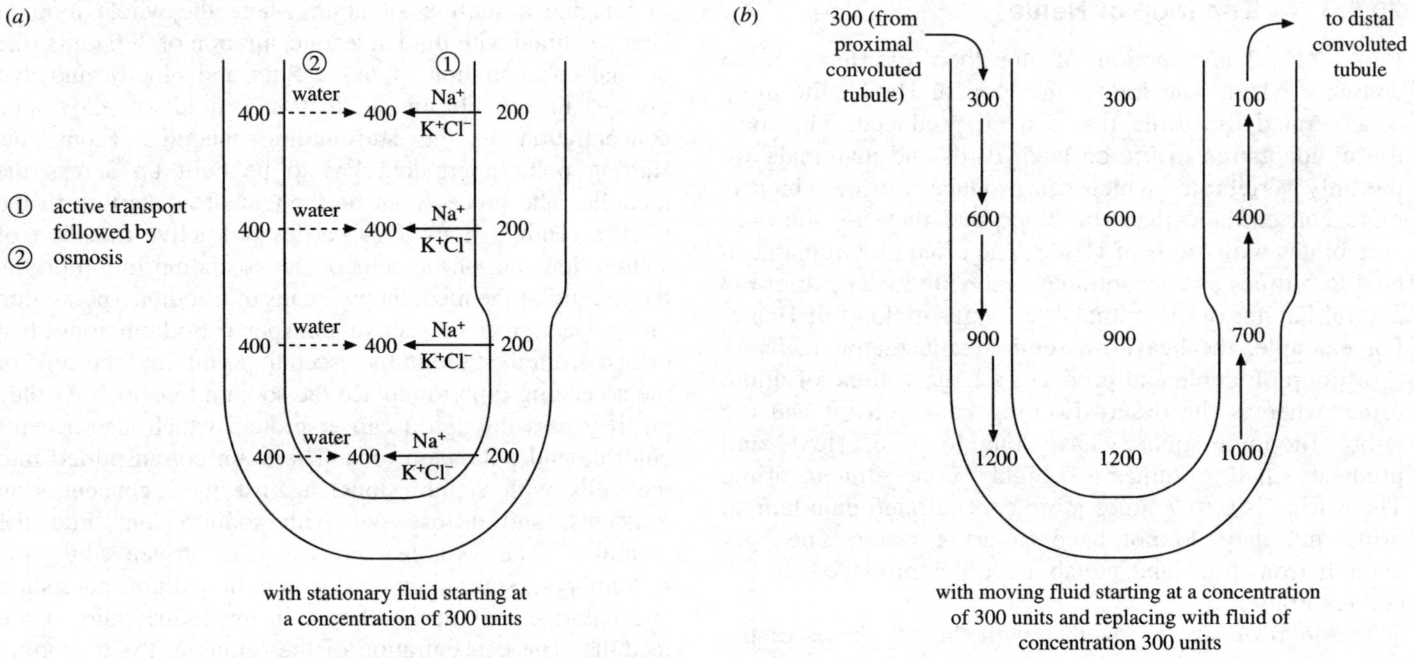

Fig 20.25 *Movement of ions and water from the loop of Henle into the medulla of the kidney.* (a) *The situation that would develop if the fluid in the loop of Henle were stationary.* (b) *The situation that develops in reality due to movement of fluid round the loop of Henle. Numbers refer to concentration of the fluid in milliosmoles per* dm^3.

The high concentration of salts in the medulla makes water leave the descending limb by osmosis, so the fluid in the descending limb gets more concentrated. If the fluid *did not move round* the loop of Henle the situation shown in fig 20.25*a* would develop. The pump cannot work fast enough to raise the concentration higher than 400 units because of leakage of ions back into the thick ascending limb. Now, imagine the real situation of fluid moving round the loop. The higher the fluid goes in the thick ascending limb the more sodium ions will have been removed, so the more dilute it gets. A gradient is established in the ascending limb. There is always a 200-unit difference between the ascending limb and the medulla due to the pump, so a similar gradient develops in the medulla. At the same time, more and more water is removed from the fluid as it passes down the descending limb, so the concentration of the fluid in the descending limb increases from top to bottom. The result of all this is the situation shown in fig 20.25*b*.

Some ions do move out of the descending limb, but water moves much more rapidly. The water does not dilute the medulla because it is carried away by the blood vessels of the vasa recta. The vasa recta follow the loops of Henle and the changes in the composition of the blood in the vasa recta are similar to those in the medulla. The flow of blood is slow, allowing it to come into equilibrium with the medulla at all levels. It does not therefore disrupt the gradient established in the medulla. The whole process is dynamic, meaning that it would soon be disrupted if the fluid stopped moving, or the pumps stopped working, or the blood in the vasa recta stopped flowing.

The loop of Henle is known as a countercurrent multiplier. The term countercurrent refers to the fact that the fluid flows in opposite directions in the two sides of the loop, down one side and up the other. The multiplier effect is seen by comparing figs 20.25*a* and *b*. In fig 20.25*b* a gradient from 300 to 1200 units is created by a pump which is only capable of maintaining a difference of 200 units between one side of the loop of Henle and the other. The effect of the pump is multiplied by constant removal of sodium and other ions from the ascending side and their replacement from the proximal convoluted tubule on the descending side.

20.5.7 The distal convoluted tubule and collecting duct

In the last two regions of the nephron, the distal convoluted tubule and the collecting duct, fine tuning of the body fluid composition is achieved. The proximal convoluted tubule always functions in the same way, removing, for example, the same proportions of water and salts all the time as described in section 20.5.5. It functions as a coarse control, reabsorbing into the blood the bulk of the substances required.

Fine control of the precise amounts of water and salts reabsorbed is important in **osmoregulation**. This is one role of the distal convoluted tubule and collecting duct. They also control blood pH. These functions are described in more detail in sections 20.6 and 20.7.

The cells of the distal tubule have a similar structure to those of the proximal tubule, with microvilli lining the inner surface to increase the surface area for reabsorption, and numerous mitochondria to supply energy for active transport. The collecting duct carries fluid from the outer region of the medulla, next to the cortex, to the pyramids (fig 20.12). As the fluid moves down the collecting duct, the tissue fluid in the medulla surrounding the duct gets more and more concentrated, as noted above. Water therefore leaves the collecting duct by osmosis. The final concentration of the urine can be as high as the medulla, about 1200 units, although the actual amount of water lost is controlled by ADH as explained in section 20.6.

A comparison of the composition of plasma and urine is shown in table 20.4. It shows that the urine has a higher concentration of all solutes except sodium ions and normally lacks protein and glucose.

20.6 Osmoregulation, antidiuretic hormone (ADH) and the formation of a concentrated or dilute urine

The body maintains the solute potential of the blood at an approximately steady state by balancing water uptake from the diet with water lost in evaporation, sweating, egestion and urine, as shown in fig 20.9. The precise control of solute potential, however, is achieved primarily by the effect of a hormone called **antidiuretic hormone** (**ADH**). Diuresis is the production of large amounts of dilute urine. Antidiuresis is therefore the opposite. ADH is antidiuretic in its effects, so has the effect of making urine more concentrated. ADH is a peptide (table 17.10). It is sometimes known as **vasopressin**.

ADH is made in the hypothalamus and passes the short distance to the posterior pituitary gland by a process called neurosecretion. This is explained in section 17.6.2.

When the blood becomes more concentrated (solute potential more negative), as in a situation where too little water has been drunk, excessive sweating has occurred or large amounts of salt have been eaten, osmoreceptors in the hypothalamus detect a fall in blood solute potential. Osmoreceptors are special receptors which are extremely sensitive to changes in blood concentration. They set up nerve impulses which pass to the posterior pituitary gland where ADH is released. ADH travels in the blood to the kidney where it increases the permeability of the distal convoluted tubule and collecting duct to water. It does this by bringing about an increase in the number of water channels in the membranes lining the tubules. Water channels are proteins, like ion channels. They are manufactured inside the cell and 'stored' in the membranes of small Golgi vesicles which accumulate in the cytoplasm. When ADH binds to its specific receptors in the cell surface membrane it acts, via cyclic AMP (the second messenger system described in sections 17.6.1, 17.6.2 and 18), to stimulate the fusion of these vesicles to the cell surface membrane. When ADH secretion is stopped, the process goes into reverse, and by a process of endocytosis the vesicles are taken back into the cell ready for recycling next time ADH is secreted.

Table 20.4 The composition of plasma and urine and changes in concentration occurring during urine formation in humans.

	Plasma %	*Urine %*	*Increase*
water	90	95	–
protein	8	0	–
glucose	0.1	0	–
urea	0.03	2	67×
uric acid	0.004	0.05	12×
creatinine	0.001	0.075	75×
Na^+	0.32	0.35	1×
NH_4^+	0.0001	0.04	400×
K^+	0.02	0.15	7×
Mg^{2+}	0.0025	0.01	4×
Cl^-	0.37	0.60	2×
PO_4^{3-}	0.009	0.27	30×
SO_4^{2-}	0.002	0.18	90×

In the presence of ADH, the increased number of water channels allows water to move from the glomerular filtrate into the cortex and medulla by osmosis, reducing the volume of the urine and making it more concentrated (fig 20.26). The water is carried away in the blood.

ADH also increases the permeability of the collecting duct to urea, which diffuses out of the urine into the tissue fluid of the medulla. Here it increases the osmotic concentration, resulting in the removal of an increased volume of water from the thin descending limb.

The opposite occurs when there is a high intake of water. The solute potential of the blood begins to get less negative. ADH release is inhibited, the walls of the distal convoluted tubule and collecting duct become impermeable to water, less water is reabsorbed as the filtrate passes through the medulla and a large volume of dilute urine is excreted (fig 20.26).

Table 20.5 shows a summary of the events involved in regulating water balance, and the control mechanisms involved in regulating water balance are shown in fig 20.27. The hypothalamus also contains a 'thirst centre'. When blood solute potential is very negative, the thirst centre stimulates the sensation of thirst.

Failure to release sufficient ADH leads to a condition known as **diabetes insipidus** in which large quantities of dilute urine are produced (diuresis). The fluid lost in the urine has to be replaced by excessive drinking.

20.7 Control of blood sodium level

The maintenance of the plasma sodium level at a steady state is controlled by the steroid hormone **aldosterone** which also influences water reabsorption. It is secreted by the cortex (outer) region of the adrenal glands. A decrease in blood sodium leads to a decrease in blood volume because less water enters the blood by osmosis.

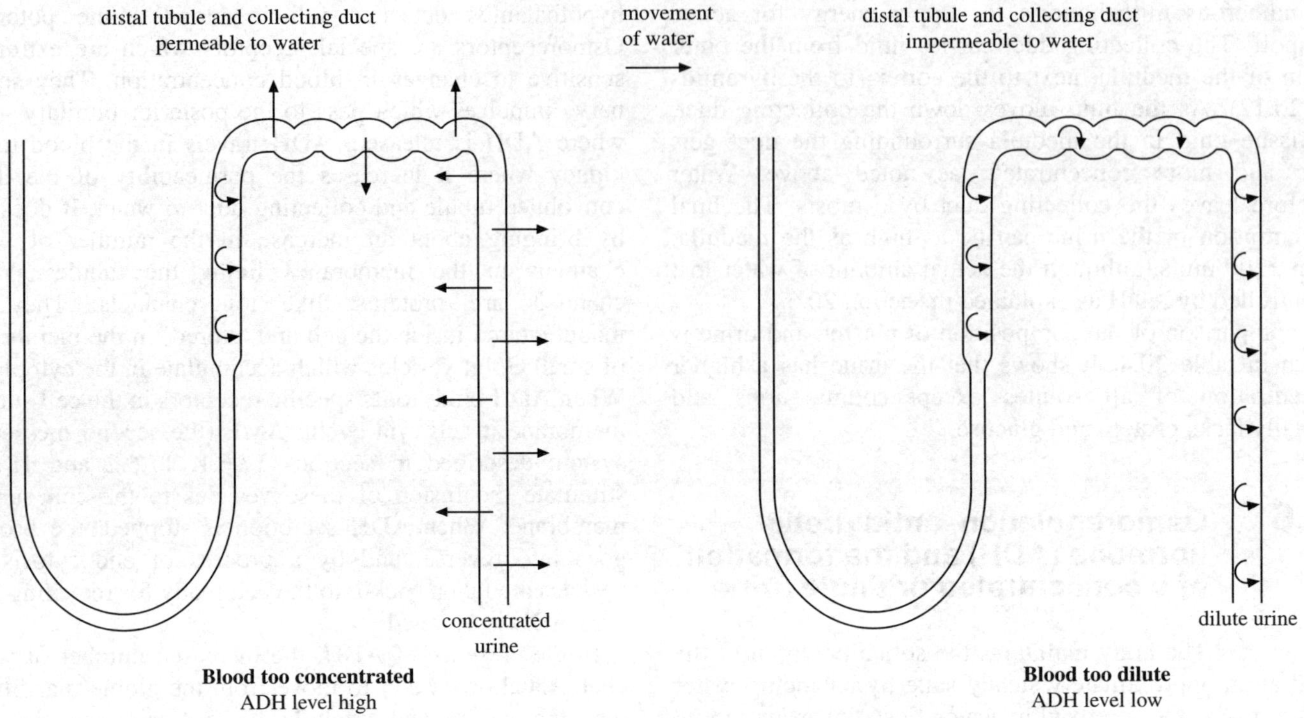

Fig 20.26 *Diagram illustrating the effect of ADH on the permeability of the distal convoluted tubule and collecting duct to water.*

Table 20.5 Summary of the changes produced in the distal convoluted tubule and collecting duct in response to ADH.

Blood concentration	*Blood solute potential*	*ADH*	*Epithelium*	*Urine*
rises	falls (more negative)	released	permeable	concentrated
falls	rises (less negative)	not released	impermeable	dilute

This in turn reduces blood pressure. This decrease in pressure and volume stimulates a group of secretory cells, the **juxtaglomerular complex**, situated between the distal convoluted tubule and the afferent arteriole (fig 20.28), to release an enzyme called **renin**. Renin activates a protein in the blood plasma, produced in the liver, to form the active hormone **angiotensin**, and this releases aldosterone from the adrenal cortex. Aldosterone travels in the blood to the distal convoluted tubule of the kidney. Here it stimulates the sodium–potassium pumps in the cells of the tubule, resulting in more sodium ions being pumped out of the distal convoluted tubule and into the blood capillaries around the tubule. Potassium moves in the opposite direction. This is an example of active transport.

Aldosterone also stimulates sodium absorption in the gut and decreases loss of sodium in sweat; both these effects tend to raise blood sodium levels. This in turn causes more water to enter the blood by osmosis, raising its volume and hence its pressure.

20.8 Control of blood pH

pH is a measure of hydrogen ion concentration. A neutral pH is 7.0, an acid pH is lower than 7 and a basic (or alkaline) pH is higher than 7. Some chemicals have the ability to resist pH changes in solution. These are called buffers. The normal pH of the blood plasma is 7.4. This must be kept to within very narrow limits. One reason is that enzymes and other proteins are easily denatured by changes in pH and this could prove fatal. Many other changes in body chemistry would also be affected by a large change in pH.

The body produces more acids than bases as a result of its chemistry, so the problem is usually one of reducing acidity. One factor which tends to increase acidity is production of carbon dioxide during cell respiration. This can dissolve to form a weak acid, carbonic acid, H_2CO_3. This dissociates into hydrogen ions, H^+, and hydrogencarbonate ions, HCO_3^-. The hydrogen ions are buffered by haemoglobin as explained in section 14.8.3. A rise in carbon dioxide concentration brings about a reflex response which causes an increased breathing rate. This helps to get

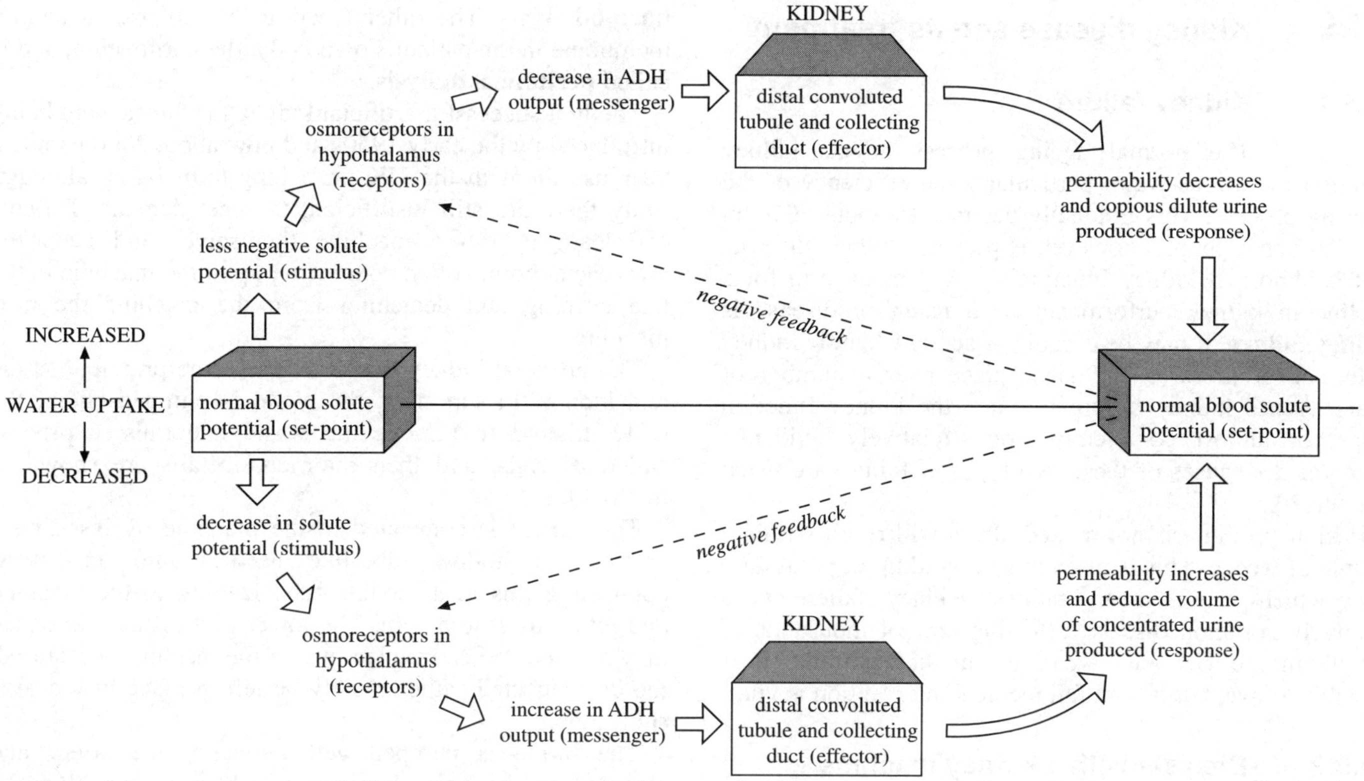

Fig 20.27 *Summary diagram of the control of blood solute potential.*

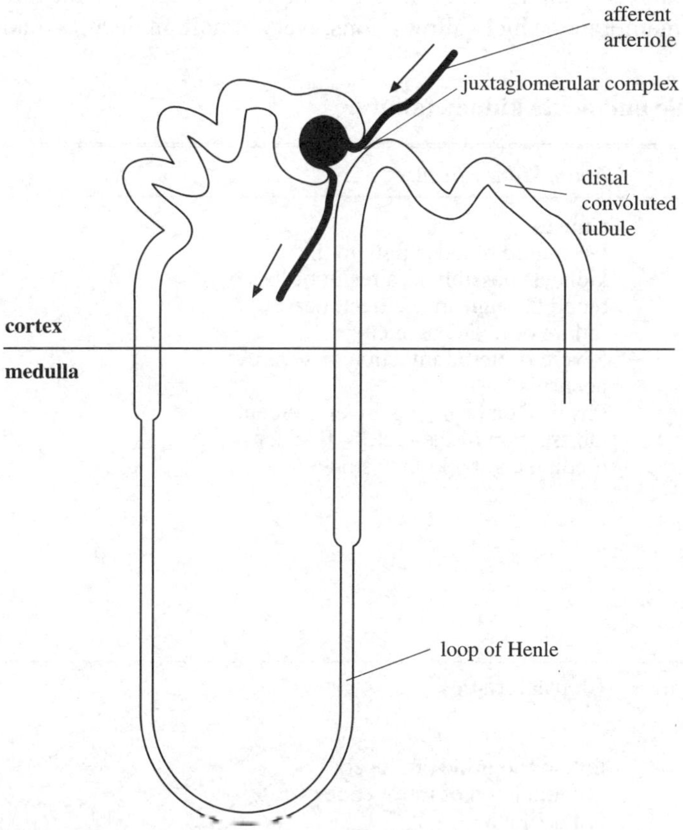

Fig 20.28 *Location of the juxtaglomerular cells in the kidney. The cells are found in the wall of the afferent arteriole and are sensitive to blood sodium level and blood pressure.*

rid of the excess carbon dioxide. The mechanism is explained in section 9.5.5. Hydrogencarbonate ions can also act as a buffer because at high concentrations of hydrogen ions, they combine with the hydrogen ions to form carbonic acid.

Hydrogencarbonate and phosphate buffers in the blood help to prevent excess hydrogen ions, produced by metabolic activities, from decreasing the pH of the blood. Changes in blood chemistry which would change the plasma pH from its normal level of 7.4 are also counteracted by the proximal and distal convoluted tubules, and the collecting duct, in two ways.

- If the blood starts to become too acidic, hydrogen ions are secreted by active transport across the cell surface membranes of the tubule or collecting duct cells from the blood into the tubules or collecting duct. If the source of hydrogen ions is carbon dioxide then the hydrogencarbonate ions also generated will return to the blood by diffusion. The reverse may happen if pH rises. The pH of the urine can vary from 4.5 to 8.5 as a result of these changes. (Note that a pH of 4.5 is 1000 times more acidic than a pH of 7.5 because the hydrogen ion concentration changes by a factor of 10 for each change in pH of 1 unit.)
- A fall in pH also stimulates the kidney cells to produce the base ion ammonium (NH_4^+) which combines with acids brought to the kidney and is then excreted as ammonium salts.

20.9 Kidney disease and its treatment

20.9.1 Kidney failure

The normal ageing process affects kidney function in various ways, particularly the efficiency of the filtering process. This gradually declines to about 50% by age 70. Some people, however, experience kidney disease, that is abnormal kidney functioning. A general term for a decline in kidney performance as a result of disease is **kidney failure**. It may be chronic or acute. **Chronic** kidney failure is progressive and takes place over a number of years. **Acute** kidney failure is when the kidney function stops, or almost completely stops, relatively suddenly. Some of the causes of these two types of failure are listed in table 20.6.

If kidney failure is not treated, death will result within a couple of weeks. This is often due to build up of potassium ions which causes heart failure. Kidney failure is a relatively common disease, affecting tens of thousands of people in the UK each year. If one kidney fails, it is possible to live, but if both fail medical intervention is vital.

20.9.2 Dialysis with a kidney machine – haemodialysis

There are two forms of dialysis. One uses an artificial membrane in a 'kidney machine' and is called **haemodialysis**. The other (section 20.9.3) uses a natural membrane in the patient's own body, the peritoneum, and is called **peritoneal dialysis**.

The first successful artificial kidney machines were being introduced by the early 1950s and now about 2500 people a year use them in the UK on a long-term basis, although sadly there are still insufficient to meet demand. Patients can learn to run a machine themselves and keep the machine at home, often connecting up to the machine in the late evening and detaching from the machine the next morning.

The artificial kidney works on the same principle as the real kidney. In summary, the blood is pumped out of the body, filtered to remove the waste materials, a process called **dialysis**, and then returned. Details are shown in fig 20.29.

The patient is connected to the machine by inserting a catheter (a hollow tube-like needle) into an artery, connecting this to a flexible tube leading to the machine and returning it to a vein. The lower part of the arm or leg may be used. When frequent use of the machine is required, the disconnected catheters may be left in place linked by a short tube.

The blood is pumped gently out of the artery and returned to the vein. Heparin is added to the blood to prevent clotting. The blood circulates slowly through dialysis tubing. This is an artificial partially permeable membrane which allows ions, very small molecules and

Table 20.6 Some common causes and characteristics of chronic and acute kidney failure.

Chronic kidney failure	*Acute kidney failure*
Causes Bacterial infection of the pelvis and surrounding tissue Nephritis – inflammation of the glomeruli due to antibodies produced against certain bacterial infections such as throat infections Damage due to high blood pressure Damage due to obstruction in the ureters, bladder or urethra, e.g. by kidney stones (these may appear anywhere in the urinary tract but are most common in the pelvis) Sugar diabetes Atherosclerosis (reduces blood supply)	**Causes** Decreased blood supply to the kidneys, possibly as a result of loss of blood through an accident, heart failure or toxic chemicals Severe bacterial infection or severe nephritis Physical damage, e.g. in an accident Obstruction of the ureters, bladder or urethra, e.g. by kidney stones
Characteristics Progressive destruction of nephrons leading to: reduced quantity of urine dilute urine dehydration salt imbalance severe high blood pressure coma and convulsions	**Characteristics** little or no urine produced accumulation of nitrogenous waste in the blood salt imbalance pain Often reversible if treated quickly

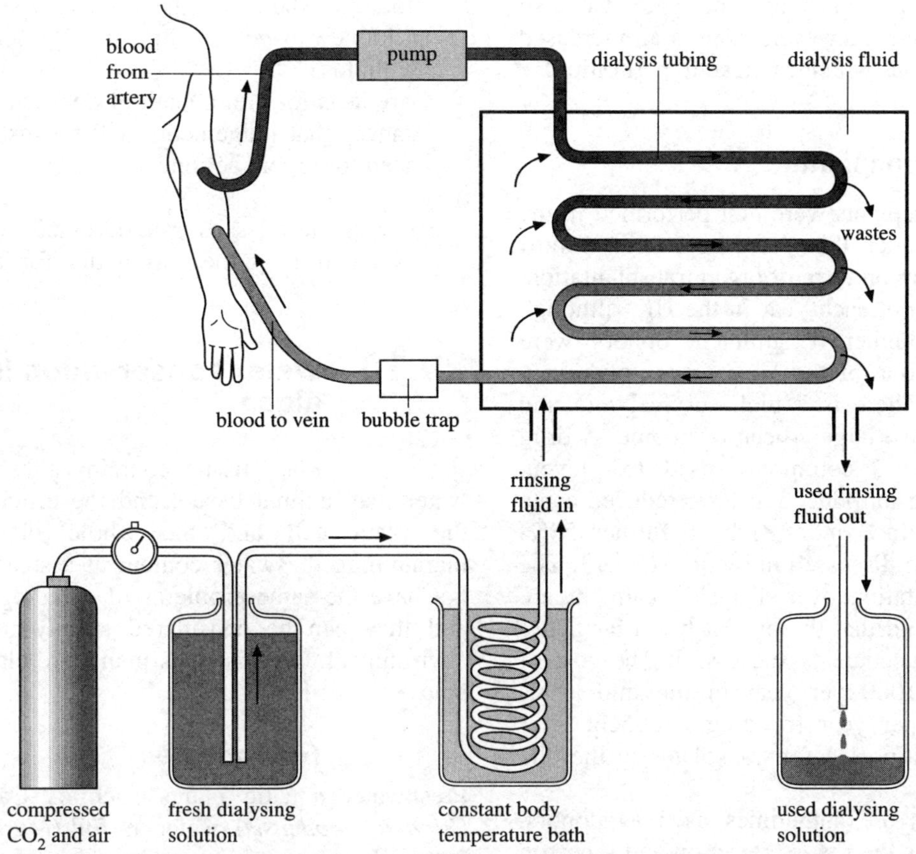

Fig 20.29 *Haemodialysis machine. Blood passes from an artery, through a coiled tube and back into a vein. The tube is a partially permeable membrane and is surrounded by dialysate. Waste products filter out into the dialysate.*

water to diffuse through it. Blood cells, platelets and protein molecules are too large to escape from the patient's blood. The tubing is bathed on the outside by a dialysing solution which has:

- the correct temperature;
- the correct ionic balance, particularly Na^+, K^+, Cl^-, Mg^{2+}, Ca^{2+}, and HCO_3^- (in the form of acetate, an organic anion);
- additional nutrients, such as glucose, which help to maintain the correct solute potential;
- the correct pH and buffering capacity.

Exchange between the blood and the dialysing solution (dialysate) takes place until an equilibrium is reached. Overall, unwanted substances are removed, particularly urea and excess sodium and potassium, and needed substances are kept. Note that the process is simpler than the real kidney because ultrafiltration does not occur and reabsorption of useful substances is not necessary.

The process takes 6 to 8 hours and is usually done at least twice a week. The solute potential of the dialysate is kept *less* negative (its water potential is higher) than that of the blood, despite the addition of glucose to the dialysate. Water would therefore tend to enter the blood. However, the blood pressure is raised to reverse this trend by squeezing the tube returning blood to the vein with a clip. The higher the pressure the greater the amount of water that will leave the blood. Regulation can therefore be achieved.

Acetate in the dialysate can be converted to hydrogencarbonate ions in the body. This restores the body's own buffering capacity which tends to get used up during periods between treatments.

20.9.3 Peritoneal dialysis – use of the peritoneum, a natural membrane

This can be carried out in hospital. A thin plastic tube is inserted into the abdominal cavity through a small slit in the abdomen wall and can be left in permanently. The peritoneal membrane, or peritoneum, which lines the abdominal cavity is the dialysing membrane and is partially permeable. Dialysis fluid is added to the abdominal cavity down the tube and left for several hours before removal. Exchange takes place between the fluid and the tissue fluid in the rest of the abdomen. The fluid can be replaced regularly, 3 or 4 times a day. In between, the patient can be mobile and free to live a relatively normal life. For this reason it is described as **continuous ambulatory peritoneal dialysis**, or **CAPD** (ambulatory means moving about). Many patients prefer this to having to be connected to a kidney machine. The method is also simpler and less expensive. However, there is an increased risk of dangerous infections which can lead to peritonitis.

20.9.4 Kidney transplants

Kidney transplants were first performed in the UK in the 1960s and since then have become the most successful and most common form of organ transplantation. About 1800 are carried out each year in the UK, although this figure would be higher if sufficient donors were available. The UK has one of the lowest rates of kidney transplants in Europe. There is a high survival rate and problems of rejection have largely been overcome. A drug called cyclosporin A was commonly used to prevent rejection, but a new drug introduced in 1996 reduced acute rejection (rejection within 3 months) by a further 50%. Problems of rejection are discussed in section 25.7.13. The advantages of transplantation over dialysis treatment are that it is a lot cheaper and that the patient has a far better quality of life if the procedure is successful. The cost of dialysis was about £20 000 per year in the mid-1990s compared with £3000 per year for drug treatment after transplantation. The waiting list for transplants in the UK was 3700 in 1989.

Close living relatives are sometimes used as donors, since this greatly reduces the risk of rejection and a person can live normally with just one kidney. National and international systems for locating suitable donors have been set up. A patient is 'tissue typed' so that their antigens can be matched as closely as possible to those of the donor. If a recently deceased donor is to be used, tests for brain death of the donor are carried out and the kidneys are then removed and packed in ice. Meanwhile tissue typing and tests for hepatitis and HIV are carried out. The kidney is used as soon as possible, and must be used within 48 hours. The recipient may receive only a few hours notice of the operation.

Many moral and ethical issues are raised by kidney transplantation and dialysis. Some of the main ones are as follows.

- How much money? How much should we pay through taxes to the National Health Service?
- Sharing the money. How should existing resources be used? Should other branches of medicine receive more money, e.g. preventive medicine?
- Who? Who should be chosen for dialysis treatment and transplantation surgery when there are waiting lists? Should any people receive priority?
- Is it fair to ask a living relative to be a donor?
- Should more be done to promote the carrying of donor cards?
- How should the problem of getting permission from recently bereaved families for using someone's kidneys be handled?
- Should there be a market in kidneys for sale? In some countries there have been many cases of people selling a kidney to get out of debt. Is there anything wrong with this?
- Are tests for brain death of donors adequate? Is there a danger that judgements will be rushed because of the need for urgent action?

Some of these issues are addressed in *SATIS 16–19* unit 7, available through the Association for Science Education.

20.10 Water conservation in plants and algae

Plant tissue contains a higher proportion of water than animal tissue, and the efficient functioning of the plant cell and the whole plant depends upon maintaining the water content at a steady state. Plants do not have the same problems of osmoregulation as animals and they can be considered simply in relation to their environment. On this basis plants are classified as outlined below.

Hydrophytes

Freshwater aquatic plants such as Canadian pondweed (*Elodea canadensis*), water milfoil (*Myriophyllum*) and the water lily (*Nymphaea*) are classed as hydrophytes and have fewer osmoregulatory problems than any of the other plant types. Plant cells in fresh water are surrounded by a solution of higher water potential and water enters the cell by osmosis. The water passes through the freely permeable cell wall and the partially permeable cell surface and tonoplast membranes. As the volume of the vacuole increases due to water uptake, it generates a **turgor pressure** (pressure potential). The cell becomes turgid and a point is reached when the water potential has increased to equal that of the surrounding water (about zero) and no further water enters (see section 13.1.5).

Halophytes

Many algae live in sea water and they are the major autotrophic organisms of the seashore. The distribution of algal species down the shore is determined by many factors, including tolerance to wave action, desiccation when exposed by tides, and the nature of their photosynthetic pigments. In all cases these species can tolerate increases in salinity and their main osmoregulatory problem is the prevention of water loss by evaporation. Channel wrack (*Pelvetia canaliculata*) occupies the highest algal zone on sheltered rocky shores surrounding the British Isles, and its tolerance of dry conditions is aided by thick cell walls, a thick covering of mucilage and a stipe shaped as a channel. Fig 20.30 shows the rate of water loss and degree of tolerance of four common British species of seaweed which are zoned according to their ability to retain water when exposed to air.

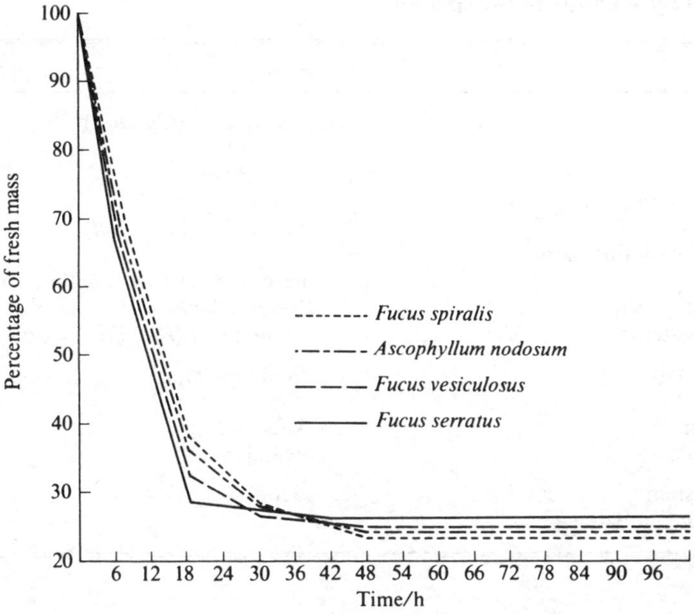

Fig 20.30 *Graph showing the comparative rates of water loss upon exposure to air for four species of algae found on the seashore. (From J. Zaneveld (1937)* J. Ecol., ***25***, 431–68.*)*

Halophytes, however, are defined as plants inhabiting areas of high salinity, such as those encountered in estuaries and salt marshes where salinity is constantly changing and may exceed that of sea water. Whilst the shoot system is not regularly exposed to high salinities, the root system must tolerate the increased salinities of the sand and mud which accompany hot windy periods when the tide is out. It was thought that these plants must tolerate periods of 'physiological drought' when water is unavailable to the tissues due to the low water potential of the environment of the roots. However, this does not seem to be the case and high transpiration rates and low water potential in root cells enable water to be taken up. Cord grass (*Spartina*) is a common halophyte found low down on estuaries and salt marshes; it has an extensive system of rhizomes for propagation, bearing adventitious roots for anchorage and purposes of water and ion uptake. Other halophytes of estuaries and salt marshes include smaller plants which store water when it is freely available. Common examples of these species are glasswort (*Salicornia*), seablite (*Suaeda maritima*) and sea purslane (*Halimione*). Some species, such as sea milkwort (*Glaux*) and *Spartina*, are able to regulate their salt content by excreting salt from glands at the margins of the leaves.

Mesophytes

The majority of angiosperm plant species are mesophytes, and they occupy habitats with adequate water supplies. They are faced with the problem of water loss by evaporation from all aerial parts. Features which help to reduce water loss are both structural (xeromorphic) and physiological, and include the presence of a cuticle, protected stomata whose diameters can be regulated, a variable leaf shape, abscission (leaf fall) and an ecological distribution based upon tolerance to dehydration.

Xerophytes

Plants adapted to life in dry regions and able to survive long periods of drought are called xerophytes. These form the typical flora of desert and semi-desert regions and are common along the strand line of the seashore and in sand dunes. Some plants respond to extreme conditions by surviving in the seed or spore stage. These are known as **drought evaders** and can germinate following rainfall and grow, flower and complete seed formation in four weeks, for example the Californian poppy (*Escholtzia*). The seeds produced lie dormant until the next rainy spell.

Drought endurers, on the other hand, show many structural (xeromorphic) and physiological adaptations enabling them to survive in extremely dry conditions. Most of the xerophytic species of the British Isles are associated with the strand line and sand dunes, such as saltwort (*Salsola*) and sea sandwort (*Honkenya*) found growing in small mounds of sand on the shore. Sand couch grass (*Agropyron*) and marram grass (*Ammophila*) (fig 13.11) are dominant species of embryo dunes and have extensive rhizome systems with adventitious roots for obtaining water from well below sand level. *Agropyron* is able to tolerate salt concentrations in the sand up to 20 times that of sea water. Both *Ammophila* and *Agropyron* are important pioneer plants in the development of sand-dune systems.

Xerophytic plant species of desert regions show several adaptations to reducing water loss and obtaining and storing water. Some of these are summarised in table 20.7.

Fig 20.31 *The 'prickly pear' cactus,* Opuntia. *This desert plant is xerophytic, meaning it is adapted for dry conditions.*

Table 20.7 Summary of methods of conserving water shown by various plant species.

Mechanism of water conservation	*Adaptation*	*Example*
reduction in transpiration rate	waxy cuticle	prickly pear (*Opuntia*)
	few stomata	
	sunken stomata	pine (*Pinus*)
	stomata open at night and closed by day	ice plant (*Mesembryanthemum*)
	surface covered with fine hairs	
	curled leaves	marram grass (*Ammophila*)
storage of water	fleshy succulent leaves	*Bryophyllum*
	fleshy succulent stems	candle plant (*Kleinia*)
	fleshy underground tuber	*Raphionacme*
water uptake	deep root system	acacia
	below water table	oleander
	shallow root system absorbing surface moisture	cactus

Chapter Twenty-one

Reproduction

Reproduction is the production of a new generation of individuals of the same species. It is one of the fundamental characteristics of living organisms. It involves the transmission of genetic material from one generation to the next, ensuring that the species survives over long periods of time, even though individual members of the species die.

Some members of a species will die before they reach reproductive age, due to factors such as predation, disease and accidental death, so that a species will only survive if each generation produces more offspring than the parental generation. Population sizes will vary according to the balance between rate of reproduction and rate of death of individuals. An increase in total numbers will occur where conditions are suitable. There are a number of different reproductive strategies, all with certain advantages and disadvantages which are described in this chapter.

A new individual normally has to go through a period of growth and development before it reaches the stage at which it can reproduce itself, and this is discussed in chapter 22.

Asexual and sexual reproduction

There are two basic types of reproduction, asexual and sexual. **Asexual reproduction** is reproduction by a single organism without production of gametes. It usually results in the production of genetically identical offspring, the only genetic variation arising as a result of random mutations among the individuals.

Sexual reproduction is the fusion of two gametes to form a zygote which develops into a new organism. It leads to genetic variation. Genetic variation is advantageous to a species because it provides the 'raw material' for natural selection, and hence evolution. Offspring showing most adaptations to the environment will have a competitive advantage over other members of the species and be more likely to survive and pass on their genes to the next generation. Over time, this process of natural selection can result in the species changing. Eventually new species may form, a process known as **speciation** (section 27.7). Increased variation can be achieved by the mixing of genes from two different individuals, a process known as genetic recombination. This is the essential feature of sexual reproduction. It occurs in almost all species, including in a primitive form in bacteria (section 2.3.3).

21.1 Asexual reproduction

Asexual reproduction is the production of offspring from a single organism without the production of gametes. The offspring are identical to the parent. Identical offspring from a single parent are referred to as a **clone**. Members of a clone only differ genetically as a result of random mutation. Most animal species do not naturally reproduce asexually, though successful attempts (section 21.1.4) have been made to clone certain species artificially.

There are several types of asexual reproduction. Examples from each of the five kingdoms of living organisms will be described below.

21.1.1 Kingdom: Prokaryotae (bacteria) and Kingdom: Protoctista

In unicellular organisms, such as bacteria and most protoctists, asexual reproduction occurs by a process called **fission**. This is the division of the cell into two or more daughter cells identical to the parent cell. The DNA replicates. In the case of protoctists, which are eukaryotes and therefore have nuclei, this is followed by nuclear division. In bacteria, and many protoctists such as *Amoeba* and *Paramecium*, two identical daughter cells are produced, a process called **binary fission** (section 2.3.3, figs 2.11 and 21.1). Under suitable conditions it results in rapid population growth, as described in section 2.3.5 for bacteria.

Multiple fission, in which repeated divisions of the parent nucleus are followed by division into many daughter cells, occurs in a group of protoctists which includes the malaria parasite *Plasmodium*. Here it occurs immediately after infection when the parasite enters the liver. About 1000 daughter cells are produced from one parent cell, each capable of invading a red blood cell and producing up to a further 24 daughter cells by multiple fission. Such enormous powers of reproduction compensate for the large losses associated with the difficulties of successful transfer from one human host to another by way of the vector organism, the mosquito.

21.1.2 Kingdom: Fungi

The typical body structure of fungi is a mass of fine tubes called **hyphae**. The whole mass of hyphae is called a **mycelium**. At the tips of the hyphae, spores can form, either enclosed in a special structure called the **sporangium**, or free. Spores are small structures containing a nucleus. They are produced in large numbers and are very light, being easily dispersed by air currents as well as by animals, particularly insects. Being small they usually have

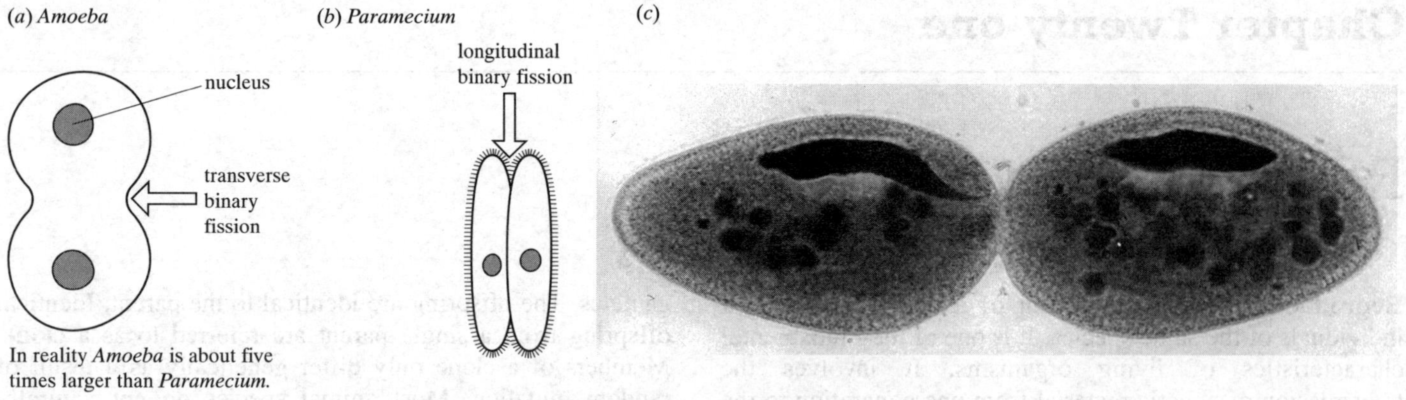

Fig 21.1 *Asexual reproduction by binary fission. (a) Transverse binary fission in* Amoeba*. (b) Longitudinal binary fission in* Paramecium*. (c)* Amoeba *dividing by binary fission.*

small food stores and there is great wastage as many fail to find a suitable place for germination. However, they allow rapid multiplication and spread of fungi.

An example of a fungus that produces sporangia is *Mucor*. This is described in section 2.5.2 and fig 2.26. An example of a fungus that produces spores directly at the tips of its hyphae is *Penicillium*, as described in section 2.5.2 and fig 2.25. Yeasts are unusual fungi because they are unicellular and do not have hyphae. Yeast cells multiply rapidly in a form of asexual reproduction called budding (fig 2.27).

Budding is a form of asexual reproduction in which a new individual is produced as an outgrowth (bud) of the parent, and is later released as an independent, identical copy of the parent. It takes place in a number of other groups of organisms, notably the cnidarians, for example *Hydra* (see animals below).

21.1.3 Plant kingdom

The most common form of asexual reproduction in plants is called **vegetative propagation**. Vegetative propagation (or vegetative reproduction) is a form of asexual reproduction in which a bud grows and develops into a new plant. At some stage the new plant becomes detached from the parent plant and starts to lead an independent existence. Specialised organs of propagation often develop, but they must all have buds, and since buds only occur on stems, they must all contain at least a small part of a stem. Examples are bulbs, corms, rhizomes, stolons and tubers. Some of these also store food and are means of surviving adverse conditions, such as cold periods or drought. The food is used for growth when conditions become suitable. Plants possessing them can therefore survive from one year to the next. The structures are called **perennating organs**, and include bulbs, corms, rhizomes and tubers. In all cases the food stored comes mainly from photosynthesis of the current year's leaves.

Some of the organs of vegetative propagation and perennation are described below. Artificial propagation is considered later in this chapter (section 21.3).

Bulb. A modified shoot, for example onion (*Allium*), daffodil (*Narcissus*) and tulip (*Tulipa*). An organ of perennation as well as vegetative propagation.

A bulb has a very short stem and fleshy storage leaves. It is surrounded by brown scaly leaves, the remains of the previous year's leaves after their food stores have been used. The bulb contains one or more buds. Each bud that grows forms a shoot which produces a new bulb at the end of the growing season. Roots are adventitious, that is they grow from the stem rather than from a main 'tap' root.

A typical bulb is illustrated in fig 21.2.

Corm. A short, swollen, vertical underground stem, as in *Crocus* and *Gladiolus*. An organ of perennation as well as vegetative propagation.

A corm consists of the swollen base of a stem surrounded by protective scale leaves; there are no fleshy leaves, unlike bulbs. Scale leaves are the remains of the previous season's foliage leaves. Roots are adventitious. At the end of the growing season contractile roots pull the new corm down into the soil. The corm contains one or more buds which may result in vegetative propagation (compare bulb).

A typical corm is illustrated in fig 21.3.

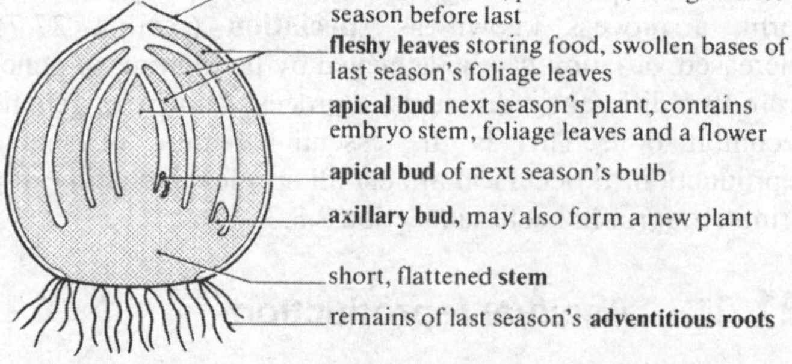

Fig 21.2 *Diagrammatic section through a dormant bulb.*

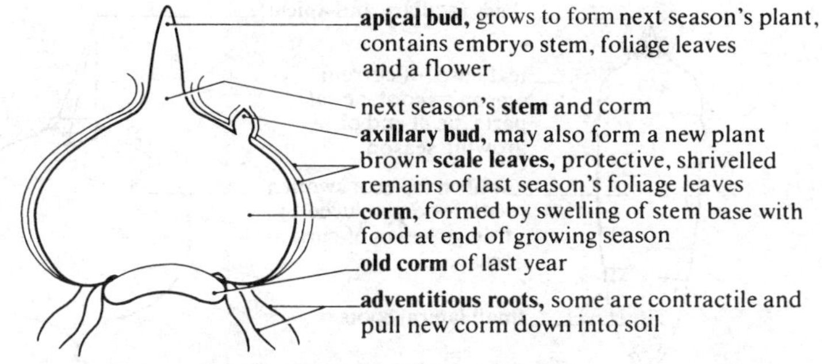

Fig 21.3 *Diagrammatic section through a dormant corm.*

Rhizome. A horizontally growing underground stem, such as in *Iris*, couch grass (*Agropyron repens*), mint (*Mentha*) and Michaelmas daisy (*Aster*). It is usually an organ of perennation as well as vegetative propagation.

A rhizome bears leaves, buds and adventitious roots. The leaves may be scale-like (small and thin, whitish or brownish in colour) as in couch grass or green foliage leaves as in *Iris*. An *Iris* rhizome is illustrated in fig 21.4.

Stolon. A creeping, horizontally growing stem that grows along the surface of the ground, for example blackberry (*Rubus*), gooseberry, blackcurrant and redcurrant (all *Ribes* spp.). It is not an organ of perennation. Roots are adventitious, growing from nodes.

A plan of a typical stolon is illustrated in fig 21.5.

Runner. A type of stolon that elongates rapidly, as in strawberry (*Fragaria*) and creeping buttercup (*Ranunculus repens*).

A runner bears scale leaves with axillary buds and the buds give rise to adventitious roots and new plants. The runners eventually decay once the new plants are established. The runner may represent the main stem or grow from one of the lower axillary buds on the main stem, as illustrated in fig 21.6. In strawberry, scale leaves and axillary buds occur at every node, but roots and foliage leaves arise only at every other node. All axillary buds may give rise to new runners.

Tuber. A tuber is an underground storage organ formed from a stem or a root, swollen with food and capable of perennation. Tubers survive only one year, and shrivel as their contents are used during the growing season. New tubers are made at the end of the growing season, but do not arise from old tubers (in contrast to corms, which arise from old corms).

Stem tubers are stem structures produced at the tips of thin rhizomes, as in potato (*Solanum tuberosum*). Their stem structure is revealed by the presence of axillary buds in the axils of scale leaves (fig 21.7). Each bud may grow into a new plant during the next growing season. Bearing in

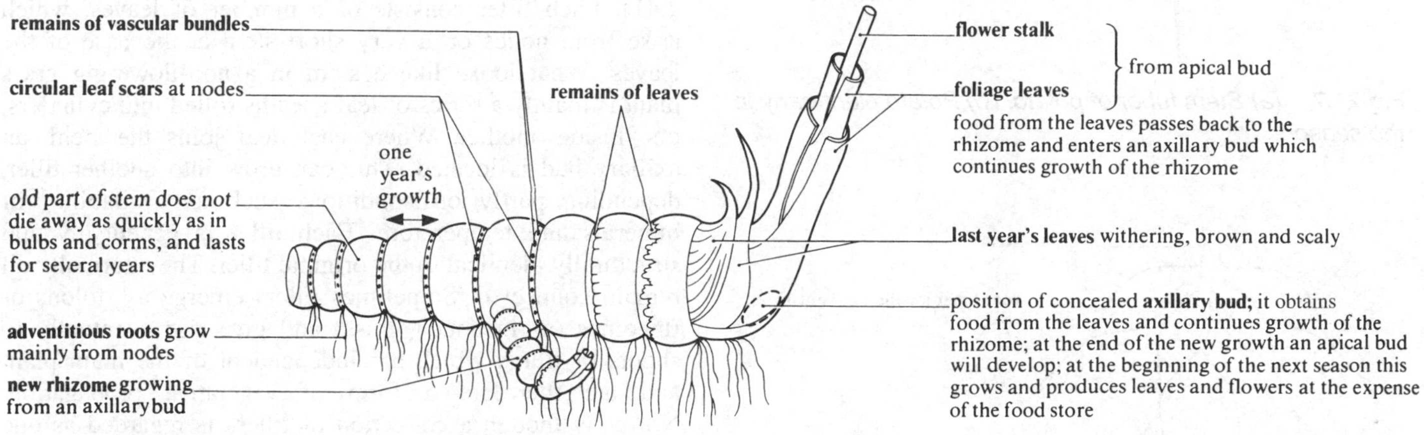

Fig 21.4 *Diagrammatic structure of an* Iris *rhizome.*

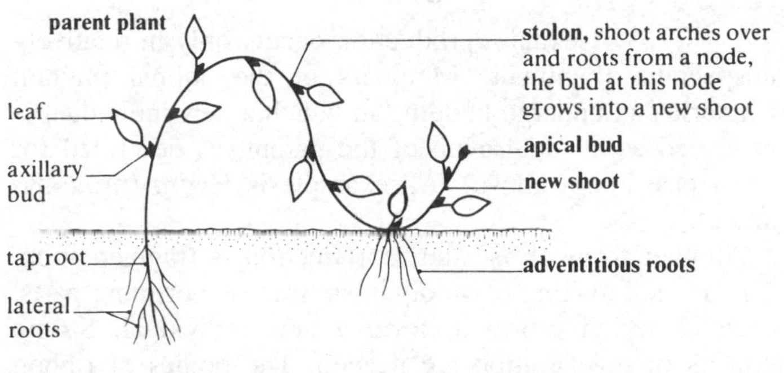

Fig 21.5 *Generalised plan of a stolon.*

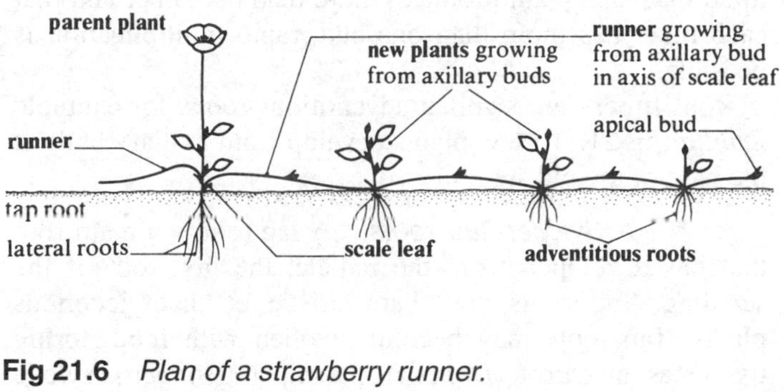

Fig 21.6 *Plan of a strawberry runner.*

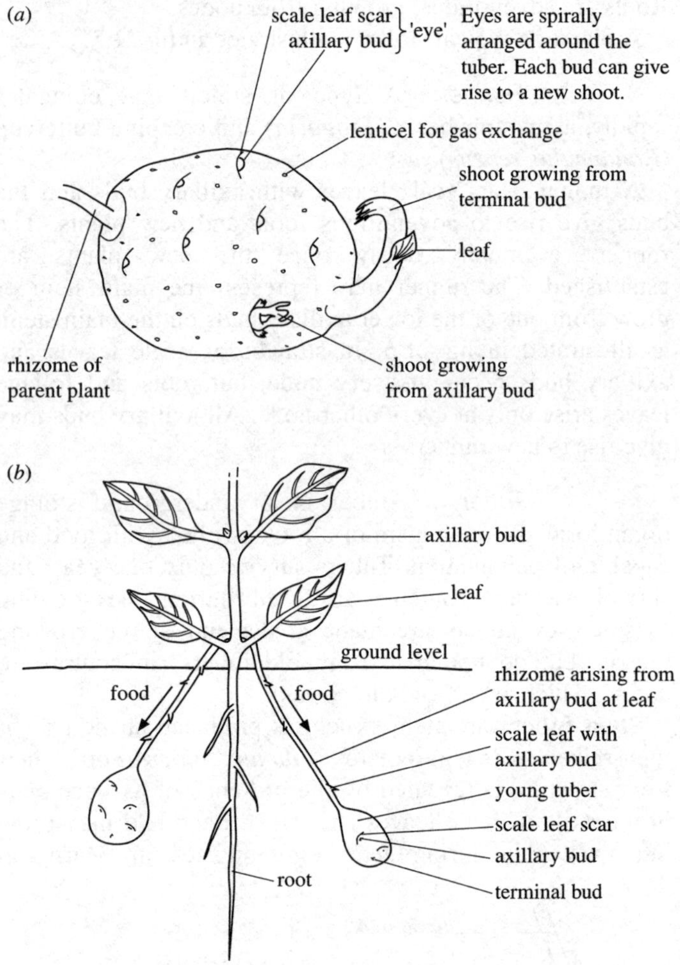

Fig 21.7 *(a) Stem tuber of potato. (b) Potato plant early in the season.*

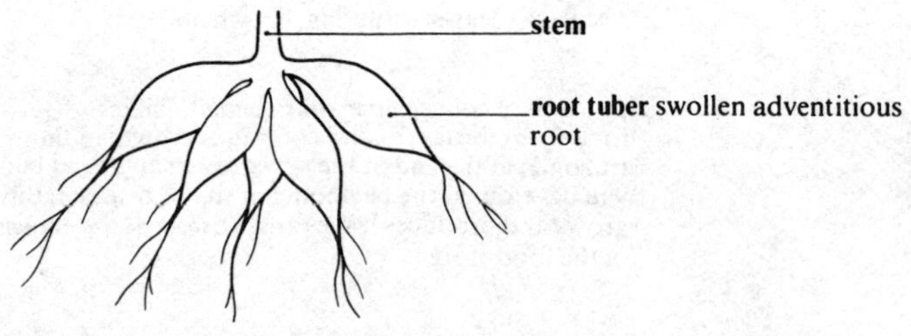

Fig 21.8 *Root tubers of* Dahlia.

mind that each plant produces more than one tuber and that each tuber has more than one bud, rapid multiplication is possible.

Root tubers are swollen adventitious roots, for example *Dahlia* (fig 21.8). New plants develop from axillary buds at the base of the old stem.

Swollen tap roots. A tap root is a main root that has developed from the radicle, the first root of the seedling. Tap roots are characteristic of dicotyledonous plants. Tap roots may become swollen with food-storing tissue, as in carrot (*Daucus*), parsnip (*Pastinaca*), swede

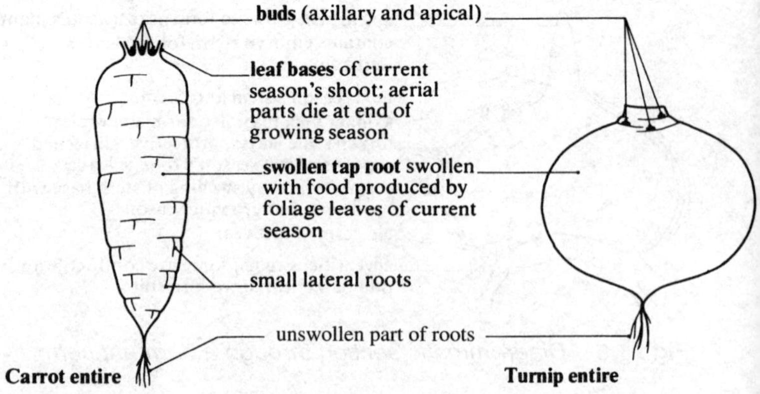

Fig 21.9 *Tap roots of carrot and turnip.*

(*Brassica napus*), turnip (*Brassica rapa*) and radish (*Raphanus sativus*). Together with buds at the base of the old stem, just above the tap root, they form organs of perennation and vegetative propagation. Two types of swollen tap root are shown in fig 21.9.

Swollen tap roots are characteristic of biennial plants, plants that grow vegetatively during the first year of growth and survive winter by means of an underground storage organ. They produce flowers and seeds during the second year of growth, at the end of which they die.

Tillers. Before they flower, grass plants consist of a collection of shoots, also known as tillers (fig 2.41). Each tiller consists of a number of leaves, which arise from nodes on a very short stem at the base of the leaves. What looks like a stem in a non-flowering grass plant is mainly a series of leaf sheaths rolled into cylinders, one inside another. Where each leaf joins the stem, an axillary bud is located. This can grow into another tiller, depending partly on conditions such as availability of minerals and temperature. Each tiller is genetically and structurally identical to the original tiller. They normally all remain connected. Sometimes tillers emerge as stolons or rhizomes (as in couch grass) and give rise eventually to shoots or tillers which are independent of the main plant and are therefore a form of vegetative propagation. Normally though a collection of tillers is regarded as one plant.

21.1.4 Animal kingdom

Asexual reproduction occurs only in relatively unspecialised animals. Members of the animal phylum Cnidaria can undergo budding in which a new individual is produced as an outgrowth of the parent, as described for yeast (see Fungi above). An example is *Hydra* (fig 2.48*a* and *g*).

Another form of asexual reproduction is fragmentation. This is the breaking of an organism into two or more parts, each of which grows to form a new individual. Strong powers of regeneration are needed. The bodies of ribbon worms, a group of simple marine worms, break up easily into small pieces, each of which can regenerate a new individual. Starfish can regenerate if accidental fragmentation occurs.

Cloning of animals

Cloning is the production of many genetically identical copies of an individual by asexual reproduction. It may occur naturally, but techniques have been developed which allow the process to be carried out artificially. The first successful cloning of a vertebrate was carried out in the late 1960s by Dr J. Gurdon at Oxford University.

The process does not occur naturally among vertebrates but, by taking a cell from the intestine or skin of a frog and introducing its nucleus into an egg cell whose own nucleus had been destroyed by ultraviolet radiation, he was able to grow a tadpole, which in turn grew into a frog identical to the parent from which the nucleus was transplanted (fig 21.10).

Experiments like these showed that differentiated (specialised) cells still contain all the information needed to make the whole organism. They are said to be **totipotent**. They also suggested that similar techniques might successfully be used in cloning more advanced vertebrates. Research in Scotland in 1996 led to the successful cloning of a sheep (Dolly) from a cell taken from the parent's udder.

Cloning of a human embryo was carried out in the USA in 1993, although the clones were only grown to the stage of a few cells to demonstrate the possibility. (The process is banned on ethical grounds in the UK.) Cloning does find a use for other animal species however. It is possible to work on animal embryos at the stage of a few cells and to deliberately split the ball of cells into identical twins. This process can be repeated many times because the cells at this stage have not yet become irreversibly specialised. In this way many identical copies of a single useful animal can be created. The embryos can then be transferred to surrogate (recipient) mothers for further growth and eventual birth of the desired animals. Cloning embryos in this way is becoming increasingly important in animal breeding and has been used, for example, for cattle, sheep and goats. It speeds up the selective breeding of animals from desirable parents but raises many difficult issues (section 25.6).

Cells may also be cloned for special purposes. This technique is called **tissue culture**. Certain cells when placed in a suitable medium can be cultured indefinitely. The use of cloned cells allows a study of the action of such chemicals as hormones, drugs, antibiotics, cosmetics and pharmaceutical products to be made on cells. Such a technique is a useful substitute for laboratory animals such as rats, cats and dogs.

21.2 Advantages and disadvantages of natural asexual reproduction

Both asexual and sexual reproduction have advantages and both forms are used by many organisms. Some organisms such as humans and many other animals rely entirely on sexual reproduction. *Amoeba* apparently relies entirely on asexual reproduction. We cannot say that one form of reproduction is better than the other. They are both successful strategies in the right situation.

Advantages

During asexual reproduction parent cells divide into genetically identical daughter cells. In eukaryotes this involves mitosis. The advantages of this process are as follows.

- **Only one parent is required**. Where sexual reproduction involves two organisms, time and energy are used in finding a mate, or, in the case of non-motile organisms such as plants, special mechanisms such as pollination are required which may be wasteful of gametes. One solution to this problem is hermaphrodite organisms which produce both male and female sex organs.
- **Genetically identical offspring**. If the organism is well adapted to its environment, the fact that the offspring are genetically identical may be an advantage. Successful combinations of genes are preserved.
- **Dispersal and spread**. The methods of asexual reproduction often enable dispersal of a species. For example, *Penicillium* and *Mucor* are common moulds which spread rapidly by means of asexually produced spores which are light and easily dispersed by air currents. This enables the fungi to find fresh sources of food. Plants that produce rhizomes, such as sea couch grass in sand dunes, bracken, and *Spartina* (cord grass) in mud flats, spread rapidly by this means.
- **Rapid multiplication**. Bacteria can divide as often as once every 20 minutes allowing numbers to build up very rapidly. Many parasites rely on one or more asexual stages where rapid multiplication compensates for large losses at other stages in the life cycle. The malaria parasite, tapeworm and liver fluke are good examples.

Disadvantages

- No genetic variation occurs among the offspring. The advantages of variation are discussed with sexual reproduction (section 21.5).
- If spores are produced, many will fail to find a suitable place for germination and so energy and materials used in their manufacture are wasted.
- If an organism spreads in one area, it may result in overcrowding and exhaustion of nutrients.

21.3 Artificial propagation of plants – cloning

A number of methods of artificial propagation of plants are used in agriculture and horticulture. The first three methods discussed below, namely cuttings, grafting

(a)

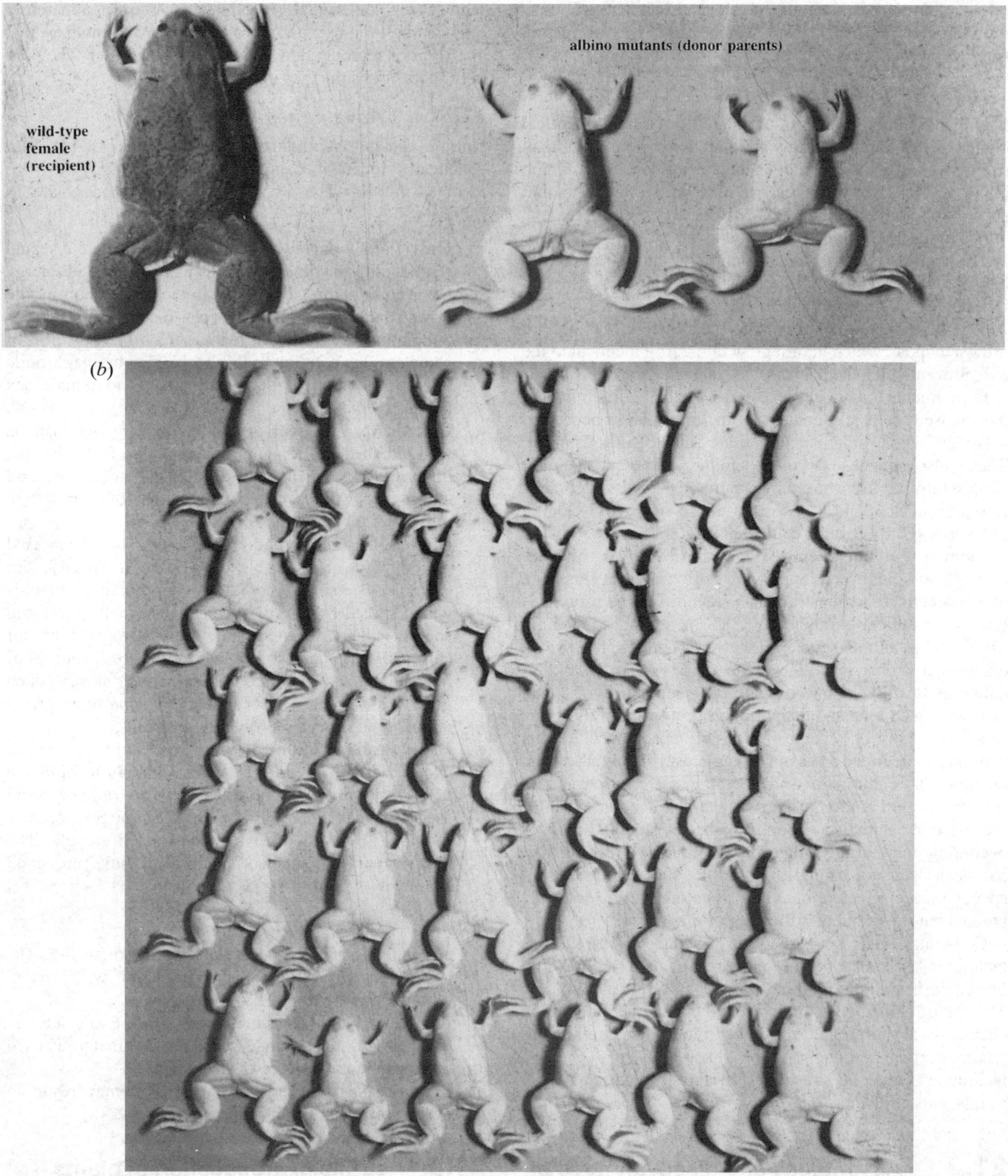

(b)

Fig 21.10 *(a) A clone of frogs (*Xenopus laevis*) produced by nuclear transplantation. An embryo was obtained from a cross between two albino mutants (donor parents). Its cells were separated and their nuclei transplanted into unfertilised eggs whose own nuclei had been destroyed with ultra-violet radiation. The eggs were from the wild-type female shown (the recipient). (*b*) The clones produced are all female and albino; this group of 30 were obtained from a total of 54 nuclear transfers.*

and layering, are traditional methods, but for commercial purposes they are gradually being replaced by modern methods involving tissue culture.

21.3.1 Cuttings

This is a simple procedure in which part of the plant is removed by cutting and placed in a suitable medium for growth. It produces roots and grows into a new plant. Rooting can be stimulated with a rooting hormone. The popular house plants *Geranium* and *Pelargonium* are commonly propagated in this way, using cuttings from their shoots. The African violet, another popular house plant, can be propagated from leaf cuttings. The process is used commercially for blackcurrant bushes in which cuttings from shoots are taken in the autumn. Chrysanthemums are also propagated by cuttings from shoots.

21.3.2 Grafting and budding

Grafting is the transfer of part of one plant, the **scion**, onto the lower part of another plant, the **stock**. This was originally done for apple trees because the plants could not be grown from cuttings, and seedlings showed too much variation because they are produced by sexual reproduction. It is now also used for some other fruit trees, such as plum and peach. The scion is chosen for its fruit and the stock for properties such as disease resistance and hardiness.

Most rose bushes are propagated by a variation of this method known as **budding** in which a bud is used as the scion rather than a shoot. New varieties are produced by *sexual* reproduction, but, like fruit trees, are not pure breeding. Artificial propagation therefore has an important role in preserving desired varieties.

21.3.3 Layering

Layering is used for plants that produce runners, such as strawberries. The runners are pegged out (layered) around the parent plant until they root, and are then cut to detach them from the parent plant.

21.3.4 Tissue culture, or micropropagation

Micropropagation is the propagation, or cloning, of plants by tissue culture. 'Micro' refers to the small size of the material used, usually isolated cells or small pieces of tissue. The material is grown in special culture solutions, so the process is also known as **tissue culture**. It developed from experiments which showed that plant tissues removed ('excised') from plants could be stimulated to grow in solution by the addition of nutrients and certain plant hormones, particularly auxins and cytokinins. The latter are needed for continued cell division. Tissue culture is now widely used for the rapid propagation of desired varieties (fig 21.11).

Totipotency

It was shown in the early 1960s that the nuclei of mature plant cells still contain all the information needed to code for the entire organism. Professor F. C. Steward of Cornell University, USA showed that mature cells removed from a carrot and placed in a suitable culture solution could be stimulated to start dividing again and to provide new carrot plants. The cells were described as **totipotent**, meaning that they retain the ability to grow into new plants in suitable conditions even when mature and specialised.

Why use tissue culture?

A summary of some of the important advantages of tissue culture is given below. Further details are given at the end of this section.

- Plants with desired characteristics can sometimes (though not always) be multiplied rapidly, producing many identical copies of the same plant. This is not easily done using conventional breeding methods which rely on sexual reproduction, particularly when plants are adapted for cross-pollination and outbreeding (see later). The technique is important for certain crops and other commercially important plants. New varieties obtained by plant breeding can be reproduced quickly. The technique therefore speeds up the development and introduction of new varieties.
- Cells can be genetically modified ('transformed') and grown into whole plants, known as **transgenic** plants, using tissue culture. Transgenic plants are discussed in section 25.2.1.
- Tissue cultures take up little space.
- Plants are grown in disease-free conditions. Viruses can be eliminated as described below.
- New methods of producing hybrids by fusing protoplasts, that is cells which have been released from their cell walls, have been developed. Hybrids between different species, such as potato and tomato, have been achieved in this way.
- The techniques may prove an efficient way of producing useful chemicals, such as pharmaceutical products, from plants.

Outline of the technique

The culture medium must contain the correct nutrients and hormones (plant growth substances, section 16.2). A typical medium will contain the inorganic ions needed for plant growth, such as a source of nitrogen, magnesium, iron and potassium (table 7.7). Also needed are sucrose as a source of energy, and vitamins. These chemicals are usually mixed with agar to form a jelly-like nutrient agar similar to that used for growing bacteria and fungi. Auxins and cytokinins are the two essential hormones. Auxins stimulate root growth and cell elongation and cytokinins stimulate shoot growth and cell division. Differences in the amount of auxin relative to cytokinin affect the way unspecialised cells develop. The tissue culture is grown on the surface of the agar in flasks or petri dishes.

Fig 21.11 *Tissue culture. (a) Auricula plant with well-developed leaves which can be cut into many small pieces for cloning. (b) Leaf fragments are transferred to agar under sterile conditions. (c) The leaf fragment has developed a callus and a new shoot. (d) The shoot has been removed from the leaf fragment and placed in deep agar to encourage roots to grow. (e) Young clones are then planted into light growing medium to strengthen root development. (f) Identical plants cloned from fragments of one leaf.*

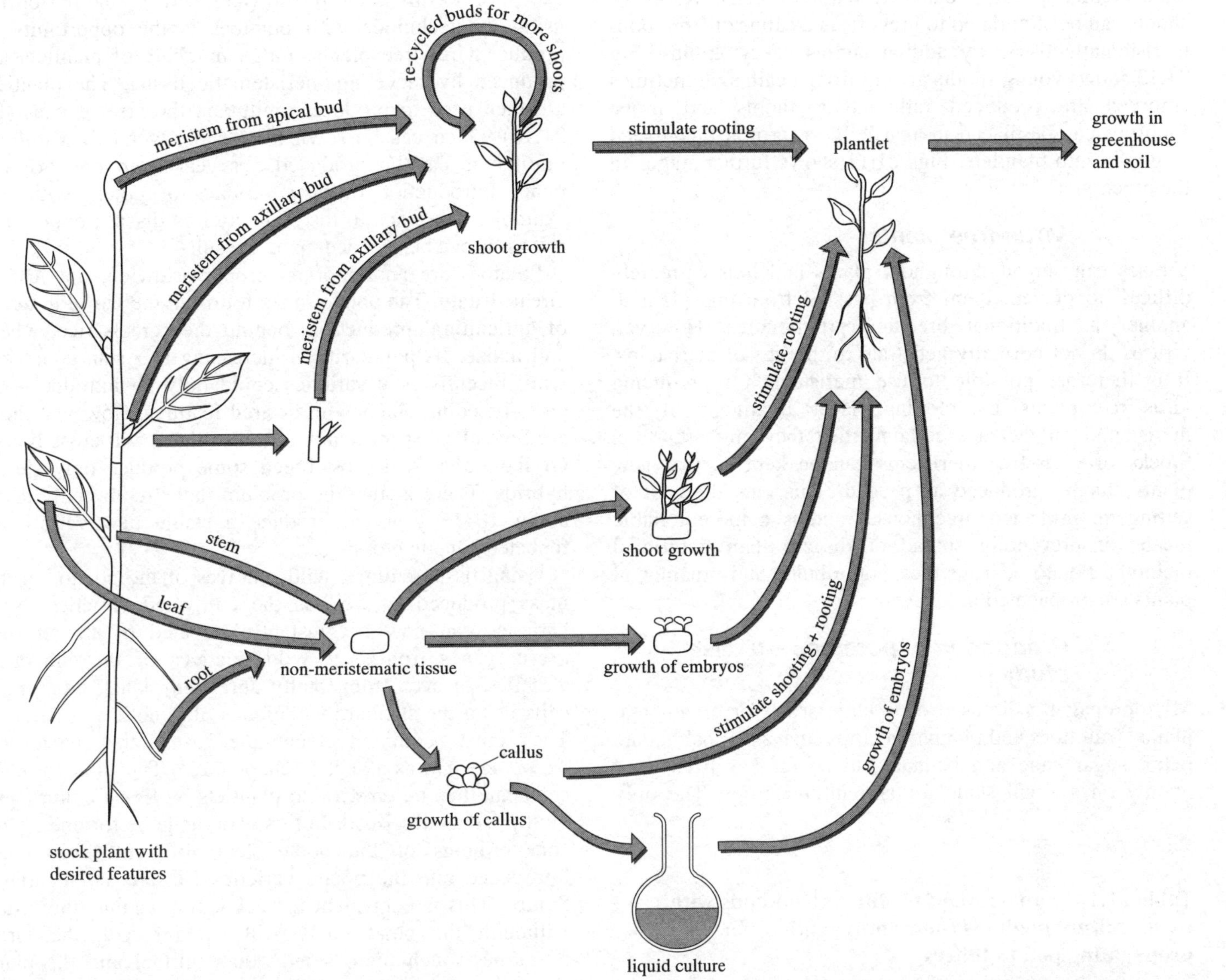

Fig 21.12 *Methods for cloning from a stock plant.*

Temperature, light intensity, light quality, daylength and humidity are all controlled by growing the cultures in special growth rooms or cabinets. All procedures must be sterile because bacteria and fungi can also grow in the cultures and would grow faster and out-compete the plants in these conditions. The plant tissues themselves are surface sterilised in a dilute bleach solution, and other materials are also sterilised before use. All apparatus must be handled under sterile conditions, as in microbiological work (section 12.3).

Fig 21.12 summarises the main methods by which new plants can be grown using tissue culture. The pieces taken from the stock plant are known as **explants**. The most common method is to use meristematic tissue from apical or axillary buds. A **meristem** is a region where cell division is still taking place. In plants, growth is confined to such regions. Another method is to produce a callus from non-meristematic tissue (fig 21.13). A callus is an

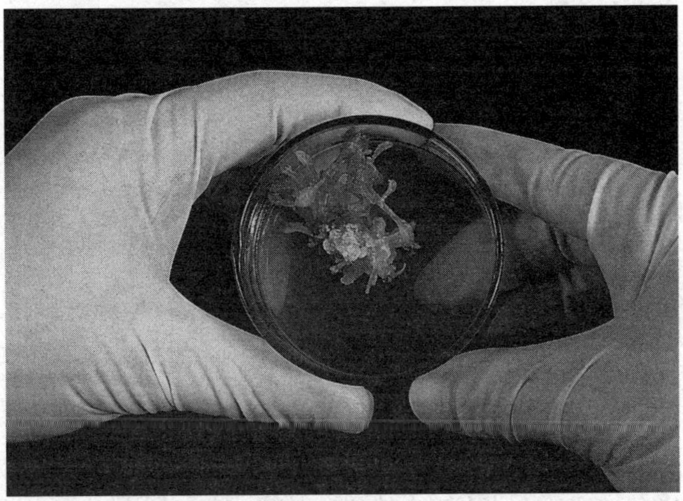

Fig 21.13 Nicotiana *plantlets growing from callus tissue culture on sterile agar.*

undifferentiated (unspecialised) mass of cells. Roots or shoots can be stimulated to grow from a callus or from non-meristematic tissue by adding auxins or cytokinins. Fig 21.13 shows young shoots arising from a callus. Sometimes embryos are produced rather than shoots and roots. Embryos can be placed in agar jelly to form artificial seed or grown into plantlets. Figs 21.11 shows further stages in the process.

Virus-free plants

Viruses can spread throughout plants and it is extremely difficult to prevent them from passing from one plant to another in traditional breeding programmes. However, viruses do not normally penetrate to the tips of meristems. It is therefore possible to use meristems for producing virus-free plants by cloning. Heat treatment of the meristems can increase certainty that they are virus-free. Stocks of virus-free meristems can be kept so that new plants can be produced as needed. This cuts the cost of setting up protected greenhouses and is a more reliable means of preventing spread of disease than traditional methods. Potatoes, fruit trees, some bulbs and ornamental plants are propagated in this way.

Production of potatoes – a case study

Micropropagation is used on a large scale for ornamental plants, fruit trees and plantation crops such as oil palm, date palm, sugar cane and banana, but so far has been used mainly on a small scale for agricultural crops. The only exception to this is the potato (table 21.1). One reason for using the technique with potatoes is the opportunity to produce virus-free plants. Large numbers of plantlets are produced by recycling meristematic tissue. The plantlets are then used to produce minitubers the size of peas (fig 21.14) which can be sown like seed. Over half a million minitubers can be produced per year from one original plant. Introduction of new varieties of potato with, for example, resistance to the serious virus disease potato leaf roll virus can be speeded up as a result.

Potatoes are an important crop worldwide, not just in Great Britain. The potato ranks fourth in the world in terms of agricultural production, behind the cereals rice, wheat and maize. Its popularity is increasing in Asian countries. Until recently new varieties could only be introduced by cross-breeding plants with desired features. However, there are several subspecies and not all of these will cross-breed. Of those that will cross-breed some produce only sterile hybrids. There is also the problem that it commonly takes about 10–15 years to produce a stable new variety by repeated genetic crossings.

Using tissue culture, wild relatives of the potato can be mass produced as well as the cultivated varieties. New varieties can now be 'instantly' created by transferring useful genes from, say, wild relatives of the cultivated varieties, or even from totally unrelated plants, into single cells using the standard techniques of genetic engineering. The most common technique uses the bacterium *Agrobacterium* as a vector (chapter 25). The 'transformed' cells can then be grown into plantlets by tissue culture and multiplied as described. In this way the gene for one of the coat proteins of the potato leaf roll virus has been introduced into the potato varieties Desiree and Pentland Squire. This is equivalent to vaccination against the virus. Although the potato may still get infected, the virus multiplies much more slowly than normal and the plant shows no, or few, signs of disease.

Another technique which has been successfully used with potatoes is to fuse two somatic (non-sex) cells from

Table 21.1 Comparison of traditional methods with tissue culture methods (micropropagation) for propagating potato tubers.

Traditional	*Micropropagation*
Year 1	
100 g tuber	100 g tuber
↓	↓
1 mature plant	one stock plant
│	↓
│	10 buds cultured
│	↓
│	shoot multiplication
↓	↓
1600 g tubers	65 000 minitubers each capable of producing 10 plants
↓	↓
Year 2	
16 mature plants	650 000 possible plants
│	5% loss for establishment in soil
│	617 500 mature plants
↓	↓
16 × 1600 g tubers = **25.6 kg**	617 500 × 500 g tubers = **308 750 kg**

(Based on table 12.3 *Molecular biotechnology*, 2nd ed., S.D. Primrose (1991), Blackwell, reproduced from Mantell *et al.* (1985).)

Fig 21.14 *Minitubers of potato look exactly like normal tubers but are only the size of peas.*

two different varieties. This involves first removing the cell walls to create naked protoplasts. The resulting **somatic hybrids** can be grown in tissue culture. Such hybrids can be made between two varieties which would not be able to be crossed using sexual reproduction due to incompatibility. Resistance to potato leaf roll virus and tolerance to cold have been bred into commercial varieties from wild potatoes in this way.

Advantages of micropropagation

- **Rapid multiplication**. When shoots are produced during cloning they produce buds in the normal way. These buds can be used to generate more shoots using the same tissue culture techniques. By constantly re-cycling buds in this way, the number of potential plants is multiplied at each stage. This can result in thousands or even millions of plants being produced from one shoot over a period of a year. This is much faster than traditional breeding methods, so new varieties can be introduced several years earlier than was possible by the old techniques.
- **Genetic uniformity**. Since plants produced are genetically identical, they all possess the desirable features of the stock plant. It is difficult to produce plants that breed true (are homozygous for the required features) when sexual reproduction is used.
- **Disease-free plants**. As explained, virus diseases can be eliminated by using meristematic tissue for propagation. Since the plants are prepared in sterile conditions they are also free from surface bacteria and fungi, some of which would cause disease.
- Tissue culture takes up relatively little space compared with growing plants in greenhouses or fields.
- Tissue culture can be carried out independently of seasonal changes in climate. One advantage of this is that plants can be produced out of season, when they can be sold for higher prices.
- Plant development is more closely controlled, guaranteeing product uniformity for customers.
- Some plants, such as banana, are sterile and have to be propagated asexually.
- The seeds of some plants, such as certain orchids, are difficult to germinate. The plants can be produced more reliably by asexual means.
- Micropropagation can be linked with genetic engineering to produce transgenic plants as described in section 25.4.
- Airfreighting is economic because the cultures are not bulky. This increases the possibility of international trade.

Disadvantages of micropropagation

- It is very labour intensive and not as convenient as sowing seed. The work is also skilled. Organisation and training of staff presents problems when carrying out procedures on a large scale. It also adds significantly to the cost of the operation. The process is normally only commercially viable for expensive products such as ornamental plants, and is not suitable for low-cost crops such as carrots.
- Sterile conditions must be maintained. This adds to the cost and makes operations much more demanding.
- Plants obtained from callus cultures sometimes undergo genetic changes. A small proportion of these changes may be commercially useful, but most are undesirable.
- Since the clones are genetically identical, crops are very susceptible to new diseases or changes in environmental conditions. Whole crops could be wiped out.

21.4 Sexual reproduction

Sexual reproduction is the production of offspring by the fusion of two **gametes** to form a diploid **zygote** which develops into the mature organism. The act of fusion is called **fertilisation**. At fertilisation the nuclei of the gametes fuse, bringing together two sets of chromosomes, one set from each parent. The presence of two sets of chromosomes is referred to as the diploid condition. Gametes are haploid, meaning that they carry one set of chromosomes.

Meiosis is an essential feature of life cycles in which sexual reproduction occurs because it provides a mechanism for reducing the amount of genetic material by half. This ensures that when gametes fuse, the diploid number of chromosomes is restored. If the chromosome number were not halved before production of gametes, the chromosome number would double with every generation (see fig 23.10).

During meiosis random segregation of chromosomes (**independent assortment**) and exchange of genetic material between homologous chromosomes (**crossing-over**) results in new combinations of genes being brought together in the gamete and this reshuffling increases genetic variation (section 23.4). The combination of two sets of chromosomes (**genetic recombination**), one set from each parent, in the zygote forms the basis of variation within species. The zygote grows and develops into the mature organism of the next generation.

Gametes are usually of two types, male and female, but in some primitive organisms they are of one type only. If the gametes are of two types, they may be produced by separate male and female parents or by a single parent bearing both male and female reproductive organs. Species that have separate male and female individuals are described as **unisexual**, such as humans and most other animals. Species capable of producing both male and female gametes within the same organism are described as **hermaphrodite**, or **bisexual**. Many protozoans, for example *Paramecium*, cnidarians such as *Obelia*, platyhelminths such as *Taenia* (tapeworm), oligochaetes such as *Lumbricus* (earthworm), crustacea such as *Balanus*

Table 21.2 Comparison of asexual reproduction with sexual reproduction (omitting bacteria).

Asexual reproduction	*Sexual reproduction*
One parent only	Usually two parents
No gametes are produced	Gametes are produced. These are haploid and nuclei of two gametes fuse (fertilisation) to form a diploid zygote
Depends on mitosis	Depends on meiosis being present at some stage in life cycle to prevent chromosome doubling in every generation
Offspring identical to parent	Offspring are *not* identical to parents. They show genetic variation as a result of genetic recombination
Commonly occurs in plants, simple animals and microorganisms. Absent in more complex animals	Occurs in almost all plant and animal species
Often results in rapid production of large numbers of offspring	Less rapid increase in numbers

(barnacle), molluscs such as *Helix* (garden snail), some fish, lizards and birds, and most flowering plants are hermaphrodite.

A summary of some of the typical features of asexual and sexual reproduction is given in table 21.2.

21.5 Sexual reproduction in flowering plants

21.5.1 The life cycle of flowering plants

Flowering plants owe much of their success to the ways in which their sexual reproduction has been adapted to dry land. This has been discussed in section 2.7.7. The major adaptations are:

- the production of seeds and fruits to nourish and protect the embryo plants and to help in their dispersal.
- the absence of swimming male gametes. Male gametes are carried inside pollen grains to the female parts of the plant, a process called **pollination**. This is followed by the production of a pollen tube carrying male nuclei to the female gamete.
- the extreme reduction of the gametophyte generation, which is poorly adapted to life on land in simpler plants like bryophytes (see below).

An outline of the life cycle of flowering plants is given in fig 21.15*a*. If you are studying a range of plants, it would

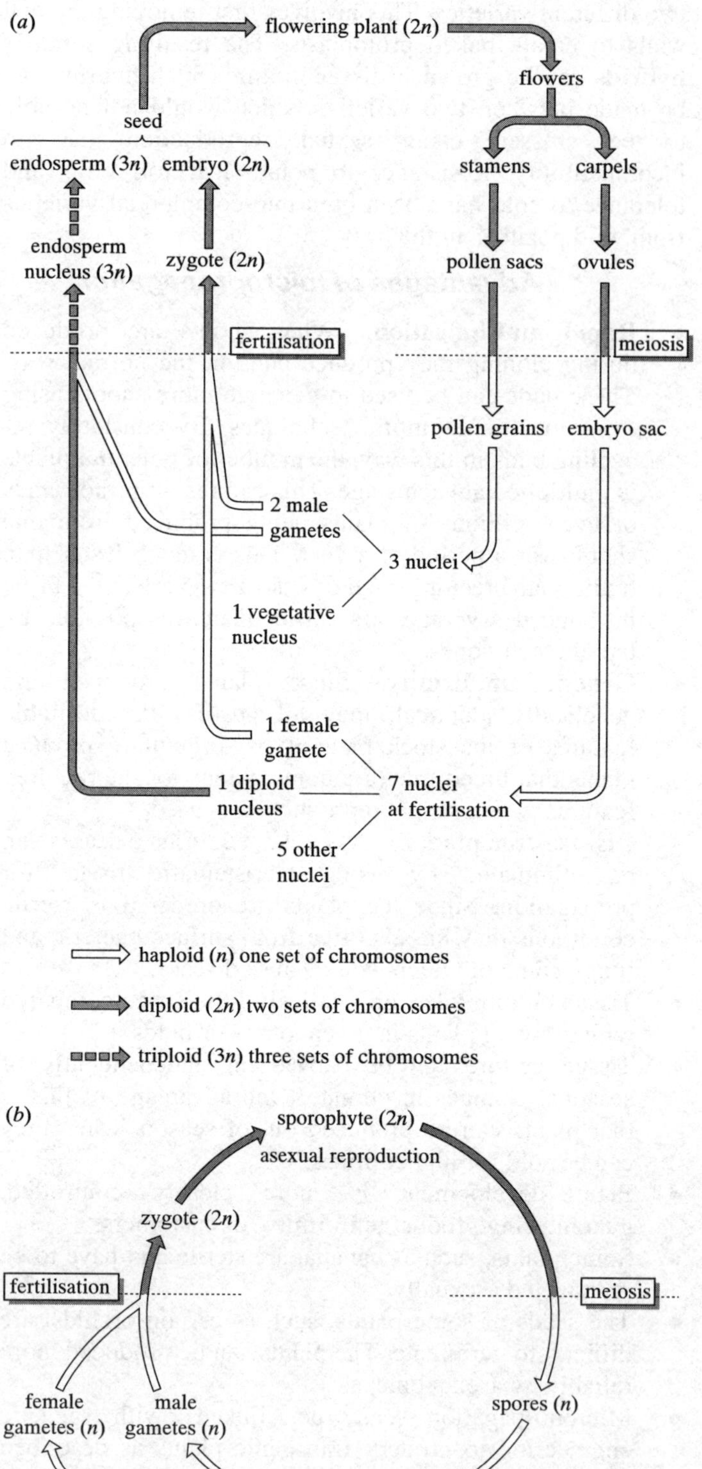

Fig 21.15 *(a) Life cycle of a flowering plant. (b) Life cycle showing alternation of generations. A diploid sporophyte generation alternates with a haploid gametophyte generation. The sporophyte generation reproduces asexually by means of spores, and the gametophyte generation reproduces sexually by means of gametes. All plants have this type of life cycle, although it is not very obvious in flowering plants.*

be useful to bear in mind that there is an alternation of generations in the life cycle, as is shown more simply in fig 21.15*b*. This is discussed in section 2.7.1. In flowering plants, the gametophyte generation is virtually non-existent and is not free-living. It would be difficult to realise that alternation of generations occurs if it were not for the comparison that can be made with more primitive ancestors.

21.5.2 The parts of a flower

The use of the term 'flowering plants' is a reference to the uniqueness of this group in producing flowers. Flowers are reproductive structures whose evolutionary origins are unclear, but which are sometimes regarded as collections of highly specialised leaves. A generalised flower is shown in fig 21.16, and some of the terms used to describe the flower parts are explained below. A collection of flowers borne on the same stalk is called an **inflorescence**. A collection of flowers may be more attractive to pollinating insects than a small solitary flower.

The **receptacle** is the top of the flower stalk **(pedicel)** from which the flower parts arise. The **perianth** consists of two whorls of structures called perianth segments. In monocotyledons the two whorls are usually similar, for instance daffodil, tulip and bluebell. In dicotyledons the two whorls are usually different, consisting of an outer whorl of **sepals** called the **calyx** and an inner whorl of **petals** called the **corolla**.

The **calyx** is the collection of sepals. Sepals are usually green and leaf-like structures that enclose and protect the flower buds. Occasionally they are brightly coloured and petal-like, serving to attract insects for pollination.

The **corolla** is the collection of petals. In insect-pollinated flowers the petals are usually large and brightly coloured, serving to attract insects. In wind-pollinated flowers the petals are usually reduced in size and green, or may be entirely absent.

The **androecium** is the collection of **stamens**, forming the male reproductive organs of the flower. Each stamen consists of an **anther** and a **filament**. The anther contains the **pollen sacs** in which **pollen** is made. The filament contains a vascular bundle that carries food and water to the anther.

The **gynoecium** is the collection of **carpels**, forming the female reproductive organs of the flower. A carpel consists of a **stigma**, **style** and **ovary**. The stigma receives the pollen grains during pollination and the style bears the stigma in a suitable position in the flower to receive the pollen. The ovary is the swollen, hollow base of the carpel and contains one or more ovules. **Ovules** are the structures in which the embryo sacs develop and which, after fertilisation, become seeds. Each is attached to the ovary wall by a short stalk called the funicle and the point of attachment is called the placenta.

The carpels of a flower may be separate and free, as in buttercup, or fused to form a single structure, as in white deadnettle. The styles of flowers with fused carpels may be fused or separate.

The receptacle and flower are described as **hypogynous** if the stamens and perianth are inserted below the gynoecium as in fig 21.16, **epigynous** if stamens and perianth are inserted above the ovary, and **perigynous** if the receptacle is flattened or cup-shaped with the gynoecium at the centre and the stamens and perianth attached round the rim. A **superior** ovary is an ovary located *above* the other flower parts on the receptacle, that is the ovary of a hypogynous flower. An **inferior** ovary is an ovary located *below* the other flower parts on the receptacle, that is the ovary of an epigynous flower.

The **nectaries** are glandular structures that secrete nectar, a sugary fluid that attracts animals for pollination, usually insects, but also birds and bats in the tropics.

The following terms are applied to whole plants and flowers.

- **hermaphrodite** (bisexual) plants – male and female sex organs borne on the same plant.
- **dioecious** (unisexual) plants – male and female sex organs borne on separate plants, that is the plants are either male or female, for example yew, willow, poplar and holly.
- **monoecious** plants – separate male and female flowers borne on the same plant, such as oak, hazel, beech and sycamore. Such plants are hermaphrodite.

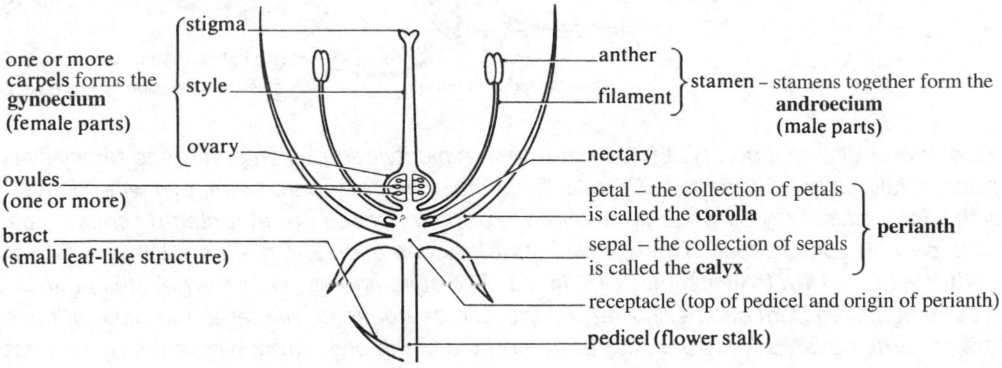

Fig 21.16 *Longitudinal section of a generalised flower.*

Representative flowers. Some representative insect-pollinated flowers are illustrated in figs 21.17–21.19. Fig 21.20 illustrates the structure of a typical grass flower, a wind-pollinated monocotyledon.

The structure of a flower is best illustrated by means of a half-flower, a view of the flower obtained by cutting the flower vertically down the middle into two equal halves and drawing one half, showing the cut surface as a continuous line.

21.5.3 Development of pollen grains

Each stamen consists of an **anther** and a **filament**. The anther contains four pollen sacs which produce pollen. The filament contains a vascular bundle supplying food and water to the anther. Fig 21.21 shows the

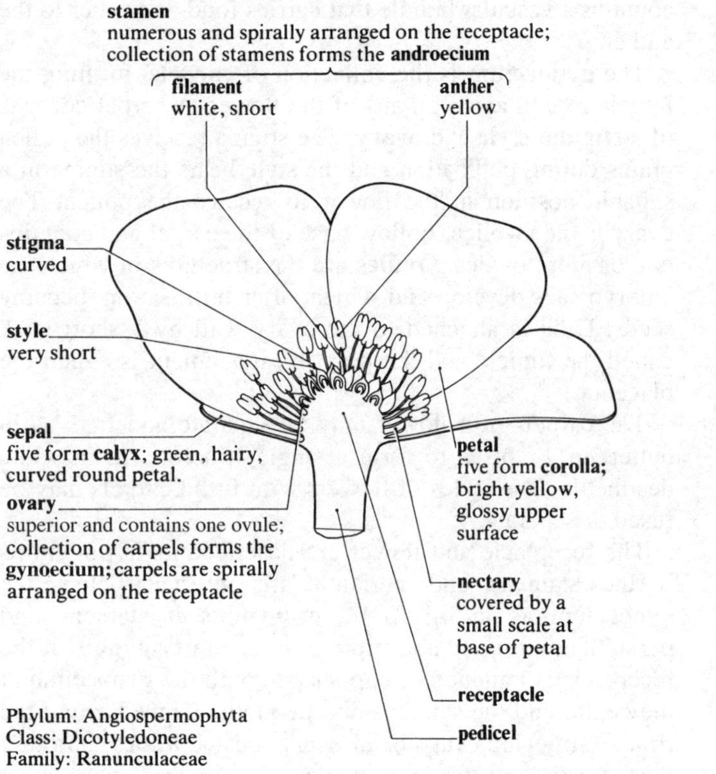

Fig 21.17 *(left) Half-flower of meadow buttercup (*Ranunculus acris*), a dicotyledon. It is a herbaceous perennial common in damp meadows and pastures. Flowers appear April to September. Pollination: insects such as flies and small bees. Fruit: each carpel contains one seed. No special dispersal mechanism.*

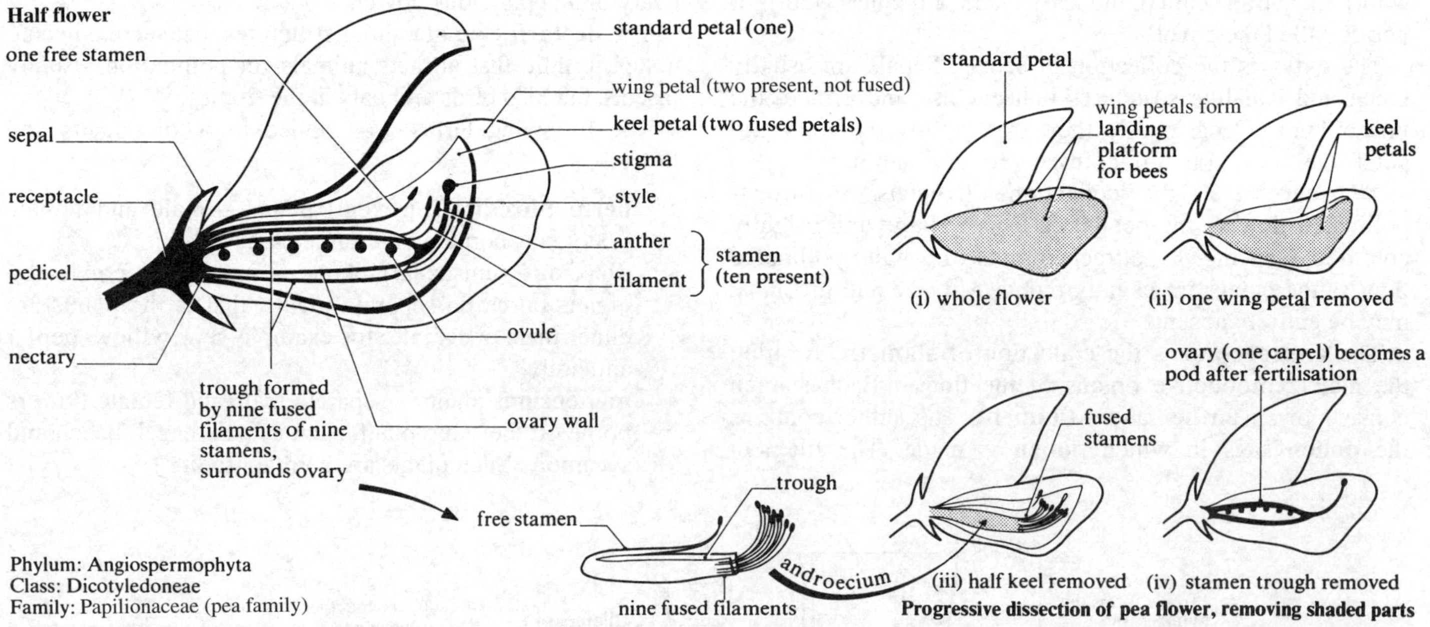

Fig 21.18 *Structure of flower of sweet pea (*Lathyrus odoratus*), a dicotyledon belonging to the family Papilionaceae (pea family). Flowers appear in July. Calyx: five sepals. Corolla: five petals, one standard petal, two wing petals, two fused keel petals interlocking with wing petals. May be white or coloured. Pollination: bees are attracted by colour, scent and nectar. The standard petal is especially conspicuous. The two wing petals act as a landing platform. When a bee lands, its weight pulls them down together with the keel to which they are linked. The style and stigma emerge, striking the undersurface of the bee that may be carrying pollen from another flower. As the bee searches for nectar at the base of the ovary with its long proboscis, the anthers may rub pollen directly onto the undersurface of the style, from where it may be passed to the bee. Self-pollination may also occur. The garden pea (*Pisum sativum*) is similar, but more commonly self-pollinated. Fruit: a pod consisting of one carpel with many seeds.*

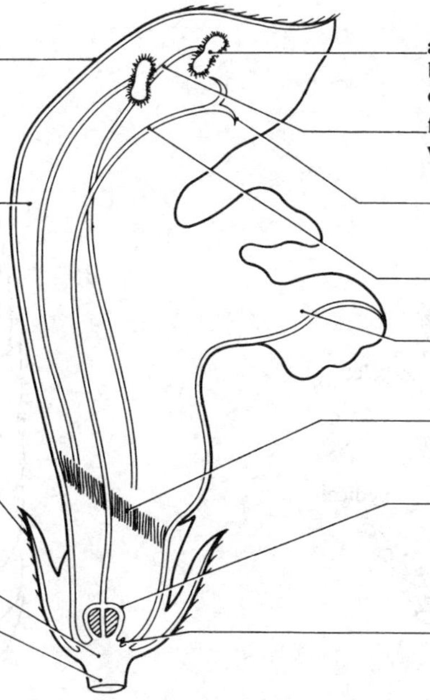

Fig 21.19 *Half-flower of white deadnettle (*Lamium album*), a dicotyledon. The plant is a herbaceous perennial common in hedgerows and on waste ground. Flowers appear in April to June and autumn. Pollination: mainly bumblebees. A bee lands on the lower lip and as it enters the flower its back, which may be carrying pollen, touches the stigma first, thus favouring cross-pollination, although self-pollination is possible.*

internal structure of an anther, with its four pollen sacs containing pollen mother cells. Each pollen mother cell undergoes meiosis to form four pollen grains as shown in fig 21.22. The appearance as seen with a microscope is shown in fig 21.23.

Immediately after meiosis the young pollen grains are seen in groups of four called **tetrads**. Each grain develops a thick wall, often with an elaborate sculptured pattern characteristic of the species or genus. The outer wall, or **exine**, is made of a waterproof substance called sporopollenin. It is one of the most resistant and long-lasting substances in nature and allows grain coats to survive unchanged over long periods of time, sometimes millions of years. This fact, together with the ease of identifying the genus or species which produced the grain, has given rise to the science of palynology or pollen analysis. By studying pollen grains from a particular time and place it is possible to determine what plants were growing and thus to gain information about, for example, the ecosystems (including animals) and climate of that time. A particularly abundant source of pollen grains is peat, which accumulates to great depths over long periods of time in peat bogs.

> **21.1** How might pollen grains be useful as indicators of (*a*) past climate, (*b*) past human activities?

The pollen grain nucleus divides into two by mitosis to form a **generative nucleus** and a **pollen tube nucleus** (fig 21.22).

21.5.4 Development of the ovule

Each carpel consists of a stigma, style and ovary. Within the ovary one or more ovules develop. Each is attached to the ovary wall at a point called the **placenta** by a short stalk or **funicle** through which food and water pass to the developing ovule.

The main body of the ovule is called the **nucellus** and is enclosed and protected by two sheaths or **integuments**. A small pore is left in the integuments at one end of the ovule, the **micropyle**.

Inside the nucellus, at the end nearest the micropyle, one spore mother cell develops, known as the **embryo sac mother cell**. This diploid cell undergoes meiosis to produce four haploid cells, only one of which develops. This forms the **embryo sac** as shown in fig 21.24. The embryo sac grows and its nucleus undergoes repeated mitosis until eight nuclei are produced, four at each end of the embryo sac. One of these is the nucleus of the female gamete.

Two nuclei migrate to the centre of the embryo sac and fuse to become a single diploid nucleus. The remaining six nuclei, three at each end, become separated by thin cell walls and only one of these, the female gamete, appears to serve any further function. The rest disintegrate.

The final appearance of the mature carpel at fertilisation is shown in fig 21.25.

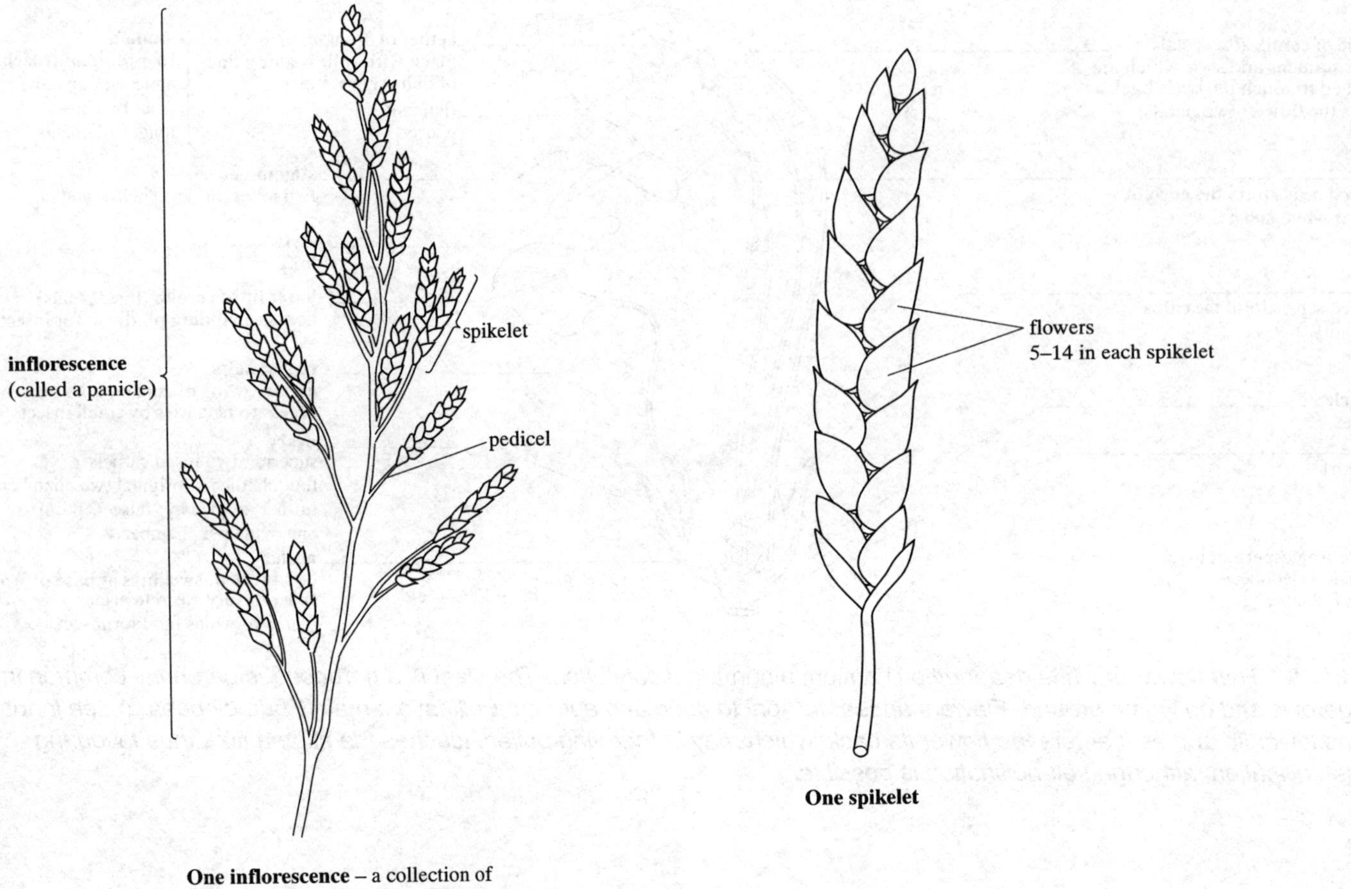

Fig 21.20 *Meadow fescue (*Festuca pratensis*), a wind-pollinated grass flower.*

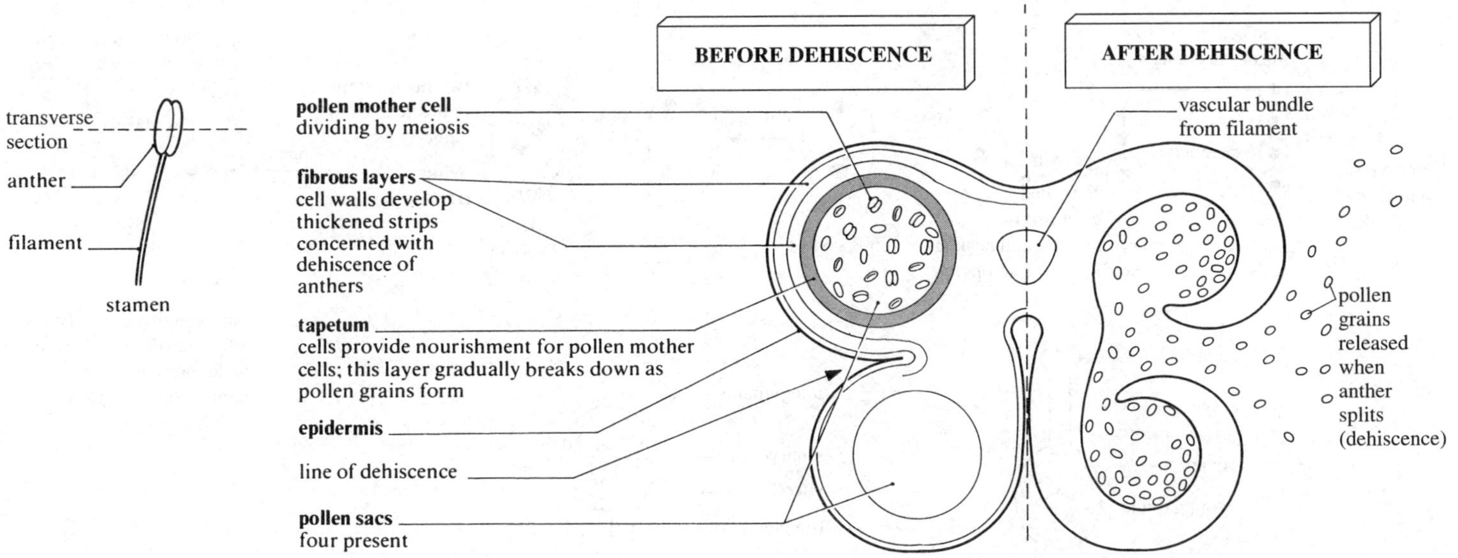

Fig 21.21 *TS mature anther before and after dehiscence.*

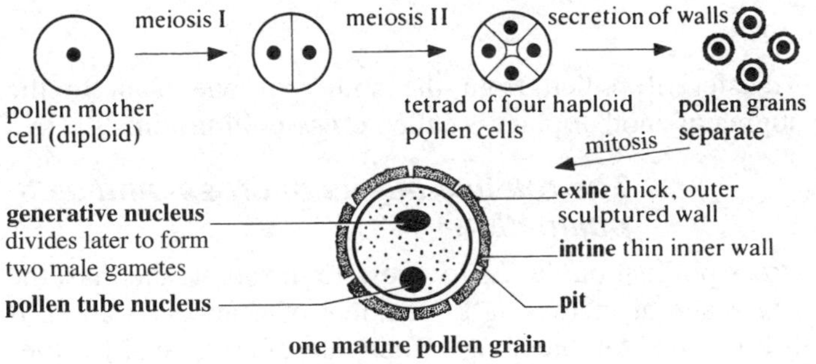

Fig 21.22 *Development of pollen grains.*

21.5.5 Pollination

After formation of pollen grains in the pollen sacs, the cells in the walls of the anther begin to dry and shrink, setting up tensions that eventually result in splitting (dehiscing) of the anther down the sides along two lines of weakness (fig 21.21). The pollen grains are thus released.

The transfer of pollen grains from an anther to a stigma is called **pollination**. (*Be careful not to confuse pollination with fertilisation.*) Pollination must be achieved if the male gametes, which develop inside the pollen grains, are to reach the female gamete. The male gametes are protected from drying out inside the pollen grain. Special mechanisms have evolved that increase the chances of successful pollination.

Transfer from an anther to a stigma of the same flower, or a flower on the same plant, is called **self-pollination**.

(*a*) (*b*)

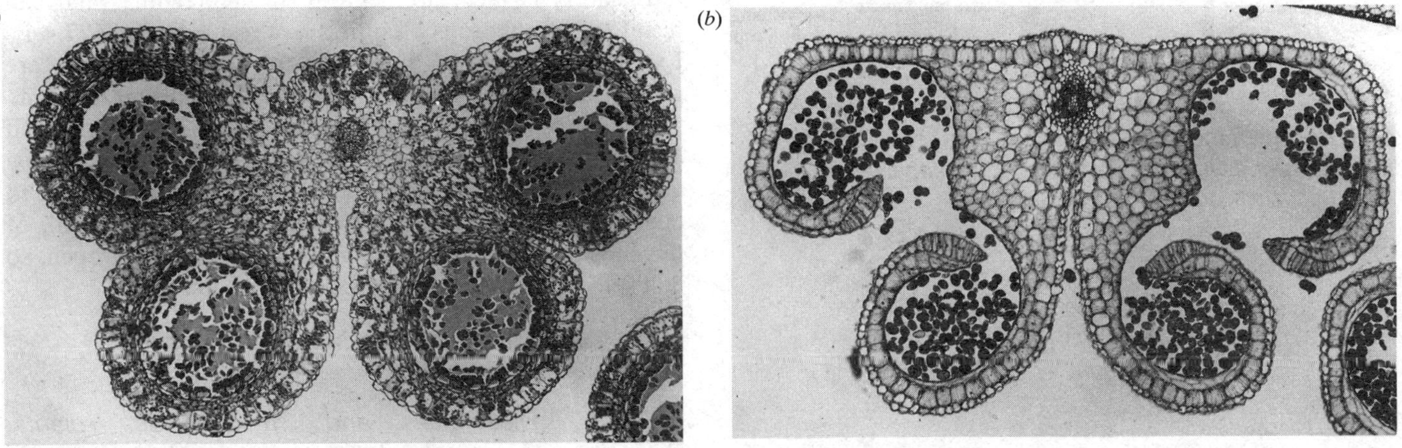

Fig 21.23 *TS anther of* Lilium *(a) before dehiscence and (b) after dehiscence.*

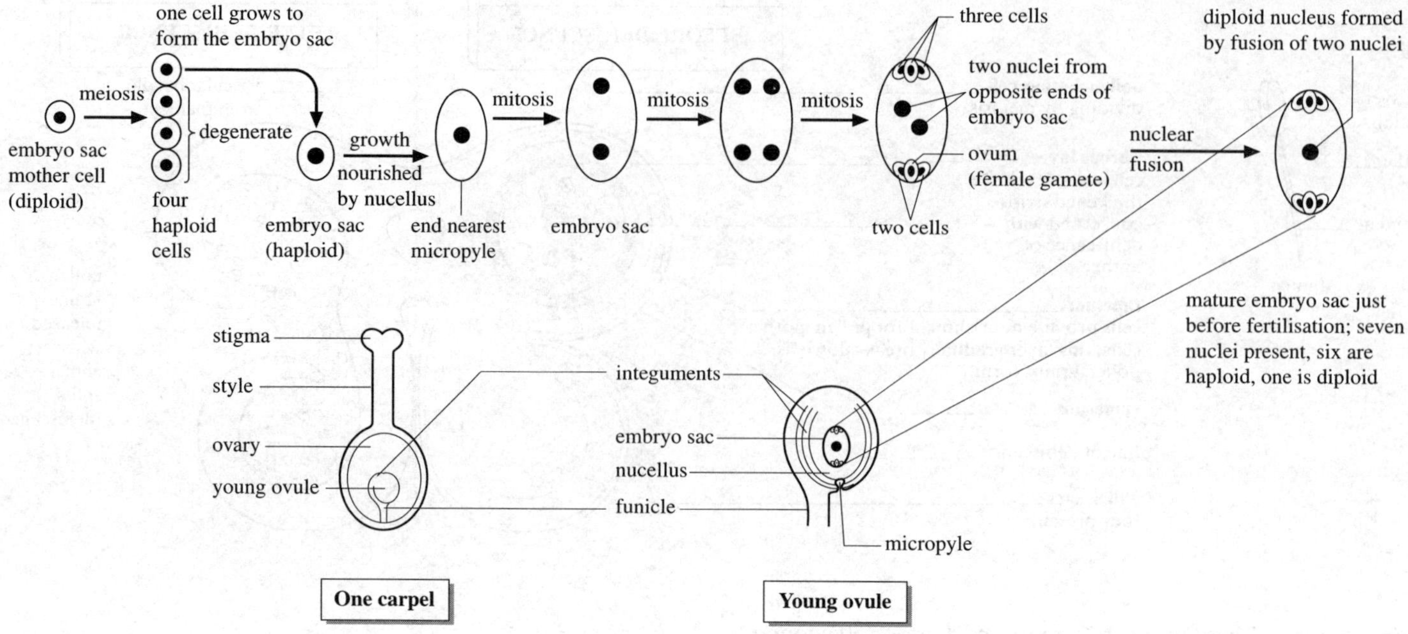

Fig 21.24 *Development of embryo sac and female gamete.*

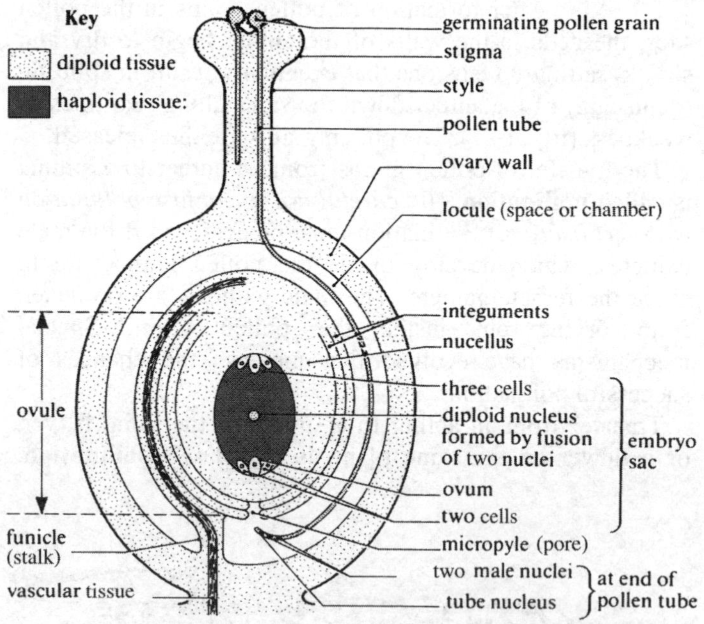

Fig 21.25 *LS carpel at fertilisation. Note that the ovule, which becomes the seed after fertilisation, contains both diploid parent tissue and haploid embryo sac tissue.*

Transfer of pollen from the anther of one plant to the stigma of another plant is called **cross-pollination.**

The relative merits of cross- and self-pollination

Cross-pollination leads to cross-fertilisation and has the advantage of increasing the amount of genetic variation. It is a form of 'outbreeding'. There are often special features to encourage it, some of which are described below. It is more wasteful of pollen, however.

Self-pollination leading to self-fertilisation has the advantage of greater reliability, particularly where members of the species are uncommon and are separated by large distances. This is because it is not dependent on an external factor, such as wind or insects, to deliver the pollen. It is also useful in harsh climates where insects are less common, such as high up on mountains. However, self-fertilisation is the extreme form of 'inbreeding' and can result in less vigorous offspring as a result (section 27.4.1). Examples of plants that rely on self-pollination are groundsel and chickweed. They produce no nectar or scent.

Both cross- and self-pollination have advantages and disadvantages and many plants balance the advantages by devices which favour cross-pollination but allow selfing to occur if crossing fails. For example, *some*, but not all, of the buds produced by violet and wood sorrel never open, so that self-pollination inside these is inevitable.

Mechanisms favouring cross-pollination

Dioecious and monoecious plants. Dioecious plant species have separate male and female plants. Self-pollination in dioecious plants is therefore impossible. Monoecious plants have separate male and female flowers on the same hermaphrodite plant. This also favours cross-pollination but selfing may also occur.

> **21.2** Dioecious plants are rare, despite the advantages of cross-pollination. Suggest two possible reasons for this.
>
> **21.3** Dioecism (separate sexes) is common in animals. Suggest why this phenomenon is more successful in animals than in flowering plants.

Protandry and protogyny. Sometimes anthers and stigmas mature at different times. If the anthers mature first, it is described as **protandry**; if the stigmas mature first it is **protogyny**. Protandrous flowers are much more common, for example white deadnettle, dandelion, and sage (*Salvia*) (fig 21.26). Protogyny occurs in bluebell and figwort (*Scrophularia*). In most cases of protandry and protogyny there is an overlapping period when both anthers and stigmas are ripe, thus allowing selfing if crossing has been unsuccessful.

Self-incompatibility (self-sterility). Even if self-pollination does occur, the pollen grain often does not develop, or develops very slowly, so preventing or reducing the chances of self-fertilisation. In all such cases there is a specific inhibition of pollen penetration of the stigma, or of pollen tube growth down the style, and this is genetically determined by 'self-incompatibility' genes.

When self-incompatibility occurs, some cross-pollinations are also incompatible. The most efficient use of the pollen will occur when a high proportion of cross-pollinations are compatible. An extreme example is clover, where all plants are self-incompatible, but cross-incompatibility occurs between less than one in 22 000 pairs.

> **21.4** Self-incompatibility is controlled by multiple alleles. Assuming (*a*) that there are three alleles, S_1, S_2 and S_3, and (*b*) that self-incompatibility occurs if the pollen grain and the style tissue have an allele in common, what proportion of the pollen grains from a plant with the genotype S_1S_2 would be capable of successfully germinating on a plant with the genotype S_2S_3?

Special floral structures. In most hermaphrodite flowers there are structural features that favour cross-pollination. In the case of insect-pollinated flowers the stigma is usually borne above the anthers, thus removing the possibility of pollen falling onto the stigma of the same flower. A visiting insect, possibly carrying pollen from another plant, will touch the stigma first as it enters the flower. Later, while the insect is seeking nectar, pollen is either brushed against it or falls onto it before it leaves the flower. This occurs in white deadnettle (fig 21.19). A more primitive mechanism may ensure that the stigma brushes against the insect as it lands, as in pea (see fig 21.18 for details). Such mechanisms are generally reinforced by protandry or protogyny and the flowers are often complex and irregular in shape, as in white deadnettle.

Flowers attract insects by providing a source of food (nectar or pollen) and stimulating the senses of sight and smell of the insects. The characteristics that enable flowers to do this are discussed below.

In the case of wind-pollinated flowers the stamens, the whole flower, or the inflorescence may hang downwards so that falling pollen will drop clear of the plant before being blown away, for example hazel catkins.

> **21.5** Fig 21.27 shows two types of primrose flower. They occur naturally in roughly equal numbers and differ in length of style and position of anthers.
> (*a*) Given that bees collect nectar from the base of the corolla tube, explain how cross-pollination between pin-eyed and thrum-eyed flowers, rather than between flowers of the same type, is favoured.
> (*b*) What is the advantage of such a system?

Although the variation in length of style described in question 21.5 and fig 21.27 apparently favours outbreeding, a much more important difference between pin-eyed and thrum-eyed primroses is a self-incompatibility mechanism which tends to restrict *cross-fertilisation* to pin-eye/thrum-eye crosses. The genes that control incompatibility, style length and anther height lie close together on the same chromosome and behave as a single inheritable unit.

Wind pollination and insect pollination

Pollen grains must be transferred to the female parts of flowers. The original agent of dispersal earlier in evolution was wind, but this is very inefficient because it relies solely on chance interception of the pollen grains by the flowers. Many flowering plants such as grasses, oak and hazel, still rely on wind, but they have to produce enormous quantities of pollen, a drain on the plant's materials and energy. Table 21.3 summarises the typical features of wind- and insect-pollinated flowers.

Insects specialised for flower-feeding appear in the fossil record at the same time as the flowering plants, and include bees, wasps, butterflies and moths. This suggests that once flowers evolved, insect pollination also rapidly evolved. The insect is a much more precise agent of dispersal than wind. It can carry a small amount of pollen from the anthers of one flower and deposit it precisely on the stigma of another flower. As a result, special relationships between flowers and insects have evolved. The reward the insects

Fig 21.26 *A bee entering a flower of meadow sage. The stamens are hinged to a plate and, as the head of the bee pushes against this, the stamens are lowered. This brushes pollen onto the bee's abdomen. The stigma of the flower increases in length as the flower ages. If a bee now enters an older flower, it will come into contact with the stigma and the pollen will be transferred from its abdomen to the stigma. This series of activities causes cross-pollination to take place.*

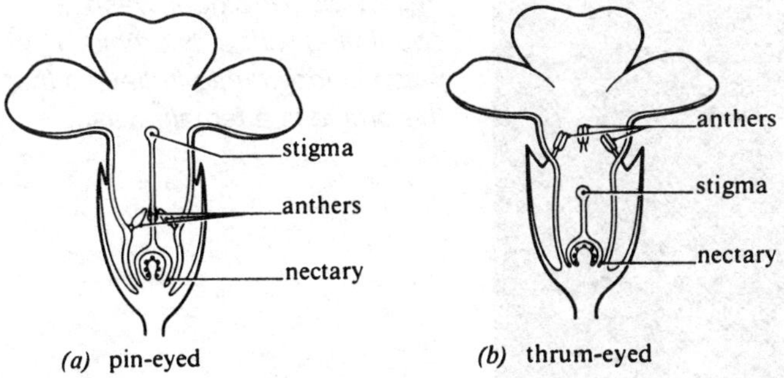

Fig 21.27 *Variation in flower structure in primrose.*

receive from the flowers is food in the form of nectar or pollen. In a few special cases the insect and the plant it pollinates are so interdependent that neither species can survive without the other, such as one species of yucca plant and its associated moth.

Table 21.3 Summary of typical differences between wind-pollinated and insect-pollinated flowers.

Typical wind-pollinated flower	*Typical insect-pollinated flower*
Small petals not brightly coloured (usually green), or petals absent; flowers therefore inconspicuous	Large coloured petals; flowers therefore conspicuous. If flowers relatively inconspicuous they may be gathered together in inflorescences
Not scented	Scented
Nectaries absent	Nectaries present
Large branched and feathery stigma hanging outside flower to trap pollen	Small stigma, sticky to hold pollen and enclosed within flower
Stamens hanging outside flower to release pollen	Stamens enclosed within flower
Anthers attached only at midpoints to tip of filament so that they swing freely in air currents	Anthers fixed at their bases or fused along their backs to the filaments so that they are immovable
Large quantities of pollen owing to high wastage	Less pollen produced
Pollen grains relatively light and small; dry, often smooth, walls	Pollen grains relatively heavy and large. Spiny walls and stickiness help attachment to insect body (fig 21.29)
Flower structure relatively simple	Complex structural modifications for particular insects often occur
Flowers borne well above foliage on long stalks (e.g. grasses) or appear before leaves (e.g. many British trees)	Position and time of appearance variable in relation to foliage, though often borne above it for increased conspicuousness

Insect pollination has the important additional advantage that it encourages cross-pollination and hence cross-fertilisation, so the modifications of flowers to encourage insect pollination described below could be added to the list of features favouring cross-pollination.

In order to attract insects, flowers generally are large, with brightly coloured or white petals or, if small, are grouped into inflorescences. Often there are markings on the petals such as lines, spots or an increased intensity of colour that guide insects to the nectaries, as in violet, pansy, orchids and foxgloves. Insects can see ultraviolet wavelengths that are invisible to humans so many flowers have markings which reflect ultraviolet that are invisible to us. More specific than colours are the scents produced by flowers, some of which, like lavender and rose, are used by humans in perfumes. Smells of rotting flesh that attract carrion-eating insects are also produced by some plants, such as the arum lily which attracts dung flies. Recognition is also helped by flower shape.

One of the most complex and strange mechanisms for ensuring cross-pollination is the sexual impersonation of female wasps by certain orchids. The flower parts mimic the shape, colourings and even the odour of the female wasp, and the impersonation is so convincing that the male wasps attempt to copulate with the flower (fig 21.28). While doing so they deposit pollen and, on leaving the flower, collect fresh pollen to take to the next flower.

An example of a wind-pollinated flower is the grass, meadow fescue, shown in fig 21.20. Examples of insect-pollinated flowers are shown in figs 21.17 (buttercup), 21.18 (sweet pea) and 21.19 (white deadnettle).

21.5.6 Fertilisation

Once a pollen grain has landed on the stigma (fig 21.29) of a compatible species, it will germinate. A sucrose solution is secreted by the epidermal cells of the stigma. This stimulates germination of the grain and possibly supplies food. A **pollen tube** emerges from one of the pores (pits) in the wall of the pollen grain and grows rapidly down the style to the ovary. Its growth is controlled by the tube nucleus of the pollen grain, which is found at the growing tip of the tube. Growth is stimulated by auxins produced by the gynoecium, and the pollen tube is directed towards the ovary by certain chemicals, an example of chemotropism. It is probably also negatively aerotropic, that is it grows away from air. Growth depends on compatibility between the pollen and the style tissue as already described.

During growth of the pollen tube the **generative nucleus** of the pollen grain divides by mitosis to produce two male nuclei that represent the **male gametes** (fig 21.25). They cannot swim, unlike the sperm of lower plants, and depend on the pollen tube to reach the female gamete which is located in the embryo sac of the ovule. The pollen tube enters the ovule through the micropyle, the tube nucleus degenerates and the tip of the tube bursts, releasing the

Fig 21.28 *A digger wasp copulating with a fly orchid. The wasp is fooled into believing that the orchid is a female wasp.*

male gametes near the embryo sac, which they enter. One nucleus fuses with the female gamete, forming a diploid **zygote**, and the other fuses with the diploid nucleus forming a triploid nucleus known as the **primary endosperm nucleus**. This double fertilisation is unique to flowering plants. It leads eventually to the two structures found in the seed, namely the **embryo** and the **endosperm** (food store).

If, as is often the case, more than one ovule is present in the gynoecium, each must be fertilised by a separate pollen grain if it is to become a seed. Thus each seed may have been fertilised by a pollen grain from a different plant.

Experiment 21.1: To investigate the growth of pollen tubes

Stigmas secrete a solution containing sucrose ranging in concentration from about 2 to 45%. This helps to stick pollen grains to the stigma and to promote their germination. The addition of borate to the experimental solution helps to prevent osmotic bursting of pollen tube tips and stimulates growth.

Materials

- microscope
- cavity slide
- flowers containing dehiscing anthers, such as deadnettle, wallflower, *Pelargonium* or *Impatiens*
- 10–20% (w/v) sucrose solution also containing sodium borate to a concentration of 0.01%
- acetocarmine or neutral red

Method

Place a drop of sucrose solution in the central depression of a cavity slide and add pollen grains by touching the drop with the surface of a dehisced anther. Observe the slide at intervals over a period of 1–2 hours. The nuclei at the tip of growing tubes may be stained by irrigating with a drop of acetocarmine or neutral red.

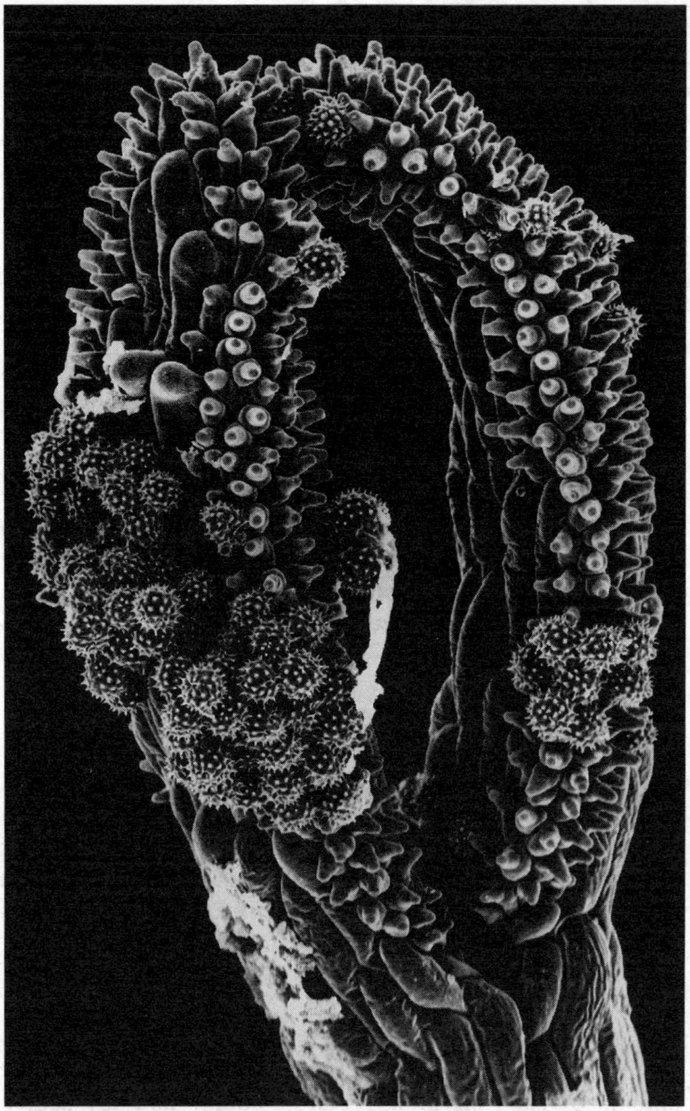

Fig 21.29 *Scanning electron micrograph of pollen grains on the stigma of a flower. The spiked surface of the grains is typical of insect-pollinated flowers.*

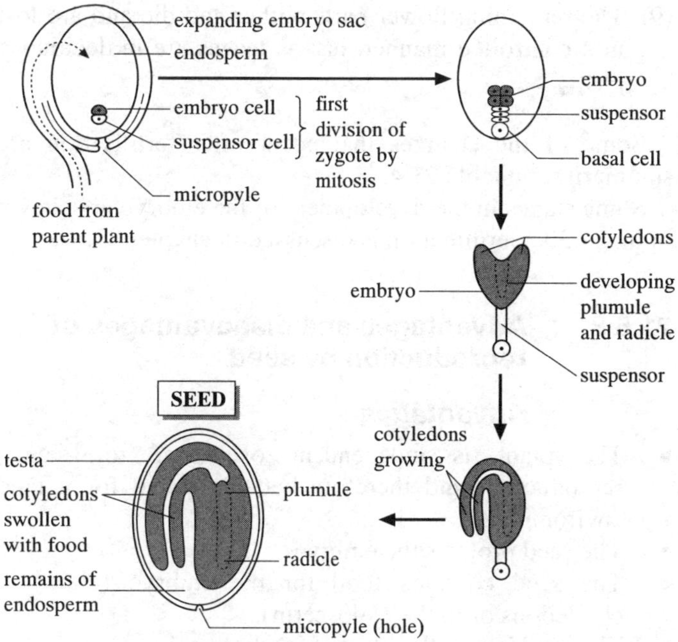

Fig 21.30 *Growth of an embryo in a non-endospermous dicotyledon seed, such as shepherd's purse (*Capsella bursa-pastoris*).*

21.5.7 Development of the seed and fruit

Immediately after fertilisation, the ovule becomes known as the **seed** and the ovary the **fruit**. The following changes take place. It will be useful to study figs 21.30 and 21.25 in order to understand these changes.

(1) The zygote grows by mitotic divisions to become a multicellular embryo which consists of a first shoot, the **plumule**, a first root, the **radicle**, and either one or two seed-leaves called **cotyledons** (one in monocotyledons and two in dicotyledons). These cotyledons are simpler in structure than the first true foliage leaves and may become swollen with food to act as storage tissue, as in the pea and broad bean. The plumule consists of a stem, the first pair of true foliage leaves and a terminal bud.

(2) The triploid primary endosperm nucleus undergoes repeated mitotic divisions to form the **endosperm**, a mass of triploid nuclei which are separated from one another by thin cell walls. In some seeds this remains as the food store, as in cereals such as wheat and maize.

(3) If the cotyledons act as a food store they grow at the expense of the endosperm, which may disappear altogether. Some seeds store food in both endosperm and cotyledons.

(4) As growth of the embryo and food store continues, the surrounding nucellus breaks down supplying nutrients for growth. Further nutrients are supplied by the vascular bundle in the stalk (funicle) of the ovule.

(5) The testa develops from the integument. It is a thin but tough protective layer.

(6) The micropyle remains a small pore in the testa through which oxygen and water will enter when the seed germinates.

(7) The final stages in development of the seed involve a reduction in the water content of the seed from the normal levels for plant tissues of about 90% by mass to about 10–15% by mass. This greatly reduces the potential for metabolic activity and is an essential step in ensuring seed dormancy.

(8) While the seeds develop, the ovary becomes a mature fruit, its wall being known as the **pericarp**. The changes that occur vary with species, but generally the fruit is adapted to protect the seeds and to help in their dispersal. The hormonal control of fruit development is discussed in section 16.3.5.

(9) The remaining flower parts wither and die and are lost in a controlled manner, just as leaves are in deciduous plants.

Some of the changes that occur after fertilisation are summarised in table 21.4.

Some stages in the development of the embryo are shown in fig 21.30. Germination is discussed in chapter 22.

21.5.8 Advantages and disadvantages of reproduction by seed

Advantages

- The plant is independent of water for sexual reproduction and therefore better adapted for a land environment.
- The seed protects the embryo.
- The seed contains food for the embryo (either in cotyledons or in the endosperm).
- The seed is usually adapted for dispersal.
- The seed can remain dormant and survive adverse conditions.
- The seed is physiologically sensitive to favourable conditions and sometimes must undergo a period of after-ripening so that it will not germinate immediately (chapter 16).

Disadvantages

- Seeds are relatively large structures because of the extensive food reserves. This makes dispersal more difficult than by spores.
- Seeds are often eaten by animals for their food reserves.
- There is a reliance on external agents such as wind, insects and water for pollination. This makes pollination (and hence fertilisation) more dependent on chance, particularly wind pollination.
- There is a large wastage of seeds because the chances of survival of a given seed are limited. The parent must therefore invest large quantities of material and energy in seed production to ensure success.
- The food supply in a seed is limited, whereas in vegetative reproduction food is available from the parent plant until the daughter plant is fully established.
- Two individuals are required in dioecious species, making the process more dependent on chance than reproduction in which only one parent is involved. However, dioecious plants are relatively rare.

The information provided above can be used to compare the advantages and disadvantages that seed-bearing plants have compared with non-seed-bearing plants, or to compare the relative merits of sexual reproduction and vegetative propagation within the seed-bearing plants.

21.6 Review of sexual reproduction in vertebrates

The evolution of the vertebrates shows a gradual adaptation to life on land. One of the major problems that had to be overcome in making the transition from an aquatic existence to a terrestrial existence involved reproduction.

Fish

The majority of fish shed their gametes directly into water. Fertilisation is external. Eggs contain a considerable amount of yolk, larval stages are common and any degree of parental care is rare.

Table 21.4 Summary of the changes that occur after fertilisation in flowering plants.

Before fertilisation		*After fertilisation*			
ovule	female gamete + male gamete	zygote → embryo	plumule; radicle; one or two cotyledons (may or may not store food)	seed – a fertilised ovule, the seed, or seeds, are contained in the fruit	fruit – the fertilised ovary
	diploid nucleus + male gamete	primary endosperm nucleus →	endosperm (may or may not persist as a food store)		
	nucellus		disappears		
	integuments		become the testa		
	ovary wall		becomes the pericarp		

Amphibia

Amphibia have to return to water to mate, and the early stages of their development take place there also. There are, however, many amphibian species that show elaborate behavioural patterns associated with parental care. For example, the male *Pipa* toad spreads the fertilised eggs over the back of the female where they stick, become 'embedded' in the skin and develop into tadpoles. After about three weeks they escape from the mother's back and lead an independent existence.

Reptiles

Reptiles were the earliest group of vertebrates to overcome the problems of fertilisation and development on land. Release of gametes on to dry land would result in their drying up, so the first requirement of totally land-dwelling organisms must have been the introduction of male gametes into the female body, that is internal fertilisation. Internal fertilisation occurs in reptiles and the increased chances of fertilisation reduces the numbers of gametes which it is necessary to produce. Once fertilised, the zygote develops within a specialised structure, the **amniote (cleidoic) egg**, which provides the embryo with a fluid-filled cavity in which it can develop on land (fig 21.31). The outer shell provides protection from mechanical damage and surrounds the four membranes which develop from the embryo. These four extra-embryonic membranes, the yolk sac, amnion, chorion and allantois, provide the embryo with protection and are necessary for many of its metabolic activities including nutrition, respiration and excretion.

The **yolk sac** develops as an outgrowth of the embryonic gut and encloses the yolk, a food supply which is gradually absorbed by the blood vessels of the yolk sac. When the yolk has been used up the yolk sac is withdrawn into the gut.

The **amnion** completely encloses the embryo in the **amniotic cavity** which becomes filled with **amniotic fluid** secreted by the cells of the amnion. This provides the embryo with a fluid environment in which the embryo can develop. All reptiles, birds and mammals have an amnion and are called **amniotes**. As the embryo grows, the amnion is pushed out until it fuses with the third embryonic membrane, the **chorion**, which lies just inside the shell and prevents excessive water loss from the amnion.

The **allantois** is an outgrowth of the embryonic hindgut and rapidly expands, in reptiles and birds, to underlie the chorion. Here it functions primarily as a 'bladder' for storing excretory products and as the gaseous exchange organ of the embryo, facilitating the transfer of respiratory gases between the environmental atmosphere and the amniotic fluid via the porous shell.

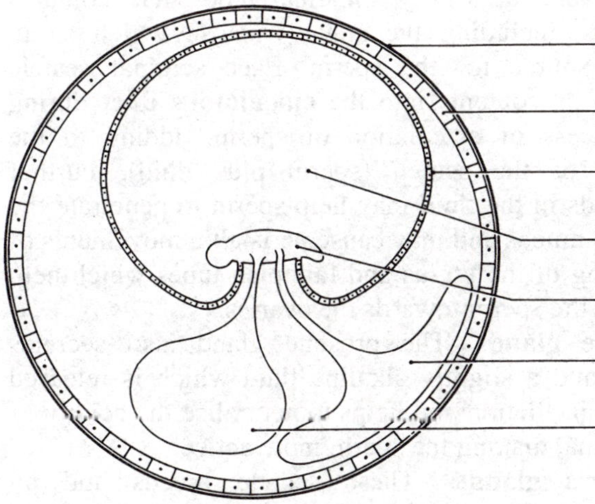

Fig 21.31 *Simplified diagram of the amniote egg.*

Birds and egg-laying mammals

Birds and the early egg-laying mammals called monotremes, such as the duck-billed platypus, *Ornithorhyncus*, all produce an amniotic egg and, whilst the shell of the egg is lost in higher mammals, the four extra-embryonic membranes are retained; two of them, the chorion and allantois, give rise to the placenta in placental mammals (fig 21.49).

21.7 Human reproductive systems

21.7.1 The male reproductive system

The structure of the male reproductive system is shown in side view in fig 21.32*a* and more diagrammatically in front view in fig 21.32*b*. The excretory system is closely associated with the reproductive system and is also shown in the diagrams. The two systems together are traditionally known as the urinogenital system.

The main structures and their functions are listed below.

- **Testes** (singular testis). The two testes are the male **gonads**, that is the sites where the male gametes, or sperm, are made. They also produce the male sex hormone **testosterone**.
- **Scrotal sac**. The testes are situated outside the abdominal cavity in a sac of skin called the scrotal sac. As a result, the sperm develop at a temperature 2–3 °C lower than the main body temperature. This is the optimum temperature for sperm production. The life of sperm is greatly reduced if the temperature is higher.
- **Seminiferous tubules**. Inside each testis are about 1000 coiled seminiferous tubules (fig 21.33). Each tubule is about 50 cm long and 200 μm in diameter. The walls of the tubules produce the sperm (also known as spermatozoa), a process called **spermatogenesis**. **Leydig cells**, also known as **interstitial cells**, between the tubules produce the male sex hormone **testosterone**.
- **Vasa efferentia** (singular vas efferens). 10 to 20 vasa efferentia collect sperm from inside the testis and transfer them to the epididymis (fig 21.33).

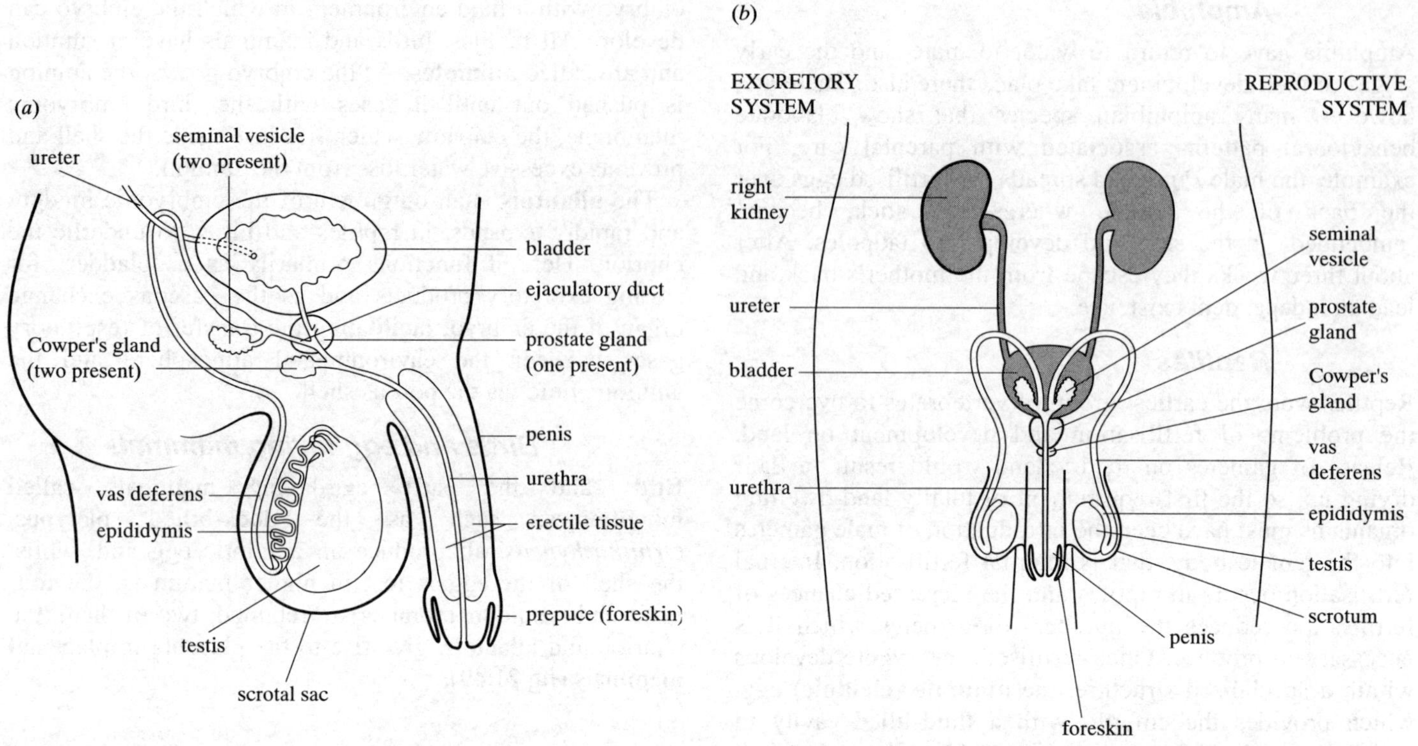

Fig 21.32 *(a) Side view of the human male reproductive system. (b) Male reproductive system seen from the front. The shaded structures are the excretory system. The urethra is part of both systems, being also connected to the bladder. The two systems together are often referred to as the urinogenital system.*

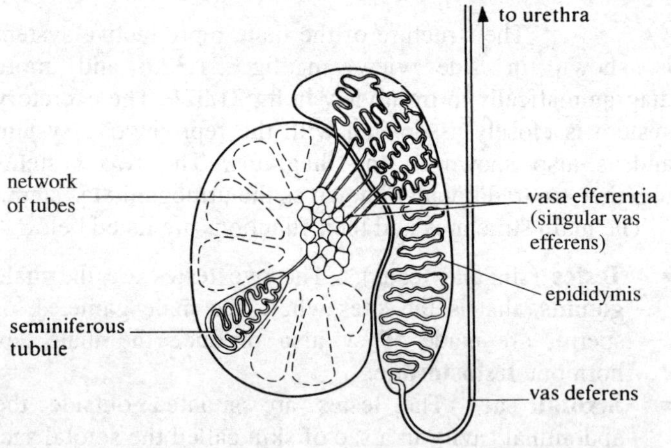

Fig 21.33 *Simplified diagram showing the structure of the human testis and tubes carrying sperm from seminiferous tubules to the urethra.*

- **Epididymis**. The epididymis is a very coiled tube, about 6 m long, pressed against the testis. Sperm take several days to pass through this tube. Sperm are concentrated here to about 5000 million per cm^3 by reabsorption of fluid secreted originally by the seminiferous tubule. They also develop the ability to swim, although they are inhibited from doing so until after ejaculation. Sperm pass to the base of the tube for a short period of storage before entering the vas deferens. Chemicals produced by the lining of the tube are essential for maturation of the sperm.
- **Vas deferens** (plural vasa deferentia). This is a straight tube about 40 cm long which carries sperm to the urethra. Most of the sperm are stored in the vas deferens.
- **Urethra**. This tube carries urine from the bladder, as well as sperm from the vasa deferentia, through the penis.
- **Penis**. The penis contains erectile tissue. When the male is sexually excited this tissue fills with blood, causing the penis to become erect. During sexual intercourse the erect penis is inserted into the vagina of the female before ejaculation of the semen.
- **Seminal vesicles**. The seminal vesicles secrete mucus and a watery alkaline fluid that contains nutrients, including the sugar fructose which is an energy source for the sperm. Each seminal vesicle empties its contents into the **ejaculatory duct** during the process of ejaculation of sperm, adding to the volume of the semen (sperm plus fluid). Further chemicals in the fluid may help sperm to penetrate the cervical mucus and may cause peristaltic movements of the lining of the uterus and fallopian tubes which help to carry the sperm towards the ovaries.
- **Prostate gland**. The prostate gland also secretes mucus and a slightly alkaline fluid which is released during ejaculation and helps to neutralise the acidity of the vagina, making the sperm more active.
- **Cowper's glands**. These secrete mucus and an alkaline fluid into the urethra. The alkaline fluid neutralises the acidity of any remaining urine.

21.7.2 The female reproductive system

The structure of the female reproductive system is shown in fig 21.34. The main structures and their functions are listed below. Note that, unlike the structures in the male, there are separate external openings to the excretory and reproductive systems.

- **Ovaries**. The two ovaries are the female gonads, the sites where the female gametes are made. The gametes are known as **eggs** or **ova**. (Biologists and the medical profession usually refer to them as eggs.) The ovaries are almond-shaped, measure about 3–5 cm long, and 2–3 cm wide and also secrete the female sex hormones **oestrogen** and **progesterone**. Usually, one egg is produced every month during the fertile years of a woman. The outermost layer of cells of the ovary is composed of germinal epithelial cells from which gamete cells are produced. The outer region of the ovary is composed of developing follicles and the middle is composed of **stroma**, which contains connective tissue, blood vessels and mature follicles.
- **Oviducts or fallopian tubes**. The tubes are about 12 cm long and carry eggs from the ovaries to the uterus. The ends of the tubes nearest the ovaries have feathery processes called **fimbriae**. They move closer to the ovaries at ovulation. Cilia lining the fimbriae beat and cause a current which draws in the ovum or egg (more precisely the secondary oocyte) after it is released from the ovary. Cilia lining the oviduct beat and smooth muscle contracts causing peristaltic movements which move the egg down the oviduct to the uterus. If fertilisation takes place it occurs in the oviduct.
- **Uterus** (womb). The uterus is about 7.5 cm long and 5 cm wide and is about the size and shape of an inverted pear. It lies behind the bladder. If fertilisation takes place, the embryo implants in the wall of the uterus and grows there until birth. The uterus grows much larger during pregnancy. The outer layer of the uterus wall, the **myometrium**, contains smooth muscle which contracts strongly during birth. The inner layer, the **endometrium**, contains glands and many blood vessels.
- **Cervix**. This is the narrow entrance to the uterus from the vagina. It is normally blocked by a plug of mucus and a ring of muscle can close it.
- **Vagina**. This is a muscular tube about 8–10 cm long whose walls contain elastic tissue. The lining is folded. It stretches during childbirth to allow passage of the baby, and during sexual intercourse when the penis is placed in it. The **clitoris** is a small structure which is equivalent to the male penis and like the penis can become erect.

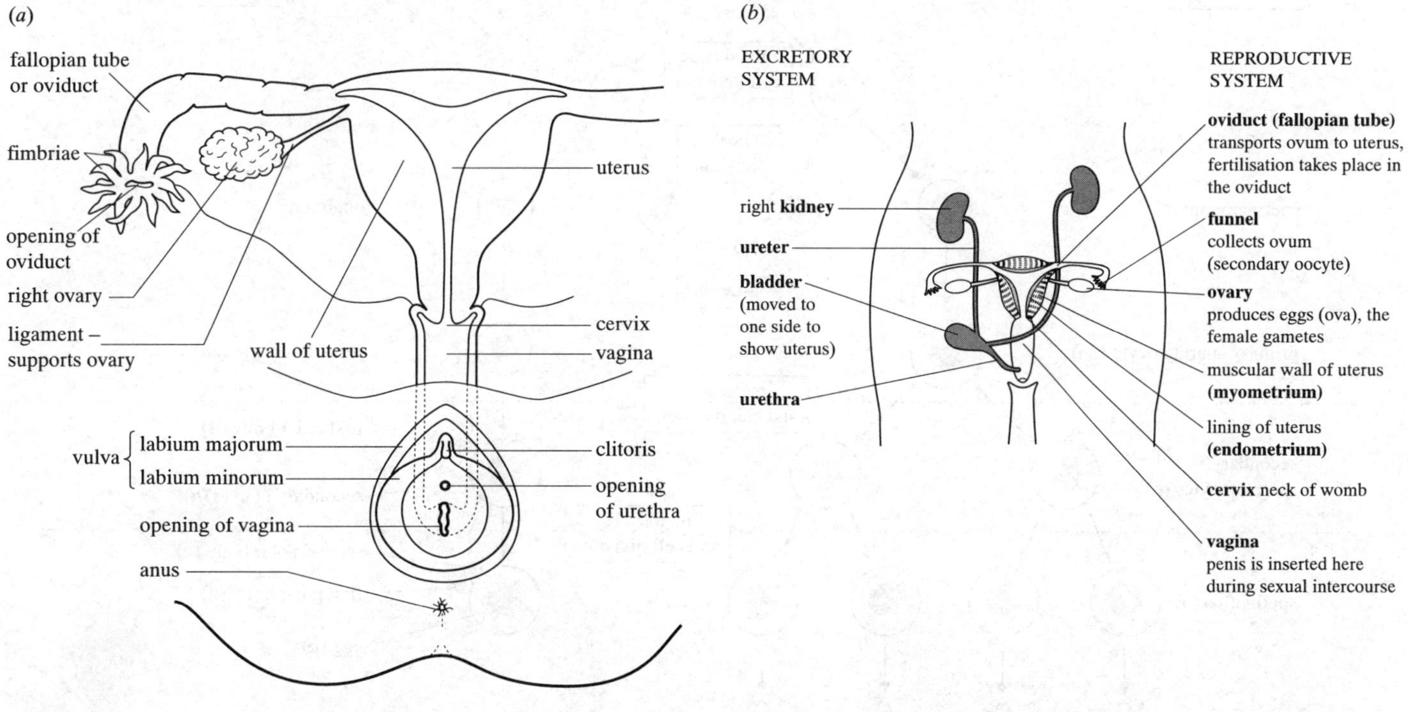

Fig 21.34 *(a) Diagram showing the human female reproductive system. The uterus and vagina are shown in section. The external genitalia and the openings of the urethra and anus are shown in surface view with the labia parted. (b) Female reproductive system seen from the front. The shaded structures are the excretory system. The two systems together are often referred to as the urinogenital system.*

21.7.3 Gametogenesis

There are three main stages to reproduction, namely gametogenesis, fertilisation and the development of the embryo. **Gametogenesis** is the production of gametes. Production of sperm is called **spermatogenesis** and production of eggs is called **oogenesis**. Both take place in the gonads, namely the testes in the male and the ovaries in the female. Both processes involve meiosis, the type of nuclear division which halves the number of chromosomes from two sets (diploid condition) to one set (haploid condition). The cells which undergo meiosis are called mother cells. Sperm mother cells are known as **spermatocytes** and egg mother cells as **oocytes**. Fig 21.35 is a diagrammatic summary of spermatogenesis and oogenesis which emphasises the main similarities and differences between the two processes. The processes are described in more detail in sections 21.7.4 and 21.7.5.

Note that both processes start with cells in the outer layer of the gonad, known as the **germinal epithelium**. In both males and females the process involves three stages, a multiplication stage, a growth stage and a maturation stage. The multiplication stage involves repeated mitotic divisions producing many spermatogonia and oogonia. Each then undergoes a period of growth in preparation for the first meiotic division and cell division. This marks the beginning of the maturation stage during which the first and second meiotic divisions occur followed by the formation of mature haploid gametes.

The gametes produced by a given individual will show variation as a result of independent assortment of chromosomes and crossing over during meiosis (section 23.4).

21.7.4 Spermatogenesis – development of sperm

Sperm are produced at the rate of about 120 million per day. Production of a given sperm takes about 70 days. The seminiferous tubule has a wall with an outer layer of **germinal epithelial cells** and about six layers of cells produced by repeated cell divisions of this layer (figs 21.36 and 21.37*a* and *b*). These represent successive stages in the development of sperm. The first divisions of the germinal epithelial cells give rise to many **spermatogonia** which increase in size to form **primary spermatocytes**.

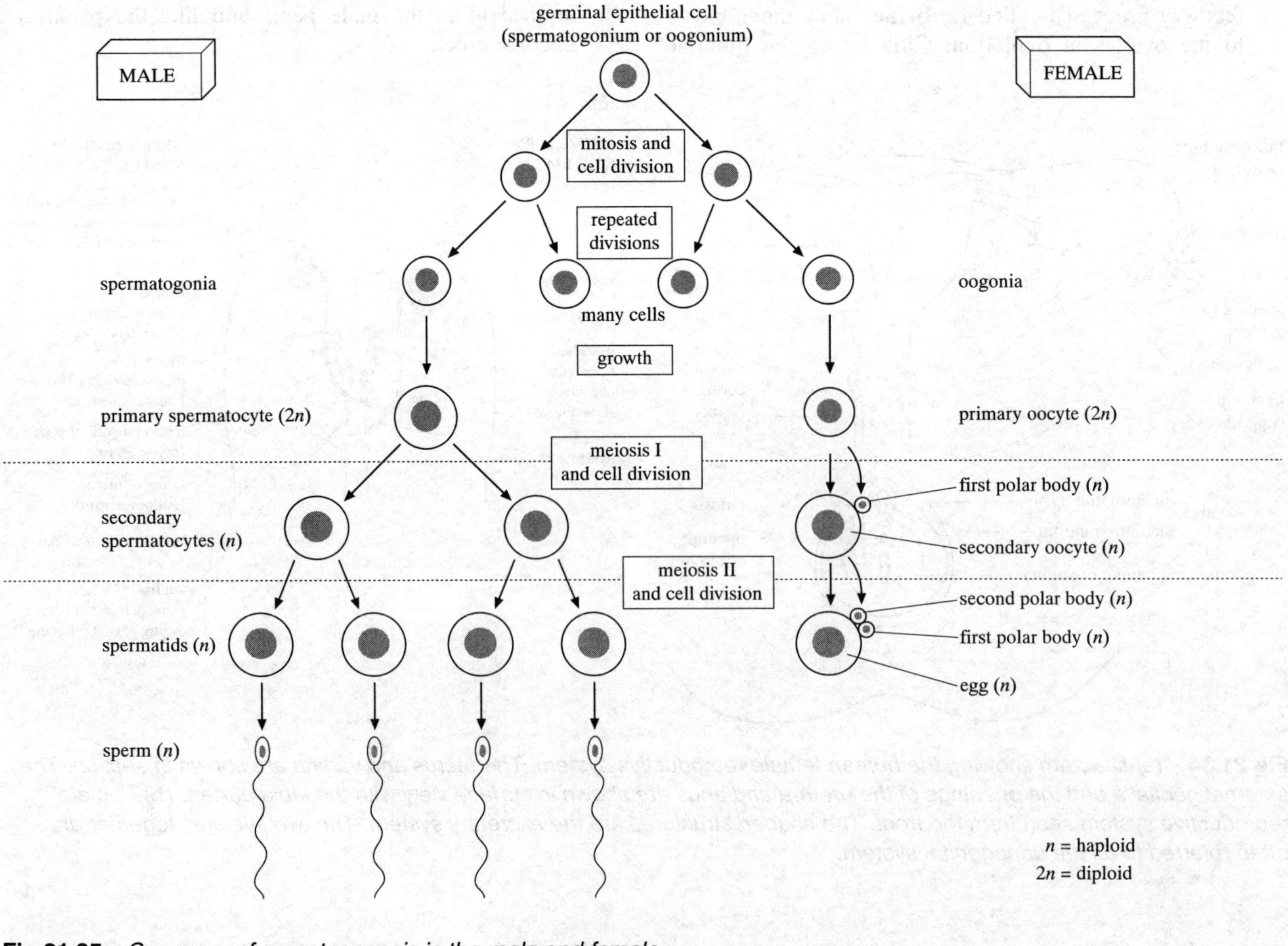

Fig 21.35 *Summary of gametogenesis in the male and female.*

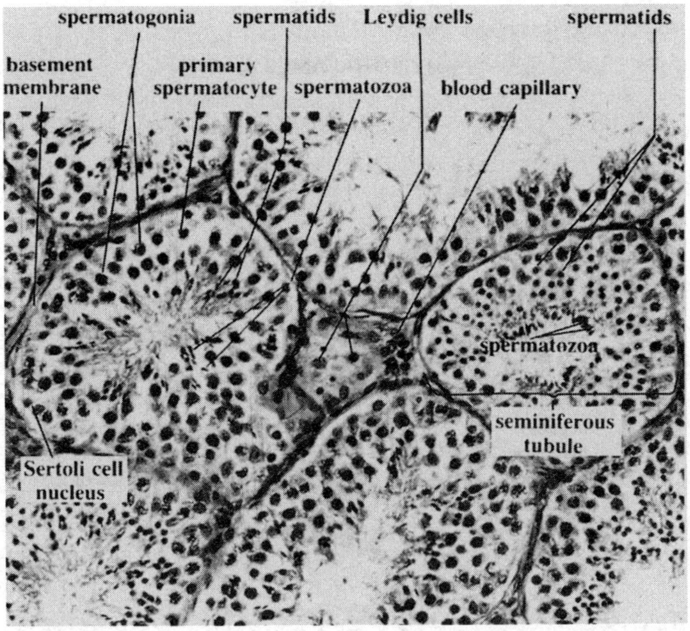

Fig 21.36 *Section through the human testis showing seminiferous tubules and Leydig (interstitial) cells.*

These undergo the first meiotic division to form haploid **secondary spermatocytes** and the second meiotic division to form **spermatids**. Between these rows of developing cells are large **Sertoli cells** stretching from the outer layer of the tubule to the lumen.

> **21.6** Sertoli cells contain abundant smooth endoplasmic reticulum, Golgi apparatus and many mitochondria and lysosomes. In view of the structure of these cells what can you suggest about their function?

Spermatocytes become embedded in the many infoldings of the cell surface membranes of the Sertoli cells and develop into spermatids before passing to the tops of the cells next to the lumen (central space) of the seminiferous tubule. Here they become mature **sperm**. The Sertoli cells carry out the remoulding of the spermatids to form sperm. Also all nutrients, oxygen and waste substances exchanged between the developing sperm and the blood vessels surrounding the tubules pass through the Sertoli cells. The fluid carrying sperm through the tubules is secreted by the Sertoli cells. The entire process from spermatogonia to sperm takes about two months.

Sperm

Sperm are extremely small cells, only 2.5 μm in diameter (compared with an average of 20 μm for animal cells) and about 50 μm long. The structure is shown in fig 21.38. The head contains the nucleus, which contains the haploid number of chromosomes (23 in humans). It also contains the **acrosome**, a large lysosome which contains hydrolytic enzymes which will be involved in the penetration of the layers of cells surrounding the egg immediately before fertilisation.

The short neck region of the sperm contains a pair of centrioles lying at right-angles to each other. The microtubules of one of the centrioles elongate during development of the sperm and run the entire length of the mature sperm forming the **axial filament** of the tail or flagellum.

The middle piece is the first part of the tail and is enlarged by the presence of many mitochondria arranged in a spiral around the axial filament. Mitochondria carry out aerobic respiration and produce ATP as a source of energy. The energy is used to bring about beating movements of the tail, allowing the sperm to swim at about 1–4 mm per minute. In transverse section the tail shows the

(a)

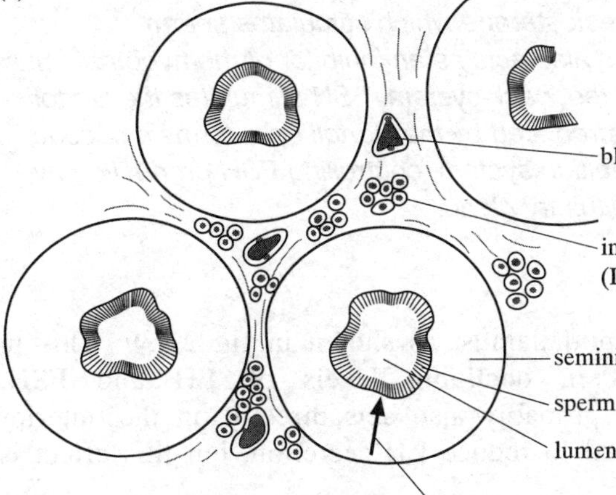

(b)

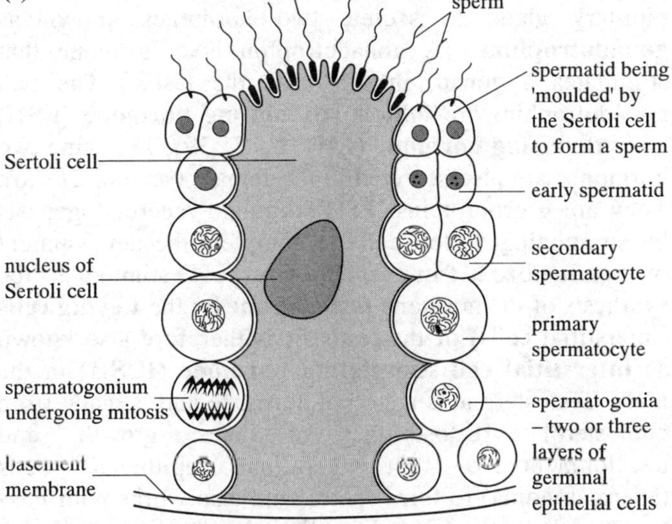

Fig 21.37 *(a) A group of seminiferous tubules seen in transverse section. (b) Diagram showing the structure of part of the wall of a seminiferous tubule. Cells in various stages of spermatogenesis are shown.*

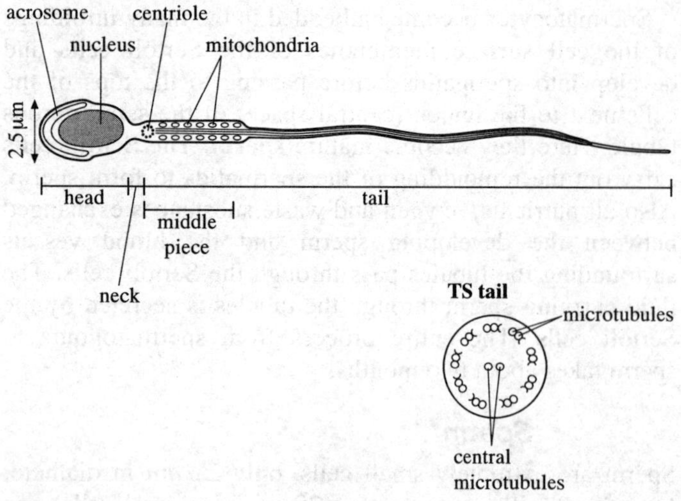

Fig 21.38 *Structure of a mature human sperm.*

characteristic arrangement found in flagella of nine pairs of peripheral microtubules surrounding a central pair of microtubules.

Activation of the tail takes place in the vagina. Movement of the tail is insufficient to cover the distance from the vagina to the site of fertilisation. However, its main function is to enable sperm to cluster around the oocyte, to orientate themselves and to help penetrate the oocyte.

Hormonal control of spermatogenesis

Spermatogenesis is controlled by the hypothalamus and anterior pituitary gland working together. The hypothalamus is part of the brain (fig 17.24) and the pituitary gland is found just below it. The hypothalamus secretes **gonadotrophin-releasing hormone (GnRH)** which travels in a small vein from the hypothalamus to the pituitary gland. GnRH in turn stimulates the anterior pituitary gland to secrete two hormones known as **gonadotrophins**. (A gonadotrophin is a hormone that stimulates a gonad, in this case the testis.) The two gonadotrophins are **follicle stimulating hormone (FSH)** and **luteinising hormone (LH)** (fig 21.39). The same two hormones are also secreted in the female (section 21.7.6). They are glycoproteins. FSH stimulates spermatogenesis by stimulating Sertoli cells to complete the development of spermatozoa from spermatids. LH stimulates the synthesis of the hormone **testosterone** by the **Leydig cells** (interstitial cells) of the testis. It is therefore also known as **interstitial cell stimulating hormone (ICSH)** in the male. Testosterone is a steroid hormone and is made from cholesterol. Testosterone stimulates growth and development of the germinal epithelial cells (spermatogonia) to form sperm, and also works with FSH to stimulate the Sertoli cells. A negative feedback mechanism operates whereby an increase in the level of testosterone results in a decrease in secretion of GnRH

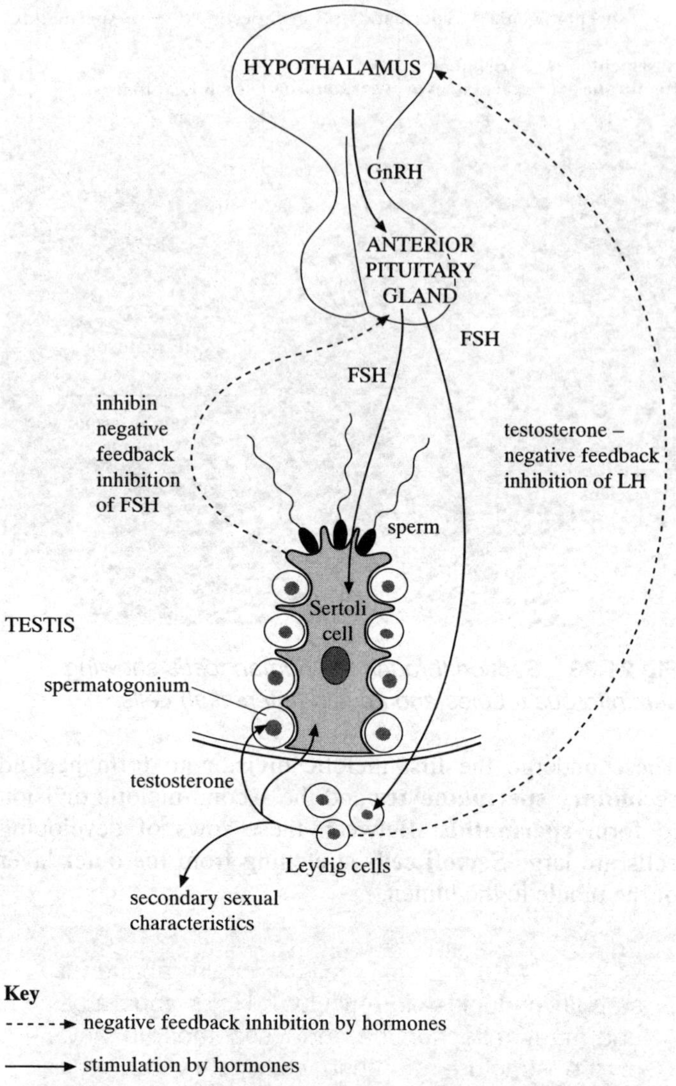

Fig 21.39 *Hormonal control of spermatogenesis. GnRH secreted by the hypothalamus stimulates the anterior pituitary gland to secrete FSH and LH. LH stimulates secretion of testosterone which stimulates sperm production but also acts as an inhibitor of the hypothalamus in a negative feedback system. FSH stimulates the Sertoli cells. Inhibin produced by the Sertoli cells forms a second feedback inhibition system, controlling FSH production by the anterior pituitary gland.*

from the hypothalamus, as shown in fig 21.39. This in turn results in declining levels of LH and FSH. Testosterone probably also acts directly on the anterior pituitary gland to reduce LH secretion, but this effect is weaker.

Role of inhibin

The Sertoli cells secrete another glycoprotein hormone called **inhibin** which is involved in the negative feedback control of sperm production. If spermatogenesis proceeds too rapidly, inhibin is released. Its target is the anterior pituitary gland where it reduces secretion of FSH (fig 21.39). It may also have a slight effect on the hypothalamus, reducing GnRH secretion. When the rate of spermatogenesis is low, inhibin is not secreted and FSH stimulates spermatogenesis.

Role of cyclic AMP

Both FSH and LH act by causing the release of cyclic AMP (adenosine monophosphate) within the cells they stimulate. Cyclic AMP is the 'second messenger' system discussed in section 17.6.1. It is released into the cytoplasm and then passes to the nucleus where it stimulates the synthesis of enzymes. In the case of LH, for example, the enzymes are involved in the synthesis of testosterone from cholesterol.

Secondary sexual characteristics

Testosterone is the main male sex hormone and it affects both primary and secondary sexual characteristics. Primary sexual characteristics are those present at birth, whereas secondary sexual characteristics are those that develop at puberty. Both testosterone and FSH are required for the successful production of sperm whereas testosterone alone controls the development of the secondary sexual characteristics during puberty and maintains these throughout adult life. These characteristics include:

- the development and enlargement of the testes, penis and glands of the reproductive tract;
- increased muscle development;
- enlargement of the larynx producing deepening of the voice;
- the growth of pubic hair and extra hair on the face, in the armpits and on the chest;
- changes in behaviour associated with courtship, mating and parental concern.

21.7.5 Oogenesis – development of eggs

Unlike the production of sperm in males, which only begins at puberty, the production of eggs in females begins before birth. An outline of the stages in oogenesis is shown in fig 21.35. During development of the fetus many oogonia are produced. These undergo mitosis and form **primary oocytes** which remain at prophase of meiosis I throughout childhood. Primary oocytes are enclosed by a single layer of cells, the **granulosa cells** (or follicle cells), and form structures known as **primordial follicles** (figs 21.40, 21.41 and 21.42). About two million of these follicles exist in the female just before birth, but only about 450 ever develop secondary oocytes which are released from the ovary during the menstrual cycle. During a woman's fertile years one primordial follicle per month develops into a mature follicle, known as a **Graafian follicle**. This is in response to the hormone FSH (follicle stimulating hormone). Within each developing follicle, a primary oocyte starts to develop into an egg. The primordial follicle first becomes a **primary follicle** as the granulosa cells multiply and form many layers of cells around the primary oocyte (figs 21.40, 21.41 and 21.42). In addition, cells from the stroma of the ovary form further layers outside these cells known collectively as the **theca**. The outer part of the theca contains blood vessels and merges with the stroma, the general 'background' material of the ovary. The inner part of the theca secretes female sex hormones, as do the granulosa cells.

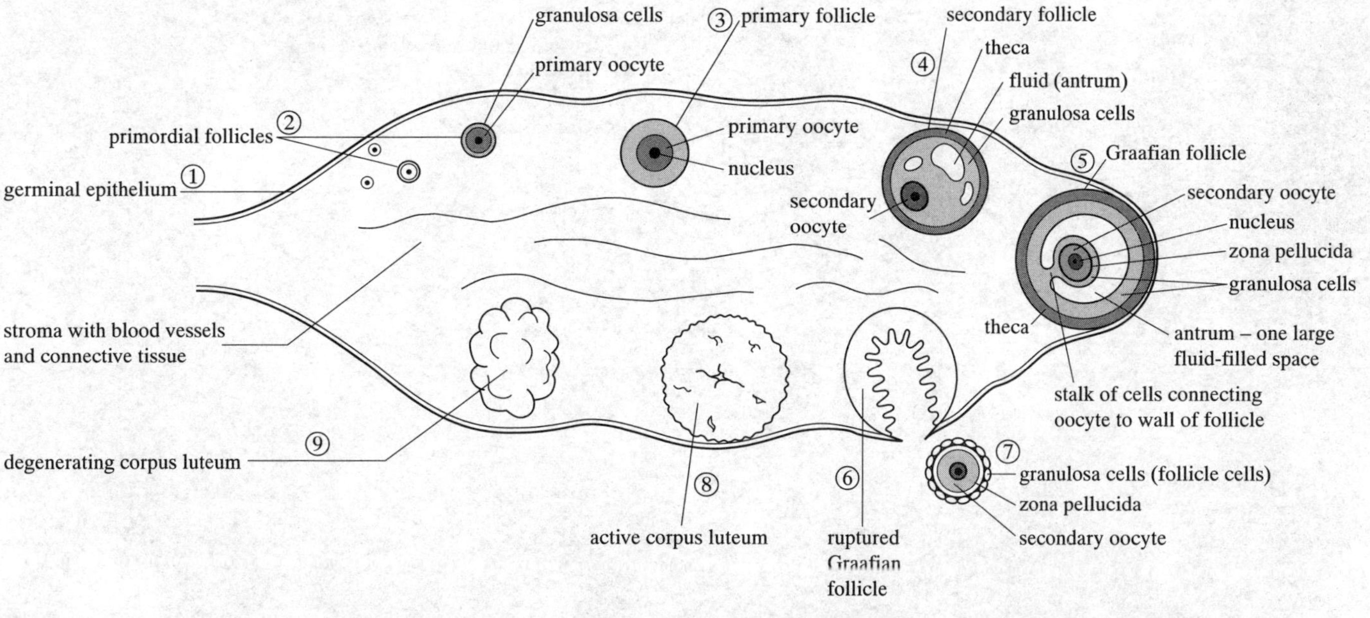

Fig 21.40 *Section through a human ovary showing the stages in the development of a Graafian follicle, ovulation and the formation and degeneration of the corpus luteum. Not all these stages would be seen together. The numbers indicate the sequence of the stages.*

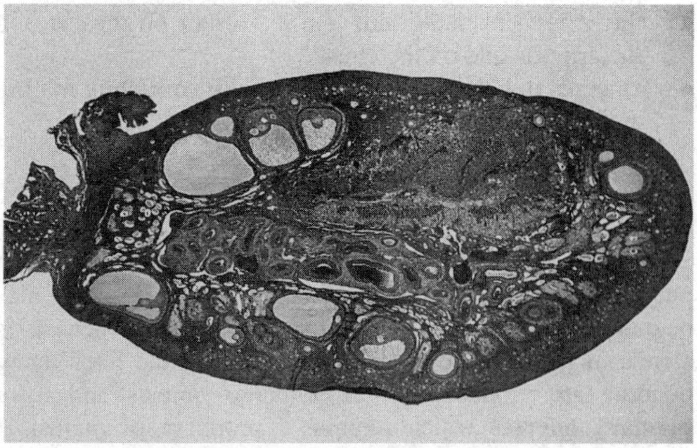

Fig 21.41 *LS human ovary showing developing follicles, X 7.*

As the primary follicle develops, a fluid is secreted by the granulosa cells which contains **oestrogen**, one of the female sex hormones. A fluid-filled space, the antrum, develops in the follicle. It is now referred to as a **secondary follicle**. Oestrogen stimulates growth of the follicle, which eventually becomes a mature follicle, also known as a **Graafian follicle**. It is about 1 cm in diameter. It contains a secondary oocyte and the first polar body, formed when the primary oocyte divides by meiosis I (fig 21.35). The secondary oocyte is haploid. The second meiotic division proceeds as far as metaphase but does not continue until a sperm fuses with the oocyte. At fertilisation the secondary oocyte undergoes the second meiotic division producing a large cell, the **ovum**, and a **second polar body**. All polar bodies are small cells. They have no role in oogenesis and they eventually degenerate. The structure of the secondary oocyte/ovum is shown in fig 21.43.

21.7 (*a*) In what important way is the structure of the egg similar to that of the sperm? (*b*) Summarise the important differences between sperm and eggs.

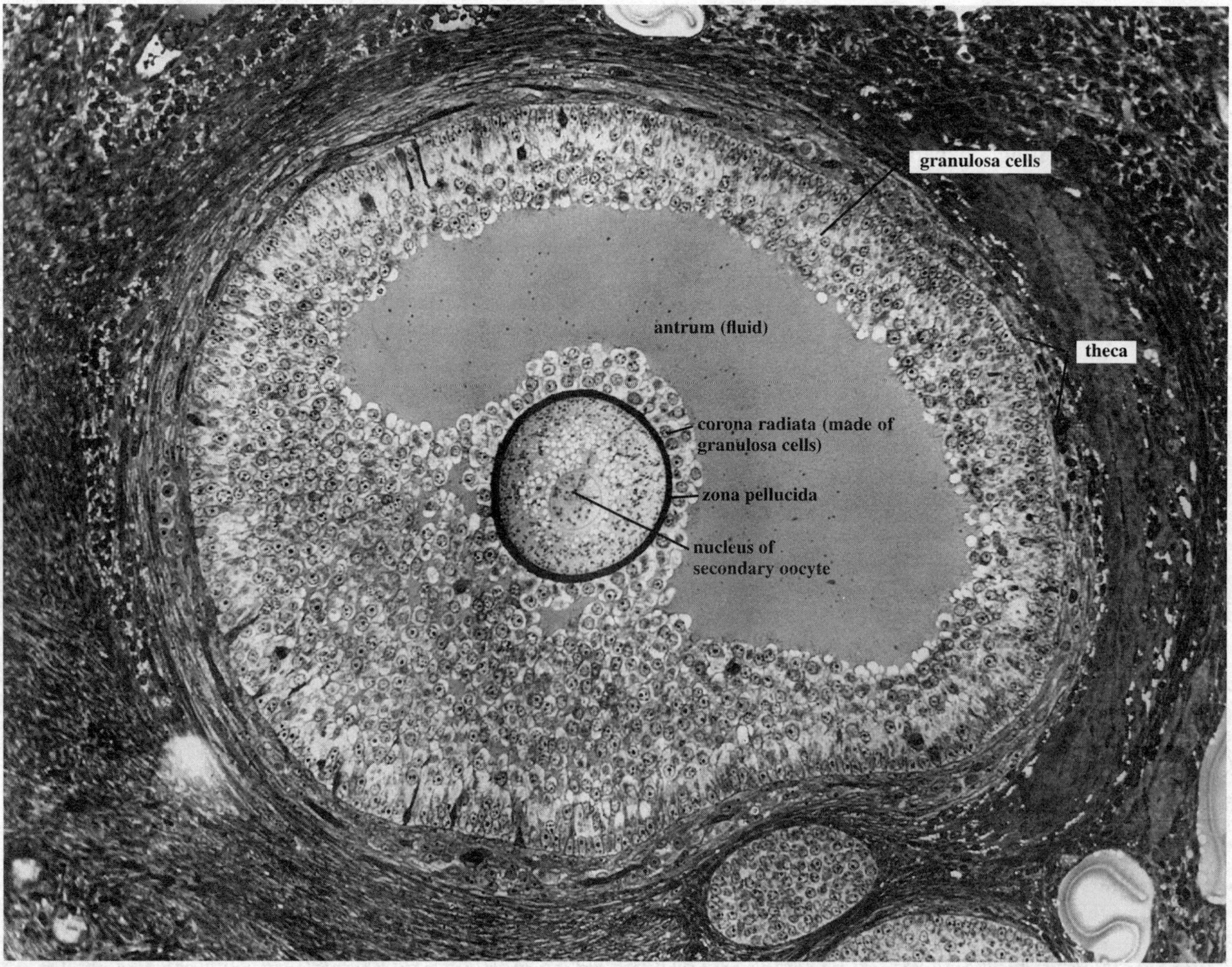

Fig 21.42 *Mature human Graafian follicle.*

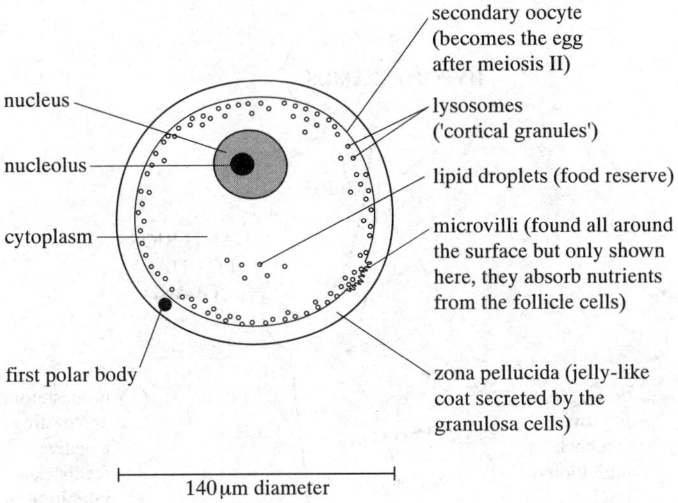

Fig 21.43 *Structure of a secondary oocyte. The structure of the ovum or egg is identical except that one more polar body will be present as a result of division of the secondary oocyte. This division occurs at fertilisation. Granulosa (follicle) cells surround the zona pellucida but are not shown.*

21.7.6 Hormonal control of oogenesis and the menstrual cycle

As in the male, the control centres for producing gametes are the hypothalamus and the pituitary gland. The hypothalamus secretes GnRH which stimulates release of FSH and LH from the pituitary gland as in the male. As explained before, FSH and LH are referred to as gonadotrophic hormones because they stimulate the gonads, in this case the female gonads or ovaries. However, in the female, hormones are not secreted constantly but in cycles. Each cycle lasts about 28 days and is referred to as the **menstrual cycle**. Normally only one egg is produced per cycle. A new cycle begins on the first day of menstruation. A summary of the cycle is given below. Fig 21.44 shows the hormonal changes associated with the cycle, as well as the events occurring in the uterus and ovary at the same time.

(1) **GnRH** stimulates the anterior pituitary gland to secrete **FSH** (follicle stimulating hormone). FSH travels in the blood to its target, the ovaries.

(2) FSH molecules fit into receptor sites in the primordial follicles. They stimulate the development of several follicles, only one of which will complete development.

(3) The granulosa cells of the developing follicle start to produce the female sex hormone **oestrogen.** Oestrogen is a steroid hormone which is produced more and more rapidly in the first half of the cycle, corresponding to the growth of the follicle (fig 21.44). Oestrogen has two targets, the uterus and the anterior pituitary gland. In the uterus it stimulates repair and development of the lining of the uterus, the endometrium. This is in preparation for the possibility of pregnancy when an embryo will implant in the endometrium. In the anterior pituitary gland oestrogen inhibits the secretion of FSH. This is an example of negative feedback. It prevents the possibility of further follicles being stimulated so that only one egg is produced at a time. At the midpoint of the cycle oestrogen levels have built up to a high level which triggers the secretion of **LH** (luteinising hormone).

(4) LH and oestrogen are released in a surge (fig 21.44). The target of LH is the ovary where it causes ovulation. **Ovulation** is the release of the secondary oocyte from the Graafian follicle. The surge ensures precise timing of ovulation. At ovulation the secondary oocyte detaches from the wall of the follicle, is released into the body cavity and passes into the fallopian tube. Usually only one oocyte is released each month by one of the ovaries so that ovulation alternates between the pair of ovaries. The ovulated oocyte consists of a cell whose nucleus is in metaphase I of meiosis surrounded by a cell layer known as the **zona pellucida** and a layer of granulosa cells known as the **corona radiata** which protects the oocyte up to fertilisation.

The remaining part of the Graafian follicle is stimulated by LH to develop into the **corpus luteum** ('yellow body').

(5) The corpus luteum continues to secrete oestrogen, as well as another hormone, **progesterone**. Like oestrogen, progesterone has two targets, the uterus and the anterior pituitary gland. It stimulates the uterus to maintain its thickening and also stimulates glandular activity. In the anterior pituitary gland it inhibits release of LH, a second example of negative feedback. Like oestrogen, it also inhibits FSH. Release of progesterone is associated with a rise in body temperature of the female just after ovulation.

(6) If fertilisation does not occur, the corpus luteum starts to degenerate about 28 days into the cycle. The cause of this is not known, although it is suspected that it is due to chemicals secreted by the corpus luteum itself. Once it starts to degenerate, levels of oestrogen and progesterone decline, so inhibition of FSH is removed. Also, the endometrium breaks down, causing **menstruation** (the 'period'). This lasts for about five days into the next cycle.

A summary of the hormonal control of oogenesis is given in fig 21.45.

Role of hormones in premenstrual tension

Premenstrual tension (PMT) is the term used to describe the 'distressing psychological or physical symptoms' which some women experience regularly towards the end of each menstrual cycle and which 'significantly regress throughout the rest of the cycle'. Tension is not the only symptom; in fact more than 150 symptoms have been attributed to PMT

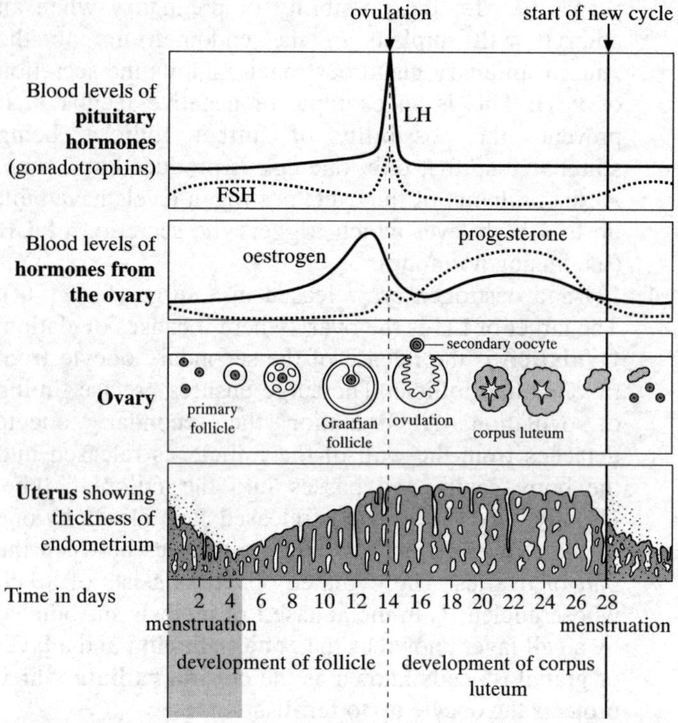

Fig 21.44 *Changes occurring during the menstrual cycle. Levels of the hormones FSH, LH, oestrogen and progesterone are shown, and the associated changes in the ovaries and uterus can be seen.*

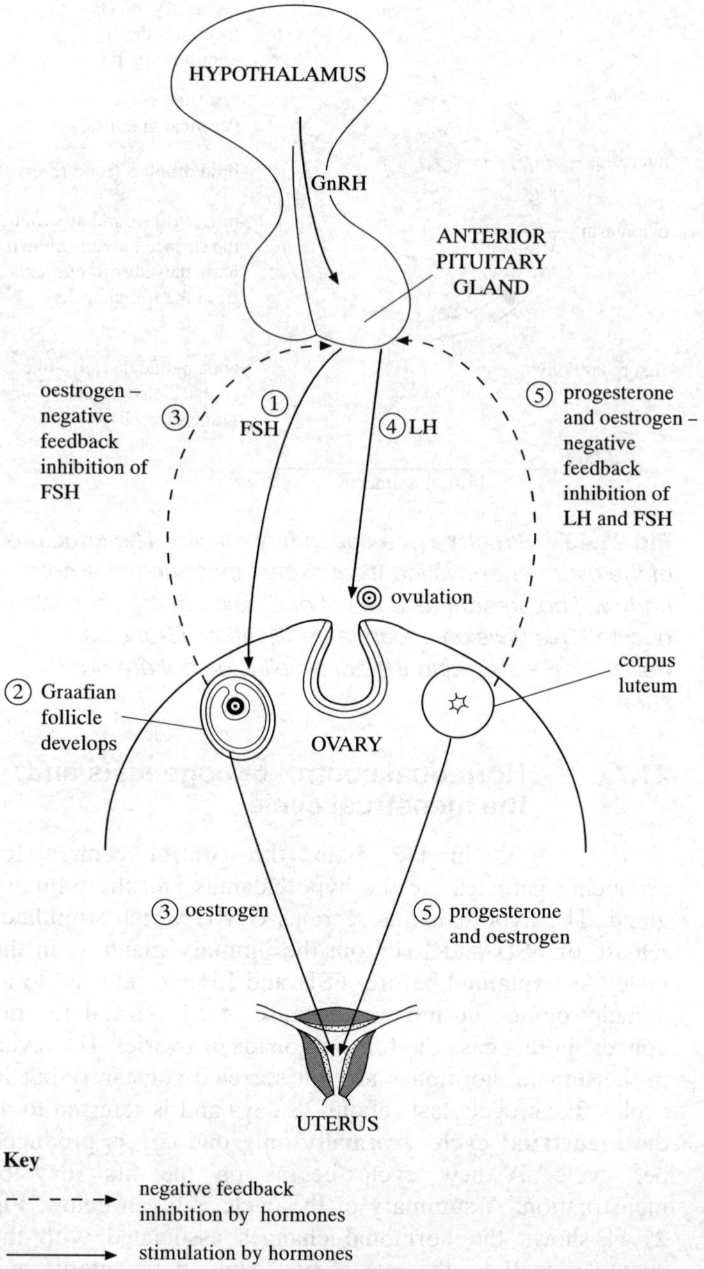

Fig 21.45 *Hormonal control of oogenesis.*

at various times, and the term premenstrual syndrome (PMS) was introduced in 1953 to acknowledge this variety of effects. Some women who experience it describe the condition as devastating and about 75% of women are said to be affected in some way. The most common symptoms are depression, changes in mood, water retention and aches and pains. Little is known about the cause, but it is usually assumed to be hormonal in origin. Since it takes place in the few days before menstruation, it could be due to changes in the balance between progesterone and oestrogen which decline at different rates at this time. It may also be due to progesterone deficiency.

There is still debate as to whether the condition really exists. In 1993, certain psychologists (not all of them men) went so far as to suggest that it may not exist at all, that it was more of a 'social construct' than a medical condition, and that it was a way of 'legitimising and expressing distress'.

Menopause and hormone replacement therapy

Menopause is the cessation of monthly periods and marks the end of a woman's fertility. The average age of menopause in the UK is 51 years. Periods usually become irregular, before finally ceasing. The cause is the gradual failure of the ovaries. The number of follicles declines and they become less sensitive to FSH so that eggs are less and less likely to be produced each month. Secretion of oestrogen declines and since oestrogen normally inhibits FSH by negative feedback, higher levels of FSH (and later LH) are typical of menopause. Many symptoms, both physical and psychological, are associated with menopause and these are mostly due to the reduced oestrogen levels (although progesterone levels also decline). The commonest symptoms are night sweats, random hot flushes during the day, and vaginal dryness. Other common symptoms are depression, irritability, fatigue, and softening of the bones due to loss of minerals, particularly calcium. Loss of calcium from the bones causes a condition known as **osteoporosis** (section 15.7.2). It is characterised by loss of bone mass; as a result the bones break more easily. It occurs because oestrogen is antagonistic to the hormone **parathormone**, which stimulates the raising of blood calcium levels. The symptoms can be prevented relatively easily by **hormone replacement therapy (HRT)**, in which oestrogen is taken either in pill form or by implants below the skin. HRT greatly reduces the rate at which calcium is lost from the bones, slowing it down to roughly the same rate as in men. 80% of female doctors in the relevant age group in the UK use HRT compared to 15% of the rest of the female population in this age group. Treatment can be short-term or continued for years, although in the long term, blood clotting and other undesirable side effects may occur. Some of these can be prevented by adding progesterone to the oestrogen.

It is important to realise that HRT is the *replacement* of *natural* hormones, unlike the Pill which adds synthetic hormones (section 21.9.1). Therefore the risks associated with taking the Pill, such as increased risk of thrombosis, are thought to be lower with HRT.

21.8 Sexual reproduction in humans

21.8.1 Sexual intercourse (copulation)

Internal fertilisation is an essential part of reproduction in terrestrial organisms. This is achieved in many organisms, including humans, by the development of a special organ, the penis, which is inserted into the vagina and releases gametes within the female reproductive tract.

Erection of the penis occurs as a result of an increase in blood content of spongy erectile tissue in the penis. Sexual stimulation involves stimulation of the parasympathetic nervous system. It results in vasoconstriction of the veins leading away from the penis and vasodilation of the arterioles entering the penis. As a result blood volume and pressure within the penis increase. In this erect state the penis can be inserted into the vagina where the friction, produced by the rhythmic movements of sexual intercourse, increases the stimulation of sensory cells at the tip of the penis. This activates neurones of the sympathetic nervous system which lead to closure of the internal sphincter of the bladder and contraction of the smooth muscle of the epididymis, vas deferens, and the male accessory glands, namely the seminal vesicles, prostate and Cowper's glands. This action releases sperm and seminal fluids into the top of the urethra where they mix to form **semen**. The increased pressure of these fluids in the urethra leads to reflex activity in the motor neurones supplying the muscles at the base of the penis. Rhythmic wave-like contractions of these muscles force semen out through the urethra during **ejaculation** which marks the climax of copulation. The other physiological and psychological sensations associated with this climax in both males and females are called **orgasm**.

Lubrication is provided during intercourse partly by a clear mucus secreted by the male Cowper's glands following erection but mainly by glands in the vagina and vulva of the female. Fluid from the blood also seeps through the vaginal epithelium.

Sexual excitement also causes the clitoris of the female to become erect. This is the female equivalent of the penis. The female orgasm involves muscular contractions of both the vagina and the uterus, and sensation can be centred on the clitoris, vagina or both.

The secretions of the male accessory glands are alkaline and decrease the normal acidity of the vagina to pH 6–6.5 which is the optimum pH for sperm motility following ejaculation. Approximately 3 cm^3 of semen is discharged during ejaculation of which only 10% is sperm. Semen contains about 100 million sperm per cm^3.

21.8.2 Passage of sperm to egg

Sperm are deposited at the top of the vagina close to the cervix. For fertilisation to take place, sperm have to travel from here through the cervix and uterus to the oviducts. The cervix is normally blocked by thick mucus. This mucus becomes thinner in the first part of the menstrual cycle, allowing the penetration of sperm to the uterus. Progesterone causes it to become thicker in the second half of the cycle.

Investigations have shown that some sperm pass from the vagina through the uterus and to the top of the oviducts within five minutes. This is faster than can be achieved by swimming and may be a result of contractions of the uterus and oviducts. These contractions could be caused by chemicals in the semen, including prostaglandins, and possibly by hormones such as oxytocin released by the female during sexual intercourse. It is more likely, however, to be due to the action of cilia which line the uterus and oviducts. It takes about 4–8 hours for most sperm to reach the oviducts. Sperm can survive in the female for 1–3 days but are only highly fertile for 12–24 h. Only a few thousand complete the journey from vagina to egg.

Capacitation

Sperm can only fertilise the secondary oocyte after spending several hours in the female genital tract, usually about seven hours, during which time they undergo an activating process known as **capacitation**. This involves a number of changes, including the removal of a layer of glycoprotein and plasma proteins from the outer surface of the sperm. Glycoprotein is originally added by the epididymis and plasma proteins from seminal fluid. They are removed by enzymes in the uterus. Cholesterol is also lost from the cell surface membrane around the sperm head, weakening the membrane. The membrane also becomes more permeable to calcium ions, which have the dual effect of increasing the beating activity of the sperm tail and promoting the acrosome reaction (see below).

During capacitation the acrosome membrane fuses with the cell surface membrane, a process which starts the release of acrosomal enzymes (see below).

Acrosome reaction

When a sperm reaches the secondary oocyte (fig 21.46), normally high up in the oviduct, the outer cell surface membrane of the sperm next to the acrosome, and the membrane of the acrosome, rupture. This enables hydrolytic (digestive) enzymes, such as **hyaluronidase** and **proteases**, stored in the acrosome to be rapidly released. These changes in the sperm head are known as the **acrosome reaction**.

21.8.3 Fertilisation

Fertilisation is the fusion of the sperm nucleus with the egg nucleus to form a diploid cell known as the zygote. It takes place in the following stages.

(1) The enzymes (particularly hyaluronidase) released by the acrosomes of the many sperm digest a path through the material that holds the granulosa (follicle) cells together.

(2) The sperm move by lashing their tails and reach the outer surface of the zona pellucida, a thick layer surrounding the secondary oocyte (see figs 21.42 and 21.43). The zona pellucida has special receptors to which the sperm heads can bind.

(3) Another acrosomal enzyme digests a path through the zona pellucida and the sperm moves through to the surface of the secondary oocyte.

(4) Here the head of a sperm will fuse with microvilli surrounding the secondary oocyte (fig 21.43) and penetrate its cytoplasm.

(5) Immediately the sperm has penetrated, the lysosomes in the outer region of the secondary oocyte, also known as **cortical granules**, release their enzymes which cause the zona pellucida to thicken and harden, forming the '**fertilisation membrane**'. This is called the **cortical reaction**. It prevents the entry of further sperm and therefore the possibility of more than one

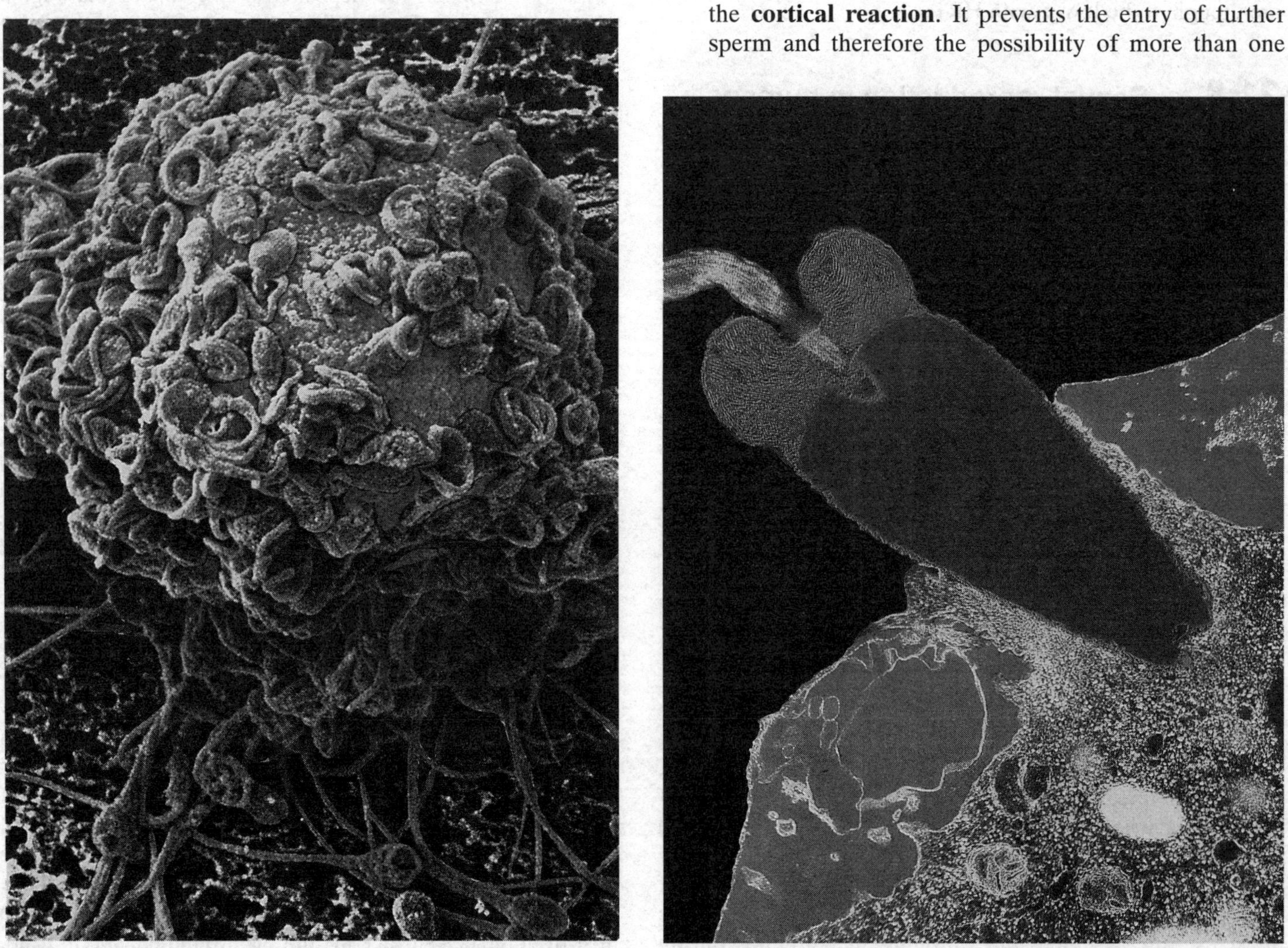

Fig 21.46 *(a) Scanning electron micrograph of human sperm clustered around a secondary oocyte. (b) The precise moment of fertilisation as a sperm penetrates the membrane of a sea urchin's egg. The dark wedge is the head of the sperm, which contains the genetic code. The grey shape behind it is the mitochondrion where energy is released that provides power for the tail. The sperm has digested the egg's surface coating of sugary protein and entered.*

sperm fertilising the same egg. The enzymes also destroy the sperm receptor sites, so sperm can no longer bind to the zona pellucida.

(6) The entry of a sperm acts as the stimulus for completion of the second meiotic division of the secondary oocyte which produces the ovum and the second polar body. The second polar body immediately degenerates and the tail of the sperm is lost within the cytoplasm of the ovum.

(7) The nucleus of the sperm swells as its chromatin becomes less tightly coiled. At this stage the nuclei of the sperm and secondary oocyte are called **pronuclei**.

(8) The male pronucleus fuses with the female pronucleus. This is the actual act of fertilisation. The new nucleus formed has two sets of chromosomes, one from the egg and one from the sperm. The cell is now diploid and is called the **zygote**. The new nucleus divides immediately by mitosis. The zygote then undergoes cytokinesis, or cell division, and produces two diploid cells.

21.8.4 The effect of fertilisation

If fertilisation occurs, the zygote develops into a ball of cells called the **blastocyst** which embeds itself into the wall of the uterus within eight days of ovulation. The outer cells of the blastocyst, the **trophoblastic cells**, then begin to secrete a hormone called **human chorionic gonadotrophin** (**HCG**), which has a similar function to LH. This function includes prevention of the breakdown of the corpus luteum. The corpus luteum therefore continues to secrete progesterone and oestrogen and these bring about increased growth of the endometrium of the uterus. Loss of the lining of the endometrium is prevented and the absence of menstruation (the 'period') is the earliest sign of pregnancy. The placenta gradually takes over from the corpus luteum from about week 10 of pregnancy when it begins to secrete most of the progesterone and oestrogen essential for a normal pregnancy. Failure of the corpus luteum before the placenta is fully established is a common cause of miscarriage at about 10–12 weeks of pregnancy.

During pregnancy HCG may be detected in the urine and this forms the basis of pregnancy testing (section 12.11.2).

21.8.5 Implantation

As the zygote passes down the oviduct it divides by successive nuclear and cell divisions into a small ball of cells by a process called **cleavage**. Cleavage involves cell division without growth in size, because the cells continue to be retained within the zona pellucida. The cells just get smaller and smaller at this stage. Nuclear division is by mitosis. The cells formed are called **blastomeres** and they form a hollow ball of cells whose central cavity is called the **blastocoel**. This fills with liquid from the oviduct. The outer layer of blastomeres is called the **trophoblast** and this thickens at one point to form a mass of cells called the **inner cell mass**. This stage is called the **blastocyst** and is reached about 4–5 days after fertilisation. The structure of the blastocyst is shown in fig 21.47.

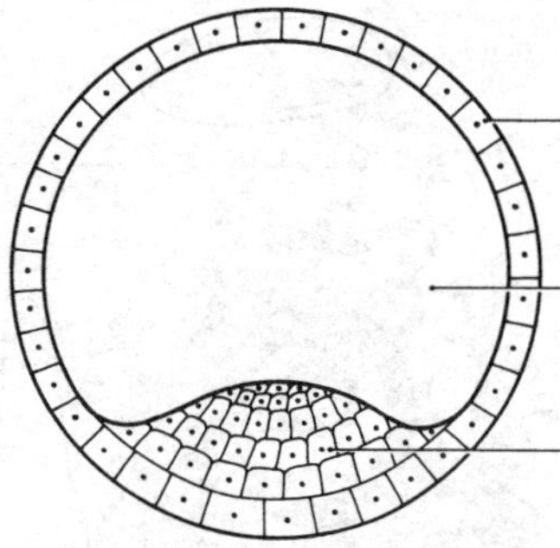

Fig 21.47 *Simplified diagram of a human blastocyst four days after fertilisation.*

When the blastocyst arrives in the uterus the zona pellucida gradually disappears over about two days, allowing the cells of the trophoblast to make contact with the cells of the endometrium. The term 'tropho-' means 'feeding' and the trophoblast starts the process of invading the uterus wall and gaining nutrients from the endometrium. The trophoblast cells multiply in the presence of these nutrients and between the sixth and ninth days after fertilisation the blastocyst becomes embedded within the endometrium. This process is called **implantation**.

The cells of the trophoblast differentiate into an inner layer and an outer layer. The outer layer is called the **chorion**, and forms the **chorionic villi**, finger-like processes which grow into the endometrium (fig 21.48). The areas of the endometrium between these villi form interconnecting spaces which give this region of the endometrium a spongy appearance. Hydrolytic enzymes released by the trophoblast cause the arterial and venous blood vessels in the endometrium to break down and blood from them fills the spaces. In the early stages of development of the blastocyst, exchange of nutrients, oxygen and excretory materials between the cells of the blastocyst and the maternal blood in the uterus wall occurs through the chorionic villi. Later in development this function is taken over by the placenta.

21.8.6 Early embryonic development and development of extra-embryonic membranes

As stated, the outer cells of the blastocyst, the trophoblast, grow and develop into an outer layer or 'membrane' called the **chorion**. This plays a major role in nourishing and removing waste from the developing embryo. Meanwhile, two cavities appear within the inner cell mass (fig 21.47) and the cells lining these give rise to two further 'membranes', the **amnion** and the **yolk sac** (fig 21.48). As with the chorion, the use of the term

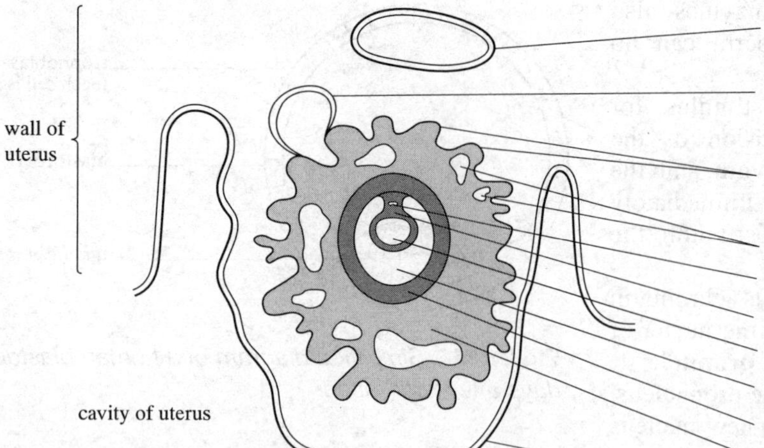

Fig 21.48 *Simplified diagram showing a recently implanted human blastocyst in the endometrium of the uterus. Enzymes produced by the outer layer of the trophoblast, the chorion, break down the blood vessels of the endometrium producing spaces containing blood which are used in the nourishment and excretion of the blastocyst.*

'membrane' here does not refer to membranes like those surrounding cells. The structures are called membranes because they are relatively thin, but they are made of cells.

The amnion is a thin membrane covering the embryo like an umbrella and has a protective function. Between the amnion and the embryo is the **amniotic fluid** which is secreted by the cells of the amnion and fills the amniotic cavity. As the embryo increases in size the amnion expands so that it is always pressed up against the uterus wall opposite the embryo. The amniotic fluid supports the embryo and protects it from mechanical shock.

The yolk sac has no significant function in humans but is important in reptiles and birds, where it absorbs food from the separate yolk and transfers food to the gut of the developing embryo.

The cells of the inner cell mass, between the early amnion and the yolk sac, form a structure called the **embryonic disc**, which gives rise to the embryo. The cells of the disc differentiate at an early stage (when the diameter is less than 2 mm) and form an outer layer of cells, the **ectoderm**, and an inner layer, the **endoderm**. At a later stage the **mesoderm** is formed and these three 'germ' layers give rise to all the tissues of the developing embryo. The development of three layers in this way is called **gastrulation** and occurs 10–11 days after fertilisation. The brain and spinal cord start to develop in the third week from a tube called the **neural tube** which arises in the ectoderm.

The tube starts as a groove in the ectoderm. Gradually the sides of the groove curve around to form a hollow tube which becomes swollen at one end to form the brain. This is the first organ to appear.

During the early stages of embryonic development exchange of materials between embryo and mother across the chorionic villi is adequate, but soon a fourth membrane, the **allantois**, develops from the embryonic hindgut. The chorion, amnion, yolk sac and allantois are called **extra-**

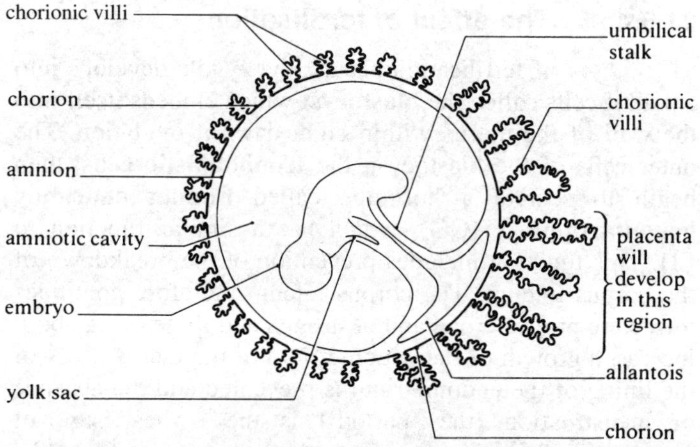

Fig 21.49 *Simplified diagram showing the relationship between the human embryo and the four extra-embryonic membranes (amnion, chorion, allantois and yolk sac) about five weeks after fertilisation. The area where the allantois meets the chorion is called the allanto-chorion and becomes the placenta.*

embryonic membranes (fig 21.49). The allantois grows outwards until it comes into contact with the chorion where it forms a structure rich in blood vessels called the **allanto-chorion**. This contributes towards the development of a more efficient and effective exchange structure, the **placenta**.

21.8.7 Development of the embryo and fetus

Some of the early stages of embryo and fetal development are summarised in fig 21.50. Times are from conception, the time of fertilisation. After 6 weeks, the embryo is recognisably human and is called a fetus. Doctors and midwives date pregnancy from the last period,

Fig 21.50 *Diary of development of the human fetus.*

Week 1
Fertilisation. Cleavage to form a blastocyst 4–5 days after fertilisation. More than 100 cells. Implantation 6–9 days after fertilisation.

Week 2
The three basic layers of the embryo develop, namely ectoderm, mesoderm and endoderm. No research allowed on human embryos beyond this stage.

Week 3
Woman will not have a period. This may be the first sign that she is pregnant. Beginnings of the backbone. Neural tube develops, the beginning of the brain and spinal cord (first organs). Embryo about 2 mm long.

Week 4
Heart, blood vessels, blood and gut start forming.
Umbilical cord developing. Embryo about 5 mm long.

Week 5
Brain developing. 'Limb buds', small swellings which are the beginnings of the arms and legs. Heart is a large tube and starts to beat, pumping blood. This can be seen on an ultrasound scan. Embryo about 8 mm long.

Week 6
Eyes and ears start to form.

Week 7
All major internal organs developing. Face forming. Eyes have some colour. Mouth and tongue. Beginnings of hands and feet. Fetus is 17 mm long.

By week 12
Fetus fully formed, with all organs, muscles, bones, toes and fingers. Sex organs well developed. Fetus is moving. For the rest of the gestation period, it is mainly growing in size. Fetus is 56 mm long from head to bottom. Pregnancy may be beginning to show.

By week 20
Hair beginning to grow, including eyebrows and eyelashes. Fingerprints developed. Fingernails and toenails growing. Firm hand grip. Between 16 and 20 weeks baby usually felt moving for first time. Baby is 160 mm long from head to bottom.

Week 24
Eyelids open. Legal limit for abortion in most circumstances.

By week 26
Has a good chance of survival if born prematurely.

By week 28
Baby moving vigorously. Responds to touch and loud noises. Swallowing amniotic fluid and urinating.

By week 30
Usually lying head down ready for birth. Baby is 240 mm from head to bottom.

40 weeks (9 months)
Birth.

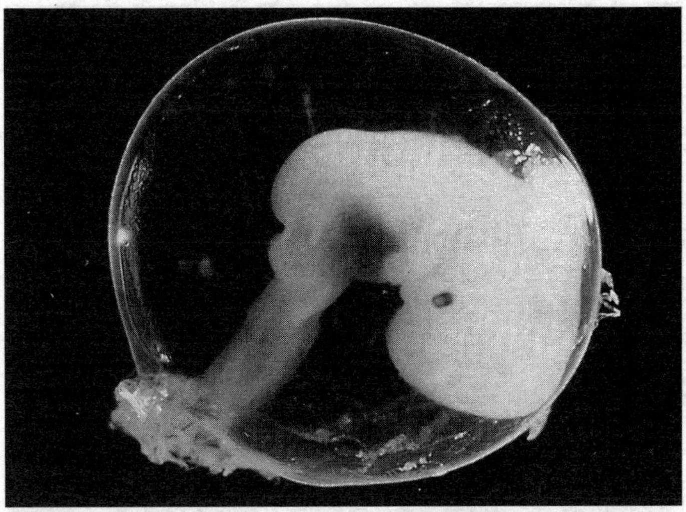

(a) Week 4: The heart is visible as the dark area near the middle of the embryo

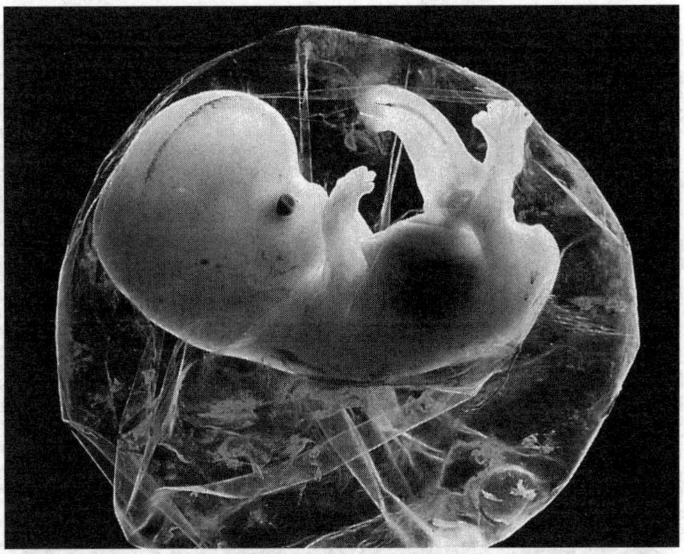

(b) Week 5–6: The arms and legs are developing.

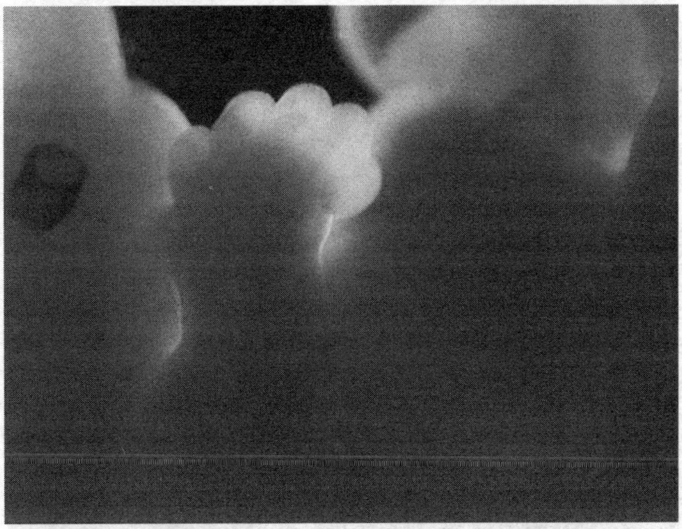

(c) Week 7–8: Close up view of hand and face showing early development of fingers and retina (dark area) of eye.

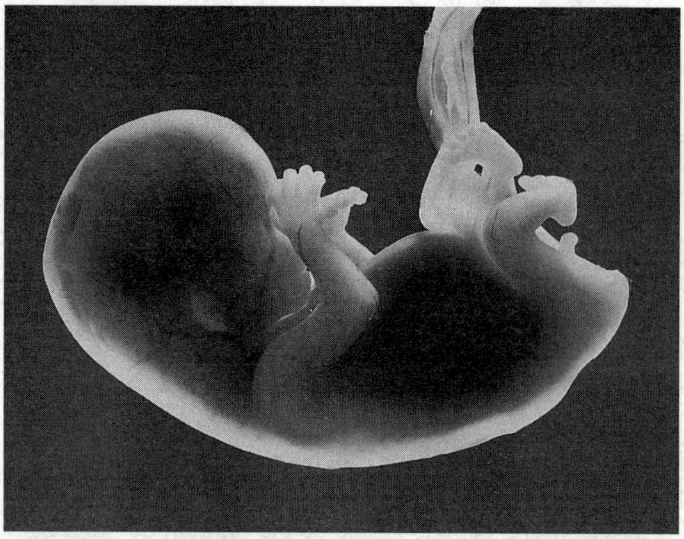

(d) *Week 11: Eyelids and ears have developed. The head is about half the total length*

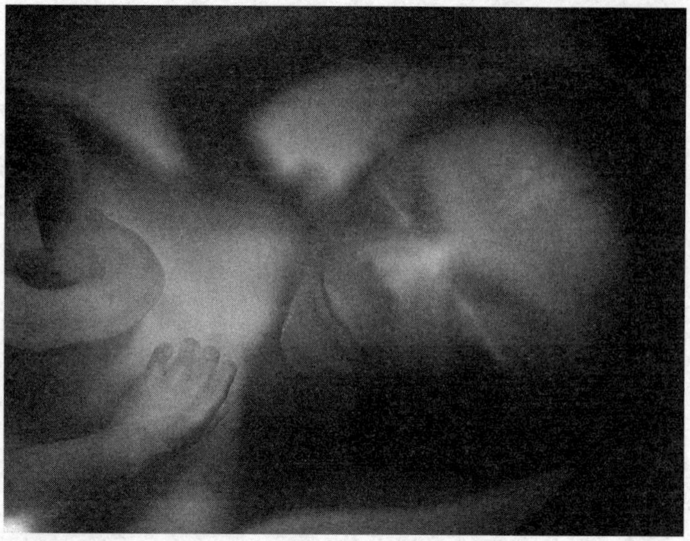

(e) *Week 12: The face looks more human with eyelids and lips visible.*

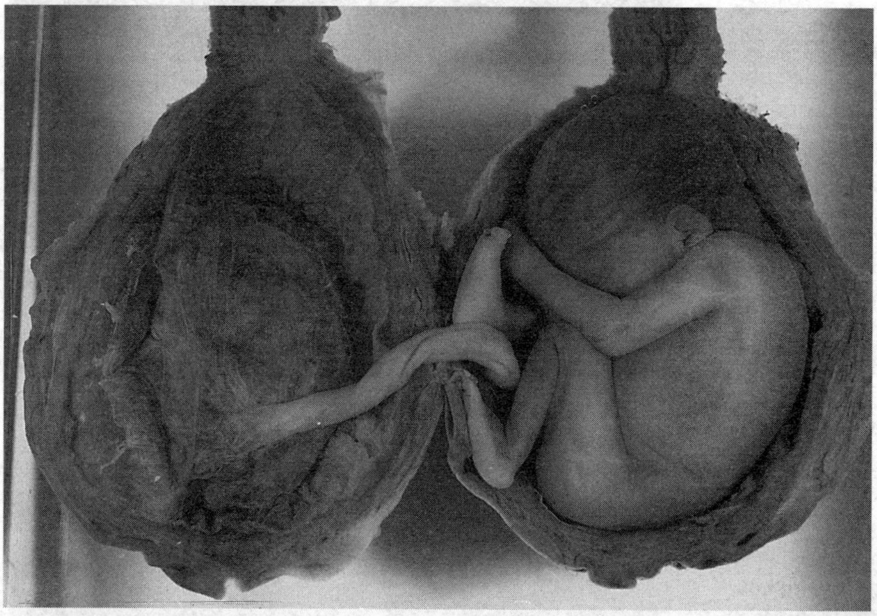

(f) *Week 20: Fetus inside embryonic membranes showing placenta in left half.*

(g) (bottom left) *Week 20: close-up view showing development of hands and fingers. The fingers will close if the palm of the hand is touched.*

(h) (below) *Week 34: Nuclear magnetic resonance image of side view of women's abdomen. The fetus, on the right of the mother's spine, is in the normal position for birth with head down.*

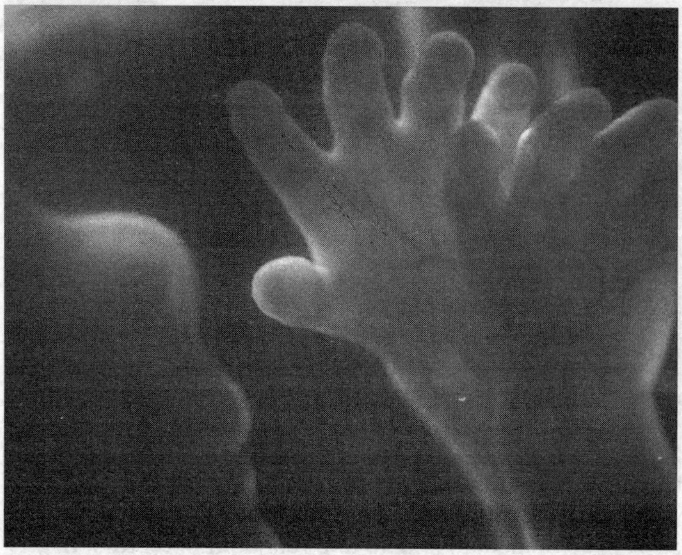

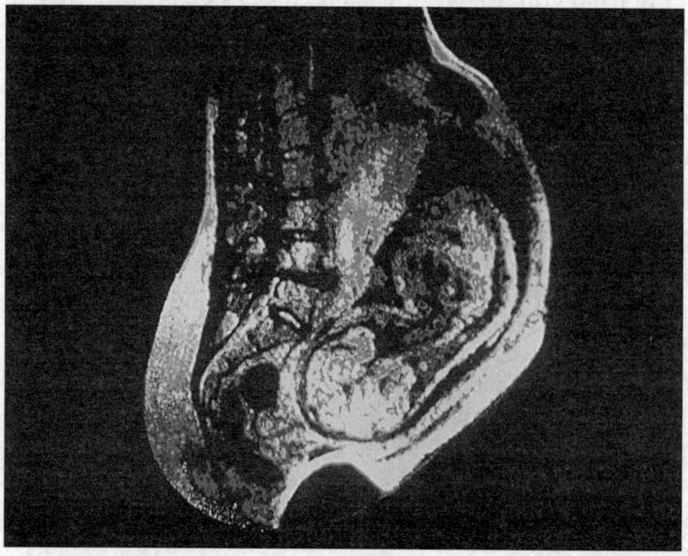

adding two weeks to the times given below. The fetus normally completes a total of about 38 weeks of development, the **gestation period**, before birth occurs. Most of the major organs are formed by the twelfth week of pregnancy, and the remainder of the gestation period is taken up by growth.

21.8.8 The placenta

The placenta is an organ found only in mammals and is the only organ in animals composed of cells derived from two different organisms, the fetus and the mother. Its function is to allow exchange of materials between mother and fetus. It takes over from the chorionic villi as the main site of exchange of materials after 12 weeks of pregnancy.

Structure of the placenta

The fetal part of the placenta consists of cells of the chorion which produce projections called **chorionic villi** (fig 21.51). These increase surface area for absorption. The chorionic villi become invaded by branches of two blood vessels of the fetus, the umbilical artery and the umbilical vein. They form capillary networks inside the villi. The blood vessels run between the fetus and the uterus wall in the **umbilical cord** which is a tough structure about 40 cm long covered by cells derived from the amnion and chorion.

The maternal part of the placenta consists of projections from the endometrium. Between these and the chorionic villi are spaces supplied with arterial blood from arterioles in the uterus wall. Blood flows through the spaces from the arterioles to venules in the uterus wall. The placenta is a relatively large structure, weighing on average about 600 g when fully formed, and measuring 15–20 cm in diameter and 3 cm thick at the centre. About 10% of the mother's blood flows through it for each circuit of blood round the body. Because there is no direct contact between the mother's blood and the fetal circulation, the fetus is not exposed to the relatively high blood pressure of the maternal circulation. Another reason for preventing mixing of the blood is that mother and fetus may be of different ABO blood groups and their blood may not be compatible. Separation at birth would also be more difficult.

Mechanisms of uptake across the placenta

The cell surface membranes of the cells in the walls of the chorionic villi have microvilli, which increase their surface

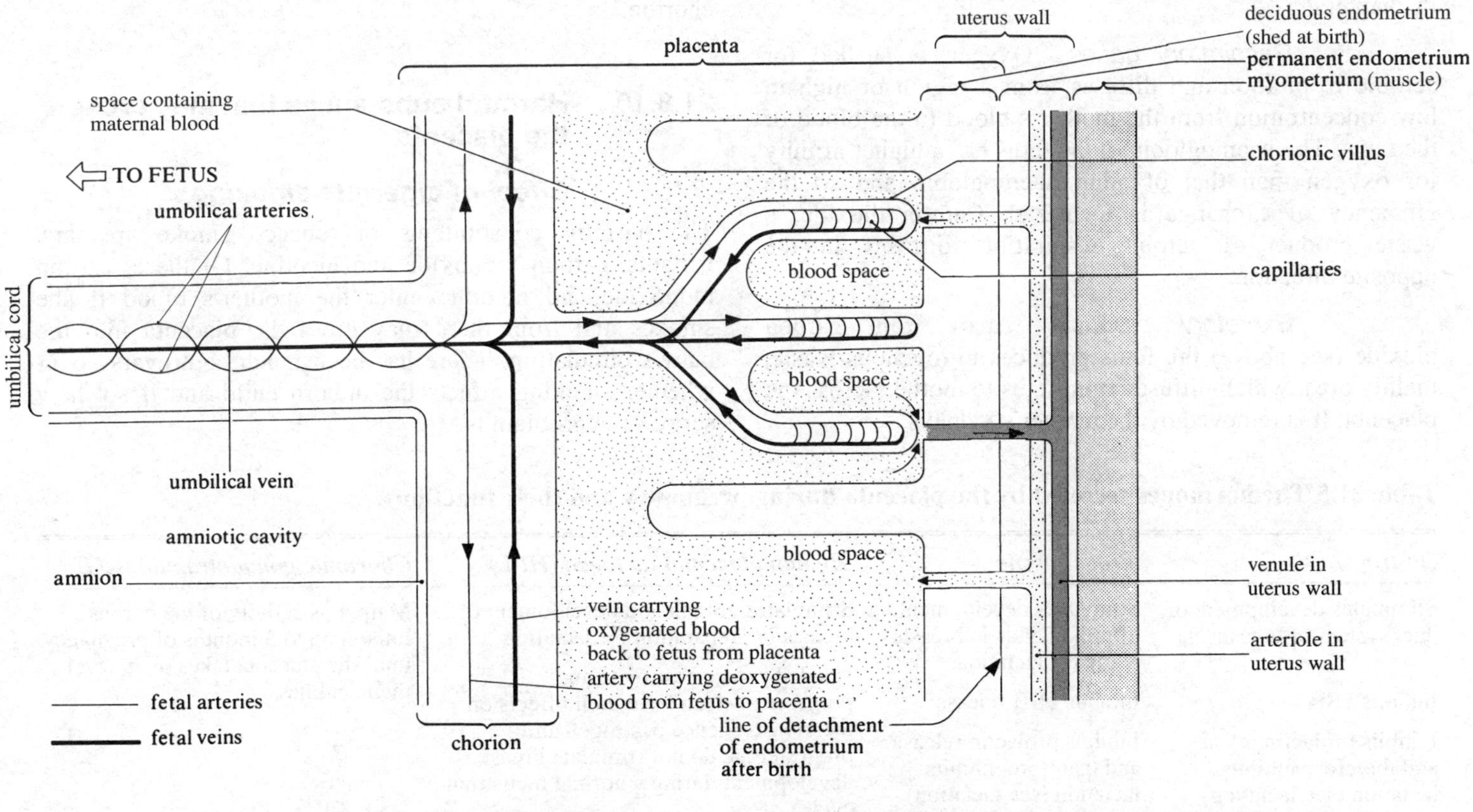

Fig 21.51 *The relationship between blood vessels in the umbilical cord, capillaries of the chorionic villi and the blood spaces of the human placenta. The placenta forms the link between the circulatory systems of the fetus and the mother.*

area for the exchange of substances by diffusion, facilitated diffusion, active transport and pinocytosis. Numerous mitochondria are found in these cells. They provide the energy for active transport and pinocytosis. The cell surface membranes contain carrier molecules used in the uptake of materials into the villi by active transport. Numerous small vesicles are found inside the cells of the villi as a result of materials being taken up from the blood by pinocytosis (section 5.9.8).

Examples of the different mechanisms of uptake are given in sections 21.8.9 and 21.8.10 below.

21.8.9 Exchange of useful substances between mother and fetus

Water. Water can cross the placenta by osmosis.

Nutrients. Glucose, amino acids, lipids, mineral salts and vitamins can all cross the placenta from the mother to the fetus. Glucose moves by facilitated diffusion through a special carrier protein as described in section 5.9.8. Ions, such as sodium, potassium and calcium, cross mainly by active transport, though some diffusion takes place. Amino acids, iron and vitamins cross by active transport. The importance of diet in pregnancy is discussed in chapter 8.

Respiratory gases. Oxygen is needed for aerobic respiration and diffuses from a region of high to low concentration from the mother's blood to the blood of the fetus. The haemoglobin of the fetus has a higher affinity for oxygen than that of adult haemoglobin and so the efficiency of exchange is increased. Carbon dioxide, a waste product of aerobic respiration, diffuses in the opposite direction.

Excretory products. Apart from carbon dioxide (see above) the fetus produces nitrogenous waste, mainly urea, which diffuses from fetus to mother across the placenta. It is removed by the mother's kidneys.

Antibodies. Antibodies can cross the placenta from mother to fetus. The fetus is therefore protected against the same diseases as the mother. This is known as **passive immunity**. It fades away gradually after birth because the immune system of the fetus has not learned to make the antibodies for itself (active immunity). Occasionally antibodies from the mother may be harmful to the fetus, as is sometimes the case with Rhesus antibodies (see below).

Endocrine organ. The placenta also functions as an endocrine organ. From the third month of pregnancy the placenta takes over completely from the corpus luteum as the main source of oestrogen and progesterone. The role of these hormones is summarised in table 21.5.

Throughout pregnancy oestrogen and progesterone are secreted in progressively greater amounts, first by the corpus luteum and then principally by the placenta. In the last three months of pregnancy oestrogen secretion increases faster than progesterone secretion and, immediately prior to birth, the progesterone level declines and the oestrogen level increases.

The placenta also secretes chorionic gonadotrophin and human placental lactogen whose functions are summarised in table 21.5. All these hormones are secreted by the chorion.

21.8.10 Harmful substances that may cross the placenta

Effect of cigarette smoking

The harmful constitutents of tobacco smoke are tars, irritants, carbon monoxide and nicotine. Of these, carbon monoxide and nicotine enter the mother's blood if she smokes and from there may cross the placenta into the baby's circulation. There has been much controversy over whether smoking affects the unborn child and if so, how serious the problem is.

Table 21.5 The hormones secreted by the placenta during pregnancy and their functions.

Oestrogen	*Progesterone*	*Human placental lactogen (HPL)*	*Chorionic gonadotrophin (CG)*
Stimulates development of duct system of breasts	Stimulates development of milk glands in breasts ready for lactation	Stimulates growth and development of breasts in preparation for lactation	Maintains activity of the corpus luteum up to 3 months of pregnancy until the placenta takes over, level then declines
Inhibits FSH release	Inhibits FSH release	Needed before oestrogen and progesterone can have their effects on the breasts (hence oestrogen and progesterone do not stimulate breast development during a normal menstrual cycle)	
Inhibits prolactin release and therefore inhibits lactation (see lactation)	Inhibits prolactin release and therefore inhibits lactation (see lactation)		
Stimulates growth of uterus, particularly muscle	Inhibits contraction of myometrium (relaxes muscle and helps to prevent miscarriage)		
Increases sensitivity of myometrium to oxytocin (see birth)	Maintains lining of uterus		

The most clearcut evidence concerns the effect on birthweight. Average birthweight is 3.40 kg (3400 g). The babies of mothers who smoke 10–20 cigarettes a day throughout pregnancy are on average 200 g (about 6%) lighter at birth. Smoking 30 cigarettes a day reduces birthweight by as much as 10%. This condition is described as **intra-uterine growth retardation (IUGR)**. IUGR can be a cause of premature birth. Babies born underweight or prematurely have a higher risk of complications if there are problems, resulting in higher rates of perinatal death (deaths just before or just after birth). Midwives and other members of the medical profession concerned with the care of new-born babies are very familiar with IUGR and claim that it is possible to tell from the umbilical cord and placenta whether the mother is a heavy smoker. This is because the blood vessels of the cord are noticeably narrower than normal and the placenta is smaller. Nicotine is known to cause vasoconstriction and therefore reduces blood flow through the placenta. Constant exposure to nicotine may therefore permanently affect development of the blood vessels in the umbilical cord and placenta. If blood flow to the fetus is restricted, oxygen and nutrients will be less freely available, possibly causing IUGR. Carbon monoxide may also have some effect since it reduces the oxygen carrying capacity of haemoglobin. There is no proof, however, that it is harmful to the fetus.

Other problems are associated with smoking in pregnancy. Premature birth, late miscarriages and perinatal deaths are all more common in smokers. A study in Great Britain of children followed from birth to age 7 years showed that children of smokers had a 30% greater chance of perinatal death and a 50% greater chance of heart abnormalities. At age 7 they were on average 6 months behind the children of non-smokers in reading age. These facts alone do not prove that smoking is the cause of all these problems. There could be some common factor. Smoking, for example, is more common among lower socioeconomic groups, where, for example, poverty and poor diet are also more common. However, carefully conducted investigations in which smokers are compared with control groups of non-smokers similar with respect to other important variables, such as weight, nutrition and socioeconomic group, show that smoking *can* probably cause effects such as premature birth and reduced birthweight.

Most women are well aware that smoking may harm the fetus, so why do some pregnant women smoke? Many women feel guilt or anxiety when they continue to smoke during pregnancy, but smoking may be physically or psychologically addictive, and it is difficult to give up if close friends or partners smoke, or if social circumstances are stressful. Sometimes women feel that if they give up smoking they will become so tense and irritable that other family members will suffer. In short, although smoking may be harmful it is not an easy issue to deal with.

Alcohol

Alcohol crosses the placenta easily. Again, there is controversy over exactly how harmful alcohol is to the fetus, and there are the same problems as with smoking over whether other factors associated with heavy drinking, such as poor diet, may also play a role.

Before considering the effects of alcohol, it is useful to be familiar with the units commonly used to measure alcohol consumption. Half a pint of beer or one glass of wine, sherry or spirit contains roughly one unit of pure alcohol (one unit is 8 g). A daily intake of this amount is estimated to increase the risk of developmental abnormalities by 1.7% and to reduce fetal growth by 1%. Women who drink more than 5 units a day are defined as heavy drinkers (7 units a day for men). It is estimated that an intake of 7.5 to 10 units a day is necessary for development of cirrhosis of the liver in the adult, a disease particularly associated with alcohol. This very high level of drinking in pregnancy can cause a condition known as **fetal alcohol syndrome (FAD)**. One or more of the following symptoms may then occur:

- mental retardation;
- microcephaly (small head/brain);
- behavioural problems such as hyperactivity and poor concentration;
- reduced growth rate, continuing after birth;
- poor muscle tone;
- flat face (poorly developed cheek bones), long thin upper lip, short upturned nose, sometimes a cleft palate.

The greatest harm is probably done during the early stages of pregnancy when the brain of the fetus is developing rapidly. Bearing in mind the particularly harmful effects on the brain, many doctors argue that, strictly speaking, there is no safe limit of alcohol consumption and recommend that women do not drink at all during pregnancy. On the other hand, it is probably more realistic to advise that the odd drink causes no harm, but that intake on any given occasion should be kept relatively low.

Apart from fetal alcohol syndrome, which is the third most common cause of mental retardation in the USA, studies from the USA indicate that drinking more than 100 g (12.5 units) of alcohol per week more than doubles the risk of delivering an underweight baby compared with women drinking less than 50 g (6.25 units) per week. Miscarriages are also more common. As mentioned, the effects of alcohol are difficult to disentangle from other sociological factors in such situations.

Drugs

The fetus is most sensitive to damage by any drugs during the phase of organ development which starts in the third week of pregnancy. At this stage the woman may only just be suspecting that she is pregnant. Pre-conceptual care and advice is therefore becoming more common, so that couples planning pregnancy can plan changes in lifestyle if necessary. Alcohol and nicotine are, strictly speaking, drugs but are described as 'socially acceptable' because they are legal. We can also consider illegal drugs and pharmaceutical products under the heading of drugs.

Illegal drugs. Of the illegal drugs, heroin and cocaine are probably of most concern. This is particularly true of the very pure form of cocaine, namely crack. If a woman is addicted to one of these drugs, her baby is also likely to become addicted and will usually have to undergo withdrawal symptoms after birth. Permanent brain damage of the fetus may occur during pregnancy, resulting in mental retardation or behavioural problems later in life. An addicted mother is less likely to maintain her own health and this in turn may result in premature birth with all the associated problems, including higher perinatal mortality. Statistics obtained in the USA show that among heroin-dependent mothers, perinatal mortality is 2.7 times higher than in control groups matched for social class and race. Babies had higher rates of jaundice (a liver problem), congenital (at birth) abnormalities, five times the risk of growth retardation, and twice the risk of being born prematurely. There were also more complications associated with birth. Both heroin and cocaine can damage the nervous system. In the USA about 300 000 babies are born each year addicted to crack.

Users of drugs that are injected are in general at more risk of catching HIV or hepatitis B through use of infected needles. LSD and marijuana may have harmful effects on the fetus. Marijuana appears to reduce the length of gestation by about 1 week.

Pharmaceutical products. Pharmaceutical products on general sale are carefully tested for harmful effects and are regarded as safe if instructions on doses are properly followed. Where possible, it is probably a wise precaution not to use drugs, especially early in pregnancy, since most will cross the placenta with ease and it is not really possible to prescribe safely in all situations.

The tragedy of the drug thalidomide resulted in increased precautions being introduced in the testing of drugs. Thalidomide was introduced in the early 1960s and prescribed to pregnant women who suffered particularly badly from 'morning sickness', a type of nausea which can become very intense in some women. An increase in numbers of deformed babies was soon traced back to the drug and it was banned. However, several thousand babies had already been affected. Deformities were characteristic. They often included missing or very stunted limbs. The bones of the limbs were much reduced or absent. The hands or feet often arose directly from the trunk of the body (fig 21.52). Either arms or legs, or sometimes both, were affected, depending on when the drug was taken during pregnancy. The drug also caused defects of the heart, gut, eyes and ears.

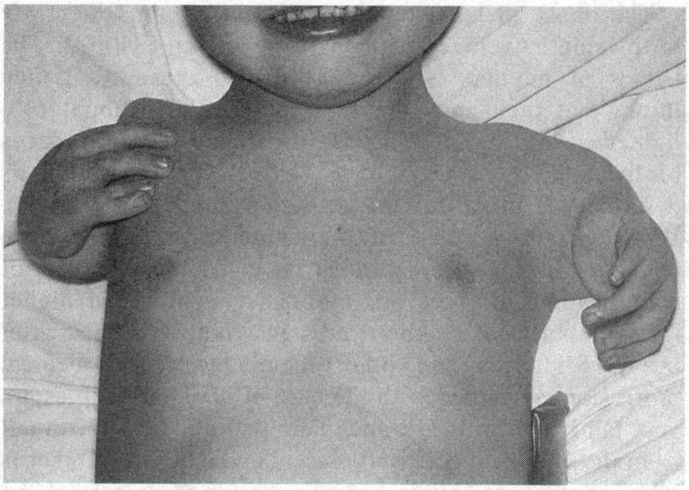

Fig 21.52 *Baby suffering from limb deformities caused by thalidomide.*

Viruses

Although most bacteria are too large to cross the placenta, most viruses are small enough to do so. If the mother catches a viral disease, she can therefore pass it to the fetus, where it may cause much greater damage, especially in the early stages of pregnancy when organs are still developing. Three examples which are of particular concern are Rubella (German measles), HIV and hepatitis B.

If the embryo or fetus is infected with Rubella it may cause miscarriage or congenital malformations. Common malformations are blindness, deafness and malformations of the heart and nervous system. Severe mental retardation may result. If contracted in the first month of pregnancy, as many as 50% of babies develop congenital malformations. A national vaccination programme, with particular care being taken to vaccinate females, has helped to reduce the risk of contracting German measles during pregnancy.

HIV and hepatitis B do not cause congenital malformations, but are life-threatening diseases.

Rhesus factor

The rhesus factor is an antigen found in the cell surface membranes of red blood cells. 84% of people possess the factor and are described as rhesus positive. Those who do not are described as rhesus negative. Presence of the rhesus antigen is genetically determined.

Red blood cells are too large to cross the placenta under normal circumstances. However, during birth, when the placenta is damaged, red blood cells of the fetus can reach the mother's blood.

A problem arises if the mother is rhesus negative and the baby is rhesus positive. If red blood cells from the fetus get into the mother's circulation, her body will recognise the rhesus (D) antigens as foreign, and make anti-rhesus (anti-D) antibodies against them. During a second pregnancy with a rhesus positive baby, the mother's immune system has already learned to make anti-D antibodies. Anti-D can cross the placenta into the blood of the fetus and cause problems which may prove fatal. A fuller discussion of the problems and how they can be prevented can be found in section 14.9.8.

21.8 Why should a Rh– woman who has not passed child-bearing age not be given Rh+ blood in a transfusion?

21.9 Why is there no problem if the baby is Rh– and the mother is Rh+?

21.8.11 Sex determination in the developing embryo

The sex of the embryo is determined by the sex chromosomes carried by the sperm, X in the case of a female and Y in the case of a male. All eggs carry one X chromosome. The resulting zygote therefore has the genotype XX (female) or XY (male) (section 24.6). In the early stages of embryonic development a pair of embryonic gonads, the **genital ridges**, develops, together with the beginnings of the female and male reproductive systems. Up to the sixth week of development, there is no structural difference between the male and the female. After that the sex chromosomes determine whether the system continues to develop as female or male.

The X chromosome carries a gene, the **Tfm gene** (**testicular feminisation gene**) which specifies the production of a particular protein molecule in the cell surface membranes of cells in the developing reproductive system. This protein acts as a testosterone receptor. Since both male and female embryos carry at least one X chromosome, this molecule is present in both sexes. The Y chromosome carries a gene called the **testis-determining gene**. This codes for the production of a protein molecule called the **H-Y antigen** which is found on the surface of all body cells of the male and is absent from the body cells of a female. It is responsible for cells of the undifferentiated genital ridges (gonads) differentiating into seminiferous tubules and interstitial cells. Without the H-Y antigen, the gonad becomes an ovary. Testosterone produced by the developing testes is released into the embryonic circulatory system and reacts with the testosterone receptor molecules in the target cells of the potential male reproductive system. The testosterone receptor/testosterone complex formed passes to the nuclei where it activates genes associated with the development of the tissues which give rise to the male reproductive system. Therefore an XY embryo will develop into a male fetus. The tissues of the potential female reproductive system are not activated and do not develop. In an XX embryo, the absence of testosterone allows the reproductive system to become female. A summary of these events is shown in fig 21.53.

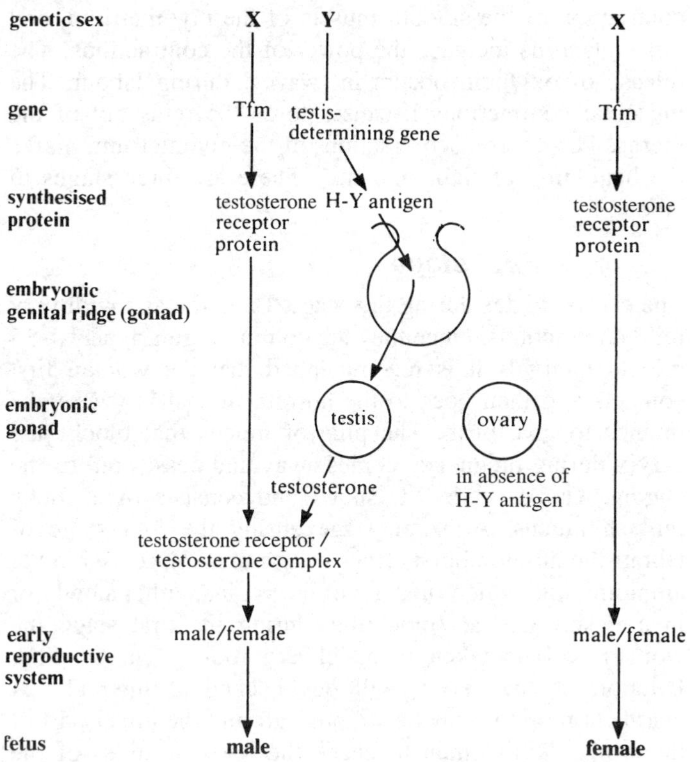

Fig 21.53 *The events involved in the differentiation of the early reproductive system of the embryo into a male or female system.*

The location of the testis-determining gene on the male chromosome has now been discovered. It is referred to as the **sex-determining region** (**SRY**) and is about 30 000 base pairs long (the total DNA in a human cell is about 3000 million base pairs).

21.8.12 Birth

During the final month of pregnancy the uterus becomes more and more sensitive to oxytocin. Oxytocin is a peptide hormone produced in the hypothalamus and released from the posterior pituitary gland. It causes contraction of the smooth muscle of the myometrium, the muscular lining of the uterus. The increased sensitivity is due partly to the synthesis of more and more oxytocin receptors in the myometrium, possibly a result of high levels of oestrogen. Oxytocin levels also rise as a result of the level of progesterone decreasing late in pregnancy. A third factor may be an influence of the fetus. The hypothalamus of the fetus releases ACTH from its pituitary gland. This stimulates the fetal adrenal gland to release corticosteroids which cross the placenta and enter the mother's circulation, causing a decrease in progesterone production and an increase in secretion of prostaglandins. Prostaglandins are secreted by the uterus and stimuiate contraction of the uterus. The reduction in progesterone level also removes the inhibitory effect of progesterone on contraction of the myometrium. Oxytocin causes contraction of the smooth muscle of the myometrium, and prostaglandins increase the power of the contractions. The release of oxytocin occurs in 'waves' during labour. The muscular contractions it causes force the fetus out of the uterus. The onset of contractions of the myometrium marks the beginning of 'labour pains'. There are three stages to labour.

First stage

The cervix dilates during this stage. Labour starts with very mild contractions. Once they are coming regularly at 10–15 minute intervals it is recommended that the woman first contacts and then goes to the hospital if that is where she intends to give birth. The plug of mucus that blocks the cervix during pregnancy comes away and passes out of the vagina. This is called a 'show' and consists of a sticky pinkish mucus. At some stage during the first stage of labour the amnion bursts (the 'waters break'), releasing the amniotic fluid which runs out of the vagina, either slowly or in a gush. Also at some time during the first stage the woman will be taken to a delivery room if in hospital. Dilation of the cervix will be checked at intervals. A heartbeat monitor may be strapped around the upper part of the woman's abdomen to check the level of stress of the fetus. Alternatively, a stethoscope may be used.

The first stage of labour can be the painful stage and usually lasts 6–12 hours for first babies. Contractions gradually get stronger and more frequent due to positive feedback control of oxytocin production. The more the uterus contracts, the greater the stimulation of stretch receptors in the uterus and cervix. These send nerve impulses via the autonomic nervous system to the myometrium, which contracts even more, and so on. Other nerve impulses pass to the hypothalamus, stimulating the release of more oxytocin from the posterior pituitary gland. Contractions spread down the uterus and are strongest from top to bottom, thus pushing the baby downwards. Prostaglandins, hormones secreted by the uterus, also cause contraction. Throughout these contractions the cervix gradually dilates.

Pain relief is usually available in the form of 'gas and air' (a mixture of air and nitrogen(I) oxide), injections of pethidine or by means of an epidural. This is a local anaesthetic applied close to the spinal cord which removes feeling below about waist level and allows the woman to be conscious even if a Caesarean operation is necessary (surgical removal of the baby by opening the abdomen).

The cervix is fully dilated when it is about 10 cm wide, wide enough to allow the baby's head to pass through.

Second stage

In the second stage the baby is born. The baby is pushed out of the uterus and down the vagina, usually head first. Once the head is born the hard work is almost over because the rest of the body follows much more easily. As soon as the baby is breathing normally the umbilical cord is clamped at two places and cut between the clamps.

Third stage

This is the 'birth' of the placenta, caused by powerful contractions of the uterus. The placenta comes away from the wall of the uterus and passes out through the vagina. This stage happens quickly and easily. Bleeding is limited by contraction of muscle fibres around the blood vessels of the uterus which supply the placenta. Average blood loss is kept to about 350 cm^3.

21.8.13 Lactation

Lactation is the production of milk. The **mammary glands**, or breasts, contain milk glands. In the glands, special epithelial cells line small sacs called alveoli. These cells secrete milk. The alveoli are surrounded by a layer of tissue containing smooth muscle fibres. When the muscle contracts it causes milk to be released. Milk enters a series of ducts (tubes), and each duct has an expanded space called a sinus which stores milk (fig 21.54). The ducts eventually pass to separate openings in the nipple.

The breasts increase in size during pregnancy due to the development of the milk glands, controlled by progesterone, and ducts, controlled by oestrogen. **Human placental lactogen**, another hormone, is also involved, as shown in table 21.5. However, for milk to be produced, the hormone **prolactin** must be present. This is secreted by the anterior pituitary gland. Throughout pregnancy the presence of oestrogen and progesterone inhibits the secretion of prolactin and therefore the formation of milk. At birth, when the oestrogen and progesterone levels fall due to loss of the placenta, prolactin is no longer inhibited and it stimulates the alveoli to secrete milk.

Suckling reflex

The ejection of milk from the nipple involves a simple reflex action, the **milk ejection reflex** (fig 21.55). The

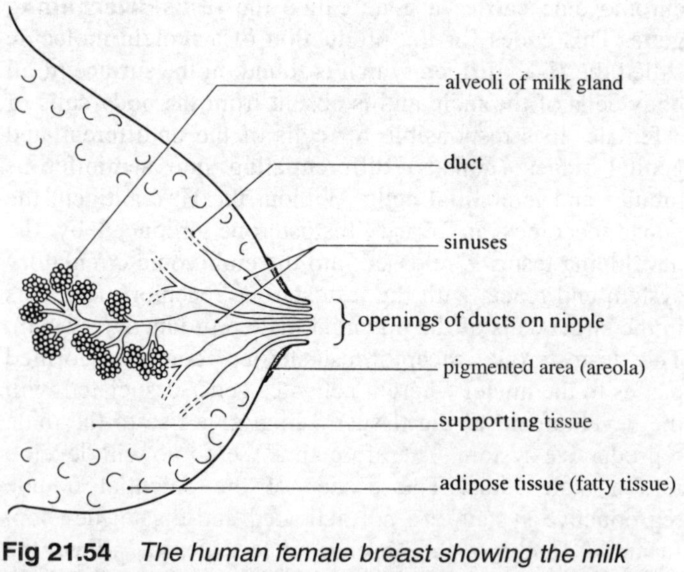

Fig 21.54 *The human female breast showing the milk glands where milk is secreted and the ducts and sinuses carrying milk to the nipple.*

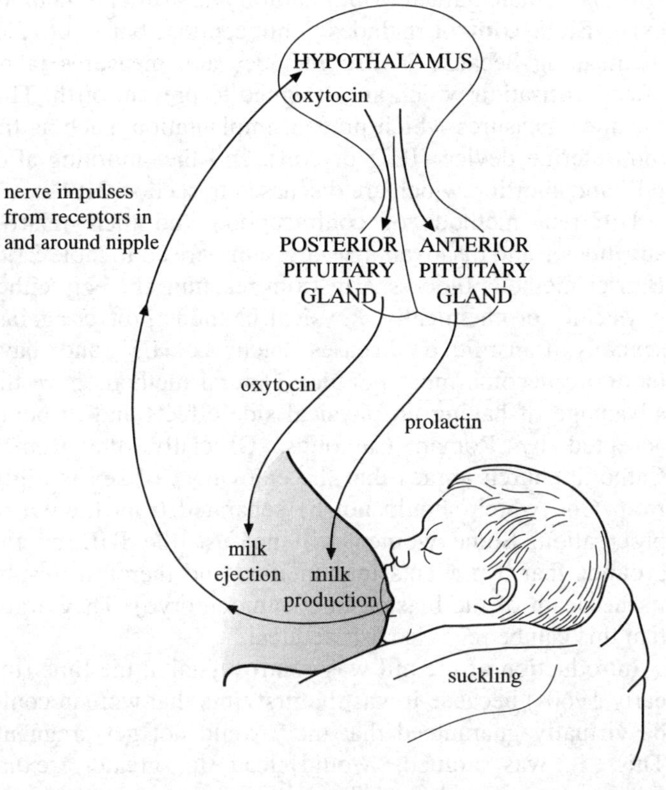

Fig 21.55 *Newborn infant suckling at the breast showing details of the suckling reflex. Lactation involves stimulation of milk secretion by prolactin and milk ejection by oxytocin.*

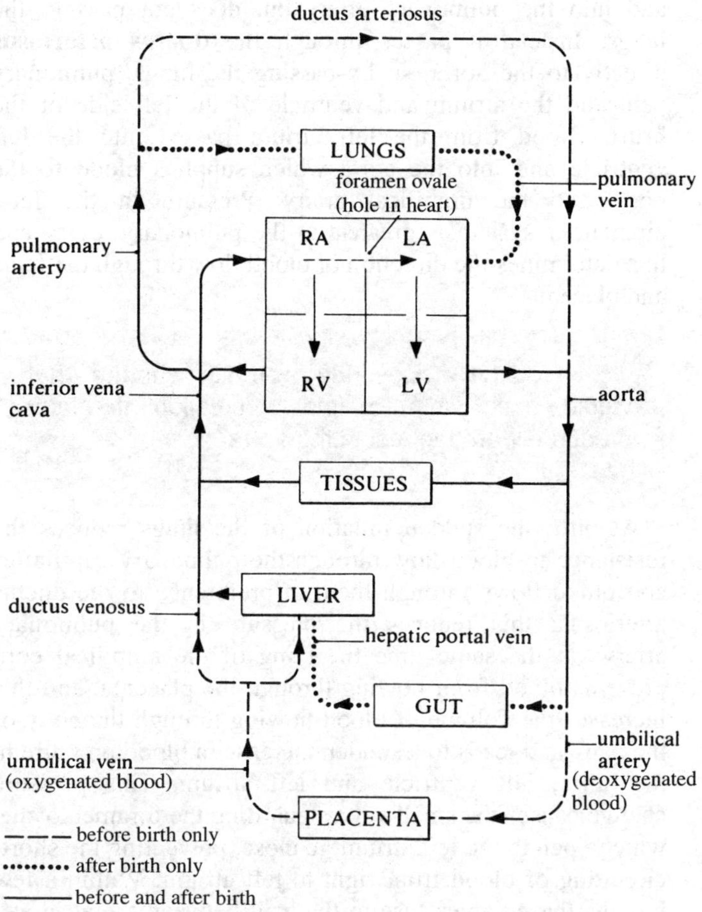

Fig 21.56 *Ante-natal (fetal) and post-natal circulatory systems. (The periods when the blood vessels are functional are indicated on the diagram.)*

sucking of the baby on the breast stimulates sensory receptors around and in the nipple. Nerve impulses pass from the receptors via the spinal cord to the hypothalamus which releases oxytocin from the posterior pituitary gland. This causes contraction of the smooth muscle fibres surrounding the alveoli and forces milk through the ducts and sinuses and out of the nipples. The stimulus of sucking also stimulates the release of prolactin by means of another reflex to the hypothalamus. Oxytocin also stimulates contraction of the muscle in the uterus, helping it to recover its normal tone after birth.

Colostrum

The first secretion of the breasts, following birth, is not milk but **colostrum**. This has a yellow colour and contains cells from the alveoli. It is rich in the protein globulin, but low in fat. It is believed to be a means of passing antibodies, particularly IgA (section 14.9.5), from mother to baby.

Milk as a food

Human milk contains fat, lactose (milk sugar) and the proteins lactalbumin and casein which are all easily digestible. The milk is made from nutrients circulating in the blood, such as lactose from glucose, protein from amino acids, and fats from fatty acids and glycerol. Milk alone is adequate to produce weight gains in the baby of 25–30 g per day.

In between breast feeds prolactin stimulates milk production for the next feed.

21.8.14 Changes in fetal circulation at birth

Throughout development in the uterus the fetal lungs and gut do not function since gaseous exchange and nutrition are provided by the mother via the placenta. Most of the oxygenated blood returning to the fetus via the umbilical vein by-passes its liver in a vessel, the **ductus venosus**, which shunts blood into the inferior vena cava and passes it to the right atrium (fig 21.56). Some blood from the umbilical vein flows directly to the liver; blood entering the right atrium, therefore, contains a mixture of oxygenated and deoxygenated blood. From here most of the blood passes through an opening in the wall separating the atria, the **foramen ovale**, into the left atrium. Some blood passes from the right atrium into the right ventricle and into the pulmonary artery but does not pass to the lungs. Instead it passes through the **ductus arteriosus** directly to the aorta, so by-passing the lungs, pulmonary vein and the atrium and ventricle of the left side of the heart. Blood from the left atrium passes into the left ventricle and into the aorta which supplies blood to the body and the umbilical artery. Pressure in the fetal circulatory system is greatest in the pulmonary artery and this determines the direction of blood flow through the fetus and placenta.

21.10 Describe a major change that would occur in the fetal circulation if blood pressure were highest in the aorta.

At birth the sudden inflation of the lungs reduces the resistance to blood flow through the pulmonary capillaries and blood flows through them in preference to the ductus arteriosus; this reduces the pressure in the pulmonary artery. At the same time the tying of the umbilical cord prevents blood from flowing through the placenta, and this increases the volume of blood flowing through the body of the baby and leads to a sudden increase in blood pressure in the aorta, left ventricle and left atrium. This pressure change causes the small valves guarding the foramen ovale, which open to the left atrium, to close, preventing the short-circuiting of blood from right to left atrium. Within a few months these valves fuse to the wall between the atria and close the foramen ovale completely. If this does not occur the baby is left with a 'hole in the heart' and will require surgery to correct the defect.

The increased pressure in the aorta and decreased pressure in the pulmonary artery force blood backwards along the ductus arteriosus into the pulmonary artery and hence to the lungs, thereby boosting its supply. After a few hours muscles in the wall of the ductus arteriosus constrict under the influence of the rising concentration of oxygen in the blood and close off this blood vessel. A similar mechanism of muscular contraction closes off the ductus venosus and increases blood flow through the liver. The mechanism of closing down the ductus venosus is not known but is essential in transforming the ante-natal (before birth) circulation into the post-natal (after birth) condition.

21.9 Human intervention in reproduction

Humans are increasingly able to intervene in their own reproduction as a result of advances in our understanding of reproductive physiology, genetics, molecular biology and improvements in medical technology. Some of the ways in which we do this are examined in this section, together with some of the social and ethical issues which are raised.

21.9.1 Contraception and birth control

Contraception is the prevention of conception, that is preventing the fusion of the male gamete with the female gamete. Both natural and artificial methods exist. **Birth control** includes contraception, but is broader in meaning because it also includes any measures taken after fertilisation which are designed to prevent birth. This includes measures which prevent implantation, such as the intra-uterine device (IUD or coil), and the 'morning-after pill' and abortion which are discussed in section 21.9.2.

Different methods of contraception and their relative advantages and disadvantages are summarised in table 21.6. Barrier methods block sperm from reaching the egg, either physically or chemically. Physical methods protect against sexually transmitted diseases such as HIV and have therefore become more popular. Natural methods have the advantage of having no physical side effects and of being accepted by Roman Catholics. Officially, the Roman Catholic church argues that the enjoyment of sex is a gift from God which should not be separated from the act of procreation. Some women will not use the IUD on the grounds that it prevents implantation and therefore results in the death of the blastocyst (young embryo). They argue that this can be regarded as unethical.

Introduction of the pill was controversial at the time (the early 1960s) because it was the first time that women could be virtually guaranteed that they would not get pregnant. This, it was argued, would lead to greater sexual permissiveness and contribute to a 'permissive society' with declining moral values. It can equally be argued that the easy availability of safe and effective contraception, together with education about its use, prevents many unwanted pregnancies and saves much emotional suffering. Some authorities argue that the relatively high level of unwanted pregnancies is evidence that there is still insufficient education about contraceptives and about the consequences of unwanted pregnancy, and that contraceptives should be more freely available. The problem is particularly acute in Britain compared with most other western European countries.

Termination of a pregnancy using a morning-after pill or, more controversially, abortion raises further issues. Use of the morning-after pill could be harmful to some women, although it is unlikely that the risks are greater than those of having a baby. Abortion is discussed more fully in section 21.9.2. It is widely used in some countries where contraceptives are not readily available as a form of birth control, as in Russia where the average woman has several abortions in her lifetime.

From a global point of view, contraception seems to be the only hope of restricting the growth of the human population. Strenuous efforts have been made to make contraception available in developing countries, with some success. For the first time it seems that we are in a position to be able to say that the total human population of the world will never again double. In developed countries

Table 21.6 Methods of contraception and birth control: their modes of action and relative advantages and disadvantages.

Method	*Basis of action*	*Notes on use*	*Approximate failure rate**	*Relative advantages*	*Relative disadvantages*
Barrier methods					
Condom	A thin, strong rubber sheath. Prevents sperm entering the vagina.	Placed over erect penis just before sexual intercourse.	10% falling to 3% with experienced use.	Cheap, easy and convenient to use. Easily obtained. Gives some protection against sexually transmitted diseases such as HIV.	May disrupt act of love-making. Not as reliable as the Pill. Relies on male. May tear or slip off after climax.
Femidom	Female equivalent of the condom – a thin rubber or polyurethane tube with a closed end which fits inside vagina. Has two flexible rings, one each end, to keep it in place. The open end stays outside the vagina, flat against the vulva.	Inserted before intercourse and removed any time later.	Relatively new so no data. Probably similar to condom.	Gives woman some control. Other advantages as male condom.	
Diaphragm/cap	A flexible rubber dome which fits over the cervix and prevents entry of sperm to uterus. Used with a spermicidal cream or jelly (a spermicide is a chemical which kills sperm).	Inserted before intercourse. Must be left in place at least 6 hours after intercourse.	3–15%	Can be inserted a few hours before intercourse.	Must first be fitted by a doctor and training is required to fit. Occasionally causes abdominal pain. Check every 6 months that cap is right size.
Spermicide	Chemical kills sperm	Placed in vagina to cover lining of vagina and cervix. Effective for about 1 hour.	10–25%	Can be quite effective when used with condom or diaphragm.	High failure rate if used on its own.
Sponge	Polyurethane sponge impregnated with spermicide. Fits over cervix. Disposable.	Fit up to 24 hours before intercourse. Leave in place at least 6 hours after intercourse.	10–25%	Easier than cap because one size fits all and no fitting required.	High failure rate.
Hormonal methods					
'Pill'	Contains the female sex hormones oestrogen and progesterone. Prevents development of eggs and ovulation by inhibiting secretion of FSH.	One taken orally each day during first 3 weeks of cycle. After week 4 menstruation starts and the Pill is started again.	1%	Very reliable (almost 100% if properly used). Woman has control. No interference with lovemaking.	Short-term side effects may include nausea, fluid retention, and weight gain. Long-term side effects not fully understood, but increased risk of blood clotting may occur in some women. Not recommended for older women, or smokers over 35.
Minipill	Progesterone only. Ovulation may occur but cervical mucus is thickened, preventing entry of sperm.	Must be taken within 3 hours of same time every day.	2%	Very reliable. Lower dose of hormones than Pill. Therefore less risk to older women.	Breakthrough bleeding (light bleeding between periods) more common than with Pill. May cause headaches. Must be taken within 3 hours of same time each day.
Preventing implantation					
IUD (intra-uterine device) or coil	Small device made of copper, plastic or stainless steel. Inserted into uterus by a doctor and left in place. Prevents implantation.		3%	Can be left in place for long periods (up to 5 years). Suitable for women who have had children (easier to fit and larger uterus).	May cause bleeding and discomfort. IUD may come out.
Natural methods					
Abstinence	Avoid sexual intercourse.		0%	Effective.	Restricts emotional development of a relationship.
Rhythm method	Avoid sexual intercourse around the time of ovulation (total abstinence for about 7 days)		20%	Natural and acceptable to Roman Catholics.	High failure rate, even higher if periods are irregular. Requires good knowledge of body and good record-keeping. Requires a period of abstinence.
Temperature method	Note rise in temperature at ovulation and avoid sexual intercourse at these times.		up to 20%	Can increase reliability of rhythm method.	as above
Billings method	Note appearance of clear, thin, stretchy mucus in vaginal secretions at ovulation and avoid sexual intercourse at these times.		up to 20%	Can increase reliability of rhythm method.	as above
Coitus interruptus (withdrawal)	Penis is withdrawn from vagina before ejaculation.		20%	Accepted by Roman Catholic church.	High failure rate. Requires much self-discipline. Penis may leak some sperm before ejaculation.
Sterilisation					
Vasectomy – male	Cut each vas deferens.		less than 1%	Very reliable. No side effects. Simple.	Very difficult to reverse.
Tying of oviducts – female	Cut both oviducts.		less than 1%	Very reliable.	Even more difficult to reverse than vasectomy.
Termination					
Morning-after Pill (section 21.9.2)	Contains RU486, an anti-progesterone.	Taken within 3 days of sexual intercourse.			For use only in emergencies. Long-term effects not known.
Abortion	Up to 24 weeks in UK	Premature termination of pregnancy by surgical intervention.		A measure of last resort.	Risk of infertility or other complications. Ethical, moral, religious issues raised. Emotionally difficult.

* % of pregnancies in first year of use.

contraception underpins our whole lifestyle and economy. It allows women to plan careers and couples to maintain a higher standard of living than would be possible with large families. It has allowed the populations of most developed countries to achieve stability. Some governments are now even concerned about declining populations, as in France.

21.9.2 Abortion

Abortion is the premature termination of a pregnancy. Medically speaking a miscarriage is a natural abortion, or **spontaneous abortion**. The theme of this section, though, is **induced abortion**, that is abortion which is deliberately brought about. The term abortion will therefore be used to mean induced abortion in this section.

Abortion is one of the most controversial issues associated with reproduction. It raises important ethical issues. There is an on-going debate within society, and in some countries such as Ireland and the USA it has become an important political issue. As in any debate, it is important not only to be able to express your own views clearly, but also to be willing to listen to and respect the views of others. People's opinions usually change slowly, after weighing up arguments over a period of time, and if you can put your own case well it will obviously have more influence on this process.

The law

Abortion was made legal in the UK by the Abortion Act, 1968. The Act was introduced because the number of illegal abortions being carried out was unacceptably high, and women's health was being put too much at risk. In the end, ethical arguments concerning the fetus played little part in the decision to legalise abortion. The Abortion Act was amended in 1991 so that at present:

- the legal age limit of the fetus for abortion is 24 weeks, the age at which a premature baby has a reasonable chance of survival using medical technology;
- in exceptional circumstances, when the fetus is shown to be suffering from severe handicap, abortions are allowed at any time up to birth (only 52 abortions were carried out on these grounds in 1991).

The grounds for abortion may be summarised as:

- risk to life, or grave permanent injury to physical or mental health, of the woman (any time);
- risk to the physical or mental health of the woman (up to 24 weeks) (these are the commonest grounds);
- risk to the physical or mental health of existing children (up to 24 weeks);
- substantial risk of the child being born seriously handicapped (any time).

Two medical practitioners must agree to the abortion and the father has no legal right to prevent an abortion.

During the 1990s, about 20% of all pregnancies in the UK were aborted.

Biological aspects of abortion

Four standard methods of abortion are described in table 21.7. The earlier the procedure is carried out, the easier and safer it is and there is a trend towards earlier abortions. Although no procedure is entirely without risk, an early abortion is in fact safer than allowing a pregnancy to proceed, so risk to the mother's health cannot be used as an argument against early abortion. Possible medical problems associated with abortion include the following.

- Occasionally infection occurs as a result of abortion. If this spreads through the uterus to the oviducts it may cause blockage of the oviducts and infertility. Such infections cause **pelvic inflammatory disease** whose symptoms include discomfort or pain in the pelvis region. About 5% of women who have abortions are made infertile as a result.
- The risk of a subsequent ectopic pregnancy is increased. This is implantation and development of the embryo in the oviducts.
- Damage to the cervix can occur which may affect the ability to carry another baby.
- Damage to the uterus may occur.
- The placenta may be retained, resulting in bleeding.
- An average of five women a year die from abortions in England and Wales.
- Emotional damage can occur. Counselling should be available after abortion as well as before. Feelings of guilt are common.

Ethical issues

The subject of abortion is a very emotional one and deserves careful consideration. Some of the commonly raised ethical issues are listed below.

- Abortion could be regarded as murder. Christians and Muslims believe that the soul is independent of the body and that it enters the body at the moment of conception. Many others also believe that from the moment of conception the new individual should have the same rights as anyone has after birth. However, the growing embryo can successfully split into identical twins at any time up to about 14 days, so it could be argued that it does not have an individual identity until at least that time.
- Even if abortion is accepted, is the scale of abortion too large? The rise in the number of abortions since it was legalised suggests that it has become more and more acceptable. Is the value of life being cheapened?
- Many abortions are being carried out on fetuses with disabilities such as Down's syndrome where the child could be expected to have a life of acceptable quality. Are we becoming intolerant of disability in general? Will this make us a less caring society and what would be the consequences? (This issue is also raised in section 25.7.)
- Many abortions are carried out beyond 12 weeks when either D and C or induced labour is necessary to

Table 21.7 Commonly used methods of abortion.

Technique	*Stage of pregnancy*	*Summary of method*	*Further notes*
Vacuum aspiration	Up to 12 weeks	Cervix is stretched and the contents of the uterus are gently sucked out using a flexible tube and a pump (aspirator). It takes about 30 minutes. Local anaesthetic used.	Most common method
Dilation and curettage (D and C)	12–16 weeks	Cervix is stretched and a curette, a spoon-shaped knife, is used to gently scrape the lining of the uterus. The contents can then be sucked out by vacuum aspiration. Local anaesthetic used.	
Prostaglandins or saline injection	16+ weeks	Prostaglandins are injected into the amniotic fluid, killing the fetus and causing contractions of the uterus. The fetus is born dead. Saline also kills the fetus and a hormone is added to cause contractions.	Fetus is too large to remove by earlier methods. More distressing than methods used at earlier stage.
RU486 (abortion pill)	Less than 10 weeks	Anti-progesterone. Embryo or fetus is rejected. Prostaglandin injection is given 1 or 2 days later to cause contractions and complete the abortion.	

achieve the abortion. All the major organs are developed by this stage. The fetus may therefore be able to feel pain. Abortion is allowed up to 24 weeks and there is evidence from hormonal changes that the fetus of 23 weeks can feel pain. We cannot know, though, the nature or extent of this pain since pain is a very subjective experience and is coloured by other feelings such as fear and anticipation.

- Is the legal limit of 24 weeks too late? Some 'Pro-Life' groups want the limit lowered, for example to 18 weeks, arguing that the fetus can feel pain at 24 weeks. The 24-week limit is based on the survival chances of the fetus if it is born prematurely. Should its life depend simply on the current state of medical technology? Some 'Pro-Life' groups are totally against abortion.
- If abortions were illegal, how should society deal with women who have become pregnant as a result of rape?
- The original argument for the legalisation of abortion is still a powerful one, namely that if it were not legal, many illegal abortions would be performed, resulting in financial exploitation of women, greater risk to health and the threat of being a social 'outcast'.
- Should the father have any right to prevent an abortion? Currently he does not.
- There is a worrying number of unwanted pregnancies among teenagers. If abortion were not available for these young women, it is likely that they would become trapped in relative poverty, unable to earn a living because of lack of child care facilities and a burden on the State through the need to claim benefits.
- 'Pro-Choice' groups, such as the National Abortion Campaign, argue that abortion should be available on demand, and therefore that there should be no legal restrictions. They also argue that there should be more National Health Service clinics carrying out the procedure rather than private practices, and that abortion should be allowed at any stage of pregnancy, otherwise women are being denied the freedom to control their 'own' bodies and futures.
- Extra children may impose severe financial stress on an existing family. Unwanted children are more likely to suffer from physical abuse, and are less likely to succeed in education, employment and marriage.

21.9.3 Infertility

Infertility cannot be defined precisely because there are varying degrees of infertility. For example, an average fertile couple have a 15% chance of conceiving each month if no contraception is used. A useful working definition is the failure to achieve pregnancy after one year of trying. In this situation the medical profession would recommend that investigations into the problem were justified.

The problem of infertility is surprisingly common. As many as 1 in 8 couples will experience problems in trying to conceive. Traditionally the finger of 'blame' tended to be pointed at the woman as the cause, but it is now known that the problem is almost as likely to be due to the man. Unfortunately, a lot less is known about male infertility because women have been studied in more detail, partly due to the traditional attitudes in medicine. For many people, having children is the most important aim in their lives, and is the thing that gives most purpose to their lives. To be deprived of the experience can cause great anguish and emotional pain. People will do almost anything to have children, and may pay large sums of money for treatment.

Those concerned with the treatment of infertile couples therefore have a great responsibility. Infertility treatment is not given high priority in the National Health Service (NHS) because it is not a life-threatening condition and because there is no shortage of people! In fact, IVF was withdrawn as a NHS treatment in 1996. However, affected couples are generally advised to visit their local GP first, and the NHS will carry out tests such as sperm counts in order to try to identify the cause of the problem. Couples will be referred to a hospital if necessary.

Some of the more common causes of infertility and their treatment are dealt with below.

Female infertility

Failure to ovulate. About 30% of female infertility is due to failure to ovulate. In 70% of such cases the cause is hormonal. Sometimes the hypothalamus or pituitary gland fail to produce hormones normally, with the result that either no follicles develop (lack of FSH) or egg release is affected (lack of LH). Alternatively, the ovaries may not be producing oestrogen and/or progesterone normally. In other cases there may be physical damage to the ovaries, they may be absent or not functioning normally, or very severe emotional upset or physical stress may prevent them from working normally.

Failure to ovulate can be cured in over 90% of cases. Hormone imbalance can be corrected using synthetic versions of natural hormones. The most commonly used drug is Clomiphene, a synthetic oestrogen-like drug which stimulates ovulation by bringing about the release of FSH from the pituitary gland. Tamoxifen is an alternative similar chemical. These drugs are taken as pills, usually for about five days soon after the menstrual cycle starts. If failure to *release* eggs is the problem, this may be overcome by using an injection of HCG (human chorionic gonadotrophin) which is chemically very similar to LH and acts as a substitute for LH. It is given at the time of the expected LH surge in the middle of the cycle. 'Fertility drugs' may also be used if other treatments fail since they are more powerful. They contain the pituitary hormones FSH and LH, or just FSH. They may, however, cause the release of several eggs in the same month and there is therefore a risk of multiple pregnancy. This risk can be reduced by checking for multiple ovulation using ultrasound, and by careful monitoring of hormone levels in the body.

An alternative strategy is to use injections of the hormone made by the hypothalamus, namely GnRH (gonadotrophin releasing hormone). This can be made artificially and administered in pulses using a small pump attached to the upper arm in order to mimic the natural activity of the hypothalamus.

Damage to the oviducts. About one-third of female infertility is due to tubal disease, that is damage to the oviducts. The tubes may be completely blocked, although this is not usually the problem. Infections may cause scarring, and hence partial or complete blockage or narrowing of the tubes, or damage the delicate linings of the oviducts. Another problem can be adhesions, where strands of tissue anchor the oviducts to other organs such as the uterus. This may be a result of the body repairing damage in that area, sometimes after operations. Adhesions can prevent the natural movements of the oviduct which are needed to collect and transport eggs.

If an infection is responsible, it is often pelvic inflammatory disease (PID) caused by bacteria which are normally harmless. PID is more common among women who have had a number of sexual partners and among women who use the coil (IUD). Other infections may occur after delivery of a baby or after a miscarriage or abortion. In a few rare cases women are born with blocked oviducts.

Endometriosis is another possible cause of damage to oviducts. In this condition, parts of the endometrium (the lining of the uterus) break away and grow elsewhere, for example in or around the ovaries or oviducts.

The best means of diagnosing this problem is to take an X-ray of the uterus and oviducts. A dye that is opaque to X-rays is injected into the uterus and will enter the tubes if they are not blocked. Laparoscopy (see IVF below) can be used to view the oviducts externally, and is particularly useful for examining adhesions.

The most common treatment is surgery. This is carried out with the aid of a microscope (microsurgery) because the tubes are extremely fine. Lasers have been used with some success. Sometimes the blocked portion of a tube is cut out and the tube rejoined. Adhesions can be removed relatively easily.

Uterus damage. About 5–10% of female infertility is caused by problems with the uterus. Here the problem is not one of getting pregnant, but of maintaining the pregnancy and preventing miscarriage. Implantation may be a problem, or problems may arise later. Occasionally the growth of fibroids, which are non-malignant tumours which grow from the walls of the uterus, can cause infertility. Similar, but smaller growths called polyps may have the same effect. Fibroids and polyps can be removed relatively easily by surgery. Adhesions within the uterus, sometimes the result of the scraping procedure of an abortion, can stick parts of the inside of the uterus together. It may be possible to remove these surgically. IUDs or infections can cause inflammation, in which case antibiotics can be prescribed. Less commonly there is a congenital abnormality (a problem already present when the woman was born), such as an absent, small or odd-shaped uterus.

Cervix damage. The cervix is the neck of the uterus. It may be damaged as a result of abortion or a difficult birth. Scar tissue may narrow it or it may stop producing the cervical mucus needed for the sperm to reach the uterus. Alternatively it may be widened, with the result that miscarriage is much more likely after 3 months of pregnancy. All these conditions can be treated by surgery.

Antibodies to sperm. In a few rare cases, women produce antibodies against their husband's sperm. These are found in the cervix, uterus and oviducts. This may be treated in various ways, including the use of drugs which suppress the immune system, but IVF is probably the best treatment (see below).

Male infertility

Absence of sperm. Absence of sperm in the semen is called **azoospermia**. About 5% of male infertility is a result of sperm not being produced at all, but even if they are produced they may not appear in the semen if the tubes between the testes and the seminal vesicles are blocked (fig 21.32). Blockage can be a result of scarring due to infection or injury. Gonorrhoea and TB (tuberculosis) are possible causes of infection. The blockage may be congenital.

Another cause of azoospermia can be a failure of the ejaculation mechanism. Retrograde ejaculation, in which sperm pass up the urethra to the bladder instead of down the urethra and out of the body, is possible for various reasons.

Failure to produce sperm may be a result of physical injury to the testes, or occasionally a result of infection by the mumps virus after puberty. Alternatively the problem may be hormonal, in which case it is very difficult to treat. Fig 21.57*a* summarises some of the causes of azoospermia.

Low sperm count. More than 90% of male infertility is due to low sperm counts. The cause is often unknown, but a number of factors have been shown to be possible causes. These are summarised in fig 21.57*b*.

Abnormal sperm. It is usual for a small proportion of sperm to be abnormal, for example having two tails, no tail, no head or abnormal shapes. However, if the proportion is high, fertility is reduced. The reasons for these abnormalities are usually not known, though they could be due to hormonal problems or infections in some cases.

Autoimmunity. About 5–10% of male infertility is probably due to an immune response by the male to his own sperm. Antibodies are made which attack the sperm and reduce the living sperm count. Reasons for this are unknown. The condition is difficult to treat; treatment with corticosteroids to reduce the immune response may result in unacceptable risk of infections. Alternatively, donor insemination may be recommended (see below).

Premature ejaculation. This is the situation where the man has an orgasm before penetration of the penis into the vagina. The problem is often overcome with experience.

Impotence. This is the inability to achieve or maintain an erection of the penis. The problem is usually psychological and counselling may help if it persists.

21.9.4 Treatment for infertility

Where environmental factors such as smoking, obesity, and stress are involved, treatment is aimed at removing or reducing the factor responsible. Surgical and hormone treatments are mentioned above where relevant and in fig 21.58. A number of other treatments are also available including in vitro fertilisation, donor insemination and artificial insemination.

In vitro fertilisation (IVF)

This is commonly known as the 'test-tube baby' technique. It was first devised by Patrick Steptoe and Robert Edwards in 1978 as a means of helping women with blocked oviducts. It is still mainly used for this problem and for patients with damaged oviducts which cannot be repaired surgically. However, it may be used where there are problems with endometriosis, low sperm counts or abnormal sperm (since fewer sperm are needed to fertilise eggs in vitro), and where the male or female are producing antibodies against the sperm. If the female cannot produce eggs it allows donor eggs to be used (see egg donation).

The technique involves fertilising one or more eggs outside the body, and then transferring the fertilised eggs, known as '**pre-embryos**', back into the uterus. This step is referred to as **embryo transfer**. The main stages of IVF are:

- stimulation of the ovaries with fertility drugs to produce several eggs;
- collecting the mature eggs;
- fertilising the eggs in the laboratory;
- culture of the pre-embryo;
- embryo transfer.

Stimulating the ovaries. The ovaries are stimulated with a fertility drug containing FSH in order to produce several eggs, thereby increasing the chances of successful egg collection and pregnancy later. Another drug, such as Clomiphene, may also be used to stimulate ovulation. The drugs are given early in the menstrual cycle. Growth of the follicles is monitored with ultrasound scans which allow the number and size of the follicles to be measured. Blood tests for oestrogen, progesterone and LH also help judge when ovulation is about to occur. An injection of LH in the form of HCG (a closely similar hormone) can help to control the timing of ovulation more precisely.

Collecting the eggs. Eggs can be collected from the follicles by laparoscopy. The laparoscope is a short telescope several centimetres long which is inserted into the abdomen, usually through the navel, in order to examine the pelvic region where the ovaries, oviducts and uterus are situated. Carbon dioxide gas is introduced to distend the abdominal cavity and separate the body wall from the gut. The procedure is carried out under general anaesthetic. A powerful light is shone down the telescope and enables the

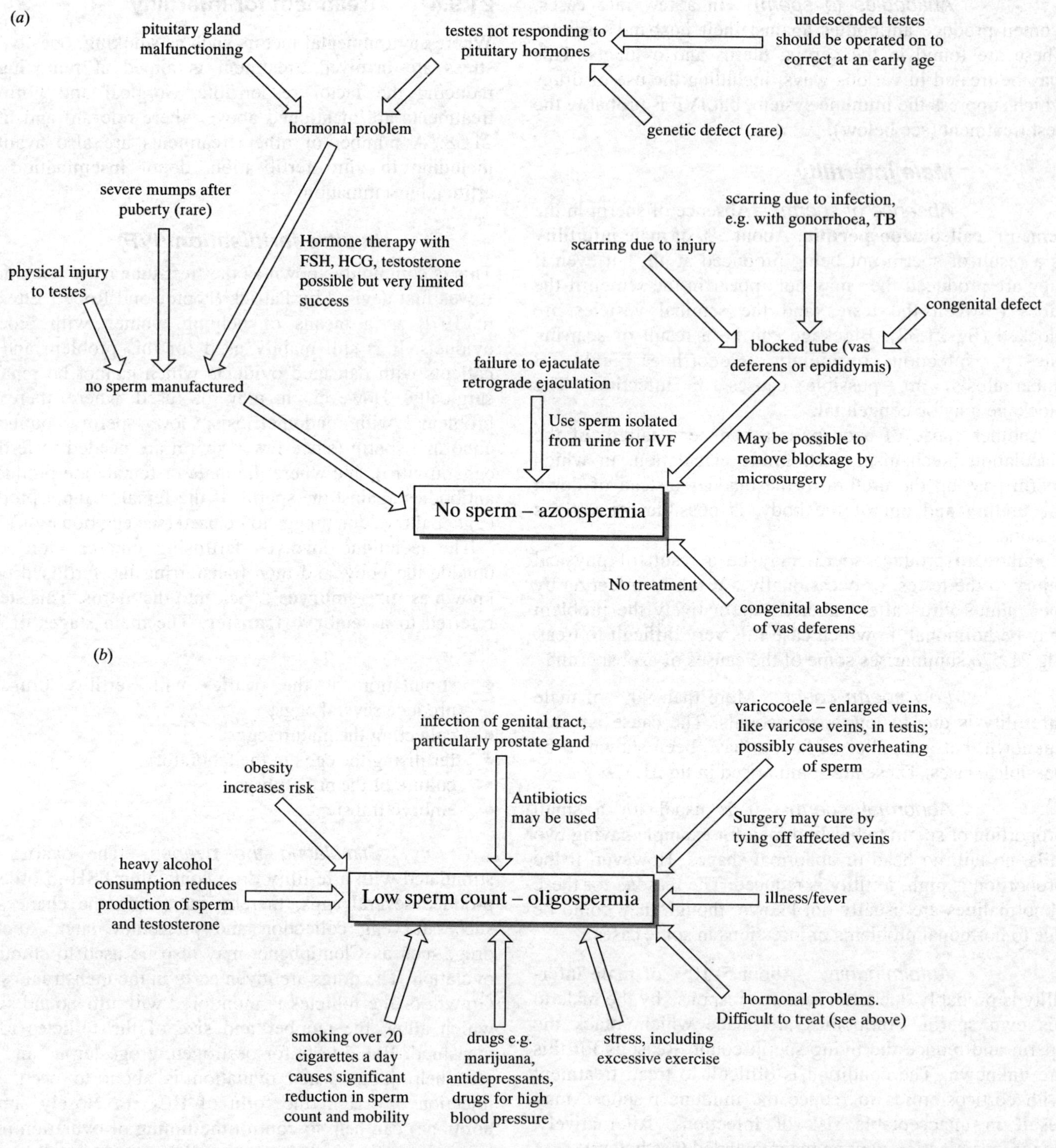

Fig 21.57 *(a) Some causes and treatments of azospermia, the inability to produce semen containing sperm. (b) Some causes and treatments of low sperm counts (oligospermia).*

surgeon to view inside the body. Eggs are collected by sucking out the fluid contents of mature follicles with a fine hollow needle, also inserted through the abdominal wall. The fluid is immediately examined with a microscope for eggs and any eggs found are placed in a special culture medium in an incubator at body temperature. Between 5 and 15 eggs are usually collected.

An alternative way of collecting eggs is by using ultrasound as a guide and collecting through the vaginal route to the uterus and oviducts. This does not require a general anaesthetic, although pain relief and a sedative are given.

Fertilisation. Sperm are collected from the male partner and washed in a culture fluid to remove seminal fluid. About 100 000 healthy sperm are added to each egg about six hours after egg collection. This is done in a glass dish or tube. The fertilised eggs are grown for about two days, after which they are usually at the 2- to 8-cell stage. They are examined with a microscope to check suitability for embryo transfer into the woman's uterus. The larger the number transferred, the greater the chance of success, but this has to be balanced against the need to avoid multiple pregnancies, particularly of three or more fetuses. It is now recommended that a maximum of three pre-embryos are transferred. Spare pre-embryos can be frozen for further attempts if necessary. If three pre-embryos are transferred the risk of having twins if the woman gets pregnant is about 23% and the risk of triplets about 5%.

Embryo transfer. Embryos are transferred in a drop of culture fluid through the cervix and into the uterus using a fine plastic tube. A syringe gently squirts the fluid into the uterus. From 10 minutes to 4 hours later the woman can leave the hospital and resume normal life, although she is usually advised to 'take it easy' for a few days. A pregnancy test is usually performed 14 days after the procedure.

Success rate. The success rate if one pre-embryo is transferred is about 10%, for two pre-embryos about 14% and for three about 25%. There is a decline in success with age of the female from an average of 22% at age 28 years to 9.5% at age 40 years. Most clinics allow about three or four attempts, but the procedure is emotionally extremely demanding and some couples give up before this if they are not successful.

Ethical considerations. A number of fundamental ethical issues are raised by IVF, as well as more everyday social issues. These are summarised below but are worth discussing further.

- Some people would object on religious or moral grounds to any unnatural intervention into human reproduction, particularly to fertilisation outside the body.
- There is a serious ethical/moral problem over what to do with spare pre-embryos. The same arguments about the right to life of embryos that have arisen in the discussion on abortion in section 21.9.2 can be raised. Should doctors have the right to dispose of spare pre-embryos? Should the parents have the right? It is estimated that at least half of all human embryos conceived naturally do not implant and are lost at the next period. Also, the embryo can split into identical twins at any point up to 14 days, making it possible to argue that there is no unique individual up to that point. Should research on spare pre-embryos be allowed? At present experiments can be carried out on pre-embryos up to 14 days after fertilisation because the nervous system does not start to form until that point and no organs have formed. Parents can opt to keep the pre-embryos frozen, possibly for later attempts at IVF or for embryo donation.
- Does the freezing of pre-embryos damage them in any way? So far there is no evidence that it does.
- Embryo research allows scientists to continue to perfect the techniques needed to improve IVF. Pre-embryos can also be used for research into infertility generally, contraception, cause of miscarriage and genetic diseases.
- Soon it will be possible to carry out extensive genetic screening of pre-embryos. Will this be a good thing, allowing us to reduce genetic disease, or will it encourage society to become less tolerant of genetic disease and disability? Will it encourage the selection of pre-embryos for particular characteristics such as sex?
- There is a possibility of multiple births, which increases the risk of miscarriage and perinatal mortality. It also increases the economic and care burden on a family.
- There is a higher risk of ectopic pregnancy with IVF (5% compared with 1% in the general population in the UK). This is linked with the fact that IVF is often used for women with damaged oviducts.
- An important problem with IVF is that it is extremely demanding emotionally and psychologically. There is a relatively high failure rate combined with desperate desire for success. Couples have to be counselled carefully before they undertake the procedure.
- The procedure is expensive, about £1000–£2000 in the UK in the early 1990s. Couples who are desperate for children may be tempted to make financial sacrifices they cannot really afford. It is sometimes difficult to judge whether the treatment should be recommended and clinics must guard against commercial exploitation of vulnerable people.
- There is no evidence that IVF babies have an increased risk of any abnormalities.

Gamete intra-fallopian transfer (GIFT)

GIFT is a variation of IVF which can be useful if the oviducts are not blocked. It is mainly used when there is no known cause for the infertility, or when the woman's cervical mucus is hostile to her partner's sperm, for example making antibodies. Sperm and eggs are added separately to the oviducts so that fertilisation takes place naturally in the woman's body rather than outside as in IVF. Laparoscopy or a more recent technique via the cervix is used. Pre-embryos move down the oviducts to the uterus where they may implant. The success rate is about 21%. Implantation is more likely to be successful with GIFT than with IVF.

Zygote intra-fallopian transfer (ZIFT)

This is another treatment for unexplained infertility. Pre-embryos (zygotes) are transferred into the oviducts rather than gametes as in GIFT. The advantage over GIFT is that fertilisation can be confirmed – with GIFT there is no means of knowing if fertilisation is taking place unless the woman becomes pregnant. The advantage over IVF is that the pre-embryos enter the uterus naturally by way of the oviducts. Greater success rates than with GIFT and IVF are anticipated.

Donor insemination (DI) or Artificial insemination by donor (AID)

If the male is infertile or has a very low sperm count, donor insemination is often a preferred option to adoption of a child. The technique may also be used by a fertile male if there is a risk of passing on an inherited disease. The technique is straightforward but requires careful counselling before being undertaken.

Potential donors are carefully screened for health, fertility and any history of genetic disease in their family. They are screened for HIV antibodies, hepatitis B and a few other diseases which may be transmitted in semen. Donors are matched to the male partner for race, height, skin, hair and eye colouring, body build and blood group. The Human Fertilisation and Embryology Act which came into force in 1991 makes it a requirement that information about donors and recipients is registered with the Human Fertilisation and Embryology Authority (HFEA). Eventually, it is planned that anyone over the age of 18 will be able to check whether they are DI offspring. However, the donor will remain legally anonymous.

In order to maximise the chances of success, the woman receiving the treatment is normally asked to keep a temperature chart to note when ovulation takes place. Clomiphene may be used to promote ovulation and to make its date more predictable. On the appropriate date the woman visits the clinic, the frozen sperm sample is selected and is gently released next to the opening of the cervix by means of a small plastic tube. The success rate is high – most women will fall pregnant within 6 months using this technique.

Ethical issues.

- The major issue is whether the child should be told the identity of the father. The genetic father (donor) is not the legal father since a change in the law in the late 1980s, so legally there are no problems with the relationship. Evidence suggests that children who are told are curious to know more about their genetic fathers but still maintain a loving relationship with the parents who have brought them up. If the child is not told, there is the risk that the information might be revealed at an inappropriate time. The father may feel very insecure about the child knowing since he will not wish to risk damaging their relationship.
- If the decision is made to tell the child, there follows the decision about when this should be done, and possibly whether family and friends should be told. Professional advice is that, on balance, the child is probably better told sooner rather than later, preferably before the teenage years.
- It may be imagined that because the legal father is not the genetic father, he may not have the same long-term commitment to the child. There is no evidence, however, that DI children are any less loved as a result of their origins. This is perhaps not surprising given the high motivation to have children needed to go through with the procedure and the counselling involved.
- Should the child have legal access to the identity of their genetic father and would this deter donors from coming forward?
- There is a slight risk that genetically related DI offspring might meet, enter a sexual relationship and have children. The number of successful donations from one donor is therefore legally restricted.
- There is some controversy over whether DI should be made available to unmarried couples, single women or lesbians. Where a stable home background is likely to be provided, some clinics consider such applicants. This is probably preferable to allowing people to select their own donors from among family, friends or acquaintances, which tends to happen if DI is denied.
- Some women and some men regard DI as a form of adultery, perhaps on religious grounds, and are therefore opposed to the method.

Artificial insemination by husband or partner (AIH)

AIH can be used in some cases of impotence and premature ejaculation. It has also been used in cases where the man is likely to face sterility as a result of surgery, for example for testicular cancer, or is living away from home for long periods of time. Some men give written consent for their widows to use their semen after their deaths. A man on active service in the forces, or someone diagnosed with a terminal illness, might do this for example. As a treatment for low sperm counts it is not very useful, although it is possible to enrich sperm samples for healthy sperm. IVF is a better option in this circumstance.

Egg donation

If the woman is unable to produce eggs, one solution is to use an egg donor. The techniques developed for IVF are followed to obtain the eggs, and after fertilisation with the male partner's sperm, the pre-embryos can be placed in the oviducts of the future mother. The technical difficulties of donating eggs are greater than donating sperm, and they cannot be successfully frozen. This makes HIV screening more difficult. One convenient source of eggs is spare eggs from women who are undergoing IVF or GIFT.

The ethical issues are similar to those for DI (see above).

Surrogacy

Surrogacy is a possibility when the woman cannot bear a child, for example through lack of a uterus. An agreement is reached with another woman to have the baby using sperm from the male partner. The egg may be supplied by the infertile woman if she can produce eggs. In this case the IVF or GIFT procedure is used and the pre-embryo replaced in the surrogate mother. Alternatively the egg of the surrogate mother can be used, in which case straightforward DI using the sperm of the male partner is possible.

Ethical issues.

- It is now illegal in the UK to enter into such an arrangement on a commercial 'Rent A Womb' basis. This is to avoid exploitation of infertile women.
- Probably the biggest potential problem is if the surrogate mother does not wish to hand over the baby at birth. It may be genetically hers. It is difficult to predict how a surrogate mother will react when the time comes to give up the baby.
- Should the child be told of its origins? (This question has already been discussed under DI.)
- If the surrogate mother suffers health damage or even death as a result of the pregnancy, what are the legal implications? Could the commissioning parties be sued? Who should decide if an abortion should take place if the child is going to be born disabled?
- If the child is born with a defect, particularly a genetic defect, the future mother may not wish to accept it.

Sub-zonal insemination (SUZI) or Micro-insemination sperm transfer (MIST)

SUZI is a newly developing technique whereby a few sperm are introduced directly next to the egg using a very fine probe while viewing with a microscope. The sperm are introduced below the zona pellucida, hence the term 'sub-zonal'. The eggs are obtained by the usual IVF procedure. The technique is not yet perfected nor widely available, but represents a promising line of research.

Chapter Twenty-two

Growth and development

22.1 What is growth?

Growth is a fundamental characteristic of all living organisms. It is often thought of simply as an **increase in size**, but if you think about this carefully you will realise that this is not an adequate definition. For example, the size of a plant cell may increase as it takes up water by osmosis, but this process may be reversible and cannot then be thought of as genuine growth. Also, when a zygote divides repeatedly to form a ball of cells, the early embryo, there is an increase in cell numbers without an increase in size (volume or mass). This is known as cleavage and is the result of cell division without subsequent increase in size of daughter cells. The process does involve development, so perhaps it should be regarded as growth despite the fact that no increase in size occurs.

The process of development is so closely linked with growth that the phrase 'growth and development' is commonly used to describe the processes which are normally thought of as growth. Development could be described as an **increase in complexity**.

Starting with an individual cell, growth of a multicellular organism can be divided into three phases:

- **cell division** – an increase in cell number as a result of mitosis and cell division;
- **cell enlargement** – an irreversible increase in cell size as a result of the uptake of water or the synthesis of living material;
- **cell differentiation** – the specialisation of cells; in its broad sense, growth also includes this phase of cell development.

However, each of these processes can occur at separate points in time. The example of cleavage has already been mentioned above. An increase in size without change in cell numbers may also occur, as in the region of cell elongation behind the root and shoot tips of plants. In the case of single-celled organisms, such as bacteria, cell division results in *reproduction* (not growth) of the individual and *growth* of the population.

It is therefore difficult to define growth. One acceptable definition is that growth is **an irreversible increase in dry mass of living material**. By specifying *dry* mass we can ignore the short-term fluctuations in water content that are particularly characteristic of plants. By specifying *living* material we can ignore non-living material such as inorganic limestone in corals.

All stages of growth involve biochemical activity. During growth the DNA message is translated and as a result specific proteins are made, including enzymes. Enzymes control cell activities. They bring about the changes which eventually result in a change in overall form and structure, both of individual organs and of the organism as a whole. This process is known as **morphogenesis** and is influenced by the environment as well as the genes.

Growth may be positive or negative. **Positive growth** occurs when synthesis of materials (anabolism) exceeds breakdown of materials (catabolism), whereas **negative growth** occurs when catabolism exceeds anabolism. For example, in the course of germination of a seed and the production of a seedling there is an increase in cell number, cell size, fresh mass, length, volume and complexity of form, while at the same time dry mass may actually *decrease* because reserves are being used up. From this definition, germination therefore includes a period of negative growth which only becomes positive when the seedling starts to photosynthesise and make its own 'food'.

22.2 Measuring growth

Fig 22.1 shows a variety of growth curves produced by plotting different parameters such as length, height, mass, surface area, volume and number against time. The shape of these curves is described as **sigmoid**, meaning S-shaped, and is typical of much growth.

A sigmoid curve can be divided into four parts as shown in fig 22.2.

- The first phase is the **lag phase** during which little growth occurs.
- This leads into the second phase, the **logarithmic** or **log phase**, during which growth proceeds exponentially. During this phase the rate of growth accelerates and at any point is proportional to the amount of material or number of cells already present. In all cases of growth the exponential increase eventually declines and the rate of growth begins to decrease. The point at which this occurs is known as the **inflexion point**. Rate of growth is at its maximum at the inflexion point (it is the point where the curve is steepest).
- The third phase is the **decelerating phase** during which growth becomes limited as a result of the effect

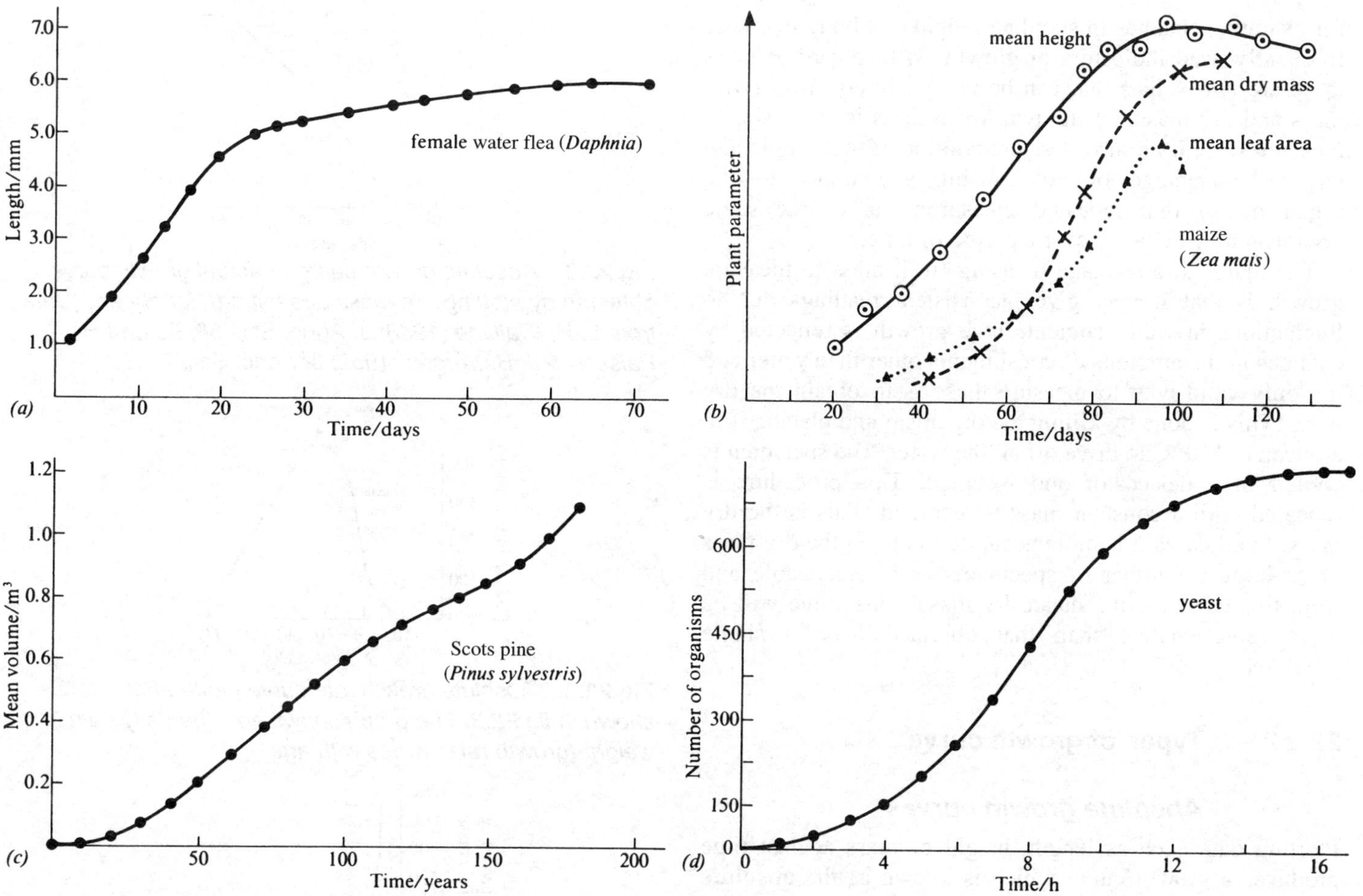

Fig 22.1 *Growth curves obtained using six different parameters and four different species. In all cases the curves are sigmoid.*

Fig 22.2 *A typical sigmoid growth curve showing the four characteristic growth phases and the inflexion point. The inflexion point is the point of maximum growth rate where the curve is steepest.*

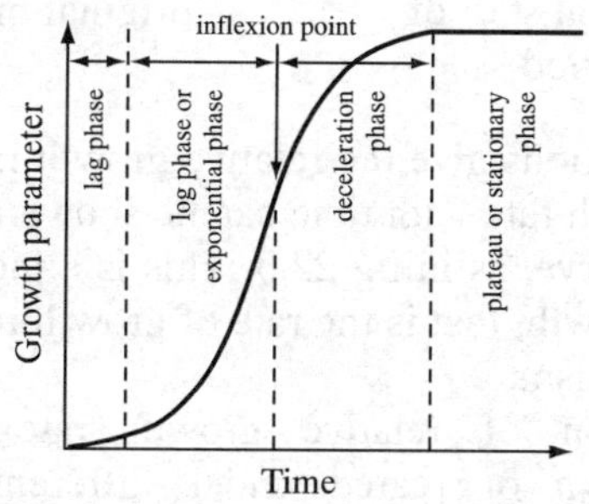

of some internal or external factor, or the interaction of both.

- The final phase is the **plateau phase** or **stationary phase**. This usually marks the period where overall growth has ceased and the parameter under consideration remains constant. The precise nature of the curve during this phase varies, depending on the species and what is being measured. In some cases the curve may continue to rise slightly until the organism dies, as is the case with monocotyledonous leaves, many non-vertebrates, fish and certain reptiles. This indicates continuing **positive growth**. In the case of certain cnidarians the curve flattens out, indicating zero growth, whilst other growth curves may tail off, indicating a period of **negative growth**. The latter pattern is characteristic of many mammals, including humans, and is a sign of physical senescence associated with increasing age.

22.2.1 Methods of measuring growth

Growth can be measured at various levels of biological organisation, such as growth of a cell, organism or population. The numbers of organisms in a population at different times can be counted and plotted against time to produce a population growth curve as shown in fig 22.1*d*. At the level of the organism there are a variety of parameters which may be measured; length, area, volume and mass are commonly used. In plants, growth curves for roots, stems, internodes and leaves are often required, and length and area are the parameters commonly chosen. In the case of growth in animals and entire plants, length and mass are two commonly measured parameters. In humans, for example, changes in standing height and body mass are frequently used indicators of growth. With regard to mass there are two values that can be used, namely fresh (wet) mass and dry mass. Of the two, fresh mass is the easier to measure since it requires less preparation of the sample and has the advantage of not causing any injury to the organism, so that repeated measurements of the same organism may be taken over a period of time.

The major disadvantage of using fresh mass to measure growth is that it may give inconsistent readings due to fluctuations in water content. True growth is reflected by changes in the amounts of constituents other than water and the only valid way to measure these is to obtain the dry mass. This is done by killing the organism and placing it in an oven at 110 °C to drive off all the water. The specimen is cooled in a desiccator and weighed. This procedure is repeated until a constant mass is recorded. This is the dry mass. In all cases it is more accurate to obtain the dry mass of as large a number of specimens as is practicable and from this calculate the mean dry mass. This value will be more representative than that obtained from a single specimen.

22.2.2 Types of growth curve

Absolute growth curves

Plotting data such as length, height or mass against time produces a growth curve which is known as the **absolute growth curve** or **actual growth curve** (fig 22.3). The usefulness of this curve is that it shows the overall growth pattern and the extent of growth. Data from this graph enable the growth curves in figs 22.4 and 22.5 to be constructed.

Absolute growth rate

An **absolute growth rate** curve shows how the *rate* of growth changes with time (fig 22.4). The rate is measured as the change in a particular parameter, such as height or mass, in a particular time. For example, it could be the increase in height of a human over a period of a year. In particular it shows the period when growth is most rapid and this corresponds to the steepest part of the absolute growth curve. The peak of the absolute growth curve marks the point of inflexion on the sigmoid curve (fig 22.2) after which the rate of growth decreases as the adult size is attained. Overall a bell-shaped absolute growth rate curve is obtained from a sigmoid absolute growth curve.

Relative growth rate

A **relative growth rate** curve takes into account existing size (fig 22.5). Thus if a 5 year old and a 10 year old human both grew 10 cm in height in one year, their absolute growth rates would be the same, but the 5 year old would be growing relatively faster and have a greater relative growth rate. Existing size is taken into account by using the following calculation:

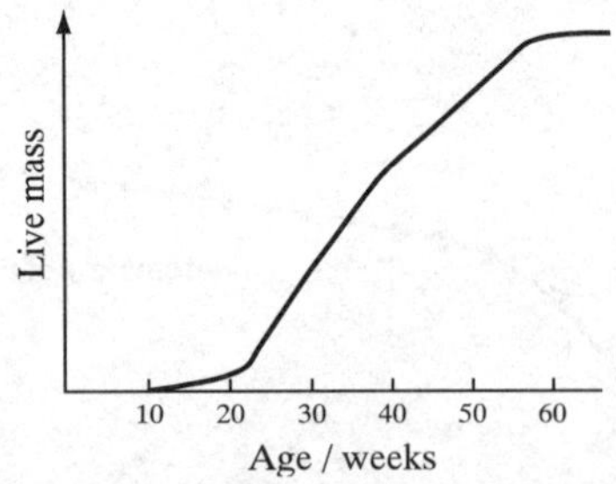

Fig 22.3 *Absolute growth curve or actual growth curve obtained by plotting live mass against age for sheep. (Data from L. R. Wallace (1948)* J. Agric. Sci., ***38****, 93 and H. Pálsson & J. B. Vergés (1952)* J. Agric. Sci., ***42****, 93.*)

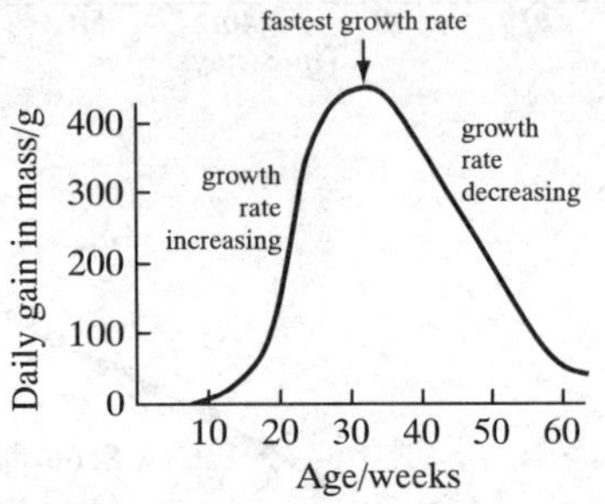

Fig 22.4 *Absolute growth rate curve plotted from data shown in fig 22.3. The graph shows how the daily gain in weight (growth rate) varies with age.*

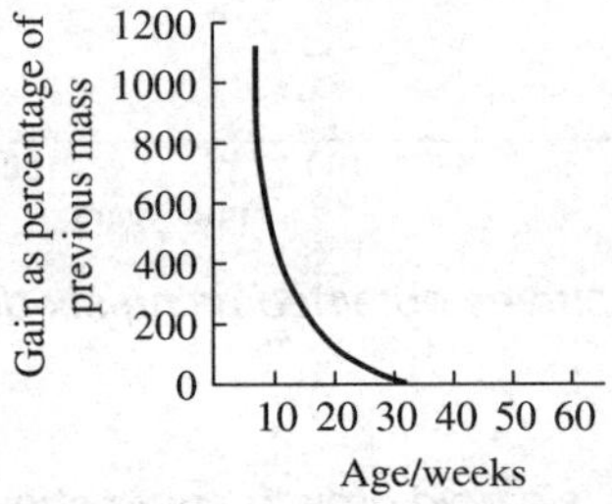

Fig 22.5 *Relative growth rate curve plotted from data shown in fig 22.3.*

$$\frac{\text{growth in given time period}}{\text{measurement at start of time period}} \quad \text{or} \quad \frac{\text{absolute growth rate}}{\text{original measurement}}$$

These calculations give the relative growth rate. Changes in relative growth rate with time can be shown on a **relative growth rate curve**, as in fig 22.5. This is a measure of the *efficiency* of growth, that is the rate of growth relative to the size of the organism.

A comparison of relative growth rate curves for organisms grown or reared under different conditions shows clearly the most favourable conditions for rapid growth and for growth over an extended period.

Growth in mammals, including humans

In the case of mammals the absolute growth curve is sigmoid but the exact shape of the curve appears to be related to the time taken to reach sexual maturity. In the rat the curve is steep and truly sigmoid since sexual maturity is reached quickly (within 12 months) whereas in humans the absolute growth curve shows four distinct phases of increased growth (fig 22.6). The absolute growth curve shows that maximum mass is achieved in adulthood. The absolute growth rate curve shows that rate of growth is greatest during infancy and adolescence, with a distinct adolescent spurt of growth being typical. The relative growth rate curve shows that relatively speaking, growth is greatest during embryological development.

22.3 Patterns of growth

Various patterns of growth occur among organisms.

22.3.1 Isometric and allometric growth

Isometric (*isos*, same; *metron*, measure) growth occurs when an organ grows at the same mean rate as the rest of the body. In this situation change in size of the organism is not accompanied by a change in shape of the organism. The relative proportions of the organs and the whole body remain the same. This type of growth pattern is seen in fish and certain insects, such as locusts (except for wings and genitalia) (fig 22.7). In such cases there is a simple relationship between linear dimension, area, volume and mass. The area increases as the square of linear dimension ($A \propto l^2$) whereas volume and mass increase as the cube of linear dimension ($V \propto l^3$ and $M \propto l^3$). An animal showing a small increase in overall dimensions with time therefore shows a marked increase in mass: for example, an increase in length of only 10% is accompanied by a 33% increase in mass.

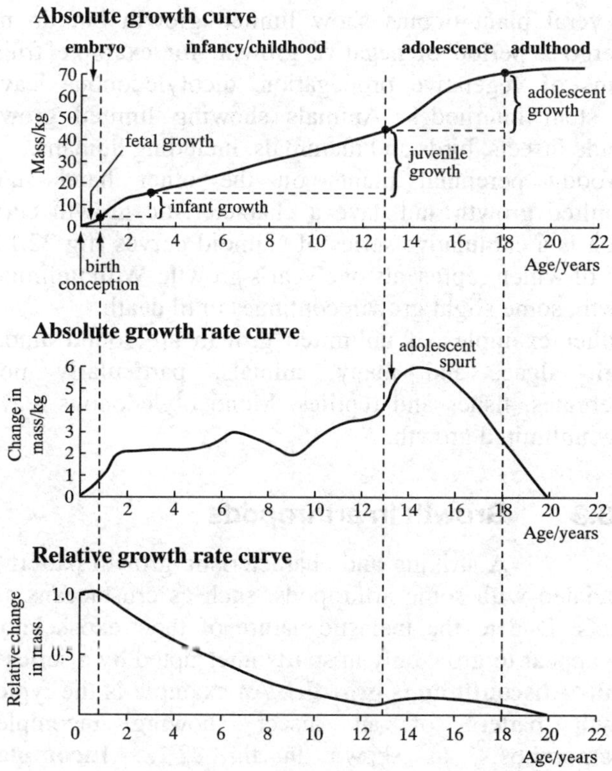

Fig 22.6 *Three types of growth curve for humans.*

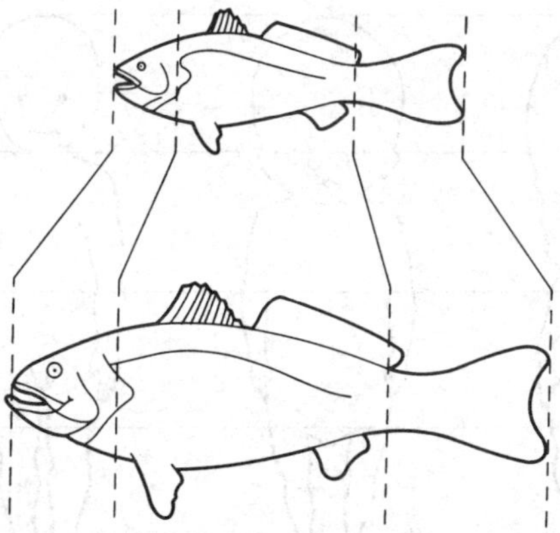

Fig 22.7 *Development in fish – an example of isometric growth. The external structures retain their shape and spatial relationships as a result of a proportional growth rate. (After Batt (1980)* Influences on animal growth and development, *Studies in Biology, No. 116, Arnold.)*

Allometric (*allos*, other; *metron*, measure) growth occurs when an organ grows at a different rate from the rest of the body. This produces a change in size of the organism which is accompanied by a change in shape of the organism. This pattern of growth is characteristic of mammals and illustrates the relationship between growth and development. Fig 22.8 shows how the relative proportions of various structures in humans change as a result of simultaneous changes in patterns of growth and development. In almost all animals the last organs to develop and differentiate are the reproductive organs. These show allometric growth and can be observed only in those organisms with external genital organs, hence they are not seen in many species of fish where growth appears to be purely isometric. Fig 22.9 shows the degree of variation in patterns of growth of different organs of a human. Again it can be seen that the last organs to develop are the reproductive organs.

The shapes of absolute growth curves for whole organisms as represented by length or mass show remarkable similarities and generally conform to the sigmoid shape described in section 22.2. However, several groups of organisms show variations on the general pattern which reflect adaptations to particular modes of life and environments as described in sections 22.3.2 and 22.3.3.

22.3.2 Limited and unlimited growth

Studies of the duration of growth in plants and animals show that there are two basic patterns. These are called **limited** (definite or determinate) growth and **unlimited** (indefinite or indeterminate) growth.

Growth in annual plants is limited and, after the plant matures and reproduces, there is a period of negative growth or senescence before the death of the plant. If the dry mass of the annual plant is plotted against time then an interesting variation on the sigmoid curve of fig 22.2 is seen, as shown in fig 22.10.

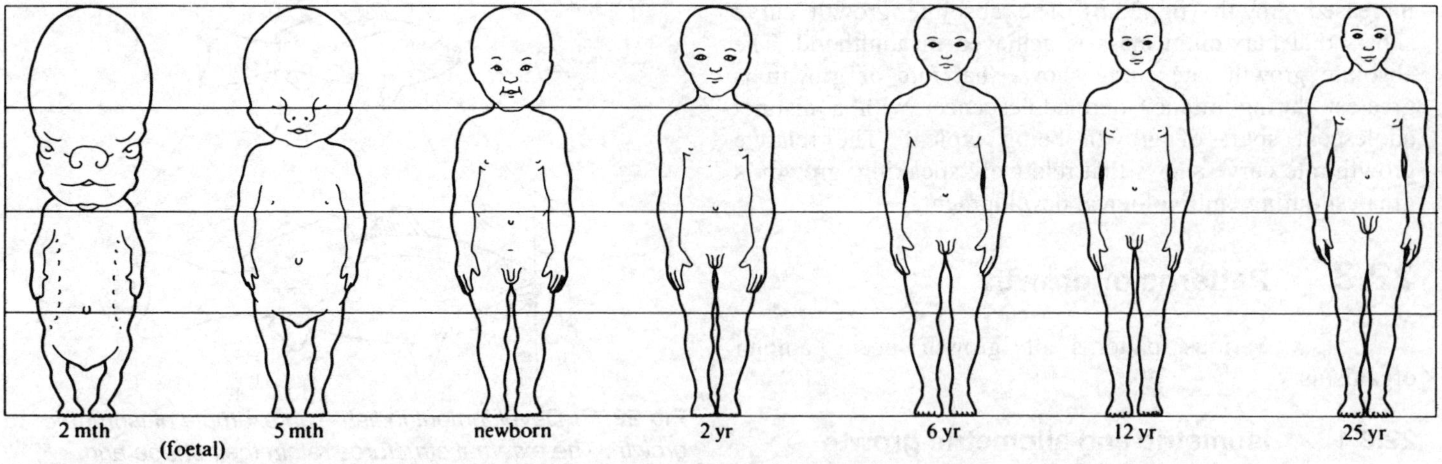

Fig 22.8 *Development in humans – an example of allometric growth. To show the relative rates of growth from the age of two months to 25 years each stage has been given a constant height. (After Stratz, cited in J. Hammond (ed.) (1955)* Progress in the physiology of farm animals, *2, 431, Butterworths.)*

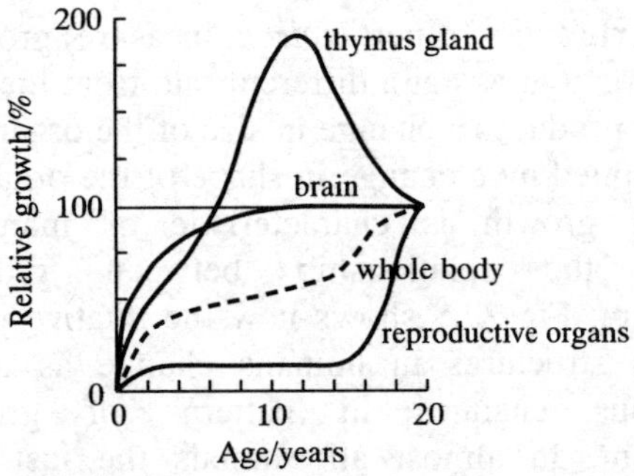

Fig 22.9 *Relative growth rates of the brain, thymus gland and reproductive organs of humans. The absolute growth curve of the whole body is also drawn for comparison. The thymus gland is involved in the early development of the immune system. (After Scammon.)*

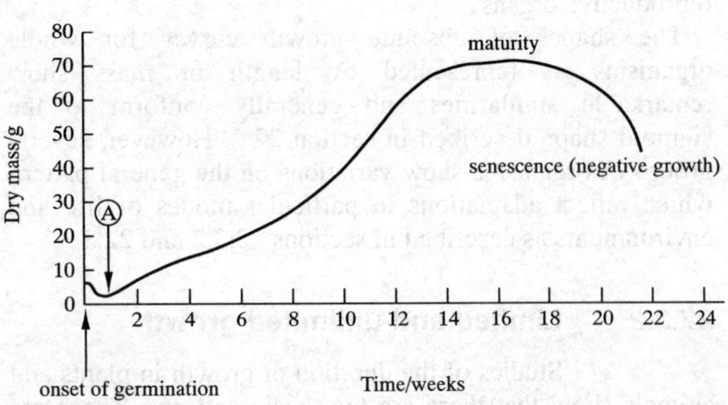

Fig 22.10 *Growth curve of a broad bean (*Vicia faba*) planted in March – an annual plant. This is an example of limited growth, where growth does not continue throughout life.*

22.1 Examine fig 22.10 and answer the following questions, based on your knowledge of the life cycle of an annual plant.
(*a*) Why is there negative growth initially during the germination of the seed?
(*b*) Describe the appearance of the seedling when positive growth occurs at A.
(*c*) What physiological process occurs here to account for positive growth?
(*d*) Why is the decrease in dry mass after 20 weeks very sudden?

Several plant organs show limited growth but do not undergo a period of negative growth, for example fruits, organs of vegetative propagation, dicotyledonous leaves and stem internodes. Animals showing limited growth include insects, birds and mammals, including humans.

Woody perennial plants on the other hand show unlimited growth and have a characteristic growth curve which is a cumulative series of sigmoid curves (fig 22.11), each of which represents one year's growth. With unlimited growth, some slight growth continues until death.

Other examples of unlimited growth are found among fungi, algae, and many animals, particularly non-vertebrates, fishes and reptiles. Monocotyledonous leaves show unlimited growth.

22.3.3 Growth in arthropods

A striking and characteristic growth pattern is associated with some arthropods, such as crustaceans and insects. Due to the inelastic nature of their exoskeletons they appear to grow only in spurts interrupted by a series of moults (**discontinuous growth**). An example is the typical growth pattern of an insect showing incomplete metamorphosis, as shown in fig 22.12. **Incomplete metamorphosis** occurs when there is a series of larval

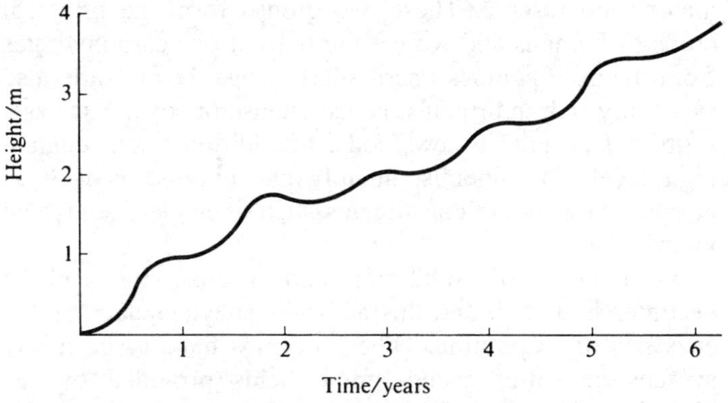

Fig 22.11 *Growth curve of a birch tree – a woody perennial. This is an example of unlimited growth, in which growth continues throughout life.*

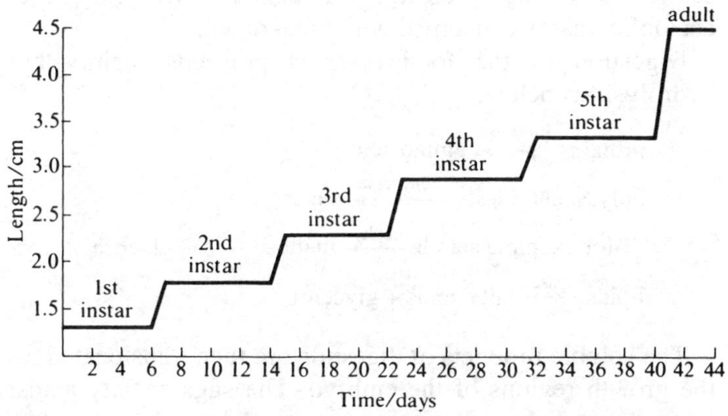

Fig 22.12 *Growth curve showing increase in length of the short-horned grasshopper. (After R. Soper & T. Smith (1979)* Modern Biology, *Macmillan.)*

stages, called instars, with each successive stage resembling the adult more closely. Moulting occurs between each stage, and growth in length is confined to the brief period before the new exoskeleton hardens. In such situations, growth curves based on length, such as that shown in fig 22.12, do not give a true reflection of growth. If a growth curve is plotted for the same insect, using dry mass as the growth parameter, a normal sigmoid curve is produced demonstrating that true growth, as represented by increase in living material, is continuous.

22.4 Growth and development in the flowering plant

22.4.1 Seed dormancy

Certain environmental conditions, namely availability of water, oxygen and a suitable temperature, must be present before the embryo of a seed will grow. However, in the presence of these factors, some seeds will not germinate and are described as **dormant**. They must undergo certain internal changes which can generally be described as **after-ripening** before they will germinate. These changes ensure that premature germination does not occur. For example, seedlings produced immediately from seeds shed in summer or autumn would probably not survive winter. In other words, mechanisms exist which ensure that germination is synchronised with the onset of a season favourable for growth. Some of the common mechanisms are discussed below.

Barrier methods

This mechanism often involves the outer layers of the seed being impervious to water or the passage of oxygen, or being physically strong enough to prevent growth of the embryo, as in many legumes. Sometimes physical damage (**scarification**) to the seed coat can remove this restriction. The process can be induced artificially by removing the testa or simply by pricking it with a pin. Under natural circumstances, bacteria or passage through the gut of an animal may have the same effect. The seeds of some species are stimulated to germinate by fire. More usually, however, the restriction is removed by some physiological change, involving the following factors.

Growth inhibitors. Many fruits or seeds contain chemical growth inhibitors which prevent germination. Abscisic acid often has this role, for instance in ash seeds. Thorough soaking of the seeds might remove the inhibitor or its effect may be overridden by an increase in a growth promoter such as gibberellin. Tomato seeds contain high levels of abscisic acid which prevent germination of the seeds inside the tomato fruit.

Light. The dormancy of some seeds is broken by light after water uptake, a phytochrome-controlled response. This stimulation of germination by light is associated with a rise in gibberellin levels within the seed. It occurs, for example, in some lettuce varieties. Less commonly, germination is inhibited by light, such as in *Phacelia* and *Nigella*. (For a relevant experiment, refer to the foot of table 16.5, section 16.4.2.) This may be an advantage if it ensures that the seed does not germinate until it is buried in a suitable medium such as soil.

Temperature. Seeds commonly require a cold period, a process known as **prechilling**, before germination will occur. This is common among cereals and members of the rose family (Rosaceae), such as plum, cherry and apple. It is associated with a rise in gibberellin activity and sometimes a reduction in growth inhibitors. It ensures that seeds must pass through a cold spell of a particular length before germination and are less likely to germinate during a warm spell in winter.

Exactly how light and cold treatments affect seeds is not clear, but increased permeability of the seed coat, as well as changes in levels of growth substances, may be involved.

22.2 Seeds which require a stimulus of light for germination are usually relatively small. What could be the significance of this?

22.3 The light which passes through leaves is enriched in green and far-red light relative to the light which strikes the leaf surface.
(*a*) Why is this?
(*b*) What ecological significance might this have in relation to seeds like lettuce, where germination is a phytochrome-controlled response? (Read section 16.4.2 if necessary.)

22.4.2 Germination

Germination is the onset of growth of the embryo, usually after a period of dormancy. The structure of the seed at germination has been described in section 21.5.

Environmental conditions needed for germination

Water. The initial uptake of water by a seed is by a process called **imbibition**. It takes place through the micropyle (a tiny hole in the testa, or seed coat) and testa and is purely a physical process caused by **adsorption** of water by substances within the seed. These include proteins, starch and cell wall materials such as hemicelluloses and pectic substances. The swelling of these substances can lead to strong imbibitional forces which are great enough to rupture the testa or pericarp (fruit coat) surrounding the seed. Water then moves from cell to cell by osmosis. Water is needed to activate the biochemical reactions associated with germination, because these take place in aqueous solution. Water is also an important reagent at this stage, being used in the hydrolysis (digestion) of food stores.

Minimum or optimum temperature. There is usually a characteristic temperature range outside which a given type of seed will not germinate. This will be related to the normal environment of the plant concerned and will be within the range 5–40 °C. Temperature influences the rate of enzyme-controlled reactions as described in section 4.3.3.

Oxygen. This is required for aerobic respiration, although such respiration can be supplemented with anaerobic respiration if necessary.

Physiology of germination

A typical seed stores carbohydrates, lipids and proteins, either in its endosperm or in the cotyledons of the embryo. Usually lipids in the form of oils form the major food reserves of the seed, though notable exceptions are the legume family (which includes peas and beans) and the grass family (which includes cereals) where starch is the major food reserve. These two groups form the principal crops of humans and we get the bulk of our carbohydrates from them. Legumes, particularly soya beans, are also especially rich in proteins; hence the use of soya beans as a source of protein in new foods. In addition, seeds contain high levels of minerals, notably phosphorus, as well as normal cytoplasmic constituents such as nucleic acids and vitamins.

As a result of imbibition and osmosis the embryo becomes hydrated, and this activates enzymes such as the enzymes of respiration. Other enzymes have to be newly synthesised, often using amino acids provided by the digestion of stored proteins.

Broadly speaking, there are two centres of activity in the germinating seed, the **storage centre** (food reserve) and the **growth centre** (embryo). The main events in the storage centre, with the exception of enzyme synthesis, are catabolic, that is concerned with breakdown.

Digestion of the food reserves proceeds mainly by hydrolysis as below:

$$\text{proteins} \xrightarrow{\text{proteases}} \text{amino acids}$$

$$\text{polysaccharides} \xrightarrow{\text{carbohydrases}} \text{sugars}$$

$$\text{for example starch} \xrightarrow{\text{amylase}} \text{maltose} \xrightarrow{\text{maltase}} \text{glucose}$$

$$\text{lipids} \xrightarrow{\text{lipases}} \text{fatty acids + glycerol}$$

The soluble products of digestion are then translocated to the growth regions of the embryo. The sugars, fatty acids and glycerol may be used to provide substrates for respiration in both the storage and the growth centres. They may also be used for anabolic reactions in the growth centre, that is, reactions concerned with synthesis. Of particular importance in these reactions are glucose and amino acids. A major use of glucose is for the synthesis of cellulose and other cell wall materials. Amino acids are used mainly for protein synthesis, proteins being important as enzymes and structural components of protoplasm. In addition, mineral salts are required for the many reasons given in tables 7.7 and 7.8.

Both storage and growth centres obtain the energy for their activities from respiration. This involves oxidation of a substrate, usually sugar, to carbon dioxide and water. A net loss in dry mass of the seed therefore occurs, since carbon dioxide is lost as a gas, and has a greater mass than the oxygen gas taken up in aerobic respiration. Water, another product of respiration, does not contribute to dry mass. This loss will continue until the seedling produces green leaves and starts to make its own food (fig 22.10).

A well-studied example of germination of a starchy seed is the barley grain, where it has been shown that the synthesis of α-amylase and other enzymes takes place in the outer layers of the endosperm in response to gibberellin secreted by the embryo. These outer layers contain stored protein which is the source of amino acids for protein synthesis. The process is described and experimentally investigated in section 16.2.6. Fig 16.20 shows an example of the role of hormones in early germination. The appearance of amylase in germinating barley grains can also be investigated by grinding them in water, filtering and centrifuging to obtain a clear extract and testing the activity of the extract on starch solution. By using samples of barley grains at different times from germination, the increase in amylase activity per grain over a period of a week can be determined.

22.4 Explain the results shown in fig 22.13.

Those seeds which store lipids convert them to fatty acids and glycerol. Each molecule of lipid yields three molecules of fatty acid and one of glycerol (section 3.3). Fatty acids are either oxidised directly in respiration or converted to sucrose, which is then translocated to the embryo.

22.5 (This question tests some basic knowledge of chemistry and of the chemistry of lipids. The latter is covered in section 3.3.)

Suppose 51.2 g dry mass of seeds, containing 50% fatty acid by mass, converted all the fatty acid to sugar in the following reaction:

$$\underset{\text{fatty acid}}{C_{16}H_{32}O_2} + 11O_2 \longrightarrow \underset{\text{sugar}}{C_{12}H_{22}O_{11}} + 4CO_2 + 5H_2O + \text{energy}$$

(*a*) Assuming that no other changes occurred which might affect dry mass, calculate the gain or loss in dry mass of the seeds.
(Relative atomic masses: C = 12, H = 1, O = 16.)
(*b*) What other important change might affect dry mass?
(*c*) Calculate the volume of carbon dioxide evolved from the seeds at STP (standard temperature and pressure).
(1 mole of gas at STP occupies 22.4 dm^3.)
(*d*) How can fatty acid be obtained from a lipid, and what would be the other component of the lipid?
(*e*) How many carbon atoms would one molecule of the parent lipid have contained if $C_{16}H_{32}O_2$ was the only fatty acid produced?
(*f*) What is the identity of the sugar formed in the reaction shown?
(*g*) How does the oxygen reach the storage tissue?

Respiration in germinating seeds

Rates of respiration in both storage tissues and embryo are high owing to the intense metabolic activity in both regions. Substrates for respiration may differ in each region and

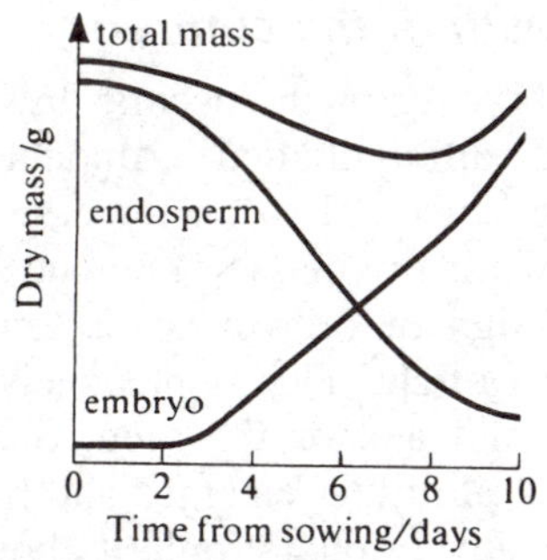

Fig 22.13 *Relative changes in dry mass of endosperm and embryo during germination of barley.*

may also change during germination. This is revealed by changes in the respiratory quotient (section 9.5.9).

22.6 When castor oil seeds were analysed for lipid and sugar content during germination in darkness, the results shown in fig 22.14 were obtained.

The RQ of the seedings was measured at day 5 and the embryo was found to have an RQ of about 1.0, while the remaining cotyledons had an RQ of about 0.4–0.5.

(*a*) Suggest as full an explanation of these results as you can (refer to section 22.4.2 for relevant information).
(*b*) What would you expect the RQ of the whole seedling to be on day 11? Explain very briefly.

22.7 The RQ of peas is normally between 2.8 and 4 during the first seven days of germination, but is 1.5–2.4 if the testas are removed. In both cases ethanol accumulates in the seeds, but in much smaller amounts when the testas are removed. Account for these observations.

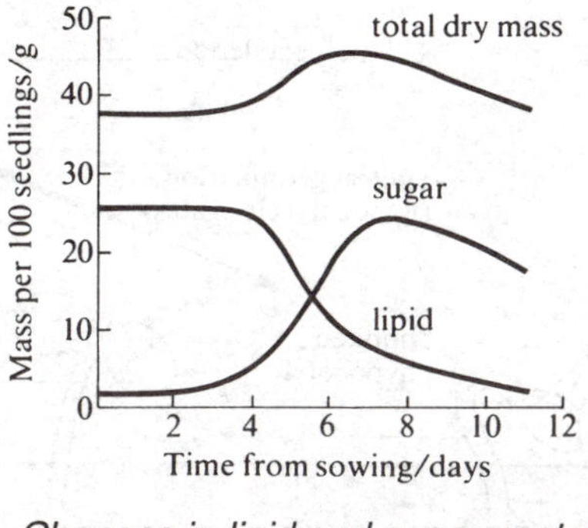

Fig 22.14 *Changes in lipid and sugar content of castor oil seeds during germination in the dark. (Based on data from R. Desveaux & M. Kogane-Charles (1952)* Annls. Inst. natn. Rech. Agron., *Paris,* ***3****, 385–416; cited by H. S. Street & H. Opik (1976)* The physiology of flowering plants *2nd ed., Arnold.)*

Growth of the embryo

Within the embryo growth occurs by cell division, enlargement and differentiation. Amounts of proteins, cellulose, nucleic acids and so on increase steadily in the growing regions while dry mass of the food store decreases. The first visible sign of growth is the emergence of the embryo root, the **radicle**. This is positively geotropic and will grow down and anchor the seed. Subsequently, the embryo shoot, the **plumule**, emerges and being negatively geotropic (and positively phototropic if above ground) will grow upwards.

There are two types of germination according to whether or not the cotyledons grow above ground or remain below it. In dicotyledons, if that part of the shoot axis, or internode, just below the cotyledons (the **hypocotyl**) elongates, then the cotyledons are carried above ground. This is **epigeal** germination. If the internode just above the cotyledons (the **epicotyl**) elongates, then the cotyledons remain below ground. This is **hypogeal** germination.

In epigeal germination, the hypocotyl remains hooked as it grows through the soil, as shown in fig 22.15*b*, thus meeting the resistance of the soil rather than the delicate plumule tip, which is further protected by being enclosed by the cotyledons. In hypogeal germination of dicotyledons the epicotyl is hooked, again protecting the plumule tip, as shown in fig 22.15*c*. In both cases the hooked structure immediately straightens on exposure to light, a phytochrome-controlled response.

In the grasses, which are monocotyledons, the plumule is protected by a sheath called the **coleoptile**, which is positively phototropic and negatively geotropic as described in section 16.1.1. The first leaf grows out through the coleoptile and unrolls in response to light.

On emerging into light a number of phytochrome-controlled responses rapidly occur in leaves, collectively known as photomorphogenesis. The overall effect is a change from etiolation (section 16.4.1) to normal growth. The major changes involved are summarised in table 16.5 and include expansion of the cotyledons or first true foliage leaves, as well as formation of chlorophyll ('greening'). At this point photosynthesis begins and net dry mass of the seedling starts to increase as it finally becomes independent of its food reserves. Once exposed to light, the shoot also shows phototropic responses although these are not phytochrome controlled.

22.4.3 Growth of the primary plant body

Meristems

In contrast to animals, growth in plants is confined to certain regions known as meristems. A **meristem** is a group of cells which retain the ability to divide by mitosis, producing daughter cells which grow and form the rest of the plant body. The daughter cells form the permanent tissue, that is, cells which have lost the ability to divide. There are three types of meristem, described in table 22.1. Two types of growth are mentioned in table 22.1, namely primary and secondary growth.

Primary growth is the first form of growth to occur. A whole plant can be built up by primary growth, and in most monocotyledonous plants and herbaceous dicotyledons it is the only type of growth. It is a result of the activity of the apical, and sometimes intercalary, meristems. The anatomy of mature primary roots and stems is dealt with in sections 13.4 and 13.5.

Some plants continue with **secondary growth** from lateral meristems. This is most notable in shrubs and trees. Plants which lack extensive secondary growth are called **herbaceous** plants or **herbs**. A few herbaceous plants show restricted amounts of secondary thickening, as in the development of additional vascular bundles in *Helianthus* (sunflower).

Apical meristems and primary growth

A typical apical meristem cell is relatively small, cuboid, with a thin cellulose cell wall and dense cytoplasmic

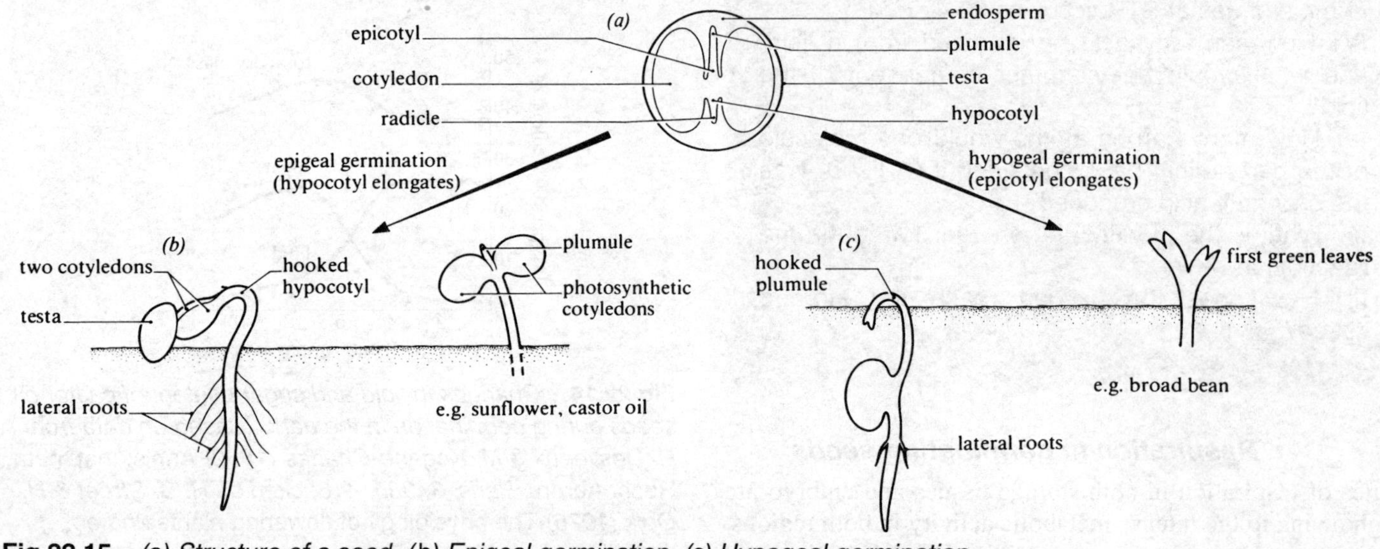

Fig 22.15 *(a) Structure of a seed. (b) Epigeal germination. (c) Hypogeal germination.*

Table 22.1 Types of meristem and their functions.

Type of meristem	*Location*	*Role*	*Effect*
Apical	Root and shoot apex	Responsible for primary growth, giving rise to primary plant body	Increase in length
Lateral (cambium)	Laterally situated in older parts of the plant parallel with the long axis of organs, e.g. the cork cambium (phellogen) and the vascular cambium	Responsible for secondary growth. The vascular cambium gives rise to secondary vascular tissue, including wood (secondary xylem); the cork cambium gives rise to the periderm, which replaces the epidermis and includes cork	Increase in girth
Intercalary	Between regions of permanent tissue, e.g. at nodes of many monocotyledons, such as bases of grass leaves	Allows growth in length to occur in regions other than tips. This is useful if the tips are susceptible to damage or destruction, e.g. eating by herbivores (grasses), wave action (kelps). Branching from the main axis is not then necessary	Increase in length

contents. It has a few small vacuoles rather than the large vacuoles characteristic of parenchyma cells, and the cytoplasm contains small, undifferentiated plastids called proplastids. Meristematic cells are packed tightly together with no obvious air spaces between the cells.

The cells are called **initials**. When they divide by mitosis one daughter cell remains in the meristem while the other increases in size and differentiates to become part of the permanent plant body.

22.4.4 Primary growth of the shoot

The structure of a typical apical shoot meristem is illustrated in figs 22.16 and 22.17. Fig 22.17 shows the approximate division of the shoot apex into regions of cell division, cell enlargement and cell differentiation. Passing back from the dome-shaped apical meristem, the cells get progressively older, so that different stages of growth can be observed simultaneously in the same shoot. Thus it is relatively easy to study developmental sequences of plant tissue.

Three basic types of meristematic tissue occur, namely the **protoderm**, which gives rise to the epidermis; the **procambium**, giving rise to the vascular tissues, including pericycle, phloem, vascular cambium and xylem; and the **ground meristem**, producing the parenchyma **ground tissues**, which in the dicotyledons are the cortex and pith. These meristematic types are produced by division of the meristematic cells (initials) in the apex. In the zone of enlargement, the daughter cells produced by the initials increase in size, mainly by osmotic uptake of water into the cytoplasm and then into the vacuoles. Increase in the length of stems and roots is mainly brought about by elongation of cells during this stage. The process is illustrated in fig 22.18.

The small vacuoles increase in size, eventually fusing to form a single large vacuole. The pressure potential developed inside the cells stretches their thin walls and the orientation of cellulose microfibrils in the walls helps to determine the final shape assumed by the cells. The final

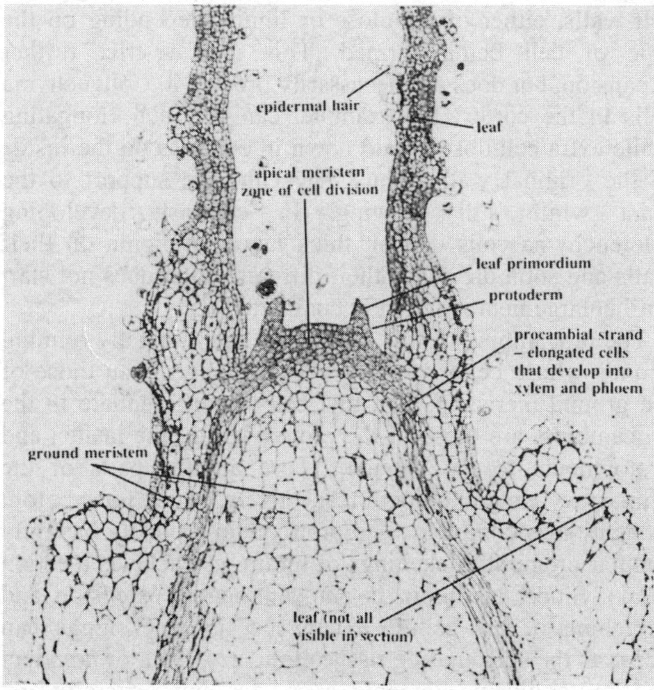

Fig 22.16 *The apical meristem of a shoot.*

Fig 22.17 *LS shoot tip of a dicotyledon showing apical meristem and regions of primary growth. For simplicity, vascular tissue to leaves and buds has been omitted.*

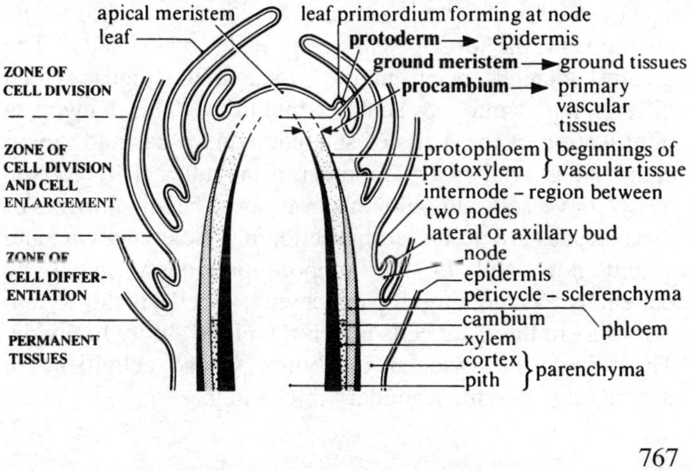

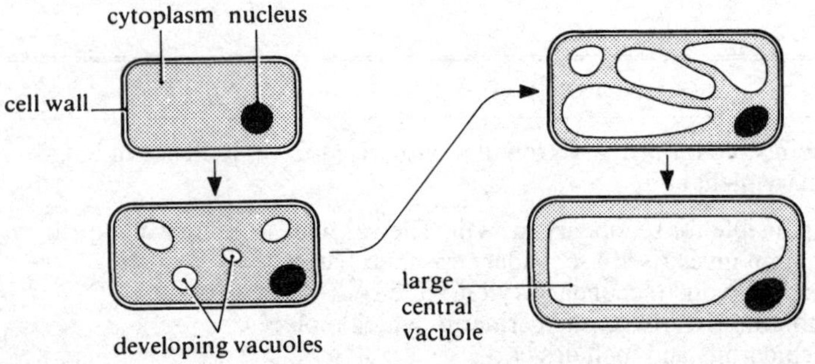

Fig 22.18 *Cell enlargement phase of growth of a meristematic cell.*

volume of cytoplasm may not be significantly greater than in the original meristematic cell, but is now confined to the cell periphery by the vacuole. As enlargement nears completion, many cells develop additional thickening of the cell walls, either of cellulose or lignin, depending on the type of cell being formed. This may restrict further expansion, but does not necessarily prevent it. Collenchyma cells in the cortex, for example, can continue elongating while extra cellulose is laid down in columns on the inside of the original walls. Thus they can give support to the plant while still growing. In contrast, developing sclerenchyma cells deposit thick layers of lignin on their walls and soon die. Thus their differentiation does not start until enlargement is virtually completed.

The procambium forms a series of longitudinally running strands whose cells are narrower and longer than those of the ground meristem. The first cells to differentiate in the procambium are those of the protoxylem to the inside, and protophloem to the outside. These are the parts of the primary xylem and phloem respectively, which form before elongation is complete. The protoxylem typically has only annular or spiral thickenings of lignin on tracheids (section 6.2.1). Since the lignin is not continuous, extension and stretching of the cellulose between the thickenings can occur as the surrounding tissue elongates. Both protoxylem and protophloem elements soon die and generally get crushed and stretched to the point of collapse as growth continues around them. Their function is taken over by later-developing xylem and phloem in the zone of differentiation.

In the zone of differentiation each cell becomes fully specialised for its own particular function, according to its position in the organ with respect to other cells. The greatest changes occur in the procambial strands, which differentiate into vascular bundles. This involves lignification of the walls of sclerenchyma fibres and xylem elements, as well as development of the tubes characteristic of xylem vessels and phloem sieve tubes. The final forms of these tissues are described in section 6.0. Sclerenchyma and xylem now add to the support previously given by collenchyma and turgid parenchyma. Between the xylem and phloem there are cells which retain the ability to divide. They form the vascular cambium, whose activities are described later with secondary thickening.

Leaf primordia and lateral buds

Development of the shoot also includes growth of leaves and lateral buds. Leaves arise as small swellings or ridges called leaf primordia, shown particularly clearly in fig 22.16. The swellings contain groups of meristematic cells and appear at regular intervals, their sites of origin being called **nodes** and the regions between **internodes**. The pattern of leaf arrangement on the stem varies and is called **phyllotaxis**. Leaves may arise in whorls with two or more leaves at each node, or singly, either in two opposite ranks or in a spiral pattern. Generally, however, they are arranged to minimise overlapping, and hence shading, when fully grown so that they form a **mosaic**.

The primordia elongate rapidly, so they soon enclose and protect the apical meristem, both physically and by the heat they generate in respiration. Later they grow and increase in area to form the blades. Cell division gradually ceases but may continue until they are about half their mature size.

Soon after the leaves start to grow, buds develop in the axils between them and the stem. These are small groups of meristematic cells which normally remain dormant, but retain the capacity to divide and grow at a later stage. They form branches or specialised structures such as flowers and underground structures such as rhizomes and tubers. They are thought to be under the control of the apical meristem (see Apical dominance, section 16.3.3).

22.4.5 Primary growth of the root

The structure of the typical apical root system is illustrated in fig 22.19.

At the very tip of the apical meristem is a **quiescent zone**, a group of **initials** (meristematic cells) from which all other cells in the root can be traced, but whose rate of cell division is much slower than their daughter cells in the apical meristem around them. To the outside, the cells of the **root cap** are formed. These become large parenchyma cells which protect the apical meristem as the root grows through the soil. They are constantly being worn away and replaced. They also have the important additional function of acting as gravity sensors, since they contain large starch grains which act as statoliths, sedimenting to the bottoms of cells in response to gravity. Their role is described in more detail in section 16.2.2.

Behind the quiescent centre, orderly rows of cells can be seen and the meristematic regions already described in the shoot, namely protoderm, ground meristem and procambium, can be distinguished (fig 22.19). In the root the term procambium is used to describe the whole central cylinder of the root, even though at maturity this contains the non-vascular tissues of the pericycle and the pith, if present.

The zone of cell division typically extends 1–2 mm back from the root tip, and overlaps slightly with the zone of cell elongation. Root tips are convenient material for observation of mitosis and a procedure for this is described

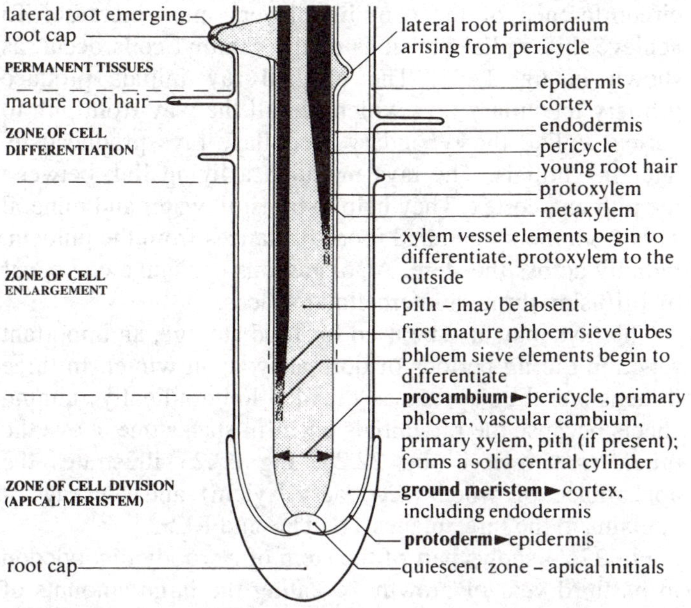

Fig 22.19 *LS apical meristem of a typical root. Xylem differentiation is shown to the right and phloem differentiation to the left. In reality xylem and phloem alternate round the root and would be on different radii. Also, in reality, the zone of enlargement would be longer.*

in section 23.0. Behind this zone, growth is mainly by cell enlargement, cells increasing in size in the manner described for the shoot and shown in fig 22.18. The zone of enlarging cells extends to a point about 10 mm behind the root tip and their increase in length forces the root tip down through the soil.

Some cell differentiation begins in the zone of cell division, with the development of the first phloem sieve tube elements (fig 22.19). In longitudinal sections, neat files of developing sieve tube elements can be seen, getting progressively more mature further back from the root tip, until they become mature sieve tubes. Development of phloem is from the outside inwards.

Further back in the zone of enlargement, the xylem vessels start to differentiate, also from the outside inwards (exarch xylem) in contrast to the stem (endarch xylem). The first-formed vessels are protoxylem vessels, as in the stem, and they show the same pattern of lignification and ability to stretch as cells around them grow. Their role is taken over by metaxylem, which develops later and matures in the zone of differentiation after enlargement has ceased. The xylem often spreads to the centre of the root, in which case no pith develops.

Development is easier to examine in roots than in shoots. In the latter, procambial strands to the leaves complicate the distribution of developing tissues. Development of xylem, in particular, is easily seen by squashing apical portions of fine roots such as those of cress seedlings and staining appropriately.

After all cells have stopped enlarging, further differentiation is completed. This includes the development of root hairs from the epidermis.

22.4.6 Lateral meristems and secondary growth

Secondary growth is that growth which occurs after primary growth as a result of the activity of lateral meristems. It results in an increase in girth. It is usually associated with deposition of large amounts of secondary xylem, called **wood**, which completely modifies the primary structure and is a characteristic feature of trees and shrubs.

There are two types of lateral meristem, the **vascular cambium** which gives rise to new vascular tissue, and the **cork cambium** or **phellogen**, which arises later to replace the ruptured epidermis of the expanding plant body.

Vascular cambium

There are two types of cell in the vascular cambium, the **fusiform initials** and the **ray initials**, illustrated in fig 22.20. Fusiform initials are narrow, elongated cells which divide by mitosis to form **secondary phloem** to the outside or **secondary xylem** to the inside. The amount of xylem produced normally exceeds the amount of phloem. Successive divisions are shown in fig 22.21. Secondary phloem contains sieve tubes, companion cells, sclerenchyma fibres and sclereids, and parenchyma.

Ray initials are almost spherical and divide by mitosis to form parenchyma cells which accumulate to form rays between the neighbouring xylem and phloem.

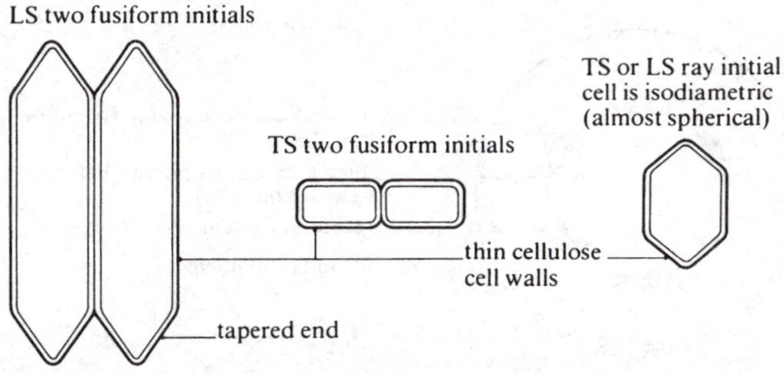

Fig 22.20 *Fusiform and ray initials.*

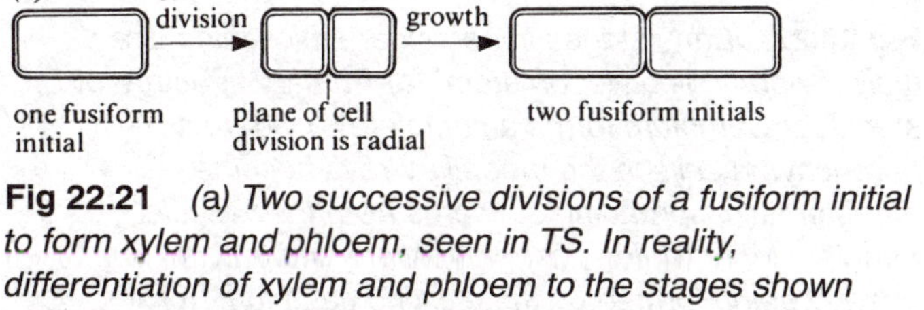

Fig 22.21 *(a) Two successive divisions of a fusiform initial to form xylem and phloem, seen in TS. In reality, differentiation of xylem and phloem to the stages shown would take some time, during which more cells would be produced. (b) Division of a fusiform initial to form a new fusiform initial, seen in TS.*

Secondary growth in woody dicotyledon stems

The vascular cambium is originally located between the primary xylem and primary phloem of the vascular bundles, its derivation from the apical meristem being shown in fig 22.17. It becomes active very soon after primary growth is complete. Fig 22.22 summarises the early stages in secondary thickening of a typical woody dicotyledon stem.

Fig 22.22*a* shows the original primary stem structure, omitting the pericycle for simplicity. Fig 22.22*b* shows the development of a complete cylinder of cambium. Fig 22.22*c* shows a complete ring of secondary thickening. Here, fusiform initials have produced large quantities of secondary xylem, and lesser quantities of secondary phloem, while the ray initials have produced rays of parenchyma. As the stem increases in thickness, so the circumference of the cambium layers must increase. To achieve this, radial divisions of the cambial cells occur, as shown in fig 22.21. The original ray initials produce primary medullary rays which run all the way from pith to cortex, unlike the secondary medullary rays produced by later ray initials. The rays maintain a living link between the pith and cortex. They help to transmit water and mineral salts from the xylem, and food substances from the phloem, radially across the stem. Also, gaseous exchange can occur by diffusion through intercellular spaces.

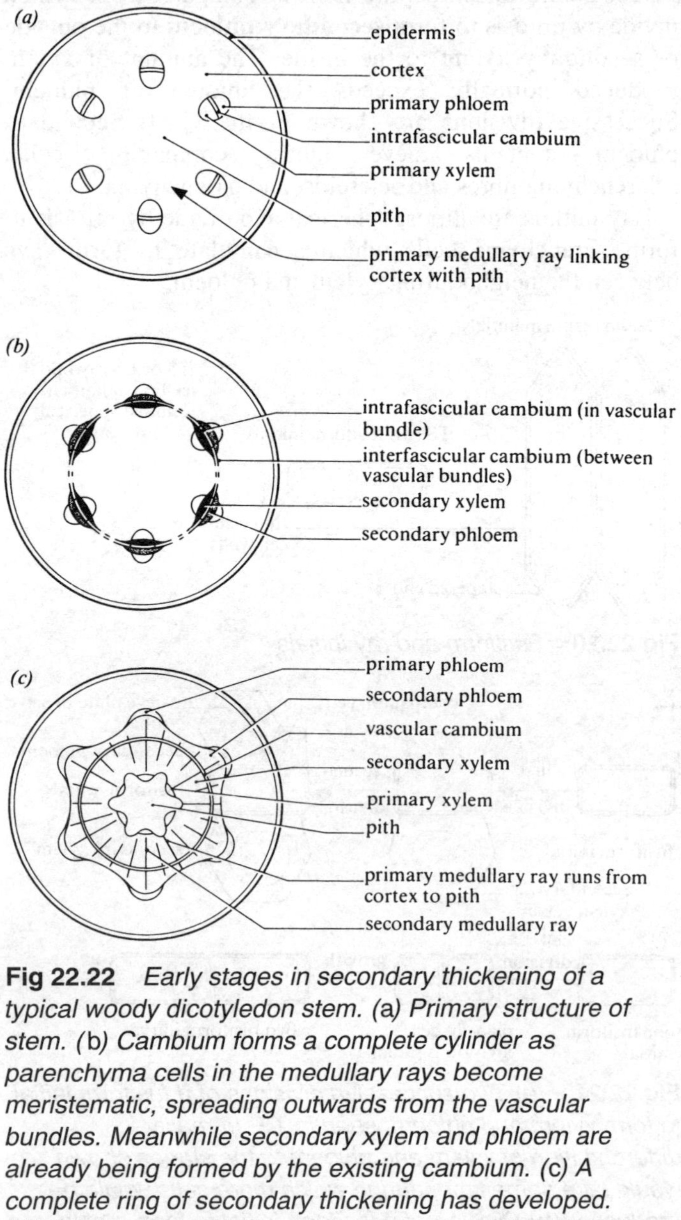

Fig 22.22 *Early stages in secondary thickening of a typical woody dicotyledon stem. (a) Primary structure of stem. (b) Cambium forms a complete cylinder as parenchyma cells in the medullary rays become meristematic, spreading outwards from the vascular bundles. Meanwhile secondary xylem and phloem are already being formed by the existing cambium. (c) A complete ring of secondary thickening has developed. Thickening is most advanced at the sites of the original vascular bundles where cambial activity first started.*

The rays may also be used for food storage, an important function during periods of dormancy, as in winter. In three dimensions they appear as radially–longitudinally running sheets because the ray initials occur in stacks one above the other, as shown in fig 22.23. Fig 22.23 illustrates the appearance of wood (secondary xylem) and the rays it contains in the three planes TS, TLS and RLS.

Fig 22.24 shows part of the stem of a woody dicotyledon in its third year of growth, revealing the large amounts of secondary xylem produced. Fig 22.25 shows photographs of a three-year-old and a five-year-old stem of *Tilia*, the lime tree.

Annual rings

Each year in temperate climates, growth resumes in the spring. The first vessels formed are wide and thin-walled, being suitable for the conduction of large quantities of water. Water is required to initiate growth, particularly the expansion of new cells, as in developing leaves. Later in the year, fewer vessels are produced and they are narrower with thicker walls. During winter the cambium remains dormant. The autumn wood produced at the end of one year, as growth ceases, will therefore be immediately next to the spring wood of the following year and will differ markedly in appearance. This contrast is seen as the **annual ring** and is clearly visible in fig 22.25. Where vessels are concentrated in the early wood it is said to be **ring porous**, as opposed to **diffuse porous** wood, where they are evenly distributed, and where it is more difficult to see annual rings. In tropical climates, seasonal droughts may induce similar fluctuations in cambial activity.

The width of an annual ring will vary partly according to climate, a favourable climate resulting in production of more wood and hence a greater distance between rings. This has been used in two areas of science, namely dendroclimatology and dendrochronology. **Dendroclimatology** is the study of climate using tree ring data. Applications vary from correlation of recent climatic records with tree growth, of possible interest in a specific locality, to investigations of more distant climatic events several hundreds or even thousands of years in the past. The oldest-known living trees, the bristlecone pines, are about 5000 years old, and fossil wood of even greater age can be found.

Dendrochronology is the dating of wood by recognition of the pattern of annual rings. This pattern can act as a 'fingerprint', pinpointing the time during which the wood

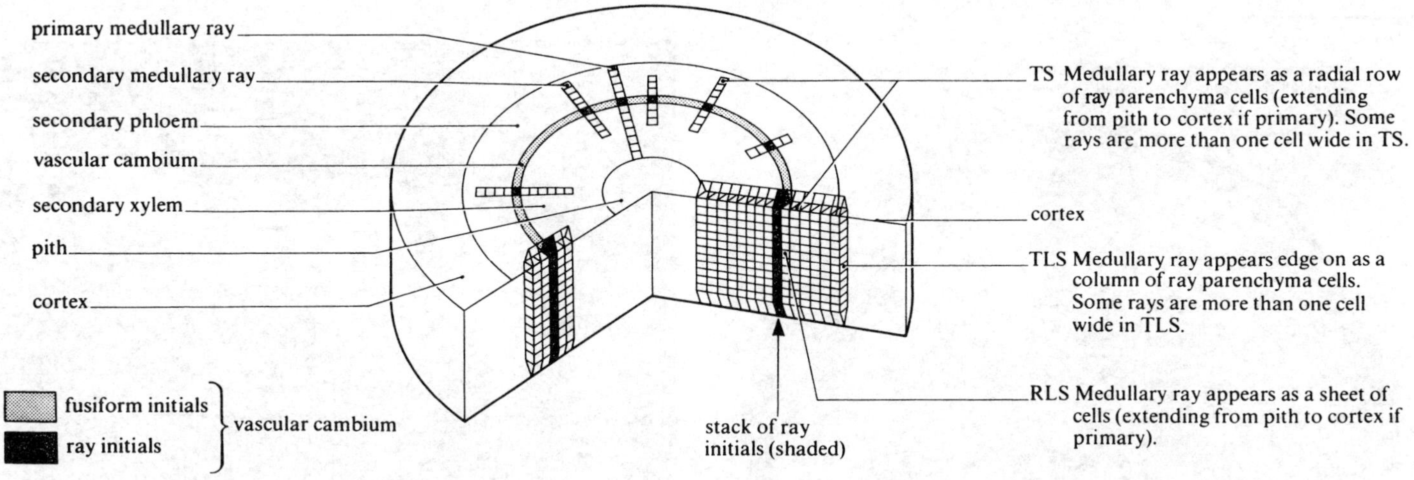

Fig 22.23 *Diagrammatic representation of primary and secondary medullary rays in a typical woody dicotyledon stem. A primary ray is shown to the right and a secondary ray to the left. (TLS, transverse longitudinal section; RLS, radial longitudinal section.)*

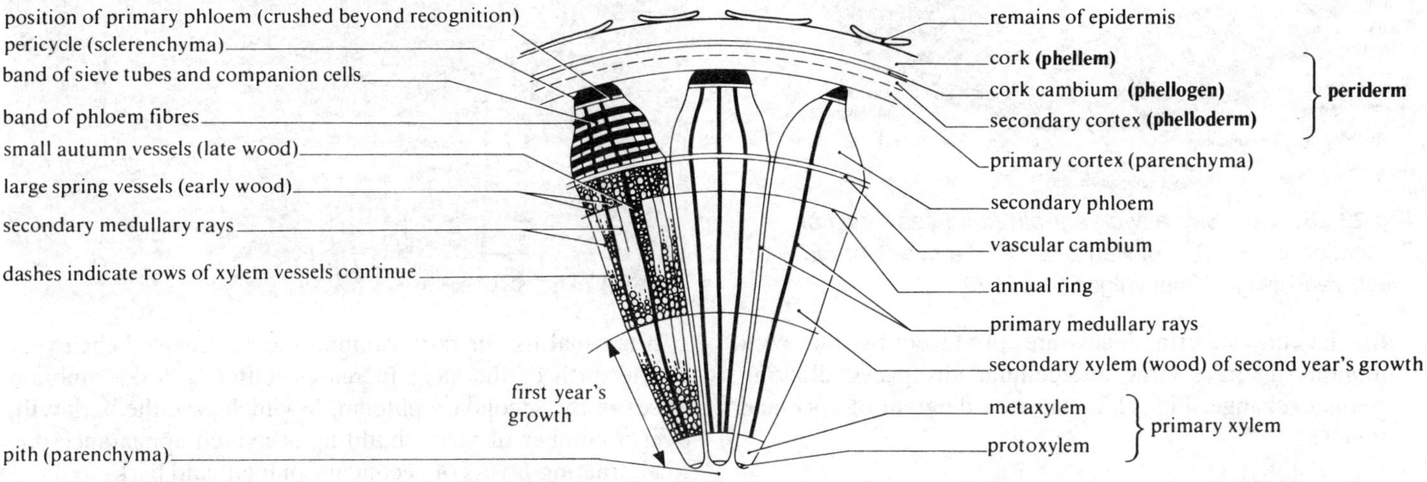

Fig 22.24 *TS of a typical woody dicotyledon stem in the third year of growth (age two years), such as* Tilia. *Details of secondary phloem, secondary xylem and secondary medullary rays are shown only in the left-hand sector.*

was growing. Dating of timbers at archaeological sites, in old buildings and ships and so on thus becomes feasible, provided enough data are available.

Heartwood and sapwood

As a tree ages, the wood at the centre may cease to serve a conducting function and become blocked with darkly staining deposits such as tannins. It is called **heartwood**, whereas the outer, wetter conducting wood is called **sapwood**.

Cork and lenticels

As the secondary xylem grows outwards, so the tissues outside it become increasingly compressed, as well as being stretched sideways by the increasing circumference. This affects the epidermis, cortex, primary phloem and all but the most recent secondary phloem. The epidermis eventually ruptures and is replaced by cork as the result of the activity of a second lateral meristem, the **cork cambium** or **phellogen**. It generally arises immediately below the epidermis. **Cork** (or **phellem**) is produced to the outside of the cork cambium, while to the inside one or two layers of parenchyma are produced. These are indistinguishable from the primary cortex and form the **phelloderm** or secondary cortex. The phellogen, cork and phelloderm together comprise the periderm (fig 22.24).

As the cork cells mature, their walls become impregnated with a fatty substance called suberin which is impermeable to water and gases. The cells gradually die and lose their living contents, becoming filled either with air or with resin or tannins. The older, dead cork cells fit together around the stem, preventing desiccation, infection and mechanical injury. They become compressed as the stem increases in girth and may eventually be lost and replaced by younger cells from beneath. If the cork layer were complete, the respiratory gases oxygen and carbon dioxide could not be exchanged between the living cells of the stem and the environment, and the cells would die. At random intervals, however, slit-like openings, or **lenticels**, develop in the cork which contain a mass of loosely packed, thin-walled dead

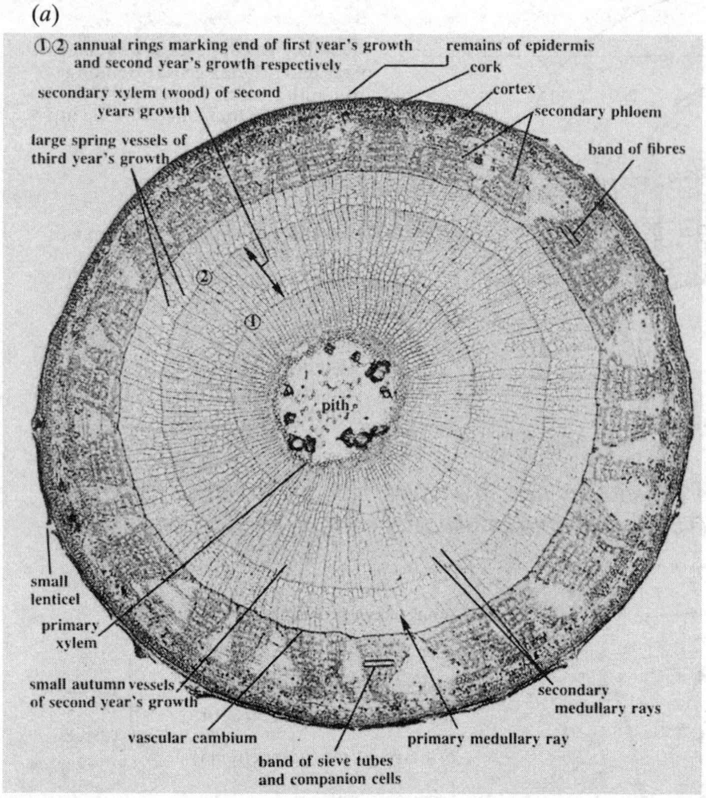

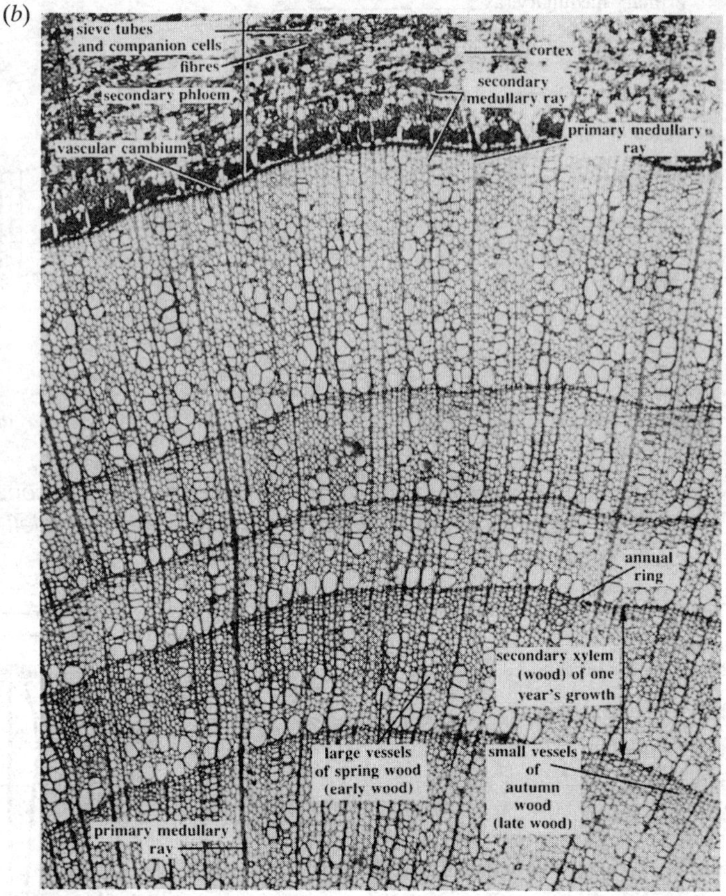

Fig 22.25 *(a) TS of a two-year-old (third year) twig of* Tilia vulgaris *(×2.2). (b) Part of a TS of a five-year-old (sixth year) twig of* Tilia vulgaris *(×11.5).*

cells, lacking suberin. They are produced by the cork cambium and have large intercellular air spaces allowing gaseous exchange. Fig 22.26 shows a diagram of cork and lenticels.

Bark

Eventually a woody stem becomes covered with a layer commonly known as **bark**. The term bark is an imprecise one which is used to refer either to all the tissues outside the vascular system, or more strictly to those tissues outside the cork cambium. Peeling bark from a tree generally strips tissues down to the vascular cambium, a thin layer of cells which is easily ruptured.

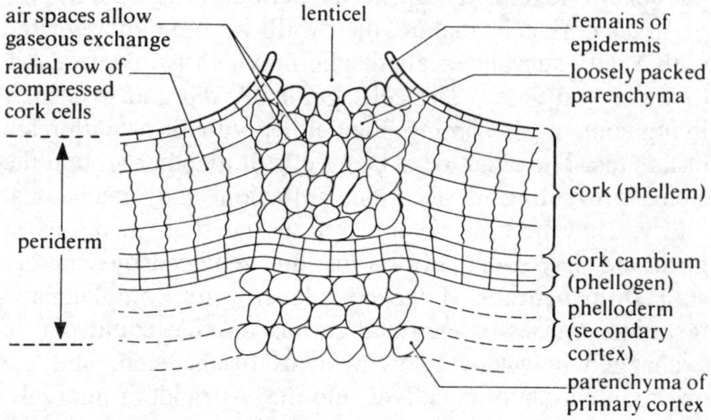

Fig 22.26 *VS lenticel (cell contents omitted).*

It is usual for the cork cambium to be renewed each year as the girth of the stem increases. Often a cork cambium arises in the secondary phloem, in which case the bark will, over a number of years, build up a layered appearance due to alternating layers of secondary phloem and bark.

22.5 Role of hormones in growth and development in humans

Human growth and development is controlled centrally by the hypothalamus and pituitary gland. The close association between these structures has already been noted in chapter 17 (section 17.6.2). The hypothalamus is the part of the brain immediately above the pituitary gland and receives information from the rest of the brain as well as from chemicals circulating in the blood. The hypothalamus controls the pituitary gland by secreting specific releasing and inhibitory factors which control the release of hormones from the pituitary gland. These in turn control other endocrine glands, which secrete hormones. In the case of growth and development these other glands include the thyroid gland, liver, adrenal cortex and gonads.

22.5.1 Pituitary gland and growth hormone

The most important hormone controlling growth and development is **growth hormone**, also known as **somatotrophin**, which is secreted by the anterior pituitary gland. It is a protein and was discovered during the early part of the twentieth century as a result of numerous experiments involving removal of the pituitary gland from animals. The rate and extent of growth in the experimental animals were greatly reduced. An extract from the pituitary gland was later isolated which, when injected into animals, caused an increase in body mass.

Secretion of human growth hormone (**hGH**) is controlled by the combined effects of two other hormones, produced by the hypothalamus (fig 22.27). These are **growth hormone releasing hormone (GHRH)**, also known as **somatocrinin**, and **growth hormone inhibitory hormone (GHIH)**, also known as **somatostatin**. Human growth hormone has a direct effect on all parts of the body, but particularly on growth of the skeleton and skeletal muscles. It also has an indirect effect by stimulating the release of small protein hormones called **somatomedins** from the liver. Somatomedins, also known as **insulin-like growth factors** or **IGFs** because they resemble insulin in structure and in some aspects of function, mediate or regulate some of the effects of human growth hormone. A summary of the regulation of the secretion of these hormones and their effects on the body is given in fig 22.27.

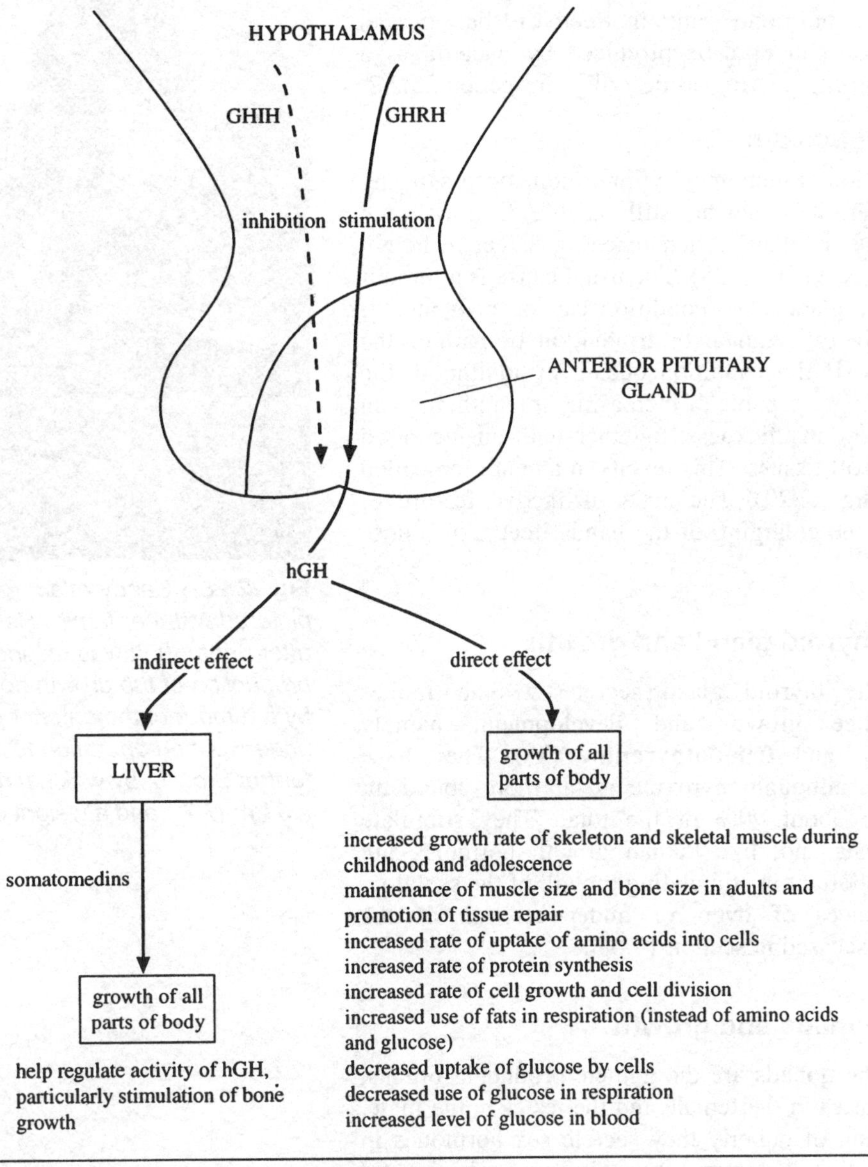

GHIH	growth hormone inhibitory hormone, also known as somatostatin
GHRH	growth hormone releasing hormone, also known as GHRF, growth hormone releasing factor or somatocrinin
hGH	human growth hormone, also known as somatotrophin or somatotropin
Somatomedins	also known as insulin-like growth factors (IGFs) because they resemble insulin in structure and in some aspects of function

Fig 22.27 *Regulation of the secretion of human growth hormone and somatomedins by the hypothalamus and anterior pituitary gland. The effects of these hormones on growth and development are also shown.*

Dwarfism

A deficiency of growth hormone results in dwarfism, more properly called pituitary dwarfism to distinguish it from other causes. Brain development and IQ are unaffected, unlike the situation with thyroid deficiency (section 17.6.4), and the body parts stay in proportion. The victim simply develops much more slowly. If the problem is due to growth hormone alone, then affected individuals do mature sexually. The dwarfism associated with African Pygmies has been discovered to be due to a deficiency of one of the somatomedins, indicating the additional importance of these growth factors.

In the past it was difficult to obtain sufficient human growth hormone to treat children with this condition (other animal growth hormones are ineffective), but human growth hormone can now be produced by bacteria as a result of genetic engineering, as described in section 25.2.2.

Gigantism

If over-production of human growth hormone occurs during childhood, when the bones are still capable of growth, the person becomes a 'giant', often reaching 8 feet in height (almost 2.5 metres) (fig 22.28). The usual cause is a tumour of the pituitary gland. The condition can be prevented if diagnosed early by removal or irradiation of part of the pituitary gland. If the condition occurs in adulthood, the bones are no longer capable of increasing in length, but can continue to grow in thickness, together with an increased growth of the soft tissues. This results in a condition called **acromegaly** (fig 22.29). The most distinctive feature of acromegaly is the enlarging of the hands, feet, skull, nose and jawbone.

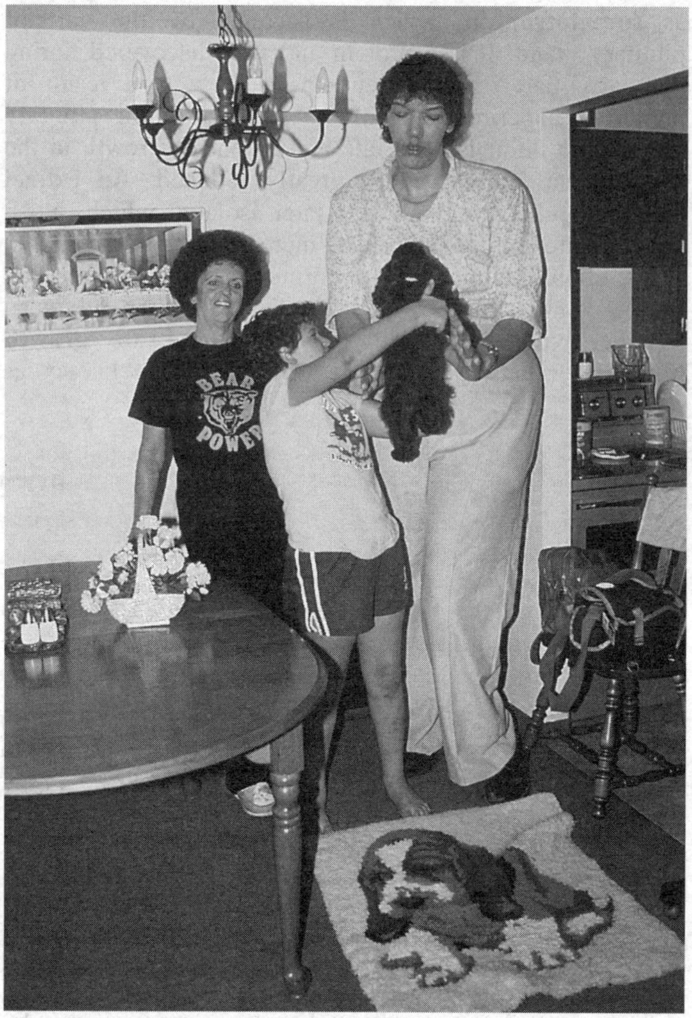

Fig 22.28 *Sandy Allen, the world's tallest woman, pictured with her family. Her abnormal growth started soon after her birth. It was diagnosed as being due to excessive production of the growth hormone somatotrophin, caused by a tumour of the anterior pituitary gland. In 1977 Allen underwent an operation to remove the gland and prevent further growth, by which time she had reached a height of 2.31 m (7'7") and a weight of 209 kg (nearly 33 stone).*

22.5.2 Thyroid gland and growth

The thyroid gland secretes two hormones which influence growth and development, namely **thyroxine (T_4)** and **triiodothyronine (T_3)**. They have similar effects, although thyroxine is far more abundant, accounting for about 90% of the total. They stimulate protein synthesis and, like human growth hormone, are particularly important in stimulating growth of the skeleton. The consequences of over- or under-secretion of the hormones is discussed in section 17.6.4.

22.5.3 Gonads and growth

The gonads are the gamete-producing organs, namely the ovaries in the female and the testes in the male. At the beginning of puberty they secrete sex hormones in response to signals from the pituitary gland and hypothalamus. The sex hormones are responsible for a fundamental change in growth and development and stimulate the development of secondary sexual characteristics. This is described in sections 21.6.4 and 21.6.6.

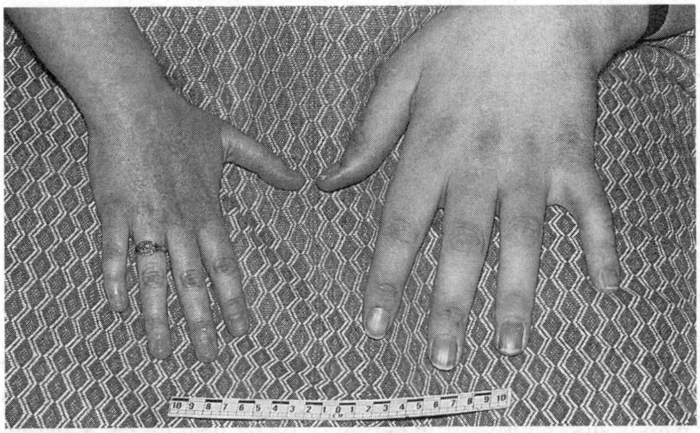
Fig 22.29 *Acromegaly – a woman's hand with a normal hand (left) for comparison. Acromegaly is a condition which results in an increase in the size of the hands, feet and face due to an excessive production of the growth hormone somatotrophin.*

22.5.4 Adrenal cortex and growth

The adrenal cortex is the outer region of each of the two adrenal glands and secretes steroid hormones. In both sexes, these include small amounts of both female and male sex hormones, oestrogens and androgens respectively. The androgens contribute to the adolescent growth spurt and development of pubic hair and underarm hair in both boys and girls. In the adult male, there is very little production of sex hormones by the adrenal glands, but small amounts of oestrogens and androgens continue to be made in females. The androgens may contribute to sexual behaviour, including sexual drive.

Chapter Twenty-three

Continuity of life

As noted at the beginning of chapter 5, one of the most important ideas in biology is the concept that the basic unit of structure and function of living organisms is the cell.

The cell theory was first proposed by Schleiden in 1838 and Schwann in 1839. Rudolph Virchow extended the theory in 1855 by declaring that new cells come only from pre-existing cells by cell division. Recognition of the continuity of life stimulated further workers throughout the later part of the nineteenth century to investigate the structure of the cell and the mechanisms involved in cell division. Improved techniques of staining and better microscopes revealed the importance of the nucleus, and in particular the chromosomes within it, as being the structures providing continuity between one generation of cells and the next. In 1879 Boveri and Flemming described the events occurring within the nucleus leading to the production of two identical cells, and in 1887 Weismann suggested that a specialised form of division occurred in the production of gametes. These two forms of division are called mitosis and meiosis respectively. It is useful to learn more about chromosomes before studying mitosis and meiosis.

23.1 Chromosomes

23.1.1 Chromosomes and karyotypes

The most important structures in the cell during division are the **chromosomes**. This is because they are responsible for the transmission of the hereditary information from one generation to the next. They do this because they contain DNA, the molecule of inheritance. Between divisions of the nucleus each chromosome contains one DNA molecule. Before the nucleus divides a copy of this DNA molecule is made so that at nuclear division the chromosome is a double structure, containing two identical DNA molecules. The two parts of the chromosome are referred to as **chromatids**. Each chromatid of a pair contains one of the two identical DNA molecules.

Although chromosomes stain intensely with certain dyes (stains), *individual* chromosomes cannot be seen very clearly in the period between divisions, known as interphase. This is because the chromosomes become very loosely coiled, long, thin threads spread throughout the nucleus. This material is referred to as **chromatin** (meaning 'coloured material'). Just before nuclear division, the chromosomes coil up into much more compact structures which are shorter, thicker and recognisable as separate structures which stain more intensely. Fig 23.1 shows a photograph of chromosomes in a human cell at this stage. You can see that each chromosome is made up of two chromatids. The chromatids are held together at a point called the **centromere** which may occur anywhere along the length of the chromosome (fig 23.2).

Each species has a characteristic number of chromosomes in each cell. In humans this is 46, as shown in fig 23.1. The number is very variable between species. For example, fruit flies have only eight chromosomes, whereas a small butterfly from Spain, *Lysandra*, has 380 chromosomes. Cats have 38 and dogs have 78. Most species have between 12 and 50 chromosomes per cell. The units of inheritance, the genes, are arranged along the chromosomes as indicated very diagrammatically in fig 23.2. In humans there are about 100 000 different genes.

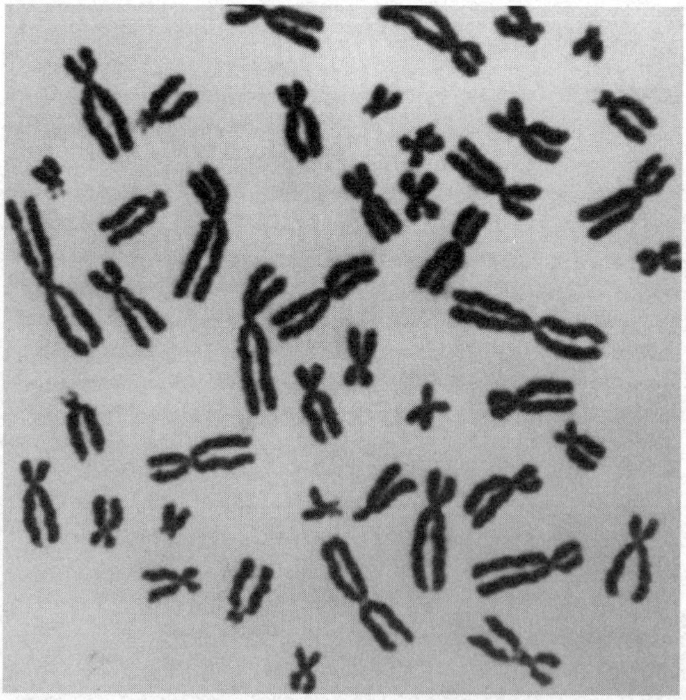

Fig 23.1 *Photograph of a set of chromosomes in a human male just before cell division. Each chromosome is made up of two chromatids held together at a point called the centromere. Forty-six chromosomes are present. Note their different sizes and positions of the centromeres.*

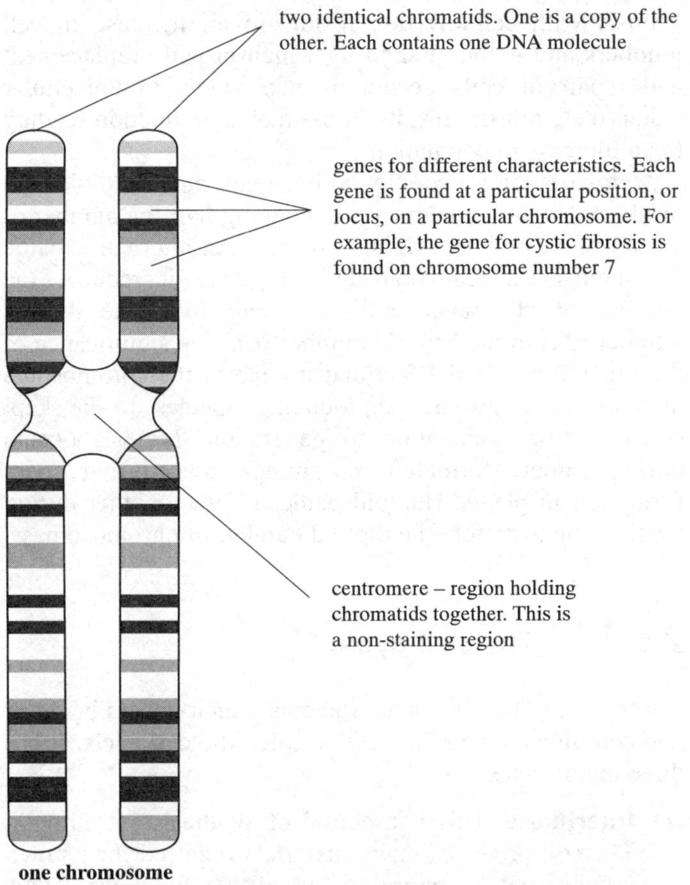

Fig 23.2 *Simplified diagram of a chromosome. In reality there would be several hundred to several thousand genes. The number and size of genes is variable.*

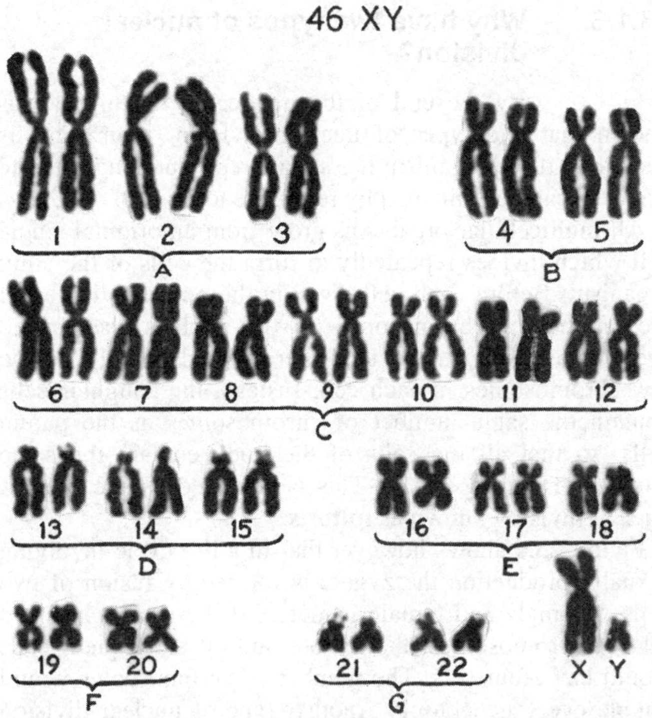

Fig 23.3 *Karyogram of a human male, prepared from fig 23.1. Non-sex chromosomes (autosomes) are placed in groups of similar size (A to G). The sex chromosomes are placed separately. X, female, Y, male. Note there are 22 pairs of autosomes and one pair of sex chromosomes. Genes on the autosomes are described as autosomal. Genes on the sex chromosomes are described as sex-linked.*

If the chromosomes are cut out from a photograph such as fig 23.1 and lined up according to size it can be seen that there are in fact *pairs* of chromosomes. These are referred to as **homologous pairs** because they are similar in structure. A photograph of such an arrangement of human chromosomes is shown in fig 23.3. Such a photograph is called a **karyogram** and the set of chromosomes is called the **karyotype**. There are 23 pairs of chromosomes in fig 23.3. The reason that there are pairs of chromosomes is that one set comes from the female parent by way of the egg, and one set comes from the male parent by way of the sperm. When the sperm fuses with the egg at fertilisation, the resulting cell, the zygote, has two sets of chromosomes.

You will notice that there is an odd pair of chromosomes in fig 23.3, labelled X and Y. These are the sex chromosomes. The male, or Y, chromosome is shorter than the female, or X, chromosome, and lacks some of the genes found on the female chromosome, as explained in section 24.6. Normally, homologous pairs of chromosomes have genes for the same characteristics. The person whose karyotype is shown in fig 23.3 is a male (XY). A female would have two X chromosomes (XX). Chromosome mutations are sometimes visible in karyograms, as discussed in chapter 24.

23.1.2 Haploid and diploid cells

Species in which there are two sets of chromosomes as described above are referred to as **diploid**, given the symbol 2*n*. The great majority of animal species and about half the plant species are diploid, with two sets of chromosomes per nucleus or cell. A few simple organisms have only one set of chromosomes and are referred to as **haploid** (symbol *n*) (see, for example, alternation of generations in chapter 2). In addition, gametes are haploid. Some organisms, including many plants, have three or more sets and are referred to as **polyploid**, but we shall ignore these in this chapter.

The advantages of possessing two sets of chromosomes are two-fold:

(1) Genetic variation is increased. Each individual will have a mixture of characteristics from both parents.
(2) If a gene on one chromosome of a pair is faulty, the second chromosome may provide a normal back-up.

23.1.3 Why have two types of nuclear division?

By the end of the nineteenth century it was known that two types of nuclear division occur. This is necessary if the organism has sexual reproduction in its life cycle, as can be explained by reference to fig 23.4.

All multicellular organisms grow from an original single cell which divides repeatedly to form the cells of the adult organism. Before each cell division the nucleus divides. If the number of chromosomes in the nucleus was halved every time a nucleus divided there would quickly be too few chromosomes in each cell. Instead, the daughter cells contain the same number of chromosomes as the parent cells, so that all the cells of the body contain the same number of chromosomes. This is achieved by the type of nuclear division known as **mitosis**.

Figure 23.4 shows however that in a life cycle involving sexual reproduction the zygote is formed by fusion of two cells, the male and female gametes. If these cells had two sets of chromosomes, the zygote and all subsequent cells would have four sets. The number of chromosomes would double every generation. Another type of nuclear division is therefore needed at some point in the life cycle to reduce the number of chromosomes. The diploid condition is then restored in the zygote. The type of nuclear division that reduces two sets of chromosomes to one set in the daughter cells is called **meiosis**. It is sometimes referred to as **reduction division**.

23.1.4 Summary

Mitosis is the process by which a cell nucleus divides to produce two daughter nuclei containing identical sets of chromosomes to the parent cell. It is usually followed immediately by division of the whole cell to form two daughter cells. This process is known as **cell division**.

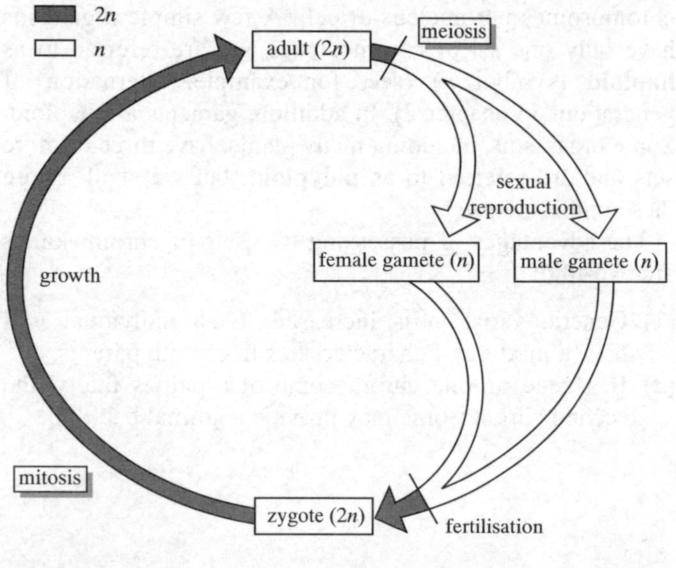

Fig 23.4 *Outline of the life cycle of an animal.*

Mitosis with cell division results in an increase in cell numbers and is the method by which growth, replacement and repair of cells occurs in eukaryotes. In unicellular eukaryotes, mitosis results in asexual reproduction leading to an increase in population size.

Meiosis is the process by which a cell nucleus divides to produce daughter nuclei each containing half the number of chromosomes of the original nucleus. An alternative name for meiosis is reduction division since it reduces the number of chromosomes in the cell from the diploid number ($2n$) to the haploid number (n). The significance of the process lies in the fact that it enables the chromosome number of a sexually reproducing species to be kept constant from generation to generation. Meiosis occurs during gamete formation in animals and during spore formation in plants. Haploid gametes fuse together during fertilisation to restore the diploid number of chromosomes.

23.2 The cell cycle

The sequence of events which occurs between one cell division and the next is called the **cell cycle**. It has three main stages.

(1) **Interphase.** This is a period of synthesis and growth. The cell produces many materials required for its own growth and for carrying out all its functions. DNA replication occurs during interphase.
(2) **Mitosis.** This is the process of nuclear division and is described later.
(3) **Cell division.** This is the process of division of the cytoplasm into two daughter cells.

The entire cycle is laid out in fig 23.5. The length of the cycle depends on the type of cell and external factors such as temperature, food and oxygen supplies. Bacteria may divide every 20 minutes, epithelial cells of the intestine wall every 8–10 hours, onion root-tip cells may take 20 hours whilst many cells of the nervous system never divide.

Experiment 23.1: To investigate the phases of mitosis using a root tip squash

Chromosomes can normally be observed only during nuclear division. The apical meristem of roots (root tips) of garlic ($2n = 16$), onion ($2n = 16$) and broad bean ($2n = 12$) provide suitable material for the experiment. The material is set up so that root development is stimulated. Later the root tips are removed, fixed, stained and macerated so that the chromosomes may be observed under the microscope.

Materials

pins	pair of fine needles
test-tube containing water	several sheets of blotting paper
scalpel	

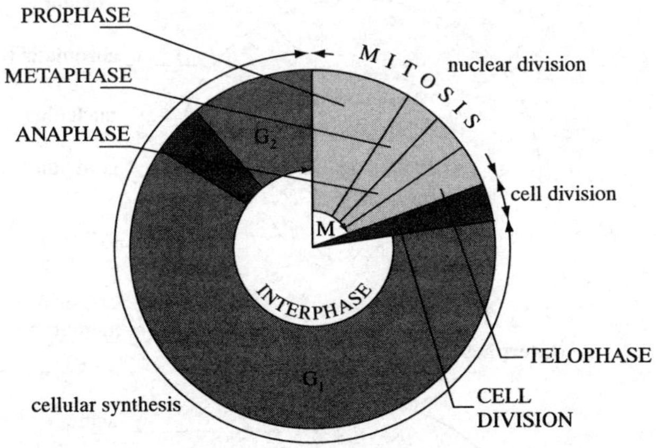

Phase	Events within cell
G_1	Intensive cellular synthesis, including new cell organelles. Cell metabolic rate high. Cell growth occurs. Substances produced to inhibit or stimulate onset of next phase as appropriate.
S	DNA replication occurs. Protein molecules called histones are synthesised and cover each DNA strand. Each chromosome becomes two chromatids. At this stage the cell is $4n$ (4 copies of each DNA molecule, 2 in each homologous chromosome).
G_2	Intensive cellular synthesis. Mitochondria and chloroplasts divide. Energy stores increase. Mitotic spindle begins to form.
M	Nuclear division occurs in four phases.
C	Equal distribution of organelles and cytoplasm into each daughter cell.

Fig 23.5 *The cell cycle.*

small corked tube
forceps
2 petri dishes
water bath and test-tube
microscope slide
cover-slip
clove of garlic
distilled water
acetic alcohol
molar hydrochloric acid
Feulgen stain

Method

(1) Place a pin through a clove of garlic and suspend in a test-tube full of water so that the base of the clove is covered with water. Leave for 3–4 days without any disturbance as this is likely to inhibit cell division temporarily.
(2) When several roots have grown 1–2 cm, remove the clove and cut off the terminal 1 cm of the roots.
(3) Transfer the roots to a small corked tube containing acetic alcohol and leave overnight at room temperature to fix the material.
(4) Remove the root tips with forceps by grasping the cut end of the root, transfer to a petri dish containing distilled water and wash for a few minutes to remove the fixative.
(5) Transfer the root tips to a test-tube containing molar hydrochloric acid which is maintained at 60 °C for 3 min (6–10 min for onion, peas and beans). This breaks down the middle lamellae holding the cells together and hydrolyses the DNA of the chromosomes to form deoxyribose aldehydes which will react with the stain.
(6) Pour the root tips and acid into a petri dish. Remove the roots into another petri dish containing distilled water and wash to remove the acid. Leave for 5 min.
(7) Transfer the roots to a small tube containing Feulgen stain and cork. Leave in a cool dark place (preferably a refrigerator) for a minimum of 2 h.
(8) Remove a root tip and place in a drop of acetic alcohol on a clean microscope slide.
(9) Cut off the terminal 1–2 mm of the root tip and discard the rest of the root.
(10) Tease out the root tip using a pair of fine needles and cover with a cover-slip. Place the slide on a flat surface, cover with several sheets of blotting paper and press down firmly over the cover-slip with the ball of the thumb. Do not allow the cover-slip to move sideways. The technique is called a 'squash'.
(11) Examine the slide under the low and high powers of the microscope and identify cells showing different phases of mitosis.
(12) Draw and label nuclei showing the various phases.

23.3 Mitosis

The events occurring within the nucleus during mitosis are usually observed in cells which have been fixed and stained. This in effect provides a series of 'snapshots' of the phases through which chromosomes pass during cell division. It must be remembered though that mitosis is a continuous process with no sharp distinction between the phases. The use of the phase-contrast microscope and time-lapse photography has enabled the events of nuclear division to be seen in the living cell as they happen. By speeding up the film, mitosis is seen as a continuous process. It can be divided for convenience into four stages. The changes occurring during these stages in an animal cell are described in fig 23.6. Photographs of mitosis in animal and plant cells are shown in figs 23.7 and 23.8.

23.3.1 Centrioles and spindle formation

Centrioles are organelles situated in the cytoplasm close to the nuclear envelope in animal and simpler plant cells. They occur in pairs and lie at right-angles to each other.

Each centriole is approximately 500 nm long and 200 nm in diameter and is composed of nine groups of microtubules arranged in triplets. Neighbouring triplets are attached to each other by fibrils (fig 23.9). **Microtubules**

Interphase
Variable duration depending on function of the cell. Just before nuclear division the DNA of each chromosome replicates. Each chromosome now exists as a pair of **chromatids** joined together by a **centromere**. At this stage the cell is 4*n* (4 copies of each DNA molecule, 2 in each chromosome of a homologous pair). During interphase chromosome material is in the form of very loosely coiled threads called **chromatin**. Centrioles have replicated.

Prophase
Usually the longest phase of division. Chromosomes shorten and thicken by coiling and tighter packaging of their components. Staining shows up the chromosomes clearly. Each chromosome is seen to consist of two chromatids held together by a centromere that does not stain. In animal cells the **centrioles** move to opposite poles of the cell. Short **microtubules** may be seen radiating from the centrioles. These are called **asters** (*astra*, a star). The **nucleoli** disappear as their DNA passes to certain chromosomes. At the end of prophase the nuclear envelope is no longer visible because it breaks up into small vesicles which disperse. A spindle is formed.

Metaphase
Chromosomes line up around the equator of the spindle, attached by their centromeres to the 'spindle fibres', which are microtubules.

Anaphase
This stage is very rapid. The centromeres split into two and the spindle fibres pull the daughter centromeres to opposite poles. The separated chromatids are pulled along behind the centromeres.

Telophase
The chromatids reach the poles of the cell, uncoil and lengthen to form chromatin again, losing the ability to be seen clearly. The spindle fibres disintegrate and the centrioles replicate. A nuclear envelope re-forms around the chromosomes at each pole and the nucleoli reappear. Telophase may lead straight into **cytokinesis** (cell division).

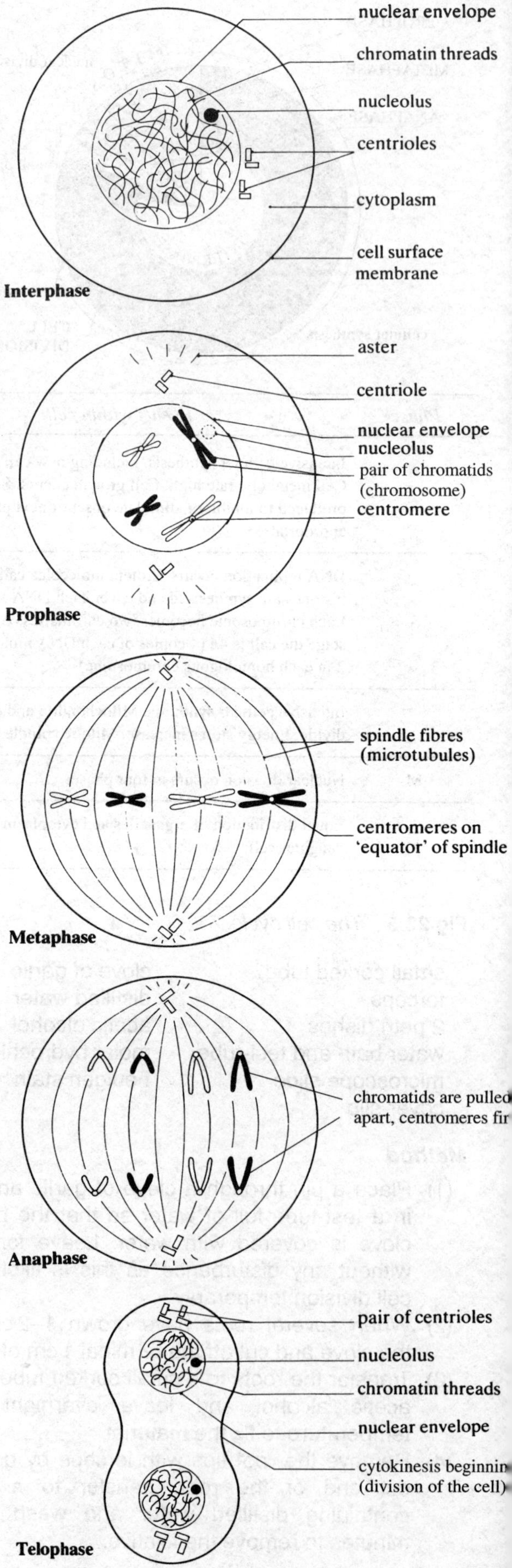

Fig 23.6 *Mitosis in an animal cell.*

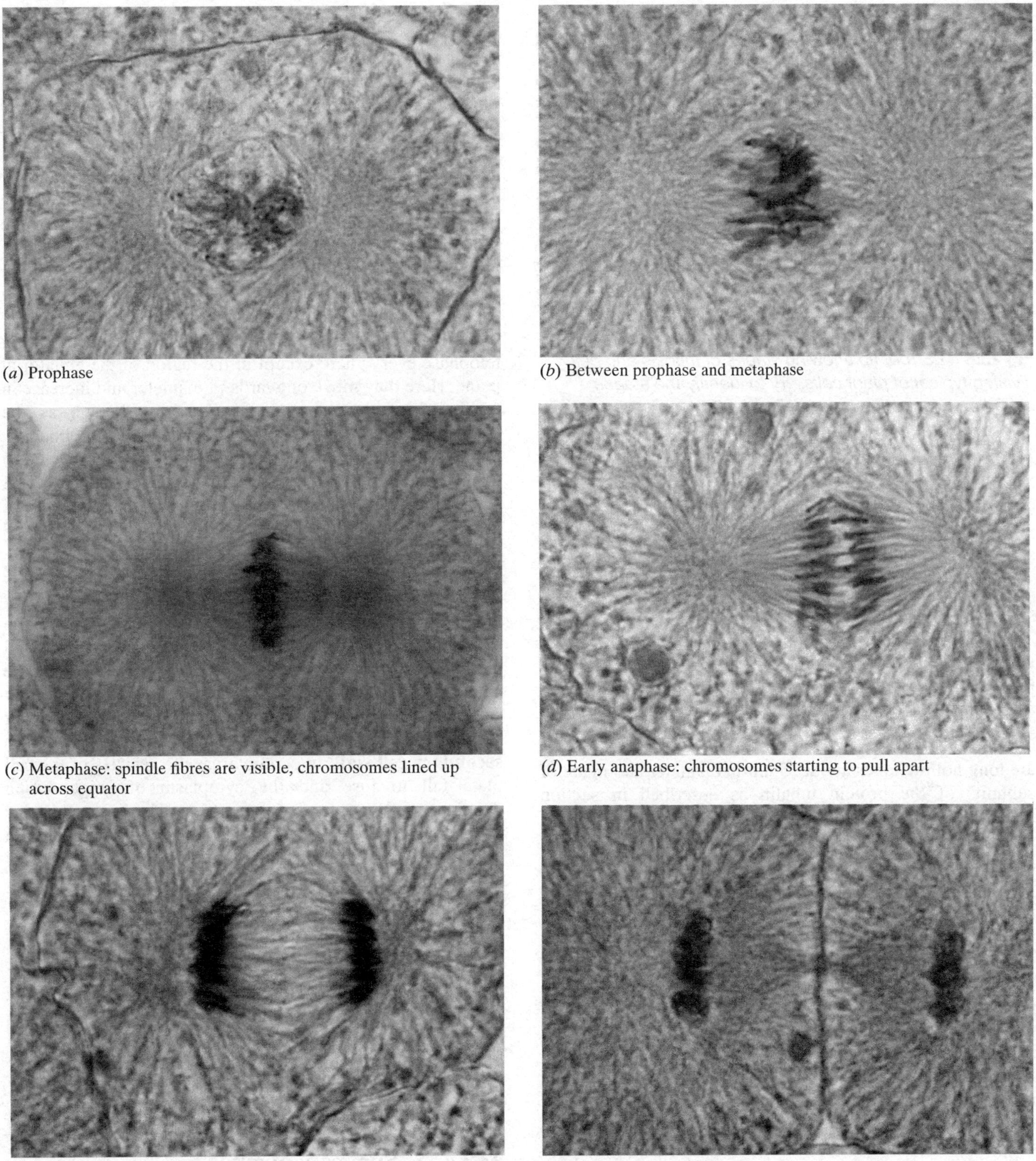

(*a*) Prophase

(*b*) Between prophase and metaphase

(*c*) Metaphase: spindle fibres are visible, chromosomes lined up across equator

(*d*) Early anaphase: chromosomes starting to pull apart

(*e*) Anaphase

(*f*) Telophase and cytokinesis

Fig 23.7 *Stages of mitosis and cell division in an animal cell.*

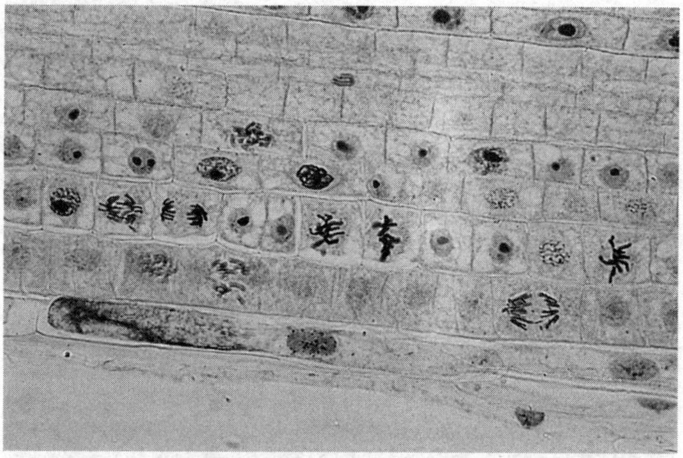

Fig 23.8 *LS root tip showing stages of mitosis and cell division typical of plant cells. Try to identify the stages based on information given in figure 23.7.*

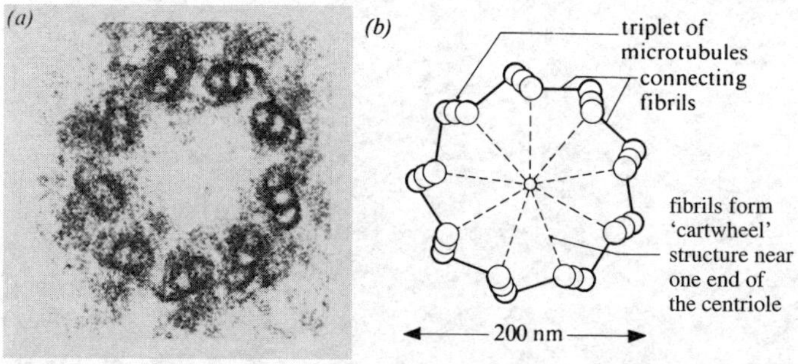

Fig 23.9 *(a) Electron micrograph of a TS of a centriole from embryonic chick pancreas. (b) Diagram of TS through a centriole.*

are long hollow tubes about 25 nm in diameter and made of subunits of the protein tubulin as described in section 5.10.7.

In all cases centrioles lie in a material of poorly defined structure which initiates the development of microtubules. This region is called the **centrosome**. It is the centrosome and not the centriole which is responsible for making the spindle because the 'spindle fibres' are in fact microtubules. This explains how plants and fungi which lack centrioles also make spindles from microtubules. The function of the centrioles in nuclear division is not clear. They may be involved in orienting the spindle, thus helping to determine in which plane the cell divides. Some spindle fibres run from pole to pole, others run from the poles to the centromeres. The shortening of these spindle fibres by removal of the tubulin subunits accounts for the movement of chromosomes and chromatids during nuclear division. They are in effect 'reeled in' by the centrosomes.

The addition of the chemical colchicine to actively dividing cells inhibits spindle formation and the chromatid pairs remain in their metaphase positions. This technique enables the number and structure of chromosomes to be examined under the microscope.

Modified centrioles also occur at the bases of cilia and flagella, where they are known as **basal bodies**.

23.3.2 Cell division

Cytokinesis is the division of the cytoplasm. This stage normally follows telophase and leads into the G_1 phase of interphase. In preparation for division, the cell organelles become evenly distributed towards the two poles of the telophase cell along with the chromosomes. In animal cells the cell surface membrane begins to invaginate during telophase towards the region previously occupied by the spindle equator. Microfilaments in the region are thought to be responsible for drawing in the cell surface membrane to form a furrow around the outside surface of the cell. The cell surface membranes in the furrow eventually join up and completely separate the two cells.

In plant cells the spindle fibres begin to disappear during telophase everywhere except in the region of the equatorial plane. Here they move outwards in diameter and increase in number to form a barrel-shaped region known as the **phragmoplast**. Microtubules, ribosomes, mitochondria, endoplasmic reticulum and Golgi apparatus are attracted to this region and the Golgi apparatus produces a number of small fluid-filled vesicles. These appear first in the centre of the cell and, guided by microtubules, fuse to form a **cell plate** which grows across the equatorial plane (fig 5.30). The contents of the vesicles contribute to the new middle lamella and cell walls of the daughter cells, whilst their membranes form the new cell surface membranes. The spreading plate eventually fuses with the parent cell wall and separates the two daughter cells. The new cell walls are called **primary cell walls** and may be thickened at a later stage by the deposition of further cellulose and other substances such as lignin and suberin to produce a **secondary cell wall**. In certain areas the vesicles of the cell plate fail to fuse and the cytoplasm of neighbouring daughter cells remains in contact. These cytoplasmic channels are lined by the cell surface membrane and form structures known as **plasmodesmata**.

23.3.3 Comparison of mitosis in animal and plant cells

The most important event occurring during mitosis is the equal distribution of duplicate chromosomes between the two daughter cells. This process is almost identical in animal and plant cells but there are a number of differences, and these are summarised in table 23.1.

23.3.4 Summary of mitosis

As a result of mitosis a parent nucleus divides into two daughter nuclei, each with the same number of chromosomes as the parent nucleus. This is followed by division of the whole cell. In order to achieve this, chromosomes first replicate during interphase. The two replicate chromosomes are known as chromatids and separate during mitosis.

Table 23.1 Differences between mitosis in plant and animal cells.

Plant	*Animal*
No centriole present	Centrioles present
No aster forms	Asters form
Cell division involves formation of a cell plate	Cell division involves furrowing and cleavage of cytoplasm
Occurs mainly at meristems	Occurs in tissues throughout the body

23.3.5 Significance of mitosis

- **Genetic stability** Mitosis produces two nuclei which have the same number of chromosomes as the parent cell. Since these chromosomes were derived from parental chromosomes by the exact replication of their DNA, they will carry the same hereditary information in their genes. Daughter cells are genetically identical to the parent cell and no variation in genetic information can therefore be introduced during mitosis. This results in genetic stability within populations of cells derived from the same parental cells.
- **Growth** The number of cells within an organism increases by mitosis and this is the basis of growth in multicellular organisms (chapter 22).
- **Cell replacement** Replacement of cells and tissues also involves mitosis. Cells are constantly dying and being replaced, an obvious example being in the skin.
- **Regeneration** Some animals are able to regenerate whole parts of the body, such as legs in crustacea and arms in starfish. Production of the new cells involves mitosis.
- **Asexual reproduction** Mitosis is the basis of asexual reproduction, the production of new individuals of a species by one parent organism. Many species undergo asexual reproduction. The various methods are described more fully in chapter 21.

23.4 Meiosis

Meiosis (*meio*, to reduce) is a form of nuclear division in which the chromosome number is halved from the diploid number ($2n$) to the haploid number (n). Like mitosis, it involves DNA replication during interphase in the parent cell, but this is followed by *two* cycles of nuclear divisions and cell divisions, known as **meiosis I** (the **first meiotic division**) and **meiosis II** (the **second meiotic division**). Thus a single diploid cell gives rise to four haploid cells as shown in outline in fig 23.10.

Meiosis occurs during the formation of sperm and eggs (gametogenesis) in animals (chapter 21) and during spore formation in plants.

Like mitosis, meiosis is a continuous process but is conveniently divided into prophase, metaphase, anaphase and telophase. These stages occur in the first meiotic division and again in the second meiotic division. The behaviour of chromosomes during these stages is illustrated in fig 23.11, which shows a nucleus containing four chromosomes ($2n = 4$), that is two homologous pairs of chromosomes.

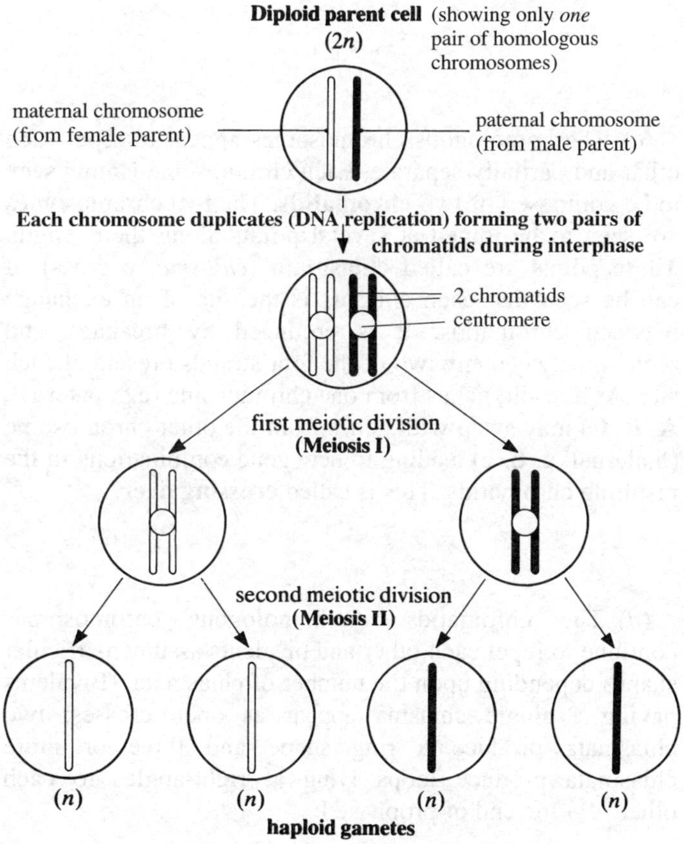

Fig 23.10 *The basic characteristics of meiosis showing one chromosome duplication followed by two nuclear and cell divisions. Note that, as for mitosis, chromosomes may be single or double structures. When double, the two parts are called chromatids.*

MEIOSIS I

Photographs of meiosis are shown in figs 23.12–14.

Prophase I

The longest phase.

(*a*) Chromosomes shorten and become visible as single structures.

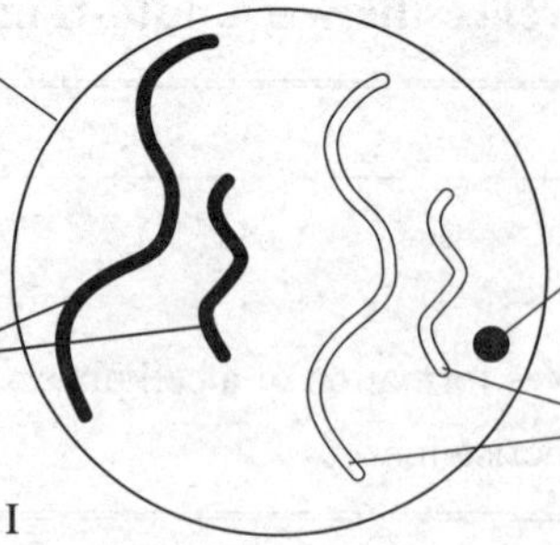

(*a*) early prophase I

(*b*) Homologous chromosomes pair up. This process is called **synapsis**. Each pair is called a **bivalent**. One of the pair comes from the male parent and one from the female parent. Each member of the pair is the same length, their centromeres are in the same positions and they usually have the same number of genes arranged in the same order. The bivalents shorten and thicken, partly by coiling. Each chromosome and its centromeres can now be seen clearly.

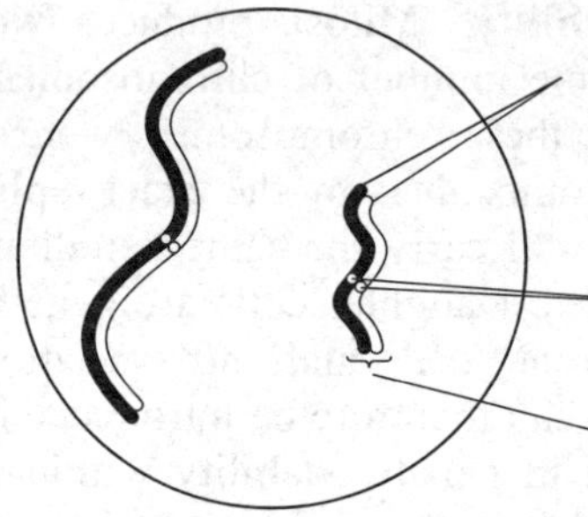

(*b*) prophase I

(*c*) The homologous chromosomes appear to repel each other and partially separate. Each chromosome is now seen to be composed of two **chromatids**. The two chromosomes are seen to be joined at several points along their length. These points are called **chiasmata** (*chiasma*, a cross). It can be seen that each chiasma is the site of an exchange between chromatids. It is produced by breakage and reunion between any two of the four strands present at each site. As a result, genes from one chromosome (e.g. paternal, **A**, **B**, **C**) may swap with genes from the other chromosome (maternal, **a**, **b**, **c**) leading to new gene combinations in the resulting chromatids. This is called **crossing over**.

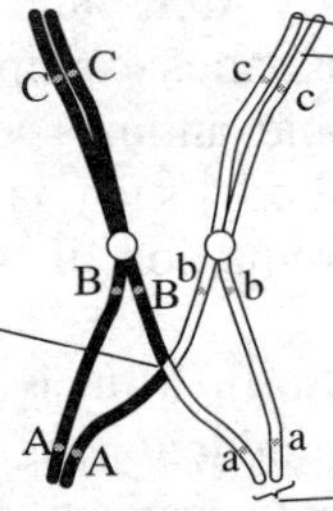

(*c*) crossing over during prophase I

(*d*) The chromatids of homologous chromosomes continue to repel each other and bivalents assume particular shapes depending upon the number of chiasmata. (Bivalents having a single chiasma appear as open crosses, two chiasmata produce a ring shape and three or more chiasmata produce loops lying at right-angles to each other.) By the end of prophase I:

- all chromosomes are fully contracted and deeply stained;
- the centrioles (if present) have migrated to the poles;
- the nucleoli and nuclear envelope have dispersed;
- lastly the spindle fibres form.

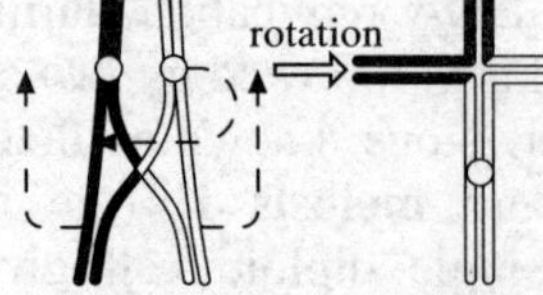

(*d*) bivalent with a single chiasma showing change in shape to an open cross as a result of rotation of the chromatids

Fig 23.11 *(a)–(k) Meiosis in an animal cell.*

Metaphase I
The bivalents become arranged around the equator of the spindle, attached by their centromeres.

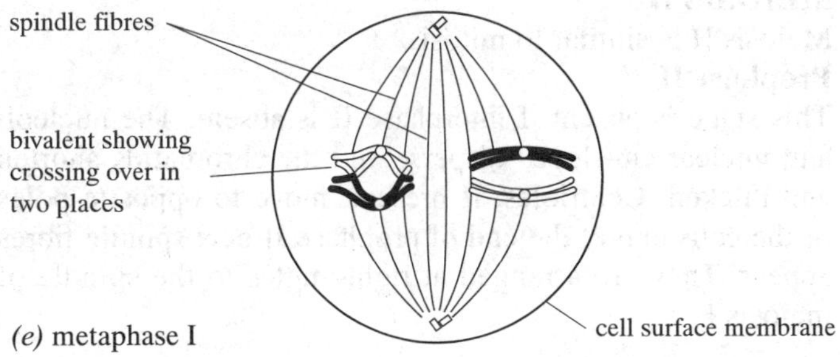

(e) metaphase I

Anaphase I
Spindle fibres pull homologous chromosomes, centromeres first, towards opposite poles of the spindle. This separates the chromosomes into two haploid sets, one set at each end of the spindle.

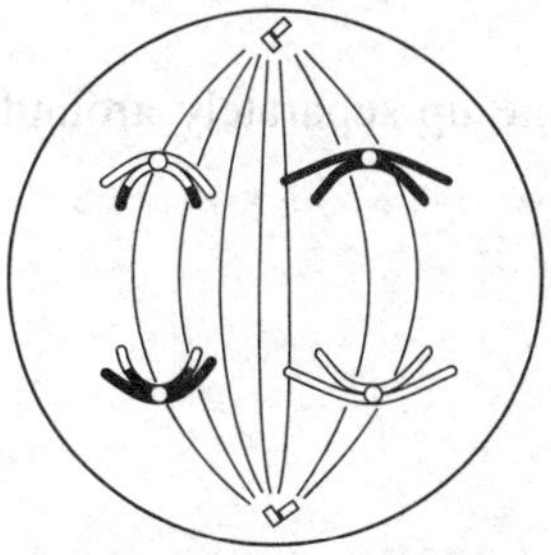
(f) anaphase I

Telophase I
The arrival of homologous chromosomes at opposite poles marks the end of meiosis I. Halving of chromosome number has occurred but the chromosomes are still composed of two chromatids.
If crossing over has occurred these chromatids are not genetically identical and must be separated in a second meiotic division. Spindles and spindle fibres usually disappear.
In animals and some plants the chromatids usually uncoil and a nuclear envelope re-forms at each pole and the nucleus enters interphase. Cleavage (animals) or cell wall formation (plants) then occurs as in mitosis. In many plants there is no telophase, cell wall formation or interphase and the cell passes straight from anaphase I into prophase of the second meiotic division.

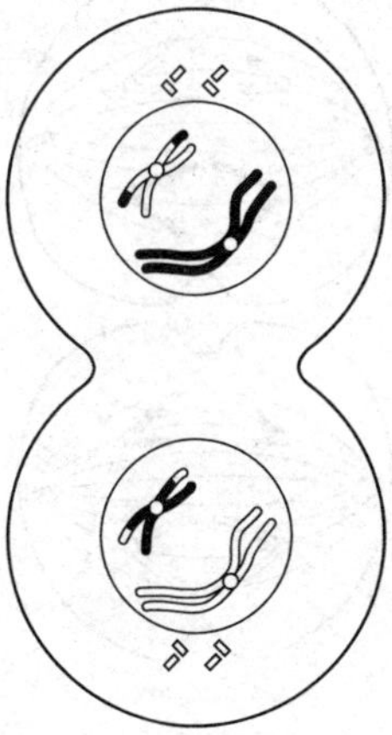
(g) telophase I in an animal cell

Interphase II
This stage is present usually only in animal cells and varies in length. No further DNA replication occurs.

Fig 23.11 *(continued)*

MEIOSIS II

Meiosis II is similar to mitosis.

Prophase II

This stage is absent if interphase II is absent. The nucleoli and nuclear envelopes disperse and the chromatids shorten and thicken. Centrioles, if present, move to opposite poles of the cells and at the end of prophase II new spindle fibres appear. They are arranged at right-angles to the spindle of meiosis I.

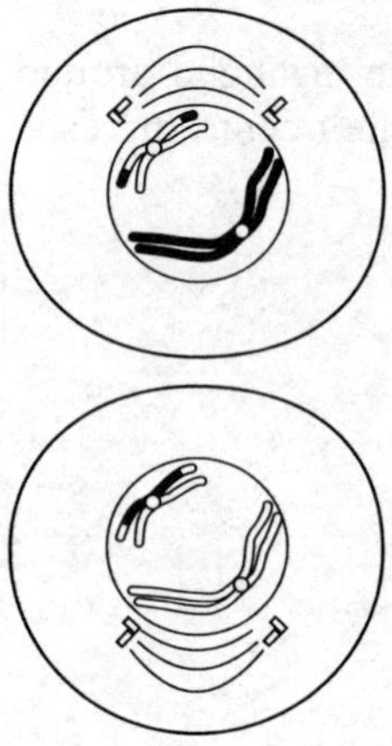

(h) prophase II

Metaphase II

Chromosomes line up separately around the equator of the spindle.

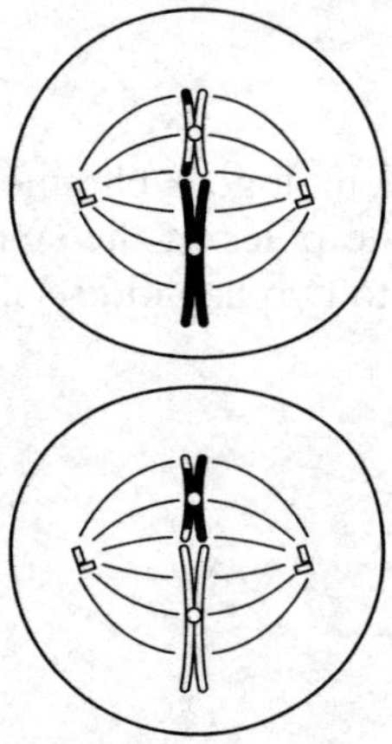

(i) metaphase II

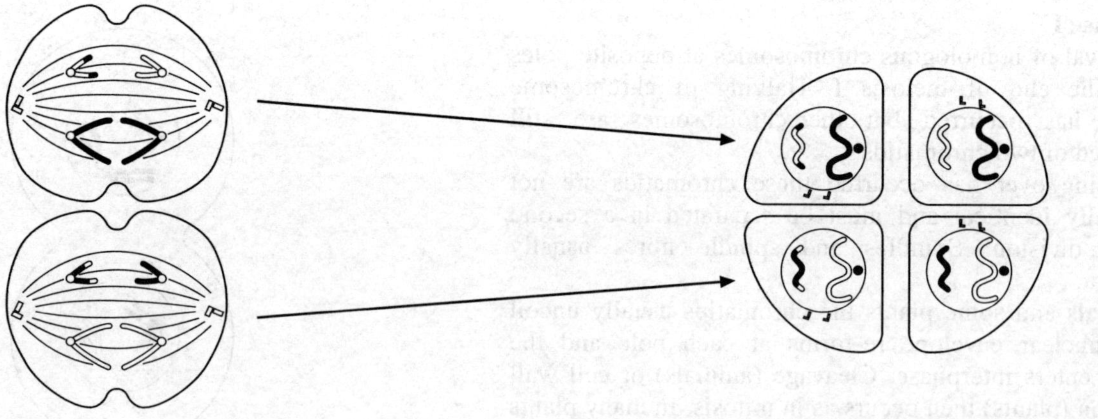

(j) anaphase II

(k) telophase II and cell cleavage in an animal cell

Anaphase II

The centromeres divide and the spindle fibres pull the chromatids to opposite poles, centromeres first.

Telophase II

As telophase in mitosis but four haploid daughter cells are formed. The chromosomes uncoil, lengthen and become very indistinct. The spindle fibres disappear and the centrioles replicate. Nuclear envelopes re-form around each nucleus which now possess half the number of chromosomes of the original parent cell (haploid). Subsequent cleavage (animals) or cell wall formation (plants) will produce four daughter cells from the original single parent cell.

Fig 23.11 (continued)

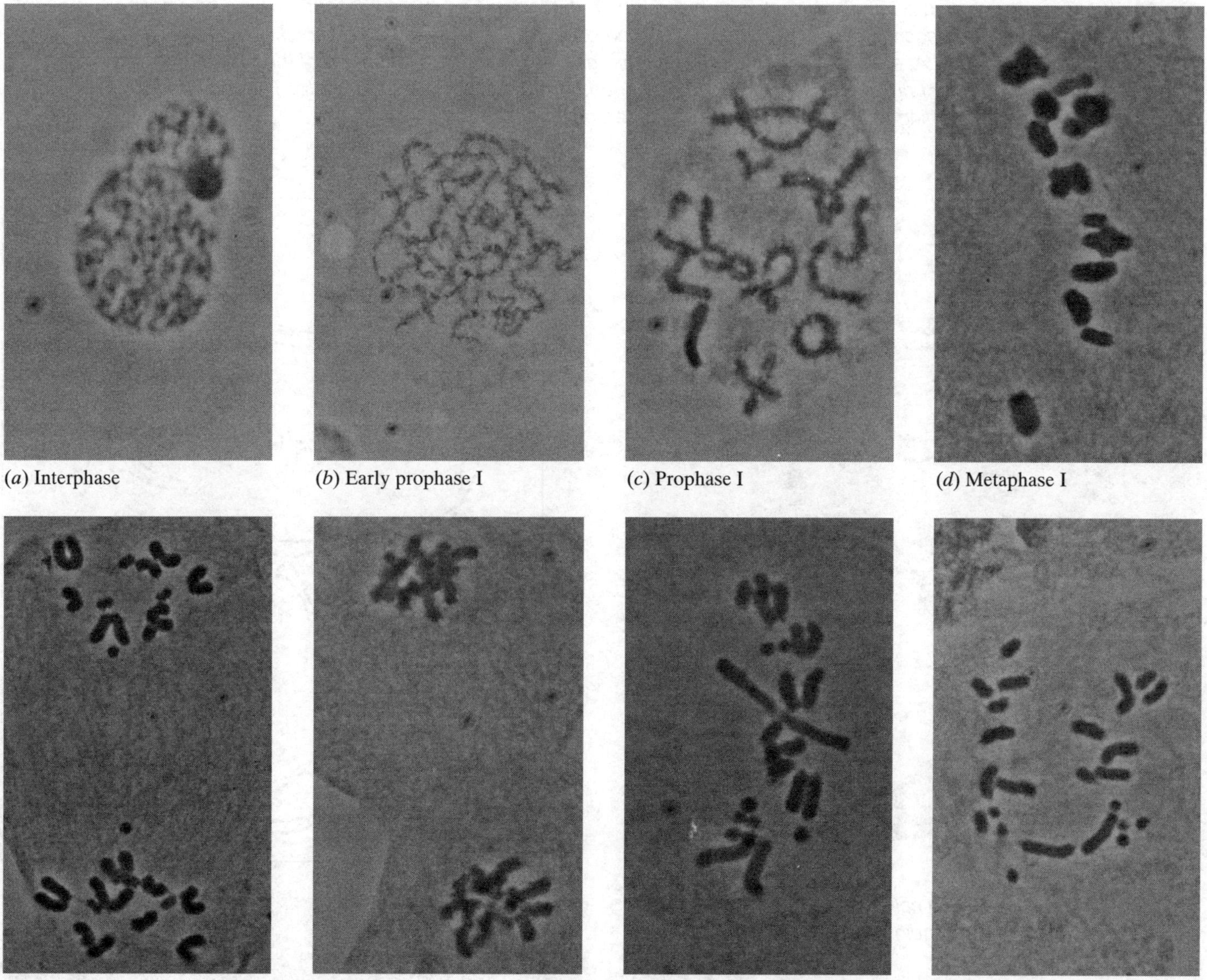

(*a*) Interphase (*b*) Early prophase I (*c*) Prophase I (*d*) Metaphase I

(*e*) Anaphase I (*f*) Telophase I (*g*) Metaphase II (*h*) Anaphase II

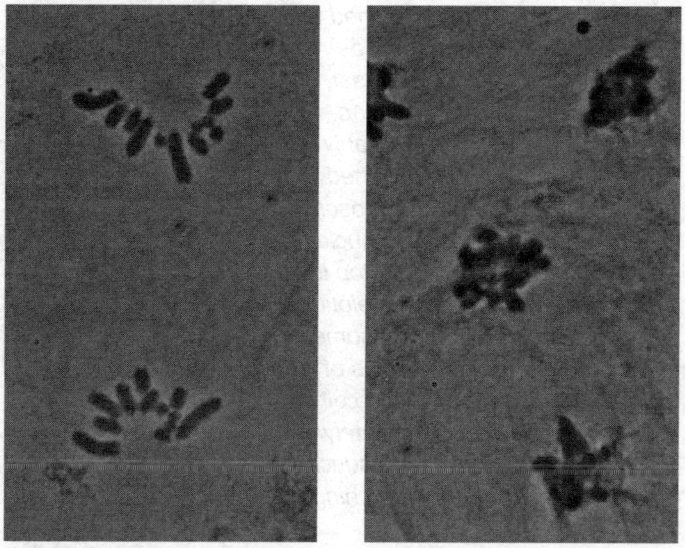

(*i*) Late anaphase II (*j*) Telophase II

Fig 23.12 *Stages of meiosis in an animal cell. Interphase is also shown.*

23.4.1 Summary of meiosis

As a result of meiosis four daughter cells are formed from one parent cell. Each daughter cell has half the number of chromosomes of the parent cell. Usually the parent cell is diploid and the daughter cells are therefore haploid.

Fig 23.13 *Crossing over in prophase I, showing chiasmata, in the locust* Locusta migratoria. *Eleven bivalents are present with one or two chiasmata each. Paternal and maternal chromosomes are represented by solid and dotted lines respectively in the drawing. At each chiasma a genetic exchange has occurred. The shape of the bivalent will vary from rod-shaped, to cross-shaped or ring-shaped, depending on the number and position of chiasmata. The unpaired X chromosome is deeply staining at this stage (Dr S. A. Henderson).*

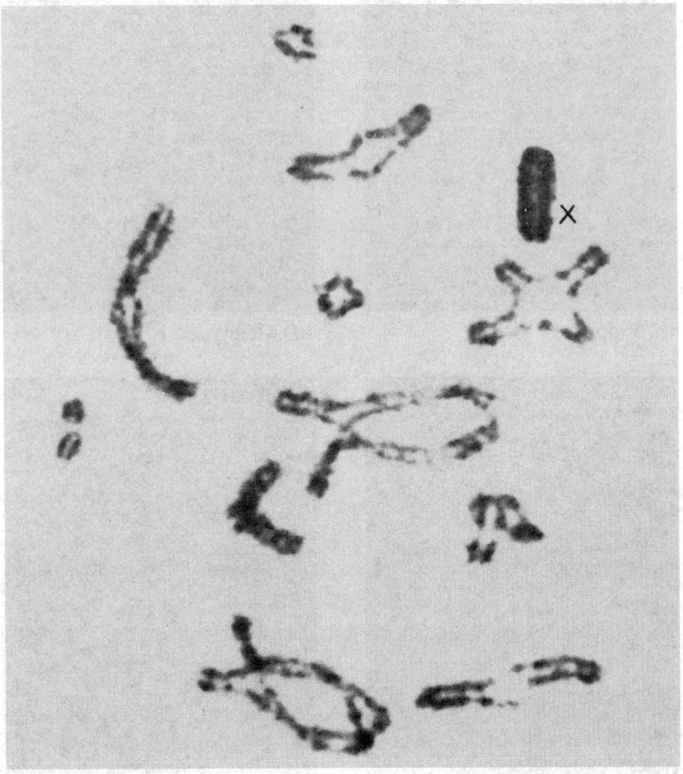

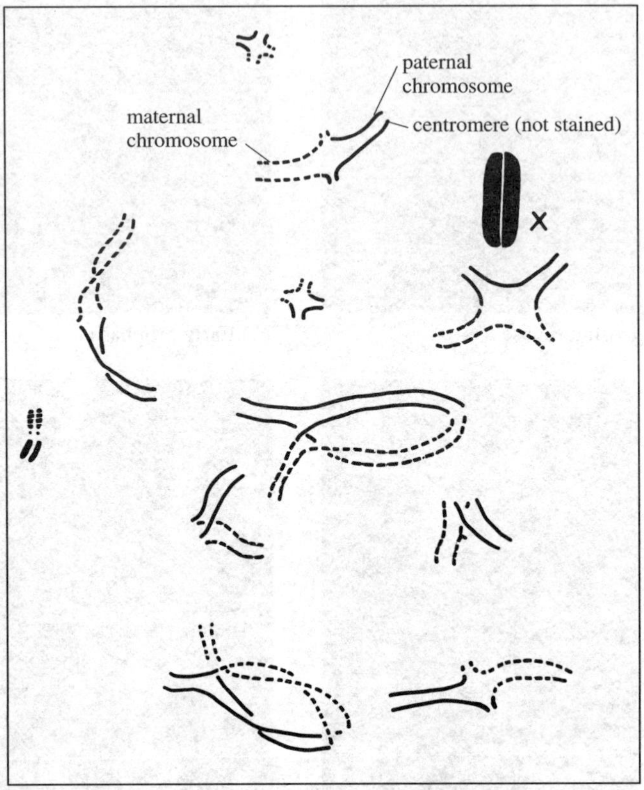

Fig 23.14 *Meiosis in living cells. Pairing and cell division at meiosis in living sperm mother cells (spermatocytes) of the locust* Locusta migratoria. *These preparations are photographed with an optical technique known as Nomarski interference contrast, which uses polarised light and produces images of remarkably 3D appearance in living, unstained cells. Two cells show chromosome pairing in early prophase I nuclei (arrowed). The two cells at top left are at the end of the first meiotic division. Groups of chromosomes have reached opposite poles of the spindle and cleavage of the cell has taken place. Two approximately equal daughter cells are produced. The fibrous structures stretching between the two groups of chromosomes are the microtubules of the spindle.*

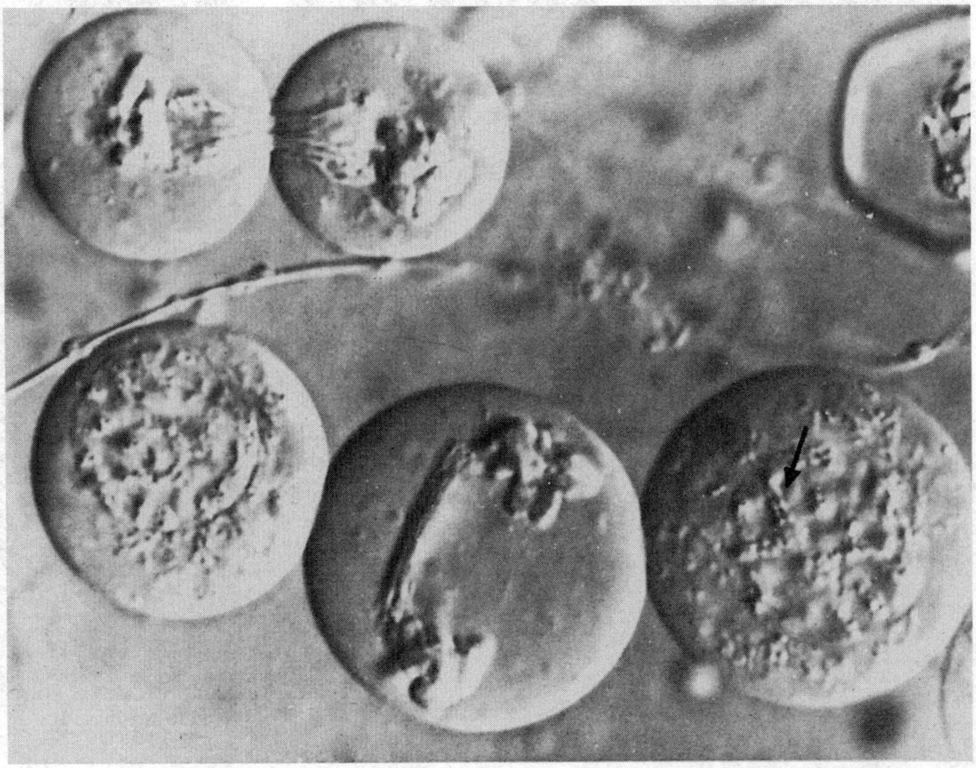

23.4.2 Significance of meiosis

- **Sexual reproduction** Meiosis occurs in all organisms carrying out sexual reproduction. During fertilisation the nuclei of the two gamete cells fuse. Each gamete has one set of chromosomes (is haploid, n). The product of fusion is a zygote which has two sets of chromosomes (the diploid condition, $2n$). If meiosis did not occur fusion of gametes would result in a doubling of the chromosomes for each successive sexually reproduced generation. (An exception to this is shown in polyploidy which is described in section 24.9.) This situation is prevented in the life cycle of all sexually reproducing organisms by the occurrence, at some stage, of cell division involving a reduction in the diploid number of chromosomes ($2n$) to the haploid number (n).

> **23.1** The amount of DNA present per cell during several nuclear divisions is represented diagrammatically in fig 23.15.
> (*a*) Which type of nuclear division is represented by fig 23.15?
> (*b*) What phases are represented by the dashed lines W, X and Y?
> (*c*) What type of cells are represented by the line Z?

- **Genetic variation** Meiosis also provides opportunities for new combinations of genes to occur in the gametes. This leads to genetic variation in the offspring produced by fusion of the gametes. Meiosis does this in two ways, namely independent assortment of chromosomes and crossing over in meiosis I.

(1) *Independent assortment of chromosomes* This is best explained by means of a diagram (fig 23.16). The orientations of bivalents at the equator of the spindle in metaphase I is random. Fig 23.16 shows a simple situation in which there are only two bivalents and therefore only two possible orientations (one in which the white chromosomes are alongside each other and one in which black and white are alongside). The more bivalents there are, the more variation is possible. Independent assortment refers to the fact that the bivalents line up independently and therefore the chromosomes in each bivalent separate (assort) independently of those in other bivalents during anaphase I. The black and white chromosomes in fig 23.16 represent maternal and paternal chromosomes. Independent assortment is the basis of Mendel's second law (section 24.1.3).

(2) *Crossing over* As a result of chiasmata, crossing over of segments of chromatids occurs between homologous chromosomes during prophase I, leading to the formation of new combinations of genes on the chromosomes of the gametes. This is shown in

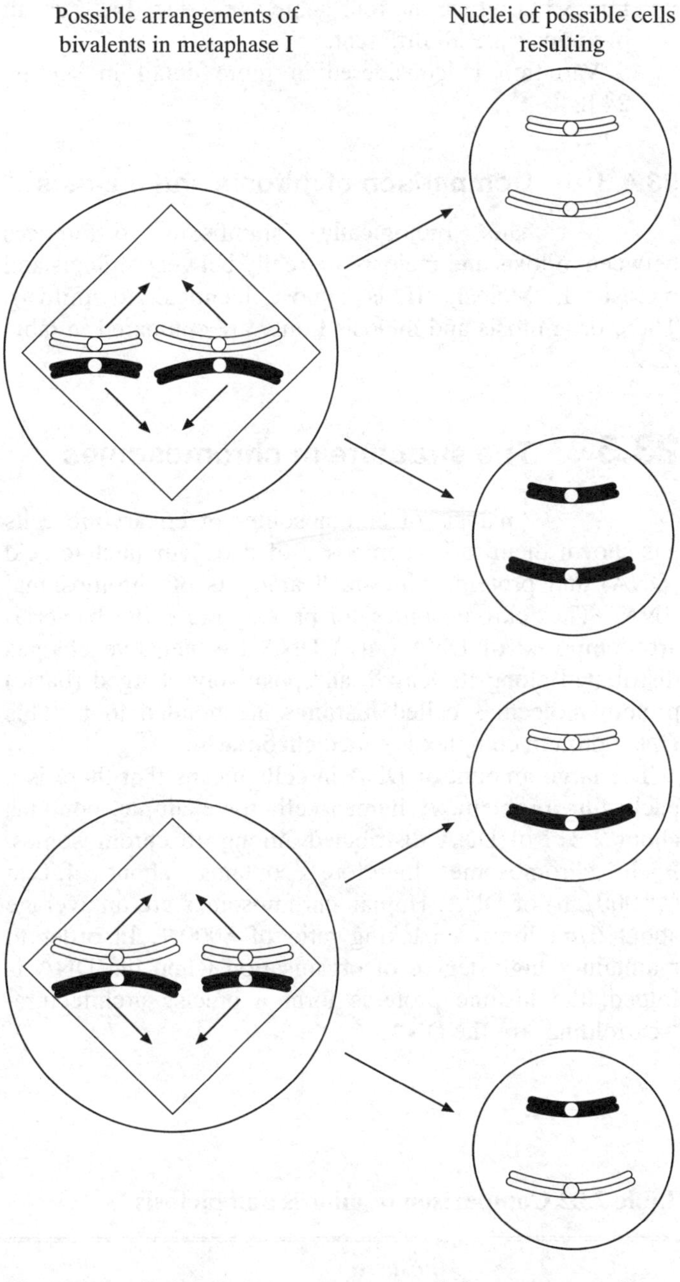

Fig 23.16 *Independent assortment of chromosomes in meiosis. Two bivalents are shown. Each bivalent lines up on the equator independently of other bivalents, so that both possibilities shown in the diagram are equally likely. This increases the potential variation in the gametes, which are also shown.*

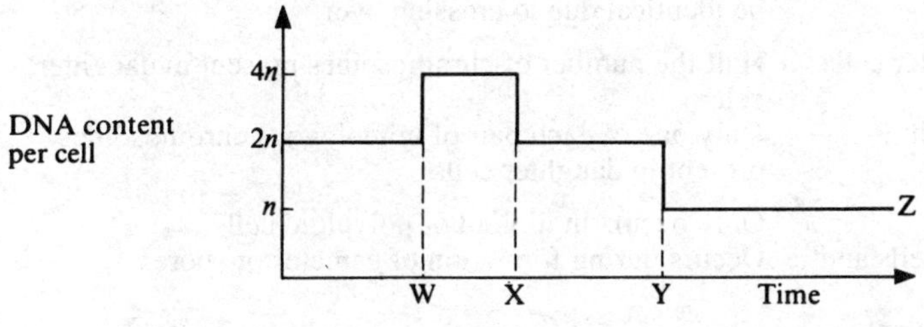

Fig 23.15 *Diagram for use in question 23.1.*

fig 23.11*k* where the four gametes produced as a result of meiosis are all different.

Variation is considered in more detail in section 24.8.4.

23.4.3 Comparison of mitosis and meiosis

The biologically significant differences between mitosis and meiosis are really between mitosis and meiosis I. Meiosis II is almost identical to mitosis. Therefore mitosis and meiosis I only are compared in table 23.2.

23.5 The structure of chromosomes

Analysis of chromosomes of eukaryotic cells has shown them to be composed of deoxyribonucleic acid (DNA) and protein, with small amounts of chromosomal RNA. (The 'chromosomes' of prokaryotic cells (bacteria) are composed of DNA only.) DNA has negative charges distributed along its length, and positively charged (basic) protein molecules called **histones** are bonded to it. This DNA–protein complex is called **chromatin**.

The large amount of DNA in cells means that there is a packaging problem. A human cell, for example, contains about 2.2 m of DNA distributed among 46 chromosomes. Each chromosome therefore contains about 4.8 cm (48 000 μm) of DNA. Human chromosomes are on average about 6 μm long, a packing ratio of 8000:1. In order to maintain a high degree of organisation when the DNA is folded, the histone proteins form a precise architectural 'scaffolding' for the DNA.

It has been shown that the DNA helix combines with groups of eight histone molecules to form structures known as **nucleosomes** having the appearance of 'beads on a string'. These nucleosomes, and the DNA strands linking them, are packed closely together to produce a 30 nm diameter helix with about six nucleosomes per turn. This is known as the 30 nm fibre, or the solenoid fibre. It has a packing ratio of about 40, that is 40 μm of DNA are packed into a 1 μm length of solenoid. The appearance of the solenoid fibres and the 'unpacked' solenoid ('beads on a string' form) are shown in fig 23.17.

Since the DNA must be even more tightly packed than this, the solenoids themselves must be folded or coiled in some way. How this is done is not yet known. The only clues at the moment come from a few examples of cells in which chromosomes have an unusual appearance. One example is in the amphibian oocyte (egg mother cell) which has 'lampbrush chromosomes', so-called because of their resemblance to brushes which were used to clean the glass of oil lamps. Electron micrographs of lampbrush chromosomes during metaphase show that each chromatid appears to be composed of a tightly coiled axis from which emerge several loops made up of a single DNA double helix (fig 23.18). These loops may represent the active DNA, that is DNA which has been exposed for the purpose of transcription (section 23.8.6).

The proposed structure of the chromosome is shown in fig 23.17.

Table 23.2 Comparison of mitosis and meiosis I.

	Mitosis	*Meiosis*
Prophase	Homologous chromosomes remain separate No formation of chiasmata No crossing over	Homologous chromosomes pair up Chiasmata form Crossing over may occur
Metaphase	Pairs of chromatids line up on the equator of the spindle	Pairs of chromosomes line up on the equator
Anaphase	Centromeres divide Chromatids separate Separating chromatids identical	Centromeres do not divide Whole chromosomes separate Separating chromosomes and their chromatids may not be identical due to crossing over
Telophase	Same number of chromosomes present in daughter cells as parent cells Both homologous chromosomes present in daughter cells if diploid	Half the number of chromosomes present in daughter cells Only one of each pair of homologous chromosomes present in daughter cells
Occurrence	May occur in haploid, diploid or polyploid cells Occurs during the formation of somatic (body) cells and some spores. Also occurs during the formation of gametes in plants	Only occurs in diploid or polyploid cells Occurs during formation of gametes or spores

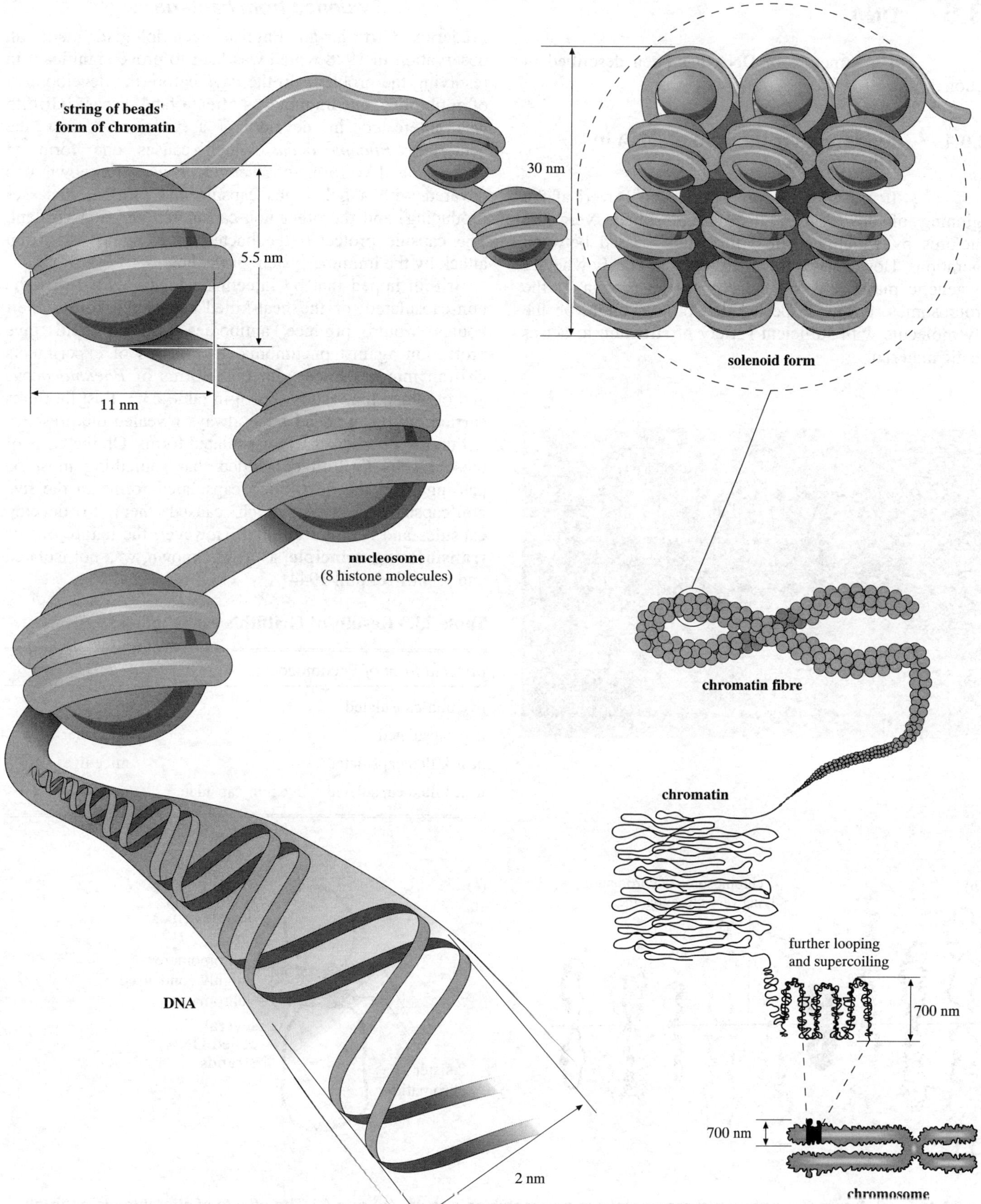

Fig 23.17 *Proposed structure of the nucleosome and its relationship to the chromosome and the DNA molecule.*

23.6 DNA

The structure of DNA has been described in section 3.6.3.

23.6.1 Evidence for the role of DNA in inheritance

It was proposed by Sutton and Boveri at the beginning of this century that chromosomes were the structures by which genetic information passed between generations. However, it took many years to clarify whether the genetic material was the DNA or the protein of the chromosomes. It was suspected that protein might be the only molecule with sufficient variety of structure to act as genetic material.

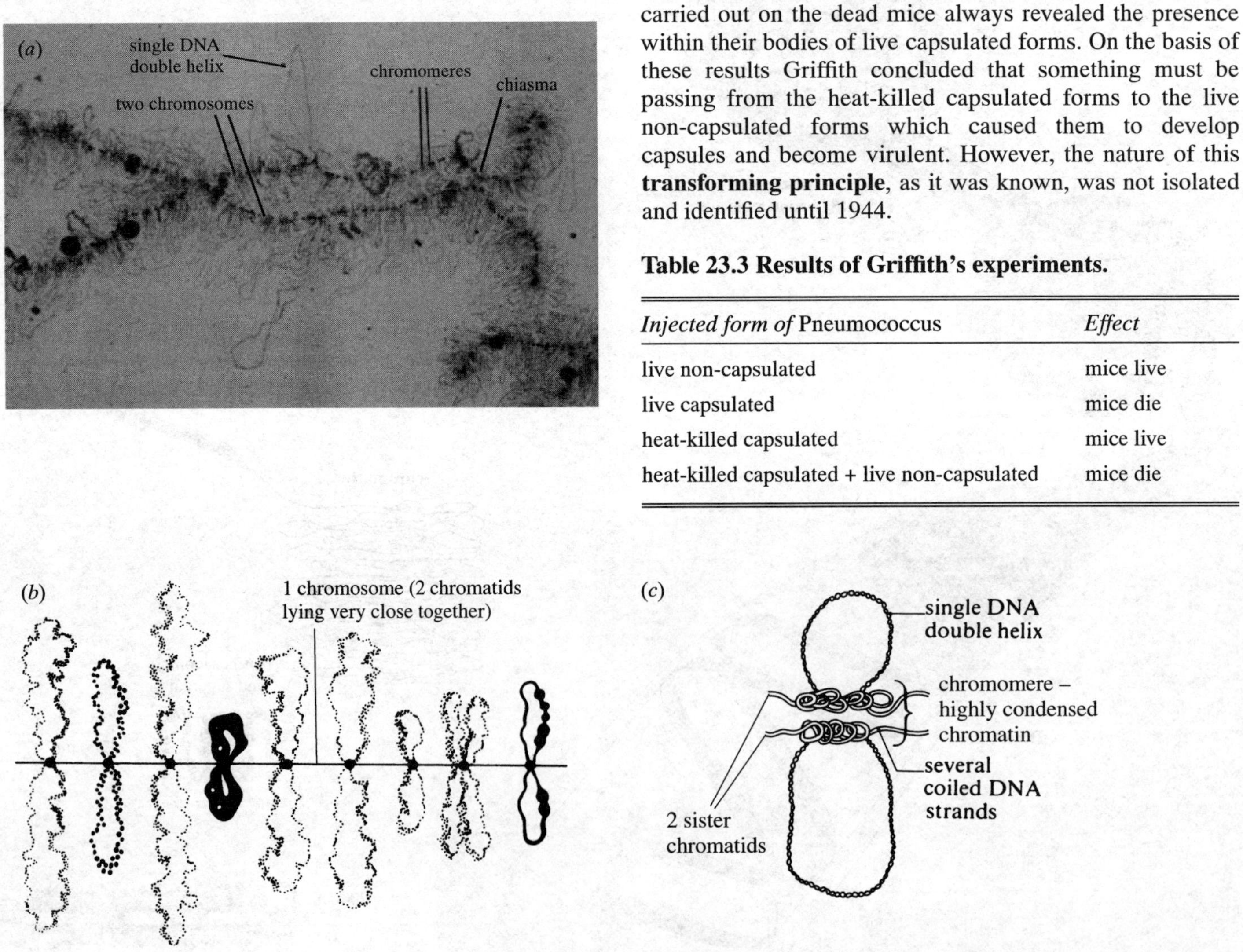

Fig 23.18 *(a) A pair of lampbrush chromosomes in an amphibian oocyte. (b) and (c) The effects of stretching lampbrush chromosomes to show the central filament of DNA and the loops of DNA where mRNA synthesis takes place. The dense regions are known as chromomeres. Each chromomere and its associated loop is thought to be associated with a specific gene. (From H. G. Callan (1963)* Int. Rev. Cytology, ***15****, 1.)*

Evidence from bacteria

Frederick Griffith, an English bacteriologist, made an observation in 1928 which was later to prove significant in resolving the problem. In the days before the development of antibiotics, pneumonia was often a fatal disease. Griffith was interested in developing a vaccine against the bacterium *Pneumococcus* which causes one form of pneumonia. Two forms of *Pneumococcus* were known, one covered with a gelatinous capsule and virulent (disease-producing) and the other non-capsulated and non-virulent. The capsule protected the bacterium in some way from attack by the immune system of the host.

Griffith hoped that by injecting patients with either the non-capsulated, or the heat-killed capsulated forms, their bodies would produce antibodies which would give protection against pneumonia. In a series of experiments Griffith injected mice with both forms of *Pneumococcus* and obtained the results shown in table 23.3. Post mortems carried out on the dead mice always revealed the presence within their bodies of live capsulated forms. On the basis of these results Griffith concluded that something must be passing from the heat-killed capsulated forms to the live non-capsulated forms which caused them to develop capsules and become virulent. However, the nature of this **transforming principle**, as it was known, was not isolated and identified until 1944.

Table 23.3 Results of Griffith's experiments.

Injected form of Pneumococcus	*Effect*
live non-capsulated	mice live
live capsulated	mice die
heat-killed capsulated	mice live
heat-killed capsulated + live non-capsulated	mice die

For ten years Avery, McCarty and McCleod analysed and purified the constituent molecules of heat-killed capsulated pneumococcal cells and tested their ability to bring about transformation in live non-capsulated cells. Removal of the polysaccharide capsule and the protein fraction from the cell extracts had no effect on transformation, but the addition of the enzyme deoxyribonuclease (DNase), which breaks down (hydrolyses) DNA, prevented transformation. The ability of extremely purified extracts of DNA from capsulated cells to bring about transformation finally demonstrated that Griffith's 'transforming principle' was in fact DNA. Despite this evidence many scientists still refused to accept that DNA, not protein, was the genetic material. In the early 1950s a wealth of additional evidence, based upon the study of viruses, eventually demonstrated that DNA is indeed the carrier of hereditary information.

Evidence from viruses

Viruses became one of the major experimental materials in genetic research in the 1940s. Virus particles have an extremely simple structure consisting of a protein coat surrounding a molecule of nucleic acid, either DNA or RNA (section 2.4.2). As such they provided ideal research material to investigate whether protein or nucleic acid is the genetic material. In 1952 Hershey and Chase began a series of experiments involving a particular type of virus which specifically attacks bacterial cells and is called a **bacteriophage**. Bacteriophage T_2 attacks the bacterium *Escherichia coli* (*E. coli*) which lives in the human gut. The phage causes *E. coli* to produce large numbers of T_2-phage particles in a very short time.

The essence of Hershey and Chase's experiment involved growing T_2-phage particles in *E. coli* which had been grown on a medium containing radioactive isotopes of either sulphur (^{35}S) or phosphorus (^{32}P). The phage protein contains sulphur but not phosphorus, and the DNA contains phosphorus but not sulphur. Therefore the phage particles formed in *E. coli* labelled with radioactive sulphur had incorporated this into their protein coats, whereas those formed in phosphorus-labelled *E. coli* contained radioactively labelled ^{32}P DNA.

The labelled T_2-phage particles were allowed to infect non-radioactively labelled *E. coli* and after a few minutes the cells were agitated in a blender or liquidiser which stripped off the phage particles from the bacterial walls. The bacteria were then incubated and examined for radioactivity. The results are shown in fig 23.19.

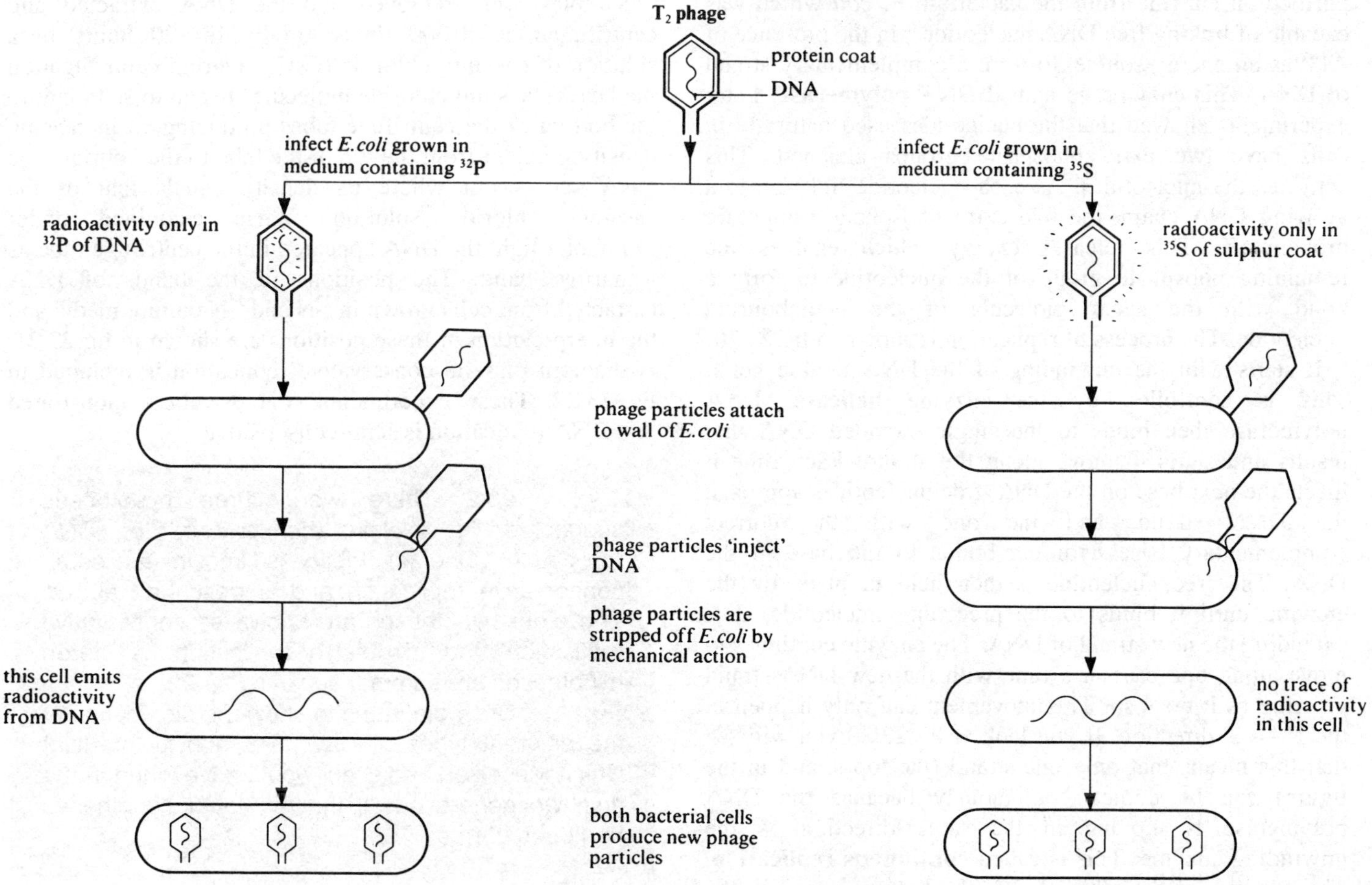

Fig 23.19 *Summary diagrams of Hershey and Chase's work on T_2 phage and E. coli.*

On the basis of these results Hershey and Chase concluded that it was the phage DNA and not the protein which entered the bacterial cell and gave rise to large numbers of phage progeny. These experiments demonstrated that DNA is the hereditary material. Confirmatory evidence that only the DNA contained within the phage is introduced into the bacterial cell has been provided by electron microscopy and increased knowledge of the life cycle of viruses. (The life cycle of virus and phage particles is described in sections 2.4.3 and 2.4.5.)

23.6.2 DNA replication

The double helical structure of DNA, as determined by Watson and Crick, is described in section 3.6.3. One of its most attractive features is that it immediately suggests a method by which replication could occur. Watson and Crick proposed that the two strands were capable of unwinding and separating, and acting as templates to which a complementary set of nucleotides would attach by base pairing. In this way each original DNA molecule would give rise to two copies with identical structures.

In 1956 Kornberg succeeded in demonstrating the synthesis of a DNA molecule in a test tube using a single strand of DNA as a template. Kornberg extracted and purified an enzyme from the bacterium *E. coli* which was capable of linking free DNA nucleotides, in the presence of ATP as an energy source, to form a complementary strand of DNA. This enzyme he named **DNA polymerase**. Later experiments showed that the nucleotides used naturally in cells have two extra phosphate groups attached. This activates the nucleotides. As each nucleotide links up to a growing DNA chain, the two extra phosphate groups are broken off. This releases energy which enables the remaining phosphate group of the nucleotide to form a bond with the sugar molecule of the neighbouring nucleotide. The process of replication is shown in fig 23.20.

It starts with the unwinding of the DNA double helix. This is controlled by the enzyme **helicase**. DNA polymerase then binds to the single stranded DNA that results and starts to move along the strand. Each time it meets the next base on the DNA, free nucleotides approach the DNA strand, and the one with the correct complementary base hydrogen-bonds to the base in the DNA. The free nucleotide is then held in place by the enzyme until it binds to the preceding nucleotide, thus extending the new strand of DNA. The enzyme continues to move along one base at a time with the new DNA strand growing as it does so. This movement can only happen in the 5' → 3' direction. If you look at fig 23.20 you will see that this means that only one strand (the top strand in the figure) can be copied continuously because the DNA polymerase is moving in the same direction as the unwinding enzyme. This is called **continuous replication**. The copying of the other strand (the bottom strand in the figure) has to keep being started again, because the DNA polymerase has to move away from the unwinding enzyme in the 5' → 3' direction. This results in small gaps being left because the DNA polymerase cannot join the 3' end of one newly synthesised piece of DNA to the 5' end of the next. Another enzyme, **DNA ligase**, is needed to close the gaps. This is called **discontinuous replication**.

Proof of semi-conservative replication

The method of DNA replication proposed by Watson and Crick and shown in fig 23.20 is known as **semi-conservative replication** since each new double helix retains (conserves) one of the two strands of the original DNA double helix. The evidence for this mechanism was provided by a series of classic experiments carried out by Meselson and Stahl in 1958. *E. coli* has a single circular chromosome, and when cultures of these cells were grown for many generations in a medium containing the heavy isotope of nitrogen ^{15}N all the DNA became labelled with ^{15}N. The cells containing DNA labelled with ^{15}N were transferred to a culture medium containing the normal isotope of nitrogen (^{14}N) and allowed to grow. The cells were then left for periods of time corresponding to the generation time for *E. coli* (50 min at 36 °C), that is the time needed for the cells to divide once and therefore for the DNA to replicate once.

Samples were removed and the DNA extracted and centrifuged at 40 000 times gravity for 20 hours in a solution of caesium chloride (CsCl). During centrifugation the heavy caesium chloride molecules began to sediment to the bottom of the centrifuge tubes producing an increasing density gradient from the top of the tube to the bottom. The DNA settles out where its density equals that of the caesium chloride solution. When examined under ultraviolet light the DNA appeared in the centrifuge tube as a narrow band. The positions of the bands of DNA extracted from cells grown in ^{15}N and ^{14}N culture media and the interpretation of these positions are shown in fig 23.21. A diagram of semi-conservative replication is included in fig 23.22. These experiments conclusively demonstrated that DNA replication is semi-conservative.

23.2 There were three hypotheses suggested to explain the process of DNA replication. One of these is known as semi-conservative replication and is described above. The other hypotheses are known as conservative replication and dispersive replication. All three hypotheses are summarised in fig 23.22.

Draw diagrams to show the distribution of the different types of DNA in a density gradient which Meselson and Stahl would have found in the first two generations if the two latter hypotheses had been correct.

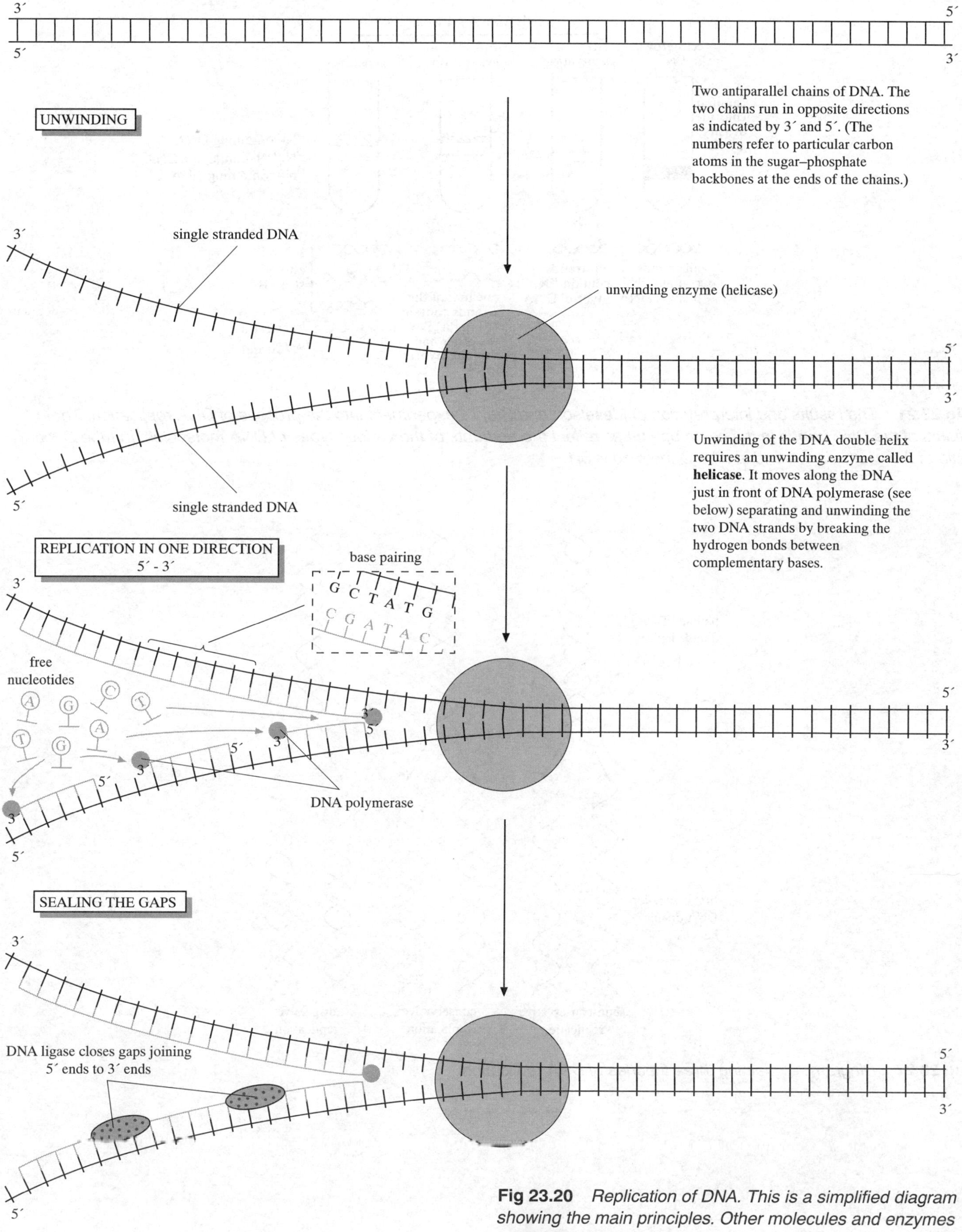

Fig 23.20 *Replication of DNA. This is a simplified diagram showing the main principles. Other molecules and enzymes are involved which are not shown.*

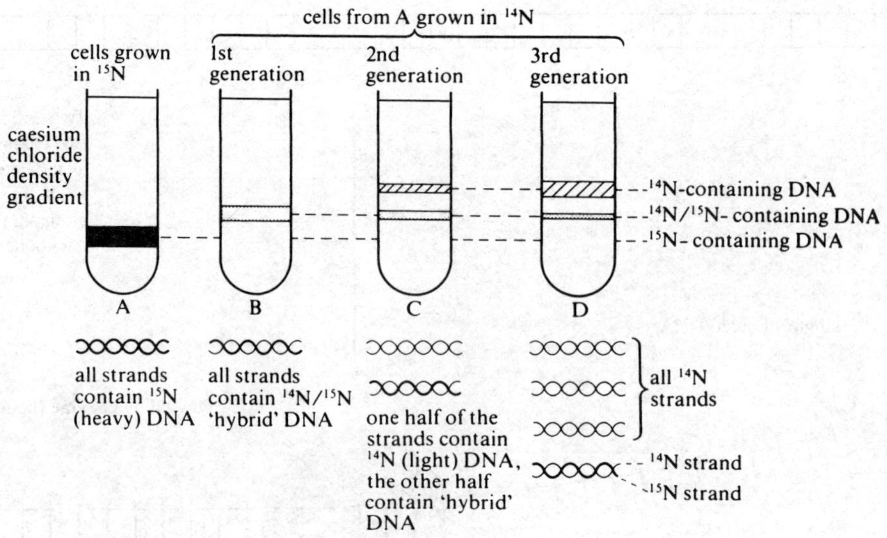

Fig 23.21 *The results and interpretation of Meselson and Stahl's experiment into the process of DNA replication. The widths of the DNA bands in the centrifuge tubes reflect the amounts of the various types of DNA molecules. In tube C the ratio of the widths is 1:1 and in tube D the ratio is 3:1.*

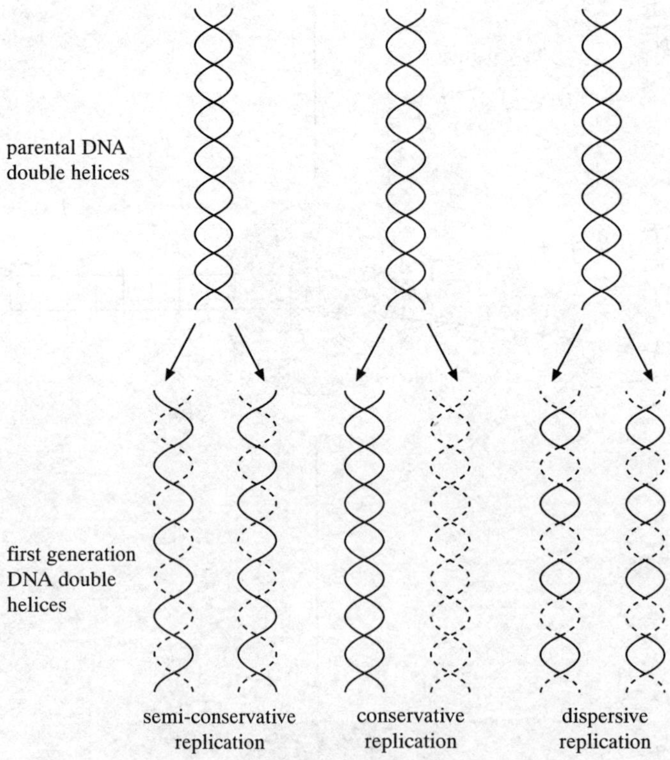

Fig 23.22 *Diagrams explaining three theories of DNA replication.*

23.7 The nature of genes

23.7.1 What are genes?

Mendel proposed in 1866 that the characteristics of organisms were determined by hereditary units which he called 'elementes'. These were later termed **genes** and shown to be located on chromosomes which transmitted them from generation to generation. Mendel would therefore have defined a gene as a **unit of inheritance**. This is a perfectly acceptable definition but it does not tell us anything about the physical nature of the gene.

Two possible ways of overcoming this objection are considered below.

A unit of recombination. From his studies of chromosome mapping in *Drosophila* (section 24.2), Morgan postulated that a gene was **the shortest segment of a chromosome which could be separated from adjacent segments by crossing-over**. This definition regards the gene as a specific region of the chromosome determining a distinct characteristic in the organism.

A unit of function. Since genes are known to determine the structural, physiological and biochemical characteristics of organisms it has been suggested that a gene can be defined in terms of its function. Originally it was proposed that a gene was **the shortest segment of a chromosome responsible for the production of a specific product**. We now know that genes are the codes for proteins. A gene could therefore be defined as **a piece of DNA which codes for a protein**. This definition can be made even more precise by stating that a gene is **the DNA code for a polypeptide** since some proteins are made up of more than one polypeptide chain and are therefore coded for by more than one gene.

23.7.2 The genetic code is a sequence of bases

When Watson and Crick proposed the double helical structure for DNA in 1953 they also suggested that the genetic information which passed from generation to generation, and which controlled the activities of the cell, might be stored in the form of the sequence of bases in the DNA molecule. Once it had been shown that DNA is a code for the production of protein molecules it became clear that the sequence of bases in the DNA must be a code for the sequence of amino acids in protein molecules. This relationship between bases and amino acids is known as the **genetic code**. The problems remaining in 1953 were to demonstrate that a base code existed, to break the code and to determine how the code is translated into the amino acid sequence of a protein molecule.

23.7.3 The code is a triplet code

There are four bases in the DNA molecule, **adenine** (A), **guanine** (G), **thymine** (T) and **cytosine** (C) (section 3.6). Each base is part of a nucleotide and the nucleotides are arranged as a polynucleotide strand. The sequence of bases in the strand can be indicated by the initial letters of the bases. This 'alphabet' of four letters is responsible for carrying the code that results in the synthesis of a potentially infinite number of different protein molecules. There are 20 common amino acids used to make proteins and that the bases in the DNA must code for. If one base determined the position of a single amino acid in the primary structure of a protein, the protein could only contain four different amino acids. If a combination of pairs of bases coded for each amino acid then 16 amino acids could be specified into the protein molecule.

23.3 Using different pairs of the bases A, G, T and C list the 16 possible combinations of bases that can be produced.

Only a code composed of three bases could incorporate all 20 amino acids into the structure of protein molecules. Such a code would produce 64 combinations of bases, more than enough. Watson and Crick therefore predicted that the code would be a triplet code.

23.4 If four bases used singly would code for four amino acids, pairs of bases code for 16 amino acids and triplets of bases code for 64 amino acids, deduce a mathematical expression to explain this.

It was later proved that the code is indeed a triplet code, meaning that three bases is the code for one amino acid.

Evidence for a triplet code

Evidence that the code is a triplet code was provided by Francis Crick in 1961. He produced mutations involving the addition or deletion of bases in T_4 phages. Adding or deleting a base changes the way in which the code is read after the point of addition or deletion, as explained in fig 23.23. The mutation is said to produce a 'frame-shift'. These frame-shifts produced base triplet sequences which failed to result in the synthesis of protein molecules with the original amino acid sequence. Only by adding a base and deleting a base at specific points could the original base sequence be restored. Restoring the original base sequence prevented the appearance of mutants in the experimental T_4 phages. Adding a single base is referred to as a (+) type mutation and deleting a base a (–) type. (+)(–) restores the

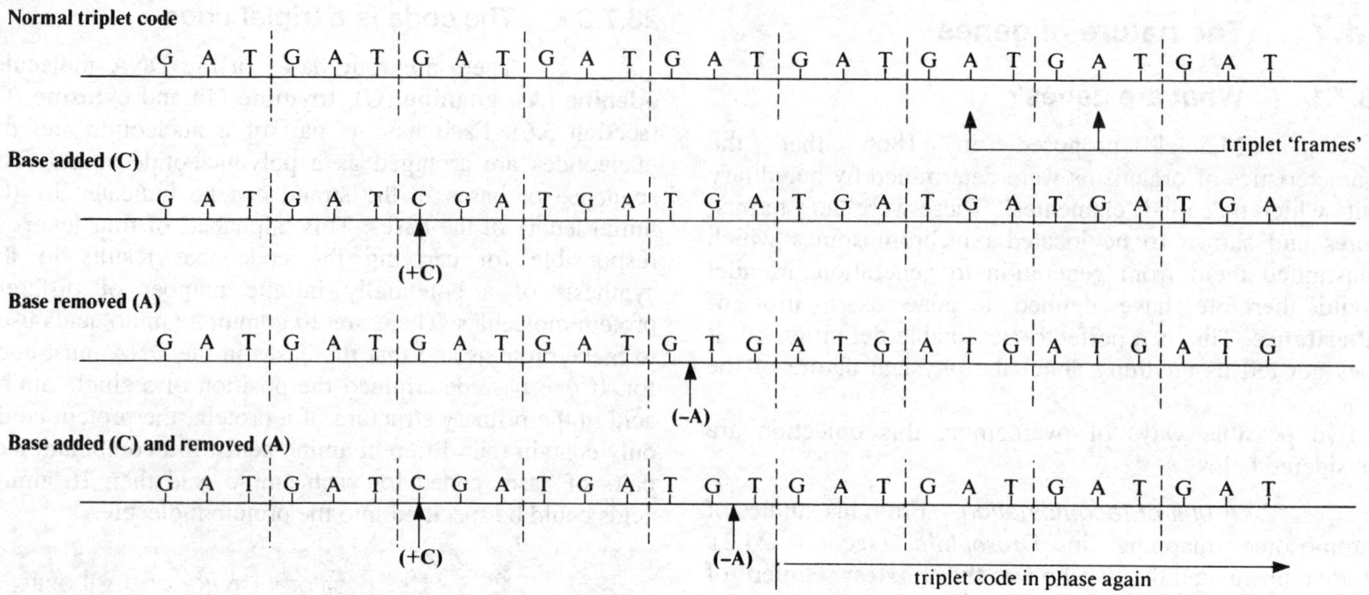

Fig 23.23 *Diagrammatic explanation of the effect of adding and deleting bases to the triplet code. The addition of the base C produces a frame-shift which makes the original message GAT, GAT, . . . read as TGA, TGA, . . . The deletion of the base A produces a frame shift changing the original message from GAT, GAT, . . . to ATG, ATG, . . . The addition of the base C at the point indicated and the deletion of the base A at the point indicated restores the original message GAT, GAT. (After F. H. C. Crick (1962)* The genetic code I, *Scientific American Offprint No. 123, Wm. Saunders & Co.)*

correct reading. The double mutants (++) or (––) also produced frame-shifts which resulted in mutants which produced faulty proteins. However (+++) or (–––) mutations usually had no effect on protein function. Crick argued that this is because such mutations do not cause frame-shifts, only the addition or deletion of one amino acid which often does not affect the performance of a protein. This implies that the code is read three bases at a time, that is in triplets.

These experiments also demonstrated that the code is **non-overlapping**, that is to say no base of a given triplet contributes to part of the code of the adjacent triplet (fig 23.24).

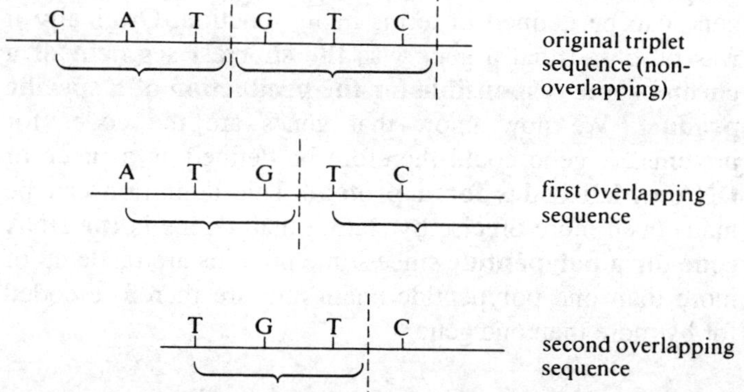

Fig 23.24 *Base sequences indicating non-overlapping and overlapping codes.*

> **23.5** Using repeated sequences of the triplet GTA and the base C show that the sequence of triplets can only be restored by adding or deleting three bases. (Set out your answer as in fig 23.24.)

23.7.4 Breaking the code

Having established that the code was a triplet code, it remained to find out which triplets coded for which amino acids, in other words to break the code. In order to understand the experimental procedures used it is necessary to appreciate, in outline, the mechanism by which the triplet code is translated into a protein molecule.

Protein synthesis involves two types of nucleic acid, deoxyribonucleic acid, DNA and ribonucleic acid, RNA. There are three kinds of RNA: messenger RNA (**mRNA**), ribosomal RNA (**rRNA**) and transfer RNA (**tRNA**). The DNA base sequence is copied (**transcribed**) on to strands of messenger RNA (mRNA) which leave the nucleus. These become attached to ribosomes in the cytoplasm where the base sequence of mRNA is **translated** into an amino acid sequence. Specific amino acids become attached to tRNA molecules which link up with complementary triplet bases on the mRNA. Adjacent amino acids brought together in this way react together to form a polypeptide chain. The process of protein synthesis, therefore, depends upon the presence of DNA, mRNA, ribosomes, tRNA, amino acids, ATP as an energy source and various enzymes and cofactors which catalyse each stage in the process.

Nirenberg used this information and various research techniques which had been developed during the late 1950s to design a series of experiments to break the code. The essence of his experiments involved using a known base sequence of mRNA as a coded message and analysing the amino acid sequence of the polypeptide chain produced from it. Nirenberg was able to synthesise a mRNA molecule that consisted of the same triplet (UUU) repeated many times. This was called **polyuridylic acid** (poly-U) and acted as a code. A series of 20 test-tubes was prepared, each containing cell-free extracts of *E. coli* including ribosomes, tRNA, ATP, enzymes and a different radioactive labelled amino acid. Poly-U was added to each test-tube and time was allowed for synthesis of polypeptides to occur. Analysis of the contents of the test-tubes showed that a polypeptide had been formed only in the test-tube containing the amino acid phenylalanine. Thus the genetic code had been partly solved. Nirenberg had shown that the base triplet of the mRNA, or **codon**, UUU determines the position of phenylalanine in a polypeptide chain. Nirenberg and his co-workers then began preparing synthetic polynucleotide molecules of all 64 possible codons and by 1964 had translated the codes for all 20 amino acids (table 23.4).

23.7.5 Features of the genetic code

The code is a triplet code

As already stated, the genetic code is a triplet code, meaning that three bases in DNA code for one amino acid in a protein. The DNA code for a protein is first copied into messenger RNA (mRNA) before a protein is made. mRNA is complementary to the DNA. The complementary triplets in the mRNA are referred to as **codons**. Each codon is therefore three bases long and is the code for one amino acid. The DNA code for each amino acid can be obtained by converting the RNA codons back into their complementary DNA triplets of bases according to the rules shown in table 23.5.

Table 23.4 The base sequences of the triplet code and the amino acids for which they code.

NB these are **codons**, i.e. base sequences of mRNA and not DNA. The DNA genetic code would have complementary bases and T would replace U.

First base	Second base U		Second base C		Second base A		Second base G		Third base
U	UUU	phe	UCU	ser	UAU	tyr	UGU	cys	U
	UUC		UCC		UAC		UGC		C
	UUA	leu	UCA		UAA	c.t.*	UGA	c.t.*	A
	UUG		UCG		UAG	c.t.*	UGG	trp	G
C	CUU	leu	CCU	pro	CAU	his	CGU	arg	U
	CUC		CCC		CAC		CGC		C
	CUA		CCA		CAA	gln	CGA		A
	CUG		CCG		CAG		CGG		G
A	AUU	ileu	ACU	thr	AAU	asn	AGU	ser	U
	AUC		ACC		AAC		AGC		C
	AUA		ACA		AAA	lys	AGA	arg	A
	AUG	met	ACG		AAG		AGG		G
G	GUU	val	GCU	ala	GAU	asp	GGU	gly	U
	GUC		GCC		GAC		GGC		C
	GUA		GCA		GAA	glu	GGA		A
	GUG		GCG		GAG		GGG		G

*c.t., chain termination codon, equivalent to a full stop in the message.

Table 23.5 The RNA bases which are complementary to those of DNA.

DNA bases	*Complementary RNA bases*
A (adenine)	U (uracil)
G (guanine)	C (cytosine)
T (thymine)	A (adenine)
C (cytosine)	G (guanine)

23.6 Write out the base sequence of mRNA formed from a DNA strand with the following sequence.

ATGTTCGAGTACCATGTAACG

The code is degenerate

Table 23.4 shows the genetic code in the form of codons. As can be seen from the table some amino acids are coded for by several codons. This type of code where the number of amino acids is less than the number of codons is termed **degenerate**. Analysis of the code also shows that for many amino acids only the first two letters appear to be significant.

The code is punctuated

Three of the codons shown in table 23.4 act as 'full stops' in determining the end of the code message. An example is UAA. They are sometimes described as 'nonsense codons' and do not code for amino acids. They presumably mark the end-point of a gene. They act as 'stop signals' for the termination of polypeptide chains during translation.

Certain codons act as 'start signals' for the initiation of polypeptide chains, such as AUG (methionine).

The code is universal

One of the remarkable features of the genetic code is that it is thought to be universal. All living organisms contain the same 20 common amino acids and the same five bases, A, G, T, C and U.

Advances in molecular biology have reached the point now where it is possible to determine the base sequences for whole genes and for whole organisms. The first organism whose complete genetic code was established was a virus, the phage ΦX174. The phage has only ten genes and its complete genetic code is 5386 bases long. The sequence was discovered by Fred Sanger, the man who first discovered the sequence of amino acids in a protein. He was awarded Nobel prizes for both these sequencing milestones. Whole genes can now be synthesised artificially, a practice which is of use in genetic engineering. Soon after the beginning of the twenty-first century it is anticipated that it will be possible to write out the entire genetic code of a human, an estimated 3000 million base pairs long, as a result of the Human Genome Project. (The genome is the total DNA in an organism.) Other organisms whose genomes are being sequenced are *E. coli*, yeast, the fruit fly *Drosophila*, a nematode worm and the laboratory mouse.

Summary

The main features of the genetic code are summarised below.

- A **triplet** of bases in the polynucleotide chain of DNA is the code for one amino acid in a polypeptide chain.
- It is **universal**: the same triplets code for the same amino acids in all organisms. (A few triplet codes in mitochondrial DNA and some ancient bacteria differ from the 'universal code'.)
- It is **degenerate**: a given amino acid may be coded for by more than one codon.
- It is **non-overlapping**: for example, an mRNA sequence beginning AUGAGCGCA is not read AUG/UGA/GAG ... (an overlap of two bases) or AUG/GAG/GCG ... (an overlap of one base). (However, overlapping of certain genes does occur in a few organisms such as the bacteriophage ΦX174. This seems likely to be exceptional and may be an economy measure when there are very few genes.)

23.8 Protein synthesis

'DNA makes RNA and RNA makes protein'

The information given so far in this chapter has shown that although DNA controls the activities of cells, the only molecules capable of being synthesised directly from DNA are proteins. These may have a structural role, such as keratin and collagen, or a functional role, such as insulin, fibrinogen and most importantly enzymes, which are responsible for controlling cell metabolism. It is the particular range of enzymes in the cell which determines what type of cell it becomes. This is the way in which DNA controls the activities of a cell.

The 'instructions' for the manufacture of enzymes and all other proteins are located in the DNA, which is found in the nucleus. However, it was shown in the early 1950s that the actual synthesis of proteins occurs in the cytoplasm at the ribosomes. Therefore a mechanism had to exist for carrying the genetic information from nucleus to cytoplasm. In 1961 two French biochemists, Jacob and Monod, suggested that the link was a specific form of RNA which they called **messenger RNA** (mRNA). This idea later proved to be correct. The sequence of events during protein synthesis was summarised by the slogan 'DNA makes RNA and RNA makes protein'.

23.8.1 The role of RNA

RNA exists as a single-stranded molecule in all living cells. It differs from DNA in possessing the pentose sugar ribose instead of deoxyribose and the pyrimidine uracil instead of thymine. Analysis of the RNA content of cells has shown the existence of three types of RNA which are all involved in the synthesis of protein molecules. These are messenger RNA (mRNA), transfer RNA (tRNA) and ribosomal RNA (rRNA). All three types are synthesised directly on DNA, and the amount of RNA in each cell is directly related to the amount of protein synthesis.

23.8.2 Messenger RNA

Analyses of cells have shown that 3–5% of the total RNA of the cell is mRNA. This is a single-stranded molecule formed on a single strand of DNA by a process known as **transcription**. In the formation of mRNA only one strand of the DNA molecule is copied. The synthesis of mRNA is described later. The base sequence of mRNA is a complementary copy of the DNA strand being copied and varies in length according to the length of the polypeptide chain for which it codes. Most mRNA exists within the cell for a short time. In the case of bacteria this may be a matter of minutes whereas in developing red blood cells the mRNA may continue to produce haemoglobin for several days.

23.8.3 Ribosomal RNA

Ribosomal RNA makes up approximately 80% of the total RNA of the cell. It is synthesised by genes present on the DNA of several chromosomes found within a region of the nucleolus known as the **nucleolar organiser**. The base sequence of rRNA is similar in all organisms from bacteria to higher plants and animals. It is found in the cytoplasm where it is associated with protein molecules which together form the cell organelles known as ribosomes (see section 5.10.4).

Ribosomes are the sites of protein synthesis. Here the mRNA 'code' is **translated** into a sequence of amino acids in a polypeptide chain.

23.8.4 Transfer RNA

The existence of transfer RNA (tRNA) was postulated by Crick and demonstrated by Hoagland in 1955. Each amino acid has its own family of tRNA molecules. tRNA transfers amino acids present in the cytoplasm to the ribosome. Consequently it acts as an intermediate molecule between the triplet code of mRNA and the amino acid sequence of the polypeptide chain. It makes up about 15% of the total RNA of the cell and, having on average 80 nucleotides per molecule, it is the smallest of all the RNAs. There are more than 20 different tRNA molecules in a given cell (60 have so far been identified) carrying specific amino acids. All tRNA molecules have the same basic structure (see fig 23.25).

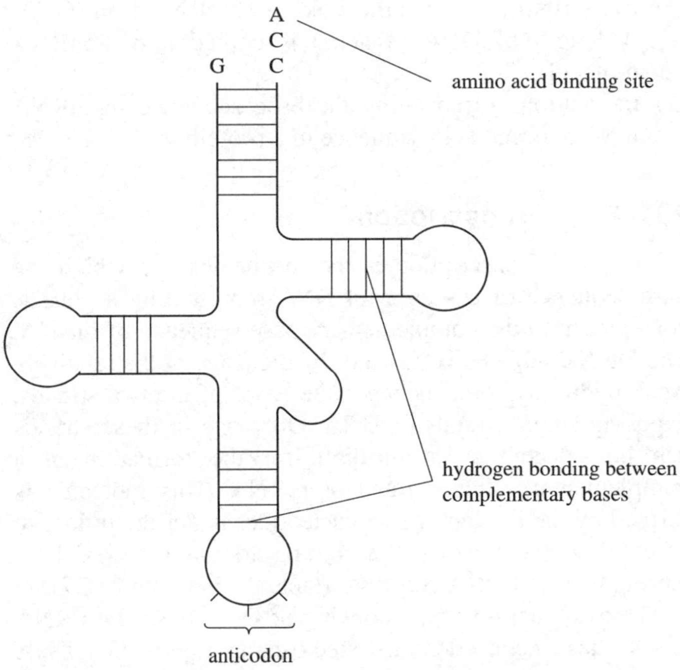

Fig 23.25 *A proposed model for the structure of transfer RNA (RNA). The whole molecule is composed of about 80 nucleotides but only 21 show complementary base pairing.*

The 5'-end of the tRNA always ends in the base guanine whilst the 3'-end always ends in the base sequence of CCA. The base sequence of the rest of the molecule is variable and may include some 'unusual bases' such as inosine (I) and pseudouracil (y). The triplet base sequence at the anticodon (fig 23.25) is directly related to the amino acid carried by that tRNA molecule. Each amino acid is attached to its specific tRNA by its own form of the enzyme **aminoacyl–tRNA synthetase**. This produces an amino acid–tRNA complex known as **aminoacyl–tRNA** with sufficient energy in the bond between the final A nucleotide of CCA and the amino acid to later form a peptide bond with the adjacent amino acid. In this way a polypeptide chain is synthesised.

23.8.5 Summary of protein synthesis

Protein synthesis is a two-stage process, as summarised in fig 23.26:

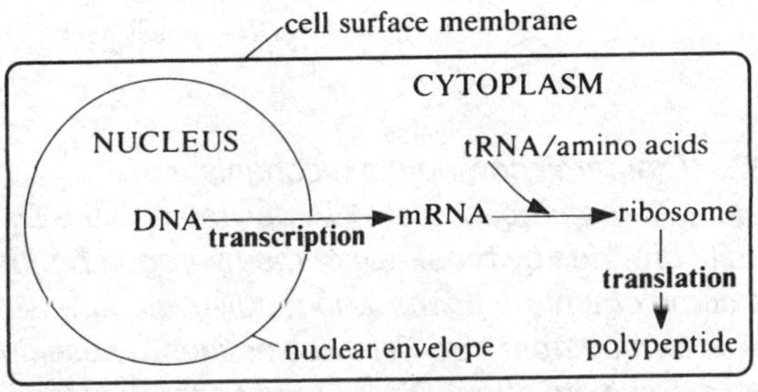

Fig 23.26 *Summary diagram of the main steps involved in protein synthesis.*

(1) transcription – this is the making of mRNA from DNA. A length of DNA (a gene) is copied into a mRNA molecule.
(2) translation – translating the base sequence in mRNA into an amino acid sequence in a protein.

23.8.6 Transcription

Transcription is the mechanism by which the base sequence of a section of DNA representing a gene is converted into the complementary base sequence of mRNA. The DNA double helix unwinds by breakage of the relatively weak hydrogen bonds between the bases of the two strands, exposing single strands of DNA. Only one of these strands can be selected as a **template** for the formation of a complementary single strand of mRNA. This molecule is formed by the linking of free nucleotides under the influence of RNA polymerase and according to the rules of base pairing between DNA and RNA (table 23.5 and fig 23.27).

The exact nature of the copying of DNA bases into RNA bases has been demonstrated using synthetic DNA composed solely of thymine nucleotides (TTT). When introduced into a cell-free system containing RNA polymerase and all four nucleotides (A, U, C and G) the messenger RNA formed was composed entirely of complementary adenine nucleotides.

When the mRNA molecules have been synthesised they leave the nucleus via the nuclear pores and carry the genetic code to the ribosomes. When sufficient numbers of mRNA molecules have been formed from the gene the RNA polymerase molecule leaves the DNA and the two strands 'zip up' again, re-forming the double helix.

23.8.7 Translation

Translation is the mechanism by which the sequence of bases in a mRNA molecule is converted into a sequence of amino acids in a polypeptide chain. It occurs on ribosomes. Several ribosomes may become attached to a molecule of mRNA like beads on a string and the whole structure is known as a polyribosome or **polysome**. These structures can be seen under the electron microscope (fig 23.28). The advantage of such an arrangement is that it allows several polypeptides to be synthesised at the same

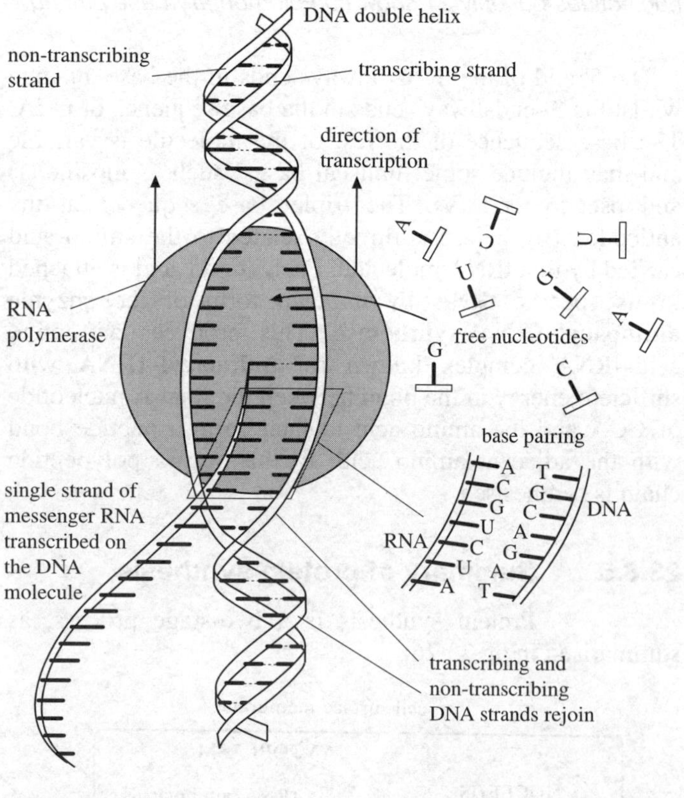

Fig 23.27 *Diagram showing the mechanism of transcription. In the presence of RNA polymerase the DNA double helix unwinds by breakage of the hydrogen bonds between complementary bases, and a polynucleotide strand of mRNA is formed from free RNA nucleotides. These line up opposite complementary DNA bases on the transcribing strand of template DNA. (Modified from E. J. Ambrose & D. M. Easty (1977)* Cell biology, *2nd ed., Nelson.)*

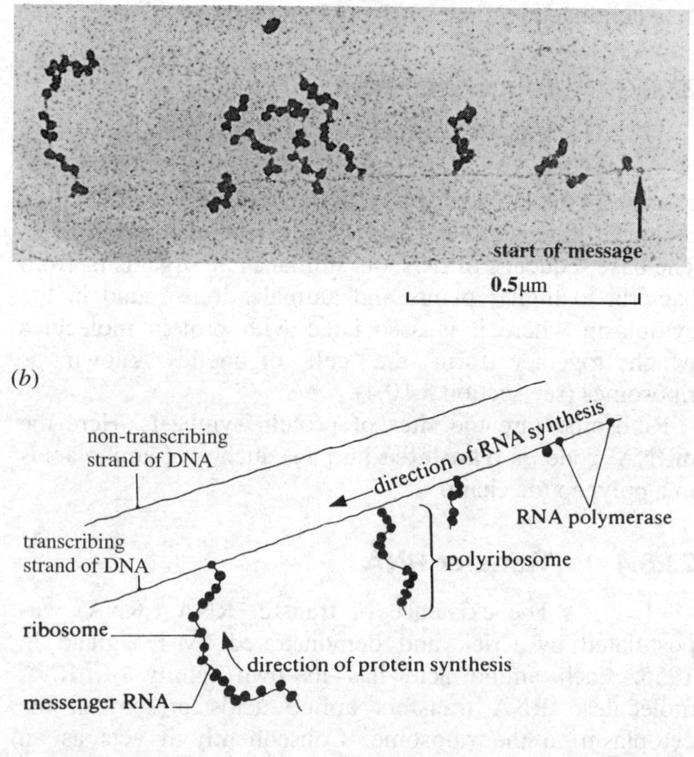

Fig 23.28 *Transcription and formation of a polysome in bacteria. Note that the mRNA does not need to leave the DNA in bacteria because there is no nucleus. (a) Electron micrograph of a piece of bacterial DNA showing stages in the development of mRNA and the attachment of ribosomes. (b) Diagrammatic representation of the structure shown in the electron micrograph in (a).*

time (section 23.8.3). Each ribosome is composed of a small and a large subunit, resembling a 'cottage loaf' (fig 5.27). The first two mRNA codons (a total of 6 bases) enter the ribosome as shown in fig 23.29*a*. The first codon binds the aminoacyl–tRNA molecule having the complementary anticodon and which is carrying the first amino acid (usually methionine) of the polypeptide being synthesised. The second codon then also attracts an aminoacyl–tRNA molecule showing the complementary anticodon (figs 23.29*a* and *b*). The function of the ribosome is to hold in position the mRNA, tRNA and the associated enzymes controlling the process until a peptide bond forms between the adjacent amino acids.

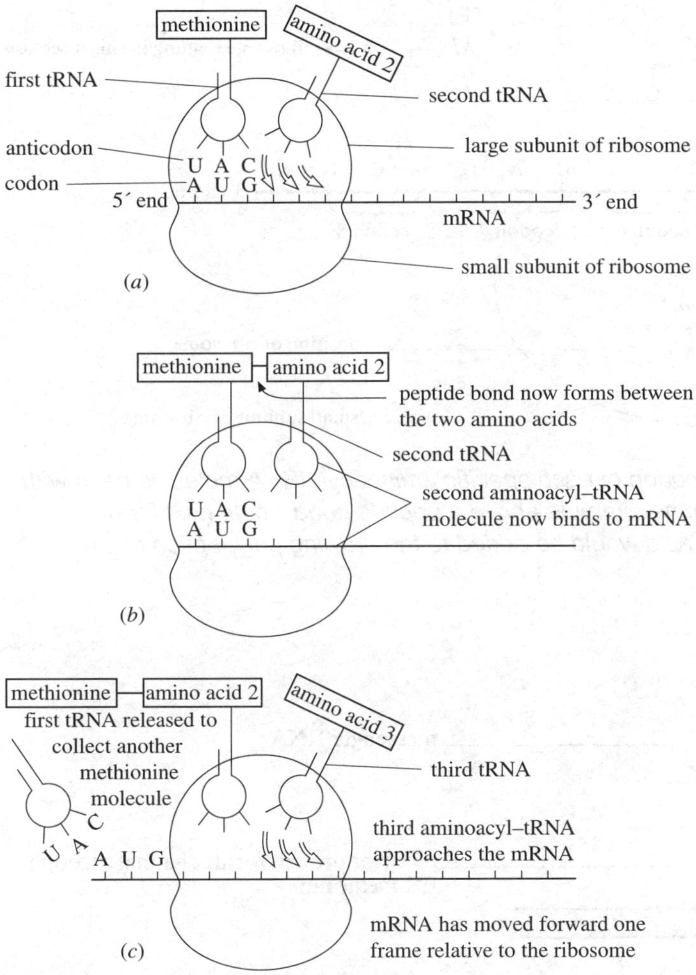

Fig 23.29 *(a) and (b) Stages in the attachment of aminoacyl–tRNA molecules by their anticodons to the codons on mRNA and the formation of a peptide bond between adjacent amino acids. (c) The relative movements of mRNA and ribosome exposing a new triplet (frame) for the attachment of the aminoacyl–tRNA molecule. The initial tRNA molecule is now released from the ribosome and cycles back into the cytoplasm to be reactivated by enzymes to form a new aminoacyl–tRNA molecule.*

Once the new amino acid has been added to the growing polypeptide chain the ribosome moves one codon along the mRNA. The tRNA molecule which was previously attached to the polypeptide chain now leaves the ribosome and passes back to the cytoplasm to be reconverted into a new aminoacyl–tRNA molecule (fig 23.29*c*).

This sequence of the ribosome 'reading' and 'translating' the mRNA code continues until it comes to a codon signalling 'stop'. These terminating codons are UAA, UAG and UGA. At this point the polypeptide chain, now with its primary structure as determined by the DNA, leaves the ribosome and translation is complete. The main steps involved in translation may be summarised under the following headings:

(1) binding of mRNA to ribosome,
(2) amino acid activation and attachment to tRNA,
(3) polypeptide chain initiation,
(4) chain elongation,
(5) chain termination,
(6) fate of mRNA,

and the process is summarised in fig 23.30.

As the polypeptide chains leave the ribosome they may immediately assume either secondary, tertiary or quaternary structures (section 3.5.3). If the ribosome is attached to ER (rough ER) the protein enters the ER to be transported. In such a situation the first part of the growing chain of amino acids consists of a 'signal sequence' of amino acids which fit a specific receptor in the ER membrane, thus binding the ribosome to the ER. The growing protein passes through the receptor into the ER as shown in fig 23.31. Once inside, the signal sequence is removed and the protein folds up into its final shape.

Evidence that it is the complementary base pairing between the mRNA codon and the tRNA anticodon which determines the incorporation of an amino acid into the polypeptide chain, and not the amino acid, was demonstrated by the following experiment. The tRNA–cysteine molecule normally pairs up, via its anticodon ACA, with the mRNA codon UGU. Exposure of this tRNA–cysteine molecule to a catalyst, Raney nickel, converted the cysteine to the amino acid alanine. When the new tRNA–alanine molecule (carrying the tRNA–cysteine anticodon) was placed in a cell-free system containing poly-UGU-mRNA the polypeptide chain formed contained only alanine. This experiment demonstrated the importance of the role of the mRNA–codon–tRNA–anticodon mechanism in translating the genetic code.

The whole sequence of protein synthesis occurs as a continuous process and is summarised in fig 23.32.

23.8.8 Non-coding DNA

Human DNA contains about 3000 million base pairs and an estimated 100 000 genes, although the number of genes can only be a rough guess at present. The problem

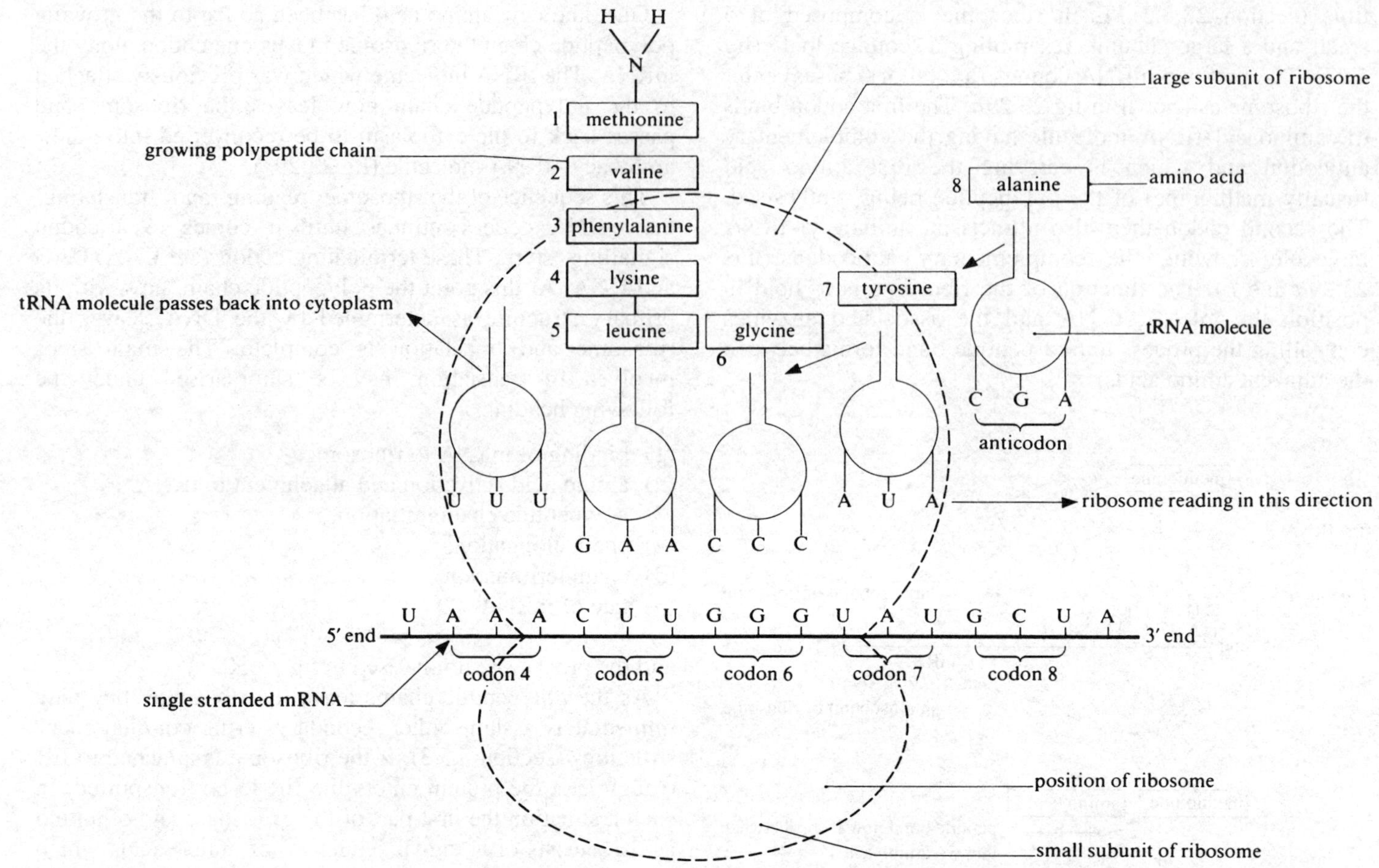

Fig 23.30 *Diagrammatic representation of translation. The anticodon of each specific aminoacyl–tRNA molecule pairs with its complementary bases of the mRNA codon in the ribosome. In the example above a peptide bond would next form between leucine and glycine and in this way an additional amino acid would be added to the growing polypeptide chain.*

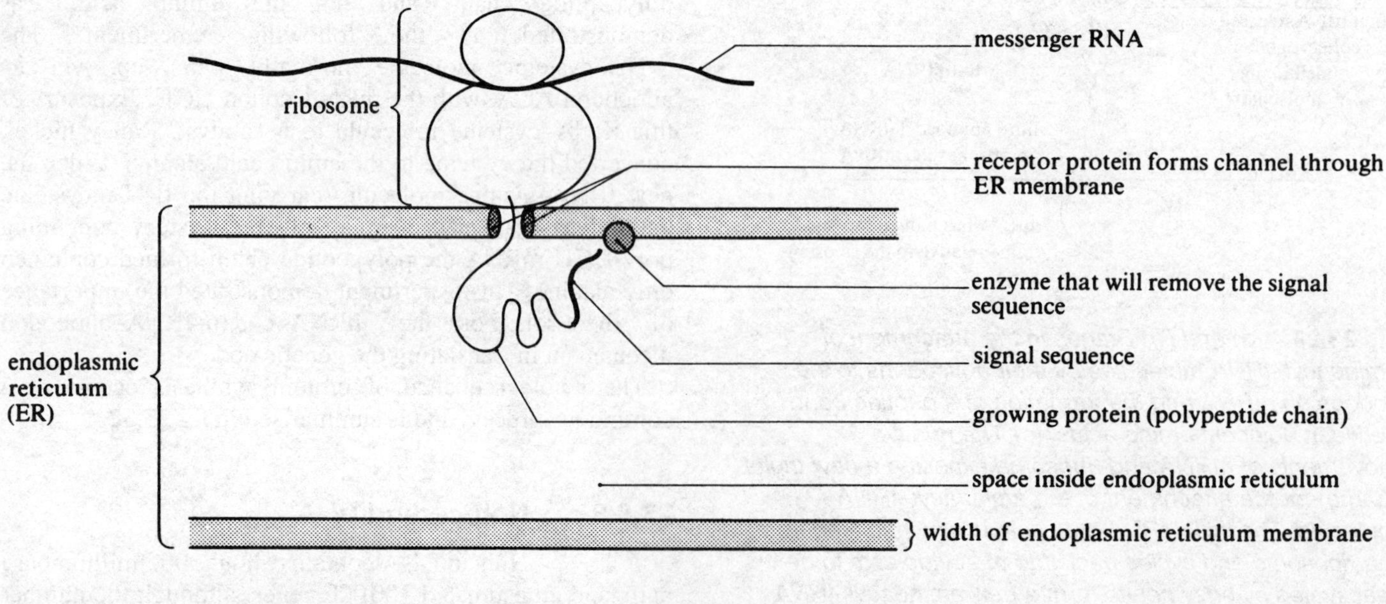

Fig 23.31 *Entry of newly synthesised protein into the endoplasmic reticulum.*

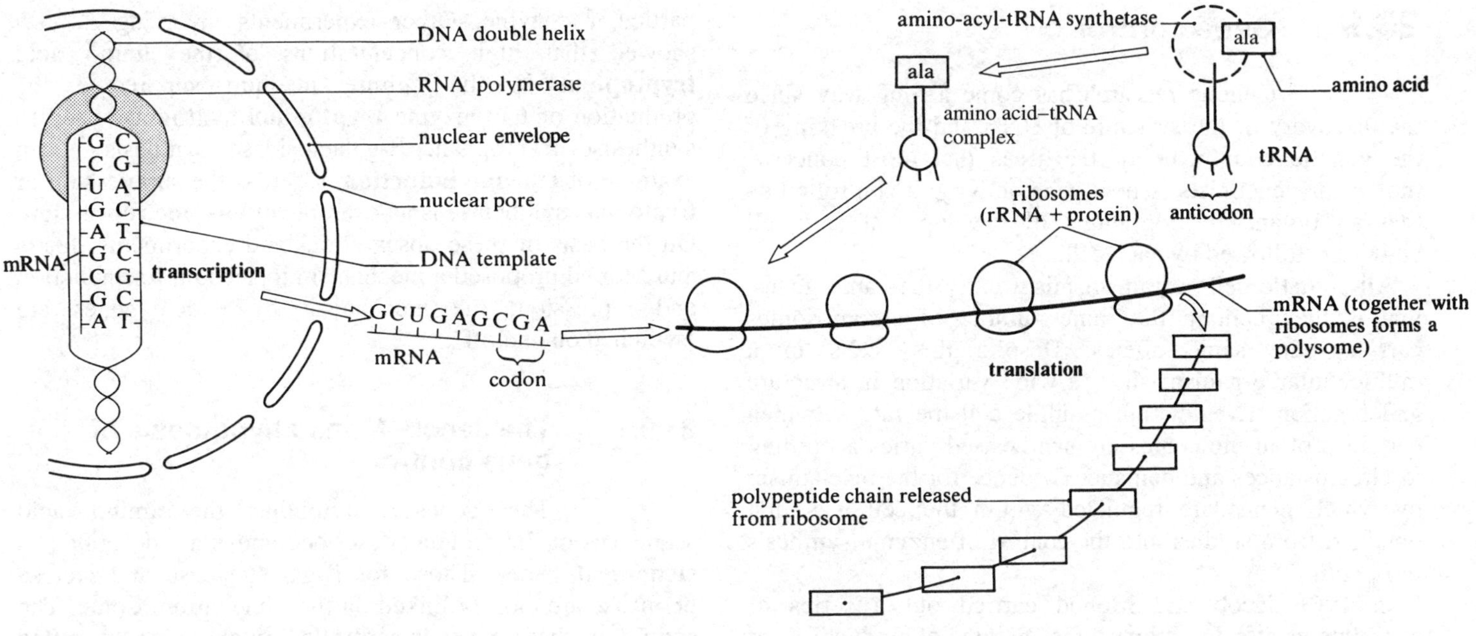

Fig 23.32 *Simplified summary diagram of the major structures and processes involved in protein synthesis in the cell.*

is that about 95% of the DNA appears to have no obvious function because it is non-coding. In other words it does not code for proteins or RNA. It has sometimes been referred to as 'junk DNA', although it is unwise to assume that something has no function just because its function is unknown. Some of it may be former genes which no longer serve any useful purpose. Some is probably structural, involved perhaps in packaging of chromosomes, for example. About 30–40% consists of short base sequences which are repeated many times. This includes the satellite DNA whose use in genetic fingerprinting is described in section 25.4.12. Some make up pieces of DNA called introns. These are discussed below.

Introns and exons

In 1977 biologists were surprised to discover that the DNA of a eukaryotic gene is longer than its corresponding mRNA. It should be the same length because the mRNA is a direct copy. It was discovered that immediately after the mRNA is made, certain sections of the molecule are cut out, before it is used in translation. The sections of the gene that code for these unused pieces of RNA are called **introns**. The remaining sections of the gene are the code for the protein and are called **exons** (fig 23.33). The size and arrangement of introns is very variable and characteristic for a particular gene. In prokaryotes there are no introns.

One possible function for introns has come with the discovery that the same mRNA may have different introns removed in different cells. The gene therefore has alternative introns and can code for different, though similar, proteins. This increases its potential use.

An example is the calcitonin gene. Two different forms of mRNA can be produced by this gene, depending on which introns are removed. One is produced in the thyroid gland and codes for the protein calcitonin, which has 32 amino acids. Calcitonin is a hormone which acts to lower calcium levels in the blood. The other is produced in the hypothalamus and codes for a protein with 37 amino acids which is similar to calcitonin and is called CGRP (calcitonin gene-related peptide). This is a powerful vasodilatory agent. It is also released from nerve endings in some parts of the peripheral nervous system.

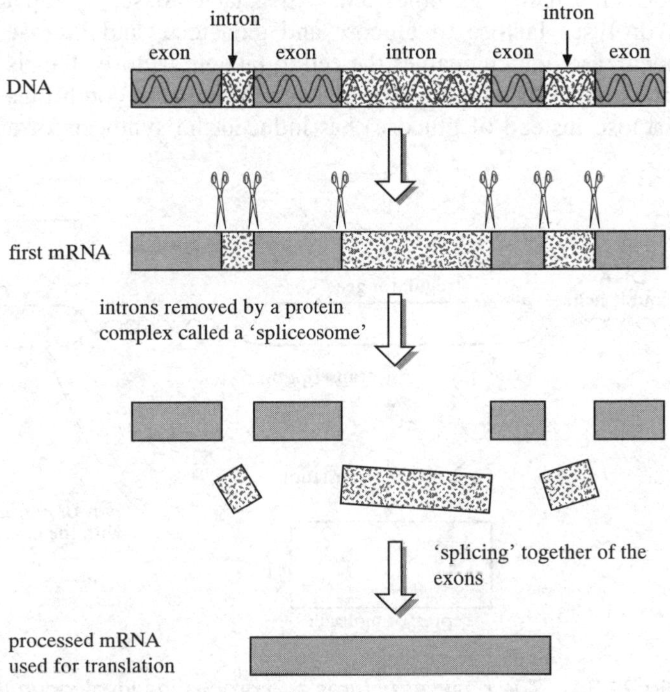

Fig 23.33 *Exons, introns and intron splicing.*

23.9 Gene control

Genetic research has come a long way since the discovery of the structure of DNA and the breaking of the genetic code. One of the areas that most concerns molecular geneticists is how gene activity is controlled so that an organised programme of development and cell activity is followed by each cell.

All somatic cells of an organism carry the same genes, that is they contain the same number of chromosomes carrying the same alleles. Despite this, cells in a multicellular organism show a wide variation in structure and function. Even within a single cell the rate at which certain protein molecules are synthesised varies according to circumstances and demand. Evidence for the mechanism by which genes are regulated within the cell was first obtained from studies into the control of enzyme synthesis in *E. coli*.

In 1961 Jacob and Monod carried out a series of experiments to investigate the nature of induction of enzyme synthesis in *E. coli*. Of the 800 enzymes thought to be synthesised by *E. coli* some are synthesised continuously and are called **constitutive enzymes**; others are synthesised only in the presence of an inducer compound, which may not be the substrate, and are called **inducible enzymes**. One of the latter enzymes is **β-galactosidase**.

E. coli will grow rapidly on a culture medium containing glucose. When transferred to a medium containing lactose instead of glucose it will not grow immediately but after a short delay begins to show the same growth rate as seen on a glucose medium. Investigations revealed that growth on the lactose medium required the presence of two substances not normally synthesised: β-galactosidase, which hydrolyses lactose to glucose and galactose, and **lactose permease**, which enables the cell to take up lactose. This is an example of where a change in environmental conditions (lactose instead of glucose) has induced the synthesis of a particular enzyme. Other experiments involving *E. coli* showed that high concentrations of the amino acid **tryptophan** in the culture medium suppressed the production of the enzyme **tryptophan synthetase** used to synthesise tryptophan. β-galactosidase synthesis is an example of **enzyme induction**, whereas the suppression of tryptophan synthetase is an example of **enzyme repression**. On the basis of these observations and experiments, Jacob and Monod proposed a mechanism to account for induction and repression, the mechanism by which genes are 'switched on and off'.

23.9.1 The Jacob–Monod hypothesis of gene control

The genes determining the amino acid sequences of the proteins described above are described as **structural genes**. Those for β-galactosidase and lactose permease are closely linked on the same chromosome. The activity of these genes is controlled by another gene called a **regulator gene** which is thought to prevent the structural genes from becoming active. This may be situated some distance from the structural genes. Evidence for the existence of a regulator gene comes from studies of mutant *E. coli* which lack this gene and as a consequence produce β-galactosidase continuously. The regulator gene carries the genetic code which results in the production of a **repressor molecule**. This prevents the structural genes from being active; it does not directly affect the structural genes but is thought to influence a gene immediately adjacent to the structural genes, the **operator gene**. The operator and structural genes are collectively known as the **operon** (fig 23.34).

The repressor molecule is thought to be a particular type of protein known as an **allosteric protein** which can either bind with the operator gene and suppress its activity ('switch it off') or not bind and permit the operator gene to become active ('switch it on'). When the operator gene is

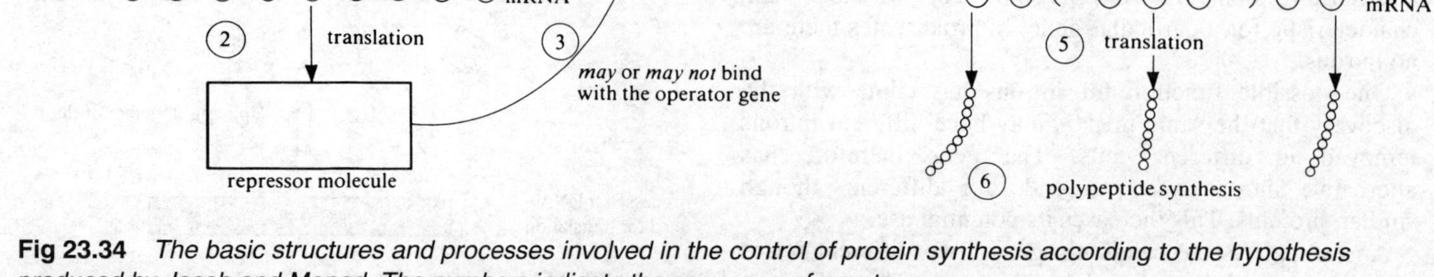

Fig 23.34 *The basic structures and processes involved in the control of protein synthesis according to the hypothesis produced by Jacob and Monod. The numbers indicate the sequence of events.*

'switched on' the structural genes carry out transcription and mRNA is formed which the ribosomes and tRNA translate into polypeptides. When the operator gene is 'switched off' no mRNA and no polypeptides are formed (fig 23.34).

The mechanism controlling whether or not the allosteric protein binds to the operator gene is simple, yet sensitive to varying intracellular conditions. It is thought that the repressor molecule has at least two active sites to which either an inducer molecule or a co-repressor molecule may become attached, depending upon their relative concentrations at any given time, as described in section 23.9.4.

23.9.2 Enzyme induction

The binding of an inducer molecule to its active site on the repressor molecule alters the tertiary structure of the repressor (allosteric effect) (section 4.4.4) so that it cannot bind with the operator gene and repress it. The operator gene becomes active and 'switches on' the structural genes.

In the case of *E. coli* grown on glucose medium, the regulator gene produces a repressor substance which combines with the operator gene and switches it 'off'. The structural genes are not activated and no β-galactosidase and lactose permease are produced. When transferred to a lactose medium the lactose is thought to act as an inducer of protein synthesis by combining with the repressor molecule and preventing it combining with the operator gene. The structural genes become active, mRNA is produced and proteins are synthesised. Lactose is thus an inducer of its own breakdown.

23.9.3 Enzyme repression

If a co-repressor molecule binds with its active site on the repressor molecule it reinforces the normal binding response of the repressor molecule with the operator gene. This inactivates the operator gene which, in effect, prevents the structural genes from being 'switched on'.

E. coli synthesises the amino acid tryptophan in the presence of the enzyme tryptophan synthetase. When the cell contains an excess of tryptophan some of it acts as a co-repressor of enzyme synthesis by combining with the repressor molecule. Co-repressor and repressor molecules combine with the operator gene and inhibit its activity. The structural genes are 'switched off', no mRNA is produced and no further tryptophan synthetase is synthesised. This is an example of feedback inhibition acting at the gene level.

23.9.4 Control of metabolic pathways

This dual mechanism of induction and repression enables the cytoplasm and nucleus to interact in a delicate control of cell metabolism. In the case of a simple metabolic pathway the initial substrate and final product can act as inducer and co-repressor respectively. This mechanism enables the cell to produce the amount of enzyme required at any given time to maintain the correct level of product. This method of metabolic control is highly economical. Negative feedback involving the inactivation of the initial enzyme by combination with the end-product would rapidly halt the pathway but would not prevent the continued synthesis of the other enzymes. In the system proposed by Jacob and Monod, the end-product, by combining with the repressor molecule to increase its repressive effect on the operator gene, would prevent the synthesis of all enzymes and halt the pathway.

23.9.5 Modification to the operon hypothesis

Since 1961, when Jacob and Monod suggested a mechanism by which genes are switched on and off, further evidence has accumulated which has helped to clarify aspects of the mechanism. Genetic evidence has suggested the existence of a **promoter gene** situated adjacent to the operator gene which acts between it and the regulator gene. It is thought to have two functions. First the promoter gene is the site to which RNA polymerase binds before moving along the DNA to begin the transcription of mRNA on the structural genes. This movement will, of course, depend upon whether the operator gene is 'operational' or not. Secondly, the base sequence of the promoter gene determines which strand of the DNA double helix attracts the RNA polymerase. In this way the promoter gene determines which strand of the DNA double helix acts as the template for mRNA transcription.

Chapter Twenty-four

Variation and genetics

Genetics may rightly be claimed to be one of the most important branches of biology. For thousands of years, humans have used the techniques of genetics in the improvement of domestic animals and crops without having any real knowledge of the mechanisms which underlie these practices. Various pieces of archaeological evidence dating back 6000 years suggest that humans understood that certain physical characteristics could be transmitted from one generation to another. By selecting particular organisms from wild stocks and interbreeding these, humans have been able to produce improved varieties of plants and animals to suit their needs.

It is only since the beginning of this century, though, that scientists have begun to appreciate fully the principles and mechanisms of heredity. Whilst advances in microscopy revealed that the sperm and the ova transmitted the hereditary characteristics from generation to generation, the problem nevertheless remained of how minute particles of biological matter could carry the vast number of characteristics that make up an individual organism.

The first really scientific advance in the study of inheritance was made by the Austrian monk Gregor Mendel who published a paper in 1866 which laid the foundations for the present-day science of genetics. He demonstrated that characteristics do not blend but pass from parents to offspring as discrete (separate) units. These units, which appear in the offspring in pairs, remain discrete and are passed on to subsequent generations by the male and female gametes which each contain a single unit. The Danish botanist Johannsen called these units **genes** in 1909, and the American geneticist Morgan, in 1912, demonstrated that they are carried on the chromosomes. Since the early 1900s the study of genetics has made great advances in explaining the nature of inheritance at both the level of the organism and at the level of the gene.

24.1 Mendel's work

Gregor Mendel was born in Moravia in 1822. In 1843 he joined an Augustinian monastery at Brünn in Austria (now Brno, in Czechoslovakia) where he took Holy Orders. From there he went to the University of Vienna where he spent two years studying natural history and mathematics before returning to the monastery in 1853. This choice of subjects undoubtedly had a significant influence on his subsequent work on inheritance in pea plants. Whilst in Vienna, Mendel had become interested in the process of hybridisation in plants and, in particular, the different forms in which hybrid offspring appear and the statistical relationships between them. This formed the basis of Mendel's scientific investigations on inheritance which he began in the summer of 1856.

Mendel's success was due, in part, to his careful choice of experimental organism, the garden pea, *Pisum sativum*. He established that it had the following advantages over other species:

(1) There were several varieties available which had quite distinct characteristics.
(2) The plants were easy to cultivate.
(3) The reproductive structures were completely enclosed by the petals so that the plant was normally self-pollinating. This led to the varieties producing the same characteristics generation after generation, a phenomenon known as **pure breeding**.
(4) Artificial cross-breeding between varieties was possible and resulting hybrids were completely fertile. From the 34 varieties of garden pea, Mendel selected 22 varieties which showed clear-cut differences in characteristics and used these in his breeding experiments. The seven basic characteristics, or **traits**, that Mendel was interested in were length of stem, shape of seed, colour of seed, shape and colour of pod, position and colour of flower.

Many scientists before Mendel had performed similar experiments on plants but none had produced results which had the accuracy and detail of Mendel's, nor were they able to explain their results in terms of a mechanism of inheritance. The reasons for Mendel's success may be taken as a model of how to carry out a scientific investigation. They may be summarised as follows:

(1) Preliminary investigations were carried out to obtain familiarity with the experimental organism.
(2) All experiments were carefully planned so that attention was focused on only one variable at any time, thus simplifying the observations to be made.
(3) Meticulous care was taken in carrying out all techniques, thus preventing the introduction of other variables (see below for details).
(4) Accurate records were kept of all the experiments and the results obtained.
(5) Sufficient data were obtained to have statistical significance.

As Mendel stated,

'The value and utility of any experiment are determined by the fitness of the material to the purpose for which it is used.'

However, it is worth stating that there was an element of luck in Mendel's choice of experimental organism. The characters chosen by Mendel lacked many of the more complex genetic features which were later discovered, such as codominance (section 24.7.1), characteristics controlled by more than one pair of genes (section 24.7.6) and linkage (section 24.3).

24.1.1 Monohybrid inheritance and the principle of segregation

Mendel's earliest experiments involved selecting plants of two varieties which had clearly different characteristics, such as flowers distributed along the main stem (axial) or flowers at the tip of the stem (terminal). These plants, showing a single pair of contrasted characteristics, were grown for a number of generations. Seeds collected from axial plants always produced plants with axial flowers, whilst those from terminal plants always produced terminal flowers. This demonstrated to Mendel that he was using pure-breeding plants. With this information he was in a position to carry out hybridisation experiments (experimental crosses) using these plants. His experimental technique involved removing the anthers from a number of plants of one variety before self-fertilisation could have occurred. These he called 'female' plants. Pollen was then transferred, by means of a brush, from the anthers of another plant of the same variety to the stigmas of the 'female' plant. The experimental flowers were then enclosed in a small bag to prevent pollen from other plants reaching their stigmas. **Reciprocal crosses** were carried out by transferring pollen grains from axial plants to terminal plants and pollen grains from terminal plants to axial plants. In all cases the seeds subsequently collected from both sets of plants gave rise to plants with axial flowers. This characteristic, 'axial flower', shown by these first generation hybrid plants (subsequently called the **first filial generation** or **F_1 generation** by Bateson and Saunders in 1902) was termed **dominant** by Mendel. None of the F_1 plants produced terminal flowers.

The F_1 plants then had their flowers enclosed in bags (to prevent cross-pollination occurring) and were left to self-pollinate. The seeds collected from these F_1 plants were counted and planted the following spring to produce the **second filial generation** or **F_2 generation**. (An F_2 generation is always the result of allowing the F_1 generation to inbreed or, as in this case, to self-pollinate.) When these plants flowered, some bore axial flowers and others terminal flowers. In other words, the characteristic 'terminal flower', which was absent in the F_1 generation, had reappeared in the F_2 generation. Mendel reasoned that the terminal characteristic must have been present in the F_1 generation but as it failed to be expressed in this generation he termed it **recessive**. Of the 858 F_2 plants that Mendel obtained, 651 had axial flowers and 207 had terminal flowers. Mendel carried out a series of similar experiments involving in each case the inheritance of a single pair of contrasting characteristics. Seven pairs of contrasting characteristics were studied and the results of the experimental crosses are shown in table 24.1. In all cases the analyses of the results revealed that the ratios of dominant to recessive characteristics in the F_2 generation were approximately 3:1.

The example quoted above is typical of all Mendel's experiments involving the inheritance of a single characteristic (**monohybrid inheritance**) and may be summarised as follows.

Observations

Parents	axial flowers × terminal flowers	
F_1	all axial flowers	
F_2	651 axial flowers	207 terminal flowers
F_2 ratio	3 :	1

On the basis of these, and similar results, Mendel drew the following conclusions.

(1) Since the original parental stocks were pure breeding, the axial variety must have possessed *two* axial factors and the terminal variety *two* terminal factors.

Table 24.1 The results of Mendel's experiments on the inheritance of seven pairs of contrasted characteristics. (The observed ratio of dominant to recessive characteristics approximates to the theoretical value of 3 : 1.)

Characteristic	*Parental appearance* (dominant)	*Parental appearance* (recessive)	*F_2 appearance* (dominant)	*F_2 appearance* (recessive)	*Ratio*
length of stem	tall	dwarf	787	277	2.84:1
shape of seed	round	wrinkled	5 474	1 850	2.96:1
colour of seed	yellow	green	6 022	2 001	3.01:1
shape of pod	inflated	constricted	882	299	2.95:1
colour of pod	green	yellow	428	152	2.82:1
position of flower	axial	terminal	651	207	3.14:1
colour of flower	red	white	705	224	3.15:1
total			14 949	5 010	2.98:1

(2) The F_1 generation possessed *one* factor from *each* parent which were carried by the gametes.
(3) These factors do not blend in the F_1 generation but retain their individuality.
(4) The axial factor is dominant to the terminal factor, which is recessive.

The separation of the pair of parental factors, so that one factor is present in each gamete, became known as **Mendel's first law**, or the **principle of segregation**. This states that:

the characteristics of an organism are determined by internal factors which occur in pairs. Only one of a pair of such factors can be represented in a single gamete.

We now know that these factors determining characteristics, such as flower position, are regions of the chromosome known as **genes**.

The experimental procedure described above which was carried out by Mendel in the investigation of the inheritance of a *single* pair of contrasted characteristics is an example of a **monohybrid cross**. This may be represented in terms of symbols and placed in a modern context of gamete formation and fertilisation. By convention, the initial letter of the dominant characteristic is used as the symbol for the gene and its capital form (e.g. **A**) represents the dominant form of the gene (the dominant allele) while the lower case (e.g. **a**) represents the recessive allele. All of the terms and symbols described above are used in genetics and are summarised in table 24.2.

Fig 24.1 shows the correct way to describe a monohybrid cross or arrive at the solution to a genetics problem involving the inheritance of a single pair of contrasted characteristics.

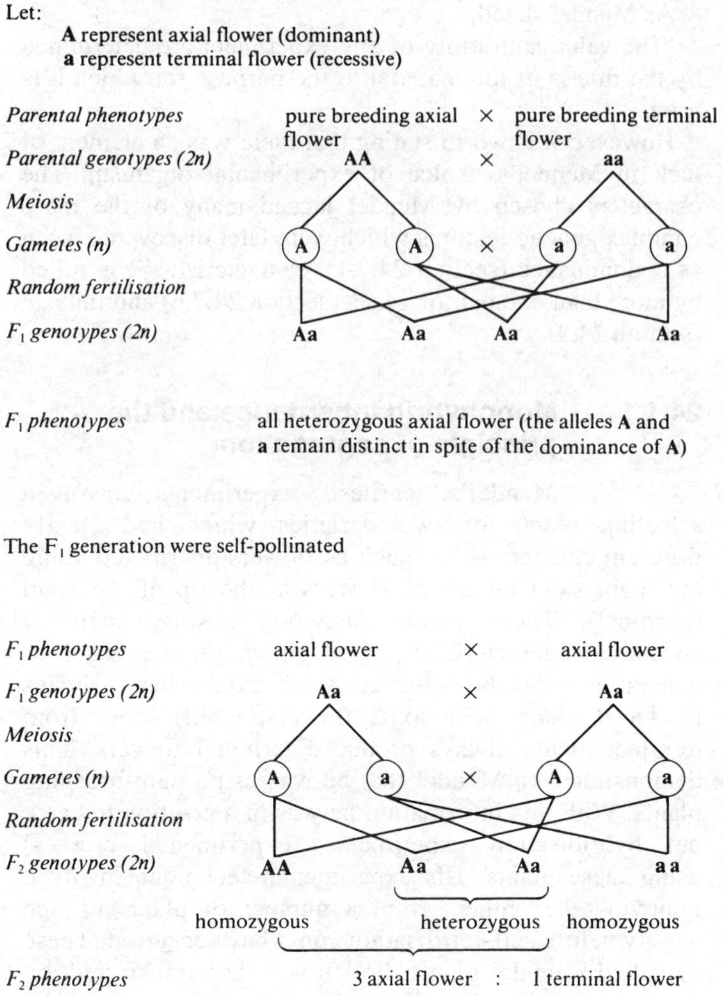

Fig 24.1 *Full genetic explanation of one of Mendel's monohybrid crosses. (2*n *represents the diploid condition,* n *represents the haploid condition; see section 23.1.2.)*

Table 24.2 Glossary of common genetic terms with examples based on fig 24.1.

Genetic term	*Explanation*	*Example*
gene	The basic unit of inheritance for a given characteristic	flower position
allele	One of a number of alternative forms of the same gene responsible for determining contrasting characteristics	**A** or **a**
locus	Position of an allele within a DNA molecule	
homozygous	The diploid condition in which the alleles at a given locus are identical	**AA** or **aa**
heterozygous	The diploid condition in which the alleles at a given locus are different	**Aa**
phenotype	The observable characteristics of an individual usually resulting from the interaction between the genotype and the environment in which development occurs	axial, terminal
genotype	The genetic constitution of an organism with respect to the alleles under consideration	**AA, Aa, aa**
dominant	The allele which influences the appearance of the phenotype even in the presence of an alternative allele	**A**
recessive	The allele which influences the appearance of the phenotype only in the presence of another identical allele	**a**
F_1 generation	The generation produced by crossing homozygous parental stocks	
F_2 generation	The generation produced by crossing two F_1 organisms	

The ratio of dominant phenotypes to recessive phenotypes of 3:1 is called the **monohybrid ratio**. Mendel's conclusions regarding the transfer of a single characteristic by each gamete and the resulting genotypes can be demonstrated by mathematical probability. The probability of a gamete cell from a heterozygous F_1 parent containing either the dominant allele **A** or the recessive allele **a** is 50% or $\frac{1}{2}$. If each gamete is represented by $\frac{1}{2}$, the number of possible combinations of F_2 genotypes is represented by $\frac{1}{2} \times \frac{1}{2} = \frac{1}{4}$. Hence there are four possible F_2 genotypes. The statistical probability of the **A** and **a** containing gametes combining by random fertilisation is shown in fig 24.2. As a result of dominance the phenotypic appearance will be 3 dominant phenotypes : 1 recessive phenotype. The results of Mendel's breeding experiments bear out this theoretical ratio as shown in table 24.1.

> **24.1** If a pure strain of mice with brown-coloured fur are allowed to breed with a pure strain of mice with grey-coloured fur they produce offspring having brown-coloured fur. If the F_1 mice are allowed to interbreed they produce an F_2 generation with fur colour in the proportion of three brown-coloured to one grey.
> (*a*) Explain these results fully.
> (*b*) What would be the result of mating a brown-coloured heterozygote from the F_2 generation with the original grey-coloured parent?

24.1.2 Test cross

The genotype of an F_1 organism, produced by the breeding of homozygous dominant and homozygous recessive parents, is heterozygous but shows the dominant phenotype. An organism displaying the recessive phenotype must have a genotype which is homozygous for the recessive allele. In the case of F_2 organisms showing the dominant phenotype the genotype may be either homozygous or heterozygous. It may be of interest to a breeder to know the genotype and the only way in which it can be determined is to carry out a breeding experiment. This involves the use of a technique known as **test cross**. By crossing an organism having an unknown genotype with a homozygous recessive organism it is possible to determine an unknown genotype within one breeding generation. For example in the fruit fly, *Drosophila*, long wing is dominant to vestigial wing. The genotype of a long wing *Drosophila* may be homozygous (**LL**) or heterozygous (**Ll**). In order to establish which is the correct genotype the fly is test crossed with a double recessive (**ll**) vestigial wing fly. If the test cross offspring are all long wing the unknown genotype is homozygous dominant. A ratio of 1 long wing : 1 vestigial wing indicates that the unknown is heterozygous (fig 24.3).

Let the probability of the alleles **A** and **a** appearing in the heterozygote (**Aa**) = 1,
therefore **A** = $\frac{1}{2}$
a = $\frac{1}{2}$
Using these values the probability of each genotype and phenotype appearing in the F_2 generation can be demonstrated as shown below:

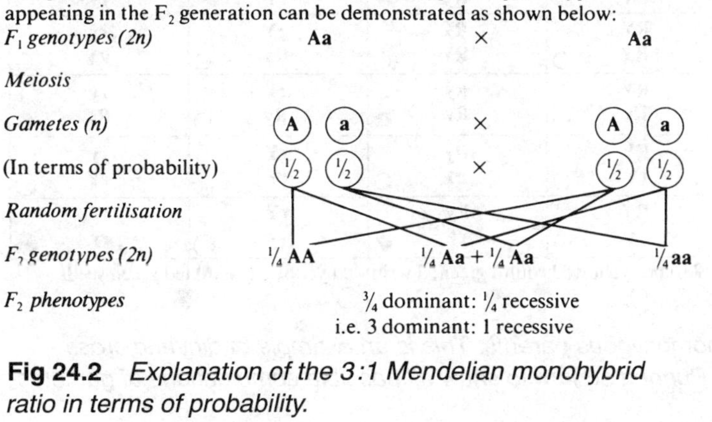

Fig 24.2 *Explanation of the 3:1 Mendelian monohybrid ratio in terms of probability.*

> **24.2** Why is it not possible to use a homozygous dominant organism (such as **TT**) in a test cross experiment to determine the genotype of an organism showing the dominant phenotype? Illustrate your answer fully using appropriate genetic symbols.

24.1.3 Dihybrid inheritance and the principle of independent assortment

Having established that it was possible to predict the outcome of breeding crosses involving a single pair of contrasted characteristics, Mendel turned his attention to the inheritance of two pairs of contrasted characteristics. Since two pairs of alleles are found in the heterozygotes, this condition is known as **dihybrid inheritance**.

In one of his experiments Mendel used pea shape and pea cotyledon colour as the characteristics. Using the same techniques as described in section 24.1.1, he crossed pure-breeding (homozygous) plants having round and yellow peas with pure-breeding plants having wrinkled and green peas. The F_1 generation seeds were round and yellow. Mendel knew that these characteristics were dominant from earlier monohybrid breeding experiments but it was the nature and number of organisms of the F_2 generation produced from the self-pollination of the F_1 plants that now interested him. He collected a total of 556 F_2 seeds from the F_1 generation which showed the following characteristics:

315 round and yellow,
101 wrinkled and yellow,
108 round and green,
32 wrinkled and green.

The proportions of each phenotype approximated to a ratio of 9 : 3 : 3 : 1. This is known as the **dihybrid ratio**. Mendel made two deductions from these observations.

(1) Two new combinations of characteristics had appeared in the F_2 generation: wrinkled and yellow, and round and green.

Let: **L** represent long wing (dominant)
l represent vestigial wing (recessive)

	homozygous long wing parent				heterozygous long wing parent			
Testcross phenotypes	long wing (homozygous)	×	vestigial wing		long wing (heterozygous)	×	vestigial wing	
Testcross genotypes (2n)	**LL**	×	**ll**		**Ll**	×	**ll**	
Meiosis								
Gametes (n)	(L) (L)	×	(l) (l)		(L) (l)	×	(l) (l)	
Random fertilisation								
Offspring genotypes (2n)	**Ll**	**Ll**	**Ll**	**Ll**	**Ll**	**Ll**	**ll**	**ll**
Offspring phenotypes	all long wing (heterozygous)				long wing (heterozygous) 1	:	vestigial wing (homozygous) 1	

Fig 24.3 *A full genetic explanation of how to determine the genotype of an organism showing a dominant characteristic. This technique is known as a test cross, and produces offspring phenotypes as shown.*

(2) The ratios of each pair of **allelomorphic** characteristics (phenotypes determined by different alleles) appeared in the monohybrid ratio of 3:1, that is 423 round to 133 wrinkled, and 416 yellow to 140 green.

On the basis of these results Mendel was able to state that the two pairs of characteristics (seed shape and colour), whilst combining in the F_1 generation, separate and behave independently from one another in subsequent generations. This forms the basis of **Mendel's second law** or the **principle of independent assortment** which states that:

any one of a pair of characteristics may combine with either one of another pair.

The above experiment can be written out in terms of our present knowledge of genetics as shown in fig 24.4*a*. As a result of separation (segregation) of alleles (**R**, **r**, **Y** and **y**) and their independent assortment (rearrangement or **recombination**), four possible arrangements of alleles can be found in each of the male and female gametes. In order to demonstrate all the possible combinations of gametes that occur during random fertilisation a **Punnett square** is used. This is a grid named after the Cambridge geneticist R. C. Punnett and its value lies in minimising the errors which can occur when listing all possible combinations of gametes. It is advisable when filling in the Punnett square to enter all the 'male' gametes first in the vertical squares and then enter all the 'female' gametes in the horizontal squares. Likewise, when determining the F_2 phenotypes, it is advisable to mark off identical phenotypes in some easily identifiable way, as shown in fig 24.4*b*. From figs 24.4*a* and *b*, which are based on Mendel's first and second laws, it can

(*a*) Let: **R** represent round seed (dominant) **Y** represent yellow seed (dominant)
r represent wrinkled seed (recessive) **y** represent green seed (recessive)

Parental phenotypes	round seed and yellow seed (homozygous)	×	wrinkled seed and green seed (homozygous)
Parental genotypes (2n)	**RRYY**	×	**rryy**
Meiosis			
Gametes (n)	all (RY)	×	all (ry)
Random fertilisation			
F_1 genotypes (2n)	all **RrYy**		
F_1 phenotypes	all heterozygous round and yellow seeds		

(*b*)

F_1 phenotypes	round and yellow seed	×	round and yellow seed
F_1 genotypes (2n)	**Rr Yy**	×	**RrYy**
Meiosis			
Gametes (n) (as shown by ♂ and ♀).			
Random fertilisation (as shown by Punnett square)			
F_2 genotypes (2n) (listed in each square)			
F_2 phenotypes	9 round yellow ○ : 3 round green ● : 3 wrinkled yellow ◌ : 1 wrinkled green seeds ●		

♀ \ ♂	**RY**	**Ry**	**rY**	**ry**
RY	**RY RY** ○	**Ry RY** ○	**rY RY** ○	**ry RY** ○
Ry	**RY Ry** ○	**Ry Ry** ●	**rY Ry** ○	**ry Ry** ●
rY	**RY rY** ○	**Ry rY** ○	**rY rY** ◌	**ry rY** ◌
ry	**RY ry** ○	**Ry ry** ●	**rY ry** ◌	**ry ry** ●

Fig 24.4 *(a) Stages in the formation of F_1 phenotypes from homozygous parents. This is an example of dihybrid cross since two characteristics are being considered. (b) Use of the Punnett square to show all possible combinations of gametes to form F_2 genotypes.*

be seen that each F_1 male and female genotype can give rise to gametes with the following combination of alleles:

R can only be present with **Y** or **y** (not **r**), that is **RY** or **Ry**,

r can only be present with **Y** or **y** (not **R**), that is **rY** or **ry**.

Thus there is a 1 in 4 chance of any gamete containing any of the four allele combinations shown above.

From a consideration of monohybrid inheritance, where $^3/_4$ of the F_2 phenotypes show the dominant allele and $^1/_4$ the recessive allele, the probability of the four alleles appearing in any F_2 phenotype is as follows:

round (dominant)	$^3/_4$
yellow (dominant)	$^3/_4$
wrinkled (recessive)	$^1/_4$
green (recessive)	$^1/_4$

Hence the probability of the following combinations of alleles appearing in the F_2 phenotypes is as follows:

round and yellow	=	$^3/_4 \times {^3/_4}$	=	$^9/_{16}$
round and green	=	$^3/_4 \times {^1/_4}$	=	$^3/_{16}$
wrinkled and yellow	=	$^1/_4 \times {^3/_4}$	=	$^3/_{16}$
wrinkled and green	=	$^1/_4 \times {^1/_4}$	=	$^1/_{16}$

The results of Mendel's breeding experiments with two pairs of contrasted characteristics approximated to the theoretical values shown above.

24.3 In the guinea pig (*Cavia*), there are two alleles for hair colour, black and white, and two alleles for hair length, short and long. In a breeding experiment all the F_1 phenotypes produced from a cross between pure-breeding, short black-haired and pure-breeding, long white-haired parents had short black hair. Explain (*a*) which alleles are dominant, and (*b*) the expected proportions of F_2 phenotypes.

24.4 Flower colour in sweet pea plants is determined by two allelomorphic pairs of genes (**R**,**r**, and **S**,**s**). If at least one dominant gene from each allelomorphic pair is present the flowers are purple. All other genotypes are white.

If two purple plants, each having the genotype **RrSs**, are crossed, what will be the phenotypic ratio of the offspring?

24.1.4 Summary of Mendel's hypotheses

The following summary includes terms taken from our present knowledge of the nature of genetics.

(1) Each characteristic of an organism is controlled by a pair of alleles.
(2) If an organism has two unlike alleles for a given characteristic, one may be expressed (the dominant allele) to the total exclusion of the other (the recessive allele).
(3) During meiosis each pair of alleles separates (segregates) and each gamete receives one of each pair of alleles (*the principle of segregation*).
(4) During gamete formation in each sex, either one of a pair of alleles may enter the same gamete cell (combine randomly) with either one of another pair (*the principle of independent assortment*).
(5) Each allele is transmitted from generation to generation as a discrete unchanging unit.
(6) Each organism inherits one allele (for each characteristic) from each parent.

NB The mechanism of dihybrid inheritance, the examples quoted in this section and the typical dihybrid ratio of 9 : 3 : 3 : 1 only apply to characteristics controlled by genes on **different** chromosomes. Genes situated on the **same** chromosome may not show this pattern of independent assortment, as described in section 24.3.

24.2 The chromosomal basis of inheritance

Mendel published his results and hypotheses in 1866 in a journal, *The Proceedings of the Brünn Natural History Society*, which was sent to most of the learned scientific societies throughout the world. In all cases they failed to appreciate the importance of his findings, possibly because scientists at the time were unable to relate them to any physical structures in the gametes by which the hereditary factors might be transmitted from parent to offspring.

By 1900, as a result of improvements in the optical properties of microscopes and advances in cytological techniques, the behaviour of chromosomes in gametes and zygotes had been observed. In 1875 Hertwig noted that during the fertilisation of sea urchin eggs two nuclei, one from the sperm and one from the egg, fused together. Boverin, in 1902, demonstrated the importance of the nucleus in controlling the development of characteristics in organisms, and in 1882 Flemming clarified the chromosomal events involved in mitosis.

In 1900 the significance of Mendel's work was realised almost simultaneously by three scientists, de Vries, Correns and Tschermak. In fact, it was Correns who summarised Mendel's conclusions in the familiar form of two principles and coined the term '**factor**', Mendel having used the term '*elemente*' to describe the hereditary unit. It was an American, William Sutton, however, who noticed the striking similarities between the behaviour of chromosomes during gamete formation and fertilisation, and the transmission of Mendel's hereditary factors. These have been summarised in table 24.3.

On the basis of the evidence suggested above, Sutton and Boveri proposed that chromosomes were the carriers of Mendel's factors, the so-called **chromosome theory of heredity**. According to this theory, each

Table 24.3 A summary of the similarities between events occurring during meiosis and fertilisation and Mendel's hypotheses.

Meiosis and fertilisation	*Mendel's hypotheses*
Diploid cells contain *pairs* of chromosomes(homologous chromosomes)	Characteristics are controlled by *pairs* of factors
Homologous chromosomes *separate* during meiosis	Pairs of factors *separate* during gamete formation
One homologous chromosome passes into each gamete cell	Each gamete receives *one* factor
Only the *nucleus* of the male gamete fuses with the egg cell nucleus	Factors are transmitted from generation to generation as *discrete units*
Homologous pairs of chromosomes are restored at fertilisation, each gamete (♂ and ♀) contributing *one* homologous chromosome	Each organism inherits *one* factor from each parent

pair of factors is carried by a pair of homologous chromosomes, with each chromosome carrying one of the factors. Since the number of characteristics of any organism vastly outnumber the chromosomes, as revealed by microscopy, each chromosome must carry many factors.

The term **factor** as the basic unit of heredity was replaced by Johannsen, in 1909, with the term **gene**. Whilst gene is used to describe the unit of heredity, it is the alternative forms of the gene or **alleles** which influence phenotypic expression. Alleles are the alternative forms in which a gene may exist and they occupy the same **loci** (singular **locus**) on **homologous chromosomes**, as shown in fig 24.5.

Mendel's principle of segregation of factors could now be explained in terms of the separation (segregation) of homologous chromosomes which occurs during anaphase I of meiosis and the random distribution of alleles into gamete cells. These events are summarised in fig 24.6.

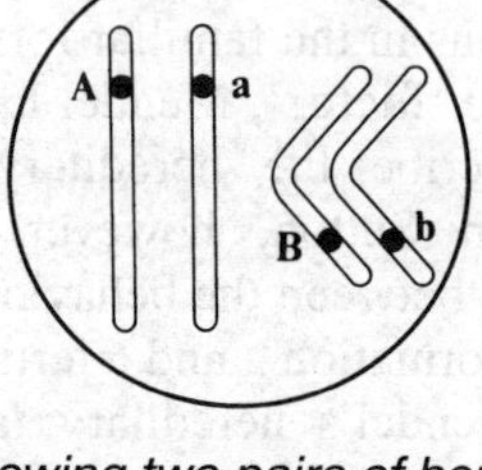

Fig 24.5 *A cell showing two pairs of homologous chromosomes. The positions of two different gene loci are indicated by circles. In this example two gene loci are shown situated on different pairs of homologous chromosomes and each gene is present as two alleles.*

24.2.1 Chromosomal explanation of independent assortment

Mendel's principle of independent assortment may also be explained in terms of the movement of chromosomes during meiosis. During gamete formation the distribution of each allele from a pair of homologous chromosomes is entirely independent of the distribution of alleles of other pairs. This situation is described in fig 24.7. It is the random alignment or assortment of homologous chromosomes on the equatorial spindle during metaphase I of meiosis, and their subsequent separation during metaphase I and anaphase I, that leads to the variety of allele recombinations in the gamete cells. It is possible to predict the number of allele combinations in either the male or female gamete using the general formula 2^n, where n = haploid number of chromosomes. In the case of humans, where $n = 23$, the possible number of different combinations is $2^{23} = 8\,388\,608$.

24.5 The deposition of starch in pollen grains in maize is controlled by the presence of one allele of a certain gene. The other allele of that gene results in no starch being deposited. Explain in terms of meiosis why half the pollen grains produced by a heterozygous maize plant contain starch.

24.6 Calculate the number of different combinations of chromosomes in the pollen grains of the crocus (*Crocus balansae*) which has a diploid number of six ($2n = 6$).

24.3 Linkage

All the situations and examples discussed so far in this chapter have dealt with the inheritance of genes situated on different chromosomes. Cytological studies have revealed that humans possess 46 chromosomes in all the somatic (body) cells. Since humans possess thousands of characteristics such as blood group, eye colour and the ability to secrete insulin, it follows that each chromosome must carry a large number of genes.

Genes situated on the same chromosome are said to be **linked**. All genes on a single chromosome form a **linkage group** and usually pass into the same gamete and are inherited together. As a result of this, genes belonging to the same linkage group usually do not show independent assortment. Since these genes do not conform to Mendel's principle of independent assortment they fail to produce the expected 9 : 3 : 3 : 1 ratio in a breeding situation involving the inheritance of two pairs of contrasted characteristics (dihybrid inheritance). In these situations a variety of ratios are produced which may be explained quite simply now that we possess a basic understanding of the mechanisms of inheritance as revealed by Mendel. (At this point it is worth

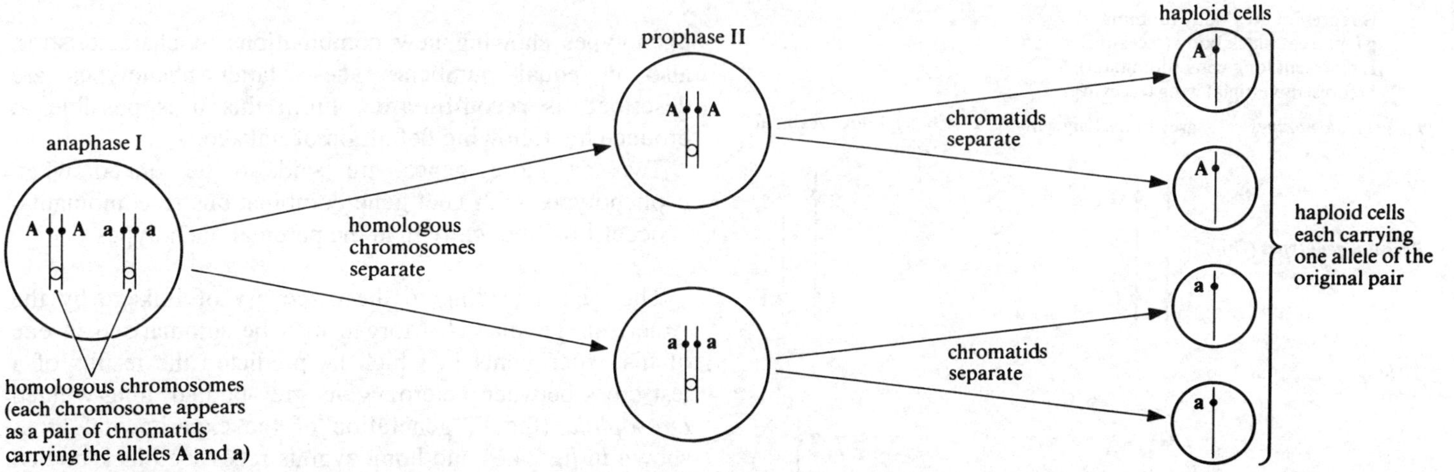

Fig 24.6 *Mendel's principle of segregation of factors (alleles)* **A** *and* **a** *described in terms of the separation of homologous chromosomes which occurs during meiosis.*

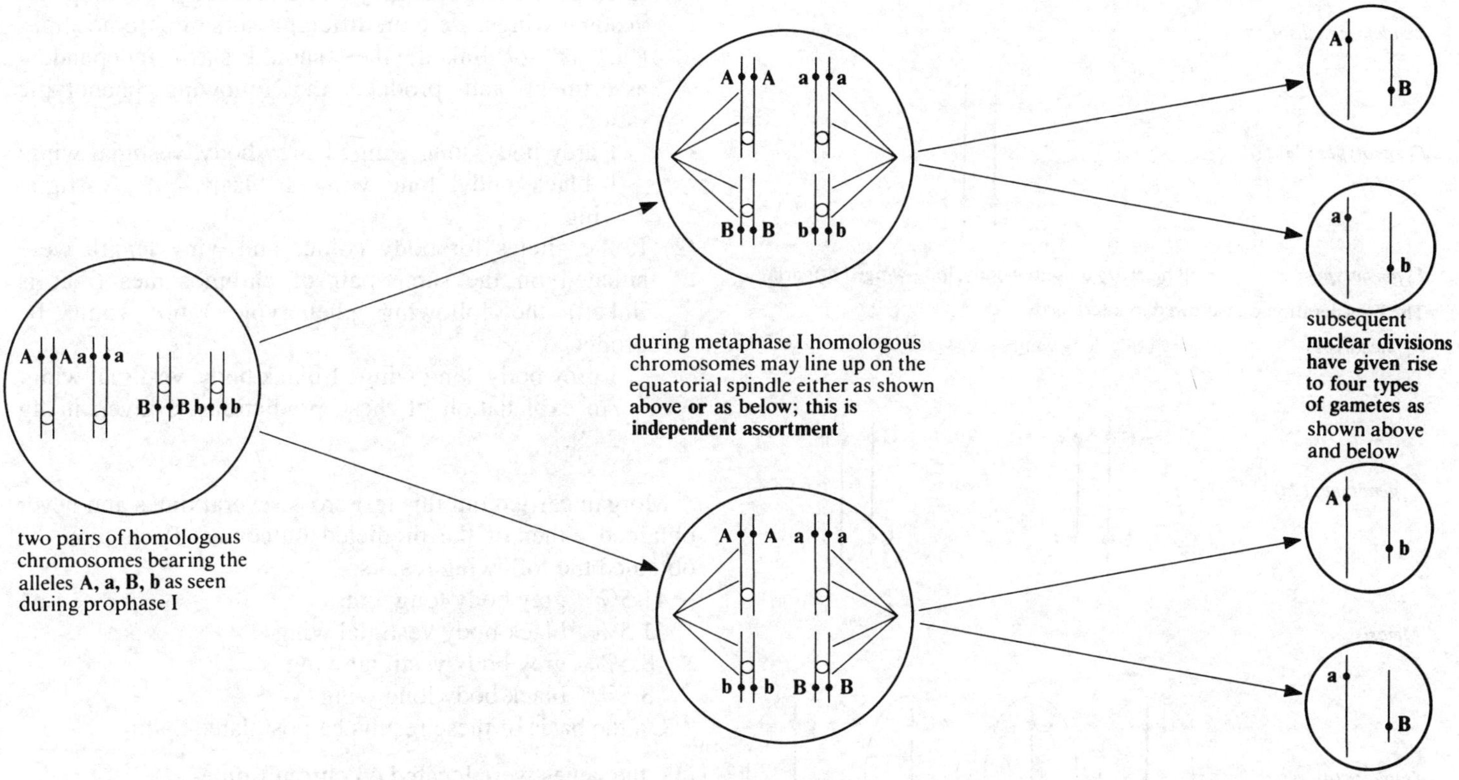

Fig 24.7 *Mendel's principle of independent assortment of factors (alleles)* **A, a, B, b**, *described in terms of the separation of homologous chromosomes which occurs during meiosis (compare fig 23.16).*

re-emphasising Mendel's good fortune in choosing to study the inheritance of pairs of characteristics located on *different* chromosomes.) In *Drosophila* the genes for body colour and wing length have the following **allelomorphs** (phenotypic characteristics determined by different alleles): grey and black body, and long and vestigial (short) wings. Grey body and long wing are dominant. If pure-breeding grey-bodied long-winged *Drosophila* are crossed with black-bodied vestigial-winged *Drosophila*, the expected F_2 phenotypic ratio would be 9 : 3 : 3 : 1. This would indicate a normal case of Mendelian dihybrid inheritance with random assortment resulting from the genes for body colour and wing length being situated on non-homologous chromosomes. However this result is not obtained. Instead the F_2 show an approximately 3 : 1 ratio of parental phenotypes. This may be explained by assuming that the genes for body colour and wing length are found on the same chromosome, that is they are linked, as shown in fig 24.8.

In practice, though, this 3 : 1 ratio is never achieved and four phenotypes are invariably produced. This is because **total** linkage is rare. Most breeding experiments involving linkage produce approximately equal numbers of the parental phenotypes and a significantly smaller number of phenotypes showing new combinations of characteristics, also in equal numbers. These latter phenotypes are described as **recombinants**. From this it is possible to produce the following definition of linkage:

Two or more genes are said to be linked when phenotypes with new gene combinations (recombinants) occur less frequently than the parental phenotypes.

The events leading to the discovery of linkage by the American Thomas H. Morgan may be summarised in one of his experiments in which he predicted the results of a test cross between heterozygous grey-bodied, long-winged *Drosophila* (the F_1 generation of the experimental cross shown in fig 24.8) and homozygous recessive black-bodied vestigial-winged *Drosophila*. The two possible outcomes were predicted as follows:

(1) If the four alleles for grey and black body, and long and vestigal wings, were on different pairs of chromosomes (that is *not* linked) they should show independent assortment and produce the following phenotypic ratios:

1 grey body, long wing : 1 grey body, vestigial wing; 1 black body, long wing : 1 black body, vestigial wing.

(2) If the alleles for body colour and wing length were situated on the same pair of chromosomes (that is linked) the following phenotypic ratio would be produced:

1 grey body, long wing : 1 black body, vestigial wing.

An explanation of these predictions is given in fig 24.9.

Morgan carried out this test cross several times and never obtained either of the predicted outcomes. Each time, he obtained the following results:

41.5% grey body long wing
41.5% black body vestigial wing
8.5% grey body vestigial wing
8.5% black body long wing

On the basis of these results he postulated that:

(1) the genes were located on chromosomes;
(2) both the genes were situated on the same chromosome, that is linked;
(3) the alleles for each gene were on homologous chromosomes;
(4) alleles were exchanged between homologous chromosomes during meiosis.

The reappearance of recombinant alleles in 17% of the offspring was explained in terms of point (4). This is known as **crossing-over**.

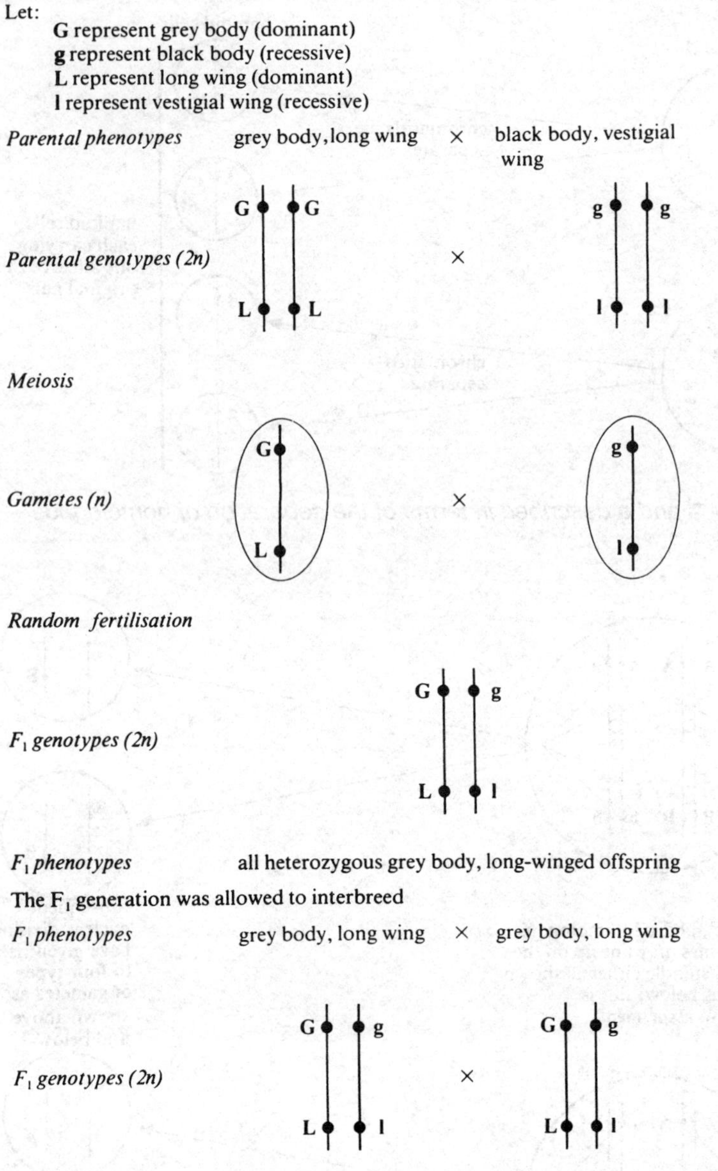

Fig 24.8 *Genetic explanation of the 3 : 1 ratio produced in F_2 phenotypes as a result of linkage.*

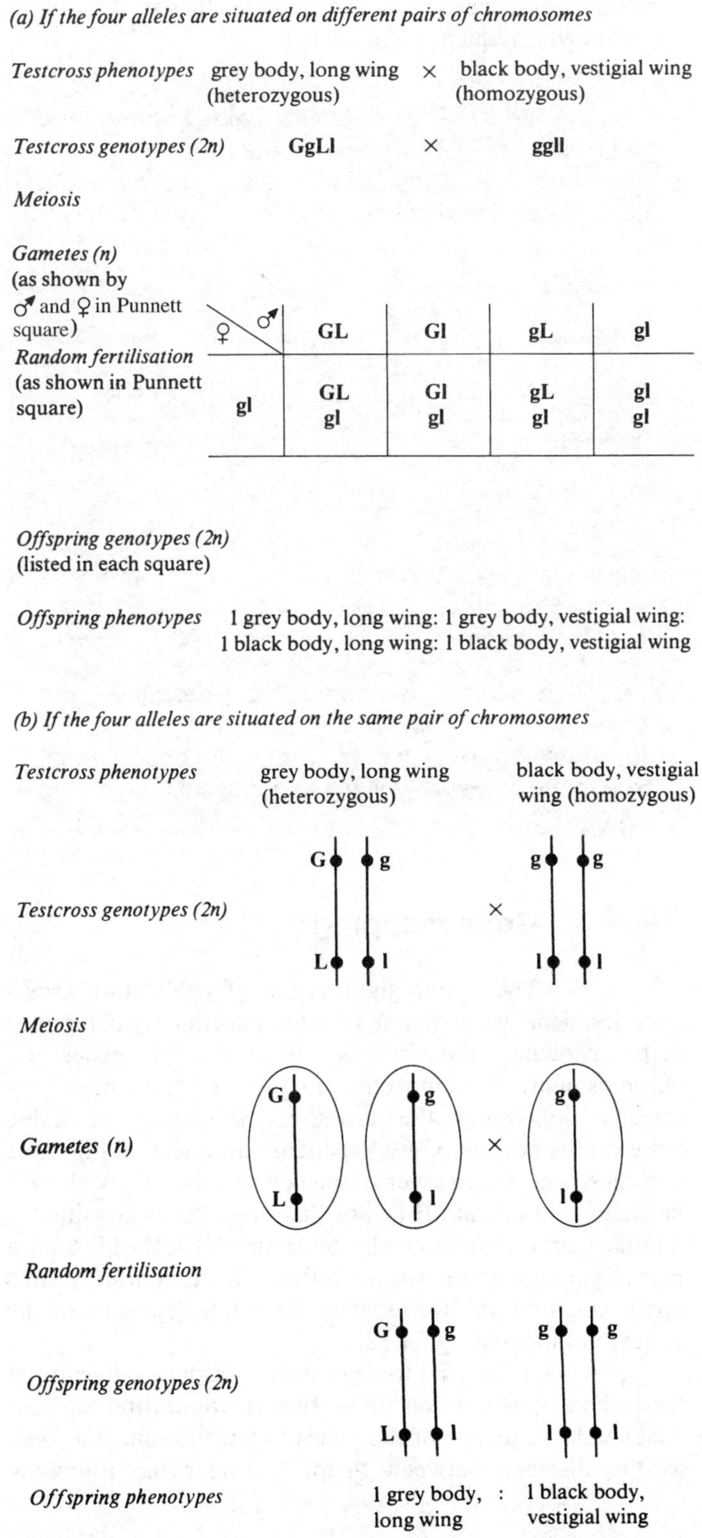

Fig 24.9 *(a) and (b) Genetic explanation of Morgan's predictions.*

> **24.7** A homozygous purple-flowered short-stemmed plant was crossed with a homozygous red-flowered long-stemmed plant and the F_1 phenotypes had purple flowers and short stems. When the F_1 generation was test crossed with a double homozygous recessive plant the following progeny were produced.
>
> 52 purple flower, short stem
> 47 purple flower, long stem
> 49 red flower, short stem
> 45 red flower, long stem
>
> Explain these results fully.

24.3.1 Crossing-over and crossover values

In 1909 the Belgian cytologist Janssens observed **chiasmata formation** during prophase I of meiosis (section 23.4). The genetic significance of this process was clarified by Morgan who proposed that crossing-over of alleles occurred as a result of the breakage and recombination of homologous chromosomes during chiasmata. Subsequent research based on the microscopic examination of cells and recombinant phenotypic ratios has confirmed that crossover of genetic material occurs between virtually all homologous chromosomes during meiosis. The alleles of parental linkage groups separate and new associations of alleles are formed in the gamete cells, a process known as **genetic recombination**. Offspring formed from these gametes showing 'new' combinations of characteristics are known as **recombinants**. Thus crossing-over is a major source of observable genetic variation within populations.

The behaviour of a pair of homologous chromosomes in *Drosophila*, carrying the alleles grey body and long wing (both dominant) and black body and vestigial wing (both recessive), during formation of chiasmata may be used to illustrate the principle of crossing-over. A cross between a male heterozygous grey-bodied long-winged *Drosophila* and a female homozygous black-bodied vestigial-winged *Drosophila* produced heterozygous F_1 offspring with grey bodies and long wings as shown in fig 24.10.

Test crossing the F_1 generation flies with homozygous double recessive flies produced the following results.

Parental phenotypes	grey body, long wing	965
	black body, vestigial wing	944
Recombinant phenotypes	black body, long wing	206
	grey body, vestigial wing	185

These results indicate that the genes for body colour and wing length are linked. (Remember that a hybrid cross between an F_1 heterozygote and a double homozygous recessive would have produced a 1 : 1 : 1 : 1 phenotypic ratio if the genes had been situated on different chromosomes and therefore had undergone random assortment.) Using

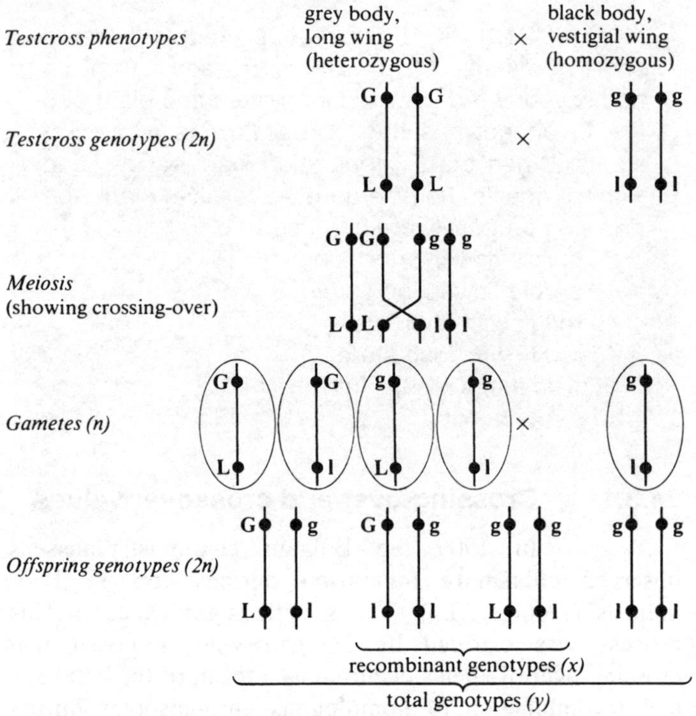

Fig 24.10 *Genetic explanation of crossing-over and the reappearance of recombinant genotypes. The recombination frequency can be calculated by counting the number of individuals showing recombination and the total number of individuals and applying the following formula: recombination frequency (%) = x/y × 100.*

the figures obtained from the above cross it is possible to calculate the recombination frequency of the genes for body colour and wing length.

The **recombination frequency** is calculated using the formula:

$$\frac{\text{number of individuals showing recombination}}{\text{number of offspring}} \times 100$$

From the example above the recombination frequency (%) is:

$$\frac{(206 + 185)}{(965 + 944) + (206 + 185)} \times 100$$

$$= \frac{391}{2300} \times 100$$

$$= 17\%$$

This value indicates the number of crossovers which have occurred during gamete formation. A. H. Sturtevant, a student of Morgan, postulated that the recombinant frequency or **crossover frequency (crossover value (COV))** demonstrated that genes are arranged linearly along the chromosome. More importantly, he suggested that the crossover frequency reflects the relative positions of genes on chromosomes because the further apart linked genes are on the chromosomes, the greater the possibility of crossing-over occurring between them, that is the greater the crossover frequency (fig 24.11).

24.8 The diagram below shows the gene loci of 12 alleles situated on a pair of chromosomes. Their relative distances from the centromere are shown.

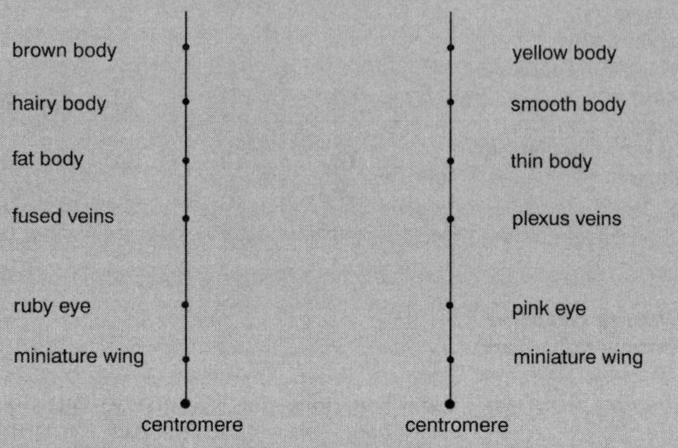

(*a*) What term is used to describe the chromosomes shown above?
(*b*) Crossing-over would probably occur most frequently between which two gene loci?

24.4 Gene mapping

The major significance of calculating cross-over frequencies is that it enables geneticists to produce maps showing the relative positions of genes on chromosomes. Chromosome maps are constructed by directly converting the crossover frequency or value between genes into hypothetical distances along the chromosome. A crossover frequency or value (COV) of 4% between genes **A** and **B** means that those genes are situated 4 units apart on the same chromosome. A COV of 9% for a pair of genes **A** and **C** would indicate that they were 9 units apart, but it would not indicate the linear sequence of the genes, as shown in fig 24.12.

In practice it is usual to determine crossover values for at least three genes at once, as this **triangulation** process enables the sequence of the genes to be determined as well as the distance between them. Consider the following

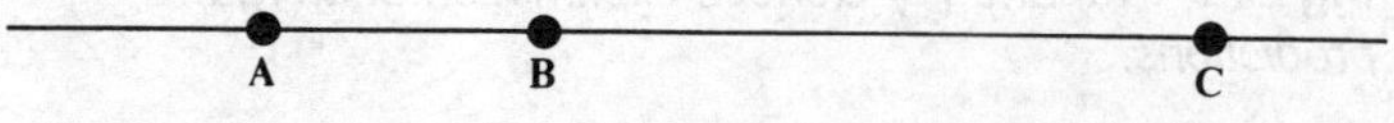

Fig 24.11 *Three gene loci represented by* **A**, **B** *and* **C** *are shown on the chromosome. Crossing-over and separation of genes is more likely to occur between* **B** *and* **C** *or* **A** *and* **B** *since the frequency of crossing-over is related to the distance between the genes.*

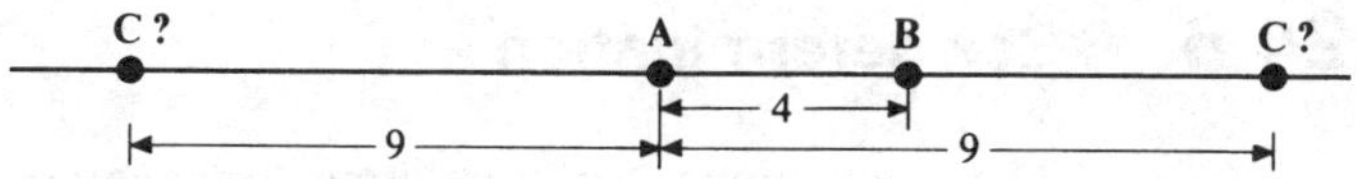

Fig 24.12 *Possible gene loci of* **A**, **B** *and* **C** *on the basis of the data presented.*

values as determined by a series of breeding experiments involving four genes **P**, **Q**, **R** and **S**:

P – Q = 24%
R – P = 14%
R – S = 8%
S – P = 6%

To calculate the sequence and distances apart of the genes, a line is drawn representing the chromosome and the following procedure carried out.

(1) Insert the positions of the genes with the least COV in the middle of the chromosome, that is **S – P** = 6% (fig 24.13*a*).
(2) Examine the next largest COV, that is **R – S** = 8%, and insert both possible positions of **R** on the chromosome, relative to **S** (fig 24.13*b*).
(3) Repeat the procedure for the next largest COV, that is **R – P** = 14%. This indicates that the right-hand position of **R** is incorrect (fig 24.13*c*).
(4) Repeat the procedure for the COV for **P – Q** = 24% (fig 24.13*d*). The position of **Q** cannot be ascertained without additional information. If, for example, the COV for **Q – R** = 10% this would confirm the left-hand position for gene **Q**.

A problem which arises in preparing chromosome maps is that of **double crossover**, particularly when considering genes which are widely separated, since the number of apparent crossovers will be less than the actual number. For example, if crossovers occur between alleles **A** and **B** and **B** and **C** in fig 24.14, **A** and **C** will still appear linked, but the chromosome will now carry the recessive allele **b**.

24.9 In maize the genes for coloured seed and full seed are dominant to the genes for colourless seed and shrunken seed. Pure-breeding strains of the double dominant variety were crossed with the double recessive variety and a test cross of the F_1 generation produced the following results.

coloured, full seed	380
colourless, shrunken seed	396
coloured, shrunken seed	14
colourless, full seed	10

Calculate the distance in units between the genes for coloured seed and seed shape on the chromosomes.

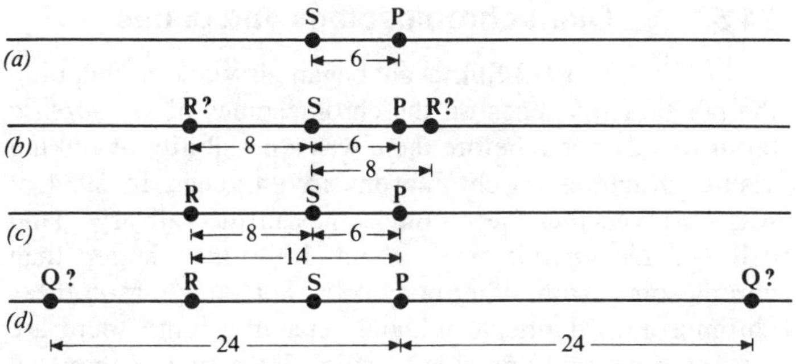

Fig 24.13 *Use of the triangulation process to establish the positions of genes* **P**, **Q**, **R** *and* **S** *on a chromosome.*

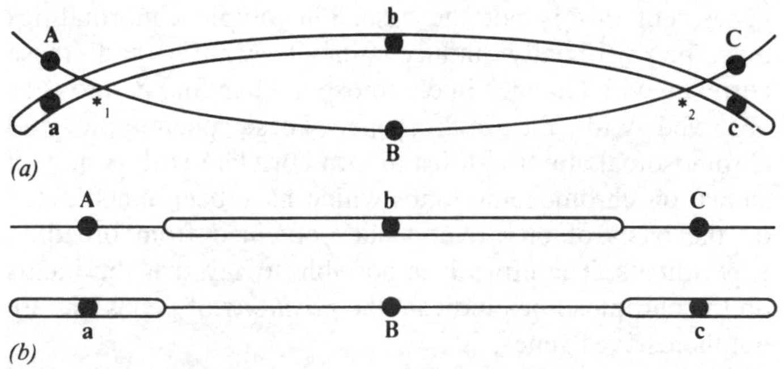

Fig 24.14 *(a) A pair of homologous chromatids, one carrying the dominant alleles* **A**, **B** *and* **C** *and the other carrying the recessive alleles* **a**, **b** *and* **c**. *Crossing-over occurs at two points* $*_1$ *and* $*_2$. (b) *The result of separation of the chromatids in which the sequences of alleles are different, although the sequences of gene loci and the distances between them remain the same.*

24.5 Linkage groups and chromosomes

Much of the evidence presented in this chapter so far has shown how our knowledge of the mechanics of inheritance has gradually increased. Most of the research into genetics in the early part of this century involved establishing the role of genes in inheritance. Morgan's research with the fruit fly (*Drosophila melanogaster*) established that the majority of phenotypic characteristics were transmitted together in four groups and these were called **linkage groups**. It was observed that the number of linkage groups corresponded to the number of pairs of chromosomes.

Studies on other organisms produced similar results. Breeding experiments using a variety of organisms revealed that some linkage groups were larger than others (that is they carried more genes). Examination of chromosomes in these organisms showed that they varied in length. Morgan demonstrated that there was a distinct relationship between these observations. This provided further confirmatory evidence that genes were located on chromosomes.

24.5.1 Giant chromosomes and genes

In 1913 Sturtevant began his work on mapping the positions of genes on the chromosomes of *Drosophila* but it was 21 years before there was a possibility of linking visible structures on chromosomes with genes. In 1934, it was observed that the chromosomes in the salivary gland cells of *Drosophila* were about 100 times larger than chromosomes from other body cells. For some reason these chromosomes duplicate without separating until there are several thousand lying side by side. When stained they can be seen with the light microscope and appear to be made up of alternating light and dark bands. Each chromosome has its own distinctive pattern of bands (fig 24.15). It was originally thought, or rather hoped, that these bands were genes, but this is not the case. Phenotypic abnormalities may be artificially induced in *Drosophila* and these correlate with changes in chromosomal banding patterns, as observed with the microscope. These phenotypic and chromosomal abnormalities in turn correlate with gene loci shown on chromosome maps which have been constructed on the basis of crossover values obtained from breeding experiments. Therefore it is possible to say that the bands on the chromosomes indicate the *positions* of genes but are not themselves genes.

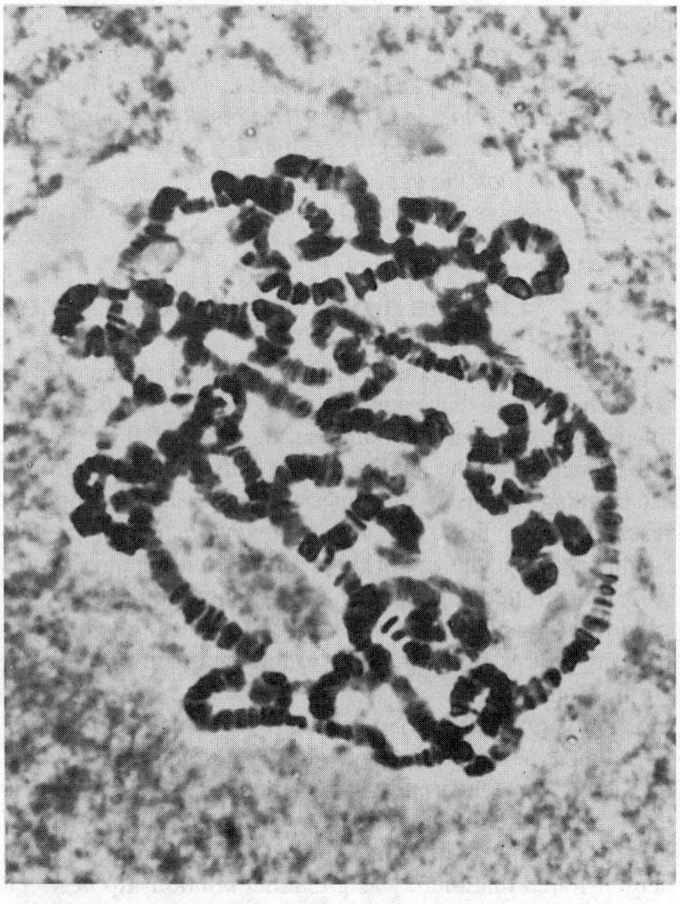

Fig 24.15 *Giant chromosomes from the salivary glands of* Drosophila melanogaster. *Four pairs of chromosomes are shown joined at their centromeres.*

24.6 Sex determination

The technique of relating phenotypic characteristics of organisms to the structure of their chromosomes, as described in earlier sections, is seen most clearly in the determination of sex. In *Drosophila* the observed phenotypic differences between the two sexes appear to be related to the differences in the size of their chromosomes, as shown in fig 24.16. Examination of the chromosome structure of a range of animals revealed that males and females showed certain chromosomal differences. Pairs of chromosomes (homologous chromosomes) are found in all cells, but one pair of chromosomes always shows differences between the sexes. These are the **sex chromosomes** or **heterosomes**. All other chromosomes are known as **autosomal chromosomes** or **autosomes**. As can be seen in fig 24.16, *Drosophila* has four pairs of chromosomes. Three pairs appear identical in both sexes (numbers II, III and IV), but the other pair, whilst appearing identical in the female, differ in the male. The chromosomes are known as X and Y chromosomes (see fig 24.17), and the genotype of the female is XX and that of the male is XY. These characteristic sex genotypes are found in most animals, including humans; but in the case of birds (including poultry), moths and butterflies the sex genotypes are reversed: the females are XY and the males are XX. In some insects, such as the grasshopper, the Y chromosome may be absent entirely and so the male has the genotype XO (see fig 23.13).

In the production of gametes the sex chromosomes segregate in typical Mendelian fashion. For example, in mammals each ovum contains an X chromosome; in males one half of the sperm contain an X chromosome and the

Fig 24.16 *Structure of chromosomes in male and female* Drosophila melanogaster. *Four pairs of chromosomes are shown. The sex chromosomes are numbered I.*

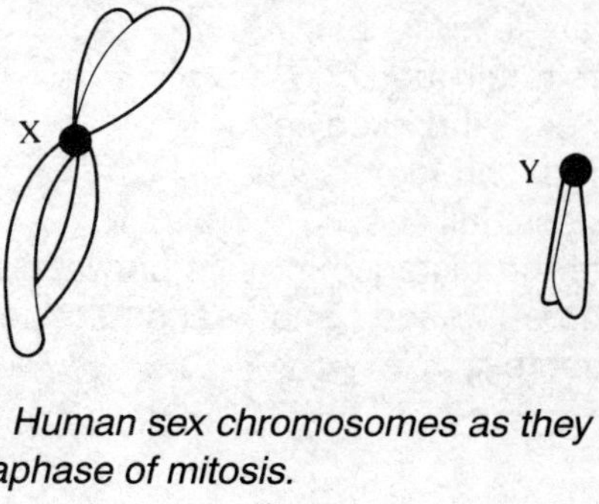

Fig 24.17 *Human sex chromosomes as they appear during metaphase of mitosis.*

other half contain a Y chromosome as shown in fig 24.18. The sex of the offspring depends upon which type of sperm fertilises the ovum. The sex having the XX genotype is described as **homogametic** as it produces gamete cells containing only X chromosomes. Organisms with the XY genotype are described as **heterogametic** since half their gametes contain the X chromosome and half the Y chromosome. In humans, the genotypic sex of an individual is determined by examining non-dividing cells. One X chromosome always appears in the active state, which has the normal appearance. If another is present, it is seen in a resting state as a tightly coiled dark-staining body called the **Barr body**. The number of Barr bodies is always one less than the number of X chromosomes present, that is male (XY) = 0, female (XX) = 1. The function of the Y chromosome appears to vary according to the species. In humans the presence of a Y chromosome controls the differentiation of the testes which subsequently influences the development of the genital organs and male characteristics (section 21.6.4). In some organisms, however, the Y chromosome does not carry genes concerned with sex. In fact it is described as genetically inert or genetically empty since it carries so few genes. In *Drosophila* it is thought that the genes determining male characteristics are carried on the autosomes and their phenotypic effects are masked by the presence of a pair of X chromosomes. Male characteristics, on the other hand, appear in the presence of a single X chromosome. This is an example of **sex-limited inheritance**, as opposed to sex-linked inheritance, and in humans is thought to cause suppression of the genes for the growth of a beard in females.

Morgan and his co-workers noticed that inheritance of eye colour in *Drosophila* was related to the sex of the parent flies. Red eye is dominant over white eye. A red-eyed male crossed with a white-eyed female produced equal numbers of F_1 red-eyed females and white-eyed males (fig 24.19*a*). A white-eyed male, however, crossed with a red-eyed female produced equal numbers of F_1 red-eyed males and females (fig 24.19*b*). Inbreeding these F_1 flies produced red-eyed females, red-eyed males and white-eyed males but *no* white-eyed females (fig 24.19*c*). The fact that male flies showed the recessive characteristic more

Parental phenotypes	female (♀)		×	male (♂)	
Parental genotypes (2n)	XX		×	XY	
Meiosis					
Gametes(n)	(X)	(X)	×	(X)	(Y)
Random fertilisation					
Offspring genotypes (2n)	XX	XY		XX	XY
Offspring phenotypes	♀	♂		♀	♂

sex ratio 1 female : 1 male

Fig 24.18 *Genetic explanation of the sex ratio in humans.*

frequently than female flies suggested that the white eye recessive allele was present on the X chromosome and that the Y chromosome lacked the eye colour gene. To test this hypothesis Morgan crossed the original white-eyed male with an F_1 red-eyed female (fig 24.19*d*). The offspring included red-eyed and white-eyed males and females. From this Morgan rightly concluded that only the X chromosome carries the gene for eye colour. There is no gene locus for eye colour on the Y chromosome. This phenomenon is known as **sex linkage**.

24.10 In *Drosophila* the genes for wing length and for eye colour are sex-linked. Normal wing and red eye are dominant to miniature wing and white eye.

(*a*) In a cross between a miniature wing, red-eyed male and a homozygous normal wing, white-eyed female, explain fully the appearance of (i) the F_1 and (ii) the F_2 generations.

(*b*) Crossing a female from the F_1 generation above with a miniature wing, white-eyed male gave the following results:

normal wing, white-eyed males and females	35
normal wing, red-eyed males and females	17
miniature wing, white-eyed males and females	18
miniature wing, red-eyed males and females	36

Account for the appearance and numbers of the phenotypes shown above.

24.6.1 Sex linkage

Genes carried on the sex chromosomes are said to be sex-linked. In the case of the heterogametic sex there is a portion of the X chromosome for which there is no homologous region of the Y chromosome (fig 24.20). Characteristics determined by genes carried on the non-homologous portion of the X chromosome therefore appear in males even if they are recessive. This special form of linkage explains the inheritance of **sex-linked traits** such as red–green colour blindness, premature balding and haemophilia. Haemophilia or 'bleeder's disease' is a sex-linked recessive condition which prevents the formation of factor VIII, an important factor in increasing the rate of blood clotting. The gene for factor VIII is carried on the non-homologous portion of the X chromosome and can appear in two allelomorphic forms: normal (dominant) and mutant (recessive). The following possible genotypes and phenotypes can occur:

genotype	*phenotype*
$X^H X^H$	normal female
$X^H X^h$	normal female (carrier)
$X^H Y$	normal male
$X^h Y$	haemophiliac male

In all sex-linked traits, females who are heterozygous are described as **carriers** of the trait. They are phenotypically

Let: **R** represent red eye (dominant)
r represent white eye (recessive)
XX represent female fly (♀)
XY represent male fly (♂)

	(a)	(b)
Parental phenotypes	red-eyed ♂ × white-eyed ♀	white-eyed ♂ × red-eyed ♀
Parental genotypes (2n)	X^RY × X^rX^r	X^rY × X^RX^R
Meiosis		
Gametes (n)	(X^R) (Y) × (X^r) (X^r)	(X^r) (Y) × (X^R) (X^R)
Random fertilisation		
F_1 genotypes (2n)	X^RX^r X^RX^r X^rY X^rY	X^rX^R X^rX^R X^RY X^RY
F_1 phenotypes	red-eyed ♀ (X^RX^r, X^RX^r); white-eyed ♂ (X^rY, X^rY)	red-eyed ♀ (X^rX^R, X^rX^R); red-eyed ♂ (X^RY, X^RY)

	(c)	(d)
F_1 phenotypes	red-eyed ♂ × red-eyed ♀	white-eyed ♂ × red-eyed ♀
F_1 genotypes (2n)	X^RY × X^rX^R	X^rY × X^RX^r
Meiosis		
Gametes (n)	(X^R) (Y) × (X^r) (X^R)	(X^r) (Y) × (X^R) (X^r)
Random fertilisation		
F_2 genotypes (2n)	X^RX^r X^RX^R X^rY X^RY	X^rX^R X^rX^r X^RY X^rY
F_2 phenotypes	red-eyed ♀ (X^RX^r, X^RX^R); ♂: white-eyed (X^rY), red-eyed (X^RY)	♀: red-eyed (X^rX^R), white-eyed (X^rX^r); ♂: red-eyed (X^RY), white-eyed (X^rY)

Fig 24.19 *(a) and (b) Morgan's reciprocal experimental crosses between red-eyed and white-eyed* Drosophila. *Note the low frequency of appearance of white eyes. (c) Morgan's confirmatory inbreeding experimental cross between an F_1 red-eyed male and an F_1 (heterozygous) red-eyed female. (d) The experimental cross between a white-eyed male and an F_1 (heterozygous) red-eyed female. Note the appearance of the white-eye characteristic only in homozygous white-eyed female flies.*

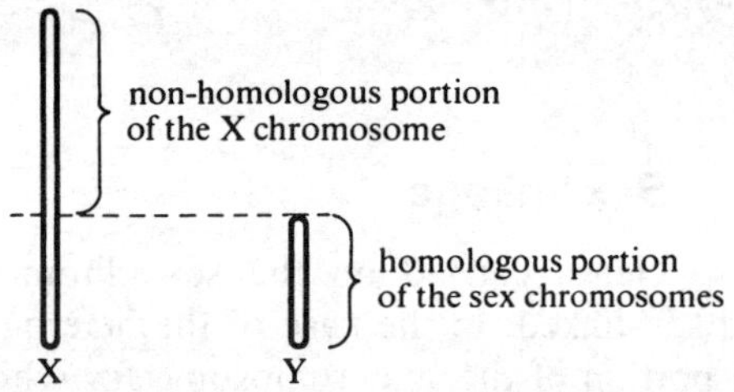

Fig 24.20 *Homologous and non-homologous regions of the sex chromosomes.*

Let: **H** represent normal allele for blood clotting (dominant)
h represent allele for haemophilia (recessive)
XX represent female chromosomes
XY represent male chromosomes

Parental phenotypes	normal female (carrier)		×	normal male
Parental genotypes (2n)	X^HX^h		×	X^HY
Meiosis				
Gametes (n)	(X^H)	(X^h)	×	(X^H) (Y)
Random fertilisation				
Offspring genotypes (2n)	X^HX^H	X^HY	X^hX^H	X^hY
Offspring phenotypes	normal female	normal male	normal female (carrier)	haemophiliac male

Fig 24.21 *Mechanism of inheritance of the sex-linked allele for haemophilia.*

normal but half their gametes carry the recessive gene. Despite the father having a normal gene there is a 50% probability (probability $^1/_2$) that sons of carrier females will show the trait. In the situation where a carrier haemophiliac female marries a normal male they may have children with phenotypes as shown in fig 24.21.

One of the best-documented examples of the inheritance of haemophilia is shown by the descendants of Queen Victoria. It is thought that the gene for haemophilia arose as a mutation in Queen Victoria or one of her parents. Fig 24.22 shows how the haemophilia gene was inherited by her descendants.

24.11 Body colour in cats and magpie moths is controlled by a sex-linked gene on the X chromosome. The following data were obtained in two breeding experiments where the homogametic sex was homozygous for body colour in the parental generation.

	Magpie moth (normal colour dominant to pale colour)	*Cat* (black colour dominant to yellow colour)
Parental phenotypes	pale male × normal female	black male × yellow female
Offspring phenotypes	1 normal male : 1 pale female	1 yellow male : 1 black female

Which is the heterogametic sex in each of these organisms?

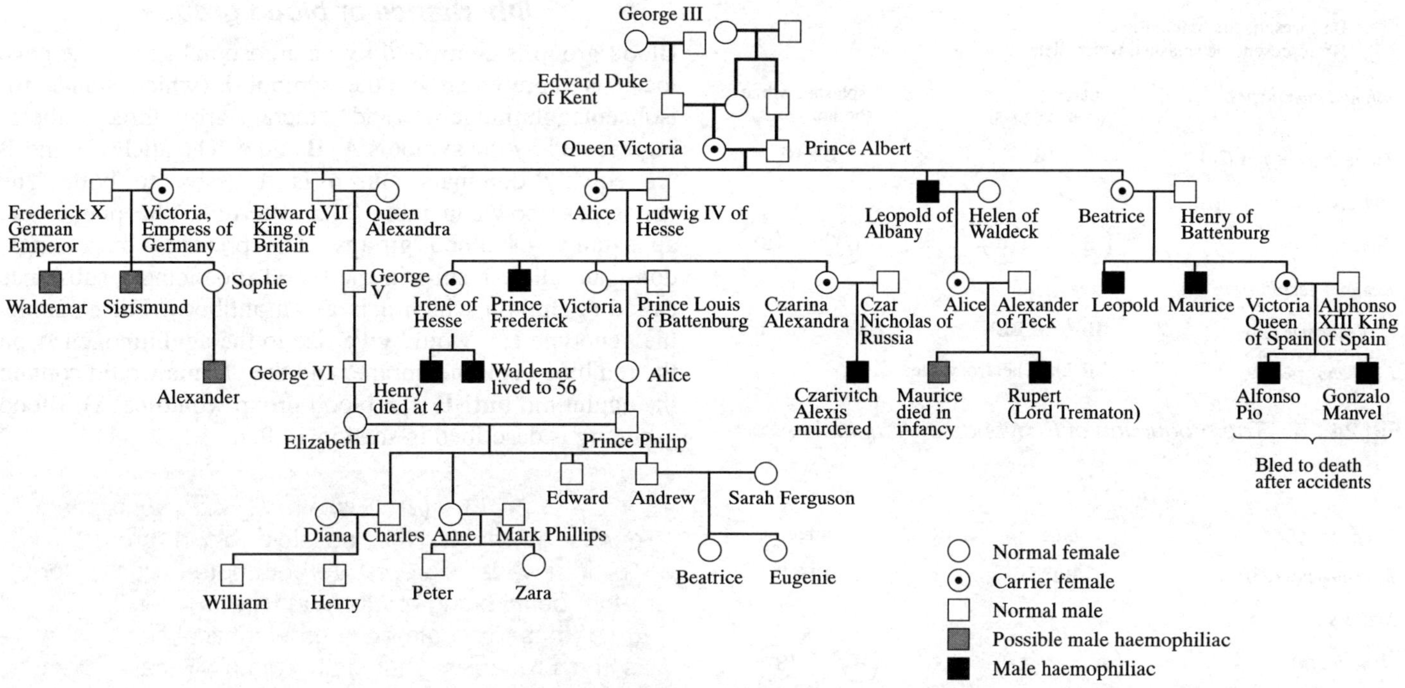

Fig 24.22 *Transmission of haemophilia in the descendants of Queen Victoria. In the diagram* ***only those descendants involved in the transmission and appearance of haemophilia have been shown.*** *The ancestry of the British Royal Family has also been given to show why haemophilia is absent from seven generations of Queen Victoria's descendants.*

24.7 Gene interactions

The topics in this chapter so far have represented the simpler aspects of genetics: dominance, monohybrid and dihybrid inheritance, linkage, sex determination and sex linkage. There are many situations in genetics where genes interact in ways other than those already described and it is probable that the majority of phenotypic characteristics in organisms result from these. Several types of gene interaction will now be considered.

24.7.1 Codominance

There are several conditions where two or more alleles do not show complete dominance or recessiveness due to the failure of any allele to be dominant in the heterozygous condition. This state of **codominance** is an exception to the situation described by Mendel in his monohybrid breeding experiments. It is fortunate that he did not select organisms which show this condition as it may have unnecessarily complicated his early work.

Codominance is found in both plants and animals. In most cases the heterozygote has a phenotype which is intermediate between the homozygous dominant and recessive conditions. An example is the production of blue Andalusian fowls by crossing pure-breeding black and splashed white parental stocks. The presence of black plumage is the result of the possession of an allele for the production of the black pigment melanin. The splashed white stock lack this allele. The heterozygotes show a partial development of melanin which produces a blue sheen in the plumage.

As there are no accepted genotypic symbols for alleles showing codominance, the importance of specifying symbols in genetic explanations is apparent. For example, in the case of the Andalusian fowl, the following genotypic symbols may be used to illustrate the alleles: black – **B**; splashed white – **b**, **W**, $\mathbf{B}^{\mathrm{W}}$ or $\mathbf{B}^{\mathrm{BW}}$. The results of a cross between black and splashed white homozygous fowl are shown in fig 24.23.

If the F_1 generation are allowed to interbreed, the F_2 generation shows a modification of the normal Mendelian phenotypic monohybrid ratio of 3:1. In this case a phenotypic ratio of 1:2:1 is produced where half the F_2 generation have the F_1 genotype (fig 24.24). This ratio of 1:2:1 is characteristic of examples of codominance. Other examples are shown in table 24.4.

24.12 In cats, the genes controlling the coat colour are carried on the X chromosomes and are codominant. A black-coat female mated with a ginger-coat male produced a litter consisting of black male and tortoiseshell female kittens. What is the expected F_2 phenotypic ratio? Explain the results.

Let:
B represent the black allele
B^W represent the splashed white allele

Parental phenotypes	black (homozygous)	×	splashed white (homozygous)	
Parental genotypes (2n)	**BB**	×	**$B^W B^W$**	
Meiosis				
Gametes (n)	B	B	×	B^W B^W
Random fertilisation				
F_1 genotypes (2n)	**BB^W**	**BB^W**	**BB^W**	**BB^W**
F_1 phenotypes	all 'blue' heterozygotes			

Fig 24.23 *The production of F_1 hybrids of Andalusian fowl.*

F_1 phenotypes	blue	×	blue	
F_1 genotypes (2n)	**BB^W**	×	**BB^W**	
Meiosis				
Gametes (n)	B B^W	×	B B^W	
Random fertilisation				
F_2 genotypes (2n)	**BB**	**BB^W**	**BB^W**	**$B^W B^W$**
F_2 phenotypes	black 1	: blue 2 :		splashed white 1

Fig 24.24 *The production of F_2 hybrids of Andalusian fowl.*

Table 24.4 Examples of codominance.

Characteristic	*Alleles*	*Heterozygous phenotype*
Antirrhinum flower (snapdragon)	red × white	pink
Mirabilis flower (four-o'clock flower)	red × white	pink
Short-horn cattle	red × white	roan
Angora and rex rabbits	long hair and short hair	intermediate silky fur

24.7.2 Multiple alleles

In all the cases studied so far, each characteristic has been controlled by a gene which may have appeared in one of two forms or alleles. There are several conditions where a single characteristic may appear in several different forms controlled by three or more alleles, of which any two may occupy the same gene loci on homologous chromosomes. This is known as the **multiple allele** (or **multiple allelomorph**) condition and it controls such characteristics as coat colour in mice, eye colour in mice and blood group in humans.

Inheritance of blood groups

Blood group is controlled by an autosomal gene. The gene locus is represented by the symbol **I** (which stands for isohaemagglutinogen) and there are three alleles represented by the symbols **A**, **B** and **o**. The alleles **A** and **B** are equally dominant and **o** is recessive to both. The genotypes shown in table 24.5 determine the phenotypic appearance of blood groups. The presence of a single dominant allele results in the blood producing a substance called agglutinin which acts as an antibody. For example, the genotype **I^AI^o** would give rise to the agglutinogen **A** on the red blood cell membrane, and the plasma would contain the agglutinin **anti-B** (the blood group would be A). Blood grouping is described in section 14.9.7.

> **24.13** (*a*) Explain, using appropriate genetic symbols, the possible blood groups of children whose parents are both heterozygous, the father being blood group A and the mother B.
> (*b*) If these parents have non-identical twins, what is the probability that both twins will have blood group A?

24.7.3 Lethal genes

There are several examples of conditions where a single gene may affect several characteristics, including mortality. In the case of humans and other mammals a certain recessive gene may lead to internal adhesion of the lungs resulting in death at birth. Another example involving a single gene affects the formation of cartilage and produces congenital deformities leading to fetal and neonatal death.

In chickens which are homozygous for an allele controlling feather structure called 'frizzled', several phenotypic effects result from the incomplete development of the feathers. These chickens lack adequate feather insulation and suffer from heat loss. To compensate for this they exhibit a range of structural and physiological adaptations, but these are largely unsuccessful and there is a high mortality rate.

The effects of a lethal gene are clearly illustrated by the inheritance of fur colour in mice. Wild mice have grey-coloured fur, a condition known as agouti. Some mice have

Table 24.5 Human blood group genotypes.

Genotype	*Blood group (phenotype)*
I^AI^A	A
I^AI^o	A
I^BI^B	B
I^BI^o	B
I^AI^B	AB
I^oI^o	O

yellow fur. Cross-breeding yellow mice produces offspring in the ratio 2 yellow fur : 1 agouti fur. These results can only be explained on the basis that yellow is dominant to agouti and that all the yellow coat mice are heterozygous. The atypical Mendelian ratio is explained by the fetal death of *homozygous* yellow coat mice (fig 24.25). Examination of the uteri of pregnant yellow mice from the above crosses revealed dead yellow fetuses. Similar examination of the uteri of crosses between yellow fur and agouti fur mice revealed no dead yellow fetuses. The explanation is that this cross would not produce homozygous yellow (**YY**) mice.

24.7.4 Gene complex

The presence of a pair of alleles occupying a given gene locus and controlling the production of a single phenotypic characteristic is true in some cases only and exceptional in most organisms. Most characteristics are determined by the interaction of several genes which form a '**gene complex**'. For example, a single characteristic may be controlled by the interaction of two or more genes situated at different loci. In the case of the inheritance of the shape of the comb in domestic fowl there are genes at two loci situated on different chromosomes which interact and give rise to four distinct phenotypes, known as pea, rose, walnut and single combs (fig 24.26). The appearance of pea comb and rose comb are each determined by the presence of their respective dominant allele (**P** or **R**) and the absence of the other dominant allele. Walnut comb results from a modified form of codominance in which at least one dominant allele for pea comb and rose comb is present (that is **PR**). Single comb appears only in the homozygous double recessive condition (that is **pprr**). These phenotypes and genotypes are shown in table 24.6.

The F_2 genotypes and F_2 phenotypic ratios resulting from crossing a pure-breeding pea-comb hen with a pure-breeding rose-comb cock are shown in fig 24.27.

Let:
Y represent yellow fur (dominant)
y represent agouti fur (recessive)

Parental phenotypes	yellow fur		×	yellow fur	
Parental genotypes (2n)	**Yy**		×	**Yy**	
Meiosis					
Gametes (n)	(Y)	(y)	×	(Y)	(y)
Random fertilisation					
Offspring genotypes (2n)	**YY**	**Yy Yy**		**yy**	
Offspring phenotypes	1 yellow fur: die before birth	2 yellow fur :		1 agouti fur	

Fig 24.25 *Genetic explanation of fur colour inheritance in mice showing the lethal genotype* **YY**.

> **24.14** In poultry, the allele for white feather (**W**) is dominant over the allele for black feather (**w**). The alleles for pea comb, **P**, and rose comb, **R**, produce the phenotypes stated. If these alleles are present together they produce a phenotype called walnut comb and if their recessive alleles are present in the homozygous condition they produce a phenotype called single comb.
>
> A cross between a black rose-comb cock and a white walnut-comb hen produced the following phenotypes:
>
> 3 white walnut-comb, 3 black walnut-comb, 3 white rose-comb, 3 black rose-comb, 1 white pea-comb, 1 black pea-comb, 1 white single-comb and 1 black single-comb.
>
> What are the parental genotypes? Show clearly how they give rise to the phenotypes described above?

24.7.5 Epistasis

A gene is said to be **epistatic** (*epi*, over) when its presence suppresses the effect of a gene at another locus. Epistatic genes are sometimes called '**inhibiting genes**' because of their effect on the other genes which are described as **hypostatic** (*hypo*, under).

Fur colour in mice is controlled by a pair of genes occupying different loci. The epistatic gene determines the presence of colour and has two alleles, coloured (dominant) and albino (white) (recessive). The hypostatic gene determines the nature of the colour and its alleles are agouti (grey) (dominant) and black (recessive). The mice may have agouti or black fur depending upon their genotypes, but this will only appear if accompanied by the allele for coloured fur. The albino condition appears in mice that are homozygous recessive for colour even if the alleles for agouti and black fur are present. Three possible phenotypes can occur and they are agouti, black and albino. A variety of phenotypic ratios can be obtained depending on the genotypes of the mating pair (fig 24.28 and table 24.7).

> **24.15** In White Leghorn fowl, plumage colour is controlled by two sets of genes, including the following:
>
> **W** (white) dominant over **w** (colour)
>
> **B** (black) dominant over **b** (brown).
>
> The heterozygous F_1 genotype **WwBb** is white. Account for this type of gene interaction and show the phenotypic ratio of the F_2 generation.

Fig 24.26 *Variation in comb shape in domestic fowl (*top left*) single comb, (*top right*) pea comb, (*bottom left*) rose comb, (*bottom right*) strawberry comb.*

Table 24.6 Phenotypes and possible genotypes associated with comb shape in poultry.

Phenotype	*Possible genotypes*
pea	**PPrr, Pprr**
rose	**RRpp, Rrpp**
walnut	**PPRR, PpRR, PPRr, PpRr**
single	**pprr**

Table 24.7 Some examples of the range of phenotypic ratios which can be produced as a result of epistatic gene interaction (see fig 24.28 for explanation of alleles).

Parental phenotypes	*Genotypes*	*Phenotypic ratios*
agouti × agouti	**AaCc × AaCc**	9 agouti : 3 black : 4 albino
agouti × black	**AaCc × aaCc**	3 agouti : 3 black : 2 albino
agouti × albino	**AaCc × Aacc**	3 agouti : 1 black : 4 albino
agouti × albino	**AaCc × aacc**	1 agouti : 1 black : 2 albino
agouti × albino	**AACc × aacc**	1 agouti : 1 albino
agouti × black	**AaCc × aaCC**	1 agouti : 1 black
albino × black	**AAcc × aaCC**	all agouti
albino × black	**AAcc × aaCc**	1 agouti : 1 albino

Fig 24.27 *Genetic explanation of comb inheritance in fowl.*

Let:
P represent presence of pea comb (dominant)
p represent absence of pea comb (recessive)
R represent presence of rose comb (dominant)
r represent absence of rose comb (recessive)

Parental phenotypes — pea comb × rose comb
Parental genotypes (2n) — **PPrr** × **RRpp**
Meiosis
Gametes (n) — (**Pr**) × (**Rp**)
Random fertilisation
F₁ genotypes (2n) — all **PpRr**
F₁ phenotypes — all walnut comb
F₁ phenotypes — walnut comb × walnut comb
F₁ genotypes (2n) — **PpRr** × **PpRr**
Meiosis
Gametes (n) (as shown by ♂ and ♀)
Random fertilisation
F₂ genotypes (2n) (shown in Punnett square)

♀ \ ♂	**PR**	**Pr**	**pR**	**pr**
PR	**PR PR** ○	**Pr PR** ○	**pR PR** ○	**pr PR** ○
Pr	**PR Pr** ○	**Pr Pr** □	**pR Pr** ○	**pr Pr** □
pR	**PR pR** ○	**Pr pR** ○	**pR pR** △	**pr pR** △
pr	**PR pr** ○	**Pr pr** □	**pR pr** △	**pr pr** ◌

F₂ phenotypes — 9 walnut comb : 3 pea comb : 3 rose comb : 1 single comb
Offspring symbols — ○ □ △ ◌

Fig 24.28 (below) *A genetic explanation of how unusual phenotypic ratios can be produced in the case of epistatic genes.*

Let:
A represent agouti fur (dominant)
a represent black fur (recessive)
C represent coloured fur (dominant)
c represent albino fur (recessive)

Parental phenotypes — agouti × albino
Parental genotypes (2n) — **AaCc** × **Aacc**
Meiosis
Gametes (n) (as shown by ♂ and ♀)
Random fertilisation
Offspring genotypes (2n) (as shown in Punnett square)

♀ \ ♂	**AC**	**Ac**	**aC**	**ac**
Ac	**AC Ac** ○	**Ac Ac** □	**aC Ac** ○	**ac Ac** □
ac	**AC ac** ○	**Ac ac** □	**aC ac** △	**ac ac** □

Offspring phenotypes — 3 agouti : 4 albino : 1 black
Offspring symbols — ○ □ △

24.7.6 Polygenic inheritance

Many of the obvious characteristics of organisms are produced by the combined effect of many different genes. These genes form a special gene complex known as a polygenic system. Whilst the effect of each gene alone is too small to make any significant impression on the phenotype, the almost infinite variety produced by the combined effect of these genes (**polygenes**) has been shown to form the genetic basis of **continuous variation**, which is described further in section 24.8.2.

24.8 Variation

The term variation describes the difference in characteristics shown by organisms belonging to the same natural population or species. It was the amazing diversity of structure within any species that caught the attention of Darwin and Wallace during their travels. The regularity and predictability with which these differences in characteristics were inherited formed the basis of Mendel's research. Whilst Darwin recognised that particular characteristics could be developed by selective breeding, as described in section 25.4.2, it was Mendel who explained the mechanism by which selected characteristics were passed on from generation to generation.

Mendel described how hereditary factors determine the genotype of an organism which in the course of development becomes expressed in the structural, physiological and biochemical characteristics of the phenotype. Whilst the phenotypic appearance of any characteristic is ultimately determined by the genes controlling that characteristic, the extent to which certain characteristics develop may be more influenced by the environment.

A study of the phenotypic differences in any large population shows that two forms of variation occur, discontinuous and continuous. Studies of variation in a character involve measuring the expression of that characteristic in a large number of organisms within the population, such as height in humans. The results are plotted as histograms or a graph which reveals the **frequency distribution** of the variations of that characteristic within the population. Typical results obtained from such studies are shown in fig 24.29 and they highlight the difference between the two forms of variation.

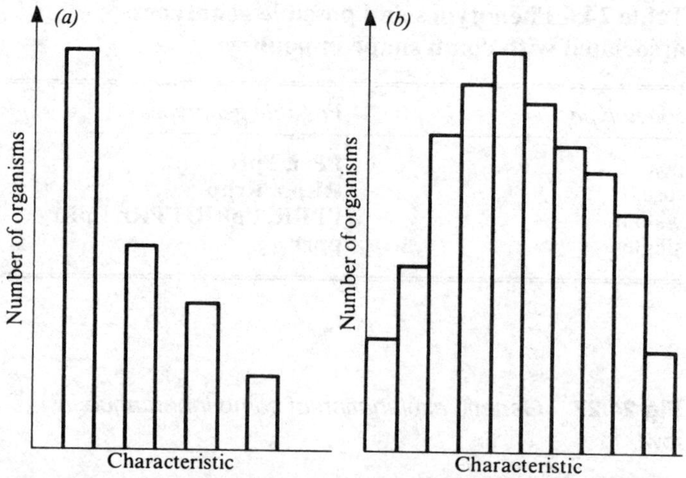

Fig 24.29 *Histograms representing frequency distribution in the case of (a) discontinuous variation and (b) continuous variation.*

24.8.1 Discontinuous variation

There are certain characteristics within a population which exhibit a limited form of variation. Variation in this case produces individuals showing clear-cut differences with no intermediates between them, such as blood groups in humans, wing lengths in *Drosophila*, melanic and light forms in *Biston betularia*, style length in *Primula* and sex in animals and plants. Characteristics showing discontinuous variation are usually controlled by one or two major genes which may have two or more allelic forms and their phenotypic expression is relatively unaffected by environmental conditions.

Since the phenotypic variation is restricted to certain clear-cut characteristics, this form of variation is alternatively known as **qualitative inheritance**, as opposed to **quantitative inheritance** which is characteristic of continuous variation.

24.8.2 Continuous variation

Many characteristics in a population show a complete gradation from one extreme to the other without any break. This is illustrated most clearly by characteristics such as mass, linear dimension, shape and colour of organs and organisms. The frequency distribution for a characteristic exhibiting continuous variation is a **normal distribution curve** (appendix section A2.3.3). Most of the organisms in the population fall in the middle of the range with approximately equal numbers showing the two extreme forms of the characteristic. Characteristics exhibiting continuous variation are produced by the combined effects of many genes (polygenes) and environmental factors.

Individually each of these genes has little effect on the phenotype but their combined effect is significant.

24.8.3 Influence of the environment

The ultimate factor determining a phenotypic characteristic is the genotype. At the moment of fertilisation the genotype of the organism is determined, but the subsequent degree of expression allowed to this genetic potential is influenced greatly by the action of environmental factors during the development of the organism. For example, Mendel's tall variety of garden pea normally attained a height of six feet. However, it would only do so if provided with adequate light, water and soil conditions. A reduction in the supply of any of these factors (**limiting factors**) would prevent the gene for height exerting its full effect. It was the Danish geneticist Johanssen who demonstrated the effect of the interaction of genotypic and environmental factors on phenotype. In a series of experiments on the mass of dwarf bean seeds he selected the heaviest and lightest seeds from each generation of self-pollinating dwarf bean plants and used these to produce the next generation. After repeating these experiments for several years he found only small differences in the mean mass of seeds from the same selected line, that is heavy or light, but large differences in mean mass of seeds from different selected lines, that is heavy and light. This suggested that both heredity and environment were influencing the phenotypic appearance of the characteristic. From these results it is possible to describe continuous phenotypic variation as being '**the cumulative effect of varying environmental factors acting on a variable genotype**'. The results also indicated that the extent to which a characteristic is inherited is determined primarily by the genotype. In the development of human characteristics such as personality, temperament and intelligence, there is evidence to suggest that both **nature** (hereditary factors) and **nurture** (environmental factors) interact to varying degrees in different individuals to influence the final appearance of the characteristic. It is these genetic and environmental differences which act to produce phenotypic differences between individuals. There is no firm evidence, as yet, to suggest that one factor is universally more influential than the other, but the environment can never increase the extent of the phenotype beyond that determined by the genotype (fig 24.30).

24.8.4 Sources of variation

It will be appreciated that, as a result of the interaction between discontinuous and continuous variations and the environment, no two organisms will possess identical phenotypes. Replication of DNA is so nearly perfect that there is little possibility of variation occurring in the genotypes of asexually reproducing organisms. Any apparent variation between these organisms is therefore almost certainly the result of environmental influences. In the case of sexually reproducing organisms there is ample opportunity for genetic variation to arise. Two processes occurring during meiosis, and the fusion of gametes during fertilisation, provide the means of introducing unlimited genetic variation into the population. These may be summarised as follows:

(1) *Crossing-over* – reciprocal crossing-over of genes between chromatids of homologous chromosomes may occur during prophase I of meiosis. This produces new linkage groups and so provides a major source of genetic recombination of alleles (sections 24.3 and 23.3).

(2) *Independent assortment* – the orientation of the chromatids of homologous chromosomes (bivalents) on

Fig 24.30 *Phenotypic variation in human height. All these children are the same age.*

the equatorial spindle during metaphase I of meiosis determines the direction in which the pairs of chromatids move during anaphase I. This orientation of the chromatids is random. During metaphase II the orientation of pairs of chromatids once more is random and determines which chromosomes migrate to opposite poles of the cell during anaphase II. These random orientations and the subsequent independent assortment (segregation) of the chromosomes give rise to a large calculable number of different chromosome combinations in the gametes (section 24.2.1).

(3) *Random fusion of gametes* – a third source of variation occurs during sexual reproduction as a result of the fact that the fusion of male and female gametes is completely random (at least in theory). Thus, any male gamete is potentially capable of fusing with any female gamete.

These sources of genetic variation account for the routine '**gene reshuffling**' which is the basis of continuous variation. The environment acts on the range of phenotypes produced and those best suited to it thrive. This leads to changes in allele and genotypic frequencies as described in chapter 27. However, these sources of variation do not generate the major changes in genotype which are necessary in order to give rise to new species as described by evolutionary theory. These changes are produced by mutations.

24.9 Mutation

A mutation is a change in the amount, arrangement or structure of the DNA of an organism. This produces a change in the genotype which may be inherited by cells derived by mitosis or meiosis from the mutant cell. A mutation may result in the change in appearance of a characteristic in a population. Mutations occurring in gamete cells are inherited, whereas those occurring in somatic cells can only be inherited by daughter cells produced by mitosis. The latter are known as **somatic mutations**.

A mutation resulting from a change in the amount or arrangement of DNA is known as **chromosomal mutation** or **chromosomal aberration**. Some forms of these affect the chromosomes to such an extent that they may be seen under the microscope. Increasingly the term mutation is being used only when describing a change in the structure of the DNA at a single locus and this is known as a **gene mutation** or **point mutation**.

The concept of mutation as the cause of the sudden appearance of a new characteristic was first proposed by the Dutch botanist Hugo de Vries in 1901, following his work on inheritance in the evening primrose *Oenothera lamarckiana*. Nine years later T. H. Morgan began a series of investigations into mutations in *Drosophila* and, with the assistance of geneticists throughout the world, identified over 500 mutations.

24.9.1 Mutation frequency and causes of mutation

Mutations occur randomly and spontaneously; that is to say any gene can undergo mutation at any time. The rates at which mutations occur vary between organisms.

As a result of the work of H. J. Muller in the 1920s it was observed that the frequency of mutation could be increased above the spontaneous level by the effects of X-rays. Since then it has been shown that the mutation rates can be significantly increased by the effects of high energy electromagnetic radiation such as ultra-violet light, X-rays and γ rays. High-energy particles, such as α and β particles, neutrons and cosmic radiation, are also **mutagenic**, that is cause mutations. A variety of chemical substances, including mustard gas, caffeine, formaldehyde, colchicine, certain constituents of tobacco and an increasing number of drugs, food preservatives and pesticides, have been shown to be mutagenic.

24.9.2 Chromosome mutations

Chromosomal mutations may be the result of changes in the number or structure of chromosomes. Certain forms of chromosomal mutation may affect several genes and have a more profound effect on the phenotype than gene mutations. Changes in the number of chromosomes are usually the result of errors occurring during meiosis but they can also occur during mitosis. These changes may involve the loss or gain of single chromosomes, a condition called **aneuploidy**, or the increase in entire haploid sets of chromosomes, a condition called **euploidy (polyploidy)**.

Aneuploidy

In this condition half the daughter cells produced have an extra chromosome $(n+1)$, $(2n+1)$ and so on, whilst the other half have a chromosome missing $(n-1)$, $(2n-1)$ and so on. Aneuploidy can arise from the failure of a pair, or pairs, of homologous chromosomes to separate during anaphase I of meiosis. If this occurs, both sets of chromosomes pass to the same pole of the cell and separation of the homologous chromosomes during anaphase II may lead to the formation of gamete cells containing either one or more chromosomes too many or too few as shown in fig 24.31. This is known as **non-disjunction**. Fusion of either of these gametes with a normal haploid gamete produces a zygote with an odd number of chromosomes.

Zygotes containing less than the diploid number of chromosomes usually fail to develop, but those with extra chromosomes may develop. In most cases where this occurs in animals it produces severe abnormalities. One of the commonest forms of chromosomal mutation in humans resulting from non-disjunction is a form of trisomy called Down's syndrome $(2n=47)$. The condition, which is named

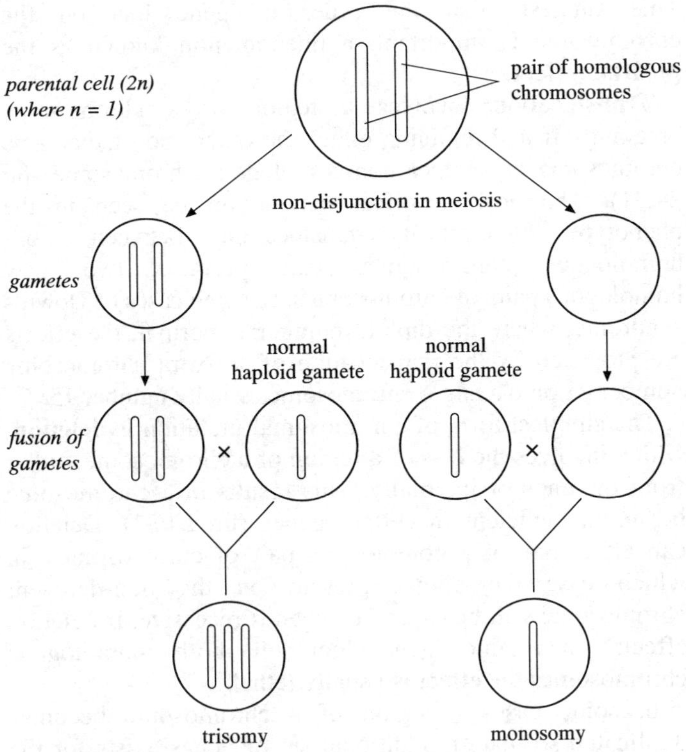

Fig 24.31 *Non-disjunction in gamete cell formation and the results of fusion of these abnormal gametes with normal haploid cells. The resulting cells may show a form of polysomy where the chromosome number may be (2n + 1) trisomy, (2n + 2) tetrasomy, (2n + 3) pentasomy etc., or monosomy (2n − 1) depending upon the number of homologous chromosomes which fail to separate normally (see also fig 25.34).*

after the doctor who first described it in 1866, is due to the non-disjunction of the two chromosomes 21. It is described in section 25.7.6.

Non-disjunction of the male and female sex chromosomes may also occur and produce aneuploidy affecting secondary sexual characteristics, fertility and, in some cases, intelligence (sections 25.7.7 and 25.7.8).

Euploidy (polyploidy)

Gamete and somatic cells containing multiples of the haploid number of chromosomes are called **polyploids**, and the prefixes tri-, tetra-, and so on, indicate the extent of polyploidy, for example $3n$ is triploid, $4n$ is tetraploid, $5n$ is pentaploid and so on. Polyploidy is much more common in plants than in animals. For example, approximately half the 300 000 known species of angiosperms are polyploid. The relatively low occurrence in animals is explained by the fact that the increased number of chromosomes in polyploids makes normal gamete formation during meiosis much more prone to error. Since most plants are capable of propagating themselves vegetatively they are able to reproduce despite being polyploid. Polyploidy is often associated with advantageous features such as increased size, hardiness and resistance to disease. This is called **hybrid vigour** (section 27.4.2). Most of our domestic plants are polyploids producing large fruits, storage organs, flowers or leaves.

There are two forms of polyploidy, autopolyploidy and allopolyploidy.

Autopolyploidy. This condition may arise naturally or artificially as a result of an increase in number of chromosomes within the same species. For example, if chromosomes undergo replication (during interphase) and the chromatids separate normally (during anaphase) but the cytoplasm fails to cleave (during cytokinesis), a **tetraploid** ($4n$) cell with a large nucleus is produced. This cell will undergo division and produce tetraploid cells. The amount of cytoplasm in these cells increases to preserve the ratio of the volumes of nucleus : cytoplasm and leads to an increase in the size of the whole plant or some part of it. Autopolyploidy can be induced by the use of a drug called **colchicine** which is extracted from the corm of the autumn crocus (*Colchicum*). Concentrations of about 0.01% inhibit spindle formation by disrupting microtubules so that the chromatids fail to separate during anaphase. Colchicine and related drugs have been used in the breeding of certain varieties of economically important crops such as tobacco, tomatoes and sugarbeet. Autopolyploids can be as fertile as diploids if they have an even number of chromosomes sets.

A modified form of polyploidy can occur in animals and gives rise to cells and tissues which are polyploid. This process is called **endomitosis** and involves chromosome replication without cell division. The giant chromosomes in the salivary glands of *Drosophila* and tetraploid cells in the human liver are produced by endomitosis.

Allopolyploidy. This condition arises when the chromosome number in a sterile hybrid becomes doubled and produces fertile hybrids. F_1 hybrids produced from different species are usually sterile since their chromosomes cannot form homologous pairs during meiosis. This is called **hybrid sterility**. However, if multiples of the original haploid number of chromosomes, for example $2(n_1 + n_2)$, $3(n_1 + n_2)$ and so on (where n_1 and n_2 are the haploid numbers of the parent species) occur, a new species is produced which is fertile with polyploids like itself but infertile with both parental species.

Most allopolyploid species have a diploid chromosome number which is the sum of the diploid numbers of their parental species; for example rice grass, *Spartina anglica* ($2n = 122$), is a fertile allopolyploid hybrid produced from a cross between *Spartina maritima* (*stricta*) ($2n = 60$) and *Spartina alterniflora* ($2n = 62$). The F_1 hybrid formed from the latter two species is sterile and is called *Spartina townsendii* ($2n = 62$). Most allopolyploid plants have different characteristics from either parental species, and include many of our most economically important plants. For example, the species of wheat used to make bread, *Triticum aestivum* ($2n = 42$), has been selectively bred over a period of more than 5000 years. By crossing a wild variety of wheat, einkorn wheat ($2n = 14$), with 'wild grass' ($2n = 14$), a different species of wheat, emmer wheat

($2n=28$), was produced. Emmer wheat was crossed with another species of wild grass ($2n=14$) to produce *Triticum aestivum* ($2n=42$) which actually represents the hexaploid condition ($6n$) of the original einkorn wheat. Another example of interspecific hybridisation involving crossing the radish and cabbage is described in section 27.9.

Allopolyploidy does not occur in animals because there are fewer instances of cross-breeding between species. Polyploidy does not add new genes to a gene pool (section 27.1.1) but gives rise to a new combination of genes.

Structural changes in chromosomes

Crossing-over during prophase I of meiosis involves the reciprocal transfer of genetic material between homologous chromosomes. This changes the allele sequence of parental linkage groups and produces recombinants, but no gene loci are lost. Similar effects to these are produced by the structural changes in chromosomes known as inversions and translocations. In other forms of change, such as deletions and duplications, the number of gene loci on chromosomes is changed, and this can have profound effects on the phenotypes. Structural changes in chromosomes resulting from inversion, deletion and duplication, and in some cases from translocation, may be observed under the microscope when homologous chromosomes attempt to pair during prophase I of meiosis. Homologous genes undergo synapsis (pairing) (section 23.3) and a loop or twist is formed in one of the homologous chromosomes as a result of the structural change. Which chromosome forms the loop and the arrangement of its genes depends upon the type of structural change.

Inversion occurs when a region of a chromosome breaks off and rotates through 180° before rejoining the chromosome. No change in genotype occurs as a result of inversion but phenotypic changes may be seen (fig 24.32).

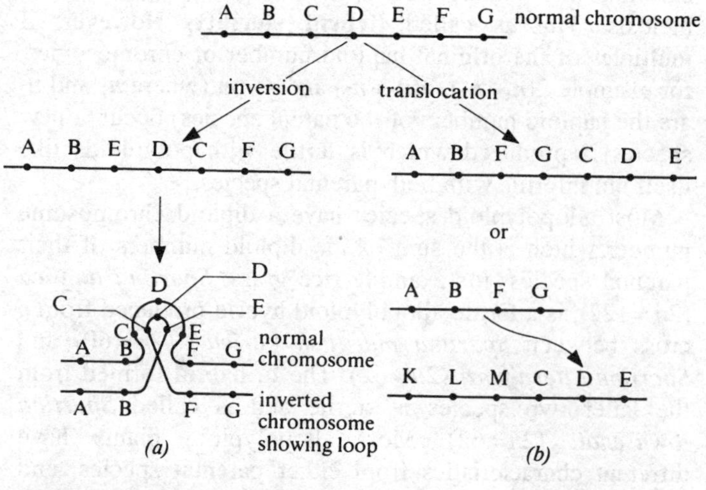

Fig 24.32 *Diagrammatic representation of inversion and translocation and their effects on the positions of genes A–G. (a) Looping in prophase due to inversion. (b) Part of the chromosome carrying genes C, D and E has broken off and become attached to the chromosome carrying genes K, L and M.*

This suggests that the order of gene loci on the chromosome is important, a phenomenon known as the **position effect**.

Translocation involves a region of a chromosome breaking off and rejoining either the other end of the same chromosome or another non-homologous chromosome (fig 24.31). The position effect may again be seen in the phenotype. Reciprocal translocation between non-homologous chromosomes can produce two new homologous pairs of chromosomes. In some cases of Down's syndrome, where the diploid number is normal, the effects are produced by the translocation of an extra chromosome number 21 onto a larger chromosome, usually number 15.

The simplest form of chromosomal mutation is **deletion**, which involves the loss of a region of a chromosome, either from the ends or internally. This results in a chromosome becoming deficient in certain genes (fig 24.33). Deletion can affect one of a homologous pair of chromosomes, in which case the alleles present on the non-deficient chromosome will be expressed even if recessive. If deletion affects the same gene loci on both homologous chromosomes the effect is usually lethal.

In some cases a region of a chromosome becomes duplicated so that an additional set of genes exists for the region of **duplication**. The additional region of genes may be incorporated within the chromosome or at one end of the chromosome, or become attached to another chromosome (fig 24.33).

24.9.3 Gene mutations

What is a gene mutation?

Sudden and spontaneous changes in phenotype, for which there are no conventional genetic explanations or any microscopic evidence of chromosomal mutation, can only be explained in terms of changes in gene structure. A **gene mutation** or **point mutation** (since it applies to a particular

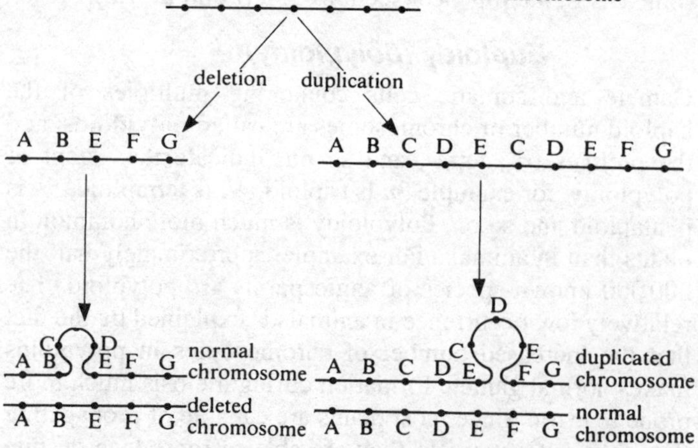

Fig 24.33 *Diagrammatic representations of deletion and duplication and their effects on the positions of genes A–G. In both cases looping can be seen.*

gene locus) is the result of a change in the nucleotide sequence of the DNA molecule in a particular region of the chromosome. Such a change in the base sequence of the gene is transmitted to mRNA during transcription and may result in a change in the amino acid sequence of the polypeptide chain produced from it during translation at the ribosomes.

Types of gene mutation

There are a variety of forms of gene mutation involving the addition, loss or rearrangement of bases in the gene. These mutations take the form of the **duplication**, **insertion**, **deletion**, **inversion** or **substitution** of bases. In all cases they change the nucleotide sequence and result in the formation of a modified polypeptide. For example, deletion causes a **frame-shift** and the implications of this are described in section 23.7.3.

Consequences of gene mutations

Gene mutations occurring during gamete formation are transmitted to all the cells of the offspring and may be significant for the future of the species. Somatic gene mutations which arise in the organism are inherited only by those cells derived from the mutant cells by mitosis. Whilst they may affect that organism, they are lost on the death of the organism. Somatic mutations are probably very common and go unnoticed, but in some cases they may produce cells with an increased rate of growth and division. These cells may give rise to a tumour which may be **benign** and not affect other tissues, or **malignant**, which live parasitically on healthy cells, a condition known as **cancer**.

The effects of gene mutation are extremely variable. Most minor gene mutations pass unnoticed in the phenotype since they are recessive, but there are several cases where a change in a single base in the genetic code can have a profound effect on the phenotype. **Sickle cell anaemia** in humans is an example of **base substitution** mutation affecting a base in one of the genes involved in the production of haemoglobin. This condition and its cause is described in more detail in section 25.7.2.

24.9.4 Implications of mutation

The effects of chromosome and gene mutations are very variable. In many cases the mutations are lethal and prevent development of the organism, for example in humans about 20% of pregnancies end in natural abortion before 12 weeks and of these about 50% exhibit a chromosome abnormality. Some forms of chromosome mutation may bring certain gene sequences together, and that combined effect may produce a 'beneficial' characteristic. Another significance of bringing certain genes closer together is that they are less likely to be separated by crossing-over and this is an advantage with beneficial genes.

Gene mutation may lead to several alleles occupying a specific locus. This increases both the heterozygosity and size of the gene pool of the population and leads to an increase in variation within the population. Gene reshuffling as a result of crossing-over, independent assortment, random fertilisation and mutations, may increase the amount of continuous variation but the evolutionary implications of this are often short-lived since the changes produced may be rapidly diluted. Certain gene mutations, on the other hand, increase discontinuous variation and this has the more profound effect on changes in the population. Most gene mutations are recessive to the 'normal' allele which has come to form genetic equilibrium with the rest of the genotype and the environment as a result of successfully withstanding selection over many generations. Being recessive the mutant alleles may remain in the population for many generations until they come together in the homozygous condition and are expressed phenotypically. Occasionally a dominant mutant allele may arise in which case it will appear immediately in the phenotype (section 27.5, *Biston betularia*).

The information provided in this chapter accounts for the origins of variation within populations and the mechanism by which characteristics are inherited, but it does not explain how the amazing diversity of living organisms described in chapter 2 may have arisen. Possible answers to this problem form the basis of the next three chapters.

Chapter Twenty-five

Applied genetics

In the second half of the twentieth century, biology has entered what some scientists have referred to as its 'Golden Age'. From the discovery of the structure of DNA in 1953 to the ability to write the genetic code for a human being by the end of the century, the relatively new science of molecular biology has combined with genetics to give us a new and powerful biotechnology. It will have applications in industry, medicine and agriculture and many other fields. Its importance is reflected in the fact that the United States now spends about half its academic research budget on the life sciences. Like all new knowledge, it can be used for the benefit of humankind, but also presents new dangers. Physicists faced the same problem in the first half of the twentieth century when knowledge of the structure of the atom led to our ability to use nuclear power, with its potential for peaceful or destructive use. In fact, both Francis Crick and Maurice Wilkins, who with James Watson won the Nobel prize for their work in discovering the structure of DNA, had been physicists and had worked on weapons research during the Second World War before moving on to the science of life. Past experience has made scientists very aware of the social and ethical implications of their research and, as we shall see later, they even slowed down their work in the late 1970s while strict regulations and guidelines for genetic engineering were worked out. As a student of biology you will share in the responsibility for discussing the new issues that will certainly arise from our expanding knowledge of molecular genetics. The more informed our opinions are, and the more people are prepared to discuss the issues, the more likely we all are to benefit.

Genetic engineering

In the first part of this chapter we shall be looking at genetic engineering. This is the most powerful technique available in applied genetics and biotechnology. It gives us the power to study and to change the genetic instructions of an organism, including ourselves. Other living organisms can be changed for the benefit of humans, and we are even beginning to manipulate our own genes to cure genetic diseases ('gene therapy'). A brief summary of the history and applications of genetic engineering is given in table 25.1.

25.1 Genetic engineering of bacteria

The basic techniques of genetic engineering were worked out in the early 1970s. It usually involves inserting a new gene into an organism. The gene may be newly synthesised or transferred from another organism. In the case of bacteria, genetic engineering turns the bacterium into a living factory for the production of whatever protein the gene codes for. Examples we shall study later are the transfer of genes for human insulin, human growth hormone and bovine somatotrophin (BST) (sections 25.2.1–25.2.3).

25.1.1 Overview

It is now a routine process to be able to obtain copies of any gene. In some cases only a single original molecule is required. Making many identical copies of a molecule is called **cloning**. Traditionally it relies on the use of plasmids or bacteriophages. **Plasmids** are small circular pieces of DNA found in certain bacteria. They are separate from the bulk of the DNA and can replicate independently of the rest of the DNA (section 2.3.1). **Bacteriophages** (known as phages for short) are viruses which can 'inject' their DNA into bacteria for replication (fig 2.21). The piece of DNA to be cloned is combined with either a plasmid or the DNA of a phage. This modified plasmid or phage DNA is called recombinant DNA. **Recombinant DNA** is the name given to DNA formed after a piece from one organism is joined to a piece from another organism. If it is inserted into a bacterium, it will replicate (clone) itself and as the bacterium multiplies, so the recombinant DNA will multiply. If desired, the cloned DNA can be separated from the plasmid or phage DNA again. This allows, for example, its base sequence to be determined. While inside the bacterium the new gene may be active and used to make a useful protein, such as human insulin, which would not normally be made in that cell. The protein can later be extracted.

The plasmid or phage is known as the '**vector**' or '**cloning vector**', because it acts as a carrier for the DNA to be cloned. The process is summarised in fig 25.1 and described in more detail below. Inserting new genes into the

Table 25.1 A brief summary of the history and development of genetic engineering. (Based on table G7, p. 418, *The encyclopaedia of molecular biology*, ed. Sir John Kendrew (1994) Blackwell Science.)

1960s–1970s	Isolation of restriction enzymes and their use to analyse DNA structure.
1972–73	DNA cloning techniques involving recombinant DNA developed. First gene cloned (bacterial).
1974	First expression in a bacterium of a gene from a different species.
1977	First complete genetic code of an organism (base sequence of a complete genome). The organism was the phage ΦX 174 and its genetic code is 5375 bases long.
1978	Bacteria produce human somatostatin from a synthetic gene. Later the same year bacteria also produce human insulin from a synthetic gene.
1982	Insulin (Eli Lilly's Humulin) is the first product made by genetically engineered bacteria to be approved for use in Britain and the USA.
1981/82	First transgenic animals (mice) produced.
1983	First transgenic plants produced.
1985	First transgenic farm animals produced (rabbits, pigs and sheep).
1986	First controlled release of genetically engineered organisms into the environment.
1989	First patented transgenic animal, the oncomouse.
1990	Human genome project started. First successful gene therapy for SCID (section 25.7.11) in USA.
1990–92	First transgenic cereal plants (maize and wheat).
1992	Regulations for deliberate release of genetically engineered organisms established in the USA and EU. First complete base sequence of a chromosome (yeast chromosome III).
1993	First human gene therapy trial in UK. Gene therapy for cystic fibrosis and SCID begun in UK.
1994	Genetically engineered tomato marketed in the USA.
1996	Genetically engineered tomato marketed in Britain.
1997	First cloned mammal produced from a single cell. The sheep, Dolly, was developed from a single udder cell.

embryos of plants or animals to create what are known as **transgenic** organisms (organisms which can pass their genes on to their offspring) is more difficult and will be discussed in sections 25.3–25.5.

Genetic engineering in bacteria can be broken down into five stages.

- Stage 1: Obtain a copy of the required gene from among all the others in the DNA of the donor organism.
- Stage 2: Place the gene in a vector.
- Stage 3: Use the vector to introduce the gene into the host cell.
- Stage 4: Select the cells which have taken up the foreign DNA (the DNA of the donor).
- Stage 5: Clone the gene.

This is the simplest order of stages. However, they may not be carried out in this order, for example a group (or library) of genes may be inserted into a vector and cloned before it is possible to isolate the required gene.

25.1.2 Stage 1: Obtaining a copy of the gene required

This is the most difficult part of the process. For example, there are around 3000 million bases and 100 000 genes in the human genome (the total DNA in a human cell). A typical gene is several thousand base pairs long, so even to find a particular gene presents a difficult problem. Three methods are used to get a copy of a gene:

- make a copy of the gene from its mRNA, using reverse transcriptase;
- synthesise the gene artificially;
- use a 'shotgun' approach, which involves chopping up the DNA with 'restriction enzymes' and searching for the piece with the required gene.

The first two methods are the most straightforward and the technique of genetic engineering will be illustrated with these first. The third method will then be examined.

Using reverse transcriptase

Although there are only two copies of each gene in a diploid cell (one on a chromosome from the female parent and one on a chromosome from the male parent), if the gene is active it usually produces thousands of mRNA molecules which are complementary to the gene (section 23.8). It is often known in which cells the gene is active. For example, the gene for insulin is active in the β cells of the pancreas. Retroviruses contain an enzyme which can make a complementary DNA copy of an RNA molecule. (Making DNA from RNA is the opposite process to normal transcription where RNA is made from DNA.) The enzyme was therefore named **reverse transcriptase**. It has come to

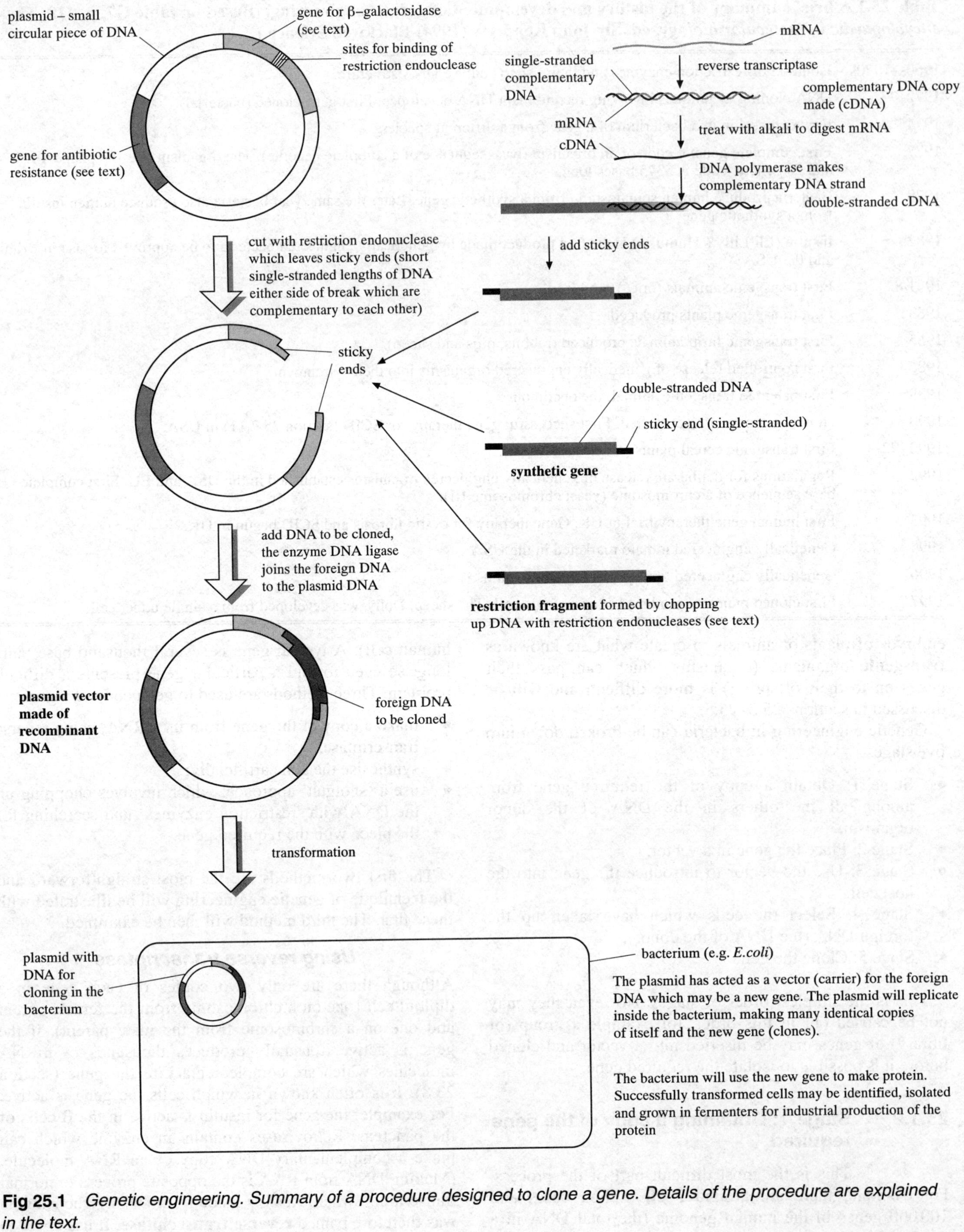

Fig 25.1 *Genetic engineering. Summary of a procedure designed to clone a gene. Details of the procedure are explained in the text.*

be a valuable tool in genetic engineering. (The virus uses it to turn its RNA genetic code into DNA so that it can infect new cells (section 2.4.5).) For some genes it is relatively easy to isolate the mRNA from that particular gene in a particular type of cell. Once this has been done, the procedure shown in fig 25.1 is used to make the gene coding for the required protein. The DNA formed in this way is called **complementary DNA**, or **cDNA**, whether it is single-stranded or double-stranded.

Synthesising a gene

The base sequence of a gene can be found directly, or a suitable base sequence can be worked out from the amino acid sequence of the protein it makes. A gene can then be constructed using nucleotides (remember each base is part of one nucleotide) and joining them together in the right order. This is only possible for short genes at present but, as techniques improve, this could become a routine possibility for any gene. It has been used for the synthesis of proinsulin and somatostatin genes. Somatostatin (otherwise known as growth hormone inhibitory hormone) is a protein hormone which contains only 14 amino acids.

The shotgun approach – using restriction enzymes

This was the original method for isolating genes and came with the discovery of enzymes called **restriction endonucleases** in the late 1960s and early 1970s. These enzymes are found in bacteria and they cut up DNA. Their function in bacteria is to cut up any invading virus DNA, thus *restricting* the multiplication of viruses in the cell. Different species of bacteria produce different restriction endonucleases. An endonuclease cuts (digests) a nucleic acid (hence 'nuclease') at specific points along the length of the molecule ('endo'- means internally, rather than attacking the molecule from the ends). The enzyme recognises a particular sequence of bases and cuts at these points. These cutting points are called **restriction sites**. Different enzymes attack different sequences. Well over 2000 different enzymes have now been isolated which attack 230 different sequences. The bacterium protects its own DNA by adding a methyl group to certain bases at the cutting sites.

Each enzyme is named after the bacterium from which it comes (fig 25.2). Note that the base sequence is often six bases long and palindromic, meaning that it reads the same in both directions. When studying fig 25.2 remember that the two complementary strands of DNA run in opposite directions. Some restriction enzymes leave a staggered cut with single-stranded ends (e.g. EcoRI, fig 25.2). These ends are described as '**sticky ends**' because they can be used to re-join fragments of DNA. They stick together by forming hydrogen bonds to complementary sticky ends from other DNA molecules cut by the same restriction enzyme (fig 25.1). For example, EcoRI produces the sticky end –TTAA. Some produce **blunt ends**, for example HindII in fig 25.2. In this way the DNA of any organism can be chopped up into pieces of different size. The pieces are known as **restriction fragments**. The different lengths of these fragments depend on the restriction enzyme used, and on where the particular base sequences that the enzyme recognises are located (fig 25.3).

Each nucleotide in a piece of DNA carries a phosphate group which is negatively charged. Thus different lengths of DNA carry different total charges. These differences can be used to separate pieces of DNA of different length by placing them in an electrical field and allowing them to migrate to the positive electrode. This is done in a gel, and the technique is known as **gel electrophoresis** (fig 25.4). The gel is made of agarose (for very large fragments) or polyacrylamide (for smaller fragments). DNA is colourless, so its final position is revealed by staining or by using radioactive DNA and carrying out **autoradiography** which involves exposure of the gel to photographic film. The radiation blackens the film, showing the location of the DNA.

Having chopped up the DNA of the donor organism, one of the fragments will hopefully, by chance, contain an entire copy of the desired gene and not much else. This technique is sometimes referred to as the shotgun approach because it is not very specific and relies on chopping up all the DNA. It leaves the problem of finding the fragment with the desired gene which is discussed in section 25.1.3 below.

Split genes

Using reverse transcriptase or gene synthesis has an advantage over the shotgun approach in that the gene that is made is not a '**split gene**'. Split genes contain one or more sections of DNA called **introns** which are not part of the code for the final protein. The function of introns is still unclear, however if a eukaryote gene containing introns is placed in a bacterium, the bacterium does not have the necessary enzymes to remove the introns from the mRNA and a useless protein will therefore be made. Introns can be removed from the mRNA as shown in fig 25.5.

25.1.3 Stage 2: Putting genes into a vector

As explained earlier, the vectors most commonly used are plasmid DNA and phage DNA. The procedure for plasmid DNA will be described first, but is the same in principle to the procedure for phage DNA.

Plasmid DNA

The circular plasmid DNA molecules in bacteria are much smaller than those of the main chromosomal DNA and can easily be separated on the basis of size. The bacterial cells are broken open and chromosomal DNA is centrifuged down, leaving the plasmid DNA in the liquid (supernatant) above the pellet. The plasmids are then purified before cutting with a restriction enzyme (fig 25.1).

If a restriction enzyme has been used to isolate the donor DNA (i.e. the shotgun approach) the same restriction

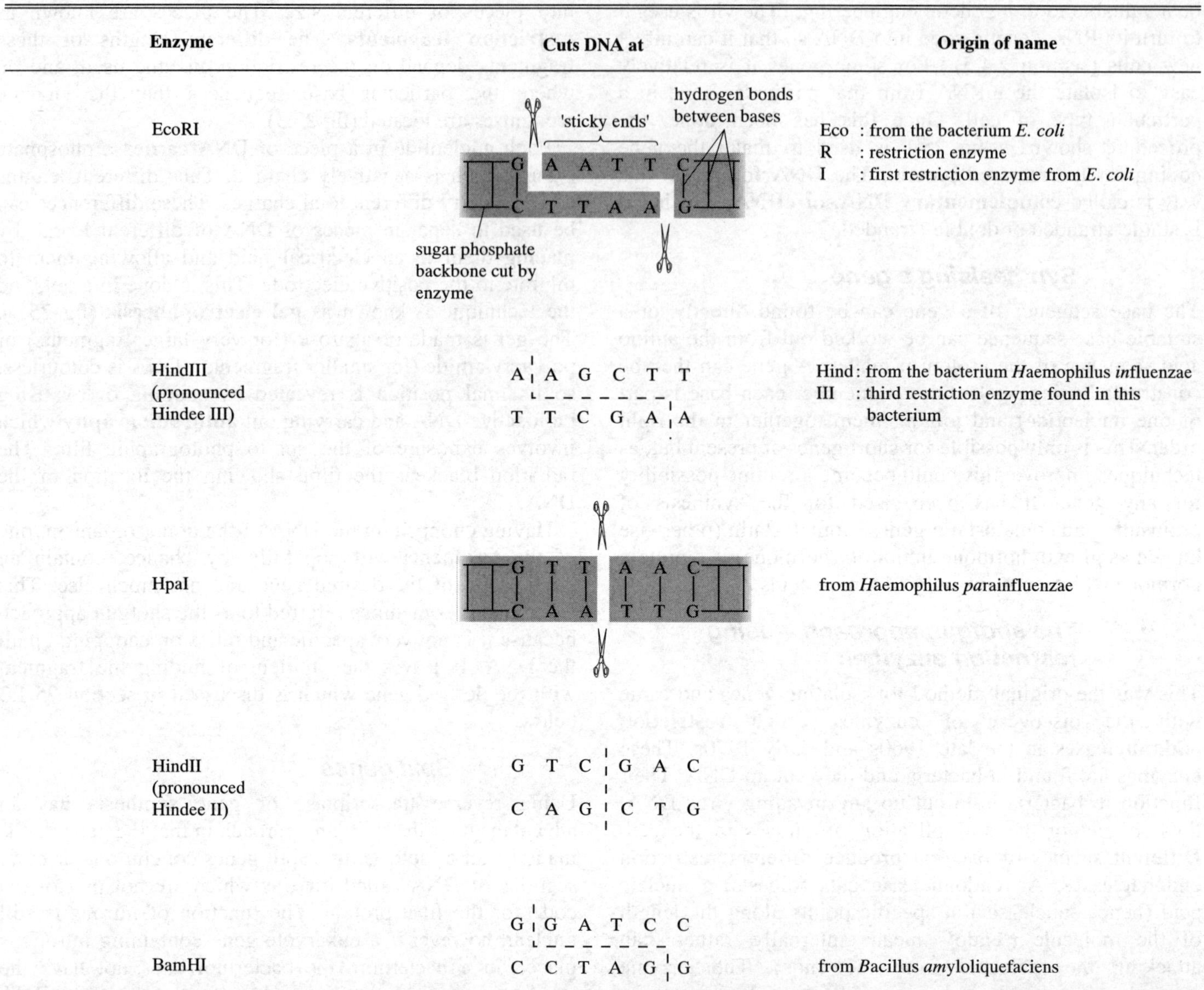

Fig 25.2 *Some commonly used restriction enzymes. EcoRI, HindIII and BamHI make staggered cuts in the DNA leaving 'sticky ends'. A sticky end produced by, say, EcoRI can join to another sticky end produced by EcoRI. HindII and HpaI leave blunt ends. Diagrams of EcoRI and HpaI are shown in more detail.*

enzyme must be used for the plasmid DNA. The restriction fragments from the donor DNA, including those containing the wanted gene, are then mixed with the plasmid DNA and joined by their sticky ends. For example, the sticky end –AATT will bind to the complementary sticky end –TTAA. The initial attraction is due to hydrogen bonding, but the sugar–phosphate backbones are then joined using an enzyme called **DNA ligase**.

If the donor DNA is cDNA or a synthetic gene (from the first two methods above), or if the restriction enzyme in the shotgun method has produced blunt ends, then the procedure shown in fig 25.6 must be used.

Phage vector

Phages are useful as vectors for larger pieces of DNA than can reliably be carried by plasmids. One phage which is commonly used is λ phage (fig 2.20). Part of the phage DNA is replaced with the DNA required for cloning. The part replaced is not needed for replication of the phage DNA inside the bacterial host cell, so cloning is unaffected. The procedure is summarised in fig 25.7. It is often used with the shotgun method of preparing DNA.

25.1.4 Stage 3: Introducing vector DNA into the host cell

The plasmid or phage vector must now be introduced into a bacterial cell which will allow the vector to multiply (clone itself and the foreign donor DNA it contains). The bacterium commonly used is *Escherichia coli*. *E. coli* is a normal inhabitant of the human gut and was chosen for this task because a great deal is known

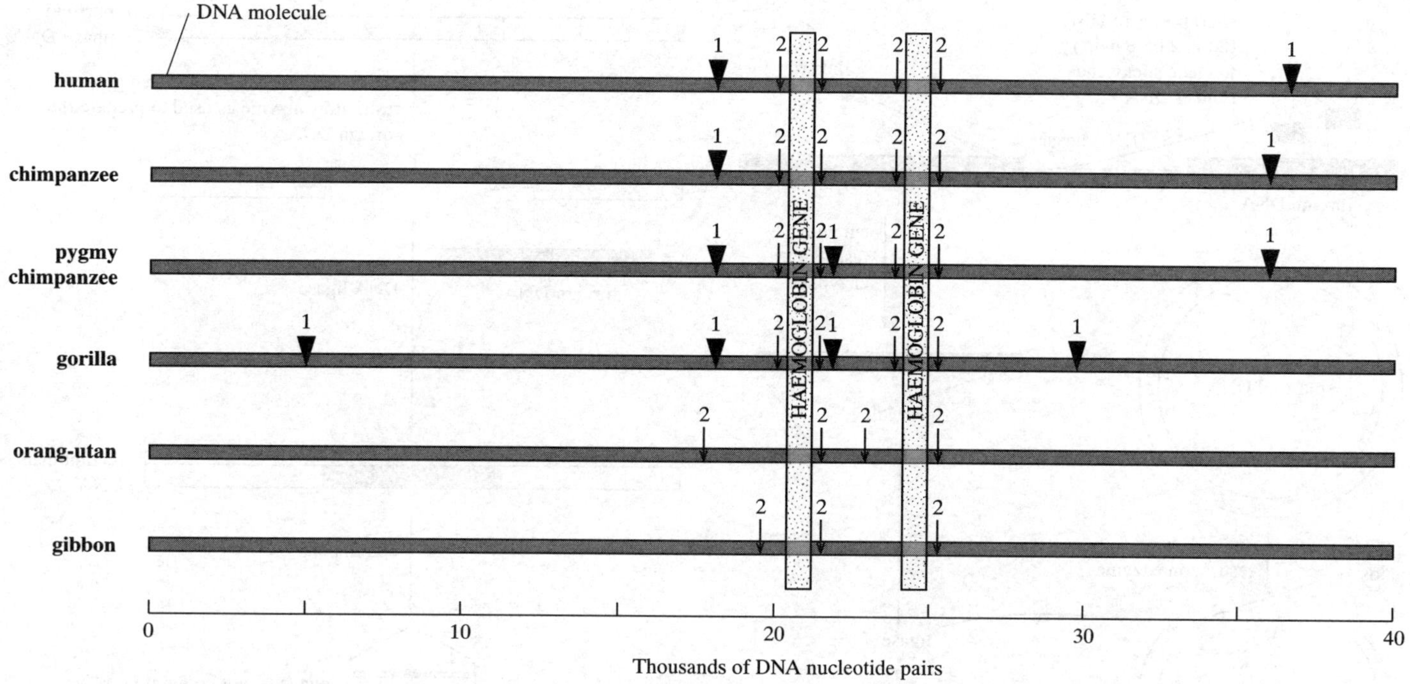

Fig 25.3 *Use of restriction enzymes to chop up DNA and to produce restriction fragments of different length. In the example a part of the DNA which contains two haemoglobin genes from a range of primates has been chopped up by two restriction enzymes. Enzyme 1 is represented by ▼ and enzyme 2 is represented by ↓. Enzyme 1 will give two fragments with human DNA and enzyme 2 will give five fragments. The lengths of these fragments, and of those obtained if both enzymes are used together, would allow the positions of the cuts relative to each other to be worked out. This produces the 'restriction map' shown in the diagram. The more restriction enzymes used, the more detailed the map becomes. Note also that the more closely related the species, the more similar is their DNA and the more similar their restriction sites (cutting points). (Based on fig 7–4, p294,* Molecular biology of the cell, *3rd ed., B. Alberts* et al. *(1994) Garland.)*

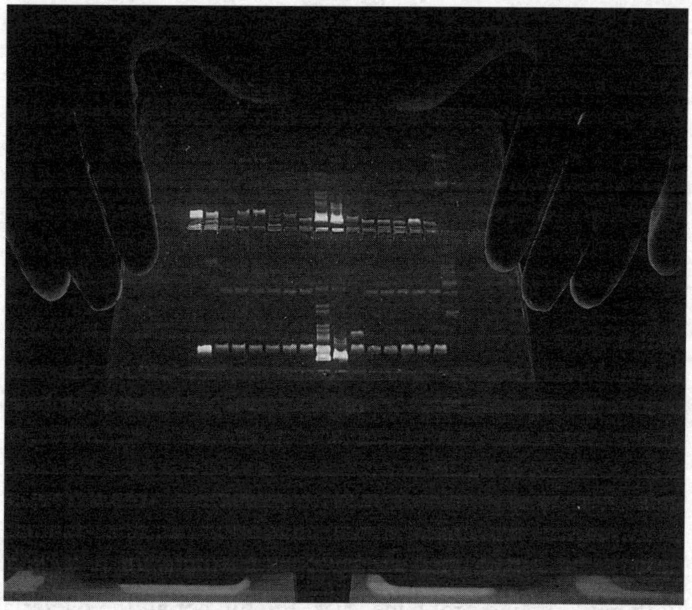

Fig 25.4 *Gel electrophoresis to separate fragments of DNA of different length. The fragments are produced by chopping up the DNA with one or more restriction enzymes. The largest fragments are the slowest moving because they have more difficulty in moving through the pores in the gel. They are nearest the top of the gel in the photo.*

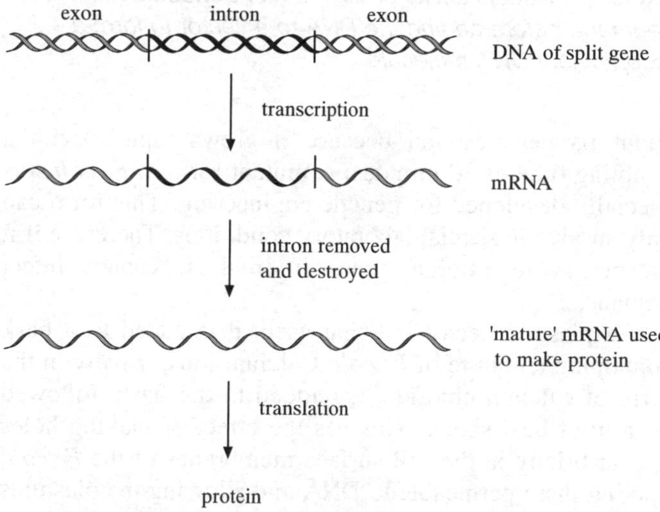

Fig 25.5 *Transcription and translation of a gene containing an intron. The regions around introns are called exons. Genes may contain many introns. Only the exons code for protein.*

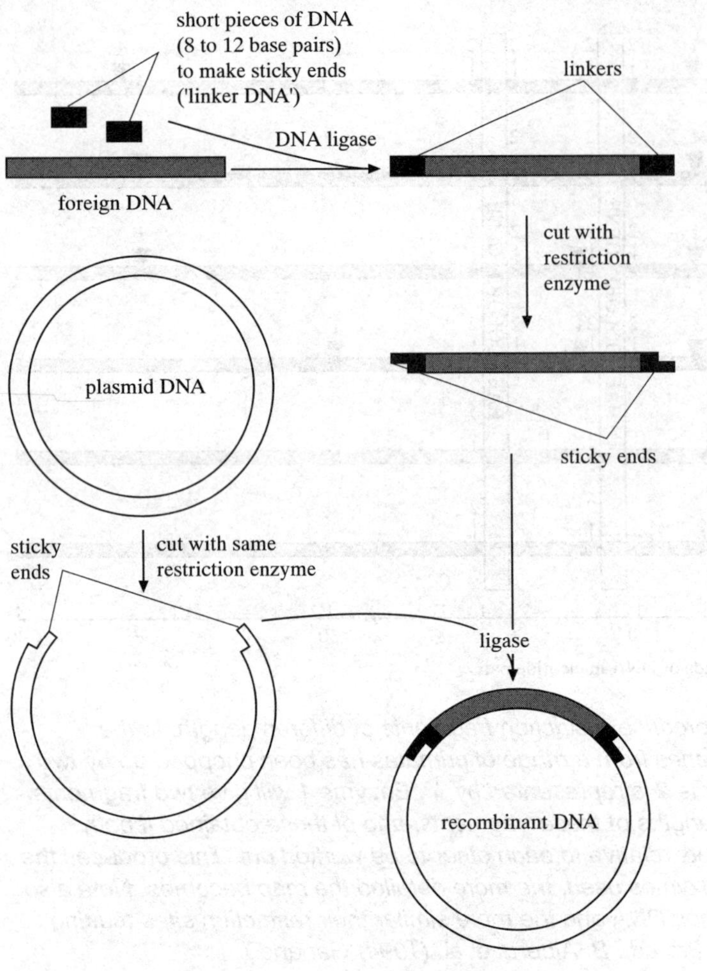

Fig 25.6 *Adding sticky ends to a blunt-ended DNA fragment, before adding the DNA to a vector to form a recombinant DNA molecule.*

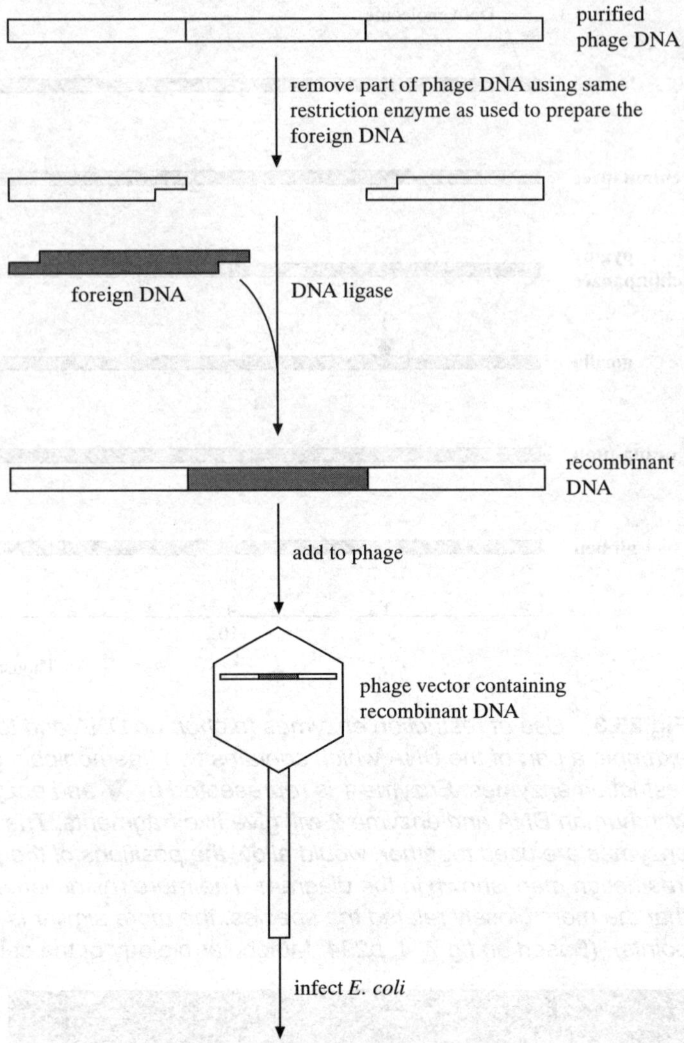

Fig 25.7 *Adding DNA to a phage vector.*

about its genetics and because it grows rapidly with a doubling time of 30 minutes. A mutant form of *E. coli* was specially developed for genetic engineering. This form can only survive in special laboratory conditions. Therefore if it escapes, with foreign genes inserted, it cannot infect humans.

If a plasmid vector is being used, it is added to a flask containing a culture of *E. coli*. Calcium ions, usually in the form of calcium chloride, are added to the flask, followed by a brief heat shock. This has the effect of making holes appear briefly in the cell surface membranes of the *E. coli*, making them permeable to DNA and allowing the plasmids to enter. The process of adding new DNA to a bacterial cell is called **transformation**

Phage vectors are introduced by infection of a bacterial lawn growing on an agar plate (section 12.8).

25.1.5 Stage 4: Cloning the DNA

A single phage containing one recombinant DNA molecule can produce more than 10^{12} identical copies of itself and the molecule in less than one day. *E. coli* cells containing plasmids are usually plated out onto nutrient agar in petri dishes. They can grow and divide once every 30 minutes, eventually forming visible colonies. This alone would produce at least as many copies of the required DNA as is obtained from phage vectors, but bacteria can also contain hundreds of copies of a plasmid and these will be copied each time the bacterium divides. Thus billions of clones are produced in a very short time with both techniques. The transformed bacteria must now be selected before further cloning.

Selecting the transformed bacteria

This will only be discussed for the situation in which plasmids have been used. When plasmid DNA is mixed with bacteria two problems arise. Firstly, not all the bacteria will be transformed (take up plasmids). Secondly, not all the plasmids will have taken up the foreign donor DNA. This problem is cleverly avoided by using plasmids which have two special features (fig 25.8):

- a gene for resistance to a particular antibiotic – if the bacteria are grown on a medium containing that region containing restriction sites (indicated by lines) – these are the points which can be cut by restriction enzymes to allow insertion of foreign DNA

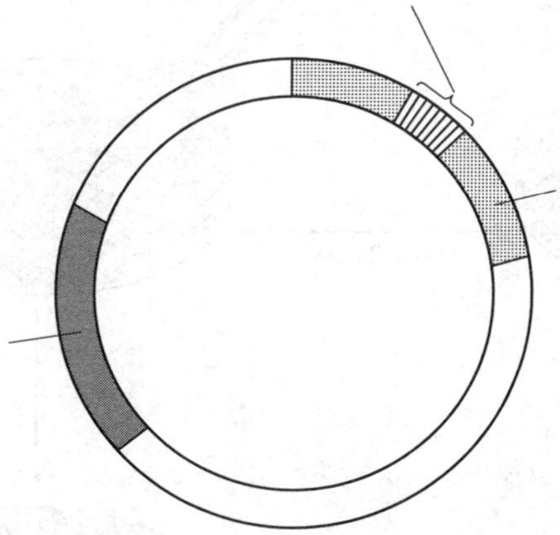

Fig 25.8 *A plasmid vector containing a gene for resistance to the antibiotic ampicillin (a close relative of penicillin). This allows cells which are carrying the plasmid to be selected by treatment with ampicillin. Cells not carrying the plasmid will die.*

antibiotic, only the transformed cells (the ones containing plasmids) will survive and multiply to form colonies.

- a gene for the enzyme β-galactosidase which has had a group of restriction sites added – these restriction sites do not affect the performance of the gene. β-galactosidase is an enzyme which breaks down lactose to galactose and glucose (it breaks down any disaccharide containing galactose). It can also break down a colourless compound called X-gal to a blue compound. If foreign DNA is inserted at a restriction site in the gene, the gene will not work. Therefore if the bacteria that survive growing on the antibiotic are then grown on a medium containing X-gal, those colonies which lack the donor DNA will appear blue. Bacteria which form colourless colonies are the ones containing the donor DNA and can be isolated for further cloning.

25.1.6 Selecting bacteria with the required gene

If the shotgun method is used in Stage 1, the bacteria which are successful in cloning donor DNA are not necessarily all cloning the DNA containing the required gene. This is because the donor DNA was a mixture of a very large number of restriction fragments (up to a million in the case of human DNA). Only one or a few of these are likely to contain the piece of DNA or gene required for cloning, yet all will have been cloned. A mixture of clones like this is called a **library**. A library will also be produced by the reverse transcriptase method of Stage 1 if a mixture of mRNAs is used. This is sometimes necessary if the desired mRNA cannot be isolated in pure form. So after Stage 4, bacterial cultures have been isolated which are libraries unless a single gene was cloned, either by synthesis or from a single type of mRNA.

Using a gene probe

The required bacteria are selected using a **gene probe** as illustrated in fig 25.9. A gene probe can be used if some or all of the base sequence of the DNA being looked for is known. Alternatively the base sequence (or one very similar) can be predicted from a knowledge of the amino acid sequence of the protein it codes for, if this is known. The DNA or RNA probe that is made is a short sequence of nucleotides which is complementary to part of the required DNA and will therefore bind to it. For example, the probe AGTCCA would find and hydrogen bond to TCAGGT. Probes can be as short as 15 to 20 nucleotides, or much longer. The probe is usually made from radioactively labelled nucleotides using the radioactive element ^{32}P. When this binds to the DNA, the radioactivity acts as a marker which can be detected by autoradiography as shown in fig 25.9.

25.2 Applications of genetically engineered bacteria

25.2.1 Human insulin

Insulin is a protein hormone made in the pancreas which plays a vital role in the regulation of blood sugar levels (section 17.6.6). Its deficiency is one of the causes of the disease diabetes mellitus (sugar diabetes) where blood sugar levels become raised with harmful consequences. At least 3% of the population is affected by diabetes mellitus. This became a treatable disease from 1921 when two Canadian workers, Banting and Best, first isolated the hormone. Before that it resulted in terrible wasting symptoms (fig 25.10) and eventual death. Now more than 2 million people worldwide use insulin and the world market is worth several hundred million pounds a year.

Daily injections of insulin isolated from the pancreases of slaughtered pigs and cattle became the standard treatment. However, due to minor differences in the amino acid composition of insulin from species different to ourselves, and to traces of impurities, some patients were allergic to animal insulin and showed damaging side effects as a result of the injections. The ideal solution became possible with the introduction of genetic engineering. The gene for human insulin is inserted into a bacterium, and the bacterium is grown in a fermenter to make large quantities of the protein. An outline of the procedure currently used for human insulin is shown in fig 25.11.

A final problem which has not been discussed so far is how to switch the gene on in the bacterium. Not all the genes in a cell are switched on at any one time. Certain regions of DNA called **promoter regions** situated next door to the genes have to be activated before a gene is expressed. If the new gene is inserted in the middle of an existing gene, the switch for that gene may be used. The gene used in *E. coli* was for β-galactosidase (fig 25.8) but is

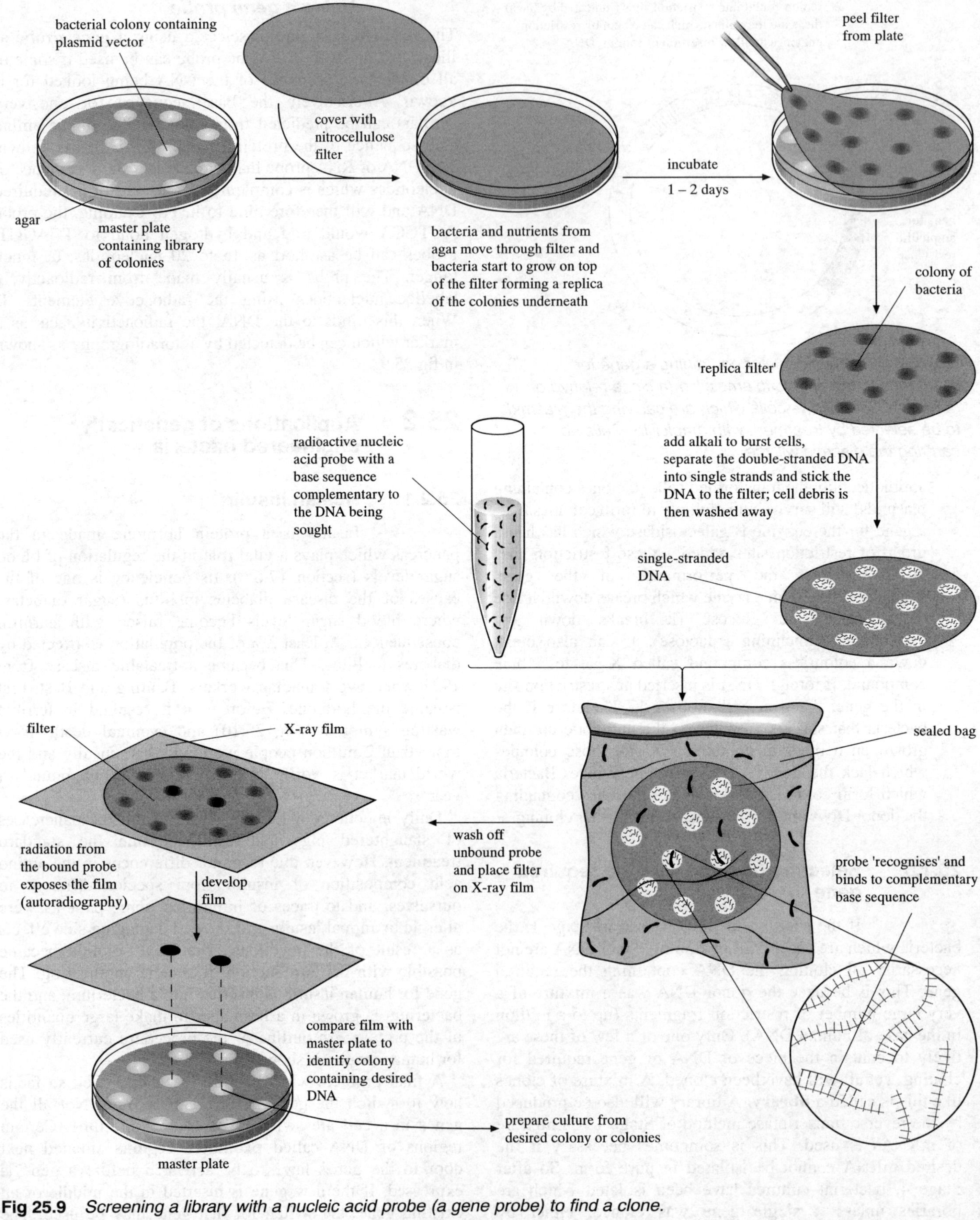

Fig 25.9 *Screening a library with a nucleic acid probe (a gene probe) to find a clone.*

Fig 25.10 *A young child who suffers from diabetes injecting herself with insulin.*

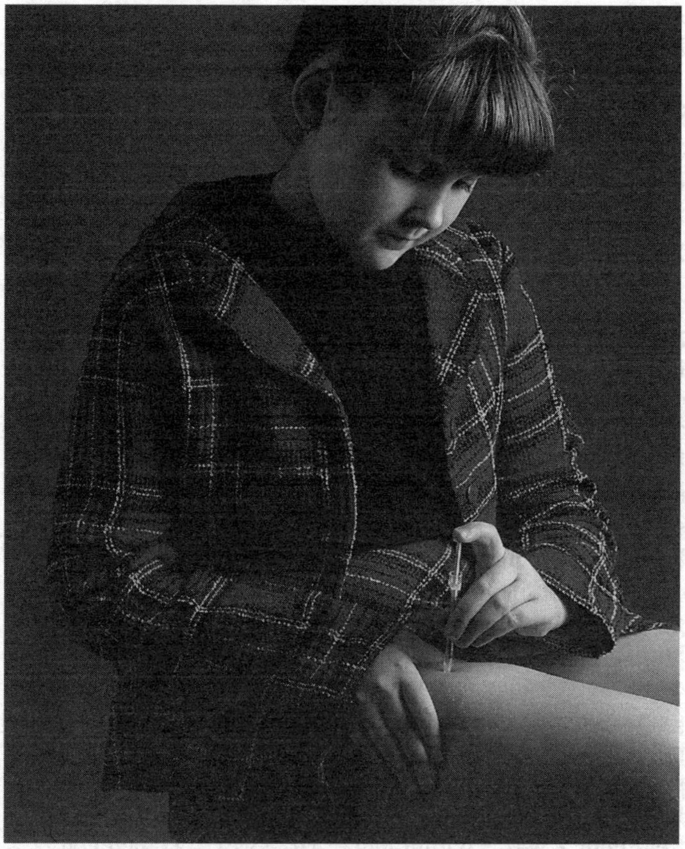

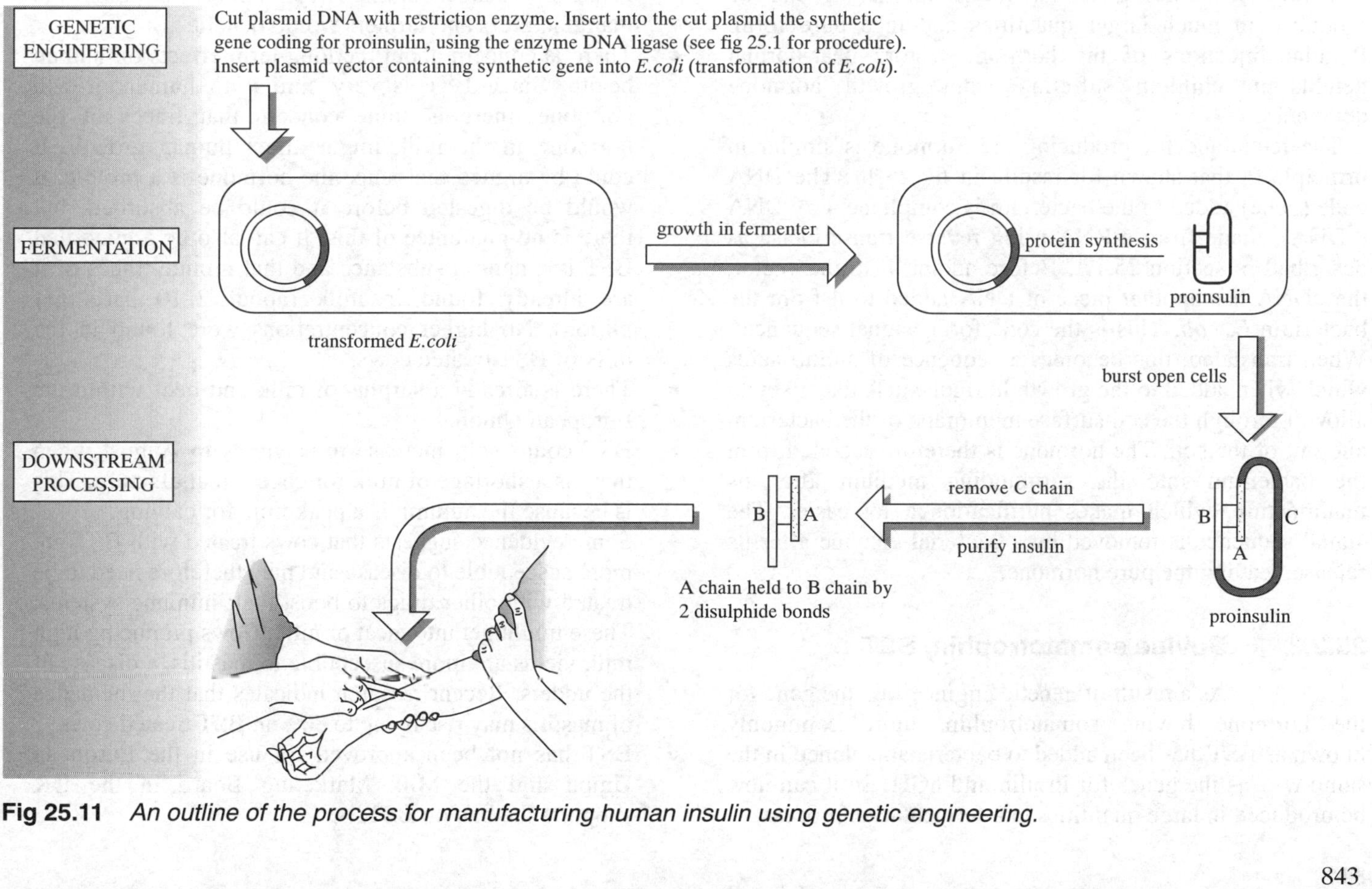

Fig 25.11 *An outline of the process for manufacturing human insulin using genetic engineering.*

now tryptophan synthetase. Its promoter is switched on if *E. coli* is grown in a medium containing lactose.

The original technique was developed by Eli Lilly and Company and in 1982 human insulin, marketed as 'humulin', became the first genetically engineered pharmaceutical product to be approved for use.

25.2.2 Human growth hormone

Growth hormone is a small protein molecule produced in the pituitary gland. It affects all the tissues of the body, causing growth of almost all those that are capable of growing. Abnormally low levels of growth hormone in childhood result in dwarfism in which the body has normal proportions but is much smaller; intelligence is unaffected. Unlike the case with insulin, where insulin from slaughtered animals will function in humans, the growth hormones of different animals work only in the species of origin. Treatment of dwarfism has therefore relied on growth hormone extracted from the pituitary glands of dead humans, and the supply was not large enough to meet demand. Another problem was that extracts from pituitary glands were occasionally contaminated with the infectious protein that causes Creutzfeldt–Jakob disease (the same protein that may cause mad cow disease). After several fatalities among people who were treated in the 1970s the treatment was withdrawn. However, Genentech, a California-based company, have produced human growth hormone (hGH) from genetically engineered bacteria which contain the human gene for the hormone. It can be produced in much larger quantities and in a pure form. Regular injections of the hormone restore near-normal heights in children suffering from growth hormone deficiency.

The technique for producing the hormone is similar in principle to that shown for insulin in fig 25.11. The DNA code (gene) added to the bacterium is complementary DNA (cDNA), made from mRNA using reverse transcriptase as described in section 25.1.1. Before adding it to the vector, the cDNA has another piece of DNA added to it from the bacterium *E. coli*. This is the code for a 'signal sequence'. When translated, this becomes a sequence of amino acids which when added to the growth hormone acts like a key to allow it through the cell surface membrane of the bacterium and out of the cell. The hormone is therefore secreted from the bacterium into the surrounding medium after its manufacture, which makes purification a lot easier. The signal sequence is removed by a bacterial enzyme after its release, leaving the pure hormone.

25.2.3 Bovine somatotrophin, BST

As a result of genetic engineering, the gene for the hormone bovine somatotrophin, more commonly known as BST, has been added to bacteria and cloned in the same way as the genes for insulin and hGH. So it can now be produced in large quantities in a fermentation process.

BST is similar to human growth hormone. Like the latter, it is a small protein made in the pituitary gland and stimulates cell division, protein synthesis and growth in most parts of the body. It is particularly important for muscle and skeletal growth. If small doses are injected into cows every 1 to 2 weeks, it increases milk production by up to 25% (up to 5 dm^3 a day) and can result in a 10–15% increase in weight of beef cattle. Although the cattle consume more food, there is a net increase in profit. The dairy farmer can either sell more milk, or if restricted by milk quotas, can produce the same amount more cheaply (with fewer cows). Preliminary trials showed no change in behaviour, health or reproduction of the cattle, according to the manufacturers of BST, Monsanto of the USA. The cattle return to normal growth and milk production soon after the last injection.

During a trial programme with the hormone in the UK in 1985, the Ministry of Agriculture, Fisheries and Food (MAFF) allowed the milk from the cows to be mixed with that from other cows and sold to the public. This was despite protests from supermarkets and consumer groups. They also refused to identify the test herds. This produced great public alarm and protests. Use of BST has been banned in the EU. Some of the issues raised are summarised below and are worth discussing as an example of how science interacts with society and the importance of having informed opinions.

- Many people do not believe that use of BST would result in cheaper milk. They believe that only the manufacturers and farmers would benefit.
- There is concern about the long-term effects on human health. Since BST is very similar to human growth hormone, there is some concern that traces of the hormone in the milk might affect human growth. It could be argued that since the hormone is a protein, it would be digested before it could be absorbed, but there is no guarantee of this. It can also be argued that BST is a natural substance and that minute traces of it are already found in milk (about 2–10 parts per billion). No higher concentrations were found in the milk of BST-treated cows.
- There is already a surplus of milk and beef within the European Union.
- BST could help increase milk yields in August when there is a shortage of milk for cheese manufacture. This is because the autumn is a peak time for calving.
- Some evidence suggests that cows treated with BST are more susceptible to disease and may therefore need to be treated with other drugs to boost their immune systems. These might get into meat or milk. Cows producing high milk yields are more susceptible to mastitis, a disease of the udders. Recent research indicates that the incidence of mastitis may rise by up to 80% in BST-treated cows.
- BST has not been approved for use in the European Union and the Milk Marketing Board in the UK supports the ban on BST.

- Since the end of 1993 BST has been approved for use in the USA.
- Laboratory rats fed relatively high doses of BST showed no ill effects.
- Long-term trials on humans have not been carried out, nor have the effects on pregnant or lactating mothers been studied.

You can decide for yourself which arguments you find most persuasive. You might like to consider the issue from the point of view of (i) the manufacturer, (ii) the farmer and (iii) the consumer. Where should we go from here?

25.2.4 Cleaning up oil spills

Examples of how microorganisms can be used to clear up waste have already been met in chapter 12 in relation to treatment of sewage, recycling, biological mining and conversion of organic wastes into useful products such as sugar, alcohol and methane. Improvements in these will be possible with genetic engineering.

Another potentially important example is the attempt to produce genetically engineered bacteria capable of cleaning up oil spills. We still have no environmentally friendly method of doing this efficiently. Trials are underway with a genetically engineered strain of *Pseudomonas*, which can break down the four main groups of hydrocarbons present in oil (xylenes, naphthalenes, octanes and camphors) and can clean up oil in an oil–water mixture. The relevant genes occur on plasmids of naturally occurring *Pseudomonas* strains, but no single strain contains all four plasmids. All four types of gene have now been introduced into a single 'superbug'.

Such bacteria might be sprayed onto surfaces polluted with oil. The bacteria only work at the oil–water interface since they need oxygen. They are therefore better suited to cleaning up thin films of oil such as might cover rocks after a pollution incident, rather than thick slicks of oil. There is also the problem of releasing genetically engineered bacteria into the environment. At present the bacteria only work very slowly at low temperatures and would therefore not be very suitable for use in cold climates, as was needed when the oil tanker *Exxon Valdez* shed its oil in Alaskan waters in 1989.

25.3 Genetic engineering in eukaryotes

It is possible to genetically engineer eukaryotic organisms as well as bacteria. Organisms that have been genetically altered using the techniques of genetic engineering are generally referred to as **transgenic**.

Because transgenic organisms offer an alternative to traditional methods of animal and plant breeding, they offer an exciting new way forward in agriculture. Improving crops or domestic animals by traditional methods is a slow process which relies a lot on chance because of crossing over in meiosis and random segregation of chromosomes during sexual reproduction. For example, it takes 7–12 years to develop a new cereal variety. Genetic engineering offers the chance to add new genes directly, without relying on sexual reproduction. It opens up the possibility of 'designer' plants and animals with desirable properties such as disease resistance. Animals and plants can become 'living factories' for useful products other than food, just like bacteria in fermenters. The greatest challenge in agriculture is to improve food production in the developing countries and hopefully some of the new techniques will be applied to regions where food shortage is greatest. Ethical issues are raised, however, such as the well-being of animals and the fact that new genes are being released into the environment (fig 25.12).

Some key aims in plant and animal breeding which might be the subject of transgenics are:

- increased **yield**.
- improved **quality of food** from the point of view of health or digestibility, for example oil, fat, and protein.
- **resistance to pests and disease** – genes for resistance can be transferred from one species to another.
- increasing **tolerance of**, or **resistance to**, **environmental stress** such as drought, cold, heat or crowded conditions; for crops, tolerance to wind damage, acid or salty soils, waterlogged soils. Genes

Fig 25.12 *Newspaper headlines from the* Times, Sun *and* Evening Standard *relating to transgenic plants and animals.*

controlling stress responses can be isolated and moved from one organism to another.

- **rate of growth**, including time from birth, or planting, to maturity – the range of some crops might be extended by shortening their growing seasons.
- **herbicide resistance** (section 25.4.4).

The advantages of transgenics can be summarised as:

- a gene for a desirable characteristic can be identified and cloned;
- all the beneficial characteristics of an existing variety can be kept and just the desired new gene can be added;
- sexual reproduction is not necessary;
- transgenics is much faster than conventional breeding.

25.4 Transgenic plants

25.4.1 Getting new genes into plants

Using Agrobacterium

The most effective method of transferring genes into plants is to use the soil bacterium *Agrobacterium tumefaciens* as a vector. This bacterium contains a plasmid which can be used to carry the desired gene. It can infect, and therefore be used for, most dicotyledonous plants and causes **crown gall disease**. It enters through wounds and stimulates host cells to multiply rapidly, forming large lumps called **galls** which are really tumours (fig 25.13), masses of undifferentiated cells which grow independently of the rest of the plant like a cancer. Normally, wounded plant tissues release chemicals that stimulate cell division and the plant produces a group of cells called a **callus** which quickly covers the wound. Chemicals released by the wounded cells also stimulate *Agrobacterium* to infect the wound. Once it has infected the plant, the plasmid in *Agrobacterium* causes development of the gall. The plasmid, known as the **T_i plasmid** (tumour-inducing), contains a short piece of DNA called **T-DNA**. This leaves the bacterium and enters the plant cells where it is inserted into the plant's own DNA. Here it brings about the unregulated growth. The bacterium itself does not enter the cells, but can live between them, feeding on new products which the T-DNA directs the plant cells to make. The plant cells are said to be **transformed**.

The plasmid genes which control infection are different from those which cause unrestricted growth to form a tumour. The latter are on the T-DNA. It is therefore possible to remove the infection genes to make room for a new gene without affecting the ability of the T-DNA to transform plant cells. Also, the plasmid will no longer cause the crown gall disease.

Originally this technique could not be used for monocotyledonous plants which was a great disadvantage because monocots include the cereals, such as maize and wheat, which are the most important group of crops worldwide. However, this problem has now been overcome. The technique also works well for tomatoes, potatoes and many trees.

Whole plants can be grown from single transformed cells using the cloning techniques described in section 21.3. The first stage involves growing the cells in a liquid culture medium to produce an undifferentiated mass of cells called a callus. The callus can be plated out on to nutrient agar, and with the correct balance of hormones will produce shoots and roots and grow into a new plant. Another technique is to use *Agrobacterium* to infect the cut edges of discs punched out from leaves and to culture the discs on nutrient agar.

Using viruses

Phages (viruses that infect bacteria) can be used as vectors in the genetic engineering of bacteria, so viruses which attack plant cells should open up the possibility of doing the same in plants. The technique is still being developed.

Using guns

A surprisingly crude but effective method of introducing foreign DNA directly into plant cells is to use guns. The required DNA is coated onto the surface of 1 mm diameter gold or tungsten beads. These are placed next to the tip of a plastic bullet in the barrel of a specially designed gun. Originally the bullet was fired in the normal way with an explosive charge (and a fire-arms licence was needed) but pressurised gas is now used. The bullet is fired at a plate containing a microscopic hole. Some of the DNA-covered particles are sprayed through this hole into target cells or tissue held in a chamber just behind the hole. The barrel and the chamber are under vacuum during the firing process so that the particles do not slow down. Particles can be found in the cytoplasm of successfully transformed cells, so presumably the cell surface membranes heal themselves immediately after being shot.

25.4.2 Pest resistance – insecticides

Insects cause enormous crop losses in agriculture and some transmit diseases to farm animals. Since the Second World War many chemicals, starting with DDT, have been used as insecticides. Gradually we have come to understand the ecological damage that can be done (section 10.8.4) and are developing ways to reduce or avoid the problem. One strategy is to rely more on biological control.

The soil bacterium *Bacillus thuringiensis*, referred to by biologists as Bt, produces a powerful protein toxin which can be used against several species of insect pest. It is 80 000 times more powerful than the organophosphate insecticides commonly sprayed on crops and is fairly selective, killing only the larvae of certain species. Different strains of Bt kill different insects, mainly the larvae of moths and butterflies (caterpillars) and of some hemipterans, such as white flies (maggots), the larvae of flies, such as mosquitoes (aquatic larvae). Some kill nematode worms,

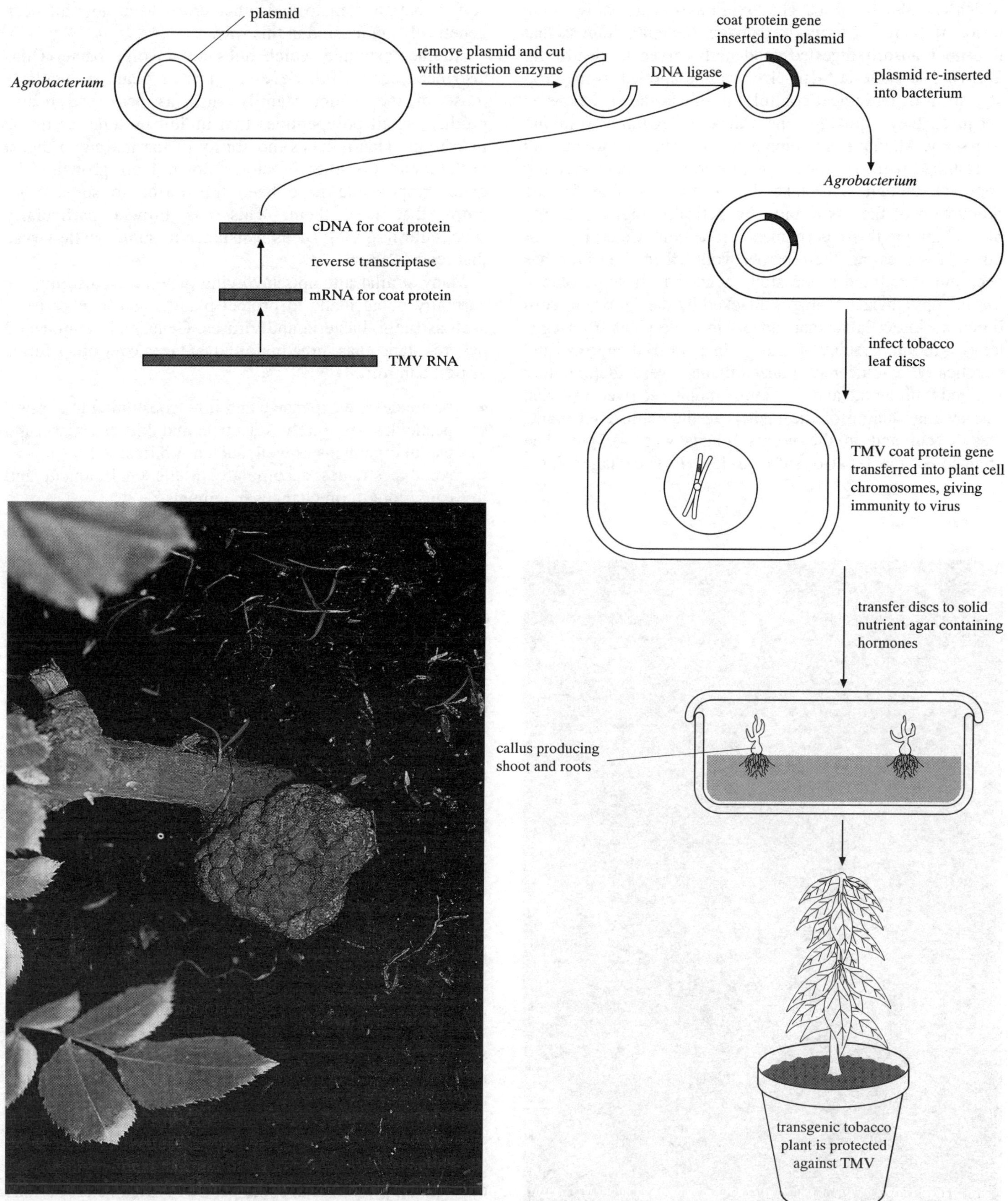

Fig 25.13 *(a) Introducing a new gene into a plant cell using* Agrobacterium. *(b) Crown-gall tumour caused by* Agrobacterium *infection of a wound.*

which can also be pests. The toxin binds specifically to the inside of the insect's gut and damages the epithelium so that it cannot absorb digested food and starves to death (fig 25.14). The bacteria themselves can be applied to the crop as a form of biological control, but this is rather expensive because they quickly die, making regular spraying necessary. Attempts are being made to isolate the toxins and to stabilise them by protein engineering. A more cunning approach though is to take the gene responsible for the production of the toxin and to genetically engineer it into plants, giving them permanent protection. Caterpillars or other larvae eating their leaves would then die. This has been attempted and successfully achieved for some plants, for example maize. Maize is attacked by the European corn borer, an insect larva that tunnels into the plant from eggs laid on the undersides of leaves. In field trials normal and genetically engineered maize plants were deliberately infested with larvae and the results monitored over 6 weeks. The average length of the tunnels in the engineered plants was 6.3 cm and in the normal plants was 40.7 cm. The engineered plants also suffered less leaf damage. Rice, cotton, potato, tomato and other crops have now all been genetically engineered in this way.

Another example which does not involve bacteria has also been successful. Several types of plant, particularly those of the legume family such as peas and beans, produce small polypeptides that inhibit proteinases in the insect gut. This reduces the ability of the insects to digest protein and prevents or slows down their growth. The genes responsible have been transferred to some other crops that lack them. This has proved particularly successful in giving seeds resistance to some beetle larvae that feed on them.

Many similar attempts involving genetic engineering are currently being made to protect plants against other pests such as fungi, bacteria and viruses. Genetically engineered pest resistance has three major advantages over other forms of pest control:

- pesticides are expensive and time consuming to apply;
- pesticides are rarely selective and kill harmless and useful organisms as well, such as pollinators;
- some pesticides accumulate in the environment and cause long-term changes in animals.

Fig 25.14 *Cabbage white caterpillar six days after feeding on a plant treated with a Bt toxin. The caterpillar is dead and decomposing from the gut.*

25.4.3 Pest resistance – viruses

Plant viruses are serious pests of crops. The first attempt to genetically engineer resistance to a virus was with tobacco plants. Tobacco is attacked by an RNA virus called tobacco mosaic virus (TMV – fig 2.18). TMV also attacks tomato plants, causing losses worth over 50 million dollars in the USA every year. *Agrobacterium* has been used to introduce a gene from TMV into tobacco plants. It codes for a virus coat protein. When deliberately infected with TMV these plants prove much more resistant than untreated control plants. Something rather similar to vaccination seems to take place (fig 25.15). Similar experiments have more recently been done to protect potato, tomato and alfalfa from virus attack.

25.4.4 Herbicide-resistant crops

An interesting application of genetic engineering in plants is the introduction of genes which give resistance to certain herbicides. The crop can then be sprayed with that herbicide and only the weeds will be killed. This solves the problem that weedkillers are not normally very selective. Weeds can reduce crop yields by over 10% even in developed countries where modern agricultural methods are used. It is estimated that genetically engineered herbicide resistance could double or even quadruple yields in some parts of Africa where serious weed problems are found. Particularly harmful are the parasitic weeds broomrape (*Orobanche*), found north of the Sahara, and witchweed (*Striga*), found in sub-Saharan Africa. They affect maize, millet, wheat, sorghum,

Fig 25.15 *Tobacco mosaic virus causes pale mottling on infected plants.*

sunflowers and legumes. Herbicide-resistant corn, wheat, sugar beet and oilseed rape have so far been produced by developed countries for their own use. These are resistant to the herbicide Basta.

25.4.5 Nitrogen fixation

A long-term goal in agriculture is to introduce the genes for nitrogen fixation into crop plants. Nitrogen fixation is the process by which atmospheric nitrogen gas is reduced to ammonia within cells so that it can be used to make protein and other organic compounds. The process can be carried out only by certain bacteria. Some nitrogen-fixing species live in the root nodules of plants, particularly plants such as peas, beans, alfalfa and clover which are legumes. These plants benefit, but artificial nitrogen fertilisers have to be added to most crops if the soil is not to become deficient in nitrogen, especially where the same crop is grown year after year in the same soil as is often the case with cereals. Over 60 million tonnes of nitrogen fertiliser were used worldwide in 1987. If plants contained their own nitrogen-fixing genes, enormous savings could be made in the time, money and energy used in making, transporting and spraying the fertilisers.

Genetic engineering of nitrogen fixation is made difficult by the fact that nitrogen fixation is a complex process involving many enzymes. About 15 genes are involved (the Nif genes). Also, part of the process is anaerobic, requiring a means of excluding oxygen. Although a great deal of work has been done, it has so far proved impossible to make the genes function properly in eukaryotes.

25.4.6 Transgenic tomatoes

Soft fruits such as tomatoes, bananas and red peppers are usually picked green and ripened artificially using ethene gas in warehouses (section 16.2.9). This means they are still hard when picked, reducing bruising and enabling the fruit to be picked mechanically and tipped into containers. It also allows controlled ripening so that the fruit will have maximum appeal to the customer. However, much of the flavour of the fruit is lost during transport and shipping due to biochemical changes. A company called Calgene in the USA, and ICI Seeds in the UK, have produced a genetically engineered tomato in which the ripening process is slowed down. This means the fruit can be left on the plant for longer, giving both increased yields and a fuller development of flavour. There is therefore a twin advantage for farmer and customer. The tomatoes first went on sale in the USA in 1995 as 'Flavr Savr' tomatoes. They are expensive, but do taste better. They were first introduced in the UK by Sainsbury in 1996, initially in tomato paste. The issue of safety and genetically engineered foods, including Flavr Savr tomatoes, is discussed in section 25.6.

25.4.7 More examples of genetic engineering in plants

- New colours, patterns and shapes of flowers are being experimented with by the horticultural industry. For example, experiments are underway to produce blue roses.

- Use of crops to produce medical drugs instead of food. This should be cheaper than using cultures of mammalian cells as at present. For example, the human enkephalin gene has been expressed in plants.
- Use of plants to produce mouse monoclonal antibodies.
- Improve the poor bread-making quality of the high-yielding British wheats. Improving the quality of protein will improve the flour quality. At least 11% of the protein needs to be of the high quality required to produce the large volume and suitable texture of a good loaf.
- Improve the nutritional qualities of plant foods, for example increase the proportion of essential amino acids. Many legumes are deficient in sulphur-containing amino acids. Genes from the brazil nut may rectify this.

25.5 Transgenic animals

25.5.1 Getting new genes into animals

One of the earliest successes in creating transgenic animals was in a mouse. A growth hormone gene from a rat was inserted into the genome of a mouse. Attached to the growth hormone gene was a powerful promoter which was stimulated by the presence of heavy metals in the mouse's diet. When these heavy metals were included in the mouse's food, the growth hormone gene was almost continually 'switched on'. This made the mouse grow at 2–3 times faster than mice without the gene. The mouse with the growth gene also finished growth at about twice as large as normal. This was achieved through genetic engineering. There are five basic methods now used in the development of transgenic animals:

- microinjection of eggs;
- use of stem cells;
- virus vectors;
- direct uptake of DNA stimulated by calcium or an electric current ('transfection');
- use of liposomes.

Microinjection of eggs

If it is desired that all the cells of an animal should contain a new gene, it must be introduced into an egg cell. This is done by firstly giving a hormone fertility drug to a female to stimulate production of extra eggs by the ovary. Fertilisation is allowed to occur and then the fertilised eggs are collected. The donor DNA is then injected directly into one of the pronuclei of a fertilised egg using a very fine needle-like pipette while viewing under a microscope. The process is described in more detail in fig 25.16. In some, though not all, cases the DNA integrates into one or more of the chromosomes. The two pronuclei later fuse and the egg becomes the zygote. The fertilised eggs are transferred to one or more foster mothers (two offspring maximum per mother if sheep or cattle) and the offspring are later screened for the presence of the new gene. The best success rates achieved so far are about one transgenic animal for every 20 eggs treated (sheep) or 100 eggs (cows). Herds of animals must therefore be kept, making it an expensive process. The first experiments on farm animals were carried out on rabbits, pigs and sheep, and these have been followed by cattle and fish.

Use of stem cells

A process which gives more control than the method described above is becoming more popular. Here a few cells (known as 'stem cells') are taken from a young embryo. These cells can be cloned indefinitely in a test tube. The new gene can be introduced into the cells by various means, including microinjection. The advantage is that the cells that are expressing the new gene (transformed cells) can be identified before adding them to the foster mother. This saves producing many unwanted non-transgenic animals. It might also be possible to introduce

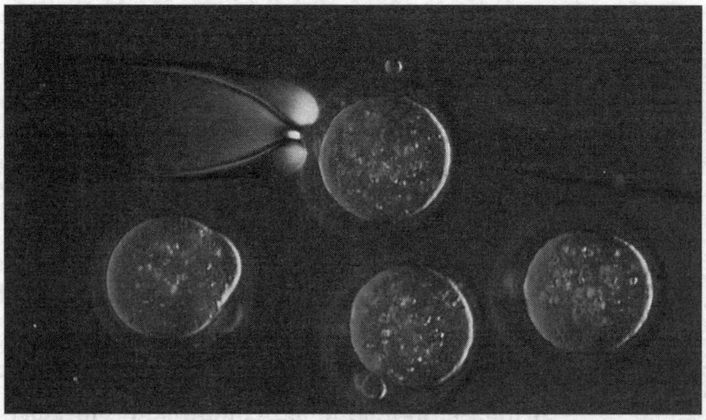

Fig 25.16 *Microinjection of DNA into an egg. The nuclei of the male and female gamete have not yet fused and at this stage are called pronuclei. They are visible at the centre of the egg. The DNA is injected into one of the pronuclei using the very fine needle-like pipette at the right of the photo. The egg cell is held steady by the larger pipette at the left. Several hundred copies of the DNA fragment are added.*

the gene into a specific region of a chromosome which may be necessary for normal expression of some genes. This ability would be particularly important if the technique is ever used with humans.

The successfully transformed cells are injected back into a normal embryo and become part of its normal development. The resulting animal is a mixture of two genetically different types of cell, some derived from the normal stem cells and some derived from the transformed stem cells. Such an animal is known as a **chimera** (after a mythological animal which had the head of a lion, the body of a goat and the tail of a serpent). The sex cells of the animal will also be mixed and some of its gametes will carry the new gene. These will give rise to completely transgenic animals in the next generation.

Virus vectors

This method is similar in principle to that used for phage vectors in bacteria. Plasmid vectors are not possible because animal cells do not contain plasmids. It is not used with egg cells but only when some of the body cells need to be transformed, as with gene therapy (section 25.7.11).

Direct uptake of DNA

Fragments of DNA can be taken up directly by phagocytosis under the right conditions. If DNA is prepared in the presence of calcium phosphate this process is stimulated, although it is not very efficient. It is best suited to gene therapy where only some of the cells in the body need to be modified. Alternatively a process called **electroporation** can be used in which the cells are stimulated with a brief shock from a weak electric current. This causes temporary holes to appear in the cell surface membrane, which becomes more permeable to DNA as a result.

Liposomes

Liposomes are small artificially created spheres (vesicles) surrounded by a phospholipid bilayer like a membrane. The required DNA is contained within the liposome. The liposomes fuse with and enter the cells (see also section 25.7.11).

25.5.2 Pharmaceutical proteins from milk

One use of transgenic animals is to produce relatively large quantities of rare and expensive proteins for use in medicine, a process sometimes referred to as 'pharming' of drugs. Such drugs cannot always be produced by bacteria in the way we have already seen for human insulin and hGH because bacteria do not always have the necessary machinery to process the proteins made. For example, the protein may have to be folded precisely or modified using mammalian cell machinery. Factor IX protein, for example, has to have a –COOH group added to some of its amino acids after production. Large-scale cell culture of the cells that produce these proteins is possible in theory but is very expensive and technically difficult.

The most successful approach so far has been to use the mammary glands to produce the protein so that it can be harvested by milking the animal. It is then relatively easy to purify it. A commercially successful example is the manufacture of **AAT** (**α-1-antitrypsin**) by PPL Pharmaceuticals, a company founded in 1987 and based in Edinburgh. They have several flocks of sheep, each producing different proteins. The first transgenic sheep to produce AAT was Tracy, although she herself is no longer used as a source of the protein. The offspring of transgenic animals continue to carry the gene, so whole flocks of transgenic sheep can eventually be built up (fig 25.17). Transgenic animals are seemingly perfectly normal, and no ill effects have been detected as a result of the treatment.

AAT is a naturally occurring protein found in human blood. A mutant form of the gene that codes for it causes a genetic disease which leads to emphysema (section 9.7.3) as a result of uninhibited elastase activity. Elastase is an enzyme produced by some white blood cells which destroys elastic fibres in the lungs as part of the normal turnover of elastic tissue. Its activity is normally regulated by AAT, which inhibits the enzyme. (Smoking is also thought to inhibit AAT, explaining one link between smoking and emphysema.) AAT is made in the liver and can be extracted from blood. However, more people need AAT than can be supplied by the usual means. The healthy human gene for AAT has now been added to sheep and the mammary gland of the sheep is used to express the gene. Sheep are used because they have a shorter generation time, lower cost and are easier to handle than cows.

The gene has a high degree of expression, meaning it is switched on most of the time. This results in nearly 50% of the milk protein being human AAT (fig 25.18). The sheep carries out the correct modification of the protein, including adding sugar to it to make it a glycoprotein. The procedure for making AAT is outlined in fig 25.19.

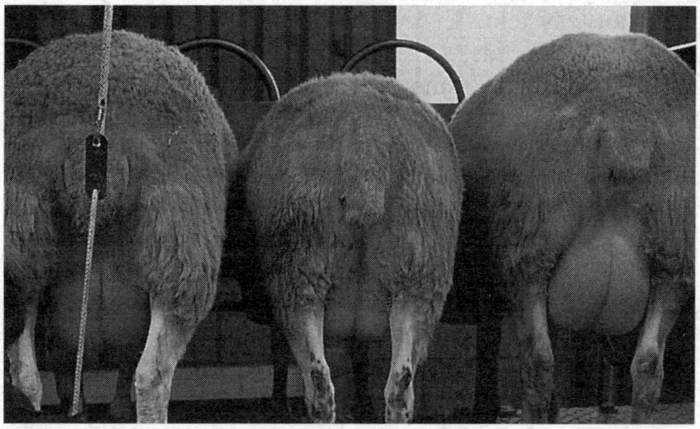

Fig 25.17 *Transgenic sheep awaiting milking. They have a human gene incorporated into their DNA which is responsible for production of the protein α-1-antitrypsin. This is produced in the mammary cells and excreted in the sheep's milk.*

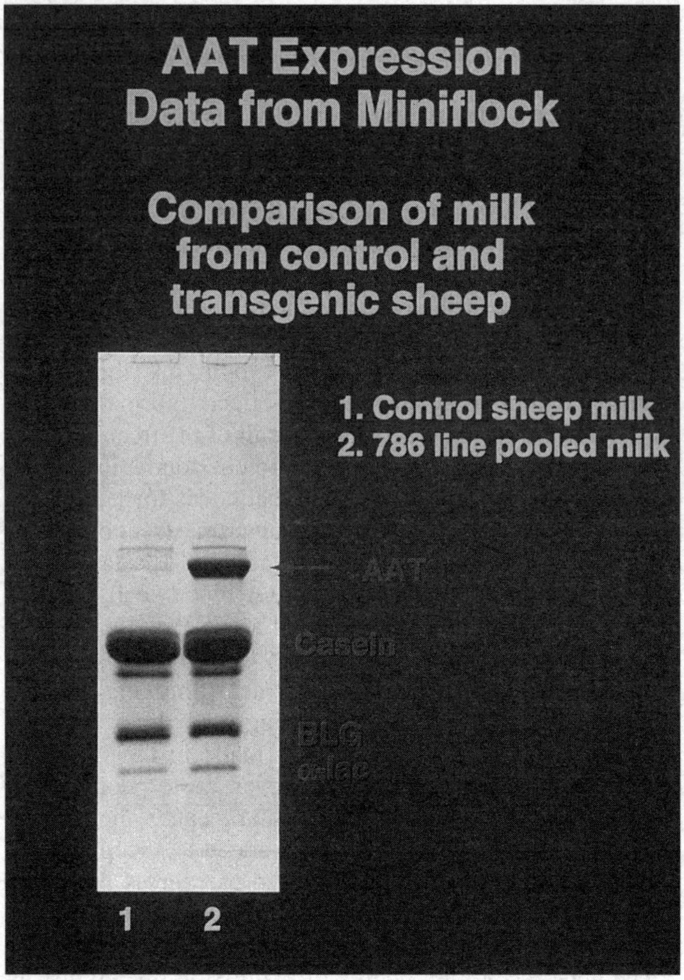

Fig 25.18 *Gel electrophoresis of milk proteins from a sheep. (1) Normal milk. (2) Milk from a transgenic sheep showing presence of a new band due to AAT.*

All the cells of the sheep contain the same gene for AAT, so how do we ensure that only the cells of the mammary gland make it? The answer is to use the body's own regulatory system. Every cell in the body contains the entire genetic code, so every cell contains the genes for milk proteins (the same is true whether you are male or female). However, for the gene to be expressed the promoter, a piece of DNA next to the gene, must be switched on. By cloning one of the sheep's own milk protein promoters and attaching it to the human DNA, we can ensure that only mammary glands express that DNA. The promoter chosen is that for β-lactoglobulin, a protein present in high concentrations of milk.

Clinical trials with AAT will have to be completed before it is released, but it should be available within a few years.

Other examples of proteins produced in this way include factor IX, a blood-clotting protein whose absence causes one type of haemophilia (section 24.6.1). Another is **tPA (tissue plasminogen activator)** which is used to dissolve blood clots in patients suffering from heart disease. PPL Pharmaceuticals are now working on fibrinogen, the major

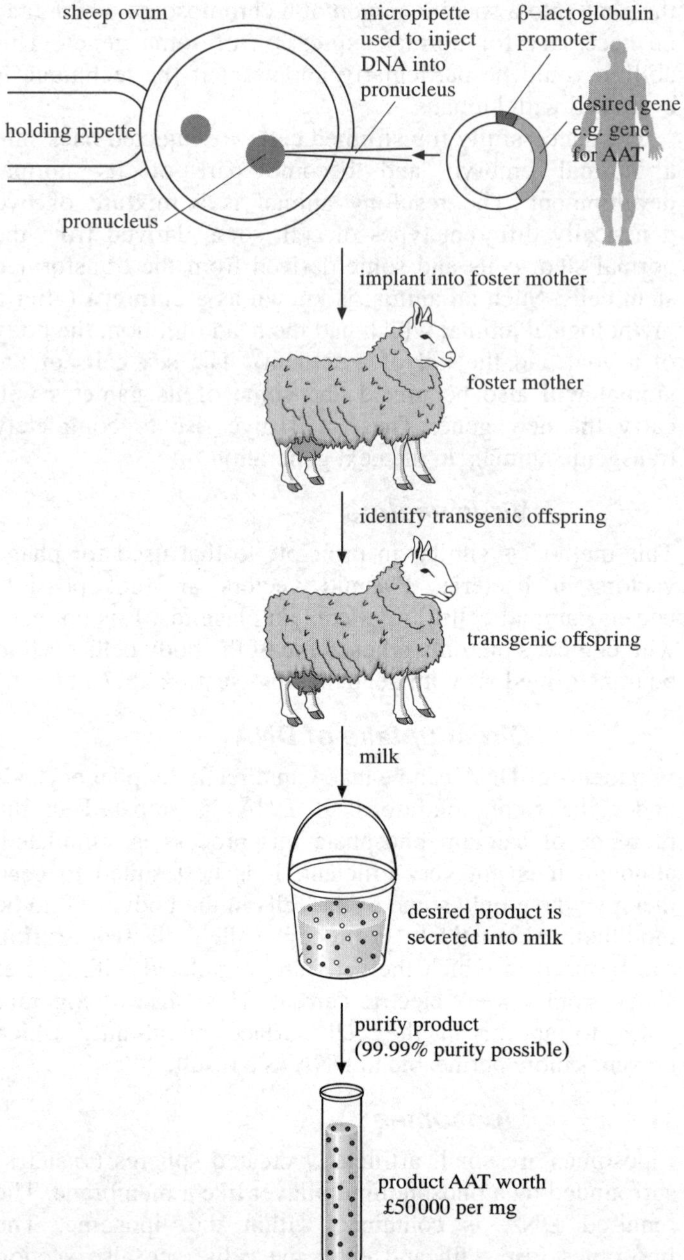

Fig 25.19 *Procedure for producing a transgenic sheep containing a human gene, e.g. the gene for AAT.*

protein involved in blood clotting, which has a very complex secondary and tertiary structure. It would be used as a tissue glue after surgery; laid on a wound, it helps healing. It has already been purified from blood and used for this purpose in the USA.

25.5.3 Growth hormone

Growth hormone genes have already been discussed in sections 25.2.2 and 25.2.3 and the controversy over BST was discussed in section 25.2.3. Similar controversy arose over the introduction of human growth hormone genes into farm animals. When this was done the normal controls over production of the hormone were avoided. Transgenic sheep which overproduce growth hormone grow leaner and put on weight more quickly, making more efficient use of their food. However, they are more prone to infection, tend to die young and the females are infertile. Similarly, transgenic pigs grow leaner meat more efficiently. However, even more side effects were noted than with sheep, including arthritis, gastric ulcers, heart and kidney disease. Until ways are found of regulating the genes more precisely, the process will not be used commercially.

More recently scientists in Canada have added a gene from another fish (the ocean pout) to salmon which activates the salmon's own growth hormone gene. The salmon grow up to 30 times their normal weight and at 10 times the normal rate. Scottish fish farmers started breeding them on a trial basis in 1996.

25.5.4 Summary

Table 25.2 summarises some of the important examples of plant and animal transgenesis which had been achieved by the mid-1990s. Most new scientific and medical procedures are still tried first on mice because they breed so rapidly and are convenient to handle. The Home Office monitors experiments done in the UK. 181 000 procedures were carried out on mice in the UK in 1994 compared with 3000 on other transgenic animals. These are still relatively small numbers compared with what is expected in the future.

25.6 Benefits and hazards – the ethical and social implications of genetic engineering

From the earliest days of genetic engineering scientists have been very aware of the need to consider the potential hazards and ethical issues associated with this new branch of biology. Original concerns in 1971 focused on plans to clone cancer genes from viruses into *E. coli*. It was argued that if the genetically altered *E. coli* escaped from the laboratory, they might spread the gene into *E. coli* that live in the human gut by transferring the plasmids, for example by conjugation (section 2.3.3). It was also argued that human DNA, and the DNA of other mammals like mice, could contain cancer-causing genes (oncogenes) and that these might inadvertently get transmitted with neighbouring pieces of DNA being used for genetic engineering. In February 1975 a group of more than 100 internationally well-known molecular biologists met in California and decided that, until the risks could be more precisely estimated, certain restrictions should be placed on genetic engineering research. This was a remarkable self-imposed brake on scientific progress, not imposed by the Government but by the scientists themselves. Work on cancer viruses was stopped. Non-scientists who had been part of the debate were invited to join the Advisory Committee that was set up by the American Government in 1976. Similar bodies were set up in Europe, including the Genetic Manipulation Advisory Group (GMAG) in the UK. After a two-year halt, during which safe procedures were established, research on cancer viruses and other work continued. Careful checks on the safety and implications of procedures remain to this day. Debate continues about how strict the rules and regulations should be. There has been more resistance to developments in Europe than in the USA.

In the 1980s an explosion of activity and interest was unleashed. Manufacturing companies quickly began investing billions of dollars into not just genetic engineering but all the 'biotechnologies' which were emerging from molecular biology. **Biotechnology**, a new industry, was born. We have examined some of the current applications of biotechnology in agriculture, medicine, industry and waste treatment in this chapter and in chapter 12, and further applications such as gene therapy and genetic fingerprinting will be discussed later in this chapter. What, then, are the major issues which face us now?

25.6.1 Human safety

The first food containing genetically engineered DNA to be approved for marketing was the 'Flavr Savr' tomato (section 25.4.6). This is a useful case study of concerns about the safety of foods.

One of the main concerns relates to the vectors used for transforming plant cells. These contain genes for antibiotic resistance, most frequently kanamycin resistance. These genes enter the transformed plants with the desired gene (fig 25.1). Flavr Savr tomatoes contain one of these antibiotic resistance genes. The concern is that when the tomato is eaten, the gene may pass from the tomato to the *E. coli* bacteria in the gut, making them resistant to kanamycin and related antibiotics. Since bacteria leave the gut in the faeces, the gene may spread to other potentially harmful bacteria in the environment which, if they infected humans, would be antibiotic resistant. In practice, the tomato gene, along with all the other DNA, would most likely be digested once eaten, and even if it were not, the chances of the gene passing through a series of organisms is extremely remote. Also, the kanamycin-resistance gene is already common in the environment. Nevertheless, scientists are trying to find ways to remove the marker genes after transformation.

In 1996 the European Union allowed genetically modified maize to be imported from the USA. The maize has a bacterial gene which increases its resistance to pests and disease, but also has a gene for resistance to the antibiotic ampicillin. Greenpeace is opposed to this introduction and is threatening legal action.

The public are extremely wary of genetically engineered products because of the publicity which has surrounded such issues as the use of growth hormones (see section

Table 25.2 Some plants and animals that have been genetically manipulated and some of the characteristics involved. (From Claire Pickering & John Beringer, *Modern Genetics*, *Biol. Sci.* Rev. 7, no. 4, March 1995, p. 34, Philip Allan.)

	Examples of characteristics															
Organism	Toxin for insect resistance	Herbicide tolerance	Antibiotic resistance	Change of flower colour	Virus resistance	Altered nutrients	Resistance to fungi	Production of pharmaceuticals	Resistance to bacteria	Prevention of ice crystal formation	Reduced accumulation of heavy metals in leaves	Delayed ripening	More rapid growth	Alterations for research on diseases	Tolerance to low temperatures	
Allegheny serviceberry	•															**Plants**
Apple	•															
Cabbage		•														
Chicory			•													
Chrysanthemum				•												
Cotton	•	•														
Cucumber					•											
Eucalyptus			•													
Flax		•														
Lettuce					•											
Lucerne		•			•	•										
Maize	•	•			•	•										
Melon family					•											
Oilseed rape	•	•	•			•	•	•								
Papaya					•											
Peanut		•			•											
Petunia				•												
Plum					•											
Poplar	•															
Potato		•			•	•		•	•							
Rice	•		•			•										
Soybean		•				•										
Strawberry		•								•						
Sugar beet		•			•											
Sunflower						•										
Tobacco	•	•	•		•	•	•			•	•					
Tomato	•	•			•	•	•			•		•				
Walnut	•		•													
Wheat		•														
Carp													•			**Animals**
Catfish													•			
Cattle								•					•			
Goat								•								
Mice								•						•		
Pigs								•					•	•		
Rats								•						•		
Salmon													•		•	
Sea bass															•	
Sheep								•								
Trout													•			

25.2.3 for a discussion of the issue of BST). Companies that have invested millions of pounds in research and development cannot afford to make mistakes, and therefore have a strong vested interest in making sure their products are safe. For example, PPL who manufacture the anti-emphysema drug AAT imported all their sheep from New Zealand to ensure that they were scrapie-free. (Scrapie causes a disease in sheep similar to mad cow disease.) Transgenic goats being used by Genzyme Transgenics for the production of monoclonal antibodies in milk are fed only food free of pesticides and herbicides. Also, no protein or animal fat additives to their food are allowed so that any possibility of transfer of disease to humans from other animals is prevented. The products are probably safer than many traditional products which we are happy to accept.

25.6.2 Safety of the environment

In section 25.4.3 the possibility of producing virus-resistant crops was described. A fear which has been raised with the method used is that a different virus might infect the crop and have its genetic code (RNA or DNA) wrapped in the protein coat of TMV instead of its own normal protein coat. It might then be able to invade all the crops that TMV can invade. It is important that extensive trials are carried out in natural conditions in all cases of crop protection to be sure that incidents like this, or unthought-of problems, do not occur.

Both North America and Europe have very strict regulations controlling the release of genetically engineered organisms (GEOs) into the environment. In the European Union each member country has its own authority which oversees all releases of GEOs. In the UK the authority is jointly held by the Department of the Environment and the Ministry of Agriculture, Fisheries and Food (MAFF). One of the early controversies concerned genetically engineered 'ice-minus' bacteria. The original bacterium lives on many crop plants and makes them susceptible to frost damage because a protein it secretes helps the formation of ice crystals on the plants. The bacterium was genetically engineered to remove the gene coding for this protein, producing the so-called 'ice-minus' bacterium. The intention was to spray this on crops such as strawberries to make them more resistant to freezing. There was a passionate legal battle about the dangers of releasing GEOs into the environment, but permission was eventually given for release. After that the rules were made clearer and less restrictive.

The first approval for *unrestricted* release of a GEO in Britain was given by a Department of the Environment Advisory Committee in 1994. The organism was produced by the Belgian company 'Plant Genetic Systems' (PGS). It was a new type of oilseed rape which contains genes for resistance to the herbicide Basta (section 25.4.4). By that time over 60 small-scale field trials of GEOs had taken place in Britain and over 1000 in Europe and North America. There are far more potential dangers once unrestricted release is granted. Rapeseed, for example, can become a weed in hedgerows and would be impossible to control with Basta. It could cross-fertilise with relatives such as wild mustard, thus spreading the resistance to wild plants. PGS claim that the environmental risks with rapeseed are negligible.

Another concern is that developing herbicide-resistant plants may encourage the use of greater amounts of herbicides, particularly Basta, although the companies concerned argue that it may lead to less herbicide spraying because it will be more effective, and that older, more harmful herbicides will be phased out. Greenpeace were one organisation which opposed the release of the rapeseed.

Other crops that are resistant to disease, drought or other types of environmental stress, might similarly spread their resistance to weeds, producing weeds that might overrun agricultural areas very rapidly. There have been many hundreds of releases of transgenic plants elsewhere in the world. In China, for example, virus-resistant tobacco is being grown commercially. None of these releases has resulted in any known harmful environmental effects.

Genetically engineered fish, such as the giant salmon mentioned in section 25.5.3, pose a serious threat. The fish are contained so that in theory they should not escape. However, young, small fish have been known to be carried away by birds and dropped in local waters, and larger fish have been known to escape. There are many examples from the past of newly introduced animals causing great ecological damage, such as the rabbit in the UK and in Australia. If the Scottish salmon escape into the sea, where they migrate as adults, there are fears that they may affect the balance of the already endangered wild salmon populations. They might also affect food chains in unpredictable ways. Already more than 90% of the salmon in some Scottish streams are descended from salmon which have escaped from fish farms in Norway.

25.6.3 Animals and ethics

Humans often think of themselves as being superior to other animals (not to mention plants, fungi, bacteria and so on) and therefore having the 'right' to exploit other organisms for their own benefit. However there has been a growing trend in recent years to challenge the human-centred (anthropocentric) view of our relationship with other species. There is particular concern about the way we exploit animals for food and for development of medical products. One aim of genetic engineering is to increase the growth rate and yield of animals like cattle, pigs and poultry. The harmful effects of unregulated production of growth hormone on the health of pigs and sheep has been described in section 25.5.3. Use of BST in dairy cattle in the United States carries increased risk of mastitis (section 25.2.3). There appears to be little concern about whether the animals are biologically 'designed' to withstand the additional stress of increased production of milk, meat, eggs and other products. An interesting case study is Hermann, a transgenic bull born in

Holland in 1990. Hermann contains a gene which, if passed on to his female offspring, will enable them to produce a human milk protein (lactoferrin) in their milk. Environmental groups threatened to boycott companies that sponsored the work and this forced a Dutch producer of baby foods to withdraw from the project.

An important motive for producing modified animals for food is commercial profit. Where there is an additional motive, such as preventing or treating disease, the issues get even more complex because the well-being of the animal has to be balanced against the well-being of humans. Medical experiments may involve a certain amount of animal suffering. An example is provided by the oncomouse which was the first animal to be patented. The oncomouse is a transgenic mouse to which an oncogene has been added, a gene that causes cancer. The mice develop tumours much more frequently than normal and are used in cancer research. Some people argue that patenting animals is itself unethical because it reduces them to the level of objects. Others argue that experiments such as those with oncomice cause suffering and should therefore be banned. In January 1993 two UK animal rights groups, the British Union for the Abolition of Vivisection (BUAV) and Compassion in World Farming (CIWF) joined with other European groups to launch an appeal against the European patent for the oncomouse, which was granted in 1992. The European Patent Office held public hearings from November 1995 but ran out of time before a judgement could be made, leaving the issue unresolved. Patenting animals, it is argued, makes producing them more profitable, so by preventing patenting the animal welfare groups hope to reduce exploitation of animals. However, there is no guarantee that this would happen. In fact, some patents that have run out have been allowed to lapse. The Cancer Research Campaign (which has reduced its animal experimentation enormously in recent years) says its policy now is 'not to patent transgenic animals after consideration of the moral, scientific and utility issues'. The utility issues may include a growing feeling that it is not commercially worth patenting the animal. Public opinion may be part of the reason.

Fig 25.20 shows part of a letter sent by BUAV to its members in August 1995. Donald Crawford of BUAV says that transgenics 'causes pain and suffering to a large number of animals. We believe that the insertion of genes from other species into laboratory animals is an ethical minefield. The directions in which that approach to life can lead opens up a Pandora's box.' In response to BUAV's claims it can be argued that the research is contributing to our understanding of diseases like cystic fibrosis, heart disease, AIDS, multiple sclerosis and cancer for which cures will only come about if their genetic basis is understood. Transgenic animals are protected by the same laws used for all laboratory animals. At the moment it seems likely that there will be a rise in the number of genetically engineered mice because of their help to us in understanding genetic disease.

At the other extreme of animal welfare are those animals that are used for pharmaceutical products such as AAT and factor IX which are probably the best cared-for farm animals in the world because they are the most valuable.

25.6.4 Patenting

Apart from the ethical aspects of patenting mentioned above there are other related issues. In 1992 an American company attempted to patent genetically engineered cotton and soya plants, however they were produced. Farmers would then have to pay royalties to sow the crop. Although the patents were granted, they have been challenged by the international community. The US National Institutes of Health tried to patent the human genome in 1991 but, after more international protest, they withdrew their application. However, the human breast cancer gene (BRCA1) was patented in the US once its base sequence had been determined and attempts are being made to patent the second breast cancer gene (BRCA2).

Some European companies tried to extend their patents on genetically engineered seed to preventing farmers from re-sowing seed from genetically engineered crops. They would therefore have had to buy new seed every year. Similarly they have tried to remove farmer's rights to breed

BUAV
campaigning to end animal experiments

Our Ref: admin/masters/geneng

British Union for the Abolition of Vivisection
16a Crane Grove,
London N7 8LB
Telephone 0171-700 4888
Fax 0171-700 0252

Dear Supporter,

Thank you for your enquiry about genetically engineered animals.

I am sending you our genetic engineering pack in which you will find detailed information about particular areas of concern, such as xenotransplantation (cross-species transplants) and animal patenting. I also enclose our general reading list which contains two books on the issue should you wish to pursue the topic further.

It is extremely worrying to note that, according to the most recent Home Office figures, the use of transgenic animals in research has increased by 33%, from 138,965 in 1993 to 184,188 in 1994 - a figure which does not include a further 202,311 animals deliberately created with a "harmful genetic defect". These statistics confirm that the genetic engineering of animals is *the* growth area of the vivisection industry. This *must* be challenged.

Fig 25.20 *Part of a letter sent by BUAV to its members in August 1995 as part of a campaign against patenting the oncomouse.*

from transgenic animals. The European Parliament and national governments have to try to balance the interests of all, including those of the farmers, the manufacturers and the consumers. In such circumstances ethical issues are not the only ones which are considered.

25.6.5 Insurance

A new UK Genetic Manipulatory Advisory Commission met for the first time in early 1997. On its first agenda was the issue of insurance companies. The insurance issue concerns how life insurance companies should use the results of genetic tests. Should they refuse insurance or raise premiums for people with an increased chance of dying from a particular disease, such as breast cancer or heart disease? Insurance companies are arguing that they *should* have access to the results of any genetic tests carried out on a person whose life is being insured.

25.6.6 Cloning

The cloning of the sheep, Dolly, in 1997 was an inevitable consequence of the progress being made in genetics and biotechnology (section 21.1.4). It raises the possibility of breeding many identical copies of animals, including transgenic animals, showing desirable features. One of the ethical concerns is that the techniques could be applied to humans, although such work is currently banned.

25.7 Human genetics

25.7.1 The scope of human genetics

About 1% of all live births produce children who suffer from some genetic disorder (more than 40 births per day in the UK). A high proportion of infant mortality is due to such disorders. Around 1 in 20 children admitted to hospital in the UK have a disorder which is entirely genetic in origin and 1 in 10 individuals will develop, sooner or later, some disorder that has been inherited. In addition, certain genes make certain diseases more likely in adulthood, in other words give a 'predisposition' to a disease. Examples are coronary heart disease, breast cancer and diabetes. Since any gene can undergo a mutation, and there are something like 100 000 human genes, there are theoretically thousands of possible genetic diseases. About 4000 have been recorded that are due to defects in single genes, but this number is increasing rapidly with modern genetic techniques. For 600 of these a known biochemical defect occurs. You yourself are probably a carrier of 4–8 different hereditary diseases which you may not suffer yourself, but which could be passed on to your children. Some mutations are fatal, some cause varying degrees of harm, generally referred to as genetic disease, and others are harmless. Some give both advantages and disadvantages, such as the gene for sickle cell anaemia which we shall study later together with a number of other diseases.

So far genetic diseases are incurable, but the study of human genetics is reaching the point where cures for some will be possible, a topic we shall consider in section 25.7.11. As we learn more about genetic disease, so the need for more specialist clinics will grow, together with the need for more genetic counsellors who will have to help people to understand and cope with the decisions that will become more complex. Genetic counselling will be examined in sections 25.7.9 and 25.7.10.

As other diseases, particularly infectious diseases, have become successfully controlled, so the relative importance of genetic diseases has grown.

25.1 Give one economic and one social argument for research into genetic disease.

Although genetic disease is one of the main reasons for studying human genetics, there are other applications. Two of these, namely genetic fingerprinting and genetic compatibility in transplant surgery will also be studied (sections 25.7.12 and 25.7.13). Sometime near the beginning of the twenty-first century our knowledge of human genetics will be based on a knowledge of the entire base sequence of human DNA and the location of all the genes on our 46 chromosomes. This is currently being worked out by teams of scientists all over the world in a cooperative project called the Human Genome Project. It is impossible to predict the eventual value of this project but it cannot fail to be of fundamental importance to further understanding of human genetics.

The study of human genetics is raising some controversial issues which will have to be discussed by society as a whole, not just by geneticists and molecular biologists. You may already have read, for example, about genes for 'intelligence', 'criminal behaviour' or hetero- and homosexual behaviour. Some of the issues relating to genetic screening, gene therapy and genetic fingerprinting will be discussed in this chapter. In order to understand this chapter you will need to have an understanding of the basic laws of genetics and of the nature of gene and chromosome mutations.

Table 25.3 shows some of the genetic diseases we will be considering in the following sections, and the nature of these diseases.

25.7.2 Sickle cell anaemia

This disease is an excellent example of how a single mutation in a gene can have devastating consequences, and also of the role of natural selection in regulating how common a gene is in a population.

In 1904, a young Chicago doctor named James Herrick examined a 20-year-old black college student who had been admitted to hospital complaining of a variety of symptoms.

Table 25.3 Some common genetic diseases.

Genetic disease/disorder	*Chromosome affected*	*Type of mutation*	*Expression of gene*	*Main symptoms*	*Defect*	*Frequency at birth*
Gene mutations						
sickle cell anaemia	11	substitution	codominant (sometimes described as recessive) autosomal	anaemia and interference with circulation	abnormal haemoglobin molecule	1 in 1600 among black people
cystic fibrosis	7	in 70% of cases is a deletion of three bases	recessive autosomal	unusually thick mucus clogs lungs, liver and pancreas	failure of chloride ion transport mechanism in cell surface membranes of epithelial cells	1 in 1800 among white people
PKU (phenylketonuria)	12	substitution	recessive autosomal	brain fails to develop normally	enzyme phenylalanine hydroxylase defective	1 in 18 000
Huntington's chorea (disease)	4	a newly discovered type of mutation – the normal gene has 10–34 repeats of CAG at one end, the HC gene has 42–100 repeats of CAG	dominant autosomal	gradual deterioration of brain tissue starting on average in middle age	brain cell metabolism is inhibited	1 in 10 000 to 1 in 20 000 worldwide
haemophilia	X	substitution	recessive sex-linked	blood does not clot	factor VIII or IX protein defective	1 in 7000
Chromosome mutations						
Down's syndrome	21	extra chromosome 21 (trisomy 21)		reduced intelligence, characteristic facial features		1 in 750
Klinefelter's syndrome	sex	extra X chromosome in male (trisomy)		feminised male		1 in 500
Turner's syndrome	sex	missing X chromosome in female (monosomy)		sterile female		1 in 2500

autosomal – affecting non-sex chromosome (autosome)
monosomy – one chromosome missing ($2n - 1$)
trisomy – one extra chromosome ($2n + 1$)
monosomy and trisomy are examples of **aneuploidy**, where the total number of chromosomes is not an exact multiple of the haploid number

These included fever, headache, weakness, dizziness and a cough. Herrick discovered other problems. The patient's lymph nodes were enlarged, his heart was abnormally large and his urine revealed kidney damage. What was particularly striking though was the appearance of the patient's red blood cells under the microscope. Herrick described them as sickle-shaped (fig 25.21). The patient's haemoglobin level was about half normal, in other words he was suffering from anaemia (lack of haemoglobin). After resting for four weeks, the patient was discharged, but it was 6 years before Herrick published details of the case. Once he did, other cases were soon reported and the search began for the cause of the disease.

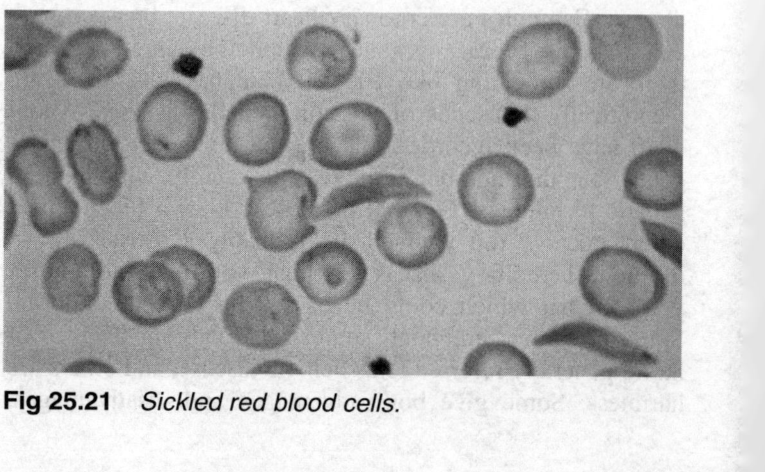

Fig 25.21 *Sickled red blood cells.*

The disease became known as **sickle cell anaemia** and was soon found to be associated with black people. Since the disease ran in families it was also soon realised that it is an inherited disease and that the gene causing it appeared to be recessive. A person is therefore only a sufferer if they have two copies of the gene, one inherited from the mother and one from the father. In other words, a sufferer is homozygous. Today about 1 in 400 black people in Britain and America suffers from the condition and about 1 in 10 is a carrier. A carrier is a person who has only one copy of the faulty gene, in other words is heterozygous. It is estimated that the disease causes about 100 000 deaths per year worldwide. It is not entirely confined to black people and is particularly common in Africa, Pakistan and India.

Symptoms

The major characteristics of the disease are anaemia and a tendency of the red blood cells to change shape (sickle) at low oxygen concentrations. The sickle cells are useless and have to be broken down. They tend to jam in capillaries and small blood vessels and prevent normal blood flow. A whole range of secondary symptoms is possible as a result (fig 25.22). The kidneys and joints are particularly affected. The blocking of blood vessels causes pain in the arms, legs, back and stomach which can be quite severe. Joints may become stiff and painful and hands and feet may swell. Affected individuals may show poor growth and development and are more prone to infections. If death occurs it usually follows an infection.

Children with sickle cell anaemia usually feel quite well for most of the time and can lead relatively normal lives. However, they will generally suffer occasional 'sickle cell crises' when a sudden worsening of pain, infection or anaemia may occur.

Cause

In 1949 a team of workers led by Linus Pauling showed that haemoglobin from sickle cell anaemia sufferers, HbS, is different from that of normal adult haemoglobin, HbA. Electrophoresis, a technique in which proteins are separated according to their overall charge, revealed that at pH 6.9 the overall charge on HbS is positive whereas it is negative on HbA. This was the first time that a disease had been shown to be caused by a faulty molecule. In 1956 Ingram showed that this difference was due to a single amino acid and since then the entire amino acid sequence of HbA and HbS has been determined. Haemoglobin is made of four polypeptide chains (fig 3.36), two α-chains which are 141 amino acids long and two β-chains which are 146 amino acids long. The fault occurs at the sixth amino acid in the β-chain. The amino acid should be glutamic acid. In HbS however it is replaced by valine. Using the codes for amino acids shown in table 23.4:

HbA	Val –	His –	Leu –	Thr –	Pro –	**Glu** –	Glu –	Lys –
HbS	Val –	His –	Leu –	Thr –	Pro –	**Val** –	Glu –	Lys –
amino acid	1	2	3	4	5	6	7	8

Glutamic acid carries a negative charge and is polar whereas valine is non-polar and hydrophobic. The presence of valine makes deoxygenated HbS less soluble. Therefore when HbS loses its oxygen the molecules come out of solution and crystallise into rigid rod-like fibres. These change the shape of the red cell, which is normally a flat circular disc. The reason for the changed amino acid is a change, or mutation, in the DNA coding for the amino acid. If you look at table 23.4 you should be able to make a prediction about what this change might have been. The possible mRNA codons for the two amino acids are as follows:

Glu: GAA GAG
Val: GUU GUC GUA GUG

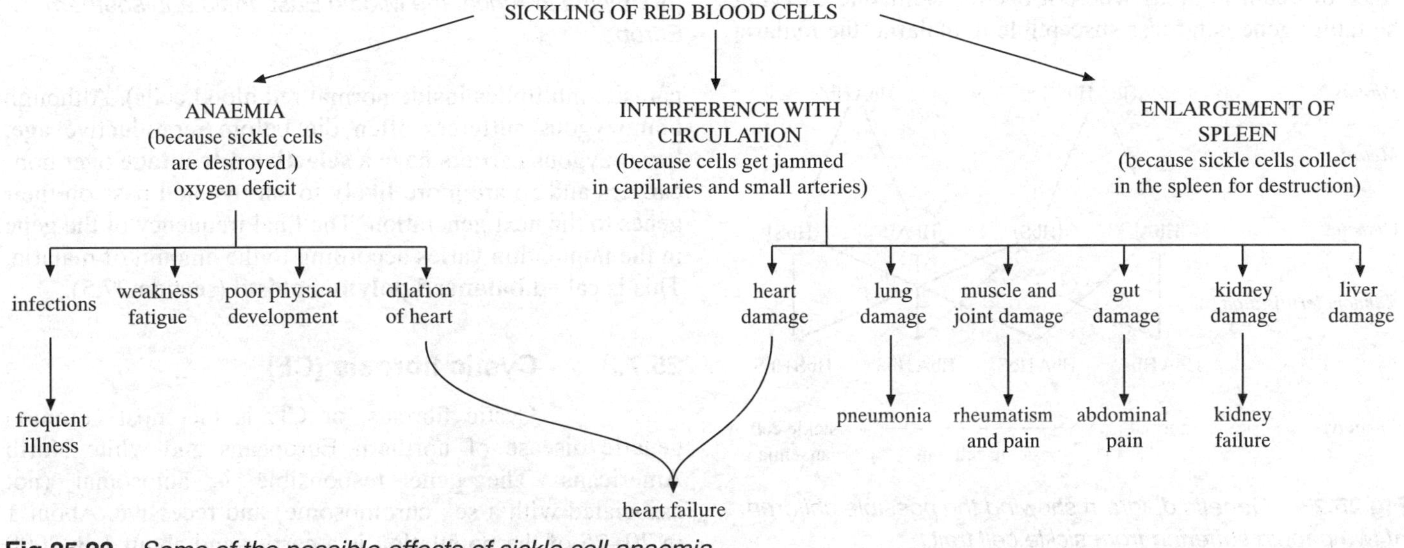

Fig 25.22 *Some of the possible effects of sickle cell anaemia.*

The complementary DNA triplet codes are therefore:

Glu:	CTT CTC
Val:	CAA CAG CAT CAC

To change the message from glu to val, T (thymine) must be replaced by A (adenine) in the second position of the triplet. Such a mutation is called a **substitution**. We now know that CTC is changed to CAC in the β-haemoglobin gene and that the gene is situated on chromosome 11.

What, then, happens in heterozygous individuals? Here, about half the molecules made are HbS and half are HbA. Strictly speaking therefore, the alleles HbA and HbS are co-dominant and the faulty gene is not recessive. Heterozygous people are unaffected except at unusually low oxygen concentrations, such as when flying in an unpressurised aircraft or climbing at high altitude. Then some of the cells sickle. The heterozygous condition is known as **sickle cell trait**.

Fig 25.23 shows that if two people suffering from sickle cell trait (carriers of sickle cell anaemia) have children, there is a 1 in 4 chance of any given child being a sufferer of sickle cell anaemia. Blood tests can be done to find out the phenotype of a given person so, if it runs in the family, people would be advised to have blood tests before having children. Prenatal diagnosis can be done using a HbS gene probe, or a particular restriction enzyme, on the DNA of the embryo or fetus. This is obtained from cells obtained by CVS or amniocentesis (section 25.7.9).

There is a final twist in this story. The faulty gene is a disadvantage and it is therefore surprising to find that it is so common. In such circumstances, geneticists suspect that the gene may also be an advantage in some circumstances. In the case of the sickle cell gene, an advantage has been found. Fig 25.24 shows the distribution of the gene worldwide and also shows the distribution of malaria. The distributions match quite closely, the gene becoming more common as malaria becomes more common. The frequency of the gene may reach 40% (40% HbS, 60% HbA genes in the population) in some parts of Africa. Malaria is a leading cause of death in areas where it occurs. Someone carrying the faulty gene is far less susceptible to malaria (the malaria parasite multiplies inside normal red blood cells). Although homozygous sufferers often die before reproductive age, heterozygous carriers have a **selective advantage** over non-carriers and so are more likely to survive and pass on their genes to the next generation. The final frequency of the gene in the population varies according to the amount of malaria. This is called **balanced polymorphism** (section 27.5).

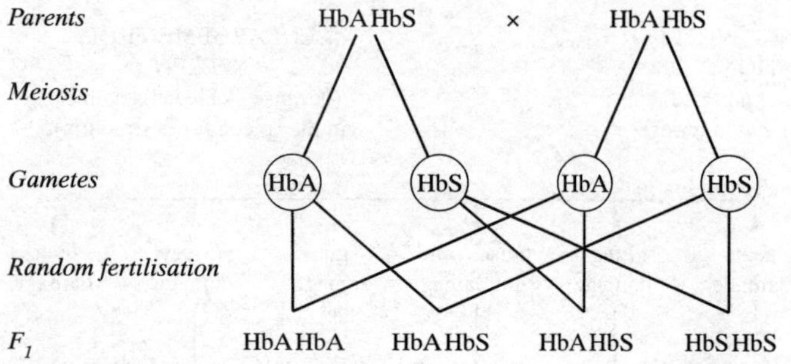

Fig 25.23 *Genetic diagram showing the possible children of two people suffering from sickle cell trait.*

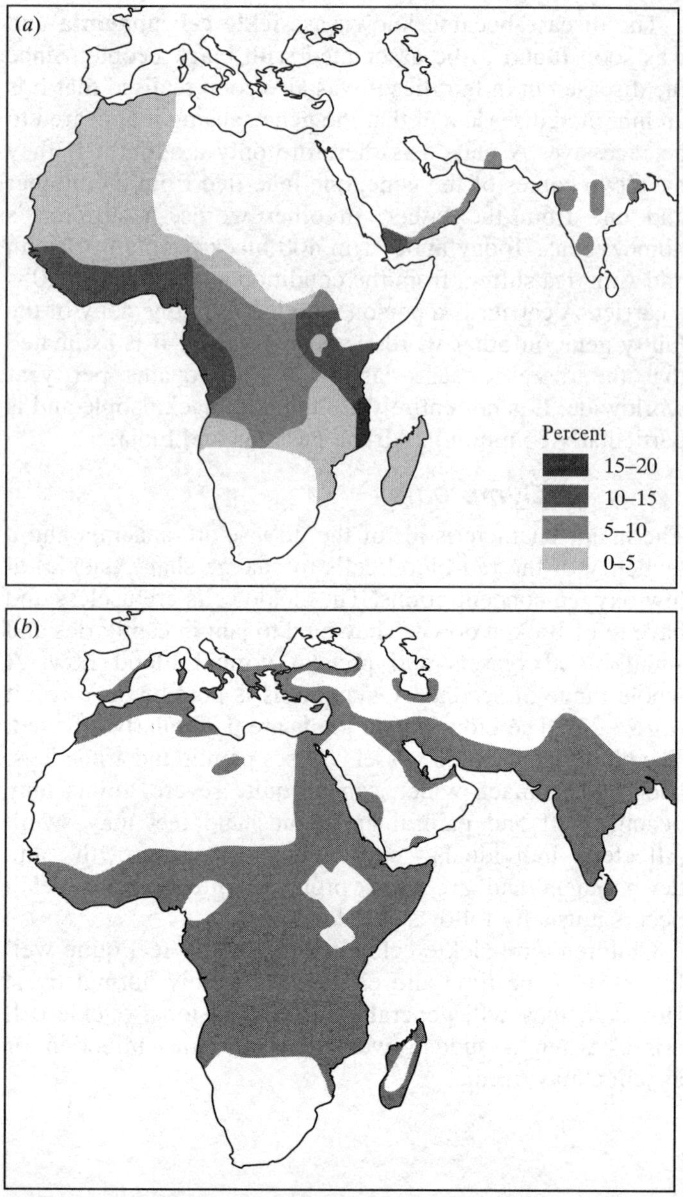

Fig 25.24 *Distribution of (a) the sickle cell gene and (b) malaria in Africa, the Middle East, India and southern Europe.*

25.7.3 Cystic fibrosis (CF)

Cystic fibrosis, or CF, is the most common genetic disease of northern Europeans and white North Americans. The gene responsible is autosomal (not associated with a sex chromosome) and recessive. About 1 in 20–25 of the population is a carrier and about 1 in 2000 is a sufferer. It occurs in fewer than 1 in 100 000 births among Africans and Asians. The disease gets its name from the fibrous cysts that appear in the pancreas. It is of particular interest, not just because it is so common, but because it is one of the first genetic diseases for which a cure has been attempted (section 25.7.11). As with sickle cell anaemia, the fact that it is so common suggests that carriers of the disease have an advantage over those with a normal genotype. It is unclear what this advantage is, but it may give increased resistance to cholera, a disease which was a common killer in Europe until the late nineteenth century.

Cause

The cause is a recessive mutation in a gene located on chromosome 7. The gene codes for a chloride channel (section 7.9.8) which is a protein, 1480 amino acids long, and known as **CFTR (cystic fibrosis transmembrane regulator)**. It allows diffusion of chloride ions into and out of epithelial cells and is located in the cell surface membranes of these cells. In CF sufferers it does not function. Since the gene is recessive, CF sufferers are homozygous and have two copies of the faulty gene.

The gene responsible for CF was cloned in 1989. This allowed the nature of the mutation to be discovered and also resulted in improved ability to detect carriers (section 25.7.9) using a simple mouthwash technique to obtain cells for DNA analysis. It also became possible to identify carriers or sufferers by prenatal diagnosis. In 70% of cases in central and western Europe the cause of the problem is the deletion of three base pairs from the gene; codon number 508 in the mRNA is therefore missing. As a result the amino acid phenylalanine (F) is missing at position 508 in the protein. The mutation is therefore called ΔF508 (Δ is the Greek letter delta, d, standing for deletion). More than 400 other mutations have been found in the same gene which also cause CF, but a further 15% of cases are caused by just five other mutations. Some result in mild forms of the disease. Some have been found in only one person.

Symptoms

One of the normal functions of the epithelial cells is to form mucus glands which secrete mucus. In CF patients this mucus becomes abnormally thick and sticky because the normal outward flow of chloride ions from the cells is prevented. Chloride ions are negatively charged, so in order to balance the negative charge which builds up in the cells more sodium ions enter. The high ion concentration inside the cell in turn prevents water from leaving the cell. The parts of the body most affected are the lungs, pancreas and liver. In the pancreas fibrous patches, called cysts, develop which give the disease its name. The thick mucus clogs up the airways of the lungs, and the branches of the pancreatic duct and the bile duct from the liver into the gut. Repeated lung infections are caused, as well as digestive problems, including poor release of pancreatic enzymes and poor absorption of digested food. The intestine may also become obstructed. In addition, males are almost always infertile and females are frequently infertile. Another characteristic symptom is that the sweat is saltier than usual because the sweat duct is relatively impermeable to chloride ions and once again sodium follows the chloride. This may explain an old saying: 'Woe is the child who tastes salty from a kiss on the brow for he is hexed and soon must die'. 95% of deaths are the result of lung complications. Average life expectancy has increased from 1 year to about 20–30 years with modern treatment methods, but about half the sufferers die by the age of 20.

Treatment

Treatment is concentrated on the lungs. It usually involves physiotherapy, possibly as much as five times a day, including slapping the back to dislodge mucus from the lungs (fig 25.25). Enzyme supplements can be given to improve food digestion and antibiotics to fight infection. In severe cases heart and lung transplants may be used. Much of the viscosity of the mucus is caused by DNA of dead infectious bacteria and dead white blood cells. Some success at reducing this has been achieved using the enzyme human DNase in the form of an aerosol to break down the DNA. The most desirable treatment though is gene therapy (section 25.7.11).

25.7.4 Phenylketonuria (PKU)

PKU occurs in about 1 in 10 000 live births among white Europeans and about 1 in 80 is a carrier. It is very rare in other races.

Cause

Like CF, PKU is a recessive, autosomal condition. It is a very distressing condition if not treated, but fortunately early diagnosis and treatment can prevent damage to health.

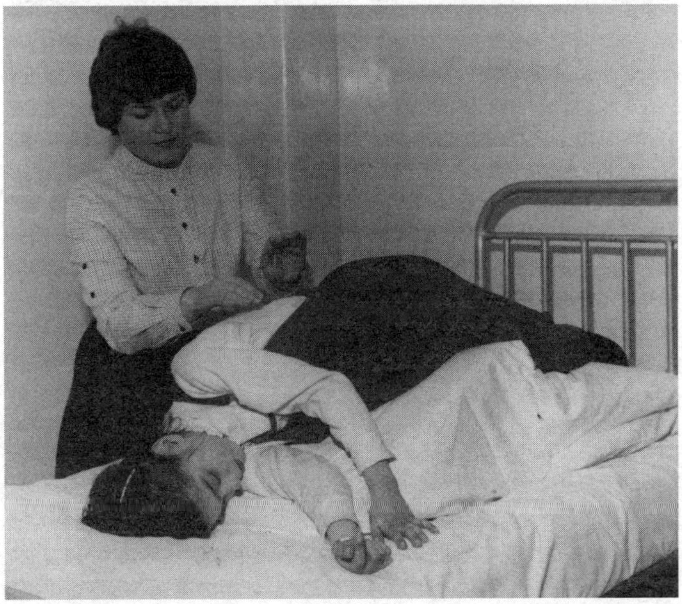

Fig 25.25 *Cystic fibrosis patient receiving physiotherapy.*

The disease is due to an inability to convert the amino acid phenylalanine to another amino acid, tyrosine:

$$\text{phenylalanine} \xrightarrow{\text{phenylalanine hydroxylase (PAH)}} \text{tyrosine}$$

The enzyme PAH is normally present in the liver, but is faulty in sufferers from PKU. The gene for this enzyme is on chromosome 12. As a result of faulty PAH, phenylalanine builds up in the body. The excess is converted to toxins which affect mental development. Affected children appear normal at birth because, while in their mother's uterus during pregnancy, excess phenylalanine moves across the placenta and is removed by the mother's liver. If not treated in infancy, harmful effects are soon noted. The most serious of these is severe mental retardation. Many untreated patients have IQs of less than 20. Before treatment became available more than 1% of all patients in mental hospitals were sufferers. Untreated sufferers rarely live beyond the age of 30. Other effects, which vary from patient to patient, include:

- hyperactive and irritable behaviour in children;
- awkward posture and walk;
- lighter skin pigmentation and fair hair (because tyrosine is normally used in the synthesis of the brown skin pigment melanin);
- dry, rough skin (eczema);
- repetitive movements of the fingers, hands or entire body;
- convulsions due to abnormal brain activity.

25.2 Examine the pattern of inheritance of PKU shown below.

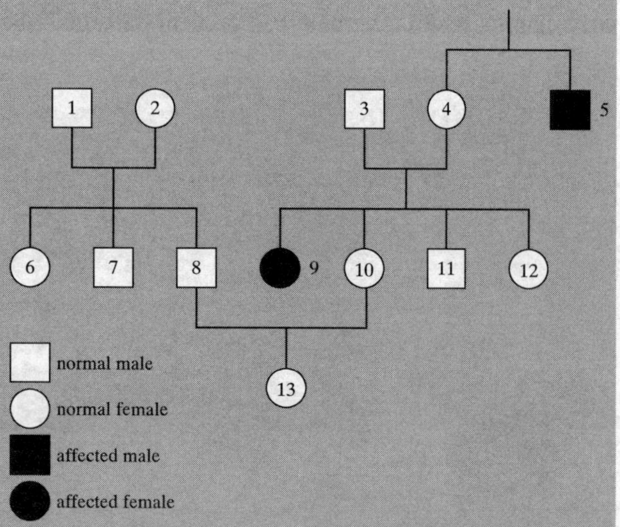

(*a*) What evidence is there that PKU is controlled by a recessive gene?

(*b*) What evidence is there that PKU is *not* sex-linked?

(*c*) Which individuals are definitely carriers (heterozygous) based on the evidence available?

(*d*) Which other individuals *could* be carriers?

(*e*) In a real situation, the individuals numbered 10, 11 and 12 may well wish to know if they are carriers since their sister suffers from PKU. If one of them asked you what were their chances of being a carrier, what would you reply. Think carefully!

Identifying PKU in newborn babies

It is important to test for PKU in babies because there are no obvious symptoms for about the first six months, and by this time irreversible brain damage will have occurred if no action has been taken. A very sensitive blood test was developed in 1963 which detects the higher than normal levels of free phenylalanine in the blood of sufferers. Levels are commonly 30–50 times higher than normal. The test is now carried out on all babies four days after birth by pricking the heel for a blood sample.

25.3 Why is the baby not tested when it is first born?

Identifying carriers and prenatal diagnosis

Population screening is a possibility since 95% of all carriers can be detected using modern methods of DNA analysis. (Carriers used to be detected using a blood test.) Prenatal diagnosis of sufferers or carriers is also now possible, using chorionic villus sampling or amniocentesis, to sample cells followed by DNA analysis. It is not normally recommended because the disease is treatable. Prenatal diagnosis was not possible before because the excess phenylalanine was removed by the mother. Early experiments on gene transfer suggest that gene therapy may be possible in the future.

Treatment

The condition is managed by reducing the amount of phenylalanine in the diet to the minimum required. Phenylalanine is an essential amino acid (meaning that it cannot be made from other amino acids so some must be present in the diet for making proteins). Since sufferers cannot make tyrosine from phenylalanine (see above), tyrosine is also an essential amino acid needed in the diet of sufferers. Blood levels are monitored for the first few years of life to check that the correct balance is being maintained. Excess phenylalanine in adulthood is not damaging, presumably because brain development has ceased, so a normal diet can then be adopted. Suitable food is rather tasteless, but can be supplemented by small amounts of other foods such as meat and products low in phenylalanine. Sufferers must avoid gluten which means avoiding bread as well as some children's favourites such as sweets and some orange juices. This makes it difficult for young children. Children will feel ill if they do not stick to the diet.

> **25.4** Why is it fortunate for PKU sufferers that phenylalanine is an essential amino acid?
>
> **25.5** Adult sufferers from PKU return to a normal diet. Suggest why it is important for a woman sufferer to go back to a phenylalanine-restricted diet before pregnancy.

25.7.5 Huntington's chorea (HC)

Huntington's chorea, or Huntington's disease, is caused by an autosomal mutation which is *dominant*, unlike the previous three examples. It affects about 1 in 10 000 people. In 1983 the gene was located on chromosome 4, but it took another 10 years of painstaking work to locate it precisely and to be able to clone it. The function of the protein it codes for is unknown, although it has been given a name, 'huntingtin'.

The disease was first described in 1872 by George Huntington. As an 8-year old boy he had witnessed the appalling consequences of the disease in two women: 'mother and daughter, both tall, thin, almost cadaverous, both bowing, twisting, grimacing.' The disease causes progressive deterioration of brain cells and gradual loss of motor control (control of voluntary muscle by motor nerves) resulting in uncontrollable shaking and dance-like movements. This accounts for the use of the term 'chorea', meaning dance, to describe the disease. Intellectual ability is lost, hallucinations, slurring of speech, mood changes, personality changes and memory loss (temporary or permanent) may all occur. The brain shrinks between 20–30% in size.

Two particularly difficult problems are associated with the disease.

- Firstly, symptoms do not usually start to appear until middle age (fig 25.26), by which time many sufferers will already have passed the disease on to their children. This is an additional source of distress for families. Life expectancy averages 15 years from the onset of symptoms and the slow decline can be emotionally very painful for close family members to witness.
- Secondly, because it is a dominant gene it will always be expressed, and on average 50% of the children will be sufferers if one parent is affected (there can be no 'carriers'). In the past, the children were unlikely to discover until they were young adults that a parent was affected and that they had a 50% chance of going through what they observed happening to the parent.

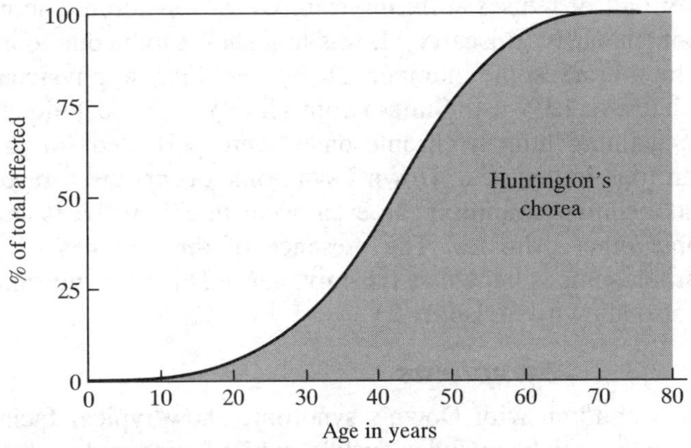

Fig 25.26 *Variation in the age of onset of Huntington's chorea.*

Predictive diagnosis

Genetic counselling of families with a high risk of the disease is a clear need and, since the cloning of the gene in 1993, has been made easier by the introduction of highly reliable predictive DNA tests for the condition. These can tell a person if they will get the disease. There is now an internationally agreed code of practice for predictive testing, which includes counselling both before and after testing. Perhaps surprisingly, many children of affected parents who stand a 50% risk say that they would rather not know if they are going to develop the disease.

> **25.6** Give two possible reasons for (*a*) wanting to know, (*b*) not wanting to know if you were going to develop HC.
>
> **25.7** There seems to have been little selection against the gene for the disease. Suggest why this is so.

25.7.6 Down's syndrome

Chromosome mutations are discussed in section 24.9.2. They are a common cause of miscarriage, probably accounting for about 50 to 60% of all miscarriages. Three well-known examples in humans are discussed in the following sections. One affects autosomes (non-sex chromosomes), namely Down's syndrome, and two affect the sex chromosomes, namely Klinefelter's syndrome and Turner's syndrome. It will help if you revise your knowledge of meiosis (section 23.4) and karyograms (section 23.1.1).

Down's syndrome is named after the nineteenth century physician John Langdon Down who worked at an asylum in Surrey, England and who in 1866 was the first to describe the condition. It affects about 1 in 750 babies at birth, but over half of fetuses suffering from Down's syndrome abort spontaneously (miscarry). It was first shown to be due to an extra chromosome number 21 by the French physician Lejeune in 1959 using microscopy (fig 25.27) as techniques for staining human chromosomes were perfected for the first time in that year. Down's syndrome occurs in all races and a similar condition can even occur in chimpanzees and some other primates. The presence of three copies of a chromosome is known as **trisomy**, hence Down's syndrome is also known as **trisomy 21**.

Symptoms

Most children with Down's syndrome show typical facial features which include eyelids which apparently slant upwards due to a fold of skin over the inner corner of the eye. The face is typically flat and rounded (fig 25.28). Other characteristics include:

- mental retardation, often severe;
- short stature and relatively small skull due to poor skeletal development;
- heart defects occur in about one-quarter of Down's children;
- increased risk of infection, particularly respiratory and ear infections;
- coarse, straight hair;
- squat hands with a characteristic crease which runs all the way across the palm;
- intestinal problems and leukaemia are slightly more common than normal.

Affected children are characteristically very friendly, cheerful and often greatly enjoy music. It is important to understand that the condition ranges widely in degree and that jobs with limited responsibility can be successfully undertaken by many Down's sufferers, and many can live independent lives. Personality can equally vary widely. In other words they should be treated and thought of as people first and Down's sufferers second.

Explanation

In 96% of cases the cause of Down's syndrome is non-disjunction of chromosome 21 during anaphase of meiosis. This can take place during production of sperm, but is more common during production of eggs (see effect of maternal age below). About 70% of non-disjunctions occur in meiosis I and 30% in meiosis II. In meiosis I, it is caused by the failure of whole chromosomes to separate, whereas

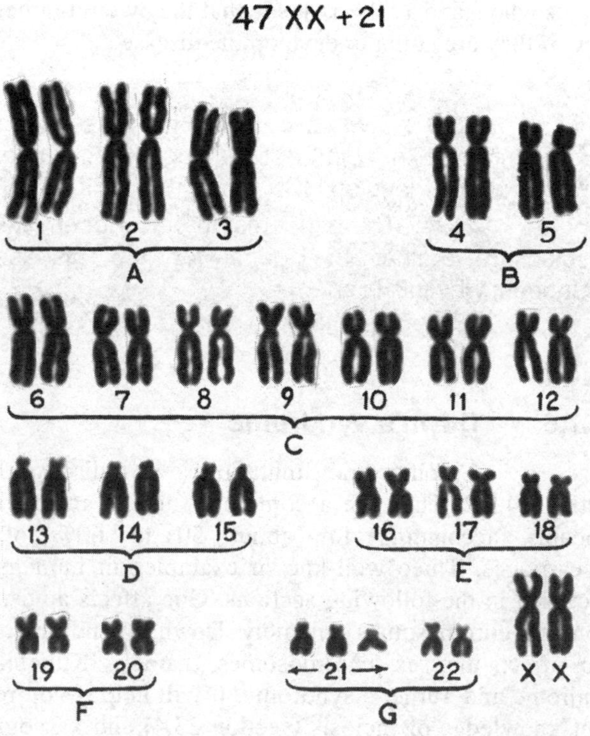

Fig 25.27 *The chromosomes of a female suffering from Down's syndrome. The non-disjunction of chromosomes 21 in one of the gametes has led to the presence of three chromosomes 21 in this female.*

Fig 25.28 *Young girl showing typical physical features characteristic of Down's Syndrome – slightly slanted eyes, round head and flat nasal bridge.*

in meiosis II chromatids fail to separate. The final effect is the same, with two chromosomes or two chromatids entering one daughter cell and none entering the other, instead of one entering each (fig 24.31). This should lead to an equal number of cases of monosomy 21 (only one chromosome 21), however this condition is fatal early in the development of the fetus like all cases of monosomy.

About 3–4% of Down's syndrome cases are due to a type of mutation known as a translocation (section 24.9.2). Chromosome 21 is translocated (moved) to chromosome 14 or, less commonly, to chromosome 22. An even less common cause is a 21 to 21 translocation.

Effect of mother's age

A correlation with the age of the mother has been shown for Down's syndrome (fig 25.29). There is no correlation with the age of the father. At age 20 the risk is 1 in 2000, at 30 it is 1 in 900, at 40 1 in 100 and at 44 1 in 40. The incidence therefore rises more and more steeply with age, shown by a straight line when a logarithmic scale is used as in fig 25.29. This is probably due to the fact that a woman's egg cells are produced while she is still an embryo. They are stored until stimulated to develop at the rate of one a month during her fertile years. Men, on the other hand, constantly produce new sperm from puberty until death.

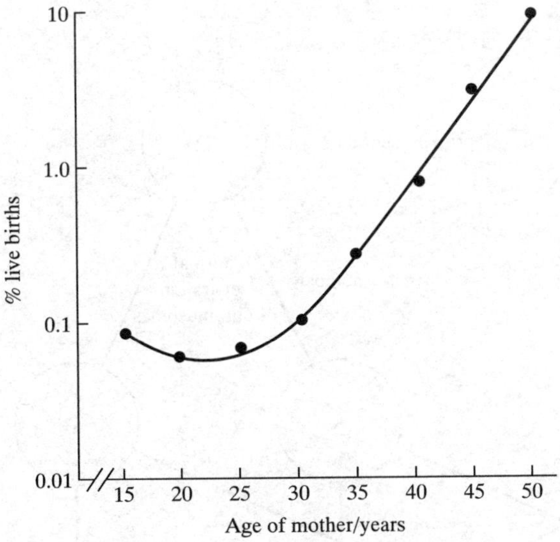

Fig 25.29 *Effect of maternal age on the incidence of Down's syndrome. Note that the vertical axis is logarithmic.*

Other trisomies involving autosomes

Most cases of trisomy involving autosomes are fatal and result in early miscarriage. Cases of trisomy involving some of the smallest chromosomes, namely chromosomes 13 and 18, sometimes survive to birth, but suffer many defects, including mental retardation. They usually die within three months because development cannot proceed normally. Other cases of trisomy surviving to birth are extremely rare.

25.7.7 Klinefelter's syndrome

In 1942, an American, Dr H. F. Klinefelter, studied nine male patients who could best be described as feminised males. Typical symptoms of these and similar patients are as follows:

- infertility – sperm are never produced, although erection and ejaculation are possible;
- usually taller than average;
- some breast development, although not necessarily very obvious;
- smaller testes than normal, although this is not necessarily obvious;
- higher than usual FSH secretion for males (FSH is follicle stimulating hormone and is produced by the pituitary gland in both men and women;
- trunk may show signs of obesity (eunuch-like appearance);
- little facial hair;
- voice pitched higher than normal;
- educational difficulties and behavioural problems are fairly common.

In 1959, it was discovered that Klinefelter's syndrome, as it became known, is due to an extra X chromosome. The genotype is therefore XXY instead of the normal XY and the sufferer has 47 chromosomes instead of 46. Like Down's syndrome, it is an example of trisomy. Fig 25.30 shows how the extra chromosome arises as a result of non-disjunction during meiosis. It may occur during spermatogenesis (sperm production) in the male parent or during oogenesis (egg production) in the female parent. The figure also shows that, as a result of non-disjunction in the male sex chromosomes, equal numbers of zygotes will contain only one X chromosome and no Y chromosome (represented as XO). This gives rise to Turner's syndrome, which is described in the next section. In the case of non-disjunction in the female, XXX and YO zygotes are also created. XXX does not produce a superwoman. In fact, there is no apparent physical difference between XXX women and XX women, apart from the fact that XXX women tend to be slightly taller. There is some evidence though that behavioural abnormalities and learning difficulties may occur more frequently in XXX women. YO zygotes do not develop because many vital genes are missing completely (remember the Y chromosome has missing genes compared with the X chromosome).

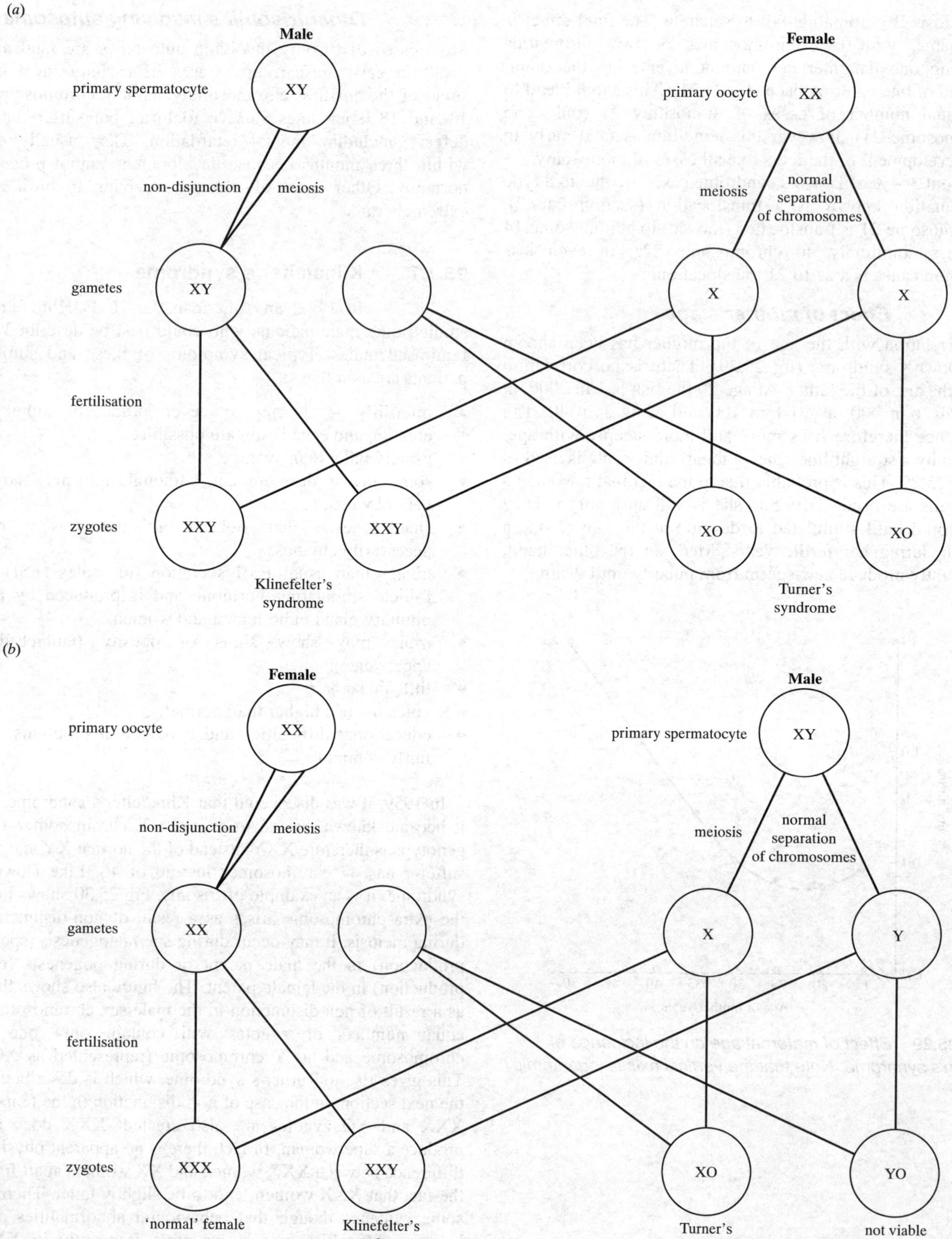

Fig 25.30 *Explanation of Klinefelter's syndrome and Turner's syndrome as a result of (a) non-disjunction of the father's sex hormones and (b) non-disjunction of the mother's sex chromosomes.*

Treatment

The condition is usually diagnosed only after puberty. Male hormones can be given. Breasts then return to normal size and a normal sex life can be led, even though sterility remains.

25.7.8 Turner's syndrome

This syndrome was first described by an American, Turner, in 1938. Patients can best be described as incompletely developed females, although there are often no obvious external differences compared with normal females. Typical symptoms are as follows:

- infertility – ovaries are absent (represented only as connective tissue);
- shortness of stature, averaging 1.5 m (less than 5 feet);
- small uterus;
- webbed neck may occur;
- puffy fingers with deep set finger nails which are more convex than normal;
- the hair line (line at which hair starts to grow) at the back of the head is lower than normal.

In 1959, Turner's syndrome was shown to be due to a missing X chromosome. The genotype is therefore XO instead of the normal XX and the sufferer has 45 chromosomes instead of 46. This is an example of monosomy. Fig 25.30 shows how Turner's syndrome can arise as a result of non-disjunction during meiosis and that it represents the 'flip-side' of Klinefelter's syndrome. In theory, equal numbers of Klinefelter's and Turner's syndrome individuals should be born. In reality, Turner's syndrome is significantly rarer, with about 1 in 2500 births compared with 1 in 500 births for Klinefelter's. This is because Turner's syndrome is far more likely to be fatal early in pregnancy. It is estimated that only about 2–3% of Turner's syndrome conceptions survive to birth. Interestingly, Turner's syndrome also seems to be responsible for a relatively high proportion of miscarriages, perhaps as great as 20%.

Treatment

From the age of puberty women can be given female sex hormones to make them develop breasts and have periods. This does not cure infertility but is done for social reasons to make the person feel more 'normal'. Growth can be stimulated with growth hormone.

Other combinations of sex chromosomes

Various other combinations of sex chromosomes are possible. Individuals with a Y chromosome are always male due to the presence of the SRY gene (sex-related Y gene), that is the gene for male sex development. Some XX males have been identified (1 in 20 000 births) but these have a piece of Y chromosome containing the SRY gene in one of their X chromosomes (a translocation mutation). They are physically like Klinefelter's patients. XXXY and XXXXY males occur and are mentally retarded. XXXX women are severely mentally retarded, although physically normal. The first XYY man was detected in 1961. Such men are usually normal and more research is needed to establish whether there are any consistent abnormal physical or mental characteristics. XXYY men have also been identified.

25.7.9 Genetic screening and prenatal diagnosis

In section 25.7.1 it was pointed out that everyone probably carries several genetic defects which can be referred to as genetic diseases or genetic disorders. More than 4000 such disorders have been identified, but with 100 000 different human genes there must be many more to find. Some common and important examples have already been discussed. Detecting mutant genes in an individual is known as **genetic screening**. Modern genetics is making this much easier than it was in the past. There are three situations where genetic screening is of particular relevance, namely prenatal diagnosis, carrier diagnosis and predictive diagnosis. These are explained below.

Prenatal diagnosis. This is the use of modern medical techniques to identify any health problems of the unborn baby. It includes the detection of genetic disease. If such a disease *is* detected, it is usually possible to provide counselling about the quality of life the child can expect and other potential problems. The parents are usually also given the option of an abortion.

Carrier diagnosis. This is the identification of people who carry a particular genetic disease, usually with no visible symptoms or harm to themselves. As stated already, we are all carriers of a few genetic diseases. Important examples are the genes for sickle cell anaemia, cystic fibrosis and PKU. Identification of carriers is becoming increasingly important and controversial as more and more genetic tests become available. There are obvious advantages. For example, if a couple considering having children both have a history of sickle cell anaemia in their families, they would be advised to be tested to discover if they were carriers. If they were, then it could be explained that they had a 1 in 4 chance of producing a child who would suffer from sickle cell anaemia. A genetic counsellor could then discuss the issues involved and possible options (section 25.7.10). As more tests become available, problems of economics arise. Can the Health Service afford mass screening programmes for cystic fibrosis, for example?

Predictive diagnosis. This is the prediction of a future disease which you are likely to suffer as a result of your genes but which has not yet produced any symptoms. The classic example of this 'genetic time bomb' is Huntington's chorea where the onset of the disease typically occurs in middle age. In this case, a single dominant gene is responsible and it is certain that if the person lives long enough they will get the disease. Many diseases are more complex than this. For example, people can be 'genetically predisposed' to suffer from heart disease, breast cancer and many other common diseases, but environmental factors, such as smoking, exercise and diet, also play a role. People can now be tested, if they wish, to find out if they have the Huntington's chorea gene, and fairly soon we may be able to carry out tests which estimate the risks of other diseases like heart disease. We therefore need to consider a number of related issues. For example, should it be illegal for insurance companies to demand that genetic testing be done before offering life insurance and should they be allowed to penalise those who are at greater risk? Also, with more testing there would be a much greater demand for genetic counselling, with funding implications. These issues are discussed at the end of this section.

How, then, is genetic screening carried out. The four most important techniques are chorionic villus sampling, amniocentesis, pre-implantation diagnosis and gene probes.

Chorionic villus sampling (CVS)

The chorion is a 'membrane' (thin layer of cells) which grows from the embryo in the earliest stages of pregnancy. Its role is to invade the mother's uterus with small finger-like processes, the **chorionic villi**. These penetrate into blood spaces in the uterus lining and allow exchange of materials such as nutrients, oxygen and waste products between the mother and the embryo. The chorion later becomes part of the placenta. The technique was developed in China in the 1970s for withdrawing a small sample of the chorion for examination. The technique is shown in fig 25.31. The abdominal route is more common (fig 25.31*a*) and involves inserting a hollow needle through the body wall and uterus wall. A local anaesthetic is used. The cervical route (fig 25.31*b*) involves the use of a narrow, flexible tube called a catheter. No anaesthetic is needed. In both cases a syringe is attached to the tube or needle and is

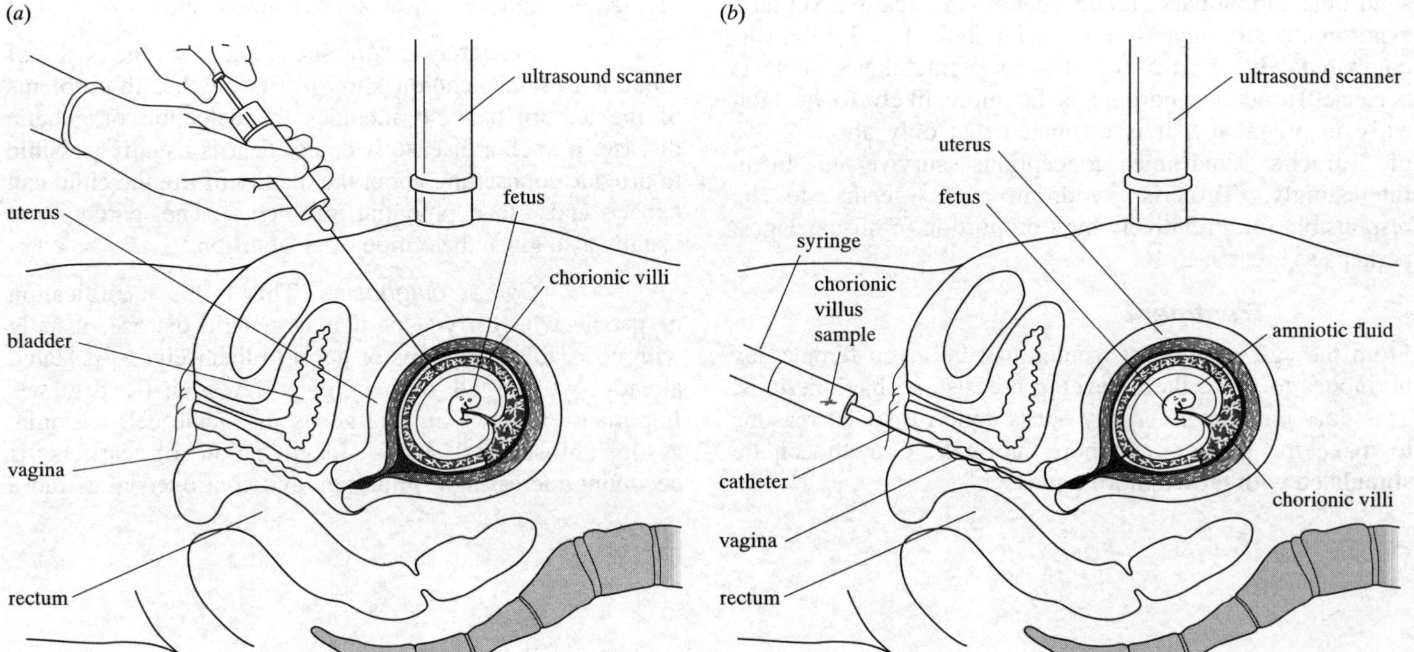

Fig 25.31 *(a) Chorionic villus sampling (CVS) through the wall of the abdomen (known as transabdominal sampling). (b) CVS through the vagina and cervix (known as transcervical sampling). The sample taken is a few millimetres across.*

used to gently suck away a few fragments of chorionic villus material. Both methods are used in association with an ultrasound scanner which shows the contents of the uterus on a screen and helps to guide the tube into place. The fetus is checked before and after the procedure for signs of life.

The cells of the chorion are all derived from the zygote, like the embryo, and so are genetically identical to the embryo. The cell sample is relatively large compared with that obtained by amniocentesis (see below). Most of the cells are also actively dividing at this stage, so it is possible to examine cells immediately for their chromosome content (chromosomes are only visible when cells are dividing). A picture of the chromosomes can be used to make a karyogram and any chromosomal abnormalities can be identified, such as Down's, Klinefelter's or Turner's syndromes. This is known as **karyotype analysis**. The sex of the child can also be seen, which may be relevant if a sex-linked disease is suspected. The cells can also be cultured in a suitable medium in a laboratory so that further tests, such as DNA analysis, can be carried out. This could identify conditions such as cystic fibrosis, Huntington's chorea or thalassemia. The cells grow quickly and results can be obtained within 5–12 days. The great advantage of CVS is that it can be carried out early in pregnancy, between 8 to 12 weeks. It does, though, carry a slightly greater risk of miscarriage than amniocentesis. It is estimated that the risk of miscarriage is about 2% higher than the rate of natural miscarriage. If as a result of tests a decision is made to abort the fetus, the abortion is less difficult and risky, both physically and emotionally, than the later abortions carried out after amniocentesis.

Amniocentesis

Amniocentesis is a more widely available and older technique than CVS, introduced in 1967. It is usually performed at 15–16 weeks of pregnancy but can be done a few weeks earlier or up to 18 weeks. It cannot be done before there is sufficient amniotic fluid or cells to analyse. It is safer than CVS, causing an increased risk of miscarriage of about 0.5% (about 1 in 200). A hollow needle is inserted through the abdomen and uterus walls using ultrasound for guidance and a local anaesthetic (fig 25.32). About 20 cm^3 of amniotic fluid is sucked out of the amniotic sac which is tough and seals itself after removal of the needle. If possible, the placenta is avoided. The fluid is normally clear and yellow like urine. Amniotic fluid is constantly being swallowed by the baby and passes through its gut. It also contains urine from the baby. It therefore contains some living cells swept from inside the body as well as any living skin cells that may flake off. It may also contain cells from the amnion which, like the chorion, is derived from the zygote and is genetically identical to the fetus. The cells are spun down in a centrifuge and are cultured to increase their numbers and to obtain dividing cells. Results therefore take about 3–4 weeks to obtain, a regrettable delay if abortion is to be carried out. The chromosomes of the dividing cells can be examined for abnormalities. The cells and the fluid from the sample can both be tested for products of faulty genes, such as **α-feto protein** (**AFP**), high levels of which indicate a higher than normal risk of certain birth defects such as spina bifida. Results from the fluid can be obtained within 7–10 days.

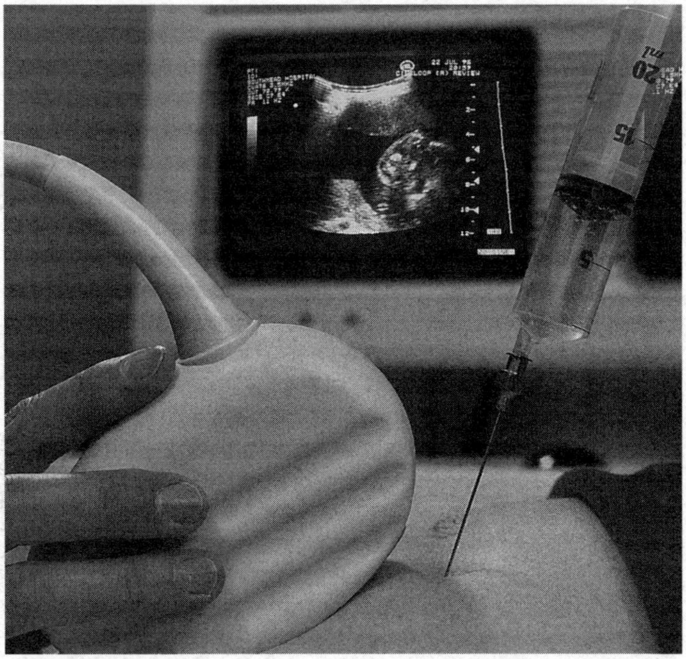

Fig 25.32 *A doctor is drawing in a sample of amniotic fluid from a pregnant woman. In the other hand the doctor holds an ultrasound transducer to detect the position of the fetus and placenta within the uterus. The image of the fetus is displayed on the screen in the background of the picture.*

Pre-implantation diagnosis

This is a form of prenatal diagnosis which is carried out on the embryo before it implants in the uterus. It requires use of the test tube baby technique (IVF) (section 21.7.4). After fertilisation outside the body, a cell can be removed from the developing embryo at about the 8-cell stage without damaging its ability to continue to develop normally. After testing, only the healthy embryos need be replaced in the woman. This avoids the need for abortion. However, the low success rate and expense of IVF mean that this cannot be used as a routine procedure.

Gene probes and DNA analysis

Genetic screening by examining a person's DNA is known as **DNA analysis**. It involves the following stages and is illustrated in fig 25.33.

(1) The DNA is first **extracted** from cells. The cells may be obtained from a fetus by CVS or amniocentesis, or from a child or adult using, for example, a blood sample or a mouthwash.
(2) The DNA is then cut up into fragments of different length using **restriction enzymes**.

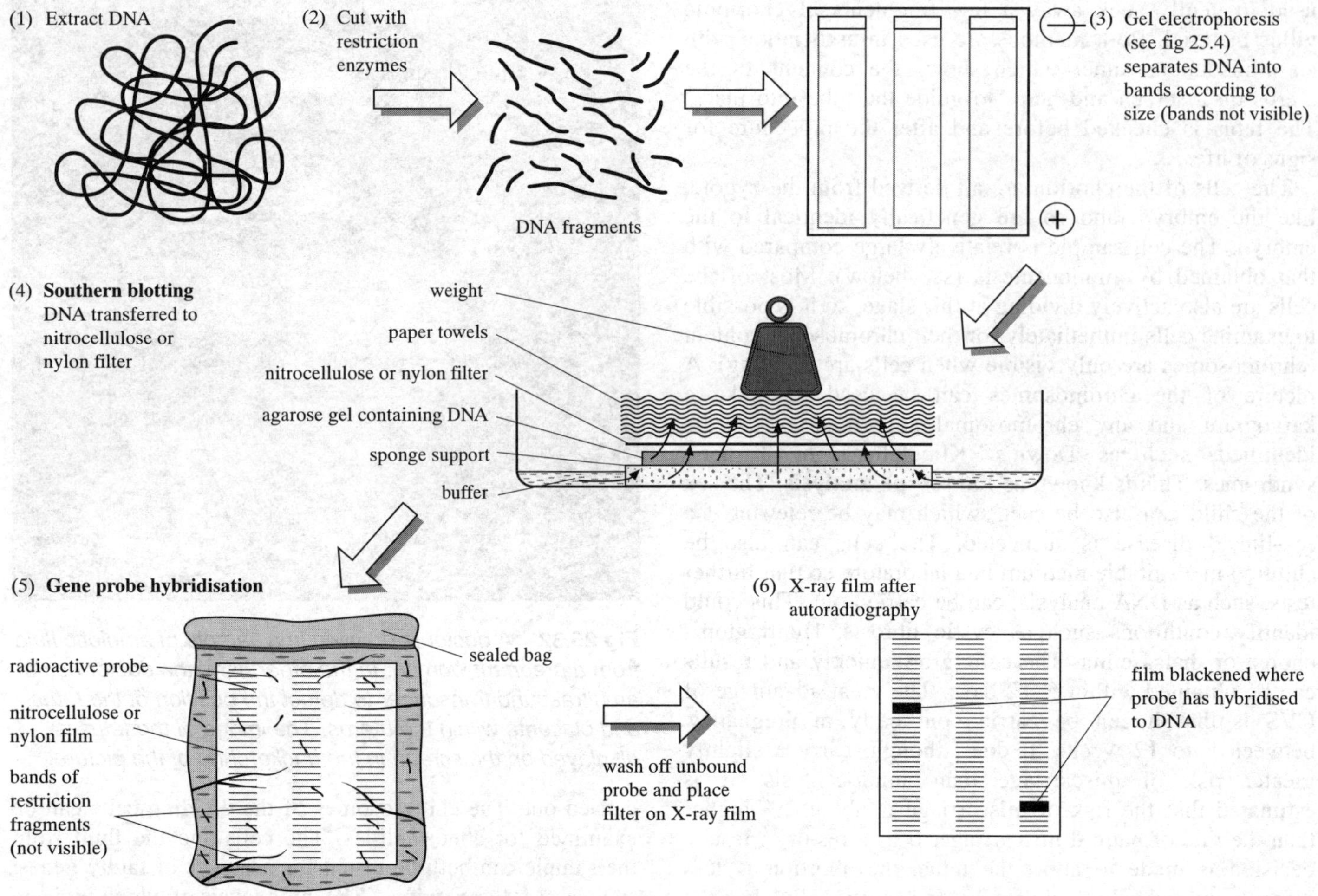

Fig 25.33 *Some stages in DNA analysis during genetic screening. During Southern blotting, denatured (single-stranded) DNA is transferred from an agarose gel to a nitrocellulose or nylon filter. Paper towels are no longer commonly used; techniques known as electroblotting or vacuum blotting are used instead.*

(3) The fragments are separated according to size using **gel electrophoresis** through an agarose gel as described earlier (fig 25.4). The smaller fragments move faster down the gel than the larger ones.

(4) The aim is now to find if a particular piece of mutant DNA is present in one of the fragments. It is looked for with a **gene probe**. The gene probe does not move very well through the gel, so a technique known as **Southern blotting** (after Edward Southern, who first developed the method in 1975) is used first. This is illustrated in fig 25.33 and is a commonly used technique in genetics (see also genetic fingerprinting, 25.7.12).

The DNA in the gel is first converted to the single-stranded form (denatured) by heating. It is then transferred to a nitrocellulose or nylon filter by a blotting action, making a precise replica of the original gel. The stack of paper towels acts as a wick which draws the buffer up through the sponge, the gel and the filter. The DNA sticks tightly to the filter.

(5) A **gene probe** is now used (section 25.1.6). This is a short, single-stranded piece of DNA whose base sequence is complementary to part of the gene being sought. Large quantities of the probe are made by cloning. It is made radioactive by labelling with ^{32}P and then added to a solution for hybridising (complementary binding) with the DNA on the nitrocellulose or nylon filter.

(6) **Autoradiography** is then carried out by placing the filter on an X-ray film. Radiation from any bands to which the probe has hybridised will blacken the film (fig 25.33). Many mutations are major chromosome deletions and so shorten the length of the restriction fragment. This makes the fragment more mobile in the gel so it will show up as a different band after radiography. One such example is the gene for sickle cell anaemia which can be detected in this way.

Other relevant techniques

Ultrasound scans can be used to detect a variety of conditions, some of which may be genetic in origin such as spina bifida.

Blood tests of the mother during pregnancy can detect raised levels of chemicals, such as AFP, associated with certain genetic conditions in the fetus.

Benefits of genetic screening

As already noted, despite the benefits of genetic screening, it also raises a number of problems. Both the possible benefits and the problems are considered below, but these are all points which need discussion.

- Genetic screening reduces suffering of both victims of genetic disease and their families. It could be argued that it is not morally right to bring a child into the world knowing it will suffer more than usual. Genetic screening gives couples a choice of whether to allow the birth of a child with a genetic defect.
- Genetic screening reduces the economic cost to the country. For example, it is estimated that it currently costs the NHS about £40 000 a year to keep alive a child who is suffering from cystic fibrosis.
- There is a public desire for tests. The cystic fibrosis test is not available on the NHS but 80–90% of people offered the test use it.
- If you know that you are genetically predisposed to a certain condition you may be able to change your lifestyle to reduce the possibility of it occurring. For example, a known predisposition to heart disease might make someone give up smoking.
- Screening programmes for diseases have been successful in reducing the incidence of the diseases when carried out with the understanding and approval of the communities involved. An example is screening for thalassemia in Cyprus. **Thalassemia** is similar to sickle cell anaemia in being a recessive gene which affects haemoglobin and gives resistance to malaria in the carrier state. It was common in Cyprus until a screening programme was introduced to identify mothers carrying affected fetuses and offer them abortions.
- Mothers over the age of 35 are usually offered free CVS or amniocentesis on the NHS because of the increased risk of Down's syndrome.

Problems associated with genetic screening

- Eventually we may be able to predict who is likely to suffer common diseases such as coronary heart disease and rheumatoid arthritis. Will insurance companies or employers demand genetic profiles of individuals and discriminate against them in some way, even though they may not even develop the disease? Will companies that do not use genetic testing be at a disadvantage? Should laws be passed which guarantee genetic privacy?
- Some genetic defects are not as serious as others. In fact, there is a continuous range of severity of disease. Who should decide then what is a severe abnormality? In the UK, abortion is allowed after the normal limit of 24 weeks for cases of 'severe abnormality'. Should abortion of an individual suffering from a treatable condition such as PKU be allowed or possibly encouraged by screening programmes?
- There is a risk that people who suffer from a genetic disease or condition will feel like social outcasts if more screening leads to fewer cases. Some genetically disabled people see the future as a nightmare. They do not necessarily view their impairment in the same way that non-sufferers do. Their values are often different and their opinions should be listened to and respected by non-disabled people. Do we have false values about what makes a life worthwhile? Should we accept that suffering is inevitable and learn the best way to cope with it?
- The eugenics movement at the beginning of the twentieth century showed the risks associated with genetic knowledge. Eugenics is the study of the possible improvement of the genetics of a species. Selective breeding is one method used. Negative eugenics is the removal of harmful genes; positive eugenics is the addition of beneficial genes. Attempts were made by the Nazis, for example, to justify the complete elimination of mentally and physically disabled people in the pursuit of a genetically 'pure' race. There have been various attempts to discriminate between races on the basis of genetics. If removal of harmful genes becomes common, will this lead to 'designer babies' with the addition of desirable genes for characteristics such as intelligence, musical ability, sports prowess, and height when techniques become available?
- Screening programmes which are designed to eliminate diseases can lead to social pressures on women or couples to have abortions when it may not be what they personally want. This has happened to some women in Cyprus as a result of the screening programme for thalassemia.
- If a disease is associated with one race, screening programmes may encourage discrimination. This happened with screening for sickle cell anaemia in the USA in the 1970s.
- What is now regarded as within the bounds of normal variation may come to be regarded as undesirable, for example low IQ.
- How accurate should tests be before they can be used? Is it acceptable, for example, to inform someone that there is an 80% probability that they will suffer a particular fatal disease?
- How should we handle the genetic counselling of people who are diagnosed as having a genetic disease, particularly one such as Huntington's chorea which carries a death sentence?

- If parents choose to go ahead with a pregnancy that they know will bring into the world a child who suffers from a genetic disease, should they have to pay the medical cost of treatment or should society pay? In 1963 Francis Crick suggested that the day might come when we should have a licence to have children. A law passed in China in 1995 requires parents to seek permission to have children if there is a significant risk of inherited defects. In the USA it has been suggested that couples should lose their medical insurance if they opt not to have a pregnancy terminated when the fetus is disabled.
- Will children sue their parents if they have allowed them to be born knowing that they will suffer? Successful lawsuits have been brought by children against their parents in the USA for other medical problems.
- Genes that cause harm later in life might be beneficial earlier in life.
- Techniques offer the possibility of selecting the sex of the baby, so that a baby of the undesired sex could be aborted.

25.7.10 Genetic counselling

Genetic counselling is the giving of information and advice about the risks of genetic disease and their outcome. Patients can be referred to genetic counsellors by their local doctors or other health professionals. Doctors have an important role to play in alerting patients to risks before problems arise. Genetic counselling is often carried out in special clinics in hospitals, and the counsellor is part of a team which includes specialist laboratory workers. Counsellors are usually doctors or paediatricians (specialists in childhood conditions) who have gone on to specialise. They must have a good understanding of medical genetics as well as being trained in sympathetic counselling techniques. It is particularly important during counselling not to try to impose one's own views on the people being counselled. By giving information and encouraging discussion people must be helped to make their own decisions.

Some of the issues which are commonly discussed are listed below.

Making a diagnosis. Making an accurate diagnosis of the problem is the first essential. This may involve a physical examination and laboratory tests. For example, a woman aged 20 who has not begun to menstruate and whose breasts have not developed could be suffering from Turner's syndrome, a hormone deficiency or simply delayed puberty. Chromosomal analysis could confirm whether or not Turner's syndrome is the cause.

Family history. Investigating the family history for previous cases of genetic diseases is commonly done. It is helped by drawing the family tree with the person being counselled.

Calculating risk. The genetic counsellor often has to calculate and explain the risk of having affected children. This is particularly likely when previous cases of genetic disease have occurred in the family. For example, if both members of a couple are carriers of a recessive gene, it can be explained that there is a 1 in 4 chance of any child being affected. This is not quite as straightforward as it sounds. One couple who had already had an affected child sued a doctor who gave this advice when later they had another affected child. They had assumed they could have three more children before another was affected. More complex situations such as sex linkage may have to be explained. Haemophilia is a sex-linked disease. Some diseases are controlled by many genes (polygenic) and here environmental factors also have an influence. Spina bifida, cleft lip and clubfoot belong in this category. Calculating risk becomes more difficult in such situations, and counsellors must rely on statistics. For example, cleft lip or cleft palate occurs in 1 in 1000 births in the general population, but is 40 times more common among the brothers and sisters of affected children. Older women may want to know the risks of having a Down's syndrome child, especially if they have already had one such child. It can be explained that chromosomal disorders have a low risk of recurrence in the same family because they are faults in cell division, not genetically inherited.

Explaining cause. The counsellor will normally try to explain the cause of any problem. For example, sickle cell anaemia is caused by sickling of red blood cells and it can be explained that this leads to anaemia and blockage of small blood vessels and that these in turn cause many other symptoms.

Quality of life. The quality and likely length of an affected child's life will also be relevant. Availability of treatment, support groups and financial help will need to be discussed. The likely effects on other members of the family if the affected child needs time-consuming and expensive care can be discussed. Many practical issues such as family holidays need to be considered.

Options. If a couple decides the risks are unacceptable, possible options must be explained. These include contraception or sterilisation, adoption, artificial insemination to avoid the husband's genes being passed on or IVF using a donor egg if the problem lies with the woman. Prenatal diagnosis is an option if the couple are willing to consider abortion of an affected fetus. CVS, amniocentesis and abortion itself must then be discussed. The reliability of any tests done, such as DNA analysis or ultra-sound scanning, must also be explained. If IVF (test tube baby technique) becomes more reliable in future, pre-implantation diagnosis (see above) may become possible. Gene therapy may also become possible (section 25.7.11).

Genetic screening. Part of the counsellor's job is to discuss the results of genetic screening with the person who has been tested. In section 25.7.9 three categories of screening were considered, namely prenatal diagnosis, carrier diagnosis, and predictive diagnosis. Genetic counselling after a positive diagnosis is very important. The possibility of receiving bad news from a test is difficult to prepare for and making decisions which have long-term consequences when one is feeling shocked and emotional is helped by the support of a counsellor. Some situations are particularly distressing, such as the case of a positive identification of Huntington's chorea. Counselling after screening is one of the most challenging tasks of a counsellor, and one which will become increasingly important as more genetic tests become available.

Responsibility. If one partner is responsible for transmission of a genetic defect, as in the case of dominant genes or sex-linked conditions, guilt and blame are common responses. The counsellor should try to emphasise sharing of responsibility and can point to the fact that we all carry some harmful genes, and in most cases we are lucky enough not to find out which ones.

25.7.11 Gene therapy

Although genetic disorders can often be prevented by genetic counselling and prenatal diagnosis, or treated with varying degrees of success, such as PKU, it has not yet been possible to cure genetic diseases. Advances in genetics have reached the point where this is beginning to be possible. The basic principle is to replace faulty genes with normal genes. An outline of the procedures used will be given in this section, together with discussion of the social and ethical issues raised.

Germ-line and somatic cell therapy

Successful experiments with mice have shown that gene therapy is possible. Genes were microinjected into fertilised eggs with a known genetic disorder and the corrected eggs were re-implanted into the mother. In this method all cells of the future mouse are normal because they are all derived from a corrected egg. This procedure is known as **germ-line therapy** and all descendants of the cured animal will be normal. At the moment such treatment is regarded as unethical in humans (see later) because the gene would be passed on to future generations. Treatment at the moment is focused on another technique called **somatic cell therapy**. This involves changing some, though not all, of the somatic cells which are the non-sex cells of the body. Changes in these cannot be inherited. The people treated will therefore be cured but they will still be able to pass the faulty gene on to their offspring.

Outline of procedure

In principle the procedure is to isolate the normal gene and clone it using methods discussed earlier in the chapter. A safe and efficient vector is then needed to introduce it into the chosen human cells. These cells may have to be isolated from the body first, corrected and then replaced. This would be relatively easy for blood diseases such as sickle cell anaemia because the cells that make blood cells can easily be removed from bone marrow and replaced. The final problem is to make sure the gene is expressed normally. If it is not switched on and off normally it could create more problems than it solves, for example by making too much product. The situation is more complex when a disease is caused by a dominant gene. Here the dominant gene must also be removed or made ineffective and techniques have not yet been developed for this.

Vectors which have been considered are:

- **viruses** – these are efficient at delivering DNA to the nucleus of the cell. The virus can be genetically engineered to remove the genes that allow it to multiply and cause disease. Retroviruses have been used successfully with mice, but it is not yet possible to control where the DNA is inserted and this can cause chromosome mutations.
- **liposomes** – their use is described below for cystic fibrosis therapy.
- **microinjection** and **electroporation** have been described in section 25.5.1. They are possible methods but are not as promising as viruses and liposomes.

Example of cystic fibrosis

Cystic fibrosis (CF) affects the epithelial cells of the body, but the life-threatening problems mainly affect the lungs. Lung and trachea epithelial cells are therefore the initial target for gene therapy. The aim is to get the gene, complete with its control sequences, into the cells so that it can make the normal protein, known as CFTR (section 25.7.3). Probably only about 10% of cells will need to be corrected to eliminate the problem.

Clinical trials have begun on human sufferers from CF using an aerosol inhaler to deliver the gene, like those used by asthmatics. The gene (a cDNA clone) is enclosed in a special liposome (section 25.5.1). Liposomes are specially designed to enter cells and release DNA into the cell so that it may bind with DNA in the nucleus. The technique was used successfully in mice in 1993. Trials with a virus vector have also begun. The virus chosen is adenovirus which normally attacks the respiratory tract. Unfortunately it does not insert its DNA into the host DNA. If the cell divides, the new DNA is not replicated at the same time so it eventually becomes diluted.

The treatment may only be effective for a few weeks until the epithelial cells die, but it will be easy to repeat the treatment at regular intervals. Hopefully a way will eventually be discovered of targetting the cells that make the epithelial cells so that a permanent cure can be given.

Example of SCID (severe combined immunodeficiency disease)

One form of this disease affects a gene coding for the enzyme ADA (adenosine deaminase). The mutant gene is recessive and makes no enzyme. Heterozygous children are unaffected because their one normal gene makes sufficient ADA. ADA is needed by the white blood cells (lymphocytes) responsible for immunity to infection. Without ADA they die, so sufferers of the disease have to live in a completely sterile environment, with no direct human contact, otherwise they normally die by the age of two (fig 25.34).

Two children, aged 4 and 9, were selected for gene therapy in the USA in 1990. Lymphocytes were isolated from the children, a normal gene introduced by means of a retrovirus vector, and the cells replaced. The treatment has to be repeated every 1 to 2 months. After a year they had shown significant improvement, were able to show an immune response and were able to start school.

Ethical issues

Few people would argue that somatic cell therapies present ethical problems, particularly those using simple techniques like the use of an inhaler. It is similar to using any other pharmaceutical product. Germline therapy, though, *is* controversial. It opens up the whole field of eugenics already discussed under genetic screening in the previous section. Exactly the same technique which can be used to cure a disease by replacing a faulty gene could be used for 'gene enhancement'. This is the addition of a desired characteristic. The American public has already demonstrated a desire for enhanced growth of normal children with demands for human growth hormone to be

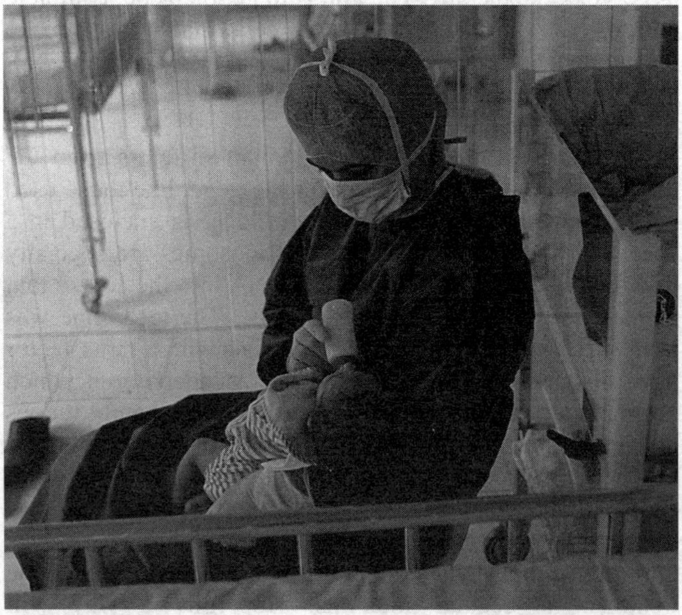

Fig 25.34 *Nurse bottle feeding a baby suffering from severe combined immunodeficiency disease (SCID). The nurse is wearing a full body gown with hair and face mask to prevent infecting the baby. Gene therapy for this condition involves inserting a gene for the enzyme into stem bone marrow cells and transplanting them into the baby where they may produce normal immune system blood cells.*

generally available. In countries like India and China the birth statistics have been distorted by abortion of girls because families want sons, although abortion on grounds of sex is now banned in India. If people are prepared to go to such lengths to select the sex of their children, why should they not make use of genetic engineering to select other inherited characteristics?

Germline therapy is also controversial because the change can be passed on to the children of the treated person and all subsequent generations. It can be argued that we should not have the right to affect future generations as well. Also, if there is even a slight possibility that the gene could be harmful in some circumstances the treatment should not be allowed. Strict regulatory procedures have to be gone through for any form of gene therapy and, on the recommendation of a Government committee in 1992, germline therapy has been made illegal for the time being in the UK.

25.7.12 Genetic fingerprinting and DNA profiling

Genetic fingerprinting was developed in 1984 by Alec Jeffreys and colleagues at the University of Leicester. A more recent and more sensitive version is known as DNA profiling. The technique has become well known to the general public through its use in criminal trials, such as the O.J. Simpson trial in the USA in 1995.

There are about 100 000 genes in the human genome (the genome being the total DNA in a cell). These code for proteins, but about 95% of the DNA is non-coding. Its function is still not clear, but some is probably structural. About 30–40% of this DNA consists of short sequences of bases which are repeated many times. Some of the repeat sequences are scattered throughout the DNA but some are found joined together in clusters, in other words in tandem. These 'tandem repeats' are known as **satellite DNA** (they were originally discovered as a separate fraction of DNA after centrifugation of the total DNA). Each cluster of repeated sequences is known as a **satellite**. Each sequence varies in the number of times it is repeated. Some satellites have only a small number of repeats and are known as **minisatellites**. It has been discovered that there is a great deal of variation between individuals in the number of repeats of these short sequences (they are 'hypervariable'). Minisatellites are therefore sometimes known as **VNTRs** (variable number tandem repeats). Each individual will have two allelic minisatellites at a particular locus, one inherited from the mother and one from the father. Genetic fingerprinting is a way of analysing the lengths of the minisatellites of a given individual.

Procedure

The DNA is extracted from cells and treated with a restriction enzyme. The enzyme chosen cuts either side of the minisatellites, leaving them intact so that their variable lengths are unaltered. Agarose gel electrophoresis is used to

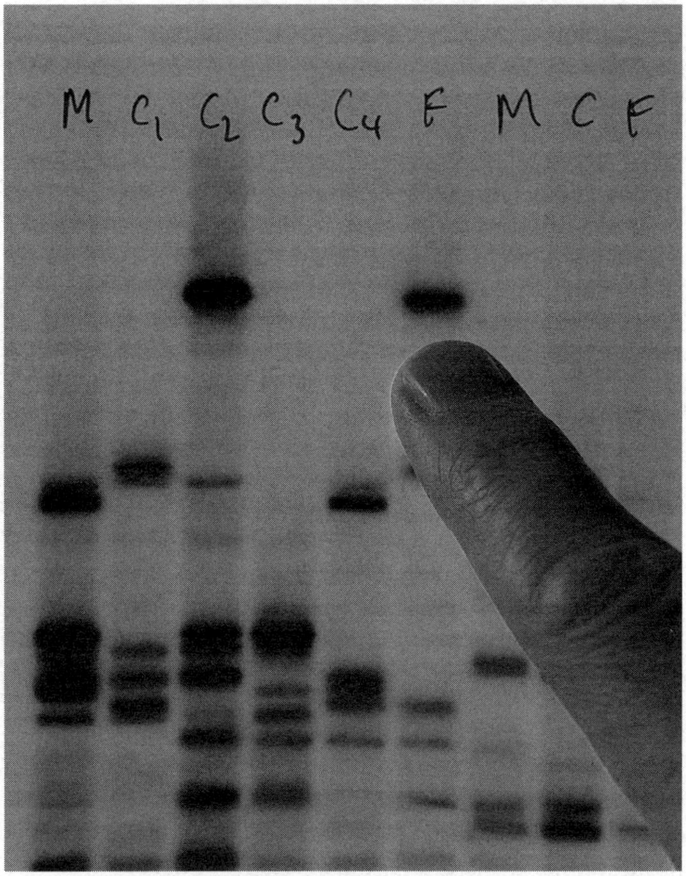

Fig 25.35 *Genetic fingerprints.*

separate the fragments, lining them up according to size as explained in section 25.1 and fig 25.4. From here the technique is as described for DNA analysis in section 25.7.9 and fig 25.33. Southern blotting is used to transfer the DNA to a nitrocellulose or nylon filter. A radioactive DNA probe with a base sequence which is complementary to part of the minisatellite repeat sequence is then hybridised to the DNA and its location found by autoradiography.

As stated, the minisatellites are different lengths in different individuals. The probe binds to the minisatellites which are found in different bands on the filter in different individuals because of their different lengths. The pattern for an individual is characteristic and therefore known as a fingerprint (fig 25.35). If the probe recognises and binds to several types of minisatellites throughout the genome it is called a **multi-locus probe** and will produce many bands on the autoradiograph. The more bands there are, the more unique the pattern. The probability of four bands matching by chance between two people is about 1 in about 250, whereas it is less than 1 in 1 million million for 20 bands (there are about 4.5 thousand million people in the world).

Multi-locus probes give best results with entire DNA. However, forensic scientists rarely work on good quality fresh material. Often it is dried samples contaminated with other materials like soil and bacteria. A more 'robust' method is often required. This can be achieved by using **single-locus probes**. These can be used on smaller pieces of DNA (such as partially decomposed DNA) and also on smaller total amounts of DNA. They recognise just a single short repeating sequence which is unique to one minisatellite and therefore found only on one particular pair of homologous chromosomes. Restriction enzymes therefore produce two characteristic fragments per individual and only two bands eventually appear on the autoradiograph, one of maternal origin and one of paternal origin (fig 25.36). If two single-locus probes are used, four bands will appear, if three are used, six will appear, and so on. By using more probes greater validity in distinguishing between two individuals can be achieved if desired (see below). This technique is called **DNA profiling** and is now the most commonly used method in forensic work.

The technique can be made even more sensitive by using a method for amplifying the amount of DNA, known as the **polymerase chain reaction** (**PCR**). This has meant that much smaller samples can be used for detection. This was demonstrated by a scientist who sent a piece of a licked stamp punched out with a hole punch to a laboratory for analysis. The saliva on the back of the stamp was diluted 200-fold and the sender was still correctly identified. (This technique is almost too sensitive for forensic work. For example, anyone who has shed dandruff or sneezed at the scene of the crime might become a suspect!)

Applications

Just as fingerprinting revolutionised forensic work in the early 1900s, so genetic fingerprinting is doing so now. DNA can be extracted from small samples of cells found at the scene of the crime, for example in traces of blood, hair roots or saliva. In cases of rape, semen may be used (fig 25.37). An even more common routine use of the technique is to settle paternity disputes. Fig 25.38 shows an example of the use of DNA profiling in such a dispute.

25.8 Look at fig 25.38. Which of the mother's children were fathered by F?

The first use of the technique for forensic work was in 1986 in the UK. In 1983 a schoolgirl had been found raped and murdered in a village near Leicester, and in 1986 a second girl was found. A man confessed to the second crime, but police believed he may also have been responsible for the 1983 crime. They asked Jeffreys at the University of Leicester to carry out DNA fingerprinting on semen samples from both crimes and on a blood sample from the suspect. These established that the man was innocent of both crimes! All the local men, a total of 1500, were then tested, but still with no positive result. The murderer was finally caught as a result of a conversation overheard by chance in a pub. DNA fingerprinting confirmed his guilt. In about 30% of the times that DNA fingerprinting is used, the person tested is shown not to match the scene-of-crime-DNA.

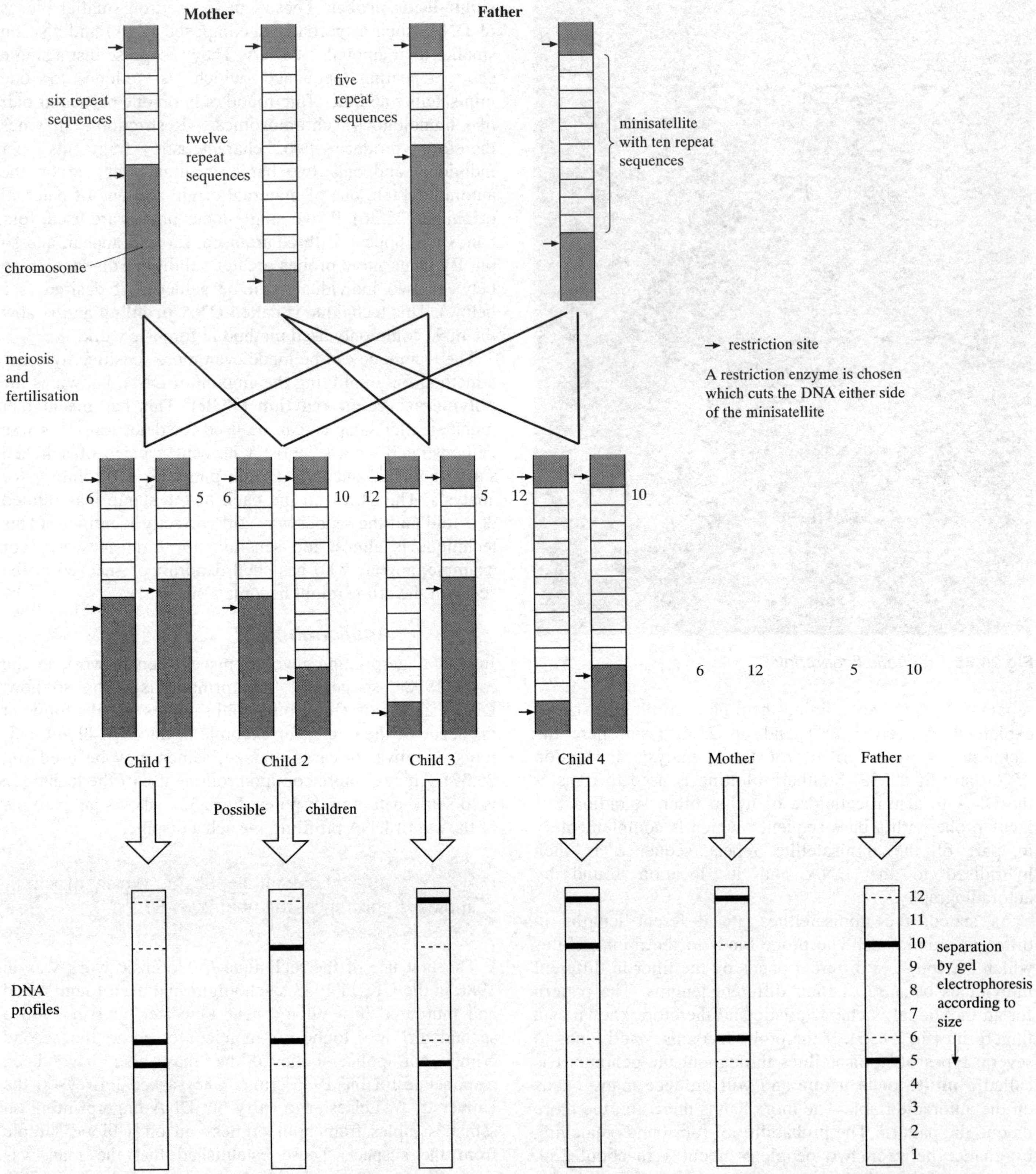

Fig 25.36 *Inheritance of minisatellites and variation in DNA profiles which result. In the population as a whole there is great variation in the lengths of the minisatellites, making the chances of two individuals producing the same DNA profile very small (see text).*

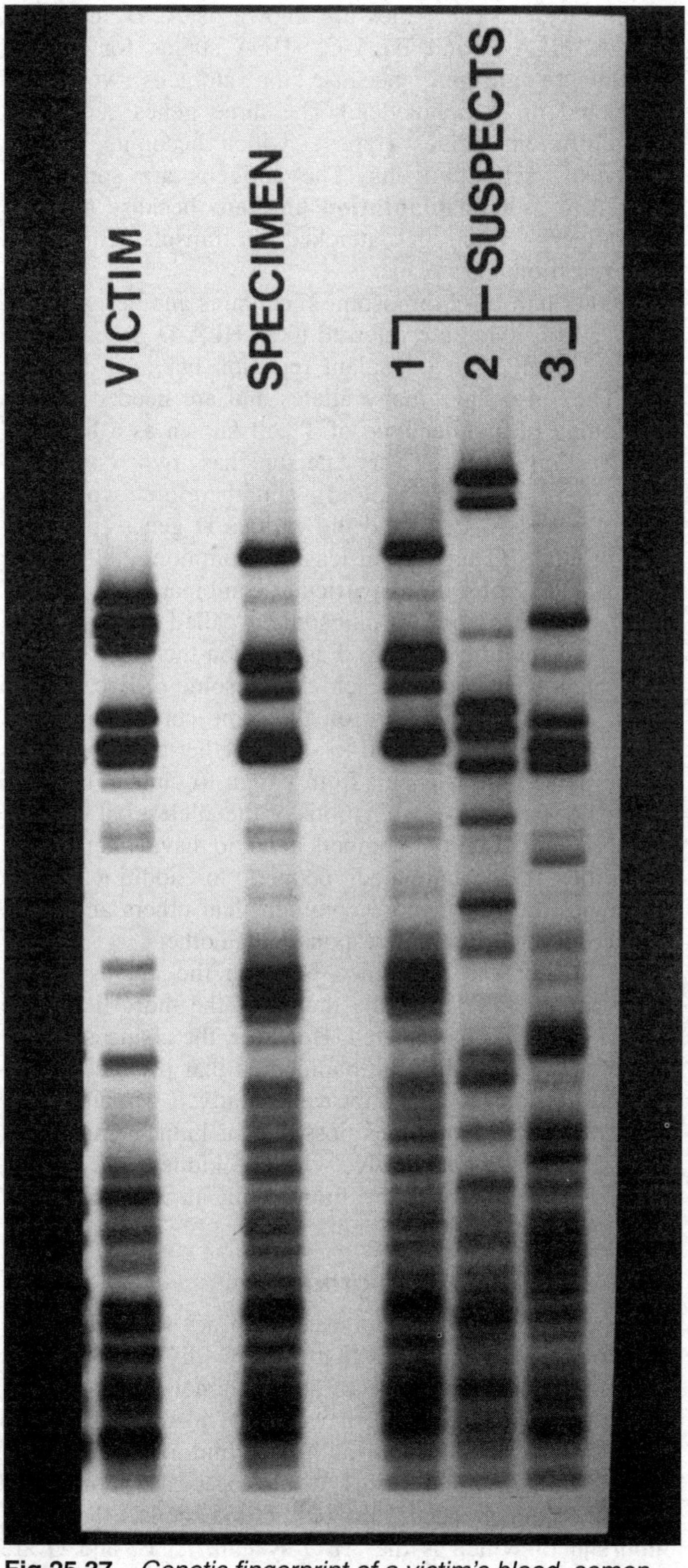

Fig 25.37 *Genetic fingerprint of a victim's blood, semen (specimen) and the blood of suspect rapists.*

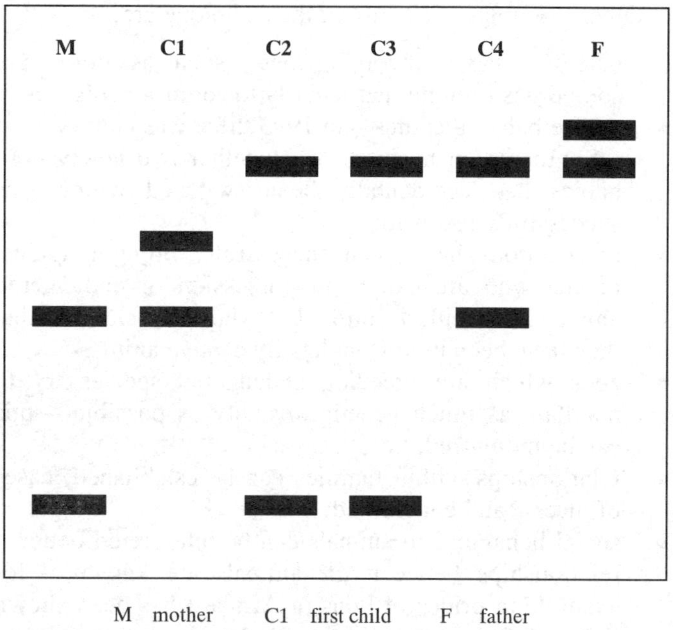

Fig 25.38 *DNA profiles of people involved in a paternity dispute.*

Reliability and validity

Reliability is being able to get the same result each time a test is carried out. For example, would two different laboratories produce the same result with the same sample? Attempts are being made to standardise techniques throughout Europe to increase reliability. This is important for international crimes, such as terrorism and drug-rings.

Validity concerns whether or not the test is appropriate. The chief concern in forensic cases is the probability of one individual having the same profile or fingerprint as another. Courtroom battles over this issue have made headlines on a number of occasions. Jeffreys calculated that there was about a one in four chance of two people sharing a band if a multi-locus probe is used. Therefore there is a 1 in 4^n chance where n = number of bands. Thus there is a 1 in 256 chance of two people sharing four bands and less than 1 in 1 million million chance of sharing 20–30 bands (the number of bands in an average DNA fingerprint). Often though, forensic scientists are working with poor quality material and degraded DNA may only give a few bands. It cannot therefore always give a definite answer. Another complication is that the odds quoted apply only to the population as a whole. Relatives will show more similarities.

It is harder to make statistical calculations for single-locus probes because each band occurs with a different frequency in the population. On average, using one such probe the bands will match in 1 in 100 at most, with two probes 1 in 10 000 and with three probes 1 in 1 million. There is variation between different ethnic groups, so separate databases have been set up for different groups. It is easier to store the data digitally for single-locus probes than with multilocus probes.

The PCR method will probably become the standard method in the future. It is fast, could be automated, would be cheaper once set up, can use samples as small as a few cells and the data is easily stored. However, new databases will have to be set up.

Other examples of the use of the technique are:

- paternity cases involving pets, such as dogs, are sometimes brought, particularly to confirm pedigrees;
- whose baby? disputes – in 1993 there was confusion at a Southampton hospital over whether two newly born babies had accidentally been switched which was successfully resolved;
- prosecutions have been successfully brought against people who are caught in possession of endangered animals, particularly birds, but who have claimed that they have been bred from legally captive animals;
- zoos which are breeding endangered species try to maintain as much genetic diversity as possible – this can be monitored;
- relationships within families can be established; cases of incest can be identified;
- social behaviour in animals can be interpreted better if relationships between the animals are known – for example in prides of lions in Africa it has been shown that not all the males mate with the females; those who do not, though, are closely related to those who do, so their genes are still likely to be passed on;
- in immigration cases disputes over family relationships can be settled;
- unusual cases sometimes occur – for example a seed pod found in a truck matched an isolated tree where a corpse had been found; this helped to convict the owner of the truck.

25.7.13 Transplant surgery and the Major Histocompatibility Complex (MHC)

A familiar problem with transplant surgery is that of rejection of the transplanted organ. The body recognises the transplanted organ as 'foreign' and mounts an immune response against it which gradually destroys it. This is known as **rejection**. The immune system must also be able to recognise 'self', so that it does not attack and destroy its own body. (There are some diseases where this happens.) The body is able to label its own cells with marker proteins rather like flags, which enable different cells to recognise each other. This helps cells to 'know' where and what they are and is important, for example, in embryonic development when tissues and organs are forming. One set of marker proteins identifies 'self'. All cells except red blood cells have these 'self' markers. They are coded for by a set of genes on chromosome 6 known as the **major histocompatibility complex (MHC)**. There is more variation among these genes than for any others in human populations, so a vast range of markers is possible. The markers act as antigens, and are attacked by a particular type of white blood cell, the killer T cell, if they are foreign. T cells are a type of lymphocyte (section 14.9). The markers are in fact glycoproteins, that is proteins with a carbohydrate tail which sit in the cell surface membranes with the tail projecting outwards.

The MHC genes that produce these markers, or antigens, are arranged at three loci quite close together on chromosome 6. The genes are known as A, B and C, or HLA-A, HLA-B and HLA-C. (HLA stands for **human lymphocyte antigen**, because the antigens were first discovered on lymphocytes.) The three genes each have many different alleles (types), which accounts for the variety of 'self' antigens. The antigens are sometimes referred to as **transplantation antigens** because they are the antigens which get attacked in transplanted organs when rejection takes place.

Another part of chromosome 6 contains another group of genes, all of which are referred to as HLA-D genes. These are also involved in transplant rejection but in a different way. They also have many alleles and are needed for the functioning of another type of T cell known as a helper T cell (section 14.9). Every person has two copies of chromosome 6 (being diploid), and therefore two alleles each of genes A, B, C and the various D genes (probably three in total). (There are at least 40 common alleles of A, 59 of B and 12 of C). The particular combination of A, B, C and D antigens on one chromosome is called the **haplotype** (short for haploid genotype). Each person therefore has two MHC haplotypes, one on each chromosome 6. Because the genes in one haplotype are on the same chromosome and not very far apart, they tend to stay together during meiosis and be passed on as a unit from parent to child. There are millions of possible combinations of the alleles, but because of haplotypes, family members tend to have many more similarities than unrelated people. In addition, some haplotypes are much more common than others and some produce stronger immune responses than others.

The greater the difference between the alleles of the donor and the alleles of the recipient, the more likely the donor organ is to be rejected. However, the chances can be improved by matching the haplotypes that produce strong immune responses, and improved greatly if a relative can be used. This is sometimes possible for kidney transplants because a person can survive with one kidney. The perfect match is only likely to be found with an identical twin, since they are genetically identical.

ABO blood groups

Fortunately, human red blood cells do not show the same variation in MHC antigens that is shown by all other cells (unlike the red blood cells of some animals such as mice and chickens). If they did, it would be just as difficult to find a suitable blood donor as it is to find a suitable organ donor, and far more deaths would occur. A number of antigen systems occur in red blood cells, the most important of which is the ABO system. A, B, and O are alleles of one gene. The A and B alleles produce A and B antigens respectively, which are glycoproteins, but the protein produced by the O allele is non-functional. The function of the system is unknown. The consequences for blood transfusions are discussed in section 14.9.9. It is just as important to match blood groups when organs are being transplanted because the A and B antigens are present on the surfaces of many cells apart from red blood cells.

Chapter Twenty-six

Evolution – history of life

The word evolution means change over a period of time. It is one of the most powerful ideas in biology that living things may have evolved from relatively simple chemicals. The study of evolution provides a focus for investigations into the nature of life itself, the origins of life, the great diversity of living things and the underlying similarities in structure and function which they have.

This chapter attempts to describe and discuss the many theories concerning the origin of life and the possible ways in which species have originated. Traditionally the study of the history of life has been fraught with allegations of indoctrination. Indoctrination may be defined as a conscious effort to inculcate an unshakeable commitment to a belief or doctrine. Such an approach is not only anti-scientific but also intellectually dishonest, and efforts to avoid this have been made in this text.

A brief outline of the main theories concerning the origin of life is presented in this chapter so that students are aware that there is a range of opinions on the subject. Much of the evidence on which these theories are based is metaphysical, that is to say it is impossible to repeat the exact events of the origin of life in any demonstrable way. This is true of both scientific and religious (theological) accounts. However, one theory, evolution, is increasingly being seen not as a single theory but as a collection of individual scientific hypotheses each of which is capable of being tested, as described in appendix section A2.1.

In this chapter, and in chapter 27, scientific facts have been selected to produce an account of the processes underlying the origins and diversity of life. Because of the necessity to be selective this account lacks absolute objectivity; indeed, this is inevitably true of any account, be it historical, scientific or metaphysical. However, by stressing the limitations and assumptions associated with the evidence presented here, this account may have a degree of objectivity and tentativeness which characterises good scientific writing. It must be stressed that the evidence presented in this chapter, and the conclusions drawn from it, represent current views. These are constantly under review and their validity is limited by the knowledge available to us at any given time.

26.1 Theories of the origin of life

Theories concerned with the origin of the Earth, and indeed the Universe, are diverse and uncertain. It may have begun as a ball of particles, exploded in a 'big bang', emerged from one of several black holes, or be the design of a Creator. Science, contrary to popular belief, cannot contradict the idea of a divine origin for the early universe. Nor do theological views necessarily dismiss the scientific hypothesis that, during the origins of life, life acquired those characteristics which are explained by the natural laws of science.

The major theories accounting for the origin of life on Earth are:

(1) life was created by a supernatural being at a particular time (**special creation**);
(2) life arose from non-living matter on numerous occasions by a process of **spontaneous generation**;
(3) life has no origin (**steady-state**);
(4) life arrived on this planet from elsewhere (**cosmozoan**);
(5) life arose according to chemical and physical laws (**biochemical evolution**).

Each of these theories is dealt with in turn below.

26.1.1 Special creation

This theory is supported by most of the world's major religions and civilisations and attributes the origin of life to a supernatural event at a particular time in the past. Archbishop Ussher of Armagh calculated in 1650 AD that God created the world in October 4004 BC, beginning on October 1st and finishing with Man at 9.00 a.m. on the morning of October 23rd. He achieved this figure by adding up the ages of all the people in the biblical genealogies from Adam to Christ (the 'begats'). Whilst the arithmetic is sound, it places Adam as having lived at a time when archaeological evidence suggests that there was already a well-established urban civilisation in the Middle East.

The traditional Judaeo-Christian account of creation, given in Genesis 1 : 1–26, has attracted, and continues to attract, controversy. Whilst all Christians would agree that the Bible is God's word to Man, there are differences of interpretation concerning the length of the 'day' mentioned in Genesis. Some believe that the world and all species were created in six days of 24 hours' duration. They reject any other possible views and rely absolutely on inspiration, meditation and divine revelation. Other Christians do not regard the Bible as a scientific textbook and see the Genesis account as the theological revelation of the Creation of all living things through the power of God, described in terms understandable to humans in all ages. For them the Creation account is concerned with answering the question 'Why?' rather than 'How?'. Whilst science broadly relies on observation and experiment to seek truth, theology draws its insights from divine revelation and faith.

> 'Faith is the substance of things hoped for, the evidence of things not seen . . . by faith . . . we understand that the universe was created by God's word, so that what can be seen was made out of what cannot be seen.' (Hebrews 11 : 1, 3)

Faith accepts things for which there is no evidence in the scientific sense. This means that logically there can be no intellectual conflict between scientific and theological accounts of creation, since they are mutually exclusive realms of thought. Scientific truth to the scientist is tentative, but theological truth to the believer is absolute.

Since the process of special creation occurred only once and therefore cannot be observed, this is sufficient to put the concept of special creation outside the framework of scientific investigation. Science concerns itself only with observable phenomena and as such will never be able to prove or disprove special creation.

26.1.2 Spontaneous generation

This theory was prevalent in ancient Chinese, Babylonian and Egyptian thought as an alternative to special creation, with which it coexisted. Aristotle (384–322 BC), often hailed as the founder of biology, believed that life arose spontaneously. On the basis of his personal observations he developed this belief further in relating all organisms to a continuum, a *scala natura* (ladder of life).

> 'For nature passes from lifeless objects to animals in such unbroken sequence, interposing between them, beings which live and yet are not animals, that scarcely any difference seems to exist between neighbouring groups, owing to their close proximity.' (Aristotle)

In stating this he reinforced the previous speculations of Empedocles on organic evolution. Aristotle's hypothesis of spontaneous generation assumed that certain 'particles' of matter contained an 'active principle' which could produce a living organism when conditions were suitable. He was correct in assuming that the active principle was present in a fertilised egg, but incorrectly extended this to the belief that sunlight, mud and decaying meat also had the active principle.

> 'Such are the facts, everything comes into being not only from the mating of animals but from the decay of earth . . . And among the plants the matter proceeds in the same way, some develop from seed, others, as it were, by spontaneous generation by natural forces; they arise from decaying earth or from certain parts of plants.' (Aristotle)

With the spread of Christianity, the spontaneous generation theory fell from favour, except among those who believed in magic and devil-worship, although it remained as a background idea for many more centuries.

Van Helmont (1557–1644), a much-acclaimed and successful scientist, described an experiment which gave rise to mice in three weeks. The raw materials for the experiment were a dirty shirt, a dark cupboard and a handful of wheat grains. The active principle in this process was thought to be human sweat.

> **26.1** What did Van Helmont omit from his experiment?

In 1688 Francesco Redi, an Italian biologist and physician living in Florence, took a more rigorous approach to the problem of the origin of life and questioned the theory of spontaneous generation. Redi observed that the little white worms seen on decaying flesh were fly larvae. By a series of experiments he produced evidence to support the idea that life can arise only from pre-existing life, the concept of **biogenesis**.

> 'Belief would be vain without the confirmation of experiment, hence in the middle of July, I put a snake, some fish, some eels of the Arno and a slice of milk-fed veal in four large, wide-mouthed flasks; having well closed and sealed them, I then filled the same number of flasks in the same way, leaving only these open.' (Redi)

Redi reported his results as follows.

> 'It was not long before the meat and the fish, in these second vessels (the unsealed ones), became wormy and the flies were seen entering and leaving at will; but in the closed flasks I did not see a worm, though many days had passed since the dead fish had been put in them.'

> **26.2** What do you consider was Redi's basic assumption?

These experiments, however, did not destroy the idea of spontaneous generation and, whilst the old theory took a set-back, it continued to be the dominant theory within the secular (non-religious) community.

Whilst Redi's experiments appeared to disprove the spontaneous generation of flies, the pioneer work in microscopy by Anton van Leeuwenhoeck appeared to reinforce the theory with regard to micro-organisms. Whilst not entering the debate between biogenesis and spontaneous generation, his observations with the microscope provided fuel for both theories and finally stimulated other scientists to design experiments to settle the question of the origin of life by spontaneous generation.

In 1765 Lazzaro Spallanzani boiled animal and vegetable broths for several hours and sealed them immediately. He then removed them from the source of heat. After being set aside for several days, none of them, on examination, showed any signs of life. He concluded from this that the high temperature had destroyed all forms of living organisms in his vessel and without their presence no life could appear.

> **26.3** Suggest another reason why Spallanzani's experiment might have prevented the growth of organisms.

In 1860 Louis Pasteur turned his attention to the problem of the origins of life. By this stage he had demonstrated the existence of bacteria and found solutions to the economic problems of the silk and wine industries. He had also shown that bacteria were ubiquitous (found in all environments) and that non-living matter could easily become contaminated by living matter if all materials were not adequately sterilised.

> **26.4** What were Pasteur's basic assumptions about the origins of life?

In a series of experiments based upon those of Spallanzani, Pasteur demonstrated the theory of biogenesis and finally disproved the theory of spontaneous generation.

The validation of biogenesis however raised another problem. Since it was now clear that a living organism was required in order to produce another living organism, where did the first living organism come from? The steady-state hypothesis has an answer for this but all the other theories imply a transition from non-living to living at some stage in the history of life. Was this a primeval spontaneous generation?

26.1.3 Steady-state theory

This theory asserts that the Earth had no origin, has always been able to support life, has changed remarkably little, if at all, and that species had no origin.

Estimates of the age of the Earth have varied greatly from the 4004 BC calculation of Archbishop Ussher to the present-day values of 5000 million years based on radioactive decay rates. Improved scientific dating techniques (appendix 4) have given increasing ages for the Earth, and extrapolation of this trend provides supporters of this theory with the hypothesis that the Earth had no origin. Whilst generally discrediting the value of geochronology in giving a precise age for the Earth, the steady-state theory uses this as a basis for supposing that the Earth has always existed. This theory proposes that species, too, never originated, they have always existed and that in the history of a species the only alternatives are for its numbers to vary, or for it to become extinct.

The theory does not accept the palaeontological evidence that presence or absence of a fossil indicates the origin or extinction of the species represented and quotes, as an example, the case of the coelacanth, *Latimeria*. Fossil evidence indicated that the coelacanths died out at the end of the Cretaceous period, 70 million years ago. The discovery of living specimens off the coast of Madagascar has altered this view. The steady-state theory claims that it is only by studying living species and comparing them with the fossil record that extinction can be assumed and then there is a high probability that this may be incorrect. The palaeontological evidence presented in support of the steady-state theory describes the fossil's appearance in ecological terms. For example, the sudden appearance of a fossil in a particular stratum would be associated with an increase in population size or movement of the organism into an area which favoured fossilisation.

26.1.4 Cosmozoan theory

This theory does not offer a mechanism to account for the origin of life but favours the idea that it could have had an extraterrestrial origin. It does not therefore, constitute a theory of origin as such, but merely shifts the problem to elsewhere in the Universe.

The theory states that life could have arisen once or several times in various parts of our Galaxy or the Universe. Its alternative name is the theory of **panspermia**. Repeated sightings of UFOs, cave drawings of rocket-like objects and 'spacemen' and reports of encounters with aliens provide the background evidence for this theory. Russian and American space probes have not yet succeeded in finding life within our Solar System but cannot comment on the nature of life outside our Solar System. Research into materials from meteorites and comets has revealed the presence of many organic molecules, such as cyanogen and hydrocyanic acid, which may have acted as 'seeds' falling on a barren Earth. There are several claims that objects bearing resemblance to primitive forms of life on Earth have been found in meteorites and in 1996 NASA scientists in the USA identified what they thought could be the remains of bacteria-like organisms in a rock from Mars (although they do not propose that life on Earth can be traced back to life on Mars). In all cases further evidence is needed.

26.1.5 Biochemical evolution

It is generally agreed by astronomers, geologists and biologists that the Earth is some 4.5–5.0 thousand million years old.

Many biologists believe that the original state of the Earth bore little resemblance to its present-day form and had the following probable appearance: it was hot (about 4000–8000 °C) and as it cooled carbon and the less volatile metals condensed and formed the Earth's core; the surface was probably barren and rugged as volcanic activity, continuous earth movements and contraction on cooling, folded and fractured the surface.

The atmosphere is believed to have been totally different in those days. The lighter gases hydrogen, helium, nitrogen, oxygen and argon would have escaped because the gravitational field of the partially condensed planet would not contain them. However, simple compounds containing these elements (amongst others) would have been retained, such as water, ammonia, carbon dioxide and methane, and until the earth cooled below 100 °C all water would have existed as vapour. The atmosphere would appear to have been a 'reducing atmosphere', as indicated by the presence of metals in their reduced form (such as iron(II)) in the oldest rocks of the Earth. More recent rocks contain metals in their oxidised form (for example iron(III)). The lack of oxygen in the atmosphere would probably be a necessity, since laboratory experiments have shown, paradoxically, that is far easier to generate organic molecules (the basis of living organisms) in a reducing atmosphere than in an oxygen-rich atmosphere.

In 1923 Alexander Oparin suggested that the atmosphere of the primeval Earth was not as we know it today but fitted the description given above. On theoretical grounds he argued that organic compounds, probably hydrocarbons, could have formed in the oceans from more simple compounds, the energy for these synthesis reactions probably being supplied from the strong solar radiation (mainly ultra-violet) which surrounded the Earth before the formation of the ozone layer, which now blocks much of it out. Oparin argued that if one considered the multitude of simple molecules present in the oceans, the surface area of the Earth, the energy available and the time scale, it was conceivable that oceans would gradually accumulate organic molecules to produce a 'primeval soup', in which life could have arisen. This was not a new idea, indeed Darwin himself expressed a similar thought in a letter he wrote in 1871:

> 'It is often said that all the conditions for the first production of a living organism are now present, which could ever have been present. But if, (and oh what a big if) we could conceive of some warm little pond, with all sorts of ammonia and phosphoric salts, light, heat, electricity, etc. present that a protein compound was chemically formed ready to undergo still more complex changes, at the present day, such matter would be constantly devoured or absorbed, which could not have been the case before living creatures were formed.'

In 1953 Stanley Miller, in a series of experiments, simulated the proposed conditions on the primitive Earth. In his experimental high-energy chamber (fig 26.1) he successfully synthesised many substances of considerable biological importance, including amino acids, adenine and simple sugars such as ribose. More recently Orgel at the Salk Institute has succeeded in synthesising nucleotides six units long (a simple nucleic acid molecule) in a similar experiment.

It has since been suggested that carbon dioxide was present in relatively high concentrations in the primeval atmosphere. Recent experiments using Miller's apparatus but containing mixtures of carbon dioxide and water and only traces of other gases have produced similar results to

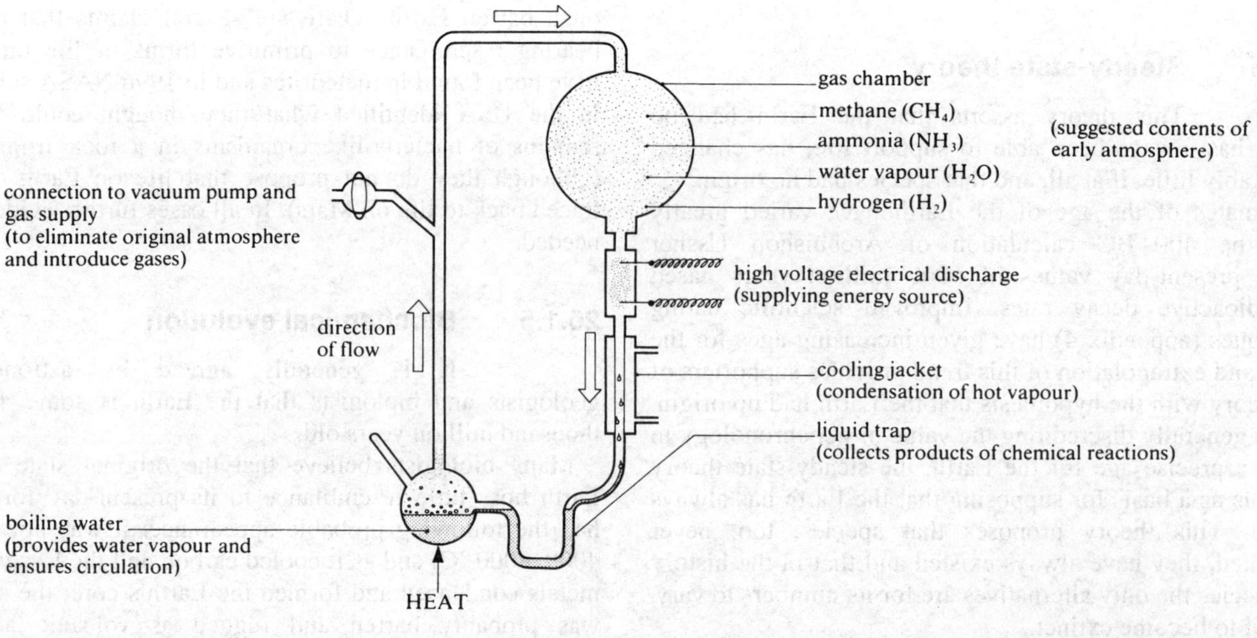

Fig 26.1 *Stanley Miller's apparatus in which he synthesised amino acids from gases under conditions thought to have been present in the primeval atmosphere. The gases and vapours were circulated under pressure and exposed to a high voltage for one week. At the end of the period the liquid products in the trap were analysed by paper chromatography. A total of 15 amino acids were isolated including glycine, alanine and aspartic acid.*

those of Miller. Oparin's theory has been widely accepted, but major problems remain in explaining the transition from complex organic molecules to living organisms. This is where the theory of a process of biochemical evolution offers a broad scheme which is acceptable to the majority of contemporary biologists. However, there is no agreement as to the precise mechanism by which it may have occurred.

Oparin considered that protein molecules were crucial to the transformation from inaminate to animate. Because of the zwitterionic nature of protein molecules they are able to form colloidal hydrophilic complexes which attract, and become surrounded by, envelopes of water molecules. These bodies may separate from the body of the liquid in which they are suspended (aqueous phase) and form a type of emulsion. Coalescence of these structures produces a separation of colloids from their aqueous phase, a process known as **coacervation** (*coacervus*, clump or heap). These colloid-rich coacervates may have been able to exchange substances with their environment and selectively concentrate compounds within them, particularly crystalloids. The colloid composition of a coacervate would depend on the composition of the medium. The varying composition of the 'soup' in different areas would lead to variation in the chemical composition of coacervates, producing the raw material for 'biochemical natural selection'.

It is suggested that substances within the coacervates may have undergone further chemical reactions and, by absorbing metal ions into the coacervates, formed enzymes. The alignment of lipid molecules (complex hydrocarbons) along the boundary between the coacervates and the external medium would have produced a primitive cell membrane which conferred stability on the coacervates. Thus the incorporation of a pre-existing molecule capable of self-replication into the coacervate, and an internal rearrangement of the lipid-coated coacervate, may have produced a primitive type of cell. Increase in size of coacervates and their fragmentation possibly led to the formation of identical coacervates which could absorb more of the medium and the cycle could continue. This possible sequence of events would have produced a primitive self-replicating heterotrophic organism feeding on an organic-rich primeval soup.

26.2 The nature of the earliest organisms

Current evidence suggests that the first organisms were heterotrophs as these were the only organisms capable of using the external supplies of available energy locked up within the complex of organic molecules present in the 'soup'. The chemical reactions involved in synthesising food substances appear to have been too complex to have arisen within the earliest forms of life.

As more complex organic molecules arose through 'biochemical evolution', it is assumed that some of these were able to harness solar radiation as an energy source and use it to synthesise new cellular materials. Incorporation of these molecules into pre-existing cells may have enabled the cells to synthesise new cellular materials without the need for them to absorb organic molecules, hence becoming autotrophic. Increasing numbers of heterotrophs would have reduced the available food resources in the primeval soup and this competition for resources would hasten the appearance of autotrophs.

The earliest photosynthetic organisms, whilst utilising solar radiation as their energy source, lacked the biochemical pathways to produce oxygen. At a later stage, it is believed that oxygen-evolving photosynthetic organisms developed, similar to existing blue-green bacteria (section 2.3), and this resulted in the gradual build-up of oxygen in the atmosphere. The increase in atmospheric oxygen and its ionisation to form the ozone layer would reduce the ultra-violet radiation striking the Earth. Whilst decreasing the rate of synthesis of new complex molecules, the decrease in radiation would confer some stability on successful forms of life. A study of the physiology of present-day organisms reveals a great diversity in biochemical pathways associated with energy capture and release, which may mirror many of Nature's early experiments with living organisms.

Despite the simplified account given above, the problem of the origin(s) of life remains. All that has been outlined is speculative and, despite tremendous advances in biochemistry, answers to the problem remain hypothetical. The above account is a simplified amalgam of present-day hypotheses. No 'ruling hypothesis' has yet achieved the status of an all-embracing theory (appendix section A2.1). Details of the transition from complex non-living materials to simple living organisms remain a mystery.

26.3 Summary of the 'theories' of the origin of life

Many of these 'theories' and the way they explain the existing diversity of species cover similar ground but with varying emphases. Scientific theories may be ultra-imaginative on the one hand and ultra-sceptical on the other. Theological considerations too, may fit into this framework, depending upon one's religious views. One of the major areas of controversy, even before the days of Darwin, was the relationship between scientific and theological views on the history of life.

Diagrams (*a*)–(*e*) in fig 26.2 represent straightforward descriptions of theories, hypotheses or beliefs on the history of life, whereas (*f*) and (*g*) represent an attempt to combine certain aspects of three theories, (*b*), (*c*) and (*d*), into an alternative acceptable to many people. The practices of science and religion are not, therefore, necessarily mutually exclusive, as witnessed by the number of scientists who hold religious beliefs.

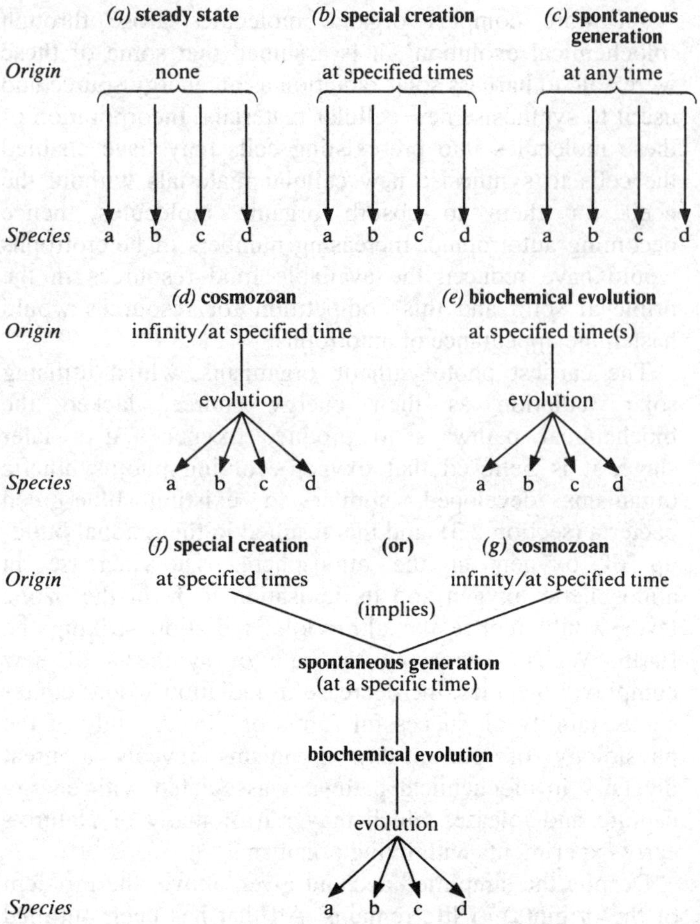

Fig 26.2 *Diagrammatic representation of various theories of the origin of life and the formation of species.*

26.4 The theory of evolution

The term 'evolution' has a special place in the study of the history of life. It has become the unifying concept which underpins the whole study of biology. Evolution implies an overall gradual development which is both ordered and sequential. In terms of living organisms it may be defined as **the development of differentiated organisms from pre-existing, less differentiated organisms over the course of time.**

The concept of evolution did not begin with Darwin and the publication of *On the Origin of Species*. Long before Darwin, attempts to explain the obvious diversity of living organisms which surround us had, paradoxically, led people to consider the basic structural and functional similarities which exist between organisms. Evolutionary hypotheses had been proposed to account for this and these ideas have themselves 'evolved' since the time of Darwin as knowledge has advanced.

The historical background to the development of the theory of evolution, as outlined in table 26.1, shows that the concept of continuity or gradual development of more complex species from pre-existing simpler forms had occurred to several philosophers and natural historians before the formal declarations of evolutionary hypotheses were put forward in the early nineteenth century.

26.4.1 Lamarckian evolution

The French biologist Lamarck proposed, in 1809, a hypothesis to account for the mechanism of evolution based on two conditions: the use and disuse of parts, and the inheritance of acquired characteristics. Changes in the environment may lead to changed patterns of behaviour which can necessitate new or increased use (or disuse) of certain organs or structures. Extensive use would lead to increased size and/or efficiency whilst disuse would lead to degeneracy and atrophy. These traits acquired during the lifetime of the individual were believed to be heritable and thus transmitted to offspring.

According to Lamarckism, as the theory came to be known, the long neck and legs of the modern giraffe were the result of generations of short-necked and short-legged ancestors feeding on leaves at progressively higher levels of trees. The slightly longer necks and legs produced in each generation were passed on to the subsequent generation, until the size of the present-day giraffe was reached. The webbed feet of aquatic birds and the shape of flat fish could be explained similarly. In aquatic birds the constant spreading of the toe bones and the skin between them in order to swim to find food and escape predators gave rise to their webbed feet. Likewise adaptations resulting from fish lying on their sides in shallow water were proposed to explain the shape of flat fish. Whilst Lamarck's theory helped prepare the way for acceptance of the concept of evolution, his views on the mechanism of change were never widely accepted.

However, Lamarck's emphasis on the role of the environment in producing phenotypic changes in the individual was correct. For example, body-building exercises will increase the size of muscles, but these acquired traits, whilst affecting the phenotype, are non-genetic, and having no influence on the genotype cannot be inherited. To demonstrate this, Weismann cut off the tails of mice over many successive generations. According to Lamarckism, the enforced disuse of tails should have led to progeny with smaller tails. This was not the case. Weismann postulated that somatic (body)-acquired characteristics (resulting in phenotypic changes) did not directly affect the germ (gamete) cells which are the means by which characteristics are passed on to the next generation.

26.4.2 Darwin, Wallace and the origin of species by natural selection

Charles Darwin was born in 1809, the son of a wealthy doctor, and like many great people he had an undistinguished academic career. In 1831 he accepted an unpaid post as naturalist on the survey ship HMS *Beagle*,

Table 26.1 The history of evolutionary thought.

Ancient Chinese	
Confucius	Life originated from a single source through a gradual unfolding and branching
Greek and Mediaeval period	
Diogenes	All things are differentiations of the same thing and are the same thing
Empedocles	Air, earth, fire and water are the four roots of all things. Life arose by the action of the forces of attraction and repulsion on the four elements. Explained origin of Universe, plants, animals and humans (produced the germ of the idea of organic evolution)
Democritus	Living things arose by spontaneous generation from the slime of the Earth
Anaxogoras	Organisms sprang from atmospheric germs
Thales (640–546 BC)	All life came from water
Anaximander	Plants, then animals and finally humans arose from the mud of the emerging Earth
Aristotle (384–322 BC)	Proposed theory of continuous and gradual evolution from lifeless matter, based on his observations of animals. Recognised a 'scala natura' for animals
Dark Ages (400–1400 AD)	All theories based on those above or acceptance of special creation
Age of speculation (1400–1790)	
John Ray (1627–1705)	Developed concept of species
Carl Linnaeus (1707–78)	Formalised 'binomial classification' system. Suggested genera were created separately and species were variants of them
Buffon (1707–88)	Suggested different types of animals had different origins at different times. Recognised influence of external environment. Believed in acquired inheritance
James Hutton (1726–97)	Theory of uniformatarianism. Gave age of Earth in millions of years
Age of formulation (1790–1900)	
Erasmus Darwin (1731–1802)	Life arose from one single 'filament' made by God. Did not accept the preformation of humans. The filament evolved by acquired characteristics
Jean-Baptiste Lamarck (1744–1829)	Inheritance of acquired characteristics. Environment acts on organisms. Phenotype changes are passed on. Concept of use and disuse of organs
Georges Cuvier (1769–1832)	Established palaeontological evidence. Fossils the results of 'catastrophes' by which new species arose
William Smith (1769–1838)	Opposed Cuvier's theory of catastrophism on basis of continuity of similar species in related strata
Charles Lyell (1797–1875)	Demonstrated the progressive history of fossil evidence
Charles Darwin (1809–82)	Influenced by Lyell and Malthus. Established a theory of evolution by means of natural selection
Alfred Russel Wallace (1823–1913)	Similar theory to Darwin, but excepted humans from his theory
Hugo de Vries (1848–1935)	Recognised existence of mutations which were heritable as a basis for discontinuous variation and regarded species as arising by mutation
August Weismann (1834–1914)	Showed that the reproductive cells of animals are distinct and therefore unaffected by the influences acting on somatic tissues
Gregor Mendel (1822–84)	Work on genetics (published 1865) only came to light after 1900. Laws of inheritance
Developments in twentieth century (neo-Darwinism)	
W. L. Johannsen	Phenotypic characteristics are determined by genotype and environmental factors
T. Henry Morgan	Developed chromosome theory of heredity on basis of cytological evidence
H. J. Muller (1927)	Genotype can be altered by X-rays; induced mutation
R. A. Fisher (1930)	No difference between change investigated by geneticists and change shown in the fossil record
G. W. Beadle and E. L. Tatum (1941)	Demonstrated the genetic basis of biochemical synthesis (following A. E. Garrod (1909) and J. B. S. Haldane (1935))
J. Lederberg and A. D. Hershey (1951)	Demonstrated value of using bacteria in studying changes in genotype
J. D. Watson and F. H. C. Crick (1953)	Proposed molecular structure of DNA and its mechanism of replication
F. Jacob and J. Monod (1961)	Proposed a mechanism for regulation of gene activity

which spent the next five years at sea charting the East Coast of South America. The *Beagle* returned to Falmouth in October 1836 via the coast of Chile, the Galapagos Islands, Tahiti, New Zealand, Tasmania and South Africa. For most of this time Darwin was concerned with studying geology, but during a five-week stay on the Galapagos Islands he was struck by the similarities shown by the flora and fauna of the islands and mainland. In particular he was intrigued by the characteristic distribution of species of tortoises and finches (section 26.7.2). He collected a great deal of biological data concerned with variation between organisms which convinced him that species were not unchangeable.

On his return home his work on the selective breeding of pigeons and other domestic animals gave him a clue to the concept of artificial selection, but he was unable to appreciate how this could operate in the wild. An earlier *Essay on the Principles of Population* by the Reverend Thomas Malthus, published in 1778, had highlighted the consequences of the reproductive potential of humans. Darwin applied this to other organisms and saw that despite this the numbers within populations remained relatively constant.

Having collated a vast amount of information he began to realise that under the intense competition of numbers in a population, any variation which favoured survival in a particular environment would increase that individual's ability to reproduce and leave fertile offspring. Less favourable variations would be at a disadvantage and organisms possessing them would therefore have their chances of successful reproduction decreased. These data provided Darwin with the framework to formulate, by 1839, a theory of evolution by natural selection, but he did not publish his findings at that time. Indeed Darwin's greatest contribution to science was not so much to show that evolution occurs but how it might occur.

In the meantime, another naturalist, Alfred Russel Wallace, who had travelled widely in South America, Malaya and the Eastern Indian archipelago, and also read Malthus, had come to the same conclusions as Darwin regarding natural selection.

In 1858, Wallace wrote a 20-page essay outlining his theory and sent it to Darwin. This stimulated and encouraged Darwin and in July 1858, Darwin and Wallace presented papers on their ideas at a meeting of the Linnean Society in London. Over a year later, in November 1859, Darwin published *On the Origin of Species by Means of Natural Selection*. All 1250 printed copies were sold on the day of publication and it is said that this book has been second only to the Bible in its impact on human thinking.

26.5 Natural selection

Darwin and Wallace proposed that natural selection is the mechanism by which new species arise from pre-existing species. This hypothesis/theory is based on three observations and two deductions which may be summarised as follows.

Observation 1: Individuals within a population produce on average more offspring than are needed to replace themselves.

Observation 2: The numbers of individuals in a population remain approximately constant.

Deduction 1: Many individuals fail to survive or reproduce. There is a 'struggle for existence' within a population.

Observation 3: Variation exists within all populations.

Deduction 2: In the 'struggle for existence' those individuals showing variations best adapted to their environment have a 'reproductive advantage' and produce more offspring than less well-adapted organisms.

Deduction 2 offers a hypothesis called natural selection which provides a mechanism accounting for evolution.

26.5.1 Evidence for natural selection

Observation 1: It was Malthus who highlighted the reproductive potential of humans and observed that human populations are able to increase exponentially (section 10.7). The capacity for reproduction is basic to all living organisms, and is a fundamental drive which ensures continuity of the species. This applies to other organisms as shown in table 26.2. If every female gamete was fertilised and developed to maturity, the Earth would be totally overcrowded in a matter of days.

Observation 2: All population sizes are limited by various environmental factors, such as food availability, space and light. Populations tend to increase in size until the environment supports no further increase and an equilibrium is reached. The population fluctuates around this equilibrium, as discussed in section 10.7.3 Hence population sizes generally remain approximately constant over a period of time related to the length of the organism's life cycle.

Deduction 1: The continuous competition between individuals for environmental resources creates a 'struggle for existence'. Whether this competition occurs within a species (**intraspecific competition**) or between members of different species (**interspecific competition**) may be irrelevant in affecting the size of the individual population (section 10.7), but it will still imply that certain organisms will fail to survive or reproduce.

Table 26.2 Reproductive potential of selected species.

Crassostrea virginica	American oyster	1.0 million eggs per season
Lycoperdon sp.	giant puff ball	7.0×10^{11} spores
Papaver rhoeas	poppy capsule	6000 seeds
Carcinus maenas	shore crab	4.0 million eggs per season

Observation 3: Darwin's study of beetles whilst an undergraduate at Cambridge, his subsequent journey in the *Beagle* and his knowledge gained through the selective breeding of certain characteristics in pigeons convinced him of the importance of variation within a species. Likewise the adaptive significance of the interspecific variation seen in Galapagos finches gave Darwin a clue to his second deduction. Data collected by Wallace in the Malayan archipelago provided further evidence of variation between populations. Darwin and Wallace, however, were unable to account for the sources of the variation. This was not to be clarified until Mendel's work on the particulate nature of inheritance.

Deduction 2: Since all individuals within a population show variation and a 'struggle for existence' has been clearly established, it follows that some individuals possessing particular variations will be more suited to survive and reproduce. The key factor in determining survival is adaptation to the environment. Any variation, however slight, be it physical, physiological or behavioural, which gives one organism an advantage over another organism will act as a **selective advantage** in the 'struggle for existence'. (The term 'selective advantage' is less emotive than that coined by the social philosopher, Herbert Spencer, who described natural selection as 'survival of the fittest'. Spencer did not mean fit in the usual sense of the word. He used 'fit' to mean well adapted to the environment. The phrase 'survival of the fittest' is often misunderstood. There is not some kind of physical contest going on between members of a species, the 'nature red in tooth and claw' spoken of by the Victorians. The phrase 'survival of the fittest' is therefore probably best not used.)

Favourable variations will be inherited by the next generation. Unfavourable variations are 'selected out' or 'selected against', their presence conferring a **selective disadvantage** on that organism. In this way natural selection leads to increased vigour within the species and ensures the survival of that species. The whole of Darwin's and Wallace's hypothesis of natural selection is summed up most succinctly in Darwin's own words:

> 'As many more individuals of each species are born than can possibly survive, and as, consequently, there is a frequently recurring struggle for existence, it follows that any being, if it vary however slightly in any manner profitable to itself, under the complex and sometimes varying conditions of life, will have a better chance of surviving and thus be naturally selected. From the strong principle of inheritance, any selected variety will tend to propagate its new and modified form.' (Darwin, 1859)

Many misconceptions have grown up around the theory of evolution as outlined by Darwin and they may be summarised as follows.

(1) Darwin made no attempt to describe how life originated on the Earth: his concern was with how new species might arise from pre-existing species.

(2) Natural selection is not simply a negative, destructive force, but can be a positive mechanism of change within a population (section 27.5). The 'struggle for existence' described by Darwin was popularised by the coiling of unfortunate terms such as 'survival of the fittest' and 'elimination of the unfit' by the philosopher Herbert Spencer and the Press of the day.

(3) The misconception that humans were 'descended from the apes' by some process of linear progression was over-sensationalised by the Press and offended both the religious and secular communities. The former saw this as an insult to their belief that 'Man' was created in the 'image of God', whilst the latter were outraged by the apparent undermining of the 'superior position' of humans within the animal kingdom.

(4) The apparent contradiction between the Genesis six-day Creation account and that of a progressive origin for species was highlighted by the meeting of the British Association for the Advancement of Science in June 1860. Bishop Samuel Wilberforce of Oxford vehemently attacked the conclusions of Darwin as outlined in *On the Origin of Species* but not being a biologist his address lacked accuracy. In concluding, he turned to Professor Thomas Henry Huxley, a supporter of Darwin's theory, and asked whether he claimed his descent from a monkey through his grandfather or grandmother. Huxley replied by explaining the more important ideas of Darwin and correcting the misconceptions of Bishop Wilberforce. In conclusion he implied that he would prefer to have a monkey for an ancestor than 'to be connected with a man who used great gifts to obscure the truth'. This unfortunate controversy has continued as the Genesis versus Evolution debate. Professor R. J. Berry has summarised the extremes of the debate as:

(*a*) those who are awed by scientists and believe that the Bible has been disproved;

(*b*) those who cling to the inspiration of Scripture and their interpretations of it, and shut their eyes to the fact that God's work can be studied by scientific methods.

26.6 Modern views on evolution

The theory of evolution as proposed by Darwin and Wallace has been modified in the light of modern evidence from genetics, molecular biology, palaeontology, ecology and ethology (the study of behaviour) and is known as **neo-Darwinism** (*neo*, new) This may be defined as *the theory of organic evolution by the natural selection of inherited characteristics*.

Different types of evidence support different aspects of the theory. In order to accept neo-Darwinian evolutionary theory it is necessary to:

(1) establish the fact that evolution (change) has taken place in the past (**past evolution**);

(2) demonstrate a mechanism which results in evolution (**natural selection of genes**);
(3) observe evolution happening today (‘**evolution in action**’).

Evidence for past evolution comes from many sources based on geology, such as fossils and stratigraphy (the study of the order and ages of rock formations). Evidence for a mechanism is found in the experimental and observational data of the natural selection of characteristics that are inherited, such as the selection of shell colour in the snail *Cepaea* (section 27.5.1), and the mechanism of inheritance demonstrated by Mendelian genetics, as in Mendel’s work on peas. Finally, evidence for the action of these processes occurring today is provided by studies of present populations, such as speciation in the herring gull (section 27.8.4), and the results of artificial selection and genetic engineering, as in the cultivation of wheat and the synthesis of genes.

There are no laws of evolution, only well-supported hypotheses which add together to form a convincing theory. However, we must guard against accepting current ideas as proven fact because that would stifle intellectual growth and the search for truth. The uncritical acceptance of evolutionary theory is a case in point. Some of the events presented as evidence for evolutionary theory can be reproduced under laboratory conditions, but that does not prove that they did take place in the past; it merely indicates the possibility that these events occurred. The debate these days is not so much about whether evolution takes place but about how it takes place, in particular whether it always takes place by natural selection of randomly generated mutations.

26.7 Evidence for the theory of evolution

Evidence of relevance to the theory of evolution is provided from many sources, the main ones being:

- palaeontology
- geographical distribution
- classification
- plant and animal breeding
- comparative anatomy
- adaptive radiation
- comparative embryology
- comparative biochemistry

These are all discussed below.

Much of the evidence presented in this chapter was unavailable to Darwin and Wallace at the time of publication of their papers on the origin of species by natural selection, although this did not prevent Darwin from using his intuition, as is typical of great scientists. This is shown by his statement:

‘In October 1838, that is, 15 months after I had begun my systematic enquiry, I happened to read, for amusement, Malthus on Population, and being well prepared to appreciate the struggle for existence which everywhere goes on from long-continued observation of the habits of animals and plants, it at once struck me that under these circumstances favourable variations would tend to be preserved, and unfavourable ones to be destroyed. The result of this would be the formation of new species. Here, then, I had at last got a theory by which to work.’

The evidence presented here largely supports the theory of evolution by natural selection as outlined in section 26.5, although remember it does not provide proof, nor does it prove that no other mechanisms are involved. It draws on data obtained from many sources, and in all cases is interpreted in terms that assume that evolution does occur. Circular arguments and exceptions to the evidence are common and alternative interpretations can be found, but the broad concept of evolution is backed up by a wealth of scientific evidence. When reading the following sections try to judge for yourself the evidence available and decide whether the conclusions drawn are justified. Try to distinguish between what is evidence for evolution, and what is evidence for natural selection being the mechanism. In the remainder of this chapter we are mainly concerned with evidence for evolution, while in chapter 27 evidence for natural selection is presented.

26.7.1 Palaeontology

Palaeontology is the study of fossils. Fossils are any form of preserved remains thought to be derived from a living organism. They may include the following: entire organisms, hard skeletal structures, moulds and casts, petrifications, impressions, imprints and coprolites (fossilised faecal pellets) (table 26.3).

Fossil evidence alone is not sufficient to prove that evolution has occurred, but it supports a theory of progressive increase in complexity of organisms. Fossils were well known before evolution was generally accepted. They were interpreted either as the remains of former creations or as artefacts inserted into the rocks by God. Most of the remains found so far can be classified into the same taxonomic groups (phyla and classes) as living species, but whether they represent the ancestors of present-day forms can only be debated, not proved.

The oldest fossil-bearing rocks contain very few types of fossilised organisms and these all have a simple structure. Younger rocks contain a greater variety of fossils with increasingly complex structures. Throughout the fossil record many species which appear at an early stratigraphic level (their level in the rock deposits) disappear at a later level. This is interpreted in evolutionary terms as

Table 26.3 Types of fossils, their formation and examples.

Fossil	*Fossilisation process*	*Examples*
Entire organism	Frozen into ice during glaciation	Woolly mammoths found in Siberian permafrost
" "	Encased in the hardened resin (amber) of coniferous trees	Insect exoskeletons found in Oligocene rocks in Baltic coast
" "	Encased in tar	'Mummies' found in asphalt lakes of California
" "	Trapped in acidic bogs: lack of bacterial and fungal activity prevents total decomposition	'Mummies' found in bogs and peat in Scandinavia
Hard skeletal materials	Trapped by sedimentary sand and clay which form sedimentary rocks, e.g. limestone, sandstone and silt	Bones, shells and teeth (very common in British Isles)
Moulds and casts	Hard materials trapped as above. Sediments harden to rock. The skeleton dissolves leaving its impression as a mould of the organism. This can be infilled with fine materials which harden to form a cast. Great detail is thus preserved	Gastropods from Portland Stone, Jurassic. Casts of giant horsetails (*Calamites*) of Carboniferous forests. Internal casts of mollusc shells showing muscle attachment points
Petrifaction	Gradual replacement by water-carried mineral deposits, such as silica, pyrites, calcium carbonate or carbon. Slow infilling as organism decomposes producing fine detail	Silica replacements of the echinoderm *Micraster*
Impressions	Impressions of remains of organisms in fine-grained sediments on which they died	Feathers of *Archaeopteryx* in Upper Jurassic. Jellyfish in Cambrian found in British Columbia. Carboniferous leaf impressions
Imprints	Footprints, trails, tracks and tunnels of various organisms made in mud are rapidly baked and filled in with sand and covered by further sediments	Dinosaur footprints and tail scrapings indicate size and posture of organisms
Coprolites	Faecal pellets prevented from decomposing, later compressed in sedimentary rock. Often contain evidence of food eaten, e.g. teeth and scales	Cenozoic mammalian remains

indicating the times at which species originated and became extinct.

Evidence suggests that geographical regions and climatic conditions have varied throughout the Earth's history. Since organisms are adapted to particular environments, the constantly changing conditions may have favoured a mechanism for evolutionary change that accounts for the progressive changes in the structures of organisms as shown by the fossil record. Ecological considerations also fit in with the fossil evidence. For example, plants appeared on land before animals, and insects appeared before insect-pollinated flowers.

One of the major criticisms of using fossil evidence in support of an evolutionary theory is the lack of a continuous fossil record. Gaps in the fossil record ('missing links') are taken as strong evidence against a theory of descent by modification. However, there are several explanations for the incompleteness of the fossil record. These include the facts that:

- dead organisms decompose rapidly;
- dead organisms are eaten by scavengers;
- soft-bodied organisms do not fossilise easily;
- only a small fraction of living organisms will have died in conditions favourable for fossilisation;
- only a fraction of fossils have been discovered.

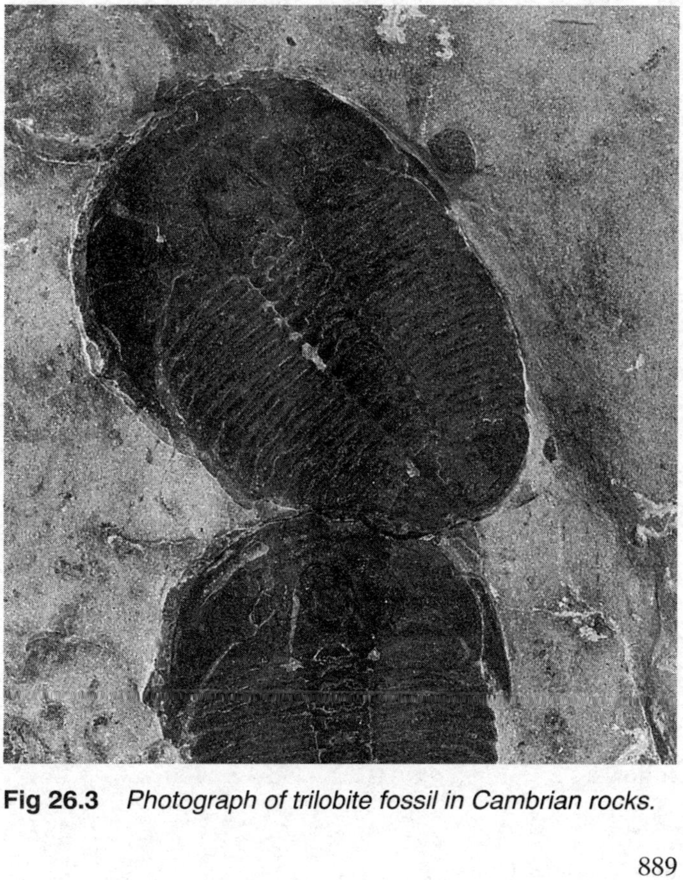

Fig 26.3 *Photograph of trilobite fossil in Cambrian rocks.*

Support for an evolutionary process increases as more and more possible 'missing links' are discovered, either as fossils, such as *Seymouria* (amphibian/reptile), *Archaeopteryx* (reptile/bird) and *Cynognathus* (reptile/mammal), or as living organisms representing groups with close structural similarities, such as *Peripatus* (fig 26.17) and *Latimeria*.

Alternatively, there exists the possibility that new species appeared so suddenly that intermediate forms in the lineage do not exist. Eldredge and Gould have proposed a process called '**punctuated equilibria**' which accounts for the sudden appearance of new species. According to this proposal species remain unchanged for long periods of time before giving rise to new species in comparatively short periods of time. This presumably would be the result of a relatively sudden and important change in environmental conditions. This process depends on the fact that evolutionary rates can vary and that some new species arise rapidly, making the fossil record appear incomplete. These apparent 'jumps' in the evolutionary sequence have given rise to the term '**saltatory evolution**' (*saltare*, to jump). Darwin himself considered this possibility and stated as much in the *Origin of Species*:

> 'I do not suppose that the process (speciation) . . . goes on continuously; it is far more probable that each form remains for long periods unaltered, and then again undergoes modification.'

Evolution, then, does not always have to be gradual.

The fossil history of the horse

The horse provides one of the best examples of evolutionary history (phylogeny) based on an almost complete fossil record found in North American sedimentary deposits from the early Eocene to the present.

The earliest recognisable odd-toed, hoofed mammals (perissodactyls) appeared about 54 million years ago and present-day perissodactyls include horses, tapirs and rhinoceroses. The oldest recognisable horse-like fossils belong to a genus called *Hyracotherium* which was widely distributed throughout North America and Europe during the early Eocene. By the beginning of the Oligocene it was extinct everywhere except North America. It was a small animal, lightly built and adapted for running. The limbs were short and slender and the feet elongated so that the digits were almost vertical. There were four digits in the forelimbs and three digits in the hindlimb. The incisors were small and the molars had low crowns with rounded cusps covered in enamel.

The probable course of development of horses from *Hyracotherium* to *Equus* involved at least twelve genera and several hundred species. The major trends seen in the development of the horse were connected with locomotion and feeding. They represent adaptations to changing environmental conditions and may be summarised as follows:

- increase in size,
- lengthening of limbs and feet,
- reduction of lateral digits,
- increase in length and thickness of the third digit,
- straightening and stiffening of the back,
- better-developed sense organs,
- increase in size and complexity of the brain associated with development of sense organs,
- increase in width of incisors,
- replacement of premolars by molars,
- increase in tooth length,
- increase in crown height of molars,
- increased lateral support of teeth by cement,
- increased surface areas of cusps by exposure of enamel ridges.

A dominant genus from each geological period of the Cenozoic has been selected to show the progressive development of the horse in fig 26.4. However, it is important to note that there is no evidence that the forms illustrated are direct relatives of each other.

The significance of the fossil sequence shown in fig 26.4 is that it supports a theory of progressive change based on homologous structures such as limbs and teeth. **Homologous structures** are structures found in different species which are believed to have a common evolutionary origin. Each of the species shown in fig 26.4 represents a stage of development which was successful for several million years (as judged by the abundance of fossils) before becoming extinct. The extinction of a species did not, however, represent the disappearance of the family line. The fossil evidence reveals that another closely related species always 'took over' after its extinction. As all the species in the sequence show structural and ecological similarities, this gives support to a theory of descent with modification. Other fossils found in the same rock strata suggest changing climatic conditions which, together with other evidence, indicates that each species was adapted to prevailing conditions.

The history of the horse does not show a gradual transition regularly spaced in time and locality, and neither is the fossil record totally complete. It would appear that several offshoots occurred from the line represented in fig 26.4, but they all became extinct. All modern horses appear to be descended from *Pliohippus*. The modern genus *Equus* arose in North America during the Pleistocene and migrated into Eurasia and Africa where it gave rise to zebras and asses as well as the modern horse. Paradoxically, having survived in North America for millions of years, the horse became extinct there several thousand years ago, at a time which coincided with the arrival of humans. Cave-paintings from other parts of the world suggest that the earliest use for the horse was as a source of food. The horse was absent from North America until its reintroduction by the Spaniards almost 500 years ago.

Epoch and age of oldest rocks	*Genus*	*Body form* (all heights are ground to shoulder)	*Bones of right forelimb*	*Mode of life, climate and structural modification*
Pleistocene 1×10^6 yr	*Equus*	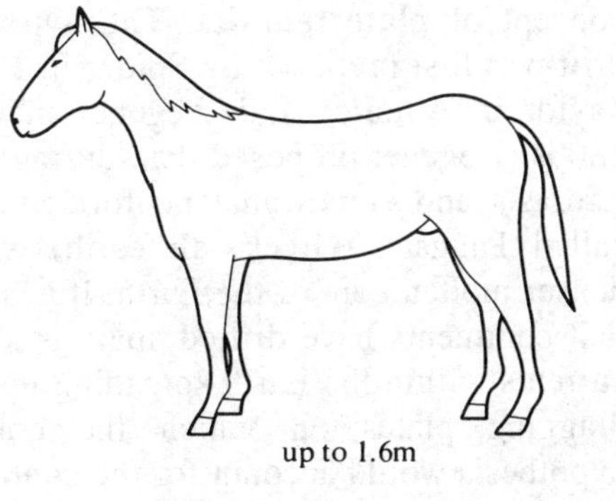 up to 1.6m	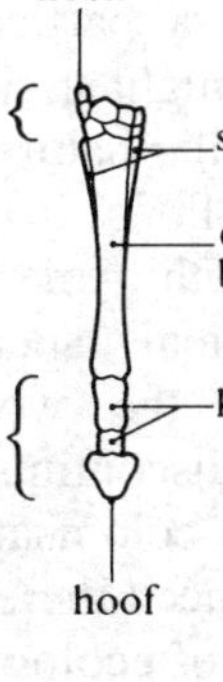	Adapted to life in dry grasslands. Very efficient at running. Metacarpals and metatarsals lengthened. Hoof formed from broadened phalanx 3 covering soft pad and all covered by claw. Teeth with large surface area. Enamel exposed where cement worn away. Premolars replaced by molars. Grind food.
Pliocene 7×10^6 yr	*Pliohippus*	 1.0m	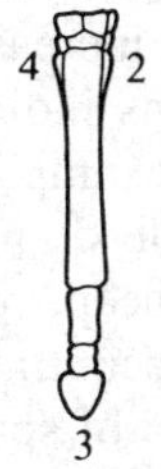	Increased reliance on speed. Digits 2 and 4 very much reduced. Thickening of metacarpals and metatarsals (hindlimb) for support. Phalanx 3 forms hoof. High-crowned teeth for eating grass.
Miocene 26×10^6 yr	*Merychippus*	 up to 1.0m	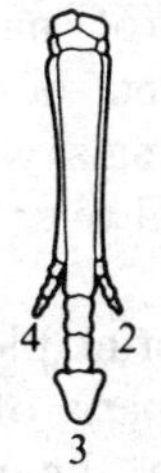	Very dry conditions: prairies. Speed more important. Reduction of digits 2 and 4. Running on digit 3. Increase in length of remaining metacarpal and metatarsal. Taller with longer neck. Teeth longer with cement on crown.
Oligocene 38×10^6 yr	*Mesohippus*	 up to 0.6m	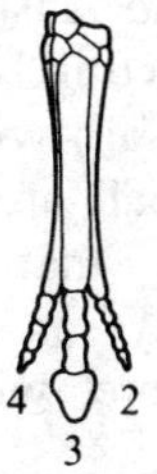	Dry conditions: forests and prairies. Speed important to escape enemies. Only three digits very obvious. Third digit much enlarged.
Eocene 54×10^6 yr	*Hyracotherium*	 about 0.4m	 metacarpals (numbered as shown)	Size of fox. Lived on soft ground near streams. Four digits in forelimb and three digits in hindlimb increase surface area for support. Low-crowned molar teeth adapted to browsing on soft lush vegetation.

Fig 26.4 *Evolution of the modern horse.*

26.7.2 Geographical distribution

All organisms are adapted to their environment to a greater or lesser extent. If the abiotic and biotic factors within a habitat are capable of supporting a particular species in one geographical area, then one might assume that the same species would be found in a similar habitat in a similar geographical area, for example lions in the savannah of Africa and the pampas of South America. This is not the case. Plant and animal species are discontinuously distributed throughout the world. Ecological factors often account for this discontinuous distribution, but evidence from the successful colonisation of habitats by plant and animal species introduced there by humans suggests that factors other than those of ecological adaptation are involved. Rabbits are not endemic (naturally occurring) species in Australia, yet their rapid increase in numbers following their introduction by humans indicates the suitability of the Australian habitat. Similar examples of this principle are illustrated by the spread of domestic animals and plants by humans, such as sheep, corn, potatoes and wheat. An explanation for the discontinuous distribution of organisms is based on the concept of species originating in a given area and then spreading (dispersing) outwards from that point. The extent of the dispersal will depend upon the success of the organisms, the efficiency of the dispersal mechanism and the existence of natural barriers such as oceans, mountain ranges and deserts. Wind-blown spores and seeds and flying animals would appear to have the best adaptations for dispersal over land and sea.

In contrast to, and despite the general principle of organisms being naturally confined to certain parts of the world, many related forms are found in widely separated regions, for example the three remaining species of lungfish are found separately in tropical areas of South America (*Lepidosiren*), Africa (*Protopterus*) and Australia (*Neoceratodus*); camels and llamas are distributed in North Africa, Asia and South America; and racoons are widely found in North and South America and a small area of south-east Asia. Fossil evidence indicates that the distribution of these organisms was not always as seen today and that in the past they were more widely distributed.

Whilst none of this evidence has any immediate significance for evolutionary theory, it does point to the fact that the distribution of land masses was not always as it is today, as explained below.

It used to be believed that the world had always been as it now is and that the present continents and oceans had never changed positions. Early geologists, such as Hutton and Lyell (table 26.1), accounted for the existence of sedimentary rocks in terms of the periodic rise and fall of the sea. Later it was suggested that there were once two large continental masses, one in the Northern Hemisphere called Laurasia and one in the Southern Hemisphere called Gondwanaland, linked by extensive land bridges across which animals and plants could migrate and disperse. Subsequent geological research has modified this idea and favours the hypothesis of **continental drift**, based on the concept of **plate tetonics**. The hypothesis of continental drift was first proposed by Snider in 1858 but developed by Taylor in America and Wegener in Germany in the late 1800s. Wegener proposed that, during Carboniferous times, Laurasia and Gondwanaland formed one large land mass called Pangaea (Greek, all earth) which floated on the denser molten core of the Earth. It is now believed, though, that continents have drifted apart as a result of convection currents within the Earth spreading upwards and outwards, dragging plates on which the continents float. This hypothesis would account for the continuous movements of land masses and the present distribution of species such as that of the lungfish (fig 26.5).

In the case of the camels and llamas (family Camelidae) it is believed that they arose from a common ancestor which fossil evidence suggests had its origin in North America. During the Pleistocene this ancestor spread southwards into South America via the Isthmus of Panana, and northwards into Asia before changes in sea level separated it from North America (fig 26.6). Throughout this time it is thought that progressive changes within the Camelidae occurred, producing the two genera *Camelus* and *Lama* at the extremes of their Pleistocene migration.

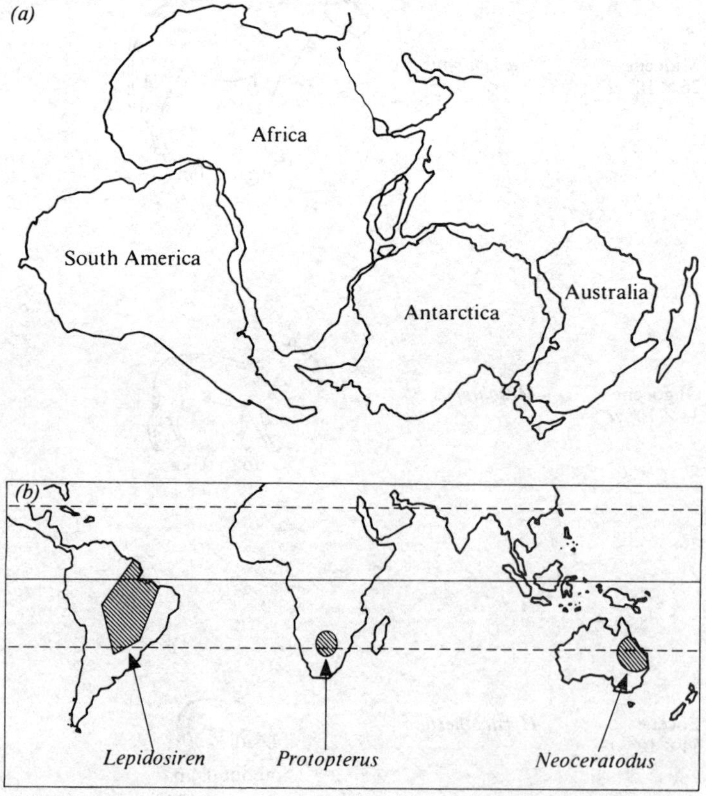

Fig 26.5 *(a) Relative positions of South America, Africa and Australia during early stages of continental drift, indicating proximity of areas where lungfish may have originated.*
(b) Present distribution of species of lungfish.

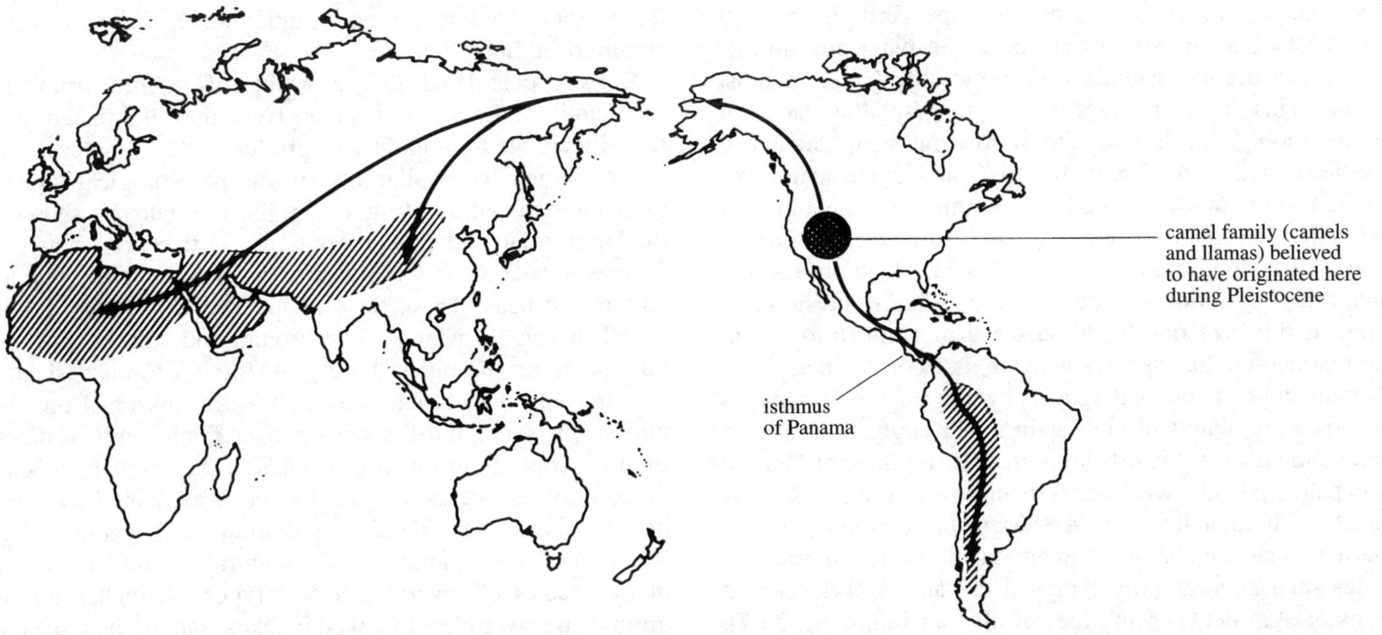

Fig 26.6 *Map of the world showing distribution of present members of the family Camelidae, the camels in North Africa and Asia and the llamas in South America. During Pleistocene times the camel family was distributed throughout North and South America and much of Asia and North Africa. This distribution is based on fossil evidence. Solid black lines indicate possible migration routes. (Based on Matthews (1939)* Climate of evolution, *Vol. 1, 2nd ed., NY Acad. of Sci.)*

Forms intermediate between the present camels and llamas exist in the fossil record throughout North America, Asia and North Africa. The fossil record indicates that other animals of the camel family in other parts of the world became extinct at the close of the last Ice Age.

Another example of discontinuous distribution as a result of geographical isolation is provided by the monotremes and marsupials of Australasia. Australasia is believed to have broken away from the other land masses during the late Jurassic, just after the appearance of primitive mammals. The mammals are divided into three orders: monotremes, marsupials and eutherians. In Australasia only the monotremes and marsupials developed. Here they coexisted and underwent adaptive radiation (modification of the same basic structures for different functions) to produce the characteristic Australasian fauna represented by the monotremes *Tachyglossus* and *Zaglossus* (the spiny anteaters) and the duck-billed platypus (*Ornithorhynchus*), and 45 genera of marsupials. Elsewhere in the world the more advanced eutherian (placental) mammals also developed. As they spread out over the continents it is believed that they ousted the more primitive monotremes and marsupials from their ecological niches, except where geographical barriers disrupted their dispersal, as into Australasia.

These points may be summarised as:

- species originated in a particular area;
- species dispersed outwards from that area;
- dispersal could only occur for most species where land masses were close enough together to permit dispersal;
- the absence of more advanced organisms from a region usually indicates the prior separation of that region from the area of origin of those organisms.

Whilst none of the evidence presented above indicates the mechanism by which species are thought to have originated, it does suggest that various groups have originated at various times and in various regions. Fossil evidence reveals the ways in which these organisms have undergone gradual modification (evolution), but again no indication of the possible mechanism.

Evidence for a possible mechanism of the origin of species by natural selection is supplied by the distribution of plants and animals on oceanic islands. Both Wallace and Darwin were struck by the amazing diversity of species found on islands such as the Hawaiian and the Galapagos groups. Geological evidence indicates that these islands were formed by oceanic volcanic activity which thrust them up above sea level, so that they have never had any direct geographical links with any land mass. Plant species must have arrived on the islands by wind dispersal as spores and seeds, or water dispersal as floating seeds and masses of vegetation. Aquatic and semi-aquatic organisms are believed to have been carried there by ocean currents, whilst terrestrial organisms may have been carried clinging to logs or floating masses of vegetation. Birds, bats and flying insects would have fewer problems of dispersal to these islands.

The Galapagos Islands are situated in the Pacific Ocean on the equator almost 1200 km west of Ecuador and form an archipelago described further in section 27.8.3. When Darwin visited the islands in 1853, he noticed the similarity of the species found there to those on the nearest mainland, a fact he had also observed on the Cape Verde Islands off the coast of West Africa. However, the plant and animal species on oceanic islands were noticeably larger in most cases. This may be accounted for by the lack of competition from larger, and more dominant, advanced species which were absent from the islands, but which co-habited with smaller related species on the mainland. For example, the giant tortoise (*Geochelone elephantopus*), nearly 2 m long and weighing 260 kg, feeding on the plentiful vegetation found on the islands presumably attained this size due to the absence of competition from various mammalian species which existed on the mainland. Darwin noticed too that iguana lizards on the Galapagos Islands were abundant and again much larger than related mainland species. Lizards are terrestrial reptiles, but on the Galapagos Islands, where two species were found, one was aquatic. The aquatic form, *Amblyrhyncus cristatus*, fed on marine algae and showed adaptations for locomotion in water such as a laterally flattened tail and well-developed webs of skin between the toes of all four limbs (fig 26.7). Competition for food, space and a mate within the terrestrial form is thought to have exerted a selection pressure on the lizards and favoured those showing variations with aquatic adaptations. This mechanism of environmental factors operating on a variable genotype is called **natural selection** and is described above. It could have been the process which gradually gave rise to the aquatic species. It was, however, the diversity of adaptive structures shown by the 13 species of finches found within the archipelago which had the greatest influence on Darwin's thinking on the mechanism of the origin of species. Only one type of finch existed on the mainland of Ecuador and its beak was adapted to crushing seeds. On the Galapagos Islands, six major beak types were found, each adapted to a particular method of feeding. The various types, their feeding methods and number of species are summarised in fig 26.8.

Fig 26.7 *Giant aquatic lizard of the Galapagos Islands underwater.*

Darwin postulated that a group of finches from the mainland colonised the islands. Here they flourished, and the inevitable competition produced by increase in numbers, and the availability of vacant ecological niches, favoured occupation of niches by those organisms showing the appropriate adaptive variations. Differences between species relate to small differences in body size, feather colour and beak shape. Several species of finch are found on all the bigger islands. The ground and warbler finches, thought to be the oldest types, are found on most islands. The tree and vegetarian/tree finches are missing from the outlying islands, and the woodpecker finches are confined to the central group of islands. The actual species distribution is interesting and has been explained by Lack on the basis of adaptive radiation and geographical isolation. For example, on the central islands there are many species of several different types of finch, such as ground, tree warbler and woodpecker, rather than several species of the same type. Even where several species of only one type of finch are present, as on the outlying islands, each species differs in its ecological requirements. This fits in with the Gaussian exclusion principle (section 10.7.5) which states that two or more closely related species will not occupy the same area unless they differ in their ecological requirements.

26.7.3 Classification

The system of classification described in chapter 2 was proposed by Linnaeus before the time of Darwin and Wallace, but has implications for the origin of species and evolutionary theory. Whilst it is possible to conceive that all species, both living and extinct, were created separately at a specific time or had no origin, the structural similarity between organisms, which forms the basis of a natural system of **phylogenetic classification**, suggests the existence of an evolutionary process. These similarities and differences between organisms may be explained as the result of progressive adaptation by organisms within each taxonomic group to particular environmental conditions over a period of time.

Numerical taxonomists, working mainly from comparative phenotypic characters have found it possible to construct a phenetic classification system which is consistent, to the extent of present knowledge, with the concept of evolution. These systems of classification are capable of standing in their own right as a basis for biological organisation, but they also strongly suggest that an evolutionary process has occurred.

26.7.4 Plant and animal breeding

One of the earliest features of human civilisation was the cultivation of plants and domestic animals from ancestral wild stocks. By selecting those

(*a*)

Types of finch	*Beak shape*	*Food source*	*Habitat*	*Number of species*
large ground finch (ancestral)	typical main land type: short and straight 	crushing seed	coastal	1
ground finches 	various, but short and straight as above 	seeds/insects 	coast/lowlands	3
cactus ground finches 	long slightly curved, split-tongue 	nectar of prickly-pear cactus	lowland	2
insectivorous tree finches 	parrot-like 	seeds/insects 	forest	3
vegetarian tree finch	curved, parrot-like 	fruit/buds/soft fruit	forest	1
warbler finch	slender 	insects in flight	forest	1
woodpecker finches 	large, straight, (uses cactus spine or stick to poke insects out of holes in wood)	larvae insect 	forest	2

members of the species which showed a favourable variation, such as increased size or improved flavour, and artificially breeding them by selective mating, selective propagation or selective pollination, the desired characteristics were perpetuated. Continued selective breeding by humans has produced the varieties of domestic animals and plants of agricultural importance seen today. It is known from archaeological remains that early humans were successful in rearing cattle, pigs and fowl, and cultivating cereal crops and certain vegetables. Until Mendel's work on genetics was revealed, the theoretical basis of inheritance and breeding was not clear, but this has not stopped humans carrying out practical breeding programmes. In terms of genetics, humans are preserving those animal or plant genes which are considered desirable and eliminating those which are undesirable for their purposes. This selection uses naturally occurring gene variation, together with any fortuitous mutations which occur from time to time.

Whilst varieties of dogs, cats, birds, fish and flowers have been produced for sporting or decorative purposes, it is economically important varieties of animals and plants that have been studied most by plant and animal breeders (fig 26.9). Some specific examples of phenotypic characteristics which have been artificially selected are shown in table 26.4. A recently developed form of artificial selection is the selection for resistance to antibiotics, pesticides and

Fig 26.8 *(a) Adaptive radiation of Darwin's finches. (After Lack.) (b) A male cactus finch (*Geospiza scandens*).*

(*b*)

(*a*)

(*b*)

Fig 26.9 *The result of selective breeding. The wild pig (a) is native to Europe, Asia and Africa but has been selectively bred to produce a variety of breeds, of which the English Large White (b) with its high quality of meat yield, is an example.*

herbicides shown respectively by pathogens, pests and weeds. A vicious circle is produced as new strains of organisms become immune to the ever-increasing number of chemical substances produced to contain and control them.

Since characteristics can be 'produced' by our ability to selectively breed, as in the case of breeds of dogs or pigeons, Darwin used this as evidence for a mechanism by which species might arise naturally. In the case of *natural* selection the environment rather than humans was believed to act as the agent of selection. Artificially selected forms probably would not have arisen in the wild; in most cases they are unable to compete successfully with closely related non-domesticated forms.

Table 26.4 Selected phenotypic characteristics and examples of them.

Phenotypic characteristic	*Example*
Hardiness	Sweetcorn grown in England
Size	Potato, cabbage
Increased yield	Milk, eggs, wool, fruit
Earlier maturity	Cereal crops (two per season)
Lengthened season	Strawberries
Taste/eating quality	Apples, seedless grapes
Harvesting ease	Peas
Length of storage	Beans/peas for freezing
Increased ecological efficiency	Protein from plants, e.g. soyabean
Resistance to disease	Rust and mildew (fungi)-resistant wheat

26.7.5 Comparative anatomy

Comparative study of the anatomy of groups of animals or plants reveals that certain structural features are basically similar. For example, the basic structure of all flowers consists of sepals, petals, stamens, stigma, style and ovary; yet the size, colour, number of parts and specific structure are different for each individual species. Similarly, the limb-bone pattern of all tetrapods (animals with four legs) from amphibia to mammals has the same structural plan: it is called the **pentadactyl limb** (fig 18.11). This basic structure has been modified in several ways as illustrated in fig 26.10. In each case, the particular structure is adapted to a certain method of locomotion in a particular environment.

Organs from different species having a similar basic form, microscopic structure, body position, and embryonic development are said to be **homologous**, a term introduced in 1843 by Richard Owen.

Homologous structures showing adaptations to different environmental conditions and modes of life are examples of adaptive radiation. The ecological significance of these processes is considered in section 26.7.6. The specific functions that these structures carry out may vary in different organisms. These differences reflect the particular ways the organisms are adapted to their environments and modes of life. Other examples of homology are given below.

Branchial arches/Ear bones. Certain bones of the jaw in fish can be traced through other vertebrates, where they are involved in jaw suspension, to mammals where they appear as the ear bones: the malleus, incus and stapes (fig 26.11).

Halteres. The hind pair of wings typical of most insects have been modified in the Diptera into little rods, the halteres, which serve as gyroscopic organs helping to maintain balance in flight.

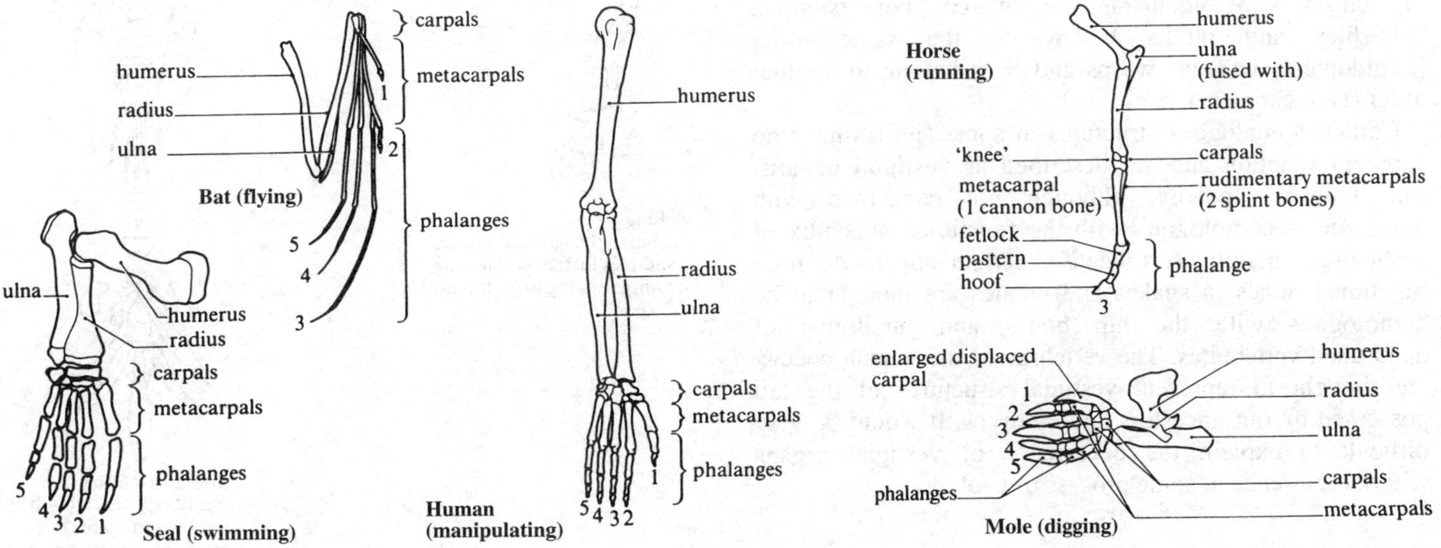

Fig 26.10 *(above)* *Adaptations of the pentadactyl limb shown by mammals.*

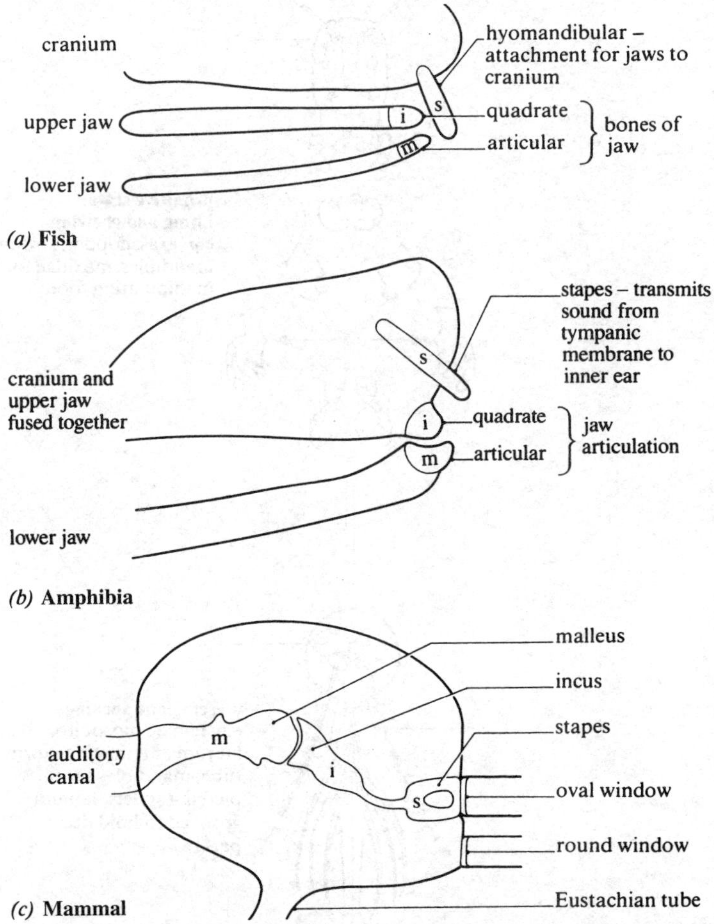

Fig 26.11 *Relative positions and functions of bones of the mammalian ear as seen in fish and amphibia.*

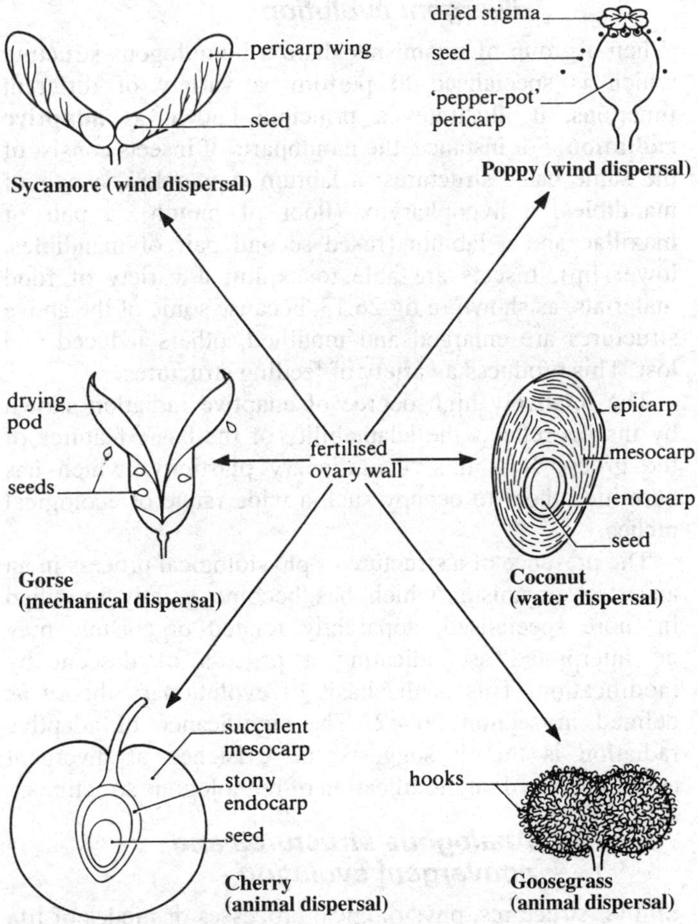

Fig 26.12 *Variation in pericarp (fertilised ovary wall) structure for different methods of seed dispersal.*

Pericarp. The ovary wall in flowering plants becomes modified, following fertilisation of the ovules, in a variety of ways to aid seed dispersal (fig 26.12).

Whilst homology does not prove that evolution has occurred, the existence of homology within a group of organisms is interpreted as evidence of their descent from a common ancestor and indicates close phylogenetic relationships.

Linnaeus used homology as the basis of his system of classification. The more exclusive the shared homologies, the closer two organisms are related. For example, butterflies and moths belong to the same order (Lepidoptera) whereas wasps and bees belong to another order (Hymenoptera).

Certain homologous structures in some species have no apparent function and are described as **vestigial organs**. The human appendix, although not concerned with digestion, is homologous with the functional appendix of herbivorous mammals. Likewise, certain apparently non-functional bones in snakes and whales are thought to be homologous with the hip bones and hindlimbs of quadruped vertebrates. The vertebrae of the human coccyx are thought to represent vestigial structures of the tail possessed by our ancestors and embryos. It would be very difficult to explain the occurrence of vestigial organs without reference to some process of evolution.

26.7.6 Adaptive radiation

Homologous structures and divergent evolution

When a group of organisms share a homologous structure which is specialised to perform a variety of different functions, it illustrates a principle known as **adaptive radiation**. For instance, the mouthparts of insects consist of the same basic structures: a labrum (upper lip), a pair of mandibles, a hypopharynx (floor of mouth), a pair of maxillae and a labium (fused second pair of mandibles, lower lip). Insects are able to exploit a variety of food materials, as shown in fig 26.13, because some of the above structures are enlarged and modified, others reduced and lost. This produces a variety of feeding structures.

The relatively high degree of adaptive radiation shown by insects reflects the adaptability of the basic features of the group. It is this 'evolutionary plasticity' which has permitted them to occupy such a wide range of ecological niches.

The presence of a structure or physiological process in an ancestral organism, which has become greatly modified in more specialised, apparently related organisms, may be interpreted as indicating a process of descent by modification. This is the basis of evolutionary theory as defined in section 26.4.2. The significance of adaptive radiation is that it suggests the existence of divergent evolution based on modification of homologous structures.

Analogous structures and convergent evolution

Similar structures, physiological processes or modes of life in organisms apparently bearing no close phylogenetic links but showing adaptations to perform the same functions are described as **analogous**. Examples of analogous structures include the eyes of vertebrates and cephalopod molluscs (squids and octopuses), the wings of insects and bats, the jointed legs of insects and vertebrates, the presence of thorns on plant stems and spines on animals, and the existence of vertebrate neuroendocrines, such as

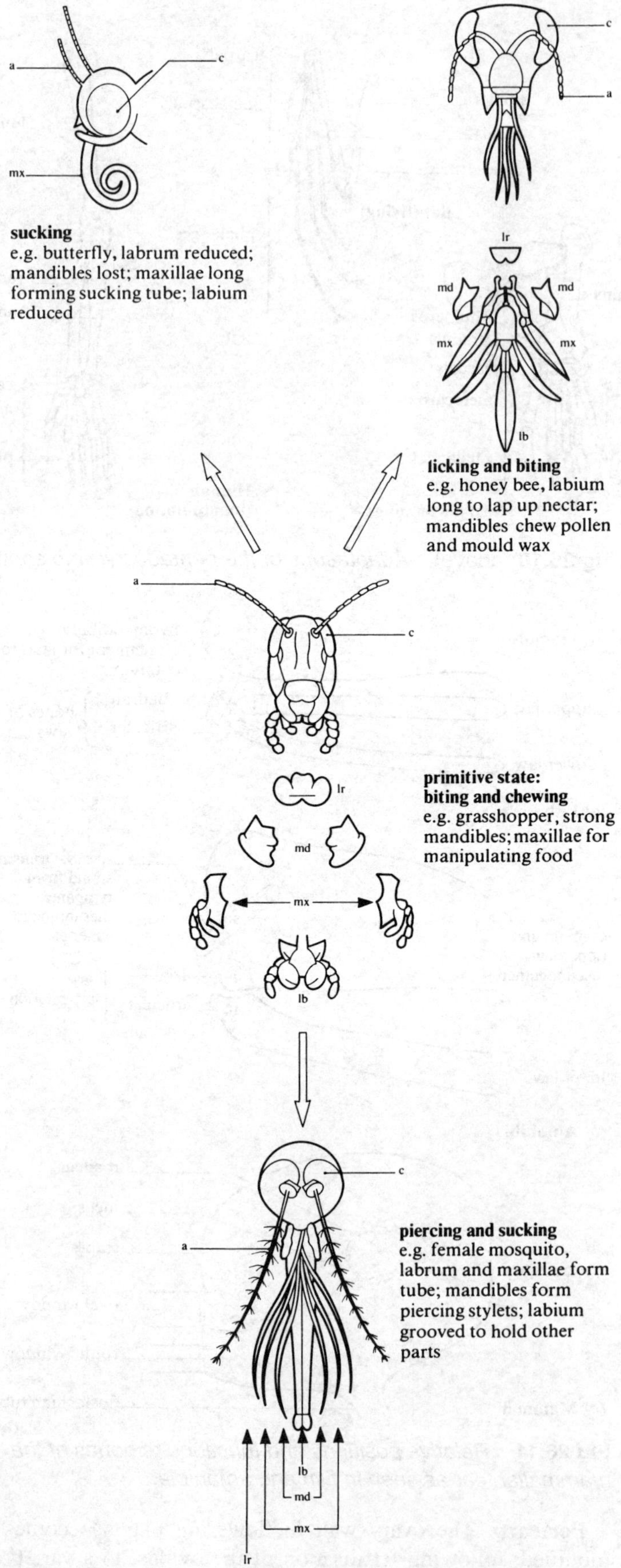

Fig 26.13 *Adaptive radiation of insect mouthparts; a, antennae; c, compound eye; lb, labium; lr, labrum; md, mandibles; mx, maxillae.*

acetylcholine, 5-hydroxytryptamine and histamine, in nettle stings. Analogous structures only bear superficial similarities. For example, the wings of insects are supported by toughened veins composed of cuticle, whereas both bats and birds have hollow bones for support. Likewise the embryological development of the cephalopod and vertebrate eye is different. The former produces an erect retina with photoreceptors facing the incoming light, whereas the latter has an inverted retina with photoreceptors separated from incoming light by their connecting neurones (fig 17.36). Thus the vertebrate eye has a blind spot which is absent in cephalopods.

The existence of analogous structures suggest the occurrence of **convergent evolution**. Convergent evolution may be explained in terms of the environment, acting through the agency of natural selection, favouring those variations which confer increased survival and reproductive potential on those organisms possessing them.

The significance of divergent evolution, suggesting an evolutionary process, and convergent evolution, suggesting an evolutionary mechanism, is highlighted by the **parallel evolution** of marsupial and placental mammals. Both groups are thought to have undergone convergent evolution and come to occupy identical ecological niches in different parts of the world (fig 26.14 and table 26.5).

26.7.7 Comparative embryology

A study of the embryonic development of the vertebrate groups by Von Baer (1792–1867) revealed striking structural similarities occurring in all the groups, particularly during cleavage, gastrulation and the early stages of differentiation. Haeckel (1834–1919) suggested that this had an evolutionary significance. He formulated the principle that 'ontogeny recapitulates phylogeny', that is the developmental stages through which an organism passes repeat the evolutionary history of the group to which it belongs. Whilst this principle over-generalises the situation, it is attractive and there is some evidence to support it. If just the embryos and fetal stages of all the vertebrate groups are examined it is impossible to identify the group to which they belong.

Fig 26.15 shows that it is only in the later stages of development that they begin to assume some similarity to their adult form. At comparable stages the vertebrate embryos all possess the following:

(1) External branchial grooves (visceral clefts) in the pharyngeal region and a series of internal paired gill pouches. These join up in fishes to form the gill slits involved in gaseous exchange. In the other vertebrate groups the only perforation that develops becomes in adults the Eustachian tube and auditory canal involved in hearing.
(2) Segmental myotomes (muscle blocks), which are evident in the tail-like structure. These are retained only in certain species.
(3) A single circulation which includes a two-chambered heart showing no separation into right and left halves, a situation retained completely only in fishes.

As development proceeds in the vertebrate embryo, changes occur which produce the characteristics of fish, amphibian, reptile, bird or mammal depending upon the embryo's parentage. The interpretation placed on these observations is that these embryos, and hence the groups to which they belong, had a common ancestor. There seems little point in an organism having developmental structures which are apparently non-functional in the adult unless they are the remaining stages of ancestral structures. However the principle of recapitulation cannot be accepted entirely since no living organisms can show all the features of their proposed evolutionary ancestors. What appears to be probable is that organisms retain the inherited development mechanisms of their ancestors. Hence at various stages in development it is likely that an organism will show structural similarities to the embryos of its ancestors. Subsequent adaptations to different environmental conditions and modes of life will modify later stages of the developmental process. Observation reveals that the closer the organisms are classified on the basis of common adult homologous structures the longer their embryological development will remain similar. Organisms showing adaptations to certain modes of life and environments not typical of the major group to which they belong show fewer similarities to other members of the group during their embryonic development. This is clearly seen in the development of the parasitic flatworms *Fasciola* and *Taenia*, where a series of larval stages showing adaptations to secondary hosts exist which do not appear in the development of the free-living flatworms, such as *Planaria*. Similarly, the terrestrial earthworm *Lumbricus* does not possess the ciliated trochopore larva which is typical of more ancestral annelids. This evidence highlights the limitations of Haeckel's principle of recapitulation.

Study of the embryological development of major groups of organisms reveals structural similarities in the embryonic and larval stages which are not apparent in the adult stages. These observations are interpreted as suggesting phylogenetic relationships between various groups of organisms and the implication underlying this is that an evolutionary process exists. On the basis of the cleavage patterns of the zygote and the fate of the blastopore, triploblastic animals may be divided into two groups, the protostomes and deuterostomes. **Protostomes** show spiral cleavage and their blastopore becomes the mouth of the adult. This pattern of development is seen in the annelids, molluscs and arthropods. **Deuterostomes** show radial cleavage and their blastopore becomes the anus of the adult. The echinoderms and chordates show this pattern of development. These differences are shown in fig 26.16. It is evidence such as this which has helped to clarify problems of the phylogenetic affinities of the echinoderms. The adult structure of echinoderms suggests that they are a

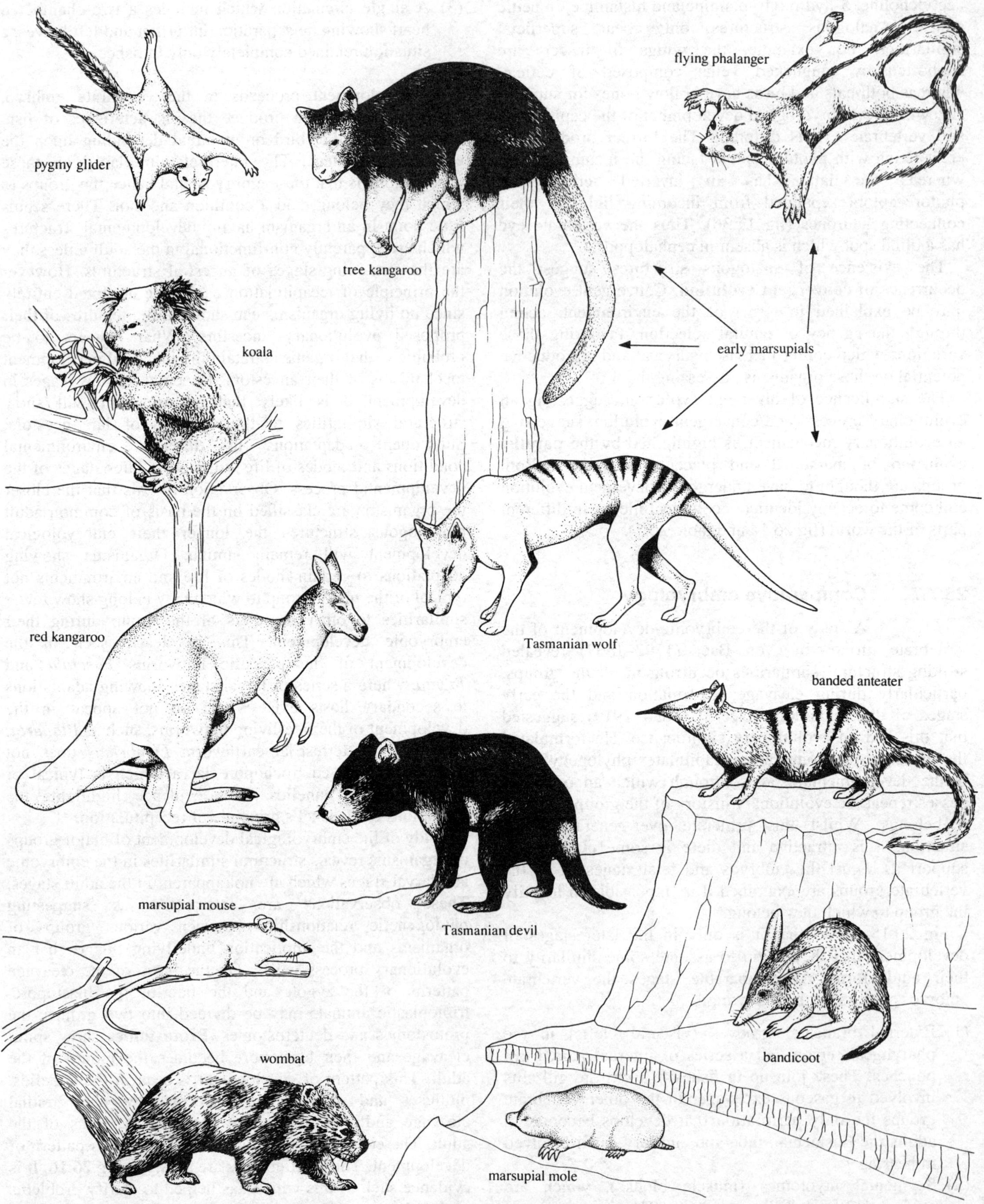

Fig 26.14 *Adaptive radiation of marsupials in Australia (from a variety of sources).*

Table 26.5 Examples of parallel evolution shown by marsupial and placental mammals.

Marsupial mammals (Australasia)	*Placental mammals (elsewhere)*
Marsupial mole	Mole
Marsupial mouse	Mouse
Banded anteater	Anteater
Wombat	Prairie dog
Kangaroo	Antelope
Bandicoot	Rabbit
Flying phalanger	Flying squirrel
Koala	Sloth
Tasmanian wolf	Hyena

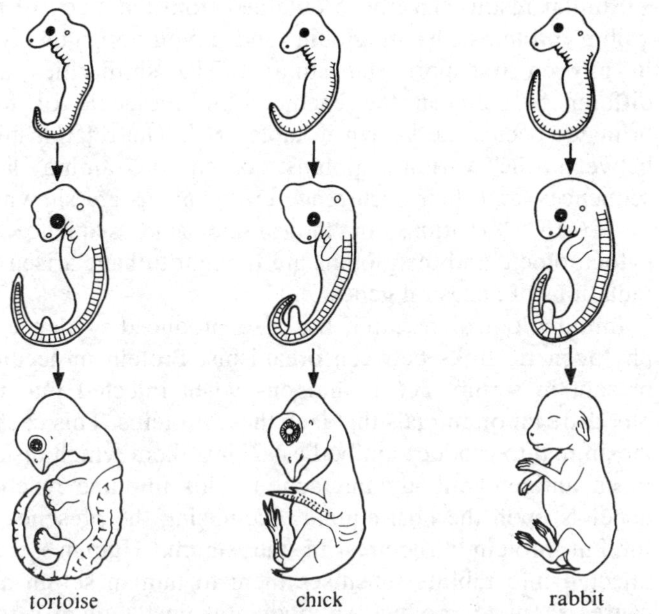

Fig 26.15 *Stages in embryological development as shown by examples from three vertebrate classes.*

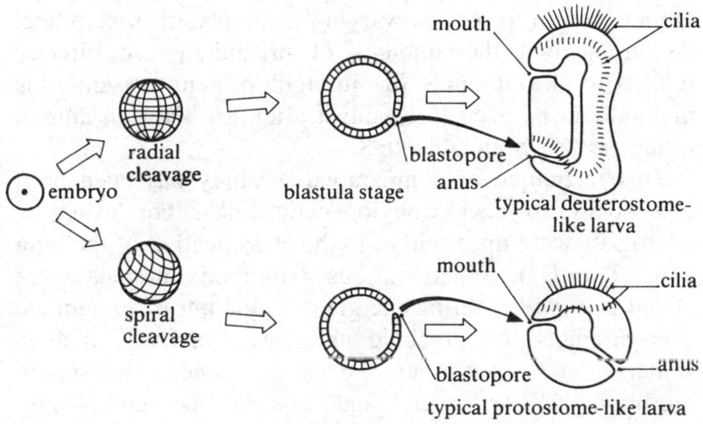

Fig 26.16 *Early development stages of deuterostomes and protostomes.*

non-vertebrate phylum, but their deuterostomic embryological development confirms their affinity with the chordate line of development. This example illustrates the principle that phylogenetic relationships should not be decided purely on the evidence of adult homologous structures.

Evidence of the progressive development of various groups on the basis of their embryology can be seen within the plant kingdom, but examples are less well documented than for the animal kingdom. The early gametophyte of mosses and ferns, as represented by the protonema produced by germination of the spores, has a similar structure, physiology and pattern of growth to the filamentous green algae from which it is therefore thought to have developed. The existence of alternation of generations in plant life cycles, and the various forms the generations take (which are adaptations to various environmental conditions) may be interpreted as examples of homology and therefore provide further evidence for evolutionary relationships between plant groups.

The cone-bearing plants represent a group which show features intermediate between those plants adapted to a terrestrial existence and those plants which still require water for the transfer of gametes. In the cone-bearing cycads the male gametophyte resembles the light dry microspore (pollen grain) of the flowering plants in that it is distributed by wind. As the male gametophyte develops, a pollen tube is formed as in flowering plants, but instead of carrying a non-motile male gamete to the archegonium, two flagellated sperm are produced in the tube which swim to the ovule to bring about fertilisation. The cycads therefore appear to represent an intermediate group between non-vascular plants and flowering plants and this suggests that a phylogenetic continuum exists within the plant kingdom.

The existence of a group of organisms possessing features common to two other groups showing different levels of complexity, or adapted to different environments, may be interpreted as suggesting phylogenetic continuity between the three organisms based on the descent of one group (such as the flowering plants) from another (the non-vascular plants) via the intermediate form (the cycads). Many of these intermediate forms are extinct and it is only by studying the fossil record that a progressive developmental sequence can be deduced. In many cases intermediate forms have not been found. These are equivalent to the 'missing links' (gaps) in the fossil record. It may be that these links do not exist, according to the hypothesis of punctuated equilibria (section 26.7.1). However, their absence may be explained by the possibility that they do not fossilise or have not yet been found. In the case of the phylogenetic link between the annelids and the arthropods there is one group of organisms, of which *Peripatus* is typical, which has features of both annelids and arthropods (fig 26.17). The annelid features include a body wall containing circular and longitudinal muscles, non-jointed parapodia-like limbs, segmental excretory tubules and a double ventral nerve chord. The arthropod features include a chitinous cuticle, spiracles and tracheae

Fig 26.17 *The primitive arthropod* Peripatus.

and an open blood system. Another 'living fossil' intermediate form is represented by the lungfish, which suggests a link between fish and amphibians.

Whilst much of this evidence suggests that some form of evolutionary process has occurred, it must be appreciated that there is no conclusive proof that it did occur.

26.7.8 Comparative biochemistry

As techniques of biochemical analysis have become more precise, this field of research has shed new light on evolutionary ideas. The occurrence of similar molecules in a complete range of organisms suggests the existence of biochemical homology in a similar way to the anatomical homology shown by organs and tissues. Again, this evidence for an evolutionary theory is supportive of other evidence rather than confirmatory in its own right. Most of the research which has been carried out on comparative biochemistry has involved analysis of the primary structure of widely distributed protein molecules, such as cytochrome *c* and haemoglobin, and more recently of nucleid acid molecules, particularly ribosomal RNA. Slight changes in the genetic code as a result of gene mutation produce subtle variations in the overall structure of a given protein or nucleic acid. This forms a basis for determining phylogenetic relationships if the following assumption is made: the fewer the differences in the molecular structure, the fewer the mutations which have occurred and the more closely related in an evolutionary sense are the organisms containing the molecule. Large differences in the molecular structure represent large differences in the DNA. Predictably, this situation exists in organisms showing fewer anatomical homologies.

Cytochromes are respiratory proteins situated in the mitochondria of cells and are responsible for the transfer of electrons along the respiratory pathway (section 9.3.5). Cytochrome *c* is one such protein from the pathway. It has an iron-containing prosthetic group surrounded by a polypeptide chain containing between 104 and 112 amino acids, depending upon the species. Modern techniques of computerised mass spectrometry have enabled the primary structure of the cytochrome *c* polypeptide chain to be worked out for a range of organisms, including bacteria, fungi, wheat, screwworm fly, silkworm, tuna, penguin, kangaroo and primates. The similarity in the cytochrome *c* amino acid sequence between 21 organisms studied in this way is surprisingly high. In 20 out of 21 organisms studied, ranging from the athlete's-foot fungus to humans, the amino acids in positions 78–88 were identical (table 26.6). The amino acid sequence for cytochrome *c* of humans and chimpanzees is identical and differs from the rhesus monkey by only one amino acid. The computer studies, based on amino acid sequences of cytochrome *c*, have produced plant and animal phylogenetic trees which show close agreement with phylogenetic trees based on anatomical homologies.

Similar results have been obtained from the study of the globin proteins, haemoglobin and myoglobin, involved in oxygen transport and storage. The similarities and differences between the haemoglobin molecule of four primate species are shown in table 26.7. The relationships between the various globins, based on amino acid sequences, and their occurrence in organisms are shown in fig 26.18. Variations in the amino acid sequence of cytochrome *c* and the globins are thought to have arisen by mutations of ancestral genes.

Immunological research has also produced evidence of phylogenetic links between organisms. Protein molecules, present in serum, act as antigens when injected into the bloodstream of animals that lack these proteins. This causes the animal to produce antibodies against them which results in an antigen/antibody interaction. This immune reaction depends upon the host animal recognising the presence of foreign protein structures in the serum. Human serum injected into rabbits sensitises them to human serum and causes them to produce antibodies against human serum proteins. After a period of time, if human serum is added to a sample of sensitised rabbit serum, antigen/antibody complexes form which settle out as a precipitate that can be measured. Adding serum from a variety of animals to samples of rabbit serum containing antibodies against human serum produces varying amounts of precipitate. Assuming that the amounts of precipitate are directly related to the amounts of 'foreign' protein present, this method can be used to establish affinities between animal groups as shown in table 26.8.

This technique of comparative serology has been used extensively to check phylogenetic links. For example, zoologists were uncertain as to the classification of the king crab (*Limulus*). When various arthropod antigens were added to *Limulus* serum the greatest amount of precipitate was produced by arachnid antigens. Arachnids include spiders and scorpions. This evidence reinforced morphological evidence, and *Limulus* is now firmly established in the class Arachnida. Similar work has clarified many phylogenetic uncertainties amongst the mammals.

Table 26.6 Cytochrome *c* amino acid sequences for 21 species.

Species	70	71	72	73	74	75	76	77	78	79	80	81	82	83	84	85	86	87	88	89	90	91	92	93	94	95
Human	D	T	L	M	E	Y	L	E	N	P	K	K	Y	I	P	G	T	K	M	I	F	V	G	I	K	K
Rhesus monkey	D	T	L	M	E	Y	L	E	N	P	K	K	Y	I	P	G	T	K	M	I	F	V	G	I	K	K
Horse	E	T	L	M	E	Y	L	E	N	P	K	K	Y	I	P	G	T	K	M	I	F	A	G	I	K	K
Pig, bovine, sheep	E	T	L	M	E	Y	L	E	N	P	K	K	Y	I	P	G	T	K	M	I	F	A	G	I	K	K
Dog	E	T	L	M	E	Y	L	E	N	P	K	K	Y	I	P	G	T	K	M	I	F	A	G	I	K	K
Grey whale	E	T	L	M	E	Y	L	E	N	P	K	K	Y	I	P	G	T	K	M	I	F	A	G	I	K	K
Rabbit	D	T	L	M	E	Y	L	E	N	P	K	K	Y	I	P	G	T	K	M	I	F	A	G	I	K	K
Kangaroo	D	T	L	M	E	Y	L	E	N	P	K	K	Y	I	P	G	T	K	M	I	F	A	G	I	K	K
Chicken, turkey	D	T	L	M	E	Y	L	E	N	P	K	K	Y	I	P	G	T	K	M	I	F	A	G	I	K	K
Penguin	D	T	L	M	E	Y	L	E	N	P	K	K	Y	I	P	G	T	K	M	I	F	A	G	I	K	K
Pekin duck	D	T	L	M	E	Y	L	E	N	P	K	K	Y	I	P	G	T	K	M	I	F	A	G	I	K	K
Snapping turtle	E	T	L	M	E	Y	L	E	N	P	K	K	Y	I	P	G	T	K	M	I	F	A	G	I	K	K
Bullfrog	D	T	L	M	E	Y	L	E	N	P	K	K	Y	I	P	G	T	K	M	I	F	A	G	I	K	K
Tuna	D	T	L	M	E	Y	L	E	N	P	K	K	Y	I	P	G	T	K	M	I	F	A	G	I	K	K
Screwworm fly	D	T	L	F	E	Y	L	E	N	P	K	K	Y	I	P	G	T	K	M	I	F	A	G	I	K	K
Silkworm moth	D	T	L	F	E	Y	L	E	N	P	K	K	Y	I	P	G	T	K	M	I	F	A	G	L	K	K
Wheat	N	T	L	Y	D	Y	L	L	N	P	K	K	Y	I	P	G	T	K	M	V	F	A	G	L	K	K
Fungus (*Neurospora*)	N	T	L	F	E	Y	L	E	N	P	K	K	Y	I	P	G	T	K	M	V	F	P	G	L	K	K
Fungus (*baker's yeast*)	N	N	M	S	E	Y	L	T	N	P	K	K	Y	I	P	G	T	K	M	A	F	G	G	L	K	K
Fungus (*Candida*)	P	T	M	S	D	Y	L	E	N	P	K	K	Y	I	P	G	T	K	M	A	F	G	G	L	K	K
Bacterium (*Rhodospirillum*)	A	N	L	A	A	Y	V	K	N	P	K	A	F	V	L	E	S	K	M	T	F	K	–	L	T	K

Key to amino acids

A	alanine	F	phenylalanine	K	lysine	P	proline	T	threonine
C	cysteine	G	glycine	L	leucine	Q	glutamine	V	valine
D	aspartic acid	H	histidine	M	methionine	R	arginine	W	tryptophan
E	glutamic acid	I	isoleucine	N	asparagine	S	serine	Y	tyrosine

After Dayhoff, M. O. and Eck, R. V. (1967–8) *Atlas of protein sequence and structure*, National Biomedical Research Foundation, Silver Spring, Md.

Table 26.7 Similarities and differences between the polypeptide chains of haemoglobin in four primate species.

Species	*Polypeptide chains* α-*haemoglobin (141 amino acids)*	β-*haemoglobin (146 amino acids)*	γ-*haemoglobin*
Human	+	+	+
Chimpanzee	+	+	1
Gorilla	1	1	1
Gibbon	3	3	2

Haemoglobin is composed of four polypeptide chains, made up of α, β, and γ polypeptides. + indicates no difference in amino acid sequence from that of human, figures indicate number of amino acid differences.

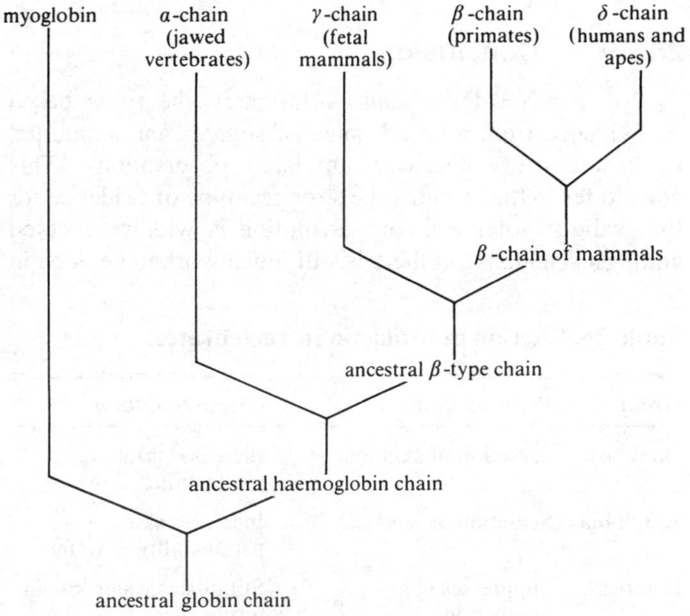

Fig 26.18 *Supposed origins of myoglobin and vertebrate globin polypeptide chains. All five types are found in humans. (After V. M. Ingram, (1963)* Haemoglobins in genetics and evolution, *Columbia University Press.)*

Table 26.8 Amounts of precipitate produced by adding serum from the following mammals to rabbit serum containing anti-human antibodies against human serum (amount of precipitate produced with human serum taken as 100%).

Human	100%
Chimpanzee	97%
Gorilla	92%
Gibbon	79%
Baboon	75%
Spider monkey	58%
Lemur	37%
Hedgehog	17%
Pig	8%

The separation of animal phyla into protostomes and deuterostomes on the basis of embryological development has been reinforced by analysis of the phosphate-containing storage molecules found in muscle and used in the synthesis of ATP. Protostomes, represented by annelids, molluscs and arthropods, contain arginine phosphate, whilst deuterostomes, represented by echinoderms and chordates, contain creatine phosphate.

A final example of biochemical homology is provided by the presence of similar or even identical hormones in vertebrates, where they carry out a range of different functions. For example, a hormone similar to mammalian prolactin occurs in all vertebrate groups where it is produced by the pituitary gland. Although it has been reported that there may be 90 distinct effects of prolactin, these can be arranged under two broad headings, reproduction and osmoregulation (table 26.9).

26.7.9 Conclusion

Neo-Darwinian evolutionary theory is based on evidence from a broad range of sources and supported by a mass of otherwise unrelated observations. This constitutes to the scientist the strongest type of evidence for the 'validity' of the theory. Evolution is widely accepted amongst scientists but there is still much work to be done in refining the theory and its application to all observed circumstances.

All scientific accounts, hypotheses and theories of the history of life are tentative and, as long as we remain objective in our search for truth, will remain so.

Since evolution forms a focal point within the study of biology it would be remiss to conclude this chapter without relating evolution to the perspective of the natural world. To do this it is fitting to quote from Darwin's final paragraph in the *Origin of Species*,

> 'There is a grandeur in this view of life, with its several powers, having been originally breathed by the Creator into a few forms or into one; and that, whilst this planet has gone cycling on according to the fixed law of gravity, from so simple a beginning endless forms most beautiful and most wonderful have been and are evolving.'

Table 26.9 Action of prolactin in vertebrates.

Group	*Reproduction*	*Osmoregulation*
Bony fish	Secretion of skin mucus	Increases urine production
Amphibia	Secretion of 'egg jelly'	Increases skin permeability to water
Reptiles	Suppresses egg production	Stimulates water loss in turtles
Birds	Production of 'crop milk'	Increases water uptake
Mammals	Mammary development and lactation	ADH-like activity

26.8 Human evolution

The course of human evolution has been followed mainly by means of the fossil record which is itself incomplete. However, the fragmentary fossil evidence recovered has enabled **palaeoanthropologists** to begin to piece together a phylogeny (evolutionary history) for primates.

The early stages of human evolution are studied by means of the comparative anatomy of fossils and the evidence of the comparison of many features, from biochemistry to behaviour, of present-day humans with other mammalian species. Later stages of human evolution are studied using the additional evidence from archaeological investigations. The existence of **artefacts** (objects made by humans) such as stone tools, pottery and fire hearths, provide us with an insight into the ways in which modern humans have developed culturally as well as biologically.

Undoubtedly the greatest problem in studying human phylogeny is finding adequate fossil remains. Some excellent remains have been found, for example those in the sediments of the **Olduvai Gorge** in northern Tanzania by Louis, Mary and Richard Leakey, but usually these consist of the skull and teeth only. These structures persist due to their extreme thickness and great hardness (fig 26.19).

Initially fossils are dated with respect to the age of the strata (layers) of rocks in which they are found and the ages of those above and below. This gives a **relative dating** for the fossils. **Absolute dating** is achieved by radioactive dating techniques as described in appendix 4. Age estimates based upon both techniques are usually preferable to either in isolation.

Whatever conclusions we reach regarding humans and their probable ancestors must be tentative and open to revision in the light of new discoveries. Despite this caution, there is a generally accepted view of human phylogeny which is presented below.

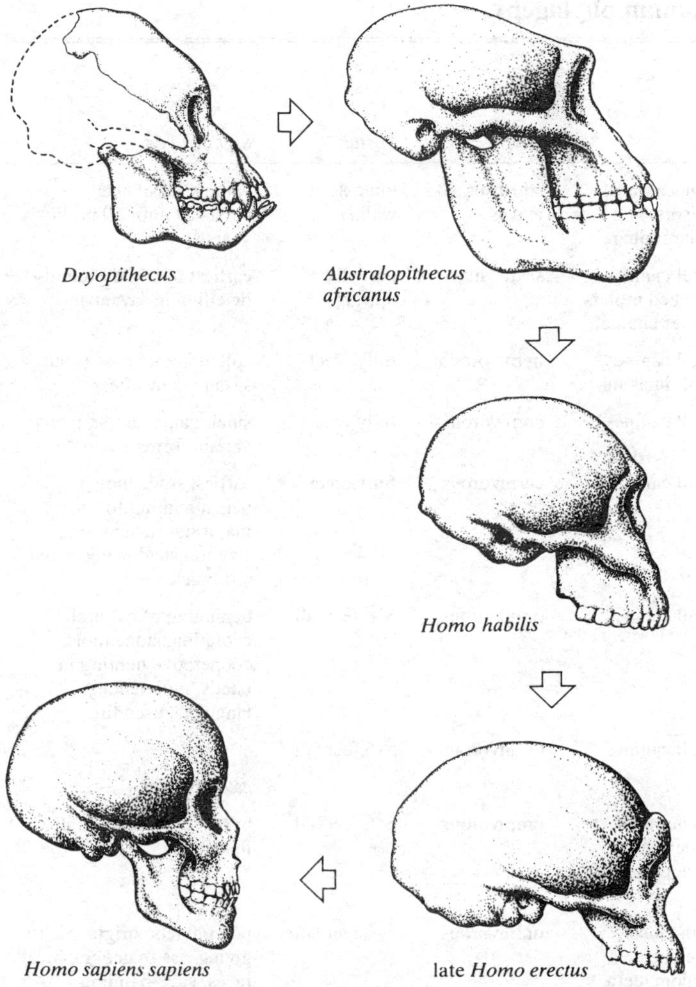

Fig 26.19 *Representative skulls showing the transition from* Dryopithecus *to* Homo sapiens.

Table 26.10 Characteristics of the order Primates.

Grasping limbs	opposable thumb with grip for power and precision
Rotating forelimb	hand can rotate through 180°
Stereoscopic vision	eyes close together on face with parallel optical axes
Visual acuity	increased numbers of rods/cones with own nerve cells
Reduced olfaction	reduced snout allowing flatter face
Enlarged skull	expanded area for cerebrum, ventral foramen magnum
Large brain	increased sensory/motor areas, deeply fissured
Few offspring	longer gestation period, increased parental care
Social dependency	corporate activities, group cohesion

26.8.1 Human phylogeny

Humans belong to an order of mammals called **primates** which also includes tarsiers, lorises, lemurs, monkeys and apes (see table 26.10). Many of the characteristics of this order are adaptations to life in a forest environment, and it was these requirements for an aboreal (tree-dwelling) existence which were essential for the later evolutionary development of our ancestors. They enabled them to exploit the new ecological niches which appeared as the lush forests of the Miocene period gave way to the drier grassland savannahs of the Pliocene period.

Within the order Primates are three groups of animals called **anthropoids**. These include the **New World monkeys** (marmosets and spider monkeys), the **Old World monkeys** (baboons and proboscis monkeys) and **hominoids** (apes and humans). Humans and their ancestors are more closely related to apes than any other anthropoid, and apes, in turn, are closer in phylogeny to Old World monkeys than to New World monkeys (see table 26.8).

It is generally accepted that the ape/human stock probably diverged from that of monkeys about 25–30 million years ago, during the Oligocene period, and the subsequent separation of apes and human ancestors occurred between 5 and 10 million years ago in the middle of the Miocene period. From that time onwards, the family **Pongidae** (fossil forms and present-day gibbons, orangutans, gorillas and chimpanzees) and the family **Hominidae** (fossil forms and modern humans) have evolved along different lines. Recent evidence, based on comparative biochemistry (section 26.7.8), has suggested that gorillas and chimpanzees may have diverged from human stock as recently as 5 million years ago. No supporting fossil evidence for this exists as yet.

Various fossil forms of relevance to human ancestry are represented in table 26.11. Four genera and six species are included. They show a transition in biological features such as skull appearance, tooth structure, brain size, upright posture and diet.

Of particular significance in the evolution of humans was the development of an upright posture (bipedalism) and the increase in brain size.

The transition from walking on four legs to walking on two legs (**bipedalism**) had implications far beyond those affecting the skeleton and muscles. It is now believed that the acquisition of an upright posture and the accompanying changes in the nervous system allowed the subsequent enlargement of the cerebral hemispheres. The common ancestors of humans and apes are likely to have used all four limbs for movement, something like chimpanzees, but with the appearance of *Ramapithecus* more time was spent in an upright posture. By about 4 million years ago our hominid ancestors were bipedal and fully erect.

Freedom of the hands from locomotion enabled them to be used for carrying objects and manipulating the environment, all vital activities preadapting hominids for later dextrous activities associated with their cultural evolution. In addition, an upright posture gave the hominids increased height and range of vision which would have had advantages for them living, as they did, in the open savannah.

Table 26.11 Summary of the main features associated with human phylogeny.

Genus	*Age of appearance/ million years ago*	*Skull*	*Brain capacity/cm*^{3}	*Teeth*	*Diet*	*Posture*	*Significance*
Dryopithecus (earliest fossil ape)	25 (Miocene)	large muzzle	?	large canines, incisors, molars square	soft fruit, leaves	knuckle walker	earliest fossil ape, persisted until 10 million years ago
Ramapithecus	15 (Miocene)	deeper jaw	?	small canines, flattened molars, thicker enamel	seeds, nuts	partially upright	earliest hominid ground-dwelling in savannah
Australopithecus afarensis ('Lucy')	4.0 (Pliocene)	large jaws	450	small canines, small incisors	herbivorous	fully erect	still at home in trees but savannah dwellers
A. africanus	2.5	ventral foramen	450	small canines	carnivorous	fully erect	small game hunter, many variant forms
Homo habilis	2.0 (Pleistocene)	lighter jaw	700	small canines	carnivorous	fully erect	earliest stone tools, began hunting for meat, major increase in brain size foreshadowing social attributes
Homo erectus ('Peking Man')	1.5	thick, low forehead brow ridges	880	small canines	omnivorous	5–6 feet tall	beginning of cultural evolution, stone tools, cooperative hunting in bands, rudimentary language, used fire
Homo sapiens	0.25			small canines	omnivorous	5–6 feet tall	
(Swanscombe)	0.25	heavy jaw	1200				cave-dweller
(Neanderthal)	0.08	face long and narrow, brow ridges, enlarged nasal cavity	1500	heavier than modern teeth, wisdom teeth	omnivorous	5–6 feet tall	buried their dead, flint flake tools
(Cro-Magnon, modern man)	0.03	vaulted cranium, shorter skull, reduced jaws	1400	teeth closer together, wisdom teeth	omnivorous	5–6 feet tall	polyphyletic origin giving rise to geographical races, cave-painting

Along with the advantages of bipedalism was the **increasing brain size** as recorded by cranial capacities. Table 26.11 shows that the cranial capacities of hominids increased from about 450 cm^3 to about 1400 cm^3. However, sheer volume alone does not give a complete picture of the brain potential which developed during human evolution. The complex infolding of the outer cortical tissue increased the surface area to give a much greater working area for the brain. This increase in effective area enabled control and coordination to be exercised over the newly developing behavioural activities such as tool-making, hunting and speech.

The course of human evolution is remarkable in that the gradual transitions in physical features (skeleton, movement, diet) were paralleled by an accelerating development in social behaviour. The process of becoming human is called **hominisation** and it is believed to have been influenced by:

- the development of *manipulative skills* and *speech;*
- changes in sexual behaviour allowing *pair bonding* and increased *parental supervision* of children;
- the establishment of *communal organisation* and *social responsibility*, arising from the principle of *food sharing*.

These biological and social changes were accompanied by changes which were transmitted from person to person by communication rather than inherited genetically. They signalled the development of *culture* which is defined as 'a store of information and set of behaviour patterns, transmitted, not by genetical inheritance, but by learning by imitation, by instruction or by example.'* Culture embraces many different aspects of the life of people including customs, rituals, shared knowledge, language, beliefs, laws, religion, food and employment. Our knowledge of early human cultural evolution is limited to artefacts which archaeologists have recovered. Most of these are stone tools but their study gives useful insights into early human activities.

*Stephen Tomkins (1984), *The Origins of Mankind*, CUP.

26.8.2 Stone tools

The increased brain size of *Homo habilis*, the dissociation of the hands from locomotion and the ability of the hands to achieve both power and precision grips led to the development of **stone tools**. *Homo habilis* (literally 'handy man') at first probably used pebble tools and sticks in much the same way as present-day chimpanzees and gorillas. Chimpanzees strip leaves from twigs before inserting the twig into termite holes. When the termites climb onto the twig it is withdrawn and the termites eaten. The earliest human artefacts were made by *Homo habilis* (2 million years ago) and these were *chopping tools*, *hammer stones*, and *percussion flakes* made from lava or quartz and used for scraping. Later artefacts produced by *Homo erectus* (1.5 million years ago) required greater skill in manufacture and included *hand-held axes* with two cutting edges leading to a point. Sophisticated tools made from flint, bone and wood, however, did not appear until the Upper Palaeolithic, 35 000 years ago. The physical ability to make tools requires sophisticated coordination of the hands and eyes. Such biological activity must be associated with the knowledge required to select materials, impart the skills to others and use these tools, the so-called cultural components of human development (fig 26.20).

The rate of progress in design, manufacture and use of hand tools from the pebbles of 2.5 million years ago to the hand-axes of 0.2 million years ago seems incredibly slow when compared to human technological achievements of the last 100 years. Since 1890 we have witnessed the origin of aircraft and sent people to the Moon, conquered most infectious diseases with vaccinations and antibiotics, transplanted organs and created artificial limbs and organs, developed computing to a sophisticated level, extended our senses with electron microscopes and radio telescopes, harnessed nuclear energy and exploited the potentials of biotechnology. This rapid increase in technology is not associated with increasing brain size but results from advances in research and development based on knowledge and skills transmitted from the previous generation. A child brought up by animals in total isolation from other humans (as in the case of the fabled Tarzan) would have no greater technological expertise than our hominid ancestors. It is through education alone, that is the transmission of culture, that humans are capable of the exponential technological advancement witnessed in the last 100 years.

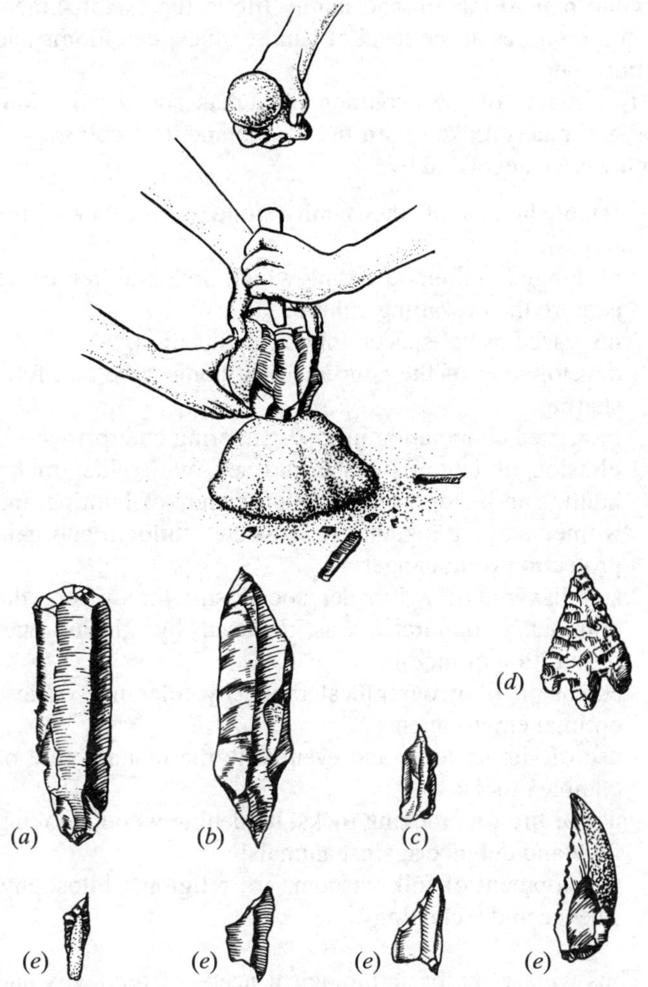

Fig 26.20 *The Upper Palaeolithic tool kit of Cro-Magnon man. Using a hammer stone and antler tine punch, long-bladed flakes were struck from a flint core placed on a stone anvil. Flakes were then retouched to make such things as (a) end scraper, (b) burin chisel, (c) microburin drill, (d) arrowhead, (e) microliths (barbs). (After S. Tomkins (1984)* The Origins of Mankind, *Cambridge University Press.)*

26.8.3 Language

Oral communication is not unique to humans. Birds sing, porpoises 'beep', bats 'chirp' and monkeys and apes chatter, grunt and howl. Humans alone have developed spoken and written languages which are used to communicate information not just about the physical world but to formulate abstract concepts of art, science, philosophy and religion. We do not know when speech began but whatever its origins, the basic anatomical structures associated with speech had to be present in our ancestors. These include the lips, tongue and larynx and three areas of the brain, the speech motor cortical area (controlling the delivery of speech) and two further areas also on the left side of the cerebrum. One of these areas stores auditory, visual and verbal information, and the other is involved in formulating statements and response, that is putting words together. Studies of imprints of blood vessels and brain convolutions present in fossil skulls (**endocasts**) show that there was a substantial development of these areas in both *Australopithecus africanus* and *Homo habilis*.

26.8.4 Social behaviour in humans

Social behaviour is developed to a greater extent in humans than in any other species and extends beyond pair formation and family life to the establishment of communities at the level of bands, tribes, chiefdoms and nation states.

The course of the evolution of human social behaviour was intimately linked with the development of culture and both were categorised by:

- establishment of the family (one partner or many wives);
- prolonged childhood during which time children could acquire the prevailing culture;
- increased use of speech for communication;
- development of the concepts of a home base and food sharing;
- increased cooperation in food-gathering enterprises;
- division of labour by age and sex, with older males hunting in bands to increase efficiency of hunting and women staying together to 'educate' children and gain protection from danger;
- stabilisation of a broader social structure where the dominance hierarchy was replaced by kinship and prohibition of incest;
- extension of geographical range by tolerance of less optimal environments;
- use of simple tools and eventually the manufacture of complex tools;
- use of fire for cracking rocks, hardening wood, cooking food and defence against animals;
- development of folk wisdom, art, religion, philosophy, science and technology.

Thus we see the basic biological needs of food, sex and safety were satisfied more efficiently by the development of group activities based on a common economic–political–sexual structure enriched and supported by the rapid development of culture.

Indeed it can be said that current human evolution is based more on cultural development than on social behaviour.

26.8.5 Art and religion

Whilst humans share many aspects of behaviour with other primates and non-primates, there are some which are unique to the species and these include art, religion and free-will.

The earliest examples of representations of animals and humans come from the Upper Palaeolithic (30 000 years ago). Some are carved in wood or ivory and some are carved on cave walls. The significance of this early art is not known, but we do know that such activities require tools, skill, observation, thought, motivation and possibly leisure. Most of the best-known *cave paintings*, such as those at Lascaux in France, are no older than 20 000 years and used earth pigments, soot and burnt animal residues. What is interesting about these paintings is the variety of abstractions and techniques which are employed and the significance of work. Were they connected with rituals, religious beliefs, simply 'art for art's sake' or an early attempt at graffiti?

In some cases the art forms depicted animals and sex and these were often associated with death and birth respectively. Whether they had religious significance is not clear, but current opinion suggests they were not associated with religious figures as we know them today. *Religion* is believed to have developed at about the same time as cave painting as evidenced by the form of burials found in various parts of the world. In many cases the dead were buried along with offerings such as food, tools and decorative ornaments. It is believed that this symbolism indicates established religious practices. Such a development requires the involvement of conscious intelligent thought, one of the most sophisticated aspects of cultural development. Religion as it is perceived today is fairly recent, the earliest shrines and temples and their accompanying artefacts being less than 10 000 years old.

Chapter Twenty-seven

Mechanisms of speciation

The previous chapter described how Darwin came to appreciate that inherited variation occurs in natural populations as well as in artificial breeding situations. He realised that this variation was significant in the process of evolution by natural selection, but could not explain the mechanism by which variations could appear. It was only with the reappearance of the work of Mendel on inheritance, and the appreciation of its importance in the understanding of evolution, that scientists began to get a deeper understanding of the mechanism. Modern explanations of variation between organisms are a blend of evolutionary theory based on the work of Darwin and Wallace and genetic theory based on principles established by Mendel. Variation, inheritance and evolutionary theory may now be studied using evidence from a branch of biology known as **population genetics**.

27.1 Population genetics

A population is a group of organisms of the same species usually found in a clearly defined geographical area. Following the rediscovery of Mendel's demonstration of the particulate nature of inheritance, the study of genes became important in the study of variation, inheritance and evolutionary change. Bateson, the scientist who introduced the term 'genetics' in 1905, saw genetics as

> 'the elucidation of the phenomena of heredity and variation'.

It is the study of population genetics which forms the basis of modern views of evolutionary theory, a theory referred to as **neo-Darwinism**.

Genes, sometimes working together with environmental factors, determine the phenotypes of organisms and are responsible for variation within populations. The theory of natural selection suggests that phenotypes adapted to the environmental conditions are '**selected for**' whereas non-adapted phenotypes are '**selected against**' and eventually eliminated. Whilst natural selection operates on *individual* organisms of a species, it is the collective genetic response of the *whole population* that determines not only the survival of the species but also the formation of new species. Only those organisms which successfully reproduce before dying contribute to the future of the species. The fate of an individual organism is relatively insignificant in the history of a species. In other words, the long-term effects of natural selection are at the level of the gene and the population rather than the individual because members of the population can interbreed, exchange genes and thereby pass on genes to the next generation and there is therefore a flow of genes between members of a population. Natural selection of the 'fittest' genes takes place.

27.1.1 Gene pool

A gene pool is the **total variety of genes and alleles present in a sexually reproducing population**. In any given population the composition of the gene pool may be constantly changing from generation to generation. New combinations of genes produce unique genotypes which, when expressed in physical terms as phenotypes, undergo environmental selection pressures which continually select and determine which genes pass on to the next generation.

A population whose gene pool shows consistent change from generation to generation is undergoing evolutionary change. A static gene pool represents a situation where genetic variation between members of the species is inadequate to bring about evolutionary change.

27.1.2 Allele frequency

The appearance of any physical characteristic, for example coat colour in mice, is determined by one or more genes. Several forms of each gene may exist and these are called alleles (table 24.2). The number of organisms in a population carrying a particular allele determines the **allele frequency** (which is sometimes, incorrectly, referred to as the gene frequency). For example, in humans the frequency of the dominant allele for the production of pigment in the skin, hair and eyes is 99%. The recessive allele, which is responsible for the lack of pigment, a condition known as **albinism**, has a frequency of 1%. This means that of the total number of alleles controlling production of the pigment, 1% result in a lack of pigment and 99% result in its presence. It is usual in studies of population genetic studies to represent gene or allele frequencies as decimals rather than percentages or fractions. Hence this dominant allele frequency is 0.99 and the recessive albino allele frequency is 0.01. Since the total population represents 100% or 1.0

it can be seen that:

dominant allele frequency	+	recessive allele frequency	=	1
0.99	+	0.01	=	1

In terms of Mendelian genetics the dominant allele would be represented by a letter, say **N** (for normal pigmentation), and the recessive allele would be represented by **n** (the albino condition). In the example above, **N** = 0.99 and **n** = 0.01.

Population genetics has borrowed two symbols from the mathematics of probability, p and q, to express the frequency with which a pair of dominant and recessive alleles appear in the gene pool of the population. Therefore,

$$p+q=1$$

where p = dominant allele frequency, and q = recessive allele frequency.

In the case of pigmentation in humans, p = 0.99 and q = 0.01, since

$$p+q = 1$$
$$0.99+0.01 = 1$$

The value of the above equation lies in the fact that if the frequency of either allele is known, the frequency of the other may be determined. For example, if the frequency of the recessive allele is 25% then q = 25% or 0.25.

Since
$$p+q = 1$$
$$p+0.25 = 1$$
$$p = 1-0.25$$
$$p = 0.75$$

That is, the frequency of the dominant allele is 0.75 or 75%.

27.1.3 Genotype frequencies

The frequencies of particular alleles in the gene pool are of importance in calculating genetic changes in the population and in determining the frequency of genotypes. Since the genotype of an organism is the major factor determining its phenotype, calculations of genotype frequency are used in predicting possible outcomes of particular matings or crosses. This has great significance in horticulture, agriculture and medicine.

The mathematical relationship between the frequencies of alleles and genotypes in populations was developed independently in 1908 by an English mathematician G. H. Hardy and a German physician W. Weinberg. The relationship, known as the **Hardy–Weinberg equilibrium**, is based upon a principle which states that

'the frequency of dominant and recessive alleles in a population will remain constant from generation to generation provided certain conditions exist.'

These conditions are:

(1) the population is large;
(2) mating is random;
(3) no mutations occur;
(4) all genotypes are equally fertile, so that no selection occurs;
(5) generations do not overlap;
(6) there is no emigration or immigration from or into the population, that is, there is no gene flow between populations.

Any changes in allele or genotype frequencies must therefore result from the introduction of one or more of the conditions above. These are the factors that are significant in producing evolutionary change, and when changes occur the **Hardy–Weinberg equation** provides a means of studying the change and of measuring its rate.

27.1.4 The Hardy–Weinberg equation

Whilst the Hardy–Weinberg equation provides a simple mathematical model of how genetic equilibrium can be maintained in a gene pool, its major application in population genetics is in calculating allele and genotype frequencies.

Starting with two homozygous organisms, one dominant for allele **A** and one recessive for allele **a**, it can be seen that all offspring will be heterozygous (**Aa**).

Let **A** = dominant allele
a = recessive allele

Parental phenotypes	homozygous dominant	×	homozygous recessive
Parental genotypes (2n)	**AA**	×	**aa**
Meiosis			
Gametes (n)	Ⓐ Ⓐ	×	ⓐ ⓐ
Random fertilisation			
F_1 genotypes (2n)	**Aa** **Aa**		**Aa** **Aa**
F_1 phenotypes		all heterozygous	

If the presence of the dominant allele **A** is represented by the symbol p and the recessive allele **a** by the symbol q, the nature and frequency of the genotypes produced by crossing the F_1 genotypes above are seen to be:

F_1 phenotypes	heterozygous	×	heterozygous
F_1 genotypes (2n)	**Aa**	×	**Aa**
Meiosis			
Gametes (n)	Ⓐ ⓐ	×	Ⓐ ⓐ

Random fertilisation

	A (p)	**a** (q)
A (p)	**AA** (p^2)	**Aa** (pq)
a (q)	**Aa** (pq)	**aa** (q^2)

F_2 genotypes (2n)	**AA** (p^2)	**2Aa** ($2pq$)	**aa** (q^2)
F_2 phenotypes	homozygous dominant	heterozygous	homozygous recessive

Since **A** is dominant, the ratio of dominant to recessive genotypes will be 3:1, the Mendelian monohybrid cross ratio. From the cross shown above it can be seen that the following genotypes can be described in terms of the symbols p and q:

p^2 = homozygous dominant
$2pq$ = heterozygous
q^2 = homozygous recessive

The distribution of possible genotypes is statistical and based on probability. Of the three possible genotypes resulting from such a cross it can be seen that they are represented in the following frequencies:

AA	**2Aa**	**aa**
$^1/_4$	$^1/_2$	$^1/_4$

In terms of genotype frequency the sum of the three genotypes presented in the above population equal one, or, expressed in terms of the symbols p and q, it can be seen that the genotypic probabilities are:

$$p^2 + 2pq + q^2 = 1$$

(In mathematical terms $p+q=1$ is the mathematical equation of probability and $p^2+2pq+q^2=1$ is the binomial expansion of that equation (that is $(p+q)^2$)).

To summarise, since

p = dominant allele frequency
q = recessive allele frequency
p^2 = homozygous dominant genotype
$2pq$ = heterozygous genotype
q^2 = homozygous recessive genotype

it is possible to calculate all allele and genotype frequencies using the expressions:

allele frequency $p+q = 1$, and
genotype frequency $p^2+2pq+q^2 = 1$.

However, in most populations it is only possible to estimate the frequency of the two alleles from the proportion of homozygous recessives, as this is the only genotype that can be identified directly from its phenotype.

For example, one person in 10 000 is albino, that is to say that the albino genotype frequency is 1 in 10 000. Since the albino condition is recessive, that person must possess the homozygous recessive genotype and in terms of probability it can be seen that

$$q^2 = \frac{1}{10\,000} = 0.0001$$

Knowing that $q^2=0.0001$ the frequencies of the albino allele (q), the dominant pigmented allele (p), the homozygous dominant genotype (p^2) and the heterozygous genotype ($2pq$) may be determined in the following manner.

Since

$$q^2 = 0.0001$$
$$q = \sqrt{0.0001} = 0.01,$$

the frequency of the albino allele in the population is 0.01 or 1%.

Since

$$p+q=1$$
$$p = 1-q = 1-0.01 = 0.99,$$

the frequency of the dominant allele in the population is 0.99 or 99%.

Since

$$p = 0.99$$
$$p^2 = (0.99)^2 = 0.9801,$$

the frequency of the homozygous dominant genotype in the population is 0.9801, or approximately 98%.

Since

$$p = 0.99 \text{ and } q = 0.01,$$
$$2pq = 2 \times (0.99) \times (0.01) = 0.0198,$$

the frequency of the heterozygous genotype is 0.0198, or approximately 2% of the population carry the albino allele either as heterozygotes or albino homozygotes.

These calculations reveal a surprisingly high value for the frequency of the recessive allele in the population considering the low number of individuals showing the homozygous recessive genotype.

Heterozygous individuals showing normal phenotypic characteristics but possessing a recessive gene capable of producing some form of metabolic disorder when present in homozygous recessives are described as **carriers**. Calculations based on the Hardy–Weinberg equation show that the frequency of carriers in a population is always higher than would be expected from estimates of the occurrence of the disorder in the phenotype. This is shown in table 27.1.

> **27.1** Cystic fibrosis occurs in the population with a frequency of 1 in 2200. Calculate the frequency of the carrier genotype.

27.1.5 Implications of the Hardy–Weinberg equation

The Hardy–Weinberg equation shows that a large proportion of the recessive alleles in a population exist in carrier heterozygotes. In fact, the heterozygous genotypes maintain a substantial potential source of genetic variability. As a result of this, very few of the recessive alleles can be eliminated from the population in each generation. Only the alleles present in the homozygous recessive organism will be expressed in the phenotype and so be exposed to environmental selection and possible elimination.

Many recessive alleles are eliminated because they confer disadvantages on the phenotype. This may result

Table 27.1 Some metabolic disorders and the frequencies of homozygous recessive and heterozygous genotypes.

Metabolic disorder	*Approximate frequency of homozygous recessive genotype* (q^2)	*Frequency of 'carrier' heterozygous genotype* ($2pq$)
albinism (lack of pigmentation in body)	1 in 10 000 (in Europe)	1 in 50
alkaptonuria (urine turns black upon exposure to air)	1 in 1 000 000	1 in 503
amaurotic family idiocy (leads to blindness and death)	1 in 40 000	1 in 100
diabetes mellitus (failure to secrete insulin)	1 in 200	1 in 7.7
phenylketonuria (may lead to mental retardation if not diagnosed)	1 in 10 000 (in Europe)	1 in 50

from the death of the organism prior to breeding or **genetic death**, that is the failure to reproduce. Not all recessive alleles, however, are disadvantageous to the population. For example, in human blood groups the commonest phenotypic characteristic in the population is blood group O, the homozygous recessive condition. This phenomenon is also clearly illustrated in the case of sickle-cell anaemia. This is a genetic disease of the blood common in certain populations in Africa, India, certain Mediterranean countries and amongst black North Americans. Homozygous recessive individuals usually die before reaching adulthood thereby eliminating two recessive alleles from the populations. Heterozygotes, on the other hand, do not suffer the same fate. Studies have revealed that the sickle-cell allele frequency has remained relatively stable in many parts of the world. In some African tribes the genotype frequency is as high as 40%, and it was thought that this figure was maintained by the appearance of new mutants. Investigations have revealed that this is not the case, and in many parts of Africa where malaria is a major source of illness and death, individuals possessing a single sickle-cell allele have increased resistance to malaria. In malaria regions of Central America the selective advantage of the heterozygous genotype maintains the sickle-cell allele in the population at frequencies between 10 and 20%.

The maintenance of a fairly constant frequency for a recessive allele which may be potentially harmful is known as **heterozygote advantage**. In the case of black North Americans who have not been exposed to the selection effect of malaria for 200–300 years the frequency of the sickle-cell allele has fallen to 5%. Some of this loss may be accounted for by increased gene flow resulting from black–white marriages, but an important factor is the removal of the selection pressure for the heterozygote due to the absence of malaria in North America. As a result of this the recessive allele is slowly being eliminated from the population. This is an example of evolutionary change in action. It clearly shows the influence of an environmental selection mechanism on changes in allele frequency, a mechanism which disrupts the genetic equilibrium predicted by the Hardy–Weinberg principle. It is mechanisms such as these that bring about the changes in populations which lead to evolutionary change.

27.2 Factors producing changes in populations

The Hardy–Weinberg principle states that given certain conditions the allele frequencies remain constant from generation to generation. Under these conditions a population will be in genetic equilibrium and there will be no evolutionary change. However the Hardy–Weinberg principle is purely theoretical. Few, if any, natural populations show the conditions necessary for equilibrium to exist (section 27.1.3).

The four major sources of genetic variation within a gene pool were described in detail in section 24.8.4, and they are crossing-over during meiosis, independent segregation during meiosis, random fertilisation and mutation. The first three sources of variation are often collectively referred to as **sexual recombination**, and they account for **gene reshuffling**. These processes however, whilst producing new genotypes and altering genotype frequencies, do not produce any changes in the existing alleles, hence the allele frequencies within the population remain constant. Many evolutionary changes, however, usually occur following the appearance of new alleles and the source of this is mutation.

Other situations in which the conditions for the Hardy–Weinberg principle do not exist are when:

- there is non-random breeding;
- the population is small and leads to genetic drift;
- genotypes are not equally fertile so there is genetic load;
- gene flow occurs between populations.

These situations are discussed below.

27.2.1 Non-random breeding

Mating in most natural populations is non-random. Sexual selection occurs whenever the presence of one or more inherited characteristics increases the likelihood of bringing about successful fertilisation of gametes. There are many structural and behavioural mechanisms in both plants and animals which prevent mating from being random. For example, flowers possessing increased size of petals and amounts of nectar are likely to attract more insects and increase the likelihood of pollination. Colour patterns in insects, fishes and birds, and behavioural patterns involving nest-building, territory possession and courtship, all increase the selective nature of breeding.

An experimental investigation with *Drosophila* illustrated the effect of non-random mating on genotype and allele frequencies. A culture of fruit flies containing equal numbers of red-eyed and white-eyed males and females was set up and within 25 generations all white-eyed fruit flies were eliminated from the population. Observation revealed that both red-eyed and white-eyed females preferred mating with red-eyed males. Thus sexual selection, as a mechanism of non-random mating, ensures that certain individuals within the population have an increased reproductive potential so their alleles are more likely to be passed on to the next generation. Organisms with less favourable characteristics have a decreased reproductive potential and the frequency of their alleles being passed on to subsequent generations is reduced.

27.2.2 Genetic drift

This refers to the fact that variation in gene frequencies within populations can occur by chance rather than by natural selection. Random genetic drift or the **Sewall Wright effect** (named after the American geneticist who realised its evolutionary significance) may be an important mechanism in evolutionary change in small or isolated populations. In a small population not all the alleles which are representative of that species may be present. Chance events such as the premature accidental death prior to mating of an organism which is the sole possessor of a particular allele would result in the elimination of that allele from the population. For example, if an allele has a frequency of 1% (that is $q = 0.01$) in a population of 1 000 000 then 10 000 individuals will possess that allele. In a population of 100 only one individual will possess that allele so the probability of losing the allele from a small population by chance is much greater.

Just as it is possible for an allele to disappear from a population, it is equally possible for it to drift to a higher frequency simply by chance. Random genetic drift, as its name implies, is unpredictable. In a small population it can lead to the extinction of the population or result in the population becoming even better adapted to the environment or more widely divergent from the parental population. In due course this may lead to the origin of a new species by natural selection. Genetic drift is thought to have been a significant factor in the origin of new species on islands and in other reproductively isolated populations.

A phenomenon associated with genetic drift is the **founder principle**. This refers to the fact that when a small population becomes split off from the parent population it may not be truly representative, in terms of alleles, of the parent population. Some alleles may be absent and others may be disproportionally represented. Continuous breeding within the **pioneer** population will produce a gene pool with allele frequencies different from that of the original parent population. Genetic drift tends to reduce the amount of genetic variation within the population, mainly as a result of the loss of those alleles which have a low frequency. Continual mating within a small population decreases the proportion of heterozygotes and increases the number of homozygotes. Examples of the founder principle were shown by studies carried out on the small populations of religious sects in America who emigrated from Germany in the eighteenth century. Some of these sects have married almost exclusively amongst their own members. In these cases they show allele frequencies which are uncharacteristic of either the German or American populations. In the case of the Dunkers, a religious sect in Pennsylvania, each community studied was made up of about 100 families, a population so small as to be likely to lead to genetic drift. Blood group analyses produced the following results:

	Blood group A
indigenous Pennsylvanian population	42%
indigenous West German population	45%
Dunker population	60%

These values would appear to be the result of genetic drift occurring within small populations.

Whilst genetic drift may lead to a reduction in variation within a population it can increase variation within the species as a whole. Small isolated populations may develop characteristics atypical of the main population which may have a selective advantage if the environment changes. In this way genetic drift can contribute to the process of speciation (evolution of new species).

27.2.3 Genetic load

The existence within the population of disadvantageous alleles in heterozygous genotypes is known as **genetic load**. As mentioned in section 27.1.5, some recessive alleles which are disadvantageous in the homozygous genotype may be carried in the heterozygous genotype and confer a selective advantage on the phenotype in certain environmental conditions, such as the sickle-cell trait in regions where malaria is endemic. Any increase in recessive alleles in a population as a result of harmful mutations will increase the genetic load of the population.

27.2.4 Gene flow

Within the gene pool of a given breeding population there is a continual interchange of alleles between organisms. Providing there are no changes in allele frequency as a result of mutation, gene reshuffling will confer genetic stability or equilibrium on the gene pool. If a mutant allele should arise, it will be distributed throughout the gene pool by random fertilisation.

Gene flow is often used loosely to describe the movement of alleles within a population as described above, but strictly speaking it refers to the movement of alleles from one population to another as a result of interbreeding between members of the two populations. The random introduction of new alleles into the **recipient** population and their removal from the **donor** population affects the allele frequency of both populations and leads to increased genetic variation. Despite introducing genetic variation into populations, gene flow has a conservative effect in terms of evolutionary change. By distributing mutant alleles throughout all populations, gene flow ensures that all populations of a given species share a common gene pool, that is it reduces differences between populations. The interruption of gene flow between populations is therefore a prerequisite for the formation of new species.

The frequency of gene flow between populations depends upon their geographical proximity, and the ease with which organisms or gametes can pass between the two populations. For example, two populations may be situated so close together that interbreeding is continuous and they may be considered in genetic terms as being one population since they share a common gene pool, such as two snail populations in adjacent gardens separated by a privet hedge.

It is relatively easy for flying animals and pollen grains to be actively or passively dispersed into new environments. Here they may interbreed or cross with the resident population, thereby introducing genetic variation into that population.

27.3 Selection

This is a mechanism that can be thought of as operating at two interrelated levels, at the level of the organism and at the level of the alleles.

Selection is the process by which those organisms which appear physically, physiologically and behaviourally better adapted to the environment survive and reproduce; those organisms not so well adapted either fail to reproduce or die. The former organisms pass on their successful characteristics to the next generation, whereas the latter do not. Selection depends upon the existence of phenotypic variation within the population and is part of the mechanism by which a species adapts to its environment.

When a population increases in size, certain environmental factors become limiting, such as food availability in animals and light in the case of plants. This produces competition for resources between members of the population. Those organisms exhibiting characteristics which give them a competitive advantage will obtain the resource, survive and reproduce. Organisms without those characteristics are at a disadvantage and may die before reproducing. Both the environment and population size operate together to produce a **selection pressure** which can vary in intensity.

Therefore, selection is the process determining which alleles are passed on to the next generation by virtue of the relative advantages they show when expressed as phenotypes. Selection pressure can then be seen as a means of increasing or decreasing the spread of an allele within the gene pool and these changes in allele frequency can lead to evolutionary change.

Major changes in genotype arise from the spread of mutant alleles through the gene pool. The extent of selection and the time it takes will depend upon the nature of the mutant allele and the degree of effect it has upon the phenotype. If the allele is dominant, it will appear in the phenotype more frequently and be selected for or against more rapidly. If the allele is recessive and has no effect in the heterozygous state, as is the case with most mutants, it will not undergo selection until it appears in the homozygous state. The chances of this occurring immediately are slight and the allele may be 'lost' from the gene pool before appearing in the homozygous condition. An allele which is recessive in a given environment may persist until changes in the environment occur where it may have an advantage. These effects would probably appear first in the heterozygote and selection would favour its spread throughout the population, as in the case of sickle-cell anaemia.

A recessive mutant allele may spread rapidly through a population if it occupies a position (locus) on a chromosome very close (linked) to a functionally important dominant allele which is strongly selected for. In this 'linked' condition the chances of the mutant allele combining with another mutant allele to produce the homozygous condition are increased (fig 27.1).

The influence of a given mutant allele can vary. Those mutations affecting alleles controlling important functions are likely to be lethal and removed from the population immediately. Evolutionary change is generally brought about by the gradual appearance of many mutant alleles which exert small progressive changes in phenotypic characteristics.

There are three types of selection process occurring in natural and artificial populations and they are described as stabilising, directional and disruptive. They may be best explained in terms of the normal distribution curve associated with the continuous phenotypic variation found in natural populations (fig 27.2).

27.3.1 Stabilising selection

This operates when phenotypic features coincide with optimal environmental conditions and competition is not severe. It occurs in all populations and

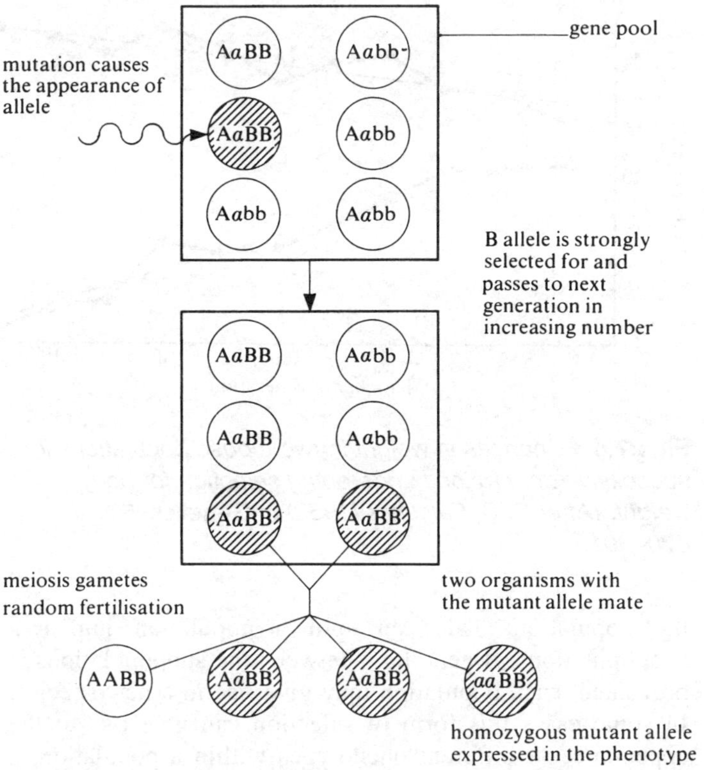

Fig 27.1 *Diagram showing the increased rate of spreading of a mutant allele (α) through a population if linked to a dominant allele (B) which is strongly selected for.*

tends to eliminate extremes from the population. For example, there is an optimum wing length for a hawk of a particular size with a certain mode of life in a given environment. Stabilising selection, operating through differences in breeding potential, will eliminate those hawks with wing spans larger or smaller than this optimum length.

Karn and Penrose carried out a study on the correlation between birth weight and post-natal mortality on 13 730 babies born in London between 1935 and 1946. Of these 614 were still-born or died within one month of birth. Fig 27.3 shows that there is an optimum birth weight of about 3.6 kg (about 8 lb). Babies heavier or lighter than this are at a selective disadvantage and have a slightly increased rate of mortality. From these results it is possible to calculate the intensity of selection pressure.

If 614 babies died at birth or within one month this represents a mortality of 4.5%. Even at the optimum birth weight 1.8% of babies died. Hence the selection pressure for weight at birth for babies of 3.6 kg is 4.5% − 1.8% = 2.7% or 0.027. At a birth weight of 1.8 kg there is a 34% mortality giving an intensity of selection pressure at this weight of approximately 30% or 0.3. It should be pointed out, however, that advances in paediatric medicine have considerably reduced post-natal mortality since 1946.

Stabilising selection pressures do not promote evolutionary change but tend to maintain phenotypic stability within the population from generation to generation.

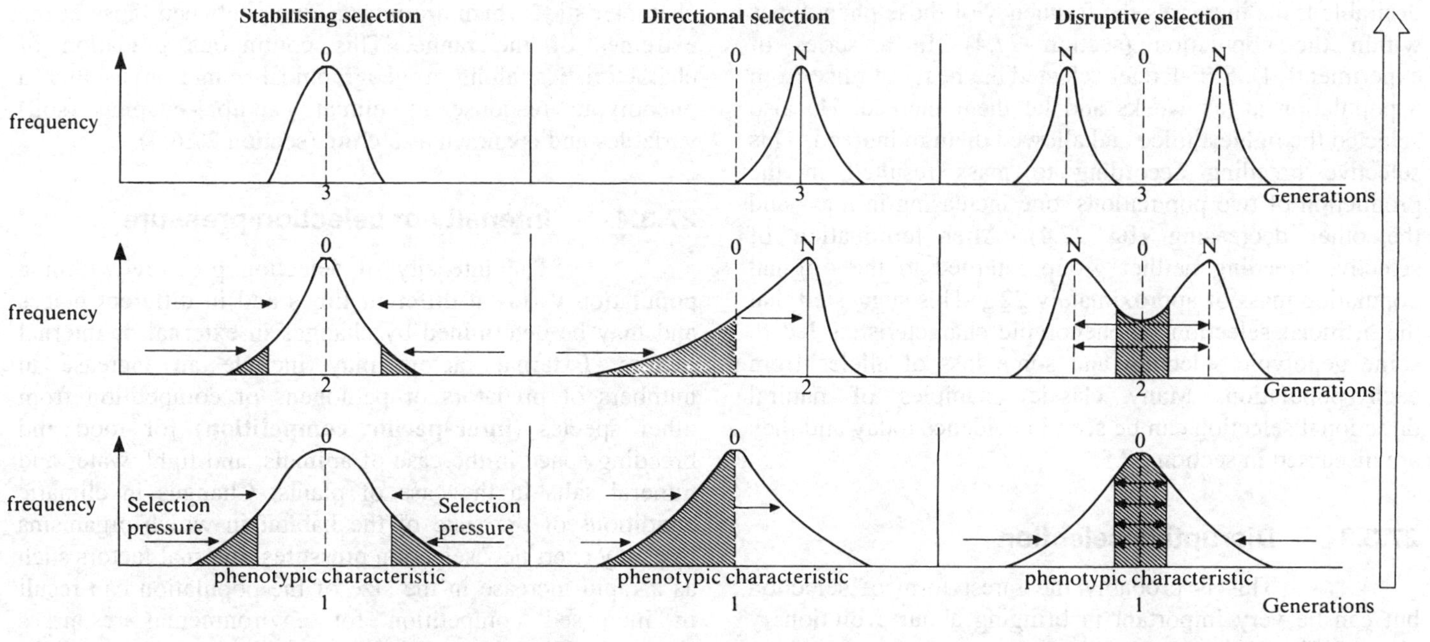

Fig 27.2 *Diagrams showing the three types of selection operating within populations. 0 indicates the original coincidence between optimum phenotype and optimum environmental conditions; N indicates the new position of coincidence of optimum phenotype and optimum environmental conditions. Organisms possessing characteristics in the shaded portions of the normal distribution are at a selective disadvantage and are eliminated by selection pressure. (The numbers 1–3 indicate the order of generations.)*

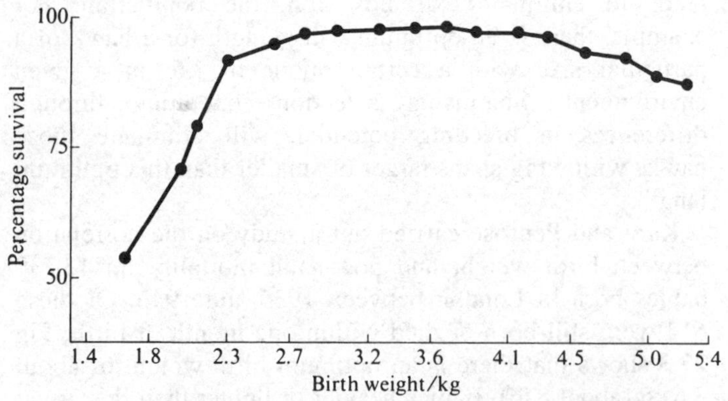

Fig 27.3 *The relationship between percentage survival and birth weight in human babies. (After M. N. Karn & L. S. Penrose (1951)* Ann. Eugen., London, **16** *147–64.)*

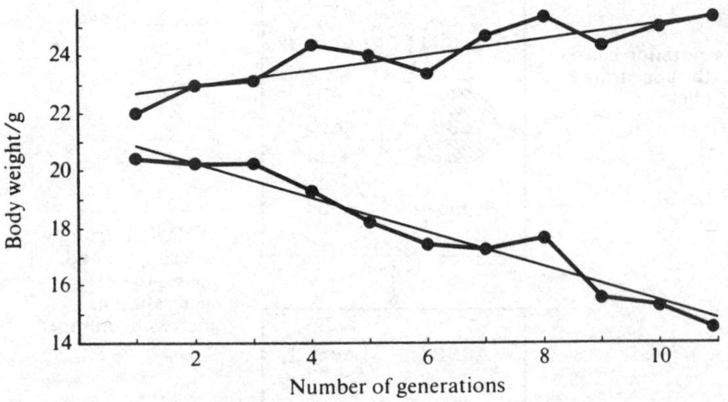

Fig 27.4 *Changes in weight in two mouse populations in successive generations undergoing selection for body weight. (After D. S. Falconer (1953)* J. Genetics, **51** *470–501.)*

27.3.2 Directional selection

This form of selection operates in response to gradual changes in environmental conditions. It operates on the range of phenotypes existing within the population and exerts selection pressure which moves the mean phenotype towards one phenotypic extreme. Once the mean phenotype coincides with the new optimum environmental conditions, stabilising selection will take over.

This kind of selection brings about evolutionary change by producing a selection pressure which favours the increase in frequency of new alleles within the population. Directional selection forms the basis of artificial selection where the selective breeding of phenotypes showing desirable traits increases the frequency of those phenotypes within the population (section 27.4). In a series of experiments, D. S. Falconer selected the heaviest mice from a population at six weeks and let them inbreed. He also selected the lightest mice and allowed them to inbreed. This selective breeding according to mass resulted in the production of two populations, one increasing in mass and the other decreasing (fig 27.4). After termination of selective breeding neither group returned to the original population mass of approximately 22 g. This suggested that the artificial selection of phenotypic characteristics led to some genotypic selection and some loss of alleles from each population. Many classic examples of natural directional selection can be seen in evidence today and they are discussed in section 27.5.

27.3.3 Disruptive selection

This is probably the rarest form of selection but can be very important in bringing about evolutionary change. Fluctuating conditions within an environment, say associated with season and climate, may favour the presence of more than one phenotype within a population. Selection pressures acting from within the population as a result of increased competition may push the phenotypes away from the population mean towards the extremes of the population. This can split a population into two subpopulations. If gene flow between the subpopulations is prevented, each population may give rise to a new species. In some cases this form of selection can give rise to the appearance of different phenotypes within a population, a phenomenon known as **polymorphism** (*poly*, many; *morphos*, form), and is discussed in section 27.5.1. Within a species organisms with different phenotypes, or **ecotypes**, may show adaptations to particular environmental conditions (section 27.6.2). When a species occupies an extremely large geographical range, organisms distributed along it may show local changes in phenotypic characteristics which are intermediate between those at the extremes of the range. This continuous gradation of characteristics along a geographical range is usually a phenotypic response to climate and/or edaphic (soil) variables and is known as a **cline** (section 27.6.3).

27.3.4 Intensity of selection pressure

The intensity of selection pressure within a population varies at different times and in different places and may be determined by changes in external or internal factors. External factors may include an increase in numbers of predators or pathogens or competition from other species (**interspecific competition**) for food and breeding space in the case of animals, and light, water and mineral salts in the case of plants. Changes in climatic conditions or the state of the habitat in which organisms live may exert new selection pressures. Internal factors such as a rapid increase in the size of the population can result in increased competition for environmental resources (**intraspecific competition**). As the population size increases, so do the numbers of parasites and predators. Pathogens, too, are more easily transmitted from organism to organism as the host population rises and diseases spread very rapidly. All of these factors may not only affect the intensity of the selection pressure but also the direction of the pressure. 'New' phenotypes (and genotypes) are selected for, and poorly adapted organisms are eliminated from the population. The organisms to be eliminated first are those at the non-adaptive extremes of the phenotypic range.

One result of increased selection pressure is that it may cause organisms to become **specialised** to certain modes of life or narrower environmental conditions. This may be a disadvantage for the future of that species. Increased uniformity and dependency by a species increases the likelihood of that species becoming extinct should environmental conditions change. The fossil record contains many extinct organisms that were bizarre and overspecialised.

> **27.2** How might a knowledge of selection pressure and mode of life be useful in the eradication of a **named** parasite?

From what has been said it can be seen that increased selection pressure is a conservative mechanism selecting for the phenotype best adapted to the prevailing environmental condition (the optimum phenotype).

A reduction in the intensity of selection pressure usually has the opposite effects to those described above. It may be produced by an absence of predators, pathogens, parasites and competing species or an increase in optimum environmental conditions. These conditions are usually found when an organism is introduced into a new environment. It is conditions such as these which are believed to have favoured the diversity of finch species found on the Galapagos Islands.

27.4 Artificial selection

Humans have practised artificial selection in the form of the domestication of animals and plants since the earliest times of civilisation. Darwin used evidence from artificial selection to account for the mechanism whereby changes in species could arise in natural populations, that is natural selection. The basis of artificial selection is the isolation of natural populations and the selective breeding of organisms showing characteristics or traits which have some usefulness to humans. In the case of cattle, the Hereford and Aberdeen Angus breeds have been selected for the quality and quantity of their meat, whereas Jersey and Guernsey cows are favoured for their milk yield. Hampshire and Suffolk sheep mature early and produce a good quality meat but lack the hardiness and foraging ability of the Cheviot and Scotch Blackface. The latter examples show that no single breed has all the characteristics necessary for the best economic yield under all conditions and therefore a planned programme of selective breeding is often practised to increase the quality of the breed and the yield.

In artificial selection humans are exerting a directional selection pressure which leads to changes in allele and genotype frequencies within the population. This is an evolutionary mechanism which gives rise to new breeds, strains, varieties, races and subspecies. In all cases these groups have isolated gene pools, but they have retained the basic gene and chromosomal structure which is characteristic of the species to which they still belong.

27.4.1 Inbreeding

This involves selective reproduction between closely related organisms, for example between offspring produced by the same parents, in order to propagate particularly desirable characteristics. Inbreeding is a particularly common practice in the breeding of 'show' animals such as cats and dogs. It was used by livestock breeders to produce cattle, pigs, poultry and sheep with high yields of milk, meat, eggs and wool respectively, but for reasons stated below inbreeding is not now widely practised.

Prolonged inbreeding can lead to a reduction in fertility and this is a particular problem in the breeding of livestock. Intensive breeding reduces the variability of the genome (the sum of all the alleles of an individual) by increasing the number of homozygous genotypes at the expense of the number of heterozygous genotypes. In order to overcome these problems breeders resort to outbreeding after several generations of inbreeding. For example, a dairy farmer may use his own bull and successive generations of his own cows to produce cows with a high milk yield. Before the cattle begin to show signs of decreased resistance to disease and reduced fertility, the farmer will use another bull or artificially inseminate his breeding cows with semen acquired from a cattle-breeding centre. This introduces new alleles into the herd, thereby increasing the heterozygosity of the breeding population.

27.4.2 Outbreeding

This is particularly useful in plant breeding, but is being used increasingly in the commercial production of meat, eggs and wool. It involves crossing individuals from genetically distinct populations. Outbreeding usually takes place between members of different varieties or strains, and in certain plants between closely related species. The progeny are known as **hybrids**, and have phenotypes showing characteristics which are superior to either of the parental stock. This phenomenon is known as **hybrid vigour** or **heterosis**. Hybrids produced from crossing homozygous parental stocks from different populations are called F_1 hybrids and show advantages such as increased fruit size and number, increased resistance to disease and earlier maturity. In maize (sweet corn), hybridisation has increased the grain yield of the F_1 hybrids by 250% over the parental stocks (fig 27.5). In the case of double-cross hybridisation, the hybrids produced by

(a) (b)

Fig 27.5 *An example of hybrid vigour. Photograph (a) shows two parental maize varieties which when interbred produce the hybrid shown in the centre of the photograph. The ear shown in the centre of the photograph (*b*) was produced by hybridisation of parental stocks with ears A and B as shown on the left and right of the photograph. (Photograph by D. F. Jones, Connecticut Agricultural Experiment Station.)*

crossing two inbred strains are themselves crossed. The resulting hybrid produces ears having the quality and yield which more than covers the costs involved in a two-year breeding programme (fig 27.6).

Increased vigour results from the increased heterozygosity which arises from gene mixing. For example, whilst each homozygous parent may possess some, but not all, of the dominant alleles for vigorous growth, the heterozygote produced will carry all the dominant alleles, as shown in fig 27.7.

Increased vigour in certain varieties may not result simply from the increased prominence of dominant alleles, but also from some form of interaction between particular combinations of alleles in the heterozygote.

If F_1 phenotypes are continually inbred the vigour will decrease as the proportion of homozygotes increases (fig 27.8).

Selective hybridisation can induce changes in chromosome number (chromosomal mutation), a phenomenon known as **polyploidy**, which can lead to the production of new species. An example of this is described in section 24.9.2.

27.4.3 Artificial selection in humans

Recent advances in human knowledge of the structure of the gene, the genetic code, the mechanisms of heredity and the prenatal diagnosis of genetic defects, have opened up the possibilities of selecting or eliminating certain characteristics in humans. The science of **eugenics** is concerned with the possibilities of 'improving' the 'quality' of the human race by the selective mating of certain individuals. This is a very emotive topic and raises all sorts of objections. Aldous Huxley in his book *Brave New World*, published in 1932, fictionalised the day when eugenics would be taken to its extreme possibilities and particular types of individuals would be produced according to the needs of society at that time. Whilst these ideas are repugnant to societies in which the freedom and rights of the individual are paramount, there are strong arguments for the exercise of limited forms of eugenic practice. In medicine, **genetic counselling** is becoming more acceptable as a means of informing couples with family histories of genetic abnormalities about the possible risks involved in having children. By applying the Hardy–Weinberg equation it is possible to calculate the number of carriers of metabolic disorders such as phenylketonuria or abnormalities of the blood, such as thalassaemia, sickle-cell anaemia or haemophilia. Known carriers can be advised as to the likelihood of marrying another carrier and the possibilities of producing offspring affected by the disorder. Such forms of preventive medicine offer advice rather than dictate policy. Any scientific advances which reduce suffering must receive sympathetic appreciation. The dangers of eugenics lie in their possible abuse.

Fig 27.6 *The phenotypes produced by double-cross hybridisation in maize. The maize crop on the right was produced by crossing the hybrids of the inbred strain (shown on the left).*

Parental genotypes (2n)	**FFgghhIIjj × FFGGHHiiJJ**
Meiosis	
Gametes (n)	**(FghIJ) × (FGHiJ)**
Random fertilisation	
F_1 genotypes (2n)	**FfGgHhIiJj**
F_1 phenotypes	This carries a dominant allele for each gene

Fig 27.7 *A simple genetic explanation of increased vigour in F_1 hybrids.*

Fig 27.8 *Maize stalks of eight generations. The seven stalks on the right demonstrate loss of hybrid vigour as a result of inbreeding from the hybrid shown on the left. The last three generations show reduced loss of vigour as a result of their becoming homozygous. (Photograph by D. F. Jones, Connecticut Agricultural Experiment Station.)*

27.5 Natural selection

Natural selection, as postulated by Darwin and Wallace, represented a hypothesis based on historical evidence. For Darwin, the time span involved in the evolutionary change of a population was such that it could not be observed directly. Recent changes accompanying the industrial, technological and medical revolutions have produced such strong directional and disruptive pressures that we can now observe the results of dramatic changes in genotypic and phenotypic characteristics of populations within days. The introduction of antibiotics in the 1940s provided a strong selection pressure for strains of bacteria that have the genetic capability of being resistant to the effects of the antibiotics. Bacteria reproduce very rapidly, producing many generations and millions of individuals each day. Random mutation may produce a resistant organism in the population which will thrive in the absence of competition from other bacteria which have been eliminated by the antibiotic. As a result, new antibiotics have to be developed to eliminate the resistant bacteria, and so the cycle continues. Other examples of the effects of chemicals in producing selection pressure have been seen with DDT on body-lice and mosquitoes and the effect of the anticoagulant warfarin on rats. Following the development of resistant strains they spread very rapidly throughout the population.

Perhaps the classic example of evolutionary change is provided by the response of moth species to the directional selection pressure produced by the atmospheric pollution which accompanied the industrial revolution. Within the last 100 years darkened forms of about 80 species of moths have appeared in varying frequencies throughout the United Kingdom. This is a phenomenon known as **industrial melanism**. Up to 1848 all reported forms of the peppered moth (*Biston betularia*) appeared creamy-white with black dots and darkly shaded areas (fig 27.9). In 1848 a black form of the moth was recorded in Manchester, and by 1895, 98% of the peppered moth population in Manchester was black. This black 'melanic' form arose by a recurring random mutation, but its phenotypic appearance had a strong selective advantage in industrial areas for reasons put forward and tested by Dr H. B. D. Kettlewell.

The moths fly by night and during the day they rest on the trunks of trees. The normal form of the moth is extremely well camouflaged as its colouration merges with that of the lichens growing on the trunks. With the spread of the industrial revolution sulphur dioxide pollution from the burning of coal killed off the lichens growing on trees in industrial areas, exposing the darker bark, which was further darkened by soot deposits (fig 27.10).

In the 1950s Kettlewell released known numbers of marked light and dark forms into two areas, one a polluted area near Birmingham where 90% of the population was the black form, and the other an unpolluted area in Dorset where the dark form was rarely found. On recapturing the moths using a light trap he obtained the following results:

Fig 27.9 *Polymorphic forms of peppered moth,* Biston betularia. *(a) The normal form,* Biston betularia typica*; (b) the melanic form,* Biston betularia carbonaria. *(From E. B. Ford (1973)* Evolution studied by observation and experiment, *Oxford Biology Readers,* **55**, *Oxford University Press.)*

	Birmingham	*Dorset*
Percentage marked dark form	34.1	6.3
Percentage marked light form	15.9	12.5

Kettlewell demonstrated using cine-film that robins and thrushes feed on the moths. This is a form of natural selection known as **selective predation**, and it acts as a selection pressure on the distribution of the melanic and non-melanic forms.

The results show that the melanic form of the moth, *Biston betularia carbonaria*, has a selective advantage in industrial areas over the lighter form, *Biston betularia typica*, whereas the lighter form has the selective advantage in non-polluted areas.

Subsequent research has demonstrated that the colouration of the dark form is due to the presence of a dominant melanic allele. Fig 27.11 shows the distribution of the two forms in the British Isles in 1958.

The presence of melanic forms in non-industrial areas of the east of England is explained by the distribution of melanic forms by prevailing westerly winds. Since the introduction of the Clean Air Act in 1956 the proportion of non-melanic forms has increased to much higher levels again as the selection pressure on these forms has been reduced in industrial areas.

(a)

(b)

Fig 27.10 *Melanic and non-melanic forms of* Biston betularia *on tree trunks in (*a*) an area near Birmingham, and (*b*) an area in Dorset. (Courtesy of Dr H. B. D. Kettlewell, Department of Zoology, University of Oxford.)*

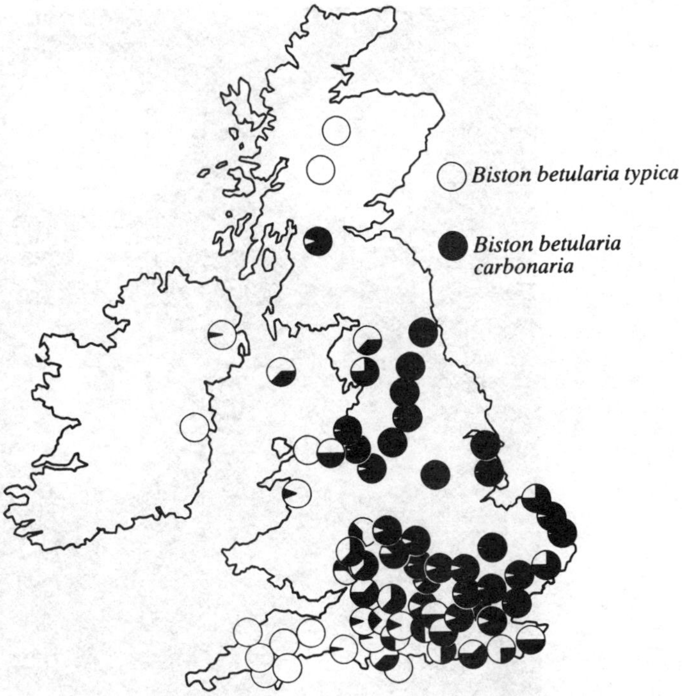

Fig 27.11 *The distribution of melanic and non-melanic forms of* Biston betularia *in the British Isles in 1958. (After H. B. D. Kettlewell (1978)* Heredity, **12**, *51–72.)*

27.5.1 Polymorphism

Polymorphism plays a significant role in the process of natural selection. It is defined as the existence of two or more forms of the same species within the same population, and can apply to biochemical, morphological and behavioural characteristics. There are two forms of polymorphism, **transient polymorphism,** and **balanced**, or **stable, polymorphism**.

Balanced polymorphism

This occurs when different forms coexist in the same population in a stable environment. It is illustrated most clearly by the existence of the two sexes in animals and plants. The genotypic frequencies of the various forms exhibit equilibrium since each form has a selective advantage of equal intensity. In humans, the existence of the A, B, AB and O blood groups are examples of balanced polymorphism. Whilst the genotypic frequencies within different populations may vary, they remain constant from generation to generation within that population. This is because none of them have a selective advantage over the other. Statistics reveal that white men of blood group O have a greater life expectancy than those of other blood groups, but, interestingly, they also have an increased risk of developing a duodenal ulcer which may perforate and lead to death. Red–green colour blindness in humans is another example of polymorphism, as is the existence of workers, drones and queens in social insects and pin-eyed and thrum-eyed forms in primroses.

A classic quantitative study of balanced polymorphism was carried out by Cain, Currey and Shepherd on the common land snail *Cepaea nemoralis*. The shells of this species may be yellow (and appear green with the living snail inside), brown, or various shades including pale fawn, pink, orange and red. The lip of the shell may be dark brown, pink or white and the whole shell may have up to five dark brown bands following the contours of the shell (fig 27.12). Both colouration and banding pattern are determined genetically. The colours are determined by multiple alleles with brown being dominant to pink and both being dominant to yellow. Banding is recessive.

Studies have revealed that the snails are predated upon by thrushes which carry the snails to a nearby stone which they use as an 'anvil' to crack open the shell; the snail inside is then eaten. By studying the proportion of types of shell found near an anvil with those in the immediate

Fig 27.12 *Colour and banding pattern variation in the shells of* Cepaea nemoralis. *The extremes of colour and banding are shown as a progression from yellow unbanded (top left) to brown banded (bottom right). Photographs* (a) *and* (b) *show top and side views of the same shells. (After Tribe, Tallan & Erant (1978)* Basic Biology Course, *Book 12, Cambridge University Press.)*

Fig 27.13 *Unbanded shells of* Cepaea nemoralis *against a background of leaf litter. The shell on the extreme right is yellow, the shell at the top of the photograph is pink and the two shells on the left are brown. (After E. B. Ford (1973)* Evolution studied by observation and experiment, *Oxford Biology Reader,* **55**, *Oxford University Press.)*

habitat, Cain, Currey and Shepherd demonstrated that selective forces were at work within the population. In areas where the background was fairly uniform, such as grass and woodland litter, the yellow and brown unbanded shells had a selective advantage as fewer of these shells were found near the anvil (fig 27.13). In areas where the ground cover was tangled and mottled, as in rough pasture or hedgerows, the darker banded shells had a selective advantage. The forms suffering the greatest predation in any area were those which were visually conspicuous to the thrushes. A large population of polymorphic snails may include several areas with a range of backgrounds. Seasonal effects also produce changes in background colour and pattern. Although predation of conspicuous forms is continuous there is no overall selective advantage for any form, hence the numbers of each form within a population remain fairly constant from year to year.

The balance in numbers of each form may not be determined purely by colour and banding pattern. There is evidence to suggest that physiological effects may help to maintain the polymorphic equilibrium. In some areas where the soil is calcareous and dry and the background cover is light, the dominant forms are not always those with the least conspicuous colour and banding pattern. The genetic basis for the polymorphism shown by *Cepaea* is thought to rely on the existence of a special form of gene linkage. The genes for colour and banding pattern are linked and form a **super-gene** which acts as a single genetic unit and is inherited as such. These genes determine characteristics which have such a selective advantage that they are maintained within the population. It is the variety of allelic forms of these genes maintained by the heterozygotes which forms the basis of the polymorphism. The added linkage of genes controlling certain physiological effects is also thought to contribute to the maintenance of the balanced polymorphism. The existence of a number of distinct inherited varieties coexisting in the same population at frequencies too great to be explained by recurrent mutations, as in the case of *Cepaea*, is called **genetic polymorphism.**

Transient polymorphism

This arises when different forms, or **morphs**, exist in a population undergoing a strong selection pressure. The frequency of the phenotypic appearance of each form is determined by the intensity of the selection pressure, such as the melanic and non-melanic forms of the peppered moth. Transient polymorphism usually applies in situations where one form is gradually being replaced by another.

27.6 The concept of species

A species represents the lowest taxonomic group which is capable of being defined with any degree of precision. It may be defined in a variety of ways and some of these are summarised in table 27.2.

Organisms belonging to a given species rarely exist naturally as a single large population. It is usual for a species to exist as small interbreeding populations, called **demes**, each with its own gene pool. These populations may occupy adjacent or widely dispersed geographical areas. Spatial separation of populations means that the species may encounter a variety of environmental conditions and degree of selection pressure. Mutation and selection within the isolated populations may produce the following degrees of phenotypic variation within the species.

27.6.1 Geographical races

Populations which are distributed over a wide geographical range or have occupied well-separated geographical habitats for a long period of time may show considerable phenotypic differences. These are usually based on adaptations to climatic factors. For example, the gypsy moth (*Hymantria dispar*) is distributed throughout the Japanese islands and eastern Asia. Over this range a variety of climatic conditions is encountered, ranging from subarctic to subtropical. Ten geographical races have been recognised which differ from each other with regard to the timing of hatching of their eggs. The northern races hatch later than the southern races. The phenotypic variations shown by the 10 races are thought to be the result of climatic factors producing changes in gene frequencies within their gene pools. The evidence that these variations are genetically controlled is shown by the fact that under identical environmental conditions the different races still hatch at different times.

Table 27.2 Alternative ways of defining a species.

Biological aspect	*Definition*
Breeding	A group of organisms capable of interbreeding and producing fertile offspring
Ecological	A group of organisms sharing the same ecological niche; no two species can share the same ecological niche
Genetic	A group of organisms showing close similarity in genetic karyotype
Evolutionary	A group of organisms sharing a unique collection of structural and functional characteristics

27.6.2 Ecological races (ecotypes)

Populations adapted to ecologically dissimilar habitats may occupy adjacent geographical areas; for example the plant species *Gilia achilleaefolia* occurs as two races along the coast of California. One race, the 'sun' race, is found on exposed southerly facing grassy slopes, whilst the 'shade' race is found in shaded oak woodlands and redwood groves. These races differ in the size of their petals, a characteristic which is determined genetically.

27.6.3 Clines

A species exhibiting a gradual change in phenotypic characteristics throughout its geographical range is referred to as a **cline**. More than one cline may be exhibited by a species and they may run in opposite directions as shown by fig 27.14.

Species exhibiting marked phenotypic variation within a population according to their degree of geographical isolation are known as **polytypic species**. One classic form of a polytypic species is illustrated by gulls belonging to the genus *Larus* (section 27.8.4).

All cases of phenotypic variation described above represent varying degrees of genetic dissimilarity which may interfere with the breeding potential of members of the populations if brought together.

27.7 Speciation

This is the process by which one or more species arise from previously existing species. A single species may give rise to new species (**intraspecific speciation**), or, as is common in many flowering plants, two different species may give rise to a new species (**interspecific hybridisation**). If intraspecific speciation occurs whilst the populations are separated it is termed **allopatric speciation**. If the process occurs whilst the populations are occupying the same geographical area it is called **sympatric speciation**.

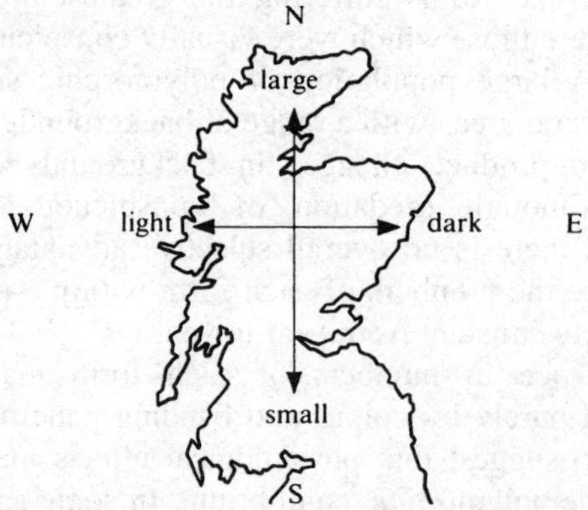

Fig 27.14 *Phenotypic variation in size and fur colour shown by the wood mouse (*Apodemus*) in Scotland.*

27.8 Intraspecific speciation

There are several factors involved in intraspecific speciation, but in all cases gene flow within populations must be interrupted. As a result of this each subpopulation becomes genetically isolated. Changes in allele and genotype frequencies within the populations, as a result of the effects of natural selection on the range of phenotypes produced by mutation and sexual recombination, lead to the formation of races and subspecies. If the genetic isolation persists over a prolonged period of time and the subspecies then come together to occupy the same area they may or may not interbreed. If the breeding is successful they may still be considered to belong to the same species. If the breeding is unsuccessful, then speciation has occurred and the subspecies may now be considered to be a separate species. This is the way in which it is believed evolutionary change can be brought about.

An initial factor in the process of speciation may be the reduction in the intensity of selection pressure within the population. This may lead to increased intraspecific variability. These new phenotypes may enable the population to increase its geographical range if the phenotypes show adaptations to environmental conditions found at the extremes of the range. Providing there is no reduction in gene flow throughout the population, the species, whilst exhibiting the localised phenotypic variation (ecotypes), will share the same gene pool and continue to exist as a single species. This is the situation found in a cline.

Speciation will only occur as a result of the formation of barriers which lead to reproductive isolation between members of the population. Reproductive isolation is brought about by some form of what the geneticist Theodosius Dobzhansky called **isolating mechanisms**.

27.8.1 Isolating mechanisms

An isolating mechanism is a means of producing and maintaining reproductive isolation within a population. This can be brought about by mechanisms acting before or after fertilisation. Dobzhansky suggested a classification of isolating mechanisms which has been modified and is shown in table 27.3.

Table 27.3 Isolating mechanisms (after Dobzhansky).

Prezygotic mechanisms (barriers to the formation of hybrids)	
Seasonal isolation	Occurs where two species mate or flower at different times of the year; for example in California *Pinus radiata* flowers in February whereas *Pinus attenuata* flowers in April
Ecological isolation	Occurs where two species inhabit similar regions but have different habitat preferences; for example *Viola arvensis* grows on calcareous soils whereas *Viola tricolor* prefers acid soils
Behavioural isolation	Occurs where animals exhibit courtship patterns, mating only results if the courtship display by one sex results in acceptance by the other sex; for example certain fish, bird and insect species
Mechanical isolation	Occurs in animals where differences in genitalia prevent successful copulation and in plants where related species of flowers are pollinated by different animals
Postzygotic mechanisms (barriers affecting hybrids)	
Hybrid inviability	Hybrids are produced but fail to develop to maturity; for example hybrids formed between northern and southern races of the leopard frog (*Rana pipiens*) in North America
Hybrid sterility	Hybrids fail to produce functional gametes; for example the mule ($2n = 63$) results from the cross between the horse (*Equus equus*, $2n = 60$) and the ass (*Equus hemionus*, $2n = 66$)
Hybrid breakdown	F_1 hybrids are fertile but the F_2 generation and backcrosses between F_1 hybrids and parental stocks fail to develop or are infertile, for example hybrids formed between species of cotton (genus *Gossypium*)

27.8.2 Allopatric speciation

Allopatric (*allos*, other; *patria*, native land) speciation is characterised by the occurrence, at some stage, of spatial separation. Geographical barriers such as mountain ranges, seas or rivers, or habitat preferences, may produce a barrier to gene flow because of spatial separation. This inability of organisms or their gametes to meet leads to reproductive isolation. Adaptations to new conditions or random genetic drift in small populations lead to changes in allele and genotype frequencies. Prolonged separation of populations may result in them becoming genetically isolated even if brought together. In this way new species may arise. For example, the variety and distribution of the finch species belonging to the family Geospizidae on the islands of the Galapagos archipelago are thought to be the result of allopatric speciation.

David Lack suggested that an original stock of finches reached the Galapagos Islands from the mainland of South America and, in the absence of competition from endemic species (representing relaxed selection pressure), adaptive radiation occurred to produce a variety of species adapted to particular ecological niches. The various species are believed to have evolved in geographical isolation to the point that when dispersal brought them together on certain islands they were able to coexist as separate species.

27.8.3 Sympatric speciation

Genetic differences may accumulate allopatrically in populations which have been geographically isolated for a much shorter period of time. If these populations are brought together, hybrids may form where these overlap. For example, both the carrion crow (*Corvus corone*) and the hooded crow (*Corvus corone cornix*) are found in the British Isles. The carrion crow is completely black and is common in England and southern Scotland. The hooded crow is black with a grey back and belly and is found in the north of Scotland. Hybrids formed from the mating of carrion and hooded crows occupying a narrow region extending across central Scotland (fig 27.15). These hybrids have reduced fertility and serve as an efficient reproductive barrier to gene flow between the populations of the carrion and hooded crows.

In time, selection against cross-breeding may occur, leading to speciation. Since such speciation occurs finally in the same geographical area, this is called **sympatric** (*sym*, together; *patria*, native land) **speciation**.

Sympatric speciation does not involve geographical separation of populations at the time at which genetic isolation occurs. It requires the development of some form of reproductive isolating mechanism which has arisen by selection within a geographically confined area. This may be structural, physiological, behavioural or genetic.

Sympatric speciation is more commonly thought of as providing an explanatory mechanism of how closely related species, which probably arose from a common ancestor by temporary isolation, can coexist as separate species within the same geographical area. For example, in the Galapagos archipelago the finch *Camarhynchus pauper* is found only on Charles Island, where it coexists with a related form *C. psittacula* which is widely distributed throughout the

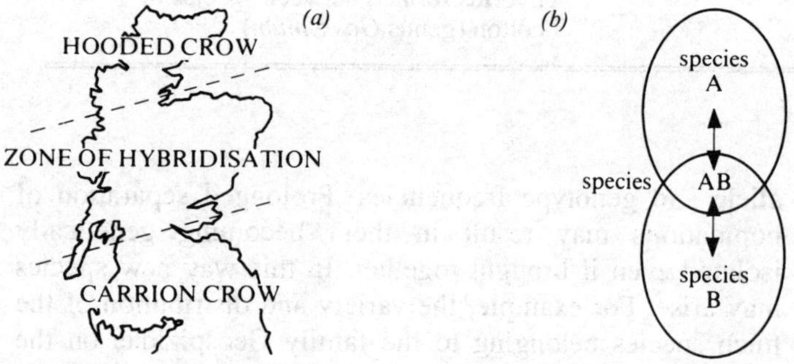

Fig 27.15 *Hybrid barrier as a means of preventing gene flow between two populations. The maintenance of the two crow species is shown to be due to the existence of a zone of hybridisation extending across Scotland as shown in (a). The existence of hybrid barriers between adjacent populations is common and functions as follows. Where the geographical ranges of A and B overlap, mating produces a hybrid with lowered fertility. A will interbreed freely with AB and AB with B but the existence of AB prevents free interbreeding of A and B populations.*

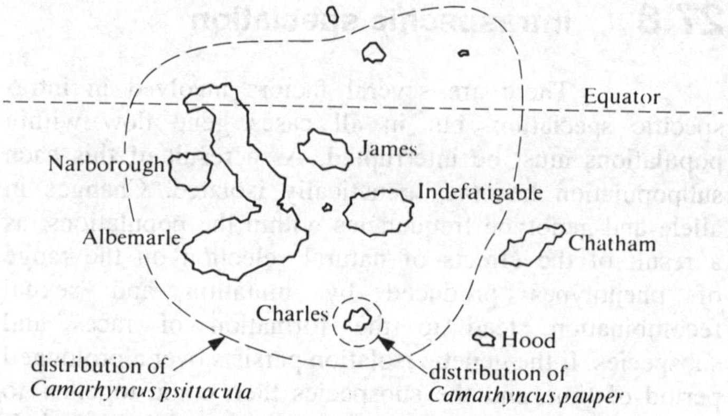

Fig 27.16 *The Galapagos Islands and the distribution of two species of finch illustrating coexistence following allopatric speciation.*

central islands (fig 27.16). The finch species appear to choose their mates on the basis of beak size. The range of beak sizes of *C. pauper* on Charles Island and *C. psittacula* on Albemarle Island are approximately equal, but on Charles Island *C. psittacula* has a longer beak. This difference is significant enough to ensure that the two species, which feed on different foods, appear unattractive to each other during the breeding season. In this way the species remain distinct and are able to coexist.

27.8.4 Ring species

This is a special form of sympatric speciation which occurs at the point where two populations at the extremes of a cline meet up and inhabit the same area, thus 'closing' the ring. For example, gulls of the genus *Larus* form a continuous population between latitudes 50–80 °N, encircling the North Pole. A ring of ten recognisable races or subspecies exist which principally differ in size and in the colour of their legs, backs and wings. Gene flow occurs freely between all races except at the point where the 'ends of the ring' meet at the British Isles. Here, at the extremes of the geographical range, the gulls behave as distinct species, that is the herring gull (*Larus argentatus*) and the lesser black-backed gull (*L. fuscus*). These have a different appearance, different tone of call, different migratory patterns and rarely interbreed. Selection against cross-breeding is said to occur sympatrically.

Sympatric speciation without geographical isolation in sexually reproducing species is unlikely. However, in asexually reproducing organisms, including vegetatively propagated angiosperms, a single mutant so different from its parent population as to be genetically isolated could give rise to a new species sympatrically. An example is polyploidy in *Spartina* (section 24.9.2).

27.3 Ten subspecies of the *Larus argentatus–fuscus* population form a continuous ring extending from the British Isles through Scandinavia, Russia, Siberia, across the Bering Straits, through Alaska and Canada and back to the British Isles. If the subspecies inhabiting the Bering Straits and Alaska was eliminated what predicted effects might this have on the population?

27.9 Interspecific hybridisation

This is a form of sympatric speciation which occurs when a new species is produced by the crossing of individuals from two unrelated species. Fertile hybrids usually appear only in cases of interspecific hybridisation as a result of a form of chromosome mutation known as **allopolyploidy** (section 24.9.2). An example of this was demonstrated by Kapechenko in the case of hybrids formed between the cabbage and the radish. The genetic changes involved in this hybridisation are shown in fig 27.17.

	cabbage *(Brassica oleracea)* $2n = 18$	×	radish *(Raphanus sativus)* $2n = 18$
Meiosis			
Gametes	$n = 9$		$n = 9$
F_1 hybrids		$2n = 18$	

During meiosis in the F_1 hybrids chromosomes from each parent cannot pair together to form homologous chromosomes. The F_1 hybrids are therefore sterile. Occasionally non-disjunction of the F_1 hybrids produces gametes with the diploid set of chromosomes ($2n = 18$).

Gametes	$2n = 18$	×	$2n = 18$
F_2 hybrids		$4n = 36$ (tetraploid) *(Rhaphanobrassica)*	

The F_2 hybrids are fertile. Homologous pairing can occur in meiosis as two sets of parental chromosomes are present. Diploid gametes ($2n = 18$), are produced which possess 9 chromosomes from the parental cabbage and 9 chromosomes from the parental radish.

Fig 27.17 *Stages involved in the hybridisation of the cabbage and the radish.*

Answers and discussion

Chapter 13

13.1 The external solution. Remember that the cell wall is freely permeable to solutions (fig 13.2).

13.2 Zero. The protoplast is not exerting pressure against the cell wall.

13.3 Prokaryotes, fungi and some protoctists such as algae. These organisms are also protected against bursting in solutions with higher water potential, or in pure water.

13.4 (*a*) Cell B (*b*) From cell B to A

(*c*) Cell A at equilibrium: $\psi_P = \psi - \psi_s$
$= -1000\,\text{kPA} - -2000\,\text{kPa}$
$= 1000\,\text{kPA}$

Cell B at equilibrium: $\psi_P = \psi - \psi_s$
$= -1000\,\text{kPa} - -1400\,\text{kPa}$
$= 400\,\text{kPa}$

13.5 −1060 kPa. For intermediate values between those shown in table 13.4, plot a graph of molarity of sucrose solution against solute potential.

13.6 Average ψ_s of beetroot cells is about −1400 kPa.

13.7 ψ beetroot is about −940 kPa.

13.8 A more accurate result can be obtained by taking the mean value of two or more replicates. Some indication of the variation that can be expected between strips is given in table 13.6.

13.9 To prevent evaporation of water, with subsequent increase in concentration of sucrose solutions, and possible drying up of beetroot strips.

13.10 $\psi_P = \psi - \psi_s$
$= -950\,\text{kPa} - -1400\,\text{kPa}$
$= 450\,\text{kPa}.$

Note that different beetroots may have different values of ψ_s and ψ.

13.11 (*a*) The cells of the intact scape are turgid and their walls are therefore tending to expand as pressure potential increases. The thick walls of the epidermal cells are less capable of stretching than the thin walls of the cortex cells and therefore exert a restraining influence on expansion of the cortical cells. The latter are under compression. Cutting the epidermis removes the restraint, each cortical cell expands slightly and there is an overall increase in volume of the cortex which causes the strip to curve outwards.

(*b*) Distilled water has a higher water potential than the scape cells. Water therefore enters the tissue from the distilled water by osmosis, inflating the cortical cells even further and causing outward curvature.

(*c*) The concentrated sucrose solution has a lower water potential than the scape cells. Water therefore leaves the tissue by osmosis, causing greater shrinkage of the cortical cells than the epidermal cells and a bending inwards of the tissue.

(*d*) The dilute sucrose solution must have the same water potential as the scape cells. There is therefore no net gain or loss of water by solution or tissue.

(*e*) Water potential. An outline of the experiment is as follows.

Prepare a dilution series of sucrose solutions from 1 M to distilled water (such as distilled water, 0.2 M, 0.4 M, 0.6 M, 0.8 M and 1.0 M). The typical curvature of freshly cut dandelion scapes should be recorded by drawing and then two pieces of scape placed in each solution in separate labelled petri dishes (two pieces are preferred so that an average can be obtained). Observe and accurately record curvatures (such as by drawing) after equilibrium has been reached (about 30 min). The solution which induces no change in curvature has the same ψ as the average dandelion scape cell immediately after the cut was made.

13.12 Outlines of two suitable experiments are as follows.

Effect of temperature. Cut cubes of fresh beetroot, wash to remove the red pigment from broken cells, and place in beakers of water at different temperatures over a range, say, from 20–100°C. The appearance of red pigment in the water would indicate destruction of the partial permeability of the tonoplast (vacuole membrane) and cell surface membrane, attended by diffusion of the pigment from the cell sap to the water. The time taken for the appearance of a standard amount of pigment would give an indication of the rapidity of breakdown of membrane structure. The colour could be measured in a colorimeter or simply by eye.

Effect of ethanol. Method as above, using a range of ethanol concentrations instead of a range of temperatures.

13.13 (*a*) Leaves contain a very large number of stomata for gaseous exchange and there is little resistance to movement of water vapour through these pores.

(*b*) Leaves have a large surface area (for trapping sunlight and exchanging gases). The greater the surface area, the greater will be the loss of water by transpiration.

13.14 Light intensity increases as the Sun rises, reaching a maximum at midday when the Sun attains its highest point in the sky. Air temperature rises similarly, but it takes about two hours for the heating effect of the Sun to be reflected in a rise in air temperature (mainly because the soil has to heat up first and then radiate heat to the air). The initial rise in transpiration rate between 3 a.m. and 6 a.m., before air temperature rises is due to opening of the stomata in the light. From 6 a.m. onwards transpiration rate is closely correlated with temperature for reasons explained in the text. It is not closely correlated with light intensity, presumably because the stomata are now fully open and any further increase in light intensity has no effect.

During the afternoon, light intensity decreases as the Sun sinks, followed by a drop in temperature with the same lag of about two hours. Transpiration rate decreases both as a result of decreasing temperature and decreasing light intensity, but it is much more closely correlated with a decrease in the latter, probably because this causes stomatal closure. By about 7.30 p.m. it is dark and the stomata are probably closed. Any remaining transpiration is probably cuticular and still influenced by temperature.

13.15 (*a*) Hollow cylinder
(*b*) Solid rod/cylinder providing support
(*c*) Solid rod/cylinder providing support
(*d*) Solid cylinder

13.16 (1) Long tubes formed by fusion of neighbouring cells, with breakdown of cross-walls between them.

(2) No living contents, so less resistance to flow.
(3) Tubes are rigid so do not collapse.
(4) Fine tubes are necessary to prevent water columns from collapsing.

13.17 ψ soil solution > root hair cell > cell C > cell B > cell A > xylem sap

13.18 (*a*) There is a rapid initial uptake of potassium (K^+) at both temperatures (during the first 10–20 min). After 20 min there is a continuous gradual uptake of K^+ at 25 °C but no further uptake at 0 °C. Uptake at 25 °C is inhibited by KCN.
(*b*) The inhibition by KCN indicates that it is dependent on respiration. The uptake at this time is therefore probably by active transport across cell surface membranes into cells.
(*c*) To flush out any existing K^+ ions from the root.

13.19 Rise in respiratory rate is accompanied by a rise in KCl uptake. Once KCl is available, it is therefore apparently taken up by active transport, the energy being supplied by an increased respiratory rate.

13.20 KCN inhibits respiration and therefore inhibits active transport of KCl into the carrot discs.

13.21 Much of the phosphate inside the root was in the apoplast and could therefore diffuse out to the water outside, reversing passive uptake.

13.22 No. The endodermis is a barrier to movement of water and solutes through the apoplast pathway (see section 13.5.2).

13.23 Autoradiography reveals the location of the ion in thin sections. Treat one plant with an inhibitor of active transport (such as low temperature or KCN) and have an untreated control plant: allow them both to take up the radioactive ion. In the treated plant ions will move only passively by way of the cell walls. Autoradiography should show that the radioactive ion tends to penetrate the root only as far as the endodermis, whereas the control should show much greater movement of ions to the tissue inside the endodermis.

13.24 2500 sieve plates per metre:

1 m	=	$10^6\,\mu$m.
400 μm	=	$4 \times 10^2\,\mu$m
$10^6/(4 \times 10^2)$	=	$10^4/4 = 2500$

Chapter 14

14.1 The majority of the pellet consists of red blood cells.

14.2 Solutes, such as Na^+ and K^+ ions, products of digestion etc., plasma proteins, gases (O_2 in red blood cells, CO_2 in rbcs and plasma).

14.3 The oxygenated blood of the systemic circulation reaches the body capillaries at a much higher pressure. This is essential for the efficient function of organs and tissue fluid formation and permits a high metabolic rate and a high body temperature to be maintained. It is essential that a much lower pressure is developed in the pulmonary artery in order to prevent rupture of the delicate pulmonary capillaries.

14.4 Local vasodilation in the wounded area enables more blood carrying oxygen and nutrients to arrive there and speed up the process of repair and replacement. Increased body blood pressure prepares the body of the animal to respond to any further stress more readily and efficiently.

14.5 **Before the race.** Adrenaline is secreted in anticipation of the race. This stimulates vasoconstriction throughout the body in all but the most vital organs. Hence blood pressure is raised. Heart rate is also increased. (Extra blood is also passed to the general circulation from the spleen.)
During the race. Increased metabolic activity takes place during the race, especially in the skeletal muscles. Increased carbon dioxide levels in these regions promote local vasodilation. The increased body temperature further enhances vasodilation. However the general increase in carbon dioxide level in blood is noted by the chemoreceptors of the aorta and carotid bodies which in turn stimulate the vasomotor centre to promote vasoconstriction. This increases blood pressure and therefore speeds up blood flow. Heart rate is also increased and a more complete emptying of the ventricles occurs. Towards the end of the race the muscles will be respiring anaerobically and producing lactic acid (section 9.3.8). Strong contractions of the muscles squeeze the veins and promote faster venous return to the heart.
Recovery. The oxygen debt is paid off and lactic acid removed from the blood system. Tissues subside in activity and the carbon dioxide level decreases. Consequently there is a return to normal of heartbeat and blood pressure.

14.6 (*a*) Increased metabolic activity increases the temperature in a part of the body. This produces a reduction in the affinity of oxygen for haemoglobin and an increased dissociation of oxygen. Thus the dissociation curve is shifted to the right. This is physiologically advantageous as more oxygen is delivered to the active regions.
(*b*) Small mammals possess a much higher metabolic rate than humans and therefore it is appropriate that oxygen should be released much more readily.

14.7 The position of the curve of the fetus relative to that of its mother means that its blood has a greater affinity for oxygen than the maternal blood. This has to be so, as the fetus must obtain all of its oxygen from its mother's blood at the placenta. So, at any given partial pressure of oxygen the fetal blood will take up oxygen from the maternal blood and will always be more saturated with oxygen than the maternal blood. This is just as true for the human fetus.

14.8 This means that the blood has a high affinity for oxygen and that it is able to combine with it at the low oxygen tensions experienced at high altitude. This is another good example of physiological adaptation.

14.9 (1) Carboxyhaemoglobin reaches the lungs and takes up oxygen and forms oxyhaemoglobin.
(2) Oxyhaemoglobin has a lower affinity for H^+ ions than haemoglobin and releases H^+ ions.
(3) H^+ ions combine with hydrogencarbonate ions in the red blood cell so forming carbonic acid.
(4) Carbonic acid dissociates into carbon dioxide and water, catalysed by the enzyme carbonic anhydrase.
(5) As a result of the loss of hydrogencarbonate ions from the red blood cell, further hydrogencarbonate ions diffuse into the red blood cell from the plasma.
(6) More carbonic acid is formed which dissociates into more carbon dioxide and water.
(7) Carbon dioxide diffuses out of the red blood cell and is eventually excreted from the body via the lungs.

Chapter 15

15.1 Some suggestions are:
- reason for test;
- recall interval;
- relevant age group;
- particular groups at risk (older women and those who have had several sexual partners are more at risk);
- cost (free);
- location of cervix;

- test procedure (internal examination involved, with possibility of slight discomfort or pain/embarrassment);
- incidence of positive results;
- treatment of test is positive;
- how to obtain further information.

15.2 Some suggestions are:
- male doctor;
- knowing the GP;
- embarrassment;
- fear of pain/discomfort;
- fear of result.

Chapter 16

16.1 Locomotion is primarily associated with the need to search for food (and is closely associated with the development of a nervous system). Plants are autotrophic, that is make their own organic requirements, so do not need to search for food.

16.2 See table 16.2(ans).

16.3 Various methods are possible. A simple experiment is illustrated in fig 16.3(ans).

16.4 (*a*) The bacteria are aerobic and positively aerotactic. Therefore they swim towards oxygen along a gradient from low oxygen concentration to high oxygen concentration. The highest oxygen concentrations are around the edges of the cover-slip, where oxygen is diffusing into the water from the atmosphere, and adjacent to the algal filament where oxygen is being released as a waste product of photosynthesis.

(*b*) Leave the slide in the dark for about 30 minutes and re-examine. All the bacteria should now be around the edges of the cover-slip because the alga cannot photosynthesise in the dark.

Table 16.2(ans).

Example	*Advantage*
Shoots and coleoptiles positively phototropic	Leaves exposed to the light which is the source of energy for photosynthesis
Roots negatively phototropic	Exposed roots more likely to grow towards soil or equivalent suitable substrate
Shoots and coleoptiles negatively geotropic	Shoots of germinating seeds will grow upwards through soil towards light
Roots positively geotropic	Roots penetrate soil
Rhizomes, runners diageotropic	Helps plants colonise new areas of soil
Dicotyledonous leaves diageotropic	Flat surface of leaf will gain maximum exposure to sunlight (at right-angles to incident radiation)
Lateral roots plagiogeotropic	Large volume of soil exploited and the arrangement of roots provide support (similar to guy-ropes supporting a tent)
Branches plagiogeotropic	Larger volume of space occupied for exploitation of light
Hyphae positively chemotropic	Grow towards food
Pollen tubes positively chemotropic	Grow towards ovule, where fertilisation takes place
Roots and pollen tubes positively hydrotropic	Water essential for all living processes
Tendrils positively haptotropic	Essential for their function of support
Sundew tentacles positively haptotropic	Enables plants to imprison insects which walk over the tentacles
Pollen tubes negatively aerotropic	Another mechanism ensuring that initial growth of the pollen tube is towards the issue of the style (away from air)

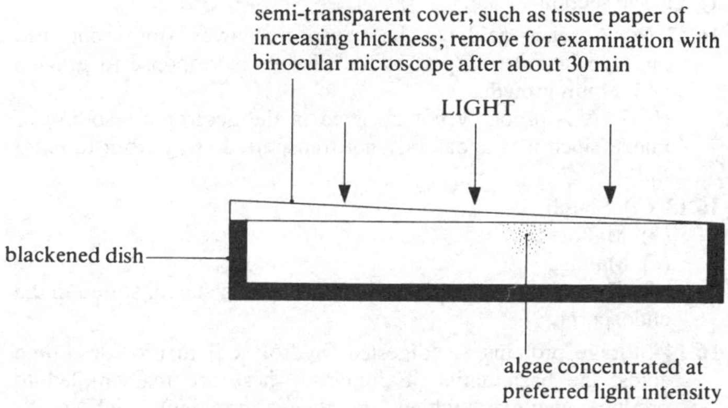

Fig 16.3(ans) *Experiment to demonstrate preferred light intensity of* Euglena *or* Chlamydomonas.

16.5 (*a*) The stimulus of light is detected by the coleoptile tip. Some kind of signal is transmitted from the tip (the receptor) to the region behind the tip (the effector).

(*b*) Experiment *c* was a check on the result from experiment *b* which could have been the result of injury to the coleoptile.

16.6 Further evidence of the existence of a signal, presumably a chemical transmitter substance (hormone), has been obtained. It cannot pass through an impermeable barrier. It moves mainly down the shaded side of the coleoptile. In experiment *b* mica prevented this movement. Light therefore either inhibits production of the hormone, causes its inactivation (stimulates its breakdown) or causes it to be redistributed laterally.

16.7 See fig 16.7(ans).

16.8 The coleoptile tip produces a chemical which diffuses into the agar. It can stimulate growth in the region behind the tip and restores normal growth (experiment *a*). There is little or no lateral transmission of the chemical (experiment *b*) under conditions of uniform illumination or darkness.

16.9 The coleoptile would have grown to the left.

16.10 A 100 ppm B 10 ppm C 1 ppm D 0.1 ppm E 0.01 ppm F zero

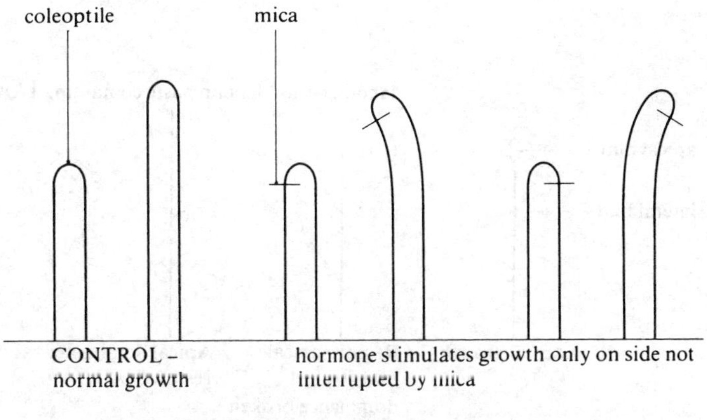

Fig 16.7(ans) *Repetition of Boysen-Jensen's experiments in uniform light. Three experiments are shown; treatment left, result right, in each case.*

16.11 See section 17.5.4.

16.12 (*a*) Abscisic acid can be transported away from root tips, undergo lateral transport in root tissues in response to gravity, and inhibit growth.

(*b*) IAA is probably not involved in the geotropic response of maize since it is apparently not transported away from the root tip.

16.13 (*a*) Starch

(*b*) Maltose

(*c*) Maltase

(*d*) The main food reserve of cereal seeds is starch, stored in the endosperm.

16.14 Storage proteins are digested (hydrolysed) to provide amino acids, the basic units of proteins. These are reassembled to produce enzymes (which are always proteins), such as α-amylase, which are then used to digest the food stores of the endosperm.

16.15 The amylase activity could be associated with micro-organisms present on the fingers or with saliva which has been transferred from mouth to fingers. Note the importance, therefore, of not handling the seeds after their surface sterilisation in this kind of experiment.

16.16 Incubate seeds with radioactive (^{14}C-labelled) amino acids. This results in production of labelled amylases. Alternatively, incubation of seeds with inhibitors of protein synthesis (such as cycloheximide) prevents synthesis of amylase and no amylase activity is then recorded.

16.17 Dissection of the seeds into aleurone and non-aleurone portions should show that the initial appearance of labelled amylase is in the aleurone layer. Alternatively, separate incubation of endosperm with aleurone layers and endosperm without aleurone layer, with starch–gibberellin agar would result in amylase production only in the former (difficult to do in practice).

16.18 One of the best bioassays for gibberellin (quick, reliable and sensitive) involves incubating embryo halves of barley grains with the substance being assayed. After two days the amount of reducing sugar present is proportional to the amount of gibberellin present.

16.19 (*a*) The amino acid is retained by the young leaf and does not move very far from the point of application. In the old leaf some of it is exported via the veins and midrib.

(*b*) The young leaf would use the amino acid to make protein in growth. The old leaf is no longer growing and so is exporting nutrients to other parts of the plant such as roots and young leaves.

(*c*) Amino acids are retained by, or move towards, tissues treated with kinetin. (The reasons for this are unknown, but presumably connected with the maintenance or stimulation of normal cell activity by kinetin.)

16.20 One solution would be to take a plant where applied gibberellin is known to affect stem growth and remove its source of auxin by removing the shoot apex. Gibberellin should then prove ineffective. It is important to demonstrate that the response can be restored by addition of auxin (such as IAA in lanolin paste) as injury might be the reason for lack of response to gibberellins, or another chemical might be involved. Such experiments do demonstrate a total dependence on auxin.

16.21 (*a*) Auxin (IAA)

(*b*) See fig 16.21(ans).

16.22 Small leaves offer less resistance to passage through the soil (leaves of grasses remain inside the coleoptile). The hooked plumule of dicotyledonous plants protects the delicate apical meristem from soil particles. Elongated internodes ensure the maximum chance of reaching light.

16.23 See chapter 7.

16.24 The graph is shown in fig 16.24(ans).

The opposite effects of red and far-red light are demonstrated. Red light exposure of 30 s, at the intensity used in the experiment, completely nullifies the inductive effect of a long night. The effectiveness of red light increases with time of exposure up to 30 s. The red light is reversed by far-red light, although a longer exposure (50 s) was needed to completely reverse the effect. These results suggest that phytochrome is the photoreceptor involved.

16.25 There are several possible methods. Fig 16.25(ans) illustrates one simple solution. Boxes represent light-proof covers, used as appropriate to give short days.

16.26 Lateral bud inhibition or apical dominance is largely controlled by auxins. (See apical dominance, section 16.3.3.)

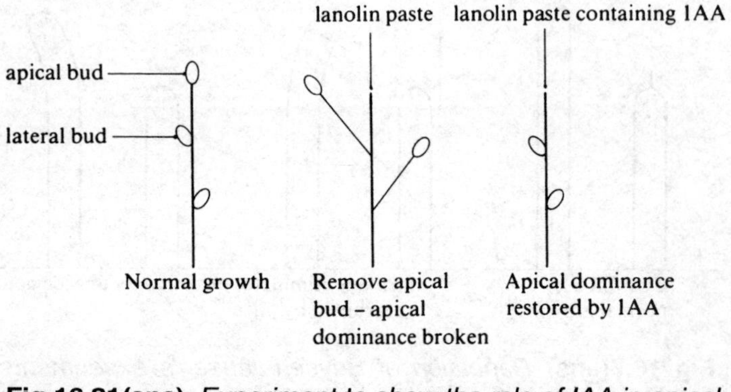

Fig 16.21(ans) *Experiment to show the role of IAA in apical dominance.*

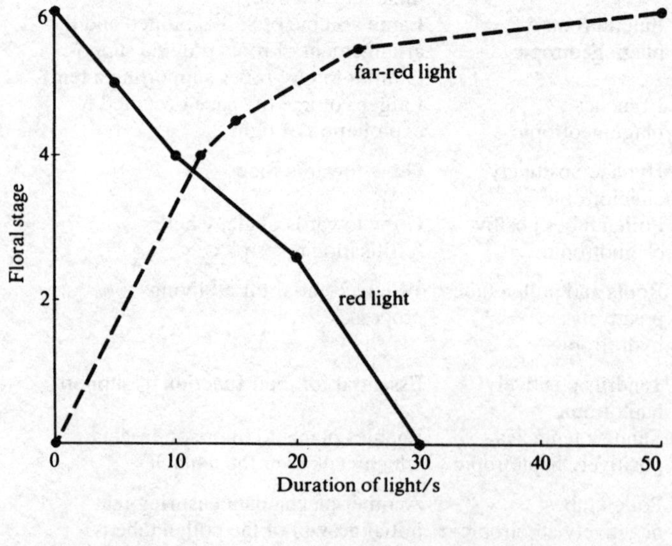

Fig 16.24(ans) *Effects of red light and red/far-red light interruption of long night on flowering of cocklebur.*

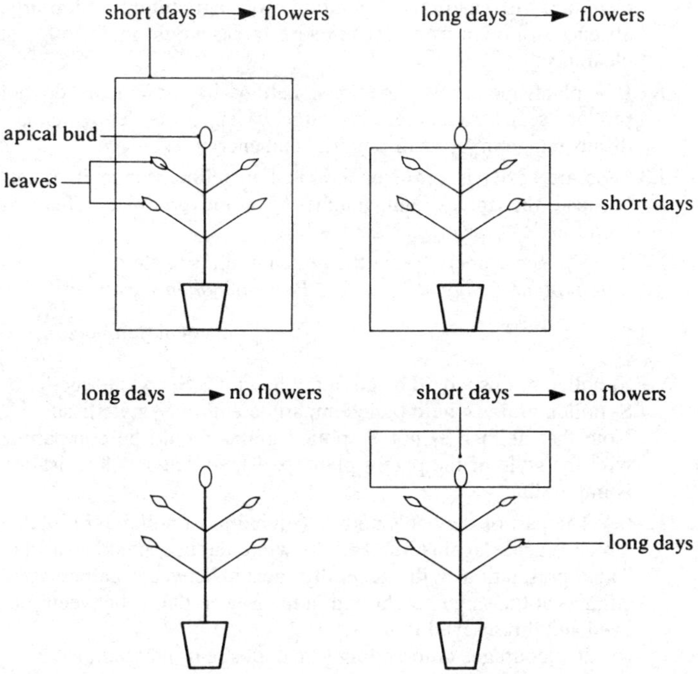

Fig 16.25(ans) *Experiment to determine whether leaves or floral apex are sensitive to the photoperiod that stimulates flowering.*

Chapter 17

17.1 (*a*) A steep concentration gradient of Na^+ ions exists between the outside and inside of the axon and Na^+ ions rapidly diffuse down this gradient.
(*b*) The relatively high negative potential within the axon encourages the inward movement of the positively charged Na^+ ions.

17.2 If the outflow of positive K^+ ions from the axon balanced the inflow of positive Na^+ ions into the axon there would be no change, or perhaps only a slight decrease, in the resting potential. Such a slight change would be insufficient to reach the threshold required to produce an action potential.

17.3 (*a*) Normal sea water
(*b*) One-half sea water
(*c*) One-third sea water
The size of the membrane potential spikes is determined by the number of Na^+ ions entering the axon from the extracellular fluid. The solutions in which (*a*), (*b*) and (*c*) were recorded contained progressively fewer Na^+ ions.

17.4 The resistance of axoplasm decreases with increasing diameter of the axon. As the resistance decreases, the length of the membrane influenced by the local circuit increases and this lengthens the distance between adjacent depolarisations and leads to an increase in conduction velocity.

17.5 The frog is a cold-blooded (poikilothermic) organism, active within the temperature range 4–25°C, whereas the cat, being warm-blooded (homeothermic), maintains a constant temperature of 35 °C. This increase in temperature increases the speed of conduction of the nerve impulses by a factor of three.

17.6 The path taken by light as it passes through the eye is as follows: conjuctiva → cornea → aqueous humour → lens → vitreous humour → retina.

17.7 Light from an object falling onto several rods which are linked to the brain by separate neurones may not have sufficient energy to produce a propagated action potential in each neurone and therefore the light may not be detected. If, however, the same light falls on three rods which are linked to the same neurone supplying the brain, the separate generator potentials produced by the rods would summate and produce an action potential which would be registered in the brain as light.

17.8 When looking directly at an object, light reflected from it passes along the optical axis of the eye and strikes the retina at the fovea which contains cones only. During daylight this will produce a detailed image in the brain due to the high light intensity activating the cones. At night the light intensity would be too low to activate the cones. By looking slightly to one side of the object the reflected light from it will not strike the fovea but a point on the retina to the side of it where there are rods. At night these will be activated by the low light intensity and an image will be produced in the brain.

17.9 The object will appear yellow. Each retina will distinguish one colour only. In one eye, green cones will be stimulated by light of 530 nm and, in the other, red cones will be stimulated by light of 620 nm. Mixing will occur in the brain due to equal stimulation by these colours and the object will appear to be the colour of the average of the combined wavelengths, that is

$$\frac{530 + 620}{2} \text{ nm} = 575 \text{ nm},$$ which corresponds to yellow.

Note that mixing light does not produce the same results as mixing pigments like paints. Blue and yellow light does not, for example, produce the sensation of green (you can work out from table 17.8 what it does produce). Failure to realise this held up theories on colour vision for a long time.

Chapter 18

18.1 (*a*) It allows free movement of the rib cage of the mammal.
(*b*) The flexible suspension enables the animal to withstand the shock sustained by the forelimbs when it lands at the end of a jump.
(*c*) The forelimbs possess a wide range of movement, which is useful for such activities as climbing, cleaning the face, manipulating food and digging.

18.2 A band remains the same length.

18.3 It allows greater movement of Ca^{2+} ions needed for muscle contraction.

18.4 Synchronous. This type of muscle has much more sarcoplasmic reticulum because it requires more nervous impulses to operate it, and each nerve impulse depends on the release of Ca^{2+} ions by the sarcoplasmic reticulum.

18.5 (*a*) Streamlined shape.
(*b*) Smooth surface
- scales overlap each other in an appropriate direction.
- mucus/oily covering thus reducing friction.

(*c*) Various types of fin to promote forward propulsion and stability during swimming.
(*d*) Highly muscular body.
(*e*) Swim bladder in bony fishes.
(*f*) Highly coordinated neuromuscular activity with segmental muscle blocks along the back.

18.6 This position increases the effective length of their limbs. Consequently each stride taken is longer and so propels the body forward over a greater distance. Assuming that the speed of movement of the limbs remains the same, the sprinter will therefore move forward at a faster pace.

Chapter 19

19.1 Rate of transpiration gets lower as atmospheric humidity increases. When humidity is high, the rate of transpiration is low and the plant cannot lose heat and reduce its temperature.

19.2 4 dm^3 sweat lost per day = 4000 cm^3
2.45 kJ of energy is lost per cm^3 of sweat.
Therefore energy lost = 4000 $\times$ 2.45 = 9800 kJ
9800/50 000 $\times$ 100 = 19.6%

19.3 During this period the subject was allowed to equilibrate with his surroundings.

19.4 There is a direct relationship between these two variables which suggests that the rate of sweating is controlled by activity of the hypothalamus.

19.5 The direct relationship between skin temperature and evaporation during the first 20 min established that an equilibrium exists between the two. As the evaporation rate falls, due to the action of the hypothalamus in response to the ingestion of iced water, latent heat of evaporation is not being lost from the skin and this accounts for the observed rise in skin temperature.

19.6 'Fever' is due to the resetting of the 'thermostat' at a higher temperature. Until the core temperature rises to that temperature, the 'normal' body temperature is too low and the body reacts as if it has been cooled. In these conditions the body responds by shivering and the body continues to *feel* cold until the core temperature reaches the temperature of the thermostat in the hypothalamus.

Chapter 20

20.1 Hepatic vein, posterior vena cava, right atrium of heart, right ventricle, pulmonary artery, lungs, pulmonary vein, left atrium of heart, left ventricle, dorsal aorta, renal artery, kidney.

20.2 Protein has 'disappeared' between blood and renal capsule. This is because these molecules are too large to be filtered into the capsule. All the other solutes pass through into the renal capsule in solution and their concentration remains unaffected.

20.3 80%. 20% remains in the tubule.

20.4 All the glucose was reabsorbed. Na^+ stayed at *d*, therefore 80% was reabsorbed (1/5 or 20% remains).

20.5 Flow rate index changed from 20 to 1. Therefore 19/20 or 95% of the remaining was reabsorbed. Therefore concentration of solutes should have increased by a factor of 20. Na^+ concentration increased from *d* to 2*d*, 20*d* – 2*d* = 18/20 reabsorbed = 90% remaining Na^+ reabsorbed. Urea concentration increased from 3*c* to 60*c* which is a 20-fold increase, therefore there was no change in the *amount* of urea in the nephron.

20.6 99% of water was reabsorbed (FRI changed from 100 to 1). 98% of the Na^+ was reabsorbed.

Chapter 21

21.1 (*a*) If the parent plants that produced the pollen grains can be identified, then certain deductions can be made about the climate that such plants would have grown in.
(*b*) Any human interference with the natural vegetation would be reflected in the pollen record. For example, pollen of weed species and agricultural plants, such as wheat, would indicate clearance of natural vegetation for agriculture. Similarly, absence of pollen from trees in some areas would indicate forest clearance.

21.2 If a plant species is dioecious, half of its individuals do not produce seeds. Also, there is a large wastage of pollen which is a disadvantage in terms of material and energy resources.

21.3 Separate sexes is more economical in animals than in plants because the males and females can move about. There is therefore less wastage of gametes.

21.4 (50%). Remembering that the pollen grain is haploid:

Parent plant genotype	*Possible pollen genotypes*	
S_1S_2	S_1 S_2	in equal numbers

S_1 pollen grains would be compatible with S_2S_3 style tissue
S_2 pollen grains would be incompatible with S_2S_3 style tissue
Note that neither S_1 nor S_2 pollen grains would be compatible with the style of the parent plant (S_1S_2), so that self-fertilisation is impossible.

21.5 (*a*) The part of the bee's body receiving most pollen will be that which brushes against the anthers while the bee is taking nectar. Thus pollination will generally occur between anthers and stigmas at the same height within the flower, that is between pin-eyed and thrum-eyed flowers.
(*b*) It encourages outbreeding (the opposite of inbreeding).

21.6 The functions of the cell organelles suggest that the cells manufacture materials for use within the cell. The raw materials for these processes come from the breakdown of materials entering the cell, using enzymes stored in the lysosomes. The synthesised products are packaged by the Golgi apparatus and stored for subsequent usage. Smooth ER is used to produce testosterone, a steroid. Mitochondria supply energy in the form of ATP.

21.7 (*a*) Both are haploid.
(*b*) Sperm: motile, small (2.5 μm diameter), no food store, produced continuously.
Egg: stationary, large (140 μm diameter), some food store, produced once per month in cycles.

21.8 The Rh antigens in the donor's blood would stimulate the mother's immune system to make Rh antibodies. These would not harm the mother, but it would then be impossible to prevent haemolytic disease of the newborn if she had a Rh^+ baby.

21.9 The baby's immune system does not start to function until after birth. Even if it did, it would not have time to react and produce antibodies to cross the placenta into the mother during the period immediately before birth when the placenta is damaged.

21.10 Blood would flow in the reverse direction along the ductus arteriosus.

Chapter 22

22.1 (*a*) There is loss of mass due to respiration of food reserves in the seed.
(*b*) Green leaves have grown and opened above the ground.
(*c*) Photosynthesis. Its rate must now be greater than respiration.
(*d*) This is due to dispersal of fruits and seeds.

22.2 Small seeds have relatively small food reserves; it is therefore important that the growing shoot reaches light quickly so that photosynthesis can start before the reserves are exhausted.

22.3 (*a*) Chlorophyll strongly absorbs red and blue light, but not green and far-red light (see chlorophyll absorption spectrum fig 7.11).

(*b*) Red light stimulates lettuce seed germination, but far-red light inhibits it (section 16.4.2). Seeds under a leaf canopy, where the light will be enriched in far-red, might therefore be inhibited from germinating until a break in the canopy ensures that they will not be too shaded for efficient photosynthesis and growth.

22.4 At the onset of germination, food reserves in the barley grain, principally starch, with some protein, are mobilised. Starch is converted to sugars, and proteins to amino acids, and these are translocated to the embryo for use in growth. Therefore, endosperm dry mass decreases while embryo dry mass increases.

At the same time there is an overall loss in dry mass during the first week. This is due to aerobic respiration, which consumes sugar, in both endosperm and embryo (though to a greater extent in the latter). At about day 7 the first leaf emerges and starts to photosynthesise. The resulting increase in dry mass more than compensates for respiration losses so that a net increase in dry mass is observed. At the same time the rate of growth of the embryo, now a seedling, increases.

22.5 (*a*) There is a gain in dry mass of 8.6 g, calculated as follows.
Mass of seeds = 51.2 g
Mass of fatty acid = 51.2/2 = 25.6 g
M_r fatty acid = 256
Therefore 1 mole = 256 g, so 25.6 g = 0.1 mole
From the equation,
0.1 mole fatty acid ⟶ 0.1 mole sugar + 0.5 mole water + 0.4 mole carbon dioxide
M_r sugar = 342
Therefore 25.6 g fatty acid ⟶ 34.2 g sugar + water + carbon dioxide
Water is not included in the dry mass and carbon dioxide is lost as a gas, therefore the gain in dry mass = (34.2 – 25.6) g = 8.6 g.
(*b*) Respiration would result in a decrease in dry mass. In reality there would still be an increase in dry mass.
(*c*) Volume of carbon dioxide evolved from the seeds = 8.96 dm^3 at STP, calculated as follows:
from the equation, 0.1 mole fatty acid ⟶ 0.4 mole carbon dioxide. 0.4 mole carbon dioxide occupies 0.4 × 22.4 dm^3 at STP = 8.96 dm^3.
(*d*) By hydrolysis, catalysed by a lipase. The other component of the lipid is glycerol.
(*e*) 51 carbon atoms (the lipid would be tripalmitin: the fatty acid is palmitic acid). Each lipid molecule comprises three fatty acid molecules, each with 16 carbon atoms, plus one glycerol molecule with three carbon atoms.
(*f*) Sucrose or maltose.
(*g*) Oxygen reaches the storage tissue by diffusion through the testa and micropyle.

22.6 (*a*) The dominant food store is lipid, which comprises about 70% of the dry mass of the seeds before germination. By day 4 the mass of lipid is starting to decrease and the mass of sugar begins to rise. Lipid is therefore being converted to sugar and translocated to the embryo. Note that no sugars can be formed by photosynthesis since germination occurs in darkness. At day 5 the RQ of the embryo = 1, indicating that the embryo is respiring the sugar derived from the lipids. At the same time, the cotyledons (RQ = 0.4–0.5) are gaining energy from the conversation of lipid to sugar, and possibly from oxidation of sugar and fatty acids.

$C_{18}H_{34}O_3 + 13O_2 \longrightarrow C_{12}H_{22}O_{11} + 6CO_2 + 6H_2O$ + energy
ricinoleic acid (fatty acid derived from a lipid) — sucrose

RQ = 6/13 = 0.46
Conversion of lipid to sugar takes place with an increase in dry mass, so total dry mass of the seedlings increases up to 6 or 7 days. Beyond this point, the lipid reserves are running low, so rate of use of sugar starts to exceed the rate of production. Net mass of sugar, and total mass of seedlings, then starts to decrease. Sugar is used in respiration and in anaerobic reactions.
(*b*) At day 11, the RQ of the whole seedlings would probably be slightly less than 1.0. It is a combination of two reactions: the main one is the oxidation of sugar in respiration, RQ = 1, but there would probably still be a small contribution from the conversion of lipid to sugar, RQ 0.4–0.5.

22.7 Normally insufficient oxygen is able to penetrate the testa to allow exclusively aerobic respiration: the RQ is a combination of the RQ for aerobic respiration (probably about 1.0) and that for anaerobic respiration, which is infinity (∞). Removal of the testa allows more rapid penetration of oxygen by diffusion, with a consequent increase in aerobic respiration and decrease in RQ. Ethanol is a product of anaerobic respiration so less accumulates when the testas are removed.

Chapter 23

23.1 (*a*) Meiosis
(*b*) W – interphase
X – telophase I
Y – telophase II
(*c*) Gamete cells

23.2 See fig 23.2(ans).

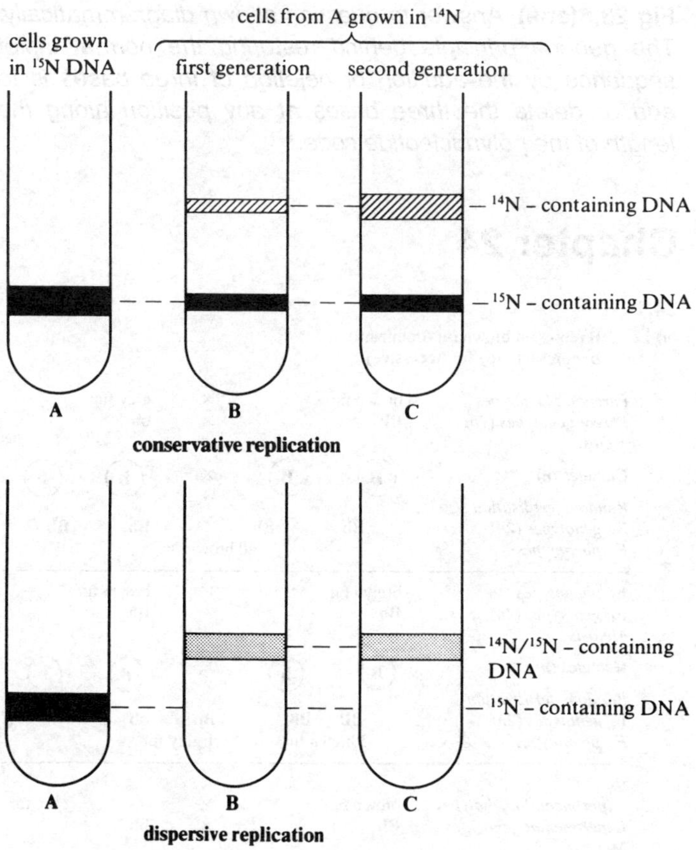

Fig 23.2(ans) *Diagrams explaining two further theories of DNA replication. The appearance of DNA in a caesium chloride density gradient according to the theories presented in fig 23.22.*

23.3

Bases	A	G	T	C
A	AA	AG	AT	AC
G	GA	GG	GT	GC
T	TA	TG	TT	TC
C	CA	CG	CT	CC

23.4 4 bases used once = $4 \times 1 = 4^1 = 4$
4 bases used twice = $4 \times 4 = 4^2 = 16$
4 bases used three times = $4 \times 4 \times 4 = 4^3 = 64$
therefore the mathematical expression is x^y where x = number of bases and y = number of bases used.

23.5 See fig 23.5(ans).

23.6 UAC AAG CUC AUG GUA CAU UGC

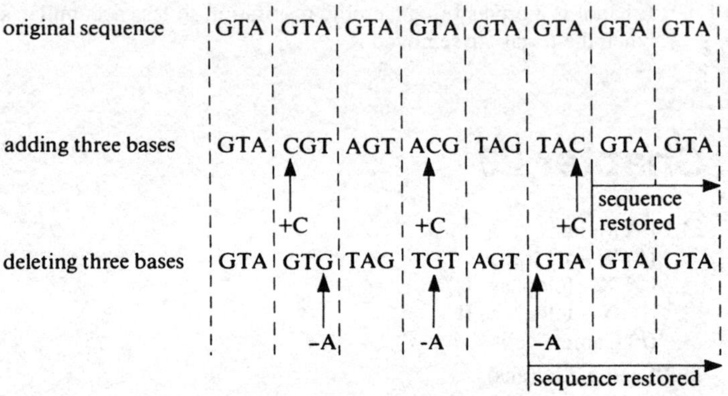

Fig 23.5(ans) *Answer to question shown diagrammatically. The general principle behind restoring the normal triplet sequence by the addition or deletion of three bases is to add or delete the three bases at any position along the length of the polynucleotide code.*

Chapter 24

24.1

(a) Let: **B** represent brown fur (dominant)
b represent grey fur (recessive)

Parental phenotypes	brown fur		×	grey fur	
Parent genotypes (2n)	**BB**		×	**bb**	
Meiosis					
Gametes (n)	(**B**)	(**B**)	×	(**b**)	(**b**)
Random fertilisation					
F_1 genotypes (2n)	**Bb**	**Bb**		**Bb**	**Bb**
F_1 phenotypes			all brown fur		

F_1 phenotypes	brown fur		×	brown fur	
F_1 genotypes (2n)	**Bb**			**Bb**	
Meiosis					
Gametes (n)	(**B**)	(**b**)	×	(**B**)	(**b**)
Random fertilisation					
F_2 genotypes (2n)	**BB**	**Bb**		**Bb**	**bb**
F_2 phenotypes	3 brown fur		:	1 grey fur	

(b)

Experimental phenotypes	brown fur		×		grey fur
Experimental genotypes (2n)	**Bb**		×	**bb**	
Meiosis					
Gametes (n)	(**B**)	(**b**)	×	(**b**)	(**b**)
Random fertilisation					
Offspring genotypes (2n)	**Bb**	**Bb**		**bb**	**bb**
Offspring phenotypes	1 brown fur		:		1 grey fur

In the case of monohybrid inheritance, the offspring from a heterozygous genotype crossed with a homozygous recessive genotype produce equal numbers of offspring showing each phenotype: in this case 50% brown fur and 50% grey fur.

24.2 If an organism having an unknown genotype is testcrossed with a homozygous dominant organism, all the offspring will show the dominant characteristics in the phenotype, as shown below.

Let: **T** represent a dominant allele
t represent a recessive allele

Testcross phenotypes	homozygous × homozygous	heterozygous × homozygous
Testcross genotypes (2n)	**TT** × **TT**	**Tt** × **TT**
Meiosis		
Gametes (n)	(**T**)(**T**) × (**T**)(**T**)	(**T**)(**t**) × (**T**)(**T**)
Random fertilisation		
Offspring genotyes (2n)	**TT TT TT TT**	**TT TT Tt Tt**
Offspring phenotypes	all tall (homozygous)	all tall (½ homozygous, ½ heterozygous)

24.3 *(a)* If short black hair appeared in the F_1 phenotypes, then short hair must be dominant to long hair and black hair must be dominant to white.

(b) Let: **B** represent black hair
b represent white hair
S represent short hair
s represent long hair

F_1 phenotypes short black hair × short black hair

F_1 genotypes (2n) **SsBb** × **SsBb**

Meiosis

Gametes (n) (as shown by ♂ and ♀)

Random fertilization (as shown by Punnett square)

F_2 genotypes (2n) (as shown in each square)

♀ \ ♂	(**SB**)	(**Sb**)	(**sB**)	(**sb**)
(**SB**)	**SB SB** ■	**Sb SB** ■	**sB SB** ■	**sb SB** ■
(**Sb**)	**SB Sb** ■	**Sb Sb** □	**sB Sb** ■	**sb Sb** □
(**sB**)	**SB sB** ■	**Sb sB** ■	**sB sB** ▮	**sb sB** ▮
(**sb**)	**SB sb** ■	**Sb sb** □	**sB sb** ▮	**sb sb** ▯

F_2 phenotypes	9 short black hair	:	3 short white hair	:	3 long black hair	:	1 long white hair
(Symbols)	■		□		▮		▯

24.4 Let:

R, r and **S, s** represent two allelomorphic pairs of genes controlling flower colour

Parental phenotypes	purple × purple
Parental genotypes (2n)	**RrSs** × **RrSs**
Meiosis	
Gametes (n) (as shown by ♂ and ♀)	
Random fertilization (as shown by Punnett square)	
F₂ genotypes (2n) (as shown in each square)	

♀ \ ♂	(RS)	(Rs)	(rS)	(rs)
(RS)	RS RS ●	Rs RS ●	rS RS ●	rs RS ●
(Rs)	RS Rs ●	Rs Rs ○	rS Rs ●	rs Rs ○
(rS)	RS rS ●	Rs rS ●	rS rS ○	rs rS ○
(rs)	RS rs ●	Rs rs ○	rS rs ○	rs rs ○

Offspring phenotypic ratio 9 purple : 7 white
(Symbols) ● ○

24.5 The two alleles segregate during metaphase I and anaphase I.

24.6 The number of different combinations of chromosomes in the pollen gamete cells is calculated using the formula 2^n, where n is the haploid number of chromosomes.

In crocus, since $2n = 6$, $n = 3$.

Therefore, combinations $= 2^3 = 8$.

24.7 The F_1 phenotypes show that purple flower and short stem are dominant and red flower and long stem are recessive. The approximate ratio of 1:1:1:1 in a dihybrid cross suggests that the two genes controlling the characteristics of flower colour and stem length are not linked and the four alleles are situated on different pairs of chromosomes (see below).

Let:

P represent purple flower
p represent red flower
S represent short stem
s represent long stem

Since the parental stocks were both homozygous for both characters the F_1 genotypes must be **PpSs**.

Testcross phenotypes	purple flower, short stem × red flower, long stem:
Testcross genotypes (2n)	**PpSs** × **ppss**
Meiosis	
Gametes (n) (as shown by ♂ and ♀)	
Random fertilisation (as shown in Punnett square)	

♀ \ ♂	(PS)	(Ps)	(pS)	(ps)
(ps)	PS ps	Ps ps	pS ps	ps ps

Offspring genotypes (2n) (listed in each square)

Offspring phenotypes 1 purple flower, short stem: 1 purple flower, long stem: 1 red flower, short stem: 1 red flower, long stem

24.8 (*a*) Homologous chromosomes
(*b*) Body colour and wing length

24.9 Out of the 800 seeds produced, only 24 show the results of crossing-over between the genes for seed colour and seed shape. In the other 776, the alleles for seed colour and seed shape have remained linked as shown by their approximate 1:1 ratio.

Hence the crossover value is $(24/800) \times 100 = 3\%$. Therefore the distance between the genes for seed colour and seed shape is 3 units.

24.10 (*a*) Let:

N represent normal wing (dominant)
n represent miniature wing (recessive)
R represent red eye (dominant)
r represent white eye (recessive)
XX represent female fly (♀)
XY represent male fly (♂)

(i) *Parental phenotypes*	miniature wing, red eye ♂	× normal wing, white eye ♀
Parental phenotypes (2n)	$X^{nR}Y$	× $X^{Nr}X^{Nr}$
Meiosis		
Gametes (n)	(X^{nR}) (Y)	(X^{Nr}) (X^{Nr})
Random fertilisation		
F_1 genotypes (2n)	$X^{nR}X^{Nr}$ $X^{nR}X^{Nr}$	$X^{Nr}Y$ $X^{Nr}Y$
F_1 phenotypes	normal wing, red eye ♀	normal wing, white eye ♂

(ii) Assuming no crossing-over between the genes for wing length and eye colour in the female, the following results are likely to appear:

F_1 phenotypes	normal wing, white eye ♂		× normal wing, red eye ♀	
F_1 genotypes (2n)	$X^{Nr}Y$		× $X^{nr}X^{Nr}$	
Meiosis				
Gametes (n)	(X^{NR})	(Y)	× (X^{nR})	(X^{Nr})
Random fertilisation				
F_2 genotypes (2n)	$X^{Nr}X^{nR}$	$X^{Nr}X^{Nr}$	$X^{nR}Y$	$X^{Nr}Y$
F_2 phenotypes	normal wing, red eye ♀	normal wing, white eye ♀	miniature wing, red eye ♂	normal wing, white eye ♂

(*b*) The lack of a 1:1:1:1 ratio of phenotypes resulting from this cross indicates crossing-over between the genes for wing length and eye colour in the female

Testcross phenotypes	normal wing, red eye ♀ × miniature wing, white eye ♂
Testcross genotypes (2n)	$X^{nR}X^{Nr}$ × $X^{nR}Y$
Meiosis	
Gametes (n) (as shown by ♀ and ♂)	
Random fertilization (as shown in Punnett square)	
Offspring genotypes (2n) (as listed in squares)	

♂ \ ♀	(X^{nR})	(X^{Nr})	(X^{nr})	(X^{NR})
(X^{nr})	$X^{nR}X^{nr}$ ♀	$X^{Nr}X^{nr}$ ♀	$X^{nr}X^{nr}$ ♀	$X^{NR}X^{nr}$ ♀
(Y)	$X^{nR}Y$ ♂	$X^{Nr}Y$ ♂	$X^{nr}Y$ ♂	$X^{NR}Y$ ♂

Offspring phenotypes	wing:	miniature	normal	miniature	normal
	eye:	red	white	white	red
Experimental results		36	35	18	17

The alleles for wing length and eye colour are shown on the two F_1 female (X) chromosomes in the explanation above. Crossing-over between the alleles will give the recombinant genotypes shown above. Out of 106 flies, 35 show recombination of alleles (18 + 17), therefore the crossover value is 35/106 = approximately 30%.

24.11 Magpie moth

Let: **N** represent normal colour (dominant)
n represent pale colour (recessive)

Parental phenotypes pale colour male × normal colour female

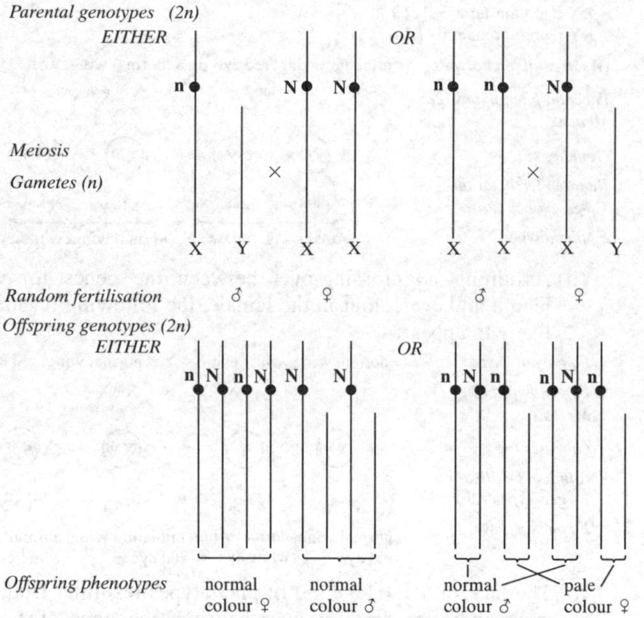

From the results for the offspring phenotypes it is seen that the heterogametic sex in the magpie moth is the female.

Cat

Let: **B** represent black colour (dominant)
n represent yellow colour (recessive)

Parental phenotypes black colour male × yellow colour female

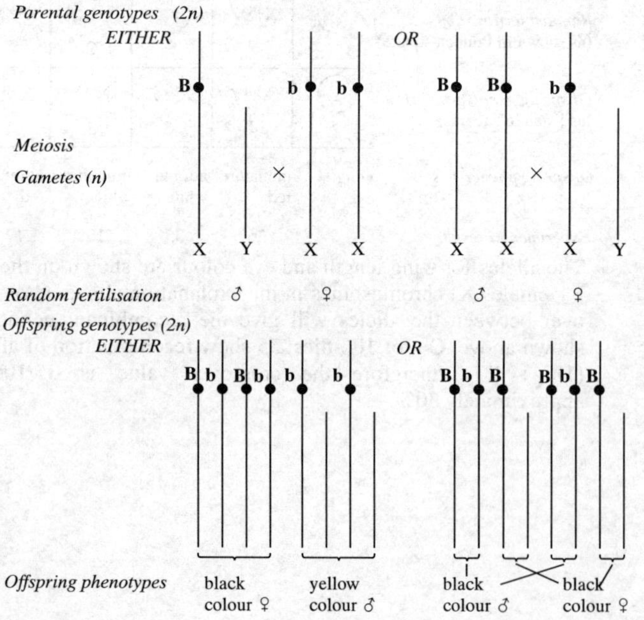

From the results for the offspring phenotypes it is seen that the heterogametic sex in the cat is the male.

24.12 Let:

B represent black coat colour
G represent ginger coat colour
XX represent female cat
XY represent male cat

Parental phenotypes	ginger-coat male	×	black-coat female
Parental genotypes (2n)	X^GY	×	X^BX^B
Meiosis			
Gametes (n)	(X^G) (Y)	×	(X^B) (X^B)
Random fertilisation			
F_1 genotypes (2n)	X^GX^B X^GX^B		X^BY X^BY
F_1 phenotypes	tortoiseshell-coat colour ♀		black coat colour ♀

(The parental female must be homozygous for black-coat colour since this is the only condition to produce a black-coat phenotype).

F_1 phenotypes	black-coat male	×	tortoiseshell-coat female	
F_1 genotypes (2n)	X^BY	×	X^GX^B	
Meiosis				
Gametes (n)	(X^B) (Y)	×	(X^G) (X^B)	
Random fertilisation				
F_1 genotypes (2n)	X^BX^G	X^BX^B	X^GY	X^BY
F_2 phenotypes	tortoiseshell-coat colour ♀	black coat colour ♀	ginger coat colour ♂	black coat colour ♂

24.13 (*a*)

Let:
I represent the gene for blood group
A represent the allele A } (equally dominant)
B represent the allele B }
o represent the allele O (recessive)

Parental phenotypes	blood group A	×	blood group B	
Parental genotypes (2n)	I^AI^o	×	I^BI^o	
Meiosis				
Gametes (n)	(I^A) (I^o)	×	(I^B) (I^o)	
Random fertilisation				
Offspring genotypes (2n)	I^AI^B	I^AI^o	I^oI^B	I^oI^o
Offspring phenotypes blood groups	AB	A	B	O

(b) There is a probability of $\frac{1}{4}$ (25%) that each child will have blood group A. So the probability that both will have blood group A is $\frac{1}{4} \times \frac{1}{4} = 1/16$ (6.25%).

24.14 Let:

P represent pea comb
R represent rose comb
a single **P** allele and a single **R** allele occurring together produce walnut comb
a double homozygous recessive genotype produces single comb
W represent white feathers (dominant)
w represent black feathers (recessive)

If eight different phenotypes are produced from the cross, each parent must possess as many heterozygous alleles as possible. Hence the genotypes are as shown below:

Parental phenotypes	black, rose-comb cock	×	white, walnut-comb hen
Parental genotypes (2n)	**wwRrpp**	×	**WwRrPp**

Meiosis

Gametes (n) (as shown by ♀ and ♂)

Random fertilisation (as shown in Punnett square)

Offspring genotypes (2n) (as shown in squares)

♂ \ ♀	(WRP)	(WRp)	(WrP)	(Wrp)	(wRP)	(wRp)	(wrP)	(wrp)
(wRp)	**WRP wRp** ○	**WRp wRp** △	**WrP wRp** ○	**Wrp wRp** △	**wRP wRp** ●	**wRp wRp** ▲	**wrP wRp** ●	**wrp wRp** ▲
(wrp)	**WRP wrp** ○	**WRp wrp** △	**WrP wrp** □	**Wrp wrp** ◌	**wRP wrp** ●	**wRp wrp** ▲	**wrP wrp** ■	**wrp wrp** ◉

Offspring phenotypes (Symbols)

3 white, walnut-comb: 3 black, walnut-comb: 3 white, rose-comb: 3 black, rose-comb: 1 white, pea-comb:
○ ● △ ▲ □
1 black, pea-comb: 1 white, single-comb: 1 black, single-comb
■ ◌ ◉

24.15 Since both dominant alleles, **W**, white, and **B**, black, are present in the heterozygous F_1 genotype, and the phenotype is white, it may be concluded that the alleles show an epistatic interaction where the white allele represents the epistatic gene. The F_2 generation is shown below.

Using the symbols given in the question

F_1 phenotypes	White cock	×	white hen
F_1 genotypes (2n)	**WwBb**	×	**WwBb**

Meiosis

Gametes (n) (as shown by ♂ and ♀)

Random fertilisation (as listed in Punnett square)

F_2 genotypes (2n) (as listed in the squares)

♀ \ ♂	(WB)	(Wb)	(wB)	(wb)
(WB)	**WB WB** ○	**Wb WB** ○	**wB WB** ○	**wb WB** ○
(Wb)	**WB Wb** ○	**Wb wB** ○	**wB Wb** ○	**wb Wb** ○
(wB)	**WB wB** ○	**Wb WB** ○	**wB wB** ●	**wb wB** ●
(wb)	**WB wb** ○	**Wb wb** ○	**wB wb** ●	**wb wb** ⊘

F_2 genotypes (Symbols)

12 white colour: 3 black colour: 1 brown colour
○ ● ⊘

Chapter 25

25.1 Economic: relieve economic burden of (i) individual families who have to care for sufferers, and (ii) the community which has to bear the cost of treatment on the National Health Service.
Social: relieve suffering of (i) individuals affected, and (ii) their families.

25.2 (*a*) Couple 3 and 4 are phenotypically normal but have an affected daughter. If the gene were dominant, at least one of the parents would be affected. The gene is unlikely to have arisen as a spontaneous mutation because it is already in the family (individual 5).
(*b*) Individual 9 is an affected woman born to phenotypically normal parents. Given that the gene is recessive, both parents must have a copy of the gene. If it were sex-linked, the father would show the symptoms of PKU because the Y chromosome only carries genes for sex.
(*c*) Individuals 3 and 4 are definitely carriers.
(*d*) Individuals 1, 2, 6, 7, 8, 10, 11, 12 and 13 *could* all be carriers. It is impossible to prove a person is not a carrier on the basis of normal breeding patterns. A biochemical test would be needed.
(*e*) A common answer to this question is 1 in 2 (or 50%) since a 1 : 2 : 1 ratio of affected : carrier : normal would be expected among the children of individuals 3 and 4. However, individuals 10, 11 and 12 know they are not PKU sufferers and so are either carriers or normal. In this situation there is a 2 in 3 chance of being a carrier (66.7%). Your advice might also include visiting a GP to enquire about the possibility of genetic counselling and being tested for being a carrier.

25.3 Its excess phenylalanine is removed by the mother while it is in the uterus. It takes a few days for the levels of phenylalanine to build up.

25.4 If it were not essential, it might be manufactured in the body. A phenylalanine-restricted diet would then be useless for controlling the condition.

25.5 A high level of phenylalanine in the mother's blood would cross the placenta and lead to restricted brain development in the fetus.

25.6 (*a*) End uncertainty/plan for future/may affect decision on whether to have children.
(*b*) Possible death sentence/would keep wondering if symptoms had started, e.g. if you forgot something/still don't know when symptoms will start even if you know you have the disease/no treatment available.

25.7 Age of onset is after it is passed on to children.

25.8 C2, C3 and C4

Chapter 26

26.1 A control experiment in which each variable was systematically eliminated.

26.2 Redi's basic assumption was that the presence of 'worms' was due to the entry of the flies through the open flasks.

26.3 Sealing the broths would prevent the entry of organisms to the vessels. Lack of air within the vessels may have deprived organisms of oxygen for respiration.

26.4 Pasteur's basic assumptions were that each generation of organisms develops from the previous generation and not spontaneously.

Chapter 27

27.1 The carrier genotype is the heterozygous genotype. The Hardy–Weinberg equation is used to calculate genotype frequencies. The equation may be represented as

$$p^2 + 2pq + q^2 = 1$$

where

p^2 = frequency of homozygous dominant genotype,
$2pq$ = frequency of heterozygous genotype,
q^2 = frequency of homozygous recessive genotype.

The incidence of cystic fibrosis in the population appears in individuals with the homozygous recessive genotype, hence q^2 is 1 in 2000 or 1/2000 = 0.0005.

Therefore $q = \sqrt{0.005}$
$= 0.0224.$
Since $p + q = 1$
$p = 1 - q$
$= 1 - 0.0224$
$= 0.9776.$

The frequency of the heterozygous genotype ($2pq$) is therefore
$2 \times (0.9776) \times (0.0224)$
$= 0.044$
1 in 23
$\approx 5\%$

Approximately 5% of the population are carriers of the recessive gene for cystic fibrosis.

27.2 *Fasciola hepatica*, the liver fluke, is a parasite which infests sheep. It has an intermediate host, the snail *Limnaea truncatula*, which lives in fresh water and damp pastures. Draining ponds and wet areas would bring about a change in environmental conditions which would exert a selection pressure tending to eliminate *Limnaea*. As the numbers of the snail fall this would reduce the numbers of available hosts which would lead to a decrease in the numbers of the parasite, *Fasciola*.

27.3 Reduced selection pressure at the extremes of each new population would favour increased variability. New phenotypes may show adaptations to the areas previously occupied by the eliminated subspecies and spread inwards to occupy the vacated ecological niche. The initial geographical separation of the cline may have initiated allopatric speciation. If the ring was reformed, gene flow may be impossible due to genetic isolation and each subpopulation would diverge genetically even further to form distinct species, as is the present case in the British Isles where the species exist sympatrically. If the genetic isolation between the two subpopulations was not too great, hybrids may form when the subpopulations were reunited. This zone of hybridisation may act as a reproductive barrier as is the case with the carrion and hooded crows.

Appendix 1
Biological chemistry

A1.1 Elementary chemistry

An **atom** is the smallest part of an element that can take part in a chemical change. An **element** is a substance which cannot be split into simpler substances by chemical means, for example the elements carbon, oxygen and nitrogen. A **compound** is a substance which contains two or more elements chemically combined, as shown below.

Compounds	*Elements*
water	hydrogen and oxygen
glucose	carbon, hydrogen and oxygen
sodium chloride	sodium and chlorine

A **molecule** is the smallest part of an element or compound which can exist alone under normal conditions, such as H_2, O_2, CO_2 and H_2O.

A1.1.1 Structure of the atom

All elements are made up of atoms. The word 'atom' comes from the Greek word *atomos* meaning indivisible.

The particles which make up atoms are protons, neutrons and electrons, details of which are given in table A1.1. Protons and neutrons have equal mass, and together make up the mass of the nucleus. The electrons have very much lower mass than the protons or neutrons and, when the mass of an atom is being considered, usually only the mass of the nucleus is taken into account.

A neutron is composed of a proton and an electron bound together, so that its charge is neutral.

Atoms are electrically neutral because the number of protons in a nucleus equals the number of electrons orbiting around it.

The number of protons in the nucleus of an atom is called the **atomic number** of that element. It also equals the number of electrons in an atom. For an individual atom, the number of protons plus the number of neutrons equals the **mass number**.

The atoms of some elements exist in different forms called **isotopes** which have different mass numbers (section A1.3). The average mass of an atom is the **relative atomic mass** (A_r) and is usually an average value for a natural mixture of the isotopes. For example, chlorine is made up of a mixture of isotopes of mass numbers 35 and 37; the proportions of the isotopes are such that naturally occurring chlorine has a relative atomic mass of 35.5.

The known elements, of which there are over 100, can be listed in order of ascending atomic number as shown in table A1.2. As indicated in the table the electrons are arranged in successive shells around the nucleus. The first shell can hold up to two electrons (being nearest the nucleus it is the smallest), the second shell can hold up to eight electrons, the third shell can hold up to 18 electrons and the fourth shell can hold up to 32 electrons.

Table A1.1 The locations, masses and charges of protons, neutrons and electrons.

Particle	*Location*	*Mass*	*Charge*
proton	the dense central core of the atom, forming the nucleus which has a diameter about 1/100 000 that of the atom	1 unit (1.7×10^{-24}g)	positive (+1)
neutron	(as above)	1 unit	neutral (0)
electron	in 'orbits' around the nucleus	1/1870 unit (9.1×10^{-28}g)	netative (–1)

Table A1.2 The first 20 elements in order of ascending atomic number.

Atomic number	*Mass number*	*Relative atomic mass**	*Element*	*Symbol*	*Arrangement of electrons*
1	1	1.0	hydrogen	H	1
2	4	4.0	helium	He	2
3	7	6.9	lithium	Li	2,1
4	9	9.0	beryllium	Be	2,2
5	11	10.8	boron	B	2,3
6	12	12.0	carbon	C	2,4
7	14	14.0	nitrogen	N	2,5
8	16	16.0	oxygen	O	2,6
9	19	19.0	fluorine	F	2,7
10	20	20.2	neon	Ne	2,8
11	23	23.0	sodium	Na	2,8,1
12	24	24.3	magnesium	Mg	2,8,2
13	27	27.0	aluminium	Al	2,8,3
14	28	28.1	silicon	Si	2,8,4
15	31	31.0	phosphorus	P	2,8,5
16	32	32.1	sulphur	S	2,8,6
17	35	35.5	chlorine	Cl	2,8,7
18	40	39.9	argon	Ar	2,8,8
19	39	39.1	potassium	K	2,8,8,1
20	40	40.1	calcium	Ca	2,8,8,2

*Relative atomic mass (A_r) was formerly atomic weight.
Figures for A_r are given to the nearest decimal place.
The symbols for some common elements are, in ascending order of atomic number, chromium (Cr), manganese (Mn), iron (Fe), cobalt (Co), nickel (Ni), copper (Cu), zinc (Zn), arsenic (As), bromine (Br), molybdenum (Mo), silver (Ag), cadmium (Cd), iodine (I), barium (Ba), platinum (Pt), mercury (Hg), lead (Pb), radium (Ra), uranium (U), plutonium (Pu).

There are also further shells in the larger atoms, but these need not be considered here. The arrangements of electrons in shells for the first 12 elements are shown in fig A1.1.

Any element with an electronic configuration in which the outermost shell is full is particularly unreactive. Hence helium and neon (table A1.2) are so unreactive that they seldom form compounds with other atoms. Thus they are called noble gases.

The tendency of all other elements is to attain full electron shells through reaction with other elements. When two atoms react to form a compound there are basically two types of bond that can form between them, ionic and covalent bonds.

A1.1.2 Ionic bonding

This is a process in which electrons are transferred from one atom to another. Consider sodium reacting with chlorine (fig A1.2*a*). The sodium atom loses an electron and therefore has an overall positive charge of +1 (its nucleus contains 11 positively charged protons and is surrounded by 10 negatively charged electrons). Similarly the chlorine atom has gained an electron and now has an overall negative charge of –1. Both have full, and therefore stable, electron shells.

These charged particles are no longer true atoms and instead are called **ions**. Hence the sodium ion is represented as Na^+ and the chloride ion as Cl^-. Positively charged ions are called **cations**, and negatively charged ions **anions**. The resulting compound is sodium chloride (formula NaCl) but no molecules of NaCl exist. Instead there is an association of sodium and chloride ions in equal numbers (ionic formula Na^+Cl^-). Compounds like this which are formed by

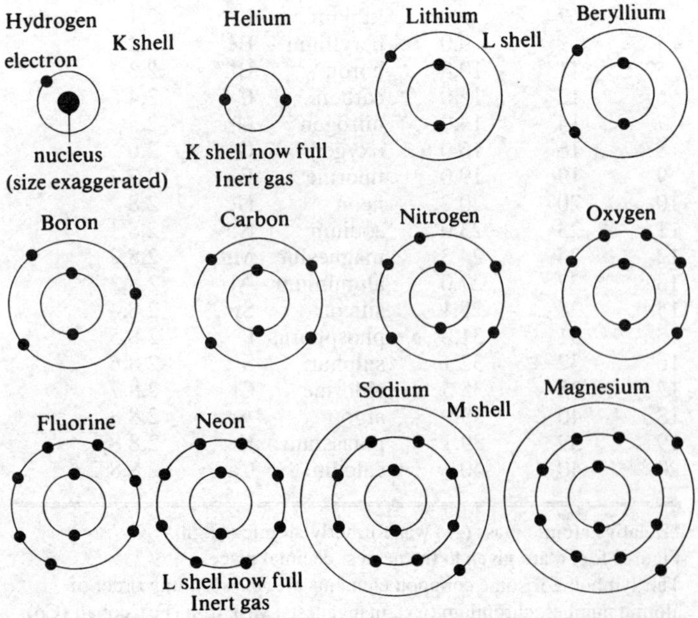

Fig A1.1 *The arrangement of electrons in shells for the first 12 elements. (The nucleus is omitted from all except hydrogen.)*

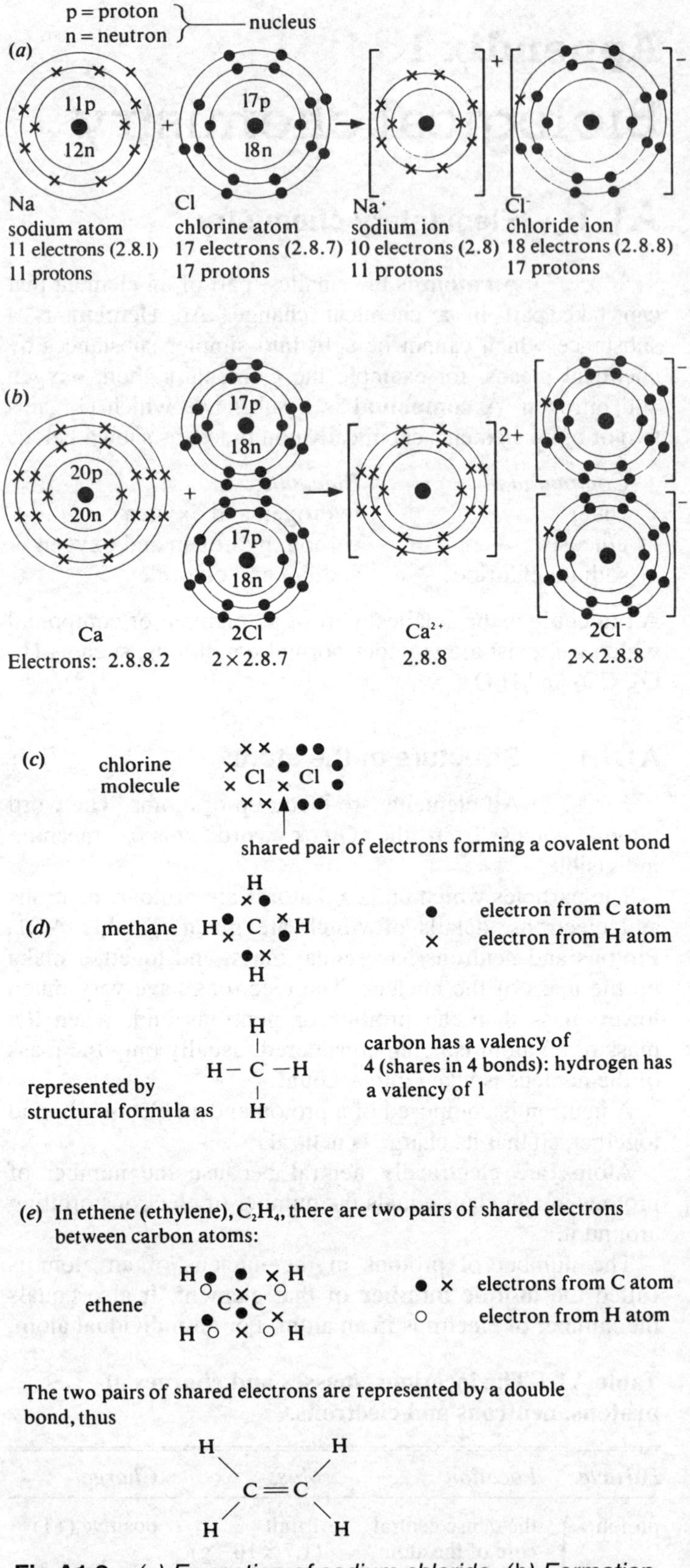

Fig A1.2 *(a) Formation of sodium chloride. (b) Formation of calcium chloride. (c) Formation of the covalent chlorine molecule. (d) Formula of methane. (e) Formula of ethene. For clarity, electrons from different atoms have been given different symbols (x, • or o). In reality, all electrons are indistinguishable. Only the outer shells of electrons are shown in (c), (d), and (e).*

the transfer of electrons are called **ionic compounds**. They are usually formed when metals react with non-metals. The metal produces a cation and the non-metal an anion. Salts are all ionic compounds.

Another typical example is calcium chloride (formula $CaCl_2$) (fig A1.2*b*). Here two electrons are lost from the calcium atom, and one gained by each chlorine atom. Therefore the calcium ion is represented as Ca^{++} or Ca^{2+}.

The number of electrons transferred (lost or gained) is the **valency**, sometimes called the **combining power**. Therefore sodium and chlorine have a valency of one and calcium has a valency of two. The number of plus or minus signs shown for an ion is therefore equal to its valency; for example, potassium and hydroxyl ions have a valency of one and so are written as K^+ and OH^-, magnesium and sulphate ions have a valency of two and so are written as Mg^{2+} and SO_4^{2-}, and aluminium has a valency of three and so is written as Al^{3+}.

Ionic formulae

Ionic compounds do not exist as molecules but as collections of ions. The ionic formula shows the ratio in which elements are combined in an ionic compound; for example, the formula for the compound aluminium oxide is Al_2O_3, meaning that Al^{3+} and O^{2-} ions are in the ratio of 2 : 3. If an ion with more than one atom is present, such as in SO_4^{2-} (sulphate), the number of that ion present, if more than one, is indicated by using brackets, for example $Al_2(SO_4)_3$; but Na_2SO_4 needs no brackets as only one sulphate ion is involved.

A1.1.3 Covalent bonding

In this type of bonding electrons are not donated or received by the atoms concerned; instead, they are shared. Consider two chlorine atoms; each has seven electrons in its outer shell (electron configuration 2, 8, 7). In covalency, two chlorine atoms contribute one electron each to a shared pair of electrons, making a chlorine molecule, formula Cl_2. In this way both atoms obtain an approximation to the noble gas configuration and molecules are produced, not ions (fig A1.2*c*). The shared pair of electrons is conventionally written as a single bond thus Cl–Cl. Chlorine is said to have a covalency of one (it shares one of its electrons). Another example is methane, CH_4. Carbon has an atomic number of six with four electrons in its outer shell (2, 4); hydrogen has an atomic number of one and has only one electron (fig A1.2*d*).

In ethene (ethylene), C_2H_4, there are two pairs of shared electrons between carbon atoms, and the two pairs are represented by a double bond (fig A1.2*e*). In some compounds there are three pairs of shared electrons, a triple bond, such as in ethyne (acetylene), C_2H_2.

Covalent compounds are far more common than ionic compounds in biological systems.

The valencies of some common elements and the charges on common ions are given in table A1.3.

Table A1.3 Valencies of some elements and charges of some ions.

(*a*) Valencies of some elements.

1	*2*	*3*	*4*
F	O	N	C
C1	S	P	Si

(*b*) Charges of some ions of single elements

−2	*−1*	*+1*	*+2*	*+3*
O^{2-}	F^-	H^+	Mg^{2+}	$A1^{3+}$
	Cl^-	Na^+	Ca^{2+}	
	Br^-	K^+	Zn^{2+}	
			Ba^{2+}	
	I^-	Cu^+ copper (I)	Cu^{2+} copper (II)	
			Fe^{2+} iron (II)	Fe^{3+} iron (III)
			Pb^{2+} lead (II)	

(*c*) Charges of some ions of more than one element

−3	*−2*	*−1*	*+1*
PO_4^{3-} phosphate (V)	SO_4^{2-} sulphate	NO_3^- nitrate	NH_4^+ ammonium
	CO_3^{2-} carbonate	NO_2^- nitrite	
		OH^- hydroxyl or hydroxide	
		HCO_3^- hydrogen carbonate (formerly bicarbonate)	

Formulae of covalent compounds

For simple covalent compounds the formula represents the number of each type of atom present in one molecule; for example CO_2 (carbon dioxide) means that one atom of carbon is combined with two atoms of oxygen.

A1.1.4 Chemical equations

When chemical equations are written, not only must the correct formulae for the chemicals be used but also the equations must be balanced, that is there must be the same number of atoms of each element on the right-hand side of the equation as on the left. This can be done in the following way.

(1) Write a word equation, for example

methane + oxygen ⟶ carbon dioxide + water

(2) Write the correct formulae.

$$CH_4 + O_2 \longrightarrow CO_2 + H_2O$$

(3) Check if the equation balances. The equation in (2) does not, as there are three oxygen atoms on the right side and only two on the left; also there are four hydrogen atoms on the left and two on the right.

(4) Balance the equation, if necessary, using large numbers in front of the relevant formulae and remembering that the formulae cannot be altered.

$CH_4 + 2O_2 \rightarrow CO_2 + 2H_2O$

($2O_2$ means two molecules of oxygen (4 atoms of oxygen); $2H_2O$ means two molecules of water (4 atoms of hydrogen, 2 atoms of oxygen).)

Ionic equations

Equations for reactions between ionic compounds can be written simply as ionic equations. Consider the following reaction:

$2NaOH$ +	H_2SO_4 →	Na_2SO_4 +	$2H_2O$
sodium hydroxide	sulphuric acid	sodium sulphate	water

(All three compounds are aqueous.)
The equation can be rewritten to show the ions present:

$$2Na^+ + 2OH^- + 2H^+ + SO_4^{2-} \rightarrow 2Na^+ + SO_4^{2-} + 2H_2O$$

Removing ions common to both sides of the equation (not involved in the reaction) gives the ionic equation:

$$2OH^- + 2H^+ \rightarrow 2H_2O$$

This is the only reaction which has taken place.

A1.1.5 Acids, bases, salts, pH and buffers

A hydrogen atom consists of one electron and one proton. If the electron is lost it leaves a proton, and a proton may therefore be regarded as a hydrogen ion, usually written H^+. An **acid** is a substance which can act as a **proton donor** and this is a substance which can ionise to form H^+ as the cation. For the purpose of this book an acid will be defined as a substance which ionises in water to give H^+ ions as the cation. A **strong** acid, such as hydrochloric acid, is one which undergoes almost complete **dissociation** (separation of its constituent ions). It is therefore a more efficient proton donor than a **weak** acid, such as ethanoic acid or carbonic acid, in which only a small proportion of the acid molecules dissociate to give hydrogen ions:

$HCl \rightleftharpoons H^+ + Cl^-$ (hydrochloric acid) $CH_3COOH \rightleftharpoons CH_3COO^- + H^+$ (ethanoic acid)

Typical properties of acids are as follows.

- Many acids react with the more reactive metals such as magnesium or zinc to produce hydrogen.
- Acids are neutralised by bases to give salts and water only.
- Almost all acids react with carbonate to give carbon dioxide.
- Acids have a sour taste in dilute solution, for example ethanoic acid (vinegar).
- Solutions of acids give a characteristic colour with **indicators**; for example, they turn blue litmus red.

A **base** is a substance which reacts with an acid to form a salt and water only (otherwise defined as a **proton acceptor**). Most bases are insoluble in water. Those that are soluble in water form solutions called **alkalis**, such as sodium hydroxide, calcium hydroxide and ammonia solutions. Other typical properties are as follows.

- Bases usually have little action on metals.
- Bases react with aqueous solutions of the salts of most metals to precipitate an insoluble hydroxide.
- Reactions with ammonium salts gives ammonia.
- Solutions of alkalis give a characteristic colour with indicators; for example, they turn red litmus blue.

A **salt** is a compound in which the replaceable hydrogen of an acid has been partly or wholly replaced by a metal. An example is sodium chloride where the hydrogen atom of hydrochloric acid has been replaced by an atom of sodium. When a salt dissolves in water its constituent ions dissociate, that is they become free ions separated from one another by water molecules.

The pH scale

The acidity or alkalinity of a solution is related to the concentration of hydrogen ions in the solution. This is expressed as its pH (p represents a mathematical operation, H represents hydrogen). The pH is defined as the logarithm to the base of 10 of the reciprocal of the hydrogen ion concentration. Pure water contains 1×10^{-7} moles of hydrogen ions per decimetre cubed (litre). The pH of water is therefore $\log(1/10^7) = 7$.

A pH of 7.0 represents a neutral solution (at room temperature). A pH of less than 7.0 represents an acidic solution, and a pH of more than 7.0 represents an alkaline solution.

The pH scale ranges from about –1 to about 15 (usually 0–14). The scale is logarithmic, so a change in pH of one unit represents a ten-fold change in hydrogen ion concentration.

Cells and tissues normally require a pH value close to 7 and fluctuations of more than one or two units from this usually cannot be tolerated. Mechanisms therefore exist to keep the pH of body fluids as constant as possible. This is partly achieved by buffers.

Buffers

A **buffer solution** is a solution containing a mixture of a weak acid and its soluble salt. It acts to resist changes in pH. Such changes can be brought about by dilution, or by addition of acid or alkali.

As acidity (hydrogen ion concentration) increases, the free anion from the salt combines more readily with free hydrogen ions, removing them from solution. As acidity decreases, the tendency to release hydrogen ions increases. Thus the buffer solution tends to maintain a constant, balanced hydrogen ion concentration. For example

$$\underset{\text{hydrogen phosphate}}{HPO_4^{2-}} + H^+ \underset{\text{high pH}}{\overset{\text{low pH}}{\rightleftharpoons}} \underset{\text{dihydrogen phosphate}}{H_2PO_4^-}$$

Some organic compounds, notably proteins, can function as buffers and they are particularly important in blood.

A1.2 Oxidation and reduction

All biological processes require energy to drive them and the biologist must be aware of the various reactions that make energy available for such processes. Chemical reactions which liberate energy are termed **exothermic** or **exergonic**, whilst those that use energy are **endothermic** or **endergonic**. Synthetic (anabolic) processes are endergonic (such as photosynthesis), whilst breakdown (catabolic) processes are exergonic (such as respiration). The sum of the catabolic and anabolic reactions of the cell occurring at any one moment represents its metabolism.

A cell obtains the majority of its energy by oxidising food molecules during the process of respiration. **Oxidation** is defined as the loss of electrons. The opposite process, in which electrons are gained, is called **reduction**. The two always occur together, electrons being transferred from the **electron donor**, which is thereby oxidised, to the **electron acceptor**, which is thereby reduced. Such reactions are called **redox** reactions, and they are widespread in the chemical processes of biological systems. Several mechanisms of oxidation and reduction exist, as described in the following sections.

A1.2.1 Oxidation

Oxidation may occur directly by the addition of molecular oxygen to a substance, which is then said to be oxidised.

$$A + O_2 \rightarrow AO_2$$

However, the most common form of biological oxidation is when hydrogen is removed from a substance (**dehydrogenation**).

$$AH_2 + B \xrightarrow[\text{dehydrogenase}]{\text{dehydrogenation}} A + BH_2$$

A has been oxidised and *B* reduced.

A cell possesses a number of substances called **hydrogen carriers** which act like B in the equation above. Each dehydrogenation is catalysed by a specific dehydrogenase enzyme and the carriers are arranged in a linear order such that their level of potential energy (section A1.6.2) decreases from one end of the line (which is where the hydrogen atoms enter) to the other. This means that each time hydrogen atoms are transferred from one carrier to another of lower potential energy, a small quantity of energy is liberated. In some cases this can be incorporated into ATP.

In some reactions, each atom of hydrogen (which can be regarded as a hydrogen ion or proton, H^+, plus one negatively charged electron, e^-) is not transferred as a whole. Here the process only involves the transfer of electrons. For example

$$\underset{\text{iron(II) chloride}}{2FeCl_2} + Cl_2 \rightleftharpoons \underset{\text{iron(III) chloride}}{2FeCl_3}$$

Iron(II) ions are oxidised to iron(III) ions by the loss of one electron per ion, or

$$\underset{\text{reduced}}{Fe^{2+}} \rightleftharpoons \underset{\text{oxidised}}{Fe^{3+} + e^-}$$

The electrons are transferred to the chlorine molecule which is thereby reduced and forms two chloride ions. So the complete ionic equation is

$$\underset{\text{reduced}}{2Fe^{2+}} + \underset{\text{oxidised}}{Cl_2} \rightleftharpoons \underset{\text{oxidised}}{2Fe^{3+}} + \underset{\text{reduced}}{2Cl^-}$$

Cytochromes, which contain iron, work in mitochondria and convey electrons (derived from hydrogen atoms which have split into hydrogen ions and electrons) along an electron transport chain. Here the electrons are passed from less electronegative atoms to more electronegative ones. The products of such reactions possess less potential energy than the reactants and the difference is liberated as energy which is utilised in one form or another. At the end of the chain is a cytochrome that also contains copper. This copper transfers its electrons directly to atmospheric oxygen and is thereby oxidised itself.

$$2Cu^+ - 2e^- \rightleftharpoons 2Cu^{2+}$$
$$2H^+ + 2e^- + \tfrac{1}{2}O_2 \rightleftharpoons H_2O$$
$$2H^+ + 2Cu^+ + \tfrac{1}{2}O_2 \rightleftharpoons Cu^{2+} + H_2O$$

A1.2.2 Reduction

Reduction occurs when molecular oxygen is removed from a substance, or hydrogen atoms are gained by a substance, or when an electron is gained by a substance.

A1.3 Isotopes

Atoms of some elements exist in more than one form, the different forms being called **isotopes** (*iso*, same; *topos*, place; same position in the periodic table of elements). All the isotopes of a given element have the same number of protons and electrons (same atomic number) and therefore have identical chemical properties. However, they differ in the neutron content of their nuclei and therefore have different masses. To distinguish between isotopes, mass number is added to the symbol of the element; for example oxygen has three naturally occurring isotopes, ^{16}O, ^{17}O and ^{18}O. One isotope is usually much commoner than the others; for example the ratio of presence of $^{16}O : ^{17}O : ^{18}O$ is 99.759% : 0.037% : 0.204%.

Some combinations of protons and neutrons give nuclei which can exist without change for a long time. These nuclei are said to be stable. Other combinations give unstable nuclei, that is they tend to break up or decay, emitting particles and radiation. Such nuclei are said to be radioactive and can easily be detected using various instruments such as Geiger–Müller tubes and counters, scintillation counters and so on. As the atomic number of the nucleus increases, so the relative number of neutrons needed for stability increases. For example, the 92 protons of uranium need 138 neutrons to be stable. Isotopes of uranium with larger numbers of neutrons are radioactive, their nuclei being unstable.

The rate of decay is often expressed as the **half-life**. This is the time during which, on average, half the atoms present will decay. For example, ^{14}C has a half-life of 5570 years.

Radioactive isotopes can emit 'rays' of particles and radiation of three kinds.

(1) α **particles**. These are identical to helium nuclei, that is they consist of two protons plus two neutrons. They have two positive charges.

An example of α particle emission is given below (see also fig A4.1). (The upper number on the left-hand side of each element's symbol is the mass number and the lower is the atomic number.)

$$^{238}_{92}U \longrightarrow ^{234}_{90}Th + ^{4}_{2}He$$

The ^{238}U nucleus ejects an α particle, thus losing four units of mass and two of charge and becoming an isotope of thorium.

(2) β **particles**. These are fast-moving electrons derived from the nucleus when a neutron changes to a proton. β particles have a single negative charge (see also fig A4.1).

An example of β particle emission is

$$^{234}_{90}Th \longrightarrow ^{234}_{91}Pa + \beta(e^-)$$

The thorium nucleus ejects an electron; one of its neutrons therefore becomes a proton. Its atomic mass is unchanged, but its atomic number (number of protons) is increased by one, and it becomes an isotope of protactinium.

(3) γ **rays**. These are very short wavelength electromagnetic waves associated with α and β decay. They have a high energy and are very difficult to stop, passing, for example, through thick sheets of lead.

α particles are easily stopped, for example by air or by a thin sheet of paper. β particles have a greater penetrating power but are stopped by a thick sheet of aluminium or a thin sheet of lead. The particles and the radiation can be harmful to living organisms if they are in close enough proximity to cells.

A1.4 Solutions and the colloidal state

Solutions have at least two parts or phases: the **continuous** (**dispersion**) phase or **solvent**, in which the **disperse** phase or **solute** is supported or dissolved.

In 1861 Graham distinguished between two types of solute which he called **crystalloids** and **colloids**. These he differentiated according to whether the solute molecules were capable of passing through a parchment (partially permeable) membrane. In fact, in biological systems there is no clear distinction between them since the biological solvent is always water and the properties of any water-based solution depend upon the size of the solute molecule and the effect of gravity. Three types of solution may be identified.

(1) **True solution**. In this, solute particles are small and comparable in size to the solvent molecules, forming a homogeneous system, and the particles do not separate out under the influence of gravity; for example salt solution and sucrose solution. Such solutions are regarded by chemists as forming one phase.
(2) **Colloidal solution**. The solute particles are large by comparison with those of the solvent, forming a heterogeneous system, but the particles still do not separate out under gravity; for example clay in water.
(3) **Suspension** or **emulsion**. The solute particles are so large that they cannot remain dispersed against gravitational force unless the suspension is stirred continuously. A suspension has solid particles whereas an emulsion has liquid particles in the disperse phase, for example a silt suspension.

The three systems above can be described as **dispersion systems** because the particles are dispersed through a medium. Dispersion systems can involve all three states of matter, namely solid, liquid and gas; for example gas in water (soda water), sodium chloride in water (salt solution) and solid in solid (copper in zinc as brass). All can be called solutions, but generally this term refers to those systems that have a liquid solvent.

Many biological systems exist as colloidal solutions which are either hydrophobic or hydrophilic: a **hydrophobic sol** is water-hating, such as clay or charcoal in water, and a **hydrophilic sol** is water-loving, such as starch, table jelly, gelatin and agar-agar. Most of the colloidal solutions occurring in organisms, such as protein solutions, are hydrophobic sols. The viscosity of a hydrophilic sol, such as table jelly, can be increased by making it more concentrated or by lowering the temperature. As viscosity increases, the sol may set and is then called a **gel**. A gel is a more or less rigid colloidal system, although there is no sharp distinction between sol and gel. Ionic composition, pH and pressure are other factors which can affect sol–gel transformations and all may be important in living cells under certain circumstances. Characteristics of the colloidal state are shown in table A1.4.

Table A1.4 Characteristics of the colloidal state.

Phenomenon	*Physical properties*	*Biological properties*
Dialysis (the separation of particles by partially permeable membranes)	Colloids cannot pass through such membranes	Colloidal cytoplasm is retained within the cell surface membrane. Large molecules cannot pass through and therefore must be changed to smaller molecules such as starch to glucose
Brownian movement	Very small particles viewed under a microscope vibrate without changing position. This movement is due to the continuous bombardment of the molecules by the solvent molecules, for example Indian ink in water	All living cytoplasm is colloidal and minute particles in the cell can be seen to exhibit Brownian movement
Filtration	The movement of particles during this process depends on the size of the molecule. The actual size of the particles can be measured by varying the size of the filter pores	
Solute potential (osmotic potential)	Hydrophobic colloids develop extremely small solute potentials in solution. Hydrophilic colloids develop small but measurable solute potentials in solution	
Precipitation	Hydrophobic colloids can be precipitated (coagulation). A positively charged colloid will precipitate a negatively charged colloid. Electrolytes have the same effect	Dilute acids or rennet coagulate casein of milk, as in cheese-making. Precipitation of pectin from cell walls occurs during jam-making. Heat irreversibly coagulates egg albumen
Surface properties	Colloidal particles present an enormous surface area to the surrounding solvent. The surface energy is considerable here and this energy can cause molecules to aggregate at the surface interface. This is called **adsorption**. For example, charcoal is used to adsorb gases in respirators, or dyes from solution. This phenomenon can be used for stabilising colloidal sols, such as in the addition of egg to mayonnaise, or soap to oil-based insecticides	Adsorption of molecules occurs in the cell colloids particularly in cells near to, or concerned in the uptake of, ions, such as cortical cells of the root
Gel to sol and reverse changes	The sol state is fluid and the gel state is solid; for example, starch in hot water is a colloidal sol but when cooled it becomes a colloidal gel. Change of pH, temperature, pressure and the presence of metallic ions can also be equally effective	Clotting of blood is a sol to gel change with the gelation of the protein fibrinogen. Heat changes egg albumen from sol to gel
Imbibition	The absorption of fluid by colloids is called **imbibition**; for example gelatin taking in water	The testa of a dry seed or cellulose in cell walls take up water by imbibition. The release of gametes from sex organs, such as antheridia, is due to imbibitional swelling

A1.5 Diffusion and osmosis

Molecules and ions in solution can move passively and spontaneously in a particular direction as a result of diffusion. Osmosis is a special type of diffusion. It is discussed in detail in section 5.9.8. Such movements in living organisms do not require the expenditure of energy, unlike active transport. Another type of movement, namely mass flow, is considered in chapter 13.

A1.5.1 Diffusion

Diffusion involves the random and spontaneous movement of individual molecules and ions. For example, if a bottle of concentrated ammonia solution is left on a bench and the stopper removed, the smell of ammonia soon penetrates the room. The process by which the ammonia molecules spread is diffusion, and although individual molecules may move in any direction, the net direction is outwards from the concentrated source to areas of lower concentration. Thus diffusion may be described as the *movement of molecules or ions from a region of their high concentration to a region of their low concentration down a concentration gradient*. In contrast to mass flow it is possible for the net diffusion of different types of molecule or ion to be in different directions at the same time, each type moving down its own concentration gradient. Thus in the lungs, oxygen diffuses into the blood at the same time as carbon dioxide diffuses out into the alveoli; mass flow of blood through the lungs, however, can be in one direction only. Also, smaller molecules and ions diffuse faster than larger ones, assuming equal concentration gradients. There is a modified form of diffusion called facilitated diffusion which is described in section 5.9.8.

A1.6 Laws of thermodynamics

All chemical changes are governed by the laws of thermodynamics. The first law, called the **law of conservation of energy**, states that for any chemical process the total energy of the system and its surroundings always remains constant. This means that energy is neither created nor destroyed, and that if the chemical system gains energy then that quantity of energy must have been provided by the surroundings of the system, and vice versa. Therefore energy may be redistributed, converted into another form or both, but never lost.

The second law states that when left to themselves, systems and their surroundings usually approach a state of maximum disorder (**entropy**). This implies that highly ordered systems will readily deteriorate unless energy is used to maintain their order. All biological processes obey, and are governed by, these two laws of thermodynamics.

A1.6.1 Energy relations in living systems

Consider the decomposition of hydrogen peroxide into water and oxygen:

$$2H_2O_2 \rightleftharpoons 2H_2 + O_2$$

Generally, pure hydrogen peroxide will persist for a long time with no significant decomposition. For decomposition to occur, molecues must, on collision, have energy greater than a certain level, called the **activation energy**, E_a. Once this energy is reached, changes in the bonding pattern of the molecules occur and the reaction may generate enough energy to proceed spontaneously. The activation energy required varies with different reactants.

Addition of heat energy is the most common way in which activation energy is reached, and most reactants require quantities far greater than that provided by normal temperatures. For example, it is only when hydrogen peroxide is heated to 150 °C that it decomposes rapidly enough to cause an explosive reaction. Water and oxygen are produced and energy is liberated. The overall energy change which occurs in this reaction is called the **free energy change** (ΔG). As the reaction is very rapid, and the products water and oxygen generally do not re-unite to form hydrogen peroxide, the energy liberated is lost from the chemical system to the environment. Therefore ΔG is negative (fig A1.3).

Obviously high temperatures would be lethal to biological systems and so enzymes are used instead. Acting as catalysts, they reduce the activation energy required by the reactants and therefore increase the rates of chemical reactions without addition of energy, such as a rise in temperature, to the system. Catalase is the enzyme that promotes rapid decomposition of aqueous solutions of hydrogen peroxide in living systems.

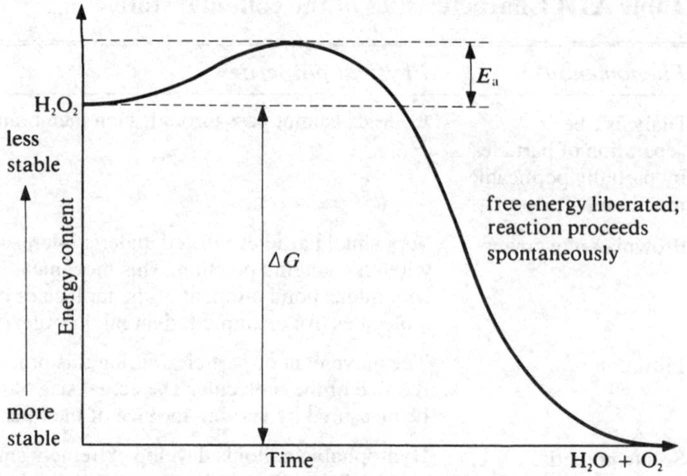

Fig A1.3 *Activation energy.*

A1.6.2 Potential energy

This is defined as the energy which a system possesses because of its position and condition. Consider a stationary ball at the top of a slope (fig A1.4). The ball possesses an amount of gravitational potential energy equal to the work done to place it there originally. When it rolls down the slope some of the potential energy of the ball is converted into kinetic energy. When the ball comes to rest at the bottom of the slope it possesses less potential energy than it had at the top. In order to restore the ball's potential energy to its original value, energy from the environment must be used to raise it once more to the top of the slope.

Potential energy for biological systems is built up by green plants during the production of sugar when photosynthesis occurs (fig A1.5). During this process, solar energy boosts certain electrons from their orbits with the result that they acquire potential energy. When oxidation of the sugar takes place during respiration, the potential energy of the electrons is used in various forms by living systems.

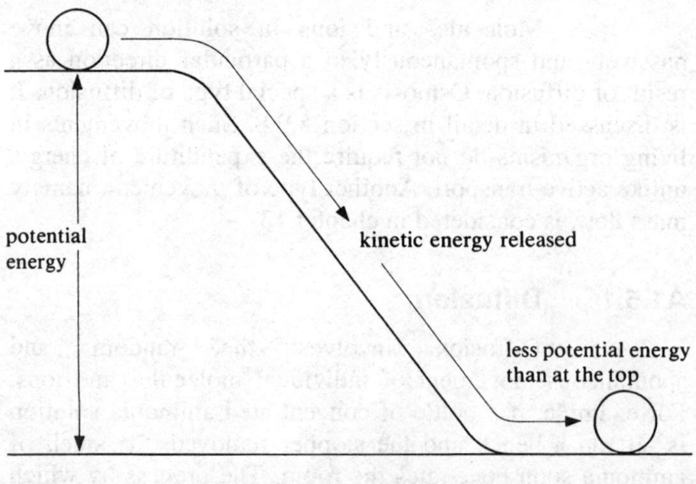

Fig A1.4 *Potential and kinetic energy.*

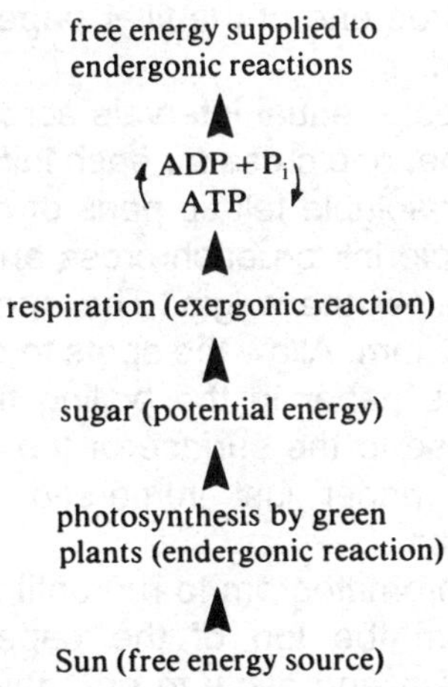

Fig A1.5 *Movement of energy through biological systems.*

A1.7 Chromatography

Chromatography is a technique used for the separating of mixtures into their components. The technique depends upon the differential movement of each component through a stationary medium under the influence of a moving solvent. For example, the green pigment in plants, when dissolved in a suitable solvent and allowed to pass through a stationary medium such as powdered chalk, separates into a number of different coloured pigments. A similar experiment is described in experiment A1.3.

There are three basic types of chromatography, depending on the nature of the stationary medium: **paper chromatography**, **adsorption column chromatography** and **thin-layer chromatography**. Paper chromatography is used in experiments A1.1–1.3 and is also described further in section 7.6.3. The various techniques of chromatography are now widely used in chemistry, biology, biochemistry and such specialist sciences as forensic medicine.

Electrophoresis

Electrophoresis is a modified form of chromatography used to separate charged molecules. An electric current is applied across a chromatographic medium such that one end has a positive charge and the other a negative charge. Individual molecules in the mixture move outwards through the medium towards the ends depending on their relative masses and charges. Electrophoresis is commonly used in the isolation and identification of amino acids, where the technique is improved further by adjusting the pH of the medium.

A1.7.1 The concept of R_f values

The movement of the solute relative to the solvent front on a chromatographic system is constant for that solute. This can be expressed in the term R_f as shown below.

$$R_f = \frac{\text{distance moved by solute}}{\text{distance moved by solvent front}}$$

If the solvent front goes off the end of the paper then it is possible to express the movement of a particular solute in comparison with the movement of another standard solute.

$$\text{Thus } R_x = \frac{\text{distance moved by solute}}{\text{distance moved by standard solute } x}$$

See fig A1.6*c*.

A1.7.2 Two-dimensional paper chromatography

Complex mixtures of solutes cannot always be separated efficiently by chromatography in one direction only. Thus a further separation must be carried out using a second solvent at right-angles to the first for a better separation of the spots (fig A1.6*d*). A square sheet of paper

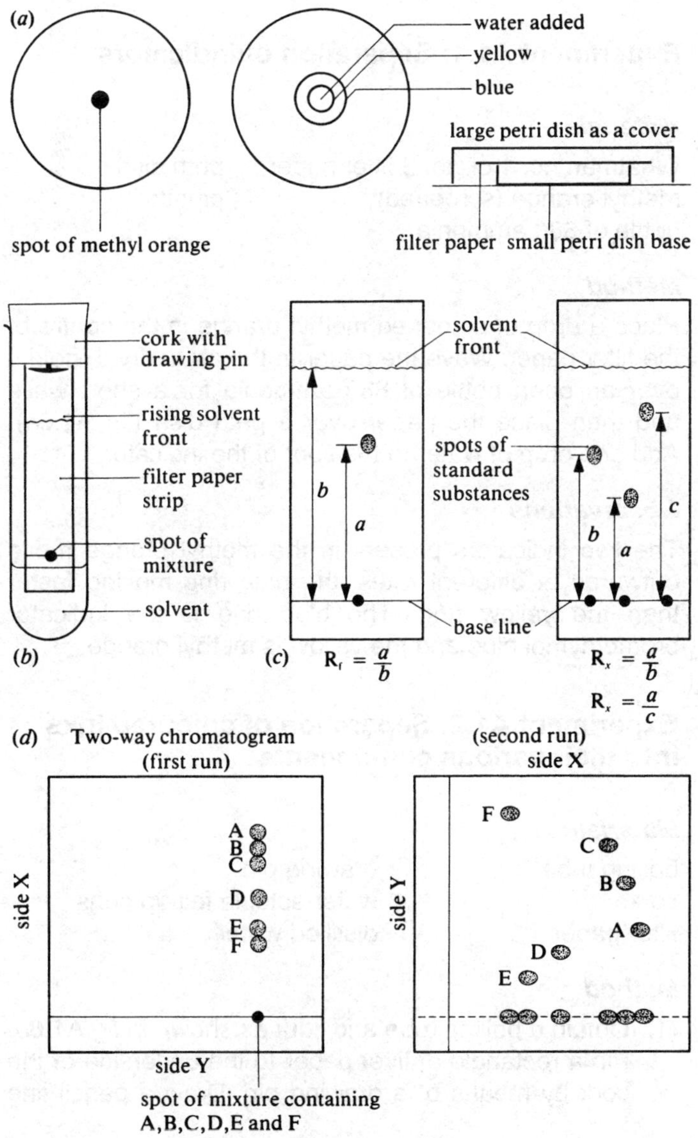

Fig A1.6 *Techniques of chromatography.*

is used. The test solution is applied to the base line near one end and the first separation is carried out. The paper is removed, dried and turned through 90° and a further chromatographic run is made with a different solvent. As a result the partially separated solutes of the first run are further separated in the second solvent which has different characteristics from the first. The paper is removed, dried, and the solutes located with a suitable reagent. The identification of a given compound can be made by comparison of its position with that of known standard compounds. This was the technique used by Calvin during his experiments to identify the initial products of the photosynthetic process (section 7.6.3).

Before running chromatograms of biological interest, it is helpful to practise the technique using coloured inks or indicators. The following experiments will show that the smaller, that is the more concentrated, the spot of origin, the better the separation. They also show that the longer the chromatogram runs, the better the separation of the samples.

Experiment A1.1: Separation of indicators

Materials

Whatman no. 1 or no. 3 filter paper	petri dish
methyl orange (screened)	pipette
bottle of 880 ammonia	

Method

Place a drop of screened methyl orange in the centre of the filter paper. Wave the paper in the air to dry it, hold it over an open bottle of 880 ammonia for a short while and then place the paper over a petri dish (fig A1.6*a*). Add one drop of water to the spot of the indicator.

Observations

The two indicators present in the methyl orange move outwards at different rates, the blue ring moving faster than the yellow ring. The blue ring is the indicator bromothymol blue and the yellow is methyl orange.

Experiment A1.2: Separation of coloured inks into their various components

Materials

boiling tube	drawing pin
cork	water-soluble felt-tip pens
filter paper	distilled water

Method

(1) Obtain a boiling tube and cork as shown in fig A1.6*b*. Pin a rectangle of filter paper to the underside of the cork by means of a drawing pin. Draw a pencil line across the free end of the filter paper about 1 cm up from the end.
(2) Mark crosses at equal intervals across the paper on the origin line, one cross for each ink being tested.
(3) Using water-soluble felt-tip pens of different colours, spot a sample ink on each cross and label the spot in pencil below the origin. The spot should be no larger than 2 mm. Allow the spots to dry.
(4) Suspend the paper in the boiling tube so that the origin is close to the surface of the solvent with the end of the paper just immersed. The solvent is distilled water.
(5) Allow the chromatogram to run until the solvent front is 1 cm from the top of the paper. Remove the chromatogram and allow to dry, having marked the end of the solvent front in pencil.
(6) If larger scale chromatography tanks are available these can be used for either ascending or descending runs.

Experiment A1.3: To separate plant pigments by paper chromatography

Materials

leaves of nettle or spinach	Buchner funnel
blender or knife	separating funnel
90% propanone (acetone)	light petroleum (BP 37.8–48.9 °C)
mortar and pestle	boiling tube and cork
small piece of capillary tube	filter paper
	drawing pin

Method

Mince some leaves of nettle (*Urtica dioica*) or spinach in a blender (or simply cut them up into small pieces by chopping with a knife). Grind up the leaves with 90% propanone (acetone) in a mortar. Filter the extract through a Buchner funnel into a separating funnel. Add an equal volume of light petroleum. Shake the mixture thoroughly. Wash through with water several times and each time discard the water layer with its contents. The solvent for running the chromatogram is 100 parts of light petroleum : 12 parts of 90% propanone. Use a boiling tube and filter paper as described in the previous experiment. In the same manner rule a pencil line about 1 cm from the bottom of the filter paper. By means of a small piece of capillary tube, spot the mixture of pigments in the centre of the pencil line. Pour the solvent into the boiling tube to a depth of about 2 cm and then fix the cork and paper into the tube. The solvent should be allowed to run until it is just below the cork. This should take about 1–2 h. The tube should be placed in dim light.

Results

The following colour bands should be shown

Colour of spot	R_f value	Pigments present
yellow	0.95	carotene
yellow-grey	0.83	phaeophytin
yellow-brown	0.71	xanthophyll (often differentiates into two spots)
blue-green	0.65	chlorophyll *a*
green	0.45	chlorophyll *b*

Appendix 2

Biological techniques

A2.1 Scientific method

Science may be defined in terms of either knowledge or method. Scientific **knowledge** is the total body of factual material which has been accumulated (by scientific **method**) relating to the events of the material world.

> 'Science is almost wholly the outgrowth of pleasurable intellectual curiosity.'
> A. N. Whitehead

In order to satisfy their curiosity, scientists must continually pose questions about the world. The secret of success in science is to ask the right questions.

> 'The formulation of a problem is often more essential than its solution, which may be mainly a matter of mathematical or experimental skill. To raise new questions, new possibilities, to regard old problems from a new angle, requires creative imagination and marks real advance in science.' Albert Einstein

Scientific investigations may begin in response to observations made by scientists or in response to some internal 'inductive' process on the part of scientists. Those aspects of knowledge which are described as scientific must, as the contemporary philosopher of science Karl Popper has stated, be capable of 'refutation'. This means that the facts of scientific knowledge must be testable and repeatable by other scientists. Thus it is essential that all scientific investigations are described fully and clearly as described in section A2.2. If investigations yield identical results under identical conditions, then the results may be accepted as valid. Knowledge which cannot be investigated as described above is not scientific and is described as 'metaphysical'.

Facts are based on **observations** obtained directly or indirectly by the senses or instruments, such as light or radio telescopes, light or electron microscope and cathode ray oscilloscopes, which act as extensions of our senses. All the facts related to a particular problem are called **data**. Observations may be **qualitative** (that is describe colour, shape, taste, presence and so on) or **quantitative**. The latter is a more precise form of observation and involves the measurement of an amount or quantity which may have been demonstrated qualitatively.

Observations provide the raw material which leads to the formulation of a hypothesis (fig A2.1). A **hypothesis** is an assumption or question based on the observations, that may provide a valid explanation of the observations. Einstein stated that a hypothesis has two functions.

(1) It should account for all the observed facts relevant to that problem.
(2) It should lead to the prediction of new information. New observations (facts, data) which support the hypothesis will strengthen it. New observations which contradict the hypothesis must result in it being modified or even rejected.

In order to assess the validity of a hypothesis it is necessary to design a series of experiments aimed at producing new observations which will support or contradict the hypothesis. In most hypotheses there are a number of factors which may influence the observation; these are called **variables**. Hypotheses are objectively tested by a series of experiments in which each one of the hypothetical variables influencing the observations is

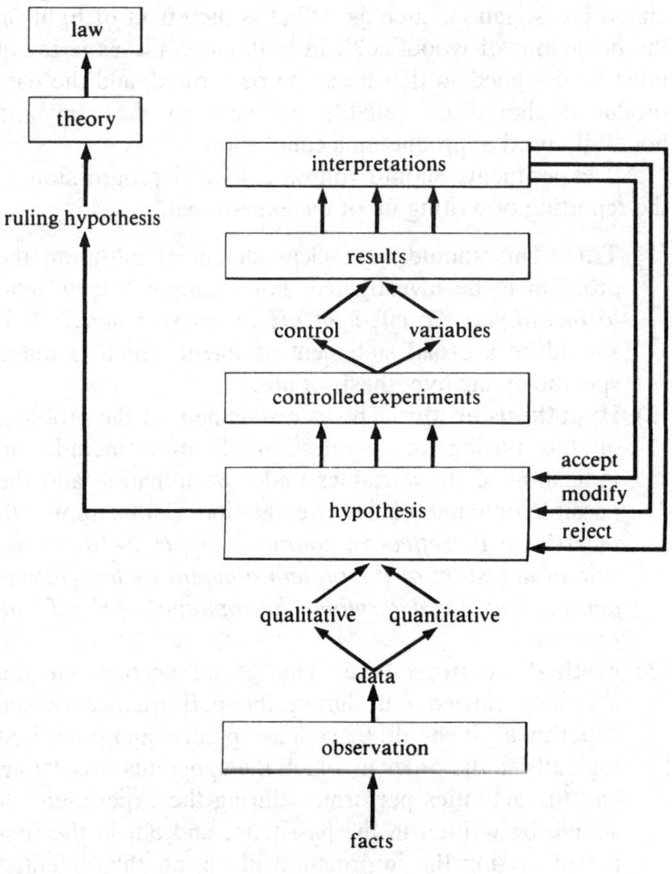

Fig A2.1 *Summary of scientific method.*

systematically eliminated. The experimental series is said to be **controlled** and this ensures that only one variable of the problem is tested at a given time.

The most successful hypothesis becomes a **ruling hypothesis**, and if it withstands attempts at falsification and continues to be successful in predicting previously unexplained facts and relationships it may become a **theory**.

The trend throughout scientific study is to achieve higher levels of predictability (probability). When a theory has proved invariable under all circumstances, or such variations as occur are systematic and predictable, then it may be accepted as a **law**.

As knowledge increases and techniques of investigation improve, hypotheses and even well-established theories may be challenged, modified and even rejected. Science is dynamic and controversial and the objective methods of science are always exposed to challenge.

A2.2 Laboratory work and writing up an experiment

Before beginning any experimental investigation, the aim of the experiment should be made clear. This may involve the testing of a hypothesis, such as 'The germination of seeds requires the presence of water, oxygen and an optimum temperature', or a more open-ended investigation, such as 'What is the effect of light on the behaviour of woodlice?'. In both cases the experiment must be designed so that it can be performed, and the data produced should be reliable, relevant to the aim and hopefully used in producing a conclusion.

All experiments should follow a logical progression in the reporting or writing up of the experiment.

(1) **Title**. This should be a clear statement outlining the problem to be investigated. For example *'Experiment to investigate the effect of pH on enzyme activity'*. It should be a broad statement of intent which is made specific by the hypothesis or aim.

(2) **Hypothesis or aim**. This is a statement of the problem or the posing of a question. It may include an indication of the variables under examination and the possible outcome of the investigation. For example *'To investigate the effect of solutions of pH 2–10 on the rate of digestion of the protein albumin by the enzyme pepsin and to determine the optimum pH of the reaction'*.

(3) **Method or procedure**. This is an account of the activities carried out during the performance of the experiment. It should be concise, precise and presented logically in the order in which the apparatus was set up and the activities performed during the experiment. It should be written in the past tense and not in the first person. Using the information given, another scientist should be able to repeat the experiment.

(4) **Results and observations**. These may be qualitative or quantitative and should be presented as clearly as possible in some appropriate form or forms, such as verbal description, tables of data, graphs, histograms, bar charts, kite diagrams and so on. If several numerical values are obtained for repeated measurements of one variable, the mean (x) of these values should be calculated and recorded.

(5) **Discussion**. This should be brief and take the form of the answer(s) to possible questions posed by the hypothesis, or confirmation of the aim. The discussion should not be a verbal repetition of the results, but an attempt to relate theoretical knowledge of the experimental variables to the results obtained.

A **conclusion** may be included if there is clear-cut verification of the stated aim. For example, for the aim given in (2) above a conclusion could state that there is 'a relationship between pH and enzyme activity and for this reaction the optimum pH is x'. The discussion of the results of this same experiment should include such theoretical aspects as the nature of the reaction and the possible chemical and physical aspects of the effects of pH on the three-dimensional structure of enzyme molecules.

A2.3 Presenting data

As a result of qualitative and quantitative investigations, observations are made and numerical data obtained. In order for the maximum amount of information to be gained from investigations, they must be planned carefully and the data must be presented comprehensively and analysed thoroughly.

A2.3.1 Tabulations

Tables form the simplest way of presenting data and consist of columns displaying the values for two or more related variables. This method gives neither an immediate nor clear indication of the relationships between the variables, but is often the first step in recording information and forms the basis for selecting some subsequent form of graphical representation.

A2.3.2 Graphical representation

A graph is a two-dimensional plot of two or more measured variables. In its simplest form a graph consists of two axes. The vertical y axis bears values called **ordinates** which show the magnitude of the **dependent** variable. This is the 'unknown quantity', that is the variable whose value is not chosen by the experimenter. The horizontal x axis bears values called **abscissae** which show the magnitude of the **independent** variable, which is the known quantity, that is the variable whose value is chosen by the experimenter.

The following stages are used in constructing a graph.

(1) The scale and intervals for each axis should match the magnitude of the variables being plotted and fill the graph paper as completely as possible.
(2) Each axis should begin at 0, but if all the values for one variable are clustered together, such as ten points lying between 6.12 and 6.68, a large scale will be required to cover these points. In this case, still begin with the axis at 0 but mark a break in the axis, marked as -/ /-, just beyond.
(3) Each axis must be labelled fully in terms of the variable, for example 'temperature/°C', and have equally spaced intervals covering the range of the interval, such as 0–60 at 12 five-unit intervals.
(4) The points plotted on the graph are called **coordinates** and represent the corresponding values of the two variables, such as when $x = a$ and $y = b$.
(5) Actual points should be marked by an **X** or ⊙ and never by a dot only.
(6) The points marked on the graph are the record of the actual observations made and may be joined by a series of straight line segments drawn with a ruler, by a smooth curve or, in some cases, a regression line (a line of best fit) (section A2.4.3). These graphs are called **line graphs**. Straight line segments and smooth curves are preferable to a regression line.
(7) The graph should have a full title, such as 'Graph showing the relationship between . . .'.
(8) Only the points on the graph represent actual data, but estimates of other values can be obtained from reading off coordinates at any point on the line. This is called **interpolation**. Similarly, coordinates outside the range of the graph may be determined by extending the line of the graph, a technique known as **extrapolation**. In both cases it must be stressed that these values are only estimates.

In graphs where the x axis is 'time', the steepness of the curve or **gradient** at any point can be calculated and this gives a measure of the rate of change of the variable under investigation. For example, in the graph shown in fig A2.2 the rate of growth is calculated by drawing a tangent to the curve at the desired point and completing a triangle with the tangent at the hypotenuse, as shown in fig A2.3. The value of the y interval is then divided by the value of the x interval and this gives the rate of change in terms of the units used in labelling the graph.

A2.3.3 Frequency distributions

Many relationships exist where each value of the dependent variable, corresponding to the independent variable, represents the number of times the latter value occurs, that is its frequency. Such relationships form a **frequency distribution** or **distribution**, for example lengths of earthworms in a population.

(a)

Time/days	0	2	4	6	8	10	12	14	16	18	20	22	24	26
Mean height/mm	1	2	4	11	24	43	73	92	105	112	117	122	124	126

(b)

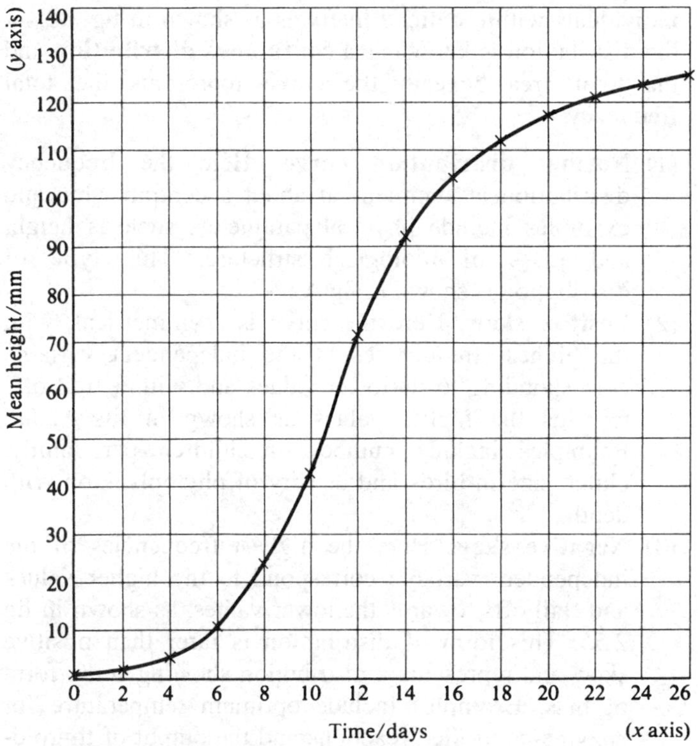

Fig A2.2 *(a) Two sets of data relating to mean heights of oat seedlings and time. (b) Graph showing the relationship between mean heights of oat seedlings and time.*

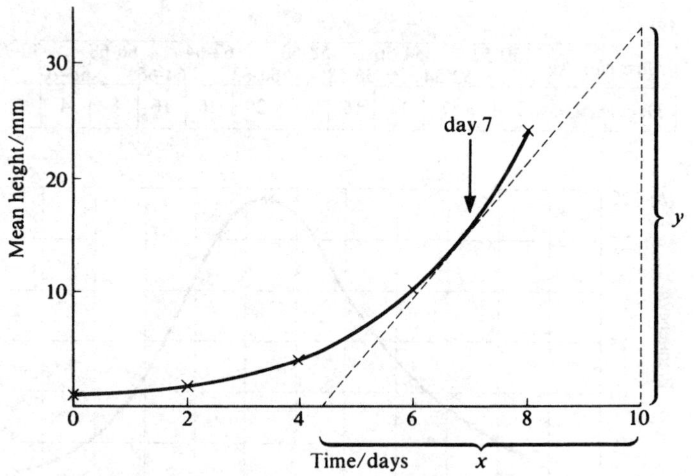

From the graph: $y = 33$ mm
$x = 5.6$ days

Therefore rate of growth $= \frac{y}{x} = \frac{33}{5.6}$

$= 5.9$ mm day^{-1}

Fig A2.3 *Method of calculating rate of change at a given point, for example day 7.*

If the value of the independent variable can assume any value within a given range, its frequency distribution can be represented by a conventional graph as described above. These graphs are called **frequency curves** and may take one of the following forms depending upon how the data are presented. If the data are presented as numbers of individuals within defined intervals as shown in fig A2.4*a*, the distribution is known as a **continuous distribution** and the total area beneath the curve represents the total frequency.

(1) **Normal distribution curve**. Here the frequency distribution is symmetrical about a central value and examples include physical parameters such as height and mass of biological structure. This type of distribution is shown in fig A2.4.

(2) **Positive skew**. Here the curve is asymmetrical, with the highest frequencies of the independent variable corresponding to its lower values and with a 'tail off' towards the higher values as shown in fig A2.5*a*. Examples include number of children per family, clutch size in birds and density of phytoplankton with depth.

(3) **Negative skew**. Here the highest frequencies of the independent variable correspond to the higher values and 'tail off' towards the lower values, as shown in fig 2.5*b*. This form of distribution is rarer than positive skew and represents a distribution showing some form of bias. Examples include optimum temperature for enzyme-controlled reactions and the output of thyroid-stimulating hormone in response to thyroxine.

(4) **Bimodal distribution**. Here there are two peaks (or modes), and it usually indicates the presence of two populations each exhibiting a partial normal distribution.

(*a*)

Mass class/kg	50-52	52-54	54-56	56-58	58-60	60-62	62-64	64-66	66-68	68-70	70-72
Frequency	4	7	11	16	24	29	26	16	8	4	2

(*b*)

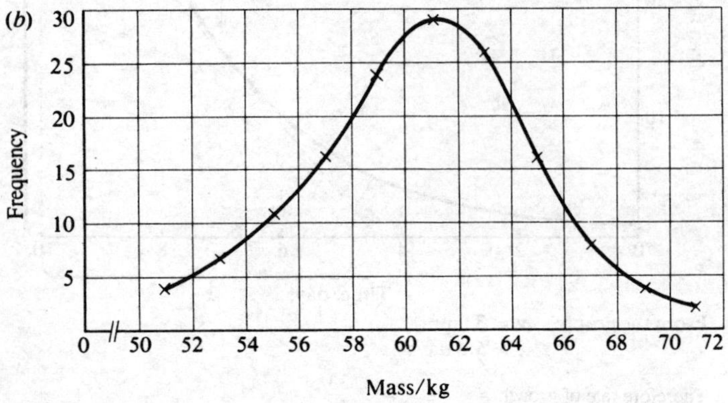

Fig A2.4 *(a) Number of 18-year-old males falling into 2 kg mass classes and represented as a table. (b) The graph representing the data from (a) forms a normal distribution curve.*

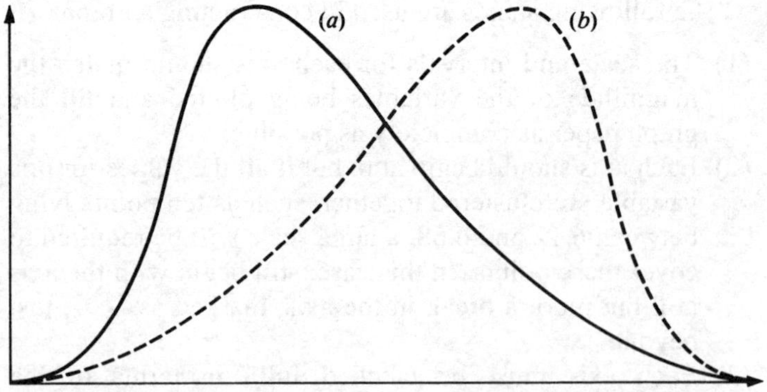

Fig A2.5 *(a) Positive skewed distribution. (b) Negative skewed distribution.*

(5) **Cumulative frequency distribution.** The data presented in fig A2.4 may be presented as in fig A2.6, where the cumulative number of individuals below certain arbitrary class boundaries are shown. Where these data are presented graphically a cumulative frequency curve is produced.

(*a*)

Mass/kg	50	52	54	56	58	60	62	64	66	68	70	72
Cumulative frequency	0	4	11	22	38	62	91	117	133	141	145	147

(*b*)

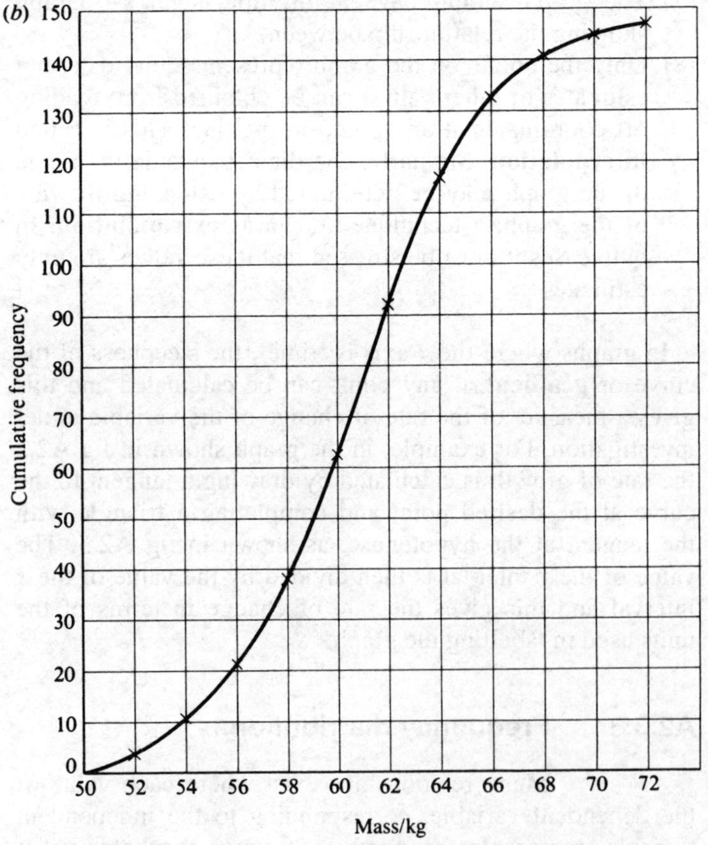

Fig A2.6 *Data (a) and graph (b) based upon fig A2.8a showing the cumulative frequency distribution of mass in 18-year-old males.*

If the values of the independent variable assume discrete values, that is whole numbers such as 3 and 5 (as in the numbers of petals of a dicotyledon), or represent physical traits such as blood groups, they exist as only discrete values and the distribution is described as **discontinuous**. In these cases it is inappropriate to plot a continuous graph and other forms of graphical representation are used as described below.

(1) **Column graph**. This shows the frequency with which distinct characteristics occur within a population, such as human blood groups (see fig A2.7*a*).

(2) **Histogram**. This represents continuous values of the independent variable which have been grouped into classes of equal widths. Where classes of equal width are chosen, for example 0–5, 5–10, 10–15, and so on, the limits of the interval are conventionally represented by the lower integer, that is 0–4.99, 5–9.99, 10–14.99 and so on. This is a useful way of representing data from a small sample and superficially resembles a column graph (fig A2.7*b*).

(3) **Bar graph**. This is a modified form of histogram usually representing the relationship between a continuous dependent variable, such as energy content, and a non-numerical independent variable, for example various foods (fig A2.7*c*). A modified form of bar graph is used in presenting ecological data and this is called a **presence–absence graph**. An example of this is shown in fig 11.22.

(4) **Kite diagram**. This is a special type of bar graph that provides an extremely clear visual display of the change in frequency of non-numerical variables which are continuously distributed within an area. A kite diagram is constructed by plotting the frequencies of each variable as a line symmetrically placed astride the *x* axis as shown in fig A2.8*a*. Once all of the frequencies have been plotted along the *x* axis, the adjacent limits of the lines are joined together by straight lines, as in a line graph, as shown in fig A2.8*b*. The enclosed area is usually shaded to present a clearer visual display. The use of kite diagrams is described in section 11.4.3.

Each one of the methods of presenting data described above is applicable to different biological situations, and all are represented in the chapters in these books. All methods have their relative merits, and the choice of which to use should be made on the basis of which will accurately and efficiently reveal relationships and patterns between variables.

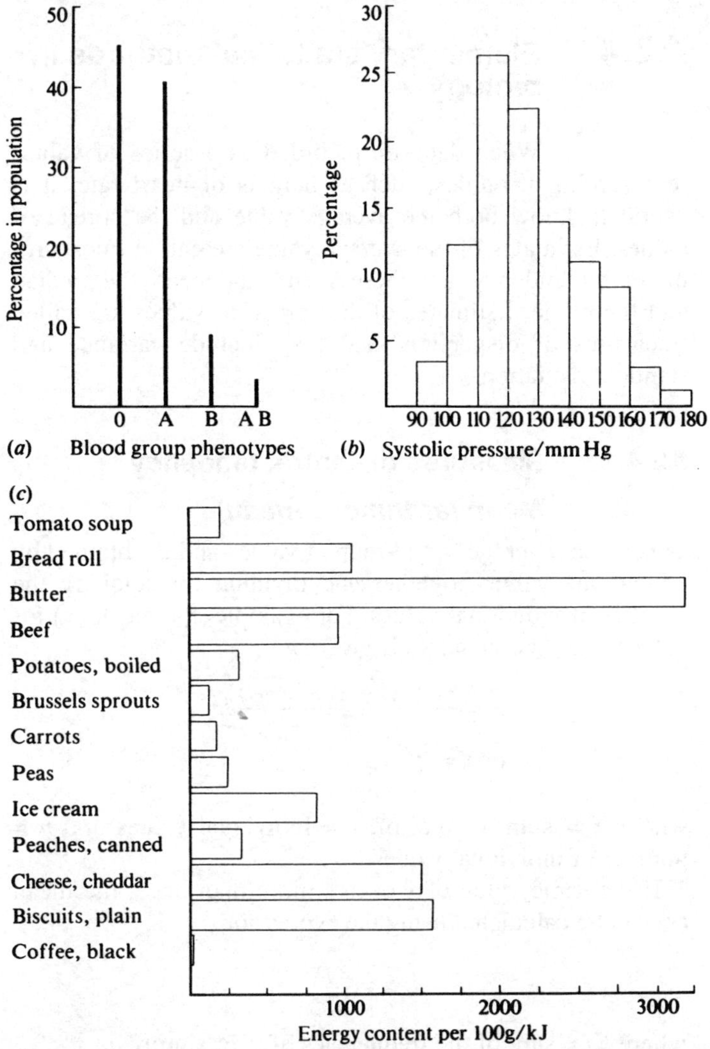

Fig A2.7 *Methods of presenting data. (a) Column graph showing frequency of blood group phenotypes in the population. (b) Histogram showing systolic blood pressure frequencies in women aged 30–9 years. (c) Bar chart showing energy content of foods in a three-course meal.*

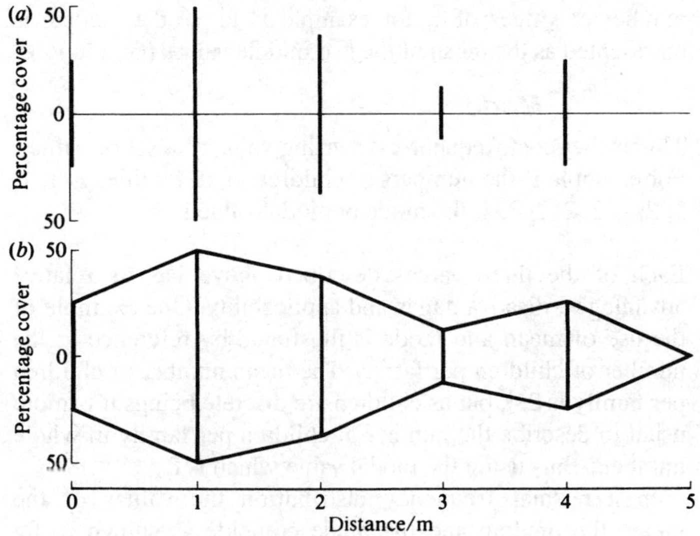

Fig A2.8 *(a) and (b) method of constructing a kite diagram.*

A2.4 Elementary statistical methods in biology

When data are recorded as a series of values representing variables, such as heights or heart rate, it is useful to know both the average value and the spread of values. Estimates of the average value are called 'measures of central tendency' and these include the mean, the median and the mode. Estimates of the spread of values are called 'measures of dispersion' and they include variance and standard deviation.

A2.4.1 Measures of central tendency

Mean (arithmetic mean)

This is the 'average' of a group of values and is obtained by adding the values together and dividing the total by the number of individual values. For example, the mean ($\bar{x}$) for values $x_1, x_2, x_3, x_4, ... x_n$ is given by

$$\bar{x} = \frac{x_1 + x_2 + x_3 ...... + x_n}{n}$$

$$\text{or } x = \frac{\Sigma x}{n}$$

where Σ = sum or total of, x = individual values and n = number of individual values.

If the same value of x occurs more than once, the mean ($\bar{x}$) can be calculated using the expression

$$\bar{x} = \frac{\Sigma fx}{\Sigma f}$$

where Σf = sum of the frequencies of x, or simply n.

Median

This represents the middle or central value of a set of values. For example, if five values of x are arranged in ascending order as x_1, x_2, x_3, x_4 and x_5, the median value would be x_3 since there are as many values above it as below it. If there are an even number of values of x, for example x_1 to x_6, the median is represented as the mean of the two middle values $((x_3 + x_4)/2)$.

Mode

This is the most frequently occurring value of a set of values. For example if the numbers of children in 10 families is 1, 1, 1, 2, 2, 2, 2, 2, 3, 4, the mode or modal value is 2.

Each of the three values described above has its relative advantages, disadvantages and applicability. One example of the use of mean and mode is illustrated by reference to the number of children per family. The mean number of children per family is 2.4, but as children are discrete beings it is more usual to describe the number of children per family in whole numbers, thus using the modal value which is 2.

In a normal frequency distribution the values of the mean, the median and the mode coincide as shown in fig A2.9*a*, whereas in cases where the frequency distribution is skewed, these values do not coincide as shown in fig A2.9*b*.

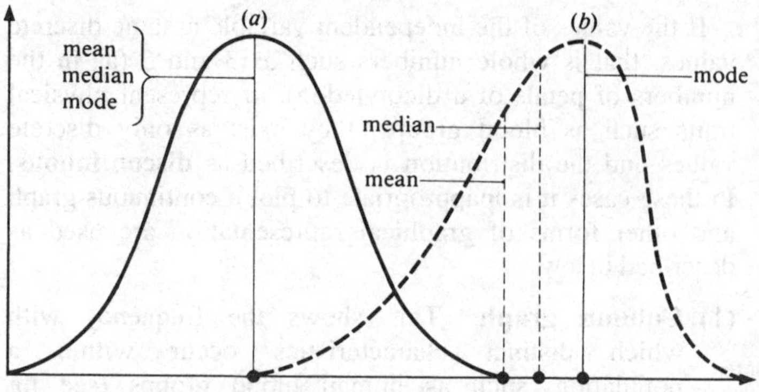

Fig A2.9 Distributions of mean, median and mode in (a) a normal distribution and (b) a skewed distribution.

A2.4.2 Measures of dispersion

Measures of dispersion are used in conjunction with measures of central tendency to give an indication of the extent to which values are 'spread' or 'clustered' around the 'average'. This is illustrated with respect to a normal distribution by the curves shown in fig A2.10. In statistical analysis, one of the most useful measures of dispersion is the root mean square deviation or standard deviation, since it may be applied both to predicting the distribution of values about the average and to determining whether two sets of data are significantly different from one another and the degree of difference between them.

Standard deviation

The standard deviation (s) of a set of values is a measure of the variation from the arithmetical mean of these values and is calculated using the expression

$$s = \sqrt{\left(\frac{\Sigma fx^2}{\Sigma f} - \bar{x}^2\right)}$$

where Σ = sum of, f = frequency of occurrence of, x = specific values, and $\bar{x}$ = mean of the specific values.

For example, a sample of ten common limpet shells (*Patella vulgaris*) from a rocky shore have the following maximum basal diameters in millimetres: 36, 34, 41, 39, 37, 43, 36, 37, 41, 39. In order to calculate the mean maximum basal diameter and the standard deviation it is necessary to calculate f, fx^2 and x^2 as shown in the following table:

x	f	fx	fx^2
34	1	34	1 156
36	2	72	2 592
37	2	74	2 738
39	2	78	3 042
41	2	82	3 362
43	1	43	1 849
	$\Sigma f = 10$	$\Sigma fx = 383$	$\Sigma fx^2 = 14\,739$

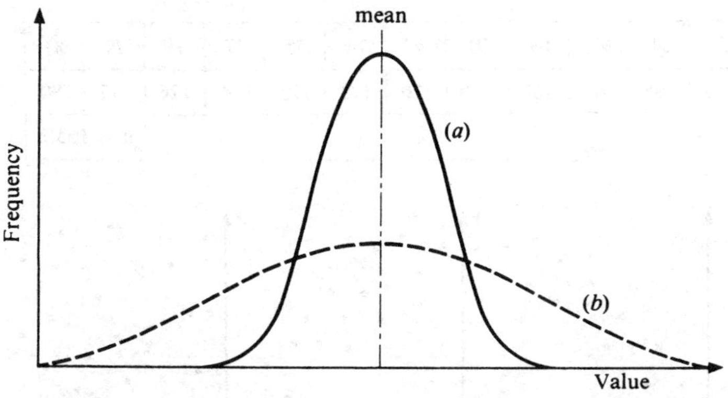

Fig A2.10 *Two normal distribution curves showing the distribution of two sets of data (possibly populations) with identical total frequencies (that is the areas under the curves are equal). Curve (a) has a restricted range and is clustered around the mean. Curve (b) represents a wider range and is not clustered around the mean.*

Therefore $\bar{x} = 38.3$ and $\bar{x}^2 = 1\,466.9$.

$$s = \sqrt{\left(\frac{\Sigma fx^2}{\Sigma f} - \bar{x}^2\right)}$$

$$s = \sqrt{\left(\frac{14739}{10} - 1466.9\right)}$$

$$= \sqrt{(1473.9 - 1466.9)}$$
$$= \sqrt{7}$$

Therefore $s = 2.65$.

In this population of the common limpet the mean maximum basal diameter of the shell is 38.3 mm, with a standard deviation of 2.7 mm (correct to one decimal place). If these values are applied to a larger population of the common limpet then it may be assumed, on statistical grounds, that approximately 68% of the population will have a basal diameter of the shell of 38.3 mm plus and minus one standard deviation (2.7 mm), that is they will lie within a range 35.6–41.0 mm; approximately 95% of the population will have a basal diameter of the shell of 38.3 plus and minus two standard deviations (5.4 mm), that is they will lie within the range 32.9–43.7 mm, and practically 100% will lie within plus and minus three standard deviations. The value of calculating the standard deviation is that it gives a measure of the spread of values from the mean. A small standard deviation indicates that there is little dispersion or variation from the mean and that the population is fairly homogeneous, as shown by curve (*a*) in fig A2.10. As the value of the standard deviation increases, the degree of variation within the population increases as shown by curve (*b*) in fig A2.10.

Variance

The **variance** is the square of the standard deviation and the variance for a set of numbers is calculated using the expression:

$$\text{variance } (s^2) = \frac{\Sigma fx^2}{\Sigma f} - \bar{x}^2$$

where f is the number of values in the set.

Variance is useful in ecological investigations involving, nutrition, reproduction and behaviour since it gives an indication of how organisms are dispersed within the population. Populations may be:

- randomly dispersed,
- aggregated into clusters, or
- regularly dispersed.

To determine the type of population dispersion within an area, the area is divided up into a number of equal-sized quadrats (section 11.2) and the number of individuals within the population, per quadrat, is counted. From these data the mean and variance are calculated using the expressions:

$$\text{mean } (\bar{x}) = \frac{\Sigma fx}{f} \qquad \text{variance } (s^2) = \frac{\Sigma fx^2}{\Sigma f} - \bar{x}^2$$

where f is the number of quadrats containing x individuals. Using the following expression:

$$\text{population dispersion} = \frac{\text{variance}}{\text{mean}}$$

the three types of dispersion can be determined as shown in fig A2.11.

A2.4.3 Relationships between variables

Data should always be presented in such a way as to reveal relationships between two or more sets of data. The simplest way of doing this is to plot a graph showing the relationship between variables, but this is only valuable if one of the variables (the independent variable) is under the control of the experimenter, as for example in the case of data shown in fig A2.2.

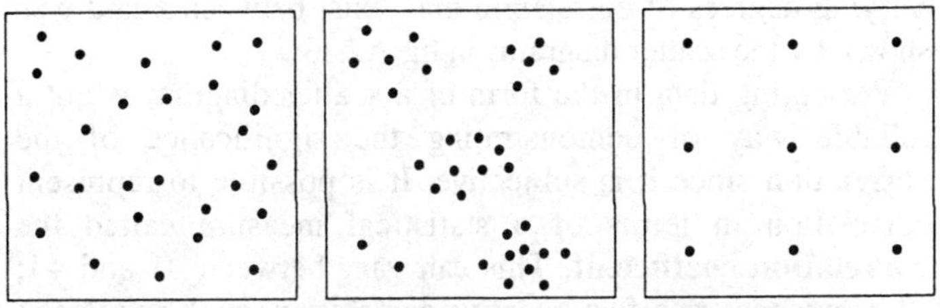

(*a*) variance = mean population dispersion = 1 random dispersion **(*b*) variance > mean population dispersion > 1 aggregated dispersion** **(*c*) variance < mean population dispersion < 1 regular dispersion**

Fig A2.11 *Types of dispersion.*

(*a*)

Mass/kg	51	51	53	55	59	60	62	60	58	64	67	69	71	68	74	75	77	79	79	81
Height/cm	154	155	156	158	158	159	161	162	163	165	166	168	169	170	172	173	174	176	177	180
													$\bar{x}$ = 65.7							$\bar{y}$ = 165.8

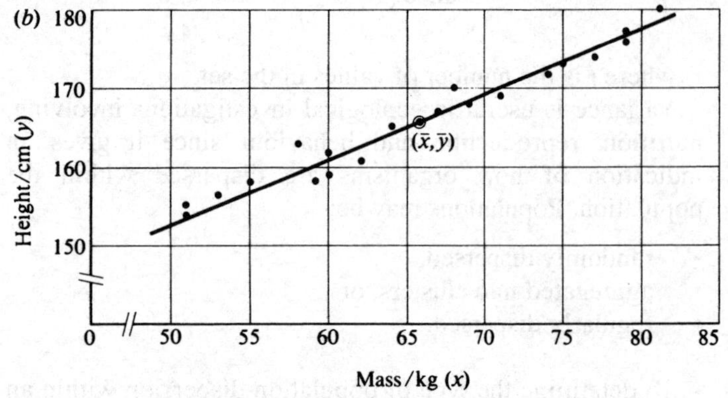

Fig A2.12 *Data showing mass and corresponding heights of 20 16-year-old male students, represented as a table (*a*) and a scatter diagram (*b*). The regression line is drawn.*

In other cases, where both variables are dependent, the value of one variable is plotted against the corresponding value of the other variable as, for example, in the case of heights and masses of 20 sixth-form students given in fig A2.12*a*. These values are plotted and shown in fig A2.12*b*, which is called a **scatter diagram**. Visual inspection shows that there is some form of relationship between the two variables but this cannot be described more accurately until a relationship, presented by a straight line, can be drawn through the points.

This single line is called a **'line of best fit'** or a **regression line** and the proximity of the points to the line gives an indication of the degree of correlation between the two variables. The position of the line of best fit should pass through the point representing the mean values of mass and height ($\bar{x}$ = 65.7 kg and $\bar{y}$ = 165.8 cm) and the distribution of points above and below the line should be approximately equal. From this line the predicted values of height corresponding to mass can be calculated.

Correlation

The relationship between the two variables, *x* and *y*, described above may be represented by a term called **correlation**. Varying degrees of correlation may exist between *x* and *y* as shown by the scatter diagrams in fig A2.13.

Presenting data in the form of a scatter diagram is not a reliable way of demonstrating the significance of the correlation since it is subjective. It is possible to represent correlation in terms of a statistical measure called the **correlation coefficient**. This can vary between –1 and +1; –1 represents a perfect negative correlation, such as oxygen tension in the atmosphere and rate of spiracle opening in insects; 0 represents no correlation, such as size of tomato fruits and number of seeds; +1 represents a perfect positive correlation, such as age and body length in the locust.

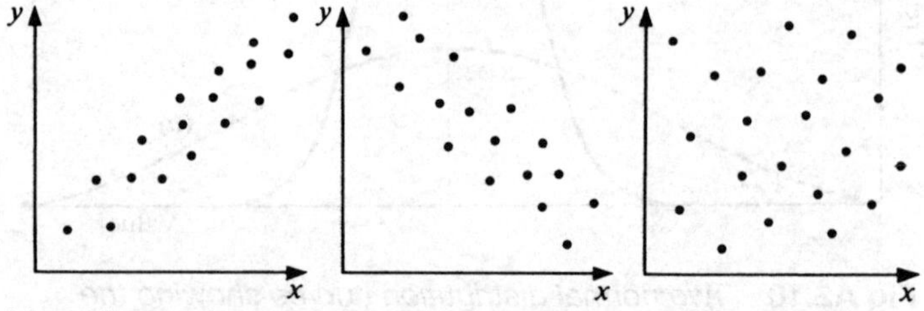

(*a*) positive correlation (*b*) negative correlation (*c*) no correlation

Fig A2.13 *Patterns of correlation; (a) positive correlation; (b) negative correlation; (c) no correlation.*

Appendix 3

Nomenclature and units

A3.1 Alphabetical list of common current names of chemicals

Old name	*New name (nc = no change)*
acetaldehyde	ethanal
acetamide	ethanamide or nc
acetic acid	ethanoic acid or nc
acetoacetic acid	3-oxobutanoic acid
acetone	propanone
acetylene	ethyne
adenine	nc
adenosine	nc
adipic acid	hexanedioic acid
alanine	2-aminopropanoic acid
alcohol (ethyl)	ethanol
alcohol (wood)	methanol
aldehyde	nc
aldol	3-hydroxybutanal
aliphatic	nc
alkyl	nc
ammonium hydroxide	ammonia solution
p-amino benzoic acid	(4-aminobenzoic acid (4-aminobenzenecarboxylic acid
aspartic acid	aminobutanedoic acid
benzaldehyde	benzenecarbaldehyde or nc
benzene	nc
benzoic acid	benzenecarboxylic acid or nc
bicarbonate	hydrogencarbonate
butyric acid	butanoic acid
camphor	nc
cane sugar	sucrose
carbon tetrachloride	tetrachloromethane
carboxylic acids	nc
chloroform	trichloromethane

citric acid	2-hydroxypropane-1,2,3-tricarboxylic acid
cobalt chloride	cobalt(II) chloride
dextrose	(+) glucose
ethyl acetate	ethyl ethanoate or nc
ethyl alcohol	ethanol
ethylene	ethene
ferric	iron(III)
ferrous	iron(II)
formaldehyde	methanal
fructose	nc
fumaric acid	*trans*-butanedioic acid
glucose	nc
glutamic acid	2-aminopentanedioic acid
glycerine/glycerol	propane-1,2,3-triol
glycine	aminoethanoic acid aminoacetic acid
glycolic acid	hydroxyethanoic acid hydroxyacetic acid
glyoxyllic acid	oxoethanoic acid oxoacetic acid
indoleacetic acid (IAA)	indolylethanoic acid
isopropyl alcohol	propan-2-ol
lactic acid	2-hydroxypropanoic acid
lactose	nc
malic acid	2-hydroxybutanedioic acid
malonic acid	propanedioic acid
maltose	nc
nitric acid	nc
oleic acid	*cis*-octadec-9-enoic acid
oxalic acid	ethanedioic acid
palmitic acid	hexadecanoic acid
phenol	nc
phosphate	phosphate(V)
phosphoric acid	phosphoric(V) acid
phosphorous acid	phosphoric acid
potassium permanganate	potassium manganate(VII)
pyridine	nc
pyrogallol	benzene-1,2,3-triol
pyruvic acid	2-oxopropanoic acid
quinol	benzene-1,4-diol
stearic acid	octadecanoic acid
succinic acid	butanedoic acid
sucrose	nc
1-tartaric acid	(-) 2,3-dihydroxybutanedioic acid
thiourea	thiocarbamide or nc
toluene	methylbenzene
urea	carbamide or nc
n-valeric acid	pentanoic acid
xylene	dimethylbenzene

A3.2 Units, symbols, abbreviations and conventional terms

absolute	abs.
adenosine 5'-pyrophosphate	ADP *use* adenosine diphosphate
adenosine 5'-triphosphate	ATP *use* adenosine triphosphate
adrenocorticotrophic hormone	ACTH
angstrom	Å *preferably* use SI units 10 Å = 1 nm
anterior	ant.
antidiuretic hormone	ADH
approximately equal	≈
basal metabolic rate	b.m.r. or BMR
calciferol	vitamin D_2 preferred for biological activity (vitamin D is the generic term)
calorie	cal *use* SI unit joule 1 cal = 4.2 J
Centigrade	*use* Celsius (°C)
central nervous system	CNS
cerebrospinal fluid	c.s.f. or CSF
chi-squared	χ^2
coenzyme A	CoA
degree Celsius	°C
deoxyribonucleic acid	DNA
endoplasmic reticulum	ER
extracellular fluid	e.c.f. or ECF
figure (diagram)	fig
Geiger–Müller tube	GM
gram	g (gramme is continental spelling)
growth hormone	GH
haemoglobin	Hb
joule	J
kilo (10^3 ×)	k
Krebs cycle	or tricarboxylic acid cycle
luteinising hormone	LH
mass	m
maximum	max.
mean value of *x* (statistics)	$\bar{x}$
minimum	min.
minute	min
molar (concentration)	M (mol dm^{-3}) molar means 'divided by amount of substance'
mole (unit of amount of substance)	mol replaces gram-molecule, gram-ion, gram-atom etc.
negative logarithm of hydrogen ion concentration	pH, plural – pH values
newton	N
normal saline	avoid *use* isosmotic saline
number of observations	*n* (or *f*)
parts per million	ppm
petroleum ether	avoid *use* light petroleum
pressure	*p*
red blood corpuscle	r.b.c. or RBC
respiratory quotient	r.q. or RQ
ribonucleic acid	RNA
solidus	/ expressed in units of
solution	soln.
species	sp. (singular), spp. (plural)
standard deviation	
(of hypothetical population)	*s*
(of observed sample)	S or s.d.
sum (statistics)	
(of hypothetical population)	Σ
(of observed sample)	S or Σ
temperature (quantity)	*T* (absolute) *t* (other scales)
thyroid stimulating hormone	TSH
time	*t*
variety (biology)	var.
volume	vol.
white blood corpuscle	w.b.c. or WBC

A3.3 SI units

A3.3.1 Names and symbols for SI units

Physical quantity	*Name of SI unit*	*Symbol*
length	metre	m
mass	kilogram	kg
time	second	s
electric current	ampere	A
thermodynamic temperature	kelvin	K
luminous intensity	candela	cd
amount of substance	mole	mol
solid angle	steradian	sr

A3.3.2 Derived units from SI units

Quantity	*SI unit*	*Symbol*	*Expressed in terms of SI units*
work, energy, quantity of heat	joule	J	$kg\ m^2\ s^{-2}$; $1\ J = 1\ N\ m$
force	newton	N	$kg\ m\ s^{-2}$; $= J\ m^{-1}$
power	watt	W	$kg\ m^2\ s^{-3}$; $= J\ s^{-1}$
quantity of electricity	coulomb	C	A s
electrical potential	volt	V	$kg\ m^2\ s^{-3}\ A^{-1}$; W A
luminous flux	lumen	lm	cd sr
illumination	lux	lx	$cd\ sr\ m^{-2}$ or $lm\ m^{-2}$
area	square metre		m^2
volume	cubic metre		m^3
density	kilogram per cubic metre		$kg\ m^{-3}$

A3.3.3 Special units still in use (should be progressively abandoned)

Quantity	*Unit name and symbol*		*Conversion factor to SI*
length	angstrom	Å	10^{-10} m = 0.1 nm
length	micron	μm	10^{-6} m = 10^{-3} mm = 1 μm
volume	litre	l	$10^{-3}\ m^3 = 1\ dm^3$
mass	tonne	t	10^3 kg = Mg
pressure	millimetres of mercury	mmHg	10^2 mmHg = 13.3 kPa

A3.3.4 Prefixes for SI units

These are used to indicate decimal fractions of the basic or derived SI units.

Multiplication factor	*Prefix*	*Symbol*
0.000 000 000 001 = 10^{-12}	pico	p
0.000 000 001 = 10^{-9}	nano	n
0.000 001 = 10^{-6}	micro	μ
0.001 = 10^{-3}	milli	m
1000 = 10^3	kilo	k
1 000 000 = 10^6	mega	M
1 000 000 000 = 10^9	giga	G
1 000 000 000 000 = 10^{12}	tera	T

Thus 1 nanometre (nm) = 1×10^{-9} m,
also 1 centimetre (cm) = 1×10^{-2} m.
Note that the kilogram is somewhat out of place in the above table since it is a basic SI unit. In the school laboratory the most convenient units are grams (g) and cubic centimetres (cm^3). Where possible the basic SI units should be used.

A3.3.5 Rules for writing SI units

- The symbol is not an abbreviation and thus a full stop is not written after the symbol except at the end of a sentence.
- There is no plural form of a unit; thus 20 kg or 30 m, not 20 kgs or 30 ms.
- Capital initial letters are never used for units except when named after famous scientists, such as N (Newton), W (Watt) and J (Joule).
- Symbols combined in a quotient can be written as, for example, metre per second or $m\ s^{-1}$. The use of the solidus (stroke, /) is restricted to indicating the unit of a variable, such as temperature /°C.
- The raised decimal point is not correct. The internationally accepted decimal sign is placed level with the feet of the numerals, for example 3.142. The comma is no longer used to separate groups of three digits but a space is left instead so that figures appear 493 645 189 not as 493,645,189.

Appendix 4
The geological time scale

The history of the Earth is divided for convenience into a series of four geological **eras** and eleven periods. The two most recent periods are further divided into seven systems or **epochs**.

The rocks of the Earth's crust are stratified, that is they lie layer upon layer. Unless disrupted by earth movements, the rocks get progressively younger towards the top of a series of layers (strata). William Smith in the eighteenth century noticed an association between fossil groups and particular strata. The sequence of fossils revealed a gradual increase in complexity of organisms from the lower strata to the highest, indicating that over geological periods of time some organisms have advanced in complexity.

Radioactive dating has established an approximate age for the oldest rocks belonging to each period. The geological time scale and the distinctive biological events associated with each period, as revealed by fossil evidence, are shown in table A4.1.

A4.1 The age of the Earth

Current estimates are that the planet Earth is about $4.6–4.9 \times 10^9$ years old. These estimates are based mainly on the dating of rocks (**geochronology**) by radioactive dating techniques.

In section A1.3 it was explained that atoms of some elements exist in a number of forms called isotopes, some of which are radioactive. Radioactive elements 'decay' at a constant rate which is independent of temperature, gravity, magnetism or any other force. The rate is measured in terms of the 'half-life'.

Three principal methods of radioactive dating are currently used, as shown in fig A4.1. Methods (1) and (2) are used for determining the ages of rocks in the Earth's crust, whereas the third method, radiocarbon dating, is used for dating fossils and has direct relevance in discussions of the history of life.

Radiocarbon dating

The normal non-radioactive isotope of carbon is ^{12}C. The radioactive isotope, ^{14}C, occurs in minute quantities (<0.1%) in air, surface waters and living organisms. It is continually being produced in the atmosphere by the action

Table A4.1 Geological time scale and history of life (age = years × 10^6).

Era	*Period*	*Epoch*	*Age*	*Animal groups*	*Plant groups*
CENOZOIC (*cenos*, recent)	Quaternary	Recent (Holocene)	0.01	Dominance of humans	
		Glacial (Pleistocene)	2	Origin of humans	
	Tertiary	Pliocene	7	Adaptive radiation of mammals	Adaptive radiation of flowering plants, especially herbaceous types
		Miocene	26		
		Oligocene	38	Dogs and bears appeared	
		Eocene	54	Apes and pigs appeared	
		Palaeocene	65	Horses, cattle and elephants	
MESOZOIC (*mesos*, middle)	Cretaceous		135	Exinction of ammonites and dinosaurs; origin of modern fish and placental mammals	Dominance of flowering plants
	Jurassic		195	Dinosaurs dominant; origin of birds and mammals; insects abundant	Origin of flowering plants
	Triassic		225	Dinosaurs appear; adaptive radiation of reptiles	Abundance of cycads and conifers
PALAEOZOIC (*palaeos*, ancient)	Permian		280	Adaptive radiation of reptiles; beetles appear; extinction of trilobites	Origin of conifers
	Carboniferous		350	Origin of reptiles and insects; adaptive radiation of amphibia	Abundance of tree-like ferns, e.g. *Lepidodendron*, forming 'coal forests'
	Devonian		400	Origin of amphibia and ammonites; spiders appear; adaptive radiation of fish (cartilaginous and bony)	Earliest mosses and ferns
	Silurian		440	Origin of jawed fish; earliest coral reefs	Earliest spore-bearing, vascular plants
	Ordovician		500	Origin of vertebrates, jawless fish; trilobites, molluscs and crustacea abundant	
	Cambrian		570	Origin of all non-vertebrate phyla and echinoderms	
ARCHEOZOIC	Pre-Cambrian			*Selected organisms*	
			1 000	primitive metazoans	
			2 000	primitive eukaryotes	
			3 000	blue-green bacteria (prokaryotes), bacteria	
			3 500?	origins of life?	
			5 000?	origin of Earth?	

of cosmic rays on nitrogen and oxygen nuclei, and there is good evidence to suggest that the rate of ^{14}C production has been constant for several thousand years. An equilibrium has been set up whereby production of ^{14}C balances ^{14}C loss by radioactive decay. ^{14}C is found occurring freely as $^{14}CO_2$ and the ratio of $^{14}C : ^{12}C$ compounds remains theoretically constant. Living organisms absorb ^{14}C either as carbon dioxide or as organic molecules throughout life. At death no more carbon is taken in and the ^{14}C continues to decay according to the rate shown by the half-life. By calculating the amount of ^{14}C in the dead organism and comparing it with the amount of ^{14}C in a living organism, the age of the dead organism can be estimated. For example, if the amount of ^{14}C in a fossil mammalian bone was found to be one-quarter that in the same bone from a recently killed mammal and the half-life is 5.6×10^3 years, the estimated age of the fossil bone would theoretically be 11.2×10^3 years. Using this technique, organic remains can be dated back, fairly accurately, for up to 10.0×10^4 years.

There are many sources of error involved in radiometric dating, so ages determined by these methods are only approximate. However, these methods have proved to be of great value in extending our knowledge of the Earth.

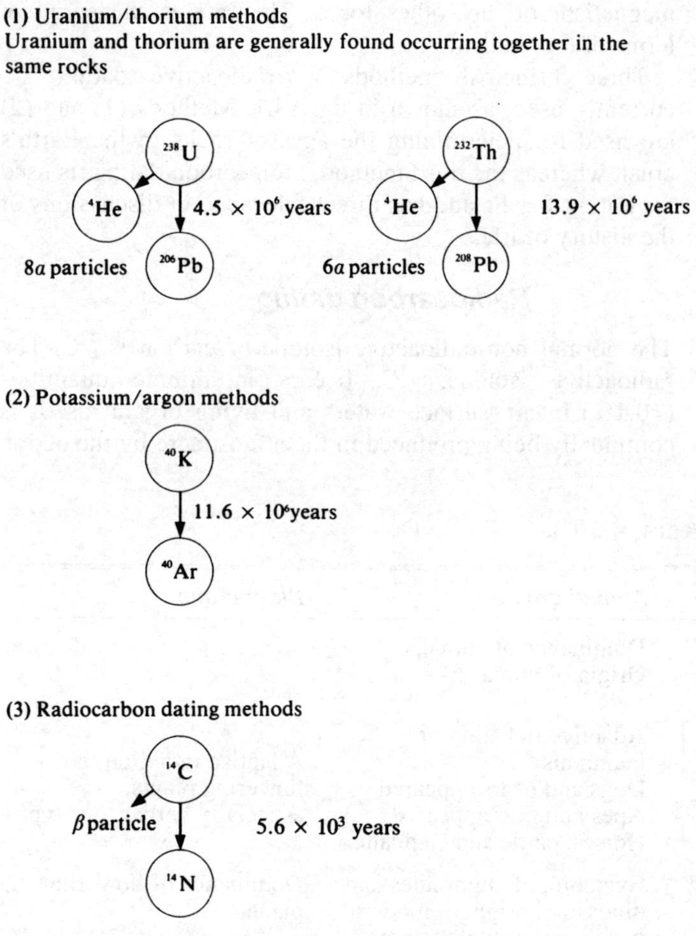

Fig A4.1 *Radiometric dating techniques.*

Index

References to figures are underlined; to tables are *underlined italics*